Your Guide to Success in

Complete Step 0 as soon as you begin your math course.

STEP 0: PLAN YOUR SEMESTER

- ☐ Register for the online part of the course (if there is one) as soon as possible.
- ☐ Fill in your Course and Contact information on this pull-out card.
- ☐ Write important dates from your syllabus on the Semester Organizer on this pull-out card.

Follow Steps 1–4 during your course. Your instructor will tell you which resources to use—and when—in the textbook or eText, *MyMathGuide* workbook, videos, and MyMathLab. Use these resources for extra help and practice.

STEP 1: LEARN THE SKILLS AND CONCEPTS

- ☐ Read the **textbook** or **eText,** listen to your instructor's lecture, and/or watch the **videos.** You can work in ***MyMathGuide*** as you do this. As you are learning:
    - ☐ Take notes, write down your questions, and save all your work (including homework solutions, quizzes, and tests) to review throughout the course.
    - ☐ Work the *Skill to Review* exercises at the beginning of each section.
    - ☐ Stop and do the *Margin* and *Guided Solution Exercises* as directed.
    - ☐ Watch the videos. Answer the *Interactive Your Turn* questions in the videos and in *MyMathGuide.*

STEP 2: CHECK YOUR UNDERSTANDING

- ☐ Answer the *Reading Checks* in the Section Exercise sets or in MyMathLab.
- ☐ Explore the concepts using the *Active Learning Figures* in MyMathLab.

STEP 3: DO YOUR HOMEWORK

- ☐ Plan to spend 2 hours studying and doing homework for every hour of class.
- ☐ Complete your assigned homework from the textbook and/or in MyMathLab.
    - ☐ When doing homework from the textbook, use the answer section to check your work.
    - ☐ When doing homework in MyMathLab, use the Learning Aids, such as Help Me Solve This and View an Example, as needed, working toward being able to complete exercises without the aids.

STEP 4: REVIEW AND TEST YOUR UNDERSTANDING

- ☐ Work the exercises in the *Mid-Chapter Review.*
- ☐ Make your own chapter study sheet by doing the *Chapter Summary and Review.*
- ☐ Take the *Chapter Test* as a practice exam. To watch an instructor solve each problem, go to the Chapter Test Prep Videos in MyMathLab or on YouTube (search "BittingerPreIntro" and click on "Channels").

Use the ***Studying for Success*** tips in the text and the MyMathLab **Study Skills modules** (with videos, tips, and activities) to help you develop effective time-management, note-taking, test-prep, and other skills.

Student Organizer

Course Information

Course Number: ______________ Name: ____

Location: ___________________________________ Days/Time: _________________________

Contact Information

Contact	Name	Email	Phone	Office Hours	Location
Instructor					
Tutor					
Math Lab					
Classmate					
Classmate					

Semester Organizer

Week	Homework	Quizzes and Tests	Other

At a Glance: Prealgebra and Introductory Algebra

Place Value

529,630,718,249.70651

Billions			Millions			Thousands			Ones							
5	2	9	6	3	0	7	1	8	2	4	9	7	0	6	5	1
Hundreds	Tens	Ones	Hundreds	Tens	Ones	Hundreds	Tens	Ones	Hundreds	Tens	Ones	Tenths	Hundredths	Thousandths	Ten-thousandths	Hundred-thousandths

Exponential Notation

Exponent

Base → $3^4 = \underbrace{3 \cdot 3 \cdot 3 \cdot 3}_{\text{4 factors of 3}}$

Large Numbers

$10^3 = 1000 = 1$ thousand

$10^6 = 1{,}000{,}000 = 1$ million

$10^9 = 1{,}000{,}000{,}000 = 1$ billion

Order of Operations

1. Do all calculations within grouping symbols before operations outside.
2. Evaluate all exponential expressions.
3. Do all multiplications and divisions in order from left to right.
4. Do all additions and subtractions in order from left to right.

Prime Factorization

$220 = 2 \cdot 2 \cdot 5 \cdot 11$ All factors are prime.

Divisibility Rules

The number is divisible

by 2: if it has a ones digit of 0, 2, 4, 6, or 8.

by 3: if the sum of its digits is divisible by 3.

by 6: if its ones digit is 0, 2, 4, 6, or 8 and the sum of its digits is divisible by 3.

by 5: if its ones digit is 0 or 5.

by 10: if its ones digit is 0.

Least Common Multiple

$12 = 2 \cdot 2 \cdot 3$ $15 = 3 \cdot 5$

The LCM of 12 and 15 $= 2 \cdot 2 \cdot 3 \cdot 5$, or 60.

Fraction Notation

$\frac{8}{1} = 8$ $\frac{8}{8} = 1$ $\frac{0}{8} = 0$ $\frac{8}{0}$ Not defined

$$\frac{5}{6} + \frac{3}{4} = \frac{5}{6} \cdot \frac{2}{2} + \frac{3}{4} \cdot \frac{3}{3} = \frac{10}{12} + \frac{9}{12} = \frac{10 + 9}{12} = \frac{19}{12}$$

$$\frac{5}{6} - \frac{3}{4} = \frac{5}{6} \cdot \frac{2}{2} - \frac{3}{4} \cdot \frac{3}{3} = \frac{10}{12} - \frac{9}{12} = \frac{10 - 9}{12} = \frac{1}{12}$$

$$\frac{5}{6} \cdot \frac{3}{4} = \frac{5 \cdot 3}{6 \cdot 4} = \frac{15}{24} = \frac{3 \cdot 5}{3 \cdot 8} = \frac{3}{3} \cdot \frac{5}{8} = 1 \cdot \frac{5}{8} = \frac{5}{8}$$

$$\frac{5}{6} \div \frac{3}{4} = \frac{5}{6} \cdot \frac{4}{3} = \frac{20}{18} = \frac{2 \cdot 10}{2 \cdot 9} = \frac{2}{2} \cdot \frac{10}{9} = 1 \cdot \frac{10}{9} = \frac{10}{9}$$

Mixed Numerals

$\frac{11}{3} = 3\frac{2}{3}$ $4\frac{1}{2} = \frac{9}{2}$

Percent Notation

$3\% = 3 \times \frac{1}{100} = \frac{3}{100}$ $3\% = 3 \times 0.01 = 0.03$

Percent Increase and Decrease

Price increased from \$480 to \$504:

$$\text{Percent increase} = \frac{\text{Change}}{\text{Original}} = \frac{504 - 480}{480} = \frac{24}{480} = 0.05 = 5\%$$

Price decreased from \$60 to \$45:

$$\text{Percent decrease} = \frac{\text{Change}}{\text{Original}} = \frac{60 - 45}{60} = \frac{15}{60} = 0.25 = 25\%$$

Interest

Simple: $I = P \cdot r \cdot t$

Compound: $A = P \cdot \left(1 + \frac{r}{n}\right)^{n \cdot t}$

Fraction, Decimal, Percent Equivalents

Fraction Notation	$\frac{1}{10}$	$\frac{1}{8}$	$\frac{1}{6}$	$\frac{1}{5}$	$\frac{1}{4}$	$\frac{3}{10}$	$\frac{1}{3}$	$\frac{3}{8}$	$\frac{2}{5}$	$\frac{1}{2}$	$\frac{3}{5}$	$\frac{5}{8}$	$\frac{2}{3}$	$\frac{7}{10}$	$\frac{3}{4}$	$\frac{4}{5}$	$\frac{5}{6}$	$\frac{7}{8}$	$\frac{9}{10}$	$\frac{1}{1}$
Decimal Notation	0.1	0.125	$0.16\overline{6}$	0.2	0.25	0.3	$0.33\overline{3}$	0.375	0.4	0.5	0.6	0.625	$0.66\overline{6}$	0.7	0.75	0.8	$0.83\overline{3}$	0.875	0.9	1
Percent Notation	10%	12.5% or $12\frac{1}{2}\%$	$16.\overline{6}\%$ or $16\frac{2}{3}\%$	20%	25%	30%	$33.\overline{3}\%$ or $33\frac{1}{3}\%$	37.5% or $37\frac{1}{2}\%$	40%	50%	60%	62.5% or $62\frac{1}{2}\%$	$66.\overline{6}\%$ or $66\frac{2}{3}\%$	70%	75%	80%	$83.\overline{3}\%$ or $83\frac{1}{3}\%$	87.5% or $87\frac{1}{2}\%$	90%	100%

Translating to Percent Equations

What is 20% of 75?
↓ ↓ ↓ ↓ ↓
$a = 20\% \cdot 75$

15 is 20% of what?
↓ ↓ ↓ ↓ ↓
$15 = 20\% \cdot b$

15 is what percent of 75?
↓ ↓ ↓ ↓ ↓
$15 = p \cdot 75$

American Units of Length

12 inches (in.) = 1 foot (ft)
36 inches = 1 yard (yd)
3 feet = 1 yard
5280 feet = 1 mile (mi)

Pythagorean Theorem

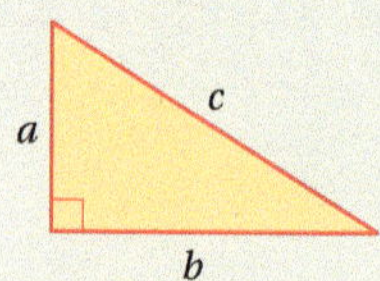

$$a^2 + b^2 = c^2$$

Formulas

Perimeter of a Square: $P = 4 \cdot s$
Perimeter of a Rectangle: $P = 2 \cdot l + 2 \cdot w$
Area of a Square: $A = s^2$
Area of a Rectangle: $A = l \cdot w$
Area of a Parallelogram: $A = b \cdot h$
Area of a Triangle: $A = \frac{1}{2} \cdot b \cdot h$
Area of a Trapezoid: $A = \frac{1}{2} \cdot h \cdot (a + b)$
Circumference of a Circle: $C = \pi \cdot d$, or $C = 2 \cdot \pi \cdot r$
Area of a Circle: $A = \pi \cdot r^2$
Volume of a Rectangular Solid: $V = l \cdot w \cdot h$
Volume of a Circular Cylinder: $V = \pi \cdot r^2 \cdot h$
Volume of a Sphere: $V = \frac{4}{3} \cdot \pi \cdot r^3$
Volume of a Cone: $V = \frac{1}{3} \cdot \pi \cdot r^2 \cdot h$

Temperature

$F = \frac{9}{5} \cdot C + 32$ $\quad$ $C = \frac{5}{9}(F - 32)$

Metric Units of Length

1 kilometer (km) = 1000 meters (m)
1 hectometer (hm) = 100 meters
1 dekameter (dam) = 10 meters
1 meter
1 decimeter (dm) = $\frac{1}{10}$ meter
1 centimeter (cm) = $\frac{1}{100}$ meter
1 millimeter (mm) = $\frac{1}{1000}$ meter

Angles

Right: 90°
Straight: 180°
Acute: Greater than 0° and less than 90°
Obtuse: Greater than 90° and less than 180°
Complementary: Two angles are complementary if the sum of their measures is 90°.
Supplementary: Two angles are supplementary if the sum of their measures is 180°.
Sum of Angle Measures of a Triangle: 180°
Sum of Angle Measures of a Polygon: $(n - 2) \cdot 180°$

Vertical Angle Property

Vertical angles are congruent.

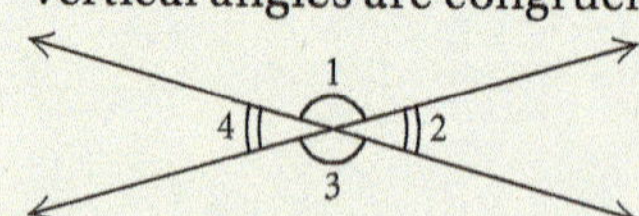

$\angle 1 \cong \angle 3, \quad \angle 2 \cong \angle 4$

Properties of Parallel Lines

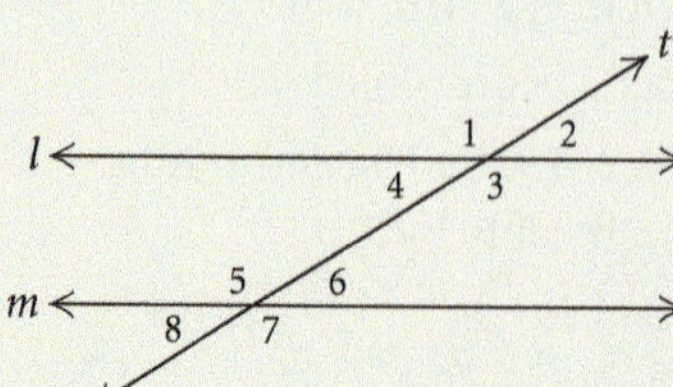

Transversal t intersects parallel lines l and m.

Corresponding angles are congruent:
$\angle 1 \cong \angle 5, \quad \angle 2 \cong \angle 6, \quad \angle 3 \cong \angle 7, \quad \angle 4 \cong \angle 8.$

Alternate interior angles are congruent:
$\angle 4 \cong \angle 6, \quad \angle 3 \cong \angle 5.$

Prealgebra and Introductory Algebra

Fourth Edition

Marvin L. Bittinger
Indiana University Purdue University Indianapolis

David J. Ellenbogen
Community College of Vermont

Judith A. Beecher

Barbara L. Johnson
Indiana University Purdue University Indianapolis

PEARSON

Boston Columbus Hoboken Indianapolis New York San Francisco
Amsterdam Cape Town Dubai London Madrid Milan Munich Paris Montréal Toronto
Delhi Mexico City São Paulo Sydney Hong Kong Seoul Singapore Taipei Tokyo

Editorial Director	Christine Hoag
Executive Editor	Cathy Cantin
Editorial Assistant	Chase Hammond
Program Management Team Lead	Karen Wernholm
Program Manager	Patty Bergin
Project Management Team Lead	Christina Lepre
Project Manager	Ron Hampton
Composition	Lumina Datamatics
Production and Editorial Services	Jane Hoover/Lifland et al., Bookmakers
Art Editor and Photo Researcher	The Davis Group, Inc.
Associate Producer	Jonathan Wooding
Executive Content Manager	Rebecca Williams (MathXL)
Senior Content Developer	John Flanagan (TestGen)
Marketing Manager	Rachel Ross
Marketing Assistant	Kelly Cross
Senior Manufacturing Buyer	Carol Melville
Text Designer	The Davis Group, Inc.
Associate Design Director	Andrea Nix
Program Design Lead	Barbara Atkinson
Cover Photograph	Shutterstock

Credits
Credits appear on pages CR-1 and CR-2.

Library of Congress Cataloging-in-Publication Data
Bittinger, Marvin L.
Prealgebra and introductory algebra. — Fourth edition / Marvin L. Bittinger, Indiana University Purdue University, Indianapolis, David J. Ellenbogen, Community College of Vermont, Judith A. Beecher, Barbara L. Johnson, Indiana University Purdue University, Indianapolis.
pages cm
ISBN: 978-0-321-99716-6
1. Algebra—Textbooks. 2. Mathematics—Textbooks. I. Ellenbogen, David. II. Beecher, Judith A. III. Johnson, Barbara L. IV. Title.
QA152.3.B59 2016
512.9—dc23 2014035497

www.pearsonhighered.com

ISBN-13: 978-0-321-99716-6
ISBN-10: 0-321-99716-6

15 2022

Contents

Index of Applications

Chemistry

Construction

Consumer

Domestic

Economics

Education

Engineering

Environment

Physics

Social Sciences

Sports/Entertainment

Statistics/Demographics

Technology

Transportation

Preface

The Bittinger Program

Math hasn't changed, but students—and the way they learn it—have.

Prealgebra and Introductory Algebra, 4th Edition, continues the Bittinger tradition of objective-based, guided learning, while integrating timely updates to the proven pedagogy. In this edition, there is a greater emphasis on guided learning and helping students get the most out of all of the course resources available with the Bittinger program, including new opportunities for mobile learning.

The program has expanded to include these comprehensive new teaching and learning resources: ***MyMathGuide*** **workbook**, **To-the-Point Objective Videos**, and enhanced, media-rich **MyMathLab** courses. Feedback from instructors and students motivated these and several other significant improvements: a new design to support guided learning, new figures and photos to help students visualize both concepts and applications, and many new and updated real-data applications to bring the math to life.

With so many resources available in so many formats, the trusted guidance of the Bittinger team on *what to do* and *when* will help today's math students stay on task. Students are encouraged to use ***Your Guide to Success in Math***, a four-step learning path and checklist available on the handy reference card in the front of this text and in MyMathLab. The guide will help students identify the resources in the textbook, supplements, and MyMathLab that support *their* learning style, as they develop and retain the skills and conceptual understanding they need to succeed in this and future courses.

In this preface, a look at the key new *and* hallmark resources and features of the *Prealgebra and Introductory Algebra* program—including the textbook/eText, video program, *MyMathGuide* workbook, and MyMathLab—is organized around ***Your Guide to Success in Math***. This will help instructors direct students to the tools and resources that will help them most in a traditional lecture, hybrid, lab-based, or online environment.

NEW AND HALLMARK FEATURES IN RELATION TO Your Guide to Success in Math

STEP 1 Learn the Skills and Concepts

Students have several options for learning, reviewing, and practicing the math concepts and skills.

Textbook/eText

☐ **Skill to Review.** At the beginning of nearly every text section, *Skill to Review* offers a just-in-time review of a previously presented skill that relates to the new material in the section. Section and objective references are included for the student's convenience, and two practice exercises are provided for review and reinforcement.

- ☐ **Margin Exercises.** For each objective, problems labeled "Do Exercise …" give students frequent opportunities to solve exercises while they learn.
- ☐ *New!* **Guided Solutions.** Nearly every section has *Guided Solution* margin exercises with fill-in blanks at key steps in the problem-solving process.
- ☐ *Enhanced!* **MyMathLab.** MyMathLab now includes *Active Learning Figures* for directed exploration of concepts; more problem types, including *Reading Checks* and *Guided Solutions;* and new, objective-based videos. (See pp. xviii–xxii for a detailed description of the features of MyMathLab.)
  - ☐ *New!* **Skills Checks.** In the Learning Path for Ready-to-Go MyMathLab, each chapter begins with a brief assessment of students' mastery of the prerequisite skills needed to learn the new material in the chapter. Based on the results of this pre-test, a personalized homework set is designed to help each student prepare for the chapter.
- ☐ *New!* **To-the-Point Objective Videos.** This is a comprehensive new program of objective-based, interactive videos that are incorporated into the Learning Path in MyMathLab and can be used hand-in-hand with the *MyMathGuide* workbook.
  - ☐ *New!* **Interactive Your Turn Exercises.** For each objective in the videos, students solve exercises and receive instant feedback on their work.
- ☐ *New!* ***MyMathGuide: Notes, Practice, and Video Path.*** This is an objective-based workbook (available printed and in MyMathLab) for guided, hands-on learning. It offers vocabulary, skill, and concept review—along with problem-solving practice—with space to show work and write notes. Incorporated in the Learning Path in MyMathLab, it can be used together with the To-the-Point Objective Video program, instructor lectures, and the textbook.

STEP 2 Check Your Understanding

Throughout the program, students have frequent opportunities to check their work and confirm that they understand each skill and concept before moving on to the next topic.

- ☐ *New!* **Reading Checks.** At the beginning of each set of section exercises in the text, students demonstrate their grasp of the skills and concepts.
- ☐ *New!* **Active Learning Figures.** In MyMathLab, Active Learning Figures guide students in exploring math concepts and reinforcing their understanding.
- ☐ **Translating for Success.** In the text and in MyMathLab, these problem sets offer students extra practice with the important first step of the process for solving applied problems.
- ☐ **Visualizing for Success.** In the text and in MyMathLab, these problem sets ask students to match an equation or an inequality with its graph by focusing on characteristics of the graph or the inequality and the corresponding attributes of the graph.

STEP 3 Do Your Homework

Prealgebra and Introductory Algebra, 4th Edition, has a wealth of proven and updated exercises. Prebuilt assignments are available for instructors in MyMathLab, and they are preassigned and incorporated into the Learning Path in the Ready-to-Go course.

- ☐ **Skill Maintenance.** In each section, these exercises offer a review of the math in the preceding text.
- ☐ **Synthesis Exercises.** To help build critical-thinking skills, these section exercises require students to use what they know and combine learning objectives from the current section with those from previous sections.

STEP 4 Review and Test Your Understanding

Students have a variety of resources to check their skills and understanding along the way and to help them prepare for tests.

- ☐ **Mid-Chapter Review.** Midway through each chapter, students work a set of exercises (*Concept Reinforcement, Guided Solutions, Mixed Review,* and *Understanding Through Discussion and Writing*) to confirm that they have grasped the skills and concepts covered in the first half before moving on to new material.
- ☐ **Summary and Review.** This resource provides an in-text opportunity for active learning and review for each chapter. *Vocabulary Reinforcement, Concept Reinforcement,* objective-based *Study Guide* (examples paired with similar exercises), *Review Exercises* (including *Synthesis* problems), and *Understanding Through Discussion and Writing* are included in these comprehensive chapter reviews.
- ☐ **Chapter Test.** Chapter Tests offer students the opportunity for comprehensive review and reinforcement prior to taking their instructor's exam. **Chapter Test Prep Videos** (in MyMathLab and on YouTube) show step-by-step solutions to the Chapter Tests.

Study Skills

Developing solid time-management, note-taking, test-taking, and other study skills is key to student success in math courses (as well as professionally and personally). Instructors can direct students to related study skills resources as needed.

- ☐ *New!* **Student Study Reference.** This pull-out card at the front of the text is perforated, three-hole-punched, and binder-ready for convenient reference. It includes **Your Guide to Success in Math** course checklist, **Student Organizer**, and **At a Glance**, a list of key information and expressions for quick reference as students work exercises and review for tests.
- ☐ *New!* **Studying for Success.** Checklists of study skills—designed to ensure that students develop the skills they need to succeed in math, school, and life—are integrated throughout the text at the beginning of selected sections.
- ☐ *New!* **Study Skills Modules.** In MyMathLab, interactive modules address common areas of weakness, including time-management, test-taking, and note-taking skills. Additional modules support career-readiness.

Learning Math in Context

- ☐ *New!* **Applications.** Throughout the text in examples and exercises, real-data applications encourage students to see and interpret the mathematics that appears every day in the world around them. Applications that use real data are drawn from business and economics, life and physical sciences, medicine, technology, and areas of general interest such as sports and daily life. New applications include "Fastest-Growing Occupations" (p. 68), "Training Regimens" (p. 272), *The Hobbit: An Unexpected Journey* (p. 393), "Mosquito Netting" (p. 532), "Cycling in Vietnam" (p. 731), "Gold Leaf on Georgia's State Capitol Dome" (p. 874), "Speed of Sea Animals" (p. 1073), and "Produce Prices" (p. 1130).

BREAKTHROUGH
To improving results

MyMathLab

Ties the Complete Learning Program Together

MyMathLab® Online Course (access code required)
MyMathLab from Pearson is the world's leading online resource in mathematics, integrating interactive homework, assessment, and media in a flexible, easy to use format. MyMathLab delivers **proven results** in helping individual students succeed. It provides **engaging experiences** that personalize, stimulate, and measure learning for each student, and it comes from an **experienced partner** with educational expertise and an eye on the future.

MyMathLab for Developmental Mathematics

Prepared to go wherever you want to take your students.

Personalized Support for Students

Homework with Built-in Support

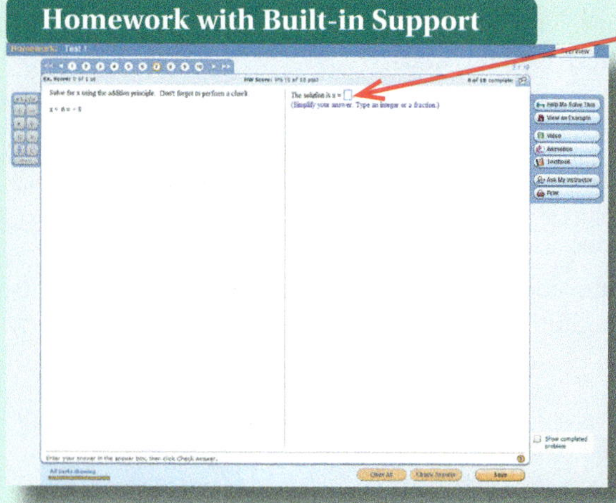

Exercises: The homework and practice exercises in MyMathLab are correlated to the exercises in the textbook, and they regenerate algorithmically to give students unlimited opportunities for practice and mastery. The software offers immediate, helpful feedback when students enter incorrect answers.

Multimedia Learning Aids: Exercises include guided solutions, sample problems, animations, videos, and eText access for extra help at point-of-use.

Expert Tutoring: Although many students describe the whole of MyMathLab as "like having your own personal tutor," students using MyMathLab do have access to live tutoring from qualified math instructors.

Personalized Homework

To help students achieve mastery, MyMathLab can generate **personalized homework** based on individual performance on tests or quizzes. Personalized homework allows students to focus on topics they have not yet mastered.

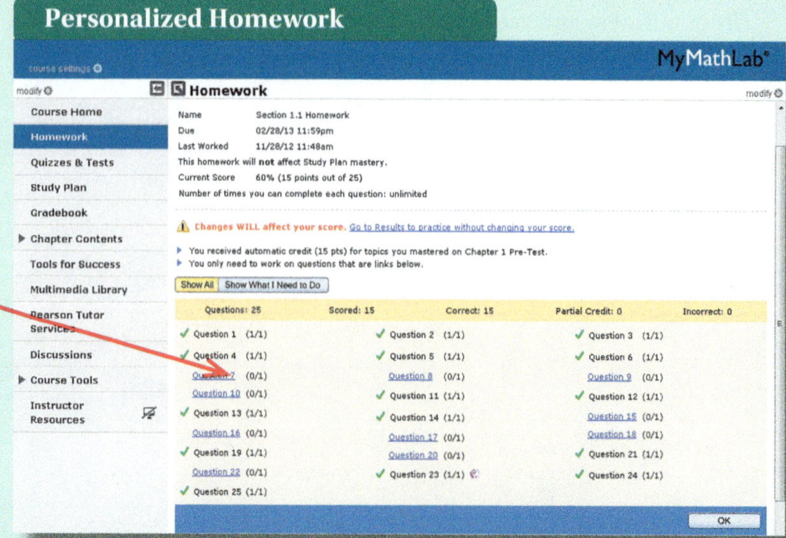

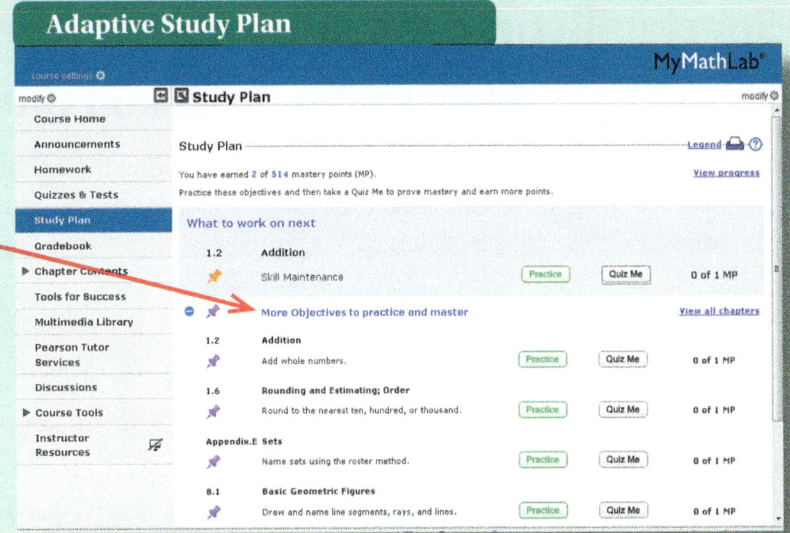

The **Adaptive Study Plan** makes studying more efficient and effective for every student. Performance and activity are assessed continually in real time. The data and analytics are used to provide personalized content—reinforcing concepts that target each student's strengths and weaknesses.

Flexible Design, Easy Start-Up, and Results for Instructors

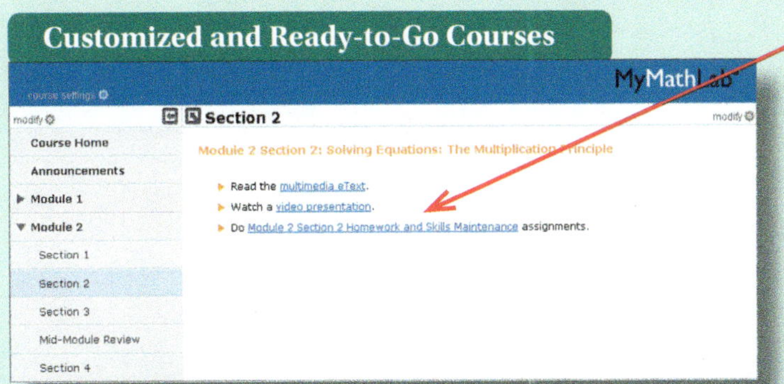

Instructors can modify the site navigation and insert their own directions on course-level landing pages; also, a **custom MyMathLab** course can be built that reorganizes and structures the course material by chapters, modules, units—whatever the need may be.

Ready-to-Go courses include preassigned homework, quizzes, and tests to make it even easier to get started. The Bittinger Ready-to-Go courses include new *Mid-Chapter Reviews* and *Reading Check Assignments*, plus a four-step Learning Path on each section-level landing page to help instructors direct students where to go and what resources to use.

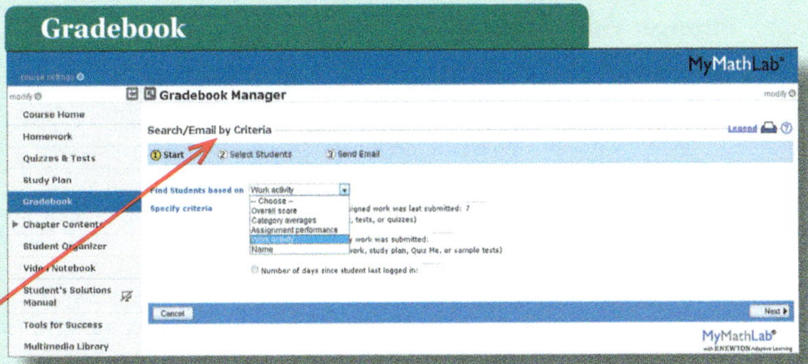

The **comprehensive online gradebook** automatically tracks students' results on tests, quizzes, and homework and in the study plan. Instructors can use the gradebook to quickly intervene if students have trouble, or to provide positive feedback on a job well done. The data within MyMathLab are easily exported to a variety of spreadsheet programs, such as Microsoft Excel. Instructors can determine which points of data to export and then analyze the results to determine success.

New features, such as **Search/Email by Criteria**, make the gradebook a powerful tool for instructors. With this feature, instructors can easily communicate with both at-risk and successful students. They can search by score on specific assignments, noncompletion of assignments within a given time frame, last login date, or overall score.

Special Bittinger Resources

in MyMathLab for Students and Instructors

In addition to robust course delivery, MyMathLab offers the full Bittinger eText, additional Bittinger Program features, and the entire set of instructor and student resources in one easy-to-access online location.

New! Active Learning Figures

In MyMathLab, Active Learning Figures guide students in exploring math concepts and reinforcing their understanding. Instructors can use Active Learning Figures in class or as media assignments in MyMathLab to guide students to explore math concepts and reinforce their understanding.

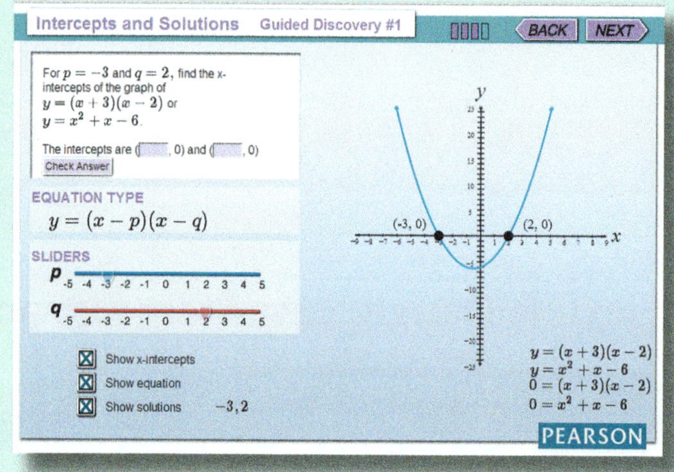

New! Four-Step Learning Path

Each of the section-level landing pages in the Ready-to-Go MyMathLab course includes a Learning Path that aligns with *Your Guide to Success in Math* to link students directly to the resources they should use when they need them. This also allows instructors to point students to the best resources to use at particular times.

New! Integrated Bittinger Video Program and *MyMathGuide* workbook

Bittinger Video Program*

The Video Program is available in MyMathLab and includes closed captioning and the following video types:

New! **To-the-Point Objective Videos.** These objective-based, interactive videos are incorporated into the Learning Path in MyMathLab and can be used along with the *MyMathGuide* workbook.

Chapter Test Prep Videos. The Chapter Test Prep Videos let students watch instructors work through step-by-step solutions to all the Chapter Test exercises from the textbook. Chapter Test Prep Videos are also available on YouTube (search using author name and book title).

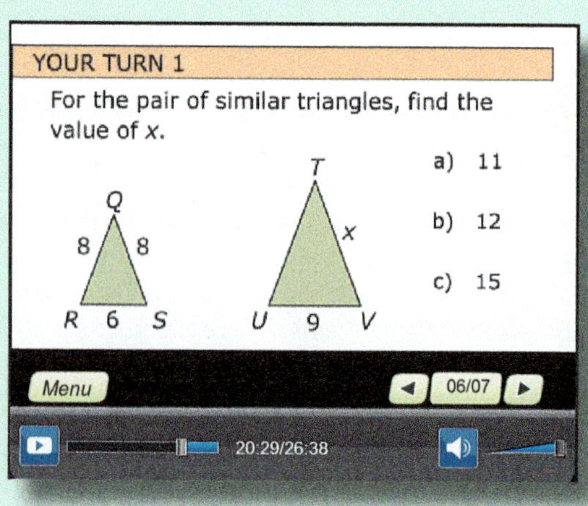

*Printed supplements are also available for separate purchase through MyMathLab, MyPearsonStore.com, or other retail outlets. They can also be value-packed with a textbook or MyMathLab code.

New! ***MyMathGuide: Notes, Practice, and Video Path* workbook***
(Printed Workbook ISBN: 978-0-13-391932-5)

This objective-based workbook for guided, hands-on learning offers vocabulary, skill, and concept review—along with problem-solving practice—with space to show work and write notes. Incorporated in the Learning Path in MyMathLab, *MyMathGuide* can be used together with the To-the-Point Objective Video program, instructor lectures, and the textbook. Instructors can assign To-the-Point Objective Videos in MyMathLab in conjunction with the *MyMathGuide* workbook.

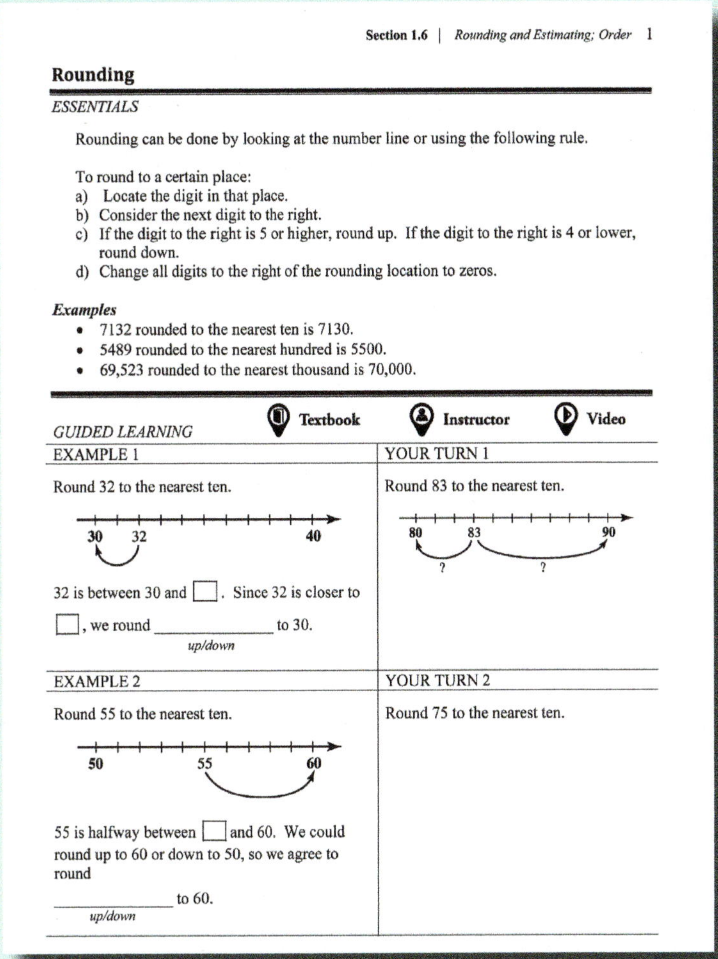
Section 1.6 | *Rounding and Estimating; Order* 1

Rounding

ESSENTIALS

Rounding can be done by looking at the number line or using the following rule.

To round to a certain place:
a) Locate the digit in that place.
b) Consider the next digit to the right.
c) If the digit to the right is 5 or higher, round up. If the digit to the right is 4 or lower, round down.
d) Change all digits to the right of the rounding location to zeros.

Examples
- 7132 rounded to the nearest ten is 7130.
- 5489 rounded to the nearest hundred is 5500.
- 69,523 rounded to the nearest thousand is 70,000.

GUIDED LEARNING Textbook Instructor Video

EXAMPLE 1	YOUR TURN 1
Round 32 to the nearest ten. 30 32 40 32 is between 30 and ☐. Since 32 is closer to ☐, we round ________ (up/down) to 30.	Round 83 to the nearest ten. 80 83 90 ? ?
EXAMPLE 2	YOUR TURN 2
Round 55 to the nearest ten. 50 55 60 55 is halfway between ☐ and 60. We could round up to 60 or down to 50, so we agree to round ________ (up/down) to 60.	Round 75 to the nearest ten.

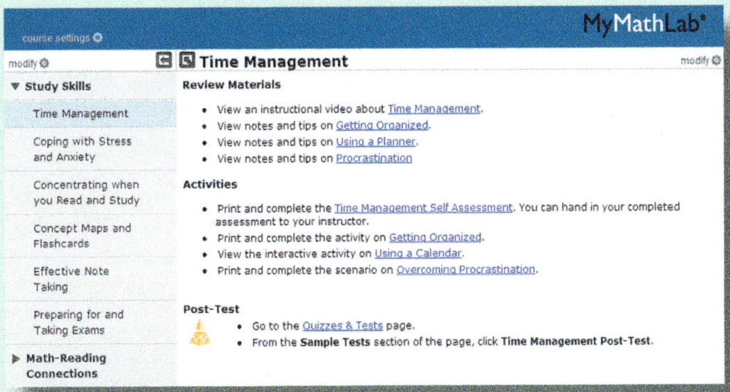

Study Skills Modules

In MyMathLab, interactive modules address common areas of weakness, including time-management, test-taking, and note-taking skills. Additional modules support career-readiness. Instructors can assign module material with a post-quiz.

*Printed supplements are also available for separate purchase through MyMathLab, MyPearsonStore.com, or other retail outlets. They can also be value-packed with a textbook or MyMathLab code.

Additional Resources in MyMathLab

For Students

Student's Solutions Manual*

(ISBN: 978-0-13-392230-8)

By Judy Penna

Contains completely worked-out annotated solutions for all the odd-numbered exercises in the text. Also includes fully worked-out annotated solutions for all the exercises (odd- and even-numbered) in the Mid-Chapter Reviews, the Summary and Reviews, the Chapter Tests, and the Cumulative Reviews.

For Instructors

Annotated Instructor's Edition**

(ISBN: 978-0-13-392231-8)

This version of the text includes answers to all exercises presented in the book, as well as helpful teaching tips.

Instructor's Resource Manual with Tests and Mini Lectures**

(download only)

By Laurie Hurley

This manual includes resources designed to help both new and experienced instructors with course preparation and classroom management. It contains support for media supplements, two multiple-choice tests per chapter, six free-response tests per chapter, and eight final exams.

Instructor's Solutions Manual**

(download only)

By Judy Penna

This manual contains detailed, worked-out solutions to all odd-numbered exercises and brief solutions to the even-numbered exercises in the exercise sets.

PowerPoint® Lecture Slides**

(download only)

Present key concepts and definitions from the text.

To learn more about how MyMathlab combines proven learning applications with powerful assessment, visit www.mymathlab.com or contact your Pearson representative.

*Printed supplements are also available for separate purchase through MyMathLab, MyPearsonStore.com, or other retail outlets. They can also be value-packed with a textbook or MyMathLab code.

**Also available for download from the Instructor Resource Center (IRC) on www.pearsonhighered.com.

Acknowledgments

Our deepest appreciation to all of you who helped to shape this edition by reviewing and spending time with us on your campuses. In particular, we would like to thank the following reviewers:

Afsheen Akbar, *Bergen Community College*
Morgan Arnold, *Central Georgia Technical College*
Gurdial Arora, *Xavier University*
Donna Beatty, *Venture College*
Connie Buller, *Metropolitan Community College*
Erin Cooke, *Gwinnett Technical College*
Kay Davis, *Del Mar College*
Edward Dillon, *Century Community and Technical College*
Beverlee Drucker, *Northern Virginia Community College*
Sabine Eggleston, *Edison State College*
Dylan Faullin, *Dodge City Community College*
Anne Fischer, *Tulsa Community College Metro Campus*
Rebecca Gubitti, *Edison State College*
Shawna Haider, *Salt Lake Community College*
Exie Hall, *Del Mar College*
Nancy Hixson, *Columbia Southern University*
Stephanie Houdek, *St. Cloud Technical Institute*
Linda Kass, *Bergen Community College*
Chauncey Keaton, *Central Georgia Technical College*
Jamie Kleine, *Southwestern Illinois College*
Edith Lester, *Volunteer State Community College*
Dorothy Marshall, *Edison State College*
Kimberley McHale, *Heartland Community College*
Arda Melkonian, *Victor Valley College*
Christian Miller, *Glendale Community College*
Christine Mirbaha, *Community College of Baltimore County–Dundalk*
Joan Monaghan, *County College of Morris*
Eli Nettles, *Nashville State Community College*
Louise Olshan, *County College of Morris*
Deborah Poetsch, *County College of Morris*
Thomas Pulver, *Waubonsee Community College*
Renee Quick, *Wallace State Community College–Hanceville*
Nimisha Raval, *Central Georgia Technical College*
Chris Schultz, *Iowa State University*
Richard Semmler, *Northern Virginia Community College*
Jane Serbousek, *Northern Virginia Community College*
Alissa Sustarsic, *Lake Sumter Community College*
Alexis Thurman, *County College of Morris*
Michelle Van Wagoner, *Nashville State Community College*
Melanie Walker, *Bergen Community College*

The endless hours of hard work by Jane Hoover, Martha Morong, and Geri Davis have led to products of which we are immensely proud. We also want to thank Judy Penna for writing the Student's and Instructor's Solutions Manuals and for her strong leadership in the preparation of the printed supplements and video lectures. Other strong support has come from Laurie Hurley for the *Instructor's Resource Manual* and for accuracy checking, along with checkers Laurie Hurley, Joanne Koratich, Holly Martinez, Judy Penna, and Mike Rosenborg, and from proofreaders David Johnson and Monroe Street. Michelle Lanosga assisted with applications research. We also wish to recognize Nelson Carter, Tom Atwater, Judy Penna, and Laurie Hurley who prepared the videos.

In addition, a number of people at Pearson have contributed in special ways to the development and production of this textbook, including the Developmental Math team: Project Manager Ron Hampton, Senior Designer Barbara Atkinson, Editorial Assistant Chase Hammond, and Associate Producer Jonathan Wooding. Executive Editor Cathy Cantin and Marketing Manager Rachel Ross encouraged our vision and provided marketing insight.

CHAPTER

1

Whole Numbers

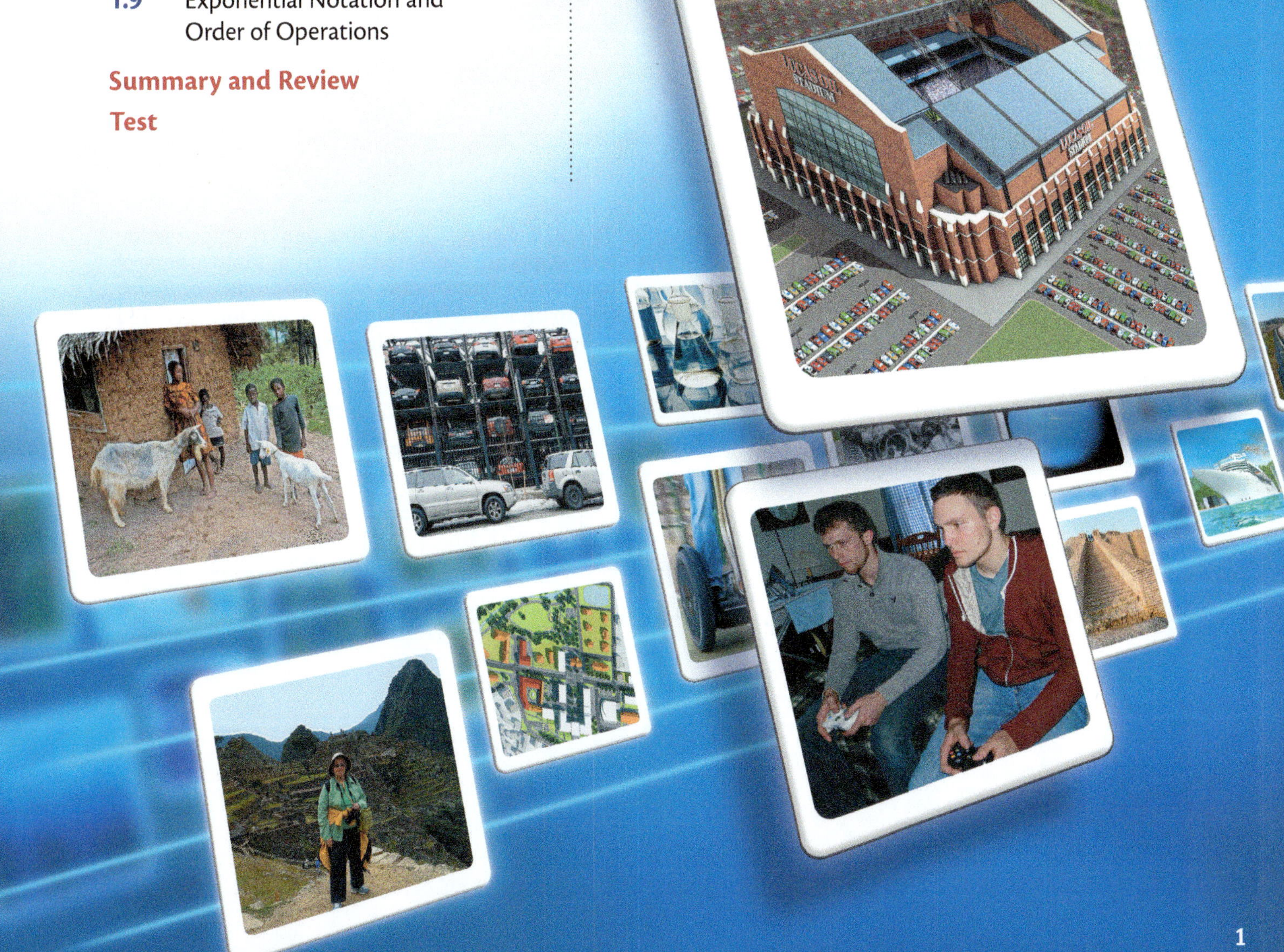

STUDYING FOR SUCCESS *Getting Off to a Good Start*

- ☐ Your syllabus for this course is extremely important. Read it carefully, noting required texts and materials.
- ☐ If there is an online component for your course, register for it as soon as possible.
- ☐ At the front of the text, you will find a Student Organizer card. This pullout card will help you keep track of important dates and useful contact information.

1.1 Standard Notation

OBJECTIVES

- **a** Give the meaning of digits in standard notation.
- **b** Convert from standard notation to expanded notation.
- **c** Convert between standard notation and word names.

We study mathematics in order to be able to solve problems. In this section, we study how numbers are named. We begin with the concept of place value.

a PLACE VALUE

The numbers of jobs available in 2010 for several professions are shown in the following table.

PROFESSION	NUMBER OF JOBS, 2010
Registered nurses	2,737,400
Radiologic technologists	219,900
Radiation therapists	16,900

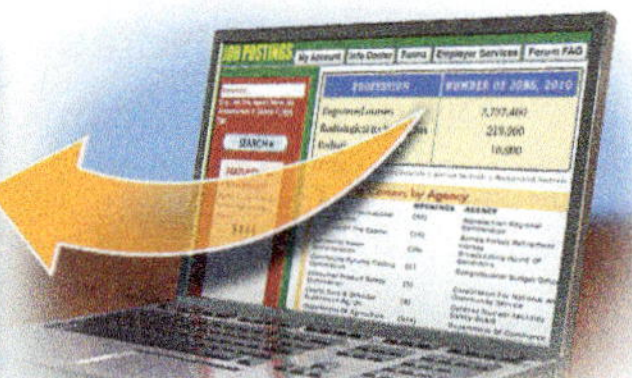

SOURCE: U.S. Department of Labor, Bureau of Labor Statistics

A **digit** is a number 0, 1, 2, 3, 4, 5, 6, 7, 8, or 9 that names a place-value location. For large numbers, digits are separated by commas into groups of three, called **periods**. Each period has a name: *ones, thousands, millions, billions, trillions,* and so on. To understand the number of jobs for registered nurses in the table above, we can use a **place-value chart**, as shown below.

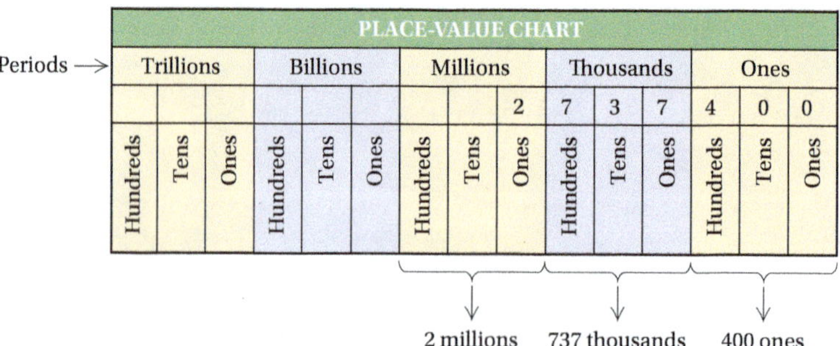

PLACE-VALUE CHART

Periods →	Trillions			Billions			Millions			Thousands			Ones		
									2	7	3	7	4	0	0
	Hundreds	Tens	Ones	Hundreds	Tens	Ones	Hundreds	Tens	Ones	Hundreds	Tens	Ones	Hundreds	Tens	Ones

2 millions 737 thousands 400 ones

EXAMPLES In each of the following numbers, what does the digit 8 mean?

1. 278,342 — 8 thousands
2. 872,342 — 8 hundred thousands
3. 28,343,399,223 — 8 billions
4. 98,413,099 — 8 millions
5. 6328 — 8 ones

Do Margin Exercises 1–6 (in the margin at right). ▶

What does the digit 2 mean in each number?

1. 526,555
2. 265,789
3. 42,789,654
4. 24,789,654
5. 8924
6. 5,643,201

EXAMPLE 6 *Charitable Organizations.* Since its founding in 1881 by Clara Barton, the American Red Cross has been the nation's best-known emergency response organization. As part of a worldwide organization, the American Red Cross also aids victims of devastating natural disasters. For the fiscal year ending June 2011, the total revenue of the American Red Cross was $3,452,960,387. What digit names the number of ten millions?
Source: charitynavigator.org

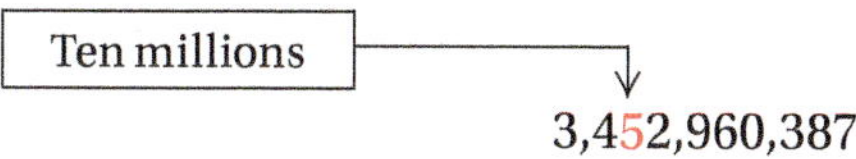

The digit 5 is in the ten millions place, so 5 names the number of ten millions.

Do Exercise 7. ▶

7. *Government Payroll.* In March 2011, the total payroll for all state employees in the United States was $19,971,861,990. What digit names the number of ten billions?
Source: *2011 Annual Survey of Public Employment and Payroll*

b CONVERTING FROM STANDARD NOTATION TO EXPANDED NOTATION

Heifer International is a charitable organization whose mission is to work with communities to end hunger and poverty and care for the earth by providing farm animals to impoverished families around the world. Consider the data in the following table.

GEOGRAPHICAL AREAS OF NEED	NUMBER OF FAMILIES ASSISTED DIRECTLY AND INDIRECTLY BY HEIFER INTERNATIONAL IN 2011
Africa	220,275
Americas	934,871
Asia, South Pacific	407,640
Central and Eastern Europe	344,945

SOURCE: *Heifer International 2011 Annual Report*

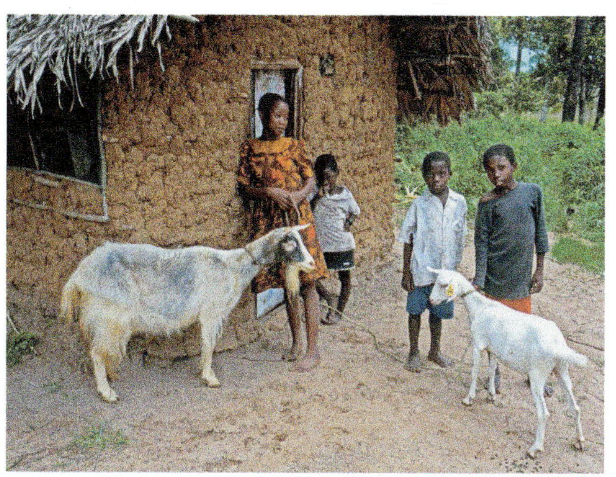

Answers

1. 2 ten thousands 2. 2 hundred thousands 3. 2 millions 4. 2 ten millions 5. 2 tens 6. 2 hundreds 7. 1

The number of families assisted in the Americas was 934,871. This number is expressed in **standard notation**. We write **expanded notation** for 934,871 as follows:

934,871 = 9 hundred thousands + 3 ten thousands
+ 4 thousands + 8 hundreds
+ 7 tens + 1 one.

EXAMPLE 7 Write expanded notation for 1815 ft, the height of the CN Tower in Toronto, Canada.

1815 = 1 thousand + 8 hundreds + 1 ten + 5 ones

EXAMPLE 8 Write expanded notation for 407,640, the number of families in Asia and the South Pacific assisted by Heifer International in 2011.

407,640 = 4 hundred thousands + 0 ten thousands
+ 7 thousands + 6 hundreds + 4 tens + 0 ones

or

4 hundred thousands + 7 thousands + 6 hundreds + 4 tens

Do Exercises 8–11.

Write expanded notation.

8. 2718 mi, the length of the Congo River in Africa **GS**

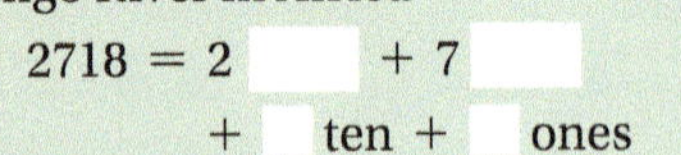

2718 = 2 ____ + 7 ____
+ ____ ten + ____ ones

9. 344,945, the number of families in Central and Eastern Europe assisted by Heifer International in 2011

10. 1670 ft, the height of the Taipei 101 Tower in Taiwan

11. 104,094 square miles, the area of Colorado

C CONVERTING BETWEEN STANDARD NOTATION AND WORD NAMES

We often use **word names** for numbers. When we pronounce a number, we are speaking its word name. Russia won 82 medals in the 2012 Summer Olympics in London, Great Britain. A word name for 82 is "eighty-two." Word names for some two-digit numbers like 36, 51, and 72 use hyphens. Others like that for 17 use only one word, "seventeen."

2012 Summer Olympics Medal Count

COUNTRY	GOLD	SILVER	BRONZE	TOTAL
United States of America	46	29	29	104
People's Republic of China	38	27	23	88
Russia	24	26	32	82
Great Britain	29	17	19	65
Germany	11	19	14	44

SOURCE: espn.go.com

Answers

8. 2 thousands + 7 hundreds + 1 ten + 8 ones
9. 3 hundred thousands + 4 ten thousands + 4 thousands + 9 hundreds + 4 tens + 5 ones
10. 1 thousand + 6 hundreds + 7 tens + 0 ones, or 1 thousand + 6 hundreds + 7 tens
11. 1 hundred thousand + 0 ten thousands + 4 thousands + 0 hundreds + 9 tens + 4 ones, or 1 hundred thousand + 4 thousands + 9 tens + 4 ones

Guided Solution:
8. thousands, hundreds, 1, 8

EXAMPLES Write a word name.

9. 46, the number of gold medals won by the United States

Forty-six

10. 19, the number of silver medals won by Germany

Nineteen

11. 104, the total number of medals won by the United States

One hundred four

Do Exercises 12–14. ▶

Write a word name. (Refer to the chart on the previous page.)

12. 65, the total number of medals won by Great Britain

13. 14, the number of bronze medals won by Germany

14. 38, the number of gold medals won by the People's Republic of China

For word names for larger numbers, we begin at the left with the largest period. The number named in the period is followed by the name of the period; then a comma is written and the next number and period are named. Note that the name of the ones period is not included in the word name for a whole number.

EXAMPLE 12 Write a word name for 46,605,314,732.

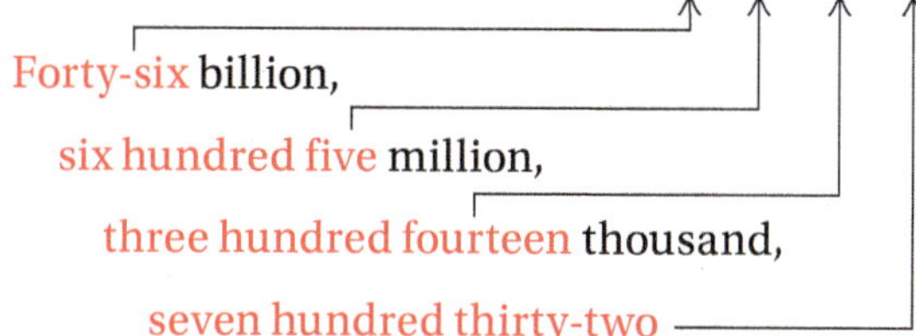

The word "and" *should not* appear in word names for whole numbers. Although we commonly hear such expressions as "two hundred *and* one," the use of "and" is not, strictly speaking, correct in word names for whole numbers. For decimal notation, it is appropriate to use "and" for the decimal point. For example, 317.4 is read as "three hundred seventeen *and* four tenths."

Do Exercises 15–18. ▶

Write a word name.

15. 204

16. $44,640, the average annual wage for athletic trainers in the United States in 2012

Source: U.S. Bureau of Labor Statistics

17. 1,879,204

One ____, eight hundred ____ thousand, two hundred ____

18. 7,052,428,785, the world population in 2012

Source: U.S. Census Bureau

EXAMPLE 13 Write standard notation.

Five hundred six million,
three hundred forty-five thousand,
two hundred twelve

Standard notation is 506,345,212.

Do Exercise 19. ▶

19. Write standard notation.

Two hundred thirteen million, one hundred five thousand, three hundred twenty-nine

Answers

12. Sixty-five **13.** Fourteen **14.** Thirty-eight **15.** Two hundred four **16.** Forty-four thousand, six hundred forty **17.** One million, eight hundred seventy-nine thousand, two hundred four **18.** Seven billion, fifty-two million, four hundred twenty-eight thousand, seven hundred eighty-five **19.** 213,105,329

Guided Solution:
17. Million, seventy-nine, four

1.1 Exercise Set

For Extra Help
MyMathLab®
MathXL® PRACTICE | WATCH | READ | REVIEW

Reading Check

Complete each statement with the correct word from the following list.

digit expanded period standard

RC1. In 983, the ________ 9 represents 9 hundreds.

RC2. In 615,702, the number 615 is in the thousands ________.

RC3. The phrase "3 hundreds + 2 tens + 9 ones" is ________ notation for 329.

RC4. The number 721 is written in ________ notation.

a What does the digit 5 mean in each number?

1. 235,888

2. 253,777

3. 1,488,526

4. 500,736

Movie Receipts. The final movie of the Harry Potter series, *Harry Potter and the Deathly Hallows: Part II*, grossed $1,328,111,219 worldwide.

Source: Nash Information Services, LLC

What digit names the number of:

5. thousands?

6. millions?

7. ten millions?

8. hundred thousands?

b Write expanded notation.

Stair-Climbing Races. The figure below shows the number of stairs in four buildings in which stair-climbing races are held. In Exercises 9–12, write expanded notation for the number of stairs in each race.

Stair-Climbing Races

SOURCE: towerrunning.com

9. 2058 steps in the International Towerthon, Kuala Lumpur, Malaysia

10. 1776 steps in the CN Tower Stair Climb, Toronto, Ontario, Canada

11. 1576 steps in the Empire State Building Run-Up, New York City, New York

12. 1081 steps in the Skytower Vertical Challenge, Auckland, New Zealand

13. 5702 **14.** 3097 **15.** 93,986 **16.** 38,453

Population. The table below shows the populations of four countries in 2012. In Exercises 17–20, write expanded notation for the population of the given country.

Four Most Populous Countries in the World

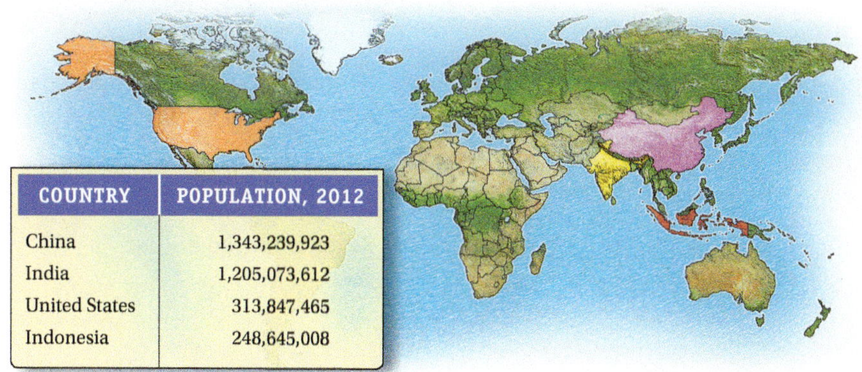

COUNTRY	POPULATION, 2012
China	1,343,239,923
India	1,205,073,612
United States	313,847,465
Indonesia	248,645,008

SOURCE: *CIA World Factbook*

17. 1,343,239,923 for China

18. 1,205,073,612 for India

19. 248,645,008 for Indonesia

20. 313,847,465 for the United States

C Write a word name.

21. 85 **22.** 48 **23.** 88,000 **24.** 45,987

25. 123,765 **26.** 111,013 **27.** 7,754,211,577 **28.** 43,550,651,808

29. *English Language Learners.* In the 2007–2008 academic year, there were 701,799 English language learners in Texas schools. Write a word name for 701,799.

Source: U.S. Department of Education

30. *College Football.* The 2012 Rose Bowl game was attended by 91,245 fans. Write a word name for 91,245.

Source: bizjournals.com

31. *Auto Racing.* Dario Franchitti, winner of the 2012 Indianapolis 500 auto race, won a prize of $2,474,280. Write a word name for 2,474,280.

Source: sbnation.com

32. *Busiest Airport.* In 2010, the world's busiest airport, Hartsfield-Jackson Atlanta International Airport, hosted 89,331,622 passengers. Write a word name for 89,331,622.

Source: Airports Council International

Write each number in standard notation.

33. Six hundred thirty-two thousand, eight hundred ninety-six

34. Three hundred fifty-four thousand, seven hundred two

35. Fifty thousand, three hundred twenty-four

36. Seventeen thousand, one hundred twelve

37. Two million, two hundred thirty-three thousand, eight hundred twelve

38. Nineteen million, six hundred ten thousand, four hundred thirty-nine

39. Eight billion

40. Seven hundred million

41. Forty million

42. Twenty-six billion

43. Thirty million, one hundred three

44. Two hundred thousand, seventeen

Write standard notation for the number in each sentence.

45. ***Pacific Ocean.*** The area of the Pacific Ocean is sixty-four million, one hundred eighty-six thousand square miles.

46. The average distance from the sun to Neptune is two billion, seven hundred ninety-three million miles.

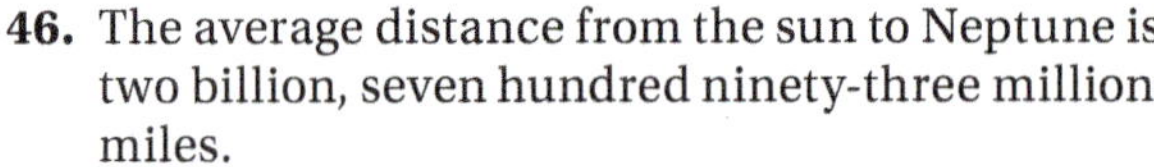

Synthesis

To the student and the instructor: The Synthesis exercises found at the end of every exercise set challenge students to combine concepts or skills studied in the section or in preceding parts of the text. Exercises marked with a 🖩 symbol are meant to be solved using a calculator.

47. How many whole numbers between 100 and 400 contain the digit 2 in their standard notation?

48. 🖩 What is the largest number that you can name on your calculator? How many digits does that number have? How many periods?

Addition

1.2

OBJECTIVES

a Add whole numbers.

b Use addition in finding perimeter.

SKILL TO REVIEW

Objective 1.1a: Give the meaning of digits in standard notation.

In each of the following numbers, what does the digit 4 mean?

1. 8342
2. 14,976

a ADDITION OF WHOLE NUMBERS

To answer questions such as "How many?", "How much?", and "How tall?", we often use whole numbers. The set, or collection, of **whole numbers** is

0, 1, 2, 3, 4, 5, 6, 7, 8, 9, 10, 11, 12, ...

The set goes on indefinitely. There is no largest whole number, and the smallest whole number is 0. Each whole number can be named using various notations. The set 1, 2, 3, 4, 5, . . . , without 0, is called the set of **natural numbers**.

Addition of whole numbers corresponds to combining things.

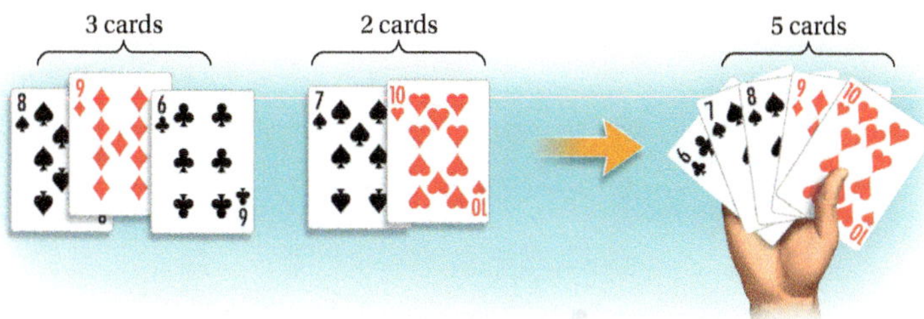

We say that the **sum** of 3 and 2 is 5. The numbers added are called **addends**. The addition that corresponds to the figure above is

$$3 + 2 = 5.$$

3 → Addend, 2 → Addend, 5 → Sum

To add whole numbers, we add the ones digits first, then the tens, then the hundreds, then the thousands, and so on.

EXAMPLE 1 Add: 878 + 995.

Place values are lined up in columns.

$$\begin{array}{r} \overset{1}{8}\ 7\ 8 \\ +\ 9\ 9\ 5 \\ \hline 3 \end{array}$$

Add ones. We get 13 ones, or 1 ten + 3 ones. Write 3 in the ones column and 1 above the tens. This is called *carrying*, or *regrouping*.

$$\begin{array}{r} \overset{1}{8}\ \overset{1}{7}\ 8 \\ +\ 9\ 9\ 5 \\ \hline 7\ 3 \end{array}$$

Add tens. We get 17 tens, so we have 10 tens + 7 tens. This is also 1 hundred + 7 tens. Write 7 in the tens column and 1 above the hundreds.

$$\begin{array}{r} \overset{1}{8}\ \overset{1}{7}\ 8 \\ 9\ 9\ 5 \\ \hline 1\ 8\ 7\ 3 \end{array}$$

Add hundreds. We get 18 hundreds.

We show you these steps for explanation. You need write only this.

$$\begin{array}{r} \overset{1}{8}\ \overset{1}{7}\ 8 \\ +\ 9\ 9\ 5 \\ \hline 1\ 8\ 7\ 3 \end{array}$$

878 and 995 ← Addends; 1873 ← Sum

Answers

Skill to Review:

1. 4 tens
2. 4 thousands

TO THE INSTRUCTOR AND THE STUDENT

This section presents a review of addition of whole numbers. Students who are successful should go on to Section 1.3. Those who have trouble should study developmental unit A near the back of this text and then repeat Section 1.2.

How do we perform an addition of three numbers, like $2 + 3 + 6$? We could do it by adding 3 and 6, and then 2. We can show this with parentheses:

$$2 + (3 + 6) = 2 + 9 = 11.$$ Parentheses tell what to do first.

We could also add 2 and 3, and then 6:

$$(2 + 3) + 6 = 5 + 6 = 11.$$

Either way the result is 11. It does not matter how we group the numbers. This illustrates the **associative law of addition**, $a + (b + c) = (a + b) + c$. We can also add whole numbers in any order. That is, $2 + 3 = 3 + 2$. This illustrates the **commutative law of addition**, $a + b = b + a$. Together, the commutative and associative laws tell us that to add more than two numbers, we can use any order and grouping we wish. Adding 0 to a number does not change the number: $a + 0 = 0 + a = a$. That is, $6 + 0 = 0 + 6 = 6$, or $198 + 0 = 0 + 198 = 198$. We say that 0 is the **additive identity**.

EXAMPLE 2 Add: $391 + 1276 + 789 + 5498$.

```
      2
    3 9 1
  1 2 7 6
    7 8 9
+ 5 4 9 8
        4
```

Add ones. We get 24, so we have 2 tens + 4 ones. Write 4 in the ones column and 2 above the tens.

```
    3 2
    3 9 1
  1 2 7 6
    7 8 9
+ 5 4 9 8
      5 4
```

Add tens. We get 35 tens, so we have 30 tens + 5 tens. This is also 3 hundreds + 5 tens. Write 5 in the tens column and 3 above the hundreds.

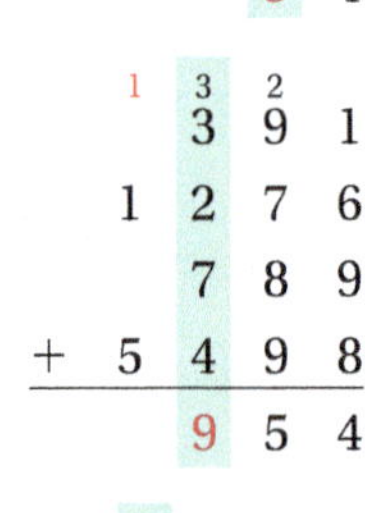

Add hundreds. We get 19 hundreds, so we have 10 hundreds + 9 hundreds. This is also 1 thousand + 9 hundreds. Write 9 in the hundreds column and 1 above the thousands.

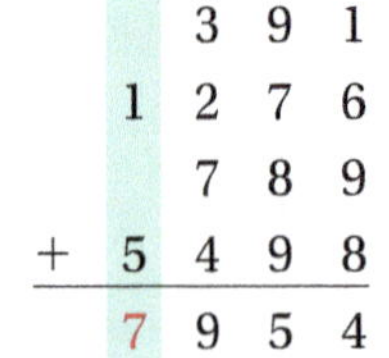

Add thousands. We get 7 thousands.

◀ Do Exercises 1–4.

Add.

1. $6203 + 3542$

2.
```
  7 9 6 8
+ 5 4 9 7
```

GS
```
  1 1 □
  7 9 6 8
+ 5 4 9 7
□ □,4 □ 5
```

3.
```
  9 8 0 4
+ 6 3 7 8
```

4.
```
  1 9 3 2
  6 7 2 3
  9 8 7 8
+ 8 9 4 1
```

Answers

1. 9745 **2.** 13,465 **3.** 16,182 **4.** 27,474

Guided Solution:

2.
```
  1 1 1
  7 9 6 8
+ 5 4 9 7
 13,4 6 5
```

CALCULATOR CORNER

Adding Whole Numbers This is the first of a series of *optional* discussions on using a calculator. A calculator is *not* a requirement for this textbook. Check with your instructor about whether you are allowed to use a calculator in the course.

There are many kinds of calculators and different instructions for their usage. Be sure to consult your user's manual.

To add whole numbers on a calculator, we use the [+] and [=] keys. After we press [=], the sum appears on the display.

EXERCISES Use a calculator to find each sum.

1. $73 + 48$ **2.** $925 + 677$ **3.** $826 + 415 + 691$ **4.** $253 + 490 + 121$

b FINDING PERIMETER

Addition can be used when finding perimeter.

PERIMETER

The distance around an object is its **perimeter**.

EXAMPLE 3 Find the perimeter of the figure.

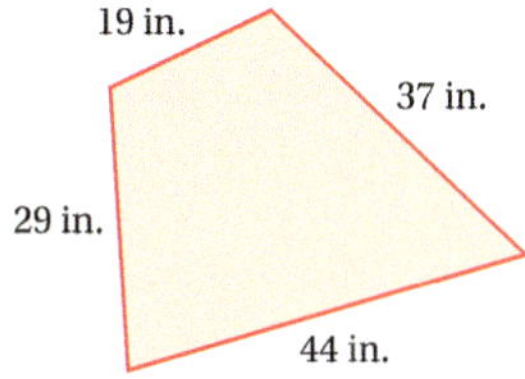

We add the lengths of the sides:

Perimeter = 29 in. + 19 in. + 37 in. + 44 in.
= 129 in.

The perimeter of the figure is 129 in. (inches).

Do Exercises 5 and 6. ▶

EXAMPLE 4 Lucas Oil Stadium in Indianapolis has a unique retractable roof. When the roof is opened (retracted) in good weather to create an open-air stadium, the opening approximates a rectangle 588 ft long and 300 ft wide. Find the perimeter of the opening.

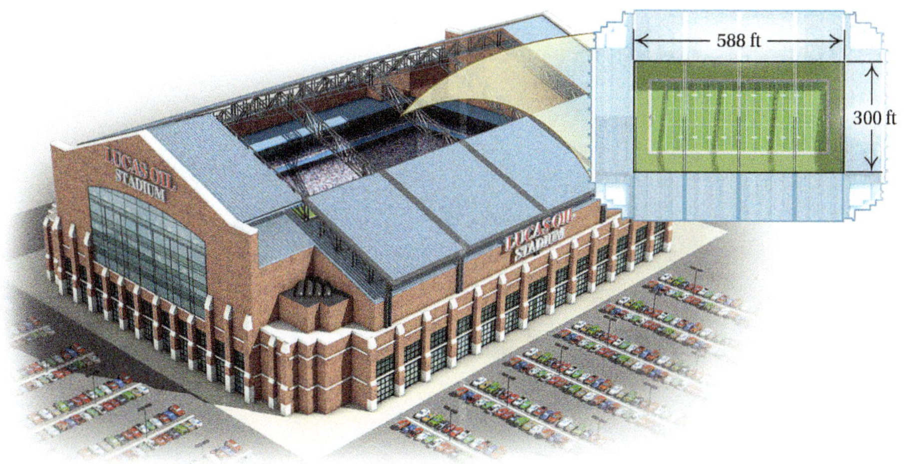

Opposite sides of a rectangle have equal lengths, so this rectangle has two sides of length 588 ft and two sides of length 300 ft.

Perimeter = 588 ft + 300 ft + 588 ft + 300 ft
= 1776 ft

The perimeter of the opening is 1776 ft.

Do Exercise 7. ▶

Find the perimeter of each figure.

GS **5.**

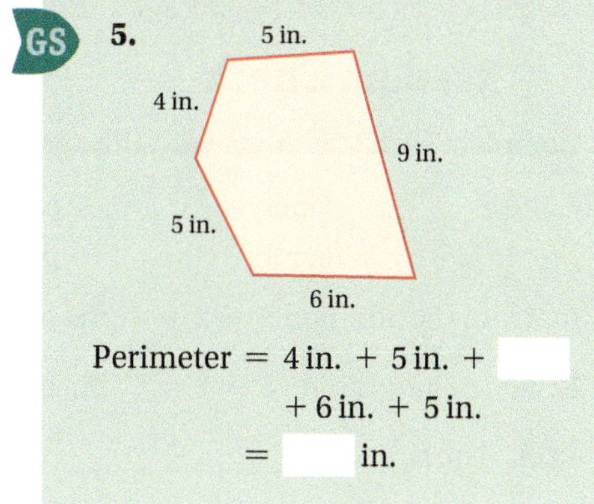

Perimeter = 4 in. + 5 in. + ____ + 6 in. + 5 in.
= ____ in.

6.

16 ft

15 ft 15 ft

16 ft

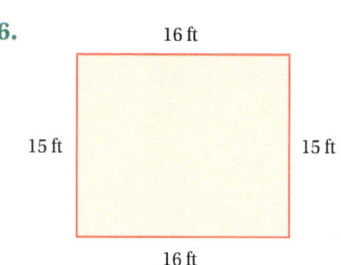

7. ***Index Cards.*** Two standard sizes for index cards are 3 in. by 5 in. and 5 in. by 8 in. Find the perimeter of each type of card.

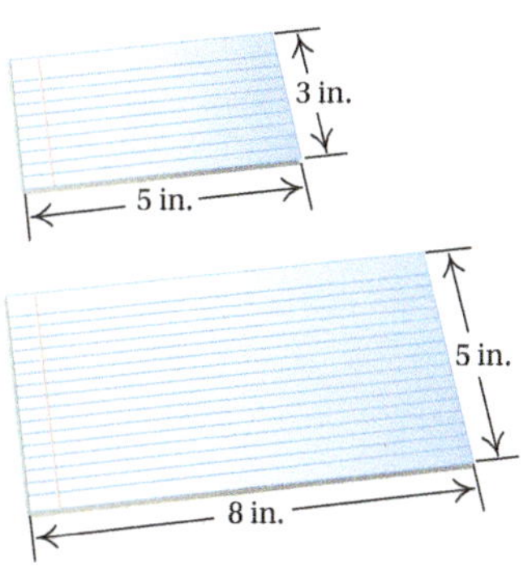

Answers

5. 29 in. **6.** 62 ft **7.** 16 in.; 26 in.

Guided Solution:
5. 9 in., 29

1.2 Exercise Set

For Extra Help MyMathLab® MathXL® PRACTICE WATCH READ REVIEW

Reading Check

Complete each statement with the appropriate word or number from the following list. Not every choice will be used.

0	addends	law	product
1	factors	perimeter	sum

RC1. In the addition $5 + 2 = 7$, the numbers 5 and 2 are ________.

RC2. In the addition $5 + 2 = 7$, the number 7 is the ________.

RC3. The sum of ________ and any number a is a.

RC4. The distance around an object is its ________.

a Add.

1. $364 + 23$

2. $1521 + 348$

3. $86 + 78$

4. $73 + 69$

5. $1716 + 3482$

6. $7503 + 2683$

7. $99 + 1$

8. $999 + 11$

9. $8113 + 390$

10. $271 + 3338$

11. $356 + 4910$

12. $280 + 34{,}902$

13. $3870 + 92 + 7 + 497$

14. $10{,}120 + 12{,}989 + 5738$

15. $4825 + 1783$

16. $3654 + 2700$

17. $23{,}443 + 10{,}989$

18. $45{,}879 + 21{,}786$

19. $77{,}543 + 23{,}767$

20. $99{,}999 + 112$

21. $45 + 25 + 36 + 44 + 80$

22. $38 + 27 + 32 + 14 + 76$

23. $12{,}070 + 2{,}954 + 3{,}400$

24. $42{,}487 + 83{,}141 + 36{,}712$

25. $4835 + 729 + 9204 + 8986 + 7931$

26. $989 + 566 + 834 + 920 + 703$

b Find the perimeter of each figure.

27.

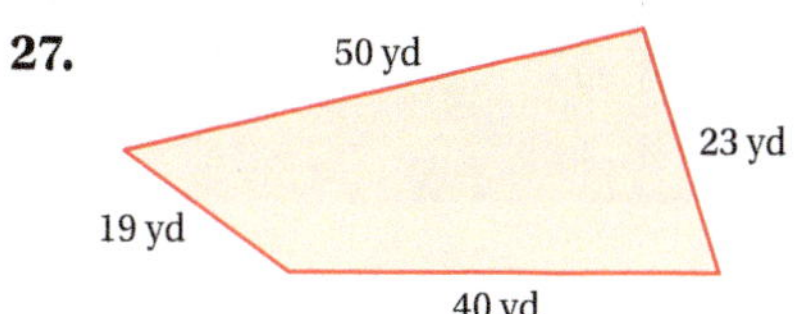

28.

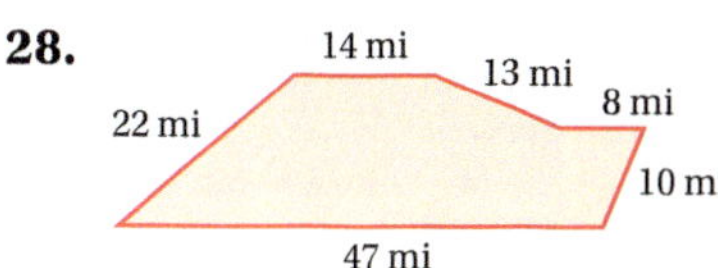

29.

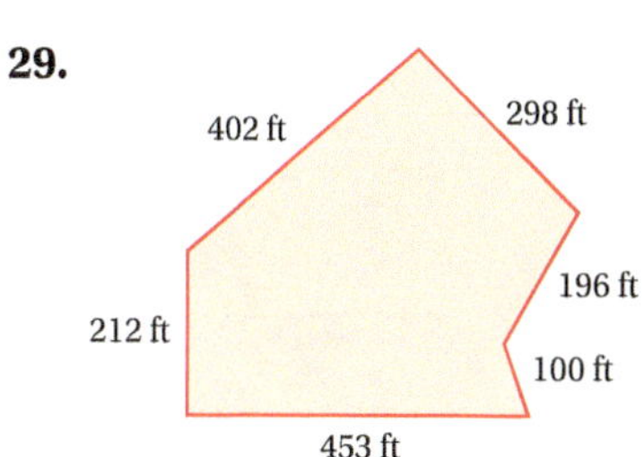

30.

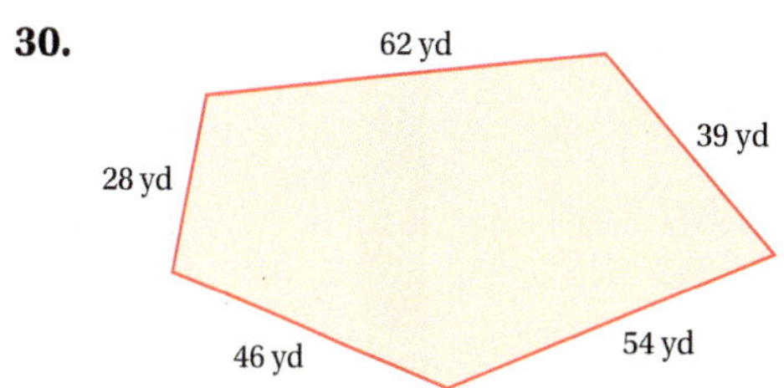

31. Find the perimeter of a standard hockey rink.

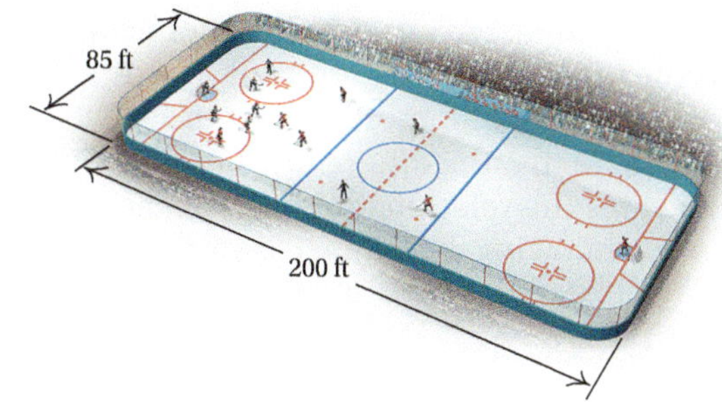

32. In Major League Baseball, how far does a batter travel when circling the bases after hitting a home run?

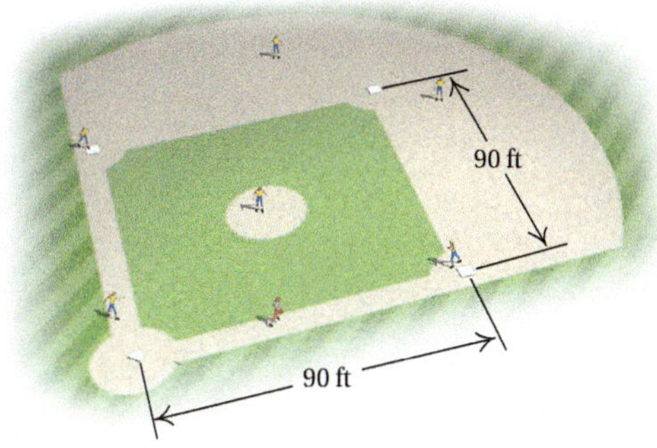

Skill Maintenance

The exercises that follow begin an important feature called *Skill Maintenance exercises.* These exercises provide an ongoing review of topics previously covered in the book. You will see them in virtually every exercise set. It has been found that this kind of continuing review can significantly improve your performance on a final examination.

33. What does the digit 8 mean in 486,205? [1.1a]

34. The population of the world is projected to be 9,346,399,468 in 2050. Write a word name for 9,346,399,468. [1.1c]

Source: U.S. Census Bureau

Synthesis

35. A fast way to add all the numbers from 1 to 10 inclusive is to pair 1 with 9, 2 with 8, and so on. Use a similar approach to add all numbers from 1 to 100 inclusive.

1.3 Subtraction

OBJECTIVE

a Subtract whole numbers.

SKILL TO REVIEW

Objective 1.1a: Give the meaning of digits in standard notation.

Consider the number 328,974.

1. What digit names the number of hundreds?
2. What digit names the number of ones?

a SUBTRACTION OF WHOLE NUMBERS

Subtraction is finding the difference of two numbers. Suppose you purchase 6 tickets for a concert and give 2 to a friend.

The subtraction that represents this situation is

$$6 - 2 = 4.$$

6 → Minuend, 2 → Subtrahend, 4 → Difference

The **minuend** is the number from which another number is being subtracted. The **subtrahend** is the number being subtracted. The **difference** is the result of subtracting the subtrahend from the minuend.

In the subtraction above, note that the difference, 4, is the number we add to 2 to get 6. This illustrates the relationship between addition and subtraction and leads us to the following definition of subtraction.

SUBTRACTION

The difference $a - b$ is that unique whole number c for which $a = c + b$.

We see that $6 - 2 = 4$ because $4 + 2 = 6$.

To subtract whole numbers, we subtract the ones digits first, then the tens digits, then the hundreds, then the thousands, and so on.

EXAMPLE 1 Subtract: $9768 - 4320$.

$$\begin{array}{r} 9\,7\,6\,8 \\ -\,4\,3\,2\,0 \\ \hline 8 \end{array}$$ Subtract ones.

$$\begin{array}{r} 9\,7\,6\,8 \\ -\,4\,3\,2\,0 \\ \hline 4\,8 \end{array}$$ Subtract tens.

$$\begin{array}{r} 9\,7\,6\,8 \\ -\,4\,3\,2\,0 \\ \hline 4\,4\,8 \end{array}$$ Subtract hundreds.

$$\begin{array}{r} 9\,7\,6\,8 \\ -\,4\,3\,2\,0 \\ \hline 5\,4\,4\,8 \end{array}$$ Subtract thousands.

We show these steps for explanation. You need write only this.

$$\begin{array}{r} 9\,7\,6\,8 \\ -\,4\,3\,2\,0 \\ \hline 5\,4\,4\,8 \end{array}$$

Answers

Skill to Review:
1. 9 2. 4

Because subtraction is defined in terms of addition, we can use addition to *check* subtraction.

Subtraction:

```
  9 7 6 8
− 4 3 2 0
  5 4 4 8
```

Check by Addition:

```
  5 4 4 8
+ 4 3 2 0
  9 7 6 8
```

?

Do Exercise 1. ▶

EXAMPLE 2 Subtract: 348 − 165.

We have

$$\begin{aligned} &3 \text{ hundreds} + 4 \text{ tens} + 8 \text{ ones} = 2 \text{ hundreds} + 14 \text{ tens} + 8 \text{ ones} \\ &-1 \text{ hundred} - 6 \text{ tens} - 5 \text{ ones} = -1 \text{ hundred} - 6 \text{ tens} - 5 \text{ ones} \\ &\qquad = 1 \text{ hundred} + 8 \text{ tens} + 3 \text{ ones} \\ &\qquad = 183. \end{aligned}$$

First, we subtract the ones.

```
  3 4 8
− 1 6 5
      3
```

Subtract ones.

We cannot subtract the tens because there is no whole number that when added to 6 gives 4. To complete the subtraction, we must *borrow* 1 hundred from 3 hundreds and regroup it with the 4 tens. Then we can do the subtraction 14 tens − 6 tens = 8 tens.

```
  2 14
  3 4 8
− 1 6 5
      3
```

Borrow one hundred. That is, 1 hundred = 10 tens, and 10 tens + 4 tens = 14 tens. Write 2 above the hundreds column and 14 above the tens.

```
  2 14
  3 4 8
− 1 6 5
  1 8 3
```

Subtract tens; subtract hundreds.

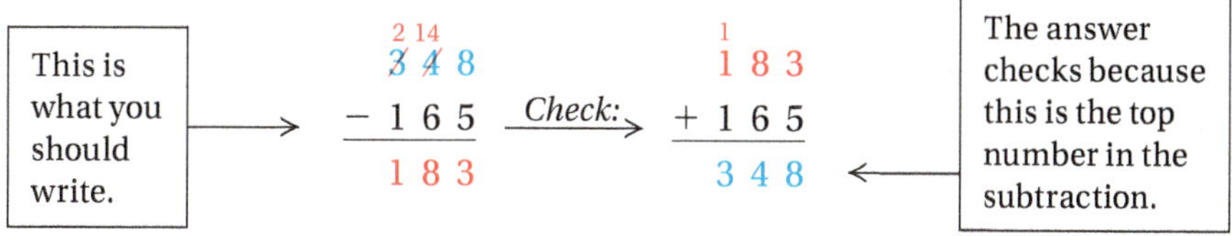

This is what you should write.
2 14
3 4 8
− 1 6 5
1 8 3
Check:
1
1 8 3
+ 1 6 5
3 4 8
The answer checks because this is the top number in the subtraction.

GS

1. Subtract. Check by adding.

```
  7 8 9 3
− 4 0 9 2
```

```
  7 8 9 3
− 4 0 9 2
  _ 8 _ 1
```

Check:

```
  _ _ _ _
+ 4 0 9 2
  _ _ _ _
```

CALCULATOR CORNER

Subtracting Whole Numbers To subtract whole numbers on a calculator, we use the [−] and [=] keys.

EXERCISES Use a calculator to perform each subtraction. Check by adding.

1. 57 − 29
2. 81 − 34
3. 145 − 78
4. 612 − 493
5. 4976 − 2848
6. 12,406 − 9813

Answer

1. 3801

Guided Solution:

1. 3, 0; 3801, 7893

TO THE INSTRUCTOR AND THE STUDENT

This section presents a review of subtraction of whole numbers. Students who are successful should go on to Section 1.4. Those who have trouble should study developmental unit S near the back of this text and then repeat Section 1.3.

EXAMPLE 3 Subtract: $6246 - 1879$.

$$\begin{array}{r} \scriptstyle 3\ 16 \\ 6\,2\,\cancel{4}\,\cancel{6} \\ -\,1\,8\,7\,9 \\ \hline 7 \end{array}$$

We cannot subtract 9 ones from 6 ones, but we can subtract 9 ones from 16 ones. We borrow 1 ten to get 16 ones.

$$\begin{array}{r} \scriptstyle 13 \\ \scriptstyle 1\ \ 3\ 16 \\ 6\,\cancel{2}\,\cancel{4}\,\cancel{6} \\ -\,1\,8\,7\,9 \\ \hline 6\,7 \end{array}$$

We cannot subtract 7 tens from 3 tens, but we can subtract 7 tens from 13 tens. We borrow 1 hundred to get 13 tens.

$$\begin{array}{r} \scriptstyle 11\,13 \\ \scriptstyle 5\ \ 1\ \ 3\ 16 \\ \cancel{6}\,\cancel{2}\,\cancel{4}\,\cancel{6} \\ -\,1\,8\,7\,9 \\ \hline 4\,3\,6\,7 \end{array}$$

We cannot subtract 8 hundreds from 1 hundred, but we can subtract 8 hundreds from 11 hundreds. We borrow 1 thousand to get 11 hundreds. Finally, we subtract the thousands.

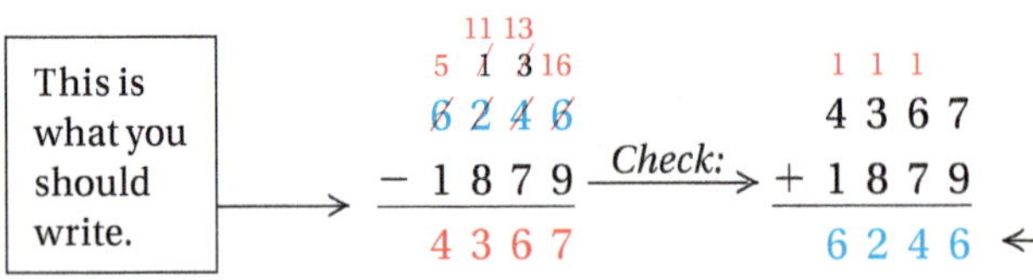
This is what you should write.
11 13
5 1 3 16
6 2 4 6
− 1 8 7 9
4 3 6 7
Check:
1 1 1
4 3 6 7
+ 1 8 7 9
6 2 4 6
The answer checks because this is the top number in the subtraction.

Subtract. Check by adding.

2. $\begin{array}{r} 8\,6\,8\,6 \\ -\,2\,3\,5\,8 \\ \hline \end{array}$ **3.** $\begin{array}{r} 7\,1\,4\,5 \\ -\,2\,3\,9\,8 \\ \hline \end{array}$

◀ **Do Exercises 2 and 3.**

EXAMPLE 4 Subtract: $902 - 477$.

$$\begin{array}{r} \scriptstyle 8\ \ 9\ 12 \\ \cancel{9}\,\cancel{0}\,\cancel{2} \\ -\,4\,7\,7 \\ \hline 4\,2\,5 \end{array}$$

We cannot subtract 7 ones from 2 ones. We have 9 hundreds, or 90 tens. We borrow 1 ten to get 12 ones. We then have 89 tens.

Subtract.

4. $\begin{array}{r} 7\,0 \\ -\,1\,4 \\ \hline \end{array}$

5. $\begin{array}{r} 5\,0\,3 \\ -\,2\,9\,8 \\ \hline \end{array}$

GS

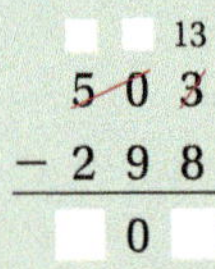

◀ **Do Exercises 4 and 5.**

EXAMPLE 5 Subtract: $8003 - 3667$.

$$\begin{array}{r} \scriptstyle 7\ \ 9\ \ 9\ 13 \\ \cancel{8}\,\cancel{0}\,\cancel{0}\,\cancel{3} \\ -\,3\,6\,6\,7 \\ \hline 4\,3\,3\,6 \end{array}$$

We have 8 thousands, or 800 tens. We borrow 1 ten to get 13 ones. We then have 799 tens.

EXAMPLES

6. Subtract: $6000 - 3762$.

$$\begin{array}{r} \scriptstyle 5\ \ 9\ \ 9\ 10 \\ \cancel{6}\,\cancel{0}\,\cancel{0}\,\cancel{0} \\ -\,3\,7\,6\,2 \\ \hline 2\,2\,3\,8 \end{array}$$

7. Subtract: $6024 - 2968$.

$$\begin{array}{r} \scriptstyle 11 \\ \scriptstyle 5\ \ 9\ \ 1\ 14 \\ \cancel{6}\,\cancel{0}\,\cancel{2}\,\cancel{4} \\ -\,2\,9\,6\,8 \\ \hline 3\,0\,5\,6 \end{array}$$

Subtract.

6. $\begin{array}{r} 7\,0\,0\,7 \\ -\,6\,3\,4\,9 \\ \hline \end{array}$ **7.** $\begin{array}{r} 6\,0\,0\,0 \\ -\,3\,1\,4\,9 \\ \hline \end{array}$

8. $\begin{array}{r} 9\,0\,3\,5 \\ -\,7\,4\,8\,9 \\ \hline \end{array}$ **9.** $\begin{array}{r} 2\,0\,0\,1 \\ -\,1\,2\,4 \\ \hline \end{array}$

◀ **Do Exercises 6–9.**

Answers

2. 6328 **3.** 4747 **4.** 56 **5.** 205
6. 658 **7.** 2851 **8.** 1546 **9.** 1877

Guided Solution:

5. $\begin{array}{r} \scriptstyle 4\ \ 9\ 13 \\ \cancel{5}\,\cancel{0}\,\cancel{3} \\ -\,2\,9\,8 \\ \hline 2\,0\,5 \end{array}$

1.3 Exercise Set

For Extra Help MyMathLab® MathXL® PRACTICE WATCH READ REVIEW

Reading Check

Match each word or phrase from the following list with the indicated part of the subtraction sentence.

difference minuend subtraction symbol subtrahend

RC1. A ______________

RC2. B ______________

RC3. C ______________

RC4. D ______________

(A) 97 (B) − (C) 51 = (D) 26

a Subtract. Check by adding.

1. $\begin{array}{r} 65 \\ -21 \\ \hline \end{array}$

2. $\begin{array}{r} 87 \\ -34 \\ \hline \end{array}$

3. $\begin{array}{r} 866 \\ -333 \\ \hline \end{array}$

4. $\begin{array}{r} 526 \\ -323 \\ \hline \end{array}$

5. 86 − 47

6. 73 − 28

7. 51 − 37

8. 64 − 19

9. $\begin{array}{r} 563 \\ -194 \\ \hline \end{array}$

10. $\begin{array}{r} 795 \\ -398 \\ \hline \end{array}$

11. $\begin{array}{r} 391 \\ -365 \\ \hline \end{array}$

12. $\begin{array}{r} 316 \\ -247 \\ \hline \end{array}$

13. 981 − 747

14. 887 − 698

15. 683 − 266

16. 342 − 217

17. $\begin{array}{r} 7769 \\ -2387 \\ \hline \end{array}$

18. $\begin{array}{r} 6431 \\ -2896 \\ \hline \end{array}$

19. $\begin{array}{r} 4512 \\ -1734 \\ \hline \end{array}$

20. $\begin{array}{r} 8364 \\ -5375 \\ \hline \end{array}$

21. 5318 − 2249

22. 9241 − 5643

23. 3947 − 2858

24. 7583 − 3641

25. $\begin{array}{r} 12,647 \\ -\ 4,899 \\ \hline \end{array}$

26. $\begin{array}{r} 16,222 \\ -\ 5,888 \\ \hline \end{array}$

27. $\begin{array}{r} 51,342 \\ -\ 47,198 \\ \hline \end{array}$

28. $\begin{array}{r} 32,194 \\ -\ 29,236 \\ \hline \end{array}$

29. $\begin{array}{r} 80 \\ -\ 24 \\ \hline \end{array}$

30. $\begin{array}{r} 90 \\ -\ 78 \\ \hline \end{array}$

31. $\begin{array}{r} 690 \\ -\ 236 \\ \hline \end{array}$

32. $\begin{array}{r} 803 \\ -\ 418 \\ \hline \end{array}$

33. $\begin{array}{r} 6808 \\ -\ 3059 \\ \hline \end{array}$

34. $\begin{array}{r} 9405 \\ -\ 258 \\ \hline \end{array}$

35. $\begin{array}{r} 2300 \\ -\ 109 \\ \hline \end{array}$

36. $\begin{array}{r} 7500 \\ -\ 3604 \\ \hline \end{array}$

37. 90,237 − 47,209

38. 84,703 − 298

39. 101,734 − 5760

40. 15,017 − 7809

41. $\begin{array}{r} 6007 \\ -\ 1589 \\ \hline \end{array}$

42. $\begin{array}{r} 8003 \\ -\ 599 \\ \hline \end{array}$

43. $\begin{array}{r} 39,000 \\ -\ 37,695 \\ \hline \end{array}$

44. $\begin{array}{r} 17,000 \\ -\ 11,598 \\ \hline \end{array}$

45. 10,008 − 19

46. 40,006 − 147

47. 50,001 − 1984

48. 30,004 − 6749

Skill Maintenance

Add. [1.2a]

49. 567 + 778

50. 901 + 23

51. 12,885 + 9807

52. 9909 + 1011

53. Write a word name for 6,375,602. [1.1c]

54. Write expanded notation for 9103. [1.1b]

Synthesis

55. Fill in the missing digits to make the subtraction true:

9,□48,621 − 2,097,□81 = 7,251,140.

56. 🖩 Subtract: 3,928,124 − 1,098,947.

Multiplication

1.4

OBJECTIVES

a Multiply whole numbers.

b Use multiplication in finding area.

SKILL TO REVIEW

Objective 1.2a: Add whole numbers.

Add.

1. $\begin{array}{r} 347 \\ +556 \\ \hline \end{array}$

2. $\begin{array}{r} 1835 \\ +4619 \\ \hline \end{array}$

a MULTIPLICATION OF WHOLE NUMBERS

Repeated Addition

The multiplication 3×5 corresponds to this repeated addition.

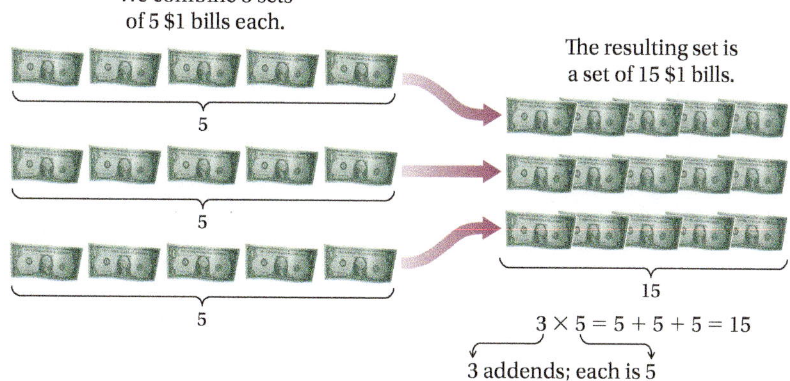
We combine 3 sets
of 5 $1 bills each.
5
5
5
The resulting set is
a set of 15 $1 bills.
15
3 × 5 = 5 + 5 + 5 = 15
3 addends; each is 5

The numbers that we multiply are called **factors**. The result of the multiplication is called a **product**.

$$3 \times 5 = 15$$

Factor Factor Product

Rectangular Arrays

Multiplications can also be thought of as rectangular arrays. Each of the following corresponds to the multiplication 3×5.

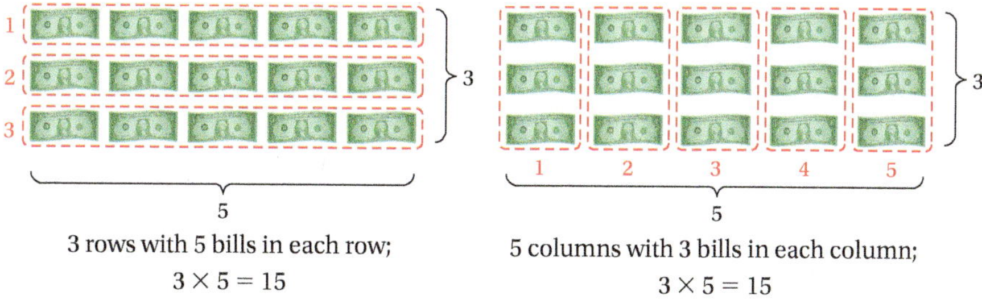
1
2
3
3
5
1 2 3 4 5
3
5

3 rows with 5 bills in each row;
$3 \times 5 = 15$

5 columns with 3 bills in each column;
$3 \times 5 = 15$

When you write a multiplication corresponding to a real-world situation, you should think of either a rectangular array or repeated addition. In some cases, it may help to think both ways.

We have used an "×" to denote multiplication. A dot " · " is also commonly used. (Use of the dot is attributed to the German mathematician Gottfried Wilhelm von Leibniz over three centuries ago.) Parentheses are also used to denote multiplication. For example,

$$3 \times 5 = 3 \cdot 5 = (3)(5) = 3(5) = 15.$$

Answers

Skill to Review:
1. 903 2. 6454

TO THE INSTRUCTOR AND THE STUDENT

This section presents a review of multiplication of whole numbers. Students who are successful should go on to Section 1.5. Those who have trouble should study developmental unit M near the back of this text and then repeat Section 1.4.

The product of 0 and any whole number is 0: $0 \cdot a = a \cdot 0 = 0$. For example, $0 \cdot 3 = 3 \cdot 0 = 0$. Multiplying a number by 1 does not change the number: $1 \cdot a = a \cdot 1 = a$. For example, $1 \cdot 3 = 3 \cdot 1 = 3$. We say that 1 is the **multiplicative identity**.

EXAMPLE 1 Multiply: 5×734.

We have

$$\begin{array}{r l} 7\,3\,4 & \\ \times \quad\; 5 & \\ \hline 2\,0 & \leftarrow \text{Multiply the 4 ones by 5: } 5 \times 4 = 20. \\ 1\,5\,0 & \leftarrow \text{Multiply the 3 tens by 5: } 5 \times 30 = 150. \\ 3\,5\,0\,0 & \leftarrow \text{Multiply the 7 hundreds by 5: } 5 \times 700 = 3500. \\ \hline 3\,6\,7\,0 & \leftarrow \text{Add.} \end{array}$$

Instead of writing each product on a separate line, we can use a shorter form.

$$\begin{array}{r} {}^{2} \\ 7\;3\;4 \\ \times \quad\; 5 \\ \hline 0 \end{array}$$

Multiply the 4 ones by 5: $5 \cdot (4 \text{ ones}) = 20$ ones $= 2$ tens $+ 0$ ones. Write 0 in the ones column and 2 above the tens.

$$\begin{array}{r} {}^{1}\;{}^{2}\;\; \\ 7\;3\;4 \\ \times \quad\; 5 \\ \hline 7\;0 \end{array}$$

Multiply the 3 tens by 5 and add 2 tens: $5 \cdot (3 \text{ tens}) = 15$ tens, 15 tens $+ 2$ tens $= 17$ tens $= 1$ hundred $+ 7$ tens. Write 7 in the tens column and 1 above the hundreds.

$$\begin{array}{r} {}^{1}\;{}^{2}\;\; \\ 7\;3\;4 \\ \times \quad\; 5 \\ \hline 3\;6\;7\;0 \end{array}$$

Multiply the 7 hundreds by 5 and add 1 hundred: $5 \cdot (7 \text{ hundreds}) = 35$ hundreds, 35 hundreds $+ 1$ hundred $= 36$ hundreds.

$$\begin{array}{r} {}^{1}\;{}^{2}\;\; \\ 7\;3\;4 \\ \times \quad\; 5 \\ \hline 3\;6\;7\;0 \end{array}$$

You should write only this.

Do Exercises 1–4.

Multiply.

1. $\begin{array}{r} 5\,8 \\ \times\; 2 \\ \hline \end{array}$

2. $\begin{array}{r} 3\,7 \\ \times\; 4 \\ \hline \end{array}$

3. $\begin{array}{r} 8\,2\,3 \\ \times \quad 6 \\ \hline \end{array}$

4. $\begin{array}{r} 1\,3\,4\,8 \\ \times \quad\; 5 \\ \hline \end{array}$

GS

$$\begin{array}{r} \square\;{}^{2}\;\square\;\; \\ 1\;3\;4\;8 \\ \times \qquad 5 \\ \hline \square\;7\;\square\;0 \end{array}$$

Multiplication of whole numbers is based on a property called the **distributive law**. It says that to multiply a number by a sum, $a \cdot (b + c)$, we can multiply each addend by a and then add like this: $(a \cdot b) + (a \cdot c)$. Thus, $a \cdot (b + c) = (a \cdot b) + (a \cdot c)$. For example, consider the following.

$4 \cdot (2 + 3) = 4 \cdot 5 = 20$ Adding first; then multiplying

$4 \cdot (2 + 3) = (4 \cdot 2) + (4 \cdot 3) = 8 + 12 = 20$ Multiplying first; then adding

The results are the same, so $4 \cdot (2 + 3) = (4 \cdot 2) + (4 \cdot 3)$.

Let's find the product 51×32. Since $32 = 2 + 30$, we can think of this product as

$$51 \times 32 = 51 \times (2 + 30) = (51 \times 2) + (51 \times 30).$$

That is, we multiply 51 by 2, then we multiply 51 by 30, and finally we add. We can write our work in columns.

Answers

1. 116 **2.** 148 **3.** 4938 **4.** 6740

Guided Solution:

4. $\begin{array}{r} {}^{1}\;{}^{2}\;{}^{4}\;\; \\ 1\;3\;4\;8 \\ \times \qquad 5 \\ \hline 6\;7\;4\;0 \end{array}$

```
      5 1
    × 3 2
    1 0 2     Multiplying by 2
  1 5 3 0     Multiplying by 30. (We write a 0 and then multiply 51 by 3.)
```

You may have learned that such a 0 need not be written. You may omit it if you wish. If you do omit it, remember, when multiplying by tens, to start writing the answer in the tens place.

We add to obtain the product.

```
      5 1
    × 3 2
    1 0 2
  1 5 3 0
  1 6 3 2     Adding to obtain the product
```

EXAMPLE 2 Multiply: 457 × 683.

```
      5 2
      6 8 3
    × 4 5 7
    4 7 8 1     Multiplying 683 by 7

      4 1
      5 2
      6 8 3
    × 4 5 7
    4 7 8 1
  3 4 1 5 0     Multiplying 683 by 50

    3 1
    4 1
    5 2
    6 8 3
  × 4 5 7
    4 7 8 1
  3 4 1 5 0
2 7 3 2 0 0     Multiplying 683 by 400. (We write 00 and then multiply 683 by 4.)
3 1 2,1 3 1     Adding
```

Do Exercises 5–8.

EXAMPLE 3 Multiply: 306 × 274.

Note that 306 = 3 hundreds + 6 ones.

```
    2 7 4
  × 3 0 6     Multiplying by 6
  1 6 4 4     Multiplying by 3 hundreds. (We write 00
8 2 2 0 0     and then multiply 274 by 3.)
8 3,8 4 4     Adding
```

Do Exercises 9–12.

CALCULATOR CORNER

Multiplying Whole Numbers To multiply whole numbers on a calculator, we use the [×] and [=] keys.

EXERCISES Use a calculator to find each product.

1. 56 × 8
2. 845 × 26
3. 5 · 1276
4. 126(314)
5. 3760 × 48
6. 5218 × 453

Multiply.

5. 45 × 23
6. 48 × 63
7. 746 × 62
8. 245 × 837

Multiply.

9. 472 × 306
10. 408 × 704
11. 2344 × 6005
12. 1006 × 703

Answers

5. 1035 6. 3024 7. 46,252 8. 205,065 9. 144,432 10. 287,232 11. 14,075,720 12. 707,218

EXAMPLE 4 Multiply: 360×274.

Note that 360 = 3 hundreds + 6 tens.

$$\begin{array}{r} 274 \\ \times\ 360 \\ \hline 16440 \\ 82200 \\ \hline 98{,}640 \end{array}$$

Multiplying by 6 tens. (We write 0 and then multiply 274 by 6.)

Multiplying by 3 hundreds. (We write 00 and then multiply 274 by 3.)

Adding

Multiply.

13. $\begin{array}{r} 472 \\ \times\ 830 \\ \hline \end{array}$ 14. $\begin{array}{r} 2344 \\ \times\ 7400 \\ \hline \end{array}$

15. 100×562 16. 1000×562

◀ **Do Exercises 13–16.**

When we multiply two numbers, we can change the order of the numbers without changing their product. For example, $3 \cdot 6 = 18$ and $6 \cdot 3 = 18$. This illustrates the **commutative law of multiplication:** $a \cdot b = b \cdot a$.

17. a) Find $23 \cdot 47$.
b) Find $47 \cdot 23$.
c) Compare your answers to parts (a) and (b).

◀ **Do Exercise 17.**

To multiply three or more numbers, we group them so that we multiply two at a time. Consider $2 \cdot 3 \cdot 4$. We can group these numbers as $2 \cdot (3 \cdot 4)$ or as $(2 \cdot 3) \cdot 4$. The parentheses tell what to do first:

$2 \cdot (3 \cdot 4) = 2 \cdot (12) = 24$. We multiply 3 and 4 and then that product by 2.

We can also multiply 2 and 3 and then that product by 4:

$$(2 \cdot 3) \cdot 4 = (6) \cdot 4 = 24.$$

Either way we get 24. It does not matter how we group the numbers. This illustrates the **associative law of multiplication:** $a \cdot (b \cdot c) = (a \cdot b) \cdot c$.

Multiply.

18. $5 \cdot 2 \cdot 4$

19. $4 \cdot 2 \cdot 6$

◀ **Do Exercises 18 and 19.**

b FINDING AREA

We can think of the area of a rectangular region as the number of square units needed to fill it. Here is a rectangle 4 cm (centimeters) long and 3 cm wide. It takes 12 square centimeters (sq cm) to fill it.

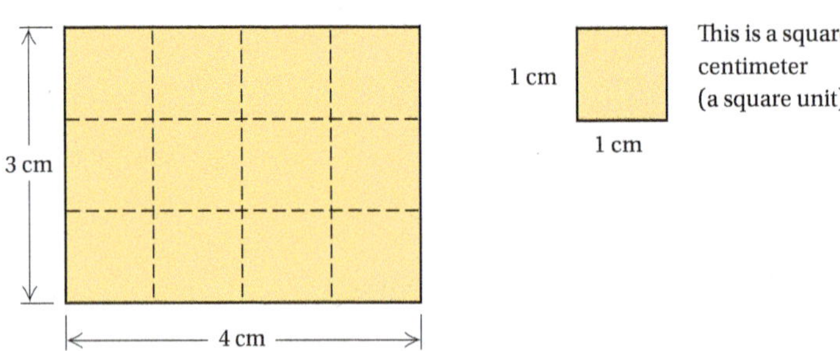

In this case, we have a rectangular array of 3 rows, each of which contains 4 squares. The number of square units is given by $3 \cdot 4$, or 12. That is, $A = l \cdot w = 3\text{ cm} \cdot 4\text{ cm} = 12$ sq cm.

Answers

13. 391,760 **14.** 17,345,600
15. 56,200 **16.** 562,000
17. **(a)** 1081; **(b)** 1081; **(c)** same
18. 40 **19.** 48

EXAMPLE 5 *Professional Pool Table.* The playing area of a pool table used in professional tournaments is 50 in. by 100 in. (There are 6-in. wide rails on the outside that are not included in the playing area.) Determine the playing area.

If we think of filling the rectangle with square inches, we have a rectangular array. The length $l = 100$ in. and the width $w = 50$ in. Thus the area A is given by the formula

$$A = l \cdot w = 100 \text{ in.} \cdot 50 \text{ in.} = 5000 \text{ sq in.}$$

Do Exercise 20.

GS **20. Painting a Room.** Ben and Elizabeth plan to paint one wall of a bedroom in a dark accent color. The wall is a rectangle 12 ft long and 8 ft high. Determine the area of the wall.

$$A = l \cdot w$$
$$= 12 \text{ ft} \cdot \underline{\quad}$$
$$= \underline{\quad} \text{ sq ft}$$

Answer

20. 96 sq ft

Guided Solution:

20. 8 ft, 96

1.4 Exercise Set

For Extra Help

MyMathLab® MathXL® PRACTICE WATCH READ REVIEW

Reading Check

Complete each statement with the appropriate word or number from the following list. Not every choice will be used.

0	addends	product
1	factors	sum

RC1. In the multiplication $4 \times 3 = 12$, 4 and 3 are ________________.

RC2. In the multiplication $4 \times 3 = 12$, 12 is the ________________.

RC3. The product of ________________ and any number a is 0.

RC4. The product of ________________ and any number a is a.

a Multiply.

1. $\begin{array}{r} 65 \\ \times \quad 8 \\ \hline \end{array}$

2. $\begin{array}{r} 87 \\ \times \quad 4 \\ \hline \end{array}$

3. $\begin{array}{r} 94 \\ \times \quad 6 \\ \hline \end{array}$

4. $\begin{array}{r} 76 \\ \times \quad 9 \\ \hline \end{array}$

5. $3 \cdot 509$

6. $7 \cdot 806$

7. $7(9229)$

8. $4(7867)$

9. $90(53)$

10. $60(78)$

11. $(47)(85)$

12. $(34)(87)$

13. $\begin{array}{r} 87 \\ \times 10 \\ \hline \end{array}$

14. $\begin{array}{r} 2340 \\ \times 1000 \\ \hline \end{array}$

15. $\begin{array}{r} 96 \\ \times 20 \\ \hline \end{array}$

16. $\begin{array}{r} 800 \\ \times 700 \\ \hline \end{array}$

17. $\begin{array}{r} 643 \\ \times \quad 72 \\ \hline \end{array}$

18. $\begin{array}{r} 777 \\ \times \quad 77 \\ \hline \end{array}$

19. $\begin{array}{r} 444 \\ \times \quad 33 \\ \hline \end{array}$

20. $\begin{array}{r} 549 \\ \times \quad 88 \\ \hline \end{array}$

21. $\begin{array}{r} 564 \\ \times 458 \\ \hline \end{array}$

22. $\begin{array}{r} 432 \\ \times 375 \\ \hline \end{array}$

23. $\begin{array}{r} 853 \\ \times 936 \\ \hline \end{array}$

24. $\begin{array}{r} 346 \\ \times 659 \\ \hline \end{array}$

25. $\begin{array}{r} 6428 \\ \times 3224 \\ \hline \end{array}$

26. $\begin{array}{r} 8928 \\ \times 3172 \\ \hline \end{array}$

27. $\begin{array}{r} 3482 \\ \times \quad 104 \\ \hline \end{array}$

28. $\begin{array}{r} 6408 \\ \times 6064 \\ \hline \end{array}$

29. $\begin{array}{r} 876 \\ \times 345 \\ \hline \end{array}$

30. $\begin{array}{r} 355 \\ \times 299 \\ \hline \end{array}$

31. $\begin{array}{r} 7889 \\ \times 6224 \\ \hline \end{array}$

32. $\begin{array}{r} 6521 \\ \times 3449 \\ \hline \end{array}$

33. $\begin{array}{r} 5608 \\ \times\ 4500 \\ \hline \end{array}$

34. $\begin{array}{r} 4506 \\ \times\ 7800 \\ \hline \end{array}$

35. $\begin{array}{r} 5006 \\ \times\ 4008 \\ \hline \end{array}$

36. $\begin{array}{r} 6009 \\ \times\ 2003 \\ \hline \end{array}$

b Find the area of each region.

37.

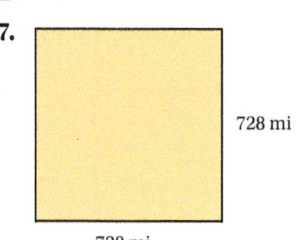

728 mi

728 mi

38.

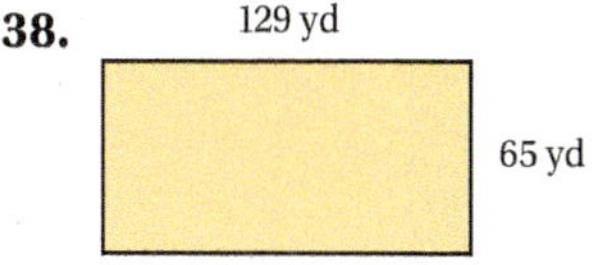

39. Find the area of the region formed by the base lines on a Major League Baseball diamond.

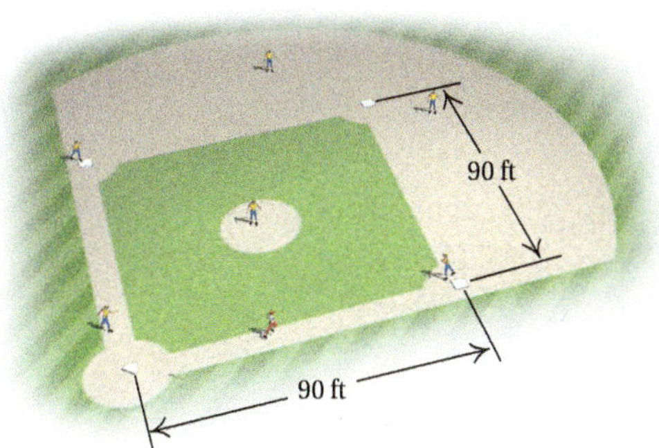

40. Find the area of a standard-sized hockey rink.

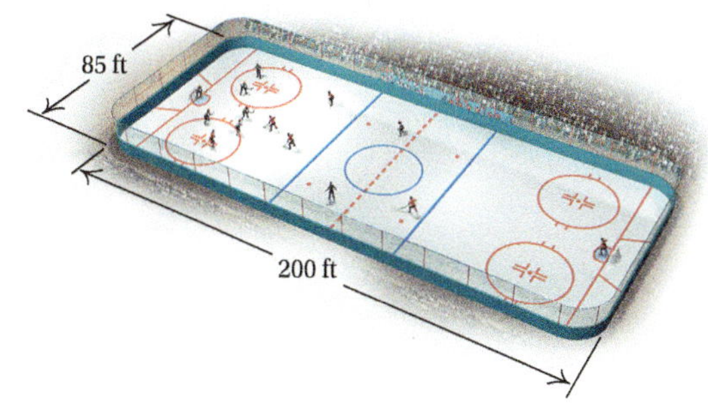

Skill Maintenance

Add. [1.2a]

41. $\begin{array}{r} 4908 \\ 5667 \\ +\ 2110 \\ \hline \end{array}$

42. $\begin{array}{r} 9876 \\ 876 \\ 76 \\ +\quad 6 \\ \hline \end{array}$

Subtract. [1.3a]

43. $\begin{array}{r} 9876 \\ -\ \ 987 \\ \hline \end{array}$

44. $\begin{array}{r} 340{,}798 \\ -\ \ 86{,}679 \\ \hline \end{array}$

45. What does the digit 4 mean in 9,482,157? [1.1a]

46. What digit in 38,026 names the number of hundreds? [1.1a]

47. Write expanded notation for 12,847. [1.1b]

48. Write a word name for 7,432,000. [1.1c]

Synthesis

49. 🖩 An 18-story office building is box-shaped. Each floor measures 172 ft by 84 ft with a 20-ft by 35-ft rectangular area lost to an elevator and a stairwell. How much area is available as office space?

1.5 Division

OBJECTIVE

a Divide whole numbers.

SKILL TO REVIEW

Objective 1.3a:
Subtract whole numbers.

Subtract.

1. $\begin{array}{r} 564 \\ -\ 397 \\ \hline \end{array}$ 2. $\begin{array}{r} 7035 \\ -\ 2944 \\ \hline \end{array}$

a DIVISION OF WHOLE NUMBERS

Repeated Subtraction

Division of whole numbers applies to two kinds of situations. The first is repeated subtraction. Suppose we have 20 doughnuts, and we want to find out how many sets of 5 there are. One way to do this is to repeatedly subtract sets of 5 as follows.

How many sets of 5 doughnuts each?

Since there are 4 sets of 5 doughnuts each, we have

$$\underset{\text{Dividend}}{20} \div \underset{\text{Divisor}}{5} = \underset{\text{Quotient}}{4}.$$

The division 20 ÷ 5 is read "20 divided by 5." The **dividend** is 20, the **divisor** is 5, and the **quotient** is 4. We divide the *dividend* by the *divisor* to get the *quotient.* We can also express the division 20 ÷ 5 = 4 as

$$\frac{20}{5} = 4 \quad \text{or} \quad 5\overline{)20}\text{ with quotient } 4.$$

Rectangular Arrays

We can also think of division in terms of rectangular arrays. Consider again the 20 doughnuts and division by 5. We can arrange the doughnuts in a rectangular array with 5 rows and ask, "How many are in each row?"

We can also consider a rectangular array with 5 doughnuts in each column and ask, "How many columns are there?" The answer is still 4.

In each case, we are asking, "What do we multiply 5 by in order to get 20?"

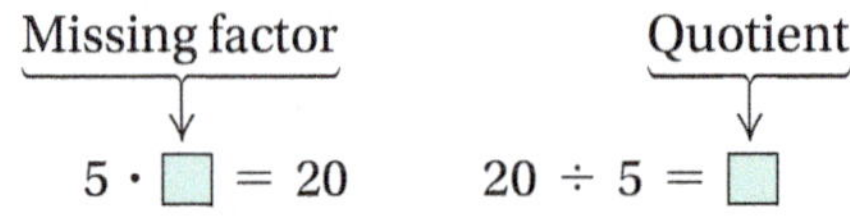

This leads us to the following definition of division.

Answers

Skill to Review:
1. 167 2. 4091

DIVISION

The quotient $a \div b$, where b is not 0, is that unique number c for which $a = b \cdot c$.

This definition shows the relation between division and multiplication. We see, for instance, that

$$20 \div 5 = 4 \text{ because } 20 = 5 \cdot 4.$$

This relation allows us to use multiplication to check division.

EXAMPLE 1 Divide. Check by multiplying.

a) $16 \div 8$ **b)** $\frac{36}{4}$ **c)** $7\overline{)56}$

We do so as follows.

a) $16 \div 8 = 2$ *Check*: $8 \cdot 2 = 16$.

b) $\frac{36}{4} = 9$ *Check*: $4 \cdot 9 = 36$.

c) $7\overline{)56}$ with quotient 8 *Check*: $7 \cdot 8 = 56$.

Do Exercises 1–3.

Divide. Check by multiplying.

1. $9\overline{)45}$
2. $27 \div 3$
3. $\frac{48}{6}$

Let's consider some basic properties of division.

DIVIDING BY 1

Any number divided by 1 is that same number: $a \div 1 = \frac{a}{1} = a$.

For example, $6 \div 1 = 6$ and $\frac{15}{1} = 15$.

DIVIDING A NUMBER BY ITSELF

Any nonzero number divided by itself is 1: $a \div a = \frac{a}{a} = 1, \quad a \neq 0$.

For example, $7 \div 7 = 1$ and $\frac{22}{22} = 1$.

DIVIDENDS OF 0

Zero divided by any nonzero number is 0: $0 \div a = \frac{0}{a} = 0, \quad a \neq 0$.

For example, $0 \div 14 = 0$ and $\frac{0}{3} = 0$.

Do Exercises 4–7.

Divide.

4. $\frac{9}{9}$
5. $\frac{8}{1}$
6. $\frac{0}{12}$
7. $\frac{28}{28}$

Answers

1. 5 2. 9 3. 8 4. 1 5. 8 6. 0 7. 1

EXCLUDING DIVISION BY 0

Division by 0 is not defined: $a \div 0$, or $\frac{a}{0}$, is **not defined**.

For example, $16 \div 0$, or $\frac{16}{0}$, is not defined.

8. Divide if possible: $0 \div 2$. If not possible, write "not defined."

$0 \div 2$ means ___ divided by ___.

Since zero divided by any nonzero number is 0, $0 \div 2 =$ ___.

9. Divide if possible: $7 \div 0$. If not possible, write "not defined."

$7 \div 0$ means ___ divided by ___.

Since division by 0 is not defined, $7 \div 0$ is ___.

Why can't we divide by 0? Suppose the number 4 could be divided by 0. Then if $\square$ were the answer, we would have

$$4 \div 0 = \square,$$

and since 0 times any number is 0, we would have

$$4 = \square \cdot 0 = 0. \quad \text{False!}$$

Thus, the only possible number that could be divided by 0 would be 0 itself. But such a division would give us any number we wish. For instance,

$$\left.\begin{aligned} 0 \div 0 &= 8 \quad \text{because} \quad 0 = 8 \cdot 0; \\ 0 \div 0 &= 3 \quad \text{because} \quad 0 = 3 \cdot 0; \\ 0 \div 0 &= 7 \quad \text{because} \quad 0 = 7 \cdot 0. \end{aligned}\right\} \quad \text{All true!}$$

We avoid the preceding difficulties by agreeing to exclude division by 0.

Do Exercises 8–9.

Division with a Remainder

Suppose everyone in a group of 22 people wants to ride a roller coaster. If each car on the ride holds 6 people, the group will fill 3 cars and there will be 4 people left over.

We can think of this situation as the following division. The people left over are the **remainder**.

$$\begin{array}{r} 3 \leftarrow \text{Quotient} \\ 6\overline{)\,22} \\ \underline{18} \\ 4 \leftarrow \text{Remainder} \end{array}$$

Answers

8. 0 **9.** Not defined

Guided Solutions:

8. 0, 2, 0

9. 7, 0, not defined

We express the result as

$$\underset{\text{Dividend}}{22} \div \underset{\text{Divisor}}{6} = \underset{\text{Quotient}}{3}\,\text{R}\,\underset{\text{Remainder}}{4}.$$

Note that

Quotient · Divisor + Remainder = Dividend.

Thus we have

	$3 \cdot 6 = 18$	Quotient · Divisor
and	$18 + 4 = 22.$	Adding the remainder. The result is the dividend.

We now show a procedure for dividing whole numbers.

EXAMPLE 2 Divide and check: $4\,\overline{)\,3\,4\,5\,7}$.

First, we try to divide the first digit of the dividend, 3, by the divisor, 4. Since $3 \div 4$ is not a whole number, we consider the first *two* digits of the dividend.

```
     8
 4 ) 3 4 5 7
     3 2
       2
```

Since $4 \cdot 8 = 32$ and 32 is smaller than 34, we write an 8 in the quotient above the 4. We also write 32 below 34 and subtract.

What if we had chosen a number other than 8 for the first digit of the quotient? Suppose we had used 7 instead of 8 and subtracted $4 \cdot 7$, or 28, from 34. The result would have been $34 - 28$, or 6. Because 6 is larger than the divisor, 4, we know that there is at least one more factor of 4 in 34, and thus 7 is too small. If we had used 9 instead of 8, then we would have tried to subtract $4 \cdot 9$, or 36, from 34. That difference is not a whole number, so we know 9 is too large. When we subtract, the difference must be smaller than the divisor.

Let's continue dividing.

```
     8 6
 4 ) 3 4 5 7
     3 2 ↓
       2 5
       2 4
         1
```

Now we bring down the 5 in the dividend and consider $25 \div 4$. Since $4 \cdot 6 = 24$ and 24 is smaller than 25, we write 6 in the quotient above the 5. We also write 24 below 25 and subtract. The difference, 1, is smaller than the divisor, so we know that 6 is the correct choice.

```
     8 6 4
 4 ) 3 4 5 7
     3 2   |
       2 5 |
       2 4 ↓
         1 7
         1 6
           1  ← The remainder is 1.
```

We bring down the 7 and consider $17 \div 4$. Since $4 \cdot 4 = 16$ and 16 is smaller than 17, we write 4 in the quotient above the 7. We also write 16 below 17 and subtract.

Check: $864 \cdot 4 = 3456$ and $3456 + 1 = 3457$.

The answer is 864 R 1.

Do Exercises 10–12. ▶

Divide and check.

10. $3\,\overline{)\,2\,3\,9}$

11. $5\,\overline{)\,5\,8\,6\,4}$

12. $6\,\overline{)\,3\,8\,5\,5}$

Answers

10. 79 R 2 **11.** 1172 R 4 **12.** 642 R 3

EXAMPLE 3 Divide: 8904 ÷ 42.

Because 42 is close to 40, we think of the divisor as 40 when we make our choices of digits in the quotient.

```
        2
42 )8 9 0 4   <— Think: 89 ÷ 40. We try 2. Multiply 42 · 2 and
    8 4 ↓        subtract. Then bring down the 0.
      5 0
```

```
      2 1
42 )8 9 0 4
    8 4
      5 0     <— Think: 50 ÷ 40. We try 1. Multiply 42 · 1 and
      4 2 ↓      subtract. Then bring down the 4.
        8 4
```

```
      2 1 2
42 )8 9 0 4
    8 4
      5 0
      4 2
        8 4   <— Think: 84 ÷ 40. We try 2. Multiply 2 · 42 and subtract.
        8 4
          0
```

The remainder is 0, so the answer is 212.

◀ **Do Exercises 13 and 14.**

Divide.

13. 45)6030

14. 52)3288

Caution!

Be careful to keep the digits lined up correctly when you divide.

CALCULATOR CORNER

Dividing Whole Numbers To divide whole numbers on a calculator, we use the [÷] and [=] keys.

When we enter 453 ÷ 15, the display reads [30.2]. Note that the result is not a whole number. This tells us that there is a remainder. The number 30.2 is expressed in decimal notation. The symbol "." is called a decimal point. Although it is possible to use the number to the right of the decimal point to find the remainder, we will not do so here.

EXERCISES Use a calculator to perform each division.

1. 19)532

2. 7)861

3. 9367 ÷ 29

4. 12,276 ÷ 341

Answers

13. 134 **14.** 63 R 12

Zeros in Quotients

EXAMPLE 4 Divide: 6341 ÷ 7.

```
     9
 7 )6 3 4 1
    6 3 ↓
        4
```

← *Think:* 63 ÷ 7 = 9. The first digit in the quotient is 9. We do not write the 0 when we find 63 − 63. Bring down the 4.

```
     9 0
 7 )6 3 4 1
    6 3   ↓
        4 1
```

← *Think:* 4 ÷ 7. If we subtract a group of 7's, such as 7, 14, 21, etc., from 4, we do not get a whole number, so the next digit in the quotient is 0. Bring down the 1.

```
     9 0 5
 7 )6 3 4 1
    6 3
        4 1
        3 5
          6
```

41 ← *Think:* 41 ÷ 7. We try 5. Multiply 7 · 5 and subtract.

6 ← The remainder is 6.

The answer is 905 R 6.

Do Exercises 15 and 16. ▶

EXAMPLE 5 Divide: 8169 ÷ 34.

Because 34 is close to 30, we think of the divisor as 30 when we make our choices of digits in the quotient.

```
         2
 3 4 )8 1 6 9
      6 8 ↓
      1 3 6
```

← *Think:* 81 ÷ 30. We try 2. Multiply 34 · 2 and subtract. Then bring down the 6.

```
         2 4
 3 4 )8 1 6 9
      6 8
      1 3 6
      1 3 6 ↓
            9
```

136 ← *Think:* 136 ÷ 30. We try 4. Multiply 34 · 4 and subtract. The difference is 0, so we do not write it. Bring down the 9.

```
         2 4 0
 3 4 )8 1 6 9
      6 8
      1 3 6
      1 3 6 ↓
            9
            0
            9
```

9 ← *Think:* 9 ÷ 34. If we subtract a group of 34's, such as 34 or 68, from 9, we do not get a whole number, so the last digit in the quotient is 0.

9 ← The remainder is 9.

The answer is 240 R 9.

Do Exercises 17 and 18. ▶

TO THE INSTRUCTOR AND THE STUDENT

This section presents a review of division of whole numbers. Students who are successful should go on to Section 1.6. Those who have trouble should study developmental unit D near the back of this text and then repeat Section 1.5.

Divide.

15. $6\,\overline{)\,4846}$

16. $7\,\overline{)\,7616}$

Divide.

17. $27\,\overline{)\,9724}$

18. $56\,\overline{)\,44,847}$

Answers

15. 807 R 4 **16.** 1088
17. 360 R 4 **18.** 800 R 47

1.5 Exercise Set

For Extra Help
MyMathLab®

MathXL® PRACTICE  WATCH  READ  REVIEW 

✓ Reading Check

Match each word from the following list with the indicated part of the division.

dividend   divisor   quotient   remainder

RC1. A ______________

RC2. B ______________

RC3. C ______________

RC4. D ______________

$$
\begin{array}{r}
29 \leftarrow (A) \\
(D) \rightarrow 8\overline{)235} \leftarrow (B) \\
\underline{16} \\
75 \\
\underline{72} \\
3 \leftarrow (C)
\end{array}
$$

a Divide, if possible. If not possible, write "not defined."

1. $72 \div 6$

2. $54 \div 9$

3. $\frac{23}{23}$

4. $\frac{37}{37}$

5. $22 \div 1$

6. $\frac{56}{1}$

7. $\frac{0}{7}$

8. $\frac{0}{32}$

9. $\frac{16}{0}$

10. $74 \div 0$

11. $\frac{48}{8}$

12. $\frac{20}{4}$

Divide.

13. $277 \div 5$

14. $699 \div 3$

15. $864 \div 8$

16. $869 \div 8$

17. $4\overline{)1228}$

18. $3\overline{)2124}$

19. $6\overline{)4521}$

20. $9\overline{)9110}$

21. $297 \div 4$

22. $389 \div 2$

23. $738 \div 8$

24. $881 \div 6$

25. $5\overline{)8515}$

26. $3\overline{)6027}$

27. $9\overline{)8888}$

28. $8\overline{)4139}$

29. 127,000 ÷ 10

30. 127,000 ÷ 100

31. 127,000 ÷ 1000

32. 4260 ÷ 10

33. $70\overline{)3692}$

34. $20\overline{)5798}$

35. $30\overline{)875}$

36. $40\overline{)987}$

37. 852 ÷ 21

38. 942 ÷ 23

39. $85\overline{)7672}$

40. $54\overline{)2729}$

41. $111\overline{)3219}$

42. $102\overline{)5612}$

43. $8\overline{)843}$

44. $7\overline{)749}$

45. $5\overline{)8047}$

46. $9\overline{)7273}$

47. $5\overline{)5036}$

48. $7\overline{)7074}$

49. 1058 ÷ 46

50. 7242 ÷ 24

51. 3425 ÷ 32

52. $48\overline{)4899}$

53. $24\overline{)8880}$ **54.** $36\overline{)7563}$ **55.** $28\overline{)17{,}067}$ **56.** $36\overline{)28{,}929}$

57. $80\overline{)24{,}320}$ **58.** $90\overline{)88{,}560}$ **59.** $285\overline{)999{,}999}$

60. $306\overline{)888{,}888}$ **61.** $456\overline{)3{,}679{,}920}$ **62.** $803\overline{)5{,}622{,}606}$

Skill Maintenance

Subtract. [1.3a]

63. $4908 - 3667$

64. $88{,}777 - 22{,}333$

Multiply. [1.4a]

65. 198×100

66. 268×35

Use the following figure for Exercises 67 and 68.

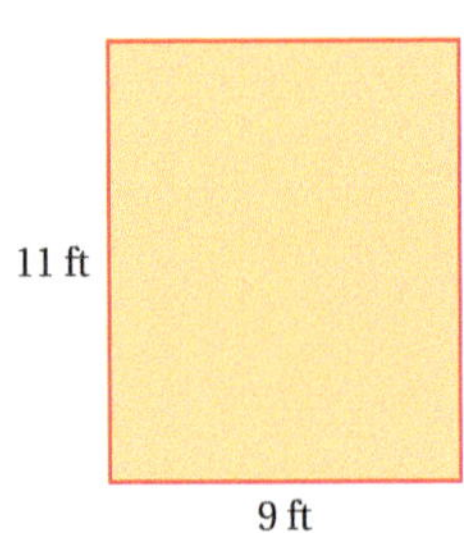

67. Find the perimeter of the figure. [1.2b]

68. Find the area of the figure. [1.4b]

Synthesis

69. Complete the following table.

a	b	$a \cdot b$	$a + b$
	68	3672	
84			117
		32	12

70. Find a pair of factors whose product is 36 and:

a) whose sum is 13.
b) whose difference is 0.
c) whose sum is 20.
d) whose difference is 9.

71. A group of 1231 college students is going to use buses to take a field trip. Each bus can hold 42 students. How many buses are needed?

72. Fill in the missing digits to make the equation true:

$34{,}584{,}132 \div 76\square = 4\square{,}386.$

Mid-Chapter Review

Concept Reinforcement

Determine whether each statement is true or false.

_____________ **1.** If $a - b = c$, then $b = a + c$. [1.3a]

_____________ **2.** We can think of the multiplication 4×3 as a rectangular array containing 4 rows with 3 items in each row. [1.4a]

_____________ **3.** We can think of the multiplication 4×3 as a rectangular array containing 3 columns with 4 items in each column. [1.4a]

_____________ **4.** The product of two whole numbers is always greater than either of the factors. [1.4a]

_____________ **5.** Zero divided by any nonzero number is 0. [1.5a]

_____________ **6.** Any number divided by 1 is the number 1. [1.5a]

Guided Solutions

GS Fill in each blank with the number that creates a correct statement or solution.

7. Write a word name for 95,406,237. [1.1c]

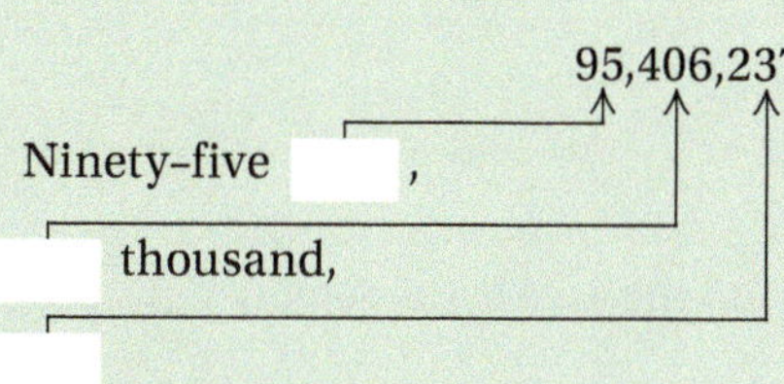

8. Subtract: $604 - 497$. [1.3a]

```
   □ □ □
   6 0 4
 - 4 9 7
 -------
   □ □ □
```

Mixed Review

In each of the following numbers, what does the digit 6 mean? [1.1a]

9. 2698

10. 61,204

11. 146,237

12. 586

Consider the number 306,458,129. What digit names the number of: [1.1a]

13. tens?

14. millions?

15. ten thousands?

16. hundreds?

Write expanded notation. [1.1b]

17. 5602

18. 69,345

Write a word name. [1.1c]

19. 136

20. 64,325

Write standard notation. [1.1c]

21. Three hundred eight thousand, seven hundred sixteen

22. Four million, five hundred sixty-seven thousand, two hundred sixteen

Add. [1.2a]

23. $\begin{array}{r} 316 \\ +\ 482 \\ \hline \end{array}$

24. $\begin{array}{r} 593 \\ +\ 437 \\ \hline \end{array}$

25. $\begin{array}{r} 2638 \\ +\ 5284 \\ \hline \end{array}$

26. $\begin{array}{r} 4617 \\ 2436 \\ +\ 481 \\ \hline \end{array}$

Subtract. [1.3a]

27. $\begin{array}{r} 786 \\ -\ 321 \\ \hline \end{array}$

28. $\begin{array}{r} 624 \\ -\ 285 \\ \hline \end{array}$

29. $\begin{array}{r} 3602 \\ -\ 1748 \\ \hline \end{array}$

30. $\begin{array}{r} 5004 \\ -\ 676 \\ \hline \end{array}$

Multiply. [1.4a]

31. $\begin{array}{r} 36 \\ \times\ 6 \\ \hline \end{array}$

32. $\begin{array}{r} 567 \\ \times\ 28 \\ \hline \end{array}$

33. $\begin{array}{r} 407 \\ \times\ 325 \\ \hline \end{array}$

34. $\begin{array}{r} 9435 \\ \times\ 602 \\ \hline \end{array}$

Divide. [1.5a]

35. $4\overline{)1012}$

36. $38\overline{)4261}$

37. $60\overline{)1399}$

38. $56\overline{)8095}$

39. Find the perimeter of the figure (m stands for "meters"). [1.2b]

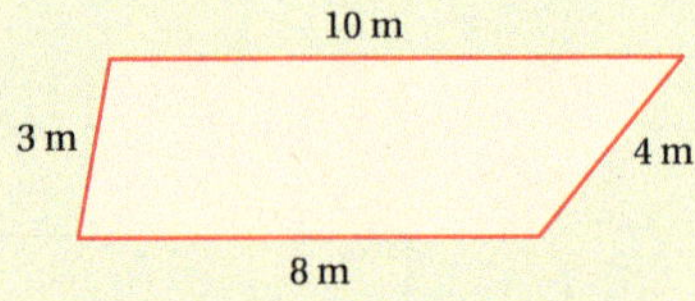

40. Find the area of the region. [1.4b]

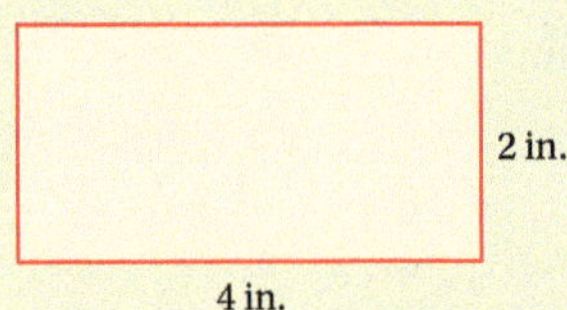

Understanding Through Discussion and Writing

To the student and the instructor: The Discussion and Writing exercises are meant to be answered with one or more sentences. They can be discussed and answered collaboratively by the entire class or by small groups.

41. Explain in your own words what the associative law of addition means. [1.2a]

42. Is subtraction commutative? That is, is there a commutative law of subtraction? Why or why not? [1.3a]

43. Describe a situation that corresponds to each multiplication: $4 \cdot \$150$; $\$4 \cdot 150$. [1.4a]

44. Suppose a student asserts that "$0 \div 0 = 0$ because nothing divided by nothing is nothing." Devise an explanation to persuade the student that the assertion is false. [1.5a]

STUDYING FOR SUCCESS *Using the Textbook*

- ☐ Study the step-by-step solutions to the examples, paying attention to the extra explanations shown in red.
- ☐ Stop and do the margin exercises when directed to do so. Answers are at the bottom of the margin, so you can check your work right away.
- ☐ The objective symbols a, b, c, and so on, allow you to refer to the appropriate place in the text whenever you need to review a topic.

Rounding and Estimating; Order 1.6

OBJECTIVES

- **a** Round to the nearest ten, hundred, or thousand.
- **b** Estimate sums, differences, products, and quotients by rounding.
- **c** Use $<$ or $>$ for □ to write a true sentence in a situation like 6 □ 10.

SKILL TO REVIEW

Objective 1.1a: Give the meaning of digits in standard notation.

In the number 145,627, what digit names the number of:

1. Tens?
2. Thousands?

a ROUNDING

We round numbers in various situations when we do not need an exact answer. For example, we might round to see if we are being charged the correct amount in a store. We might also round to check if an answer to a problem is reasonable or to check a calculation done by hand or on a calculator.

To understand how to round, we first look at some examples using the number line. The number line displays numbers at equally spaced intervals.

EXAMPLE 1 Round 47 to the nearest ten.

47 is between 40 and 50. Since 47 is closer to 50, we round up to 50.

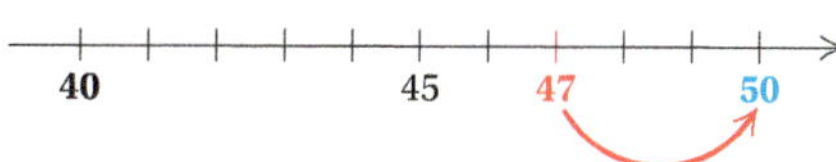

EXAMPLE 2 Round 42 to the nearest ten.

42 is between 40 and 50. Since 42 is closer to 40, we round down to 40.

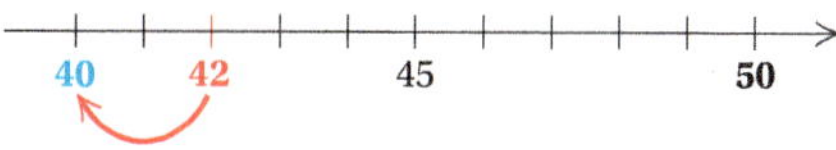

EXAMPLE 3 Round 45 to the nearest ten.

45 is halfway between 40 and 50. We could round 45 down to 40 or up to 50. We agree to round up to 50.

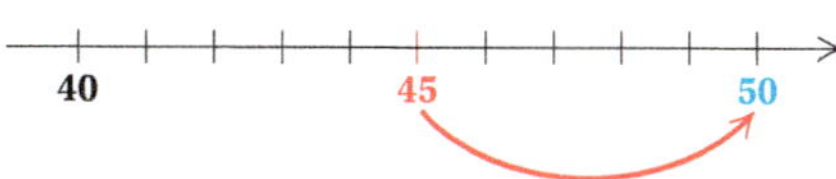

> When a number is halfway between rounding numbers, round up.

Do Margin Exercises 1–7. ▶

Based on these examples, we can state a rule for rounding whole numbers.

Round to the nearest ten.

1. 37

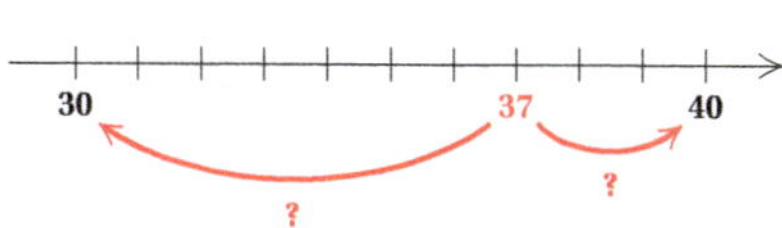

2. 52
3. 35
4. 73
5. 75
6. 88
7. 64

Answers

Skill to Review:
1. 2 2. 5

Margin Exercises:
1. 40 2. 50 3. 40 4. 70
5. 80 6. 90 7. 60

ROUNDING WHOLE NUMBERS

To round to a certain place:

a) Locate the digit in that place.

b) Consider the next digit to the right.

c) If the digit to the right is 5 or higher, round up. If the digit to the right is 4 or lower, round down.

d) Change all digits to the right of the rounding location to zeros.

EXAMPLE 4 Round 6485 to the nearest ten.

a) Locate the digit in the tens place, 8.

6 4 8 5
↑

b) Consider the next digit to the right, 5.

6 4 8 5

c) Since that digit, 5, is 5 or higher, round 8 tens up to 9 tens.

d) Change all digits to the right of the tens digit to zeros.

6 4 9 0 ← This is the answer.

Do Exercises 8–11.

Round to the nearest ten.

8. 137 **9.** 473

10. 235 **11.** 285

EXAMPLE 5 Round 6485 to the nearest hundred.

a) Locate the digit in the hundreds place, 4.

6 4 8 5
↑

b) Consider the next digit to the right, 8.

6 4 8 5

c) Since that digit, 8, is 5 or higher, round 4 hundreds up to 5 hundreds.

d) Change all digits to the right of hundreds to zeros.

6 5 0 0 ← This is the answer.

Do Exercises 12–15.

Round to the nearest hundred.

12. 641 **13.** 759

14. 1871 **15.** 9325

Caution!

It is incorrect in Example 6 to round from the ones digit over, as follows:

6485 → 6490 → 6500 → 7000.

Note that 6485 is closer to 6000 than it is to 7000.

EXAMPLE 6 Round 6485 to the nearest thousand.

a) Locate the digit in the thousands place, 6.

6 4 8 5
↑

b) Consider the next digit to the right, 4.

6 4 8 5

c) Since that digit, 4, is 4 or lower, round down, meaning that 6 thousands stays as 6 thousands.

d) Change all digits to the right of thousands to zeros.

6 0 0 0 ← This is the answer.

Do Exercises 16–19.

Round to the nearest thousand.

16. 7896 **17.** 8459

18. 19,343 **19.** 68,500

Answers

8. 140 **9.** 470 **10.** 240 **11.** 290
12. 600 **13.** 800 **14.** 1900 **15.** 9300
16. 8000 **17.** 8000 **18.** 19,000 **19.** 69,000

Sometimes rounding involves changing more than one digit in a number.

EXAMPLE 7 Round 78,595 to the nearest ten.

a) Locate the digit in the tens place, 9.

7 8 , 5 9 5
↑

b) Consider the next digit to the right, 5.

7 8 , 5 9 5

c) Since that digit, 5, is 5 or higher, round 9 tens to 10 tens. We think of 10 tens as 1 hundred + 0 tens and increase the hundreds digit by 1, to get 6 hundreds + 0 tens. We then write 6 in the hundreds place and 0 in the tens place.

d) Change the digit to the right of the tens digit to zero.

7 8 , 6 0 0 ← This is the answer.

Note that if we round this number to the nearest hundred, we get the same answer.

Do Exercises 20 and 21. ▶

b ESTIMATING

Estimating can be done in many ways. In general, an estimate rounded to the nearest ten is more accurate than one rounded to the nearest hundred, and an estimate rounded to the nearest hundred is more accurate than one rounded to the nearest thousand, and so on.

EXAMPLE 8 Estimate this sum by first rounding to the nearest ten:

78 + 49 + 31 + 85.

We round each number to the nearest ten. Then we add.

```
   78      80
   49      50
   31      30
 + 85    + 90
          250 ← Estimated answer
```

Do Exercises 22 and 23. ▶

EXAMPLE 9 Estimate the difference by first rounding to the nearest thousand: 9324 − 2849.

We have

```
   9324     9000
 − 2849   − 3000
            6000 ← Estimated answer
```

Do Exercises 24 and 25. ▶

20. Round 48,968 to the nearest ten, hundred, and thousand.

21. Round 269,582 to the nearest ten, hundred, and thousand.

22. Estimate the sum by first rounding to the nearest ten. Show your work.

```
   74
   23
   35
 + 66
```

23. Estimate the sum by first rounding to the nearest hundred. Show your work.

```
   650
   685
   238
 + 168
```

24. Estimate the difference by first rounding to the nearest hundred. Show your work.

```
   9285
 − 6739
```

25. Estimate the difference by first rounding to the nearest thousand. Show your work.

```
   23,278
 − 11,698
```

Answers

20. 48,970; 49,000; 49,000
21. 269,580; 269,600; 270,000
22. 70 + 20 + 40 + 70 = 200
23. 700 + 700 + 200 + 200 = 1800
24. 9300 − 6700 = 2600
25. 23,000 − 12,000 = 11,000

In the sentence $7 - 5 = 2$, the equals sign indicates that $7 - 5$ is the *same* as 2. When we round to make an estimate, the outcome is rarely the same as the exact result. Thus we cannot use an equals sign when we round. Instead, we use the symbol $\approx$. This symbol means **"is approximately equal to."** In Example 9, for instance, we could have written

$$9324 - 2849 \approx 6000.$$

EXAMPLE 10 Estimate the following product by first rounding to the nearest ten and then to the nearest hundred: 683×457.

Nearest ten:

$$\begin{array}{r} 680 \\ \times\ \ 460 \\ \hline 40800 \\ 272000 \\ \hline 312{,}800 \end{array} \qquad \begin{array}{l} 683 \approx 680 \\ 457 \approx 460 \end{array}$$

Nearest hundred:

$$\begin{array}{r} 700 \\ \times\ \ 500 \\ \hline 350{,}000 \end{array} \qquad \begin{array}{l} 683 \approx 700 \\ 457 \approx 500 \end{array}$$

Exact:

$$\begin{array}{r} 683 \\ \times\ \ 457 \\ \hline 4781 \\ 34150 \\ 273200 \\ \hline 312{,}131 \end{array}$$

We see that rounding to the nearest ten gives a better estimate than rounding to the nearest hundred.

Do Exercise 26.

EXAMPLE 11 Estimate the following quotient by first rounding to the nearest ten and then to the nearest hundred: $12{,}238 \div 175$.

Nearest ten:

$$\begin{array}{r} 68 \\ 180\overline{)12{,}240} \\ \underline{1080} \\ 1440 \\ \underline{1440} \\ 0 \end{array}$$

Nearest hundred:

$$\begin{array}{r} 61 \\ 200\overline{)12{,}200} \\ \underline{1200} \\ 200 \\ \underline{200} \\ 0 \end{array}$$

The exact answer is 69 R 163. Again we see that rounding to the nearest ten gives a better estimate than rounding to the nearest hundred.

Do Exercise 27.

26. Estimate the product by first rounding to the nearest ten and then to the nearest hundred. Show your work.

GS

$$\begin{array}{r} 837 \\ \times 245 \\ \hline \end{array}$$

Nearest ten:

$$\begin{array}{r} 840 \\ \times\ \ \square \\ \hline 42000 \\ \square\square\square 000 \\ \hline \square\square\square{,}000 \end{array}$$

Nearest hundred:

$$\begin{array}{r} 800 \\ \times\ \ \square \\ \hline \square\square 0{,}000 \end{array}$$

27. Estimate the quotient by first rounding to the nearest hundred. Show your work.

$$64{,}534 \div 349$$

Answers

26. $840 \times 250 = 210{,}000$;
$800 \times 200 = 160{,}000$
27. $64{,}500 \div 300 = 215$

Guided Solution:
26. Nearest ten: 250, 1, 6, 8, 2, 1, 0;
Nearest hundred: 200, 1, 6

The next two examples show how estimating can be used in making financial decisions.

EXAMPLE 12 *Tuition.* Ellen plans to take 12 credit hours of classes next semester. If she takes the courses on campus, the cost per credit hour is $248. Estimate, by rounding to the nearest ten, the total cost of tuition.

We have

$$\begin{array}{r} 250 \\ \times \quad 10 \\ \hline 2500. \end{array}$$

The tuition will cost about $2500.

Do Exercise 28. ▶

28. *Tuition.* If Ellen takes courses online, the cost per credit hour is $198. Estimate, by rounding to the nearest ten, the total cost of 12 credit hours of classes.

EXAMPLE 13 *Purchasing a New Car.* Jon and Joanna are shopping for a new car. They are considering buying a Ford Focus S sedan. The base price of the car is $16,200. A 6-speed automatic transmission package can be added to this, as well as several other options, as shown in the chart below. Jon and Joanna want to stay within a budget of $18,000.

Estimate, by rounding to the nearest hundred, the cost of the Focus with the automatic transmission package and all other options and determine whether this will fit within their budget.

FORD FOCUS S SEDAN	PRICE
Base price	$16,200
6-speed automatic transmission	$1,095
Remote start system (requires purchase of automatic transmission)	$445
SYNC basic (includes Bluetooth and USB input jacks)	$295
Cargo management	$115
Graphics package	$375
Exterior protection package	$245

SOURCE: motortrend.com

First, we list the base price of the car and then the cost of each of the options. We then round each number to the nearest hundred and add.

$$\begin{array}{rr} 16,200 & 16,200 \\ 1,095 & 1,100 \\ 445 & 400 \\ 295 & 300 \\ 115 & 100 \\ 375 & 400 \\ +\quad 245 & +\quad 200 \\ \hline & 18,700 \end{array} \leftarrow \text{Estimated cost}$$

The estimated cost is $18,700. This exceeds Jon and Joanna's budget of $18,000, so they will have to forgo at least one option.

Do Exercises 29 and 30. ▶

Refer to the chart in Example 13 to do Margin Exercises 29 and 30.

29. By eliminating at least one option, determine how Jon and Joanna can buy a Focus S sedan and stay within their budget. Keep in mind that purchasing the remote-start system requires purchasing the automatic transmission.

30. Elizabeth and C.J. are also considering buying a Focus S sedan. Estimate, by rounding to the nearest hundred, the cost of this car with automatic transmission, SYNC basic, and exterior protection package.

Answers

28. $2000 **29.** Eliminate either the automatic transmission and the remote-start system or the remote-start system and the graphics package. There are other correct answers as well. **30.** $17,800

C ORDER

A sentence like $8 + 5 = 13$ is called an **equation**. It is a *true* equation. The equation $4 + 8 = 11$ is a *false* equation.

A sentence like $7 < 11$ is called an **inequality**. The sentence $7 < 11$ is a *true* inequality. The sentence $23 > 69$ is a *false* inequality.

Some common **inequality symbols** follow.

INEQUALITY SYMBOLS

$<$ means "is less than"

$>$ means "is greater than"

$\neq$ means "is not equal to"

We know that 2 is not the same as 5. We express this by the sentence $2 \neq 5$. We also know that 2 is less than 5. We symbolize this by the expression $2 < 5$. We can see this order on the number line: 2 is to the left of 5. The number 0 is the smallest whole number.

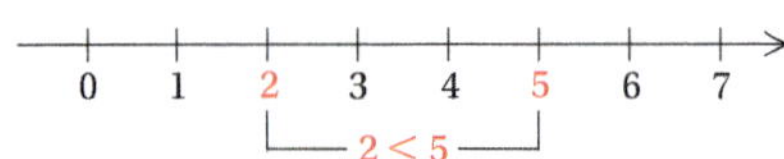

ORDER OF WHOLE NUMBERS

For any whole numbers a and b:

1. $a < b$ (read "a is less than b") is true when a is to the left of b on the number line.
2. $a > b$ (read "a is greater than b") is true when a is to the right of b on the number line.

EXAMPLE 14 Use $<$ or $>$ for $\square$ to write a true sentence: $7 \square 11$.

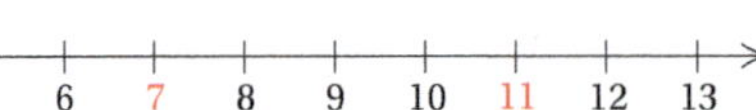

Since 7 is to the left of 11 on the number line, $7 < 11$.

EXAMPLE 15 Use $<$ or $>$ for $\square$ to write a true sentence: $92 \square 87$.

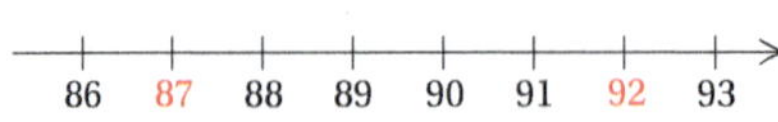

Since 92 is to the right of 87 on the number line, $92 > 87$.

Do Exercises 31–36.

Use $<$ or $>$ for $\square$ to write a true sentence. Draw the number line if necessary.

31. $8 \square 12$ (GS)

Since 8 is to the ______ of 12 on the number line, 8 ______ 12.

32. $12 \square 8$

33. $76 \square 64$

34. $64 \square 76$

35. $217 \square 345$

36. $345 \square 217$

Answers

31. $<$ 32. $>$ 33. $>$ 34. $<$
35. $<$ 36. $>$

Guided Solution:
31. left, $<$

1.6 Exercise Set

For Extra Help
MyMathLab®
MathXL® PRACTICE | WATCH | READ | REVIEW

✓ Reading Check

Determine whether each statement is true or false.

_______________ **RC1.** When rounding to the nearest hundred, if the digit in the tens place is 5 or higher, we round up.

_______________ **RC2.** When rounding 3500 to the nearest thousand, we should round down.

_______________ **RC3.** An estimate made by rounding to the nearest thousand is more accurate than an estimate made by rounding to the nearest ten.

_______________ **RC4.** Since 78 rounded to the nearest ten is 80, we can write 78 $\approx$ 80.

a Round to the nearest ten.

1. 48 **2.** 532 **3.** 463 **4.** 8945

5. 731 **6.** 54 **7.** 895 **8.** 798

Round to the nearest hundred.

9. 146 **10.** 874 **11.** 957 **12.** 650

13. 9079 **14.** 4645 **15.** 32,839 **16.** 198,402

Round to the nearest thousand.

17. 5876 **18.** 4500 **19.** 7500 **20.** 2001

21. 45,340 **22.** 735,562 **23.** 373,405 **24.** 6,713,255

b Estimate each sum or difference by first rounding to the nearest ten. Show your work.

25.
$$\begin{array}{r} 78 \\ +\ 92 \\ \hline \end{array}$$

26.
$$\begin{array}{r} 62 \\ 97 \\ 46 \\ +\ 81 \\ \hline \end{array}$$

27.
$$\begin{array}{r} 8074 \\ -\ 2347 \\ \hline \end{array}$$

28.
$$\begin{array}{r} 673 \\ -\ \ 28 \\ \hline \end{array}$$

Estimate each sum by first rounding to the nearest ten. State if the given sum seems to be incorrect when compared to the estimate.

29.
$$\begin{array}{r} 45 \\ 77 \\ 25 \\ +\ 56 \\ \hline 343 \end{array}$$

30.
$$\begin{array}{r} 41 \\ 21 \\ 55 \\ +\ 60 \\ \hline 177 \end{array}$$

31.
$$\begin{array}{r} 622 \\ 78 \\ 81 \\ +\ 111 \\ \hline 932 \end{array}$$

32.
$$\begin{array}{r} 836 \\ 374 \\ 794 \\ +\ 938 \\ \hline 3947 \end{array}$$

Estimate each sum or difference by first rounding to the nearest hundred. Show your work.

33. $\begin{array}{r} 7348 \\ +\,9247 \\ \hline \end{array}$

34. $\begin{array}{r} 568 \\ 472 \\ 938 \\ +\,402 \\ \hline \end{array}$

35. $\begin{array}{r} 6852 \\ -\,1748 \\ \hline \end{array}$

36. $\begin{array}{r} 9438 \\ -\,2787 \\ \hline \end{array}$

Estimate each sum by first rounding to the nearest hundred. State if the given sum seems to be incorrect when compared to the estimate.

37. $\begin{array}{r} 216 \\ 84 \\ 745 \\ +\,595 \\ \hline 1640 \end{array}$

38. $\begin{array}{r} 481 \\ 702 \\ 623 \\ +\,1043 \\ \hline 1849 \end{array}$

39. $\begin{array}{r} 750 \\ 428 \\ 63 \\ +\,205 \\ \hline 1446 \end{array}$

40. $\begin{array}{r} 326 \\ 275 \\ 758 \\ +\,943 \\ \hline 2302 \end{array}$

Estimate each sum or difference by first rounding to the nearest thousand. Show your work.

41. $\begin{array}{r} 9643 \\ 4821 \\ 8943 \\ +\,7004 \\ \hline \end{array}$

42. $\begin{array}{r} 7648 \\ 9348 \\ 7842 \\ +\,2222 \\ \hline \end{array}$

43. $\begin{array}{r} 92{,}149 \\ -\,22{,}555 \\ \hline \end{array}$

44. $\begin{array}{r} 84{,}890 \\ -\,11{,}110 \\ \hline \end{array}$

Estimate each product by first rounding to the nearest ten. Show your work.

45. $\begin{array}{r} 45 \\ \times\,67 \\ \hline \end{array}$

46. $\begin{array}{r} 51 \\ \times\,78 \\ \hline \end{array}$

47. $\begin{array}{r} 34 \\ \times\,29 \\ \hline \end{array}$

48. $\begin{array}{r} 63 \\ \times\,54 \\ \hline \end{array}$

Estimate each product by first rounding to the nearest hundred. Show your work.

49. $\begin{array}{r} 876 \\ \times\,345 \\ \hline \end{array}$

50. $\begin{array}{r} 355 \\ \times\,299 \\ \hline \end{array}$

51. $\begin{array}{r} 432 \\ \times\,199 \\ \hline \end{array}$

52. $\begin{array}{r} 789 \\ \times\,434 \\ \hline \end{array}$

Estimate each quotient by first rounding to the nearest ten. Show your work.

53. $347 \div 73$

54. $454 \div 87$

55. $8452 \div 46$

56. $1263 \div 29$

Estimate each quotient by first rounding to the nearest hundred. Show your work.

57. 1165 ÷ 236 **58.** 3641 ÷ 571 **59.** 8358 ÷ 295 **60.** 32,854 ÷ 748

Planning a Vacation. Most cruise ships offer a choice of rooms at varying prices, as well as additional packages and shore excursions at each port. The table below lists room prices for a seven-day Mediterranean cruise, as well as prices for several additional packages and excursions.

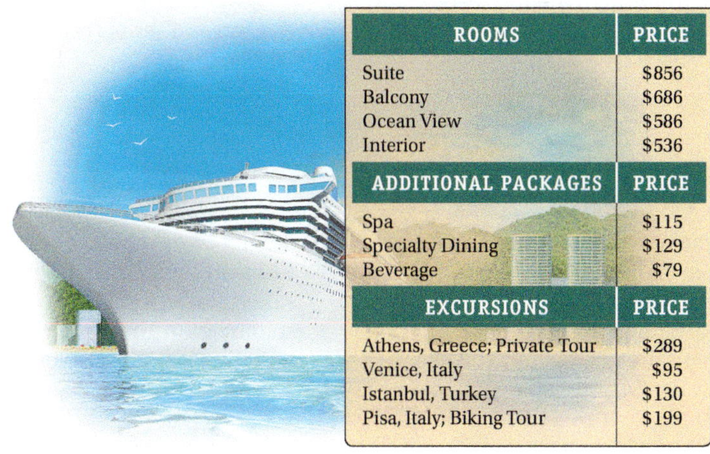

ROOMS	PRICE
Suite	$856
Balcony	$686
Ocean View	$586
Interior	$536
ADDITIONAL PACKAGES	**PRICE**
Spa	$115
Specialty Dining	$129
Beverage	$79
EXCURSIONS	**PRICE**
Athens, Greece; Private Tour	$289
Venice, Italy	$95
Istanbul, Turkey	$130
Pisa, Italy; Biking Tour	$199

61. Estimate the total price of a cruise with an ocean view room, a spa package, and an Istanbul excursion. Round each price to the nearest hundred dollars.

62. Estimate the total price of a cruise with a balcony room, no additional packages, a Venice excursion, and a biking tour in Pisa. Round each price to the nearest hundred dollars.

63. Antonio has a budget of $1000 for a Mediterranean cruise. He would like a balcony room, a specialty dining package, a private tour of Athens, and a tour of Venice. Estimate the total price of this cruise by rounding each price to the nearest hundred dollars. Can he afford his choices?

64. Alyssa has a budget of $1400 for a Mediterranean cruise. She is planning to book an interior room and would like to go on all the excursions listed. Estimate the total price of this cruise by rounding each price to the nearest hundred dollars. Does her budget cover her choices?

65. If you were going on a Mediterranean cruise and had a budget of $1500, what options would you choose? Decide on the options you would like and estimate the total price by rounding each price to the nearest hundred dollars. Could you afford all your chosen options?

66. If you were going on a Mediterranean cruise and had a budget of $1200, what options would you choose? Decide on the options you would like and estimate the total price by rounding each price to the nearest hundred dollars. Could you afford all your chosen options?

67. *Mortgage Payments.* To pay for their new home, Tim and Meribeth will make 360 payments of $751.55 each. In addition, they must add an escrow amount of $112.67 to each payment for insurance and taxes.

a) Estimate the total amount they will pay by rounding the number of payments, the amount of each payment, and the escrow amount to the nearest ten.

b) Estimate the total amount they will pay by rounding the number of payments, the amount of each payment, and the escrow amount to the nearest hundred.

68. *Conference Expenses.* The cost to attend a three-day teachers' conference is $245, and a hotel room costs $169 a night. One year, 489 teachers attended the conference, and 315 rooms were rented for two nights each.

a) Estimate the total amount spent by the teachers by rounding the cost of attending the conference, the cost of a hotel room, the number of teachers, and the number of rooms to the nearest ten.

b) Estimate the total amount spent by the teachers by rounding the cost of attending the conference, the cost of a hotel room, the number of teachers, and the number of rooms to the nearest hundred.

69. *Banquet Attendance.* Tickets to the annual awards banquet for the Riviera Swim Club cost $28 each. Ticket sales for the banquet totaled $2716. Estimate the number of people who attended the banquet by rounding the cost of a ticket to the nearest ten and the total sales to the nearest hundred.

70. *School Fundraiser.* For a school fundraiser, Charlotte sells trash bags at a price of $11 per roll. If her sales totaled $2211, estimate the number of rolls she sold by rounding the price per roll to the nearest ten and the total sales to the nearest hundred.

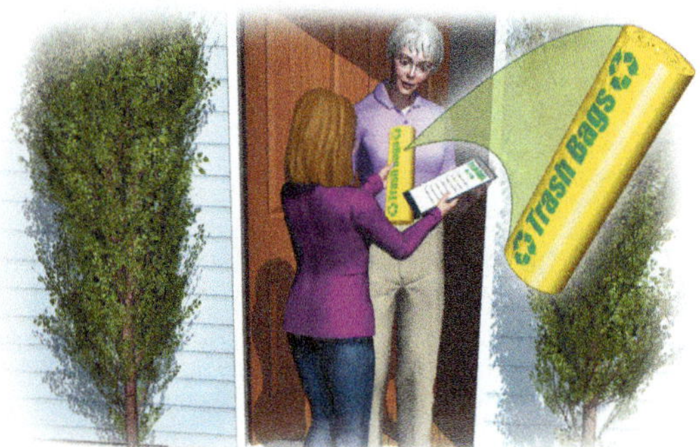

C Use $<$ or $>$ for □ to write a true sentence. Draw the number line if necessary.

71. 0 □ 17

72. 32 □ 0

73. 34 □ 12

74. 28 □ 18

75. 1000 □ 1001

76. 77 □ 117

77. 133 □ 132

78. 999 □ 997

79. 460 □ 17

80. 345 □ 456

81. 37 □ 11

82. 12 □ 32

New Book Titles. The number of new book titles published in the United States in each of three recent years is shown in the table below. Use this table to do Exercises 83 and 84.

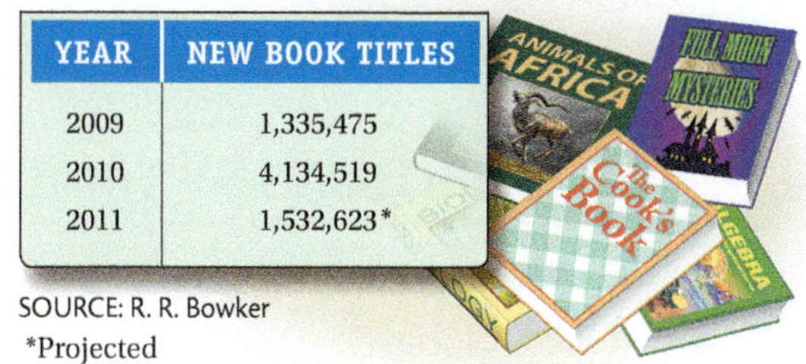

YEAR	NEW BOOK TITLES
2009	1,335,475
2010	4,134,519
2011	1,532,623*

SOURCE: R. R. Bowker
*Projected

83. Write an inequality to compare the numbers of new titles published in 2009 and in 2010.

84. Write an inequality to compare the numbers of new titles published in 2010 and in 2011.

85. *Public Schools.* The number of public schools in the United States increased from 97,382 in 2006 to 98,817 in 2010. Write an inequality to compare these numbers of schools.

Public Schools

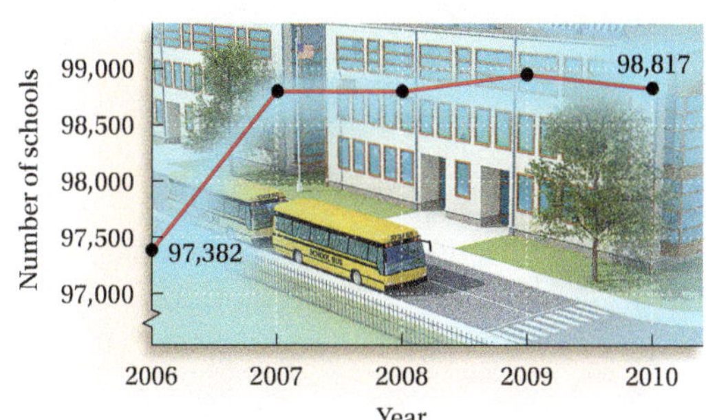

SOURCE: National Center for Education Statistics

86. *Life Expectancy.* The life expectancy of a female in the United States in 2020 is predicted to be about 82 years and that of a male about 77 years. Write an inequality to compare these life expectancies.

Life Expectancy in the United States

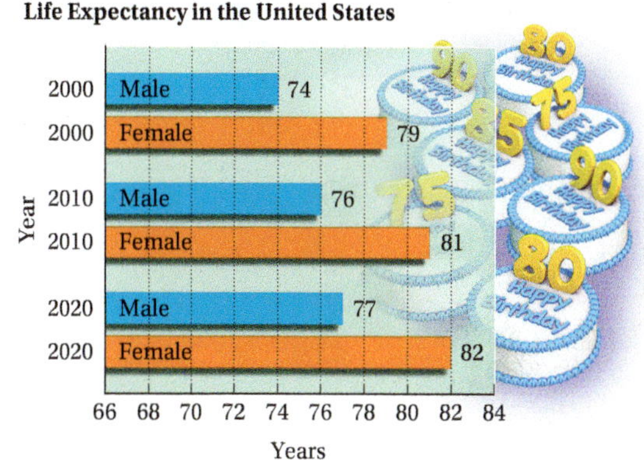

SOURCE: U.S. Census Bureau

Skill Maintenance

Add. [1.2a]

87. $\begin{array}{r} 67,789 \\ +\ 18,965 \\ \hline \end{array}$

88. $\begin{array}{r} 9002 \\ +\ 4587 \\ \hline \end{array}$

Subtract. [1.3a]

89. $\begin{array}{r} 67,789 \\ -\ 18,965 \\ \hline \end{array}$

90. $\begin{array}{r} 9002 \\ -\ 4587 \\ \hline \end{array}$

Multiply. [1.4a]

91. $\begin{array}{r} 46 \\ \times\ 37 \\ \hline \end{array}$

92. $\begin{array}{r} 306 \\ \times\ 58 \\ \hline \end{array}$

Divide. [1.5a]

93. $328 \div 6$

94. $4784 \div 23$

Synthesis

95.–98. Use a calculator to find the sums and the differences in each of Exercises 41–44. Then compare your answers with those found using estimation. Even when using a calculator it is possible to make an error if you press the wrong buttons, so it is a good idea to check by estimating.

1.7 Solving Equations

OBJECTIVES

a Solve simple equations by trial.

b Solve equations like $t + 28 = 54$, $28 \cdot x = 168$, and $98 \cdot 2 = y$.

SKILL TO REVIEW

Objective 1.5a: Divide whole numbers.

Divide.

1. $1008 \div 36$
2. $675 \div 15$

Find a number that makes each sentence true.

1. $8 = 1 + \square$
2. $\square + 2 = 7$

3. Determine whether 7 is a solution of $\square + 5 = 9$.
4. Determine whether 4 is a solution of $\square + 5 = 9$.

Solve by trial.

5. $n + 3 = 8$
6. $x - 2 = 8$
7. $45 \div 9 = y$
8. $10 + t = 32$

Answers

Skill to Review:
1. 28 2. 45

Margin Exercises:
1. 7 2. 5 3. No 4. Yes 5. 5
6. 10 7. 5 8. 22

a SOLUTIONS BY TRIAL

Let's find a number that we can put in the blank to make this sentence true:

$$9 = 3 + \square.$$

We are asking "9 is 3 plus what number?" The answer is 6.

$$9 = 3 + 6$$

◀ **Do Margin Exercises 1 and 2.**

A sentence with = is called an **equation**. A **solution** of an equation is a number that makes the sentence true. Thus, 6 is a solution of

$9 = 3 + \square$ because $9 = 3 + 6$ is true.

However, 7 is not a solution of

$9 = 3 + \square$ because $9 = 3 + 7$ is false.

◀ **Do Exercises 3 and 4.**

We can use a letter in an equation instead of a blank:

$$9 = 3 + n.$$

We call n a **variable** because it can represent any number. If a replacement for a variable makes an equation true, the replacement is a solution of the equation.

SOLUTIONS OF AN EQUATION

A **solution of an equation** is a replacement for the variable that makes the equation true. When we find all the solutions, we say that we have **solved** the equation.

EXAMPLE 1 Solve $y + 12 = 27$ by trial.

We replace y with several numbers.

If we replace y with 13, we get a false equation: $13 + 12 = 27$.
If we replace y with 14, we get a false equation: $14 + 12 = 27$.
If we replace y with 15, we get a true equation: $15 + 12 = 27$.

No other replacement makes the equation true, so the solution is 15. ■

EXAMPLES Solve.

2. $7 + n = 22$
(7 plus what number is 22?)
The solution is 15.

3. $63 = 3 \cdot x$
(63 is 3 times what number?)
The solution is 21.

◀ **Do Exercises 5–8.**

b SOLVING EQUATIONS

We now begin to develop more efficient ways to solve certain equations. When an equation has a variable alone on one side and a calculation on the other side, we can find the solution by carrying out the calculation.

EXAMPLE 4 Solve: $x = 245 \times 34$.

To solve the equation, we carry out the calculation.

$$\begin{array}{r} 245 \\ \times\ 34 \\ \hline 980 \\ 7350 \\ \hline 8330 \end{array} \qquad \begin{aligned} x &= 245 \times 34 \\ x &= 8330 \end{aligned}$$

The solution is 8330.

Do Exercises 9–12. ▶

Solve.

9. $346 \times 65 = y$

10. $x = 2347 + 6675$

11. $4560 \div 8 = t$

12. $x = 6007 - 2346$

Look at the equation

$$x + 12 = 27.$$

We can get x alone by subtracting 12 *on both sides*. Thus,

$$\begin{aligned} x + 12 - 12 &= 27 - 12 && \text{Subtracting 12 on both sides} \\ x + 0 &= 15 && \text{Carrying out the subtraction} \\ x &= 15. \end{aligned}$$

SOLVING $x + a = b$

To solve $x + a = b$, subtract a on both sides.

If we can get an equation in a form with the variable alone on one side, we can "see" the solution.

EXAMPLE 5 Solve: $t + 28 = 54$.

We have

$$\begin{aligned} t + 28 &= 54 \\ t + 28 - 28 &= 54 - 28 && \text{Subtracting 28 on both sides} \\ t + 0 &= 26 \\ t &= 26. \end{aligned}$$

To check the answer, we substitute 26 for t in the original equation.

Check: $t + 28 = 54$

$26 + 28 \ ? \ 54$

54 | TRUE   Since $54 = 54$ is true, 26 checks.

The solution is 26.

Do Exercises 13 and 14. ▶

Solve. Be sure to check.

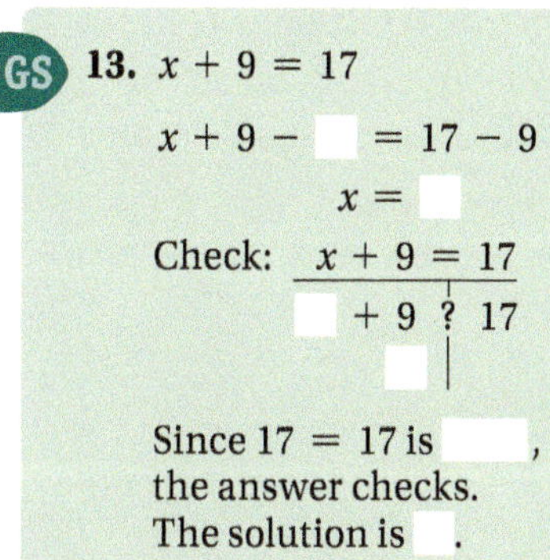

GS
13. x + 9 = 17
x + 9 − = 17 − 9
x =
Check: x + 9 = 17
+ 9 ? 17
Since 17 = 17 is ,
the answer checks.
The solution is .

14. $77 = m + 32$

Answers

9. 22,490 **10.** 9022 **11.** 570 **12.** 3661 **13.** 8 **14.** 45

Guided Solution:
13. 9, 8, 8, 17; true; 8

EXAMPLE 6 Solve: $182 = 65 + n$.

We have

$$\begin{aligned} 182 &= 65 + n \\ 182 - 65 &= 65 + n - 65 \quad \text{Subtracting 65 on both sides} \\ 117 &= 0 + n \quad \text{65 plus } n \text{ minus 65 is } 0 + n. \\ 117 &= n. \end{aligned}$$

Check:

$$\begin{array}{c|l} 182 = 65 + n & \\ \hline 182 \; ? & 65 + 117 \\ & 182 \qquad \text{TRUE} \end{array}$$

The solution is 117.

15. Solve: $155 = t + 78$. Be sure to check.

Do Exercise 15.

EXAMPLE 7 Solve: $7381 + x = 8067$.

We have

$$\begin{aligned} 7381 + x &= 8067 \\ 7381 + x - 7381 &= 8067 - 7381 \quad \text{Subtracting 7381 on both sides} \\ x &= 686. \end{aligned}$$

Check:

$$\begin{array}{r|l} 7381 + x = & 8067 \\ \hline 7381 + 686 \; ? & 8067 \\ 8067 & \qquad \text{TRUE} \end{array}$$

The solution is 686.

Solve. Be sure to check.

16. $4566 + x = 7877$

17. $8172 = h + 2058$

Do Exercises 16 and 17.

We now learn to solve equations like $8 \cdot n = 96$. Look at

$8 \cdot n = 96$.

We can get n alone by dividing by 8 *on both sides*. Thus,

$$\frac{8 \cdot n}{8} = \frac{96}{8} \quad \text{Dividing by 8 on both sides}$$

$$n = 12. \quad 8 \text{ times } n \text{ divided by 8 is } n.$$

To check the answer, we substitute 12 for n in the original equation.

Check:

$$\begin{array}{r|l} 8 \cdot n = & 96 \\ \hline 8 \cdot 12 \; ? & 96 \\ 96 & \qquad \text{TRUE} \end{array}$$

Since $96 = 96$ is a true equation, 12 is the solution of the equation.

SOLVING $a \cdot x = b$

To solve $a \cdot x = b$, divide by a on both sides.

Answers

15. 77 16. 3311 17. 6114

EXAMPLE 8 Solve: $10 \cdot x = 240$.

We have

$$10 \cdot x = 240$$
$$\frac{10 \cdot x}{10} = \frac{240}{10} \quad \text{Dividing by 10 on both sides}$$
$$x = 24.$$

Check: $10 \cdot x = 240$

$10 \cdot 24 \ ? \ 240$

240 | TRUE

The solution is 24.

EXAMPLE 9 Solve: $5202 = 9 \cdot t$.

We have

$$5202 = 9 \cdot t$$
$$\frac{5202}{9} = \frac{9 \cdot t}{9} \quad \text{Dividing by 9 on both sides}$$
$$578 = t.$$

Check: $5202 = 9 \cdot t$

$5202 \ ? \ 9 \cdot 578$

| 5202 TRUE

The solution is 578.

Do Exercises 18–20. ▶

EXAMPLE 10 Solve: $14 \cdot y = 1092$.

We have

$$14 \cdot y = 1092$$
$$\frac{14 \cdot y}{14} = \frac{1092}{14} \quad \text{Dividing by 14 on both sides}$$
$$y = 78.$$

The check is left to the student. The solution is 78.

EXAMPLE 11 Solve: $n \cdot 56 = 4648$.

We have

$$n \cdot 56 = 4648$$
$$\frac{n \cdot 56}{56} = \frac{4648}{56} \quad \text{Dividing by 56 on both sides}$$
$$n = 83.$$

The check is left to the student. The solution is 83.

Do Exercises 21 and 22. ▶

Solve. Be sure to check.

18. $8 \cdot x = 64$

GS **19.** $144 = 9 \cdot n$

$$\frac{144}{9} = \frac{9 \cdot n}{\square}$$
$$\square = n$$

Check: $144 = 9 \cdot n$

$144 \ ? \ 9 \cdot \square$

| $\square$

Since 144 = 144 is ____,
the answer checks.
The solution is ____.

20. $5152 = 8 \cdot t$

Solve. Be sure to check.

21. $18 \cdot y = 1728$

22. $n \cdot 48 = 4512$

Answers

18. 8 **19.** 16 **20.** 644 **21.** 96 **22.** 94

Guided Solution:
19. 9, 16, 16, 144; true; 16

1.7 Exercise Set

For Extra Help MyMathLab® MathXL® PRACTICE WATCH READ REVIEW

✓ Reading Check

Match each word with its definition from the list on the right.

RC1. Equation ________

RC2. Solution ________

RC3. Solved ________

RC4. Variable ________

a) A replacement for the variable that makes an equation true

b) A letter that can represent any number

c) A sentence containing $=$

d) An equation for which we have found all solutions

a Solve by trial.

1. $x + 0 = 14$

2. $x - 7 = 18$

3. $y \cdot 17 = 0$

4. $56 \div m = 7$

b Solve. Be sure to check.

5. $x = 12{,}345 + 78{,}555$

6. $t = 5678 + 9034$

7. $908 - 458 = p$

8. $9007 - 5667 = m$

9. $16 \cdot 22 = y$

10. $34 \cdot 15 = z$

11. $t = 125 \div 5$

12. $w = 256 \div 16$

13. $13 + x = 42$

14. $15 + t = 22$

15. $12 = 12 + m$

16. $16 = t + 16$

17. $10 + x = 89$

18. $20 + x = 57$

19. $61 = 16 + y$

20. $53 = 17 + w$

21. $3 \cdot x = 24$

22. $6 \cdot x = 42$

23. $112 = n \cdot 8$

24. $162 = 9 \cdot m$

25. $3 \cdot m = 96$

26. $4 \cdot y = 96$

27. $715 = 5 \cdot z$

28. $741 = 3 \cdot t$

29. $8322 + 9281 = x$

30. $9281 - 8322 = y$

31. $47 + n = 84$

32. $56 + p = 92$

33. $45 \cdot 23 = x$

34. $23 \cdot 78 = y$

35. $x + 78 = 144$

36. $z + 67 = 133$

37. $6 \cdot p = 1944$

38. $4 \cdot w = 3404$

39. $567 + x = 902$

40. $438 + x = 807$

41. $234 \cdot 78 = y$

42. $10{,}534 \div 458 = q$

43. $18 \cdot x = 1872$

44. $19 \cdot x = 6080$

45. $40 \cdot x = 1800$

46. $20 \cdot x = 1500$

47. $2344 + y = 6400$

48. $9281 = 8322 + t$

49. $m = 7006 - 4159$

50. $n = 3004 - 1745$

51. $165 = 11 \cdot n$

52. $660 = 12 \cdot n$

53. $58 \cdot m = 11{,}890$

54. $233 \cdot x = 22{,}135$

55. $491 - 34 = y$

56. $512 - 63 = z$

Skill Maintenance

Divide. [1.5a]

57. $1283 \div 9$

58. $1278 \div 9$

59. $17\overline{)5678}$

60. $17\overline{)5689}$

Use $>$ or $<$ for $\square$ to write a true sentence. [1.6c]

61. $123 \square 789$

62. $342 \square 339$

63. $688 \square 0$

64. $0 \square 11$

65. Round 6,375,602 to the nearest thousand. [1.6a]

66. Round 6,375,602 to the nearest ten. [1.6a]

Synthesis

Solve.

67. ▦ $23{,}465 \cdot x = 8{,}142{,}355$

68. ▦ $48{,}916 \cdot x = 14{,}332{,}388$

1.8 Applications and Problem Solving

OBJECTIVE

a Solve applied problems involving addition, subtraction, multiplication, or division of whole numbers.

SKILL TO REVIEW

Objective 1.6b: Estimate sums, differences, products, and quotients by rounding.

Estimate each sum or difference by first rounding to the nearest thousand. Show your work.

1. $\begin{array}{r} 367{,}982 \\ +\ \ 43{,}495 \\ \hline \end{array}$

2. $\begin{array}{r} 9287 \\ -\ 3502 \\ \hline \end{array}$

Answers

Skill to Review:
1. 368,000 + 43,000 = 411,000
2. 9000 − 4000 = 5000

a A PROBLEM-SOLVING STRATEGY

One of the most important ways in which we use mathematics is as a tool in solving problems. To solve a problem, we use the following five-step strategy.

FIVE STEPS FOR PROBLEM SOLVING

1. **Familiarize** yourself with the problem situation.
2. **Translate** the problem to an equation using a variable.
3. **Solve** the equation.
4. **Check** to see whether your possible solution actually fits the problem situation and is thus really a solution of the problem.
5. **State** the answer clearly using a complete sentence and appropriate units.

The first of these five steps, becoming familiar with the problem, is probably the most important.

THE FAMILIARIZE STEP

- If the problem is presented in words, read and reread it carefully until you understand what you are being asked to find.
- Make a drawing, if it makes sense to do so.
- Write a list of the known facts and a list of what you wish to find out.
- Assign a letter, or *variable*, to the unknown.
- Organize the information in a chart or a table.
- Find further information. Look up a formula, consult a reference book or an expert in the field, or do research on the Internet.
- Guess or estimate the answer and check your guess or estimate.

EXAMPLE 1 *Video Game Platforms.* The following table shows the total number of units of eight popular video game platforms sold worldwide, as of December 8, 2012. Three of these are made by PlayStation®. Find the total number of PlayStation units sold.

PLATFORM	GLOBAL SALES
Nintendo 3DS	682,396
Xbox 360	614,353
Nintendo Wii U	610,384
PlayStation® 3	576,565
Nintendo Wii	252,484
Nintendo DS	178,209
PlayStation® Vita	152,049
PlayStation® Portable	54,677

Source: www.vgchartz.com

1. **Familiarize.** First, we assign a letter, or variable, to the number we wish to find. We let $p =$ the total number of video game units sold by PlayStation. Since we are combining numbers, we will add.
2. **Translate.** We translate to an equation:

Number of PlayStation 3 units	plus	Number of PlayStation Vita units	plus	Number of PlayStation Portable units	is	Total number of PlayStation units
↓	↓	↓	↓	↓	↓	↓
576,565	+	152,049	+	54,677	=	p

3. **Solve.** We solve the equation by carrying out the addition.

$$576{,}565 + 152{,}049 + 54{,}677 = p$$
$$783{,}291 = p$$

$$\begin{array}{r} 576{,}565 \\ 152{,}049 \\ +\quad 54{,}677 \\ \hline 783{,}291 \end{array}$$

4. **Check.** We check our result by rereading the original problem and seeing if 783,291 answers the question. Since we are looking for a total, we could repeat the addition calculation. We could also check whether the answer is reasonable. In this case, since the total is greater than any of the three separate sales numbers, the result seems reasonable. Another way to check is to estimate the expected result and compare the estimate with the calculated result. If we round each PlayStation sales number to the nearest ten thousand and add, we have

$$580{,}000 + 150{,}000 + 50{,}000 = 780{,}000.$$

Since $780{,}000 \approx 783{,}291$, our result again seems reasonable.

5. **State.** The total number of the three PlayStation units sold worldwide as of December 8, 2012, is 783,291.

Do Exercises 1–3. ▶

Refer to the table in Example 1 to do Margin Exercises 1–3.

1. Find the total number of Nintendo units sold.
2. Find the total number of units sold for the four most popular game platforms listed in the table.
3. Find the total number of units sold for all the game platforms listed in the table.

Answers

1. 1,723,473 units 2. 2,483,698 units
3. 3,121,117 units

EXAMPLE 2 *Travel Distance.* Abigail is driving from Indianapolis to Salt Lake City to attend a family reunion. The distance from Indianapolis to Salt Lake City is 1634 mi. In the first two days, she travels 1154 mi to Denver. How much farther must she travel?

1. **Familiarize.** We first make a drawing or at least visualize the situation. We let $d =$ the remaining distance to Salt Lake City.

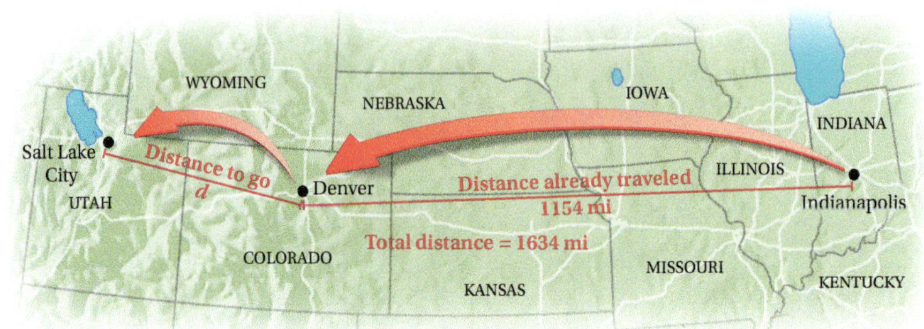

2. **Translate.** We want to determine how many more miles Abigail must travel. We translate to an equation:

Distance already traveled	plus	Distance to go	is	Total distance of trip
↓	↓	↓	↓	↓
1154	+	d	=	1634.

3. **Solve.** To solve the equation, we subtract 1154 on both sides.

$$1154 + d = 1634$$
$$1154 + d - 1154 = 1634 - 1154$$
$$d = 480$$

$$\begin{array}{r} \overset{5\ 13}{1\,\not{6}\,\not{3}\,4} \\ -\ 1\,1\,5\,4 \\ \hline 4\,8\,0 \end{array}$$

4. **Check.** We check our answer of 480 mi in the original problem. This number should be less than the total distance, 1634 mi, and it is. We can add the distance traveled, 1154, and the distance left to go, 480: $1154 + 480 = 1634$. We can also estimate:

$$1634 - 1154 \approx 1600 - 1200$$
$$= 400 \approx 480.$$

The answer, 480 mi, checks.

5. **State.** Abigail must travel 480 mi farther to Salt Lake City.

◀ Do Exercise 4.

4. Reading Assignment. William has been assigned 234 pages of reading for his history class. He has read 86 pages. How many more pages does he have to read? GS

1. **Familiarize.** Let $p =$ the number of pages William still has to read.
2. **Translate.**

Pages already read	plus	Number of pages to read	is	Total number of pages
↓	↓	↓	↓	↓
86	+	☐	=	☐

3. **Solve.**

$$86 + p = 234$$
$$86 + p - \square = 234 - 86$$
$$p = \square$$

4. **Check.** If William reads 148 more pages, he will have read a total of 86 + 148 pages, or ☐ pages.
5. **State.** William has ☐ more pages to read.

Answer

4. 148 pages

Guided Solution:

4. p, 234; 86, 148; 234; 148

EXAMPLE 3 *Total Cost of Chairs.* What is the total cost of 6 Adirondack chairs if each one costs $169?

1. **Familiarize.** We make a drawing and let C = the cost of 6 chairs.

2. **Translate.** We translate to an equation:

Number of chairs	times	Cost of each chair	is	Total cost
↓	↓	↓	↓	↓
6	×	169	=	C.

3. **Solve.** This sentence tells us what to do. We multiply.

$$6 \times 169 = C$$
$$1014 = C$$

$$\begin{array}{r} 1\ 6\ 9 \\ \times \quad 6 \\ \hline 1\ 0\ 1\ 4 \end{array}$$

4. **Check.** We have an answer, 1014, that is greater than the cost of any one chair, which is reasonable. We can also check by estimating:

$$6 \times 169 \approx 6 \times 170 = 1020 \approx 1014.$$

The answer checks.

5. **State.** The total cost of 6 chairs is $1014.

Do Exercise 5. ▶

5. *Total Cost of Gas Grills.* What is the total cost of 14 gas grills, each with 520 sq in. of total cooking surface, if each one costs $398?

EXAMPLE 4 *Area of an Oriental Rug.* The dimensions of the oriental rug in the Fosters' front hallway are 42 in. by 66 in. What is the area of the rug?

1. **Familiarize.** We let A = the area of the rug and use the formula for the area of a rectangle, $A = \text{length} \cdot \text{width} = l \cdot w$. Since we usually consider length to be larger than width, we will let $l = 66$ in. and $w = 42$ in.

2. **Translate.** We substitute in the formula:

$$A = l \cdot w = 66 \cdot 42.$$

Answer

5. $5572

3. Solve. We carry out the multiplication.

$$A = 66 \cdot 42 \qquad \begin{array}{r} 66 \\ \times\ 42 \\ \hline 132 \\ 2640 \\ \hline 2772 \end{array}$$

$$A = 2772$$

4. Check. We can repeat the calculation. We can also round and estimate:

$$66 \times 42 \approx 70 \times 40 = 2800 \approx 2772.$$

The answer checks.

5. State. The area of the rug is 2772 sq in.

6. *Bed Sheets.* The dimensions of a flat sheet for a queen-size bed are 90 in. by 102 in. What is the area of the sheet?

◀ **Do Exercise 6.**

EXAMPLE 5 *Packages of Gum.* A candy company produces 3304 sticks of gum. How many 12-stick packages can be filled? How many sticks will be left over?

1. Familiarize. We make a drawing to visualize the situation and let $n =$ the number of 12-stick packages that can be filled. The problem can be considered as repeated subtraction, taking successive sets of 12 sticks and putting them into n packages.

2. Translate. We translate to an equation:

Number of sticks	divided by	Number in each package	is	Number of packages
↓	↓	↓	↓	↓
3304	$\div$	12	$=$	n.

3. Solve. We solve the equation by carrying out the division.

$$3304 \div 12 = n$$

$$275\,\text{R}\,4 = n$$

$$\begin{array}{r} 275 \\ 12\overline{)3304} \\ \underline{24} \\ 90 \\ \underline{84} \\ 64 \\ \underline{60} \\ 4 \end{array}$$

Answer

6. 9180 sq in.

4. **Check.** We can check by multiplying the number of packages by 12 and adding the remainder, 4:

$$12 \cdot 275 = 3300,$$
$$3300 + 4 = 3304.$$

5. **State.** Thus, 275 twelve-stick packages of gum can be filled. There will be 4 sticks left over.

Do Exercise 7. ▶

7. ***Packages of Gum.*** The candy company in Example 5 also produces 6-stick packages. How many 6-stick packages can be filled with 2269 sticks of gum? How many sticks will be left over?

EXAMPLE 6 ***Automobile Mileage.*** A 2013 Toyota Matrix gets 26 miles per gallon (mpg) in city driving. How many gallons will it use in 4758 mi of city driving?

Source: Toyota

1. **Familiarize.** We make a drawing and let $g =$ the number of gallons of gasoline used in 4758 mi of city driving.

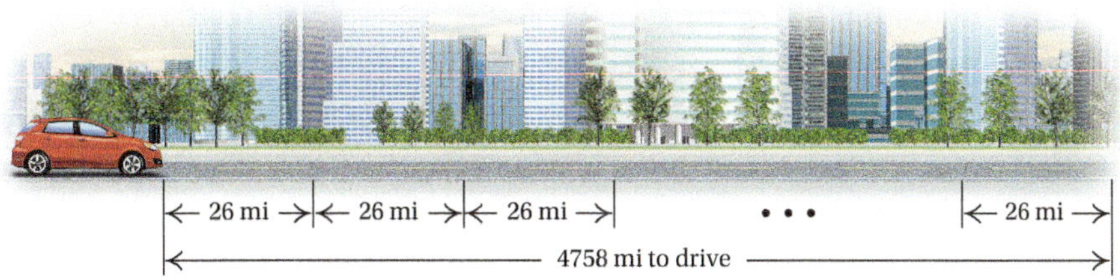

2. **Translate.** Repeated addition or multiplication applies here.

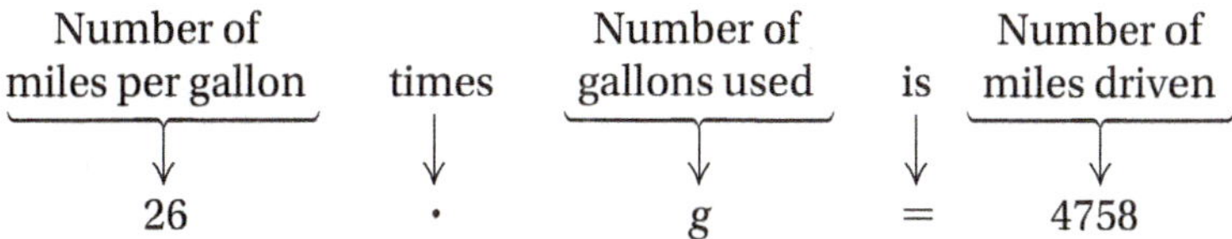

3. **Solve.** To solve the equation, we divide by 26 on both sides.

$$26 \cdot g = 4758$$
$$\frac{26 \cdot g}{26} = \frac{4758}{26}$$
$$g = 183$$

$$\begin{array}{r} 183 \\ 26\overline{)4758} \\ \underline{26} \\ 215 \\ \underline{208} \\ 78 \\ \underline{78} \\ 0 \end{array}$$

4. **Check.** To check, we multiply 183 by 26.

$$\begin{array}{r} 183 \\ \times\ \ 26 \\ \hline 1098 \\ 3660 \\ \hline 4758 \end{array}$$

The answer checks.

5. **State.** The Toyota Matrix will use 183 gal of gasoline.

Do Exercise 8. ▶

8. ***Automobile Mileage.*** A 2013 Toyota Matrix gets 32 miles per gallon (mpg) in highway driving. How many gallons will it use in 2528 mi of highway driving?

Source: Toyota

Answers

7. 378 packages with 1 stick left over

8. 79 gal

Multistep Problems

Sometimes we must use more than one operation to solve a problem, as in the following example.

EXAMPLE 7 *Weight Loss.* To lose one pound, you must burn about 3500 calories in excess of what you already burn doing your regular daily activities. The following chart shows how long a person must engage in several types of exercise in order to burn 100 calories. For how long would a person have to run at a brisk pace in order to lose one pound?

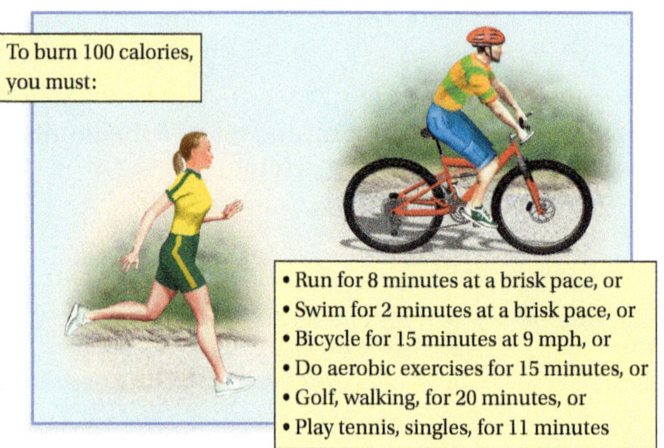

To burn 100 calories, you must:
• Run for 8 minutes at a brisk pace, or
• Swim for 2 minutes at a brisk pace, or
• Bicycle for 15 minutes at 9 mph, or
• Do aerobic exercises for 15 minutes, or
• Golf, walking, for 20 minutes, or
• Play tennis, singles, for 11 minutes

1. **Familiarize.** This is a multistep problem. We will first find how many hundreds are in 3500. This will tell us how many times a person must run for 8 min in order to lose one pound. Then we will find the total number of minutes required for the weight loss.

   We let $x =$ the number of hundreds in 3500 and $t =$ the time it takes to lose one pound.

2. **Translate.** We translate to two equations.

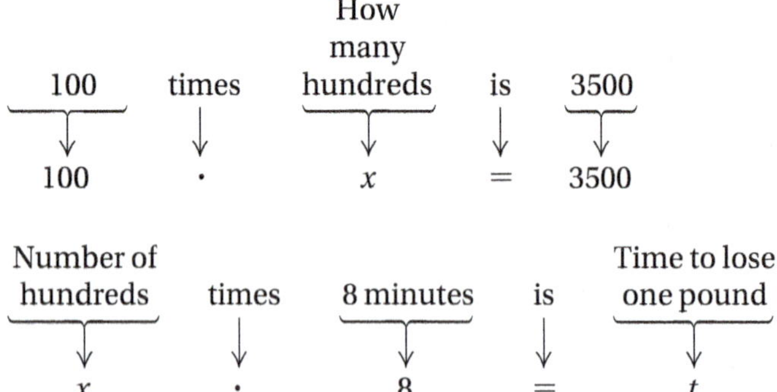

3. **Solve.** We divide by 100 on both sides of the first equation to find x.

$$100 \cdot x = 3500$$
$$\frac{100 \cdot x}{100} = \frac{3500}{100}$$
$$x = 35$$

$$\begin{array}{r} 35 \\ 100\overline{)3500} \\ \underline{300} \\ 500 \\ \underline{500} \\ 0 \end{array}$$

Then we use the fact that $x = 35$ to find t.

$$x \cdot 8 = t$$
$$35 \cdot 8 = t$$
$$280 = t$$

$$\begin{array}{r} 35 \\ \times\ \ 8 \\ \hline 280 \end{array}$$

4. **Check.** Suppose you run for 280 min. For every 8 min of running, you burn 100 calories. Since $280 \div 8 = 35$, there are 35 groups of 8 min in 280 min, so you will burn $35 \times 100 = 3500$ calories.
5. **State.** You must run for 280 min, or 4 hr 40 min, at a brisk pace in order to lose one pound.

The key words, phrases, and concepts in the following table are useful when translating the problems to equations.

Key Words, Phrases, and Concepts

ADDITION (+)	SUBTRACTION (−)	MULTIPLICATION (×)	DIVISION (÷)
add	subtract	multiply	divide
added to	subtracted from	multiplied by	divided by
sum	difference	product	quotient
total	minus	times	repeated subtraction
plus	less than	of	missing factor
more than	decreased by	repeated addition	finding equal quantities
increased by	take away	rectangular arrays	
	how much more		

The following tips are also helpful in problem solving.

PROBLEM-SOLVING TIPS

1. Look for patterns when solving problems.
2. When translating in mathematics, consider the dimensions of the variables and constants in the equation. The variables that represent length should all be in the same unit, those that represent money should all be in dollars or all in cents, and so on.
3. Make sure that units appear in the answer whenever appropriate and that you completely answer the original problem.

9. **Weight Loss.** Use the information in Example 7 to determine how long an individual must swim at a brisk pace in order to lose one pound.

1. **Familiarize.** Let $x =$ the number of hundreds in 3500. Let $t =$ the time it takes to lose one pound.
2. **Translate.**

   $100 \cdot x = \_\_\_\_$

   $x \cdot \_\_\_\_ = t$

3. **Solve.** From Example 7, we know that $x = \_\_\_\_$.

   $x \cdot 2 = t$

   $\_\_\_\_ \cdot 2 = t$

   $\_\_\_\_ = t$

4. **Check.** Since $70 \div 2 = 35$, there are ____ groups of 2 min in 70 min. Thus, you will burn $35 \times 100 = \_\_\_\_$ calories.
5. **State.** You must swim for ____ min, or 1 hr ____ min, in order to lose one pound.

Answer

9. 70 min, or 1 hr 10 min

Guided Solution:
9. 3500, 2; 35, 35, 70, 35, 3500, 70, 10

Translating for Success

The goal of these matching questions is to practice step (2), Translate, of the five-step problem-solving process. Translate each word problem to an equation and select a correct translation from equations A–O.

A. $8 \cdot 52 = n$

B. $69 \cdot n = 76$

C. $73 - 76 - 69 = n$

D. $423 + 73 - 76 - 69 = n$

E. $30 \cdot 10{,}860 = n$

F. $15 \cdot n = 195$

G. $69 + n = 423$

H. $n = 10{,}860 - 300$

I. $n = 423 \div 69$

J. $30 \cdot n = 10{,}860$

K. $15 \cdot 195 = n$

L. $n = 52 - 8$

M. $69 + n = 76$

N. $15 \div 195 = n$

O. $52 + n = 60$

Answers on page A-2

1. *Brick-Mason Expense.* A commercial contractor is building 30 two-unit condominiums in a retirement community. The brick-mason expense for each building is \$10,860. What is the total cost of bricking the buildings?

2. *Heights.* Dean's sons are on the high school basketball team. Their heights are 73 in., 69 in., and 76 in. How much taller is the tallest son than the shortest son?

3. *Account Balance.* James has \$423 in his checking account. Then he deposits \$73 and uses his debit card for purchases of \$76 and \$69. How much is left in the account?

4. *Purchasing a Computer.* A computer is on sale for \$423. Jenny has only \$69. How much more does she need to buy the computer?

5. *Purchasing Coffee Makers.* Sara purchases 8 coffee makers for the newly remodeled bed-and-breakfast that she manages. If she pays \$52 for each coffee maker, what is the total cost of her purchase?

6. *Hourly Rate.* Miller Auto Repair charges \$52 per hour for labor. Jackson Auto Care charges \$60 per hour. How much more does Jackson charge than Miller?

7. *College Band.* A college band with 195 members marches in a 15-row formation in the homecoming halftime performance. How many members are in each row?

8. *Shoe Purchase.* A college football team purchases 15 pairs of shoes at \$195 a pair. What is the total cost of this purchase?

9. *Loan Payment.* Kendra's uncle loans her \$10,860, interest free, to buy a car. The loan is to be paid off in 30 payments. How much is each payment?

10. *College Enrollment.* At the beginning of the fall term, the total enrollment in Lakeview Community College was 10,860. By the end of the first two weeks, 300 students had withdrawn. How many students were then enrolled?

1.8 Exercise Set

For Extra Help
MyMathLab®
MathXL® PRACTICE | WATCH | READ | REVIEW

✓ Reading Check

List the steps of the problem-solving strategy in order, using the choices given below. The last step is already listed.

Check   Familiarize   Solve   Translate

RC1. 1. ____________.

RC2. 2. ____________.

RC3. 3. ____________.

RC4. 4. ____________.

5. State.

Towers Never Built. The buildings shown in the figure below were designed but never completed. Use the information to do Exercises 1–4.

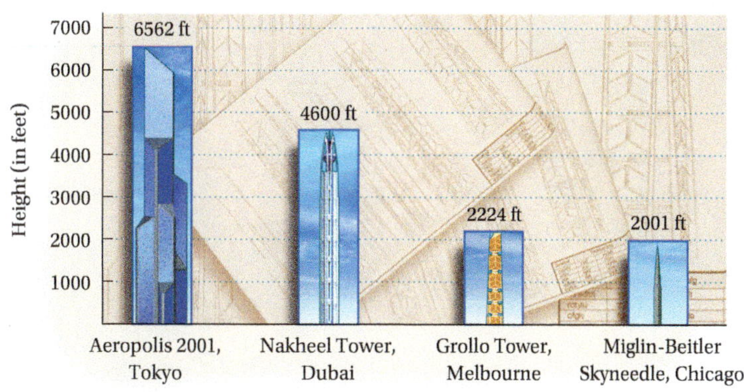

SOURCE: http://en.wikipedia.org/wiki/Proposed_tall_buildings_and_structures

1. How much taller would the Aeropolis 2001 have been than the Nakheel Tower?

2. How much taller would the Grollo Tower have been than the Miglin-Beitler Skyneedle?

3. The Willis Tower (formerly the Sears Tower) is the tallest building in Chicago. If the Miglin-Beitler Skyneedle had been built, it would have been 551 ft higher than the Willis Tower. What is the height of the Willis Tower?

4. The Burj Khalifa is the tallest building in Dubai. If the Nakheel Tower had been built, it would have been 1883 ft higher than the Burj Khalifa. What is the height of the Burj Khalifa?

5. *Caffeine Content.* An 8-oz serving of Red Bull energy drink contains 76 milligrams of caffeine. An 8-oz serving of brewed coffee contains 19 more milligrams of caffeine than the energy drink. How many milligrams of caffeine does the 8-oz serving of coffee contain?

Source: The Mayo Clinic

6. *Caffeine Content.* Hershey's 6-oz milk chocolate almond bar contains 25 milligrams of caffeine. A 20-oz bottle of Coca-Cola has 32 more milligrams of caffeine than the Hershey bar. How many milligrams of caffeine does the 20-oz bottle of Coca-Cola have?

Source: *National Geographic*, "Caffeine," by T. R. Reid, January 2005

7. A carpenter drills 216 holes in a rectangular array to construct a pegboard. There are 12 holes in each row. How many rows are there?

8. Lou arranges 504 entries on a spreadsheet in a rectangular array that has 36 rows. How many entries are in each row?

9. ***Olympics.*** There were 302 events in the 2012 Summer Olympics in London, England. This was 259 more events than there were in the first modern Olympic games in Athens, Greece, in 1896. How many events were there in 1896?

Sources: *USA Today* research; infoplease.com

10. ***Drilling Activity.*** In 2011, there were 984 rotary rigs drilling for crude oil in the United States. This was 687 more rigs than were active in 2007. Find the number of active rotary oil rigs in 2007.

Source: Energy Information Administration

11. ***Boundaries between Countries.*** The boundary between mainland United States and Canada including the Great Lakes is 3987 mi long. The length of the boundary between the United States and Mexico is 1933 mi. How much longer is the Canadian border?

Source: U.S. Geological Survey

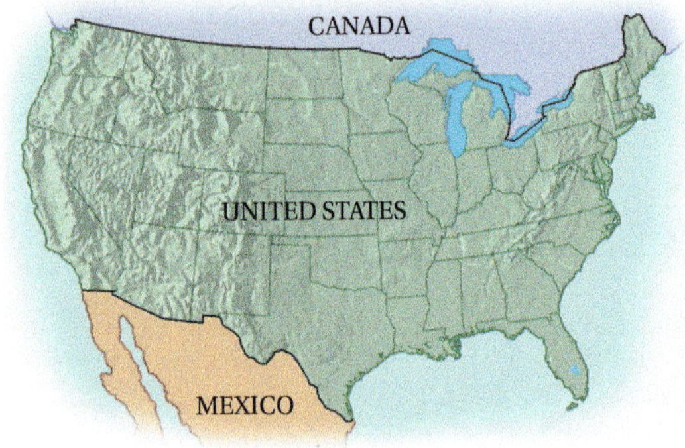

12. ***Longest Rivers.*** The longest river in the world is the Nile in Africa at about 4135 mi. The longest river in the United States is the Missouri–Mississippi at about 3860 mi. How much longer is the Nile?

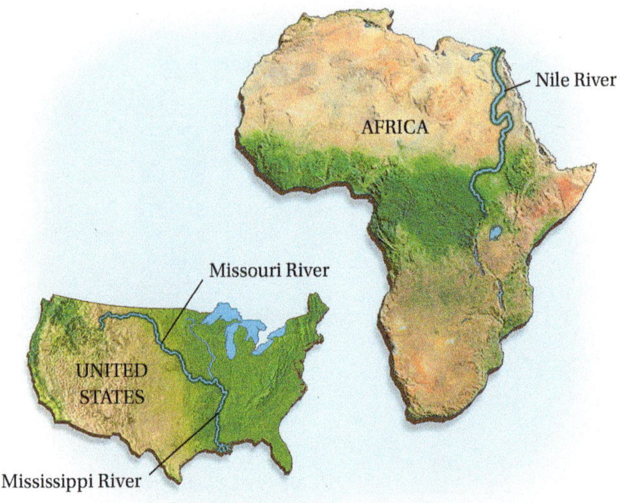

13. ***Pixels.*** A high-definition television (HDTV) screen consists of small rectangular dots called *pixels*. How many pixels are there on a screen that has 1080 rows with 1920 pixels in each row?

14. ***Crossword Puzzle.*** The *USA Today* crossword puzzle is a rectangle containing 15 rows with 15 squares in each row. How many squares does the puzzle have altogether?

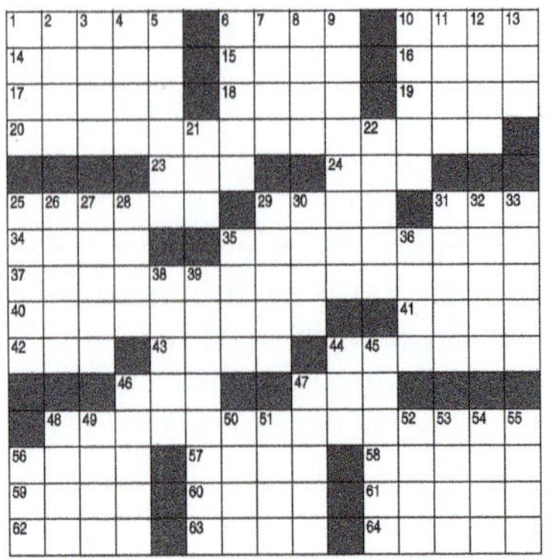

15. There are 24 hr in a day and 7 days in a week. How many hours are there in a week?

16. There are 60 min in an hour and 24 hr in a day. How many minutes are there in a day?

Housing Costs. The graph below shows the average monthly rent for a one-bedroom apartment in several cities in November 2012. Use this graph to do Exercises 17–22.

Average Rent for a One-Bedroom Apartment

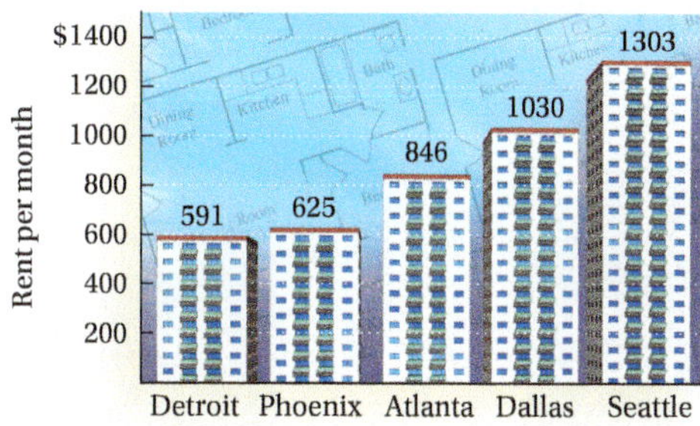

SOURCE: rentjungle.com

17. How much higher is the average monthly rent in Seattle than in Dallas?

18. How much lower is the average monthly rent in Phoenix than in Atlanta?

19. Phil, Scott, and Julio plan to rent a one-bedroom apartment in Detroit immediately after graduation, sharing the rent equally. What average monthly rent can each of them expect to pay?

20. Maria and her sister Theresa plan to share a one-bedroom apartment in Dallas, dividing the monthly rent equally between them. About how much can each of them expect to pay?

21. On average, how much rent would a tenant pay for a one-bedroom apartment in Detroit during a 12-month period?

22. On average, how much rent would a tenant pay for a one-bedroom apartment in Seattle during a 6-month period?

23. *Colonial Population.* Before the establishment of the U.S. Census in 1790, it was estimated that the colonial population in 1780 was 2,780,400. This was an increase of 2,628,900 from the population in 1680. What was the colonial population in 1680?

Source: *Time Almanac*

24. *Interstate Speed Limits.* The speed limit for passenger cars on interstate highways in rural areas in Montana is 75 mph. This is 10 mph faster than the speed limit for trucks on the same roads. What is the speed limit for trucks?

25. *Yard-Sale Profit.* Ruth made \$312 at her yard sale and divided the money equally among her four grandchildren. How much did each child receive?

26. *Paper Measures.* A quire of paper consists of 25 sheets, and a ream of paper consists of 500 sheets. How many quires are in a ream?

27. *Parking Rates.* The most expensive parking in the United States is found in midtown New York City, where the average rate is \$585 per month. This is \$545 per month more than in the city with the least expensive rate, Bakersfield, California. What is the average monthly parking rate in Bakersfield?

Source: Colliers International

28. *Trade Balance.* In 2011, international visitors spent \$153,000,000,000 traveling in the United States, while Americans spent \$110,200,000,000 traveling abroad. How much more was spent by visitors to the United States than by Americans traveling abroad?

Source: U.S. Office of Travel and Tourism Industries

29. *Refrigerator Purchase.* Gourmet Deli has a chain of 24 restaurants. It buys a commercial refrigerator for each store at a cost of \$1019 each. Determine the total cost of the purchase.

30. *Microwave Purchase.* Each room in the new dorm at Bridgeway College has a small kitchen. To furnish the kitchens, the college buys 96 microwave ovens at \$88 each. Determine the total cost of the purchase.

31. *"Seinfeld."* A local television station plans to air the 177 episodes of the long-running comedy series "Seinfeld." If the station airs 5 episodes per week, how many full weeks will pass before it must begin re-airing previously shown episodes? How many unaired episodes will be shown the following week before the previously aired episodes are rerun?

32. *"Everybody Loves Raymond."* The popular television comedy series "Everybody Loves Raymond" had 208 scripted episodes and 2 additional episodes consisting of clips from previous shows. A local television station plans to air the 208 scripted episodes, showing 5 episodes per week. How many full weeks will pass before it must begin re-airing episodes? How many unaired episodes will be shown the following week before the previously aired episodes are rerun?

33. *Crossword Puzzle.* The *Los Angeles Times* crossword puzzle is a rectangle containing 441 squares arranged in 21 rows. How many columns does the puzzle have?

34. *Mailing Labels.* A box of mailing labels contains 750 labels on 25 sheets. How many labels are on each sheet?

35. ***Automobile Mileage.*** The 2013 Hyundai Tucson GLS gets 30 miles per gallon (mpg) in highway driving. How many gallons will it use in 7080 mi of highway driving?

Source: Hyundai

36. ***Automobile Mileage.*** The 2013 Volkswagen Jetta (5 cylinder) gets 24 miles per gallon (mpg) in city driving. How many gallons will it use in 3960 mi of city driving?

Source: Volkswagen of America, Inc.

37. ***High School Court.*** The standard basketball court used by high school players has dimensions of 50 ft by 84 ft.

   a) What is its area?
   b) What is its perimeter?

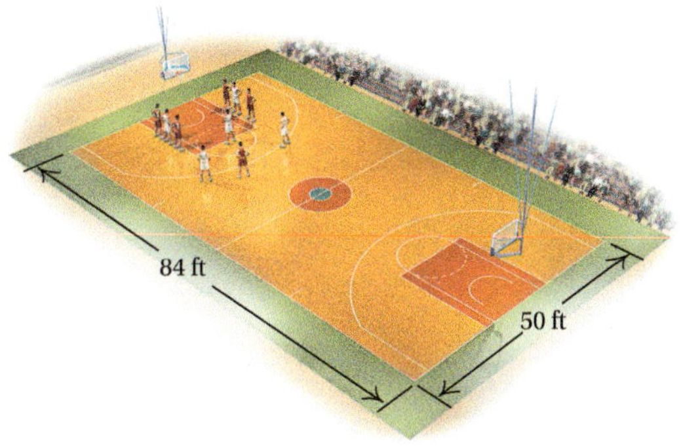

38. ***College Court.*** The standard basketball court used by college players has dimensions of 50 ft by 94 ft.

   a) What is its area?
   b) What is its perimeter?
   c) How much greater is the area of a college court than a high school court? (See Exercise 37.)

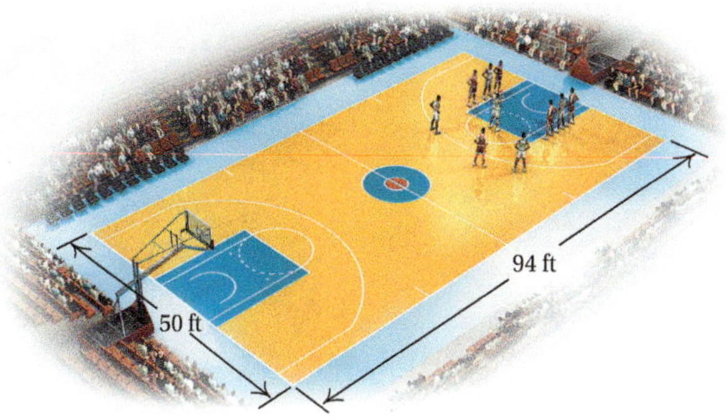

39. Copies of this book are usually shipped from the warehouse in cartons containing 24 books each. How many cartons are needed to ship 1344 books?

40. The H. J. Heinz Company ships 16-oz bottles of ketchup in cartons containing 12 bottles each. How many cartons are needed to ship 528 bottles of ketchup?

41. ***Map Drawing.*** A map has a scale of 215 mi to the inch. How far apart *in reality* are two cities that are 3 in. apart on the map? How far apart *on the map* are two cities that, in reality, are 1075 mi apart?

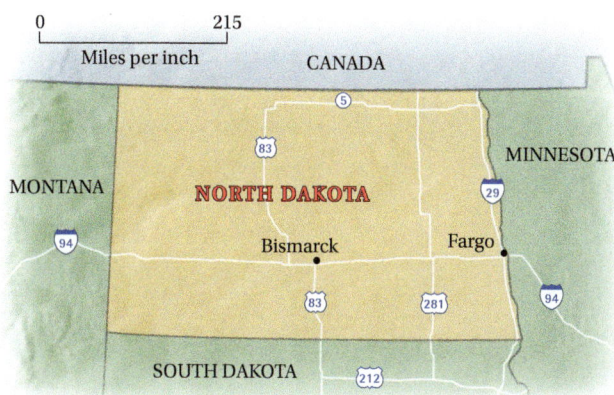

42. ***Map Drawing.*** A map has a scale of 288 mi to the inch. How far apart *on the map* are two cities that, in reality, are 2016 mi apart? How far apart *in reality* are two cities that are 8 in. apart on the map?

43. *Loan Payments.* Dana borrows $5928 for a used car. The loan is to be paid off in 24 equal monthly payments. How much is each payment (excluding interest)?

44. *Home Improvement Loan.* The Van Reken family borrows $7824 to build a detached garage next to their home. The loan is to be paid off in equal monthly payments of $163 (excluding interest). How many months will it take to pay off the loan?

Refer to the information in Example 7 to do Exercises 45 and 46.

45. For how long must you do aerobic exercises in order to lose one pound?

46. For how long must you bicycle at 9 mph in order to lose one pound?

New Jobs. Many of the fastest-growing occupations in the United States require education beyond a high school diploma. The following table lists some of these and gives the projected numbers of new jobs expected to be created between 2010 and 2020. Use the information in the table for Exercises 47 and 48.

New Jobs Created, 2010–2020

JOB	NUMBER
Registered nurse	711,900
Postsecondary teacher	305,700
Nursing aide	302,000
Elementary school teacher	248,800
Accountant	190,700
Licensed practical nurse	168,500

SOURCE: U.S. Bureau of Labor Statistics

47. The U.S. Bureau of Labor Statistics predicts that between 2010 and 2020, there will be 1,014,100 more new jobs created for registered nurses, nursing aides, and licensed practical nurses than there will be for physicians. How many new jobs will be created for physicians between 2010 and 2020?

48. The U.S. Bureau of Labor Statistics predicts that between 2010 and 2020, there will be 484,600 more new jobs created for postsecondary teachers and elementary teachers than there will be for secondary teachers. How many new jobs will be created for secondary teachers between 2010 and 2020?

49. *Seating Configuration.* The seats in the Boeing 737-500 airplanes in United Airlines' North American fleet are configured with 2 rows of 4 seats across in first class and 16 rows of 6 seats across in economy class. Determine the total seating capacity of one of these planes.

Source: United Airlines

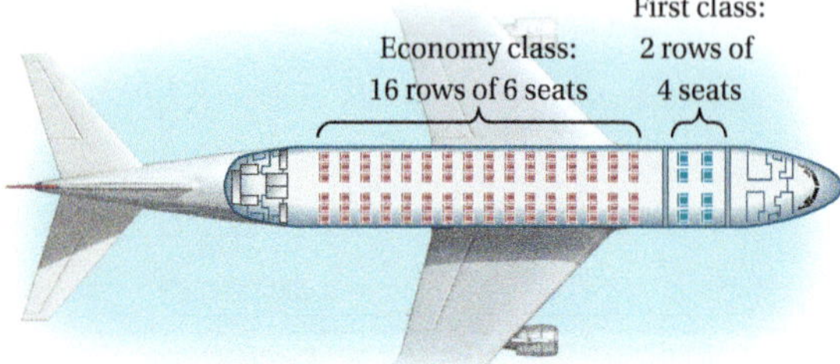

50. *Seating Configuration.* The seats in the Airbus 320 airplanes in United Airlines' North American fleet are configured with 3 rows of 4 seats across in first class and 21 rows of 6 seats across in economy class. Determine the total seating capacity of one of these planes.

Source: United Airlines

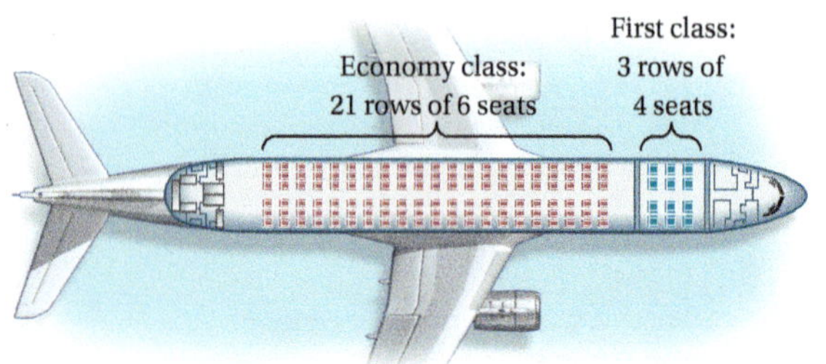

51. Elena buys 5 video games at $64 each and pays for them with $10 bills. How many $10 bills does it take?

52. Pedro buys 5 video games at $64 each and pays for them with $20 bills. How many $20 bills does it take?

53. The balance in Meg's bank account is $568. She uses her debit card for purchases of $46, $87, and $129. Then she deposits $94 in the account after returning a textbook. How much is left in her account?

54. The balance in Dylan's bank account is $749. He uses his debit card for purchases of $34 and $65. Then he makes a deposit of $123 from his paycheck. What is the new balance?

55. *Bones in the Hands and Feet.* There are 27 bones in each human hand and 26 bones in each human foot. How many bones are there in all in the hands and feet?

56. An office for adjunct instructors at a community college has 6 bookshelves, each of which is 3 ft wide. The office is moved to a new location that has dimensions of 16 ft by 21 ft. Is it possible for the bookshelves to be put side by side on the 16-ft wall?

Skill Maintenance

57. Add: [1.2a]

$$\begin{array}{r} 6\,2\,5\,4 \\ 1\,5\,3\,7 \\ +\ \ 4\,8\,2 \\ \hline \end{array}$$

58. Subtract: [1.3a]

$$\begin{array}{r} 9\,6\,0\,2 \\ -\ 1\,8\,4\,3 \\ \hline \end{array}$$

59. Multiply: [1.4a]

$$\begin{array}{r} 3\,4\,0\,5 \\ \times\ 2\,3\,7 \\ \hline \end{array}$$

60. Divide: [1.5a]

$$32\overline{)4708}$$

61. Find the perimeter of the figure. [1.2b]

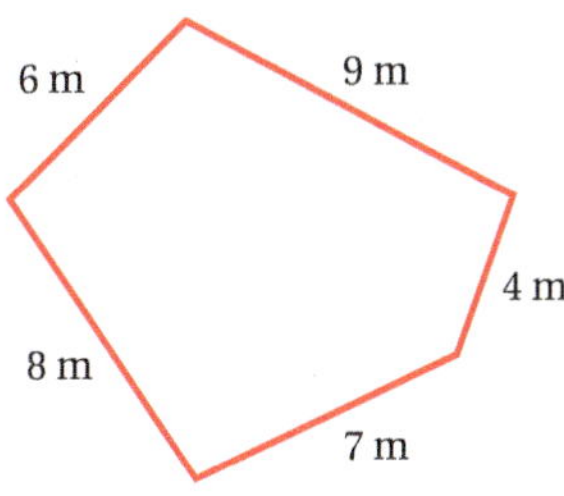

62. Find the area of the region. [1.4b]

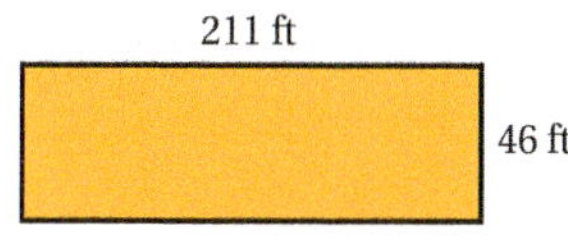

63. Estimate 238×596 by rounding to the nearest hundred. [1.6b]

64. Solve: $x + 15 = 81$. [1.7b]

Synthesis

65. *Speed of Light.* Light travels about 186,000 miles per second (mi/sec) in a vacuum such as in outer space. In ice it travels about 142,000 mi/sec, and in glass it travels about 109,000 mi/sec. In 18 sec, how many more miles will light travel in a vacuum than in ice? than in glass?

66. Carney Community College has 1200 students. Each instructor teaches 4 classes, and each student takes 5 classes. There are 30 students and 1 instructor in each classroom. How many instructors are there at Carney Community College?

1.9 Exponential Notation and Order of Operations

OBJECTIVES

a Write exponential notation for products such as $4 \cdot 4 \cdot 4$.

b Evaluate exponential notation.

c Simplify expressions using the rules for order of operations.

d Remove parentheses within parentheses.

SKILL TO REVIEW

Objective 1.4a: Multiply whole numbers.

Multiply.

1. $5 \times 5 \times 5$

2. $2 \times 2 \times 2 \times 2 \times 2$

a WRITING EXPONENTIAL NOTATION

Consider the product $3 \cdot 3 \cdot 3 \cdot 3$. Such products occur often enough that mathematicians have found it convenient to create a shorter notation, called **exponential notation**, for them. For example,

$$\underbrace{3 \cdot 3 \cdot 3 \cdot 3}_{4 \text{ factors}} \text{ is shortened to } 3^4.$$

(4 is the exponent; 3 is the base.)

We read exponential notation as follows.

NOTATION	WORD DESCRIPTION
3^4	"three to the fourth power," or "the fourth power of three"
5^3	"five cubed," or "the cube of five," or "five to the third power," or "the third power of five"
7^2	"seven squared," or "the square of seven," or "seven to the second power," or "the second power of seven"

The wording "seven squared" for 7^2 is derived from the fact that a square with side s has area A given by $A = s^2$.

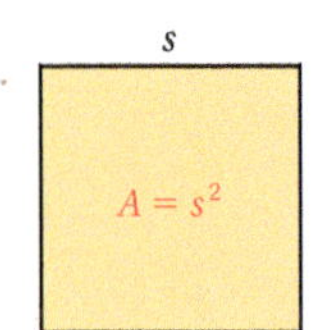

An expression like $3 \cdot 5^2$ is read "three times five squared," or "three times the square of five."

EXAMPLE 1 Write exponential notation for $10 \cdot 10 \cdot 10 \cdot 10 \cdot 10$.

Exponential notation is 10^5. 5 is the *exponent.* 10 is the *base.*

EXAMPLE 2 Write exponential notation for $2 \cdot 2 \cdot 2$.

Exponential notation is 2^3.

◀ Do Margin Exercises 1–4.

Write exponential notation.

1. $5 \cdot 5 \cdot 5 \cdot 5$

2. $5 \cdot 5 \cdot 5 \cdot 5 \cdot 5 \cdot 5$

3. $10 \cdot 10$

4. $10 \cdot 10 \cdot 10 \cdot 10$

Answers

Skill to Review:
1. 125 **2.** 32

Margin Exercises:
1. 5^4 **2.** 5^6 **3.** 10^2 **4.** 10^4

b EVALUATING EXPONENTIAL NOTATION

We evaluate exponential notation by rewriting it as a product and then computing the product.

EXAMPLE 3 Evaluate: 10^3.

$10^3 = 10 \cdot 10 \cdot 10 = 1000$

Caution!

10^3 does not mean $10 \cdot 3$.

EXAMPLE 4 Evaluate: 5^4.

$5^4 = 5 \cdot 5 \cdot 5 \cdot 5 = 625$

Do Exercises 5–8.

GS **5.** Evaluate: 10^4.

$10^4 = \square \cdot \square \cdot \square \cdot \square = \square$

Evaluate.

6. 10^2

7. 8^3

8. 2^5

CALCULATOR CORNER

Exponential Notation Many calculators have a y^x or ^ key for raising a base to a power. To find 16^3, for example, we press 1 6 y^x 3 = or 1 6 ^ 3 =. The result is 4096.

EXERCISES Use a calculator to find each of the following.

1. 3^5
2. 5^6
3. 12^4
4. 2^{11}

c SIMPLIFYING EXPRESSIONS

Suppose we have a calculation like the following:

$3 + 4 \cdot 8.$

How do we find the answer? Do we add 3 to 4 and then multiply by 8, or do we multiply 4 by 8 and then add 3? In the first case, the answer is 56. In the second, the answer is 35. We agree to compute as in the second case:

$3 + 4 \cdot 8 = 3 + 32 = 35.$

The following rules are an agreement regarding the order in which we perform operations. These are the rules that computers and most scientific calculators use to do computations.

RULES FOR ORDER OF OPERATIONS

1. Do all calculations within parentheses ( ), brackets [ ], or braces { } before operations outside.
2. Evaluate all exponential expressions.
3. Do all multiplications and divisions in order from left to right.
4. Do all additions and subtractions in order from left to right.

EXAMPLE 5 Simplify: $16 \div 8 \cdot 2$.

There are no parentheses or exponents, so we begin with the third step.

$16 \div 8 \cdot 2 = 2 \cdot 2$
$= 4$
} Doing all multiplications and divisions in order from left to right

EXAMPLE 6 Simplify: $7 \cdot 14 - (12 + 18)$.

$7 \cdot 14 - (12 + 18) = 7 \cdot 14 - 30$ Carrying out operations inside parentheses

$= 98 - 30$ Doing all multiplications and divisions

$= 68$ Doing all additions and subtractions

Do Exercises 9–12.

Simplify.

9. $93 - 14 \cdot 3$

10. $104 \div 4 + 4$

11. $25 \cdot 26 - (56 + 10)$

12. $75 \div 5 + (83 - 14)$

Answers

5. 10,000 **6.** 100 **7.** 512 **8.** 32 **9.** 51 **10.** 30 **11.** 584 **12.** 84

Guided Solution:
5. 10, 10, 10, 10, 10,000

EXAMPLE 7 Simplify and compare: $23 - (10 - 9)$ and $(23 - 10) - 9$.

We have

$$23 - (10 - 9) = 23 - 1 = 22;$$
$$(23 - 10) - 9 = 13 - 9 = 4.$$

We can see that $23 - (10 - 9)$ and $(23 - 10) - 9$ represent different numbers. Thus subtraction is not associative.

Do Exercises 13 and 14.

Simplify and compare.

13. $64 \div (32 \div 2)$ and $(64 \div 32) \div 2$

14. $(28 + 13) + 11$ and $28 + (13 + 11)$

EXAMPLE 8 Simplify: $7 \cdot 2 - (12 + 0) \div 3 - (5 - 2)$.

$7 \cdot 2 - (12 + 0) \div 3 - (5 - 2)$

$= 7 \cdot 2 - 12 \div 3 - 3$ Carrying out operations inside parentheses

$= 14 - 4 - 3$ Doing all multiplications and divisions in order from left to right

$= 10 - 3$

$= 7$ Doing all additions and subtractions in order from left to right

Do Exercise 15.

15. Simplify:

GS

$9 \times 4 - (20 + 4) \div 8 - (6 - 2)$.

$9 \times 4 - (20 + 4) \div 8 - (6 - 2)$

$= 9 \times 4 - \square \div 8 - \square$

$= \square - 24 \div 8 - 4$

$= 36 - \square - 4$

$= \square - 4$

$= \square$

EXAMPLE 9 Simplify: $15 \div 3 \cdot 2 \div (10 - 8)$.

$15 \div 3 \cdot 2 \div (10 - 8)$

$= 15 \div 3 \cdot 2 \div 2$ Carrying out operations inside parentheses

$= 5 \cdot 2 \div 2$

$= 10 \div 2$

$= 5$ Doing all multiplications and divisions in order from left to right

Do Exercises 16–18.

Simplify.

16. $5 \cdot 5 \cdot 5 + 26 \cdot 71 - (16 + 25 \cdot 3)$

17. $30 \div 5 \cdot 2 + 10 \cdot 20 + 8 \cdot 8 - 23$

18. $95 - 2 \cdot 2 \cdot 2 \cdot 5 \div (24 - 4)$

EXAMPLE 10 Simplify: $4^2 \div (10 - 9 + 1)^3 \cdot 3 - 5$.

$4^2 \div (10 - 9 + 1)^3 \cdot 3 - 5$

$= 4^2 \div (1 + 1)^3 \cdot 3 - 5$ Subtracting inside parentheses

$= 4^2 \div 2^3 \cdot 3 - 5$ Adding inside parentheses

$= 16 \div 8 \cdot 3 - 5$ Evaluating exponential expressions

$= 2 \cdot 3 - 5$

$= 6 - 5$ Doing all multiplications and divisions in order from left to right

$= 1$ Subtracting

Do Exercises 19–21.

Simplify.

19. $5^3 + 26 \cdot 71 - (16 + 25 \cdot 3)$

20. $(1 + 3)^3 + 10 \cdot 20 + 8^2 - 23$

21. $81 - 3^2 \cdot 2 \div (12 - 9)$

Answers

13. 4; 1 **14.** 52; 52 **15.** 29 **16.** 1880 **17.** 253 **18.** 93 **19.** 1880 **20.** 305 **21.** 75

Guided Solution:

15. 24, 4, 36, 3, 33, 29

EXAMPLE 11 Simplify: $2^9 \div 2^6 \cdot 2^3$.

$$2^9 \div 2^6 \cdot 2^3 = 512 \div 64 \cdot 8$$ Since there are no parentheses, we evaluate the exponential expressions.

$$= 8 \cdot 8$$
$$= 64$$
Doing all multiplications and divisions in order from left to right

Do Exercise 22.

22. Simplify: $2^3 \cdot 2^8 \div 2^9$.

Averages

In order to find the average of a set of numbers, we use addition and then division. For example, the average of 2, 3, 6, and 9 is found as follows.

$$\text{Average} = \frac{2 + 3 + 6 + 9}{4} = \frac{20}{4} = 5$$

The number of addends is 4.

Divide by 4.

The fraction bar acts as a grouping symbol, so

$$\frac{2 + 3 + 6 + 9}{4} \text{ is equivalent to } (2 + 3 + 6 + 9) \div 4.$$

Thus we are using order of operations when we compute an average.

AVERAGE

The **average** of a set of numbers is the sum of the numbers divided by the number of addends.

EXAMPLE 12 *National Parks.* Since 1995, four national parks have been established in the United States. The sizes of these parks are shown in the figure below. Determine the average size of the four parks.

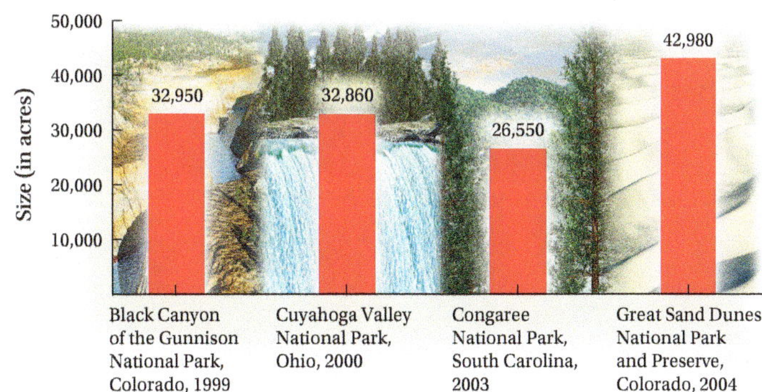

SOURCE: us-national-parks.findthedata.org

CALCULATOR CORNER

Order of Operations We can test whether a calculator is programmed to follow the rules for order of operations by entering a calculation such as $3 + 4 \cdot 2$. Using the rules for order of operations, the value of this expression is 11. If a calculator returns the result 14 instead of 11, that calculator performs operations as they are entered rather than following the rules for order of operations. When using such a calculator, we have to enter the operations in the order in which we want them performed.

EXERCISES Simplify.

1. $84 - 5 \cdot 7$

2. $80 + 50 \div 10$

3. $3^2 + 9^2 \div 3$

4. $4^4 \div 64 - 4$

5. $15 \cdot 7 - (23 + 9)$

6. $(4 + 3)^2$

Answer

22. 4

The average is given by

$$\frac{32{,}950 + 32{,}860 + 26{,}550 + 42{,}980}{4} = \frac{135{,}340}{4} = 33{,}835.$$

The average size of the four national parks is 33,835 acres.

Do Exercise 23.

d REMOVING PARENTHESES WITHIN PARENTHESES

When parentheses occur within parentheses, we can make them different shapes, such as [ ] (called "brackets") and { } (called "braces"). All of these have the same meaning. When parentheses occur within parentheses, computations in the innermost ones are to be done first.

EXAMPLE 13 Simplify: $[25 - (4 + 3) \cdot 3] \div (11 - 7)$.

$[25 - (4 + 3) \cdot 3] \div (11 - 7)$	
$= [25 - 7 \cdot 3] \div (11 - 7)$	Doing the calculations in the innermost parentheses first
$= [25 - 21] \div (11 - 7)$	Doing the multiplication in the brackets
$= 4 \div 4$	Subtracting
$= 1$	Dividing

EXAMPLE 14 Simplify: $16 \div 2 + \{40 - [13 - (4 + 2)]\}$.

$16 \div 2 + \{40 - [13 - (4 + 2)]\}$	
$= 16 \div 2 + \{40 - [13 - 6]\}$	Doing the calculations in the innermost parentheses first
$= 16 \div 2 + \{40 - 7\}$	Again, doing the calculations in the innermost brackets
$= 16 \div 2 + 33$	Subtracting inside the braces
$= 8 + 33$	Dividing
$= 41$	Adding

Do Exercises 24 and 25.

23. ***Average Number of Career Hits.*** The numbers of career hits of five Hall of Fame baseball players are given in the graph below. Find the average number of career hits of all five.

Career Hits

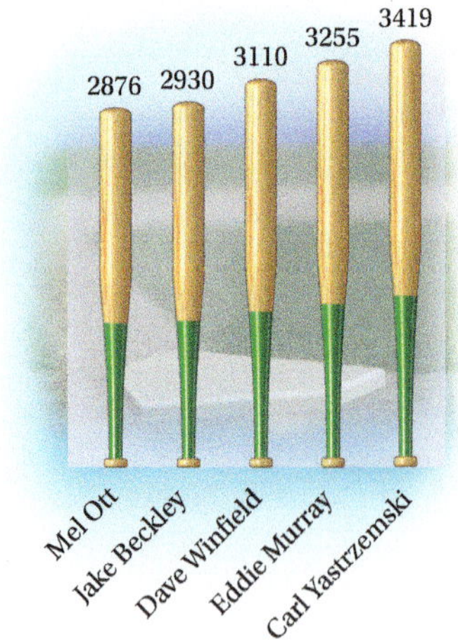

SOURCES: Associated Press; Major League Baseball

Simplify.

24. $9 \times 5 + \{6 \div [14 - (5 + 3)]\}$

25. $[18 - (2 + 7) \div 3] - (31 - 10 \times 2)$ GS

$= [18 - \square \div 3] - (31 - 10 \times 2)$

$= [18 - \square] - (31 - \square)$

$= \square - \square$

$= \square$

Answers

23. 3118 hits **24.** 46 **25.** 4

Guided Solution:

25. 9, 3, 20, 15, 11, 4

1.9 Exercise Set

For Extra Help
MyMathLab®
MathXL® PRACTICE WATCH READ REVIEW

Reading Check

Complete each statement by choosing the correct word or number from below the blank.

RC1. In the expression 5^3, the number 3 is the ______________.
base/exponent

RC2. The expression 9^2 can be read "nine ______________."
cubed/squared

RC3. To calculate $10 - 4 \cdot 2$, we perform the ______________ first.
multiplication/subtraction

RC4. To find the average of 7, 8, and 9, we add the numbers and divide the sum by ______________.
2/3

a Write exponential notation.

1. $3 \cdot 3 \cdot 3 \cdot 3$

2. $2 \cdot 2 \cdot 2 \cdot 2 \cdot 2$

3. $5 \cdot 5$

4. $13 \cdot 13 \cdot 13$

5. $7 \cdot 7 \cdot 7 \cdot 7 \cdot 7$

6. $9 \cdot 9$

7. $10 \cdot 10 \cdot 10$

8. $1 \cdot 1 \cdot 1 \cdot 1$

b Evaluate.

9. 7^2

10. 5^3

11. 9^3

12. 8^2

13. 12^4

14. 10^5

15. 3^5

16. 2^6

c Simplify.

17. $12 + (6 + 4)$

18. $(12 + 6) + 18$

19. $52 - (40 - 8)$

20. $(52 - 40) - 8$

21. $1000 \div (100 \div 10)$

22. $(1000 \div 100) \div 10$

23. $(256 \div 64) \div 4$

24. $256 \div (64 \div 4)$

25. $(2 + 5)^2$

26. $2^2 + 5^2$

27. $(11 - 8)^2 - (18 - 16)^2$

28. $(32 - 27)^3 + (19 + 1)^3$

29. $16 \cdot 24 + 50$

30. $23 + 18 \cdot 20$

31. $83 - 7 \cdot 6$

32. $10 \cdot 7 - 4$

33. $10 \cdot 10 - 3 \cdot 4$

34. $90 - 5 \cdot 5 \cdot 2$

35. $4^3 \div 8 - 4$

36. $8^2 - 8 \cdot 2$

37. $17 \cdot 20 - (17 + 20)$

38. $1000 \div 25 - (15 + 5)$

39. $6 \cdot 10 - 4 \cdot 10$

40. $3 \cdot 8 + 5 \cdot 8$

41. $300 \div 5 + 10$

42. $144 \div 4 - 2$

43. $3 \cdot (2 + 8)^2 - 5 \cdot (4 - 3)^2$

44. $7 \cdot (10 - 3)^2 - 2 \cdot (3 + 1)^2$

45. $4^2 + 8^2 \div 2^2$

46. $6^2 - 3^4 \div 3^3$

47. $10^3 - 10 \cdot 6 - (4 + 5 \cdot 6)$

48. $7^2 + 20 \cdot 4 - (28 + 9 \cdot 2)$

49. $6 \cdot 11 - (7 + 3) \div 5 - (6 - 4)$

50. $8 \times 9 - (12 - 8) \div 4 - (10 - 7)$

51. $120 - 3^3 \cdot 4 \div (5 \cdot 6 - 6 \cdot 4)$

52. $80 - 2^4 \cdot 15 \div (7 \cdot 5 - 45 \div 3)$

53. $2^3 \cdot 2^8 \div 2^6$

54. $2^7 \div 2^5 \cdot 2^4 \div 2^2$

55. Find the average of $64, $97, and $121.

56. Find the average of four test grades of 86, 92, 80, and 78.

57. Find the average of 320, 128, 276, and 880.

58. Find the average of $1025, $775, $2062, $942, and $3721.

d Simplify.

59. $8 \times 13 + \{42 \div [18 - (6 + 5)]\}$

60. $72 \div 6 - \{2 \times [9 - (4 \times 2)]\}$

61. $[14 - (3 + 5) \div 2] - [18 \div (8 - 2)]$

62. $[92 \times (6 - 4) \div 8] + [7 \times (8 - 3)]$

63. $(82 - 14) \times [(10 + 45 \div 5) - (6 \cdot 6 - 5 \cdot 5)]$

64. $(18 \div 2) \cdot \{[(9 \cdot 9 - 1) \div 2] - [5 \cdot 20 - (7 \cdot 9 - 2)]\}$

65. $4 \times \{(200 - 50 \div 5) - [(35 \div 7) \cdot (35 \div 7) - 4 \times 3]\}$

66. $15(23 - 4 \cdot 2)^3 \div (3 \cdot 25)$

67. $\{[18 - 2 \cdot 6] - [40 \div (17 - 9)]\} + \{48 - 13 \times 3 + [(50 - 7 \cdot 5) + 2]\}$

68. $(19 - 2^4)^5 - (141 \div 47)^2$

Skill Maintenance

Solve. [1.7b]

69. $x + 341 = 793$

70. $4197 + x = 5032$

71. $7 \cdot x = 91$

72. $1554 = 42 \cdot y$

73. $6000 = 1102 + t$

74. $10{,}000 = 100 \cdot t$

Solve. [1.8a]

75. *Colorado.* The state of Colorado is roughly the shape of a rectangle that is 273 mi by 382 mi. What is its area?

76. On a long four-day trip, a family bought the following amounts of gasoline for their motor home: 23 gal, 24 gal, 26 gal, and 25 gal. How much gasoline did they buy in all?

Synthesis

Each of the answers in Exercises 77–79 is incorrect. First find the correct answer. Then place as many parentheses as needed in the expression in order to make the incorrect answer correct.

77. $1 + 5 \cdot 4 + 3 = 36$

78. $12 \div 4 + 2 \cdot 3 - 2 = 2$

79. $12 \div 4 + 2 \cdot 3 - 2 = 4$

80. Use one occurrence each of 1, 2, 3, 4, 5, 6, 7, 8, and 9, in order, and any of the symbols $+$, $-$, $\cdot$, $\div$, and ( ) to represent 100.

CHAPTER

1 Summary and Review

Vocabulary Reinforcement

In each of Exercises 1–8, fill in the blank with the correct term from the given list. Some of the choices may not be used and some may be used more than once.

1. The distance around an object is its _______________. [1.2b]
2. The _______________ is the number from which another number is being subtracted. [1.3a]
3. For large numbers, _______________ are separated by commas into groups of three, called _______________. [1.1a]
4. In the sentence $28 \div 7 = 4$, the _______________ is 28. [1.5a]
5. In the sentence $10 \times 1000 = 10{,}000$, 10 and 1000 are called _______________ and 10,000 is called the _______________. [1.4a]
6. The number 0 is called the _______________ identity. [1.2a]
7. The sentence $3 \times (6 \times 2) = (3 \times 6) \times 2$ illustrates the _______________ law of multiplication. [1.4a]
8. We can use the following statement to check division: quotient · _______________ + _______________ = _______________. [1.5a]

associative
commutative
addends
factors
area
perimeter
minuend
subtrahend
product
digits
periods
additive
multiplicative
dividend
quotient
remainder
divisor

Concept Reinforcement

Determine whether each statement is true or false.

_______________ 1. $a > b$ is true when a is to the right of b on the number line. [1.6c]

_______________ 2. Any nonzero number divided by itself is 1. [1.5a]

_______________ 3. For any whole number a, $a \div 0 = 0$. [1.5a]

_______________ 4. Every equation is true. [1.7a]

_______________ 5. The rules for order of operations tell us to multiply and divide before adding and subtracting. [1.9c]

_______________ 6. The average of three numbers is the middle number. [1.9c]

Study Guide

Objective 1.1a Give the meaning of digits in standard notation.

Example What does the digit 7 mean in 2,379,465?

2,3 **7** 9,4 6 5

7 means 7 ten thousands.

Practice Exercise

1. What does the digit 2 mean in 432,079?

Objective 1.2a Add whole numbers.

Example Add: 7368 + 3547.

$$\begin{array}{r} {\scriptstyle 1\ 1\ \ \ } \\ 7\,3\,6\,8 \\ +\ 3\,5\,4\,7 \\ \hline 1\,0{,}9\,1\,5 \end{array}$$

Practice Exercise

2. Add: 36,047 + 29,255.

Objective 1.3a Subtract whole numbers.

Example Subtract: 8045 − 2897.

$$\begin{array}{r} {\scriptstyle 13\ \ \ } \\ {\scriptstyle 7\ 9\ \not{3}\ 15} \\ \not{8}\,\not{0}\,\not{4}\,\not{5} \\ -\ 2\,8\,9\,7 \\ \hline 5\,1\,4\,8 \end{array}$$

Practice Exercise

3. Subtract: 4805 − 1568.

Objective 1.4a Multiply whole numbers.

Example Multiply: 57 × 315.

$$\begin{array}{rl} {\scriptstyle 2\ \ } & \\ {\scriptstyle 1\ 3\ \ } & \\ 3\,1\,5 & \\ \times\ \ 5\,7 & \\ \hline 2\,2\,0\,5 & \leftarrow 315 \times 7 \\ 1\,5\,7\,5\,0 & \leftarrow 315 \times 50 \\ \hline 1\,7{,}9\,5\,5 & \end{array}$$

Practice Exercise

4. Multiply: 329 × 684.

Objective 1.5a Divide whole numbers.

Example Divide: 6463 ÷ 26.

$$\begin{array}{r} 2\,4\,8 \\ 2\,6\,\overline{)6\,4\,6\,3} \\ \underline{5\,2} \\ 1\,2\,6 \\ \underline{1\,0\,4} \\ 2\,2\,3 \\ \underline{2\,0\,8} \\ 1\,5 \end{array}$$

The answer is 248 R 15.

Practice Exercise

5. Divide: 8519 ÷ 27.

Objective 1.6a Round to the nearest ten, hundred, or thousand.

Example Round to the nearest thousand.

6 4 7 1
↑

The digit 6 is in the thousands place. We consider the next digit to the right. Since the digit, 4, is 4 or lower, we round down, meaning that 6 thousands stays as 6 thousands. Change all digits to the right of the thousands digit to zeros. The answer is 6000.

Practice Exercise

6. Round 36,468 to the nearest thousand.

Objective 1.6c Use < or > for □ to write a true sentence in a situation like 6 □ 10.

Example Use < or > for □ to write a true sentence:

34 □ 29.

Since 34 is to the right of 29 on the number line,

$34 > 29.$

Practice Exercise

7. Use < or > for □ to write a true sentence:

78 □ 81.

Objective 1.7b Solve equations like $t + 28 = 54$, $28 \cdot x = 168$, and $98 \cdot 2 = y$.

Example Solve: $y + 12 = 27$.

$$y + 12 = 27$$
$$y + 12 - 12 = 27 - 12$$
$$y + 0 = 15$$
$$y = 15$$

The solution is 15.

Practice Exercise

8. Solve: $24 \cdot x = 864$.

Objective 1.9b Evaluate exponential notation.

Example Evaluate: 5^4.

$$5^4 = 5 \cdot 5 \cdot 5 \cdot 5 = 625$$

Practice Exercise

9. Evaluate: 6^3.

Review Exercises

The review exercises that follow are for practice. Answers are given at the back of the book. If you miss an exercise, restudy the objective indicated in red next to the exercise or on the direction line that precedes it.

1. What does the digit 8 mean in 4,678,952? [1.1a]

2. In 13,768,940, what digit tells the number of millions? [1.1a]

Write expanded notation. [1.1b]

3. 2793

4. 56,078

5. 4,007,101

Write a word name. [1.1c]

6. 67,819

7. 2,781,427

Write standard notation. [1.1c]

8. Four hundred seventy-six thousand, five hundred eighty-eight

9. *Subway Ridership.* Ridership on the New York City Subway system totaled one billion, six hundred forty million in 2011.

Source: Metropolitan Transit Authority

Add. [1.2a]

10. $7304 + 6968$

11. $27{,}609 + 38{,}415$

12. $2703 + 4125 + 6004 + 8956$

13.
$$\begin{array}{r} 91{,}426 \\ +\ 7{,}495 \\ \hline \end{array}$$

Subtract. [1.3a]

14. $8045 - 2897$

15. $9001 - 7312$

16. $6003 - 3729$

17.
$$\begin{array}{r} 37{,}405 \\ -\ 19{,}648 \\ \hline \end{array}$$

Multiply. [1.4a]

18. $17{,}000 \cdot 300$

19. $7846 \cdot 800$

20. $726 \cdot 698$

21. $587 \cdot 47$

22.
$$\begin{array}{r} 8305 \\ \times\ 642 \\ \hline \end{array}$$

Divide. [1.5a]

23. $63 \div 5$

24. $80 \div 16$

25. $7\overline{)6394}$

26. $3073 \div 8$

27. $60\overline{)286}$

28. $4266 \div 79$

29. $38\overline{)17{,}176}$

30. $14\overline{)70{,}112}$

31. $52{,}668 \div 12$

Round 345,759 to the nearest: [1.6a]

32. Hundred.

33. Ten.

34. Thousand.

35. Hundred thousand.

Use $<$ or $>$ for $\square$ to write a true sentence. [1.6c]

36. $67 \square 56$

37. $1 \square 23$

Estimate each sum, difference, or product by first rounding to the nearest hundred. Show your work. [1.6b]

38. $41{,}348 + 19{,}749$

39. $38{,}652 - 24{,}549$

40. $396 \cdot 748$

Solve. [1.7b]

41. $46 \cdot n = 368$

42. $47 + x = 92$

43. $1 \cdot y = 58$

44. $24 = x + 24$

45. Write exponential notation: $4 \cdot 4 \cdot 4$. [1.9a]

Evaluate. [1.9b]

46. 10^4

47. 6^2

Simplify. [1.9c, d]

48. $8 \cdot 6 + 17$

49. $10 \cdot 24 - (18 + 2) \div 4 - (9 - 7)$

50. $(80 \div 16) \times [(20 - 56 \div 8) + (8 \cdot 8 - 5 \cdot 5)]$

51. Find the average of 157, 170, and 168. [1.9c]

Solve. [1.8a]

52. *Computer Purchase.* Natasha has \$196 and wants to buy a computer for \$698. How much more does she need?

53. Toni has \$406 in her checking account. She is paid \$78 for a part-time job and deposits that in her checking account. How much is then in her account?

54. *Lincoln-Head Pennies.* In 1909, the first Lincoln-head pennies were minted. Seventy-three years later, these pennies were first minted with a decreased copper content. In what year was the copper content reduced?

55. A beverage company packed 228 cans of soda into 12-can cartons. How many cartons were filled?

56. An apartment builder bought 13 gas stoves at $425 each and 13 refrigerators at $620 each. What was the total cost?

57. An apple farmer keeps bees in her orchard to help pollinate the apple blossoms. The bees from an average beehive can pollinate 30 surrounding trees during one growing season. A farmer has 420 trees. How many beehives does she need to pollinate all of them?

Source: Jordan Orchards, Westminster, PA

58. *Olympic Trampoline.* Shown below is an Olympic trampoline. Determine the area and the perimeter of the trampoline. [1.2b], [1.4b]

Source: International Trampoline Industry Association, Inc.

59. A chemist has 2753 mL of alcohol. How many 20-mL beakers can be filled? How much will be left over?

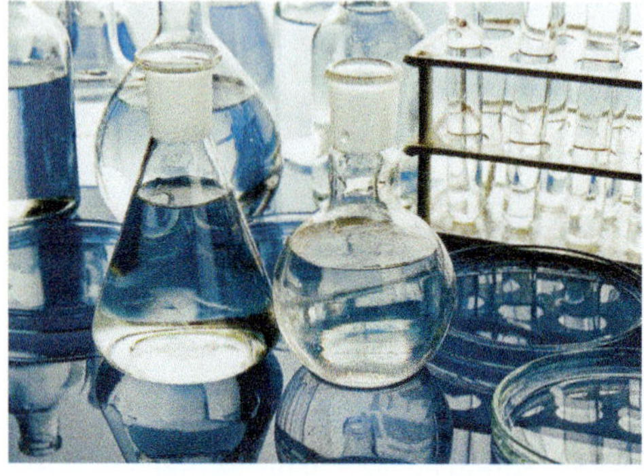

60. A family budgeted $7825 a year for food and clothing and $2860 for entertainment. The yearly income of the family was $38,283. How much of this income remained after these two allotments?

61. Simplify: $7 + (4 + 3)^2$. [1.9c]

A. 32 B. 56
C. 151 D. 196

62. Simplify: $7 + 4^2 + 3^2$. [1.9c]

A. 32 B. 56
C. 130 D. 196

63. $[46 - (4 - 2) \cdot 5] \div 2 + 4$ [1.9d]

A. 6 B. 20
C. 114 D. 22

Synthesis

64. Determine the missing digit d. [1.4a]

$$\begin{array}{r} 9\,d \\ \times\ \ d\,2 \\ \hline 8\,0\,3\,6 \end{array}$$

65. Determine the missing digits a and b. [1.5a]

$$\begin{array}{r} 9\ a\ 1 \\ 2\ b\ 1\,\overline{)\,2\ 3\ 6,4\ 2\ 1} \end{array}$$

66. A mining company estimates that a crew must tunnel 2000 ft into a mountain to reach a deposit of copper ore. Each day, the crew tunnels about 500 ft. Each night, about 200 ft of loose rocks roll back into the tunnel. How many days will it take the mining company to reach the copper deposit? [1.8a]

Understanding Through Discussion and Writing

1. Is subtraction associative? Why or why not? [1.3a]

2. Explain how estimating and rounding can be useful when shopping for groceries. [1.6b]

3. Write a problem for a classmate to solve. Design the problem so that the solution is "The driver still has 329 mi to travel." [1.8a]

4. Consider the expressions $9 - (4 \cdot 2)$ and $(3 \cdot 4)^2$. Are the parentheses necessary in each case? Explain. [1.9c]

CHAPTER 1

Test

For Extra Help For step-by-step test solutions, access the Chapter Test Prep Videos in MyMathLab® or on YouTube (search "BittingerPreIntro" and click on Channels).

1. In the number 546,789, which digit tells the number of hundred thousands?

2. Write expanded notation: 8843.

3. Write a word name: 38,403,277.

Add.

4. $\begin{array}{r} 6811 \\ +\ 3178 \\ \hline \end{array}$

5. $\begin{array}{r} 45{,}889 \\ +\ 17{,}902 \\ \hline \end{array}$

6. $\begin{array}{r} 1239 \\ 843 \\ 301 \\ +\ \ 782 \\ \hline \end{array}$

7. $\begin{array}{r} 6203 \\ +\ 4312 \\ \hline \end{array}$

Subtract.

8. $\begin{array}{r} 7983 \\ -\ 4353 \\ \hline \end{array}$

9. $\begin{array}{r} 2974 \\ -\ 1935 \\ \hline \end{array}$

10. $\begin{array}{r} 8907 \\ -\ 2059 \\ \hline \end{array}$

11. $\begin{array}{r} 23{,}067 \\ -\ 17{,}892 \\ \hline \end{array}$

Multiply.

12. $\begin{array}{r} 4568 \\ \times\ \ \ \ 9 \\ \hline \end{array}$

13. $\begin{array}{r} 8876 \\ \times\ 600 \\ \hline \end{array}$

14. $\begin{array}{r} 65 \\ \times\ 37 \\ \hline \end{array}$

15. $\begin{array}{r} 678 \\ \times\ 788 \\ \hline \end{array}$

Divide.

16. $15 \div 4$

17. $420 \div 6$

18. $89\overline{)8633}$

19. $44\overline{)35{,}428}$

Round 34,528 to the nearest:

20. Thousand.

21. Ten.

22. Hundred.

Estimate each sum, difference, or product by first rounding to the nearest hundred. Show your work.

23. $\begin{array}{r} 23{,}649 \\ +\ 54{,}746 \\ \hline \end{array}$

24. $\begin{array}{r} 54{,}751 \\ -\ 23{,}649 \\ \hline \end{array}$

25. $\begin{array}{r} 824 \\ \times\ 489 \\ \hline \end{array}$

Use $<$ or $>$ for $\square$ to write a true sentence.

26. $34 \ \square\ 17$

27. $117 \ \square\ 157$

Solve.

28. $28 + x = 74$

29. $169 \div 13 = n$

30. $38 \cdot y = 532$

31. $381 = 0 + a$

Solve.

32. *Calorie Content.* An 8-oz serving of whole milk contains 146 calories. This is 63 calories more than the number of calories in an 8-oz serving of skim milk. How many calories are in an 8-oz serving of skim milk?

Source: *American Journal of Clinical Nutrition*

33. A box contains 5000 staples. How many staplers can be filled from the box if each stapler holds 250 staples?

34. *Largest States.* The following table lists the five largest states in terms of their land area. Find the total land area of these states.

STATE	AREA (in square miles)
Alaska	571,951
Texas	261,797
California	155,959
Montana	145,552
New Mexico	121,356

Sources: U.S. Department of Commerce; U.S. Census Bureau

35. *Pool Tables.* The Bradford™ pool table made by Brunswick Billiards comes in three sizes of playing area, 50 in. by 100 in., 44 in. by 88 in., and 38 in. by 76 in.

Source: Brunswick Billiards

a) Determine the perimeter and the playing area of each table.

b) By how much does the area of the largest table exceed the area of the smallest table?

36. *Hostess Ding Dongs®.* Hostess packages its Ding Dong snack cakes in 12-packs. How many 12-packs can it fill with 22,231 cakes? How many will be left over?

37. *Office Supplies.* Morgan manages the office of a small graphics firm. He buys 3 black inkjet cartridges at \$15 each and 2 photo inkjet cartridges at \$25 each. How much does the purchase cost?

38. Write exponential notation: $12 \cdot 12 \cdot 12 \cdot 12$.

Evaluate.

39. 7^3

40. 10^5

Simplify.

41. $35 - 1 \cdot 28 \div 4 + 3$

42. $10^2 - 2^2 \div 2$

43. $(25 - 15) \div 5$

44. $2^4 + 24 \div 12$

45. $8 \times \{(20 - 11) \cdot [(12 + 48) \div 6 - (9 - 2)]\}$

46. Find the average of 97, 99, 87, and 89.

A. 93 **B.** 124 **C.** 186 **D.** 372

Synthesis

47. An open cardboard container is 8 in. wide, 12 in. long, and 6 in. high. How many square inches of cardboard are used?

48. Use trials to find the single-digit number a for which

$$359 - 46 + a \div 3 \times 25 - 7^2 = 339.$$

49. Cara spends \$229 a month to repay her student loan. If she has already paid \$9160 on the 10-year loan, how many payments remain?

CHAPTER

2

Introduction to Integers and Algebraic Expressions

STUDYING FOR SUCCESS *Doing Your Homework*

- ☐ Prepare for doing your homework by reading explanations of concepts and by following the step-by-step solutions of examples in the text.
- ☐ Include all the steps when solving a problem. This will keep you organized, help you avoid computational errors, and give you a study guide for an exam.
- ☐ Try to do your homework as soon as possible after each class. Avoid waiting until a deadline is near to begin work on an assignment.

2.1 Integers and the Number Line

OBJECTIVES

a. State the integer that corresponds to a real-world situation.

b. Form a true sentence using $<$ or $>$.

c. Find the absolute value of any integer.

d. Find the opposite of any integer.

SKILL TO REVIEW

Objective 1.6c: Use $<$ or $>$ for $\square$ to write a true sentence in a situation like $6 \square 10$.

Use $<$ or $>$ for $\square$ to write a true sentence.

1. $0 \square 10$ **2.** $51 \square 15$

Answers

Skill to Review:

1. $<$ 2. $>$

In this section, we extend the set of whole numbers to form the set of *integers*. You have probably already used negative numbers. For example, the outside temperature could drop to *negative five* degrees, or a credit card statement could indicate activity of *negative forty-eight* dollars.

To describe integers, we start with the whole numbers, 0, 1, 2, 3, and so on. For each number 1, 2, 3, and so on, we obtain a new number that is the same number of units to the left of zero on the number line.

For the number 1, there is the *opposite* number -1 (negative 1).

For the number 2, there is the *opposite* number -2 (negative 2).

For the number 3, there is the *opposite* number -3 (negative 3), and so on.

The **integers** consist of the whole numbers and these new numbers. We picture them on the number line as follows.

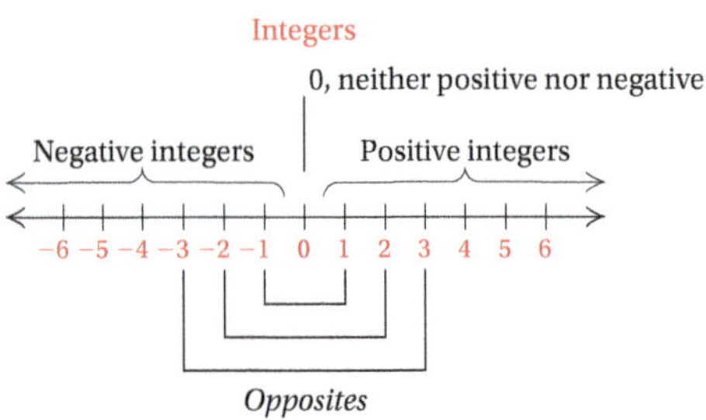

We call the integers to the left of zero **negative integers**. Those to the right of zero are called **positive integers**. Zero is neither positive nor negative. We call -1 and 1 **opposites** of each other. Similarly, -2 and 2 are opposites, -3 and 3 are opposites, -100 and 100 are opposites, and 0 is its own opposite.

INTEGERS

The **integers**: $\ldots, -5, -4, -3, -2, -1, 0, 1, 2, 3, 4, 5, \ldots$

a INTEGERS AND THE REAL WORLD

Integers correspond to many real-world problems and situations. The following examples will help you get ready to translate problem situations to mathematical language.

EXAMPLE 1 Tell which integer corresponds to this situation: The temperature is 4 degrees below zero.

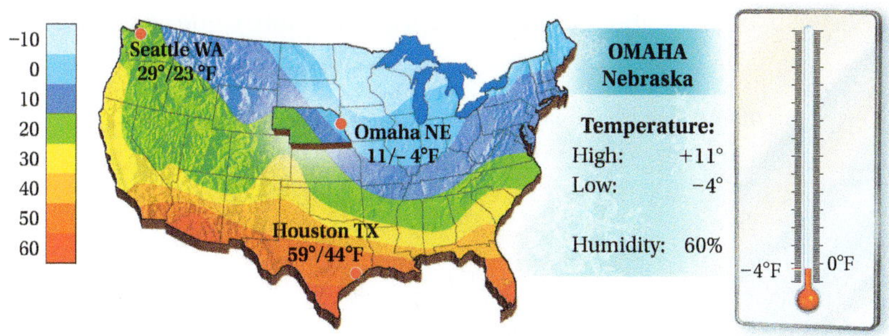

The integer -4 corresponds to the situation. The temperature is $-4°$.

EXAMPLE 2 *Stock Price Change.* Tell which integers correspond to this situation: Hal owns a stock whose price decreased from \$27 per share to \$11 per share over a recent time period. He owns another stock whose price increased from \$20 per share to \$22 per share over the same time period.

The integer -16 corresponds to the decrease in the value of the first stock. The integer 2 represents the increase in the value of the second stock.

Do Exercises 1–5.

Tell which integers correspond to each situation.

1. ***High and Low Temperatures.*** As of 2010, the highest recorded temperature in Illinois was 117°F on July 14, 1954, in East St. Louis. The lowest recorded temperature in Illinois was 36°F below zero on January 5, 1999, in Congerville.
Source: National Climate Data Center, NESDIS, NOAA, U.S. Dept. of Commerce

2. ***Stock Decrease.*** The price of a stock decreased from \$41 per share to \$38 per share over a recent period.

3. At 10 sec before liftoff, ignition occurs. At 148 sec after liftoff, the first stage is detached from the rocket.

4. The halfback gained 8 yd on first down. The quarterback was sacked for a 5-yd loss on second down.

5. A submarine dived 120 ft, rose 50 ft, and then dived 80 ft.

b ORDER ON THE NUMBER LINE

Numbers are written in order on the number line, increasing as we move from left to right. For any two numbers on the line, the one to the left is *less than* the one to the right.

Since the symbol $<$ means "is less than," the sentence $-5 < 9$ is read "-5 is less than 9." The symbol $>$ means "is greater than," so the sentence $-4 > -8$ is read "-4 is greater than -8."

EXAMPLES Use either $<$ or $>$ for □ to form a true sentence.

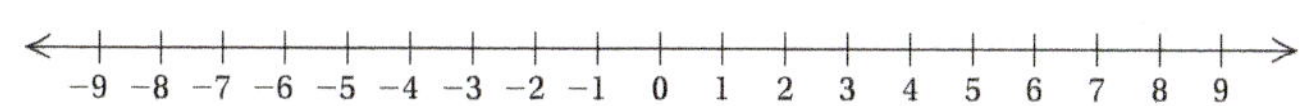

3. -9 □ 2 Since -9 is to the left of 2, we have $-9 < 2$.

4. 4 □ -6 Since 4 is to the right of -6, we have $4 > -6$.

5. -8 □ -1 Since -8 is to the left of -1, we have $-8 < -1$.

Do Exercises 6–9.

Use either $<$ or $>$ for □ to write a true sentence.

6. -3 □ 7

7. -8 □ -5

8. 7 □ -10

9. -4 □ -20

c ABSOLUTE VALUE

From the number line, we see that some integers, like 4 and -4, are the same distance from zero. We call the distance of a number from zero the **absolute value** of the number. Since distance is always a nonnegative number, absolute value is always nonnegative.

Answers

1. 117; -36 **2.** The integer -3 corresponds to the decrease in the value of the stock. **3.** -10; 148 **4.** 8; -5 **5.** -120; 50; -80 **6.** $<$ **7.** $<$ **8.** $>$ **9.** $>$

ABSOLUTE VALUE

The **absolute value** of a number is its distance from zero on the number line. We use the symbol $|x|$ to represent the absolute value of a number x.

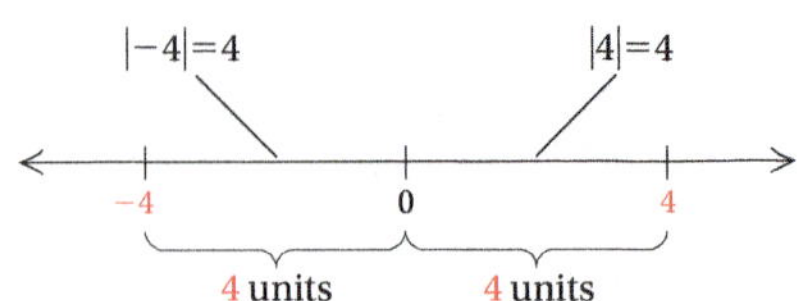

The distance of −4 from 0 is 4.
The absolute value of −4 is 4.
|−4|=4
The distance of 4 from 0 is 4.
The absolute value of 4 is 4.
|4|=4
−4
0
4
4 units
4 units

To find the absolute value of a number:

a) If a number is negative, its absolute value is its opposite.

b) If a number is positive or zero, its absolute value is the same as the number.

Find the absolute value.

10. $|18|$

The distance of 18 from 0 is ____, so $|18| =$ ____.

11. $|-9|$

12. $|-29|$

13. $|52|$

EXAMPLES Find the absolute value of each number.

6. $|-3|$ The distance of -3 from 0 is 3, so $|-3| = 3$.

7. $|25|$ The distance of 25 from 0 is 25, so $|25| = 25$.

8. $|0|$ The distance of 0 from 0 is 0, so $|0| = 0$.

Do Exercises 10–13.

d OPPOSITES

Given a number on one side of 0 on the number line, we can get a number on the other side by *reflecting* the number across zero. For example, the *reflection* of 2 is -2. We can read -2 as "negative 2" or "the opposite of 2."

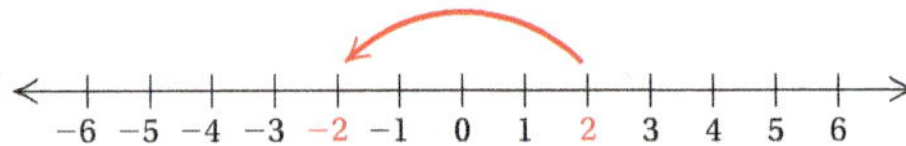

−6 −5 −4 −3 −2 −1 0 1 2 3 4 5 6

NOTATION FOR OPPOSITES

The **opposite** of a number x is written $-x$ (read "the opposite of x").

The opposite of a number is also called its *additive inverse*.

EXAMPLE 9 If x is -3, find $-x$.

To find the opposite of x when x is -3, we reflect -3 to the other side of 0.

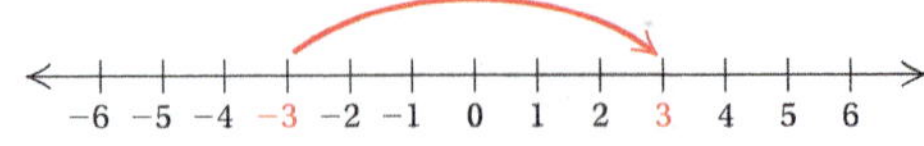

−6 −5 −4 −3 −2 −1 0 1 2 3 4 5 6

When $x = -3$, $-x = -(-3)$. We have $-(-3) = 3$. The opposite of -3 is 3.

Answers

10. 18 11. 9 12. 29 13. 52

Guided Solution:

10. 18, 18

When we replace a variable with a number and find the value of an expression we say that we are **evaluating** the expression.

EXAMPLE 10 Evaluate $-x$ when x is 4.

To find the opposite of x when x is 4, we reflect 4 to the other side of 0.

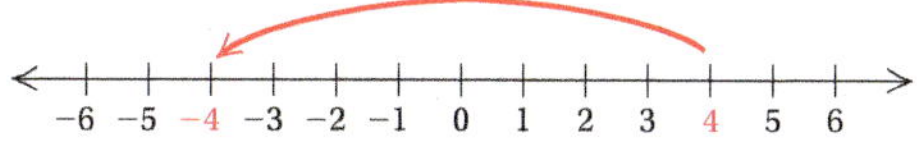

We have $-(4) = -4$. The opposite of 4 is -4.

EXAMPLE 11 Evaluate $-x$ when x is 0.

$-x = 0$ when x is 0. The opposite of 0 is 0; $-0 = 0$.

Do Exercises 14–16.

In each case, draw a number line, if necessary.

14. Find $-x$ when x is 1.

15. Find $-x$ when x is 0.

16. Evaluate $-x$ when x is -2.

A negative number is sometimes said to have a negative *sign*. A positive number is said to have a positive sign, even though it is rarely written. Thus, for example, -7 has a negative sign and 23 has a positive sign. Replacing a number with its opposite, or additive inverse, is sometimes called *changing the sign.*

EXAMPLES Change the sign (find the opposite, or additive inverse) of each number.

12. -6 $\quad -(-6) = 6$

13. -10 $\quad -(-10) = 10$

14. 0 $\quad -(0) = 0$

15. 14 $\quad -(14) = -14$

Do Exercises 17–20.

Change the sign. (Find the opposite, or additive inverse.)

17. -4 **18.** -13

19. 39 **20.** 0

EXAMPLE 16 If x is 2, find $-(-x)$.

If $x = 2$, then $-(-x) = -(-2) = 2$. The opposite of the opposite of 2 is 2.

EXAMPLE 17 Evaluate $-(-x)$ for $x = -4$.

If $x = -4$, then $-(-x) = -(-(-4)) = -(4) = -4$. The opposite of the opposite of -4 is -4.

When we change a number's sign twice, we return to the original number.

Do Exercises 21 and 22.

21. If x is 7, find $-(-x)$.

GS **22.** Evaluate $-(-x)$ for $x = -2$.

$-(-x) = -(-(\quad))$
$= -(\quad) = \quad$

It is important not to confuse parentheses with absolute-value symbols.

EXAMPLE 18 Evaluate $-|-x|$ for $x = 2$.

If $x = 2$, then $-|-x| = -|-2| = -2$. The absolute value of -2 is 2.

Note that $-(-2) = 2$, whereas $-|-2| = -2$.

Do Exercises 23 and 24.

23. Find $-|-7|$.

24. Find $-|-39|$.

Answers

14. -1 **15.** 0 **16.** 2 **17.** 4 **18.** 13 **19.** -39 **20.** 0 **21.** 7 **22.** -2 **23.** -7 **24.** -39

Guided Solution:
22. $-2, 2, -2$

2.1 Exercise Set

For Extra Help
MyMathLab®
MathXL® PRACTICE | WATCH | READ | REVIEW

Reading Check

Use the number line below, on which the letters name numbers, for Exercises RC1–RC8.

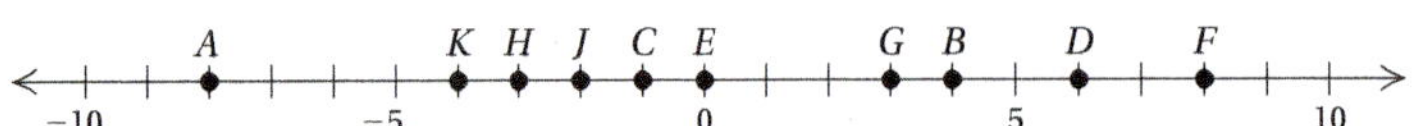

Determine whether each statement is true or false.

______ **RC1.** $K < B$ ______ **RC2.** $H < B$ ______ **RC3.** $E < C$ ______ **RC4.** $J > D$

______ **RC5.** $|K| = 4$ ______ **RC6.** $|H| = |B|$ ______ **RC7.** $A = -F$ ______ **RC8.** $-G = H$

a Tell which integers correspond to each situation.

1. At tax time, Janine received an $820 refund while David owed $541.

2. A student deposited her tax refund of $750 in a savings account. Two weeks later, she withdrew $125 to pay technology fees.

3. *Temperature Extremes.* The highest temperature ever created on Earth was 950,000,000°F. The lowest temperature ever created was approximately 460°F below zero.

Source: *The Guinness Book of Records*

4. *Shipwreck.* There are numerous shipwrecks to explore near Bermuda. One of the most frequently visited wrecks is *L'Herminie*, a French warship that sank in 1837. This wreck is 35 ft below the surface.

Source: www./10best.com/interests/adventure/scuba-diving-in-pirate-territory/

5. *Empire State Building.* The Empire State Building has a total height above ground level, including the lightning rod at the top, of 1454 ft. The foundation depth is 55 ft below ground level.

Source: www.empirestatebuildingfacts.com

6. *Extreme Climate.* Verkhoyansk, a river port in northeast Siberia, has the most extreme climate on the planet. Its average monthly winter temperature is 59°F below zero, and its average monthly summer temperature is 57°F.

Source: *The Guinness Book of Records*

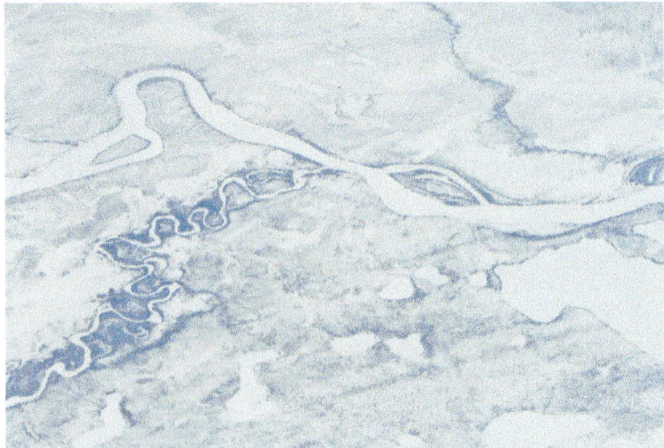

b Use either < or > for □ to form a true sentence.

7. $-8 \square 0$ **8.** $7 \square 0$ **9.** $9 \square 0$ **10.** $-7 \square 0$ **11.** $8 \square -8$

12. $6 \square -6$ **13.** $-6 \square -4$ **14.** $-1 \square -7$ **15.** $-8 \square -5$ **16.** $-5 \square -3$

17. $-13 \square -9$ **18.** $-5 \square -11$ **19.** $-3 \square -4$ **20.** $-6 \square -5$

c Find the absolute value.

21. $|57|$ **22.** $|11|$ **23.** $|0|$ **24.** $|-4|$ **25.** $|-24|$

26. $|-36|$ **27.** $|53|$ **28.** $|54|$ **29.** $|-8|$ **30.** $|-79|$

d Find $-x$ when x is each of the following.

31. -7 **32.** -6 **33.** 7 **34.** 6 **35.** 0 **36.** -1

Change the sign. (Find the opposite, or additive inverse.)

37. -21 **38.** -67 **39.** 53 **40.** 0 **41.** -1 **42.** 16

Evaluate $-(-x)$ when x is each of the following.

43. 7 **44.** -8 **45.** -9 **46.** 3 **47.** -17 **48.** -19

49. 23 **50.** 0 **51.** -1 **52.** 73 **53.** 85 **54.** -37

Evaluate $-|-x|$ when x is each of the following.

55. 345 **56.** 729 **57.** 0 **58.** 1 **59.** -8 **60.** -3

Skill Maintenance

61. Add: $327 + 498$. [1.2a]

62. Evaluate: 5^3. [1.9b]

63. Multiply: $209 \cdot 34$. [1.4a]

64. Solve: $300 \cdot x = 1200$. [1.7b]

65. Simplify: $7(9 - 3)$. [1.9c]

66. Simplify: $9^2 - 3[2 + (10 - 8)]$. [1.9d]

Synthesis

Use either $<$, $>$, or $=$ for $\square$ to write a true sentence.

67. $|-5| \square |-2|$

68. $|4| \square |-7|$

69. $|-8| \square |8|$

Solve. Consider only integer replacements.

70. $|x| = 7$

71. $|x| < 2$

72. Simplify $-(-x)$, $-(-(-x))$, and $-(-(-(-x)))$.

73. List these integers in order from least to greatest.
2^{10}, -5, $|-6|$, 4, $|3|$, -100, 0, 2^7, 7^2, 10^2

2.2 Addition of Integers

OBJECTIVE

a Add integers without using the number line.

Add using the number line.

1. $1 + (-4)$

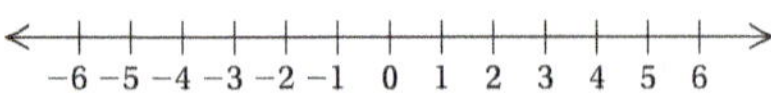

2. $-3 + (-2)$

-6 -5 -4 -3 -2 -1 0 1 2 3 4 5 6

3. $-3 + 7$

-6 -5 -4 -3 -2 -1 0 1 2 3 4 5 6

4. $-5 + 5$

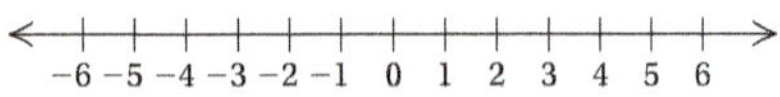

For each illustration, write a corresponding addition sentence.

5.

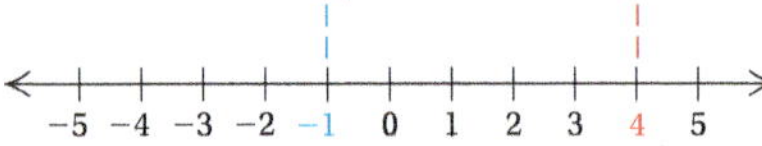

6.

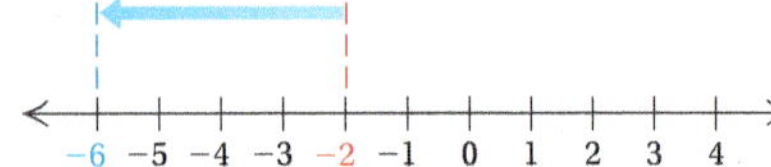

7.

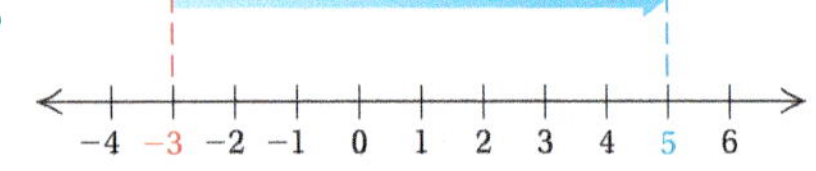

a ADDITION

To explain addition of integers, we can use the number line. Once our understanding is developed, we will streamline our approach.

Addition on the Number Line

To find $a + b$, we start at a and then move according to b.

a) If b is positive, we move from a to the right.

b) If b is negative, we move from a to the left.

c) If b is 0, we stay at a.

EXAMPLE 1 Add: $2 + (-5)$.

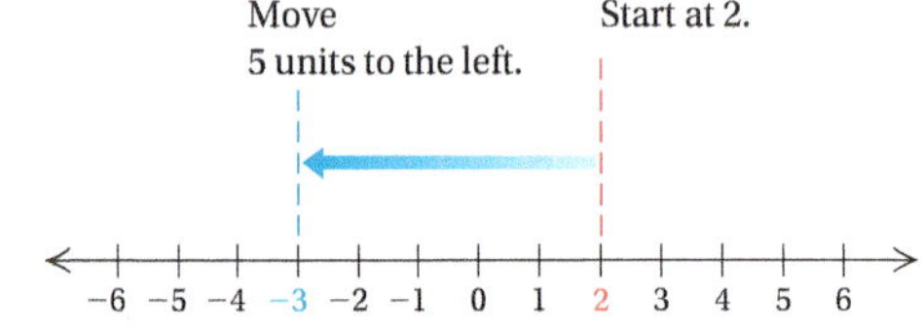

$2 + (-5) = -3$

EXAMPLE 2 Add: $-1 + (-3)$.

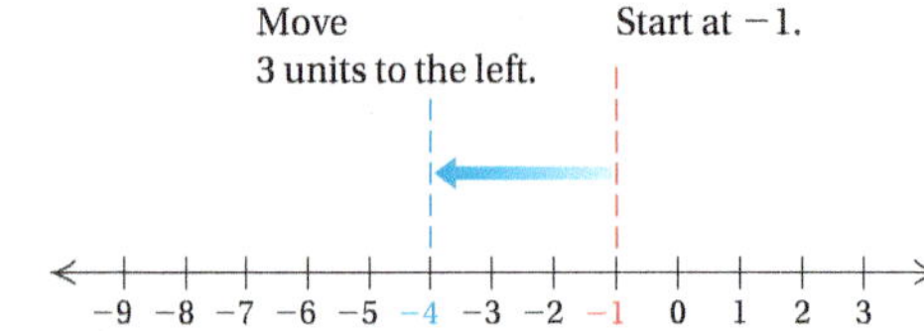

$-1 + (-3) = -4$

EXAMPLE 3 Add: $-4 + 9$.

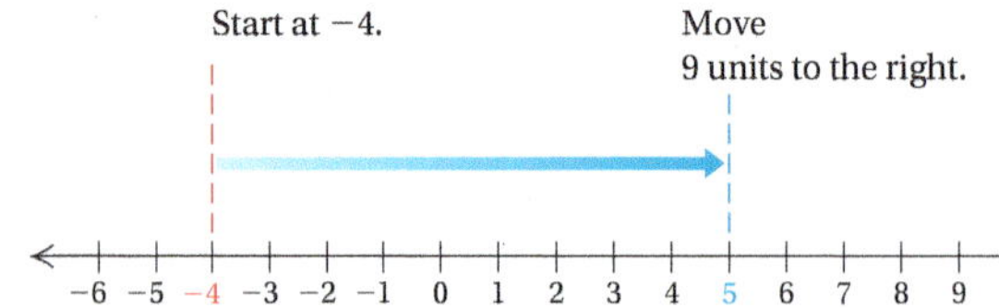

$-4 + 9 = 5$

Do Exercises 1–7.

You may have noticed a pattern in Example 2 and Margin Exercises 2 and 6. When two negative integers are added, the result is negative.

ADDING NEGATIVE INTEGERS

To add two negative integers, add their absolute values and change the sign (making the answer negative).

Answers

1. -3 2. -5 3. 4 4. 0
5. $4 + (-5) = -1$ 6. $-2 + (-4) = -6$
7. $-3 + 8 = 5$

EXAMPLES Add.

4. $-5 + (-7) = -12$ *Think*: Add the absolute values: $5 + 7 = 12$. Make the answer negative, -12.

5. $-8 + (-2) = -10$

Do Exercises 8–11. ▶

Add. Do not use the number line except as a check.

8. $-5 + (-6)$

9. $-9 + (-3)$

10. $-20 + (-14)$

11. $-11 + (-11)$

When the number 0 is added to any number, that number remains unchanged. For this reason, the number 0 is referred to as the **additive identity**.

EXAMPLES Add.

6. $-4 + 0 = -4$ 7. $0 + (-9) = -9$ 8. $17 + 0 = 17$

Do Exercises 12–14. ▶

Add.

12. $0 + (-17)$

13. $49 + 0$

14. $-56 + 0$

When we add a positive integer and a negative integer, as in Examples 1 and 3, the sign of the number with the greater absolute value is the sign of the answer.

ADDING POSITIVE AND NEGATIVE INTEGERS

To add a positive integer and a negative integer, find the difference of their absolute values.

a) If the negative integer has the greater absolute value, the answer is negative.

b) If the positive integer has the greater absolute value, the answer is positive.

c) If the integers have the same absolute value, the answer is 0.

EXAMPLES Add.

9. $3 + (-5) = -2$ *Think*: The absolute values are 3 and 5. The difference is 2. Since the negative number has the larger absolute value, the answer is *negative*, -2.

10. $11 + (-8) = 3$ *Think*: The absolute values are 11 and 8. The difference is 3. The positive number has the larger absolute value, so the answer is *positive*, 3.

11. $-7 + 4 = -3$

12. $-6 + 10 = 4$

13. $9 + (-9) = 0$

Do Exercises 15–19. ▶

Add, using the number line only as a check.

15. $-4 + 6$

16. $-7 + 3$

17. $5 + (-7)$

18. $10 + (-7)$

GS **19.** $-12 + 12$

-12 and 12 have the same  value. The answer is ___.

Sometimes $-a$ is referred to as the **additive inverse** of a because adding any number to its additive inverse always results in the additive identity, 0.

ADDING OPPOSITES

For any integer a,

$$a + (-a) = -a + a = 0.$$

(The sum of any number and its additive inverse, or opposite, is 0.)

Answers

8. -11 **9.** -12 **10.** -34 **11.** -22
12. -17 **13.** 49 **14.** -56 **15.** 2
16. -4 **17.** -2 **18.** 3 **19.** 0

Guided Solution:
19. absolute, 0

Add, using the number line only as a check.

20. $5 + (-5)$

21. $-6 + 6$

22. $89 + (-89)$

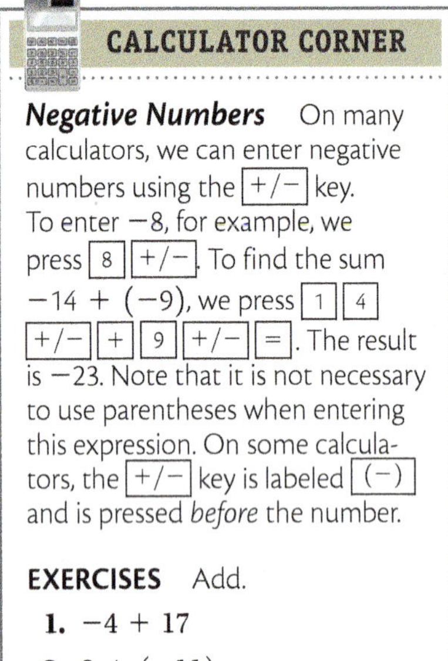

CALCULATOR CORNER

Negative Numbers On many calculators, we can enter negative numbers using the [+/−] key. To enter -8, for example, we press [8] [+/−]. To find the sum $-14 + (-9)$, we press [1] [4] [+/−] [+] [9] [+/−] [=]. The result is -23. Note that it is not necessary to use parentheses when entering this expression. On some calculators, the [+/−] key is labeled [(−)] and is pressed *before* the number.

EXERCISES Add.

1. $-4 + 17$

2. $3 + (-11)$

EXAMPLES Add.

14. $-8 + 8 = 0$

15. $14 + (-14) = 0$

◀ **Do Exercises 20–22.**

In summary, to add integers, look first at the signs of the numbers you are adding. This tells you whether you should add or subtract to find the sum.

RULES FOR ADDITION OF INTEGERS

1. *Positive numbers:* Add the same way as you add arithmetic numbers. The answer is positive.
2. *Negative numbers:* Add absolute values. The answer is negative.
3. *A positive and a negative number:* Subtract absolute values.
   a) If the positive number has the greater absolute value, the answer is positive.
   b) If the negative number has the greater absolute value, the answer is negative.
   c) If the numbers have the same absolute value, they are additive inverses and the answer is 0.
4. *One number is zero:* The sum is the other number.

Add.

23. $-5 + (-10)$

24. $18 + (-11)$

25. $-13 + 13$

26. $-20 + 7$

EXAMPLES Add.

16. $-12 + (-7) = -19$ Two negative numbers; add absolute values. The answer is negative.

17. $-20 + 36 = 16$ A positive and a negative number; subtract absolute values. The answer is positive.

◀ **Do Exercises 23–26.**

Suppose we wish to add several numbers, some positive and some negative, as in $15 + (-2) + 7 + 14 + (-5) + (-12)$. Because of the commutative and associative laws for addition, we can group the positive numbers together and the negative numbers together and add them separately. Then we add the two results.

GS

Add.

27. $-15 + (-5) + 25 + (-9) + 10 + (-14)$

Add the positive numbers:
$25 + \_\_\_ = \_\_\_$.

Add the negative numbers:
$-15 + (-5) + (-9) + (\_\_\_) = \_\_\_$.

Finally, add the results:
$35 + (\_\_\_) = \_\_\_$.

EXAMPLE 18 Add: $15 + (-2) + 7 + 14 + (-5) + (-12)$.

First add the positive numbers: $15 + 7 + 14 = 36$.
Then add the negative numbers: $-2 + (-5) + (-12) = -19$.
Finally, add the results: $36 + (-19) = 17$.

We can also add in any other order we wish, say, from left to right:

$$\begin{aligned} 15 + (-2) + 7 + 14 + (-5) + (-12) &= 13 + 7 + 14 + (-5) + (-12) \\ &= 20 + 14 + (-5) + (-12) \\ &= 34 + (-5) + (-12) \\ &= 29 + (-12) \\ &= 17. \end{aligned}$$

◀ **Do Exercise 27.**

Answers

20. 0 **21.** 0 **22.** 0 **23.** -15 **24.** 7
25. 0 **26.** -13 **27.** -8

Guided Solution:
27. 10, 35, -14, -43, -43, -8

2.2 Exercise Set

For Extra Help
MyMathLab®
MathXL® PRACTICE WATCH READ REVIEW

Reading Check

Fill in each blank with either "left" or "right" so that the statement describes the steps for adding the given numbers on the number line.

RC1. To add $7 + 2$, start at 7 and then move 2 units ________________. The sum is 9.

RC2. To add $-3 + (-5)$, start at -3 and then move 5 units ________________. The sum is -8.

RC3. To add $4 + (-6)$, start at 4 and then move 6 units ________________. The sum is -2.

RC4. To add $-8 + 3$, start at -8 and then move 3 units ________________. The sum is -5.

a Add, using the number line.

1. $-7 + 2$

2. $1 + (-5)$

3. $-9 + 5$

4. $8 + (-3)$

5. $-3 + 9$

6. $9 + (-9)$

7. $-7 + 7$

8. $-8 + (-5)$

9. $-3 + (-1)$

10. $-2 + (-9)$

Add. Use the number line only as a check.

11. $-3 + (-9)$

12. $-3 + (-7)$

13. $-6 + (-5)$

14. $-10 + (-14)$

15. $-15 + 0$

16. $0 + (-11)$

17. $0 + 42$

18. $27 + 0$

19. $9 + (-4)$

20. $-7 + 8$

21. $-10 + 6$

22. $6 + (-13)$

23. $5 + (-5)$

24. $10 + (-10)$

25. $-2 + 2$

26. $-3 + 3$

27. $-4 + (-5)$

28. $10 + (-12)$

29. $13 + (-6)$

30. $-3 + 14$

31. $-25 + 25$

32. $40 + (-40)$

33. $63 + (-18)$

34. $85 + (-65)$

35. $-11 + 8$

36. $0 + (-34)$

37. $-19 + 19$

38. $-10 + 3$

39. $-16 + 6$

40. $-15 + 5$

41. $-17 + (-7)$

42. $-15 + (-5)$

43. $11 + (-16)$

44. $-8 + 14$

45. $-15 + (-6)$

46. $-8 + 8$

47. $-15 + (-15)$

48. $-25 + (-25)$

49. $-11 + 17$

50. $19 + (-19)$

51. $-15 + (-7) + 1$

52. $23 + (-5) + 4$

53. $30 + (-10) + 5$

54. $40 + (-8) + 5$

55. $-23 + (-9) + 15$

56. $-25 + 25 + (-9)$

57. $40 + (-40) + 6$

58. $63 + (-18) + 12$

59. $12 + (-65) + (-12)$

60. $-35 + (-63) + 35$

61. $-24 + (-37) + (-19) + (-45) + (-35)$

62. $75 + (-14) + (-17) + (-5)$

63. $28 + (-44) + 17 + 31 + (-94)$

64. $27 + (-54) + (-32) + 65 + 46$

65. $-19 + 73 + (-23) + 19 + (-73)$

66. $35 + (-51) + 29 + 51 + (-35)$

Skill Maintenance

67. Add: $587 + 6094$. [1.2a]

68. Subtract: $3046 - 2973$. [1.3a]

69. Write in expanded notation: 39,417. [1.1b]

70. Multiply: $42 \cdot 56$. [1.4a]

71. Divide: $288 \div 9$. [1.5a]

72. Round to the nearest ten: 3496. [1.6a]

Synthesis

Add.

73. $-|27| + (-|-13|)$

74. $|-32| + (-|15|)$

75. 🖩 $-3496 + (-2987)$

76. 🖩 $497 + (-3028)$

77. For what numbers x is $-x$ positive?

78. For what numbers x is $-x$ negative?

Tell whether each sum is positive, negative, or zero.

79. If n is positive and m is negative, then $-n + m$ is ________________.

80. If $n = m$ and n is negative, then $-n + (-m)$ is ________________.

81. If n is negative and m is less than n, then $n + m$ is ________________.

82. If n is positive and m is greater than n, then $n + m$ is ________________.

Subtraction of Integers

2.3

a SUBTRACTION

We now consider subtraction of integers. To find the difference $a - b$, we look for a number to add to b that gives us a.

THE DIFFERENCE

The difference $a - b$ is the number that when added to b gives a.

For example, $45 - 17 = 28$ because $28 + 17 = 45$. Let's consider an example in which the answer is a negative number.

EXAMPLE 1 Subtract: $5 - 8$.

Think: $5 - 8$ is the number that when added to 8 gives 5. What number can we add to 8 to get 5? The number must be negative. The number is -3:

$$5 - 8 = -3.$$

That is, $5 - 8 = -3$ because $8 + (-3) = 5$.

Do Exercises 1–3. ▶

The definition of $a - b$ above does not always provide the most efficient way to subtract. To understand a faster way to subtract, consider finding $5 - 8$ using the number line. We start at 5. Then we move 8 units to the *left* to do the subtracting. Note that this is the same as adding the opposite of 8, or -8, to 5.

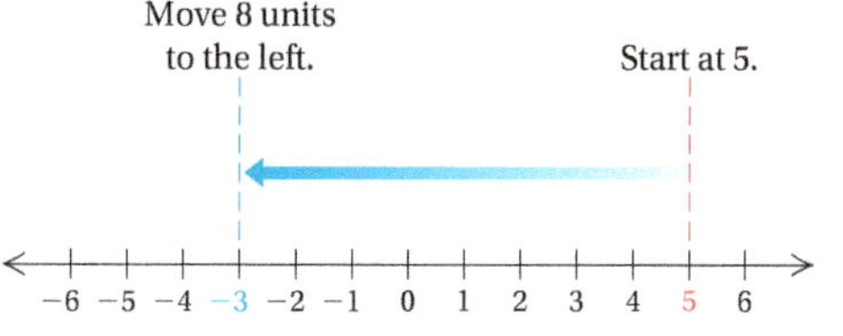

$5 - 8 = -3$

Look for a pattern in the following table.

SUBTRACTIONS	ADDING AN OPPOSITE
$5 - 8 = -3$	$5 + (-8) = -3$
$-6 - 4 = -10$	$-6 + (-4) = -10$
$-7 - (-10) = 3$	$-7 + 10 = 3$
$-7 - (-2) = -5$	$-7 + 2 = -5$

Do Exercises 4–7. ▶

Perhaps you have noticed that we can subtract by adding the opposite of the number being subtracted. This can always be done.

OBJECTIVES

a Subtract integers, and simplify combinations of additions and subtractions.

b Solve applied problems involving addition and subtraction of integers.

Subtract.

1. $-6 - 4$
   Think: What number can be added to 4 to get -6:
   $\square + 4 = -6$?

2. $-7 - (-10)$
   Think: What number can be added to -10 to get -7:
   $\square + (-10) = -7$?

3. $-7 - (-2)$
   Think: What number can be added to -2 to get -7:
   $\square + (-2) = -7$?

Complete the addition and compare with the subtraction.

4. $4 - 6 = -2$;
   $4 + (-6) =$ ______

5. $-3 - 8 = -11$;
   $-3 + (-8) =$ ______

6. $-5 - (-9) = 4$;
   $-5 + 9 =$ ______

7. $-5 - (-3) = -2$;
   $-5 + 3 =$ ______

Answers

1. -10 2. 3 3. -5 4. -2 5. -11
6. 4 7. -2

SUBTRACTING BY ADDING THE OPPOSITE

To subtract, add the opposite, or additive inverse, of the number being subtracted:

$$a - b = a + (-b).$$

This is the method generally used for quick subtraction of integers.

EXAMPLES Write each subtraction as a corresponding addition. Then write the equation in words.

2. $-12 - 30$

$-12 - 30 = -12 + (-30)$ Adding the opposite of 30

Negative twelve minus thirty is negative twelve plus negative thirty.

3. $-20 - (-17)$

$-20 - (-17) = -20 + 17$ Adding the opposite of -17

Negative twenty minus negative seventeen is negative twenty plus seventeen.

Write each subtraction as a corresponding addition. Then write the equation in words.

8. $3 - 10$

9. $-12 - (-9)$

10. $-12 - 10$

Do Exercises 8–10.

EXAMPLES Subtract.

4. $2 - 6 = 2 + (-6)$ The opposite of 6 is -6. We change the subtraction to addition and add the opposite. Instead of subtracting 6, we add -6.

$= -4$

5. $4 - (-9) = 4 + 9$ The opposite of -9 is 9. We change the subtraction to addition and add the opposite. Instead of subtracting -9, we add 9.

$= 13$

6. $-4 - 8 = -4 + (-8)$ We change the subtraction to addition and add the opposite. Instead of subtracting 8, we add -8.

$= -12$

7. $10 - 7 = 10 + (-7)$ We change the subtraction to addition and add the opposite. Instead of subtracting 7, we add -7.

$= 3$

8. $-4 - (-9) = -4 + 9$ Instead of subtracting -9, we add 9.

$= 5$ To check, note that $5 + (-9) = -4$.

9. $-7 - (-3) = -7 + 3$ Instead of subtracting -3, we add 3.

$= -4$ *Check*: $-4 + (-3) = -7$.

Subtract.

11. $2 - 8$
$= 2 + \square = \square$ GS

12. $-6 - 10$

13. $13 - 8$

14. $-7 - (-9)$

15. $-8 - (-2)$

16. $5 - (-8)$

Do Exercises 11–16.

Answers

8. $3 - 10 = 3 + (-10)$; three minus ten is three plus negative ten. **9.** $-12 - (-9) = -12 + 9$; negative twelve minus negative nine is negative twelve plus nine. **10.** $-12 - 10 = -12 + (-10)$; negative twelve minus ten is negative twelve plus negative ten. **11.** -6 **12.** -16 **13.** 5 **14.** 2 **15.** -6 **16.** 13

Guided Solution:
11. $(-8), -6$

When several additions and subtractions occur together, we can make them all additions. The commutative law for addition can then be used.

EXAMPLE 10 Simplify: $-3 - (-5) - 9 + 4 - (-6)$.

$$
\begin{aligned}
-3 - (-5) - 9 + 4 - (-6) &= -3 + 5 + (-9) + 4 + 6 && \text{Adding opposites} \\
&= -3 + (-9) + 5 + 4 + 6 && \text{Using a commutative law} \\
&= -12 + 15 \\
&= 3
\end{aligned}
$$

Do Exercises 17 and 18. ▶

Simplify.

GS **17.** $-6 - (-2) - (-4) - 12 + 3$
$= -6 + 2 + \square + (-12) + 3$
$= -6 + (-12) + 2 + \square + 3$
$= -18 + \square$
$= \square$

18. $9 - (-6) + 7 - 9 - 8 - (-20)$

b APPLICATIONS AND PROBLEM SOLVING

We use addition and subtraction of real numbers to solve a variety of applied problems.

EXAMPLE 11 *Surface Temperatures on Mars.* Surface temperatures on Mars vary from $-128°$C during the polar night to 27°C at the equator at midday when Mars is at the point in its orbit closest to the sun. Find the difference between the highest value and the lowest value in this temperature range.
Source: Mars Institute

We let $D =$ the difference in the temperatures. Then the problem translates to the following subtraction:

Difference in temperature	is	Highest temperature	minus	Lowest temperature
↓	↓	↓	↓	↓
D	$=$	27	$-$	(-128).

We then solve the equation: $D = 27 - (-128) = 27 + 128 = 155$.

The difference in the temperatures is 155°C.

Do Exercise 19. ▶

19. *Temperature Extremes.* The highest temperature ever recorded in the United States was 134°F in Greenland Ranch, California, on July 10, 1913. The lowest temperature ever recorded was $-80°$F in Prospect Creek, Alaska, on January 23, 1971. How much higher was the temperature in Greenland Ranch than the temperature in Prospect Creek?
Source: National Oceanographic and Atmospheric Administration

Answers

17. -9 **18.** 25 **19.** 214°F
Guided Solution:
17. 4, 4, 9, -9

2.3 Exercise Set

For Extra Help MyMathLab® MathXL® PRACTICE WATCH READ REVIEW

Reading Check

Match each expression with the expression from the list at the right that names the same number.

RC1. $18 - 6$ ______ **a)** $18 + 6$

RC2. $-18 - (-6)$ ______ **b)** $-18 + 6$

RC3. $-18 - 6$ ______ **c)** $18 + (-6)$

RC4. $18 - (-6)$ ______ **d)** $-18 + (-6)$

a Subtract.

1. $3 - 7$

2. $5 - 10$

3. $0 - 7$

4. $0 - 8$

5. $-8 - (-2)$

6. $-6 - (-8)$

7. $-10 - (-10)$

8. $-8 - (-8)$

9. $12 - 16$

10. $14 - 19$

11. $20 - 27$

12. $26 - 7$

13. $-9 - (-3)$

14. $-6 - (-9)$

15. $-11 - (-11)$

16. $-14 - (-14)$

17. $7 - 7$

18. $9 - 9$

19. $7 - (-7)$

20. $4 - (-4)$

21. $8 - (-3)$

22. $-7 - 4$

23. $-6 - 8$

24. $6 - (-10)$

25. $-4 - (-9)$

26. $-14 - 2$

27. $2 - 9$

28. $2 - 8$

29. $-6 - (-5)$

30. $-4 - (-3)$

31. $8 - (-10)$

32. $5 - (-6)$

33. $0 - 10$

34. $0 - 23$

35. $-5 - (-2)$

36. $-3 - (-1)$

37. $-7 - 14$

38. $-9 - 16$

39. $0 - (-5)$

40. $0 - (-1)$

41. $-8 - 0$

42. $-9 - 0$

43. $7 - (-5)$

44. $7 - (-4)$

45. $6 - 25$

46. $18 - 63$

47. $-42 - 26$

48. $-18 - 63$

49. $-72 - 9$

50. $-49 - 3$

51. $24 - (-92)$

52. $48 - (-73)$

53. $-50 - (-50)$

54. $-70 - (-70)$

55. $-30 - (-85)$

56. $-25 - (-15)$

Simplify.

57. $7 - (-5) + 4 - (-3)$

58. $-5 - (-8) + 3 - (-7)$

59. $-31 + (-28) - (-14) - 17$

60. $-43 - (-19) - (-21) + 25$

61. $-34 - 28 + (-33) - 44$

62. $39 + (-88) - 29 - (-83)$

63. $-93 - (-84) - 41 - (-56)$

64. $84 + (-99) + 44 - (-18) - 43$

65. $-5 - (-30) + 30 + 40 - (-12)$

66. $14 - (-50) + 20 - (-32)$

67. $132 - (-21) + 45 - (-21)$

68. $81 - (-20) - 14 - (-50) + 53$

Solve.

69. ***"Flipping" Houses.*** Buying run-down houses, fixing them up, and reselling them is referred to as "flipping" houses. Charlie and Sophia flipped four houses in a recent year. The profits and losses are shown in the following bar graph. Find the sum of the profits and losses.

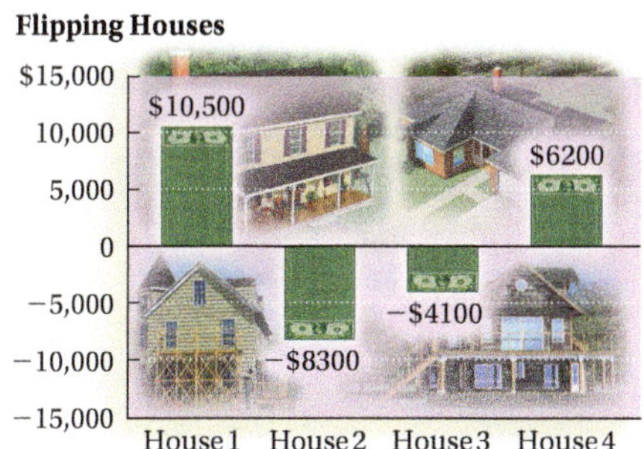

70. ***Elevations in Asia.*** The elevation of the highest point in Asia, Mt. Everest, on the border between Nepal and Tibet, is 29,035 ft. The lowest elevation, at the Dead Sea, between Israel and Jordan, is −1348 ft. What is the difference in the elevations of the two locations?

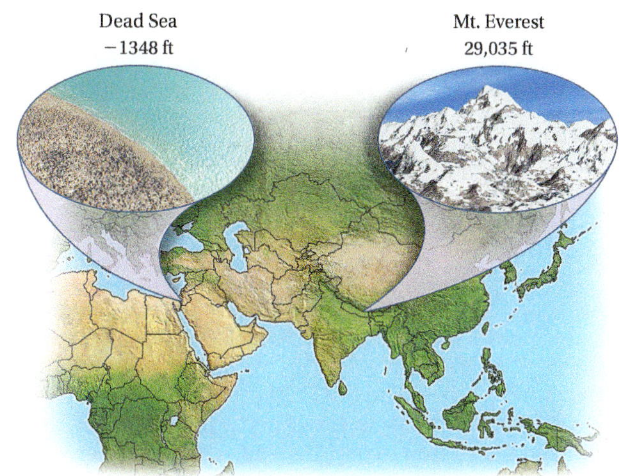

71. ***Difference in Elevation.*** The highest elevation in Japan is 3776 m above sea level at Fujiyama. The lowest elevation in Japan is 4 m below sea level at Hachirogata. Find the difference in the elevations.

Source: *The CIA World Factbook 2012*

72. ***Copy-Center Account.*** Rachel's copy-center bill for July was \$327. She made a payment of \$200 and then made \$48 worth of copies in August. How much did she then owe on her account?

73. Through exercise, Rod went from 8 lb above his "ideal" body weight to 9 lb below it. How many pounds did Rod lose?

74. Laura has a charge of \$477 on her credit card, but she then returns a sweater that cost \$129. How much does she now owe on her credit card?

75. *Temperature Changes.* One day the temperature in Lawrence, Kansas, is 32° at 6:00 A.M. It rises 15° by noon, but falls 50° by midnight after a cold front moves in. What is the final temperature?

76. *Stock Price Changes.* On a recent day, the price of a stock opened at \$61. It rose \$5, dropped \$7, and rose \$4. Find the price of the stock at the end of the day.

77. *Points per Game Differential.* A basketball team's point differential is the difference between that team's score in a game and its opponent's score. The points per game differential is the sum of a team's point differentials divided by the number of games it has played. In January 2014, the Indiana Pacers had a points per game differential of +9, and the Milwaukee Bucks had a points per game differential of −9. How much higher was the Pacers' points per game differential?

78. *Golf.* As a result of coaching, Cedric's average golf score improved from 3 over par to 2 under. By how many strokes did his score change?

79. *Offshore Oil.* In 1998, the elevation of the world's deepwater drilling record was −7718 ft. In 2013, the deepwater drilling record was 2693 ft deeper. What was the elevation of the deepwater drilling record in 2013?

Source: www.deepwater.com

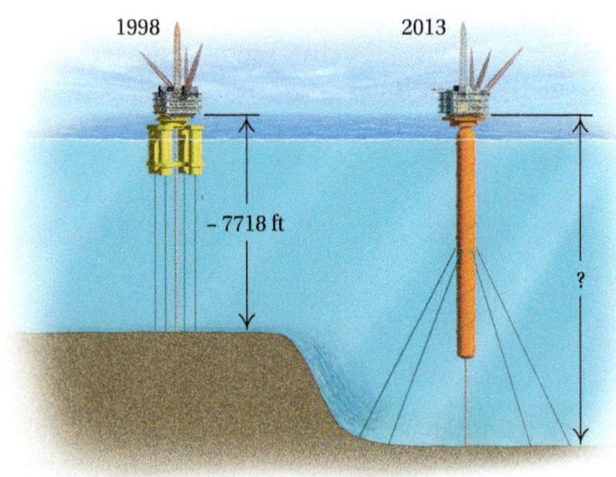

80. *Oceanography.* The deepest point in the Pacific Ocean is the Marianas Trench, with a depth of 11,033 m. The deepest point in the Atlantic Ocean is the Puerto Rico Trench, with a depth of 8648 m. What is the difference in the elevation of the two trenches?

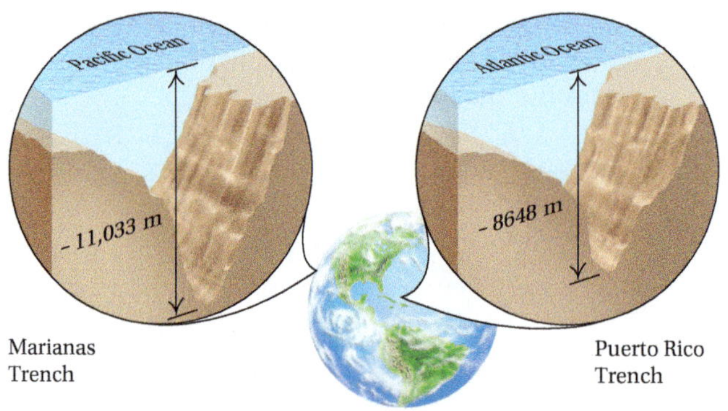

81. *Toll Roads.* The E-Z Pass program allows drivers in the Northeast to travel certain toll roads without having to stop to pay. Instead, a transponder attached to the vehicle is scanned as the vehicle rolls through a toll booth. Recently the Ramones began a trip to New York City with a balance of \$12 in their E-Z Pass account. They accumulated \$15 in tolls on their trip, and because they overspent their balance, the Ramones had to pay \$80 in administrative fees. By how much were the Ramones in debt as a result of their travel on toll roads?

Source: State of New Jersey

82. *Toll Roads.* The Murrays began a trip with \$13 in their E-Z Pass account (see Exercise 81). They accumulated \$20 in tolls and had to pay \$80 in administrative fees. By how much were the Murrays in debt as a result of their travel on toll roads?

Skill Maintenance

Evaluate.

83. 4^3 [1.9b]

84. $68 \cdot 72$ [1.4a]

85. 1^7 [1.9b]

86. $143 \cdot 29$ [1.4a]

87. How many 12-oz cans of soda can be filled with 96 oz of soda? [1.8a]

88. A case of soda contains 24 bottles. If each bottle contains 12 oz, how many ounces of soda are in the case? [1.8a]

Simplify. [1.9c]

89. $5 + 4^2 + 2 \cdot 7$

90. $45 \div (2^2 + 11)$

91. $(9 + 7)(9 - 7)$

92. $(13 - 2)(13 + 2)$

Synthesis

Subtract.

93. 🖩 $123{,}907 - 433{,}789$

94. 🖩 $23{,}011 - (-60{,}432)$

For Exercises 95–100, tell whether each statement is true or false for all integers a and b. If false, give an example to show why.

95. $a - 0 = 0 - a$

96. $0 - a = a$

97. If $a \neq b$, then $a - b \neq 0$.

98. If $a = -b$, then $a + b = 0$.

99. If $a + b = 0$, then a and b are opposites.

100. If $a - b = 0$, then $a = -b$.

101. If $a - 54$ is -37, find the value of a.

102. If $x - 48$ is -15, find the value of x.

103. Maureen kept track of the weekly changes in the stock market over a period of 5 weeks. By how many points (pts) had the market risen or fallen over this time?

WEEK 1	WEEK 2	WEEK 3	WEEK 4	WEEK 5
Down 13 pts	Down 16 pts	Up 36 pts	Down 11 pts	Up 19 pts

104. *Blackjack Counting System.* Players of the casino game of blackjack make use of many card-counting systems in which a negative count means that a player has an advantage. One such system, called *High-Low*, was developed by Harvey Dubner in 1963. Each card already showing counts as −1, 0, or 1 as follows:

2, 3, 4, 5, 6	count as +1;
7, 8, 9	count as 0;
10, J, Q, K, A	count as −1.

Source: Patterson, Jerry L. *Casino Gambling.* New York: Perigee, 1982

a) Find the total count on this sequence of cards:

K, A, 2, 4, 5, 10, J, 8, Q, K, 5.

b) Does the player have an advantage?

2.4 Multiplication of Integers

OBJECTIVES

a Multiply integers.

b Find products of three or more integers, and simplify powers of integers.

a MULTIPLICATION

Multiplication of integers is like multiplication of whole numbers. The difference is that we must determine whether the answer is positive or negative.

Multiplication of a Positive Integer and a Negative Integer

To see how to multiply a positive integer and a negative integer, consider the following pattern.

This number decreases by 1 each time. This number decreases by 5 each time.

$$\begin{aligned} 4 \cdot 5 &= 20 \\ 3 \cdot 5 &= 15 \\ 2 \cdot 5 &= 10 \\ 1 \cdot 5 &= 5 \\ 0 \cdot 5 &= 0 \\ -1 \cdot 5 &= -5 \\ -2 \cdot 5 &= -10 \\ -3 \cdot 5 &= -15 \end{aligned}$$

Do Exercise 1.

1. Complete, as in the example in the text.

$$\begin{aligned} 4 \cdot 10 &= 40 \\ 3 \cdot 10 &= 30 \\ 2 \cdot 10 &= \\ 1 \cdot 10 &= \\ 0 \cdot 10 &= \\ -1 \cdot 10 &= \\ -2 \cdot 10 &= \\ -3 \cdot 10 &= \end{aligned}$$

According to this pattern, it looks as though the product of a negative integer and a positive integer is negative. To confirm this, use repeated addition:

$$-1 \cdot 5 = 5 \cdot (-1) = -1 + (-1) + (-1) + (-1) + (-1) = -5$$
$$-2 \cdot 5 = 5 \cdot (-2) = -2 + (-2) + (-2) + (-2) + (-2) = -10$$
$$-3 \cdot 5 = 5 \cdot (-3) = -3 + (-3) + (-3) + (-3) + (-3) = -15$$

MULTIPLYING A POSITIVE INTEGER AND A NEGATIVE INTEGER

To multiply a positive integer and a negative integer, multiply their absolute values. The answer is negative.

EXAMPLES Multiply.

1. $8(-5) = -40$ **2.** $50(-1) = -50$ **3.** $-7 \cdot 6 = -42$

Do Exercises 2–4.

Multiply.

2. $-3 \cdot 6$ **3.** $20 \cdot (-5)$

4. $12(-1)$

Multiplication of Two Negative Integers

How do we multiply two negative integers? Again we look for a pattern.

This number decreases by 1 each time. This number increases by 5 each time.

$$\begin{aligned} 4 \cdot (-5) &= -20 \\ 3 \cdot (-5) &= -15 \\ 2 \cdot (-5) &= -10 \\ 1 \cdot (-5) &= -5 \\ 0 \cdot (-5) &= 0 \\ -1 \cdot (-5) &= 5 \\ -2 \cdot (-5) &= 10 \\ -3 \cdot (-5) &= 15 \end{aligned}$$

Do Exercise 5.

5. Complete, as in the example in the text.

$$\begin{aligned} 3 \cdot (-10) &= -30 \\ 2 \cdot (-10) &= -20 \\ 1 \cdot (-10) &= \\ 0 \cdot (-10) &= \\ -1 \cdot (-10) &= \\ -2 \cdot (-10) &= \\ -3 \cdot (-10) &= \end{aligned}$$

Answers

1. 20; 10; 0; −10; −20; −30 **2.** −18 **3.** −100 **4.** −12 **5.** −10; 0; 10; 20; 30

According to the pattern, the product of two negative integers is positive. This leads to the second part of the rule for multiplying integers.

MULTIPLYING TWO NEGATIVE INTEGERS

To multiply two negative integers, multiply their absolute values. The answer is positive.

EXAMPLES Multiply.

4. $(-2)(-4) = 8$

5. $(-10)(-7) = 70$

Do Exercises 6–8. ▶

The following is another way to state the rules for multiplication.

To multiply two nonzero integers:

a) Multiply the absolute values.
b) If the signs are the same, the answer is positive.
c) If the signs are different, the answer is negative.

EXAMPLES Multiply.

6. $(-3)(-5) = 15$

7. $8(-10) = -80$

Do Exercises 9–12. ▶

Multiplication by 0

No matter how many times 0 is added to itself, the answer is 0.

The product of 0 and any integer is 0:

$$a \cdot 0 = 0.$$

EXAMPLES Multiply.

8. $-19 \cdot 0 = 0$

9. $0(-7) = 0$

Do Exercises 13 and 14. ▶

b MULTIPLICATION OF MORE THAN TWO INTEGERS

When multiplying more than two integers, we can choose order and grouping as we please, using the commutative and associative laws.

EXAMPLES Multiply.

10. a) $-8 \cdot 2(-3) = -16(-3)$ Multiplying the first two numbers
$= 48$ Multiplying the results

b) $-8 \cdot 2(-3) = 24 \cdot 2$ Multiplying the negative numbers
$= 48$ The result is the same as above.

Multiply.

6. $(-3)(-4)$

7. $-9(-5)$

8. $(-1)(-6)$

Multiply.

GS **9.** $3(-6)$

Multiply absolute values:

$3 \cdot 6 =$ ____.

The signs are different, so the answer is ____.

$3(-6) =$ ____

10. $(-6)(-5)$

11. $(-1)(-50)$

12. $(-8) \cdot 11$

Multiply.

13. $0(-5)$

14. $-23 \cdot 0$

Answers

6. 12 **7.** 45 **8.** 6 **9.** −18 **10.** 30 **11.** 50 **12.** −88 **13.** 0 **14.** 0

Guided Solution:
9. 18, negative, −18

Multiply.

15. $5 \cdot (-3) \cdot 2$

16. $-2 \cdot (-5) \cdot (-4) \cdot (-3)$

17. $(-4)(-5)(-2)(-3)(-1)$
$= \square \cdot 6 \cdot (-1)$
$= \square \cdot (-1)$
$= \square$

GS

18. $(-1)(-1)(-2)(-3)(-1)(-1)$

Simplify.

19. $(-2)^3$

20. $(-9)^2$

21. $(-1)^9$

22. 2^5

Answers

15. −30 16. 120 17. −120 18. 6
19. −8 20. 81 21. −1 22. 32

Guided Solution:
17. 20, 120, −120

11. $7(-1)(-4)(-2) = (-7)8$ Multiplying the first two numbers and the last two numbers
$= -56$

12. $-5 \cdot (-2) \cdot (-3) \cdot (-6) = 10 \cdot 18$ Each pair of negative numbers gives a positive product.
$= 180$

13. $(-3)(-5)(-2)(-3)(-1) = 15 \cdot 6 \cdot (-1)$
$= 90(-1) = -90$

We can see the following pattern in the results of Examples 11–13.

> The product of an even number of negative integers is positive.
> The product of an odd number of negative integers is negative.

Do Exercises 15–18.

Powers of Integers

A positive number raised to any power is positive. When a negative number is raised to a power, the sign of the result depends upon whether the exponent is even or odd.

EXAMPLES Simplify.

14. $(-7)^2 = (-7)(-7) = 49$ The result is positive.

15. $(-4)^3 = (-4)(-4)(-4)$
$= 16(-4)$
$= -64$ The result is negative.

16. $(-3)^4 = (-3)(-3)(-3)(-3)$
$= 9 \cdot 9$
$= 81$ The result is positive.

17. $(-2)^5 = (-2)(-2)(-2)(-2)(-2)$
$= 4 \cdot 4 \cdot (-2)$
$= 16(-2)$
$= -32$ The result is negative.

Perhaps you noted the following.

> When a negative number is raised to an even exponent, the result is positive.
> When a negative number is raised to an odd exponent, the result is negative.

Do Exercises 19–22.

When an integer is multiplied by -1, the result is the opposite of that integer.

> For any integer a,
> $$-1 \cdot a = -a.$$

EXAMPLE 18 Simplify: -7^2.

In the expression -7^2, the base is 7, not -7. We can regard -7^2 as $-1 \cdot 7^2$:

$$-7^2 = -1 \cdot 7^2$$ The rules for order of operations tell us to square first: $7^2 = 7 \cdot 7$.

$$= -1 \cdot 49$$

$$= -49.$$

Compare Examples 14 and 18 and note that $(-7)^2 \neq -7^2$.

Do Exercises 23 and 24. ▶

23. Simplify: -5^2.

24. Simplify: $(-5)^2$.

Answers

23. -25 24. 25

CALCULATOR CORNER

Exponential Notation When using a calculator to calculate expressions like $(-39)^4$ or -39^4, it is important to use the correct sequence of keystrokes.

Calculators with [+/−] key: To calculate $(-39)^4$, we press [3] [9] [+/−] [x^y] [4] [=].

To calculate -39^4, we must first raise 39 to the power 4. Then the sign of the result must be changed. This can be done with the keystrokes [3] [9] [x^y] [4] [=] [+/−], or by multiplying 39^4 by -1 using the keystrokes [3] [9] [x^y] [4] [=] [×] [1] [+/−] [=].

Calculators with [(−)] key: On some calculators, the [(−)] key is pressed before a number to indicate that the number is negative. This is similar to the way the expression is written on paper. With these calculators, $(-39)^4$ is found by pressing [(] [(−)] [3] [9] [)] [^] [4] [ENTER], and -39^4 is found by pressing [(−)] [3] [9] [^] [4] [ENTER].

You can either experiment or consult a user's manual if you are unsure of the proper keystrokes for your calculator.

EXERCISES Use a calculator to determine each of the following.

1. $(-23)^6$
2. $(-17)^5$
3. $(-104)^3$
4. $(-4)^{10}$
5. -9^6
6. -7^6
7. -6^5
8. -3^9

2.4 Exercise Set

For Extra Help

MyMathLab®

PRACTICE

WATCH

READ

REVIEW

✓ Reading Check

Fill in each blank with either "positive" or "negative."

RC1. To multiply a positive integer and a negative integer, we multiply their absolute values. The answer is ____________.

RC2. To multiply two negative integers, we multiply their absolute values. The answer is ____________.

RC3. The product of an even number of negative integers is ____________.

RC4. The product of an odd number of negative integers is ____________.

a Multiply.

1. $-2 \cdot 8$
2. $-7 \cdot 3$
3. $10 \cdot (-6)$
4. $12 \cdot (-2)$
5. $8 \cdot (-6)$
6. $8 \cdot (-3)$
7. $-10 \cdot 3$
8. $-9 \cdot 8$
9. $-3(-5)$
10. $-8 \cdot (-2)$
11. $-9 \cdot (-2)$
12. $(-8)(-9)$
13. $(-6)(-7)$
14. $-8 \cdot (-3)$
15. $-10(-2)$
16. $-9(-8)$
17. $12(-10)$
18. $15(-8)$
19. $-23 \cdot 0$
20. $-38 \cdot 0$
21. $(-72)(-1)$
22. $41(-3)$
23. $(-20)17$
24. $(-1)(-43)$
25. $-8(-50)$
26. $(-25) \cdot 8$
27. $0(-14)$
28. $0(-38)$

b Multiply.

29. $3 \cdot (-8) \cdot (-1)$
30. $(-7) \cdot (-4) \cdot (-1)$
31. $7(-4)(-3)5$
32. $9(-2)(-6)7$
33. $-2(-5)(-7)$
34. $(-2)(-5)(-3)(-5)$
35. $(-5)(-2)(-3)(-1)$
36. $-6(-5)(-9)$
37. $(-15)(-29) \cdot 0 \cdot 8$
38. $19(-7)(-8) \cdot 0 \cdot 6$
39. $(-7)(-1)(7)(-6)$
40. $(-5)6(-4)5$

Simplify.

41. $(-6)^2$

42. $(-8)^2$

43. $(-5)^3$

44. $(-2)^4$

45. $(-10)^4$

46. $(-1)^5$

47. -2^4

48. $(-2)^6$

49. $(-3)^5$

50. -10^4

51. $(-1)^{12}$

52. $(-1)^{13}$

53. -11^2

54. -2^6

55. -4^3

56. -2^5

Skill Maintenance

57. Round 532,451 to the nearest hundred. [1.6a]

58. Write standard notation for sixty million. [1.1c]

59. Divide: $2880 \div 36$. [1.5a]

60. Multiply: 75×34. [1.4a]

61. Simplify: $10 - 2^3 + 6 \div 2$. [1.9c]

62. Simplify: $2 \cdot 5^2 - 3 \cdot 2^3 \div (3 + 3^2)$. [1.9c]

63. A rectangular rug measures 5 ft by 8 ft. What is the area of the rug? [1.4b], [1.8a]

64. How many 12-egg cartons can be filled with 2880 eggs? [1.8a]

65. A ferry can accommodate 12 cars and 53 cars are waiting. How many trips will be required to ferry them all? [1.8a]

66. An elevator can hold 16 people and 50 people are waiting to go up. How many trips will be required to transport all of them? [1.8a]

Synthesis

Simplify.

67. $(-3)^5(-1)^{379}$

68 $(-2)^3 \cdot [(-1)^{29}]^{46}$

69. $-9^4 + (-9)^4$

70. $-5^2(-1)^{29}$

71. $|(-2)^5 + 3^2| - (3 - 7)^2$

72. $|-12(-3)^2 - 5^3 - 6^2 - (-5)^2|$

73. 🖩 -47^2

74. 🖩 -53^2

75. 🖩 $(-19)^4$

76. 🖩 $(-23)^4$

77. 🖩 $(73 - 86)^3$

78. 🖩 $(-49 + 34)^3$

79. 🖩 $-935(238 - 243)^3$

80. 🖩 $(-17)^4(129 - 133)^5$

81. Jo had a balance of \$68 in her account and wrote seven checks for \$13 each. What was her balance after writing the checks?

82. After diving 95 m below the surface, a diver rises at a rate of 7 meters per minute for 9 min. What is the diver's new elevation?

83. What must be true of m and n if $[(-5)^m]^n$ is to be **(a)** negative? **(b)** positive?

84. What must be true of m and n if $-mn$ is to be **(a)** positive? **(b)** zero? **(c)** negative?

2.5 Division of Integers and Order of Operations

OBJECTIVES

a Divide integers.

b Use the rules for order of operations with integers.

SKILL TO REVIEW

Objective 1.9c: Simplify expressions using the rules for order of operations.

Simplify.

1. $5^2 - 2(10 - 3)$
2. $2[21 - (11 - 3)]$

We now consider division of integers. Because of the way in which division is defined, its rules are similar to those for multiplication.

a DIVISION OF INTEGERS

DIVISION

The quotient $\frac{a}{b}$ (or $a \div b$, or a/b) is the number, if there is one, that when multiplied by b gives a.

Let's use the definition to divide integers.

EXAMPLES Divide, if possible. Check each answer.

1. $14 \div (-7) = -2$ — *Think*: What number multiplied by -7 gives 14? The number is -2. *Check*: $(-2)(-7) = 14$.
2. $\frac{-32}{-4} = 8$ — *Think*: What number multiplied by -4 gives -32? The number is 8. *Check*: $8(-4) = -32$.
3. $-21 \div 7 = -3$ — *Think*: What number multiplied by 7 gives -21? The number is -3. *Check*: $(-3) \cdot 7 = -21$.
4. $\frac{0}{-5} = 0$ — *Think*: What number multiplied by -5 gives 0? The number is 0. *Check*: $0(-5) = 0$.
5. $\frac{-5}{0}$ is **not defined**. — *Think*: What number multiplied by 0 gives -5? There is no such number because the product of 0 and *any* number is 0.

The rules for determining the sign of a quotient are the same as those for determining the sign of a product. We state them together.

To multiply or divide two integers:

a) Multiply or divide the absolute values.
b) If the signs are the same, the answer is positive.
c) If the signs are different, the answer is negative.

Do Margin Exercises 1–6.

Dividing by 0

Recall that, in general, $a \div b$ and $b \div a$ are different numbers. In Example 4, we divided *into* 0. In Example 5, we attempted to divide *by* 0. Since any number times 0 gives 0, not -5, we say that $-5 \div 0$ is **not defined** or is **undefined**. Also, since *any* number times 0 gives 0, $0 \div 0$ is also not defined.

Divide.

1. $6 \div (-2)$
   Think: What number multiplied by -2 gives 6?
2. $\frac{-15}{-3}$
   Think: What number multiplied by -3 gives -15?
3. $-24 \div 8$
   Think: What number multiplied by 8 gives -24?
4. $\frac{0}{-4}$
5. $\frac{30}{-5}$
6. $\frac{-45}{9}$

Answers

Skill to Review:
1. 11 **2.** 26

Margin Exercises:
1. -3 **2.** 5 **3.** -3 **4.** 0
5. -6 **6.** -5

EXCLUDING DIVISION BY 0

Division by 0 is not defined:

$a \div 0$, or $\frac{a}{0}$, is undefined for all real numbers a.

Dividing 0 by Other Numbers

Note that $0 \div 8 = 0$ because $0 = 0 \cdot 8$.

DIVIDENDS OF 0

Zero divided by any nonzero real number is 0:

$$\frac{0}{a} = 0, \quad a \neq 0.$$

EXAMPLES Divide.

6. $0 \div (-6) = 0$

7. $\frac{0}{12} = 0$

8. $\frac{-3}{0}$ is undefined.

Do Exercises 7–9. ▶

Divide, if possible.

7. $34 \div 0$

8. $0 \div (-4)$

9. $-52 \div 0$

b RULES FOR ORDER OF OPERATIONS

When more than one operation appears in a calculation or problem, we apply the rules for order of operations.

RULES FOR ORDER OF OPERATIONS

1. Do all calculations within grouping symbols, including parentheses, brackets, braces, and absolute-value symbols, and within numerators or denominators.
2. Evaluate all exponential expressions.
3. Do all multiplications and divisions in order from left to right.
4. Do all additions and subtractions in order from left to right.

EXAMPLE 9 Simplify: $17 - 10 \div 2 \cdot 4$.

There are no grouping symbols or exponents, so we begin with the third rule.

$$\begin{aligned} 17 - 10 \div 2 \cdot 4 &= 17 - 5 \cdot 4 \\ &= 17 - 20 \\ &= -3 \end{aligned}$$

Carrying out all multiplications and divisions in order from left to right

Answers

7. Undefined **8.** 0 **9.** Undefined

EXAMPLES Simplify.

10. $|(-2)^3 \div 4| - 5(-2)$

We first simplify within the absolute-value symbols.

$$\begin{aligned} |(-2)^3 \div 4| - 5(-2) &= |-8 \div 4| - 5(-2) && (-2)(-2)(-2) = -8 \\ &= |-2| - 5(-2) && \text{Dividing} \\ &= 2 - 5(-2) && |-2| = 2 \\ &= 2 - (-10) && \text{Multiplying} \\ &= 12 && \text{Subtracting; } 2 - (-10) = 2 + 10 \end{aligned}$$

11. $2^4 + 51 \cdot 4 - (37 + 23 \cdot 2)$

$$\begin{aligned} &2^4 + 51 \cdot 4 - (37 + 23 \cdot 2) \\ &= 2^4 + 51 \cdot 4 - (37 + 46) && \text{Carrying out all operations inside parentheses first, following the rules for order of operations within the parentheses} \\ &= 2^4 + 51 \cdot 4 - 83 && \text{Adding inside parentheses} \\ &= 16 + 51 \cdot 4 - 83 && \text{Evaluating exponential expressions} \\ &= 16 + 204 - 83 && \text{Doing all multiplications} \\ &= 220 - 83 && \text{Doing all additions and subtractions in order from left to right} \\ &= 137 \end{aligned}$$

Always regard a fraction bar as a grouping symbol. It separates any calculations in the numerator from those in the denominator.

EXAMPLE 12 Simplify: $\dfrac{5 - (-3)^2}{8 - 10}$.

$$\begin{aligned} \frac{5 - (-3)^2}{8 - 10} &= \frac{5 - 9}{-2} \\ &= \frac{-4}{-2} && \text{Calculating within the numerator and within the denominator: } (-3)^2 = (-3)(-3) = 9,\ 8 - 10 = -2, \text{ and } 5 - 9 = -4 \\ &= 2 && \text{Dividing} \end{aligned}$$

Do Exercises 10–13.

Simplify.

10. $5 - (-7)(-3)^2$

11. $(-2) \cdot |3 - 2^2| + 5$
$= (-2) \cdot |3 - \square| + 5$
$= (-2) \cdot |\square| + 5$
$= (-2) \cdot \square + 5$
$= \square + 5$
$= \square$

GS

12. $52 \cdot 5 + 5^3 - (4^2 - 48 \div 4)$

13. $\dfrac{(-5)(-9)}{1 - 2 \cdot 2}$

Answers

10. 68 **11.** 3 **12.** 381 **13.** −15

Guided Solution:
11. 4, −1, 1, −2, 3

CALCULATOR CORNER

Grouping Symbols On calculators, grouping symbols may appear as [( ] and [ )] or [[(...] and [...)]]. We often need grouping symbols when simplifying expressions written in fraction form. For example, the fraction bar in the expression

$$\frac{38 + 142}{2 - 47}$$

acts as a grouping symbol. To simplify, we add parentheses around the numerator and around the denominator, pressing [( ] [3] [8] [+] [1] [4] [2] [ )] [÷] [( ] [2] [−] [4] [7] [ )] [=]. The result is −4. Without the grouping symbols, we would be simplifying a different expression:

$$38 + \frac{142}{2} - 47.$$

EXERCISES Use a calculator with grouping symbols to simplify each of the following.

1. $\dfrac{38 - 178}{5 + 30}$ **2.** $\dfrac{311 - 17^2}{2 - 13}$ **3.** $785 - \dfrac{285 - 5^4}{17 + 3 \cdot 51}$

2.5 Exercise Set

For Extra Help MyMathLab® MathXL® PRACTICE WATCH READ REVIEW

✓ Reading Check

Tell whether each division results in 0 or is undefined.

RC1. $0 \div 17$

RC2. $(-3) \div 0$

RC3. $\frac{132}{0}$

RC4. $\frac{0}{-74}$

a Divide, if possible, and check each answer by multiplying. If an answer is undefined, state so.

1. $36 \div (-6)$

2. $\frac{42}{-7}$

3. $\frac{26}{-2}$

4. $24 \div (-12)$

5. $\frac{-16}{8}$

6. $-22 \div (-2)$

7. $\frac{-48}{-12}$

8. $-72 \div (-9)$

9. $\frac{-72}{8}$

10. $\frac{-50}{25}$

11. $-100 \div (-50)$

12. $\frac{-400}{8}$

13. $-108 \div 9$

14. $\frac{-128}{8}$

15. $\frac{200}{-25}$

16. $-651 \div (-31)$

17. $\frac{-56}{0}$

18. $\frac{0}{-5}$

19. $\frac{88}{-11}$

20. $\frac{-145}{-5}$

21. $-\frac{276}{12}$

22. $-\frac{217}{7}$

23. $\frac{0}{-2}$

24. $\frac{-13}{0}$

25. $\frac{19}{-1}$

26. $\frac{-17}{1}$

27. $-41 \div 1$

28. $23 \div (-1)$

b Simplify, if possible. If an answer is undefined, state so.

29. $8 - 2 \cdot 3 - 9$

30. $8 - (2 \cdot 3 - 9)$

31. $(8 - 2 \cdot 3) - 9$

32. $(8 - 2)(3 - 9)$

33. $16 \cdot (-24) + 50$

34. $10 \cdot 20 - 15 \cdot 24$

35. $40 - 3^2 - 2^3$

36. $2^4 + 2^2 - 10$

37. $4 \cdot (6 + 8)/(4 + 3)$

38. $4^3 + 10 \cdot 20 + 8^2 - 23$

39. $4 \cdot 5 - 2 \cdot 6 + 4$

40. $5^3 + 4 \cdot 9 - (8 + 9 \cdot 3)$

41. $1 - (-2)^2 \cdot 3 \div 6$

42. $-6 + (-3)^2 + 6 \div (-2)$

43. $18 - (-3)^3 - 3^2 \cdot 5$

44. $9 - (-2)^3 - 50 \div 2$

45. $\dfrac{9^2 - 1}{1 - 3^2}$

46. $\dfrac{100 - 6^2}{(-5)^2 - 3^2}$

47. $8(-7) + 6(-5)$

48. $10(-5) \div 1(-1)$

49. $20 \div 5(-3) + 3$

50. $14 \div 2(-6) + 7$

51. $18 - 0(3^2 - 5^2 \cdot 7 - 4)$

52. $9 \cdot 0 \div 5 \cdot 4$

53. $-4^2 + 6$

54. $-5^2 + 7$

55. $-8^2 - 3$

56. $-9^2 - 11$

57. $4 \cdot 5^2 \div 10$

58. $(2 - 5)^2 \div (-9)$

59. $(3 - 8)^2 \div (-1)$

60. $3 - 3^2$

61. $12 - 20^3$

62. $20 + 4^3 \div (-8)$

63. $2 \times 10^3 - 5000$

64. $-7(3^4) + 18$

65. $6[9 - (3 - 4)]$

66. $8[(6 - 13) - 11]$

67. $-1000 \div (-100) \div 10$

68. $256 \div (-32) \div (-4)$

69. $-7 - 3[-80 \div (2 - 10)]$

70. $-1 - 5[3 - (7 - 4^2)]$

71. $-2[3 - (7 - 9)^3]$

72. $-10[(2 - 8)^2 - 6]$

73. $8 - |7 - 9| \cdot 3$

74. $|8 - 7 - 9| \cdot 2 + 1$

75. $9 - |7 - 3^2|$

76. $9 - |5 - 7|^3$

77. $\dfrac{6^3 - 7 \cdot 3^4 - 2^5 \cdot 9}{(1 - 2^3)^3 + 7^3}$

78. $\dfrac{6 \div 2 \cdot 4^2 - 7^2 + 1}{(7 - 4)^3 - 2 \cdot 5 - 4}$

79. $\dfrac{2 \cdot 3^2 \div (3^2 - (2 + 1))}{5^2 - 6^2 - 2^2(-3)}$

80. $\dfrac{5 \cdot 6^2 \div (2^2 \cdot 5) - 7^2}{3^2 - 4^2 - (-2)^3 - 2}$

81. $\dfrac{(-5)^3 + 17}{10(2 - 6) - 2(5 + 2)}$

82. $\dfrac{(3 - 5)^2 - (7 - 13)}{(2 - 5)3 + 2 \cdot 4}$

83. $\dfrac{2 \cdot 4^3 - 4 \cdot 32}{19^3 - 17^4}$

84. $\dfrac{-16 \cdot 28 \div 2^2}{5 \cdot 25 - 5^3}$

Skill Maintenance

85. Fabrikant Fine Diamonds ran a 4-in. by 7-in. advertisement in the *New York Times*. Find the area of the ad. [1.4b], [1.8a]

86. A hotel has 4 floors with 62 rooms on each floor. How many rooms are there in the hotel? [1.8a]

87. Cindi's Ford Focus gets 32 mpg (miles per gallon). How many gallons will it take to travel 384 mi? [1.8a]

88. Craig's Chevy Blazer gets 14 mpg. How many gallons will it take to travel 378 mi? [1.8a]

89. A 7-oz bag of tortilla chips contains 1050 calories. How many calories are in a 1-oz serving? [1.8a]

90. A 7-oz bag of tortilla chips contains 8 g (grams) of fat per ounce. How many grams of fat are in a carton containing 12 bags of chips? [1.8a]

91. There are 18 sticks in a large pack of Trident gum. If 4 people share the pack equally, how many whole pieces will each person receive? How many extra pieces will remain? [1.8a]

92. A bag of Ricola throat lozenges contains 24 cough drops. If 5 people share the bag equally, how many lozenges will each person receive? How many extra lozenges will remain? [1.8a]

Synthesis

Simplify, if possible.

93. $\dfrac{9 - 3^2}{2 \cdot 4^2 - 5^2 \cdot 9 + 8^2 \cdot 7}$

94. $\dfrac{7^3 \cdot 9 - 6^2 \cdot 8 + 4^3 \cdot 6}{5^2 - 25}$

95. $\dfrac{(25 - 4^2)^3}{17^2 - 16^2} \cdot ((-6)^2 - 6^2)$

96. $\dfrac{(7 - 8)^{37}}{7^2 - 8^2} \cdot (98 - 7^2 \cdot 2)$

97. 🖩 $\dfrac{19 - 17^2}{13^2 - 34}$

98. 🖩 $\dfrac{195 + (-15)^3}{195 - 7 \cdot 5^2}$

99. 🖩 $28^2 - 36^2/4^2 + 17^2$

100. 🖩 $9^3 - 36^3/12^2 + 9^2$

101. 🖩 Write down the keystrokes needed to calculate $\dfrac{15^2 - 5^3}{3^2 + 4^2}$.

102. 🖩 Write down the keystrokes needed to calculate $\dfrac{16^2 - 24 \cdot 23}{3 \cdot 4 + 5^2}$.

103. Evaluate the expression for which the keystrokes are as follows: [4] [−] [1] [0] [÷] [2] [+] [6].

104. Evaluate the expression for which the keystrokes are [4] [−] [1] [6] [÷] [(] [2] [+] [6] [)].

Determine the sign of each expression if m is negative and n is positive.

105. $\dfrac{-n}{m}$

106. $\dfrac{-n}{-m}$

107. $-\left(\dfrac{-n}{m}\right)$

108. $-\left(\dfrac{n}{-m}\right)$

109. $-\left(\dfrac{-n}{-m}\right)$

Mid-Chapter Review

Concept Reinforcement

Determine whether each statement is true or false.

__________ **1.** Every integer is either positive or negative. [2.1a]

__________ **2.** If $a > b$, then a lies to the left of b on the number line. [2.1b]

__________ **3.** The absolute value of a number is always nonnegative. [2.1c]

Guided Solutions

Fill in each blank with the number that creates a correct statement or solution.

4. Evaluate $-x$ and $-(-x)$ when $x = -4$. [2.1d]

$-x = -(\quad) = \quad$;

$-(-x) = -(-(\quad)) = -(\quad) = \quad$

Subtract. [2.3a]

5. $5 - 13 = 5 + (\quad) = \quad$

6. $-6 - (-7) = -6 + \quad = \quad$

Mixed Review

7. State the integers that correspond to this situation.
Jerilyn deposited $450 in her checking account. Later that week she used her debit card for a $79 purchase. [2.1a]

8. Change the sign of 9. [2.1d]

Use either $<$ or $>$ for $\square$ to write a true sentence. [2.1b]

9. $-6 \square 6$

10. $-5 \square -3$

11. $-10 \square 0$

12. $-20 \square -30$

Find the absolute value. [2.1c]

13. $|38|$

14. $|-18|$

15. $|0|$

16. $|-12|$

Find the opposite, or additive inverse, of the number. [2.1d]

17. -56

18. 3

19. 0

20. -49

21. Find $-x$ when x is -19. [2.1d]

22. Evaluate $-(-x)$ when x is 23. [2.1d]

Compute and simplify. [2.2a], [2.3a], [2.4a], [2.5a], [2.5b]

23. $7 + (-9)$

24. $-6 + (-10)$

25. $36 + (-36)$

26. $-8 + (-9)$

27. $-9 + 10$

28. $19 + (-17)$

29. $2 - 28$

30. $-8 - (-4)$

31. $-3 - 10$

32. $5 - (-11)$

33. $0 - (-6)$

34. $12 - 24$

35. $-12 \cdot 3$

36. $6(-9)$

37. $(-13)(-2)$

38. $(-2)(-41)$

39. $(-9)^2$

40. -9^2

41. $-75 \div (-3)$

42. $-20 \div 4$

43. $17 - (-25) + 15 - (-18)$

44. $-9 + (-3) + 16 - (-10)$

45. $(-7)(-2)(-1)(-3)$

46. $3 - 6 \cdot 5 - 11$

47. $-5^2 + 6[1 - (3 - 4)]$

48. $\dfrac{6^2 - 3(5 - 9)}{7^2 - (-5)^2}$

Solve. [2.3b]

49. ***Temperature Change.*** In a chemistry lab, Ben works with a substance whose initial temperature is 25°C. During an experiment, the temperature falls to −8°C. Find the difference between the two temperatures.

50. ***Stock Price Change.*** The price of a stock opened at $56. During the day, it dropped $6, then rose $2, and dropped $8. Find the value of the stock at the end of the day.

Understanding Through Discussion and Writing

51. A student states "−45 is bigger than −21." What mistake do you think the student is making? [2.1b]

52. Is subtraction of integers associative? Why or why not? [2.3a]

53. Explain in your own words why the sum of two negative numbers is always negative. [2.2a]

54. If a negative number is subtracted from a positive number, will the result always be positive? Why or why not? [2.3a]

STUDYING FOR SUCCESS *Your Textbook as a Resource*

- ☐ Study any drawings. Observe details in any sketches or graphs that accompany the explanations.
- ☐ Note the careful use of color to indicate substitutions and highlight steps in a multi-step solution.

2.6 Introduction to Algebra and Expressions

OBJECTIVES

a Evaluate an algebraic expression by substitution.

b Use the distributive law to find equivalent expressions.

SKILL TO REVIEW

Objective 1.9b: Evaluate exponential notation.

Evaluate.

1. 8^2 **2.** 5^4

In this section, we will write *equivalent expressions* by making use of the *distributive law.*

a EVALUATING ALGEBRAIC EXPRESSIONS

In arithmetic, we work with expressions such as

$$37 + 86, \quad 7 \cdot 8, \quad 19 - 7, \quad \text{and} \quad \frac{3}{8}.$$

In algebra, we use both numbers and letters and work with *algebraic expressions* such as

$$x + 86, \quad 7 \cdot t, \quad 19 - y, \quad \text{and} \quad \frac{a}{b}.$$

When a letter can stand for various numbers, we call the letter a **variable**. A number or a letter that stands for just one number is called a **constant**. Let $b =$ your birth year. Then b is a constant. Let $a =$ your age. Then a is a variable since the value of a changes from year to year.

An **algebraic expression** consists of variables, constants, numerals, and operation signs. When we replace a variable with a number, we say that we are **substituting** for the variable. Carrying out the **operations** of addition, subtraction, and so on, is called **evaluating the expression**.

EXAMPLE 1 Evaluate $x + y$ for $x = 37$ and $y = 29$.

We substitute 37 for x and 29 for y and carry out the addition:

$$x + y = 37 + 29 = 66.$$

The number 66 is called the **value** of the expression.

Algebraic expressions involving multiplication can be written in several ways. For example, "8 times a" can be written as $8 \times a$, $8 \cdot a$, $8(a)$, or simply $8a$. Two letters written together without an operation symbol, such as ab, also indicate multiplication.

EXAMPLE 2 Evaluate $3y$ for $y = -14$.

$$3y = 3(-14) = -42 \qquad \text{Parentheses are required here.}$$

◀ Do Margin Exercises 1–3.

1. Evaluate $a + b$ for $a = 38$ and $b = 26$.
2. Evaluate $x - y$ for $x = 57$ and $y = 29$.
3. Evaluate $5t$ for $t = -14$.

Algebraic expressions involving division can also be written several ways. For example, "8 divided by t" can be written as $8 \div t$, $8/t$, or $\frac{8}{t}$.

Answers

Skill to Review:
1. 64 2. 625

Margin Exercises:
1. 64 2. 28 3. −70

EXAMPLE 3 Evaluate $\frac{a}{b}$ and $\frac{-a}{-b}$ for $a = 35$ and $b = 7$.

We substitute 35 for a and 7 for b:

$$\frac{a}{b} = \frac{35}{7} = 5; \quad \frac{-a}{-b} = \frac{-35}{-7} = 5.$$

EXAMPLE 4 Evaluate $-\frac{a}{b}$, $\frac{-a}{b}$, and $\frac{a}{-b}$ for $a = 15$ and $b = 3$.

We substitute 15 for a and 3 for b:

$$-\frac{a}{b} = -\frac{15}{3} = -5; \quad \frac{-a}{b} = \frac{-15}{3} = -5; \quad \frac{a}{-b} = \frac{15}{-3} = -5.$$

Examples 3 and 4 illustrate the following.

> $\frac{-a}{-b}$ and $\frac{a}{b}$ represent the same number.
>
> $-\frac{a}{b}$, $\frac{-a}{b}$, and $\frac{a}{-b}$ all represent the same number.

Do Exercises 4–7.

EXAMPLE 5 Evaluate $\frac{9C}{5} + 32$ for $C = 20$.

This expression can be used to find the Fahrenheit temperature that corresponds to 20 degrees Celsius.

$$\frac{9C}{5} + 32 = \frac{9 \cdot 20}{5} + 32 = \frac{180}{5} + 32 = 36 + 32 = 68.$$

Do Exercise 8.

EXAMPLE 6 Evaluate $5x^2$ for $x = 3$ and $x = -3$.

The rules for order of operations specify that the replacement for x be squared. That result is then multiplied by 5.

$$5x^2 = 5(3)^2 = 5(9) = 45;$$
$$5x^2 = 5(-3)^2 = 5(9) = 45$$

You can always use parentheses when substituting. They are usually necessary when substituting a negative number.

Do Exercises 9 and 10.

EXAMPLE 7 Evaluate $(-x)^2$ and $-x^2$ for $x = 7$.

$(-x)^2 = (-7)^2 = (-7)(-7) = 49$ — Substitute 7 for x. Then evaluate the power.

To evaluate $-x^2$, we again substitute 7 for x. We must recall that taking the opposite of a number is the same as multiplying that number by -1.

$-x^2 = -1 \cdot x^2$ — $-a = -1 \cdot a$

$-7^2 = -1 \cdot 7^2$ — Substituting 7 for x

$= -1 \cdot 49 = -49$ — Using the rules for order of operations

Do Exercises 11–13.

For each expression, find two equivalent expressions with the negative sign in different places.

GS **4.** $\frac{-6}{x}$

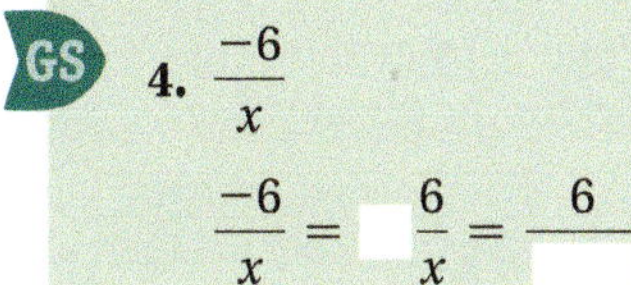

$\frac{-6}{x} = \square\frac{6}{x} = \frac{6}{\square}$

5. $-\frac{m}{n}$

6. $\frac{r}{-4}$

7. Evaluate $\frac{a}{-b}$, $\frac{-a}{b}$, and $-\frac{a}{b}$ for $a = 28$ and $b = 4$.

8. Find the Fahrenheit temperature that corresponds to 10 degrees Celsius. (See Example 5.)

9. Evaluate $3x^2$ for $x = 4$ and $x = -4$.

10. Evaluate a^4 for $a = 3$ and $a = -3$.

11. Evaluate $(-x)^2$ and $-x^2$ for $x = 3$.

12. Evaluate $(-x)^2$ and $-x^2$ for $x = 2$.

13. Evaluate x^5 for $x = 2$ and $x = -2$.

Answers

4. $-\frac{6}{x}$; $\frac{6}{-x}$ **5.** $\frac{-m}{n}$; $\frac{m}{-n}$ **6.** $\frac{-r}{4}$; $-\frac{r}{4}$
7. -7; -7; -7 **8.** 50 **9.** 48; 48 **10.** 81; 81
11. 9; -9 **12.** 4; -4 **13.** 32; -32

Guided Solution:
4. $-$, $-x$

Examples 6 and 7 illustrate the following.

$$(-x)^2 = (x)^2$$
$$(-x)^2 \neq -x^2$$

b EQUIVALENT EXPRESSIONS AND THE DISTRIBUTIVE LAW

Some pairs of algebraic expressions are *equivalent*.

EXAMPLE 8 Evaluate $x + x$ and $2x$ for $x = 3$ and $x = -5$.

We substitute 3 for x in $x + x$ and again in $2x$:

$$x + x = 3 + 3 = 6; \qquad 2x = 2 \cdot 3 = 6.$$

Next we repeat the procedure, substituting -5 for x:

$$x + x = -5 + (-5) = -10; \qquad 2x = 2(-5) = -10.$$

The results can be shown in a table. It appears that $x + x$ and $2x$ represent the same number.

	$x + x$	$2x$
$x = 3$	6	6
$x = -5$	-10	-10

◀ Do Exercises 14 and 15.

Complete each table by evaluating each expression for the given values.

14.

	$3x + 2x$	$5x$
$x = 4$		
$x = -2$		
$x = 0$		

15.

	$4x - x$	$3x$
$x = 2$		
$x = -2$		
$x = 0$		

Example 8 suggests that $x + x$ and $2x$ represent the same number for any replacement of x. This is in fact true, so we can say that $x + x$ and $2x$ are **equivalent expressions.**

EQUIVALENT EXPRESSIONS

Two expressions that have the same value for all allowable replacements are called **equivalent**.

In Examples 3 and 7 we saw that the expressions $\frac{-a}{-b}$ and $\frac{a}{b}$ are equivalent but that the expressions $(-x)^2$ and $-x^2$ are *not* equivalent.

An important concept, known as the **distributive law**, is useful for finding equivalent algebraic expressions. The distributive law involves two operations: multiplication and either addition or subtraction.

To understand how the distributive law works, consider the following:

$$\begin{array}{r l} 45 & \\ \times\ 7 & \\ \hline 35 & \leftarrow \text{This is } 7 \cdot 5. \\ 280 & \leftarrow \text{This is } 7 \cdot 40. \\ \hline 315 & \leftarrow \text{This is the sum } 7 \cdot 40 + 7 \cdot 5. \end{array}$$

To carry out the multiplication, we actually added two products. That is,

$$7 \cdot 45 = 7(40 + 5) = 7 \cdot 40 + 7 \cdot 5.$$

The distributive law says that if we want to multiply a sum of several numbers by a number, we can either add within the grouping symbols and then multiply, or multiply each of the terms separately and then add.

CALCULATOR CORNER

Evaluating Powers To evaluate an expression like $-x^3$ for $x = -14$ with a calculator, we must keep in mind the rules for order of operations. On some calculators, this expression is evaluated by pressing [1] [4] [+/−] [x^y] [3] [=] [+/−]. Other calculators use the keystrokes [(−)] [(] [(−)] [1] [4] [)] [^] [3] [ENTER]. The result should be 2744. Consult your owner's manual or an instructor, or simply experiment if your calculator behaves differently.

EXERCISES Evaluate.

1. $-a^5$ for $a = -3$
2. $-x^5$ for $x = -4$
3. $-x^5$ for $x = 2$
4. $-x^5$ for $x = 5$

Answers

14. 20, 20; −10, −10; 0, 0
15. 6, 6; −6, −6; 0, 0

THE DISTRIBUTIVE LAW

For any numbers a, b, and c,

$$a(b + c) = ab + ac.$$

EXAMPLE 9 Evaluate $a(b + c)$ and $ab + ac$ for $a = 3$, $b = 4$, and $c = 2$.

We have

$$a(b + c) = 3(4 + 2) = 3 \cdot 6 = 18$$

and $$ab + ac = 3 \cdot 4 + 3 \cdot 2 = 12 + 6 = 18.$$

The parentheses in the statement of the distributive law tell us to multiply both b and c by a. Without the parentheses, we would have $ab + c$.

EXAMPLE 10 Use the distributive law to write an expression equivalent to $2(l + w)$.

$2(l + w) = 2 \cdot l + 2 \cdot w$ — Note that the + sign between l and w now appears between $2 \cdot l$ and $2 \cdot w$.

$= 2l + 2w.$ — Try to go directly to this step.

Do Exercises 16 and 17. ▶

Since subtraction can be regarded as addition of the opposite, it follows that the distributive law is true for subtraction as well as addition.

EXAMPLE 11 Use the distributive law to write an expression equivalent to each of the following:

a) $9(x - 5)$; **b)** $(a - 7)b$; **c)** $-4(x - 2y + 3z)$; **d)** $-(2x - 3y)$.

a) $9(x - 5) = 9x - 9(5)$

$= 9x - 45$ — Try to go directly to this step.

b) $(a - 7)b = b(a - 7)$ — Using a commutative law

$= b \cdot a - b \cdot 7$ — Using the distributive law

$= ab - 7b$ — Using a commutative law to write ba alphabetically and $b \cdot 7$ with the constant first

c) $-4(x - 2y + 3z) = -4 \cdot x - (-4)(2y) + (-4)(3z)$ — Using the distributive law

$= -4x - (-4 \cdot 2)y + (-4 \cdot 3)z$ — Using an associative law (twice)

$= -4x - (-8y) + (-12z)$

$= -4x + 8y - 12z$

d) $-(2x - 3y) = -1(2x - 3y)$ — The opposite of a number is the same as multiplying by -1.

$= -1 \cdot 2x - (-1)(3y)$ — Using the distributive law

$= -2x - (-3y)$ — Using an associative law (twice)

$= -2x + 3y$

Do Exercises 18–22. ▶

Use the distributive law to write an equivalent expression.

16. $5(a + b)$

GS **17.** $6(x + y + z)$

$= 6 \cdot \square + 6 \cdot \square + 6 \cdot \square$

$= 6x + \square + 6z$

Use the distributive law to write an equivalent expression.

18. $4(x - y)$

19. $3(a - b + c)$

20. $(m - 4)6$

21. $-8(2a - b + 3c)$

22. $-(c - 8d)$

Answers

16. $5a + 5b$ **17.** $6x + 6y + 6z$ **18.** $4x - 4y$ **19.** $3a - 3b + 3c$ **20.** $6m - 24$ **21.** $-16a + 8b - 24c$ **22.** $-c + 8d$

Guided Solution:
17. $x, y, z, 6y$

2.6 Exercise Set

For Extra Help
MyMathLab® MathXL® PRACTICE WATCH READ REVIEW

Reading Check

Classify each algebraic expression as involving either multiplication or division.

RC1. $3/q$ ________

RC2. $3q$ ________

RC3. $3 \cdot q$ ________

RC4. $\frac{3}{q}$ ________

a Evaluate.

1. $10n$, for $n = 2$
(The cost, in cents, of sending 2 text messages)

2. $99n$, for $n = 2$
(The cost, in cents, of downloading 2 songs)

3. $\frac{x}{y}$, for $x = 6$ and $y = -3$

4. $\frac{m}{n}$, for $m = 18$ and $n = 2$

5. $\frac{2d}{c}$, for $c = 6$ and $d = 3$

6. $\frac{5y}{z}$, for $y = 15$ and $z = -25$

7. $\frac{72}{r}$, for $r = 4$
(The approximate doubling time, in years, for an investment earning 4% interest per year)

8. $\frac{72}{i}$, for $i = 2$
(The approximate doubling time, in years, for an investment earning 2% interest per year)

9. $3 - 5 \cdot x$, for $x = 2$

10. $9 - 2 \cdot x$, for $x = 5$

11. $2l + 2w$, for $l = 3$ and $w = 4$
(The perimeter, in feet, of a 3-ft by 4-ft rectangle)

12. $3(a + b)$, for $a = 2$ and $b = 4$

13. $2(l + w)$, for $l = 3$ and $w = 4$
(The perimeter, in feet, of a 3-ft by 4-ft rectangle)

14. $3a + 3b$, for $a = 2$ and $b = 4$

15. $7a - 7b$, for $a = -1$ and $b = 2$

16. $4x - 4y$, for $x = -5$ and $y = 1$

17. $7(a - b)$, for $a = -1$ and $b = 2$

18. $4(x - y)$, for $x = -5$ and $y = 1$

19. $16t^2$, for $t = 5$
(The distance, in feet, that an object falls in 5 sec)

20. $\frac{49t^2}{10}$, for $t = 10$
(The distance, in meters, that an object falls in 10 sec)

21. $a + (b - a)^2$, for $a = 6$ and $b = 10$

22. $(x + y)^2$, for $x = 2$ and $y = 10$

23. $9a + 9b$, for $a = 13$ and $b = -13$

24. $8x + 8y$, for $x = 17$ and $y = -17$

25. $\frac{n^2 - n}{2}$, for $n = 9$
(For determining the number of handshakes possible among 9 people)

26. $\frac{5(F - 32)}{9}$, for $F = 50$
(For converting 50 degrees Fahrenheit to degrees Celsius)

27. $1 - x^2$, for $x = -2$

28. $4 - y^2$, for $y = -1$

29. $m^2 - n^2$, for $m = 6$ and $n = 5$

30. $a^2 + b^2$, for $a = 3$ and $b = 4$

31. $a^3 - a^2$, for $a = -10$

32. $x^2 - x^3$, for $x = -10$

For each expression, write two equivalent expressions with the negative sign in different places.

33. $-\frac{5}{t}$

34. $\frac{7}{-x}$

35. $\frac{-n}{b}$

36. $-\frac{3}{r}$

37. $\frac{9}{-p}$

38. $\frac{-u}{5}$

39. $\frac{-14}{w}$

40. $\frac{-23}{m}$

Evaluate $\frac{-a}{b}$, $\frac{a}{-b}$, and $-\frac{a}{b}$ for the given values.

41. $a = 45, b = 9$

42. $a = 40, b = 2$

43. $a = 81, b = 3$

44. $a = 56, b = 7$

Evaluate.

45. $(-3x)^2$ and $-3x^2$, for $x = 2$

46. $(-2x)^2$ and $-2x^2$, for $x = 3$

47. $5x^2$, for $x = 3$ and $x = -3$

48. $2x^2$, for $x = 5$ and $x = -5$

49. x^3, for $x = 6$ and $x = -6$

50. x^6, for $x = 2$ and $x = -2$

51. x^8, for $x = 1$ and $x = -1$

52. x^5, for $x = 3$ and $x = -3$

53. a^5, for $a = 2$ and $a = -2$

54. a^7, for $a = 1$ and $a = -1$

b Use the distributive law to write an equivalent expression.

55. $5(a + b)$

56. $7(x + y)$

57. $4(x + 1)$

58. $6(a + 1)$

59. $2(b + 5)$

60. $3(x - 6)$

61. $7(1 - t)$

62. $4(1 - y)$

63. $6(5x - 2)$

64. $9(6m - 7)$

65. $8(x + 7 + 6y)$

66. $4(5x + 8 + 3p)$

67. $-7(y - 2)$

68. $-9(y - 7)$

69. $(x + 2)3$

70. $(x + 4)2$

71. $-4(x - 3y - 2z)$

72. $8(2x - 5y - 8z)$

73. $8(a - 3b + c)$

74. $-6(a + 2b - c)$

75. $4(x - 3y - 7z)$

76. $5(9x - y + 8z)$

77. $(4a - 5b + c - 2d)5$

78. $(9a - 4b + 3c - d)7$

79. $-1(3m + 2n)$

80. $-1(6a + 7b)$

81. $-1(2a - 3b + 4)$

82. $-1(7x - 8y - 9)$

83. $-(x - y - z)$

84. $-(a - b - c)$

Skill Maintenance

85. Write a word name for 23,043,921. [1.1c]

86. Multiply: $17 \cdot 53$. [1.4a]

87. Estimate by rounding to the nearest ten. Show your work. [1.6b]

$$\begin{array}{r} 5\,2\,8\,3 \\ -\ 2\,4\,7\,5 \\ \hline \end{array}$$

88. Divide: $2982 \div 3$. [1.5a]

89. On March 9, it snowed 12 in., but on March 10, the sun melted 7 in. How much snow remained? [1.8a]

90. For Tania's graduation party, her husband ordered three buckets of chicken wings at $12 apiece and 3 trays of nachos at $9 a tray. How much did he pay for the wings and nachos? [1.8a]

Synthesis

91. A car's catalytic converter works most efficiently after it is heated to about 370°C. To what Fahrenheit temperature does this correspond? (*Hint*: See Example 5.)

92. Evaluate $\frac{9C}{5} + 32$ for $C = 10$ and for $C = 20$. (See Example 5.) When the Celsius temperature is doubled, is the corresponding Fahrenheit temperature also doubled?

Evaluate.

93. ▦ $a - b^3 + 17a$, for $a = 19$ and $b = -16$

94. ▦ $x^2 - 23y + y^3$, for $x = 18$ and $y = -21$

95. ▦ $r^3 + r^2t - rt^2$, for $r = -9$ and $t = 7$

96. ▦ $a^3b - a^2b^2 + ab^3$, for $a = -8$ and $b = -6$

97. $a^{1996} - a^{1997}$, for $a = -1$

98. $x^{1493} - x^{1492}$, for $x = -1$

99. $(m^3 - mn)^m$, for $m = 4$ and $n = 6$

100. $5a^{3a-4}$, for $a = 2$

Replace each blank with $\boxed{+}$, $\boxed{-}$, $\boxed{\times}$, or $\boxed{\div}$ to make each statement true.

101. ▦ $-32\square(88\square 29) = -1888$

102. ▦ $59\square 17\square 59\square 8 = 1475$

Classify each statement as true or false. If false, write an example showing why.

103. For any choice of x, $x^2 = (-x)^2$.

104. For any choice of x, $x^3 = -x^3$.

105. For any choice of x, $x^6 + x^4 = (-x)^6 + (-x)^4$.

106. For any choice of x, $(-3x)^2 = 9x^2$.

2.7 Like Terms and Perimeter

OBJECTIVES

a Combine like terms.

b Determine the perimeter of a polygon.

One common way in which equivalent expressions are formed is by *combining like terms.*

a COMBINING LIKE TERMS

A **term** is a number, a variable, a product of numbers and/or variables, or a quotient of numbers and/or variables. Terms are separated by addition signs. If there are subtraction signs, we can find an equivalent expression that uses addition signs.

EXAMPLE 1 What are the terms of $3xy - 4y + \frac{2}{z}$?

$$3xy - 4y + \frac{2}{z} = 3xy + (-4y) + \frac{2}{z}$$ Separating parts with + signs

The terms are $3xy$, $-4y$, and $\frac{2}{z}$.

Do Exercises 1 and 2.

What are the terms of each expression?

1. $5x - 4y + 3$
2. $-4y - 2x + \frac{x}{y}$

Terms in which the variable factors are exactly the same, such as $9x$ and $-4x$, are called **like**, or **similar, terms**. For example, $3y^2$ and $7y^2$ are like terms, but $5x$ and $6x^2$ are not. Constants, like 7 and 3, are also like terms.

EXAMPLES Identify the like terms.

2. $7x + 5x^2 + 2x + 8 + 5x^3 + 1$

$7x$ and $2x$ are like terms; 8 and 1 are like terms.

3. $5ab + a^3 - a^2b - 2ab + 7a^3$

$5ab$ and $-2ab$ are like terms; a^3 and $7a^3$ are like terms.

Do Exercises 3 and 4.

Identify the like terms.

3. $9a^3 + 4ab + a^3 + 3ab + 7$
4. $3xy - 5x^2 + y^2 - 4xy + y$

When an algebraic expression contains like terms, an equivalent expression can be formed by **combining**, or **collecting, like terms.** To combine like terms, we use the distributive law.

EXAMPLE 4 Combine like terms to form an equivalent expression.

a) $4x + 3x$ **b)** $6mn - 7mn$

c) $7y - 2 - 6y + 5$ **d)** $2a^5 + 9ab + 3 + a^5 - 7 - 4ab$

a) $4x + 3x = (4 + 3)x$ Using the distributive law (in "reverse")

$= 7x$ We usually go directly to this step.

b) $6mn - 7mn = (6 - 7)mn$ Try to do this mentally.

$= -1mn$, or simply $-mn$

c) $7y - 2 - 6y + 5 = 7y + (-2) + (-6y) + 5$ Rewriting as addition

$= 7y + (-6y) + (-2) + 5$ Using a commutative law

$= 1y + 3$, or simply $y + 3$

Answers

1. $5x$; $-4y$; 3 2. $-4y$; $-2x$; $\frac{x}{y}$
3. $9a^3$ and a^3; $4ab$ and $3ab$ 4. $3xy$ and $-4xy$

d) $2a^5 + 9ab + 3 + a^5 - 7 - 4ab$

$= 2a^5 + 9ab + 3 + a^5 + (-7) + (-4ab)$

$= 2a^5 + a^5 + 9ab + (-4ab) + 3 + (-7)$ Rearranging terms

$= 3a^5 + 5ab + (-4)$ Think of a^5 as $1a^5$; $2a^5 + a^5 = 3a^5$

$= 3a^5 + 5ab - 4$

b PERIMETER

PERIMETER OF A POLYGON

A **polygon** is a closed geometric figure with three or more sides. The **perimeter** of a polygon is the distance around it, or the sum of the lengths of its sides.

EXAMPLE 5 Find the perimeter of this polygon.

A polygon with five sides is called a *pentagon*.

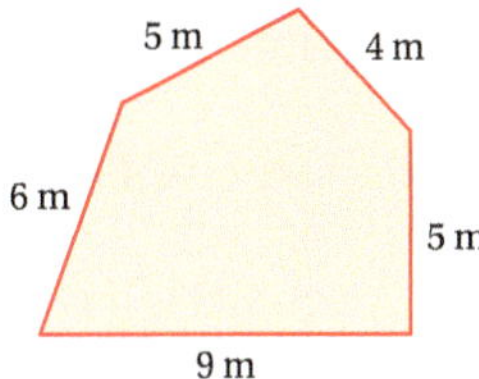

We add the lengths of all sides. Since all the units are the same, we are effectively combining like terms.

Perimeter $= 6\text{ m} + 5\text{ m} + 4\text{ m} + 5\text{ m} + 9\text{ m}$

$= (6 + 5 + 4 + 5 + 9)\text{ m}$ Using the distributive law

$= 29\text{ m}$ Try to go directly to this step.

Caution!

When units of measurement are given in the statement of a problem, as in Example 5, the solution should also contain units of measurement.

Do Exercises 8 and 9.

A **rectangle** is a polygon with four sides and four 90° angles. Opposite sides of a rectangle have the same measure. The symbol ⌈ or ⌉ indicates a 90° angle. A 90° angle is often referred to as a **right angle**.

EXAMPLE 6 Find the perimeter of a rectangle that is 3 cm by 4 cm.

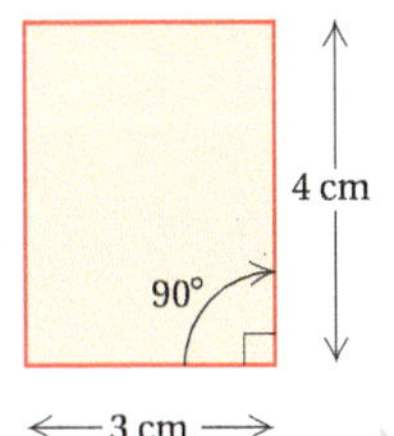

Perimeter $= 3\text{ cm} + 3\text{ cm} + 4\text{ cm} + 4\text{ cm}$

$= (3 + 3 + 4 + 4)\text{ cm}$

$= 14\text{ cm}$

Do Exercise 10.

Combine like terms to form an equivalent expression.

5. $2a + 7a$

6. $5x^2 + 9 - 4x^2 + 3$

7. $4m - 2n^2 + 5 + n^2 + m - 9$

The like terms are

$4m$ and ____,

$-2n^2$ and ____,

5 and ____.

$4m + (-2n^2) + 5 + n^2 + m + (-9)$

$= 4m + m + (-2n^2) +$ ____ $+ 5 + (-9)$

$= 5m + ($ ____ $) + (-4)$

$= 5m -$ ____ $- 4$

Find the perimeter of each polygon.

8.

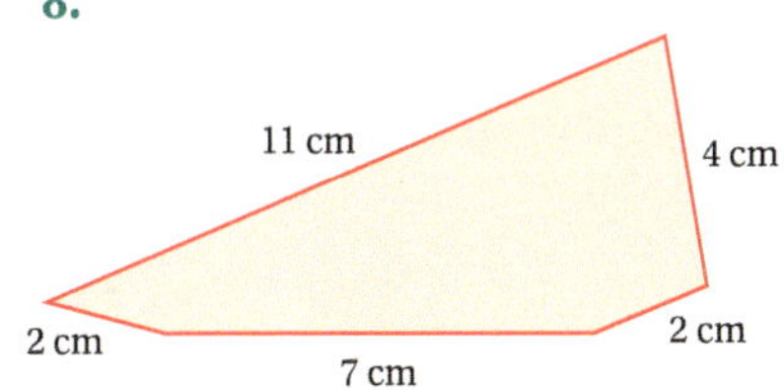

9.

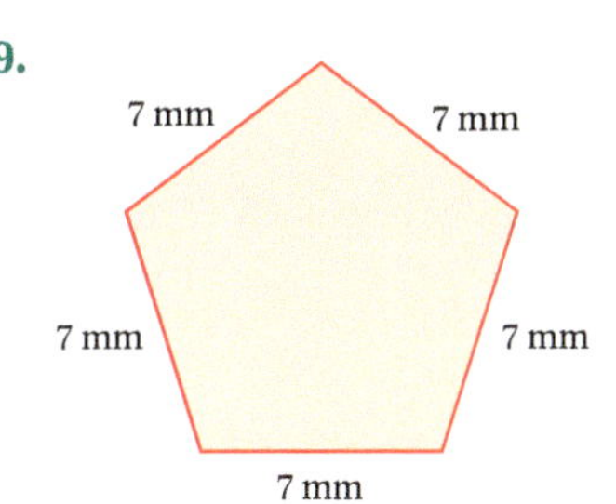

10. Find the perimeter of a rectangle that is 4 cm by 2 cm.

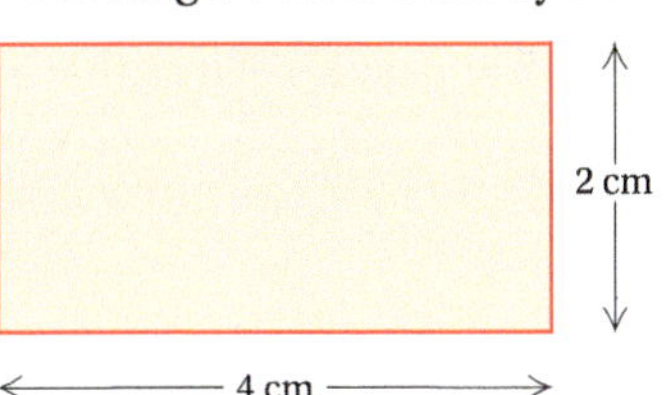

Answers

5. $9a$ **6.** $x^2 + 12$ **7.** $5m - n^2 - 4$

8. 26 cm **9.** 35 mm **10.** 12 cm

Guided Solution:

7. $m, n^2, -9, n^2, -n^2, n^2$

The perimeter of the rectangle in Example 6 is $2 \cdot 3\text{ cm} + 2 \cdot 4\text{ cm}$, or equivalently $2(3\text{ cm} + 4\text{ cm})$. This can be generalized, as follows.

PERIMETER OF A RECTANGLE

The **perimeter P of a rectangle** of length l and width w is given by

$$P = 2l + 2w, \quad \text{or} \quad P = 2 \cdot (l + w).$$

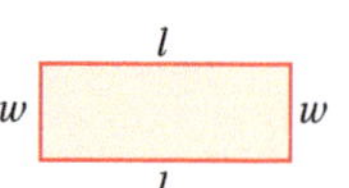

EXAMPLE 7 A common door size is 7 ft by 3 ft. Find the perimeter of such a door.

$$\begin{aligned} P &= 2l + 2w && \text{We could also use } P = 2(l + w). \\ &= 2 \cdot 7\text{ ft} + 2 \cdot 3\text{ ft} \\ &= (2 \cdot 7)\text{ ft} + (2 \cdot 3)\text{ ft} && \text{Try to do this mentally.} \\ &= 14\text{ ft} + 6\text{ ft} \\ &= 20\text{ ft} && \text{Combining like terms} \end{aligned}$$

The perimeter of the door is 20 ft.

11. Find the perimeter of a 4-ft by 8-ft sheet of plywood.

Do Exercise 11.

A **square** is a rectangle in which all sides have the same length.

EXAMPLE 8 Find the perimeter of a square with sides of length 9 mm.

12. Find the perimeter of a square with sides of length 10 km.

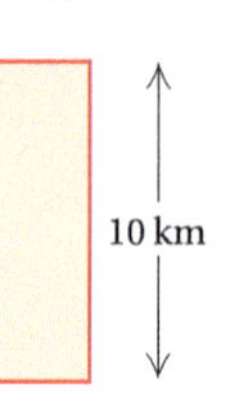

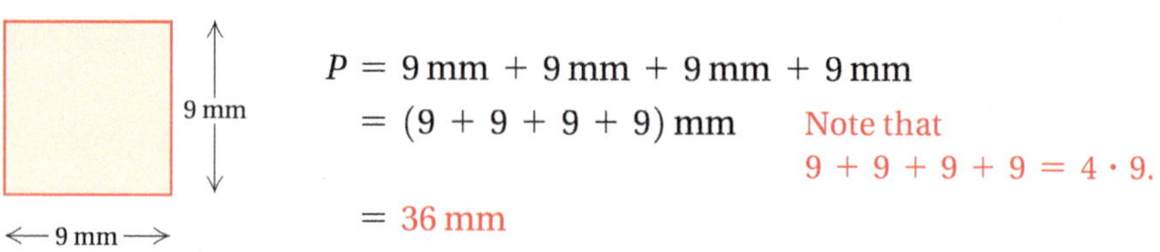

$$\begin{aligned} P &= 9\text{ mm} + 9\text{ mm} + 9\text{ mm} + 9\text{ mm} \\ &= (9 + 9 + 9 + 9)\text{ mm} && \text{Note that } 9 + 9 + 9 + 9 = 4 \cdot 9. \\ &= 36\text{ mm} \end{aligned}$$

Do Exercise 12.

PERIMETER OF A SQUARE

The **perimeter P of a square** is four times s, the length of a side:

$$\begin{aligned} P &= s + s + s + s \\ &= 4s. \end{aligned}$$

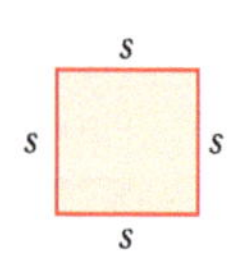

13. Find the perimeter of a square book with sides of length 9 in. GS

$$\begin{aligned} P &= 4s \\ &= 4 \cdot \square\text{ in.} \\ &= \square\text{ in.} \end{aligned}$$

EXAMPLE 9 Find the perimeter of a square room with sides of length 12 ft.

$$\begin{aligned} P &= 4s \\ &= 4 \cdot 12\text{ ft} \\ &= 48\text{ ft} \end{aligned}$$

The perimeter of the room is 48 ft.

Do Exercise 13.

Answers

11. 24 ft **12.** 40 km **13.** 36 in.

Guided Solution:
13. 9, 36

Translating for Success

1. *Wood Costs.* It costs \$8 for the wood for each birdhouse that Annette builds. If she used \$120 worth of wood, how many birdhouses did she build?

2. *Elevation.* Genine started hiking a 10-mi trail at an elevation that was 150 ft below sea level. At the end of the trail, she was 75 ft above sea level. How many feet higher was she at the end of the trail than at the beginning?

3. *Community Service.* In order to fulfill the requirements for a sociology class, Glen must log 120 hr of community service. So far, he has spent 75 hr volunteering at a youth center. How many more hours must he serve?

4. *Disaster Relief.* Each package that is prepared for a disaster relief effort contains 15 meal bars. How many packages can be filled from a donation of 750 bars?

5. *Perimeter.* A rectangular building lot is 75 ft wide and 150 ft long. What is the perimeter of the lot?

The goal of these matching questions is to practice step (2), Translate, of the five-step problem-solving process. Translate each word problem to an equation and select a correct translation from equations A–O.

A. $75 + x = 120$

B. $15 \div 750 = x$

C. $-10 - 15 = x$

D. $8 \cdot 120 = x$

E. $150 - 75 = x$

F. $15 \cdot 750 = x$

G. $75 - (-150) = x$

H. $2 \cdot 150 + 2 \cdot 75 = x$

I. $-10 - (-15) = x$

J. $8 \cdot x = 120$

K. $750 \div 15 = x$

L. $15 - (-10) = x$

M. $75 + 120 = x$

N. $75 - 150 = x$

O. $75 = 120 + x$

Answers on page A-4

6. *Account Balance.* Lorenzo had \$75 in his checking account. He then wrote a check for \$150. What was the balance in his account?

7. *Laptop Computers.* Great Graphics purchased a laptop computer for each of its 15 employees. If each laptop cost \$750, how much did the computers cost?

8. *Basketball.* A basketball team scored 75 points in one game. In the next game, the team scored a record 120 points. How many points did the team score in the two games?

9. *Pizza Sales.* A youth club sold 120 pizzas for a fundraiser. If each pizza sold for \$8, how much money was taken in?

10. *Temperature.* The temperature in Fairbanks was $-10°$ at 6:00 P.M. and fell another 15° by midnight. What was the temperature at midnight?

2.7 Exercise Set

For Extra Help
MyMathLab®
MathXL® PRACTICE · WATCH · READ · REVIEW

Reading Check

Complete each statement with the correct word from the following list. A word may be used more than once or not at all.

closed	perimeter	rectangle
open	polygon	square

RC1. A polygon is a(n) ________________ figure with three or more sides.

RC2. The distance around a polygon is its ________________.

RC3. The formula $P = 2l + 2w$ gives the ________________ of a rectangle.

RC4. The perimeter of a(n) ________________ is given by the formula $P = 4s$.

a List the terms of each expression.

1. $2a + 5b - 7c$

2. $4x - 6y + 7z$

3. $mn - 6n + 8$

4. $7rs - s - 5$

5. $3x^2y - 4y^2 - 2z^3$

6. $4a^3b + ab^2 - 9b^3$

Combine like terms to form an equivalent expression.

7. $5x + 9x$

8. $9a + 7a$

9. $10a - 15a$

10. $-17x + x$

11. $2x + 6y + x$

12. $3t - y + 7t$

13. $27a + 70 - 40a - 8$

14. $42x - 6 - x + 2$

15. $9 + 5t + 7y - t - y - 13$

16. $8 - 4a + 9b + 5a - 3b - 15$

17. $a + 3b + 5a - 2 + b$

18. $x + 7y + 5 - 2y + 3x$

19. $-8 + 11a - 5b - 10a - 7b + 7$

20. $8x - 5x + 6 + 3y - y - 4$

21. $8x^2 + 3y - x^2$

22. $8y^3 - 3z + 4y^3$

23. $11x^4 + 2y^3 - 4x^4 - y^3$

24. $13a^5 + 9b^4 - 2a^5 - 4b^4$

25. $9a^2 - 4a + a - 3a^2$

26. $3a^2 + 7a^3 - a^2 + 5 + a^3$

27. $x^3 - 5x^2 + 2x^3 - 3x^2 + 4$

28. $9xy + 4y^2 - 2xy + 2y^2 - 1$

29. $9x^3y + 4xy^3 - 5xy^3 + 3xy$

30. $8a^2b - 3ab^2 - 7a^2b + 2ab$

31. $3a^6 - b^4 + 2a^6b^4 - 7a^6 - 2b^4$

32. $3x^4 - 2y^4 + 8x^4y^4 - 7x^4 + y^4$

b Find the perimeter of each polygon.

33.

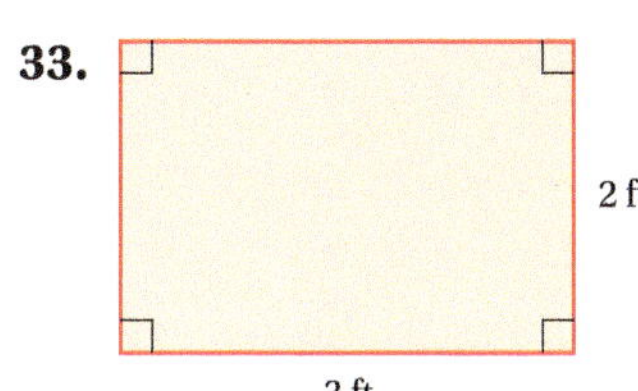

34.

5 in.

5 in.

35.

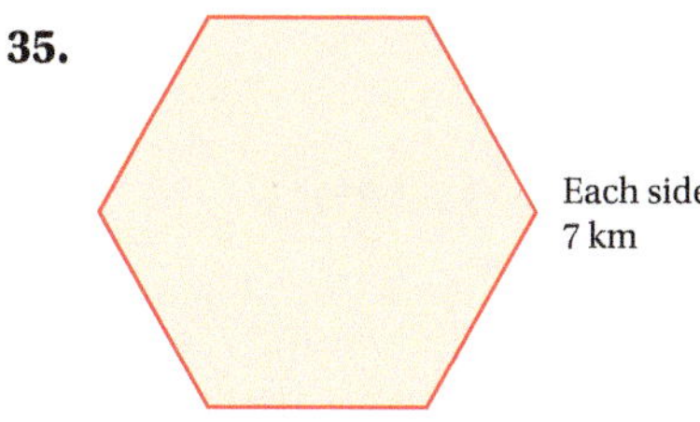

36.

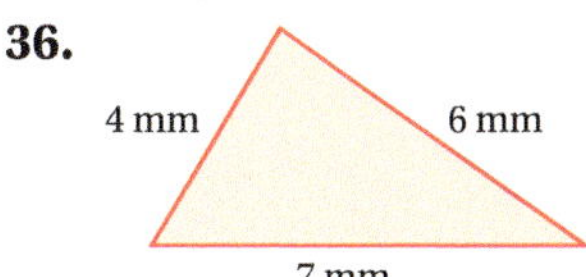

37.

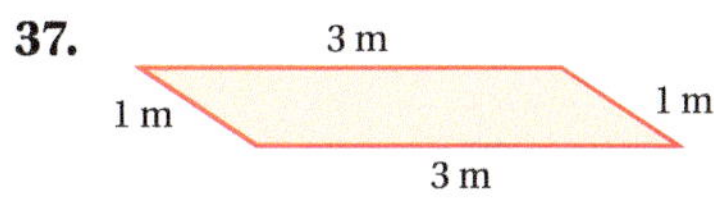

38.

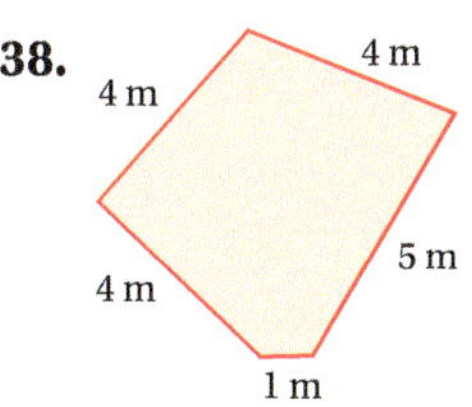

Tennis Court. A tennis court contains many rectangles. Use the diagram of a regulation tennis court to calculate the perimeters in Exercises 39–42.

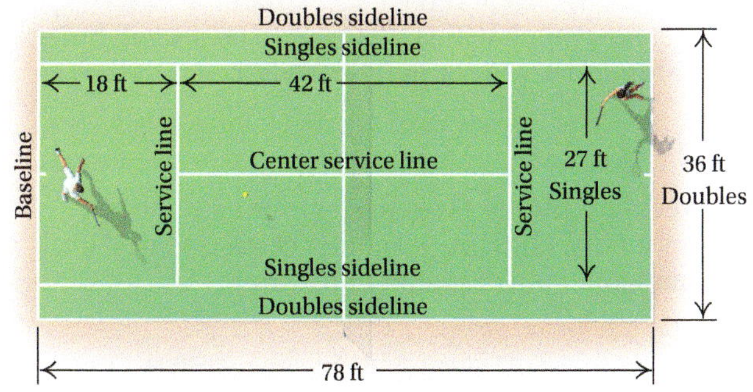

39. The perimeter of a singles court

40. The perimeter of a doubles court

41. The perimeter of the rectangle formed by the service lines and the singles sidelines

42. The perimeter of the rectangle formed by a service line, a baseline, and the singles sidelines

43. Find the perimeter of a rectangular 8-ft by 10-ft bedroom.

44. Find the perimeter of a rectangular 3-ft by 4-ft doghouse.

45. Find the perimeter of a checkerboard that is 14 in. on each side.

46. Find the perimeter of a square skylight that is 2 m on each side.

47. Find the perimeter of a square frame that is 65 cm on each side.

48. Find the perimeter of a square garden that is 12 yd on each side.

49. Find the perimeter of a 12-ft by 20-ft rectangular deck.

50. Find the perimeter of a 40-ft by 35-ft rectangular backyard.

Skill Maintenance

51. A box of Shaw's Corn Flakes contains 510 grams (g) of corn flakes. A serving of corn flakes weighs 30 g. How many servings are in one box? [1.8a]

52. Estimate the difference by rounding to the nearest ten. [1.6b]

$$\begin{array}{r} 704 \\ -\ 486 \\ \hline \end{array}$$

Solve. [1.7b]

53. $25 = t + 9$

54. $19 = x + 6$

55. $45 = 3x$

56. $2t = 50$

57. $25 + n = 400$

58. $25 \cdot n = 400$

Synthesis

Simplify. (Multiply and then combine like terms.)

59. $5(x + 3) + 2(x - 7)$

60. $3(a - 7) + 7(a + 4)$

61. $2(3 - 4a) + 5(a - 7)$

62. $7(2 - 5x) + 3(x - 8)$

63. $-5(2 + 3x + 4y) + 7(2x - y)$

64. $3(4 - 2x) + 5(9x - 3y + 1)$

65. In order to save energy, Andrea plans to run a bead of caulk sealant around 3 exterior doors and 13 windows. Each window measures 3 ft by 4 ft, each door measures 3 ft by 7 ft, and there is no need to caulk the bottom of each door. If each cartridge of caulk seals 56 ft and costs $9, how much will it cost Andrea to seal the windows and doors?

66. Jacie is adding a border to small rectangular bulletin boards that are 5 ft by 7 ft and to larger bulletin boards that are 7 ft by 9 ft. If the border material costs $5 per yard, how much will Jacie spend to do 4 small bulletin boards and 3 large bulletin boards?

67. A square wooden rack is used to store the 15 numbered pool balls as well as the cue ball. If a pool ball has a diameter of 57 mm, find the inside perimeter of the storage rack.

68. A rectangular box is used to store six Christmas ornaments. Find the perimeter of such a box if each ornament has a diameter of 72 mm.

Solving Equations

2.8

A *solution* of an equation is a replacement for the variable that makes the equation true. To *solve* an equation means to find all of its solutions.

a THE ADDITION PRINCIPLE

The solution of the equation $x = 10$ is 10. The solution of the equation $x + 5 = 15$ is also 10, since $10 + 5 = 15$ is true. Because their solutions are identical, $x = 10$ and $x + 5 = 15$ are said to be **equivalent equations**.

EQUIVALENT EQUATIONS

Equations with the same solutions are called **equivalent equations**.

It is important to be able to distinguish between equivalent *expressions* and equivalent *equations*.

- $6a$ and $4a + 2a$ are equivalent *expressions* because, for any replacement of a, both expressions represent the same number.
- $3x = 15$ and $4x = 20$ are equivalent *equations* because any solution of one equation is also a solution of the other equation.

EXAMPLE 1 Classify each pair as either equivalent equations or equivalent expressions:

a) $5x + 1;\quad 2x - 4 + 3x + 5$ **b)** $x = -7;\quad x + 2 = -5.$

a) First note that these are expressions, not equations. To see if they are equivalent, we combine like terms in the second expression:

$$2x - 4 + 3x + 5 = (2 + 3)x + (-4 + 5)$$ Regrouping and using the distributive law

$$= 5x + 1.$$

We see that $2x - 4 + 3x + 5$ and $5x + 1$ are *equivalent expressions.*

b) Both $x = -7$ and $x + 2 = -5$ are equations. The solution of $x = -7$ is -7. We substitute to see if -7 is also the solution of $x + 2 = -5$:

$$x + 2 = -5$$
$$-7 + 2 = -5 \quad \text{TRUE}$$

Since $x = -7$ and $x + 2 = -5$ have the same solution, they are *equivalent equations.*

Do Margin Exercises 1 and 2.

We now begin to consider principles that enable us to start with one equation and create an equivalent equation similar to $x = 15$, for which the solution is obvious. One such principle, the *addition principle*, tells us that we can add the same number to both sides of an equation without changing the solutions of the equation.

OBJECTIVES

- **a** Use the addition principle to solve equations.
- **b** Use the division principle to solve equations.
- **c** Decide which principle should be used to solve an equation.
- **d** Solve equations that require use of both the addition principle and the division principle.

SKILL TO REVIEW

Objective 1.7b: Solve equations like $t + 28 = 54$, $28 \cdot x = 168$, and $98 \cdot 2 = y$.

Solve.

1. $x + 13 = 87$
2. $17 \cdot y = 357$

Classify each pair as equivalent expressions or equivalent equations.

1. $a - 5 = -3;\quad a = 2$
2. $a - 9 + 6a;\quad 7a - 9$

Answers

Skill to Review:
1. 74 2. 21

Margin Exercises:
1. Equivalent equations
2. Equivalent expressions

THE ADDITION PRINCIPLE

For any numbers a, b, and c,

$$a = b \quad \text{is equivalent to} \quad a + c = b + c.$$

EXAMPLE 2 Solve: $x - 7 = -2$.

We have

$x - 7 = -2$

$x - 7 + 7 = -2 + 7$ — Using the addition principle: adding 7 to both sides

$x + 0 = 5$ — Adding 7 "undoes" the subtraction of 7.

$x = 5.$ — This equation has the same solution as $x - 7 = -2$.

The solution appears to be 5. To check, we use the original equation.

Check: $x - 7 = -2$

$5 - 7 \ ? \ -2$

$-2 \mid$ TRUE

The solution is 5.

Solve.

3. $x - 5 = 19$

4. $x - 9 = -12$

Do Exercises 3 and 4.

We can subtract by adding the opposite of the number being subtracted. Because of this, the addition principle allows us to subtract the same number from both sides of an equation.

EXAMPLE 3 Solve: $23 = t + 7$.

We have

$23 = t + 7$

$23 - 7 = t + 7 - 7$ — Using the addition principle to add -7 or to subtract 7 on both sides

$16 = t + 0$ — Subtracting 7 "undoes" the addition of 7.

$16 = t.$ — The solution of $23 = t + 7$ is also 16.

The solution is 16. The check is left to the student.

To visualize the addition principle, think of a jeweler's balance. When both sides of the balance hold equal amounts of weight, the balance is level. If weight is added or removed, equally, on both sides, the balance remains level.

Solve.

5. $42 = x + 17$

6. $a + 8 = -6$

Do Exercises 5 and 6.

Answers

3. 24 **4.** -3 **5.** 25 **6.** -14

b THE DIVISION PRINCIPLE

We can solve $8n = 96$ by dividing both sides by 8:

$$8 \cdot n = 96$$

$$\frac{8 \cdot n}{8} = \frac{96}{8} \qquad \text{Dividing both sides by 8}$$

$$n = 12. \qquad 8 \text{ times } n \text{, divided by 8, is } n.\ 96 \div 8 \text{ is } 12.$$

Both $8n = 96$ and $n = 12$ have the solution 12. We can divide both sides of an equation by any nonzero number in order to find an equivalent equation.

THE DIVISION PRINCIPLE

For any numbers a, b, and c $(c \neq 0)$,

$$a = b \quad \text{is equivalent to} \quad \frac{a}{c} = \frac{b}{c}.$$

After we have discussed multiplication of fractions, we can use an equivalent form of this principle: the multiplication principle.

EXAMPLE 4 Solve: $9x = 63$.

We have

$$9x = 63$$

$$\frac{9x}{9} = \frac{63}{9} \qquad \text{Using the division principle to divide both sides by 9}$$

$$x = 7.$$

Check:

$9x = 63$		Checking in the original equation
$9 \cdot 7$? 63		
63		TRUE

The solution is 7.

Do Exercises 7 and 8. ▶

Solve.

7. $7x = 42$

8. $-24 = 3t$

EXAMPLE 5 Solve: $48 = -8n$.

To undo multiplication by -8, we use the division principle:

$$48 = -8n$$

$$\frac{48}{-8} = \frac{-8n}{-8} \qquad \text{Dividing both sides by } -8$$

$$-6 = n.$$

Check:

$48 = -8n$	
48 ? $-8(-6)$	
48	TRUE

The solution is -6.

Do Exercises 9 and 10. ▶

Solve.

9. $63 = -7n$

10. $-6x = 72$

Answers

7. 6 8. -8 9. -9 10. -12

Our goal in equation solving is to have the variable by itself on one side of the equation. In an equation like $-x = 7$, the variable is not by itself—there is a negative sign in front of x. We can solve an equation like $-x = 7$ by dividing or by multiplying both sides of the equation by -1.

EXAMPLE 6 Solve: $-x = 7$.

One way to solve this equation is to note that $-x = -1 \cdot x$. Then we can divide both sides by -1:

$$-x = 7$$
$$-1 \cdot x = 7$$
$$\frac{-1 \cdot x}{-1} = \frac{7}{-1} \quad \text{Using the division principle}$$
$$x = -7.$$

Check: $-x = 7$ — Be sure to check in the original equation.

$-(-7) \; ? \; 7$

7 | TRUE

The solution is -7.

Another way to solve the equation in Example 6 is to remember that when an expression is multiplied or divided by -1, its sign is changed. Here we multiply on both sides by -1 to change the sign of $-x$:

$$-x = 7$$
$$(-1)(-x) = (-1) \cdot 7 \quad \text{Multiplying both sides by } -1$$
$$x = -7. \quad \text{Note that } (-1)(-x) \text{ is the same as } (-1)(-1)x.$$

Do Exercises 11 and 12.

C SELECTING THE CORRECT APPROACH

You can determine which principle should be used to solve a particular equation by thinking of undoing an operation.

EXAMPLE 7 Solve: $39 = -3 + t$.

Note that -3 is added to t. To undo addition of -3, we subtract -3 or simply add 3 on both sides:

$$3 + 39 = 3 + (-3) + t \quad \text{Using the addition principle}$$
$$42 = 0 + t$$
$$42 = t.$$

Check: $39 = -3 + t$

$39 \; ? \; -3 + 42$

| 39 TRUE

The solution is 42.

It is important to distinguish between addition of a negative number, as in $39 = -3 + t$, and multiplication by a negative number, as in $39 = -3t$. We undo addition by subtracting on both sides, and we undo multiplication by dividing on both sides.

Solve.

11. $-x = 23$

12. $-t = -3$ (GS)

$\square \cdot t = -3$

$\frac{-1 \cdot t}{\square} = \frac{-3}{\square}$

$t = \square$

Answers

11. -23 12. 3

Guided Solution:

12. $-1, -1, -1, 3$

EXAMPLE 8 Solve: $39 = -3t$.

Here t is multiplied by -3. To undo multiplication by -3, we divide by -3 on both sides:

$$39 = -3t$$

$$\frac{39}{-3} = \frac{-3t}{-3} \qquad \text{Using the division principle}$$

$$-13 = t.$$

Check:

$39 = -3t$	
$39 \ ? \ -3(-13)$	
$\quad 39$	TRUE

The solution is -13.

Do Exercises 13–15. ▶

Solve.

13. $-2x = -52$

14. $-2 + x = -52$

15. $x \cdot 7 = -28$

d USING THE PRINCIPLES TOGETHER

Suppose we want to determine whether 7 is the solution of $5x - 8 = 27$. To check, we replace x with 7 and simplify.

Check:

$5x - 8 = 27$	
$5 \cdot 7 - 8 \ ? \ 27$	
$35 - 8$	
27	TRUE

This shows that 7 is the solution.

Do Exercises 16 and 17. ▶

16. Determine whether -9 is the solution of $7x + 8 = -55$.

17. Determine whether -6 is the solution of $4x + 3 = -25$.

In the check above, note that the rules for order of operations require that we multiply before we subtract (or add). Thus, to evaluate $5x - 8$,

we *select* a value: x

then *multiply* by 5: $5x$

and then *subtract* 8: $5x - 8$.

In Example 9, which follows, these steps are reversed to solve for x:

we *add* 8: $5x - 8 + 8$

then *divide* by 5: $\frac{5x}{5}$

and then *isolate* x: x.

In general, the *last* step performed when calculating is the *first* step to be reversed when finding a solution.

When x is by itself on one side of an equation, we say that it is *isolated*. Our goal in the remaining examples in this section will be to isolate the term containing the variable (using the addition principle) and then to isolate the variable itself (using the division principle).

Answers

13. 26 **14.** -50 **15.** -4 **16.** Yes **17.** No

EXAMPLE 9 Solve: $5x - 8 = 27$.

We first note that the term containing x is $5x$. To isolate $5x$, we add 8 on both sides:

$$
\begin{aligned}
5x - 8 &= 27 \\
5x - 8 + 8 &= 27 + 8 && \text{Using the addition principle} \\
5x + 0 &= 35 && \text{Try to do this step mentally.} \\
5x &= 35. && \text{We have isolated } 5x.
\end{aligned}
$$

Next, we isolate x by dividing by 5 on both sides:

$$
\begin{aligned}
5x &= 35 \\
\frac{5x}{5} &= \frac{35}{5} && \text{Using the division principle} \\
1x &= 7 && \text{Try to do this step mentally.} \\
x &= 7. && \text{We have isolated } x.
\end{aligned}
$$

The check was performed on the previous page. The solution is 7.

18. Solve: $2x - 9 = 43$.

GS

$$
\begin{aligned}
2x - 9 &= 43 \\
2x - 9 + \square &= 43 + \square \\
2x &= \square \\
\frac{2x}{\square} &= \frac{52}{\square} \\
x &= \square
\end{aligned}
$$

◀ **Do Exercise 18.**

EXAMPLE 10 Solve: $38 = -9t + 2$.

We first isolate $-9t$ by subtracting 2 on both sides:

$$
\begin{aligned}
38 &= -9t + 2 \\
38 - 2 &= -9t + 2 - 2 && \text{Subtracting 2 (or adding } -2\text{) on both sides} \\
36 &= -9t + 0 && \text{Try to do this step mentally.} \\
36 &= -9t.
\end{aligned}
$$

Now that we have isolated $-9t$ on one side of the equation, we can divide by -9 to isolate t:

$$
\begin{aligned}
36 &= -9t \\
\frac{36}{-9} &= \frac{-9t}{-9} && \text{Dividing both sides by } -9 \\
-4 &= t. && \text{Simplifying}
\end{aligned}
$$

Check:

$$
\begin{array}{c|l}
38 & = -9t + 2 \\
\hline
38 & ?\ -9 \cdot (-4) + 2 \\
 & 36 + 2 \\
 & 38 \qquad \text{TRUE}
\end{array}
$$

The solution is -4.

19. Solve: $-3n + 2 = 47$.

◀ **Do Exercise 19.**

Answers

18. 26 **19.** -15

Guided Solution:

18. 9, 9, 52, 2, 2, 26

2.8 Exercise Set

For Extra Help MyMathLab® MathXL® PRACTICE WATCH READ REVIEW

Reading Check

Match each equation with the correct first step for solving it from the list on the right.

RC1. $7x = 28$ ______

RC2. $x - 7 = 13$ ______

RC3. $-7x = -14$ ______

RC4. $x + 7 = 1$ ______

a) Add 7 to both sides.

b) Add -7 to both sides.

c) Divide both sides by 7.

d) Divide both sides by -7.

a Classify each pair as either equivalent expressions or equivalent equations.

1. $x = -1;\ x + 5 = 4$

2. $4x + 1;\ 6 + 4x - 5$

3. $7a - 3;\ 13 + 7a - 16$

4. $t = -2;\ 5t = -10$

5. $4r + 3;\ 9 - r + 5r - 6$

6. $2r - 7;\ r - 10 + r + 3$

7. $x - 9 = 8;\ x = 20 - 3$

8. $t + 4 = 19;\ t = 9 + 6$

9. $3(t + 2);\ 5 + 3t + 1$

10. $2x = -14;\ x - 2 = -9$

11. $x + 4 = -8;\ 2x = -24$

12. $4(x - 7);\ 3x - 28 + x$

Solve.

13. $x - 6 = -9$

14. $x - 5 = -7$

15. $x - 4 = -12$

16. $x - 7 = 5$

17. $a + 7 = 25$

18. $x + 9 = -3$

19. $-8 = n + 7$

20. $38 = a + 12$

21. $24 = t - 8$

22. $-9 = x + 3$

23. $-12 = x + 5$

24. $17 = n - 6$

25. $-5 + a = 12$

26. $3 = 17 + x$

27. $-8 = -8 + t$

28. $-7 + t = -7$

b Solve.

29. $6x = 60$

30. $-8t = 40$

31. $-3t = 42$

32. $3x = 24$

33. $-7n = -35$

34. $64 = -2t$

35. $0 = 6x$

36. $-5n = -65$

37. $55 = -5t$

38. $-x = 83$

39. $-x = 56$

40. $-2x = 0$

41. $n(-4) = -48$

42. $-x = -475$

43. $-x = -390$

44. $n(-7) = 42$

c Solve.

45. $t - 6 = -2$

46. $3t = -45$

47. $6x = -54$

48. $x + 9 = -15$

49. $15 = -x$

50. $-13 = x - 4$

51. $-21 = x + 5$

52. $-42 = -x$

53. $35 = -7t$

54. $7 + t = -18$

55. $-17x = 68$

56. $-34 = x - 10$

57. $12 + t = -160$

58. $-48 = t(-12)$

59. $-27 = x - 23$

60. $-135 = -9t$

d Solve.

61. $5x - 1 = 34$

62. $7x - 3 = 25$

63. $4t + 2 = 14$

64. $3t + 5 = 26$

65. $6a + 1 = -17$

66. $8a + 3 = -37$

67. $2x - 9 = -23$

68. $3x - 5 = -35$

69. $-2x + 1 = 17$

70. $-4t + 3 = -17$

71. $-8t - 3 = -67$

72. $-7x - 4 = -46$

73. $-x + 9 = -15$

74. $-x - 6 = 8$

75. $7 = 2x - 5$

76. $9 = 4x - 7$

77. $13 = 3 + 2x$

78. $33 = 5 - 4x$

79. $13 = 5 - x$

80. $12 = 7 - x$

Skill Maintenance

Simplify. [1.9c]

81. $5 + 3 \cdot 2^3$

82. $(9 - 7)^4 - 3^2$

83. $12 \div 3 \cdot 2$

84. $27 \div 3(2 + 1)$

85. $15 - 3 \cdot 2 + 7$

86. $30 - 4^2 \div 8 \cdot 2$

87. $3(8 - 6) - (9 - 4)$

88. $2(10 - 6)^2 - (120 \div 6 - 3 \cdot 4)$

Synthesis

Solve.

89. $2x - 7x = -40$

90. $9 + x - 5 = 23$

91. $2x - 7 + x = 5 - 12$

92. $3 - 6x + 5 = 2(4)$

93. $n + n = -2 - 3 \cdot 6 \div 2 + 1$

94. $10 \div 5 \cdot 2 + 3 = 5t - 4t$

95. $17 - 3^2 = 4 + t - 5^2$

96. $(-9)^2 = 2^3t + (3 \cdot 6 + 1)t$

97. $(-7)^2 - 5 = t + 4^3$

98. 🖩 $(-42)^3 = 14^2t$

99. 🖩 $x - (19)^3 = -18^3$

100. 🖩 $23^2 = x + 22^2$

101. 🖩 $35^3 = -125t$

102. 🖩 $248 = 24 - 32x$

103. 🖩 $529 - 143x = -1902$

CHAPTER

2 Summary and Review

Vocabulary Reinforcement

Complete each statement with the appropriate word or phrase from the column on the right. Some of the choices may not be used.

1. The ______________ are . . . , −3, −2, −1, 0, 1, 2, 3, ... [2.1a]

2. The ______________ value of a number is its distance from zero on the number line. [2.1c]

3. The number 3 is the ______________ of −3. [2.1d]

4. Division by 0 is ______________. [2.5a]

5. When we replace a variable with a number, we say that we are ______________ for the variable. [2.6a]

6. A letter that stands for just one number is called a ______________. [2.6a]

7. The ______________ states that for any numbers a, b, and c, $a(b + c) = ab + ac$. [2.6b]

8. The ______________ for solving equations states that for any real numbers a, b, and c, $a = b$ is equivalent to $a + c = b + c$. [2.8a]

9. Equations with the same solutions are called ______________ equations. [2.8a]

constant
variable
opposite
equivalent
distributive law
integers
substituting
absolute
not defined
zero
addition principle
division principle
distributive law

Concept Reinforcement

Determine whether each statement is true or false.

______________ **1.** The opposite of the opposite of a number is the original number. [2.1d]

______________ **2.** The product of an even number of negative numbers is positive. [2.4b]

______________ **3.** The expression $2(x + 3)$ is equivalent to the expression $2 \cdot x + 3$. [2.6b]

______________ **4.** $3y$ and $3y^2$ are like terms. [2.7a]

Study Guide

Objective 2.1c Find the absolute value of any integer.

Example Find the absolute value.

a) $|-97|$ **b)** $|35|$

a) The number is negative, so we write the opposite.

$|-97| = 97$

b) The number is positive, so the absolute value is the same as the number.

$|35| = 35$

Practice Exercise

1. Find the absolute value.

a) $|-17|$ **b)** $|300|$

Objective 2.2a Add integers without using the number line.

Example Add without using the number line: $-15 + 9$.

To add a negative number and a positive number, subtract the absolute values: $15 - 9 = 6$. The negative number has the larger absolute value, so the answer is negative.

$$-15 + 9 = -6$$

Example Add without using the number line: $-8 + (-9)$.

To add two negative numbers, add the absolute values: $8 + 9 = 17$. The answer is negative.

$$-8 + (-9) = -17$$

Practice Exercise

2. Add without using the number line: $37 + (-16)$.

Objective 2.3a Subtract integers, and simplify combinations of additions and subtractions.

Example Subtract: $8 - 12$.

We add the opposite of the number being subtracted.

$$\begin{aligned} 8 - 12 &= 8 + (-12) \\ &= -4 \end{aligned}$$

Practice Exercise

3. Subtract: $6 - (-8)$.

Objective 2.4a Multiply integers.

Example Multiply: $-6(-4)$.

The signs are the same, so the answer is positive.

$$-6(-4) = 24$$

Example Multiply: $-4(3)$.

The signs are different, so the answer is negative.

$$-4(3) = -12$$

Practice Exercise

4. Multiply: $6(-15)$.

Objective 2.5a Divide integers.

Example Divide: $-36 \div (-4)$.

The signs are the same, so the answer is positive.

$$-36 \div (-4) = 9$$

Example Divide: $-30 \div 10$.

The signs are different, so the answer is negative.

$$-30 \div 10 = -3$$

Practice Exercise

5. Divide: $99 \div (-9)$.

Objective 2.5b Use the rules for order of operations with integers.

Example Simplify: $3^2 - 24 \div 2 - (4 + 2 \cdot 8)$.

$$\begin{aligned} &3^2 - 24 \div 2 - (4 + 2 \cdot 8) \\ &\quad = 3^2 - 24 \div 2 - (4 + 16) \\ &\quad = 3^2 - 24 \div 2 - 20 \\ &\quad = 9 - 24 \div 2 - 20 \\ &\quad = 9 - 12 - 20 \\ &\quad = -3 - 20 \\ &\quad = -23 \end{aligned}$$

Practice Exercise

6. Simplify: $4 - 8^2 \div (10 - 6)$.

Objective 2.6b Use the distributive law to find equivalent expressions.

Example Use the distributive law to write an expression equivalent to $-2(3x - 4)$.

$$\begin{aligned} -2(3x - 4) &= -2(3x) - (-2)(4) \\ &= -6x - (-8) \\ &= -6x + 8 \end{aligned}$$

Practice Exercise

7. Use the distributive law to write an expression equivalent to $5(6x - 8y - z)$.

Objective 2.7a Combine like terms.

Example Combine like terms to form an expression equivalent to $6 - 30x + 12x - 7$.

$$\begin{aligned} 6 - 30x + 12x - 7 &= 6 + (-30x) + 12x + (-7) \\ &= -30x + 12x + 6 + (-7) \\ &= -18x - 1 \end{aligned}$$

Practice Exercise

8. Combine like terms to form an expression equivalent to $8a - b + 9a - 6b$.

Objective 2.8d Solve equations that require use of both the addition principle and the division principle.

Example Solve: $-3x + 1 = 16$.

$$\begin{aligned} -3x + 1 &= 16 \\ -3x + 1 - 1 &= 16 - 1 \\ -3x &= 15 \\ \frac{-3x}{-3} &= \frac{15}{-3} \\ x &= -5 \end{aligned}$$

The solution is -5.

Practice Exercise

9. Solve: $-19 = 5x + 11$.

Review Exercises

1. Tell which integers correspond to this situation: David has a debt of \$45 and Joe has \$72 in his savings account. [2.1a]

Use either $<$ or $>$ for $\square$ to form a true statement. [2.1b]

2. $0 \square -5$ **3.** $-7 \square 6$ **4.** $-4 \square -19$

Find the absolute value. [2.1c]

5. $|-39|$ **6.** $|23|$ **7.** $|0|$

8. Find $-x$ when $x = -72$. [2.1d]

9. Find $-(-x)$ when $x = 59$. [2.1d]

Compute and simplify.

10. $-14 + 5$ [2.2a]

11. $-5 + (-6)$ [2.2a]

12. $14 + (-8)$ [2.2a]

13. $0 + (-24)$ [2.2a]

14. $17 - 29$ [2.3a]

15. $9 - (-14)$ [2.3a]

16. $-8 - (-7)$ [2.3a]

17. $-3 - (-3)$ [2.3a]

18. $-3 + 7 + (-8)$ [2.3a]

19. $8 - (-9) - 7 + 2$ [2.3a]

20. $-23 \cdot (-4)$ [2.4a]

21. $7(-12)$ [2.4a]

22. $2(-4)(-5)(-1)$ [2.4b]

23. $15 \div (-5)$ [2.5a]

24. $\frac{-55}{11}$ [2.5a]

25. $\frac{0}{7}$ [2.5a]

Simplify. [2.5b]

26. $625 \div (-25) \div 5$

27. $-16 \div 4 - 30 \div (-5)$

28. $9[(7 - 14) - 13]$

29. $(-3)|4 - 3^2| - 5$

30. $[-12(-3) - 2^3] - (-9)(-10)$

31. Evaluate $3a + b$ for $a = 4$ and $b = -5$. [2.6a]

32. Evaluate $\frac{-x}{y}$, $\frac{x}{-y}$, and $-\frac{x}{y}$ for $x = 30$ and $y = 5$. [2.6a]

Use the distributive law to write an equivalent expression. [2.6b]

33. $4(5x + 9)$

34. $3(2a - 4b + 5)$

35. $-10(2x + y)$

Combine like terms. [2.7a]

36. $5a + 12a$

37. $-7x + 13x$

38. $9m + 14 - 12m - 8$

39. Find the perimeter of a rectangular frame that is 8 in. by 10 in. [2.7b]

40. Find the perimeter of a square pane of glass that is 25 cm on each side. [2.7b]

Solve. [2.8a, b, c, d]

41. $x - 9 = -17$

42. $-4t = 36$

43. $13 = -x$

44. $56 = 6x - 10$

45. $-x + 3 = -12$

46. $18 = 4 - 2x$

Solve.

47. On the first, second, and third downs, a football team had these gains and losses: 5-yd gain, 12-yd loss, and 15-yd gain, respectively. Find the total gain (or loss). [2.3b]

48. Kaleb's total assets are $170. He borrows $300. What are his total assets now? [2.3b]

49. Evaluate $-|-x|$ when $x = -10$. [2.1c], [2.1d]

A. -10
B. -20
C. 100
D. 10

50. Simplify: $-3 \cdot 4 - 12 \div 4$. [2.5b]

A. -16
B. -15
C. 0
D. 6

Synthesis

51. The sum of two numbers is 800. The difference is 6. Find the numbers. [2.3b]

52. The following are examples of consecutive integers: 4, 5, 6, 7, 8; $-13, -12, -11, -10$. [2.2a], [2.4b]

a) Express the number 8 as the sum of 16 consecutive integers.

b) Find the product of the 16 consecutive integers in part (a).

Simplify. [2.5b]

53. ▦ $87 \div 3 \cdot 29^3 - (-6)^6 + 1957$

54. ▦ $1969 + (-8)^5 - 17 \cdot 15^3$

55. ▦ $\dfrac{113 - 17^3}{15 + 8^3 - 507}$

56. For what values of x will $8 + x^3$ be negative? [2.6a]

57. For what values of x is $|x| > x$? [2.1b, c]

Understanding Through Discussion and Writing

1. What rule have we developed that would tell you the sign of $(-7)^8$ and $(-7)^{11}$ without doing the computations? Explain. [2.4b]

2. Does $-x$ always represent a negative number? Why or why not? [2.1d]

3. Jake enters $18/2 \cdot 3$ on his calculator and expects the result to be 3. What mistake is he making? [2.5b]

4. Does $-x^2$ always represent a negative number? Why or why not? [2.6a]

CHAPTER

2 Test

For Extra Help For step-by-step test solutions, access the Chapter Test Prep Videos in MyMathLab® or on YouTube (search "BittingerPreIntro" and click on Channels).

1. Tell which integers correspond to this situation: The Tee Shop sold 542 fewer muscle shirts than expected in January and 307 more than expected in February.

2. Use either $<$ or $>$ for $\square$ to form a true statement.
$-14 \square -21$

3. Find the absolute value: $|-739|$.

4. Find $-(-x)$ when $x = -19$.

Compute and simplify.

5. $6 + (-17)$

6. $-9 + (-12)$

7. $-8 + 17$

8. $0 - 12$

9. $7 - 22$

10. $-5 - 19$

11. $-8 - (-27)$

12. $31 - (-3) - 5 + 9$

13. $(-4)^3$

14. $27(-10)$

15. $-9 \cdot 0$

16. $-72 \div (-9)$

17. $\frac{-56}{7}$

18. $8 \div 2 \cdot 2 - 3^2$

19. $29 - (3 - 5)^2$

20. ***Antarctica Highs and Lows.*** The continent of Antarctica, which lies in the southern hemisphere, experiences winter in July. The average high temperature is $-67°F$ and the average low temperature is $-81°F$. How much higher is the average high than the average low?

Source: National Climatic Data Center

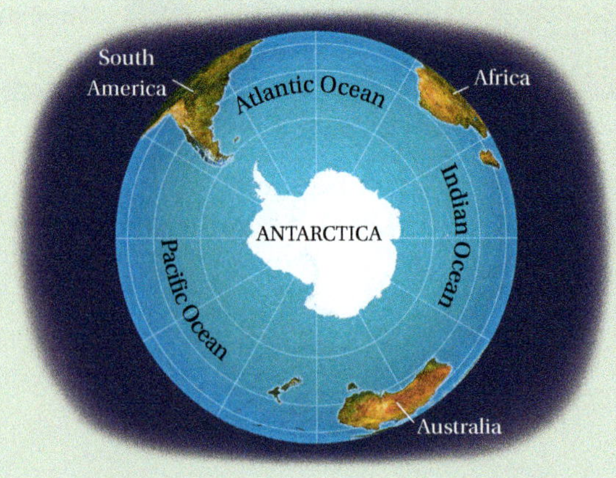

21. Evaluate $\frac{a-b}{6}$ for $a = -8$ and $b = 10$.

22. Use the distributive law to write an expression equivalent to $7(2x + 3y - 1)$.

23. Combine like terms.
$9x - 14 - 5x - 3$

24. Find the perimeter of a square garden that is 5 ft on each side.

Solve.

25. $-7x = -35$

26. $a + 9 = -3$

27. $95 = -x$

28. $3t - 7 = 5$

29. Use the distributive law to write an expression equivalent to $-2(n - 6m)$.
A. $-2n - 6m$ **B.** $-2n - 12m$
C. $-2n + 12m$ **D.** $2n + 12m$

Synthesis

30. Monty plans to attach trim around the doorway and along the base of the walls in a 12-ft by 14-ft room. If the doorway is 3 ft by 7 ft, how many feet of trim are needed? (Only three sides of a doorway get trim.)

Simplify.

31. $9 - 5[x + 2(3 - 4x)] + 14$

32. $15x + 3(2x - 7) - 9(4 + 5x)$

33. 🖩 $49 \cdot 14^3 \div 7^4 + 1926^2 \div 6^2$

34. 🖩 $3487 - 16 \div 4 \cdot 4 \div 2^8 \cdot 14^4$

CHAPTER

3

Fraction Notation: Multiplication and Division

STUDYING FOR SUCCESS *Learning Resources on Campus*

- ☐ There may be a learning lab or a tutoring center for drop-in tutoring.
- ☐ There may be group tutoring sessions for this specific course.
- ☐ The math department may have a bulletin board or network that provides contact information for private tutors.

3.1 Multiples and Divisibility

OBJECTIVES

a Find some multiples of a number, and determine whether a number is divisible by another number.

b Test to see if a number is divisible by 2, 3, 5, 6, 9, or 10.

SKILL TO REVIEW

Objective 1.5a: Divide whole numbers.

Divide.

1. $329 \div 8$
2. $23\overline{)1081}$

In this section, we discuss *multiples* and *divisibility* in order to be able to simplify fractions like $\frac{117}{225}$.

a MULTIPLES

A **multiple** of a number is a product of that number and an integer. For example, some multiples of 2 are

2 (because $2 = 1 \cdot 2$);
4 (because $4 = 2 \cdot 2$);
6 (because $6 = 3 \cdot 2$);
8 (because $8 = 4 \cdot 2$).

We can also find multiples of 2 by counting by twos: 2, 4, 6, 8, and so on.

EXAMPLE 1 Show that each of the numbers 3, 6, 9, and 15 is a multiple of 3.

$3 = 1 \cdot 3$; $6 = 2 \cdot 3$; $9 = 3 \cdot 3$; $15 = 5 \cdot 3$.

Do Margin Exercises 1 and 2.

EXAMPLE 2 Multiply by 1, 2, 3, and so on, to find eight multiples of six.

$1 \cdot 6 = 6$	$5 \cdot 6 = 30$
$2 \cdot 6 = 12$	$6 \cdot 6 = 36$
$3 \cdot 6 = 18$	$7 \cdot 6 = 42$
$4 \cdot 6 = 24$	$8 \cdot 6 = 48$

Do Exercise 3.

1. Show that each of the numbers 5, 45, and 100 is a multiple of 5.
2. Show that each of the numbers 10, 60, and 110 is a multiple of 10.
3. Multiply by 1, 2, 3, and so on, to find ten multiples of 5.

DIVISIBILITY

A number b is said to be **divisible** by another number a if b is a multiple of a.

Thus,

6 is divisible by 2 because 6 is a multiple of 2 ($6 = 3 \cdot 2$), and
100 is divisible by 25 because 100 is a multiple of 25 ($100 = 4 \cdot 25$).

Answers

Skill to Review:
1. 41 R 1 2. 47

Margin Exercises:
1. $5 = 1 \cdot 5$; $45 = 9 \cdot 5$; $100 = 20 \cdot 5$
2. $10 = 1 \cdot 10$; $60 = 6 \cdot 10$; $110 = 11 \cdot 10$
3. 5, 10, 15, 20, 25, 30, 35, 40, 45, 50

Saying that b is divisible by a means that if we divide b by a, the remainder is 0. When this happens, we sometimes say that a divides b "evenly."

EXAMPLE 3 Determine **(a)** whether 45 is divisible by 9 and **(b)** whether 45 is divisible by 4.

a) We divide 45 by 9.

$$\begin{array}{r} 5 \\ 9\overline{)45} \\ \underline{45} \\ 0 \end{array} \leftarrow \text{Remainder is 0.}$$

Because the remainder is 0, 45 is divisible by 9.

b) We divide 45 by 4.

$$\begin{array}{r} 11 \\ 4\overline{)45} \\ \underline{4} \\ 5 \\ \underline{4} \\ 1 \end{array} \leftarrow \text{Not 0}$$

Since the remainder is not 0, 45 is not divisible by 4.

Do Exercises 4–6. ▶

4. Determine whether 16 is divisible by 2.

$$\begin{array}{r} 8 \\ \square\overline{)16} \\ \underline{16} \\ \square \end{array}$$

Since the remainder is ___, 16 ___ divisible by 2.
is/is not

5. Determine whether 125 is divisible by 5.

6. Determine whether 125 is divisible by 6.

b TESTS FOR DIVISIBILITY

We now look at quick ways of checking for divisibility by 2, 3, 5, 6, 9, and 10 without actually performing long division. Tests do exist for divisibility by 4, 7, and 8, but they can be as difficult to perform as the actual long division.

To test for divisibility by 2, 5, and 10, we examine the ones digit.

Divisibility by 2

All even numbers are divisible by 2.

BY 2

A number is **divisible by 2** (is *even*) if it has a ones digit of 0, 2, 4, 6, or 8 (that is, it has an even ones digit).

To see why this test works, start counting by twos: 2, 4, 6, 8, 10, 12, 14, 16, 18, 20, 22, ... Note that the ones digit will always be 0, 2, 4, 6, or 8, no matter how high we count.

EXAMPLES Determine whether each number is divisible by 2.

4. 355 *is not* divisible by 2 because 5 is not even.
5. 4786 *is* divisible by 2 because 6 is even.
6. 8990 *is* divisible by 2 because 0 is even.
7. 4261 *is not* divisible by 2 because 1 is not even.

Do Exercises 7–10. ▶

CALCULATOR CORNER

Divisibility We can use a calculator to determine whether one number is divisible by another number. For example, to determine whether 387 is divisible by 9, we press [3] [8] [7] [÷] [9] [=]. The display is [43].

Since 43 contains no digits to the right of the decimal point, we know that 387 is divisible by 9. On the other hand, since $387 \div 10 = 38.7$, we know that 387 is *not* divisible by 10.

EXERCISES For each pair of numbers, determine whether the first number is divisible by the second number.

1. 731; 17
2. 1502; 79
3. 1053; 36
4. 4183; 47

Determine whether each number is divisible by 2.

7. 84 **8.** 59

9. 998 **10.** 2225

Answers

4. Yes **5.** Yes **6.** No **7.** Yes
8. No **9.** Yes **10.** No

Guided Solution:
4. 2, 0; 0, is

Divisibility by 5

To determine the test for divisibility by 5, we start counting by fives: 5, 10, 15, 20, 25, 30, 35, ... Note that the ones digit will always be 5 or 0, no matter how high we count.

BY 5

A number is **divisible by 5** if its ones digit is 0 or 5.

EXAMPLES Determine whether each number is divisible by 5.

8. 220 *is* divisible by 5 because the ones digit is 0.

9. 475 *is* divisible by 5 because the ones digit is 5.

10. 6514 *is not* divisible by 5 because the ones digit is neither 0 nor 5.

Do Exercises 11–14.

Determine whether each number is divisible by 5.

11. 5780 **12.** 3427

13. 34,678 **14.** 7775

Divisibility by 10

BY 10

A number is **divisible by 10** if its ones digit is 0.

We know that this test works because the product of 10 and *any* number has a ones digit of 0.

EXAMPLES Determine whether each number is divisible by 10.

11. 3440 *is* divisible by 10 because the ones digit is 0.

12. 3447 *is not* divisible by 10 because the ones digit is not 0.

Do Exercises 15–18.

Determine whether each number is divisible by 10.

15. 305 **16.** 847

17. 300 **18.** 8760

Divisibility by 3

To test for divisibility by 3, we examine the sum of a number's digits.

BY 3

A number is **divisible by 3** if the sum of its digits is divisible by 3.

An explanation of why this test works is outlined in Exercise 69 at the end of this section.

EXAMPLES Determine whether each number is divisible by 3.

13. 18 $1 + 8 = 9$
14. 93 $9 + 3 = 12$
15. 201 $2 + 0 + 1 = 3$

Each *is* divisible by 3 because the sum of its digits *is* divisible by 3.

16. 256 $2 + 5 + 6 = 13$ The sum of the digits, 13, *is not* divisible by 3, so 256 *is not* divisible by 3.

Do Exercises 19–22.

Determine whether each number is divisible by 3.

19. 111 **20.** 1111

21. 309

22. 17,216

GS

Add the digits:

$1 + 7 + \_\_ + 1 + 6 = \_\_$.

Since 17 ____ (is/is not) divisible by 3, the number 17,216 ____ (is/is not) divisible by 3.

Answers

11. Yes **12.** No **13.** No **14.** Yes
15. No **16.** No **17.** Yes **18.** Yes
19. Yes **20.** No **21.** Yes **22.** No

Guided Solution:
22. 2, 17; is not, is not

Divisibility by 9

The test for divisibility by 9 is similar to the test for divisibility by 3.

BY 9

A number is **divisible by 9** if the sum of its digits is divisible by 9.

EXAMPLES Determine whether each number is divisible by 9.

17. 6984

Because $6 + 9 + 8 + 4 = 27$ and 27 is divisible by 9, 6984 *is* divisible by 9.

18. 322

Because $3 + 2 + 2 = 7$ and 7 is not divisible by 9, 322 *is not* divisible by 9.

Do Exercises 23–26.

Determine whether each number is divisible by 9.

23. 16 **24.** 117

25. 309 **26.** 29,223

Divisibility by 6

A number divisible by 6 is a multiple of 6. But $6 = 2 \cdot 3$, so the number is also a multiple of 2 and 3. Since 2 and 3 have no factors in common, a number is divisible by 6 if it is divisible by 2 *and* by 3.

BY 6

A number is **divisible by 6** if its ones digit is 0, 2, 4, 6, or 8 (is even) and the sum of its digits is divisible by 3.

EXAMPLES Determine whether each number is divisible by 6.

19. 720

Because 720 is even, it is divisible by 2. Also, $7 + 2 + 0 = 9$ and 9 is divisible by 3, so 720 is divisible by 3. Thus, 720 *is* divisible by 6.

720 $7 + 2 + 0 = 9$

↑ Even ↑ Divisible by 3

20. 73

73 *is not* divisible by 6 because it is not even.

21. 256

Although 256 is even, it *is not* divisible by 6 because the sum of its digits, $2 + 5 + 6$, or 13, is not divisible by 3.

Do Exercises 27–30.

Determine whether each number is divisible by 6.

27. 420 **28.** 106

29. 321 **30.** 444

Answers

23. No **24.** Yes **25.** No **26.** Yes
27. Yes **28.** No **29.** No **30.** Yes

3.1 Exercise Set

For Extra Help
MyMathLab®

MathXL® PRACTICE  WATCH  READ  REVIEW 

Reading Check

Match the beginning of each divisibility test with the appropriate ending from the list at the right.

RC1. A number is divisible by 2 if ______________.

RC2. A number is divisible by 3 if ______________.

RC3. A number is divisible by 5 if ______________.

RC4. A number is divisible by 6 if ______________.

RC5. A number is divisible by 9 if ______________.

RC6. A number is divisible by 10 if ______________.

a) the sum of its digits is divisible by 3

b) the sum of its digits is divisible by 9

c) it has an even ones digit

d) its ones digit is 0 or 5

e) its ones digit is 0

f) it has an even ones digit and the sum of its digits is divisible by 3

a Multiply by 1, 2, 3, and so on, to find ten multiples of each number.

1. 7 **2.** 4 **3.** 20 **4.** 50 **5.** 3 **6.** 8

7. 12 **8.** 15 **9.** 10 **10.** 11 **11.** 25 **12.** 100

13. Determine whether 83 is divisible by 3.

14. Determine whether 29 is divisible by 2.

15. Determine whether 525 is divisible by 7.

16. Determine whether 346 is divisible by 8.

17. Determine whether 8127 is divisible by 9.

18. Determine whether 4144 is divisible by 4.

b For Exercises 19–30, answer "Yes" or "No" and give a reason based on the tests for divisibility.

19. Determine whether 84 is divisible by 3.

20. Determine whether 467 is divisible by 9.

21. Determine whether 5553 is divisible by 5.

22. Determine whether 2004 is divisible by 6.

23. Determine whether 671,500 is divisible by 10.

24. Determine whether 6120 is divisible by 5.

25. Determine whether 1773 is divisible by 9.

26. Determine whether 3286 is divisible by 3.

27. Determine whether 21,687 is divisible by 2.

28. Determine whether 64,091 is divisible by 10.

29. Determine whether 32,109 is divisible by 6.

30. Determine whether 9840 is divisible by 2.

For Exercises 31–38, test each number for divisibility by 2, 3, 5, 6, 9, and 10.

31. 6825

32. 12,600

33. 119,117

34. 2916

35. 127,575

36. 25,088

37. 9360

38. 143,507

To answer Exercises 39–44, consider the following numbers. Use the tests for divisibility.

46	300	85	256
224	36	711	8064
19	45,270	13,251	1867
555	4444	254,765	21,568

39. Which of the above are divisible by 3?

40. Which of the above are divisible by 2?

41. Which of the above are divisible by 10?

42. Which of the above are divisible by 5?

43. Which of the above are divisible by 6?

44. Which of the above are divisible by 9?

To answer Exercises 45–50, consider the following numbers.

56	200	75	35
324	42	812	402
784	501	2345	111,111
55,555	3009	2001	1005

45. Which of the above are divisible by 2?

46. Which of the above are divisible by 3?

47. Which of the above are divisible by 5?

48. Which of the above are divisible by 10?

49. Which of the above are divisible by 9?

50. Which of the above are divisible by 6?

Skill Maintenance

Solve.

51. $16 \cdot t = 848$ [1.7b], [2.8b]

52. $m + 9 = 14$ [1.7b], [2.8a]

53. $23 + x = 15$ [1.7b], [2.8a]

54. $24 \cdot m = -576$ [1.7b], [2.8b]

Solve. [1.8a]

55. Marty's automobile has a 5-speed transmission and gets 33 mpg in city driving. How many gallons of gas will it use in 1485 mi of city driving?

56. There are 60 min in 1 hr. How many minutes are there in 72 hr?

Evaluate. [1.9b], [2.4b]

57. 5^3

58. $(-2)^4$

Write in exponential notation. [1.9a]

59. $9 \cdot 9 \cdot 9$

60. $3 \cdot 3 \cdot 3 \cdot 3 \cdot 3 \cdot 3$

Synthesis

61. Find the largest five-digit number that is divisible by 47.

62. Find the largest six-digit number that is divisible by 53.

Find the smallest number that is simultaneously a multiple of the given numbers.

63. 2, 3, and 5

64. 3, 5, and 7

65. 6, 10, and 14

66. 17, 43, and 85

67. 30, 70, and 120

68. 25, 100, and 175

69. To help see why the tests for division by 3 and 9 work, note that any four-digit number *abcd* can be rewritten as $1000 \cdot a + 100 \cdot b + 10 \cdot c + d$, or $999a + 99b + 9c + a + b + c + d$.

a) Explain why $999a + 99b + 9c$ is divisible by both 9 and 3 for all choices of a, b, c, and d.

b) Explain why the four-digit number *abcd* is divisible by 9 if $a + b + c + d$ is divisible by 9 and is divisible by 3 if $a + b + c + d$ is divisible by 3.

70. A passenger in a taxicab asks for the cab number. The driver says abruptly, "Sure—it's the smallest multiple of 11 that, when divided by 2, 3, 4, 5, or 6, has a remainder of 1." What is the number?

Following are the tests for divisibility by 4, 8, 7, and 11. Use these for Exercises 71 and 72.

NUMBER	TEST	TEST 23,904,328 FOR DIVISIBILITY
4	A number is divisible by 4 if the number named by its last two digits is divisible by 4.	The number named by the last two digits is 28. Since 28 is divisible by 4, 23,904,328 is divisible by 4.
8	A number is divisible by 8 if the number named by its last three digits is divisible by 8.	The number named by the last three digits is 328. Since 328 is divisible by 8, 23,904,328 is divisible by 8.
7	Divide the number into groups of three digits, starting at the right. Start with the group at the right, subtract the next group to the left, add the next group to the left, and so on. If the resulting number is divisible by 7, the original number is divisible by 7.	The number is already divided into groups of three by commas. Calculate $328 - 904 + 23 = -553$. Since -553 is divisible by 7, 23,904,328 is divisible by 7.
11	Find the sum of the odd-numbered digits (the 1st digit plus the 3rd plus the 5th and so on). Find the sum of the even-numbered digits. Subtract these sums. If the difference is divisible by 11, then the original number is divisible by 11.	Add the odd-numbered digits: $2 + 9 + 4 + 2 = 17$. Add the even-numbered digits: $3 + 0 + 3 + 8 = 14$. Subtract: $17 - 14 = 3$. Since 3 is *not* divisible by 11, 23,904,328 is not divisible by 11.

71. Test 332,986,412 for divisibility by 4, 8, 7, and 11.

72. Test 6,637,105,860 for divisibility by 4, 8, 7, and 11.

Factorizations

3.2

When we express a number as a product, we say that we have *factored* the original number. The numbers in the product are called *factors*. Thus "factor" can be used as either a noun or a verb. The ability to factor is an important skill needed for a solid understanding of fractions.

a FACTORS AND FACTORIZATIONS

From the equation $3 \cdot 4 = 12$, we can say that 3 and 4 are **factors** of 12. Since $12 = 12 \cdot 1$, we know that 12 and 1 are also factors of 12.

FACTORS AND FACTORIZATIONS

- In the product $a \cdot b$, a and b are called **factors.**
- A number c is a **factor** of a if a is divisible by c.
- A **factorization** of a expresses a as a product of two or more numbers.

Note that saying that 3 and 4 are factors of 12 is the same as saying that 12 is a multiple of 3 and that 12 is a multiple of 4. Each of the following gives a factorization of 12.

$12 = 4 \cdot 3$ ← This factorization shows that 4 and 3 are factors of 12.
$12 = 12 \cdot 1$ ← This factorization shows that 12 and 1 are factors of 12.
$12 = 6 \cdot 2$ ← This factorization shows that 6 and 2 are factors of 12.
$12 = 2 \cdot 3 \cdot 2$ ← This factorization shows that 2 and 3 are factors of 12.

Thus, 1, 2, 3, 4, 6, and 12 are all factors of 12. Note that since $n = n \cdot 1$, every number has a factorization, and every number has itself and 1 as factors.

EXAMPLE 1 Find all the factors of 18.

Beginning at 1, we check all positive integers to see if they are factors of 18. If they are, we write the factorization. We stop when we have already included the next integer in a factorization.

1 is a factor of every number.	$1 \cdot 18$
2 is a factor of 18.	$2 \cdot 9$
3 is a factor of 18.	$3 \cdot 6$
4 is *not* a factor of 18.	
5 is *not* a factor of 18.	

The next integer is 6, but we have already listed 6 as a factor in the product $3 \cdot 6$. We need check no additional numbers, because any integer greater than 6 must be paired with a factor less than 6.

We now write the factors of 18 beginning with 1, going down the list of factorizations writing each first factor, then up the list of factorizations writing each second factor:

1, 2, 3, 6, 9, 18.

Do Margin Exercises 1–4. ▶

OBJECTIVES

a Find the factors of a number.

b Given a number from 1 to 100, tell whether it is prime, composite, or neither.

c Find the prime factorization of a composite number.

SKILL TO REVIEW

Objective 1.9a: Write exponential notation for products such as $4 \cdot 4 \cdot 4$.

Write in exponential notation.

1. $2 \cdot 2 \cdot 2$ **2.** $5 \cdot 5 \cdot 5 \cdot 5$

Find all the factors of each number.

1. 10 **2.** 62 **3.** 24

4. 45

1 is a factor of 45. $1 \cdot 45$
2 ____ a factor of 45.
3 ____ a factor of 45. $3 \cdot$ ____
4 ____ a factor of 45.
5 ____ a factor of 45. $5 \cdot$ ____
6 is not a factor of 45.
7 is not a factor of 45.
8 is not a factor of 45.

Factors of 45: 1, 3, 5, ____, ____, 45.

Answers

Skill to Review:
1. 2^3 **2.** 5^4

Margin Exercises:
1. 1, 2, 5, 10 **2.** 1, 2, 31, 62
3. 1, 2, 3, 4, 6, 8, 12, 24 **4.** 1, 3, 5, 9, 15, 45

Guided Solution:
4. is not; is, 15; is not; is, 9; 9, 15

b PRIME AND COMPOSITE NUMBERS

PRIME AND COMPOSITE NUMBERS

A natural number that has exactly two different factors, only itself and 1, is called a **prime number**.

- The number 1 is *not* prime.
- A natural number, other than 1, that is not prime is **composite**.

EXAMPLE 2 Determine which of the numbers 1, 2, 7, 8, 9, 11, 18, 27, 39, 43, 56, 59, and 77 are prime, which are composite, and which are neither.

The number 1 is not prime. It does not have *two* different factors.

The number 2 is prime. It has only the factors 2 and 1.

The numbers 7, 11, 43, and 59 are prime. Each has only two factors, itself and 1.

The number 8 is not prime. It has the factors 1, 2, 4, and 8 and is composite.

The numbers 9, 18, 27, 39, 56, and 77 are composite. Each has more than two factors.

Thus we have

Prime: 2, 7, 11, 43, 59;
Composite: 8, 9, 18, 27, 39, 56, 77;
Neither: 1.

We can make several observations about prime numbers.

- Because 0 is not a natural number, it is neither prime nor composite.
- The number 1 is not prime because it does not have two different factors.
- The number 2 is the smallest prime and the only even prime, since 2 is a factor of all even numbers.
- To determine whether an odd number is prime, check divisibility by prime numbers beginning with 3 and 5. If you reach a point where the quotient is less than the divisor and none of the primes up to that point are factors, the number you are checking is prime.

5. Classify each number as prime, composite, or neither.
1, 2, 6, 12, 13, 23, 41, 65, 73, 99

Do Exercise 5.

The following is a table of the prime numbers from 2 to 157. Although you need not memorize the entire list, remembering at least the first nine or ten is important.

A TABLE OF PRIMES FROM 2 TO 157

2, 3, 5, 7, 11, 13, 17, 19, 23, 29, 31, 37, 41, 43, 47, 53, 59, 61, 67, 71, 73, 79, 83, 89, 97, 101, 103, 107, 109, 113, 127, 131, 137, 139, 149, 151, 157

There are infinitely many prime numbers. It takes extensive computer operations to determine whether very large numbers are prime. In 2014, the largest known prime was $2^{57885161} - 1$. This number has over 17 million digits! (**Source:** primes.utm.edu/largest.html#largest.)

Answer

5. 2, 13, 23, 41, 73 are prime; 6, 12, 65, 99 are composite; 1 is neither

C PRIME FACTORIZATIONS

When we factor a composite number into a product of primes, we find the **prime factorization** of the number. We can do this by making a series of successive divisions or by using a *factor tree.*

To use division, we consider the prime numbers 2, 3, 5, 7, 11, 13, and so on, and determine whether a given number is divisible by the primes.

EXAMPLE 3 Find the prime factorization of 39.

We check for divisibility by the first prime, 2. Since 39 is not even, 2 is not a factor of 39. Since the sum of the digits in 39 is 12 and 12 is divisible by 3, we know that 39 is divisible by 3. We then perform the division.

$$3\overline{)39} \quad 13 \quad R = 0$$

A remainder of 0 confirms that 3 is a factor of 39.

Because 13 is prime, we can now write the prime factorization:

$$39 = 3 \cdot 13.$$

EXAMPLE 4 Find the prime factorization of 220.

We consider the first prime, 2. Since 220 is even, it must have 2 as a factor.

$$2\overline{)220} \quad 110 \quad R = 0 \qquad 220 = 2 \cdot 110$$

Because 110 is also even, we divide again by 2.

$$2\overline{)110} \quad 55 \quad R = 0 \qquad 220 = 2 \cdot 2 \cdot 55$$

Since 55 is odd, it is not divisible by 2. We move to the next prime, 3. The sum of the digits is 10 and 10 is not divisible by 3, so we move to the next prime, 5. Since the ones digit of 55 is 5, we divide by 5.

$$5\overline{)55} \quad 11 \quad R = 0 \qquad 220 = 2 \cdot 2 \cdot 5 \cdot 11$$

Because 11 is prime, we are finished. The prime factorization is

$$220 = 2 \cdot 2 \cdot 5 \cdot 11.$$

We abbreviate our procedure as follows.

$$\begin{array}{r} 11 \\ 5\overline{)55} \\ 2\overline{)110} \\ 2\overline{)220} \end{array}$$

$$220 = 2 \cdot 2 \cdot 5 \cdot 11$$

A factorization like $2 \cdot 2 \cdot 5 \cdot 11$ could also be expressed as $5 \cdot 2 \cdot 2 \cdot 11$ or $2 \cdot 5 \cdot 11 \cdot 2$ or $2^2 \cdot 5 \cdot 11$ or $11 \cdot 2^2 \cdot 5$. The prime factors are the same in each case. For this reason, we agree that any of these may be considered "the" prime factorization of 220.

Every composite number has just one (unique) prime factorization.

This result is sometimes called the Fundamental Theorem of Arithmetic.

Caution!

Finding the prime factorization of a number is not the same as finding all of its factors. A prime factorization of a number is a product; the factors of a number are a list. Compare the following for the number 72:

Prime factorization of 72:

$2 \cdot 2 \cdot 2 \cdot 3 \cdot 3$;

Factors of 72:

1, 2, 3, 4, 6, 8, 9, 12, 18, 24, 36, 72.

EXAMPLE 5 Find the prime factorization of 187.

We check for divisibility by 2, 3, and 5, and find that 187 is not divisible by any of these numbers. The next prime number, 7, does not divide 187 evenly. However, when we divide by 11, the remainder is 0, so 11 is a factor of 187.

$$11\overline{)187}\ \text{ quotient } 17$$

We can write $187 = 11 \cdot 17$.

Because 17 is prime, we can factor no further. The complete factorization is

$187 = 11 \cdot 17$. All factors are prime.

EXAMPLE 6 Find the prime factorization of 72.

3 ← 3 is prime, so we stop dividing.

$3\overline{)9}$

$2\overline{)18}$

$2\overline{)36}$

$2\overline{)72}$ ← Begin here and work upward.

$72 = 2 \cdot 2 \cdot 2 \cdot 3 \cdot 3$

Another way to find the prime factorization of 72 is to use a **factor tree** as follows. Begin by determining any factorization you can, and then continue factoring until all of the factors are prime numbers. Each of the following trees gives the same prime factorization.

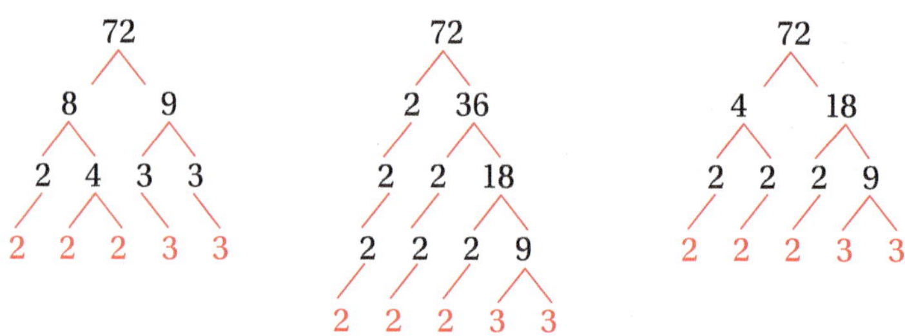

Find the prime factorization of each number.

6. 6 **7.** 12

8. 45

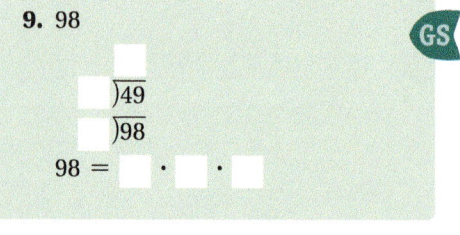

10. 126 **11.** 144

12. 91 **13.** 1925

EXAMPLE 7 Find the prime factorization of 130.

We can use a string of divisions or a factor tree.

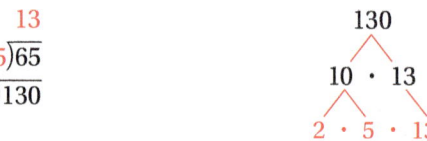

$130 = 2 \cdot 5 \cdot 13$

◀ **Do Exercises 6–13.**

Finding a number's prime factorization can be quite challenging, especially when the prime factors themselves are large. This difficulty is used worldwide as a way of securing transactions over the Internet.

Answers

6. $2 \cdot 3$ **7.** $2 \cdot 2 \cdot 3$ **8.** $3 \cdot 3 \cdot 5$
9. $2 \cdot 7 \cdot 7$ **10.** $2 \cdot 3 \cdot 3 \cdot 7$
11. $2 \cdot 2 \cdot 2 \cdot 2 \cdot 3 \cdot 3$ **12.** $7 \cdot 13$
13. $5 \cdot 5 \cdot 7 \cdot 11$

Guided Solution:

9. $7\overline{)49}$ (quotient 7), $2\overline{)98}$; 2, 7, 7

3.2 Exercise Set

For Extra Help

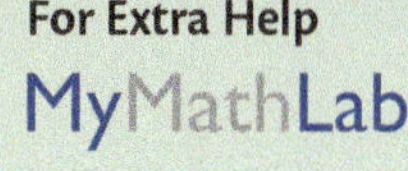

✓ Reading Check

Determine whether each statement is true or false.

________ **RC1.** One factorization of 20 is $4 \cdot 5$.

________ **RC2.** One factor of 15 is 30.

________ **RC3.** A prime number has exactly two different factors.

________ **RC4.** The smallest prime number is 1.

________ **RC5.** The prime factorization of 30 is $3 \cdot 10$.

a Determine whether the second number is a factor of the first.

1. 52; 14

2. 52; 13

3. 625; 25

4. 680; 16

List all the factors of each number.

5. 18

6. 16

7. 54

8. 48

9. 9

10. 4

11. 13

12. 11

13. 98

14. 100

15. 255

16. 120

b Classify each number as prime, composite, or neither.

17. 19

18. 24

19. 22

20. 31

21. 48

22. 43

23. 53

24. 54

25. 1

26. 2

27. 81

28. 37

29. 47

30. 51

31. 29

32. 49

c Find the prime factorization of each number.

33. 27

34. 16

35. 14

36. 15

37. 80

38. 32

39. 25

40. 75

41. 62

42. 169

43. 100 **44.** 110 **45.** 143 **46.** 40 **47.** 121

48. 170 **49.** 273 **50.** 675 **51.** 175 **52.** 196

53. 209 **54.** 217 **55.** 1200 **56.** 1800 **57.** 693

58. 675 **59.** 2884 **60.** 484 **61.** 1122 **62.** 6435

Skill Maintenance

Multiply.

63. $-2 \cdot 13$ [2.4a]

64. $(-8)(-32)$ [2.4a]

Add.

65. $-17 + 25$ [2.2a]

66. $-9 + (-14)$ [2.2a]

Divide.

67. $53 \div 53$ [1.5a]

68. $-98 \div 1$ [2.5a]

69. $0 \div 22$ [1.5a]

70. $0 \div (-42)$ [2.5a]

Synthesis

Find the prime factorization of each number.

71. 136,097

72. 102,971

73. 473,073,361

74. Describe an arrangement of 54 objects that corresponds to the factorization $54 = 6 \times 9$.

75. Describe an arrangement of 24 objects that corresponds to the factorization $24 = 2 \cdot 3 \cdot 4$.

76. Two numbers are **relatively prime** if there is no prime number that is a factor of both numbers. For example, 10 and 21 are relatively prime but 15 and 18 are not. List five pairs of composite numbers that are relatively prime.

77. *Factors and Sums.* In the table below, the top number in each column can be factored in such a way that the sum of the factors is the bottom number in the column. For example, in the first column, 56 has been factored as $7 \cdot 8$, and $7 + 8 = 15$, the bottom number. (Such thinking will be important in understanding the meaning of a factor and in algebra.) Fill in the missing numbers in the table.

PRODUCT	56	63	36	72	140	96	48	168	110	90	432	63
FACTOR	7											
FACTOR	8											
SUM	15	16	20	38	24	20	14	29	21	19	42	24

Fractions and Fraction Notation

3.3

OBJECTIVES

a Identify the numerator and the denominator of a fraction, and write fraction notation for part of an object or part of a set of objects and as a ratio.

b Simplify fraction notation like n/n to 1, $0/n$ to 0, and $n/1$ to n.

SKILL TO REVIEW

Objective 1.5a: Divide whole numbers.

Divide.

1. $36 \div 36$ **2.** $50 \div 1$

The study of arithmetic begins with the set of whole numbers. We also need to be able to use fractional parts of numbers such as halves and thirds.

Households in auto-dependent locations spend about $\frac{1}{4}$ of their income on transportation costs, while location-efficient households (those with easy access to public transportation) can hold transportation costs to $\frac{1}{10}$ of their income.

Auto-Dependent Households **Location-Efficient Households**

$\frac{1}{4}$ Transportation $\frac{3}{4}$ Remaining $\frac{1}{10}$ Transportation $\frac{9}{10}$ Remaining

SOURCE: Based on data from U.S. Department of Transportation, Federal Highway Administration

a IDENTIFYING NUMERATORS AND DENOMINATORS

Expressions like those below are written in **fraction notation**. The top number is called the **numerator** and the bottom number is called the **denominator**.

$$\frac{1}{2}, \quad \frac{7}{8}, \quad \frac{-8}{5}, \quad \frac{x}{y}, \quad -\frac{4}{25}, \quad \frac{2a}{7b}$$

EXAMPLE 1 Identify the numerator and the denominator.

$\frac{7}{8}$ 7 ← Numerator, 8 ← Denominator

Do Margin Exercises 1–4. ▶

Fractions as a Partition of an Object Divided into Equal Parts

Consider a candy bar divided into 5 equal sections. If you eat 2 sections, you have eaten $\frac{2}{5}$ of the candy bar. The denominator 5 tells us the unit, $\frac{1}{5}$. The numerator 2 tells us the number of equal parts we are considering, 2.

For each fraction, identify the numerator and the denominator.

1. About $\frac{1}{5}$ of people age 5 and older in the United States speak a language other than English at home.

   Source: 2010 American Community Survey

2. About $\frac{4}{5}$ of the parts of a Toyota Camry were produced in the United States.

   Source: "Made in America: Which Car Creates the Most Jobs?" by David Muir and Sharyn Alfonsi, on abcnews.go.com

3. $\frac{5a}{7m}$

4. $\frac{-22}{3}$

Answers

Skill to Review:

1. 1 **2.** 50

Margin Exercises:

1. Numerator: 1; denominator: 5
2. Numerator: 4; denominator: 5
3. Numerator: $5a$; denominator: $7m$
4. Numerator: -22; denominator: 3

EXAMPLE 2 What part is shaded?

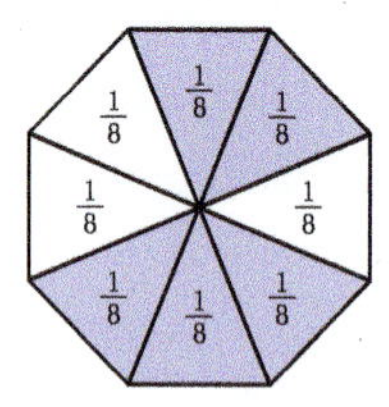

There are 8 equal parts. This tells us the unit, $\frac{1}{8}$. The *denominator* is 8. We have 5 of the units shaded. This tells us the *numerator,* 5. Thus,

$\frac{5}{8}$ 5 ← 5 units are shaded. 8 ← The unit is $\frac{1}{8}$.

is shaded.

EXAMPLE 3 What part of the measuring cup is filled?

The measuring cup is divided into 3 parts of the same size, and 2 of them are filled. This is $2 \cdot \frac{1}{3}$, or $\frac{2}{3}$. Thus, $\frac{2}{3}$ (read *two-thirds*) of the cup is filled.

Do Exercises 5–8.

What part is shaded?

5.

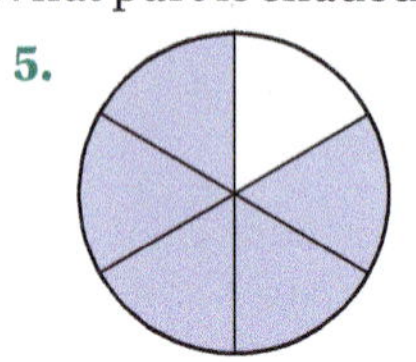

6.

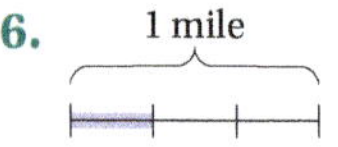

7.

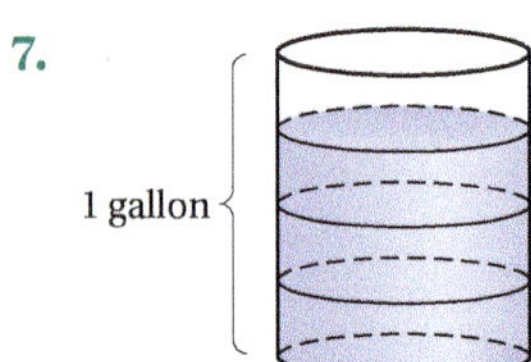

8.

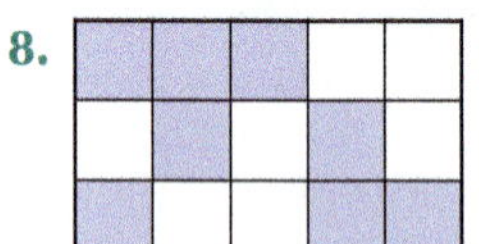

The markings on a ruler use fractions.

EXAMPLE 4 What part of an inch is indicated?

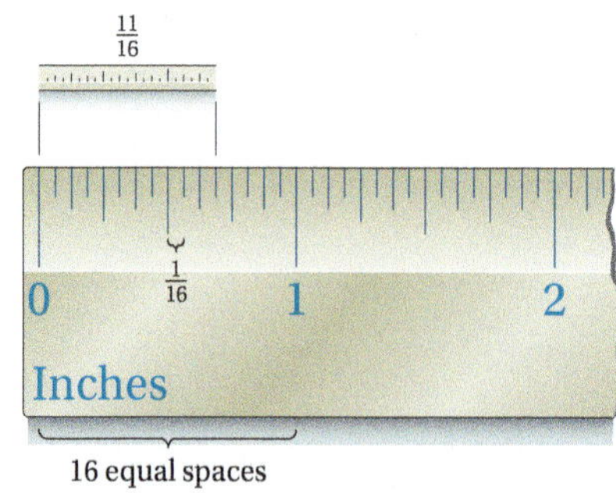

Each inch on the ruler shown above is divided into 16 equal parts. The marked section extends to the 11th mark. Thus, $\frac{11}{16}$ (read *eleven-sixteenths*) of an inch is indicated.

Do Exercise 9.

9. What part of an inch is indicated?

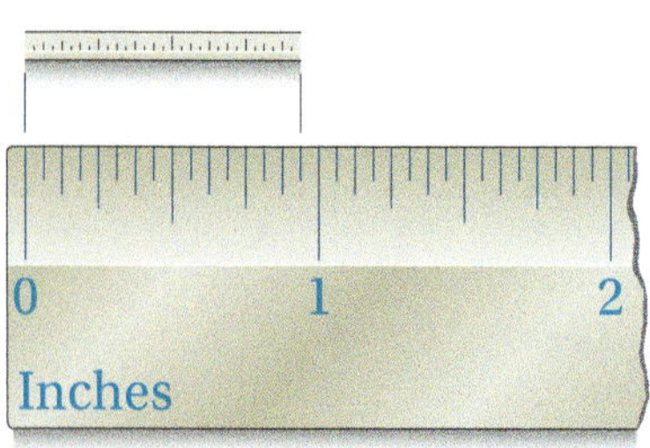

Answers

5. $\frac{5}{6}$ **6.** $\frac{1}{3}$ **7.** $\frac{3}{4}$ **8.** $\frac{8}{15}$ **9.** $\frac{15}{16}$

Fractions greater than or equal to 1, such as $\frac{24}{24}$, $\frac{10}{3}$, and $\frac{5}{4}$, correspond to situations like the following.

EXAMPLE 5 What part is shaded?

a)

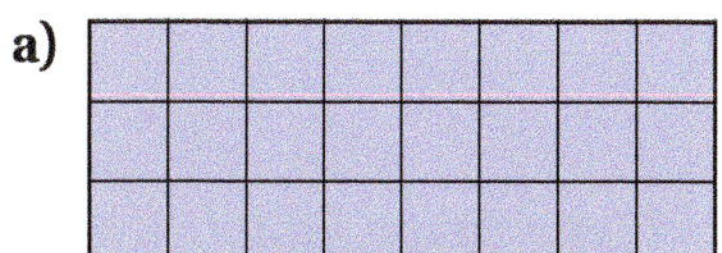

The rectangle is divided into 24 equal parts. Thus the unit is $\frac{1}{24}$. The denominator is 24. All 24 equal parts are shaded. This tells us that the numerator is 24. Thus, $\frac{24}{24}$ is shaded.

b)

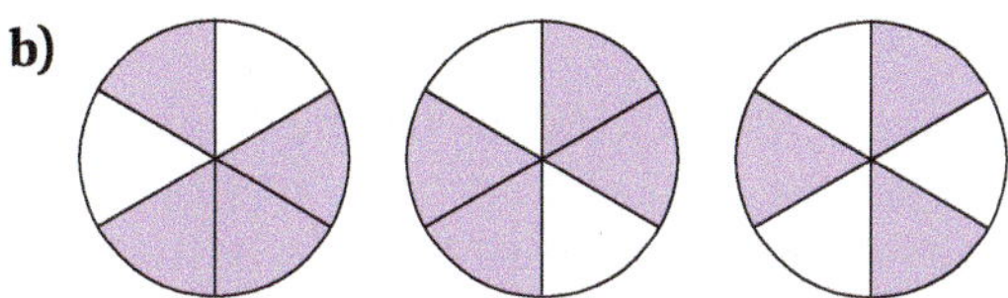

Each circle is divided into 6 parts. Thus the unit is $\frac{1}{6}$. The denominator is 6. We see that 11 of the equal units are shaded. This tells us that the numerator is 11. Thus, $\frac{11}{6}$ is shaded.

EXAMPLE 6 *Ice-Cream Roll-up Cake.* What part of an ice-cream roll-up cake is shaded?

Each cake is divided into 6 equal slices. The unit is $\frac{1}{6}$. The *denominator* is 6. We see that 13 of the slices are shaded. This tells us that the *numerator* is 13. Thus, $\frac{13}{6}$ is shaded.

Do Exercises 10–12.

What part is shaded?

10.

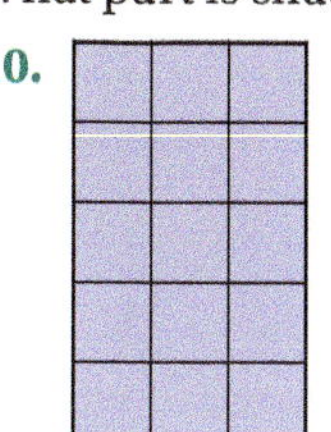

11.

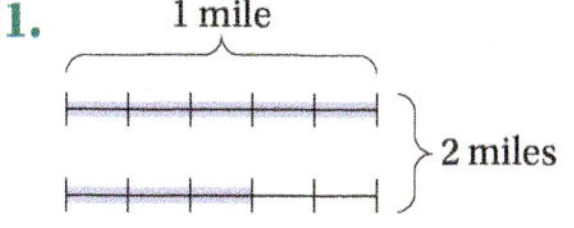

GS **12.**

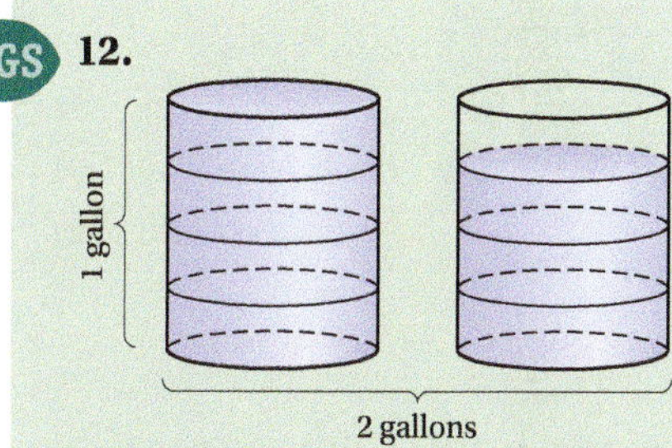

Each gallon is divided into ☐ equal parts.

The unit is $\frac{1}{\square}$.

There are ☐ equal units shaded.

The part that is shaded is $\frac{\square}{\square}$.

Fractions larger than or equal to 1, such as $\frac{13}{6}$ or $\frac{9}{9}$, are sometimes referred to as "improper" fractions. We will not use this terminology because notation such as $\frac{27}{8}$, $\frac{11}{3}$, or $\frac{4}{4}$ is quite "proper" and very common in algebra.

Fractions as Ratios

A **ratio** is a quotient of two quantities. We can express a ratio with fraction notation. (We will consider ratios in more detail in Chapter 7.)

Answers

10. $\frac{15}{15}$ 11. $\frac{8}{5}$ 12. $\frac{7}{4}$

Guided Solution:

12. 4, 4, 7, $\frac{7}{4}$

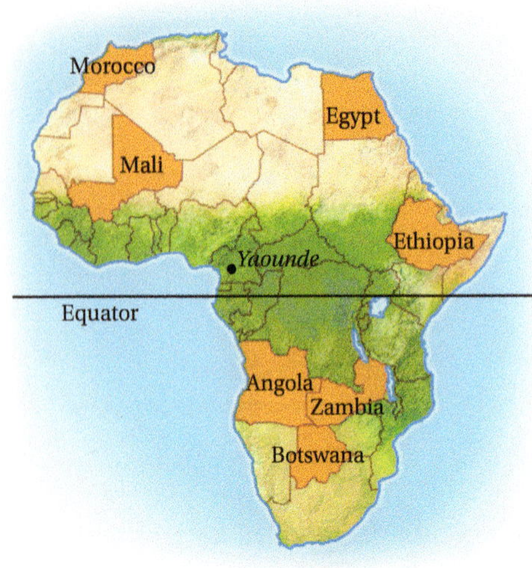

EXAMPLE 7 *Countries of Africa.* What part of this group of countries is north of the equator? south of the equator?

Angola	Mali
Botswana	Morocco
Egypt	Zambia
Ethiopia	

There are 7 countries in the set, and 4 of them, Egypt, Ethiopia, Mali, and Morocco, are north of the equator. Thus, 4 of 7, or $\frac{4}{7}$, are north of the equator. The 3 remaining countries are south of the equator. Thus, $\frac{3}{7}$ are south of the equator.

◀ Do Exercise 13.

13. What part of the set of countries in Example 7 is west of Yaounde?

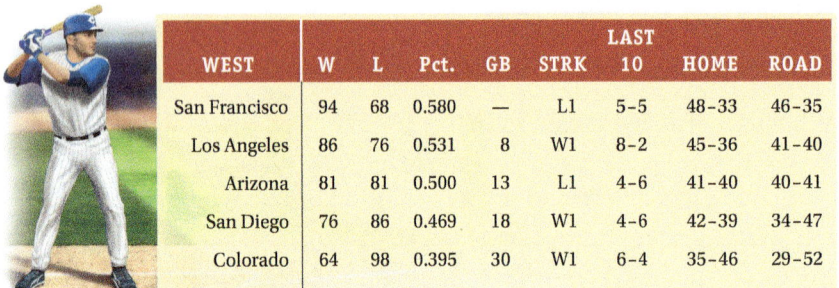

EXAMPLE 8 *Baseball Standings.* The following are the final standings in the National League West for 2012, when the San Francisco Giants won the division. Find the ratio of Giants wins to losses, wins to total games, and losses to total games.

WEST	W	L	Pct.	GB	STRK	LAST 10	HOME	ROAD
San Francisco	94	68	0.580	—	L1	5-5	48-33	46-35
Los Angeles	86	76	0.531	8	W1	8-2	45-36	41-40
Arizona	81	81	0.500	13	L1	4-6	41-40	40-41
San Diego	76	86	0.469	18	W1	4-6	42-39	34-47
Colorado	64	98	0.395	30	W1	6-4	35-46	29-52

SOURCE: espn.go.com

The Giants won 94 games and lost 68 games. They played a total of 94 + 68, or 162 games. Thus we have the following.

The ratio of wins to losses is $\frac{94}{68}$.

The ratio of wins to total games is $\frac{94}{162}$.

The ratio of losses to total games is $\frac{68}{162}$.

◀ Do Exercise 14.

14. *Baseball Standings.* Refer to the table in Example 8. The Arizona Diamondbacks finished third in the National League West in 2012. Find the ratio of Diamondback wins to losses, wins to total games, and losses to total games.

b SOME FRACTION NOTATION FOR WHOLE NUMBERS

Fraction Notation for 1

The number 1 corresponds to situations like those shown here. If we divide an object into n parts and take n of them, we get all of the object (1 whole object).

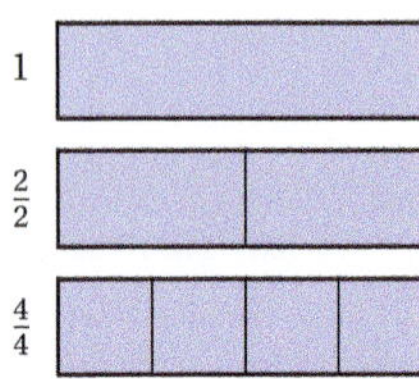

Since a negative number divided by a negative number is a positive number, we state the following for all nonzero integers.

THE NUMBER 1 IN FRACTION NOTATION

$\frac{n}{n} = 1$, for any integer n that is not 0.

Answers

13. $\frac{2}{7}$ **14.** $\frac{81}{81}, \frac{81}{162}, \frac{81}{162}$

EXAMPLES Simplify. Assume any variables are nonzero.

9. $\frac{5}{5} = 1$ **10.** $\frac{-9}{-9} = 1$ **11.** $\frac{23x}{23x} = 1$

Do Exercises 15–20. ▶

Fraction Notation for 0

Consider the fraction $\frac{0}{4}$. This corresponds to dividing an object into 4 parts and taking none of them. We get 0.

THE NUMBER 0 IN FRACTION NOTATION

$\frac{0}{n} = 0$, for any integer n that is not 0.

EXAMPLES Simplify.

12. $\frac{0}{1} = 0$ **13.** $\frac{0}{9} = 0$

14. $\frac{0}{-23} = 0$ **15.** $\frac{0}{5a} = 0$, assuming that $a \neq 0$

Fraction notation with a denominator of 0, such as $n/0$, does not represent a number because we cannot speak of an object being divided into *zero* parts. (If it is not divided at all, then we say it is undivided and remains in one part.)

A DENOMINATOR OF 0

$\frac{n}{0}$ is not defined. We say that $\frac{n}{0}$ is *undefined.*

Do Exercises 21–26. ▶

Other Integers

Consider the fraction $\frac{4}{1}$. This corresponds to taking 4 objects and dividing each into 1 part. (In other words, we do not divide them.) We have 4 objects.

ANY INTEGER IN FRACTION NOTATION

Any integer divided by 1 is the original integer. That is,

$\frac{n}{1} = n$, for any integer n.

EXAMPLES Simplify.

16. $\frac{2}{1} = 2$ **17.** $\frac{-9}{1} = -9$ **18.** $\frac{3x}{1} = 3x$

Do Exercises 27–30. ▶

Simplify.

15. $\frac{1}{1}$ **16.** $\frac{4}{4}$

17. $\frac{a}{a}$ **18.** $\frac{-100}{-100}$

19. $\frac{2347}{2347}$ **20.** $\frac{54n}{54n}$

Simplify, if possible. Assume that $x \neq 0$.

21. $\frac{0}{1}$ **22.** $\frac{0}{-6}$

23. $\frac{0}{4x}$

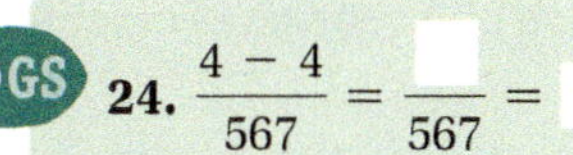

GS **24.** $\frac{4-4}{567} = \frac{\square}{567} = \square$

25. $\frac{15}{0}$ **26.** $\frac{-4}{3-3}$

Simplify.

27. $\frac{8}{1}$ **28.** $\frac{-10}{1}$

29. $\frac{-346}{1}$ **30.** $\frac{24-1}{23-22}$

Answers

15. 1 **16.** 1 **17.** 1 **18.** 1 **19.** 1 **20.** 1 **21.** 0 **22.** 0 **23.** 0 **24.** 0 **25.** Undefined **26.** Undefined **27.** 8 **28.** −10 **29.** −346 **30.** 23

Guided Solution:
24. 0, 0

3.3 Exercise Set

For Extra Help
MyMathLab® MathXL® PRACTICE WATCH READ REVIEW

Reading Check

Match each expression with the appropriate description or value from the list at the right.

RC1. The 3 in $\frac{3}{4}$ ______ **a)** a denominator

RC2. The 4 in $\frac{3}{4}$ ______ **b)** a numerator

RC3. The fraction $\frac{3}{4}$ ______ **c)** a ratio

RC4. $\frac{0}{1}$ ______ **d)** n

RC5. $\frac{n}{0}$ ______ **e)** 0

RC6. $\frac{n}{1}$ ______ **f)** undefined

a Identify the numerator and the denominator.

1. $\frac{3}{4}$ **2.** $\frac{-9}{10}$ **3.** $\frac{7}{-9}$ **4.** $\frac{15}{8}$ **5.** $\frac{2x}{3y}$ **6.** $\frac{9a}{2b}$

What part of each object or set of objects is shaded?

7.

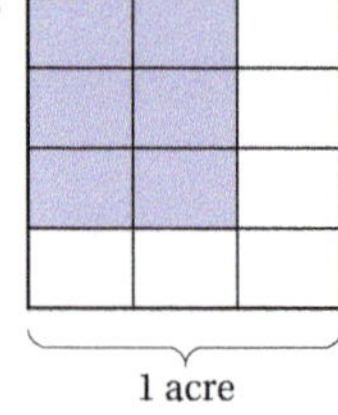

8.

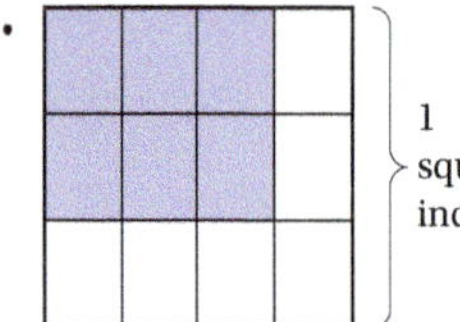

9.

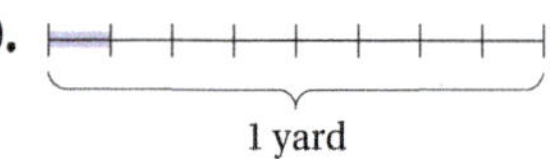

10.

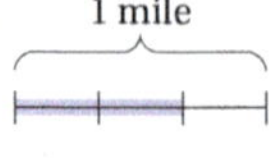

11.

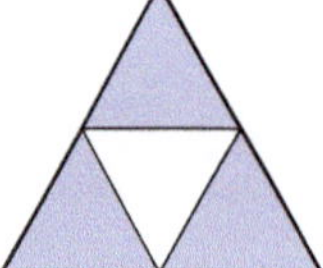

12.

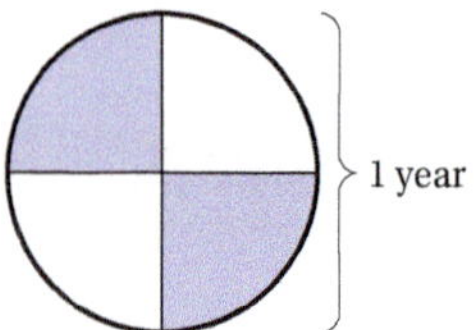

13.

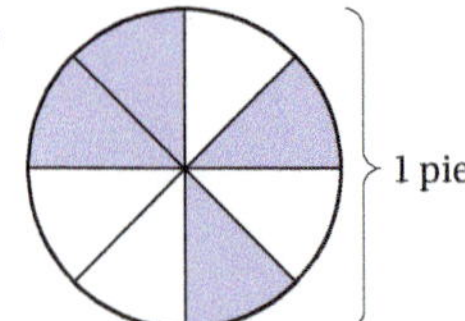

14.

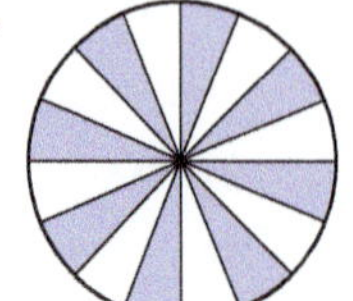

15.

16. 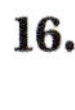

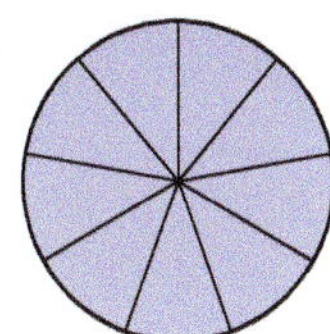

17.

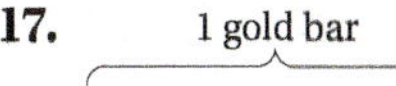

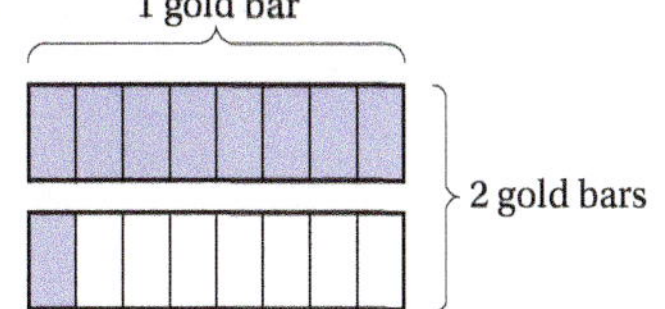

18.

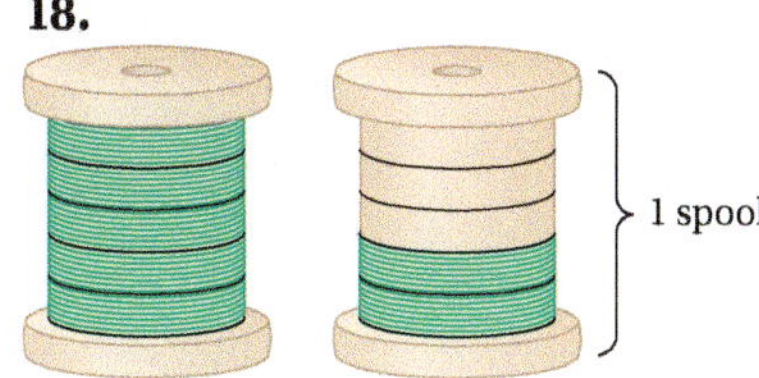

19.

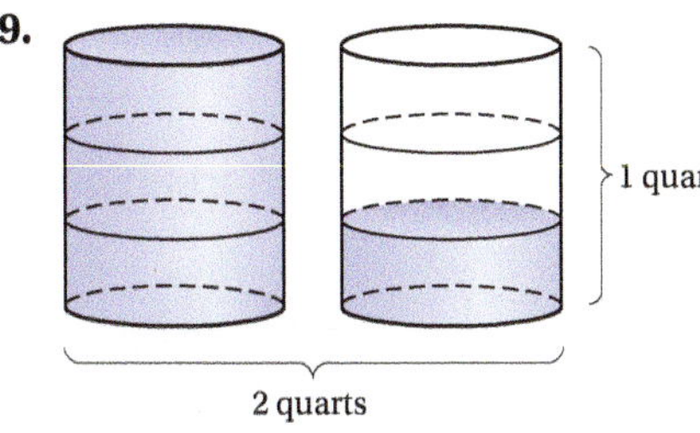

20.

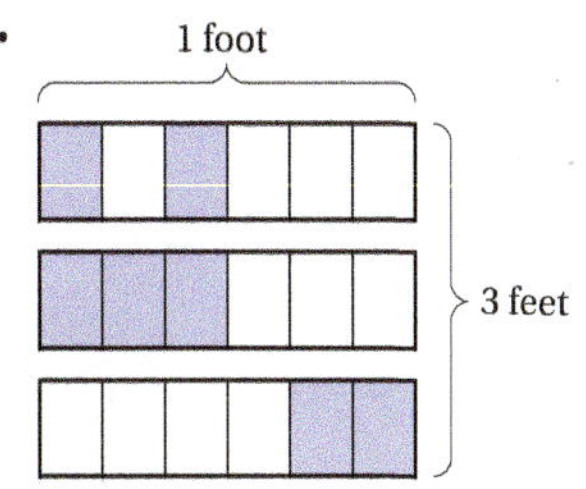

21.

22.

23.

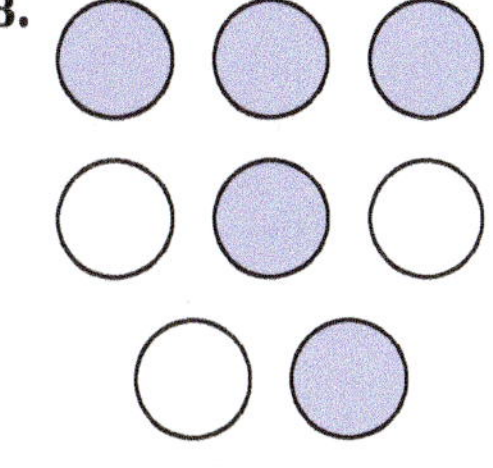

24.

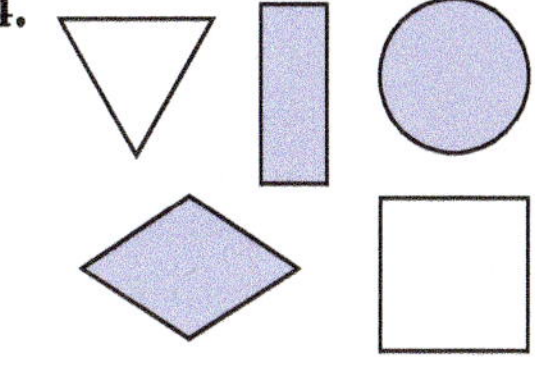

25.

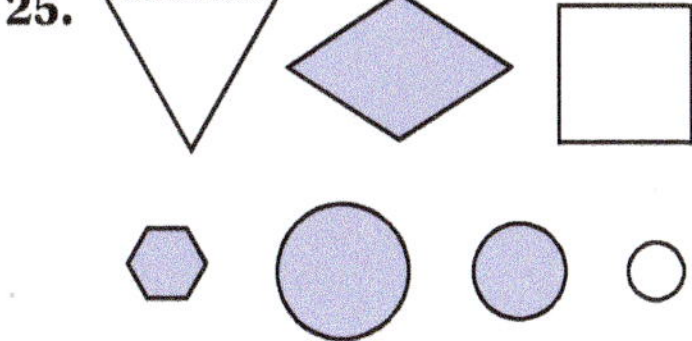

26.

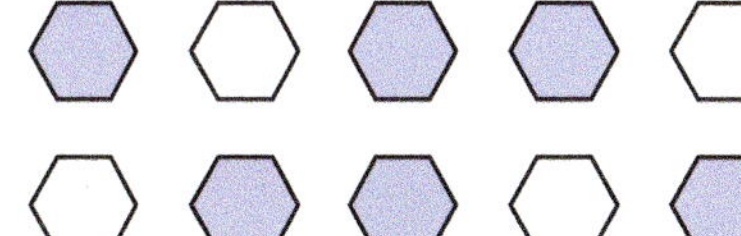

What part of an inch is indicated?

27.

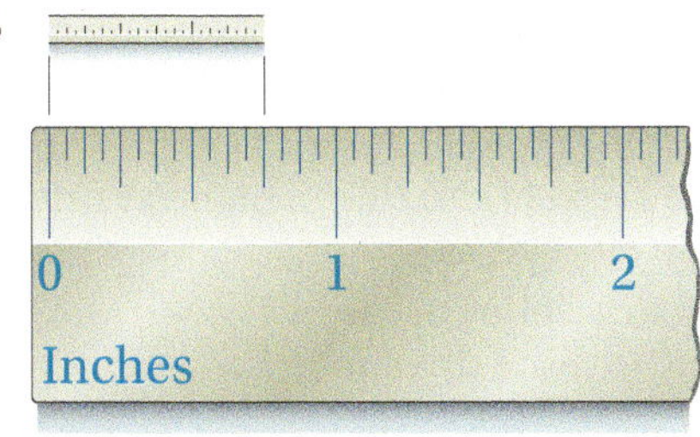

28.

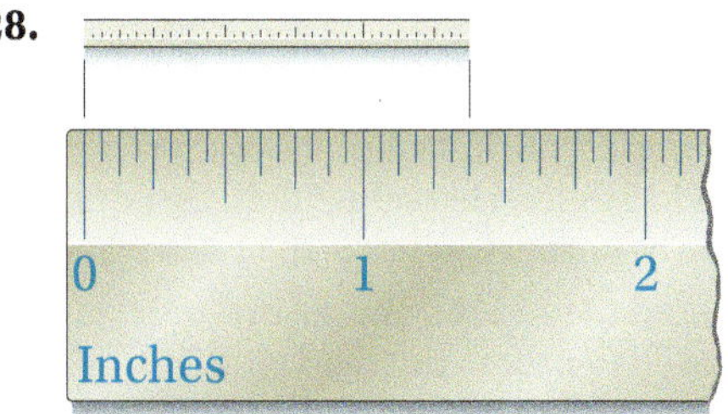

29. 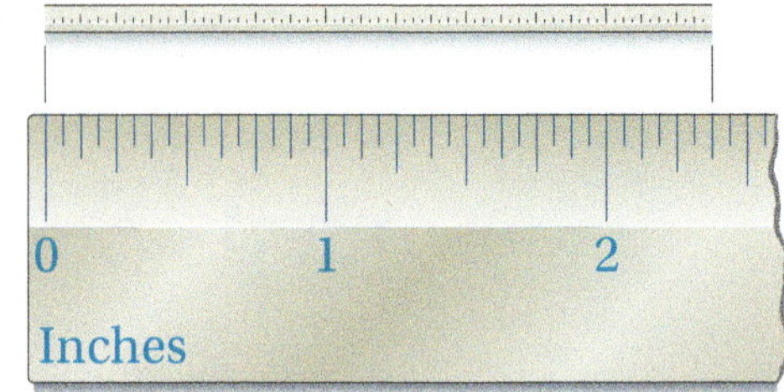

30. 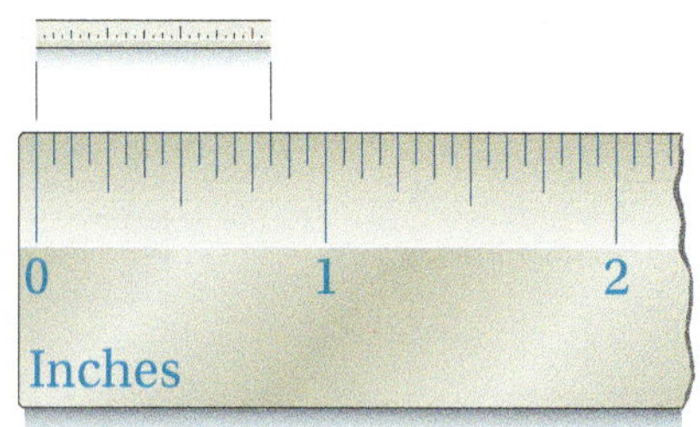

For each of Exercises 31–34, give fraction notation for the amount of gas **(a)** in the tank and **(b)** used from a full tank.

31. 

32. 

33. 

34. 

35. For the following set of animals, what is the ratio of:

a) puppies to the total number of animals?
b) puppies to kittens?
c) kittens to the total number of animals?
d) kittens to puppies?

36. For the following set of sports equipment, what is the ratio of:

a) basketballs to footballs?
b) footballs to basketballs?
c) basketballs to the total number of balls?
d) total number of balls to basketballs?

37. Bryce delivers car parts to auto service centers. On Thursday he had 15 deliveries scheduled. By noon he had delivered only 4 orders. What is the ratio of:

a) orders delivered to total number of orders?
b) orders delivered to orders not delivered?
c) orders not delivered to total number of orders?

38. *Gas Mileage.* A Volkswagen Passat TDI® SE will travel 473 mi on 11 gal of gasoline in highway driving. What is the ratio of:

a) miles driven to gasoline used?
b) gasoline used to miles driven?

Source: vw.com

For Exercises 39 and 40, use the following bar graph, which shows the number of registered nurses per 100,000 residents in each of twelve states or districts.

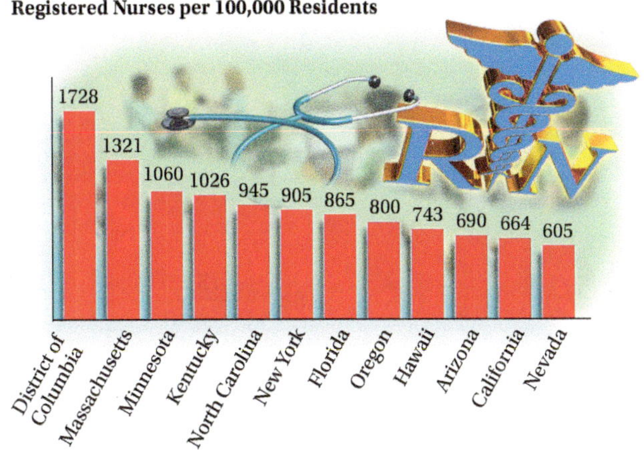

SOURCES: www.statehealthfacts.org; Kaiser Family Foundation, 2011

39. What is the ratio of registered nurses to 100,000 residents in the given state or district?

a) Minnesota
b) Hawaii
c) Florida
d) Kentucky
e) New York
f) District of Columbia

40. What is the ratio of registered nurses to 100,000 residents in the given state?

a) North Carolina
b) California
c) Massachusetts
d) Nevada
e) Arizona
f) Oregon

For Exercises 41 and 42, use the following set of states, as illustrated on the map.

Alabama Nebraska West Virginia
Arkansas South Dakota Wisconsin
Illinois

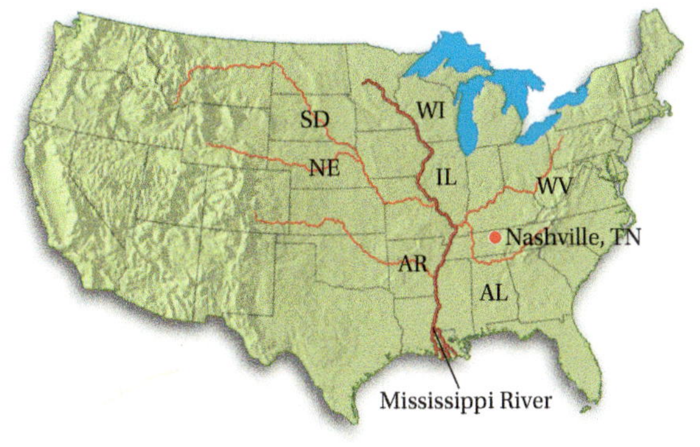

41. What part of this group of states is east of the Mississippi River?

42. What part of this group of states is north of Nashville, Tennessee?

b Simplify. Assume that all variables are nonzero.

43. $\frac{0}{8}$

44. $\frac{8}{8}$

45. $\frac{8-1}{9-8}$

46. $\frac{16}{1}$

47. $\frac{20}{20}$

48. $\frac{-20}{1}$

49. $\frac{-45}{-45}$

50. $\frac{11-1}{10-9}$

51. $\frac{0}{-238}$

52. $\frac{19x}{19x}$

53. $\frac{19x}{1}$

54. $\frac{0}{-16}$

55. $\frac{13t}{13t}$

56. $\frac{-27m}{1}$

57. $\frac{-87}{1}$

58. $\frac{-98}{-98}$

59. $\frac{0}{2a}$

60. $\frac{0}{18}$

61. $\frac{52}{0}$

62. $\frac{8-8}{1247}$

63. $\frac{7n}{1}$

64. $\frac{1317}{0}$

65. $\frac{5}{6-6}$

66. $\frac{13}{10-10}$

Skill Maintenance

Multiply.

67. $-7(30)$ [2.4a]

68. $23 \cdot (-14)$ [2.4a]

69. $(-71)(-12)0$ [2.4b]

70. $32(-29)0$ [2.4b]

Solve. [1.8a]

71. *Museums.* At the end of 2011, there were 3415 museums in China. By the end of 2012, the number of museums had increased to 3866. How many museums opened in China in 2012?
Source: China Museums Association

72. *Gas Mileage.* The 2014 Subaru XV Cross gets 33 mpg in highway driving. How many gallons will it use in 2343 mi of highway driving?
Source: www.kbb.com

Synthesis

What part of each object is shaded?

73.

74. 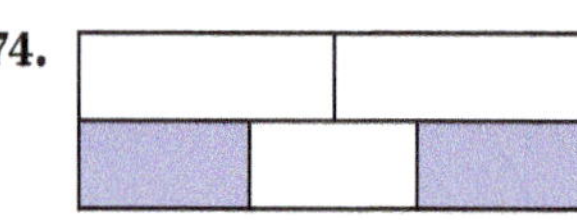

75. 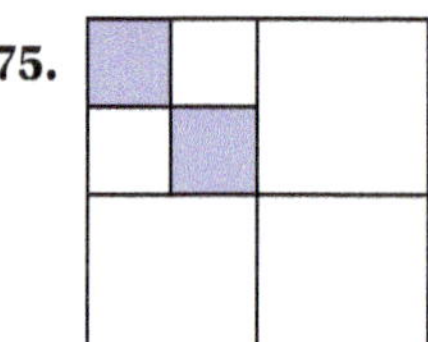

76. 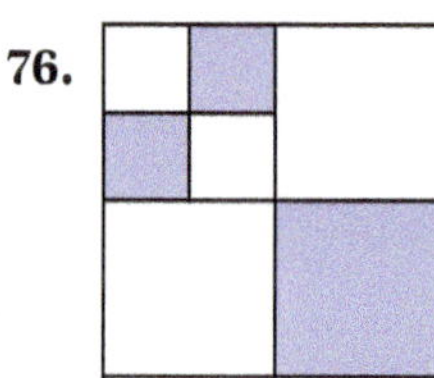

Shade or mark each figure to show $\frac{3}{5}$.

77. 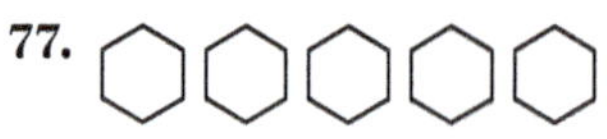

78. 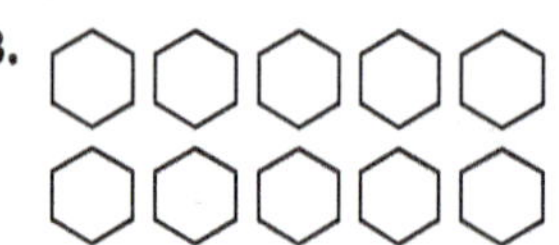

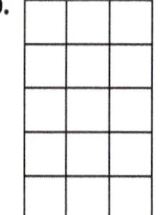

79. 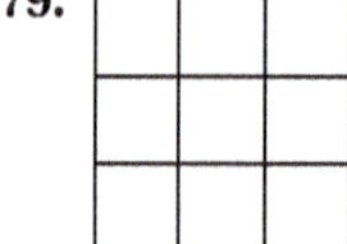

80. 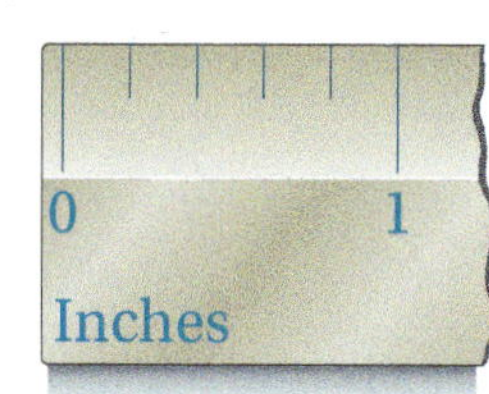

81. The year 2013 began on a Tuesday. What fractional part of the days in 2013 were Wednesdays?

82. The year 2014 began on a Wednesday. What fractional part of the days in 2014 were Wednesdays?

83. The surface of Earth is 3 parts water and 1 part land. What fractional part of Earth is water? land?

84. A couple had 3 sons, each of whom had 3 daughters. If each daughter gave birth to 3 sons, what fractional part of the couple's descendants is female?

Multiplication and Applications

3.4

a MULTIPLICATION BY AN INTEGER

We can find $3 \cdot \frac{1}{4}$ by thinking of repeated addition. We add three $\frac{1}{4}$'s.

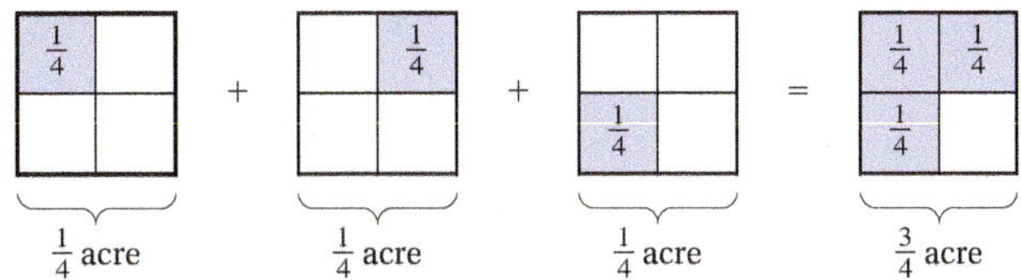

We see that $3 \cdot \frac{1}{4} = \frac{1}{4} + \frac{1}{4} + \frac{1}{4} = \frac{3}{4}$.

Do Margin Exercises 1 and 2.

To multiply a fraction by an integer,

a) multiply the top number (the numerator) by the integer and

b) keep the same denominator.

$$6 \cdot \frac{4}{5} = \frac{6 \cdot 4}{5} = \frac{24}{5}$$

EXAMPLES Multiply.

1. $5 \times \frac{3}{8} = \frac{5 \times 3}{8} = \frac{15}{8}$ We generally replace the $\times$ symbol with $\cdot$.

Skip this step when you feel comfortable doing so.

2. $\frac{2}{5} \cdot 13 = \frac{2 \cdot 13}{5} = \frac{26}{5}$

3. $-10 \cdot \frac{1}{3} = \frac{-10}{3}$, or $-\frac{10}{3}$ Recall that $\frac{-a}{b} = -\frac{a}{b}$.

4. $a \cdot \frac{4}{7} = \frac{4a}{7}$ Recall that $a \cdot 4 = 4 \cdot a$.

Do Exercises 3–6.

OBJECTIVES

a Multiply an integer and a fraction.

b Multiply using fraction notation.

c Solve problems involving multiplication of fractions.

SKILL TO REVIEW

Objective 1.4a: Multiply whole numbers.

Multiply.

1. $24 \cdot 17$ **2.** $5(13)$

1. Find $2 \cdot \frac{1}{3}$.

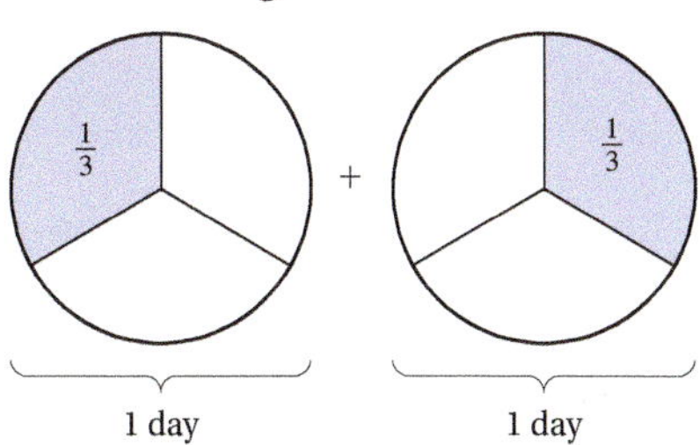

2. Find $5 \cdot \frac{1}{8}$.

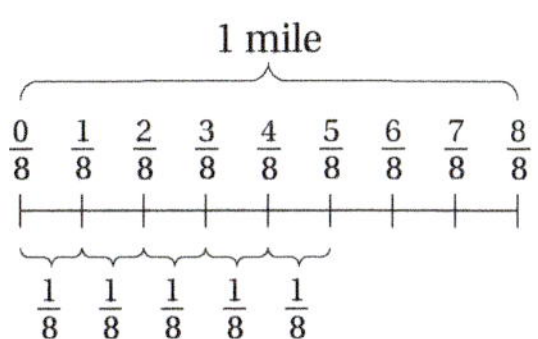

Multiply.

3. $7 \times \frac{2}{3}$

4. $(-11) \cdot \frac{3}{10}$

5. $34 \cdot \frac{2}{5}$

6. $x \cdot \frac{4}{9}$

Answers

Skill to Review:
1. 408 **2.** 65

Margin Exercises:
1. $\frac{2}{3}$ **2.** $\frac{5}{8}$ **3.** $\frac{14}{3}$ **4.** $-\frac{33}{10}$, or $\frac{-33}{10}$
5. $\frac{68}{5}$ **6.** $\frac{4x}{9}$

b MULTIPLICATION USING FRACTION NOTATION

7. Draw diagrams like the ones at right to illustrate $\frac{1}{4}$ and $\frac{1}{2} \cdot \frac{1}{4}$.

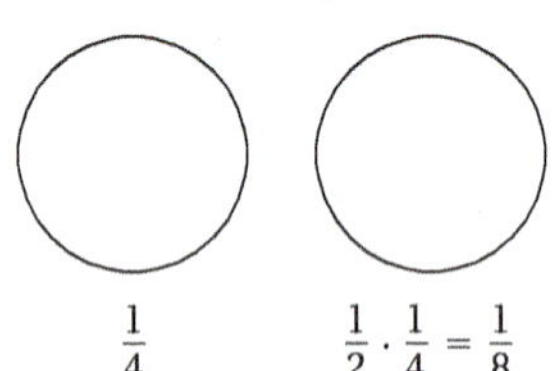

To illustrate the meaning of an expression like $\frac{1}{2} \cdot \frac{1}{3}$, we first represent $\frac{1}{3}$ and then shade half of that region. Note that $\frac{1}{2} \cdot \frac{1}{3}$ is the same as $\frac{1}{2}$ of $\frac{1}{3}$.

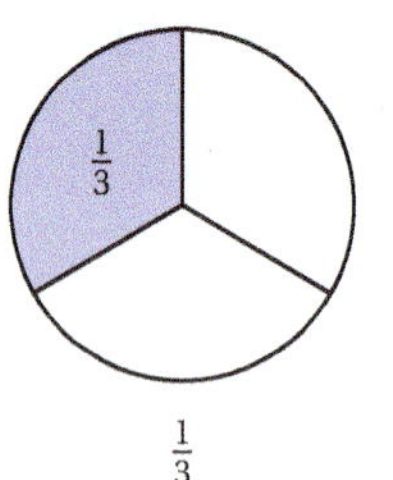

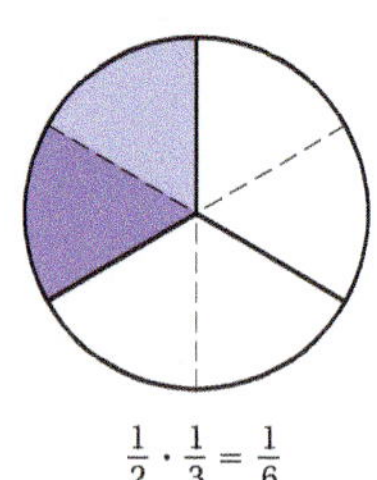

◀ **Do Exercise 7.**

To visualize $\frac{2}{5} \cdot \frac{3}{4}$, we first represent $\frac{3}{4}$. This is shown by the shading on the left below. Next, we divide the shaded area into 5 equal parts (using horizontal lines) and take 2 of them. That is shown by the darker shading on the right below. Note that $\frac{2}{5} \cdot \frac{3}{4}$ is the same as $\frac{2}{5}$ of $\frac{3}{4}$.

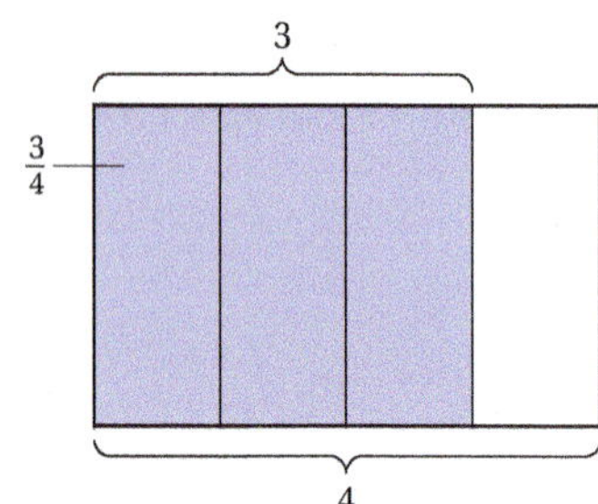

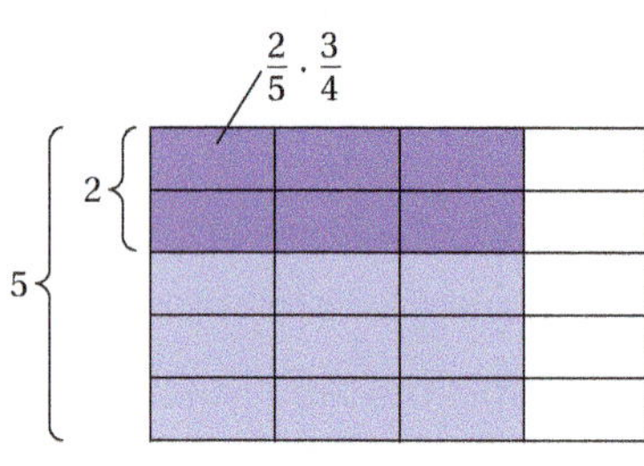

8. Draw diagrams like the ones at right to illustrate $\frac{1}{3}$ and $\frac{4}{5} \cdot \frac{1}{3}$.

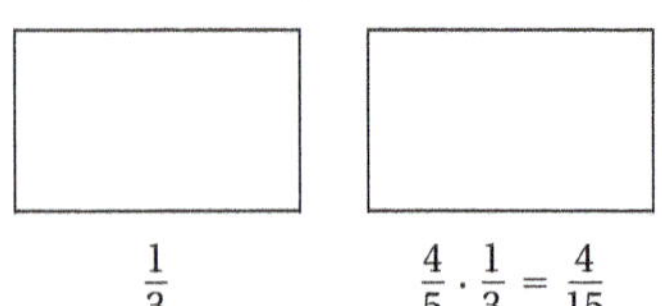

$\frac{1}{3}$ $\qquad \frac{4}{5} \cdot \frac{1}{3} = \frac{4}{15}$

The entire object has now been divided into 20 parts, and we have shaded 6 of them twice. Thus,

$$\frac{2}{5} \cdot \frac{3}{4} = \frac{6}{20}.$$

← This is the product of the numerators.
← This is the product of the denominators.

◀ **Do Exercise 8.**

Notice that the product of two fractions is the product of the numerators over the product of the denominators.

> To multiply a fraction by a fraction,
>
> **a)** multiply the numerators and
>
> $$\frac{9}{7} \cdot \frac{3}{4} = \frac{9 \cdot 3}{7 \cdot 4} = \frac{27}{28}$$
>
> **b)** multiply the denominators.

EXAMPLES Multiply.

5. $\frac{5}{6} \cdot \frac{7}{4} = \frac{5 \cdot 7}{6 \cdot 4} = \frac{35}{24}$

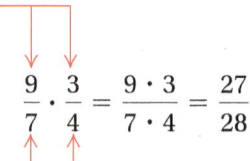

Skip this step when you feel comfortable doing so.

6. $\frac{3}{5} \cdot \frac{7}{8} = \frac{3 \cdot 7}{5 \cdot 8} = \frac{21}{40}$

Answers

7.

8.

7. $\left(-\frac{4}{x}\right)\left(-\frac{y}{7}\right) = \frac{4y}{7x}$ A negative times a negative is positive.

8. $(-6)\left(-\frac{4}{5}\right) = \frac{-6}{1} \cdot \frac{-4}{5} = \frac{24}{5}$ We can always write n as $\frac{n}{1}$.

Do Exercises 9–12.

Multiply.

9. $\frac{3}{8} \cdot \frac{5}{7} = \frac{3 \cdot 5}{8 \cdot \square}$

$= \frac{\square}{\square}$

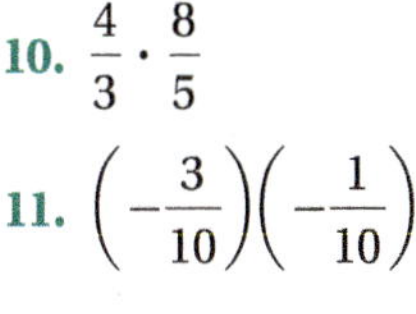

10. $\frac{4}{3} \cdot \frac{8}{5}$

11. $\left(-\frac{3}{10}\right)\left(-\frac{1}{10}\right)$

12. $(-7)\frac{a}{b}$

C APPLICATIONS AND PROBLEM SOLVING

Many problems that can be solved by multiplying fractions can be thought of in terms of rectangular arrays.

EXAMPLE 9 A real estate developer owns a plot of land and plans to use $\frac{4}{5}$ of the plot for a small strip mall and parking lot. Of this, $\frac{2}{3}$ will be needed for the parking lot. What part of the plot will be used for parking?

1. **Familiarize.** We first make a drawing to help familiarize ourselves with the problem. The land may not be rectangular, but we can think of it as a rectangle. The strip mall, including the parking lot, uses $\frac{4}{5}$ of the plot. We shade $\frac{4}{5}$ as shown on the left below. The parking lot alone uses $\frac{2}{3}$ of the part we just shaded. We shade that as shown on the right below.

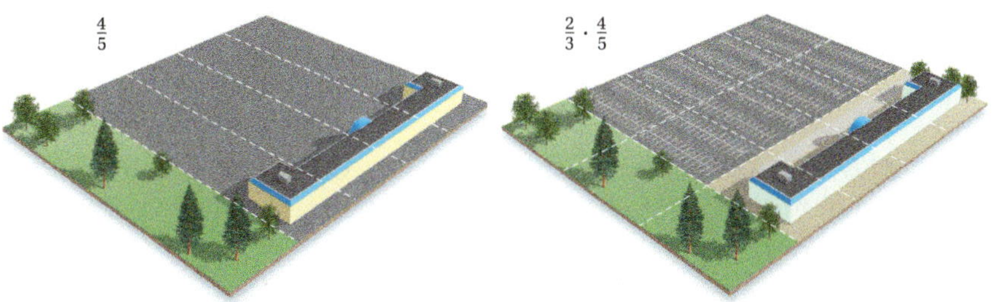

2. **Translate.** We let n = the part of the plot that is used for parking. We are taking "two-thirds of four-fifths." The word "of" corresponds to multiplication. Thus the following multiplication sentence corresponds to the situation:

$$\frac{2}{3} \cdot \frac{4}{5} = n.$$

3. **Solve.** The number sentence tells us what to do. We multiply:

$$\frac{2}{3} \cdot \frac{4}{5} = \frac{2 \cdot 4}{3 \cdot 5} = \frac{8}{15}.$$

Thus, $\frac{8}{15} = n$.

4. **Check.** We can do a partial check by noting that the answer is a fraction less than 1, which we expect since the developer is using only part of the original plot of land. Thus, $\frac{8}{15}$ is a reasonable answer. We can also check this in the figure above, where we see that 8 of 15 parts represent the parking lot.

5. **State.** The parking lot takes up $\frac{8}{15}$ of the plot of land.

Do Exercise 13.

13. A developer plans to set aside $\frac{3}{4}$ of the land in a housing development as open (undeveloped) space. Of this, $\frac{1}{2}$ will be green (natural) space. What part of the land will be green space?

Answers

9. $\frac{15}{56}$ 10. $\frac{32}{15}$ 11. $\frac{3}{100}$

12. $-\frac{7a}{b}$, or $\frac{-7a}{b}$ 13. $\frac{3}{8}$

Guided Solution:

9. 7, $\frac{15}{56}$

Caution!

When stating the solution of an applied problem, be sure to include appropriate units. Recall that units for area are square inches, square meters, and so on.

The area of a rectangular region is found by multiplying length by width. That is true whether length and width are whole numbers or not. Remember, the area of a rectangular region is given by the formula

$$A = l \cdot w \quad (\textit{Area} = \textit{length} \cdot \textit{width}).$$

EXAMPLE 10 *Area of a Cranberry Bog.* The length of a rectangular cranberry bog is $\frac{9}{16}$ mi. The width is $\frac{3}{8}$ mi. What is the area of the bog?

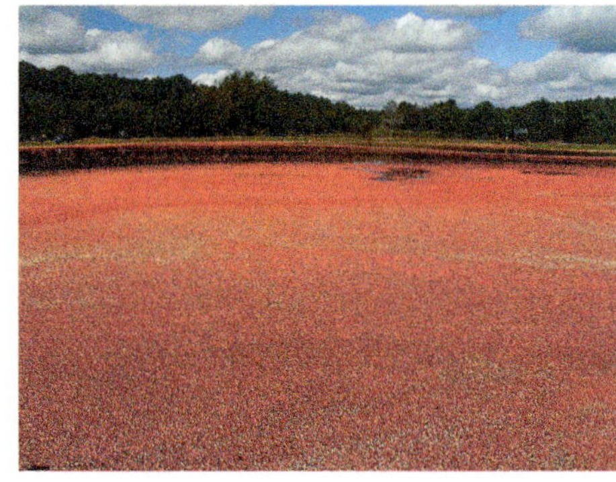

1. **Familiarize.** Recall that area is length times width. We let A = the area of the cranberry bog.
2. **Translate.** Next, we translate:

Area	is	Length	times	Width
↓	↓	↓	↓	↓
A	$=$	$\frac{9}{16}$	$\times$	$\frac{3}{8}$.

3. **Solve.** The sentence tells us what to do. We multiply:

$$A = \frac{9}{16} \cdot \frac{3}{8} = \frac{9 \cdot 3}{16 \cdot 8} = \frac{27}{128}.$$

4. **Check.** We check by repeating the calculation. This is left to the student.
5. **State.** The area is $\frac{27}{128}$ square mile (mi^2).

◀ Do Exercise 14.

14. *Area of a Ceramic Tile.* The length of a rectangular ceramic tile inlaid on a countertop is $\frac{4}{9}$ ft. The width is $\frac{2}{9}$ ft. What is the area of the tile?

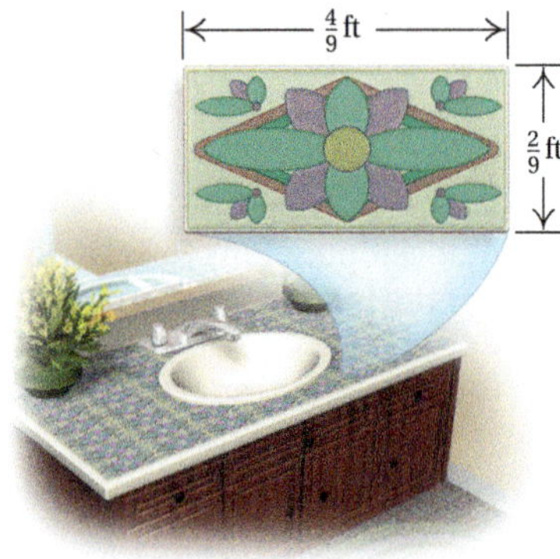

EXAMPLE 11 A recipe for oatmeal chocolate chip cookies calls for $\frac{3}{4}$ cup of rolled oats. Monica is making $\frac{1}{2}$ of the recipe. How much oats should she use?

1. **Familiarize.** We first make a drawing or at least visualize the situation. We let n = the amount of oats that Monica should use.

$\frac{3}{4}$ cup in the recipe $\qquad$ $\frac{1}{2} \cdot \frac{3}{4}$ cup in $\frac{1}{2}$ the recipe

2. **Translate.** We are finding $\frac{1}{2}$ of $\frac{3}{4}$, so the multiplication sentence $\frac{1}{2} \cdot \frac{3}{4} = n$ corresponds to the situation.
3. **Solve.** We carry out the multiplication:

$$\frac{1}{2} \cdot \frac{3}{4} = \frac{1 \cdot 3}{2 \cdot 4} = \frac{3}{8}.$$

Thus, $\frac{3}{8} = n$.

4. **Check.** We check by repeating the calculation. This is left to the student.
5. **State.** Monica should use $\frac{3}{8}$ cup of oats.

◀ Do Exercise 15.

15. Of the students at Overton Junior College, $\frac{1}{8}$ participate in sports and $\frac{3}{5}$ of these play football. What fractional part of the students play football?

Answers

14. $\frac{8}{81}\ \text{ft}^2$ **15.** $\frac{3}{40}$

3.4 Exercise Set

For Extra Help MyMathLab® MathXL® PRACTICE WATCH READ REVIEW

Reading Check

Determine whether each statement is true or false.

_______________ **RC1.** Multiplying $\frac{1}{2} \cdot \frac{3}{5}$ is the same as finding $\frac{1}{2}$ of $\frac{3}{5}$.

_______________ **RC2.** When we multiply two fractions, the new numerator is the product of the numerators in the two fractions.

_______________ **RC3.** The whole number 6 can be written $\frac{6}{1}$.

_______________ **RC4.** The product of two fractions can be smaller than either of the two fractions.

a Multiply.

1. $3 \cdot \frac{1}{8}$

2. $2 \cdot \frac{1}{5}$

3. $(-5) \times \frac{1}{6}$

4. $(-4) \times \frac{1}{7}$

5. $\frac{2}{3} \cdot 7$

6. $\frac{2}{5} \cdot 7$

7. $(-1)\frac{7}{9}$

8. $(-1)\frac{4}{11}$

9. $\frac{5}{6} \cdot x$

10. $\left(-\frac{7}{8}\right)y$

11. $\frac{2}{5}(-3)$

12. $\frac{3}{5}(-4)$

13. $a \cdot \frac{2}{7}$

14. $b \cdot \frac{3}{8}$

15. $-17 \times \frac{m}{6}$

16. $\frac{n}{7} \cdot 30$

17. $-3 \cdot \frac{-2}{5}$

18. $-4 \cdot \frac{-5}{7}$

19. $-\frac{2}{7}(-x)$

20. $-\frac{3}{4}(-a)$

b Multiply.

21. $\frac{2}{5} \cdot \frac{2}{3}$

22. $\frac{3}{4} \cdot \frac{3}{5}$

23. $\left(-\frac{1}{4}\right) \times \frac{1}{10}$

24. $\left(-\frac{1}{3}\right) \times \frac{1}{10}$

25. $\frac{2}{3} \times \frac{1}{5}$

26. $\frac{3}{5} \times \frac{1}{5}$

27. $\frac{2}{y} \cdot \frac{x}{9}$

28. $\left(-\frac{3}{4}\right)\left(-\frac{3}{5}\right)$

29. $\left(-\frac{3}{4}\right)\left(-\frac{3}{4}\right)$

30. $\frac{3}{b} \cdot \frac{a}{7}$

31. $\frac{2}{3} \cdot \frac{7}{13}$

32. $\frac{3}{11} \cdot \frac{4}{5}$

33. $\frac{1}{10}\left(\frac{-3}{5}\right)$

34. $\frac{3}{10}\left(\frac{-7}{5}\right)$

35. $\frac{7}{8} \cdot \frac{a}{8}$

36. $\frac{4}{5} \cdot \frac{7}{x}$

37. $\frac{1}{y} \cdot 100$

38. $\frac{b}{10} \cdot 13$

39. $\frac{-21}{4} \cdot \frac{7}{5}$

40. $\frac{-8}{3} \cdot \frac{20}{9}$

c Solve.

41. *Hair Bows.* It takes $\frac{5}{3}$ yd of ribbon to make a hair bow. How much ribbon is needed to make 8 bows?

42. *Gasoline Can Capacity.* A gasoline can holds $\frac{5}{2}$ gal. How much will the can hold when it is $\frac{1}{2}$ full?

43. ***Basketball: High School to Pro.*** One of 35 high school basketball players plays college basketball. One of 75 college players plays professional basketball. What fractional part of high school basketball players play professional basketball?

Source: National Basketball Association

44. ***Football: High School to Pro.*** One of 42 high school football players plays college football. One of 85 college players plays professional football. What fractional part of high school football players play professional football?

Source: National Football League

45. ***Serving of Cheesecake.*** At the Cheesecake Factory, a piece of cheesecake is $\frac{1}{12}$ of a cheesecake. How much of the cheesecake is $\frac{1}{2}$ piece?

Source: The Cheesecake Factory

46. ***Tossed Salad.*** The recipe for a tossed salad calls for $\frac{3}{4}$ cup of sliced almonds. How much is needed to make $\frac{1}{2}$ of the recipe?

47. ***Floor Tiling.*** The floor of a room is being covered with tile. An area $\frac{3}{5}$ of the length and $\frac{3}{4}$ of the width is covered. What fraction of the floor has been tiled?

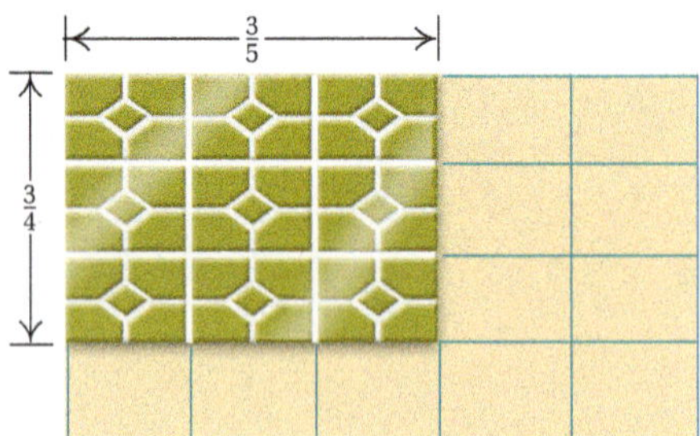

48. ***Table Top Size.*** A rectangular table top measures $\frac{4}{5}$ m long by $\frac{3}{5}$ m wide. What is its area?

Skill Maintenance

Divide. [1.5a], [2.5a]

49. $7140 \div 35$

50. $32{,}200 \div 46$

51. $-65 \div (-5)$

52. $540 \div (-6)$

Simplify. [1.9c]

53. $8 \cdot 12 - (63 \div 9 + 13 \cdot 3)$

54. $(10 - 3)^4 + 10^3 \cdot 4 - 10 \div 5$

Synthesis

Multiply. Write each answer using fraction notation.

55. $\frac{341}{517} \cdot \frac{209}{349}$

56. $\left(\frac{57}{61}\right)^3$

57. $\left(\frac{2}{5}\right)^3\left(-\frac{7}{9}\right)$

58. $\left(-\frac{1}{2}\right)^5\left(\frac{3}{5}\right)$

59. A chain saw holds $\frac{1}{5}$ gal of fuel. Chain saw fuel is $\frac{1}{16}$ two-cycle oil and $\frac{15}{16}$ unleaded gasoline. How much two-cycle oil is in a freshly filled chain saw?

60. Evaluate $-\frac{2}{3}xy$ for $x = \frac{2}{5}$ and $y = -\frac{1}{7}$.

Simplifying

3.5

a MULTIPLYING BY 1

Recall the following:

$$1 = \frac{1}{1} = \frac{2}{2} = \frac{3}{3} = \frac{4}{4} = \frac{-13}{-13} = \frac{45}{45} = \frac{100}{100} = \frac{n}{n}.$$

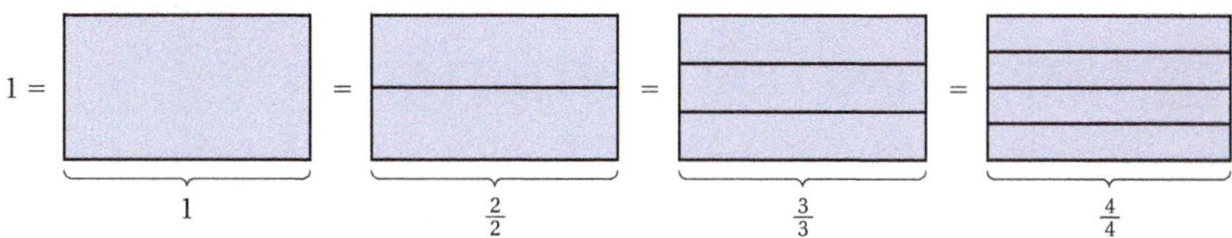

Recall that for any whole number a, we have $1 \cdot a = a \cdot 1 = a$. Since any nonzero number divided by itself is 1, we can state the multiplicative identity using fraction notation.

MULTIPLICATIVE IDENTITY FOR FRACTIONS

When we multiply a number by 1, we get the same number:

$$a = a \cdot 1 = a \cdot \frac{n}{n}.$$

For example, $\frac{3}{5} = \frac{3}{5} \cdot 1 = \frac{3}{5} \cdot \frac{4}{4} = \frac{12}{20}$.

Since $\frac{3}{5} = \frac{12}{20}$, we know that $\frac{3}{5}$ and $\frac{12}{20}$ are two names for the same number. We also say that $\frac{3}{5}$ and $\frac{12}{20}$ are **equivalent fractions**.

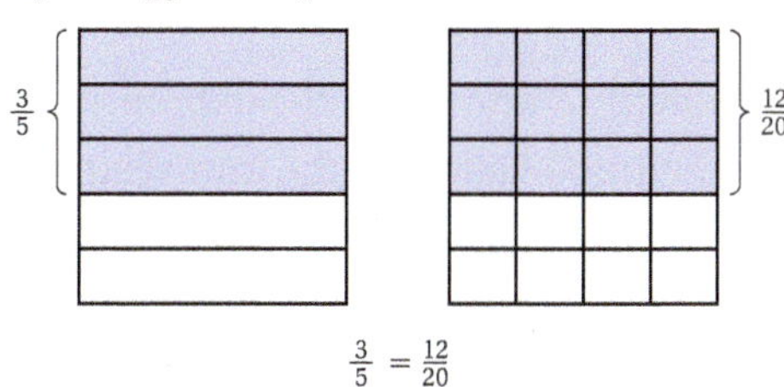

Do Margin Exercises 1–4.

Suppose we want to rename $\frac{2}{3}$, using a denominator of 15. We can multiply by 1 to find a number equivalent to $\frac{2}{3}$. We choose $\frac{5}{5}$ for 1 because $15 \div 3 = 5$.

$$\frac{2}{3} = \frac{2}{3} \cdot \frac{5}{5} = \frac{2 \cdot 5}{3 \cdot 5} = \frac{10}{15}.$$

EXAMPLE 1 Find a number equivalent to $\frac{1}{4}$ with a denominator of 24.

Since $24 \div 4 = 6$, we multiply by 1, using $\frac{6}{6}$:

$$\frac{1}{4} = \frac{1}{4} \cdot \frac{6}{6} = \frac{1 \cdot 6}{4 \cdot 6} = \frac{6}{24}.$$

OBJECTIVES

a Multiply by 1 to find an equivalent expression using a different denominator.

b Simplify fraction notation.

c Test to determine whether two fractions are equivalent.

SKILL TO REVIEW

Objective 3.2c: Find the prime factorization of a composite number.

Find the prime factorization.

1. 84 **2.** 2250

Multiply.

1. $\frac{1}{2} \cdot \frac{8}{8}$ **2.** $\frac{3}{7} \cdot \frac{a}{a}$

3. $-\frac{8}{25} \cdot \frac{4}{4}$ **4.** $\frac{8}{3}\left(\frac{-2}{-2}\right)$

Answers

Skill to Review:

1. $2 \cdot 2 \cdot 3 \cdot 7$ **2.** $2 \cdot 3 \cdot 3 \cdot 5 \cdot 5 \cdot 5$

Margin Exercises:

1. $\frac{8}{16}$ **2.** $\frac{3a}{7a}$ **3.** $-\frac{32}{100}$ **4.** $\frac{-16}{-6}$

Find an equivalent expression for each number, using the denominator indicated. Use multiplication by 1.

GS

5. $\frac{4}{3} = \frac{?}{15}$

$\frac{4}{3} = \frac{4}{3} \cdot \frac{\square}{\square}$

$= \frac{4 \cdot 5}{3 \cdot 5}$

$= \frac{\square}{15}$

6. $\frac{3}{4} = \frac{?}{-24}$

7. $\frac{9}{10} = \frac{?}{10x}$

8. $\frac{15}{1} = \frac{?}{2}$

9. $\frac{-8}{7} = \frac{?}{49}$

10. $\frac{1}{-2} = \frac{?}{-6}$

EXAMPLE 2 Find a number equivalent to $\frac{2}{5}$ with a denominator of -35.

Since $-35 \div 5 = -7$, we multiply by 1, using $\frac{-7}{-7}$:

$$\frac{2}{5} = \frac{2}{5}\left(\frac{-7}{-7}\right) = \frac{2(-7)}{5(-7)} = \frac{-14}{-35}.$$

EXAMPLE 3 Find an expression equivalent to $\frac{9}{8}$ with a denominator of $8a$.

Since $8a \div 8 = a$, we multiply by 1, using $\frac{a}{a}$:

$$\frac{9}{8} \cdot \frac{a}{a} = \frac{9a}{8a}.$$

◀ Do Exercises 5–10.

b SIMPLIFYING FRACTION NOTATION

All of the following are names for three-fourths:

$$\frac{3}{4}, \quad \frac{-3}{-4}, \quad \frac{6}{8}, \quad \frac{30}{40}, \quad \frac{-15}{-20}.$$

We say that $\frac{3}{4}$ is **simplest** because it has the smallest positive denominator. Note that 3 and 4 have no factor in common other than 1.

To simplify fraction notation, we reverse the process of multiplying by 1:

$$\frac{12}{18} = \frac{2 \cdot 6}{3 \cdot 6} \qquad \begin{array}{l}\leftarrow \text{Factoring the numerator} \\ \leftarrow \text{Factoring the denominator}\end{array}$$

$$= \frac{2}{3} \cdot \frac{6}{6} \qquad \text{Factoring the fraction}$$

$$= \frac{2}{3} \cdot 1 \qquad \frac{6}{6} = 1$$

$$= \frac{2}{3} \qquad \text{Removing the factor 1: } \frac{2}{3} \cdot 1 = \frac{2}{3}$$

SIMPLIFYING FRACTION NOTATION

1. Factor the numerator and factor the denominator.
2. Identify any common factors in the numerator and denominator.
3. Use the common factors to remove a factor equal to 1.

EXAMPLES Simplify.

4. $\frac{-8}{20} = \frac{-2 \cdot 4}{5 \cdot 4} = \frac{-2}{5} \cdot \frac{4}{4} = \frac{-2}{5}$ Removing a factor equal to 1: $\frac{4}{4} = 1$

5. $\frac{2}{6} = \frac{1 \cdot 2}{3 \cdot 2} = \frac{1}{3} \cdot \frac{2}{2} = \frac{1}{3}$

Writing 1 allows for pairing of factors in the numerator and the denominator.

6. $\frac{30}{6} = \frac{5 \cdot 6}{1 \cdot 6} = \frac{5}{1} \cdot \frac{6}{6} = \frac{5}{1} = 5$

We could also simplify $\frac{30}{6}$ by doing the division $30 \div 6$. That is, $\frac{30}{6} = 30 \div 6 = 5$.

7. $-\frac{15}{10} = -\frac{3 \cdot 5}{2 \cdot 5} = -\frac{3}{2} \cdot \frac{5}{5} = -\frac{3}{2}$ Removing a factor equal to 1: $\frac{5}{5} = 1$

Answers

5. $\frac{20}{15}$ 6. $\frac{-18}{-24}$ 7. $\frac{9x}{10x}$
8. $\frac{30}{2}$ 9. $\frac{-56}{49}$ 10. $\frac{3}{-6}$

Guided Solution:

5. $\frac{5}{5}$, 20

8. $\dfrac{4x}{15x} = \dfrac{4 \cdot x}{15 \cdot x} = \dfrac{4}{15} \cdot \dfrac{x}{x} = \dfrac{4}{15}$ Removing a factor equal to 1: $\dfrac{x}{x} = 1$

Do Exercises 11–16.

The use of prime factorizations can be helpful for simplifying.

EXAMPLE 9 Simplify: $\dfrac{90}{84}$.

$\dfrac{90}{84} = \dfrac{2 \cdot 3 \cdot 3 \cdot 5}{2 \cdot 2 \cdot 3 \cdot 7}$ Factoring the numerator and the denominator into primes

$= \dfrac{2 \cdot 3 \cdot 3 \cdot 5}{2 \cdot 3 \cdot 2 \cdot 7}$ Changing the order so that like primes are above and below each other

$= \dfrac{2}{2} \cdot \dfrac{3}{3} \cdot \dfrac{3 \cdot 5}{2 \cdot 7}$ Factoring the fraction

$= \dfrac{3 \cdot 5}{2 \cdot 7}$ Removing factors of 1

$= \dfrac{15}{14}$

EXAMPLE 10 Simplify: $\dfrac{105}{135}$.

Since both 105 and 135 end in 5, we know that 5 is a factor of both the numerator and the denominator:

$\dfrac{105}{135} = \dfrac{21 \cdot 5}{27 \cdot 5} = \dfrac{21}{27} \cdot \dfrac{5}{5} = \dfrac{21}{27}$. To find the 21, we divided 105 by 5. To find the 27, we divided 135 by 5.

A fraction is not "simplified" if common factors of the numerator and the denominator remain. Because 21 and 27 are both divisible by 3, we must simplify further:

$\dfrac{105}{135} = \dfrac{21}{27} = \dfrac{7 \cdot 3}{9 \cdot 3} = \dfrac{7}{9} \cdot \dfrac{3}{3} = \dfrac{7}{9}$. To find the 7, we divided 21 by 3. To find the 9, we divided 27 by 3.

EXAMPLE 11 Simplify: $\dfrac{322}{434}$.

Since 322 and 434 are both even, we know that 2 is a common factor:

$\dfrac{322}{434} = \dfrac{2 \cdot 161}{2 \cdot 217} = \dfrac{2}{2} \cdot \dfrac{161}{217} = \dfrac{161}{217}$ Removing a factor equal to 1: $\dfrac{2}{2} = 1$

$= \dfrac{7 \cdot 23}{7 \cdot 31} = \dfrac{7}{7} \cdot \dfrac{23}{31} = \dfrac{23}{31}$. 7 is also a common factor; removing a factor equal to 1

We found the common factor 7 by focusing first on 161. After determining that 7 is a factor of 161, we checked to see if 7 is also a factor of 217.

Do Exercises 17–23.

Simplify.

11. $\dfrac{8}{14}$ **12.** $\dfrac{-10}{12}$

13. $\dfrac{40}{8}$ **14.** $\dfrac{4a}{3a}$

15. $-\dfrac{50}{30}$ **16.** $\dfrac{x}{3x}$

Simplify.

17. $\dfrac{-35}{40}$ **18.** $\dfrac{801}{702}$

19. $\dfrac{24}{21}$ **20.** $\dfrac{429}{561}$

21. $\dfrac{280}{960}$ **22.** $\dfrac{1332}{2880}$

23. Simplify each fraction in this circle graph.

What College Graduates Will Do If They Are Not Offered a Full-Time Job in Their First-Choice Field

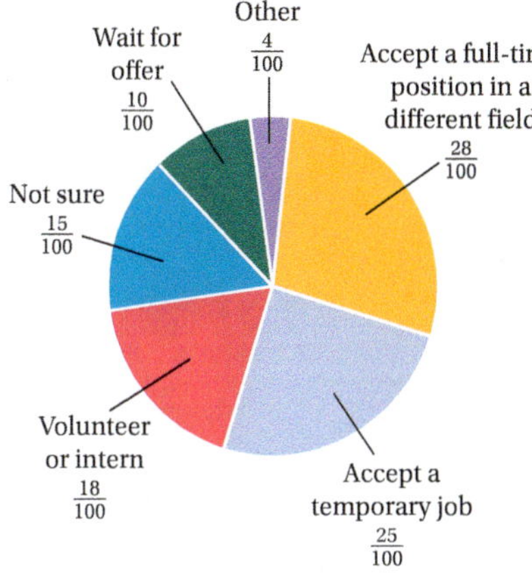

SOURCE: Yahoo Hotjobs survey

Answers

11. $\frac{4}{7}$ 12. $\frac{-5}{6}$ 13. 5 14. $\frac{4}{3}$ 15. $-\frac{5}{3}$
16. $\frac{1}{3}$ 17. $\frac{-7}{8}$ 18. $\frac{89}{78}$ 19. $\frac{8}{7}$ 20. $\frac{13}{17}$
21. $\frac{7}{24}$ 22. $\frac{37}{80}$ 23. $\frac{28}{100} = \frac{7}{25}$; $\frac{25}{100} = \frac{1}{4}$; $\frac{18}{100} = \frac{9}{50}$; $\frac{15}{100} = \frac{3}{20}$; $\frac{10}{100} = \frac{1}{10}$; $\frac{4}{100} = \frac{1}{25}$

CALCULATOR CORNER

Simplifying Fraction Notation Fraction calculators are equipped with a key, often labeled $\boxed{a\,b/c}$, that allows numbers to be entered using fraction notation. To simplify $\frac{208}{256}$ with such a fraction calculator, press

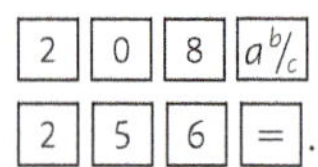

2 0 8 $a\,b/c$ 2 5 6 = .

The display, 13 ⌟ 16 , or 13/16 , represents the simplified fraction notation $\frac{13}{16}$.

On many graphing calculators, the Frac option accessed by pressing MATH will convert a number to simplified fraction notation. To simplify $\frac{208}{256}$ with such a calculator, press

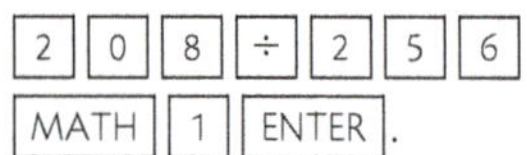

2 0 8 ÷ 2 5 6 MATH 1 ENTER .

Simplified fraction notation is $\frac{13}{16}$.

Caution!

It is still important for you to understand how to work with fractions by hand. For this reason, your instructor may not allow the use of calculators.

EXERCISES Use a fraction calculator to simplify each of the following.

1. $\dfrac{84}{90}$ **2.** $\dfrac{35}{40}$

3. $\dfrac{690}{835}$ **4.** $\dfrac{42}{150}$

Canceling

Canceling is a shortcut that you may have used for removing a factor that equals 1 when working with fraction notation. With concern, we mention it as a possibility for speeding up your work. Canceling may be done only when removing common factors in numerators and denominators. Each common factor allows us to remove a factor equal to 1 in a product.

Generally, slashes are used to indicate factors equal to 1 that have been removed. For instance, Example 11 might have been done faster as follows:

$$\frac{322}{434} = \frac{2 \cdot 161}{2 \cdot 217} \quad \text{Factoring the numerator and the denominator}$$

$$= \frac{\not{2} \cdot 161}{\not{2} \cdot 217} \quad \text{When a factor equal to 1 is noted, it is "canceled" as shown: } \frac{2}{2} = 1.$$

$$= \frac{161}{217} = \frac{\not{7} \cdot 23}{\not{7} \cdot 31} = \frac{23}{31}.$$

Caution!

The problem with canceling is that it is often applied incorrectly in situations like the following:

$$\frac{\not{2} + 3}{\not{2}} = 3; \qquad \frac{\not{4} + 1}{\not{4} + 2} = \frac{1}{2}; \qquad \frac{1\not{5}}{\not{5}4} = \frac{1}{4}.$$

Wrong! Wrong! Wrong!

The correct answers are

$$\frac{2+3}{2} = \frac{5}{2}; \qquad \frac{4+1}{4+2} = \frac{5}{6}; \qquad \frac{15}{54} = \frac{\not{3} \cdot 5}{\not{3} \cdot 18} = \frac{5}{18}.$$

In each incorrect case, the numbers canceled did not form a factor equal to 1. Factors are parts of products. For example, in $2 \cdot 3$, the numbers 2 and 3 are factors, but in $2 + 3$, 2 and 3 are terms, not factors.

- **If you cannot factor, do not cancel! If in doubt, do not cancel!**
- **Only factors can be canceled, and factors are never separated by + or − signs.**

C A TEST FOR EQUALITY

When denominators are the same, we say that fractions have a **common denominator**. One way to compare fractions like $\frac{2}{4}$ and $\frac{3}{6}$ is to find a common denominator and compare numerators. We can multiply each fraction by 1, using the other denominator to write 1.

The denominator is 6.

$$\frac{3}{6} = \frac{3}{6} \cdot \frac{4}{4} = \frac{3 \cdot 4}{6 \cdot 4} = \frac{12}{24}$$

$$\frac{2}{4} = \frac{2}{4} \cdot \frac{6}{6} = \frac{2 \cdot 6}{4 \cdot 6} = \frac{12}{24}$$

Both denominators are 24.

The denominator is 4.

Because $\dfrac{12}{24} = \dfrac{12}{24}$ is true, it follows that $\dfrac{3}{6} = \dfrac{2}{4}$ is also true.

The "key" to the above work is that $3 \cdot 4$ and $2 \cdot 6$ are equal. Had these products differed, we would have shown that $\frac{3}{6}$ and $\frac{2}{4}$ were *not* equal.

A TEST FOR EQUALITY

Two fractions are equal if their cross products are equal.

We multiply these two numbers: $3 \cdot 4 = 12$.

We multiply these two numbers: $6 \cdot 2 = 12$.

$$\frac{3}{6} \square \frac{2}{4}$$

We call $3 \cdot 4$ and $6 \cdot 2$ **cross products**. Since the cross products are the same—that is, $3 \cdot 4 = 6 \cdot 2$—we know that $\frac{3}{6} = \frac{2}{4}$.

In the sentence $a \neq b$, the symbol $\neq$ means "is not equal to."

EXAMPLE 12 Use $=$ or $\neq$ for $\square$ to write a true sentence:

$$\frac{6}{7} \square \frac{7}{8}.$$

We multiply these two numbers: $6 \cdot 8 = 48$.

We multiply these two numbers: $7 \cdot 7 = 49$.

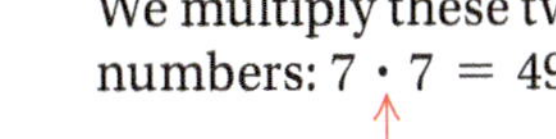

Because $48 \neq 49$, $\frac{6}{7}$ and $\frac{7}{8}$ do not name the same number. Thus, $\frac{6}{7} \neq \frac{7}{8}$. ■

EXAMPLE 13 Use $=$ or $\neq$ for $\square$ to write a true sentence:

$$\frac{6}{10} \square \frac{3}{5}.$$

We multiply these two numbers: $6 \cdot 5 = 30$.

We multiply these two numbers: $10 \cdot 3 = 30$.

$$\frac{6}{10} \square \frac{3}{5}$$

Because the cross products are the same, we have $\frac{6}{10} = \frac{3}{5}$. ■

Remembering that $\frac{-a}{b}$, $\frac{a}{-b}$, and $-\frac{a}{b}$ all represent the same number can be helpful when checking for equality.

EXAMPLE 14 Use $=$ or $\neq$ for $\square$ to write a true sentence:

$$\frac{-6}{8} \square -\frac{9}{12}.$$

We rewrite $-\frac{9}{12}$ as $\frac{-9}{12}$ and then check cross products:

$-6 \cdot 12 = -72$

$8(-9) = -72$

$$\frac{-6}{8} \square \frac{-9}{12}$$

Because the cross products are the same, we have $\frac{-6}{8} = -\frac{9}{12}$.

Do Exercises 24–26. ▶

Use $=$ or $\neq$ for $\square$ to write a true sentence.

24. $\frac{2}{6} \square \frac{3}{9}$

GS **25.** $\frac{2}{3} \square \frac{14}{20}$

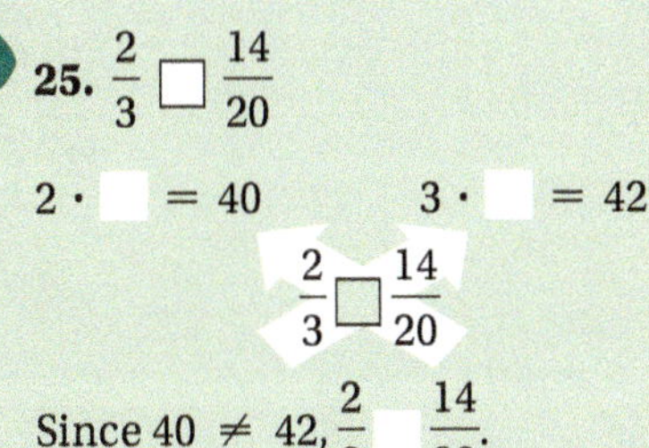

$2 \cdot \square = 40$
$3 \cdot \square = 42$
$\frac{2}{3} \square \frac{14}{20}$
Since $40 \neq 42$, $\frac{2}{3} \square \frac{14}{20}$.

26. $-\frac{10}{15} \square \frac{8}{-12}$

Answers

24. $=$ **25.** $\neq$ **26.** $=$

Guided Solution:

25. 20, 14; $\neq$

For Extra Help

MyMathLab®

MathXL® PRACTICE  WATCH  READ  REVIEW 

Reading Check

Complete each statement with the appropriate the word from the following list.

common   cross   equivalent   simplify

RC1. ______________ fractions name the same number.

RC2. To ______________ a fraction, we find a fraction that names the same number and that has a numerator and a denominator with no common factor.

RC3. The fractions $\frac{2}{7}$ and $\frac{4}{7}$ have a ______________ denominator.

RC4. Two fractions are equal if their ______________ products are equal.

a Find an equivalent expression for each number, using the denominator indicated. Use multiplication by 1.

1. $\frac{1}{2} = \frac{?}{10}$

2. $\frac{1}{6} = \frac{?}{12}$

3. $\frac{3}{4} = \frac{?}{-48}$

4. $\frac{2}{9} = \frac{?}{-18}$

5. $\frac{7}{10} = \frac{?}{50}$

6. $\frac{3}{8} = \frac{?}{48}$

7. $\frac{11}{5} = \frac{?}{5t}$

8. $\frac{5}{3} = \frac{?}{3a}$

9. $\frac{5}{1} = \frac{?}{4}$

10. $\frac{7}{1} = \frac{?}{5}$

11. $-\frac{17}{18} = -\frac{?}{54}$

12. $-\frac{11}{16} = -\frac{?}{256}$

13. $\frac{3}{-8} = \frac{?}{-40}$

14. $\frac{7}{-8} = \frac{?}{-32}$

15. $\frac{-7}{22} = \frac{?}{132}$

16. $\frac{-10}{21} = \frac{?}{126}$

17. $\frac{1}{8} = \frac{?}{8x}$

18. $\frac{1}{3} = \frac{?}{3a}$

19. $\frac{-10}{7} = \frac{?}{7a}$

20. $\frac{-4}{3} = \frac{?}{3n}$

21. $\frac{4}{9} = \frac{?}{9ab}$

22. $\frac{8}{11} = \frac{?}{11xy}$

23. $\frac{4}{9} = \frac{?}{27b}$

24. $\frac{8}{11} = \frac{?}{55y}$

b Simplify.

25. $\frac{2}{4}$

26. $\frac{3}{6}$

27. $-\frac{6}{9}$

28. $\frac{-9}{12}$

29. $\frac{10}{25}$

30. $\frac{8}{10}$

31. $\frac{24}{8}$

32. $\frac{36}{9}$

33. $\frac{27}{36}$

34. $\frac{30}{40}$

35. $-\frac{24}{14}$

36. $-\frac{16}{10}$

37. $\frac{3n}{4n}$

38. $\frac{7x}{8x}$

39. $\frac{-17}{51}$

40. $-\frac{13}{26}$

41. $\frac{-100}{20}$

42. $\frac{-150}{25}$

43. $\frac{420}{480}$ 44. $\frac{180}{240}$ 45. $\frac{-540}{810}$ 46. $\frac{-1000}{1080}$ 47. $\frac{12x}{30x}$ 48. $\frac{54n}{90n}$

49. $\frac{153}{136}$ 50. $\frac{117}{91}$ 51. $\frac{132}{143}$ 52. $\frac{91}{259}$ 53. $\frac{221}{247}$ 54. $\frac{299}{403}$

55. $\frac{3ab}{8ab}$ 56. $\frac{6xy}{7xy}$ 57. $\frac{9xy}{6x}$ 58. $\frac{10ab}{15a}$ 59. $\frac{-18a}{20ab}$ 60. $\frac{-19x}{38xy}$

c Use = or ≠ for □ to write a true sentence.

61. $\frac{3}{4}$ □ $\frac{9}{12}$ 62. $\frac{4}{8}$ □ $\frac{3}{6}$ 63. $\frac{1}{5}$ □ $\frac{2}{9}$ 64. $\frac{1}{4}$ □ $\frac{2}{9}$

65. $\frac{3}{8}$ □ $\frac{6}{16}$ 66. $\frac{2}{6}$ □ $\frac{6}{18}$ 67. $\frac{2}{5}$ □ $\frac{2}{7}$ 68. $\frac{3}{10}$ □ $\frac{3}{11}$

69. $\frac{-3}{10}$ □ $\frac{-4}{12}$ 70. $\frac{-2}{9}$ □ $\frac{-8}{36}$ 71. $-\frac{12}{9}$ □ $\frac{-8}{6}$ 72. $\frac{-8}{7}$ □ $-\frac{16}{14}$

73. $\frac{5}{-2}$ □ $-\frac{17}{7}$ 74. $-\frac{10}{3}$ □ $\frac{24}{-7}$ 75. $\frac{305}{145}$ □ $\frac{122}{58}$ 76. $\frac{425}{165}$ □ $\frac{130}{66}$

Skill Maintenance

Use < or > for □ to write a true sentence. [1.6c]

77. 0 □ 23 78. 34 □ 43 79. 124 □ 98 80. 999 □ 1001

Solve. [1.7b]

81. $5280 = 1760 + t$ 82. $10{,}947 = 123 \cdot y$ 83. $8797 = y + 2299$ 84. $x \cdot 74 = 6290$

Synthesis

Simplify. Use a list of prime numbers (see p. 158).

85. $\frac{391}{667}$ 86. $\frac{209ab}{247ac}$ 87. $-\frac{1073x}{555y}$ 88. 🖩 $\frac{3473}{3197}$

89. Sociologists have found that 4 of 10 people are shy. Write fraction notation for **(a)** the part of the population that is shy and **(b)** the part that is not shy. Simplify.

90. Sociologists estimate that 3 of 20 people are left-handed. In a crowd of 460 people, how many would you expect to be left-handed?

91. 🖩 *Batting Averages.* For the 2013 season, Michael Cuddyer of the Colorado Rockies won the National League batting title with 162 hits in 489 times at bat. Miguel Cabrera of the Detroit Tigers won the American League title with 193 hits in 555 times at bat. Did they have the same fraction of hits in times at bat (batting average)? Why or why not?

Source: Major League Baseball

92. 🖩 On a test with 82 questions, Taylor got 63 correct. On another test with 100 questions, she got 77 correct. Did she get the same portion of each test correct? Why or why not?

Mid-Chapter Review

Concept Reinforcement

Determine whether each statement is true or false.

______ **1.** A number a is divisible by another number b if b is factor of a. [3.2a]

______ **2.** If a number is not divisible by 6, then it is not divisible by 3. [3.1b]

______ **3.** The fraction $\frac{9}{4}$ is equal to the fraction $\frac{13}{6}$. [3.5c]

______ **4.** The number 1 is not prime. [3.2b]

Guided Solutions

Fill in each blank with the number that creates a correct statement or solution.

5. $\frac{25}{\square} = 1$ [3.3b] **6.** $\frac{\square}{9} = 0$ [3.3b] **7.** $\frac{8}{\square} = 8$ [3.3b] **8.** $\frac{6}{13} = \frac{\square}{39}$ [3.5a]

9. Simplify: $\frac{70}{225}$. [3.5b]

$$\frac{70}{225} = \frac{2 \cdot \square \cdot 7}{\square \cdot 3 \cdot 5 \cdot \square}$$ Factoring the numerator / Factoring the denominator

$$= \frac{5}{5} \cdot \frac{\square \cdot 7}{3 \cdot \square \cdot 5}$$ Factoring the fraction

$$= \square \cdot \frac{\square}{45}$$ $\frac{5}{5} = 1$

$$= \frac{\square}{\square}$$ Removing the factor 1

Mixed Review

To answer Exercises 10–14, consider the following numbers. [3.1b]

84	132	594	350
300	500	120	14,850
17,576	180	1125	504
224	351	495	1632

10. Which of the above are divisible by 2 but not by 10?

11. Which of the above are divisible by 4 but not by 8?

12. Which of the above are divisible by 4 but not by 6?

13. Which of the above are divisible by 3 but not by 9?

14. Which of the above are divisible by 4, 5, and 6?

Determine whether each number is prime, composite, or neither. [3.2b]

15. 61 **16.** 2 **17.** 91 **18.** 1

Find all the factors of each composite number. Then find the prime factorization of the number. [3.2a], [3.2c]

19. 160 **20.** 222 **21.** 98 **22.** 315

What part of each object or set of objects is shaded? [3.3a]

23.

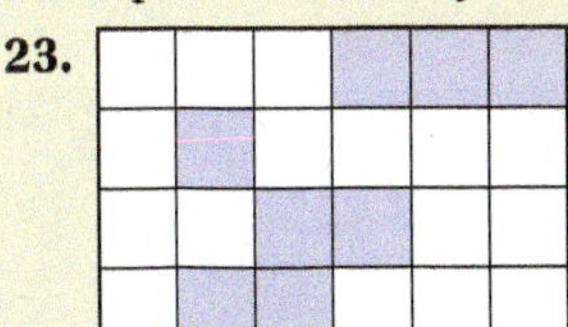

24.

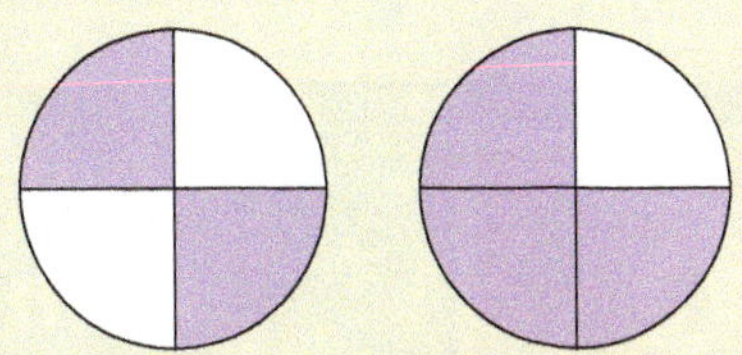

Multiply. [3.4a], [3.4b]

25. $7 \cdot \frac{1}{9}$

26. $\frac{4}{15} \cdot \frac{2}{3}$

27. $\frac{5}{11}(-8)$

Simplify. [3.3b], [3.5b]

28. $\frac{24}{60}$

29. $\frac{220n}{60n}$

30. $\frac{17x}{17x}$

31. $\frac{0}{-23}$

32. $\frac{54}{186}$

33. $\frac{-36}{20}$

34. $\frac{75}{630}$

35. $\frac{315}{435}$

36. $\frac{14}{0}$

Use = or ≠ for □ to write a true sentence. [3.5c]

37. $\frac{3}{-7} \;\square\; \frac{-39}{91}$

38. $\frac{19}{3} \;\square\; \frac{95}{18}$

39. ***Job Applications.*** Of every 200 online job applications started, only 25 are reviewed by a hiring manager. What is the ratio of applications reviewed to applications started? [3.3a], [3.5b]

Source: Talent Function Group LLC, in "Your Résumé vs. Oblivion," wsj.com, 1/24/12

40. ***Area of an Ice-Skating Rink.*** The length of a rectangular ice-skating rink in the atrium of a shopping mall is $\frac{7}{100}$ mi. The width is $\frac{3}{100}$ mi. What is the area of the rink? [3.4c]

Understanding Through Discussion and Writing

41. Explain a method for finding a composite number that contains exactly two factors other than itself and 1. [3.2b]

42. Which of the years from 2000 to 2020, if any, also happen to be prime numbers? Explain at least two ways in which you might go about solving this problem. [3.2b]

43. Explain in your own words when it *is* possible to "cancel" and when it *is not* possible to "cancel." [3.5b]

44. Can fraction notation be simplified if the numerator and the denominator are two different prime numbers? Why or why not? [3.5b]

STUDYING FOR SUCCESS *A Valuable Resource—Your Instructor*

- ☐ Don't be afraid to ask questions in class. Other students probably have the same questions you do.
- ☐ Visit your instructor during office hours if you need additional help.
- ☐ Many instructors welcome e-mails from students who have questions.

3.6 Multiplying, Simplifying, and More with Area

OBJECTIVES

a Multiply and simplify using fraction notation.

b Solve applied problems involving multiplication of fractions.

SKILL TO REVIEW

Objective 3.1b: Test to see if a number is divisible by 2, 3, 5, 6, 9, or 10.

Determine whether each number is divisible by 9.

1. 486 **2.** 129

a SIMPLIFYING WHEN MULTIPLYING

It is often possible to simplify after we multiply. To make such simplifying easier, it is usually best not to carry out the products in the numerator and the denominator immediately, but to factor and simplify first. Consider the product

$$\frac{5}{6} \cdot \frac{14}{15}.$$

We proceed as follows:

$$\frac{5}{6} \cdot \frac{14}{15} = \frac{5 \cdot 14}{6 \cdot 15}$$ We write the products in the numerator and the denominator, but we do not yet carry out the multiplication.

$$= \frac{5 \cdot 2 \cdot 7}{2 \cdot 3 \cdot 5 \cdot 3}$$ Factoring and identifying common factors

$$= \frac{5 \cdot 2}{5 \cdot 2} \cdot \frac{7}{3 \cdot 3}$$ Factoring the fraction

$$= 1 \cdot \frac{7}{3 \cdot 3}$$

$$= \frac{7}{3 \cdot 3}$$ Removing a factor equal to 1: $\frac{5 \cdot 2}{5 \cdot 2} = 1$

$$= \frac{7}{9}.$$

To multiply and simplify:

a) Write the products in the numerator and the denominator, but do not carry out the multiplication.

b) Factor the numerator and the denominator.

c) Factor the fraction to remove a factor equal to 1, if possible.

d) Carry out the remaining products.

EXAMPLES Multiply and simplify.

1. $\frac{2}{3} \cdot \frac{9}{4} = \frac{2 \cdot 9}{3 \cdot 4} = \frac{2 \cdot 3 \cdot 3}{3 \cdot 2 \cdot 2} = \frac{2 \cdot 3}{2 \cdot 3} \cdot \frac{3}{2} = 1 \cdot \frac{3}{2} = \frac{3}{2}$

Answers

Skill to Review:
1. Yes **2.** No

2. $\dfrac{6}{7} \cdot \dfrac{-5}{3} = \dfrac{3 \cdot 2 \cdot (-5)}{7 \cdot 3}$ Note that 3 is a common factor of 6 and 3.

$= \dfrac{3}{3} \cdot \dfrac{2(-5)}{7} = \dfrac{-10}{7}$, or $-\dfrac{10}{7}$ Removing a factor equal to 1: $\dfrac{3}{3} = 1$

3. $\dfrac{10}{21} \cdot \dfrac{14a}{15} = \dfrac{5 \cdot 2 \cdot 7 \cdot 2 \cdot a}{7 \cdot 3 \cdot 5 \cdot 3}$ Note that 5 is a common factor of 10 and 15. Note that 7 is a common factor of 21 and $14a$.

$= \dfrac{5 \cdot 7}{5 \cdot 7} \cdot \dfrac{2 \cdot 2 \cdot a}{3 \cdot 3}$

$= \dfrac{4a}{9}$ Removing a factor equal to 1: $\dfrac{5 \cdot 7}{5 \cdot 7} = 1$

4. $32 \cdot \dfrac{7}{8} = \dfrac{32}{1} \cdot \dfrac{7}{8} = \dfrac{8 \cdot 4 \cdot 7}{8 \cdot 1}$ Note that 8 is a common factor of 32 and 8.

$= \dfrac{8}{8} \cdot \dfrac{4 \cdot 7}{1} = 28$ Removing a factor equal to 1: $\dfrac{8}{8} = 1$

Caution!

Canceling can be used as follows for these examples.

1. $\dfrac{2}{3} \cdot \dfrac{9}{4} = \dfrac{\cancel{2} \cdot \cancel{3} \cdot 3}{\cancel{3} \cdot \cancel{2} \cdot 2} = \dfrac{3}{2}$ Removing a factor equal to 1: $\dfrac{2 \cdot 3}{2 \cdot 3} = 1$

2. $\dfrac{6}{7} \cdot \dfrac{-5}{3} = \dfrac{\cancel{3} \cdot 2(-5)}{7 \cdot \cancel{3}} = \dfrac{-10}{7}$ Removing a factor equal to 1: $\dfrac{3}{3} = 1$

3. $\dfrac{10}{21} \cdot \dfrac{14a}{15} = \dfrac{\cancel{5} \cdot 2 \cdot \cancel{7} \cdot 2 \cdot a}{\cancel{7} \cdot 3 \cdot \cancel{5} \cdot 3} = \dfrac{4a}{9}$ Removing a factor equal to 1: $\dfrac{5 \cdot 7}{5 \cdot 7} = 1$

4. $32 \cdot \dfrac{7}{8} = \dfrac{\cancel{8} \cdot 4 \cdot 7}{\cancel{8} \cdot 1} = 28$ Removing a factor equal to 1: $\dfrac{8}{8} = 1$

Remember, only factors can be canceled!

Do Exercises 1–4.

Multiply and simplify.

1. $\dfrac{2}{3} \cdot \dfrac{7}{8} = \dfrac{2 \cdot 7}{3 \cdot \square}$

$= \dfrac{2 \cdot 7}{3 \cdot 2 \cdot 2 \cdot \square}$

$= \dfrac{2}{\square} \cdot \dfrac{7}{3 \cdot 2 \cdot 2}$

$= \square \cdot \dfrac{7}{3 \cdot 2 \cdot 2}$

$= \dfrac{7}{\square}$

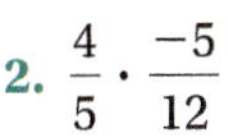

2. $\dfrac{4}{5} \cdot \dfrac{-5}{12}$

3. $16 \cdot \dfrac{3}{8}$

4. $\dfrac{5}{2x} \cdot 4$

b SOLVING PROBLEMS

EXAMPLE 5 *Landscaping.* Celina's Landscaping uses $\frac{2}{3}$ lb of peat moss when planting a rosebush. How much will be needed to plant 21 rosebushes?

1. **Familiarize.** We let n = the number of pounds of peat moss needed. Each rosebush requires $\frac{2}{3}$ lb of peat moss, so repeated addition, or multiplication, applies.
2. **Translate.** The problem translates to the following equation:

$$n = 21 \cdot \frac{2}{3}.$$

3. **Solve.** To solve the equation, we carry out the multiplication:

$n = 21 \cdot \dfrac{2}{3} = \dfrac{21}{1} \cdot \dfrac{2}{3} = \dfrac{21 \cdot 2}{1 \cdot 3}$ Multiplying

$= \dfrac{3 \cdot 7 \cdot 2}{1 \cdot 3} = \dfrac{3}{3} \cdot \dfrac{7 \cdot 2}{1} = 14.$ Removing the factor $\frac{3}{3}$ and simplifying

Answers

1. $\frac{7}{12}$ 2. $-\frac{1}{3}$ 3. 6 4. $\frac{10}{x}$

Guided Solution:

1. 8, 2, 2, 1, 12

5. *Candy.* Chocolate Delight sells $\frac{4}{5}$-lb boxes of truffles. How many pounds of truffles will be needed to fill 85 boxes?

85 boxes

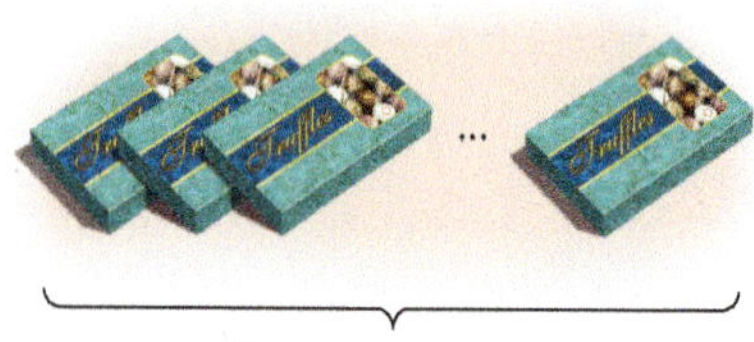

$\frac{4}{5}$ pound of truffles in each box

4. Check. We check by repeating the calculation. (This is left to the student.) We can also ask if the answer seems reasonable. We are putting less than a pound of peat moss on each bush, so the answer should be less than 21. Since 14 is less than 21, we have a partial check.

A second partial check can be performed using the units:

$$21 \text{ bushes} \cdot \frac{2}{3} \text{pounds per bush}$$
$$= 21 \cdot \frac{2}{3} \cdot \text{bushes} \cdot \frac{\text{pounds}}{\text{bush}}$$
$$= 14 \text{ pounds}.$$

Since the resulting unit is pounds, we have another partial check.

5. State. Celina's Landscaping will need 14 lb of peat moss to plant 21 rosebushes.

Do Exercise 5.

Area

We multiply to find the area of a triangle. Consider a triangle with a base of length b and a height of h. A rectangle can be formed by splitting and inverting a copy of this triangle.

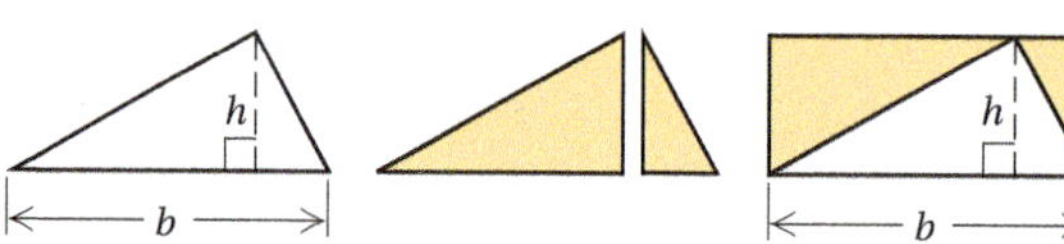

The rectangle's area, $b \cdot h$, is exactly twice the area of the triangle. We have the following result.

AREA OF A TRIANGLE

The **area A of a triangle** is half the length of the base b times the height h:

$$A = \frac{1}{2} \cdot b \cdot h.$$

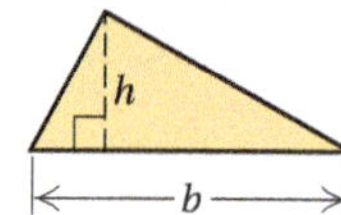

EXAMPLE 6 Find the area of this triangle.

$$A = \frac{1}{2} \cdot b \cdot h$$
$$= \frac{1}{2} \cdot 9 \text{ yd} \cdot 6 \text{ yd}$$
$$= \frac{9 \cdot 6}{2} \text{yd}^2$$

$$= 27 \text{ yd}^2, \text{ or } 27 \text{ square yards}$$

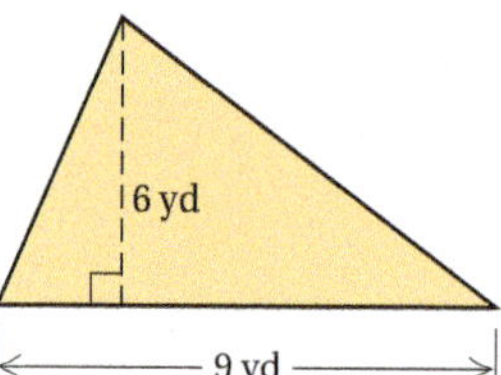

Caution!

Use square units for units of area.

Answer

5. 68 lb

EXAMPLE 7 Find the area of this triangle.

$$A = \frac{1}{2} \cdot b \cdot h$$
$$= \frac{1}{2} \cdot \frac{10}{3}\text{ cm} \cdot 4\text{ cm}$$
$$= \frac{1 \cdot 10 \cdot 4}{2 \cdot 3}\text{ cm}^2$$
$$= \frac{1 \cdot 2 \cdot 5 \cdot 4}{2 \cdot 3}\text{ cm}^2 \qquad \text{Removing a factor equal to 1: } \frac{2}{2} = 1$$
$$= \frac{20}{3}\text{ cm}^2$$

4 cm

$\frac{10}{3}$ cm

Do Exercises 6 and 7.

EXAMPLE 8 Find the area of this kite.

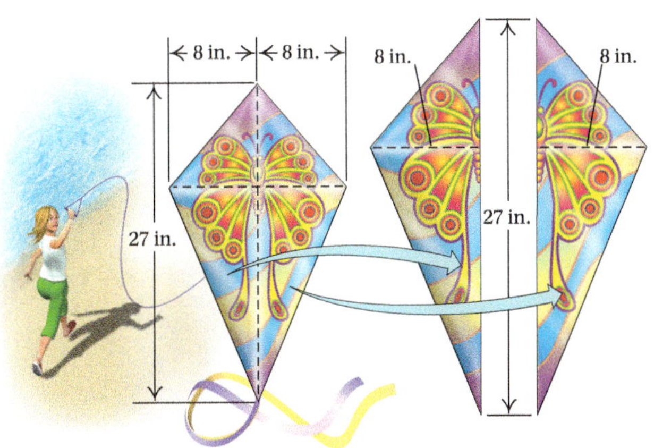

1. **Familiarize.** We look for figures with areas we can calculate using area formulas that we already know. We let $K =$ the kite's area.
2. **Translate.** The kite consists of two triangles, each with a base of 27 in. and a height of 8 in. We can apply the formula $A = \frac{1}{2} \cdot b \cdot h$ for the area of a triangle and then multiply by 2.

Rephrase:	Kite's area	is	twice		Area of long triangle
	↓	↓	↓		↓
Translate:	K	$=$	2	$\cdot$	$\frac{1}{2}(27\text{ in.}) \cdot (8\text{ in.})$

3. **Solve.** We have

$$K = 2 \cdot \frac{1}{2} \cdot (27\text{ in.}) \cdot (8\text{ in.})$$
$$= 1 \cdot 27\text{ in.} \cdot 8\text{ in.} = 216\text{ in}^2.$$

4. **Check.** We can check by repeating the calculations. The unit, in^2, is appropriate for area.
5. **State.** The area of the kite is 216 in^2.

Do Exercise 8.

Find the area.

6.

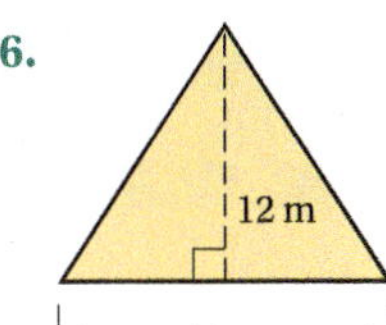

7. GS

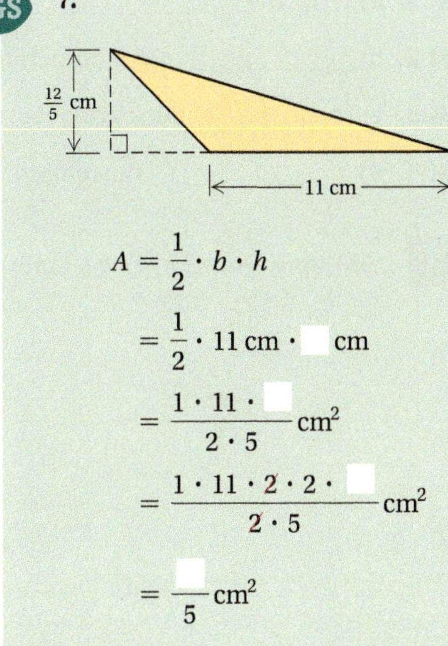

$$A = \frac{1}{2} \cdot b \cdot h$$
$$= \frac{1}{2} \cdot 11\text{ cm} \cdot \square\text{ cm}$$
$$= \frac{1 \cdot 11 \cdot \square}{2 \cdot 5}\text{ cm}^2$$
$$= \frac{1 \cdot 11 \cdot 2 \cdot 2 \cdot \square}{2 \cdot 5}\text{ cm}^2$$
$$= \frac{\square}{5}\text{ cm}^2$$

8. Find the area. (*Hint*: The figure is made up of a rectangle and a triangle.)

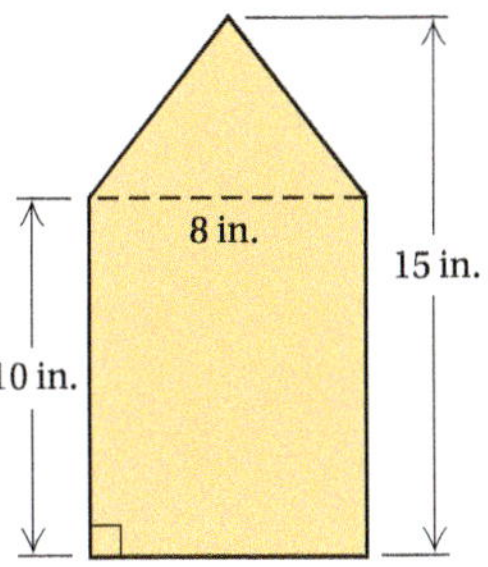

Answers

6. 96 m^2 **7.** $\frac{66}{5}\text{ cm}^2$
8. Rectangle: $(10\text{ in.}) \cdot (8\text{ in.}) = 80\text{ in}^2$; triangle: $\frac{1}{2}(8\text{ in.}) \cdot (5\text{ in.}) = 20\text{ in}^2$; $80\text{ in}^2 + 20\text{ in}^2 = 100\text{ in}^2$

Guided Solution:
7. $\frac{12}{5}$, 12, 3, 66

3.6 Exercise Set

For Extra Help
MyMathLab®
MathXL®

PRACTICE

WATCH

READ

REVIEW

Reading Check

Complete each step in the process for multiplying and simplifying using fraction notation.

RC1. a) Write the _______________ in the numerator and the denominator, but do not carry out the multiplication.

RC2. b) _______________ the numerator and the denominator.

RC3. c) Factor the fraction to remove a factor equal to _______________, if possible.

RC4. d) _______________ the remaining products.

a Multiply. Don't forget to simplify, if possible.

1. $\frac{2}{3} \cdot \frac{1}{2}$

2. $\frac{4}{5} \cdot \frac{1}{4}$

3. $\frac{7}{8} \cdot \frac{-1}{7}$

4. $\frac{5}{6} \cdot \frac{-1}{5}$

5. $\frac{2}{3} \cdot \frac{6}{7}$

6. $\frac{2}{5} \cdot \frac{3}{10}$

7. $\frac{2}{9} \cdot \frac{3}{10}$

8. $\frac{3}{5} \cdot \frac{10}{9}$

9. $\frac{9}{-5} \cdot \frac{12}{8}$

10. $\frac{16}{-15} \cdot \frac{5}{4}$

11. $\frac{5x}{9} \cdot \frac{4}{5}$

12. $\frac{25}{4a} \cdot \frac{4}{3}$

13. $9 \cdot \frac{1}{9}$

14. $4 \cdot \frac{1}{4}$

15. $\frac{7}{10} \cdot \frac{10}{7}$

16. $\frac{8}{9} \cdot \frac{9}{8}$

17. $\frac{1}{4} \cdot 12$

18. $\frac{1}{6} \cdot 12$

19. $21 \cdot \frac{1}{3}$

20. $18 \cdot \frac{1}{2}$

21. $-16\left(-\frac{3}{4}\right)$

22. $-24\left(-\frac{5}{6}\right)$

23. $\frac{3}{8} \cdot 8a$

24. $\frac{2}{9} \cdot 9x$

25. $\left(-\frac{3}{8}\right)\left(-\frac{8}{3}\right)$

26. $\left(-\frac{7}{9}\right)\left(-\frac{9}{7}\right)$

27. $\frac{a}{b} \cdot \frac{b}{a}$

28. $\frac{n}{m} \cdot \frac{m}{n}$

29. $\frac{4}{10} \cdot \frac{5}{10}$

30. $\frac{11}{24} \cdot \frac{3}{5}$

31. $\frac{8}{10} \cdot \frac{45}{100}$

32. $\frac{3}{10} \cdot \frac{8}{10}$

33. $\frac{1}{6} \cdot 360n$

34. $\frac{1}{3} \cdot 12y$

35. $20\left(\frac{1}{-6}\right)$

36. $35\left(\frac{1}{-10}\right)$

37. $-8x \cdot \frac{1}{-8x}$

38. $-5a \cdot \frac{1}{-5a}$

39. $\frac{2x}{9} \cdot \frac{27}{2x}$

40. $\frac{10a}{3} \cdot \frac{3}{5a}$

41. $\frac{7}{10} \cdot \frac{34}{150}$

42. $\frac{15}{22} \cdot \frac{4}{7}$

43. $\frac{36}{85} \cdot \frac{25}{-99}$

44. $\frac{-70}{45} \cdot \frac{50}{49}$

45. $\frac{-98}{99} \cdot \frac{27a}{175a}$

46. $\frac{70}{-49} \cdot \frac{63}{300x}$

47. $\frac{110}{33} \cdot \frac{-24}{25x}$

48. $\frac{-19}{130} \cdot \frac{65}{38x}$

49. $\left(-\frac{11}{24}\right)\frac{3}{5}$

50. $\left(-\frac{15}{22}\right)\frac{4}{7}$

51. $\frac{10a}{21} \cdot \frac{3}{8b}$

52. $\frac{17}{21y} \cdot \frac{3x}{5}$

 Solve.

The *pitch* of a screw is the distance between its threads. With each complete rotation, the screw goes in or out a distance equal to its pitch. Use this information to answer Exercises 53 and 54.

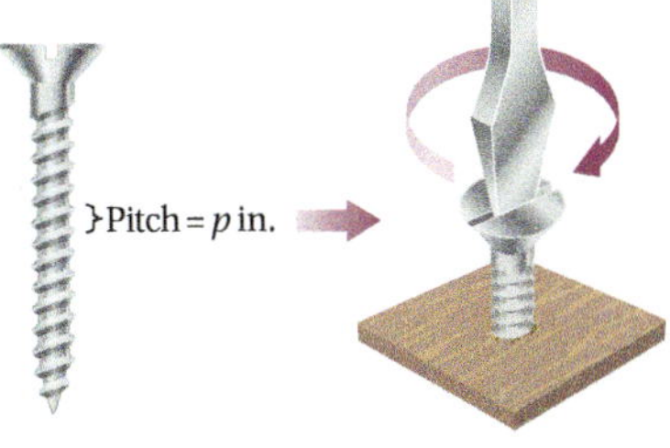
}Pitch = p in.
Each rotation moves the screw in or out p in.

53. The pitch of a screw is $\frac{1}{16}$ in. How far will it go into a piece of oak when it is turned 10 complete rotations clockwise?

54. The pitch of a screw is $\frac{3}{32}$ in. How far will it come out of a piece of plywood when it is turned 10 complete rotations counterclockwise?

55. *World Silver Supply.* The total world supply of silver for new uses in 2011 was about 1040 million ounces. Of this, approximately $\frac{1}{4}$ was scrap silver being reused or repurposed. How many ounces of silver for new uses was supplied as scrap silver?

Source: Based on information from the *World Silver Survey 2012*

56. *Substitute Teaching.* After Jack completes 60 hr of teacher training in college, he can earn $75 for working a full day as a substitute teacher. How much will he receive for working $\frac{3}{5}$ of a day?

57. *Mailing-List Changes.* The United States Postal Service estimates that $\frac{4}{25}$ of the addresses on a mailing list will change in one year. A business has a mailing list of 3000 people. After one year, how many addresses on that list will be incorrect?

Source: Based on information from usps.com

58. *Shy People.* Sociologists have determined that $\frac{2}{5}$ of the people in the world are shy. A sales manager has 650 applicants for a sales position that requires an outgoing personality. How many of these people might be shy?

59. A recipe for piecrust calls for $\frac{2}{3}$ cup of flour. A chef is making $\frac{1}{2}$ of the recipe. How much flour should the chef use?

60. Of the students in the freshman class, $\frac{4}{5}$ have digital cameras; $\frac{1}{4}$ of these students also join the college photography club. What fraction of the students in the freshman class join the photography club?

61. *Assessed Value.* A house worth $154,000 is assessed for $\frac{3}{4}$ of its value. What is the assessed value of the house?

62. Roxanne's tuition was $4600. A loan was obtained for $\frac{3}{4}$ of the tuition. How much was the loan?

63. *Map Scaling.* On a map, 1 in. represents 240 mi. How much does $\frac{2}{3}$ in. represent?

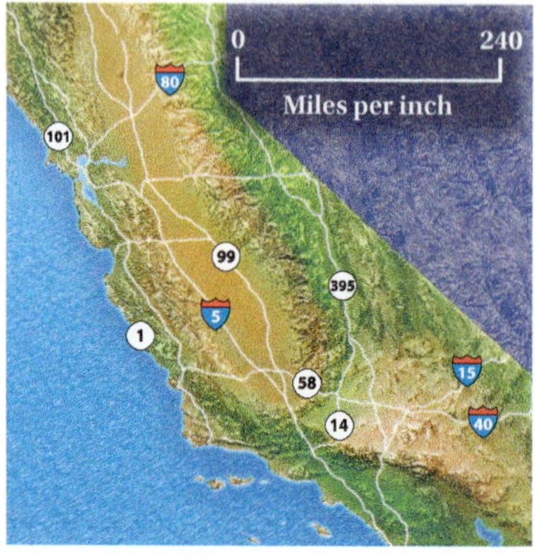

64. *Map Scaling.* On a map, 1 in. represents 120 mi. How much does $\frac{3}{4}$ in. represent?

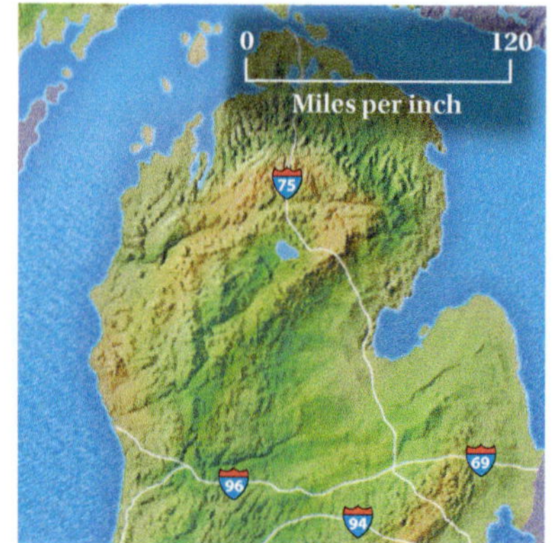

65. ***Household Budgets.*** A family has an annual income of $42,000. Of this, $\frac{1}{5}$ is spent for food, $\frac{1}{4}$ for housing, $\frac{1}{10}$ for clothing, $\frac{1}{14}$ for savings, $\frac{1}{5}$ for taxes, and the rest for other expenses. How much is spent for each?

66. ***Household Budgets.*** A family has an annual income of $28,140. Of this, $\frac{1}{5}$ is spent for food, $\frac{1}{4}$ for housing, $\frac{1}{10}$ for clothing, $\frac{1}{14}$ for savings, $\frac{1}{5}$ for taxes, and the rest for other expenses. How much is spent for each?

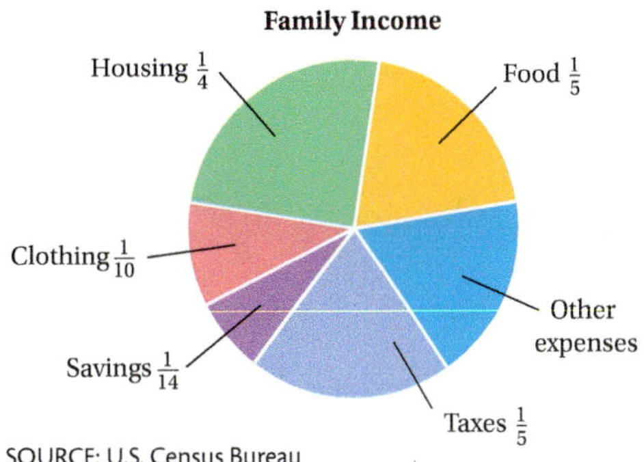

Find the area.

67.

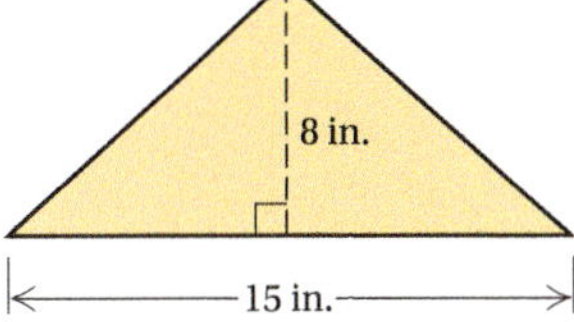

68.

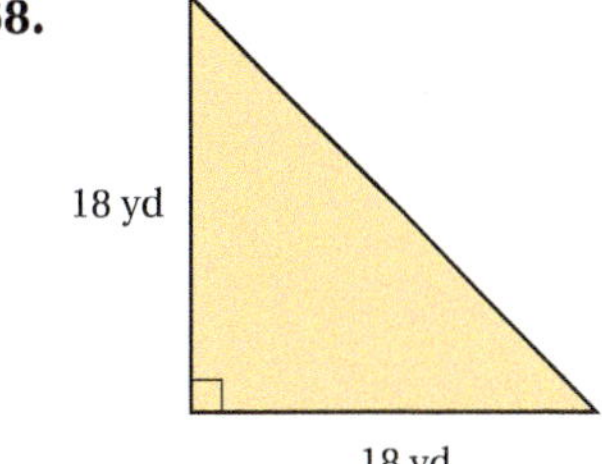

69.

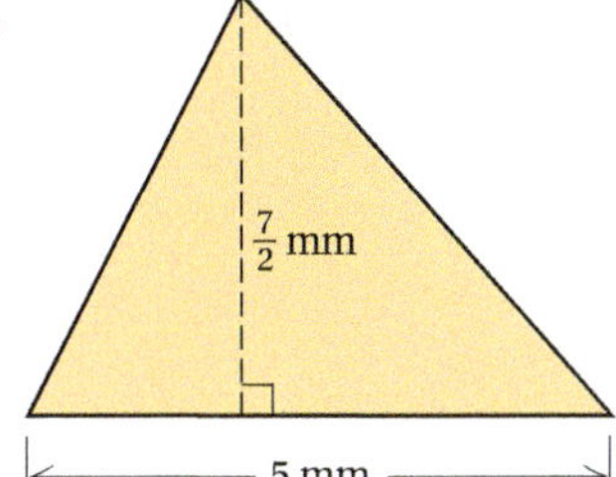

70.

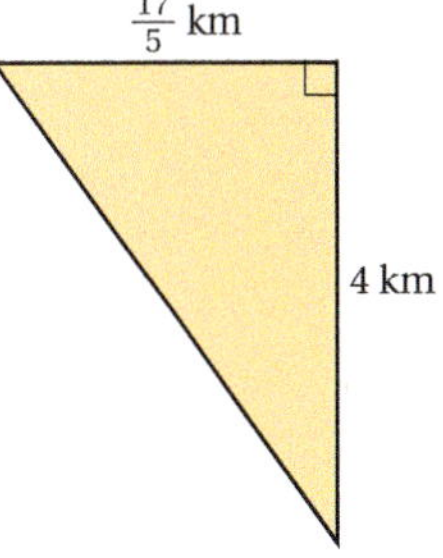

71.

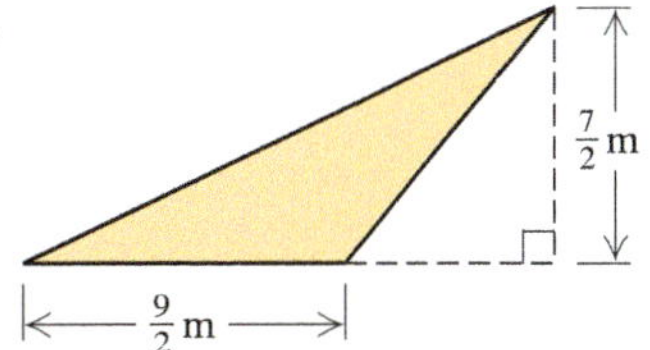

72.

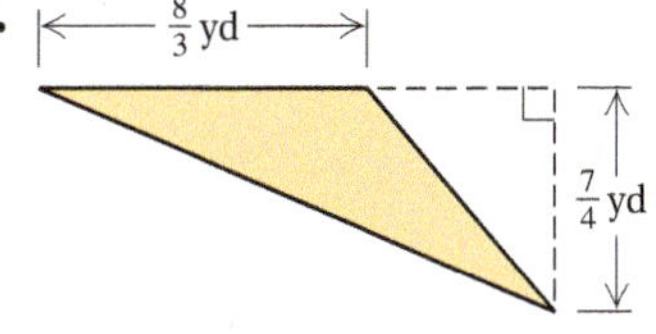

73.

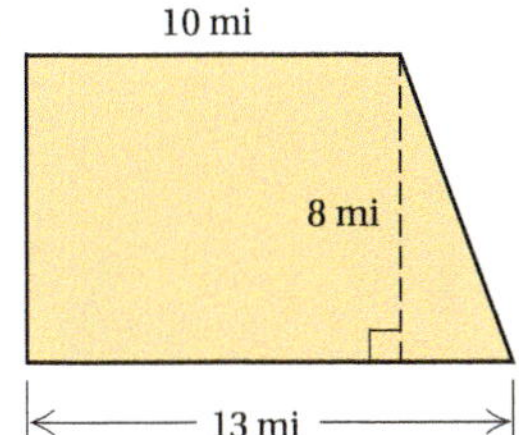

74.

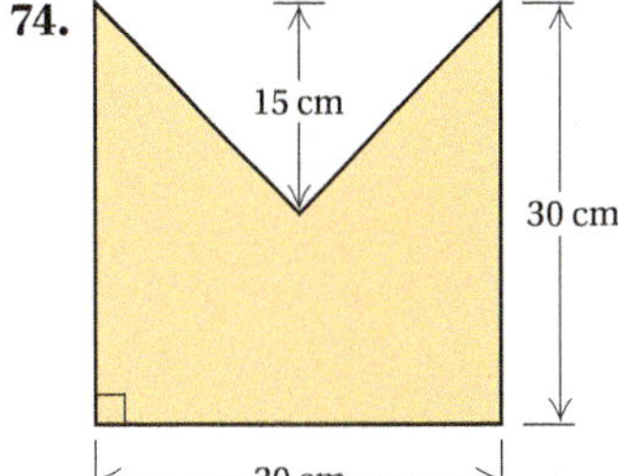

75. *Quilting.* Michelle is designing a quilt using hexagons. Each hexagon is made using six kite-shaped pieces of cloth, as shown below. Find the area of one of the kite-shaped pieces.

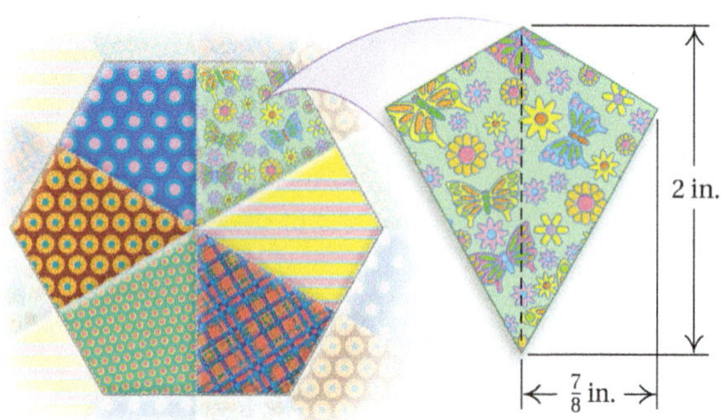

76. *Construction.* Find the total area of the sides and ends of the town office building shown. Do not subtract for any windows, doors, or steps.

Skill Maintenance

Solve. [1.7b]

77. $48 \cdot t = 1680$

78. $74 \cdot x = 6290$

79. $3125 = 25 \cdot t$

80. $2880 = 24 \cdot y$

81. $t + 28 = 5017$

82. $456 + x = 9002$

83. $8797 = y + 98$

84. $10{,}000 = 3593 + m$

Synthesis

Simplify. Use the list of prime numbers on p. 158.

85. $\frac{201}{535} \cdot \frac{4601}{6499}$

86. $\frac{667}{899} \cdot \frac{558}{621}$

87. *College Profile.* Of students entering a college, $\frac{7}{8}$ have completed high school and $\frac{2}{3}$ are older than 20. If $\frac{1}{7}$ of all students are left-handed, what fraction of students entering the college are left-handed high school graduates over the age of 20?

88. *College Profile.* Refer to the information in Exercise 87. If 480 students are entering the college, how many of them are left-handed high school graduates 20 years old or younger?

89. *Manufacturing.* A candy box is triangular at each end, as shown below. Find the surface area of the box.

90. *Painting.* Shoreline Painting needs to determine the surface area of an octagonal steeple. Find the total area, if the dimensions are as shown below.

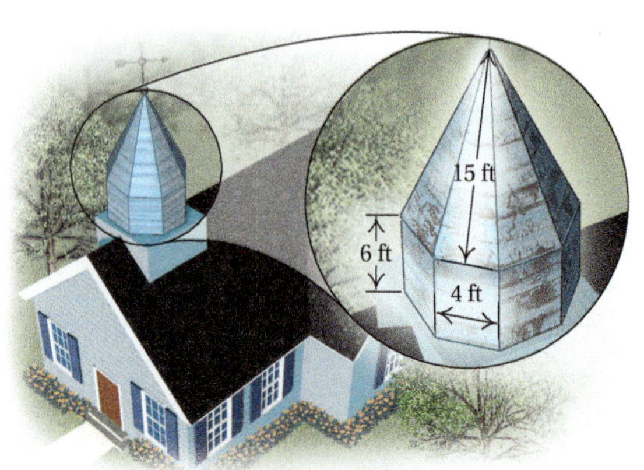

Reciprocals and Division

3.7

OBJECTIVES

a Find the reciprocal of a number.

b Divide and simplify using fraction notation.

SKILL TO REVIEW

Objective 2.5a: Divide integers.

Divide.

1. $\dfrac{26}{-2}$ 2. $\dfrac{-100}{-25}$

a RECIPROCALS

Each of the following products is 1:

$$8 \cdot \frac{1}{8} = \frac{8}{8} = 1; \qquad \frac{-2}{3} \cdot \frac{3}{-2} = \frac{-6}{-6} = 1.$$

RECIPROCALS

If the product of two numbers is 1, we say that they are **reciprocals** of each other.* To find the reciprocal of a fraction, interchange the numerator and the denominator.

Number: $\dfrac{3}{4}$ ⟶ Reciprocal: $\dfrac{4}{3}$

EXAMPLES Find the reciprocal.

1. The reciprocal of $\frac{4}{5}$ is $\frac{5}{4}$. $\quad \frac{4}{5} \cdot \frac{5}{4} = \frac{20}{20} = 1$

2. The reciprocal of $\frac{a}{b}$ is $\frac{b}{a}$. $\quad \frac{a}{b} \cdot \frac{b}{a} = \frac{ab}{ba} = 1$

3. The reciprocal of 8 is $\frac{1}{8}$. $\quad$ Think of 8 as $\frac{8}{1}$: $\frac{8}{1} \cdot \frac{1}{8} = \frac{8}{8} = 1$.

4. The reciprocal of $\frac{1}{3}$ is 3. $\quad \frac{1}{3} \cdot 3 = \frac{3}{3} = 1$

5. The reciprocal of $-\frac{5}{9}$ is $-\frac{9}{5}$. $\quad$ Negative numbers have negative reciprocals: $\left(-\frac{5}{9}\right)\left(-\frac{9}{5}\right) = \frac{45}{45} = 1$.

Do Margin Exercises 1–5. ▶

Find the reciprocal.

1. $\dfrac{2}{5}$ 2. $\dfrac{-6}{x}$

3. 9 4. $\dfrac{1}{5}$

5. $-\dfrac{3}{10}$

Does 0 have a reciprocal? If it did, it would have to be a number x such that $0 \cdot x = 1$. But 0 times any number is 0. Thus we have the following.

0 HAS NO RECIPROCAL

The number 0, or $\frac{0}{n}$, has no reciprocal. $\left(\text{Recall that } \frac{n}{0} \text{ is not defined.}\right)$

b DIVISION

Consider the division $\frac{3}{4} \div \frac{1}{8}$. We are asking how many $\frac{1}{8}$'s are in $\frac{3}{4}$. From the figure at the right, we see that there are six $\frac{1}{8}$'s in $\frac{3}{4}$. Thus,

$$\frac{3}{4} \div \frac{1}{8} = 6.$$

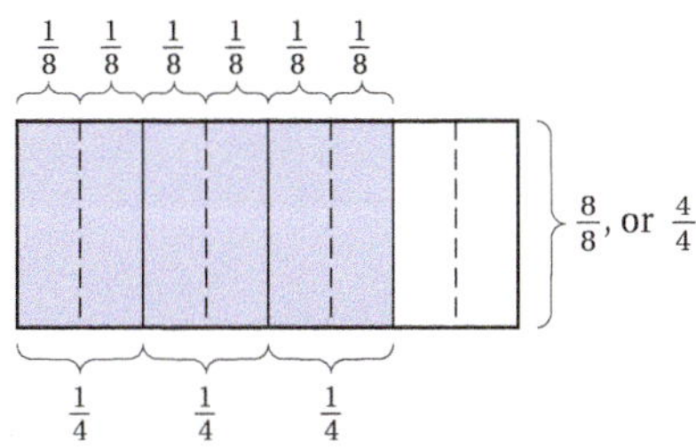

*A reciprocal is also called a *multiplicative inverse*.

Answers

Skill to Review:
1. −13 2. 4

Margin Exercises:
1. $\frac{5}{2}$ 2. $\frac{x}{-6}$ 3. $\frac{1}{9}$ 4. 5 5. $-\frac{10}{3}$

CALCULATOR CORNER

Multiplying and Dividing Fractions On a fraction calculator, multiplication or division of fractions is entered the way it is written. To perform the division $\frac{4}{9} \div \frac{6}{5}$ with such a calculator, the following keystrokes can be used:

[4] [a b/c] [9] [÷]
[6] [a b/c] [5] [=].

The simplified fraction notation is $\frac{10}{27}$.

On calculators without an [a b/c] key, parentheses must be used around the second fraction when dividing. The following keystrokes can be used to perform the division and convert the result to fraction notation:

[4] [÷] [9] [÷] [(] [6] [÷]
[5] [)] [MATH] [1] [ENTER].

Again, the result is $\frac{10}{27}$.

EXERCISES Use a calculator to perform the following operations.

1. $\frac{5}{8} \cdot \frac{4}{15}$

2. $\frac{10}{12} \div \frac{3}{8}$

3. $\frac{-6}{7} \div \frac{2}{3}$

4. $\frac{-9}{10} \div \frac{-3}{5}$

We can check this by multiplying:

$$6 \cdot \frac{1}{8} = \frac{6}{1} \cdot \frac{1}{8} = \frac{6}{8} = \frac{2 \cdot 3}{2 \cdot 4} = \frac{2}{2} \cdot \frac{3}{4} = \frac{3}{4}.$$

Here is a faster way to do this division:

$$\frac{3}{4} \div \frac{1}{8} = \frac{3}{4} \cdot \frac{8}{1} = \frac{3 \cdot 8}{4 \cdot 1} = \frac{24}{4} = 6.$$ Multiplying by the reciprocal of the divisor

DIVISION OF FRACTIONS

To divide by a fraction, multiply by its reciprocal:

$$\frac{a}{b} \div \frac{c}{d} = \frac{a}{b} \cdot \frac{d}{c}.$$ Multiply by the reciprocal of the divisor.

EXAMPLES Divide and simplify.

6. $\frac{5}{6} \div \frac{2}{3} = \frac{5}{6} \cdot \frac{3}{2}$ Multiplying by the reciprocal of the divisor

$= \frac{5 \cdot 3}{3 \cdot 2 \cdot 2}$ Factoring and identifying a common factor

$= \frac{3}{3} \cdot \frac{5}{2 \cdot 2}$ Removing a factor equal to 1: $\frac{3}{3} = 1$

$= \frac{5}{4}$

7. $\frac{-3}{5} \div \frac{1}{2} = \frac{-3}{5} \cdot 2$ The reciprocal of $\frac{1}{2}$ is 2.

$= \frac{-3 \cdot 2}{5} = \frac{-6}{5}$, or $-\frac{6}{5}$

8. $\frac{2a}{5} \div 7 = \frac{2a}{5} \cdot \frac{1}{7}$ The reciprocal of 7 is $\frac{1}{7}$.

$= \frac{2a \cdot 1}{5 \cdot 7} = \frac{2a}{35}$

9. $\dfrac{\frac{7}{10}}{-\frac{14}{15}} = \frac{7}{10} \div \left(-\frac{14}{15}\right)$ The fraction bar indicates division.

$= \frac{7}{10} \cdot \left(-\frac{15}{14}\right)$ Multiplying by the reciprocal of the divisor

$= \frac{7 \cdot 5(-3)}{2 \cdot 5 \cdot 7 \cdot 2}$ Factoring and identifying common factors

$= \frac{7 \cdot 5}{7 \cdot 5} \cdot \frac{-3}{4}$ Removing a factor equal to 1: $\frac{7 \cdot 5}{7 \cdot 5} = 1$

$= \frac{-3}{4}$, or $-\frac{3}{4}$

Caution!

Canceling can be used as follows for Examples 6 and 9.

6. $\frac{5}{6} \div \frac{2}{3} = \frac{5}{6} \cdot \frac{3}{2} = \frac{5 \cdot 3}{6 \cdot 2} = \frac{5 \cdot \cancel{3}}{\cancel{3} \cdot 2 \cdot 2} = \frac{5}{2 \cdot 2} = \frac{5}{4}$

Removing a factor equal to 1: $\frac{3}{3} = 1$

9. $\frac{7}{10} \div \left(-\frac{14}{15}\right) = \frac{7}{10} \cdot \left(-\frac{15}{14}\right) = \frac{\cancel{7} \cdot \cancel{5}(-3)}{2 \cdot \cancel{5} \cdot \cancel{7} \cdot 2} = \frac{-3}{4}$, or $-\frac{3}{4}$

Removing a factor equal to 1: $\frac{7 \cdot 5}{7 \cdot 5} = 1$

Remember, if you can't factor, you can't cancel!

Divide and simplify.

GS **6.** $\frac{6}{7} \div \frac{3}{4} = \frac{6}{7} \cdot \frac{\square}{\square}$

$= \frac{6 \cdot 4}{7 \cdot 3}$

$= \frac{2 \cdot 3 \cdot 2 \cdot \square}{7 \cdot 3}$

$= \frac{3}{\square} \cdot \frac{2 \cdot 2 \cdot 2}{7}$

$= \frac{2 \cdot 2 \cdot 2}{7}$

$= \frac{\square}{7}$

7. $\left(-\frac{2}{3}\right) \div \frac{1}{4}$

8. $\frac{4}{5} \div 8$

9. $60 \div \frac{3a}{5}$

10. $\frac{\frac{-6}{7}}{\frac{3}{5}}$

Do Exercises 6–10. ▶

Why do we multiply by a reciprocal when dividing? To see why, consider $\frac{2}{3} \div \frac{7}{5}$. We will multiply by 1 to find an equivalent expression. To write 1, we use $(5/7)/(5/7)$, since $\frac{5}{7}$ is the reciprocal of $\frac{7}{5}$.

$$\frac{2}{3} \div \frac{7}{5} = \frac{\frac{2}{3}}{\frac{7}{5}}$$ Writing fraction notation for the division

$$= \frac{\frac{2}{3}}{\frac{7}{5}} \cdot 1$$ Multiplying by 1

$$= \frac{\frac{2}{3}}{\frac{7}{5}} \cdot \frac{\frac{5}{7}}{\frac{5}{7}}$$ Multiplying by 1; $\frac{5}{7}$ is the reciprocal of $\frac{7}{5}$ and $\frac{\frac{5}{7}}{\frac{5}{7}} = 1$.

$$= \frac{\frac{2}{3} \cdot \frac{5}{7}}{\frac{7}{5} \cdot \frac{5}{7}}$$ Multiplying the numerators and the denominators

$$= \frac{\frac{2}{3} \cdot \frac{5}{7}}{1} = \frac{2}{3} \cdot \frac{5}{7} = \frac{10}{21}$$ After we multiplied, we got 1 for the denominator. The numerator (in color) shows the multiplication by the reciprocal of $\frac{7}{5}$.

Thus,

$$\frac{2}{3} \div \frac{7}{5} = \frac{2}{3} \cdot \frac{5}{7} = \frac{10}{21}.$$

Expressions of the form $\frac{\frac{a}{b}}{\frac{c}{d}}$ are examples of *complex fractions*:

$$\frac{\frac{a}{b}}{\frac{c}{d}} = \frac{a}{b} \div \frac{c}{d}.$$

11. To remember *why* fractions are divided as they are, multiply by 1 to perform the following division, using the reciprocal of $\frac{4}{5}$ to write 1.

$$\frac{\frac{6}{7}}{\frac{4}{5}}$$

Do Exercise 11. ▶

Answers

6. $\frac{8}{7}$ 7. $-\frac{8}{3}$ 8. $\frac{1}{10}$ 9. $\frac{100}{a}$
10. $\frac{-10}{7}$, or $-\frac{10}{7}$ 11. $\frac{15}{14}$

Guided Solution:

6. $\frac{4}{3}$, 2, 3, 8

3.7 Exercise Set

For Extra Help

MyMathLab®

MathXL® PRACTICE  WATCH  READ  REVIEW

✓ Reading Check

Determine whether each statement is true or false.

__________ **RC1.** The numbers $\frac{1}{7}$ and 7 are reciprocals.

__________ **RC2.** The number 1 has no reciprocal.

__________ **RC3.** To divide by a fraction, we multiply by its reciprocal.

__________ **RC4.** We can divide by $\frac{1}{2}$ by multiplying by 2.

a Find the reciprocal.

1. $\frac{7}{3}$

2. $\frac{6}{5}$

3. 9

4. 3

5. $\frac{1}{7}$

6. $\frac{1}{4}$

7. $-\frac{8}{9}$

8. $-\frac{12}{5}$

9. $\frac{a}{c}$

10. $\frac{x}{y}$

11. $\frac{-3n}{m}$

12. $\frac{8t}{-7r}$

13. $\frac{8}{-15}$

14. $\frac{-6}{25}$

15. $7m$

16. $5n$

17. $\frac{1}{4a}$

18. $\frac{1}{9t}$

19. $-\frac{1}{3z}$

20. $-\frac{1}{2x}$

b Divide. Don't forget to simplify when possible. Assume that all variables are nonzero.

21. $\frac{3}{7} \div \frac{3}{4}$

22. $\frac{2}{3} \div \frac{3}{4}$

23. $\frac{3}{5} \div \frac{9}{4}$

24. $\frac{6}{7} \div \frac{3}{5}$

25. $\frac{4}{3} \div \frac{1}{3}$

26. $\frac{10}{9} \div \frac{1}{2}$

27. $\left(-\frac{1}{3}\right) \div \frac{1}{6}$

28. $\left(-\frac{1}{4}\right) \div \frac{1}{5}$

29. $\left(-\frac{10}{21}\right) \div \left(-\frac{2}{15}\right)$

30. $-\frac{15}{28} \div \left(-\frac{9}{20}\right)$

31. $\frac{3}{8} \div 3$

32. $\frac{5}{6} \div 5$

33. $\frac{12}{7} \div 16$

34. $\frac{18}{5} \div 27$

35. $(-12) \div \frac{3}{2}$

36. $(-24) \div \frac{3}{8}$

37. $\frac{x}{8} \div \frac{1}{4}$

38. $\frac{3}{4} \div \frac{2}{y}$

39. $\frac{2}{3} \div (6x)$

40. $\frac{12}{5} \div (4x)$

41. $28 \div \frac{4}{5a}$

42. $40 \div \frac{2}{3m}$

43. $\left(-\frac{5}{8}\right) \div \left(-\frac{5}{8}\right)$

44. $\left(-\frac{2}{5}\right) \div \left(-\frac{2}{5}\right)$

45. $\frac{-8}{15} \div \frac{4}{5}$

46. $\frac{6}{-13} \div \frac{3}{26}$

47. $\frac{77}{64} \div \frac{49}{18}$

48. $\frac{81}{42} \div \frac{33}{56}$

49. $120a \div \frac{45}{14}$

50. $360n \div \frac{27n}{8}$

51. $\dfrac{\frac{2}{5}}{\frac{3}{7}}$

52. $\dfrac{\frac{5}{6}}{\frac{2}{7}}$

53. $\dfrac{-\frac{7}{20}}{-\frac{8}{5}}$

54. $\dfrac{-\frac{8}{21}}{-\frac{6}{5}}$

55. $\dfrac{-\frac{15}{8}}{\frac{9}{10}}$

56. $\dfrac{-\frac{27}{10}}{\frac{21}{20}}$

Skill Maintenance

57. Evaluate $-(-x)$ when $x = -9$. [2.1d]

58. Evaluate $-x^2$ when $x = -9$. [2.6a]

59. Evaluate $3x^2$ when $x = 5$. [2.6a]

60. Evaluate $(-x)^2$ when $x = -9$. [2.6a]

61. Evaluate $\frac{-a}{b}$ for $a = 20$ and $b = -10$. [2.6a]

62. Evaluate $-\frac{a}{b}$ for $a = 20$ and $b = -10$. [2.6a]

Use the distributive law to write an equivalent expression. [2.6b]

63. $3(2x + 3y - 1)$

64. $-4(a - 3b + 2c)$

Synthesis

Simplify.

65. $\left(\frac{4}{15} \div \frac{2}{25}\right)^2$

66. $\left(\frac{9}{10} \div \frac{12}{25}\right)^2$

67. $\left(\frac{9}{10} \div \frac{2}{5} \div \frac{3}{8}\right)^2$

68. $\dfrac{\left(-\frac{3}{7}\right)^2 \div \frac{12}{5}}{\left(\frac{-2}{9}\right)\left(\frac{9}{2}\right)}$

69. $\left(\frac{14}{15} \div \frac{49}{65} \cdot \frac{77}{260}\right)^2$

70. $\left(\frac{10}{9}\right)^2 \div \frac{35}{27} \cdot \frac{49}{44}$

Simplify. Use the list of prime numbers on p. 158.

71. 🖩 $\frac{711}{1957} \div \frac{10{,}033}{13{,}081}$

72. 🖩 $\frac{8633}{7387} \div \frac{485}{581}$

73. 🖩 $\frac{451}{289} \div \frac{123}{340}$

74. 🖩 $\frac{530}{490} \div \frac{1060}{980}$

3.8 Solving Equations: The Multiplication Principle

OBJECTIVES

a Use the multiplication principle to solve equations.

b Solve problems by using the multiplication principle.

SKILL TO REVIEW

Objective 2.8b: Use the division principle to solve equations.

Solve.

1. $-8t = 32$ **2.** $-x = -9$

With fraction notation, we can solve equations like $a \cdot x = b$ by using multiplication.

a THE MULTIPLICATION PRINCIPLE

To divide by a fraction, we multiply by the reciprocal of that fraction. This suggests that we restate the division principle in its more common form—the multiplication principle.

THE MULTIPLICATION PRINCIPLE

For any numbers a, b, and c, with $c \neq 0$,

$a = b$ is equivalent to $a \cdot c = b \cdot c$.

EXAMPLE 1 Solve: $\frac{3}{4}x = 15$.

We multiply both sides of the equation by the reciprocal of $\frac{3}{4}$.

$$\frac{3}{4}x = 15$$

$$\frac{4}{3} \cdot \frac{3}{4}x = \frac{4}{3} \cdot 15$$ Using the multiplication principle; note that $\frac{4}{3}$ is the reciprocal of $\frac{3}{4}$.

$$\left(\frac{4}{3} \cdot \frac{3}{4}\right)x = \frac{4 \cdot 15}{3}$$ Using an associative law; try to do this mentally.

$$1x = 20$$ Multiplying; note that $\frac{4 \cdot 15}{3} = \frac{4 \cdot \cancel{3} \cdot 5}{\cancel{3}}$.

$$x = 20$$ Remember that $1x$ is x.

To confirm that 20 is the solution, we perform a check.

Check: $\frac{3}{4}x = 15$

$$\frac{3}{4} \cdot 20 \ ? \ 15$$

$$\frac{3 \cdot \cancel{4} \cdot 5}{\cancel{4}}$$ Removing a factor equal to 1: $\frac{4}{4} = 1$

$$3 \cdot 5 \mid 15$$ TRUE

The solution is 20.

Do Margin Exercises 1 and 2.

In an expression like $\frac{3}{4}x$, the constant factor—in this case, $\frac{3}{4}$—is called the **coefficient**. In Example 1, we multiplied on both sides by $\frac{4}{3}$, the reciprocal of the coefficient of x. After we carried out the multiplication, the coefficient of x was 1. Note that using the multiplication principle to multiply by $\frac{4}{3}$ on both sides is the same as using the division principle to divide by $\frac{3}{4}$ on both sides.

Solve.

1. $\frac{2}{3}x = 8$ GS

$$\frac{\square}{\square} \cdot \frac{2}{3}x = \frac{\square}{\square} \cdot 8$$

$$\square\, x = \frac{3 \cdot 8}{2}$$

$$x = \frac{3 \cdot 2 \cdot \square}{2}$$

$$x = \square$$

2. $\frac{2}{7}a = -6$

Answers

Skill to Review:
1. −4 2. 9

Margin Exercises:
1. 12 2. −21

Guided Solution:
1. $\frac{3}{2}, \frac{3}{2}$, 1, 4, 12

EXAMPLE 2 Solve: $5a = -\frac{7}{3}$.

We have

$$5a = -\frac{7}{3}$$

$$\frac{1}{5} \cdot 5a = \frac{1}{5} \cdot \left(-\frac{7}{3}\right) \qquad \text{Multiplying both sides by } \tfrac{1}{5}, \text{ the reciprocal of 5}$$

$$a = -\frac{1 \cdot 7}{5 \cdot 3} = -\frac{7}{15}.$$

We leave the check to the student. The solution is $-\frac{7}{15}$.

EXAMPLE 3 Solve: $\frac{10}{3} = -\frac{4}{9}x$.

We have

$$\frac{10}{3} = -\frac{4}{9}x$$

$$-\frac{9}{4} \cdot \frac{10}{3} = -\frac{9}{4} \cdot \left(-\frac{4}{9}\right)x \qquad \text{Multiplying both sides by } -\tfrac{9}{4}, \text{ the reciprocal of } -\tfrac{4}{9}$$

$$-\frac{\cancel{3} \cdot 3 \cdot \cancel{2} \cdot 5}{\cancel{2} \cdot 2 \cdot \cancel{3}} = x$$

$$-\frac{15}{2} = x. \qquad \text{Removing a factor equal to 1: } \frac{3 \cdot 2}{2 \cdot 3} = 1$$

We leave the check to the student. The solution is $-\frac{15}{2}$.

Do Exercises 3 and 4.

Solve.

3. $-\frac{9}{8} = 4x$

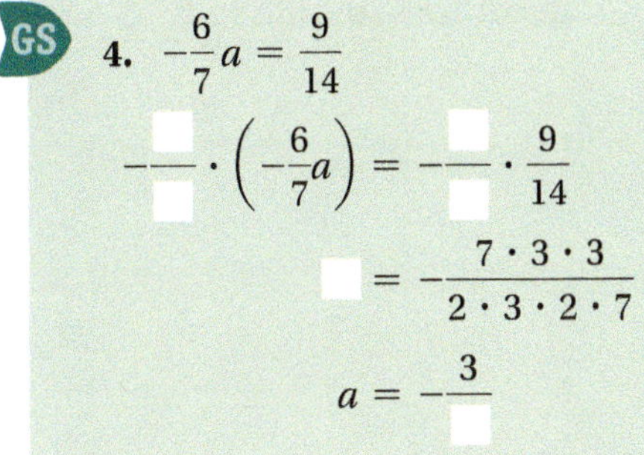

GS **4.** $-\frac{6}{7}a = \frac{9}{14}$

$$-\frac{\square}{\square} \cdot \left(-\frac{6}{7}a\right) = -\frac{\square}{\square} \cdot \frac{9}{14}$$

$$\square = -\frac{7 \cdot 3 \cdot 3}{2 \cdot 3 \cdot 2 \cdot 7}$$

$$a = -\frac{3}{\square}$$

b APPLICATIONS AND PROBLEM SOLVING

EXAMPLE 4 ***Doses of an Antibiotic.*** How many doses, each containing $\frac{15}{4}$ milliliters (mL), can be obtained from a bottle of a children's antibiotic that contains 60 mL?

1. **Familiarize.** We make a drawing, as shown to the right, and let $n =$ the number of doses in the bottle.

2. **Translate.** The problem can be translated to the equation

$$\frac{15}{4} \cdot n = 60.$$

3. **Solve.** To solve the equation, we use the multiplication principle.

$$\frac{4}{15} \cdot \frac{15}{4} \cdot n = \frac{4}{15} \cdot 60 \qquad \text{Multiplying both sides by } \frac{4}{15}$$

$$1n = \frac{4 \cdot 60}{15}$$

$$n = \frac{2 \cdot 2 \cdot 2 \cdot 2 \cdot 3 \cdot 5}{1 \cdot 3 \cdot 5} = \frac{3 \cdot 5}{3 \cdot 5} \cdot \frac{2 \cdot 2 \cdot 2 \cdot 2}{1} = 16$$

$\frac{15}{4}$ milliliter in each dose

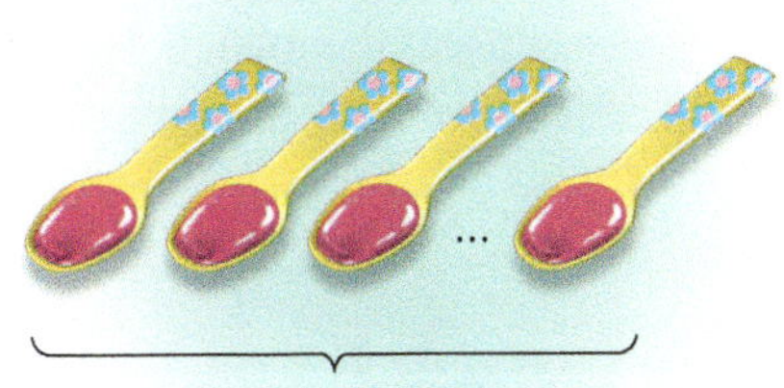

n doses in the bottle

Answers

3. $-\frac{9}{32}$ **4.** $-\frac{3}{4}$

Guided Solution:

4. $\frac{7}{6}, \frac{7}{6}, a, 4$

5. Each loop in a spring uses $\frac{21}{8}$ in. of wire. How many loops can be made from 210 in. of wire?

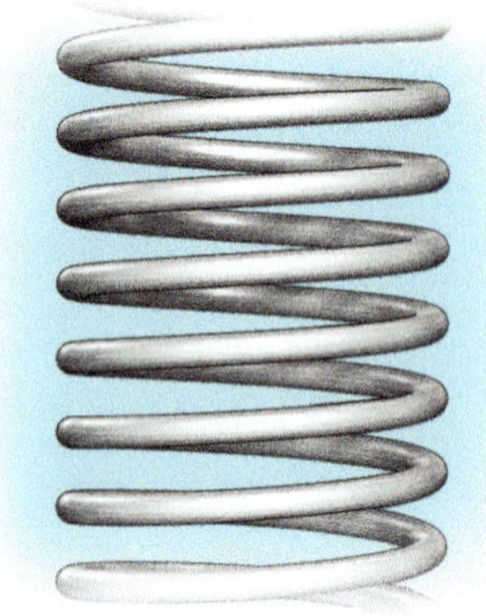

6. *Servings of Cereal.* A box contains 12 cups of cereal. The nutrition label on the box states that the serving size is $\frac{3}{4}$ cup. How many servings of cereal are in the box?

4. Check. We check by multiplying the number of doses by the size of the dose: $16 \cdot \frac{15}{4} = 60$. Note too that

$$\text{doses} \cdot \frac{\text{mL}}{\text{dose}} = \text{mL},$$

so the units also check.

5. State. There are 16 doses of the antibiotic in the 60-mL bottle.

◀ **Do Exercises 5 and 6.**

EXAMPLE 5 *Bicycle Paths.* The city of Indianapolis has adopted the *Indianapolis Bicycle Master Plan* as a strategy for creating an environment where bicycling is a safe, practical, and enjoyable transportation choice. After the city finished constructing 60 mi of bike paths and on-road bike lanes, the master plan was $\frac{3}{10}$ complete. What is the total number of miles of bicycling surface that the city of Indianapolis plans to construct?

Source: *Indianapolis Bicycle Master Plan*, June 2012

1. Familiarize. We ask: "60 mi is $\frac{3}{10}$ of what length?" We make a drawing or at least visualize the problem. We let $b =$ the total number of miles of bicycling surface in the master plan.

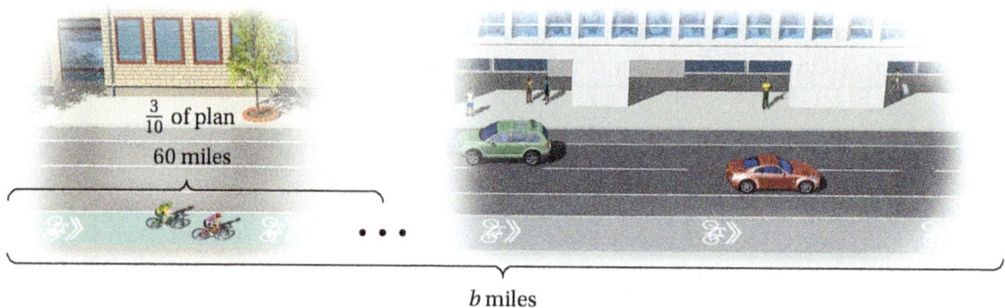

2. Translate. We translate to an equation:

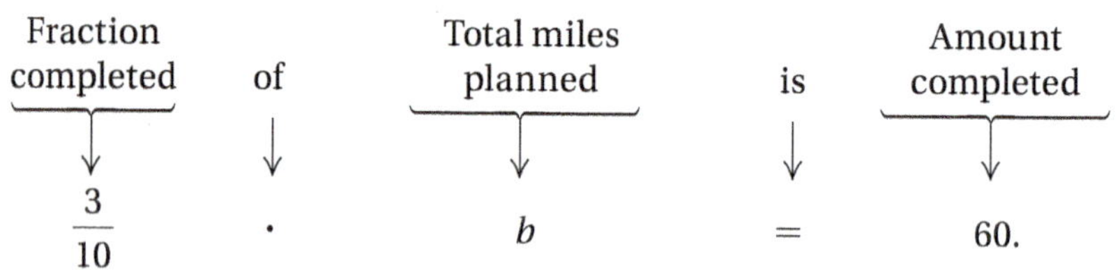

$$\frac{3}{10} \cdot b = 60.$$

3. Solve. We divide by $\frac{3}{10}$ on both sides and carry out the division:

$$b = 60 \div \frac{3}{10} = \frac{60}{1} \cdot \frac{10}{3} = \frac{60 \cdot 10}{1 \cdot 3} = \frac{3 \cdot 20 \cdot 10}{1 \cdot 3} = \frac{3}{3} \cdot \frac{20 \cdot 10}{1} = 200.$$

4. Check. We determine whether $\frac{3}{10}$ of 200 is 60: $\frac{3}{10} \cdot 200 = 60$. The answer, 200, checks.

5. State. The *Indianapolis Bicycle Master Plan* calls for 200 mi of bicycling surface.

7. *Sales Trip.* John Penna sells soybean seeds to seed companies. After he had driven 210 mi, $\frac{5}{6}$ of his sales trip was completed. How long was the total trip?

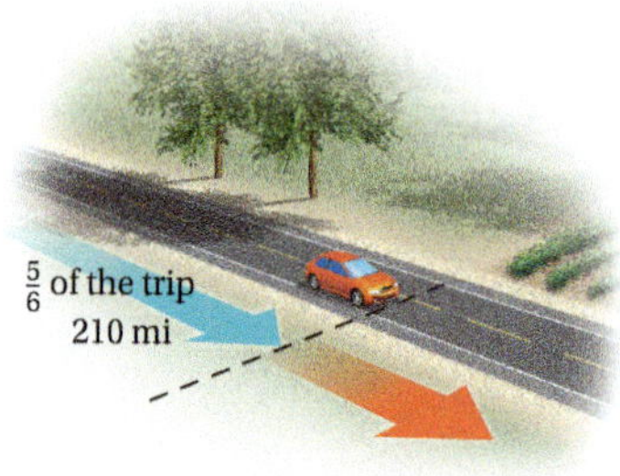

◀ **Do Exercise 7.**

Answers

5. 80 loops **6.** 16 servings **7.** 252 mi

Translating for Success

1. *Valentine Boxes.* Jane's Fudge Shop is preparing Valentine boxes. How many pounds of fudge will be needed to fill 80 boxes if each box contains $\frac{5}{16}$ lb?

2. *Gallons of Gasoline.* On the third day of a business trip, a sales representative used $\frac{4}{5}$ of a tank of gasoline. If the tank is a 20-gal tank, how many gallons were used on the third day?

3. *Purchasing a Shirt.* Tom received $36 for his birthday. If he spends $\frac{3}{4}$ of the gift on a new shirt, what is the cost of the shirt?

4. *Checkbook Balance.* The balance in Sam's checking account is $1456. He writes a check for $28 and makes a deposit of $52. What is the new balance?

5. *Valentine Boxes.* Jane's Fudge Shop prepared 80 lb of fudge for Valentine boxes. If each box contains $\frac{5}{16}$ lb, how many boxes can be filled?

The goal of these matching questions is to practice step (2), Translate, of the five-step problem-solving process. Translate each problem to an equation and select a correct translation from equations A–O.

A. $x = \frac{3}{4} \cdot 36$

B. $28 \cdot x = 52$

C. $x = 80 \cdot \frac{5}{16}$

D. $x = 1456 \div 28$

E. $\frac{5}{4} \cdot x = 20$

F. $20 = \frac{4}{5} \cdot x$

G. $x = 12 \cdot 28$

H. $x = \frac{4}{5} \cdot 20$

I. $\frac{3}{4} \cdot x = 36$

J. $x = 1456 - 52 - 28$

K. $x \div 28 = 1456$

L. $x = 52 - 28$

M. $x = 52 \cdot 28$

N. $x = 1456 - 28 + 52$

O. $\frac{5}{16} \cdot x = 80$

Answers on page A-6

6. *Gasoline Tank.* A gasoline tank contains 20 gal when it is $\frac{4}{5}$ full. How many gallons can it hold when full?

7. *Knitting a Scarf.* It takes Rachel 36 hr to knit a scarf. She can knit only $\frac{3}{4}$ hr per day because she is taking 16 hr of college classes. How many days will it take her to knit the scarf?

8. *Bicycle Trip.* On a recent 52-mi bicycle trip, David stopped to make a cell-phone call after completing 28 mi. How many more miles did he bicycle after the call?

9. *Crème de Menthe Thins.* Andes Candies L.P. makes Crème de Menthe Thins. How many 28-piece packages can be filled with 1456 pieces?

10. *Cereal Donations.* The Williams family donates 28 boxes of cereal weekly to the local Family in Crisis Center. How many boxes does this family donate in one year?

3.8 Exercise Set

For Extra Help
MyMathLab® MathXL® PRACTICE WATCH READ REVIEW

Reading Check

From the list on the right, choose the fraction to use in solving each equation by multiplying on both sides.

RC1. $\frac{3}{4}x = 10$ _____ **a)** $\frac{3}{4}$

RC2. $-\frac{4}{9}x = \frac{3}{4}$ _____ **b)** $\frac{4}{3}$

RC3. $-7 = \frac{9}{4}x$ _____ **c)** $-\frac{9}{4}$

RC4. $\frac{4}{9} = \frac{4}{3}x$ _____ **d)** $\frac{4}{9}$

a Use the multiplication principle to solve each equation. Don't forget to check!

1. $\frac{4}{5}x = 12$

2. $\frac{4}{3}x = 20$

3. $\frac{7}{3}a = 21$

4. $\frac{4}{5}a = 24$

5. $\frac{2}{9}y = -10$

6. $\frac{3}{8}y = -21$

7. $6t = \frac{12}{17}$

8. $3t = \frac{15}{14}$

9. $\frac{1}{4}x = \frac{3}{5}$

10. $\frac{1}{6}x = \frac{2}{7}$

11. $\frac{3}{2}t = -\frac{8}{7}$

12. $\frac{4}{3}t = -\frac{5}{2}$

13. $\frac{4}{5} = -10a$

14. $\frac{6}{5} = -12a$

15. $x \cdot \frac{9}{5} = \frac{3}{10}$

16. $x \cdot \frac{10}{3} = \frac{8}{15}$

17. $-\frac{1}{10}x = 8$

18. $-\frac{1}{11}x = -5$

19. $a \cdot \frac{9}{7} = -\frac{3}{14}$

20. $a\left(-\frac{9}{4}\right) = -\frac{3}{10}$

21. $-x = \frac{7}{13}$

22. $-x = \frac{7}{11}$

23. $-x = -\frac{27}{31}$

24. $-x = -\frac{35}{39}$

25. $7t = 6$

26. $-6t = 1$

27. $-24 = -10a$

28. $-18 = -20a$

29. $-\frac{14}{9} = \frac{10}{3}t$

30. $-\frac{15}{7} = \frac{3}{2}t$

31. $n \cdot \frac{4}{15} = \frac{12}{25}$

32. $n \cdot \frac{5}{16} = \frac{15}{14}$

33. $-\frac{7}{20}x = -\frac{21}{10}$

34. $-\frac{7}{15}x = -\frac{21}{10}$

35. $-\frac{25}{17} = -\frac{35}{34}a$

36. $-\frac{49}{45} = -\frac{28}{27}a$

Solve.

37. *Extension Cords.* An electrical supplier sells rolls of SJO 14-3 cable to a company that makes extension cords. It takes $\frac{7}{3}$ ft of cable to make each cord. How many extension cords can be made with a roll of cable containing 2240 ft of cable?

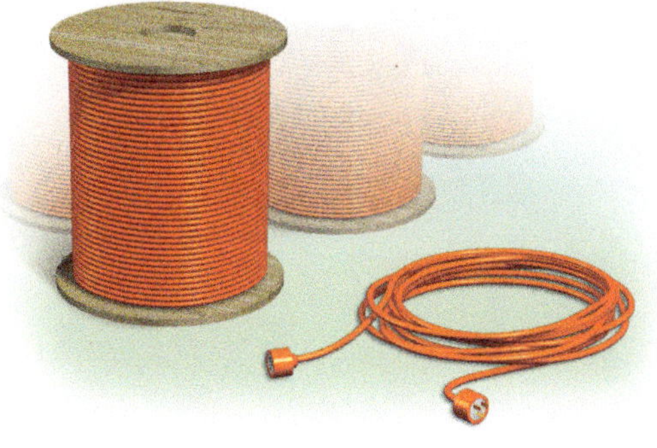

38. Benny uses $\frac{2}{5}$ gram (g) of toothpaste each time he brushes his teeth. If Benny buys a 30-g tube, how many times will he be able to brush his teeth?

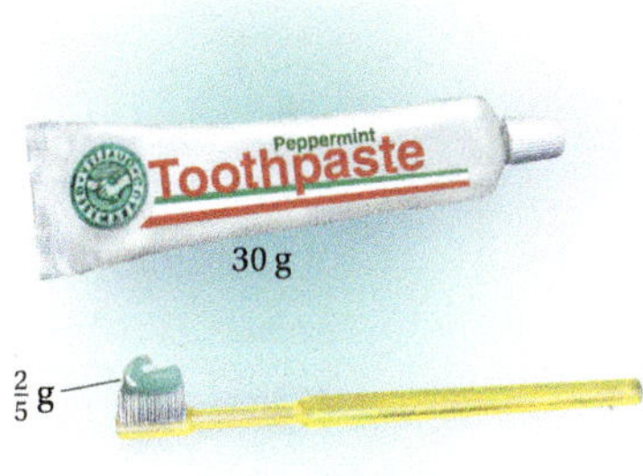

39. *Sewing.* A pair of basketball shorts requires $\frac{3}{4}$ yd of nylon. How many pairs of shorts can be made from 24 yd of nylon?

40. *Sewing.* A child's baseball shirt requires $\frac{5}{6}$ yd of fabric. How many shirts can be made from 25 yd of fabric?

41. How many $\frac{2}{3}$-cup sugar bowls can be filled from 16 cups of sugar?

42. For a party, Kyrsten makes an 8-ft submarine sandwich. If one serving is $\frac{2}{3}$ ft, how many servings does Kyrsten's sub contain?

43. A bucket had 12 L of water in it when it was $\frac{3}{4}$ full. How much could it hold when full?

44. A tank had 20 L of gasoline in it when it was $\frac{4}{5}$ full. How much could it hold when full?

45. ***Packaging.*** The South Shore Co-op prepackages cheddar cheese in $\frac{3}{4}$-lb packages. How many packages can be made from a 15-lb wheel of cheese?

46. ***Meal Planning.*** Ian purchased 6 lb of cold cuts for a luncheon. If Ian is to allow $\frac{3}{8}$ lb per person, how many people can attend the luncheon?

47. ***Art Supplies.*** The Ferristown School District purchased $\frac{3}{4}$ T (ton) of clay. The clay is to be shared equally among the district's 6 art departments. How much will each art department receive?

48. ***Gardening.*** The Bingham community garden is to be split into 16 equally sized plots. If the garden occupies $\frac{3}{4}$ acre of land, how large will each plot be?

49. ***Honey.*** A worker bee will produce $\frac{1}{12}$ tsp (teaspoon) of honey in her lifetime. How many worker bees does it take to produce $\frac{3}{4}$ tsp of honey?

Source: www.pbs.org/wgbh/nova/bees/buzz.html

50. ***Gardening.*** Large quantities of soil, gravel, or mulch are normally sold by the *yard* (yd). Although technically the unit for volume is *cubic yard* (yd^3), in this context only the word *yard* is used. Green Season Gardening uses about $\frac{2}{3}$ yd of bark mulch per customer every spring. How many customers can they accommodate with one 30-yd load of bark mulch?

51. ***Writing.*** On Monday, Ayesha wrote 450 words of an essay that was due on Friday. She then calculated that she had written $\frac{3}{16}$ of the minimum number of words required. What was the minimum number of words required for the essay?

52. ***Landscaping.*** As part of a semester project for an urban planning course, Fridrik planted 240 perennials in the median strip of an avenue. If Fridrik planted $\frac{5}{32}$ of the total number of perennials planted, how many perennials were planted in all?

53. ***Running.*** Chad runs 12 mi three days a week. In bad weather, he runs on an indoor track that is $\frac{3}{8}$ mi long. How many laps must he complete in order to run 12 mi?

54. ***Herbal Tea.*** At Perfect Tea, Donna fills each bag of Sereni-Tea with $\frac{3}{5}$ g (gram) of chamomile. If she begins with 51 g of chamomile, how many tea bags can she fill?

55. Yoshi Teramoto sells tools to hardware stores. After driving 180 kilometers (km), he has completed $\frac{5}{8}$ of a sales trip. How long is the total trip? How many kilometers are left to drive?

56. A piece of coaxial cable $\frac{4}{5}$ meter (m) long is to be cut into 8 pieces of the same length. What is the length of each piece?

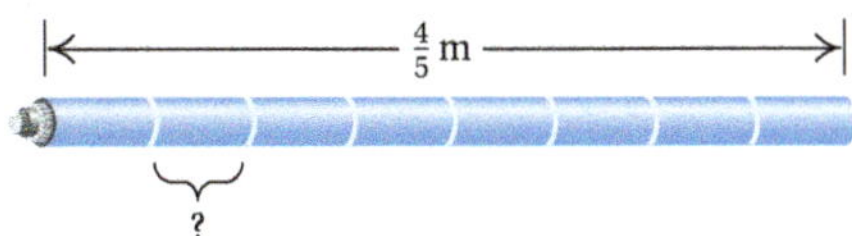

Pitch of a Screw. The pitch of a screw is the distance between its threads. With each complete rotation, the screw goes in or out a distance equal to its pitch. Use this information to do Exercises 57 and 58.

57. After a screw has been turned 8 complete rotations, it is extended $\frac{1}{2}$ in. into a piece of wallboard. What is the pitch of the screw?

58. The pitch of a screw is $\frac{3}{32}$ in. How many complete rotations are necessary to drive the screw $\frac{3}{4}$ in. into a piece of pine wood?

Skill Maintenance

Simplify.

59. $-23 + 49$ [2.2a]

60. $-69 + 27$ [2.2a]

61. $-38 - 29$ [2.3a]

62. $-47 - 18$ [2.3a]

63. $36 \div (-3)^2 \times (7 - 2)$ [2.5b]

64. $(-37 - 12 + 1) \div (-2)^3$ [2.5b]

Form an equivalent expression by combining like terms. [2.7a]

65. $13x + 4x$

66. $9a - 5a$

67. $2a + 3 + 5a$

68. $3x - 7 + x$

Synthesis

Solve.

69. $2x - 7x = -\frac{10}{9}$

70. $\left(-\frac{4}{7}\right)^2 = \left(\frac{2^3 - 9}{3}\right)^3 x$

Solve using the five-step problem-solving approach.

71. A package of coffee beans weighed $\frac{21}{32}$ lb when it was $\frac{3}{4}$ full. How much could the package hold when completely filled?

72. After swimming $\frac{2}{7}$ mi, Katie had swum $\frac{3}{4}$ of the race. How long a race was Katie competing in?

73. A block of Swiss cheese is 12 in. long. How many slices will it yield if half of the brick is cut by a slicer set for $\frac{3}{32}$-in. slices and half is cut by a slicer set for $\frac{5}{32}$-in. slices?

74. If $\frac{1}{3}$ of a number is $\frac{1}{4}$, what is $\frac{1}{2}$ of the number?

75. See Exercise 49. There are 3 teaspoons in a tablespoon and 4 tablespoons in $\frac{1}{4}$ cup. How many worker bees does it take to produce $\frac{1}{2}$ cup of honey?

76. See Exercise 50. If each customer of Green Season Gardening uses $\frac{3}{4}$ yd of bark mulch and Green Season charges each customer \$45 for the mulch, how much will Green Season receive from a 25-yd load of mulch?

CHAPTER 3

Summary and Review

Vocabulary Reinforcement

Fill in each blank with the correct term from the list at the right. Some of the choices may not be used.

1. For any number a, $a \cdot 1 = a$. The number 1 is the ______________ identity. [3.5a]
2. In the product $10 \cdot \frac{3}{4}$, 10 and $\frac{3}{4}$ are called ______________. [3.2a]
3. A natural number that has exactly two different factors, only itself and 1, is called a(n) ______________ number. [3.2b]
4. In the fraction $\frac{4}{17}$, we call 17 the ______________. [3.3a]
5. Since $\frac{2}{5}$ and $\frac{6}{15}$ are two names for the same number, we say that $\frac{2}{5}$ and $\frac{6}{15}$ are ______________ fractions. [3.5a]
6. The product of 6 and $\frac{1}{6}$ is 1. We say that 6 and $\frac{1}{6}$ are ______________. [3.7a]
7. Since $20 = 4 \cdot 5$, we say that $4 \cdot 5$ is a ______________ of 20. [3.2a]
8. Since $20 = 4 \cdot 5$, we say that 20 is a ______________ of 5. [3.1a]

equivalent
additive
multiplicative
reciprocals
factors
prime
composite
numerator
denominator
factorization
variables
multiple

Concept Reinforcement

Determine whether each statement is true or false.

______________ 1. For any natural number n, $\frac{n}{n} > \frac{0}{n}$. [3.3b]

______________ 2. A number is divisible by 10 if its ones digit is 0 or 5. [3.1b]

______________ 3. If a number is divisible by 9, then it is also divisible by 3. [3.1b]

______________ 4. The fraction $\frac{-6}{8}$ is equivalent to the fraction $\frac{15}{-20}$. [3.5c]

Study Guide

Objective 3.2a Find the factors of a number.

Example Find the factors of 84.

We find as many "two-factor" factorizations as we can.

$1 \cdot 84$ $4 \cdot 21$
$2 \cdot 42$ $6 \cdot 14$
$3 \cdot 28$ $7 \cdot 12$ ← Since 8, 9, 10, and 11 are not factors, we are finished.

The factors are 1, 2, 3, 4, 6, 7, 12, 14, 21, 28, 42, and 84.

Practice Exercise

1. Find the factors of 104.

Objective 3.2c Find the prime factorization of a composite number.

Example Find the prime factorization of 84.

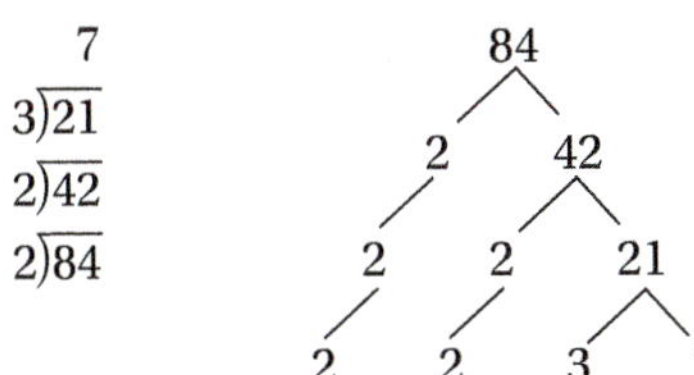

Thus, $84 = 2 \cdot 2 \cdot 3 \cdot 7$.

Practice Exercise

2. Find the prime factorization of 104.

Objective 3.3b Simplify fraction notation like n/n to 1, $0/n$ to 0, and $n/1$ to n.

Example Simplify $\frac{6}{6}$, $\frac{0}{6}$, and $\frac{6}{1}$.

$$\frac{6}{6} = 1, \quad \frac{0}{6} = 0, \quad \frac{6}{1} = 6$$

Practice Exercise

3. Simplify $\frac{0}{18}$, $\frac{18}{18}$, and $\frac{18}{1}$.

Objective 3.5b Simplify fraction notation.

Example Simplify: $\frac{315}{1650}$.

Using the test for divisibility by 5, we see that both the numerator and the denominator are divisible by 5:

$$\frac{315}{1650} = \frac{5 \cdot 63}{5 \cdot 330} = \frac{5}{5} \cdot \frac{63}{330} = 1 \cdot \frac{63}{330}$$

$$= \frac{63}{330} = \frac{3 \cdot 21}{3 \cdot 110} = \frac{3}{3} \cdot \frac{21}{110} = 1 \cdot \frac{21}{110} = \frac{21}{110}.$$

Practice Exercise

4. Simplify: $\frac{100}{280}$.

Objective 3.5c Test to determine whether two fractions are equivalent.

Example Use $=$ or $\neq$ for $\square$ to write a true sentence:

$$\frac{10}{54} \square \frac{15}{81}.$$

We find the cross products. We multiply 10 and 81: $10 \cdot 81 = 810$. Then we multiply 54 and 15: $54 \cdot 15 = 810$. Because the cross products are the same, we have

$$\frac{10}{54} = \frac{15}{81}.$$

If the cross products had been different, the fractions would not be equal.

Practice Exercise

5. Use $=$ or $\neq$ for $\square$ to write a true sentence:

$$\frac{8}{48} \square \frac{6}{44}.$$

Objective 3.6a Multiply and simplify using fraction notation.

Example Multiply and simplify: $\frac{7}{16} \cdot \frac{40}{49}$.

$$\frac{7}{16} \cdot \frac{40}{49} = \frac{7 \cdot 40}{16 \cdot 49} = \frac{7 \cdot 2 \cdot 2 \cdot 2 \cdot 5}{2 \cdot 2 \cdot 2 \cdot 2 \cdot 7 \cdot 7}$$

$$= \frac{2 \cdot 2 \cdot 2 \cdot 7}{2 \cdot 2 \cdot 2 \cdot 7} \cdot \frac{5}{2 \cdot 7} = 1 \cdot \frac{5}{14} = \frac{5}{14}$$

Practice Exercise

6. Multiply and simplify: $\frac{80}{3} \cdot \frac{21}{72}$.

Objective 3.7b Divide and simplify using fraction notation.

Example Divide and simplify: $\frac{9}{20} \div \frac{18}{25}$.

$$\frac{9}{20} \div \frac{18}{25} = \frac{9}{20} \cdot \frac{25}{18} = \frac{9 \cdot 25}{20 \cdot 18} = \frac{3 \cdot 3 \cdot 5 \cdot 5}{2 \cdot 2 \cdot 5 \cdot 2 \cdot 3 \cdot 3}$$

$$= \frac{3 \cdot 3 \cdot 5}{3 \cdot 3 \cdot 5} \cdot \frac{5}{2 \cdot 2 \cdot 2} = 1 \cdot \frac{5}{8} = \frac{5}{8}$$

Practice Exercise

7. Divide and simplify: $\frac{9}{4} \div \frac{45}{14}$.

Objective 3.8b Solve problems by using the multiplication principle.

Example A rental car had 18 gal of gasoline when its gas tank was $\frac{6}{7}$ full. How much could the tank hold when full?

The equation that corresponds to the situation is

$$\frac{6}{7} \cdot g = 18.$$

We multiply both sides by $\frac{7}{6}$:

$$\frac{7}{6} \cdot \frac{6}{7} \cdot g = \frac{7}{6} \cdot 18$$

$$g = \frac{7}{6} \cdot \frac{18}{1} = \frac{7 \cdot 3 \cdot \cancel{6}}{\cancel{6} \cdot 1} = 21.$$

The rental car can hold 21 gal of gasoline.

Practice Exercise

8. A flower vase has $\frac{7}{4}$ cups of water in it when it is $\frac{3}{4}$ full. How much can it hold when full?

Review Exercises

1. Multiply by 1, 2, 3, and so on, to find ten multiples of 8. [3.1a]

Use the tests for divisibility to answer Exercises 2–6. [3.1b]

2. Determine whether 3920 is divisible by 6.

3. Determine whether 68,537 is divisible by 3.

4. Determine whether 673 is divisible by 5.

5. Determine whether 4936 is divisible by 2.

6. Determine whether 5238 is divisible by 9.

Find all the factors of each number. [3.2a]

7. 60 **8.** 176

Classify each number as prime, composite, or neither. [3.2b]

9. 37 **10.** 1 **11.** 91

Find the prime factorization of each number. [3.2c]

12. 70 **13.** 72

14. 45 **15.** 150

16. 648 **17.** 1200

18. Identify the numerator and the denominator of $\frac{9}{7}$. [3.3a]

What part is shaded? [3.3a]

19. **20.**

21. For a committee in the United States Senate that consists of 3 Democrats and 5 Republicans, what is the ratio of: [3.3a]

a) Democrats to Republicans?
b) Republicans to Democrats?
c) Democrats to the total number of members of the committee?

Simplify, if possible. Assume that all variables are nonzero.

22. $\frac{0}{6}$ [3.3b]
23. $\frac{74}{74}$ [3.3b]
24. $\frac{48}{1}$ [3.3b]

25. $\frac{7x}{7x}$ [3.3b]
26. $-\frac{10}{15}$ [3.5b]
27. $\frac{7}{28}$ [3.5b]

28. $\frac{-42}{42}$ [3.5b]
29. $\frac{9m}{12m}$ [3.5b]
30. $\frac{-12}{-30}$ [3.5b]

31. $\frac{-27}{0}$ [3.3b]
32. $\frac{140}{490}$ [3.5b]
33. $\frac{288}{2025}$ [3.5b]

Find an equivalent expression for each number, using the denominator indicated. Use multiplication by 1. [3.5a]

34. $\frac{5}{7} = \frac{?}{21}$
35. $\frac{-6}{11} = \frac{?}{55}$

36. Simplify, if possible, the fractions on this circle graph. [3.5b]

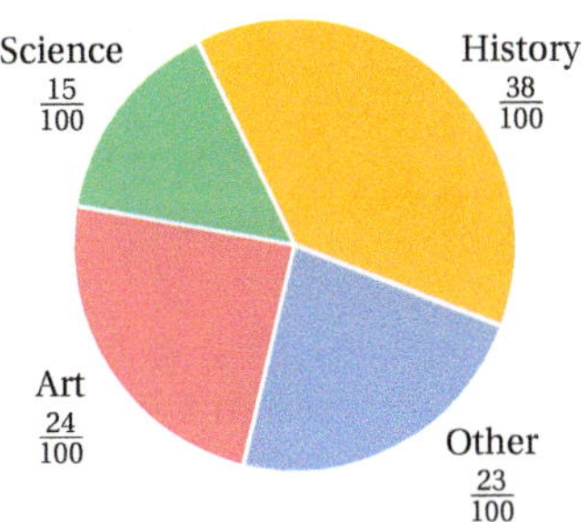

Use = or ≠ for □ to write a true sentence. [3.5c]

37. $\frac{3}{5} \square \frac{4}{6}$
38. $\frac{4}{-7} \square \frac{-8}{14}$

39. $\frac{4}{5} \square \frac{5}{6}$
40. $\frac{4}{3} \square \frac{28}{21}$

Find the reciprocal of each number. [3.7a]

41. $\frac{2}{13}$
42. -7

43. $\frac{1}{8}$
44. $\frac{3x}{5y}$

Perform the indicated operation and, if possible, simplify.

45. $\frac{2}{9} \cdot \frac{7}{5}$ [3.4b]
46. $\frac{3}{x} \cdot \frac{y}{7}$ [3.4b]

47. $\frac{3}{4} \cdot \frac{8}{9}$ [3.6a]
48. $-10 \cdot \frac{7}{5}$ [3.6a]

49. $\frac{11}{3} \cdot \frac{30}{77}$ [3.6a]
50. $\frac{4a}{7} \cdot \frac{7}{4a}$ [3.6a]

51. $\frac{6}{5} \cdot 20x$ [3.6a]
52. $\frac{3}{14} \div \frac{6}{7}$ [3.7b]

53. $20 \div \frac{3}{4}$ [3.7b]
54. $-\frac{5}{36} \div \left(-\frac{25}{12}\right)$ [3.7b]

55. $21 \div \frac{7}{2a}$ [3.7b]
56. $-\frac{23}{25} \div \frac{23}{25}$ [3.7b]

57. $\frac{\frac{21}{30}}{\frac{14}{15}}$ [3.7b]
58. $\frac{-\frac{2}{3}}{-\frac{3}{2}}$ [3.7b]

Solve. [3.8a]

59. $\frac{2}{3}x = 160$
60. $\frac{3}{8} = -\frac{5}{4}t$

61. $-\frac{1}{7}n = -4$
62. $y \cdot \frac{1}{2} = \frac{1}{3}$

Find the area. [3.6b]

63.

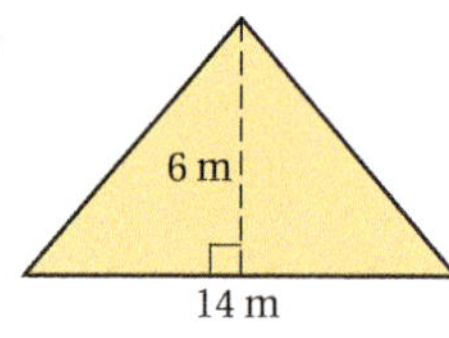

64.

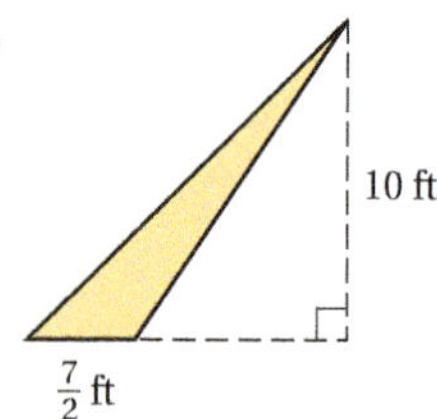

Solve. [3.8b]

65. A road crew repaves $\frac{1}{12}$ mi of road each day. How long will it take the crew to repave a $\frac{3}{4}$-mi stretch of road?

66. *Level of Education and Median Income.* The median yearly income of individuals with an associate's degree is approximately $\frac{3}{4}$ of the median income of individuals with a bachelor's degree. If the median income for those with bachelor's degrees is \$42,780, what is the median income of those with associate's degrees?

Source: U.S. Census Bureau

67. After driving 600 km, the Buxton family has completed $\frac{3}{5}$ of their vacation. How long is the total trip?

68. Molly is making a pepper steak recipe that calls for $\frac{2}{3}$ cup of green bell peppers. How much would be needed to make $\frac{1}{2}$ recipe?

69. The Winchester swim team has 4 swimmers in a $\frac{2}{3}$-mi relay race. If each swims the same distance, how far will each person swim?

70. A book bag requires $\frac{4}{5}$ yd of fabric. How many bags can be made from 48 yd?

71. Solve: $\frac{2}{13} \cdot x = \frac{1}{2}$. [3.8a]

A. $\frac{1}{13}$ **B.** 13 **C.** $\frac{4}{13}$ **D.** $\frac{13}{4}$

72. Multiply and simplify: $\frac{15}{26} \cdot \frac{13}{90}$. [3.6a]

A. $\frac{195}{234}$ **B.** $\frac{1}{12}$ **C.** $\frac{3}{36}$ **D.** $\frac{13}{156}$

Synthesis

73. Simplify: $\frac{15x}{14z} \cdot \frac{17yz}{35xy} \div \left(-\frac{3}{7}\right)^2$. [3.6a], [3.7b]

74. What digit(s) could be inserted in the ones place to make 574 _____ divisible by 6? [3.1b]

75. 🖩 In the division below, find a and b. [3.7b]

$$\frac{19}{24} \div \frac{a}{b} = \frac{187{,}853}{268{,}224}$$

76. A prime number that remains a prime number when its digits are reversed is called a **palindrome prime**. For example, 17 is a palindrome prime because both 17 and 71 are primes. Which of the following numbers are palindrome primes? [3.2b]

13, 91, 16, 11, 15, 24, 29, 101, 201, 37

Understanding Through Discussion and Writing

1. A student incorrectly insists that $\frac{2}{5} \div \frac{3}{4}$ is $\frac{15}{8}$. What mistake is he probably making? [3.7b]

2. Use the number 9432 to explain why the test for divisibility by 9 works. [3.1b]

3. A student claims that "taking $\frac{1}{2}$ of a number is the same as dividing by $\frac{1}{2}$." Explain the error in this reasoning. [3.7b]

4. On p. 174 we explained, using words and pictures, why $\frac{2}{5} \cdot \frac{3}{4}$ equals $\frac{6}{20}$. Present a similar explanation of why $\frac{2}{3} \cdot \frac{4}{7}$ equals $\frac{8}{21}$. [3.4b]

5. Without performing the division, explain why $5 \div \frac{1}{7}$ is a greater number than $5 \div \frac{2}{3}$. [3.5c], [3.7b]

6. If a fraction's numerator and denominator have no factors (other than 1) in common, can the fraction be simplified? Why or why not? [3.5b]

CHAPTER 3

Test

For Extra Help — For step-by-step test solutions, access the Chapter Test Prep Videos in MyMathLab® or on YouTube (search "BittingerPreIntro" and click on Channels).

1. Determine whether 5682 is divisible by 3. Do not use long division.

2. Determine whether 7018 is divisible by 5. Do not use long division.

3. Find all the factors of 90.

4. Determine whether 93 is prime, composite, or neither.

Find the prime factorization of each number.

5. 36

6. 60

7. Identify the numerator and the denominator of $\frac{4}{9}$.

8. What part is shaded?

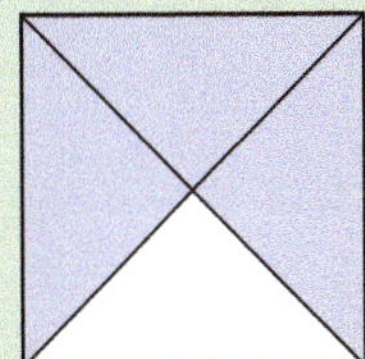

9. What part of the set is shaded?

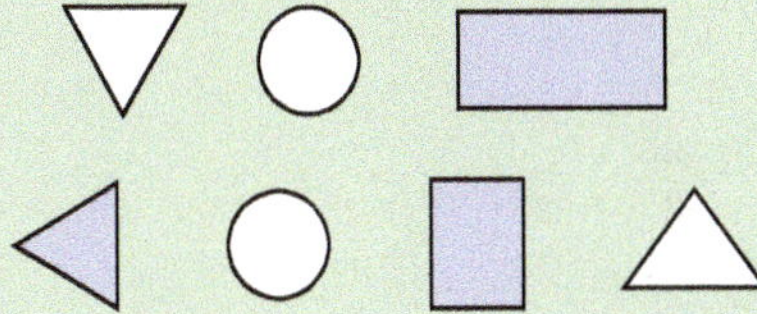

10. *Cholesterol.* Morrison's cholesterol test showed that his total cholesterol level was 180 mg/dL, his HDL level was 47 mg/dL, and his LDL level was 93 mg/dL.

a) What was the ratio of total cholesterol to HDL cholesterol?
b) What was the ratio of HDL cholesterol to LDL cholesterol?

Simplify, if possible. Assume that all variables are nonzero.

11. $\frac{32}{1}$

12. $\frac{-12}{-12}$

13. $\frac{0}{16}$

14. $\frac{-8}{24}$

15. $\frac{42}{7}$

16. $\frac{9x}{45x}$

17. $\frac{-62}{0}$

18. $\frac{72}{108}$

Use $=$ or $\neq$ for $\square$ to write a true sentence.

19. $\frac{3}{4} \square \frac{6}{8}$

20. $\frac{5}{4} \square \frac{9}{7}$

21. Find an equivalent expression for $\frac{3}{8}$ with a denominator of 40.

Find the reciprocal.

22. $\frac{a}{42}$

23. -9

Perform the indicated operation. Simplify, if possible.

24. $\frac{2}{3} \cdot \frac{15}{4}$

25. $\frac{2}{11} \div \frac{3}{4}$

26. $3 \cdot \frac{x}{8}$

27. $\dfrac{\frac{4}{7}}{-\frac{8}{3}}$

28. $12 \div \frac{2}{3}$

29. $\frac{22c}{15} \cdot \frac{5}{33c}$

Solve.

30. A $\frac{3}{4}$-lb slab of cheese is shared equally by 5 people. How much does each person receive?

31. Monroe weighs $\frac{5}{7}$ of his dad's weight. If his dad weighs 175 lb, how much does Monroe weigh?

32. $\frac{7}{8} \cdot x = 56$

33. $\frac{7}{10} = \frac{-2}{5} \cdot t$

34. Find the area.

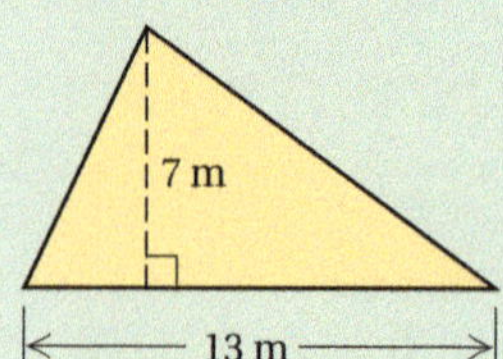

35. In which figure does the shaded part represent $\frac{7}{6}$ of the figure?

A.

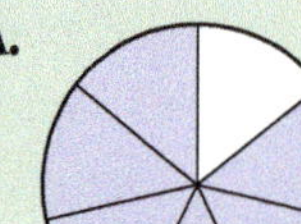

B.

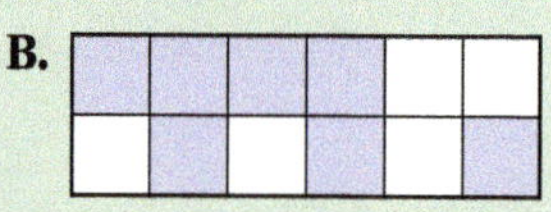

C.

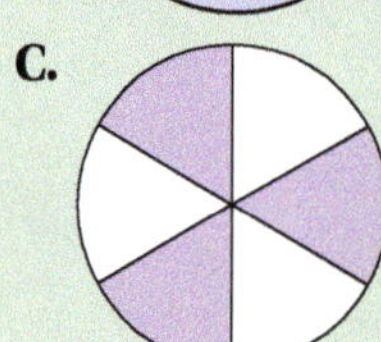

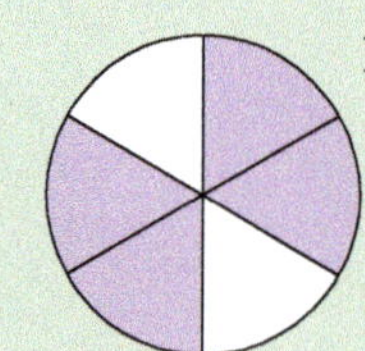

D.

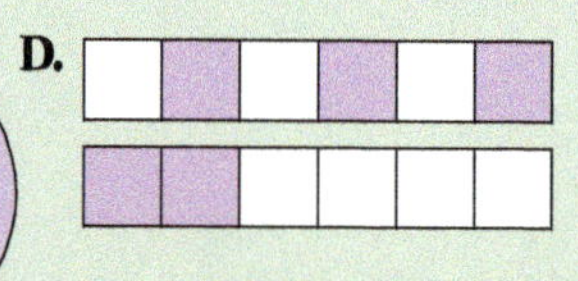

Synthesis

36. Grandma Shelby left $\frac{2}{3}$ of her $\frac{7}{8}$-acre apple farm to Karl. Karl gave $\frac{1}{4}$ of his share to his oldest daughter, Shannon. How much land did Shannon receive?

37. Simplify: $\left(-\frac{3}{8}\right)^2 \div \frac{6}{7} \cdot \frac{2}{9} \div (-5)$.

CHAPTER

4

Fraction Notation: Addition, Subtraction, and Mixed Numerals

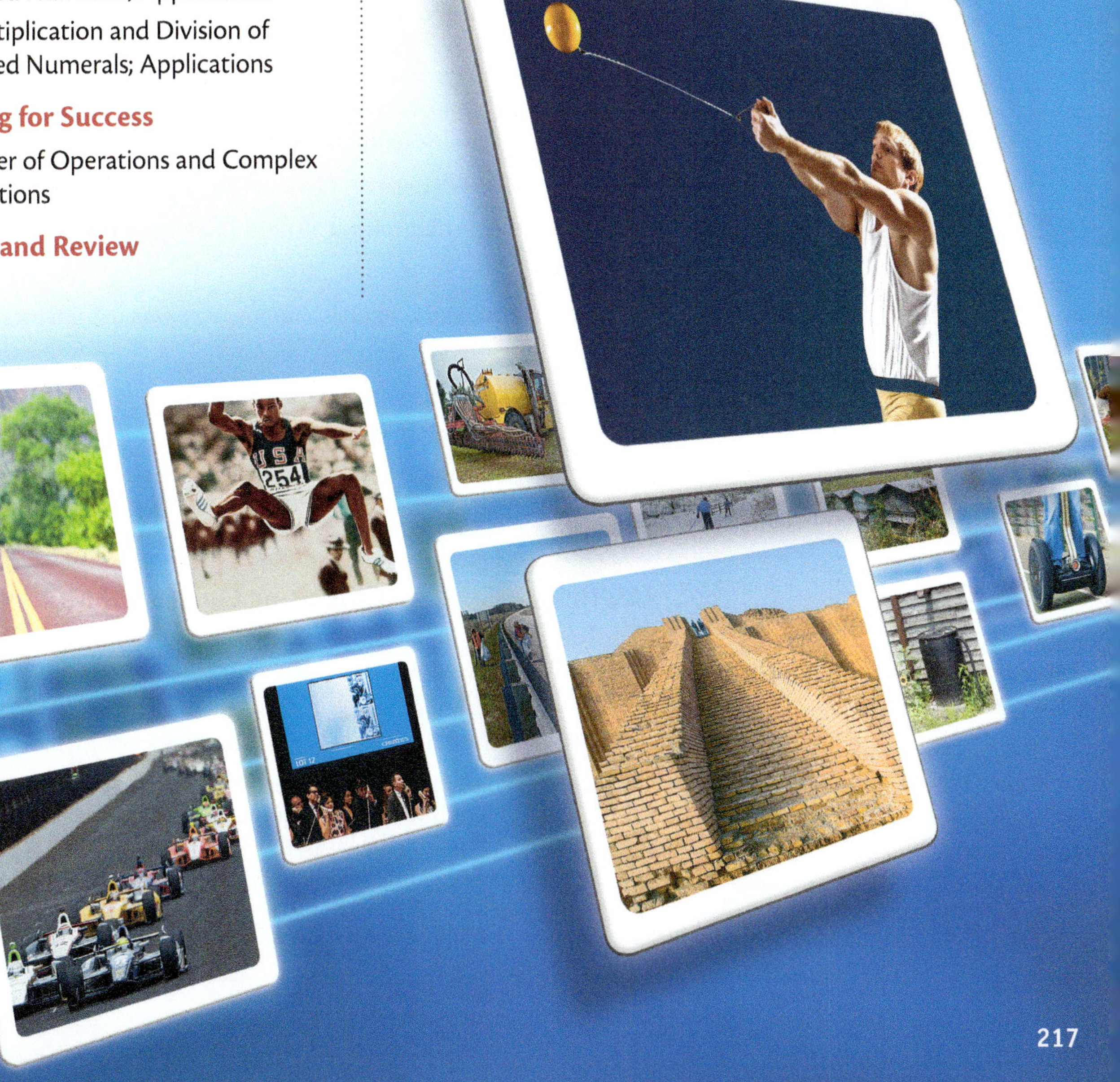

STUDYING FOR SUCCESS *Preparing for a Test*

- ☐ Make up your own test questions as you study.
- ☐ Do an overall review of the chapter focusing on the objectives and the examples.
- ☐ Do the exercises in the Mid-Chapter Review and in the end-of-chapter Review.
- ☐ Take the chapter test at the end of the chapter.

4.1 Least Common Multiples

OBJECTIVE

a Find the least common multiple, or LCM, of two or more numbers.

SKILL TO REVIEW

Objective 3.1a: Find some multiples of a number.

Multiply by 1, 2, 3, and so on, to find six multiples of each number.

1. 8 **2.** 25

1. Find the LCM of 9 and 15 by examining lists of multiples.

2. Find the LCM of 8 and 14 by examining lists of multiples.

Answers

Skill to Review:
1. 8, 16, 24, 32, 40, 48
2. 25, 50, 75, 100, 125, 150

Margin Exercises:
1. 45 **2.** 56

In order to add or subtract fractions, all fractions must share a common denominator. Before discussing common denominators, we look at how to find the **least common multiple** of two or more numbers.

a FINDING LEAST COMMON MULTIPLES

LEAST COMMON MULTIPLE, LCM

The **least common multiple**, or **LCM**, of two natural numbers is the smallest number that is a multiple of both numbers.

EXAMPLE 1 Find the LCM of 20 and 30.

First, we list some multiples of 20 by multiplying 20 by 1, 2, 3, and so on:

20, 40, 60, 80, 100, 120, 140, 160, 180, 200, 220, 240, ...

Then we list some multiples of 30 by multiplying 30 by 1, 2, 3, and so on:

30, 60, 90, 120, 150, 180, 210, 240, ...

Now we determine the smallest number *common* to both lists. The LCM of 20 and 30 is 60.

◀ Do Margin Exercises 1 and 2.

Next we develop two more efficient methods for finding LCMs. You may choose to learn only one method. (Consult your instructor.) However, if you intend to study algebra, you should definitely learn method 2.

Method 1: Finding LCMs Using One List of Multiples

The first method for finding LCMs works especially well when the numbers are relatively small.

Method 1. To find the LCM of a set of numbers using a list of multiples:

1. Determine whether the largest number is a multiple of the others. If it is, it is the LCM. That is, if the largest number has the others as factors, the LCM is that number.
2. If not, check multiples of the largest number until you get one that is a multiple of each of the others.

EXAMPLE 2 Find the LCM of 8 and 10.

1. 10 is the larger number, but it is not a multiple of 8.
2. Check multiples of 10:

$2 \cdot 10 = 20,$ Not a multiple of 8
$3 \cdot 10 = 30,$ Not a multiple of 8
$4 \cdot 10 = 40.$ A multiple of both 8 and 10

The LCM $= 40$.

EXAMPLE 3 Find the LCM of 4 and 14.

1. 14 is the larger number, but it is not a multiple of 4.
2. Check multiples of 14:

$2 \cdot 14 = 28.$ A multiple of 4

The LCM $= 28$.

EXAMPLE 4 Find the LCM of 8 and 32.

1. 32 is the larger number and 32 is a multiple of 8, so it is the LCM.

The LCM $= 32$.

Do Exercises 3–6. ▶

Find each LCM using one list of multiples.

3. 6, 9 **4.** 12, 15

5. 9, 36 **6.** 3, 20

To find the least common multiple of three numbers, we find the LCM of two of the numbers and then find the LCM of that number and the third number.

EXAMPLE 5 Find the LCM of 4, 10, and 15. We can start by finding the LCM of any two of the numbers. Let's use 10 and 15:

1. 15 is the larger number, but it is not a multiple of 10.
2. Check multiples of 15:

$2 \cdot 15 = 30.$ A multiple of 10. The LCM of 10 and 15 is 30.

Note now that any multiple of 30 will automatically be a multiple of 10 and 15. Thus, to find the LCM of 10, 15, and 4, we need only find the LCM of 30 and 4:

1. 30 is not a multiple of 4.
2. Check multiples of 30:

$2 \cdot 30 = 60.$ Since it is a multiple of 4, we know that 60 is the LCM of 30 and 4.

The LCM of 4, 10, and 15 is 60.

Do Exercises 7 and 8. ▶

Find each LCM using lists of multiples.

7. 30, 40, 50 **8.** 10, 20, 40

Method 2: Finding LCMs Using Prime Factorizations

A second method for finding LCMs uses prime factorizations and is usually the best method when the numbers are large. Consider again 20 and 30. Their prime factorizations are $20 = 2 \cdot 2 \cdot 5$ and $30 = 2 \cdot 3 \cdot 5$. Let's look at these prime factorizations in order to find the LCM. Any multiple of 20 will have to have *two* 2's as factors and *one* 5 as a factor. Any multiple of 30 will need to have *one* 2, *one* 3, and *one* 5 as factors.

Answers

3. 18 **4.** 60 **5.** 36 **6.** 60 **7.** 600 **8.** 40

The smallest number satisfying these conditions is

$$2 \cdot 2 \cdot 3 \cdot 5.$$

Two 2's, one 5; $2 \cdot 2 \cdot 3 \cdot 5$ is a multiple of 20.

One 2, one 3, one 5; $2 \cdot 2 \cdot 3 \cdot 5$ is a multiple of 30.

Thus, the LCM of 20 and 30 is $2 \cdot 2 \cdot 3 \cdot 5$, or 60. It has all the factors of 20 and all the factors of 30, but the factors are not repeated when they are common to both numbers.

Note that each prime factor is used the greatest number of times that it occurs in either of the individual factorizations.

Method 2. To find the LCM of two numbers using prime factorizations:

1. Write the prime factorization of each number.
2. Select one of the factorizations and see whether it contains the other.
   a) If it does, it is the LCM.
   b) If it does not, multiply that factorization by those prime factors of the other number that it lacks. The final product is the LCM.
3. As a check, make sure that the LCM includes each factor the greatest number of times that it occurs in either factorization.

EXAMPLE 6 Find the LCM of 18 and 21.

1. We begin by writing the prime factorization of each number:

   $18 = 2 \cdot 3 \cdot 3$ and $21 = 3 \cdot 7.$

2. We select the factorization of 18: $2 \cdot 3 \cdot 3$.
   a) We note that $2 \cdot 3 \cdot 3$ does not contain the other factorization, $3 \cdot 7$.
   b) To find the LCM of 18 and 21, we multiply $2 \cdot 3 \cdot 3$ by the factor of 21 that it lacks, 7:

   $$\text{LCM} = 2 \cdot 3 \cdot 3 \cdot 7.$$

   18 is a factor.

   21 is a factor.

3. The greatest number of times that 2 occurs as a factor of 18 or 21 is **one** time; the greatest number of times that 3 occurs as a factor of 18 or 21 is **two** times; and the greatest number of times that 7 occurs as a factor of 18 or 21 is **one** time. To check, note that the LCM has exactly **one** 2, **two** 3's, and **one** 7. The LCM is $2 \cdot 3 \cdot 3 \cdot 7$, or 126.

EXAMPLE 7 Find the LCM of 7 and 21.

1. Because 7 is prime, we think of $7 = 7$ as a "factorization":

   $7 = 7$ and $21 = 3 \cdot 7.$

2. One factorization, $3 \cdot 7$, contains the other. Thus, the LCM is $3 \cdot 7$, or 21.

EXAMPLE 8 Find the LCM of 24 and 36.

1. We write the prime factorization of each number:

$$24 = 2 \cdot 2 \cdot 2 \cdot 3 \quad \text{and} \quad 36 = 2 \cdot 2 \cdot 3 \cdot 3.$$

2. We select the factorization of 24.
   a) The factorization of 24 does not contain the factorization of 36.
   b) To find the LCM of 24 and 36, we multiply the factorization of 24 by any prime factors of 36 that it lacks. We need another factor of 3.

24 is a factor.

$$\text{LCM} = 2 \cdot 2 \cdot 2 \cdot 3 \cdot 3.$$

36 is a factor.

3. Note that the LCM includes 2 and 3 the greatest number of times that each appears as a factor of either 24 or 36. The LCM is $2 \cdot 2 \cdot 2 \cdot 3 \cdot 3$, or 72.

Do Exercises 9–12.

Exponential notation is often helpful when writing least common multiples. Let's reconsider Example 8 using exponents. The largest exponents indicate the greatest number of times that 2 and 3 occur as factors.

$24 = 2 \cdot 2 \cdot 2 \cdot 3 = 2^3 \cdot 3^1$ — 2^3 is the greatest power of 2

$36 = 2 \cdot 2 \cdot 3 \cdot 3 = 2^2 \cdot 3^2$ — 3^2 is the greatest power of 3

$\text{LCM} = 2 \cdot 2 \cdot 2 \cdot 3 \cdot 3 = 2^3 \cdot 3^2$, or 72.

Note that the greatest power of each factor is used to construct the LCM.

Lining up the different prime numbers in the factorizations can help us construct the LCM. This method also works well when finding the LCM of more than two numbers.

EXAMPLE 9 Find the LCM of 27, 90, and 84.

We find the prime factorization of each number and write the factorizations in exponential notation.

$$27 = 3 \cdot 3 \cdot 3 = 3^3$$
$$90 = 2 \cdot 3 \cdot 3 \cdot 5 = 2 \cdot 3^2 \cdot 5$$
$$84 = 2 \cdot 2 \cdot 3 \cdot 7 = 2^2 \cdot 3 \cdot 7$$

No one factorization contains the others. The prime numbers 2, 3, 5, and 7 appear as factors.

We write the factorizations, lining up all the powers of 2, the powers of 3, and so on.

$$27 = \qquad 3^3$$
$$90 = 2 \cdot 3^2 \cdot 5$$
$$84 = 2^2 \cdot 3 \cdot \qquad 7$$

The LCM is formed by choosing the greatest power of each factor:

$$2^2 \cdot 3^3 \cdot 5 \cdot 7 = 3780.$$

The LCM of 27, 90, and 84 is 3780.

Do Exercises 13 and 14.

Use prime factorizations to find the LCM.

9. 8, 10 **10.** 5, 30

11. 12, 48

12. 18, 40

1. $18 = 2 \cdot 3 \cdot \square$

   $40 = 2 \cdot 2 \cdot 2 \cdot \square$

2. Select the factorization of 40:

   $2 \cdot 2 \cdot 2 \cdot 5.$

   This is not a multiple of 18. We need two factors of $\square$.

3. $\text{LCM} = 2 \cdot 2 \cdot 2 \cdot 5 \cdot \square \cdot \square$
   $= \square$

Find the LCM.

13. 8, 18, 30 **14.** 10, 20, 25

Answers

9. 40 **10.** 30 **11.** 48 **12.** 360 **13.** 360 **14.** 100

Guided Solution:
12. 3, 5; 3; 3, 3, 360

EXAMPLE 10 Find the LCM of 8 and 25.

We write the prime factorization of each number in exponential notation.

$$8 = 2 \cdot 2 \cdot 2 = 2^3$$
$$25 = 5 \cdot 5 = 5^2$$

The prime numbers 2 and 5 appear as factors. We write the factorizations as

$$8 = 2^3$$
$$25 = \quad 5^2.$$

Note that the two numbers, 8 and 25, have no common prime factor. When this is the case, the LCM is just the product of the two numbers. Thus, the LCM is $2^3 \cdot 5^2 = 8 \cdot 25 = 200$.

Find the LCM.

15. 4, 9 **16.** 5, 6, 7

Do Exercises 15 and 16.

The same method works perfectly with variables.

EXAMPLE 11 Find the LCM of $7a^2b$ and ab^3.

We have the following factorizations:

$$7a^2b = 7 \cdot a \cdot a \cdot b \quad \text{and} \quad ab^3 = a \cdot b \cdot b \cdot b.$$

No one factorization contains the other.

Consider the factorization of $7a^2b$, which is $7 \cdot a \cdot a \cdot b$. Since ab^3 contains two more factors of b, we multiply the factorization of $7a^2b$ by $b \cdot b$.

$$7 \cdot a \cdot a \cdot b \cdot b \cdot b$$

$7a^2b$ is a factor.

ab^3 is a factor.

Find the LCM.

17. xy, yz

18. $5a^2, a^3b$

GS

$5a^2 = 5 \cdot$ ☐

$a^3b = \quad a^3 \cdot$ ☐

LCM $= 5 \cdot a^3 \cdot$ ☐, or ☐

As a second approach, we find the greatest power of each factor using exponential notation.

$$7a^2b = 7 \cdot a^2 \cdot b$$
$$ab^3 = \quad a \cdot b^3$$

The LCM is $7 \cdot a \cdot a \cdot b \cdot b \cdot b$, or $7a^2b^3$.

Do Exercises 17 and 18.

EXAMPLE 12 Find the LCM of $12x^2y^3z$ and $18x^4z^3$.

We write the factorizations using exponential notation:

$$12x^2y^3z = 2^2 \cdot 3 \cdot x^2 \cdot y^3 \cdot z \quad \text{and} \quad 18x^4z^3 = 2 \cdot 3^2 \cdot x^4 \cdot z^3.$$

No one factorization contains the other.

We form the LCM using the greatest power of each factor.

$$12x^2y^3z = 2^2 \cdot 3 \cdot x^2 \cdot y^3 \cdot z$$
$$18x^4z^3 = 2 \cdot 3^2 \cdot x^4 \cdot \quad z^3$$

The LCM is $2^2 \cdot 3^2 \cdot x^4 \cdot y^3 \cdot z^3$, or $36x^4y^3z^3$.

19. Find the LCM of $8a^3b^2$ and $10a^2c^4$.

Do Exercise 19.

Answers

15. 36 **16.** 210 **17.** xyz **18.** $5a^3b$
19. $40a^3b^2c^4$

Guided Solution:
18. $a^2, b; b, 5a^3b$

4.1 Exercise Set

For Extra Help MyMathLab® MathXL®  PRACTICE  WATCH  READ  REVIEW

Reading Check

Determine whether each statement is true or false.

_______________ **RC1.** Any two numbers have more than one common multiple.

_______________ **RC2.** If one number is a multiple of a second number, the larger number is the LCM of the two numbers.

_______________ **RC3.** If two numbers have no common prime factor, then the LCM of the numbers is their product.

_______________ **RC4.** LCMs cannot be found using prime factorizations.

a Find the LCM of each set of numbers or expressions.

1. 2, 4 **2.** 3, 15 **3.** 10, 25 **4.** 10, 15 **5.** 20, 40

6. 8, 12 **7.** 18, 27 **8.** 9, 11 **9.** 30, 50 **10.** 8, 36

11. 30, 40 **12.** 21, 27 **13.** 18, 24 **14.** 12, 18 **15.** 60, 70

16. 35, 45 **17.** 16, 36 **18.** 24, 32 **19.** 18, 20 **20.** 36, 48

21. 2, 3, 7 **22.** 2, 5, 9 **23.** 3, 6, 15 **24.** 6, 12, 18 **25.** 24, 36, 12

26. 8, 16, 22 **27.** 5, 12, 15 **28.** 12, 18, 40 **29.** 9, 12, 6 **30.** 8, 16, 12

31. 180, 100, 450 **32.** 18, 30, 50, 48 **33.** 8, 48 **34.** 16, 32 **35.** 10, 21

36. 14, 15 **37.** 75, 100 **38.** 81, 90 **39.** 12, 15, 60 **40.** 24, 36, 72

41. ab, bc

42. $7x, xy$

43. $3x, 9x^2$

44. $10x^4, 5x^3$

45. $4x^3, x^2y$

46. $6ab^2, 9a^3b$

47. $6r^3st^4, 8rs^2t$

48. $3m^2n^4p^5, 9mn^2p^4$

49. a^3b, b^2c, ac^2

50. x^2z^3, x^3y, y^2z

Applications of LCMs: Planet Orbits. Jupiter, Saturn, and Uranus all revolve around the sun. Jupiter takes 12 yr, Saturn 30 yr, and Uranus 84 yr to make a complete revolution. On a certain night, you look at Jupiter, Saturn, and Uranus and wonder how many years it will take before they have the same position again. (*Hint*: The number of years is the LCM of 12, 30, and 84.)

Source: *The Handy Science Answer Book*

51. How often will Jupiter and Saturn appear in the same direction in the night sky as seen from the earth?

52. How often will Jupiter and Uranus appear in the same direction in the night sky as seen from the earth?

53. How often will Saturn and Uranus appear in the same direction in the night sky as seen from the earth?

54. How often will Jupiter, Saturn, and Uranus appear in the same direction in the night sky as seen from the earth?

Skill Maintenance

Perform the indicated operation and, if possible, simplify.

55. $-38 + 52$ [2.2a]

56. $-18 \div \left(\frac{2}{3}\right)$ [3.7b]

57. $23 \cdot 345$ [1.4a]

58. $\frac{4}{5} \cdot \frac{10}{12}$ [3.6a]

59. $\frac{4}{5} \div \left(-\frac{7}{10}\right)$ [3.7b]

60. $382 - 549$ [2.3a]

Synthesis

Use a calculator and the multiples method to find the LCM of each pair of numbers.

61. 288; 324

62. 2700; 7800

63. 7719; 18,011

64. 17,385; 24,339

65. The tables at a flea market are either 6 ft long or 8 ft long. Each row consists entirely of 6-ft tables or entirely of 8-ft tables, and all rows are the same length. What is the shortest possible length of the rows at the flea market?

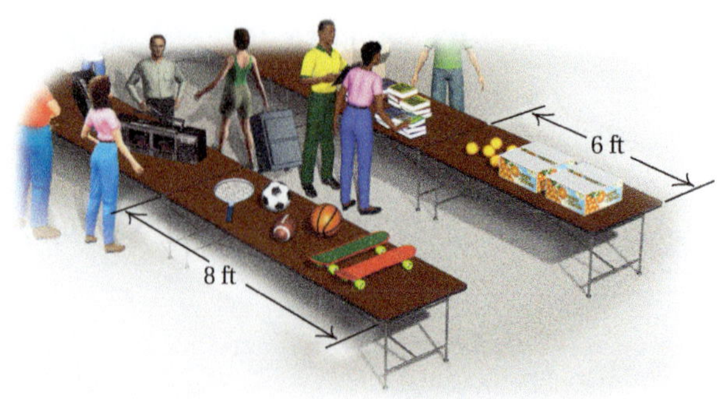

African Artistry. In southern Africa, the design of every woven handbag, or *gipatsi* (plural *sipatsi*), is created by repeating two or more geometric patterns. Each pattern encircles the bag, sharing the strands of fabric with any pattern above or below. The length, or period, of each pattern is the number of strands required to construct the pattern. For a gipatsi to be considered beautiful, each individual pattern must fit a whole number of times around the bag.

Source: Gerdes, Paulus. *Women, Art and Geometry in Southern Africa.* Asmara, Eritrea: Africa World Press, Inc., p. 5.

66. A weaver is using two patterns to create a gipatsi. Pattern A is 10 strands long, and pattern B is 3 strands long. What is the smallest number of strands that can be used to complete the gipatsi?

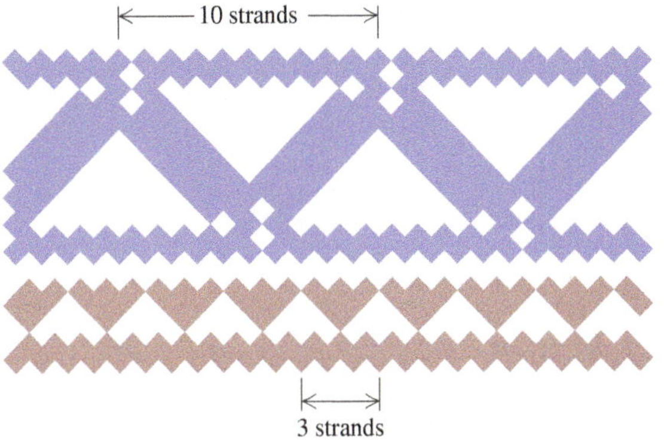

67. A weaver is using a four-strand pattern, a six-strand pattern, and an eight-strand pattern. What is the smallest number of strands that can be used to complete the gipatsi?

68. *Prescriptions.* Prescriptions for a 30-day supply of simvastatin and a 14-day supply of pain medication are filled at a pharmacy. Assuming the prescriptions are refilled regularly, how long will it be until they are both refilled on the same day?

69. Consider a^3b^2 and a^2b^5. Determine whether each of the following is the LCM of a^3b^2 and a^2b^5. Tell why or why not.

a) a^3b^3
b) a^2b^5
c) a^3b^5

70. Use Example 9 to help find the LCM of 27, 90, 84, 210, 108, and 50.

71. Use Examples 6 and 8 to help find the LCM of 18, 21, 24, 36, 63, 56, and 20.

72. Find three different pairs of numbers for which 56 is the LCM. Do not use 56 itself in any of the pairs.

73. Find three different pairs of numbers for which 54 is the LCM. Do not use 54 itself in any of the pairs.

4.2 Addition, Order, and Applications

OBJECTIVES

a Add using fraction notation when denominators are the same.

b Add using fraction notation when denominators are different.

c Use < or > to form a true statement with fraction notation.

d Solve problems involving addition with fraction notation.

SKILL TO REVIEW

Objective 3.2c: Find the prime factorization of a composite number.

Find the prime factorization of each number.

1. 96 **2.** 1400

a LIKE DENOMINATORS

Addition using fraction notation corresponds to combining or putting like things together, just as when we combined like terms. For example,

We combine two sets, each of which consists of equally sized parts of one object.

This is the resulting set.

1 1 1

$\frac{2}{8}$ + $\frac{3}{8}$ = $\frac{5}{8}$

2 eighths + 3 eighths = 5 eighths,

or $2 \cdot \frac{1}{8} + 3 \cdot \frac{1}{8} = 5 \cdot \frac{1}{8}$, or $\frac{2}{8} + \frac{3}{8} = \frac{5}{8}$.

Do Margin Exercise 1.

To add when denominators are the same,

a) add the numerators,

b) keep the denominator, and

c) simplify, if possible.

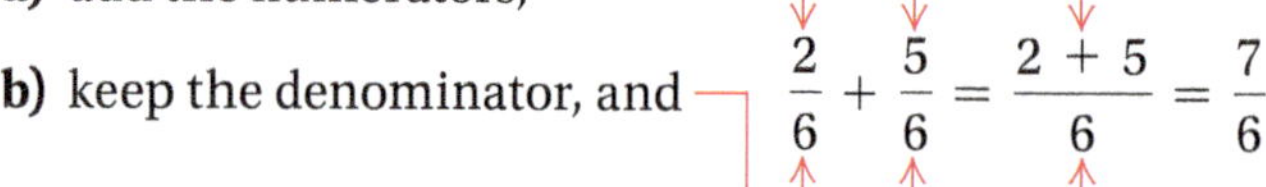
$$\frac{2}{6} + \frac{5}{6} = \frac{2+5}{6} = \frac{7}{6}$$

EXAMPLES Add and, if possible, simplify.

1. $\frac{2}{4} + \frac{1}{4} = \frac{2+1}{4} = \frac{3}{4}$ No simplifying is possible.

2. $\frac{3}{12} + \frac{5}{12} = \frac{3+5}{12} = \frac{8}{12}$ Adding numerators; the denominator remains unchanged.

$= \frac{4}{4} \cdot \frac{2}{3} = \frac{2}{3}$ Simplifying by removing a factor equal to 1: $\frac{4}{4} = 1$

3. $\frac{-11}{6} + \frac{3}{6} = \frac{-11+3}{6} = \frac{-8}{6}$

$= \frac{2}{2} \cdot \frac{-4}{3} = \frac{-4}{3}$, or $-\frac{4}{3}$ Removing a factor equal to 1: $\frac{2}{2} = 1$

4. $-\frac{2}{a} + \left(-\frac{3}{a}\right) = \frac{-2}{a} + \frac{-3}{a}$ Recall that $-\frac{m}{n} = \frac{-m}{n}$.

$= \frac{-2 + (-3)}{a} = \frac{-5}{a}$, or $-\frac{5}{a}$

Do Exercises 2–5.

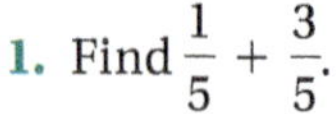
1. Find $\frac{1}{5} + \frac{3}{5}$.

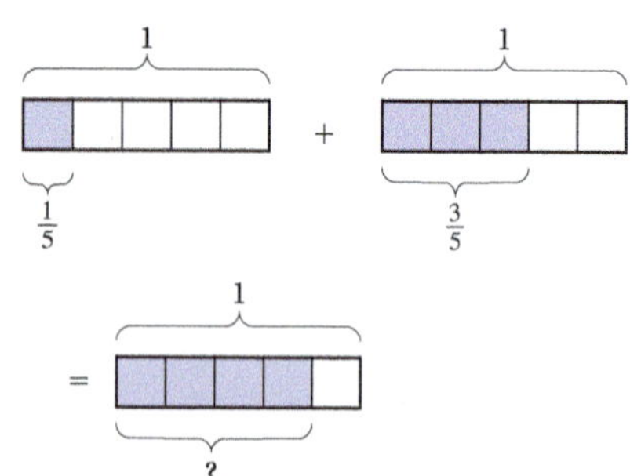

Add and, if possible, simplify.

2. $\frac{5}{13} + \frac{9}{13}$ **3.** $\frac{1}{3} + \frac{2}{3}$

4. $\frac{-5}{12} + \left(\frac{-1}{12}\right)$ **5.** $\frac{3}{x} + \frac{-7}{x}$

Answers

See next page.

We may need to add fractions when combining like terms.

EXAMPLE 5 Simplify by combining like terms: $\frac{2}{7}x + \frac{3}{7}x$.

$$\begin{aligned}\frac{2}{7}x + \frac{3}{7}x &= \left(\frac{2}{7} + \frac{3}{7}\right)x \quad \text{Try to do this step mentally.}\\ &= \frac{5}{7}x\end{aligned}$$

Do Exercises 6 and 7. ▶

Simplify by combining like terms.

6. $\frac{3}{10}a + \frac{1}{10}a$ **7.** $-\frac{3}{4}x + \frac{1}{4}x$

b DIFFERENT DENOMINATORS

We cannot add $\frac{1}{2} + \frac{1}{3}$ by simply adding numerators. However, by rewriting $\frac{1}{2}$ as $\frac{1}{2} \cdot \frac{3}{3} = \frac{3}{6}$ and $\frac{1}{3}$ as $\frac{1}{3} \cdot \frac{2}{2} = \frac{2}{6}$, we can determine the sum.

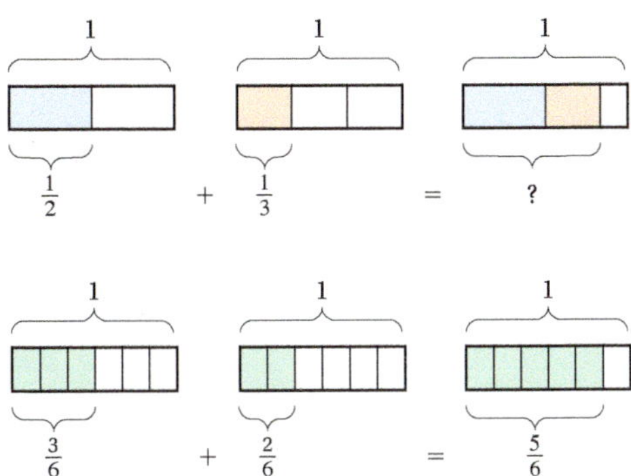

When denominators are different, we can find a common denominator by multiplying by 1. Consider the addition $\frac{3}{4} + \frac{1}{6}$ using two different denominators.

A. We use 24 as a common denominator:

$$\begin{aligned}\frac{3}{4} + \frac{1}{6} &= \frac{3}{4} \cdot \frac{6}{6} + \frac{1}{6} \cdot \frac{4}{4}\\ &= \frac{18}{24} + \frac{4}{24} = \frac{22}{24} = \frac{11}{12}.\end{aligned}$$

B. We use 12 as a common denominator:

$$\begin{aligned}\frac{3}{4} + \frac{1}{6} &= \frac{3}{4} \cdot \frac{3}{3} + \frac{1}{6} \cdot \frac{2}{2}\\ &= \frac{9}{12} + \frac{2}{12} = \frac{11}{12}.\end{aligned}$$

We had to simplify at the end of (A), but not in (B). In (B), we used the *least* common multiple of the denominators, 12, as the common denominator. That number is called the **least common denominator**, or **LCD**. We may still need to simplify when using the LCD, but it is usually easier than when we use a larger denominator.

> To add when denominators are different:
>
> **a)** Find the least common multiple of the denominators. That number is the least common denominator, LCD.
>
> **b)** Multiply by 1, writing 1 in the form of n/n, to express each fraction in an equivalent form that contains the LCD.
>
> **c)** Add the numerators, keeping the same denominator.
>
> **d)** Simplify, if possible.

Answers

Skill to Review:

1. $2 \cdot 2 \cdot 2 \cdot 2 \cdot 2 \cdot 3$
2. $2 \cdot 2 \cdot 2 \cdot 5 \cdot 5 \cdot 7$

Margin Exercises:

1. $\frac{4}{5}$ **2.** $\frac{14}{13}$ **3.** 1 **4.** $-\frac{1}{2}$ **5.** $-\frac{4}{x}$
6. $\frac{2}{5}a$ **7.** $-\frac{1}{2}x$

EXAMPLE 6 Add: $\frac{1}{8} + \frac{3}{4}$.

a) Since 4 is a factor of 8, the LCM of 4 and 8 is 8. Thus, the LCD is 8.

b) We need to find a fraction equivalent to $\frac{3}{4}$ with a denominator of 8:

$$\frac{1}{8} + \frac{3}{4} = \frac{1}{8} + \frac{3}{4} \cdot \frac{2}{2}.$$

Think: $4 \times \square = 8$. The answer is 2, so we multiply by 1, using $\frac{2}{2}$.

c) We add: $\frac{1}{8} + \frac{6}{8} = \frac{7}{8}$. Adding numerators

d) No simplifying is possible. The sum is $\frac{7}{8}$.

In Examples 7–10, we follow the same steps without labeling them.

Add using the least common denominator.

8. $\frac{2}{3} + \frac{1}{6}$

9. $\frac{3}{8} + \frac{5}{6}$

The LCD is ____.

$$\frac{3}{8} + \frac{5}{6} = \frac{3}{8} \cdot 1 + \frac{5}{6} \cdot 1$$
$$= \frac{3}{8} \cdot \frac{3}{3} + \frac{5}{6} \cdot \frac{\square}{\square}$$
$$= \frac{\square}{24} + \frac{\square}{24}$$
$$= \frac{\square}{24}$$

EXAMPLE 7 Add: $\frac{5}{6} + \frac{1}{9}$.

The LCD is 18. $6 = 2 \cdot 3$ and $9 = 3 \cdot 3$, so the LCM of 6 and 9 is $2 \cdot 3 \cdot 3$, or 18.

$$\frac{5}{6} + \frac{1}{9} = \frac{5}{6} \cdot \frac{3}{3} + \frac{1}{9} \cdot \frac{2}{2}$$

Think: $9 \times \square = 18$. The answer is 2, so we multiply by 1, using $\frac{2}{2}$.

Think: $6 \times \square = 18$. The answer is 3, so we multiply by 1, using $\frac{3}{3}$.

$$= \frac{15}{18} + \frac{2}{18} = \frac{17}{18}$$

Do Exercises 8 and 9.

EXAMPLE 8 Add: $\frac{3}{-5} + \frac{11}{10}$.

$$\frac{3}{-5} + \frac{11}{10} = \frac{-3}{5} + \frac{11}{10}$$

Recall that $\frac{m}{-n} = \frac{-m}{n}$. We generally avoid negative signs in the denominator. The LCD is 10.

$$= \frac{-3}{5} \cdot \frac{2}{2} + \frac{11}{10}$$
$$= \frac{-6}{10} + \frac{11}{10}$$
$$= \frac{5}{10} = \frac{1}{2}$$

We may still have to simplify, but simplifying is almost always easier if the LCD has been used.

10. Add: $\frac{1}{-6} + \frac{7}{18}$.

Do Exercise 10.

EXAMPLE 9 Add: $\frac{5}{8} + 2$.

$$\frac{5}{8} + 2 = \frac{5}{8} + \frac{2}{1}$$ Rewriting 2 in fraction notation

$$= \frac{5}{8} + \frac{2}{1} \cdot \frac{8}{8}$$ The LCD is 8.

$$= \frac{5}{8} + \frac{16}{8} = \frac{21}{8}$$

11. Add: $7 + \frac{3}{5}$.

Do Exercise 11.

Answers

8. $\frac{5}{6}$ **9.** $\frac{29}{24}$ **10.** $\frac{2}{9}$ **11.** $\frac{38}{5}$

Guided Solution:

9. 24; $\frac{4}{4}$, 9, 20, 29

EXAMPLE 10 Add: $\frac{9}{70} + \frac{11}{21} + \frac{-4}{15}$.

We need to determine the LCM of 70, 21, and 15:

$$70 = 2 \cdot 5 \cdot 7, \quad 21 = 3 \cdot 7, \quad 15 = 3 \cdot 5.$$

The LCM is $2 \cdot 3 \cdot 5 \cdot 7$, or 210.

$$\frac{9}{70} + \frac{11}{21} + \frac{-4}{15} = \frac{9}{70} \cdot \frac{3}{3} + \frac{11}{21} \cdot \frac{2 \cdot 5}{2 \cdot 5} + \frac{-4}{15} \cdot \frac{7 \cdot 2}{7 \cdot 2}$$

In each case, we multiply by 1 to obtain the LCD. To form 1, look at the prime factorization of the LCD and use the factor(s) missing from each denominator.

$$= \frac{9 \cdot 3}{70 \cdot 3} + \frac{11 \cdot 10}{21 \cdot 10} + \frac{-4 \cdot 14}{15 \cdot 14}$$

$$= \frac{27}{210} + \frac{110}{210} + \frac{-56}{210}$$

$$= \frac{137 + (-56)}{210} = \frac{81}{210}$$

$$= \frac{3 \cdot 3 \cdot 3 \cdot 3}{2 \cdot 3 \cdot 5 \cdot 7} = \frac{3}{3} \cdot \frac{3 \cdot 3 \cdot 3}{2 \cdot 5 \cdot 7} = \frac{27}{70}$$

Do Exercises 12 and 13.

Add.

12. $\frac{4}{10} + \frac{1}{100} + \frac{3}{1000}$

13. $\frac{7}{10} + \frac{-2}{21} + \frac{1}{7}$

C ORDER

When two fractions share a common denominator, the larger number can be found by comparing numerators. For example, 4 is greater than 3, so $\frac{4}{5}$ is greater than $\frac{3}{5}$.

$$\frac{4}{5} > \frac{3}{5}$$

$\frac{4}{5}$

$\frac{3}{5}$

Similarly, because -6 is less than -2, we have

$$\frac{-6}{7} < \frac{-2}{7}, \quad \text{or} \quad -\frac{6}{7} < -\frac{2}{7}.$$

Do Exercises 14–16.

Use $<$ or $>$ for $\square$ to form a true sentence.

14. $\frac{3}{8} \square \frac{5}{8}$

15. $\frac{7}{10} \square \frac{6}{10}$

16. $\frac{-2}{9} \square \frac{-5}{9}$

EXAMPLE 11 Use $<$ or $>$ for $\square$ to form a true sentence:

$$\frac{5}{8} \square \frac{2}{3}.$$

You can confirm that the LCD is 24. We multiply by 1 to find two fractions equivalent to $\frac{5}{8}$ and $\frac{2}{3}$ with denominators the same:

$$\frac{5}{8} \cdot \frac{3}{3} = \frac{15}{24}; \quad \frac{2}{3} \cdot \frac{8}{8} = \frac{16}{24}. \quad \frac{5}{8} \square \frac{2}{3} \text{ is equivalent to } \frac{15}{24} \square \frac{16}{24}.$$

Since $15 < 16$, it follows that $\frac{15}{24} < \frac{16}{24}$. Thus,

$$\frac{5}{8} < \frac{2}{3}.$$

Answers

12. $\frac{413}{1000}$ 13. $\frac{157}{210}$ 14. $<$ 15. $>$ 16. $>$

EXAMPLE 12 Use < or > for $\square$ to form a true sentence:

$$-\frac{89}{100} \square -\frac{9}{10}.$$

We rewrite $-\frac{9}{10}$ with a denominator of 100: $-\frac{9}{10} \cdot \frac{10}{10} = -\frac{90}{100}$. We then have

$$\frac{-89}{100} \square \frac{-90}{100}. \qquad \text{Recall that } -\frac{m}{n} = \frac{-m}{n}.$$

Since $-89 > -90$, it follows that $-\frac{89}{100} > -\frac{90}{100}$, so

$$-\frac{89}{100} > -\frac{9}{10}.$$

◀ **Do Exercises 17–19.**

Use < or > for $\square$ to form a true sentence.

17. $\frac{2}{3} \square \frac{3}{4}$ **18.** $\frac{-3}{4} \square \frac{-8}{12}$

19. $\frac{7}{8} \square \frac{5}{6}$

d APPLICATIONS AND PROBLEM SOLVING

EXAMPLE 13 *Construction.* A contractor uses two layers of subflooring under a ceramic tile floor. First, she installs a $\frac{3}{4}$-in. layer of oriented strand board (OSB). Then a $\frac{1}{2}$-in. sheet of cement board is mortared to the OSB. The mortar is $\frac{1}{8}$-in. thick. What is the total thickness of the two installed subfloors?

1. Familiarize. We first make a drawing. We let $T =$ the total thickness of the subfloors.

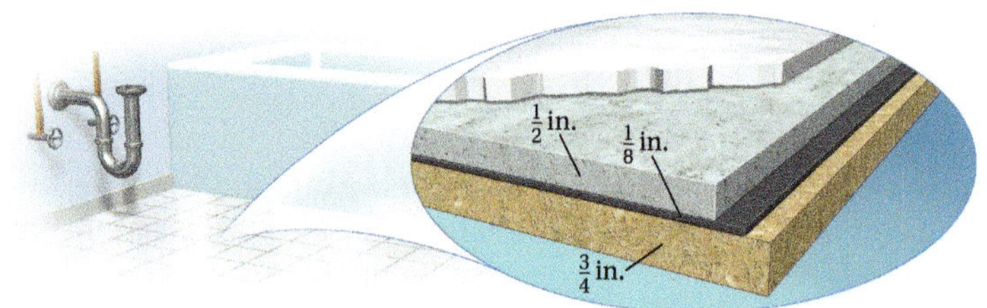

2. Translate. The problem can be translated to an equation as follows.

OSB	plus	mortar	plus	cement board	is	total thickness
↓	↓	↓	↓	↓	↓	↓
$\frac{3}{4}$	$+$	$\frac{1}{8}$	$+$	$\frac{1}{2}$	$=$	T

3. Solve. To solve the equation, we carry out the addition.

$$\frac{3}{4} + \frac{1}{8} + \frac{1}{2} = T \qquad \text{The LCM of the denominators is 8.}$$

$$\frac{3}{4} \cdot \frac{2}{2} + \frac{1}{8} + \frac{1}{2} \cdot \frac{4}{4} = T \qquad \text{Multiplying by 1 to obtain the LCD}$$

$$\frac{6}{8} + \frac{1}{8} + \frac{4}{8} = T$$

$$\frac{11}{8} = T$$

4. Check. We check by repeating the calculation.

5. State. The total thickness of the installed subfloors is $\frac{11}{8}$ in.

◀ **Do Exercise 20.**

20. *Catering.* A caterer prepares a mixed berry salad with $\frac{7}{8}$ qt of strawberries, $\frac{3}{4}$ qt of raspberries, and $\frac{5}{16}$ qt of blueberries. How many quarts of berries are in the salad? GS

1. **Familiarize.** Let $T =$ the total amount of berries in the salad.
2. **Translate.** To find the total amount, we add.

$$\frac{7}{\square} + \frac{3}{\square} + \frac{5}{\square} = T$$

3. **Solve.** The LCD is $\square$.

$$\frac{7}{8} \cdot \frac{2}{2} + \frac{3}{4} \cdot \frac{\square}{\square} + \frac{5}{16} = T$$

$$\frac{\square}{16} + \frac{\square}{16} + \frac{5}{16} = T$$

$$\frac{\square}{16} = T$$

4. **Check.** The answer is reasonable because it is larger than any of the individual amounts.
5. **State.** There are $\square$ qt of berries in the salad.

Answers

17. < **18.** < **19.** > **20.** $\frac{31}{16}$ qt

Guided Solution:

20. 8, 4, 16; 16; $\frac{4}{4}$, 14, 12, 31; $\frac{31}{16}$

4.2 Exercise Set

For Extra Help MyMathLab® MathXL® PRACTICE WATCH READ REVIEW

Reading Check

Determine whether each statement is true or false.

_______________ **RC1.** Before we can add two fractions, they must have the same denominator.

_______________ **RC2.** To add fractions, we add numerators and add denominators.

_______________ **RC3.** If we use the LCD to add fractions, we never need to simplify the result.

_______________ **RC4.** Adding fractions with different denominators involves multiplying at least one fraction by 1.

a, **b** Add and, if possible, simplify.

1. $\frac{4}{9}+\frac{1}{9}$

2. $\frac{3}{11}+\frac{5}{11}$

3. $\frac{4}{7}+\frac{3}{7}$

4. $\frac{7}{8}+\frac{1}{8}$

5. $\frac{7}{10}+\frac{3}{-10}$

6. $\frac{1}{-6}+\frac{5}{6}$

7. $\frac{9}{a}+\frac{4}{a}$

8. $\frac{2}{t}+\frac{3}{t}$

9. $\frac{-1}{4}+\frac{-1}{4}$

10. $\frac{7}{12}+\frac{-5}{12}$

11. $\frac{2}{9}x+\frac{5}{9}x$

12. $\frac{3}{11}a+\frac{2}{11}a$

13. $\frac{3}{32}t+\frac{13}{32}t$

14. $\frac{3}{25}x+\frac{12}{25}x$

15. $-\frac{2}{x}+\left(-\frac{7}{x}\right)$

16. $-\frac{7}{a}+\frac{5}{a}$

17. $\frac{1}{8}+\frac{1}{6}$

18. $\frac{1}{9}+\frac{1}{6}$

19. $\frac{-4}{5}+\frac{7}{10}$

20. $\frac{-3}{4}+\frac{-1}{12}$

21. $\frac{7}{12}+\frac{3}{8}$

22. $\frac{7}{8}+\frac{1}{16}$

23. $\frac{3}{20}+4$

24. $\frac{2}{15}+3$

25. $\frac{5}{-8}+\frac{5}{6}$

26. $\frac{5}{-6}+\frac{7}{9}$

27. $\frac{3}{10}x+\frac{7}{100}x$

28. $\frac{9}{20}a+\frac{3}{40}a$

29. $\frac{5}{12}+\frac{8}{15}$

30. $\frac{3}{16}+\frac{1}{12}$

31. $\frac{7}{8} + \frac{0}{1}$

32. $\frac{0}{6} + \frac{5}{3}$

33. $\frac{-7}{10} + \frac{-29}{100}$

34. $\frac{-3}{10} + \frac{-27}{100}$

35. $-\frac{1}{10}x + \frac{1}{15}x$

36. $-\frac{1}{6}x + \frac{1}{4}x$

37. $-5t + \frac{2}{7}t$

38. $-4x + \frac{3}{5}x$

39. $-\frac{5}{12} + \frac{7}{-24}$

40. $-\frac{1}{18} + \frac{5}{-12}$

41. $\frac{3}{16} + \frac{5}{16} + \frac{4}{16}$

42. $\frac{3}{8} + \frac{1}{8} + \frac{2}{8}$

43. $\frac{4}{10} + \frac{3}{100} + \frac{7}{1000}$

44. $\frac{7}{10} + \frac{2}{100} + \frac{9}{1000}$

45. $\frac{3}{10} + \frac{5}{12} + \frac{8}{15}$

46. $\frac{1}{2} + \frac{3}{8} + \frac{1}{4}$

47. $\frac{5}{6} + \frac{25}{52} + \frac{7}{4}$

48. $\frac{15}{24} + \frac{7}{36} + \frac{91}{48}$

49. $\frac{2}{9} + \frac{7}{10} + \frac{-4}{15}$

50. $\frac{5}{12} + \frac{-3}{8} + \frac{1}{10}$

51. $-\frac{3}{4} + \frac{1}{5} + \frac{-7}{10}$

52. $\frac{1}{3} + \frac{-7}{9} + \frac{-1}{2}$

C Use $<$ or $>$ for $\square$ to form a true sentence.

53. $\frac{3}{8} \square \frac{2}{8}$

54. $\frac{7}{9} \square \frac{5}{9}$

55. $\frac{2}{3} \square \frac{5}{6}$

56. $\frac{11}{18} \square \frac{5}{9}$

57. $\frac{-2}{7} \square \frac{-5}{7}$

58. $\frac{-4}{5} \square \frac{-3}{5}$

59. $\frac{9}{15} \square \frac{7}{10}$

60. $\frac{5}{14} \square \frac{8}{21}$

61. $\frac{3}{4} \square -\frac{1}{5}$

62. $\frac{3}{8} \square -\frac{13}{16}$

63. $\frac{-7}{20} \square \frac{-6}{15}$

64. $\frac{-7}{12} \square \frac{-9}{16}$

Arrange each group of fractions from smallest to largest.

65. $\frac{3}{10}, \frac{5}{12}, \frac{4}{15}$

66. $\frac{5}{6}, \frac{19}{21}, \frac{11}{14}$

d Solve.

67. *Riding a Segway.*® Tate rode a Segway® Personal Transporter $\frac{5}{6}$ mi to the library, then $\frac{3}{4}$ mi to class, and then $\frac{3}{2}$ mi to his part-time job. How far did he ride his Segway®?

68. *Volunteering.* For a community project, an earth science class volunteered one hour per day for three days to join the state highway beautification project. The students collected trash along a $\frac{4}{5}$-mi stretch of highway the first day, a $\frac{5}{8}$-mi stretch the second day, and a $\frac{1}{2}$-mi stretch the third day. How many miles along the highway did they clean?

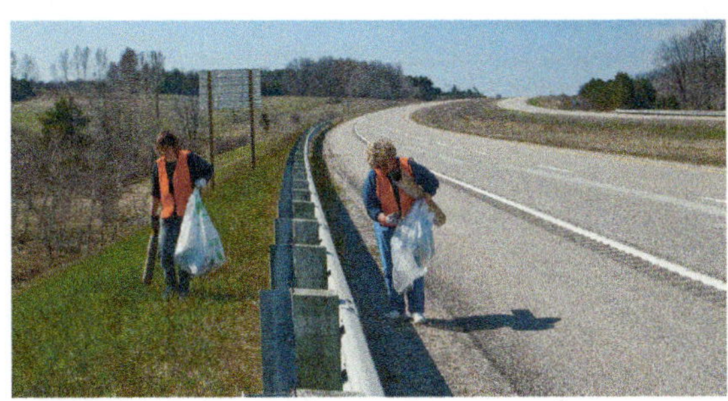

69. *Caffeine.* To cut back on caffeine intake, Michelle and Gerry mix caffeinated and decaffeinated coffee beans before grinding for a customized mix. They mix $\frac{3}{16}$ lb of decaffeinated beans with $\frac{5}{8}$ lb of caffeinated beans. What is the total amount of coffee beans in the mixture?

70. *Purchasing Tea.* Alyse bought $\frac{1}{3}$ lb of orange pekoe tea and $\frac{1}{2}$ lb of English cinnamon tea. How many pounds of tea did she buy?

71. *Culinary Arts.* The campus culinary arts department is preparing brownies for the international student reception. Students in the catering program iced the $\frac{11}{16}$-in. ($\frac{11}{16}''$) brownies with a $\frac{5}{32}$-in. ($\frac{5}{32}''$) layer of butterscotch icing. What is the thickness of the iced brownies?

72. *Carpentry.* A carpenter glues two kinds of plywood together. He glues a $\frac{1}{4}$-in. ($\frac{1}{4}''$) piece of cherry plywood to a $\frac{3}{8}$-in. ($\frac{3}{8}''$) piece of less expensive plywood. What is the total thickness of these pieces?

73. *Baking.* A baker used $\frac{1}{2}$ lb of flour for rolls, $\frac{1}{4}$ lb for donuts, and $\frac{1}{3}$ lb for cookies. How much flour was used?

74. *Baking.* A recipe for muffins calls for $\frac{1}{2}$ qt (quart) of buttermilk, $\frac{1}{3}$ qt of skim milk, and $\frac{1}{16}$ qt of oil. How many quarts of liquid ingredients does the recipe call for?

75. *Meteorology.* On April 15, it rained $\frac{1}{2}$ in. in the morning and $\frac{3}{8}$ in. in the afternoon. How much did it rain altogether?

76. *Medication.* Janine took $\frac{1}{5}$ g of ibuprofen before lunch and $\frac{1}{2}$ g after lunch. How much did she take altogether?

77. A park naturalist hikes $\frac{3}{5}$ mi to a lookout, another $\frac{3}{10}$ mi to an osprey's nest, and finally $\frac{3}{4}$ mi to a campsite. How far does the naturalist hike?

78. A triathlete runs $\frac{7}{8}$ mi, canoes $\frac{1}{3}$ mi, and swims $\frac{1}{6}$ mi. How many miles does the triathlete cover?

79. *Punch Recipe.* A recipe for strawberry punch calls for $\frac{1}{5}$ qt of ginger ale and $\frac{3}{5}$ qt of strawberry soda. How much liquid is needed? If the recipe is doubled, how much liquid is needed? If the recipe is halved, how much liquid is needed?

80. *Concrete Mix.* A cubic meter of concrete mix contains 420 kg (kilograms) of cement, 150 kg of stone, and 120 kg of sand. What is the total weight of a cubic meter of the mix? What fractional part is cement? stone? sand? Add these fractional amounts. What is the result?

Skill Maintenance

Subtract. [2.3a]

81. $-7 - 6$

82. $-5 - (-9)$

83. $9 - 17$

84. $-8 - 23$

Evaluate. [2.6a]

85. $\frac{x - y}{3}$, for $x = 7$ and $y = -3$

86. $3(x + y)$ and $3x + 3y$, for $x = 5$ and $y = 9$

Solve.

87. $48 \cdot t = 1680$ [1.7b]

88. $10{,}000 = m + 3593$ [1.7b]

89. $3x - 8 = 25$ [2.8d]

90. $5x + 9 = 24$ [2.8d]

91. $\frac{2}{3}x = \frac{6}{7}$ [3.8a]

92. $-\frac{5}{8}t = \frac{1}{4}$ [3.8a]

Synthesis

Add and, if possible, simplify.

93. $\frac{3}{10}t + \frac{2}{7} + \frac{2}{15}t + \frac{3}{5}$

94. $\frac{2}{9} + \frac{4}{21}x + \frac{4}{15} + \frac{3}{14}x$

95. $5t^2 + \frac{6}{a}t + 2t^2 + \frac{3}{a}t$

Use $<$, $>$, or $=$ for $\square$ to form a true sentence.

96. 🖩 $\frac{10}{97} + \frac{67}{137} \square \frac{8123}{13{,}289}$

97. 🖩 $\frac{12}{169} + \frac{53}{103} \square \frac{10{,}192}{17{,}407}$

98. 🖩 $\frac{37}{157} + \frac{20}{107} \square \frac{6942}{16{,}799}$

99. A guitarist's band is booked for Friday and Saturday nights at a local club. The guitarist's group is a trio on Friday and expands to a quintet on Saturday. Thus, the guitarist is paid one-third of one-half the weekend's pay for Friday and one-fifth of one-half the weekend's pay for Saturday. What fractional part of the total pay did the guitarist receive for the weekend's work? If the band was paid $1200, how much did the guitarist receive?

100. 🖩 Consider only the numbers 2, 3, 4, and 5. Assume each is placed in a blank in the following.

$$\frac{\square}{\square} + \frac{\square}{\square} = ?$$

What placement of the numbers in the blanks yields the largest sum?

101. 🖩 In the sum below, a and b are digits (so $1b$ is a two-digit number and $35a$ is a three-digit number). Find a and b. (*Hint:* $a < 4$ and $b > 6$.)

$$\frac{a}{17} + \frac{1b}{23} = \frac{35a}{391}$$

102. 🖩 Use a standard calculator. Arrange the following in order from smallest to largest.

$$\frac{3}{4}, \frac{17}{21}, \frac{13}{15}, \frac{7}{9}, \frac{15}{17}, \frac{13}{12}, \frac{19}{22}$$

Subtraction, Equations, and Applications

4.3

OBJECTIVES

a Subtract using fraction notation.

b Solve equations of the type $x + a = b$ and $a + x = b$, where a and b may be fractions.

c Solve applied problems involving subtraction with fraction notation.

SKILL TO REVIEW

Objective 2.8a: Use the addition principle to solve equations.

Solve.

1. $-2 + x = -8$

2. $y - 3 = -3$

a SUBTRACTION

Like Denominators

Let's consider the difference $\frac{4}{8} - \frac{3}{8}$.

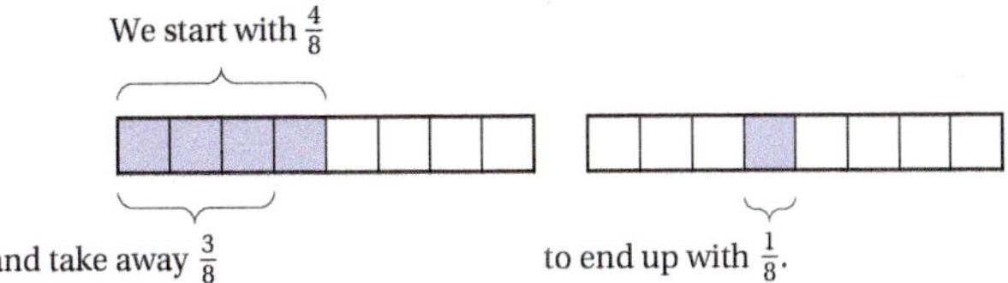

We start with $\frac{4}{8}$
and take away $\frac{3}{8}$
to end up with $\frac{1}{8}$.

We start with 4 eighths and take away 3 eighths:

4 eighths − 3 eighths = 1 eighth,

or $4 \cdot \frac{1}{8} - 3 \cdot \frac{1}{8} = \frac{1}{8}$, or $\frac{4}{8} - \frac{3}{8} = \frac{1}{8}$.

> To subtract when denominators are the same,
>
> **a)** subtract the numerators,
>
> **b)** keep the denominator, and
>
> **c)** simplify, if possible.
>
> $$\frac{7}{10} - \frac{4}{10} = \frac{7-4}{10} = \frac{3}{10}$$

EXAMPLES Subtract and, if possible, simplify.

1. $\frac{8}{13} - \frac{3}{13} = \frac{8-3}{13} = \frac{5}{13}$

2. $\frac{3}{35} - \frac{13}{35} = \frac{3-13}{35} = \frac{-10}{35} = \frac{5}{5} \cdot \frac{-2}{7} = \frac{-2}{7}$, or $-\frac{2}{7}$ Removing a factor equal to 1: $\frac{5}{5} = 1$

3. $\frac{13}{2a} - \frac{5}{2a} = \frac{13-5}{2a} = \frac{8}{2a} = \frac{2}{2} \cdot \frac{4}{a} = \frac{4}{a}$ Removing a factor equal to 1: $\frac{2}{2} = 1$

4. $-\frac{7}{t} - \frac{2}{t} = \frac{-7-2}{t} = \frac{-9}{t}$, or $-\frac{9}{t}$

Do Margin Exercises 1–4. ▶

Subtract and, if possible, simplify.

1. $\frac{7}{8} - \frac{3}{8}$

2. $\frac{5}{9a} - \frac{1}{9a}$

3. $\frac{7}{10} - \frac{13}{10}$

4. $-\frac{2}{x} - \frac{4}{x}$

Different Denominators

> To subtract when denominators are different:
>
> **a)** Find the least common multiple of the denominators. That number is the least common denominator, LCD.
>
> **b)** Multiply by 1, writing 1 in the form n/n, to express each fraction in an equivalent form that contains the LCD.
>
> **c)** Subtract the numerators, keeping the same denominator.
>
> **d)** Simplify, if possible.

Answers

Skill to Review:
1. −6 **2.** 0

Margin Exercises:
1. $\frac{1}{2}$ **2.** $\frac{4}{9a}$ **3.** $-\frac{3}{5}$ **4.** $-\frac{6}{x}$

EXAMPLE 5 Subtract: $\frac{2}{5} - \frac{3}{8}$.

a) The LCM of 5 and 8 is 40, so the LCD is 40.

b) We need to find numbers equivalent to $\frac{2}{5}$ and $\frac{3}{8}$ with denominators of 40:

$$\frac{2}{5} - \frac{3}{8} = \frac{2}{5} \cdot \frac{8}{8} - \frac{3}{8} \cdot \frac{5}{5}.$$

Think: $8 \times \square = 40$. The answer is 5, so we multiply by 1, using $\frac{5}{5}$.

Think: $5 \times \square = 40$. The answer is 8, so we multiply by 1, using $\frac{8}{8}$.

c) We subtract: $\frac{16}{40} - \frac{15}{40} = \frac{16 - 15}{40} = \frac{1}{40}$.

d) Since $\frac{1}{40}$ cannot be simplified, we are finished. The answer is $\frac{1}{40}$.

Do Exercise 5.

5. Subtract: $\frac{3}{4} - \frac{2}{3}$.

EXAMPLE 6 Subtract: $\frac{7}{12} - \frac{5}{6}$.

Since 12 is a multiple of 6, the LCM of 6 and 12 is 12. The LCD is 12.

$$\frac{7}{12} - \frac{5}{6} = \frac{7}{12} - \frac{5}{6} \cdot \frac{2}{2}$$

Think: $6 \times \square = 12$. The answer is 2, so we multiply by 1, using $\frac{2}{2}$.

$$= \frac{7}{12} - \frac{10}{12}$$

$$= \frac{7 - 10}{12} = \frac{-3}{12}$$

$7 - 10 = 7 + (-10) = -3$

$$= \frac{3}{3} \cdot \frac{-1}{4} = \frac{-1}{4}, \text{ or } -\frac{1}{4}$$

Simplifying by removing a factor equal to 1: $\frac{3}{3} = 1$

EXAMPLE 7 Subtract: $\frac{17}{24} - \frac{4}{15}$.

We need to find the LCM of 24 and 15:

$$24 = 2 \cdot 2 \cdot 2 \cdot 3, \quad 15 = 3 \cdot 5.$$

The LCM is $2 \cdot 2 \cdot 2 \cdot 3 \cdot 5$, or 120.

$$\frac{17}{24} - \frac{4}{15} = \frac{17}{24} \cdot \frac{5}{5} - \frac{4}{15} \cdot \frac{8}{8}$$

Multiplying by 1 to obtain the LCD. To form 1, use the factors of the LCM that each denominator lacks. Note that $2 \cdot 2 \cdot 2 = 8$.

$$= \frac{85}{120} - \frac{32}{120} = \frac{85 - 32}{120} = \frac{53}{120}$$

Do Exercises 6–9.

Subtract.

6. $\frac{5}{6} - \frac{1}{9}$

The LCD is $\square$.

$$\frac{5}{6} - \frac{1}{9} = \frac{5}{6} \cdot \frac{3}{3} - \frac{1}{9} \cdot \frac{\square}{\square}$$

$$= \frac{\square}{18} - \frac{\square}{18}$$

$$= \frac{\square}{18}$$

7. $\frac{2}{5} - \frac{7}{10}$

8. $\frac{2}{3} - \frac{5}{6}$

9. $\frac{11}{28} - \frac{5}{16}$

EXAMPLE 8 Simplify by combining like terms: $\frac{7}{8}x - \frac{3}{4}x$.

$$\frac{7}{8}x - \frac{3}{4}x = \left(\frac{7}{8} - \frac{3}{4}\right)x$$

Try to do this step mentally.

$$= \left(\frac{7}{8} - \frac{6}{8}\right)x = \frac{1}{8}x$$

Multiplying $\frac{3}{4}$ by $\frac{2}{2}$ and subtracting

Do Exercise 10.

10. Simplify: $\frac{9}{10}x - \frac{3}{5}x$.

Answers

5. $\frac{1}{12}$ 6. $\frac{13}{18}$ 7. $-\frac{3}{10}$ 8. $-\frac{1}{6}$ 9. $\frac{9}{112}$ 10. $\frac{3}{10}x$

Guided Solution:

6. 18; $\frac{2}{2}$, 15, 2, 13

b SOLVING EQUATIONS

We can use the addition principle to solve equations containing fractions.

EXAMPLE 9 Solve: $x - \frac{1}{3} = -\frac{2}{7}$.

$$x - \frac{1}{3} = -\frac{2}{7}$$

$$x - \frac{1}{3} + \frac{1}{3} = -\frac{2}{7} + \frac{1}{3}$$ Using the addition principle: adding $\frac{1}{3}$ to both sides

$$x + 0 = -\frac{2}{7} + \frac{1}{3}$$ Adding $\frac{1}{3}$ "undid" the subtraction of $\frac{1}{3}$ on the left-hand side of the equation.

$$x = -\frac{2}{7} \cdot \frac{3}{3} + \frac{1}{3} \cdot \frac{7}{7}$$ Multiplying by 1 to obtain the LCD, 21

$$x = -\frac{6}{21} + \frac{7}{21} = \frac{1}{21}$$ The solution appears to be $\frac{1}{21}$.

Check:

$$\begin{array}{c|c} x - \frac{1}{3} = -\frac{2}{7} & \\ \hline \frac{1}{21} - \frac{1}{3} \;?\; & -\frac{2}{7} \\ \frac{1}{21} - \frac{1}{3} \cdot \frac{7}{7} & \\ \frac{1}{21} - \frac{7}{21} & \\ \frac{-6}{21} & \\ -\frac{2}{7} \cdot \frac{3}{3} & \\ -\frac{2}{7} & \text{TRUE} \end{array}$$

Our answer checks. The solution is $\frac{1}{21}$.

Recall that we can also create equivalent equations if we subtract the same number on both sides of an equation.

EXAMPLE 10 Solve: $x + \frac{1}{4} = \frac{3}{5}$.

$$x + \frac{1}{4} - \frac{1}{4} = \frac{3}{5} - \frac{1}{4}$$ Using the addition principle: adding $-\frac{1}{4}$ to, or subtracting $\frac{1}{4}$ from, both sides

$$x + 0 = \frac{3}{5} \cdot \frac{4}{4} - \frac{1}{4} \cdot \frac{5}{5}$$ The LCD is 20. We multiply by 1 to get the LCD.

$$x = \frac{12}{20} - \frac{5}{20} = \frac{7}{20}$$

The solution is $\frac{7}{20}$. We leave the check to the student.

Do Exercises 11–13.

Solve.

11. $x - \frac{2}{5} = \frac{1}{5}$

12. $x + \frac{2}{3} = \frac{5}{6}$

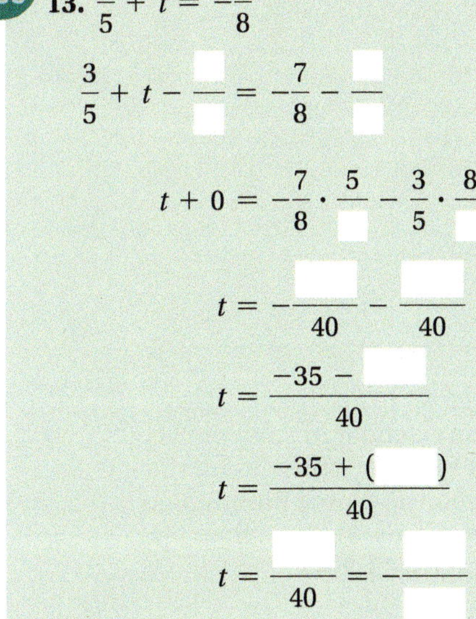

GS
13. $\frac{3}{5} + t = -\frac{7}{8}$
$\frac{3}{5} + t - \frac{\square}{\square} = -\frac{7}{8} - \frac{\square}{\square}$
$t + 0 = -\frac{7}{8} \cdot \frac{5}{\square} - \frac{3}{5} \cdot \frac{8}{\square}$
$t = -\frac{\square}{40} - \frac{\square}{40}$
$t = \frac{-35 - \square}{40}$
$t = \frac{-35 + (\quad)}{40}$
$t = \frac{\square}{40} = -\frac{\square}{\square}$

Answers

11. $\frac{3}{5}$ 12. $\frac{1}{6}$ 13. $-\frac{59}{40}$

Guided Solution:

13. $\frac{3}{5}, \frac{3}{5}$, 5, 8, 35, 24, 24, −24, −59, $\frac{59}{40}$

C APPLICATIONS AND PROBLEM SOLVING

11 12 13 14 15 16 17
1/5 1/4 9/25 First quarter 1/2 3/5 7/10 4/5
18 19 20 21 22 23 24
17/20 19/20 99/100 Full moon 1 49/50 19/20 17/20
25 26 27 28 29 30 31
4/5 7/10 3/5 Last quarter 1/2 2/5 1/3 6/25

SOURCE: Astronomical Applications Department, U.S. Naval Observatory, Washington, DC 20392-5420

EXAMPLE 11 *Phases of the Moon.* The moon rotates in such a way that the same side always faces the earth. Throughout a lunar cycle, the portion of the moon that appears illuminated increases from nearly none (new moon) to nearly all (full moon), then decreases back to nearly none. These *phases* of the moon can be described by fractions between 0 and 1, indicating the portion of the moon illuminated. The partial calendar from August 2013 shows the fraction of the moon illuminated at midnight, Eastern Standard Time, for each day.

How much more of the moon appeared illuminated on August 18, 2013, than on August 15, 2013?

1. **Familiarize.** From the calendar, we see that $\frac{3}{5}$ of the moon was illuminated on August 15 and that $\frac{17}{20}$ of the moon was illuminated on August 18. We let $m =$ the additional part of the moon that appeared illuminated on August 18.

2. **Translate.** We translate to an equation.

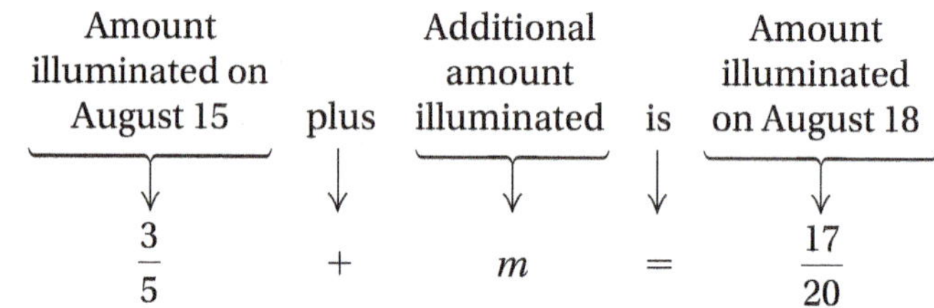

$$\frac{3}{5} + m = \frac{17}{20}$$

3. **Solve.** To solve the equation, we subtract $\frac{3}{5}$ on both sides:

$$\frac{3}{5} + m = \frac{17}{20}$$

$$\frac{3}{5} + m - \frac{3}{5} = \frac{17}{20} - \frac{3}{5} \qquad \text{Subtracting } \frac{3}{5} \text{ on both sides}$$

$$m + 0 = \frac{17}{20} - \frac{3}{5} \cdot \frac{4}{4} \qquad \text{The LCD is 20. We multiply by 1 to obtain the LCD.}$$

$$m = \frac{17}{20} - \frac{12}{20} = \frac{5}{20} = \frac{5 \cdot 1}{5 \cdot 4} = \frac{5}{5} \cdot \frac{1}{4} = \frac{1}{4}.$$

4. **Check.** To check, we add:

$$\frac{3}{5} + \frac{1}{4} = \frac{3}{5} \cdot \frac{4}{4} + \frac{1}{4} \cdot \frac{5}{5} = \frac{12}{20} + \frac{5}{20} = \frac{17}{20}.$$

This is the amount of the moon illuminated on August 18, so the answer checks.

5. **State.** On August 18, $\frac{1}{4}$ more of the moon was illuminated than on August 15.

◀ **Do Exercise 14.**

14. *Phases of the Moon.* Use the calendar in Example 11 to find how much less of the moon appeared illuminated on August 31, 2013, than on August 23, 2013.

Answer

14. $\frac{71}{100}$

Translating for Success

1. *Bubble Wrap.* One-Stop Postal Center orders bubble wrap in 64-yd rolls. On average, $\frac{3}{4}$ yd is used per small package. How many small packages can be prepared with 2 rolls of bubble wrap?

2. *Distance from College.* The post office is $\frac{7}{9}$ mi from the community college. The medical clinic is $\frac{2}{5}$ as far from the college as the post office is. How far is the clinic from the college?

3. *Swimming.* Andrew swims $\frac{7}{9}$ mi every day. One day, he swims $\frac{2}{5}$ mi by 11:00 A.M. How much farther must Andrew swim to reach his daily goal?

4. *Tuition.* The average tuition at Waterside University is $12,000. If a loan is obtained for $\frac{1}{3}$ of the tuition, how much is the loan?

5. *Thermos Bottle Capacity.* A thermos bottle holds $\frac{11}{12}$ gal. How much is in the bottle when it is $\frac{4}{7}$ full?

The goal of these matching questions is to practice step (2), Translate, of the five-step problem-solving process. Translate each word problem to an equation and select a correct translation from equations A–O.

A. $\frac{3}{4} \cdot 64 = x$

B. $\frac{1}{3} \cdot 12{,}000 = x$

C. $\frac{1}{3} + \frac{2}{5} = x$

D. $\frac{2}{5} + x = \frac{7}{9}$

E. $\frac{2}{5} \cdot \frac{7}{9} = x$

F. $\frac{3}{4} \cdot x = 64$

G. $\frac{4}{7} = x + \frac{11}{12}$

H. $\frac{2}{5} = x + \frac{7}{9}$

I. $\frac{4}{7} \cdot \frac{11}{12} = x$

J. $\frac{3}{4} \cdot x = 2 \cdot 64$

K. $\frac{1}{3} \cdot x = 12{,}000$

L. $\frac{1}{3} + \frac{2}{5} + x = 1$

M. $\frac{2}{5} = \frac{7}{9}x$

N. $\frac{4}{7} + x = \frac{11}{12}$

O. $\frac{1}{3} + x = \frac{2}{5}$

Answers on page A-7

6. *Cutting Rope.* A piece of rope $\frac{11}{12}$ yd long is cut into two pieces. One piece is $\frac{4}{7}$ yd long. How long is the other piece?

7. *Planting Corn.* Each year, Prairie State Farm plants 64 acres of corn. With good weather, $\frac{3}{4}$ of the planting can be completed by April 20. How many acres can be planted by April 20 with good weather?

8. *Painting Trim.* A painter used $\frac{1}{3}$ gal of white paint for the trim in the library and $\frac{2}{5}$ gal for the trim in the family room. How much paint was used for the trim in the two rooms?

9. *Lottery Winnings.* Sally won $12,000 in a state lottery and decided to give the net amount after taxes to three charities. One received $\frac{1}{3}$ of the money, and a second received $\frac{2}{5}$. What fractional part did the third charity receive?

10. *Reading Assignment.* When Lowell had read 64 pages of his political science assignment, he had completed $\frac{3}{4}$ of his required reading. How many total pages were assigned?

4.3 Exercise Set

For Extra Help

MyMathLab®

MathXL®

PRACTICE

WATCH

READ

REVIEW

Reading Check

Complete each statement with the appropriate word or words from the following list. A word may be used more than once or not at all.

denominator numerator
denominators numerators

RC1. To subtract fractions with like denominators, we subtract the ______ and keep the ______.

RC2. Before we can subtract fractions, the ______ must be the same.

RC3. To subtract fractions when denominators are different, we find the LCM of the ______.

RC4. To subtract fractions when denominators are different, we multiply one or both fractions by 1 to make the ______ the same.

a Subtract and, if possible, simplify.

1. $\frac{5}{6} - \frac{1}{6}$ **2.** $\frac{7}{5} - \frac{2}{5}$ **3.** $\frac{9}{16} - \frac{13}{16}$ **4.** $\frac{5}{12} - \frac{7}{12}$ **5.** $\frac{8}{a} - \frac{6}{a}$

6. $\frac{4}{t} - \frac{9}{t}$ **7.** $-\frac{3}{8} - \frac{1}{8}$ **8.** $-\frac{3}{10} - \frac{1}{10}$ **9.** $\frac{3}{5a} - \frac{7}{5a}$ **10.** $\frac{2}{7t} - \frac{10}{7t}$

11. $\frac{10}{3t} - \frac{4}{3t}$ **12.** $\frac{9}{2a} - \frac{5}{2a}$ **13.** $\frac{7}{8} - \frac{1}{16}$ **14.** $\frac{4}{3} - \frac{5}{6}$ **15.** $\frac{7}{15} - \frac{4}{5}$

16. $\frac{3}{28} - \frac{3}{4}$ **17.** $\frac{3}{4} - \frac{1}{20}$ **18.** $\frac{3}{4} - \frac{4}{16}$ **19.** $\frac{2}{15} - \frac{5}{12}$ **20.** $\frac{11}{16} - \frac{9}{10}$

21. $\frac{7}{10} - \frac{23}{100}$ **22.** $\frac{9}{10} - \frac{3}{100}$ **23.** $\frac{7}{15} - \frac{3}{25}$ **24.** $\frac{18}{25} - \frac{4}{35}$ **25.** $\frac{-41}{100} - \frac{3}{10}$

26. $\frac{-13}{100} - \frac{7}{20}$ **27.** $\frac{2}{3} - \frac{1}{8}$ **28.** $\frac{3}{4} - \frac{1}{2}$ **29.** $-\frac{3}{10} - \frac{7}{25}$ **30.** $-\frac{5}{18} - \frac{2}{27}$

31. $\frac{3}{8} - \frac{5}{12}$ **32.** $\frac{2}{9} - \frac{7}{12}$ **33.** $\frac{-5}{18} - \frac{7}{24}$ **34.** $\frac{-7}{25} - \frac{2}{15}$ **35.** $\frac{13}{90} - \frac{17}{120}$

36. $\frac{8}{25} - \frac{29}{150}$

37. $\frac{2}{3}x - \frac{4}{9}x$

38. $\frac{7}{4}x - \frac{5}{12}x$

39. $\frac{2}{5}a - \frac{3}{4}a$

40. $\frac{4}{7}a - \frac{1}{3}a$

b Solve.

41. $x - \frac{4}{9} = \frac{3}{9}$

42. $x - \frac{3}{11} = \frac{7}{11}$

43. $a + \frac{2}{11} = \frac{6}{11}$

44. $a + \frac{4}{15} = \frac{13}{15}$

45. $y + \frac{1}{30} = \frac{1}{10}$

46. $y + \frac{1}{3} = \frac{5}{6}$

47. $a - \frac{3}{8} = \frac{3}{4}$

48. $x - \frac{3}{10} = \frac{2}{5}$

49. $\frac{2}{3} + x = \frac{4}{5}$

50. $\frac{4}{5} + x = \frac{6}{7}$

51. $\frac{3}{8} + a = \frac{1}{12}$

52. $\frac{5}{6} + a = \frac{2}{9}$

53. $n - \frac{3}{10} = -\frac{1}{6}$

54. $n - \frac{3}{4} = -\frac{5}{12}$

55. $x + \frac{3}{4} = -\frac{1}{2}$

56. $x + \frac{5}{6} = -\frac{11}{12}$

c Solve.

57. For a research paper, Kaitlyn spent $\frac{3}{4}$ hr searching the Internet on google.com and $\frac{1}{3}$ hr on yahoo.com. How much longer did she spend on google.com than on yahoo.com?

58. As part of a fitness program, Deb swims $\frac{1}{2}$ mi every day. One day she had already swum $\frac{1}{5}$ mi. How much farther did Deb need to swim?

59. The tread depth of an IRL Indy Car Series tire is $\frac{3}{32}$ in. Tires for a normal car have a tread depth of $\frac{5}{16}$ in. when new and are considered bald at $\frac{1}{16}$ in. How much deeper is the tread depth of an Indy Car tire than that of a bald tire for a normal car?

$\frac{3}{32}$ in.

Sources: Indy500.com; *Consumer Reports*

60. Gerry uses $\frac{1}{3}$ lb of fresh mozzarella cheese and $\frac{1}{4}$ lb of grated Parmesan cheese on a homemade margherita pizza. How much more mozzarella cheese does he use than Parmesan cheese?

61. From a $\frac{4}{5}$-lb wheel of cheese, a $\frac{1}{4}$-lb piece was served. How much cheese remained on the wheel?

62. A baker has a dispenser containing $\frac{15}{16}$ cup of icing and puts $\frac{1}{12}$ cup on a cinnamon roll. How much icing remains in the dispenser?

63. Jorge's $\frac{3}{4}$-hr drive to a job was part city and part country driving. If $\frac{2}{5}$ hr was city driving, how much time was spent on country driving?

64. Keri exercises $\frac{5}{6}$ of an hour every day. She jogs for $\frac{7}{12}$ of an hour and spends the remaining time warming up and cooling down. How much time does she spend warming up and cooling down?

65. *Woodworking.* Natalie is replacing a $\frac{3}{4}$-in.-thick shelf in her bookcase. If her replacement board is $\frac{15}{16}$ in. thick, how much must it be planed down before the repair can be completed?

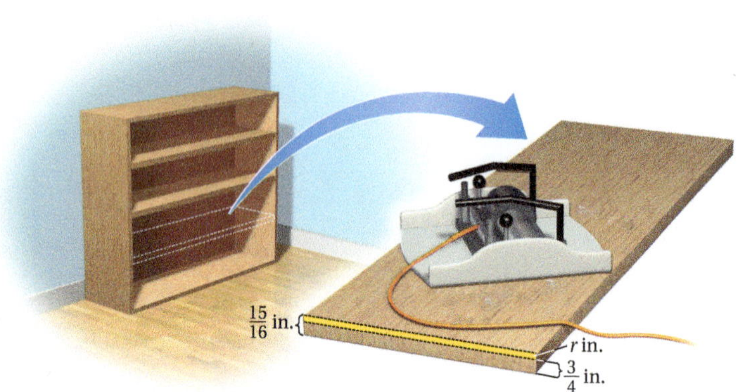

66. *Furniture Cleaner.* A $\frac{2}{3}$-cup mixture of lemon juice and olive oil makes an excellent cleaner for wood furniture. If the mixture contains $\frac{1}{4}$ cup of lemon juice, how much olive oil is in the cleaner?

67. Blake used $\frac{1}{3}$ cup of maple syrup in preparing the batter for a batch of maple oatbran muffins. Sheila pointed out that the recipe actually calls for $\frac{5}{8}$ cup of syrup. How much more syrup should Blake add to the batter?

68. Amber added $\frac{1}{3}$ qt of two-cycle oil to a fuel mixture for her lawn mower. She then noticed that the owner's manual indicates $\frac{1}{2}$ qt should have been added. How much more two-cycle oil should Amber add to the mixture?

Skill Maintenance

Divide, if possible. If not possible, write "Not defined." [1.5a], [3.3b]

69. $\frac{38}{38}$ **70.** $\frac{38}{0}$ **71.** $\frac{124}{0}$ **72.** $\frac{124}{31}$

Divide and simplify. [3.7b]

73. $\frac{3}{7} \div \frac{9}{4}$ **74.** $\frac{9}{10} \div \frac{3}{5}$ **75.** $7 \div \frac{1}{3}$ **76.** $\frac{1}{4} \div 8$

Multiply and simplify. [3.6a]

77. $18 \cdot \frac{2}{3}$ **78.** $\frac{5}{12} \cdot 6$ **79.** $-\frac{17}{25} \cdot \frac{15}{34}$ **80.** $\frac{7}{20} \cdot \left(-\frac{45}{49}\right)$

Synthesis

Simplify.

81. $\frac{7}{8} - \frac{3}{4} - \frac{1}{16}$

82. $\frac{9}{10} - \frac{1}{2} - \frac{2}{15}$

83. $\frac{2}{5} - \frac{1}{6}(-3)^2$

84. $\frac{7}{8} - \frac{1}{10}\left(-\frac{5}{6}\right)^2$

85. $-4 \cdot \frac{3}{7} - \frac{1}{7} \cdot \frac{4}{5}$

86. $\left(\frac{5}{6}\right)^2 - \left(\frac{3}{4}\right)^2$

87. $\left(-\frac{2}{5}\right)^3 - \left(-\frac{3}{10}\right)^3$

88. $\frac{3}{17} - \frac{2}{19} - \left(\frac{3}{17} - \frac{2}{19}\right)$

89. At a party, three friends, Ashley, Cole, and Lauren, shared a big tub of popcorn. Within 30 min, the tub was empty. Ashley ate $\frac{7}{12}$ of the tub while Lauren ate only $\frac{1}{6}$ of the tub. How much did Cole eat?

90. A small community garden was divided among four local residents. Based on the time they could spend on their garden sections and their individual crop plans, each resident received a different-size plot to tend. One received $\frac{1}{4}$ of the garden, the second $\frac{1}{16}$, and the third $\frac{3}{8}$ of the garden. How much did the fourth gardener receive?

91. The circle graph below shows how long shoppers stay when visiting a mall. What portion of shoppers stay for 0–2 hr?

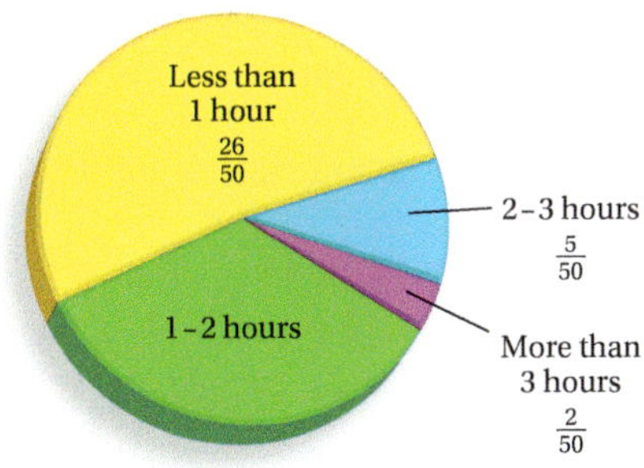

92. Four business partners plan to upgrade their computer system at a cost of $12,600. Celia will pay $\frac{1}{2}$ of the cost, Reba will pay $\frac{1}{4}$ of the cost, Jon will pay $\frac{1}{6}$ of the cost, and Karl will pay the rest.

a) How much will Karl pay?

b) What fractional part will Karl pay?

93. Mark Romano owns $\frac{7}{12}$ of Romano-Chrenka Chevrolet and Lisa Romano owns $\frac{1}{6}$. If Paul and Ella Chrenka own the remaining share of the dealership equally, what fractional piece does Paul own?

94. As part of a rehabilitation program, an athlete must swim and then walk a total of $\frac{9}{10}$ km each day. If one lap in the swimming pool is $\frac{3}{80}$ km, how far must the athlete walk after swimming 10 laps?

95. Solve: $1x + \frac{16}{323} = \frac{10}{187}$.

96. Determine what whole number a must be in order for the following to be true:

$$\frac{10 + a}{23} = \frac{330}{391} - \frac{a}{17}.$$

97. *Microsoft Interview.* The following is a question taken from an employment interview with Microsoft. "Given a gold bar that can be cut exactly twice and a contractor who must be paid one-seventh of a gold bar every day for seven days, how should the bar be cut?"

Source: *Fortune Magazine,* January 22, 2001

4.4 Solving Equations: Using the Principles Together

OBJECTIVES

a Solve equations that involve fractions and require use of both the addition principle and the multiplication principle.

b Solve equations by using the multiplication principle to clear fractions.

SKILL TO REVIEW

Objective 3.8a: Use the multiplication principle to solve equations.

Solve.

1. $\frac{2}{3}x = \frac{5}{6}$

2. $18 = -\frac{2}{3}t$

We have used the multiplication principle to solve equations like

$$\frac{2}{3}x = \frac{5}{6} \quad \text{and} \quad 7 = \frac{5}{4}t,$$

and we have used the addition principle to solve equations like

$$\frac{4}{5} + x = \frac{1}{2} \quad \text{and} \quad \frac{7}{3} = t - \frac{2}{9}.$$

We are now ready to solve equations in which both principles are required.

a USING THE PRINCIPLES TOGETHER

Recall that we use the addition and multiplication principles to write equivalent equations. In the following steps, all five equations are equivalent:

$$5x - 2 = 43 \qquad \text{We first isolate } 5x.$$

$$5x - 2 + 2 = 43 + 2 \qquad \text{Using the addition principle}$$

$$5x = 45 \qquad \text{We now isolate } x.$$

$$\frac{1}{5} \cdot 5x = \frac{1}{5} \cdot 45 \qquad \text{Using the multiplication principle}$$

$$x = 9. \qquad \text{The solution of } x = 9 \text{ is the solution of } 5x - 2 = 43.$$

As a check, note that $5 \cdot 9 - 2 = 45 - 2 = 43$, as desired. The solution is 9.

EXAMPLE 1 Solve: $\frac{3}{4}x - \frac{1}{8} = \frac{1}{2}$.

We first isolate $\frac{3}{4}x$ by adding $\frac{1}{8}$ to both sides:

$$\frac{3}{4}x - \frac{1}{8} = \frac{1}{2}$$

$$\frac{3}{4}x - \frac{1}{8} + \frac{1}{8} = \frac{1}{2} + \frac{1}{8} \qquad \text{Using the addition principle}$$

$$\frac{3}{4}x + 0 = \frac{4}{8} + \frac{1}{8} \qquad \text{Writing with a common denominator}$$

$$\frac{3}{4}x = \frac{5}{8}.$$

Next, we isolate x by multiplying both sides by $\frac{4}{3}$:

$$\frac{3}{4}x = \frac{5}{8} \qquad \text{Note that the reciprocal of } \tfrac{3}{4} \text{ is } \tfrac{4}{3}.$$

$$\frac{4}{3} \cdot \frac{3}{4}x = \frac{4}{3} \cdot \frac{5}{8} \qquad \text{Using the multiplication principle}$$

$$1x = \frac{20}{24}, \text{ or } \frac{5}{6}. \qquad \text{Simplifying; the solution appears to be } \tfrac{5}{6}.$$

Answers

Skill to Review:

1. $\frac{5}{4}$ 2. -27

Check:

$$\begin{array}{c|c} \frac{3}{4}x - \frac{1}{8} & \frac{1}{2} \\ \hline \frac{3}{4}\cdot\frac{5}{6} - \frac{1}{8} & ?\ \frac{1}{2} \\ \frac{\not{3}\cdot 5}{4\cdot 2\cdot \not{3}} - \frac{1}{8} & \\ \frac{5}{8} - \frac{1}{8} & \\ \frac{1}{2} & \end{array}$$

Removing a factor equal to 1: $\frac{3}{3} = 1$

TRUE

The solution is $\frac{5}{6}$.

Do Exercises 1 and 2.

EXAMPLE 2 Solve: $5 + \frac{9}{2}t = -\frac{7}{2}$.

We first isolate $\frac{9}{2}t$ by subtracting 5 from both sides:

$$5 + \frac{9}{2}t = -\frac{7}{2}$$

$$5 + \frac{9}{2}t - 5 = -\frac{7}{2} - 5 \qquad \text{Subtracting 5 from both sides}$$

$$\frac{9}{2}t = -\frac{7}{2} - \frac{10}{2} \qquad \text{Writing 5 as } \tfrac{10}{2} \text{ to use the LCD}$$

$$\frac{9}{2}t = -\frac{17}{2} \qquad \text{Note that the reciprocal of } \tfrac{9}{2} \text{ is } \tfrac{2}{9}.$$

$$\frac{2}{9}\cdot\frac{9}{2}t = \frac{2}{9}\cdot\left(-\frac{17}{2}\right) \qquad \text{Multiplying both sides by } \tfrac{2}{9}$$

$$1t = -\frac{\not{2}\cdot 17}{9\cdot \not{2}} \qquad \text{Removing a factor equal to 1: } \tfrac{2}{2} = 1$$

$$t = -\frac{17}{9}.$$

Check:

$$\begin{array}{c|c} 5 + \frac{9}{2}t & -\frac{7}{2} \\ \hline 5 + \frac{\not{9}}{2}\left(-\frac{17}{\not{9}}\right) & ?\ -\frac{7}{2} \\ 5 + \left(-\frac{17}{2}\right) & \\ \frac{10}{2} + \left(\frac{-17}{2}\right) & \\ \frac{10-17}{2} & \\ \frac{-7}{2} & \end{array}$$

Removing a factor equal to 1: $\frac{9}{9} = 1$

TRUE

The solution is $-\frac{17}{9}$.

Do Exercises 3 and 4.

Sometimes the variable appears on the right side of the equation. The strategy for solving the equation remains the same.

Solve.

1. $\frac{3}{5}t - \frac{8}{15} = \frac{2}{15}$

2. $\frac{1}{2}x - \frac{1}{5} = \frac{7}{10}$

$$\frac{1}{2}x - \frac{1}{5} + \frac{\square}{\square} = \frac{7}{10} + \frac{1}{5}$$

$$\frac{1}{2}x = \frac{7}{10} + \frac{\square}{10}$$

$$\frac{1}{2}x = \frac{\square}{10}$$

$$\square\cdot\frac{1}{2}x = 2\cdot\frac{9}{10}$$

$$1x = \frac{\not{2}\cdot 3\cdot 3}{\not{2}\cdot\square}$$

$$x = \frac{\square}{\square}$$

Solve.

3. $3 + \frac{14}{5}t = -\frac{21}{5}$

4. $2x + 4 = \frac{1}{2}$

Answers

1. $\frac{10}{9}$ 2. $\frac{9}{5}$ 3. $-\frac{18}{7}$ 4. $-\frac{7}{4}$

Guided Solution:

2. $\frac{1}{5}$, 2, 9, 2, 5, $\frac{9}{5}$

EXAMPLE 3 Solve: $20 = 6 - \frac{2}{3}x$.

Our plan is to first use the addition principle to isolate $-\frac{2}{3}x$ and then use the multiplication principle to isolate x.

$$20 = 6 - \frac{2}{3}x$$

$$20 - 6 = 6 - \frac{2}{3}x - 6 \qquad \text{Subtracting 6 (or adding } -6\text{) on both sides}$$

$$14 = -\frac{2}{3}x$$

$$\left(-\frac{3}{2}\right)14 = \left(-\frac{3}{2}\right)\left(-\frac{2}{3}x\right) \qquad \text{Multiplying both sides by } -\tfrac{3}{2}$$

$$-\frac{3 \cdot 14}{2} = 1x$$

$$-\frac{3 \cdot 7 \cdot \cancel{2}}{\cancel{2}} = 1x \qquad \text{Removing a factor equal to 1: } \tfrac{2}{2} = 1$$

$$-21 = x$$

Check:

$$20 = 6 - \frac{2}{3}x$$

$$20 \;?\; 6 - \frac{2}{3}(-21)$$

$$6 + \frac{42}{3}$$

$$6 + 14$$

$$20 \qquad \text{TRUE}$$

The solution is -21.

5. Solve: $9 - \frac{3}{4}x = 21$.

Do Exercise 5.

b CLEARING FRACTIONS

We now look at an alternative approach for solving Examples 1–3. Key to this approach is using the multiplication principle in the *first* step to produce an equivalent equation that is "cleared of fractions."

To "clear fractions," we identify the LCM of the denominators and use the multiplication principle. Because the LCM is a common multiple of the denominators, when both sides of the equation are multiplied by the LCM, the resulting terms can all be simplified. An equivalent equation can then be written without using fractions. We demonstrate this approach by solving Examples 1 and 2 by clearing fractions.

Caution!

We can "clear fractions" in equations, not in expressions. Do not multiply to clear fractions when simplifying an expression.

Either of the methods discussed in this section can be used to solve equations that contain fractions, but it is important for students planning to continue in algebra to thoroughly understand *both* methods.

Answer

5. -16

EXAMPLE 4 Solve Example 1 by clearing fractions:

$$\frac{3}{4}x - \frac{1}{8} = \frac{1}{2}.$$

The LCM of the denominators is 8, so we begin by multiplying both sides of the equation by 8:

$$\frac{3}{4}x - \frac{1}{8} = \frac{1}{2}$$

$$8\left(\frac{3}{4}x - \frac{1}{8}\right) = 8 \cdot \frac{1}{2}$$ Multiplying both sides by 8. We use parentheses when we are multiplying more than one term.

$$\frac{8 \cdot 3}{4}x - 8 \cdot \frac{1}{8} = \frac{8}{2}$$

Caution!

Use the distributive law carefully! Here we multiply every term inside the parentheses by 8.

$$\frac{\not{4} \cdot 2 \cdot 3}{\not{4}}x - 1 = 4$$ Factoring and simplifying

$$6x - 1 = 4$$ The equation is now cleared of fractions. This is the advantage of using this method.

$$6x - 1 + 1 = 4 + 1$$ Adding 1 to both sides

$$6x = 5$$

$$\frac{6x}{6} = \frac{5}{6}$$ Dividing both sides by 6 (or multiplying both sides by $\frac{1}{6}$)

$$x = \frac{5}{6}.$$ Simplifying

Since $\frac{5}{6}$ was the solution in Example 1, we have a check. The solution is $\frac{5}{6}$.

EXAMPLE 5 Solve Example 2 by clearing fractions:

$$5 + \frac{9}{2}t = -\frac{7}{2}.$$

The LCM of the denominators is 2, so we begin by multiplying both sides of the equation by 2:

$$2\left(5 + \frac{9}{2}t\right) = 2\left(-\frac{7}{2}\right)$$ Using the multiplication principle

$$2 \cdot 5 + \frac{2 \cdot 9}{2}t = -\frac{2 \cdot 7}{2}$$ Using the distributive law; multiplying every term by 2

$$10 + 9t = -7$$ Simplifying and removing a factor equal to 1: $\frac{2}{2} = 1$. The equation is now cleared of fractions.

$$10 + 9t - 10 = -7 - 10$$ Subtracting 10 from both sides

$$9t = -17$$ Simplifying

$$\frac{9t}{9} = -\frac{17}{9}$$ Dividing both sides by 9 (or multiplying both sides by $\frac{1}{9}$)

$$t = -\frac{17}{9}.$$ Simplifying

Since the solution in Example 2 is also $-\frac{17}{9}$, we have a check. The solution is $-\frac{17}{9}$.

Do Exercises 6 and 7.

GS **6.** Solve Example 3 by clearing fractions:

$$20 = 6 - \frac{2}{3}x.$$

$$\square(20) = 3\left(6 - \frac{2}{3}x\right)$$

$$\square = 3 \cdot \square - \frac{\not{3} \cdot 2}{\not{3}}x$$

$$60 = \square - \square$$

$$60 - \square = 18 - 2x - 18$$

$$\square = -2x$$

$$\frac{42}{\square} = \frac{-2x}{-2}$$

$$\square = x$$

7. Solve Margin Exercise 1 by clearing fractions:

$$\frac{3}{5}t - \frac{8}{15} = \frac{2}{15}.$$

Answers

6. -21 **7.** $\frac{10}{9}$

Guided Solution:

6. 3, 60, 6, 18, $2x$, 18, 42, -2, -21

4.4 Exercise Set

For Extra Help MyMathLab® MathXL® PRACTICE WATCH READ REVIEW

Reading Check

Determine whether each statement is true or false.

________ **RC1.** We cannot use both the additon principle and the multiplication principle in the solution of an equation.

________ **RC2.** To solve for a variable, it must appear on the left side of an equation.

________ **RC3.** The first step in the solution of an equation always involves the addition principle.

________ **RC4.** We clear fractions by multiplying both sides of an equation by the LCM of the denominators.

a Solve using the addition principle and/or the multiplication principle. Don't forget to check!

1. $6x - 3 = 15$

2. $7x - 6 = 22$

3. $5x + 7 = 10$

4. $19 = 2x + 4$

5. $8 = 3x + 11$

6. $2a + 9 = -7$

7. $\frac{2}{3}y - 8 = 1$

8. $\frac{3}{10}y - 7 = 3$

9. $\frac{3}{2}t - \frac{1}{4} = \frac{1}{2}$

10. $\frac{1}{4}t + \frac{1}{8} = \frac{1}{2}$

11. $\frac{1}{5}x + \frac{3}{10} = \frac{3}{5}$

12. $\frac{4}{3}x - \frac{2}{15} = \frac{2}{15}$

13. $5 - \frac{3}{4}x = 6$

14. $3 - \frac{2}{5}x = 6$

15. $-1 + \frac{2}{5}t = -\frac{4}{5}$

16. $-2 + \frac{1}{6}t = -\frac{7}{4}$

17. $12 = 8 + \frac{7}{2}t$

18. $7 = 5 + \frac{3}{2}t$

19. $-11 = \frac{2}{3}x - 7$

20. $-10 = \frac{2}{5}x - 4$

21. $7 = a + \frac{14}{5}$

22. $9 = a + \frac{47}{10}$

23. $\frac{2}{5}t - 1 = \frac{7}{5}$

24. $-\frac{53}{4} = \frac{3}{2}a + 2$

25. $\frac{39}{8} = \frac{11}{4} - \frac{1}{2}x$

26. $\frac{7}{2} = \frac{13}{2} - \frac{1}{7}y$

27. $-\frac{13}{3}s + \frac{11}{2} = \frac{35}{4}$

28. $-\frac{11}{5}t + \frac{36}{5} = \frac{7}{2}$

b Solve by using the multiplication principle to clear fractions.

29. $\frac{1}{2}x - \frac{1}{4} = \frac{1}{2}$

30. $\frac{1}{3}x - \frac{1}{6} = \frac{2}{3}$

31. $7 = \frac{4}{9}t + 5$

32. $5 = \frac{4}{7}t + 3$

33. $-3 = \frac{3}{4}t - \frac{1}{2}$

34. $-2 = \frac{4}{3}t - \frac{5}{6}$

35. $\frac{4}{3} - \frac{5}{6}x = \frac{3}{2}$

36. $\frac{3}{2} - \frac{5}{3}x = \frac{5}{6}$

37. $-\frac{3}{4} = -\frac{5}{6} - \frac{1}{2}x$

38. $-\frac{1}{4} = -\frac{2}{3} - \frac{1}{6}x$

39. $\frac{4}{3} - \frac{1}{5}t = \frac{3}{4}$

40. $\frac{2}{5} - \frac{3}{4}t = \frac{4}{3}$

Skill Maintenance

Divide. [2.5a]

41. $39 \div (-3)$

42. $56 \div (-7)$

43. $(-72) \div (-4)$

44. $(-81) \div (-3)$

Solve. [2.3b]

45. Jeremy withdraws \$200 from his bank account, makes a \$90 deposit, and then withdraws another \$40. How much has Jeremy's account balance changed?

46. Animal Instinct, a pet supply store, makes a profit of \$850 on Friday and a profit of \$375 on Saturday, but suffers a loss of \$45 on Sunday. Find the total profit or loss for the three days.

Divide and simplify. [3.7b]

47. $\frac{10}{7} \div (2m)$

48. $45n \div \frac{9}{4}$

Synthesis

Solve.

49. 🖩 $\frac{553}{2451}a - \frac{13}{57} = \frac{29}{43}$

50. 🖩 $\frac{1081}{3599}x - \frac{17}{61} = \frac{19}{59}$

51. 🖩 $\frac{11}{17} = \frac{13}{41} - \frac{23}{29}t$

52. $-\frac{a}{5} + \frac{31}{4} = \frac{16}{3}$

53. $\frac{47}{5} - \frac{a}{4} = \frac{44}{7}$

54. $\frac{49}{8} + \frac{2x}{9} = 4$

55. The perimeter of the figure shown is 15 cm. Solve for x.

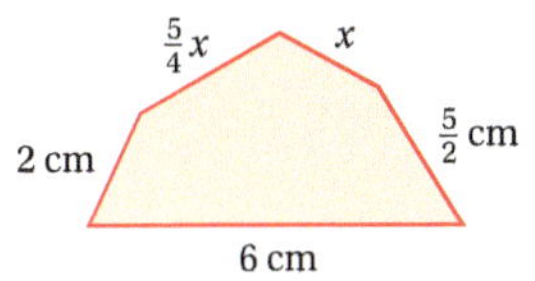

56. The perimeter of the figure is 15 cm. Solve for n.

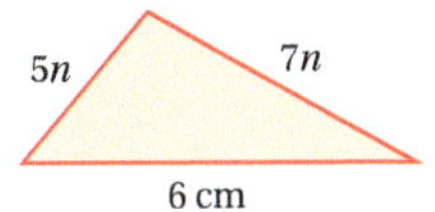

4.5 Mixed Numerals

OBJECTIVES

a Convert between mixed numerals and fraction notation.

b Divide, writing the quotient as a mixed numeral.

SKILL TO REVIEW

Objective 1.5a: Divide whole numbers.

Divide.

1. 735 ÷ 16

2. $23\overline{)6023}$

Convert to a mixed numeral.

1. $1 + \frac{2}{3} = \square\frac{\square}{\square}$

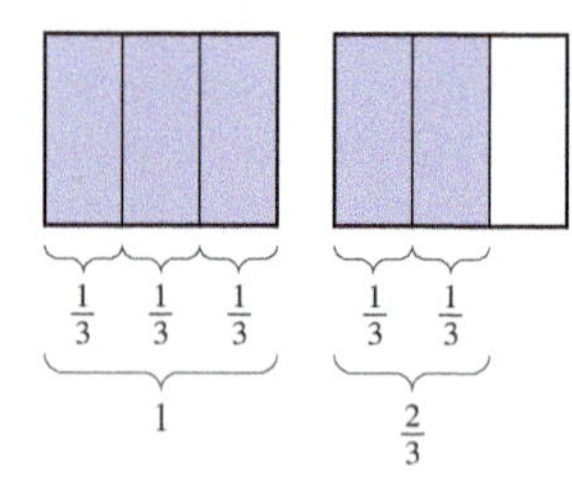

2. $2 + \frac{3}{4} = \square\frac{\square}{\square}$

3. $12 + \frac{2}{7}$

a MIXED NUMERALS

The following figure illustrates the use of a **mixed numeral**. The bolt shown is $2\frac{3}{8}$ in. long. The length is given as a whole-number part, 2, and a fractional part less than 1, $\frac{3}{8}$. We can represent the measurement as $\frac{19}{8}$, but $2\frac{3}{8}$ makes the length easier to visualize and is thus more descriptive.

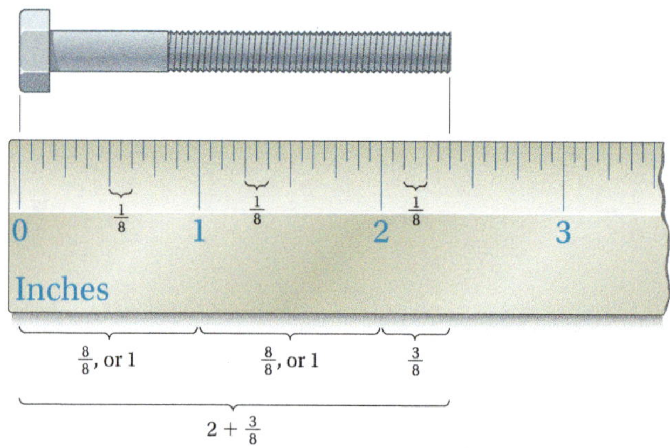

A mixed numeral $2\frac{3}{8}$ represents a sum:

$$2\frac{3}{8} \quad \text{means} \quad 2 + \frac{3}{8}.$$

This is a whole number. This is a fraction less than 1.

EXAMPLES Convert to a mixed numeral.

1. $7 + \frac{2}{5} = 7\frac{2}{5}$

2. $4 + \frac{3}{10} = 4\frac{3}{10}$

◀ **Do Margin Exercises 1–3.**

The notation $2\frac{3}{4}$ has a plus sign left out. To aid in understanding, we sometimes write the missing plus sign: $2 + \frac{3}{4}$. Similarly, the notation $-5\frac{2}{3}$ has a minus sign left out, since $-5\frac{2}{3} = -(5 + \frac{2}{3}) = -5 - \frac{2}{3}$.

Mixed numerals can be displayed on the number line, as shown here.

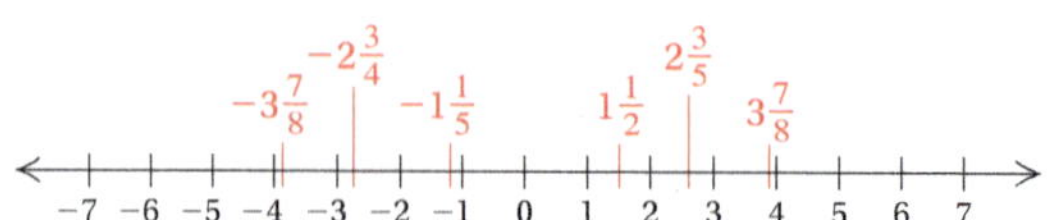

Answers

Skill to Review:
1. 45 R 15 **2.** 261 R 20

Margin Exercises:
1. $1\frac{2}{3}$ **2.** $2\frac{3}{4}$ **3.** $12\frac{2}{7}$

EXAMPLES Convert to fraction notation.

3. $2\frac{3}{4} = 2 + \frac{3}{4}$ — Inserting the missing plus sign

$= \frac{2}{1} + \frac{3}{4}$ — $2 = \frac{2}{1}$

$= \frac{2}{1} \cdot \frac{4}{4} + \frac{3}{4}$ — Finding a common denominator

$= \frac{8}{4} + \frac{3}{4}$

$= \frac{11}{4}$ — Adding

4. $4\frac{3}{10} = 4 + \frac{3}{10} = \frac{4}{1} + \frac{3}{10} = \frac{4}{1} \cdot \frac{10}{10} + \frac{3}{10} = \frac{40}{10} + \frac{3}{10} = \frac{43}{10}$

Do Exercises 4 and 5. ▶

Using Example 4, we can develop a faster way to convert.

> To convert from a mixed numeral like $4\frac{3}{10}$ to fraction notation:
>
> (a) Multiply the whole number by the denominator: $4 \cdot 10 = 40$.
>
> (b) Add the result to the numerator: $40 + 3 = 43$.
>
> (c) Keep the denominator.
>
> $4\frac{3}{10} = \frac{43}{10}$ (a: ×, b: +, c: keep 10)

EXAMPLES Convert to fraction notation.

5. $6\frac{2}{3} = \frac{20}{3}$ — $6 \cdot 3 = 18,\ 18 + 2 = 20$

6. $8\frac{2}{9} = \frac{74}{9}$ — $8 \cdot 9 = 72,\ 72 + 2 = 74$

7. $10\frac{7}{8} = \frac{87}{8}$ — $10 \cdot 8 = 80,\ 80 + 7 = 87$

Do Exercises 6–9. ▶

To find the opposite of the number in Example 5, we can write either $-6\frac{2}{3}$ or $-\frac{20}{3}$. Thus, to convert a negative mixed numeral to fraction notation, we remove the negative sign for purposes of computation and then include it in the answer.

EXAMPLES Convert to fraction notation.

8. $-5\frac{1}{3} = -\left(5 + \frac{1}{3}\right) = -\frac{16}{3}$ — $5 \cdot 3 = 15;\ 15 + 1 = 16;$ include the negative sign

9. $-7\frac{5}{6} = -\left(7 + \frac{5}{6}\right) = -\frac{47}{6}$ — $7 \cdot 6 = 42;\ 42 + 5 = 47$

Do Exercises 10 and 11. ▶

Convert to fraction notation.

4. $4\frac{2}{5}$ **5.** $6\frac{1}{10}$

Convert to fraction notation. Use the faster method.

GS **6.** $4\frac{5}{6}$

$4 \cdot 6 = \square$

$24 + \square = 29$

$4\frac{5}{6} = \frac{\square}{6}$

7. $9\frac{1}{4}$ **8.** $20\frac{2}{3}$

9. $1\frac{9}{13}$

Convert to fraction notation.

10. $-6\frac{2}{5}$ **11.** $-7\frac{2}{9}$

Answers

4. $\frac{22}{5}$ 5. $\frac{61}{10}$ 6. $\frac{29}{6}$ 7. $\frac{37}{4}$ 8. $\frac{62}{3}$ 9. $\frac{22}{13}$ 10. $-\frac{32}{5}$ 11. $-\frac{65}{9}$

Guided Solution:
6. 24, 5, 29

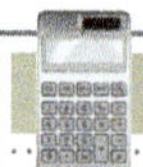

CALCULATOR CORNER

Writing Quotients as Mixed Numerals To calculate 36 ÷ 5 on a fraction calculator, we press [3] [6] [$a^b/_c$] [5] [=]. The display [7 ⌟ 1 ⌟ 5] or [7 ⊔ 1/5] represents $7\frac{1}{5}$.

Another approach, which works on all calculators, is to press 36 ÷ 5 and then examine the result, 7.2. The whole-number part of our answer is 7. To find the remainder, we multiply the whole-number part of our answer, 7, by the divisor, 5, and subtract that result from 36:

$$36 - 7 \times 5 = 1.$$

The remainder is 1, and so the desired fraction is $\frac{1}{5}$. Thus,

$$36 \div 5 = 7\tfrac{1}{5}.$$

EXERCISES Use a calculator to divide. Write each result as a mixed numeral.

1. $7\overline{)39}$
2. $5\overline{)42}$
3. $6\overline{)8857}$
4. $9\overline{)6088}$
5. $45\overline{)6033}$
6. $908\overline{)11{,}234}$

Writing Mixed Numerals

We can find a mixed numeral for $\frac{5}{3}$ as follows:

$$\frac{5}{3} = \frac{3}{3} + \frac{2}{3} = 1 + \frac{2}{3} = 1\frac{2}{3}.$$

In terms of objects, we can think of $\frac{5}{3}$ as $\frac{3}{3}$, or 1, plus $\frac{2}{3}$, as shown below.

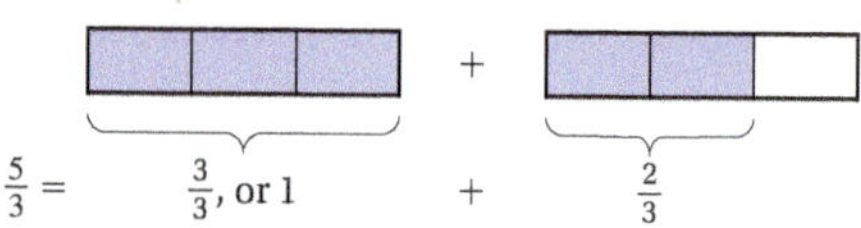

$\frac{5}{3} = \frac{3}{3}$, or 1 $+$ $\frac{2}{3}$

Fraction symbols like $\frac{5}{3}$ also indicate division; $\frac{5}{3}$ means 5 ÷ 3. Let's divide the numerator by the denominator.

$$\begin{array}{r} 1 \\ 3\overline{)5} \\ \underline{3} \\ 2 \end{array} \leftarrow 2 \div 3 = \frac{2}{3}$$

Thus, $\frac{5}{3} = 1\frac{2}{3}$.

To convert from fraction notation to a mixed numeral, divide.

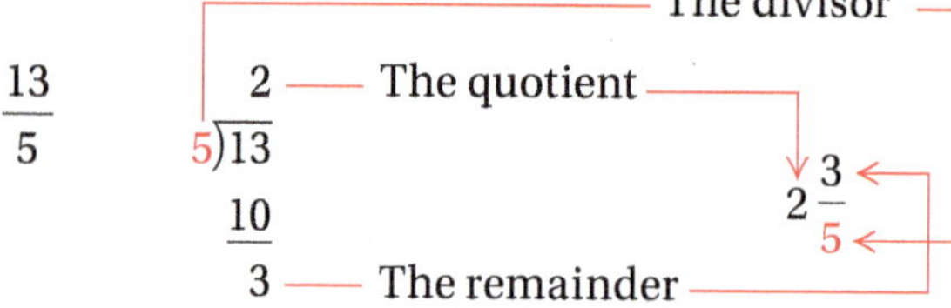

$$\frac{13}{5} \qquad \begin{array}{r} 2 \\ 5\overline{)13} \\ \underline{10} \\ 3 \end{array} \qquad 2\frac{3}{5}$$

The divisor; The quotient; The remainder

EXAMPLES Convert to a mixed numeral.

10. $\frac{69}{10}$ $\qquad \begin{array}{r} 6 \\ 10\overline{)69} \\ \underline{60} \\ 9 \end{array} \qquad \frac{69}{10} = 6\frac{9}{10}$

11. $\frac{122}{8}$ $\qquad \begin{array}{r} 15 \\ 8\overline{)122} \\ \underline{8} \\ 42 \\ \underline{40} \\ 2 \end{array} \qquad \frac{122}{8} = 15\frac{2}{8} = 15\frac{1}{4}$

Simplify the fraction part of a mixed numeral, if possible.

◀ **Do Exercises 12–15.**

Convert to a mixed numeral.

12. $\frac{7}{3}$ **13.** $\frac{11}{10}$

14. $\frac{110}{6}$ **15.** $\frac{231}{18}$

A fraction larger than 1, such as $\frac{27}{8}$, is sometimes referred to as an "improper" fraction. However, the use of notation such as $\frac{27}{8}$, $\frac{11}{9}$, and $\frac{89}{10}$ is quite "proper" and very common in algebra.

Answers

12. $2\frac{1}{3}$ 13. $1\frac{1}{10}$ 14. $18\frac{1}{3}$ 15. $12\frac{5}{6}$

The same procedure also works with negative numbers. Of course, the result will be a negative mixed numeral.

EXAMPLE 12 Convert $\frac{-9}{4}$ to a mixed numeral.

Since $4\overline{)9}$ gives quotient 2, $\frac{8}{1}$, we have $\frac{9}{4} = 2\frac{1}{4}$. Thus, $\frac{-9}{4} = -2\frac{1}{4}$.

Do Exercises 16 and 17. ▶

Convert to a mixed numeral.

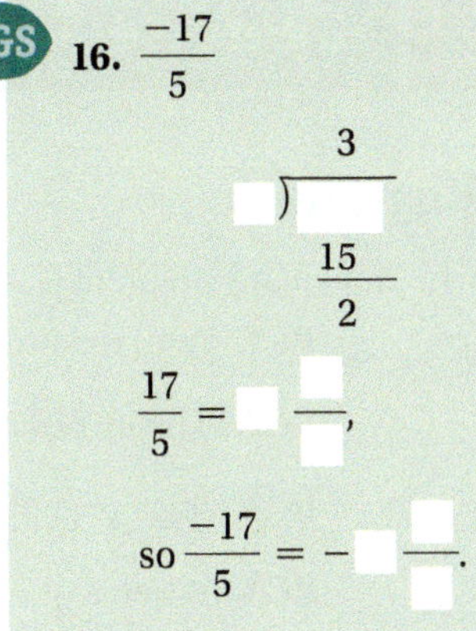

16. $\frac{-17}{5}$

$5\overline{)17}$ quotient 3, $\frac{15}{2}$

$\frac{17}{5} = \square\frac{\square}{\square}$,

so $\frac{-17}{5} = -\square\frac{\square}{\square}$.

17. $-\frac{134}{12}$

b FINDING QUOTIENTS AND AVERAGES

It is quite common when performing long division to express the quotient as a mixed numeral. As in Examples 10–12, the remainder becomes the numerator of the fraction part of the mixed numeral.

EXAMPLE 13 Divide. Write a mixed numeral for the quotient.

$7\overline{)6341}$

We first divide as usual.

$$\begin{array}{r} 905 \\ 7\overline{)6341} \\ \underline{6300} \\ 41 \\ \underline{35} \\ 6 \end{array} \qquad \frac{6341}{7} = 905\frac{6}{7}$$

The answer is 905 R 6, or $905\frac{6}{7}$. Using fraction notation, we write $\frac{6341}{7} = 905\frac{6}{7}$.

Do Exercises 18 and 19. ▶

Divide. Write a mixed numeral for the answer.

18. $6\overline{)4846}$

19. $45\overline{)6053}$

EXAMPLE 14 *Dietary Fiber.* Each of the following five fruits is a good source of dietary fiber. This list gives the amount of fiber, in grams, contained in each serving of fruit. How much fiber is contained, on average, in these fruits?

Source: www.NationalFiberCouncil.org

Raspberries (1 cup)	8 g
Pear (1 medium)	6 g
Blueberries (1 cup)	4 g
Apple (1 medium)	3 g
Banana (1 medium)	3 g

To find the *average* of a set of values, we add the values and divide the sum by the number of values being added.

$$\text{Average grams of fiber} = \frac{8\text{ g} + 6\text{ g} + 4\text{ g} + 3\text{ g} + 3\text{ g}}{5} = \frac{24\text{ g}}{5} = 4\frac{4}{5}\text{ g}$$

On average, these fruits contain $4\frac{4}{5}$ g of fiber per serving.

Do Exercise 20. ▶

20. *Fitness.* Shelby recently added jogging to her exercise routine. The following list gives the total amount of time she jogged for each of the first four weeks of her new routine. Find the average number of minutes she jogged per week.

Week 1: 24 min
Week 2: 30 min
Week 3: 27 min
Week 4: 32 min

Answers

16. $-3\frac{2}{5}$ 17. $-11\frac{1}{6}$ 18. $807\frac{2}{3}$
19. $134\frac{23}{45}$ 20. $28\frac{1}{4}$ min

Guided Solution:
16. 5, 17, $3\frac{2}{5}$, $3\frac{2}{5}$

4.5 Exercise Set

For Extra Help

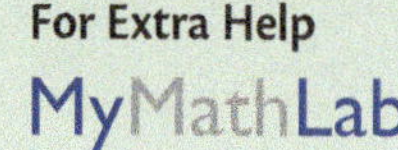

Reading Check

Determine whether each statement is true or false.

_______________ **RC1.** A mixed numeral consists of a whole-number part and a fraction less than 1.

_______________ **RC2.** The mixed numeral $5\frac{1}{4}$ represents $5 + \frac{1}{4}$.

_______________ **RC3.** It is never appropriate to use fraction notation such as $\frac{33}{25}$.

_______________ **RC4.** When a quotient is written as a mixed numeral, the divisor is the denominator of the fraction part (assuming that the fraction has not been simplified).

a Convert to fraction notation.

1. $7\frac{2}{3}$ **2.** $6\frac{2}{5}$ **3.** $6\frac{1}{4}$ **4.** $8\frac{1}{2}$ **5.** $-20\frac{1}{8}$

6. $-10\frac{1}{3}$ **7.** $5\frac{1}{10}$ **8.** $8\frac{1}{10}$ **9.** $20\frac{3}{5}$ **10.** $30\frac{4}{5}$

11. $-33\frac{1}{3}$ **12.** $-66\frac{2}{3}$ **13.** $1\frac{5}{8}$ **14.** $1\frac{3}{5}$ **15.** $-12\frac{3}{4}$

16. $-15\frac{2}{3}$ **17.** $5\frac{7}{10}$ **18.** $7\frac{3}{100}$ **19.** $-5\frac{7}{100}$ **20.** $-6\frac{4}{15}$

Convert to a mixed numeral.

21. $\frac{16}{3}$ **22.** $\frac{19}{8}$ **23.** $\frac{45}{6}$ **24.** $\frac{30}{9}$ **25.** $\frac{57}{10}$ **26.** $\frac{-89}{10}$

27. $\frac{65}{9}$ **28.** $\frac{65}{8}$ **29.** $\frac{-33}{6}$ **30.** $\frac{-50}{8}$ **31.** $\frac{46}{4}$ **32.** $\frac{39}{9}$

33. $\frac{-12}{8}$ **34.** $-\frac{57}{6}$ **35.** $\frac{307}{5}$ **36.** $\frac{227}{4}$ **37.** $-\frac{413}{50}$ **38.** $\frac{467}{100}$

b Divide. Write a mixed numeral for the answer.

39. $8\overline{)869}$ **40.** $3\overline{)2126}$ **41.** $7\overline{)6345}$ **42.** $9\overline{)9110}$ **43.** $21\overline{)852}$

44. $85\overline{)7670}$ **45.** $-302 \div 15$ **46.** $-475 \div 13$ **47.** $471 \div (-21)$ **48.** $545 \div (-25)$

Education. Since 2001, the non-profit organization School on Wheels has provided one-on-one tutoring and educational advocacy for school-aged children experiencing homelessness in Indianapolis. The table below shows the number of children who received services each year, as well as other related information.

Sources: www.indysow.org; School on Wheels annual reports

SCHOOL YEAR	NUMBER OF HOMELESS CHILDREN SERVED	NUMBER OF VOLUNTEER TUTOR HOURS	NUMBER OF BACKPACKS WITH SCHOOL SUPPLIES GIVEN TO CHILDREN
2012–2013	359	5511	278
2011–2012	407	6018	310
2010–2011	415	6708	320
2009–2010	373	5965	242
2008–2009	365	4796	303
2007–2008	322	4027	201
2006–2007	363	4083	101
2005–2006	388	3385	105
2004–2005	375	3131	87
2003–2004	240	1298	92
2002–2003	256	338	46
2001–2002	50	158	63

49. What is the average number of backpacks with school supplies given to children for the school years 2007–2008 through 2012–2013?

50. What is the average number of homeless children served for the school years 2007–2008 through 2012–2013?

51. What is the average number of volunteer tutor hours for the school years 2007–2008 through 2012–2013?

52. What is the average number of volunteer tutor hours for the school years 2001–2002 through 2006–2007?

Skill Maintenance

Use the distributive law to write an equivalent expression. [2.6b]

53. $6(3x + y - 4)$

54. $-3(a - 2b + 4c)$

Synthesis

Write a mixed numeral for each number or sum listed.

55. 🖩 $\dfrac{128{,}236}{541}$ **56.** 🖩 $\dfrac{103{,}676}{349}$ **57.** $\dfrac{56}{7} + \dfrac{2}{3}$ **58.** $\dfrac{72}{12} + \dfrac{5}{6}$ **59.** $\dfrac{12}{5} + \dfrac{19}{15}$

60. There are $\frac{366}{7}$ weeks in a leap year.

61. There are $\frac{365}{7}$ weeks in a year.

62. *Athletics.* At a track and field meet, the hammer that is thrown has a wire length ranging from 3 ft $10\frac{1}{4}$ in. to 3 ft $11\frac{3}{4}$ in., a $4\frac{1}{8}$-in. grip, and a 16-lb ball with a diameter of $4\frac{3}{8}$ in. to $5\frac{1}{8}$ in. Give specifications for the wire length and diameter of an "average" hammer.

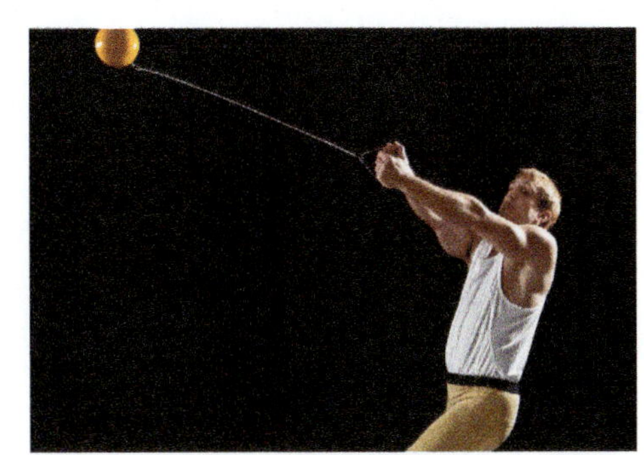

Mid-Chapter Review

Concept Reinforcement

Determine whether each statement is true or false.

_______________ **1.** If $\frac{a}{b} > \frac{c}{b}$, $b \neq 0$, then $a > c$. [4.2c]

_______________ **2.** The mixed numeral $1\frac{1}{2}$ and the fraction $\frac{6}{4}$ represent the same number. [4.5a]

_______________ **3.** The least common multiple of two natural numbers is the smallest number that is a factor of both. [4.1a]

_______________ **4.** To add fractions when denominators are the same, we keep the numerator and add the denominators. [4.2a]

Guided Solutions

 Fill in each blank with the number that creates a correct solution.

5. Subtract: $\frac{11}{42} - \frac{3}{35}$. [4.3a]

$$\frac{11}{42} - \frac{3}{35} = \frac{11}{2 \cdot \square \cdot 7} - \frac{3}{\square \cdot 7}$$ Factoring the denominators

$$= \frac{11}{2 \cdot 3 \cdot 7} \cdot \left(\frac{\square}{\square}\right) - \frac{3}{5 \cdot 7} \cdot \left(\frac{2 \cdot 3}{2 \cdot 3}\right)$$ Multiplying by 1 to get the LCD

$$= \frac{11 \cdot \square}{2 \cdot 3 \cdot 7 \cdot \square} - \frac{3 \cdot 2 \cdot 3}{5 \cdot 7 \cdot 2 \cdot 3}$$ Multiplying

$$= \frac{\square}{2 \cdot 3 \cdot 5 \cdot 7} - \frac{\square}{2 \cdot 3 \cdot 5 \cdot 7}$$ Simplifying

$$= \frac{\square - \square}{2 \cdot 3 \cdot 5 \cdot 7} = \frac{\square}{\square}$$ Subtracting and simplifying

6. Solve: $x + \frac{1}{8} = \frac{2}{3}$. [4.3b]

$$x + \frac{1}{8} = \frac{2}{3}$$

$$x + \frac{1}{8} - \square = \frac{2}{3} - \square$$ Subtracting on both sides

$$x + 0 = \frac{2}{3} \cdot \frac{\square}{\square} - \frac{1}{8} \cdot \frac{\square}{\square}$$ Multiplying by 1 to get the LCD

$$x = \frac{\square}{24} - \frac{\square}{24}$$ Simplifying and multiplying

$$x = \frac{\square}{\square}$$ Subtracting

The solution is $\frac{\square}{\square}$.

Mixed Review

7. Match each set of numbers in the first column with its least common multiple in the second column by drawing connecting lines. [4.1a]

45 and 50	120
50 and 80	720
30 and 24	400
18, 24, and 80	450
30, 45, and 50	

Calculate and simplify. [4.2b], [4.3a]

8. $\frac{1}{5} + \frac{7}{45}$

9. $\frac{5}{6} + \frac{2}{3} + \frac{7}{12}$

10. $\frac{2}{9} - \frac{1}{6}$

11. $\frac{1}{15} - \frac{5}{18}$

12. $\frac{19}{48} - \frac{11}{30}$

13. $-\frac{3}{8}x + \frac{1}{12}x$

14. $\frac{-3}{40} + \frac{-5}{24}$

15. $\frac{8}{65} - \frac{2}{35}$

Solve.

16. Miguel jogs for $\frac{4}{5}$ mi, rests, and then jogs for another $\frac{2}{3}$ mi. How far does he jog in all? [4.2d]

17. One weekend, Kirby spent $\frac{39}{5}$ hr playing two iPod games—Brain Challenge: Cerebral Burn and Scrabble: Go for a Triple Word Score. She spent $\frac{11}{4}$ hr playing Scrabble. How many hours did she spend playing Brain Challenge? [4.3c]

18. Arrange in order from smallest to largest:

$\frac{4}{9}, \frac{3}{10}, \frac{2}{7}$, and $\frac{1}{5}$. [4.2c]

19. Solve: $\frac{2}{5} + x = \frac{9}{16}$. [4.3b]

20. Solve: $\frac{3}{4}x + 1 = \frac{1}{3}$. [4.4a], [4.4b]

21. Divide: $15\overline{)263}$. Write a mixed numeral for the answer. [4.5b]

22. Fraction notation for $9\frac{3}{8}$ is which of the following? [4.5a]

A. $\frac{27}{8}$ **B.** $\frac{93}{8}$ **C.** $\frac{75}{8}$ **D.** $\frac{80}{3}$

23. Mixed numeral notation for $-\frac{39}{4}$ is which of the following? [4.5a]

A. $-35\frac{1}{4}$ **B.** $-\frac{4}{39}$ **C.** $-9\frac{3}{4}$ **D.** $-36\frac{3}{4}$

Understanding Through Discussion and Writing

24. Is the LCM of two numbers always larger than either number? Why or why not? [4.1a]

25. Explain the role of multiplication when adding using fraction notation with different denominators. [4.2b]

26. A student made the following error:

$$\frac{8}{5} - \frac{8}{2} = \frac{8}{3}.$$

Find at least two ways to convince him of the mistake. [4.3a]

27. Are the numbers $2\frac{1}{3}$ and $2 \cdot \frac{1}{3}$ equal? Why or why not? [4.5a]

STUDYING FOR SUCCESS *Taking a Test*

- ☐ Read each question carefully. Know what the question is asking before you answer it.
- ☐ Try to do all the questions the first time through, marking those to recheck if you have time at the end.
- ☐ Pace yourself.
- ☐ Write your answers in a neat and orderly manner.

4.6 Addition and Subtraction Using Mixed Numerals; Applications

OBJECTIVES

a Add using mixed numerals.

b Subtract using mixed numerals.

c Solve applied problems involving addition and subtraction with mixed numerals.

d Add and subtract using negative mixed numerals.

SKILL TO REVIEW

Objective 3.5b: Simplify fraction notation.

Simplify.

1. $\frac{18}{32}$ 2. $\frac{78}{117}$

Add.

1. $2\frac{3}{10} + 5\frac{1}{10}$

2. $8\frac{2}{5} + 3\frac{7}{10}$

Answers

Skill to Review:

1. $\frac{9}{16}$ 2. $\frac{2}{3}$

Margin Exercises:

1. $7\frac{2}{5}$ 2. $12\frac{1}{10}$

a ADDITION USING MIXED NUMERALS

To add mixed numerals, we first add the fractions. Then we add the whole numbers.

EXAMPLE 1 Add: $1\frac{5}{8} + 3\frac{1}{8}$. Write a mixed numeral for the answer.

$$\begin{array}{rl} 1\frac{5}{8} & = \\ +\,3\frac{1}{8} & = \\ \hline \frac{6}{8} & \end{array} \qquad \begin{array}{r} 1\frac{5}{8} \\ +\,3\frac{1}{8} \\ \hline 4\frac{6}{8} = 4\frac{3}{4} \end{array}$$

Add the fractions. Add the whole numbers. Simplifying: $\frac{6}{8} = \frac{3}{4}$

Sometimes we must write the fractional parts with a common denominator before we can add.

EXAMPLE 2 Add: $5\frac{2}{3} + 3\frac{5}{6}$. Write a mixed numeral for the answer.

The LCD is 6.

$$\begin{array}{rcr} 5\frac{2}{3}\cdot\frac{2}{2} & = & 5\frac{4}{6} \\ +\,3\frac{5}{6} & = & +\,3\frac{5}{6} \\ \hline & & 8\frac{9}{6} \end{array}$$

$$8\frac{9}{6} = 8 + \frac{9}{6}$$
$$= 8 + 1\frac{1}{2} \quad \text{Writing } \tfrac{9}{6} \text{ as a mixed numeral; } \tfrac{9}{6} = 1\tfrac{3}{6} = 1\tfrac{1}{2}$$
$$= 9\frac{1}{2}$$

Do Margin Exercises 1 and 2.

The fractional part of a mixed numeral should always be less than 1.

EXAMPLE 3 Add: $10\frac{5}{6} + 7\frac{3}{8}$.

The LCD is 24.

$$\begin{array}{rcr} 10\frac{5}{6} \cdot \frac{4}{4} & = & 10\frac{20}{24} \\ +\,7\frac{3}{8} \cdot \frac{3}{3} & = & +\,7\frac{9}{24} \\ \hline & & 17\frac{29}{24} \end{array}$$

$$17\frac{29}{24} = 17 + \frac{29}{24}$$

$$= 17 + 1\frac{5}{24}$$ Writing $\frac{29}{24}$ as a mixed numeral, $1\frac{5}{24}$

$$= 18\frac{5}{24}$$

Do Exercise 3. ▶

3. Add.

$$\begin{array}{r} 9\frac{3}{4} \\ +\,3\frac{5}{6} \\ \hline \end{array}$$

b SUBTRACTION USING MIXED NUMERALS

EXAMPLE 4 Subtract: $7\frac{3}{4} - 2\frac{1}{4}$.

$$\begin{array}{rcr} 7\frac{3}{4} & = & 7\frac{3}{4} \\ -\,2\frac{1}{4} & = & -\,2\frac{1}{4} \\ \hline \frac{2}{4} & & 5\frac{2}{4} = 5\frac{1}{2} \end{array}$$ ← Simplifying: $\frac{2}{4} = \frac{1}{2}$

↑ Subtract the fractions. ↑ Subtract the whole numbers.

Subtract.

4.
$$\begin{array}{r} 10\frac{7}{8} \\ -\,9\frac{3}{8} \\ \hline \end{array}$$

GS **5.**
$$\begin{array}{rcr} 8\frac{2}{3} & = & 8\frac{\square}{6} \\ -\,5\frac{1}{2} & = & -\,5\frac{\square}{6} \\ \hline & & 3\frac{\square}{6} \end{array}$$

EXAMPLE 5 Subtract: $9\frac{4}{5} - 3\frac{1}{2}$.

The LCD is 10.

$$\begin{array}{rcr} 9\frac{4}{5} \cdot \frac{2}{2} & = & 9\frac{8}{10} \\ -\,3\frac{1}{2} \cdot \frac{5}{5} & = & -\,3\frac{5}{10} \\ \hline & & 6\frac{3}{10} \end{array}$$

Do Exercises 4 and 5. ▶

GS **6.** Subtract: $5 - 1\frac{1}{3}$.

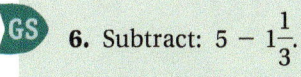

$$\begin{array}{rcr} 5 & = & 4\frac{\square}{\square} \\ -\,1\frac{1}{3} & = & -\,1\frac{1}{3} \\ \hline & & 3\frac{\square}{3} \end{array}$$

EXAMPLE 6 Subtract: $12 - 9\frac{3}{8}$.

$$\begin{array}{rcr} 12 & = & 11\frac{8}{8} \\ -\,9\frac{3}{8} & = & -\,9\frac{3}{8} \\ \hline & & 2\frac{5}{8} \end{array}$$ $12 = 11 + 1 = 11 + \frac{8}{8} = 11\frac{8}{8}$

Do Exercise 6. ▶

Answers

3. $13\frac{7}{12}$ **4.** $1\frac{1}{2}$ **5.** $3\frac{1}{6}$ **6.** $3\frac{2}{3}$

Guided Solutions:

5. 4, 3, 1 **6.** $\frac{3}{3}$, 2

EXAMPLE 7 Subtract: $7\frac{1}{6} - 2\frac{1}{4}$.

$$\begin{array}{rcl} 7\frac{1}{6}\cdot\frac{2}{2} & = & 7\frac{2}{12} \\ -\ 2\frac{1}{4}\cdot\frac{3}{3} & = & -\ 2\frac{3}{12} \end{array}$$

← The LCD is 12.

$$\begin{array}{rcl} 7\frac{2}{12} & = & 6\frac{14}{12} \\ -\ 2\frac{3}{12} & = & -\ 2\frac{3}{12} \\ \hline & & 4\frac{11}{12} \end{array}$$

We cannot subtract $\frac{3}{12}$ from $\frac{2}{12}$.
We borrow 1, or $\frac{12}{12}$, from 7:
$7\frac{2}{12} = 6 + 1 + \frac{2}{12} = 6 + \frac{12}{12} + \frac{2}{12} = 6\frac{14}{12}$.

7. Subtract.

$$\begin{array}{r} 8\frac{1}{9} \\ -4\frac{5}{6} \\ \hline \end{array}$$

◀ Do Exercise 7.

EXAMPLE 8 Combine like terms: $9\frac{3}{4}x + 4\frac{1}{2}x$.

$$9\frac{3}{4}x + 4\frac{1}{2}x = \left(9\frac{3}{4} + 4\frac{1}{2}\right)x$$ Using the distributive law; this is often done mentally.

$$= \left(9\frac{3}{4} + 4\frac{2}{4}\right)x$$ The LCD is 4.

$$= 13\frac{5}{4}x = 14\frac{1}{4}x$$

Combine like terms.

8. $7\frac{1}{12}t + 1\frac{5}{12}t$

9. $7\frac{11}{12}x - 5\frac{2}{3}x$

10. $5\frac{11}{15}x + 8\frac{3}{10}x$

◀ Do Exercises 8–10.

C APPLICATIONS AND PROBLEM SOLVING

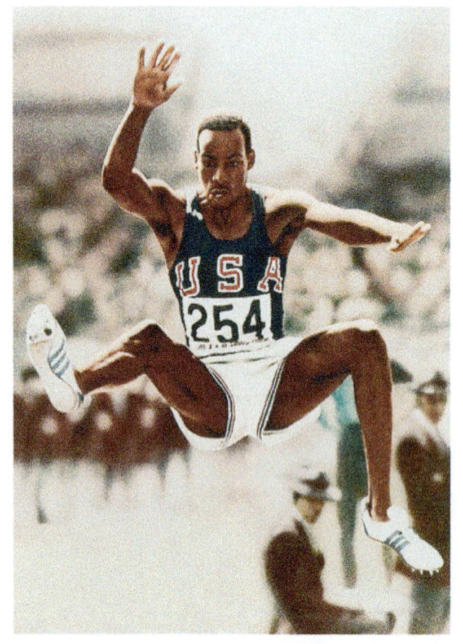

EXAMPLE 9 ***Men's Long-Jump World Records.*** On October 18, 1968, Bob Beamon set a world record of $29\frac{3}{16}$ ft for the long jump, a record that was not broken for nearly 23 years. This record-setting jump was significantly longer than the previous one of $27\frac{19}{48}$ ft, accomplished on May 29, 1965, by Ralph Boston. How much longer was Beamon's jump than Boston's?

Source: "Long jump world record progression," en.wikipedia.org

1. **Familiarize.** The phrase "how much longer" indicates subtraction. We let w = the difference in the world records.
2. **Translate.** We translate as follows:

Beamon's jump	−	Boston's jump	=	Difference in length
$29\frac{3}{16}$	−	$27\frac{19}{48}$	=	w.

3. **Solve.** To solve the equation, we subtract. The LCD is 48.

$$\begin{array}{rcccccl} 29\frac{3}{16} & = & 29\frac{3}{16}\cdot\frac{3}{3} & = & 29\frac{9}{48} & = & 28\frac{57}{48} \\ -\ 27\frac{19}{48} & = & -\ 27\frac{19}{48} & = & -\ 27\frac{19}{48} & = & -\ 27\frac{19}{48} \\ \hline & & & & & & 1\frac{38}{48} = 1\frac{19}{24} \end{array}$$

Thus, $w = 1\frac{19}{24}$.

Answers

7. $3\frac{5}{18}$ 8. $8\frac{1}{2}t$ 9. $2\frac{1}{4}x$ 10. $14\frac{1}{30}x$

4. **Check.** To check, we add the difference, $1\frac{19}{24}$, to Boston's jump.

Since $27\frac{19}{48} + 1\frac{19}{24} = 29\frac{3}{16}$, the answer checks.

5. **State.** Beamon's jump was $1\frac{19}{24}$ ft longer than Boston's.

Do Exercise 11. ▶

11. *Travel Distance.* On a two-day business trip, Paul drove $213\frac{7}{10}$ mi the first day and $107\frac{5}{8}$ mi the second day. What was the total distance that Paul drove?

EXAMPLE 10 *Locks of Love.* Locks of Love, Inc., is a non-profit organization that provides hairpieces for children who have lost their hair as a result of medical conditions. The minimum length for hair that is donated is 10 in. Karissa and Cayla allowed their hair to grow in order to donate it. The length cut from Karissa's hair was $15\frac{1}{4}$ in., and the length cut from Cayla's hair was $14\frac{1}{2}$ in. After the hair was cut, it was discovered that the ends of each lock had highlighting that needed to be trimmed. Because of the highlighting, $1\frac{1}{2}$ in. was cut from Karissa's lock of hair and $2\frac{3}{4}$ in. was cut from Cayla's. In all, what was the total usable length of hair that Karissa and Cayla donated?

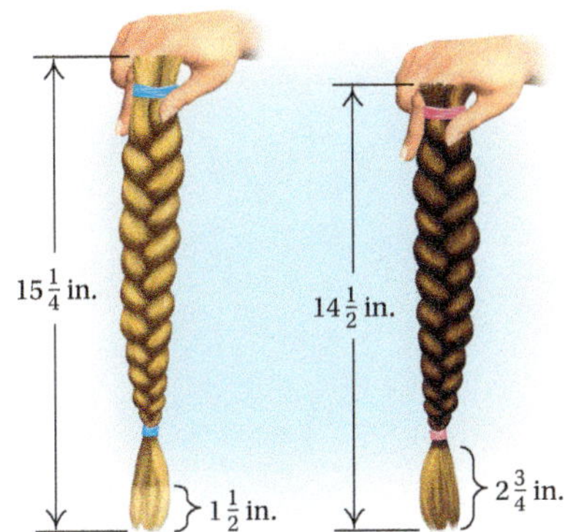

1. **Familiarize.** We let l = the total usable length of hair that Karissa and Cayla donated.
2. **Translate.** The length l is the sum of the lengths that were cut, minus the sum of the lengths that were trimmed from the locks. Thus we have

$$l = (15\tfrac{1}{4} + 14\tfrac{1}{2}) - (1\tfrac{1}{2} + 2\tfrac{3}{4}).$$

3. **Solve.** This is a three-step problem.

a) We first add the two lengths $15\frac{1}{4}$ and $14\frac{1}{2}$.

$$\begin{array}{rcr} 15\frac{1}{4} & = & 15\frac{1}{4} \\ +\ 14\frac{1}{2} & = & +\ 14\frac{2}{4} \\ \hline & & 29\frac{3}{4} \end{array}$$

b) Next, we add the two lengths $1\frac{1}{2}$ and $2\frac{3}{4}$.

$$\begin{array}{rcr} 1\frac{1}{2} & = & 1\frac{2}{4} \\ +\ 2\frac{3}{4} & = & +\ 2\frac{3}{4} \\ \hline & & 3\frac{5}{4} = 4\frac{1}{4} \end{array}$$

c) Finally, we subtract $4\frac{1}{4}$ from $29\frac{3}{4}$.

$$\begin{array}{r} 29\frac{3}{4} \\ -\ 4\frac{1}{4} \\ \hline 25\frac{2}{4} = 25\frac{1}{2} \end{array}$$

Thus, $l = 25\frac{1}{2}$.

4. **Check.** We can check by doing the problem a different way. We can subtract the trimmed length from each lock, then add the adjusted lengths together.

Karissa's lock	Cayla's lock	Sum of lengths
$15\frac{1}{4} = 14\frac{5}{4}$	$14\frac{1}{2} = 13\frac{6}{4}$	$13\frac{3}{4}$
$-\ 1\frac{1}{2} = -\ 1\frac{2}{4}$	$-\ 2\frac{3}{4} = -\ 2\frac{3}{4}$	$+\ 11\frac{3}{4}$
$13\frac{3}{4}$	$11\frac{3}{4}$	$24\frac{6}{4} = 25\frac{1}{2}$

We obtained the same answer, so our answer checks.

5. **State.** The sum of the usable lengths of hair donated was $25\frac{1}{2}$ in.

Do Exercise 12. ▶

12. *Liquid Fertilizer.* There is $283\frac{5}{8}$ gal of liquid fertilizer in a fertilizer application tank. After applying $178\frac{2}{3}$ gal to a soybean field, the farmer requests that Braden's Farm Supply deliver an additional 250 gal to the tank. How many gallons of fertilizer are in the tank after the delivery?

Answers

11. $321\frac{13}{40}$ mi **12.** $354\frac{23}{24}$ gal

d NEGATIVE MIXED NUMERALS

Consider the numbers $5\frac{3}{4}$ and $-5\frac{3}{4}$ on the number line.

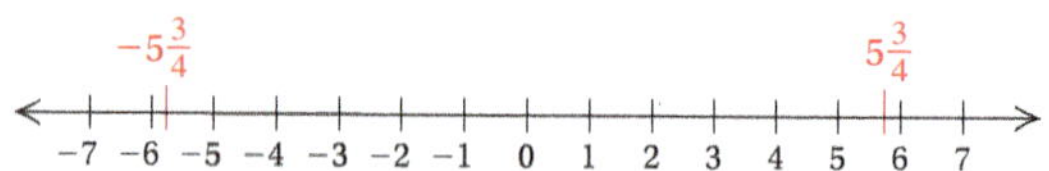

Note that just as $5\frac{3}{4}$ means $5 + \frac{3}{4}$, we can regard $-5\frac{3}{4}$ as $-5 - \frac{3}{4}$.

Consider the subtraction $4 - 4\frac{1}{2}$. We know that if we have \$4 and make a $\$4\frac{1}{2}$ purchase, we will owe half a dollar. Thus,

$$4 - 4\frac{1}{2} = -\frac{1}{2}.$$

We can subtract by rewriting the subtraction as addition:

$$4 - 4\frac{1}{2} = 4 + \left(-4\frac{1}{2}\right).$$

Because $-4\frac{1}{2}$ has the greater absolute value, the answer will be negative. The difference in absolute value is $4\frac{1}{2} - 4 = \frac{1}{2}$, so

$$4 - 4\frac{1}{2} = -\frac{1}{2}.$$

Another way to see this is to convert to fraction notation and subtract:

$$4 - 4\frac{1}{2} = \frac{8}{2} - \frac{9}{2} = \frac{8}{2} + \left(-\frac{9}{2}\right) = -\frac{1}{2}.$$

EXAMPLE 11 Subtract: $3\frac{2}{7} - 4\frac{2}{5}$.

Since $4\frac{2}{5}$ is greater than $3\frac{2}{7}$, the answer will be negative. We can also see this by rewriting the subtraction as $3\frac{2}{7} + \left(-4\frac{2}{5}\right)$. The difference in absolute values is

$$\begin{array}{rcrcr} 4\frac{2}{5} & = & 4\,\frac{2}{5}\cdot\frac{7}{7} & = & 4\frac{14}{35} \\ -\ 3\frac{2}{7} & = & -\ 3\,\frac{2}{7}\cdot\frac{5}{5} & = & -\ 3\frac{10}{35} \\ \hline & & & & 1\frac{4}{35}. \end{array}$$

Because $-4\frac{2}{5}$ has the larger absolute value, we make the answer negative.

Thus, $3\frac{2}{7} - 4\frac{2}{5} = -1\frac{4}{35}$.

◀ Do Exercises 13–15.

Subtract.

13. $7 - 7\frac{3}{4}$

14. $5\frac{1}{2} - 9\frac{3}{4}$

15. $4\frac{2}{3} - 7\frac{1}{6}$

Answers

13. $-\frac{3}{4}$ 14. $-4\frac{1}{4}$ 15. $-2\frac{1}{2}$

EXAMPLE 12 Subtract: $-6\frac{4}{5} - \left(-9\frac{3}{10}\right)$.

We write the subtraction as addition:

$$-6\frac{4}{5} - \left(-9\frac{3}{10}\right) = -6\frac{4}{5} + 9\frac{3}{10}.$$

Instead of subtracting, we add the opposite.

Since $9\frac{3}{10}$ has the greater absolute value, the answer will be positive. The difference in absolute values is

$$\begin{array}{rcrcrcr} 9\frac{3}{10} & = & 9\frac{3}{10} & = & 9\frac{3}{10} & = & 8\frac{13}{10} \\ -6\frac{4}{5} & = & -6\frac{4}{5}\cdot\frac{2}{2} & = & -6\frac{8}{10} & = & -6\frac{8}{10} \\ \hline & & & & & & 2\frac{5}{10} = 2\frac{1}{2}. \end{array}$$

Thus, $-6\frac{4}{5} - \left(-9\frac{3}{10}\right) = 2\frac{1}{2}$.

We can check by converting to fraction notation and redoing the calculation:

$$-6\frac{4}{5} - \left(-9\frac{3}{10}\right) = -\frac{34}{5} - \left(-\frac{93}{10}\right)$$

$$= -\frac{34}{5} + \frac{93}{10}$$ Adding the opposite

$$= -\frac{68}{10} + \frac{93}{10}$$ Writing with a common denominator

$$= \frac{25}{10} = \frac{5}{2} = 2\frac{1}{2}.$$ $-68 + 93 = 25$

Do Exercises 16 and 17. ▶

Subtract.

16. $-7\frac{1}{3} - \left(-5\frac{1}{2}\right)$

17. $-4\frac{1}{10} - \left(-7\frac{2}{5}\right)$

To add two negative numbers we add absolute values and make the answer negative.

EXAMPLE 13 Add: $-4\frac{1}{6} + \left(-5\frac{2}{9}\right)$.

We add the absolute values and make the answer negative.

$$\begin{array}{rcrcr} 4\frac{1}{6} & = & 4\frac{1}{6}\cdot\frac{3}{3} & = & 4\frac{3}{18} \\ +\,5\frac{2}{9} & = & +\,5\frac{2}{9}\cdot\frac{2}{2} & = & +\,5\frac{4}{18} \\ \hline & & & & 9\frac{7}{18} \end{array}$$

Adding absolute values

Thus, $-4\frac{1}{6} + \left(-5\frac{2}{9}\right) = -9\frac{7}{18}$.

Do Exercise 18. ▶

18. Add: $-7\frac{1}{10} + \left(-6\frac{2}{15}\right)$.

Answers

16. $-1\frac{5}{6}$ 17. $3\frac{3}{10}$ 18. $-13\frac{7}{30}$

4.6 Exercise Set

For Extra Help
MyMathLab®

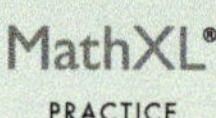

Reading Check

Match each addition or subtraction with the correct first step from the following list.

a) Add the fractions.

b) Write the fractional parts with a common denominator.

c) Rename 5 as $4\frac{9}{9}$.

d) Borrow 1 from 5 and add it to $\frac{1}{9}$.

_____ **RC1.** $5\frac{1}{9} - 3\frac{4}{9}$

_____ **RC2.** $5\frac{4}{9} + 3\frac{1}{9}$

_____ **RC3.** $5\frac{4}{9} + 3\frac{1}{18}$

_____ **RC4.** $5 - 3\frac{1}{9}$

a, b Perform the indicated operation. Write a mixed numeral for each answer.

1. $6 + 5\frac{2}{5}$

2. $3 + 6\frac{5}{7}$

3. $2\frac{7}{8} + 6\frac{5}{8}$

4. $2\frac{5}{6} + 5\frac{5}{6}$

5. $4\frac{1}{4} + 1\frac{1}{12}$

6. $4\frac{1}{12} + 5\frac{1}{6}$

7. $7\frac{3}{4} + 5\frac{5}{6}$

8. $4\frac{3}{8} + 6\frac{5}{12}$

9. $3\frac{2}{5} + 8\frac{7}{10}$

10. $5\frac{1}{2} + 3\frac{7}{10}$

11. $6\frac{3}{8} + 10\frac{5}{6}$

12. $\frac{5}{8} + 1\frac{5}{6}$

13. $18\frac{4}{5} + 2\frac{7}{10}$

14. $15\frac{5}{8} + 11\frac{3}{4}$

15. $14\frac{5}{8} + 13\frac{1}{4}$

16. $16\frac{1}{4} + 15\frac{7}{8}$

17. $8\frac{9}{10} - 1\frac{7}{10}$

18. $6\frac{7}{8} - 5\frac{3}{8}$

19. $9\frac{3}{5} - 3\frac{1}{2}$

20. $8\frac{2}{3} - 7\frac{1}{2}$

21. $4\frac{1}{5} - 2\frac{3}{5}$

22. $5\frac{1}{8} - 2\frac{3}{8}$

23. $19 - 5\frac{3}{4}$

24. $17 - 3\frac{7}{8}$

25. $34 - 18\frac{5}{8}$

26. $\begin{array}{r} 23 \\ -\ 19\frac{3}{4} \\ \hline \end{array}$

27. $\begin{array}{r} 21\frac{1}{6} \\ -\ 13\frac{3}{4} \\ \hline \end{array}$

28. $\begin{array}{r} 42\frac{1}{10} \\ -\ 23\frac{7}{12} \\ \hline \end{array}$

29. $\begin{array}{r} 25\frac{1}{9} \\ -\ 13\frac{5}{6} \\ \hline \end{array}$

30. $\begin{array}{r} 23\frac{5}{16} \\ -\ 14\frac{7}{12} \\ \hline \end{array}$

Combine like terms.

31. $1\frac{1}{8}t + 7\frac{5}{8}t$

32. $6\frac{1}{4}x + 8\frac{3}{4}x$

33. $9\frac{1}{2}x - 7\frac{1}{2}x$

34. $7\frac{3}{4}x - 2\frac{1}{4}x$

35. $5\frac{9}{10}y + 2\frac{2}{5}y$

36. $9\frac{3}{4}t + 2\frac{3}{8}t$

37. $37\frac{5}{9}t - 25\frac{2}{3}t$

38. $23\frac{1}{6}t - 19\frac{2}{3}t$

39. $2\frac{5}{6}x + 7\frac{3}{8}x$

40. $7\frac{3}{20}t + 1\frac{2}{15}t$

41. $11a - 8\frac{2}{3}a$

42. $6a - 3\frac{7}{10}a$

C Solve.

43. ***Widening a Driveway.*** Sherry and Woody are widening their existing $17\frac{1}{4}$-ft driveway by adding $5\frac{9}{10}$ ft on one side. What is the new width of the driveway?

44. ***Plumbing.*** A plumber uses two pipes, each of length $51\frac{5}{16}$ in., and one pipe of length $34\frac{3}{4}$ in. when installing a shower. How much pipe is used in all?

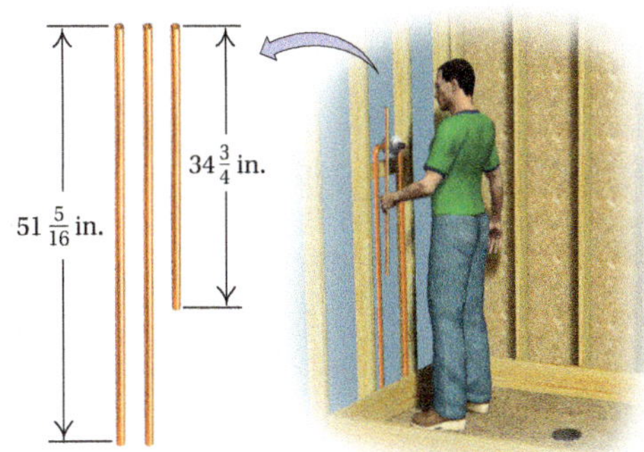

45. ***Upholstery Fabric.*** Executive Car Care sells 45-in. upholstery fabric for car restoration. Art bought $9\frac{1}{4}$ yd and $10\frac{5}{6}$ yd for two car projects. How many yards did Art buy?

46. ***Painting.*** A painter used $1\frac{3}{4}$ gal of paint for the Garcias' living room and $1\frac{1}{3}$ gal for their family room. How much paint was used in all?

47. ***Height.*** Casey's beagle is $14\frac{1}{4}$ in. from shoulder to floor, and her basset hound is $13\frac{5}{16}$ in. from shoulder to floor. How much shorter is her basset hound?

48. ***Winterizing a Swimming Pool.*** To winterize their swimming pool, the Jablonskis are draining the water into a nearby field. The distance to the field is $103\frac{1}{2}$ ft. Because their only hose measures $62\frac{3}{4}$ ft, they need to buy an additional hose. How long must the new hose be?

Find the perimeter of (distance around) each figure.

49.

50.

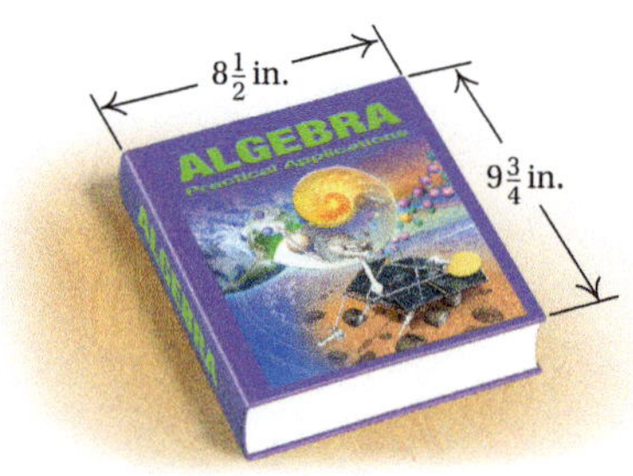

51.

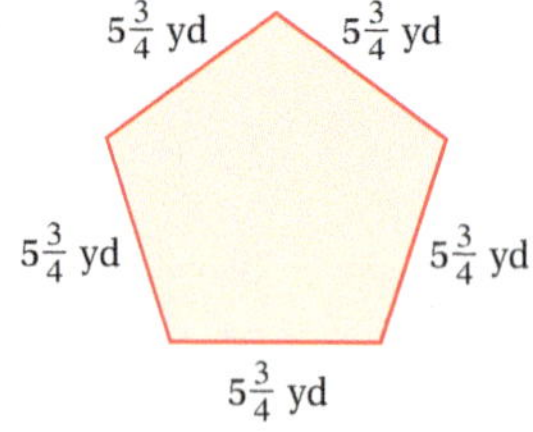

52.

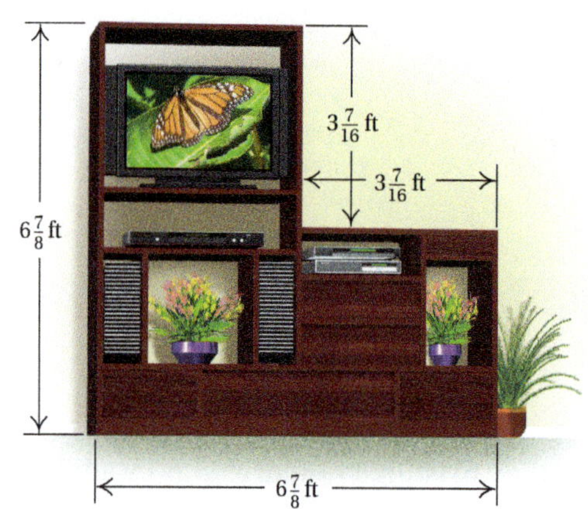

53.

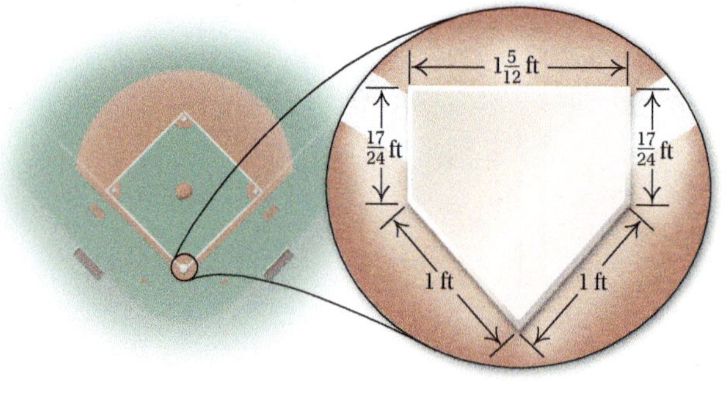

54.

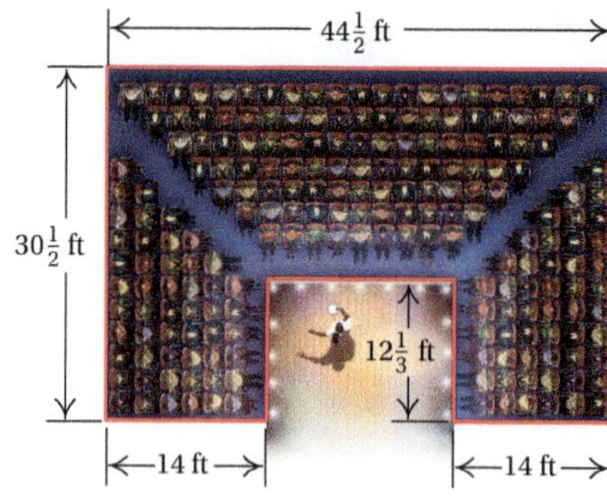

Find the length d in each figure.

55.

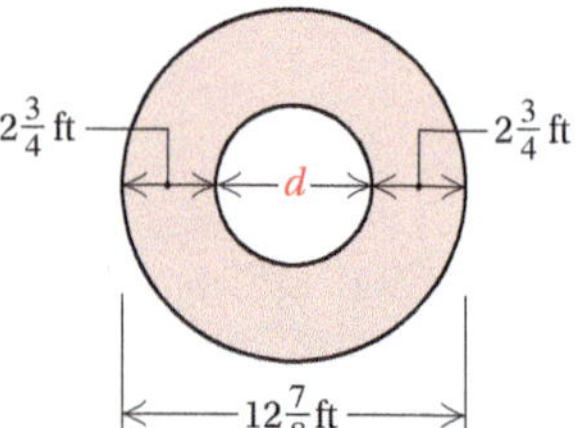

56.

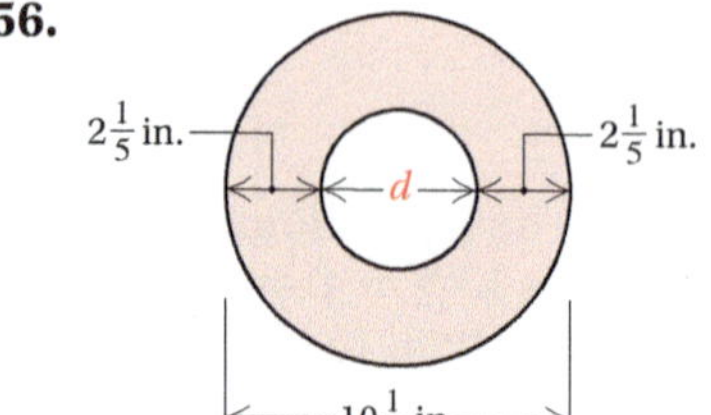

57. ***Stone Bench.*** Baytown Village Stone Creations is making a custom stone bench as shown below. The recommended height for the bench is 18 in. The depth of the stone bench is $3\frac{3}{8}$ in. Each of the two supporting legs is made up of three stacked stones. Two of the stones measure $3\frac{1}{2}$ in. and $5\frac{1}{4}$ in. How much must the third stone measure?

58. ***Window Dimensions.*** The Sanchez family is replacing a window in their home. The original window measures $4\frac{5}{6}$ ft $\times$ $8\frac{1}{4}$ ft. The new window is $2\frac{1}{3}$ ft wider. What are the dimensions of the new window?

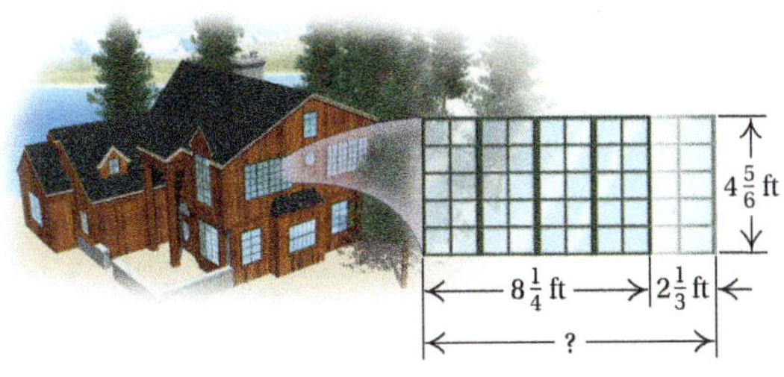

59. ***Carpentry.*** When cutting wood with a saw, a carpenter must take into account the thickness of the saw blade. Suppose that from a piece of wood 36 in. long, a carpenter cuts a $15\frac{3}{4}$-in. length with a saw blade that is $\frac{1}{8}$ in. thick. How long is the piece that remains?

60. ***Cutco Cutlery.*** The Essentials 5-piece set sold by Cutco contains three knives with different blade lengths: $7\frac{5}{8}''$ Petite Chef, $6\frac{3}{4}''$ Petite Carver, and $2\frac{3}{4}''$ Paring Knife. How much larger is the blade of the Petite Chef than that of the Petite Carver? than that of the Paring Knife?

Source: Cutco Cutlery Corporation

61. ***Interior Design.*** Eric worked $10\frac{1}{2}$ hr over a three-day period on an interior design project. If he worked $2\frac{1}{2}$ hr on the first day and $4\frac{1}{5}$ hr on the second, how many hours did Eric work on the third day?

62. ***Painting.*** Geri had $3\frac{1}{2}$ gal of paint. It took $2\frac{3}{4}$ gal to paint the family room. She estimated that it would take $2\frac{1}{4}$ gal to paint the living room. How much more paint did Geri need?

63. ***Fly Fishing.*** Bryan is putting together a fly fishing line and uses $58\frac{5}{8}$ ft of slow-sinking fly line and $8\frac{3}{4}$ ft of leader line. He uses $\frac{3}{8}$ ft of the slow-sinking fly line to connect the two lines. The knot used to connect the fly to the leader line uses $\frac{1}{6}$ ft of the leader line. How long is the finished fly fishing line?

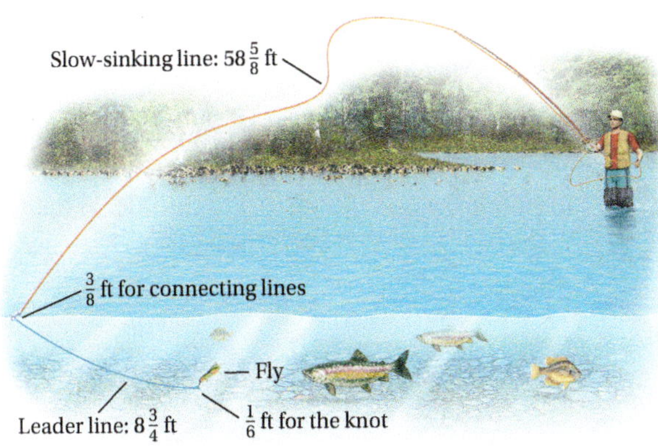

64. Find the smallest length of a bolt that will pass through a piece of tubing with an outside diameter of $\frac{1}{2}$ in., a washer $\frac{1}{16}$ in. thick, a piece of tubing with a $\frac{3}{4}$-in. outside diameter, another washer, and a nut $\frac{3}{16}$ in. thick.

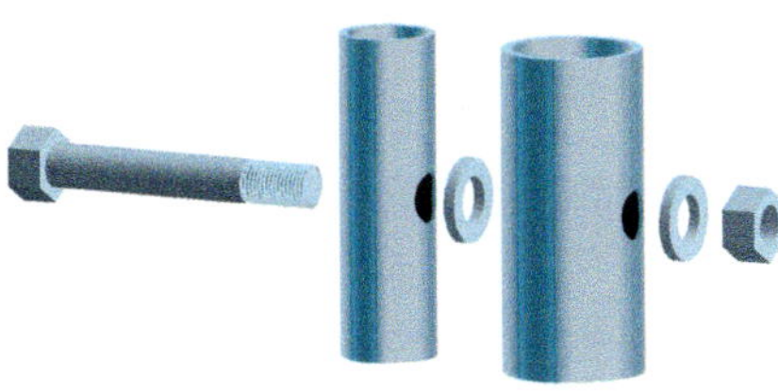

d Perform the indicated operation. Write a mixed numeral for each answer.

65. $9 - 9\frac{2}{5}$

66. $8 - 8\frac{3}{7}$

67. $3\frac{1}{2} - 6\frac{3}{4}$

68. $5\frac{1}{2} - 7\frac{3}{4}$

69. $3\frac{4}{5} + \left(-7\frac{2}{3}\right)$

70. $2\frac{3}{7} + \left(-5\frac{1}{2}\right)$

71. $-3\frac{1}{5} - 4\frac{2}{5}$

72. $-5\frac{3}{8} - 4\frac{1}{8}$

73. $-4\frac{1}{12} + 6\frac{2}{3}$

74. $-2\frac{3}{4} + 5\frac{3}{8}$

75. $-6\frac{1}{9} - \left(-4\frac{2}{9}\right)$

76. $-2\frac{3}{5} - \left(-1\frac{1}{5}\right)$

Skill Maintenance

77. Write expanded notation for 38,125. [1.1b]

78. Write a word name for 2,005,689. [1.1c]

79. Write exponential notation for $9 \cdot 9 \cdot 9 \cdot 9$. [1.9a]

80. Evaluate: 3^4. [1.9b]

Determine whether the first number is divisible by the second. [3.1a], [3.1b]

81. 9993 by 3

82. 9993 by 9

83. 2345 by 9

84. 2345 by 5

85. 2335 by 10

86. 7764 by 6

87. 18,888 by 2

88. 18,888 by 6

89. Multiply and simplify: $\frac{15}{9} \cdot \frac{18}{39}$. [3.6a]

90. Divide and simplify: $\frac{12}{25} \div \frac{24}{5}$. [3.7b]

Synthesis

Calculate each of the following. Write the result as a mixed numeral.

91. 🖩 $3289\frac{1047}{1189} + 5278\frac{32}{41}$

92. 🖩 $4230\frac{19}{73} - 5848\frac{17}{29}$

Solve.

93. $35\frac{2}{3} + n = 46\frac{1}{4}$

94. $42\frac{7}{9} = x - 13\frac{2}{5}$

95. $-15\frac{7}{8} = 12\frac{1}{2} + t$

96. A post for a pier is 29 ft long. Half of the post extends above the water's surface and $8\frac{3}{4}$ ft of the post is buried in mud. How deep is the water at that location?

97. An algebra text is $1\frac{1}{8}$ in. thick, $9\frac{3}{4}$ in. long, and $8\frac{1}{2}$ in. wide. If the front, back, and spine of the book were unfolded, they would form a rectangle. What would the perimeter of that rectangle be?

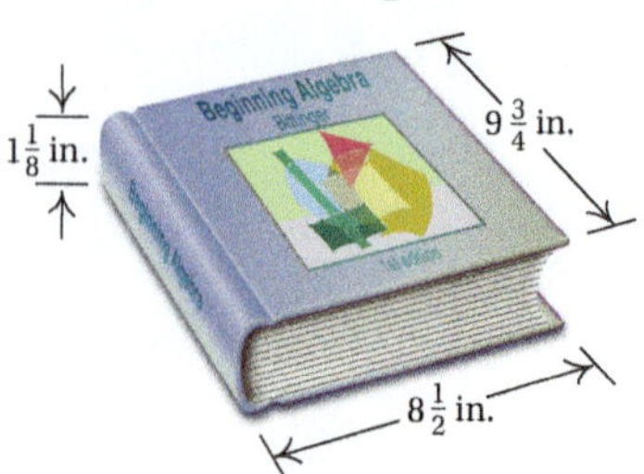

Multiplication and Division of Mixed Numerals; Applications

4.7

OBJECTIVES

a Multiply using mixed numerals.

b Divide using mixed numerals.

c Evaluate expressions using mixed numerals.

d Solve applied problems involving multiplication and division with mixed numerals.

SKILL TO REVIEW

Objective 3.7b: Divide and simplify using fraction notation.

Divide and simplify.

1. $85 \div \frac{17}{5}$ 2. $\frac{7}{65} \div \frac{21}{25}$

Carrying out addition and subtraction with mixed numerals is usually easier if the numbers are left as mixed numerals. With multiplication and division, however, it is easier to convert the numbers to fraction notation first.

a MULTIPLICATION

MULTIPLICATION USING MIXED NUMERALS

To multiply using mixed numerals, first convert to fraction notation and multiply. Then convert the answer to a mixed numeral, if appropriate.

EXAMPLE 1 Multiply: $6 \cdot 2\frac{1}{2}$.

$$6 \cdot 2\frac{1}{2} = \frac{6}{1} \cdot \frac{5}{2} = \frac{6 \cdot 5}{1 \cdot 2} = \frac{\cancel{2} \cdot 3 \cdot 5}{\cancel{2} \cdot 1} = 15$$

Removing a factor equal to 1: $\frac{2}{2} = 1$

Here we write fraction notation.

EXAMPLE 2 Multiply: $3\frac{1}{2} \cdot \frac{3}{4}$.

$$3\frac{1}{2} \cdot \frac{3}{4} = \frac{7}{2} \cdot \frac{3}{4} = \frac{21}{8} = 2\frac{5}{8}$$

Recall that common denominators are *not* required when multiplying fractions.

Do Margin Exercises 1 and 2.

Multiply.

1. $8 \cdot 3\frac{1}{2}$

2. $5\frac{1}{2} \cdot \frac{3}{7}$

EXAMPLE 3 Multiply: $-10 \cdot 5\frac{2}{3}$.

$$-10 \cdot 5\frac{2}{3} = -\frac{10}{1} \cdot \frac{17}{3} = -\frac{170}{3} = -56\frac{2}{3}$$

EXAMPLE 4 Multiply: $2\frac{1}{4} \cdot 5\frac{2}{3}$.

$$2\frac{1}{4} \cdot 5\frac{2}{3} = \frac{9}{4} \cdot \frac{17}{3} = \frac{9 \cdot 17}{4 \cdot 3} = \frac{\cancel{3} \cdot 3 \cdot 17}{2 \cdot 2 \cdot \cancel{3}} = \frac{51}{4} = 12\frac{3}{4}$$

Caution!

$2\frac{1}{4} \cdot 5\frac{2}{3} \neq 10\frac{2}{12}$. A common error is to multiply the whole numbers and then the fractions. The correct answer, $12\frac{3}{4}$, is found only after converting to fraction notation.

Do Exercises 3 and 4.

Multiply.

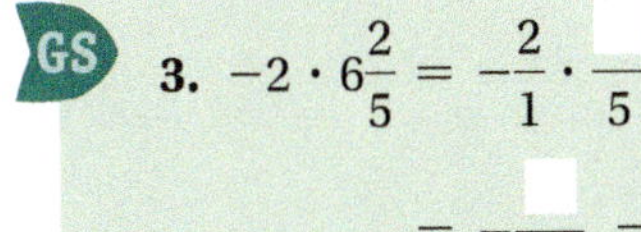

GS 3. $-2 \cdot 6\frac{2}{5} = -\frac{2}{1} \cdot \frac{\square}{5}$

$= -\frac{\square}{5} = -\square\frac{\square}{\square}$

4. $3\frac{1}{3} \cdot 2\frac{1}{2}$

Answers

Skill to Review:

1. 25 2. $\frac{5}{39}$

Margin Exercises:

1. 28 2. $2\frac{5}{14}$ 3. $-12\frac{4}{5}$ 4. $8\frac{1}{3}$

Guided Solution:

3. 32, 64, $12\frac{4}{5}$

b DIVISION

The division $1\frac{1}{2} \div \frac{1}{6}$ is shown here. This division means "How many $\frac{1}{6}$'s are in $1\frac{1}{2}$?" We see that the answer is 9.

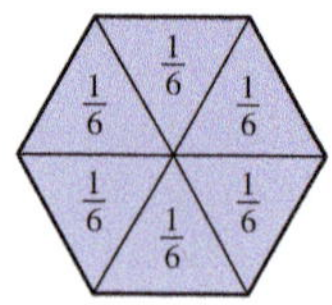

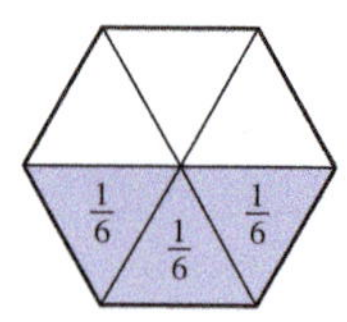

$\frac{1}{6}$ goes into $1\frac{1}{2}$ nine times.

When we divide using mixed numerals, we convert to fraction notation first.

$$1\frac{1}{2} \div \frac{1}{6} = \frac{3}{2} \div \frac{1}{6}$$ We convert $1\frac{1}{2}$ to fraction notation.

$$= \frac{3}{2} \cdot \frac{6}{1}$$ To divide by $\frac{1}{6}$, multiply by the reciprocal, $\frac{6}{1}$.

$$= \frac{3 \cdot 6}{2 \cdot 1} = \frac{3 \cdot 3 \cdot 2}{2 \cdot 1} = \frac{3 \cdot 3}{1} \cdot \frac{2}{2} = \frac{3 \cdot 3}{1} \cdot 1 = 9$$

DIVISION USING MIXED NUMERALS

To divide using mixed numerals, first write fraction notation and divide. Then convert the answer to a mixed numeral, if appropriate.

EXAMPLE 5 Divide: $32 \div 3\frac{1}{5}$.

$$32 \div 3\frac{1}{5} = \frac{32}{1} \div \frac{16}{5}$$ Converting to fraction notation

$$= \frac{32}{1} \cdot \frac{5}{16} = \frac{32 \cdot 5}{1 \cdot 16} = \frac{2 \cdot 16 \cdot 5}{1 \cdot 16} = 10$$ Removing a factor equal to 1: $\frac{16}{16} = 1$

Remember to multiply by the reciprocal of the divisor.

Caution!

The reciprocal of $3\frac{1}{5}$ is neither $5\frac{1}{3}$ nor $3\frac{5}{1}$!

◀ **Do Exercise 5.**

EXAMPLE 6 Divide: $2\frac{1}{3} \div 1\frac{3}{4}$.

$$2\frac{1}{3} \div 1\frac{3}{4} = \frac{7}{3} \div \frac{7}{4} = \frac{7}{3} \cdot \frac{4}{7} = \frac{7 \cdot 4}{7 \cdot 3} = \frac{4}{3} = 1\frac{1}{3}$$ Removing a factor equal to 1: $\frac{7}{7} = 1$

EXAMPLE 7 Divide: $-1\frac{3}{5} \div (-3\frac{1}{3})$.

$$-1\frac{3}{5} \div \left(-3\frac{1}{3}\right) = -\frac{8}{5} \div \left(-\frac{10}{3}\right) = \frac{8}{5} \cdot \frac{3}{10}$$ The product or quotient of two negatives is positive.

$$= \frac{2 \cdot 4 \cdot 3}{5 \cdot 2 \cdot 5} = \frac{12}{25}$$ Removing a factor equal to 1: $\frac{2}{2} = 1$

◀ **Do Exercises 6 and 7.**

5. Divide: $63 \div 5\frac{1}{4}$.

Divide.

6. $2\frac{1}{4} \div 1\frac{1}{5}$

$= \frac{\square}{4} \div \frac{\square}{5}$

$= \frac{9}{4} \cdot \frac{5}{\square}$

$= \frac{3 \cdot 3 \cdot 5}{2 \cdot 2 \cdot 2 \cdot \square}$

$= \frac{\square}{\square} \cdot \frac{3 \cdot 5}{2 \cdot 2 \cdot 2}$

$= \frac{15}{\square}$

$= \square\frac{\square}{\square}$

GS

7. $1\frac{3}{4} \div \left(-2\frac{1}{2}\right)$

Answers

5. 12 6. $1\frac{7}{8}$ 7. $-\frac{7}{10}$

Guided Solution:

6. 9, 6, 6, 3, $\frac{3}{3}$, 8, $1\frac{7}{8}$

C EVALUATING EXPRESSIONS

Mixed numerals can appear in algebraic expressions.

EXAMPLE 8 A train traveling r miles per hour for t hours travels a total of rt miles. (*Remember*: Distance = Rate · Time.)

a) Find the distance traveled by a train moving at 60 mph in $2\frac{3}{4}$ hr.

b) Find the distance traveled if the speed of the train is $26\frac{1}{2}$ mph and the time is $2\frac{2}{3}$ hr.

a) We evaluate rt for $r = 60$ and $t = 2\frac{3}{4}$:

$$rt = 60 \cdot 2\frac{3}{4}$$
$$= \frac{60}{1} \cdot \frac{11}{4}$$
$$= \frac{15 \cdot 4 \cdot 11}{1 \cdot 4} = 165.$$

Removing a factor equal to 1: $\frac{4}{4} = 1$

In $2\frac{3}{4}$ hr, a train moving at 60 mph travels 165 mi.

b) We evaluate rt for $r = 26\frac{1}{2}$ and $t = 2\frac{2}{3}$:

$$rt = 26\frac{1}{2} \cdot 2\frac{2}{3}$$
$$= \frac{53}{2} \cdot \frac{8}{3} = \frac{53 \cdot 2 \cdot 4}{2 \cdot 3}$$
$$= \frac{212}{3} = 70\frac{2}{3}.$$

Removing a factor equal to 1: $\frac{2}{2} = 1$

In $2\frac{2}{3}$ hr, a train moving at $26\frac{1}{2}$ mph travels $70\frac{2}{3}$ mi.

EXAMPLE 9 Evaluate $x + yz$ for $x = 7\frac{1}{3}$, $y = \frac{1}{3}$, and $z = 5$.

We substitute and follow the rules for order of operations:

$$x + yz = 7\frac{1}{3} + \frac{1}{3} \cdot 5$$
$$= 7\frac{1}{3} + \frac{1}{3} \cdot \frac{5}{1}$$

Multiply first; then add.

$$= 7\frac{1}{3} + \frac{5}{3}$$
$$= 7\frac{1}{3} + 1\frac{2}{3}$$
$$= 8\frac{3}{3} = 9.$$

Adding mixed numerals

Do Exercises 8–10.

Evaluate.

8. rt, for $r = 78$ and $t = 2\frac{1}{4}$

9. $7xy$, for $x = 9\frac{2}{5}$ and $y = 2\frac{3}{7}$

10. $x - y \div z$, for $x = 5\frac{7}{8}$, $y = \frac{1}{4}$, and $z = 2$

Answers

8. $175\frac{1}{2}$ **9.** $159\frac{4}{5}$ **10.** $5\frac{3}{4}$

d APPLICATIONS AND PROBLEM SOLVING

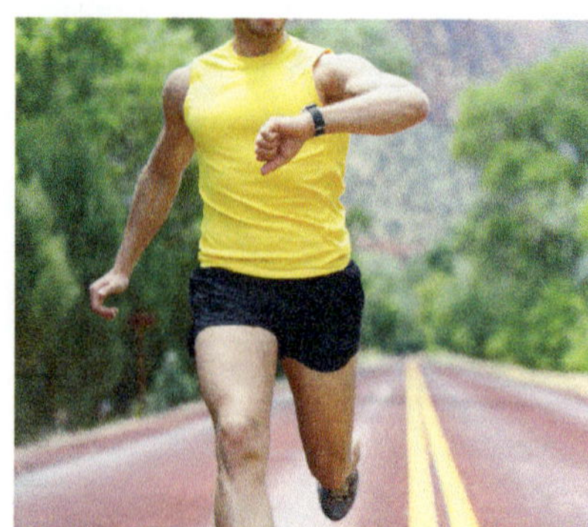

EXAMPLE 10 *Training Regimens.* Fitness trainers suggest training regimens for athletes who are preparing to run marathons and mini-marathons. One suggested twelve-week regimen combines days of short, easy running with other days of cross-training, rest, and long-distance running. During week nine, this regimen calls for a long-distance run of 10 mi, which is $2\frac{1}{2}$ times the length of the long-distance run recommended for week one. What is the length of the long-distance run recommended for week one?

Source: shape.com

1. **Familiarize.** We ask the question "10 is $2\frac{1}{2}$ times what number?" We let $r =$ the length of the long-distance run recommended for week one.

2. **Translate.** The problem can be translated to an equation.

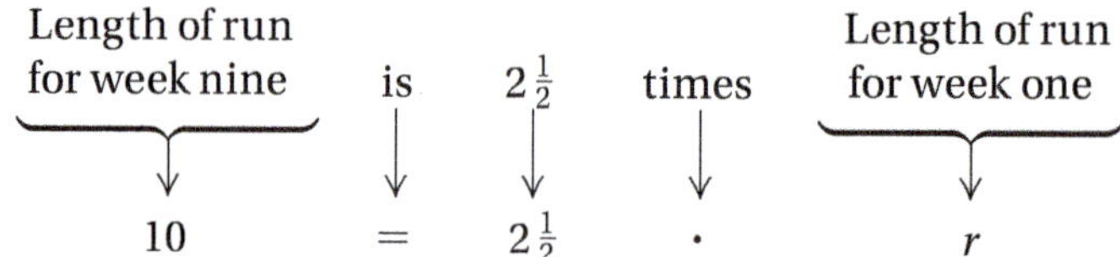

3. **Solve.** To solve the equation, we multiply on both sides.

$$10 = 2\frac{1}{2} \cdot r$$

$$10 = \frac{5}{2} \cdot r \qquad \text{Converting } 2\frac{1}{2} \text{ to fraction notation}$$

$$\frac{2}{5} \cdot 10 = \frac{2}{5} \cdot \frac{5}{2}r \qquad \text{Using the multiplication principle}$$

$$\frac{2 \cdot 10}{5} = 1 \cdot r \qquad \text{Multiplying}$$

$$4 = r \qquad \text{Simplifying: } \frac{2 \cdot 10}{5} = \frac{20}{5} = 4$$

4. **Check.** If the length of the long-distance run recommended for week one is 4 mi, we find the length of the run recommended for week nine by multiplying 4 by $2\frac{1}{2}$.

$$2\frac{1}{2} \cdot 4 = \frac{5}{2} \cdot 4 = \frac{20}{2} = 10$$

The answer checks.

5. **State.** The regimen recommends a long-distance run of 4 mi for week one.

Do Exercises 11 and 12.

Solve.

11. Kyle's pickup truck travels on an interstate highway at 65 mph for $3\frac{1}{2}$ hr. How far does it travel?

12. Holly's minivan traveled 302 mi on $15\frac{1}{10}$ gal of gas. How many miles per gallon did it get?

EXAMPLE 11 *Flooring.* Ann and Tony plan to redo the floor of their living room. Since part of the room will be covered by a rug, they want to lay a hardwood floor only on the part of the room not under the rug. If the room is $22\frac{1}{2}$ ft by $15\frac{1}{2}$ ft and the rug is 9 ft by 12 ft, how much hardwood flooring do they need? How much hardwood flooring would it take to cover the entire floor of the room?

Answers

11. $227\frac{1}{2}$ mi **12.** 20 mpg

1. **Familiarize.** We draw a diagram and let $B =$ the area of the room, $R =$ the area of the rug, and $H =$ the area to be covered by hardwood flooring.
2. **Translate.** This is a multistep problem. We first find the area of the room, B, and the area of the rug, R. Then $H = B - R$. We find each area using the formula for the area of a rectangle: $A = l \times w$.
3. **Solve.** We carry out the calculations.

$$B = \text{length} \times \text{width}$$
$$= 22\frac{1}{2} \cdot 15\frac{1}{2}$$
$$= \frac{45}{2} \cdot \frac{31}{2}$$
$$= \frac{1395}{4} = 348\frac{3}{4}\text{ sq ft}$$

$$R = \text{length} \times \text{width}$$
$$= 12 \cdot 9$$
$$= 108\text{ sq ft}$$

Then $H = B - R$

$$= 348\frac{3}{4}\text{ sq ft} - 108\text{ sq ft}$$
$$= 240\frac{3}{4}\text{ sq ft}$$

4. **Check.** We can perform a check by repeating the calculations.
5. **State.** Ann and Tony will need $240\frac{3}{4}$ sq ft of hardwood flooring. It would take $348\frac{3}{4}$ sq ft of hardwood flooring to cover the entire floor of the room.

Do Exercise 13. ▶

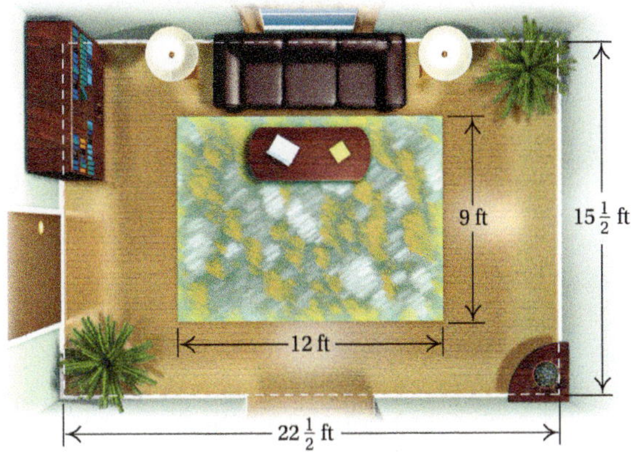

13. *Koi Pond.* Colleen designed a koi fish pond for her backyard. Using the dimensions shown in the diagram below, determine the area of Colleen's backyard remaining after the pond was completed.

Sources: en.wikipedia.org; pondliner.com

Answer

13. 945 sq yd

CALCULATOR CORNER

Operations on Fractions and Mixed Numerals Fraction calculators can add, subtract, multiply, and divide fractions and mixed numerals. The [$a\,b/c$] key is used to enter fractions and mixed numerals. The fraction $\frac{3}{4}$ is entered by pressing [3] [$a\,b/c$] [4], and it appears on the display as [3 ⌋ 4]. The mixed numeral $1\frac{5}{16}$ is entered by pressing [1] [$a\,b/c$] [5] [$a\,b/c$] [1] [6], and it is displayed as [1 ⌋ 5 ⌋ 16]. Fraction results that are greater than 1 are always displayed as mixed numerals. To express the result for $1\frac{5}{16}$ as a fraction, we press [SHIFT] [d/c]. We get [21 ⌋ 16], or $\frac{21}{16}$. Some calculators display fractions and mixed numerals in the way in which we write them.

EXERCISES Perform each calculation. Give the answer in fraction notation.

1. $\frac{1}{3} + \frac{1}{4}$
2. $\frac{7}{5} - \frac{3}{10}$
3. $\frac{15}{4} \cdot \frac{7}{12}$
4. $-\frac{4}{5} \div \frac{8}{3}$

Perform each calculation. Give the answer as a mixed numeral.

5. $4\frac{1}{3} + 5\frac{4}{5}$
6. $9\frac{2}{7} - 8\frac{1}{4}$
7. $-2\frac{1}{3} \cdot \left(-4\frac{3}{5}\right)$
8. $10\frac{7}{10} \div 3\frac{5}{6}$

Translating for Success

1. ***Raffle Tickets.*** At the Happy Hollow Camp Fall Festival, Rico and Becca, together, spent \$270 on raffle tickets that sell for $\$\frac{9}{20}$ each. How many tickets did they buy?

2. ***Irrigation Pipe.*** Jed uses two pipes, one of which measures $5\frac{1}{3}$ ft, to repair the irrigation system in the Aguilars' lawn. The total length of the two pipes is $8\frac{7}{12}$ ft. How long is the other pipe?

3. ***Vacation Days.*** Together, Oscar and Claire have 36 vacation days a year. Oscar has 22 vacation days per year. How many does Claire have?

4. ***Enrollment in Japanese Classes.*** Last year at Lakeside Community College, 225 students enrolled in basic mathematics. This number is $4\frac{1}{2}$ times as many as the number who enrolled in Japanese. How many enrolled in Japanese?

5. ***Bicycling.*** Cole rode his bicycle $5\frac{1}{3}$ mi on Saturday and $8\frac{7}{12}$ mi on Sunday. How far did he ride on the weekend?

The goal of these matching questions is to practice step (2), Translate, of the five-step problem-solving process. Translate each word problem to an equation and select a correct translation from equations A–O.

A. $13\frac{11}{12} = x + 5\frac{1}{3}$

B. $\frac{3}{4} \cdot x = 1\frac{2}{3}$

C. $270 - \frac{20}{9} = x$

D. $225 = 4\frac{1}{2} \cdot x$

E. $98 \div 2\frac{1}{3} = x$

F. $22 + x = 36$

G. $x = 4\frac{1}{2} \cdot 225$

H. $x = 5\frac{1}{3} + 8\frac{7}{12}$

I. $22 \cdot x = 36$

J. $x = \frac{3}{4} \cdot 1\frac{2}{3}$

K. $5\frac{1}{3} + x = 8\frac{7}{12}$

L. $\frac{9}{20} \cdot 270 = x$

M. $1\frac{2}{3} + \frac{3}{4} = x$

N. $98 - 2\frac{1}{3} = x$

O. $\frac{9}{20} \cdot x = 270$

Answers on page A-8

6. ***Deli Order.*** For a promotional open house for contractors last year, the Bayside Builders Association ordered 225 turkey sandwiches. Due to increased registrations this year, $4\frac{1}{2}$ times as many sandwiches are needed. How many sandwiches should be ordered?

7. ***Dog Ownership.*** In Sam's community, $\frac{9}{20}$ of the households own at least one dog. There are 270 households. How many own dogs?

8. ***Magic Tricks.*** Samantha has 98 ft of rope and needs to cut it into $2\frac{1}{3}$-ft pieces to be used in a magic trick. How many pieces can be cut from the rope?

9. ***Painting.*** Laura needs $1\frac{2}{3}$ gal of paint to paint the ceiling of the exercise room and $\frac{3}{4}$ gal of the same paint for the bathroom. How much paint does Laura need?

10. ***Chocolate Fudge Bars.*** A recipe for chocolate fudge bars that serves 16 includes $1\frac{2}{3}$ cups of sugar. How much sugar is needed for $\frac{3}{4}$ of this recipe?

4.7 Exercise Set

For Extra Help
MyMathLab® MathXL® PRACTICE WATCH READ REVIEW

Reading Check

Determine whether each statement is true or false.

__________ **RC1.** To multiply using mixed numerals, we first convert to fraction notation.

__________ **RC2.** To divide using mixed numerals, we first convert to fraction notation.

__________ **RC3.** The product of mixed numerals is generally written as a mixed numeral, unless it is a whole number or less than 1.

__________ **RC4.** To divide fractions, we multiply by the reciprocal of the divisor.

a Multiply. Write a mixed numeral for each answer.

1. $8 \cdot 2\frac{5}{6}$

2. $10 \cdot 3\frac{3}{4}$

3. $6\frac{2}{3} \cdot \frac{1}{4}$

4. $-\frac{1}{3} \cdot 5\frac{2}{5}$

5. $20\left(-2\frac{5}{6}\right)$

6. $6\frac{3}{8} \cdot 4\frac{1}{3}$

7. $3\frac{1}{2} \cdot 4\frac{2}{3}$

8. $4\frac{1}{5} \cdot 5\frac{1}{4}$

9. $-2\frac{3}{10} \cdot 4\frac{2}{5}$

10. $4\frac{7}{10} \cdot 5\frac{3}{10}$

11. $\left(-6\frac{3}{10}\right)\left(-5\frac{7}{10}\right)$

12. $-20\frac{1}{2} \cdot \left(-10\frac{1}{5}\right)$

b Divide. Write a mixed numeral for each answer whenever possible.

13. $20 \div 3\frac{1}{5}$

14. $18 \div 2\frac{1}{4}$

15. $8\frac{2}{5} \div 7$

16. $3\frac{3}{8} \div 3$

17. $6\frac{1}{4} \div 3\frac{3}{4}$

18. $5\frac{4}{5} \div 2\frac{1}{2}$

19. $-1\frac{7}{8} \div 1\frac{2}{3}$

20. $-4\frac{3}{8} \div 2\frac{5}{6}$

21. $5\frac{1}{10} \div 4\frac{3}{10}$

22. $4\frac{1}{10} \div 2\frac{1}{10}$

23. $-20\frac{1}{4} \div (-90)$

24. $-12\frac{1}{2} \div (-50)$

c Evaluate.

25. lw, for $l = 2\frac{3}{5}$ and $w = 9$

26. mv, for $m = 7$ and $v = 3\frac{2}{5}$

27. rs, for $r = 5$ and $s = 3\frac{1}{7}$

28. rt, for $r = 5\frac{2}{3}$ and $t = 4\frac{1}{5}$

29. mt, for $m = 6\frac{2}{9}$ and $t = -4\frac{3}{8}$

30. $M \div NP$, for $M = 2\frac{1}{4}$, $N = -5$, and $P = 2\frac{1}{3}$

31. $R \cdot S \div T$, for $R = 4\frac{2}{3}$, $S = 1\frac{3}{7}$, and $T = -5$

32. $a - bc$, for $a = 18$, $b = 2\frac{1}{5}$, and $c = 3\frac{3}{4}$

33. $r + ps$, for $r = 5\frac{1}{2}$, $p = 3$, and $s = 2\frac{1}{4}$

34. $s + rt$, for $s = 3\frac{1}{2}$, $r = 5\frac{1}{2}$, and $t = 7\frac{1}{2}$

35. $m + n \div p$, for $m = 7\frac{2}{5}$, $n = 4\frac{1}{2}$, and $p = 6$

36. $x - y \div z$, for $x = 9$, $y = 2\frac{1}{2}$, and $z = 3\frac{3}{4}$

d Solve.

37. *Art Prices.* A 1966 Andy Warhol portrait of Marlon Brando on his motorcycle sold for \$5 million at Christie's in 2003. On November 14, 2012, the work was sold again by Christie's for about $4\frac{4}{5}$ times the amount paid in 2003. How much was paid for the portrait in 2012?

Source: Businessweek.com

38. *Spreading Grass Seed.* Emily seeds lawns for Sam's Superior Lawn Care. When she walks at a rapid pace, the wheel on the broadcast spreader completes $150\frac{2}{3}$ revolutions per minute. How many revolutions does the wheel complete in 15 min?

39. *Population.* The population of Alabama is $6\frac{2}{3}$ times that of Alaska. The population of Alaska is approximately 720,000. What is the population of Alabama?

Source: U.S. Census Bureau

40. *Population.* The population of New York is $32\frac{1}{2}$ times the population of Wyoming. The population of Wyoming is approximately 600,000. What is the population of New York?

Source: U.S. Census Bureau

41. *Apple Net Income.* Apple, Inc., reported net income of about \$8,000,000,000 in fiscal year 2009. In fiscal year 2012, Apple's net income was about $5\frac{1}{4}$ times that amount. What was Apple's net income for fiscal year 2012?

42. *Median Income.* Median household income in the United States was about \$12,000 in 1975. By 2011, median household income was $4\frac{1}{6}$ times that amount. What was the median household income in 2011?

Source: U.S. Census Bureau

43. *Average Speed in Indianapolis 500.* Tony Kanaan won the Indianapolis 500 in 2013 with the highest average speed—about $187\frac{1}{2}$ mph. This record speed is about $2\frac{1}{2}$ times the average speed of the first winner, Ray Harroun, in 1911. What was the average speed in the first Indianapolis 500?

Source: Indianapolis Motor Speedway

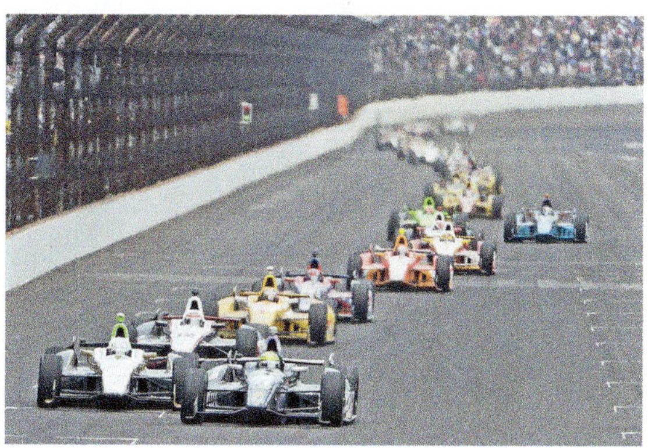

44. *Population.* The population of Cleveland is about $1\frac{1}{3}$ times the population of Cincinnati. In 2013, the population of Cincinnati was approximately 294,750. What was the population of Cleveland in 2013?

Source: U.S. Census Bureau

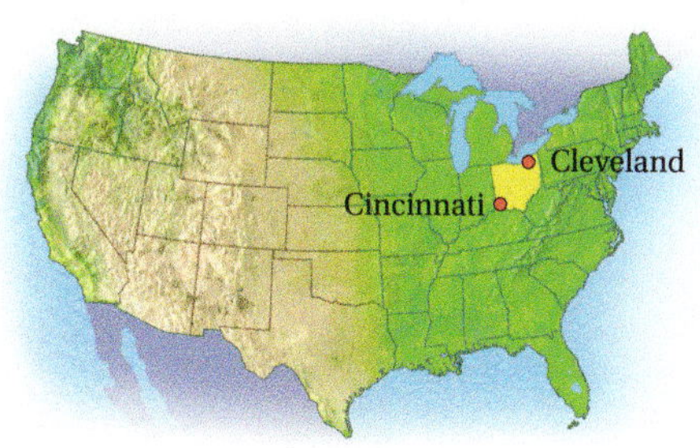

45. *Sidewalk.* A sidewalk alongside a garden at the conservatory is to be $14\frac{2}{5}$ yd long. Rectangular stone tiles that are each $1\frac{1}{8}$ yd long are used to form the sidewalk. How many tiles are used?

46. *Aeronautics.* Most space shuttles orbit the earth once every $1\frac{1}{2}$ hr. How many orbits are made every 24 hr?

47. *Doubling a Recipe.* The chef of a five-star hotel is doubling a recipe for chocolate cake. The original recipe requires $2\frac{3}{4}$ cups of flour and $1\frac{1}{3}$ cups of sugar. How much flour and sugar will she need?

48. *Half of a Recipe.* A caterer is following a salad dressing recipe that calls for $1\frac{7}{8}$ cups of mayonnaise and $1\frac{1}{6}$ cups of sugar. How much mayonnaise and sugar will he need if he prepares $\frac{1}{2}$ of the amount of salad dressing?

49. *Mileage.* A car traveled 213 mi on $14\frac{2}{10}$ gal of gas. How many miles per gallon did it get?

50. *Mileage.* A car traveled 385 mi on $15\frac{4}{10}$ gal of gas. How many miles per gallon did it get?

51. *Mural.* A student artist painted a mural on the wall under a bridge. The dimensions of the mural are $6\frac{2}{3}$ ft by $9\frac{3}{8}$ ft. What is the area of the mural?

52. *Weight of Water.* The weight of water is $62\frac{1}{2}$ lb per cubic foot. What is the weight of $2\frac{1}{4}$ cubic feet of water?

53. *Weight of Water.* The weight of water is $62\frac{1}{2}$ lb per cubic foot. How many cubic feet would be occupied by 25,000 lb of water?

54. *Weight of Water.* The weight of water is $8\frac{1}{3}$ lb per gallon. Harry rolls his lawn with an 800-lb capacity roller. Express the water capacity of the roller in gallons.

55. *Servings of Salmon.* A serving of filleted fish is generally considered to be about $\frac{1}{3}$ lb. How many servings can be prepared from $5\frac{1}{2}$ lb of salmon fillet?

56. *Servings of Tuna.* A serving of fish steak (cross section) is generally $\frac{1}{2}$ lb. How many servings can be prepared from a cleaned $18\frac{3}{4}$-lb tuna?

57. *Landscaping.* The previous owners of Ashley's new home had a large L-shaped vegetable garden consisting of a rectangle that was $15\frac{1}{2}$ ft by 20 ft adjacent to one that was $10\frac{1}{2}$ ft by $12\frac{1}{2}$ ft. Ashley wants to cover the garden with sod. What is the total area of the sod she must purchase?

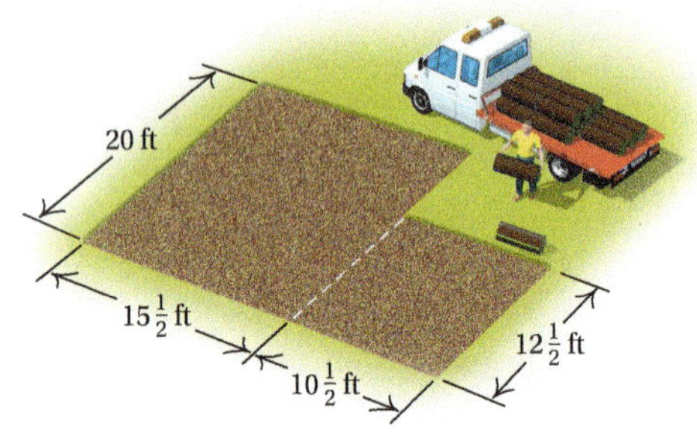

58. *Home Furnishings.* An L-shaped sunroom consists of a rectangle that is $9\frac{1}{2}$ ft by 12 ft adjacent to one that is $9\frac{1}{2}$ ft by 8 ft. What is the total area of a carpet that covers the floor?

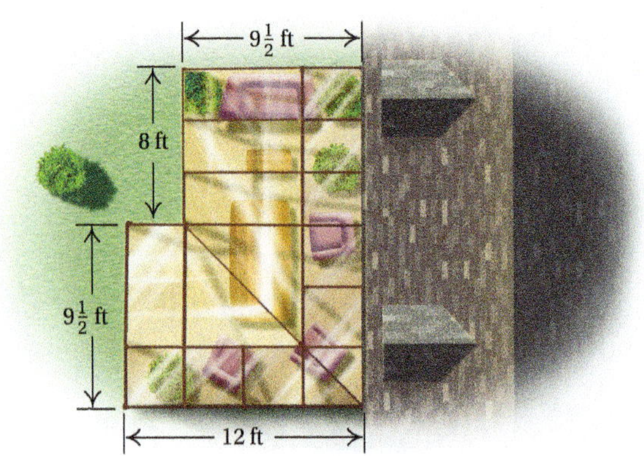

Find the area of each shaded region.

59.

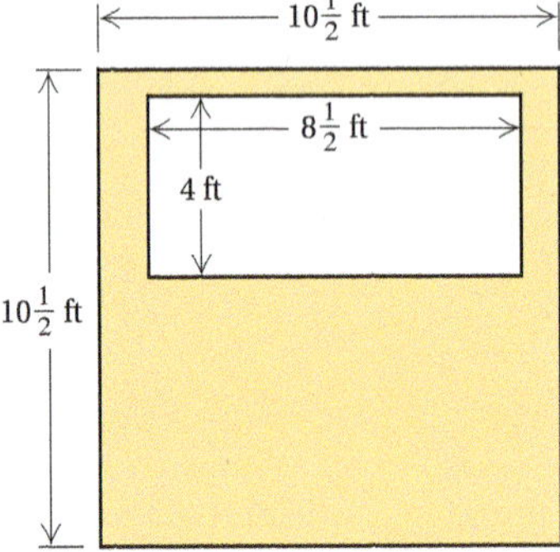

60.

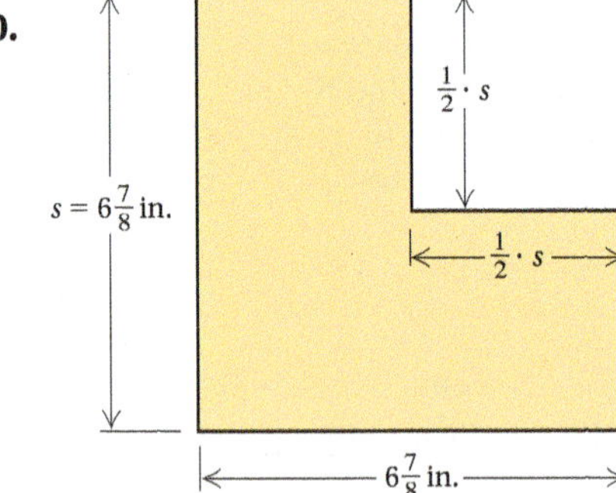

61.

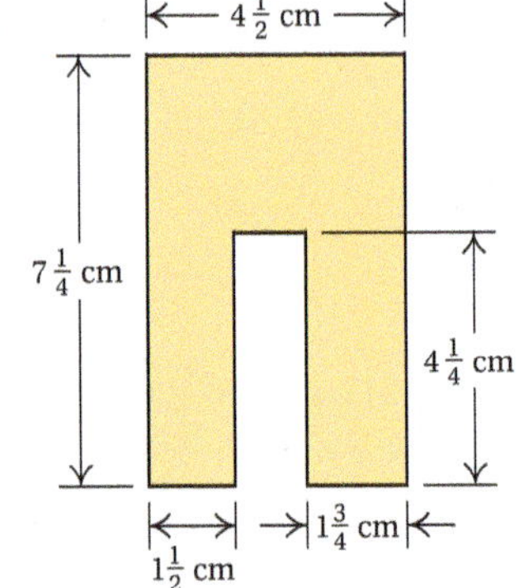

62.

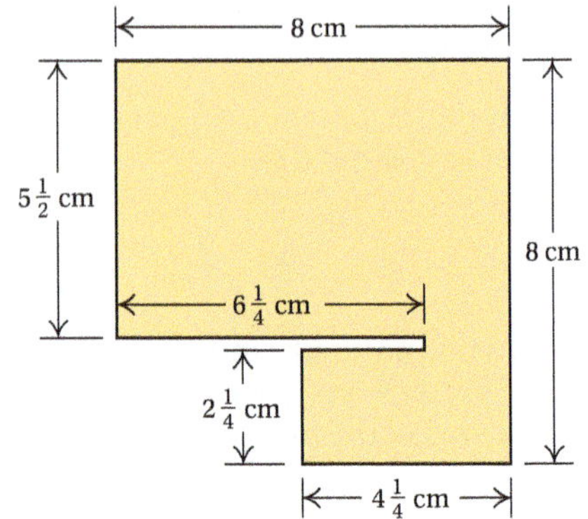

63. *Temperature.* Fahrenheit temperature can be obtained from Celsius (Centigrade) temperature by multiplying by $1\frac{4}{5}$ and adding 32°. What Fahrenheit temperature corresponds to a Celsius temperature of 20°?

64. *Temperature.* Fahrenheit temperature can be obtained from Celsius (Centigrade) temperature by multiplying by $1\frac{4}{5}$ and adding 32°. What Fahrenheit temperature corresponds to the Celsius temperature of boiling water, 100°?

65. *Building a Ziggurat.* The dimensions of all of the square bricks that King Nebuchadnezzar used over 2500 years ago to build ziggurats were $13\frac{1}{4}$ in. × $13\frac{1}{4}$ in. × $3\frac{1}{4}$ in. What are the perimeter and the area of the $13\frac{1}{4}$ in. × $13\frac{1}{4}$ in. side? of the $13\frac{1}{4}$ in. × $3\frac{1}{4}$ in. side?

Source: www.eartharchitecture.org

66. *Word Processing.* For David's design report, he needs to create a table containing two columns, each $1\frac{1}{2}$ in. wide, and five columns, each $\frac{3}{4}$ in. wide. Will this table fit on a piece of standard paper that is $8\frac{1}{2}$ in. wide? If so, how wide will each side margin be if the margins on each side are to be of equal width?

Skill Maintenance

67. Find the perimeter of the figure. [1.2b]

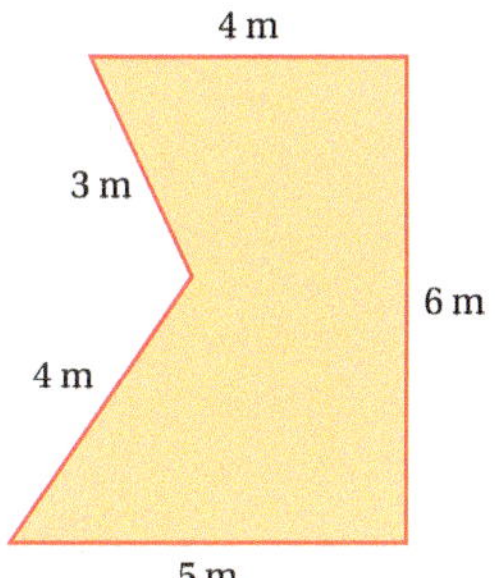

68. Find the area of the region. [1.4b]

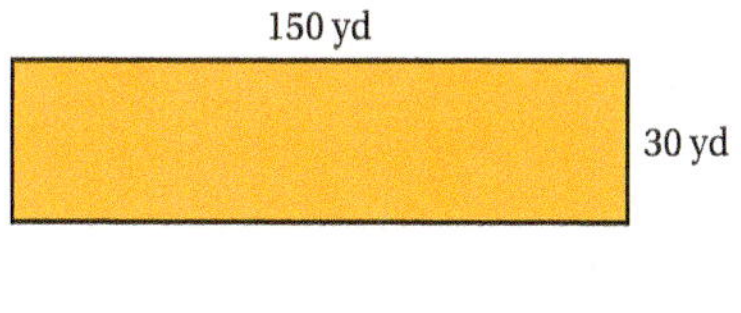

Synthesis

Simplify. Write each answer as a mixed numeral whenever possible.

69. $-8 \div \frac{1}{2} + \frac{3}{4} + \left(-5 - \frac{5}{8}\right)^2$

70. $\left(\frac{5}{9} - \frac{1}{4}\right)(-12) + \left(-4 - \frac{3}{4}\right)^2$

71. $\frac{1}{3} \div \left(\frac{1}{2} - \frac{1}{5}\right) \times \frac{1}{4} + \frac{1}{6}$

72. $\frac{7}{8} - 1\frac{1}{8} \times \frac{2}{3} + \frac{9}{10} \div \frac{3}{5}$

73. Find r if

$$\frac{1}{r} = \frac{1}{40} + \frac{1}{60} + \frac{1}{80}.$$

74. *Heights.* Find the average height of the following NBA players:

LeBron James	6 ft 8 in.
Dwayne Wade	6 ft 4 in.
Dwight Howard	6 ft 11 in.
Dirk Nowitzki	7 ft 0 in.
Al Jefferson	6 ft 10 in.

75. *Water Consumption.* According to the U.S. Department of Energy, washing one load of clothes uses $2\frac{2}{3}$ times the amount of hot water required for the average shower. If the average shower uses 12 gal of hot water, how much hot water will two showers and two loads of wash require?

4.8 Order of Operations and Complex Fractions

OBJECTIVES

a Simplify expressions containing fraction notation using the rules for order of operations.

b Simplify complex fractions.

SKILL TO REVIEW

Objective 1.9c: Simplify expressions using the rules for order of operations.

Simplify.

1. $22 - 3 \cdot 4$

2. $6(4 + 1)^2 - 3^4 \div 3$

Simplify.

1. $\frac{2}{5} \cdot \frac{5}{8} + \frac{1}{4}$

2. $\frac{1}{3} \cdot \frac{3}{4} \div \frac{5}{8} - \frac{1}{10}$ GS

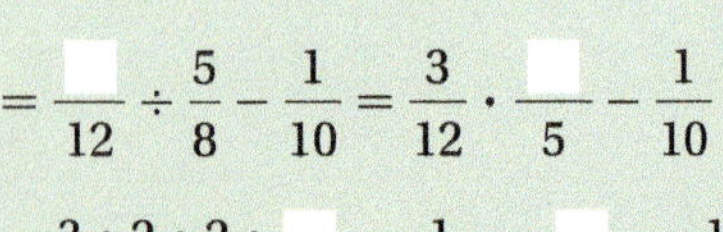

$= \frac{\square}{12} \div \frac{5}{8} - \frac{1}{10} = \frac{3}{12} \cdot \frac{\square}{5} - \frac{1}{10}$

$= \frac{3 \cdot 2 \cdot 2 \cdot \square}{3 \cdot 2 \cdot 2 \cdot 5} - \frac{1}{10} = \frac{\square}{5} - \frac{1}{10}$

$= \frac{\square}{10} - \frac{1}{10} = \frac{\square}{10}$

a ORDER OF OPERATIONS

Like expressions containing integers, expressions containing fraction notation follow the rules for order of operations.

RULES FOR ORDER OF OPERATIONS

1. Do all calculations within parentheses ( ), brackets [ ], braces { }, absolute-value symbols, numerators, or denominators.
2. Evaluate all exponential expressions.
3. Do all multiplications and divisions in order from left to right.
4. Do all additions and subtractions in order from left to right.

EXAMPLE 1 Simplify: $\frac{1}{6} + \frac{2}{3} \div \frac{1}{2} \cdot \frac{5}{8}$.

$\frac{1}{6} + \frac{2}{3} \div \frac{1}{2} \cdot \frac{5}{8} = \frac{1}{6} + \frac{2}{3} \cdot \frac{2}{1} \cdot \frac{5}{8}$ — Doing the division first by multiplying by the reciprocal of $\frac{1}{2}$

$= \frac{1}{6} + \frac{2 \cdot 2 \cdot 5}{3 \cdot 1 \cdot 8}$ — Doing the multiplications in order from left to right

$= \frac{1}{6} + \frac{2 \cdot 2 \cdot 5}{3 \cdot 1 \cdot 2 \cdot 2 \cdot 2}$ — Factoring

$= \frac{1}{6} + \frac{5}{6}$ — Removing a factor equal to 1: $\frac{2 \cdot 2}{2 \cdot 2} = 1$; simplifying

$= \frac{6}{6}$, or 1 — Doing the addition

◀ Do Margin Exercises 1 and 2.

EXAMPLE 2 Simplify: $\frac{2}{3} \cdot 24 - 11\frac{1}{2}$.

$\frac{2}{3} \cdot 24 - 11\frac{1}{2} = \frac{2 \cdot 24}{3 \cdot 1} - 11\frac{1}{2}$ — Doing the multiplication first

$= \frac{2 \cdot 3 \cdot 8}{3 \cdot 1} - 11\frac{1}{2}$ — Factoring

$= 2 \cdot 8 - 11\frac{1}{2}$ — Removing a factor equal to 1: $\frac{3}{3} = 1$

$= 16 - 11\frac{1}{2}$ — Multiplying

$= 15\frac{2}{2} - 11\frac{1}{2}$

$= 4\frac{1}{2}$, or $\frac{9}{2}$ — Subtracting

Answers

Skill to Review:
1. 10 **2.** 123

Margin Exercises:
1. $\frac{1}{2}$ **2.** $\frac{3}{10}$

Guided Solution:
2. 3, 8, 2, 2, 4, 3

Do Exercise 3.

3. Simplify: $\frac{3}{4} \cdot 16 + 8\frac{2}{3}$.

EXAMPLE 3 Simplify.

a) $\frac{1}{2} - \frac{2}{3}\left(-\frac{1}{2}\right)^2$ **b)** $1 - \frac{2}{3}\left(\frac{1}{3} - \frac{1}{2}\right)$

a) The parentheses here are not grouping symbols, so we begin by evaluating the exponential expression.

$$\frac{1}{2} - \frac{2}{3}\left(-\frac{1}{2}\right)^2 = \frac{1}{2} - \frac{2}{3} \cdot \frac{1}{4} \qquad \left(-\frac{1}{2}\right)^2 = \left(-\frac{1}{2}\right)\cdot\left(-\frac{1}{2}\right) = \frac{1}{4}$$

$$= \frac{1}{2} - \frac{2 \cdot 1}{3 \cdot 2 \cdot 2} \qquad \text{Multiplying and factoring}$$

$$= \frac{1}{2} - \frac{1}{6} \qquad \text{Simplifying}$$

$$= \frac{3}{6} - \frac{1}{6} \qquad \text{Writing with the LCD, 6, in order to subtract}$$

$$= \frac{4}{6} = \frac{2}{3} \qquad \text{Subtracting and simplifying}$$

b) We first subtract within the parentheses.

$$1 - \frac{2}{3}\left(\frac{1}{3} - \frac{1}{2}\right) = 1 - \frac{2}{3}\left(\frac{2}{6} - \frac{3}{6}\right) \qquad \text{Writing with the LCD, 6, in order to subtract}$$

$$= 1 - \frac{2}{3}\left(-\frac{1}{6}\right) \qquad \text{Subtracting within the parentheses: } \frac{2}{6} - \frac{3}{6} = \frac{2}{6} + \left(-\frac{3}{6}\right) = -\frac{1}{6}$$

$$= 1 - \left(-\frac{2 \cdot 1}{3 \cdot 2 \cdot 3}\right) \qquad \text{There are no exponential expressions, so we multiply and factor.}$$

$$= 1 - \left(-\frac{1}{9}\right) \qquad \text{Simplifying}$$

$$= \frac{9}{9} + \frac{1}{9} = \frac{10}{9} \qquad \text{Writing with the LCD, 9, and adding the opposite of } -\frac{1}{9}$$

Do Exercises 4–6.

Simplify.

4. $\left(\frac{3}{4}\right)^2 - \frac{1}{2} \div \left(-\frac{4}{5}\right)$

5. $1 - \left(\frac{2}{3} - \frac{3}{4}\right)^2$

6. $\left(\frac{2}{3} + \frac{3}{4}\right) \div 2\frac{1}{3} - \left(\frac{1}{2}\right)^3$

b COMPLEX FRACTIONS

A **complex fraction** is a fraction in which the numerator and/or denominator contain one or more fractions. The following are some examples of complex fractions.

$$\frac{\frac{2}{3}}{7} \leftarrow \text{The numerator contains a fraction.}$$

$$\frac{-\frac{1}{5}}{-\frac{9}{10}} \qquad \begin{matrix}\leftarrow \text{The numerator contains a fraction.} \\ \leftarrow \text{The denominator contains a fraction.}\end{matrix}$$

Since a fraction bar represents division, complex fractions can be rewritten using the division symbol $\div$.

Answers

3. $20\frac{2}{3}$, or $\frac{62}{3}$ 4. $\frac{19}{16}$, or $1\frac{3}{16}$ 5. $\frac{143}{144}$ 6. $\frac{27}{56}$

EXAMPLE 4 Simplify.

a) $\dfrac{\frac{2}{3}}{7}$

b) $\dfrac{-\frac{1}{5}}{-\frac{9}{10}}$

a) $\dfrac{\frac{2}{3}}{7} = \dfrac{2}{3} \div 7$ Rewriting using a division symbol

$= \dfrac{2}{3} \cdot \left(\dfrac{1}{7}\right)$ Multiplying by the reciprocal of the divisor

$= \dfrac{2 \cdot 1}{3 \cdot 7}$ Multiplying numerators and multiplying denominators

$= \dfrac{2}{21}$ This expression cannot be simplified.

b) $\dfrac{-\frac{1}{5}}{-\frac{9}{10}} = -\dfrac{1}{5} \div \left(-\dfrac{9}{10}\right)$ Rewriting using a division symbol

$= -\dfrac{1}{5} \cdot \left(-\dfrac{10}{9}\right)$ Multiplying by the reciprocal of the divisor; the product will be positive.

$= \dfrac{1 \cdot 2 \cdot \not{5}}{\not{5} \cdot 3 \cdot 3}$ Multiplying numerators and multiplying denominators; factoring

$= \dfrac{2}{9}$ Removing a factor equal to 1 and simplifying

◀ **Do Exercises 7 and 8.**

Simplify.

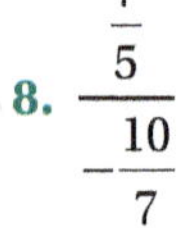

7. $\dfrac{10}{\frac{5}{8}} = 10 \div \dfrac{\square}{\square}$

$= 10 \cdot \dfrac{\square}{\square}$

$= \dfrac{10 \cdot \square}{5}$

$= \dfrac{\square \cdot \not{5} \cdot 8}{\not{5} \cdot 1}$

$= \square$

8. $\dfrac{\frac{7}{5}}{-\frac{10}{7}}$

Complex fractions may contain variables.

EXAMPLE 5 Simplify: $\dfrac{\frac{x}{24}}{\frac{15}{28}}$.

$\dfrac{\frac{x}{24}}{\frac{15}{28}} = \dfrac{x}{24} \div \dfrac{15}{28}$ Rewriting using a division symbol

$= \dfrac{x}{24} \cdot \dfrac{28}{15}$ Multiplying by the reciprocal of the divisor

$= \dfrac{x \cdot \not{2} \cdot \not{2} \cdot 7}{\not{2} \cdot \not{2} \cdot 2 \cdot 3 \cdot 3 \cdot 5}$ Multiplying numerators and multiplying denominators; factoring

$= \dfrac{x \cdot 7}{2 \cdot 3 \cdot 3 \cdot 5}$ Removing a factor equal to 1: $\dfrac{2 \cdot 2}{2 \cdot 2} = 1$

$= \dfrac{7x}{90}$ Multiplying

◀ **Do Exercises 9 and 10.**

Simplify.

9. $\dfrac{\frac{x}{20}}{\frac{3}{10}}$

10. $\dfrac{-\frac{7}{12}}{-\frac{x}{18}}$

Answers

7. 16 8. $-\dfrac{49}{50}$ 9. $\dfrac{x}{6}$ 10. $\dfrac{21}{2x}$

Guided Solution:

7. $\dfrac{5}{8}, \dfrac{8}{5}$, 8, 2, 16

When the numerator or denominator of a complex fraction consists of more than one term, first simplify the numerator and/or denominator separately.

EXAMPLE 6 Simplify: $\dfrac{\frac{1}{2}-\frac{2}{3}}{1\frac{7}{8}}$.

$$\frac{\frac{1}{2}-\frac{2}{3}}{1\frac{7}{8}} = \frac{\frac{3}{6}-\frac{4}{6}}{\frac{15}{8}}$$ Writing the fractions in the numerator with a common denominator; Writing the mixed numeral in the denominator as a fraction

$$= \frac{-\frac{1}{6}}{\frac{15}{8}}$$ Subtracting in the numerator of the complex fraction

$$= -\frac{1}{6} \div \frac{15}{8}$$ Rewriting using a division symbol

$$= -\frac{1}{6} \cdot \frac{8}{15}$$ Multiplying by the reciprocal of the divisor

$$= -\frac{1 \cdot \cancel{2} \cdot 2 \cdot 2}{\cancel{2} \cdot 3 \cdot 3 \cdot 5}$$ Multiplying numerators and multiplying denominators; factoring

$$= -\frac{4}{45}$$ Removing a factor equal to 1: $\frac{2}{2} = 1$

Do Exercises 11 and 12. ▶

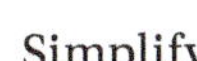

Simplify.

11. $\dfrac{\frac{7}{12}+\frac{5}{6}}{\frac{4}{9}}$

12. $\dfrac{-\frac{3}{5}}{\frac{2}{3}-\frac{7}{10}}$

EXAMPLE 7 ***Harvesting Walnut Trees.*** A woodland owner decided to harvest five walnut trees in order to improve the growing conditions of the remaining trees. The logs she sold measured $7\frac{5}{8}$ ft, $8\frac{1}{4}$ ft, $8\frac{3}{4}$ ft, $9\frac{1}{8}$ ft, and $10\frac{1}{2}$ ft. What is the average length of the logs?

Recall that to compute an average, we add the numbers and then divide the sum by the number of addends. We have

$$\frac{7\frac{5}{8}+8\frac{1}{4}+8\frac{3}{4}+9\frac{1}{8}+10\frac{1}{2}}{5} = \frac{7\frac{5}{8}+8\frac{2}{8}+8\frac{6}{8}+9\frac{1}{8}+10\frac{4}{8}}{5}$$

$$= \frac{42\frac{18}{8}}{5} = \frac{42\frac{9}{4}}{5} = \frac{\frac{177}{4}}{5}$$ Adding, simplifying, and converting to fraction notation

$$= \frac{177}{4} \div 5$$ Rewriting using a division symbol

$$= \frac{177}{4} \cdot \frac{1}{5}$$ Multiplying by the reciprocal of 5

$$= \frac{177}{20} = 8\frac{17}{20}.$$ Converting to a mixed numeral

The average length of the logs is $8\frac{17}{20}$ ft.

Do Exercises 13–15. ▶

13. Rachel has triplets. Their birth weights are $3\frac{1}{2}$ lb, $2\frac{3}{4}$ lb, and $3\frac{1}{8}$ lb. What is the average weight of her babies?

14. Find the average of $\frac{1}{2}$, $\frac{1}{3}$, and $\frac{5}{6}$.

15. Find the average of $\frac{3}{4}$ and $\frac{4}{5}$.

Answers

11. $3\frac{3}{16}$ **12.** 18 **13.** $3\frac{1}{8}$ lb **14.** $\frac{5}{9}$
15. $\frac{31}{40}$

4.8 Exercise Set

For Extra Help
MyMathLab®
MathXL®
PRACTICE  WATCH  READ  REVIEW 

✓ Reading Check

Match the beginning of each statement with the correct ending from the list at the right so that the rules for order of operations are listed in the correct order.

RC1. Do all ______.

RC2. Evaluate all ______.

RC3. Do all ______.

RC4. Do all ______.

a) multiplications and divisions in order from left to right.

b) additions and subtractions in order from left to right.

c) calculations within parentheses, brackets, braces, and so forth.

d) exponential expressions.

a Simplify.

1. $\frac{1}{8} + \frac{1}{4} \cdot \frac{2}{3}$

2. $\frac{2}{5} - \frac{4}{5} \div \frac{2}{3}$

3. $-\frac{1}{6} - 3\left(-\frac{5}{9}\right)$

4. $1 + \frac{2}{3}\left(-\frac{6}{25}\right)$

5. $\frac{9}{10} - \left(\frac{2}{5} - \frac{3}{8}\right)$

6. $-\frac{5}{9} + \left(\frac{1}{3} - \frac{5}{6}\right)$

7. $\frac{5}{8} \div \frac{1}{4} - \frac{2}{3} \cdot \frac{4}{5}$

8. $\frac{4}{7} \cdot \frac{7}{15} + \frac{2}{3} \div 8$

9. $\frac{7}{8} \div \frac{1}{2} \cdot \frac{1}{4}$

10. $\frac{7}{10} \cdot \frac{4}{5} \div \frac{2}{3}$

11. $\frac{3}{4} - \frac{2}{3} \cdot \left(\frac{1}{2} + \frac{2}{5}\right)$

12. $\frac{3}{4} \div \frac{1}{2} \cdot \left(\frac{8}{9} - \frac{2}{3}\right)$

13. $\frac{4}{5} \div \left(\frac{2}{9} \cdot \frac{1}{2}\right) \cdot \left(-\frac{5}{6}\right)$

14. $-\frac{4}{9} \cdot \left(\frac{3}{8} \div \frac{1}{2}\right) \div \left(-\frac{2}{3}\right)$

15. $\left(\frac{2}{3}\right)^2 - \frac{1}{3} \cdot 1\frac{1}{4}$

16. $-1\frac{3}{5} - \frac{9}{10} + \left(-\frac{1}{2}\right)^2$

17. $-\frac{12}{25}\left(\frac{3}{4} - \frac{1}{2}\right)^2$

18. $\frac{2}{3}\left(\frac{1}{3} - \frac{1}{5}\right)^2$

19. $-\frac{3}{4} \div \left(\frac{2}{3} - \frac{1}{6}\right) + \frac{1}{2}$

20. $-6 \div \left(\frac{1}{2} - \frac{2}{3}\right) + \frac{1}{3}$

21. $\left(-\frac{3}{2}\right)^2 - 2\left(\frac{1}{4} - \frac{3}{2}\right)$

22. $\left(-\frac{1}{2}\right)^3 - \frac{3}{4}\left(\frac{1}{3} - \frac{1}{6}\right)$

23. $\frac{1}{2} - \left(\frac{1}{2}\right)^2 + \left(\frac{1}{2}\right)^3$

24. $1 + \frac{1}{4} + \left(\frac{1}{4}\right)^2 - \left(\frac{1}{4}\right)^3$

25. $\left(\frac{3}{5} - \frac{1}{2}\right) \div \left(\frac{3}{4} - \frac{3}{10}\right)$

26. $\left(\frac{2}{3} + \frac{3}{4}\right) \div \left(\frac{5}{6} - \frac{1}{3}\right)$

b Simplify.

27. $\dfrac{\frac{3}{8}}{\frac{11}{8}}$

28. $\dfrac{-\frac{1}{8}}{\frac{3}{4}}$

29. $\dfrac{-4}{\frac{6}{7}}$

30. $\dfrac{-\frac{3}{8}}{-12}$

31. $\dfrac{\frac{1}{40}}{-\frac{1}{50}}$

32. $\dfrac{\frac{7}{9}}{\frac{3}{9}}$

33. $\dfrac{-\frac{1}{10}}{-10}$

34. $\dfrac{28}{-\frac{7}{4}}$

35. $\dfrac{\frac{5}{18}}{-1\frac{2}{3}}$

36. $\dfrac{-2\frac{1}{5}}{\frac{7}{10}}$

37. $\dfrac{\frac{x}{28}}{\frac{5}{8}}$

38. $\dfrac{\frac{x}{15}}{\frac{3}{4}}$

39. $\dfrac{\frac{n}{14}}{-\frac{2}{3}}$

40. $\dfrac{-\frac{t}{10}}{\frac{4}{45}}$

41. $\dfrac{-\frac{3}{35}}{-\frac{x}{10}}$

42. $\dfrac{\frac{7}{40}}{-\frac{6}{x}}$

43. $\dfrac{-\frac{5}{8}}{\left(\frac{3}{2}\right)^2}$

44. $\dfrac{\left(-\frac{2}{3}\right)^2}{\left(\frac{9}{10}\right)^2}$

45. $\dfrac{\frac{1}{6}-\frac{5}{9}}{\frac{2}{3}}$

46. $\dfrac{\frac{7}{12}}{\frac{1}{4}-\frac{5}{8}}$

47. $\dfrac{\frac{1}{4}-\frac{3}{8}}{\frac{1}{2}-\frac{7}{8}}$

48. $\dfrac{\frac{1}{2}-\frac{3}{5}}{\frac{2}{5}-\frac{1}{2}}$

49. Find the average of $\frac{2}{3}$ and $\frac{7}{8}$.

50. Find the average of $\frac{1}{4}$ and $\frac{1}{5}$.

51. Find the average of $\frac{1}{6}$, $\frac{1}{8}$, and $\frac{3}{4}$.

52. Find the average of $\frac{4}{5}$, $\frac{1}{2}$, and $\frac{1}{10}$.

53. Find the average of $3\frac{1}{2}$ and $9\frac{3}{8}$.

54. Find the average of $10\frac{2}{3}$ and $24\frac{5}{6}$.

55. *Hiking the Appalachian Trail.* Ellen camped and hiked for three consecutive days along a section of the Appalachian Trail. The distances she hiked on the three days were $15\frac{5}{32}$ mi, $20\frac{3}{16}$ mi, and $12\frac{7}{8}$ mi. Find the average of these distances.

56. *Vertical Leaps.* Eight-year-old Zachary registered vertical leaps of $12\frac{3}{4}$ in., $13\frac{3}{4}$ in., $13\frac{1}{2}$ in., and 14 in. Find his average vertical leap.

57. ***Black Bear Cubs.*** Black bears typically have two cubs. In January 2007 in northern New Hampshire, a black bear sow gave birth to a litter of 5 cubs. This is so rare that Tom Sears, a wildlife photographer, spent 28 hr per week for six weeks watching for the perfect opportunity to photograph this family of six. At the time of this photo, an observer estimated that the cubs weighed $7\frac{1}{2}$ lb, 8 lb, $9\frac{1}{2}$ lb, $10\frac{5}{8}$ lb, and $11\frac{3}{4}$ lb. What was the average weight of the cubs?

Source: Andrew Timmins, New Hampshire Fish and Game Department, *Northcountry News*, Warren, NH; Tom Sears, photographer

58. ***Acceleration.*** The results of a *Road & Track* road acceleration test for five cars are given in the graph below. The test measures the time in seconds required to go from 0 mph to 60 mph. What was the average time?

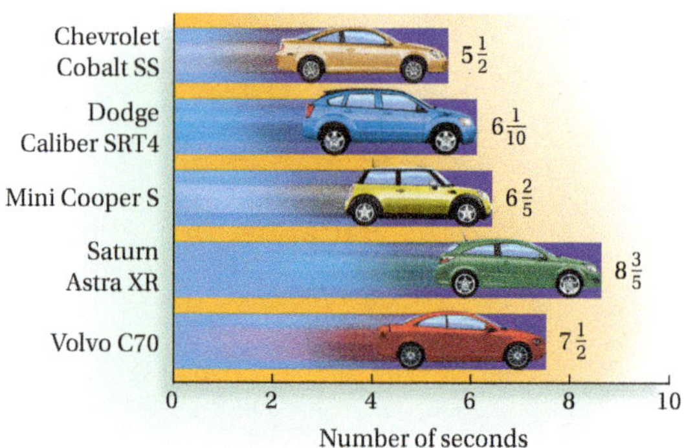

SOURCE: *Road & Track*, October 2008, pp.156–157

Skill Maintenance

Simplify.

59. $12 + 30 \div 3 - 2$ [1.9c]

60. $5 \cdot 2^2 \div 10$ [1.9c]

61. $10^2 - [3 \cdot 2^4 \div (10 - 2) + 5 \cdot 2]$ [1.9d]

62. $(10 + 3 \cdot 4 \div 6)^2 - 11 \cdot 2^2$ [1.9c]

63. List all the factors of 42. [3.2a]

64. Determine whether 114 is divisible by 7. [3.1a]

65. Classify the given numbers as prime, composite, or neither. [3.2b]

1, 5, 7, 9, 14, 23, 43

66. Find the prime factorization of 150. [3.2c]

Synthesis

Simplify.

67. $\left(1\frac{1}{2} - 1\frac{1}{3}\right)^2 \cdot 144 - \frac{9}{10} \div 4\frac{1}{5}$

68. 🖩 $\left(3\frac{1}{2} - 2\frac{1}{3}\right)^3 - 30 \cdot 2\frac{1}{2} \div (-2)^5$

69. $\dfrac{\frac{2}{3}x - \frac{1}{2}x}{\left(1\frac{1}{4} + \frac{1}{2}\right)^2}$

70. $\dfrac{\frac{x}{24}}{\frac{x}{48}}$

Estimate each of the following as 0, $\frac{1}{2}$, or 1.

71. $\frac{2}{99}$

72. $\frac{19}{20}$

73. $\frac{13}{27}$

74. $\frac{101}{100}$

75. $\frac{215}{429}$

76. $\frac{1}{1000}$

CHAPTER 4

Summary and Review

Vocabulary Reinforcement

Complete each statement with the correct term from the given list. Some of the choices may not be used and some may be used more than once.

1. The _______________ of two numbers is the smallest number that is a multiple of both numbers. [4.1a]
2. A _______________ represents a sum of a whole number and a fraction less than 1. [4.5a]
3. To multiply using mixed numerals, we first convert to _______________ notation. [4.7a]
4. A _______________ contains a fraction in its numerator and/or denominator. [4.8b]
5. To add fractions, the _______________ of the fractions must be the same. [4.2a]
6. The least common denominator of two fractions is the _______________ of the denominators of the fractions. [4.2b]
7. When finding the LCM of a set of numbers using prime factorizations, we use each prime number the _______________ number of times that it appears in any one factorization. [4.1a]
8. To compare two fractions with a common denominator, we compare their _______________. [4.2c]

greatest
least
numerators
denominators
fraction
decimal
mixed numeral
complex fraction
least common multiple
greatest common factor

Concept Reinforcement

Determine whether each statement is true or false.

_______________ 1. The mixed numeral $5\frac{2}{3}$ can be represented by the sum $5 \cdot \frac{3}{3} + \frac{2}{3}$. [4.5a]

_______________ 2. The least common multiple of two numbers is always larger than or equal to the larger number. [4.1a]

_______________ 3. To clear fractions in an equation, multiply both sides by the LCM of all denominators in the equation. [4.4b]

_______________ 4. The product of any two mixed numerals is greater than 1. [4.7a]

Study Guide

Objective 4.1a Find the least common multiple, or LCM, of two or more numbers.

Example Find the LCM of 105 and 90.

$$105 = 3 \cdot 5 \cdot 7,$$
$$90 = 2 \cdot 3 \cdot 3 \cdot 5;$$
$$\text{LCM} = 2 \cdot 3 \cdot 3 \cdot 5 \cdot 7 = 630$$

Practice Exercise

1. Find the LCM of 52 and 78.

Objective 4.2b Add using fraction notation when denominators are different.

Example Add: $\frac{5}{24} + \frac{7}{45}$.

$$\frac{5}{24} + \frac{7}{45} = \frac{5}{2 \cdot 2 \cdot 2 \cdot 3} \cdot \frac{3 \cdot 5}{3 \cdot 5} + \frac{7}{3 \cdot 3 \cdot 5} \cdot \frac{2 \cdot 2 \cdot 2}{2 \cdot 2 \cdot 2}$$
$$= \frac{75}{360} + \frac{56}{360} = \frac{131}{360}$$

Practice Exercise

2. Add: $\frac{19}{60} + \frac{11}{36}$.

Objective 4.2c Use < or > to form a true statement with fraction notation.

Example Use < or > for □ to write a true sentence:

$$\frac{5}{12} \square \frac{9}{16}.$$

The LCD is 48. Thus, we have

$$\frac{5}{12} \cdot \frac{4}{4} \square \frac{9}{16} \cdot \frac{3}{3}, \quad \text{or} \quad \frac{20}{48} \square \frac{27}{48}.$$

Since $20 < 27$, $\frac{20}{48} < \frac{27}{48}$ and thus $\frac{5}{12} < \frac{9}{16}$.

Practice Exercise

3. Use < or > for □ to write a true sentence:

$$\frac{3}{13} \square \frac{5}{12}.$$

Objective 4.3a Subtract using fraction notation.

Example Subtract: $\frac{7}{12} - \frac{11}{60}$.

$$\frac{7}{12} - \frac{11}{60} = \frac{7}{12} \cdot \frac{5}{5} - \frac{11}{60} = \frac{35}{60} - \frac{11}{60}$$
$$= \frac{35 - 11}{60} = \frac{24}{60} = \frac{2 \cdot \cancel{12}}{5 \cdot \cancel{12}} = \frac{2}{5}$$

Practice Exercise

4. Subtract: $\frac{29}{35} - \frac{5}{7}$.

Objective 4.4a Solve equations that involve fractions and require use of both the addition principle and the multiplication principle.

Example Solve: $\frac{1}{2}x + \frac{1}{6} = \frac{5}{8}$.

$$\frac{1}{2}x + \frac{1}{6} = \frac{5}{8}$$
$$\frac{1}{2}x + \frac{1}{6} - \frac{1}{6} = \frac{5}{8} - \frac{1}{6}$$
$$\frac{1}{2}x = \frac{5}{8} \cdot \frac{3}{3} - \frac{1}{6} \cdot \frac{4}{4}$$
$$\frac{1}{2}x = \frac{15}{24} - \frac{4}{24}$$
$$\frac{1}{2}x = \frac{11}{24}$$
$$\frac{2}{1}\left(\frac{1}{2}x\right) = \frac{2}{1}\left(\frac{11}{24}\right)$$
$$x = \frac{\cancel{2} \cdot 11}{1 \cdot \cancel{2} \cdot 12} = \frac{11}{12}$$

The solution is $\frac{11}{12}$.

Practice Exercise

5. Solve: $\frac{2}{9} + \frac{2}{3}x = \frac{1}{6}$.

Objective 4.5a Convert between mixed numerals and fraction notation.

Example Convert $2\frac{5}{13}$ to fraction notation: $2\frac{5}{13} = \frac{31}{13}$.

Example Convert $\frac{40}{9}$ to a mixed numeral: $\frac{40}{9} = 4\frac{4}{9}$.

Practice Exercises

6. Convert $8\frac{2}{3}$ to fraction notation.

7. Convert $\frac{47}{6}$ to a mixed numeral.

Objective 4.6b Subtract using mixed numerals.

Example Subtract: $3\frac{3}{8} - 1\frac{4}{5}$.

$$\begin{array}{rcrcr} 3\frac{3}{8} & = & 3\frac{15}{40} & = & 2\frac{55}{40} \\ -1\frac{4}{5} & = & -1\frac{32}{40} & = & -1\frac{32}{40} \\ \hline & & & & 1\frac{23}{40} \end{array}$$

Practice Exercise

8. Subtract: $10\frac{5}{7} - 2\frac{3}{4}$.

Objective 4.7a Multiply using mixed numerals.

Example Multiply: $7\frac{1}{4} \cdot 5\frac{3}{10}$. Write a mixed numeral for the answer.

$$7\frac{1}{4} \cdot 5\frac{3}{10} = \frac{29}{4} \cdot \frac{53}{10}$$
$$= \frac{1537}{40} = 38\frac{17}{40}$$

Practice Exercise

9. Multiply: $4\frac{1}{5} \cdot 3\frac{7}{15}$.

Objective 4.7d Solve applied problems involving multiplication and division with mixed numerals.

Example The population of New York is $3\frac{1}{3}$ times that of Missouri. The population of New York is approximately 19,000,000. What is the population of Missouri?

Translate:

Population of New York	is	$3\frac{1}{3}$	times	population of Missouri
↓	↓	↓	↓	↓
19,000,000	=	$3\frac{1}{3}$	$\cdot$	x.

Solve:

$$19{,}000{,}000 = \frac{10}{3} \cdot x$$

$$\frac{19{,}000{,}000}{\frac{10}{3}} = \frac{\frac{10}{3} \cdot x}{\frac{10}{3}}$$

$$19{,}000{,}000 \cdot \frac{3}{10} = x$$

$$5{,}700{,}000 = x.$$

The population of Missouri is about 5,700,000.

Practice Exercise

10. The population of Louisiana is $2\frac{1}{2}$ times the population of West Virginia. The population of West Virginia is approximately 1,800,000. What is the population of Louisiana?

Objective 4.8a Simplify expressions containing fraction notation using the rules for order of operations.

Example Simplify: $\left(\frac{4}{5}\right)^2 - \frac{1}{5} \cdot 2\frac{1}{8}$.

$$\left(\frac{4}{5}\right)^2 - \frac{1}{5} \cdot 2\frac{1}{8} = \frac{16}{25} - \frac{1}{5} \cdot \frac{17}{8}$$
$$= \frac{16}{25} - \frac{17}{40} = \frac{16}{25} \cdot \frac{8}{8} - \frac{17}{40} \cdot \frac{5}{5}$$
$$= \frac{128}{200} - \frac{85}{200} = \frac{43}{200}$$

Practice Exercise

11. Simplify: $\frac{3}{2} \cdot 1\frac{1}{3} \div \left(\frac{2}{3}\right)^2$.

Review Exercises

Find the LCM. [4.1a]

1. 12 and 18

2. 18 and 45

3. 3, 6, and 30

4. 26, 36, and 54

Perform the indicated operation and, if possible, simplify. [4.2a, b], [4.3a]

5. $\frac{2}{9} + \frac{5}{9}$

6. $\frac{7}{x} + \frac{2}{x}$

7. $-\frac{6}{5} + \frac{11}{15}$

8. $\frac{5}{16} + \frac{3}{24}$

9. $\frac{7}{9} - \frac{5}{9}$

10. $\frac{1}{4} - \frac{3}{8}$

11. $\frac{10}{27} - \frac{2}{9}$

12. $\frac{5}{6} - \frac{7}{9}$

Use < or > for □ to form a true sentence. [4.2c]

13. $\frac{4}{7} \square \frac{5}{9}$

14. $-\frac{8}{9} \square -\frac{11}{13}$

Solve. [4.3b], [4.4a]

15. $x + \frac{2}{5} = \frac{7}{8}$

16. $\frac{1}{2}a - 3 = \frac{5}{2}$

17. $5 + \frac{16}{3}x = \frac{5}{9}$

18. $\frac{22}{5} = \frac{16}{5} + \frac{5}{2}x$

Solve by using the multiplication principle to clear fractions. [4.4b]

19. $\frac{5}{3}x + \frac{5}{6} = \frac{3}{2}$

Convert to fraction notation. [4.5a]

20. $7\frac{1}{2}$

21. $8\frac{3}{8}$

22. $4\frac{1}{3}$

23. $-1\frac{5}{7}$

Convert to a mixed numeral. [4.5a]

24. $\frac{7}{3}$

25. $\frac{-27}{4}$

26. $\frac{63}{5}$

27. $\frac{7}{2}$

28. Divide. Write a mixed numeral for the answer.
$7896 \div (-9)$ [4.5b]

29. Gina's golf scores were 80, 82, and 85. What was her average score? [4.5b]

Perform the indicated operation. Write a mixed numeral for each answer. [4.6a, b, d]

30. $\begin{array}{r} 7\frac{3}{5} \\ +\ 2\frac{4}{5} \\ \hline \end{array}$

31. $\begin{array}{r} 6\frac{1}{3} \\ +\ 5\frac{2}{5} \\ \hline \end{array}$

32. $-3\frac{5}{6} + \left(-5\frac{1}{6}\right)$

33. $-2\frac{3}{4} + 4\frac{1}{2}$

34. $\begin{array}{r} 14 \\ -\ 6\frac{2}{9} \\ \hline \end{array}$

35. $\begin{array}{r} 9\frac{3}{5} \\ -\ 4\frac{13}{15} \\ \hline \end{array}$

36. $4\frac{5}{8} - 9\frac{3}{4}$

37. $-7\frac{1}{2} - 6\frac{3}{4}$

Combine like terms. [4.2b], [4.6b]

38. $\frac{4}{9}x + \frac{1}{3}x$

39. $8\frac{3}{10}a - 5\frac{1}{8}a$

Perform the indicated operation. Write a mixed numeral or integer for each answer, unless the answer is less than 1. [4.7a, b]

40. $6 \cdot 2\frac{2}{3}$

41. $-5\frac{1}{4} \cdot \frac{2}{3}$

42. $2\frac{1}{5} \cdot 1\frac{1}{10}$

43. $2\frac{2}{5} \cdot 2\frac{1}{2}$

44. $-54 \div 2\frac{1}{4}$

45. $2\frac{2}{5} \div \left(-1\frac{7}{10}\right)$

46. $3\frac{1}{4} \div 26$

47. $4\frac{1}{5} \div 4\frac{2}{3}$

Evaluate. [4.7c]

48. $5x - y$, for $x = 3\frac{1}{5}$ and $y = 2\frac{2}{7}$

49. $2a \div b$, for $a = 5\frac{2}{11}$ and $b = 3\frac{4}{5}$

Solve. [4.2d], [4.3c], [4.6c], [4.7d]

50. *Sewing.* Kim wants to make slacks and a jacket. She needs $1\frac{5}{8}$ yd of 60-in. fabric for the slacks and $2\frac{5}{8}$ yd for the jacket. How many yards in all does Kim need to make the outfit?

51. The San Diaz drama club had $\frac{3}{8}$ of a vegetarian pizza, $1\frac{1}{2}$ cheese pizzas, and $1\frac{1}{4}$ pepperoni pizzas remaining after a cast party. How many pizzas remained altogether?

52. *Turkey Servings.* Turkey contains $1\frac{1}{3}$ servings per pound. How many pounds are needed for 32 servings?

53. *Humane Society Pie Sale.* Green River's Humane Society recently hosted its annual dessert social. Each of the 83 pies donated was cut into 6 pieces. At the end of the evening, 382 pieces of pie had been sold. How many pies were sold? How many were left over? Express your answers in mixed numerals.

54. *Running.* Janelle has mapped a $1\frac{1}{2}$-mi running route in her neighborhood. One Saturday, she ran this route $2\frac{1}{2}$ times. How many miles did she run?

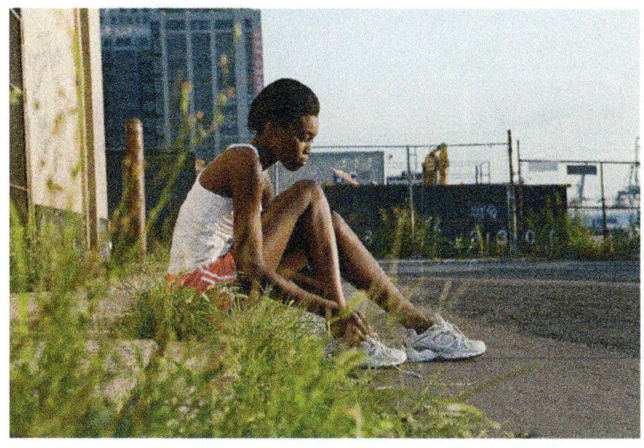

55. What is the sum of the areas in the figure below?

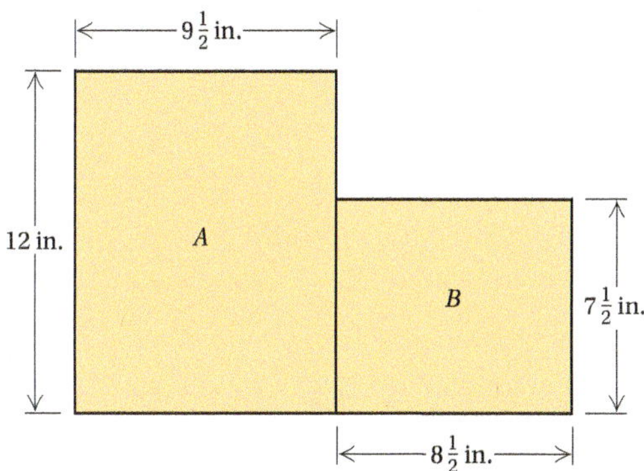

56. In the figure above, how much larger is the area of rectangle A than the area of rectangle B?

57. *Painting a Border.* Katie hired an artist to paint a decorative border around the top of her son's bedroom. The artist charges $20 per foot. The room measures $11\frac{3}{4}$ ft $\times$ $9\frac{1}{2}$ ft. What is Katie's cost for the project?

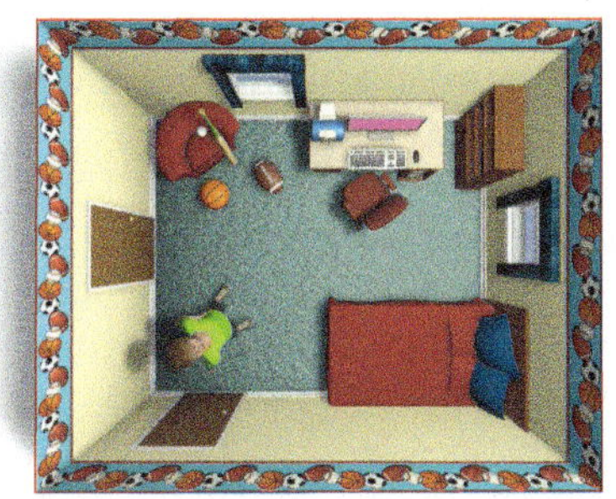

58. *NCAA Football Goalposts.* In college football, the distance between goalposts was reduced from $23\frac{1}{3}$ ft to $18\frac{1}{2}$ ft. By how much was it reduced?

Source: NCAA

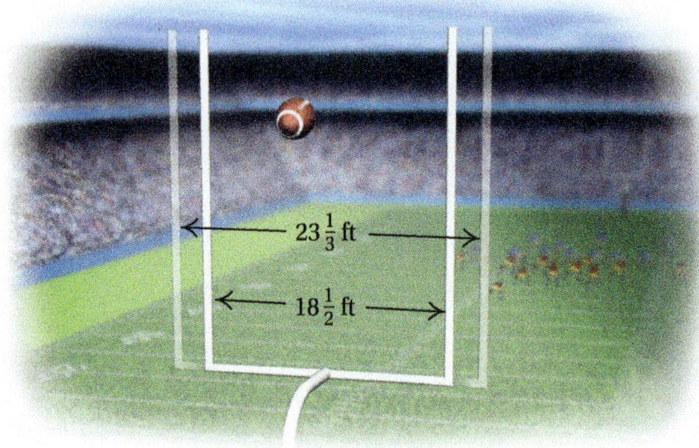

Simplify each expression using the rules for order of operations. [4.8a]

59. $\frac{1}{2} + \frac{1}{8} \div \frac{1}{4}$

60. $\frac{4}{5} - \frac{1}{2} \cdot \left(\frac{1}{4} - \frac{3}{5}\right)$

61. $20\frac{3}{4} - 1\frac{1}{2} \times 12 + \left(\frac{1}{2}\right)^2$

62. Find the average of $\frac{1}{2}, \frac{1}{4}, \frac{1}{3}$, and $\frac{1}{5}$. [4.4a]

Simplify. [4.8b]

63. $\dfrac{\frac{1}{3}}{\frac{5}{8} - 1}$

64. $\dfrac{-\frac{x}{7}}{\frac{3}{14}}$

65. Simplify: $\frac{1}{4} + \frac{2}{5} \div 5^2$. [4.8a]

A. $\frac{133}{500}$ **B.** $\frac{3}{500}$

C. $\frac{117}{500}$ **D.** $\frac{5}{2}$

66. Solve: $x + \frac{2}{3} = 5$. [4.3b]

A. $\frac{15}{2}$ **B.** $5\frac{2}{3}$

C. $\frac{10}{3}$ **D.** $4\frac{1}{3}$

Synthesis

67. Find r if

$$\frac{1}{r} = \frac{1}{100} + \frac{1}{150} + \frac{1}{200}. \quad [4.2b], [4.4b]$$

68. Place the numbers 3, 4, 5, and 6 in the boxes in order to make a true equation: [4.5a]

$$\frac{\square}{\square} + \frac{\square}{\square} = 3\frac{1}{4}.$$

69. Find the largest integer for which each fraction is greater than 1. [4.2c]

a) $\frac{7}{\square}$ **b)** $\frac{11}{\square}$

c) $\frac{\square}{-27}$ **d)** $\frac{\square}{-\frac{1}{2}}$

Understanding Through Discussion and Writing

1. Is the sum of two mixed numerals always a mixed numeral? Why or why not? [4.6a]

2. Write a problem for a classmate to solve. Design the problem so that its solution is found by performing the multiplication $4\frac{1}{2} \cdot 33\frac{1}{3}$. [4.7d]

3. A student insists that $3\frac{2}{5} \cdot 1\frac{3}{7} = 3\frac{6}{35}$. What mistake is he making and how should he have proceeded? [4.7a]

4. Discuss the role of least common multiples in adding and subtracting with fraction notation. [4.2b], [4.3a]

5. Find a real-world situation that fits this equation:

$$2 \cdot 15\frac{3}{4} + 2 \cdot 28\frac{5}{8} = 88\frac{3}{4}. \quad [4.6c], [4.7d]$$

6. A student insists that $5 \cdot 3\frac{2}{7} = (5 \cdot 3) \cdot (5 \cdot \frac{2}{7})$. What mistake is she making and how should she have proceeded? [4.7a]

CHAPTER

4 Test

For Extra Help For step-by-step test solutions, access the Chapter Test Prep Videos in MyMathLab® or on YouTube (search "BittingerPreIntro" and click on Channels).

1. Find the LCM of 12 and 16.

Perform the indicated operation and, if possible, simplify.

2. $\frac{1}{2} + \frac{5}{2}$

3. $-\frac{7}{8} + \frac{2}{3}$

4. $\frac{5}{t} - \frac{3}{t}$

5. $\frac{5}{6} - \frac{3}{4}$

6. $\frac{5}{8} - \frac{17}{24}$

Solve.

7. $x + \frac{2}{3} = \frac{11}{12}$

8. $-5x - 3 = 9$

9. $\frac{3}{4} = \frac{1}{2} + \frac{5}{3}x$

10. Use $<$ or $>$ for $\square$ to form a true sentence.

$\frac{6}{7} \square \frac{21}{25}$

Convert to fraction notation.

11. $3\frac{1}{2}$

12. $-9\frac{3}{8}$

13. Convert to a mixed numeral:

$-\frac{74}{9}$.

14. Divide. Write a mixed numeral for the answer.

$11\overline{)1789}$

Perform the indicated operation. Write a mixed numeral for each answer.

15. $\begin{array}{r} 6\frac{2}{5} \\ +\ 7\frac{4}{5} \\ \hline \end{array}$

16. $\begin{array}{r} 3\frac{1}{4} \\ +\ 9\frac{1}{6} \\ \hline \end{array}$

17. $\begin{array}{r} 10\frac{1}{6} \\ -\ 5\frac{7}{8} \\ \hline \end{array}$

18. $14 + \left(-5\frac{3}{7}\right)$

19. $3\frac{4}{5} - 9\frac{1}{2}$

Combine like terms.

20. $\frac{3}{8}x - \frac{1}{2}x$

21. $5\frac{2}{11}a - 3\frac{1}{5}a$

Perform the indicated operation.

22. $9 \cdot 4\frac{1}{3}$

23. $6\frac{3}{4} \cdot \left(-2\frac{2}{3}\right)$

24. $33 \div 5\frac{1}{2}$

25. $2\frac{1}{3} \div 1\frac{1}{6}$

Evaluate.

26. $\frac{2}{3}ab$, for $a = 7$ and $b = 4\frac{1}{5}$

27. $4 + mn$, for $m = 7\frac{2}{5}$ and $n = 3\frac{1}{4}$

Solve.

28. *Weightlifting.* In 2002, Hossein Rezazadeh of Iran completed a clean and jerk of 263 kg. This amount was about $2\frac{1}{2}$ times his body weight. How much did Rezazadeh weigh?

Source: *The Guinness Book of Records*, 2005

29. *Book Order.* An order of books for a math course weighs 220 lb. Each book weighs $2\frac{3}{4}$ lb. How many books are in the order?

30. *Carpentry.* The following diagram shows a middle drawer support guide for a cabinet drawer. Find each of the following.

a) The short length a across the top
b) The length b across the bottom

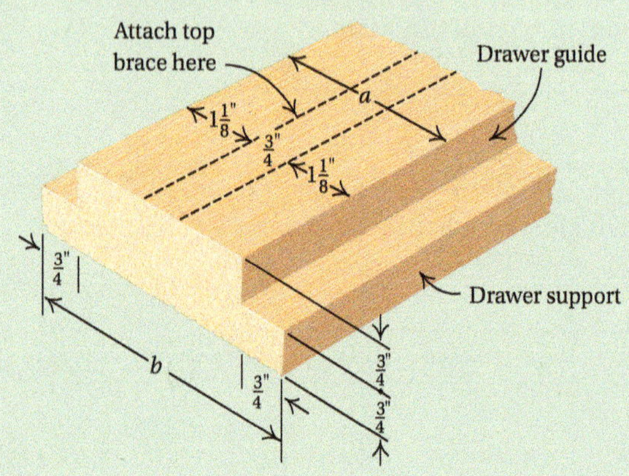

31. *Carpentry.* In carpentry, some pieces of plywood that are called "$\frac{3}{4}$-inch" plywood are actually $\frac{11}{16}$ in. thick. How much thinner is such a piece than its name indicates?

32. *Women's Dunks.* The first three women in the history of college basketball able to dunk a basketball are listed below. Their names, heights, and universities are

Michelle Snow, $6\frac{5}{12}$ ft, Tennessee;
Charlotte Smith, $5\frac{11}{12}$ ft, North Carolina;
Georgeann Wells, $6\frac{7}{12}$ ft, West Virginia.

Find the average height of these women.

Source: *USA Today*, 11/30/00, p. 3C

Simplify.

33. $\frac{2}{3} + 1\frac{1}{3} \cdot 2\frac{1}{8}$

34. $-1\frac{1}{2} - \frac{1}{2}\left(\frac{1}{2} \div \frac{1}{4}\right) + \left(\frac{1}{2}\right)^2$

35. $\dfrac{\frac{1}{3} - \frac{7}{9}}{\frac{1}{2} + \frac{1}{6}}$

36. Find the LCM of 12, 36, and 60.

A. 6
B. 12
C. 60
D. 180

Synthesis

37. The students in a math class can be organized into study groups of 8 each so that no students are left out. The same class of students can also be organized into groups of 6 so that no students are left out.

a) Find some class sizes for which this will work.
b) Find the smallest such class size.

38. Rebecca walks 17 laps at her health club. Trent walks 17 laps at his health club. If the track at Rebecca's health club is $\frac{1}{7}$ mi long and the track at Trent's is $\frac{1}{8}$ mi long, who walks farther? How much farther?

CHAPTER

5

Decimal Notation

STUDYING FOR SUCCESS *Time Management*

- ☐ As a rule, budget 2–3 hours for doing homework and studying for every hour you spend in class.
- ☐ Balance work and study carefully. Be honest with yourself about how much time you have available to work, attend class, and study.
- ☐ Make an hour-by-hour schedule of your week, planning time for leisure as well as work and study.
- ☐ Use your syllabus to help you plan your time. Transfer project deadlines and test dates to your calendar.

5.1 Decimal Notation, Order, and Rounding

OBJECTIVES

- **a** Given decimal notation, write a word name.
- **b** Convert between decimal notation and fraction notation.
- **c** Given a pair of numbers in decimal notation, tell which is larger.
- **d** Round decimal notation to the nearest thousandth, hundredth, tenth, one, ten, hundred, or thousand.

SKILL TO REVIEW

Objective 1.6a: Round to the nearest ten, hundred, or thousand.

Round 4735 to the nearest

1. Ten.

2. Hundred.

The set of **rational numbers** consists of the **integers** . . . , −3, −2, −1, 0, 1, 2, 3, . . . , and fractions like $\frac{1}{2}$, $\frac{2}{3}$, $\frac{-7}{8}$, $\frac{17}{-10}$, and so on. In this chapter, we will use *decimal notation* to represent rational numbers. Using decimal notation, we can write 0.875 for $\frac{7}{8}$, for example, or 48.97 for $48\frac{97}{100}$. A number written in decimal notation is often simply referred to as a *decimal*.

The word *decimal* comes from the Latin word *decima*, meaning a *tenth part*. Since our usual counting system is based on tens, decimal notation is a natural extension of an already familiar system.

a DECIMAL NOTATION AND WORD NAMES

One model of the Magellan GPS navigation system sells for \$249.98. The dot in \$249.98 is called a **decimal point**. Since \$0.98, or 98¢, is $\frac{98}{100}$ of a dollar, it follows that

$$\$249.98 = \$249 + \$0.98.$$

Also, since \$0.98, or 98¢, has the same value as 9 dimes + 8 cents and 1 dime is $\frac{1}{10}$ of a dollar and 1 cent is $\frac{1}{100}$ of a dollar, we can write

$$249.98 = 2 \cdot 100 + 4 \cdot 10 + 9 \cdot 1 + 9 \cdot \frac{1}{10} + 8 \cdot \frac{1}{100}.$$

This is an extension of expanded notation for whole numbers. The place values are 100, 10, 1, $\frac{1}{10}$, $\frac{1}{100}$, and so on. We can see this on a **place-value chart**. The value of each place is $\frac{1}{10}$ as large as that of the one to its left.

Let's consider decimal notation using a place-value chart to represent 2.0677 min, the women's 200-meter backstroke world record held by Missy Franklin from the United States.

PLACE-VALUE CHART							
Hundreds	Tens	Ones	Tenths	Hundredths	Thousandths	Ten-Thousandths	Hundred-Thousandths
100	10	1	$\frac{1}{10}$	$\frac{1}{100}$	$\frac{1}{1000}$	$\frac{1}{10,000}$	$\frac{1}{100,000}$
		2 .	0	6	7	7	

Answers

Skill to Review:
1. 4740 2. 4700

The decimal notation 2.0677 means

2 ones + 0 tenths + 6 hundredths + 7 thousandths + 7 ten-thousandths,

or

$$2 + \frac{0}{10} + \frac{6}{100} + \frac{7}{1000} + \frac{7}{10{,}000}, \quad \text{or} \quad 2\frac{677}{10{,}000}.$$

We read both 2.0677 and $2\frac{677}{10{,}000}$ as

"Two *and* six hundred seventy-seven ten-thousandths."

We read the decimal point as "and." Note that the place values to the right of the decimal point always end in *th*. We can also read 2.0677 as "Two *point* zero six seven seven."

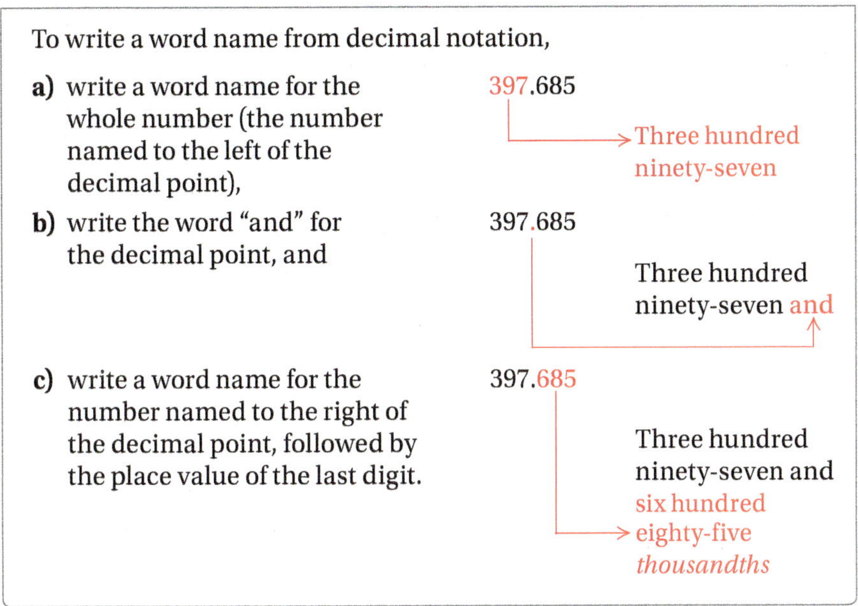

To write a word name from decimal notation,

a) write a word name for the whole number (the number named to the left of the decimal point),

397.685 → Three hundred ninety-seven

b) write the word "and" for the decimal point, and

397.685 → Three hundred ninety-seven and

c) write a word name for the number named to the right of the decimal point, followed by the place value of the last digit.

397.685 → Three hundred ninety-seven and six hundred eighty-five *thousandths*

EXAMPLE 1 ***Median Age.*** The median age of residents in Maine is 42.7 years. The median age of residents in Utah is 29.2 years. Write word names for 42.7 and 29.2.

Source: U.S. Census Bureau

Forty-two and seven tenths

Twenty-nine and two tenths

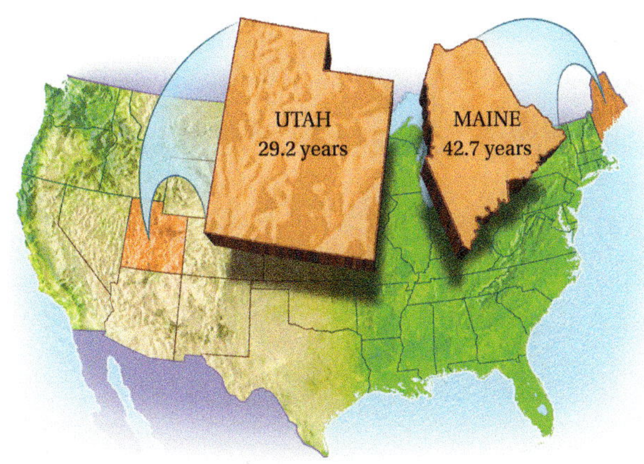

1. *Life Expectancy.* The life expectancy at birth in Kenya in 2011 was 61.3 years for males and 64.2 years for females. Write word names for 61.3 and 64.2.

Source: World Health Organization

2. *10,000-Meter Record.* Wang Junxia of China holds the women's world record for the 10,000-meter run: 29.5297 min. Write a word name for 29.5297.

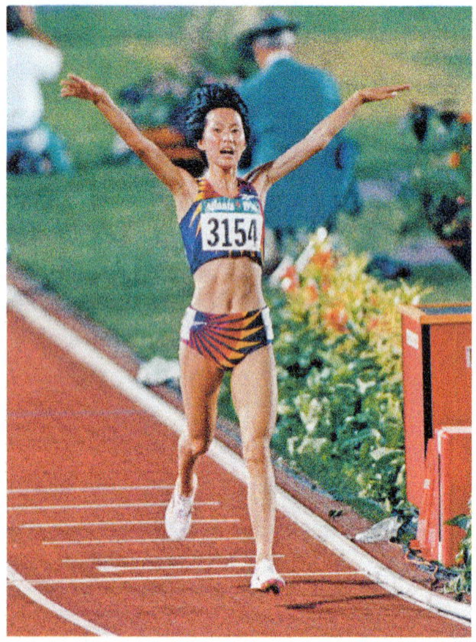

Write a word name for each number.

3. 245.89

4. 34.0064

5. 31,079.756

Answers

1. Sixty-one and three tenths; sixty-four and two tenths
2. Twenty-nine and five thousand two hundred ninety-seven ten-thousandths
3. Two hundred forty-five and eighty-nine hundredths
4. Thirty-four and sixty-four ten-thousandths
5. Thirty-one thousand, seventy-nine and seven hundred fifty-six thousandths

EXAMPLE 2 Write a word name for 410.87.

Four hundred ten and eighty-seven hundredths

EXAMPLE 3 *Record Price.* Edvard Munch's painting *The Scream* sold for \$119.9225 million at Sotheby's in New York City on May 2, 2012. This price set a world record as the highest paid for a work of art sold at auction. Write a word name for 119.9225.

Source: Sotheby's

One hundred nineteen and nine thousand two hundred twenty-five ten-thousandths

EXAMPLE 4 Write a word name for 1788.045.

One thousand, seven hundred eighty-eight and forty-five thousandths

Do Exercises 1–5.

b CONVERTING BETWEEN DECIMAL NOTATION AND FRACTION NOTATION

Given decimal notation, we can convert to fraction notation as follows:

$$\begin{aligned} 9.875 &= 9 + \frac{8}{10} + \frac{7}{100} + \frac{5}{1000} \\ &= 9 \cdot \frac{1000}{1000} + \frac{8}{10} \cdot \frac{100}{100} + \frac{7}{100} \cdot \frac{10}{10} + \frac{5}{1000} \\ &= \frac{9000}{1000} + \frac{800}{1000} + \frac{70}{1000} + \frac{5}{1000} = \frac{9875}{1000}. \end{aligned}$$

Decimal notation → 9.875 ← 3 decimal places

Fraction notation → $\frac{9875}{1000}$ ← 3 zeros

To convert from decimal notation to fraction notation,

a) count the number of decimal places, 4.98 → 2 places

b) move the decimal point that many places to the right, and 4.98. Move 2 places.

c) write the answer over a denominator of 1 followed by that number of zeros. $\frac{498}{100}$ 2 zeros

For a number like 0.876, we write a 0 to call attention to the presence of the decimal point.

EXAMPLE 5 Write fraction notation for 0.876. Do not simplify.

0.876 0.876. $0.876 = \frac{876}{1000}$

3 places 3 zeros

Decimals greater than 1 or less than −1 can be written either as fractions or as mixed numerals.

EXAMPLE 6 Write 56.23 as a fraction and as a mixed numeral.

56.23 56.23. $56.23 = \frac{5623}{100}$, and $56.23 = 56\frac{23}{100}$

2 places 2 zeros

As a check, note that both 56.23 and $56\frac{23}{100}$ are read as "fifty-six and twenty-three hundredths."

EXAMPLE 7 Write −2.6073 as a fraction and as a mixed numeral.

$-2.6073. = -\frac{26{,}073}{10{,}000}$ and $-2.6073 = -2\frac{6073}{10{,}000}$

4 places 4 zeros

Do Exercises 6–9.

To write $\frac{5328}{10}$ as a decimal, we can first divide to find an equivalent mixed numeral.

$$\frac{5328}{10} = 532\frac{8}{10}$$

```
      5 3 2
1 0 ) 5 3 2 8
      5 0
        3 2
        3 0
          2 8
          2 0
            8
```

Next note that

$$532\frac{8}{10} = 532 + \frac{8}{10}$$
$$= 532.8.$$

Write fraction notation. Do not simplify.

6. 0.5491

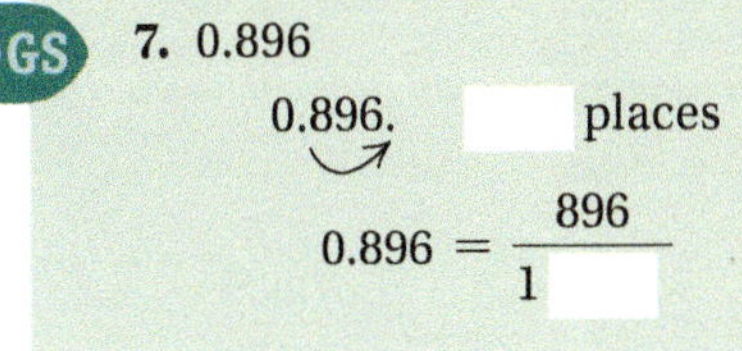
GS **7.** 0.896

0.896. ☐ places

$0.896 = \frac{896}{1☐}$

Write as a fraction and as a mixed numeral.

8. 75.069 **9.** −312.9

Answers

6. $\frac{5491}{10{,}000}$ **7.** $\frac{896}{1000}$ **8.** $\frac{75{,}069}{1000}$; $75\frac{69}{1000}$
9. $-\frac{3129}{10}$; $-312\frac{9}{10}$

Guided Solution:
7. 3; 000

Thus, if fraction notation has a denominator that is a power of ten, such as 10, 100, 1000, and so on, we convert to decimal notation by reversing the procedure that we just used in Examples 5–7.

> To convert from fraction notation to decimal notation when the denominator is 10, 100, 1000, and so on,
>
> **a)** count the number of zeros and $\frac{8679}{1000}$ — 3 zeros
>
> **b)** move the decimal point that number of places to the left. Leave off the denominator. 8.679. Move 3 places.
>
> $\frac{8679}{1000} = 8.679$

EXAMPLE 8 Write decimal notation for $\frac{47}{10}$.

$\frac{47}{10}$ 1 zero; 4.7. 1 place; $\frac{47}{10} = 4.7$ The decimal point is moved to the left.

EXAMPLE 9 Write decimal notation for $\frac{123,067}{10,000}$.

$\frac{123,067}{10,000}$ 4 zeros; 12.3067. 4 places; $\frac{123,067}{10,000} = 12.3067$

To move the decimal point to the left, we may need to write extra 0's.

EXAMPLE 10 Write decimal notation for $-\frac{9}{100}$.

$-\frac{9}{100}$ 2 zeros; −0.09. 2 places; $-\frac{9}{100} = -0.09$

Do Exercises 10–13.

Write decimal notation for each number.

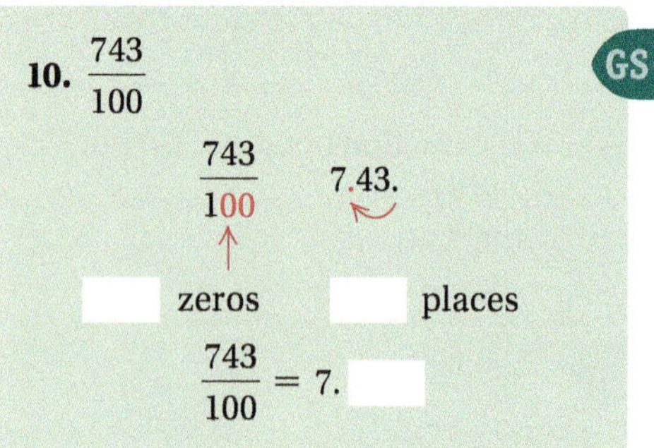

10. $\frac{743}{100}$

$\frac{743}{100}$ ____ zeros; 7.43. ____ places

$\frac{743}{100} = 7.$ ____

11. $-\frac{73}{1000}$ **12.** $\frac{67,089}{10,000}$

13. $-\frac{9}{10}$

To convert fractions with denominators other than 10, 100, and so on to decimal notation, we will usually perform long division.

If a mixed numeral has a fraction part with a denominator that is a power of ten, such as 10, 100, or 1000, and so on, we first write the mixed numeral as a sum of a whole number and a fraction. Then we convert to decimal notation.

EXAMPLE 11 Write decimal notation for $23\frac{59}{100}$.

$$23\frac{59}{100} = 23 + \frac{59}{100} = 23 \text{ and } \frac{59}{100} = 23.59$$

Do Exercises 14 and 15.

Write decimal notation for each number.

14. $-7\frac{3}{100}$ **15.** $23\frac{47}{1000}$

Answers

10. 7.43 **11.** −0.073 **12.** 6.7089 **13.** −0.9 **14.** −7.03 **15.** 23.047

Guided Solution:
10. 2, 2; 43

C ORDER

To understand how to compare numbers in decimal notation, consider 0.85 and 0.9. First note that $0.9 = 0.90$ because $\frac{9}{10} = \frac{90}{100}$. Since $0.85 = \frac{85}{100}$, it follows that $\frac{85}{100} < \frac{90}{100}$ and $0.85 < 0.9$. This leads us to a quick way to compare two numbers in decimal notation.

> To compare two positive numbers in decimal notation, start at the left and compare corresponding digits, moving from left to right. When two digits differ, the number with the larger digit is the larger of the two numbers. Extra zeros can be written to the right of the last decimal place.

EXAMPLE 12 Which is larger: 2.109 or 2.1?

2.109	2.109	2.109	2.109
↕ The same	↕ The same	↕ The same	↕ Different; 9 is larger than 0.
2.1	2.1	2.10	2.100

Thus, 2.109 is larger than 2.1. In symbols, $2.109 > 2.1$.

EXAMPLE 13 Which is larger: 0.09 or 0.108?

0.09	0.09
↕ The same	↕ Different; 1 is larger than 0.
0.108	0.108

Thus, 0.108 is larger than 0.09. In symbols, $0.108 > 0.09$.

Do Exercises 16–21. ▶

Which number is larger?

16. 2.04, 2.039

17. 0.06, 0.008

18. 0.5, 0.58

19. 1, 0.9999

20. 0.8989, 0.09898

21. 21.006, 21.05

We can use the number line to visualize order. We illustrate Examples 12 and 13 below. Larger numbers are always to the right.

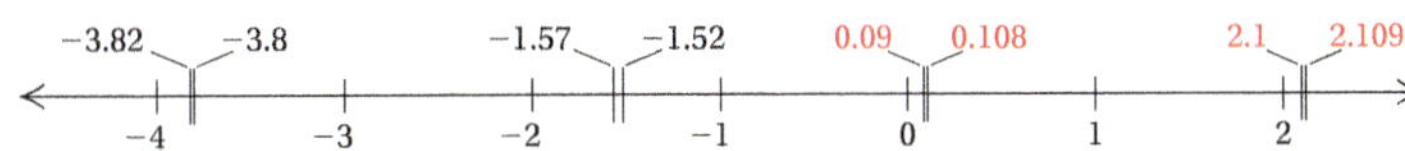

Note from the number line that $-2 < -1$. Similarly, $-1.57 < -1.52$.

> To compare two negative numbers in decimal notation, start at the left and compare corresponding digits, moving from left to right. When two digits differ, the number with the smaller digit is the larger of the two numbers.

EXAMPLE 14 Which is larger: −3.8 or −3.82?

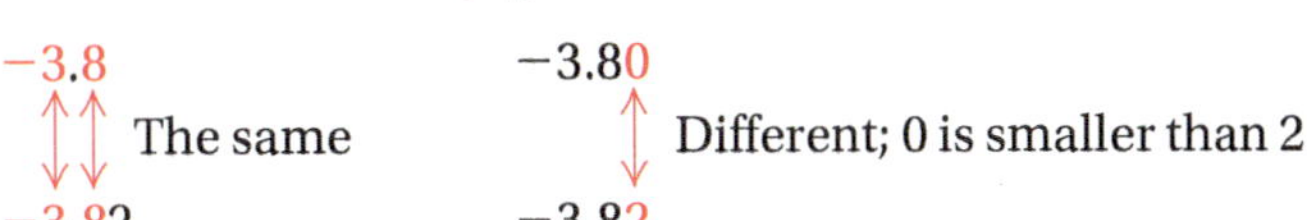

Thus, −3.8 is larger than −3.82. In symbols, $-3.8 > -3.82$. (See the number line above.)

Do Exercises 22 and 23. ▶

Which number is larger?

22. −34.01, −34.008

23. −9.12, −8.98

Answers

16. 2.04 **17.** 0.06 **18.** 0.58 **19.** 1 **20.** 0.8989 **21.** 21.05 **22.** −34.008 **23.** −8.98

d ROUNDING

We round decimals in much the same way that we round whole numbers. To see how, we use the number line.

EXAMPLE 15 Round 0.37 to the nearest tenth.

Here is part of the number line, magnified.

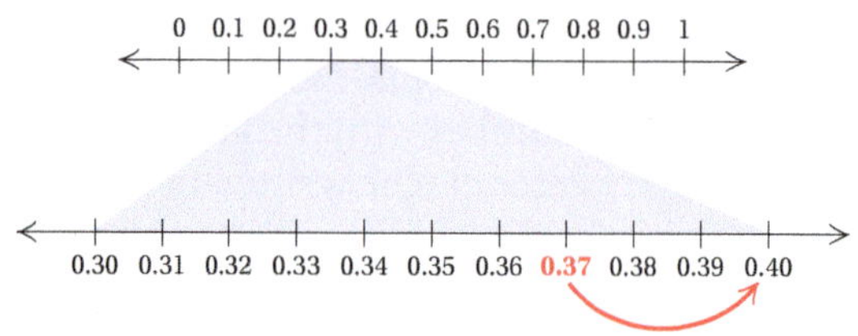

We see that 0.37 is closer to 0.40 than to 0.30. Thus, 0.37 rounded to the nearest tenth is 0.4.

> To round to a certain place:
>
> **a)** Locate the digit in that place.
>
> **b)** Consider the next digit to the right.
>
> **c)** If the digit to the right is 5 or greater, add 1 to the original digit. If the digit to the right is 4 or less, the original digit does not change. In either case, drop all numbers to the right of the original digit.

EXAMPLE 16 Round 72.3846 to the nearest hundredth.

a) Locate the digit in the hundredths place, 8.

7 2 . 3 8 4 6

b) Consider the next digit to the right, 4.

c) Since that digit, 4, is less than 5, the original digit does not change. The rounded number is 72.38.

Caution!

72.39 is not a correct answer to Example 16. It is *incorrect* to round sequentially from right to left as follows: 72.3846, 72.385, 72.39.

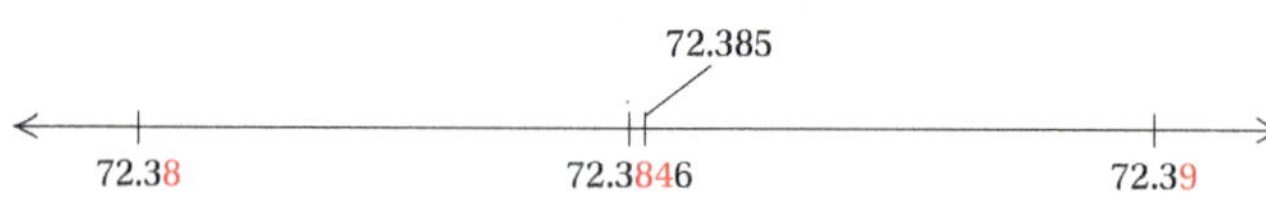

72.3846 is closer to 72.38 than to 72.39.

EXAMPLE 17 Round −0.064 to the nearest tenth.

a) Locate the digit in the tenths place, 0.

−0 . 0 6 4

b) Consider the next digit to the right, 6.

c) Since that digit, 6, is greater than 5, round from −0.064 to −0.1.

The answer is −0.1. Since $-0.1 < -0.064$, we actually rounded *down*.

Do Exercises 24–42.

Round to the nearest tenth.

24. 2.76 **25.** 13.85

26. −234.448 **27.** 7.009

Round to the nearest hundredth.

28. 0.6362 **29.** −7.8348

30. 34.67514 **31.** −0.02521

Round to the nearest thousandth.

32. 0.94347 **33.** −8.00382

34. −43.111943 **35.** 37.400526

Round 7459.3548 to the nearest

36. Thousandth.

37. Hundredth.

38. Tenth. **39.** One.

40. Ten. (*Caution:* "Tens" are not "tenths.")

41. Hundred. **42.** Thousand.

Answers

24. 2.8 **25.** 13.9 **26.** −234.4 **27.** 7.0 **28.** 0.64 **29.** −7.83 **30.** 34.68 **31.** −0.03 **32.** 0.943 **33.** −8.004 **34.** −43.112 **35.** 37.401 **36.** 7459.355 **37.** 7459.35 **38.** 7459.4 **39.** 7459 **40.** 7460 **41.** 7500 **42.** 7000

5.1 Exercise Set

For Extra Help
MyMathLab®
MathXL® PRACTICE | WATCH | READ | REVIEW

Reading Check

Name the digit that represents each place value in the number 436.81205.

RC1. Hundred-thousandths ________

RC2. Thousandths ________

RC3. Tens ________

RC4. Ten-thousandths ________

RC5. Tenths ________

RC6. Hundreds ________

RC7. Hundredths ________

RC8. Ones ________

a

1. *Currency Conversion.* One Japanese yen was worth about $0.0119 in U.S. currency recently. Write a word name for 0.0119.

Source: finance.yahoo.com

2. *Currency Conversion.* One U.S. dollar was worth 0.9949 Canadian dollars recently. Write a word name for 0.9949.

Source: finance.yahoo.com

3. *Birth Rate.* There were 137.6 triplet births per 100,000 live births in a recent year in the United States. Write a word name for 137.6.

Source: U.S. Centers for Disease Control and Prevention

4. *Coffee Consumption.* Annual per capita coffee consumption in Finland is 26.7 lb. In Spain, annual per capita coffee consumption is only 9.4 lb. Write word names for 26.7 and 9.4.

Source: International Coffee Organization

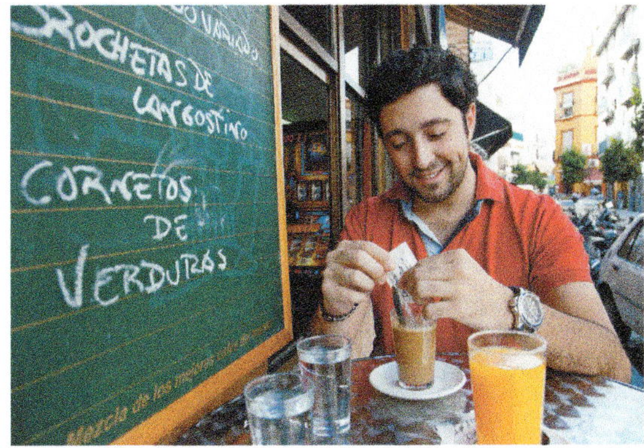

5. *Stock Price.* Apple, Inc.'s stock price was recently $519.22 per share. Write a word name for 519.22.

Source: NASDAQ

6. *Indianapolis 500.* Ryan Hunter-Reay won the 2014 Indianapolis 500 race with a time of 2 hr 40 min 48.2305 sec. Write a word name for 48.2305.

Source: Indianapolis Motor Speedway

Write a word name for the number in each sentence.

7. One gallon of paint is equal to 3.785 L of paint.

8. *Water Weight.* One gallon of water weighs 8.35 lb.

b Write each number as a fraction and, if possible, as a mixed numeral. Do not simplify.

9. 7.3 **10.** 4.9 **11.** 21.67 **12.** −57.32 **13.** −2.703

14. 0.079 **15.** 0.0109 **16.** 1.0008 **17.** −4.0003 **18.** −9.012

19. −0.0207 **20.** −0.00104 **21.** 70.00105 **22.** 60.0403

Write decimal notation for each number.

23. $\frac{3}{10}$ **24.** $\frac{73}{10}$ **25.** $-\frac{59}{100}$ **26.** $-\frac{67}{100}$ **27.** $\frac{3798}{1000}$

28. $\frac{780}{1000}$ **29.** $\frac{78}{10{,}000}$ **30.** $\frac{56{,}788}{100{,}000}$ **31.** $\frac{-18}{100{,}000}$ **32.** $\frac{-2347}{100}$

33. $\frac{486{,}197}{1{,}000{,}000}$ **34.** $\frac{8{,}953{,}074}{1{,}000{,}000}$ **35.** $7\frac{13}{1000}$ **36.** $4\frac{909}{1000}$ **37.** $-8\frac{431}{1000}$

38. $-49\frac{32}{1000}$ **39.** $2\frac{1739}{10{,}000}$ **40.** $9243\frac{1}{10}$ **41.** $8\frac{953{,}073}{1{,}000{,}000}$ **42.** $2256\frac{3059}{10{,}000}$

c Which number is larger?

43. 0.06, 0.58 **44.** 0.008, 0.8 **45.** 0.403, 0.410 **46.** 42.06, 42.1

47. −5.046, −5.043 **48.** −324.19, −325.19 **49.** 234.07, 235.07 **50.** 0.99999, 1

51. 0.007, $\frac{7}{100}$ **52.** $\frac{73}{10}$, 0.73 **53.** −0.872, −0.873 **54.** −0.8437, −0.84384

d Round to the nearest tenth.

55. 0.23 **56.** 0.85 **57.** −0.372 **58.** −0.261

59. 2.951 **60.** 7.532 **61.** −327.2347 **62.** −8.7493

Round to the nearest hundredth.

63. 0.893 **64.** 0.675 **65.** −0.6666 **66.** −7.5252

67. 0.9952 **68.** 207.9976 **69.** −0.03488 **70.** −9.27481

Round to the nearest thousandth.

71. 0.5724 **72.** 0.6666 **73.** 17.0015 **74.** 123.4562

75. −20.20202 **76.** −0.10346 **77.** 9.98487 **78.** 67.100602

Round 809.47321 to the nearest

79. Tenth. **80.** Thousandth. **81.** Hundredth. **82.** One.

Skill Maintenance

Add or subtract, as indicated.

83. $\begin{array}{r} 681 \\ +\ 149 \\ \hline \end{array}$ [1.2a]

84. $\frac{681}{1000} + \frac{149}{1000}$ [4.2a]

85. $\begin{array}{r} 267 \\ -\ 85 \\ \hline \end{array}$ [1.3a]

86. $\frac{267}{100} - \frac{85}{100}$ [4.3a]

87. $\frac{37}{55} - \frac{49}{55}$ [4.3a]

88. $-\frac{29}{34} + \frac{14}{34}$ [4.2a]

89. $\begin{array}{r} 34{,}903 \\ -\ 1{,}945 \\ \hline \end{array}$ [1.3a]

90. $\begin{array}{r} 4937 \\ +\ 5789 \\ \hline \end{array}$ [1.2a]

Synthesis

91. Arrange the following numbers in order from smallest to largest.

−0.989, −0.898, −1.009, −1.09, −0.098

92. Arrange the following numbers in order from smallest to largest.

−2.018, −2.1, −2.109, −2.0119, −2.108

Truncating. There are other methods of rounding decimal notation. A computer often uses a method called **truncating**. To truncate we drop off decimal places right of the rounding place, which is the same as changing all digits to the right of the rounding place to zeros. For example, rounding 6.78093456285102 to the ninth decimal place, using truncating, gives us 6.780934562. Use truncating to round each of the following to the fifth decimal place, that is, the hundred-thousandth place.

93. 6.78346123 **94.** 6.783461902 **95.** 99.999999999 **96.** 0.030303030303

Temperature Change. The graph below is based on the average global temperature from 1951 through 1980. Each bar indicates, in Celsius degrees, how much above or below average the temperature was for the year. Use the graph for Exercises 97–100.

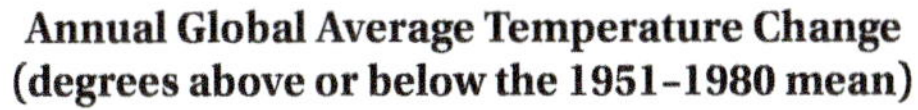

Annual Global Average Temperature Change
(degrees above or below the 1951–1980 mean)

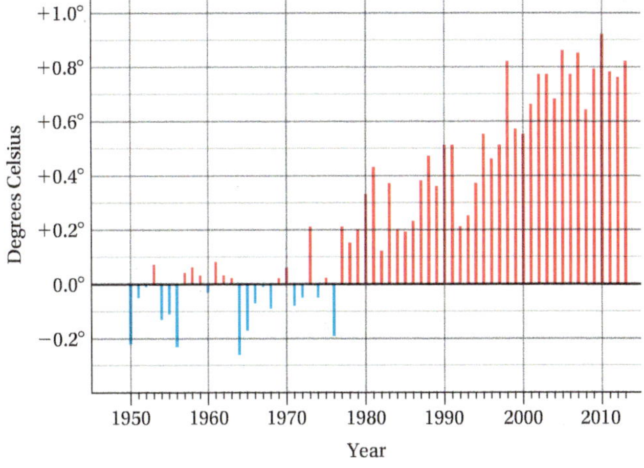

SOURCE: Goddard Institute for Space Studies

97. For what year(s) was the yearly temperature more than 0.8 degree above average?

98. What was the last year in which the yearly temperature was more than 0.2 degree below average?

99. What was the last year in which the yearly temperature was below average?

100. For what year(s) was the yearly temperature more than 0.9 degree above average?

5.2 Addition and Subtraction of Decimals

OBJECTIVES

a Add using decimal notation.

b Subtract using decimal notation.

c Add and subtract negative decimals.

d Combine like terms with decimal coefficients.

SKILLS TO REVIEW

Objective 1.2a: Add whole numbers.
Objective 1.3a: Subtract whole numbers.

Add or subtract.

1. $\begin{array}{r} 387 \\ +\,254 \\ \hline \end{array}$

2. $\begin{array}{r} 4023 \\ -\,1667 \\ \hline \end{array}$

Add.

1. $\begin{array}{r} 0.847 \\ +\,10.07 \\ \hline \end{array}$

2. $\begin{array}{r} 2.1 \\ 0.73 \\ +\,31.368 \\ \hline \end{array}$

Add.

3. 0.02 + 4.3 + 0.649

4. 0.12 + 3.006 + 0.4357

5. 0.4591 + 0.2374 + 8.70894

Answers

Skills to Review:
1. 641 2. 2356

Margin Exercises:
1. 10.917 2. 34.198 3. 4.969
4. 3.5617 5. 9.40544

a ADDITION

Adding with decimal notation is similar to adding whole numbers. First, we line up the decimal points so that we can add corresponding place-value digits. Then we add digits from the right. For example, we add the thousandths, then the hundredths, and so on, carrying if necessary. If desired, we can write extra zeros to the right of the decimal point so that the number of places is the same in all of the addends.

EXAMPLE 1 Add: 56.314 + 17.78.

$\begin{array}{r} 56.314 \\ +\,17.780 \\ \hline \end{array}$ Lining up the decimal points in order to add; Writing an extra zero to the right of the decimal point

$\begin{array}{r} 56.314 \\ +\,17.780 \\ \hline 4 \end{array}$ Adding thousandths

$\begin{array}{r} 56.314 \\ +\,17.780 \\ \hline 94 \end{array}$ Adding hundredths

$\begin{array}{r} \overset{1}{6}\ \\ 56.314 \\ +\,17.780 \\ \hline .094 \end{array}$ Adding tenths. We get 10 tenths = 1 one + 0 tenths, so we carry the 1 to the ones column. Writing a decimal point in the answer

$\begin{array}{r} 56.314 \\ +\,17.780 \\ \hline 4.094 \end{array}$ Adding ones. We get 14 ones = 1 ten + 4 ones, so we carry the 1 to the tens column.

$\begin{array}{r} 56.314 \\ +\,17.780 \\ \hline 74.094 \end{array}$ Adding tens

◀ Do Margin Exercises 1–2.

EXAMPLE 2 Add: 3.42 + 0.237 + 14.1.

$\begin{array}{r} 3.420 \\ 0.237 \\ +\,14.100 \\ \hline 17.757 \end{array}$ Lining up the decimal points and writing extra zeros; Adding

◀ Do Exercises 3–5.

Now we consider the addition 3456 + 19.347. Keep in mind that any whole number has an "unwritten" decimal point at the right that can be followed by zeros. For example, 3456 can also be written 3456.000. When adding, we can always write in the decimal point and extra zeros if desired.

EXAMPLE 3 Add: 3456 + 19.347.

```
     1
  3 4 5 6 . 0 0 0    Writing in the decimal point and extra zeros
+     1 9 . 3 4 7    Lining up the decimal points
  3 4 7 5 . 3 4 7    Adding
```

Do Exercises 6 and 7. ▶

b SUBTRACTION

Subtracting with decimal notation is similar to subtracting whole numbers. First, we line up the decimal points so that we can subtract corresponding place-value digits. Then we subtract digits from the right. For example, we subtract the thousandths, then the hundredths, the tenths, and so on, borrowing if necessary.

EXAMPLE 4 Subtract: 56.314 − 17.78.

```
  5 6 . 3 1 4    Lining up the decimal points in order to subtract
− 1 7 . 7 8 0    Writing an extra 0

  5 6 . 3 1 4
− 1 7 . 7 8 0
            4    Subtracting thousandths

        2 11
  5 6 . 3 1 4    Borrowing tenths to subtract hundredths
− 1 7 . 7 8 0
          3 4    Subtracting hundredths

       12
    5   2 11
  5 6 . 3 1 4    Borrowing ones to subtract tenths
− 1 7 . 7 8 0
        . 5 3 4  Subtracting tenths; writing a decimal point

    15 12
  4  5  2 11
  5 6 . 3 1 4    Borrowing tens to subtract ones
− 1 7 . 7 8 0
      8 . 5 3 4  Subtracting ones

    15 12
  4  5  2 11
  5 6 . 3 1 4
− 1 7 . 7 8 0
  3 8 . 5 3 4    Subtracting tens
```

Check by adding:

```
  1 1   1
  3 8 . 5 3 4
+ 1 7 . 7 8 0
  5 6 . 3 1 4  ←
```

The answer checks because this is the top number in the subtraction.

Do Exercises 8 and 9. ▶

Add.

6. 789 + 123.67

7. 45.78 + 2467 + 1.993

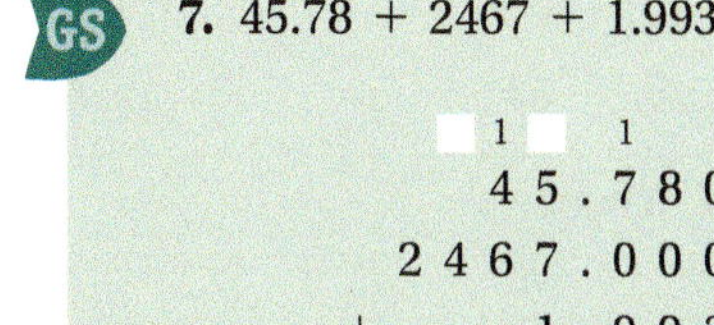

Subtract.

8. 37.428 − 26.674

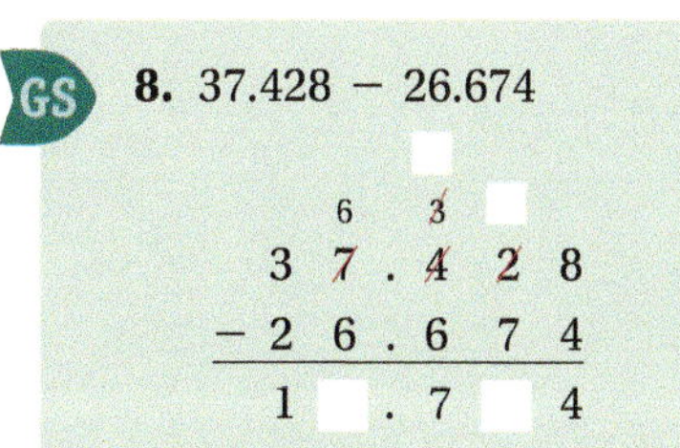

9.
```
  0 . 3 4 7
− 0 . 0 0 8
```

Answers

6. 912.67 **7.** 2514.773 **8.** 10.754 **9.** 0.339

Guided Solutions:
7. 1, 1; 5, 7 **8.** 13, 12; 0, 5

Subtract.

10. 2.9 − 0.36

11. 0.43 − 0.18762

12. 5.27 − 0.00008

Subtract.

13. 1277 − 82.78

14. 5 − 0.0089

```
        10
 5.0 0 0 0
−0.0 0 8 9
 4.     9 1
```

Add.

15. 7.42 + (−9.38)

16. −4.201 + 7.36

17. Add: −7.49 + (−5.8).

Answers

10. 2.54 **11.** 0.24238 **12.** 5.26992
13. 1194.22 **14.** 4.9911 **15.** −1.96
16. 3.159 **17.** −13.29

Guided Solution:
14. 4999; 9, 1

EXAMPLE 5 Subtract: 23.08 − 5.0053.

```
  1 13    7 9 10
  2 3.0 8 0 0      Writing two extra zeros to the right of the last digit
−    5.0 0 5 3
  1 8.0 7 4 7      Subtracting
```

◀ **Do Exercises 10–12.**

When subtraction involves an integer, the "unwritten" decimal point can be written in. Extra zeros can then be written in to the right of the decimal point.

EXAMPLE 6 Subtract: 456 − 2.467.

```
      5   9 9 10
  4 5 6.0 0 0      Writing in the decimal point and extra zeros
−     2.4 6 7
  4 5 3.5 3 3      Subtracting
```

◀ **Do Exercises 13 and 14.**

C ADDING AND SUBTRACTING WITH NEGATIVES

Negative decimals are added or subtracted like negative integers.

> To add a negative number and a positive number:
>
> **a)** Determine the sign of the number with the greater absolute value.
>
> **b)** Subtract the smaller absolute value from the larger one.
>
> **c)** The answer is the difference from part (b) with the sign from part (a).

EXAMPLE 7 Add: −13.82 + 4.69.

a) Since $|-13.82| > |4.69|$, the sign of the number with the greater absolute value is negative.

b)
```
     7 12
  1 3.8 2
−   4.6 9      Finding the difference of the absolute values
    9.1 3
```

c) The answer is negative: −13.82 + 4.69 = −9.13.

◀ **Do Exercises 15 and 16.**

> To add two negative numbers:
>
> **a)** Add the absolute values.
>
> **b)** Make the answer negative.

EXAMPLE 8 Add: −2.306 + (−3.125).

a)
```
        1
   2.3 0 6
+  3.1 2 5     |−2.306| = 2.306 and |−3.125| = 3.125
   5.4 3 1     Adding the absolute values
```

b) −2.306 + (−3.125) = −5.431   The answer is negative.

◀ **Do Exercise 17.**

To subtract, we add the opposite of the number being subtracted.

EXAMPLE 9 Subtract: $-3.1 - 4.8$.

$$-3.1 - 4.8 = -3.1 + (-4.8) \quad \text{Adding the opposite of 4.8}$$
$$= -7.9 \quad \text{The sum of two negative numbers is negative.}$$

EXAMPLE 10 Subtract: $-7.9 - (-8.5)$.

$$-7.9 - (-8.5) = -7.9 + 8.5 \quad \text{Adding the opposite of } -8.5$$
$$= 0.6 \quad \text{Subtracting absolute values. The answer is positive since 8.5 has the larger absolute value.}$$

Do Exercises 18–21.

Subtract.

18. $9.25 - 13.41$

19. $-5.72 - 4.19$

20. $9.8 - (-2.6)$

21. $-5.9 - (-3.2)$

d COMBINING LIKE TERMS

Recall that like, or similar, terms have exactly the same variable factors. To combine like terms, we add or subtract coefficients to form an equivalent expression.

EXAMPLE 11 Combine like terms: $3.2x + 4.6x$.

These are the coefficients.

$$3.2x + 4.6x = (3.2 + 4.6)x \quad \text{Using the distributive law—try to do this step mentally.}$$
$$= 7.8x \quad \text{Adding}$$

A similar procedure is used when subtracting like terms.

EXAMPLE 12 Combine like terms: $4.13a - 7.56a$.

$$4.13a - 7.56a = (4.13 - 7.56)a \quad \text{Using the distributive law}$$
$$= (4.13 + (-7.56))a \quad \text{Adding the opposite of 7.56}$$
$$= -3.43a \quad \text{Subtracting absolute values. The coefficient is negative since } |-7.56| > |4.13|.$$

When more than one pair of like terms is present, we can rearrange the terms and then simplify.

EXAMPLE 13 Combine like terms: $5.7x - 3.9y - 2.4x + 4.5y$.

$$5.7x - 3.9y - 2.4x + 4.5y$$
$$= 5.7x + (-3.9y) + (-2.4x) + 4.5y \quad \text{Rewriting as addition}$$
$$= 5.7x + (-2.4x) + (-3.9y) + 4.5y \quad \text{Using the commutative law to rearrange}$$
$$= 3.3x + 0.6y \quad \text{Combining like terms}$$

With practice, you will be able to perform many of the above steps mentally.

Do Exercises 22–24.

CALCULATOR CORNER

Addition and Subtraction with Decimal Notation To use a calculator to add and subtract with decimal notation, we use the [·], [+], [−], and [=] keys. To find $47.046 - 28.193$, for example, we press [4] [7] [·] [0] [4] [6] [−] [2] [8] [·] [1] [9] [3] [=]. The display reads [18.853], so $47.046 - 28.193 = 18.853$.

EXERCISES Use a calculator to perform the indicated operations.

1. $274.159 + 43.486$

2. $3.4 + 45 + 0.68$

3. $52.34 - 18.51$

4. $6.09 - 5.1$

5. $246 + (-3.07)$

6. $-12.7 - (-1.008)$

Combine like terms.

22. $5.8x - 2.1x$

23. $-5.9a + 7.6a$

24. $-4.8y + 7.5 + 2.1y - 2.1$

Answers

18. -4.16 **19.** -9.91 **20.** 12.4 **21.** -2.7 **22.** $3.7x$ **23.** $1.7a$ **24.** $-2.7y + 5.4$

5.2 Exercise Set

For Extra Help
MyMathLab®

MathXL®  PRACTICE  WATCH  READ  REVIEW

Reading Check

Complete each subtraction and its check by selecting a number from the list at the right.

RC1.
$$\begin{array}{r} 23.7 \\ -\ 1.876 \\ \hline \square \end{array}$$
Check:
$$\begin{array}{r} 21.824 \\ +\ 1.876 \\ \hline \square \end{array}$$

RC2.
$$\begin{array}{r} 187.623 \\ -\ 40.9 \\ \hline \square \end{array}$$
Check:
$$\begin{array}{r} 146.723 \\ +\ \square \\ \hline 187.623 \end{array}$$

23.7
187.623
21.824
40.9
1.876
146.723

a Add.

1. $\begin{array}{r} 426.25 \\ +\ 38.12 \\ \hline \end{array}$

2. $\begin{array}{r} 641.823 \\ +\ 14.915 \\ \hline \end{array}$

3. $\begin{array}{r} 659.403 \\ +\ 916.612 \\ \hline \end{array}$

4. $\begin{array}{r} 875.795 \\ +\ 324.862 \\ \hline \end{array}$

5. $\begin{array}{r} 9.104 \\ +\ 123.456 \\ \hline \end{array}$

6. $\begin{array}{r} 3.409 \\ +\ 81.001 \\ \hline \end{array}$

7. $2.006 + 5.817$

8. $0.8096 + 0.7856$

9. $20.7 + 30.0124$

10. $0.263 + 0.8$

11. $1.06 + 9$

12. $12 + 18.08$

13. $0.34 + 3.5 + 0.127 + 768$

14. $2.3 + 0.729 + 23$

15. $17 + 3.24 + 0.256 + 0.3689$

16. $\begin{array}{r} 47.8 \\ 219.852 \\ 43.59 \\ +\ 666.713 \\ \hline \end{array}$

17. $\begin{array}{r} 2.703 \\ 78.33 \\ 28.0009 \\ +\ 118.4341 \\ \hline \end{array}$

18. $\begin{array}{r} 13.72 \\ 9.112 \\ 6542.7908 \\ +\ 23.901 \\ \hline \end{array}$

b Subtract.

19. $\begin{array}{r} 47.596 \\ -\ 6.215 \\ \hline \end{array}$

20. $\begin{array}{r} 11.345 \\ -\ 2.105 \\ \hline \end{array}$

21. $\begin{array}{r} 51.31 \\ -\ 2.29 \\ \hline \end{array}$

22. $\begin{array}{r} 37.45 \\ -\ 6.32 \\ \hline \end{array}$

23. $\begin{array}{r} 3.6 \\ -\ 0.036 \\ \hline \end{array}$

24. $\begin{array}{r} 28.0 \\ -\ 0.28 \\ \hline \end{array}$

25. $\begin{array}{r} 92.341 \\ -\ 6.42 \\ \hline \end{array}$

26. $\begin{array}{r} 0.346 \\ -\ 0.0346 \\ \hline \end{array}$

27. $\begin{array}{r} 3.0074 \\ -\ 1.3408 \\ \hline \end{array}$

28. $\begin{array}{r} 32.7978 \\ -\ 0.0592 \\ \hline \end{array}$

29. $\begin{array}{r} 6.07 \\ -\ 2.0078 \\ \hline \end{array}$

30. $\begin{array}{r} 1.0 \\ -\ 0.9999 \\ \hline \end{array}$

31. $30.24 - 0.241$

32. $100.12 - 0.112$

33. $34.07 - 30.7$

34. $36.2 - 16.28$

35. $8.45 - 7.405$

36. $3.801 - 2.81$

37. $6.003 - 2.3$

38. $1 - 0.0098$

39. $2 - 1.0908$

40. $100 - 0.34$

41. $624 - 18.79$

42. $7.48 - 2.6$

43. $57.803 - 4.6$

44. $25.008 - 12.4$

45. $263.7 - 102.08$

46. $19 - 1.198$

47. $45 - 0.999$

48. $10.05 - 0.392$

c Add or subtract, as indicated.

49. $-5.02 + 1.73$

50. $-4.31 + 7.66$

51. $12.9 - 15.4$

52. $27.2 - 31.9$

53. $-2.9 + (-4.3)$

54. $-7.49 - 1.82$

55. $-4.301 + 7.68$

56. $-5.952 + 7.98$

57. $-12.9 - 3.7$

58. $-8.7 - 12.4$

59. $-2.1 - (-4.6)$

60. $-4.3 - (-2.5)$

61. $14.301 + (-17.82)$

62. $13.45 + (-18.701)$

63. $7.201 - (-2.4)$

64. $2.901 - (-5.7)$

65. $96.9 + (-21.4)$

66. $43.2 + (-10.9)$

67. $-3 - (-12.7)$

68. $-4.5 - (-7)$

69. $-4.9 - 5.392$

70. $89.3 - 100$

71. $14.7 - 15$

72. $-7.201 - 1.9$

d Combine like terms.

73. $1.8x + 3.9x$

74. $7.9x + 1.3x$

75. $17.59a - 12.73a$

76. $23.28a - 15.79a$

77. $15.2t + 7.9 + 5.9t$

78. $29.5t - 4.8 + 7.6t$

79. $5.217x - 8.134x$

80. $6.317t - 9.429t$

81. $4.906y - 7.1 + 3.2y$

82. $9.108y + 4.2 + 3.7y$

83. $4.8x + 1.9y - 5.7x + 1.2y$

84. $3.2r - 4.1t - 5.6t + 1.9r$

85. $4.9 - 3.9t - 6 - 4.5t$

86. $5 + 9.7x - 7.2 - 12.8x$

Skill Maintenance

Multiply. [3.4b]

87. $\frac{3}{5} \cdot \frac{4}{7}$

88. $\frac{2}{9} \cdot \frac{7}{5}$

89. $\frac{3}{10} \cdot \frac{21}{100}$

Evaluate. [2.6a]

90. $8 - 2x^2$, for $x = 3$

91. $5 - 3x^2$, for $x = -2$

92. $7 + 2x^2 \div 3$, for $x = 6$

Synthesis

Combine like terms.

93. 🖩 $-3.928 - 4.39a + 7.4b - 8.073 + 2.0001a - 9.931b - 9.8799a + 12.897b$

94. 🖩 $79.02x + 0.0093y - 53.14z - 0.02001y - 37.987z - 97.203x - 0.00987y$

95. 🖩 $39.123a - 42.458b - 72.457a + 31.462b - 59.491 + 37.927a$

96. Ryan presses the wrong key when using a calculator and adds 235.7 instead of subtracting it. The incorrect answer is 817.2. What is the correct answer?

97. Alicia presses the wrong key when using a calculator and subtracts 349.2 instead of adding it. The incorrect answer is -836.9. What is the correct answer?

98. 🖩 Find the errors, if any, in the balances in this checkbook.

20___ RECORD ALL CHARGES OR CREDITS THAT AFFECT YOUR ACCOUNT

DATE	CHECK NUMBER	TRANSACTION DESCRIPTION	√ T	(−) PAYMENT/ DEBIT	(+ OR −) OTHER	(+) DEPOSIT/ CREDIT	BALANCE FORWARD
							2767 73
8/16	432	Burch Laundry		23 56			2744 16
8/19	433	Rogers TV		20 49			2764 65
8/20		Deposit				85 00	2848 65
8/21	434	Galaxy Records		48 60			2801 05
8/22	435	Electric Works		267 95			2533 09

Find ***a***.

99.
$$\begin{array}{r} 93.\boldsymbol{a}43 \\ -\ 87.969 \\ \hline 5.274 \end{array}$$

100.
$$\begin{array}{r} 481.\boldsymbol{a}24 \\ -\ 72.978 \\ \hline 408.346 \end{array}$$

5.3 Multiplication of Decimals

OBJECTIVES

a Multiply using decimal notation.

b Convert from notation like 45.7 million to standard notation and convert between dollars and cents.

c Evaluate algebraic expressions using decimal notation.

SKILL TO REVIEW

Objective 1.4a: Multiply whole numbers.

Multiply.

1. $\begin{array}{r} 42 \\ \times\ 63 \\ \hline \end{array}$ **2.** $\begin{array}{r} 716 \\ \times\ \ 58 \\ \hline \end{array}$

a MULTIPLICATION

To develop an understanding of decimal multiplication, consider 2.3×1.12:

$$2.3 \times 1.12 = \frac{23}{10} \times \frac{112}{100} = \frac{2576}{1000} = 2.576.$$

Note that the number of decimal places in the product is the sum of the numbers of decimal places in the factors.

$$\begin{array}{rl} 1.12 & (2 \text{ decimal places}) \\ \times\ \ 2.3 & (1 \text{ decimal place}) \\ \hline 2.576 & (3 \text{ decimal places}) \end{array}$$

Now consider 0.02×3.412:

$$0.02 \times 3.412 = \frac{2}{100} \times \frac{3412}{1000} = \frac{6824}{100{,}000} = 0.06824.$$

Again, note that the number of decimal places in the product is the sum of the numbers of decimal places in the factors.

$$\begin{array}{rl} 3.412 & (3 \text{ decimal places}) \\ \times\ \ \ 0.02 & (2 \text{ decimal places}) \\ \hline 0.06824 & (5 \text{ decimal places}) \end{array}$$

The 0 after the decimal point is necessary.

To multiply using decimal notation: 0.8×0.43

a) Ignore the decimal points, for the moment, and multiply as though both factors were integers.

$$\begin{array}{r} \overset{2}{0.43} \\ \times\ \ 0.8 \\ \hline 344 \end{array}$$

Ignore the decimal points for now.

b) Place the decimal point in the result. The number of decimal places in the product is the sum of the numbers of places in the factors.

$$\begin{array}{rl} 0.43 & (2 \text{ decimal places}) \\ \times\ \ 0.8 & (1 \text{ decimal place}) \\ \hline 0.344 & (3 \text{ decimal places}) \end{array}$$

Count the number of decimal places by starting at the far right and moving the decimal point to the left.

EXAMPLE 1 Multiply: 8.3×74.6.

a) Ignore the decimal points and multiply as if both factors were integers.

$$\begin{array}{r} \overset{3\,4}{\overset{1\,1}{74.6}} \\ \times\ \ \ 8.3 \\ \hline 2238 \\ 59680 \\ \hline 61918 \end{array}$$

We are not yet finished.

Answers

Skill to Review:
1. 2646 **2.** 41,528

b) Place the decimal point in the result. The number of decimal places in the product is the sum, 1 + 1, of the numbers of decimal places in the factors.

```
     7 4.6     (1 decimal place)
×      8.3     (1 decimal place)
   2 2 3 8
 5 9 6 8 0
 6 1 9.1 8     (2 decimal places)
```

Do Exercise 1.

1. Multiply:

```
   7 6.3
×    8.2
```

EXAMPLE 2 Multiply: 0.0032×2148.

```
       2 1 4 8     (0 decimal places)
×  0.0 0 3 2       (4 decimal places)
       4 2 9 6
     6 4 4 4 0
     6.8 7 3 6     (4 decimal places)
```

EXAMPLE 3 Multiply: -0.104×0.86.

Multiplying the absolute values, we have

```
       0.8 6     (2 decimal places)
×    0.1 0 4     (3 decimal places)
       3 4 4
     8 6 0 0
 0.0 8 9 4 4     (5 decimal places)   We write an extra zero.
```

Since the product of a negative number and a positive number is negative, the answer is -0.08944.

Do Exercises 2–4.

Multiply.

2.

```
       4 2 1 3
×  0.0 0 5 1
```

GS 3. 42.65×0.804

```
       4 2.6 5
×      0.8 0 4
     1 ___ 0 6 0
 3 4 1 2 0 ___ 0
 3 4. ___ 9 0 ___ 0
```

4. $5.2014 \times (-2.41)$

Multiplying by 0.1, 0.01, 0.001, and So On

Now let's consider some special kinds of products. The first involves multiplying by a tenth, hundredth, thousandth, ten-thousandth, and so on. We can see a pattern in the following products.

$$0.1 \times 38 = \frac{1}{10} \times 38 = \frac{38}{10} = 3.8$$

$$0.01 \times 38 = \frac{1}{100} \times 38 = \frac{38}{100} = 0.38$$

$$0.001 \times 38 = \frac{1}{1000} \times 38 = \frac{38}{1000} = 0.038$$

$$0.0001 \times 38 = \frac{1}{10{,}000} \times 38 = \frac{38}{10{,}000} = 0.0038$$

Note in each case that the product is *smaller* than 38. That is, the decimal point in each product is farther to the left than the unwritten decimal point in 38. Note also that each product can be obtained from 38 by moving the decimal point.

Answers

1. 625.66 2. 21.4863 3. 34.2906 4. −12.535374

Guided Solution:

3. 7; 0; 2, 6

To multiply any number by 0.1, 0.01, 0.001, and so on:

a) count the number of decimal places in the tenth, hundredth, or thousandth, and so on, and

0.001×34.45678 → 3 places

b) move the decimal point in the other number that many places to the left. Use zeros as placeholders when necessary.

$0.001 \times 34.45678 = 0.034.45678$

Move 3 places to the left.

$0.001 \times 34.45678 = 0.03445678$

EXAMPLES Multiply.

4. $0.1 \times 45 = 4.5$ Moving the decimal point one place to the left

5. $0.01 \times 243.7 = 2.437$ Moving the decimal point two places to the left

6. $0.001 \times (-8.2) = -0.0082$ Moving the decimal point three places to the left. This requires writing two extra zeros.

7. $0.0001 \times 536.9 = 0.05369$ Moving the decimal point four places to the left. This requires writing one extra zero.

Multiply.

5. 0.1×746

6. 0.001×732.4

7. $(-0.01) \times 6.2$

8. 0.0001×723.6

Do Exercises 5–8.

Multiplying by 10, 100, 1000, and So On

Next we consider multiplying by 10, 100, 1000, and so on. We see a pattern in the following.

```
   5.2 3 7        5.2 3 7          5.2 3 7
 ×     1 0     ×    1 0 0     ×    1 0 0 0
   0 0 0 0        0 0 0 0          0 0 0 0
 5 2 3 7        0 0 0 0          0 0 0 0
 5 2.3 7 0    5 2 3 7          0 0 0 0
              5 2 3.7 0 0    5 2 3 7
                             5 2 3 7.0 0 0
```

Note in each case that the product is *larger* than 5.237. That is, the decimal point in each product is farther to the right than the decimal point in 5.237. Also, each product can be obtained from 5.237 by moving the decimal point.

To multiply any number by 10, 100, 1000, and so on:

a) count the number of zeros and

1000×34.45678 → 3 zeros

b) move the decimal point in the other number that many places to the right. Use zeros as placeholders when necessary.

$1000 \times 34.45678 = 34.456.78$

Move 3 places to the right.

$1000 \times 34{,}456.78$ — $1000 \times 34.45678 = 34{,}456.78$

CALCULATOR CORNER

Multiplication with Decimal Notation To use a calculator to multiply with decimal notation, we use the [·], [×], and [=] keys. To find 4.78 × 0.34, for example, we press [4] [·] [7] [8] [×] [·] [3] [4] [=]. The display reads [1.6252], so 4.78 × 0.34 = 1.6252.

EXERCISES Use a calculator to multiply.

1. 17.63×8.1

2. $0.04 \times (-12.69)$

3. 586.4×13.5

4. $-4.003 \times (-5.1)$

Answers

5. 74.6 6. 0.7324
7. −0.062 8. 0.07236

EXAMPLES Multiply.

8. $10 \times 32.98 = 329.8$ — Moving the decimal point one place to the right

9. $100 \times 4.7 = 470$ — Moving the decimal point two places to the right. The 0 in 470 is a placeholder.

10. $1000 \times (-2.4167) = -2416.7$ — Moving the decimal point three places to the right

11. $10{,}000 \times 7.52 = 75{,}200$ — Moving the decimal point four places to the right and using two zeros as placeholders

Do Exercises 9–12. ▶

Multiply.

9. 10×53.917

10. $100 \times (-62.417)$

11. 1000×83.9

12. $10{,}000 \times 57.04$

b APPLICATIONS USING MULTIPLICATION WITH DECIMAL NOTATION

Naming Large Numbers

We often see notation like the following in newspapers and magazines and on television and the Internet.

- In 2013, wildfires burned over 4.3 million acres in the United States.
  Source: National Interagency Fire Center
- The number of valid U.S. passports in circulation in 2011 exceeded 113.4 million.
  Source: U.S. State Department
- Each day, about 144.8 billion e-mails are sent worldwide.
  Source: Mashable.com
- At one point in 2014, the U.S. national debt was approximately \$17.478 trillion. The national debt has increased, on average, by \$3.81 billion per day since 2007.
  Source: www.usdebtclock.org

To understand such notation, consider the information in the following table.

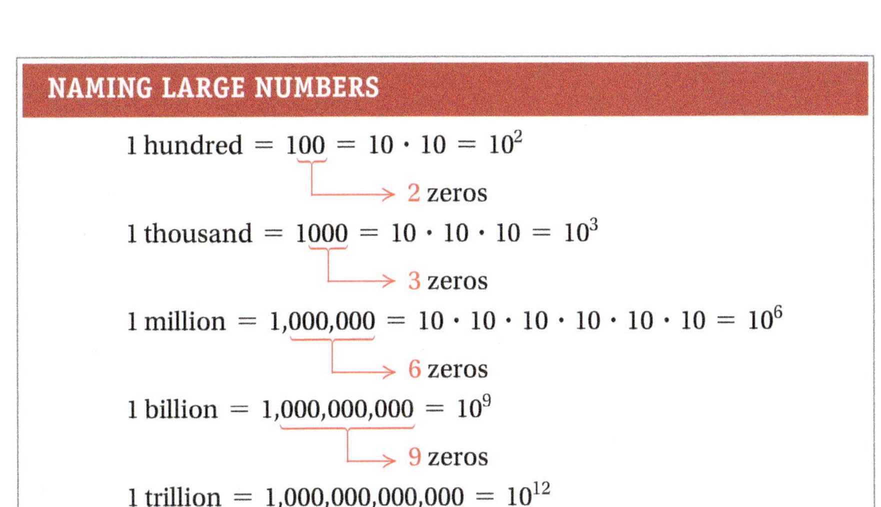

NAMING LARGE NUMBERS

1 hundred $= 100 = 10 \cdot 10 = 10^2$ → 2 zeros

1 thousand $= 1000 = 10 \cdot 10 \cdot 10 = 10^3$ → 3 zeros

1 million $= 1{,}000{,}000 = 10 \cdot 10 \cdot 10 \cdot 10 \cdot 10 \cdot 10 = 10^6$ → 6 zeros

1 billion $= 1{,}000{,}000{,}000 = 10^9$ → 9 zeros

1 trillion $= 1{,}000{,}000{,}000{,}000 = 10^{12}$ → 12 zeros

Answers

9. 539.17 **10.** −6241.7
11. 83,900 **12.** 570,400

To convert a large number to standard notation, we proceed as follows.

EXAMPLE 12 *Text Messages.* Worldwide, 8.6 trillion text messages are sent each year. Convert 8.6 trillion to standard notation.

Source: CNNTech, Portio Research

$$\begin{aligned} 8.6 \text{ trillion} &= 8.6 \times 1 \text{ trillion} \\ &= 8.6 \times \underbrace{1{,}000{,}000{,}000{,}000}_{12 \text{ zeros}} \\ &= 8{,}600{,}000{,}000{,}000 \quad \text{Moving the decimal point 12 places to the right} \end{aligned}$$

◀ Do Exercises 13 and 14.

Convert the number in each sentence to standard notation.

13. The largest building in the world is the Pentagon, which has 3.7 million square feet of floor space.

14. *Text Messages.* In the United States, 2.2 trillion text messages are sent each year.
Source: CNNTech, Forrester Research

Money Conversion

Converting from dollars to cents is like multiplying by 100. To see why, consider \$19.43.

$$\begin{aligned} \$19.43 &= 19.43 \times \$1 && \text{We think of \$19.43 as } 19.43 \times 1 \text{ dollar, or } 19.43 \times \$1. \\ &= 19.43 \times 100¢ && \text{Substituting 100¢ for \$1: \$1} = 100¢ \\ &= 1943¢ && \text{Multiplying} \end{aligned}$$

DOLLARS TO CENTS

To convert from dollars to cents, move the decimal point two places to the right and change the \$ sign in front to a ¢ sign at the end.

EXAMPLES Convert from dollars to cents.

13. \$189.64 = 18,964¢

14. \$0.75 = 75¢

◀ Do Exercises 15 and 16.

Convert from dollars to cents.

15. \$15.69 GS

\$15.69 = 15.69 × \$1
= 15.69 × ____ ¢
= ____ ¢

16. \$0.17

Converting from cents to dollars is like multiplying by 0.01. To see why, consider 65¢.

$$\begin{aligned} 65¢ &= 65 \times 1¢ && \text{We think of 65¢ as } 65 \times 1 \text{ cent, or } 65 \times 1¢. \\ &= 65 \times \$0.01 && \text{Substituting \$0.01 for 1¢: 1¢} = \$0.01 \\ &= \$0.65 && \text{Multiplying} \end{aligned}$$

CENTS TO DOLLARS

To convert from cents to dollars, move the decimal point two places to the left and change the ¢ sign at the end to a \$ sign in front.

EXAMPLES Convert from cents to dollars.

15. 395¢ = \$3.95

16. 8503¢ = \$85.03

◀ Do Exercises 17 and 18.

Convert from cents to dollars.

17. 35¢

18. 577¢

Answers

13. 3,700,000 **14.** 2,200,000,000,000
15. 1569¢ **16.** 17¢ **17.** \$0.35
18. \$5.77

Guided Solution:
15. 100, 1569

C EVALUATING

Algebraic expressions are often evaluated using numbers written in decimal notation.

EXAMPLE 17 Evaluate Prt for $P = 780$, $r = 0.12$, and $t = 0.5$.

This product can be used to determine the interest paid on \$780, borrowed at 12 percent simple interest, for half a year. We substitute as follows.

$$Prt = 780 \cdot 0.12 \cdot 0.5 = 780 \cdot 0.06 = 46.8$$ This would represent \$46.80.

Do Exercise 19. ▶

19. Evaluate lwh for $l = 3.2$, $w = 2.6$, and $h = 0.8$. (This is the formula for the volume of a rectangular box.)

EXAMPLE 18 Find the perimeter of a stamp that is 3.25 cm long and 2.5 cm wide.

Recall that the perimeter, P, of a rectangle of length l and width w is given by the formula

$$P = 2l + 2w.$$

Thus, we evaluate $2l + 2w$ for $l = 3.25$ and $w = 2.5$.

$$\begin{aligned} 2l + 2w &= 2 \cdot 3.25 + 2 \cdot 2.5 \\ &= 6.5 + 5.0 \\ &= 11.5 \end{aligned}$$

Remember the rules for order of operations.

The perimeter is 11.5 cm.

Do Exercise 20. ▶

20. Find the area of the stamp in Example 18.

EXAMPLE 19 *Multiple Births.* The expression $2.67t + 65.02$ can be used to predict the number of twin births, in thousands, in the United States t years after 1980. Predict the number of twin births in 2015.

Source: Based on information from the Centers for Disease Control

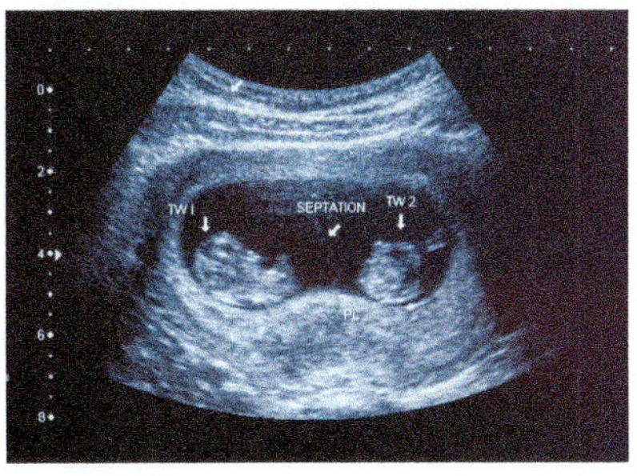

2015 is 35 years after 1980, so we evaluate $2.67t + 65.02$ for $t = 35$.

$$\begin{aligned} 2.67t + 65.02 &= 2.67 \cdot 35 + 65.02 \\ &= 93.45 + 65.02 \\ &= 158.47 \end{aligned}$$

In 2015, there will be approximately 158.47 thousand, or 158,470, twin births.

Do Exercise 21. ▶

21. Evaluate $6.28rh + 3.14r^2$ for $r = 1.5$ and $h = 5.1$. (This is the formula for the surface area of an open can.)

Answers

19. 6.656 **20.** 8.125 sq cm **21.** 55.107

5.3 Exercise Set

For Extra Help
MyMathLab® MathXL® PRACTICE WATCH READ REVIEW

Reading Check

Match each expression with an equivalent expression from the list at the right.

RC1. 0.001×38 ______ **a)** 3800¢

RC2. 1000×38 ______ **b)** 380

RC3. 38¢ ______ **c)** 0.038

RC4. \$38 ______ **d)** \$3.80

RC5. 380¢ ______ **e)** 38,000

RC6. 10×38 ______ **f)** \$0.38

a Multiply.

1. 6.8×7

2. 5.7×0.9

3. 0.84×8

4. 7.3×0.6

5. 6.3×0.04

6. 7.8×0.09

7. 87×0.006

8. 25.9×0.005

9. 10×42.63

10. 100×2.8793

11. -1000×783.686852

12. -0.34×1000

13. -7.8×100

14. $0.00238 \times (-10)$

15. 0.1×79.18

16. 0.01×789.235

17. 0.001×97.68

18. 8976.23×0.001

19. $28.7 \times (-0.01)$

20. $0.0325 \times (-0.1)$

21. 2.73×16

22. 8.27×5.4

23. 0.984×0.031

24. 7.489×1.7

25. $(-37.4)(-2.4)$

26. $569(-1.05)$

27. $749(-0.43)$

28. $(-0.876)(-0.0204)$

29. 0.87×64

30. 7.25×60

31. 46.50×75

32. 8.24×703

33. $(-0.231)(-0.5)$

34. $(-12.3)(-1.08)$

35. $9.42 \times (-1000)$

36. -7.6×1000

37. $-95.3 \times (-0.0001)$

38. $-4.23 \times (-0.001)$

 Convert from dollars to cents.

39. $57.06

40. $49.85

41. $0.95

42. $0.49

43. $0.01

44. $0.09

Convert from cents to dollars.

45. 72¢

46. 52¢

47. 2¢

48. 5¢

49. 6399¢

50. 5238¢

51. ***Farming Area.*** China has 3.48 million hectares of land devoted to farming. This area is approximately 24% of China's total area. Convert 3.48 million to standard notation.

Source: Viking Cruises

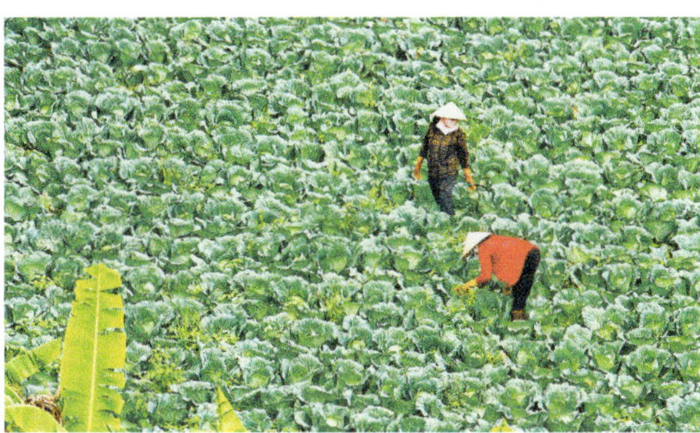

52. ***Doll Sales.*** In the United States, spending on dolls totaled $2.7 billion in 2011. Convert 2.7 billion to standard notation.

Source: The NPD Group

53. ***Spending on Pets.*** In 2011, Americans spent approximately $50.96 billion on their pets, of which $13.4 billion was for veterinary care. Convert 50.96 billion and 13.41 billion to standard notation.

Source: American Pet Association

54. ***Library of Congress.*** The Library of Congress is the largest public library in the United States. It holds about 33.5 million books. Convert 33.5 million to standard notation.

Source: American Library Association

55. ***Safe Water.*** Worldwide, about 2.2 million people die each year because of diseases caused by unsafe water. Convert 2.2 million to standard notation.

Source: United Nations Environment Programme

56. ***Overdraft Charges.*** Revenue of U.S. banks from checking account overdraft charges rose to $31.5 billion in the fiscal year ending in June 2012. Convert 31.5 billion to standard notation.

Sources: Moebs Services; abcNews, *Consumer Report*, September 27, 2012

c Evaluate.

57. $P + Prt$, for $P = 10{,}000$, $r = 0.04$, and $t = 2.5$
(*A formula for adding interest*)

58. $6.28r(h + r)$, for $r = 10$ and $h = 17.2$
(*Surface area of a cylinder*)

59. $vt + 0.5at^2$, for $v = 10$, $t = 1.5$, and $a = 9.8$
(*A physics formula*)

60. $4lh + 2h^2$, for $l = 3.5$ and $h = 1.2$
(*Surface area of a rectangular prism*)

Find **(a)** the perimeter and **(b)** the area of a rectangular room with the given dimensions.

61. 12.5 ft long, 9.5 ft wide

62. 10.25 ft long, 8 ft wide

63. 8.4 m wide, 10.5 m long

64. 8.2 yd long, 6.4 yd wide

Nursing. The expression $0.0375t + 2.2$ can be used to predict the number of registered nurses, in millions, in the United States t years after 2000. Predict the number of registered nurses in the United States in the year indicated.

Source: Based on information from the Bureau of Labor Statistics, U.S. Dept. of Labor

65. 2015

66. 2020

Skill Maintenance

Divide.

67. $-162 \div 6$ [2.5a]

68. $-216 \div (-6)$ [2.5a]

69. $-1035 \div (-15)$ [2.5a]

70. $-423 \div 3$ [2.5a]

71. $17\overline{)20,006}$ [1.5a]

72. $675 \div (-25)$ [2.5a]

Synthesis

Consider the following names for large numbers in addition to those already discussed in this section:

$$1 \text{ quadrillion} = 1{,}000{,}000{,}000{,}000{,}000 = 10^{15};$$
$$1 \text{ quintillion} = 1{,}000{,}000{,}000{,}000{,}000{,}000 = 10^{18};$$
$$1 \text{ sextillion} = 1{,}000{,}000{,}000{,}000{,}000{,}000{,}000 = 10^{21};$$
$$1 \text{ septillion} = 1{,}000{,}000{,}000{,}000{,}000{,}000{,}000{,}000 = 10^{24}.$$

Find each of the following. Express the answer with a name that is a power of 10.

73. (1 trillion) · (1 billion)

74. (1 million) · (1 billion)

75. (1 trillion) · (1 trillion)

76. Is a billion millions the same as a million billions? Explain.

77. In Great Britain, France, and Germany, a billion means a million millions. Write standard notation for the British number 6.6 billion.

78. One light-year (LY) is 9.46×10^{12} km. The star Regulus is 85 LY from the earth. How many billions of kilometers (km) from the earth is Regulus?

Source: *The Cambridge Factfinder*, 4th ed

Evaluate using a calculator.

79. $d + vt + at^2$, for $d = 79.2$, $v = 3.029$, $t = 7.355$, and $a = 4.9$ (*A physics formula for distance traveled*)

80. $0.5(b_1 + b_2)h$, for $b_1 = 9.7$ cm, $b_2 = 13.4$ cm, and $h = 6.32$ cm (*A geometry formula for the area of a trapezoid*)

81. *Electric Bills* Recently, electric bills from the Central Vermont Public Service Corporation consisted of a "customer charge" of \$0.374 per day plus an "energy charge" of \$0.1174 per kilowatt-hour (kWh) for the first 250 kWh used and \$0.09079 per kilowatt-hour for each kilowatt-hour in excess of 250. From April 20 to May 20, the Coy-Bergers used 480 kWh of electricity. What was their bill for the period?

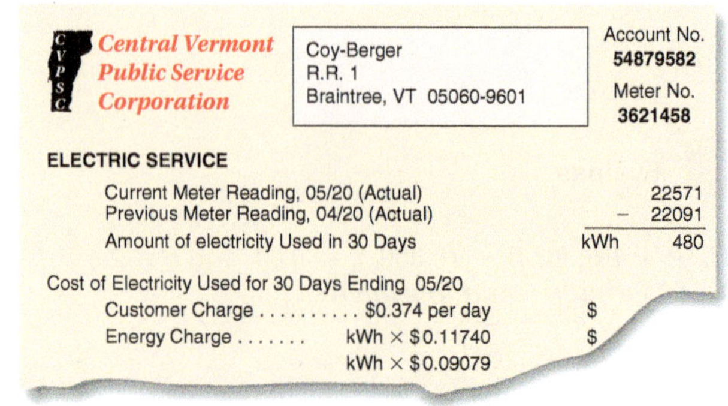

Central Vermont Public Service Corporation

Coy-Berger
R.R. 1
Braintree, VT 05060-9601

Account No. 54879582
Meter No. 3621458

ELECTRIC SERVICE

Current Meter Reading, 05/20 (Actual)		22571
Previous Meter Reading, 04/20 (Actual)		− 22091
Amount of electricity Used in 30 Days	kWh	480

Cost of Electricity Used for 30 Days Ending 05/20

Customer Charge	\$0.374 per day	\$
Energy Charge	kWh × \$0.11740	\$
	kWh × \$0.09079	

Division of Decimals

5.4

OBJECTIVES

a Divide using decimal notation.

b Simplify expressions using the rules for order of operations.

SKILL TO REVIEW

Objective 1.5a: Divide whole numbers.

Divide.

1. $5\overline{)245}$

2. $23\overline{)1978}$

a DIVISION

Whole-Number Divisors

Now that we have studied multiplication of decimals, we can develop a procedure for division. The following divisions are justified by the multiplication in each *check*:

This is the dividend. → 651; This is the divisor. → 7; This is the quotient. → 93

$$\frac{651}{7} = 93 \qquad \textit{Check}: 7 \cdot 93 = 651$$

$$\frac{65.1}{7} = 9.3 \qquad \textit{Check}: 7 \cdot 9.3 = 65.1$$

$$\frac{6.51}{7} = 0.93 \qquad \textit{Check}: 7 \cdot 0.93 = 6.51$$

$$\frac{0.651}{7} = 0.093 \qquad \textit{Check}: 7 \cdot 0.093 = 0.651$$

Note that the number of decimal places in each quotient is the same as the number of decimal places in the dividend.

> To perform long division of a decimal number by a whole number:
>
> **a)** place the decimal point directly above the decimal point in the dividend, and
>
> **b)** divide as though dividing whole numbers.
>
> $$\begin{array}{r} 0.84 \leftarrow \text{Quotient} \\ 7\overline{)5.88} \leftarrow \text{Dividend} \\ \underline{5\,6\,0} \\ 2\,8 \\ \underline{2\,8} \\ 0 \leftarrow \text{Remainder} \end{array}$$
>
> ↑ Divisor (7)

EXAMPLE 1 Divide: $82.08 \div 24$.

Place the decimal point.

$$\begin{array}{r} 3.42 \\ 24\overline{)82.08} \\ \underline{72} \\ 10\,0 \\ \underline{9\,6} \\ 48 \\ \underline{48} \\ 0 \end{array}$$

Divide as though dividing whole numbers.

Since $(3.42)(24) = 82.08$, the answer checks.

Do Margin Exercises 1–3. ▶

Divide.

1. $9\overline{)5.4}$

2. $15\overline{)22.5}$

3. $82\overline{)38.54}$

Answers

Skill to Review:
1. 49 **2.** 86

Margin Exercises:
1. 0.6 **2.** 1.5 **3.** 0.47

CALCULATOR CORNER

Division with Decimal Notation To use a calculator to divide with decimal notation, we use the [.], [÷], and [=] keys. To find 237.12 ÷ 5.2, for example, we press [2] [3] [7] [.] [1] [2] [÷] [5] [.] [2] [=]. The display reads [45.6], so 237.12 ÷ 5.2 = 45.6.

EXERCISES Use a calculator to divide.

1. $12.4\overline{)177.32}$
2. $49\overline{)125.44}$
3. $-1.6 \div 25$
4. $518.472 \div 6.84$

We can think of a whole-number dividend as having a decimal point at the end with as many 0's as we wish after the decimal point. For example, 12 = 12. = 12.0 = 12.00 = 12.000, and so on. We can also add 0's after the last digit in the decimal portion of a number: 3.6 = 3.60 = 3.600, and so on.

EXAMPLE 2 Divide: 30 ÷ 8.

$$\begin{array}{r} 3.\\ 8\overline{)30.}\\ \underline{24}\\ 6 \end{array}$$

Place the decimal point and divide to find how many ones.

$$\begin{array}{r} 3.\\ 8\overline{)30.0}\\ \underline{24}\\ 60 \end{array}$$

Write an extra zero to the right of the decimal point. This does not change the number.

$$\begin{array}{r} 3.7\\ 8\overline{)30.0}\\ \underline{24}\\ 60\\ \underline{56}\\ 4 \end{array}$$

Divide to find how many tenths.

$$\begin{array}{r} 3.7\\ 8\overline{)30.00}\\ \underline{24}\\ 60\\ \underline{56}\\ 40 \end{array}$$

Write another zero.

$$\begin{array}{r} 3.75\\ 8\overline{)30.00}\\ \underline{24}\\ 60\\ \underline{56}\\ 40\\ \underline{40}\\ 0 \end{array}$$

Divide to find how many hundredths.

Check:
$$\begin{array}{r} \overset{6\ 4}{3.75}\\ \underline{\times\quad 8}\\ 30.00 \end{array}$$

Divide.

4. $25\overline{)8}$

5. $-23 \div 4$

6. 2.15 ÷ 86

$$\begin{array}{r} 0.\square 2\square\\ 86\overline{)2.150}\\ \underline{172}\\ 43\square\\ \underline{430}\\ 0 \end{array}$$

EXAMPLE 3 Divide: −4.5 ÷ 250.

We first consider 4.5 ÷ 250.

$$\begin{array}{r} 0.018\\ 250\overline{)4.500}\\ \underline{250}\\ 2000\\ \underline{2000}\\ 0 \end{array}$$

← Since the remainder is 0, we are finished.

Since a negative number divided by a positive number is negative, the answer is −0.018. To check, note that (−0.018)(250) = −4.5.

Do Exercises 4–6.

Answers

4. 0.32 5. −5.75 6. 0.025

Guided Solution:

6. 0, 5; 0

Divisors That Are Not Whole Numbers

Consider the division

$0.24\overline{)8.208}$

We write the division as $\frac{8.208}{0.24}$. Then we multiply by 1 to change to a whole-number divisor:

$$\frac{8.208}{0.24} = \frac{8.208}{0.24} \times \frac{100}{100} = \frac{820.8}{24}.$$

The division $0.24\overline{)8.208}$ is the same as $24\overline{)820.8}$.

The divisor is now a whole number.

To divide when the divisor is not a whole number:

a) move the decimal point (multiply by 10, 100, and so on) to make the divisor a whole number,

$0.24\overline{)8.208}$

Move 2 places to the right.

b) move the decimal point in the dividend the same number of places (multiply the same way), and

$0.24\overline{)8.208}$

Move 2 places to the right.

c) place the decimal point for the answer directly above the new decimal point in the dividend and divide as though dividing whole numbers.

```
          3 4.2
0.2 4 ) 8.2 0^8
        7 2
        1 0 0
          9 6
            4 8
            4 8
              0
```

(The new decimal point in the dividend is indicated by a caret.)

EXAMPLE 4 Divide: $5.848 \div 8.6$.

$8.6\overline{)5.848}$

Multiply the divisor by 10. (Move the decimal point 1 place.) Multiply the same way in the dividend. (Move 1 place.)

```
       0.6 8
8.6 ) 5.8^4 8
      5 1 6
        6 8 8
        6 8 8
            0
```

Place a decimal point above the new decimal point in the dividend and then divide.

Note: $\frac{5.848}{8.6} = \frac{5.848}{8.6} \cdot \frac{10}{10} = \frac{58.48}{86}$.

Check: $(0.68)(8.6) = 5.848$

Do Exercises 7–9.

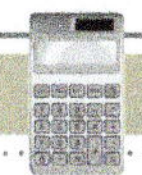

CALCULATOR CORNER

Finding Remainders We can find a whole-number remainder by multiplying the decimal portion of a quotient by the divisor. For example, to find the quotient and the whole-number remainder for $567 \div 13$, we can use a calculator to find that

$567 \div 13 \approx 43.61538462.$

To isolate the portion to the right of the decimal point, we can subtract 43. When the decimal part of the quotient is multiplied by the divisor, we have

$0.61538462 \times 13 = 8.00000006.$

The rounding error in the result may vary, depending on the calculator used. We see that $567 \div 13 = 43$ R 8.

EXERCISES Find the quotient and the whole-number remainder for each of the following.

1. $478 \div 17$
2. $815 \div 7$
3. $824 \div 11$
4. $7888 \div 19$

Divide.

7. $0.375 \div 0.25$

$$\frac{0.375}{0.25} = \frac{0.375}{0.25} \times \frac{\square}{100}$$

$$= \frac{37.5}{\square}$$

```
             1.□
0.2 5 ) 0.3 7^5
          2 5
          1 2 □
          1 2 5
              0
```

8. $4.067 \div (-0.83)$

9. $-44.8 \div (-3.5)$

Answers

7. 1.5 **8.** −4.9 **9.** 12.8

Guided Solution:

7. 100, 25; 5, 5

EXAMPLE 5 Divide: $-12 \div (-0.64)$.

Note first that a negative number divided by a negative number is positive. To find the quotient, we consider $12 \div 0.64$.

$0.64\overline{)12.}$ — Place a decimal point at the end of the whole number.

$0.64\overline{)12.00}$ — Multiply the divisor by 100. (Move the decimal point 2 places.) Multiply the same way in the dividend. (Move 2 places after adding extra zeros.)

$$
\begin{array}{r}
18.75 \\
0.64\overline{)12.00{}_\wedge00} \\
\underline{64} \\
560 \\
\underline{512} \\
480 \\
\underline{448} \\
320 \\
\underline{320} \\
0
\end{array}
$$

Place a decimal point above the new decimal point in the dividend and then divide.

We have $-12 \div (-0.64) = 18.75$.

Divide.

10. $1.6\overline{)2.5}$

11. $-9 \div 0.03$

Do Exercises 10 and 11.

Dividing by 10, 100, 1000, and So On

We can divide quickly by a ten, hundred, or thousand, or by a tenth, hundredth, or thousandth. Each procedure we use is based on multiplying by 1. Consider the following example:

$$\frac{23.789}{1000} = \frac{23.789}{1000} \cdot \frac{0.001}{0.001} = \frac{0.023789}{1} = 0.023789.$$

We are dividing by a number greater than 1: The result is *smaller* than 23.789.

> To divide by 10, 100, 1000, and so on,
>
> **a)** count the number of zeros in the divisor, and $\frac{713.49}{100}$ 2 zeros
>
> **b)** write the quotient by moving the decimal point in the dividend that number of places to the left.
>
> $$\frac{713.49}{100} = \frac{713.49}{100.} = \frac{7.1349}{1.00} = 7.1349$$
>
> 2 places to the left to change 100 to 1

EXAMPLE 6 Divide: $\frac{0.0104}{10}$.

$$\frac{0.0104}{10} = \frac{0.0104}{10.} = \frac{0.00104}{1.0} = 0.00104$$

1 zero — 1 place to the left to change 10 to 1

Answers

10. 1.5625 11. −300

EXAMPLE 7 Divide −213.75 by 100.

$$\frac{-213.75}{100} = \frac{-213.75}{100.} = \frac{-2.1375}{1.00} = -2.1375$$

2 zeros   2 places to the left

The answer is −2.1375.

Do Exercises 12 and 13. ▶

Divide.

12. $\frac{0.176}{100}$

13. $\frac{-98.47}{1000}$

Dividing by 0.1, 0.01, 0.001, and So On

Now consider the following example:

$$\frac{23.789}{0.01} = \frac{23.789}{0.01} \cdot \frac{100}{100} = \frac{2378.9}{1} = 2378.9.$$

We are dividing by a number less than 1: The result is *larger* than 23.789.

> To divide by 0.1, 0.01, 0.001, and so on,
>
> **a)** count the number of decimal places in the divisor, and
>
> $\frac{713.49}{0.001}$ 3 places
>
> **b)** write the quotient by moving the decimal point in the dividend that number of places to the right.
>
> $$\frac{713.49}{0.001} = \frac{713.49}{0.001} = \frac{713{,}490}{1} = 713{,}490$$
>
> 3 places to the right to change 0.001 to 1

EXAMPLE 8 Divide: $\frac{67.8}{0.1}$.

$$\frac{67.8}{0.1} = \frac{67.8}{0.1} = \frac{678}{1} = 678$$

1 place   1 place to the right to change 0.1 to 1

The answer is 678.

Do Exercises 14 and 15. ▶

Divide.

14. $\frac{-6.7832}{0.1}$

15. $\frac{12.78}{0.01}$

b ORDER OF OPERATIONS: DECIMAL NOTATION

The rules for order of operations apply when simplifying expressions involving decimal notation.

> **RULES FOR ORDER OF OPERATIONS**
>
> 1. Do all calculations within grouping symbols first.
> 2. Evaluate all exponential expressions.
> 3. Do all multiplications and divisions in order from left to right.
> 4. Do all additions and subtractions in order from left to right.

Answers

12. 0.00176 13. −0.09847 14. −67.832
15. 1278

EXAMPLE 9 Simplify: $2.5 \times 25 \div 25{,}000 \times 250$.

$$\begin{aligned} 2.5 \times 25 \div 25{,}000 \times 250 &= 62.5 \div 25{,}000 \times 250 \\ &= 0.0025 \times 250 \\ &= 0.625. \end{aligned}$$

Doing all multiplications and divisions in order from left to right

EXAMPLE 10 Simplify: $(5 - 0.06) \div 2 + 3.42 \times 0.1$.

$$(5 - 0.06) \div 2 + 3.42 \times 0.1 = 4.94 \div 2 + 3.42 \times 0.1$$ Working inside the parentheses

$$= 2.47 + 0.342$$ Multiplying and dividing in order from left to right

$$= 2.812$$

Simplify.

16. $625 \div 62.5 \times 25 \div 6250$

$= \square \times 25 \div 6250$

$= \square \div 6250$

$= 0.04$

17. $0.25 \cdot (1 + 0.08) - 0.0274$

18. $[(19.7 - 17.2)^2 + 3] \div (-1.25)$

EXAMPLE 11 Simplify: $13 - [5.4(1.3^2 + 0.21) \div 0.6]$.

$$\begin{aligned} &13 - [5.4(1.3^2 + 0.21) \div 0.6] \\ &= 13 - [5.4(1.69 + 0.21) \div 0.6] \\ &= 13 - [5.4 \times 1.9 \div 0.6] \end{aligned}$$ Working in the innermost parentheses first

$$= 13 - [10.26 \div 0.6]$$ Multiplying

$$= 13 - 17.1$$ Dividing

$$= -4.1$$

Do Exercises 16–18.

EXAMPLE 12 *Population Density.* The table below shows the number of residents per square mile in the six New England states. Find the average number of residents per square mile for this group of states.

Source: 2010 U.S. Census

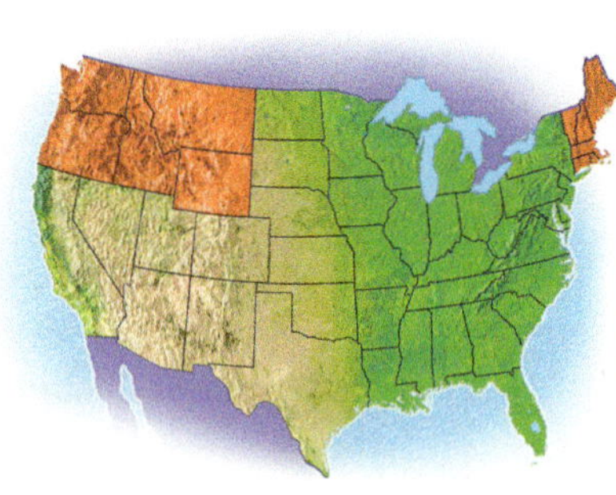

STATE	NUMBER OF RESIDENTS PER SQUARE MILE
Maine	43.1
New Hampshire	147.0
Vermont	67.9
Massachusetts	839.4
Rhode Island	1018.1
Connecticut	738.1

The **average** of a set of numbers is the sum of the numbers divided by the number of addends. We find the sum of the population densities per square mile and divide it by the number of addends, 6:

$$\frac{43.1 + 147 + 67.9 + 839.4 + 1018.1 + 738.1}{6} = \frac{2853.6}{6} = 475.6.$$

Thus the average number of residents per square mile for these six states is 475.6.

Do Exercise 19.

19. *Population Density.* The table below shows the number of residents per square mile in five northwestern states. Find the average number of residents per square mile for this group of states.

Source: 2010 U.S. Census

STATE	RESIDENTS PER SQUARE MILE
Washington	101.2
Oregon	39.9
Idaho	19.0
Montana	6.8
Wyoming	5.8

Answers

16. 0.04 **17.** 0.2426 **18.** −7.4 **19.** 34.54

Guided Solution:

16. 10, 250

5.4 Exercise Set

For Extra Help MyMathLab® MathXL® PRACTICE WATCH READ REVIEW

Reading Check

Name the operation that should be performed first in evaluating each expression. Do not calculate.

RC1. $(2 - 0.04) \div 4 + 8.5$ ________________

RC2. $0.02 + 2.06 \div 0.01$ ________________

RC3. $5 \times 2.1 + 0.1 - 8^3$ ________________

RC4. $18.2 - (4.1 + 6.9)$ ________________

RC5. $16 - 9 \div 3 + 7.3$ ________________

RC6. $4(10 - 5) \times 14.2$ ________________

a Divide.

1. $2\overline{)5.98}$

2. $5\overline{)13.5}$

3. $4\overline{)95.12}$

4. $8\overline{)25.92}$

5. $12\overline{)84.96}$

6. $23\overline{)25.07}$

7. $15\overline{)18}$

8. $30\overline{)54}$

9. $5.4 \div (-6)$

10. $3.6 \div (-4)$

11. $-30 \div 0.005$

12. $-100 \div 0.0002$

13. $0.06\overline{)8.4}$

14. $0.04\overline{)1.68}$

15. $2.6\overline{)104}$

16. $3.2\overline{)192}$

17. $1.8 \div (-12)$

18. $6 \div (-15)$

19. $8.5\overline{)27.2}$

20. $6.2\overline{)46.5}$

21. $-31.59 \div 8.1$

22. $-39.06 \div 4.2$

23. $-5 \div (-8)$

24. $-7 \div (-8)$

25. $0.47\overline{)0.1222}$

26. $0.54\overline{)0.27}$

27. $0.032\overline{)0.07488}$

28. $0.017\overline{)1.581}$

29. $-24.969 \div 82$

30. $-25.221 \div 42$

31. $\dfrac{213.4567}{100}$

32. $\dfrac{769.3265}{1000}$

33. $\dfrac{-23.59}{10}$

34. $\dfrac{-83.57}{10}$

35. $\dfrac{1.0237}{0.001}$

36. $\dfrac{3.4029}{0.001}$

37. $\dfrac{-92.36}{0.01}$

38. $\dfrac{-56.78}{0.001}$

39. $\dfrac{0.8172}{10}$

40. $\dfrac{0.5678}{1000}$

41. $\dfrac{0.97}{0.1}$

42. $\dfrac{0.97}{0.001}$

43. $\dfrac{52.7}{-1000}$

44. $\dfrac{8.9}{-100}$

45. $\dfrac{75.3}{-0.001}$

46. $\dfrac{63.47}{-0.1}$

47. $\dfrac{-75.3}{1000}$

48. $\dfrac{23{,}001}{100}$

b Simplify.

49. $14 \times (82.6 + 67.9)$

50. $(26.2 - 14.8) \times 12$

51. $0.003 + 3.03 \div (-0.01)$

52. $42 \times (10.6 + 0.024)$

53. $(4.9 - 18.6) \times 13$

54. $4.2 \times 5.7 + 0.7 \div 3.5$

55. $210.3 - 4.24 \times 1.01$

56. $-7.32 + 0.04 \div 0.1^2$

57. $0.04 \times 0.1 \div 0.4 \times 50$

58. $30 \div 0.2 \times 0.4 \div 10$

59. $12 \div (-0.03) - 12 \times 0.03^2$

60. $(5 - 0.04)^2 \div 4 + 8.7 \times 0.4$

61. $(4 - 2.5)^2 \div 100 + 0.1 \times 6.5$

62. $4 \div 0.4 - 0.1 \times 5 + 0.1^2$

63. $6 \times 0.9 - 0.1 \div 4 + 0.2^3$

64. $5.5^2 \times [(6 - 7.8) \div 0.06 + 0.12]$

65. $12^2 \div (12 + 2.4) - [(2 - 2.4) \div 0.8]$

66. $0.01 \times \{[(4 - 0.25) \div 2.5] - (4.5 - 4.025)\}$

67. ***Mountain Peaks in Colorado.*** The elevations of four mountain peaks in Colorado are listed in the table below. Find the average elevation of these peaks.

MOUNTAIN PEAK	ELEVATION (in feet)
Mount Elbert	14,440
Mount Evans	14,271
Pikes Peak	14,115
Crested Butte	12,168

68. ***Driving Costs.*** The table below shows the cost per mile when specific types of vehicles are driven 15,000 miles in a year. Find the average cost per mile for the listed vehicles.

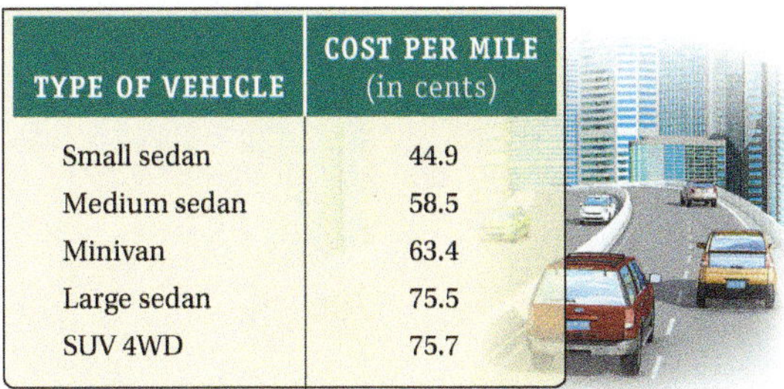

TYPE OF VEHICLE	COST PER MILE (in cents)
Small sedan	44.9
Medium sedan	58.5
Minivan	63.4
Large sedan	75.5
SUV 4WD	75.7

SOURCE: AAA

69. ***Camping in National Parks.*** The graph below shows the numbers of overnight camping stays in Park Service campgrounds in the National Park system from 2009 to 2013. Find the average number of stays per year during this period.

Camping in Park Service Campgrounds

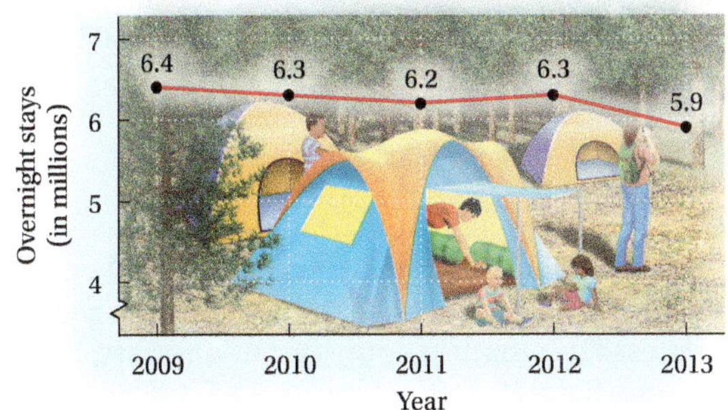

SOURCE: U.S. National Park Service

70. ***Life Expectancy.*** The information in the following bar graph shows the six highest life expectancies in the world. Determine the average life expectancy for these countries.

Life Expectancy at Birth as of 2013

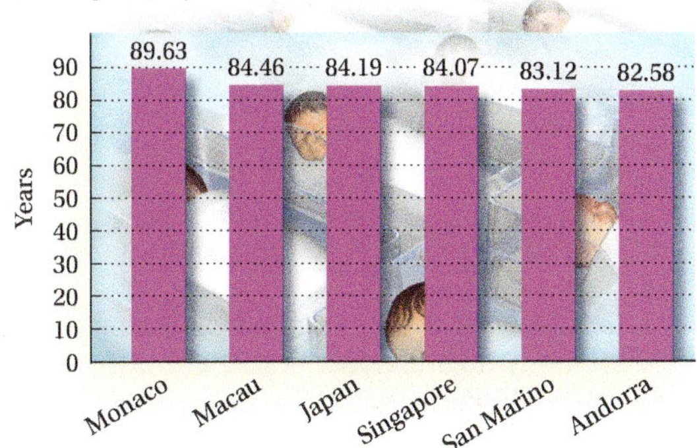

SOURCE: *CIA World Factbook*

The following table lists the lengths of the longest railway tunnels in the world. Use the table for Exercises 71 and 72.

TUNNEL	Gotthard Base, Switzerland	Brenner Base, Austria	Sei-Kan, Japan	Eurotunnel, England-France	Lötschberg, Switzerland
LENGTH, IN MILES	35.5	34.4	33.5	31.3	21.5
LENGTH, IN KILOMETERS	57.1	55.4	53.9	50.5	34.6

SOURCE: lotsberg.net/data/rail

71. Find the average length of the tunnels, in miles.

72. Find the average length of the tunnels, in kilometers.

Skill Maintenance

Simplify. [3.5b]

73. $\frac{38}{146}$

74. $\frac{4r}{20r}$

Find the prime factorization. [3.2c]

75. 684

76. 2005

77. Add: $10\frac{1}{2} + 4\frac{5}{8}$. [4.6a]

78. Subtract: $10\frac{1}{2} - 4\frac{5}{8}$. [4.6b]

Evaluate. [1.9b]

79. 7^3

80. 2^6

Solve. [1.7b]

81. $235 = 5 \cdot z$

82. $q + 31 = 72$

Synthesis

Calculate each of the following.

83. $7.434 \div (-1.2) \times 9.5 + 1.47^2$

84. $-9.46 \times 2.1^2 \div 3.5 + 4.36$

85. $9.0534 - 2.041^2 \times 0.731 \div 1.043^2$

86. $23.042(7 - 4.037 \times 1.46 - 0.932^2)$

Solve.

87. $439.57 \times 0.01 \div 1000 \cdot x = 4.3957$

88. $5.2738 \div 0.01 \times 1000 \div t = 52.738$

89. $0.0329 \div 0.001 \times 10^4 \div x = 3290$

90. $-4.302 \times 0.1^2 \div 0.001 \cdot t = -430.2$

91. *Television Ratings.* A television rating point represents 1,150,000 households. The 2014 Super Bowl was viewed in approximately 53.7 million households. How many rating points did the game receive? Round to the nearest tenth.

Source: Nielsen Media Research

92. *Size of Country.* The world's largest country, Russia, has an area of approximately 6.6 million square miles. The smallest country, Vatican City, has an area of approximately 0.2 square mile. How many times larger is Russia?

Source: U.S. Bureau of the Census, International Data Base

Electric Bills. Recently, electric bills from the Central Vermont Public Service Corporation consisted of a "customer charge" of \$0.374 per day plus an "energy charge" of \$0.1174 per kilowatt-hour (kWh) for the first 250 kWh used and \$0.09079 per kilowatt-hour for each kilowatt-hour in excess of 250.

93. From August 20 to September 20, the Kaufmans' bill was \$59.10. How many kilowatt-hours of electricity did they use? (Round to the nearest kilowatt-hour.)

94. From July 20 to August 20, the McGuires' bill was \$70. How many kilowatt-hours of electricity did they use? (Round to the nearest kilowatt-hour.)

Mid-Chapter Review

Concept Reinforcement

Determine whether each statement is true or false.

__________ **1.** In the number 308.00567, the digit 6 names the tens place. [5.1a]

__________ **2.** When writing a word name for decimal notation, we write the word "and" for the decimal point. [5.1a]

__________ **3.** On the number line, −2.3 is to the left of −2.2. [5.1c]

Guided Solutions

Fill in each blank with the number that creates a correct statement or solution.

4. Evaluate $P(1 + r)$ for $P = 5000$ and $r = 0.045$. [5.3c]

$P(1 + r) = \square(1 + \square)$ Substituting

$= 5000(\square)$ Adding within parentheses

$= \square$ Multiplying

5. Simplify: $5.6 + 4.3 \times (6.5 - 0.25)^2$. [5.4b]

$5.6 + 4.3 \times (6.5 - 0.25)^2 = 5.6 + 4.3 \times (\square)^2$ Carrying out the operation inside parentheses

$= 5.6 + 4.3 \times \square$ Evaluating the exponential expression

$= 5.6 + \square$ Multiplying

$= \square$ Adding

Mixed Review

6. ***Mile Run Record.*** The difference between the men's record for the mile run, held by Hicham El Guerrouj of Morocco (with a time of 3 min 43.13 sec) and the women's record for the mile run, held by Svetlana Masterkova of Russia (with a time of 4 min 12.56 sec) is 29.43 sec. Write a word name for 29.43. [5.1a]

Source: International Association of Athletics Federations (iaaf.org)

7. ***Skin Allergies.*** Skin allergies are the most common allergies among children. In 2010, 9.4 million children in the United States suffered from skin allergies. Convert 9.4 million to standard notation. [5.3b]

Source: CDC National Center for Health Statistics

Write each number as a fraction and, if possible, as a mixed numeral. [5.1b]

8. 4.53

9. 0.287

Which number is larger? [5.1c]

10. 0.07, 0.13

11. −5.2, −5.09

Write decimal notation. [5.1b]

12. $\frac{7}{10}$

13. $\frac{639}{100}$

14. $-35\frac{67}{100}$

15. $8\frac{2}{1000}$

Round 28.4615 to the nearest [5.1d]

16. Thousandth.

17. Hundredth.

18. Tenth.

19. One.

Add. [5.2a], [5.2c]

20. $\begin{array}{r} 47.638 \\ +\ \ 2.457 \\ \hline \end{array}$

21. $\begin{array}{r} 15.6 \\ 234.729 \\ 3.08 \\ +961.453 \\ \hline \end{array}$

22. $-10.5 + 0.27$

23. $16 + 0.34 + 1.9$

Subtract. [5.2b], [5.2c]

24. $\begin{array}{r} 321.57 \\ -\ \ 49.38 \\ \hline \end{array}$

25. $\begin{array}{r} 5.6 \\ -0.007 \\ \hline \end{array}$

26. $34.3 - 18.75$

27. $-6.9 - 13$

Multiply. [5.3a]

28. $\begin{array}{r} 4.6 \\ \times 0.9 \\ \hline \end{array}$

29. $\begin{array}{r} 15.3 \\ \times\ \ 6.07 \\ \hline \end{array}$

30. 100×81.236

31. $0.1 \times (-0.483)$

Divide. [5.4a]

32. $-20.24 \div (-4)$

33. $21.76 \div 6.8$

34. $76.3 \div 0.1$

35. $914.036 \div 1000$

36. Convert \$20.45 to cents. [5.3b]

37. Convert 147¢ to dollars. [5.3b]

38. Combine like terms: $3.08x - 7.1 - 4.3x$. [5.2d]

39. Evaluate $2(l + w)$ for $l = 1.3$ and $w = 0.8$. [5.3c]

Simplify. [5.4b]

40. $6.594 + 0.5318 \div 0.01$

41. $7.3 \times 4.6 - 0.8 \div 3.2$

Understanding Through Discussion and Writing

42. A classmate rounds 236.448 to the nearest one and gets 237. Explain the possible error. [5.1d]

43. Explain the error in the following:
Subtract.
$73.089 - 5.0061 = 2.3028$ [5.2b]

44. Explain why $10 \div 0.2 = 100 \div 2$. [5.4a]

45. Kayla made these two computational mistakes:
$0.247 \div 0.1 = 0.0247$; $0.247 \div 10 = 2.47$.
In each case, how could you convince her that a mistake has been made? [5.4a]

STUDYING FOR SUCCESS *As You Study*

- ☐ Find a quiet place to study.
- ☐ Be disciplined about your use of video games, the Internet, and television. Study first!
- ☐ Pace yourself. It is usually better to study for shorter periods several times a week than to study in one marathon session each week.

5.5 Using Fraction Notation with Decimal Notation

OBJECTIVES

a Use division to convert fraction notation to decimal notation.

b Round numbers named by repeating decimals.

c Convert certain fractions to decimal notation by using equivalent fractions.

d Simplify expressions that contain both fraction notation and decimal notation.

SKILL TO REVIEW

Objective 5.4a: Divide using decimal notation.

Divide.

1. $3 \div 4$ **2.** $25 \div 8$

a USING DIVISION TO FIND DECIMAL NOTATION

Recall that $\frac{a}{b}$ means $a \div b$. Thus, using division, we can express *any* fraction as a decimal. This means that any *rational* number (ratio of integers) can be written as a decimal.

EXAMPLE 1 Find decimal notation for $\frac{3}{20}$.

Because $\frac{3}{20}$ means $3 \div 20$, we can perform long division.

$$\begin{array}{r} 0.15 \\ 20\overline{)3.00} \\ \underline{2\,0} \\ 1\,00 \\ \underline{1\,00} \\ 0 \end{array}$$

We are finished when the remainder is 0.

We have $\frac{3}{20} = 0.15$.

EXAMPLE 2 Find decimal notation for $\frac{-7}{8}$.

Since $\frac{-7}{8}$ means $-7 \div 8$ and a negative number divided by a positive number is negative, we know that the decimal will be negative. We divide 7 by 8 and make the result negative.

$$\begin{array}{r} 0.875 \\ 8\overline{)7.000} \\ \underline{6\,4} \\ 60 \\ \underline{56} \\ 40 \\ \underline{40} \\ 0 \end{array}$$

Thus, $\frac{-7}{8} = -0.875$.

Do Margin Exercises 1 and 2.

Find decimal notation.

1. $\frac{2}{5}$ **2.** $\frac{-5}{8}$

When division with decimals ends with a remainder of 0, or *terminates*, as in Examples 1 and 2, the result is called a **terminating decimal**. If the division does not terminate, the result will be a **repeating decimal**.

Answers

Skill to Review:
1. 0.75 **2.** 3.125

Margin Exercises:
1. 0.4 **2.** −0.625

Find decimal notation.

3. $\frac{1}{6}$

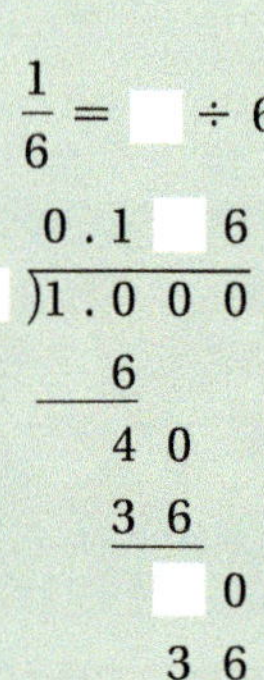

$\frac{1}{6} = 0.1666\ldots = 0.1\overline{6}$

4. $\frac{2}{3}$

Find decimal notation.

5. $\frac{5}{11}$

6. $-\frac{12}{11}$

Answers

3. $0.1\overline{6}$ **4.** $0.\overline{6}$ **5.** $0.\overline{45}$
6. $-1.\overline{09}$

Guided Solution:
3. 1, 6, 6, 4, 4

EXAMPLE 3 Find decimal notation for $\frac{5}{6}$.

Since $\frac{5}{6}$ means $5 \div 6$, we have

```
     0.8 3 3
 6 ) 5.0 0 0
     4 8
       2 0
       1 8
       → 2 0
         1 8
         → 2
```

Since 2 keeps reappearing as a remainder, the digits repeat and will continue to do so; therefore,

$$\frac{5}{6} = 0.83333\ldots.$$

The dots indicate an endless sequence of repeating digits in the quotient.

When there is a repeating pattern, we often use an overbar to indicate the repeating part, in this case, only the 3.

$$\frac{5}{6} = 0.8\overline{3}$$

Do Exercises 3 and 4.

EXAMPLE 4 Find decimal notation for $-\frac{4}{11}$.

Since $-\frac{4}{11}$ is negative, we divide 4 by 11 and make the result negative.

```
        0.3 6 3 6
 1 1 ) 4.0 0 0 0
       3 3
         7 0
         6 6
         → 4 0
           3 3
           → 7 0
             6 6
             → 4
```

Since 7 and 4 keep repeating as remainders, the sequence of digits "36" repeats in the quotient, and

$$\frac{4}{11} = 0.363636\ldots, \quad \text{or} \quad 0.\overline{36}.$$

Thus, $-\frac{4}{11} = -0.\overline{36}$.

Do Exercises 5 and 6.

When a fraction is written in simplified form, we can tell from the denominator whether its decimal notation will repeat or terminate.

For a fraction in simplified form,

- if the denominator has a prime factor other than 2 or 5, the decimal notation repeats.
- if the denominator has no prime factor other than 2 or 5, the decimal notation terminates.

EXAMPLE 5 Find decimal notation for $\frac{3}{7}$.

Because 7 is not a product of 2's and/or 5's, we expect a repeating decimal.

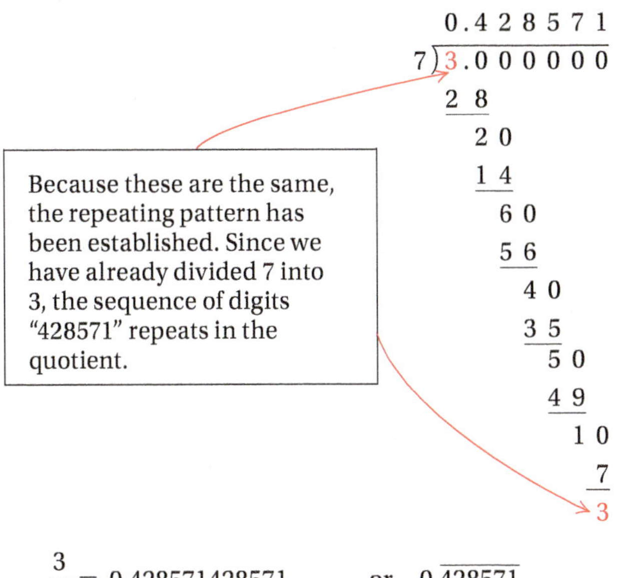

Because these are the same, the repeating pattern has been established. Since we have already divided 7 into 3, the sequence of digits "428571" repeats in the quotient.

$$\frac{3}{7} = 0.428571428571\ldots, \quad \text{or} \quad 0.\overline{428571}$$

Do Exercise 7. ▶

b ROUNDING REPEATING DECIMALS

The repeating part of a decimal can be so long that it will not fit on a calculator display. For example, when $\frac{5}{97}$ is written as a decimal, its repeating part is 96 digits long! Most calculators round repeating decimals to 9 or 10 decimal places.

In applied problems, repeating decimals are rounded to get approximate answers. To round a repeating decimal, we can extend the decimal notation at least one place past the rounding digit and then round as before.

EXAMPLES Round each to the nearest tenth, hundredth, and thousandth.

	Nearest tenth	*Nearest hundredth*	*Nearest thousandth*
6. $0.8\overline{3} = 0.83333\ldots$	0.8	0.83	0.833
7. $3.\overline{09} = 3.090909\ldots$	3.1	3.09	3.091
8. $-4.1\overline{763} = -4.1763763\ldots$	-4.2	-4.18	-4.176

Do Exercises 8–11. ▶

EXAMPLE 9 *Gas Mileage.* A car travels 457 mi on 16.4 gal of gasoline. The ratio of number of miles driven to amount of gasoline used is *gas mileage.* Find the gas mileage, and convert the ratio to decimal notation rounded to the nearest tenth.

$$\frac{\text{Miles driven}}{\text{Gasoline used}} = \frac{457}{16.4} \approx 27.86 \quad \text{Dividing to 2 decimal places}$$

$$\approx 27.9 \quad \text{Rounding to 1 decimal place}$$

The gas mileage is 27.9 miles per gallon, or 27.9 mpg.

Do Exercise 12. ▶

CALCULATOR CORNER

Recognizing Repeating Decimals Short repeating patterns can usually be recognized on a calculator's display. For example, to convert $\frac{1}{11}$ to decimal notation, press [1] [÷] [1] [1] [=]. The result for a 10-digit display is [.0909090909], so the decimal notation is $0.\overline{09}$.

For $-\frac{12}{11}$, the same calculator displays [−1.090909091]. Note that the last digit has been rounded from 0 to 1. The decimal notation for $-\frac{12}{11}$ is $-1.\overline{09}$.

EXERCISES Convert to decimal notation.

1. $-\frac{1}{6}$ **2.** $\frac{7}{11}$

3. $\frac{19}{3}$ **4.** $-\frac{514}{9}$

---

7. Find decimal notation for $\frac{5}{7}$.

Round each to the nearest tenth, hundredth, and thousandth.

8. $0.\overline{6}$ **9.** $0.\overline{608}$

10. $-7.3\overline{49}$ **11.** $2.6\overline{891}$

12. *Gas Mileage.* A car travels 380 mi on 15.7 gal of gasoline. Find the gasoline mileage, and convert the ratio to decimal notation rounded to the nearest tenth.

Answers

7. $0.\overline{714285}$ **8.** 0.7; 0.67; 0.667 **9.** 0.6; 0.61; 0.608 **10.** −7.3; −7.35; −7.349 **11.** 2.7; 2.69; 2.689 **12.** 24.2 miles per gallon

CALCULATOR CORNER

The Number π Many geometric applications of decimal notation involve the number π.

Most calculators now have a key that provides an approximation of π. Often a key labeled [SHIFT] or [2nd] must be pressed first.

To calculate the value of an expression like $2\pi(7.5)$, on most calculators we press [2] [×] [2nd] [π] [×] [7] [.] [5] and then [=] or [ENTER]. The result is 47.1238898.

EXERCISES

1. Calculate: $4\pi(9.8)$.
2. Evaluate: $2\pi r$ for $r = 8.37$.

c MORE WITH CONVERSIONS

Fractions like $\frac{3}{10}$ or $-\frac{71}{1000}$ can be converted to decimal notation without using long division. When a denominator is a factor of 10, 100, and so on, we can convert to decimal notation by finding (perhaps mentally) an equivalent fraction in which the denominator is a power of 10.

EXAMPLE 10 Find decimal notation for $-\frac{7}{500}$.

Since $500 \cdot 2 = 1000$, and 1000 is a power of 10, we use $\frac{2}{2}$ as an expression for 1.

$$-\frac{7}{500} = -\frac{7}{500} \cdot \frac{2}{2} = -\frac{14}{1000} = -0.014$$

Think: $1000 \div 500 = 2$, and $7 \cdot 2 = 14$.

EXAMPLE 11 Find decimal notation for $\frac{9}{25}$.

$$\frac{9}{25} = \frac{9}{25} \cdot \frac{4}{4} = \frac{36}{100} = 0.36$$

Using $\frac{4}{4}$ for 1 to get a denominator of 100

As a check, we can divide.

$$\begin{array}{r} 0.36 \\ 25\overline{)9.00} \\ \underline{75} \\ 150 \\ \underline{150} \\ 0 \end{array}$$

Note that multiplication by 1 is much faster.

EXAMPLE 12 Find decimal notation for $\frac{7}{4}$.

$$\frac{7}{4} = \frac{7}{4} \cdot \frac{25}{25} = \frac{175}{100} = 1.75$$

Using $\frac{25}{25}$ for 1 to get a denominator of 100. You might also note that 7 quarters is \$1.75.

Multiply by a form of 1 to find decimal notation for each number.

13. $\frac{4}{5}$
14. $-\frac{9}{20}$
15. $\frac{7}{200}$
16. $\frac{33}{25}$

Do Exercises 13–16.

d CALCULATIONS WITH FRACTION AND DECIMAL NOTATION TOGETHER

Fraction notation and decimal notation can occur together in a calculation. In such cases, there are at least three ways in which we might proceed.

EXAMPLE 13 Calculate: $\frac{2}{3} \times 0.576$.

Method 1: Perhaps the quickest method is to treat 0.576 as $\frac{0.576}{1}$. Then we multiply 0.576 by 2 and divide the result by 3.

$$\begin{aligned} \frac{2}{3} \times 0.576 &= \frac{2}{3} \times \frac{0.576}{1} \\ &= \frac{2 \times 0.576}{3} = \frac{1.152}{3} \\ &= 0.384 \end{aligned}$$

$$\begin{array}{r} 0.384 \\ 3\overline{)1.152} \\ \underline{9} \\ 25 \\ \underline{24} \\ 12 \\ \underline{12} \\ 0 \end{array}$$

Answers

13. 0.8 14. −0.45 15. 0.035 16. 1.32

Method 2: A second way to do this calculation is to convert the fraction notation to decimal notation so that both numbers are in decimal notation. Since $\frac{2}{3}$ converts to repeating decimal notation, it is first rounded to some chosen decimal place. We choose three decimal places because 0.576 has three decimal places. Then, using decimal notation, we multiply.

$$\frac{2}{3} \times 0.576 = 0.\overline{6} \times 0.576 \approx 0.667 \times 0.576 = 0.384192$$

Method 3: A third method is to convert the decimal notation to fraction notation so that both numbers are in fraction notation. The answer can be left in fraction notation and simplified, or we can convert back to decimal notation and, if appropriate, round.

$$\frac{2}{3} \times 0.576 = \frac{2}{3} \cdot \frac{576}{1000} = \frac{2 \cdot 576}{3 \cdot 1000}$$

$$= \frac{2 \cdot 2 \cdot 2 \cdot 2 \cdot 2 \cdot 2 \cdot 2 \cdot 3 \cdot 3}{2 \cdot 2 \cdot 2 \cdot 3 \cdot 5 \cdot 5 \cdot 5} \qquad \text{Factoring}$$

$$= \frac{2 \cdot 2 \cdot 2 \cdot 3}{2 \cdot 2 \cdot 2 \cdot 3} \cdot \frac{2 \cdot 2 \cdot 2 \cdot 2 \cdot 3}{5 \cdot 5 \cdot 5} \qquad \text{Removing a factor equal to 1: } \frac{2 \cdot 2 \cdot 2 \cdot 3}{2 \cdot 2 \cdot 2 \cdot 3} = 1$$

$$= \frac{2 \cdot 2 \cdot 2 \cdot 2 \cdot 3}{5 \cdot 5 \cdot 5} = \frac{48}{125}, \text{ or } 0.384$$

Note that we get an exact answer with methods 1 and 3, but method 2 gives an approximation since we rounded decimal notation for $\frac{2}{3}$.

Do Exercises 17 and 18. ▶

EXAMPLE 14 ***Boating.*** A triangular sail on a single-sail day cruiser is 3.4 m wide and 4.2 m tall. Find the area of the sail.

1. **Familiarize.** We make a drawing and recall that the formula for the area, A, of a triangle with base b and height h is $A = \frac{1}{2}bh$.
2. **Translate.** We substitute 3.4 for b and 4.2 for h.

$$A = \frac{1}{2}bh$$
$$= \frac{1}{2}(3.4)(4.2) \qquad \text{Substituting}$$

3. **Solve.** We simplify as follows.

$$A = \frac{1}{2}(3.4)(4.2)$$
$$= \frac{3.4}{2}(4.2) \qquad \text{Multiplying } \tfrac{1}{2} \text{ and } \tfrac{3.4}{1}$$
$$= 1.7(4.2) \qquad \text{Dividing}$$
$$= 7.14 \qquad \text{Multiplying}$$

4. **Check.** To check, we repeat the calculations using the commutative law. We also rewrite $\frac{1}{2}$ as 0.5.

$$\frac{1}{2}(3.4)(4.2) = 0.5(4.2)(3.4) = (2.1)(3.4) = 7.14$$

Our answer checks.

5. **State.** The area of the sail is 7.14 m^2 (square meters).

Do Exercise 19. ▶

Calculate.

17. $\frac{3}{4} \times 0.62$.

Method 1:

$$\frac{3}{4} \times 0.62 = \frac{3}{4} \times \frac{0.62}{\square}$$
$$= \frac{\square}{4} = 0.465$$

Method 2:

$$\frac{3}{4} \times 0.62 = \square \times 0.62$$
$$= 0.465$$

Method 3:

$$\frac{3}{4} \times 0.62 = \frac{3}{4} \cdot \frac{62}{\square}$$
$$= \frac{\square}{400}$$
$$= \frac{\square}{200} = 0.465$$

18. $\frac{1}{3} \times 0.384 + \frac{5}{8} \times 0.6784$

19. Find the area of a triangular window that is 3.25 ft wide and 2.6 ft tall.

Answers

17. $\frac{93}{200}$, or 0.465 **18.** 0.552 **19.** 4.225 ft^2

Guided Solution:

17. 1, 1.86; 0.75; 100, 186, 93

5.5 Exercise Set

For Extra Help

MyMathLab®

MathXL® PRACTICE 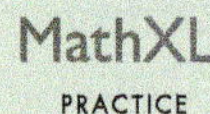 WATCH  READ  REVIEW 

✓ Reading Check

Determine whether the decimal notation for each fraction is terminating or repeating.

RC1. $\frac{4}{9}$ ________________

RC2. $\frac{3}{32}$ ________________

RC3. $\frac{39}{40}$ ________________

RC4. $\frac{7}{12}$ ________________

RC5. $\frac{2}{11}$ ________________

RC6. $\frac{80}{125}$ ________________

a, **c** Find decimal notation for each number.

1. $\frac{3}{8}$

2. $\frac{3}{5}$

3. $\frac{-1}{2}$

4. $\frac{-1}{4}$

5. $\frac{3}{25}$

6. $\frac{7}{20}$

7. $\frac{9}{40}$

8. $\frac{3}{40}$

9. $\frac{13}{25}$

10. $\frac{17}{25}$

11. $\frac{-17}{20}$

12. $\frac{-13}{20}$

13. $-\frac{9}{16}$

14. $-\frac{5}{16}$

15. $\frac{7}{5}$

16. $\frac{3}{2}$

17. $\frac{28}{25}$

18. $\frac{31}{20}$

19. $\frac{11}{-8}$

20. $\frac{17}{-10}$

21. $-\frac{39}{40}$

22. $-\frac{17}{40}$

23. $\frac{121}{200}$

24. $\frac{32}{125}$

25. $\frac{8}{15}$

26. $\frac{7}{9}$

27. $\frac{1}{3}$

28. $\frac{1}{9}$

29. $\frac{-4}{3}$

30. $\frac{-8}{9}$

31. $\frac{7}{6}$

32. $\frac{7}{11}$

33. $-\frac{14}{11}$

34. $-\frac{7}{11}$

35. $\frac{-5}{12}$

36. $\frac{-11}{12}$

37. $\frac{127}{500}$

38. $\frac{83}{500}$

39. $\frac{4}{33}$

40. $\frac{5}{33}$

41. $\frac{-12}{55}$

42. $\frac{-5}{22}$

43. $\frac{4}{7}$

44. $\frac{2}{7}$

b Round each to the nearest tenth, hundredth, and thousandth.

45. $0.\overline{18}$

46. $0.\overline{83}$

47. $0.2\overline{7}$

48. $3.5\overline{4}$

For Exercises 49–60, round the decimal notation for each number to the nearest tenth, hundredth, and thousandth.

49. $\frac{4}{11}$

50. $\frac{3}{11}$

51. $-\frac{5}{3}$

52. $-\frac{19}{16}$

53. $\frac{-8}{17}$

54. $\frac{-7}{13}$

55. $\frac{7}{12}$

56. $\frac{2}{15}$

57. $\frac{29}{-150}$

58. $\frac{37}{-150}$

59. $\frac{7}{-9}$

60. $\frac{5}{-13}$

61. For this set of people, what is the ratio, in decimal notation rounded to the nearest thousandth, where appropriate, of:

a) women to the total number of people?
b) women to men?
c) men to the total number of people?
d) men to women?

62. For this set of pennies and quarters, what is the ratio, in decimal notation rounded to the nearest thousandth, where appropriate, of:

a) pennies to quarters?
b) quarters to pennies?
c) pennies to total number of coins?
d) total number of coins to pennies?

Gas Mileage. In each of Exercises 63–66, find the gas mileage rounded to the nearest tenth.

63. 285 mi; 18 gal

64. 396 mi; 17 gal

65. 324.8 mi; 18.2 gal

66. 264.8 mi; 12.7 gal

d Calculate and write the result as a decimal.

67. $\frac{7}{8} \times 12.64$

68. $\frac{4}{5} \times 384.8$

69. $6.84 \div 2\frac{1}{2}$

70. $8\frac{1}{2} \div 2.125$

71. $\frac{47}{9}(-79.95)$

72. $\frac{7}{11}(-2.7873)$

73. $\frac{1}{2} - 0.5$

74. $3\frac{1}{8} - 2.75$

75. $\left(\frac{1}{6}\right)0.0765 + \left(\frac{3}{4}\right)0.1124$

76. $\left(\frac{2}{5}\right)6384.1 - \left(\frac{5}{8}\right)156.56$

77. $\frac{3}{4} \times 2.56 - \frac{7}{8} \times 3.94$

78. $\frac{2}{5} \times 3.91 - \frac{7}{10} \times 4.15$

79. $5.2 \times 1\frac{7}{8} \div 0.4$

80. $4\frac{3}{4} \times 0.5 \div 0.1$

Solve.

81. Find the area of a triangular shawl that is 1.8 m long and 1.2 m wide.

82. Find the area of a triangular sign that is 1.5 m wide and 1.5 m tall.

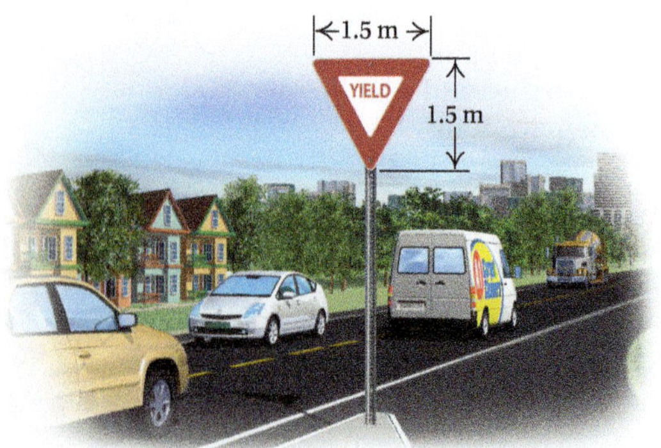

83. Find the area of a triangular stamp that is 3.4 cm wide and 3.4 cm tall.

84. Find the area of a triangular reflector that is 7.4 cm wide and 9.1 cm tall.

85. Find the area of the kite shown on the left.

86. Find the area of the kite shown on the right.

Skill Maintenance

Calculate.

87. $9 \cdot 2\frac{1}{3}$ [4.7a]

88. $16\frac{1}{10} - 14\frac{3}{5}$ [4.6b]

89. $84 \div 8\frac{2}{5}$ [4.7b]

90. $14\frac{3}{5} + 16\frac{1}{10}$ [4.6a]

Solve. [4.6c]

91. A recipe for bread calls for $\frac{2}{3}$ cup of water, $\frac{1}{4}$ cup of milk, and $\frac{1}{8}$ cup of oil. How many cups of liquid ingredients does the recipe call for?

92. A board $\frac{7}{10}$ in. thick is glued to a board $\frac{3}{5}$ in. thick. The glue is $\frac{3}{100}$ in. thick. How thick is the result?

Synthesis

93. Find decimal notation for $\frac{1}{7}, \frac{2}{7}, \frac{3}{7}, \frac{4}{7}$, and $\frac{5}{7}$. Observe the pattern and predict the decimal notation for $\frac{6}{7}$. Check your answer on a calculator.

94. Find decimal notation for $\frac{1}{9}, \frac{1}{99}$, and $\frac{1}{999}$. Observe the pattern and predict the decimal notation for $\frac{1}{9999}$. Check your answer on a calculator.

The formula $A = \pi r^2$ is used to find the area, A, of a circle with radius r. For Exercises 95 and 96, find the area of a circle with the given radius, using $\frac{22}{7}$ for π. For Exercises 97 and 98, use 3.14 for π or a calculator with a π key.

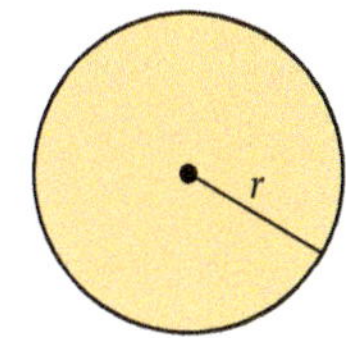

95. $r = 2.1$ cm

96. $r = 1.4$ cm

97. $r = \frac{3}{4}$ ft

98. $r = 4\frac{1}{2}$ yd

5.6 Estimating

OBJECTIVE

a Estimate sums, differences, products, and quotients.

SKILL TO REVIEW

Objective 1.6b: Estimate sums, differences, products, and quotients by rounding.

Estimate by first rounding to the nearest ten.

1. $\begin{array}{r} 467 \\ -\ 284 \\ \hline \end{array}$

2. $\begin{array}{r} 54 \\ \times\ 29 \\ \hline \end{array}$

1. Estimate by rounding to the nearest ten the total cost of one grill and one camera. Which of the following is an appropriate estimate?

a) $43 b) $400
c) $410 d) $430

2. About how much more does the TV cost than the grill? Estimate by rounding to the nearest ten. Which of the following is an appropriate estimate?

a) $100 b) $150
c) $160 d) $300

Answers

Skill to Review:
1. 190 2. 1500

Margin Exercises:
1. (d) 2. (c)

a ESTIMATING SUMS, DIFFERENCES, PRODUCTS, AND QUOTIENTS

Estimating has many uses. It can be done before a problem is even attempted and it can be done afterward as a check, even when we are using a calculator. In many situations, an estimate is all we need. We usually estimate by rounding the numbers so that there are one or two nonzero digits. Consider the following prices for Examples 1–3.

$289.95
Four Burner Gas Grill
8.2 Megapixel Digital Camera
$139.97
19" Flat-Panel HDTV
$449.99

EXAMPLE 1 Estimate by rounding to the nearest ten the total cost of one grill and one TV.

We are estimating the sum

$289.95 + $449.99 = Total cost.

The estimate found by rounding the addends to the nearest ten is

$290 + $450 = $740. (Estimated total cost)

◀ Do Margin Exercise 1.

EXAMPLE 2 About how much more does the TV cost than the camera? Estimate by rounding to the nearest ten.

We are estimating the difference

$449.99 − $139.97 = Price difference.

The estimate found by rounding each price to the nearest ten is

$450 − $140 = $310. (Estimated price difference)

◀ Do Exercise 2.

EXAMPLE 3 Estimate the total cost of 4 cameras.

We are estimating the product

$4 \times \$139.97 =$ Total cost.

The estimate is found by rounding 139.97 to the nearest ten:

$4 \times \$140 = \$560.$

Do Exercise 3. ▶

EXAMPLE 4 A student government group is planning an event for first-year students. Since the local weather is often rainy, the group decides to purchase umbrellas bearing the university logo for door prizes. About how many umbrellas that cost $39.99 each can be purchased for $356?

We are estimating the quotient

$\$356 \div \$39.99.$

Since we want a whole-number estimate, we need to round appropriately. Rounding $39.99 to the nearest one, we get $40. Since $356 is close to $360, which is a multiple of $40, we estimate

$\$360 \div \$40 = 9.$

The answer is about 9 umbrellas.

Do Exercise 4. ▶

When estimating, we usually look for numbers that are easy to work with. For example, if multiplying, we might round 0.43 to 0.5 and 8.9 to 10, because 0.5 and 10 are convenient numbers to multiply.

EXAMPLE 5 Estimate: 4.8×52. Do not find the actual product. Which of the following is an appropriate estimate?

a) 25 **b)** 250 **c)** 2500 **d)** 360

We round 4.8 to the nearest one and 52 to the nearest ten:

$5 \times 50 = 250.$ (Estimated product)

Thus, an approximate estimate is (b).

Other estimates that we might have used in Example 5 are

$5 \times 52 = 260$ or $4.8 \times 50 = 240.$

The estimate in Example 5, $5 \times 50 = 250$, is the easiest to do because the factors have the fewest nonzero digits. You could probably do it mentally. In general, we try to round so that a computation has as few nonzero digits as possible while still keeping the estimated value close to the original value.

Do Exercises 5–10. ▶

EXAMPLE 6 Estimate: $82.08 \div 24$. Which of the following is an appropriate estimate?

a) 400 **b)** 16 **c)** 40 **d)** 4

This is about $80 \div 20$, so the answer is about 4. Thus, an appropriate estimate is (d).

3. Estimate the total cost of 6 TVs. Which of the following is an appropriate estimate?

a) $450 **b)** $2700
c) $4500 **d)** $27,000

4. Refer to the umbrella price in Example 4. About how many umbrellas can be purchased for $675?

Estimate each product. Do not find the actual product. Which of the following is an appropriate estimate?

5. 2.4×8
a) 16 **b)** 34
c) 125 **d)** 5

6. 24×0.6
a) 200 **b)** 5
c) 110 **d)** 20

7. 0.86×0.432
a) 0.04 **b)** 0.4
c) 1.1 **d)** 4

8. 0.82×0.1
a) 800 **b)** 8
c) 0.08 **d)** 80

9. 0.12×18.248
a) 180 **b)** 1.8
c) 0.018 **d)** 18

10. 24.234×5.2
a) 200 **b)** 120
c) 12.5 **d)** 234

Answers

3. (b) 4. 17 umbrellas 5. (a) 6. (d) 7. (b) 8. (c) 9. (b) 10. (b)

Estimate each quotient. Which of the following is an appropriate estimate?

11. $59.78 \div 29.1$

a) 200 **b)** 20
c) 2 **d)** 0.2

12. $82.08 \div 2.4$

a) 40 **b)** 4.0
c) 400 **d)** 0.4

13. $0.1768 \div 0.08$

a) 8 **b)** 10
c) 2 **d)** 20

EXAMPLE 7 Estimate: $94.18 \div 3.2$. Which of the following is an appropriate estimate?

a) 30 **b)** 300 **c)** 3 **d)** 60

This is about $90 \div 3$, so the answer is about 30. Thus, an appropriate estimate is (a).

EXAMPLE 8 Estimate: $0.0156 \div 1.3$. Which of the following is an appropriate estimate?

a) 0.2 **b)** 0.002 **c)** 0.02 **d)** 20

This is about $0.02 \div 1$, so the answer is about 0.02. Thus, an appropriate estimate is (c).

Do Exercises 11–13.

In some cases, it is easier to estimate a quotient by checking products rather than by rounding the divisor and the dividend.

EXAMPLE 9 Estimate: $0.0074 \div 0.23$. Which of the following is an appropriate estimate?

a) 0.3 **b)** 0.03 **c)** 300 **d)** 3

We estimate 3 for a quotient. We check by multiplying.

$$0.23 \times 3 = 0.69$$

We make the estimate smaller. We estimate 0.3 and check by multiplying.

$$0.23 \times 0.3 = 0.069$$

We make the estimate smaller. We estimate 0.03 and check by multiplying.

$$0.23 \times 0.03 = 0.0069$$

This is about 0.0074, so the quotient is about 0.03. Thus, an appropriate estimate is (b).

Do Exercise 14.

14. Estimate: $0.0069 \div 0.15$. Which of the following is an appropriate estimate?

a) 0.5 **b)** 50
c) 0.05 **d)** 0.004

Answers

11. (c) 12. (a) 13. (c) 14. (c)

5.6 Exercise Set

For Extra Help
MyMathLab®

MathXL® PRACTICE 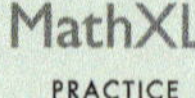 WATCH  READ  REVIEW 

Reading Check

Match each calculation with the most appropriate estimate from the list at the right.

RC1. 0.1003×0.8 ________ **a)** 0.8

RC2. $38.41 + 41.777$ ________ **b)** 0.08

RC3. 0.00152×4025 ________ **c)** 8

RC4. $1632 \div 1.9$ ________ **d)** 80

RC5. $9.054 - 8.3111$ ________ **e)** 800

RC6. $162,105 \times 0.0496$ ________ **f)** 8000

a For Exercises 1–8, use the prices shown below to estimate each sum, difference, product, or quotient. Indicate which of the choices is an appropriate estimate.

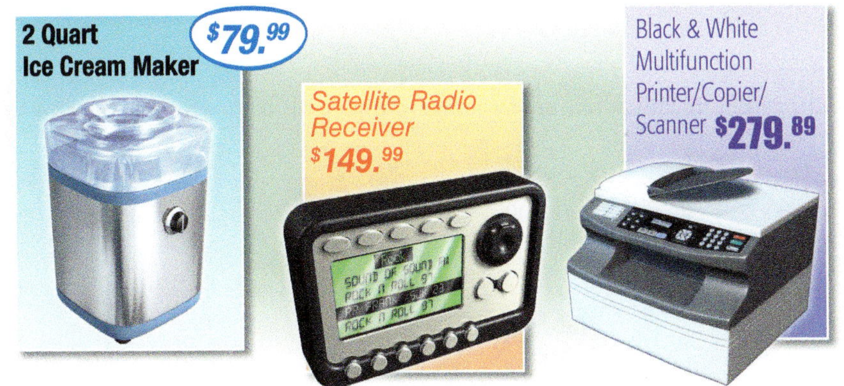

1. Estimate the total cost of one printer and one satellite radio.

a) \$43 **b)** \$4300 **c)** \$360 **d)** \$430

2. Estimate the total cost of one satellite radio and one ice cream maker.

a) \$230 **b)** \$23 **c)** \$2300 **d)** \$400

3. About how much more does the printer cost than the satellite radio?

a) \$1300 **b)** \$200 **c)** \$130 **d)** \$13

4. About how much more does the satellite radio cost than the ice cream maker?

a) \$7000 **b)** \$70 **c)** \$130 **d)** \$700

5. Estimate the total cost of 6 ice cream makers.

a) \$480 **b)** \$48 **c)** \$240 **d)** \$4800

6. Estimate the total cost of 4 printers.

a) \$1200 **b)** \$1120 **c)** \$11,200 **d)** \$600

7. About how many ice cream makers can be purchased for \$830?

a) 120 **b)** 100 **c)** 10 **d)** 1000

8. About how many printers can be purchased for \$5627?

a) 200 **b)** 20 **c)** 1800 **d)** 2000

Estimate by rounding as directed.

9. $0.02 + 1.31 + 0.34$; nearest tenth

10. $0.88 + 2.07 + 1.54$; nearest one

11. $6.03 + 0.007 + 0.214$; nearest one

12. $1.11 + 8.888 + 99.94$; nearest one

13. $52.367 + 1.307 + 7.324$; nearest one

14. $12.9882 + 1.2115$; nearest tenth

15. 2.678 − 0.445; nearest tenth

16. 12.9882 − 1.0115; nearest one

17. 198.67432 − 24.5007; nearest ten

Estimate. Choose a rounding digit that gives one or two nonzero digits. Indicate which of the choices is an appropriate estimate.

18. 234.12321 − 200.3223
a) 600 **b)** 60
c) 300 **d)** 30

19. 49 × 7.89
a) 400 **b)** 40
c) 4 **d)** 0.4

20. 7.4 × 8.9
a) 95 **b)** 63
c) 124 **d)** 6

21. 98.4 × 0.083
a) 80 **b)** 12
c) 8 **d)** 0.8

22. 78 × 5.3
a) 400 **b)** 800
c) 40 **d)** 8

23. 3.6 ÷ 4
a) 10 **b)** 1
c) 0.1 **d)** 0.01

24. 0.0713 ÷ 1.94
a) 3.5 **b)** 0.35
c) 0.035 **d)** 35

25. 74.68 ÷ 24.7
a) 9 **b)** 3
c) 12 **d)** 120

26. 914 ÷ 0.921
a) 10 **b)** 100
c) 1000 **d)** 1

27. *Fence Posts.* Melanie needs to build a fence around a new horse pasture. The perimeter of the area to be fenced is 1760 ft. Estimate the number of wooden fence posts needed if the posts are placed 8.625 ft apart.

28. *Ticketmaster.* Recently, Ticketmaster stock sold for $22.25 per share. Estimate how many shares can be purchased for $4400.

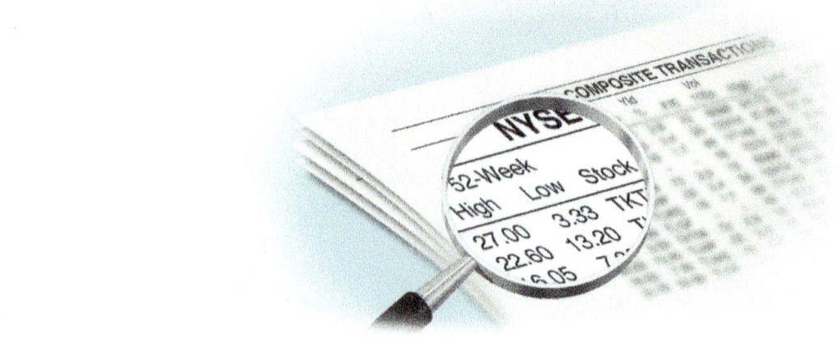

29. ***Day-Care Supplies.*** Helen wants to buy 12 boxes of crayons at \$1.89 per box for the day care center that she runs. Estimate the total cost of the crayons.

30. ***Batteries.*** Oscar buys 6 packages of AAA batteries at \$5.29 per package. Estimate the total cost of the purchase.

Skill Maintenance

31. ***Honey Production.*** In 2011, 176,462,000 lb of honey was produced in the United States. The two states with the greatest honey production were North Dakota and California. North Dakota produced 46,410,000 lb of honey, and California produced 27,470,000 lb. How many more pounds of honey were produced in North Dakota than in California? [1.8a]

Source: U.S. Department of Agriculture

32. ***Jewelry.*** Christie has designed a pair of kite-shaped earrings, as shown below. Determine the surface area of the front of one earring. [3.6b]

33. About $\frac{9}{25}$ of all pizzas that Americans order have pepperoni as a topping. If Americans eat 350 slices of pizza every second, how many of those slices are topped with pepperoni? [3.6b]

Source: inventors.about.com

34. A batch of fudge requires $\frac{3}{4}$ cup of sugar. How much sugar is needed to make 12 batches? [3.6b]

35. After her company was restructured, Meghan's pay was $\frac{9}{10}$ of what it had been. If she is now making \$32,850 a year, what was she making before the reorganization? [3.8b]

36. Rick's Market sells Swiss cheese in $\frac{3}{4}$-lb packages. How many packages can be made from a 12-lb slab of cheese? [3.8b]

Synthesis

The following were done on a calculator. Estimate to determine whether the decimal point was placed correctly.

37. $178.9462 \times 61.78 = 11{,}055.29624$

38. $14{,}973.35 \div 298.75 = 501.2$

39. $19.7236 - 1.4738 \times 4.1097 = 1.366672414$

40. $28.46901 \div 4.9187 - 2.5081 = 3.279813473$

41. ▦ Use one of $+, -, \times$, and $\div$ in each blank to make a true sentence.

a) $(0.37 \; \square \; 18.78) \; \square \; 2^{13} = 156{,}876.8$

b) $2.56 \; \square \; 6.4 \; \square \; 51.2 \; \square \; 17.4 = 312.84$

5.7 Solving Equations

OBJECTIVES

a Solve equations containing decimals and one variable term.

b Solve equations containing decimals and two or more variable terms.

SKILL TO REVIEW

Objective 2.8d: Solve equations that require use of both the addition principle and the division principle.

Solve.

1. $5x + 7 = -3$
2. $3x - 5 - x = 4 + x$

We now use the addition and division principles to solve equations involving decimals.

a EQUATIONS WITH ONE VARIABLE TERM

Recall that equations like $5x + 7 = -3$ are normally solved by first "undoing" the addition and then "undoing" the multiplication. This reverses the order of operations in which we add last and multiply first.

EXAMPLE 1 Solve: $0.5x + 5 = 8$.

$$0.5x + 5 = 8$$

$$0.5x + 5 - 5 = 8 - 5 \quad \text{Subtracting 5 from both sides}$$

$$0.5x = 3 \quad \text{Simplifying}$$

$$\frac{0.5x}{0.5} = \frac{3}{0.5} \quad \text{Dividing both sides by 0.5}$$

$$x = 6 \quad \text{Simplifying}$$

Check:

$0.5x + 5 = 8$	
$0.5(6) + 5$	? 8
$3 + 5$	
8	8 TRUE

The solution is 6.

◀ Do Margin Exercises 1 and 2.

Solve.

1. $6x + 7.4 = 11$
2. $0.2 - 0.1x = 1.4$

EXAMPLE 2 Solve: $4.2x + 3.7 = -26.12$.

$$4.2x + 3.7 = -26.12$$

$$4.2x + 3.7 - 3.7 = -26.12 - 3.7 \quad \text{Subtracting 3.7 from both sides}$$

$$4.2x = -29.82 \quad \text{Simplifying}$$

$$\frac{4.2x}{4.2} = \frac{-29.82}{4.2} \quad \text{Dividing both sides by 4.2}$$

$$x = -7.1 \quad \text{Simplifying}$$

Check:

$4.2x + 3.7 = -26.12$	
$4.2(-7.1) + 3.7$	? -26.12
$-29.82 + 3.7$	
-26.12	-26.12 TRUE

The solution is -7.1.

◀ Do Exercises 3 and 4.

Solve.

3. $7.4t + 1.25 = 27.89$
4. $-5.7 + 4.8x = -14.82$

Answers

Skill to Review:
1. -2 2. 9

Margin Exercises:
1. 0.6 2. -12 3. 3.6 4. -1.9

b EQUATIONS WITH TWO OR MORE VARIABLE TERMS

Some equations have variable terms on both sides. To solve such an equation, we use the addition principle to get all variable terms on one side of the equation and all constant terms on the other side.

EXAMPLE 3 Solve: $10x - 7 = 2x + 13$.

We begin by subtracting $2x$ from (or adding $-2x$ to) each side. This will group all variable terms on one side of the equation.

$$10x - 7 - 2x = 2x + 13 - 2x \qquad \text{Adding } -2x \text{ to both sides}$$
$$8x - 7 = 13 \qquad \text{Combining like terms}$$

We use the addition principle to isolate all constant terms on one side.

$$8x - 7 = 13$$
$$8x - 7 + 7 = 13 + 7 \qquad \text{Adding 7 to both sides}$$
$$8x = 20 \qquad \text{Simplifying (combining like terms)}$$
$$\frac{8x}{8} = \frac{20}{8} \qquad \text{Dividing both sides by 8}$$
$$x = 2.5$$

Check:

$10x - 7$	$= 2x + 13$
$10(2.5) - 7$	? $2(2.5) + 13$
$25 - 7$	$5 + 13$
18	18 TRUE

The solution is 2.5.

Sometimes it may be easier to combine all variable terms on the right side and all constant terms on the left side.

EXAMPLE 4 Solve: $11 - 3t = 7t + 8$.

We can combine all variable terms on the right side by adding $3t$ to both sides.

$$11 - 3t = 7t + 8$$
$$11 - 3t + 3t = 7t + 8 + 3t \qquad \text{Adding } 3t \text{ to both sides}$$
$$11 = 10t + 8 \qquad \text{Combining like terms}$$
$$11 - 8 = 10t + 8 - 8 \qquad \text{Subtracting 8 from both sides}$$
$$3 = 10t$$
$$\frac{3}{10} = \frac{10t}{10} \qquad \text{Dividing both sides by 10}$$
$$0.3 = t$$

Check:

$11 - 3t$	$= 7t + 8$
$11 - 3(0.3)$	? $7(0.3) + 8$
$11 - 0.9$	$2.1 + 8$
10.1	10.1 TRUE

The solution is 0.3.

CALCULATOR CORNER

Checking Solutions To check the solution of Example 2 with a calculator with a [+/−] key, we press the following sequence of keys:

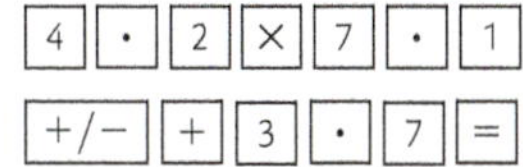

On a calculator with a [(−)] key, we press

[4] [.] [2] [×] [(−)] [7] [.]
[1] [+] [3] [.] [7] [ENTER].

In both cases, the result, −26.12, shows that −7.1 *is* a solution of $4.2x + 3.7 = -26.12$.

EXERCISES

1. Use a calculator to check the solutions of Margin Exercises 2 and 3.
2. Use a calculator to show that −3.6 is *not* a solution of $7.4t + 1.25 = 27.89$ (Margin Exercise 3).
3. Use a calculator to show that 1.9 is *not* a solution of $-5.7 + 4.8x = -14.82$ (Margin Exercise 4).

Solve.

5. $10t - 3 = 4t + 18$

6. $8 + 4x = 9x - 3$ GS

$$
\begin{aligned}
8 + 4x - 4x &= 9x - 3 - \square\\
8 &= \square - 3\\
8 + 3 &= 5x - 3 + \square\\
\square &= 5x\\
\frac{11}{5} &= \frac{5x}{\square}\\
\square &= x
\end{aligned}
$$

7. $2.1x - 45.3 = 17.3x + 23.1$

8. Solve: $3(x + 5) = 20 - x.$ GS

$$
\begin{aligned}
3(x + 5) &= 20 - x\\
3x + \square &= 20 - x\\
3x + 15 + x &= 20 - x + \square\\
\square + 15 &= 20\\
4x + 15 - \square &= 20 - 15\\
4x &= \square\\
\frac{4x}{4} &= \frac{5}{\square}\\
x &= \square
\end{aligned}
$$

9. Solve: $8(x - 2) - 15 = 4x + 2.$

Answers

5. 3.5 6. 2.2 7. −4.5

8. 1.25 9. 8.25

Guided Solutions:

6. $4x, 5x, 3, 11, 5, 2.2$

8. $15, x, 4x, 15, 5, 4, 1.25$

Note that in Example 4 the variable appears on the right side of the last equation. It does not matter whether the variable is isolated on the right or left side. What is important is that you have a clear direction to your work as you proceed from step to step.

◀ **Do Exercises 5–7.**

EXAMPLE 5 Solve: $5(x + 1) = 7x + 12.$

$$
\begin{aligned}
5(x + 1) &= 7x + 12\\
5 \cdot x + 5 \cdot 1 &= 7x + 12 && \text{Using the distributive law to remove parentheses}\\
5x + 5 &= 7x + 12 && \text{Simplifying}\\
5x + 5 - 7x &= 7x + 12 - 7x && \text{Subtracting } 7x \text{ from both sides}\\
-2x + 5 &= 12 && \text{Simplifying}\\
-2x + 5 - 5 &= 12 - 5 && \text{Subtracting 5 from both sides}\\
-2x &= 7\\
\frac{-2x}{-2} &= \frac{7}{-2} && \text{Dividing both sides by } -2\\
x &= -3.5
\end{aligned}
$$

Check: Check in the original equation.

$5(x + 1) = 7x + 12$	
$5(-3.5 + 1)$?	$7(-3.5) + 12$
$5(-2.5)$	$-24.5 + 12$
-12.5	-12.5 TRUE

The solution is -3.5.

◀ **Do Exercise 8.**

EXAMPLE 6 Solve: $9(x - 3) + 7 = 5x - 47.$

We use the distributive law and combine like terms before using the addition and division principles.

$$
\begin{aligned}
9(x - 3) + 7 &= 5x - 47\\
9x - 27 + 7 &= 5x - 47 && \text{Using the distributive law}\\
9x - 20 &= 5x - 47 && \text{Simplifying}\\
9x - 20 - 5x &= 5x - 47 - 5x && \text{Subtracting } 5x \text{ from both sides}\\
4x - 20 &= -47 && \text{Simplifying}\\
4x - 20 + 20 &= -47 + 20 && \text{Adding 20 to both sides}\\
4x &= -27 && \text{Simplifying}\\
\frac{4x}{4} &= -\frac{27}{4} && \text{Dividing both sides by 4}\\
x &= -6.75
\end{aligned}
$$

Check:

$9(x - 3) + 7 = 5x - 47$	
$9(-6.75 - 3) + 7$?	$5(-6.75) - 47$
$9(-9.75) + 7$	$-33.75 - 47$
$-87.75 + 7$	-80.75
-80.75	-80.75 TRUE

The solution is -6.75.

◀ **Do Exercise 9.**

5.7 Exercise Set

For Extra Help
MyMathLab®
MathXL® PRACTICE WATCH READ REVIEW

Reading Check

Determine whether each statement is true or false.

______ **RC1.** Equations involving decimals never have integer solutions.

______ **RC2.** An equation that does not involve decimals may have a decimal solution.

______ **RC3.** In the final step of a solution of an equation, the variable may appear on the right side or on the left side of the equation.

______ **RC4.** You should be able to give a reason for each step in your solution.

a Solve. Remember to check.

1. $5x = 27$

2. $4x = 75$

3. $16 \cdot y = 3.2$

4. $36 \cdot y = 14.76$

5. $-1.5t = 9.36$

6. $-1.2t = -11.4$

7. $x + 15.7 = 3.1$

8. $x + 13.9 = 4.2$

9. $1.25 = 3.8 + x$

10. $-1.3 = 4.05 + x$

11. $x - 0.37 = 1.6$

12. $x - 0.4 = 12$

13. $y - 9.8 = -1.42$

14. $y - 6 = -0.2$

15. $-9 = t - 3.1$

16. $-23 = t - 0.01$

17. $5x - 8 = 22$

18. $4x - 7 = 13$

19. $6.9x - 8.4 = 4.02$

20. $7.1x - 9.3 = 8.45$

21. $21.6 + 4.1t = 6.43$

22. $12.4 + 3.7t = 2.04$

23. $-26.25 = 7.5x + 9$

24. $-43.72 = 8.7x + 5$

25. $-4.2x + 3.04 = -4.1$

26. $-2.9x - 2.24 = -17.9$

27. $-3.05 = 7.24 - 3.5t$

28. $-4.62 = 5.68 - 2.5t$

29. $3 - 1.2y = -2.4$

30. $6 - 3.5y = 12.3$

b Solve. Remember to check.

31. $9x - 2 = 5x + 34$

32. $8x - 5 = 6x + 9$

33. $2x + 6 = 7x - 10$

34. $3x + 4 = 11x - 6$

35. $5y - 3 = 4 + 9y$

36. $6y - 5 = 8 + 10y$

37. $5.9x + 67 = 7.6x + 16$

38. $2.1x + 42 = 5.2x - 20$

39. $7.8a + 2 = 2.4a + 19.28$

40. $7.5a - 5.16 = 3.1a + 12$

41. $6(x + 2) = 4x + 30$

42. $5(x + 3) = 3x + 23$

43. $5(x + 3) = 15x - 6$

44. $2(x + 3) = 4x - 11$

45. $7a - 9 = 15(a - 3)$

46. $2a - 7 = 12(a - 3)$

47. $1.5(y - 6) = 1.3 - y$

48. $2.3(5 - y) = 0.2y + 1.7$

49. $2.9(x + 8.1) = 7.8x - 3.95$

50. $2(x + 7.3) = 6x - 0.83$

51. $-6.21 - 4.3t = 9.8(t + 2.1)$

52. $-7.37 - 3.2t = 4.9(t + 6.1)$

53. $4(x - 2) - 9 = 2x + 9$

54. $9(x - 4) + 13 = 4x - 23$

55. $2(4y - 1.8) + 0.4 = 8(2y - 0.4)$

56. $3(1.2y + 5) = 2(2.5y - 7) + 1$

57. $43(7 - 2x) + 34 = 50(x - 4.1) + 744$

58. $34(5 - 3.5x) = 12(3x - 8) + 653.5$

Skill Maintenance

Find the area of each figure. [3.6b]

59.

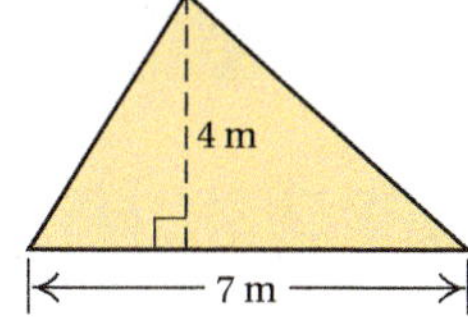

60.

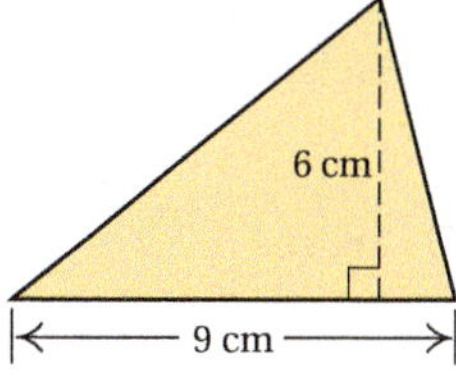

61.

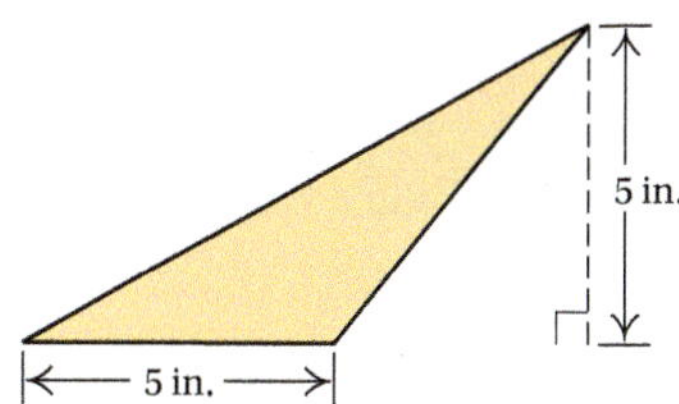

62.

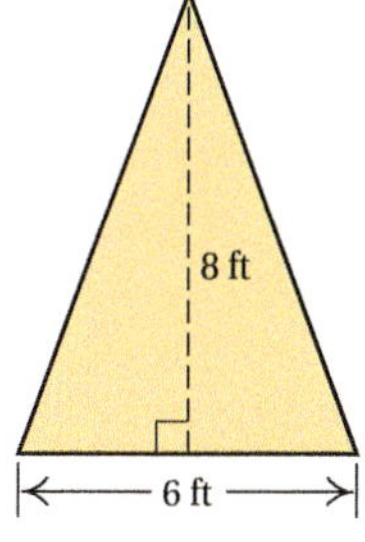

63.

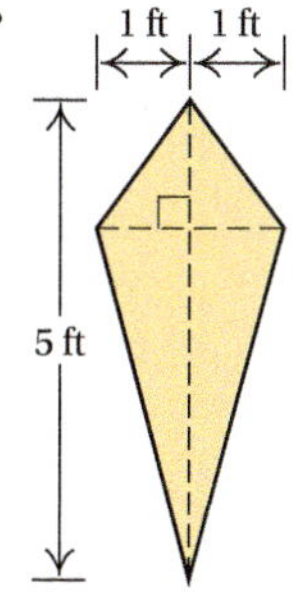

64.

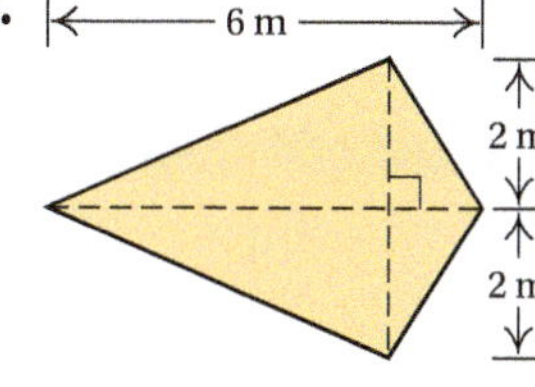

65. Subtract: $\frac{3}{25} - \frac{7}{10}$. [4.3a]

66. Simplify: $\frac{0}{-18}$. [3.3b]

67. Add: $-17 + 24 + (-9)$. [2.2a]

68. Solve: $3x - 10 = 14$. [2.8d]

Synthesis

Solve.

69. 🖩 $7.035(4.91x - 8.21) + 17.401 = 23.902x - 7.372815$

70. 🖩 $8.701(3.4 - 5.1x) - 89.321 = 5.401x + 74.65787$

71. $5(x - 4.2) + 3[2x - 5(x + 7)] = 39 + 2(7.5 - 6x) + 3x$

72. $14(2.5x - 3) + 9x + 5 = 4(3.25 - x) + 2[5x - 3(x + 1)]$

73. 🖩 $3.5(4.8x - 2.9) + 4.5 = 9.4x - 3.4(x - 1.9)$

74. 🖩 $4.19 - 1.8(4.5x - 6.4) = 3.1(9.8 + x)$

5.8 Applications and Problem Solving

OBJECTIVES

a Translate key phrases to algebraic expressions.

b Solve applied problems involving decimals.

SKILL TO REVIEW

Objective 1.8a: Solve applied problems involving addition, subtraction, multiplication, or division of whole numbers.

Solve.

1. A piece of pecan pie has 502 calories. This is 186 calories more than a piece of pumpkin pie contains. How many calories does a piece of pumpkin pie have?

a TRANSLATING TO ALGEBRAIC EXPRESSIONS

To translate problems to equations, we need to be able to translate phrases to algebraic expressions. Certain key words in phrases help direct the translation.

KEY WORDS	SAMPLE PHRASE OR SENTENCE	TRANSLATION
Addition (+)		
added to	350 lb was added to the car's weight.	$w + 350$
sum of	The sum of a number and 10	$n + 10$
plus	13 plus some number	$13 + x$
more than	176 more than last year's enrollment	$r + 176$
increased by	The original estimate, increased by 50	$y + 50$
Subtraction (−)		
subtracted from	2 oz was subtracted from the bag's weight.	$w - 2$
difference of	The difference of two prices	$p - q$
minus	A construction crew of size *c*, minus 4 workers	$c - 4$
less than	18 less than the number of volunteers last month	$v - 18$
decreased by	An essay's grade, decreased by 5 points	$g - 5$
Multiplication (·)		
multiplied by	The number of mints in a box, multiplied by 5	$5 \cdot m$
product of	The product of two numbers	$a \cdot b$
times	10 times Jon's age	$10j$
twice	Twice the wholesale price	$2w$
of	$\frac{1}{2}$ of the number of pages assigned	$\frac{1}{2}p$
Division (÷)		
divided by	A 16-oz bag of almonds, divided by 5	$16 \div 5$
quotient of	The quotient of 60 and 3	$60 \div 3$
divided into	6 divided into the cost of the meal	$c \div 6$
ratio of	The ratio of 456 to the number of gallons of gasoline	$456/g$

It is helpful to choose a descriptive variable to represent the unknown. For example, *w* suggests weight and *g* suggests the number of gallons of gasoline.

Answer

Skill to Review:

1. 316 calories

The following tips are helpful in translating phrases to algebraic expressions.

TIPS FOR TRANSLATING PHRASES TO ALGEBRAIC EXPRESSIONS

- Replace the unknown in the statement with a specific number before translating using a variable.
- Write down what each variable represents.
- Check the translation using another number to see if it matches the phrase.
- Be especially careful with order when subtracting and dividing.

EXAMPLE 1 Translate each phrase to an algebraic expression.

a) A number added to 5

b) A number subtracted from 5

a) Let n represent the number. Then the phrase "a number added to 5" can be translated $5 + n$, or $n + 5$.

b) Let n represent the number. Then the phrase "a number subtracted from 5" can be translated $5 - n$.

Do Exercises 1 and 2. ▶

Translate to an algebraic expression.

1. Twenty more than a number

2. Twenty less than a number

Caution!

Note that in Example 1, there are two correct translations for part (a) and only one correct translation for part (b). Recall that numbers can be added in either order, so $5 + n$ and $n + 5$ represent the same value. However, order is important in subtraction, so $5 - n$ and $n - 5$ do NOT represent the same value.

EXAMPLE 2 Translate each phrase to an algebraic expression.

a) Than's age increased by six

b) Half of some number

c) Twice the cost

d) Seven more than twice the weight

e) Fifteen divided by a number

f) A number divided by fifteen

g) Six less than the product of two numbers

h) Nine times the sum of two numbers

	Phrase	*Variable(s)*	*Algebraic Expression*
a)	Than's age increased by six	Let $a =$ Than's age.	$a + 6$, or $6 + a$
b)	Half of some number	Let $n =$ the number.	$\frac{1}{2}n$, or $\frac{n}{2}$, or $n \div 2$
c)	Twice the cost	Let $c =$ the cost.	$2c$
d)	Seven more than twice the weight	Let $w =$ the weight.	$2w + 7$, or $7 + 2w$

Answers

1. Let $n =$ the number; $n + 20$, or $20 + n$
2. Let $n =$ the number; $n - 20$

e) Fifteen divided by a number	Let $x =$ the number.	$15 \div x$, or $\dfrac{15}{x}$
f) A number divided by fifteen	Let $x =$ the number.	$x \div 15$, or $\dfrac{x}{15}$
g) Six less than the product of two numbers	Let m and $n =$ the numbers.	$mn - 6$
h) Nine times the sum of two numbers	Let a and $b =$ the numbers.	$9(a + b)$

Translate to an algebraic expression.

3. The height of a tree, decreased by 50
4. Four less than ten times a number
5. Fourteen more than the product of the hourly rate and the number of hours worked

Do Exercises 3–5.

b SOLVING APPLIED PROBLEMS

EXAMPLE 3 *Canals.* The Panama Canal in Panama is 50.7 mi long. The Suez Canal in Egypt is 119.9 mi long. How much longer is the Suez Canal?

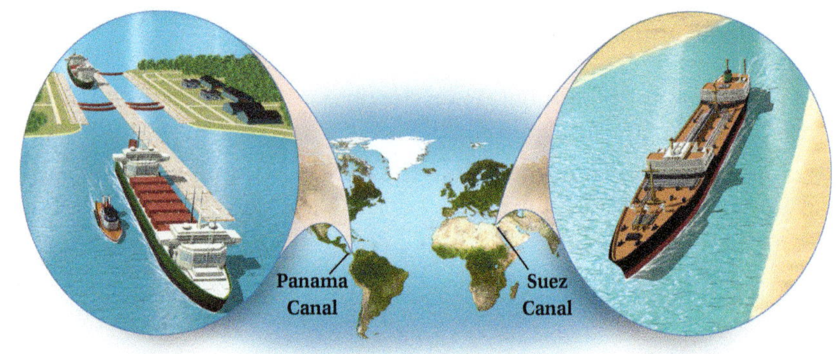

1. **Familiarize.** We let $l =$ the distance in miles that the length of the longer canal differs from the length of the shorter canal.
2. **Translate.** We translate as follows, using the given information:

Length of Panama Canal, the shorter canal	plus	Additional length	is	Length of Suez Canal, the longer canal
↓	↓	↓	↓	↓
50.7 mi	$+$	l	$=$	119.9 mi.

3. **Solve.** We solve the equation by subtracting 50.7 mi on both sides:

$$50.7 + l = 119.9$$
$$50.7 + l - 50.7 = 119.9 - 50.7$$
$$l = 69.2.$$

4. **Check.** We can check by adding.

$$\begin{array}{r} 50.7 \\ +\ 69.2 \\ \hline 119.9 \end{array}$$

The answer checks.

5. **State.** The Suez Canal is 69.2 mi longer than the Panama Canal.

Do Exercise 6.

6. *Debit-Card Transactions.* U.S. debit-card transactions in 2009 totaled 38.5 billion. The number of transactions in 2012 was 52.6 billion. How many more debit-card transactions were there in 2012 than in 2009?

Source: *The Nilson Report*

Answers

3. Let $h =$ the height of a tree; $h - 50$
4. Let $x =$ the number; $10x - 4$
5. Let $r =$ the hourly rate and $w =$ the number of hours worked; $rw + 14$, or $14 + rw$
6. 14.1 billion transactions

EXAMPLE 4 *iPad Purchase.* A large architectural firm spent \$10,399.74 on 26 iPads for its architects. How much did each iPad cost?

1. **Familiarize.** We let $c =$ the cost of each iPad.
2. **Translate.** We translate as follows:

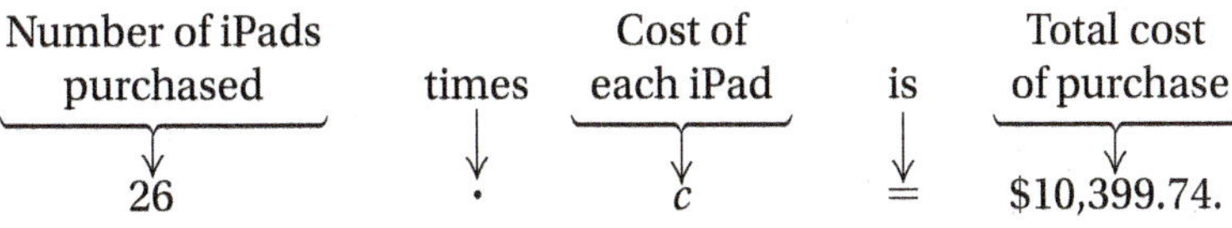

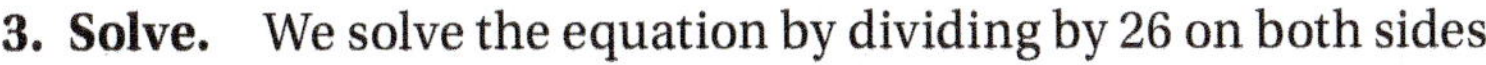

$$26 \cdot c = \$10,399.74.$$

3. **Solve.** We solve the equation by dividing by 26 on both sides.

$$\frac{26 \cdot c}{26} = \frac{10,399.74}{26}$$
$$c = \frac{10,399.74}{26}$$
$$c = 399.99$$

4. **Check.** We check by estimating:

$$10,399.74 \div 26 \approx 10,000 \div 25 = 400.$$

Since 400 is close to 399.99, the answer is probably correct.

5. **State.** The cost of each iPad was \$399.99.

Do Exercise 7. 

7. *Mileage Rates.* For a recent year, the Internal Revenue Service allowed a tax deduction of 23.5¢ per mile driven for medical or moving purposes. What deduction, in dollars, would be allowed for driving 1862 mi during a move?

Source: Internal Revenue Service

Multistep Problems

EXAMPLE 5 *Tracking a Bank Balance.* Revenue of U.S. banks from checking account overdraft charges was \$31.5 billion in a recent fiscal year. (**Source:** Moebs Services) To avoid overdrafts and to track her spending, Maggie keeps a running account of her banking transactions. She checks her balance online, but because of pending amounts that post later, she keeps a separate record. Maggie had \$2432.27 in her account. She used her debit card to pay her rent of \$835 and make purchases of \$14.13, \$38.60, and \$205.98. She then deposited her weekly pay of \$748.35. What was her balance after these transactions?

1. **Familiarize.** We first find the total of the debits. Then we find how much is left in the account after the debits are deducted. Finally, we add to this amount the deposit to find the balance in the account after all the transactions.

2, 3. **Translate and Solve.** We let $d =$ the total amount of the debits. We are combining amounts: $\$835 + \$14.13 + \$38.60 + \$205.98 = d.$ To solve the equation, we add.

835.00	First debit
14.13	Second debit
38.60	Third debit
+ 205.98	Fourth debit
1093.71	Total debits

Thus, $d = 1093.71.$

Answer

7. \$437.57

Now let a = the amount in the account after the debits are deducted. We subtract: $2432.27 − $1093.71 = a.

2432.27	Original amount
− 1093.71	Total debits
1338.56	New amount

Thus, $a = 1338.56$.

Finally, we let f = the amount in the account after the paycheck is deposited.

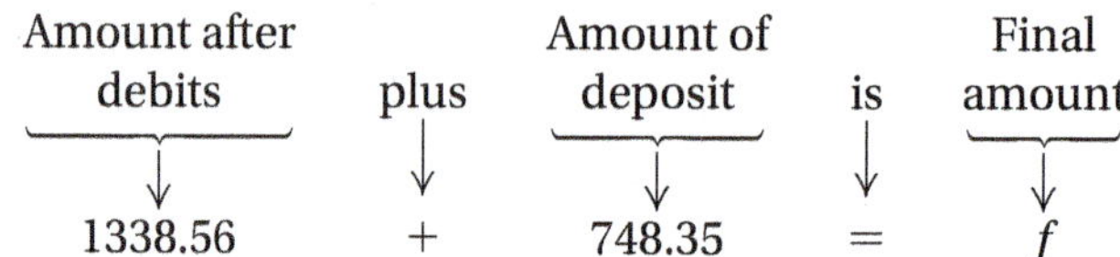

Amount after debits
plus
Amount of deposit
is
Final amount
1338.56
+
748.35
=
f

To solve the equation, we add.

1338.56	Balance after debits
+ 748.35	Paycheck deposit
2086.91	Final amount

4. **Check.** We repeat the computations.

5. **State.** Maggie had $2086.91 in her account after all the transactions.

8. ***Bank Balance.*** Stephen had $915.22 in his checking account. He used his debit card to pay a charge card minimum payment of $36 and to make purchases of $67.50, $178.23, and $429.05. He then deposited his weekly pay of $570.91. How much was in his account after these transactions?

Do Exercise 8.

EXAMPLE 6 ***Gas Mileage.*** Ava filled her gas tank and noted that the odometer read 67,507.8. After the next fill-up, the odometer read 68,006.1. It took 16.5 gal to fill the tank. How many miles per gallon did Ava's car get?

1. **Familiarize.** We first make a drawing.

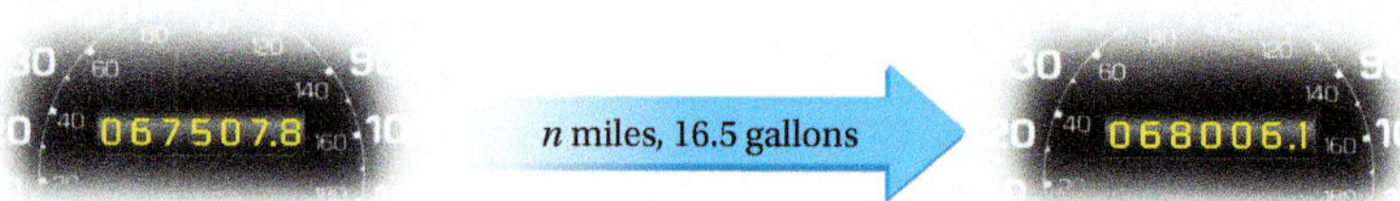

This is a two-step problem. First, we find the number of miles that have been driven between fill-ups. We let n = the number of miles driven and m = the number of miles per gallon.

2, 3. **Translate and Solve.** We translate and solve as follows:

First reading	plus	Number of miles driven	is	Second reading
↓	↓	↓	↓	↓
67,507.8	+	n	=	68,006.1.

To solve the equation, we subtract 67,507.8 on both sides:

$$n = 68{,}006.1 - 67{,}507.8$$
$$= 498.3.$$

$$\begin{array}{r} 68{,}006.1 \\ -\ 67{,}507.8 \\ \hline 498.3 \end{array}$$

Next, we divide the total number of miles driven by the number of gallons. This gives us m, that is, the gas mileage. The division that corresponds to the situation is $498.3 \div 16.5 = m$.

$$\begin{array}{r} 30.2 \\ 16.5\overline{)498.30} \\ 495 \\ \hline 330 \\ 330 \\ \hline 0 \end{array}$$

Thus, $m = 30.2$.

Answer

8. $775.35

4. **Check.** To check, we first multiply the number of miles per gallon by the number of gallons to find the number of miles driven:

$$16.5 \times 30.2 = 498.3.$$

Then we add 498.3 to 67,507.8 to find the new odometer reading:

$$67{,}507.8 + 498.3 = 68{,}006.1.$$

The gas mileage of 30.2 checks.

5. **State.** Ava's car got 30.2 miles per gallon.

Do Exercise 9. ▶

9. ***Gas Mileage.*** John filled his gas tank and noted that the odometer read 38,320.8. After the next fill-up, the odometer read 38,735.5. It took 14.5 gal to fill the tank. How many miles per gallon did John get?

Example 7 involves a formula giving the area of a circle.

In any circle, a **diameter** is a segment that passes through the center of the circle with endpoints on the circle. A **radius** is a segment with one endpoint on the center and the other endpoint on the circle. The area, A, of a circle with radius of length r is given by

$$A = \pi \cdot r^2,$$

where $\pi \approx 3.14$.

The length r of a radius of a circle is half the length d of a diameter:

$$r = \frac{1}{2}d.$$

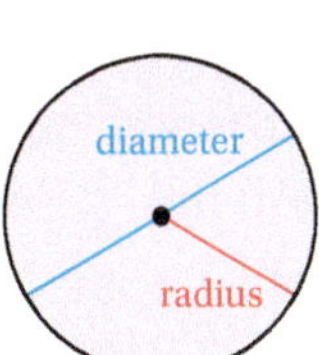

EXAMPLE 7 The Northfield Tap and Die Company stamps 6-cm-wide discs out of metal squares that are 6 cm by 6 cm. How much metal remains after the disc has been punched out?

1. **Familiarize.** We make, and label, a drawing. We let a = the amount of metal remaining, in square centimeters, and list the relevant formulas.

For a square with sides of length s, *Area* $= s^2$.

For a circle with radius of length r, *Area* $= \pi \cdot r^2$, where $\pi \approx 3.14$.

The radius r of a circle is half of its diameter: $r = \frac{1}{2}d$.

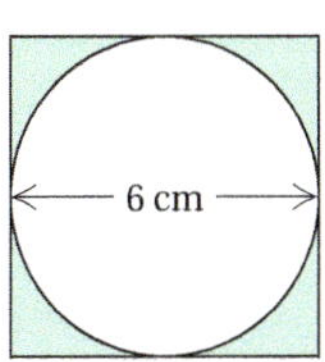

The shaded area is the amount of metal remaining.

We also list the known information about the square and the disc:

Square: The side $s = 6$ cm.

Disc: The diameter $d = 6$ cm.

The radius $r = \frac{1}{2}(6\text{ cm}) = 3$ cm.

2. **Translate.** To find the amount left over, we subtract the area of the disc from the area of the square.

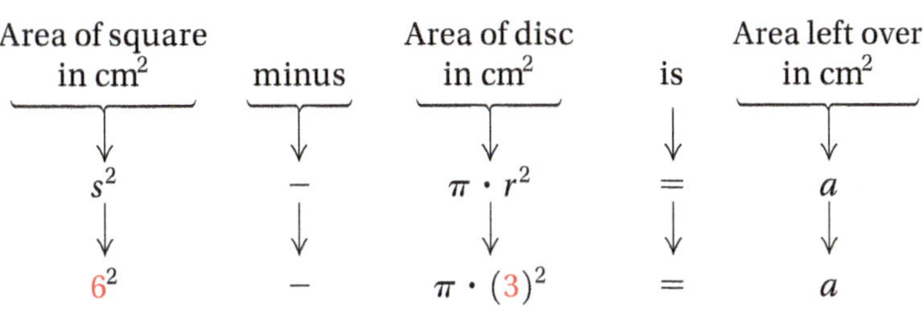

Answer

9. 28.6 mpg

3. **Solve.** We simplify as follows.

$$6^2 - 3.14(3)^2 = a$$
$$36 - 3.14 \cdot 9 = a$$
$$36 - 28.26 = a$$
$$7.74 = a$$

4. **Check.** We can repeat our calculation as a check. Note that 7.74 is less than the area of the disc, which in turn is less than the area of the square. This agrees with the impression given by our drawing.
5. **State.** The amount of material left over is 7.74 cm^2.

10. Suppose that an 8-in.-wide disc is punched out of an 8-in. by 8-in. sheet of metal. How much material is left over?

◀ **Do Exercise 10.**

EXAMPLE 8 ***Photo Books.*** Users of Apple's iPhoto application can create their own photo books. The price of a small 20-page book is $24.99. Additional pages are 79 cents per page. Marta has $35 to spend on a book. What is the greatest number of pages she can put in the book?

Source: www.apple.com

1. **Familiarize.** Suppose that Marta put 30 pages in the book. She would have to pay an additional price per page for 30 − 20, or 10, pages. The price would be the base price plus the charge for extra pages, or

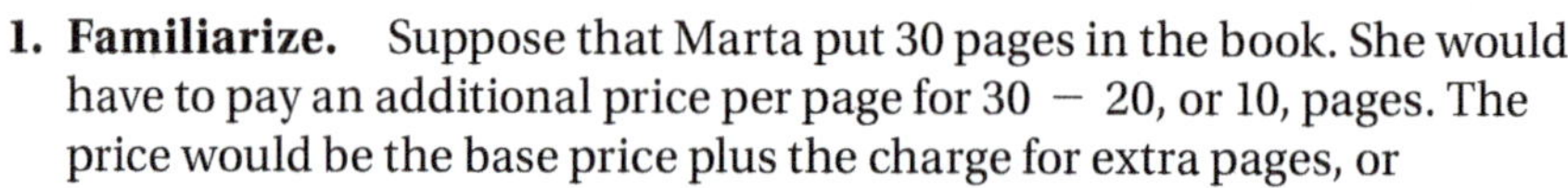

$$\$24.99 + \$0.79 \cdot 10 = \$24.99 + \$7.90 = \$32.89.$$

This familiarizes us with the way in which price is calculated. Note that we convert 79 cents to $0.79 so that only one unit, dollars, is used. Note also that Marta can put more than 30 pages in the book.

We let p = the number of extra pages Marta adds to the 20-page book. Note that the total number of pages in the book is then $20 + p$.

2. **Translate.** The problem can be rephrased and translated as follows.

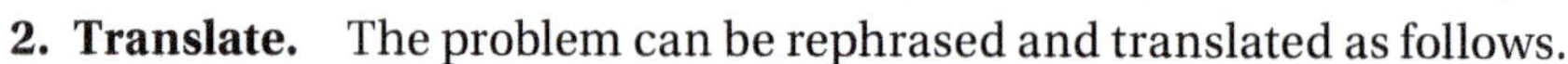

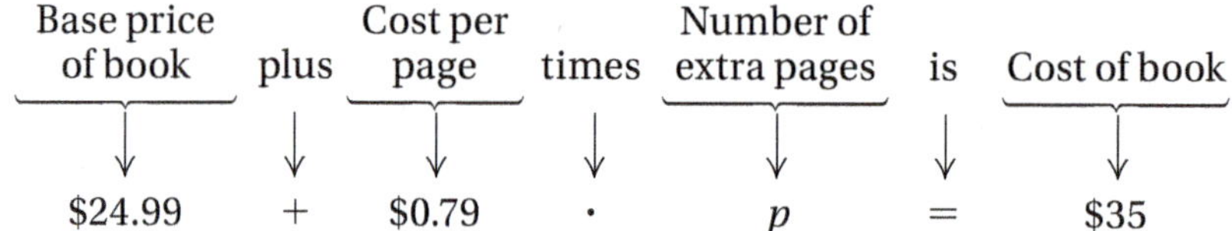

Base price of book	plus	Cost per page	times	Number of extra pages	is	Cost of book
↓	↓	↓	↓	↓	↓	↓
$24.99	+	$0.79	·	p	=	$35

3. **Solve.** We solve the equation.

$$24.99 + 0.79p = 35$$
$$0.79p = 10.01 \quad \text{Subtracting 24.99 from both sides}$$
$$p = \frac{10.01}{0.79} \quad \text{Dividing both sides by 0.79}$$
$$p \approx 12.7 \quad \text{Rounding}$$

4. **Check.** We check in the original problem. Since Marta cannot pay for parts of a page, we must round the answer *down* to 12 in order to keep the cost under $35. This makes the total length of the book 20 + 12, or 32, pages. If Marta adds 12 pages to the book, the additional cost will be 12 times $0.79, or $9.48. If we add $9.48 to the base price of $24.99, we get $34.47, which is just under the $35 Marta has to spend.
5. **State.** With $35, Marta can make a 32-page book.

11. **Bike Rentals.** Mike's Bikes rents mountain bikes. The shop charges $4.00 insurance for each rental plus $6.00 per hour. For how many hours can a person rent a bike with $25.00?

◀ **Do Exercise 11.**

Answers

10. 13.76 in^2 11. 3.5 hr

Problems with More Than One Unknown

EXAMPLE 9 *Multi-sport Recreation.* Around the Bend Expeditions offers a two-day canyon trip combining biking and canoeing. Those participating will ride mountain bikes for 12 miles longer than the distance paddled. The total length of the trip is 43 miles. How long is the biking portion of the trip, and how long is the canoeing portion of the trip?

1. **Familiarize.** We first list the quantities we are asked to find:

   Length of biking portion, in miles
   Length of canoeing portion, in miles

   We will want to represent both quantities using only one variable. To do so, we use the second sentence in the problem: *Those participating will ride mountain bikes for 12 miles longer than the distance paddled.* Since the biking portion of the trip is described in terms of the canoeing portion, we let $x =$ the length of the canoeing portion, in miles. Then the length of the biking portion, also in miles, is $x + 12$.

   Length of biking portion, in miles: $x + 12$
   Length of canoeing portion, in miles: x

2. **Translate.** We use the total length of the trip to translate to an equation. Note that all lengths are in miles.

Length of biking portion	plus	Length of canoeing portion	is	Total length of trip
↓	↓	↓	↓	↓
$(x + 12)$	$+$	x	$=$	43

3. **Solve.** We solve the equation.

$$(x + 12) + x = 43$$
$$x + 12 + x = 43 \quad \text{Removing parentheses}$$
$$2x + 12 = 43 \quad \text{Combining like terms}$$
$$2x = 31 \quad \text{Subtracting 12 from both sides}$$
$$x = 15.5 \quad \text{Dividing both sides by 2}$$

   Recall that we are looking for two quantities. We will use the value of x to find both of them. We return to the list of unknowns in the *Familiarize* step.

   Length of biking portion, in miles: $x + 12 = 15.5 + 12 = 27.5$
   Length of canoeing portion, in miles: $x = 15.5$

4. **Check.** There are two statements in the problem to verify. First, since $27.5 - 15.5 = 12$, the biking portion is 12 miles longer than the canoeing portion. Second, since $27.5 + 15.5 = 43$, the total length of the trip is 43 miles.
5. **State.** The biking portion of the trip is 27.5 miles, and the canoeing portion is 15.5 miles.

Do Exercise 12. ▶

Caution!

Do not skip the *Familiarize* step, particularly in problems involving more than one unknown. Describing the unknowns using one variable is often the most challenging part of the solution process.

12. Holly and Lorenzo worked together on a project for a psychology class. Holly spent twice as much time on the project as Lorenzo did. They spent a total of 10.5 hours on the project. How much time did each person work on the project?

Answer

12. Holly: 7 hr; Lorenzo: 3.5 hr

Translating for Success

The goal of these matching questions is to practice step (2), Translate, of the five-step problem-solving process. Translate each word problem to an equation and select a correct translation from equations A–O.

1. *Gas Mileage.* Art filled his SUV's gas tank and noted that the odometer read 38,271.8. At the next fill-up, the odometer read 38,677.9. It took 28.4 gal to fill the tank. How many miles per gallon did the SUV get?

2. *Dimensions of a Parking Lot.* A store's parking lot is a rectangle that measures 85.2 ft by 52.3 ft. What is the area of the parking lot?

3. *Game Snacks.* Three students pay $18.40 for snacks at a football game. What is each student's share of the cost?

4. *Electrical Wiring.* An electrician needs 1314 ft of wiring cut into $2\frac{1}{2}$-ft pieces. How many pieces will she have?

5. *College Tuition.* Wayne needs $4638 for the fall semester's tuition. On the day of registration, he has only $3092. How much does he need to borrow?

A. $2\frac{1}{2} \cdot n = 1314$

B. $18.4 \times 3.87 = n$

C. $n = 85.2 \times 52.3$

D. $19 - (-4) = n$

E. $3 \times 18.40 = n$

F. $2\frac{1}{2} \cdot 1314 = n$

G. $3092 + n = 4638$

H. $18.4 \cdot n = 3.87$

I. $\frac{406.1}{28.4} = n$

J. $52.3 \cdot n = 85.2$

K. $n = 19 + (-4)$

L. $52.3 + n = 85.2$

M. $3092 + 4638 = n$

N. $3 \cdot n = 18.40$

O. $85.2 + 52.3 = n$

Answers on page A-10

6. *Cost of Gasoline.* What is the cost, in dollars, of 18.4 gal of gasoline at $3.87 per gallon?

7. *Temperature.* At noon, the temperature in Pierre was 19°F. At midnight, the temperature had fallen to −4°F. By how many degrees had the temperature fallen?

8. *Acres Planted.* This season Sam planted 85.2 acres of corn and 52.3 acres of soybeans. Find the total number of acres that he planted.

9. *Amount Inherited.* Tara inherited $2\frac{1}{2}$ times as much as her cousin. Her cousin received $1314. How much did Tara receive?

10. *Travel Funds.* The athletic department needs travel funds of $4638 for the tennis team and $3092 for the golf team. What is the total amount needed for travel?

5.8 Exercise Set

For Extra Help
MyMathLab®
MathXL®

PRACTICE WATCH READ REVIEW

Reading Check

Complete each step in the five-step problem-solving strategy with the correct word from the following list.

Solve Familiarize State Translate Check

Five-Step Problem-Solving Strategy

RC1. ________ yourself with the problem situation.

RC2. ________ the problem to an equation.

RC3. ________ the equation.

RC4. ________ the solution.

RC5. ________ the answer using a complete sentence.

a Translate to an algebraic expression. Choice of variables used may vary.

1. Five more than Ron's age

2. The product of four and t

3. 6 more than b

4. 7 more than Jen's weight

5. 9 less than c

6. 4 less than d

7. A number decreased by 16

8. A number increased by 20

9. 8 times Nate's speed

10. The ratio of a number and 100

11. x divided by 17

12. 100 divided by the hourly rate

13. 20 added to half of a number

14. 18 subtracted from twice a number

15. 20 less than 4 times a number

16. 35 more than 10 times a number

17. The sum of the box's length and width

18. The Cessna's speed minus the wind speed

19. 10 more than the product of the rate and the time

20. 50 less than the product of the length and width

21. The sum of 10 times a number and the number

22. The difference of a number and twice the number

23. 5 times the difference of two numbers

24. One fourth of the sum of two numbers

Hurricane Damage. The amounts of damage caused by the five most costly Atlantic hurricanes in the United States are shown in the table below. Use this table to do Exercises 25 and 26.

Most Costly Hurricanes

RANK	HURRICANE	YEAR	COST IN 2010 DOLLARS (in billions)
1	Katrina	2005	$105.8
2	Andrew	1992	45.6
3	Ike	2008	27.8
4	Wilma	2005	20.6
5	Ivan	2004	19.8

SOURCE: National Hurricane Center

25. How much more costly was Hurricane Katrina than Hurricane Andrew?

26. What was the total cost of the two hurricanes that occurred in 2005?

27. *Gasoline Cost.* What is the cost, in dollars, of 20.4 gal of gasoline at 324.9 cents per gallon? (324.9 cents = $3.249) Round the answer to the nearest cent.

28. *Gasoline Cost.* What is the cost, in dollars, of 15.3 gal of gasoline at 389.9 cents per gallon? (389.9 cents = $3.899) Round the answer to the nearest cent.

29. *Cost of Bottled Water.* The cost of a year's supply of a popular brand of bottled water, based on consumption of 64 oz per day, at the supermarket price of $3.99 for a six-pack of half-liter bottles is $918.82. This is $918.31 more than the cost of drinking the same amount of tap water for a year. What is the cost of drinking tap water for a year?

Source: American Water Works Association

30. *Body Temperature.* Normal body temperature is 98.6°F. During an illness, a patient's temperature rose 4.2°. What was the new temperature?

31. *Lottery Winnings.* The largest lottery jackpot in the United States totaled $656,000,000 and was shared equally by 3 winners. How much was each winner's share? Round to the nearest cent.

Source: Mega Millions

32. *Lunch Costs.* A group of 4 students pays $47.84 for lunch and splits the cost equally. What is each person's share?

33. *100-Meter Record.* The fastest speed clocked for a cheetah running a distance of 100 m is 5.95 sec. The men's world record for the 100-m dash is held by Jamaican Usain Bolt. His time was 3.63 sec more than the cheetah's. What is the men's 100-m record held by Usain Bolt?

Source: "Cheetahs on the Edge," by Roff Smith, *National Geographic*, November 2012.

34. *Record Movie Openings.* The movie *Marvel's The Avengers* took in $207.4 million on its first weekend. This topped the previous record for opening-weekend revenue, set by *Harry Potter and the Deathly Hallows: Part II*, by $38.2 million. How much did *Harry Potter and the Deathly Hallows: Part II* take in on its opening weekend?

Source: Nash Information Services

35. *Odometer Reading.* The Binford family's odometer reads 22,456.8 at the beginning of a trip. The family's online driving directions tell them that they will be driving 234.7 mi. What will the odometer read at the end of the trip?

36. *Miles Driven.* Petra bought gasoline when the odometer read 14,296.3. At the next gasoline purchase, the odometer read 14,515.8. How many miles had been driven?

37. *Gas Mileage.* Peggy filled her van's gas tank and noted that the odometer read 26,342.8. After the next filling, the odometer read 26,736.7. It took 19.5 gal to fill the tank. How many miles per gallon did the van get?

38. *Gas Mileage.* Henry filled his Honda's gas tank and noted that the odometer read 18,943.2. After the next filling, the odometer read 19,306.2. It took 13.2 gal to fill the tank. How many miles per gallon did the car get?

39. Andrew bought a DVD of the movie *Horton Hears a Who* for his nephew for $23.99 plus $1.68 sales tax. He paid for it with a $50 bill. How much change did he receive?

40. Claire bought a copy of the book *Make Way for Ducklings* for her daughter for $16.95 plus $0.85 sales tax. She paid for it with a $20 bill. How much change did she receive?

41. *Health Care.* Phil injects 38 units of insulin each day for a week. Each unit is 0.01 cc (cubic centimeter). How many cc's of insulin does he use in a week?

42. *Chemistry.* The water in a filled tank weighs 748.45 lb. One cubic foot of water weighs 62.5 lb. How many cubic feet of water does the tank hold?

43. *Egg Costs.* A restaurant owner bought 20 dozen eggs for $25.80. Find the cost of each egg to the nearest tenth of a cent (thousandth of a dollar).

44. *Weight Loss.* A person weighing 170 lb burns 8.6 calories per minute while mowing a lawn. One must burn about 3500 calories in order to lose 1 lb. How many pounds would be lost by mowing for 2 hr? Round to the nearest tenth.

45. *Loose Change.* In 2012, passengers at Miami International Airport left $39,613 in loose change at airport checkpoints. This was $12,712.79 more than passengers left that year at Las Vegas McCarran International Airport. How much change was left at the Las Vegas airport?

Source: Transportation Security Administration

46. *Online Ad Spending.* It is projected that $132.1 billion will be spent for online advertising in 2015. This is $63.7 billion more than in 2010. What was spent on online advertising in 2010?

Source: eMarketer, June 2011

47. *Pole Vault Pit.* Find the area and the perimeter of the landing area of the pole vault pit shown here.

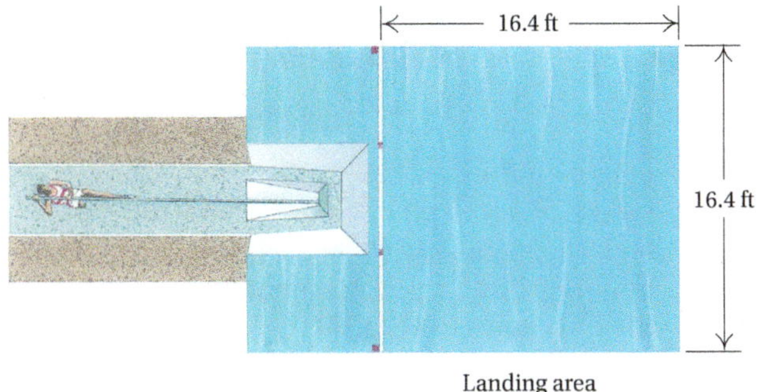

48. *Stamp.* Find the area and the perimeter of the stamp shown here.

49. *Study Cards.* An instructor allows her students to bring one 7.6-cm by 12.7-cm index card to the final exam, with notes of any sort written on the card. If both sides of the card are used, how much area is available for notes?

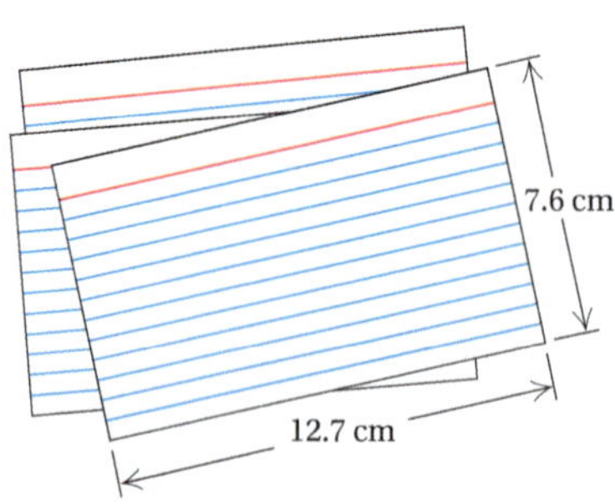

50. *Stamps.* Find the total area of the stamps shown.

Find the length d in each figure.

51.

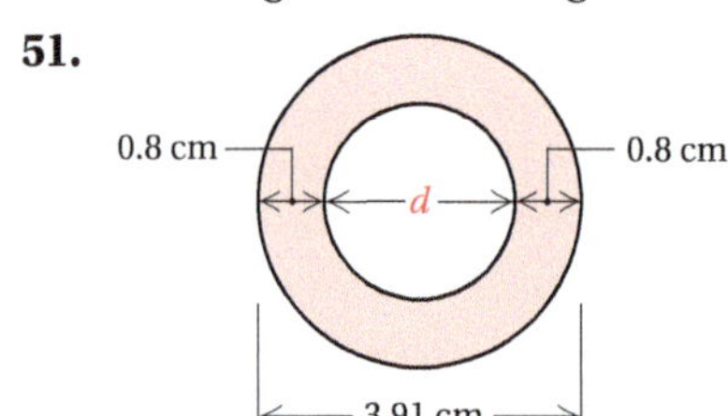

52.

0.9 cm — d — 0.9 cm

4.52 cm

53. *Carpentry.* A round, 6-ft-wide hot tub is being built into a 12-ft by 30-ft rectangular deck. How much decking is needed for the surface of the deck?

54. *Travel Poster.* Sam is decorating his dorm room with a travel poster. The dimensions of the poster are as shown. How much area is not devoted to the painting?

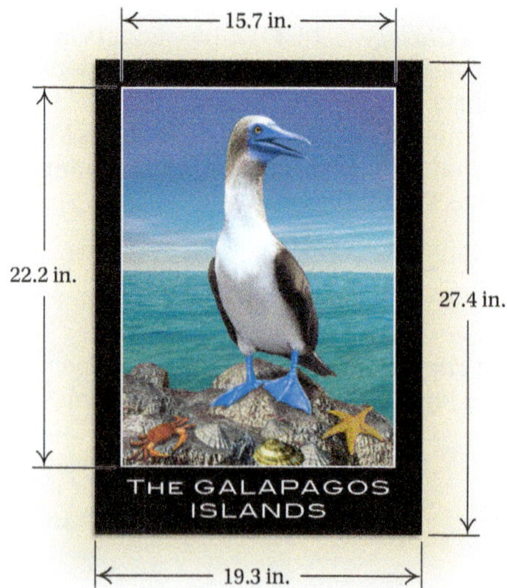

55. *Web Space.* Penn State charges \$8 a month for Internet use plus 7 cents for each gigabyte over 10 gigabytes of Web space used. Nikki's bill for one month was \$16.75. How many additional gigabytes did she pay for?

Source: Penn State Information Technology Services

56. *Service Calls.* JoJo's Service Center charges \$30 for a house call plus \$37.50 for each hour the job takes. For how long has a repairperson worked on a house call if the bill comes to \$123.75?

57. Lot A measures 250.1 ft by 302.7 ft. Lot B measures 389.4 ft by 566.2 ft. What is the total area of the two lots?

58. A 4-ft by 4-ft tablecloth is cut from a round tablecloth that is 6 ft wide. Find the area of the cloth left over.

59. Frank has been sent to the store with $40 to purchase 6 lb of cheese at $4.79 a pound and as many bottles of seltzer, at $0.64 a bottle, as possible. How many bottles of seltzer should Frank buy?

60. Janice has been sent to the store with $30 to purchase 5 pt of strawberries at $2.49 a pint and as many bags of chips, at $1.39 a bag, as possible. How many bags of chips should Janice buy?

61. *Endangered Species.* Part of the giant panda's habitat is protected by the Chinese government. There are 0.4 million more acres unprotected than there are protected acres. The total size of the giant panda's habitat is 5.4 million acres. How many acres are protected and how many are not protected?

Source: www.worldwildlife.org

62. *Hours of Daylight.* On December 22, Fairbanks, Alaska, has 16.6 more hours of darkness than it has daylight. How many hours of daylight and how many hours of darkness are there in Fairbanks on that day? [*Hint:* There are 24 hours in a day.]

Source: "Alaska's Winter Daylight," by Kimi Ross, at www.bellaonline.com

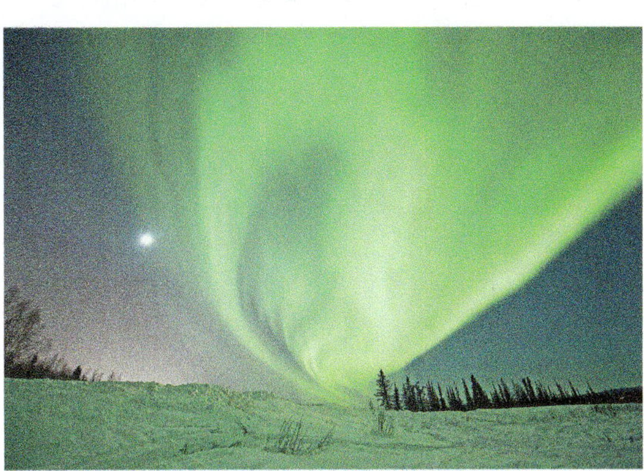

63. *Vacation Spending.* Emily spent three times as much on lodging as she did for food on a recent vacation. She spent a total of $261.20 for food and lodging. How much did she spend for each?

64. *Homework.* Ian spends twice as long on his science homework as he does on his history homework. One week, he spent a total of 13.5 hours on science and history homework. How much time did he spend on each subject?

65. *Construction Pay.* A construction worker is paid $18.50 per hour for the first 40 hr of work, and time and a half, or $27.75 per hour, for any overtime exceeding 40 hr per week. One week she works 46 hr. How much is her pay?

66. *Summer Work.* Zachary worked 53 hr during a week one summer. He earned $7.50 per hour for the first 40 hr and $11.25 per hour for overtime (hours exceeding 40). How much did Zachary earn during the week?

67. *Loan Payment.* In order to make money on loans, financial institutions are paid back more money than they loan. Suppose you borrow $120,000 to buy a house and agree to make monthly payments of $880.52 for 30 years. How much do you pay back altogether? How much more do you pay back than the amount of the loan?

68. *Loan Payment.* In order to make money on loans, financial institutions are paid back more money than they loan. Suppose you borrow $270,000 to buy a house and agree to make monthly payments of $1105.73 for 30 years. How much do you pay back altogether? How much more do you pay back than the amount of the loan?

Skill Maintenance

Simplify, if possible.

69. $\frac{0}{-13}$ [3.3b]

70. $\frac{12}{0}$ [3.3b]

71. $\frac{-76}{-76}$ [3.3b]

72. Add: $-\frac{4}{5} + \frac{7}{10}$. [4.2b]

73. Subtract: $\frac{8}{11} - \frac{4}{3}$. [4.3a]

74. Add: $4\frac{1}{3} + 2\frac{1}{2}$. [4.6a]

Synthesis

75. A poster that is 61.8 cm by 73.2 cm includes a 2-cm border. What is the area of the poster inside the border?

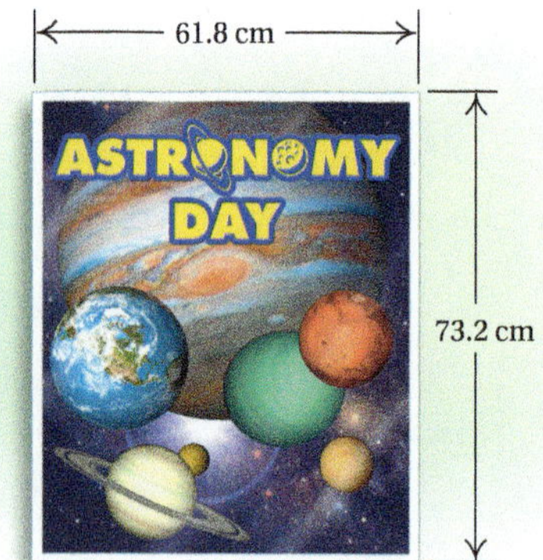

76. A French press coffeepot requires no filters, but costs $34.95. Kenny could buy a plastic drip cone for $4.49, but the cone requires filters which cost $0.04 per pot dripped. How many pots of coffee must Kenny make for the French press pot to be the more economical purchase?

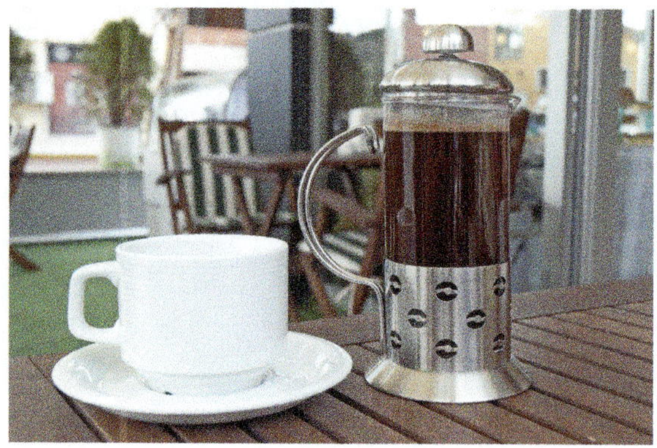

77. If the daily rental for a car is $18.90 plus a certain price per mile, and Lindsey must drive 190 mi in one day and still stay within a $55.00 budget, what is the highest price per mile that Lindsey can afford?

78. A 25-ft by 30-ft yard contains an 8-ft-wide, round fountain. How many 1-lb bags of grass seed should be purchased to seed the lawn if 1 lb of seed covers 300 ft^2?

79. Find the shaded area. What assumptions must you make?

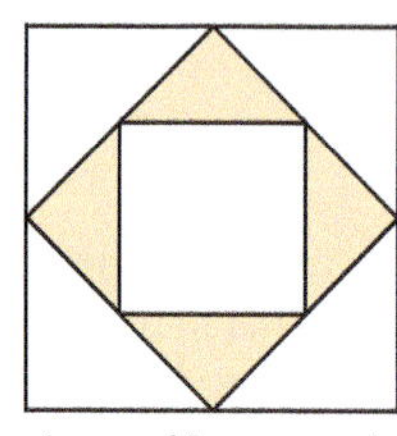

80. You can drive from home to work using either of two routes:

Route A: Via interstate highway, 7.6 mi, with a speed limit of 65 mph.

Route B: Via a country road, 5.6 mi, with a speed limit of 50 mph.

Assuming you drive at the posted speed limit, how much time can you save by taking the faster route?

CHAPTER 5

Summary and Review

Vocabulary Reinforcement

Complete each statement with the correct word from the list at the right.

1. A ______________ decimal occurs when we convert a fraction to a decimal and the denominator of the fraction has at least one factor other than 2 or 5. [5.5a]
2. A ______________ decimal occurs when we convert a fraction to a decimal and the denominator of the fraction has only 2's or 5's, or both, as factors. [5.5a]
3. One ______________ = 1,000,000,000. [5.3b]
4. One ______________ = 1,000,000. [5.3b]
5. One ______________ = 1,000,000,000,000. [5.3b]
6. The ______________ consist of the integers, . . . , −3, −2, −1, 0, 1, 2, 3, . . . , and fractions like $\frac{-1}{2}$, $\frac{4}{5}$, and $\frac{31}{25}$. [5.1a]

trillion
million
billion
rational numbers
repeating
terminating

Concept Reinforcement

Determine whether each statement is true or false.

______________ 1. One thousand billion is one trillion. [5.3b]

______________ 2. The number of decimal places in the product of two numbers is the product of the numbers of places in the factors. [5.3a]

______________ 3. When we divide a positive number by 0.1, 0.01, 0.001, and so on, the quotient is larger than the dividend. [5.4a]

______________ 4. For a fraction with a factor other than 2 or 5 in its denominator, decimal notation terminates. [5.5a]

______________ 5. An estimate found by rounding to the nearest ten is usually more accurate than one found by rounding to the nearest hundred. [5.6a]

Study Guide

Objective 5.1b Convert between decimal notation and fraction notation.

Example Write fraction notation for 5.347.

5.347 ↑ 3 decimal places

5.347. Move 3 places to the right.

$\frac{5347}{1000}$ ↑ 3 zeros

Practice Exercise

1. Write fraction notation for 50.93.

Example Write decimal notation for $\frac{29}{1000}$.

$\frac{29}{1000}$ ↑ 3 zeros

0.029. Move 3 places to the left.

$\frac{29}{1000} = 0.029$

Practice Exercise

2. Write decimal notation for $\frac{817}{10}$.

Example Write decimal notation for $4\frac{63}{100}$.

$$4\frac{63}{100} = 4 + \frac{63}{100} = 4 \text{ and } \frac{63}{100} = 4.63$$

Practice Exercise

3. Write decimal notation for $42\frac{159}{1000}$.

Objective 5.1d Round decimal notation to the nearest thousandth, hundredth, tenth, one, ten, hundred, or thousand.

Example Round 19.7625 to the nearest hundredth.

Locate the digit in the hundredths place, 6. Consider the next digit to the right, 2. Since that digit, 2, is 4 or lower, the original digit does not change.

19.7625
↓
19.76

Practice Exercise

4. Round 153.346 to the nearest hundredth.

Objective 5.2a Add using decimal notation.

Example Add: 14.26 + 63.589.

$$\begin{array}{r} \scriptstyle 1 \\ 14.260 \\ +\,63.589 \\ \hline 77.849 \end{array}$$

Writing an extra zero

Practice Exercise

5. Add: 5.54 + 33.071.

Objective 5.2b Subtract using decimal notation.

Example Subtract: 67.345 − 24.28.

$$\begin{array}{r} \scriptstyle 2\ 14 \\ 67.\cancel{3}\cancel{4}5 \\ -\,24.280 \\ \hline 43.065 \end{array}$$

Writing an extra zero

Practice Exercise

6. Subtract: 221.04 − 13.192.

Objective 5.3a Multiply using decimal notation.

Example Multiply: 1.8 × 0.04.

$$\begin{array}{rl} 1.8 & \text{(1 decimal place)} \\ \times\,0.04 & \text{(2 decimal places)} \\ \hline 0.072 & \text{(3 decimal places)} \end{array}$$

Practice Exercise

7. Multiply: 5.46 × 3.5.

Example Multiply: 0.001 × 87.1.

0.001 × 87.1   0.087.1

3 decimal places   Move 3 places to the left. We write an extra zero.

0.001 × 87.1 = 0.0871

Practice Exercise

8. Multiply: 17.6 × 0.01.

Example Multiply: 63.4 × 100.

63.4 × 100   63.40.

2 zeros   Move 2 places to the right. We write an extra zero.

63.4 × 100 = 6340

Practice Exercise

9. Multiply: 1000 × 60.437.

Objective 5.4a Divide using decimal notation.

Example Divide: $21.35 \div 6.1$.

$$\begin{array}{r} 3.5 \\ 6.1\overline{)\,21.35} \\ \underline{18\,3} \\ 3\,05 \\ \underline{3\,05} \\ 0 \end{array}$$

Practice Exercise

10. Divide: $26.64 \div 3.6$.

Example Divide: $\frac{16.7}{1000}$.

$$\frac{16.7}{1000} = \frac{016.7}{1\,000.} = \frac{0.0167}{1.0} = 0.0167$$

3 zeros — Move 3 places to the left to change 1000 to 1.

$$\frac{16.7}{1000} = 0.0167$$

Practice Exercise

11. Divide: $\frac{4.7}{100}$.

Example Divide: $\frac{42.93}{0.001}$.

$$\frac{42.93}{0.001} = \frac{42.930}{0.001} = \frac{42{,}930}{1.} = 42{,}930$$

3 decimal places — Move 3 places to the right to change 0.001 to 1.

$$\frac{42.93}{0.001} = 42{,}930$$

Practice Exercise

12. Divide: $\frac{156.9}{0.01}$.

Review Exercises

Convert the number in each sentence to standard notation. [5.3b]

1. Russia has the largest total area of any country in the world, at 6.59 million square miles.
2. Americans eat more than 3.1 billion lb of chocolate each year.

Source: Chocolate Manufacturers' Association

Write a word name. [5.1a]

3. 3.47 **4.** 0.031

5. 27.0001 **6.** 0.9

Write in fraction notation and, if possible, as a mixed numeral. [5.1b]

7. 0.09 **8.** -4.561

9. -0.089 **10.** 3.0227

Write in decimal notation. [5.1b]

11. $\frac{34}{1000}$ **12.** $\frac{42{,}603}{10{,}000}$

13. $27\frac{91}{100}$ **14.** $-867\frac{6}{1000}$

Which number is larger? [5.1c]

15. 0.034, 0.0185 **16.** -0.91, -0.19

17. 0.741, 0.6943 **18.** 1.038, 1.041

Round 17.4287 to the nearest [5.1d]

19. Tenth.

20. Hundredth.

21. Thousandth.

22. One.

Perform the indicated operation.

23. $\begin{array}{r} 236.231 \\ 263.4 \\ +\quad 0.198 \\ \hline \end{array}$ [5.2a]

24. $\begin{array}{r} 37.645 \\ -\ 8.497 \\ \hline \end{array}$ [5.2b]

25. $219.3 + 2.8 + 7$ [5.2a]

26. $745.0109 - 59.959$ [5.2b]

27. $-37.8 + (-19.5)$ [5.2c]

28. $-7.52 - (-9.89)$ [5.2c]

29. $\begin{array}{r} 48 \\ \times\ 0.27 \\ \hline \end{array}$ [5.3a]

30. $-3.7\,(0.29)$ [5.3a]

31. $\begin{array}{r} 24.68 \\ \times\ 1000 \\ \hline \end{array}$ [5.3a]

32. $25\overline{)80}$ [5.4a]

33. $11.52 \div (-7.2)$ [5.4a]

34. $\frac{276.3}{1000}$ [5.4a]

35. 3.056×0.001 [5.3a]

36. $\frac{-4.38}{0.001}$ [5.4a]

Combine like terms. [5.2d]

37. $3.7x - 5.2y - 1.5x - 3.9y$

38. $7.94 - 3.89a + 4.63 + 1.05a$

39. Evaluate: $P - Prt$ for $P = 1000$, $r = 0.05$, and $t = 1.5$. (*A formula for depreciation*) [5.3c]

40. Simplify: $9 - 3.2(-1.5) + 5.2^2$. [5.4b]

41. Convert 1549 cents to dollars. [5.3b]

42. Round $248.\overline{27}$ to the nearest hundredth. [5.5b]

Multiply by a form of 1 to find decimal notation for each number. [5.5c]

43. $\frac{13}{5}$

44. $\frac{32}{25}$

Use division to find decimal notation for each number. [5.5a]

45. $\frac{13}{4}$

46. $-\frac{7}{6}$

47. Calculate: $\frac{4}{15} \times 79.05$. [5.5d]

Solve. Remember to check.

48. $t - 4.3 = -7.5$ [5.7a]

49. $4.1x + 5.6 = -6.7$ [5.7a]

50. $6x - 11 = 8x + 4$ [5.7b]

51. $3(x + 2) = 5x - 7$ [5.7b]

Solve. [5.8b]

52. ***Record Movie Revenue.*** *Avatar* took in \$2.78 billion over its entire run in movie theaters. This is \$0.59 billion more than *Titanic* took in. How much did *Titanic* take in during its entire run?

Source: boxofficemojo.com/alltime/world

53. Stacia, a coronary intensive care nurse, earned \$620.74 during a recent 40-hr week. What was her hourly wage? Round to the nearest cent.

54. ***Landscaping.*** A rectangular yard is 20 ft by 15 ft. The yard is covered with grass except for a circular flower garden with an 8-ft diameter. Find the area of grass in the yard.

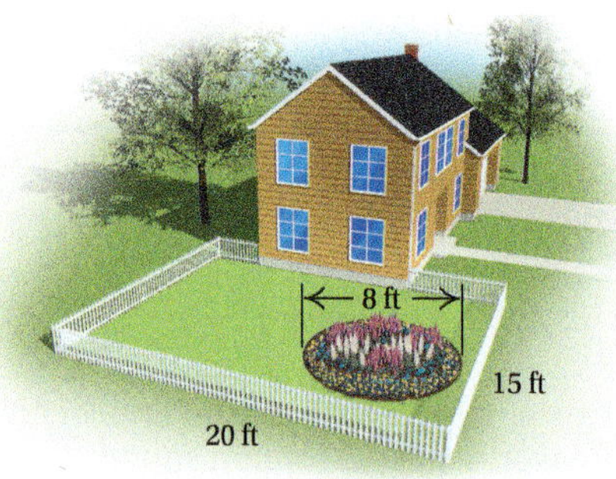

55. Derek had \$1034.46 in his bank account. He used his debit card to buy a Wii system for \$249.99. How much was left in his account?

56. ***Credit Card Processing.*** Pay Right charges \$150 for software, \$8.95 a month for service, and 21¢ per online transaction. Timeless Treasures paid \$178.90 for its first month of service. How many transactions did the store process that month?

57. ***Recycling.*** Lisa volunteers at a local recycling center. One Saturday, she processed 130 more pounds of newspaper than of glass. She processed a total of 261.4 pounds of newspaper and glass. How many pounds of each did she process?

58. *Gas Mileage.* Inge wants to estimate gas mileage per gallon. With an odometer reading 36,057.1, she fills up. At 36,217.6 mi, the tank is refilled with 11.1 gal. Find the mileage per gallon. Round to the nearest tenth.

59. *Books in Libraries.* The table below lists the numbers of books, in millions, held in the five largest public libraries in the United States. Find the average number of books per library. Round to the nearest tenth.

LIBRARY	NUMBER OF BOOKS (in millions)
Library of Congress	33.5
Boston Public Library	24.1
New York Public Library	16.6
Harvard University	16.6
University of Illinois–Urbana	12.8

SOURCE: American Library Association

60. *Scanning Posters.* A high school club needs to scan posters designed by students and load them onto a flash drive. The copy center charges \$12.99 for the flash drive and \$1.09 per square foot for scanning. If the club needs to scan 13 posters at 3 sq ft per poster, what will the total cost be?

61. One pound of lean boneless ham contains 4.5 servings. It costs \$5.99 per pound. What is the cost per serving? Round to the nearest cent.

62. *Construction.* A rectangular room measures 14.5 ft by 16.25 ft. How many feet of crown molding are needed to go around the top of the room? How many square feet of bamboo tiles are needed for the floor of the room?

63. Estimate the quotient $82.304 \div 17.287$ by rounding to the nearest ten. [5.6a]

A. 0.4 **B.** 4
C. 40 **D.** 400

64. Translate to an algebraic expression: 15 less than twice the price. [5.8a]

A. $2p - 15$ **B.** $15 - 2p$
C. $2 + p - 15$ **D.** $2(p - 15)$

Synthesis

65. In each of the following, use $+, -, \times,$ or $\div$ in each blank to make a true sentence. [5.4b]

a) $2.56 - 6.4 \square 51.2 - 17.4 + 89.7 = 119.66$
b) $(11.12 \square 0.29)3^4 = 877.23$

66. Arrange from smallest to largest:

$$-\frac{2}{3}, -\frac{15}{19}, -\frac{11}{13}, \frac{-5}{7}, \frac{-13}{15}, \frac{-17}{20}.$$

[5.1c], [5.5a]

67. *Automobile Leases.* Quentin leases a Ford Fusion Energi for \$279 a month. He must pay an additional 20 cents per mile for all miles over 10,000 in one year. In 2014, his total bill for leasing the car was \$6600. How many miles did he drive the car in 2014? [5.8b]

68. Use the fact that $\frac{1}{3} = 0.\overline{3}$ to find repeating decimal notation for 1. Explain how you got your answer. [5.5a]

69. Sal's sells Sicilian pizza as a 17-in. by 20-in. pie for \$15 or as an 18-in.-diameter round pie for \$14. Which is a better buy and why? [5.8b]

Understanding Through Discussion and Writing

1. Describe in your own words a procedure for converting from decimal notation to fraction notation. [5.1b]

2. A student insists that $346.708 \times 0.1 = 3467.08$. How could you convince him that a mistake had been made without checking on a calculator? [5.3a]

3. When is long division *not* the fastest way to convert from fraction notation to decimal notation? [5.5a]

4. Consider finding decimal notation for $\frac{44}{125}$. Discuss as many ways as you can for finding such notation and give the answer. [5.5a]

CHAPTER

5 Test

For Extra Help For step-by-step test solutions, access the Chapter Test Prep Videos in MyMathLab® or on YouTube (search "BittingerPreIntro" and click on Channels).

1. In a recent year, 2.6 billion ducks were killed for food worldwide. Convert 2.6 billion to standard notation.

Sources: FAO; *National Geographic*, May 2011, Nigel Holmes

2. Write a word name for 123.0047.

Write in fraction notation and, if possible, as a mixed numeral.

3. -0.91

4. 2.769

Write in decimal notation.

5. $\frac{74}{1000}$

6. $-\frac{37{,}047}{10{,}000}$

7. $756\frac{9}{100}$

8. $91\frac{703}{1000}$

Which number is larger?

9. 0.07, 0.162

10. 8.049, 8.0094

11. -0.09, -0.9

Round 5.6783 to the nearest

12. One.

13. Hundredth.

14. Thousandth.

15. Tenth.

Perform the indicated operation.

16. $\begin{array}{r} 402.3 \\ 2.81 \\ +\ 0.109 \\ \hline \end{array}$

17. $\begin{array}{r} 0.125 \\ \times\ 0.24 \\ \hline \end{array}$

18. $\begin{array}{r} 213.45 \\ \times\ 0.001 \\ \hline \end{array}$

19. $\begin{array}{r} 52.091 \\ -\ 7.345 \\ \hline \end{array}$

20. $342.9 + 8.1 + 5.37$

21. $-9.5 + 7.3$

22. $2 - 0.0054$

23. 1000×73.962

24. $4\overline{)19}$

25. $3.3\overline{)100.32}$

26. $\frac{-346.82}{1000}$

27. $\frac{346.82}{0.01}$

28. Convert \$179.82 to cents.

29. Combine like terms:
$4.1x + 5.2 - 3.9y + 5.7x - 9.8.$

30. Evaluate: $2l + 4w + 2h$ for $l = 2.4$, $w = 1.3$, and $h = 0.8$. (*The total girth of a postal package*)

31. Simplify: $20 \div 5(-2)^2 - 8.4$.

Multiply by a form of 1 to find decimal notation for each number.

32. $\frac{8}{5}$

33. $\frac{21}{4}$

Use division to find decimal notation for each number.

34. $-\frac{7}{16}$

35. $\frac{14}{9}$

36. Round $1.\overline{5}$ to the nearest hundredth.

Calculate.

37. $3 \div (-0.3) \cdot 2 - 1.5^2$

38. $(8 - 1.23) \div 4 + 5.6 \times 0.02$

39. $\frac{3}{8} \times 45.6 - \frac{1}{5} \times 36.9$

Solve. Remember to check.

40. $17y - 3.12 = -58.2$

41. $9t - 4 = 6t + 26$

42. $4 + 2(x - 3) = 7x - 9$

43. *Scanning Blueprints.* A building contractor needs to scan and load blueprints onto a flash drive. The copy center charges \$10.99 for the flash drive and \$1.19 per square foot for scanning. If the contractor needs to scan 5 blueprints at 6 sq ft per print, what will the total cost be?

44. *Gas Mileage.* Tina wants to estimate the gas mileage in her economy car. At 76,843 mi, she fills the tank with 14.3 gal of gasoline. At 77,310 mi, she fills the tank with 16.5 gal of gasoline. Find the mileage per gallon. Round to the nearest tenth.

45. *Checking Account Balance.* Nicholas had a balance of \$820 in his checking account before making purchases of \$123.89, \$56.68, and \$46.98 with his debit card. What was the balance after the purchases had been made?

46. The office manager for the Drake, Smith, and Hartner law firm buys 7 cases of copy paper at \$41.99 per case. What is the total cost?

47. ***Life Expectancy.*** Life expectancies at birth for seven Asian countries are listed in the table below. Find the average life expectancy for this group of countries. Round to the nearest tenth.

COUNTRY	LIFE EXPECTANCY (in years)
Japan	82.25
South Korea	79.05
People's Republic of China	73.47
India	69.9
Russia	66.03
North Korea	63.81
Afghanistan	44.64

SOURCE: *CIA Factbook* 2011

48. About how many gallons of gasoline, at $2.749 per gallon, can be bought with $20?

A. 1 gallon **B.** 3 gallons
C. 7 gallons **D.** 12 gallons

Synthesis

49. Use one of the words *sometimes*, *never*, or *always* to complete each of the following.

a) The product of two numbers greater than 0 and less than 1 is ______________ less than 1.

b) The product of two numbers greater than 1 is ______________ less than 1.

c) The product of a number greater than 1 and a number less than 1 is ______________ equal to 1.

d) The product of a number greater than 1 and a number less than 1 is ______________ equal to 0.

50. Silver's Health Club charges a $79 membership fee and $42.50 a month. Allise has a coupon that will allow her to join the club for $299 for six months. How much will Allise save if she uses the coupon?

51. ***Travel Costs.*** Roundtrip airfare between Burlington, VT, and Newark, NJ, often costs $359. One estimate of the true cost of driving is $1.35 per mile. Is it more economical to fly or drive the 600 mi for **(a)** an individual; **(b)** a couple; **(c)** a family of 4?

Source: commutesolutions.org

CHAPTER

6

Percent Notation

STUDYING FOR SUCCESS *Working with Others*

- ☐ Try being a tutor for a fellow student. You may find that you understand concepts better after you explain them to someone else.
- ☐ Consider forming a study group.
- ☐ Verbalize the math. Often simply talking about a concept with a classmate can help clarify the material.

6.1 Ratio and Proportion

OBJECTIVES

- **a** Find fraction notation for ratios.
- **b** Give the ratio of two different measures as a rate.
- **c** Determine whether two pairs of numbers are proportional.
- **d** Solve proportions.
- **e** Solve applied problems involving proportions.

SKILL TO REVIEW

Objective 3.5b: Simplify fraction notation.

Simplify.

1. $\frac{16}{64}$ 2. $\frac{40}{24}$

1. Find the ratio of 5 to 11.

2. Find the ratio of 57.3 to 86.1.

3. Find the ratio of $6\frac{3}{4}$ to $7\frac{2}{5}$.

Answers

Skill to Review:

1. $\frac{1}{4}$ 2. $\frac{5}{3}$

Margin Exercises:

1. $\frac{5}{11}$, or 5 : 11 2. $\frac{57.3}{86.1}$, or 57.3 : 86.1

3. $\frac{6\frac{3}{4}}{7\frac{2}{5}}$, or $6\frac{3}{4} : 7\frac{2}{5}$

a RATIOS

RATIO

A **ratio** is the quotient of two quantities.

The average wind speed in Chicago is 10.4 mph. The average wind speed in Boston is 12.5 mph. The *ratio* of average wind speed in Chicago to average wind speed in Boston is written using colon notation,

Chicago wind speed → 10.4 : 12.5, ← Boston wind speed

or fraction notation,

$\frac{10.4}{12.5}$. ← Chicago wind speed (numerator); ← Boston wind speed (denominator)

We read both forms of notation as "the ratio of 10.4 to 12.5."

RATIO NOTATION

The **ratio** of a to b is expressed by the fraction notation $\frac{a}{b}$, where a is the numerator and b is the denominator, or by the colon notation $a : b$.

EXAMPLE 1 Find the ratio of 31.4 to 100.

The ratio is $\frac{31.4}{100}$, or 31.4 : 100.

◀ Do Margin Exercises 1–3.

In most of our work, we will use fraction notation for ratios.

EXAMPLE 2 *Media Usage.* In 2012, Americans spent an average of 6 hr/month on Facebook and 147 hr/month watching TV. Find the ratio of average time spent on Facebook to average time spent watching TV.

Sources: comScore; Nielsen

The ratio is $\frac{6}{147}$. ■

EXAMPLE 3 Refer to the triangle below.

a) What is the ratio of the length of the longest side to the length of the shortest side?

$\frac{5}{3}$ ← Longest side / ← Shortest side

b) What is the ratio of the length of the shortest side to the length of the longest side?

$\frac{3}{5}$ ← Shortest side / ← Longest side

Do Exercises 4–6. ▶

EXAMPLE 4 *Grammy Awards.* From 1959 to 2011, a total of 798 musical artists won Grammy awards for Record of the Year, Album of the Year, and Best New Artist. The following bar graph shows the numbers of award winners from four music genres.

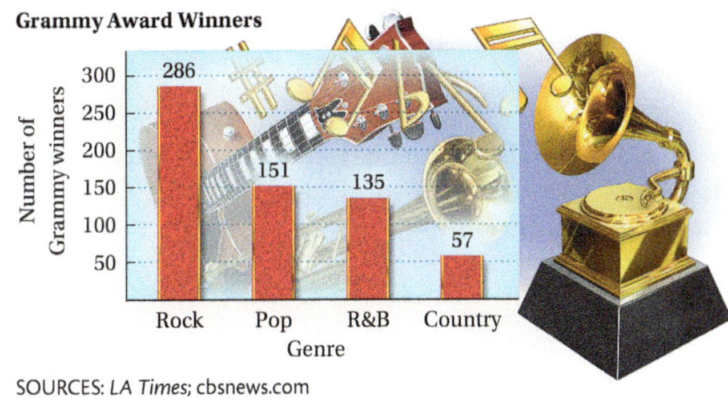

SOURCES: *LA Times*; cbsnews.com

a) What is the ratio of the number of country winners to the number of rock winners?

b) What is the ratio of the number of pop, R&B, and country winners to the number of rock winners?

c) What is the ratio of the number of Grammy winners from genres other than rock, pop, R&B, and country to the total number of Grammy winners?

4. *Record Snowfall.* The greatest snowfall recorded in North America during a 24-hr period was 76 in. in Silver Lake, Colorado, on April 14–15, 1921. What is the ratio of amount of snowfall, in inches, to time, in hours?

Source: U.S. Army Corps of Engineers

5. *Frozen Fruit Drinks.* A Berries & Kreme Chiller from Krispy Kreme contains 960 calories, while Smoothie King's MangoFest drink contains 258 calories. What is the ratio of the number of calories in the Krispy Kreme drink to the number of calories in the Smoothie King drink? of the number of calories in the Smoothie King drink to the number of calories in the Krispy Kreme drink?

Source: Physicians Committee for Responsible Medicine

6. In the triangle below, what is the ratio of the length of the shortest side to the length of the longest side?

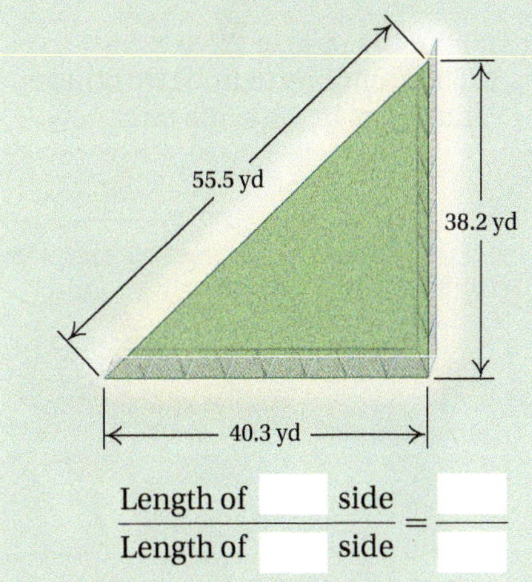

$\frac{\text{Length of } \_\_\_ \text{ side}}{\text{Length of } \_\_\_ \text{ side}} = \frac{\_\_\_}{\_\_\_}$

Answers

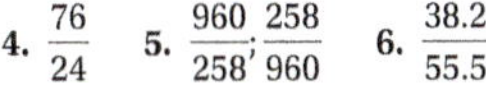

4. $\frac{76}{24}$ 5. $\frac{960}{258}$; $\frac{258}{960}$ 6. $\frac{38.2}{55.5}$

Guided Solution:

6. $\frac{\text{shortest}}{\text{longest}}$, $\frac{38.2}{55.5}$

7. ***NBA Playoffs.*** In the final game of the 2012 NBA Playoffs, the Miami Heat made a total of 67 baskets. Of these, 27 were free throws, 26 were two-point field goals, and the remainder were three-point field goals.
Source: nba.com

a) What was the ratio of the number of two-point field goals to the number of free throws?

b) What was the ratio of the number of three-point field goals to the total number of baskets?

8. Find the ratio of 3.6 to 12. Then simplify to find two other numbers in the same ratio. 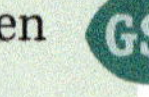

Ratio of 3.6 to 12: $\frac{\square}{\square}$

Simplifying:

$$\frac{3.6}{12} \cdot \frac{10}{\square} = \frac{\square}{120} = \frac{\square \cdot 3}{12 \cdot 10}$$

$$= \frac{\square}{12} \cdot \frac{3}{10} = \frac{\square}{10}$$

9. Find the ratio of 1.2 to 1.5. Then simplify to find two other numbers in the same ratio.

a) The ratio of the number of country winners to the number of rock winners is

$$\frac{57}{286}. \quad \begin{matrix} \leftarrow \text{Country winners} \\ \leftarrow \text{Rock winners} \end{matrix}$$

b) The total number of pop, R&B, and country winners is

$$151 + 135 + 57 = 343.$$

The ratio of the number of these winners to the number of rock winners is

$$\frac{343}{286}. \quad \begin{matrix} \leftarrow \text{Pop, R\&B, and country winners} \\ \leftarrow \text{Rock winners} \end{matrix}$$

c) The total number of rock, pop, R&B, and country Grammy winners is

$$286 + 151 + 135 + 57 = 629.$$

Thus the number of Grammy winners from other genres is

$$798 - 629 = 169.$$

The ratio of the number of Grammy winners from other genres to the total number of Grammy winners is

$$\frac{169}{798}. \quad \begin{matrix} \leftarrow \text{Winners from other genres} \\ \leftarrow \text{Total number of winners} \end{matrix}$$

◀ **Do Exercise 7.**

EXAMPLE 5 Find the ratio of 2.4 to 10. Then simplify to find two other numbers in the same ratio.

We first write the ratio in fraction notation. Next, we multiply by 1 to clear the decimal from the numerator. Then we simplify.

$$\frac{2.4}{10} = \frac{2.4}{10} \cdot \frac{10}{10} = \frac{24}{100} = \frac{4 \cdot 6}{4 \cdot 25} = \frac{4}{4} \cdot \frac{6}{25} = \frac{6}{25}$$

Thus, 2.4 is to 10 as 6 is to 25.

◀ **Do Exercises 8 and 9.**

EXAMPLE 6 An HDTV screen that measures approximately 46 in. diagonally has a width of 40 in. and a height of $22\frac{1}{2}$ in. Find the ratio of width to height and simplify.

$$\text{The ratio is } \frac{40}{22\frac{1}{2}} = \frac{40}{22.5} = \frac{40}{22.5} \cdot \frac{10}{10} = \frac{400}{225}$$

$$= \frac{25 \cdot 16}{25 \cdot 9} = \frac{25}{25} \cdot \frac{16}{9} = \frac{16}{9}.$$

Thus we can say that the ratio of width to height is 16 to 9, which can also be expressed as 16 : 9.

◀ **Do Exercise 10 on the following page.**

Answers

7. (a) $\frac{26}{27}$; (b) $\frac{14}{67}$
8. 3.6 is to 12 as 3 is to 10.
9. 1.2 is to 1.5 as 4 is to 5.

Guided Solution:

8. $\frac{3.6}{12}$; 10, 36, 12, 12, 3

b RATES

A 2013 Honda Odyssey can travel 504 mi on 18 gal of gasoline. Let's consider the ratio of miles to gallons:

Source: Kia Motors America, Inc.

$$\frac{504 \text{ mi}}{18 \text{ gal}} = \frac{504}{18}\frac{\text{miles}}{\text{gallon}} = \frac{28}{1}\frac{\text{miles}}{\text{gallon}}$$

$$= 28 \text{ miles per gallon} = 28 \text{ mpg.}$$

"per" means "division," or "for each."

The ratio

$$\frac{504 \text{ mi}}{18 \text{ gal}}, \quad \text{or} \quad \frac{504}{18}\frac{\text{mi}}{\text{gal}}, \quad \text{or} \quad 28 \text{ mpg},$$

is called a **rate**.

RATE

When a ratio is used to compare two different kinds of measure, we call it a **rate**.

Suppose David's car travels 475.4 mi on 16.8 gal of gasoline. Is the gas mileage (mpg) of his car better than that of the Honda Odyssey above? To determine this, it helps to convert the ratio to decimal notation and perhaps round. Thus we have

$$\frac{475.4 \text{ miles}}{16.8 \text{ gallons}} = \frac{475.4}{16.8} \text{mpg} \approx 28.298 \text{ mpg.}$$

Since $28.298 > 28$, David's car gets better gas mileage than the Honda Odyssey does.

EXAMPLE 7 It takes 60 oz of grass seed to seed 3000 sq ft of lawn. What is the rate in ounces per square foot?

$$\frac{60 \text{ oz}}{3000 \text{ sq ft}} = \frac{1}{50}\frac{\text{oz}}{\text{sq ft}}, \quad \text{or} \quad 0.02\frac{\text{oz}}{\text{sq ft}}$$

EXAMPLE 8 Martina bought 5 lb of organic russet potatoes for \$4.99. What was the rate in cents per pound?

$$\frac{\$4.99}{5 \text{ lb}} = \frac{499 \text{ cents}}{5 \text{ lb}} = 99.8¢/\text{lb}$$

EXAMPLE 9 *Hourly Wage.* In 2011, Walmart employees earned, on average, \$470 for a 40-hr work week. What was the rate of pay per hour?

Source: "Living Wage Policies and Big-Box Retail," UC Berkeley Center for Labor Research and Education, Ken Jacobs, Dave Graham-Squire, and Stephanie Luce, at http://laborcenter.berkeley.edu/retail/bigbox_livingwage_policies11.pdf

10. An HDTV screen that measures 44 in. diagonally has a width of 38.4 in. and a height of 21.6 in. Find the ratio of height to width and simplify.

Answer

10. $\frac{9}{16}$

A ratio of distance traveled to time is called a *speed*. What is the rate, or speed, in miles per hour?

11. 45 mi, 9 hr

12. 120 mi, 10 hr

What is the rate, or speed, in feet per second?

13. 2200 ft, 2 sec

14. 52 ft, 13 sec GS

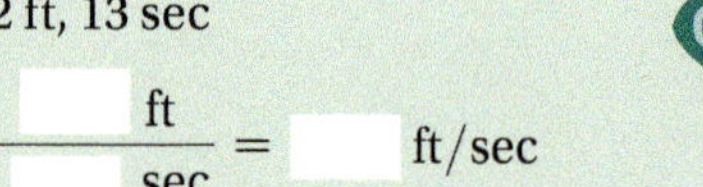

15. ***Babe Ruth.*** In his baseball career, Babe Ruth had 1330 strikeouts and 714 home runs. What was his home-run to strikeout rate?

Source: Major League Baseball

Determine whether the two pairs of numbers are proportional.

16. 3, 4 and 6, 8

17. 1, 4 and 10, 39

18. 1, 2 and 20, 39 GS

We compare cross products.

Since 39 ≠ 40, the numbers ______ proportional.
are/are not

Determine whether the two pairs of numbers are proportional.

19. 6.4, 12.8 and 5.3, 10.6

20. 6.8, 7.4 and 3.4, 4.2

Answers

11. 5 mi/hr, or 5 mph **12.** 12 mi/hr, or 12 mph **13.** 1100 ft/sec **14.** 4 ft/sec **15.** $\frac{714}{1330}$ home run per strikeout ≈ 0.537 home run per strikeout **16.** Yes **17.** No **18.** No **19.** Yes **20.** No

Guided Solution:

14. $\frac{52\text{ ft}}{13\text{ sec}} = 4\text{ ft/sec}$ **18.** 39, 39, 20, 40; are not

The rate of pay is the ratio of money earned to length of time worked, or

$$\frac{\$470}{40\text{ hr}} = \frac{470}{40}\frac{\text{dollars}}{\text{hr}} = 11.75\frac{\text{dollars}}{\text{hr}}, \quad \text{or} \quad \$11.75 \text{ per hr.}$$

◀ Do Exercises 11–15.

C PROPORTIONS

When two pairs of numbers, such as 3, 2 and 6, 4, have the same ratio, we say that they are **proportional**. The equation

$$\frac{3}{2} = \frac{6}{4}$$

states that the pairs 3, 2 and 6, 4 are proportional. Such an equation is called a **proportion**. We sometimes read $\frac{3}{2} = \frac{6}{4}$ as "3 is to 2 as 6 is to 4."

Since ratios can be written using fraction notation, we can use the test for equality of fractions to determine whether two ratios are the same.

A TEST FOR EQUALITY OF FRACTIONS

Two fractions are equal if their cross products are equal.

EXAMPLE 10 Determine whether 1, 2 and 3, 6 are proportional.

We can use cross products.

$$1 \cdot 6 = 6 \qquad \frac{1}{2} \stackrel{?}{=} \frac{3}{6} \qquad 2 \cdot 3 = 6$$

Since the cross products are the same, $6 = 6$, we know that $\frac{1}{2} = \frac{3}{6}$, so the numbers are proportional.

EXAMPLE 11 Determine whether 2, 5 and 4, 7 are proportional.

We can use cross products.

$$2 \cdot 7 = 14 \qquad \frac{2}{5} \stackrel{?}{=} \frac{4}{7} \qquad 5 \cdot 4 = 20$$

Since the cross products are not the same, $14 \neq 20$, we know that $\frac{2}{5} \neq \frac{4}{7}$, so the numbers are not proportional.

◀ Do Exercises 16–18.

EXAMPLE 12 Determine whether 3.2, 4.8 and 0.16, 0.24 are proportional.

We can use cross products.

$$3.2 \times 0.24 = 0.768 \qquad \frac{3.2}{4.8} \stackrel{?}{=} \frac{0.16}{0.24} \qquad 4.8 \times 0.16 = 0.768$$

Since the cross products are the same, $0.768 = 0.768$, we know that $\frac{3.2}{4.8} = \frac{0.16}{0.24}$, so the numbers are proportional.

◀ Do Exercises 19 and 20.

EXAMPLE 13 Determine whether $4\frac{2}{3}, 5\frac{1}{2}$ and $8\frac{7}{8}, 16\frac{1}{3}$ are proportional.

We can use cross products:

$$4\frac{2}{3} \cdot 16\frac{1}{3} = \frac{14}{3} \cdot \frac{49}{3} = \frac{686}{9} = 76\frac{2}{9}; \qquad \frac{4\frac{2}{3}}{5\frac{1}{2}} \stackrel{?}{=} \frac{8\frac{7}{8}}{16\frac{1}{3}} \qquad 5\frac{1}{2} \cdot 8\frac{7}{8} = \frac{11}{2} \cdot \frac{71}{8} = \frac{781}{16} = 48\frac{13}{16}.$$

Since the cross products are not the same, $76\frac{2}{9} \neq 48\frac{13}{16}$, we know that the numbers are not proportional.

Do Exercise 21. ▶

21. Determine whether $4\frac{2}{3}, 5\frac{1}{2}$ and $14, 16\frac{1}{2}$ are proportional.

d SOLVING PROPORTIONS

One way to solve a proportion is to use cross products. Then we can divide on both sides to get the variable alone.

EXAMPLE 14 Solve the proportion $\frac{x}{3} = \frac{4}{6}$.

$$\frac{x}{3} = \frac{4}{6}$$

$$x \cdot 6 = 3 \cdot 4 \qquad \text{Equating cross products (finding cross products and setting them equal)}$$

$$\frac{x \cdot 6}{6} = \frac{3 \cdot 4}{6} \qquad \text{Dividing by 6 on both sides}$$

$$x = \frac{3 \cdot 4}{6} = \frac{12}{6} = 2$$

We can check that 2 is the solution by replacing x with 2 and finding cross products:

$$2 \cdot 6 = 12 \qquad \frac{2}{3} \stackrel{?}{=} \frac{4}{6} \qquad 3 \cdot 4 = 12.$$

Since the cross products are the same, it follows that $\frac{2}{3} = \frac{4}{6}$. Thus the pairs of numbers 2, 3 and 4, 6 are proportional, and 2 is the solution of the proportion.

SOLVING PROPORTIONS

To solve $\frac{x}{a} = \frac{c}{d}$ for x, equate *cross products* and divide on both sides to get x alone.

Do Exercise 22. ▶

22. Solve: $\frac{x}{63} = \frac{2}{9}$.

Answers

21. Yes 22. 14

23. Solve: $\frac{x}{9} = \frac{5}{4}$. Write a mixed numeral for the answer.

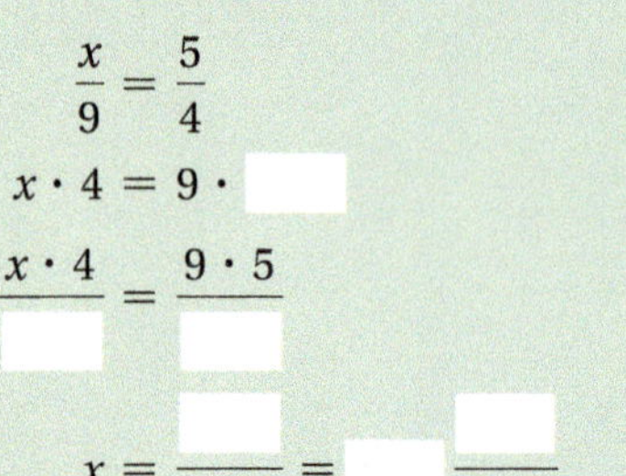

EXAMPLE 15 Solve: $\frac{x}{7} = \frac{5}{3}$. Write a mixed numeral for the answer.

We have

$$\frac{x}{7} = \frac{5}{3}$$

$$x \cdot 3 = 7 \cdot 5 \qquad \text{Equating cross products}$$

$$\frac{x \cdot 3}{3} = \frac{7 \cdot 5}{3} \qquad \text{Dividing by 3}$$

$$x = \frac{7 \cdot 5}{3} = \frac{35}{3}, \text{ or } 11\frac{2}{3}.$$

The solution is $11\frac{2}{3}$.

◀ **Do Exercise 23.**

EXAMPLE 16 Solve: $\frac{7.7}{15.4} = \frac{y}{2.2}$.

We have

$$\frac{7.7}{15.4} = \frac{y}{2.2}$$

$$7.7 \times 2.2 = 15.4 \times y \qquad \text{Equating cross products}$$

$$\frac{7.7 \times 2.2}{15.4} = \frac{15.4 \times y}{15.4} \qquad \text{Dividing by 15.4}$$

$$\frac{7.7 \times 2.2}{15.4} = y$$

$$\frac{16.94}{15.4} = y \qquad \text{Multiplying}$$

$$1.1 = y. \qquad \text{Dividing: } 15.4\overline{)16.94} = 1.1$$

Dividing: 1 5.4) 1 6.9 4 → 1.1
1 5 4
1 5 4
1 5 4
0

The solution is 1.1.

EXAMPLE 17 Solve: $\frac{8}{x} = \frac{5}{3}$. Write decimal notation for the answer.

We have

$$\frac{8}{x} = \frac{5}{3}$$

$$8 \cdot 3 = x \cdot 5 \qquad \text{Equating cross products}$$

$$\frac{8 \cdot 3}{5} = x \qquad \text{Dividing by 5}$$

$$\frac{24}{5} = x \qquad \text{Multiplying}$$

$$4.8 = x. \qquad \text{Simplifying}$$

The solution is 4.8.

◀ **Do Exercises 24 and 25.**

24. Solve: $\frac{21}{5} = \frac{n}{2.5}$.

25. Solve: $\frac{6}{x} = \frac{25}{11}$. Write decimal notation for the answer.

Answers

23. $11\frac{1}{4}$ **24.** 10.5 **25.** 2.64

Guided Solution:
23. 5, 4, 4, 45, 11, 1

EXAMPLE 18 Solve: $\frac{3.4}{4.93} = \frac{10}{n}$.

We have

$$\frac{3.4}{4.93} = \frac{10}{n}$$

$$3.4 \times n = 4.93 \times 10 \quad \text{Equating cross products}$$

$$\frac{3.4 \times n}{3.4} = \frac{4.93 \times 10}{3.4} \quad \text{Dividing by 3.4}$$

$$n = \frac{4.93 \times 10}{3.4}$$

$$n = \frac{49.3}{3.4} \quad \text{Multiplying}$$

$$n = 14.5. \quad \text{Dividing}$$

The solution is 14.5.

Do Exercise 26. ▶

26. Solve: $\frac{0.4}{0.9} = \frac{4.8}{t}$.

e APPLICATIONS AND PROBLEM SOLVING

Proportions have applications in such diverse fields as business, chemistry, health sciences, and home economics, as well as in many areas of daily life. Proportions are useful in making predictions.

EXAMPLE 19 *Predicting Total Distance.* Donna drives her delivery van 800 mi in 3 days. At this rate, how far will she drive in 15 days?

1. **Familiarize.** We let d = the distance traveled in 15 days.
2. **Translate.** We translate to a proportion. We make each side the ratio of distance to time, with distance in the numerator and time in the denominator.

$$\text{Distance in 15 days} \rightarrow \frac{d}{15} = \frac{800}{3} \leftarrow \text{Distance in 3 days}$$
$$\text{Time} \rightarrow 15 \qquad 3 \leftarrow \text{Time}$$

It may help to verbalize the proportion above as "the unknown distance d is to 15 days as the known distance 800 mi is to 3 days."

3. **Solve.** Next, we solve the proportion:

$$d \cdot 3 = 15 \cdot 800 \quad \text{Equating cross products}$$

$$\frac{d \cdot 3}{3} = \frac{15 \cdot 800}{3} \quad \text{Dividing by 3 on both sides}$$

$$d = \frac{15 \cdot 800}{3}$$

$$d = 4000. \quad \text{Multiplying and dividing}$$

4. **Check.** We substitute into the proportion and check cross products:

$$\frac{4000}{15} = \frac{800}{3};$$

$$4000 \cdot 3 = 12{,}000; \qquad 15 \cdot 800 = 12{,}000.$$

The cross products are the same.

5. **State.** Donna will drive 4000 mi in 15 days.

Do Exercise 27. ▶

27. *Burning Calories.* The readout on Mary's treadmill indicates that she burns 108 calories when she walks for 24 min. How many calories will she burn if she walks at the same rate for 30 min?

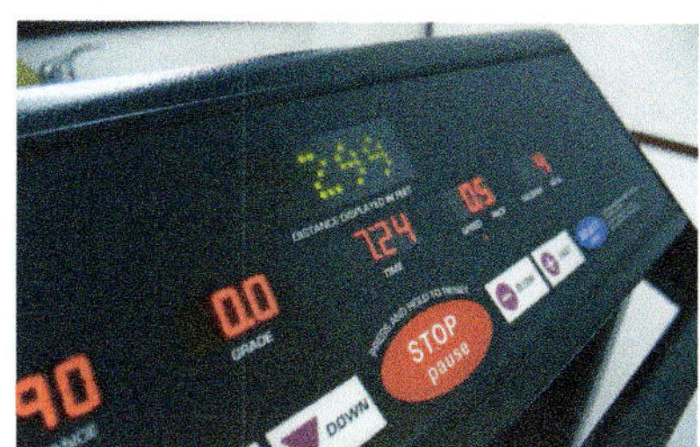

Answers

26. 10.8 **27.** 135 calories

28. Determining Paint Needs. Lowell and Chris run a painting company during the summer to pay for their college expenses. They can paint 1600 ft^2 of clapboard with 4 gal of paint. How much paint would be needed for a building with 6000 ft^2 of clapboard?

GS

1. **Familiarize.** Let p = the amount of paint needed, in gallons.
2. **Translate.**

$$\frac{4}{1600} = \frac{\square}{\square}$$

3. **Solve.**

$$4 \cdot 6000 = 1600 \cdot \square$$
$$\square = p$$

4. **Check.** The cross products are the same.
5. **State.** For 6000 ft^2, they would need ____ gal of paint.

Answer

28. 15 gal

Guided Solution:

28. $\frac{p}{6000}$; p, 15; 15

EXAMPLE 20 *Recommended Dosage.* To control a fever, a doctor suggests that a child who weighs 28 kg be given 320 mg of a liquid pain reliever. If the dosage is proportional to the child's weight, how much of the medication is recommended for a child who weighs 35 kg?

1. **Familiarize.** We let t = the number of milligrams of the liquid pain reliever recommended for a child who weighs 35 kg.
2. **Translate.** We translate to a proportion, keeping the amount of medication in the numerators.

$$\begin{matrix}\text{Medication suggested} \rightarrow \\ \text{Child's weight} \rightarrow\end{matrix} \frac{320}{28} = \frac{t}{35} \begin{matrix}\leftarrow \text{Medication suggested} \\ \leftarrow \text{Child's weight}\end{matrix}$$

3. **Solve.** Next, we solve the proportion:

$$320 \cdot 35 = 28 \cdot t \quad \text{Equating cross products}$$
$$\frac{320 \cdot 35}{28} = \frac{28 \cdot t}{28} \quad \text{Dividing by 28 on both sides}$$
$$\frac{320 \cdot 35}{28} = t$$
$$400 = t. \quad \text{Multiplying and dividing}$$

4. **Check.** We substitute into the proportion and check cross products:

$$\frac{320}{28} = \frac{400}{35};$$
$$320 \cdot 35 = 11{,}200; \qquad 28 \cdot 400 = 11{,}200.$$

The cross products are the same.

5. **State.** The dosage for a child who weighs 35 kg is 400 mg.

◀ Do Exercise 28.

EXAMPLE 21 *Purchasing Tickets.* Carey bought 8 tickets to an international food festival for \$52. How many tickets could she purchase with \$90?

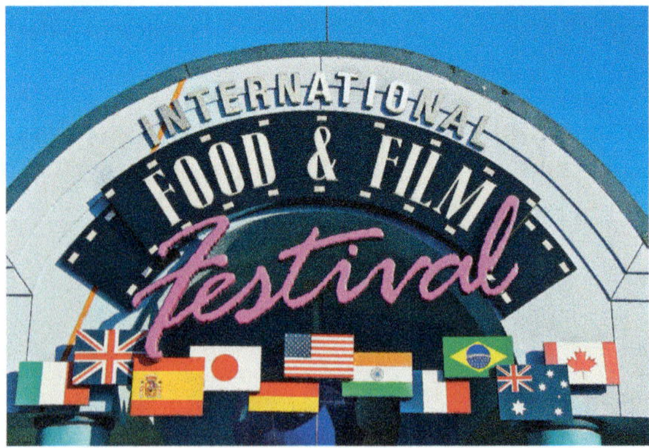

1. **Familiarize.** We let n = the number of tickets that can be purchased with \$90.
2. **Translate.** We translate to a proportion, keeping the number of tickets in the numerators.

$$\begin{matrix}\text{Tickets} \rightarrow \\ \text{Cost} \rightarrow\end{matrix} \frac{8}{52} = \frac{n}{90} \begin{matrix}\leftarrow \text{Tickets} \\ \leftarrow \text{Cost}\end{matrix}$$

3. **Solve.** Next, we solve the proportion:

$$52 \cdot n = 8 \cdot 90 \quad \text{Equating cross products}$$

$$\frac{52 \cdot n}{52} = \frac{8 \cdot 90}{52} \quad \text{Dividing by 52 on both sides}$$

$$n = \frac{8 \cdot 90}{52}$$

$$n \approx 13.8. \quad \text{Multiplying and dividing}$$

Because it is impossible to buy a fractional part of a ticket, we must round our answer *down* to 13.

4. **Check.** As a check, we use a different approach: We find the cost per ticket and then divide \$90 by that price. Since $52 \div 8 = 6.50$ and $90 \div 6.50 \approx 13.8$, we have a check.

5. **State.** Carey could purchase 13 tickets with \$90.

Do Exercise 29.

29. *Purchasing Shirts.* If 2 shirts can be bought for \$47, how many shirts can be bought with \$200?

EXAMPLE 22 *Construction Plans.* Architects make blueprints for construction projects. These are scale drawings in which lengths are in proportion to actual sizes. The Hennesseys are adding a rectangular deck to their house. The architect's blueprint is rendered such that $\frac{3}{4}$ in. on the drawing is actually 2.25 ft on the deck. The width of the deck on the drawing is 4.3 in. How wide is the deck in reality?

1. **Familiarize.** We let w = the width of the deck.

2. **Translate.** Then we translate to a proportion, using 0.75 for $\frac{3}{4}$ in.

Measure on drawing → 0.75 / Measure on deck → 2.25 = 4.3 ← Width on drawing / w ← Width on deck

$$\frac{0.75}{2.25} = \frac{4.3}{w}$$

3. **Solve.** Next, we solve the proportion:

$$0.75 \times w = 2.25 \times 4.3 \quad \text{Equating cross products}$$

$$\frac{0.75 \times w}{0.75} = \frac{2.25 \times 4.3}{0.75} \quad \text{Dividing by 0.75 on both sides}$$

$$w = \frac{2.25 \times 4.3}{0.75}$$

$$w = 12.9.$$

4. **Check.** We substitute into the proportion and check cross products:

$$\frac{0.75}{2.25} = \frac{4.3}{12.9};$$

$$0.75 \times 12.9 = 9.675; \qquad 2.25 \times 4.3 = 9.675.$$

The cross products are the same.

5. **State.** The width of the deck is 12.9 ft.

Do Exercise 30.

30. *Construction Plans.* In Example 22, the length of the actual deck is 28.5 ft. What is the length of the deck on the blueprint?

Answers

29. 8 shirts 30. 9.5 in.

EXAMPLE 23 ***Estimating a Wildlife Population.*** Scientists often use proportions to estimate the size of a wildlife population. They begin by collecting and marking, or tagging, a portion of the population. This tagged sample is released and mingles with the entire population. At a later date, the scientists collect a second sample from the population. The proportion of tagged individuals in the second sample is estimated to be the same as the proportion of tagged individuals in the entire population.

The marking can be done by using actual tags or by identifying individuals in other ways. For example, marine biologists can identify an individual whale by the patterns on its tail. Recently, scientists have begun using DNA to identify individuals in populations. For example, to identify individual bears in the grizzly bear population of the Northern Continental Divide ecosystem in Montana, geneticists use DNA from fur samples left on branches near the bears' feeding areas.

In one recent large-scale study in this ecosystem, biologists identified 545 individual grizzly bears. If later a sample of 30 bears contained 25 of the previously identified individuals, estimate the total number of bears in the ecosystem.

Source: Based on information from the Northern Divide Grizzly Bear Project

1. **Familiarize.** We let $B =$ the total number of bears in the ecosystem. We assume that the ratio of the number of identified bears to the total number of bears in the ecosystem is the same as the ratio of the number of identified bears in the later sample to the total number of bears in the later sample.
2. **Translate.** We translate to a proportion as follows.

$$\begin{matrix}\text{Identified bears} \rightarrow \\ \text{Total number of bears} \rightarrow\end{matrix} \frac{545}{B} = \frac{25}{30} \begin{matrix}\leftarrow \text{Identified bears in sample} \\ \leftarrow \text{Number of bears in sample}\end{matrix}$$

3. **Solve.** Next, we solve the proportion:

$$545 \cdot 30 = B \cdot 25 \quad \text{Equating cross products}$$

$$\frac{545 \cdot 30}{25} = \frac{B \cdot 25}{25} \quad \text{Dividing by 25 on both sides}$$

$$\frac{545 \cdot 30}{25} = B$$

$$654 = B. \quad \text{Multiplying and dividing}$$

4. **Check.** We substitute into the proportion and check cross products:

$$\frac{545}{654} = \frac{25}{30};$$

$$545 \cdot 30 = 16{,}350; \qquad 654 \cdot 25 = 16{,}350.$$

The cross products are the same.

5. **State.** We estimate that there are 654 bears in the ecosystem.

◀ **Do Exercise 31.**

31. Estimating a Deer Population. GS To determine the number of deer in a forest, a conservationist catches 153 deer, tags them, and releases them. Later, 62 deer are caught, and it is found that 18 of them are tagged. Estimate how many deer are in the forest.

1. **Familiarize.** Let $D =$ the number of deer in the forest.
2. **Translate.**

$$\frac{153}{D} = \frac{\square}{\square}$$

3. **Solve.**

$$153 \cdot 62 = D \cdot \square$$

$$\square = D$$

4. **Check.** The cross products are the same.
5. **State.** There are about ____ deer in the forest.

Answer

31. 527 deer

Guided Solution:

31. $\frac{18}{62}$; 18, 527; 527

6.1 Exercise Set

For Extra Help MyMathLab® MathXL® PRACTICE WATCH READ REVIEW

Reading Check

Complete each statement with the appropriate word from the following list.

cross products proportion proportional ratio

RC1. The quotient of two quantities is their ________________.

RC2. Two pairs of numbers that have the same ratio are ________________.

RC3. A(n) ________________ states that two pairs of numbers have the same ratio.

RC4. For the equation $\frac{2}{x} = \frac{3}{y}$, the ________________ are $2y$ and $3x$.

a Find fraction notation for each ratio. You need not simplify.

1. 178 to 572

2. 3 to 2

3. $8\frac{3}{4}$ to $9\frac{5}{6}$

4. 456.2 to 333.1

Shark Attacks. Of the 75 unprovoked shark attacks recorded worldwide in 2011, 29 occurred in U.S. waters. The following bar graph shows the breakdown by state. Use the graph for Exercises 5 and 6.

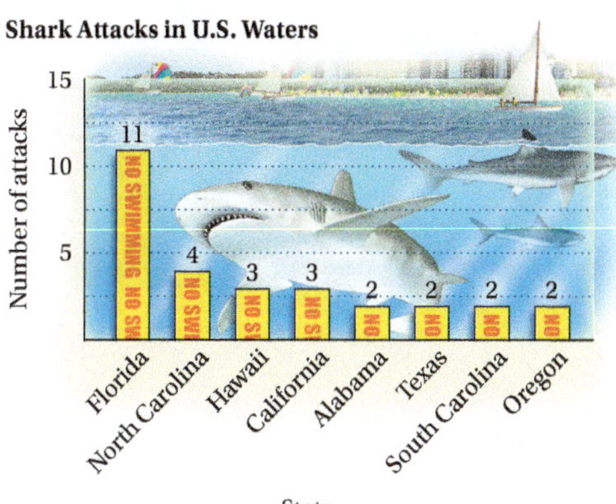

5. What is the ratio of the number of shark attacks in U.S. waters to the number of shark attacks worldwide?

6. What is the ratio of the total number of shark attacks in states bordering the Pacific Ocean (Hawaii, California, and Oregon) to the total number of shark attacks in the other five states?

7. *Tax Freedom Day.* Of the 366 days in 2012 (a leap year), the average American worked 107 days to pay his or her federal, state, and local taxes. Find the ratio of the number of days worked to pay taxes in 2012 to the number of days in the year.

Source: Tax Foundation

8. *Careers in Medicine.* The number of jobs for nurses is expected to increase by 711,900 between 2010 and 2020. During the same decade, the number of jobs for physicians is expected to increase by 168,300. Find the ratio of the increase in jobs for physicians to the increase in jobs for nurses.

Source: Bureau of Labor Statistics

Find the ratio of the first number to the second number and simplify.

9. 4 to 6

10. 28 to 36

11. 2.8 to 3.6

12. 5.6 to 10

b In Exercises 13–16, find each rate, or speed, as a ratio of distance to time. Round to the nearest hundredth where appropriate.

13. 120 km, 3 hr

14. 18 mi, 9 hr

15. 217 mi, 29 sec

16. 443 m, 48 sec

17. ***Mazda3—Highway Driving.*** A 2013 Mazda3 i SV will travel 643.5 mi on 19.5 gal of gasoline in highway driving. What is the rate in miles per gallon?

Source: mazdausa.com

18. ***Chevrolet Malibu LS—Highway Driving.*** A 2013 Chevrolet Malibu LS will travel 499.5 mi on 13.5 gal of gasoline in highway driving. What is the rate in miles per gallon?

Source: Chevrolet

19. ***Broadway Musicals.*** In the 17 years from 1987 through 2003, the musical *Les Misérables* was performed on Broadway 6680 times. What was the average rate of performances per year?

Source: broadwaymusicalhome.com

20. ***Employment Growth.*** In the 10 years from 2010 to 2020, the number of jobs for interpreters and translators is expected to grow by 24,600. What is the expected average rate of growth in jobs per year?

Source: U.S. Bureau of Labor Statistics

21. ***Speed of Light.*** Light travels 186,000 mi in 1 sec. What is its rate, or speed, in miles per second?

Source: *The Handy Science Answer Book*

22. ***Speed of Sound.*** Sound travels 1100 ft in 1 sec. What is its rate, or speed, in feet per second?

Source: *The Handy Science Answer Book*

23. ***Lawn Watering.*** Watering a lawn adequately requires 623 gal of water for every 1000 ft^2. What is the rate in gallons per square foot?

24. A car is driven 200 km on 40 L of gasoline. What is the rate in kilometers per liter?

25. ***Elephant Heart Rate.*** The heart of an elephant, at rest, beats an average of 1500 beats in 60 min. What is the rate in beats per minute?

Source: *The Handy Science Answer Book*

26. ***Human Heart Rate.*** The heart of a human, at rest, beats an average of 4200 beats in 60 min. What is the rate in beats per minute?

Source: *The Handy Science Answer Book*

c Determine whether the two pairs of numbers are proportional.

27. 5, 6 and 7, 9

28. 7, 5 and 6, 4

29. 1, 2 and 10, 20

30. 7, 3 and 21, 9

31. 2.4, 3.6 and 1.8, 2.7

32. 4.5, 3.8 and 6.7, 5.2

33. $5\frac{1}{3}, 8\frac{1}{4}$ and $2\frac{1}{5}, 9\frac{1}{2}$

34. $2\frac{1}{3}, 3\frac{1}{2}$ and 14, 21

d Solve.

35. $\frac{x}{8} = \frac{9}{6}$

36. $\frac{8}{10} = \frac{n}{5}$

37. $\frac{2}{5} = \frac{8}{n}$

38. $\frac{10}{6} = \frac{5}{x}$

39. $\frac{16}{12} = \frac{24}{x}$

40. $\frac{8}{12} = \frac{20}{x}$

41. $\frac{12}{9} = \frac{x}{7}$

42. $\frac{x}{20} = \frac{16}{15}$

43. $\frac{1.2}{4} = \frac{x}{9}$

44. $\frac{x}{11} = \frac{7.1}{2}$

45. $\frac{8}{2.4} = \frac{6}{y}$

46. $\frac{3}{y} = \frac{5}{4.5}$

47. $\frac{y}{\frac{3}{5}} = \frac{\frac{7}{12}}{\frac{14}{15}}$

48. $\frac{\frac{5}{8}}{\frac{5}{4}} = \frac{y}{\frac{3}{2}}$

49. $\frac{x}{1\frac{3}{5}} = \frac{2}{15}$

50. $\frac{1}{7} = \frac{x}{4\frac{1}{2}}$

51. $\frac{0.5}{n} = \frac{2.5}{3.5}$

52. $\frac{6.3}{0.9} = \frac{0.7}{n}$

53. $\frac{\frac{1}{5}}{\frac{1}{10}} = \frac{\frac{1}{10}}{x}$

54. $\frac{\frac{1}{4}}{\frac{1}{2}} = \frac{\frac{1}{2}}{x}$

e Solve.

55. *Movies.* If *The Hobbit: An Unexpected Journey* is played at the rate preferred by director Peter Jackson, a moviegoer sees 600 frames in $12\frac{1}{2}$ sec. How many frames does a moviegoer see in 160 sec?

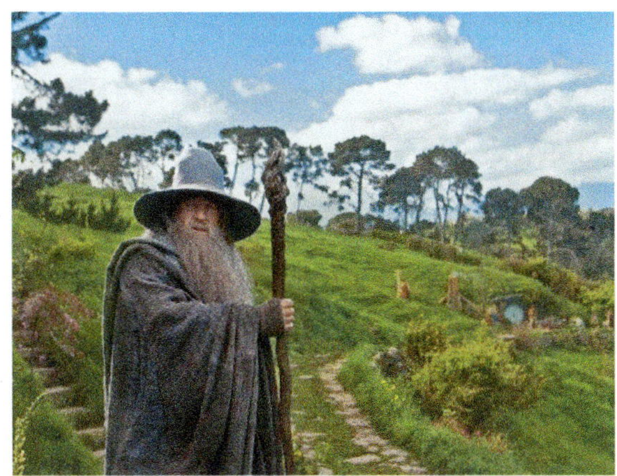

56. *Metallurgy.* In Ethan's white gold ring, the ratio of nickel to gold is 3 to 13. If the ring contains 4.16 oz of gold, how much nickel does it contain?

57. *Lefties.* In a class of 40 students, on average, 6 will be left-handed. If a class includes 9 "lefties," how many students would you estimate are in the class?

58. *Sugaring.* When 20 gal of maple sap are boiled down, the result is $\frac{1}{2}$ gal of maple syrup. How much sap is needed to produce 9 gal of syrup?

Source: University of Maine

59. *Currency Exchange.* On January 7, 2013, 1 U.S. dollar was worth about 0.76 euro.

a) How much were 50 U.S. dollars worth in euros on that day?

b) How much would a car that costs 8640 euros cost in U.S. dollars?

60. *Gas Mileage.* A 2013 Volkswagen Beetle will travel 495 mi on 16.5 gal of premium gasoline in highway driving.

a) How many gallons of gasoline will it take to drive 1650 mi from Pittsburgh to Albuquerque?

b) How far can the car be driven on 130 gal of gasoline?

Source: Volkswagen of America, Inc.

61. *Estimating a Whale Population.* To determine the number of humpback whales in a population, a marine biologist, using tail markings, identifies 27 individual whales. Several weeks later, 40 whales from the population are randomly sighted. Of the 40 sighted, 12 are among the 27 originally identified. Estimate the number of whales in the population.

62. *Home Runs.* After playing 118 games in the 2012 Major League Baseball season, Andrew McCutchen of the Pittsburgh Pirates had 24 home runs.

a) At this rate, how many games would it take him to hit 30 home runs?

b) At this rate, how many home runs would McCutchen hit in the entire 162-game season?

Source: Major League Baseball

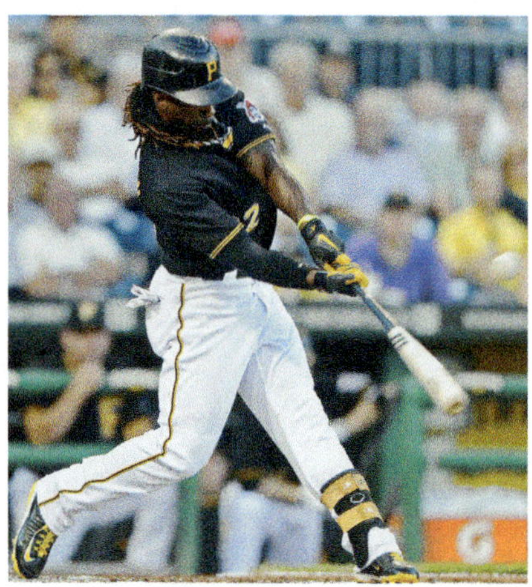

63. *Overweight Americans.* A recent study determined that of every 100 American adults, 69 are overweight or obese. It is estimated that the U.S. population will be about 337 million in 2020. At the given rate, how many Americans will be considered overweight or obese in 2020?

Source: U.S. Centers for Disease Control and Prevention

64. *Prevalence of Diabetes.* A recent study determined that of every 1000 Americans age 65 or older, 269 have been diagnosed with diabetes. It is estimated that there will be about 55 million Americans in this age group in 2020. At the given rate, how many in this age group will be diagnosed with diabetes in 2020?

Source: American Diabetes Association

65. *Grass-Seed Coverage.* It takes 60 oz of grass seed to seed 3000 ft^2 of lawn. At this rate, how much would be needed for 5000 ft^2 of lawn?

66. *Coffee Production.* Coffee beans from 18 trees are required to produce enough coffee each year for a person who drinks 2 cups of coffee per day. Jared brews 15 cups of coffee each day for himself and his coworkers. How many coffee trees are required for this?

67. ***Rice Krispies® Cereal.*** The nutritional chart on the side of a box of Kellogg's Rice Krispies® cereal states that there are 130 calories in a $1\frac{1}{4}$-cup serving. How many calories are there in 5 cups of the cereal?

Nutrition Facts

Serving Size $1\frac{1}{4}$ Cups (33g/1.2oz)

Amount Per Serving	Cereal	Cereal with 1/2 Cup Vitamins A&D Fat Free Milk
Calories	130	170
Calories from Fat	0	0
	% Daily Value	
Total Fat 0g	0%	0%
Saturated Fat 1g	0%	0%
Trans Fat 0g		
Cholesterol 0mg	0%	0%
Sodium 220mg	9%	12%
Potassium 30mg	1%	7%
Total Carbohydrate 29g	10%	11%
Dietary Fiber 0g	0%	0%
Sugars 4g		
Other Carbohydrate 25g		
Protein 2g		

68. ***Class Size.*** A college advertises that its student-to-faculty ratio is 27 to 2. If 81 students register for Introductory Spanish, how many sections of the course would you expect to see offered?

69. ***Map Scaling.*** On a road atlas map, 1 in. represents 16.6 mi. If two cities are 3.5 in. apart on the map, how far apart are they in reality?

70. ***Bicycling.*** Roy bicycled 234 mi in 14 days. At this rate, how far would Roy bicycle in 42 days?

71. ***Painting.*** Helen can paint 950 ft^2 with 2 gal of paint. How many 1-gal cans does she need in order to paint a 30,000-ft^2 wall?

72. ***Snow to Water.*** Under typical conditions, $1\frac{1}{2}$ ft of snow will melt to 2 in. of water. How many inches of water will result when $5\frac{1}{2}$ ft of snow melts?

Skill Maintenance

Solve.

73. $12 \cdot x = 1944$ [1.7b]

74. $6807 = m + 2793$ [1.7b]

75. $t - 4.25 = 8.7$ [5.7a]

76. $112.5 \cdot p = 45$ [5.7a]

77. $3.7 + y = 18$ [5.7a]

78. $0.078 = -0.3t$ [5.7a]

79. $c - \frac{4}{5} = -\frac{9}{10}$ [4.3b]

80. $\frac{5}{6} = \frac{2}{3} \cdot x$ [4.4b]

Synthesis

Fertilizer. Exercises 81 and 82 refer to a common lawn fertilizer known as "5, 10, 15." This mixture contains 5 parts of potassium for every 10 parts of phosphorus and 15 parts of nitrogen. (This is often denoted 5 : 10 : 15.)

81. Simplify the ratio 5 : 10 : 15.

82. Find and simplify the ratio of potassium to nitrogen and of nitrogen to phosphorus.

6.2 Percent Notation

OBJECTIVES

a Write three kinds of notation for a percent.

b Convert between percent notation and decimal notation.

SKILL TO REVIEW

Objective 5.3a: Multiply using decimal notation.

Multiply.

1. 68.3×0.01

2. 3013×2.4

a UNDERSTANDING PERCENT NOTATION

Today almost half of the world's population lives in cities. It is estimated that by 2050, 70% of the population will live in cities. What does 70% mean?

70 of 100 squares are shaded.

70% or $\frac{70}{100}$ or 0.70 of the large square is shaded.

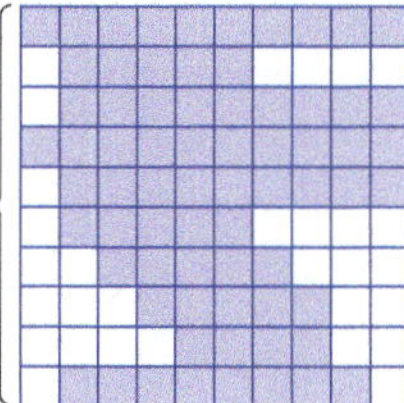

It means that of every 100 people on Earth in 2050, 70 will live in cities. Thus, 70% is a ratio of 70 to 100, or $\frac{70}{100}$.

Sources: DuPont, 2011; *National Geographic*, December 2011

We encounter percent notation frequently. Here are some examples.

- The 1964–2013 Washington quarter is 91.67% copper and 8.33% nickel. The 1946–2013 Jefferson nickel is 75% copper and 25% nickel.
  Source: Coinflation.com
- A blood alcohol level of 0.08% is the standard used by most states as the legal limit for drunk driving.
  Source: The National Highway Safety Administration
- Educational loans of $20,000 or more per year are obtained by 16% of college students.
  Source: TRU Survey
- Of senior citizens aged 76 and older, 31% own a laptop.
  Source: Internet & American Life Project

Percent notation is often represented using a circle graph, or pie chart, to show how the parts of a quantity are related. For example, the circle graph at left illustrates the percentages of coffee drinkers who get their coffee at selected locations.

Where Coffee Drinkers Get Their Coffee

Home 29%
Starbucks 25%
McDonald's 16%
Other 19%
Dunkin' Donuts 11%

SOURCE: "Starbucks' Big Mug," *TIME*, June 25, 2012

PERCENT NOTATION

The notation ***n*%** means "*n* per hundred."

Answers

Skill to Review:

1. 0.683 **2.** 7231.2

This definition leads us to the following equivalent ways of defining percent notation.

NOTATION FOR $n\%$

Percent notation, $n\%$, can be expressed using:

ratio → $n\% =$ the ratio of n to 100 $= \frac{n}{100}$,

fraction notation → $n\% = n \times \frac{1}{100}$, or

decimal notation → $n\% = n \times 0.01$.

EXAMPLE 1 Write three kinds of notation for 67.8%.

Using ratio: $67.8\% = \frac{67.8}{100}$ A ratio of 67.8 to 100

Using fraction notation: $67.8\% = 67.8 \times \frac{1}{100}$ Replacing % with $\times \frac{1}{100}$

Using decimal notation: $67.8\% = 67.8 \times 0.01$ Replacing % with $\times$ 0.01

Do Exercises 1–4.

During 2011, 42.2% of electricity in the United States was generated using coal.

Source: U.S. Energy Information Administration

Write three kinds of notation for each percent, as in Example 1.

1. 70% **2.** 23.4%

3. 100% **4.** 0.6%

b CONVERTING BETWEEN PERCENT NOTATION AND DECIMAL NOTATION

Consider 78%. To convert to decimal notation, we can think of percent notation as a ratio and write

$$78\% = \frac{78}{100}$$ Using the definition of percent as a ratio

$$= 0.78.$$ Dividing

Similarly,

$$4.9\% = \frac{4.9}{100} = 0.049.$$

We can also convert 78% to decimal notation by replacing "%" with "× 0.01" and writing

$$78\% = 78 \times 0.01$$ Replacing % with × 0.01

$$= 0.78.$$ Multiplying

Similarly,

$$4.9\% = 4.9 \times 0.01 = 0.049.$$

Dividing by 100 amounts to moving the decimal point two places to the left, which is the same as multiplying by 0.01. This leads us to a quick way to convert from percent notation to decimal notation: We drop the percent symbol and move the decimal point two places to the left.

It is thought that the Roman emperor Augustus began percent notation by taxing goods sold at a rate of $\frac{1}{100}$. In time, the symbol "%" evolved by interchanging the parts of the symbol "100" to "0/0" and then to "%."

Answers

1. $\frac{70}{100}$; $70 \times \frac{1}{100}$; 70×0.01

2. $\frac{23.4}{100}$; $23.4 \times \frac{1}{100}$; 23.4×0.01

3. $\frac{100}{100}$; $100 \times \frac{1}{100}$; 100×0.01

4. $\frac{0.6}{100}$; $0.6 \times \frac{1}{100}$; 0.6×0.01

CALCULATOR CORNER

Converting from Percent Notation to Decimal Notation Many calculators have a [%] key that can be used to convert from percent notation to decimal notation. This is often the second operation associated with a particular key and is accessed by first pressing a [2nd] or [SHIFT] key. To convert 57.6% to decimal notation, for example, you might press [5] [7] [·] [6] [2nd] [%] or [5] [7] [·] [6] [SHIFT] [%]. The display would read [0.576], so 57.6% = 0.576.

EXERCISES Use a calculator to find decimal notation.

1. 14% **2.** 0.069%

3. 43.8% **4.** 125%

Find decimal notation.

5. 34% **6.** 78.9%

Find decimal notation for the percent notation(s) in each sentence.

7. ***Energy Use.*** The United States consumes 19% of the world's energy. Russia consumes only 6% of the world's energy.

Source: U.S. Energy Information Administration

8. ***Blood Alcohol Level.*** A blood alcohol level of 0.08% is the standard used by the most states as the legal limit for drunk driving.

Answers

5. 0.34 **6.** 0.789
7. 0.19; 0.06 **8.** 0.0008

To convert from percent notation to decimal notation,	36.5%	
a) replace the percent symbol % with × 0.01, and	36.5 × 0.01	
b) multiply by 0.01, which means move the decimal point two places to the left.	0.36.5	Move 2 places to the left.
	36.5% = 0.365	

EXAMPLE 2 Find decimal notation for 99.44%.

a) Replace the percent symbol with × 0.01. 99.44 × 0.01

b) Move the decimal point two places to the left. 0.99.44

Thus, 99.44% = 0.9944.

EXAMPLE 3 The interest rate on a $2\frac{1}{2}$-year certificate of deposit is $6\frac{3}{8}\%$. Find decimal notation for $6\frac{3}{8}\%$.

a) Convert $6\frac{3}{8}$ to decimal notation and replace the percent symbol with × 0.01. $6\frac{3}{8}\%$; 6.375 × 0.01

b) Move the decimal point two places to the left. 0.06.375

Thus, $6\frac{3}{8}\% = 0.06375$.

Do Exercises 5–8.

To convert 0.38 to percent notation, we can first write fraction notation, as follows:

$$0.38 = \frac{38}{100} \quad \text{Converting to fraction notation}$$
$$= 38\%. \quad \text{Using the definition of percent as a ratio}$$

Note that 100% = 100 × 0.01 = 1. Thus to convert 0.38 to percent notation, we can multiply by 1, using 100% as a symbol for 1.

$$0.38 = 0.38 \times 1$$
$$= 0.38 \times 100\%$$
$$= 0.38 \times 100 \times 0.01 \quad \text{Replacing 100\% with } 100 \times 0.01$$
$$= (0.38 \times 100) \times 0.01 \quad \text{Using the associative law of multiplication}$$
$$= 38 \times 0.01$$
$$= 38\% \quad \text{Replacing } \times 0.01 \text{ with \%}$$

Even more quickly, since 0.38 = 0.38 × 100%, we can simply multiply 0.38 by 100 and write the % symbol.

To convert from decimal notation to percent notation, we multiply by 100%. That is, we move the decimal point two places to the right and write a percent symbol.

To convert from decimal notation to percent notation, multiply by 100%. That is,	$0.675 = 0.675 \times 100\%$
a) move the decimal point two places to the right and	0.67.5 Move 2 places to the right.
b) write a % symbol.	67.5% $0.675 = 67.5\%$

EXAMPLE 4 Of the time off that employees take as sick leave, 0.21 is actually used for family issues. Find percent notation for 0.21.

Source: CCH Inc.

a) Move the decimal point two places to the right. 0.21.

b) Write a % symbol. 21%

Thus, $0.21 = 21\%$.

EXAMPLE 5 Find percent notation for 5.6.

a) Move the decimal point two places to the right, adding an extra zero. 5.60.

b) Write a % symbol. 560%

Thus, $5.6 = 560\%$.

EXAMPLE 6 Of those who play golf, 0.149 play 8–24 rounds per year. Find percent notation for 0.149.

Source: U.S. Golf Association

a) Move the decimal point two places to the right. 0.14.9

b) Write a % symbol. 14.9%

Thus, $0.149 = 14.9\%$.

Do Exercises 9–14.

Find percent notation.

9. 0.24 **10.** 3.47

11. 1 **12.** 0.05

Find percent notation for the decimal notation(s) in each sentence.

13. ***Women in Congress.*** In 2012, 0.19 of the members of the United States Congress were women.

Source: *Wall Street Journal*

14. ***Soccer.*** Of Americans in the 18–24 age group, 0.311 have played soccer. Of those in the 12–17 age group, 0.396 have played soccer.

Source: ESPN Sports Poll, a service of TNS Sport

Answers

9. 24% **10.** 347% **11.** 100% **12.** 5%
13. 19% **14.** 31.1%; 39.6%

6.2 Exercise Set

For Extra Help

MyMathLab®

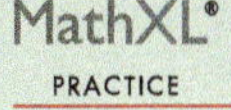

Reading Check

Find percent notation for the shaded area in each figure.

RC1.

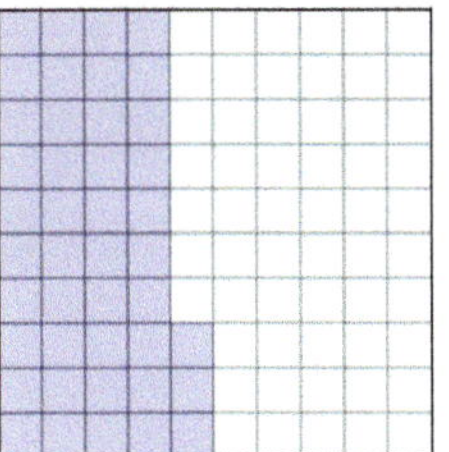

RC2.

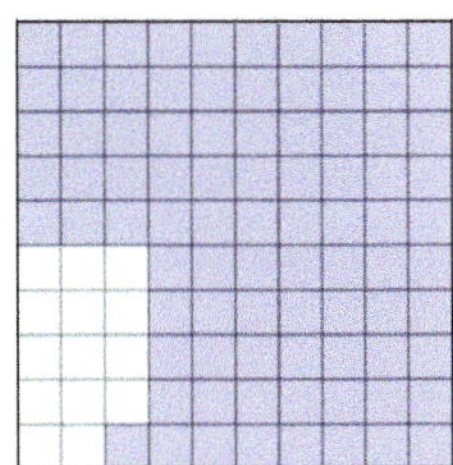

RC3.

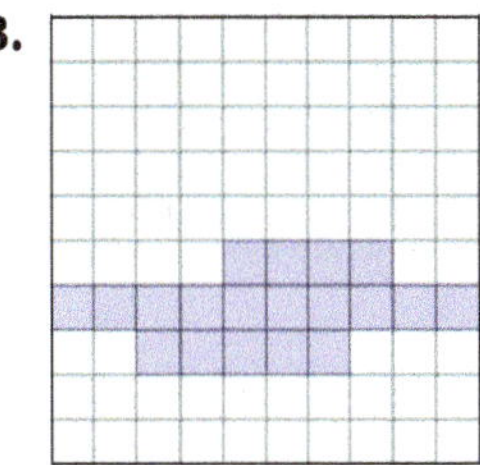

RC4.

a Write three kinds of notation, as in Example 1 on p. 397.

1. 90%

2. 58.7%

3. 12.5%

4. 130%

b Find decimal notation.

5. 67%

6. 17%

7. 45.6%

8. 76.3%

9. 59.01%

10. 30.02%

11. 10%

12. 80%

13. 1%

14. 100%

15. 200%

16. 300%

17. 0.1%

18. 0.4%

19. 0.09%

20. 0.12%

21. 0.18%

22. 5.5%

23. 23.19%

24. 87.99%

25. $14\frac{7}{8}\%$

26. $93\frac{1}{8}\%$

27. $56\frac{1}{2}\%$

28. $61\frac{3}{4}\%$

Find decimal notation for the percent notation(s) in each sentence.

29. ***Daily Calories.*** In 2006, American adults got about 13% of their daily calories from fast food. This percentage decreased to 11% in 2010.

Source: *Indianapolis Star*, February 21, 2013, Nanci Hellmich

30. ***Bachelor's Degrees.*** In 1970, 1% of taxi drivers had a bachelor's degree. By 2010, 15% of taxi drivers had a bachelor's degree.

Source: *USA Today*, January, 28, 2013, Mary Beth Marklein

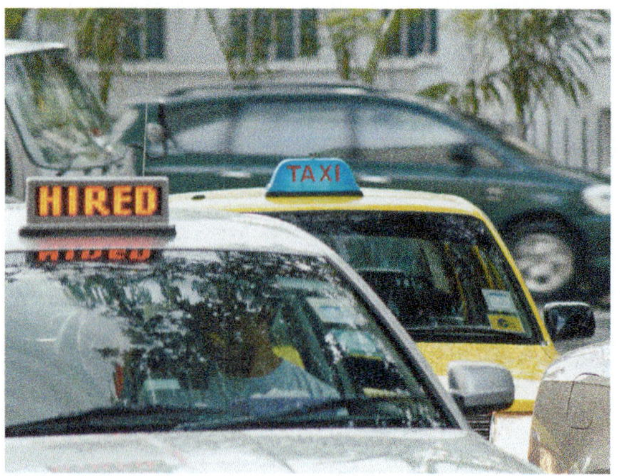

31. ***Fuel Efficiency.*** Speeding up by only 5 mph on the highway cuts fuel efficiency by approximately 7% to 8%.

Source: *Wall Street Journal*, "Pain Relief," by A. J. Miranda, September 15, 2008

32. ***Foreign-Born Population.*** In 2010, the U.S. foreign-born population was 12.9%, the highest since 1920.

Source: U.S. Census Bureau

33. ***Credit-Card Debt.*** In 2012, approximately 13.9% of American households had credit-card debt that exceeded 40% of their income.

Source: statisticbrain.com

34. ***Eating Out.*** On a given day, 58% of all Americans eat meals and snacks away from home.

Source: U.S. Department of Agriculture

Find percent notation.

35. 0.47 **36.** 0.87 **37.** 0.03 **38.** 0.01 **39.** 8.7

40. 4 **41.** 0.334 **42.** 0.889 **43.** 0.75 **44.** 0.99

45. 0.4 **46.** 0.5 **47.** 0.006 **48.** 0.008 **49.** 0.017

50. 0.024 **51.** 0.2718 **52.** 0.8911 **53.** 0.0239 **54.** 0.00073

Find percent notation for the decimal notation(s) in each sentence.

55. ***Recycling Aluminum Cans.*** Over 0.651 of all aluminum cans are recycled.

Source: earth911.com

56. ***Wasting Food.*** Americans waste an estimated 0.27 of the food available for consumption. The waste occurs in restaurants, supermarkets, cafeterias, and household kitchens.

Source: *New York Times*, "One Country's Table Scraps, Another Country's Meal," by Andrew Martin, May 18, 2008

57. ***Dining Together.*** In 2012, 0.34 of American families dined together four or five times per week.

Source: unitedfamiliesinternational.wordpress.com

58. ***Ages 65 and Older.*** In Alaska, 0.057 of the residents are ages 65 and older. In Florida, 0.176 are ages 65 and older.

Source: U.S. Census Bureau

59. ***Residents Ages 15 or Younger.*** In Haiti, 0.359 of the residents are ages 15 or younger. In the United States, 0.2 of the residents are ages 15 or younger.

Source: *The World Almanac 2012*

60. ***Graduation Rates.*** In 2010, the high school graduation rate in the United States was 0.742. The dropout rate was 0.034.

Source: National Center for Educational Statistics

Find decimal notation for each percent notation in the graph.

61.

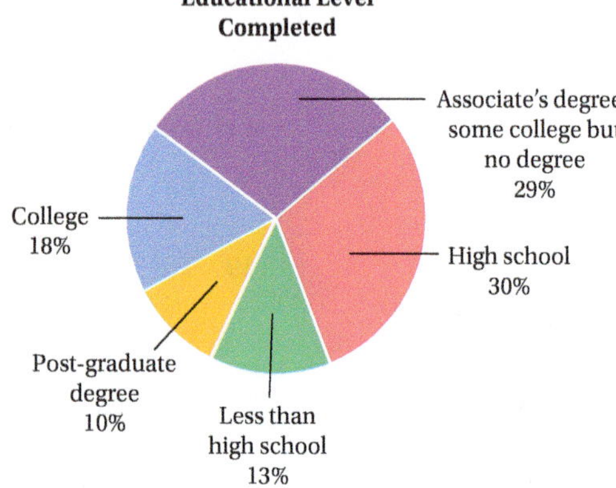

SOURCE: U.S. Census Bureau

62.

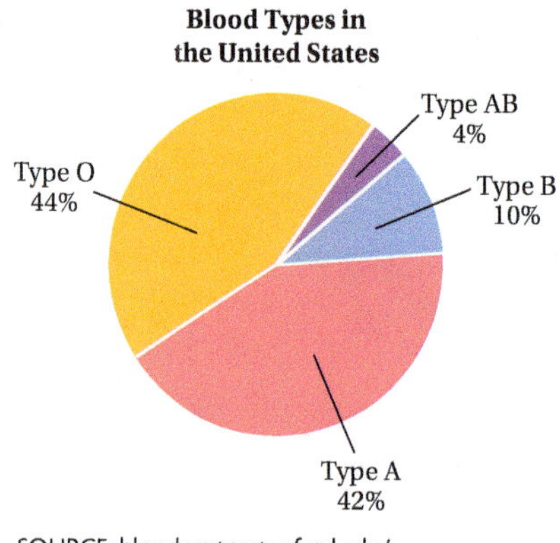

SOURCE: bloodcenter.stanford.edu/about_blood/blood_types.html

Skill Maintenance

Find all the factors of each number. [3.2a]

63. 84

64. 620

65. Find the LCM of 18 and 60. [4.1a]

66. Find the prime factorization of 90. [3.2c]

67. Solve: $\frac{5}{8} + y = \frac{13}{16}$. [4.3b]

68. Simplify: $(12 - 3)^2 - 9 + 5^2$. [1.9c]

Synthesis

Find percent notation. (*Hint*: Multiply by a form of 1 and obtain a denominator of 100.)

69. $\frac{1}{2}$

70. $\frac{3}{4}$

71. $\frac{7}{10}$

72. $\frac{2}{5}$

Find percent notation for each shaded area.

73. | | ■ | | | |
|---|---|---|---|---|

74. | ■ | | | ■ | |
|---|---|---|---|---|
| | ■ | ■ | | ■ |

Percent Notation and Fraction Notation

6.3

OBJECTIVES

a Convert from fraction notation to percent notation.

b Convert from percent notation to fraction notation.

SKILL TO REVIEW

Objective 5.5a: Use division to convert fraction notation to decimal notation.

Find decimal notation.

1. $\frac{11}{16}$ 2. $\frac{5}{9}$

a CONVERTING FROM FRACTION NOTATION TO PERCENT NOTATION

Consider the fraction notation $\frac{7}{8}$. To convert to percent notation, we use two skills that we already have. We first find decimal notation by dividing: $7 \div 8$.

$$\begin{array}{r} 0.875 \\ 8\overline{)7.000} \\ \underline{64} \\ 60 \\ \underline{56} \\ 40 \\ \underline{40} \\ 0 \end{array} \qquad \frac{7}{8} = 0.875$$

Then we convert the decimal notation to percent notation. We move the decimal point two places to the right

0.87.5

and write a % symbol:

$$\frac{7}{8} = 87.5\%, \text{ or } 87\tfrac{1}{2}\%. \qquad 0.5 = \tfrac{1}{2}$$

To convert from fraction notation to percent notation,	$\frac{3}{5}$	Fraction notation
a) find decimal notation by division, and	$\begin{array}{r} 0.6 \\ 5\overline{)3.0} \\ \underline{30} \\ 0 \end{array}$	
b) convert the decimal notation to percent notation.	$0.6 = 0.60 = 60\%$ $\frac{3}{5} = 60\%$	Percent notation

EXAMPLE 1 Find percent notation for $\frac{1}{6}$.

a) We first find decimal notation by division.

$$\begin{array}{r} 0.166 \\ 6\overline{)1.000} \\ \underline{6} \\ 40 \\ \underline{36} \\ 40 \\ \underline{36} \\ 4 \end{array}$$

We get a repeating decimal; $0.16\overline{6}$.

CALCULATOR CORNER

Converting from Fraction Notation to Percent Notation A calculator can be used to convert from fraction notation to percent notation. We simply perform the division on the calculator and then use the percent key. To convert $\frac{17}{40}$ to percent notation, for example, we press [1] [7] [÷] [4] [0] [2nd] [%], or [1] [7] [÷] [4] [0] [SHIFT] [%]. The display reads [42.5], so $\frac{17}{40} = 42.5\%$.

EXERCISES Use a calculator to find percent notation. Round to the nearest hundredth of a percent.

1. $\frac{13}{25}$ 2. $\frac{5}{13}$

3. $\frac{43}{39}$ 4. $\frac{12}{7}$

5. $\frac{217}{364}$ 6. $\frac{2378}{8401}$

Answers

Skill to Review:
1. 0.6875 2. $0.\overline{5}$

b) Next, we convert the decimal notation to percent notation. We move the decimal point two places to the right and write a % symbol.

$0.16.\overline{6}$

$\frac{1}{6} = 16.\overline{6}\%$, or $16\frac{2}{3}\%$ $\quad 0.\overline{6} = \frac{2}{3}$

Don't forget the % symbol.

Find percent notation.

1. $\frac{5}{6}$

2. $\frac{1}{4}$

Do Exercises 1 and 2.

EXAMPLE 2 *First Language.* The first language of approximately $\frac{3}{16}$ of the world's population is Chinese. Find percent notation for $\frac{3}{16}$.

Sources: *National Geographic*, "Languages at Risk," Virginia W. Mason, July 2012; U.S. Census Bureau; *The CIA World Factbook 2012*

a) Find decimal notation by division.

$$\begin{array}{r} 0.1875 \\ 16\,\overline{)\,3.0000} \\ \underline{1\,6} \\ 1\,40 \\ \underline{1\,28} \\ 120 \\ \underline{112} \\ 80 \\ \underline{80} \\ 0 \end{array}$$

$\frac{3}{16} = 0.1875$

b) Convert the answer to percent notation.

$0.18.75$

$\frac{3}{16} = 18.75\%$, or $18\frac{3}{4}\%$

Do Exercises 3 and 4.

3. Water is the single most abundant chemical in the human body. The body is about $\frac{2}{3}$ water. Find percent notation for $\frac{2}{3}$.

4. Find percent notation: $\frac{5}{8}$.

Answers

1. $83.\overline{3}\%$, or $83\frac{1}{3}\%$ **2.** 25%
3. $66.\overline{6}\%$, or $66\frac{2}{3}\%$ **4.** 62.5%

In some cases, division is not the fastest way to convert a fraction to percent notation. The following are some optional ways in which the conversion might be done.

EXAMPLE 3 Find percent notation for $\frac{69}{100}$.

We use the definition of percent as a ratio.

$$\frac{69}{100} = 69\%$$

EXAMPLE 4 Find percent notation for $\frac{17}{20}$.

We want to multiply by 1 to get 100 in the denominator. We think of what we must multiply 20 by in order to get 100. That number is 5, so we multiply by 1 using $\frac{5}{5}$.

$$\frac{17}{20} \cdot \frac{5}{5} = \frac{85}{100} = 85\%$$

Note that this shortcut works only when the denominator is a factor of 100.

EXAMPLE 5 Find percent notation for $\frac{18}{25}$.

$$\frac{18}{25} = \frac{18}{25} \cdot \frac{4}{4} = \frac{72}{100} = 72\%$$

Do Exercises 5–8.

Find percent notation.

5. $\frac{57}{100}$

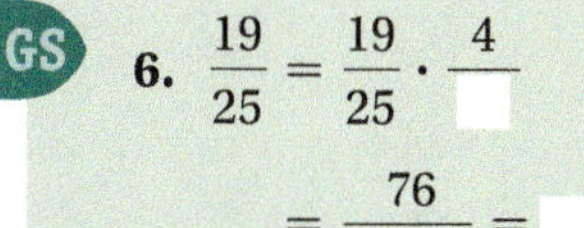

6. $\frac{19}{25} = \frac{19}{25} \cdot \frac{4}{\square}$

$= \frac{76}{\square} = \square\%$

7. $\frac{7}{10}$

8. $\frac{1}{4}$

b CONVERTING FROM PERCENT NOTATION TO FRACTION NOTATION

To convert from percent notation to fraction notation,	30%	Percent notation
a) use the definition of percent as a ratio, and	$\frac{30}{100}$	
b) simplify, if possible.	$\frac{3}{10}$	Fraction notation

EXAMPLE 6 Find fraction notation for 75%.

$$75\% = \frac{75}{100} \quad \text{Using the definition of percent}$$

$$= \frac{3 \cdot 25}{4 \cdot 25} = \frac{3}{4} \cdot \frac{25}{25} = \frac{3}{4} \quad \text{Simplifying}$$

Answers

5. 57% **6.** 76% **7.** 70% **8.** 25%

Guided Solution:
6. 4, 100, 76

EXAMPLE 7 Find fraction notation for 62.5%.

$$62.5\% = \frac{62.5}{100} \qquad \text{Using the definition of percent}$$

$$= \frac{62.5}{100} \times \frac{10}{10} \qquad \text{Multiplying by 1 to eliminate the decimal point in the numerator}$$

$$= \frac{625}{1000}$$

$$= \frac{5 \cdot 125}{8 \cdot 125} = \frac{5}{8} \cdot \frac{125}{125} \qquad \text{Simplifying}$$

$$= \frac{5}{8}$$

Find fraction notation.

9. 60%

10. 3.25%

$$= \frac{3.25}{\square} = \frac{3.25}{100} \times \frac{\square}{100}$$

$$= \frac{325}{\square} = \frac{13 \times \square}{400 \times 25}$$

$$= \frac{13}{400} \times \frac{25}{25} = \frac{\square}{400}$$

11. $66\frac{2}{3}\%$

12. $12\frac{1}{2}\%$

EXAMPLE 8 Find fraction notation for $16\frac{2}{3}\%$.

$$16\frac{2}{3}\% = \frac{50}{3}\% \qquad \text{Converting from the mixed numeral to fraction notation}$$

$$= \frac{50}{3} \times \frac{1}{100} \qquad \text{Using the definition of percent}$$

$$= \frac{50 \cdot 1}{3 \cdot 50 \cdot 2} = \frac{1}{3 \cdot 2} \cdot \frac{50}{50} \qquad \text{Simplifying}$$

$$= \frac{1}{6}$$

Do Exercises 9–12.

The following table lists fraction, decimal, and percent equivalents that are used so often it would speed up your work if you memorized them. For example, $\frac{1}{3} = 0.\overline{3}$, so we say that the **decimal equivalent** of $\frac{1}{3}$ is $0.\overline{3}$, or that $0.\overline{3}$ has the **fraction equivalent** $\frac{1}{3}$. This table also appears on the inside back cover of the book.

FRACTION, DECIMAL, AND PERCENT EQUIVALENTS

FRACTION NOTATION	$\frac{1}{10}$	$\frac{1}{8}$	$\frac{1}{6}$	$\frac{1}{5}$	$\frac{1}{4}$	$\frac{3}{10}$	$\frac{1}{3}$	$\frac{3}{8}$	$\frac{2}{5}$	$\frac{1}{2}$
DECIMAL NOTATION	0.1	0.125	$0.16\overline{6}$	0.2	0.25	0.3	$0.33\overline{3}$	0.375	0.4	0.5
PERCENT NOTATION	10%	12.5%, or $12\frac{1}{2}\%$	$16.\overline{6}\%$, or $16\frac{2}{3}\%$	20%	25%	30%	$33.\overline{3}\%$, or $33\frac{1}{3}\%$	37.5%, or $37\frac{1}{2}\%$	40%	50%

FRACTION NOTATION	$\frac{3}{5}$	$\frac{5}{8}$	$\frac{2}{3}$	$\frac{7}{10}$	$\frac{3}{4}$	$\frac{4}{5}$	$\frac{5}{6}$	$\frac{7}{8}$	$\frac{9}{10}$	$\frac{1}{1}$
DECIMAL NOTATION	0.6	0.625	$0.66\overline{6}$	0.7	0.75	0.8	$0.83\overline{3}$	0.875	0.9	1
PERCENT NOTATION	60%	62.5%, or $62\frac{1}{2}\%$	$66.\overline{6}\%$, or $66\frac{2}{3}\%$	70%	75%	80%	$83.\overline{3}\%$, or $83\frac{1}{3}\%$	87.5%, or $87\frac{1}{2}\%$	90%	100%

Find fraction notation.

13. $33.\overline{3}\%$

14. $83.\overline{3}\%$

EXAMPLE 9 Find fraction notation for $16.\overline{6}\%$.

We can use the table above or recall that $16.\overline{6}\% = 16\frac{2}{3}\% = \frac{1}{6}$. We can also recall from our work with repeating decimals in Chapter 3 that $0.\overline{6} = \frac{2}{3}$. Then we have $16.\overline{6}\% = 16\frac{2}{3}\%$ and can proceed as in Example 8.

Do Exercises 13 and 14.

Answers

9. $\frac{3}{5}$ **10.** $\frac{13}{400}$ **11.** $\frac{2}{3}$
12. $\frac{1}{8}$ **13.** $\frac{1}{3}$ **14.** $\frac{5}{6}$

Guided Solution:
10. 100, 100, 10,000, 25, 13

6.3 Exercise Set

For Extra Help

MyMathLab®

MathXL® PRACTICE 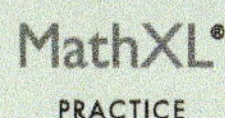  WATCH  READ  REVIEW

Reading Check

Match each fraction with the equivalent decimal notation from the list on the right. Some choices will not be used.

RC1. $\frac{1}{8}$ ______ **RC2.** $\frac{3}{8}$ ______

RC3. $\frac{5}{8}$ ______ **RC4.** $\frac{7}{8}$ ______

RC5. $\frac{1}{5}$ ______ **RC6.** $\frac{2}{5}$ ______

RC7. $\frac{3}{5}$ ______ **RC8.** $\frac{4}{5}$ ______

a) 0.875
b) 0.2
c) 0.125
d) 0.4
e) 0.375
f) 0.1
g) 0.625
h) 0.6
i) 0.8
j) 0.675

a Find percent notation.

1. $\frac{41}{100}$ **2.** $\frac{36}{100}$ **3.** $\frac{5}{100}$ **4.** $\frac{1}{100}$ **5.** $\frac{2}{10}$ **6.** $\frac{7}{10}$

7. $\frac{3}{10}$ **8.** $\frac{9}{10}$ **9.** $\frac{1}{2}$ **10.** $\frac{3}{4}$ **11.** $\frac{7}{8}$ **12.** $\frac{1}{8}$

13. $\frac{4}{5}$ **14.** $\frac{2}{5}$ **15.** $\frac{2}{3}$ **16.** $\frac{1}{3}$ **17.** $\frac{1}{6}$ **18.** $\frac{5}{6}$

19. $\frac{3}{16}$ **20.** $\frac{11}{16}$ **21.** $\frac{13}{16}$ **22.** $\frac{7}{16}$ **23.** $\frac{4}{25}$ **24.** $\frac{17}{25}$

25. $\frac{1}{20}$ **26.** $\frac{31}{50}$ **27.** $\frac{17}{50}$ **28.** $\frac{3}{20}$

Find percent notation for the fraction notation in each sentence.

29. ***Heart Transplants.*** In the United States in 2006, $\frac{2}{25}$ of the organ transplants were heart transplants and $\frac{59}{100}$ were kidney transplants.

Source: 2007 OPTN/SRTR Annual Report, Table 1.7

30. ***Car Colors.*** The four most popular colors for 2006 compact/sports cars were silver, gray, black, and red. Of all cars in this category, $\frac{9}{50}$ were silver, $\frac{3}{20}$ gray, $\frac{3}{20}$ black, and $\frac{3}{20}$ red.

Sources: Ward's Automotive Group; DuPont Automotive Products

In Exercises 31–36, write percent notation for the fractions in the following pie chart.

How Food Dollars Are Spent

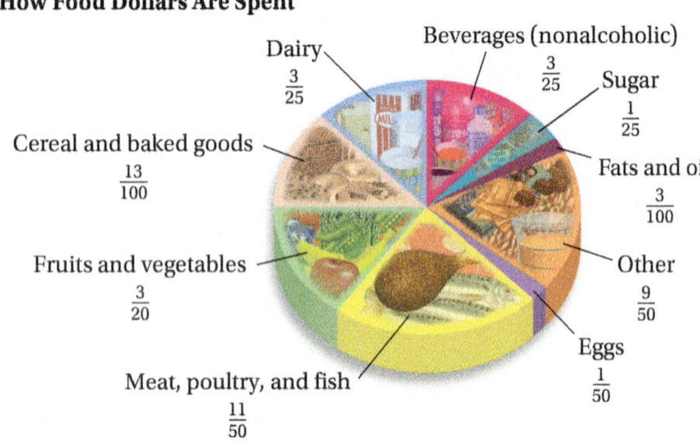

SOURCES: U.S. Bureau of Labor Statistics; Consumer Price Index; *The Hoosier Farmer*, Summer 2008

31. $\frac{11}{50}$

32. $\frac{9}{50}$

33. $\frac{3}{25}$

34. $\frac{1}{25}$

35. $\frac{3}{20}$

36. $\frac{13}{100}$

b Find fraction notation. Simplify.

37. 85%

38. 55%

39. 62.5%

40. 12.5%

41. $33\frac{1}{3}\%$

42. $83\frac{1}{3}\%$

43. $16.\overline{6}\%$

44. $66.\overline{6}\%$

45. 7.25%

46. 4.85%

47. 0.8%

48. 0.2%

49. $25\frac{3}{8}\%$

50. $48\frac{7}{8}\%$

51. $78\frac{2}{9}\%$

52. $16\frac{5}{9}\%$

53. $64\frac{7}{11}\%$

54. $73\frac{3}{11}\%$

55. 150%

56. 110%

57. 0.0325%

58. 0.419%

59. $33.\overline{3}\%$

60. $83.\overline{3}\%$

In Exercises 61–66, find fraction notation for the percent notations in the following table.

U.S. POPULATION BY SELECTED AGE CATEGORIES
(Data have been rounded to the nearest percent.)

AGE CATEGORY	PERCENT OF POPULATION
5–17 years	18%
18–24 years	10
15–44 years	41
18 years and older	76
65 years and older	13
75 years and older	6

SOURCES: U.S. Census Bureau; 2010 American Community Survey

61. 6%

62. 18%

63. 13%

64. 41%

65. 76%

66. 10%

Find fraction notation for the percent notation in each sentence.

67. A $\frac{3}{4}$-cup serving of Post Selects Great Grains cereal with $\frac{1}{2}$ cup of fat-free milk satisfies 15% of the minimum daily requirement for calcium.

Source: Kraft Foods Global, Inc.

68. A 1.8-oz serving of Frosted Mini-Wheats®, Blueberry Muffin, with $\frac{1}{2}$ cup of fat-free milk satisfies 35% of the minimum daily requirement for Vitamin B_{12}.

Source: Kellogg, Inc.

69. In 2006, 20.9% of Americans age 18 and older smoked cigarettes.

Sources: *Washington Post*, March 9, 2006; U.S. Centers for Disease Control and Prevention

70. In 2010, 11.9% of Californians age 18 and older smoked cigarettes.

Source: California Department of Public Health

Complete each table.

71.

Fraction Notation	Decimal Notation	Percent Notation
$\frac{1}{8}$		12.5%, or $12\frac{1}{2}\%$
$\frac{1}{6}$		
		20%
	0.25	
		$33.\overline{3}\%$, or $33\frac{1}{3}\%$
		37.5%, or $37\frac{1}{2}\%$
		40%
$\frac{1}{2}$		

72.

Fraction Notation	Decimal Notation	Percent Notation
$\frac{3}{5}$		
	0.625	
$\frac{2}{3}$		
	0.75	75%
$\frac{4}{5}$		
$\frac{5}{6}$		$83.\overline{3}\%$, or $83\frac{1}{3}\%$
$\frac{7}{8}$		87.5%, or $87\frac{1}{2}\%$
		100%

73.

Fraction Notation	Decimal Notation	Percent Notation
	0.5	
$\frac{1}{3}$		
		25%
		$16.\overline{6}\%$, or $16\frac{2}{3}\%$
	0.125	
$\frac{3}{4}$		
	$0.8\overline{3}$	
$\frac{3}{8}$		

74.

Fraction Notation	Decimal Notation	Percent Notation
		40%
		62.5%, or $62\frac{1}{2}\%$
	0.875	
$\frac{1}{1}$		
	0.6	
	$0.\overline{6}$	
$\frac{1}{5}$		

Skill Maintenance

Solve.

75. $13 \cdot x = 910$ [1.7b]

76. $15 \cdot y = 75$ [1.7b]

77. $0.05b = 20$ [5.7a]

78. $3 = -0.16b$ [5.7a]

79. $\frac{24}{37} = \frac{15}{x}$ [6.1d]

80. $\frac{17}{18} = \frac{x}{27}$ [6.1d]

81. $\frac{9}{10} = \frac{x}{5}$ [6.1d]

82. $\frac{7}{x} = \frac{4}{5}$ [6.1d]

Convert to a mixed numeral. [4.5a]

83. $-\frac{75}{4}$

84. $\frac{67}{9}$

Convert to fraction notation. [4.5a]

85. $101\frac{1}{2}$

86. $-20\frac{9}{10}$

Synthesis

Write percent notation.

87. $2.\overline{574631}$

88. $\frac{54}{999}$

Write decimal notation.

89. $\frac{729}{7}\%$

90. $\frac{19}{12}\%$

91. Arrange the following numbers from smallest to largest.

$$16\frac{1}{6}\%,\ 1.6,\ \frac{1}{6}\%,\ \frac{1}{2},\ 0.2,\ 1.6\%,\ 1\frac{1}{6}\%,\ 0.5\%,\ \frac{2}{7}\%,\ 0.\overline{54}$$

Solving Percent Problems Using Percent Equations

6.4

a TRANSLATING TO EQUATIONS

To solve a problem involving percents, it is helpful to translate first to an equation. To distinguish the method discussed in this section from that of Section 6.5, we will call these *percent equations*.

KEY WORDS IN PERCENT TRANSLATIONS

"**Of**" translates to "$\cdot$" or "$\times$".
"**What**" translates to any letter.
"**Is**" translates to "$=$".
"**%**" translates to "$\times \frac{1}{100}$" or "$\times$ 0.01".

EXAMPLES Translate each of the following.

1. 23% of 5 is what?

$$23\% \cdot 5 = a$$

This is a *percent equation.*

2. What is 11% of 49?

$$a = 11\% \cdot 49$$

Any letter can be used.

Do Margin Exercises 1 and 2.

EXAMPLES Translate each of the following.

3. 3 is 10% of what?

$$3 = 10\% \cdot b$$

4. 45% of what is 23?

$$45\% \times b = 23$$

Do Margin Exercises 3 and 4.

EXAMPLES Translate each of the following.

5. 10 is what percent of 20?

$$10 = p \times 20$$

6. What percent of 50 is 7?

$$p \cdot 50 = 7$$

Do Margin Exercises 5 and 6.

OBJECTIVES

a Translate percent problems to percent equations.

b Solve basic percent problems.

SKILL TO REVIEW

Objective 5.7a: Solve equations containing decimals and one variable term.

Solve.

1. $0.05x = 830$

2. $8 \cdot y = 40.648$

Translate to an equation. Do not solve.

1. 12% of 50 is what?

2. What is 40% of 60?

Translate to an equation. Do not solve.

3. 45 is 20% of what?

4. 120% of what is 60?

Translate to an equation. Do not solve.

5. 16 is what percent of 40?

6. What percent of 84 is 10.5?

Answers

Skill to Review:
1. 16,600 **2.** 5.081

Margin Exercises:
1. $12\% \times 50 = a$ **2.** $a = 40\% \times 60$
3. $45 = 20\% \times b$ **4.** $120\% \times b = 60$
5. $16 = p \times 40$ **6.** $p \times 84 = 10.5$

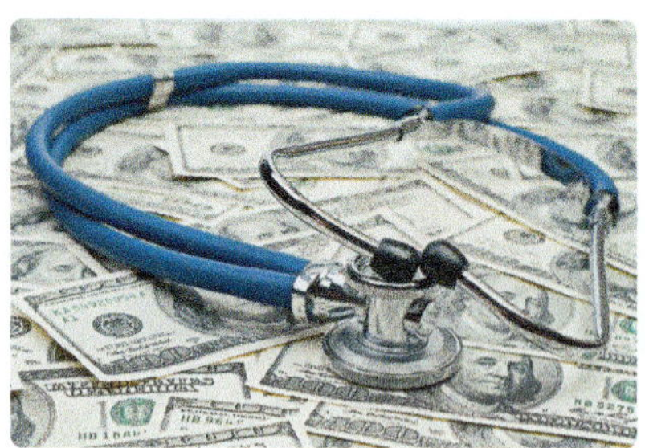

The U.S. gross domestic product for 2010 was approximately \$14.66 trillion. Of that amount, Americans spent 17.9% on health care. What was spent on health care? (See Example 7.)

Sources: *EconPost*, April 25, 2011; kaiseredu.org

b SOLVING PERCENT PROBLEMS

In solving percent problems, we use the *Translate* and *Solve* steps in the problem-solving strategy used throughout this text.

Percent problems are actually of three different types. Although the method we present does *not* require that you be able to identify which type you are solving, it is helpful to know them. Each of the three types of percent problems depends on which of the three pieces of information is missing.

1. **Finding the *amount* (the result of taking the percent)**

   Example: What is 25% of 60?

   Translation: $a = 25\% \cdot 60$

2. **Finding the *base* (the number you are taking the percent of)**

   Example: 15 is 25% of what?

   Translation: $15 = 25\% \cdot b$

3. **Finding the *percent number* (the percent itself)**

   Example: 15 is what percent of 60?

   Translation: $15 = p \cdot 60$

Finding the Amount

EXAMPLE 7 What is 17.9% of \$14,660,000,000,000?

Translate: $a = 17.9\% \times 14{,}660{,}000{,}000{,}000$.

Solve: The letter is by itself. To solve the equation, we convert 17.9% to decimal notation and multiply:

$$a = 17.9\% \times 14{,}660{,}000{,}000{,}000$$
$$= 0.179 \times 14{,}660{,}000{,}000{,}000 = 2{,}624{,}140{,}000{,}000.$$

Thus, \$2,624,140,000,000 is 17.9% of \$14,660,000,000,000. The answer is \$2,624,140,000,000, or about \$2.6 trillion.

7. Solve:

   What is 12% of \$50?

◀ Do Exercise 7.

EXAMPLE 8 120% of 42 is what?

Translate: $120\% \times 42 = a$.

Solve: The letter is by itself. To solve the equation, we carry out the calculation:

$$a = 120\% \times 42$$
$$a = 1.2 \times 42 \qquad 120\% = 1.2$$
$$a = 50.4.$$

Thus, 120% of 42 is 50.4. The answer is 50.4.

8. Solve:

   64% of 55 is what?

◀ Do Exercise 8.

Answers

7. \$6 8. 35.2

Finding the Base

EXAMPLE 9 8% of what is 32?

Translate: $8\% \times b = 32$.

Solve: This time the letter is *not* by itself. To solve the equation, we divide by 8% on both sides:

$$\frac{8\% \times b}{8\%} = \frac{32}{8\%} \qquad \text{Dividing by 8\% on both sides}$$

$$b = \frac{32}{0.08} \qquad 8\% = 0.08$$

$$b = 400.$$

Thus, 8% of 400 is 32. The answer is 400.

EXAMPLE 10 \$3 is 16% of what?

Translate: \$3 is 16% of what?

$$3 = 16\% \times b$$

Solve: To solve the equation, we divide by 16% on both sides:

$$\frac{3}{16\%} = \frac{16\% \times b}{16\%} \qquad \text{Dividing by 16\% on both sides}$$

$$\frac{3}{0.16} = b \qquad 16\% = 0.16$$

$$18.75 = b.$$

Thus, \$3 is 16% of \$18.75. The answer is \$18.75.

Do Exercises 9 and 10. ▶

Finding the Percent Number

In solving these problems, you *must* remember to convert to percent notation after you have solved the equation.

EXAMPLE 11 414,000 is what percent of 621,000?

Translate: 414,000 is what percent of 621,000?

$$414{,}000 = p \times 621{,}000$$

Solve: To solve the equation, we divide by 621,000 on both sides and convert the result to percent notation:

$$p \times 621{,}000 = 414{,}000$$

$$\frac{p \times 621{,}000}{621{,}000} = \frac{414{,}000}{621{,}000} \qquad \text{Dividing by 621,000 on both sides}$$

$$p = 0.666\ldots \qquad \text{Converting to decimal notation}$$

$$p = 66.\overline{6}\%, \text{ or } 66\tfrac{2}{3}\%. \qquad \text{Converting to percent notation}$$

Thus, 414,000 is $66\frac{2}{3}\%$ of 621,000. The answer is $66\frac{2}{3}\%$.

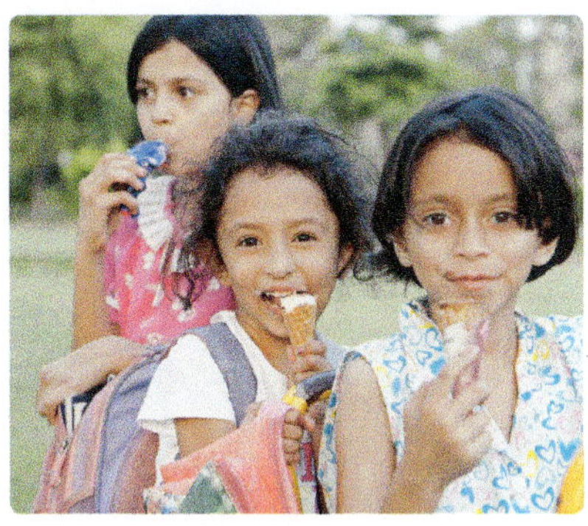

A survey of a group of people found that 8% of the group, or 32 people, chose cookies and cream as their favorite ice cream flavor. How many people were surveyed? (See Example 9.)

Source: Rasmussen Reports Survey

Solve.

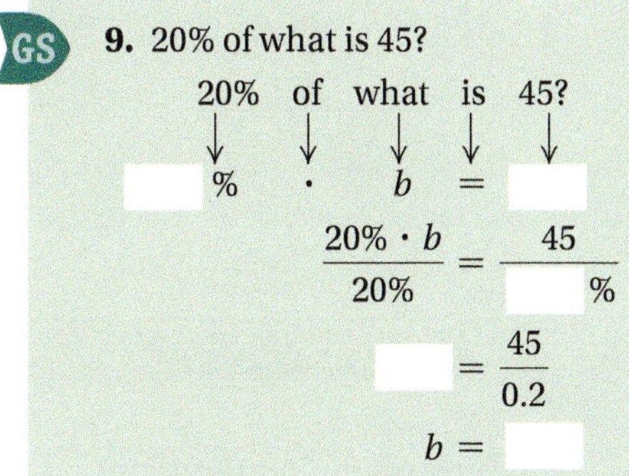

GS **9.** 20% of what is 45?

20% of what is 45?

$\square\% \cdot b = \square$

$$\frac{20\% \cdot b}{20\%} = \frac{45}{\square\%}$$

$$\square = \frac{45}{0.2}$$

$$b = \square$$

10. \$60 is 120% of what?

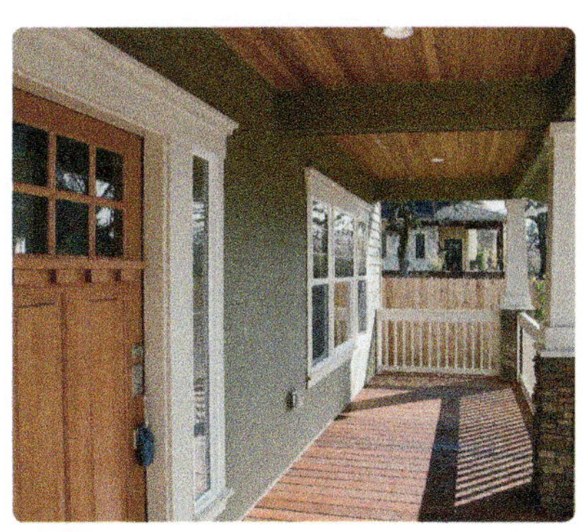

In 2011, 621,000 new housing permits were issued in the United States. Of those new houses, 414,000 have porches. What percent of the houses have porches? (See Example 11.)

Source: U.S. Census Bureau

Answers

9. 225 **10.** \$50

Guided Solution:

9. 20, 45, 20, *b*, 225

Solve.

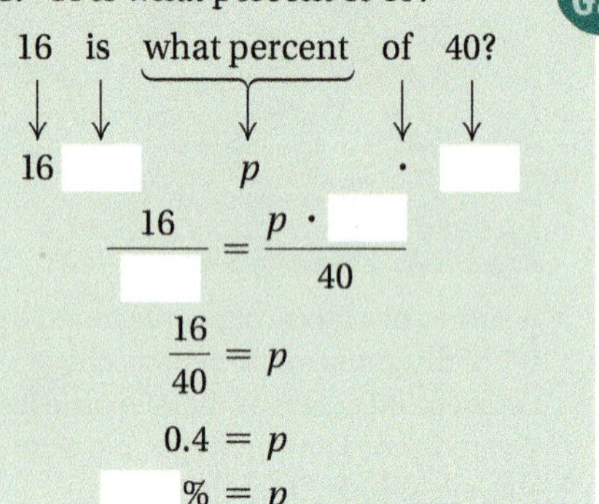

11. 16 is what percent of 40?

12. What percent of $84 is $10.50?

EXAMPLE 12 What percent of $50 is $16?

Translate: What percent of $50 is $16?

$$p \times 50 = 16$$

Solve: To solve the equation, we divide by 50 on both sides and convert the answer to percent notation:

$$\frac{p \times 50}{50} = \frac{16}{50} \quad \text{Dividing by 50 on both sides}$$

$$p = \frac{16}{50}$$

$$p = 0.32$$

$$p = 32\%. \quad \text{Converting to percent notation}$$

Thus, 32% of $50 is $16. The answer is 32%.

Do Exercises 11 and 12.

Caution!

When a question asks "what percent?", be sure to give the answer in percent notation.

CALCULATOR CORNER

Using Percents in Computations Many calculators have a [%] key that can be used in computations. (See the Calculator Corner on page 398.) For example, to find 11% of 49, we press [1] [1] [2nd] [%] [×] [4] [9] [=], or [4] [9] [×] [1] [1] [SHIFT] [%]. The display reads [5.39], so 11% of 49 is 5.39.

In Example 9, we performed the computation 32/8%. To use the [%] key in this computation, we press [3] [2] [÷] [8] [2nd] [%] [=], or [3] [2] [÷] [8] [SHIFT] [%]. The result is 400.

We can also use the [%] key to find the percent number in a problem. In Example 11, for instance, we answered the question "414,000 is what percent of 621,000?" On a calculator, we press [4] [1] [4] [0] [0] [0] [÷] [6] [2] [1] [0] [0] [0] [2nd] [%] [=], or [4] [1] [4] [0] [0] [0] [÷] [6] [2] [1] [0] [0] [0] [SHIFT] [%]. The result is $66.\overline{6}$, so 414,000 is $66.\overline{6}\%$ of 621,000.

EXERCISES Use a calculator to find each of the following.

1. What is 12.6% of $40?
2. 0.04% of 28 is what?
3. 8% of what is 36?
4. $45 is 4.5% of what?
5. 23 is what percent of 920?
6. What percent of $442 is $53.04?

Answers

11. 40% 12. 12.5%

Guided Solution:
11. =, 40, 40, 40, 40

6.4 Exercise Set

For Extra Help MyMathLab® MathXL® PRACTICE WATCH READ REVIEW

Reading Check

Match each question with the correct translation from the list at the right.

RC1. 18 is 40% of what? ________

RC2. What percent of 45 is 18? ________

RC3. What is 40% of 45? ________

RC4. 0.5% of 1200 is what? ________

RC5. 6 is what percent of 1200? ________

RC6. 6 is 0.5% of what? ________

a) $6 = 0.5\% \cdot b$

b) $6 = p \cdot 1200$

c) $18 = 40\% \cdot b$

d) $0.5\% \cdot 1200 = a$

e) $p \cdot 45 = 18$

f) $a = 40\% \cdot 45$

a Translate to an equation. Do not solve.

1. What is 32% of 78?

2. 98% of 57 is what?

3. 89 is what percent of 99?

4. What percent of 25 is 8?

5. 13 is 25% of what?

6. 21.4% of what is 20?

b Translate to an equation and solve.

7. What is 85% of 276?

8. What is 74% of 53?

9. 150% of 30 is what?

10. 100% of 13 is what?

11. What is 6% of \$300?

12. What is 4% of \$45?

13. 3.8% of 50 is what?

14. $33\frac{1}{3}\%$ of 480 is what?
(*Hint*: $33\frac{1}{3}\% = \frac{1}{3}$.)

15. \$39 is what percent of \$50?

16. \$16 is what percent of \$90?

17. 20 is what percent of 10?

18. 60 is what percent of 20?

19. What percent of \$300 is \$150?

20. What percent of \$50 is \$40?

21. What percent of 80 is 100?

22. What percent of 60 is 15?

23. 20 is 50% of what?

24. 57 is 20% of what?

25. 40% of what is $16?

26. 100% of what is $74?

27. 56.32 is 64% of what?

28. 71.04 is 96% of what?

29. 70% of what is 14?

30. 70% of what is 35?

31. What is $62\frac{1}{2}\%$ of 10?

32. What is $35\frac{1}{4}\%$ of 1200?

33. What is 8.3% of $10,200?

34. What is 9.2% of $5600?

35. 2.5% of what is 30.4?

36. 8.2% of what is 328?

Skill Maintenance

Write fraction notation. [5.1b]

37. 0.9375

38. −0.125

Write decimal notation. [5.1b]

39. $-\frac{3}{10}$

40. $\frac{17}{1000}$

Simplify. [1.9c]

41. $3 + (8 - 6) \cdot 2$

42. $2 \cdot 7 - (5 + 1)$

Synthesis

Solve.

43. $2496 is 24% of what amount?
Estimate ______
Calculate ______

44. What is 38.2% of $52,345.79?
Estimate ______
Calculate ______

45. 40% of $18\frac{3}{4}\%$ of $25,000 is what?

Solving Percent Problems Using Proportions*

6.5

OBJECTIVES

a Translate percent problems to proportions.

b Solve basic percent problems.

SKILL TO REVIEW

Objective 6.1d: Solve proportions.

Solve.

1. $\frac{3}{100} = \frac{27}{b}$

2. $\frac{4.3}{20} = \frac{N}{100}$

a TRANSLATING TO PROPORTIONS

A percent is a ratio of some number to 100. For example, 5% is the ratio $\frac{5}{100}$. The numbers 7,700,000 and 154,000,000 have the same ratio as 5 and 100.

$$\frac{5}{100} = \frac{7{,}700{,}000}{154{,}000{,}000}$$

5, 100 → 5%; 7,700,000, 154,000,000

To solve a percent problem using a proportion, we translate as follows:

$$\frac{N}{100} = \frac{a}{b}$$

Number → N; 100 → 100; a ← Amount; b ← Base

You might find it helpful to read this as "part is to whole as part is to whole."

For example, 60% of 25 is 15 translates to

$$\frac{60}{100} = \frac{15}{25}$$

15 ← Amount; 25 ← Base

A clue for translating is that the base, b, corresponds to 100 and usually follows the wording "percent of." Also, N% always translates to $N/100$. Another aid in translating is to make a comparison drawing. To do this, we start with the percent side and list 0% at the top and 100% near the bottom. Then we estimate where the specified percent—in this case, 60%—is located. The corresponding quantities are then filled in. The base—in this case, 25—always corresponds to 100%, and the amount—in this case, 15—corresponds to the specified percent.

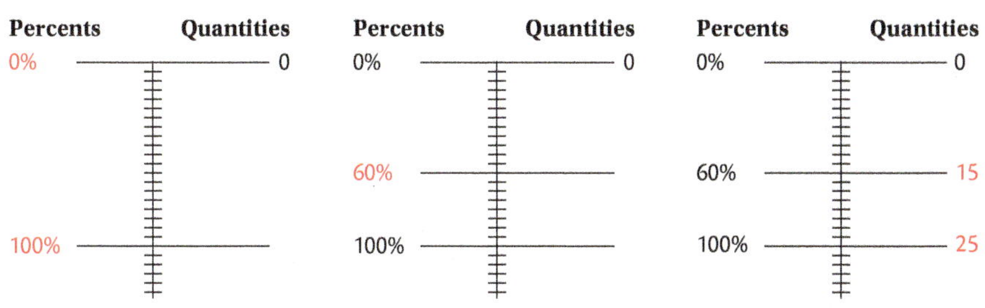

The proportion can then be read easily from the drawing: $\frac{60}{100} = \frac{15}{25}$.

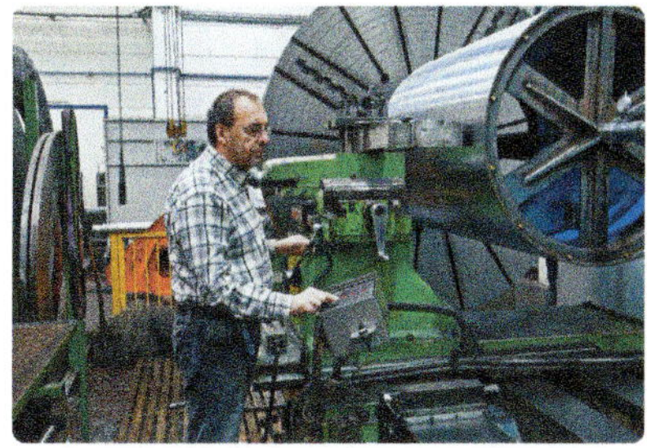

In the United States, 5% of the labor force is ages 65 or older. In 2012, there were approximately 154,000,000 people in the labor force. This means that about 7,700,000 workers were ages 65 or older.

Sources: U.S. Department of Labor; U.S. Bureau of Labor Statistics

*Note: This section presents an alternative method for solving basic percent problems. You can use either equations or proportions to solve percent problems, but you might prefer one method over the other, or your instructor may direct you to use one method over the other.

Answers

Skill to Review:
1. 900 2. 21.5

EXAMPLE 1 Translate to a proportion.

23% of 5 is what?

$$\frac{23}{100} = \frac{a}{5}$$

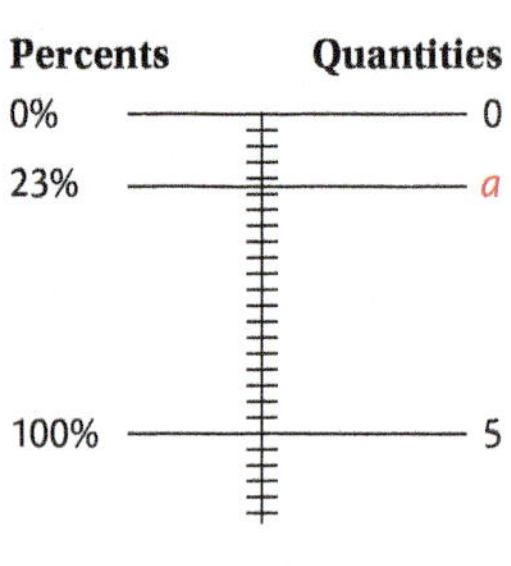

EXAMPLE 2 Translate to a proportion.

What is 124% of 49?

$$\frac{124}{100} = \frac{a}{49}$$

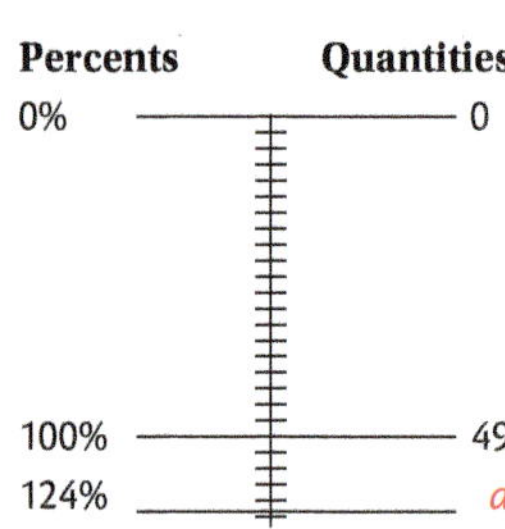

Translate to a proportion. Do not solve.

1. 12% of 50 is what?

2. What is 40% of 60?

3. 130% of 72 is what?

Do Exercises 1–3.

EXAMPLE 3 Translate to a proportion.

3 is 10% of what?

$$\frac{10}{100} = \frac{3}{b}$$

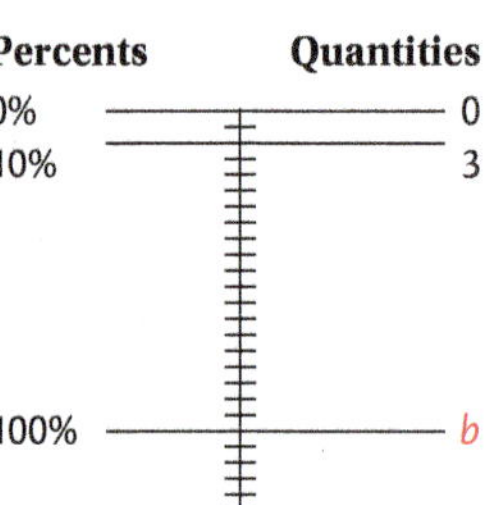

EXAMPLE 4 Translate to a proportion.

45% of what is 23?

$$\frac{45}{100} = \frac{23}{b}$$

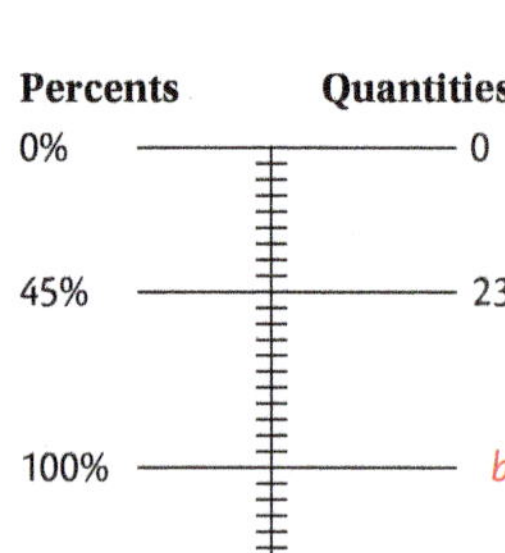

Translate to a proportion. Do not solve.

4. 45 is 20% of what?

5. 120% of what is 60?

Do Exercises 4 and 5.

EXAMPLE 5 Translate to a proportion.

10 is what percent of 20?

$$\frac{N}{100} = \frac{10}{20}$$

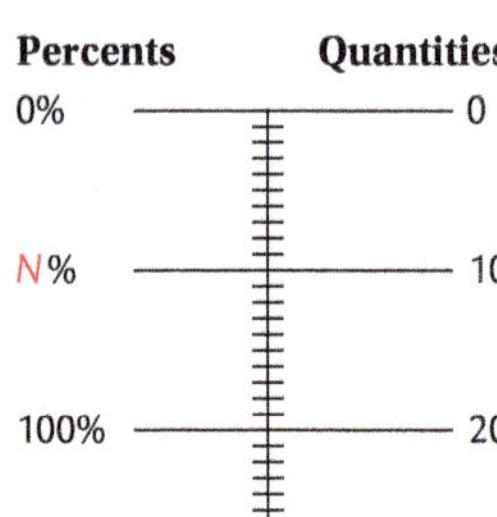

Answers

1. $\frac{12}{100} = \frac{a}{50}$ 2. $\frac{40}{100} = \frac{a}{60}$ 3. $\frac{130}{100} = \frac{a}{72}$
4. $\frac{20}{100} = \frac{45}{b}$ 5. $\frac{120}{100} = \frac{60}{b}$

EXAMPLE 6 Translate to a proportion.

What percent of 50 is 7?

$$\frac{N}{100} = \frac{7}{50}$$

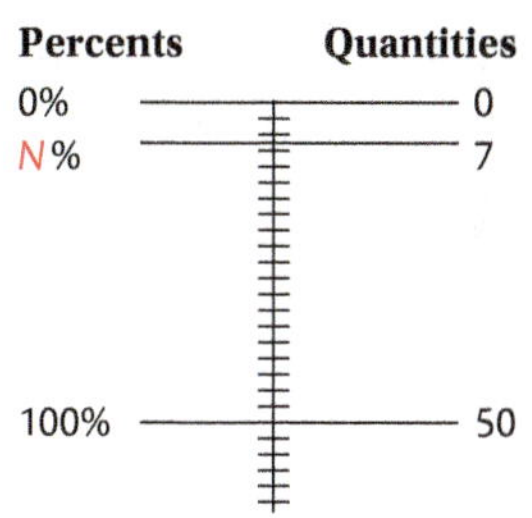

Do Exercises 6 and 7. ▶

Translate to a proportion. Do not solve.

6. 16 is what percent of 40?

7. What percent of 84 is 10.5?

b SOLVING PERCENT PROBLEMS

After a percent problem has been translated to a proportion, we solve as in Section 6.1.

EXAMPLE 7 5% of what is \$20?

Translate: $\frac{5}{100} = \frac{20}{b}$

Solve: $5 \cdot b = 100 \cdot 20$ Equating cross products

$\frac{5 \cdot b}{5} = \frac{100 \cdot 20}{5}$ Dividing by 5

$b = \frac{2000}{5}$

$b = 400$ Simplifying

Thus, 5% of \$400 is \$20. The answer is \$400.

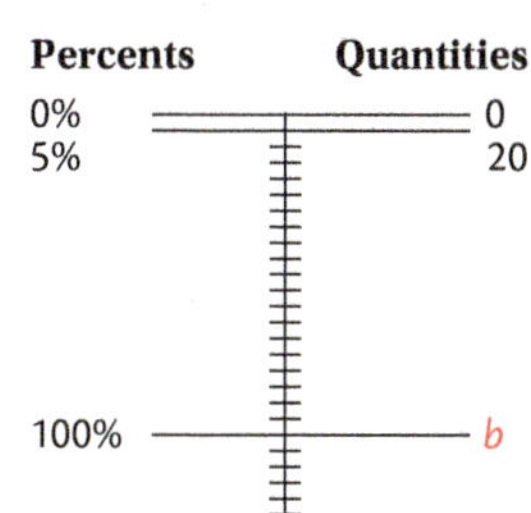

Do Exercise 8. ▶

GS 8. Solve: 20% of what is \$45?

$\frac{20}{\square} = \frac{\square}{b}$

$20 \cdot b = 100 \cdot \square$

$\frac{20b}{20} = \frac{100 \cdot 45}{\square}$

$b = \frac{4500}{20}$

$b = \square$

EXAMPLE 8 120% of 42 is what?

Translate: $\frac{120}{100} = \frac{a}{42}$

Solve: $120 \cdot 42 = 100 \cdot a$ Equating cross products

$\frac{120 \cdot 42}{100} = \frac{100 \cdot a}{100}$ Dividing by 100

$\frac{5040}{100} = a$

$50.4 = a$ Simplifying

Thus, 120% of 42 is 50.4. The answer is 50.4.

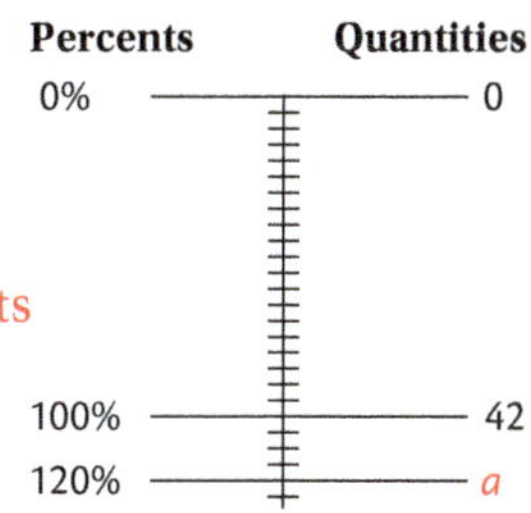

Do Exercises 9 and 10. ▶

GS 9. Solve: 64% of 55 is what?

$\frac{64}{100} = \frac{a}{\square}$

$\square \cdot 55 = 100 \cdot a$

$\frac{64 \cdot 55}{100} = \frac{100 \cdot a}{\square}$

$\frac{\square}{100} = a$

$\square = a$

10. What is 12% of 50?

Answers

6. $\frac{N}{100} = \frac{16}{40}$ 7. $\frac{N}{100} = \frac{10.5}{84}$ 8. \$225
9. 35.2 10. 6

Guided Solutions:
8. 100, 45, 45, 20, 225
9. 55, 64, 100, 3520, 35.2

EXAMPLE 9 210 is $10\frac{1}{2}\%$ of what?

Translate: $\frac{210}{b} = \frac{10.5}{100}$ $\qquad 10\frac{1}{2}\% = 10.5\%$

Solve:
$210 \cdot 100 = b \cdot 10.5$ Equating cross products

$\frac{210 \cdot 100}{10.5} = \frac{b \cdot 10.5}{10.5}$ Dividing by 10.5

$\frac{21{,}000}{10.5} = b$ Multiplying and simplifying

$2000 = b$ Dividing

Thus, 210 is $10\frac{1}{2}\%$ of 2000. The answer is 2000.

11. Solve:
60 is 120% of what?

◀ **Do Exercise 11.**

EXAMPLE 10 \$10 is what percent of \$20?

Translate: $\frac{10}{20} = \frac{N}{100}$

Solve:
$10 \cdot 100 = 20 \cdot N$ Equating cross products

$\frac{10 \cdot 100}{20} = \frac{20 \cdot N}{20}$ Dividing by 20

$\frac{1000}{20} = N$ Multiplying and simplifying

$50 = N$ Dividing

Note when solving percent problems using proportions that N is a percent and need not be converted.

Thus, \$10 is 50% of \$20. The answer is 50%.

12. Solve: GS
\$12 is what percent of \$40?

$\frac{12}{40} = \frac{N}{\square}$

$\square \cdot 100 = 40 \cdot N$

$\frac{12 \cdot 100}{\square} = \frac{40 \cdot N}{40}$

$\frac{\square}{40} = N$

$30 = N$

Thus, \$12 is 30 $\square$ of \$40.

◀ **Do Exercise 12.**

EXAMPLE 11 What percent of 50 is 16?

Translate: $\frac{N}{100} = \frac{16}{50}$

Solve:
$50 \cdot N = 100 \cdot 16$ Equating cross products

$\frac{50 \cdot N}{50} = \frac{100 \cdot 16}{50}$ Dividing by 50

$N = \frac{1600}{50}$ Multiplying and simplifying

$N = 32$ Dividing

Thus, 32% of 50 is 16. The answer is 32%.

13. Solve:
What percent of 84 is 10.5?

◀ **Do Exercise 13.**

Answers
11. 50 **12.** 30% **13.** 12.5%
Guided Solution:
12. 100, 12, 40, 1200; %

6.5 Exercise Set

For Extra Help
MyMathLab®
MathXL® PRACTICE  WATCH  READ  REVIEW 

Reading Check

Match each question with the correct translation from the list at the right.

RC1. 70 is 35% of what? ______

RC2. 70 is what percent of 200? ______

RC3. What is 35% of 200? ______

RC4. 74.8 is 110% of what? ______

RC5. What percent of 68 is 74.8? ______

RC6. 110% of 68 is what? ______

a) $\frac{110}{100} = \frac{a}{68}$

b) $\frac{70}{b} = \frac{35}{100}$

c) $\frac{a}{200} = \frac{35}{100}$

d) $\frac{74.8}{68} = \frac{N}{100}$

e) $\frac{70}{200} = \frac{N}{100}$

f) $\frac{74.8}{b} = \frac{110}{100}$

a Translate to a proportion. Do not solve.

1. What is 37% of 74?

2. 66% of 74 is what?

3. 4.3 is what percent of 5.9?

4. What percent of 6.8 is 5.3?

5. 14 is 25% of what?

6. 133% of what is 40?

b Translate to a proportion and solve.

7. What is 76% of 90?

8. What is 32% of 70?

9. 70% of 660 is what?

10. 80% of 920 is what?

11. What is 4% of 1000?

12. What is 6% of 2000?

13. 4.8% of 60 is what?

14. 63.1% of 80 is what?

15. $24 is what percent of $96?

16. $14 is what percent of $70?

17. 102 is what percent of 100?

18. 103 is what percent of 100?

19. What percent of $480 is $120?

20. What percent of $80 is $60?

21. What percent of 160 is 150?

22. What percent of 33 is 11?

23. \$18 is 25% of what?

24. \$75 is 20% of what?

25. 60% of what is 54?

26. 80% of what is 96?

27. 65.12 is 74% of what?

28. 63.7 is 65% of what?

29. 80% of what is 16?

30. 80% of what is 10?

31. What is $62\frac{1}{2}\%$ of 40?

32. What is $43\frac{1}{4}\%$ of 2600?

33. What is 9.4% of \$8300?

34. What is 8.7% of \$76,000?

35. 80.8 is $40\frac{2}{5}\%$ of what?

36. 66.3 is $10\frac{1}{5}\%$ of what?

Skill Maintenance

Solve. [6.1d]

37. $\frac{x}{188} = \frac{2}{47}$

38. $\frac{15}{x} = \frac{3}{800}$

39. $\frac{75}{100} = \frac{n}{20}$

40. $\frac{612}{t} = \frac{72}{244}$

Solve.

41. A recipe for muffins calls for $\frac{1}{2}$ qt of buttermilk, $\frac{1}{3}$ qt of skim milk, and $\frac{1}{16}$ qt of oil. How many quarts of liquid ingredients does the recipe call for? [4.2d]

42. The Ferristown School District purchased $\frac{3}{4}$ ton (T) of clay. If the clay is to be shared equally among the district's 6 art departments, how much will each art department receive? [3.8b]

Synthesis

Solve.

43. ▦ What is 8.85% of \$12,640?

Estimate ______________
Calculate ______________

44. ▦ 78.8% of what is 9809.024?

Estimate ______________
Calculate ______________

Mid-Chapter Review

Concept Reinforcement

Determine whether each statement is true or false.

______ **1.** If $\frac{x}{t} = \frac{y}{s}$, then $xy = ts$. [6.1d]

______ **2.** When converting decimal notation to percent notation, move the decimal point two places to the right and write a percent symbol. [6.2b]

______ **3.** The symbol % is equivalent to $\times$ 0.10. [6.2a]

______ **4.** Of the numbers $\frac{1}{10}$, 1%, 0.1%, and $\frac{1}{100}$, the smallest number is 0.1%. [6.2b], [6.3a, b]

Guided Solutions

Fill in each blank with the number that creates a correct statement or solution. [6.2b], [6.3a,b]

5. $\frac{1}{2}\% = \frac{1}{2} \cdot \frac{1}{\square} = \frac{1}{\square}$

6. $\frac{80}{1000} = \frac{\square}{100} = \square\%$

7. $5.5\% = \frac{\square}{100} = \frac{\square}{1000} = \frac{11}{\square}$

8. $0.375 = \frac{\square}{1000} = \frac{\square}{100} = \square\%$

9. Solve: 15 is what percent of 80? [6.4b]

$15 = p \times \square$ Translating

$\frac{15}{\square} = \frac{p \times \square}{\square}$ Dividing on both sides

$\frac{15}{\square} = p$ Simplifying

$\square = p$ Dividing

$\square\% = p$ Converting to percent notation

10. Solve: $\frac{x}{4} = \frac{3}{6}$. [6.1d]

$\frac{x}{4} = \frac{3}{6}$

$x \cdot \square = \square \cdot 3$ Equating cross products

$\frac{x \cdot 6}{\square} = \frac{4 \cdot 3}{\square}$ Dividing on both sides

$x = \square$ Simplifying

Mixed Review

Find the ratio of the first number to the second number and simplify. [6.1a]

11. 25 to 75

12. 2.4 to 8.4

Find each rate, or speed, as a ratio of distance to time. Round to the nearest hundredth where appropriate. [6.1b]

13. 146 km, 3 hr

14. 243 mi, 4 hr

Solve. [6.1d]

15. $\frac{x}{24} = \frac{30}{18}$

16. $\frac{12}{y} = \frac{20}{15}$

17. $\frac{0.24}{0.02} = \frac{y}{0.36}$

18. $\frac{\frac{1}{4}}{x} = \frac{\frac{1}{8}}{\frac{1}{4}}$

Find decimal notation. [6.2b]

19. 28%

20. 0.15%

21. $5\frac{3}{8}\%$

22. 240%

Find percent notation. [6.2b], [6.3a]

23. 0.71

24. $\frac{9}{100}$

25. 0.3891

26. $\frac{3}{16}$

27. 0.005

28. $\frac{37}{50}$

29. 6

30. $\frac{5}{6}$

Find fraction notation. Simplify. [6.3b]

31. 85%

32. 0.048%

33. $22\frac{3}{4}\%$

34. $16.\overline{6}\%$

Write percent notation for the shaded area. [6.3a]

35.

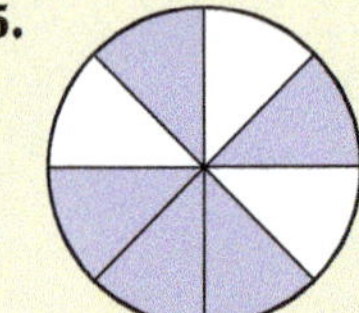

36.

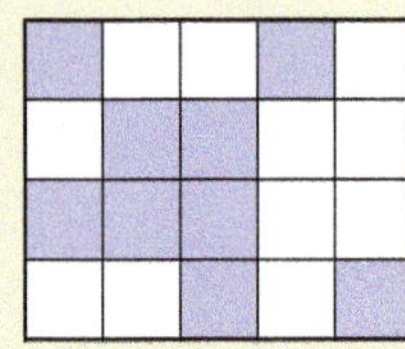

Solve. [6.1e]

37. ***Record Snowfall.*** The greatest recorded snowfall in a single storm occurred in the Mt. Shasta Ski Bowl in California, when 189 in. fell during a seven-day storm in 1959. What is the rate in inches per day?

Source: U.S. Army Corps of Engineers

Solve. [6.4b], [6.5b]

38. 25% of what is 14.5?

39. 220 is what percent of 1320?

40. What is 3.2% of 80,000?

41. \$17.50 is 35% of what?

42. Arrange the following numbers from smallest to largest. [6.2b], [6.3a, b]

$\frac{1}{2}\%$, 5%, 0.275, $\frac{13}{100}$, 1%, 0.1%, 0.05%, $\frac{3}{10}$, $\frac{7}{20}$, 10%

43. Solve: \$102,000 is what percent of \$3.6 million? [6.4b], [6.5b]

A. $\$2.8\overline{3}$ million

B. $2\frac{5}{6}\%$

C. $0.028\overline{3}\%$

D. $28.\overline{3}\%$

Understanding Through Discussion and Writing

44. What do the following have in common? Explain. [6.2b], [6.3a, b]

$\frac{23}{16}$, $1\frac{875}{2000}$, 1.4375, $\frac{207}{144}$, $1\frac{7}{16}$, 143.75%, $1\frac{4375}{10{,}000}$

45. Suppose we know that 40% of 92 is 36.8. What is a quick way to find 4% of 92? 400% of 92? Explain. [6.4b]

STUDYING FOR SUCCESS *Quiz and Test Follow-up*

- ☐ Immediately after completing a chapter quiz or test, write out a step-by-step solution for each question that you missed.
- ☐ Meet with your instructor or tutor for help with problems that are still giving you trouble.
- ☐ Keep your tests and quizzes along with their corrections to use as a study guide for the final exam.

Applications of Percent 6.6

OBJECTIVES

a Solve applied problems involving percent.

b Solve applied problems involving percent increase or percent decrease.

SKILL TO REVIEW

Objective 5.4a: Divide using decimal notation.

Divide.

1. $345 \div 57.5$

2. $111.87 \div 9.9$

a APPLIED PROBLEMS INVOLVING PERCENT

Applied problems involving percent are not always stated in a manner easily translated to an equation. In such cases, it is helpful to rephrase the problem before translating. Sometimes it also helps to make a drawing.

EXAMPLE 1 *Transportation to Work.* In the United States, there are about 154,000,000 workers who are 16 years old or older. Approximately 76.1% of these workers drive to work alone. How many workers drive to work alone?

Transportation to Work in the United States

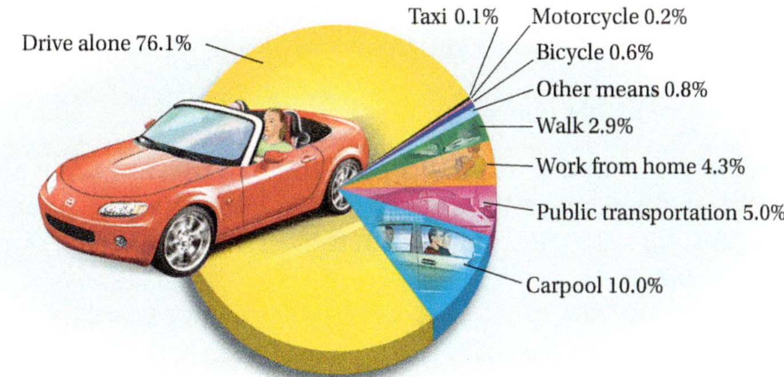

SOURCE: U.S. Bureau of Labor Statistics

1. Familiarize. We can simplify the pie chart shown above to help familiarize ourselves with the problem. We let a = the total number of workers who drive to work alone.

Transportation to Work in the United States

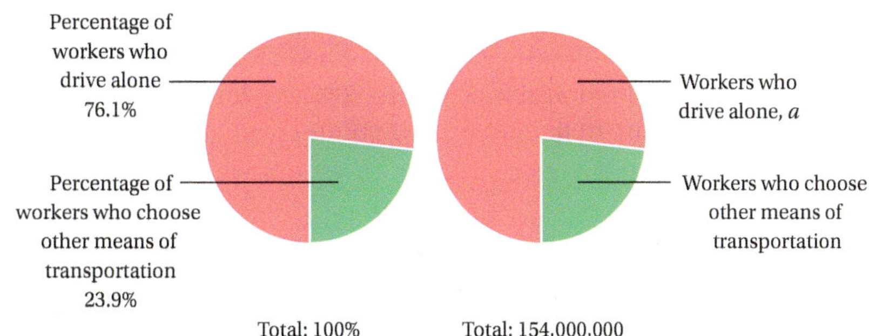

Answers

Skill to Review:
1. 6 **2.** 11.3

2. **Translate.** There are two ways in which we can translate this problem.

Percent equation (see Section 6.4):

$$\underbrace{\text{What number}}_{a} \; \underbrace{\text{is}}_{=} \; \underbrace{76.1\%}_{76.1\%} \; \underbrace{\text{of}}_{\cdot} \; \underbrace{154{,}000{,}000?}_{154{,}000{,}000}$$

Proportion (see Section 6.5):

$$\frac{76.1}{100} = \frac{a}{154{,}000{,}000}$$

3. **Solve.** We now have two ways in which to solve the problem.

Percent equation (see Section 6.4):

$$a = 76.1\% \cdot 154{,}000{,}000$$

We convert 76.1% to decimal notation and multiply:

$$a = 0.761 \times 154{,}000{,}000 = 117{,}194{,}000.$$

Proportion (see Section 6.5):

$$\frac{76.1}{100} = \frac{a}{154{,}000{,}000}$$

$$76.1 \times 154{,}000{,}000 = 100 \cdot a \quad \text{Equating cross products}$$

$$\frac{76.1 \cdot 154{,}000{,}000}{100} = \frac{100 \cdot a}{100} \quad \text{Dividing by 100}$$

$$\frac{11{,}719{,}400{,}000}{100} = a$$

$$117{,}194{,}000 = a \quad \text{Simplifying}$$

4. **Check.** To check, we can repeat the calculations. We also can do a partial check by estimating. Since 76.1% is about 75%, or, $\frac{3}{4}$, and $\frac{3}{4}$ of 154,000,000 is 115,500,000, which is close to 117,194,000, our answer is reasonable.

5. **State.** The number of workers who drive to work alone is 117,194,000.

◀ **Do Exercise 1.**

1. ***Transportation to Work.*** There are about 154,000,000 workers ages 16 and older in the United States. Approximately 10.0% of them carpool to work. How many workers carpool to work?

Sources: U.S. Census Bureau; American Community Survey

EXAMPLE 2 ***Extinction of Mammals.*** According to a study conducted for the International Union for the Conservation of Nature (IUCN), the world's mammals are in danger of an extinction crisis. Of the 5501 species of mammals on Earth, 1139 are on IUCN "vulnerable," "endangered," or "critically endangered" lists. What percent of all mammals are threatened with extinction?

1. **Familiarize.** The question asks for a percent of the world's mammals that are in danger of extinction. We note that 5501 is approximately 5500 and 1139 is approximately 1100. Since 1100 is $\frac{1100}{5500}$, or $\frac{1}{5}$, or 20% of 5500, our answer should be close to 20%. We let p = the percent of mammals that are in danger of extinction.

2. **Translate.** There are two ways in which we can translate this problem.

Percent equation:

$$\underbrace{1139}_{1139} \; \underbrace{\text{is}}_{=} \; \underbrace{\text{what percent}}_{p} \; \underbrace{\text{of}}_{\cdot} \; \underbrace{5501?}_{5501}$$

Answer

1. 15,400,000

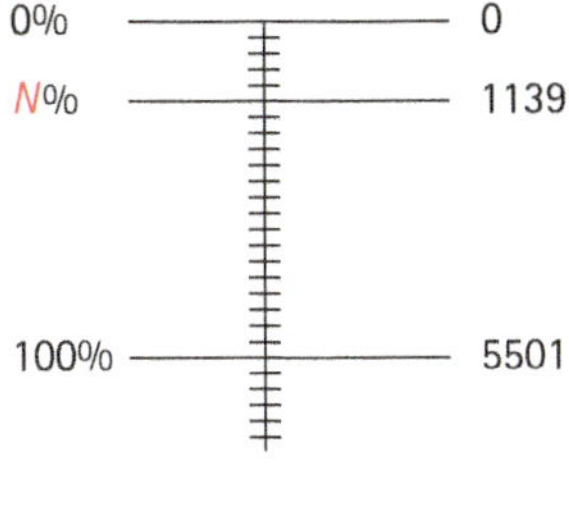

Mammals that are in danger of extinction include those shown here: Barbary macaque, black rhino, Galápagos seal, Malayan tapir, Cuvier's gazelle, Darwin's fox, and indri (a lemur).

Proportion:

$$\frac{N}{100} = \frac{1139}{5501}$$

For proportions, $N\% = p$.

Percents	Quantities
0%	0
N%	1139
100%	5501

3. **Solve.** We now have two ways in which to solve the problem.

Percent equation:

$$1139 = p \cdot 5501$$

$$\frac{1139}{5501} = \frac{p \cdot 5501}{5501}$$ Dividing by 5501 on both sides

$$\frac{1139}{5501} = p$$

$$0.207 \approx p$$ Finding decimal notation and rounding to the nearest thousandth

$$20.7\% = p$$ Remember to find percent notation.

Note here that the solution, p, includes the % symbol.

Proportion:

$$\frac{N}{100} = \frac{1139}{5501}$$

$$N \cdot 5501 = 100 \cdot 1139$$ Equating cross products

$$\frac{N \cdot 5501}{5501} = \frac{113{,}900}{5501}$$ Dividing by 5501 on both sides

$$N = \frac{113{,}900}{5501}$$

$$N \approx 20.7$$ Dividing and rounding to the nearest tenth

We use the solution of the proportion to express the answer to the problem as 20.7%. Note that in the proportion method, $N\% = p$.

4. **Check.** To check, we note that the answer 20.7% is close to 20%, as estimated in the *Familiarize* step.

5. **State.** About 20.7% of the world's mammals are threatened with extinction.

Do Exercise 2. ▶

2. ***Presidential Assassinations in Office.*** Of the 43 different U.S. presidents, 4 have been assassinated while in office. These were James A. Garfield, William McKinley, Abraham Lincoln, and John F. Kennedy. What percent have been assassinated while in office?

Answer

2. About 9.3%

b PERCENT INCREASE OR DECREASE

Percent is often used to state an increase or a decrease. Let's consider an example of each, using the price of a car as the original number.

Percent Increase

One year a car sold for \$20,455. The manufacturer decides to raise the price of the following year's model by 6%. The increase is $0.06 \times \$20,455$, or \$1227.30. The new price is $\$20,455 + \$1227.30$, or \$21,682.30. Note that the *new* price is 106% of the *former* price.

New price: \$21,682.30

Increase ↑

Former price: \$20,455

The increase, \$1227.30, is 6% of the *former* price, \$20,455. The *percent increase* is 6%.

Percent Decrease

Abigail buys the car listed above for \$20,455. After one year, the car depreciates in value by 25%. The decrease is $0.25 \times \$20,455$, or \$5113.75. This lowers the value of the car to $\$20,455 - \$5113.75$, or \$15,341.25. Note that the new value is 75% of the original price. If Abigail decides to sell the car after one year, \$15,341.25 might be the most she could expect to get for it.

Original price: \$20,455

Decrease ↓

New value: \$15,341.25

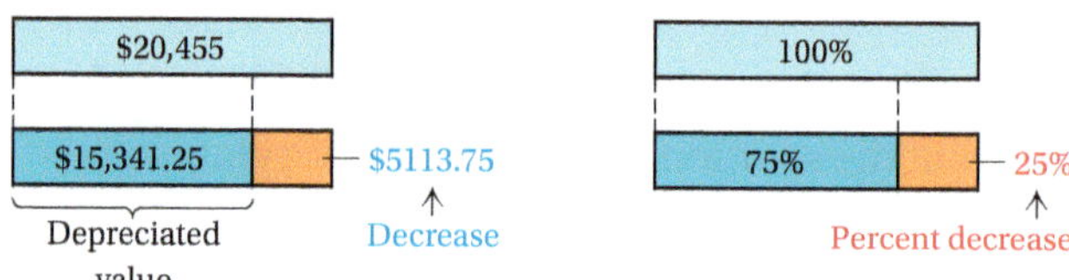

The decrease, \$5113.75, is 25% of the *original* price, \$20,455. The *percent decrease* is 25%.

◀ **Do Exercises 3 and 4.**

3. *Percent Increase.* The price of a car is \$36,875. The price is increased by 4%.
   a) How much is the increase?
   b) What is the new price?

4. *Percent Decrease.* The value of a car is \$36,875. The car depreciates in value by 25% after one year.
   a) How much is the decrease?
   b) What is the depreciated value of the car?

When a quantity is decreased by a certain percent, we say that this is a **percent decrease**.

EXAMPLE 3 *Dow Jones Industrial Average.* The Dow Jones Industrial Average (DJIA) plunged from 11,143 to 10,365 on September 29, 2008. This was the largest one-day drop in its history. What was the percent decrease?

Sources: *Nightly Business Reports*, September 29, 2008; DJIA

Answers

3. (a) \$1475; (b) \$38,350
4. (a) \$9218.75; (b) \$27,656.25

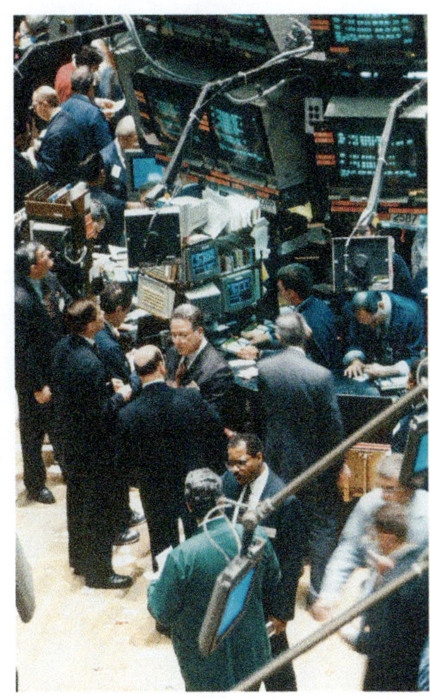

1. **Familiarize.** We first determine the amount of decrease and then make a drawing.

$$\begin{array}{rl} 11{,}143 & \text{Opening average} \\ -\,10{,}365 & \text{Closing average} \\ \hline 778 & \text{Decrease} \end{array}$$

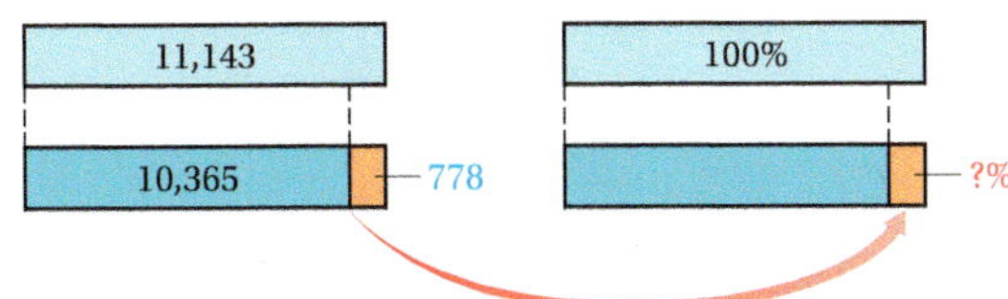

We are asking this question: The decrease is what percent of the opening average? We let p = the percent decrease.

2. **Translate.** There are two ways in which we can translate this problem.

Percent equation:

$$\underbrace{778}\ \text{is}\ \underbrace{\text{what percent}}\ \text{of}\ \underbrace{11{,}143?}$$
$$778 = p \times 11{,}143$$

Proportion:

$$\frac{N}{100} = \frac{778}{11{,}143}$$

For proportion, $N\% = p$.

Percents | Quantities
0% — 0
N% — 778
100% — 11,143

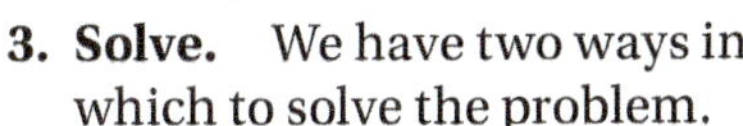

3. **Solve.** We have two ways in which to solve the problem.

Percent equation:

$$778 = p \times 11{,}143$$
$$\frac{778}{11{,}143} = \frac{p \times 11{,}143}{11{,}143} \qquad \text{Dividing by 11,143 on both sides}$$
$$\frac{778}{11{,}143} = p$$
$$0.07 \approx p$$
$$7\% = p \qquad \text{Converting to percent notation}$$

Proportion:

$$\frac{N}{100} = \frac{778}{11{,}143}$$
$$11{,}143 \times N = 100 \times 778 \qquad \text{Equating cross products}$$
$$\frac{11{,}143 \times N}{11{,}143} = \frac{100 \times 778}{11{,}143} \qquad \text{Dividing by 11,143 on both sides}$$
$$N = \frac{77{,}800}{11{,}143}$$
$$N \approx 7$$

We use the solution of the proportion to express the answer to the problem as 7%.

5. ***Volume of Mail.*** The volume of U.S mail decreased from about 102,379 million pieces of mail in 2002 to 68,696 million pieces in 2012. What was the percent decrease?

Source: U.S. Postal Service

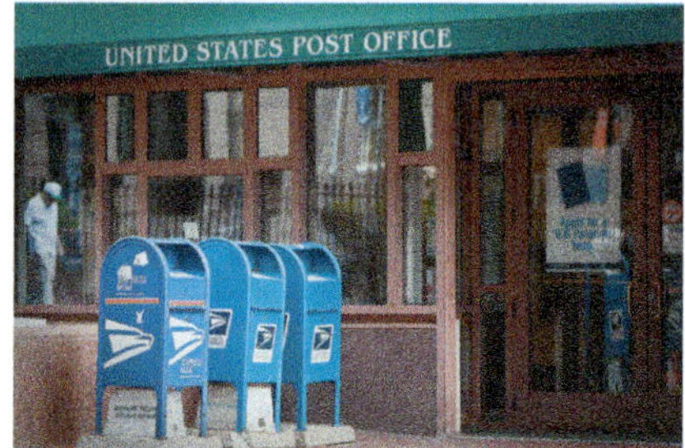

4. **Check.** To check, we note that, with a 7% decrease, the closing Dow average should be 93% of the opening average. Since

$$93\% \times 11{,}143 = 0.93 \times 11{,}143 \approx 10{,}363,$$

and 10,363 is close to 10,365, our answer checks. (Remember that we rounded to get 7%.)

5. **State.** The percent decrease in the DJIA was approximately 7%.

Do Exercise 5.

When a quantity is increased by a certain percent, we say that this is a **percent increase**.

EXAMPLE 4 ***Costs for Moviegoers.*** The average cost of a movie ticket was \$5.80 in 2002. The cost rose to \$8.12 in 2012. What was the percent increase in the cost of a movie ticket?

Sources: National Association of Theatre Owners; theaterseatstore.com

1. **Familiarize.** We first determine the increase in the cost and then make a drawing.

\$8.12	Cost in 2012
− 5.80	Cost in 2002
\$2.32	Increase

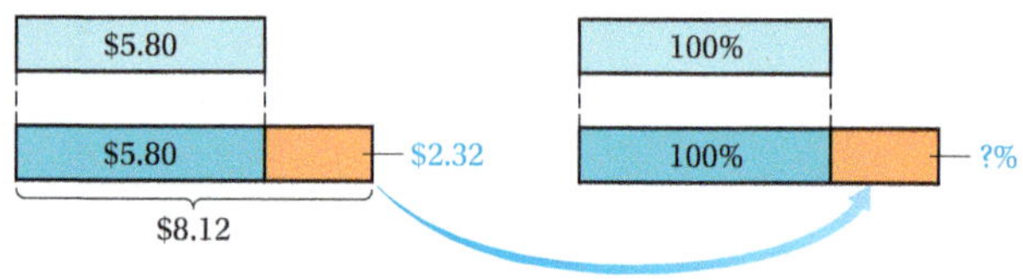

We are asking this question: The increase is what percent of the *original* cost? We let p = the percent increase.

2. **Translate.** There are two ways in which we can translate this problem.

Percent equation:

2.32 is what percent of 5.80?

$$2.32 = p \cdot 5.80$$

Proposition:

$$\frac{N}{100} = \frac{2.32}{5.80}$$

For proportions, $N\% = p$.

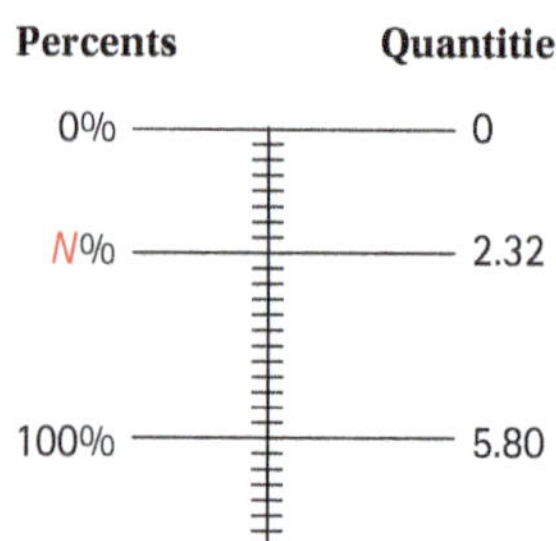

Answer

5. About 32.9%

3. Solve. We have two ways in which to solve the problem.

Percent equation:

$$2.32 = p \times 5.80$$

$$\frac{2.32}{5.80} = \frac{p \times 5.80}{5.80} \quad \text{Dividing by 5.80 on both sides}$$

$$\frac{2.32}{5.80} = p$$

$$0.4 = p$$

$$40\% = p \quad \text{Converting to percent notation}$$

Proportion:

$$\frac{N}{100} = \frac{2.32}{5.80}$$

$$5.80 \times N = 100 \times 2.32 \quad \text{Equating cross products}$$

$$\frac{5.80 \times N}{5.80} = \frac{100 \times 2.32}{5.80} \quad \text{Dividing by 5.80 on both sides}$$

$$N = \frac{232}{5.80}$$

$$N = 40$$

We use the solution of the proportion to express the answer to the problem as 40%.

4. Check. To check, we take 40% of 5.80:

$$40\% \times 5.80 = 0.40 \times 5.80 = 2.32.$$

5. State. The percent increase in the cost of a movie ticket was 40%.

Do Exercise 6. ▶

6. ***Centenarians.*** As of April 1, 2010, the number of centenarians in the United States was 53,364. It is projected that this number will increase to 601,000 by 2050. What is the projected percent increase?

Sources: *National Geographic*, November 2011; Population Reference Bureau, U.S. Census Bureau

Answer

6. About 1026%

Translating for Success

1. *Distance Walked.* After a knee replacement, Alex walked $\frac{1}{8}$ mi each morning and $\frac{1}{5}$ mi each afternoon. How much farther did he walk in the afternoon?

2. *Stock Prices.* A stock sold for \$5 per share on Monday and only \$2.125 per share on Friday. What was the percent decrease from Monday to Friday?

3. *SAT Score.* After attending a class called *Improving Your SAT Scores,* Jacob raised his total score from 884 to 1040. What was the percent increase?

4. *Change in Population.* The population of a small farming community decreased from 1040 to 884. What was the percent decrease?

5. *Lawn Mowing.* During the summer, brothers Steve and Rob earned money for college by mowing lawns. The largest lawn that they mowed was $2\frac{1}{8}$ acres. Steve can mow $\frac{1}{5}$ acre per hour, and Rob can mow only $\frac{1}{8}$ acre per hour. Working together, how many acres did they mow per hour?

The goal of these matching questions is to practice step (2), Translate, of the five-step problem-solving process. Translate each word problem to an equation and select a correct translation from equations A–O.

A. $x + \frac{1}{5} = \frac{1}{8}$

B. $250 = x \cdot 1040$

C. $884 = x \cdot 1040$

D. $\frac{250}{16.25} = \frac{1000}{x}$

E. $156 = x \cdot 1040$

F. $16.25 = 250 \cdot x$

G. $\frac{1}{5} + \frac{1}{8} = x$

H. $2\frac{1}{8} = x \cdot 5$

I. $5 = 2.875 \cdot x$

J. $\frac{1}{8} + x = \frac{1}{5}$

K. $1040 = x \cdot 884$

L. $\frac{250}{16.25} = \frac{x}{1000}$

M. $2.875 = x \cdot 5$

N. $x \cdot 884 = 156$

O. $x = 16.25 \cdot 250$

Answers on page A-12

6. *Land Sale.* Cole sold $2\frac{1}{8}$ acres of the 5 acres he inherited from his uncle. What percent of his land did he sell?

7. *Travel Expenses.* A magazine photographer is reimbursed 16.25¢ per mile for business travel, up to 1000 mi per week. In a recent week, he traveled 250 mi. What was the total reimbursement for travel?

8. *Trip Expenses.* The total expenses for Claire's recent business trip were \$1040. She put \$884 on her credit card and paid the balance in cash. What percent did she place on her credit card?

9. *Cost of Copies.* During the first summer session at a community college, the campus copy center advertised 250 copies for \$16.25. At this rate, what is the cost of 1000 copies?

10. *Cost of Insurance.* Following a rise in the cost of health insurance, 250 of a company's 1040 employees canceled their insurance. What percent of the employees canceled their insurance?

6.6 Exercise Set

For Extra Help
MyMathLab®
MathXL® PRACTICE WATCH READ REVIEW

Reading Check

Complete the table by filling in the missing numbers.

	Original Price	New Price	Change	Percent Increase or Decrease
RC1.	\$50	\$40	\$ ______	$\frac{\text{Change}}{\text{Original}} = \frac{\$}{\$} =$ ____%
RC2.	\$60	\$75	\$ ______	$\frac{\text{Change}}{\text{Original}} = \frac{\$}{\$} =$ ____%
RC3.	\$360	\$480	\$ ______	$\frac{\text{Change}}{\text{Original}} = \frac{\$}{\$} =$ ____%
RC4.	\$4000	\$2400	\$ ______	$\frac{\text{Change}}{\text{Original}} = \frac{\$}{\$} =$ ____%

a Solve.

1. *Winnings from Gambling.* Pre-tax gambling winnings were 417 billion worldwide in 2012. Approximately, 25.1% of the winnings were in the United States and 5.9% were in Italy. About how much, in dollars, were the gambling winnings in each country?

Source: H2 Gambling Capital (h2gc.com)

2. *Mississippi River.* The Mississippi River, which extends from its source, at Lake Itasca in Minnesota, to the Gulf of Mexico, is 2348 mi long. Approximately 77% of the river is navigable. How many miles of the river are navigable?

Source: National Oceanic and Atmospheric Administration

3. A person earns \$43,200 one year and receives an 8% raise in salary. What is the new salary?

4. A person earns \$28,600 one year and receives a 5% raise in salary. What is the new salary?

5. *Test Results.* On a test, Juan got 85%, or 119, of the items correct. How many items were on the test?

6. *Test Results.* On a test, Maj Ling got 86%, or 81.7, of the items correct. (There was partial credit on some items.) How many items were on the test?

7. *Farmland.* In Kansas, 47,000,000 acres are farmland. About 5% of all the farm acreage in the United States is in Kansas. What is the total number of acres of farmland in the United States?

Sources: U.S. Department of Agriculture; National Agricultural Statistics Service

8. *World Population.* World population is increasing by 1.2% each year. In 2008, it was 6.68 billion. What will the population be in 2015?

Sources: U.S. Census Bureau; International Data Base

9. *Car Depreciation.* A car generally depreciates 25% of its original value in the first year. A car is worth $27,300 after the first year. What was its original cost?

10. *Car Depreciation.* Given normal use, an American-made car will depreciate 25% of its original cost the first year and 14% of its remaining value in the second year. What is the value of a car at the end of the second year if its original cost was $36,400? $28,400? $26,800?

11. *Test Results.* On a test of 80 items, Pedro got 93% correct. (There was partial credit on some items.) How many items did he get correct? incorrect?

12. *Test Results.* On a test of 40 items, Christina got 91% correct. (There was partial credit on some items.) How many items did she get correct? incorrect?

13. *Olympic Team.* The 2012 U.S. Summer Olympics team consisted of 529 members. Approximately 49.3% of the athletes were men. How many men were on the team?

Source: United States Olympic Committee

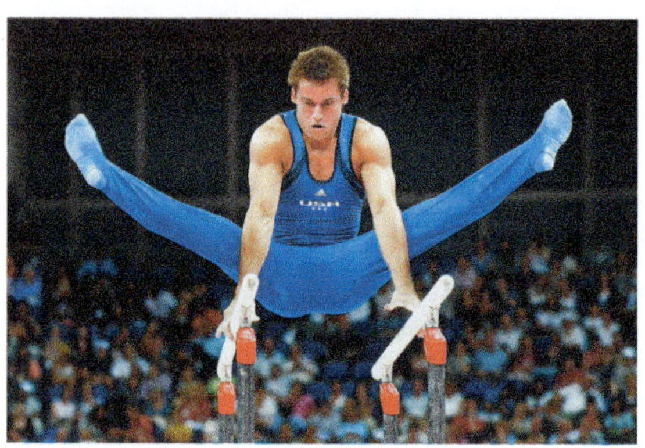

14. *Murder Case Costs.* The cost (for both trial and incarceration) of a life-without-parole case is about 30% of the cost of a death-penalty case. The average cost of a death-penalty case is $505,773. What is the average cost of a life-without-parole case?

Source: Indiana Legislative Services Agency, *2010 Study of Murder Trials*

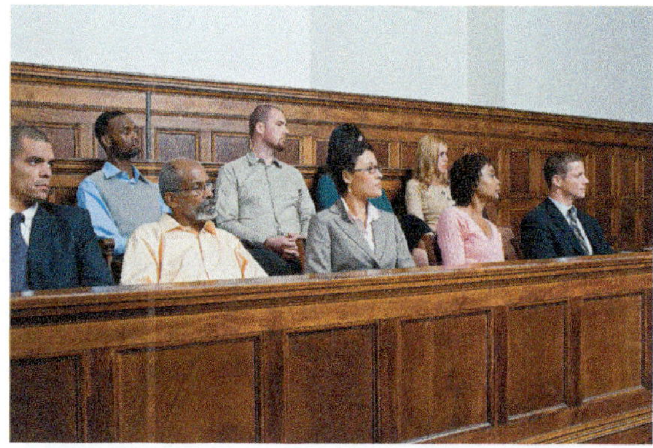

15. *Spending on Pets.* In 2011, Americans spent approximately $50.8 billion on their pets. Of this amount, $14.1 billion was for veterinarian bills. What percent of the total was spent on veterinary care?

Source: American Pet Products Association

16. *Health Workers' Salaries.* In 2010, the median annual salary of registered nurses was $64,690. The median annual salary of physicians and surgeons was $111,570. What percent of the physician and surgeon's median annual salary is the nurses' median annual salary?

Source: Bureau of Labor Statistics

17. *Tipping.* For a party of 8 or more, some restaurants add an 18% tip to the bill. What is the total amount charged for a party of 10 if the cost of the meal, without tip, is $195?

18. *Tipping.* Diners frequently add a 15% tip when charging a meal to a credit card. What is the total amount charged to a card if the cost of the meal, without tip, is $18? $34? $49?

19. *Wasting Food.* As world population increases and the number of acres of farmland decreases, improvements in food production and packaging must be implemented. Also, wealthy countries need to waste less food. In the United States, consumers waste 39 lb of every 131 lb of fruit purchased. What percent is wasted?

Sources: United States Department of Agriculture; *National Geographic*, July 2011

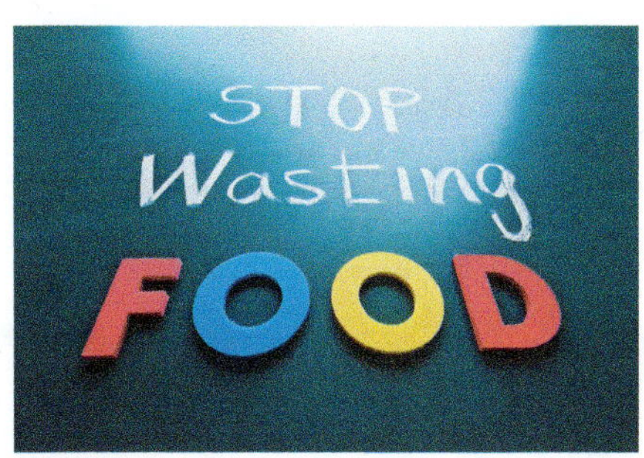

20. *Credit-Card Debt.* Michael has disposable monthly income of $3400. Each month, he pays $470 toward his credit-card debt. What percent of his disposable income is allotted to paying off credit-card debt?

Monthly Disposable Income: $3400

21. A lab technician has 540 mL of a solution of alcohol and water; 8% is alcohol. How many milliliters are alcohol? water?

22. A lab technician has 680 mL of a solution of water and acid; 3% is acid. How many milliliters are acid? water?

23. *U.S. Armed Forces.* There were 1,384,000 people in the United States in active military service in 2006. The numbers in the four armed services are listed in the following table. What percent of the total does each branch represent? Round the answers to the nearest tenth of a percent.

U.S. ARMED FORCES: 2006

TOTAL	1,384,000*
AIR FORCE	349,000
ARMY	505,000
NAVY	350,000
MARINES	180,000

*Includes National Guard, Reserve, and retired regular personnel on extended or continuous active duty. Excludes Coast Guard.

SOURCES: U.S. Department of Defense; U.S. Census Bureau

24. *Living Veterans.* There were 23,977,000 living veterans in the United States in 2006. Numbers in various age groups are listed in the following table. What percent of the total does each age group represent? Round the answers to the nearest tenth of a percent.

LIVING VETERANS BY AGE: 2006

TOTAL	23,977,000
UNDER 35 YEARS OLD	1,949,000
35–44 YEARS OLD	2,901,000
45–54 YEARS OLD	3,846,000
55–64 YEARS OLD	6,081,000
65 YEARS OLD AND OLDER	9,200,000

SOURCES: U.S. Department of Defense; U.S. Census Bureau

b Solve.

25. *Mortgage Payment Increase.* A monthly mortgage payment increases from $840 to $882. What is the percent increase?

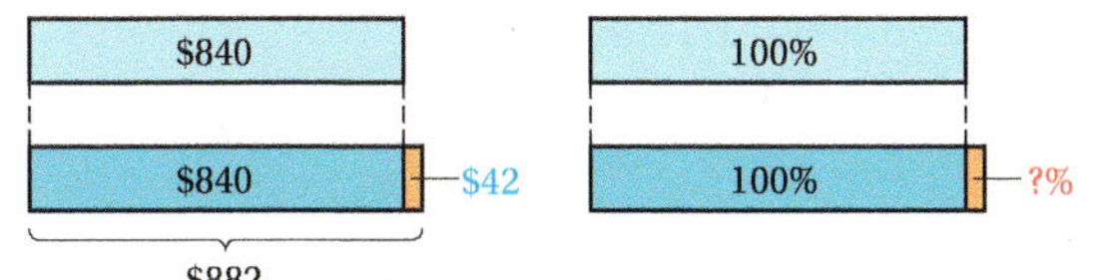

26. *Savings Increase.* The amount in a savings account increased from $200 to $216. What was the percent increase?

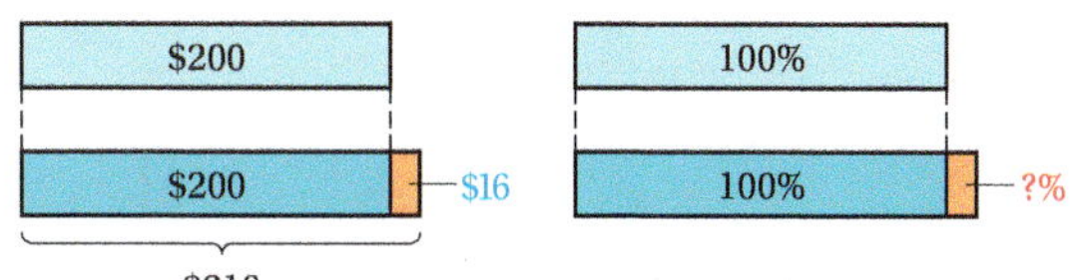

27. A person on a diet goes from a weight of 160 lb to a weight of 136 lb. What is the percent decrease?

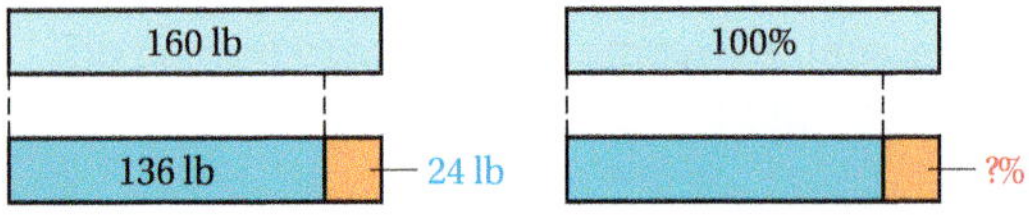

28. During a sale, a dress decreased in price from $90 to $72. What was the percent decrease?

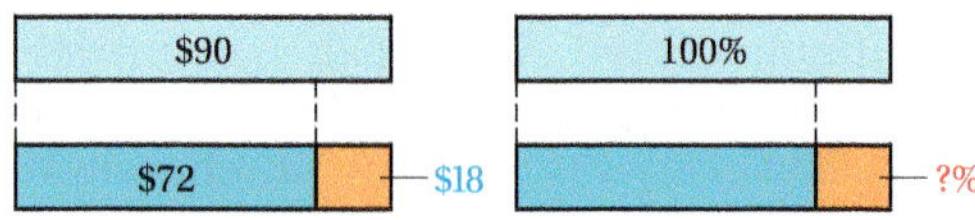

29. *Insulation.* A roll of unfaced fiberglass insulation has a retail price of $23.43. For two weeks, it is on sale for $15.31. What is the percent decrease?

30. *Set of Weights.* A 300-lb weight set retails for $199.95. For its grand opening, a sporting goods store reduced the price to $154.95. What is the percent decrease?

31. *Birds Killed.* In 2009, approximately 440,000 birds were killed by wind turbines in the United States. It is estimated that this number will increase to 1,000,000 per year by 2020. What would the percent increase be?

Sources: National Wind Coordinating Collaborative; American Bird Conservancy; U.S. Fish and Wildlife Service

32. *Miles of Railroad Track.* The greatest combined length of U.S.-owned operating railroad track was 254,037 mi in 1916, when industrial activity increased during World War I. The total length has decreased ever since. By 2006, the number of miles of track had decreased to 140,490 mi. What was the percent decrease from 1916 to 2006?

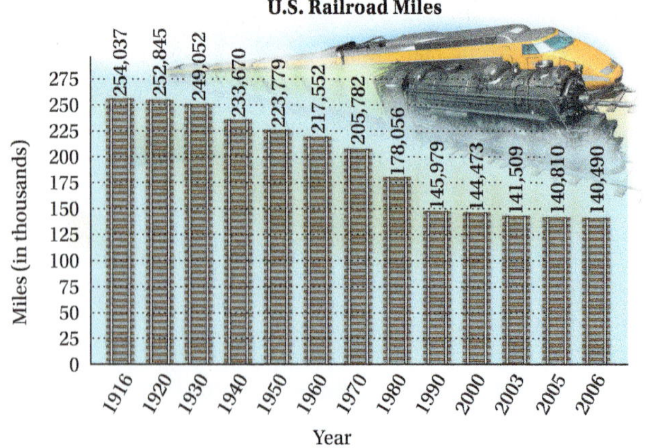

NOTE: The lengths exclude yard tracks, sidings, and parallel tracks.
SOURCE: Association of American Railroads

33. *Overdraft Fees.* Consumers are paying record amounts of fees for overdrawing their bank accounts. In 2007, banks, thrift institutions, and credit unions collected \$45.6 billion in overdraft fees, which is \$15.1 billion more than in 2001. What was the percent increase?

Source: Moebs Services

34. *Credit-Card Debt.* In 2013, the average credit-card debt per household in the United States was \$15,162. In 1990, the average credit-card debt was \$2966. What was the percent increase?

Sources: nerdwallet.com

35. *Population over 1 Million.* In 1950, 74 cities in the world had populations of 1 million or more. In 2010, 442 cities had populations of 1 million or more. What was the percent increase?

Sources: *World Cities*, George Modelski; United Nations

36. *Prescription Drug Sales.* Total spending on prescription drugs in the United States was \$329.2 billion in 2011. This amount dropped to \$325.8 billion in 2012. What was the percent decrease?

Source: IMS Institute for Healthcare Informatics

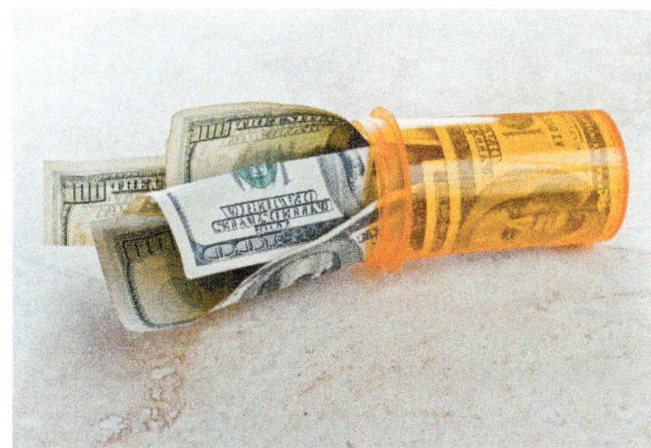

37. *Patents Issued.* The U.S. Patent and Trademark Office (USPTO) issued a total of 157,284 utility patents in 2007. This number of patents was down from 173,794 in 2006. What was the percent decrease?

Source: IFI Patent Intelligence

38. *Highway Fatalities.* In 2007, there were 41,059 highway fatalities in the United States, which was 1649 fewer than the number in 2006. What was in the percent decrease?

Source: National Highway Traffic Safety Administration

39. *Two-by-Four.* A cross-section of a standard, or nominal, "two-by-four" actually measures $1\frac{1}{2}$ in. by $3\frac{1}{2}$ in. The rough board is 2 in. by 4 in. but is planed and dried to the finished size. What percent of the wood is removed in planing and drying?

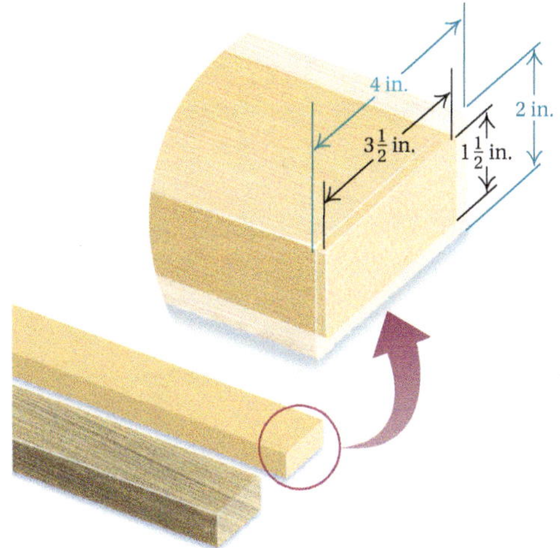

40. *Strike Zone.* In baseball, the *strike zone* is normally a 17-in. by 30-in. rectangle. Some batters give the pitcher an advantage by swinging at pitches thrown out of the strike zone. By what percent is the area of the strike zone increased if a 2-in. border is added to the outside?

Source: Major League Baseball

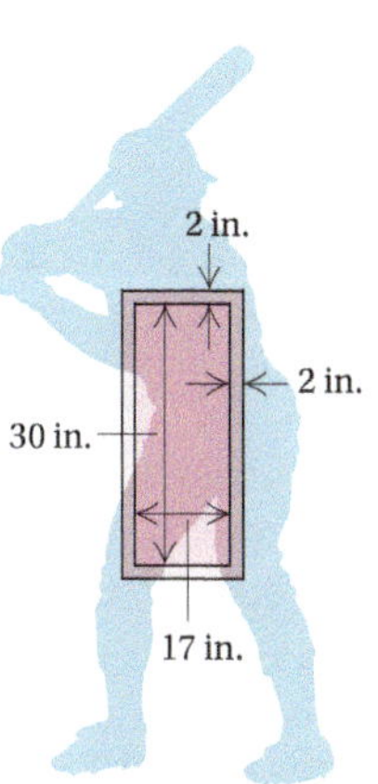

Population Increase. The following table provides data showing how the populations of various states increased from 2000 to 2010. Complete the table by filling in the missing numbers. Round percents to the nearest tenth of a percent.

	State	Population in 2000	Population in 2010	Change	Percent Change
41.	Vermont	608,827	625,741		
42.	Wisconsin	5,363,675		323,311	
43.	Arizona		6,392,017	1,261,385	
44.	Virginia		8,001,024	922,509	
45.	Idaho	1,293,953		273,629	
46.	Georgia	8,186,453	9,535,483		

SOURCE: U.S. Census Bureau

47. *Increase in Population.* Between 2000 and 2010, the population of Utah increased from 2,233,169 to 2,763,885. What was the percent increase?

Sources: U.S. Census Bureau; U.S. Department of Commerce

48. *Decrease in Population.* Between 2000 and 2010, the population of Michigan decreased from 9,938,444 to 9,883,640. What was the percent decrease?

Sources: U.S. Census Bureau; U.S. Department of Commerce

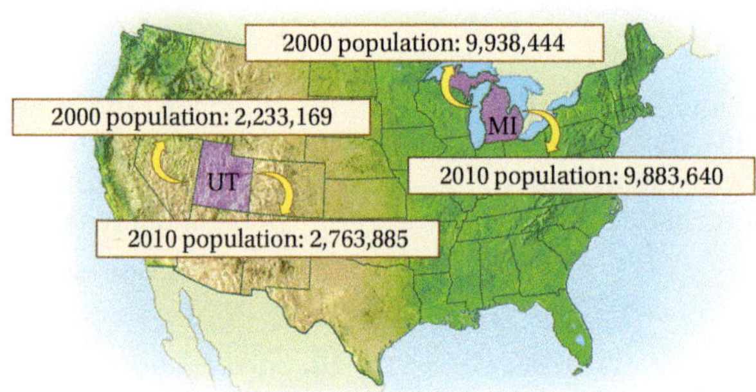

Skill Maintenance

Convert to decimal notation. [5.1b], [5.5a]

49. $-\frac{25}{11}$ **50.** $\frac{11}{25}$ **51.** $\frac{27}{8}$ **52.** $-\frac{43}{9}$

Simplify. [3.5b]

53. $\frac{18}{102}$ **54.** $\frac{135}{510}$ **55.** $\frac{192}{1000}$ **56.** $\frac{70}{406}$

Synthesis

57. A coupon allows a couple to have \$10 subtracted from their dinner bill. Before subtracting \$10, however, the restaurant adds a tip of 20%. If the couple is presented with a bill for \$40.40, how much would the dinner (without tip) have cost without the coupon?

58. If p is 120% of q, then q is what percent of p?

Sales Tax, Commission, and Discount

6.7

OBJECTIVES

a Solve applied problems involving sales tax and percent.

b Solve applied problems involving commission and percent.

c Solve applied problems involving discount and percent.

a SALES TAX

Sales tax computations represent a special type of percent increase problem. The sales tax rate in Colorado is 2.9%. This means that the tax is 2.9% of the purchase price. Suppose the purchase price of a canoe is \$749.95. The sales tax is then 2.9% of \$749.95, or $0.029 \times \$749.95$, or \$21.74855, which is about \$21.75.

BILL:

Purchase price	=	\$749.95
Sales tax (2.9% of \$749.95)	=	+ 21.75
Total price		\$771.70

The total that you would pay for the canoe is the purchase price plus the sales tax:

\$749.95 + \$21.75, or \$771.70.

SALES TAX

Sales tax = Sales tax rate × Purchase price

Total price = Purchase price + Sales tax

EXAMPLE 1 *Maine Sales Tax.* The sales tax rate in Maine is 5%. How much tax is charged on the purchase of 3 shrubs at \$42.99 each? What is the total price?

a) We first find the purchase price of the 3 shrubs. It is

$3 \times \$42.99 = \$128.97.$

b) The sales tax on items costing \$128.97 is

$$\underbrace{\text{Sales tax rate}}_{5\%} \times \underbrace{\text{Purchase price}}_{128.97},$$

or 0.05×128.97, or 6.4485. Thus the tax is \$6.45 (rounded to the nearest cent).

c) The total price is given by the purchase price plus the sales tax:

\$128.97 + \$6.45, or \$135.42.

To check, note that the total price is the purchase price plus 5% of the purchase price. Thus the total price is 105% of the purchase price. Since $1.05 \times 128.97 \approx 135.42$, we have a check. The sales tax is \$6.45, and the total price is \$135.42.

1. *Texas Sales Tax.* The sales tax rate in Texas is 6.25%. In Texas, how much tax is charged on the purchase of an ultrasound toothbrush that sells for $139.95? What is the total price?

2. **Wyoming Sales Tax.** In her hometown, Laramie, Wyoming, Samantha buys 4 copies of *It's All Good* by Gwyneth Paltrow for $18.95 each. The sales tax rate in Wyoming is 4%. How much sales tax will Samantha be charged? What is the total price?

GS

Sales tax = ___ % × 4 × $ ___
= 0.04 × $ ___
= $3.032
≈ $ ___
Total price = $75.80 + $ ___
= $ ___

3. The sales tax on the purchase of a set of holiday dishes that costs $449 is $26.94. What is the sales tax rate?

4. The sales tax on the purchase of a pair of designer jeans is $4.84 and the sales tax rate is 5.5%. Find the purchase price (the price before the tax is added).

Answers

1. $8.75; $148.70 2. $3.03, $78.83
3. 6% 4. $88

Guided Solution:
2. 4, 18.95, 75.80, 3.03; 3.03, 78.83

◀ **Do Exercises 1 and 2.**

EXAMPLE 2 The sales tax on the purchase of an eReader that costs $199 is $13.93. What is the sales tax rate?

We rephrase and translate as follows:

Rephrase: Sales tax is what percent of purchase price?

Translate: $13.93 = r × 199.

To solve the equation, we divide by 199 on both sides:

$$\frac{13.93}{199} = \frac{r \times 199}{199}$$

$$\frac{13.93}{199} = r$$

$$0.07 = r$$

$$7\% = r.$$

The sales tax rate is 7%.

◀ **Do Exercise 3.**

EXAMPLE 3 The sales tax on the purchase of a stone-top firepit is $12.74 and the sales tax rate is 8%. Find the purchase price (the price before the tax is added).

We rephrase and translate as follows:

Rephrase: Sales tax is 8% of what?

Translate: 12.74 = 8% × b,

or 12.74 = 0.08 × b.

To solve, we divide by 0.08 on both sides:

$$\frac{12.74}{0.08} = \frac{0.08 \times b}{0.08}$$

$$\frac{12.74}{0.08} = b$$

$$159.25 = b.$$

The purchase price is $159.25.

◀ **Do Exercise 4.**

b COMMISSION

When you work for a **salary**, you receive the same amount of money each week or month. When you work for a **commission**, you are paid a percentage of the total sales for which you are responsible.

COMMISSION

Commission = Commission rate × Sales

EXAMPLE 4 *Membership Sales.* A membership salesperson's commission rate is 3%. What is the commission on the sale of \$8300 worth of fitness club memberships?

$$\begin{aligned} \textit{Commission} &= \textit{Commission rate} \times \textit{Sales} \\ C &= 3\% \times 8300 \\ C &= 0.03 \times 8300 \\ C &= 249 \end{aligned}$$

The commission is \$249.

Do Exercise 5. ▶

5. Aniyah's commission rate is 15%. What commission does she earn on the sale of \$9260 worth of exercise equipment?

EXAMPLE 5 *Earth-Moving Equipment Sales.* Gavin earns a commission of \$20,800 for selling \$320,000 worth of earth-moving equipment. What is the commission rate?

$$\begin{aligned} \textit{Commission} &= \textit{Commission rate} \times \textit{Sales} \\ 20{,}800 &= r \times 320{,}000 \end{aligned}$$

Answer

5. \$1389

To solve this equation, we divide by 320,000 on both sides:

$$\frac{20{,}800}{320{,}000} = \frac{r \times 320{,}000}{320{,}000}$$

$$0.065 = r$$

$$6.5\% = r.$$

The commission rate is 6.5%.

Do Exercise 6.

6. Zion earns a commission of $2040 for selling $17,000 worth of concert tickets. What is the commission rate?

EXAMPLE 6 *Cruise Vacations.* Valentina's commission rate is 5.6%. She received a commission of $2457 on cruise vacation packages that she sold in November. How many dollars worth of cruise vacations did she sell?

$$\textit{Commission} = \textit{Commission rate} \times \textit{Sales}$$

$$2457 = 5.6\% \times S, \quad \text{or}$$

$$2457 = 0.056 \times S$$

To solve this equation, we divide by 0.056 on both sides:

$$\frac{2457}{0.056} = \frac{0.056 \times S}{0.056}$$

$$\frac{2457}{0.056} = S$$

$$43{,}875 = S.$$

Valentina sold $43,875 worth of cruise vacation packages.

Do Exercise 7.

7. Nathan's commission rate is 7.5%. He receives a commission of $2970 from the sale of winter ski passes. How many dollars worth of ski passes did he sell?

GS

$$\$2970 = \underline{\quad}\% \times S$$

$$\$2970 = 0.075 \times S$$

$$\frac{\$2970}{\underline{\quad}} = \frac{0.075 \times S}{0.075}$$

$$\$\underline{\quad} = S$$

C DISCOUNT

Suppose that the regular price of a rug is $60, and the rug is on sale at 25% off. Since 25% of $60 is $15, the sale price is $60 − $15, or $45. We call $60 the **original**, or **marked**, **price**, 25% the **rate of discount**, $15 the **discount**, and $45 the **sale price**. Note that discount problems are a type of percent decrease problem.

DISCOUNT AND SALE PRICE

Discount = Rate of discount × Original price

Sale price = Original price − Discount

Answers

6. 12% 7. $39,600

Guided Solution:

7. 7.5, 0.075, 39,600

EXAMPLE 7 A leather sofa marked \$2379 is on sale at $33\frac{1}{3}\%$ off. What is the discount? the sale price?

a) $$\begin{aligned} \textit{Discount} &= \textit{Rate of discount} \times \textit{Original price} \\ D &= 33\frac{1}{3}\% \times 2379 \\ D &= \frac{1}{3} \times 2379 \\ D &= \frac{2379}{3} = 793 \end{aligned}$$

b) $$\begin{aligned} \textit{Sale price} &= \textit{Original price} - \textit{Discount} \\ S &= 2379 - 793 \\ S &= 1586 \end{aligned}$$

The discount is \$793, and the sale price is \$1586.

Do Exercise 8. ▶

8. A computer marked \$660 is on sale at $16\frac{2}{3}\%$ off. What is the discount? the sale price?

EXAMPLE 8 The price of a snowblower is marked down from \$950 to \$779. What is the rate of discount?

We first find the discount by subtracting the sale price from the original price:

$$950 - 779 = 171.$$

The discount is \$171.

Next, we use the equation for discount:

$$\begin{aligned} \textit{Discount} &= \textit{Rate of discount} \times \textit{Original price} \\ 171 &= r \times 950. \end{aligned}$$

To solve, we divide by 950 on both sides:

$$\frac{171}{950} = \frac{r \times 950}{950}$$

$$\frac{171}{950} = r$$

$$0.18 = r$$

$$18\% = r.$$

The rate of discount is 18%.

To check, note that an 18% rate of discount means that the buyer pays 82% of the original price:

$$0.82 \times \$950 = \$779.$$

Do Exercise 9. ▶

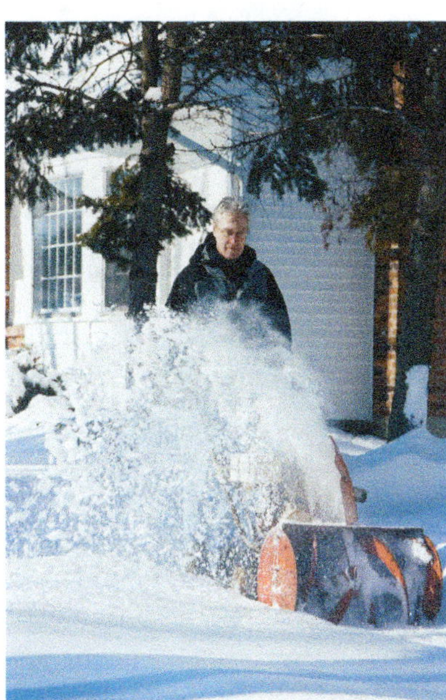

9. The price of a winter coat is reduced from \$75 to \$60. Find the rate of discount.

Answers

8. \$110; \$550 **9.** 20%

6.7 Exercise Set

For Extra Help
MyMathLab®

MathXL® PRACTICE   WATCH  READ  REVIEW

 Reading Check

Complete each definition with the word *price*, *rate*, or *tax*.

RC1. Commission = Commission ________ × Sales

RC2. Discount = ________ of discount × Original price

RC3. Sale price = Original ________ − Discount

RC4. Sales tax = Sales ________ rate × Purchase price

RC5. Total price = Purchase price + Sales ________

a Solve.

1. *Wyoming Sales Tax.* The sales tax rate in Wyoming is 4%. How much sales tax would be charged on a fireplace screen with doors that costs $239?

2. *Kansas Sales Tax.* The sales tax rate in Kansas is 6.3%. How much sales tax would be charged on a fireplace screen with doors that costs $239?

3. *New Mexico Sales Tax.* The sales tax rate in New Mexico is 5.125%. How much sales tax is charged on a camp stove that sells for $129.95?

4. *Ohio Sales Tax.* The sales tax rate in Ohio is 5.5%. How much sales tax is charged on a pet carrier that sells for $39.99?

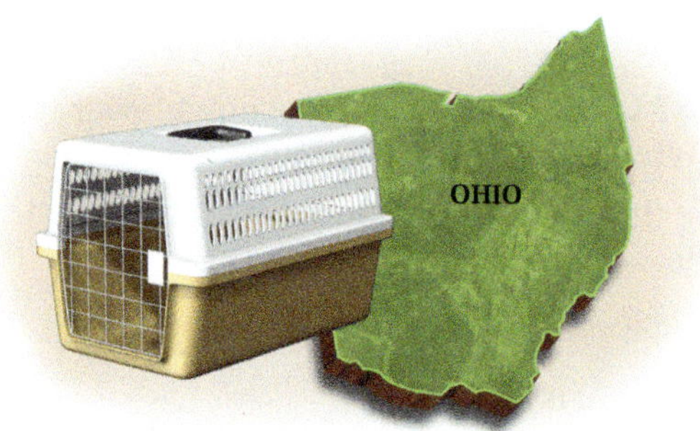

5. *California Sales Tax.* The sales tax rate in California is 7.5%. How much sales tax is charged on a purchase of 4 contour foam travel pillows at $39.95 each? What is the total price?

6. *Illinois Sales Tax.* The sales tax rate in Illinois is 6.25%. How much sales tax is charged on a purchase of 3 wet-dry vacs at $60.99 each? What is the total price?

7. The sales tax is $30 on the purchase of a diamond ring that sells for $750. What is the sales tax rate?

8. The sales tax is $48 on the purchase of a dining room set that sells for $960. What is the sales tax rate?

9. The sales tax on the purchase of a new fishing boat is $112 and the sales tax rate is 2%. What is the purchase price (the price before tax is added)?

10. The sales tax on the purchase of a used car is $100 and the sales tax rate is 5%. What is the purchase price?

11. The sales tax rate in New York City is 4.375% for the city plus 4% for the state. Find the total amount paid for 6 boxes of chocolates at $17.95 each.

12. The sales tax rate in Nashville, Tennessee, is 2.25% for Davidson County plus 7% for the state. Find the total amount paid for 2 ladders at $39 each.

13. The sales tax rate in Seattle, Washington, is 2.5% for King County plus 6.5% for the state. Find the total amount paid for 3 ceiling fans at $84.49 each.

14. The sales tax rate in Miami, Florida, is 1% for Dade County plus 6% for the state. Find the total amount paid for 2 tires at $49.95 each.

15. The sales tax rate in Atlanta, Georgia, is 1% for the city, 3% for Fulton County, and 4% for the state. Find the total amount paid for 6 basketballs at $29.95 each.

16. The sales tax rate in Dallas, Texas, is 1% for the city, 1% for Dallas County, and 6.25% for the state. Find the total amount paid for 5 flash drives at $19.95 each.

b Solve.

17. Benjamin's commission rate is 21%. What commission does he earn on the sale of $12,500 worth of windows?

18. Olivia's commission rate is 6%. What commission does she earn on the sale of $45,000 worth of lawn irrigation systems?

19. Alyssa earns $408 for selling $3400 worth of shoes. What is the commission rate?

20. Joshua earns $120 for selling $2400 worth of television sets. What is the commission rate?

21. *Real Estate Commission.* A real estate agent's commission rate is 7%. She receives a commission of $12,950 from the sale of a home. How much did the home sell for?

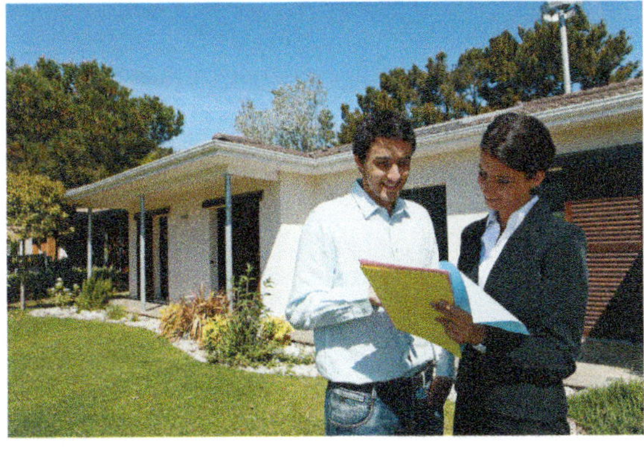

22. *Clothing Consignment Commission.* A clothing consignment shop's commission rate is 40%. The shop receives a commission of $552. How many dollars worth of clothing were sold?

23. David earns $1147.50 for selling $7650 worth of car parts. What is the commission rate?

24. Jayla earns $280.80 for selling $2340 worth of tee shirts. What is the commission rate?

25. Laila's commission is increased according to how much she sells. She receives a commission of 4% for the first $1000 of sales and 7% for the amount over $1000. What is her total commission on sales of $5500?

26. Malik's commission is increased according to how much he sells. He receives a commission of 5% for the first $2000 of sales and 8% for the amount over $2000. What is his total commission on sales of $6200?

c Complete the following table by filling in the missing numbers.

	Marked Price	Rate of Discount	Discount	Sale Price
27.	$300	10%		
28.	$2000	40%		
29.	$17	15%		
30.	$20	25%		
31.		10%	$12.50	
32.		15%	$65.70	
33.	$600		$240	
34.	$12,800		$1920	

35. Find the marked price and the rate of discount for the surfboard in this ad.

36. Find the discount and the rate of discount for the amaryllis in this ad.

Skill Maintenance

Solve. [6.1d]

37. $\frac{x}{12} = \frac{24}{16}$

38. $\frac{7}{2} = \frac{11}{x}$

Convert to standard notation. [5.3b]

39. 4.03 trillion

40. 5.8 million

Synthesis

41. Sara receives a 10% commission on the first $5000 in sales and 15% on all sales beyond $5000. If Sara receives a commission of $2405, how much did she sell? Use a calculator and trial and error if you wish.

42. Elijah collects baseball memorabilia. He bought two autographed plaques, but then became short of funds and had to sell them quickly for $200 each. On one, he made a 20% profit, and on the other, he lost 20%. Did he make or lose money on the sale?

Simple Interest and Compound Interest; Credit Cards

6.8

a SIMPLE INTEREST

Suppose you put $1000 into an investment for 1 year. The $1000 is called the **principal**. If the **interest rate** is 5%, in addition to the principal, you get back 5% of the principal, which is

5% of $1000, or $0.05 \times \$1000$, or $50.00.

The $50.00 is called **simple interest**. It is, in effect, the price that a financial institution pays for the use of the money over time.

SIMPLE INTEREST FORMULA

The **simple interest I** on principal P, invested for t years at interest rate r, is given by

$$I = P \cdot r \cdot t.$$

EXAMPLE 1 What is the simple interest on $2500 invested at an interest rate of 6% for 1 year?

We use the formula $I = P \cdot r \cdot t$:

$$\begin{aligned} I = P \cdot r \cdot t &= \$2500 \times 6\% \times 1 \\ &= \$2500 \times 0.06 \\ &= \$150. \end{aligned}$$

The simple interest for 1 year is $150.

Do Margin Exercise 1. ▶

EXAMPLE 2 What is the simple interest on a principal of $2500 invested at an interest rate of 6% for 3 months?

We use the formula $I = P \cdot r \cdot t$ and express 3 months as a fraction of a year:

$$\begin{aligned} I = P \cdot r \cdot t = \$2500 \times 6\% \times \frac{3}{12} &= \$2500 \times 6\% \times \frac{1}{4} \\ &= \frac{\$2500 \times 0.06}{4} = \$37.50. \end{aligned}$$

The simple interest for 3 months is $37.50.

Do Margin Exercise 2. ▶

When time is given in days, we generally divide it by 365 to express the time as a fractional part of a year.

EXAMPLE 3 To pay for a shipment of lawn furniture, Patio by Design borrows $8000 at $9\frac{3}{4}\%$ for 60 days. Find **(a)** the amount of simple interest that is due and **(b)** the total amount that must be paid after 60 days.

OBJECTIVES

a Solve applied problems involving simple interest.

b Solve applied problems involving compound interest.

c Solve applied problems involving interest rates on credit cards.

SKILL TO REVIEW

Objective 6.2b: Convert between percent notation and decimal notation.

Find decimal notation.

1. $34\frac{5}{8}\%$ **2.** $5\frac{1}{4}\%$

1. What is the simple interest on $4300 invested at an interest rate of 4% for 1 year?

$$\begin{aligned} I &= P \cdot r \cdot t \\ &= \$4300 \times \underline{\quad}\% \times 1 \\ &= \$\underline{\quad} \times 0.04 \times 1 \\ &= \$\underline{\quad} \end{aligned}$$

2. What is the simple interest on a principal of $4300 invested at an interest rate of 4% for 9 months?

Answers

Skill to Review:
1. 0.34625 **2.** 0.0525

Margin Exercises:
1. $172 **2.** $129

Guided Solution:
1. 4, 4300, 172

3. The Glass Nook borrows $4800 at $5\frac{1}{2}\%$ for 30 days. Find **(a)** the amount of simple interest due and **(b)** the total amount that must be paid after 30 days.

a) $I = P \cdot r \cdot t$

$= \$4800 \times 5\frac{1}{2}\% \times \frac{\square}{365}$

$= \$4800 \times 0.055 \times \frac{30}{365}$

$\approx \$\square$

b) Total amount

$= \$4800 + \square$

$= \square$

4. Find the amount in an account if $2000 is invested at 2%, compounded annually, for 2 years.

Answers

3. (a) $21.70; (b) $4821.70 4. $2080.80

Guided Solution:

3. (a) 30, 21.70; (b) $21.70, $4821.70

a) We express 60 days as a fractional part of a year:

$$I = P \cdot r \cdot t = \$8000 \times 9\frac{3}{4}\% \times \frac{60}{365}$$

$$= \$8000 \times 0.0975 \times \frac{60}{365}$$

$$\approx \$128.22.$$

The interest due for 60 days is $128.22.

b) The total amount to be paid after 60 days is the principal plus the interest:

$$\$8000 + \$128.22 = \$8128.22.$$

The total amount due is $8128.22.

◀ **Do Exercise 3.**

b COMPOUND INTEREST

When interest is paid *on interest,* we call it **compound interest**. This is the type of interest usually paid on investments. Suppose you have $5000 in a savings account at 6%. In 1 year, the account will contain the original $5000 plus 6% of $5000. Thus the total in the account after 1 year will be

106% of $5000, or $1.06 \times \$5000$, or $5300.

Now suppose that the total of $5300 remains in the account for another year. At the end of this second year, the account will contain the $5300 plus 6% of $5300. The total in the account would thus be

106% of $5300, or $1.06 \times \$5300$, or $5618.

Note that in the second year, interest is also earned on the first year's interest. When this happens, we say that interest is **compounded annually**.

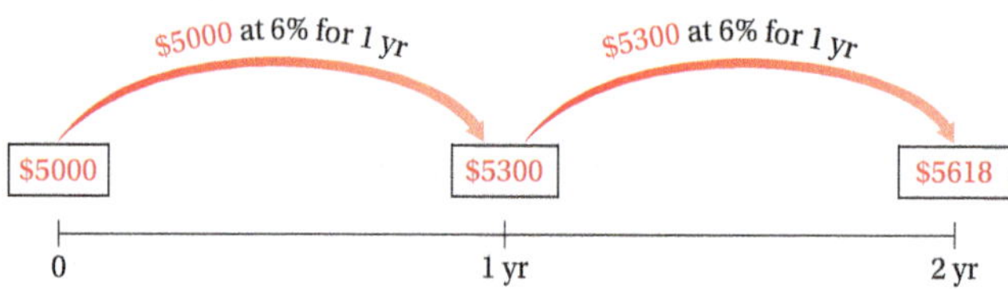

EXAMPLE 4 Find the amount in an account if $2000 is invested at 8%, compounded annually, for 2 years.

a) After 1 year, the account will contain 108% of $2000:

$$1.08 \times \$2000 = \$2160.$$

b) At the end of the second year, the account will contain 108% of $2160:

$$1.08 \times \$2160 = \$2332.80.$$

The amount in the account after 2 years is $2332.80.

◀ **Do Exercise 4.**

Suppose that the interest in Example 4 were **compounded semi-annually**—that is, every half year. Interest would then be calculated twice a year at a rate of 8% $\div$ 2, or 4% each time. The approach used in Example 4 can then be adapted, as follows.

After the first $\frac{1}{2}$ year, the account will contain 104% of \$2000:

$$1.04 \times \$2000 = \$2080.$$

After a second $\frac{1}{2}$ year (1 full year), the account will contain 104% of \$2080:

$$1.04 \times \$2080 = \$2163.20.$$

After a third $\frac{1}{2}$ year ($1\frac{1}{2}$ full years), the account will contain 104% of \$2163.20:

$$1.04 \times \$2163.20 = \$2249.728$$
$$\approx \$2249.73. \qquad \text{Rounding to the nearest cent}$$

Finally, after a fourth $\frac{1}{2}$ year (2 full years), the account will contain 104% of \$2249.73:

$$1.04 \times \$2249.73 = \$2339.7192$$
$$\approx \$2339.72. \qquad \text{Rounding to the nearest cent}$$

Let's summarize our results and look at them another way:

End of 1st $\frac{1}{2}$ year $\rightarrow 1.04 \times 2000 = 2000 \times (1.04)^1$;

End of 2nd $\frac{1}{2}$ year $\rightarrow 1.04 \times (1.04 \times 2000) = 2000 \times (1.04)^2$;

End of 3rd $\frac{1}{2}$ year $\rightarrow 1.04 \times (1.04 \times 1.04 \times 2000) = 2000 \times (1.04)^3$;

End of 4th $\frac{1}{2}$ year $\rightarrow 1.04 \times (1.04 \times 1.04 \times 1.04 \times 2000) = 2000 \times (1.04)^4$.

Note that each multiplication was by 1.04 and that

$$\$2000 \times 1.04^4 \approx \$2339.72. \qquad \text{Using a calculator and rounding to the nearest cent}$$

We have illustrated the following result.

COMPOUND INTEREST FORMULA

If a principal P has been invested at interest rate r, compounded n times a year, in t years it will grow to an amount A given by

$$A = P \cdot \left(1 + \frac{r}{n}\right)^{n \cdot t}.$$

Let's apply this formula to confirm our preceding discussion, where the amount invested is $P = \$2000$, the number of years is $t = 2$, and the number of compounding periods each year is $n = 2$. Substituting into the compound interest formula, we have

$$A = P \cdot \left(1 + \frac{r}{n}\right)^{n \cdot t} = 2000 \cdot \left(1 + \frac{8\%}{2}\right)^{2 \cdot 2}$$
$$= \$2000 \cdot \left(1 + \frac{0.08}{2}\right)^4 = \$2000(1.04)^4$$
$$= \$2000 \times 1.16985856 \approx \$2339.72.$$

If you were using a calculator, you could perform this computation in one step.

EXAMPLE 5 The Ibsens invest \$4000 in an account paying $3\frac{5}{8}\%$, compounded quarterly. Find the amount in the account after $2\frac{1}{2}$ years.

The compounding is quarterly, so n is 4. We substitute \$4000 for P, $3\frac{5}{8}\%$, or 0.03625, for r, 4 for n, and $2\frac{1}{2}$, or $\frac{5}{2}$, for t and compute A:

$$A = P \cdot \left(1 + \frac{r}{n}\right)^{n \cdot t} = \$4000 \cdot \left(1 + \frac{3\frac{5}{8}\%}{4}\right)^{4 \cdot 5/2}$$

$$= \$4000 \cdot \left(1 + \frac{0.03625}{4}\right)^{10}$$

$$= \$4000(1.0090625)^{10}$$

$$\approx \$4377.65.$$

The amount in the account after $2\frac{1}{2}$ years is \$4377.65.

Do Exercise 5.

5. A couple invests \$7000 in an account paying $6\frac{3}{8}\%$, compounded semiannually. Find the amount in the account after $1\frac{1}{2}$ years.

CALCULATOR CORNER

Compound Interest A calculator is useful in computing compound interest. Not only does it perform computations quickly but it also eliminates the need to round until the computation is completed. This minimizes round-off errors that occur when rounding is done at each stage of the computation. We must keep order of operations in mind when computing compound interest.

To find the amount due on a \$20,000 loan made for 25 days at 11% interest, compounded daily, we compute $20{,}000\left(1 + \frac{0.11}{365}\right)^{25}$. To do this on a calculator, we press [2] [0] [0] [0] [0] [×] [(] [1] [+] [.] [1] [1] [÷] [3] [6] [5] [)] [y^x] (or [^]) [2] [5] [=]. The result is \$20,151.23, rounded to the nearest cent.

Some calculators have business keys that allow such computations to be done more quickly.

EXERCISES

1. Find the amount due on a \$16,000 loan made for 62 days at 13% interest, compounded daily.
2. An investment of \$12,500 is made for 90 days at 8.5% interest, compounded daily. How much is the investment worth after 90 days?

C CREDIT CARDS

According to nerdwallet.com, the average credit-card debt among U.S. households with such debt was \$15,162 per household in 2012. According to the Aggregate Revolving Consumer Debt Survey in 2012, the average credit-card debt carried by undergraduate college students is \$3137.

The money you obtain through the use of a credit card is not "free" money. There is a price (interest) to be paid for the convenience of using a credit card. A balance carried on a credit card is a type of loan. Comparing interest rates is essential if one is to become financially responsible. A small change in an interest rate can make a large difference in the cost of a loan. When you make a payment on a credit-card balance, do you know how much of that payment is interest and how much is applied to reducing the principal?

Answer

5. \$7690.94

EXAMPLE 6 *Credit Cards.* After the holidays, Addison has a balance of $3216.28 on a credit card with an annual percentage rate (APR) of 19.7%. She decides not to make additional purchases with this card until she has paid off the balance.

a) Many credit cards require a minimum monthly payment of 2% of the balance. At this rate, what is Sarah's minimum payment on a balance of $3216.28? Round the answer to the nearest dollar.

b) Find the amount of interest and the amount applied to reduce the principal in the minimum payment found in part (a).

c) If Addison had transferred her balance to a card with an APR of 12.5%, how much of her first payment would be interest and how much would be applied to reduce the principal?

d) Compare the amounts for 12.5% from part (c) with the amounts for 19.7% from part (b).

We solve as follows.

a) We multiply the balance of $3216.28 by 2%:

$$0.02 \times \$3216.28 = \$64.3256.$$

Addison's minimum payment, rounded to the nearest dollar, is $64.

b) The amount of interest on $3216.28 at 19.7% for one month* is given by

$$I = P \cdot r \cdot t = \$3216.28 \times 0.197 \times \frac{1}{12} \approx \$52.80.$$

We subtract to find the portion of the first payment applied to reduce the principal:

$$\begin{aligned} \text{Amount applied to reduce the principal} &= \text{Minimum payment} - \text{Interest for the month} \\ &= \$64 - \$52.80 \\ &= \$11.20. \end{aligned}$$

Thus the principal of $3216.28 is decreased by only $11.20 with the first payment. (Addison still owes $3205.08.)

c) The amount of interest on $3216.28 at 12.5% for one month is

$$I = P \cdot r \cdot t = \$3216.28 \times 0.125 \times \frac{1}{12} \approx \$33.50.$$

We subtract to find the amount applied to reduce the principal in the first payment:

$$\begin{aligned} \text{Amount applied to reduce the principal} &= \text{Minimum payment} - \text{Interest for the month} \\ &= \$64 - \$33.50 \\ &= \$30.50. \end{aligned}$$

Thus the principal of $3216.28 would have decreased by $30.50 with the first payment. (Addison would still owe $3185.78.)

---

*Actually, the interest on a credit card is computed daily with a rate called a daily percentage rate (DPR). The DPR for Example 6 would be $19.7\%/365 = 0.054\%$. When no payments or additional purchases are made during a month, the difference in total interest for the month is minimal and we will not deal with it here.

d) Let's organize the information for both rates in the following table.

BALANCE BEFORE FIRST PAYMENT	FIRST MONTH'S PAYMENT	%APR	AMOUNT OF INTEREST	AMOUNT APPLIED TO PRINCIPAL	BALANCE AFTER FIRST PAYMENT
$3216.28	$64	19.7%	$52.80	$11.20	$3205.08
3216.28	64	12.5	33.50	30.50	3185.78

Difference in balance after first payment ⟶ $19.30

At 19.7%, the interest is $52.80 and the principal is decreased by $11.20. At 12.5%, the interest is $33.50 and the principal is decreased by $30.50. Thus the interest at 19.7% is $52.80 − $33.50, or $19.30, greater than the interest at 12.5%. Thus the principal is decreased by $30.50 − $11.20, or $19.30, more with the 12.5% rate than with the 19.7% rate.

Do Exercise 6.

Even though the mathematics of the information in the following table is beyond the scope of this text, it is interesting to compare how long it takes to pay off the balance of Example 6 if Addison continues to pay $64 each month, compared to how long it takes if she pays double that amount, or $128, each month. Financial consultants frequently tell clients that if they want to take control of their debt, they should pay double the minimum payment.

RATE	PAYMENT	NUMBER OF PAYMENTS TO PAY OFF DEBT	TOTAL PAID BACK	ADDITIONAL COST OF PURCHASES
19.7%	$64	107, or 8 yr 11 mo	$6848	$3631.72
19.7	128	33, or 2 yr 9 mo	4224	1007.72
12.5	64	72, or 6 yr	4608	1391.72
12.5	128	29, or 2 yr 5 mo	3712	495.72

As with most loans, if you pay an extra amount toward the principal with each payment, the length of the loan can be greatly reduced. Note that at the rate of 19.7%, it will take Addison almost 9 years to pay off her debt if she pays only $64 per month and does not make additional purchases. If she transfers her balance to a card with a 12.5% rate and pays $128 per month, she can eliminate her debt in approximately $2\frac{1}{2}$ years. You can see how debt can get out of control if you continue to make purchases and pay only the minimum payment each month. The debt will never be eliminated.

6. ***Credit Card.*** After the holidays, Logan has a balance of $4867.59 on a credit card with an annual percentage rate (APR) of 21.3%. He decides not to make additional purchases with this card until he has paid off the balance.
   a) Many credit cards require a minimum monthly payment of 2% of the balance. What is Logan's minimum payment on a balance of $4867.59? Round the answer to the nearest dollar.
   b) Find the amount of interest and the amount applied to reduce the principal in the minimum payment found in part (a).
   c) If Logan had transferred his balance to a card with an APR of 13.6%, how much of his first payment would be interest and how much would be applied to reduce the principal?
   d) Compare the amounts for 13.6% from part (c) with the amounts for 21.3% from part (b).

Answers

6. (a) $97; (b) interest: $86.40; amount applied to principal: $10.60; (c) interest: $55.17; amount applied to principal: $41.83; (d) At 13.6%, the principal was reduced by $31.23 more than at the 21.3% rate. The interest at 13.6% is $31.23 less than at 21.3%.

6.8 Exercise Set

For Extra Help MyMathLab®

MathXL®  PRACTICE  WATCH  READ  REVIEW

 Reading Check

In the simple interest formula, $I = P \cdot r \cdot t$, t must be expressed in years. Convert each length of time to years. Use 1 year = 12 months = 365 days.

RC1. 6 months

RC2. 40 days

RC3. 285 days

RC4. 9 months

RC5. 3 months

RC6. 4 months

a Find the simple interest.

	Principal	Rate of Interest	Time	Simple Interest
1.	\$200	4%	1 year	
2.	\$200	7.7%	1 year	
3.	\$4300	10.56%	$\frac{1}{4}$ year	
4.	\$80,000	$6\frac{3}{4}\%$	$\frac{1}{12}$ year	
5.	\$20,000	$4\frac{5}{8}\%$	6 months	
6.	\$8000	9.42%	2 months	
7.	\$50,000	$5\frac{3}{8}\%$	3 months	
8.	\$100,000	$3\frac{1}{4}\%$	9 months	

Solve. Assume that simple interest is being calculated in each case.

9. Mia's Boutique borrows \$10,000 at 9% for 60 days. Find **(a)** the amount of interest due and **(b)** the total amount that must be paid after 60 days.

10. Mason's Drywall borrows \$8000 at 10% for 90 days. Find **(a)** the amount of interest due and **(b)** the total amount that must be paid after 90 days.

11. Animal Instinct, a pet supply shop, borrows \$6500 at $5\frac{1}{4}\%$ for 90 days. Find **(a)** the amount of interest due and **(b)** the total amount that must be paid after 90 days.

12. Andante's Cafe borrows \$4500 at $12\frac{1}{2}\%$ for 60 days. Find **(a)** the amount of interest due and **(b)** the total amount that must be paid after 60 days.

13. Cameron's Garage borrows \$5600 at 10% for 30 days. Find **(a)** the amount of interest due and **(b)** the total amount that must be paid after 30 days.

14. Shear Delights Hair Salon borrows \$3600 at 4% for 30 days. Find **(a)** the amount of interest due and **(b)** the total amount that must be paid after 30 days.

b Interest is compounded annually. Find the amount in the account after the given length of time. Round to the nearest cent.

	Principal	Rate of Interest	Time	Amount in the Account
15.	\$400	5%	2 years	
16.	\$450	4%	2 years	
17.	\$2000	8.8%	4 years	
18.	\$4000	7.7%	4 years	
19.	\$4300	10.56%	6 years	
20.	\$8000	9.42%	6 years	
21.	\$20,000	$6\frac{5}{8}\%$	25 years	
22.	\$100,000	$5\frac{7}{8}\%$	30 years	

Interest is compounded semiannually. Find the amount in the account after the given length of time. Round to the nearest cent.

	Principal	Rate of Interest	Time	Amount in the Account
23.	\$4000	6%	1 year	
24.	\$1000	5%	1 year	
25.	\$20,000	8.8%	4 years	
26.	\$40,000	7.7%	4 years	
27.	\$5000	10.56%	6 years	
28.	\$8000	9.42%	8 years	
29.	\$20,000	$7\frac{5}{8}\%$	25 years	
30.	\$100,000	$4\frac{7}{8}\%$	30 years	

Solve.

31. A family invests \$4000 in an account paying 6%, compounded monthly. How much is in the account after 5 months?

32. A couple invests \$2500 in an account paying 3%, compounded monthly. How much is in the account after 6 months?

33. A couple invests \$1200 in an account paying 10%, compounded quarterly. How much is in the account after 1 year?

34. The O'Hares invest \$6000 in an account paying 8%, compounded quarterly. How much is in the account after 18 months?

C Solve.

35. *Credit Cards.* Amelia has a balance of \$1278.56 on a credit card with an annual percentage rate (APR) of 19.6%. The minimum payment required in the current statement is \$25.57. Find the amount of interest and the amount applied to reduce the principal in this payment and the balance after this payment.

36. *Credit Cards.* Lawson has a balance of \$1834.90 on a credit card with an annual percentage rate (APR) of 22.4%. The minimum payment required in the current statement is \$36.70. Find the amount of interest and the amount applied to reduce the principal in this payment and the balance after this payment.

37. *Credit Cards.* Antonio has a balance of \$4876.54 on a credit card with an annual percentage rate (APR) of 21.3%.
   a) Many credit cards require a minimum monthly payment of 2% of the balance. What is Antonio's minimum payment on a balance of \$4876.54? Round the answer to the nearest dollar.
   b) Find the amount of interest and the amount applied to reduce the principal in the minimum payment found in part (a).
   c) If Antonio had transferred his balance to a card with an APR of 12.6%, how much of his payment would be interest and how much would be applied to reduce the principal?
   d) Compare the amounts for 12.6% from part (c) with the amounts for 21.3% from part (b).

38. *Credit Cards.* Becky has a balance of \$5328.88 on a credit card with an annual percentage rate (APR) of 18.7%.
   a) Many credit cards require a minimum monthly payment of 2% of the balance. What is Becky's minimum payment on a balance of \$5328.88? Round the answer to the nearest dollar.
   b) Find the amount of interest and the amount applied to reduce the principal in the minimum payment found in part (a).
   c) If Becky had transferred her balance to a card with an APR of 13.2%, how much of her payment would be interest and how much would be applied to reduce the principal?
   d) Compare the amounts for 13.2% from part (c) with the amounts for 18.7% from part (b).

Skill Maintenance

39. Find the LCM of 32 and 50. [4.1a]

40. Find the prime factorization of 228. [3.2c]

Divide and simplify. [3.7b]

41. $-\frac{6}{125} \div \frac{8}{15}$

42. $\frac{16}{105} \div \frac{5}{14}$

Multiply and simplify. [3.6a]

43. $\frac{4}{15} \cdot \frac{3}{20}$

44. $\frac{8}{21}\left(-\frac{49}{800}\right)$

45. Simplify: $4^3 - 6^2 \div 2^2$. [1.9c]

46. Solve: $x - \frac{2}{5} = -\frac{3}{10}$. [4.3b]

Synthesis

Effective Yield. The *effective yield* is the yearly rate of simple interest that corresponds to a rate for which interest is compounded two or more times a year. For example, if P is invested at 12%, compounded quarterly, we multiply P by $(1 + 0.12/4)^4$, or 1.03^4. Since $1.03^4 \approx 1.126$, the 12% compounded quarterly corresponds to an effective yield of approximately 12.6%. In Exercises 47 and 48, find the effective yield for the indicated account.

47. The account pays 9%, compounded monthly.

48. The account pays 10%, compounded daily.

CHAPTER 6

Summary and Review

Vocabulary Reinforcement

Complete each statement with the correct word or phrase from the column on the right. Some of the choices will not be used.

1. When a quantity is decreased by a certain percent, we say that this is a(n) ______________. [6.6b]
2. The ______________ interest I on principal P, invested for t years at interest rate r, is given by $I = P \cdot r \cdot t$. [6.8a]
3. A(n) ______________ is a ratio used to compare two different kinds of measure. [6.1b]
4. A(n) ______________ states that two pairs of numbers have the same ratio. [6.1c]
5. Sale price = Original price − ______________. [6.7c]
6. Commission = Commission rate × ______________. [6.7b]

proportion
discount
rate
sales
commission
price
percent increase
percent decrease
simple
compound

Concept Reinforcement

Determine whether each statement is true or false.

______________ 1. When we simplify a ratio like $\frac{8}{12}$, we find two other numbers in the same ratio. [6.1a]

______________ 2. The proportion $\frac{a}{b} = \frac{c}{d}$ can also be written as $\frac{c}{a} = \frac{d}{b}$. [6.1c]

______________ 3. A fixed principal invested for 4 years will earn more interest when interest is compounded quarterly than when interest is compounded semiannually. [6.8b]

______________ 4. Of the numbers 0.5%, $\frac{5}{1000}\%$, $\frac{1}{2}\%$, $\frac{1}{5}$, and $0.\overline{1}$, the largest number is $0.\overline{1}$. [6.2b], [6.3a, b]

______________ 5. If principal A equals principal B and principal A is invested for 2 years at 4%, compounded quarterly, while principal B is invested for 4 years at 2%, compounded semiannually, the interest earned from each investment is the same. [6.8b]

Study Guide

Objective 6.1a Find fraction notation for ratios.

Example Find the ratio of 7 to 18.

Write a fraction with a numerator of 7 and a denominator of 18: $\frac{7}{18}$.

Practice Exercise

1. Find the ratio of 17 to 3.

Objective 6.1b Give the ratio of two different measures as a rate.

Example A driver travels 156 mi on 6.5 gal of gas. What is the rate in miles per gallon?

$$\frac{156 \text{ mi}}{6.5 \text{ gal}} = \frac{156}{6.5}\,\frac{\text{mi}}{\text{gal}} = 24\,\frac{\text{mi}}{\text{gal}}, \text{ or } 24 \text{ mpg}$$

Practice Exercise

2. A student earned $120 for working 16 hr. What was the rate of pay per hour?

Objective 6.1c Determine whether two pairs of numbers are proportional.

Example Determine whether 3, 4 and 7, 9 are proportional.

$$3 \cdot 9 = 27 \qquad \frac{3}{4} \stackrel{?}{=} \frac{7}{9} \qquad 4 \cdot 7 = 28$$

Since the cross products are not the same $(27 \neq 28)$, $\frac{3}{4} \neq \frac{7}{9}$ and the numbers are not proportional.

Practice Exercise

3. Determine whether 7, 9 and 21, 27 are proportional.

Objective 6.1d Solve proportions.

Example Solve: $\frac{3}{4} = \frac{y}{7}$.

$3 \cdot 7 = 4 \cdot y$ Equating cross products

$\frac{3 \cdot 7}{4} = \frac{4 \cdot y}{4}$ Dividing by 4 on both sides

$\frac{21}{4} = y$

The solution is $\frac{21}{4}$.

Practice Exercise

4. Solve: $\frac{9}{x} = \frac{8}{3}$.

Objective 6.1e Solve applied problems involving proportions.

Example Martina bought 3 tickets to a campus theater production for \$16.50. How much would 8 tickets cost?

We translate to a proportion.

$$\begin{array}{r} \text{Tickets} \rightarrow \\ \text{Cost} \rightarrow \end{array} \frac{3}{16.50} = \frac{8}{c} \begin{array}{l} \leftarrow \text{Tickets} \\ \leftarrow \text{Cost} \end{array}$$

$3 \cdot c = 16.50 \cdot 8$ Equating cross products

$c = \frac{16.50 \cdot 8}{3}$

$c = 44$

Eight tickets would cost \$44.

Practice Exercise

5. On a map, $\frac{1}{2}$ in. represents 50 mi. If two cities are $1\frac{3}{4}$ in. apart on the map, how far apart are they in reality?

Objective 6.2b Convert between percent notation and decimal notation.

Example Find percent notation for 1.3.

We move the decimal point two places to the right and write a percent symbol:

$1.3 = 130\%$.

Practice Exercise

6. Find percent notation for 0.082.

Example Find decimal notation for $12\frac{3}{4}\%$.

We convert $12\frac{3}{4}$ to a decimal and move the decimal point two places to the left:

$12\frac{3}{4}\% = 12.75\% = 0.1275$.

Practice Exercise

7. Find decimal notation for $62\frac{5}{8}\%$.

Objective 6.3a Convert from fraction notation to percent notation.

Example Find percent notation for $\frac{5}{12}$.

$$\begin{array}{r} 0.416 \\ 12\overline{)5.000} \\ \underline{48} \\ 20 \\ \underline{12} \\ 80 \\ \underline{72} \\ 8 \end{array}$$

$\frac{5}{12} = 0.41\overline{6} = 41.\overline{6}\%$, or $41\frac{2}{3}\%$

Practice Exercise

8. Find percent notation for $\frac{7}{11}$.

Objective 6.3b Convert from percent notation to fraction notation.

Example Find fraction notation for 9.5%.

$$9.5\% = \frac{9.5}{100} = \frac{95}{1000} = \frac{5 \cdot 19}{5 \cdot 200} = \frac{5}{5} \cdot \frac{19}{200} = \frac{19}{200}$$

Practice Exercise

9. Find fraction notation for 6.8%.

Objective 6.4b Solve basic percent problems.

Example 165 is what percent of 3300?

We have

$165 = p \cdot 3300$ Translating to a percent equation

$\frac{165}{3300} = p$ Dividing by 3300 on both sides

$0.05 = p$

$5\% = p.$

Thus, 165 is 5% of 3300.

Practice Exercise

10. 12 is what percent of 288?

Objective 6.5b Solve basic percent problems.

Example 18% of what is 1296?

$\frac{18}{100} = \frac{1296}{b}$ Translating to a proportion

$18 \cdot b = 100 \cdot 1296$

$b = 7200.$ Dividing by 18 on both sides

Thus, 18% of 7200 is 1296.

Practice Exercise

11. 3% of what is 300?

Objective 6.6b Solve applied problems involving percent increase or percent decrease.

Example The total cost for 16 basic grocery items in the second quarter of 2008 averaged $46.67 nationally. The total cost of these 16 items in the second quarter of 2007 averaged $42.95. What was the percent increase?

Source: American Farm Bureau Federation

Practice Exercise

12. In Indiana, the cost for 16 basic grocery items increased from $40.07 in the second quarter of 2007 to $46.20 in the second quarter of 2008. What was the percent increase from 2007 to 2008?

Objective 6.6b (continued)

We first determine the amount of increase:

$\$46.67 - \$42.95 = \$3.72.$

Then we translate to a percent equation or a proportion and solve.

Rewording: $3.72 is what percent of $42.95?

Percent Equation:

$$3.72 = p \cdot 42.95$$
$$\frac{3.72}{42.95} = \frac{p \cdot 42.95}{42.95}$$
$$\frac{3.72}{42.95} = p$$
$$0.087 \approx p$$
$$8.7\% = p$$

Proportion:

$$\frac{N}{100} = \frac{3.72}{42.95}$$
$$42.95N = 100 \cdot 3.72$$
$$\frac{42.95N}{42.95} = \frac{100 \cdot 3.72}{42.95}$$
$$N = \frac{372}{42.95}$$
$$N \approx 8.7$$

The percent increase was about 8.7%.

Objective 6.7a Solve applied problems involving sales tax and percent.

Example The sales tax is $34.23 on the purchase of a flat-screen high-definition television that costs $489. What is the sales tax rate?

Rephrase: Sales tax is what percent of purchase price?

Translate: $34.23 = r \times 489$

Solve:
$$\frac{34.23}{489} = \frac{r \times 489}{489}$$
$$\frac{34.23}{489} = r$$
$$0.07 = r$$
$$7\% = r$$

The sales tax rate is 7%.

Practice Exercise

13. The sales tax is $1102.20 on the purchase of a new car that costs $18,370. What is the sales tax rate?

Objective 6.7b Solve applied problems involving commission and percent.

Example A real estate agent's commission rate is $6\frac{1}{2}\%$. She received a commission of $17,160 on the sale of a home. For how much did the home sell?

Rephrase: Commission is $6\frac{1}{2}\%$ of what selling price?

Translate: $17{,}160 = 6\frac{1}{2}\% \times S$

Solve:
$$17{,}160 = 0.065 \times S$$
$$\frac{17{,}160}{0.065} = \frac{0.065 \times S}{0.065}$$
$$264{,}000 = S$$

The home sold for $264,000.

Practice Exercise

14. A real estate agent's commission rate is 7%. He received a commission of $12,950 on the sale of a home. For how much did the home sell?

Objective 6.8a Solve applied problems involving simple interest.

Example To meet its payroll, a business borrows \$5200 at $4\frac{1}{4}\%$ for 90 days. Find the amount of simple interest that is due and the total amount that must be paid after 90 days.

$$I = P \cdot r \cdot t = \$5200 \times 4\tfrac{1}{4}\% \times \frac{90}{365}$$
$$= \$5200 \times 0.0425 \times \frac{90}{365}$$
$$\approx \$54.49$$

The interest due for 90 days = \$54.49.

The total amount due = \$5200 + \$54.49 = \$5254.49.

Practice Exercise

15. A student borrows \$2500 for tuition at $5\frac{1}{2}\%$ for 60 days. Find the amount of simple interest that is due and the total amount that must be paid after 60 days.

Objective 6.8b Solve applied problems involving compound interest.

Example Find the amount in an account if \$3200 is invested at 5%, compounded semiannually, for $1\frac{1}{2}$ years.

$$A = P \cdot \left(1 + \frac{r}{n}\right)^{n \cdot t}$$
$$= \$3200\left(1 + \frac{0.05}{2}\right)^{2 \cdot \frac{3}{2}}$$
$$= \$3200(1.025)^3$$
$$= \$3446.05$$

The amount in the account after $1\frac{1}{2}$ years is \$3446.05.

Practice Exercise

16. Find the amount in an account if \$6000 is invested at $4\frac{3}{4}\%$, compounded quarterly, for 2 years.

Review Exercises

Write fraction notation for each ratio. Do not simplify. [6.1a]

1. 47 to 84

2. 46 to 1.27

3. 83 to 100

4. 0.72 to 197

5. At Preston Seafood Market, 12,480 lb of tuna and 16,640 lb of salmon were sold one year. [6.1a]

a) Write fraction notation for the ratio of tuna sold to salmon sold.

b) Write fraction notation for the ratio of salmon sold to the total number of pounds of both kinds of fish sold.

Find the ratio of the first number to the second number and simplify. [6.1a]

6. 9 to 12

7. 3.6 to 6.4

8. *Gas Mileage.* The Chrysler 200 will travel 406 mi on 14.5 gal of gasoline in highway driving. What is the rate in miles per gallon? [6.1b]

Source: Chrysler Motor Corporation

9. *Flywheel Revolutions.* A certain flywheel makes 472,500 revolutions in 75 min. What is the rate of spin in revolutions per minute? [6.1b]

10. A lawn requires 319 gal of water for every 500 ft^2. What is the rate in gallons per square foot? [6.1b]

Determine whether the two pairs of numbers are proportional. [6.1c]

11. 9, 15 and 36, 60

12. 24, 37 and 40, 46.25

Solve. [6.1d]

13. $\frac{8}{9} = \frac{x}{36}$

14. $\frac{6}{x} = \frac{48}{56}$

15. $\frac{120}{\frac{3}{7}} = \frac{7}{x}$

16. $\frac{4.5}{120} = \frac{0.9}{x}$

Solve. [6.1e]

17. ***Quality Control.*** A factory manufacturing computer circuits found 3 defective circuits in a lot of 65 circuits. At this rate, how many defective circuits can be expected in a lot of 585 circuits?

18. ***Exchanging Money.*** On January 7, 2013, 1 U.S. dollar was worth about 0.99 Canadian dollar.

a) How much were 250 U.S. dollars worth in Canada on that day?

b) While traveling in Canada that day, Jamal saw a sweatshirt that cost 50 Canadian dollars. How much would it cost in U.S. dollars?

19. A train travels 448 mi in 7 hr. At this rate, how far will it travel in 13 hr?

20. ***Gasoline Consumption.*** The United States consumed about 89 billion gallons of gasoline in the first 8 months of 2011. At this rate, how much was consumed in 12 months?

21. ***Trash Production.*** A study shows that 5 people generate 23 lb of trash each day. The population of Austin, Texas, is 820,611. How many pounds of trash are produced in Austin in one day?

Sources: U.S. Environmental Protection Agency; U.S. Census Bureau

22. ***Calories Burned.*** Kevin burned 200 calories while playing ultimate frisbee for $\frac{3}{4}$ hr. How many calories would he burn if he played for $1\frac{1}{5}$ hr?

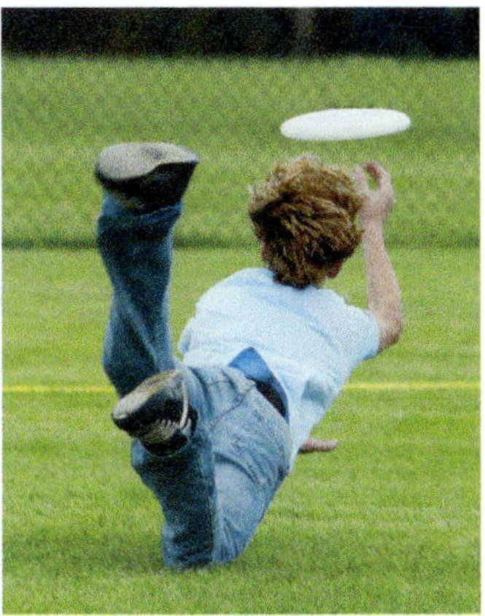

23. ***Lawyers in Chicago.*** In Illinois, there are about 4.6 lawyers for every 1000 people. The population of Chicago is 2,707,120. How many lawyers would you expect there to be in Chicago?

Sources: American Bar Association; U.S. Census Bureau

Find decimal notation for the percent notations in each sentence. [6.2b]

24. In the 2006–2007 school year, about 4% of the 15 million college students in the United States were foreign students. Approximately 14.4% of the foreign students were from India.

Source: Institute of International Education

25. Poland is 62.1% urban; Sweden is 84.2% urban.

Source: *The World Almanac*, 2008

Find percent notation. [6.2b]

26. 1.7

27. 0.065

Find percent notation. [6.3a]

28. $\frac{3}{8}$

29. $\frac{1}{3}$

Find fraction notation. [6.3b]

30. 24%

31. 6.3%

Translate to a percent equation. Then solve. [6.4a, b]

32. 30.6 is what percent of 90?

33. 63 is 84% of what?

34. What is $38\frac{1}{2}\%$ of 168?

Translate to a proportion. Then solve. [6.5a, b]

35. 24% of what is 16.8?

36. 42 is what percent of 30?

37. What is 10.5% of 84?

Solve. [6.6a, b]

38. *Favorite Ice Creams.* According to a survey, 8.9% of those interviewed chose chocolate as their favorite ice cream flavor and 4.2% chose butter pecan. At this rate, of the 2000 students in a freshman class, how many would choose chocolate as their favorite ice cream? butter pecan?

Source: International Ice Cream Association

39. *Prescriptions.* Of the 305 million people in the United States, 140.3 million take at least one type of prescription drug per day. What percent take at least one type of prescription drug per day?

Source: William N. Kelly, *Pharmacy: What It Is and How It Works*, 2nd ed., CRC Press Pharmaceutical Education, 2006

40. *Water Output.* The average person loses 200 mL of water per day by sweating. This is 8% of the total output of water from the body. How much is the total output of water?

Source: Elaine N. Marieb, *Essentials of Human Anatomy and Physiology*, 6th ed. Boston: Addison Wesley Longman, Inc., 2000

41. *Test Scores.* After Sheila got a 75 on a math test, she was allowed to go to the math lab and take a retest. She increased her score to 84. What was the percent increase?

42. *Test Scores.* James got an 80 on a math test. By taking a retest in the math lab, he increased his score by 15%. What was his new score?

Solve. [6.7a, b, c]

43. A state charges a meals tax of $7\frac{1}{2}\%$. What is the meals tax charged on a dinner party costing $320?

44. In a certain state, a sales tax of $453.60 is collected on the purchase of a used car for $7560. What is the sales tax rate?

45. Kim earns $753.50 for selling $6850 worth of televisions. What is the commission rate?

46. What is the rate of discount of this stepladder?

47. An air conditioner has a marked price of $350. It is placed on sale at 12% off. What are the discount and the sale price?

48. The price of a fax machine is marked down from $305 to $262.30. What is the rate of discount?

49. An insurance salesperson receives a 7% commission. If $42,000 worth of life insurance is sold, what is the commission?

Solve. [6.8a, b, c]

50. What is the simple interest on $1800 at 6% for $\frac{1}{3}$ year?

51. The Dress Shack borrows $24,000 at 10% simple interest for 60 days. Find **(a)** the amount of interest due and **(b)** the total amount that must be paid after 60 days.

52. What is the simple interest on a principal of $2200 at an interest rate of 5.5% for 1 year?

53. The Armstrongs invest $7500 in an investment account paying an annual interest rate of 4%, compounded monthly. How much is in the account after 3 months?

54. Find the amount in an investment account if $8000 is invested at 9%, compounded annually, for 2 years.

55. *Credit Cards.* At the end of her junior year of college, Kasha has a balance of $6428.74 on a credit card with an annual percentage rate (APR) of 18.7%. She decides not to make additional purchases with this card until she has paid off the balance.

a) Many credit cards require a minimum payment of 2% of the balance. At this rate, what is Kasha's minimum payment on a balance of $6428.74? Round the answer to the nearest dollar.

b) Find the amount of interest and the amount applied to reduce the principal in the minimum payment found in part (a).

c) If Kasha had transferred her balance to a card with an APR of 13.2%, how much of her payment would be interest and how much would be applied to reduce the principal?

d) Compare the amounts for 13.2% from part (c) with the amounts for 18.7% from part (b).

56. If 3 dozen eggs cost $5.04, how much will 5 dozen eggs cost? [6.1e]

A. $6.72 **B.** $6.96
C. $8.40 **D.** $10.08

57. Find the amount in a money market account if $10,500 is invested at 6%, compounded semiannually, for $1\frac{1}{2}$ years. [6.8b]

A. $11,139.45 **B.** $12,505.67
C. $11,473.63 **D.** $10,976.03

Synthesis

58. A worker receives raises of 3%, 6%, and then 9%. By what percent has the original salary increased? [6.6a]

59. Shine-and-Glo Painters uses 2 gal of finishing paint for every 3 gal of primer. Each gallon of finishing paint covers 450 ft^2. If a surface of 4950 ft^2 needs both primer and finishing paint, how many gallons of each should be purchased? [6.1e]

Understanding Through Discussion and Writing

1. Which is the better deal for a consumer and why: a discount of 40% or a discount of 20% followed by another of 22%? [6.7c]

2. If you were a college president, which would you prefer: a low or high faculty-to-student ratio? Why? [6.1a]

3. Ollie buys a microwave oven during a 10%-off sale. The sale price that Ollie paid was $162. To find the original price, Ollie calculates 10% of $162 and adds that to $162. Is this correct? Why or why not? [6.7c]

4. Which is better for a wage earner, and why: a 10% raise followed by a 5% raise a year later, or a 5% raise followed by a 10% raise a year later? [6.6a]

5. You take 40% of 50% of a number. What percent of the number could you take to obtain the same result making only one multiplication? Explain your answer. [6.6a]

6. A firm must choose between borrowing $5000 at 10% for 30 days and borrowing $10,000 at 8% for 60 days. Give arguments in favor of and against each option. [6.8a]

CHAPTER 6

Test

For Extra Help For step-by-step test solutions, access the Chapter Test Prep Videos in MyMathLab® or on YouTube (search "BittingerPreIntro" and click on "Channels").

Write fraction notation for each ratio. Do not simplify.

1. 85 to 97

2. 0.34 to 124

Find the ratio of the first number to the second number and simplify.

3. 18 to 20

4. 0.75 to 0.96

5 *Ham Servings.* A 12-lb shankless ham contains 16 servings. What is the rate in servings per pound?

6. *Gas Mileage.* Jeff's convertible will travel 464 mi on 14.5 gal of gasoline in highway driving. What is the rate in miles per gallon?

Determine whether the two pairs of numbers are proportional.

7. 7, 8 and 63, 72

8. 1.3, 3.4 and 5.6, 15.2

Solve.

9. $\frac{68}{y} = \frac{17}{25}$

10. $\frac{150}{2.5} = \frac{x}{6}$

11. *Map Scaling.* On a map, 3 in. represents 225 mi. If two cities are 7 in. apart on the map, how far are they apart in reality?

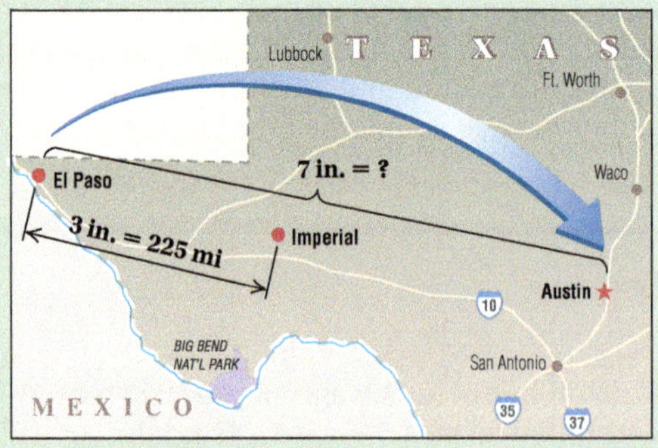

12. *Charity Work.* Kayla is crocheting hats for a charity. She can make 8 hats from 12 packages of yarn.

a) How many hats can she make from 20 packages of yarn?

b) How many packages of yarn does she need to make 20 hats?

Solve.

13. *Distance Traveled.* An ocean liner traveled 432 km in 12 hr. At this rate, how far would the boat travel in 42 hr?

14. *Thanksgiving Dinner.* A traditional turkey dinner for 8 people cost about $33.81 in a recent year. How much would it cost to serve a turkey dinner for 14 people?

Source: American Farm Bureau Federation

15. ***Households Owning Pets.*** In 2011, about 61.3% of the households in Colorado owned pets. Find decimal notation for 61.3%.

Source: American Veterinary Medical Association

16. ***Gravity.*** The gravity of Mars is 0.38 as strong as Earth's. Find percent notation for 0.38.

Source: www.marsinstitute.info/epo/mermarsfacts.html

17. Find percent notation for $\frac{11}{8}$.

18. Find fraction notation for 65%.

19. Translate to a percent equation. Then solve.
What is 40% of 55?

20. Translate to a proportion. Then solve. What percent of 80 is 65?

Solve.

21. ***Organ Transplants.*** In 2011, there were 28,535 organ transplants in the United States. The following pie chart shows the percentages for the main transplants. How many kidney transplants were there in 2011? liver transplants? heart transplants?

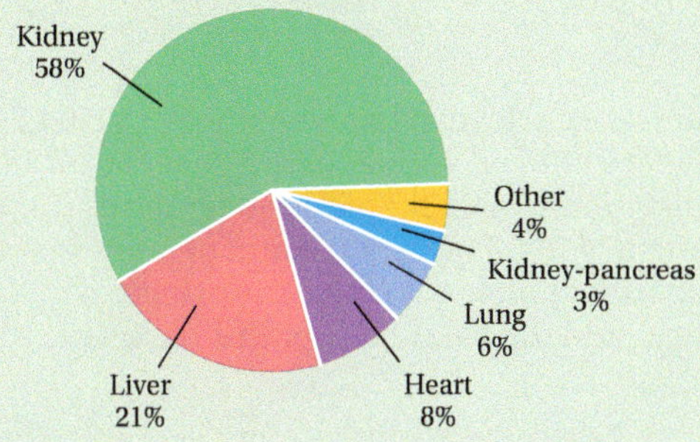

SOURCE: Milliman Research Report

22. ***Batting Average.*** Garrett Atkins, third baseman for the Colorado Rockies, got 175 hits during the 2008 baseball season. This was about 28.64% of his at-bats. How many at-bats did he have?

Source: Major League Baseball

23. ***Foreign Adoptions.*** The number of foreign children adopted by Americans declined from 20,679 in 2006 to 19,292 in 2007. Find the percent decrease.

Source: U.S. State Department, *USA TODAY*, August 13, 2008

24. There are about 6,603,000,000 people living in the world today, and approximately 4,002,000,000 live in Asia. What percent of people live in Asia?

Source: Population Division/International Programs Center, U.S. Census Bureau, U.S. Dept. of Commerce

25. ***Oklahoma Sales Tax.*** The sales tax rate in Oklahoma is 4.5%. How much tax is charged on a purchase of $560? What is the total price?

26. Noah's commission rate is 15%. What is his commission on the sale of $4200 worth of merchandise?

27. The marked price of a DVD player is $200 and the item is on sale at 20% off. What are the discount and the sale price?

28. What is the simple interest on a principal of $120 at the interest rate of 7.1% for 1 year?

29. A city orchestra invests $5200 at 6% simple interest. How much is in the account after $\frac{1}{2}$ year?

30. Find the amount in an account if $1000 is invested at $5\frac{3}{8}\%$ compounded annually, for 2 years.

31. The Suarez family invests $10,000 at an annual interest rate of 4.9%, compounded monthly. How much is in the account after 3 years?

32. *Job Opportunities.* The following table lists job opportunities in 2006 and projected increases for 2016. Complete the table by filling in the missing numbers.

Occupation	Total Employment in 2006	Projected Employment in 2016	Change	Percent of Increase
Dental assistant	280,000	362,000	82,000	29.3%
Plumber	705,000		52,000	
Veterinary assistant	71,000	100,000		
Motorcycle repair technician		24,000	3000	
Fitness professional		298,000		26.8%

SOURCE: EarnMyDegree.com

33. Find the discount and the rate of discount of the television in this ad.

19" LCD HDTV

$299.99

was $349.99

34. *Credit Cards.* Jayden has a balance of $2704.27 on a credit card with an annual percentage rate of 16.3%. The minimum payment required on the current statement is $54. Find the amount of interest and the amount applied to reduce the principal in this payment and the balance after this payment.

35. 0.75% of what number is 300?

A. 2.25 **B.** 40,000 **C.** 400 **D.** 225

36. Lucita walks $4\frac{1}{2}$ mi in $1\frac{1}{2}$ hr. What is her rate in miles per hour?

A. $\frac{1}{3}$ mph **B.** $1\frac{1}{2}$ mph

C. 3 mph **D.** $4\frac{1}{2}$ mph

Synthesis

37. By selling a home without using a realtor, Juan and Marie can avoid paying a 7.5% commission. They receive an offer of $180,000 from a potential buyer. In order to give a comparable offer, for what price would a realtor need to sell the house? Round to the nearest hundred.

38. Karen's commission rate is 16%. She invests her commission from the sale of $15,000 worth of merchandise at an interest rate of 12%, compounded quarterly. How much is Karen's investment worth after 6 months?

CHAPTER

7

Data, Graphs, and Statistics

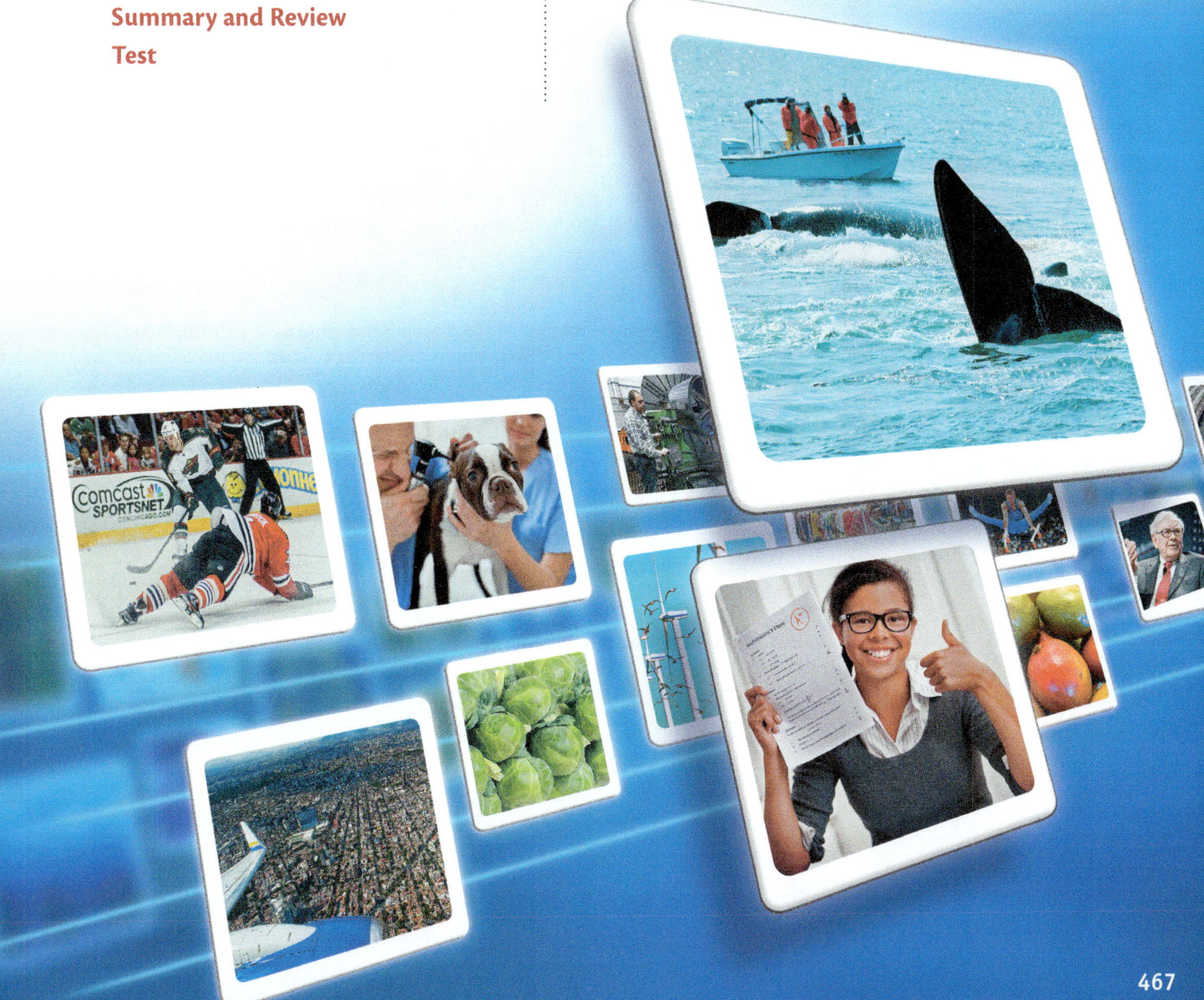

STUDYING FOR SUCCESS *Applications*

- ☐ If you find applied problems challenging, don't give up! Your skill will improve with each problem that you solve.
- ☐ Make applications real! Look for applications of the math that you are studying in newspapers and magazines.

7.1 Averages, Medians, and Modes

OBJECTIVES

a Find the average of a set of numbers and solve applied problems involving averages.

b Find the median of a set of numbers and solve applied problems involving medians.

c Find the mode of a set of numbers and solve applied problems involving modes.

SKILL TO REVIEW

Objectives 1.9c and 5.4b: Simplify expressions using the rules for order of operations.

1. Find the average of 282, 137, 5280, and 193.
2. Find the average of $23.40, $89.15, and $148.17 to the nearest cent.

a AVERAGES

A **statistic** is a number describing a set of data. One statistic is a *center point,* or *measure of central tendency,* that characterizes the data. The most common kind of center point is the **arithmetic** (pronounced ăr′ĭth-mĕt′-ĭk) **mean**, or simply the **mean**. This center point is often referred to as the *average.*

AVERAGE

To find the **average** of a set of numbers, add the numbers and then divide by the number of items of data.

EXAMPLE 1 On a 4-day trip, a car was driven the following number of miles: 240, 302, 280, 320. What was the average number of miles per day?

$$\frac{240 + 302 + 280 + 320}{4} = \frac{1142}{4}, \quad \text{or} \quad 285.5$$

The car was driven an average of 285.5 mi per day. Had the car been driven exactly 285.5 mi each day, the same total distance (1142 mi) would have been traveled.

EXAMPLE 2 *Gas Mileage.* The 2013 Volkswagen Jetta TDI is estimated to travel 546 mi on the highway on 13 gal of diesel fuel. What is the expected average number of miles per gallon (mpg)—that is, what is the fuel mileage for highway driving?
Source: vw.com

We divide the total number of miles, 546, by the total number of gallons, 13:

$$\frac{546 \text{ mi}}{13 \text{ gal}} = 42 \text{ mpg}.$$

The Jetta's expected average is 42 mi per gallon for highway driving.

Answers

Skill to Review:
1. 1473 **2.** $86.91

Do Exercises 1–4. ▶

In a *weighted average*, more importance, or *weight*, is assigned to some values than to others. For example, a course syllabus may include the following description:

COURSE COMPONENT	WEIGHT FOR GRADE
Quizzes	20
Homework	30
Tests	50

If Allison has scored 70% on quizzes, 100% on homework, and 92% on tests, she cannot calculate her course grade by averaging 70, 100, and 92, because each category is weighted differently. Instead, she must multiply each percentage by its weight, add the results, and divide by the total of the weights:

$$\text{Course grade} = \frac{70 \cdot 20 + 100 \cdot 30 + 92 \cdot 50}{20 + 30 + 50}$$

$$= \frac{9000}{100} = 90.$$

Allison's course grade is 90%.

A grade point average is another example of a weighted average.

EXAMPLE 3 *Grade Point Average.* In many schools, students are assigned grade point values for grades obtained. The **grade point average**, or **GPA**, is the average of the grade point values for each credit hour taken. At Meg's college, grade point values are assigned as follows:

A: 4.0 B: 3.0 C: 2.0 D: 1.0 F: 0.0.

Meg earned the following grades for one semester. What was her grade point average?

COURSE	GRADE	NUMBER OF CREDIT HOURS IN COURSE
Colonial History	B	3
Basic Mathematics	A	4
English Literature	A	3
French	C	4
Time Management	D	1

To find the GPA, we first multiply the grade point value for each grade by the number of credit hours in the course to determine the number of *quality points,* and then add. Here each grade is weighted by the number of credit hours in the course.

Colonial History	$3.0 \cdot 3 =$	9
Basic Mathematics	$4.0 \cdot 4 =$	16
English Literature	$4.0 \cdot 3 =$	12
French	$2.0 \cdot 4 =$	8
Time Management	$1.0 \cdot 1 =$	1
		46 (Total)

Find the average.

1. 14, 175, 36

2. 75, 36.8, 95.7, 12.1

3. In the first five games of the season, a basketball player scored 26, 21, 13, 14, and 23 points. Find the average number of points scored per game.

4. *Home-Run Batting Average.* Babe Ruth hit 714 home runs in 22 seasons in the major leagues. What was his average number of home runs per season? Round to the nearest tenth.

Source: Major League Baseball

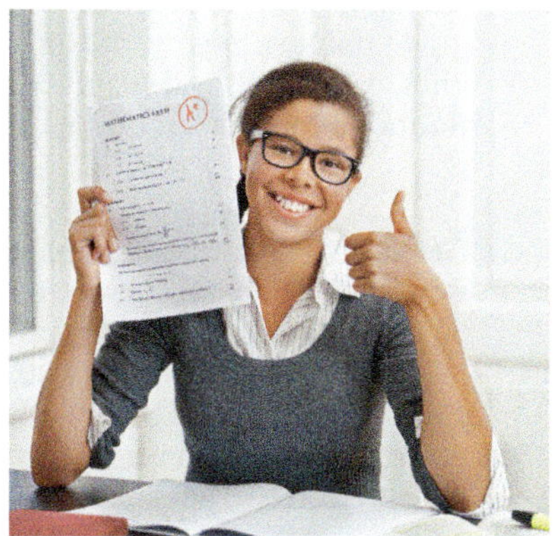

Answers

1. 75 2. 54.9 3. 19.4 points per game
4. 32.5 home runs per season

5. Soha's sociology professor included the following in the course syllabus:

COURSE COMPONENT	WEIGHT FOR GRADE
Participation	15
Book reports	25
Research paper	40

Soha received 88% on her research paper and 92% on her book reports, and she anticipates a score of 100% for participation. What is her course grade?

Course grade

$$= \frac{100 \cdot 15 + 92 \cdot \square + 88 \cdot \square}{15 + 25 + 40}$$

$$= \frac{7320}{\square} = \square$$

Soha's course grade is ____ %.

6. *Grade Point Average.* Alex earned the following grades one semester.

GRADE	NUMBER OF CREDIT HOURS IN COURSE
B	3
C	4
C	4
A	2

What was Alex's grade point average? Assume that the grade point values are 4.0 for an A, 3.0 for a B, and so on. Round to the nearest tenth.

7. *Grading.* To get an A in math, Rosa must score an average of 90 on four tests. On the first three tests, her scores were 80, 100, and 86. What is the lowest score that Rosa can get on the last test and still get an A?

Answers

5. 91.5% **6.** 2.5 **7.** 94

Guided Solution:

5. 25, 40, 80, 91.5; 91.5

The total number of credit hours taken is $3 + 4 + 3 + 4 + 1$, or 15. We divide the number of quality points, 46, by the number of credit hours, 15, and round to the nearest tenth:

$$\text{GPA} = \frac{46}{15} \approx 3.1.$$

Meg's grade point average was 3.1.

Do Exercises 5 and 6.

EXAMPLE 4 *Grading.* To get a B in math, Geraldo must score an average of 80 on five tests. On the first four tests, his scores were 79, 88, 64, and 78. What is the lowest score that Geraldo can get on the last test and still get a B?

We can find the total of the five scores needed as follows:

$$80 + 80 + 80 + 80 + 80 = 5 \cdot 80, \quad \text{or} \quad 400.$$

The total of the scores on the first four tests is

$$79 + 88 + 64 + 78 = 309.$$

Thus Geraldo needs to get at least

$$400 - 309, \quad \text{or} \quad 91$$

in order to get a B. We can check this as follows:

$$\frac{79 + 88 + 64 + 78 + 91}{5} = \frac{400}{5}, \quad \text{or} \quad 80.$$

Do Exercise 7.

b MEDIANS

Another type of center-point statistic is the *median.* Medians are useful when we wish to de-emphasize unusually extreme numbers. For example, suppose a small class scored as follows on an exam.

Jae:	78	Pat:	56
Jill:	81	Carmen:	84
Matt:	82		

Let's first list the scores in order from smallest to largest:

56, 78, 81, 82, 84.

↑ Middle score (81)

The middle score—in this case, 81—is called the **median**. Note that because of the extremely low score of 56, the average of the scores is 76.2. In this example, the median may be a more appropriate center-point statistic.

EXAMPLE 5 What is the median of this set of numbers?

99, 870, 91, 98, 106, 90, 98

We first rearrange the numbers in order from smallest to largest. Then we locate the middle number, 98.

90, 91, 98, 98, 99, 106, 870

↑ Middle number

The median is 98.

Do Exercises 8–10.

Find the median.

8. 17, 13, 18, 14, 19

9. 20, 14, 13, 19, 16, 18, 17

10. 78, 81, 83, 91, 103, 102, 122, 119, 88

MEDIAN

Once a set of data is listed in order, from smallest to largest, the **median** is the middle number if there is an odd number of data items. If there is an even number of items, the median is the number that is the average of the two middle numbers.

EXAMPLE 6 What is the median of this set of numbers?

69, 80, 61, 63, 62, 65

We first rearrange the numbers in order from smallest to largest. There is an even number of numbers. We look for the middle two, which are 63 and 65. The median is halfway between 63 and 65, the number 64.

61, 62, 63, 65, 69, 80 The average of the middle numbers is $\frac{63 + 65}{2} = \frac{128}{2}$, or 64.

The median is 64.

EXAMPLE 7 *Salaries.* The following are the salaries of the four highest-paid players in the National Hockey League. What is the median of the salaries?

PLAYER	SALARY
Sidney Crosby	$8,700,000
Alexander Ovechkin	9,500,000
Evgeni Malkin	8,700,000
Eric Staal	8,250,000

SOURCE: Forbes.com

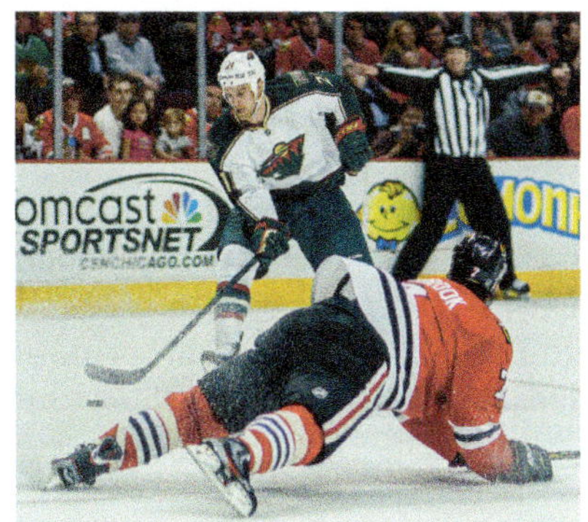

We rearrange the numbers in order from smallest to largest:

$8,250,000, $8,700,000, $8,700,000, $9,500,000.

The two middle numbers are $8,700,000 and $8,700,000. Since they are the same number, their average is $8,700,000. Thus the median salary is $8,700,000.

Do Exercises 11 and 12.

Find the median.

11. *Salaries of Part-Time Typists.* $3300, $4000, $3900, $3600, $3800, $3400

GS **12.** 68, 34, 67, 69, 34, 70

Rearrange the numbers in order from smallest to largest:

34, 34, ____, 68, ____, 70.

The middle numbers are ____ and 68.

The average of 67 and 68 is ____.

The median is ____.

Answers

8. 17 **9.** 17 **10.** 91 **11.** $3700 **12.** 67.5

Guided Solution:

12. 67, 69; 67, 67.5, 67.5

C MODES

The final type of center-point statistic we will consider is the *mode.*

MODE

The **mode** of a set of data is the number or numbers that occur most often. If each number occurs the same number of times, there is *no* mode.

EXAMPLE 8 Find the mode of these data.

17, 13, 18, 17, 14, 19

To find the mode, it is helpful to first rearrange the numbers in order from smallest to largest.

13, 14, 17, 17, 18, 19

The number that occurs most often is 17. Thus the mode is 17.

EXAMPLE 9 Find the mode of these data.

5, 5, 11, 11, 13, 13

The numbers in this set of data are 5, 11, and 13. Each occurs twice, so all the numbers are equally represented. There is *no mode.*

A set of data has just one average (mean) and just one median, but it can have more than one mode.

EXAMPLE 10 Find the modes of these data.

33, 34, 34, 34, 35, 36, 37, 37, 37, 38, 39, 40

There are two numbers that occur most often, 34 and 37. Thus the modes are 34 and 37.

Do Exercises 13–16.

Which statistic is best for a particular situation? If someone is bowling, the *average* from several games is a good indicator of that person's ability. If someone is applying for a job, the *median* salary at that business is often most indicative of what people are earning there because although executives tend to make a lot more money, there are fewer of them. For similar reasons, the selling price of homes is usually reported as a *median* price. Finally, if someone is reordering stock for a clothing store, the *mode* of the sizes sold is probably the most important statistic.

Find the modes of these data.

13. 23, 45, 45, 45, 78

14. 34, 34, 67, 67, 68, 70

15. 24, 89, 13, 28, 67, 27 GS

Rearrange the numbers in order from smallest to largest.

13, 24, ____, 28, 67, ____.

Each number occurs ____ time.

There is no mode.

16. In a lab, Gina determined the mass, in grams, of each of five eggs:

15 g, 19 g, 19 g, 14 g, 18 g.

a) What is the mean?

b) What is the median?

c) What is the mode?

Answers

13. 45 **14.** 34, 67 **15.** No mode exists.
16. **(a)** 17 g; **(b)** 18 g; **(c)** 19 g

Guided Solution:
15. 27, 89; one

7.1 Exercise Set

For Extra Help
MyMathLab® MathXL® PRACTICE WATCH READ REVIEW

✓ Reading Check

Complete each sentence with the appropriate word from the list on the right. Not all choices will be used.

RC1. A(n) ________________ is a number describing a set of data.

RC2. To find the ________________ of a set of numbers, add the numbers and then divide by the number of items of data.

RC3. To find the weighted average of a set of numbers, multiply each number by its ________________, add the results, and divide by the total of the weights.

RC4. The ________________ of a set of numbers is the number or numbers that occur most often.

average
median
mode
statistic
weight

a, b, c For each set of numbers, find the average, the median, and any modes that exist.

1. *Great Smoky Mountains National Park.* More people visit Great Smoky Mountains National Park each year than any other U.S. national park. The following bar graph shows the numbers of visitors the park had for 2005 to 2012. What is the average number of visitors for the 8 years? the median? the mode?

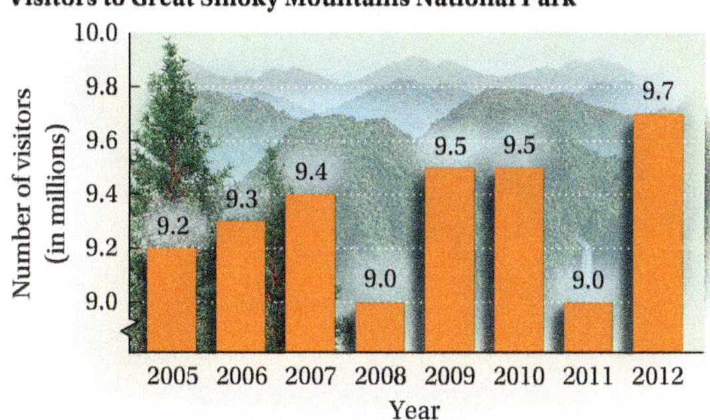

2. *Congestion.* The following bar graph shows the annual number of hours of traffic delay per auto commuter for 8 U.S. cities. What is the average delay time? the median? the mode?

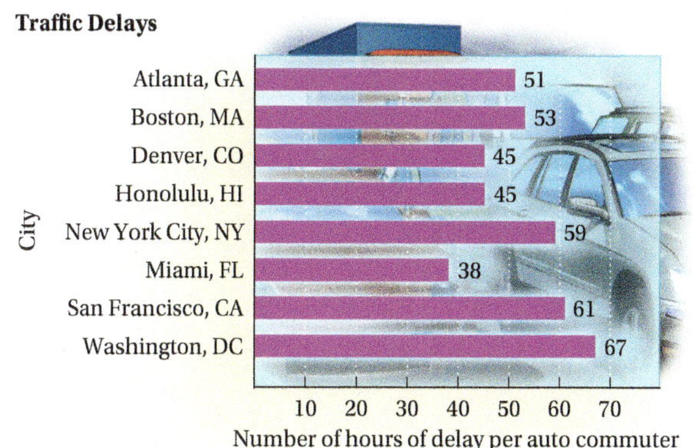

SOURCE: *2012 Annual Urban Mobility Report*, Texas A&M Transportation Institute

3. 17, 19, 29, 18, 14, 29

4. 72, 83, 85, 88, 92

5. 5, 37, 20, 20, 35, 5, 25

6. 13, 32, 25, 27, 13

7. 4.3, 7.4, 1.2, 5.7, 8.3

8. 13.4, 13.4, 12.6, 42.9

9. 234, 228, 234, 229, 234, 278

10. $29.95, $28.79, $30.62, $28.79, $29.95

11. ***Gas Mileage.*** The 2013 Kia Rio does 396 mi of highway driving on 11 gal of gasoline. What is the average number of miles expected per gallon—that is, what is the gas mileage?

Source: fueleconomy.gov

12. ***Gas Mileage.*** When using gas only, the 2013 Chevrolet Volt does 315 mi of city driving on 9 gal of gasoline. What is the average number of miles expected per gallon—that is, what is the gas mileage?

Source: fueleconomy.gov

Grade Point Average. The tables in Exercises 13 and 14 show the grades of a student for one semester. In each case, find the grade point average. Assume that the grade point values are 4.0 for an A, 3.0 for a B, and so on. Round to the nearest tenth.

13.

GRADE	NUMBER OF CREDIT HOURS IN COURSE
B	4
A	5
D	3
C	4

14.

GRADE	NUMBER OF CREDIT HOURS IN COURSE
A	5
C	4
F	3
B	5

15. ***Brussels Sprouts.*** The following prices per stalk of Brussels sprouts were found at five farmers' markets:

$3.99, $4.49, $4.99, $3.99, $3.49.

What was the average price per stalk? the median price? the mode?

16. ***Mangoes.*** The most popular fruit in the world is the mango, which is grown in over 2000 varieties. The following prices per pound of mangoes were found at five supermarkets:

$2.49, $1.59, $2.29, $2.49, $2.29.

What was the average price per pound? the median price? the mode?

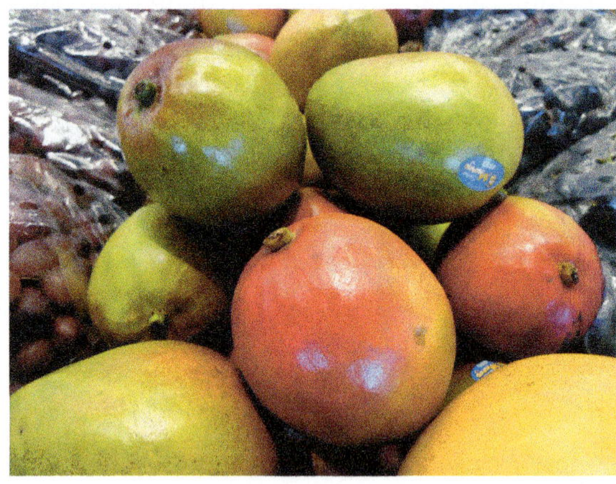

17. ***Grading.*** To get a B in math, Rich must score an average of 80 on five tests. His scores on the first four tests were 80, 74, 81, and 75. What is the lowest score that Rich can get on the last test and still receive a B?

18. ***Grading.*** To get an A in math, Cybil must score an average of 90 on five tests. Her scores on the first four tests were 90, 91, 81, and 92. What is the lowest score that Cybil can get on the last test and still receive an A?

19. ***Length of Pregnancy.*** Marta was pregnant 270 days, 259 days, and 272 days for her first three pregnancies. In order for Marta's average pregnancy to equal the worldwide average of 266 days, how long must her fourth pregnancy last?

Source: Vardaan Hospital, Dr. Rekha Khandelwal, M.S.

20. ***Male Height.*** Jason's brothers are 174 cm, 180 cm, 179 cm, and 172 cm tall. The average male is 176.5 cm tall. How tall is Jason if he and his brothers have an average height of 176.5 cm?

21. *Median Home Prices.* The following table lists the selling prices of homes in two counties during one month.

a) Find the median home price for each county.
b) Which county had the lower median home price?

JEFFERSON COUNTY	HAMILTON COUNTY
$122,587	$387,262
138,291	146,989
121,103	262,105
768,407	253,289
532,194	112,681
129,683	127,092
278,104	131,612
110,329	

22. *Median Salaries.* The following table lists salaries for two small companies.

a) Find the median salary for each company.
b) Which company has the higher median salary?

VALUE SERVICES	DEPENDABLE CARE
$ 48,267	$18,242
32,193	21,607
189,607	98,322
56,189	87,212
28,394	56,812
152,693	42,394
42,681	50,112
	52,987

23. *Movie Ticket Sales.* The following table lists the number of movie tickets sold annually, in billions, from 2002 to 2012.

a) Find the average number of tickets sold for the 8 years from 2002 to 2009.
b) Find the average number of tickets sold for the 8 years from 2005 to 2012.
c) On average, were more tickets sold per year from 2002 to 2009 or from 2005 to 2012?

YEAR	NUMBER OF MOVIE TICKETS SOLD (in billions)	YEAR	NUMBER OF MOVIE TICKETS SOLD (in billions)
2002	1.58	2008	1.39
2003	1.55	2009	1.42
2004	1.49	2010	1.33
2005	1.40	2011	1.30
2006	1.41	2012	1.37
2007	1.40		

SOURCE: the-numbers.com

24. *Movies Released.* The following table lists the number of movies released to U.S. theaters annually from 2005 to 2012.

a) Find the average number of movies released for the 5 years from 2005 to 2009.
b) Find the average number of movies released for the 5 years from 2008 to 2012.
c) On average, were more movies released from 2005 to 2009 or from 2008 to 2012?

YEAR	NUMBER OF MOVIES RELEASED	YEAR	NUMBER OF MOVIES RELEASED
2005	507	2009	557
2006	594	2010	563
2007	611	2011	609
2008	638	2012	677

SOURCE: mpaa.org

Skill Maintenance

Multiply.

25. 12.86×-17.5 [5.3a]

26. 222×0.5678 [5.3a]

27. $\left(-\frac{4}{5}\right)\left(-\frac{3}{28}\right)$ [3.6a]

28. $-\frac{28}{45} \cdot \frac{3}{2}$ [3.6a]

Synthesis

29. The ordered set of data 18, 21, 24, a, 36, 37, b has a median of 30 and an average of 32. Find a and b.

30. *Hank Aaron.* Hank Aaron averaged $34\frac{7}{22}$ home runs per year over a 22-year career. After 21 years, Aaron had averaged $35\frac{10}{21}$ home runs per year. How many home runs did Aaron hit in his final year?

31. *Price Negotiations.* Amy offers $6400 for a used Ford Taurus advertised at $8000. The first offer from Jim, the car's owner, is to "split the difference" and sell the car for $(6400 + 8000) \div 2$, or $7200. Amy's second offer is to split the difference between Jim's offer and her first offer. Jim's second offer is to split the difference between Amy's second offer and his first offer. If this pattern continues and Amy accepts Jim's third (and final) offer, how much will she pay for the car?

7.2 Interpreting Data from Tables and Graphs

OBJECTIVES

a Extract and interpret data from tables.

b Extract and interpret data from graphs.

c Extract and interpret data from histograms.

SKILL TO REVIEW

Objective 4.7a: Multiply using mixed numerals.

Multiply.

1. $3\frac{1}{2} \times 800$

2. $1\frac{3}{4} \times 3020$

We use tables and graphs to display data and to communicate information about the data. For example, the following table and graphs display data on the resting heart rate for several mammals. Examine each method of presentation. Which method do you like best, and why?

Table

	MOUSE	GIRAFFE	CAT	HUMAN	HORSE	ELEPHANT
Average Resting Heart Rate (in beats per minute)	500	170	130	70	35	28

SOURCES: elephantnaturepark.org; vetmedicine.about.com; giraffeconservation.org; learningabouthorses.com

Pictograph

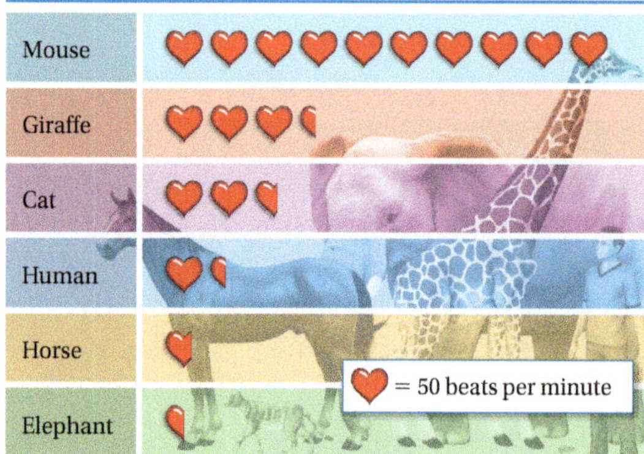

Bar Graph

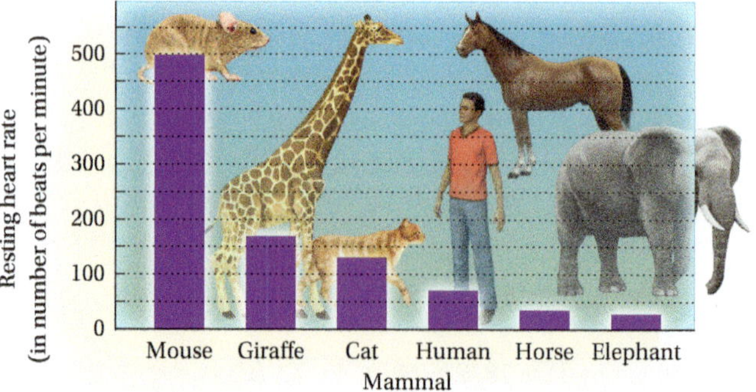

Answers

Skill to Review:
1. 2800 **2.** 5285

Comparing the table and the graphs reveals that the exact data values are most easily read in a table. The fastest and slowest heart rates can be determined easily from the graphs. The following graph communicates additional information about the data. The size of each mammal in this graph indicates the heart rate, not the actual size of the animal. The unexpected relative sizes of the mammals in the graph emphasize the fact that many smaller animals have faster heart rates than larger animals.

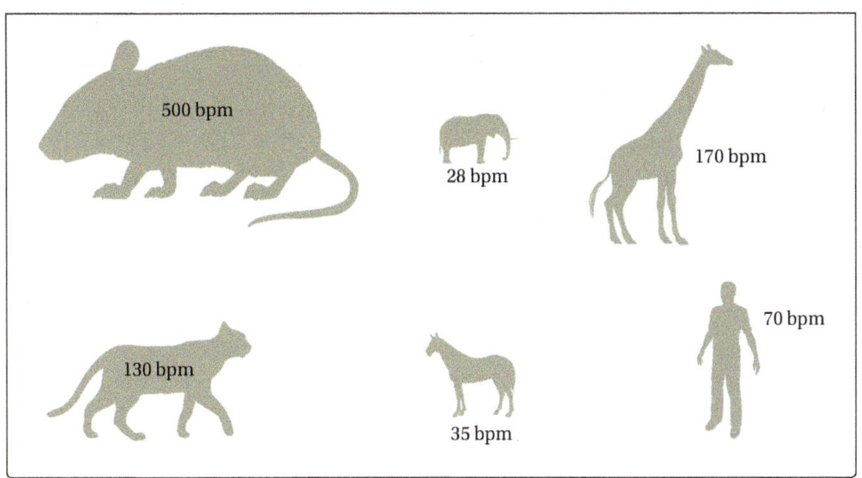

a READING AND INTERPRETING TABLES

A **table** is often used to present data in rows and columns.

EXAMPLE 1 *Population Density.* The following table lists populations and land areas of 10 countries.

COUNTRY	LAND AREA (in square miles)	POPULATION		POPULATION DENSITY (per square mile)	
		2008	2012	2008	2012
Australia	2,941,299	20,434,176	22,015,576	7	7
Brazil	3,265,077	190,010,647	199,321,413	58	61
China	3,600,947	1,321,851,888	1,343,239,923	367	373
Finland	117,558	5,238,460	5,262,930	45	45
Germany	134,836	82,400,996	81,305,856	611	603
India	1,147,955	1,129,866,154	1,205,073,612	984	1050
Japan	144,689	127,433,494	127,368,088	881	880
Kenya	219,789	36,913,721	43,013,341	168	196
Mexico	742,490	108,700,891	114,975,406	146	155
United States	3,537,439	301,139,947	313,847,465	85	89

SOURCES: *World Almanac 2008*; U.S. Census Bureau

a) Which country had the largest population in 2012?

b) In which country or countries did the population decrease from 2008 to 2012?

c) What was the percent increase in population density in India from 2008 to 2012?

d) Find the average land area of the four largest countries in the table.

Careful examination of the table allows us to answer the questions.

a) Note that the column head "Population" actually refers to two table columns. We look down the Population column headed "2012" and find the largest number. That number is 1,343,239,923. Then we look across that row to find the name of the country: China.

b) Comparing the Population columns headed "2008" and "2012," we see that the population decreased in the fifth row (from 82,400,996 to 81,305,856) and in the seventh row (from 127,433,494 to 127,368,088). Looking across these rows, we find the countries: Germany and Japan.

c) We look down the column headed "Country" and find India. Then we look across that row to the columns headed "Population Density." The population density of India in 2008 was 984 people per square mile. This increased to 1050 people per square mile in 2012. To find the percent increase, we find the amount of increase and divide by the population density in 2008.

Amount of Increase: $1050 - 984 = 66$

Percent Increase: $\dfrac{66}{984} \approx 0.067 = 6.7\%$

The population density of India increased by 6.7% from 2008 to 2012.

d) By looking down the column headed "Land Area," we determine that the four largest countries in the table are Australia, Brazil, China, and the United States. We find the average land area of these countries:

$$\frac{2{,}941{,}299 + 3{,}265{,}077 + 3{,}600{,}947 + 3{,}537{,}439}{4} = \frac{13{,}344{,}762}{4}$$
$$= 3{,}336{,}190.5 \text{ sq mi.}$$

Do Exercises 1–5.

Use the table in Example 1 to answer Margin Exercises 1–5.

1. Which country has the smallest land area?

2. What was the population of Mexico in 2012?

3. What was the percent decrease in population density in Germany from 2008 to 2012?

GS

The amount of the decrease in population density is

$611 - 603 = \square$.

The percent decrease is

$\dfrac{8}{\square} \approx 0.013$, or $\square$%.

4. Which country had the greatest increase in population from 2008 to 2012?

5. Find the median population density of these countries in 2012.

b READING AND INTERPRETING GRAPHS

Pictographs (or *picture graphs*) are another way to show information. Instead of actually listing the amounts to be considered, a **pictograph** uses symbols to represent the amounts. A pictograph includes a *key* that tells what each symbol represents.

Answers

1. Finland
2. 114,975,406
3. About 1.3%
4. India, with an increase of 75,207,458
5. 175.5 people per square mile

Guided Solution:

3. 8; 611, 1.3

EXAMPLE 2 *Roller Coasters.* The following pictograph shows the number of roller coasters listed in the Roller Coaster Data Base for six continents. Below the graph is a key that tells you that each represents 100 roller coasters.

Roller Coasters of the World

a) Which continent has the greatest number of roller coasters?

b) About how many roller coasters are there in Australia?

c) How many more roller coasters are there in Europe than in North America?

We can determine the answers by reading the pictograph.

a) The continent with the most symbols is Asia, so Asia has the greatest number of roller coasters.

b) The pictograph shows about $\frac{1}{4}$ symbol for Australia. Since each symbol represents 100 roller coasters, there are about $\frac{1}{4} \times 100$, or 25, roller coasters in Australia.

c) From the graph, we see that there are 8×100, or 800, roller coasters in Europe and about $7\frac{1}{2} \times 100$, or 750, roller coasters in North America. Thus there are $800 - 750$, or 50, more roller coasters in Europe than in North America. We could also estimate this difference by noting that Europe has $\frac{1}{2}$ of a symbol more than North America does, and $\frac{1}{2} \times 100 = 50$.

Do Exercises 6–8. ▶

When representing data with graphs, we must be sure that the areas of regions of the graph are proportional to the numbers that the regions represent. For example, in pictographs, each symbol is the same size, and the number of symbols is proportional to the actual data values. Thus the total area of the symbols is proportional to the data values. This area principle is illustrated in the following example.

Use the pictograph in Example 2 to answer Margin Exercises 6–8.

6. Which continent has the smallest number of roller coasters?

7. About how many roller coasters are there in Asia?

8. How many more roller coasters are there in South America than in Africa?

The graph shows $1\frac{1}{2}$ symbols for South America.

This represents ______ roller coasters.

The graph shows $\frac{1}{2}$ symbol for Africa.

This represents ______ roller coasters.

There are ______ more roller coasters in South America than in Africa.

Answers

6. Australia
7. 1400 roller coasters
8. About 100 more roller coasters

Guided Solution:
8. 150, 50, 100

EXAMPLE 3 *Electricity Generation.* The following graph illustrates the different methods used in the United States to generate electricity. Each yellow circle represents the amount of electricity generated, in terawatt-hours per year (TWh). Some methods of electricity generation are more efficient than others. The larger circle surrounding each yellow circle represents the amount of electricity that could have been generated if the method were 100% efficient.

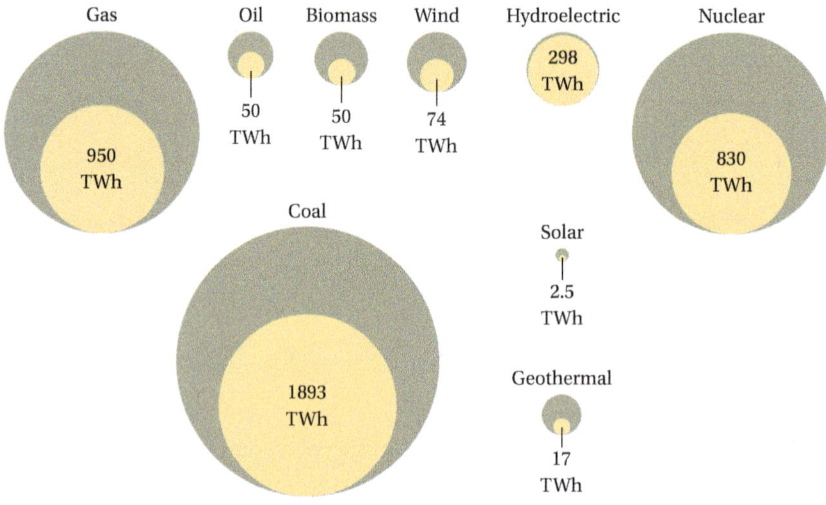

SOURCES: International Energy Agency; Eurelectric

a) How much electricity is generated annually with wind?

b) Which method generates the least electricity?

c) Which method of electricity generation is the most efficient?

d) Is generation of electricity from oil or from biomass more efficient?

We use the information in the graph to answer the questions. Note that the area of the larger circle that is not covered by the yellow circle represents the amount of energy lost or wasted during the generation process.

a) From the yellow circle labeled "Wind," we see that 74 TWh of electricity is generated annually with wind.

b) The smallest yellow circle is labeled "Solar," so the least electricity is generated by solar methods.

c) The yellow circle that most nearly fills its outer circle is labeled "Hydroelectric," so hydroelectric generation is the most efficient.

d) The same amount of electricity, 50 TWh, is generated using oil and biomass. Since the outer circle corresponding to oil is smaller than that corresponding to biomass, generation of electricity from oil is more efficient.

Do Exercises 9–12.

Use the graph in Example 3 to answer Margin Exercises 9–12.

9. How much electricity is generated annually from gas?

10. Which method generates the most electricity?

11. Solar, geothermal, hydroelectric, and wind are all considered renewable energy sources. How much electricity is generated annually from renewable sources?

12. Is nuclear generation of electricity more or less efficient than generation of electricity from gas?

Answers

9. 950 TWh
10. Coal
11. 391.5 TWh
12. Nuclear generation of electricity is less efficient than generation from gas.

C HISTOGRAMS

A **histogram** is a special kind of graph that shows how often certain numbers appear in a set of data. Such *frequencies* are usually considered in terms of ranges of values.

EXAMPLE 4 *Fuel Economy.* Listed below are the fuel economy ratings, in miles per gallon, for combined city and highway driving for all 2013 midsize car models sold in the United States.

23, 20, 21, 21, 28, 24, 21, 22, 20, 20, 21, 19, 28, 26, 24, 23, 24, 20, 17, 26,
19, 17, 16, 14, 13, 29, 29, 21, 21, 22, 24, 22, 22, 20, 23, 23, 22, 16, 14, 21,
21, 19, 22, 21, 29, 31, 30, 33, 30, 29, 27, 33, 31, 30, 28, 26, 24, 30, 28, 22,
23, 24, 32, 29, 27, 22, 17, 19, 21, 18, 17, 23, 24, 31, 32, 27, 13, 13, 47, 29,
28, 26, 26, 25, 28, 43, 43, 29, 28, 25, 25, 22, 30, 32, 32, 32, 31, 32, 30, 30,
20, 19, 18, 29, 21, 22, 20, 20, 21, 19, 18, 23, 26, 28, 28, 29, 29, 30, 26, 26,
23, 21, 31, 24, 40, 19, 18, 19, 18, 20, 26, 25, 21, 22, 45, 21, 25, 24, 31, 32,
21, 22, 23, 20, 19, 23, 22, 23, 22, 26, 25, 31, 25, 22, 30, 34, 33, 34, 24, 27,
20, 50, 50, 24, 28, 40, 41, 40, 25, 25, 34, 23, 35, 26, 25, 21, 23

It is difficult to make sense of the 177 numbers in this data set, so the data are displayed below in a histogram.

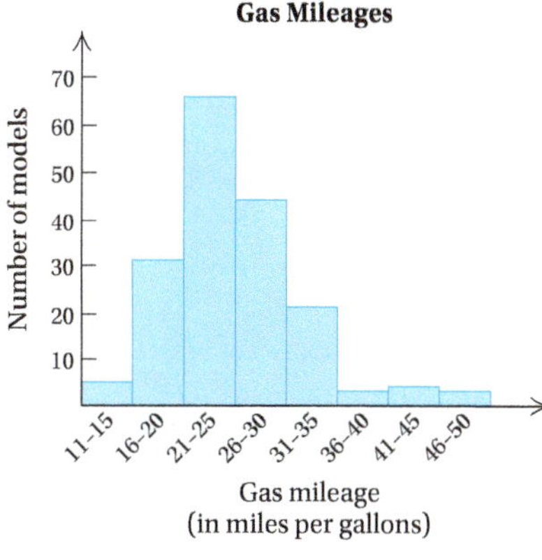

SOURCE: www.fueleconomy.gov

a) In which range of gas mileages did the greatest number of midsize models fall?

b) About how many midsize models had gas mileages that were less than 16 mpg?

c) About how many more midsize models had gas mileages between 26 mpg and 30 mpg than between 31 mpg and 35 mpg?

We use the histogram to answer the questions.

a) The tallest rectangle in the histogram is above the range 21–25, so the range 21 mpg to 25 mpg included the greatest number of midsize models.

b) The rectangle corresponding to 11–15 is 5 units high, so 5 midsize models had gas mileages that were less than 16 mpg.

c) From the histogram, we estimate that about 44 midsize models had gas mileages in the 26–30 range and about 21 models had gas mileages in the 31–35 range. Thus about 44 − 21, or 23, more midsize models had gas mileages between 26 mpg and 30 mpg than between 31 mpg and 35 mpg.

Do Exercises 13–15. ▶

The following histogram illustrates test grades for a class of 100 students. Use the histogram for Margin Exercises 13–15.

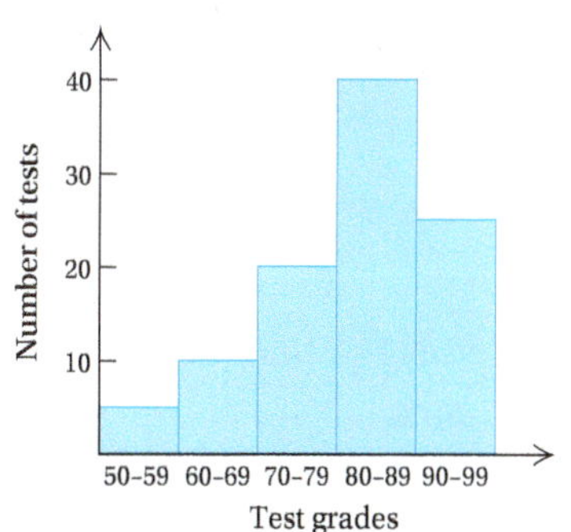

13. Which range of grades included the greatest number of students?

14. About how many students received a test grade between 90 and 99?

15. About how many more students received a grade between 90 and 99 than a grade between 50 and 59?

Answers

13. 80–89
14. 25 students
15. 20 students

7.2 Exercise Set

For Extra Help
MyMathLab® MathXL® PRACTICE WATCH READ REVIEW

Reading Check

Determine whether each statement is true or false.

RC1. There is only one correct way to represent a set of data.

RC2. It is usually easy to read exact amounts from a pictograph.

RC3. Histograms show frequencies.

RC4. If the same data were displayed in a table and in a pictograph, we would have to use the pictograph to determine a maximum or a minimum.

Heat Index. In warm weather, a person can feel hot because of reduced heat loss from the skin caused by higher humidity. The **temperature–humidity index**, or **apparent temperature**, is what the temperature would have to be with no humidity in order to give the same heat effect. The following table lists the apparent temperatures for various actual temperatures and relative humidities. Use this table for Exercises 1–12.

ACTUAL TEMPERATURE (°F)	RELATIVE HUMIDITY									
	10%	20%	30%	40%	50%	60%	70%	80%	90%	100%
	APPARENT TEMPERATURE (°F)									
75°	75	77	79	80	82	84	86	88	90	92
80°	80	82	85	87	90	92	94	97	99	102
85°	85	88	91	94	97	100	103	106	108	111
90°	90	93	97	100	104	107	111	114	118	121
95°	95	99	103	107	111	115	119	123	127	131
100°	100	105	109	114	118	123	127	132	137	141
105°	105	110	115	120	125	131	136	141	146	151

In Exercises 1–4, find the apparent temperature for the given actual temperature and humidity combinations.

1. 80°, 60%

2. 90°, 70%

3. 85°, 90%

4. 95°, 80%

5. Which temperature–humidity combinations give an apparent temperature of 100°?

6. Which temperature–humidity combinations give an apparent temperature of 111°?

7. At a relative humidity of 50%, what actual temperatures give an apparent temperature above 100°?

8. At a relative humidity of 90%, what actual temperatures give an apparent temperature above 100°?

9. At an actual temperature of 95°, what relative humidities give an apparent temperature above 100°?

10. At an actual temperature of 85°, what relative humidities give an apparent temperature above 100°?

11. At an actual temperature of 85°, what is the difference in humidities required to raise the apparent temperature from 94° to 108°?

12. At an actual temperature of 80°, what is the difference in humidities required to raise the apparent temperature from 87° to 102°?

Planets. Use the following table, which lists information about the planets, for Exercises 13–20.

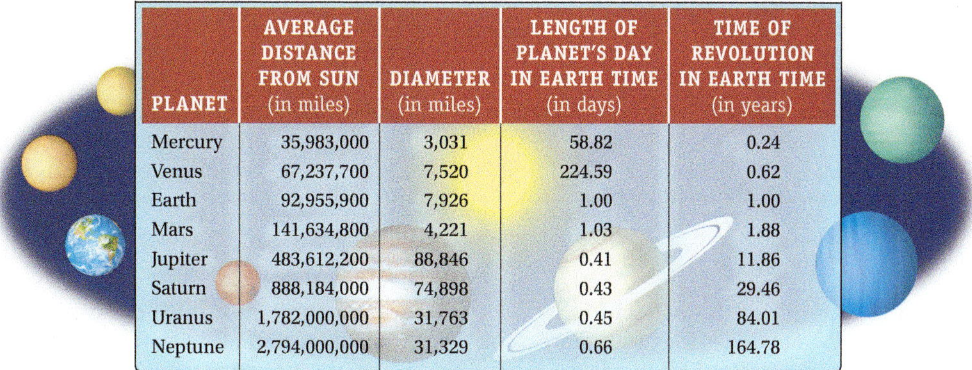

PLANET	AVERAGE DISTANCE FROM SUN (in miles)	DIAMETER (in miles)	LENGTH OF PLANET'S DAY IN EARTH TIME (in days)	TIME OF REVOLUTION IN EARTH TIME (in years)
Mercury	35,983,000	3,031	58.82	0.24
Venus	67,237,700	7,520	224.59	0.62
Earth	92,955,900	7,926	1.00	1.00
Mars	141,634,800	4,221	1.03	1.88
Jupiter	483,612,200	88,846	0.41	11.86
Saturn	888,184,000	74,898	0.43	29.46
Uranus	1,782,000,000	31,763	0.45	84.01
Neptune	2,794,000,000	31,329	0.66	164.78

SOURCE: *The Handy Science Answer Book*, Gale Research, Inc.

13. Find the average distance from the sun to Jupiter.

14. How long is a day on Venus?

15. Which planet has a time of revolution of 164.78 years?

16. Which planet has a diameter of 4221 mi?

17. About how many Earth diameters would it take to equal one Jupiter diameter?

18. How much longer is the longest time of revolution than the shortest?

19. What are the average, the median, and the mode of the diameters of the planets?

20. What are the average, the median, and the mode of the average distances from the sun of the planets?

Nutrition Facts. Most foods are required by law to provide factual information regarding nutrition, like that in the following table of nutrition facts from a box of Frosted Flakes cereal. Use the nutrition data for Exercises 21–26 on the next page.

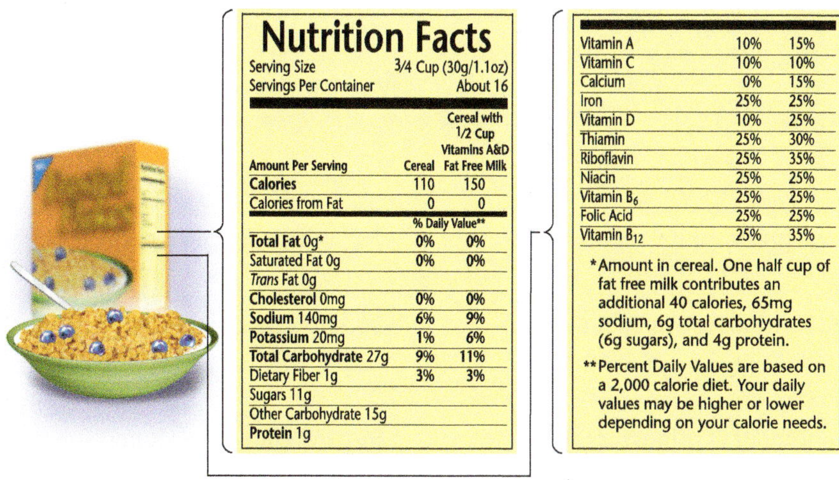

Nutrition Facts

Serving Size 3/4 Cup (30g/1.1oz)
Servings Per Container About 16

Amount Per Serving	Cereal	Cereal with 1/2 Cup Vitamins A&D Fat Free Milk
Calories	110	150
Calories from Fat	0	0
	% Daily Value**	
Total Fat 0g*	0%	0%
Saturated Fat 0g	0%	0%
Trans Fat 0g		
Cholesterol 0mg	0%	0%
Sodium 140mg	6%	9%
Potassium 20mg	1%	6%
Total Carbohydrate 27g	9%	11%
Dietary Fiber 1g	3%	3%
Sugars 11g		
Other Carbohydrate 15g		
Protein 1g		

Vitamin A	10%	15%
Vitamin C	10%	10%
Calcium	0%	15%
Iron	25%	25%
Vitamin D	10%	25%
Thiamin	25%	30%
Riboflavin	25%	35%
Niacin	25%	25%
Vitamin B_6	25%	25%
Folic Acid	25%	25%
Vitamin B_{12}	25%	35%

*Amount in cereal. One half cup of fat free milk contributes an additional 40 calories, 65mg sodium, 6g total carbohydrates (6g sugars), and 4g protein.

**Percent Daily Values are based on a 2,000 calorie diet. Your daily values may be higher or lower depending on your calorie needs.

SOURCE: © 2013 Kellogg North America Company

21. Suppose your morning bowl of cereal consists of $1\frac{1}{2}$ cups of Frosted Flakes with 1 cup of fat-free milk. How many calories do you consume?

22. Suppose your morning bowl of cereal consists of $1\frac{1}{2}$ cups of Frosted Flakes with 1 cup of fat-free milk. What percent of the daily value of dietary fiber do you consume?

23. A nutritionist recommends that you look for foods that provide 10% or more of the daily value of vitamin C. Do you get that with 1 serving of Frosted Flakes and $\frac{1}{2}$ cup of fat-free milk?

24. Suppose you are trying to limit your daily caloric intake to 2000 calories. How many servings of cereal alone would it take to exceed 2000 calories?

25. Suppose your morning bowl of cereal consists of $1\frac{1}{2}$ cups of Frosted Flakes with 1 cup of fat-free milk. How much sodium do you consume? (*Hint*: Use the data listed in the first footnote below the table of nutrition facts.)

26. Suppose your morning bowl of cereal consists of $1\frac{1}{2}$ cups of Frosted Flakes with 1 cup of fat-free milk. How much protein do you consume? (*Hint*: Use the data listed in the first footnote below the table of nutrition facts.)

Rhino Population. The rhinoceros is considered one of the world's most endangered animals. The worldwide total number of rhinoceroses is approximately 20,700. The following pictograph shows the populations of the five remaining rhino species. Located in the graph is a key that tells you that each symbol represents 300 rhinos. Use the pictograph for Exercises 27–32.

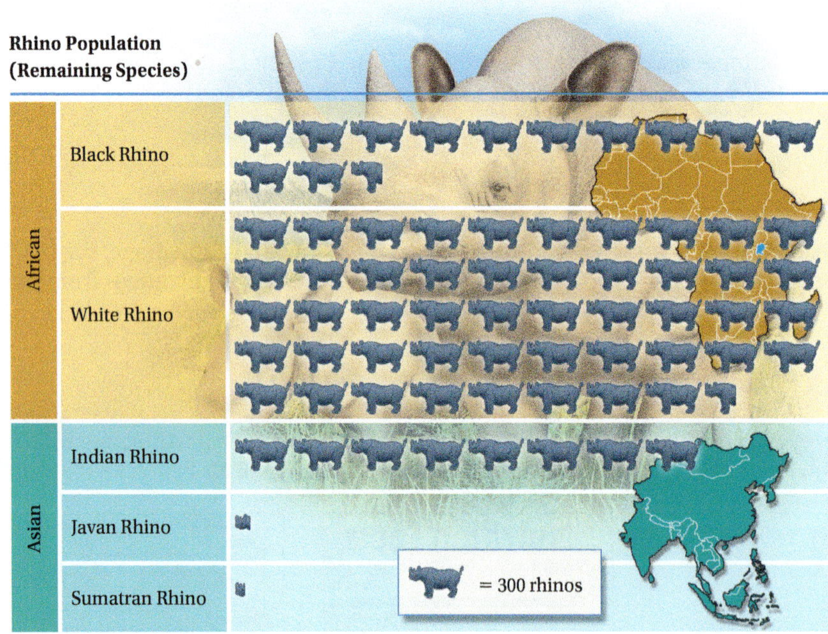

SOURCE: World Wildlife Fund, 2008

27. Which species has the greatest number of rhinos?

28. Which species has the least number of rhinos?

29. How many more black rhinos are there than Indian rhinos?

30. How many more white rhinos are there than black rhinos?

31. What is the average number of rhinos for the five species?

32. How does the white rhino population compare with the Indian rhino population?

Personal Consumption Expenditures. The following graph shows the amounts of personal consumption expenditures, in dollars per person per year, in the United States, for four years. The graph also shows the amounts spent on food and on financial services and insurance for those years, labeled as percents of the personal consumption expenditures. Use the graph for Exercises 33–40.

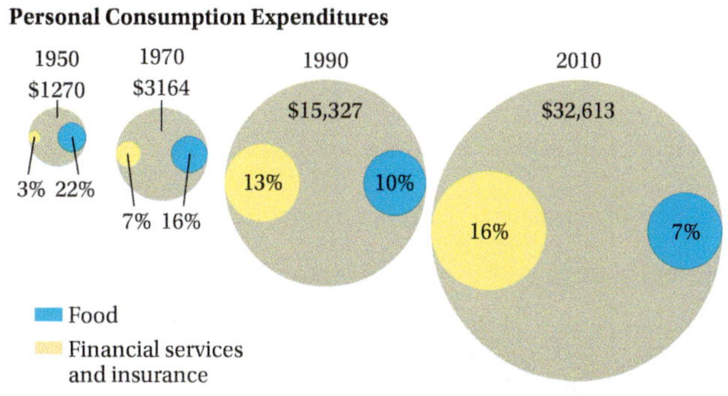

33. How much were personal consumption expenditures per person in 1950?

34. How much were personal consumption expenditures per person in 2010?

35. For which of the years shown was more spent on food than on financial services and insurance?

36. For which of the years shown was more spent on financial services and insurance than on food?

37. How much per person was spent on food in 1990?

38. How much per person was spent on financial services and insurance in 1970?

39. **a)** How much less, as a percent of personal consumption expenditures, was spent on food in 2010 than in 1950?
b) How much more, in dollars, was spent on food in 2010 than in 1950?

40. **a)** How much more, as a percent of personal consumption expenditures, was spent on financial services and insurance in 2010 than in 1950?
b) How much more, in dollars, was spent on financial services and insurance in 2010 than in 1950?

C

Basketball. The following histogram illustrates the number of points scored per game by the Los Angeles Lakers during the 2012–2013 regular basketball season. Use the graph for Exercises 41–44.

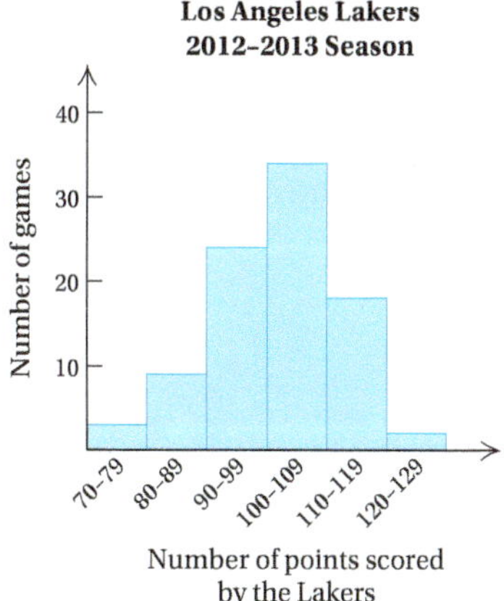

41. In how many games did the Lakers score 90–99 points?

42. In what point range did the highest number of Laker scores lie?

43. In what point range did the lowest number of Laker scores lie?

44. In how many more games did the Lakers score 100–109 points than 90–99 points?

Mid-Chapter Review

Concept Reinforcement

Determine whether each statement is true or false.

__________ **1.** A set of data has just one average and just one median, but it can have more than one mode. [7.1a, b, c]

__________ **2.** It is possible for the average, the median, and the mode of a set of data to be the same number. [7.1a, b, c]

__________ **3.** If there is an even number of items in a set of data, the middle number is the median. [7.1b]

Guided Solutions

Fill in each blank with the number that creates a correct solution.

4. The average of 60, 45, 115, 15, and 35 is

$$\frac{60 + 45 + \square + 15 + 35}{\square} = \frac{\square}{5} = \square. \quad [7.1a]$$

5. Find the median of this set of numbers:

2.1, 11.3, 8.7, 6.3, 14.5, 4.8. [7.1b]

We first arrange the numbers from smallest to largest:

$\square$, $\square$, 6.3, $\square$, 11.3, $\square$.

There is an even number of items. The median is the average of

$\square$ and $\square$.

We find that average:

$$\frac{\square + \square}{2} = \frac{\square}{2} = \square.$$

The median is $\square$.

Mixed Review

For each set of numbers, find the average, the median, and any modes that exist. [7.1a, b, c]

6. 56, 29, 45, 240, 175, 7, 29

7. 2.12, 18.42, 9.37, 43.89

8. $\frac{5}{9}, \frac{1}{9}, \frac{8}{9}, \frac{2}{9}, \frac{4}{9}$

9. 160, 102, 102, 116, 160, 116

10. \$4.96, \$5.24, \$4.96, \$10.05, \$5.24

11. $\frac{1}{2}, \frac{3}{4}, \frac{7}{8}, \frac{5}{4}$

12. 2, 5, 7, 7, 8, 5, 5, 7, 8

13. 38.2, 38.2, 38.2, 38.2

Downsizing. Companies sometimes downsize their products. That is, they charge the same price for a package that contains less product. The following table lists products that have been downsized. Use this table for Exercises 14–18. [7.2a]

PRODUCT	SIZE (in ounces)		PERCENT SMALLER
	OLD	NEW	
Breyer's ice cream	56	48	14%
Hellmann's mayonnaise	32	30	6
Hershey's Special Dark chocolate bar	8	6.8	15
Iams cat food	6	5.5	8
Nabisco Chips Ahoy cookies	16	15.25	5
Skippy creamy peanut butter	18	16.3	9
Tropicana orange juice	96	89	7

SOURCE: *Consumer Reports*

14. How much less ice cream is in the new Breyer's ice cream package than in the old package?

15. By what percent has the size of a jar of Hellmann's mayonnaise changed in the downsizing process?

16. Which product in the table showed the greatest percent decrease?

17. How much less orange juice is in the new Tropicana orange juice package than in the old package?

18. Which product in the table showed the smallest percent decrease?

Touchdown Passes. The following pictograph shows the career-high number of touchdown passes in one season for seven quarterbacks in the National Football League. Use the pictograph for Exercises 19–22. [7.2b]

Touchdown Passes (Career high for quarterback)

	PLAYER	YEAR	TOUCHDOWN PASSES
Retired	John Elway	1997	
	Dan Marino	1984	
	Joe Montana	1987	
	Roger Staubach	1978	
Active	Tom Brady	2007	
	Aaron Rodgers	2011	
	Peyton Manning	2004	

SOURCE: National Football League

= 10 touchdown passes

19. Which quarterback threw the greatest number of touchdown passes in one season?

20. About how many touchdown passes did Aaron Rodgers throw in 2011?

21. How many more touchdown passes does Peyton Manning have as his career high than John Elway?

22. What is the average career-high number of touchdown passes in one season for the seven quarterbacks?

Understanding Through Discussion and Writing

23. Is it possible for a driver to average 20 mph on a 30-mi trip and still receive a ticket for driving 75 mph? Why or why not? [7.1a]

24. You are applying for an entry-level job at a large firm. You can be informed of the mean, median, or mode salary. Which of the three figures would you request? Why? [7.1a, b, c]

STUDYING FOR SUCCESS *Making Positive Choices*

- ☐ Choose to improve your attitude and raise your goals.
- ☐ Choose to make a strong commitment to learning.
- ☐ Choose to take the primary responsibility for learning.
- ☐ Choose to allocate the proper amount of time to learn.

7.3 Interpreting and Drawing Bar Graphs and Line Graphs

OBJECTIVES

- **a** Extract and interpret data from bar graphs.
- **b** Draw bar graphs.
- **c** Extract and interpret data from line graphs.
- **d** Draw line graphs.

SKILL TO REVIEW

Objective 5.1c: Given a pair of numbers in decimal notation, tell which is larger.

Which number is larger?

1. 0.078, 0.1
2. 36.4, 9.875

a READING AND INTERPRETING BAR GRAPHS

A **bar graph** is convenient for showing comparisons because you can tell at a glance which quantity is the largest or smallest. A *scale* is usually included with a bar graph so that estimates of values can be made with some accuracy. Bar graphs may be drawn horizontally or vertically, and sometimes a double bar graph is used to make comparisons.

EXAMPLE 1 *Coffee and Tea Consumption.* The following horizontal bar graph is a double bar graph, showing per capita consumption, in pounds per person per year, of both coffee and tea for several countries.

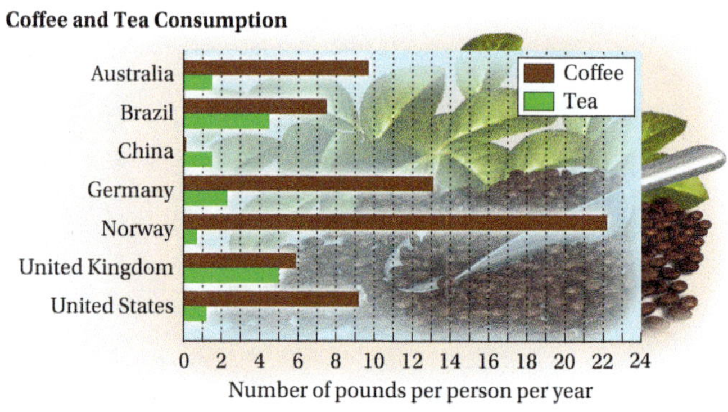

a) Which country has the highest per capita coffee consumption?

b) What is the per capita tea consumption in Brazil?

c) In which country do people consume about the same number of pounds of coffee and pounds of tea per year?

d) In which two countries do people consume about the same amount of tea?

e) In which countries is per capita coffee consumption greater than 10 pounds per year?

We use the graph to answer the questions.

a) The longest brown bar is for Norway. Thus Norway has the highest coffee consumption per capita.

b) We look to the right along the green bar associated with Brazil. Since it ends halfway between 4 and 5, we estimate Brazil's per capita tea consumption to be 4.5 pounds per year.

c) The brown and green bars representing data for the United Kingdom are about the same length, so people in the United Kingdom consume about the same number of pounds of coffee and pounds of tea per year.

Answers

Skill to Review:
1. 0.1 2. 36.4

d) The green bars are the same length for Australia and China; both countries have a per capita consumption of 1.5 pounds of tea. Thus people in Australia and China, on average, consume about the same amount of tea per year.

e) We move across the horizontal scale to 10. From there we move up, noting any brown bars that are longer than 10 units. We see that per capita coffee consumption is greater than 10 pounds per year in Germany and Norway.

Do Exercises 1–3. ▶

Use the bar graph in Example 1 to answer Margin Exercises 1–3.

1. What is the per capita coffee consumption in the United Kingdom?

2. In which countries is per capita tea consumption less than 2 pounds per year?

3. How many more pounds of coffee are consumed per person in the United States than pounds of tea?

b DRAWING BAR GRAPHS

EXAMPLE 2 *Population by Age.* Listed below are U.S. population data for selected age groups. Make a vertical bar graph of the data.

AGE GROUP	PERCENT OF POPULATION
5–17 years	17%
18 years and older	76
10–49 years	54
16–64 years	66
55 years and older	25
65 years and older	13
85 years and older	2

SOURCE: U.S. Census Bureau

First, we indicate the age groups in seven equally spaced intervals on the horizontal scale and give the horizontal scale the title "Age category." (See Figure 1 below.)

Next, we scale the vertical axis. To do so, we look over the data and note that it ranges from 2% to 76%. We start the vertical scaling at 0, labeling the marks by 10's from 0 to 80. We give the vertical scale the title "Percent of population" and the graph the overall title "U.S. Population by Age."

Finally, we draw vertical bars to show the various percents, as shown in Figure 2.

U.S. Population by Age

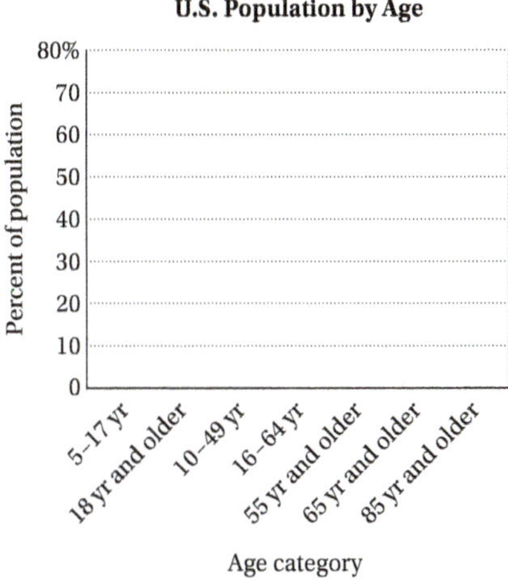

FIGURE 1

U.S. Population by Age

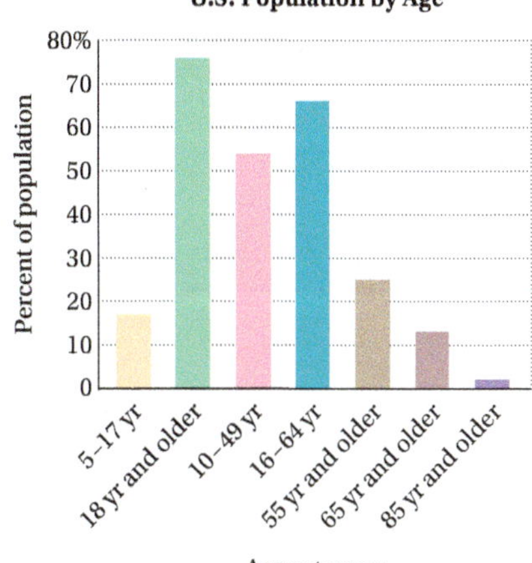

FIGURE 2

Do Exercise 4. ▶

4. *Planetary Moons.* Make a horizontal bar graph to show the numbers of moons orbiting the various planets.

PLANET	MOONS
Earth	1
Mars	2
Jupiter	63
Saturn	60
Uranus	27
Neptune	13

SOURCE: National Aeronautics and Space Administration

Answers

1. About 5.9 pounds per year
2. Australia, China, Norway, and the United States
3. About 8 pounds per person
4. Planetary Moons

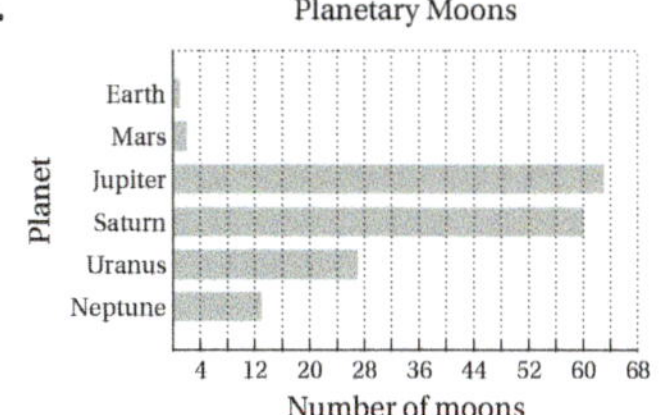

C READING AND INTERPRETING LINE GRAPHS

Line graphs are often used to show a change over time as well as to indicate patterns or trends.

EXAMPLE 3 *Gold.* The following line graph shows the average price of gold, in dollars per ounce, for various years from 1970 to 2012.

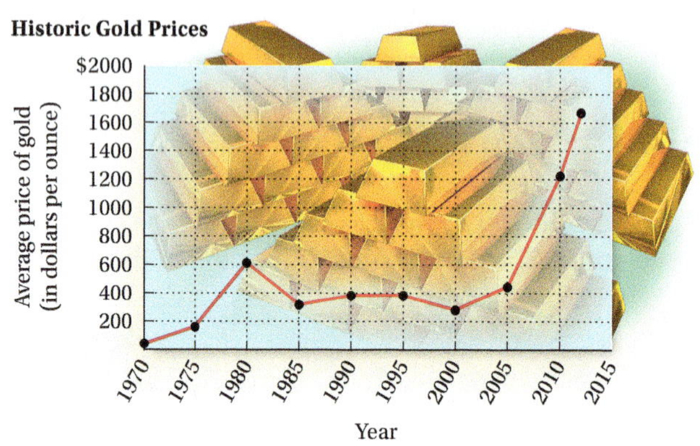

SOURCE: kitco.com

a) For which year before 2000 was the average price of gold the highest?

b) Between which year did the average price of gold decrease?

c) For which year was the average price of gold about $450 per ounce?

d) By how much did the average price of gold increase from 2010 to 2012?

We look at the graph to answer the questions.

a) Before 2000, the highest point on the graph corresponds to 1980. The highest average price of gold was about $610 per ounce in 1980.

b) Reading the graph from left to right, we see that the average price of gold decreased from 1980 to 1985 and from 1995 to 2000.

c) We look from left to right along a line at $450 per ounce. We see that the average price of gold was about $450 per ounce in 2005.

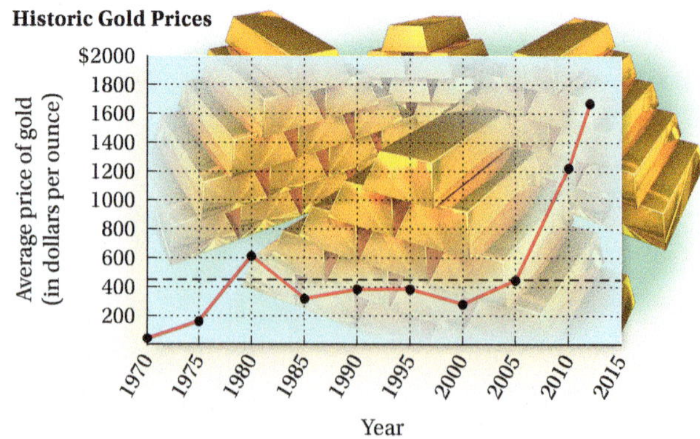

SOURCE: kitco.com

d) The graph shows that the average price of gold was about $1225 per ounce in 2010 and about $1675 per ounce in 2012. Thus the average price of gold increased by $\$1675 - \$1225 = \$450$ per ounce.

Do Exercises 5–7.

Use the line graph in Example 3 to answer Margin Exercises 5–7.

5. For which year after 1980 was the average price of gold the lowest?

6. Between which years did the average price of gold increase by about $800 per ounce?

7. For which years was the average price of gold less than $400 per ounce? GS

   We look from left to right along a line at $ ____ per ounce. The points on the graph that are below this line correspond to the years 1970, 1975, 1985, 1990, 1995, and ____ .

Answers

5. 2000
6. Between 2005 and 2010
7. 1970, 1975, 1985, 1990, 1995, 2000

Guided Solution:

7. 400; 2000

d DRAWING LINE GRAPHS

EXAMPLE 4 *Temperature in Enclosed Vehicle.* The temperature inside an enclosed vehicle increases rapidly with time. Listed in the following table are the inside temperatures of an enclosed vehicle for specified elapsed times when the outside temperature is 80°F. Make a line graph of the data.

ELAPSED TIME	TEMPERATURE IN ENCLOSED VEHICLE WITH OUTSIDE TEMPERATURE 80°F
10 min	99°
20 min	109°
30 min	114°
40 min	118°
50 min	120°
60 min	123°

SOURCES: General Motors; Jan Null, Golden Gate Weather Services

First, we indicate the 10-min elapsed time intervals on the horizontal scale and give the horizontal scale the title "Elapsed time (in minutes)." See the figure on the left below. Next, we scale the vertical axis by 10's beginning with 80 to show the number of degrees and give the vertical scale the title "Temperature (in degrees)." The jagged line at the base of the vertical scale indicates that an unused portion of the scale has been omitted. We also give the graph the overall title "Temperature in Enclosed Vehicle with Outside Temperature 80°F."

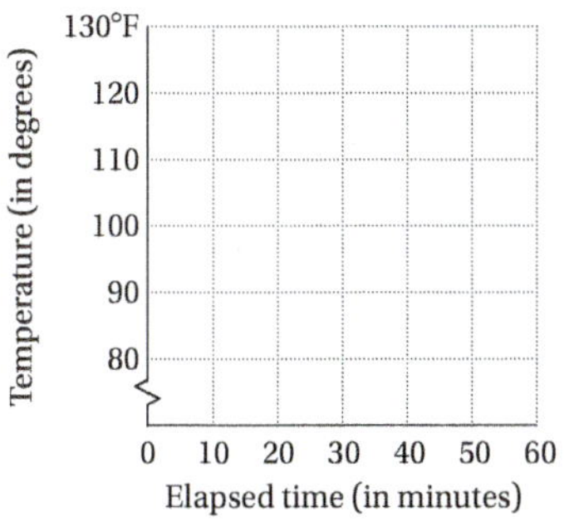

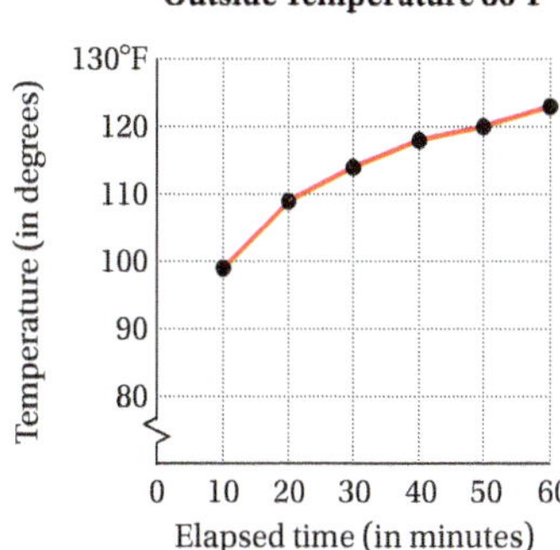

Next, we mark the temperature at the appropriate level above each elapsed time. (See the figure on the right above.) Then we draw line segments connecting the points. The rapid change in temperature can be observed easily from the graph.

Do Exercise 8. ▶

8. *Military Technologies.* Listed below are the numbers of bachelor's degrees in military technologies earned in the United States for the years 2003–2009. Make a line graph of the data.

YEAR	NUMBER OF DEGREES EARNED
2003	6
2004	10
2005	40
2006	33
2007	168
2008	39
2009	55

SOURCE: U.S. Census Bureau

Answer

8.

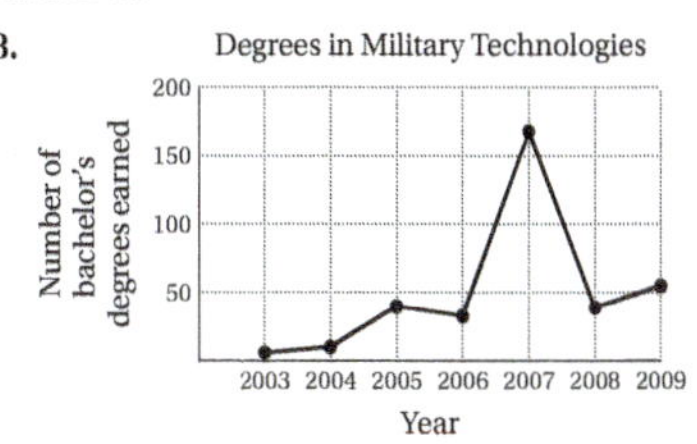

7.3 Exercise Set

For Extra Help MyMathLab® MathXL®

PRACTICE WATCH READ REVIEW

Reading Check

Determine whether each statement is true or false.

RC1. Bar graphs may be drawn horizontally or vertically.

RC2. A double bar graph indicates two amounts for each category.

RC3. A line graph is always used to show trends over time.

RC4. Some data could be illustrated using either a line graph or a bar graph.

a

Bearded Irises. A gardener planted six varieties of bearded iris in a new garden on campus. Students from the horticulture department were assigned to record data on the range of heights for each variety. The following vertical bar graph shows their results. The length of the light green shaded portion of each bar and the blossom illustrates the range of heights for a variety. For example, the range of heights for the miniature dwarf bearded iris is 2 in. to 9 in.

1. Which variety of iris has a minimum height of 17 in.?

2. Which variety of iris has a maximum height of 28 in.?

3. What is the range of heights for the border bearded iris?

4. What is the range of heights for the standard dwarf bearded iris?

5. Which variety of iris has the smallest range in heights?

6. Which irises have a maximum height less than 16 in.?

7. What is the difference between the maximum heights of the tallest iris and the shortest iris?

8. Which irises have a range in heights less than 10 in.?

SOURCE: www.irises.org/classification.htm

Chocolate Desserts. The following horizontal bar graph shows the average caloric content of various kinds of chocolate desserts. Use the bar graph for Exercises 9–16.

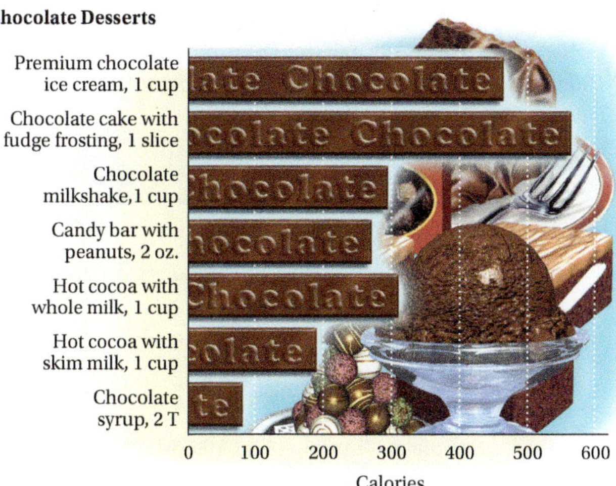

9. Estimate how many calories there are in 1 cup of hot cocoa with skim milk.

10. Estimate how many calories there are in a 2-oz candy bar with peanuts.

11. Which dessert has the highest caloric content?

12. Which dessert has the lowest caloric content?

13. Which dessert contains about 460 calories?

14. Which desserts contain about 300 calories?

15. How many more calories are there in 1 cup of hot cocoa made with whole milk than in 1 cup of hot cocoa made with skim milk?

16. If Emily drinks a 4-cup chocolate milkshake, how many calories does she consume?

Bachelor's Degrees. The graph at right provides data on the numbers of bachelor's degrees conferred on men and on women in selected years. Use the bar graph for Exercises 17–20.

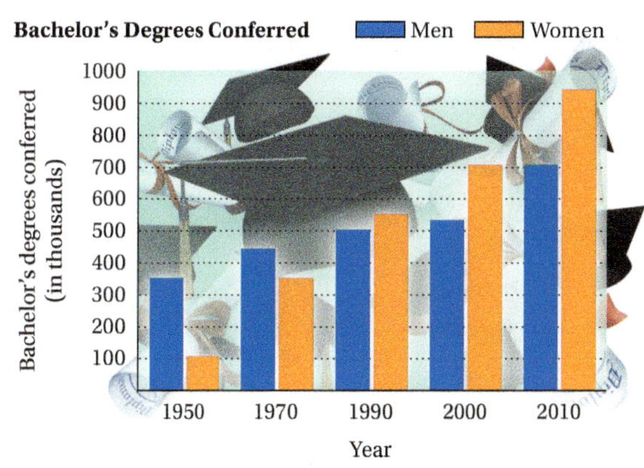

SOURCE: National Center for Education Statistics, U.S. Department of Education

17. In which years were more bachelor's degrees conferred on men than on women?

18. How many more bachelor's degrees were conferred on women in 2010 than in 1970?

19. How many more bachelor's degrees were conferred on women than on men in 2000?

20. In which years were the numbers of bachelor's degrees conferred on men and on women each greater than 500,000?

b

21. *Cost of Living Index.* The following table lists the cost of living index for several cities. The national average of this index is 100. An index greater than 100 indicates that the cost of living is higher than average, and an index less than 100 indicates that the cost of living is lower than average. Make a horizontal bar graph to illustrate the data.

CITY	COST OF LIVING INDEX
Chicago	116.9
Denver	99.4
New York City	185.8
Juneau	136.5
Indianapolis	87.2
San Diego	132.3
Salt Lake City	100.6

SOURCE: U.S. Census Bureau

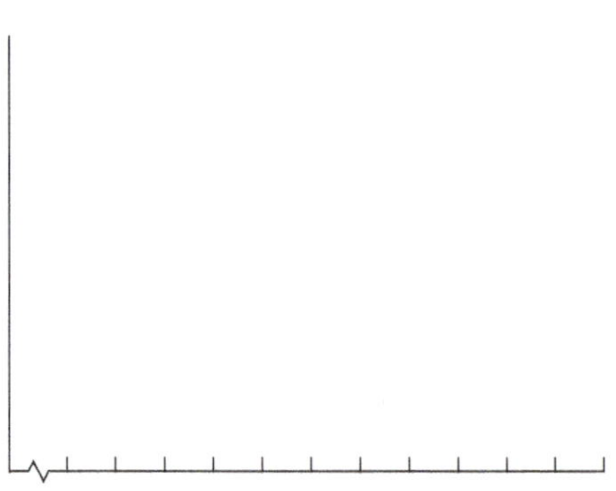

Use the data and the bar graph in Exercise 21 to do Exercises 22–25.

22. Which city has the highest cost of living index?

23. In which cities is the cost of living index less than 100?

24. In which cities is the cost of living approximately the national average?

25. How much higher is the cost of living index in New York City than in Chicago?

26. *Commuting Time.* The following table lists the average commuting time in six metropolitan areas with more than 1 million people. Make a vertical bar graph to illustrate the data.

CITY	COMMUTING TIME (in minutes)
New York City	38.3
Los Angeles	29.0
Phoenix	24.5
Houston	25.8
Indianapolis	21.7
Chicago	33.2

SOURCE: U.S. Census Bureau

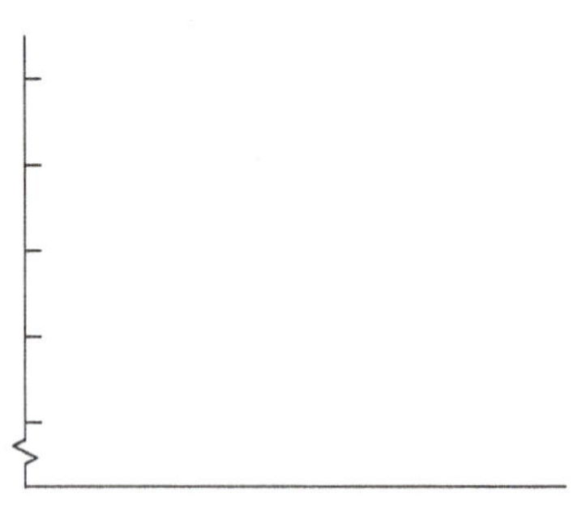

Use the data and the bar graph in Exercise 26 to do Exercises 27–30.

27. Which city has the longest commuting time?

28. Which city has the shortest commuting time?

29. What was the median commuting time for all six cities?

30. What was the average commuting time for the six cities?

C

Facebook Stock. The following line graph shows the price per share of Facebook stock when it was first offered in May 2012 and at the beginning of each month for the remainder of that year. Use the graph for Exercises 31–34.

31. Estimate the opening price per share of Facebook stock in May 2012.

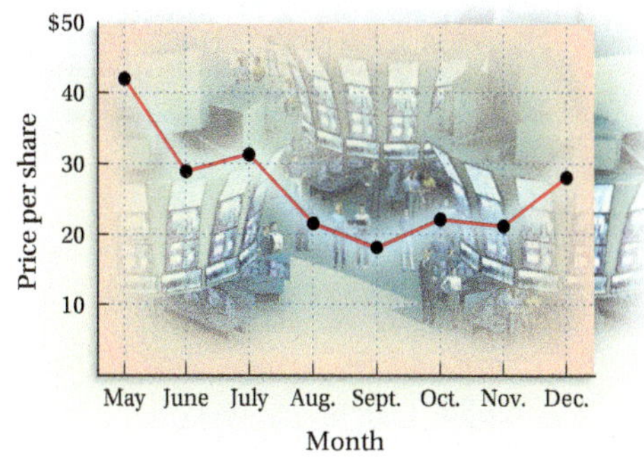

SOURCE: finance.yahoo.com

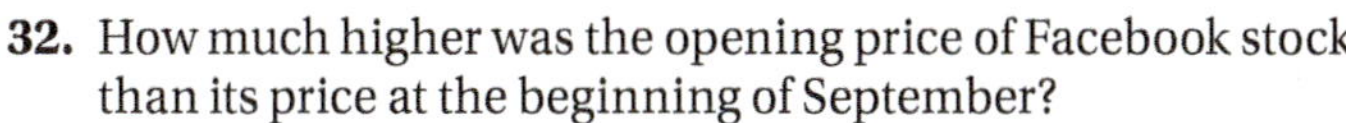

32. How much higher was the opening price of Facebook stock than its price at the beginning of September?

33. Between which months did the price of Facebook stock increase?

34. Between which months was the decrease in the price of Facebook stock the greatest?

Monthly Loan Payment. Suppose you borrow $110,000 at an interest rate of $5\frac{1}{2}\%$ to buy a condominium. The following graph shows the monthly payment required to pay off the loan, depending on the length of the loan. Use the graph for Exercises 35–42.

35. Estimate the monthly payment for a loan of 15 years.

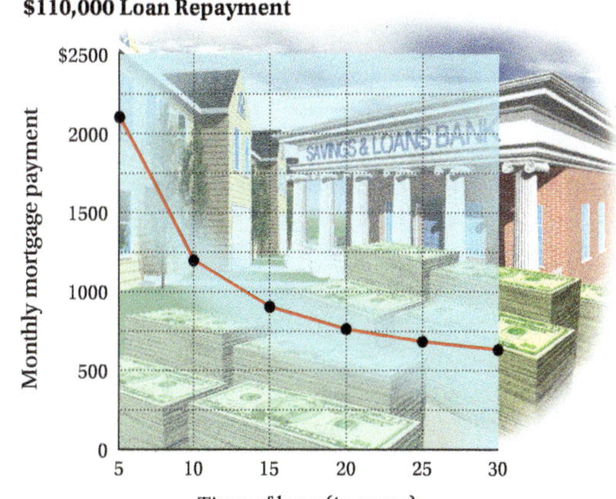

36. Estimate the monthly payment for a loan of 25 years.

37. What time period corresponds to a monthly payment of about $760?

38. What time period corresponds to a monthly payment of about $625?

39. By how much does the monthly payment decrease when the loan period is increased from 10 years to 20 years?

40. By how much does the monthly payment decrease when the loan period is increased from 5 years to 20 years?

41. For a 10-year loan, there are 120 monthly payments. In all, how much will you pay back for a 10-year loan?

42. For a 20-year loan, there are 240 monthly payments. In all, how much will you pay back for a 20-year loan?

d

43. *Longevity Beyond Age 65.* The data in the following table indicate how many years beyond age 65 a male who is 65 in the given year could expect to live. Draw a line graph using the horizontal axis to scale "Year."

YEAR	AVERAGE NUMBER OF YEARS MEN ARE ESTIMATED TO LIVE BEYOND AGE 65
1980	14
1990	15
2000	15.9
2010	16.4
2020	16.9
2030	17.5

SOURCE: 2000 Social Security Report

44. What was the percent increase in longevity (years beyond 65) between 1980 and 2000?

45. What is the expected percent increase in longevity between 1980 and 2030?

46. What is the expected percent increase in longevity between 2020 and 2030?

47. What is the expected percent increase in longevity between 2000 and 2030?

Skill Maintenance

Solve.

48. $32 + n = 115$ [1.7b]

49. $x \cdot \frac{2}{3} = -\frac{8}{9}$ [3.8a]

50. $y + \frac{5}{8} = \frac{11}{12}$ [4.3b]

51. $5 \cdot x = 11.3$ [5.7a]

52. $t + 4.752 = 11.1$ [5.7a]

53. $\frac{9}{10} = \frac{x}{8}$ [6.1d]

54. 51.2 is 64% of what? [6.4a, b], [6.5a, b]

55. What is $4\frac{1}{2}\%$ of 20? [6.4a, b], [6.5a, b]

56. 120 is what percent of 80? [6.4a, b], [6.5a, b]

Calculate.

57. $3 \times [11 + (18 - 10) \div 2^3 - 5]$ [1.9d]

58. $2.56 \div (4 - 3.84) + 6.3 \times 0.2$ [5.4b]

59. $\frac{9}{10} \div \frac{1}{2} \cdot \frac{1}{3} - \left(\frac{1}{4} - \frac{1}{6}\right)$ [4.8a]

60. $6.25 \times 7\frac{1}{5}$ [5.5d]

Interpreting and Drawing Circle Graphs

7.4

OBJECTIVES

a Extract and interpret data from circle graphs.

b Draw circle graphs.

SKILL TO REVIEW

Objective 6.3a: Convert from fraction notation to percent notation.

Find percent notation.

1. $\frac{7}{100}$ 2. $\frac{81}{100}$

We often use **circle graphs**, also called **pie charts**, to show the percent of a quantity in each of several categories. Circle graphs can also be used very effectively to show visually the *ratio* of one category to another. In either case, it is quite often necessary to use mathematics to find the actual amounts represented for each specific category.

a READING AND INTERPRETING CIRCLE GRAPHS

EXAMPLE 1 *Endangered Species.* According to the International Union for Conservation of Nature, seven species of whales are endangered or near-threatened. The following circle graph shows the approximate percentage of the entire population of endangered or near-threatened whales that each species represents.

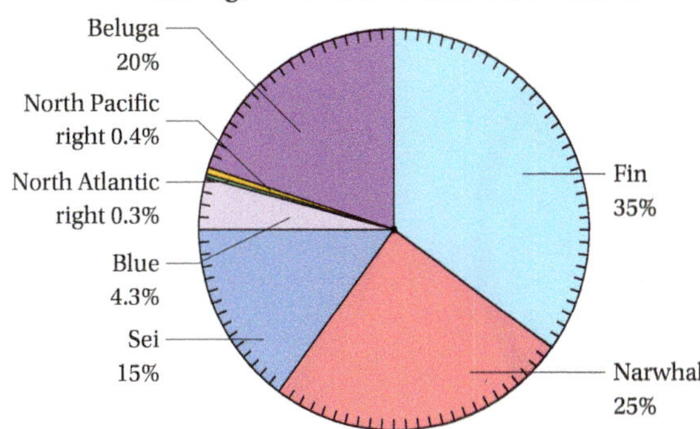

a) Which species has the greatest population?

b) Which species accounts for 25% of the entire population of endangered or near-threatened whales?

c) The total number of whales in these seven species is about 300,000. How many blue whales are there?

d) What percent of the population of endangered or near-threatened whales are right whales?

We look at the sections of the graph to find the answers.

a) The largest section (or *sector*) of the graph represents 35% of the population and corresponds to fin whales.

b) The narwhal accounts for 25% of the endangered or near-threatened whales.

c) The section representing blue whales is 4.3% of the circle. Since 4.3% of 300,000 is 12,900, there are approximately 12,900 blue whales.

d) There are two kinds of right whales represented on the graph: North Pacific right whales and North Atlantic right whales. We add the percents corresponding to these whales:

$$0.4\% + 0.3\% = 0.7\%.$$

Do Margin Exercises 1–3.

Use the circle graph in Example 1 to answer Margin Exercises 1–3.

1. Which species accounts for 20% of the entire population of endangered or near-threatened whales?

2. What percent of the population of endangered or near-threatened whales are fin whales or sei whales?

3. The total number of whales in these seven species is about 300,000. How many fin whales are there?

Answers

Skill to Review:
1. 7% 2. 81%

Margin Exercises:
1. Beluga whales 2. 50% 3. 105,000 whales

b DRAWING CIRCLE GRAPHS

EXAMPLE 2 *Education.* The following list shows the percents of students in the United States enrolled in different levels and types of schools. Use this information to draw a circle graph.

Source: *The 2012 Statistical Abstract*, U.S. Census Bureau

Grades K–8, Public:	46%
Grades K–8, Private:	5%
Grades 9–12, Public:	19%
Grades 9–12, Private:	2%
College, Public:	20%
College, Private:	8%

Using a circle with 100 equally spaced tick marks, we start with the 46% of students who are in grades K–8 in public schools. We draw a line from the center of the circle to any tick mark. Then we count off 46 ticks and draw another line. We label the wedge as shown in the figure on the left below. To distinguish the sectors, we can use different colors. We choose blue for this first sector.

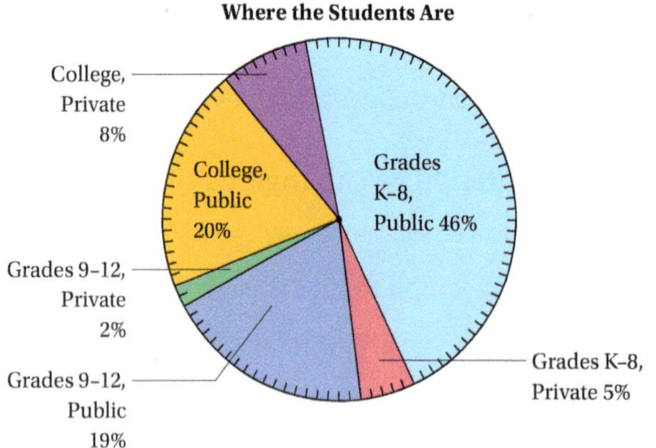

To draw a wedge for the 5% of students who are in grades K–8 in private schools, we start at one side of the wedge for 46%, count off 5 ticks, and draw another line. We label this second wedge as shown in the figure on the right above and shade it using a different color. Continuing in this manner, we obtain the final graph, at the lower middle above, to which we give the overall title "Where the Students Are."

◀ **Do Exercise 4.**

4. *Lengths of Engagement of Married Couples.* The following table lists the percents of married couples who were engaged for certain periods of time before marriage. Use this information to draw a circle graph.

ENGAGEMENT PERIOD	PERCENT
Less than 1 year	24%
1–2 years	21
More than 2 years	35
Never engaged	20

SOURCE: Bruskin Goldring Research

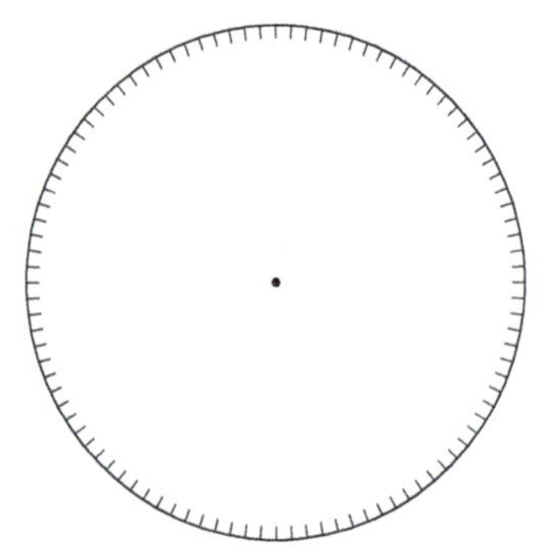

Answer

4.

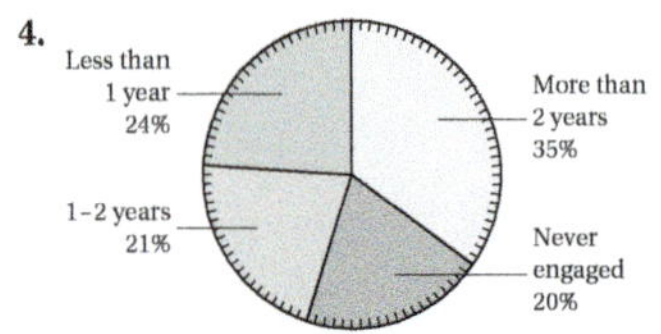

Translating for Success

1. *Vacation Miles.* The Saenz family drove their new van 13,640.8 mi in the first year. Of this total, 2018.2 mi were driven while on vacation. How many nonvacation miles did they drive?

2. *Rail Miles.* Of the recent $15\frac{1}{2}$ million passenger miles on a rail passenger line, 80% were transportation-to-work miles. How many rail miles, in millions, were to and from work?

3. *Sales Tax Rate.* The sales tax on the purchase of 10 bath towels that cost \$129.50 is \$8.42. What is the sales tax rate?

4. *Water Level.* During heavy rains in early spring, the water level in a pond rose 0.5 in. every 35 min. How much did the water rise in 90 min?

5. *Marathon Training.* At one point in his daily training routine for a marathon, Rocco had run $15\frac{1}{2}$ mi. This was 80% of the distance he intended to run that day. How far did Rocco plan to run?

The goal of these matching questions is to practice step (2), Translate, of the five-step problem-solving process. Translate each word problem to an equation and select a correct translation from equations A–O.

A. $8.42 \cdot x = 129.50$

B. $x = 80\% \cdot 15\frac{1}{2}$

C. $x = \dfrac{84 - 68}{84}$

D. $2018.2 + x = 13{,}640.8$

E. $\dfrac{5}{100} = \dfrac{x}{3875}$

F. $2018.2 = x \cdot 13{,}640.8$

G. $4\frac{1}{6} \cdot 73 = x$

H. $\dfrac{x}{5} = \dfrac{100}{3875}$

I. $15\frac{1}{2} = 80\% \cdot x$

J. $8.42 = x \cdot 129.50$

K. $\dfrac{0.5}{35} = \dfrac{x}{90}$

L. $x \cdot 4\frac{1}{6} = 73$

M. $x = \dfrac{84 - 68}{68}$

N. $x = 8.42\% \cdot 129.50$

O. $0.5 \times 35 = 90 \cdot x$

Answers on page A-13

6. *Vacation Miles.* The Ning family drove 2018.2 mi on their summer vacation. If they put a total of 13,640.8 mi on their new van during that year, what percent were vacation miles?

7. *Sales Tax.* The sales tax rate is 8.42%. Salena purchased 10 pillows at \$12.95 each. How much tax was charged on this purchase?

8. *Charity Donations.* Rachel donated \$5 to her favorite charity for each \$100 she earned. One month, she earned \$3875. How much did she donate that month?

9. *Tuxedos.* Emil Tailoring Company purchased 73 yd of fabric for a new line of tuxedos. How many tuxedos can be produced if it takes $4\frac{1}{6}$ yd of fabric for each tuxedo?

10. *Percent Increase.* In a calculus-based physics course, Mime got 68% on the first exam and 84% on the second. What was the percent increase in her score?

7.4 Exercise Set

For Extra Help MyMathLab® MathXL® PRACTICE WATCH READ REVIEW

Reading Check

The following statements refer to the graph at right. Determine whether each statement is true or false.

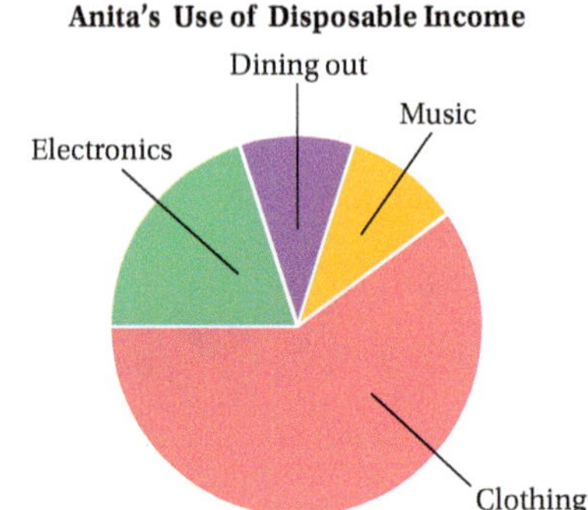

RC1. The graph is an example of a circle graph, or pie chart.

RC2. Anita spent 100% of her disposable income on music, clothing, electronics, and dining out.

RC3. Anita spent more than half of her disposable income on clothing.

RC4. Anita spent about $\frac{1}{4}$ of her disposable income on music.

RC5. Anita spent about the same amount on electronics as she spent on dining out and music combined.

RC6. If Anita has $100 in disposable income, she spends about $50 on electronics.

a *Foreign Students.* The following circle graph shows the foreign countries sending the most students to the United States to attend colleges and universities. Use this graph for Exercises 1–6.

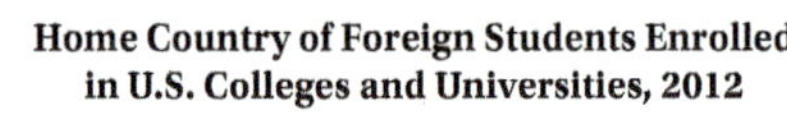

Home Country of Foreign Students Enrolled in U.S. Colleges and Universities, 2012

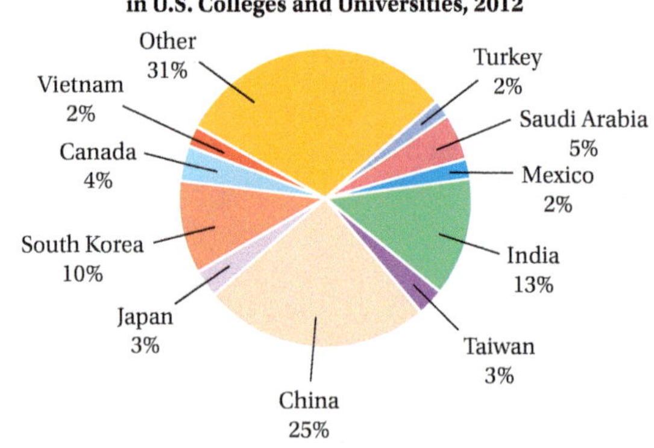

SOURCE: Institute of International Education

1. What percent of foreign students are from South Korea?

2. Together, what percent of foreign students are from China and Taiwan?

3. In 2012, there were approximately 760,000 foreign students studying at colleges and universities in the United States. According to the data in the graph, how many were from India?

4. In 2012, there were approximately 760,000 foreign students studying in the United States. How many were from Saudi Arabia?

5. Which country accounted for 4% of the foreign students?

6. Which country accounted for 13% of the foreign students?

b In Exercises 7–10, use the given information to complete a circle graph. Note that each circle is divided into 100 sections.

7. *Fruit Juice Sales.* The following table lists the percentages of various kinds of fruit juice sold.

FRUIT JUICE	PERCENT
Apple	14%
Orange	56
Blends	6
Grape	5
Grapefruit	4
Prune	1
Other	14

SOURCE: Beverage Marketing Corporation

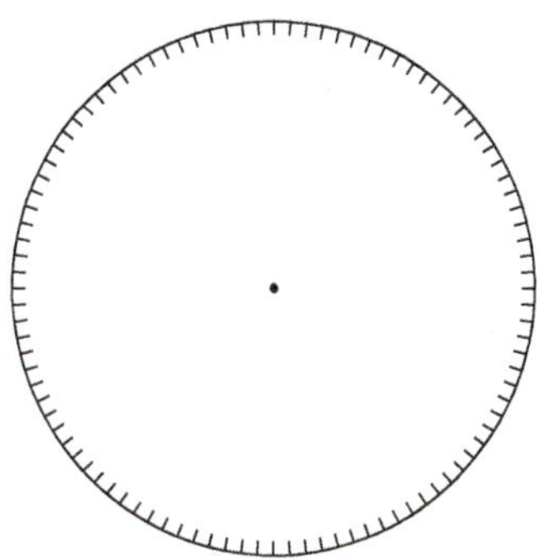

8. *Population of Continents.* The following table lists the percentage of the world population on each continent.

CONTINENT	PERCENT
Africa	15%
Asia	60
Europe	10
Oceania, includes Australia	1
North America	9
South America	5

SOURCE: Population Division/International Programs Center, U.S. Census Bureau

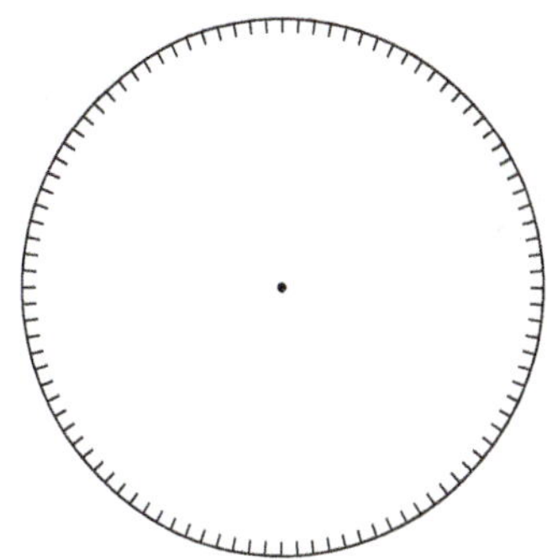

9. *Substance Abuse.* The following table lists the types of substances abused by those ages 12 and older who were admitted to substance abuse programs in the United States.

PRIMARY SUBSTANCE(S)	PERCENT
Drugs only	38%
Alcohol only	24
Alcohol with one drug	23
Alcohol with two drugs	14
No primary substance	1

SOURCE: Substance Abuse and Mental Health Services Administration

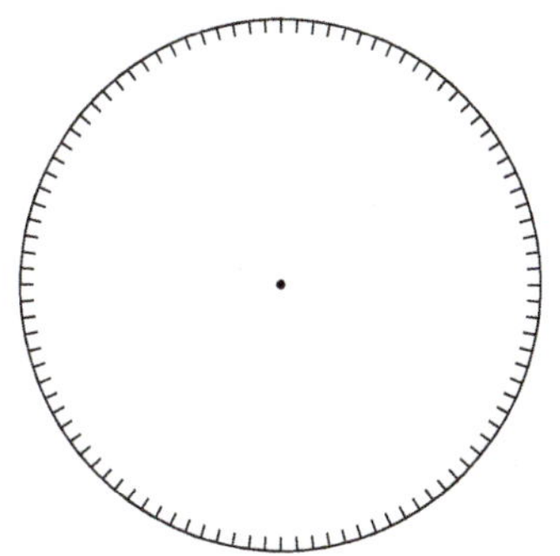

10. *Causes of Spinal Cord Injuries.* The following table lists the causes of spinal cord injury.

CAUSES	PERCENT
Motor vehicle accidents	44%
Acts of violence	24
Falls	22
Sports	8
Other	2

SOURCE: National Spinal Cord Injury Association

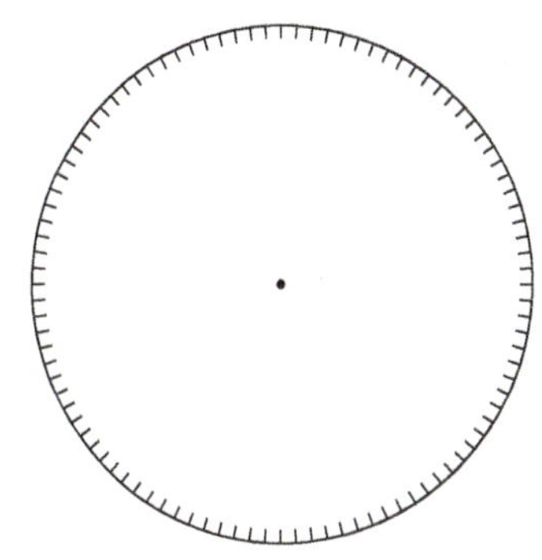

CHAPTER 7

Summary and Review

Vocabulary Reinforcement

Complete each statement with the correct term from the list on the right. Some of the choices may not be used.

1. A(n) ______________ presents data in rows and columns. [7.2a]
2. A(n) ______________ illustrates category percentages using different sized sectors or wedges. [7.4a]
3. A(n) ______________ uses symbols to represent amounts. [7.2b]
4. The ______________ of a set of data is the number or numbers that occur most often. [7.1c]
5. The ______________ of a set of data is the sum of the numbers in the set divided by the number of items of data. [7.1a]
6. The ______________ of an ordered set of data is the middle number, or the average of the middle numbers if there is an even number of items of data. [7.1b]

statistic
average
median
mode
table
pictograph
histogram
bar graph
circle graph
line graph

Concept Reinforcement

Determine whether each statement is true or false.

______________ **1.** To find the average of a set of numbers, add the numbers and then multiply by the number of items of data. [7.1a]

______________ **2.** If each number in a set of data occurs the same number of times, there is no mode. [7.1c]

______________ **3.** If there is an odd number of items in a set of data, the middle number is the median. [7.1b]

Study Guide

Objectives 7.1a, b, c Find the average, the median, and the mode of a set of numbers.

Example Find the average, the median, and the mode of this set of numbers:

2.6, 3.5, 61.8, 10.4, 3.5, 21.6, 10.4, 3.5.

Average: We add the numbers and divide by the number of data items:

$$\frac{2.6 + 3.5 + 61.8 + 10.4 + 3.5 + 21.6 + 10.4 + 3.5}{8}$$

$$= 14.6625.$$

Median: We first rearrange the numbers from smallest to largest:

2.6, 3.5, 3.5, 3.5, 10.4, 10.4, 21.6, 61.8.

The median is halfway between the middle two, which are 3.5 and 10.4. The average of these middle numbers is 6.95.

Mode: The number that occurs most often is 3.5, so it is the mode.

Practice Exercise

1. Find the average, the median, and the mode of this set of numbers:

8, 13, 1, 4, 8, 7, 15.

Objective 7.2a Extract and interpret data from tables.

Example The following table lists comparative information for oatmeal sold by six companies.

PRODUCT	PER PACKET (instant) OR SERVING (longer-cooking)					
	Cost	Calories	Fat (g)	Fiber (g)	Sugars (g)	Sodium (mg)
Quaker Quick-1 Minute	0.19	150	3.0	4	1	0
Market Pantry Maple & Brown Sugar	0.17	160	2.0	3	13	240
365 Organic Maple Spice	0.42	150	1.5	3	13	200
Kashi Heart to Heart Golden Brown Maple	0.44	160	2.0	5	12	100
McCann's Irish Maple & Brown Sugar	0.45	160	2.0	3	13	240
Quaker Organic Maple & Brown Sugar	0.54	150	2.0	3	12	95
Nature's Path Organic Maple Nut	0.47	200	4.0	4	12	105

SOURCE: *Consumer Reports*, November 2008

a) Which oatmeal has the greatest number of calories?
b) How much sodium is in a serving of Market Pantry Maple & Brown Sugar?

An examination of the table will give the answers.

a) We look down the column headed "Calories" and find the largest number. That number is 200. Then we look left across that row to find the name of the oatmeal: Nature's Path Organic Maple Nut.
b) We look down the column of products and find Market Pantry. Then we move right across that row to the column headed "Sodium" and find the amount of sodium: 240 mg.

Practice Exercises

Use the table in the example shown above for Exercises 2 and 3.

2. Which oatmeal has the greatest cost per serving? What is that cost?

3. How many grams of sugar are in the Kashi oatmeal?

Objective 7.3a Extract and interpret data from bar graphs.

Example The following horizontal bar graph shows the building costs of selected stadiums. When comparing the costs, note the year in which each stadium was built.

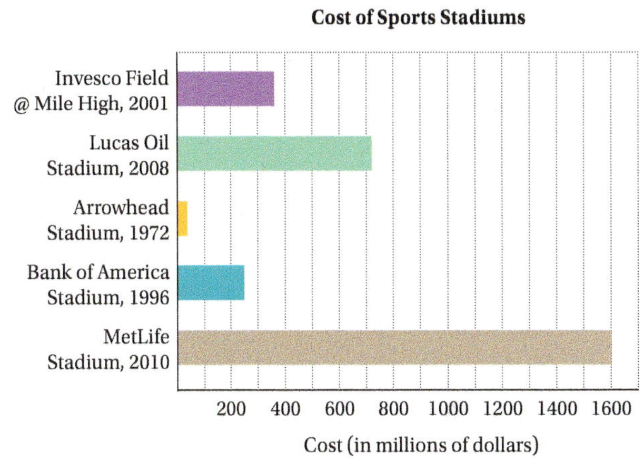

SOURCE: National Football League

a) Estimate how much more Lucas Oil Stadium cost than Invesco Field did.
b) Which stadium cost approximately $250 million to build?

We look at the graph to answer the questions.

a) We move to the right along the bars for Lucas Oil Stadium and Invesco Field and move down to the horizontal scale to estimate the costs: about $720 million for Lucas Oil Stadium and $360 million for Invesco Field. The difference in cost is about $720 million − $360 million, or $360 million. Thus Lucas Oil cost about $360 million more than Invesco Field.
b) We locate the lines representing $200 million and $300 million and go up until we reach a bar that ends close to $250 million. Then we go to the left and read the name of the stadium: Bank of America Stadium.

Practice Exercises

Use the bar graph in the example shown above for Exercises 4 and 5.

4. Which stadium cost less than $100 million?

5. Estimate how much more MetLife Stadium cost than Bank of America Stadium did.

Objective 7.4a Extract and interpret data from circle graphs.

Example The following circle graph shows the percentages of the population of the United States in various age groups.

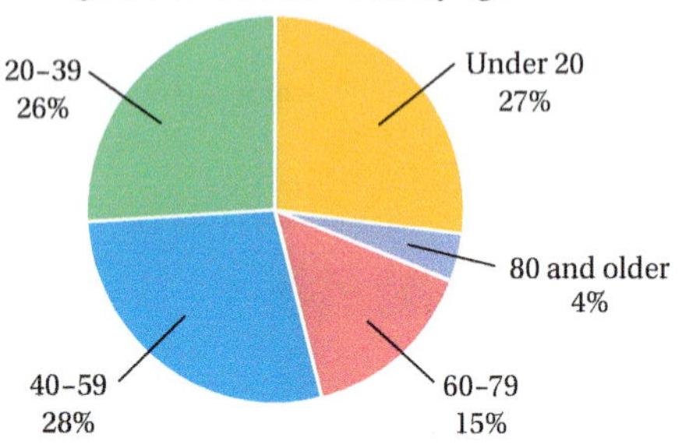

a) What percent of the population is 40–59 years old?

b) How much more of the population is in the 20–39 age group than in the 60–79 age group?

The graph gives us the answers.

a) The graph shows us that the segment of the population that is 40–59 years old is 28% of the total population.

b) From the graph, we see that 26% of the population is 20–39 years old and only 15% is 60–79 years old. We subtract: 26% − 15% = 11%. Thus the percent of the population that is 20–39 years old is 11% greater than the percentage that is 60–79 years old.

Practice Exercises

Use the circle graph at left to answer Exercises 6 and 7.

6. Which age group has the fewest people?

7. What percent of the population is under 20 years old?

Review Exercises

Find the average. [7.1a]

1. 26, 34, 43, 51

2. 11, 14, 17, 18, 7

3. 0.2, 1.7, 1.9, 2.4

4. 700, 2700, 3000, 900, 1900

5. $2, $14, $17, $17, $21, $29

6. 20, 190, 280, 470, 470, 500

7. To get an A in math, Naomi must score an average of 90 on four tests. Her scores on the first three tests were 94, 78, and 92. What is the lowest score she can make on the last test and still get an A? [7.1a]

8. ***Gas Mileage.*** A 2012 Mazda Miata does 336 mi of highway driving on 12 gal of gasoline. What is the gas mileage? [7.1a]

9. ***Grade Point Average.*** Find the grade point average for one semester given the following grades. Assume the grade point values are 4.0 for A, 3.0 for B, and so on. Round to the nearest tenth. [7.1a]

COURSE	GRADE	NUMBER OF CREDIT HOURS IN COURSE
Math	A	5
English	B	3
Computer Science	C	4
Spanish	B	3
College Skills	B	1

Find the median. [7.1b]

10. 26, 34, 43, 51

11. 7, 11, 14, 17, 18

12. 0.2, 1.7, 1.9, 2.4

13. 700, 900, 1900, 2700, 3000

14. \$2, \$17, \$21, \$29, \$14, \$17

15. 470, 20, 190, 280, 470, 500

16. One summer, a student worked part time as a veterinary assistant. She earned the following weekly amounts over a six-week period: \$360, \$192, \$240, \$216, \$420, and \$132. What was the average amount earned per week? the median? [7.1a, b]

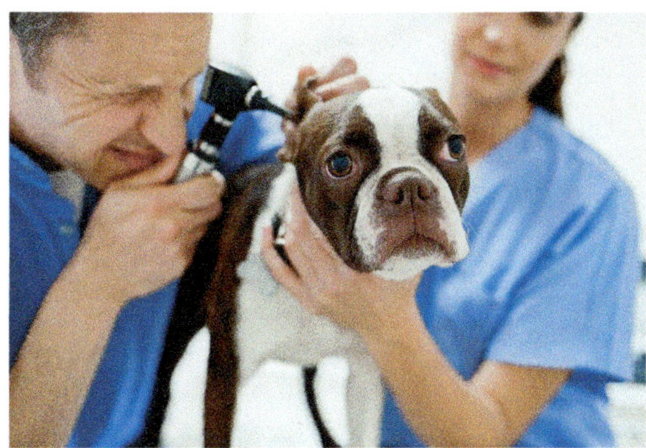

Find the mode. [7.1c]

17. 26, 34, 43, 26, 51

18. 17, 7, 11, 11, 14, 17, 18

19. 0.2, 0.2, 1.7, 1.9, 2.4, 0.2

20. 700, 700, 800, 2700, 800

21. \$14, \$17, \$21, \$29, \$17, \$2

22. 20, 20, 20, 20, 20, 500

Smartphone and Tablet Ownership. The following table lists the percents of the populations of several countries who own a smartphone and who own a tablet. Use this table for Exercises 23–25. [7.2a]

COUNTRY	PERCENT OWNING SMARTPHONE	PERCENT OWNING TABLET
Singapore	56%	18%
China	52	30
United States	35	17
Mexico	15	2
Germany	14	3
Philippines	8	5

SOURCE: Based on information from mastercard.com

23. What percent of China's population owns a smartphone?

24. In what countries does less than 10% of the population own a tablet?

25. In which country do approximately twice as many people own a smartphone as own a tablet?

Major League World Series. Except for four years, the World Series of Major League Baseball has been a best-of-seven series. In 1903, 1919, 1920, and 1921, the championship was a best-of-nine series. The championships have all been decided in 4, 5, 6, 7, or 8 games. The following pictograph shows the number of times the series has extended to each number of games. Use this graph for Exercises 26–28. [7.2b]

Number of Games Needed to Decide the World Series

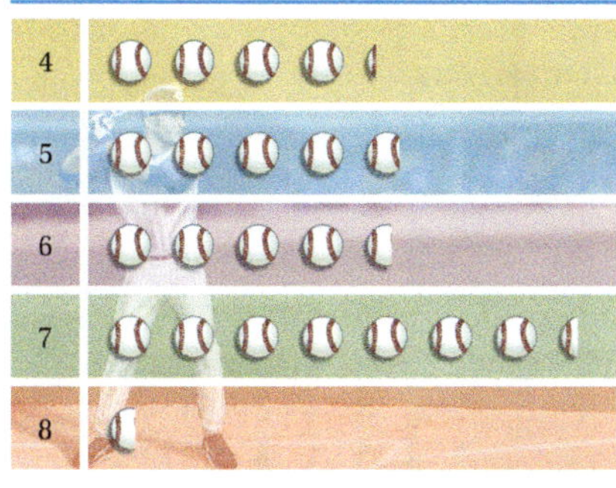

26. How many World Series were decided in 4 games?

27. In what number of games were the most World Series decided?

28. How many more World Series were decided in 7 games than were decided in 4 games?

Governors' Salaries. The following histogram shows the numbers of state governors in the United States who receive annual salaries in the given ranges. Use the graph for Exercises 29–31. [7.2c]

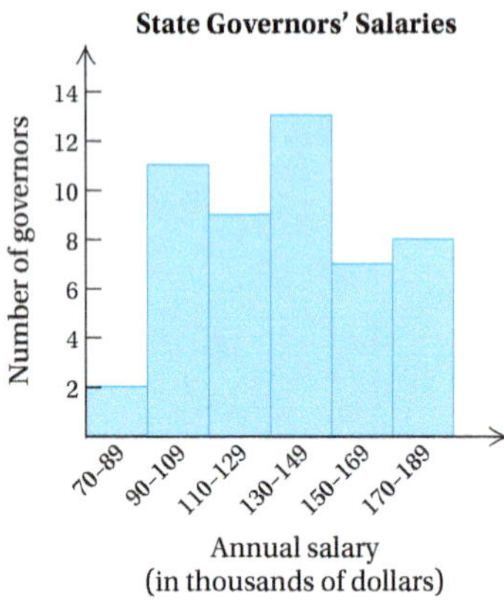

SOURCE: knowledgecenter.csg.org

29. Which salary range has the smallest number of governors?

30. How many more governors make between \$130,000 and \$149,000 than make between \$90,000 and \$109,000?

31. How many governors make less than \$130,000?

Tornadoes. The following bar graph shows the total number of tornadoes that occurred in the United States from 2010 through 2012, by month. Use the graph for Exercises 32–35. [7.3a]

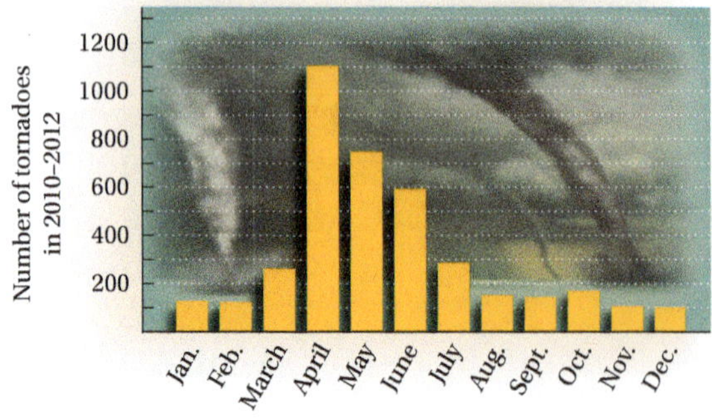

32. Which month had the greatest number of tornadoes?

33. How many tornadoes occurred in August?

34. How many more tornadoes occurred in May than in June?

35. Do more tornadoes occur in the winter or in the spring?

Homelessness. The following line graph shows the average number of homeless children in New York City's shelter system each night for various years. Use the graph for Exercises 36–40. [7.3c]

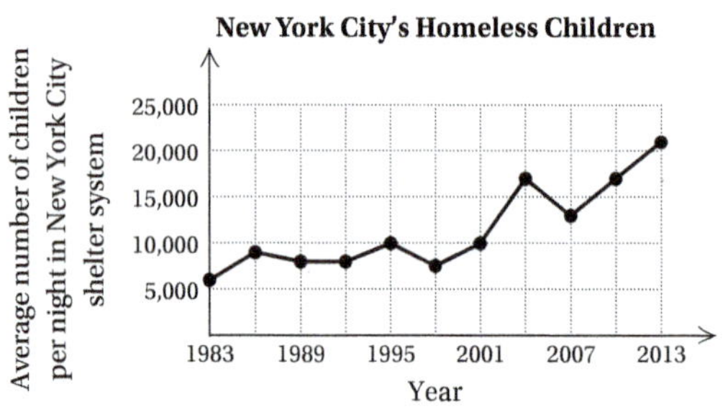

SOURCE: NYC Department of Homeless Services and Human Resources Administration and NYC Stat, shelter census reports

36. During which year after 1990 were there the fewest children in the shelter system?

37. How many children were in the shelter system each night in 2001?

38. In which years were there about 17,000 children each night in the shelter system?

39. Between which years did the number of children in the shelter system decrease?

40. By how much did the number of children in the shelter system each night increase from 2010 to 2013?

College Costs. The following circle graph shows the various cost categories for a full-time resident student at an Oklahoma regional university and the percentage of the total college cost represented by each category. Use this graph for Exercises 41–44. [7.4a]

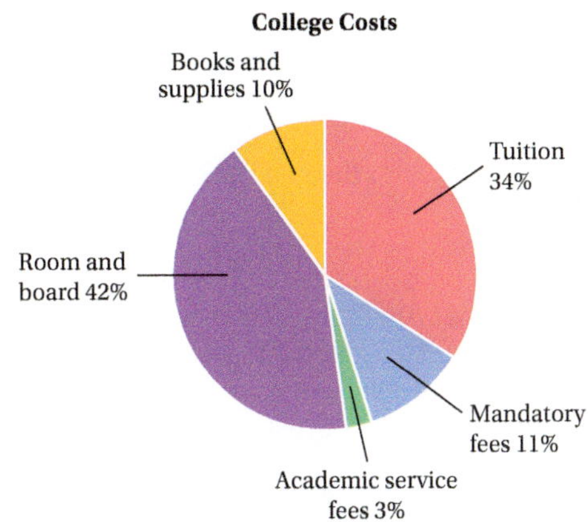

SOURCE: okcollegestart.org

41. What percent of college costs is tuition?

42. Which category accounts for the greatest part of the total college costs?

43. What percent of college costs are fees?

44. In a recent year, the total college cost for a full-time resident student at an Oklahoma regional university was $11,500. How much did a student pay for room and board?

45. Find the mode(s) of this set of data.

6, 9, 6, 8, 8, 5, 10, 5, 9, 10 [7.1c]

A. 8 **B.** 5, 6, 8, 9, 10
C. 9 **D.** No mode exists.

46. What is the average of this set of data?

$\frac{1}{2}, \frac{1}{3}, \frac{1}{4}, \frac{1}{5}$ [7.1a]

A. $\frac{77}{240}$ **B.** $\frac{1}{3}$ **C.** $\frac{7}{24}$ **D.** $\frac{1}{4}$

First-Class Postage. The following table lists the cost of first-class postage in various years. Use the table for Exercises 47 and 48.

YEAR	FIRST-CLASS POSTAGE
2001	34¢
2002	37
2006	39
2007	41
2008	42
2009	44
2012	45
2013	46

SOURCE: U.S. Postal Service

47. Make a vertical bar graph of the data. [7.3b]

48. Make a line graph of the data. [7.3d]

49. Construct a circle graph showing the governors' salaries discussed in Exercises 29–31 as percentages of the total:

$70,000–$89,000: 4%; $90,000–$109,000: 22%;

$110,000–$129,000: 18%; $130,000–$149,000: 26%;

$150,000–$169,000: 14%; $170,000–189,000: 16%. [7.4b]

Synthesis

50. The ordered set of data 298, 301, 305, a, 323, b, 390 has a median of 316 and an average of 326. Find a and b. [7.1a, b]

Understanding Through Discussion and Writing

1. Find a real-world situation that fits this equation:

$$T = \frac{20{,}500 + 22{,}800 + 23{,}400 + 26{,}000}{4}.$$ [7.1a]

2. Can bar graphs always, sometimes, or never be converted to line graphs? Why? [7.3b, d]

3. Discuss the advantages of being able to read a circle graph. [7.4a]

4. Compare bar graphs and line graphs. Discuss why you might use one rather than the other to graph a particular set of data. [7.3b, d]

5. Compare and contrast averages, medians, and modes. Discuss why you might use one over the others to analyze a set of data. [7.1a, b, c]

6. Compare circle graphs to bar graphs. [7.3a], [7.4a]

CHAPTER

7 Test

For Extra Help For step-by-step test solutions, access the Chapter Test Prep Videos in MyMathLab® or on YouTube (search "BittingerPreIntro" and click on Channels).

Find the average.

1. 45, 49, 52, 52

2. 1, 1, 3, 5, 3

3. 3, 17, 17, 18, 18, 20

Find the median and the mode.

4. 45, 49, 52, 53

5. 1, 1, 3, 5, 3

6. 3, 17, 17, 18, 18, 20

7. *Grades.* To get a C in chemistry, Ted must score an average of 70 on four tests. His scores on the first three tests were 68, 71, and 65. What is the lowest score he can get on the last test and still get a C?

8. *Grade Point Average.* Find the grade point average for one semester given the following grades. Assume the grade point values are 4.0 for A, 3.0 for B, and so on. Round to the nearest tenth.

COURSE	GRADE	NUMBER OF CREDIT HOURS IN COURSE
Introductory Algebra	B	3
English	A	3
Business	C	4
Spanish	B	3
Typing	B	2

Desirable Body Weights. The following tables list the desirable body weights for men and women over age 25. Use the tables for Exercises 9–12.

DESIRABLE WEIGHT OF MEN			
HEIGHT	SMALL FRAME (in pounds)	MEDIUM FRAME (in pounds)	LARGE FRAME (in pounds)
5 ft 7 in.	138	152	166
5 ft 9 in.	146	160	174
5 ft 11 in.	154	169	184
6 ft 1 in.	163	179	194
6 ft 3 in.	172	188	204

DESIRABLE WEIGHT OF WOMEN			
HEIGHT	SMALL FRAME (in pounds)	MEDIUM FRAME (in pounds)	LARGE FRAME (in pounds)
5 ft 1 in.	105	113	122
5 ft 3 in.	111	120	130
5 ft 5 in.	118	128	139
5 ft 7 in.	126	137	147
5 ft 9 in.	134	144	155

SOURCE: U.S. Department of Agriculture

9. What is the desirable weight for a 6 ft 1 in. man with a medium frame?

10. What size woman has a desirable weight of 120 lb?

11. How much more should a 5 ft 3 in. woman with a medium frame weigh than one with a small frame?

12. How much more should a 6 ft 3 in. man with a large frame weigh than one with a small frame?

Waste Generated. The number of pounds of waste generated per person per year varies greatly among countries around the world. In the pictograph at right, each symbol represents approximately 100 lb of waste. Use the pictograph for Exercises 13–16.

SOURCE: OECD, Key Environmental Indicators 2008

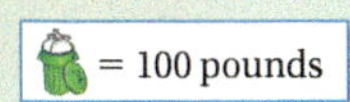

13. In which country does each person generate 1300 lb of waste per year?

14. In which countries does each person generate more than 1500 lb of waste per year?

15. How many pounds of waste per person per year are generated in Canada?

16. How many more pounds of waste per person per year are generated in the United States than in Mexico?

Hurricanes. The following line graph shows the numbers of Atlantic hurricanes for the years 2000–2012. Use the graph for Exercises 17–22.

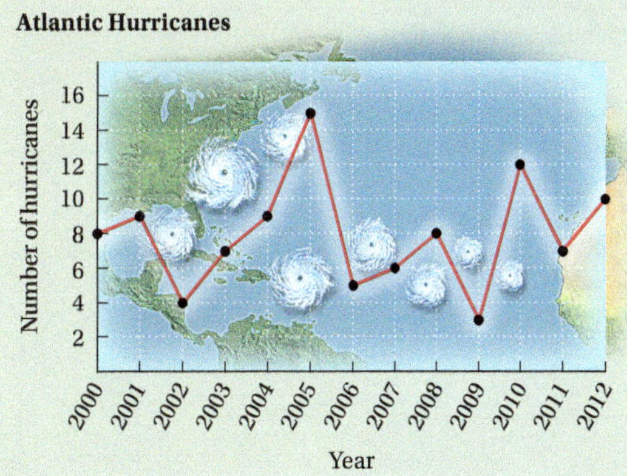

17. What year had the greatest number of Atlantic hurricanes?

18. In what year were there 3 Atlantic hurricanes?

19. How many hurricanes were there in 2012?

20. How many more hurricanes were there in 2005 than in 2006?

21. Find the average number of hurricanes per year for the years 2008–2012.

22. In what years were there 10 or more hurricanes?

Book Circulation. The following table lists the average number of books checked out per day of the week for a branch library. Use this table for Exercises 23 and 24.

DAY	NUMBER OF BOOKS CHECKED OUT
Sunday	210
Monday	160
Tuesday	240
Wednesday	270
Thursday	310
Friday	275
Saturday	420

23. Make a vertical bar graph of the data.

24. Make a line graph of the data.

25. *Food Budget.* The following table lists the percents of a family's food budget spent on selected food categories. Construct a circle graph representing these data.

FOOD CATEGORY	PERCENT OF BUDGET
Meat, poultry, fish, and eggs	23%
Fruits and vegetables	17
Cereals and bakery products	13
Dairy products	11
Other	36

SOURCE: Consumer Expenditure Survey

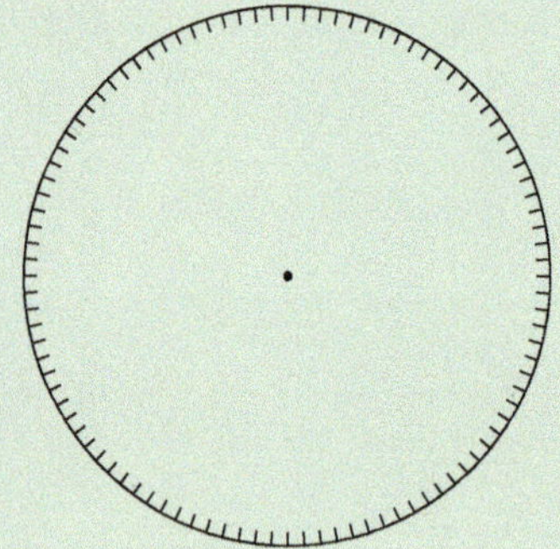

26. Referring to Exercise 25, consider a family that spends $664 per month on food. Using the percents from the table and the circle graph, find the amount of money spent on cereals and bakery products.

A. $8.63 **B.** $21.58 **C.** $86.32 **D.** $577.68

Synthesis

27. The ordered set of data 69, 71, 73, a, 78, 98, b has a median of 74 and a mean of 82. Find a and b.

CHAPTER

8

Geometry

STUDYING FOR SUCCESS *Make the Most of Your Homework*

- ☐ Double-check that you have copied an exercise correctly.
- ☐ Check that your answer is reasonable.
- ☐ Include correct units with answers.

8.1 Basic Geometric Figures

OBJECTIVES

- **a** Draw and name segments, rays, and lines. Also, identify endpoints, if they exist.
- **b** Name an angle in five different ways, and measure an angle with a protractor.
- **c** Classify an angle as right, straight, acute, or obtuse.
- **d** Identify perpendicular lines.
- **e** Classify a triangle as equilateral, isosceles, or scalene and as right, obtuse, or acute. Given a polygon of twelve, ten, or fewer sides, classify it as a dodecagon, a decagon, and so on.
- **f** Given a polygon of n sides, find the sum of its angle measures using the formula $(n - 2) \cdot 180°$.

In geometry we study sets of points. A **geometric figure** (or *figure*) is simply a set of points.

a SEGMENTS, RAYS, AND LINES

A **segment** is a geometric figure consisting of two points, called *endpoints*, and all points between them. The segment whose endpoints are A and B is shown below. It can be named $\overline{AB}$ or $\overline{BA}$.

◀ Do Exercise 1.

1. a) Draw a segment.
   b) Label its endpoints E and F.
   c) Name this segment in two ways.

We get an idea of a geometric figure called a ray by thinking of a ray of light. A **ray** consists of a segment, say $\overline{AB}$, and all points X such that B is between A and X, that is, $\overline{AB}$ and all points "beyond" B.

A ray is usually drawn as shown below. It has just one endpoint. The arrow indicates that it extends forever in one direction.

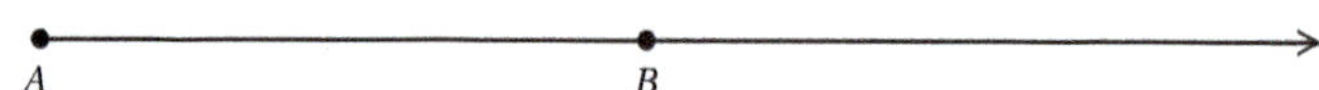

A ray is named $\overrightarrow{AB}$, where B is some point on the ray other than A. The endpoint is always listed first. Thus rays $\overrightarrow{AB}$ and $\overrightarrow{BA}$ are different.

◀ Do Exercises 2–5 on the following page.

Two rays such as $\overrightarrow{PQ}$ and $\overrightarrow{QP}$ make up what is known as a **line**. A line can be named with a smaller letter m, as shown below, or it can be named by two points P and Q on the line as $\overleftrightarrow{PQ}$.

◀ Do Exercises 6–11 on the following page.

Answer

1. (a) and (b) ; (c) $\overline{EF}$, $\overline{FE}$

Lines in the same plane are called **coplanar**. Coplanar lines that do not intersect are called **parallel**. For example, lines l and m below are *parallel* ($l \parallel m$).

The following figure shows two lines that cross. Their *intersection* is D. They are also called **intersecting lines**.

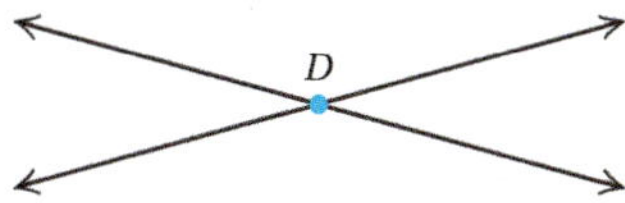

b MEASURING ANGLES

We see a real-world application of *angles* of various types in the different back postures of the bicycle riders illustrated below.

Style of Biking Determines Cycling Posture

Riders prefer a more aerodynamic flat-back position.

Riders prefer a semi-upright position to help lift the front wheel over obstacles.

Riders prefer an upright position that lessens stress on the lower back and neck.

SOURCE: USA TODAY research

An **angle** is a set of points consisting of two **rays**, or half-lines, with a common endpoint. The endpoint is called the **vertex** of the angle. The rays are called the **sides** of the angle.

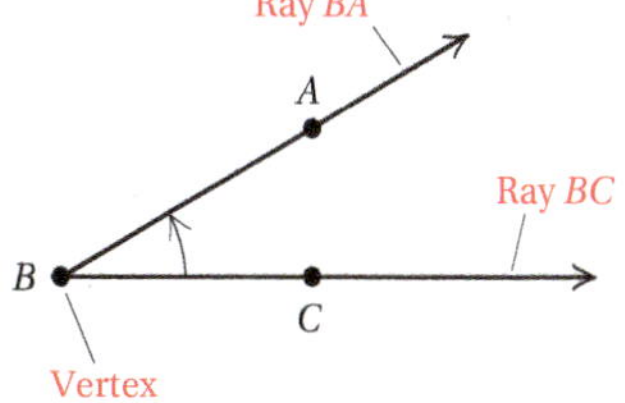

The angle shown above can be named

angle ABC, angle CBA, $\angle ABC$, $\angle CBA$, or $\angle B$.

Note that the vertex is written in the middle of the name. If there is only one angle with a given vertex in a drawing, the angle may be named using simply its vertex.

2. Draw two points P and Q.

3. Draw $\overline{PQ}$.

4. Draw $\overrightarrow{PQ}$. What is its endpoint?

5. Use a colored pencil to draw $\overrightarrow{QP}$. What is its endpoint?

6. Draw two points R and S.

7. Draw $\overline{RS}$. What are its endpoints?

8. Draw $\overrightarrow{RS}$. What is its endpoint?

9. Draw $\overrightarrow{SR}$. What is its endpoint?

10. Draw $\overleftrightarrow{RS}$. What are its endpoints?

11. Name this line in seven different ways.

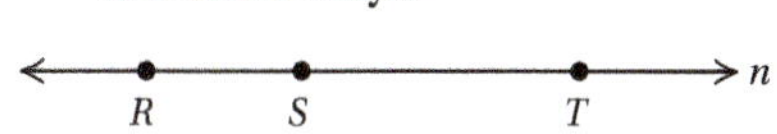

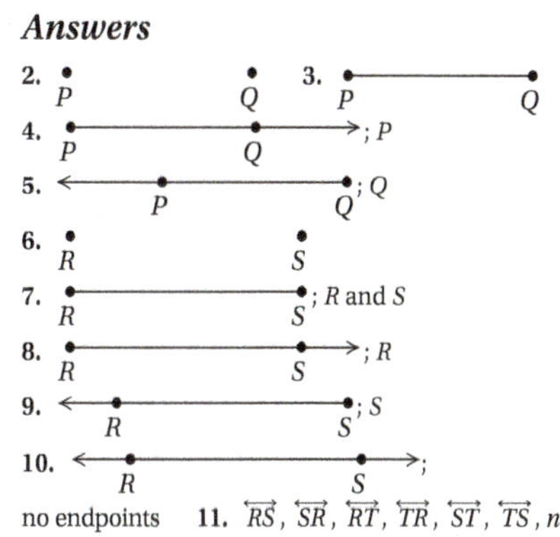

Name the angle in five different ways.

12.

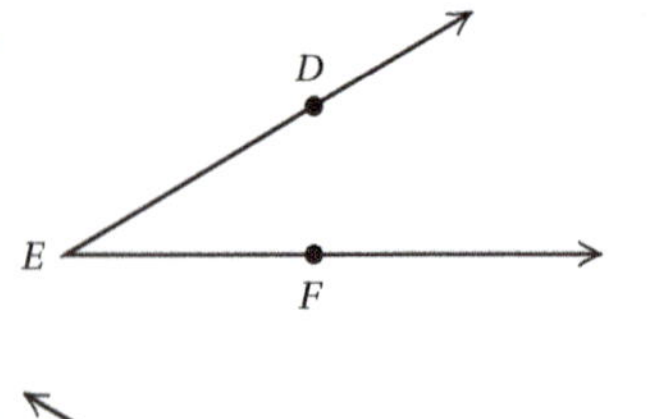

13.

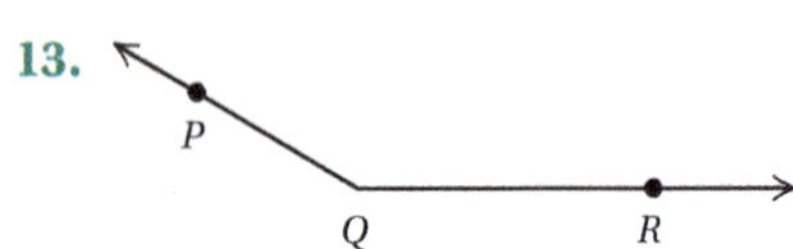

Do Exercises 12 and 13.

To measure angles, we start with some arbitrary angle and assign to it a measure of 1. We call it a *unit angle*. Suppose that $\angle U$, shown below, is a unit angle. Let's measure $\angle DEF$. If we made 3 copies of $\angle U$, they would "fill up" $\angle DEF$. Thus the measure of $\angle DEF$ would be 3.

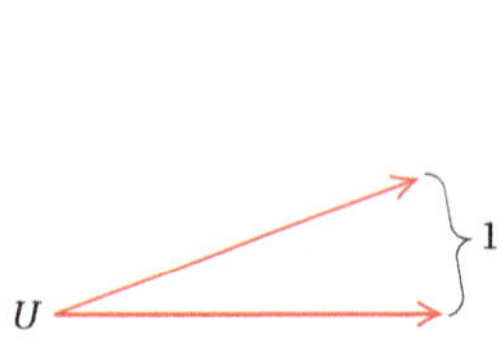

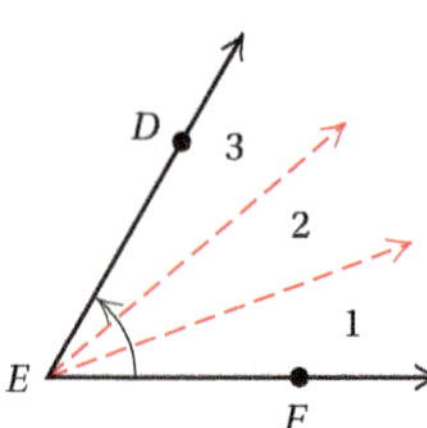

The unit most commonly used for angle measure is the degree. Below is such a unit angle. Its measure is 1 degree, or 1°. There are 360 degrees in a circle.

A 1° angle:

Here are some other angles with their degree measures.

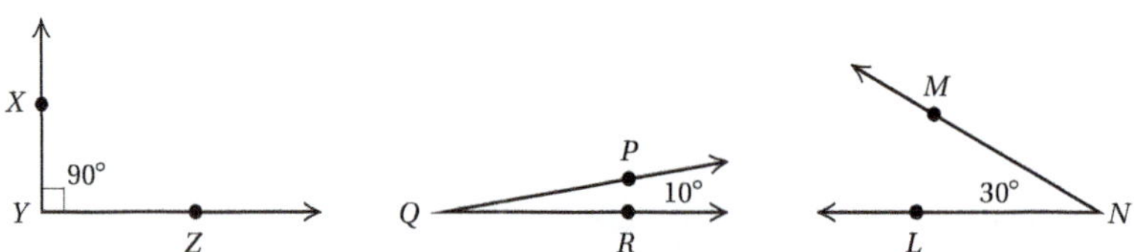

To indicate the *measure* of $\angle XYZ$, we write $m\,\angle XYZ = 90°$. Recall that the symbol ⌝ is sometimes drawn on a figure to indicate a 90° angle.

A device called a **protractor** is used to measure angles. Protractors have two scales (inside and outside). To measure an angle like $\angle Q$ below, we place the protractor's ▲ at the vertex and line up one of the angle's sides at 0°. Then we check where the angle's other side crosses the scale. In the following figure, the side $\overrightarrow{QR}$ lines up with 0° on the *inside* scale, so we check where the angle's other side, $\overrightarrow{QP}$, crosses the *inside* scale. We see that $m\,\angle Q = 145°$.

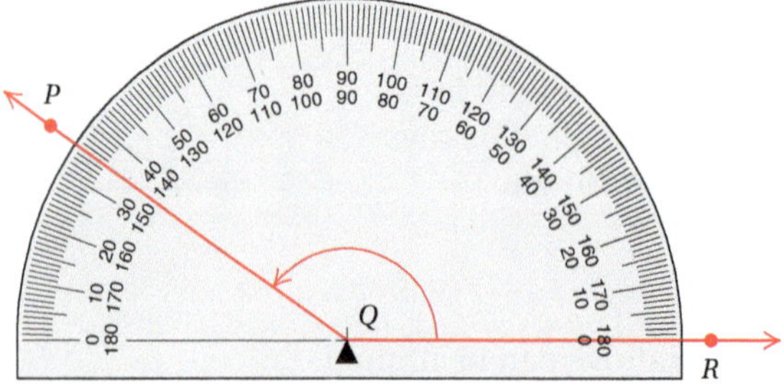

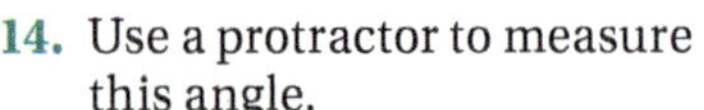

14. Use a protractor to measure this angle.

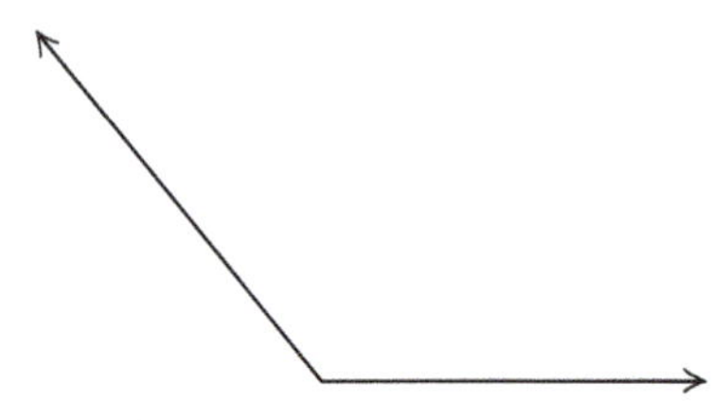

Do Exercise 14.

Answers

12. Angle *DEF*, angle *FED*, $\angle DEF$, $\angle FED$, or $\angle E$
13. Angle *PQR*, angle *RQP*, $\angle PQR$, $\angle RQP$, or $\angle Q$
14. 127°

Let's find the measure of $\angle ABC$. This time we line up one of the angle's sides, $\overrightarrow{BC}$, with 0° on the *outside* scale. Then we check where the angle's other side, $\overrightarrow{BA}$, crosses the *outside* scale. We see that $m\ \angle ABC = 42°$.

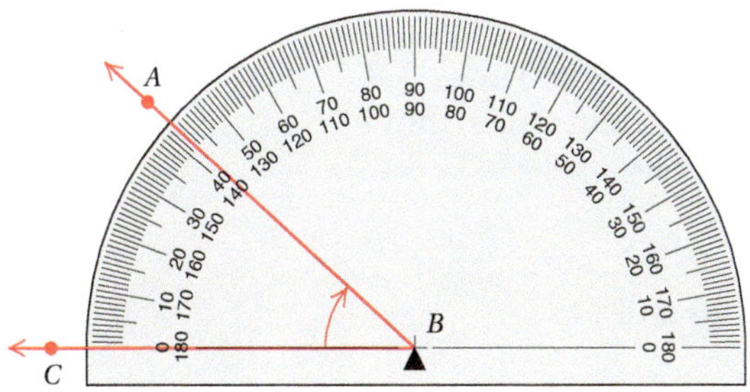

Do Exercise 15. ▶

15. Use a protractor to measure this angle.

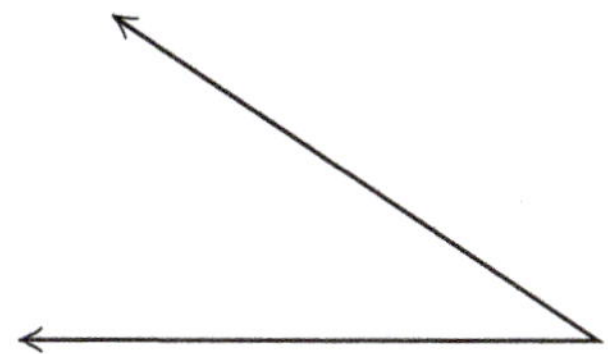

C CLASSIFYING ANGLES

The following are ways in which we classify angles.

> **TYPES OF ANGLES**
>
> **Right angle:** An angle whose measure is 90°.
>
> **Straight angle:** An angle whose measure is 180°.
>
> **Acute angle:** An angle whose measure is greater than 0° and less than 90°.
>
> **Obtuse angle:** An angle whose measure is greater than 90° and less than 180°.

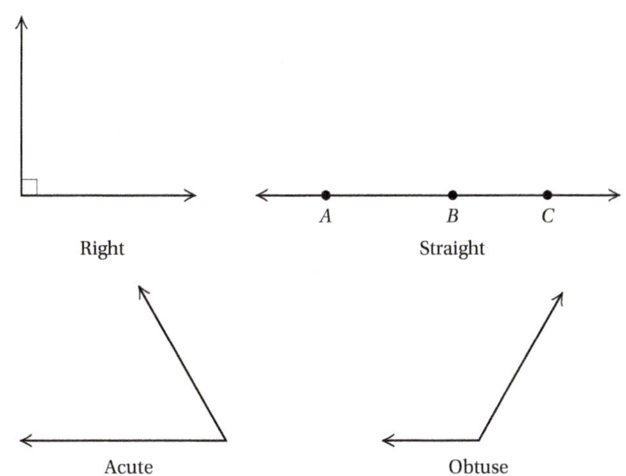

Do Exercises 16–19. ▶

Classify each angle as right, straight, acute, or obtuse. Use a protractor if necessary.

16.

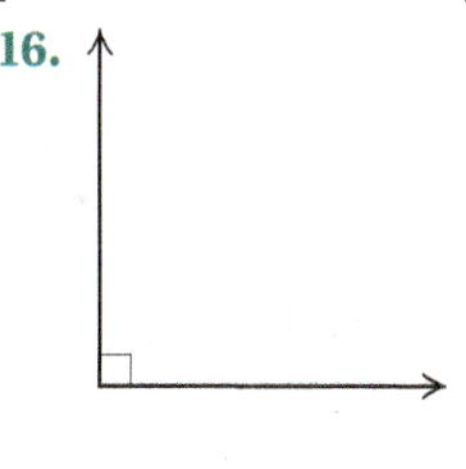

17.

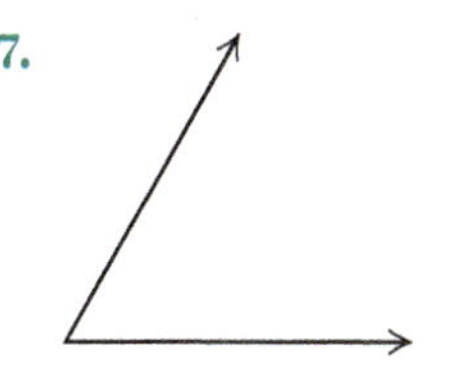

18.

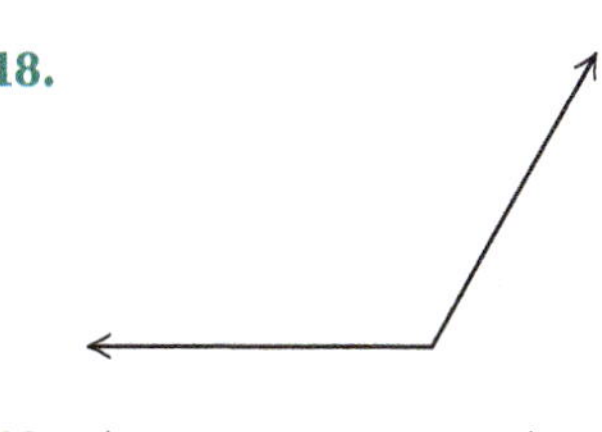

19.

Answers

15. 33° **16.** Right **17.** Acute **18.** Obtuse **19.** Straight

Determine whether the pair of lines is perpendicular. Use a protractor.

20.

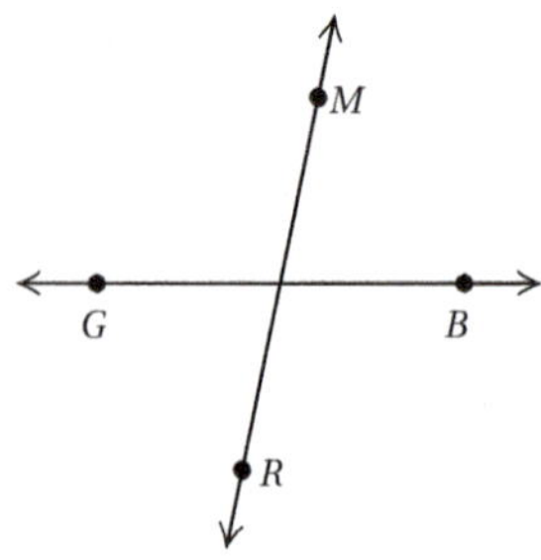

21.

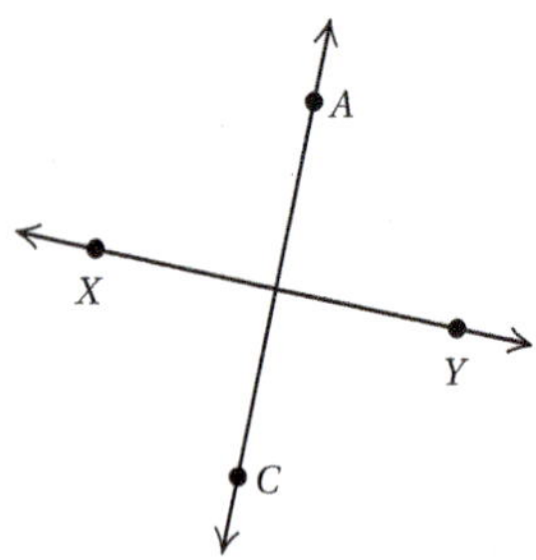

d PERPENDICULAR LINES

Two lines are **perpendicular** if they intersect to form a right angle.

To say that $\overleftrightarrow{AB}$ is perpendicular to $\overleftrightarrow{RS}$, we write $\overleftrightarrow{AB} \perp \overleftrightarrow{RS}$. If two lines intersect to form one right angle, they form four right angles.

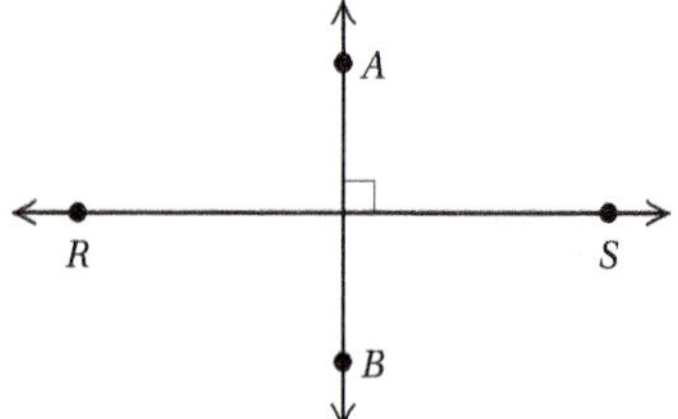

◀ Do Exercises 20 and 21.

e POLYGONS

The following figures are examples of **polygons**.

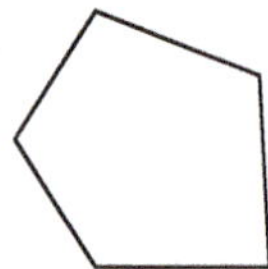
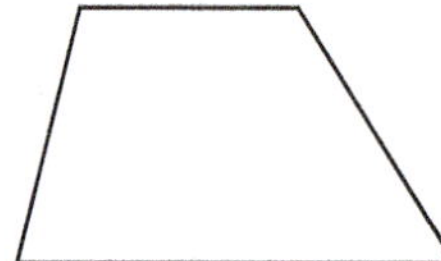
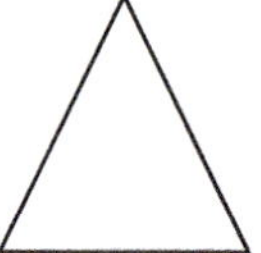

A **triangle** is a polygon made up of three segments, or sides. Consider these triangles. The triangle with vertices A, B, and C can be named $\triangle ABC$.

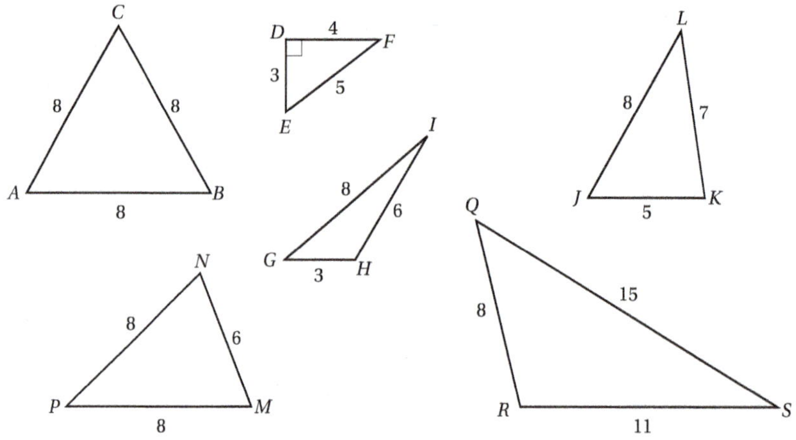

We can classify triangles according to sides and according to angles.

TYPES OF TRIANGLES

Equilateral triangle: All sides are the same length.
Isosceles triangle: Two or more sides are the same length.
Scalene triangle: All sides are of different lengths.
Right triangle: One angle is a right angle.
Obtuse triangle: One angle is an obtuse angle.
Acute triangle: All three angles are acute.

◀ Do Exercises 22–25.

22. Which triangles shown at right are:
a) equilateral?
b) isosceles?
c) scalene?

23. Are all equilateral triangles isosceles?

24. Are all isosceles triangles equilateral?

25. Which triangles shown at right are:
a) right triangles?
b) obtuse triangles?
c) acute triangles?

Answers

20. Not perpendicular **21.** Perpendicular
22. **(a)** $\triangle ABC$; **(b)** $\triangle ABC$, $\triangle MPN$; **(c)** $\triangle DEF$, $\triangle GHI$, $\triangle JKL$, $\triangle QRS$ **23.** Yes **24.** No
25. **(a)** $\triangle DEF$; **(b)** $\triangle GHI$, $\triangle QRS$; **(c)** $\triangle ABC$, $\triangle MPN$, $\triangle JKL$

We can further classify polygons as follows.

NUMBER OF SIDES	POLYGON	NUMBER OF SIDES	POLYGON
4	Quadrilateral	8	Octagon
5	Pentagon	9	Nonagon
6	Hexagon	10	Decagon
7	Heptagon	12	Dodecagon

Do Exercises 26–31. ▶

f SUM OF THE ANGLE MEASURES OF A POLYGON

The sum of the angle measures of a triangle is 180°. To see this, note that we can think of cutting apart a triangle as shown on the left below. If we reassemble the pieces, we see that a straight angle is formed.

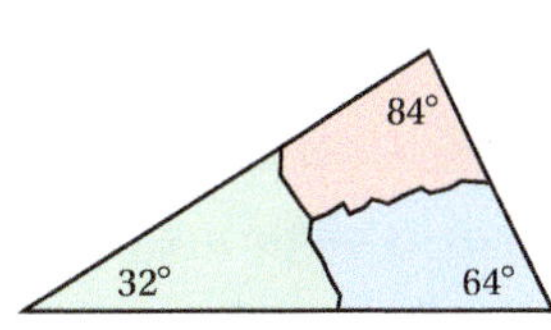

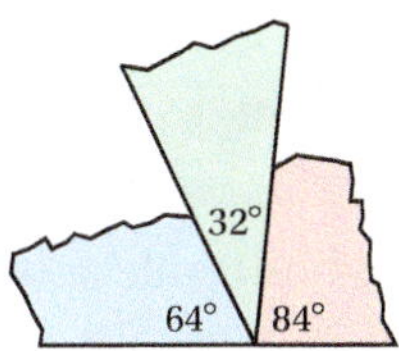

$64° + 32° + 84° = 180°$

SUM OF THE ANGLE MEASURES OF A TRIANGLE

In any $\triangle ABC$, the sum of the measures of the angles is 180°:

$$m\angle A + m\angle B + m\angle C = 180°.$$

Do Exercise 32. ▶

If we know the measures of two angles of a triangle, we can calculate the measure of the third angle.

EXAMPLE 1 Find the missing angle measure.

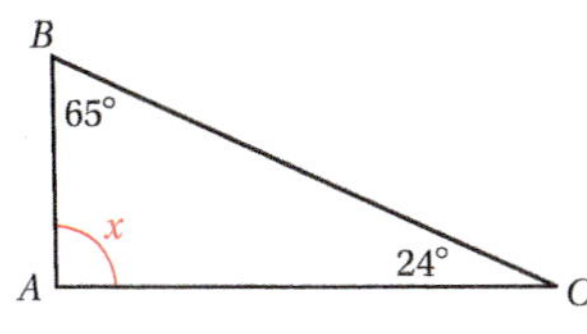

$$\begin{aligned} m\angle A + m\angle B + m\angle C &= 180° \\ x + 65° + 24° &= 180° \\ x + 89° &= 180° \\ x &= 180° - 89° \\ x &= 91° \end{aligned}$$

Thus, $m\angle A = 91°$.

Classify the polygon by name.

26.

27.

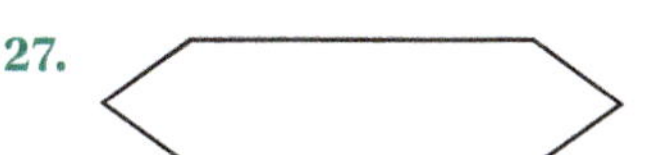

28.

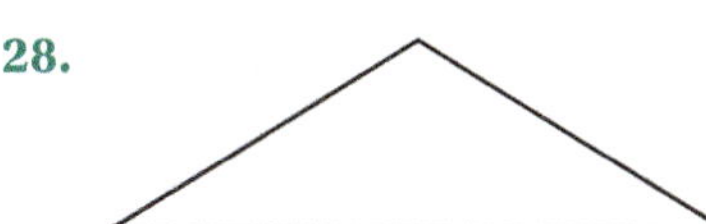

29.

30.

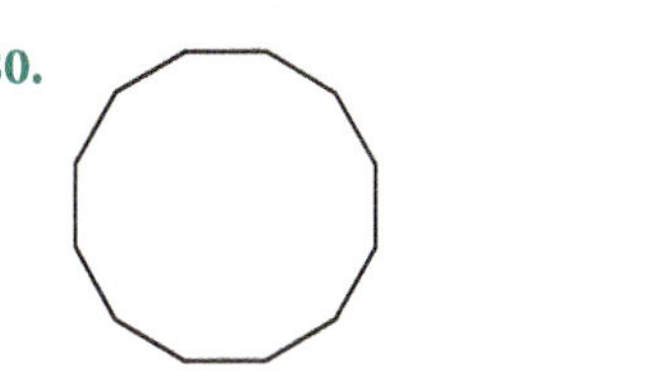

31.

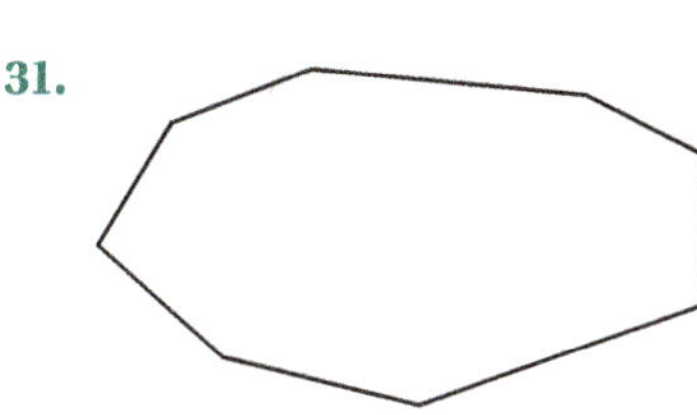

32. Find $m\angle P + m\angle Q + m\angle R$.

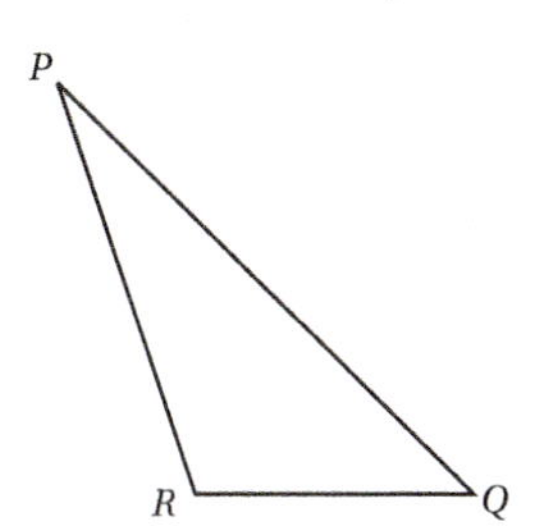

Answers

26. Quadrilateral **27.** Hexagon
28. Triangle **29.** Quadrilateral
30. Dodecagon **31.** Octagon **32.** 180°

33. Find the missing angle measure.

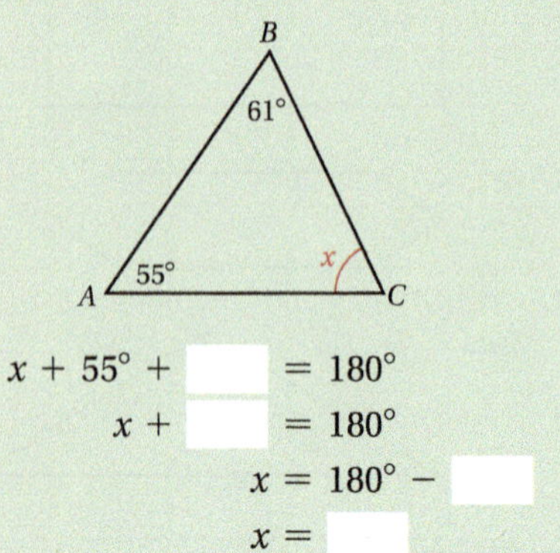

$x + 55° + \square = 180°$
$x + \square = 180°$
$x = 180° - \square$
$x = \square$

◀ **Do Exercise 33.**

Now let's use this idea to find the sum of the measures of the angles of a polygon of n sides. First, let's consider a four-sided figure.

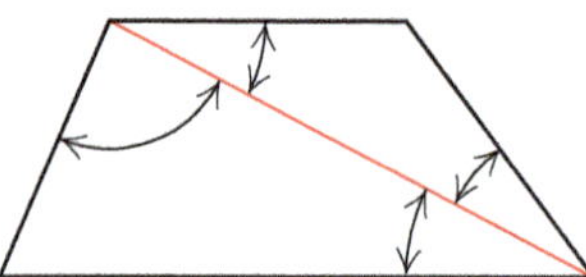

We can divide the figure into two triangles. The sum of the angle measures of each triangle is 180°. We have two triangles, so the sum of the angle measures of the figure is $2 \cdot 180°$, or 360°.

◀ **Do Exercise 34.**

If a polygon has n sides, it can be divided into $n - 2$ triangles, each having 180° as the sum of its angle measures. Thus the sum of the angle measures of the polygon is $(n - 2) \cdot 180°$.

SUM OF ANGLE MEASURES

If a polygon has n sides, then the sum of its angle measures is $(n - 2) \cdot 180°$.

34. Consider a five-sided figure:

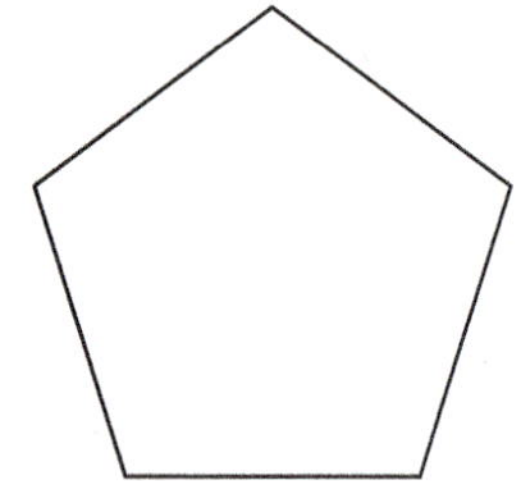

Complete.

a) The figure can be divided into __________ triangles.

b) The sum of the angle measures of each triangle is __________.

c) The sum of the angle measures of the polygon is __________ $\cdot$ 180°, or __________.

EXAMPLE 2 What is the sum of the angle measures of a hexagon?

A hexagon has 6 sides. We use the formula $(n - 2) \cdot 180°$:

$$(n - 2) \cdot 180° = (6 - 2) \cdot 180°$$
$$= 4 \cdot 180°$$
$$= 720°.$$

◀ **Do Exercises 35 and 36.**

35. What is the sum of the angle measures of an octagon?

36. What is the sum of the angle measures of a 25-sided figure?

Answers

33. 64° 34. **(a)** 3; **(b)** 180°; **(c)** 3, 540°
35. 1080° 36. 4140°

Guided Solution:
33. 61°, 116°, 116°, 64°

8.1 Exercise Set

For Extra Help MyMathLab® MathXL® PRACTICE WATCH READ REVIEW

Reading Check

Match each definition with the correct term from the list on the right.

RC1. ____ An angle whose measure is 90°

RC2. ____ An angle whose measure is 180°

RC3. ____ An angle whose measure is greater than 0° and less than 90°

RC4. ____ An angle whose measure is greater than 90° and less than 180°

RC5. ____ A triangle with three sides of the same length

RC6. ____ A triangle with all sides of different lengths

RC7. ____ A triangle with two or more sides of the same length

RC8. ____ A triangle containing a 90° angle

a) acute angle
b) equilateral triangle
c) isosceles triangle
d) obtuse angle
e) right angle
f) right triangle
g) scalene triangle
h) straight angle

1. Draw the segment whose endpoints are G and H. Name the segment in two ways.

G H

2. Draw the segment whose endpoints are C and D. Name the segment in two ways.

C D

3. Draw the ray with endpoint Q. Name the ray.

Q D

4. Draw the ray with endpoint D. Name the ray.

Q D

Name the line in seven different ways.

5. l — D E F

6. m — J K T

b Name each angle in five different ways.

7.

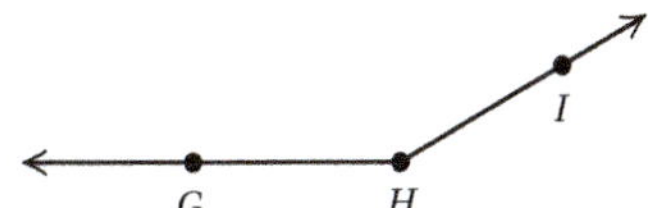

8.

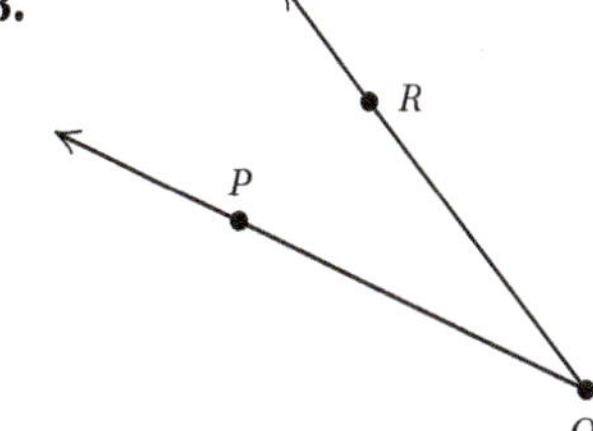

Use a protractor to measure each angle.

9.

10.

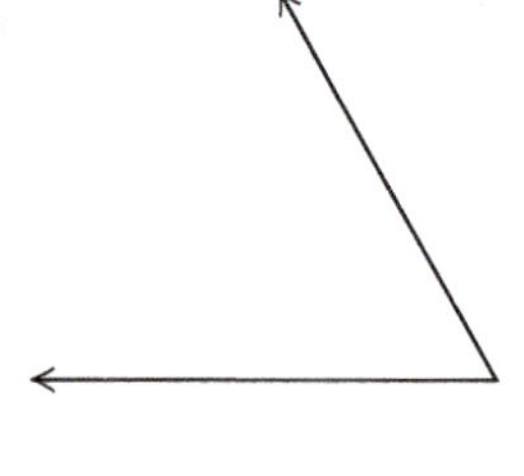

11.

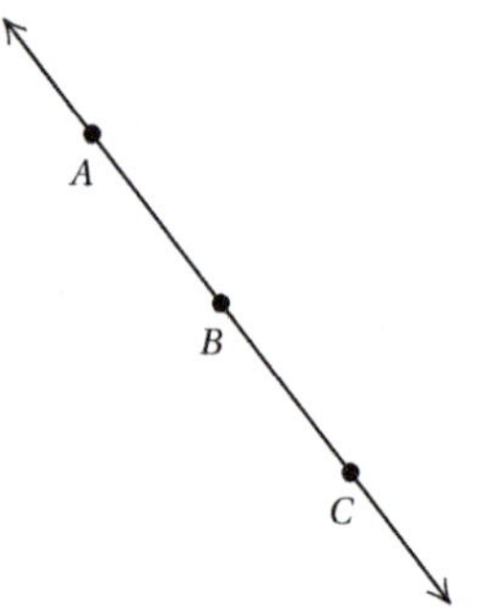

12.

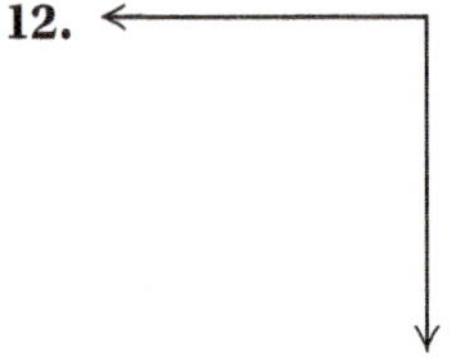

13.

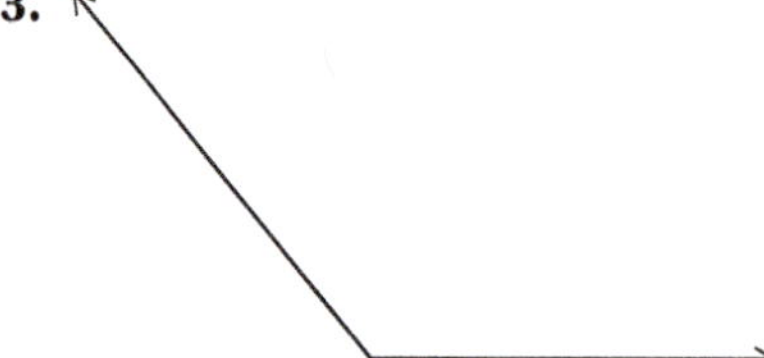

14.

c

15.–22. Classify each of the angles in Exercises 7–14 as right, straight, acute, or obtuse.

23.–26. Classify each of the angles in Margin Exercises 12–15 as right, straight, acute, or obtuse.

d Determine whether the pair of lines is perpendicular. Use a protractor.

27.

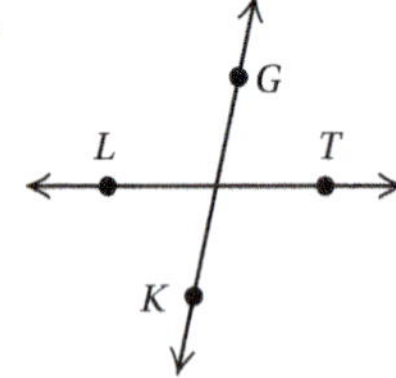

28.

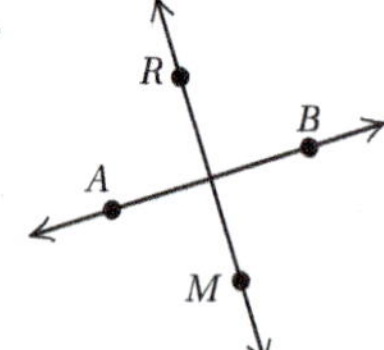

29.

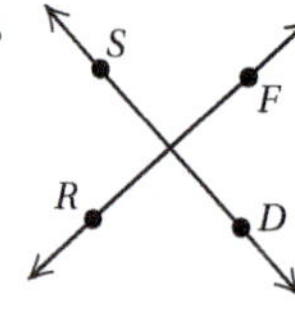

30.

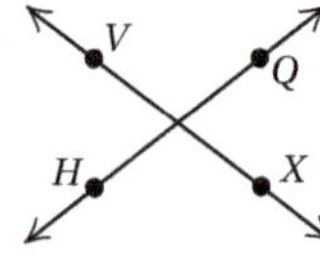

e Classify the triangle as equilateral, isosceles, or scalene. Then classify it as right, obtuse, or acute.

31.

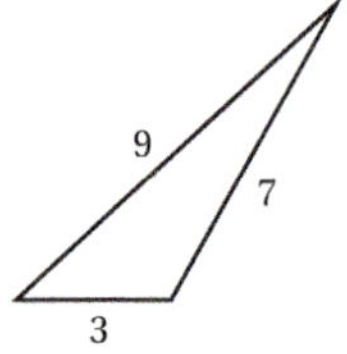

32.

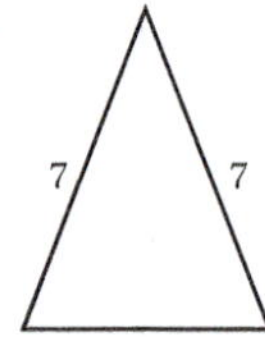

33.

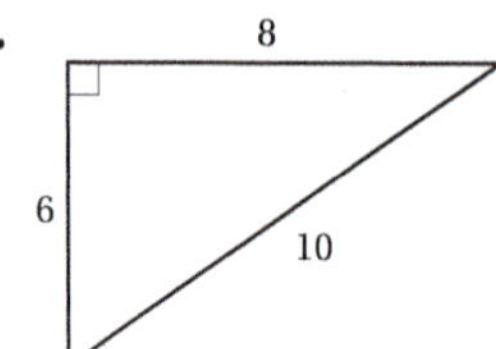

34.

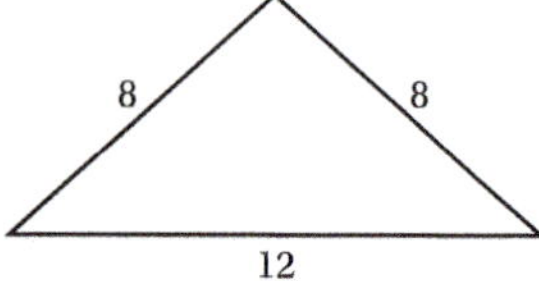

35. 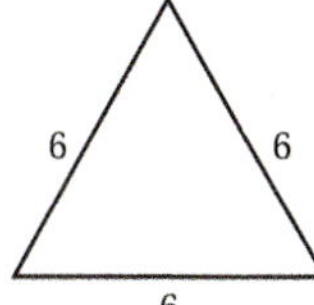

36. 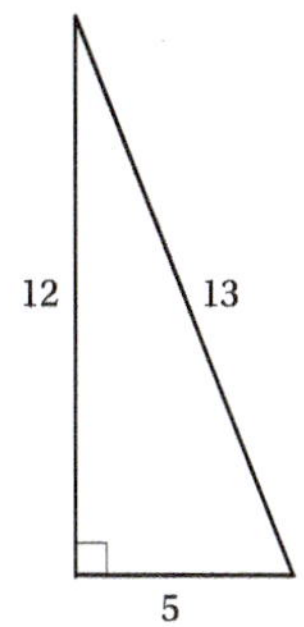

37. 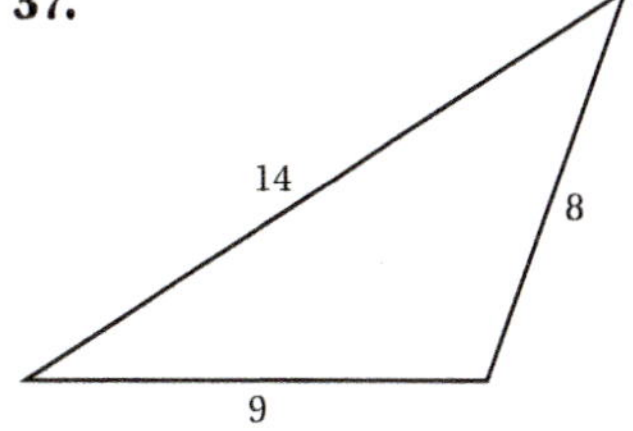

38. 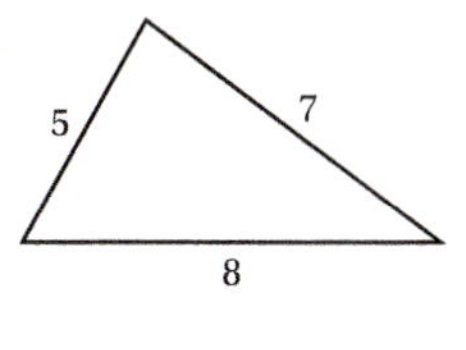

Classify each polygon by name.

39. 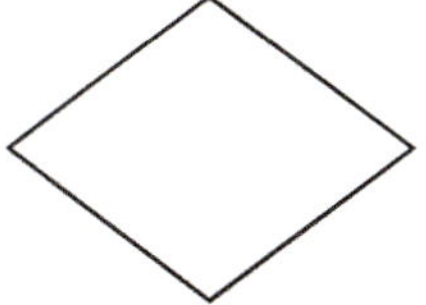

40. 

41. 

42. 

43. 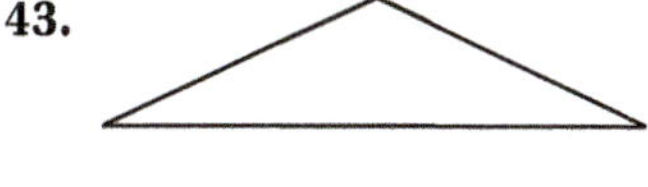

44. 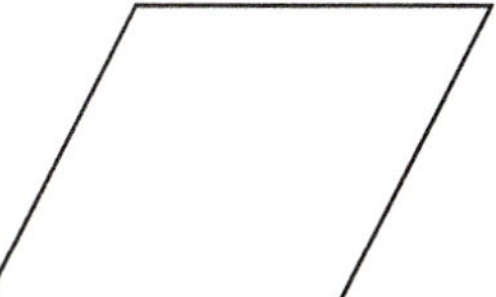

45. 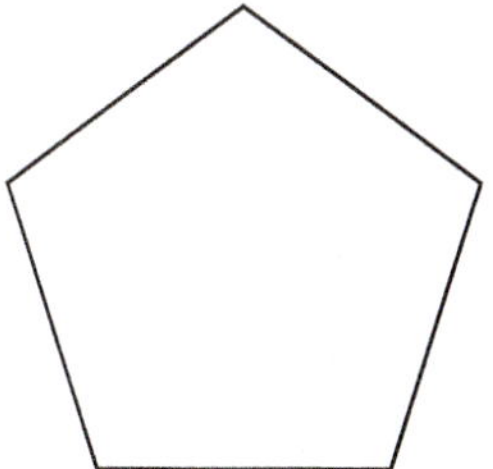

46. 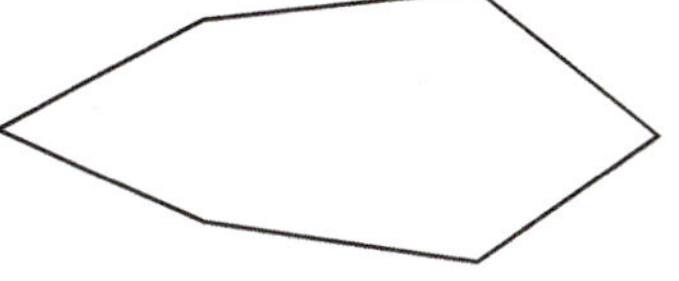

47. 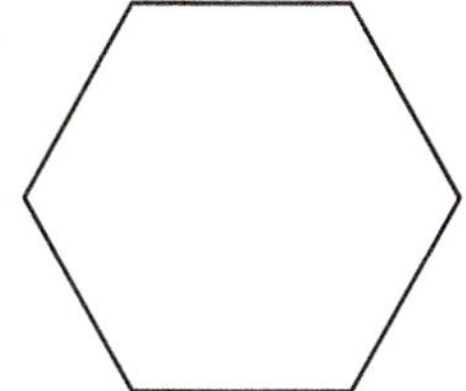

48. 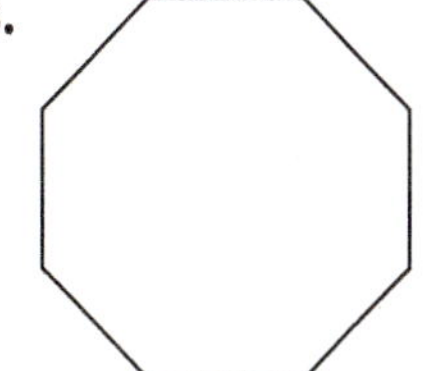

f Find the missing angle measure.

49.

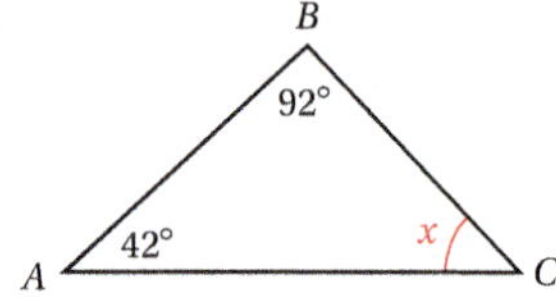

50.

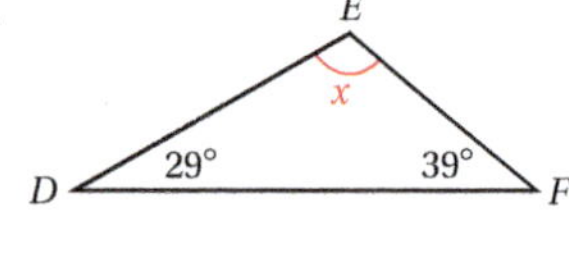

51.

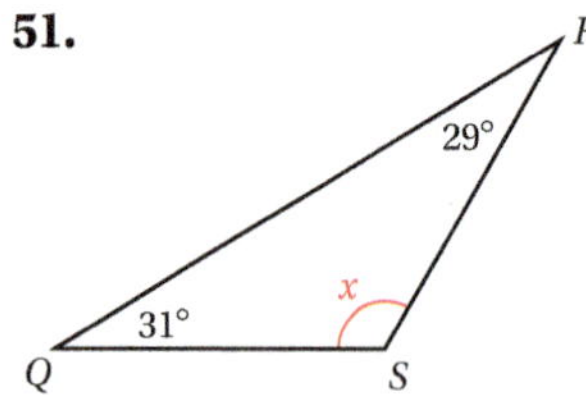

52.

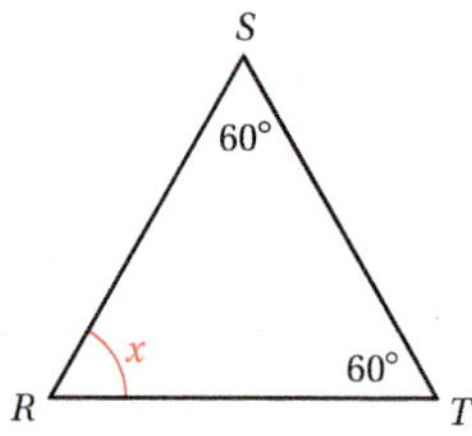

53. In $\triangle RST$, $m(\angle S) = 58°$ and $m(\angle T) = 79°$. Find $m(\angle R)$.

54. In $\triangle KNP$, $m(\angle K) = 137°$ and $m(\angle P) = 12°$. Find $m(\angle N)$.

Find the sum of the angle measures of each of the following.

55. A decagon

56. A quadrilateral

57. A heptagon

58. A nonagon

59. A 14-sided polygon

60. A 17-sided polygon

61. A 20-sided polygon

62. A 32-sided polygon

Skill Maintenance

Perform the indicated operation and simplify.

63. Subtract: $3.8 - 1.0875$. [5.2b]

64. Add: $2\frac{1}{3} + 5\frac{3}{4}$. [4.6a]

65. Add: $-\frac{3}{10} + \frac{5}{12}$. [4.2b]

66. Multiply: $\frac{1}{4} \cdot 2\frac{2}{3}$. [4.7a]

67. Divide: $-18 \div \frac{2}{3}$. [3.7b]

68. Divide: $16.8 \div 0.02$. [5.4a]

Solve.

69. $\frac{2}{5} + t = \frac{7}{10}$ [4.3b]

70. $\frac{2}{3}y - \frac{1}{9} = \frac{1}{8}$ [4.4a]

Synthesis

71. Find $m\angle ACB$, $m\angle CAB$, $m\angle EBC$, $m\angle EBA$, $m\angle AEB$, and $m\angle ADB$ in the rectangle shown below.

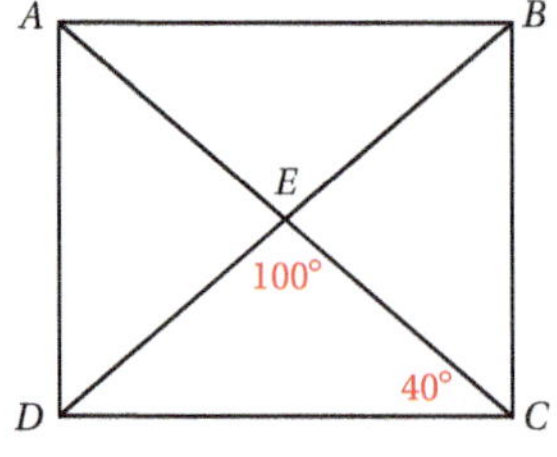

72. In the figure, $m\angle 2 = 42.17°$ and $m\angle 3 = 81.9°$. Find $m\angle 1$, $m\angle 4$, $m\angle 5$, and $m\angle 6$.

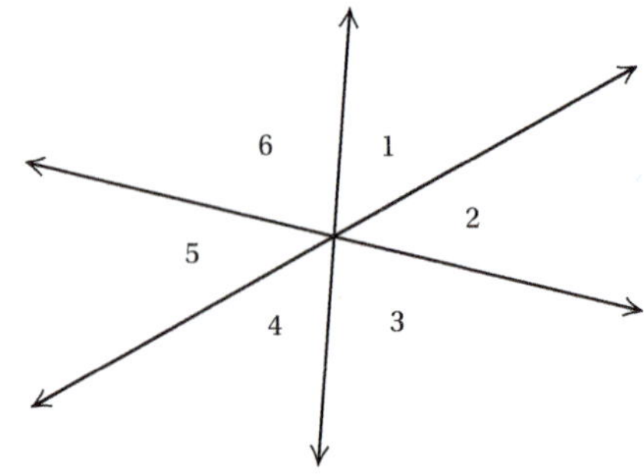

Perimeter

8.2

a FINDING PERIMETERS

PERIMETER OF A POLYGON

A **polygon** is a closed geometric figure with three or more sides. The **perimeter of a polygon** is the distance around it, or the sum of the lengths of its sides.

EXAMPLE 1 Find the perimeter of this polygon.

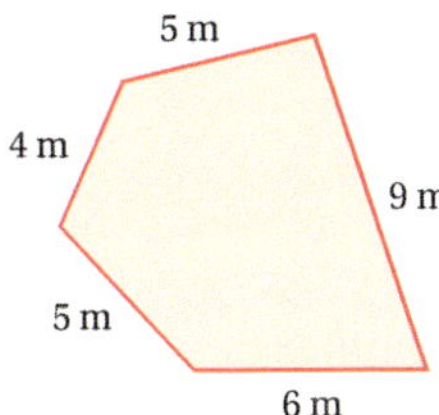

We add the lengths of the sides. Since all units are the same, we add the numbers, keeping meters (m) as the unit.

$$\begin{aligned} \text{Perimeter} &= 6\text{ m} + 5\text{ m} + 4\text{ m} + 5\text{ m} + 9\text{ m} \\ &= (6 + 5 + 4 + 5 + 9)\text{ m} \\ &= 29\text{ m} \end{aligned}$$

Do Margin Exercises 1 and 2.

A **rectangle** is a polygon with four sides and four 90° angles.

EXAMPLE 2 Find the perimeter of a rectangle that is 3 cm by 4 cm. The symbol ⏋ in the corner indicates an angle of 90°.

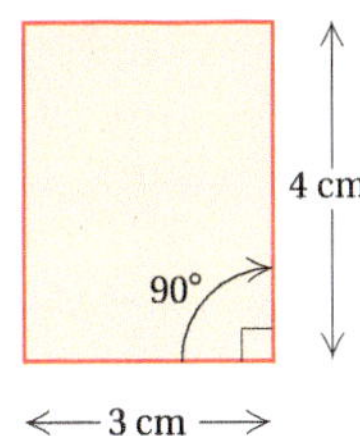

$$\begin{aligned} \text{Perimeter} &= 3\text{ cm} + 4\text{ cm} + 3\text{ cm} + 4\text{ cm} \\ &= (3 + 4 + 3 + 4)\text{ cm} \\ &= 14\text{ cm} \end{aligned}$$

OBJECTIVES

a Find the perimeter of a polygon.

b Solve applied problems involving perimeter.

SKILL TO REVIEW

Objective 4.7a: Multiply using mixed numerals.

Multiply.

1. $2 \times 8\frac{1}{3}$ **2.** $4 \times 6\frac{2}{5}$

Find the perimeter of each polygon.

1.

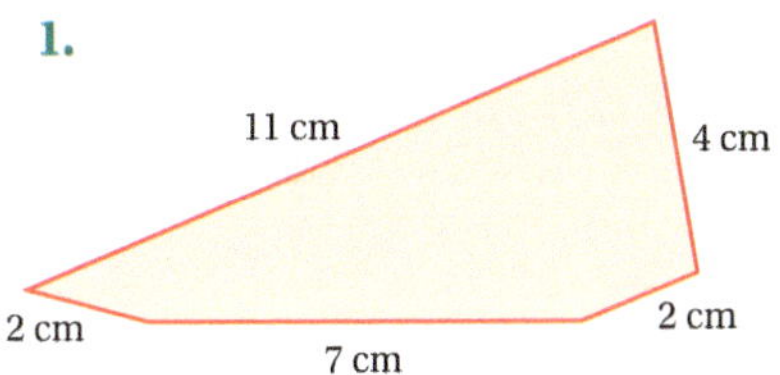

2.

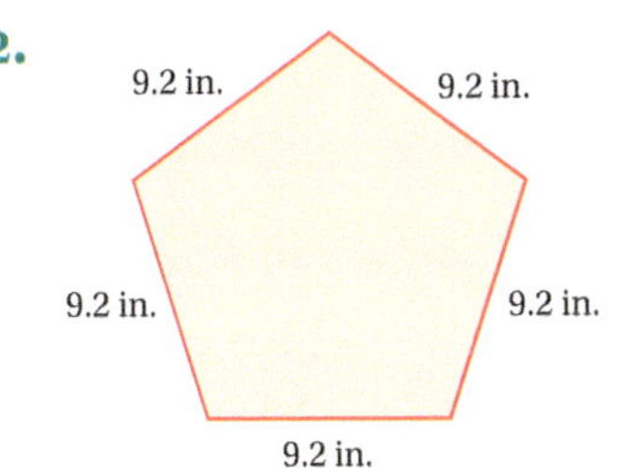

Answers

Skill to Review:
1. $\frac{50}{3}$, or $16\frac{2}{3}$ **2.** $\frac{128}{5}$, or $25\frac{3}{5}$

Margin Exercises:
1. 26 cm **2.** 46 in.

3. Find the perimeter of a rectangle that is 2 cm by 4 cm.

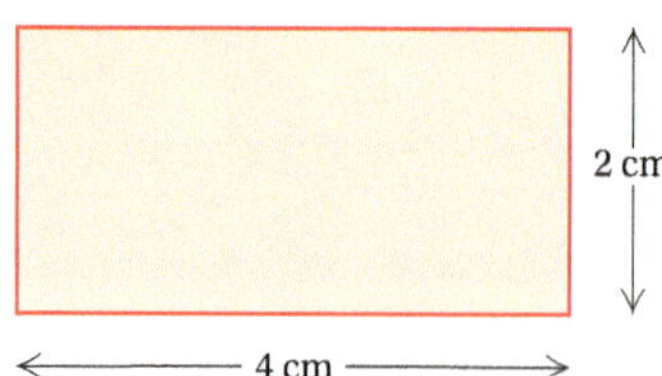

4. Find the perimeter of a rectangle that is 5.25 yd by 3.5 yd.

5. Find the perimeter of a rectangle that is $8\frac{1}{4}$ in. by 5 in.

GS

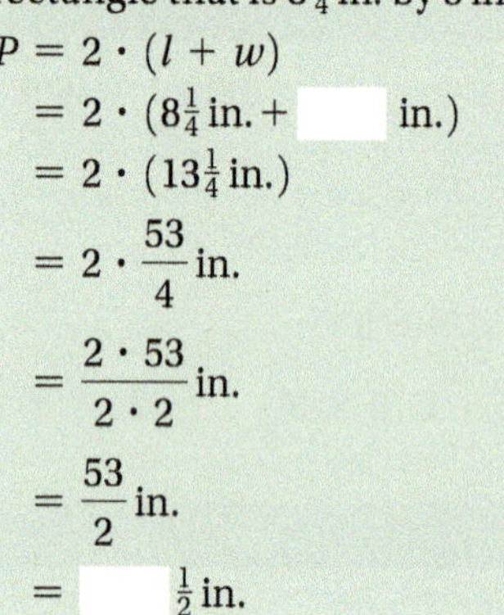

$P = 2 \cdot (l + w)$

$= 2 \cdot (8\frac{1}{4}\text{ in.} + \underline{\qquad}\text{ in.})$

$= 2 \cdot (13\frac{1}{4}\text{ in.})$

$= 2 \cdot \frac{53}{4}\text{ in.}$

$= \frac{2 \cdot 53}{2 \cdot 2}\text{ in.}$

$= \frac{53}{2}\text{ in.}$

$= \underline{\qquad}\frac{1}{2}\text{ in.}$

PERIMETER OF A RECTANGLE

The **perimeter of a rectangle** is twice the sum of the length and the width, or 2 times the length plus 2 times the width:

$$P = 2 \cdot (l + w), \quad \text{or} \quad P = 2 \cdot l + 2 \cdot w.$$

EXAMPLE 3 Find the perimeter of a rectangle that is 7.8 ft by 4.3 ft.

$P = 2 \cdot (l + w)$

$= 2 \cdot (7.8\text{ ft} + 4.3\text{ ft})$

$= 2 \cdot (12.1\text{ ft})$

$= 24.2\text{ ft}$

◀ **Do Exercises 3–5.**

A **square** is a rectangle with all sides the same length.

EXAMPLE 4 Find the perimeter of a square whose sides are 9 mm long.

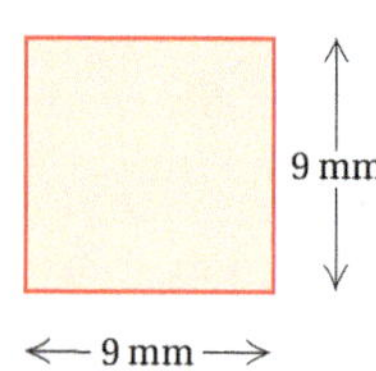

$P = 9\text{ mm} + 9\text{ mm} + 9\text{ mm} + 9\text{ mm}$

$= (9 + 9 + 9 + 9)\text{ mm}$

$= 36\text{ mm}$

◀ **Do Exercise 6.**

6. Find the perimeter of a square with sides of length 10 km.

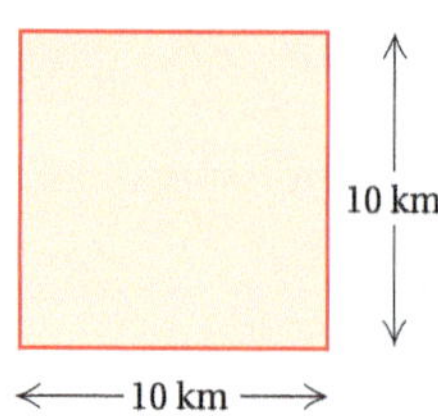

PERIMETER OF A SQUARE

The **perimeter of a square** is four times the length of a side:

$$P = 4 \cdot s.$$

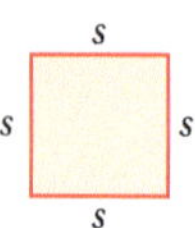

Answers

3. 12 cm **4.** 17.5 yd **5.** $26\frac{1}{2}$ in. **6.** 40 km

Guided Solution:
5. 5, 26

EXAMPLE 5 Find the perimeter of a square whose sides are $20\frac{1}{8}$ in. long.

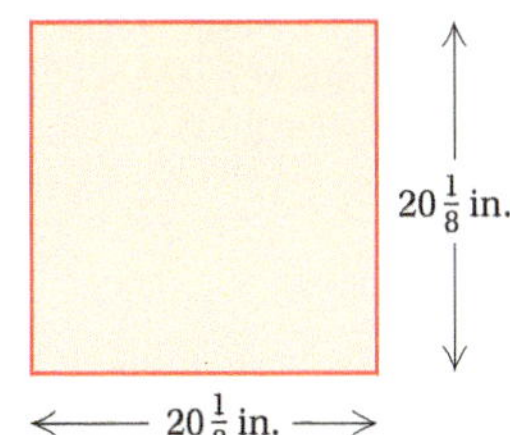

$$P = 4 \cdot s = 4 \cdot 20\frac{1}{8}\text{ in.}$$

$$= 4 \cdot \frac{161}{8}\text{ in.} = \frac{4 \cdot 161}{4 \cdot 2}\text{ in.}$$

$$= \frac{4}{4} \cdot \frac{161}{2}\text{ in.} = 80\frac{1}{2}\text{ in.}$$

Do Exercises 7 and 8. ▶

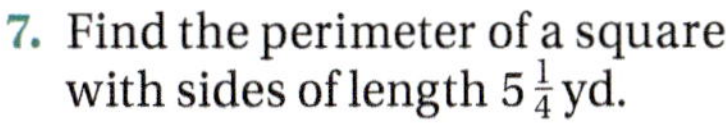

7. Find the perimeter of a square with sides of length $5\frac{1}{4}$ yd.

8. Find the perimeter of a square with sides of length 7.8 km.

$$P = 4 \cdot s$$
$$= 4 \cdot \square \text{ km}$$
$$= \square \text{ km}$$

b SOLVING APPLIED PROBLEMS

EXAMPLE 6 Jaci is adding crown molding along the top of the walls of her rectangular dining room, which measures 14 ft by 12 ft. How many feet of molding will be needed? If the molding sells for $3.25 per foot, what will its total cost be?

1. **Familiarize.** We make a drawing and let $P =$ the perimeter.

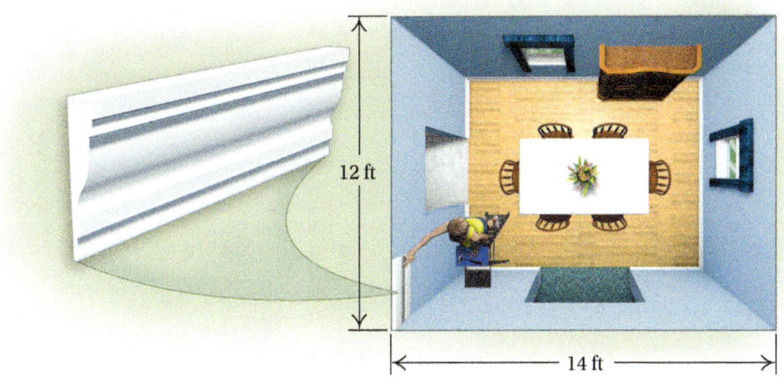

2. **Translate.** The perimeter of the room is given by

$$P = 2 \cdot (l + w) = 2 \cdot (14\text{ ft} + 12\text{ ft}).$$

3. **Solve.** We calculate the perimeter as follows:

$$P = 2 \cdot (14\text{ ft} + 12\text{ ft}) = 2 \cdot (26\text{ ft}) = 52\text{ ft}.$$

Then we multiply by $3.25 to find the cost of the crown molding:

$$\text{Cost} = \$3.25 \times \text{Perimeter} = \$3.25 \times 52\text{ ft} = \$169.$$

4. **Check.** The check is left to the student.
5. **State.** The 52 ft of crown molding that is needed will cost $169.

Do Exercise 9. ▶

9. A fence is to be built around a vegetable garden that measures 20 ft by 15 ft. How many feet of fence will be needed? If fencing sells for $2.95 per foot, what will the fencing cost?

Answers

7. 21 yd **8.** 31.2 km **9.** 70 ft; $206.50

Guided Solution:
8. 7.8, 31.2

8.2 Exercise Set

For Extra Help MyMathLab® MathXL® PRACTICE WATCH READ REVIEW

Reading Check

Complete each statement with the correct word from the following list. A word may be used more than once or not at all.

closed	perimeter	rectangle
open	polygon	square

RC1. A polygon is a(n) ________________ figure with three or more sides.

RC2. The distance around a polygon is its ________________.

RC3. The formula $P = 2 \cdot l + 2 \cdot w$ gives the ________________ of a rectangle.

RC4. The perimeter of a(n) ________________ is given by the formula $P = 4 \cdot s$.

a Find the perimeter of each polygon.

1.

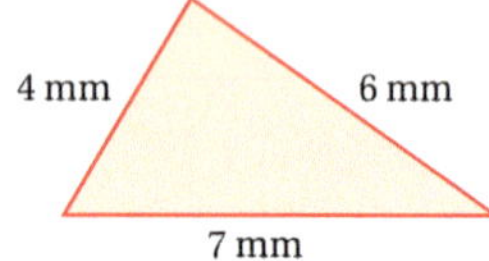

2.

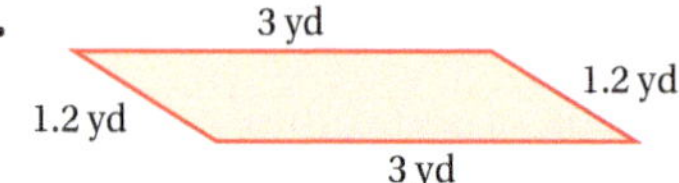

3.

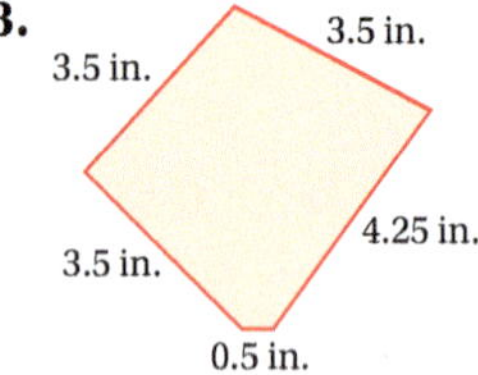

4.

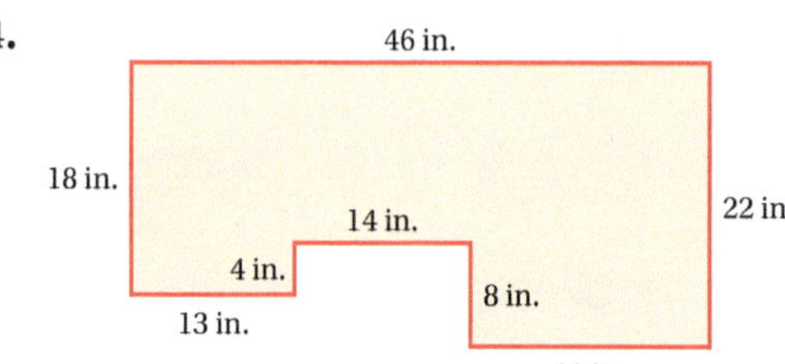

5.

6.

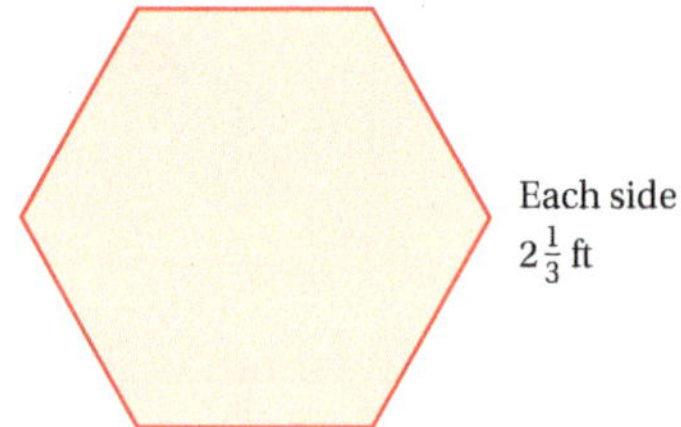

Find the perimeter of each rectangle.

7. 5 ft by 10 ft

8. 2.5 m by 100 m

9. $3\frac{1}{2}$ yd by $4\frac{1}{2}$ yd

10. 34.67 cm by 4.9 cm

Find the perimeter of each square.

11. 22 ft on a side

12. 56.9 km on a side

13. 45.5 mm on a side

14. $3\frac{1}{8}$ yd on a side

b Solve.

15. Most billiard tables are twice as long as they are wide. What is the perimeter of a billiard table that measures 4.5 ft by 9 ft?

16. A rectangular posterboard is 61.8 cm by 87.9 cm. What is the perimeter of the board?

17. A piece of flooring tile is a square with sides of length 30.5 cm. What is the perimeter of the piece of tile?

18. The Plaza de Balcarce in Balcarce, Argentina, is a public square with sides of length 300 m. What is the perimeter of the square?

19. A rain gutter is to be installed around the office building shown in the following figure.

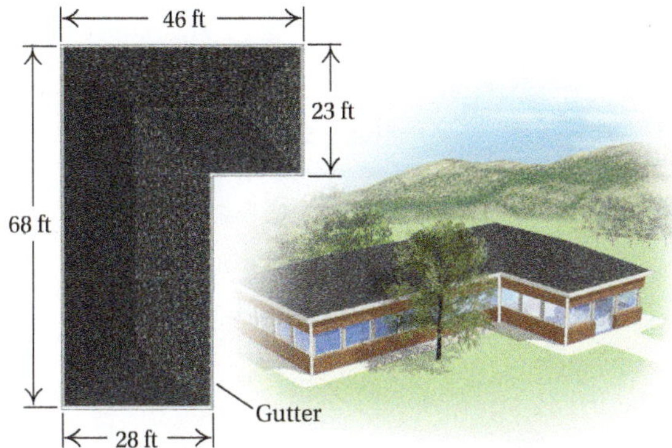

a) Find the perimeter of the office building.

b) If the gutter costs \$4.59 per foot, what is the total cost of the gutter?

20. Robbin plans to string lights around the lower level of the roof of the gazebo shown in the following figure.

a) If all sides of the roof are the same length, find the perimeter of the roof.

b) How many 6-ft strands of lights will Robbin need to buy?

Skill Maintenance

21. Find the simple interest on \$600 at 6.4% for $\frac{1}{2}$ year. [6.8a]

22. Find the simple interest on \$600 at 8% for 2 years. [6.8a]

Evaluate. [1.9b]

23. 10^3

24. 11^3

25. 15^2

26. 22^2

27. 7^2

28. 4^3

Solve.

29. ***Sales Tax.*** In a certain state, a sales tax of \$878 is collected when a car is purchased for \$17,560. What is the sales tax rate? [6.7a]

30. ***Commission Rate.*** Rich earns \$1854.60 selling \$16,860 worth of cell phones. What is the commission rate? [6.7b]

Synthesis

31. If it takes 18 in. to make the bow, how much ribbon is needed for the entire package shown here?

32. A carpenter is to build a fence around a 9-m by 12-m garden.

a) The posts are 3 m apart. How many posts will be needed?

b) The posts cost \$8.65 each. How much will the posts cost?

c) The fence will surround all but 3 m of the garden, which will be a gate. How long will the fence be?

d) The fencing costs \$3.85 per meter. What will the cost of the fencing be?

e) The gate costs \$69.95. What is the total cost of the materials?

8.3 Area

OBJECTIVES

a Find the area of a rectangle and a square.

b Find the area of a parallelogram, a triangle, and a trapezoid.

c Solve applied problems involving areas of rectangles, squares, parallelograms, triangles, and trapezoids.

SKILL TO REVIEW

Objective 5.5d: Simplify expressions that contain both fraction and decimal notation.

Calculate.

1. $\frac{1}{2} \times 16.243$

2. $0.5 \times \frac{3}{8}$

1. What is the area of this region? Count the number of square centimeters.

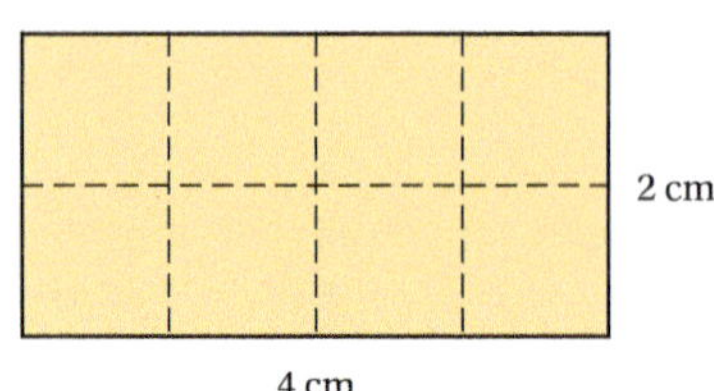

2. Find the area of a rectangle that is 7 km by 8 km.

3. Find the area of a rectangle that is $5\frac{1}{4}$ yd by $3\frac{1}{2}$ yd.

Answers

Skill to Review:

1. 8.1215 2. 0.1875, or $\frac{3}{16}$

Margin Exercises:

1. 8 cm^2 2. 56 km^2 3. $18\frac{3}{8}$ yd^2

a RECTANGLES AND SQUARES

A polygon and its interior form a plane region. We can find the area of a *rectangular region*, or *rectangle*, by filling it in with square units. Two such units, a *square inch* and a *square centimeter*, are shown below.

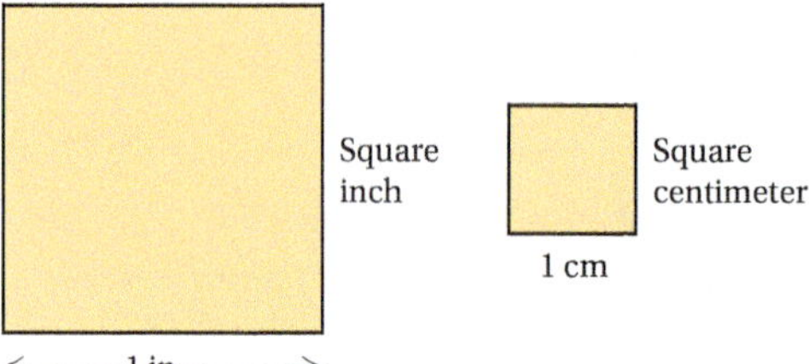

EXAMPLE 1 What is the area of this region?

We have a rectangular array. Since the region is filled with 12 square centimeters, its area is 12 square centimeters (sq cm), or 12 cm^2. The number of units is 3×4, or 12.

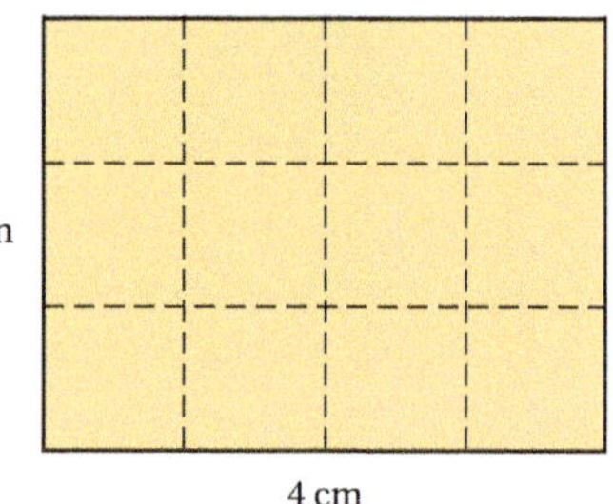

◀ **Do Margin Exercise 1.**

AREA OF A RECTANGLE

The **area of a rectangle** is the product of the length l and the width w:

$$A = l \cdot w.$$

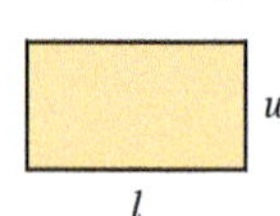

EXAMPLE 2 Find the area of a rectangle that is 7 yd by 4 yd.

We have

$$A = l \cdot w = 7\text{ yd} \cdot 4\text{ yd}$$
$$= 7 \cdot 4 \cdot \text{yd} \cdot \text{yd} = 28\text{ yd}^2.$$

We think of yd · yd as $(\text{yd})^2$ and denote it yd^2. Thus we read "28 yd^2" as "28 square yards."

◀ **Do Margin Exercises 2 and 3.**

EXAMPLE 3 Find the area of a square with sides of length 9 mm.

$$A = (9\text{ mm}) \cdot (9\text{ mm})$$
$$= 9 \cdot 9 \cdot \text{mm} \cdot \text{mm}$$
$$= 81\text{ mm}^2$$

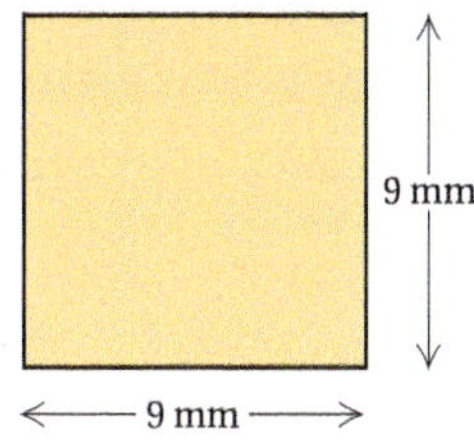

Do Exercise 4.

4. Find the area of a square with sides of length 12 km.

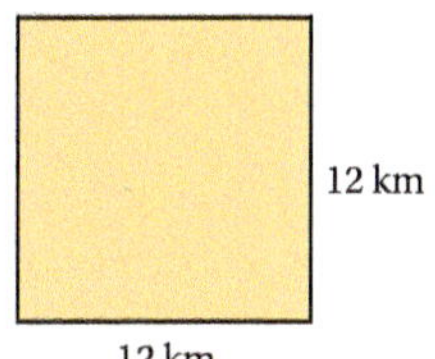

AREA OF A SQUARE

The **area of a square** is the square of the length of a side:

$$A = s \cdot s, \quad \text{or} \quad A = s^2.$$

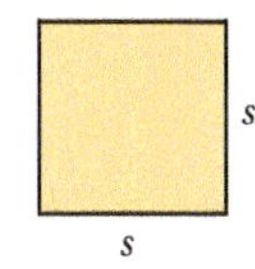

5. Find the area of a square with sides of length 10.9 m.

GS **6.** Find the area of a square with sides of length $3\frac{1}{2}$ yd.

$$A = s \cdot s$$
$$= 3\frac{1}{2}\text{ yd} \times \underline{\quad}\text{ yd}$$
$$= \frac{7}{2}\text{yd} \times \frac{7}{2}\text{yd}$$
$$= \frac{49}{4}\text{yd}^2$$
$$= \underline{\quad}\frac{1}{4}\text{yd}^2$$

EXAMPLE 4 Find the area of a square with sides of length 20.3 m.

$$A = s \cdot s = 20.3\text{ m} \times 20.3\text{ m} = 20.3 \times 20.3 \times \text{m} \times \text{m} = 412.09\text{ m}^2$$

Do Exercises 5 and 6.

b FINDING OTHER AREAS

Parallelograms

A **parallelogram** is a four-sided figure with two pairs of parallel sides, as shown below.

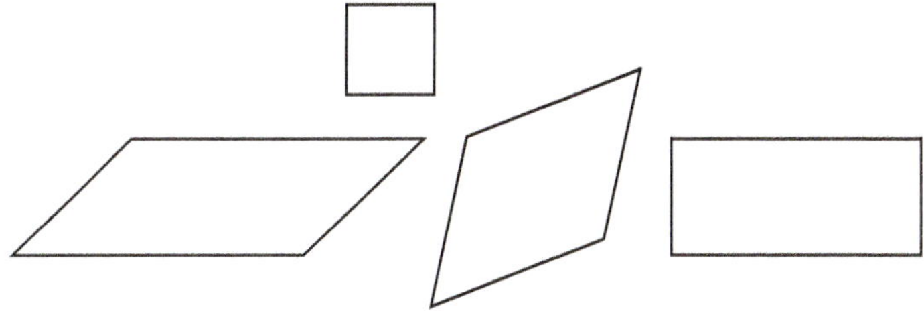

To find the area of a parallelogram, consider the one below.

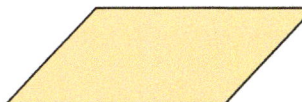

If we cut off a piece and move it to the other end, we get a rectangle.

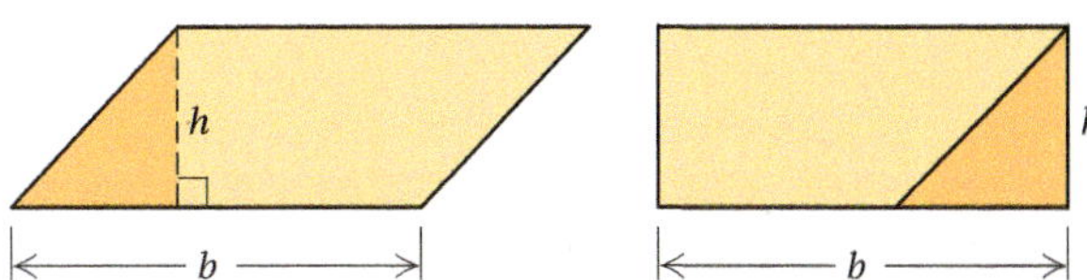

We can find the area by multiplying the length b, called a **base**, by h, called the **height**.

Answers

4. 144 km^2 **5.** 118.81 m^2 **6.** $12\frac{1}{4}\text{yd}^2$

Guided Solution:
6. $3\frac{1}{2}$, 12

AREA OF A PARALLELOGRAM

The **area of a parallelogram** is the product of the length of the base b and the height h:

$$A = b \cdot h.$$

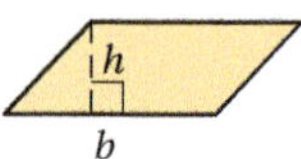

EXAMPLE 5 Find the area of this parallelogram.

$A = b \cdot h$
$= 7\text{ km} \cdot 5\text{ km}$
$= 35\text{ km}^2$

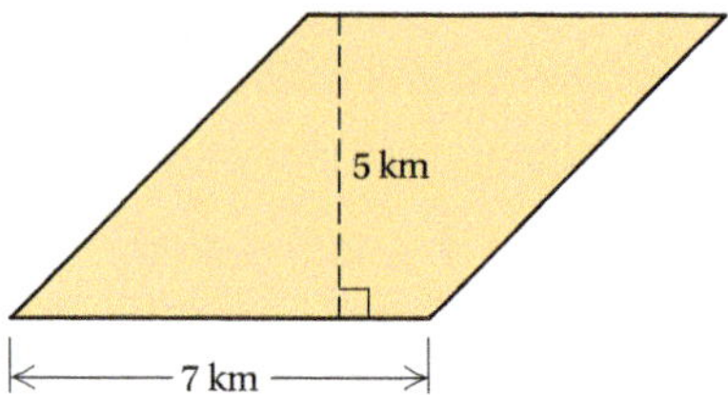

Find the area.

7.

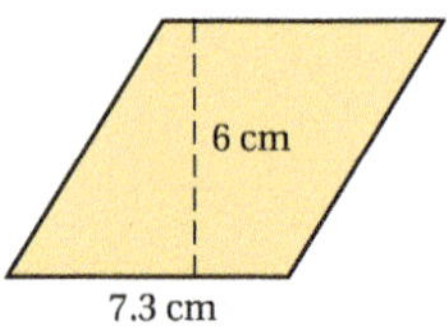

8.

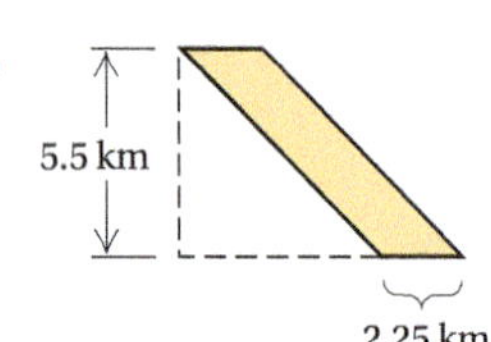

EXAMPLE 6 Find the area of this parallelogram.

$A = b \cdot h$
$= 1.2\text{ m} \times 6\text{ m}$
$= 7.2\text{ m}^2$

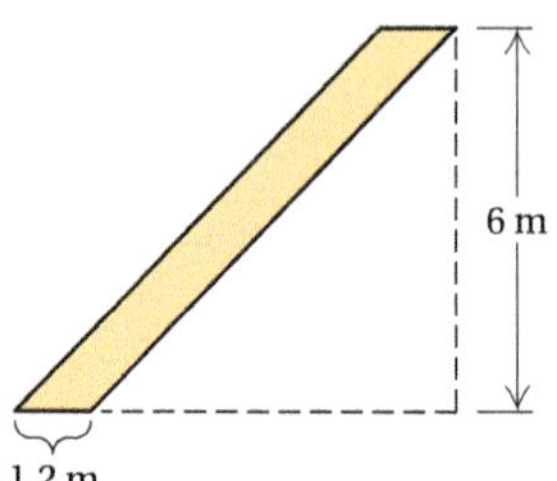

Do Exercises 7 and 8.

Triangles

A **triangle** is a polygon with three sides. To find the area of a triangle like the one shown on the left below, think of cutting out another just like it and placing it as shown on the right below.

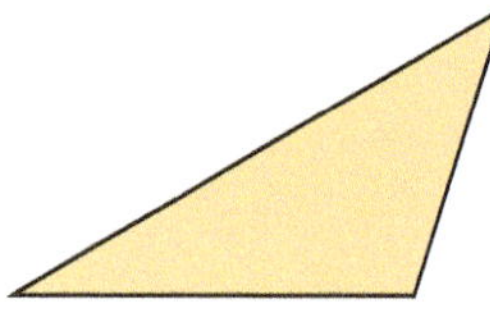

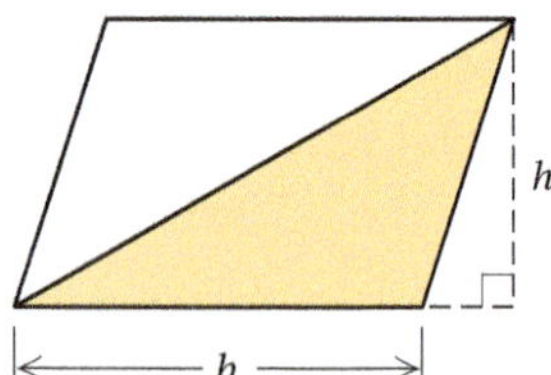

The resulting figure is a parallelogram whose area is

$$b \cdot h.$$

The triangle we began with has half the area of the parallelogram, or

$$\frac{1}{2} \cdot b \cdot h.$$

AREA OF A TRIANGLE

The **area of a triangle** is half the length of the base times the height:

$$A = \frac{1}{2} \cdot b \cdot h.$$

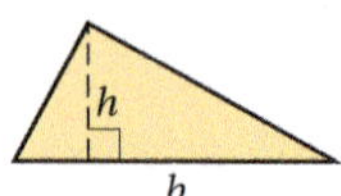

Answers

7. 43.8 cm^2 8. 12.375 km^2

EXAMPLE 7 Find the area of this triangle.

$$A = \frac{1}{2} \cdot b \cdot h$$
$$= \frac{1}{2} \cdot 9\text{ m} \cdot 6\text{ m}$$
$$= \frac{9 \cdot 6}{2}\text{ m}^2$$
$$= 27\text{ m}^2$$

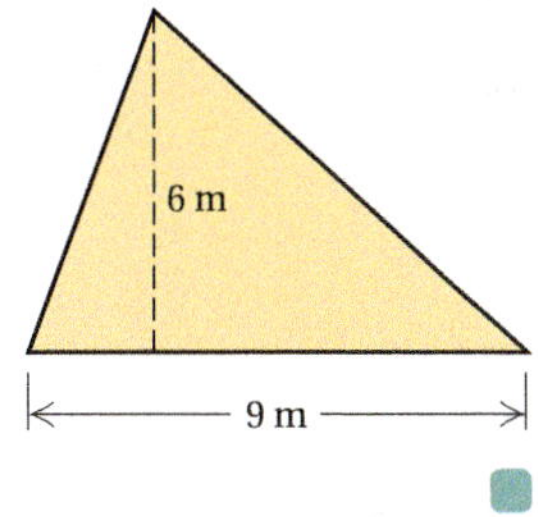

Find the area.

9.

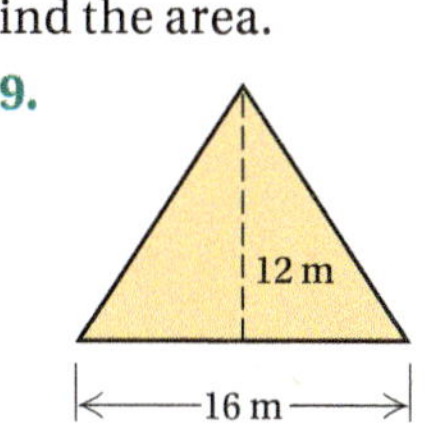

EXAMPLE 8 Find the area of this triangle.

$$A = \frac{1}{2} \cdot b \cdot h$$
$$= \frac{1}{2} \times 6.25\text{ cm} \times 5.5\text{ cm}$$
$$= 0.5 \times 6.25 \times 5.5\text{ cm}^2$$
$$= 17.1875\text{ cm}^2$$

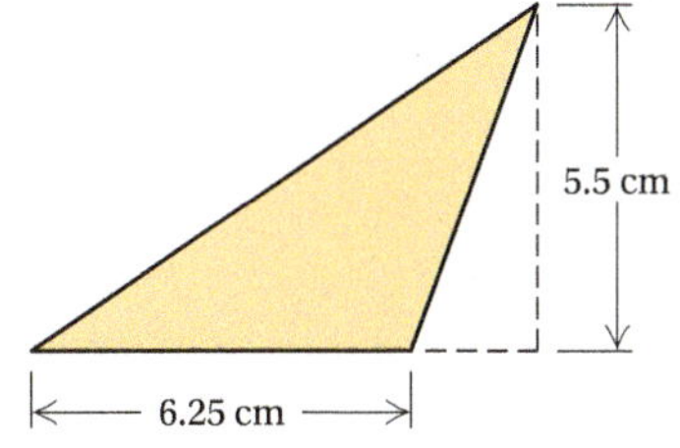

Do Exercises 9 and 10.

GS 10.

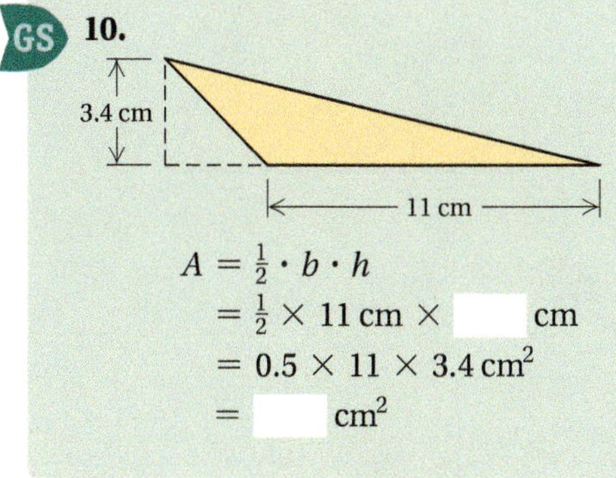

$A = \frac{1}{2} \cdot b \cdot h$

$= \frac{1}{2} \times 11\text{ cm} \times$ ____ cm

$= 0.5 \times 11 \times 3.4\text{ cm}^2$

$=$ ____ cm^2

Trapezoids

A **trapezoid** is a polygon with four sides, two of which, the **bases**, are parallel to each other.

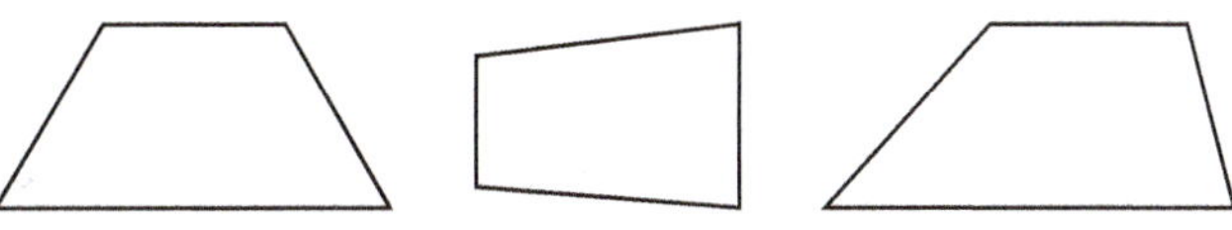

To find the area of a trapezoid, think of cutting out another just like it.

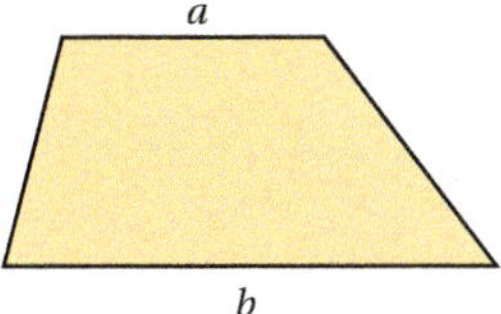

Then place the second one like this.

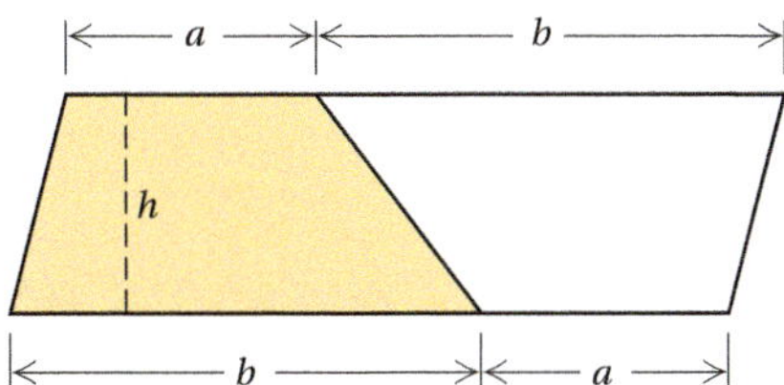

The resulting figure is a parallelogram whose area is

$h \cdot (a + b)$. The base is $a + b$.

The trapezoid we began with has half the area of the parallelogram, or

$$\frac{1}{2} \cdot h \cdot (a + b).$$

Answers

9. 96 m^2 **10.** 18.7 cm^2

Guided Solution:
10. 3.4, 18.7

AREA OF A TRAPEZOID

The **area of a trapezoid** is half the product of the height and the sum of the lengths of the parallel sides (bases):

$$A = \frac{1}{2} \cdot h \cdot (a + b), \quad \text{or} \quad A = \frac{a + b}{2} \cdot h.$$

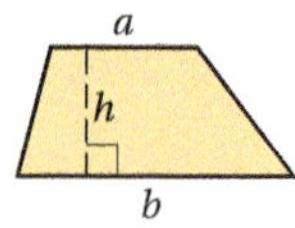

Find the area.

11.

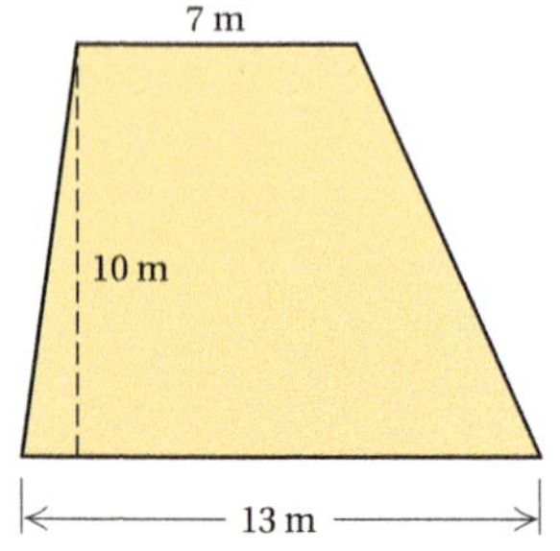

12.

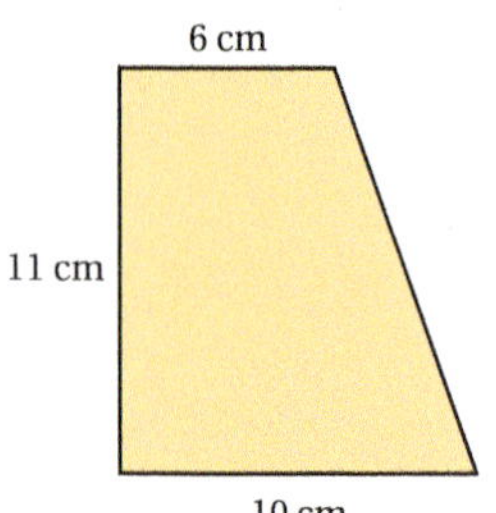

EXAMPLE 9 Find the area of this trapezoid.

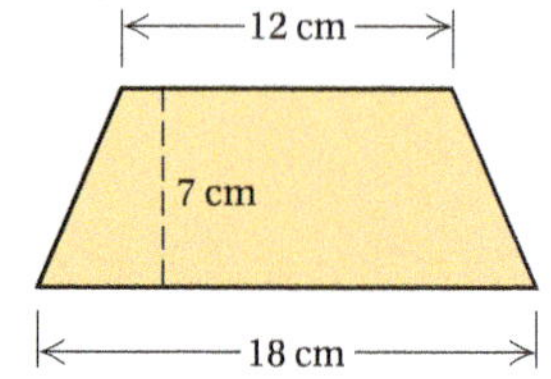

$$
\begin{aligned}
A &= \frac{1}{2} \cdot h \cdot (a + b) \\
&= \frac{1}{2} \cdot 7\text{ cm} \cdot (12 + 18)\text{ cm} \\
&= \frac{7 \cdot 30}{2} \cdot \text{cm}^2 = \frac{7 \cdot 15 \cdot 2}{1 \cdot 2}\text{cm}^2 \\
&= \frac{2}{2} \cdot \frac{7 \cdot 15}{1}\text{cm}^2 \\
&= 105\text{ cm}^2
\end{aligned}
$$

Do Exercises 11 and 12.

C SOLVING APPLIED PROBLEMS

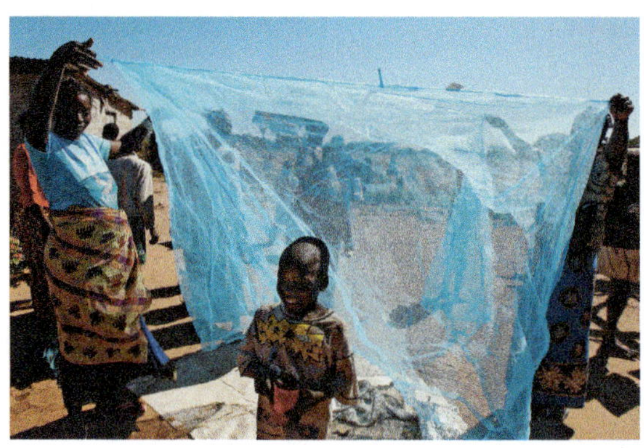

EXAMPLE 10 ***Mosquito Netting.*** Malaria is the leading cause of death among children in Africa. Bed nets prevent malaria transmission by creating a protective barrier against mosquitoes at night. In November 2006, the United Nations Foundation, the United Methodist Church, and the National Basketball Association launched the Nothing But Nets campaign to distribute mosquito netting in Africa. In the next six years, nets were sent to more than 25 countries in Africa. A medium-sized net measures approximately 9.843 ft by 8.2025 ft. A large-sized net measures approximately 13.124 ft by 8.2025 ft. Find the area of each net. How much larger is the area of the large net than that of the medium net?

Source: www.nothingbutnets.net

We find the area of each net using the area formula $A = l \cdot w$ and substituting values for l and w. Then we subtract the area of the medium net from the area of the large net.

Area of Medium Net	*Area of Large Net*
$A = l \times w$	$A = l \times w$
$A \approx 9.843\text{ ft} \times 8.2025\text{ ft}$	$A \approx 13.124\text{ ft} \times 8.2025\text{ ft}$
$A \approx 80.74\text{ ft}^2$	$A \approx 107.65\text{ ft}^2$

$$
\begin{aligned}
\text{Area of Large Net} - \text{Area of Medium Net} &= 107.65\text{ ft}^2 - 80.74\text{ ft}^2 \\
&= 26.91\text{ ft}^2
\end{aligned}
$$

The area of the large net is approximately 26.91 ft^2 larger than that of the medium net.

Answers

11. 100 m^2 12. 88 cm^2

EXAMPLE 11 *Lucas Oil Stadium.* The retractable roof of Lucas Oil Stadium, the home of the Indianapolis Colts football team, divides lengthwise. Each half measures 588 ft by 160 ft. The roof opens and closes in approximately 9–11 min. The opening measures 300 ft across. What is the total area of the retractable roof? What is the area of the opening?

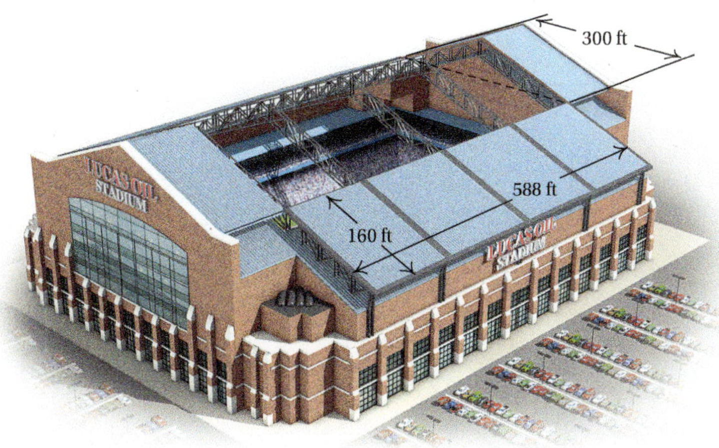

Each half of the retractable roof is a rectangle that measures 588 ft by 160 ft. The area of a rectangle is length times width, so we have

$$\begin{aligned} A &= l \cdot w \\ &= 588 \text{ ft} \times 160 \text{ ft} \\ &= 94{,}080 \text{ ft}^2. \end{aligned}$$

The total area of the two halves of the retractable roof is

$$\begin{aligned} \text{Total area} &= 2 \times 94{,}080 \text{ ft}^2 \\ &= 188{,}160 \text{ ft}^2. \end{aligned}$$

When the retractable roof is open, the dimensions of the opening are 588 ft by 300 ft. The area of this rectangle is

$$\begin{aligned} A &= l \cdot w \\ &= 588 \text{ ft} \times 300 \text{ ft} \\ &= 176{,}400 \text{ ft}^2. \end{aligned}$$

When the roof is open, the area of the opening is 176,400 ft^2.

Do Exercise 13.

13. Find the area of this kite.

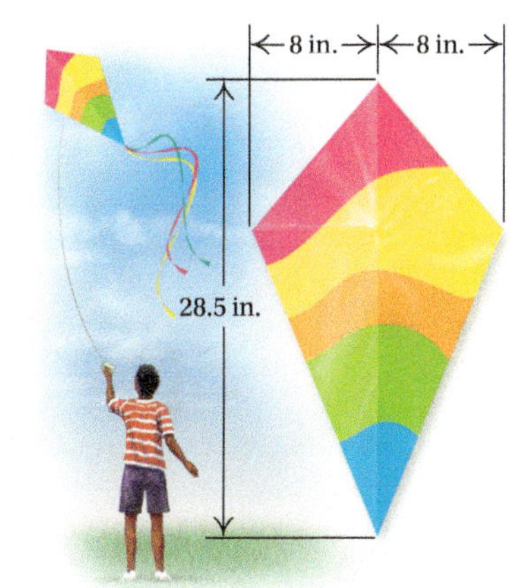

Answer

13. 228 in^2

8.3 Exercise Set

For Extra Help
MyMathLab® MathXL® PRACTICE  WATCH  READ  REVIEW

Reading Check

Complete each statement with the correct phrase from the list on the right.

RC1. The area of a square is ________.

RC2. The area of a rectangle is ________.

RC3. The area of a triangle is ________.

RC4. The area of a trapezoid is ________.

a) half the length of the base times the height

b) the square of the length of a side

c) the product of the length and the width

d) half the product of the height and the sum of the lengths of the bases

a Find the area.

1.

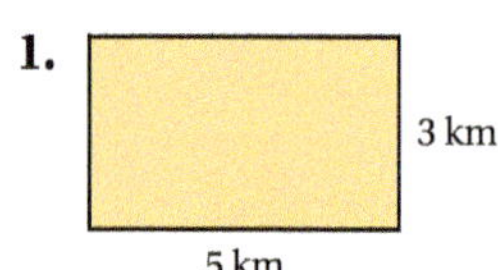

2.

1.5 ft

1.5 ft

3.

2 in.

0.7 in.

4.

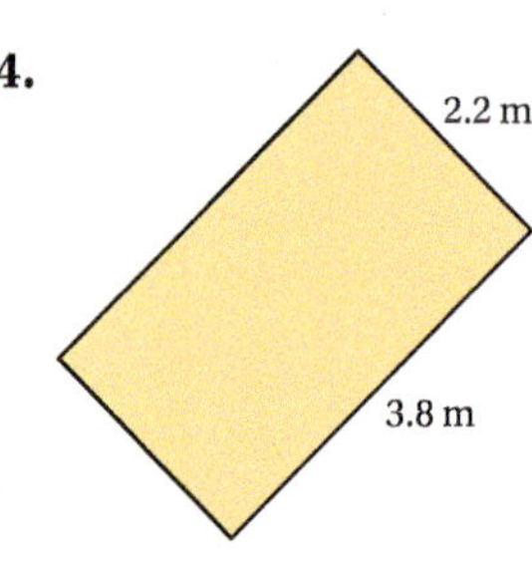

5.

$2\frac{1}{2}$ yd

$2\frac{1}{2}$ yd

6.

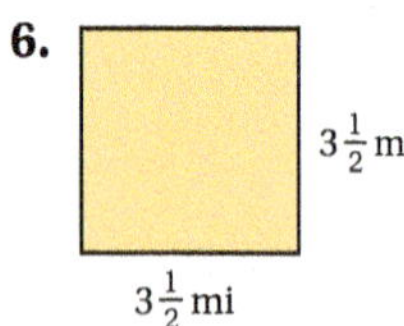

7.

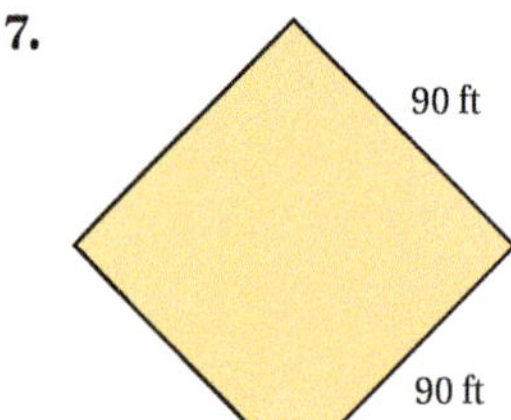

8.

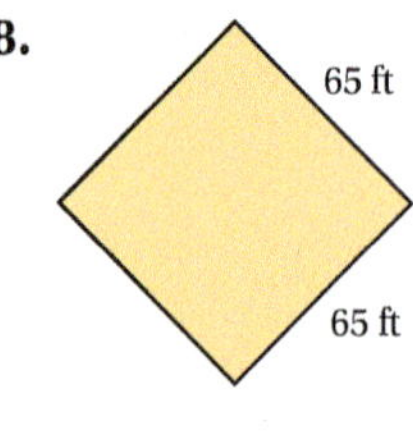

Find the area of each rectangle.

9. 5 ft by 10 ft

10. 14 yd by 8 yd

11. 34.67 cm by 4.9 cm

12. 2.45 km by 100 km

13. $4\frac{2}{3}$ in. by $8\frac{5}{6}$ in.

14. $10\frac{1}{3}$ mi by $20\frac{2}{3}$ mi

Find the area of the square.

15. 22 ft on a side

16. 18 yd on a side

17. 56.9 km on a side

18. 45.5 m on a side

19. $5\frac{3}{8}$ yd on a side

20. $7\frac{2}{3}$ ft on a side

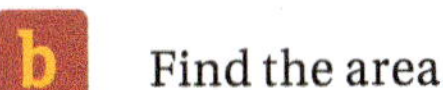

b Find the area.

21.

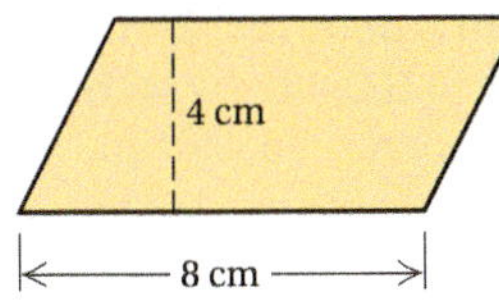

22.

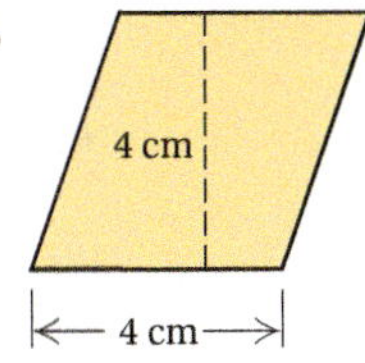

23.

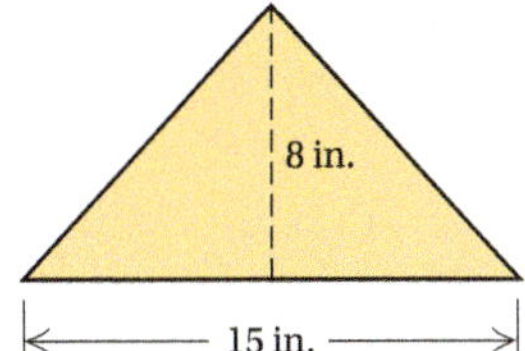

24.

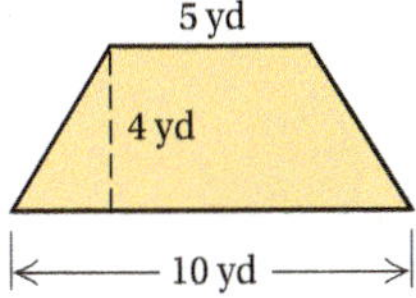

25.

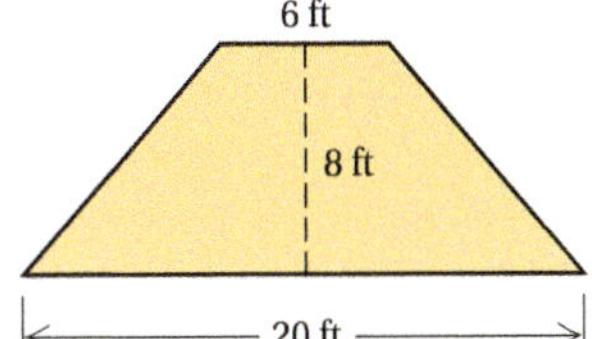

26.

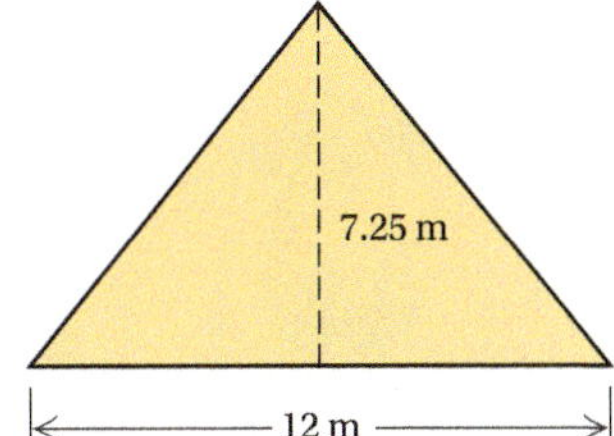

27.

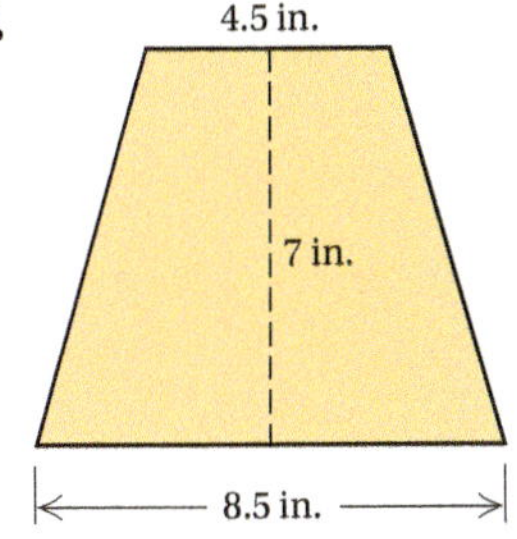

28.

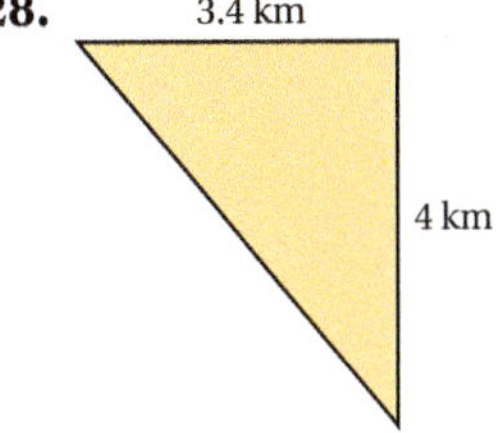

29.

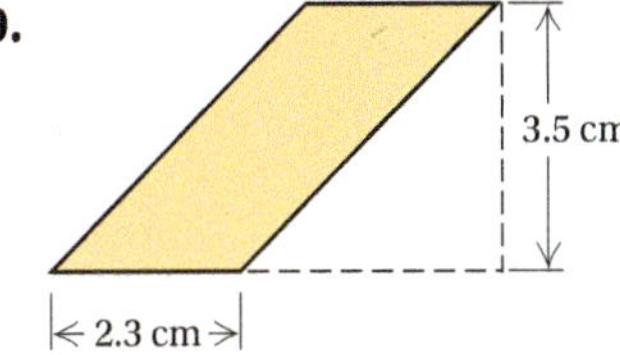

30.

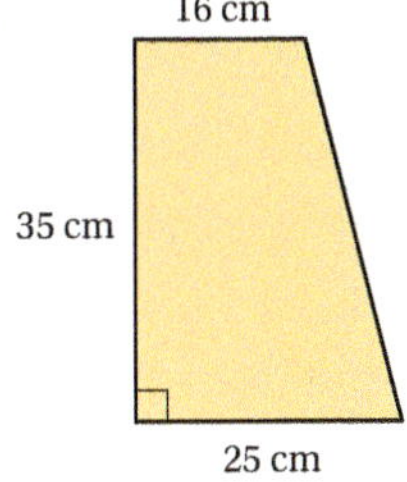

31.

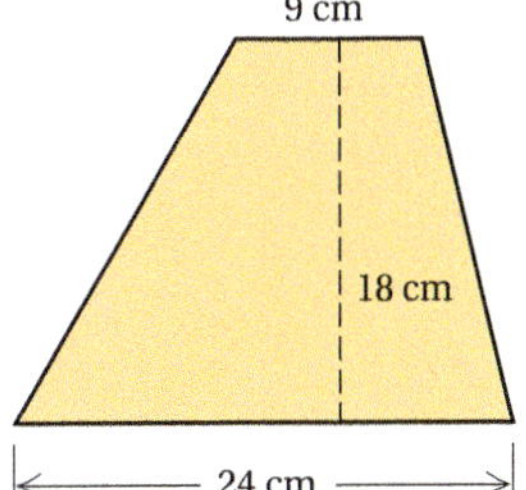

32.

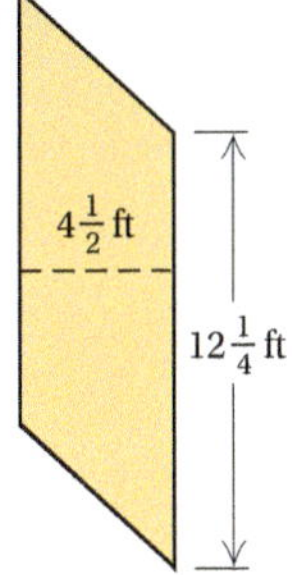

33.

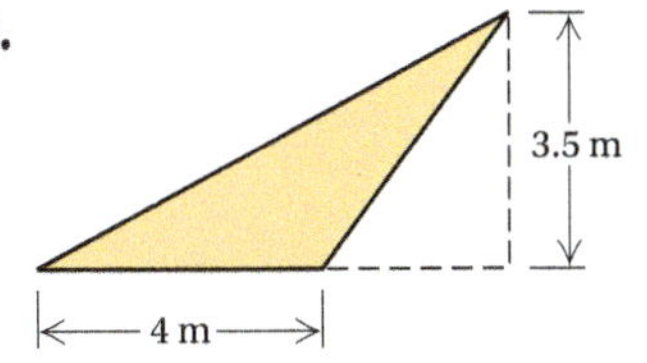

34.

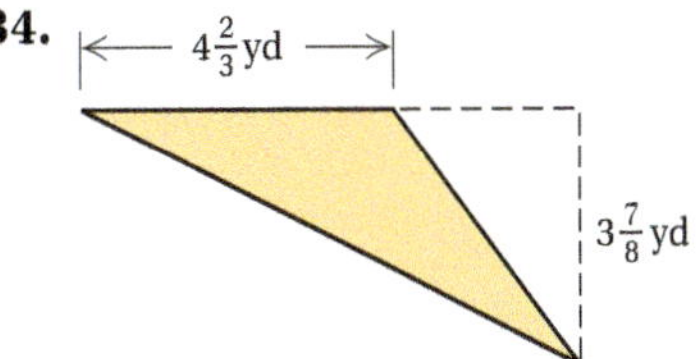

C Solve.

35. *Area of a Lawn.* A lot is 40 m by 36 m. A house 27 m by 9 m is built on the lot. How much area is left over for a lawn?

36. *Area of a Field.* A field is 240.8 m by 450.2 m. A rectangular area that measures 160.4 m by 90.6 m is paved for a parking lot. How much area is unpaved?

37. For a performance at an outdoor event, a folk music group rented a triangular tent. The base of the triangular-shaped floor of the tent was 20 ft and its height was $17\frac{1}{2}$ ft, and the tent was placed in a corner of a small park that measured 200 ft by 200 ft. Approximately how much of the park space was left for the audience?

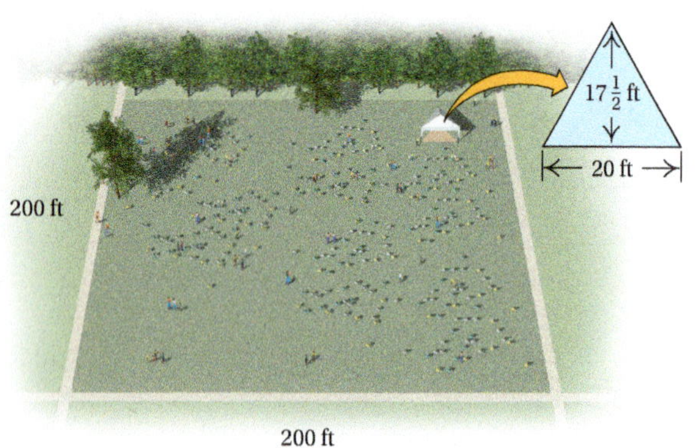

38. Becky's rectangular swimming pool measures 27 ft by 14.6 ft. She likes to relax while floating on an inflatable mattress, which measures 6.5 ft by 2.75 ft. What area of the pool is left around her for her nieces and nephews to play in?

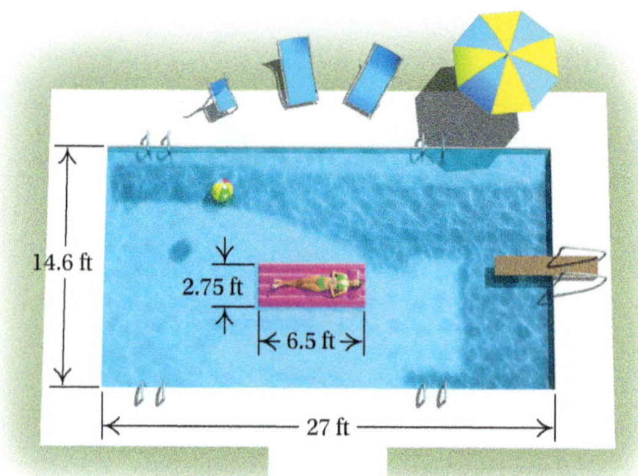

39. *Area of a Sidewalk.* Franklin Construction Company builds a sidewalk around two sides of a new library, as shown in the figure.

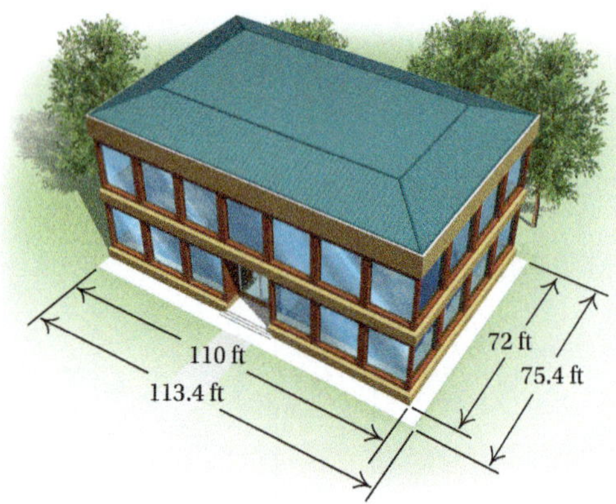

a) What is the area of the sidewalk?

b) The concrete for the sidewalk will cost the library $12.50 per square foot. How much will the concrete for the project cost?

40. Maravene is planning a wildflower border around three sides of her backyard, as shown in the figure. She will use wildflower mats to seed the border. Each mat covers 7.5 ft^2 and costs $4.99.

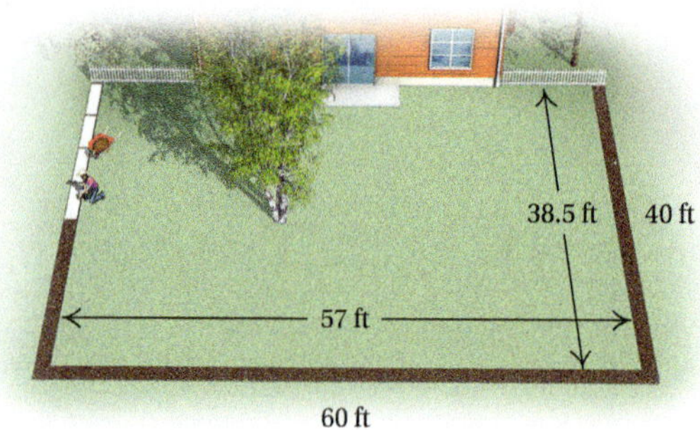

a) What is the area of the border?

b) How many wildflower mats will Maravene need to complete the job?

c) What will be the total cost of the wildflower mats?

41. *Painting Costs.* A room is 15 ft by 20 ft. The ceiling is 8 ft above the floor. There are two windows in the room, each 3 ft by 4 ft. The door is $2\frac{1}{2}$ ft by $6\frac{1}{2}$ ft.

a) What is the total area of the walls and the ceiling?

b) A gallon of paint will cover 360.625 ft^2. How many gallons of paint are needed for the room, including the ceiling?

c) Paint costs $34.95 a gallon. How much will it cost to paint the room?

42. *Carpeting Costs.* A restaurant owner wants to carpet a 15-yd by 20-yd room.

a) How many square yards of carpeting are needed?

b) The carpeting she wants is $28.50 per square yard, including installation. How much will it cost to carpet the room?

Find the area of the shaded region in each figure.

43.

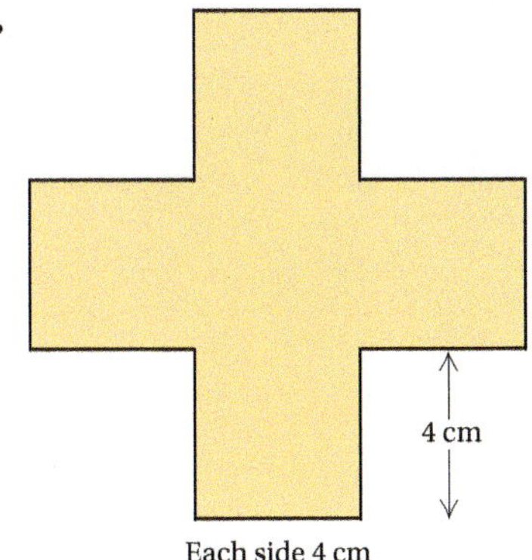

44.

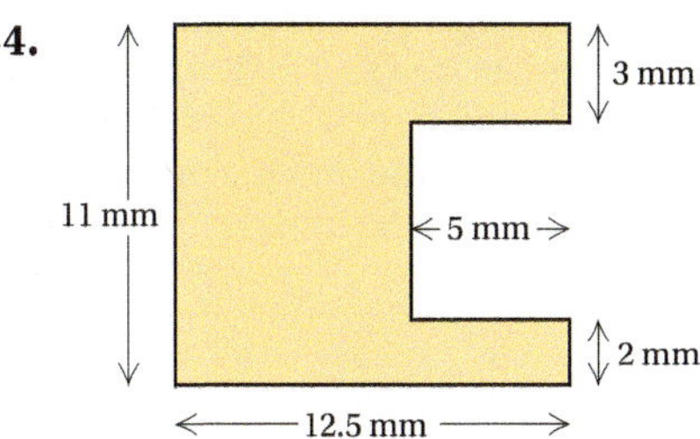

45.

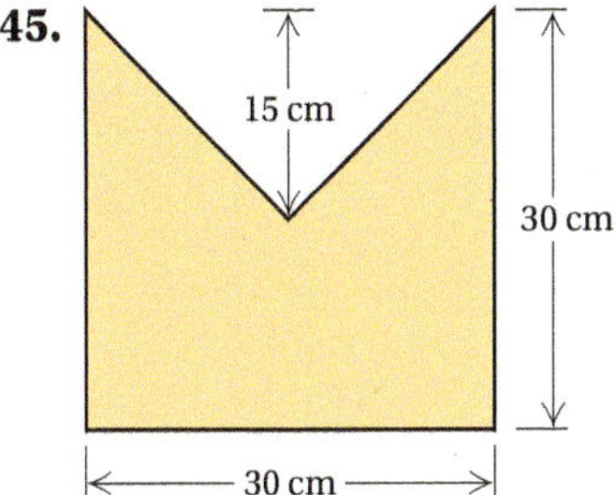

46.

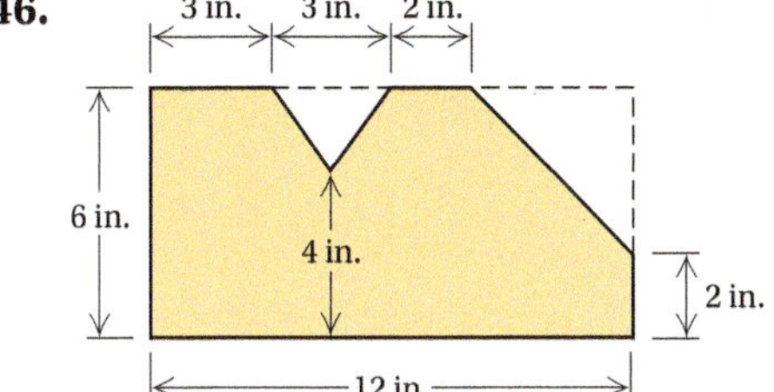

47.

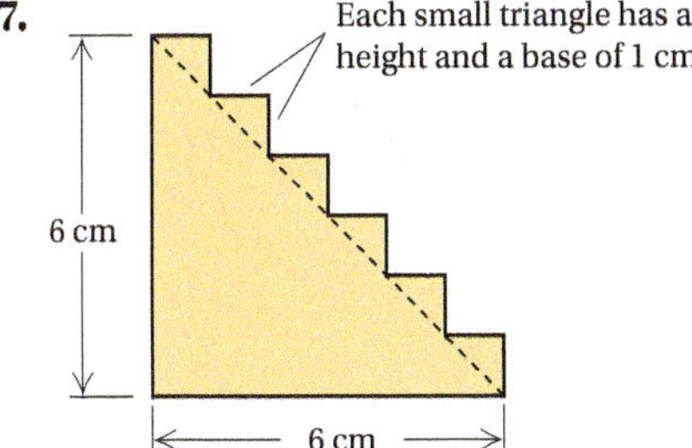

48.

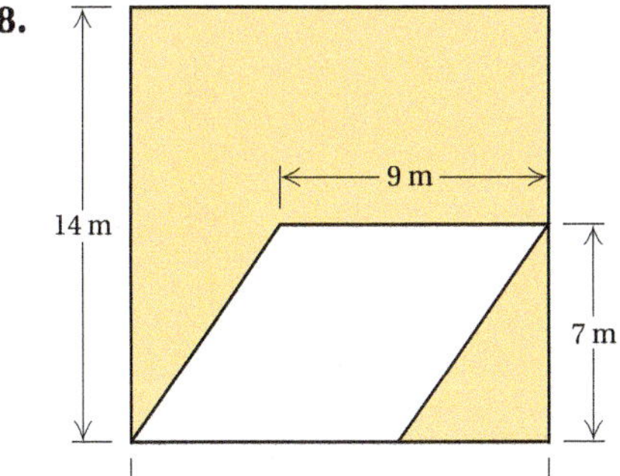

49. ***Triangular Sail.*** Jane's Custom Sails is making a custom sail for a laser sailboat. From a rectangular piece of dacron sailcloth that measures 18 ft by 12 ft, Jane cuts out a right triangular area plus a rectangular extension on each side for the hems, with the dimensions shown below. How much fabric (area) is left over?

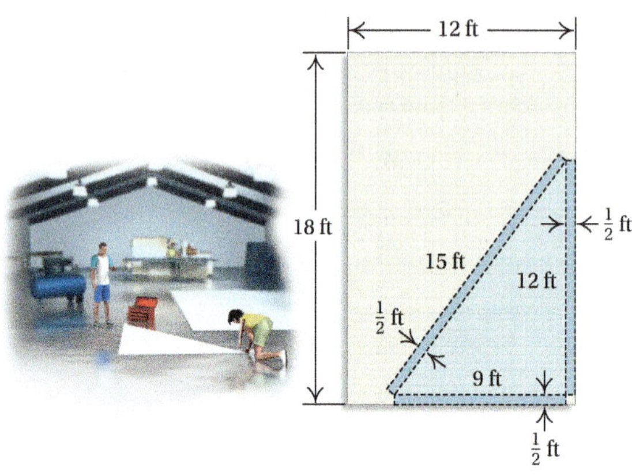

50. ***Building Area.*** Find the total area of the sides and the ends of the building.

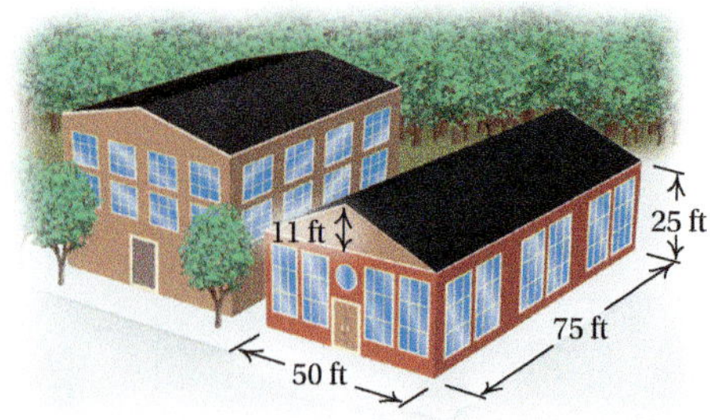

Skill Maintenance

Simplify. [3.5b]

51. $\frac{350}{1450}$

52. $\frac{1000}{1{,}000{,}000}$

53. $-\frac{165}{264}$

54. $\frac{1050}{420}$

Solve. [6.4b]

55. What percent of \$80 is \$8?

56. \$16 is what percent of 96?

57. $33\frac{1}{3}\%$ of 273 is what?

58. 40% of what is 144?

Interest is compounded semiannually. Find the amount in the account after the given length of time. Round to the nearest cent. [6.8b]

	Principal	Rate of Interest	Time	Amount in the Account
59.	\$25,000	4%	5 years	
60.	\$150,000	$6\frac{7}{8}\%$	15 years	
61.	\$150,000	7.4%	20 years	
62.	\$160,000	5%	20 years	

Synthesis

63. Find the area, in square inches, of the shaded region.

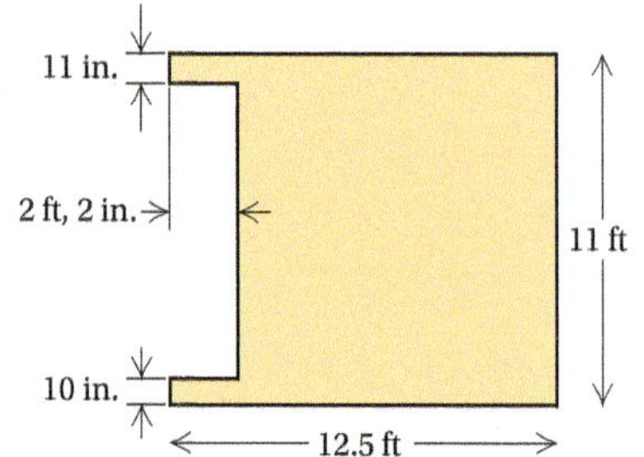

64. Find the area, in square feet, of the shaded region.

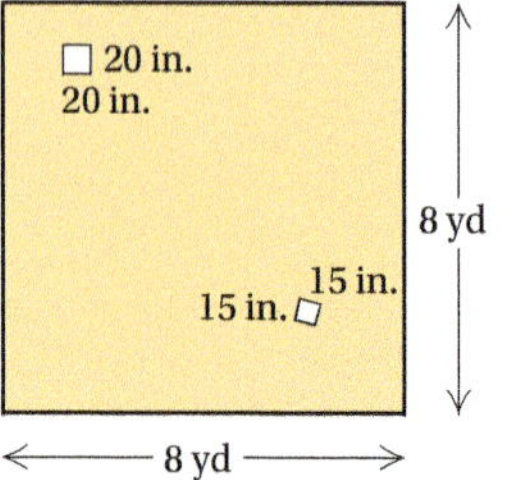

Circles

8.4

OBJECTIVES

a Find the length of a radius of a circle given the length of a diameter, and find the length of a diameter given the length of a radius.

b Find the circumference of a circle given the length of a diameter or a radius.

c Find the area of a circle given the length of a diameter or a radius.

d Solve applied problems involving circles.

SKILL TO REVIEW

Objective 3.4a: Multiply an integer and a fraction.

Multiply and simplify.

1. $2 \cdot \frac{1}{6}$ **2.** $\frac{22}{7} \cdot 21$

a RADIUS AND DIAMETER

Shown below is a circle with center O. Segment $\overline{AC}$ is a *diameter*. A **diameter** is a segment that passes through the center of the circle and has endpoints on the circle. Segment $\overline{OB}$ is called a *radius*. A **radius** is a segment with one endpoint on the center and the other endpoint on the circle.

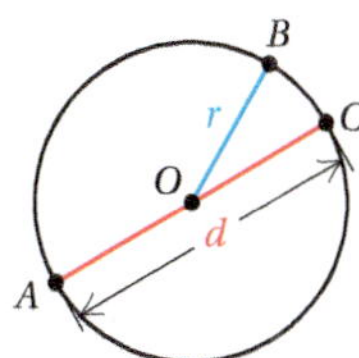

DIAMETER AND RADIUS

Suppose that d is the length of a diameter of a circle and r is the length of a radius. Then

$$\boldsymbol{d = 2 \cdot r} \quad \text{and} \quad \boldsymbol{r = \frac{d}{2}}.$$

EXAMPLE 1 Find the length of a radius of this circle.

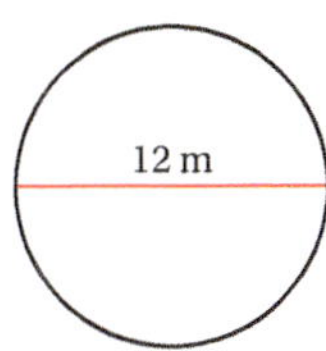

$$\begin{aligned} r &= \frac{d}{2} \\ &= \frac{12\text{ m}}{2} \\ &= 6\text{ m} \end{aligned}$$

The radius is 6 m.

EXAMPLE 2 Find the length of a diameter of this circle.

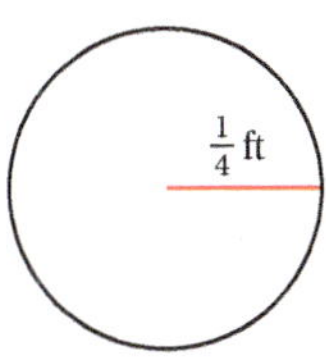

$$\begin{aligned} d &= 2 \cdot r \\ &= 2 \cdot \frac{1}{4}\text{ ft} \\ &= \frac{1}{2}\text{ ft} \end{aligned}$$

The diameter is $\frac{1}{2}$ ft.

Do Margin Exercises 1 and 2.

1. Find the length of a radius.

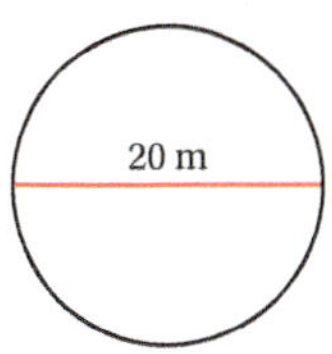

2. Find the length of a diameter.

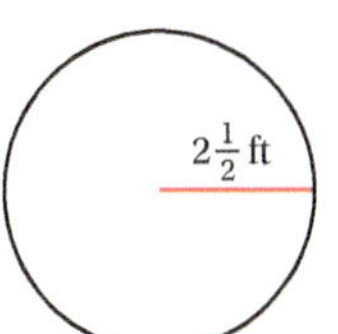

Answers

Skill to Review:

1. $\frac{1}{3}$ **2.** 66

Margin Exercises:

1. 10 m **2.** 5 ft

b CIRCUMFERENCE

The **circumference** of a circle is the distance around it. Calculating the circumference is similar to finding the perimeter of a polygon.

To find a formula for the circumference C of any circle given its diameter d, we consider the ratio C/d. Take a dinner plate and measure the circumference C with a tape measure. Also measure the diameter d. The results for a specific plate are shown in the following figure.

Finding the ratio, we have

$$\frac{C}{d} = \frac{33.7 \text{ in.}}{10.75 \text{ in.}} \approx 3.1.$$

Suppose we do this with plates and circles of several sizes. We get different values for C and d, but always a number close to 3.1 for C/d. For any circle, if we divide the circumference C by the diameter d, we get the same number. We call this number π (pi). The *exact* value of the ratio C/d is π; 3.14 and 22/7 are approximations of π. If $C/d = \pi$, then $C = \pi \cdot d$.

3. Find the circumference of the circle. Use 3.14 for π.

GS

$C = \pi \cdot d$

$\approx 3.14 \times$ ____ in.

$=$ ____ in.

CIRCUMFERENCE AND DIAMETER

The circumference C of a circle of diameter d is given by

$$\boldsymbol{C = \pi \cdot d.}$$

The number π is about 3.14, or about $\frac{22}{7}$.

EXAMPLE 3 Find the circumference of this circle. Use 3.14 for π.

$$\begin{aligned} C &= \pi \cdot d \\ &\approx 3.14 \times 6 \text{ cm} \\ &= 18.84 \text{ cm} \end{aligned}$$

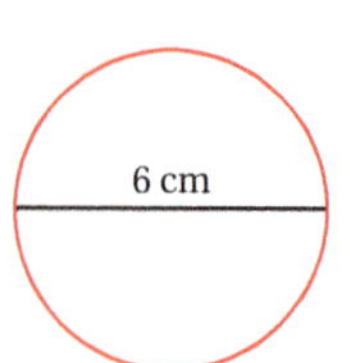

The circumference is about 18.84 cm.

◀ **Do Exercise 3.**

Answer

3. 56.52 in.

Guided Solution:

3. 18, 56.52

Since $d = 2 \cdot r$, where r is the length of a radius, it follows that

$$C = \pi \cdot d = \pi \cdot (2 \cdot r), \quad \text{or} \quad 2 \cdot \pi \cdot r.$$

CIRCUMFERENCE AND RADIUS

The circumference C of a circle of radius r is given by

$$C = 2 \cdot \pi \cdot r.$$

EXAMPLE 4 Find the circumference of this circle. Use $\frac{22}{7}$ for π.

$$C = 2 \cdot \pi \cdot r$$
$$\approx 2 \cdot \frac{22}{7} \cdot 70 \text{ in.}$$
$$= 2 \cdot 22 \cdot \frac{70}{7} \text{ in.}$$
$$= 44 \cdot 10 \text{ in.}$$
$$= 440 \text{ in.}$$

70 in.

The circumference is about 440 in.

EXAMPLE 5 Find the perimeter of this figure. Use 3.14 for π.

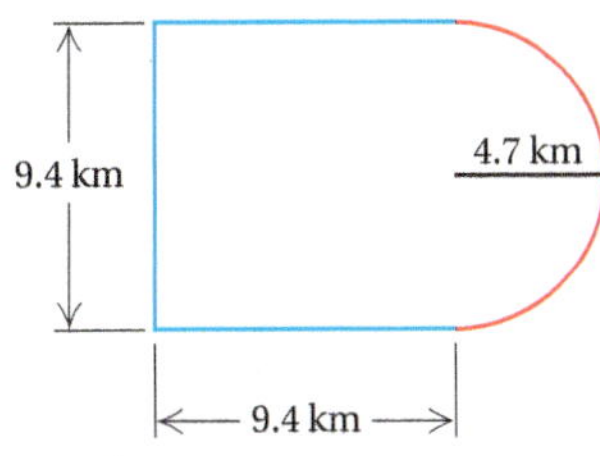

We let $P =$ the perimeter. We see that we have half a circle attached to a square. Thus we add half the circumference of the circle to the lengths of the three sides of the square.

$$P = \text{Length of three sides of the square} + \text{Half of the circumference of the circle}$$
$$= 3 \times 9.4 \text{ km} + \frac{1}{2} \times 2 \times \pi \times 4.7 \text{ km}$$
$$\approx 28.2 \text{ km} + 3.14 \times 4.7 \text{ km}$$
$$= 28.2 \text{ km} + 14.758 \text{ km}$$
$$= 42.958 \text{ km}$$

The perimeter is about 42.958 km.

Do Exercises 4 and 5. ▶

CALCULATOR CORNER

The $\boxed{\pi}$ Key Many calculators have a $\boxed{\pi}$ key that can be used to enter the value of π in a computation. Results obtained using the $\boxed{\pi}$ key might be slightly different from those obtained when 3.14 is used for the value of π in a computation.

When we use the $\boxed{\pi}$ key to find the circumference of the circle in Example 3, we get a result of approximately 18.85. Note that this is slightly different from the result found using 3.14 for the value of π.

EXERCISES

1. Use a calculator with a $\boxed{\pi}$ key to perform the computations in Examples 4 and 5.
2. Use a calculator with a $\boxed{\pi}$ key to perform the computations in Margin Exercises 3–5.

4. Find the circumference of this circle. Use $\frac{22}{7}$ for π.

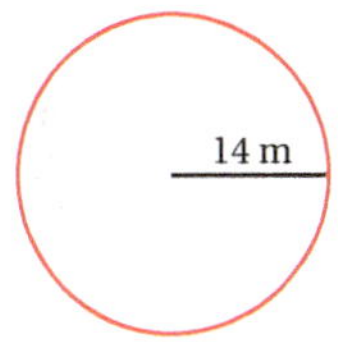

5. Find the perimeter of this figure. Use 3.14 for π.

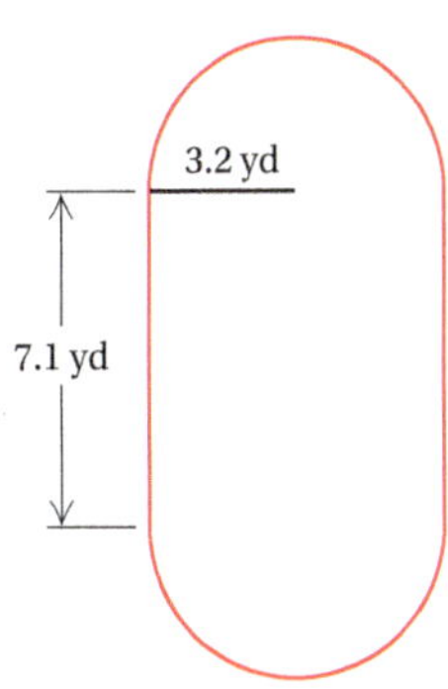

Answers

4. 88 m **5.** 34.296 yd

C AREA

To find the area of a circle of radius r, think of cutting half the circular region into small pieces and arranging them as shown below.

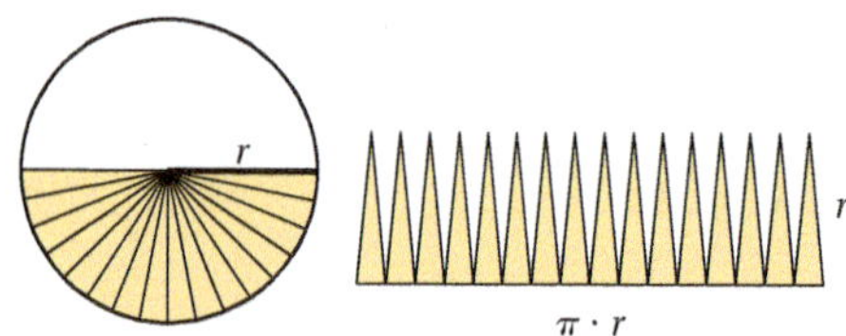

Then imagine cutting the other half of the circular region and arranging the pieces in with the others as shown below.

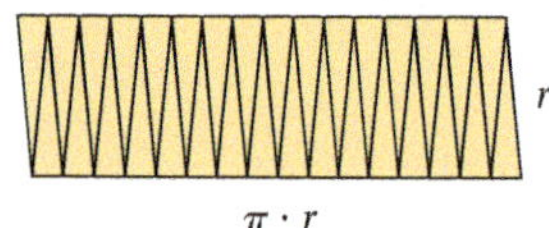

This is almost a parallelogram. The base has length $\frac{1}{2} \cdot 2 \cdot \pi \cdot r$, or $\pi \cdot r$ (half the circumference), and the height is r. Thus the area is

$$(\pi \cdot r) \cdot r.$$

Caution!

Remember that circumference is always measured in linear units like ft, m, cm, yd, and so on. But area is measured in square units like ft^2, m^2, cm^2, yd^2, and so on.

AREA OF A CIRCLE

The **area of a circle** with radius of length r is given by

$$A = \pi \cdot r \cdot r, \quad \text{or} \quad A = \pi \cdot r^2.$$

EXAMPLE 6 Find the area of this circle. Use $\frac{22}{7}$ for π.

$$\begin{aligned} A &= \pi \cdot r \cdot r \\ &\approx \frac{22}{7} \cdot 14\text{ cm} \cdot 14\text{ cm} \\ &= \frac{22}{7} \cdot 196\text{ cm}^2 \\ &= 616\text{ cm}^2 \end{aligned}$$

14 cm

The area is about 616 cm^2.

EXAMPLE 7 Find the area of this circle. Use 3.14 for π. Round to the nearest hundredth.

The diameter is 4.2 m; the radius is $4.2\text{ m} \div 2$, or 2.1 m.

$$\begin{aligned} A &= \pi \cdot r \cdot r \\ &\approx 3.14 \times 2.1\text{ m} \times 2.1\text{ m} \\ &= 3.14 \times 4.41\text{ m}^2 \\ &= 13.8474\text{ m}^2 \\ &\approx 13.85\text{ m}^2 \end{aligned}$$

4.2 m

The area is about 13.85 m^2.

◀ **Do Exercises 6 and 7.**

6. Find the area of this circle. Use $\frac{22}{7}$ for π. GS

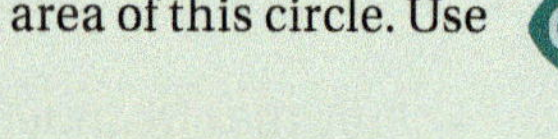

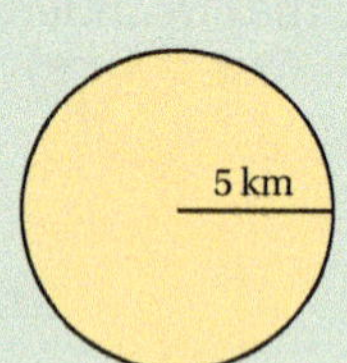

$$\begin{aligned} A &= \pi \cdot r \cdot r \\ &\approx \tfrac{22}{7} \cdot 5\text{ km} \cdot \underline{\quad}\text{ km} \\ &= \tfrac{22}{7} \cdot 25\text{ km}^2 \\ &= \tfrac{550}{7}\text{ km}^2 \\ &= \underline{\quad}\tfrac{4}{7}\text{ km}^2 \end{aligned}$$

7. Find the area of this circle. Use 3.14 for π. Round to the nearest hundredth.

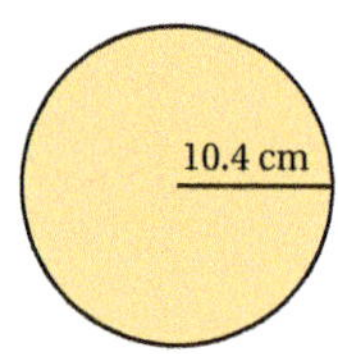

Answers

6. $78\frac{4}{7}\text{ km}^2$ 7. 339.62 cm^2

Guided Solution:

6. 5, 78

d SOLVING APPLIED PROBLEMS

EXAMPLE 8 *Areas of Cake Pans.* Tyler can make either a 9-in. round cake or a 9-in. square cake for a party. If he makes the square cake, how much more area will he have on the top for decorations?

The area of the square is

$$A = s \cdot s$$
$$= 9 \text{ in.} \times 9 \text{ in.} = 81 \text{ in}^2.$$

The diameter of the circle is 9 in., so the radius is 9 in./2, or 4.5 in. The area of the circle is

$$A = \pi \cdot r \cdot r$$
$$\approx 3.14 \times 4.5 \text{ in.} \times 4.5 \text{ in.} = 63.585 \text{ in}^2.$$

The area of the square cake is larger by about

$$81 \text{ in}^2 - 63.585 \text{ in}^2, \quad \text{or} \quad 17.415 \text{ in}^2.$$

Do Exercise 8. ▶

8. Which is larger and by how much: a 10-ft-square flower bed or a 12-ft-diameter round flower bed?

Answer

8. 12-ft-diameter round flower bed, by about 13.04 ft^2

8.4 Exercise Set

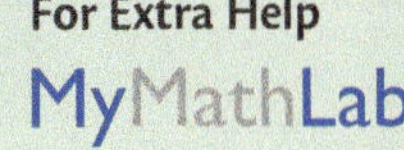

Reading Check

Complete each statement with the correct word from the following list. A word may be used more than once or not at all.

area circumference diameter radius

RC1. The ______________ of a circle is half the length of its diameter.

RC2. The ______________ of a circle is found by multiplying its diameter by π.

RC3. The ______________ of a circle is found by multiplying its radius by 2π.

RC4. The ______________ of a circle is found by multiplying the square of its radius by π.

For each circle, find the length of a diameter, the circumference, and the area. Use $\frac{22}{7}$ for π.

1. 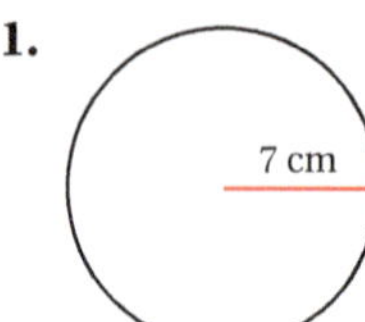

2. 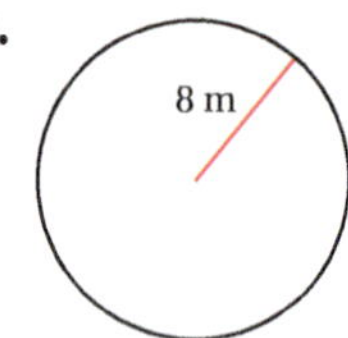

3. 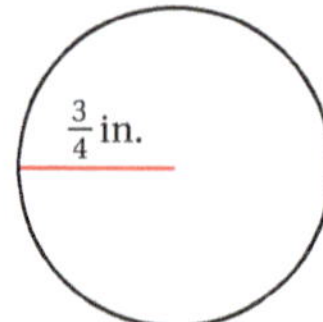

4. 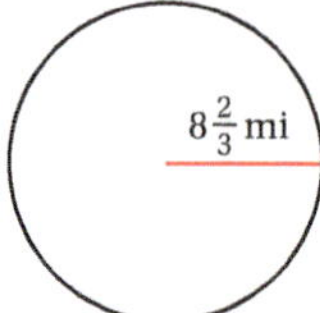

For each circle, find the length of a radius, the circumference, and the area. Use 3.14 for π.

5. 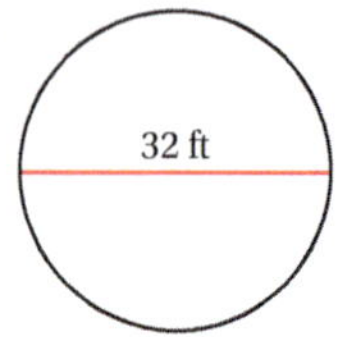

6. 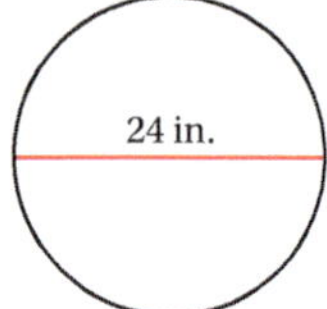

7. 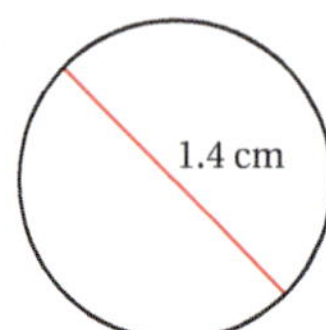

8. 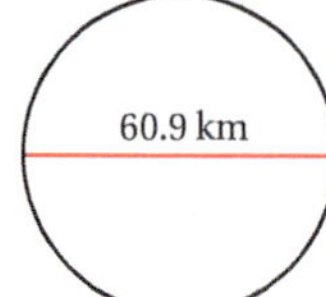

d Solve. Use 3.14 for π.

9. ***Pond Edging.*** Quiet Designs plans to incorporate a circular pond with a diameter of 30 ft in a landscape design. The pond will be edged using stone pavers. How many feet of pavers will be needed?

10. ***Gypsy-Moth Tape.*** To protect an elm tree in your backyard, you decide to attach gypsy moth caterpillar tape around the trunk. The tree has a 1.1-ft diameter. What length of tape is needed?

11. ***Areas of Pizza Pans.*** How much larger is a pizza made in a 16-in.-square pizza pan than a pizza made in a 16-in. diameter circular pan?

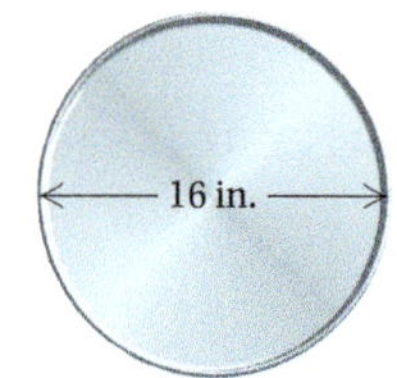

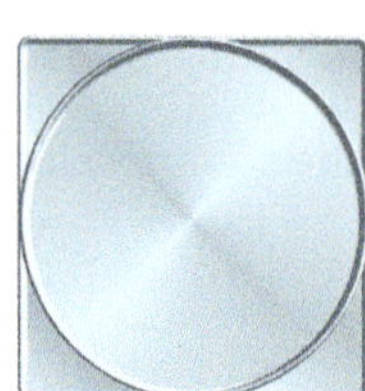

12. ***Penny.*** A penny has a 1-cm radius. What is its diameter? its circumference? its area?

13. *Earth.* The circumference of the earth at the equator is 24,901 mi. What is the diameter of the earth at the equator? the radius?

14. *Dimensions of a Quarter.* The circumference of a quarter is 7.85 cm. What is the diameter? the radius? the area?

15. *Circumference of a Baseball Bat.* In Major League baseball, the diameter of the barrel of a bat cannot be more than $2\frac{3}{4}$ in., and the diameter of the bat handle cannot be less than $\frac{16}{19}$ in. Find the maximum circumference of the barrel of a bat and the minimum circumference of the bat handle. Use $\frac{22}{7}$ for π.

Source: Major League Baseball

Maximum diameter of barrel of bat: $2\frac{3}{4}$ in.

Minimum diameter of bat handle: $\frac{16}{19}$ in.

16. *Trampoline.* The standard backyard trampoline has a diameter of 14 ft. What is its area?

Source: International Trampoline Industry Association, Inc.

17. *Swimming-Pool Walk.* You want to install a 1-yd wide walk around a circular swimming pool. The diameter of the pool is 20 yd. What is the area of the walk?

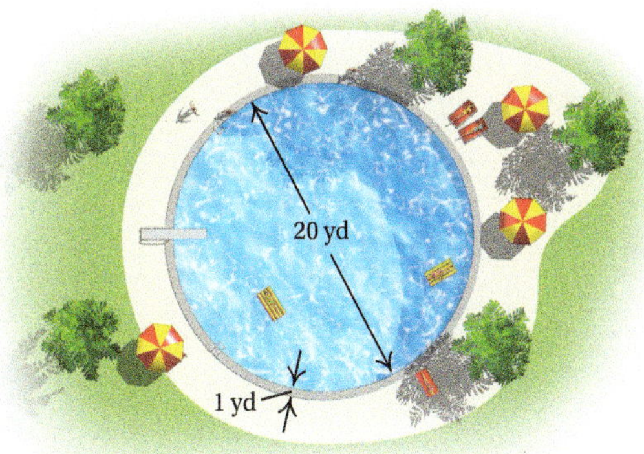

18. *Roller-Rink Floor.* A roller-rink floor is shown below. Each end is a semicircle. What is its area? If hardwood flooring costs \$32.50 per square meter, how much will the flooring cost?

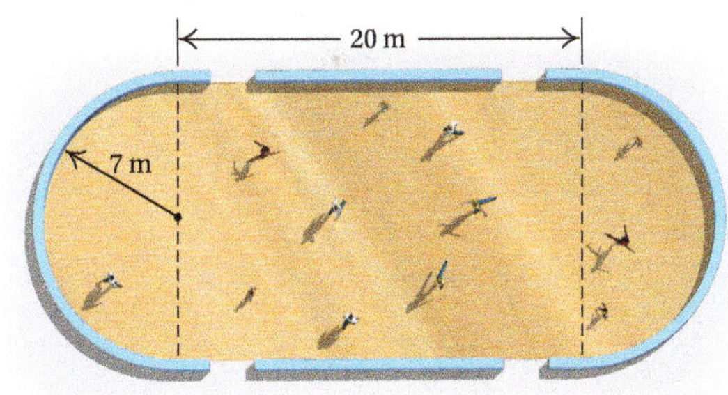

Find the perimeter of each figure. Use 3.14 for π.

19.

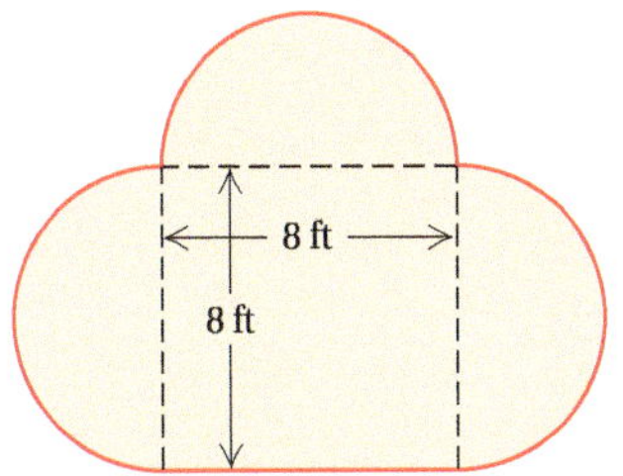

20.

4 cm

4 cm

4 cm

21.

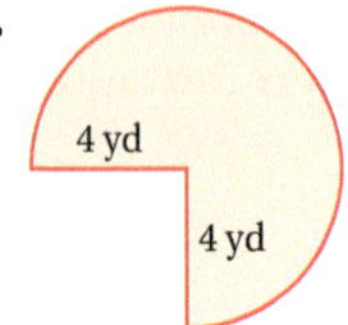

22.

8 in. 8 in. 8 in. 8 in.

23.

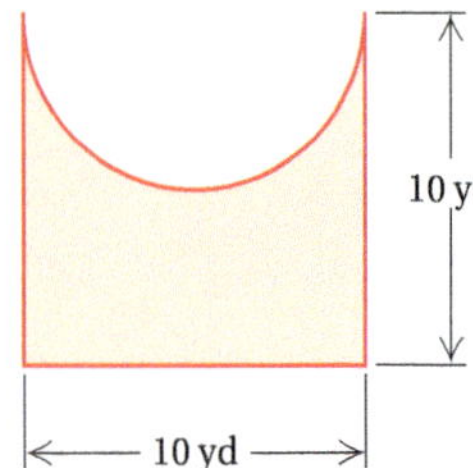

24.

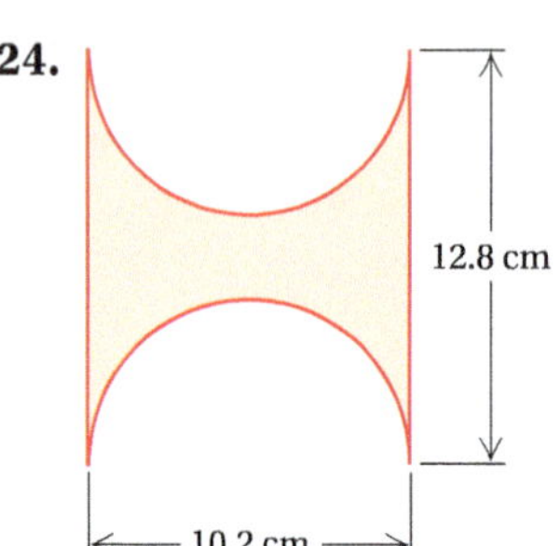

Find the area of the shaded region in each figure. Use 3.14 for π.

25.

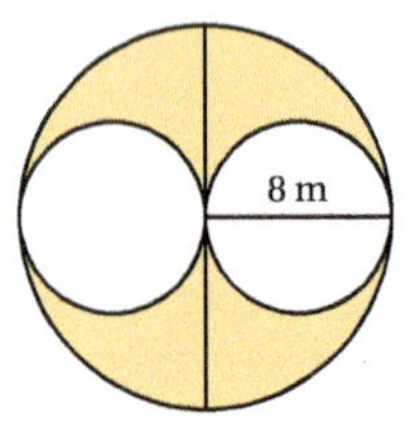

26.

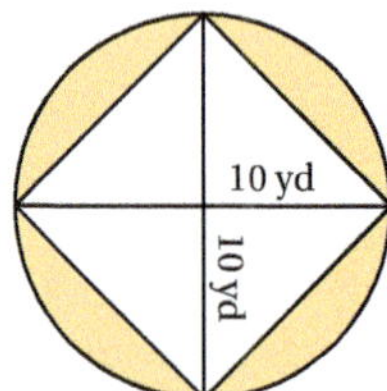

27.

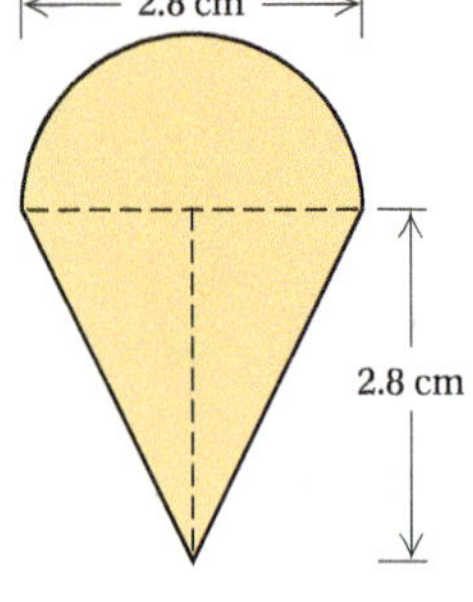

28.

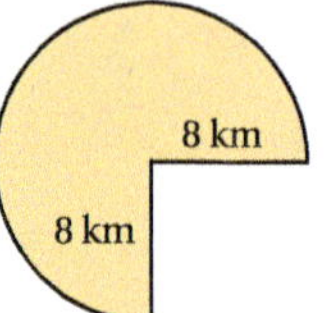

29.

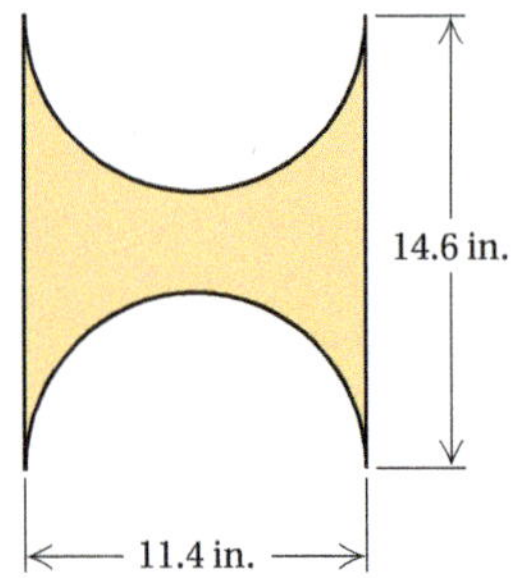

30.

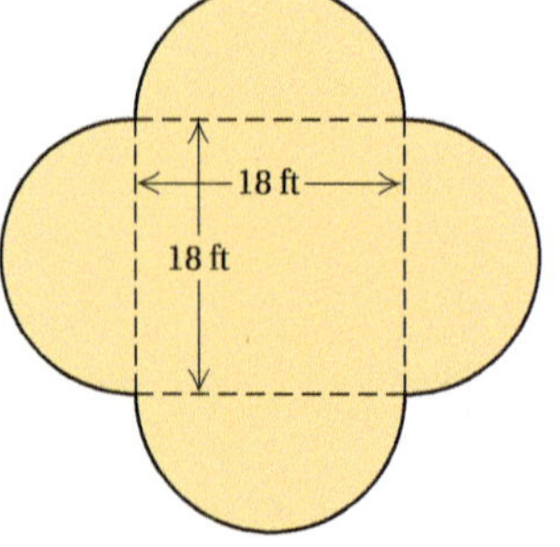

Skill Maintenance

31. A Ford C-Max Hybrid can travel 611 mi on 13 gal of gasoline. What is the rate in miles per gallon? [6.1b]

Source: edmunds.com

32. The ratio of gold to other metals in 18K gold is 18 to 6. If an 18K gold ring contains 1.2 oz of gold, what amount of other metals does the ring contain? [6.1e]

33. The weight of a human brain is 2.5% of total body weight. A person weighs 200 lb. What does the person's brain weigh? [6.6a]

34. Jack's commission is increased according to how much he sells. He receives a commission of 6% for the first $3000 and 10% on the amount over $3000. What is the total commission on sales of $8500? [6.7b]

Synthesis

Comparing Perimeters and Fencing Costs. An **acre** is a unit of area that is defined to be 43,560 ft^2. A farmer needs to fence an acre of land. She is using 32-in. fencing that costs $149.99 for a 330-ft roll. Complete the following table for Exercises 35–39 and then use the data to answer Exercise 40. Use 3.14 for π.

	Figure	Area	Perimeter or Circumference	Cost of Fencing
35.	Rectangle: 580.8 ft by 75 ft			
36.	Rectangle: 435.6 ft by 100 ft			
37.	Circle: radius 117.83 ft			
38.	Square: 208.71 ft by 208.71 ft			
39.	Rectangle: 242 ft by 180 ft			

40. Which dimensions of the acre yield the fence with **(a)** the shortest perimeter? **(b)** the least area? **(c)** the lowest cost and the largest area?

Mid-Chapter Review

Concept Reinforcement

Determine whether each statement is true or false.

______________ **1.** The area of a parallelogram with base 8 cm and height 5 cm is the same as the area of a rectangle with length 8 cm and width 5 cm. [8.3a, b]

______________ **2.** The area of a square that is 4 in. on a side is less than the area of a circle whose radius is 4 in. [8.3a], [8.4c]

______________ **3.** The perimeter of a rectangle that is 6 ft by 3 ft is greater than the circumference of a circle whose radius is 3 ft. [8.2a], [8.4b]

______________ **4.** The exact value of the ratio C/d is π. [8.4b]

Guided Solutions

GS Fill in each blank with the number that creates a correct solution.

5. Find the area. [8.3b]

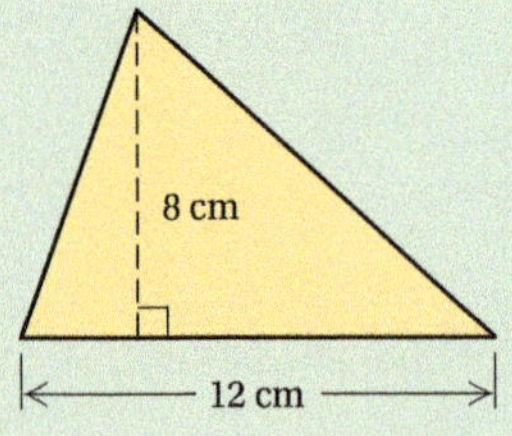

$$A = \frac{1}{2} \cdot b \cdot h$$
$$A = \frac{1}{2} \cdot \square \text{ cm} \cdot \square \text{ cm}$$
$$A = \frac{\square \cdot \square}{2} \text{ cm}^{\square}$$
$$A = \frac{\square}{2} \text{ cm}^{\square}, \text{ or } \square \text{ cm}^{\square}$$

6. Find the circumference and the area. Use 3.14 for π. [8.4b, c]

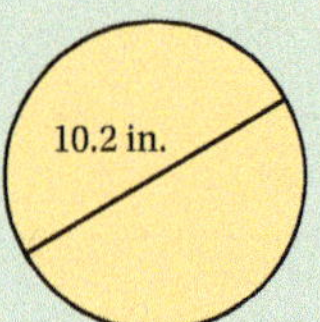

$$C = \pi \cdot d$$
$$C \approx \square \cdot \square \text{ in.}$$
$$C = \square \text{ in.}$$

$$A = \pi \cdot r \cdot r$$
$$A \approx \square \cdot \square \text{ in.} \cdot \square \text{ in.}$$
$$A = \square \text{ in}^{\square}$$

Mixed Review

7. Find the sum of the angle measures of a 19-sided polygon. [8.1f]

8. Find the missing angle measure. [8.1f]

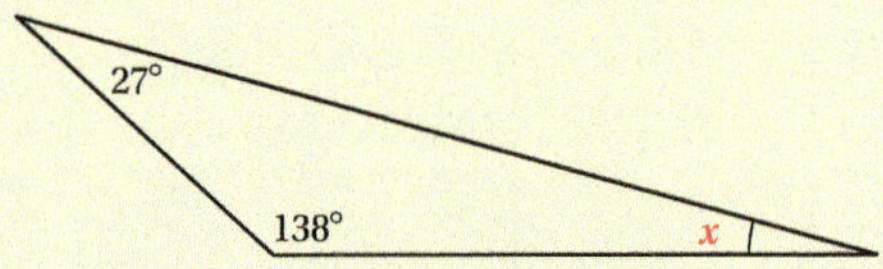

9. Classify the polygon by name. [8.1e]

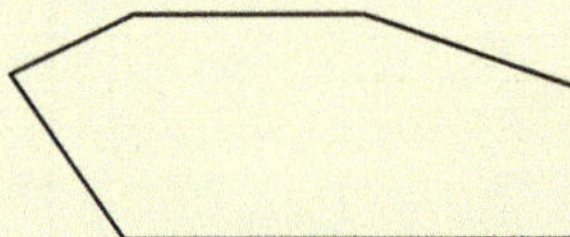

Classify each triangle as equilateral, isosceles, or scalene. Then classify it as right, obtuse, or acute. [8.1e]

10.

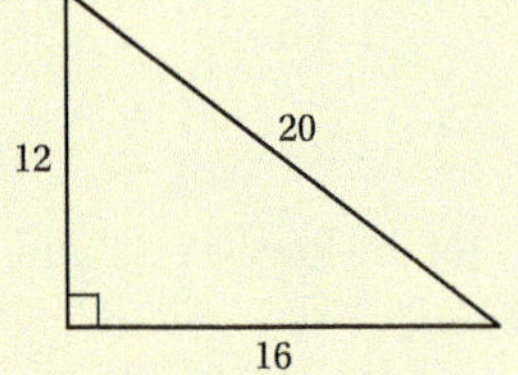

11.

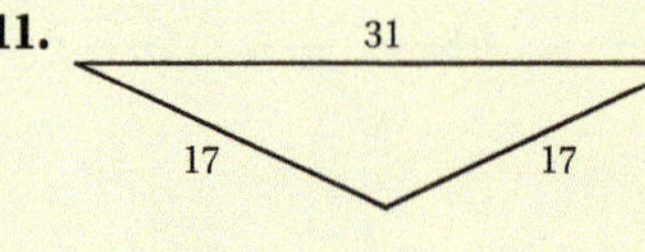

12.

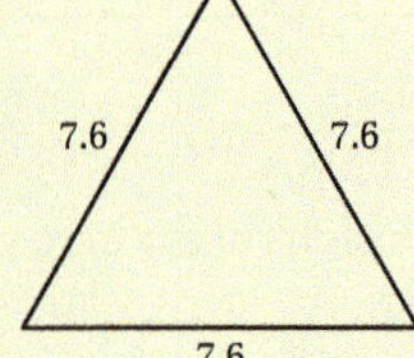

13. Find the perimeter. [8.2a]

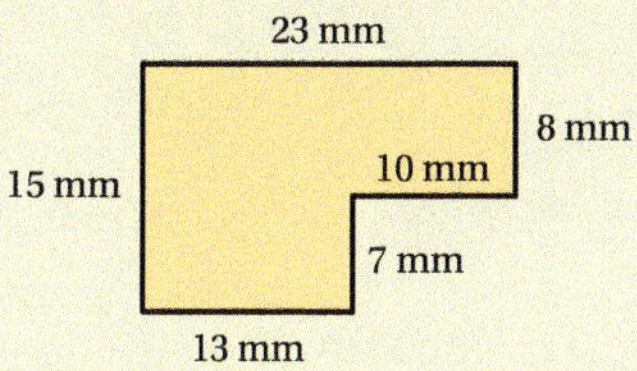

14. Find the perimeter and the area. [8.2a], [8.3a]

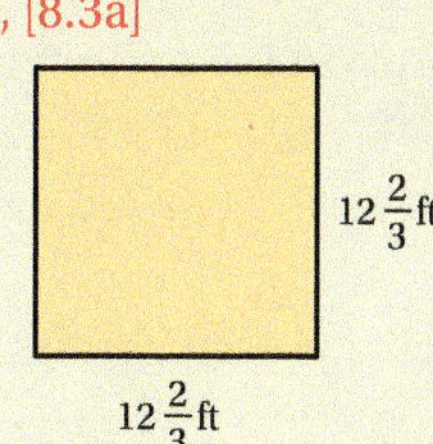

15. Find the area. [8.3b]

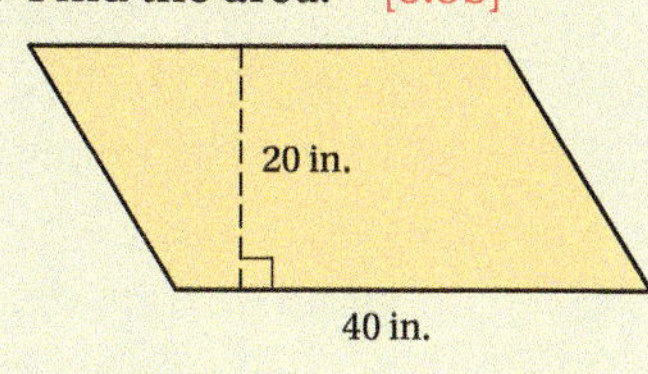

Find the area. [8.3b]

16.

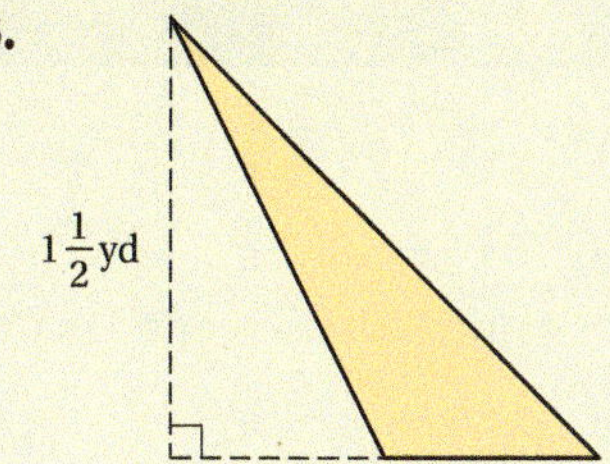

17.

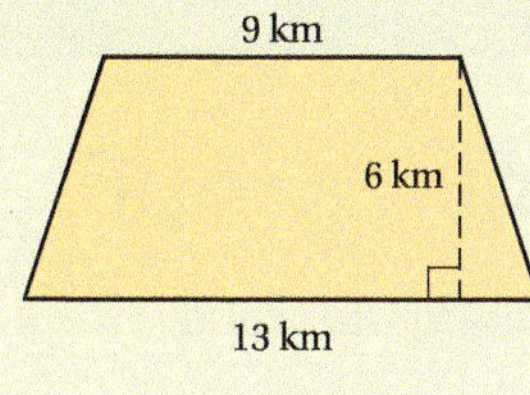

Find the circumference and the area. Use 3.14 for π. [8.4b, c]

18.

7 in.

19.

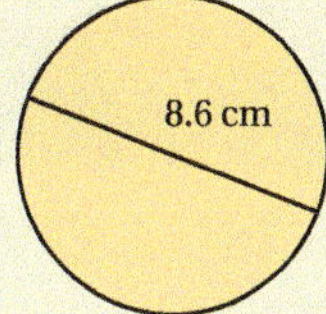

20. *Matching.* Match each item in the first column with the appropriate item in the second column by drawing connecting lines. Some expressions in the second column might be used more than once. Some expressions might not be used. [8.2a], [8.3a, b], [8.4b, c]

Area of a circle with radius 4 ft	24 ft
Area of a square with side 4 ft	16 ft
Circumference of a circle with radius 4 ft	$16 \cdot \pi\,\text{ft}^2$
Area of a rectangle with length 8 ft and width 4 ft	$8 \cdot \pi\,\text{ft}^2$
Area of a triangle with base 4 ft and height 8 ft	$32\,\text{ft}^2$
Perimeter of a square with side 4 ft	$4 \cdot \pi$ ft
Perimeter of a rectangle with length 8 ft and width 4 ft	$8 \cdot \pi$ ft
	64 ft
	$16\,\text{ft}^2$

Understanding Through Discussion and Writing

21. Explain why a 16-in.-diameter pizza that costs \$16.25 is a better buy than a 10-in.-diameter pizza that costs \$7.85. [8.4d]

22. The length and the width of one rectangle are each three times the length and the width of another rectangle. Is the area of the first rectangle three times the area of the other rectangle? Why or why not? [8.3a]

23. The length of a side of a square is $\frac{1}{2}$ the length of a side of another square. Is the perimeter of the first square $\frac{1}{2}$ the perimeter of the other square? Why or why not? [8.2a]

24. For a fellow student, develop the formula for the perimeter of a rectangle:

$$P = 2 \cdot (l + w) = 2 \cdot l + 2 \cdot w. \quad [8.2a]$$

25. Explain how the area of a triangle can be found by considering the area of a parallelogram. [8.3b]

26. The radius of one circle is twice the length of that of another circle. Is the area of the first circle twice the area of the other circle? Why or why not? [8.4c]

STUDYING FOR SUCCESS *Make the Most of Your Time in Class*

- ☐ Before each class, try to at least glance at the section in your text that will be covered, so you can concentrate on the instruction in class.
- ☐ Get a great seat! Sitting near the front will help you hear instruction more clearly and avoid distractions.
- ☐ Let your instructor know in advance if you must miss class, and do your best to keep up with any work that you miss.

8.5 Volume and Surface Area

OBJECTIVES

- **a** Find the volume and the surface area of a rectangular solid.
- **b** Given the radius and the height, find the volume of a circular cylinder.
- **c** Given the radius, find the volume of a sphere.
- **d** Given the radius and the height, find the volume of a circular cone.
- **e** Solve applied problems involving volumes of rectangular solids, circular cylinders, spheres, and cones.

a RECTANGULAR SOLIDS

The **volume** of a **rectangular solid** is the number of unit cubes needed to fill it.

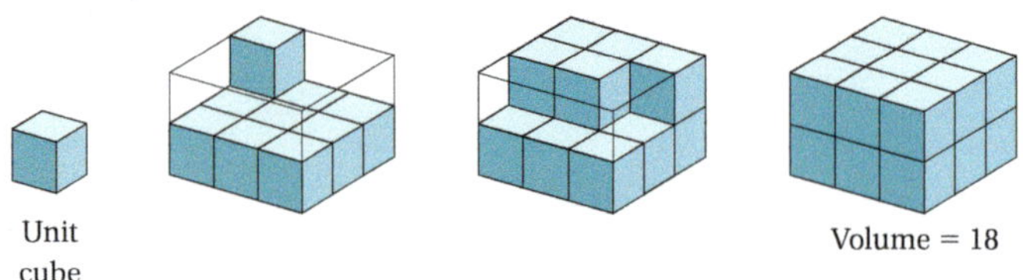

Two unit cubes used to measure volume are shown below.

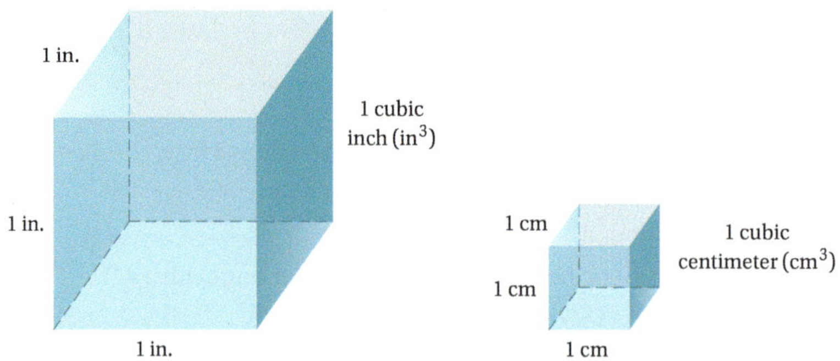

EXAMPLE 1 Find the volume.

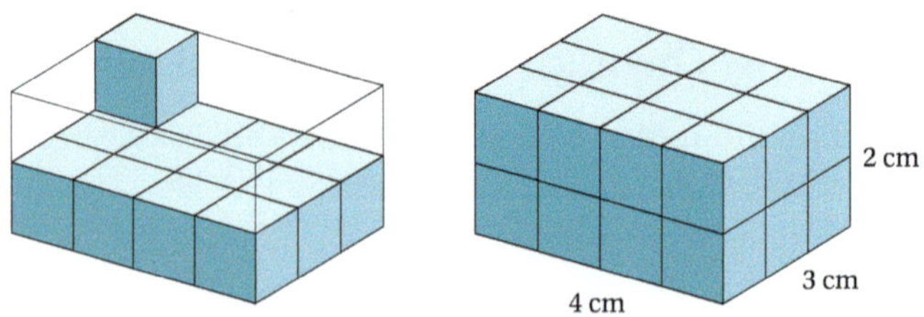

The figure is made up of 2 layers of 12 cubes each, so its volume is 24 cubic centimeters (cm^3).

Do Exercise 1.

1. Find the volume.

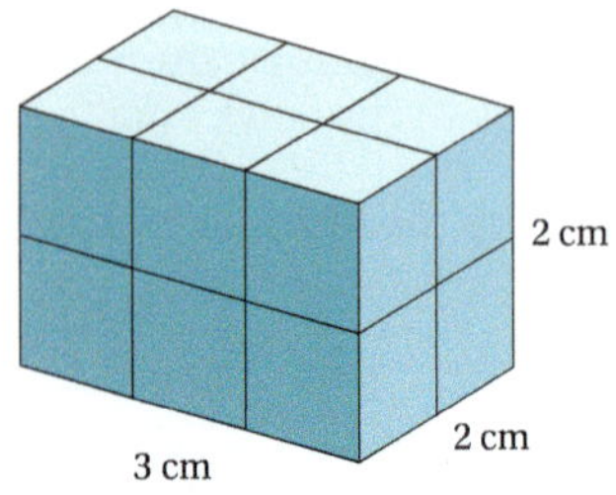

Answer

1. $12\ cm^3$

VOLUME OF A RECTANGULAR SOLID

The **volume of a rectangular solid** is found by multiplying length by width by height:

$$V = l \cdot w \cdot h.$$

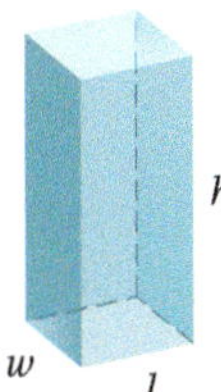

EXAMPLE 2 *Volume of a Safety Deposit Box.* Tricia rents a safety deposit box at her bank. The dimensions of the box are 18 in. × 10.5 in. × 5 in. Find the volume of this rectangular solid.

$$V = l \cdot w \cdot h$$
$$= 18 \text{ in.} \times 10.5 \text{ in.} \times 5 \text{ in.}$$
$$= 945 \text{ in}^3$$

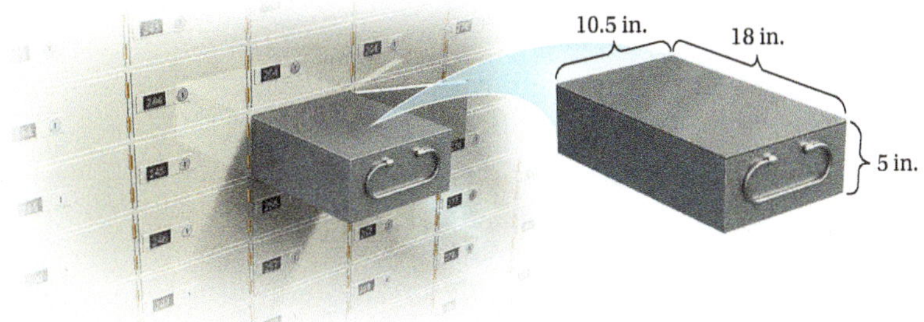

Do Exercises 2 and 3. ▶

The **surface area** of a rectangular solid is the total area of the six rectangles that form the surface of the solid. For the rectangular solid below, we can show the six rectangles with a diagram.

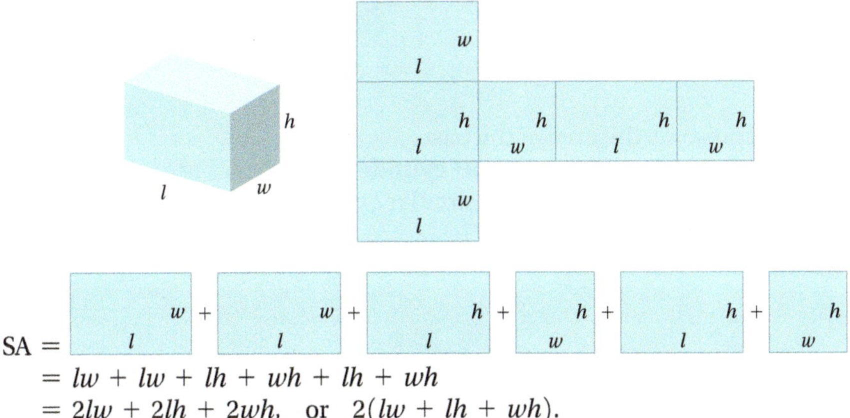

$$SA = lw + lw + lh + wh + lh + wh$$
$$= 2lw + 2lh + 2wh, \quad \text{or} \quad 2(lw + lh + wh).$$

2. *Carry-on Luggage.* The largest piece of luggage that you can carry on an airplane measures 23 in. by 10 in. by 13 in. Find the volume of this solid.

3. *Cord of Wood.* A cord of wood measures 4 ft by 4 ft by 8 ft. What is the volume of a cord of wood?

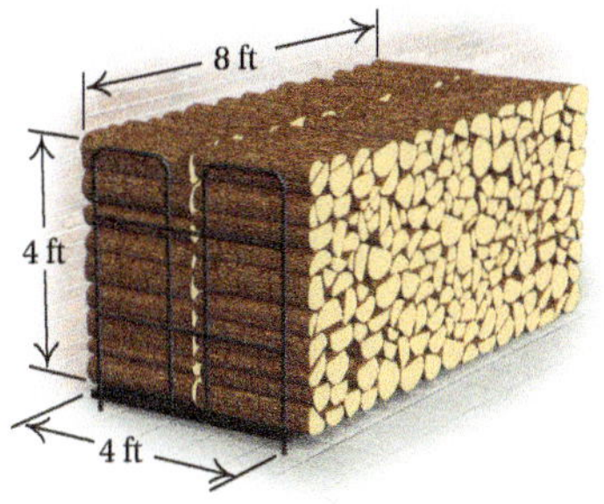

Answers

2. 2990 in^3 3. 128 ft^3

SURFACE AREA OF A RECTANGULAR SOLID

The surface area of a rectangular solid with length l, width w, and height h is given by the formula

$$\text{SA} = 2lw + 2lh + 2wh, \quad \text{or} \quad 2(lw + lh + wh).$$

EXAMPLE 3 Find the surface area of this rectangular solid.

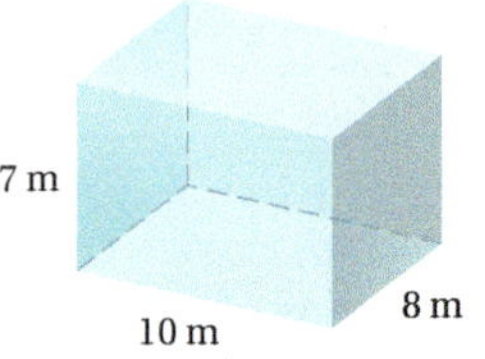

$$\begin{aligned} \text{SA} &= 2lw + 2lh + 2wh \\ &= 2 \cdot 10\text{ m} \cdot 8\text{ m} + 2 \cdot 10\text{ m} \cdot 7\text{ m} + 2 \cdot 8\text{ m} \cdot 7\text{ m} \\ &= 160\text{ m}^2 + 140\text{ m}^2 + 112\text{ m}^2 \\ &= 412\text{ m}^2 \end{aligned}$$

The units used for area are square units.
The units used for volume are cubic units.

Find the volume and the surface area of the rectangular solid.

4.

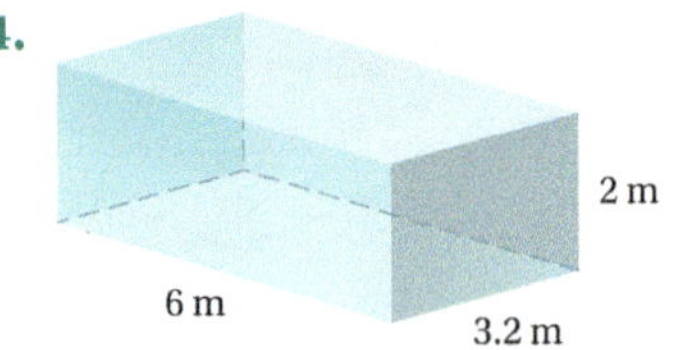

5.

$2\frac{1}{2}$ ft
1 ft
$\frac{3}{4}$ ft

Do Exercises 4 and 5.

b CYLINDERS

A rectangular solid is shown below. Note that we can think of the volume as the product of the area of the base times the height:

$$\begin{aligned} V &= l \cdot w \cdot h \\ &= (l \cdot w) \cdot h \\ &= (\text{Area of the base}) \cdot h \\ &= B \cdot h, \end{aligned}$$

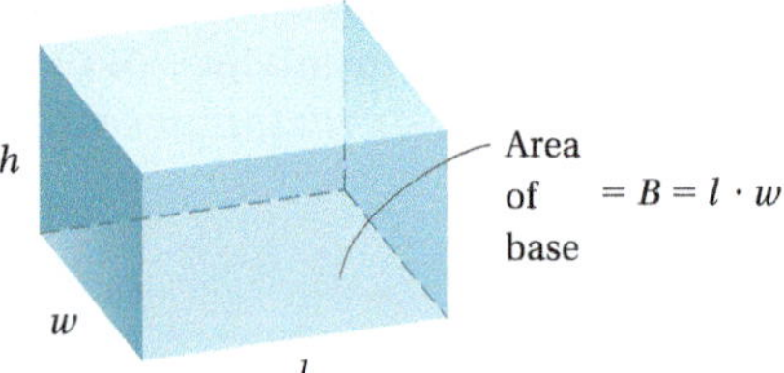

where B represents the area of the base.

Like rectangular solids, **circular cylinders** have bases of equal area that lie in parallel planes. The bases of circular cylinders are circular regions.

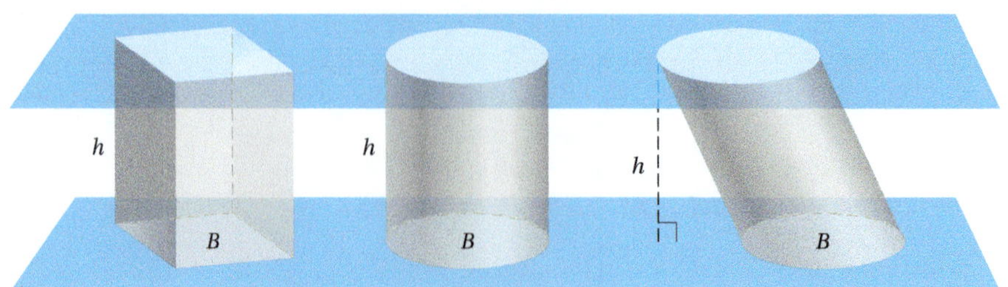

Answers

4. 38.4 m^3; 75.2 m^2 **5.** $1\frac{7}{8}\text{ ft}^3$; $10\frac{1}{4}\text{ ft}^2$

The volume of a circular cylinder is found in a manner similar to the way the volume of a rectangular solid is found. The volume is the product of the area of the base times the height. The height is always measured perpendicular to the base.

VOLUME OF A CIRCULAR CYLINDER

The **volume of a circular cylinder** is the product of the area of the base B and the height h:

$$V = B \cdot h, \quad \text{or} \quad V = \pi \cdot r^2 \cdot h.$$

EXAMPLE 4 Find the volume of this circular cylinder. Use 3.14 for π.

$$\begin{aligned} V &= B \cdot h = \pi \cdot r^2 \cdot h \\ &\approx 3.14 \times 4\text{ cm} \times 4\text{ cm} \times 12\text{ cm} \\ &= 602.88\text{ cm}^3 \end{aligned}$$

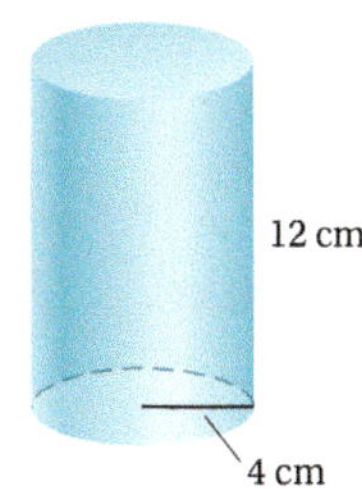

EXAMPLE 5 Find the volume of this circular cylinder. Use $\frac{22}{7}$ for π.

$$\begin{aligned} V &= B \cdot h = \pi \cdot r^2 \cdot h \\ &\approx \frac{22}{7} \times 6.8\text{ yd} \times 6.8\text{ yd} \times 11.2\text{ yd} \\ &= 1627.648\text{ yd}^3 \end{aligned}$$

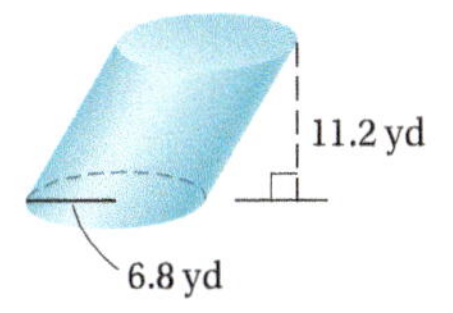

Do Exercises 6 and 7.

GS **6.** Find the volume of the cylinder. Use 3.14 for π.

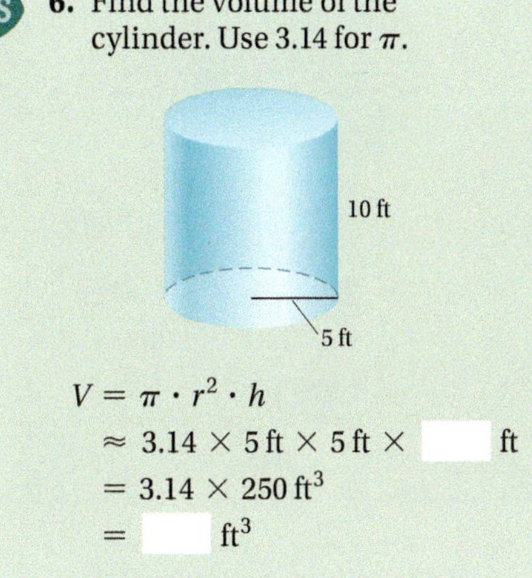

$$\begin{aligned} V &= \pi \cdot r^2 \cdot h \\ &\approx 3.14 \times 5\text{ ft} \times 5\text{ ft} \times \underline{\quad}\text{ ft} \\ &= 3.14 \times 250\text{ ft}^3 \\ &= \underline{\quad}\text{ ft}^3 \end{aligned}$$

7. Find the volume of the cylinder. Use $\frac{22}{7}$ for π.

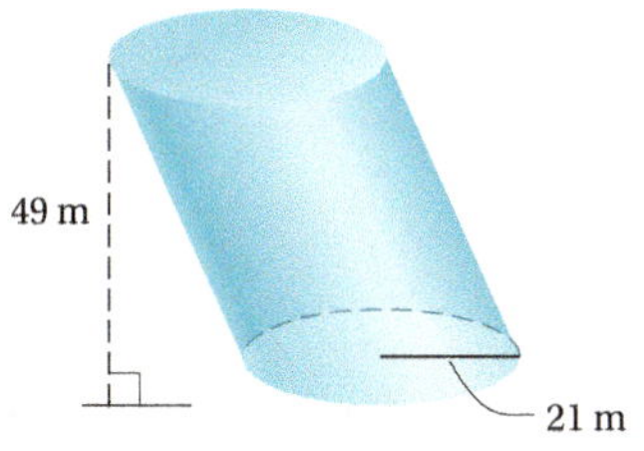

C SPHERES

A **sphere** is the three-dimensional counterpart of a circle. It is the set of all points in space that are a given distance (the radius) from a given point (the center).

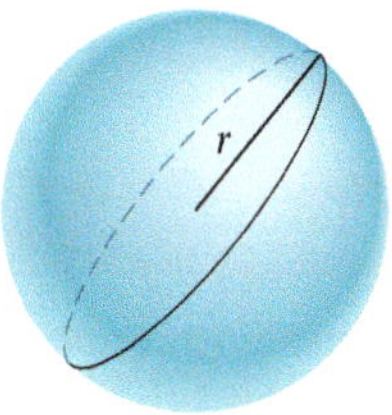

We find the volume of a sphere as follows.

VOLUME OF A SPHERE

The **volume of a sphere** of radius r is given by

$$V = \frac{4}{3} \cdot \pi \cdot r^3.$$

Answers

6. 785 ft^3 **7.** 67,914 m^3

Guided Solution:

6. 10, 785

8. Find the volume of the sphere. Use $\frac{22}{7}$ for π.

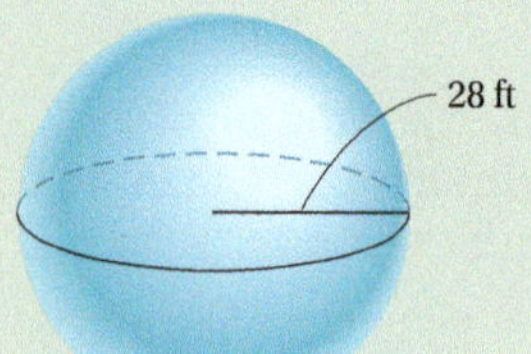

$$V = \frac{4}{3} \cdot \pi \cdot r^3$$
$$\approx \frac{4}{3} \times \frac{22}{7} \times (\quad \text{ft})^3$$
$$= \frac{4}{3} \times \frac{22}{7} \times \quad \text{ft}^3$$
$$= \frac{275{,}968}{3} \text{ ft}^3$$
$$= \quad \frac{1}{3} \text{ ft}^3$$

9. The radius of a standard-sized golf ball is 2.1 cm. Find its volume. Use 3.14 for π.

10. Find the volume of this cone. Use 3.14 for π.

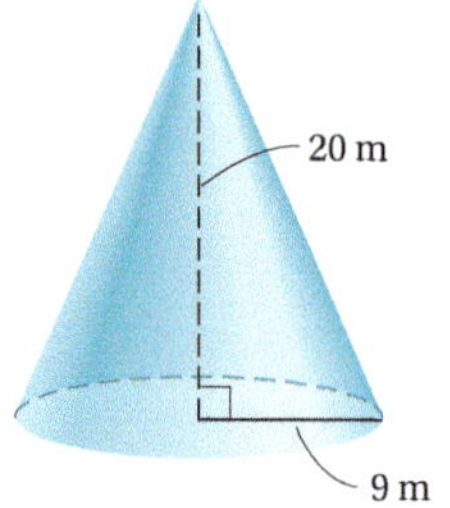

11. Find the volume of this cone. Use $\frac{22}{7}$ for π.

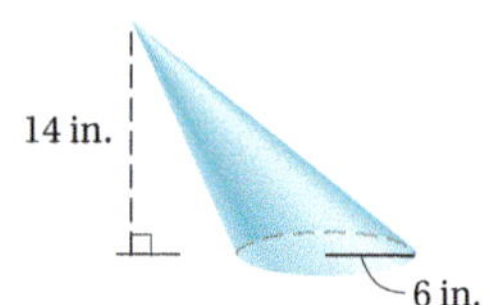

Answers

8. $91{,}989\frac{1}{3}$ ft^3 9. 38.77272 cm^3
10. 1695.6 m^3 11. 528 in^3

Guided Solution:
8. 28, 21,952, 91,989

EXAMPLE 6 ***Bowling Ball.*** The radius of a standard-sized bowling ball is 4.2915 in. Find the volume of a standard-sized bowling ball. Round to the nearest hundredth of a cubic inch. Use 3.14 for π.

We have

$$V = \frac{4}{3} \cdot \pi \cdot r^3 \approx \frac{4}{3} \times 3.14 \times (4.2915 \text{ in.})^3$$
$$\approx 330.90 \text{ in}^3. \qquad \text{Using a calculator}$$

Do Exercises 8 and 9.

d CONES

Consider a circle in a plane and choose any point P not in the plane. The circular region, together with the set of all segments connecting P to a point on the circle, is called a **circular cone**. The height of the cone is measured perpendicular to the base.

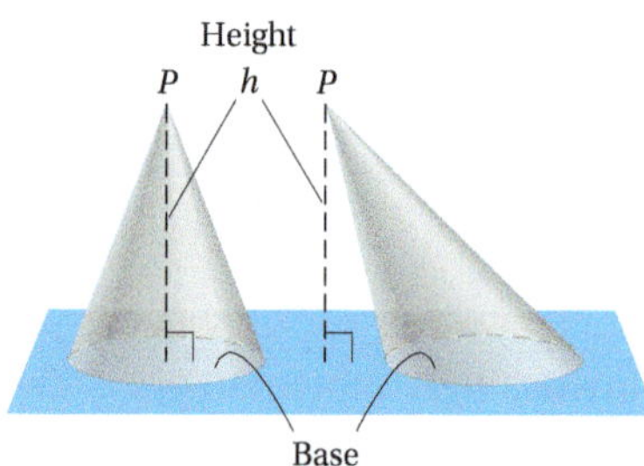

We find the volume of a cone as follows.

VOLUME OF A CIRCULAR CONE

The **volume of a circular cone** with base radius r is one-third the product of the area of the base and the height:

$$V = \frac{1}{3} \cdot B \cdot h = \frac{1}{3} \cdot \pi \cdot r^2 \cdot h.$$

EXAMPLE 7 Find the volume of this circular cone. Use 3.14 for π.

$$V = \frac{1}{3} \cdot \pi \cdot r^2 \cdot h$$
$$\approx \frac{1}{3} \times 3.14 \times 3 \text{ cm} \times 3 \text{ cm} \times 7 \text{ cm}$$
$$= 65.94 \text{ cm}^3$$

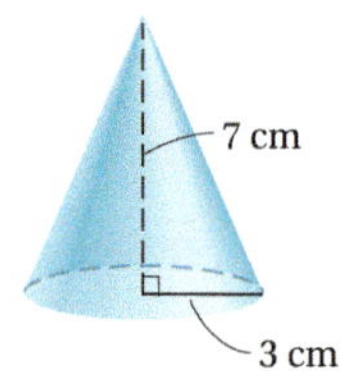

Do Exercises 10 and 11.

e SOLVING APPLIED PROBLEMS

EXAMPLE 8 *Propane Gas Tank.* A propane gas tank is shaped like a circular cylinder with half of a sphere at each end. Find the volume of the tank if the cylindrical section is 5 ft long with a 4-ft diameter. Use 3.14 for π.

1. **Familiarize.** We first make a drawing.

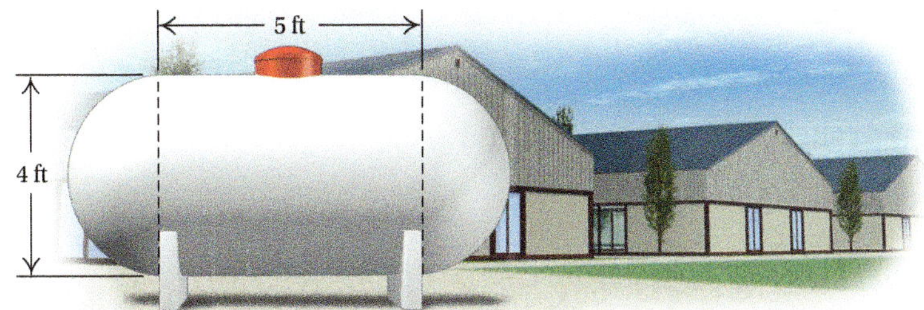

2. **Translate.** This is a two-step problem. We first find the volume of the cylindrical portion. Then we find the volume of the two ends and add. Note that together the two ends make a sphere with a radius of 2 ft. We have

Total volume	is	Volume of the cylinder	plus	Volume of the sphere
V	$=$	$\pi \cdot r^2 \cdot h$	$+$	$\frac{4}{3} \cdot \pi \cdot r^3,$

where V is the total volume. Then

$$V \approx 3.14 \cdot (2\text{ ft})^2 \cdot 5\text{ ft} + \frac{4}{3} \cdot 3.14 \cdot (2\text{ ft})^3.$$

3. **Solve.** The volume of the cylinder is approximately

$$\begin{aligned} 3.14 \cdot (2\text{ ft})^2 \cdot 5\text{ ft} &= 3.14 \cdot 2\text{ ft} \cdot 2\text{ ft} \cdot 5\text{ ft} \\ &= 62.8\text{ ft}^3. \end{aligned}$$

The volume of the two ends is approximately

$$\begin{aligned} \frac{4}{3} \cdot 3.14 \cdot (2\text{ ft})^3 &= \frac{4}{3} \cdot 3.14 \cdot 2\text{ ft} \cdot 2\text{ ft} \cdot 2\text{ ft} \\ &\approx 33.5\text{ ft}^3. \end{aligned}$$

The total volume is about

$$62.8\text{ ft}^3 + 33.5\text{ ft}^3 = 96.3\text{ ft}^3.$$

4. **Check.** We can repeat the calculations. The answer checks.
5. **State.** The volume of the tank is about 96.3 ft^3.

Do Exercise 12. ▶

12. *Medicine Capsule.* A cold capsule is 8 mm long and 4 mm in diameter. Find the volume of the capsule. Use 3.14 for π. (*Hint*: First find the length of the cylindrical section.)

Answer

12. $83.7\overline{3}$ mm^3

8.5 Exercise Set

For Extra Help
MyMathLab®

Reading Check

Match each formula with the correct phrase from the list on the right.

RC1. ________ $V = l \cdot w \cdot h$

RC2. ________ $V = \pi \cdot r^2 \cdot h$

RC3. ________ $V = \frac{4}{3} \cdot \pi \cdot r^3$

RC4. ________ $V = \frac{1}{3} \cdot \pi \cdot r^2 \cdot h$

a) the volume of a cylinder

b) the volume of a rectangular solid

c) the volume of a sphere

d) the volume of a cone

a Find the volume and the surface area of the rectangular solid.

1.

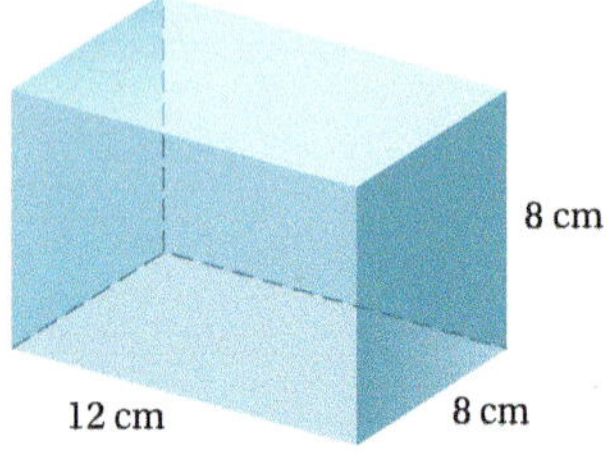

2.

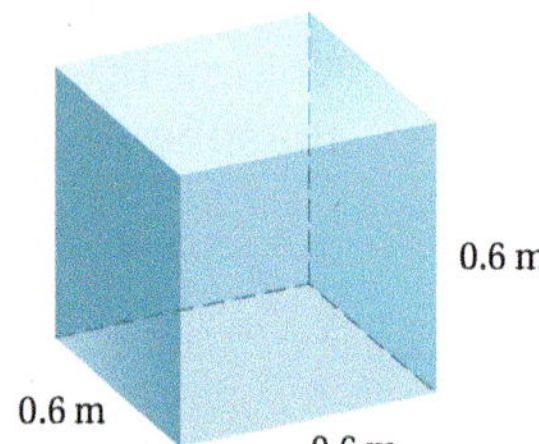

3.

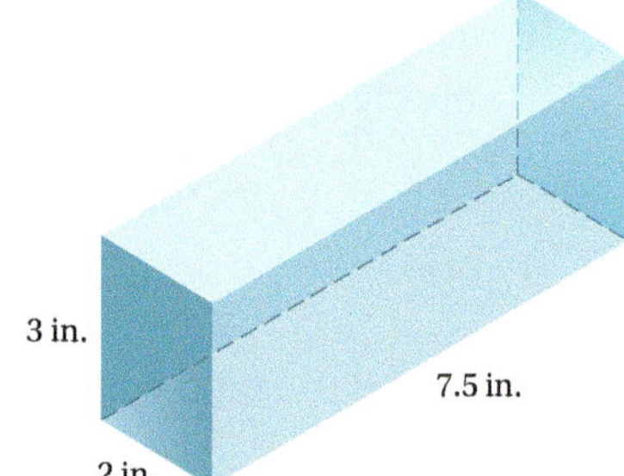

4.

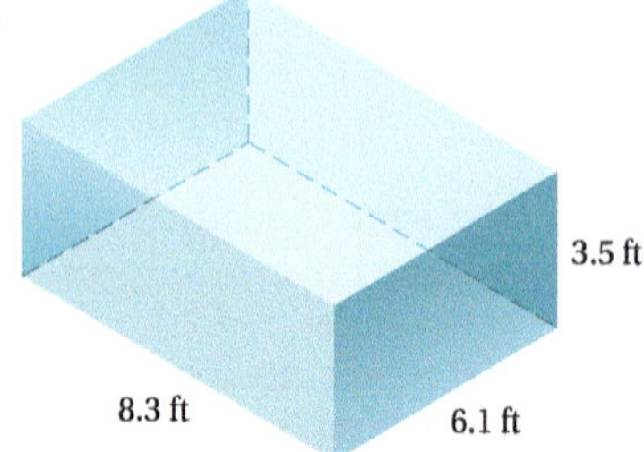

5.

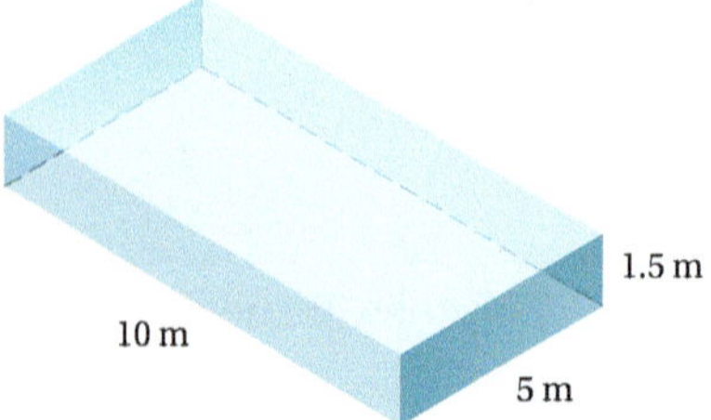

6.

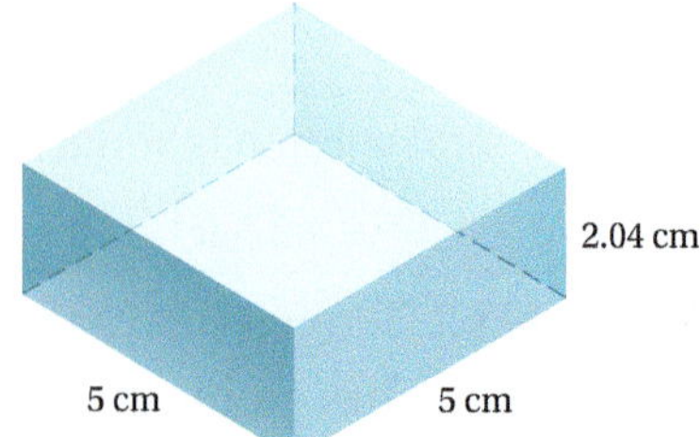

7.

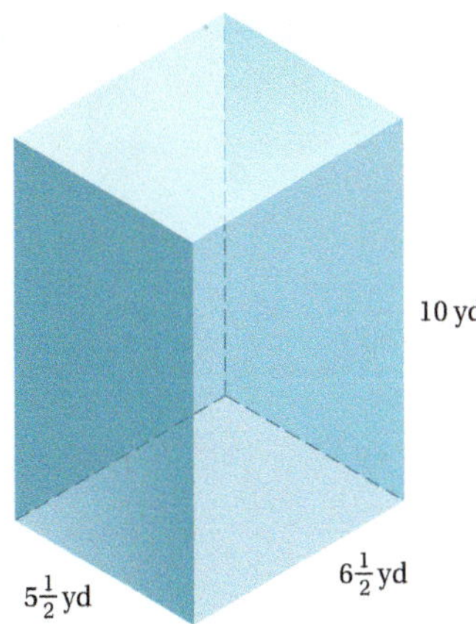

8.

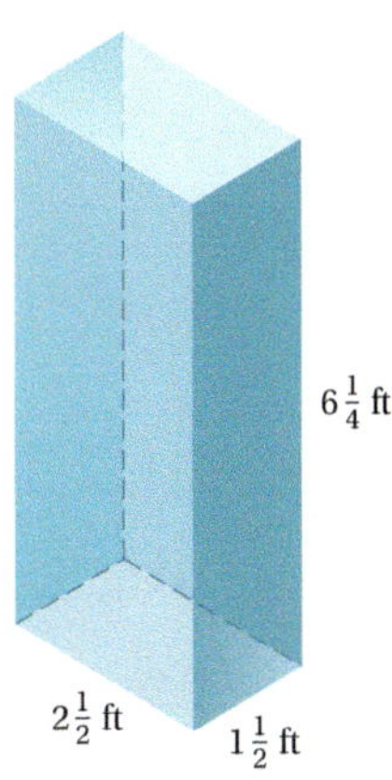

b Find the volume of the circular cylinder. Use 3.14 for π in Exercises 9–12. Use $\frac{22}{7}$ for π in Exercises 13 and 14.

9.

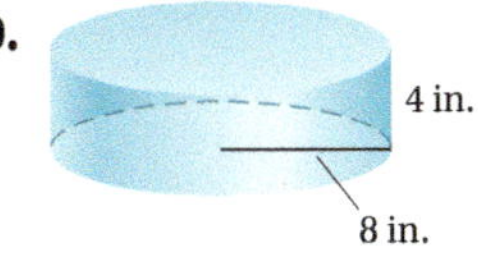

10.

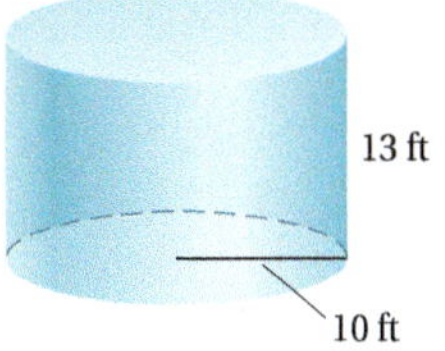

11.

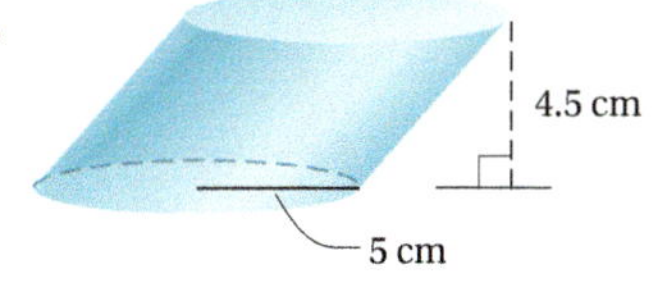

12.

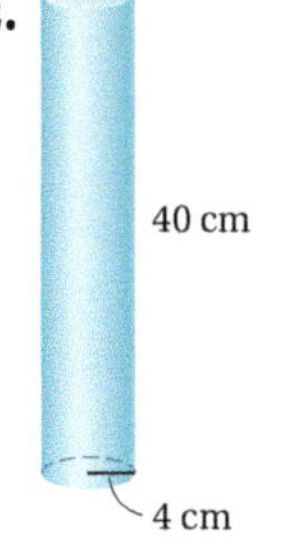

13.

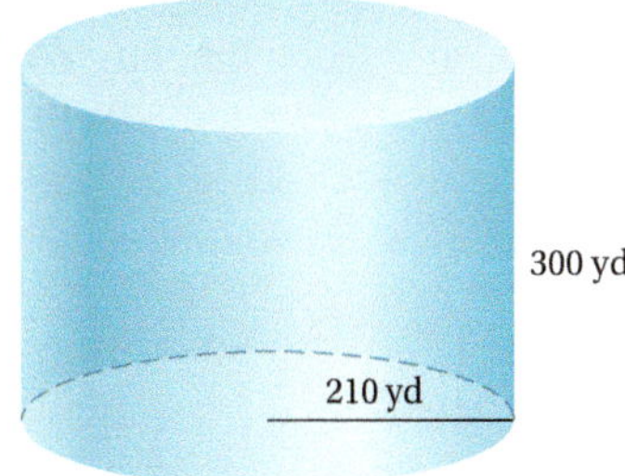

14.

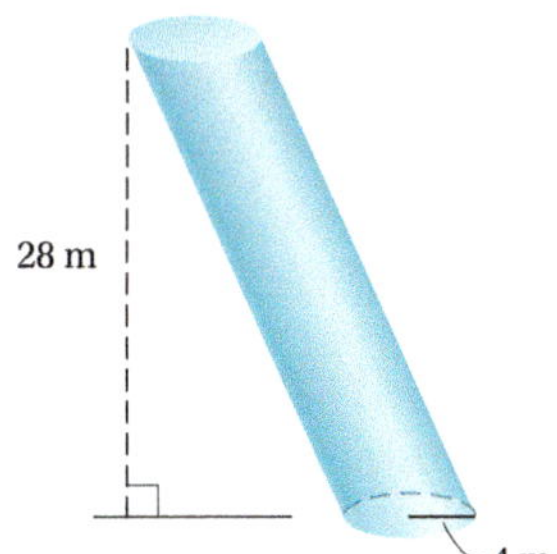

c Find the volume of the sphere. Use 3.14 for π in Exercises 15–18 and round to the nearest hundredth in Exercises 17 and 18. Use $\frac{22}{7}$ for π in Exercises 19 and 20.

15.

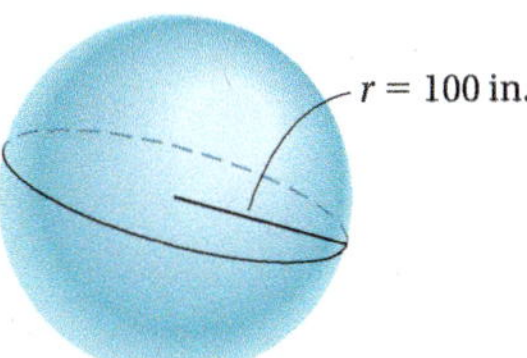

16.

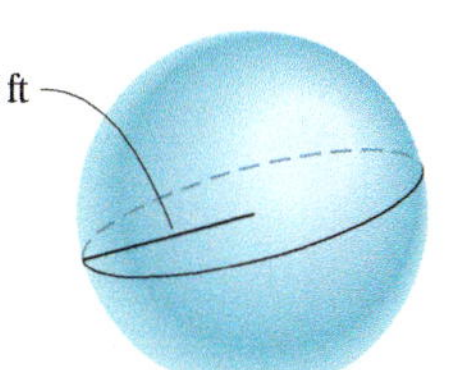

17.

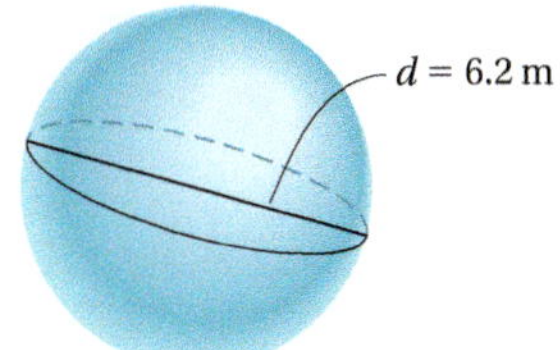

18.

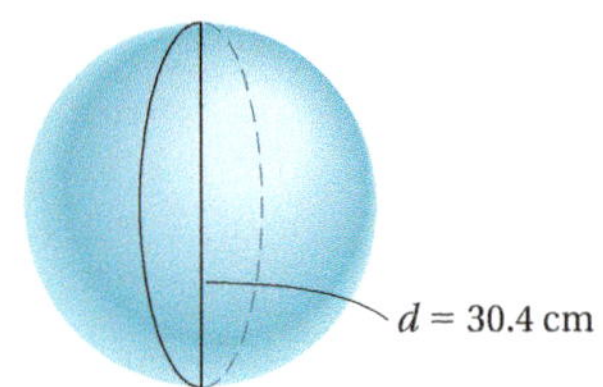

19.

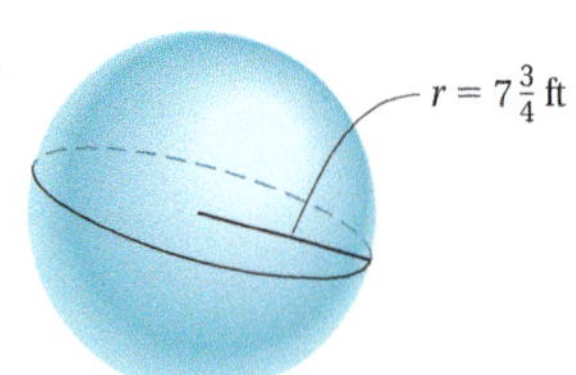

20.

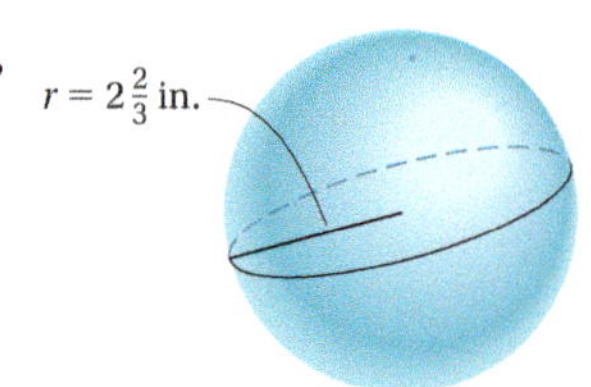

d Find the volume of the circular cone. Use 3.14 for π in Exercises 21, 22, and 26. Use $\frac{22}{7}$ for π in Exercises 23, 24, and 25.

21.

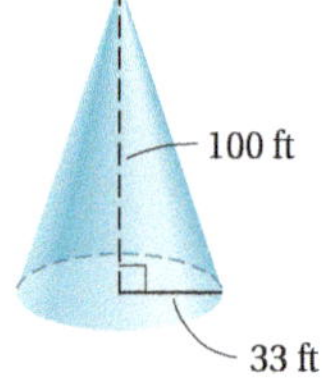

22.

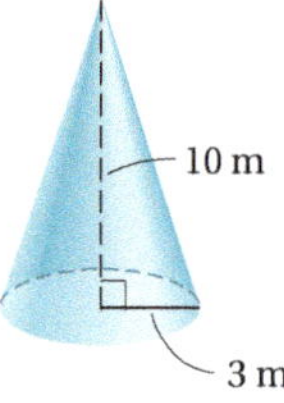

23.

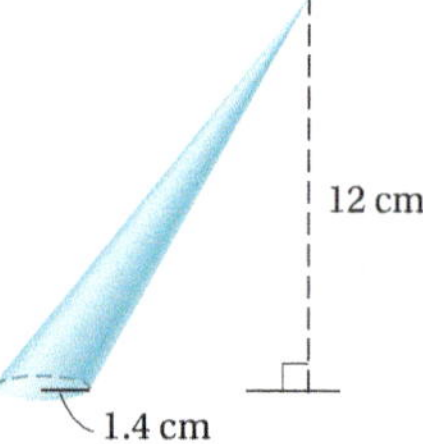

24.

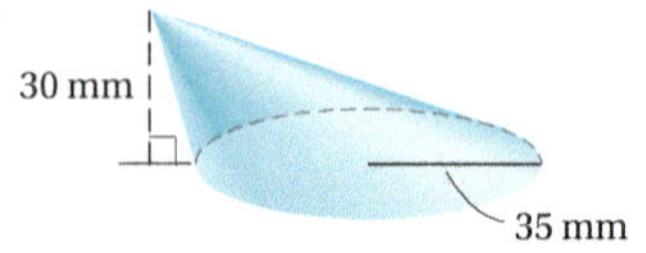

25.

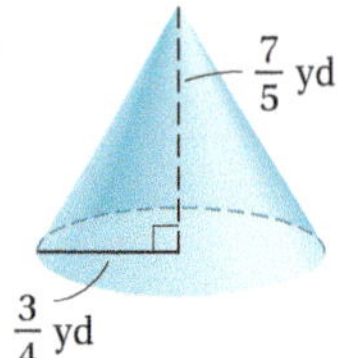

26.

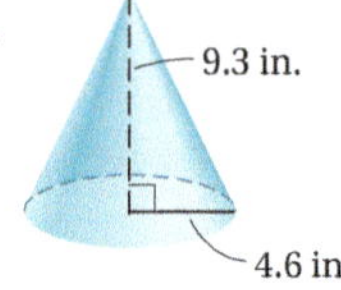

e Solve.

27. ***Oak Log.*** An oak log has a diameter of 12 cm and a length (height) of 42 cm. Find the volume. Use 3.14 for π.

28. ***Ladder Rung.*** A rung of a ladder is 2 in. in diameter and 16 in. long. Find the volume. Use 3.14 for π.

29. ***Architecture.*** The largest sphere in the Oriental Pearl TV tower in Shanghai, China, measures 50 m in diameter. Find the volume of the sphere. Use 3.14 for π, and round to the nearest cubic meter.

Source: www.emporis.com

30. ***Gas Pipeline.*** The 638-mi Rockies Express–East pipeline from Colorado to Ohio is constructed with 80-ft sections of 42-in., or $3\frac{1}{2}$-ft, steel gas pipeline. Find the volume of one section. Use $\frac{22}{7}$ for π.

Source: Rockies Express Pipeline

31. *Volume of a Candle.* Find the approximate volume of a candle that is a circular cone. The diameter of the base of the candle is 4.875 in., and the height is 12.5 in. Use 3.14 for π.

32. *Culinary Arts.* Raena often makes individual soufflés in cylindrical baking dishes called *ramekins*. The diameter of each ramekin is 3.5 in., and the height is 1.75 in. Find the approximate volume of a ramekin. Use 3.14 for π.

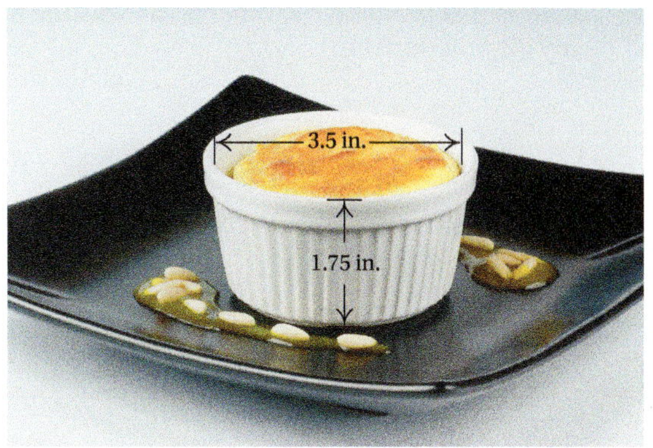

33. *Precious Metals.* If all the gold in the world could be gathered together, it would form a cube about 20.7 m on a side. Find the volume of the world's gold.

34. *Containers.* Medical Hope collects and ships medical supplies to areas of the world that have been damaged by storms or earthquakes. The containers in which the supplies are shipped measure 6 ft by 5 ft by 4.8 ft. Find the volume of a container.

35. *Architecture.* The Westhafen Tower in Frankfort, Germany, is a cylindrical building with a height of 110 m and a radius of 21 m. Find the volume of the building. Use 3.14 for π. Round to the nearest cubic meter.

36. *Roof of a Turret.* The roof of a turret is often in the shape of a circular cone. Find the volume of this circular cone structure if the radius is 2.5 m and the height is 4.6 m. Use 3.14 for π.

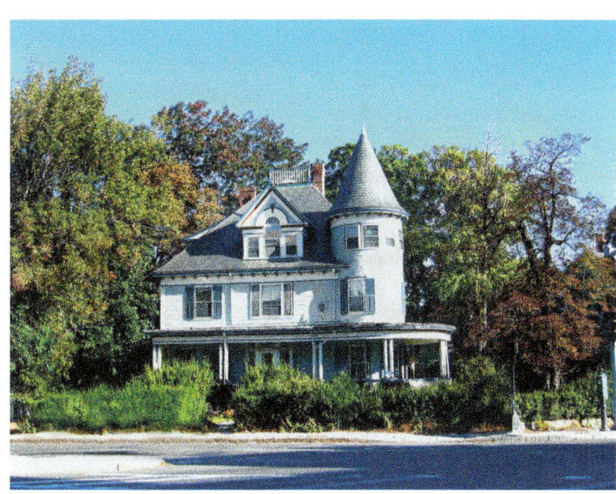

37. *Volume of Earth.* The diameter of the earth is about 7926 mi. Find the volume of the earth. Use 3.14 for π. Round to the nearest ten thousand cubic miles.

38. *Astronomy.* The diameter of the largest moon of Uranus is about 1578 km. Find the volume of this satellite. Use $\frac{22}{7}$ for π. Round to the nearest ten thousand cubic kilometers.

39. The volume of a ball is 36π cm^3. Find the dimensions of a rectangular box that is just large enough to hold the ball.

40. *Oceanography.* A research submarine is capsule-shaped. Find the volume of the submarine if it has a length of 10 m and a diameter of 8 m. Use 3.14 for π and round the answer to the nearest hundredth. (*Hint*: First find the length of the cylindrical section.)

41. *Toys.* Toy stores often sell capsules that dissolve in water allowing a toy inside the capsule to expand. One such capsule is 40 mm long with a diameter of 8 mm.

a) What is the volume of the capsule? Use 3.14 for π.

b) The manufacturer claims that the toy in the capsule will expand 600%. What is the volume of the toy after expansion?

42. *Golf-Ball Packaging.* The box shown is just big enough to hold 3 golf balls. If the radius of a golf ball is 2.1 cm, how much air surrounds the three balls? Use 3.14 for π.

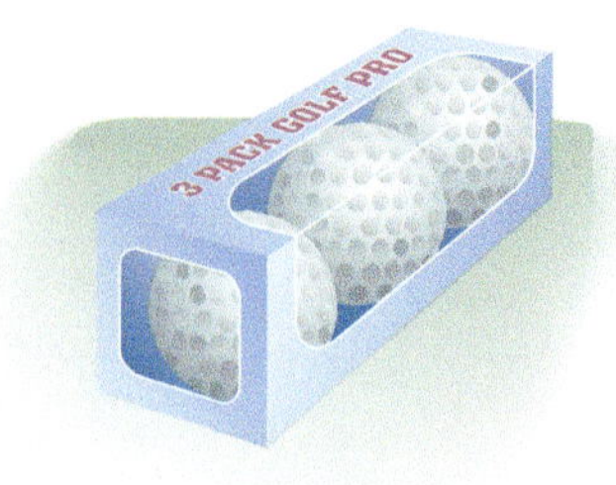

Skill Maintenance

Great Lakes. The Great Lakes contain about 5500 mi^3 (23,000 km^3) of water that covers a total area of about 94,000 mi^2 (244,000 km^2). The Great Lakes are the largest system of fresh surface water on Earth. Use the following table for Exercises 43–48.

FEATURE	UNITS	GREAT LAKE				
		SUPERIOR	MICHIGAN	HURON	ERIE	ONTARIO
Average depth	ft	483	279	195	62	283
Volume	mi^3	2,900	1,180	850	116	393
Water area	mi^2	31,700	22,300	23,000	9,910	7,340

Lake Superior, Lake Huron, Lake Ontario, Lake Michigan, Lake Erie

SOURCES: http://www.epa.gov/glnpo/factsheet.html; http://earth1.epa.gov/glnpo/statrefs.html

43. How much greater is the volume of water in Lake Michigan than in Lake Erie? [1.8a], [7.2a]

44. How much less is the water area of Lake Ontario than that of Lake Superior? [1.8a], [7.2a]

45. Find the average of the average depths of the five Great Lakes. [7.1a], [7.2a]

46. Find the average volume of water in the five Great Lakes. [7.1a], [7.2a]

Synthesis

Use the data in the table above for Exercises 47 and 48. Use 1 mi = 1.609 km.

47. Convert the volume of water in Lake Huron from cubic miles to cubic kilometers. Round to the nearest hundredth of a cubic kilometer.

48. Convert the water area of Lake Superior from square miles to square kilometers. Round to the nearest hundredth of a square kilometer.

49. A sphere with diameter 1 m is circumscribed by a cube. How much greater is the volume of the cube than the volume of the sphere? Use 3.14 for π.

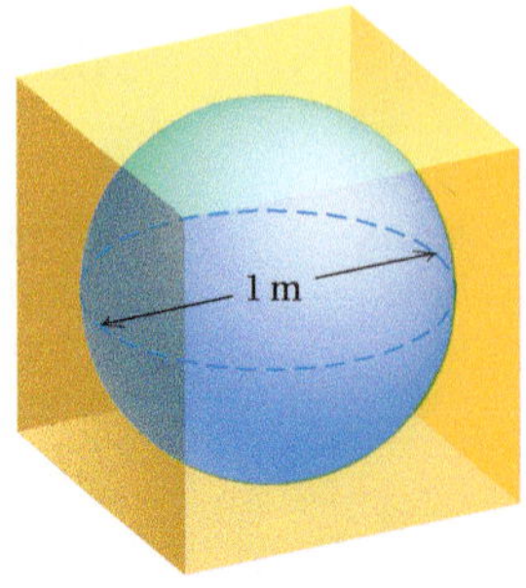

Relationships Between Angle Measures

8.6

a COMPLEMENTARY ANGLES AND SUPPLEMENTARY ANGLES

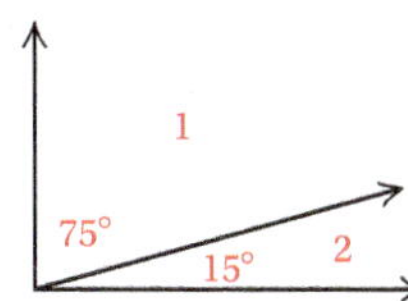

$\angle 1$ and $\angle 2$ above are **complementary** angles.

$$m\angle 1 + m\angle 2 = 90^\circ$$
$$75^\circ + 15^\circ = 90^\circ$$

COMPLEMENTARY ANGLES

Two angles are **complementary** if the sum of their measures is 90°. Each angle is called a **complement** of the other.

If two angles are complementary, each is an acute angle. When complementary angles are adjacent to each other, they form a right angle.

EXAMPLE 1 Identify each pair of complementary angles.

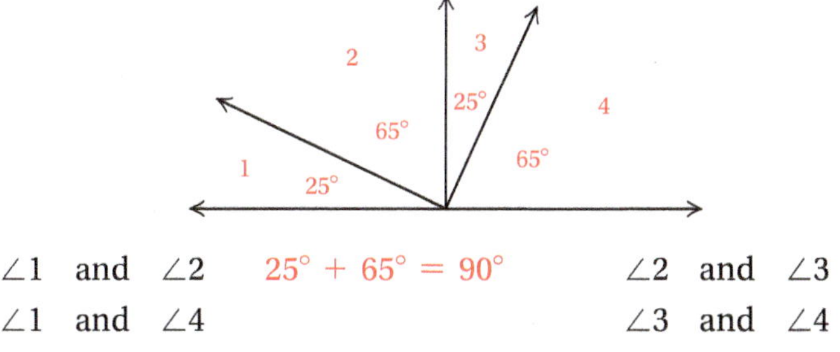

$\angle 1$ and $\angle 2$ $\quad 25^\circ + 65^\circ = 90^\circ \quad$ $\angle 2$ and $\angle 3$

$\angle 1$ and $\angle 4$ $\quad\quad$ $\angle 3$ and $\angle 4$

EXAMPLE 2 Find the measure of a complement of an angle of 39°.

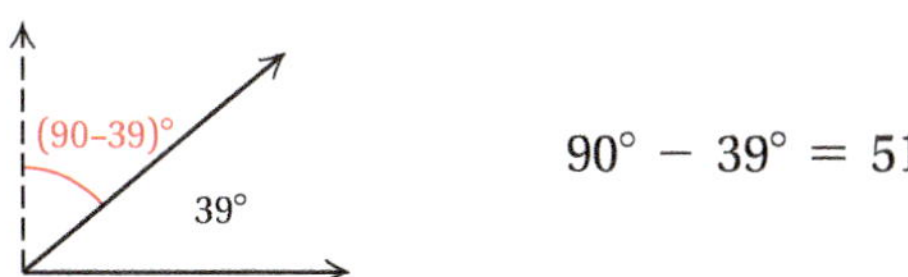

$$90^\circ - 39^\circ = 51^\circ$$

The measure of a complement is 51°.

Do Exercises 1–5. ▶

OBJECTIVES

a Identify complementary angles and supplementary angles and find the measure of a complement or a supplement of a given angle.

b Determine whether segments are congruent and whether angles are congruent.

c Use the Vertical Angle Property to find measures of angles.

d Identify pairs of corresponding angles, interior angles, and alternate interior angles and apply properties of transversals and parallel lines to find measures of angles.

1. Identify each pair of complementary angles.

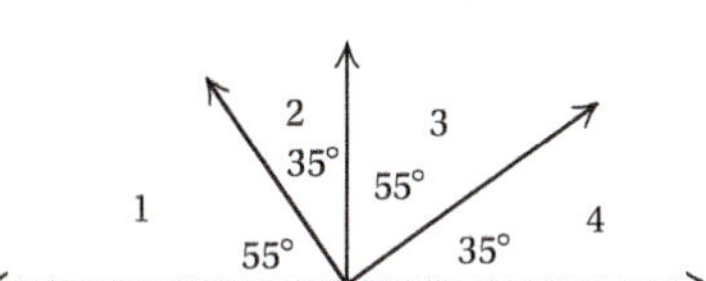

Find the measure of a complement of each angle.

2. 45° **3.** 18°

4. 85°

GS **5.** 67°

$90^\circ - 67^\circ = \square^\circ$

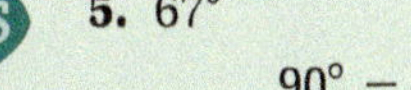

Answers

1. $\angle 1$ and $\angle 2$; $\angle 1$ and $\angle 4$; $\angle 2$ and $\angle 3$; $\angle 3$ and $\angle 4$ **2.** 45° **3.** 72° **4.** 5° **5.** 23°

Guided Solution:
5. 23

Next, consider $\angle 1$ and $\angle 2$ as shown below. Because the sum of their measures is 180°, $\angle 1$ and $\angle 2$ are said to be **supplementary**. Note that when supplementary angles are adjacent, they form a straight angle.

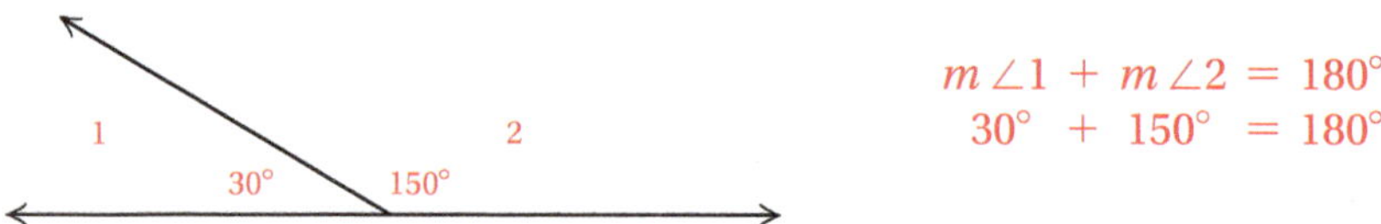

SUPPLEMENTARY ANGLES

Two angles are **supplementary** if the sum of their measures is 180°. Each angle is called a **supplement** of the other.

EXAMPLE 3 Identify each pair of supplementary angles.

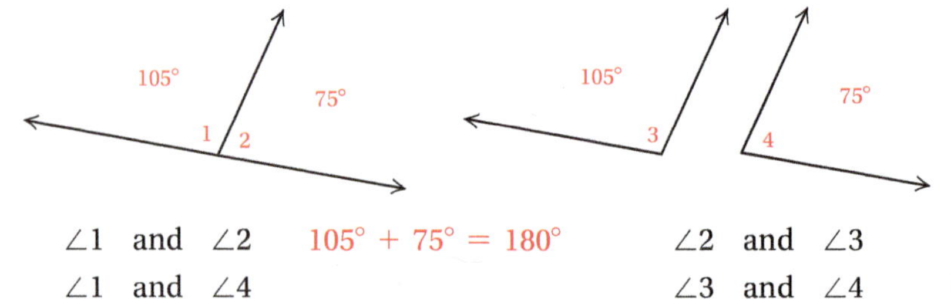

$\angle 1$ and $\angle 2$ $\quad 105° + 75° = 180°$ $\quad \angle 2$ and $\angle 3$

$\angle 1$ and $\angle 4$ $\quad \angle 3$ and $\angle 4$

EXAMPLE 4 Find the measure of a supplement of an angle of 112°.

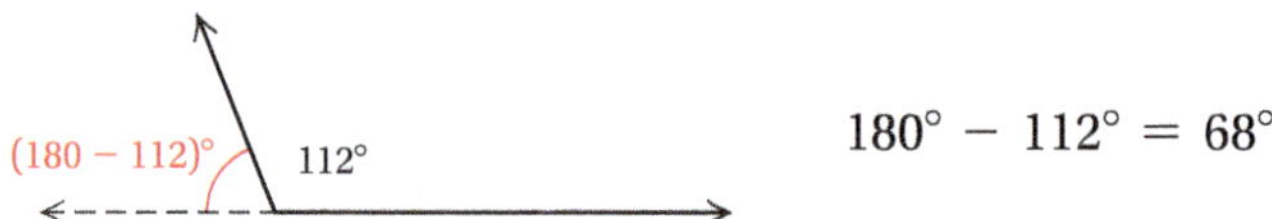

$$180° - 112° = 68°$$

The measure of a supplement is 68°.

Do Exercises 6–10.

b CONGRUENT SEGMENTS AND ANGLES

Congruent figures have the same size and shape. They fit together exactly.

CONGRUENT SEGMENTS

Two segments are **congruent** if and only if they have the same length.

---

6. Identify each pair of supplementary angles.

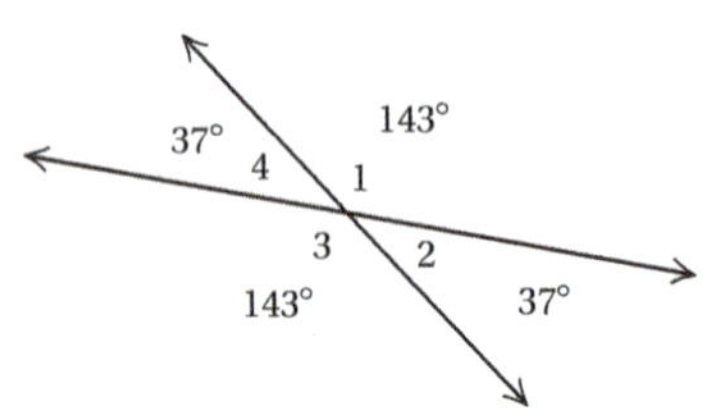

Find the measure of a supplement of each angle.

7. 38° **8.** 157°

9. 90°

10. 71°

GS $180° - 71° = \_\_\_\_°$

Answers

6. $\angle 1$ and $\angle 2$; $\angle 1$ and $\angle 4$; $\angle 2$ and $\angle 3$; $\angle 3$ and $\angle 4$ **7.** 142° **8.** 23° **9.** 90° **10.** 109°

Guided Solution:
10. 109

EXAMPLE 5 Use a ruler to show that $\overline{PQ}$ and $\overline{RS}$ are congruent.

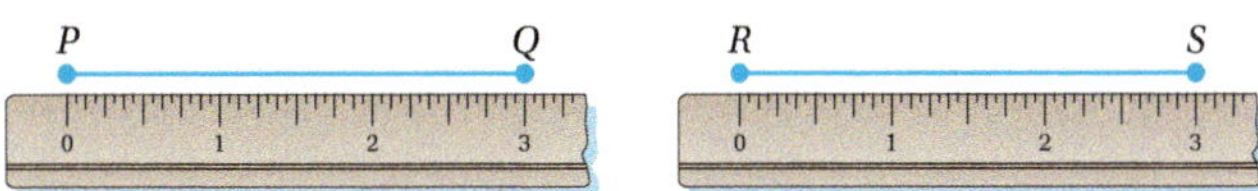

Since both segments have the same length, $\overline{PQ}$ and $\overline{RS}$ are congruent. To say that $\overline{PQ}$ and $\overline{RS}$ are congruent, we write $\overline{PQ} \cong \overline{RS}$.

EXAMPLE 6 Which pairs of segments are congruent? Use a ruler.

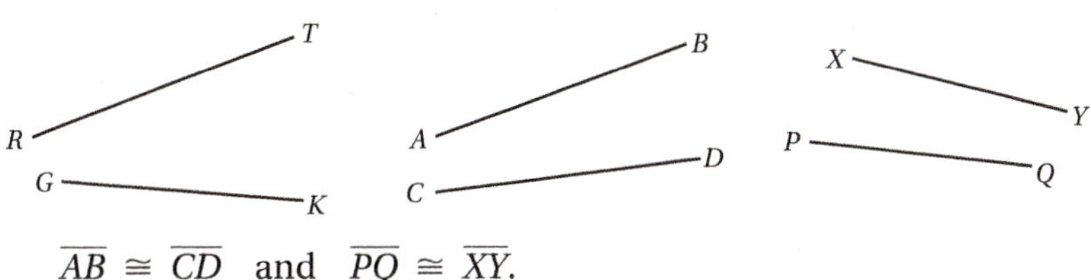

$$\overline{AB} \cong \overline{CD} \quad \text{and} \quad \overline{PQ} \cong \overline{XY}.$$

Do Exercises 11 and 12. ▶

Determine whether the given segments are congruent. Use a ruler.

11. **12.**

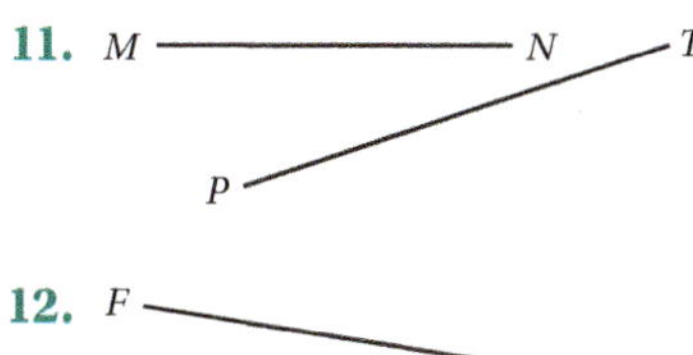

CONGRUENT ANGLES

Two angles are **congruent** if and only if they have the same measure.

EXAMPLE 7 Use a protractor to show that $\angle P$ and $\angle Q$ are congruent.

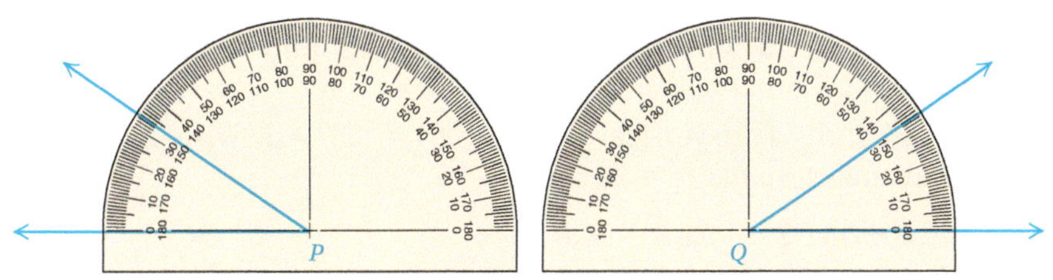

Since $m\angle P = m\angle Q = 34°$, $\angle P$ and $\angle Q$ are congruent. To say that $\angle P$ and $\angle Q$ are congruent, we write $\angle P \cong \angle Q$.

EXAMPLE 8 Which pairs of angles are congruent? Use a protractor.

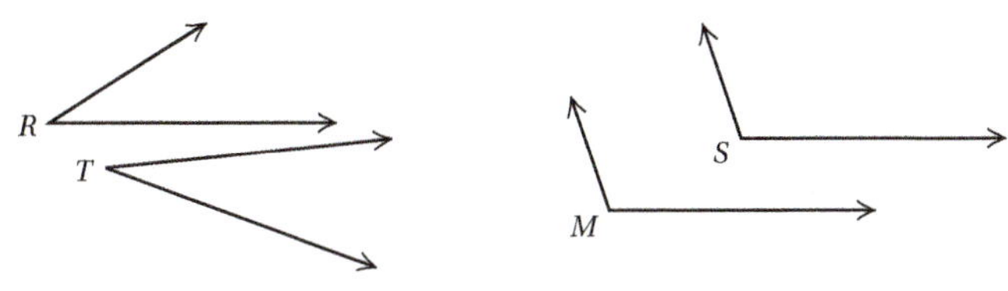

$$\angle M \cong \angle S \quad \text{since} \quad m\angle M = m\angle S = 108°.$$

Do Exercises 13 and 14. ▶

Determine whether the pair of angles is congruent. Use a protractor.

13. **14.**

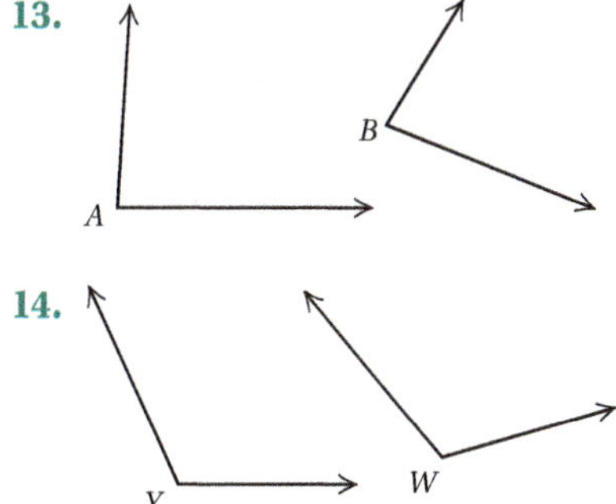

If two angles are congruent, then their supplements are congruent and their complements are congruent.

Answers

11. Not congruent **12.** Congruent
13. Not congruent **14.** Congruent

C VERTICAL ANGLES

When $\overleftrightarrow{RT}$ intersects $\overleftrightarrow{SQ}$ at P, four angles are formed:

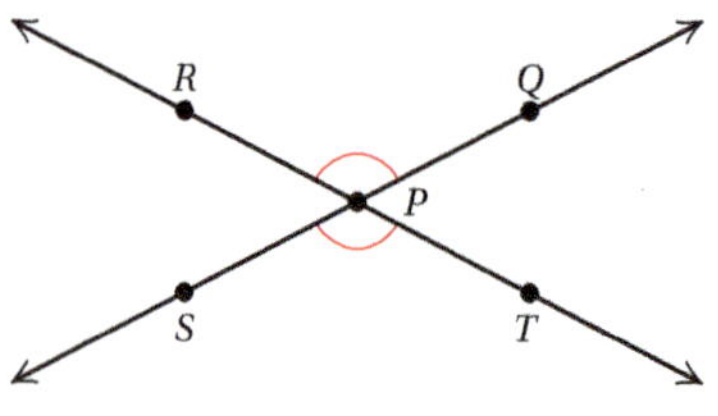

$\angle SPT$

$\angle RPQ$

$\angle SPR$

$\angle QPT$

Pairs of angles such as $\angle RPQ$ and $\angle SPT$ are called **vertical angles**.

> **VERTICAL ANGLES**
>
> Two nonstraight angles are **vertical angles** if and only if their sides form two pairs of opposite rays.

Vertical angles are supplements of the same angle. Thus they are congruent.

> **THE VERTICAL ANGLE PROPERTY**
>
> Vertical angles are congruent.

EXAMPLE 9 In the following figure, $m\angle 1 = 23°$ and $m\angle 3 = 34°$. Find $m\angle 2$, $m\angle 4$, $m\angle 5$, and $m\angle 6$.

Since $\angle 1$ and $\angle 4$ are vertical angles, $m\angle 4 = 23°$. Likewise, $\angle 3$ and $\angle 6$ are vertical angles, so $m\angle 6 = 34°$.

$$m\angle 1 + m\angle 2 + m\angle 3 = 180$$
$$23 + m\angle 2 + 34 = 180 \quad \text{Substituting}$$
$$m\angle 2 = 180 - 57$$
$$m\angle 2 = 123°$$

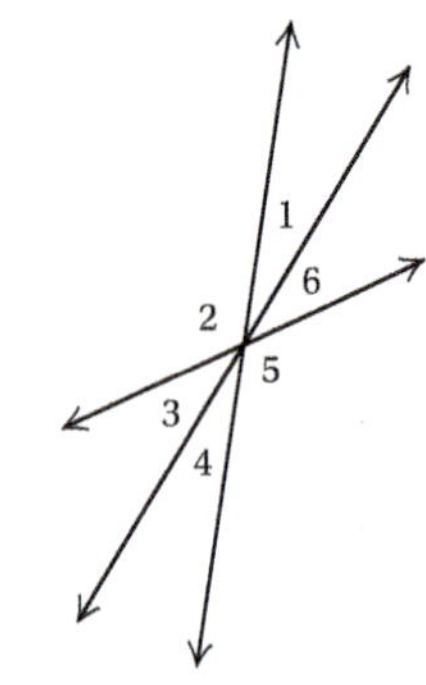

Since $\angle 2$ and $\angle 5$ are vertical angles, $m\angle 5 = 123°$.

◀ Do Exercise 15.

15. In the following figure, $m\angle 2 = 41°$ and $m\angle 4 = 10°$. Find $m\angle 1$, $m\angle 3$, $m\angle 5$, and $m\angle 6$.

Answer

15. $m\angle 1 = 10°$; $m\angle 3 = 129°$; $m\angle 5 = 41°$; $m\angle 6 = 129°$

d TRANSVERSALS AND ANGLES

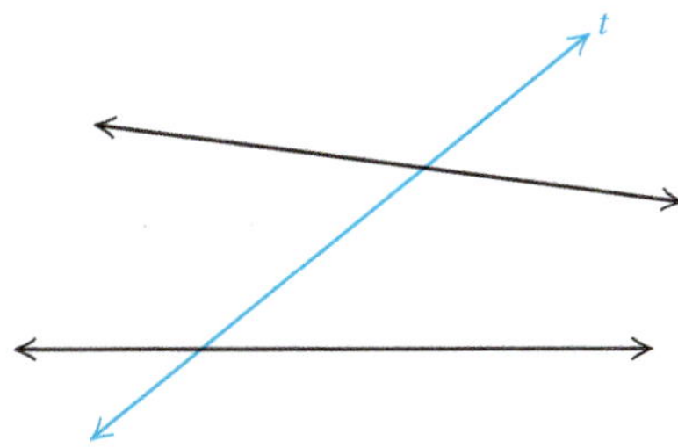

TRANSVERSAL

A **transversal** is a line that intersects two or more coplanar lines in different points.

When a transversal intersects a pair of lines, eight angles are formed. Certain pairs of these angles have special names.

Corresponding Angles

$\angle 2$ and $\angle 6$
$\angle 3$ and $\angle 7$
$\angle 1$ and $\angle 5$
$\angle 4$ and $\angle 8$

Interior Angles

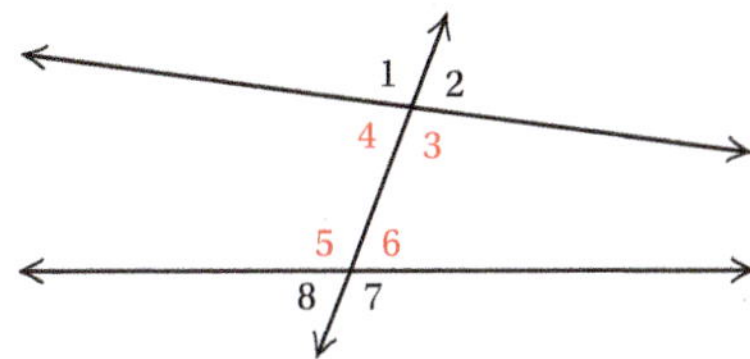

$\angle 3$, $\angle 4$, $\angle 5$, and $\angle 6$

Alternate Interior Angles

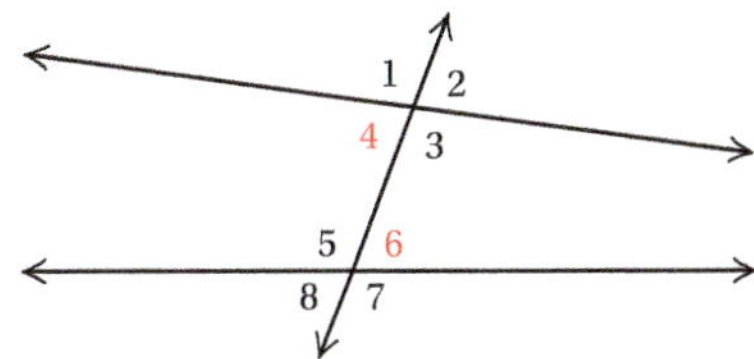

$\angle 4$ and $\angle 6$
$\angle 3$ and $\angle 5$

Use the following figure to answer Margin Exercises 16–18.

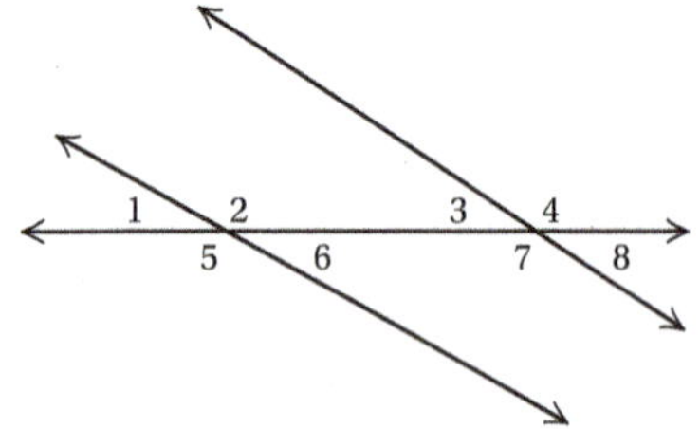

16. Identify all pairs of corresponding angles.

17. Identify all interior angles.

18. Identify all pairs of alternate interior angles.

EXAMPLE 10 Identify all pairs of corresponding angles, all interior angles, and all pairs of alternate interior angles.

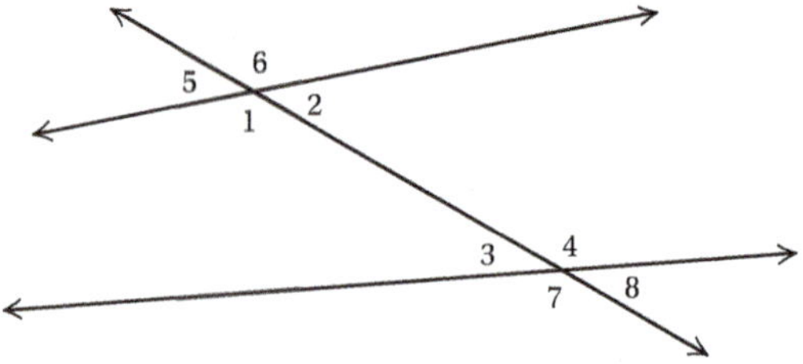

Corresponding angles:	$\angle 6$ and $\angle 4$, $\angle 2$ and $\angle 8$, $\angle 5$ and $\angle 3$, $\angle 1$ and $\angle 7$
Interior angles:	$\angle 1$, $\angle 2$, $\angle 3$, $\angle 4$
Alternate interior angles:	$\angle 1$ and $\angle 4$, $\angle 2$ and $\angle 3$

Do Exercises 16–18.

Given a line l and a point P not on l, there is only one line that contains P and is parallel to l.

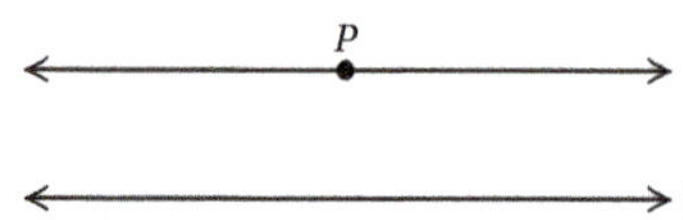

If two lines are parallel, the following relations hold.

Properties of Parallel Lines

1. If a transversal intersects two parallel lines, then the corresponding angles are congruent.

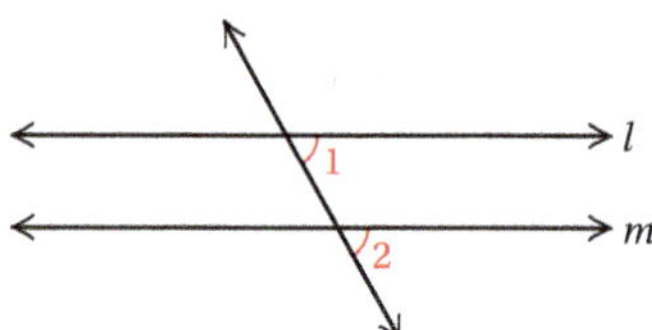

If $l \parallel m$, then $\angle 1 \cong \angle 2$.

2. If a transversal intersects two parallel lines, then the alternate interior angles are congruent.

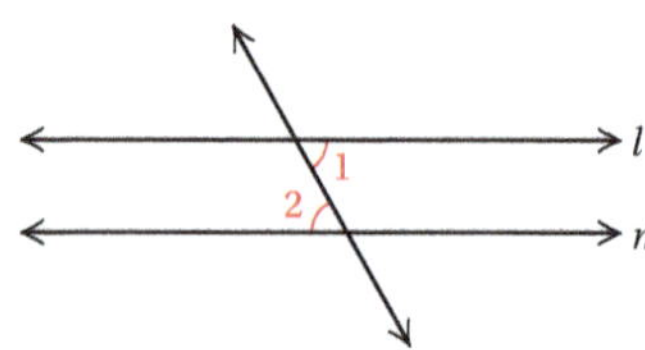

If $l \parallel m$, then $\angle 1 \cong \angle 2$.

Answers

16. $\angle 1$ and $\angle 3$, $\angle 2$ and $\angle 4$, $\angle 5$ and $\angle 7$, $\angle 6$ and $\angle 8$ **17.** $\angle 2$, $\angle 3$, $\angle 6$, and $\angle 7$
18. $\angle 2$ and $\angle 7$, $\angle 6$ and $\angle 3$

3. In a plane, if two lines are parallel to a third line, then the two lines are parallel to each other.

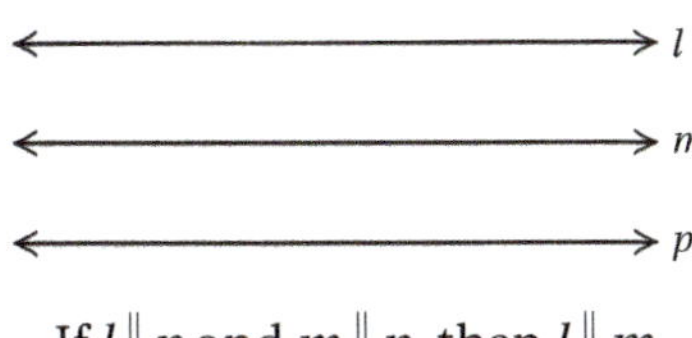

If $l \parallel p$ and $m \parallel p$, then $l \parallel m$.

4. If a transversal intersects two parallel lines, then the interior angles on the same side of the transversal are supplementary.

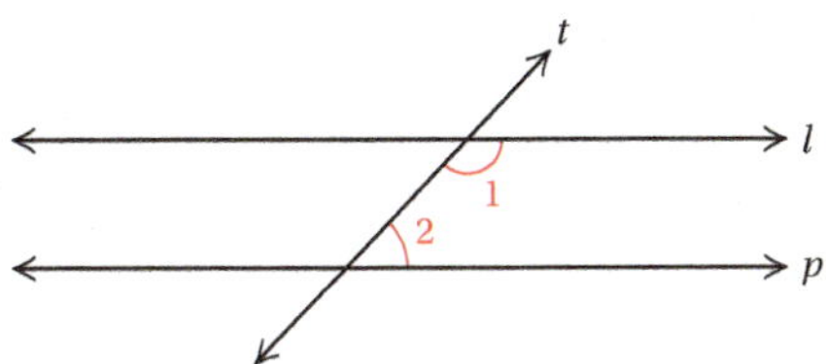

If $l \parallel p$, then $m\angle 1 + m\angle 2 = 180°$.

5. If a transversal is perpendicular to one of two parallel lines, then it is perpendicular to the other.

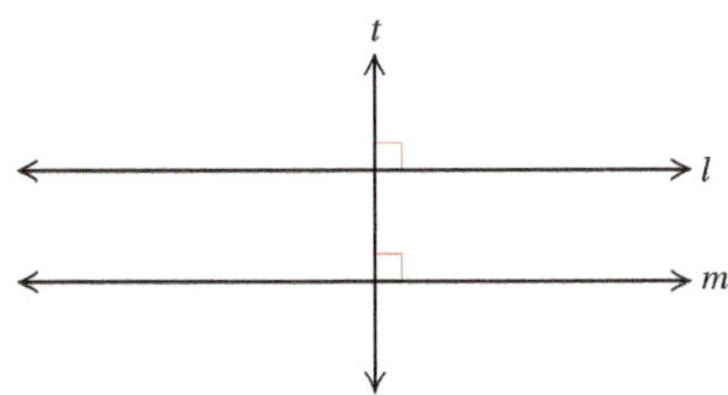

If $l \parallel m$ and $t \perp l$, then $t \perp m$.

EXAMPLE 11 If $l \parallel m$ and $m\angle 1 = 40°$, what are the measures of the other angles?

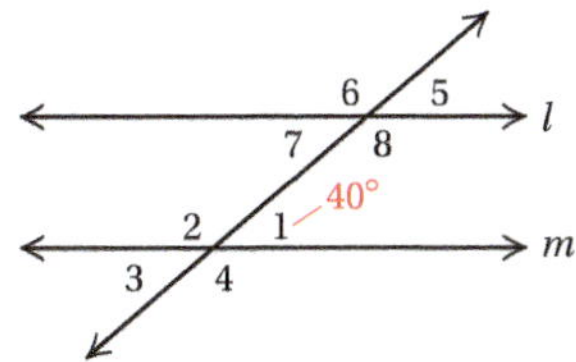

$m\angle 7 = 40°$	Using Property 2
$m\angle 5 = 40°$	Using Property 1
$m\angle 8 = 140°$	Using Property 4
$m\angle 3 = 40°$	$\angle 1$ and $\angle 3$ are vertical angles
$m\angle 4 = 140°$	Using Property 1 and $m\angle 8 = 140°$
$m\angle 2 = 140°$	$\angle 2$ and $\angle 4$ are vertical angles and $m\angle 4 = 140°$
$m\angle 6 = 140°$	$\angle 6$ and $\angle 8$ are vertical angles and $m\angle 8 = 140°$

Do Exercise 19. ▶

19. If $l \parallel m$ and $m\angle 3 = 51°$, what are the measures of the other angles?

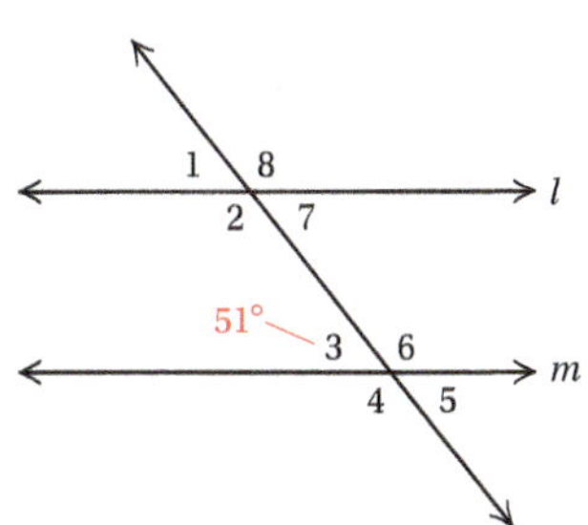

Answer

19. $m\angle 7 = m\angle 1 = m\angle 5 = 51°$; $m\angle 8 = m\angle 2 = m\angle 6 = m\angle 4 = 129°$

EXAMPLE 12 If $\overline{PT} \parallel \overline{SR}$, which pairs of angles are congruent?

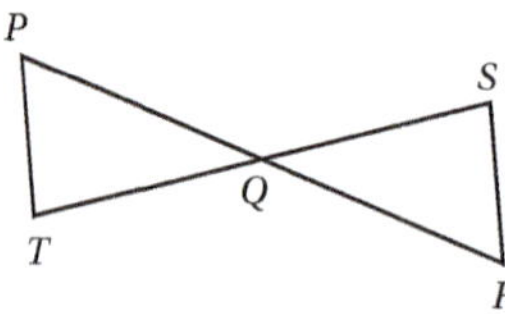

$\angle TPQ \cong \angle SRQ$ and $\angle PTQ \cong \angle RSQ$ Using Property 2

$\angle PQT \cong \angle RQS$ and $\angle PQS \cong \angle RQT$ Vertical angles

◀ **Do Exercise 20.**

EXAMPLE 13 If $\overline{DE} \parallel \overline{BC}$, which pairs of angles are congruent?

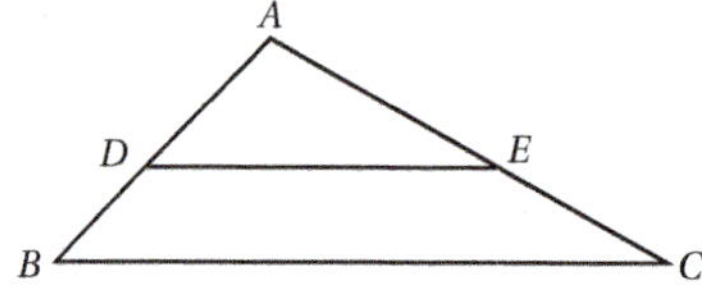

$\angle ADE \cong \angle ABC$ and $\angle AED \cong \angle ACB$ Using Property 1

◀ **Do Exercise 21.**

20. If $\overline{AB} \parallel \overline{CD}$, which pairs of angles are congruent?

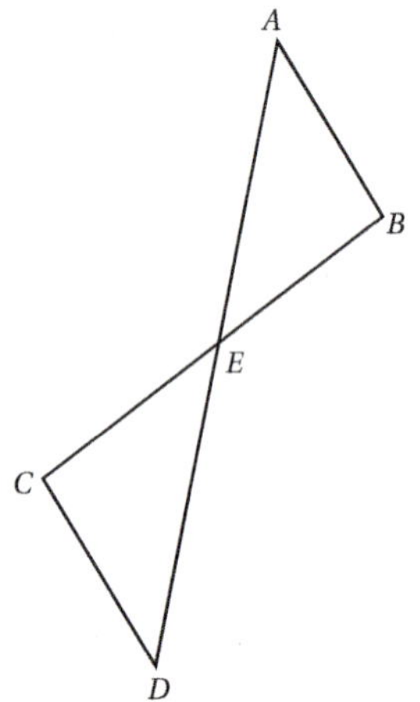

21. If $\overline{PQ} \parallel \overline{RS}$, which pairs of angles are congruent?

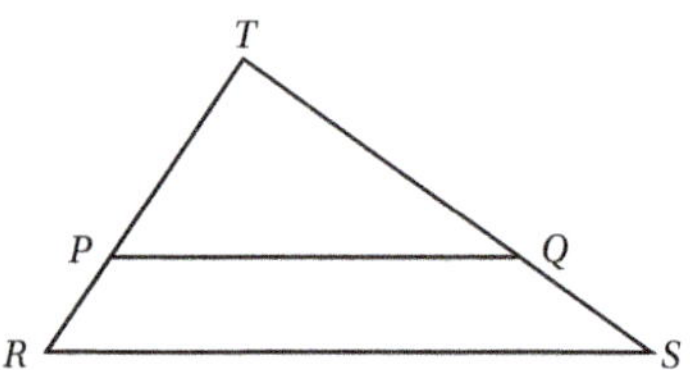

Answers

20. $\angle CED \cong \angle BEA$, $\angle ECD \cong \angle EBA$, $\angle EDC \cong \angle EAB$, $\angle CEA \cong \angle BED$
21. $\angle TPQ \cong \angle TRS$, $\angle TQP \cong \angle TSR$

8.6 Exercise Set

For Extra Help
MyMathLab®

☑ Reading Check

Determine whether each statement is true or false.

RC1. If a transversal intersects two parallel lines, then the alternate interior angles are congruent.

RC2. Two angles are complementary if the sum of their measures is 180°.

RC3. If a transversal intersects two parallel lines, then the interior angles on the same side of the transversal are complementary.

RC4. Two angles are congruent if and only if they have the same measure.

RC5. Vertical angles are congruent.

RC6. Two angles are supplementary if the sum of their measures is 180°.

a Find the measure of a complement of each angle.

1. 11°
2. 83°
3. 67°
4. 5°

5. 58°
6. 32°
7. 29°
8. 54°

Find the measure of a supplement of each angle.

9. 3°
10. 54°
11. 139°
12. 13°

13. 85°
14. 129°
15. 102°
16. 45°

b Determine whether each pair of segments is congruent. Use a ruler.

17.

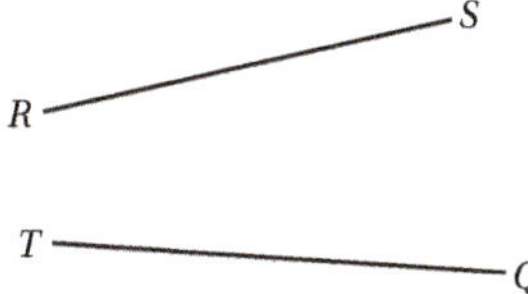

18.

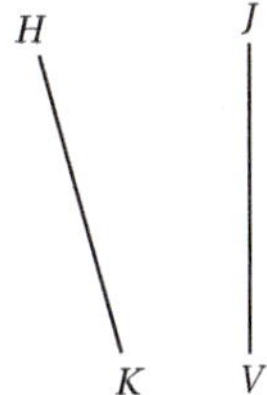

Determine whether each pair of angles is congruent. Use a protractor.

19.

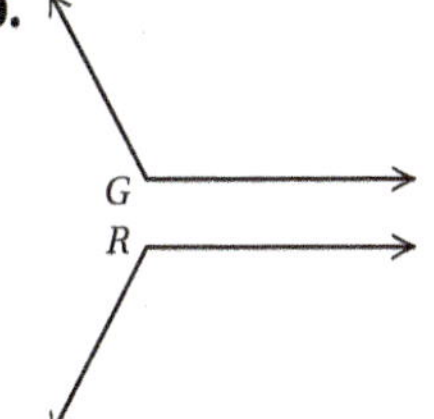

20.

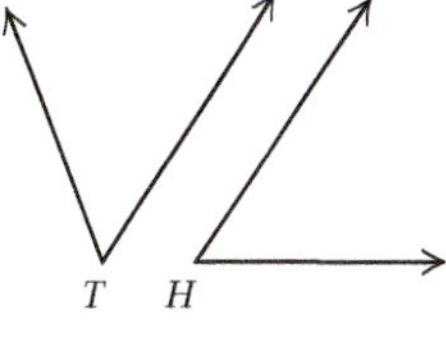

c

21. In the figure, $m\angle 1 = 80°$ and $m\angle 5 = 67°$. Find $m\angle 2$, $m\angle 3$, $m\angle 4$, and $m\angle 6$.

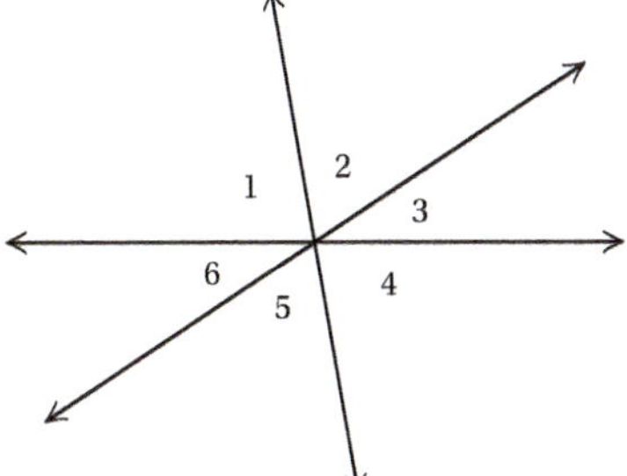

22. In the figure, $m\angle 2 = 42°$ and $m\angle 4 = 56°$. Find $m\angle 1$, $m\angle 3$, $m\angle 5$, and $m\angle 6$.

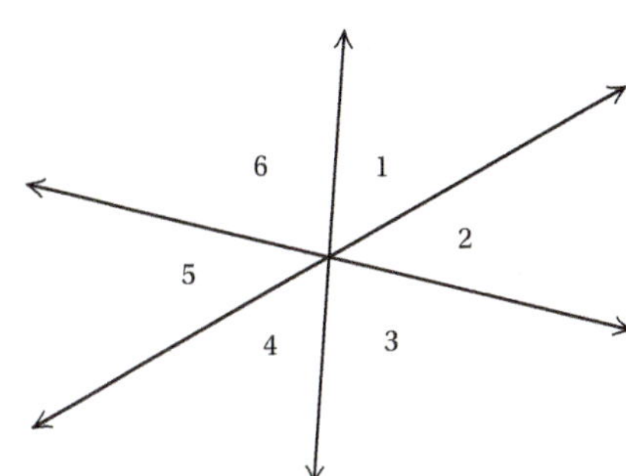

d In Exercises 23 and 24, **(a)** identify all pairs of corresponding angles, **(b)** identify all interior angles, and **(c)** identify all pairs of alternate interior angles.

23.

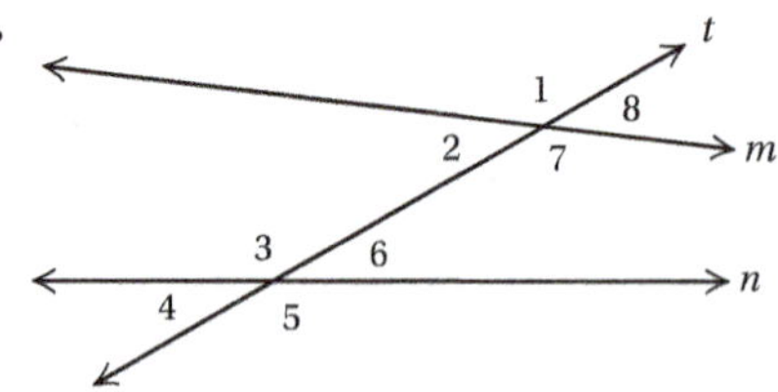

Lines m and n
Transversal t

24.

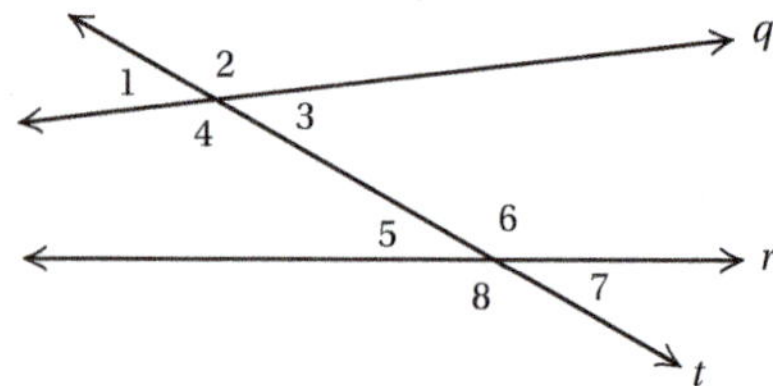

Lines q and r
Transversal t

25. If $m \parallel n$ and $m\angle 4 = 125°$, what are the measures of the other angles?

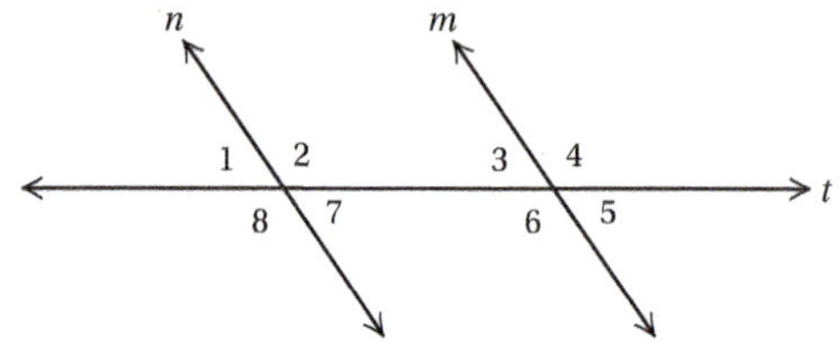

26. If $m \parallel n$ and $m\angle 8 = 34°$, what are the measures of the other angles?

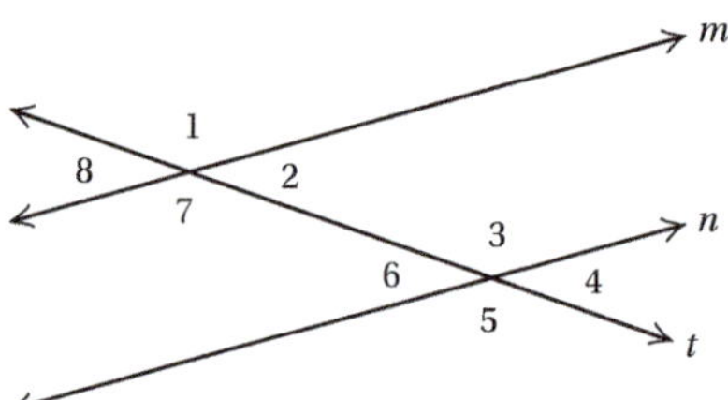

In each figure, $\overline{AB} \parallel \overline{CD}$. Identify pairs of congruent angles. Where possible, give the measures of the angles.

27.

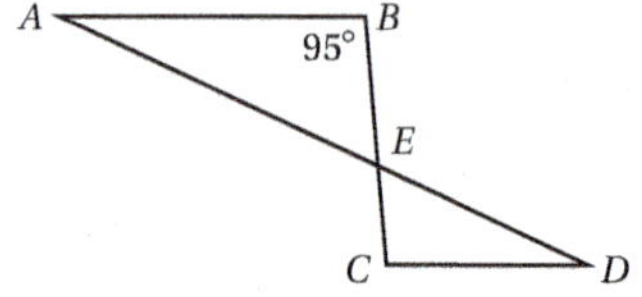

28.

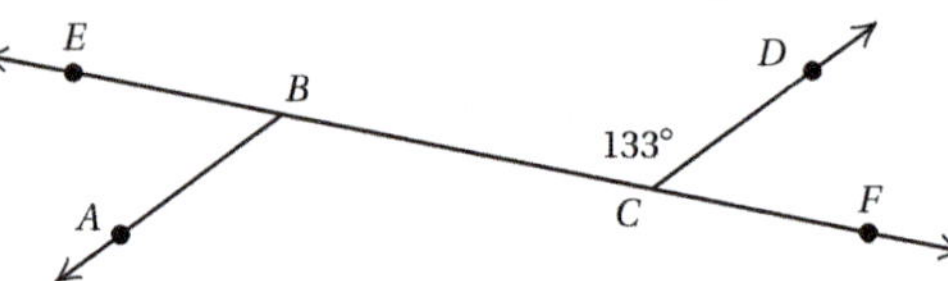

29.

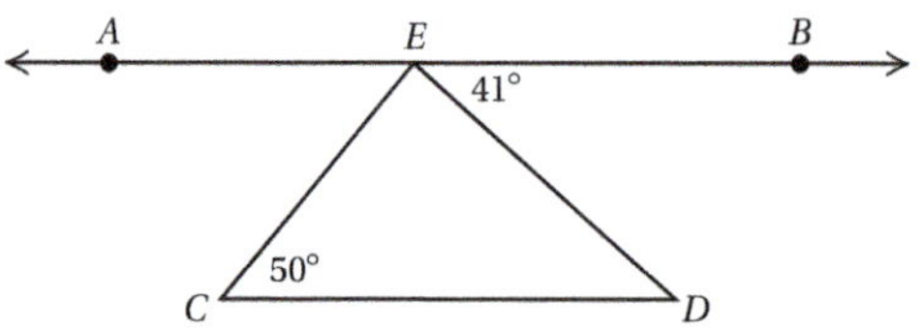

30.

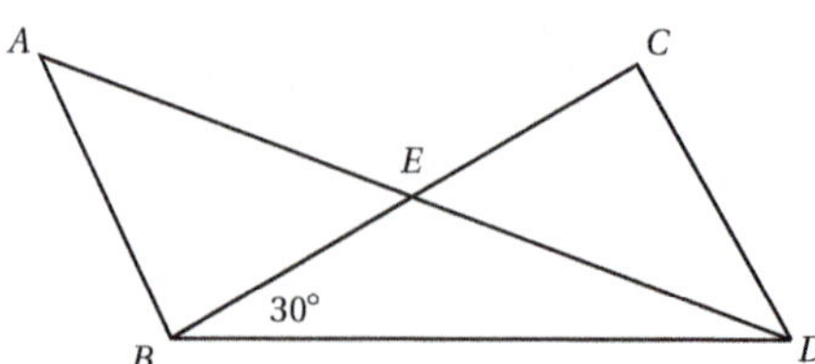

Skill Maintenance

Multiply. [4.7a]

31. $6 \times 1\frac{7}{8}$

32. $-10\frac{3}{4} \times 1\frac{1}{2}$

33. $\left(-8\frac{3}{7}\right)\left(-14\right)$

34. $2\frac{2}{3} \times 5\frac{1}{2}$

Congruent Triangles and Properties of Parallelograms

8.7

OBJECTIVES

a Identify the corresponding parts of congruent triangles and show why triangles are congruent using SAS, SSS, and ASA.

b Use properties of parallelograms to find lengths of sides and measures of angles of parallelograms.

a CONGRUENT TRIANGLES

Triangles can be classified by their angles.

Acute:	All angles acute
Right:	One right angle
Obtuse:	One obtuse angle
Equiangular:	All angles congruent

Triangles can also be classified by their sides.

Equilateral:	All sides congruent
Isosceles:	At least two sides congruent
Scalene:	No sides congruent

We know that congruent figures fit together exactly.

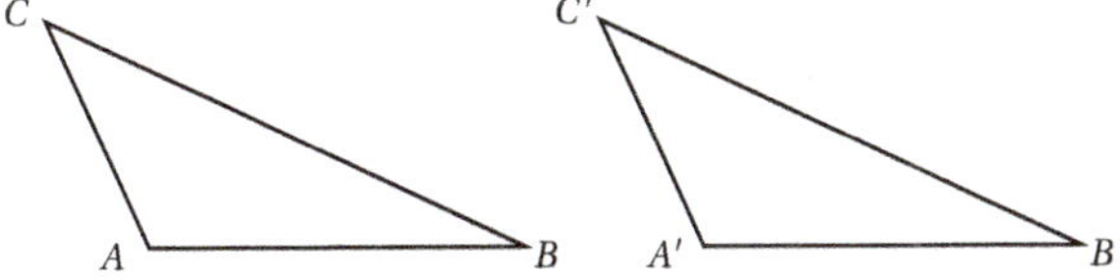

B' is read "B prime."

These triangles will fit together exactly if we match A with A', B with B', and C with C'. On the other hand, if we match A with B', B with C', and C with A', the triangles will not fit together exactly. The matching of vertices determines corresponding sides and angles.

EXAMPLES Consider $\triangle ABC$ and $\triangle A'B'C'$ above.

1. If we match A with A', B with B', and C with C', what are the corresponding sides? What are the corresponding angles?

$\overline{AB} \leftrightarrow \overline{A'B'}$ $\quad\angle A \leftrightarrow \angle A'$

$\overline{BC} \leftrightarrow \overline{B'C'}$ $\quad\angle B \leftrightarrow \angle B'$

$\overline{AC} \leftrightarrow \overline{A'C'}$ $\quad\angle C \leftrightarrow \angle C'$

$\leftrightarrow$ means "corresponds to."

If $A \leftrightarrow A'$, $B \leftrightarrow B'$, and $C \leftrightarrow C'$, then we write $ABC \leftrightarrow A'B'C'$.

2. If we match A with B', B with C', and C with A', what are the corresponding sides? What are the corresponding angles?

$\overline{AB} \leftrightarrow \overline{B'C'}$ $\quad\overline{BC} \leftrightarrow \overline{C'A'}$ $\quad\overline{AC} \leftrightarrow \overline{B'A'}$

$\angle A \leftrightarrow \angle B'$ $\quad\angle B \leftrightarrow \angle C'$ $\quad\angle C \leftrightarrow \angle A'$

CONGRUENT TRIANGLES

Two triangles are **congruent** if and only if their vertices can be matched so that the corresponding angles and sides are congruent.

The corresponding sides and angles of two congruent triangles are called *corresponding parts* of congruent triangles. Corresponding parts of congruent triangles are always congruent.

We write $\triangle ABC \cong \triangle A'B'C'$ to say that $\triangle ABC$ and $\triangle A'B'C'$ are congruent. We agree that this symbol also tells us the way in which the vertices are matched.

$$\triangle ABC \cong \triangle A'B'C'$$

$\triangle ABC \cong \triangle A'B'C'$ means that

$$\angle A \cong \angle A' \quad \text{and} \quad \overline{AB} \cong \overline{A'B'}$$
$$\angle B \cong \angle B' \qquad \overline{AC} \cong \overline{A'C'}$$
$$\angle C \cong \angle C' \qquad \overline{BC} \cong \overline{B'C'}.$$

EXAMPLE 3 Suppose that $\triangle PQR \cong \triangle STV$. What are the congruent corresponding parts?

Angles	*Sides*
$\angle P \cong \angle S$	$\overline{PQ} \cong \overline{ST}$
$\angle Q \cong \angle T$	$\overline{PR} \cong \overline{SV}$
$\angle R \cong \angle V$	$\overline{QR} \cong \overline{TV}$

Do Exercise 1.

1. Suppose that $\triangle ABC \cong \triangle DEF$. What are the congruent corresponding parts?

EXAMPLE 4 Name the corresponding parts of these congruent triangles.

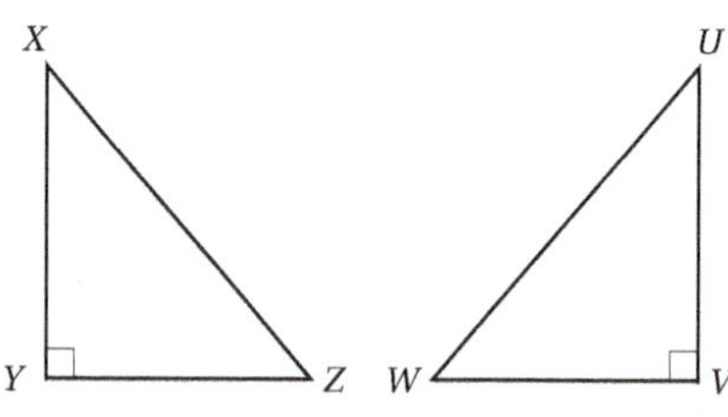

Angles	*Sides*
$\angle X \cong \angle U$	$\overline{XY} \cong \overline{UV}$
$\angle Y \cong \angle V$	$\overline{YZ} \cong \overline{VW}$
$\angle Z \cong \angle W$	$\overline{ZX} \cong \overline{WU}$

Do Exercise 2.

2. Name the corresponding parts of these congruent triangles.

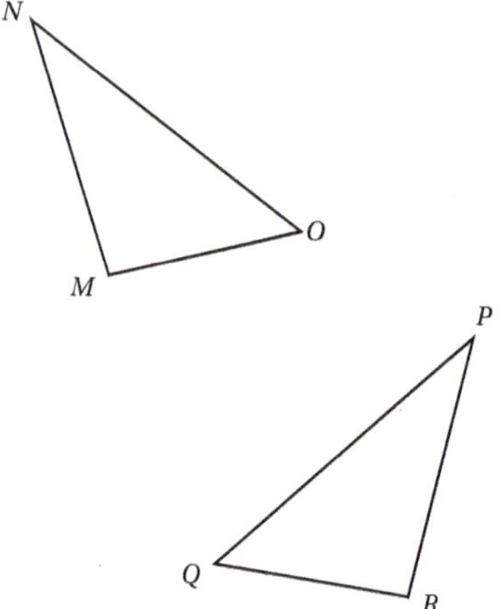

Sometimes we can show that triangles are congruent without already knowing that all six corresponding parts are congruent.

On a full sheet of paper, draw $\triangle ABC$. On another sheet of paper, make a copy of $\angle A$. Label the copy $\angle D$. On the sides of $\angle D$, copy $\overline{AB}$ and $\overline{AC}$. Label the copy $\overline{DE}$ and $\overline{DF}$. Draw $\overline{EF}$. Cut out $\triangle DEF$ and $\triangle ABC$ and place them together. What do you conclude?

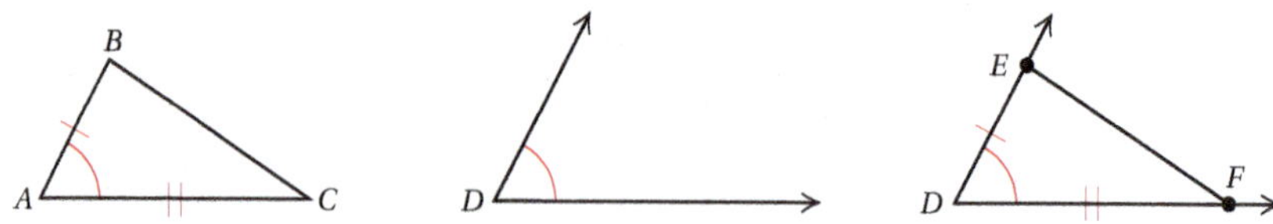

Answers

1. $\angle A \cong \angle D, \angle B \cong \angle E, \angle C \cong \angle F;$ $\overline{AB} \cong \overline{DE}, \overline{AC} \cong \overline{DF}, \overline{BC} \cong \overline{EF}$
2. $\angle N \cong \angle P, \angle M \cong \angle R, \angle O \cong \angle Q;$ $\overline{NM} \cong \overline{PR}, \overline{NO} \cong \overline{PQ}, \overline{MO} \cong \overline{RQ}$

THE SIDE-ANGLE-SIDE (SAS) PROPERTY

Two triangles are congruent if two sides and the included angle of one triangle are congruent to two sides and the included angle of the other triangle.

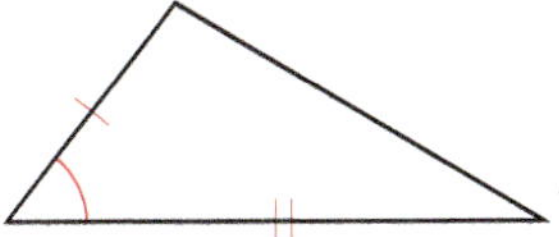
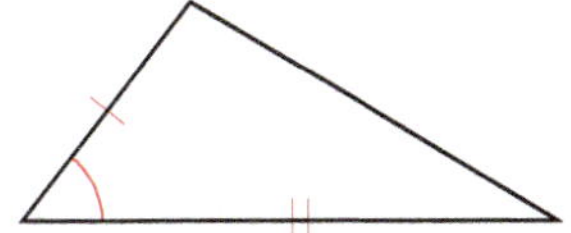

EXAMPLE 5 Which pairs of triangles are congruent by the SAS property?

a)

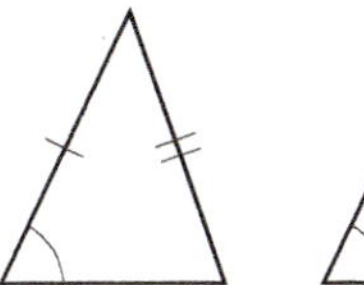
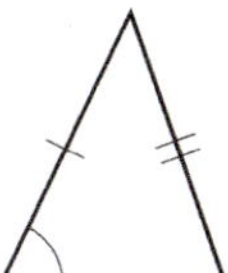

b)

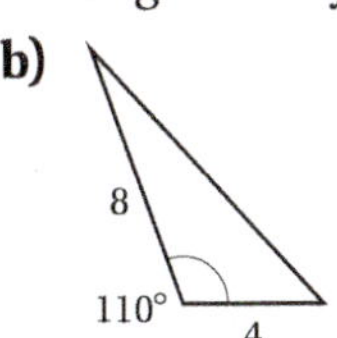

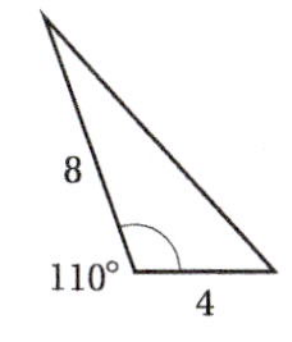

c)

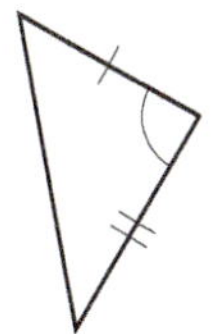
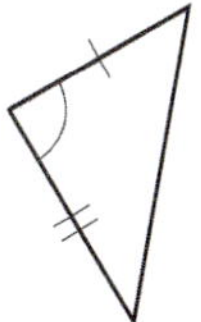

d)

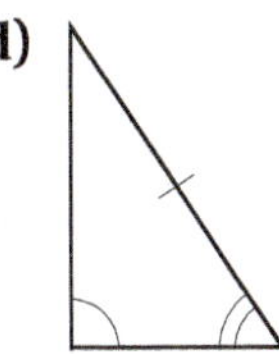
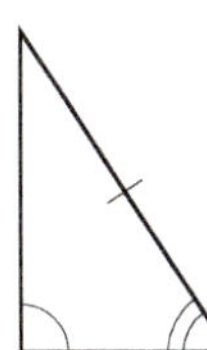

Pairs (b) and (c) are congruent by the SAS property.

Do Exercise 3. ▶

On a sheet of paper, draw a triangle. Then copy this triangle by copying each of its sides. Cut both triangles out and place them together. This suggests the following property.

THE SIDE-SIDE-SIDE (SSS) PROPERTY

If three sides of one triangle are congruent to three sides of another triangle, then the triangles are congruent.

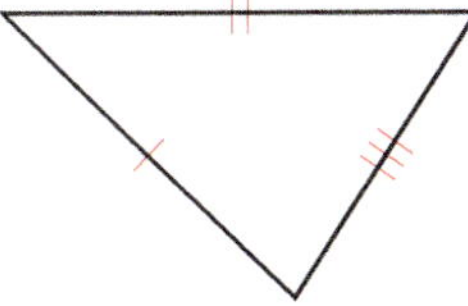
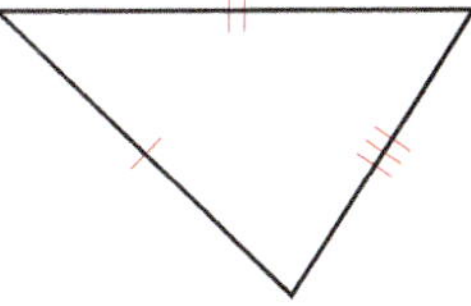

3. Which pairs of triangles are congruent by the SAS property?

a)

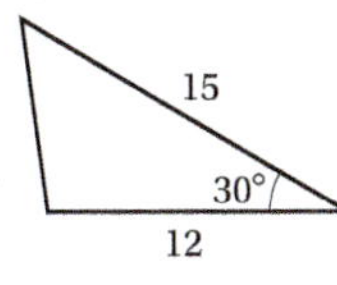

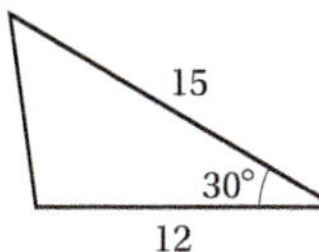

b)

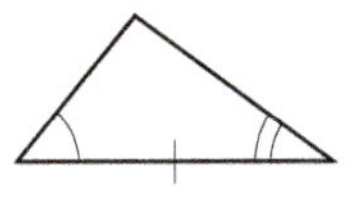
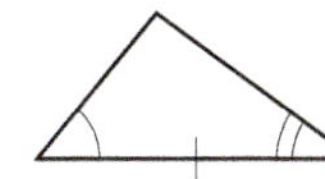

c)

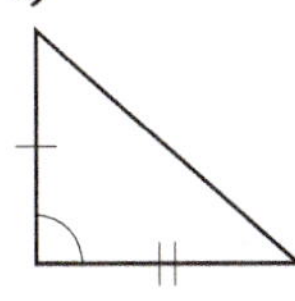
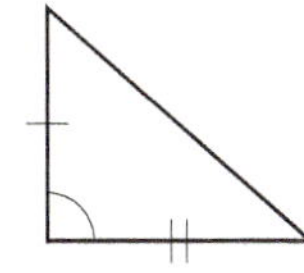

d)

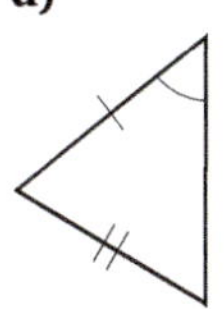
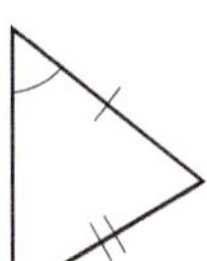

Answer

3. (a), (c)

EXAMPLE 6 Which pairs of triangles are congruent by the SSS property?

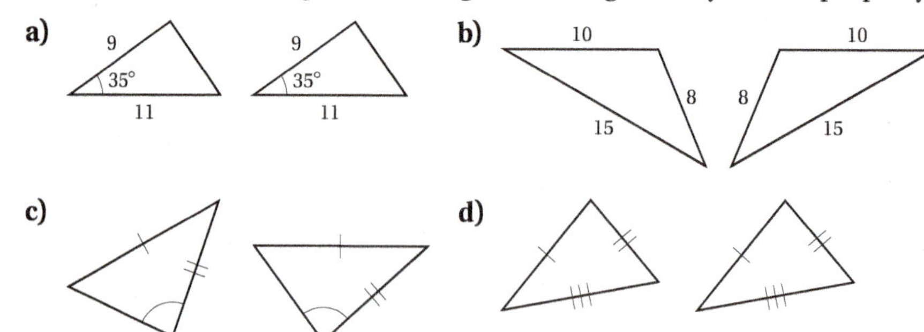

Pairs (b) and (d) are congruent by the SSS property.

◀ **Do Exercise 4.**

4. Which pairs of triangles are congruent by the SSS property?

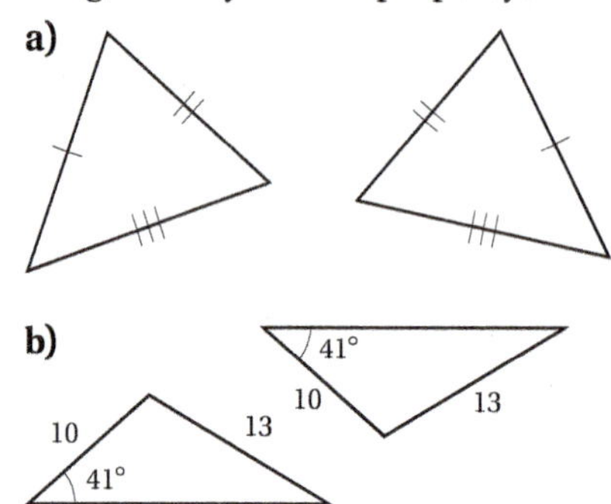

We have shown triangles to be congruent using SAS and SSS. A third way to show congruence is shown below.

On a full sheet of paper, draw a triangle, $\triangle ABC$. On another sheet of paper, draw a segment $\overline{DE}$ so that $DE = AB$*. At D, make a copy of $\angle A$. At E, make a copy of $\angle B$. Label the third vertex of the copy F. Cut out $\triangle ABC$ and $\triangle DEF$ and place them together. What do you conclude?

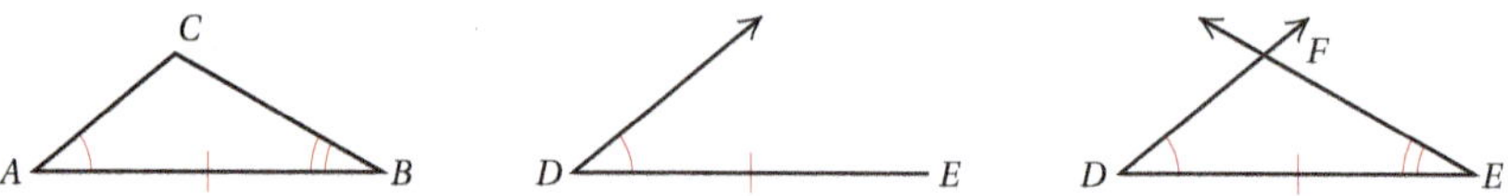

THE ANGLE-SIDE-ANGLE (ASA) PROPERTY

If two angles and the included side of a triangle are congruent to two angles and the included side of another triangle, then the triangles are congruent.

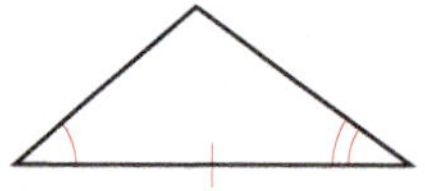

EXAMPLE 7 Which pairs of triangles are congruent by the ASA property?

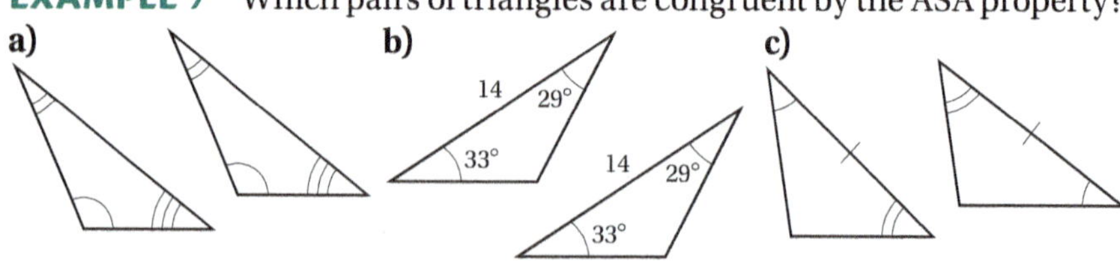

Pairs (b) and (c) are congruent by the ASA property.

◀ **Do Exercise 5.**

5. Which pairs of triangles are congruent by the ASA property?

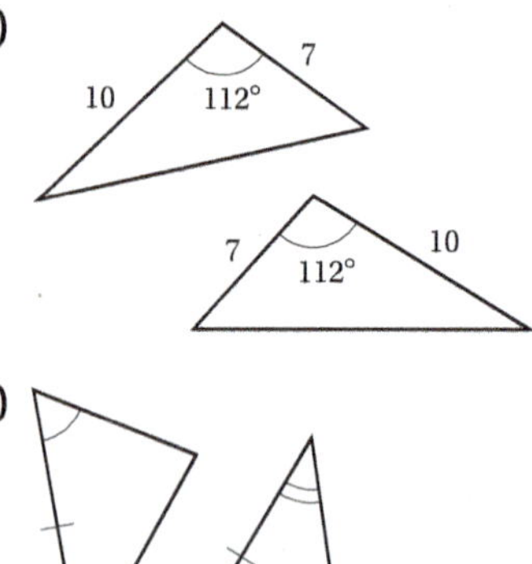

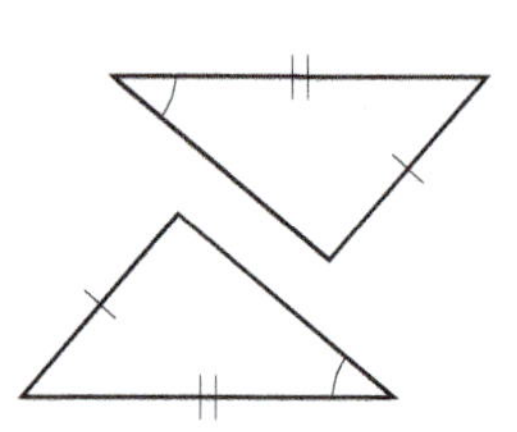

*$\overline{DE}$ denotes the segment with endpoints D and E. DE denotes the length of $\overline{DE}$.

Answers

4. (a) **5.** (b)

EXAMPLES Which property (if any) should be used to show that these pairs of triangles are congruent?

8.

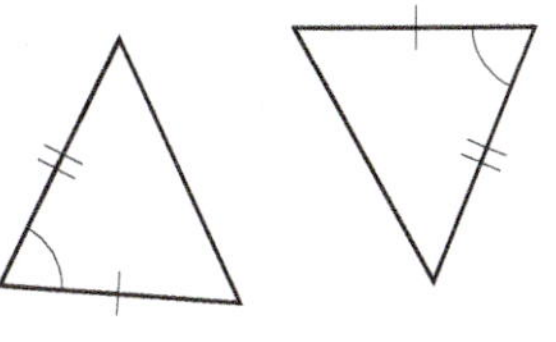

Use SAS.

9.

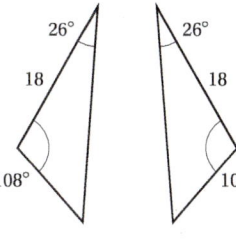

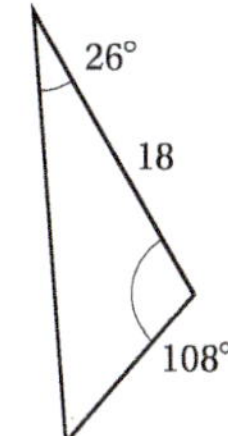

Use ASA.

10.

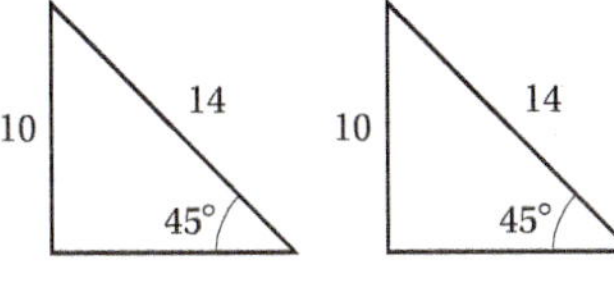

None.

11.

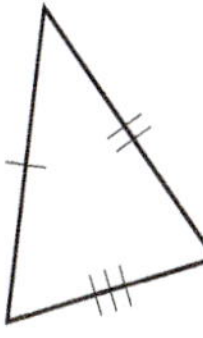

Use SSS.

Do Exercises 6–9. ▶

It is important to be able to explain why triangles are congruent.

EXAMPLE 12 In $\triangle ABC$ and $\triangle DEF$, $\overline{AB} \cong \overline{DE}$, $\overline{AC} \cong \overline{DF}$, and $\angle A \cong \angle D$. Explain why the triangles are congruent.

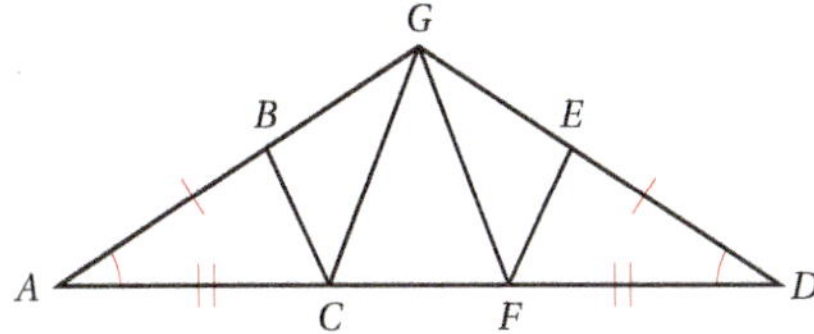

We have two sides and an included angle of $\triangle ABC$ congruent to the corresponding parts of $\triangle DEF$. Thus, $\triangle ABC \cong \triangle DEF$ by SAS. ■

EXAMPLE 13 In $\triangle CPD$ and $\triangle EQD$, $\overline{CP} \perp \overline{QP}$ and $\overline{EQ} \perp \overline{QP}$. Also, $\angle QDE \cong \angle PDC$ and D is the midpoint of $\overline{QP}$. Explain why $\triangle CPD \cong \triangle EQD$.

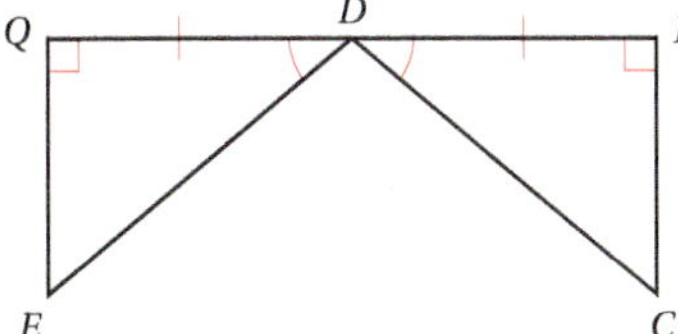

The perpendicular sides form right angles, which are congruent. Since D is the midpoint of $\overline{QP}$, we know that $\overline{QD} \cong \overline{PD}$. With $\angle QDE \cong \angle PDC$, we have $\triangle CPD \cong \triangle EQD$ by ASA.

Do Exercise 10. ▶

Which property (if any) should be used to show that the pair of triangles is congruent?

6.

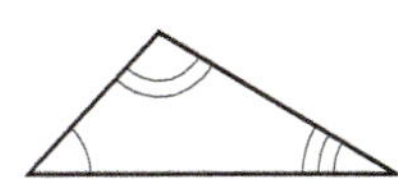

7.

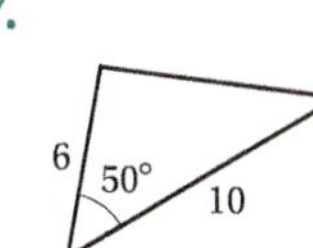
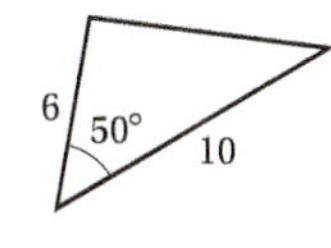

8.

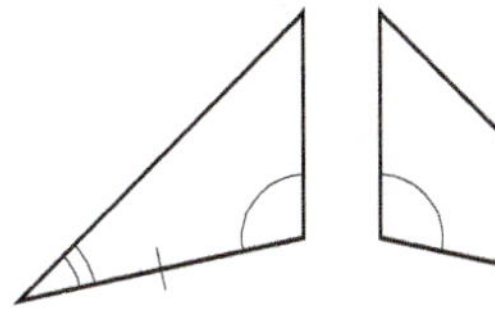

9.

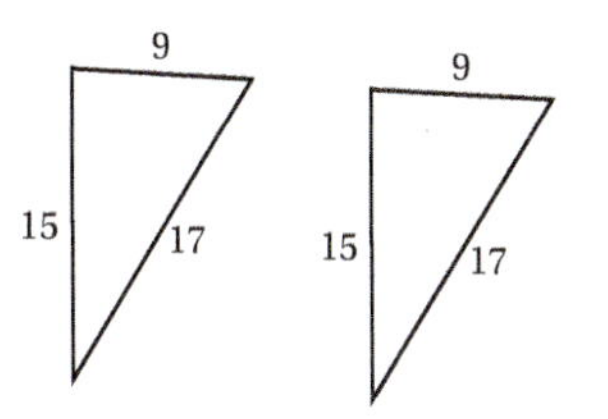

10. In this figure, $\overline{AB} \perp \overline{ED}$ and B is the midpoint of $\overline{ED}$. Explain why $\triangle ABD \cong \triangle ABE$.

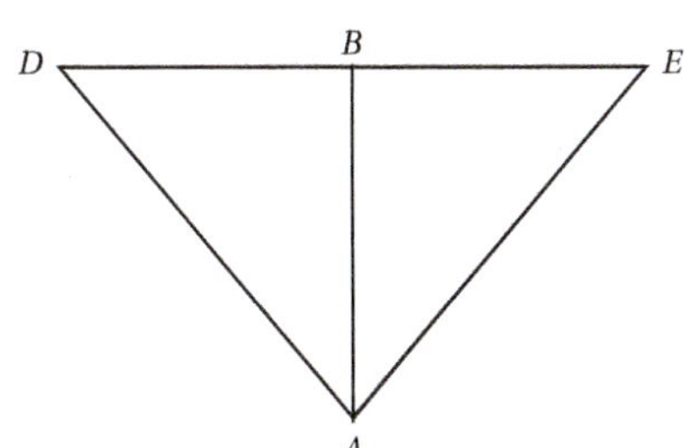

Answers

6. None **7.** SAS **8.** ASA
9. SSS **10.** SAS

Sometimes we can conclude that angles and segments are congruent by first showing that triangles are congruent.

EXAMPLE 14 $\overline{AB} \cong \overline{BC}$ and $\overline{EB} \cong \overline{DB}$. What can you conclude?

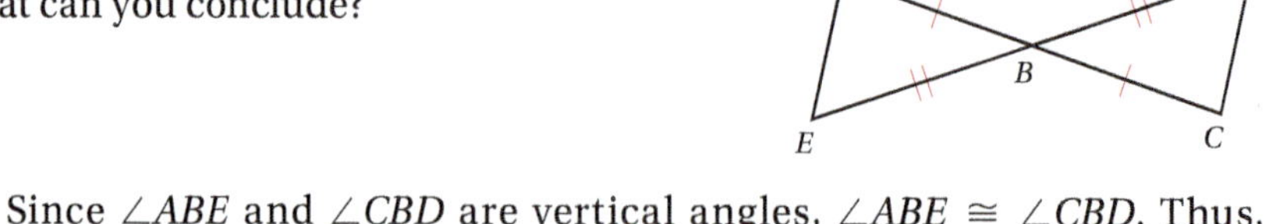

11. $\angle R \cong \angle T$, $\angle W \cong \angle V$, and $\overline{RW} \cong \overline{TV}$. What can you conclude about this figure?

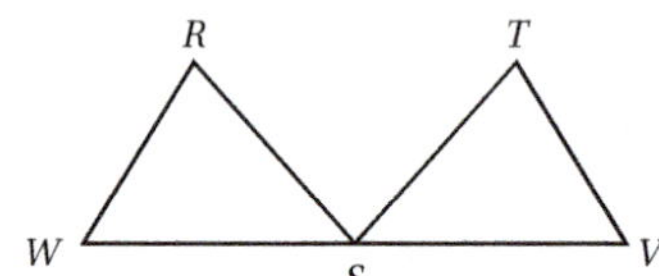

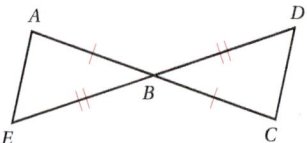

Since $\angle ABE$ and $\angle CBD$ are vertical angles, $\angle ABE \cong \angle CBD$. Thus, $\triangle ABE \cong \triangle CBD$ by SAS. As corresponding parts, $\overline{AE} \cong \overline{CD}$, $\angle A \cong \angle C$, and $\angle E \cong \angle D$.

◀ **Do Exercise 11.**

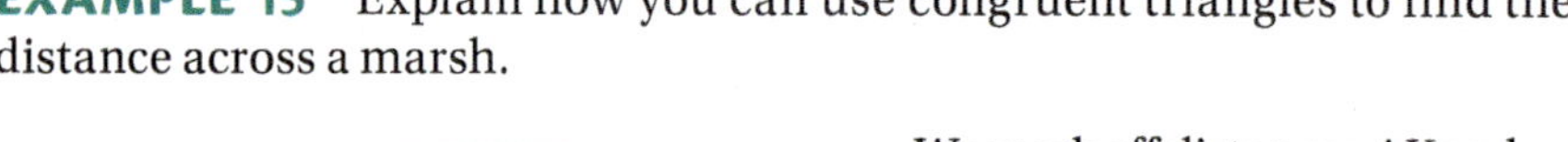

EXAMPLE 15 Explain how you can use congruent triangles to find the distance across a marsh.

12. On a pair of pinking shears, the indicated angles and sides are congruent. How do you know that P is the midpoint of $\overline{GR}$?

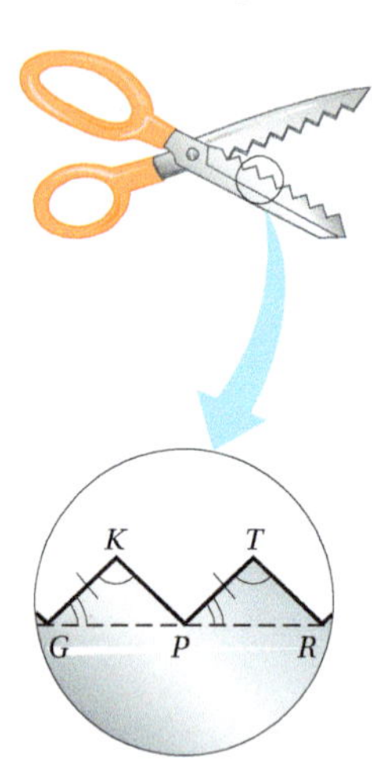

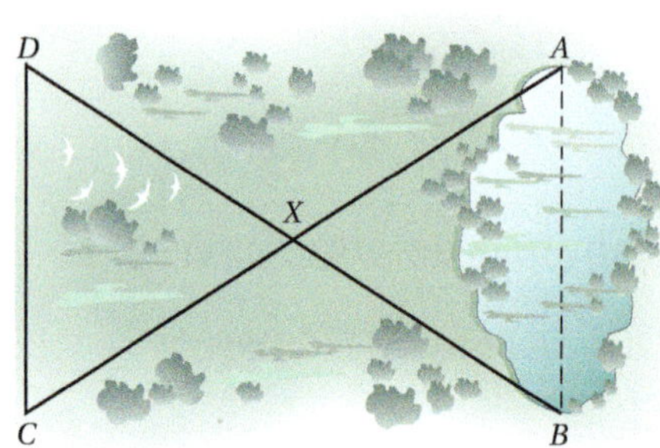

We mark off distances AX and BX and then extend $\overline{AX}$ and $\overline{BX}$ so that point X becomes the midpoint of $\overline{AC}$ and $\overline{BD}$. Then $\triangle ABX \cong \triangle CDX$ by SAS. Thus, $\overline{DC} \cong \overline{AB}$ as corresponding parts. Then we can measure $\overline{DC}$ knowing that $DC = AB$.

◀ **Do Exercise 12.**

b PROPERTIES OF PARALLELOGRAMS

A quadrilateral is a polygon with four sides. A **diagonal** of a quadrilateral is a segment that joins two opposite vertices.

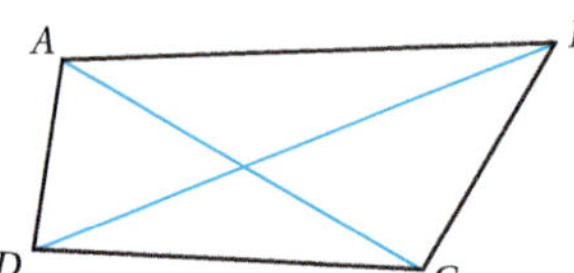

$\overline{AC}$ and $\overline{BD}$ are diagonals.

The sum of the measures of the angles of a quadrilateral is 360°.

A parallelogram is a quadrilateral with two pairs of parallel sides.

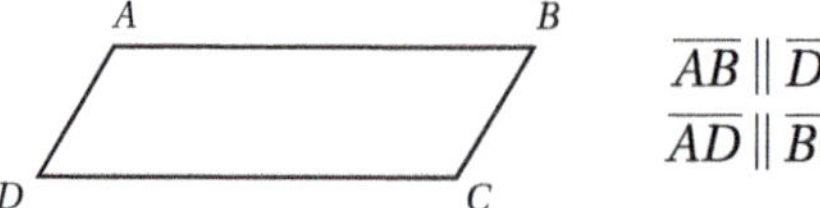

$\overline{AB} \parallel \overline{DC}$
$\overline{AD} \parallel \overline{BC}$

We draw two pairs of parallel lines to form parallelogram $ABCD$. Compare the lengths of opposite sides. Compare the measures of opposite angles. Compare the measures of consecutive angles. Then, we draw diagonal $\overline{AC}$. How are $\triangle ADC$ and $\triangle CBA$ related? Next, we draw diagonal $\overline{BD}$, intersecting $\overline{AC}$ at point E. What is special about point E?

Answers

11. $\triangle SRW \cong \triangle STV$ by ASA; $\angle RSW \cong \angle TSV$, $\overline{RS} \cong \overline{TS}$, $\overline{SW} \cong \overline{SV}$
12. $\triangle GKP \cong \triangle PTR$ by ASA. Thus corresponding parts $\overline{GP}$ and $\overline{PR}$ are congruent, and P is the midpoint of $\overline{GR}$.

Using the comparisons and the fact that corresponding parts of congruent triangles are congruent, we can list the following properties of parallelograms.

PROPERTIES OF PARALLELOGRAMS

1. A diagonal of a parallelogram determines two congruent triangles.
2. The opposite angles of a parallelogram are congruent.
3. The opposite sides of a parallelogram are congruent.
4. Consecutive angles of a parallelogram are supplementary.
5. The diagonals of a parallelogram bisect each other.

EXAMPLE 16 If $m\angle A = 120°$, find the measures of the other angles of parallelogram *ABCD*.

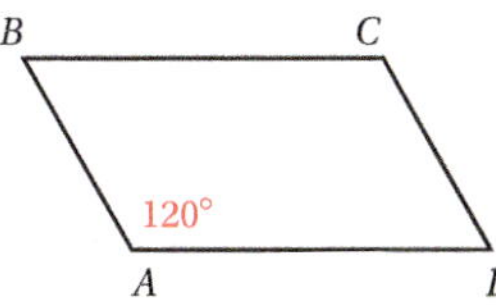

$m\angle C = 120°$ Using Property 2

$m\angle B = 60°$ Using Property 4

$m\angle D = 60°$ Using Property 2

Do Exercises 13 and 14.

EXAMPLE 17 Find *AB* and *BC*.

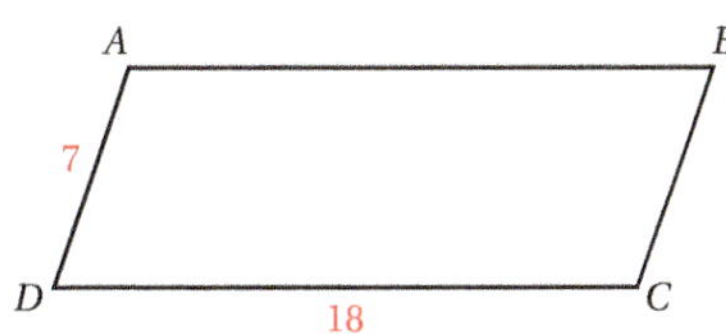

$AB = 18$ and $BC = 7$ Using Property 3

Do Exercises 15 and 16.

Find the measure of each angle.

13.

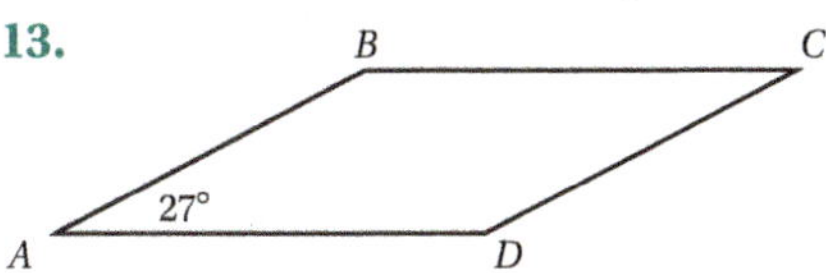

14.

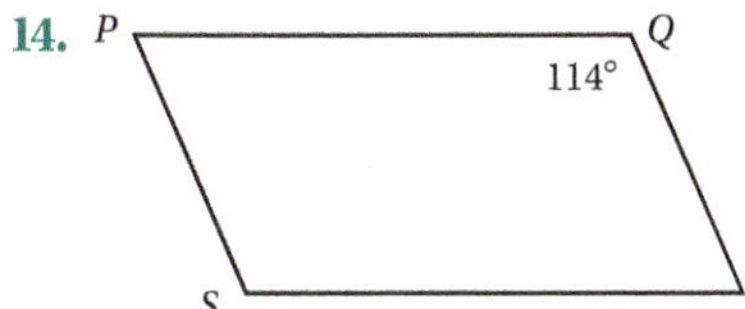

Find the length of each side.

15.

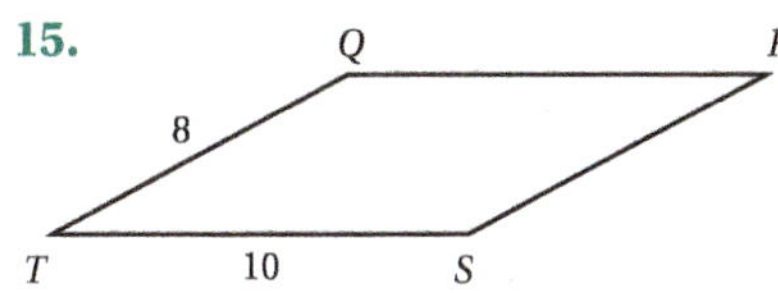

16. The perimeter of $\square DEFG$ is 68.

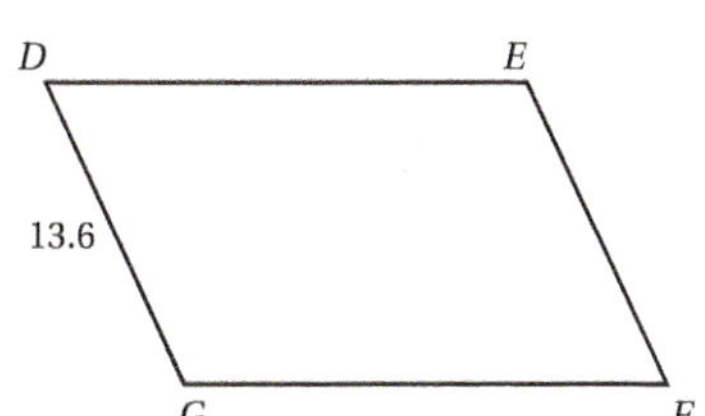

Answers

13. $m\angle C = 27°$, $m\angle B = m\angle D = 153°$
14. $m\angle S = 114°$, $m\angle P = m\angle R = 66°$
15. $QR = 10$, $SR = 8$
16. $EF = 13.6$, $DE = GF = 20.4$

8.7 Exercise Set

For Extra Help
MyMathLab®

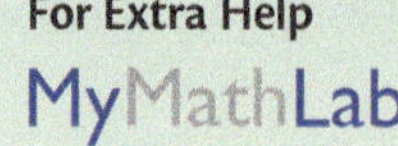

PRACTICE WATCH READ REVIEW

Reading Check

Determine whether each statement is true or false.

RC1. Consecutive angles of a parallelogram are complementary.

RC2. Two triangles are congruent if two sides and the included angle of one triangle are congruent to two sides and the included angle of the other triangle.

RC3. The diagonals of a parallelogram bisect each other.

RC4. If two angles of a triangle are congruent to two angles of another triangle, then the triangles are congruent.

a Name the corresponding parts of the congruent triangles.

1. $\triangle ABC \cong \triangle RST$

2. $\triangle MNQ \cong \triangle HJK$

3. $\triangle DEF \cong \triangle GHK$

4. $\triangle ABC \cong \triangle ABC$

5. $\triangle XYZ \cong \triangle UVW$

6. $\triangle ABC \cong \triangle ACB$

Name the corresponding parts of the congruent triangles.

7.

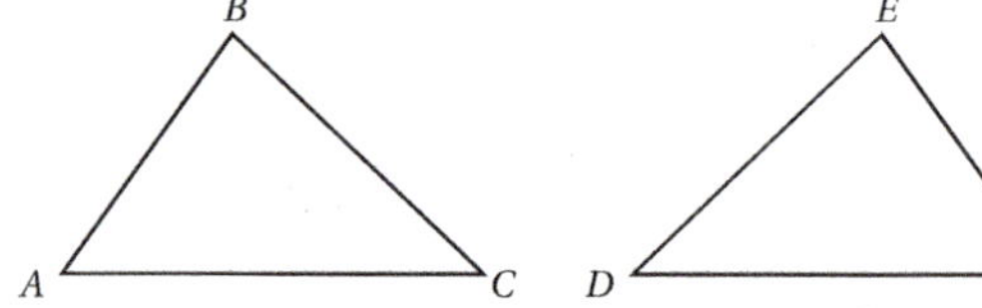

8.

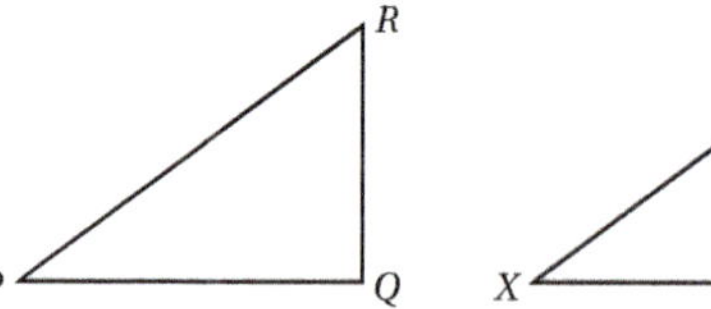

9.

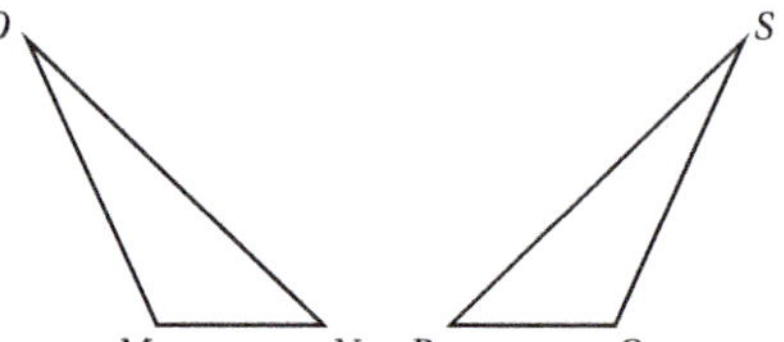

10.

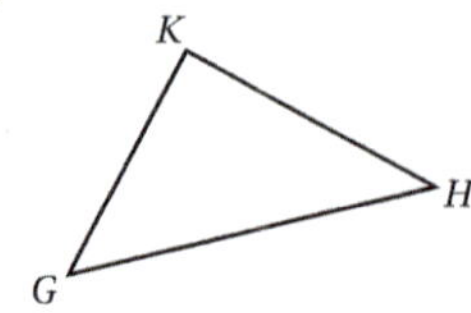

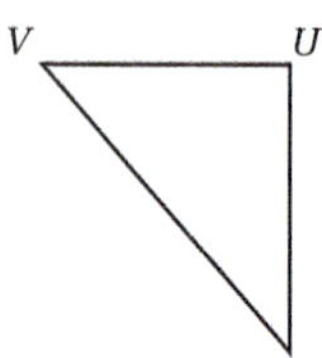

Determine whether each pair of triangles is congruent by the SAS property.

11.

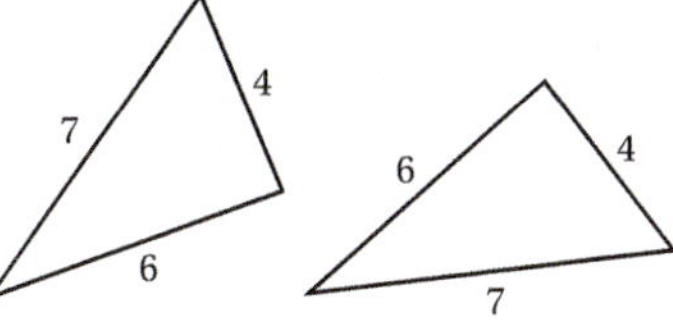

12.

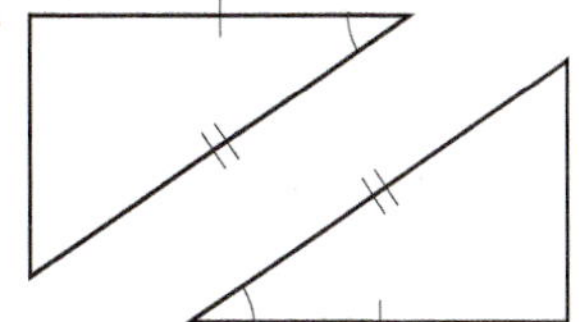

13.

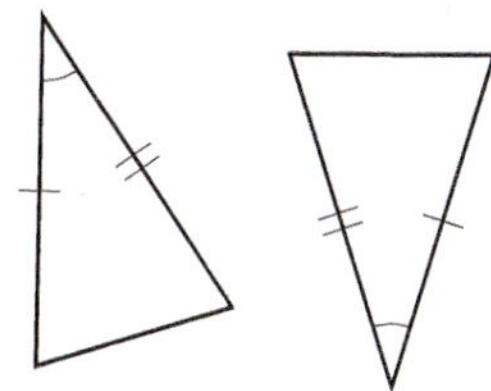

14.

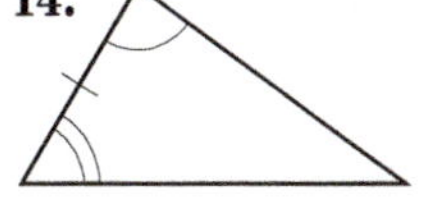

15.

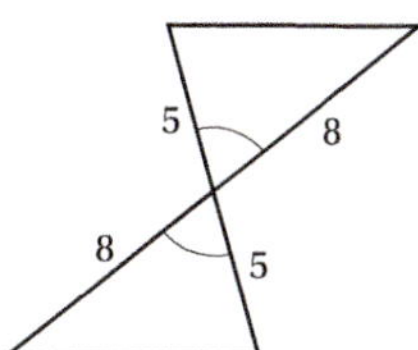

16.

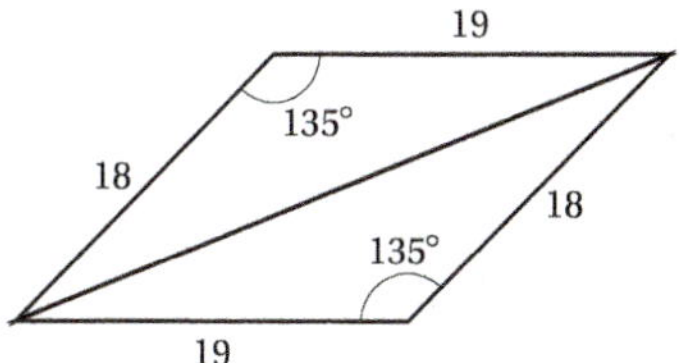

Determine whether each pair of triangles is congruent by the SSS property.

17.

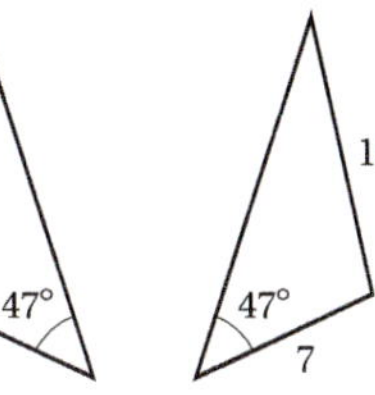

18.

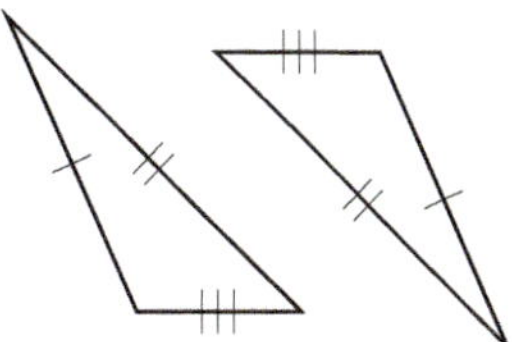

19.

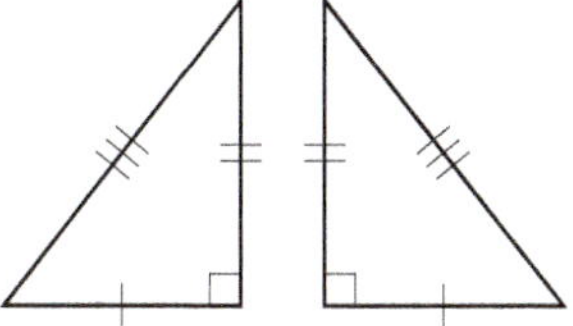

20.

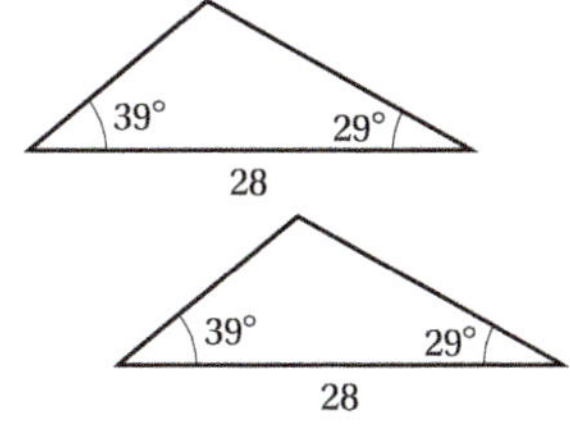

21.

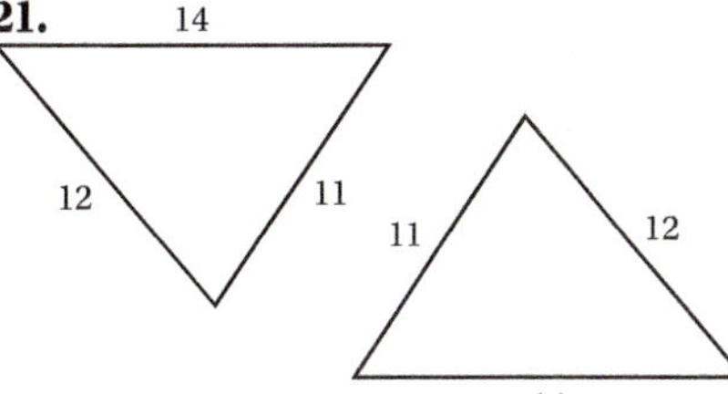

22.

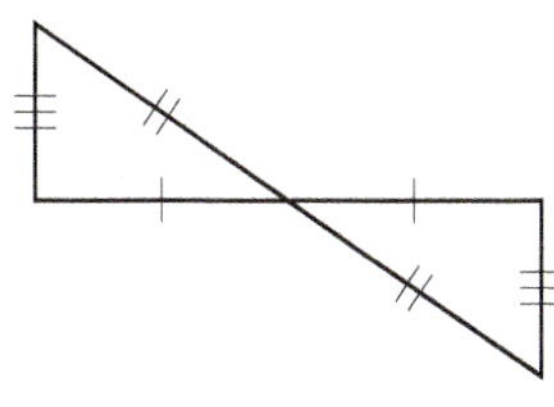

Determine whether each pair of triangles is congruent by the ASA property.

23.

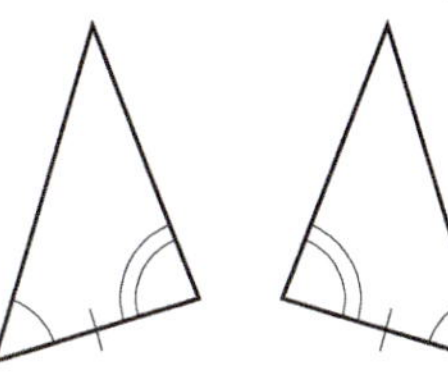

24.

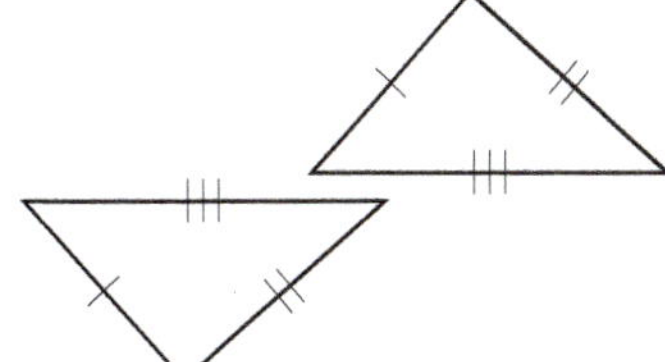

25.

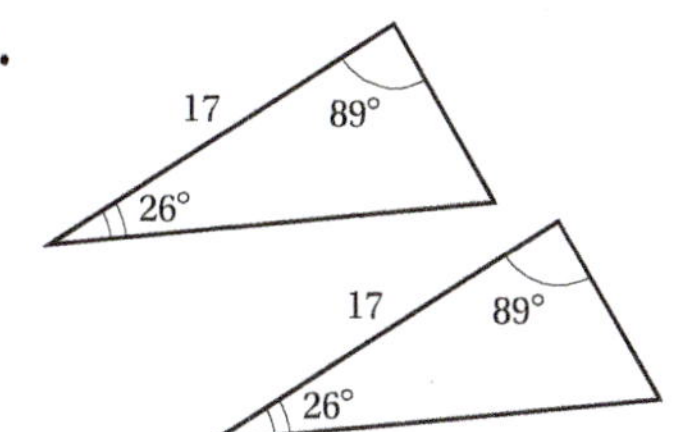

26. 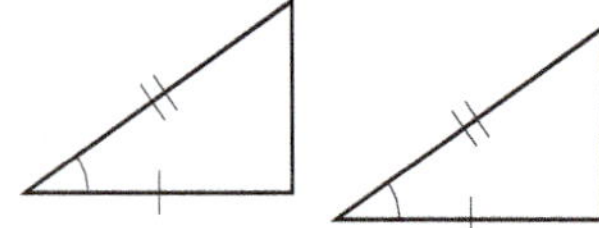

27. 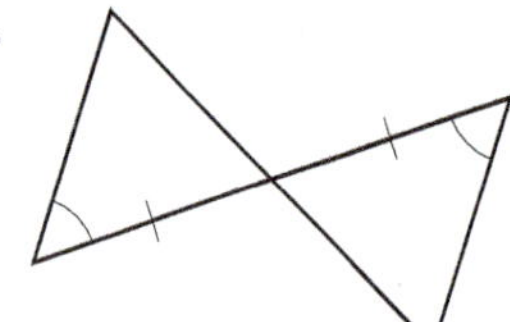

28. 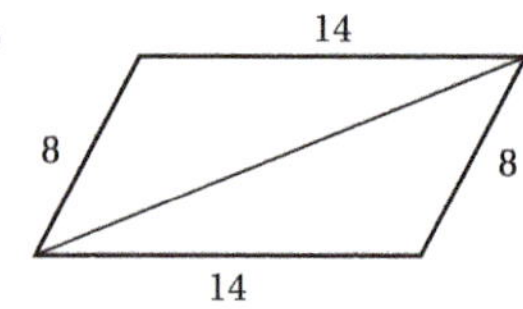

Which property (if any) should be used to show that the pair of triangles is congruent?

29. 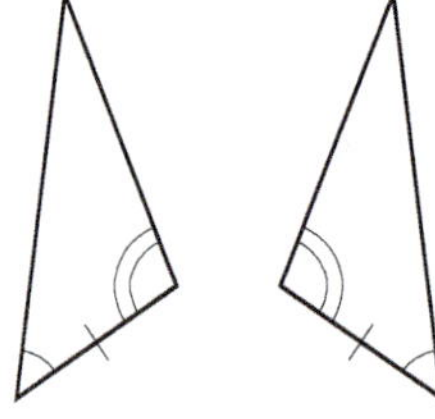

30. 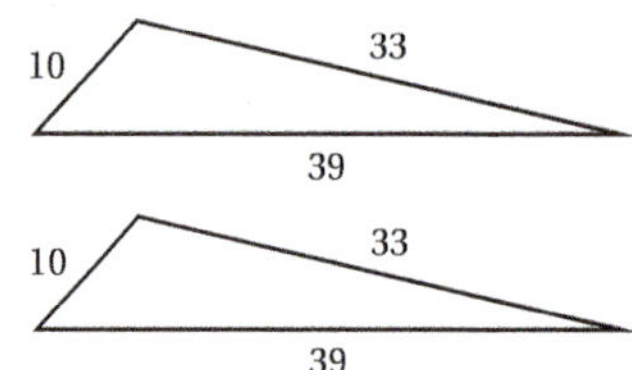

31. 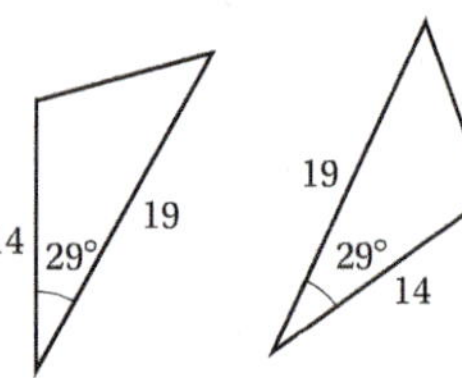

32. 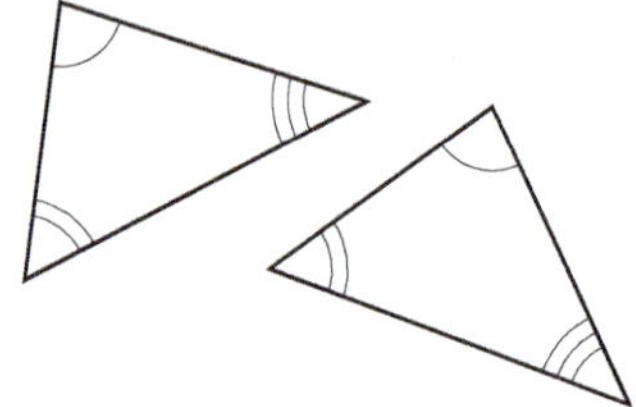

33. 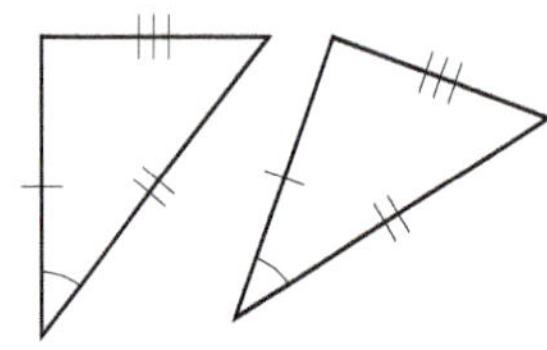

34. 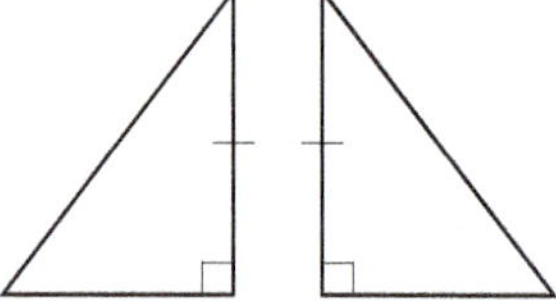

Explain why the triangles indicated in parentheses are congruent.

35. R is the midpoint of both $\overline{PT}$ and $\overline{QS}$. ($\triangle PRQ \cong \triangle TRS$)

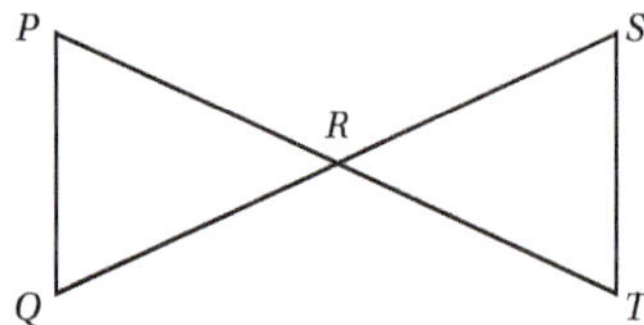

36. $\angle 1$ and $\angle 2$ are right angles, X is the midpoint of $\overline{AY}$, and $\overline{XB} \cong \overline{YZ}$. ($\triangle ABX \cong \triangle XZY$)

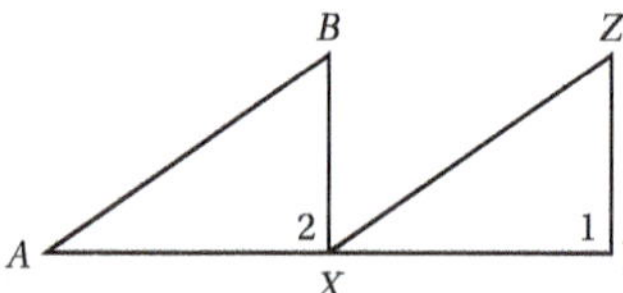

37. L is the midpoint of $\overline{KM}$ and $\overline{GL} \perp \overline{KM}$. ($\triangle KLG \cong \triangle MLG$)

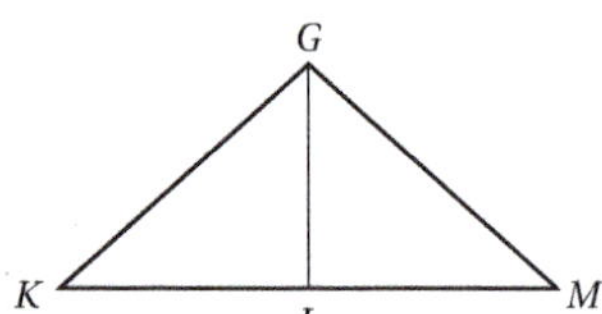

38. X is the midpoint of $\overline{QS}$ and $\overline{RP}$ with $RQ = SP$. ($\triangle RQX \cong \triangle PSX$)

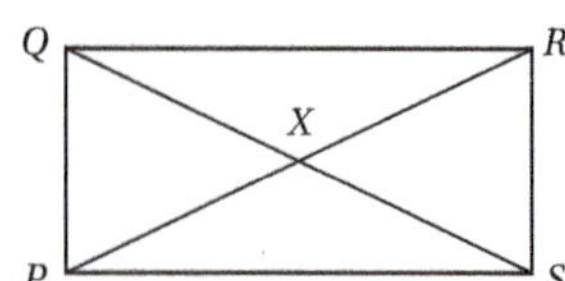

39. $\triangle AEB$ and $\triangle CDB$ are isosceles with $\overline{AE} \cong \overline{AB} \cong \overline{CB} \cong \overline{CD}$. Also, B is the midpoint of $\overline{ED}$. ($\triangle AEB \cong \triangle CDB$)

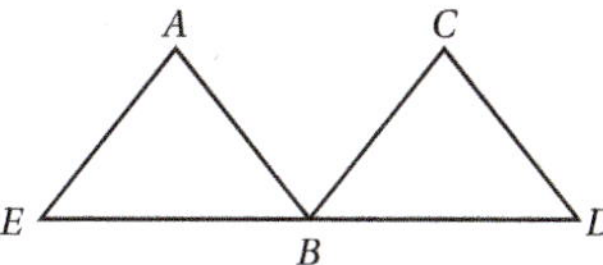

40. $\overline{AB} \perp \overline{BE}$ and $\overline{DE} \perp \overline{BE}$. $\overline{AB} \cong \overline{DE}$ and $\angle BAC \cong \angle EDC$. ($\triangle ABC \cong \triangle DEC$)

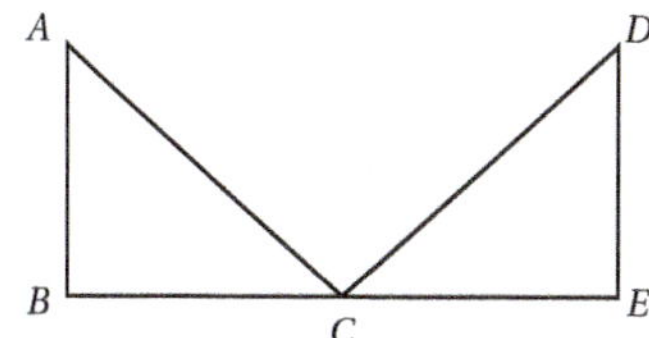

What can you conclude about each figure using the given information?

41. $\overline{GK} \perp \overline{LJ}$, $\overline{HK} \cong \overline{KJ}$, and $\overline{GK} \cong \overline{LK}$

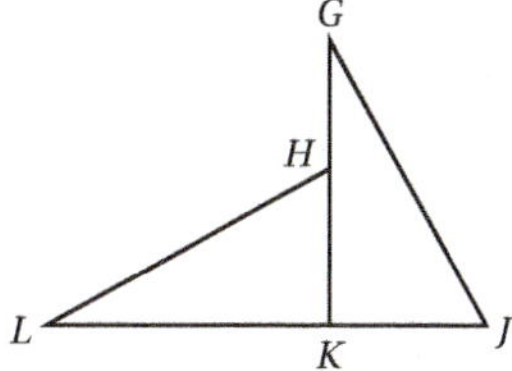

42. $\overline{AB} \cong \overline{DC}$ and $\angle BAC \cong \angle DCA$

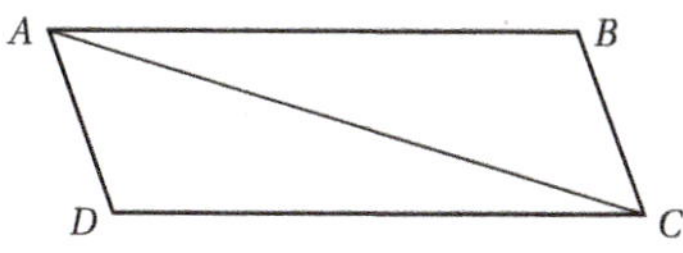

Use corresponding parts to solve Exercises 43 and 44.

43. On this national flag, the indicated segments and angles are congruent. Explain why P is the midpoint of $\overline{EF}$.

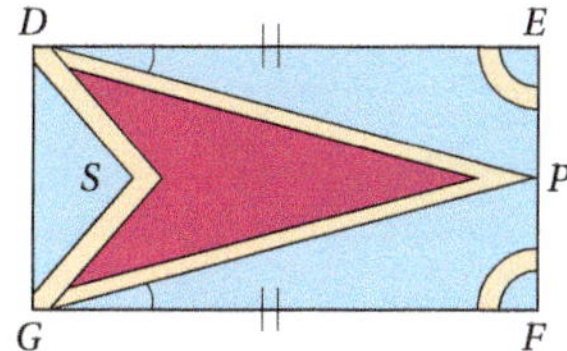

44. The indicated sides of a kite are congruent. Explain how you know that $\angle 1 \cong \angle 2$.

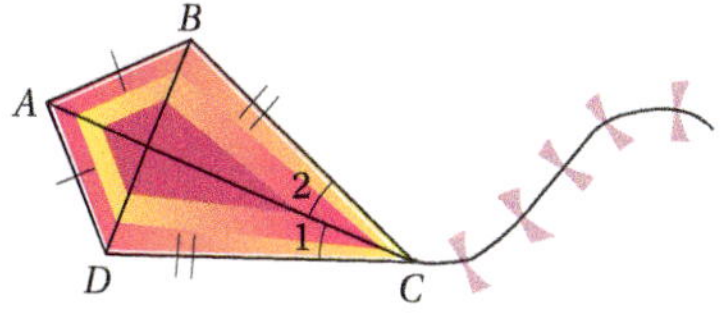

b Find the measures of the angles of each parallelogram.

45.

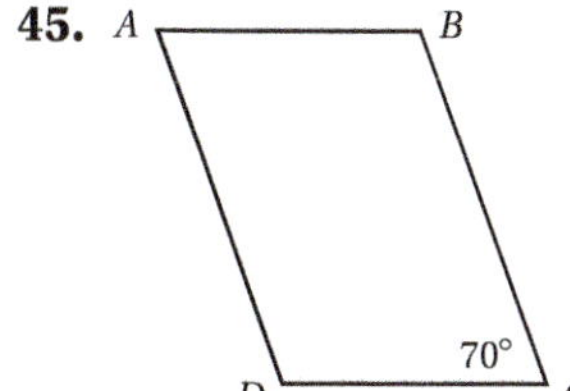

46.

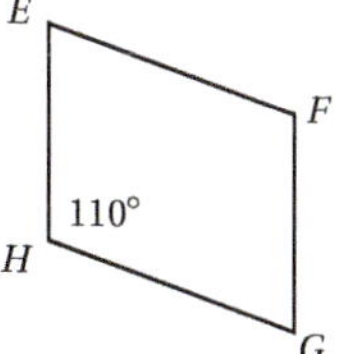

47.

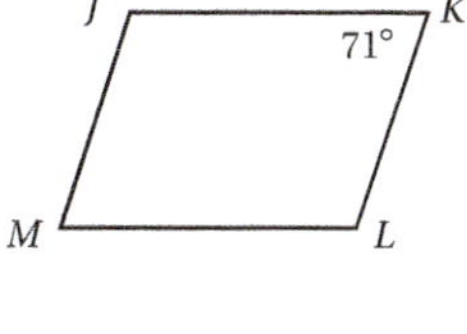

48.

P

S 105°

Q

R

Find the lengths of the sides of each parallelogram.

49.

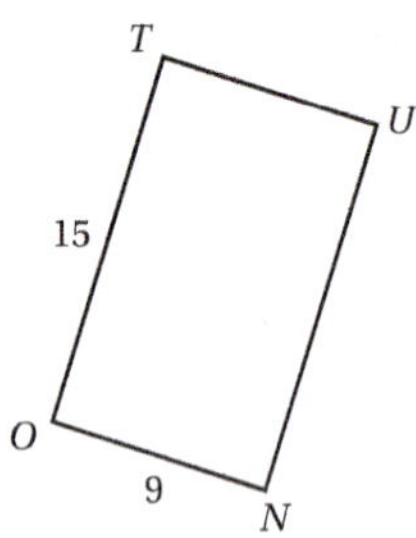

50.

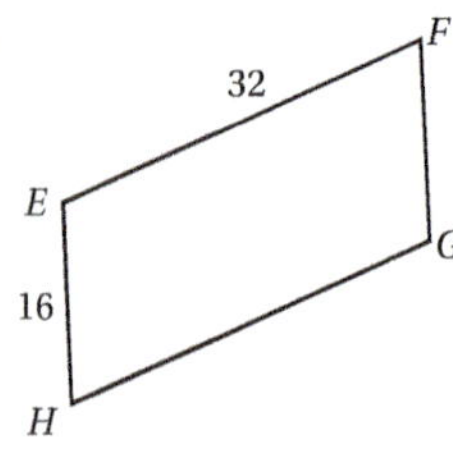

51. The perimeter of ▱*JKLM* is 22.

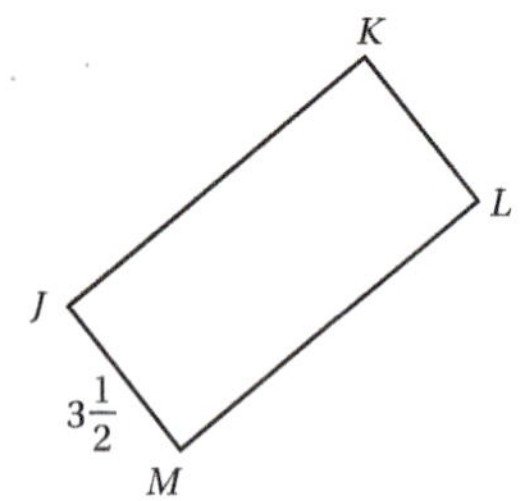

52. The perimeter of ▱*WXYZ* is 248.

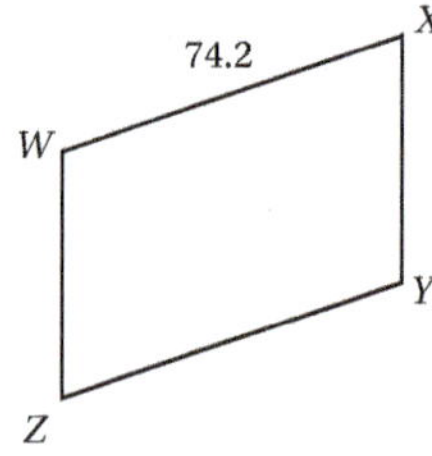

53. $AB = 14$ and $BD = 19$. Find the length of each diagonal.

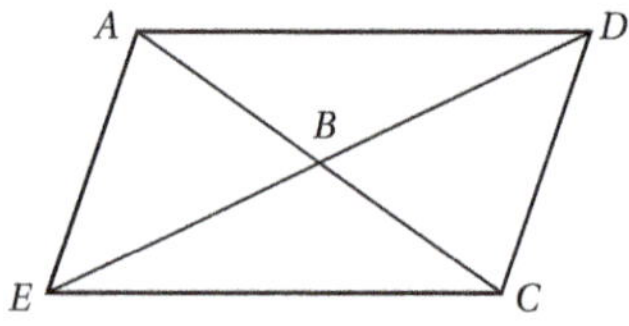

54. $EJ = 23$ and $GJ = 13$. Find the length of each diagonal.

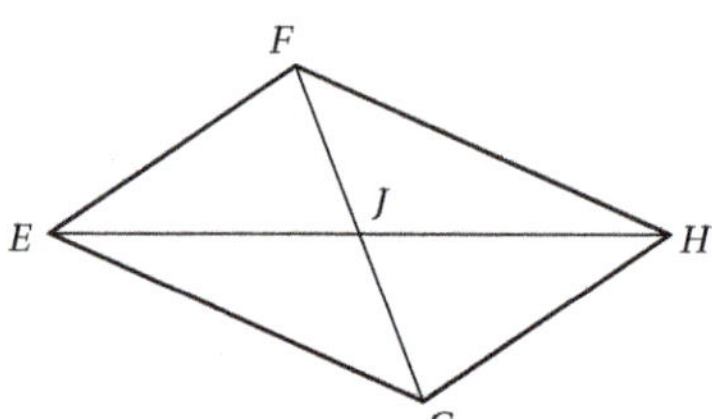

Skill Maintenance

Convert to percent notation. [6.2b], [6.3a]

55. 0.452

56. $\frac{1}{3}$

57. $\frac{11}{20}$

58. $\frac{22}{25}$

59. ***Tourist Spending.*** Foreign tourists spend $13.1 billion in this country annually. The most money, $2.7 billion, is spent in Florida. What is the ratio of amount spent in Florida to total amount spent? What is the ratio of total amount spent to amount spent in Florida? [6.1a]

60. One person in four plays a musical instrument. In a given group of people, what is the ratio of those who play an instrument to total number of people? What is the ratio of those who do not play an instrument to total number of people? [6.1a]

Divide. Find decimal notation for the answer. [5.4a]

61. $21 \div 12$

62. $-23.4 \div (-10)$

63. $23.4 \div (-100)$

64. $23.4 \div 1000$

65. Multiply 3.14×4.41. Round to the nearest hundredth. [5.3a], [5.1d]

Similar Triangles

8.8

OBJECTIVES

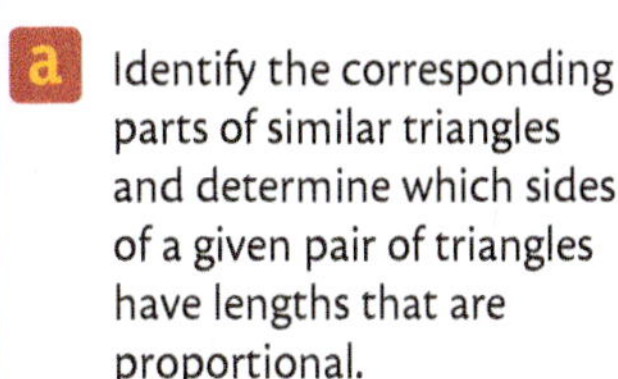

a Identify the corresponding parts of similar triangles and determine which sides of a given pair of triangles have lengths that are proportional.

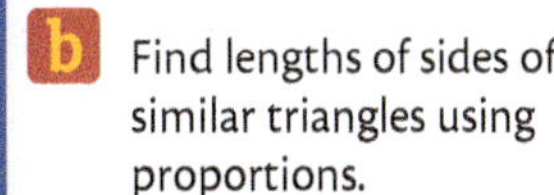

b Find lengths of sides of similar triangles using proportions.

a PROPORTIONS AND SIMILAR TRIANGLES

We know that congruent figures have the same shape and size. *Similar figures* have the same shape, but are not necessarily the same size.

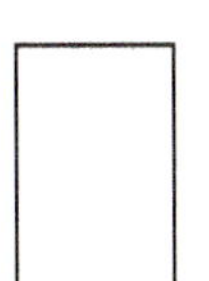
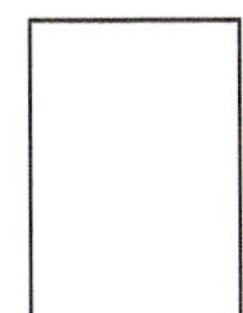
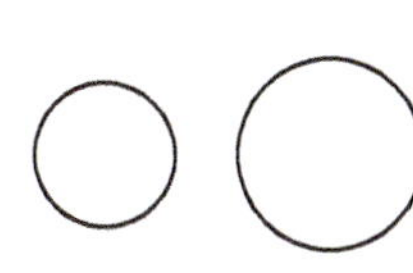
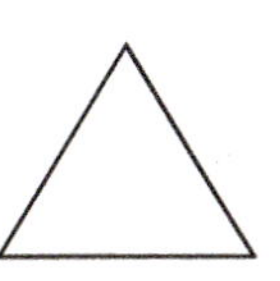
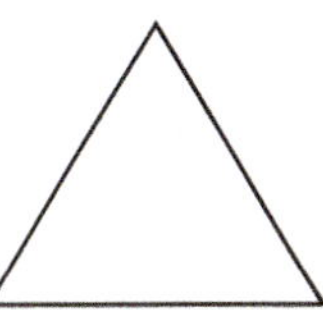

Similar figures

EXAMPLE 1 Which pairs of triangles appear to be similar?

a)

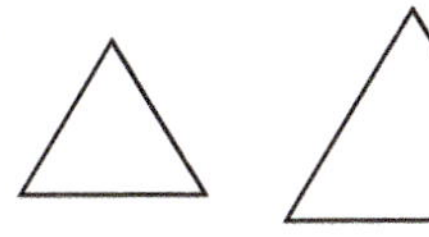

b)

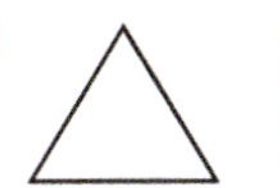
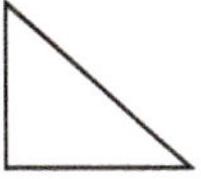

c)

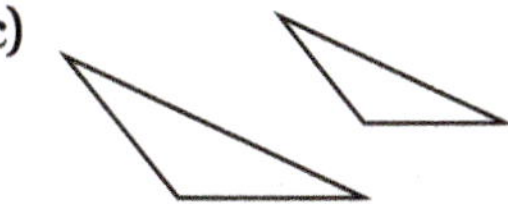

d)

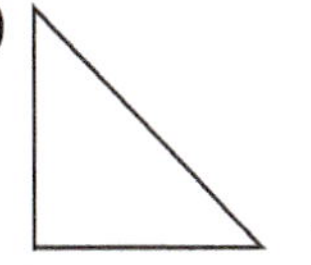
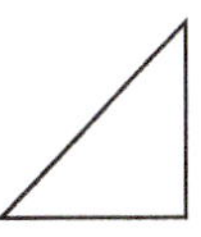

Pairs (a), (c), and (d) appear to be similar.

Do Exercise 1. ▶

Similar triangles have corresponding sides and angles.

EXAMPLE 2 $\triangle ABC$ and $\triangle DEF$ are similar. Name their corresponding sides and angles.

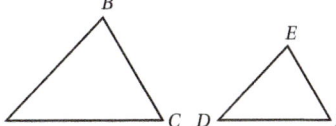

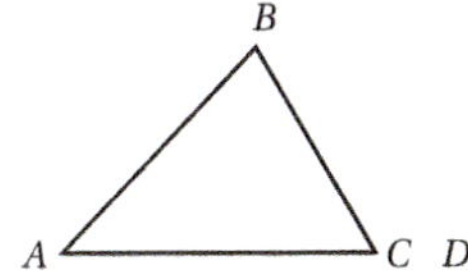

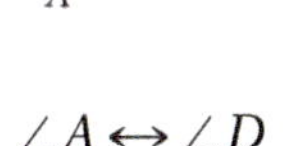

$\overline{AB} \leftrightarrow \overline{DE}$ $\quad \angle A \leftrightarrow \angle D$

$\overline{AC} \leftrightarrow \overline{DF}$ $\quad \angle B \leftrightarrow \angle E$

$\overline{BC} \leftrightarrow \overline{EF}$ $\quad \angle C \leftrightarrow \angle F$

Do Exercise 2. ▶

1. Which pairs of triangles appear to be similar?

a)

b)

c)

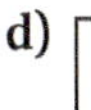

d)

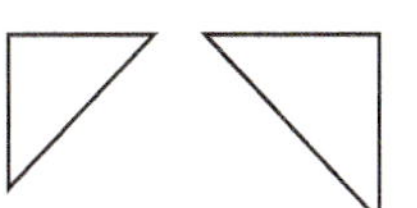

2. $\triangle PQR$ and $\triangle GHK$ are similar. Name their corresponding sides and angles.

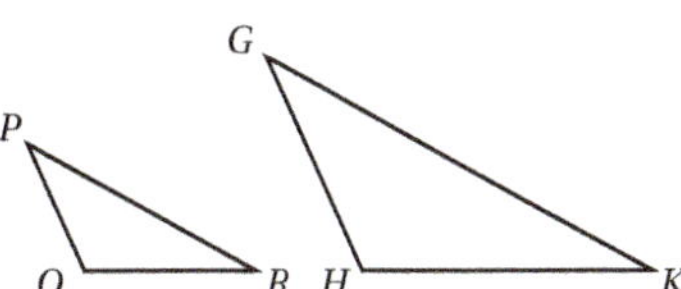

Answers

1. (a), (b), (d)
2. $\overline{PQ} \leftrightarrow \overline{GH}$, $\overline{QR} \leftrightarrow \overline{HK}$, $\overline{PR} \leftrightarrow \overline{GK}$, $\angle P \leftrightarrow \angle G$, $\angle Q \leftrightarrow \angle H$, $\angle R \leftrightarrow \angle K$

SIMILAR TRIANGLES

Two triangles are **similar** if and only if their vertices can be matched so that the corresponding angles are congruent and the lengths of corresponding sides are proportional.

To say that $\triangle ABC$ and $\triangle DEF$ are similar, we write "$\triangle ABC \sim \triangle DEF$." We will agree that this symbol also tells us the way in which the vertices are matched.

$$\triangle ABC \sim \triangle DEF$$

Thus, $\triangle ABC \sim \triangle DEF$ means that

$$\angle A \cong \angle D, \quad \angle B \cong \angle E, \quad \angle C \cong \angle F \quad \text{and} \quad \frac{AB}{DE} = \frac{AC}{DF} = \frac{BC}{EF}.$$

EXAMPLE 3 Suppose that $\triangle PQR \sim \triangle STV$. Which angles are congruent? Which sides are proportional?

$$\angle P \cong \angle S, \quad \angle Q \cong \angle T, \quad \angle R \cong \angle V \quad \text{and} \quad \frac{PQ}{ST} = \frac{PR}{SV} = \frac{QR}{TV}.$$

3. Suppose that $\triangle JKL \sim \triangle ABC$. Which angles are congruent? Which sides are proportional?

◀ **Do Exercise 3.**

EXAMPLE 4 These triangles are similar. Which sides are proportional?

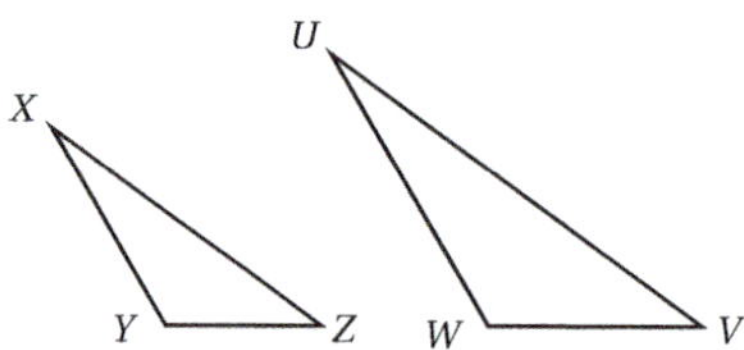

It appears that if we match X with U, Y with W, and Z with V, the corresponding angles will be congruent. Thus,

$$\frac{XY}{UW} = \frac{XZ}{UV} = \frac{YZ}{WV}.$$

4. These triangles are similar. Which sides are proportional?

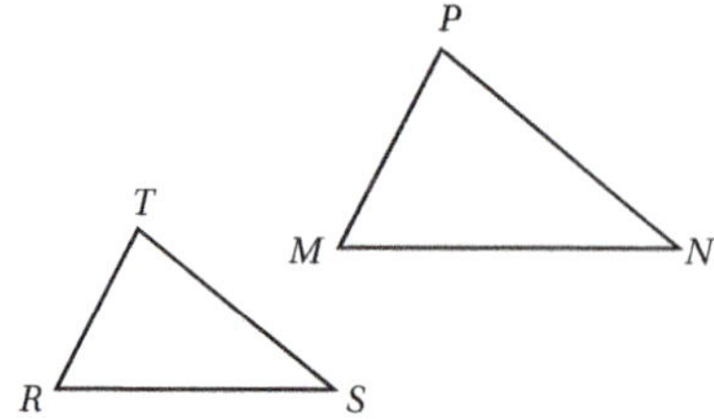

◀ **Do Exercise 4.**

Answers

3. $\angle J \cong \angle A, \angle K \cong \angle B, \angle L \cong \angle C$; $\frac{JK}{AB} = \frac{JL}{AC} = \frac{KL}{BC}$

4. $\frac{PN}{TS} = \frac{PM}{TR} = \frac{MN}{RS}$

PROPORTIONS AND SIMILAR TRIANGLES

We can find lengths of sides in similar triangles.

EXAMPLE 5 If $\triangle RAE \sim \triangle GQL$, find QL and GL.

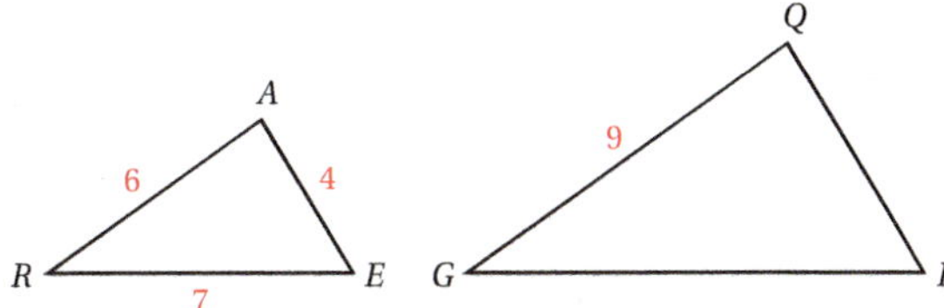

Since $\triangle RAE \sim \triangle GQL$, the corresponding sides are proportional. Thus,

$$\frac{6}{9} = \frac{4}{QL}$$

$$6(QL) = 9 \cdot 4 \quad \text{Equating cross products}$$

$$6(QL) = 36$$

$$QL = 6 \quad \text{Dividing by 6 on both sides}$$

and

$$\frac{6}{9} = \frac{7}{GL}$$

$$6(GL) = 9 \cdot 7$$

$$6(GL) = 63$$

$$GL = 10\tfrac{1}{2}.$$

Do Exercise 5. ▶

EXAMPLE 6 If $\overline{AB} \parallel \overline{CD}$, find CD.

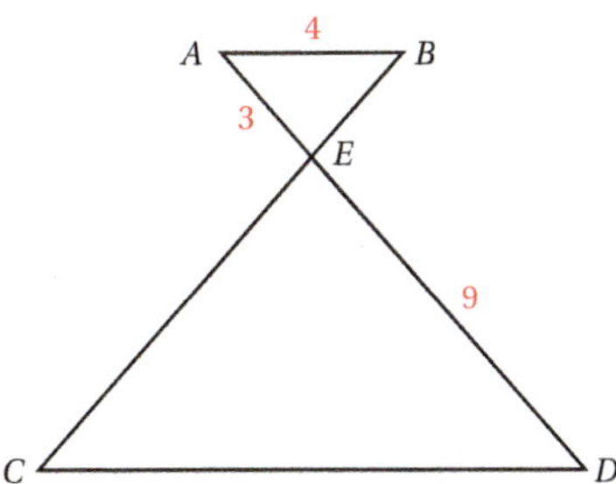

Recall that if a transversal intersects two parallel lines, then the alternate interior angles are congruent (Section 8.6). Thus,

$$\angle A \cong \angle D \quad \text{and} \quad \angle C \cong \angle B,$$

because they are pairs of alternate interior angles. Since $\angle AEB$ and $\angle DEC$ are vertical angles, they are congruent. Thus by definition

$$\triangle AEB \sim \triangle DEC$$

and the lengths of the corresponding sides are proportional. Thus,

$$\frac{AE}{DE} = \frac{AB}{CD}.$$

GS **5.** If $\triangle WNE \sim \triangle CBT$, find BT and CT.

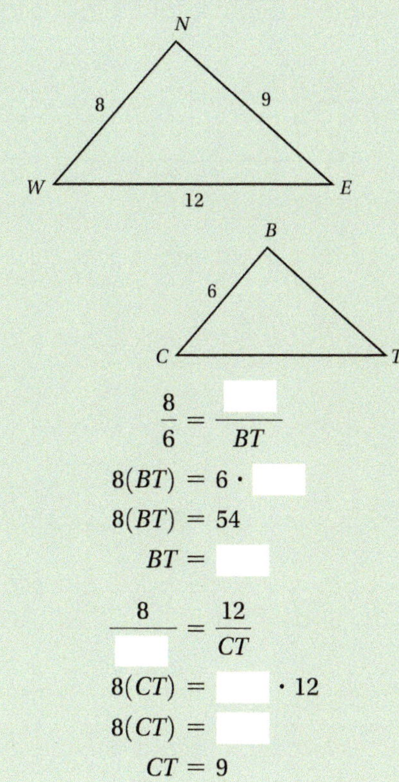

$$\frac{8}{6} = \frac{\square}{BT}$$

$$8(BT) = 6 \cdot \square$$

$$8(BT) = 54$$

$$BT = \square$$

$$\frac{8}{\square} = \frac{12}{CT}$$

$$8(CT) = \square \cdot 12$$

$$8(CT) = \square$$

$$CT = 9$$

Answer

5. $BT = 6\frac{3}{4}$, $CT = 9$

Guided Solution:

5. 9, 9, $6\frac{3}{4}$; 6, 6, 72

6. If $\overline{QR} \parallel \overline{ST}$, find QR.

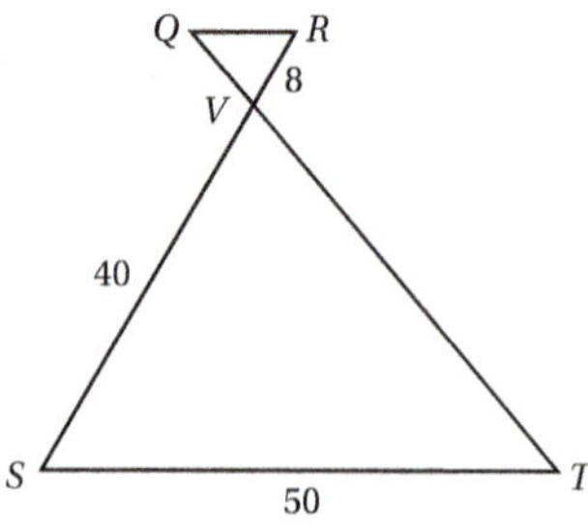

Solve: $\frac{3}{9} = \frac{4}{CD}$ Substituting

$3(CD) = 9 \cdot 4$ Equating cross products

$3(CD) = 36$

$CD = 12.$ Dividing by 3 on both sides

◀ **Do Exercise 6.**

Similar triangles and proportions can often be used to find lengths that would ordinarily be difficult to measure. For example, we could find the height of a flagpole without climbing it or the distance across a river without crossing it.

EXAMPLE 7 How high is a flagpole that casts a 56-ft shadow at the same time that a 6-ft man casts a 5-ft shadow?

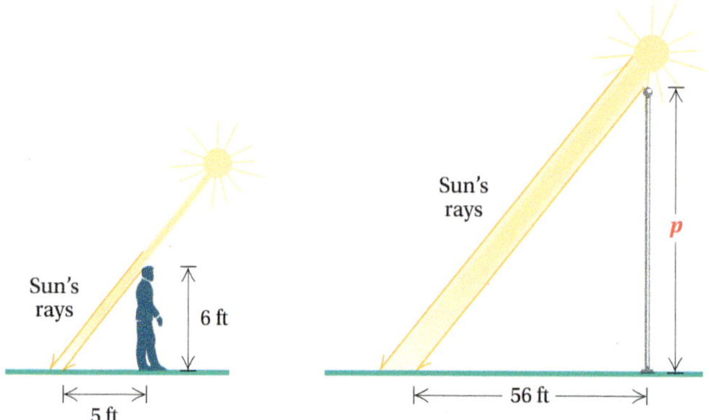

If we use the sun's rays to represent the third side of the triangle in our drawing of the situation, we see that we have similar triangles. Let $p =$ the height of the flagpole. The ratio of 6 to p is the same as the ratio of 5 to 56. Thus we have the proportion

Height of man → $\frac{6}{p} = \frac{5}{56}$. ← Length of shadow of man
Height of pole → ← Length of shadow of pole

Solve: $6 \cdot 56 = p \cdot 5$ Equating cross products

$\frac{6 \cdot 56}{5} = \frac{p \cdot 5}{5}$ Dividing by 5 on both sides

$\frac{6 \cdot 56}{5} = p$ Simplifying

$67.2 = p.$

The height of the flagpole is 67.2 ft.

◀ **Do Exercise 7.**

7. How high is a flagpole that casts a 45-ft shadow at the same time that a 5.5-ft woman casts a 10-ft shadow?

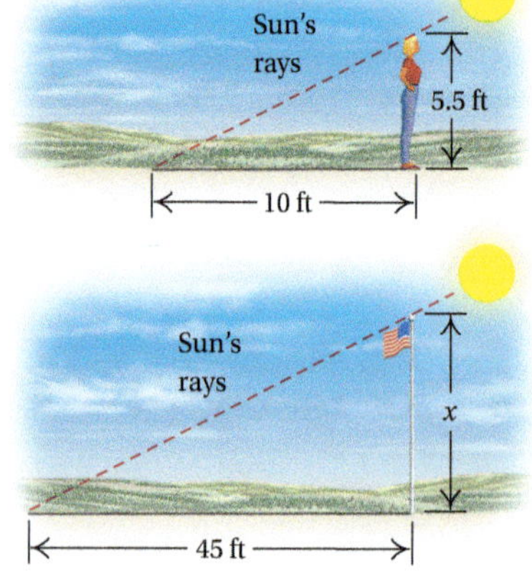

Answers

6. $QR = 10$ 7. 24.75 ft

EXAMPLE 8 ***Rafters of a House.*** Carpenters use similar triangles to determine the length of rafters for a house. They first choose the pitch of the roof, or the ratio of the rise over the run. Then using a triangle with that ratio, they calculate the length of the rafters needed for the house. Loren is making rafters for a roof with a 6/12 pitch on a house that is 30 ft wide. Using a rafter guide, Loren knows that the rafter length corresponding to the 6/12 pitch is 13.4. Find the length x of the rafters for this house to the nearest tenth of a foot.

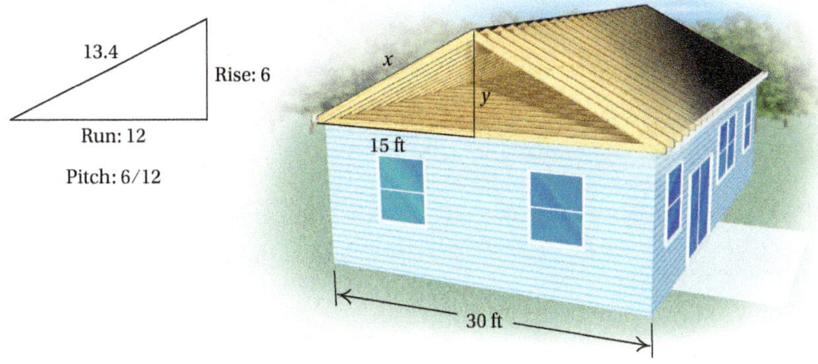

We have the proportion

Length of rafter in 6/12 triangle → ; Length of rafter on the house → ; ← Run in 6/12 triangle ; ← Run in similar triangle on the house

$$\frac{13.4}{x} = \frac{12}{15}.$$

Solve: $13.4 \cdot 15 = x \cdot 12$ — Equating cross products

$$\frac{13.4 \cdot 15}{12} = \frac{x \cdot 12}{12}$$ Dividing by 12 on both sides

$$\frac{13.4 \cdot 15}{12} = x$$

$16.8 \text{ ft} \approx x.$ — Rounding to the nearest tenth of a foot

The length x of the rafters for the house is about 16.8 ft.

Do Exercise 8. ▶

8. ***Rafters of a House.*** Referring to Example 8, find the length y of the rise of the rafters of the house to the nearest tenth of a foot.

Answer

8. 7.5 ft

Translating for Success

1. *Servings of Pork.* An 8-lb pork roast contains 37 servings of meat. How many pounds of pork would be needed for 55 servings?

2. *Height of a Ladder.* A 14.5-ft ladder leans against a house. The bottom of the ladder is 9.4 ft from the house. How high is the top of the ladder?

3. *Cruise Cost.* A group of 6 college students pays \$4608 for a spring break cruise. What is each person's share?

4. *Sales Tax Rate.* The sales tax is \$14.95 when a new ladder is purchased for \$299. What is the sales tax rate?

5. *Volume of a Sphere.* Find the volume of a sphere whose radius is 7.2 cm.

The goal of these matching questions is to practice step (2), Translate, *of the five-step problem-solving process. Translate each word problem to an equation and select a correct translation from equations A–O.*

A. $x = \frac{1}{3}\pi \cdot 6^2 \cdot (7.2)$

B. $6 \cdot x = \$4608$

C. $x = \frac{4}{3} \cdot \pi \cdot 6^2 \cdot (7.2)$

D. $x = \pi \cdot \left(5\frac{1}{2} \div 2\right)^2 \cdot 7$

E. $x = 6\% \times 5 \times \$14.95$

F. $x = \frac{1}{3}\pi\left(5\frac{1}{2}\right)^2$

G. $(9.4)^2 + x^2 = (14.5)^2$

H. $\$14.95 = x \cdot \$299$

I. $x = 2(14.5 + 9.4)$

J. $(9.4 + 14.5)^2 = x$

K. $\frac{8}{37} = \frac{x}{55}$

L. $x = 4(14.5 + 9.4)$

M. $x = 6 \cdot \$4608$

N. $8 \cdot 37 = 55 \cdot x$

O. $x = \frac{4}{3}\pi(7.2)^3$

Answers on page A-15

6. *Inheritance.* Six children each inherit \$4608 from their mother's estate. What is the total amount of the inheritance?

7. *Sales Tax.* Erica buys 5 pairs of earrings at \$14.95 each. The sales tax rate is 6%. How much sales tax will be charged?

8. *Volume of a Cone.* Find the volume of a circular cone with a base radius of 6 cm and a height of 7.2 cm.

9. *Volume of a Storage Tank.* The diameter of a cylindrical grain-storage tank is $5\frac{1}{2}$ yd. Its height is 7 yd. Find its volume.

10. *Perimeter of a Photo.* A rectangular photo is 14.5 cm by 9.4 cm. What is the perimeter of the photo?

8.8 Exercise Set

For Extra Help
MyMathLab® MathXL® PRACTICE WATCH READ REVIEW

Reading Check

Complete each proportion based on the following similar triangles.

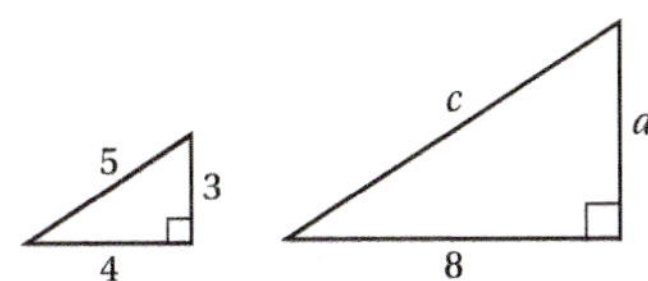

RC1. $\frac{a}{3} = \frac{c}{\square}$

RC2. $\frac{\square}{4} = \frac{c}{5}$

RC3. $\frac{3}{\square} = \frac{5}{c}$

RC4. $\frac{8}{4} = \frac{a}{\square}$

a For each pair of similar triangles, name the corresponding sides and angles.

1.

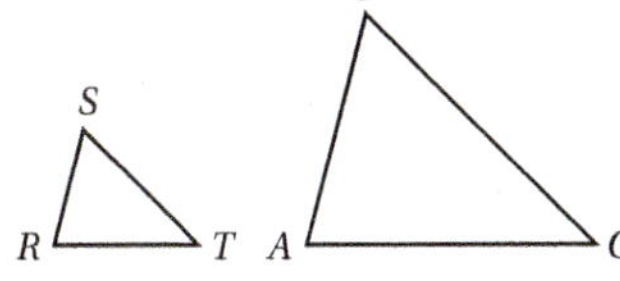

2.

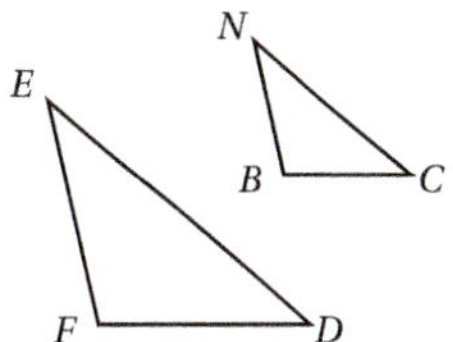

3.

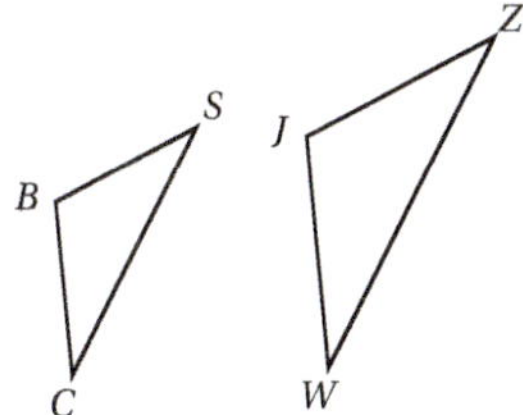

4.

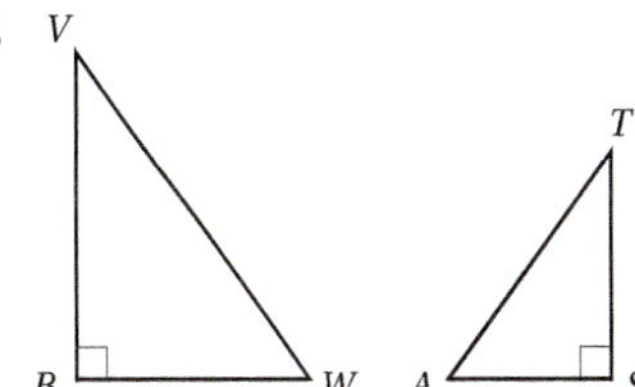

For each pair of similar triangles, name the congruent angles and proportional sides.

5. $\triangle ABC \sim \triangle RST$

6. $\triangle PQR \sim \triangle STV$

7. $\triangle MES \sim \triangle CLF$

8. $\triangle SMH \sim \triangle WLK$

Name the proportional sides in these similar triangles.

9.

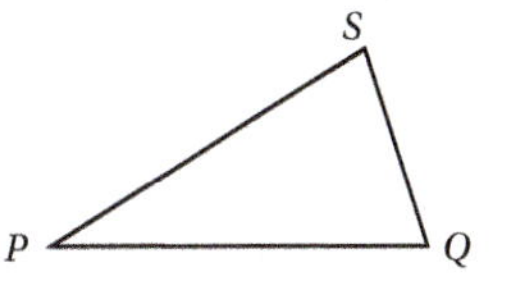

10.

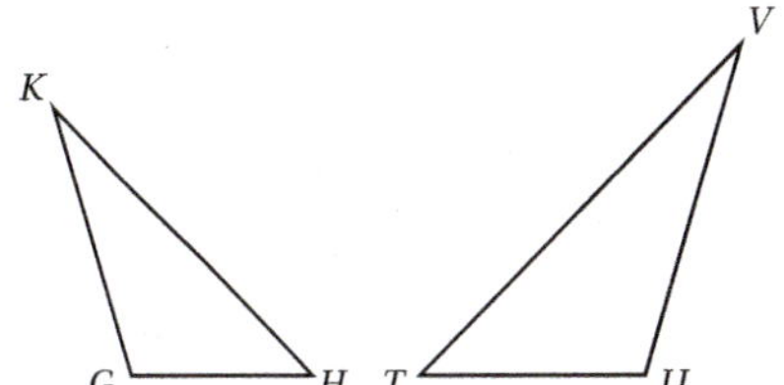

b Find the missing lengths.

11. If $\triangle ABC \sim \triangle PQR$, find QR and PR.

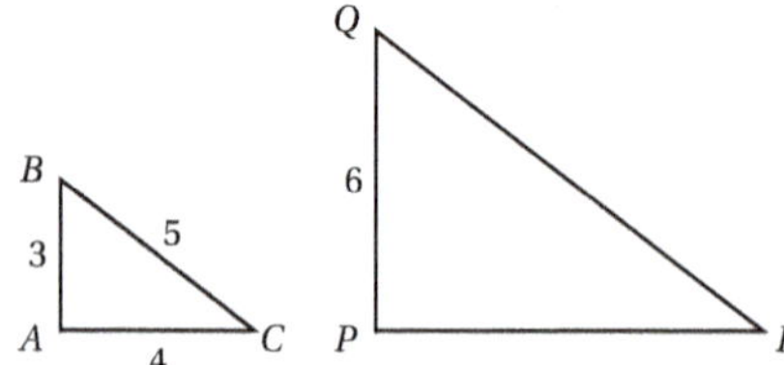

12. If $\triangle MAC \sim \triangle GET$, find AM and GT.

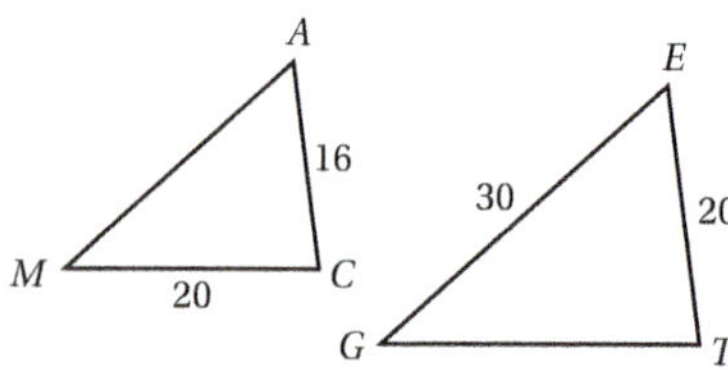

13. If $\overline{AD} \parallel \overline{CB}$, find EC.

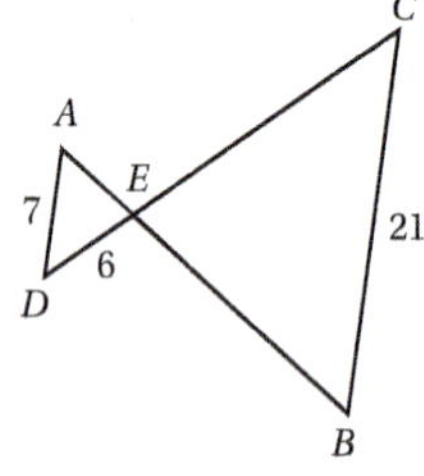

14. If $\overline{LN} \parallel \overline{PM}$, find QM.

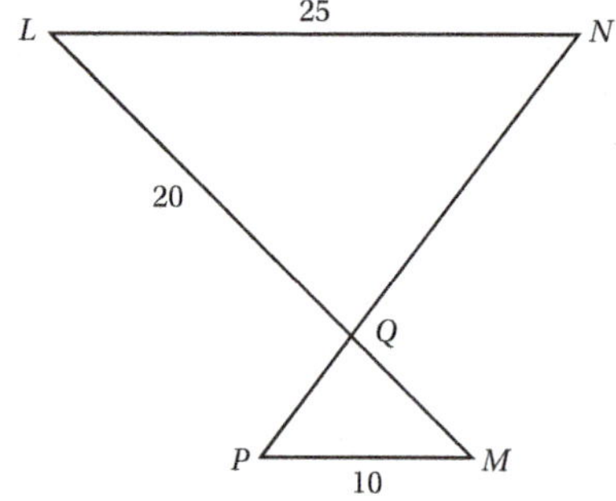

15. How high is a tree that casts a 27-ft shadow at the same time that a 4-ft fence post casts a 3-ft shadow?

16. How high is a flagpole that casts a 42-ft shadow at the same time that a $5\frac{1}{2}$-ft woman casts a 7-ft shadow?

17. Find the distance across the river. Assume that the ratio of d to 25 ft is the same as the ratio of 40 ft to 10 ft.

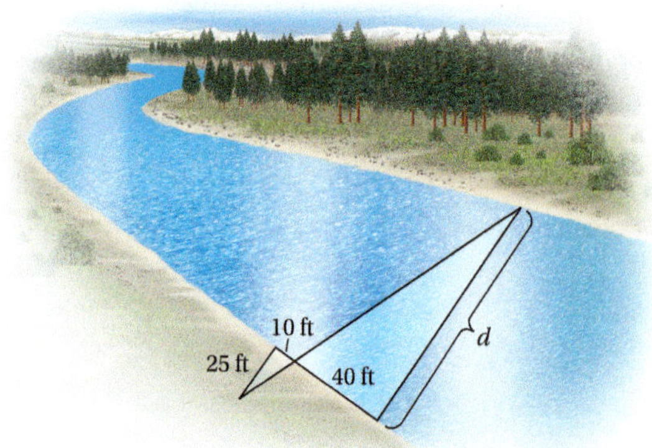

18. To measure the height of a hill, a string is stretched from level ground to the top of the hill. A 3-ft stick is placed under the string, touching it at point P, a distance of 5 ft from point G, where the string touches the ground. The string is then detached and found to be 120 ft long. How high is the hill?

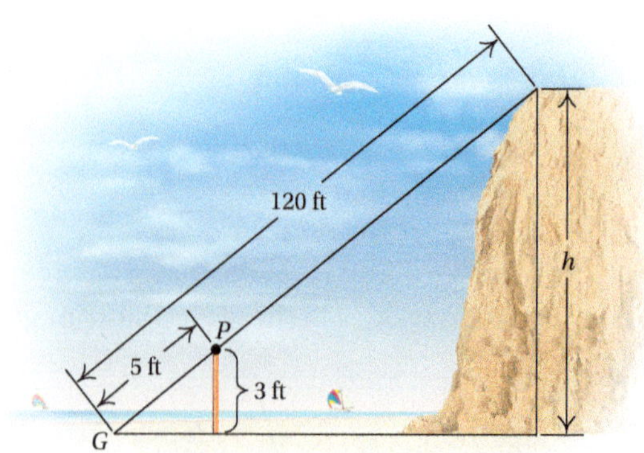

Skill Maintenance

Multiply.

19. $\left(-2\frac{4}{5}\right)\left(-10\frac{1}{2}\right)$ [4.7a]

20. 3.05×0.08 [5.3a]

21. $8 \times 9\frac{3}{4}$ [4.7a]

22. -10.01×6.11 [5.3a]

CHAPTER

8 Summary and Review

Formulas

Perimeter of a Rectangle:	$P = 2 \cdot (l + w)$, or $P = 2 \cdot l + 2 \cdot w$
Perimeter of a Square:	$P = 4 \cdot s$
Area of a Rectangle:	$A = l \cdot w$
Area of a Square:	$A = s \cdot s$, or $A = s^2$
Area of a Parallelogram:	$A = b \cdot h$
Area of a Triangle:	$A = \frac{1}{2} \cdot b \cdot h$
Area of a Trapezoid:	$A = \frac{1}{2} \cdot h \cdot (a + b)$
Radius and Diameter of a Circle:	$d = 2 \cdot r$, or $r = \dfrac{d}{2}$
Circumference of a Circle:	$C = \pi \cdot d$, or $C = 2 \cdot \pi \cdot r$
Area of a Circle:	$A = \pi \cdot r \cdot r$, or $A = \pi \cdot r^2$
Volume of a Rectangular Solid:	$V = l \cdot w \cdot h$
Surface Area of a Rectangular Solid:	$SA = 2lw + 2lh + 2wh$, or $2(lw + lh + wh)$
Volume of a Circular Cylinder:	$V = \pi \cdot r^2 \cdot h$
Volume of a Sphere:	$V = \frac{4}{3} \cdot \pi \cdot r^3$
Volume of a Cone:	$V = \frac{1}{3} \cdot \pi \cdot r^2 \cdot h$
Sum of Angle Measures of a Triangle:	$m\angle A + m\angle B + m\angle C = 180°$
Sum of Angle Measures of a Polygon:	$(n - 2) \cdot 180°$

Vocabulary Reinforcement

Complete each statement with the correct term from the column on the right. Some of the choices may not be used and some may be used more than once.

1. A parallelogram is a four-sided figure with two pairs of ______________ sides. [8.3b]
2. The ______________ of a polygon is the sum of the lengths of its sides. [8.2a]
3. The ______________ of a circle is half the length of its diameter. [8.4a]
4. Two angles are ______________ if the sum of their measures is 180°. [8.6a]
5. A(n) ______________ triangle has all sides of different lengths. [8.1e]
6. Similar triangles have the same ______________. [8.8a]

circumference
radius
perimeter
isosceles
scalene
parallel
perpendicular
shape
complementary
supplementary

Concept Reinforcement

Determine whether each statement is true or false.

______________ 1. The acute angles of a right triangle are complementary. [8.1e], [8.6a]

______________ 2. Two angles are supplementary if the sum of their measures is between 90° and 180°. [8.6a]

______________ 3. The number π is greater than 3.14 and $\frac{22}{7}$. [8.4b]

______________ 4. The volume of a sphere with diameter 6 ft is less than the volume of a rectangular solid that measures 6 ft by 6 ft by 6 ft. [8.5a, c]

______________ 5. The measure of any obtuse angle is larger than the measure of any acute angle. [8.1c]

Study Guide

Objective 8.1b Name an angle in five different ways, and measure an angle with a protractor.

Example Name this angle in five different ways and measure it with a protractor.

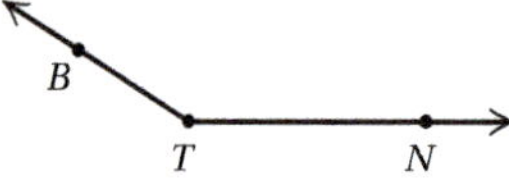

We can name this angle as

angle BTN, angle NTB, $\angle BTN$, $\angle NTB$, or $\angle T$.

The measure of the angle is 148°.

Practice Exercise

1. Name this angle in five different ways and measure it with a protractor.

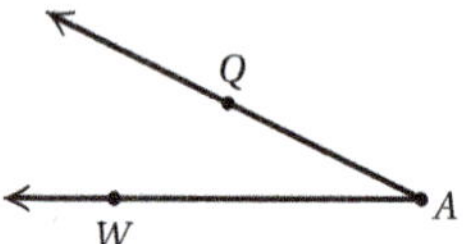

Objective 8.1c Classify an angle as right, straight, acute, or obtuse.

Example Classify each angle as right, straight, acute, or obtuse.

a)

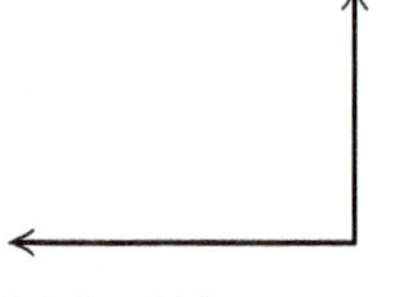

Right: 90°

b)

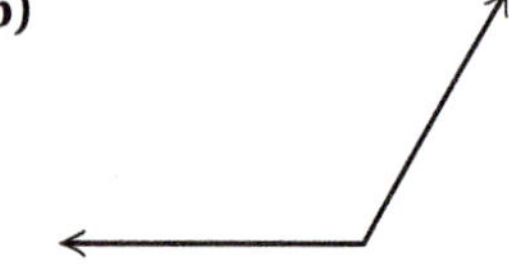

Obtuse: Greater than 90° and less than 180°

c)

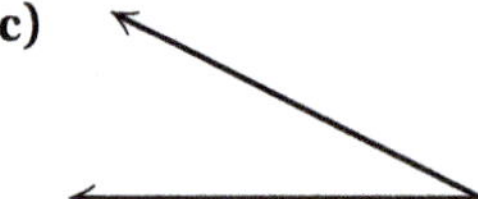

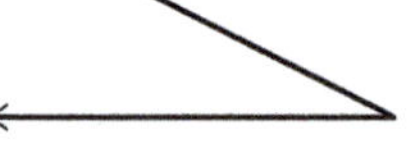

Acute: Greater than 0° and less than 90°

d) C D E

Straight: 180°

Practice Exercise

2. Classify each angle as right, straight, acute, or obtuse.

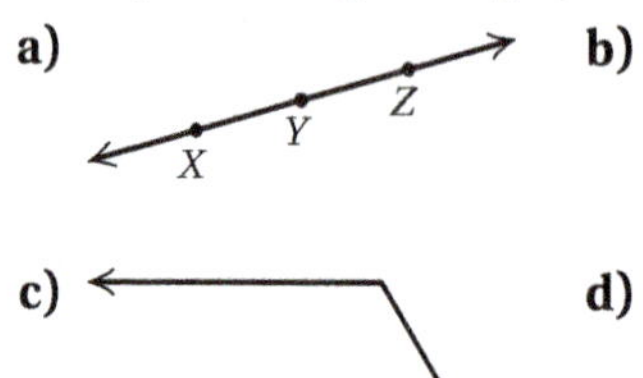

b)

d)

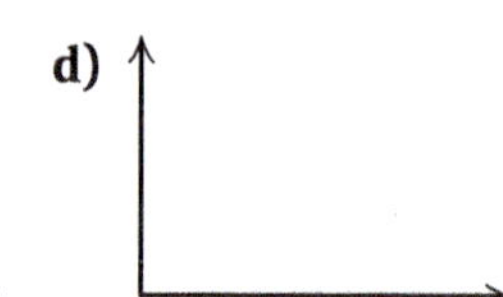

Objective 8.1e Classify a triangle as equilateral, isosceles, or scalene and as right, obtuse, or acute.

Example Classify each triangle as equilateral, isosceles, or scalene. Then classify it as right, obtuse, or acute.

a)

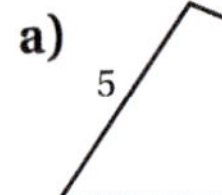

5, 11, 13

Scalene; obtuse

b)

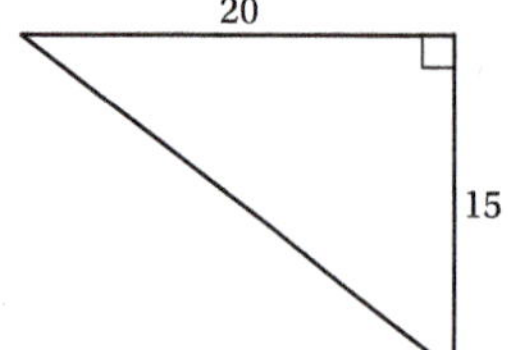

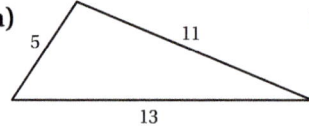

Scalene; right

c) 9, 12, 12

Isosceles; acute

d)

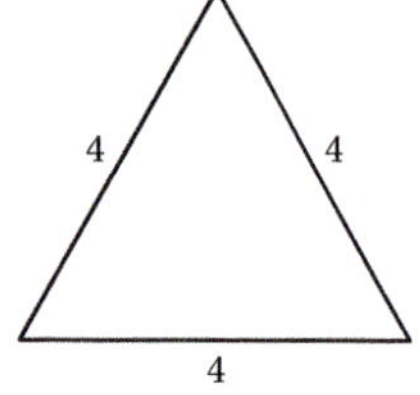

Equilateral; acute

Practice Exercise

3. Classify each triangle as equilateral, isosceles, or scalene. Then classify it as right, obtuse or acute.

a)

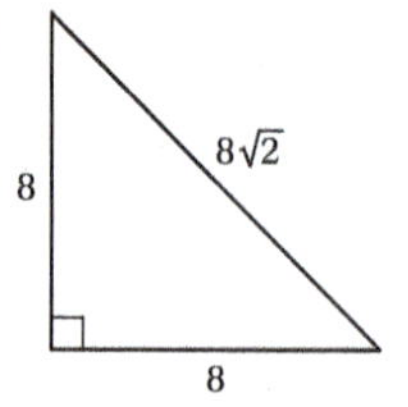

b)

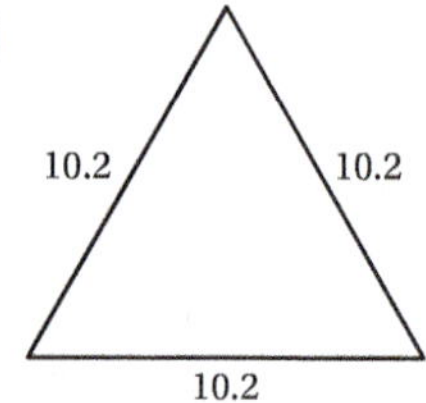

c)

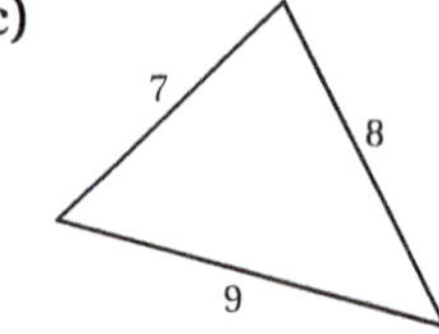

d) 10, 15, 22

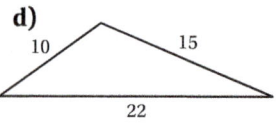

Objective 8.1f Given two of the angle measures of a triangle, find the third.

Example Find the missing angle measure.

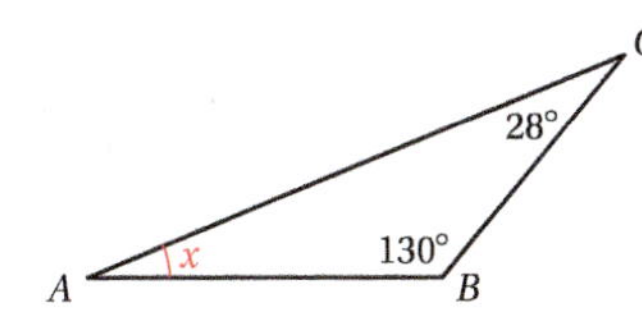

$$m\angle A + m\angle B + m\angle C = 180°$$
$$x + 130° + 28° = 180°$$
$$x + 158° = 180°$$
$$x = 180° - 158°$$
$$x = 22°$$

The measure of $\angle A$ is 22°.

Practice Exercise

4. Find the missing angle measure.

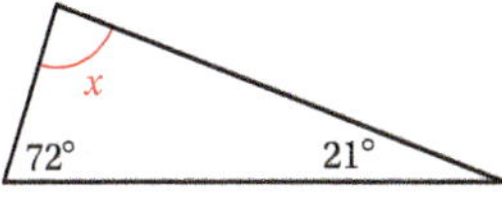

Objective 8.1f Given a polygon of n sides, find the sum of its angle measures using the formula $(n - 2) \cdot 180°$.

Example Find the sum of the angle measures of a 12-sided (dodecagon) polygon.

$$(n - 2) \cdot 180° = (12 - 2) \cdot 180° = 10 \cdot 180°$$
$$= 1800°$$

Practice Exercise

5. Find the sum of the angle measures of a 9-sided (nonagon) polygon.

Objectives 8.2a and 8.3a Find the perimeter of a polygon; find the area of a rectangle and a square.

Example Find the perimeter and the area of this rectangle.

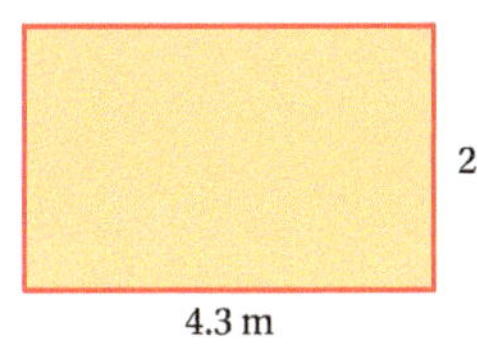

$$P = 2 \cdot (l + w)$$
$$= 2 \cdot (4.3\text{ m} + 2.7\text{ m})$$
$$= 2 \cdot (7\text{ m}) = 14\text{ m}$$
$$A = l \cdot w$$
$$= 4.3\text{ m} \cdot 2.7\text{ m} = 11.61\text{ m}^2$$

Practice Exercise

6. Find the perimeter and the area of this rectangle.

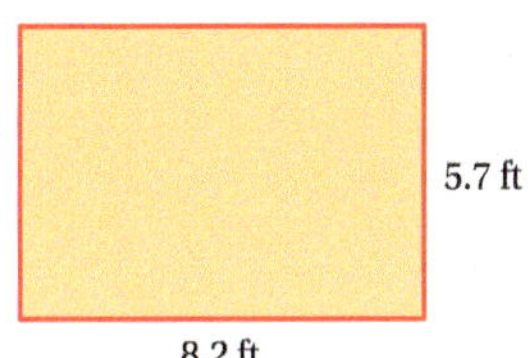

Objective 8.3b Find the area of a parallelogram, a triangle, and a trapezoid.

Examples Find the area of this parallelogram.

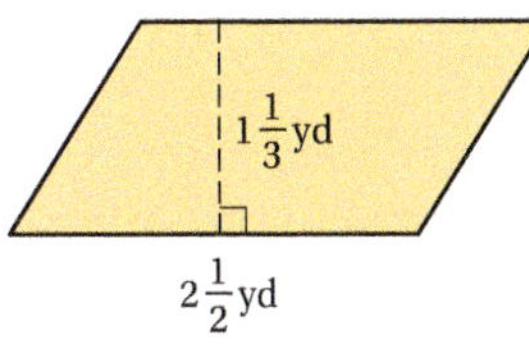

$$A = b \cdot h$$
$$= 2\frac{1}{2}\text{ yd} \cdot 1\frac{1}{3}\text{ yd}$$
$$= \frac{5}{2} \cdot \frac{4}{3} \cdot \text{yd} \cdot \text{yd}$$
$$= \frac{20}{6}\text{ yd}^2 = \frac{10}{3}\text{ yd}^2,$$
$$\text{or } 3\frac{1}{3}\text{ yd}^2$$

Practice Exercises

7. Find the area of this parallelogram.

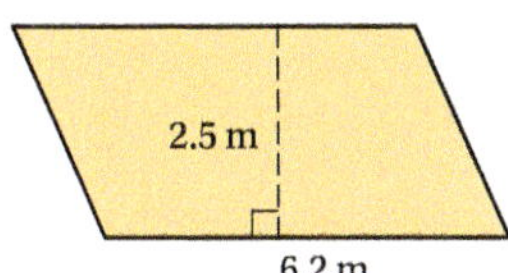

Objective 8.3b (*continued*)

Find the area of this triangle.

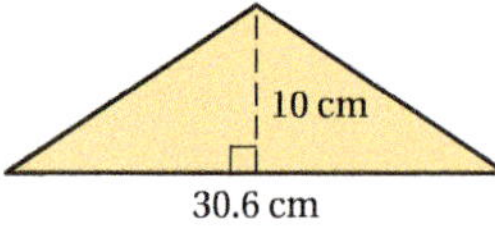

$$A = \frac{1}{2} \cdot b \cdot h$$
$$= \frac{1}{2} \cdot 30.6 \text{ cm} \cdot 10 \text{ cm}$$
$$= \frac{1}{2} \cdot 30.6 \cdot 10 \cdot \text{cm}^2$$
$$= 153 \text{ cm}^2$$

Find the area of this trapezoid.

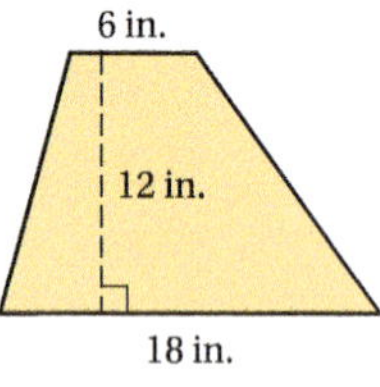

$$A = \frac{1}{2} \cdot h \cdot (a + b)$$
$$= \frac{1}{2} \times 12 \text{ in.} \times (6 \text{ in.} + 18 \text{ in.})$$
$$= \frac{1}{2} \times 12 \text{ in.} \times (24 \text{ in.})$$
$$= \frac{12 \times 24}{2} \text{in}^2 = 144 \text{ in}^2$$

8. Find the area of this triangle.

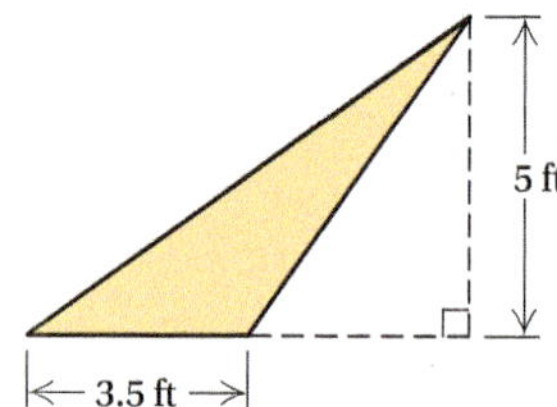

9. Find the area of this trapezoid.

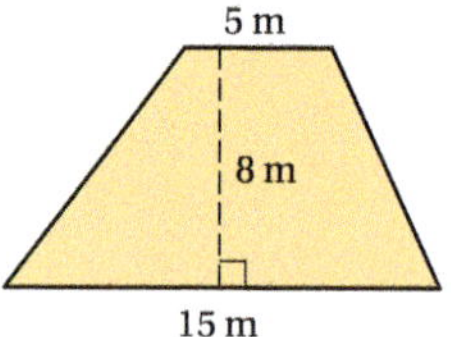

Objective 8.4b Find the circumference of a circle given the length of a diameter or a radius.

Example Find the circumference of this circle. Use 3.14 for π.

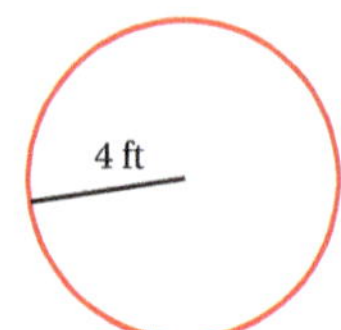

$$C = \pi \cdot d, \quad \text{or} \quad 2 \cdot \pi \cdot r$$
$$\approx 2 \times 3.14 \times 4 \text{ ft}$$
$$= 25.12 \text{ ft}$$

Practice Exercise

10. Find the circumference of this circle. Use 3.14 for π.

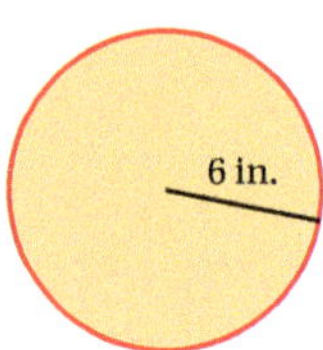

Objective 8.4c Find the area of a circle given the length of a diameter or a radius.

Example Find the area of this circle. Use $\frac{22}{7}$ for π.

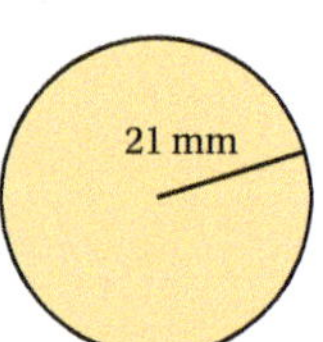

$$A = \pi \cdot r \cdot r, \quad \text{or} \quad \pi \cdot r^2$$
$$\approx \frac{22}{7} \cdot 21 \text{ mm} \cdot 21 \text{ mm}$$
$$= \frac{22 \cdot 21 \cdot 21}{7} \text{mm}^2 = 1386 \text{ mm}^2$$

Practice Exercise

11. Find the area of this circle. Use $\frac{22}{7}$ for π.

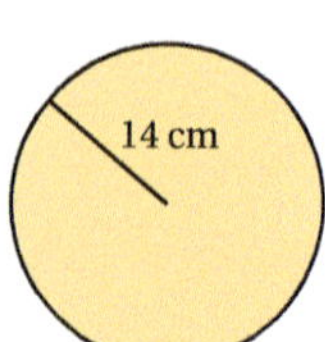

Objective 8.5a Find the volume of a rectangular solid.

Example Find the volume of this rectangular solid.

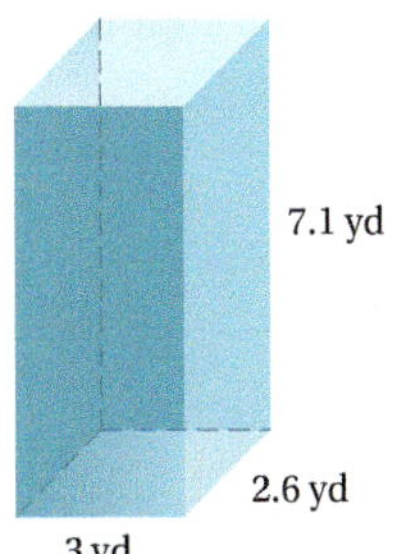

$$V = l \cdot w \cdot h$$
$$= 3\text{ yd} \times 2.6\text{ yd} \times 7.1\text{ yd}$$
$$= 3 \times 2.6 \times 7.1\text{ yd}^3$$
$$= 55.38\text{ yd}^3$$

Practice Exercise

12. Find the volume of this rectangular solid.

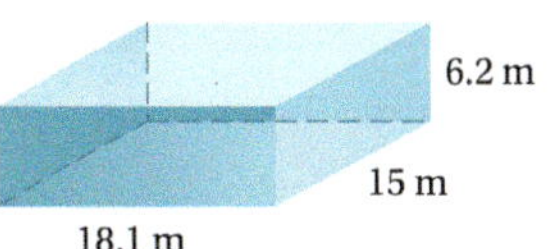

Objective 8.5b Given the radius and the height, find the volume of a circular cylinder.

Example Find the volume of this circular cylinder. Use 3.14 for π.

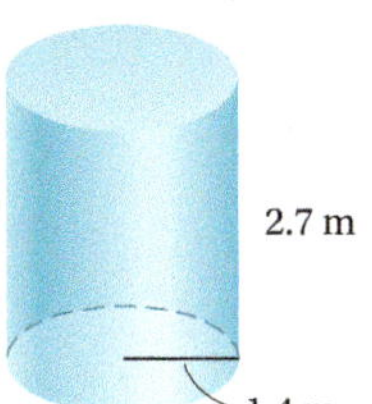

$$V = B \cdot h, \quad \text{or} \quad \pi \cdot r^2 \cdot h$$
$$\approx 3.14 \times 1.4\text{ m} \times 1.4\text{ m} \times 2.7\text{ m}$$
$$= 16.61688\text{ m}^3$$

Practice Exercise

13. Find the volume of this circular cylinder. Use $\frac{22}{7}$ for π.

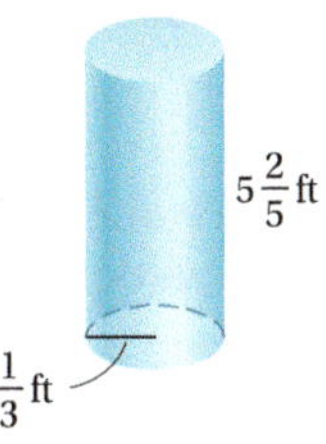

Objective 8.5c Given the radius, find the volume of a sphere.

Example Find the volume of this sphere. Use $\frac{22}{7}$ for π.

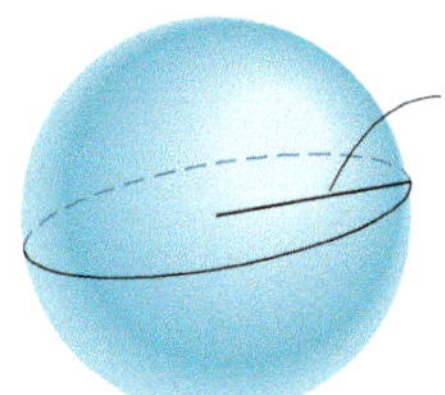

$$V = \frac{4}{3} \cdot \pi \cdot r^3$$
$$\approx \frac{4}{3} \times \frac{22}{7} \times 7\text{ in.} \times 7\text{ in.} \times 7\text{ in.}$$
$$= 1437\frac{1}{3}\text{ in}^3$$

Practice Exercise

14. Find the volume of this sphere. Use 3.14 for π.

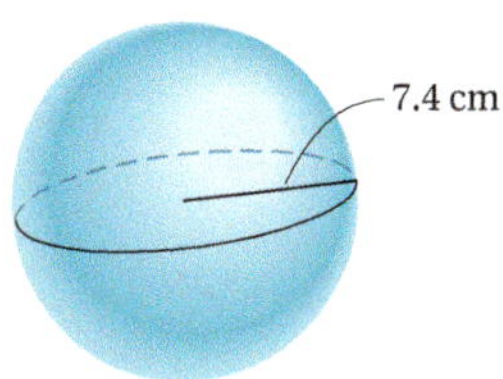

Objective 8.5d Given the radius and the height, find the volume of a circular cone.

Example Find the volume of this circular cone. Use 3.14 for π.

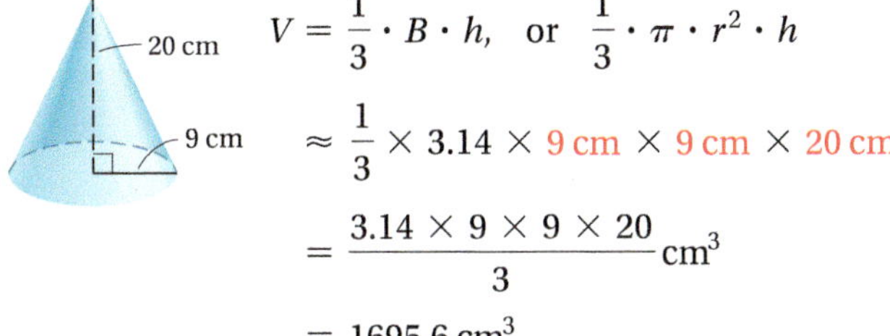

$$V = \frac{1}{3} \cdot B \cdot h, \quad \text{or} \quad \frac{1}{3} \cdot \pi \cdot r^2 \cdot h$$
$$\approx \frac{1}{3} \times 3.14 \times 9\text{ cm} \times 9\text{ cm} \times 20\text{ cm}$$
$$= \frac{3.14 \times 9 \times 9 \times 20}{3}\text{ cm}^3$$
$$= 1695.6\text{ cm}^3$$

Practice Exercise

15. Find the volume of this circular cone. Use 3.14 for π.

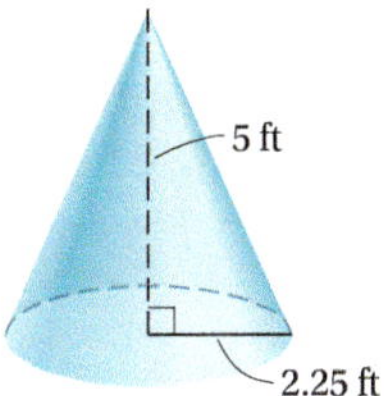

Objective 8.6a Find the measure of a complement or a supplement of a given angle.

Example Find the measure of a complement and a supplement of an angle that measures 65°.

The measure of the complement of an angle of 65° is $90° - 65°$, or 25°.

The measure of the supplement of an angle of 65° is $180° - 65°$, or 115°.

Practice Exercise

16. Find the measure of a complement and a supplement of an angle that measures 38°.

Objective 8.6c Use the Vertical Angle Property to find measures of angles.

Example In the following figure, $m\angle 2 = 110°$ and $m\angle 4 = 41°$. Find $m\angle 1$, $m\angle 3$, $m\angle 5$, and $m\angle 6$.

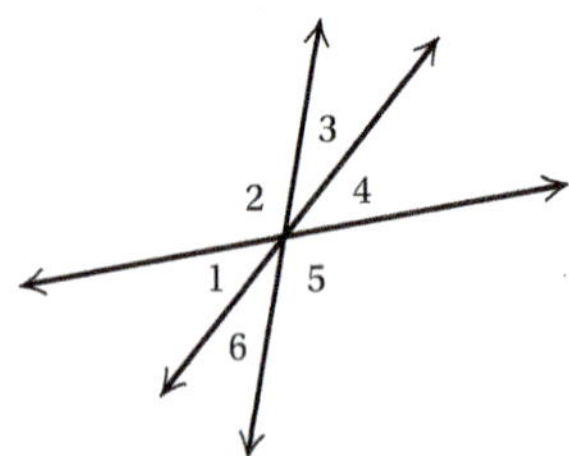

Since $\angle 2$ and $\angle 5$ are vertical angles, $m\angle 5 = 110°$. Likewise, $\angle 4$ and $\angle 1$ are vertical angles, so $m\angle 1 = 41°$.

$$m\angle 1 + m\angle 2 + m\angle 3 = 180°$$
$$41° + 110° + m\angle 3 = 180°$$
$$m\angle 3 = 180° - 151°$$
$$m\angle 3 = 29°$$

Since $\angle 3$ and $\angle 6$ are vertical angles, $m\angle 6 = 29°$.

Practice Exercise

17. In the following figure, $m\angle 8 = 65°$ and $m\angle 12 = 55°$. Find $m\angle 7$, $m\angle 9$, $m\angle 10$, and $m\angle 11$.

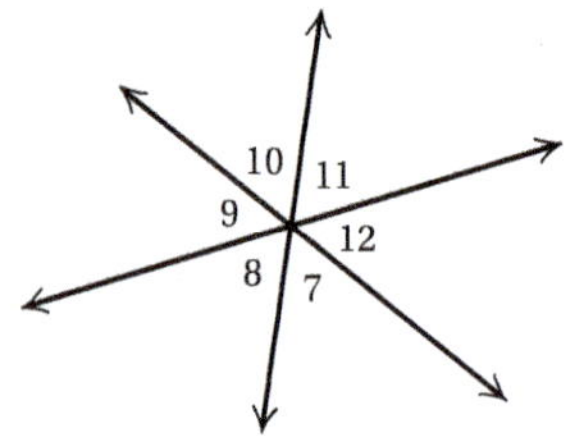

Objective 8.6d Apply properties of transversals and parallel lines to find measures of angles.

Example If $a \parallel b$ and $m\angle 3 = 37°$, what are the measures of the other angles?

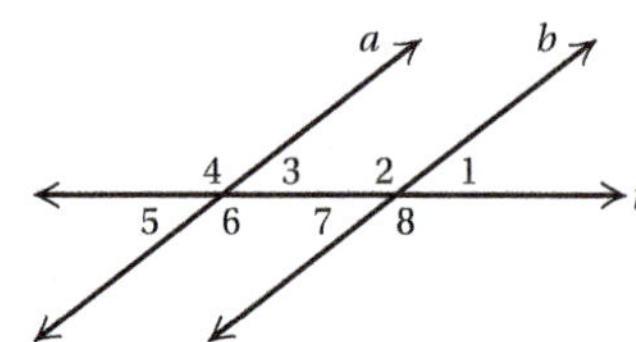

Using Property 2, $m\angle 7 = 37°$. Using the Vertical Angle Property, $m\angle 5 = 37°$ and $m\angle 1 = 37°$. Using Property 4, angles 3 and 2 are supplementary and angles 6 and 7 are supplementary. Thus, $m\angle 2 = 180° - 37° = 143°$ and $m\angle 6 = 180° - 37° = 143°$. Then using the Vertical Angle Property, $m\angle 4 = m\angle 6 = 143°$ and $m\angle 8 = m\angle 2 = 143°$.

Practice Exercise

18. In the following figure, if $c \parallel d$ and $m\angle 5 = 105°$, what are the measures of the other angles?

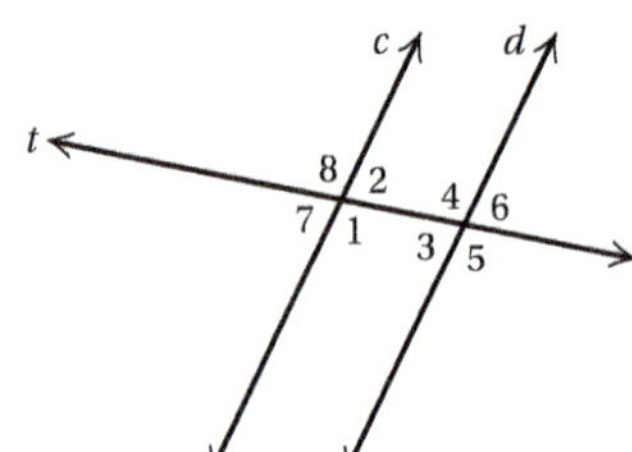

Objective 8.7a Show why triangles are congruent using SAS, SSS, and ASA.

Example Which property (if any) should be used to show that the pair of triangles is congruent?

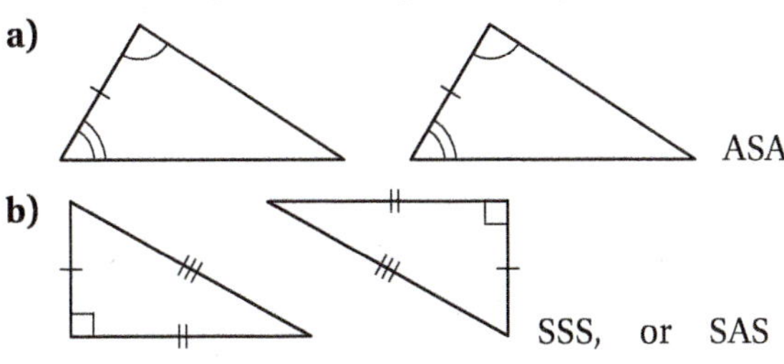

a) ASA

b) SSS, or SAS

Practice Exercise

19. Which property (if any) should be used to show that the pair of triangles is congruent?

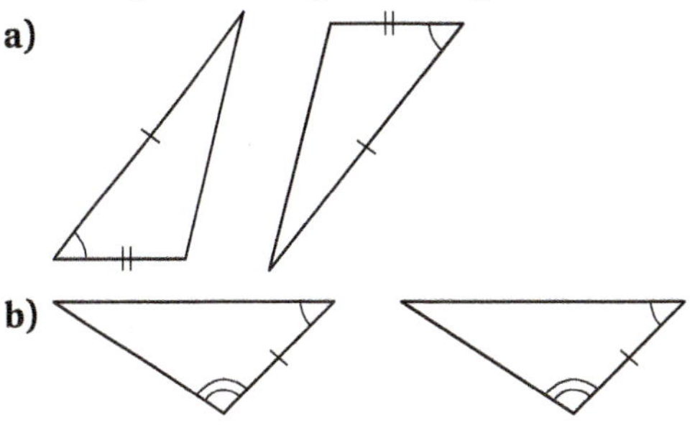

a)

b)

Objective 8.7b Use properties of parallelograms to find lengths of sides and measures of angles of parallelograms.

Example The perimeter of ▱*MNOP* is 57. Find the length of the sides of the parallelogram and the measures of the angles of the parallelogram.

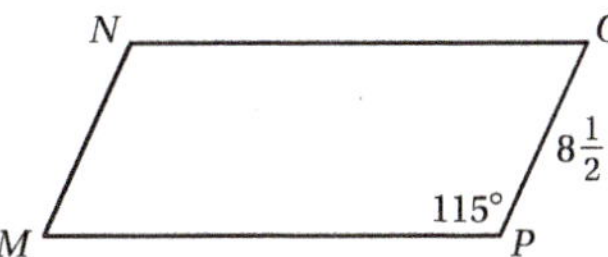

The opposite sides of a parallelogram are congruent. Since $PO = 8\frac{1}{2}$, $MN = 8\frac{1}{2}$. Then Perimeter $= MN + PO + MP + NO = 8\frac{1}{2} + 8\frac{1}{2} + MP + NO = 57$, or $MP + NO = 40$. Since $MP = NO$, we know that $MP = 20$ and $NO = 20$.

The opposite angles of a parallelogram are congruent. Thus, $m\angle P = 115°$ and $m\angle N = 115°$. Since consecutive angles of a parallelogram are supplementary, $m\angle M = m\angle O = 180° - 115° = 65°$.

Practice Exercise

20. The perimeter of ▱*ABCD* is 237. Find the lengths of the sides of the parallelogram and the measures of the angles of the parallelogram.

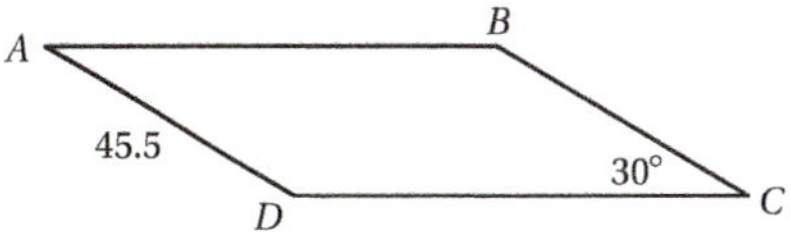

Objective 8.8b Find lengths of sides of similar triangles using proportions.

Example If $\triangle FDE \sim \triangle TZW$, find TZ and WZ.

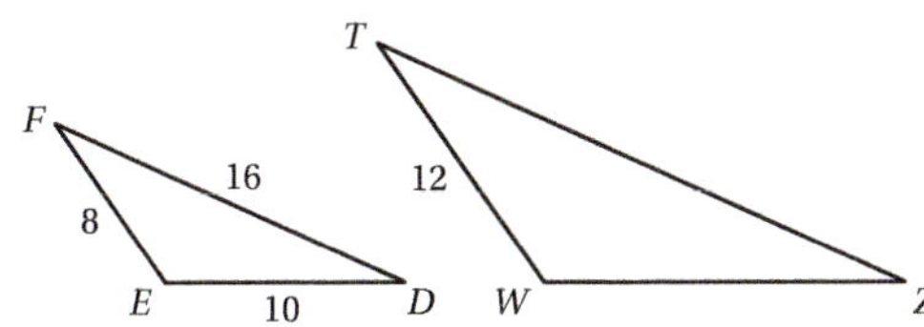

Since $\triangle FDE \sim \triangle TZW$, the corresponding sides are proportional. Thus,

$$\frac{8}{12} = \frac{16}{TZ} \qquad \frac{8}{12} = \frac{10}{WZ}$$

$$8(TZ) = 12 \cdot 16 \qquad 8(WZ) = 12 \cdot 10$$

$$8(TZ) = 192 \qquad 8(WZ) = 120$$

$$TZ = 24; \qquad WZ = 15.$$

Practice Exercise

21. If $\triangle QBX \sim \triangle ZAT$, find ZA and AT.

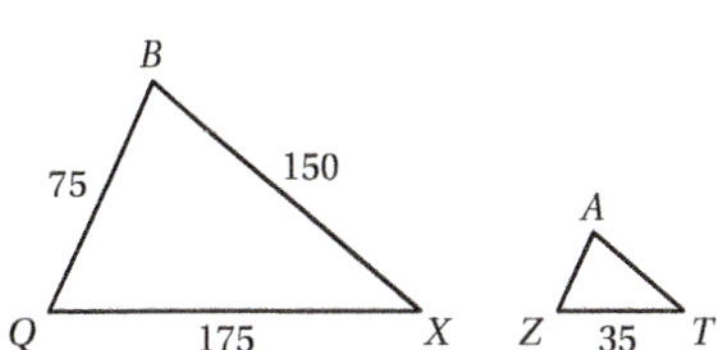

Review Exercises

Use a protractor to measure each angle. [8.1b]

1. **2.**

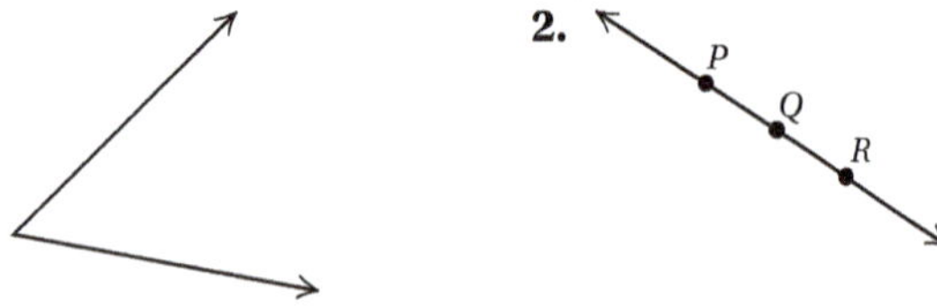

3. **4.**

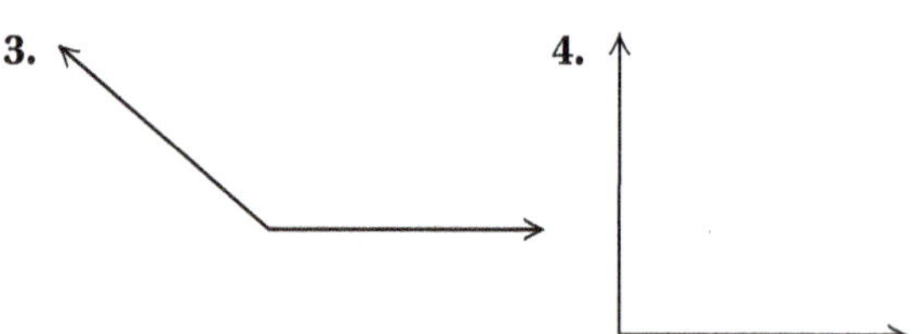

5.–8. Classify each of the angles in Exercises 1–4 as right, straight, acute, or obtuse. [8.1c]

Use the following triangle for Exercises 9–11.

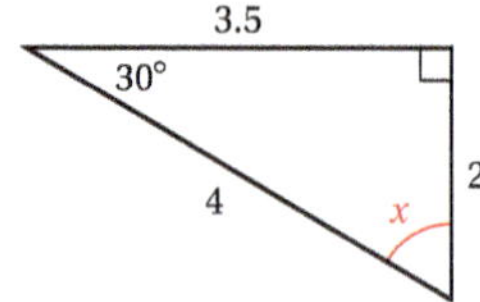

9. Find the missing angle measure. [8.1f]

10. Classify the triangle as equilateral, isosceles, or scalene. [8.1e]

11. Classify the triangle as right, obtuse, or acute. [8.1e]

12. Find the sum of the angle measures of a hexagon. [8.1f]

Find the perimeter. [8.2a]

13.

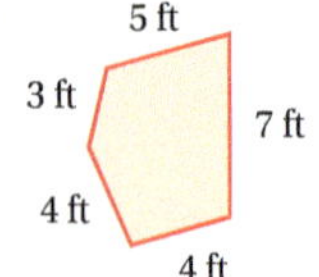

14.

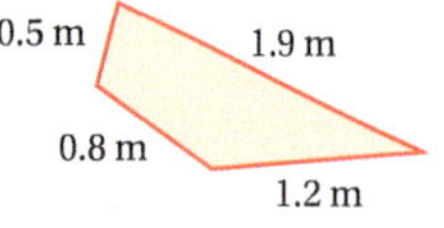

15. *Tennis Court.* The dimensions of a standard-sized tennis court are 78 ft by 36 ft. Find the perimeter and the area of the tennis court. [8.2b], [8.3c]

Find the perimeter and the area. [8.2a], [8.3a]

16.

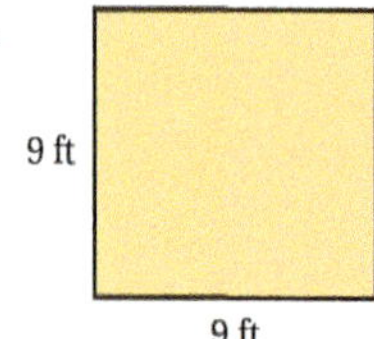

17.

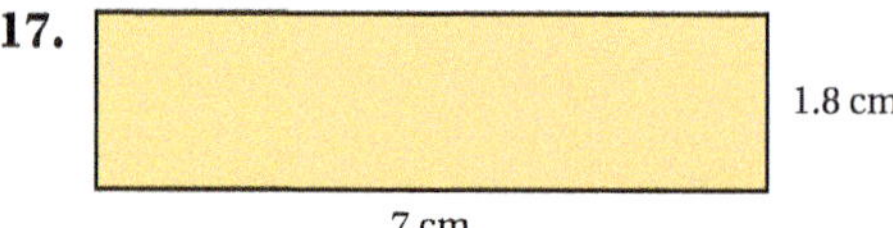

Find the area. [8.3b]

18.

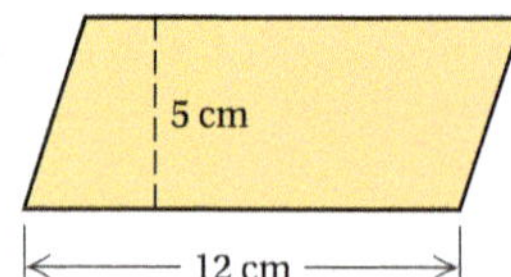

19.

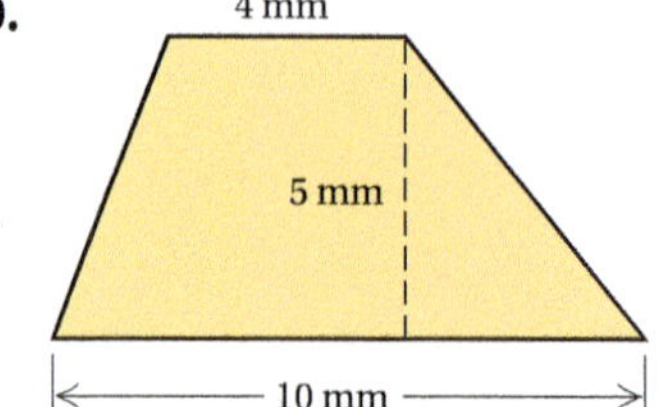

20.

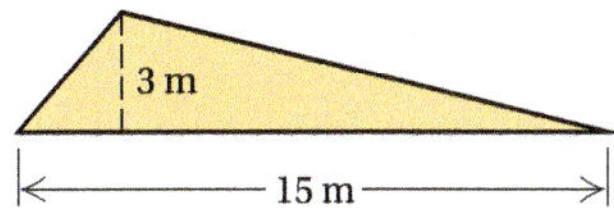

21.

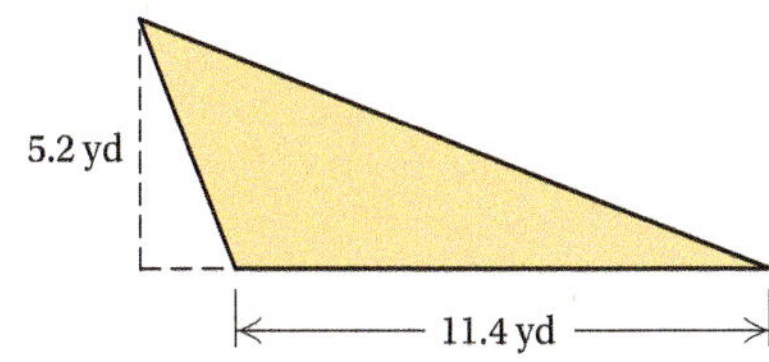

22.

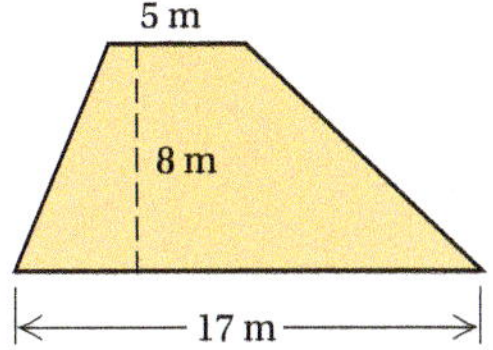

23.

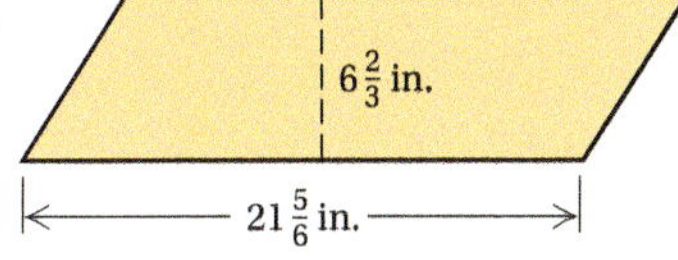

24. ***Seeded Area.*** A grassy area around three sides of a building has equal width on the three sides, as shown below, and is going to be reseeded. What is the total area to be reseeded? [8.3c]

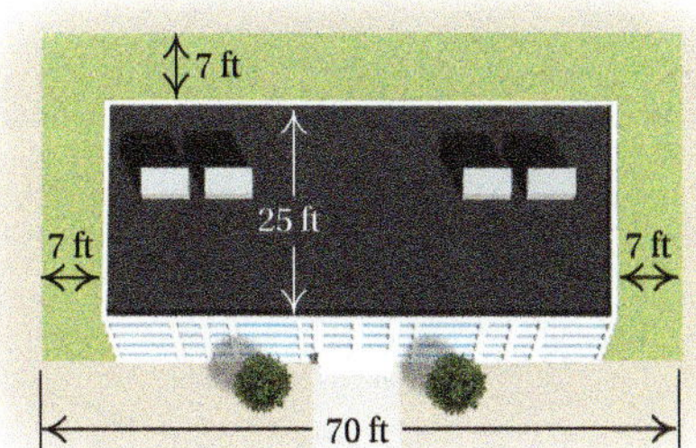

Find the length of a radius of each circle. [8.4a]

25.

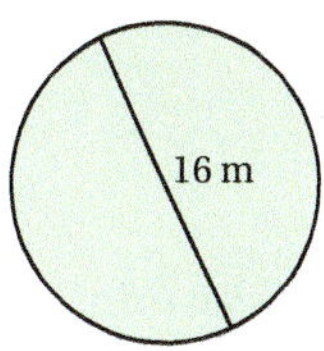

26.

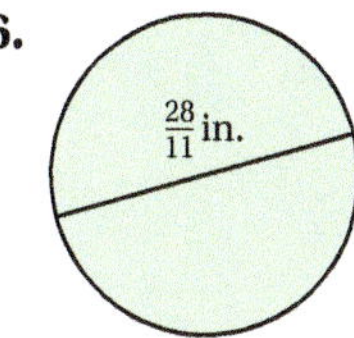

Find the length of a diameter of each circle. [8.4a]

27.

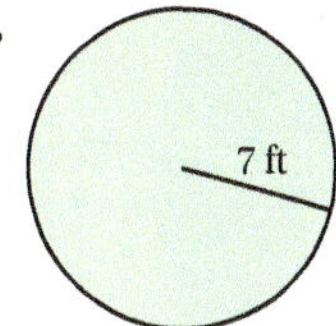

28.

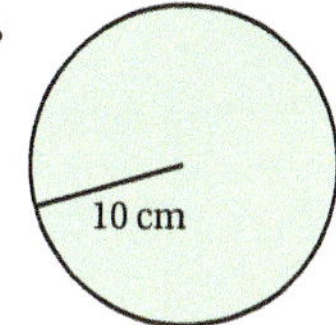

29. Find the circumference of the circle in Exercise 25. Use 3.14 for π. [8.4b]

30. Find the circumference of the circle in Exercise 26. Use $\frac{22}{7}$ for π. [8.4b]

31. Find the area of the circle in Exercise 25. Use 3.14 for π. [8.4c]

32. Find the area of the circle in Exercise 26. Use $\frac{22}{7}$ for π. [8.4c]

33. Find the area of the shaded region. Use 3.14 for π. [8.4d]

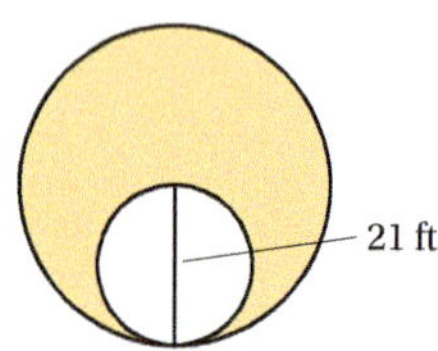

34. A Norman window is designed with dimensions as shown. Find its area and its perimeter. Use 3.14 for π. [8.4d]

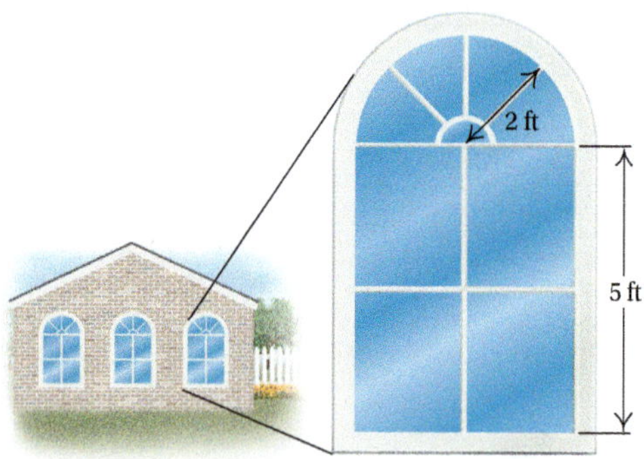

Find the volume and the surface area. [8.5a]

35.

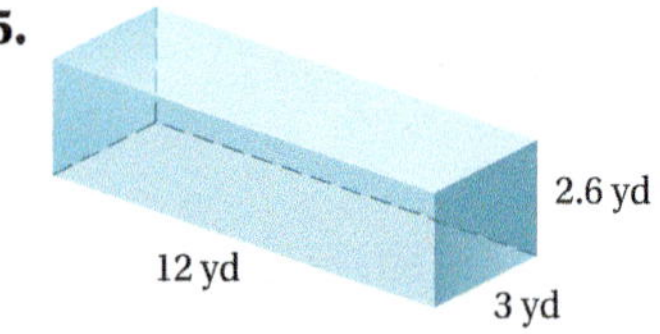

36.

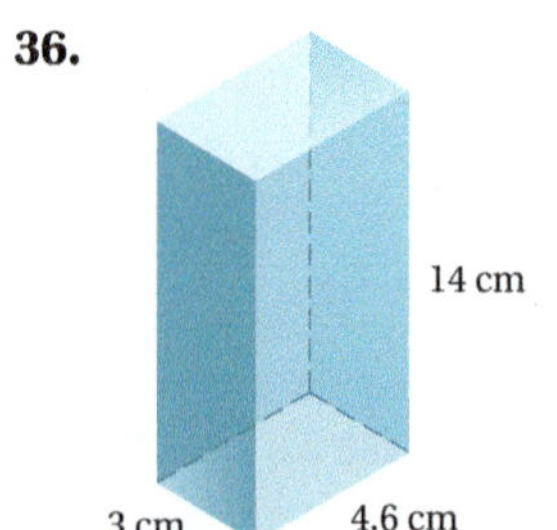

Find the volume. Use 3.14 for π.

37. [8.5b]

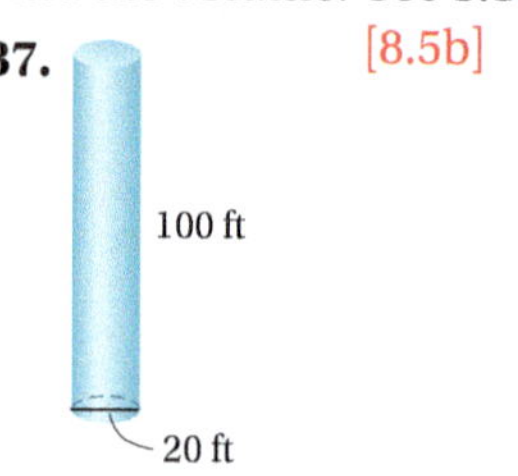

38. [8.5c]

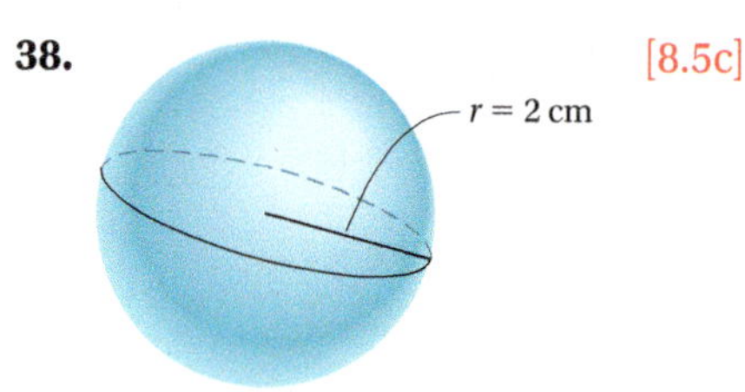

39. [8.5d]

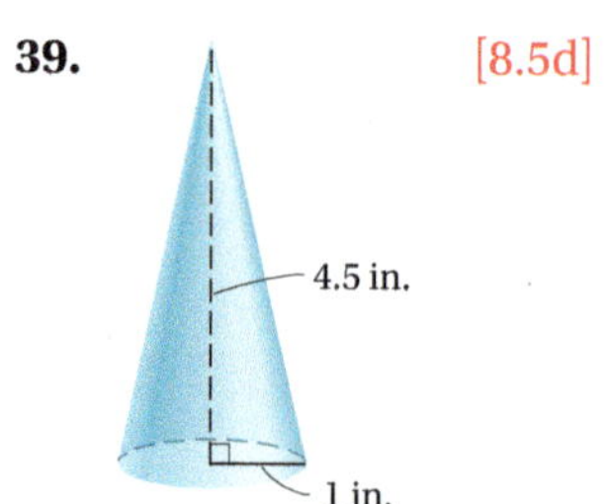

40. [8.5b]

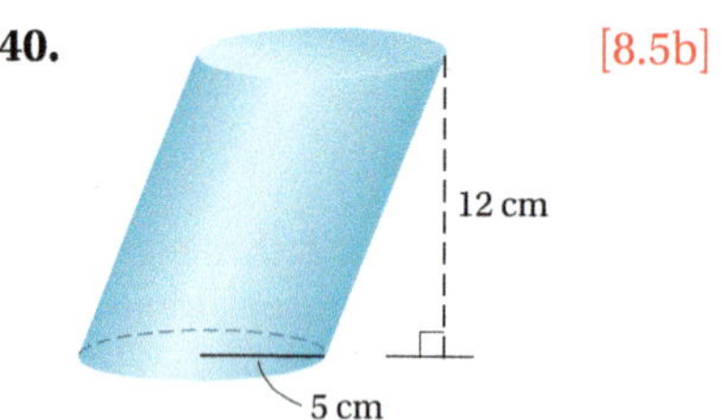

41. Find the measure of a complement of $\angle BAC$. [8.6a]

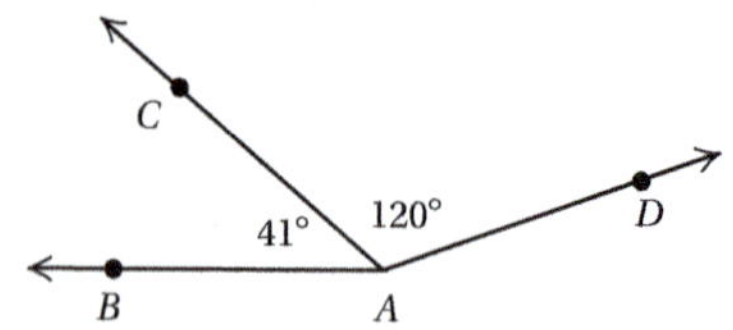

Find the measure of a complement of an angle with the given measure. [8.6a]

42. 82°

43. 5°

Find the measure of a supplement of an angle with the given measure. [8.6a]

44. 33°

45. 133°

46. In this figure, $m\angle 1 = 38°$ and $m\angle 5 = 105°$. Find $m\angle 2$, $m\angle 3$, $m\angle 4$, and $m\angle 6$. [8.6c]

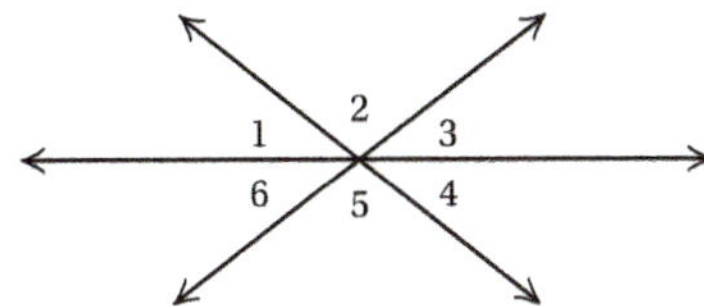

47. In this figure, identify **(a)** all pairs of corresponding angles, **(b)** all interior angles, and **(c)** all pairs of alternate interior angles. [8.6d]

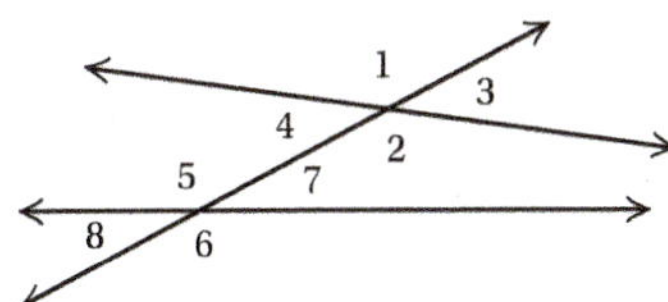

48. If $m \parallel n$ and $m\angle 4 = 135°$, what are the measures of the other angles? [8.6d]

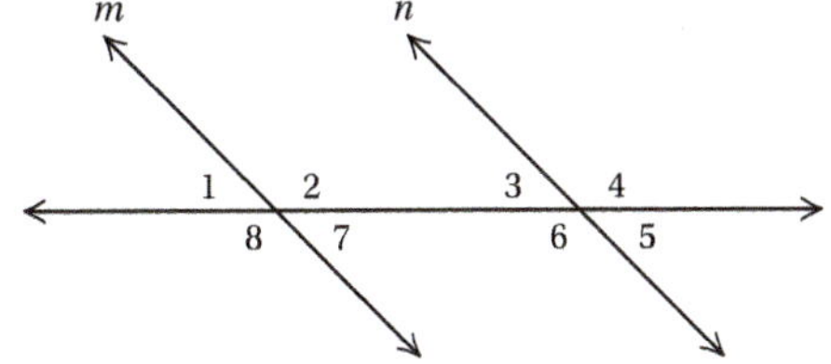

Name the corresponding parts of these congruent triangles. [8.7a]

49. $\triangle DHJ \cong \triangle RZK$

50.

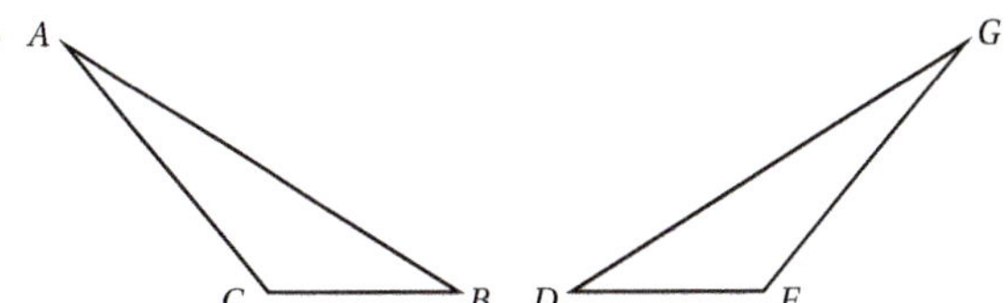

Which property (if any) should be used to show that the pair of triangles is congruent? [8.7a]

51.

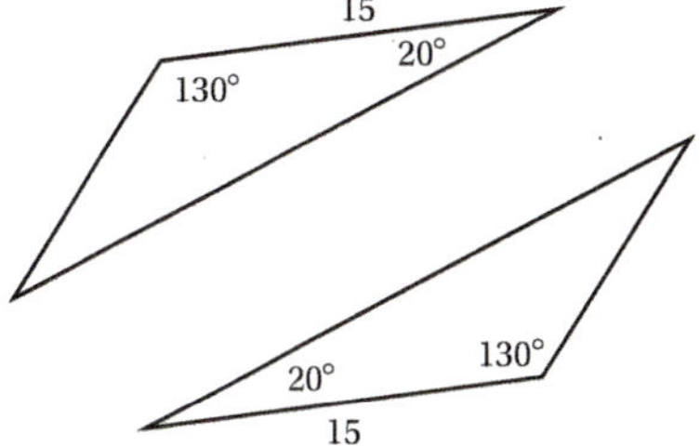

52.

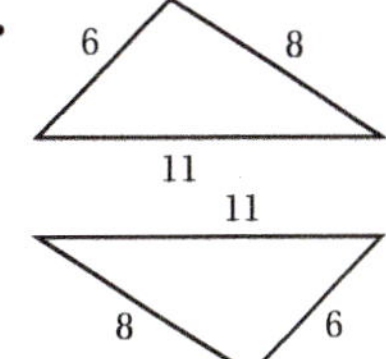

53.

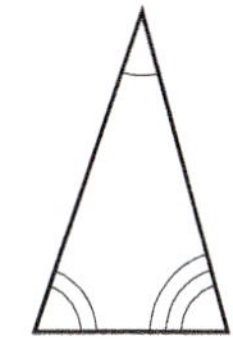

54. J is the midpoint of $\overline{IK}$ and $\overline{HI} \parallel \overline{KL}$. Explain why $\triangle JIH \cong \triangle JKL$. [8.7a]

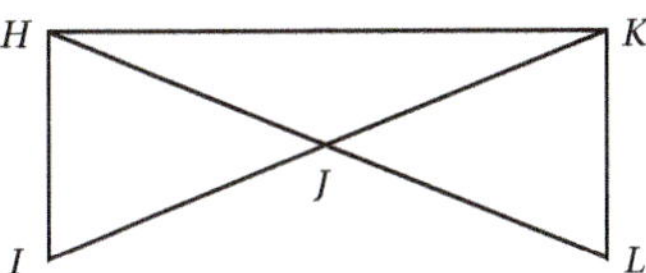

55. Find the measures of the angles and the lengths of the sides of this parallelogram. [8.7b]

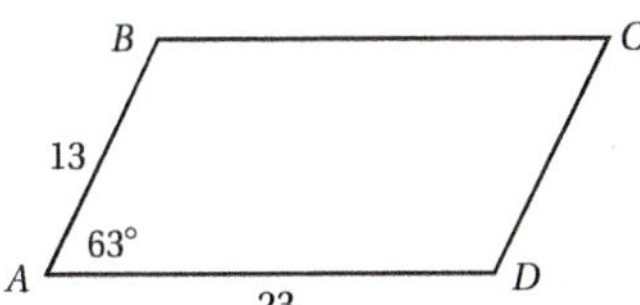

56. If $\triangle CQW \sim \triangle FAS$, name the congruent angles and the proportional sides. [8.8a]

57. If $\triangle NMO \sim \triangle STR$, find MO. [8.8b]

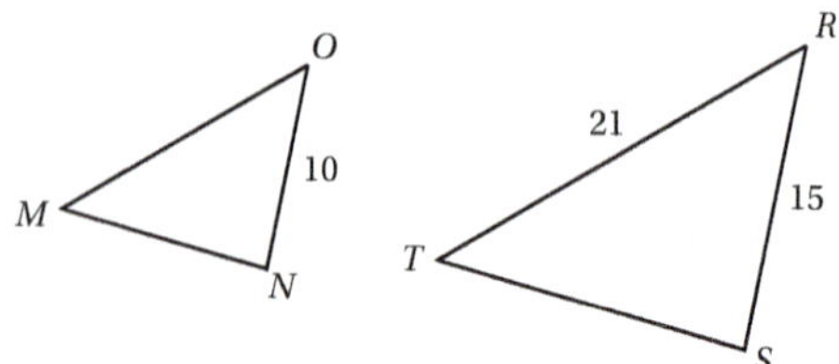

58. Find the measure of a supplement of a $20\frac{3}{4}°$ angle. [8.6a]

A. $339\frac{1}{4}°$ **B.** $159\frac{1}{4}°$

C. $69\frac{1}{4}°$ **D.** $70\frac{1}{4}°$

59. Find the area of a circle whose diameter is $\frac{7}{9}$ in. Use $\frac{22}{7}$ for π. [8.4c]

A. $\frac{11}{9}\text{ in}^2$ **B.** $\frac{77}{162}\text{ in}^2$

C. $\frac{22}{9}\text{ in}^2$ **D.** $\frac{154}{81}\text{ in}^2$

Synthesis

60. A square is cut in half so that the perimeter of the resulting rectangle is 30 ft. Find the area of the original square. [8.2a], [8.3a]

61. Find the area, in square meters, of the shaded region. [8.3c]

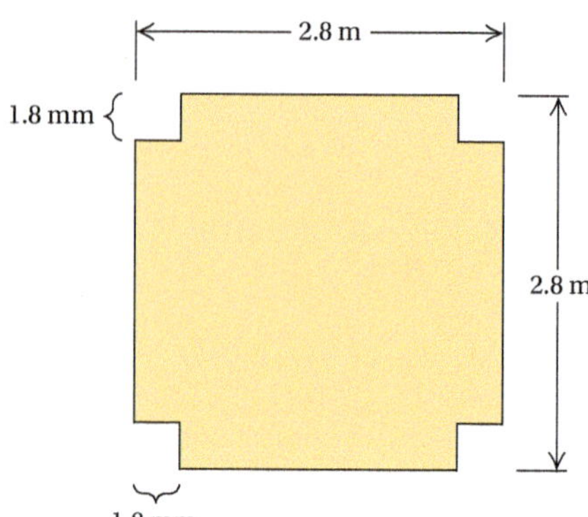

62. Find the area, in square centimeters, of the shaded region. [8.3c]

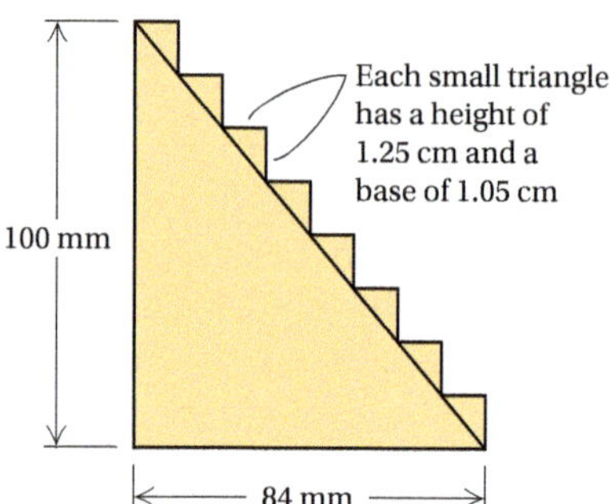

Understanding Through Discussion and Writing

1. Explain a procedure that could be used to determine the measure of an angle's supplement from the measure of the angle's complement. [8.6a]

2. How could you use the volume formulas given in Section 8.5 to help estimate the volume of an egg? [8.5a, b, c, e]

3. Describe the differences among linear, area, and volume units of measure. [8.2a], [8.3a], [8.5a]

4. Explain how you might use triangles to find the sum of the angle measures of this figure. [8.1f]

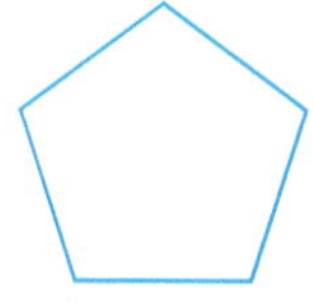

5. The design of a home includes a cylindrical tower that will be capped with either a 10-ft high dome (half of a sphere) or a 10-ft high cone. Which type of cap would be more energy-efficient and why? [8.5c, d]

6. Which occupies more volume: two spheres, each with radius r, or one sphere with radius $2r$? Explain why. [8.5c]

CHAPTER **8**

Test

For Extra Help For step-by-step test solutions, access the Chapter Test Prep Videos in MyMathLab® or on YouTube (search "BittingerPreIntro" and click on Channels).

Use a protractor to measure each angle.

1.

2.

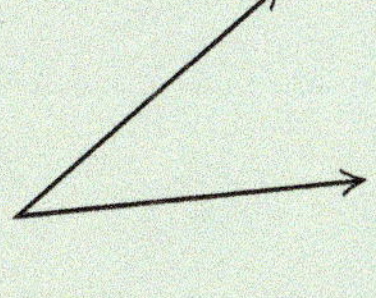

3.

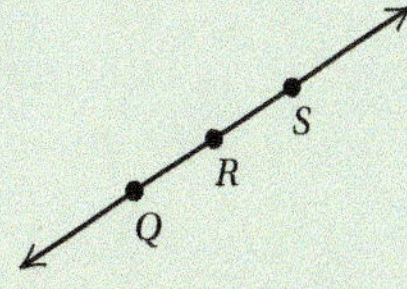

4.

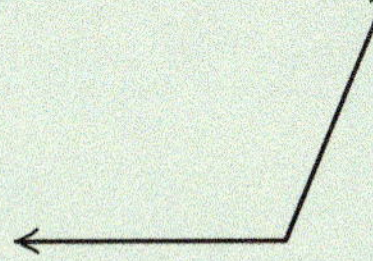

5.–8. Classify each of the angles in Exercises 1–4 as right, straight, acute, or obtuse.

Use the following triangle for Exercises 9–11.

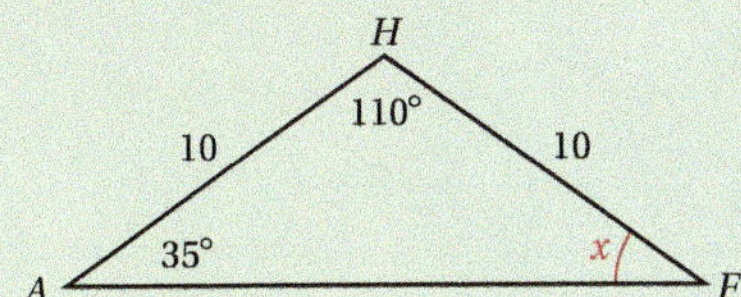

9. Find the missing angle measure.

10. Classify the triangle as equilateral, isosceles, or scalene.

11. Classify the triangle as right, obtuse, or acute.

12. Find the sum of the angle measures of a pentagon.

Find the perimeter and the area.

13. 7.01 cm; 9.4 cm

14.

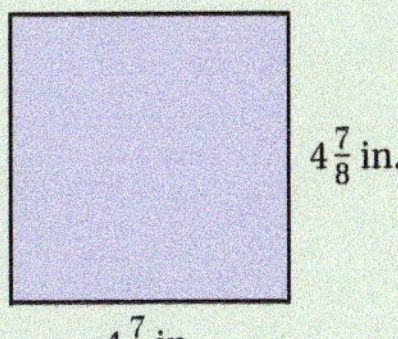

Find the area.

15.

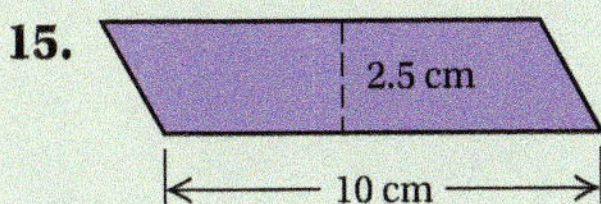

16.

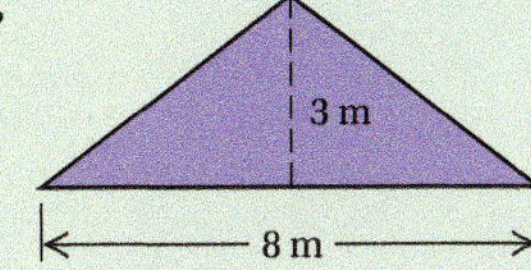

17.

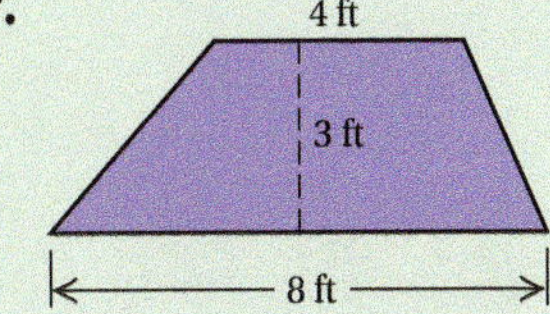

18. Find the length of a diameter of this circle.

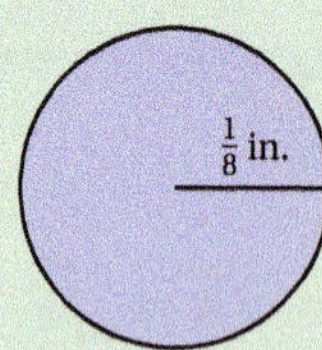

19. Find the length of a radius of this circle.

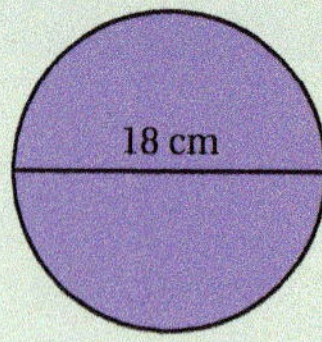

20. Find the circumference of the circle in Exercise 18. Use $\frac{22}{7}$ for π.

21. Find the area of the circle in Exercise 19. Use 3.14 for π.

22. Find the perimeter and the area of the shaded region. Use 3.14 for π.

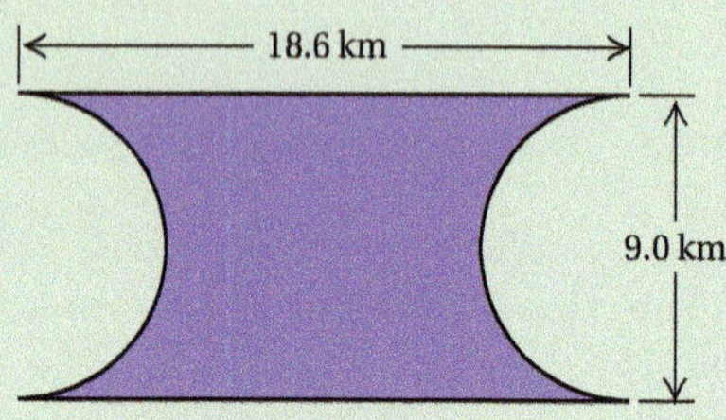

23. Find the volume and the surface area.

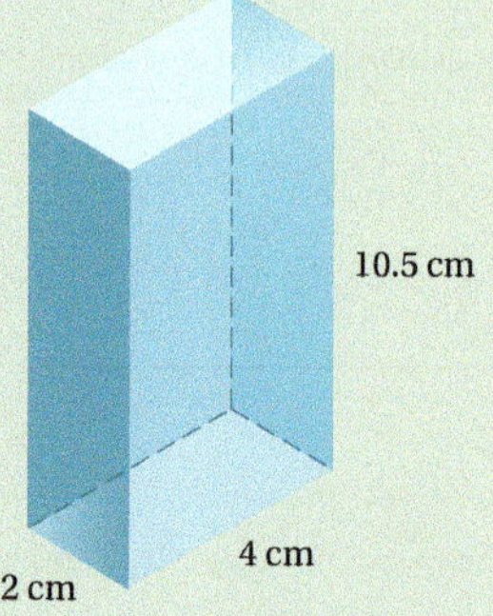

24. A twelve-box rectangular carton of 12-oz juice boxes measures $10\frac{1}{2}$ in. by 8 in. by 5 in. What is the volume of the carton?

Find the volume. Use 3.14 for π.

25.

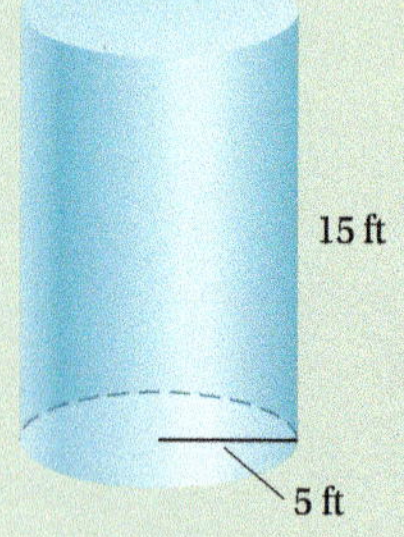

26.

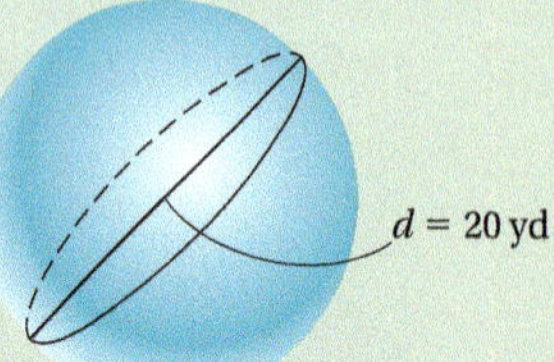

27.

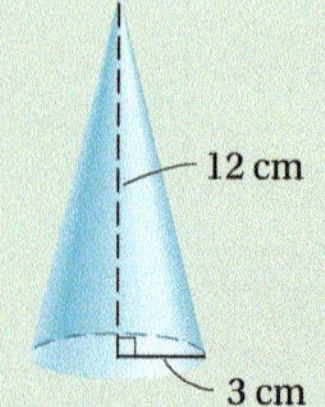

28. Find the measure of a complement and a supplement of $\angle CAD$.

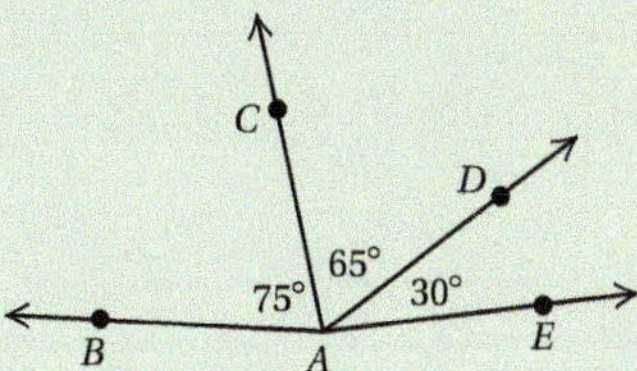

29. In the figure, $m\angle 1 = 62°$ and $m\angle 5 = 110°$. Find $m\angle 2$, $m\angle 3$, $m\angle 4$, and $m\angle 6$.

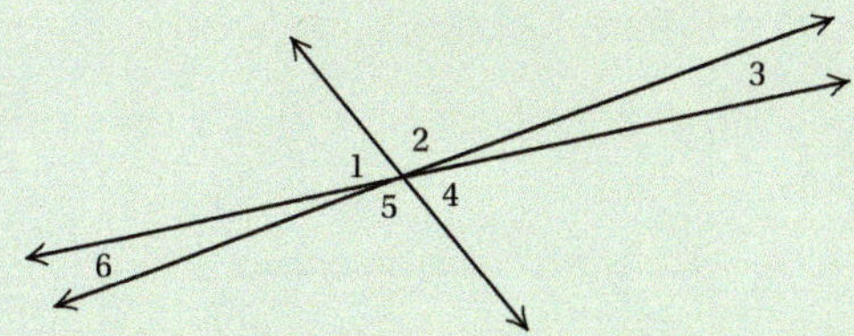

30. If $m \parallel n$ and $m\angle 4 = 120°$, what are the measures of the other angles?

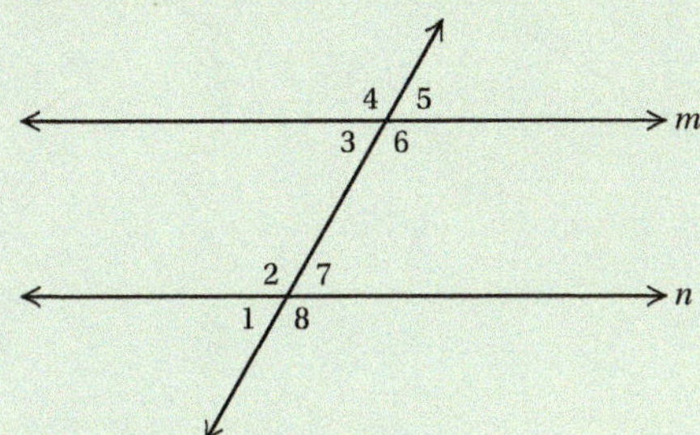

31. Name the corresponding parts of these congruent triangles: $\triangle CWS \cong \triangle ATZ$.

Which property (if any) would you use to show that $\triangle RST \cong \triangle DEF$ with the given information?

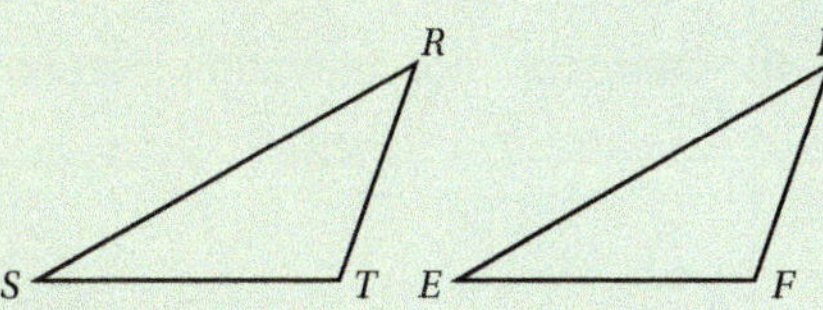

32. $\overline{RS} \cong \overline{DE}$, $\overline{RT} \cong \overline{DF}$, and $\angle R \cong \angle D$

33. $\angle R \cong \angle D$, $\angle S \cong \angle E$, and $\angle T \cong \angle F$

34. $\overline{RS} \cong \overline{DE}$, $\angle R \cong \angle D$, and $\angle S \cong \angle E$

35. $\angle R \cong \angle D$, $\overline{RT} \cong \overline{DF}$, and $\overline{ST} \cong \overline{EF}$

36. The perimeter of $\square DEFG$ is 62. Find the measures of the angles and the lengths of the sides.

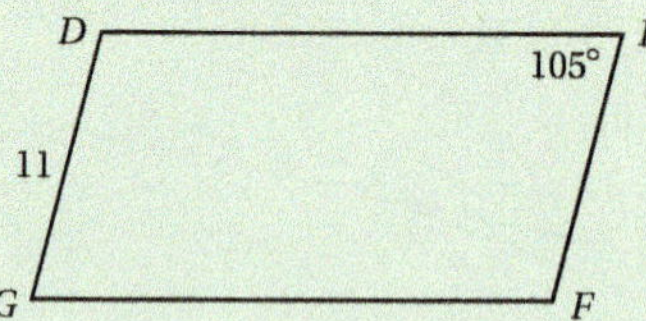

37. In $\square JKLM$, $JN = 3.2$ and $KN = 3$. Find the lengths of the diagonals, $\overline{LJ}$ and $\overline{KM}$.

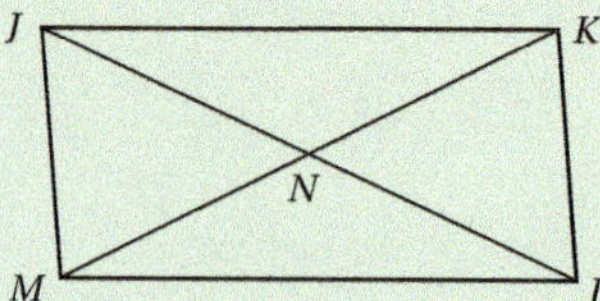

38. If $\triangle ERS \sim \triangle TGF$, name the congruent angles and the proportional sides.

39. If $\triangle GTR \sim \triangle ZEK$, find EK and ZK.

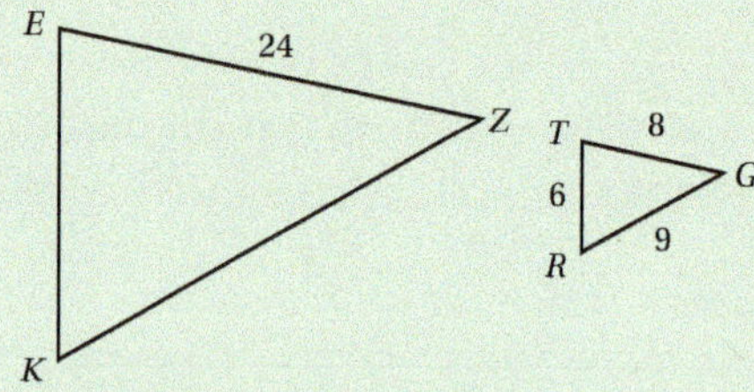

40. Find the volume of a sphere whose diameter is 42 cm. Use $\frac{22}{7}$ for π.

A. 310,464 cm^3 **B.** 9702 cm^3
C. 1848 cm^3 **D.** 38,808 cm^3

Synthesis

Find the area of the shaded region. (Note that the figures are not drawn in perfect proportion.) Give the answer in square feet.

41.

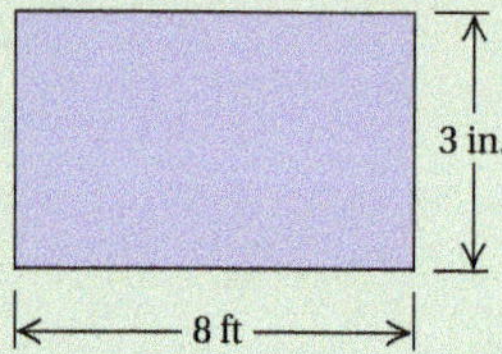

42.

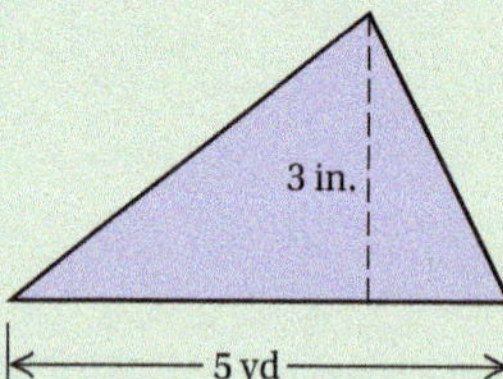

Find the volume of the solid. (Note that the solids are not drawn in perfect proportion.) Give the answer in cubic feet. Use 3.14 for π and round to the nearest thousandth in Exercises 44 and 45.

43.

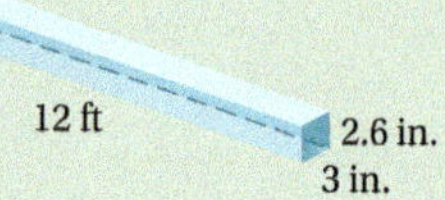

44.

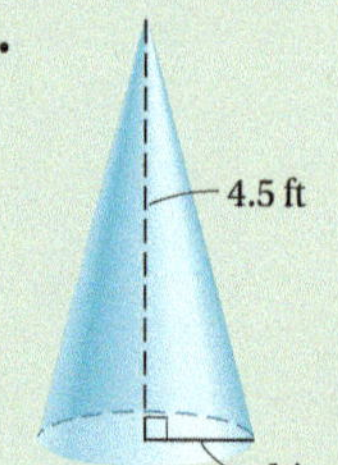

45.

18 ft

$\frac{3}{4}$ in.

CHAPTER

9

Introduction to Real Numbers and Algebraic Expressions

STUDYING FOR SUCCESS *Getting Off to a Good Start*

- ☐ Your syllabus for this course is extremely important. Read it carefully, noting required texts and materials.
- ☐ If you have an online component in your course, register for it as soon as possible.
- ☐ At the front of the text, you will find a Student Organizer card. This pullout card will help you keep track of important dates and useful contact information.

9.1 Introduction to Algebra

OBJECTIVES

- **a** Evaluate algebraic expressions by substitution.
- **b** Translate phrases to algebraic expressions.

SKILL TO REVIEW

Objective 3.5b: Simplify fraction notation.

Simplify.

1. $\frac{100}{20}$ **2.** $\frac{78}{3}$

The study of algebra involves the use of equations to solve problems. Equations are constructed from algebraic expressions.

a EVALUATING ALGEBRAIC EXPRESSIONS

In arithmetic, you have worked with expressions such as

$$49 + 75, \quad 8 \times 6.07, \quad 29 - 14, \quad \text{and} \quad \frac{5}{6}.$$

In algebra, we can use letters to represent numbers and work with *algebraic expressions* such as

$$x + 75, \quad 8 \times y, \quad 29 - t, \quad \text{and} \quad \frac{a}{b}.$$

Sometimes a letter can represent various numbers. In that case, we call the letter a **variable**. Let a = your age. Then a is a variable since a changes from year to year. Sometimes a letter can stand for just one number. In that case, we call the letter a **constant**. Let b = your date of birth. Then b is a constant.

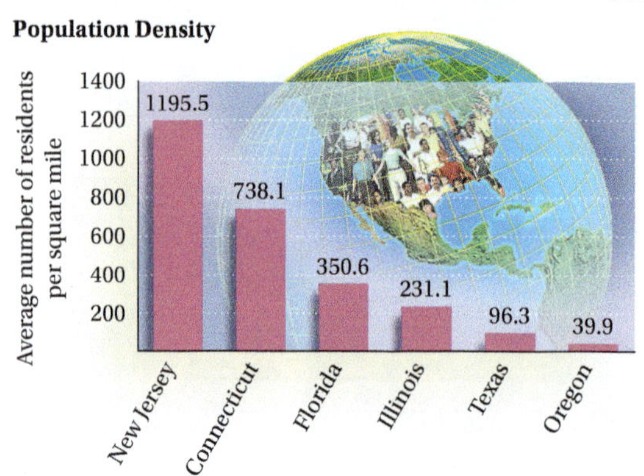

SOURCE: 2010 U.S. Census

Where do algebraic expressions occur? Most often we encounter them when we are solving applied problems. For example, consider the bar graph shown at left, one that we might find in a book or a magazine. Suppose we want to know how much greater the average population density per square mile is in New Jersey than in Illinois. Using arithmetic, we might simply subtract. But let's see how we can determine this using algebra. We translate the problem into a statement of equality, an equation. It could be done as follows:

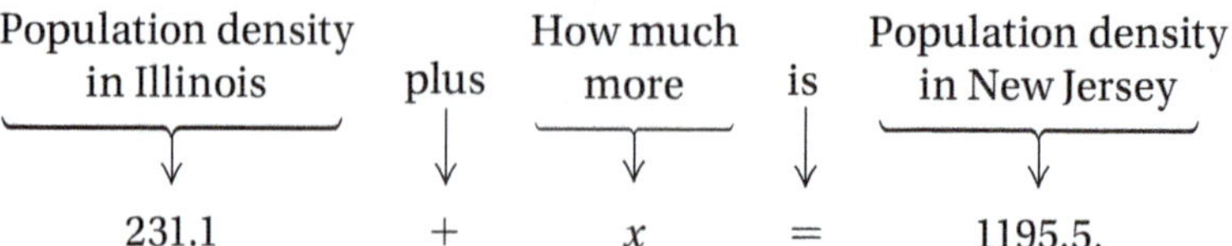

Population density in Illinois
plus
How much more
is
Population density in New Jersey
231.1
+
x
=
1195.5.

Note that we have an algebraic expression, $231.1 + x$, on the left of the equals sign. To find the number x, we can subtract 231.1 on both sides of the equation:

$$231.1 + x = 1195.5$$
$$231.1 + x - 231.1 = 1195.5 - 231.1$$
$$x = 964.4.$$

This value of x gives the answer, 964.4 residents per square mile.

Answers

Skill to Review:
1. 5 **2.** 26

We call $231.1 + x$ an *algebraic expression* and $231.1 + x = 1195.5$ an *algebraic equation*. Note that there is no equals sign, =, in an algebraic expression.

Do Margin Exercise 1. ▶

An **algebraic expression** consists of variables, constants, numerals, operation signs, and/or grouping symbols. When we replace a variable with a number, we say that we are **substituting** for the variable. When we replace all of the variables in an expression with numbers and carry out the operations in the expression, we are **evaluating the expression**.

EXAMPLE 1 Evaluate $x + y$ when $x = 37$ and $y = 29$.

We substitute 37 for x and 29 for y and carry out the addition:

$$x + y = 37 + 29 = 66.$$

The number 66 is called the **value** of the expression when $x = 37$ and $y = 29$. ■

Algebraic expressions involving multiplication can be written in several ways. For example, "8 times a" can be written as

$$8 \times a, \quad 8 \cdot a, \quad 8(a), \quad \text{or simply} \quad 8a.$$

Two letters written together without an operation symbol, such as ab, also indicate a multiplication.

EXAMPLE 2 Evaluate $3y$ when $y = 14$.

$$3y = 3(14) = 42$$

Do Exercises 2–4. ▶

EXAMPLE 3 *Area of a Rectangle.* The area A of a rectangle of length l and width w is given by the formula $A = lw$. Find the area when l is 24.5 in. and w is 16 in.

We substitute 24.5 in. for l and 16 in. for w and carry out the multiplication:

$$\begin{aligned} A = lw &= (24.5 \text{ in.})(16 \text{ in.}) \\ &= (24.5)(16)(\text{in.})(\text{in.}) \\ &= 392 \text{ in}^2, \text{ or } 392 \text{ square inches.} \end{aligned}$$

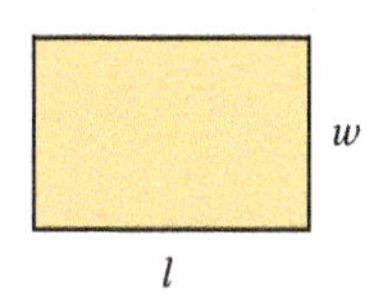

Do Exercise 5. ▶

Algebraic expressions involving division can also be written in several ways. For example, "8 divided by t" can be written as

$$8 \div t, \quad \frac{8}{t}, \quad 8/t, \quad \text{or} \quad 8 \cdot \frac{1}{t},$$

where the fraction bar is a division symbol.

EXAMPLE 4 Evaluate $\frac{a}{b}$ when $a = 63$ and $b = 9$.

We substitute 63 for a and 9 for b and carry out the division:

$$\frac{a}{b} = \frac{63}{9} = 7.$$ ■

1. Translate this problem to an equation. Then solve the equation.

Population Density. The average number of residents per square mile in six U.S. states is shown in the bar graph on the preceding page. How much greater is the population density in Connecticut than in Oregon?

2. Evaluate $a + b$ when $a = 38$ and $b = 26$.

3. Evaluate $x - y$ when $x = 57$ and $y = 29$.

4. Evaluate $4t$ when $t = 15$.

GS **5.** Find the area of a rectangle when l is 24 ft and w is 8 ft.

$A = lw$

$A = (24 \text{ ft})(\quad)$

$= (24)(\quad)(\text{ft})(\text{ft})$

$= 192 \quad$, or

192 square feet

Answers

1. $39.9 + x = 738.1$; 698.2 residents per square mile **2.** 64 **3.** 28 **4.** 60 **5.** 192 ft^2

Guided Solution:

5. 8 ft, 8, ft^2

6. Evaluate a/b when $a = 200$ and $b = 8$.

7. Evaluate $10p/q$ when $p = 40$ and $q = 25$.

8. *Commuting via Bicycle.* Find the time it takes to bike 22 mi if the speed is 16 mph.

EXAMPLE 5 Evaluate $\frac{12m}{n}$ when $m = 8$ and $n = 16$.

$$\frac{12m}{n} = \frac{12 \cdot 8}{16} = \frac{96}{16} = 6$$

◀ **Do Exercises 6 and 7.**

EXAMPLE 6 *Commuting Via Bicycle.* Commuting to work via bicycle has increased in popularity with the emerging concept of sharing bicycles. Bikes are picked up and returned at docking stations. The payment is approximately $1.50 per 30 min. Richard bicycles 18 mi to work. The time t, in hours, that it takes to bike 18 mi is given by

$$t = \frac{18}{r},$$

where r is the speed. Find the time for Richard to commute to work if his speed is 15 mph.

We substitute 15 for r and carry out the division:

$$t = \frac{18}{r} = \frac{18}{15} = 1.2 \text{ hr.}$$

◀ **Do Exercise 8.**

b TRANSLATING TO ALGEBRAIC EXPRESSIONS

We translate problems to equations. The different parts of an equation are translations of word phrases to algebraic expressions. It is easier to translate if we know that certain words often translate to certain operation symbols.

Key Words, Phrases, and Concepts

ADDITION (+)	SUBTRACTION (−)	MULTIPLICATION (·)	DIVISION (÷)
add	subtract	multiply	divide
added to	subtracted from	multiplied by	divided by
sum	difference	product	quotient
total	minus	times	
plus	less than	of	
more than	decreased by		
increased by	take away		

EXAMPLE 7 Translate to an algebraic expression:

Twice (or two times) some number.

Think of some number, say, 8. We can write 2 times 8 as 2×8, or $2 \cdot 8$. We multiplied by 2. Do the same thing using a variable. We can use any variable we wish, such as x, y, m, or n. Let's use y to represent some number. If we multiply by 2, we get an expression

$$y \times 2, \quad 2 \times y, \quad 2 \cdot y, \quad \text{or} \quad 2y.$$

Answers

6. 25 7. 16 8. 1.375 hr

EXAMPLE 8 Translate to an algebraic expression:

Thirty-eight percent of some number.

Let $n =$ the number. The word "of" translates to a multiplication symbol, so we could write any of the following expressions as a translation:

$38\% \cdot n, \quad 0.38 \times n, \quad \text{or} \quad 0.38n.$

EXAMPLE 9 Translate to an algebraic expression:

Seven less than some number.

We let x represent the number. If the number were 10, then 7 less than 10 is $10 - 7$, or 3. If we knew the number to be 34, then 7 less than the number would be $34 - 7$. Thus if the number is x, then the translation is

$x - 7.$

Caution!

Note that $7 - x$ is *not* a correct translation of the expression in Example 9. The expression $7 - x$ is a translation of "seven minus some number" or "some number less than seven."

EXAMPLE 10 Translate to an algebraic expression:

Eighteen more than a number.

We let $t =$ the number. If the number were 6, then the translation would be $6 + 18$, or $18 + 6$. If we knew the number to be 17, then the translation would be $17 + 18$, or $18 + 17$. Thus if the number is t, then the translation is

$t + 18, \quad \text{or} \quad 18 + t.$

EXAMPLE 11 Translate to an algebraic expression:

A number divided by 5.

We let $m =$ the number. If the number were 7, then the translation would be $7 \div 5$, or $7/5$, or $\frac{7}{5}$. If the number were 21, then the translation would be $21 \div 5$, or $21/5$, or $\frac{21}{5}$. If the number is m, then the translation is

$$m \div 5, \quad m/5, \quad \text{or} \quad \frac{m}{5}.$$

EXAMPLE 12 Translate each phrase to an algebraic expression.

PHRASE	ALGEBRAIC EXPRESSION
Five more than some number	$n + 5$, or $5 + n$
Half of a number	$\frac{1}{2}t$, $\frac{t}{2}$, or $t/2$
Five more than three times some number	$3p + 5$, or $5 + 3p$
The difference of two numbers	$x - y$
Six less than the product of two numbers	$mn - 6$
Seventy-six percent of some number	$76\%z$, or $0.76z$
Four less than twice some number	$2x - 4$

Do Exercises 9–17. ▶

Translate each phrase to an algebraic expression.

9. Eight less than some number

10. Eight more than some number

11. Four less than some number

12. One-third of some number

13. Six more than eight times some number

14. The difference of two numbers

15. Fifty-nine percent of some number

16. Two hundred less than the product of two numbers

17. The sum of two numbers

Answers

9. $x - 8$ **10.** $y + 8$, or $8 + y$ **11.** $m - 4$
12. $\frac{1}{3} \cdot p$, or $\frac{p}{3}$ **13.** $8x + 6$, or $6 + 8x$
14. $a - b$ **15.** $59\%x$, or $0.59x$ **16.** $xy - 200$
17. $p + q$

9.1 Exercise Set

For Extra Help

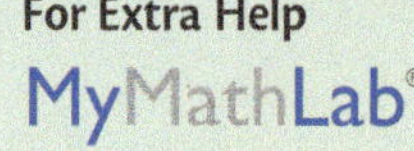

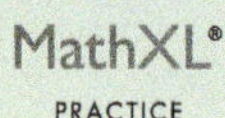

Reading Check

Classify each expression as an algebraic expression involving either multiplication or division.

RC1. $3/q$ **RC2.** $3q$ **RC3.** $3 \cdot q$ **RC4.** $\dfrac{3}{q}$

a Substitute to find values of the expressions in each of the following applied problems.

1. *Commuting Time.* It takes Abigail 24 min less time to commute to work than it does Jayden. Suppose that the variable x stands for the time it takes Jayden to get to work. Then $x - 24$ stands for the time it takes Abigail to get to work. How long does it take Abigail to get to work if it takes Jayden 56 min? 93 min? 105 min?

2. *Enrollment Costs.* At Mountain View Community College, it costs \$600 to enroll in the 8 A.M. section of Elementary Algebra. Suppose that the variable n stands for the number of students who enroll. Then $600n$ stands for the total amount of tuition collected for this course. How much is collected if 34 students enroll? 78 students? 250 students?

3. *Distance Traveled.* A driver who drives at a constant speed of r miles per hour for t hours will travel a distance of d miles given by $d = rt$ miles. How far will a driver travel at a speed of 65 mph for 4 hr?

4. *Simple Interest.* The simple interest I on a principal of P dollars at interest rate r for time t, in years, is given by $I = Prt$. Find the simple interest on a principal of \$4800 at 3% for 2 years.

5. *Wireless Internet Sign.* The U.S. Department of Transportation has designed a new sign that indicates the availability of wireless internet. The square sign measures 24 in. on each side. Find its area.

Source: *Manual on Uniform Traffic Control Devices*, U.S. Department of Transportation, Federal Highway Administration

6. *Yield Sign.* The U.S. Department of Transportation has designed a new yield sign. Each side of the triangular sign measures 30 in., and the height of the triangle is 26 in. Find its area. The area of a triangle with base b and height h is given by $A = \frac{1}{2}bh$.

Source: *Manual on Uniform Traffic Control Devices*, U.S. Department of Transportation, Federal Highway Administration

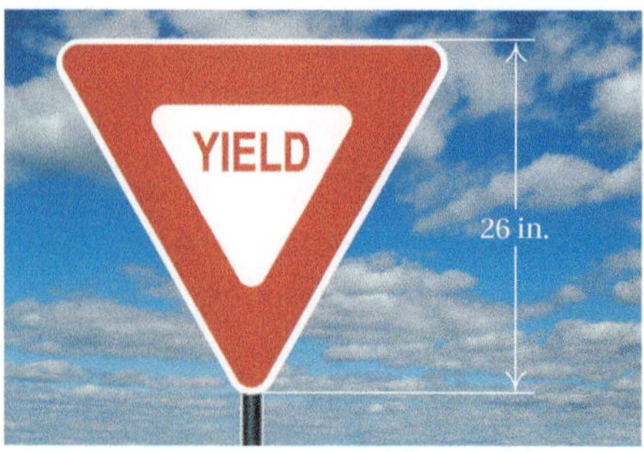

7. ***Area of a Triangle.*** The area A of a triangle with base b and height h is given by $A = \frac{1}{2}bh$. Find the area when $b = 45$ m (meters) and $h = 86$ m.

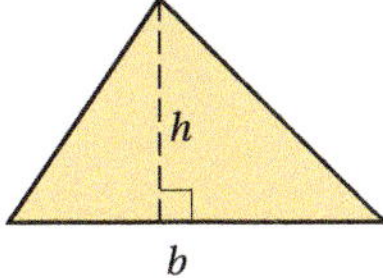

8. ***Area of a Parallelogram.*** The area A of a parallelogram with base b and height h is given by $A = bh$. Find the area of the parallelogram when the height is 15.4 cm (centimeters) and the base is 6.5 cm.

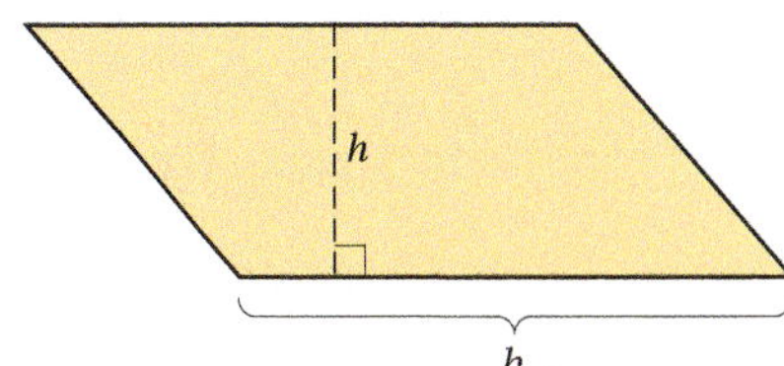

Evaluate.

9. $8x$, when $x = 7$

10. $6y$, when $y = 7$

11. $\frac{c}{d}$, when $c = 24$ and $d = 3$

12. $\frac{p}{q}$, when $p = 16$ and $q = 2$

13. $\frac{3p}{q}$, when $p = 2$ and $q = 6$

14. $\frac{5y}{z}$, when $y = 15$ and $z = 25$

15. $\frac{x + y}{5}$, when $x = 10$ and $y = 20$

16. $\frac{p + q}{2}$, when $p = 2$ and $q = 16$

17. $\frac{x - y}{8}$, when $x = 20$ and $y = 4$

18. $\frac{m - n}{5}$, when $m = 16$ and $n = 6$

b Translate each phrase to an algebraic expression. Use any letter for the variable(s) unless directed otherwise.

19. Seven more than some number

20. Some number increased by thirteen

21. Twelve less than some number

22. Fourteen less than some number

23. b more than a

24. c more than d

25. x divided by y

26. c divided by h

27. x plus w

28. s added to t

29. m subtracted from n

30. p subtracted from q

31. Twice some number

32. Three times some number

33. Three multiplied by some number

34. The product of eight and some number

35. Six more than four times some number

36. Two more than six times some number

37. Eight less than the product of two numbers

38. The product of two numbers minus seven

39. Five less than twice some number

40. Six less than seven times some number

41. Three times some number plus eleven

42. Some number times 8 plus 5

43. The sum of four times a number plus three times another number

44. Five times a number minus eight times another number

45. Your salary after a 5% salary increase if your salary before the increase was s

46. The price of a chain saw after a 30% reduction if the price before the reduction was P

47. Aubrey drove at a speed of 65 mph for t hours. How far did she travel? (See Exercise 3.)

48. Liam drove his pickup truck at 55 mph for t hours. How far did he travel? (See Exercise 3.)

49. Lisa had $50 before spending x dollars on pizza. How much money remains?

50. Juan has d dollars before spending $820 on four new tires for his truck. How much did Juan have after the purchase?

51. Sid's part-time job pays $12.50 per hour. How much does he earn for working n hours?

52. Meredith pays her babysitter $10 per hour. What does it cost her to hire the sitter for m hours?

Skill Maintenance

Find the prime factorization. [3.2c]

53. 108

54. 192

Add. [4.2b]

55. $\frac{3}{8} + \frac{5}{14}$

56. $\frac{11}{27} + \frac{1}{6}$

Multiply. [5.3a]

57. 0.05×1.03

58. 43.5×1000

Find the LCM. [4.1a]

59. 16, 24, 32

60. 18, 36, 44

Synthesis

Evaluate.

61. $\frac{a - 2b + c}{4b - a}$, when $a = 20$, $b = 10$, and $c = 5$

62. $\frac{x}{y} - \frac{5}{x} + \frac{2}{y}$, when $x = 30$ and $y = 6$

63. $\frac{12 - c}{c + 12b}$, when $b = 1$ and $c = 12$

64. $\frac{2w - 3z}{7y}$, when $w = 5$, $y = 6$, and $z = 1$

The Real Numbers

9.2

A **set** is a collection of objects. For our purposes, we will most often be considering sets of numbers. One way to name a set uses what is called **roster notation**. For example, roster notation for the set containing the numbers 0, 2, and 5 is $\{0, 2, 5\}$.

Sets that are part of other sets are called **subsets**. In this section, we become acquainted with the set of *real numbers* and its various subsets.

Two important subsets of the real numbers are listed below using roster notation.

NATURAL NUMBERS

The set of **natural numbers** $= \{1, 2, 3, \ldots\}$. These are the numbers used for counting.

WHOLE NUMBERS

The set of **whole numbers** $= \{0, 1, 2, 3, \ldots\}$. This is the set of natural numbers and 0.

We can represent these sets on the number line. The natural numbers are to the right of zero. The whole numbers are the natural numbers and zero.

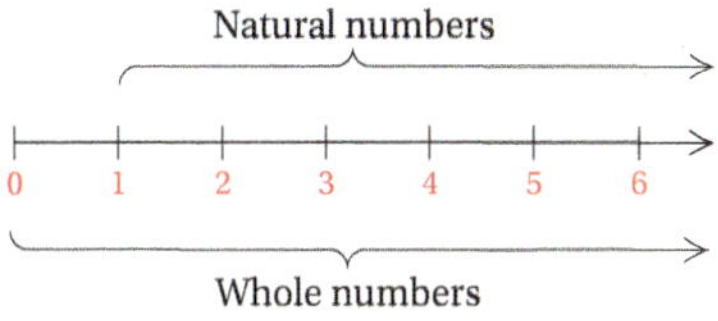

We create a new set, called the *integers*, by starting with the whole numbers, 0, 1, 2, 3, and so on. For each natural number 1, 2, 3, and so on, we obtain a new number to the left of zero on the number line:

For the number 1, there will be an *opposite* number -1 (negative 1).

For the number 2, there will be an *opposite* number -2 (negative 2).

For the number 3, there will be an *opposite* number -3 (negative 3), and so on.

The **integers** consist of the whole numbers and these new numbers.

INTEGERS

The set of **integers** $= \{\ldots, -5, -4, -3, -2, -1, 0, 1, 2, 3, 4, 5, \ldots\}$.

OBJECTIVES

a State the integer that corresponds to a real-world situation.

b Graph rational numbers on the number line.

c Convert from fraction notation for a rational number to decimal notation.

d Determine which of two real numbers is greater and indicate which, using $<$ or $>$. Given an inequality like $a > b$, write another inequality with the same meaning. Determine whether an inequality like $-3 \leq 5$ is true or false.

e Find the absolute value of a real number.

SKILL TO REVIEW

Objective 5.5a: Use division to convert fraction notation to decimal notation.

Convert to decimal notation.

1. $\frac{5}{8}$ **2.** $\frac{7}{9}$

Answers

Skill to Review:
1. 0.625 **2.** $0.\overline{77}$

We picture the integers on the number line as follows.

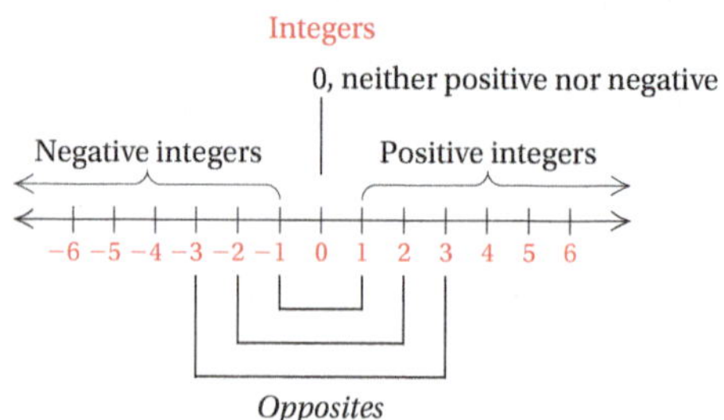

We call the integers to the left of zero **negative integers**. The natural numbers are also called **positive integers**. Zero is neither positive nor negative. We call -1 and 1 **opposites** of each other. Similarly, -2 and 2 are opposites, -3 and 3 are opposites, -100 and 100 are opposites, and 0 is its own opposite. Pairs of opposite numbers like -3 and 3 are the same distance from zero. The integers extend infinitely on the number line to the left and right of zero.

a INTEGERS AND THE REAL WORLD

Integers correspond to many real-world problems and situations. The following examples will help you get ready to translate problem situations that involve integers to mathematical language.

EXAMPLE 1 Tell which integer corresponds to this situation: The temperature is 4 degrees below zero.

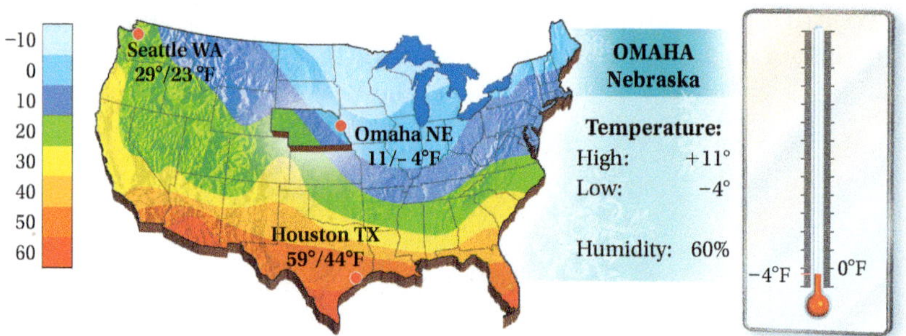

The integer -4 corresponds to the situation. The temperature is $-4°$.

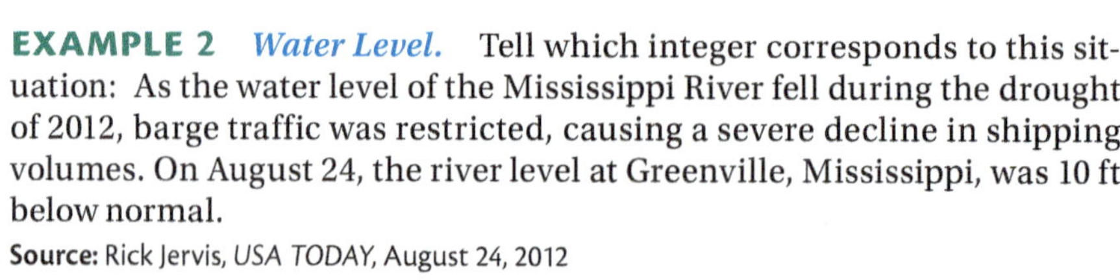

EXAMPLE 2 *Water Level.* Tell which integer corresponds to this situation: As the water level of the Mississippi River fell during the drought of 2012, barge traffic was restricted, causing a severe decline in shipping volumes. On August 24, the river level at Greenville, Mississippi, was 10 ft below normal.

Source: Rick Jervis, *USA TODAY*, August 24, 2012

The integer -10 corresponds to the drop in water level.

EXAMPLE 3 ***Stock Price Change.*** Tell which integers correspond to this situation: Hal owns a stock whose price decreased \$16 per share over a recent period. He owns another stock whose price increased \$2 per share over the same period.

The integer -16 corresponds to the decrease in the value of the first stock. The integer 2 represents the increase in the value of the second stock.

Do Exercises 1–5. ▶

b THE RATIONAL NUMBERS

We created the set of integers by obtaining a negative number for each natural number and also including 0. To create a larger number system, called the set of **rational numbers**, we consider quotients of integers with nonzero divisors. The following are some examples of rational numbers:

$$\frac{2}{3}, \quad -\frac{2}{3}, \quad \frac{7}{1}, \quad 4, \quad -3, \quad 0, \quad \frac{23}{-8}, \quad 2.4, \quad -0.17, \quad 10\frac{1}{2}.$$

The number $-\frac{2}{3}$ (read "negative two-thirds") can also be named $\frac{-2}{3}$ or $\frac{2}{-3}$; that is,

$$-\frac{a}{b} = \frac{-a}{b} = \frac{a}{-b}.$$

The number 2.4 can be named $\frac{24}{10}$ or $\frac{12}{5}$, and -0.17 can be named $-\frac{17}{100}$. We can describe the set of rational numbers as follows.

RATIONAL NUMBERS

The set of **rational numbers** = the set of numbers $\frac{a}{b}$, where a and b are integers and b is not equal to 0 ($b \neq 0$).

Note that this new set of numbers, the rational numbers, contains the whole numbers, the integers, the arithmetic numbers (also called the nonnegative rational numbers), and the negative rational numbers.

We picture the rational numbers on the number line as follows.

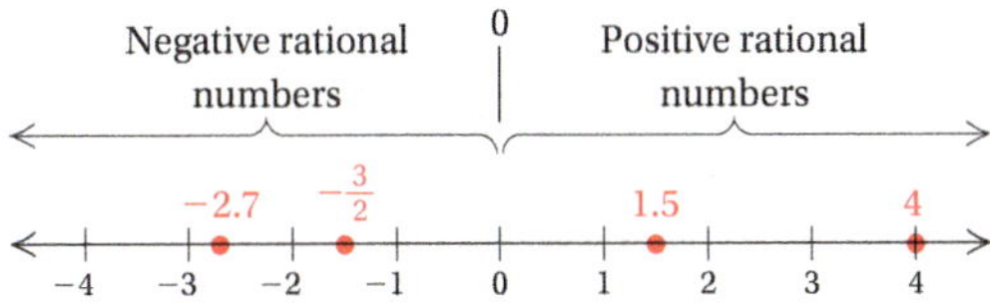

To **graph** a number means to find and mark its point on the number line. Some rational numbers are graphed in the preceding figure.

Tell which integers correspond to each situation.

1. ***Temperature High and Low.*** The highest recorded temperature in Illinois is 117°F on July 14, 1954, in East St. Louis. The lowest recorded temperature in Illinois is 36°F below zero on January 5, 1999, in Congerville.

Source: National Climate Data Center, NESDIS, NOAA, U.S. Department of Commerce (through 2010)

2. ***Stock Decrease.*** The price of a stock decreased \$3 per share over a recent period.

3. At 10 sec before liftoff, ignition occurs. At 148 sec after liftoff, the first stage is detached from the rocket.

4. The halfback gained 8 yd on first down. The quarterback was sacked for a 5-yd loss on second down.

5. A submarine dove 120 ft, rose 50 ft, and then dove 80 ft.

Answers

1. 117; −36 2. −3 3. −10; 148
4. 8; −5 5. −120; 50; −80

Graph each number on the number line.

6. $-\frac{7}{2}$

7. 1.4

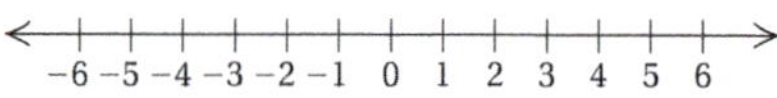

8. $-\frac{11}{4}$

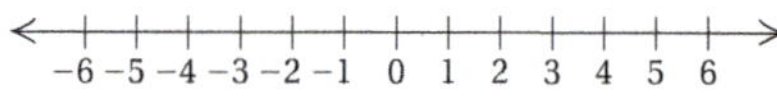

EXAMPLES Graph each number on the number line.

4. -3.2 The graph of -3.2 is $\frac{2}{10}$ of the way from -3 to -4.

5. $\frac{13}{8}$ The number $\frac{13}{8}$ can also be named $1\frac{5}{8}$, or 1.625. The graph is $\frac{5}{8}$ of the way from 1 to 2.

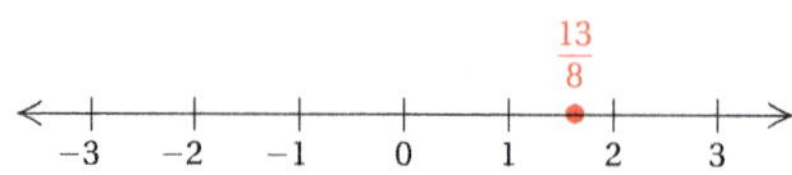

◀ Do Exercises 6–8.

C NOTATION FOR RATIONAL NUMBERS

Each rational number can be named using fraction notation or decimal notation.

EXAMPLE 6 Convert to decimal notation: $-\frac{5}{8}$.

We first find decimal notation for $\frac{5}{8}$. Since $\frac{5}{8}$ means $5 \div 8$, we divide.

$$
\begin{array}{r}
0.625 \\
8\,\overline{)\,5.000} \\
\underline{4\,8} \\
20 \\
\underline{16} \\
40 \\
\underline{40} \\
0
\end{array}
$$

Thus, $\frac{5}{8} = 0.625$, so $-\frac{5}{8} = -0.625$.

Decimal notation for $-\frac{5}{8}$ is -0.625. We consider -0.625 to be a **terminating decimal**. Decimal notation for some numbers repeats.

EXAMPLE 7 Convert to decimal notation: $\frac{7}{11}$.

$$
\begin{array}{r}
0.6363\ldots \\
11\,\overline{)\,7.0000} \\
\underline{6\,6} \\
40 \\
\underline{33} \\
70 \\
\underline{66} \\
40 \\
\underline{33} \\
7
\end{array}
$$

Dividing

We can abbreviate **repeating decimal** notation by writing a bar over the repeating part—in this case, we write $0.\overline{63}$. Thus, $\frac{7}{11} = 0.\overline{63}$.

Answers

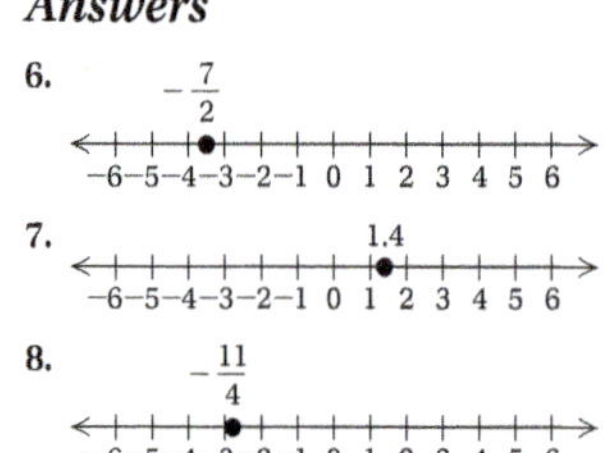

> Each rational number can be expressed in either terminating decimal notation or repeating decimal notation.

The following are other examples showing how rational numbers can be named using fraction notation or decimal notation:

$$0 = \frac{0}{8}, \quad \frac{27}{100} = 0.27, \quad -8\frac{3}{4} = -8.75, \quad -\frac{13}{6} = -2.1\overline{6}.$$

Do Exercises 9–11. ▶

Find decimal notation.

9. $-\frac{3}{8}$

10. $-\frac{6}{11}$

11. $\frac{4}{3}$

d THE REAL NUMBERS AND ORDER

Every rational number has a point on the number line. However, there are some points on the line for which there is no rational number. These points correspond to what are called **irrational numbers**.

What kinds of numbers are irrational? One example is the number π, which is used in finding the area and the circumference of a circle: $A = \pi r^2$ and $C = 2\pi r$.

Another example of an irrational number is the square root of 2, named $\sqrt{2}$. It is the length of the diagonal of a square with sides of length 1. It is also the number that when multiplied by itself gives 2—that is, $\sqrt{2} \cdot \sqrt{2} = 2$. There is no rational number that can be multiplied by itself to get 2. But the following are rational *approximations*:

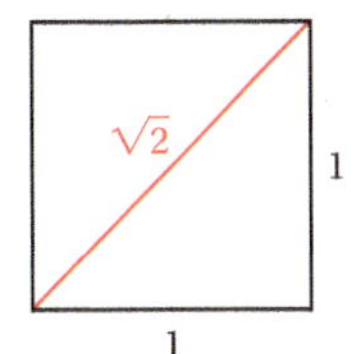

1.4 is an approximation of $\sqrt{2}$ because $(1.4)^2 = 1.96$;

1.41 is a better approximation because $(1.41)^2 = 1.9881$;

1.4142 is an even better approximation because $(1.4142)^2 = 1.99996164$.

We can find rational approximations for square roots using a calculator.

Approximating Square Roots and π Square roots are found by pressing 2ND $\sqrt{}$. ($\sqrt{}$ is the second operation associated with the x^2 key.)

To find an approximation for $\sqrt{48}$, we press 2ND $\sqrt{}$ 4 8 .

The number π is used widely enough to have its own key. (π is the second operation associated with the ^ key.) To approximate π, we press 2ND π ENTER.

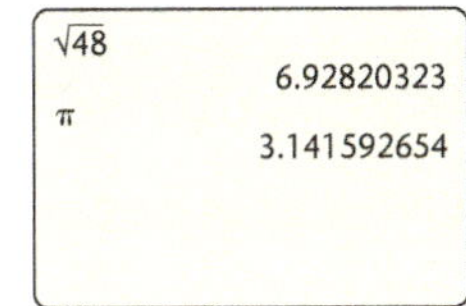

EXERCISES: Approximate.

1. $\sqrt{76}$

2. $\sqrt{317}$

3. $15 \cdot \sqrt{20}$

4. $29 + \sqrt{42}$

5. π

6. $29 \cdot \pi$

7. $\pi \cdot 13^2$

8. $5 \cdot \pi + 8 \cdot \sqrt{237}$

> Decimal notation for rational numbers *either* terminates *or* repeats.
> Decimal notation for irrational numbers *neither* terminates *nor* repeats.

Some other examples of irrational numbers are $\sqrt{3}$, $-\sqrt{8}$, $\sqrt{11}$, and 0.121221222122221. . . . Whenever we take the square root of a number that is not a perfect square, we will get an irrational number.

The rational numbers and the irrational numbers together correspond to all the points on the number line and make up what is called the **real-number system**.

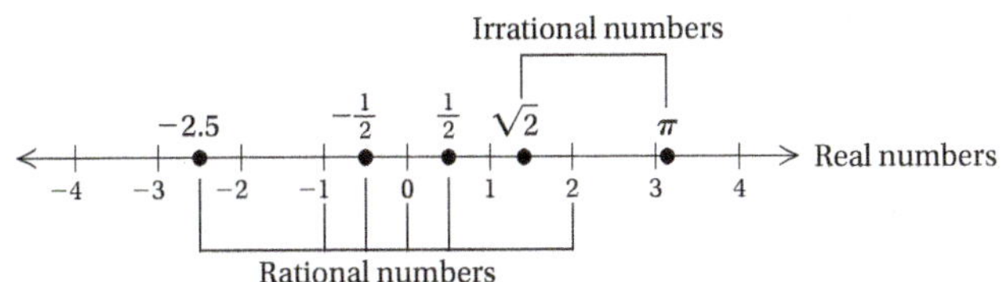

Answers

9. −0.375 **10.** $-0.\overline{54}$ **11.** $1.\overline{3}$

REAL NUMBERS

The set of **real numbers** = The set of all numbers corresponding to points on the number line.

The real numbers consist of the rational numbers and the irrational numbers. The following figure shows the relationships among various kinds of numbers.

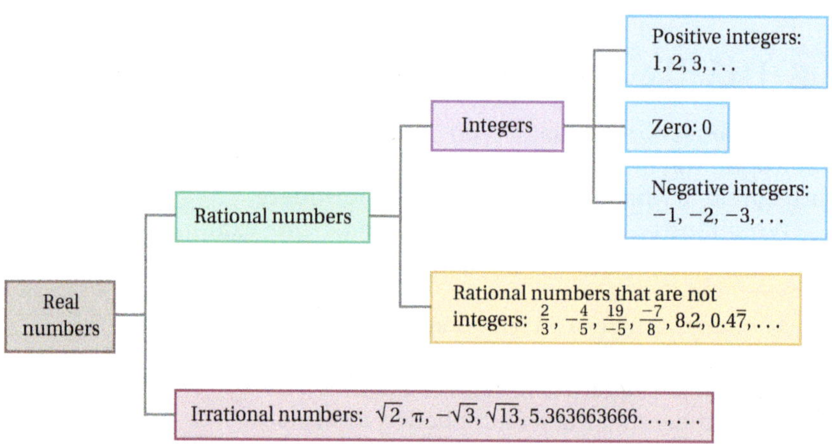

Order

Real numbers are named in order on the number line, increasing as we move from left to right. For any two numbers on the line, the one on the left is less than the one on the right.

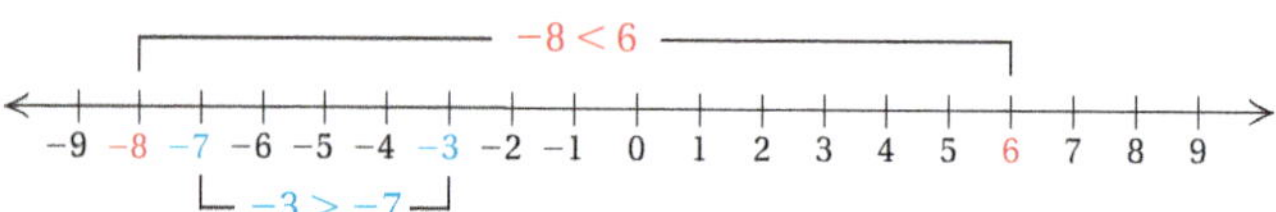

We use the symbol **<** to mean **"is less than."** The sentence $-8 < 6$ means "-8 is less than 6." The symbol **>** means **"is greater than."** The sentence $-3 > -7$ means "-3 is greater than -7." The sentences $-8 < 6$ and $-3 > -7$ are **inequalities**.

EXAMPLES Use either < or > for □ to write a true sentence.

8. 2 □ 9 — Since 2 is to the left of 9, 2 is less than 9, so $2 < 9$.

9. −7 □ 3 — Since -7 is to the left of 3, we have $-7 < 3$.

10. 6 □ −12 — Since 6 is to the right of -12, then $6 > -12$.

11. −18 □ −5 — Since -18 is to the left of -5, we have $-18 < -5$.

12. −2.7 □ $-\frac{3}{2}$ — The answer is $-2.7 < -\frac{3}{2}$.

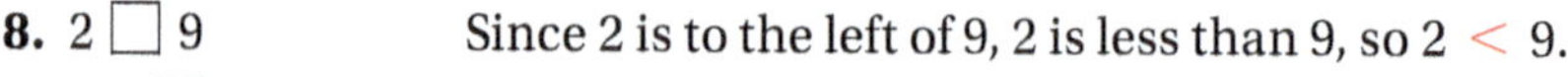

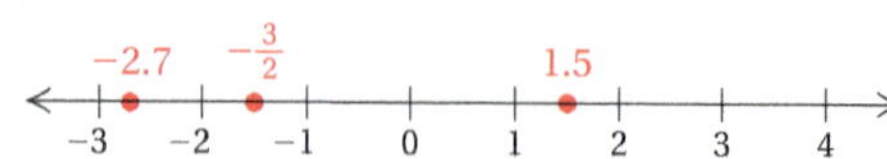

13. 1.5 □ −2.7 — The answer is $1.5 > -2.7$.

14. 1.38 □ 1.83 — The answer is $1.38 < 1.83$.

CALCULATOR CORNER

Negative Numbers on a Calculator; Converting to Decimal Notation We use the opposite key (−) to enter negative numbers on a graphing calculator. Note that this is different from the subtraction key, −.

To convert $-\frac{5}{8}$ to decimal notation, we press (−) 5 ÷ 8 ENTER. The result is −0.625.

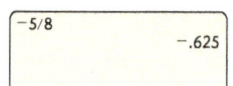

EXERCISES: Convert to decimal notation.

1. $-\frac{3}{4}$
2. $-\frac{9}{20}$
3. $-\frac{1}{8}$
4. $-\frac{9}{5}$
5. $-\frac{27}{40}$
6. $-\frac{11}{16}$
7. $-\frac{7}{2}$
8. $-\frac{19}{25}$

15. $-3.45 \square 1.32$ The answer is $-3.45 < 1.32$.

16. $-4 \square 0$ The answer is $-4 < 0$.

17. $5.8 \square 0$ The answer is $5.8 > 0$.

18. $\frac{5}{8} \square \frac{7}{11}$ We convert to decimal notation: $\frac{5}{8} = 0.625$ and $\frac{7}{11} = 0.6363\ldots$. Thus, $\frac{5}{8} < \frac{7}{11}$.

19. $-\frac{1}{2} \square -\frac{1}{3}$ The answer is $-\frac{1}{2} < -\frac{1}{3}$.

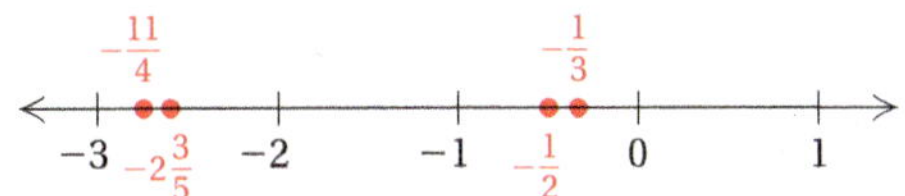

20. $-2\frac{3}{5} \square -\frac{11}{4}$ The answer is $-2\frac{3}{5} > -\frac{11}{4}$.

Do Exercises 12–19. ▶

Use either $<$ or $>$ for $\square$ to write a true sentence.

12. $-3 \square 7$

13. $-8 \square -5$

14. $7 \square -10$

15. $3.1 \square -9.5$

16. $-4.78 \square -5.01$

17. $-\frac{2}{3} \square -\frac{5}{9}$

18. $-\frac{11}{8} \square \frac{23}{15}$

19. $0 \square -9.9$

Note that both $-8 < 6$ and $6 > -8$ are true. Every true inequality yields another true inequality when we interchange the numbers or the variables and reverse the direction of the inequality sign.

ORDER; $>$, $<$

$a < b$ also has the meaning $b > a$.

EXAMPLES Write another inequality with the same meaning.

21. $-3 > -8$ The inequality $-8 < -3$ has the same meaning.

22. $a < -5$ The inequality $-5 > a$ has the same meaning.

A helpful mental device is to think of an inequality sign as an "arrow" with the arrowhead pointing to the smaller number.

Do Exercises 20 and 21. ▶

Write another inequality with the same meaning.

20. $-5 < 7$

21. $x > 4$

Note that all positive real numbers are greater than zero and all negative real numbers are less than zero.

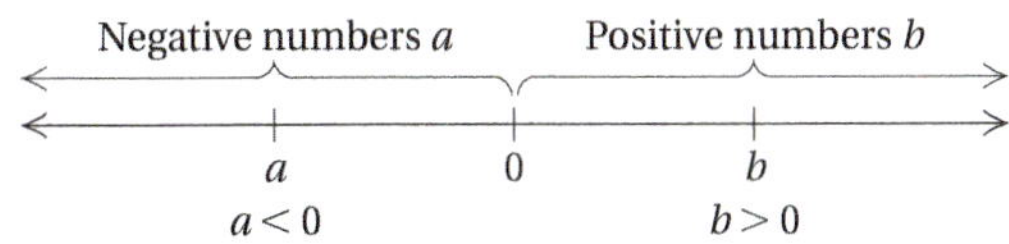

If b is a positive real number, then $b > 0$.
If a is a negative real number, then $a < 0$.

Answers

12. $<$ 13. $<$ 14. $>$ 15. $>$ 16. $>$
17. $<$ 18. $<$ 19. $>$ 20. $7 > -5$
21. $4 < x$

Expressions like $a \le b$ and $b \ge a$ are also inequalities. We read $\boldsymbol{a \le b}$ as **"a is less than or equal to b."** We read $\boldsymbol{a \ge b}$ as **"a is greater than or equal to b."**

EXAMPLES Write true or false for each statement.

23. $-3 \le 5.4$ True since $-3 < 5.4$ is true

24. $-3 \le -3$ True since $-3 = -3$ is true

25. $-5 \ge 1\frac{2}{3}$ False since neither $-5 > 1\frac{2}{3}$ nor $-5 = 1\frac{2}{3}$ is true

◀ **Do Exercises 22–24.**

Write true or false for each statement.

22. $-4 \le -6$

23. $7.8 \ge 7.8$

24. $-2 \le \frac{3}{8}$

CALCULATOR CORNER

Absolute Value Finding absolute value is the first item in the Catalog on the T1-84 Plus graphing calculator. To find $|-7|$, we first press 2ND CATALOG ENTER to copy "abs(" to the home screen. (CATALOG is the second operation associated with the 0 numeric key.) Then we press (−) 7) ENTER. The result is 7.

To find $|-\frac{1}{2}|$ and express the result as a fraction, we press 2ND CATALOG ENTER (−) 1 ÷ 2) MATH 1 ENTER. The result is $\frac{1}{2}$.

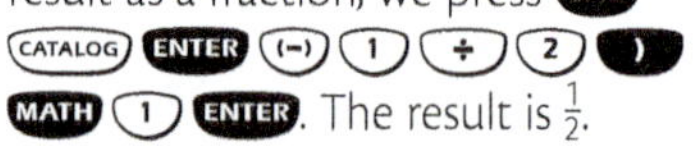

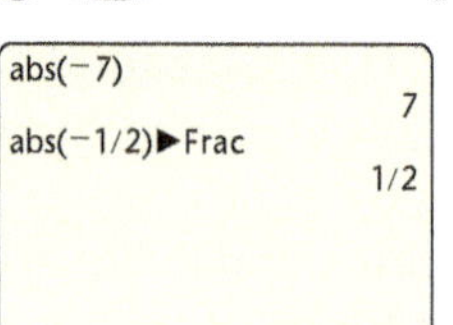
abs(−7)
7
abs(−1/2)▶Frac
1/2

EXERCISES: Find the absolute value.

1. $|-5|$ **2.** $|17|$

3. $|0|$ **4.** $|6.48|$

5. $|-12.7|$ **6.** $|-0.9|$

7. $\left|-\frac{5}{7}\right|$ **8.** $\left|\frac{4}{3}\right|$

e ABSOLUTE VALUE

From the number line, we see that numbers like 4 and -4 are the same distance from zero. Distance is always a nonnegative number. We call the distance of a number from zero on the number line the **absolute value** of the number.

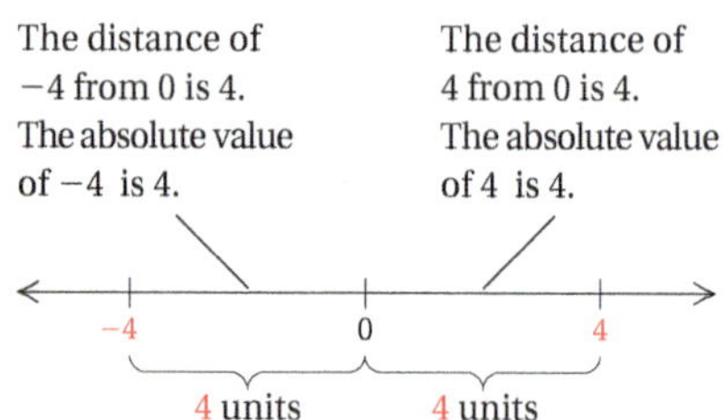
The distance of
−4 from 0 is 4.
The absolute value
of −4 is 4.
The distance of
4 from 0 is 4.
The absolute value
of 4 is 4.
−4
0
4
4 units
4 units

ABSOLUTE VALUE

The **absolute value** of a number is its distance from zero on the number line. We use the symbol $|x|$ to represent the absolute value of a number x.

FINDING ABSOLUTE VALUE

a) If a number is negative, its absolute value is its opposite.

b) If a number is positive or zero, its absolute value is the same as the number.

EXAMPLES Find the absolute value.

26. $|-7|$ The distance of -7 from 0 is 7, so $|-7| = 7$.

27. $|12|$ The distance of 12 from 0 is 12, so $|12| = 12$.

28. $|0|$ The distance of 0 from 0 is 0, so $|0| = 0$.

29. $\left|\frac{3}{2}\right| = \frac{3}{2}$

30. $|-2.73| = 2.73$

◀ **Do Exercises 25–28.**

Find the absolute value.

25. $|8|$ **26.** $|-9|$

27. $\left|-\frac{2}{3}\right|$ **28.** $|5.6|$

Answers

22. False **23.** True **24.** True **25.** 8 **26.** 9 **27.** $\frac{2}{3}$ **28.** 5.6

9.2 Exercise Set

For Extra Help
MyMathLab®
MathXL®

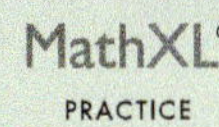

PRACTICE WATCH READ REVIEW

Reading Check

Use the number line below for Exercises RC1–RC10.

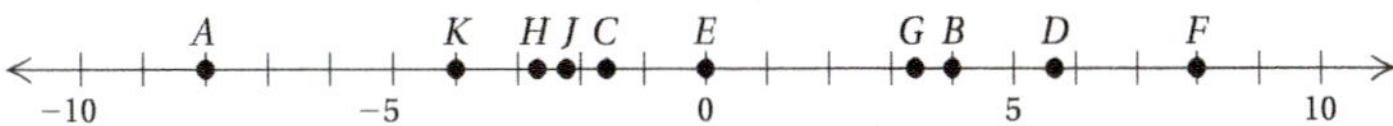

Match each number with its graph.

RC1. $-2\frac{5}{7}$ **RC2.** $\left|\frac{0}{-8}\right|$ **RC3.** -2.25 **RC4.** $\frac{17}{3}$ **RC5.** $|-4|$ **RC6.** $3.\overline{4}$

Write true or false. The letters name numbers on the number line shown above.

RC7. $K < B$ **RC8.** $H < B$ **RC9.** $E < C$ **RC10.** $J > D$

a State the integers that correspond to each situation.

1. On Wednesday, the temperature was 24° above zero. On Thursday, it was 2° below zero.

2. A student deposited her tax refund of $750 in a savings account. Two weeks later, she withdrew $125 to pay technology fees.

3. ***Temperature Extremes.*** The highest temperature ever created in a lab is 7,200,000,000,000°F. The lowest temperature ever created is approximately 460°F below zero.

Sources: *Live Science; Guinness Book of World Records*

4. ***Extreme Climate.*** Verkhoyansk, a river port in northeast Siberia, has the most extreme climate on the planet. Its average monthly winter temperature is 58.5°F below zero, and its average monthly summer temperature is 56.5°F.

Source: *Guinness Book of World Records*

5. ***Empire State Building.*** The Empire State Building has a total height, including the lightning rod at the top, of 1454 ft. The foundation depth is 55 ft below ground level.

Source: www.empirestatebuildingfacts.com

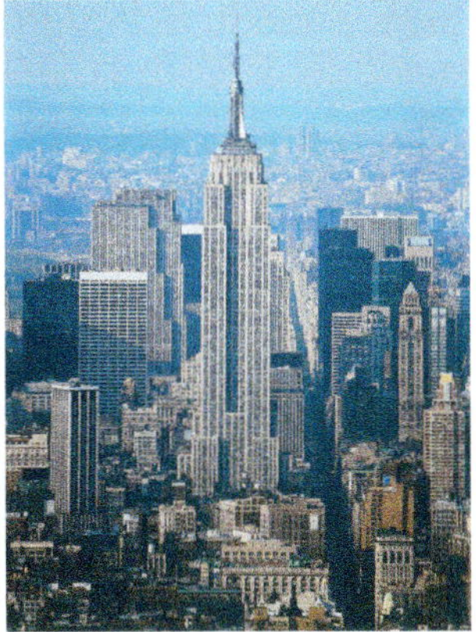

6. ***Shipwreck.*** There are numerous shipwrecks to explore near Bermuda. One of the most frequently visited sites is L'Herminie, a French warship that sank in 1838. This ship is 35 ft below the surface.

Source: www.10best.com/interest/adventures/scuba-diving-in-pirate-territory

b Graph the number on the number line.

7. $\frac{10}{3}$ 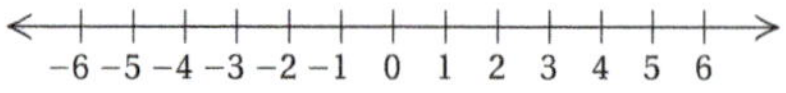

8. $-\frac{17}{4}$ 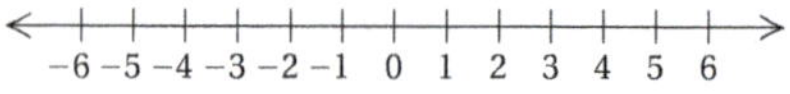

9. -5.2 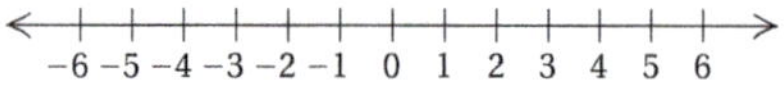

10. 4.78 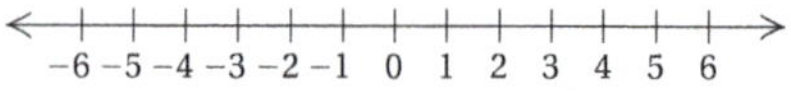

11. $-4\frac{2}{5}$ 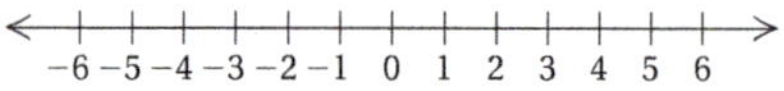

12. $2\frac{6}{11}$ 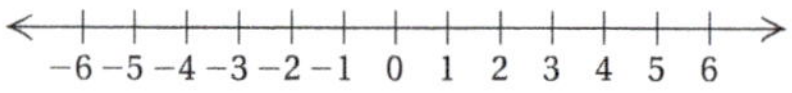

c Convert to decimal notation.

13. $-\frac{7}{8}$

14. $-\frac{3}{16}$

15. $\frac{5}{6}$

16. $\frac{5}{3}$

17. $-\frac{7}{6}$

18. $-\frac{5}{12}$

19. $\frac{2}{3}$

20. $-\frac{11}{9}$

21. $\frac{1}{10}$

22. $\frac{1}{4}$

23. $-\frac{1}{2}$

24. $\frac{9}{8}$

25. $\frac{4}{25}$

26. $-\frac{7}{20}$

d Use either $<$ or $>$ for $\square$ to write a true sentence.

27. $8 \square 0$

28. $3 \square 0$

29. $-8 \square 3$

30. $6 \square -6$

31. $-8 \square 8$

32. $0 \square -9$

33. $-8 \square -5$

34. $-4 \square -3$

35. $-5 \square -11$

36. $-3 \square -4$

37. $2.14 \square 1.24$

38. $-3.3 \square -2.2$

39. $-12.88 \square -6.45$

40. $17.2 \square -1.67$

41. $-\frac{1}{2} \square -\frac{2}{3}$

42. $-\frac{5}{4} \square -\frac{3}{4}$

43. $-\frac{2}{3} \square \frac{1}{3}$

44. $\frac{3}{4} \square -\frac{5}{4}$

45. $\frac{5}{12} \square \frac{11}{25}$

46. $-\frac{13}{16} \square -\frac{5}{9}$

Write an inequality with the same meaning.

47. $-6 > x$

48. $x < 8$

49. $-10 \leq y$

50. $12 \geq t$

Write true or false.

51. $-5 \leq -6$

52. $-7 \geq -10$

53. $4 \geq 4$

54. $7 \leq 7$

55. $-3 \geq -11$

56. $-1 \leq -5$

57. $0 \geq 8$

58. $-5 \leq 7$

e Find the absolute value.

59. $|-3|$

60. $|-6|$

61. $|11|$

62. $|0|$

63. $\left|-\frac{2}{3}\right|$

64. $|325|$

65. $\left|\frac{0}{4}\right|$

66. $|14.8|$

67. $|-2.65|$

68. $\left|-3\frac{5}{8}\right|$

Skill Maintenance

Convert to decimal notation. [6.2b]

69. 110%

70. $23\frac{4}{5}\%$

Convert to percent notation. [6.3a]

71. $\frac{13}{25}$

72. $\frac{19}{32}$

Evaluate. [1.9b]

73. 3^4

74. 5^0

Simplify. [1.9c]

75. $3(7 + 2^3)$

76. $48 \div 8 - 6$

Synthesis

List in order from the least to the greatest.

77. $\frac{2}{3}, -\frac{1}{7}, \frac{1}{3}, -\frac{2}{7}, -\frac{2}{3}, \frac{2}{5}, -\frac{1}{3}, -\frac{2}{5}, \frac{9}{8}$

78. $-8\frac{7}{8}, 7^1, -5, |-6|, 4, |3|, -8\frac{5}{8}, -100, 0, 1^7, \frac{7}{2}, -\frac{67}{8}$

Given that $0.\overline{3} = \frac{1}{3}$ and $0.\overline{6} = \frac{2}{3}$, express each of the following as a quotient or a ratio of two integers.

79. $0.\overline{9}$

80. $0.\overline{1}$

81. $5.\overline{5}$

9.3 Addition of Real Numbers

OBJECTIVES

a Add real numbers without using the number line.

b Find the opposite, or additive inverse, of a real number.

c Solve applied problems involving addition of real numbers.

SKILL TO REVIEW

Objective 4.2b: Add using fraction notation when denominators are different.

Add.

1. $\frac{1}{6} + \frac{2}{9}$ **2.** $\frac{5}{8} + \frac{7}{30}$

In this section, we consider addition of real numbers. First, to gain an understanding, we add using the number line. Then we consider rules for addition.

ADDITION ON THE NUMBER LINE

To do the addition $a + b$ on the number line, start at 0, move to a, and then move according to b.

a) If b is positive, move from a to the right.

b) If b is negative, move from a to the left.

c) If b is 0, stay at a.

EXAMPLE 1 Add: $3 + (-5)$.

We start at 0 and move to 3. Then we move 5 units left since -5 is negative.

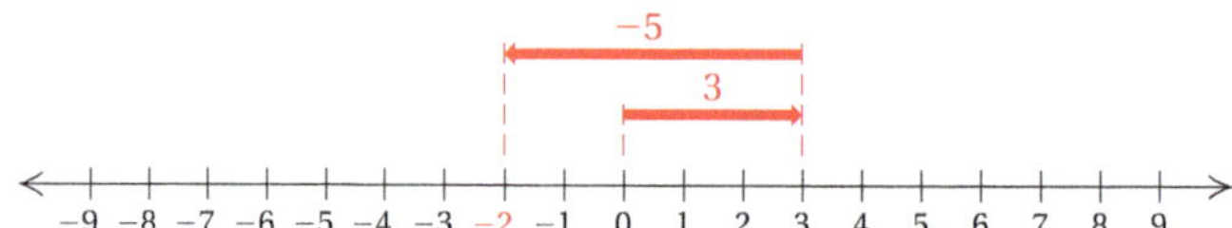

$3 + (-5) = -2$

EXAMPLE 2 Add: $-4 + (-3)$.

We start at 0 and move to -4. Then we move 3 units left since -3 is negative.

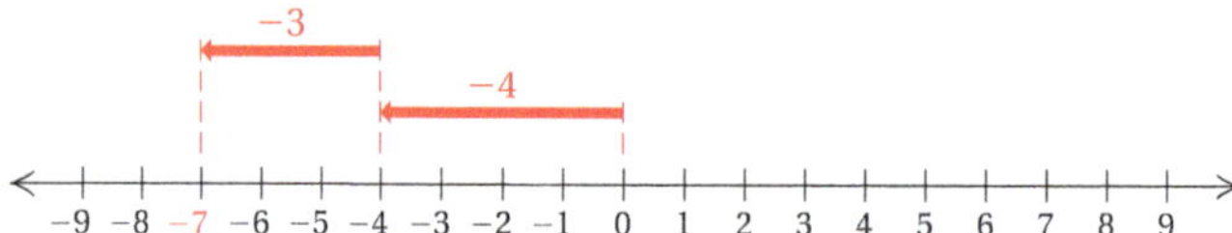

$-4 + (-3) = -7$

EXAMPLE 3 Add: $-4 + 9$.

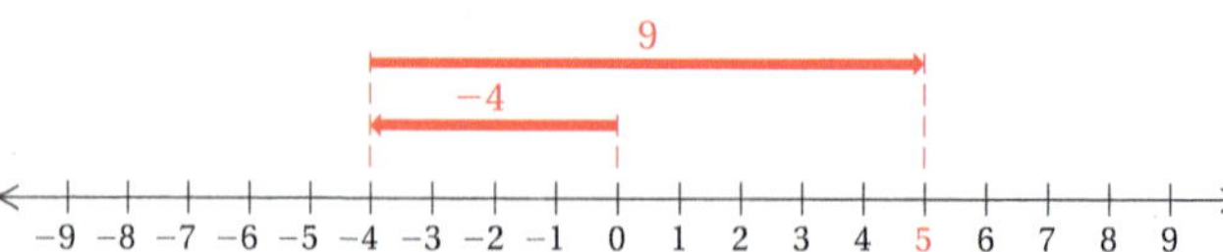

$-4 + 9 = 5$

Answers

Skill to Review:

1. $\frac{7}{18}$ **2.** $\frac{103}{120}$

EXAMPLE 4 Add: $-5.2 + 0$.

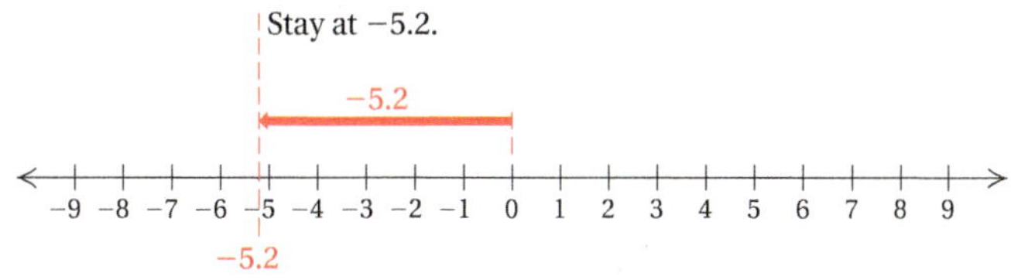

$-5.2 + 0 = -5.2$

Do Exercises 1–6. ▶

Add using the number line.

1. $0 + (-3)$

−6 −5 −4 −3 −2 −1 0 1 2 3 4 5 6

2. $1 + (-4)$

−6 −5 −4 −3 −2 −1 0 1 2 3 4 5 6

3. $-3 + (-2)$

−6 −5 −4 −3 −2 −1 0 1 2 3 4 5 6

4. $-3 + 7$

−6 −5 −4 −3 −2 −1 0 1 2 3 4 5 6

5. $-2.4 + 2.4$

−6 −5 −4 −3 −2 −1 0 1 2 3 4 5 6

6. $-\frac{5}{2} + \frac{1}{2}$

−6 −5 −4 −3 −2 −1 0 1 2 3 4 5 6

a ADDING WITHOUT THE NUMBER LINE

You may have noticed some patterns in the preceding examples. These lead us to rules for adding without using the number line that are more efficient for adding larger numbers.

RULES FOR ADDITION OF REAL NUMBERS

1. ***Positive numbers:*** Add the same as arithmetic numbers. The answer is positive.
2. ***Negative numbers:*** Add absolute values. The answer is negative.
3. ***A positive number and a negative number:***
   - If the numbers have the same absolute value, the answer is 0.
   - If the numbers have different absolute values, subtract the smaller absolute value from the larger. Then:
     - **a)** If the positive number has the greater absolute value, the answer is positive.
     - **b)** If the negative number has the greater absolute value, the answer is negative.
4. ***One number is zero:*** The sum is the other number.

Rule 4 is known as the **identity property of 0.** It says that for any real number a, $a + 0 = a$.

EXAMPLES Add without using the number line.

5. $-12 + (-7) = -19$ Two negatives. Add the absolute values: $|-12| + |-7| = 12 + 7 = 19$. Make the answer *negative*: -19.

6. $-1.4 + 8.5 = 7.1$ One negative, one positive. Find the absolute values: $|-1.4| = 1.4$; $|8.5| = 8.5$. Subtract the smaller absolute value from the larger: $8.5 - 1.4 = 7.1$. The *positive* number, 8.5, has the larger absolute value, so the answer is *positive*: 7.1.

7. $-36 + 21 = -15$ One negative, one positive. Find the absolute values: $|-36| = 36$; $|21| = 21$. Subtract the smaller absolute value from the larger: $36 - 21 = 15$. The *negative* number, -36, has the larger absolute value, so the answer is *negative*: -15.

Answers

1. −3 **2.** −3 **3.** −5
4. 4 **5.** 0 **6.** −2

Add without using the number line.

7. $-5 + (-6)$
8. $-9 + (-3)$
9. $-4 + 6$
10. $-7 + 3$
11. $5 + (-7)$
12. $-20 + 20$
13. $-11 + (-11)$
14. $10 + (-7)$
15. $-0.17 + 0.7$
16. $-6.4 + 8.7$
17. $-4.5 + (-3.2)$
18. $-8.6 + 2.4$
19. $\frac{5}{9} + \left(-\frac{7}{9}\right)$

GS

20. $-\frac{1}{5} + \left(-\frac{3}{4}\right)$
$= -\frac{4}{20} + \left(-\frac{\square}{20}\right)$
$= -\frac{19}{\square}$

Add.

21. $(-15) + (-37) + 25 + 42 + (-59) + (-14)$
22. $42 + (-81) + (-28) + 24 + 18 + (-31)$
23. $-2.5 + (-10) + 6 + (-7.5)$
24. $-35 + 17 + 14 + (-27) + 31 + (-12)$

Answers

7. -11 8. -12 9. 2 10. -4
11. -2 12. 0 13. -22 14. 3
15. 0.53 16. 2.3 17. -7.7 18. -6.2
19. $-\frac{2}{9}$ 20. $-\frac{19}{20}$ 21. -58 22. -56
23. -14 24. -12

Guided Solution:
20. 15, 20

8. $1.5 + (-1.5) = 0$ — The numbers have the same absolute value. The sum is 0.

9. $-\frac{7}{8} + 0 = -\frac{7}{8}$ — One number is zero. The sum is $-\frac{7}{8}$.

10. $-9.2 + 3.1 = -6.1$

11. $-\frac{3}{2} + \frac{9}{2} = \frac{6}{2} = 3$

12. $-\frac{2}{3} + \frac{5}{8} = -\frac{16}{24} + \frac{15}{24} = -\frac{1}{24}$

Do Exercises 7–20.

Suppose we want to add several numbers, some positive and some negative, as follows. How can we proceed?

$$15 + (-2) + 7 + 14 + (-5) + (-12)$$

We can change grouping and order as we please when adding. For instance, we can group the positive numbers together and the negative numbers together and add them separately. Then we add the two results.

EXAMPLE 13 Add: $15 + (-2) + 7 + 14 + (-5) + (-12)$.

a) $15 + 7 + 14 = 36$ — Adding the positive numbers
b) $-2 + (-5) + (-12) = -19$ — Adding the negative numbers
$36 + (-19) = 17$ — Adding the results in (a) and (b)

We can also add the numbers in any other order we wish—say, from left to right—as follows:

$$\begin{aligned} 15 + (-2) + 7 + 14 + (-5) + (-12) &= 13 + 7 + 14 + (-5) + (-12) \\ &= 20 + 14 + (-5) + (-12) \\ &= 34 + (-5) + (-12) \\ &= 29 + (-12) \\ &= 17 \end{aligned}$$

Do Exercises 21–24.

b OPPOSITES, OR ADDITIVE INVERSES

Suppose we add two numbers that are **opposites**, such as 6 and -6. The result is 0. When opposites are added, the result is always 0. Opposites are also called **additive inverses**. Every real number has an opposite, or additive inverse.

OPPOSITES, OR ADDITIVE INVERSES

Two numbers whose sum is 0 are called **opposites**, or **additive inverses**, of each other.

EXAMPLES Find the opposite, or additive inverse, of each number.

14. 34 The opposite of 34 is -34 because $34 + (-34) = 0$.

15. -8 The opposite of -8 is 8 because $-8 + 8 = 0$.

16. 0 The opposite of 0 is 0 because $0 + 0 = 0$.

17. $-\frac{7}{8}$ The opposite of $-\frac{7}{8}$ is $\frac{7}{8}$ because $-\frac{7}{8} + \frac{7}{8} = 0$.

Do Exercises 25–30.

Find the opposite, or additive inverse, of each number.

25. -4 **26.** 8.7

27. -7.74 **28.** $-\frac{8}{9}$

29. 0 **30.** 12

To name the opposite, we use the symbol $-$, as follows.

SYMBOLIZING OPPOSITES

The opposite, or additive inverse, of a number a can be named $-a$ (read "the opposite of a," or "the additive inverse of a").

Note that if we take a number, say, 8, and find its opposite, -8, and then find the opposite of the result, we will have the original number, 8, again.

THE OPPOSITE OF AN OPPOSITE

The **opposite of the opposite** of a number is the number itself. (The additive inverse of the additive inverse of a number is the number itself.) That is, for any number a,

$$-(-a) = a.$$

EXAMPLE 18 Evaluate $-x$ and $-(-x)$ when $x = 16$.

If $x = 16$, then $-x = -16$. The opposite of 16 is -16.

If $x = 16$, then $-(-x) = -(-16) = 16$. The opposite of the opposite of 16 is 16.

EXAMPLE 19 Evaluate $-x$ and $-(-x)$ when $x = -3$.

If $x = -3$, then $-x = -(-3) = 3$.

If $x = -3$, then $-(-x) = -(-(-3)) = -(3) = -3$.

Note that in Example 19 we used a second set of parentheses to show that we are substituting the negative number -3 for x. Symbolism like $--x$ is not considered meaningful.

Do Exercises 31–34.

Evaluate $-x$ and $-(-x)$ when:

31. $x = 14$.

GS **32.** $x = -1.6$.

$-x = -(\quad) = 1.6;$

$-(-x) = -(-(\quad))$

$= -(\quad) = -1.6$

33. $x = \frac{2}{3}$. **34.** $x = -\frac{9}{8}$.

A symbol such as -8 is usually read "negative 8." It could be read "the additive inverse of 8," because the additive inverse of 8 is negative 8. It could also be read "the opposite of 8," because the opposite of 8 is -8. Thus a symbol like -8 can be read in more than one way. It is never correct to read -8 as "minus 8."

Caution!

A symbol like $-x$, which has a variable, should be read "the opposite of x" or "the additive inverse of x" and *not* "negative x," because we do not know whether x represents a positive number, a negative number, or 0. You can check this in Examples 18 and 19.

Answers

25. 4 **26.** -8.7 **27.** 7.74 **28.** $\frac{8}{9}$
29. 0 **30.** -12 **31.** -14; 14
32. 1.6; -1.6 **33.** $-\frac{2}{3}$; $\frac{2}{3}$ **34.** $\frac{9}{8}$; $-\frac{9}{8}$

Guided Solution:
32. -1.6; -1.6, 1.6

We can use the symbolism $-a$ to restate the definition of opposite, or additive inverse.

OPPOSITES, OR ADDITIVE INVERSES

For any real number a, the **opposite**, or **additive inverse**, of a, denoted $-a$, is such that

$$a + (-a) = (-a) + a = 0.$$

Signs of Numbers

A negative number is sometimes said to have a "negative sign." A positive number is said to have a "positive sign." When we replace a number with its opposite, we can say that we have "changed its sign."

EXAMPLES Find the opposite. (Change the sign.)

20. -3 $\quad -(-3) = 3$

21. $-\frac{2}{13}$ $\quad -\left(-\frac{2}{13}\right) = \frac{2}{13}$

22. 0 $\quad -(0) = 0$

23. 14 $\quad -(14) = -14$

◀ **Do Exercises 35–38.**

Find the opposite. (Change the sign.)

35. -4

36. -13.4

37. 0

38. $\frac{1}{4}$

C APPLICATIONS AND PROBLEM SOLVING

Addition of real numbers occurs in many real-world situations.

EXAMPLE 24 *Banking Transactions.* On August 1st, Martias checks his bank account balance on his phone and sees that it is \$54. During the next week, the following transactions were recorded: a debit-card purchase of \$71, an overdraft fee of \$29, a direct deposit of \$160, and an ATM withdrawal of \$80. What is Martias's balance at the end of the week?

We let B = the ending balance of the bank account. Then the problem translates to the following:

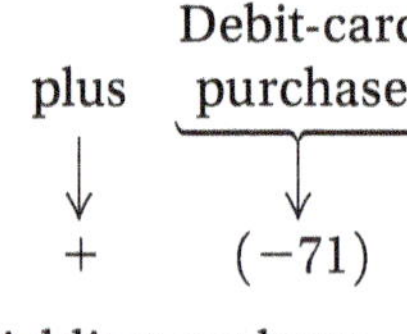
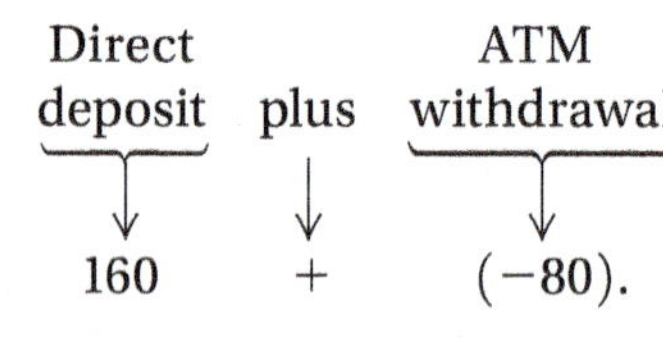

Ending balance	is	Beginning balance	plus	Debit-card purchase	plus	Overdraft fee	plus	Direct deposit	plus	ATM withdrawal
↓	↓	↓	↓	↓	↓	↓	↓	↓	↓	↓
B	$=$	54	$+$	(-71)	$+$	(-29)	$+$	160	$+$	(-80).

Adding, we have

$$B = 54 + (-71) + (-29) + 160 + (-80)$$
$$= 214 + (-180) \quad \text{Adding the positive numbers and adding the negative numbers}$$
$$= 34.$$

Martias's balance at the end of the week was \$34.

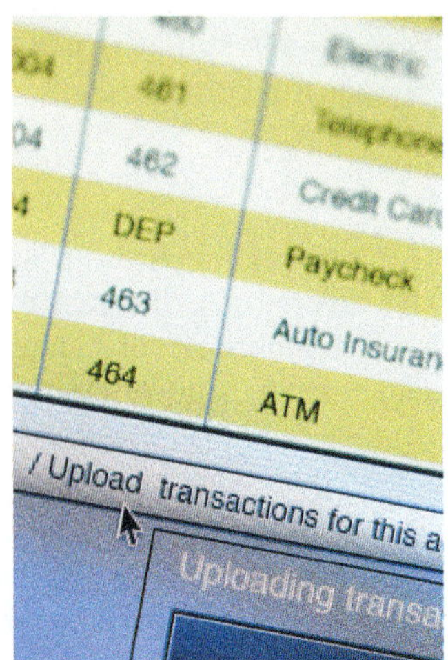

◀ **Do Exercise 39.**

39. *Change in Class Size.* During the first two weeks of the semester in Jim's algebra class, 4 students withdrew, 8 students enrolled late, and 6 students were dropped as "no shows." By how many students had the class size changed at the end of the first two weeks?

Answers

35. 4 **36.** 13.4 **37.** 0
38. $-\frac{1}{4}$ **39.** -2 students

9.3 Exercise Set

For Extra Help MyMathLab® MathXL® PRACTICE WATCH READ REVIEW

✓ Reading Check

Fill in each blank with either "left" or "right" so that the statements describe the steps when adding numbers with the number line.

RC1. To add $7 + 2$, start at 0, move ____________ to 7, and then move 2 units ____________. The sum is 9.

RC2. To add $-3 + (-5)$, start at 0, move ____________ to -3, and then move 5 units ____________. The sum is -8.

RC3. To add $4 + (-6)$, start at 0, move ____________ to 4, and then move 6 units ____________. The sum is -2.

RC4. To add $-8 + 3$, start at 0, move ____________ to -8, and then move 3 units ____________. The sum is -5.

a Add. Do not use the number line except as a check.

1. $2 + (-9)$

2. $-5 + 2$

3. $-11 + 5$

4. $4 + (-3)$

5. $-6 + 6$

6. $8 + (-8)$

7. $-3 + (-5)$

8. $-4 + (-6)$

9. $-7 + 0$

10. $-13 + 0$

11. $0 + (-27)$

12. $0 + (-35)$

13. $17 + (-17)$

14. $-15 + 15$

15. $-17 + (-25)$

16. $-24 + (-17)$

17. $18 + (-18)$

18. $-13 + 13$

19. $-28 + 28$

20. $11 + (-11)$

21. $8 + (-5)$

22. $-7 + 8$

23. $-4 + (-5)$

24. $10 + (-12)$

25. $13 + (-6)$

26. $-3 + 14$

27. $-25 + 25$

28. $50 + (-50)$

29. $53 + (-18)$

30. $75 + (-45)$

31. $-8.5 + 4.7$

32. $-4.6 + 1.9$

33. $-2.8 + (-5.3)$

34. $-7.9 + (-6.5)$

35. $-\frac{3}{5} + \frac{2}{5}$

36. $-\frac{4}{3}+\frac{2}{3}$

37. $-\frac{2}{9}+\left(-\frac{5}{9}\right)$

38. $-\frac{4}{7}+\left(-\frac{6}{7}\right)$

39. $-\frac{5}{8}+\frac{1}{4}$

40. $-\frac{5}{6}+\frac{2}{3}$

41. $-\frac{5}{8}+\left(-\frac{1}{6}\right)$

42. $-\frac{5}{6}+\left(-\frac{2}{9}\right)$

43. $-\frac{3}{8}+\frac{5}{12}$

44. $-\frac{7}{16}+\frac{7}{8}$

45. $-\frac{1}{6}+\frac{7}{10}$

46. $-\frac{11}{18}+\left(-\frac{3}{4}\right)$

47. $\frac{7}{15}+\left(-\frac{1}{9}\right)$

48. $-\frac{4}{21}+\frac{3}{14}$

49. $76+(-15)+(-18)+(-6)$

50. $29+(-45)+18+32+(-96)$

51. $-44+\left(-\frac{3}{8}\right)+95+\left(-\frac{5}{8}\right)$

52. $24+3.1+(-44)+(-8.2)+63$

b Find the opposite, or additive inverse.

53. 24

54. −64

55. −26.9

56. 48.2

Evaluate $-x$ when:

57. $x = 8$.

58. $x = -27$.

59. $x = -\frac{13}{8}$.

60. $x = \frac{1}{236}$.

Evaluate $-(-x)$ when:

61. $x = -43$.

62. $x = 39$.

63. $x = \frac{4}{3}$.

64. $x = -7.1$.

Find the opposite. (Change the sign.)

65. −24

66. −12.3

67. $-\frac{3}{8}$

68. 10

c Solve.

69. ***Tallest Mountain.*** The tallest mountain in the world, when measured from base to peak, is Mauna Kea (White Mountain) in Hawaii. From its base 19,684 ft below sea level in the Hawaiian Trough, it rises 33,480 ft. What is the elevation of the peak above sea level?

Source: *The Guinness Book of Records*

70. ***Copy Center Account.*** Rachel's copy-center bill for July was \$327. She made a payment of \$200 and then made \$48 worth of copies in August. How much did she then owe on her account?

71. *Temperature Changes.* One day, the temperature in Lawrence, Kansas, is 32°F at 6:00 A.M. It rises 15° by noon, but falls 50° by midnight when a cold front moves in. What is the final temperature?

72. *Stock Changes.* On a recent day, the price of a stock opened at a value of $61.38. During the day, it rose $4.75, dropped $7.38, and rose $5.13. Find the value of the stock at the end of the day.

73. *"Flipping" Houses.* Buying run-down houses, fixing them up, and reselling them is referred to as "flipping" houses. Charlie and Sophia bought and sold four houses in a recent year. The profits and losses are shown in the following bar graph. Find the sum of the profits and losses.

74. *Football Yardage.* In a college football game, the quarterback attempted passes with the following results. Find the total gain or loss.

ATTEMPT	GAIN OR LOSS
1st	13-yd gain
2nd	12-yd loss
3rd	21-yd gain

75. *Credit-Card Bills.* On August 1, Lyle's credit-card bill shows that he owes $470. During the month of August, Lyle makes a payment of $45 to the credit-card company, charges another $160 in merchandise, and then pays off another $500 of his bill. What is the new amount that Lyle owes at the end of August?

76. *Account Balance.* Emma has $460 in a checking account. She uses her debit card for a purchase of $530, makes a deposit of $75, and then writes a check for $90. What is the balance in her account?

Skill Maintenance

Convert to decimal notation. [6.2b]

77. 71.3%

78. $92\frac{7}{8}\%$

Convert to percent notation. [6.3a]

79. $\frac{1}{8}$

80. $\frac{13}{32}$

81. Divide and simplify: $\frac{2}{3} \div \frac{5}{12}$. [3.7b]

82. Subtract and simplify: $\frac{2}{3} - \frac{5}{12}$. [4.3a]

Synthesis

83. For what numbers x is $-x$ negative?

84. For what numbers x is $-x$ positive?

85. If a is positive and b is negative, then $-a + b$ is which of the following?

A. Positive
B. Negative
C. 0
D. Cannot be determined without more information

86. If $a = b$ and a and b are negative, then $-a + (-b)$ is which of the following?

A. Positive
B. Negative
C. 0
D. Cannot be determined without more information

9.4 Subtraction of Real Numbers

OBJECTIVES

a Subtract real numbers and simplify combinations of additions and subtractions.

b Solve applied problems involving subtraction of real numbers.

a SUBTRACTION

We now consider subtraction of real numbers.

SUBTRACTION

The difference $a - b$ is the number c for which $a = b + c$.

Consider, for example, $45 - 17$. *Think*: What number can we add to 17 to get 45? Since $45 = 17 + 28$, we know that $45 - 17 = 28$. Let's consider an example whose answer is a negative number.

EXAMPLE 1 Subtract: $3 - 7$.

Think: What number can we add to 7 to get 3? The number must be negative. Since $7 + (-4) = 3$, we know the number is -4: $3 - 7 = -4$. That is, $3 - 7 = -4$ because $7 + (-4) = 3$.

◀ Do Exercises 1–3.

The definition above does not provide the most efficient way to do subtraction. We can develop a faster way to subtract. As a rationale for the faster way, let's compare $3 + 7$ and $3 - 7$ on the number line.

To find $3 + 7$ on the number line, we start at 0, move to 3, and then move 7 units farther to the right since 7 is positive.

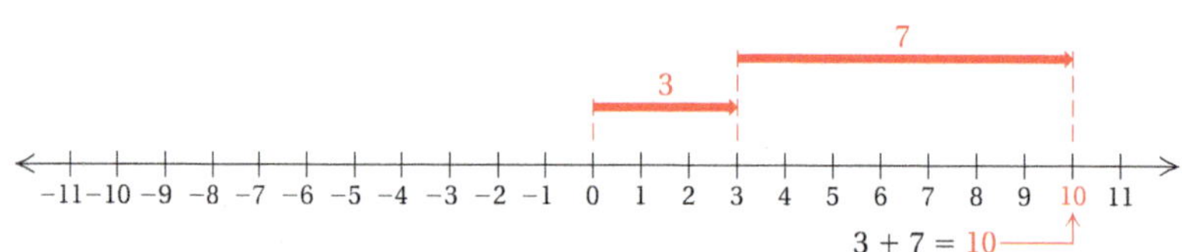

To find $3 - 7$, we do the "opposite" of adding 7: We move 7 units to the *left* to do the subtracting. This is the same as *adding* the opposite of 7, -7, to 3.

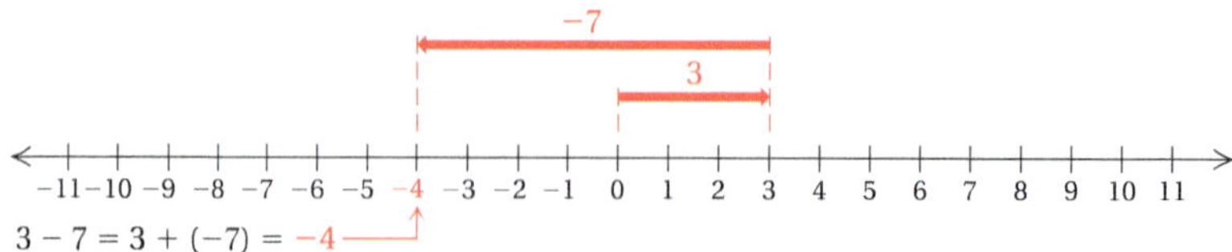

◀ Do Exercises 4–6.

Look for a pattern in the examples shown at right.

SUBTRACTING	ADDING AN OPPOSITE
$5 - 8 = -3$	$5 + (-8) = -3$
$-6 - 4 = -10$	$-6 + (-4) = -10$
$-7 - (-2) = -5$	$-7 + 2 = -5$

Subtract.

1. $-6 - 4$

Think: What number can be added to 4 to get -6:

$\square + 4 = -6?$

2. $-7 - (-10)$

Think: What number can be added to -10 to get -7:

$\square + (-10) = -7?$

3. $-7 - (-2)$

Think: What number can be added to -2 to get -7:

$\square + (-2) = -7?$

Subtract. Use the number line, doing the "opposite" of addition.

4. $5 - 9$

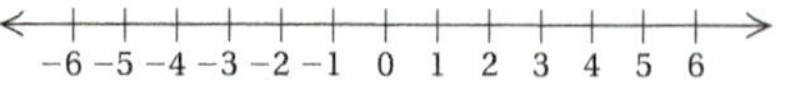

5. $-3 - 2$

6. $-4 - (-3)$

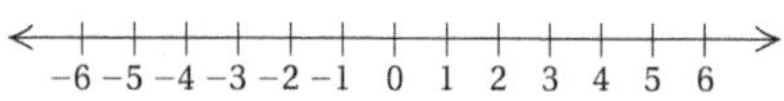

Answers

1. -10 **2.** 3 **3.** -5 **4.** -4
5. -5 **6.** -1

Do Exercises 7–10. ▶

Perhaps you have noticed that we can subtract by adding the opposite of the number being subtracted. This can always be done.

> **SUBTRACTING BY ADDING THE OPPOSITE**
>
> For any real numbers a and b,
>
> $$a - b = a + (-b).$$
>
> (To subtract, add the opposite, or additive inverse, of the number being subtracted.)

This is the method generally used for quick subtraction of real numbers.

EXAMPLES Subtract.

2. $2 - 6 = 2 + (-6) = -4$

The opposite of 6 is -6. We change the subtraction to addition and add the opposite.
Check: $-4 + 6 = 2$.

3. $4 - (-9) = 4 + 9 = 13$

The opposite of -9 is 9. We change the subtraction to addition and add the opposite.
Check: $13 + (-9) = 4$.

4. $-4.2 - (-3.6) = -4.2 + 3.6 = -0.6$

Adding the opposite.
Check: $-0.6 + (-3.6) = -4.2$.

5. $-\frac{1}{2} - \left(-\frac{3}{4}\right) = -\frac{1}{2} + \frac{3}{4}$

$= -\frac{2}{4} + \frac{3}{4} = \frac{1}{4}$

Adding the opposite.
Check: $\frac{1}{4} + \left(-\frac{3}{4}\right) = -\frac{1}{2}$.

Do Exercises 11–16. ▶

EXAMPLES Subtract by adding the opposite of the number being subtracted.

6. $3 - 5$ *Think*: "Three minus five is three plus the opposite of five"

$3 - 5 = 3 + (-5) = -2$

7. $\frac{1}{8} - \frac{7}{8}$ *Think*: "One-eighth minus seven-eighths is one-eighth plus the opposite of seven-eighths"

$\frac{1}{8} - \frac{7}{8} = \frac{1}{8} + \left(-\frac{7}{8}\right) = -\frac{6}{8}$, or $-\frac{3}{4}$

8. $-4.6 - (-9.8)$ *Think*: "Negative four point six minus negative nine point eight is negative four point six plus the opposite of negative nine point eight"

$-4.6 - (-9.8) = -4.6 + 9.8 = 5.2$

9. $-\frac{3}{4} - \frac{7}{5}$ *Think*: "Negative three-fourths minus seven-fifths is negative three-fourths plus the opposite of seven-fifths"

$-\frac{3}{4} - \frac{7}{5} = -\frac{3}{4} + \left(-\frac{7}{5}\right) = -\frac{15}{20} + \left(-\frac{28}{20}\right) = -\frac{43}{20}$

Do Exercises 17–21. ▶

Complete the addition and compare with the subtraction.

7. $4 - 6 = -2$;
$4 + (-6) =$ ______

8. $-3 - 8 = -11$;
$-3 + (-8) =$ ______

9. $-5 - (-9) = 4$;
$-5 + 9 =$ ______

10. $-5 - (-3) = -2$;
$-5 + 3 =$ ______

Subtract.

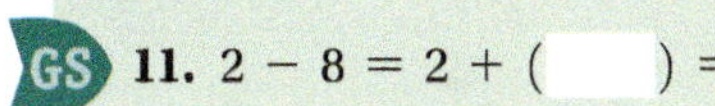

GS 11. $2 - 8 = 2 + ($ ____ $) =$ ____

12. $-6 - 10$

13. $12.4 - 5.3$

14. $-8 - (-11)$

15. $-8 - (-8)$

16. $\frac{2}{3} - \left(-\frac{5}{6}\right)$

Subtract by adding the opposite of the number being subtracted.

17. $3 - 11$

18. $12 - 5$

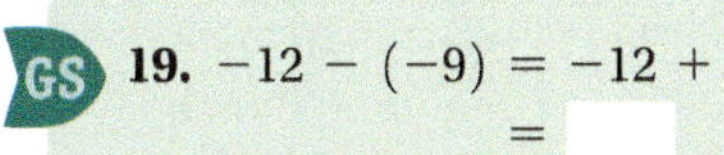

GS 19. $-12 - (-9) = -12 +$ ____
$=$ ____

20. $-12.4 - 10.9$

21. $-\frac{4}{5} - \left(-\frac{4}{5}\right)$

Answers

7. -2 **8.** -11 **9.** 4 **10.** -2 **11.** -6
12. -16 **13.** 7.1 **14.** 3 **15.** 0 **16.** $\frac{3}{2}$
17. -8 **18.** 7 **19.** -3 **20.** -23.3 **21.** 0

Guided Solutions:
11. $-8, -6$ **19.** $9, -3$

When several additions and subtractions occur together, we can make them all additions.

EXAMPLES Simplify.

10. $8 - (-4) - 2 - (-4) + 2 = 8 + 4 + (-2) + 4 + 2$ Adding the opposite
$= 16$

11. $8.2 - (-6.1) + 2.3 - (-4) = 8.2 + 6.1 + 2.3 + 4 = 20.6$

12. $\frac{3}{4} - \left(-\frac{1}{12}\right) - \frac{5}{6} - \frac{2}{3} = \frac{9}{12} + \frac{1}{12} + \left(-\frac{10}{12}\right) + \left(-\frac{8}{12}\right)$
$= \frac{9 + 1 + (-10) + (-8)}{12}$
$= \frac{-8}{12} = -\frac{8}{12} = -\frac{2}{3}$

Simplify.

22. $-6 - (-2) - (-4) - 12 + 3$

23. $\frac{2}{3} - \frac{4}{5} - \left(-\frac{11}{15}\right) + \frac{7}{10} - \frac{5}{2}$

24. $-9.6 + 7.4 - (-3.9) - (-11)$

Do Exercises 22–24.

b APPLICATIONS AND PROBLEM SOLVING

Let's now see how we can use subtraction of real numbers to solve applied problems.

EXAMPLE 13 *Surface Temperatures on Mars.* Surface temperatures on Mars vary from −128°C during polar night to 27°C at the equator during midday at the closest point in orbit to the sun. Find the difference between the highest value and the lowest value in this temperature range.
Source: Mars Institute

We let $D =$ the difference in the temperatures. Then the problem translates to the following subtraction:

Difference in temperature	is	Highest temperature	minus	Lowest temperature
↓	↓	↓	↓	↓
D	$=$	27	$-$	(-128)

$D = 27 + 128 = 155.$

The difference in the temperatures is 155°C.

Do Exercise 25.

25. *Temperature Extremes.* The highest temperature ever recorded in the United States is 134°F in Greenland Ranch, California, on July 10, 1913. The lowest temperature ever recorded is −80°F in Prospect Creek, Alaska, on January 23, 1971. How much higher was the temperature in Greenland Ranch than the temperature in Prospect Creek?
Source: National Oceanographic and Atmospheric Administration

Answers

22. −9 **23.** $-\frac{6}{5}$ **24.** 12.7 **25.** 214°F

9.4 Exercise Set

For Extra Help
MyMathLab® MathXL® PRACTICE WATCH READ REVIEW

Reading Check

Match the expression with an expression from the column on the right that names the same number.

RC1. $18 - 6$

RC2. $-18 - (-6)$

RC3. $-18 - 6$

RC4. $18 - (-6)$

a) $18 + 6$

b) $-18 + 6$

c) $18 + (-6)$

d) $-18 + (-6)$

a Subtract.

1. $2 - 9$

2. $3 - 8$

3. $-8 - (-2)$

4. $-6 - (-8)$

5. $-11 - (-11)$

6. $-6 - (-6)$

7. $12 - 16$

8. $14 - 19$

9. $20 - 27$

10. $30 - 4$

11. $-9 - (-3)$

12. $-7 - (-9)$

13. $-40 - (-40)$

14. $-9 - (-9)$

15. $7 - (-7)$

16. $4 - (-4)$

17. $8 - (-3)$

18. $-7 - 4$

19. $-6 - 8$

20. $6 - (-10)$

21. $-4 - (-9)$

22. $-14 - 2$

23. $-6 - (-5)$

24. $-4 - (-3)$

25. $8 - (-10)$

26. $5 - (-6)$

27. $-5 - (-2)$

28. $-3 - (-1)$

29. $-7 - 14$

30. $-9 - 16$

31. $0 - (-5)$

32. $0 - (-1)$

33. $-8 - 0$

34. $-9 - 0$

35. $7 - (-5)$

36. $7 - (-4)$

37. $2 - 25$

38. $18 - 63$

39. $-42 - 26$

40. $-18 - 63$

41. $-71 - 2$

42. $-49 - 3$

43. $24 - (-92)$

44. $48 - (-73)$

45. $-50 - (-50)$

46. $-70 - (-70)$

47. $-\frac{3}{8} - \frac{5}{8}$

48. $\frac{3}{9} - \frac{9}{9}$

49. $\frac{3}{4} - \frac{2}{3}$

50. $\frac{5}{8} - \frac{3}{4}$

51. $-\frac{3}{4} - \frac{2}{3}$

52. $-\frac{5}{8} - \frac{3}{4}$

53. $-\frac{5}{8} - \left(-\frac{3}{4}\right)$

54. $-\frac{3}{4} - \left(-\frac{2}{3}\right)$

55. $6.1 - (-13.8)$

56. $1.5 - (-3.5)$

57. $-2.7 - 5.9$

58. $-3.2 - 5.8$

59. $0.99 - 1$

60. $0.87 - 1$

61. $-79 - 114$

62. $-197 - 216$

63. $0 - (-500)$

64. $500 - (-1000)$

65. $-2.8 - 0$

66. $6.04 - 1.1$

67. $7 - 10.53$

68. $8 - (-9.3)$

69. $\frac{1}{6} - \frac{2}{3}$

70. $-\frac{3}{8} - \left(-\frac{1}{2}\right)$

71. $-\frac{4}{7} - \left(-\frac{10}{7}\right)$

72. $\frac{12}{5} - \frac{12}{5}$

73. $-\frac{7}{10} - \frac{10}{15}$

74. $-\frac{4}{18} - \left(-\frac{2}{9}\right)$

75. $\frac{1}{5} - \frac{1}{3}$

76. $-\frac{1}{7} - \left(-\frac{1}{6}\right)$

77. $\frac{5}{12} - \frac{7}{16}$

78. $-\frac{1}{35} - \left(-\frac{9}{40}\right)$

79. $-\frac{2}{15} - \frac{7}{12}$

80. $\frac{2}{21} - \frac{9}{14}$

Simplify.

81. $18 - (-15) - 3 - (-5) + 2$

82. $22 - (-18) + 7 + (-42) - 27$

83. $-31 + (-28) - (-14) - 17$

84. $-43 - (-19) - (-21) + 25$

85. $-34 - 28 + (-33) - 44$

86. $39 + (-88) - 29 - (-83)$

87. $-93 - (-84) - 41 - (-56)$

88. $84 + (-99) + 44 - (-18) - 43$

89. $-5.4 - (-30.9) + 30.8 + 40.2 - (-12)$

90. $14.9 - (-50.7) + 20 - (-32.8)$

91. $-\frac{7}{12} + \frac{3}{4} - \left(-\frac{5}{8}\right) - \frac{13}{24}$

92. $-\frac{11}{16} + \frac{5}{32} - \left(-\frac{1}{4}\right) + \frac{7}{8}$

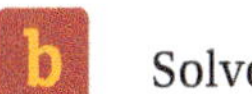

Solve.

93. ***Elevations in Asia.*** The elevation of the highest point in Asia, Mt. Everest, Nepal–Tibet, is 29,035 ft. The lowest elevation, at the Dead Sea, Israel–Jordan, is −1348 ft. What is the difference in the elevations of the two locations?

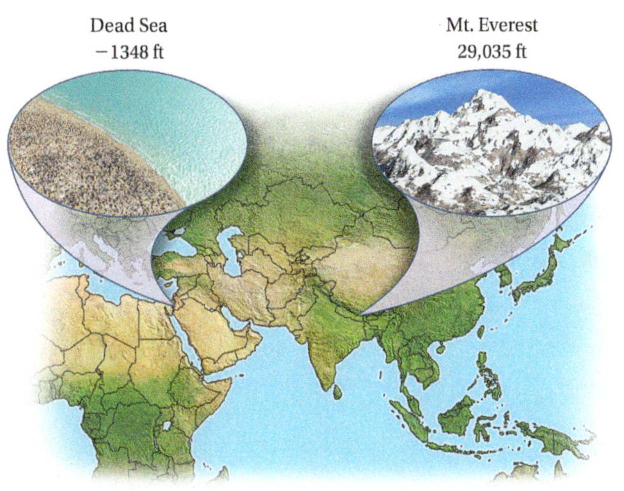

94. ***Ocean Depth.*** The deepest point in the Pacific Ocean is the Marianas Trench, with a depth of 10,924 m. The deepest point in the Atlantic Ocean is the Puerto Rico Trench, with a depth of 8605 m. What is the difference in the elevation of the two trenches?

Source: *The World Almanac and Book of Facts*

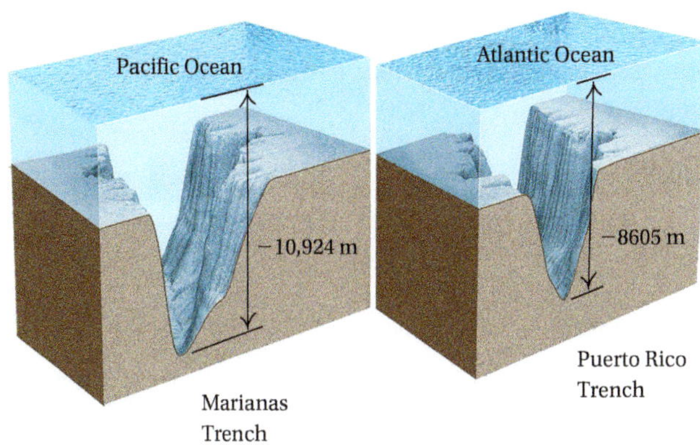

95. Francisca has a charge of $476.89 on her credit card, but she then returns a sweater that cost $128.95. How much does she now owe on her credit card?

96. Jacob has $825 in a checking account. What is the balance in his account after he has written a check for $920 to pay for a laptop?

97. *Difference in Elevation.* The highest elevation in Japan is 3776 m above sea level at Fujiyama. The lowest elevation in Japan is 4 m below sea level at Hachirogata. Find the difference in the elevations.

Source: *Information Please Almanac*

98. *Difference in Elevation.* The lowest elevation in North America—Death Valley, California—is 282 ft below sea level. The highest elevation in North America—Mount McKinley, Alaska—is 20,320 ft above sea level. Find the difference in elevation between the highest point and the lowest point.

Source: National Geographic Society

99. *Low Points on Continents.* The lowest point in Africa is Lake Assal, which is 512 ft below sea level. The lowest point in South America is the Valdes Peninsula, which is 131 ft below sea level. How much lower is Lake Assal than the Valdes Peninsula?

Source: National Geographic Society

100. *Temperature Records.* The greatest recorded temperature change in one 24-hr period occurred between January 14 and January 15, 1972, in Loma, Montana, where the temperature rose from $-54°F$ to $49°F$. By how much did the temperature rise?

Source: *The Guinness Book of Records*

101. *Surface Temperature on Mercury.* Surface temperatures on Mercury vary from 840°F on the equator when the planet is closest to the sun to $-290°F$ at night. Find the difference between these two temperatures.

102. *Run Differential.* In baseball, the difference between the number of runs that a team scores and the number of runs that it allows its opponents to score is called the *run differential.* That is,

$$\text{Run differential} = \begin{matrix}\text{Number of}\\ \text{runs scored}\end{matrix} - \begin{matrix}\text{Number of}\\ \text{runs allowed.}\end{matrix}$$

Teams strive for a positive run differential.

Source: Major League Baseball

a) In a recent season, the Kansas City Royals scored 676 runs and allowed 746 runs to be scored on them. Find the run differential.

b) In a recent season, the Atlanta Braves scored 700 runs and allowed 600 runs to be scored on them. Find the run differential.

Skill Maintenance

Translate to an algebraic expression. [5.8a], [9.1b]

103. 7 more than y

104. 41 less than t

105. h subtracted from a

106. The product of 6 and c

107. r more than s

108. x less than y

Synthesis

Determine whether each statement is true or false for all integers a and b. If false, give an example to show why. Examples may vary.

109. $a - 0 = 0 - a$

110. $0 - a = a$

111. If $a \neq b$, then $a - b \neq 0$.

112. If $a = -b$, then $a + b = 0$.

113. If $a + b = 0$, then a and b are opposites.

114. If $a - b = 0$, then $a = -b$.

Mid-Chapter Review

Concept Reinforcement

Determine whether each statement is true or false.

______________ **1.** All rational numbers can be named using fraction notation. [9.2c]

______________ **2.** If $a > b$, then a lies to the left of b on the number line. [9.2d]

______________ **3.** The absolute value of a number is always nonnegative. [9.2e]

______________ **4.** We can translate "7 less than y" as $7 - y$. [9.1b]

Guided Solutions

Fill in each blank with the number that creates a correct statement or solution.

5. Evaluate $-x$ and $-(-x)$ when $x = -4$. [9.3b]

$-x = -(\quad) = \quad$;

$-(-x) = -(-(\quad)) = -(\quad) = \quad$

Subtract. [9.4a]

6. $5 - 13 = 5 + (\quad) = \quad$

7. $-6 - 7 = -6 + (\quad) = \quad$

Mixed Review

Evaluate. [9.1a]

8. $\dfrac{3m}{n}$, when $m = 8$ and $n = 6$

9. $\dfrac{a + b}{2}$, when $a = 5$ and $b = 17$

Translate each phrase to an algebraic expression. Use any letter for the variable. [9.1b]

10. Three times some number

11. Five less than some number

12. State the integers that correspond to this situation: Jerilyn deposited \$450 in her checking account. Later that week, she wrote a check for \$79. [9.2a]

13. Graph -3.5 on the number line. [9.2b]

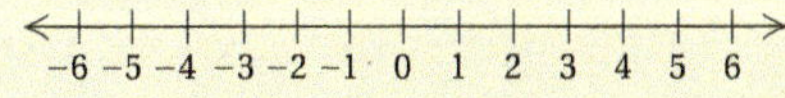

Convert to decimal notation. [9.2c]

14. $-\dfrac{4}{5}$

15. $\dfrac{7}{3}$

Use either < or > for □ to write a true sentence. [9.2d]

16. $-5 \;\square\; -3$

17. $-9.9 \;\square\; -10.1$

Write true or false. [9.2d]

18. $-8 \geq -5$

19. $-4 \leq -4$

Write an inequality with the same meaning. [9.2d]

20. $y < 5$

21. $-3 \geq t$

Find the absolute value. [9.2e]

22. $|15.6|$

23. $|-18|$

24. $|0|$

25. $\left|-\frac{12}{5}\right|$

Find the opposite, or additive inverse, of each number. [9.3b]

26. -5.6

27. $\frac{7}{4}$

28. 0

29. -49

30. Evaluate $-x$ when x is -19. [9.3b]

31. Evaluate $-(-x)$ when x is 2.3. [9.3b]

Compute and simplify. [9.3a], [9.4a]

32. $7 + (-9)$

33. $-\frac{3}{8} + \frac{1}{4}$

34. $3.6 + (-3.6)$

35. $-8 + (-9)$

36. $\frac{2}{3} + \left(-\frac{9}{8}\right)$

37. $-4.2 + (-3.9)$

38. $-14 + 5$

39. $19 + (-21)$

40. $-4.1 - 6.3$

41. $5 - (-11)$

42. $-\frac{1}{4} - \left(-\frac{3}{5}\right)$

43. $12 - 24$

44. $-8 - (-4)$

45. $-\frac{1}{2} - \frac{5}{6}$

46. $12.3 - 14.1$

47. $6 - (-7)$

48. $16 - (-9) - 20 - (-4)$

49. $-4 + (-10) - (-3) - 12$

50. $17 - (-25) + 15 - (-18)$

51. $-9 + (-3) + 16 - (-10)$

Solve. [9.3c], [9.4b]

52. ***Temperature Change.*** In chemistry lab, Ben works with a substance whose initial temperature is 25°C. During an experiment, the temperature falls to −8°C. Find the difference between the two temperatures.

53. ***Stock Price Change.*** The price of a stock opened at \$56.12. During the day, it dropped \$1.18, then rose \$1.22, and then dropped \$1.36. Find the value of the stock at the end of the day.

Understanding Through Discussion and Writing

54. Give three examples of rational numbers that are not integers. Explain. [9.2b]

55. Give three examples of irrational numbers. Explain the difference between an irrational number and a rational number. [9.2b, d]

56. Explain in your own words why the sum of two negative numbers is always negative. [9.3a]

57. If a negative number is subtracted from a positive number, will the result always be positive? Why or why not? [9.4a]

STUDYING FOR SUCCESS *Using the Textbook*

- ☐ Study the step-by-step solutions to the examples, using the red comments for extra explanations.
- ☐ Stop and do the margin exercises as directed. Answers are at the bottom of the margin, so you can check your work right away.
- ☐ The objective symbols a, b, c, and so on, allow you to refer to the appropriate place in the text whenever you need to review a topic.

Multiplication of Real Numbers

9.5

OBJECTIVES

a Multiply real numbers.

b Solve applied problems involving multiplication of real numbers.

SKILL TO REVIEW

Objective 3.4b: Multiply using fraction notation.

Multiply.

1. $\frac{5}{3} \cdot \frac{4}{7}$ **2.** $\frac{1}{8} \cdot \frac{3}{11}$

a MULTIPLICATION

Multiplication of real numbers is very much like multiplication of arithmetic numbers. The only difference is that we must determine whether the product is positive or negative.

Multiplication of a Positive Number and a Negative Number

To see how to multiply a positive number and a negative number, consider the pattern of the following.

This number decreases by 1 each time. This number decreases by 5 each time.

$$\begin{aligned} 4 \cdot 5 &= 20 \\ 3 \cdot 5 &= 15 \\ 2 \cdot 5 &= 10 \\ 1 \cdot 5 &= 5 \\ 0 \cdot 5 &= 0 \\ -1 \cdot 5 &= -5 \\ -2 \cdot 5 &= -10 \\ -3 \cdot 5 &= -15 \end{aligned}$$

Do Exercise 1. ▶

1. Complete, as in the example.

$$\begin{aligned} 4 \cdot 10 &= 40 \\ 3 \cdot 10 &= 30 \\ 2 \cdot 10 &= \\ 1 \cdot 10 &= \\ 0 \cdot 10 &= \\ -1 \cdot 10 &= \\ -2 \cdot 10 &= \\ -3 \cdot 10 &= \end{aligned}$$

According to this pattern, it looks as though the product of a negative number and a positive number is negative. That is the case, and we have the first part of the rule for multiplying real numbers.

THE PRODUCT OF A POSITIVE NUMBER AND A NEGATIVE NUMBER

To multiply a positive number and a negative number, multiply their absolute values. The product is negative.

EXAMPLES Multiply.

1. $8(-5) = -40$

2. $-\frac{1}{3} \cdot \frac{5}{7} = -\frac{5}{21}$

3. $(-7.2)5 = -36$

Do Exercises 2–5 on the following page. ▶

Answers

Skill to Review:

1. $\frac{20}{21}$ **2.** $\frac{3}{88}$

Margin Exercise:

1. 20; 10; 0; −10; −20; −30

Multiply.

2. $-3 \cdot 6$

3. $20 \cdot (-5)$

4. $-\frac{2}{3} \cdot \frac{5}{6}$

5. $-4.23(7.1)$

6. Complete, as in the example.

$$\begin{aligned} 3 \cdot (-10) &= -30 \\ 2 \cdot (-10) &= -20 \\ 1 \cdot (-10) &= \\ 0 \cdot (-10) &= \\ -1 \cdot (-10) &= \\ -2 \cdot (-10) &= \\ -3 \cdot (-10) &= \end{aligned}$$

Multiply.

7. $-9 \cdot (-3)$

8. $-16 \cdot (-2)$

9. $-7 \cdot (-5)$

10. $-\frac{4}{7}\left(-\frac{5}{9}\right)$

11. $-\frac{3}{2}\left(-\frac{4}{9}\right)$

12. $-3.25(-4.14)$

Answers

2. -18 3. -100 4. $-\frac{5}{9}$ 5. -30.033
6. $-10; 0; 10; 20; 30$ 7. 27 8. 32 9. 35
10. $\frac{20}{63}$ 11. $\frac{2}{3}$ 12. 13.455

Multiplication of Two Negative Numbers

How do we multiply two negative numbers? Again, we look for a pattern.

This number decreases by 1 each time. This number increases by 5 each time.

$$\begin{aligned} 4 \cdot (-5) &= -20 \\ 3 \cdot (-5) &= -15 \\ 2 \cdot (-5) &= -10 \\ 1 \cdot (-5) &= -5 \\ 0 \cdot (-5) &= 0 \\ -1 \cdot (-5) &= 5 \\ -2 \cdot (-5) &= 10 \\ -3 \cdot (-5) &= 15 \end{aligned}$$

Do Exercise 6.

According to the pattern, it appears that the product of two negative numbers is positive. That is actually so, and we have the second part of the rule for multiplying real numbers.

THE PRODUCT OF TWO NEGATIVE NUMBERS

To multiply two negative numbers, multiply their absolute values. The product is positive.

Do Exercises 7–12.

The following is another way to consider the rules that we have for multiplication.

To multiply two nonzero real numbers:

a) Multiply the absolute values.
b) If the signs are the same, the product is positive.
c) If the signs are different, the product is negative.

Multiplication by Zero

The only case that we have not considered is multiplying by zero. As with nonnegative numbers, the product of any real number and 0 is 0.

THE MULTIPLICATION PROPERTY OF ZERO

For any real number a,

$$a \cdot 0 = 0 \cdot a = 0.$$

(The product of 0 and any real number is 0.)

EXAMPLES Multiply.

4. $(-3)(-4) = 12$

5. $-1.6(2) = -3.2$

6. $-19 \cdot 0 = 0$

7. $\left(-\frac{5}{6}\right)\left(-\frac{1}{9}\right) = \frac{5}{54}$

8. $0 \cdot (-452) = 0$

9. $23 \cdot 0 \cdot \left(-8\frac{2}{3}\right) = 0$

Do Exercises 13–18.

Multiply.

13. $5(-6)$

14. $(-5)(-6)$

15. $(-3.2) \cdot 10$

16. $\left(-\frac{4}{5}\right)\left(\frac{10}{3}\right)$

17. $0 \cdot (-34.2)$

18. $-\frac{5}{7} \cdot 0 \cdot \left(-4\frac{2}{3}\right)$

Multiplying More Than Two Numbers

When multiplying more than two real numbers, we can choose order and grouping as we please.

EXAMPLES Multiply.

10. $-8 \cdot 2(-3) = -16(-3)$ Multiplying the first two numbers
$= 48$

11. $-8 \cdot 2(-3) = 24 \cdot 2$ Multiplying the negative numbers. Every pair of negative numbers gives a positive product.
$= 48$

12. $-3(-2)(-5)(4) = 6(-5)(4)$ Multiplying the first two numbers
$= (-30)4$
$= -120$

13. $\left(-\frac{1}{2}\right)(8)\left(-\frac{2}{3}\right)(-6) = (-4)4$ Multiplying the first two numbers and the last two numbers
$= -16$

14. $-5 \cdot (-2) \cdot (-3) \cdot (-6) = 10 \cdot 18 = 180$

15. $(-3)(-5)(-2)(-3)(-6) = (-30)(18) = -540$

Considering that the product of a pair of negative numbers is positive, we see the following pattern.

> The product of an even number of negative numbers is positive.
> The product of an odd number of negative numbers is negative.

Do Exercises 19–24.

Multiply.

19. $5 \cdot (-3) \cdot 2$

20. $-3 \times (-4.1) \times (-2.5)$

21. $-\frac{1}{2} \cdot \left(-\frac{4}{3}\right) \cdot \left(-\frac{5}{2}\right)$

22. $-2 \cdot (-5) \cdot (-4) \cdot (-3)$

23. $(-4)(-5)(-2)(-3)(-1)$

24. $(-1)(-1)(-2)(-3)(-1)(-1)$

EXAMPLE 16 Evaluate $2x^2$ when $x = 3$ and when $x = -3$.

$2x^2 = 2(3)^2 = 2(9) = 18;$
$2x^2 = 2(-3)^2 = 2(9) = 18$

Let's compare the expressions $(-x)^2$ and $-x^2$.

EXAMPLE 17 Evaluate $(-x)^2$ and $-x^2$ when $x = 5$.

$(-x)^2 = (-5)^2 = (-5)(-5) = 25;$ Substitute 5 for x. Then evaluate the power.

$-x^2 = -(5)^2 = -(25) = -25$ Substitute 5 for x. Evaluate the power. Then find the opposite.

In Example 17, we see that the expressions $(-x)^2$ and $-x^2$ are *not* equivalent. That is, they do not have the same value for every allowable replacement of the variable by a real number. To find $(-x)^2$, we take the opposite and then square. To find $-x^2$, we find the square and then take the opposite.

Answers

13. -30 14. 30 15. -32 16. $-\frac{8}{3}$
17. 0 18. 0 19. -30 20. -30.75
21. $-\frac{5}{3}$ 22. 120 23. -120 24. 6

25. Evaluate $3x^2$ when $x = 4$ and when $x = -4$.

26. Evaluate $(-x)^2$ and $-x^2$ when $x = 2$.

27. Evaluate $(-x)^2$ and $-x^2$ when $x = -3$.

EXAMPLE 18 Evaluate $(-a)^2$ and $-a^2$ when $a = -4$.

To make sense of the substitutions and computations, we introduce extra grouping symbols into the expressions.

$$(-a)^2 = [-(-4)]^2 = [4]^2 = 16;$$
$$-a^2 = -(-4)^2 = -(16) = -16$$

◀ **Do Exercises 25–27.**

b APPLICATIONS AND PROBLEM SOLVING

We now consider multiplication of real numbers in real-world applications.

EXAMPLE 19 ***Mine Rescue.*** The San Jose copper and gold mine near Copiapó, Chile, collapsed on August 5, 2010, trapping 33 miners. Each miner was safely brought out of the mine with a specially designed capsule that could be lowered into the mine at −137 feet per minute. It took approximately 15 minutes to lower the capsule to the miners' location. Determine how far below the surface of the earth the miners were trapped.

Source: Reuters News

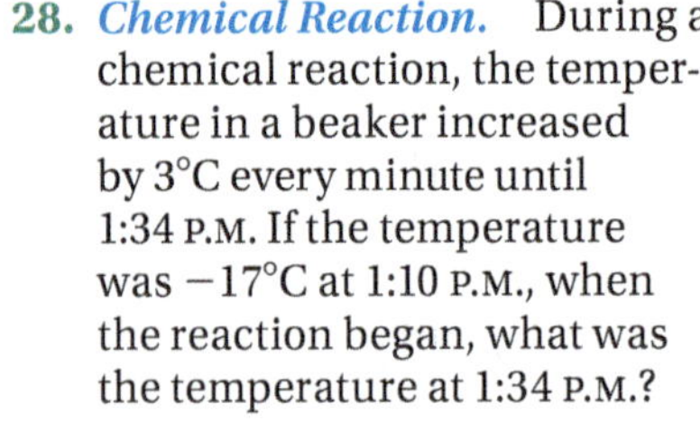

Since the capsule moved −137 feet per minute and it took 15 minutes to reach the miners, we have the depth d given by

$$d = 15 \cdot (-137) = -2055.$$

Thus the miners were trapped at −2055 ft.

◀ **Do Exercise 28.**

28. ***Chemical Reaction.*** During a chemical reaction, the temperature in a beaker increased by 3°C every minute until 1:34 P.M. If the temperature was −17°C at 1:10 P.M., when the reaction began, what was the temperature at 1:34 P.M.?

Answers

25. 48; 48 26. 4; −4 27. 9; −9 28. 55°C

9.5 Exercise Set

For Extra Help

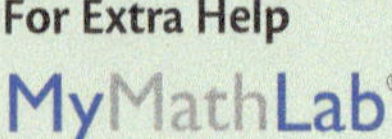
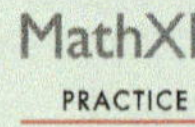

✓ Reading Check

Fill in the blank with either "positive" or "negative."

RC1. To multiply a positive number and a negative number, multiply their absolute values. The answer is ____________.

RC2. To multiply two negative numbers, multiply their absolute values. The answer is ____________.

RC3. The product of an even number of negative numbers is ____________.

RC4. The product of an odd number of negative numbers is ____________.

Evaluate.

RC5. -3^2

RC6. $(-3)^2$

RC7. $-\left(\frac{1}{2}\right)^2$

RC8. $-\left(-\frac{1}{2}\right)^2$

a Multiply.

1. $-4 \cdot 2$
2. $-3 \cdot 5$
3. $8 \cdot (-3)$
4. $9 \cdot (-5)$
5. $-9 \cdot 8$
6. $-10 \cdot 3$
7. $-8 \cdot (-2)$
8. $-2 \cdot (-5)$
9. $-7 \cdot (-6)$
10. $-9 \cdot (-2)$
11. $15 \cdot (-8)$
12. $-12 \cdot (-10)$
13. $-14 \cdot 17$
14. $-13 \cdot (-15)$
15. $-25 \cdot (-48)$
16. $39 \cdot (-43)$
17. $-3.5 \cdot (-28)$
18. $97 \cdot (-2.1)$
19. $9 \cdot (-8)$
20. $7 \cdot (-9)$
21. $4 \cdot (-3.1)$
22. $3 \cdot (-2.2)$
23. $-5 \cdot (-6)$
24. $-6 \cdot (-4)$
25. $-7 \cdot (-3.1)$
26. $-4 \cdot (-3.2)$
27. $\frac{2}{3} \cdot \left(-\frac{3}{5}\right)$
28. $\frac{5}{7} \cdot \left(-\frac{2}{3}\right)$
29. $-\frac{3}{8} \cdot \left(-\frac{2}{9}\right)$
30. $-\frac{5}{8} \cdot \left(-\frac{2}{5}\right)$
31. -6.3×2.7
32. -4.1×9.5
33. $7 \cdot (-4) \cdot (-3) \cdot 5$
34. $9 \cdot (-2) \cdot (-6) \cdot 7$
35. $-\frac{2}{3} \cdot \frac{1}{2} \cdot \left(-\frac{6}{7}\right)$
36. $-\frac{1}{8} \cdot \left(-\frac{1}{4}\right) \cdot \left(-\frac{3}{5}\right)$
37. $-3 \cdot (-4) \cdot (-5)$
38. $-2 \cdot (-5) \cdot (-7)$
39. $-2 \cdot (-5) \cdot (-3) \cdot (-5)$
40. $-3 \cdot (-5) \cdot (-2) \cdot (-1)$
41. $-4 \cdot (-1.8) \cdot 7$
42. $-8 \cdot (-1.3) \cdot (-5)$
43. $-\frac{1}{9}\left(-\frac{2}{3}\right)\left(\frac{5}{7}\right)$
44. $-\frac{7}{2}\left(-\frac{5}{7}\right)\left(-\frac{2}{5}\right)$
45. $4 \cdot (-4) \cdot (-5) \cdot (-12)$
46. $-2 \cdot (-3) \cdot (-4) \cdot (-5)$

47. $0.07 \cdot (-7) \cdot 6 \cdot (-6)$

48. $80 \cdot (-0.8) \cdot (-90) \cdot (-0.09)$

49. $\left(-\frac{5}{6}\right)\left(\frac{1}{8}\right)\left(-\frac{3}{7}\right)\left(-\frac{1}{7}\right)$

50. $\left(\frac{4}{5}\right)\left(-\frac{2}{3}\right)\left(-\frac{15}{7}\right)\left(\frac{1}{2}\right)$

51. $(-14) \cdot (-27) \cdot 0$

52. $7 \cdot (-6) \cdot 5 \cdot (-4) \cdot 3 \cdot (-2) \cdot 1 \cdot 0$

53. $(-8)(-9)(-10)$

54. $(-7)(-8)(-9)(-10)$

55. $(-6)(-7)(-8)(-9)(-10)$

56. $(-5)(-6)(-7)(-8)(-9)(-10)$

57. $(-1)^{12}$

58. $(-1)^9$

59. Evaluate $(-x)^2$ and $-x^2$ when $x = 4$ and when $x = -4$.

60. Evaluate $(-x)^2$ and $-x^2$ when $x = 10$ and when $x = -10$.

61. Evaluate $(-y)^2$ and $-y^2$ when $y = \frac{2}{5}$ and when $y = -\frac{2}{5}$.

62. Evaluate $(-w)^2$ and $-w^2$ when $w = \frac{1}{10}$ and when $w = -\frac{1}{10}$.

63. Evaluate $-(-t)^2$ and $-t^2$ when $t = 3$ and when $t = -3$.

64. Evaluate $-(-s)^2$ and $-s^2$ when $s = 1$ and when $s = -1$.

65. Evaluate $(-3x)^2$ and $-3x^2$ when $x = 7$ and when $x = -7$.

66. Evaluate $(-2x)^2$ and $-2x^2$ when $x = 3$ and when $x = -3$.

67. Evaluate $5x^2$ when $x = 2$ and when $x = -2$.

68. Evaluate $2x^2$ when $x = 5$ and when $x = -5$.

69. Evaluate $-2x^3$ when $x = 1$ and when $x = -1$.

70. Evaluate $-3x^3$ when $x = 2$ and when $x = -2$.

b Solve.

71. ***Chemical Reaction.*** The temperature of a chemical compound was 0°C at 11:00 A.M. During a reaction, it dropped 3°C per minute until 11:08 A.M. What was the temperature at 11:08 A.M.?

72. ***Chemical Reaction.*** The temperature of a chemical compound was −5°C at 3:20 P.M. During a reaction, it increased 2°C per minute until 3:52 P.M. What was the temperature at 3:52 P.M.?

73. *Weight Loss.* Dave lost 2 lb each week for a period of 10 weeks. Express his total weight change as an integer.

74. *Stock Loss.* Each day for a period of 5 days, the value of a stock that Lily owned dropped \$3. Express Lily's total loss as an integer.

75. *Stock Price.* The price of a stock began the day at \$23.75 per share and dropped \$1.38 per hour for 8 hr. What was the price of the stock after 8 hr?

76. *Population Decrease.* The population of Bloomtown was 12,500. It decreased 380 each year for 4 years. What was the population of the town after 4 years?

77. *Diver's Position.* After diving 95 m below sea level, a diver rises at a rate of 7 m/min for 9 min. Where is the diver in relation to the surface at the end of the 9-min period?

78. *Bank Account Balance.* Karen had \$68 in her bank account. After she used her debit card to make seven purchases at \$13 each, what was the balance in her bank account?

79. *Drop in Temperature.* The temperature in Osgood was 62°F at 2:00 P.M. It dropped 6°F per hour for the next 4 hr. What was the temperature at the end of the 4-hr period?

80. *Juice Consumption.* Oliver bought a 64-oz container of cranberry juice and drank 8 oz per day for a week. How much juice was left in the container at the end of the week?

Skill Maintenance

81. Find the LCM of 36 and 60 [4.1a]

82. Find the prime factorization of 4608. [3.2c]

Simplify [3.5b]

83. $\frac{26}{39}$

84. $\frac{48}{54}$

85. $\frac{264}{484}$

86. $\frac{1025}{6625}$

87. $\frac{275}{800}$

88. $\frac{111}{201}$

89. $\frac{11}{264}$

90. $\frac{78}{13}$

Synthesis

91. If a is positive and b is negative, then $-ab$ is which of the following?

A. Positive
B. Negative
C. 0
D. Cannot be determined without more information

92. If a is positive and b is negative, then $(-a)(-b)$ is which of the following?

A. Positive
B. Negative
C. 0
D. Cannot be determined without more information

93. Of all possible quotients of the numbers 10, $-\frac{1}{2}$, -5, and $\frac{1}{5}$, which two produce the largest quotient? Which two produce the smallest quotient?

9.6 Division of Real Numbers

OBJECTIVES

a Divide integers.

b Find the reciprocal of a real number.

c Divide real numbers.

d Solve applied problems involving division of real numbers.

SKILL TO REVIEW

Objective 3.7b: Divide and simplify using fraction notation.

Divide and simplify.

1. $\frac{6}{5} \div \frac{9}{2}$ 2. $30 \div \frac{5}{6}$

We now consider division of real numbers. The definition of division results in rules for division that are the same as those for multiplication.

a DIVISION OF INTEGERS

DIVISION

The quotient $a \div b$, or $\frac{a}{b}$, where $b \neq 0$, is that unique real number c for which $a = b \cdot c$.

Let's use the definition to divide integers.

EXAMPLES Divide, if possible. Check your answer.

1. $14 \div (-7) = -2$ — *Think*: What number multiplied by -7 gives 14? That number is -2. *Check*: $(-2)(-7) = 14$.

2. $\frac{-32}{-4} = 8$ — *Think*: What number multiplied by -4 gives -32? That number is 8. *Check*: $8(-4) = -32$.

3. $\frac{-10}{7} = -\frac{10}{7}$ — *Think*: What number multiplied by 7 gives -10? That number is $-\frac{10}{7}$. *Check*: $-\frac{10}{7} \cdot 7 = -10$.

4. $\frac{-17}{0}$ is **not defined.** — *Think*: What number multiplied by 0 gives -17? There is no such number because the product of 0 and *any* number is 0.

The rules for division are the same as those for multiplication.

To multiply or divide two real numbers (where the divisor is nonzero):

a) Multiply or divide the absolute values.

b) If the signs are the same, the answer is positive.

c) If the signs are different, the answer is negative.

Do Margin Exercises 1–6.

Divide.

1. $6 \div (-3)$

Think: What number multiplied by -3 gives 6?

2. $\frac{-15}{-3}$

Think: What number multiplied by -3 gives -15?

3. $-24 \div 8$

Think: What number multiplied by 8 gives -24?

4. $\frac{-48}{-6}$

5. $\frac{30}{-5}$

6. $\frac{30}{-7}$

Excluding Division by 0

Example 4 shows why we cannot divide -17 by 0. We can use the same argument to show why we cannot divide any nonzero number b by 0. Consider $b \div 0$. We look for a number that when multiplied by 0 gives b. There is no such number because the product of 0 and any number is 0. Thus we cannot divide a nonzero number b by 0.

On the other hand, if we divide 0 by 0, we look for a number c such that $0 \cdot c = 0$. But $0 \cdot c = 0$ for any number c. Thus it appears that $0 \div 0$ could be any number we choose. Getting any answer we want when we divide 0 by 0 would be very confusing. Thus we agree that division by 0 is not defined.

Answers

Skill to Review:

1. $\frac{4}{15}$ 2. 36

Margin Exercises:

1. -2 2. 5 3. -3 4. 8 5. -6 6. $-\frac{30}{7}$

EXCLUDING DIVISION BY 0

Division by 0 is not defined.

$a \div 0$, or $\frac{a}{0}$, is not defined for all real numbers a.

Dividing 0 by Other Numbers

Note that

$$0 \div 8 = 0 \text{ because } 0 = 0 \cdot 8; \qquad \frac{0}{-5} = 0 \text{ because } 0 = 0 \cdot (-5).$$

DIVIDENDS OF 0

Zero divided by any nonzero real number is 0:

$$\frac{0}{a} = 0; \qquad a \neq 0.$$

EXAMPLES Divide.

5. $0 \div (-6) = 0$

6. $\frac{0}{12} = 0$

7. $\frac{-3}{0}$ is not defined.

Do Exercises 7 and 8. ▶

Divide, if possible.

7. $\frac{-5}{0}$

8. $\frac{0}{-3}$

b RECIPROCALS

When two numbers like $\frac{1}{2}$ and 2 are multiplied, the result is 1. Such numbers are called **reciprocals** of each other. Every nonzero real number has a reciprocal, also called a **multiplicative inverse**.

RECIPROCALS

Two numbers whose product is 1 are called **reciprocals**, or **multiplicative inverses**, of each other.

EXAMPLES Find the reciprocal.

8. $\frac{7}{8}$ The reciprocal of $\frac{7}{8}$ is $\frac{8}{7}$ because $\frac{7}{8} \cdot \frac{8}{7} = 1$.

9. -5 The reciprocal of -5 is $-\frac{1}{5}$ because $-5\left(-\frac{1}{5}\right) = 1$.

10. 3.9 The reciprocal of 3.9 is $\frac{1}{3.9}$ because $3.9\left(\frac{1}{3.9}\right) = 1$.

11. $-\frac{1}{2}$ The reciprocal of $-\frac{1}{2}$ is -2 because $\left(-\frac{1}{2}\right)(-2) = 1$.

12. $-\frac{2}{3}$ The reciprocal of $-\frac{2}{3}$ is $-\frac{3}{2}$ because $\left(-\frac{2}{3}\right)\left(-\frac{3}{2}\right) = 1$.

13. $\frac{3y}{8x}$ The reciprocal of $\frac{3y}{8x}$ is $\frac{8x}{3y}$ because $\left(\frac{3y}{8x}\right)\left(\frac{8x}{3y}\right) = 1$.

Answers

7. Not defined **8.** 0

Find the reciprocal.

9. $\frac{2}{3}$

10. $-\frac{5}{4}$

11. -3

12. $-\frac{1}{5}$

13. 1.3

14. $\frac{a}{6b}$

RECIPROCAL PROPERTIES

For $a \neq 0$, the reciprocal of a can be named $\frac{1}{a}$ and the reciprocal of $\frac{1}{a}$ is a.

The reciprocal of a nonzero number $\frac{a}{b}$ can be named $\frac{b}{a}$.

The number 0 has no reciprocal.

Do Exercises 9–14.

The reciprocal of a positive number is also a positive number, because the product of the two numbers must be the positive number 1. The reciprocal of a negative number is also a negative number, because the product of the two numbers must be the positive number 1.

THE SIGN OF A RECIPROCAL

The reciprocal of a number has the same sign as the number itself.

Caution!

It is important *not* to confuse *opposite* with *reciprocal*. Keep in mind that the opposite, or additive inverse, of a number is what we add to the number to get 0. The reciprocal, or multiplicative inverse, is what we multiply the number by to get 1.

Compare the following.

NUMBER	OPPOSITE (Change the sign.)	RECIPROCAL (Invert but do not change the sign.)
$-\frac{3}{8}$	$\frac{3}{8}$	$-\frac{8}{3}$
19	-19	$\frac{1}{19}$
$\frac{18}{7}$	$-\frac{18}{7}$	$\frac{7}{18}$
-7.9	7.9	$-\frac{1}{7.9}$, or $-\frac{10}{79}$
0	0	None

$\left(-\frac{3}{8}\right)\left(-\frac{8}{3}\right) = 1$

$-\frac{3}{8} + \frac{3}{8} = 0$

Do Exercise 15.

15. Complete the following table.

NUMBER	OPPOSITE	RECIPROCAL
$\frac{2}{9}$		
$-\frac{7}{4}$		
0		
1		
-8		
-4.7		

Answers

9. $\frac{3}{2}$ **10.** $-\frac{4}{5}$ **11.** $-\frac{1}{3}$ **12.** -5 **13.** $\frac{1}{1.3}$, or $\frac{10}{13}$ **14.** $\frac{6b}{a}$ **15.** $-\frac{2}{9}$ and $\frac{9}{2}$; $\frac{7}{4}$ and $-\frac{4}{7}$; 0 and none; -1 and 1; 8 and $-\frac{1}{8}$; 4.7 and $-\frac{1}{4.7}$, or $-\frac{10}{47}$

C DIVISION OF REAL NUMBERS

We know that we can subtract by adding an opposite. Similarly, we can divide by multiplying by a reciprocal.

> **RECIPROCALS AND DIVISION**
>
> For any real numbers a and b, $b \neq 0$,
>
> $$a \div b = \frac{a}{b} = a \cdot \frac{1}{b}.$$
>
> (To divide, multiply by the reciprocal of the divisor.)

EXAMPLES Rewrite each division as a multiplication.

14. $-4 \div 3$ $-4 \div 3$ is the same as $-4 \cdot \frac{1}{3}$

15. $\frac{6}{-7}$ $\frac{6}{-7} = 6\left(-\frac{1}{7}\right)$

16. $\frac{3}{5} \div \left(-\frac{9}{7}\right)$ $\frac{3}{5} \div \left(-\frac{9}{7}\right) = \frac{3}{5}\left(-\frac{7}{9}\right)$

17. $\frac{x+2}{5}$ $\frac{x+2}{5} = (x+2)\frac{1}{5}$ Parentheses are necessary here.

18. $\frac{-17}{1/b}$ $\frac{-17}{1/b} = -17 \cdot b$

Do Exercises 16–20. ▶

Rewrite each division as a multiplication.

16. $\frac{4}{7} \div \left(-\frac{3}{5}\right)$

17. $\frac{5}{-8}$

18. $\frac{a-b}{7}$

19. $\frac{-23}{1/a}$

20. $-5 \div 7$

When actually doing division calculations, we sometimes multiply by a reciprocal and we sometimes divide directly. With fraction notation, it is usually better to multiply by a reciprocal. With decimal notation, it is usually better to divide directly.

EXAMPLES Divide by multiplying by the reciprocal of the divisor.

19. $\frac{2}{3} \div \left(-\frac{5}{4}\right) = \frac{2}{3} \cdot \left(-\frac{4}{5}\right) = -\frac{8}{15}$

20. $-\frac{5}{6} \div \left(-\frac{3}{4}\right) = -\frac{5}{6} \cdot \left(-\frac{4}{3}\right) = \frac{20}{18} = \frac{10 \cdot 2}{9 \cdot 2} = \frac{10}{9} \cdot \frac{2}{2} = \frac{10}{9}$

Caution!

Be careful *not* to change the sign when taking a reciprocal!

21. $-\frac{3}{4} \div \frac{3}{10} = -\frac{3}{4} \cdot \left(\frac{10}{3}\right) = -\frac{30}{12} = -\frac{5 \cdot 6}{2 \cdot 6} = -\frac{5}{2} \cdot \frac{6}{6} = -\frac{5}{2}$

Do Exercises 21 and 22. ▶

Divide by multiplying by the reciprocal of the divisor.

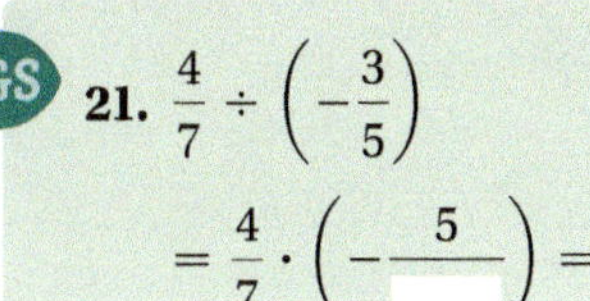

GS **21.** $\frac{4}{7} \div \left(-\frac{3}{5}\right)$

$= \frac{4}{7} \cdot \left(-\frac{5}{\square}\right) = \square$

22. $-\frac{12}{7} \div \left(-\frac{3}{4}\right)$

Answers

16. $\frac{4}{7} \cdot \left(-\frac{5}{3}\right)$ **17.** $5 \cdot \left(-\frac{1}{8}\right)$ **18.** $(a-b) \cdot \frac{1}{7}$
19. $-23 \cdot a$ **20.** $-5 \cdot \left(\frac{1}{7}\right)$ **21.** $-\frac{20}{21}$
22. $\frac{16}{7}$

Guided Solution:
21. $3, -\frac{20}{21}$

With decimal notation, it is easier to carry out long division than to multiply by the reciprocal.

EXAMPLES Divide.

22. $-27.9 \div (-3) = \frac{-27.9}{-3} = 9.3$ Do the long division $3\overline{)27.9}$ (quotient 9.3). The answer is positive.

23. $-6.3 \div 2.1 = -3$ Do the long division $2.1\overline{)6.3}$ (quotient 3.). The answer is negative.

Divide.

23. $21.7 \div (-3.1)$

24. $-20.4 \div (-4)$

Do Exercises 23 and 24.

Consider the following:

1. $\frac{2}{3} = \frac{2}{3} \cdot 1 = \frac{2}{3} \cdot \frac{-1}{-1} = \frac{2(-1)}{3(-1)} = \frac{-2}{-3}$. Thus, $\frac{2}{3} = \frac{-2}{-3}$.

(A negative number divided by a negative number is positive.)

2. $-\frac{2}{3} = -1 \cdot \frac{2}{3} = \frac{-1}{1} \cdot \frac{2}{3} = \frac{-1 \cdot 2}{1 \cdot 3} = \frac{-2}{3}$. Thus, $-\frac{2}{3} = \frac{-2}{3}$.

(A negative number divided by a positive number is negative.)

3. $\frac{-2}{3} = \frac{-2}{3} \cdot 1 = \frac{-2}{3} \cdot \frac{-1}{-1} = \frac{-2(-1)}{3(-1)} = \frac{2}{-3}$. Thus, $-\frac{2}{3} = \frac{2}{-3}$.

(A positive number divided by a negative number is negative.)

We can use the following properties to make sign changes in fraction notation.

SIGN CHANGES IN FRACTION NOTATION

For any numbers a and b, $b \neq 0$:

1. $\frac{-a}{-b} = \frac{a}{b}$

(The opposite of a number a divided by the opposite of another number b is the same as the quotient of the two numbers a and b.)

2. $\frac{-a}{b} = \frac{a}{-b} = -\frac{a}{b}$

(The opposite of a number a divided by another number b is the same as the number a divided by the opposite of the number b, and both are the same as the opposite of *a divided by b*.)

Find two equal expressions for each number with negative signs in different places.

25. $\frac{-5}{6} = \frac{5}{\square} = -\frac{\square}{6}$ GS

26. $-\frac{8}{7}$

27. $\frac{10}{-3}$

Do Exercises 25–27.

d APPLICATIONS AND PROBLEM SOLVING

EXAMPLE 24 *Chemical Reaction.* During a chemical reaction, the temperature in a beaker decreased every minute by the same number of degrees. The temperature was 56°F at 10:10 A.M. By 10:42 A.M., the temperature had dropped to −12°F. By how many degrees did it change each minute?

Answers

23. −7 24. 5.1 25. $\frac{5}{-6}; -\frac{5}{6}$ 26. $\frac{8}{-7}; \frac{-8}{7}$
27. $\frac{-10}{3}; -\frac{10}{3}$

Guided Solution:
25. −6, 5

We first determine by how many degrees d the temperature changed altogether. We subtract -12 from 56:

$$d = 56 - (-12) = 56 + 12 = 68.$$

The temperature changed a total of 68°. We can express this as $-68°$ since the temperature dropped.

The amount of time t that passed was $42 - 10$, or 32 min. Thus the number of degrees T that the temperature dropped each minute is given by

$$T = \frac{d}{t} = \frac{-68}{32} = -2.125.$$

The change was $-2.125°$F per minute.

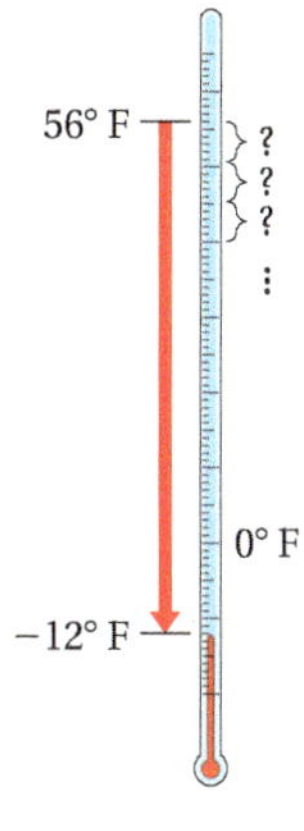

Do Exercise 28. ▶

28. ***Chemical Reaction.*** During a chemical reaction, the temperature in a beaker decreased every minute by the same number of degrees. The temperature was 71°F at 2:12 P.M. By 2:37 P.M., the temperature had changed to −14°F. By how many degrees did it change each minute?

Answer

28. −3.4°F per minute

CALCULATOR CORNER

Operations on the Real Numbers To perform operations on the real numbers on a graphing calculator, recall that negative numbers are entered using the opposite key, (−), rather than the subtraction operation key, −. Consider the sum $-5 + (-3.8)$. On a graphing calculator, the parentheses are not necessary. The result is -8.8. Note that it is not incorrect to enter the parentheses. The result will be the same if this is done. We can also subtract, multiply, and divide real numbers. At right, we see $10 - (-17)$, $-5 \cdot (-7)$, and $45 \div (-9)$. Again, it is not necessary to use parentheses.

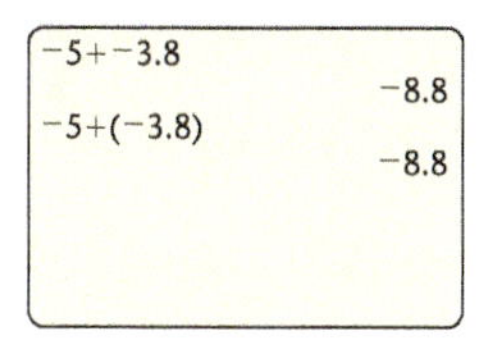

```
-5+-3.8
                -8.8
-5+(-3.8)
                -8.8
```

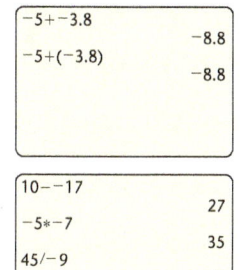

```
10--17
                  27
-5*-7
                  35
45/-9
                  -5
```

EXERCISES: Use a calculator to perform each operation.

1. $-8 + 4$

2. $-7 - (-5)$

3. $-8 \cdot 4$

4. $-7 \div (-5)$

5. $1.2 - (-1.5)$

6. $-7.6 + (-1.9)$

7. $1.2 \div (-1.5)$

8. $-7.6 \cdot (-1.9)$

9.6 Exercise Set

For Extra Help
MyMathLab®

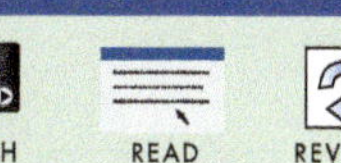

Reading Check

Choose the word or the number below the blank that will make the sentence true.

RC1. The numbers 4 and −4 are called ______________ of each other.
opposites/reciprocals

RC2. The multiplicative inverse, or reciprocal, of a number is what we multiply the number by to get ____.
0/1

RC3. The additive inverse, or opposite, of a number is what we add to the number to get ____.
0/1

RC4. The numbers $-\frac{9}{4}$ and $-\frac{4}{9}$ are called ______________ of each other.
opposites/reciprocals

a Divide, if possible. Check each answer.

1. $48 \div (-6)$

2. $\frac{42}{-7}$

3. $\frac{28}{-2}$

4. $24 \div (-12)$

5. $\frac{-24}{8}$

6. $-18 \div (-2)$

7. $\frac{-36}{-12}$

8. $-72 \div (-9)$

9. $\frac{-72}{9}$

10. $\frac{-50}{25}$

11. $-100 \div (-50)$

12. $\frac{-200}{8}$

13. $-108 \div 9$

14. $\frac{-63}{-7}$

15. $\frac{200}{-25}$

16. $-300 \div (-16)$

17. $\frac{75}{0}$

18. $\frac{0}{-5}$

19. $\frac{0}{-2.6}$

20. $\frac{-23}{0}$

b Find the reciprocal.

21. $\frac{15}{7}$

22. $\frac{3}{8}$

23. $-\frac{47}{13}$

24. $-\frac{31}{12}$

25. 13

26. -10

27. -32

28. 15

29. $\frac{1}{-7.1}$

30. $\frac{1}{-4.9}$

31. $\frac{1}{9}$

32. $\frac{1}{16}$

33. $\frac{1}{4y}$

34. $\frac{-1}{8a}$

35. $\frac{2a}{3b}$

36. $\frac{-4y}{3x}$

C Rewrite each division as a multiplication.

37. $4 \div 17$

38. $5 \div (-8)$

39. $\frac{8}{-13}$

40. $-\frac{13}{47}$

41. $\frac{13.9}{-1.5}$

42. $-\frac{47.3}{21.4}$

43. $\frac{2}{3} \div \left(-\frac{4}{5}\right)$

44. $\frac{3}{4} \div \left(-\frac{7}{10}\right)$

45. $\frac{x}{\frac{1}{y}}$

46. $\frac{13}{\frac{1}{x}}$

47. $\frac{3x + 4}{5}$

48. $\frac{4y - 8}{-7}$

Divide.

49. $\frac{3}{4} \div \left(-\frac{2}{3}\right)$

50. $\frac{7}{8} \div \left(-\frac{1}{2}\right)$

51. $-\frac{5}{4} \div \left(-\frac{3}{4}\right)$

52. $-\frac{5}{9} \div \left(-\frac{5}{6}\right)$

53. $-\frac{2}{7} \div \left(-\frac{4}{9}\right)$

54. $-\frac{3}{5} \div \left(-\frac{5}{8}\right)$

55. $-\frac{3}{8} \div \left(-\frac{8}{3}\right)$

56. $-\frac{5}{8} \div \left(-\frac{6}{5}\right)$

57. $-\frac{5}{6} \div \frac{2}{3}$

58. $-\frac{7}{16} \div \frac{3}{8}$

59. $-\frac{9}{4} \div \frac{5}{12}$

60. $-\frac{3}{5} \div \frac{7}{10}$

61. $\frac{-11}{-13}$

62. $\frac{-21}{-25}$

63. $-6.6 \div 3.3$

64. $-44.1 \div (-6.3)$

65. $\frac{48.6}{-3}$

66. $\frac{-1.9}{20}$

67. $\frac{-12.5}{5}$

68. $\frac{-17.8}{3.2}$

69. $11.25 \div (-9)$

70. $-9.6 \div (-6.4)$

71. $\frac{-9}{17 - 17}$

72. $\frac{-8}{-5 + 5}$

d To determine percent increase or decrease, we first determine the change by subtracting the new amount from the original amount. Then we divide the change, which is *positive* if an increase, or *negative* if a decrease, by the original amount. Finally, we convert the decimal answer to percent notation.

73. *Passports.* In 2006, approximately 71 million valid passports were in circulation in the United States. This number increased to approximately 102 million in 2011. What is the percent increase?

Source: U.S. State Department

74. *Super Bowl Spending.* The average amount spent during the Super Bowl per television viewer increased from \$59.33 in 2011 to \$63.87 in 2012. What is the percent increase?

Source: Retail Advertising and Marketing Association, Super Bowl Consumer Intentions and Actions Survey

75. *Pieces of Mail.* The number of pieces of mail handled by the U.S. Postal Service decreased from 212 billion in 2007 to 168 billion in 2011. What is the percent decrease?

Source: U.S. Postal Service

76. *Beef Consumption.* Beef consumption per capita in the United States decreased from 67.5 lb in 1990 to 57.4 in 2011. What is the percent decrease?

Source: U.S. Department of Agriculture

Skill Maintenance

Simplify. [1.9c]

77. $2^3 - 5 \cdot 3 + 8 \cdot 10 \div 2$

78. $16 \cdot 2^3 - 5 \cdot 3 + 80 \div 10 \cdot 2$

79. $1000 \div 100 \div 10$

80. $216 \cdot 6^3 \div 6^2$

81. Simplify: $\frac{264}{468}$. [3.5b]

82. Convert to decimal notation: 47.7% [6.2b]

83. Convert to percent notation: $\frac{7}{8}$. [6.3a]

84. Simplify: $\frac{40}{60}$. [3.5b]

85. Divide and simplify: $\frac{12}{25} \div \frac{32}{75}$. [3.7b]

86. Multiply and simplify: $\frac{12}{25} \cdot \frac{32}{75}$. [3.6a]

Synthesis

87. Find the reciprocal of -10.5. What happens if you take the reciprocal of the result?

88. Determine those real numbers a for which the opposite of a is the same as the reciprocal of a.

Determine whether each expression represents a positive number or a negative number when a and b are negative.

89. $\frac{-a}{b}$

90. $\frac{-a}{-b}$

91. $-\left(\frac{a}{-b}\right)$

92. $-\left(\frac{-a}{b}\right)$

93. $-\left(\frac{-a}{-b}\right)$

Properties of Real Numbers

9.7

a EQUIVALENT EXPRESSIONS

In solving equations and doing other kinds of work in algebra, we manipulate expressions in various ways. For example, instead of $x + x$, we might write $2x$, knowing that the two expressions represent the same number for any allowable replacement of x. In that sense, the expressions $x + x$ and $2x$ are **equivalent**, as are $\frac{3}{x}$ and $\frac{3x}{x^2}$, even though 0 is not an allowable replacement because division by 0 is not defined.

EQUIVALENT EXPRESSIONS

Two expressions that have the same value for all allowable replacements are called **equivalent**.

The expressions $x + 3x$ and $5x$ are *not* equivalent, as we see in Margin Exercise 2.

Do Margin Exercises 1 and 2. ▶

In this section, we will consider several laws of real numbers that will allow us to find equivalent expressions. The first two laws are the *identity properties of 0 and 1.*

THE IDENTITY PROPERTY OF 0

For any real number a,

$$a + 0 = 0 + a = a.$$

(The number 0 is the *additive identity.*)

THE IDENTITY PROPERTY OF 1

For any real number a,

$$a \cdot 1 = 1 \cdot a = a.$$

(The number 1 is the *multiplicative identity.*)

We often refer to the use of the identity property of 1 as "multiplying by 1." We can use this method to find equivalent fraction expressions. Recall from arithmetic that to multiply with fraction notation, we multiply the numerators and multiply the denominators.

EXAMPLE 1 Write a fraction expression equivalent to $\frac{2}{3}$ with a denominator of $3x$:

$$\frac{2}{3} = \frac{\square}{3x}.$$

OBJECTIVES

a Find equivalent fraction expressions and simplify fraction expressions.

b Use the commutative laws and the associative laws to find equivalent expressions.

c Use the distributive laws to multiply expressions like 8 and $x - y$.

d Use the distributive laws to factor expressions like $4x - 12 + 24y$.

e Collect like terms.

SKILL TO REVIEW

Objective 2.2a: Add real numbers.

Add.

1. $-16 + 5$
2. $29 + (-23)$

Complete the table by evaluating each expression for the given values.

1.

VALUE	$x + x$	$2x$
$x = 3$		
$x = -6$		
$x = 4.8$		

2.

VALUE	$x + 3x$	$5x$
$x = 2$		
$x = -6$		
$x = 4.8$		

Answers

Skill to Review:
1. -11 **2.** 6

Margin Exercises:
1. 6, 6; -12, -12; 9.6, 9.6 **2.** 8, 10; -24, -30; 19.2, 24

3. Write a fraction expression equivalent to $\frac{3}{4}$ with a denominator of 8:

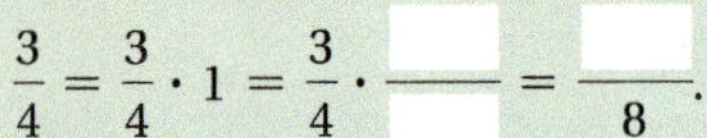

$$\frac{3}{4} = \frac{3}{4} \cdot 1 = \frac{3}{4} \cdot \frac{\square}{\square} = \frac{\square}{8}.$$

4. Write a fraction expression equivalent to $\frac{3}{4}$ with a denominator of $4t$:

$$\frac{3}{4} = \frac{3}{4} \cdot 1 = \frac{3}{4} \cdot \frac{\square}{\square} = \frac{\square}{4t}.$$

GS

Note that $3x = 3 \cdot x$. We want fraction notation for $\frac{2}{3}$ that has a denominator of $3x$, but the denominator 3 is missing a factor of x. Thus we multiply by 1, using x/x as an equivalent expression for 1:

$$\frac{2}{3} = \frac{2}{3} \cdot 1 = \frac{2}{3} \cdot \frac{x}{x} = \frac{2x}{3x}.$$

The expressions $2/3$ and $2x/(3x)$ are equivalent. They have the same value for any allowable replacement. Note that $2x/3x$ is not defined for a replacement of 0, but for all nonzero real numbers, the expressions $2/3$ and $2x/(3x)$ have the same value.

Do Exercises 3 and 4.

In algebra, we consider an expression like $2/3$ to be "simplified" from $2x/(3x)$. To find such simplified expressions, we use the identity property of 1 to remove a factor of 1.

Simplify.

5. $\frac{3y}{4y}$

6. $-\frac{16m}{12m}$

7. $\frac{5xy}{40y}$

EXAMPLE 2 Simplify: $-\frac{20x}{12x}$.

$-\frac{20x}{12x} = -\frac{5 \cdot 4x}{3 \cdot 4x}$ We look for the largest factor common to both the numerator and the denominator and factor each.

$= -\frac{5}{3} \cdot \frac{4x}{4x}$ Factoring the fraction expression

$= -\frac{5}{3} \cdot 1$ $\frac{4x}{4x} = 1$

$= -\frac{5}{3}$ Removing a factor of 1 using the identity property of 1

8. $$\frac{18p}{24pq} = \frac{6p \cdot 3}{6p \cdot \square}$$
$$= \frac{6p}{6p} \cdot \frac{\square}{4q}$$
$$= 1 \cdot \frac{3}{4q} = \frac{3}{4q}$$

GS

EXAMPLE 3 Simplify: $\frac{14ab}{56a}$.

$$\frac{14ab}{56a} = \frac{14a \cdot b}{14a \cdot 4} = \frac{14a}{14a} \cdot \frac{b}{4} = 1 \cdot \frac{b}{4} = \frac{b}{4}$$

Do Exercises 5–8.

b THE COMMUTATIVE LAWS AND THE ASSOCIATIVE LAWS

The Commutative Laws

Let's examine the expressions $x + y$ and $y + x$, as well as xy and yx.

9. Evaluate $x + y$ and $y + x$ when $x = -2$ and $y = 3$.

10. Evaluate xy and yx when $x = -2$ and $y = 5$.

EXAMPLE 4 Evaluate $x + y$ and $y + x$ when $x = 4$ and $y = 3$.

We substitute 4 for x and 3 for y in both expressions:

$$x + y = 4 + 3 = 7; \qquad y + x = 3 + 4 = 7.$$

EXAMPLE 5 Evaluate xy and yx when $x = 3$ and $y = -12$.

We substitute 3 for x and -12 for y in both expressions:

$$xy = 3 \cdot (-12) = -36; \qquad yx = (-12) \cdot 3 = -36.$$

Do Exercises 9 and 10.

Answers

3. $\frac{6}{8}$ **4.** $\frac{3t}{4t}$ **5.** $\frac{3}{4}$ **6.** $-\frac{4}{3}$ **7.** $\frac{x}{8}$
8. $\frac{3}{4q}$ **9.** 1; 1 **10.** -10; -10

Guided Solutions:
3. $\frac{2}{2}$, 6 **4.** $\frac{t}{t}$, $3t$ **8.** $4q$, 3

The expressions $x + y$ and $y + x$ have the same values no matter what the variables stand for. Thus they are equivalent. Therefore, when we add two numbers, the order in which we add does not matter. Similarly, the expressions xy and yx are equivalent. They also have the same values, no matter what the variables stand for. Therefore, when we multiply two numbers, the order in which we multiply does not matter.

The following are examples of general patterns or laws.

THE COMMUTATIVE LAWS

Addition. For any numbers a and b,

$$a + b = b + a.$$

(We can change the order when adding without affecting the answer.)

Multiplication. For any numbers a and b,

$$ab = ba.$$

(We can change the order when multiplying without affecting the answer.)

Using a commutative law, we know that $x + 2$ and $2 + x$ are equivalent. Similarly, $3x$ and $x(3)$ are equivalent. Thus, in an algebraic expression, we can replace one with the other and the result will be equivalent to the original expression.

EXAMPLE 6 Use the commutative laws to write an equivalent expression: **(a)** $y + 5$; **(b)** mn; **(c)** $7 + xy$.

a) An expression equivalent to $y + 5$ is $5 + y$ by the commutative law of addition.

b) An expression equivalent to mn is nm by the commutative law of multiplication.

c) An expression equivalent to $7 + xy$ is $xy + 7$ by the commutative law of addition. Another expression equivalent to $7 + xy$ is $7 + yx$ by the commutative law of multiplication. Another equivalent expression is $yx + 7$.

Do Exercises 11–13. ▶

Use a commutative law to write an equivalent expression.

11. $x + 9$

12. pq

13. $xy + t$

The Associative Laws

Now let's examine the expressions $a + (b + c)$ and $(a + b) + c$. Note that these expressions involve the use of parentheses as *grouping* symbols, and they also involve three numbers. Calculations within parentheses are to be done first.

EXAMPLE 7 Calculate and compare: $3 + (8 + 5)$ and $(3 + 8) + 5$.

$3 + (8 + 5) = 3 + 13$ Calculating within parentheses first; adding 8 and 5

$= 16;$

$(3 + 8) + 5 = 11 + 5$ Calculating within parentheses first; adding 3 and 8

$= 16$

Answers

11. $9 + x$ **12.** qp
13. $t + xy$, or $yx + t$, or $t + yx$

The two expressions in Example 7 name the same number. Moving the parentheses to group the additions differently does not affect the value of the expression.

14. Calculate and compare:
$8 + (9 + 2)$ and $(8 + 9) + 2$.

15. Calculate and compare:
$10 \cdot (5 \cdot 3)$ and $(10 \cdot 5) \cdot 3$.

EXAMPLE 8 Calculate and compare: $3 \cdot (4 \cdot 2)$ and $(3 \cdot 4) \cdot 2$.

$$3 \cdot (4 \cdot 2) = 3 \cdot 8 = 24; \qquad (3 \cdot 4) \cdot 2 = 12 \cdot 2 = 24$$

◀ **Do Exercises 14 and 15.**

You may have noted that when only addition is involved, numbers can be grouped in any way we please without affecting the answer. When only multiplication is involved, numbers can also be grouped in any way we please without affecting the answer.

THE ASSOCIATIVE LAWS

Addition. For any numbers a, b, and c,

$$a + (b + c) = (a + b) + c.$$

(Numbers can be grouped in any manner for addition.)

Multiplication. For any numbers a, b, and c,

$$a \cdot (b \cdot c) = (a \cdot b) \cdot c.$$

(Numbers can be grouped in any manner for multiplication.)

EXAMPLE 9 Use an associative law to write an equivalent expression: **(a)** $(y + z) + 3$; **(b)** $8(xy)$.

a) An expression equivalent to $(y + z) + 3$ is $y + (z + 3)$ by the associative law of addition.

b) An expression equivalent to $8(xy)$ is $(8x)y$ by the associative law of multiplication.

Use an associative law to write an equivalent expression.

16. $r + (s + 7)$

17. $9(ab)$

◀ **Do Exercises 16 and 17.**

The associative laws say that numbers can be grouped in any way we please when only additions or only multiplications are involved. Thus we often omit the parentheses. For example,

$$x + (y + 2) \text{ means } x + y + 2, \quad \text{and} \quad (lw)h \text{ means } lwh.$$

Using the Commutative Laws and the Associative Laws Together

EXAMPLE 10 Use the commutative laws and the associative laws to write at least three expressions equivalent to $(x + 5) + y$.

a) $(x + 5) + y = x + (5 + y)$ Using the associative law first and then using the commutative law
$= x + (y + 5)$

b) $(x + 5) + y = y + (x + 5)$ Using the commutative law twice
$= y + (5 + x)$

c) $(x + 5) + y = (5 + x) + y$ Using the commutative law first and then the associative law
$= 5 + (x + y)$

Answers

14. 19; 19 15. 150; 150 16. $(r + s) + 7$
17. $(9a)b$

EXAMPLE 11 Use the commutative laws and the associative laws to write at least three expressions equivalent to $(3x)y$.

a) $(3x)y = 3(xy)$ Using the associative law first and then using the commutative law
$= 3(yx)$

b) $(3x)y = y(3x)$ Using the commutative law twice
$= y(x \cdot 3)$

c) $(3x)y = (x \cdot 3)y$ Using the commutative law, and then the associative law, and then the commutative law again
$= x(3y)$
$= x(y \cdot 3)$

Do Exercises 18 and 19. ▶

Use the commutative laws and the associative laws to write at least three equivalent expressions.

18. $4(tu)$

19. $r + (2 + s)$

C THE DISTRIBUTIVE LAWS

The *distributive laws* are the basis of many procedures in both arithmetic and algebra. They are probably the most important laws that we use to manipulate algebraic expressions. The distributive law of multiplication over addition involves two operations: addition and multiplication.

Let's begin by considering a multiplication problem from arithmetic:

```
    4 5
  ×   7
  -----
    3 5  ← This is 7 · 5.
  2 8 0  ← This is 7 · 40.
  -----
  3 1 5  ← This is the sum 7 · 5 + 7 · 40.
```

To carry out the multiplication, we actually added two products. That is,

$$7 \cdot 45 = 7(5 + 40) = 7 \cdot 5 + 7 \cdot 40.$$

Let's examine this further. If we wish to multiply a sum of several numbers by a factor, we can either add and then multiply, or multiply and then add.

EXAMPLE 12 Compute in two ways: $5 \cdot (4 + 8)$.

a) $5 \cdot (4 + 8)$ Adding within parentheses first, and then multiplying
$= 5 \cdot 12$
$= 60$

b) $5 \cdot (4 + 8) = (5 \cdot 4) + (5 \cdot 8)$ Distributing the multiplication to terms within parentheses first and then adding
$= 20 + 40$
$= 60$

Do Exercises 20–22. ▶

Compute.

20. a) $7 \cdot (3 + 6)$
b) $(7 \cdot 3) + (7 \cdot 6)$

21. a) $2 \cdot (10 + 30)$
b) $(2 \cdot 10) + (2 \cdot 30)$

22. a) $(2 + 5) \cdot 4$
b) $(2 \cdot 4) + (5 \cdot 4)$

THE DISTRIBUTIVE LAW OF MULTIPLICATION OVER ADDITION

For any numbers a, b, and c,

$$a(b + c) = ab + ac.$$

Answers

18. $(4t)u$, $(tu)4$, $t(4u)$; answers may vary
19. $(2 + r) + s$, $(r + s) + 2$, $s + (r + 2)$; answers may vary
20. (a) $7 \cdot 9 = 63$; **(b)** $21 + 42 = 63$
21. (a) $2 \cdot 40 = 80$; **(b)** $20 + 60 = 80$
22. (a) $7 \cdot 4 = 28$; **(b)** $8 + 20 = 28$

In the statement of the distributive law, we know that in an expression such as $ab + ac$, the multiplications are to be done first according to the rules for order of operations. So, instead of writing $(4 \cdot 5) + (4 \cdot 7)$, we can write $4 \cdot 5 + 4 \cdot 7$. However, in $a(b + c)$, we cannot omit the parentheses. If we did, we would have $ab + c$, which means $(ab) + c$. For example, $3(4 + 2) = 3(6) = 18$, but $3 \cdot 4 + 2 = 12 + 2 = 14$.

Another distributive law relates multiplication and subtraction. This law says that to multiply by a difference, we can either subtract and then multiply, or multiply and then subtract.

Calculate.

23. **a)** $4(5 - 3)$
b) $4 \cdot 5 - 4 \cdot 3$

24. **a)** $-2 \cdot (5 - 3)$
b) $-2 \cdot 5 - (-2) \cdot 3$

25. **a)** $5 \cdot (2 - 7)$
b) $5 \cdot 2 - 5 \cdot 7$

THE DISTRIBUTIVE LAW OF MULTIPLICATION OVER SUBTRACTION

For any numbers a, b, and c,

$$a(b - c) = ab - ac.$$

We often refer to "*the* distributive law" when we mean *either* or *both* of these laws.

Do Exercises 23–25.

What do we mean by the *terms* of an expression? **Terms** are separated by addition signs. If there are subtraction signs, we can find an equivalent expression that uses addition signs.

EXAMPLE 13 What are the terms of $3x - 4y + 2z$?

We have

$$3x - 4y + 2z = 3x + (-4y) + 2z. \quad \text{Separating parts with + signs}$$

The terms are $3x$, $-4y$, and $2z$.

What are the terms of each expression?

26. $5x - 8y + 3$

27. $-4y - 2x + 3z$

Do Exercises 26 and 27.

The distributive laws are a basis for **multiplying** algebraic expressions. In an expression like $8(a + 2b - 7)$, we multiply each term inside the parentheses by 8:

$$8(a + 2b - 7) = 8 \cdot a + 8 \cdot 2b - 8 \cdot 7 = 8a + 16b - 56.$$

EXAMPLES Multiply.

14. $9(x - 5) = 9 \cdot x - 9 \cdot 5$ Using the distributive law of multiplication over subtraction

$= 9x - 45$

15. $\frac{2}{3}(w + 1) = \frac{2}{3} \cdot w + \frac{2}{3} \cdot 1$ Using the distributive law of multiplication over addition

$= \frac{2}{3}w + \frac{2}{3}$

16. $\frac{4}{3}(s - t + w) = \frac{4}{3}s - \frac{4}{3}t + \frac{4}{3}w$ Using both distributive laws

Do Exercises 28–30.

Multiply.

28. $3(x - 5)$

29. $5(x + 1)$

30. $\frac{3}{5}(p + q - t)$

Answers

23. **(a)** $4 \cdot 2 = 8$; **(b)** $20 - 12 = 8$
24. **(a)** $-2 \cdot 2 = -4$; **(b)** $-10 + 6 = -4$
25. **(a)** $5(-5) = -25$; **(b)** $10 - 35 = -25$
26. $5x, -8y, 3$ **27.** $-4y, -2x, 3z$ **28.** $3x - 15$
29. $5x + 5$ **30.** $\frac{3}{5}p + \frac{3}{5}q - \frac{3}{5}t$

EXAMPLE 17 Multiply: $-4(x - 2y + 3z)$.

$$-4(x - 2y + 3z) = -4 \cdot x - (-4)(2y) + (-4)(3z)$$ Using both distributive laws

$$= -4x - (-8y) + (-12z)$$ Multiplying

$$= -4x + 8y - 12z$$

We can also do this problem by first finding an equivalent expression with all plus signs and then multiplying:

$$-4(x - 2y + 3z) = -4[x + (-2y) + 3z]$$
$$= -4 \cdot x + (-4)(-2y) + (-4)(3z)$$
$$= -4x + 8y - 12z.$$

Do Exercises 31–33. ▶

Multiply.

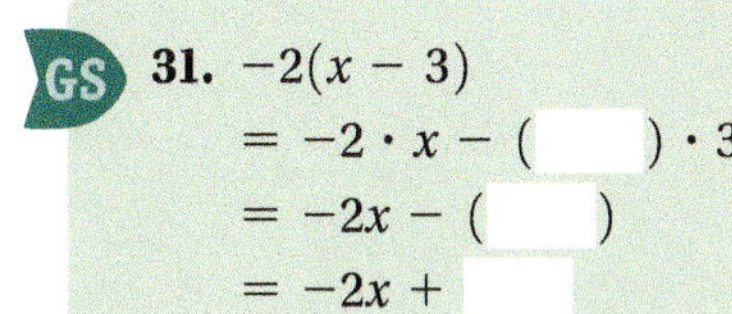

GS **31.** $-2(x - 3)$
$= -2 \cdot x - (\quad) \cdot 3$
$= -2x - (\quad)$
$= -2x + \quad$

32. $5(x - 2y + 4z)$

33. $-5(x - 2y + 4z)$

EXAMPLES Name the property or the law illustrated by each equation.

Equation	*Property*
18. $5x = x(5)$	Commutative law of multiplication
19. $a + (8.5 + b) = (a + 8.5) + b$	Associative law of addition
20. $0 + 11 = 11$	Identity property of 0
21. $(-5s)t = -5(st)$	Associative law of multiplication
22. $\frac{3}{4} \cdot 1 = \frac{3}{4}$	Identity property of 1
23. $12.5(w - 3) = 12.5w - 12.5(3)$	Distributive law of multiplication over subtraction
24. $y + \frac{1}{2} = \frac{1}{2} + y$	Commutative law of addition

Do Exercises 34–40. ▶

Name the property or the law illustrated by each equation.

34. $(-8a)b = -8(ab)$

35. $p \cdot 1 = p$

36. $m + 34 = 34 + m$

37. $2(t + 5) = 2t + 2(5)$

38. $0 + k = k$

39. $-8x = x(-8)$

40. $x + (4.3 + b) = (x + 4.3) + b$

d FACTORING

Factoring is the reverse of multiplying. To factor, we can use the distributive laws in reverse:

$$ab + ac = a(b + c) \quad \text{and} \quad ab - ac = a(b - c).$$

FACTORING

To **factor** an expression is to find an equivalent expression that is a product.

To factor $9x - 45$, for example, we find an equivalent expression that is a product: $9(x - 5)$. This reverses the multiplication that we did in Example 14. When all the terms of an expression have a factor in common, we can "factor it out" using the distributive laws. Note the following.

$9x$ has the factors $9, -9, 3, -3, 1, -1, x, -x, 3x, -3x, 9x, -9x$;
-45 has the factors $1, -1, 3, -3, 5, -5, 9, -9, 15, -15, 45, -45$

Answers

31. $-2x + 6$ **32.** $5x - 10y + 20z$ **33.** $-5x + 10y - 20z$ **34.** Associative law of multiplication **35.** Identity property of 1 **36.** Commutative law of addition **37.** Distributive law of multiplication over addition **38.** Identity property of 0 **39.** Commutative law of multiplication **40.** Associative law of addition

Guided Solution:
31. $-2, -6, 6$

We generally remove the largest common factor. In this case, that factor is 9. Thus,

$$9x - 45 = 9 \cdot x - 9 \cdot 5$$
$$= 9(x - 5).$$

Remember that an expression has been factored when we have found an equivalent expression that is a product. Above, we note that $9x - 45$ and $9(x - 5)$ are equivalent expressions. The expression $9x - 45$ is the difference of $9x$ and 45; the expression $9(x - 5)$ is the product of 9 and $(x - 5)$.

EXAMPLES Factor.

25. $5x - 10 = 5 \cdot x - 5 \cdot 2$ Try to do this step mentally.
$= 5(x - 2)$ You can check by multiplying.

26. $ax - ay + az = a(x - y + z)$

27. $9x + 27y - 9 = 9 \cdot x + 9 \cdot 3y - 9 \cdot 1 = 9(x + 3y - 1)$

Note in Example 27 that you might, at first, just factor out a 3, as follows:

$$9x + 27y - 9 = 3 \cdot 3x + 3 \cdot 9y - 3 \cdot 3$$
$$= 3(3x + 9y - 3).$$

At this point, the mathematics is correct, but the answer is not because there is another factor of 3 that can be factored out, as follows:

$$3 \cdot 3x + 3 \cdot 9y - 3 \cdot 3 = 3(3x + 9y - 3)$$
$$= 3(3 \cdot x + 3 \cdot 3y - 3 \cdot 1)$$
$$= 3 \cdot 3(x + 3y - 1)$$
$$= 9(x + 3y - 1).$$

We now have a correct answer, but it took more work than we did in Example 27. Thus it is better to look for the *greatest common factor* at the outset.

EXAMPLES Factor. Try to write just the answer, if you can.

28. $5x - 5y = 5(x - y)$

29. $-3x + 6y - 9z = -3(x - 2y + 3z)$

We generally factor out a negative factor when the first term is negative. The way we factor can depend on the situation in which we are working. We might also factor the expression in Example 29 as follows:

$$-3x + 6y - 9z = 3(-x + 2y - 3z).$$

30. $18z - 12x - 24 = 6(3z - 2x - 4)$

31. $\frac{1}{2}x + \frac{3}{2}y - \frac{1}{2} = \frac{1}{2}(x + 3y - 1)$

Remember that you can always check factoring by multiplying. Keep in mind that an expression is factored when it is written as a product.

Do Exercises 41–46.

Factor.

41. $6x - 12$

42. $3x - 6y + 9$

43. $bx + by - bz$

44. $16a - 36b + 42$ GS
$= 2 \cdot 8a - \_\_\_ \cdot 18b + 2 \cdot 21$
$= \_\_\_ (8a - 18b + 21)$

45. $\frac{3}{8}x - \frac{5}{8}y + \frac{7}{8}$

46. $-12x + 32y - 16z$

Answers

41. $6(x - 2)$ **42.** $3(x - 2y + 3)$
43. $b(x + y - z)$ **44.** $2(8a - 18b + 21)$
45. $\frac{1}{8}(3x - 5y + 7)$ **46.** $-4(3x - 8y + 4z)$, or $4(-3x + 8y - 4z)$

Guided Solution:
44. 2, 2

e COLLECTING LIKE TERMS

Terms such as $5x$ and $-4x$, whose variable factors are exactly the same, are called **like terms**. Similarly, numbers, such as -7 and 13, are like terms. Also, $3y^2$ and $9y^2$ are like terms because the variables are raised to the same power. Terms such as $4y$ and $5y^2$ are not like terms, and $7x$ and $2y$ are not like terms.

The process of **collecting like terms** is also based on the distributive laws. We can apply a distributive law when a factor is on the right-hand side because of the commutative law of multiplication.

Later in this text, terminology like "collecting like terms" and "combining like terms" will also be referred to as "simplifying."

EXAMPLES Collect like terms. Try to write just the answer, if you can.

32. $4x + 2x = (4 + 2)x = 6x$ Factoring out x using a distributive law

33. $2x + 3y - 5x - 2y = 2x - 5x + 3y - 2y$
$= (2 - 5)x + (3 - 2)y = -3x + 1y = -3x + y$

34. $3x - x = 3x - 1x = (3 - 1)x = 2x$

35. $x - 0.24x = 1 \cdot x - 0.24x = (1 - 0.24)x = 0.76x$

36. $x - 6x = 1 \cdot x - 6 \cdot x = (1 - 6)x = -5x$

37. $4x - 7y + 9x - 5 + 3y - 8 = 13x - 4y - 13$

38. $\frac{2}{3}a - b + \frac{4}{5}a + \frac{1}{4}b - 10 = \frac{2}{3}a - 1 \cdot b + \frac{4}{5}a + \frac{1}{4}b - 10$
$= (\frac{2}{3} + \frac{4}{5})a + (-1 + \frac{1}{4})b - 10$
$= (\frac{10}{15} + \frac{12}{15})a + (-\frac{4}{4} + \frac{1}{4})b - 10$
$= \frac{22}{15}a - \frac{3}{4}b - 10$

Do Exercises 47–53. ▶

Collect like terms.

47. $6x - 3x$ **48.** $7x - x$

49. $x - 9x$ **50.** $x - 0.41x$

51. $5x + 4y - 2x - y$

GS **52.** $3x - 7x - 11 + 8y + 4 - 13y$
$= (3 - \quad)x + (8 - 13)y + (\quad + 4)$
$= \quad x + (\quad)y + (\quad)$
$= -4x - 5y - 7$

53. $-\frac{2}{3} - \frac{3}{5}x + y + \frac{7}{10}x - \frac{2}{9}y$

Answers

47. $3x$ **48.** $6x$ **49.** $-8x$ **50.** $0.59x$ **51.** $3x + 3y$ **52.** $-4x - 5y - 7$ **53.** $\frac{1}{10}x + \frac{7}{9}y - \frac{2}{3}$

Guided Solution:
52. 7, −11, −4, −5, −7

9.7 Exercise Set

For Extra Help MyMathLab® MathXL®

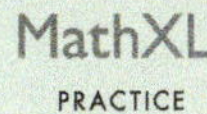
PRACTICE

WATCH

READ

REVIEW

✓ Reading Check

Choose from the column on the right an equation that illustrates the property or law.

RC1. Associative law of multiplication

RC2. Identity property of 1

RC3. Distributive law of multiplication over subtraction

RC4. Commutative law of addition

RC5. Identity property of 0

RC6. Commutative law of multiplication

RC7. Associative law of addition

a) $3 \cdot 5 = 5 \cdot 3$

b) $8 + (\frac{1}{2} + 9) = (8 + \frac{1}{2}) + 9$

c) $5(6 + 3) = 5 \cdot 6 + 5 \cdot 3$

d) $3 + 0 = 3$

e) $3 + 5 = 5 + 3$

f) $5(6 - 3) = 5 \cdot 6 - 5 \cdot 3$

g) $8 \cdot \left(\frac{1}{2} \cdot 9\right) = \left(8 \cdot \frac{1}{2}\right) \cdot 9$

h) $\frac{6}{5} \cdot 1 = \frac{6}{5}$

a Find an equivalent expression with the given denominator.

1. $\frac{3}{5} = \frac{\square}{5y}$
2. $\frac{5}{8} = \frac{\square}{8t}$
3. $\frac{2}{3} = \frac{\square}{15x}$
4. $\frac{6}{7} = \frac{\square}{14y}$
5. $\frac{2}{x} = \frac{\square}{x^2}$
6. $\frac{4}{9x} = \frac{\square}{9xy}$

Simplify.

7. $-\frac{24a}{16a}$
8. $-\frac{42t}{18t}$
9. $-\frac{42ab}{36ab}$
10. $-\frac{64pq}{48pq}$
11. $\frac{20st}{15t}$
12. $\frac{21w}{7wz}$

b Write an equivalent expression. Use a commutative law.

13. $y + 8$
14. $x + 3$
15. mn
16. yz
17. $9 + xy$
18. $11 + ab$
19. $ab + c$
20. $rs + t$

Write an equivalent expression. Use an associative law.

21. $a + (b + 2)$
22. $3(vw)$
23. $(8x)y$
24. $(y + z) + 7$
25. $(a + b) + 3$
26. $(5 + x) + y$
27. $3(ab)$
28. $(6x)y$

Use the commutative laws and the associative laws to write three equivalent expressions.

29. $(a + b) + 2$
30. $(3 + x) + y$
31. $5 + (v + w)$
32. $6 + (x + y)$
33. $(xy)3$
34. $(ab)5$
35. $7(ab)$
36. $5(xy)$

c Multiply.

37. $2(b + 5)$
38. $4(x + 3)$
39. $7(1 + t)$
40. $4(1 + y)$
41. $6(5x + 2)$
42. $9(6m + 7)$
43. $7(x + 4 + 6y)$
44. $4(5x + 8 + 3p)$

45. $7(x - 3)$

46. $15(y - 6)$

47. $-3(x - 7)$

48. $1.2(x - 2.1)$

49. $\frac{2}{3}(b - 6)$

50. $\frac{5}{8}(y + 16)$

51. $7.3(x - 2)$

52. $5.6(x - 8)$

53. $-\frac{3}{5}(x - y + 10)$

54. $-\frac{2}{3}(a + b - 12)$

55. $-9(-5x - 6y + 8)$

56. $-7(-2x - 5y + 9)$

57. $-4(x - 3y - 2z)$

58. $8(2x - 5y - 8z)$

59. $3.1(-1.2x + 3.2y - 1.1)$

60. $-2.1(-4.2x - 4.3y - 2.2)$

List the terms of each expression.

61. $4x + 3z$

62. $8x - 1.4y$

63. $7x + 8y - 9z$

64. $8a + 10b - 18c$

d Factor. Check by multiplying.

65. $2x + 4$

66. $5y + 20$

67. $30 + 5y$

68. $7x + 28$

69. $14x + 21y$

70. $18a + 24b$

71. $14t - 7$

72. $25m - 5$

73. $8x - 24$

74. $10x - 50$

75. $18a - 24b$

76. $32x - 20y$

77. $-4y + 32$

78. $-6m + 24$

79. $5x + 10 + 15y$

80. $9a + 27b + 81$

81. $16m - 32n + 8$

82. $6x + 10y - 2$

83. $12a + 4b - 24$

84. $8m - 4n + 12$

85. $8x + 10y - 22$

86. $9a + 6b - 15$

87. $ax - a$

88. $by - 9b$

89. $ax - ay - az$

90. $cx + cy - cz$

91. $-18x + 12y + 6$

92. $-14x + 21y + 7$

93. $\frac{2}{3}x - \frac{5}{3}y + \frac{1}{3}$

94. $\frac{3}{5}a + \frac{4}{5}b - \frac{1}{5}$

95. $36x - 6y + 18z$

96. $8a - 4b + 20c$

e Collect like terms.

97. $9a + 10a$

98. $12x + 2x$

99. $10a - a$

100. $-16x + x$

101. $2x + 9z + 6x$

102. $3a - 5b + 7a$

103. $7x + 6y^2 + 9y^2$

104. $12m^2 + 6q + 9m^2$

105. $41a + 90 - 60a - 2$

106. $42x - 6 - 4x + 2$

107. $23 + 5t + 7y - t - y - 27$

108. $45 - 90d - 87 - 9d + 3 + 7d$

109. $\frac{1}{2}b + \frac{1}{2}b$

110. $\frac{2}{3}x + \frac{1}{3}x$

111. $2y + \frac{1}{4}y + y$

112. $\frac{1}{2}a + a + 5a$

113. $11x - 3x$

114. $9t - 17t$

115. $6n - n$

116. $100t - t$

117. $y - 17y$

118. $3m - 9m + 4$

119. $-8 + 11a - 5b + 6a - 7b + 7$

120. $8x - 5x + 6 + 3y - 2y - 4$

121. $9x + 2y - 5x$

122. $8y - 3z + 4y$

123. $11x + 2y - 4x - y$

124. $13a + 9b - 2a - 4b$

125. $2.7x + 2.3y - 1.9x - 1.8y$

126. $6.7a + 4.3b - 4.1a - 2.9b$

127. $\frac{13}{2}a + \frac{9}{5}b - \frac{2}{3}a - \frac{3}{10}b - 42$

128. $\frac{11}{4}x + \frac{2}{3}y - \frac{4}{5}x - \frac{1}{6}y + 12$

Skill Maintenance

Find the LCM. [4.1a]

129. 16, 18

130. 18, 24

131. 16, 18, 24

132. 12, 15, 20

133. 16, 32

134. 24, 72

135. 15, 45, 90

136. 18, 54, 108

137. Evaluate $9w$ when $w = 20$. [9.1a]

138. Find the absolute value: $\left|-\frac{4}{13}\right|$. [9.2e]

Write true or false. [9.2d]

139. $-43 < -40$

140. $-3 \geq 0$

141. $-6 \leq -6$

142. $0 > -4$

Synthesis

Determine whether the expressions are equivalent. Explain why if they are. Give an example if they are not. Examples may vary.

143. $3t + 5$ and $3 \cdot 5 + t$

144. $4x$ and $x + 4$

145. $5m + 6$ and $6 + 5m$

146. $(x + y) + z$ and $z + (x + y)$

147. Factor: $q + qr + qrs + qrst$.

148. Collect like terms:
$21x + 44xy + 15y - 16x - 8y - 38xy + 2y + xy$.

9.8 Simplifying Expressions; Order of Operations

OBJECTIVES

a Find an equivalent expression for an opposite without parentheses, where an expression has several terms.

b Simplify expressions by removing parentheses and collecting like terms.

c Simplify expressions with parentheses inside parentheses.

d Simplify expressions using the rules for order of operations.

SKILL TO REVIEW

Objective 9.7c: Use the distributive laws to multiply expressions like 8 and $x - y$.

Multiply.

1. $4(x + 5)$
2. $-7(a + b)$

We now expand our ability to manipulate expressions by first considering opposites of sums and differences. Then we simplify expressions involving parentheses.

a OPPOSITES OF SUMS

What happens when we multiply a real number by -1? Consider the following products:

$$-1(7) = -7, \qquad -1(-5) = 5, \qquad -1(0) = 0.$$

From these examples, it appears that when we multiply a number by -1, we get the opposite, or additive inverse, of that number.

THE PROPERTY OF -1

For any real number a,

$$-1 \cdot a = -a.$$

(Negative one times a is the opposite, or additive inverse, of a.)

The property of -1 enables us to find expressions equivalent to opposites of sums.

EXAMPLES Find an equivalent expression without parentheses.

1. $-(3 + x) = -1(3 + x)$ — Using the property of -1

$= -1 \cdot 3 + (-1)x$ — Using a distributive law, multiplying each term by -1

$= -3 + (-x)$ — Using the property of -1

$= -3 - x$

2. $-(3x + 2y + 4) = -1(3x + 2y + 4)$ — Using the property of -1

$= -1(3x) + (-1)(2y) + (-1)4$ — Using a distributive law

$= -3x - 2y - 4$ — Using the property of -1

◀ **Do Margin Exercises 1 and 2.**

Find an equivalent expression without parentheses.

1. $-(x + 2)$
2. $-(5x + 2y + 8)$

Suppose we want to remove parentheses in an expression like

$$-(x - 2y + 5).$$

We can first rewrite any subtractions inside the parentheses as additions. Then we take the opposite of each term:

$$-(x - 2y + 5) = -[x + (-2y) + 5]$$
$$= -x + 2y + (-5) = -x + 2y - 5.$$

The most efficient method for removing parentheses is to replace each term in the parentheses with its opposite ("change the sign of every term"). Doing so for $-(x - 2y + 5)$, we obtain $-x + 2y - 5$ as an equivalent expression.

Answers

Skill to Review:
1. $4x + 20$ 2. $-7a - 7b$

Margin Exercises:
1. $-x - 2$ 2. $-5x - 2y - 8$

EXAMPLES Find an equivalent expression without parentheses.

3. $-(5 - y) = -5 + y$ Changing the sign of each term
4. $-(2a - 7b - 6) = -2a + 7b + 6$
5. $-(-3x + 4y + z - 7w - 23) = 3x - 4y - z + 7w + 23$

Do Exercises 3–6. ▶

Find an equivalent expression without parentheses. Try to do this in one step.

3. $-(6 - t)$
4. $-(x - y)$
5. $-(-4a + 3t - 10)$
6. $-(18 - m - 2n + 4z)$

b REMOVING PARENTHESES AND SIMPLIFYING

When a sum is added to another expression, as in $5x + (2x + 3)$, we can simply remove, or drop, the parentheses and collect like terms because of the associative law of addition:

$$5x + (2x + 3) = 5x + 2x + 3 = 7x + 3.$$

On the other hand, when a sum is subtracted from another expression, as in $3x - (4x + 2)$, we cannot simply drop the parentheses. However, we can subtract by adding an opposite. We then remove parentheses by changing the sign of each term inside the parentheses and collecting like terms.

EXAMPLE 6 Remove parentheses and simplify.

$3x - (4x + 2) = 3x + [-(4x + 2)]$ Adding the opposite of $(4x + 2)$

$= 3x + (-4x - 2)$ Changing the sign of each term inside the parentheses

$= 3x - 4x - 2$

$= -x - 2$ Collecting like terms

Caution!

Note that $3x - (4x + 2) \neq 3x - 4x + 2$. You cannot simply drop the parentheses.

Do Exercises 7 and 8. ▶

Remove parentheses and simplify.

7. $5x - (3x + 9)$
8. $5y - 2 - (2y - 4)$

In practice, the first three steps of Example 6 are usually combined by changing the sign of each term in parentheses and then collecting like terms.

EXAMPLES Remove parentheses and simplify.

7. $5y - (3y + 4) = 5y - 3y - 4$ Removing parentheses by changing the sign of every term inside the parentheses

$= 2y - 4$ Collecting like terms

8. $3x - 2 - (5x - 8) = 3x - 2 - 5x + 8$

$= -2x + 6$

9. $(3a + 4b - 5) - (2a - 7b + 4c - 8)$

$= 3a + 4b - 5 - 2a + 7b - 4c + 8$

$= a + 11b - 4c + 3$

Do Exercises 9–11. ▶

Remove parentheses and simplify.

9. $6x - (4x + 7)$
10. $8y - 3 - (5y - 6)$
11. $(2a + 3b - c) - (4a - 5b + 2c)$

Answers

3. $-6 + t$ 4. $-x + y$ 5. $4a - 3t + 10$ 6. $-18 + m + 2n - 4z$ 7. $2x - 9$ 8. $3y + 2$ 9. $2x - 7$ 10. $3y + 3$ 11. $-2a + 8b - 3c$

Next, consider subtracting an expression consisting of several terms multiplied by a number other than 1 or -1.

EXAMPLE 10 Remove parentheses and simplify.

$$\begin{aligned} x - 3(x + y) &= x + [-3(x + y)] && \text{Adding the opposite of } 3(x + y) \\ &= x + [-3x - 3y] && \text{Multiplying } x + y \text{ by } -3 \\ &= x - 3x - 3y \\ &= -2x - 3y && \text{Collecting like terms} \end{aligned}$$

EXAMPLES Remove parentheses and simplify.

11. $3y - 2(4y - 5) = 3y - 8y + 10$ Multiplying each term in the parentheses by -2

$= -5y + 10$

12. $(2a + 3b - 7) - 4(-5a - 6b + 12)$
$= 2a + 3b - 7 + 20a + 24b - 48 = 22a + 27b - 55$

13. $2y - \frac{1}{3}(9y - 12) = 2y - 3y + 4 = -y + 4$

14. $6(5x - 3y) - 2(8x + y) = 30x - 18y - 16x - 2y = 14x - 20y$

◀ **Do Exercises 12–16.**

Remove parentheses and simplify.

12. $y - 9(x + y)$

GS **13.** $5a - 3(7a - 6)$
$= 5a - \square + \square$
$= \square + 18$

14. $4a - b - 6(5a - 7b + 8c)$

15. $5x - \frac{1}{4}(8x + 28)$

16. $4.6(5x - 3y) - 5.2(8x + y)$

C PARENTHESES WITHIN PARENTHESES

In addition to parentheses, some expressions contain other grouping symbols such as brackets [ ] and braces { }.

> When more than one kind of grouping symbol occurs, do the computations in the innermost ones first. Then work from the inside out.

EXAMPLES Simplify.

15. $2[3 - (7 + 3)] = 2[3 - 10] = 2[-7] = -14$

16.
$$\begin{aligned} 8 - [9 - (12 + 5)] &= 8 - [9 - 17] && \text{Computing } 12 + 5 \\ &= 8 - [-8] && \text{Computing } 9 - 17 \\ &= 8 + 8 = 16 \end{aligned}$$

17.
$$\begin{aligned} \left[-4 - 2\left(-\tfrac{1}{2}\right)\right] \div \tfrac{1}{4} &= [-4 + 1] \div \tfrac{1}{4} && \text{Working within parentheses} \\ &= -3 \div \tfrac{1}{4} && \text{Computing } -4 + 1 \\ &= -3 \cdot 4 = -12 \end{aligned}$$

18.
$$\begin{aligned} &4(2 + 3) - \{7 - [4 - (8 + 5)]\} \\ &= 4(5) - \{7 - [4 - 13]\} && \text{Working with the innermost parentheses first} \\ &= 4(5) - \{7 - [-9]\} && \text{Computing } 4 - 13 \\ &= 4(5) - 16 && \text{Computing } 7 - [-9] \\ &= 20 - 16 = 4 \end{aligned}$$

◀ **Do Exercises 17–20.**

Simplify.

17. $12 - (8 + 2)$

GS **18.** $9 - [10 - (13 + 6)]$
$= 9 - [10 - (\square)]$
$= 9 - [\square]$
$= 9 + \square$
$= 18$

19. $[24 \div (-2)] \div (-2)$

20. $5(3 + 4) - \{8 - [5 - (9 + 6)]\}$

Answers

12. $-9x - 8y$ **13.** $-16a + 18$
14. $-26a + 41b - 48c$ **15.** $3x - 7$
16. $-18.6x - 19y$ **17.** 2 **18.** 18 **19.** 6
20. 17

Guided Solutions:
13. $21a, 18, -16a$ **18.** $19, -9, 9$

EXAMPLE 19 Simplify.

$$[5(x + 2) - 3x] - [3(y + 2) - 7(y - 3)]$$
$$= [5x + 10 - 3x] - [3y + 6 - 7y + 21]$$ Working with the innermost parentheses first
$$= [2x + 10] - [-4y + 27]$$ Collecting like terms within brackets
$$= 2x + 10 + 4y - 27$$ Removing brackets
$$= 2x + 4y - 17$$ Collecting like terms

Do Exercise 21. ▶

21. Simplify:
$[3(x + 2) + 2x] -$
$[4(y + 2) - 3(y - 2)]$.

d ORDER OF OPERATIONS

When several operations are to be done in a calculation or a problem, we apply the following.

RULES FOR ORDER OF OPERATIONS

1. Do all calculations within grouping symbols before operations outside.
2. Evaluate all exponential expressions.
3. Do all multiplications and divisions in order from left to right.
4. Do all additions and subtractions in order from left to right.

These rules are consistent with the way in which most computers and scientific calculators perform calculations.

EXAMPLE 20 Simplify: $-34 \cdot 56 - 17$.

There are no parentheses or powers, so we start with the third step.

$$-34 \cdot 56 - 17 = -1904 - 17$$ Doing all multiplications and divisions in order from left to right
$$= -1921$$ Doing all additions and subtractions in order from left to right

EXAMPLE 21 Simplify: $25 \div (-5) + 50 \div (-2)$.

There are no calculations inside parentheses and no powers. The parentheses with (-5) and (-2) are used only to represent the negative numbers. We begin by doing all multiplications and divisions.

$$25 \div (-5) + 50 \div (-2)$$
$$= -5 + (-25)$$ Doing all multiplications and divisions in order from left to right
$$= -30$$ Doing all additions and subtractions in order from left to right

Do Exercises 22–24. ▶

Simplify.

22. $23 - 42 \cdot 30$

23. $32 \div 8 \cdot 2$

24. $-24 \div 3 - 48 \div (-4)$

Answers

21. $5x - y - 8$ **22.** -1237 **23.** 8 **24.** 4

EXAMPLE 22 Simplify: $-2^4 + 51 \cdot 4 - (37 + 23 \cdot 2)$.

$$-2^4 + 51 \cdot 4 - (37 + 23 \cdot 2)$$

$= -2^4 + 51 \cdot 4 - (37 + 46)$	Following the rules for order of operations within the parentheses first
$= -2^4 + 51 \cdot 4 - 83$	Completing the addition inside parentheses
$= -16 + 51 \cdot 4 - 83$	Evaluating exponential expressions. Note that $-2^4 \neq (-2)^4$.
$= -16 + 204 - 83$	Doing all multiplications
$= 188 - 83$	Doing all additions and subtractions in order from left to right
$= 105$	

A fraction bar can play the role of a grouping symbol.

EXAMPLE 23 Simplify: $\dfrac{-64 \div (-16) \div (-2)}{2^3 - 3^2}$.

An equivalent expression with brackets as grouping symbols is

$$[-64 \div (-16) \div (-2)] \div [2^3 - 3^2].$$

This shows, in effect, that we do the calculations in the numerator and then in the denominator, and divide the results:

$$\frac{-64 \div (-16) \div (-2)}{2^3 - 3^2} = \frac{4 \div (-2)}{8 - 9} = \frac{-2}{-1} = 2.$$

◀ **Do Exercises 25 and 26.**

Simplify.

25. $-4^3 + 52 \cdot 5 + 5^3 - (4^2 - 48 \div 4)$
$= \_\_\_ + 52 \cdot 5 + 125 - (\_\_\_ - 48 \div 4)$
$= -64 + 52 \cdot 5 + 125 - (16 - \_\_\_)$
$= -64 + 52 \cdot 5 + 125 - 4$
$= -64 + \_\_\_ + 125 - 4$
$= \_\_\_ + 125 - 4$
$= 321 - 4$
$= 317$

GS

26. $\dfrac{5 - 10 - 5 \cdot 23}{2^3 + 3^2 - 7}$

Answers

25. 317 **26.** −12

Guided Solution:
25. −64, 16, 12, 260, 196

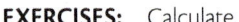

CALCULATOR CORNER

Order of Operations and Grouping Symbols Parentheses are necessary in some calculations. To simplify $-5(3 - 6) - 12$, we must use parentheses. The result is 3. Without parentheses, the computation is $-5 \cdot 3 - 6 - 12$, and the result is -33.

When a negative number is raised to an even power, parentheses also must be used. To find -3 raised to the fourth power, for example, we must use parentheses. The result is 81. Without parentheses, the computation is $-3^4 = -1 \cdot 3^4 = -1 \cdot 81 = -81$.

To simplify an expression like $\dfrac{49 - 104}{7 + 4}$, we must enter it as $(49 - 104) \div (7 + 4)$.

The result is -5.

```
-5(3-6)-12
                3
-5*3-6-12
              -33
```

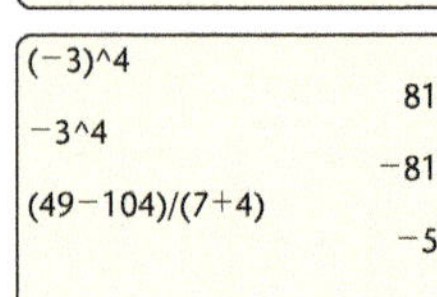

EXERCISES: Calculate.

1. $-8 + 4(7 - 9)$

2. $-3[2 + (-5)]$

3. $(-7)^6$

4. $(-17)^5$

5. -7^6

6. -17^5

7. $\dfrac{38 - 178}{5 + 30}$

8. $\dfrac{311 - 17^2}{2 - 13}$

9.8 Exercise Set

For Extra Help
MyMathLab®
MathXL® PRACTICE WATCH READ REVIEW

Reading Check

In each of Exercises 1–6, name the operation that should be performed first. Do not calculate.

RC1. $10 - 4 \cdot 2 + 5$

RC2. $10 - 4(2 + 5)$

RC3. $(10 - 4) \cdot 2 + 5$

RC4. $5[2(10 \div 5) - 3]$

RC5. $5(10 \div 2 + 5 - 3)$

RC6. $5 \cdot 2 - 4 \cdot 8 \div 2$

a Find an equivalent expression without parentheses.

1. $-(2x + 7)$

2. $-(8x + 4)$

3. $-(8 - x)$

4. $-(a - b)$

5. $-(4a - 3b + 7c)$

6. $-(x - 4y - 3z)$

7. $-(6x - 8y + 5)$

8. $-(4x + 9y + 7)$

9. $-(3x - 5y - 6)$

10. $-(6a - 4b - 7)$

11. $-(-8x - 6y - 43)$

12. $-(-2a + 9b - 5c)$

b Remove parentheses and simplify.

13. $9x - (4x + 3)$

14. $4y - (2y + 5)$

15. $2a - (5a - 9)$

16. $12m - (4m - 6)$

17. $2x + 7x - (4x + 6)$

18. $3a + 2a - (4a + 7)$

19. $2x - 4y - 3(7x - 2y)$

20. $3a - 9b - 1(4a - 8b)$

21. $15x - y - 5(3x - 2y + 5z)$

22. $4a - b - 4(5a - 7b + 8c)$

23. $(3x + 2y) - 2(5x - 4y)$

24. $(-6a - b) - 5(2b + a)$

25. $(12a - 3b + 5c) - 5(-5a + 4b - 6c)$

26. $(-8x + 5y - 12) - 6(2x - 4y - 10)$

c Simplify.

27. $9 - 2(5 - 4)$

28. $6 - 5(8 - 4)$

29. $8[7 - 6(4 - 2)]$

30. $10[7 - 4(7 - 5)]$

31. $[4(9 - 6) + 11] - [14 - (6 + 4)]$

32. $[7(8 - 4) + 16] - [15 - (7 + 8)]$

33. $[10(x + 3) - 4] + [2(x - 1) + 6]$

34. $[9(x + 5) - 7] + [4(x - 12) + 9]$

35. $[7(x + 5) - 19] - [4(x - 6) + 10]$

36. $[6(x + 4) - 12] - [5(x - 8) + 14]$

37. $3\{[7(x - 2) + 4] - [2(2x - 5) + 6]\}$

38. $4\{[8(x - 3) + 9] - [4(3x - 2) + 6]\}$

39. $4\{[5(x - 3) + 2] - 3[2(x + 5) - 9]\}$

40. $3\{[6(x - 4) + 5] - 2[5(x + 8) - 3]\}$

d Simplify.

41. $8 - 2 \cdot 3 - 9$

42. $8 - (2 \cdot 3 - 9)$

43. $(8 - 2) \div (3 - 9)$

44. $(8 - 2) \div 3 - 9$

45. $[(-24) \div (-3)] \div \left(-\frac{1}{2}\right)$

46. $[32 \div (-2)] \div \left(-\frac{1}{4}\right)$

47. $16 \cdot (-24) + 50$

48. $10 \cdot 20 - 15 \cdot 24$

49. $2^4 + 2^3 - 10$

50. $40 - 3^2 - 2^3$

51. $5^3 + 26 \cdot 71 - (16 + 25 \cdot 3)$

52. $4^3 + 10 \cdot 20 + 8^2 - 23$

53. $4 \cdot 5 - 2 \cdot 6 + 4$

54. $4 \cdot (6 + 8)/(4 + 3)$

55. $4^3/8$

56. $5^3 - 7^2$

57. $8(-7) + 6(-5)$

58. $10(-5) + 1(-1)$

59. $19 - 5(-3) + 3$

60. $14 - 2(-6) + 7$

61. $9 \div (-3) + 16 \div 8$

62. $-32 - 8 \div 4 - (-2)$

63. $-4^2 + 6$

64. $-5^2 + 7$

65. $-8^2 - 3$

66. $-9^2 - 11$

67. $12 - 20^3$

68. $20 + 4^3 \div (-8)$

69. $2 \cdot 10^3 - 5000$

70. $-7(3^4) + 18$

71. $6[9 - (3 - 4)]$

72. $8[3(6 - 13) - 11]$

73. $-1000 \div (-100) \div 10$

74. $256 \div (-32) \div (-4)$

75. $8 - (7 - 9)$

76. $(16 - 6) \cdot \frac{1}{2} + 9$

77. $\frac{10 - 6^2}{9^2 + 3^2}$

78. $\frac{5^2 - 4^3 - 3}{9^2 - 2^2 - 1^5}$

79. $\frac{3(6 - 7) - 5 \cdot 4}{6 \cdot 7 - 8(4 - 1)}$

80. $\frac{20(8 - 3) - 4(10 - 3)}{10(2 - 6) - 2(5 + 2)}$

81. $\frac{|2^3 - 3^2| + |12 \cdot 5|}{-32 \div (-16) \div (-4)}$

82. $\frac{|3 - 5|^2 - |7 - 13|}{|12 - 9| + |11 - 14|}$

Skill Maintenance

Evaluate. [9.1a]

83. $\frac{x - y}{y}$, when $x = 38$ and $y = 2$

84. $a - 3b$, when $a = 50$ and $b = 5$

Find the absolute value. [9.2e]

85. $|-0.4|$

86. $\left|\frac{15}{2}\right|$

Find the reciprocal. [9.6b]

87. -9

88. $\frac{7}{3}$

Subtract. [9.4a]

89. $5 - 30$

90. $-5 - 30$

91. $-5 - (-30)$

92. $5 - (-30)$

Synthesis

Simplify.

93. $x - [f - (f - x)] + [x - f] - 3x$

94. $x - \{x - 1 - [x - 2 - (x - 3 - \{x - 4 - [x - 5 - (x - 6)]\})]\}$

95. Use your calculator to do the following.

a) Evaluate $x^2 + 3$ when $x = 7$, when $x = -7$, and when $x = -5.013$.

b) Evaluate $1 - x^2$ when $x = 5$, when $x = -5$, and when $x = -10.455$.

96. Express $3^3 + 3^3 + 3^3$ as a power of 3.

Find the average.

97. $-15, 20, 50, -82, -7, -2$

98. $-1, 1, 2, -2, 3, -8, -10$

CHAPTER

9 Summary and Review

Vocabulary Reinforcement

Complete each statement with the correct term from the column on the right. Some of the choices may not be used.

1. The set of ______________ is $\{\ldots, -5, -4, -3, -2, -1, 0, 1, 2, 3, 4, 5, \ldots\}$. [9.2a]
2. Two numbers whose sum is 0 are called ______________ of each other. [9.3b]
3. The ______________ of addition says that $a + b = b + a$ for any real numbers a and b. [9.7b]
4. The ______________ states that for any real number a, $a \cdot 1 = 1 \cdot a = a$. [9.7a]
5. The ______________ of multiplication says that $a(bc) = (ab)c$ for any real numbers a, b, and c. [9.7b]
6. Two numbers whose product is 1 are called ______________ of each other. [9.6b]
7. The equation $y + 0 = y$ illustrates the ______________. [9.7a]

natural numbers
whole numbers
integers
real numbers
multiplicative inverses
additive inverses
commutative law
associative law
distributive law
identity property of 0
identity property of 1
property of -1

Concept Reinforcement

Determine whether each statement is true or false.

______________ 1. Every whole number is also an integer. [9.2a]

______________ 2. The product of an even number of negative numbers is positive. [9.5a]

______________ 3. The product of a number and its multiplicative inverse is -1. [9.6b]

______________ 4. $a < b$ also has the meaning $b \geq a$. [9.2d]

Study Guide

Objective 9.1a Evaluate algebraic expressions by substitution.

Example Evaluate $y - z$ when $y = 5$ and $z = -7$.

$y - z = 5 - (-7) = 5 + 7 = 12$

Practice Exercise

1. Evaluate $2a + b$ when $a = -1$ and $b = 16$.

Objective 9.2d Determine which of two real numbers is greater and indicate which, using < or >.

Example Use < or > for □ to write a true sentence:

$-5 \,\square\, -12$.

Since -5 is to the right of -12 on the number line, we have $-5 > -12$.

Practice Exercise

2. Use < or > for □ to write a true sentence:

$-6 \,\square\, -3$.

Objective 9.2e Find the absolute value of a real number.

Example Find the absolute value: **(a)** $|21|$; **(b)** $|-3.2|$; **(c)** $|0|$.

a) The number is positive, so the absolute value is the same as the number.

$$|21| = 21$$

b) The number is negative, so we make it positive.

$$|-3.2| = 3.2$$

c) The number is 0, so the absolute value is the same as the number.

$$|0| = 0$$

Practice Exercise

3. Find the absolute value: $\left|-\frac{5}{4}\right|$.

Objective 9.3a Add real numbers without using the number line.

Example Add without using the number line: **(a)** $-13 + 4$; **(b)** $-2 + (-3)$.

a) We have a negative number and a positive number. The absolute values are 13 and 4. The difference is 9. The negative number has the larger absolute value, so the answer is negative.

$$-13 + 4 = -9$$

b) We have two negative numbers. The sum of the absolute values is $2 + 3$, or 5. The answer is negative.

$$-2 + (-3) = -5$$

Practice Exercise

4. Add without using the number line: $-5.6 + (-2.9)$.

Objective 9.4a Subtract real numbers.

Example Subtract: $-4 - (-6)$.

$$-4 - (-6) = -4 + 6 = 2$$

Practice Exercise

5. Subtract: $7 - 9$.

Objective 9.5a Multiply real numbers.

Example Multiply: **(a)** $-1.9(4)$; **(b)** $-7(-6)$.

a) The signs are different, so the answer is negative.

$$-1.9(4) = -7.6$$

b) The signs are the same, so the answer is positive.

$$-7(-6) = 42$$

Practice Exercise

6. Multiply: $-8(-7)$.

Objective 9.6a Divide integers.

Example Divide: **(a)** $15 \div (-3)$; **(b)** $-72 \div (-9)$.

a) The signs are different, so the answer is negative.

$$15 \div (-3) = -5$$

b) The signs are the same, so the answer is positive.

$$-72 \div (-9) = 8$$

Practice Exercise

7. Divide: $-48 \div 6$.

Objective 9.6c Divide real numbers.

Example Divide: **(a)** $-\frac{1}{4} \div \frac{3}{5}$; **(b)** $-22.4 \div (-4)$.

a) We multiply by the reciprocal of the divisor:

$$-\frac{1}{4} \div \frac{3}{5} = -\frac{1}{4} \cdot \frac{5}{3} = -\frac{5}{12}.$$

b) We carry out the long division. The answer is positive.

$$\begin{array}{r} 5.6 \\ 4\overline{)22.4} \\ \underline{20} \\ 2\,4 \\ \underline{2\,4} \\ 0 \end{array}$$

Practice Exercise

8. Divide: $-\frac{3}{4} \div \left(-\frac{5}{3}\right)$.

Objective 9.7a Simplify fraction expressions.

Example Simplify: $-\frac{18x}{15x}$.

$$-\frac{18x}{15x} = -\frac{6 \cdot 3x}{5 \cdot 3x} \quad \text{Factoring the numerator and the denominator}$$

$$= -\frac{6}{5} \cdot \frac{3x}{3x} \quad \text{Factoring the fraction expression}$$

$$= -\frac{6}{5} \cdot 1 \quad \frac{3x}{3x} = 1$$

$$= -\frac{6}{5} \quad \text{Removing a factor of 1}$$

Practice Exercise

9. Simplify: $\frac{45y}{27y}$.

Objective 9.7c Use the distributive laws to multiply expressions like 8 and $x - y$.

Example Multiply: $3(4x - y + 2z)$.

$$3(4x - y + 2z) = 3 \cdot 4x - 3 \cdot y + 3 \cdot 2z$$
$$= 12x - 3y + 6z$$

Practice Exercise

10. Multiply: $5(x + 3y - 4z)$.

Objective 9.7d Use the distributive laws to factor expressions like $4x - 12 + 24y$.

Example Factor: $12a - 8b + 4c$.

$$12a - 8b + 4c = 4 \cdot 3a - 4 \cdot 2b + 4 \cdot c$$
$$= 4(3a - 2b + c)$$

Practice Exercise

11. Factor: $27x + 9y - 36z$.

Objective 9.7e Collect like terms.

Example Collect like terms: $3x - 5y + 8x + y$.

$$3x - 5y + 8x + y = 3x + 8x - 5y + y$$
$$= 3x + 8x - 5y + 1 \cdot y$$
$$= (3 + 8)x + (-5 + 1)y$$
$$= 11x - 4y$$

Practice Exercise

12. Collect like terms: $6a - 4b - a + 2b$.

Objective 9.8b Simplify expressions by removing parentheses and collecting like terms.

Example Remove parentheses and simplify:

$5x - 2(3x - y)$.

$5x - 2(3x - y) = 5x - 6x + 2y = -x + 2y$

Practice Exercise

13. Remove parentheses and simplify:

$8a - b - (4a + 3b)$.

Objective 9.8d Simplify expressions using the rules for order of operations.

Example Simplify: $12 - (7 - 3 \cdot 6)$.

$$\begin{aligned} 12 - (7 - 3 \cdot 6) &= 12 - (7 - 18) \\ &= 12 - (-11) \\ &= 12 + 11 \\ &= 23 \end{aligned}$$

Practice Exercise

14. Simplify: $75 \div (-15) + 24 \div 8$.

Review Exercises

1. Evaluate $\frac{x - y}{3}$ when $x = 17$ and $y = 5$. [9.1a]

2. Translate to an algebraic expression: [9.1b]

Nineteen percent of some number.

3. Tell which integers correspond to this situation: [9.2a]

Josh earned $620 for one week's work. While driving to work one day, he received a speeding ticket for $125.

Find the absolute value. [9.2e]

4. $|-38|$

5. $|126|$

Graph the number on the number line. [9.2b]

6. -2.5

7. $\frac{8}{9}$

Use either $<$ or $>$ for $\square$ to write a true sentence. [9.2d]

8. $-3 \square 10$

9. $-1 \square -6$

10. $0.126 \square -12.6$

11. $-\frac{2}{3} \square -\frac{1}{10}$

12. Write another inequality with the same meaning as $-3 < x$. [9.2d]

Write true or false. [9.2d]

13. $-9 \leq 9$

14. $-11 \geq -3$

Find the opposite. [9.3b]

15. 3.8

16. $-\frac{3}{4}$

Find the reciprocal. [9.6b]

17. $\frac{3}{8}$

18. -7

19. Evaluate $-x$ when $x = -34$. [9.3b]

20. Evaluate $-(-x)$ when $x = 5$. [9.3b]

Compute and simplify.

21. $4 + (-7)$ [9.3a]

22. $6 + (-9) + (-8) + 7$ [9.3a]

23. $-3.8 + 5.1 + (-12) + (-4.3) + 10$ [9.3a]

24. $-3 - (-7) + 7 - 10$ [9.4a]

25. $-\frac{9}{10} - \frac{1}{2}$ [9.4a]

26. $-3.8 - 4.1$ [9.4a]

27. $-9 \cdot (-6)$ [9.5a]

28. $-2.7(3.4)$ [9.5a]

29. $\frac{2}{3} \cdot \left(-\frac{3}{7}\right)$ [9.5a]

30. $3 \cdot (-7) \cdot (-2) \cdot (-5)$ [9.5a]

31. $35 \div (-5)$ [9.6a]

32. $-5.1 \div 1.7$ [9.6c]

33. $-\frac{3}{11} \div \left(-\frac{4}{11}\right)$ [9.6c]

Simplify. [9.8d]

34. $2(-3.4 - 12.2) - 8(-7)$

35. $\frac{-12(-3) - 2^3 - (-9)(-10)}{3 \cdot 10 + 1}$

36. $-16 \div 4 - 30 \div (-5)$

37. $\frac{-4[7 - (10 - 13)]}{|-2(8) - 4|}$

Solve.

38. On the first, second, and third downs, a football team had these gains and losses: 5-yd gain, 12-yd loss, and 15-yd gain, respectively. Find the total gain (or loss). [9.3c]

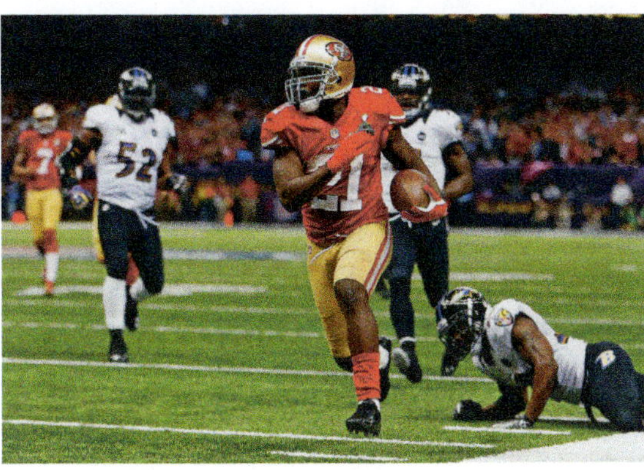

39. Chang's total assets are \$2140. He borrows \$2500. What are his total assets now? [9.4b]

40. ***Stock Price.*** The value of EFX Corp. stock began the day at \$17.68 per share and dropped \$1.63 per hour for 8 hr. What was the price of the stock after 8 hr? [9.5b]

41. ***Bank Account Balance.*** Yuri had \$68 in his bank account. After using his debit card to buy seven equally priced tee shirts, the balance in his account was $-\$64.65$. What was the price of each shirt? [9.6d]

Multiply. [9.7c]

42. $5(3x - 7)$

43. $-2(4x - 5)$

44. $10(0.4x + 1.5)$

45. $-8(3 - 6x)$

Factor. [9.7d]

46. $2x - 14$

47. $-6x + 6$

48. $5x + 10$

49. $-3x + 12y - 12$

Collect like terms. [9.7e]

50. $11a + 2b - 4a - 5b$

51. $7x - 3y - 9x + 8y$

52. $6x + 3y - x - 4y$

53. $-3a + 9b + 2a - b$

Remove parentheses and simplify.

54. $2a - (5a - 9)$ [9.8b]

55. $3(b + 7) - 5b$ [9.8b]

56. $3[11 - 3(4 - 1)]$ [9.8c]

57. $2[6(y - 4) + 7]$ [9.8c]

58. $[8(x + 4) - 10] - [3(x - 2) + 4]$ [9.8c]

59. $5\{[6(x - 1) + 7] - [3(3x - 4) + 8]\}$ [9.8c]

60. Factor out the greatest common factor: $18x - 6y + 30$. [9.7d]

A. $2(9x - 2y + 15)$ **B.** $3(6x - 2y + 10)$
C. $6(3x + 5)$ **D.** $6(3x - y + 5)$

61. Which expression is *not* equivalent to $mn + 5$? [9.7b]

A. $nm + 5$ **B.** $5n + m$
C. $5 + mn$ **D.** $5 + nm$

Synthesis

Simplify. [9.2e], [9.4a], [9.6a], [9.8d]

62. $-\left|\frac{7}{8} - \left(-\frac{1}{2}\right) - \frac{3}{4}\right|$

63. $(|2.7 - 3| + 3^2 - |-3|) \div (-3)$

64. $2000 - 1990 + 1980 - 1970 + \cdots + 20 - 10$

65. Find a formula for the perimeter of the figure below. [9.7e]

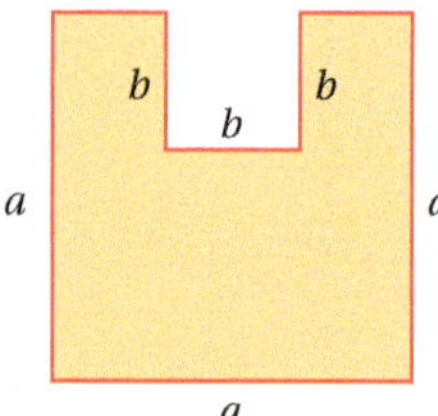

Understanding Through Discussion and Writing

1. Without actually performing the addition, explain why the sum of all integers from -50 to 50 is 0. [9.3b]

2. What rule have we developed that would tell you the sign of $(-7)^8$ and of $(-7)^{11}$ without doing the computations? Explain. [9.5a]

3. Explain how multiplication can be used to justify why a negative number divided by a negative number is positive. [9.6c]

4. Explain how multiplication can be used to justify why a negative number divided by a positive number is negative. [9.6c]

5. The distributive law was introduced before the discussion on collecting like terms. Why do you think this was done? [9.7c, e]

6. Jake keys in $18/2 \cdot 3$ on his calculator and expects the result to be 3. What mistake is he making? [9.8d]

CHAPTER

9 Test

For Extra Help For step-by-step test solutions, access the Chapter Test Prep Videos in MyMathLab® or on YouTube (search "BittingerPreIntro" and click on Channels).

1. Evaluate $\frac{3x}{y}$ when $x = 10$ and $y = 5$.

2. Translate to an algebraic expression: Nine less than some number.

Use either $<$ or $>$ for $\square$ to write a true sentence.

3. $-3 \square -8$

4. $-\frac{1}{2} \square -\frac{1}{8}$

5. $-0.78 \square -0.87$

6. Write an inequality with the same meaning as $x < -2$.

7. Write true or false: $-13 \leq -3$.

Simplify.

8. $|-7|$

9. $\left|\frac{9}{4}\right|$

10. $|-2.7|$

Find the opposite.

11. $\frac{2}{3}$

12. -1.4

Find the reciprocal.

13. -2

14. $\frac{4}{7}$

15. Evaluate $-x$ when $x = -8$.

Compute and simplify.

16. $3.1 - (-4.7)$

17. $-8 + 4 + (-7) + 3$

18. $-\frac{1}{5} + \frac{3}{8}$

19. $2 - (-8)$

20. $3.2 - 5.7$

21. $\frac{1}{8} - \left(-\frac{3}{4}\right)$

22. $4 \cdot (-12)$

23. $-\frac{1}{2} \cdot \left(-\frac{3}{8}\right)$

24. $-45 \div 5$

25. $-\frac{3}{5} \div \left(-\frac{4}{5}\right)$

26. $4.864 \div (-0.5)$

27. $-2(16) - |2(-8) - 5^3|$

28. $-20 \div (-5) + 36 \div (-4)$

29. Isabella kept track of the changes in the stock market over a period of 5 weeks. By how many points had the market risen or fallen over this time?

WEEK 1	WEEK 2	WEEK 3	WEEK 4	WEEK 5
Down 13 pts	Down 16 pts	Up 36 pts	Down 11 pts	Up 19 pts

30. ***Difference in Elevation.*** The lowest elevation in Australia, Lake Eyre, is 15 m below sea level. The highest elevation in Australia, Mount Kosciuszko, is 2229 m. Find the difference in elevation between the highest point and the lowest point.

Source: *The CIA World Factbook*, 2012

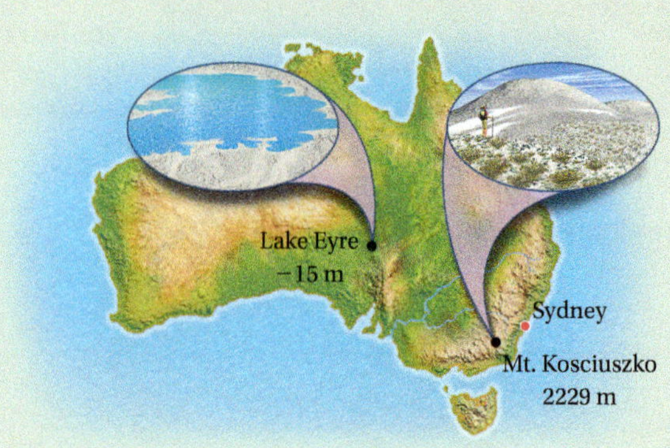

31. ***Population Decrease.*** The population of Stone City was 18,600. It dropped 420 each year for 6 years. What was the population of the city after 6 years?

32. ***Chemical Experiment.*** During a chemical reaction, the temperature in a beaker decreased every minute by the same number of degrees. The temperature was 16°C at 11:08 A.M. By 11:52 A.M., the temperature had dropped to −17°C. By how many degrees did it change each minute?

Multiply.

33. $3(6 - x)$

34. $-5(y - 1)$

Factor.

35. $12 - 22x$

36. $7x + 21 + 14y$

Simplify.

37. $6 + 7 - 4 - (-3)$

38. $5x - (3x - 7)$

39. $4(2a - 3b) + a - 7$

40. $4\{3[5(y - 3) + 9] + 2(y + 8)\}$

41. $256 \div (-16) \div 4$

42. $2^3 - 10[4 - 3(-2 + 18)]$

43. Which of the following is *not* a true statement?

A. $-5 \le -5$
B. $-5 < -5$
C. $-5 \ge -5$
D. $-5 = -5$

Synthesis

Simplify.

44. $|-27 - 3(4)| - |-36| + |-12|$

45. $a - \{3a - [4a - (2a - 4a)]\}$

46. Find a formula for the perimeter of the figure shown here.

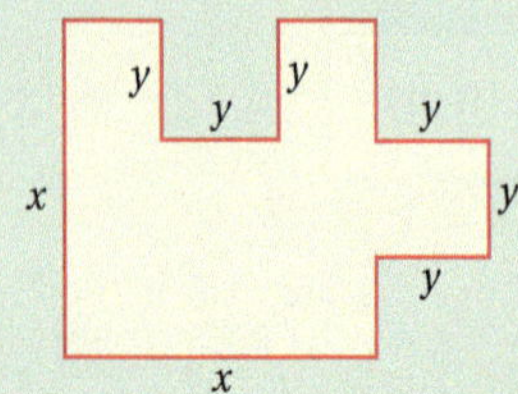

CHAPTER

10

Solving Equations and Inequalities

STUDYING FOR SUCCESS *Learning Resources on Campus*

- ☐ There may be a learning lab or a tutoring center for drop-in tutoring.
- ☐ There may be group tutoring sessions for this specific course.
- ☐ The mathematics department may have a bulletin board or a network for locating private tutors.

10.1 Solving Equations: The Addition Principle

OBJECTIVES

a Determine whether a given number is a solution of a given equation.

b Solve equations using the addition principle.

SKILL TO REVIEW

Objective 9.1a: Evaluate algebraic expressions by substitution.

1. Evaluate $x - 7$ when $x = 5$.

2. Evaluate $2x + 3$ when $x = -1$.

Determine whether each equation is true, false, or neither.

1. $5 - 8 = -4$

2. $12 + 6 = 18$

3. $x + 6 = 7 - x$

Answers

Skill to Review:
1. -2 **2.** 1

Margin Exercises:
1. False **2.** True **3.** Neither

a EQUATIONS AND SOLUTIONS

In order to solve problems, we must learn to solve equations.

EQUATION

An **equation** is a number sentence that says that the expressions on either side of the equals sign, =, represent the same number.

Here are some examples of equations:

$$3 + 2 = 5, \quad 14 - 10 = 1 + 3, \quad x + 6 = 13, \quad 3x - 2 = 7 - x.$$

Equations have expressions on each side of the equals sign. The sentence "$14 - 10 = 1 + 3$" asserts that the expressions $14 - 10$ and $1 + 3$ name the same number.

Some equations are true. Some are false. Some are neither true nor false.

EXAMPLES Determine whether each equation is true, false, or neither.

1. $3 + 2 = 5$ The equation is *true*.

2. $7 - 2 = 4$ The equation is *false*.

3. $x + 6 = 13$ The equation is *neither* true nor false, because we do not know what number x represents.

◀ **Do Margin Exercises 1–3.**

SOLUTION OF AN EQUATION

Any replacement for the variable that makes an equation true is called a **solution** of the equation. To solve an equation means to find *all* of its solutions.

One way to determine whether a number is a solution of an equation is to evaluate the expression on each side of the equals sign by substitution. If the values are the same, then the number is a solution.

EXAMPLE 4 Determine whether 7 is a solution of $x + 6 = 13$.

We have

$$\begin{array}{c|l} x + 6 = 13 & \text{Writing the equation} \\ 7 + 6 \ ?\ 13 & \text{Substituting 7 for } x \\ 13 & \text{TRUE} \end{array}$$

Since the left-hand side and the right-hand side are the same, 7 is a solution. No other number makes the equation true, so the only solution is the number 7.

EXAMPLE 5 Determine whether 19 is a solution of $7x = 141$.

We have

$$\begin{array}{c|l} 7x = 141 & \text{Writing the equation} \\ 7(19) \ ?\ 141 & \text{Substituting 19 for } x \\ 133 & \text{FALSE} \end{array}$$

Since the left-hand side and the right-hand side are not the same, 19 is not a solution of the equation.

Do Exercises 4–7.

Determine whether the given number is a solution of the given equation.

4. $8;\ x + 4 = 12$

5. $0;\ x + 4 = 12$

6. $-3;\ 7 + x = -4$

7. $-\frac{3}{5};\ -5x = 3$

b USING THE ADDITION PRINCIPLE

Consider the equation

$$x = 7.$$

We can easily see that the solution of this equation is 7. If we replace x with 7, we get

$$7 = 7, \quad \text{which is true.}$$

Now consider the equation of Example 4: $x + 6 = 13$. In Example 4, we discovered that the solution of this equation is also 7, but the fact that 7 is the solution is not as obvious. We now begin to consider principles that allow us to start with an equation like $x + 6 = 13$ and end up with an *equivalent equation,* like $x = 7$, in which the variable is alone on one side and for which the solution is easier to find.

EQUIVALENT EQUATIONS

Equations with the same solutions are called **equivalent equations**.

One of the principles that we use in solving equations involves addition. An equation $a = b$ says that a and b stand for the same number. Suppose this is true, and we add a number c to the number a. We get the same answer if we add c to b, because a and b are the same number.

THE ADDITION PRINCIPLE FOR EQUATIONS

For any real numbers a, b, and c,

$$a = b \quad \text{is equivalent to} \quad a + c = b + c.$$

Answers

4. Yes 5. No 6. No 7. Yes

8. Solve $x + 2 = 11$ using the addition principle.

GS

$$x + 2 = 11$$
$$x + 2 + (-2) = 11 + (\quad)$$
$$x + \quad = 9$$
$$x = \quad$$

9. Solve using the addition principle, subtracting 5 on both sides:

$$x + 5 = -8.$$

10. Solve: $t - 3 = 19$.

Answers

8. 9 **9.** -13 **10.** 22

Guided Solution:

8. $-2, 0, 9$

Let's solve the equation $x + 6 = 13$ using the addition principle. We want to get x alone on one side. To do so, we use the addition principle, choosing to add -6 because $6 + (-6) = 0$:

$$x + 6 = 13$$

$$x + 6 + (-6) = 13 + (-6) \qquad \text{Using the addition principle: adding } -6 \text{ on both sides}$$

$$x + 0 = 7 \qquad \text{Simplifying}$$

$$x = 7. \qquad \text{Identity property of 0: } x + 0 = x$$

The solution of $x + 6 = 13$ is 7.

◀ **Do Exercise 8.**

When we use the addition principle, we sometimes say that we "add the same number on both sides of the equation." This is also true for subtraction, since we can express every subtraction as an addition. That is, since

$$a - c = b - c \quad \text{is equivalent to} \quad a + (-c) = b + (-c),$$

the addition principle tells us that we can "subtract the same number on both sides of the equation."

EXAMPLE 6 Solve: $x + 5 = -7$.

We have

$$x + 5 = -7$$

$$x + 5 - 5 = -7 - 5 \qquad \text{Using the addition principle: adding } -5 \text{ on both sides or subtracting 5 on both sides}$$

$$x + 0 = -12 \qquad \text{Simplifying}$$

$$x = -12. \qquad \text{Identity property of 0}$$

The solution of the original equation is -12. The equations $x + 5 = -7$ and $x = -12$ are *equivalent.*

◀ **Do Exercise 9.**

Now we use the addition principle to solve an equation that involves a subtraction.

EXAMPLE 7 Solve: $a - 4 = 10$.

We have

$$a - 4 = 10$$

$$a - 4 + 4 = 10 + 4 \qquad \text{Using the addition principle: adding 4 on both sides}$$

$$a + 0 = 14 \qquad \text{Simplifying}$$

$$a = 14. \qquad \text{Identity property of 0}$$

Check:

$$\begin{array}{c|c} a - 4 & 10 \\ \hline 14 - 4 & ?\ 10 \\ 10 & \end{array} \qquad \text{TRUE}$$

The solution is 14.

◀ **Do Exercise 10.**

EXAMPLE 8 Solve: $-6.5 = y - 8.4$.

We have

$$-6.5 = y - 8.4$$

$$-6.5 + 8.4 = y - 8.4 + 8.4$$ Using the addition principle: adding 8.4 on both sides to eliminate -8.4 on the right

$$1.9 = y.$$

Check:

$$\begin{array}{c|c} -6.5 = y - 8.4 & \\ \hline -6.5 \; ? \; 1.9 - 8.4 & \\ -6.5 & \text{TRUE} \end{array}$$

The solution is 1.9.

Note that equations are reversible. That is, if $a = b$ is true, then $b = a$ is true. Thus when we solve $-6.5 = y - 8.4$, we can reverse it and solve $y - 8.4 = -6.5$ if we wish.

Do Exercises 11 and 12.

Solve.

11. $8.7 = n - 4.5$

12. $y + 17.4 = 10.9$

EXAMPLE 9 Solve: $-\frac{2}{3} + x = \frac{5}{2}$.

We have

$$-\frac{2}{3} + x = \frac{5}{2}$$

$$\frac{2}{3} - \frac{2}{3} + x = \frac{2}{3} + \frac{5}{2}$$ Adding $\frac{2}{3}$ on both sides

$$x = \frac{2}{3} + \frac{5}{2}$$

$$x = \frac{2}{3} \cdot \frac{2}{2} + \frac{5}{2} \cdot \frac{3}{3}$$ Multiplying by 1 to obtain equivalent fraction expressions with the least common denominator 6

$$x = \frac{4}{6} + \frac{15}{6}$$

$$x = \frac{19}{6}.$$

Check:

$$\begin{array}{c|c} -\frac{2}{3} + x = \frac{5}{2} & \\ \hline -\frac{2}{3} + \frac{19}{6} \; ? \; \frac{5}{2} & \\ -\frac{4}{6} + \frac{19}{6} & \\ \frac{15}{6} & \\ \frac{5}{2} & \text{TRUE} \end{array}$$

The solution is $\frac{19}{6}$.

Do Exercises 13 and 14.

Solve.

13. $x + \frac{1}{2} = -\frac{3}{2}$

14. $t - \frac{13}{4} = \frac{5}{8}$

Answers

11. 13.2 **12.** -6.5 **13.** -2 **14.** $\frac{31}{8}$

10.1 Exercise Set

For Extra Help MyMathLab® MathXL® PRACTICE WATCH READ REVIEW

✓ Reading Check

Choose from the column on the right the most appropriate first step in solving each equation.

RC1. $9 = x - 4$

RC2. $3 + x = -15$

RC3. $x - 3 = 9$

RC4. $x + 4 = 3$

a) Add -4 on both sides.
b) Add 15 on both sides.
c) Subtract 3 on both sides.
d) Subtract 9 on both sides.
e) Add 3 on both sides.
f) Add 4 on both sides.

a Determine whether the given number is a solution of the given equation.

1. 15; $x + 17 = 32$

2. 35; $t + 17 = 53$

3. 21; $x - 7 = 12$

4. 36; $a - 19 = 17$

5. -7; $6x = 54$

6. -9; $8y = -72$

7. 30; $\frac{x}{6} = 5$

8. 49; $\frac{y}{8} = 6$

9. 20; $5x + 7 = 107$

10. 9; $9x + 5 = 86$

11. -10; $7(y - 1) = 63$

12. -5; $6(y - 2) = 18$

b Solve using the addition principle. Don't forget to check!

13. $x + 2 = 6$
Check: $x + 2 = 6$?

14. $y + 4 = 11$
Check: $y + 4 = 11$?

15. $x + 15 = -5$
Check: $x + 15 = -5$?

16. $t + 10 = 44$
Check: $t + 10 = 44$?

17. $x + 6 = -8$
Check: $x + 6 = -8$?

18. $z + 9 = -14$

19. $x + 16 = -2$

20. $m + 18 = -13$

21. $x - 9 = 6$

22. $x - 11 = 12$

23. $x - 7 = -21$

24. $x - 3 = -14$

25. $5 + t = 7$

26. $8 + y = 12$

27. $-7 + y = 13$

28. $-8 + y = 17$

29. $-3 + t = -9$

30. $-8 + t = -24$

31. $x + \frac{1}{2} = 7$

32. $24 = -\frac{7}{10} + r$

33. $12 = a - 7.9$

34. $2.8 + y = 11$

35. $r + \frac{1}{3} = \frac{8}{3}$

36. $t + \frac{3}{8} = \frac{5}{8}$

37. $m + \frac{5}{6} = -\frac{11}{12}$

38. $x + \frac{2}{3} = -\frac{5}{6}$

39. $x - \frac{5}{6} = \frac{7}{8}$

40. $y - \frac{3}{4} = \frac{5}{6}$

41. $-\frac{1}{5} + z = -\frac{1}{4}$

42. $-\frac{1}{8} + y = -\frac{3}{4}$

43. $7.4 = x + 2.3$

44. $8.4 = 5.7 + y$

45. $7.6 = x - 4.8$

46. $8.6 = x - 7.4$

47. $-9.7 = -4.7 + y$

48. $-7.8 = 2.8 + x$

49. $5\frac{1}{6} + x = 7$

50. $5\frac{1}{4} = 4\frac{2}{3} + x$

51. $q + \frac{1}{3} = -\frac{1}{7}$

52. $52\frac{3}{8} = -84 + x$

Skill Maintenance

53. Divide: $\frac{2}{3} \div \left(-\frac{4}{9}\right)$. [3.7b], [9.6c]

54. Add: $-8.6 + 3.4$. [5.2c], [9.3a]

55. Subtract: $-\frac{2}{3} - \left(-\frac{5}{8}\right)$. [4.3a], [9.4a]

56. Multiply: $(-25.4)(-6.8)$. [5.3a], [9.5a]

Translate to an algebraic expression. [5.8a], [9.1b]

57. Jane had \$83 before paying x dollars for a pair of tennis shoes. How much does she have left?

58. Justin drove his S-10 pickup truck 65 mph for t hours. How far did he drive?

Synthesis

Solve.

59. $x + \frac{4}{5} = -\frac{2}{3} - \frac{4}{15}$

60. $x + x = x$

61. $16 + x - 22 = -16$

62. $x + 4 = 5 + x$

63. $x + 3 = 3 + x$

64. $|x| + 6 = 19$

10.2 Solving Equations: The Multiplication Principle

OBJECTIVE

a Solve equations using the multiplication principle.

SKILL TO REVIEW

Objective 9.6b: Find the reciprocal of a real number.

Find the reciprocal.

1. 5

2. $-\frac{5}{4}$

a USING THE MULTIPLICATION PRINCIPLE

Suppose that $a = b$ is true, and we multiply a by some number c. We get the same number if we multiply b by c, because a and b are the same number.

THE MULTIPLICATION PRINCIPLE FOR EQUATIONS

For any real numbers a, b, and c, $c \neq 0$,

$$a = b \quad \text{is equivalent to} \quad a \cdot c = b \cdot c.$$

When using the multiplication principle, we sometimes say that we "multiply on both sides of the equation by the same number."

EXAMPLE 1 Solve: $5x = 70$.

To get x alone on one side, we multiply by the *multiplicative inverse,* or *reciprocal*, of 5. Then we get the *multiplicative identity* 1 times x, or $1 \cdot x$, which simplifies to x. This allows us to eliminate 5 on the left.

$5x = 70$	The reciprocal of 5 is $\frac{1}{5}$.
$\frac{1}{5} \cdot 5x = \frac{1}{5} \cdot 70$	Multiplying by $\frac{1}{5}$ to get $1 \cdot x$ and eliminate 5 on the left
$1 \cdot x = 14$	Simplifying
$x = 14$	Identity property of 1: $1 \cdot x = x$

Check:

$$\begin{array}{c|c} 5x & 70 \\ \hline 5 \cdot 14 \ ? & 70 \\ 70 & \end{array} \quad \text{TRUE}$$

The solution is 14.

The multiplication principle also tells us that we can "divide on both sides of the equation by the same nonzero number." This is because dividing is the same as multiplying by a reciprocal. That is,

$$\frac{a}{c} = \frac{b}{c} \quad \text{is equivalent to} \quad a \cdot \frac{1}{c} = b \cdot \frac{1}{c}, \quad \text{when } c \neq 0.$$

In an expression like $5x$ in Example 1, the number 5 is called the **coefficient**. Example 1 could be done as follows, dividing by 5 on both sides, the coefficient of x.

EXAMPLE 2 Solve: $5x = 70$.

$5x = 70$	
$\frac{5x}{5} = \frac{70}{5}$	Dividing by 5 on both sides
$1 \cdot x = 14$	Simplifying
$x = 14$	Identity property of 1. The solution is 14.

Answers

Skill to Review:

1. $\frac{1}{5}$ 2. $-\frac{4}{5}$

Do Exercises 1 and 2.

EXAMPLE 3 Solve: $-4x = 92$.

We have

$$-4x = 92$$

$$\frac{-4x}{-4} = \frac{92}{-4}$$ Using the multiplication principle. Dividing by -4 on both sides is the same as multiplying by $-\frac{1}{4}$.

$$1 \cdot x = -23$$ Simplifying

$$x = -23.$$ Identity property of 1

Check:

$$\begin{array}{c|c} -4x & 92 \\ \hline -4(-23) \overset{?}{} & 92 \\ 92 & \end{array}$$ TRUE

The solution is -23.

Do Exercise 3.

EXAMPLE 4 Solve: $-x = 9$.

We have

$$-x = 9$$

$$-1 \cdot x = 9$$ Using the property of -1: $-x = -1 \cdot x$

$$\frac{-1 \cdot x}{-1} = \frac{9}{-1}$$ Dividing by -1 on both sides: $-1/(-1) = 1$

$$1 \cdot x = -9$$

$$x = -9.$$

Check:

$$\begin{array}{c|c} -x & 9 \\ \hline -(-9) \overset{?}{} & 9 \\ 9 & \end{array}$$ TRUE

The solution is -9.

Do Exercise 4.

We can also solve the equation $-x = 9$ by multiplying as follows.

EXAMPLE 5 Solve: $-x = 9$.

We have

$$-x = 9$$

$$-1 \cdot (-x) = -1 \cdot 9$$ Multiplying by -1 on both sides

$$-1 \cdot (-1) \cdot x = -9$$ $-x = (-1) \cdot x$

$$1 \cdot x = -9$$ $-1 \cdot (-1) = 1$

$$x = -9.$$

The solution is -9.

Do Exercise 5.

GS **1.** Solve $6x = 90$ by multiplying on both sides.

$$6x = 90$$

$$\frac{1}{6} \cdot 6x = \square \cdot 90$$

$$1 \cdot x = 15$$

$$\square = 15$$

Check:

$$\begin{array}{c|c} 6x & 90 \\ \hline 6 \cdot \square \overset{?}{} & 90 \\ 90 & \end{array}$$ TRUE

2. Solve $4x = -7$ by dividing on both sides.

$$4x = -7$$

$$\frac{4x}{4} = \frac{-7}{\square}$$

$$1 \cdot x = -\frac{7}{4}$$

$$\square = -\frac{7}{4}$$

Don't forget to check.

3. Solve: $-6x = 108$.

4. Solve. Divide on both sides.

$$-x = -10$$

5. Solve. Multiply on both sides.

$$-x = -10$$

Answers

1. 15 **2.** $-\frac{7}{4}$ **3.** -18 **4.** 10 **5.** 10

Guided Solutions:

1. $\frac{1}{6}$, x, 15 **2.** 4, x

In practice, it is generally more convenient to divide on both sides of the equation if the coefficient of the variable is in decimal notation or is an integer. If the coefficient is in fraction notation, it is usually more convenient to multiply by a reciprocal.

EXAMPLE 6 Solve: $\frac{3}{8} = -\frac{5}{4}x$.

$$\frac{3}{8} = -\frac{5}{4}x$$

The reciprocal of $-\frac{5}{4}$ is $-\frac{4}{5}$. There is no sign change.

$$-\frac{4}{5} \cdot \frac{3}{8} = -\frac{4}{5} \cdot \left(-\frac{5}{4}x\right)$$ Multiplying by $-\frac{4}{5}$ to get $1 \cdot x$ and eliminate $-\frac{5}{4}$ on the right

$$-\frac{12}{40} = 1 \cdot x$$

$$-\frac{3}{10} = 1 \cdot x$$ Simplifying

$$-\frac{3}{10} = x$$ Identity property of 1

Check: $\frac{3}{8} = -\frac{5}{4}x$

$\frac{3}{8} \;?\; -\frac{5}{4}\left(-\frac{3}{10}\right)$

$\quad\quad \frac{3}{8}$ TRUE

The solution is $-\frac{3}{10}$.

As noted in Section 10.1, if $a = b$ is true, then $b = a$ is true. Thus we can reverse the equation $\frac{3}{8} = -\frac{5}{4}x$ and solve $-\frac{5}{4}x = \frac{3}{8}$ if we wish.

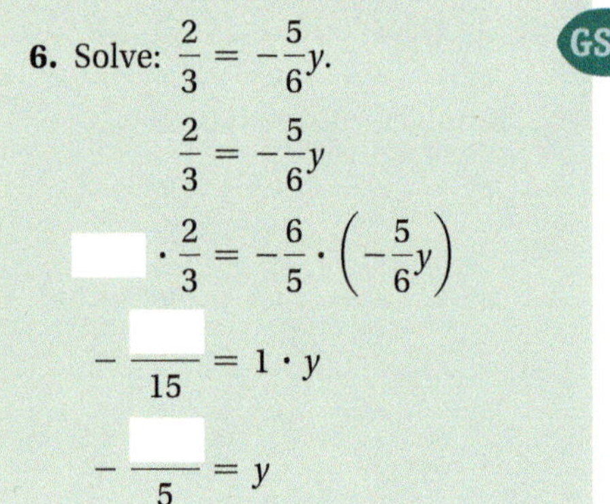

6. Solve: $\frac{2}{3} = -\frac{5}{6}y$.
$\frac{2}{3} = -\frac{5}{6}y$
$\square \cdot \frac{2}{3} = -\frac{6}{5} \cdot \left(-\frac{5}{6}y\right)$
$-\frac{\square}{15} = 1 \cdot y$
$-\frac{\square}{5} = y$
GS

Do Exercise 6.

EXAMPLE 7 Solve: $1.16y = 9744$.

$$1.16y = 9744$$

$$\frac{1.16y}{1.16} = \frac{9744}{1.16}$$ Dividing by 1.16 on both sides

$$y = \frac{9744}{1.16}$$

$$y = 8400$$ Simplifying

Check: $1.16y = 9744$

$1.16(8400) \;?\; 9744$

$\quad 9744$ TRUE

The solution is 8400.

Do Exercises 7 and 8.

Solve.

7. $1.12x = 8736$

8. $6.3 = -2.1y$

Answers

6. $-\frac{4}{5}$ 7. 7800 8. -3

Guided Solution:

6. $-\frac{6}{5}$, 12, 4

Now we use the multiplication principle to solve an equation that involves division.

EXAMPLE 8 Solve: $\frac{-y}{9} = 14$.

$$\frac{-y}{9} = 14$$

$$9 \cdot \frac{-y}{9} = 9 \cdot 14 \quad \text{Multiplying by 9 on both sides}$$

$$-y = 126$$

$$-1 \cdot (-y) = -1 \cdot 126 \quad \text{Multiplying by } -1 \text{ on both sides}$$

$$y = -126$$

Check:

$$\frac{-y}{9} = 14$$

$$\frac{-(-126)}{9} \ ? \ 14$$

$$\frac{126}{9}$$

$$14 \quad \text{TRUE}$$

The solution is -126.

There are other ways to solve the equation in Example 8. One is by multiplying by -9 on both sides as follows:

$$-9 \cdot \frac{-y}{9} = -9 \cdot 14$$

$$\frac{9y}{9} = -126$$

$$y = -126.$$

Do Exercise 9. ▶

9. Solve: $-14 = \frac{-y}{2}$.

Answer

9. 28

10.2 Exercise Set

For Extra Help

MyMathLab®

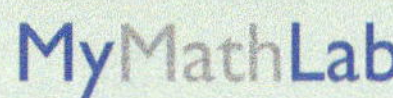

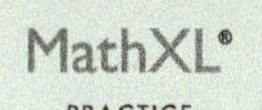

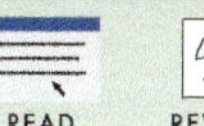

Reading Check

Choose from the column on the right the most appropriate first step in solving each equation.

RC1. $3 = -\frac{1}{12}x$

RC2. $-6x = 12$

RC3. $12x = -6$

RC4. $\frac{1}{6}x = 12$

a) Divide by 12 on both sides.
b) Multiply by 6 on both sides.
c) Multiply by 12 on both sides.
d) Divide by -6 on both sides.
e) Divide by 6 on both sides.
f) Multiply by -12 on both sides.

a Solve using the multiplication principle. Don't forget to check!

1. $6x = 36$

Check: $6x = 36$ | ?

2. $3x = 51$

Check: $3x = 51$ | ?

3. $5y = 45$

Check: $5y = 45$ | ?

4. $8y = 72$

Check: $8y = 72$ | ?

5. $84 = 7x$

6. $63 = 9x$

7. $-x = 40$

8. $-x = 53$

9. $-1 = -z$

10. $-47 = -t$

11. $7x = -49$

12. $8x = -56$

13. $-12x = 72$

14. $-15x = 105$

15. $-21w = -126$

16. $-13w = -104$

17. $\frac{t}{7} = -9$

18. $\frac{y}{5} = -6$

19. $\frac{n}{-6} = 8$

20. $\frac{y}{-8} = 11$

21. $\frac{3}{4}x = 27$

22. $\frac{4}{5}x = 16$

23. $-\frac{2}{3}x = 6$

24. $-\frac{3}{8}x = 12$

25. $\frac{-t}{3} = 7$

26. $\frac{-x}{6} = 9$

27. $-\frac{m}{3} = \frac{1}{5}$

28. $\frac{1}{8} = -\frac{y}{5}$

29. $-\frac{3}{5}r = \frac{9}{10}$

30. $-\frac{2}{5}y = \frac{4}{15}$

31. $-\frac{3}{2}r = -\frac{27}{4}$

32. $-\frac{3}{8}x = -\frac{15}{16}$

33. $6.3x = 44.1$

34. $2.7y = 54$

35. $-3.1y = 21.7$

36. $-3.3y = 6.6$

37. $38.7m = 309.6$

38. $29.4m = 235.2$

39. $-\frac{2}{3}y = -10.6$

40. $-\frac{9}{7}y = 12.06$

41. $\frac{-x}{5} = 10$

42. $\frac{-x}{8} = -16$

43. $-\frac{t}{2} = 7$

44. $\frac{m}{-3} = 10$

Skill Maintenance

Collect like terms. [9.7e]

45. $3x + 4x$

46. $6x + 5 - 7x$

47. $-4x + 11 - 6x + 18x$

48. $8y - 16y - 24y$

Remove parentheses and simplify. [9.8b]

49. $3x - (4 + 2x)$

50. $2 - 5(x + 5)$

51. $8y - 6(3y + 7)$

52. $-2a - 4(5a - 1)$

Translate to an algebraic expression. [9.1b]

53. Patty drives her van for 8 hr at a speed of r miles per hour. How far does she drive?

54. A triangle has a height of 10 meters and a base of b meters. What is the area of the triangle?

Synthesis

Solve.

55. $-0.2344m = 2028.732$

56. $0 \cdot x = 0$

57. $0 \cdot x = 9$

58. $4|x| = 48$

59. $2|x| = -12$

Solve for x.

60. $ax = 5a$

61. $3x = \frac{b}{a}$

62. $cx = a^2 + 1$

63. $\frac{a}{b}x = 4$

64. A student makes a calculation and gets an answer of 22.5. On the last step, she multiplies by 0.3 when she should have divided by 0.3. What is the correct answer?

10.3 Using the Principles Together

OBJECTIVES

a Solve equations using both the addition principle and the multiplication principle.

b Solve equations in which like terms may need to be collected.

c Solve equations by first removing parentheses and collecting like terms; solve equations with an infinite number of solutions and equations with no solutions.

SKILL TO REVIEW

Objective 9.7e: Collect like terms.

Collect like terms.

1. $q + 5t - 1 + 5q - t$

2. $7d + 16 - 11w - 2 - 10d$

a APPLYING BOTH PRINCIPLES

Consider the equation $3x + 4 = 13$. It is more complicated than those we discussed in the preceding two sections. In order to solve such an equation, we first isolate the x-term, $3x$, using the addition principle. Then we apply the multiplication principle to get x by itself.

EXAMPLE 1 Solve: $3x + 4 = 13$.

$$3x + 4 = 13$$

$$3x + 4 - 4 = 13 - 4$$ Using the addition principle: subtracting 4 on both sides

First isolate the x-term. → $$3x = 9$$ Simplifying

$$\frac{3x}{3} = \frac{9}{3}$$ Using the multiplication principle: dividing by 3 on both sides

Then isolate x. → $$x = 3$$ Simplifying

Check:

$3x + 4 = 13$	
$3 \cdot 3 + 4$?	13
$9 + 4$	
13	TRUE

We use the rules for order of operations to carry out the check. We find the product $3 \cdot 3$. Then we add 4.

The solution is 3.

◀ Do Margin Exercise 1.

1. Solve: $9x + 6 = 51$.

EXAMPLE 2 Solve: $-5x - 6 = 16$.

$$-5x - 6 = 16$$

$$-5x - 6 + 6 = 16 + 6$$ Adding 6 on both sides

$$-5x = 22$$

$$\frac{-5x}{-5} = \frac{22}{-5}$$ Dividing by -5 on both sides

$$x = -\frac{22}{5}, \text{ or } -4\frac{2}{5}$$ Simplifying

Check:

$-5x - 6 = 16$	
$-5\left(-\frac{22}{5}\right) - 6$?	16
$22 - 6$	
16	TRUE

The solution is $-\frac{22}{5}$.

◀ Do Margin Exercises 2 and 3.

Solve.

2. $8x - 4 = 28$

3. $-\frac{1}{2}x + 3 = 1$

Answers

Skill to Review:
1. $6q + 4t - 1$
2. $-3d + 14 - 11w$

Margin Exercises:
1. 5 **2.** 4 **3.** 4

EXAMPLE 3 Solve: $45 - t = 13$.

$$45 - t = 13$$
$$-45 + 45 - t = -45 + 13 \quad \text{Adding } -45 \text{ on both sides}$$
$$-t = -32$$
$$-1(-t) = -1(-32) \quad \text{Multiplying by } -1 \text{ on both sides}$$
$$t = 32$$

The number 32 checks and is the solution.

Do Exercise 4. ▶

GS **4.** Solve: $-18 - m = -57$.

$$-18 - m = -57$$
$$18 - 18 - m = \square - 57$$
$$\square = -39$$
$$\square(-m) = -1(-39)$$
$$\square = 39$$

EXAMPLE 4 Solve: $16.3 - 7.2y = -8.18$.

$$16.3 - 7.2y = -8.18$$
$$-16.3 + 16.3 - 7.2y = -16.3 + (-8.18) \quad \text{Adding } -16.3 \text{ on both sides}$$
$$-7.2y = -24.48$$
$$\frac{-7.2y}{-7.2} = \frac{-24.48}{-7.2} \quad \text{Dividing by } -7.2 \text{ on both sides}$$
$$y = 3.4$$

Check:

$$\begin{array}{c|c} \multicolumn{2}{c}{16.3 - 7.2y = -8.18} \\ \hline 16.3 - 7.2(3.4) \; ? & -8.18 \\ 16.3 - 24.48 & \\ -8.18 & \end{array} \quad \text{TRUE}$$

The solution is 3.4.

Do Exercises 5 and 6. ▶

Solve.

5. $-4 - 8x = 8$

6. $41.68 = 4.7 - 8.6y$

b COLLECTING LIKE TERMS

If there are like terms on one side of the equation, we collect them before using the addition principle or the multiplication principle.

EXAMPLE 5 Solve: $3x + 4x = -14$.

$$3x + 4x = -14$$
$$7x = -14 \quad \text{Collecting like terms}$$
$$\frac{7x}{7} = \frac{-14}{7} \quad \text{Dividing by 7 on both sides}$$
$$x = -2$$

The number -2 checks, so the solution is -2.

Do Exercises 7 and 8. ▶

Solve.

7. $4x + 3x = -21$

8. $x - 0.09x = 728$

If there are like terms on opposite sides of the equation, we get them on the same side by using the addition principle. Then we collect them. In other words, we get all the terms with a variable on one side of the equation and all the terms without a variable on the other side.

Answers

4. 39 **5.** $-\frac{3}{2}$ **6.** -4.3 **7.** -3 **8.** 800

Guided Solution:
4. $18, -m, -1, m$

EXAMPLE 6 Solve: $2x - 2 = -3x + 3$.

$$\begin{aligned} 2x - 2 &= -3x + 3 \\ 2x - 2 + 2 &= -3x + 3 + 2 && \text{Adding 2} \\ 2x &= -3x + 5 && \text{Collecting like terms} \\ 2x + 3x &= -3x + 3x + 5 && \text{Adding } 3x \\ 5x &= 5 && \text{Simplifying} \\ \frac{5x}{5} &= \frac{5}{5} && \text{Dividing by 5} \\ x &= 1 && \text{Simplifying} \end{aligned}$$

Check:

$$\begin{array}{r|l} 2x - 2 & = -3x + 3 \\ \hline 2 \cdot 1 - 2 & ? \ -3 \cdot 1 + 3 \\ 2 - 2 & -3 + 3 \\ 0 & 0 \quad \text{TRUE} \end{array}$$

Substituting in the original equation

The solution is 1.

Solve.

9. $7y + 5 = 2y + 10$

10. $5 - 2y = 3y - 5$

◀ **Do Exercises 9 and 10.**

In Example 6, we used the addition principle to get all the terms with an x on one side of the equation and all the terms without an x on the other side. Then we collected like terms and proceeded as before. If there are like terms on one side at the outset, they should be collected first.

EXAMPLE 7 Solve: $6x + 5 - 7x = 10 - 4x + 3$.

$$\begin{aligned} 6x + 5 - 7x &= 10 - 4x + 3 \\ -x + 5 &= 13 - 4x && \text{Collecting like terms} \\ 4x - x + 5 &= 13 - 4x + 4x && \text{Adding } 4x \text{ to get all terms with a variable on one side} \\ 3x + 5 &= 13 && \text{Simplifying; that is, collecting like terms} \\ 3x + 5 - 5 &= 13 - 5 && \text{Subtracting 5} \\ 3x &= 8 && \text{Simplifying} \\ \frac{3x}{3} &= \frac{8}{3} && \text{Dividing by 3} \\ x &= \frac{8}{3} && \text{Simplifying} \end{aligned}$$

The number $\frac{8}{3}$ checks, so it is the solution.

Solve.

11. $7x - 17 + 2x = 2 - 8x + 15$ GS

$\_\_\_ \cdot x - 17 = 17 - 8x$

$8x + 9x - 17 = 17 - 8x + \_\_\_$

$\_\_\_ \cdot x - 17 = 17$

$17x - 17 + 17 = 17 + \_\_\_$

$17x = 34$

$\frac{17x}{17} = \frac{34}{\_\_\_}$

$\_\_\_ = 2$

12. $3x - 15 = 5x + 2 - 4x$

◀ **Do Exercises 11 and 12.**

Clearing Fractions and Decimals

In general, equations are easier to solve if they do not contain fractions or decimals. Consider, for example, the equations

$$\frac{1}{2}x + 5 = \frac{3}{4} \quad \text{and} \quad 2.3x + 7 = 5.4.$$

Answers

9. 1 **10.** 2 **11.** 2 **12.** $\frac{17}{2}$

Guided Solution:

11. 9, $8x$, 17, 17, 17, x

If we multiply by 4 on both sides of the first equation and by 10 on both sides of the second equation, we have

$$4\left(\frac{1}{2}x + 5\right) = 4 \cdot \frac{3}{4} \quad \text{and} \quad 10(2.3x + 7) = 10 \cdot 5.4$$

$$4 \cdot \frac{1}{2}x + 4 \cdot 5 = 4 \cdot \frac{3}{4} \quad \text{and} \quad 10 \cdot 2.3x + 10 \cdot 7 = 10 \cdot 5.4$$

$$2x + 20 = 3 \quad \text{and} \quad 23x + 70 = 54.$$

The first equation has been "cleared of fractions" and the second equation has been "cleared of decimals." Both resulting equations are equivalent to the original equations and are easier to solve. *It is your choice* whether to clear fractions or decimals, but doing so often eases computations.

The easiest way to clear an equation of fractions is to multiply *every term on both sides* by the **least common multiple of all the denominators**.

EXAMPLE 8 Solve: $\frac{2}{3}x - \frac{1}{6} + \frac{1}{2}x = \frac{7}{6} + 2x$.

The denominators are 3, 6, and 2. The number 6 is the least common multiple of all the denominators. We multiply by 6 on both sides of the equation.

$$6\left(\frac{2}{3}x - \frac{1}{6} + \frac{1}{2}x\right) = 6\left(\frac{7}{6} + 2x\right) \qquad \text{Multiplying by 6 on both sides}$$

$$6 \cdot \frac{2}{3}x - 6 \cdot \frac{1}{6} + 6 \cdot \frac{1}{2}x = 6 \cdot \frac{7}{6} + 6 \cdot 2x \qquad \text{Using the distributive law (\textit{Caution!} Be sure to multiply \textit{all} the terms by 6.)}$$

$$4x - 1 + 3x = 7 + 12x \qquad \text{Simplifying. Note that the fractions are cleared.}$$

$$7x - 1 = 7 + 12x \qquad \text{Collecting like terms}$$

$$7x - 1 - 12x = 7 + 12x - 12x \qquad \text{Subtracting } 12x$$

$$-5x - 1 = 7 \qquad \text{Collecting like terms}$$

$$-5x - 1 + 1 = 7 + 1 \qquad \text{Adding 1}$$

$$-5x = 8 \qquad \text{Collecting like terms}$$

$$\frac{-5x}{-5} = \frac{8}{-5} \qquad \text{Dividing by } -5$$

$$x = -\frac{8}{5}$$

Check:

$$\frac{2}{3}x - \frac{1}{6} + \frac{1}{2}x = \frac{7}{6} + 2x$$

$\frac{2}{3}\left(-\frac{8}{5}\right) - \frac{1}{6} + \frac{1}{2}\left(-\frac{8}{5}\right)$?	$\frac{7}{6} + 2\left(-\frac{8}{5}\right)$
$-\frac{16}{15} - \frac{1}{6} - \frac{8}{10}$	$\frac{7}{6} - \frac{16}{5}$
$-\frac{32}{30} - \frac{5}{30} - \frac{24}{30}$	$\frac{35}{30} - \frac{96}{30}$
$-\frac{61}{30}$	$-\frac{61}{30}$ TRUE

The solution is $-\frac{8}{5}$.

Caution!

Check the possible solution in the *original* equation rather than in the equation that has been cleared of fractions.

CALCULATOR CORNER

Checking Possible Solutions There are several ways to check the possible solutions of an equation on a calculator. One of the most straightforward methods is to substitute and carry out the calculations on each side of the equation just as we do when we check by hand. To check the possible solution, 1, in Example 6, for instance, we first substitute 1 for x in the expression on the left side of the equation. We get 0. Next, we substitute 1 for x in the expression on the right side of the equation. Again we get 0. Since the two sides of the equation have the same value when x is 1, we know that 1 is the solution of the equation.

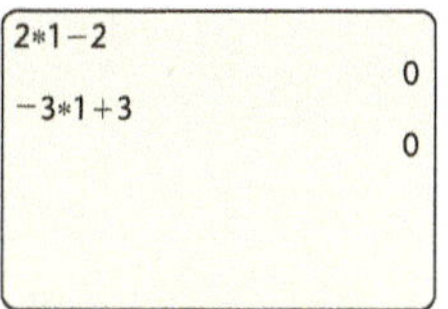

EXERCISE:

1. Use substitution to check the solutions found in Examples 1–5.

13. Solve: $\frac{7}{8}x - \frac{1}{4} + \frac{1}{2}x = \frac{3}{4} + x$.

GS

$$\frac{7}{8}x - \frac{1}{4} + \frac{1}{2}x = \frac{3}{4} + x$$
$$8 \cdot \left(\frac{7}{8}x - \frac{1}{4} + \frac{1}{2}x\right) = \square \cdot \left(\frac{3}{4} + x\right)$$
$$8 \cdot \frac{7}{8}x - \square \cdot \frac{1}{4} + 8 \cdot \frac{1}{2}x = 8 \cdot \frac{3}{4} + \square \cdot x$$
$$\square x - \square + 4x = 6 + 8x$$
$$\square x - 2 = 6 + 8x$$
$$11x - 2 - 8x = 6 + 8x - \square$$
$$3x - 2 = \square$$
$$3x - 2 + \square = 6 + 2$$
$$3x = \square$$
$$\frac{3x}{3} = \frac{8}{\square}$$
$$x = \frac{8}{3}$$

◀ **Do Exercise 13.**

To illustrate clearing decimals, we repeat Example 4, but this time we clear the equation of decimals first. Compare the methods.

To clear an equation of decimals, we count the greatest number of decimal places in any one number. If the greatest number of decimal places is 1, we multiply every term on both sides by 10; if it is 2, we multiply by 100; and so on.

EXAMPLE 9 Solve: $16.3 - 7.2y = -8.18$.

The greatest number of decimal places in any one number is *two*. Multiplying by 100, which has *two* 0's, will clear all decimals.

$100(16.3 - 7.2y) = 100(-8.18)$	Multiplying by 100 on both sides
$100(16.3) - 100(7.2y) = 100(-8.18)$	Using the distributive law
$1630 - 720y = -818$	Simplifying
$1630 - 720y - 1630 = -818 - 1630$	Subtracting 1630
$-720y = -2448$	Collecting like terms
$\frac{-720y}{-720} = \frac{-2448}{-720}$	Dividing by -720
$y = \frac{17}{5}$, or 3.4	

The number $\frac{17}{5}$, or 3.4, checks, as shown in Example 4, so it is the solution.

14. Solve: $41.68 = 4.7 - 8.6y$.

◀ **Do Exercise 14.**

C EQUATIONS CONTAINING PARENTHESES

To solve certain kinds of equations that contain parentheses, we first use the distributive laws to remove the parentheses. Then we proceed as before.

EXAMPLE 10 Solve: $8x = 2(12 - 2x)$.

$8x = 2(12 - 2x)$	
$8x = 24 - 4x$	Using the distributive laws to multiply and remove parentheses
$8x + 4x = 24 - 4x + 4x$	Adding $4x$ to get all the x-terms on one side
$12x = 24$	Collecting like terms
$\frac{12x}{12} = \frac{24}{12}$	Dividing by 12
$x = 2$	

The number 2 checks, so the solution is 2.

Solve.

15. $2(2y + 3) = 14$

16. $5(3x - 2) = 35$

◀ **Do Exercises 15 and 16.**

Answers

13. $\frac{8}{3}$ **14.** $-\frac{43}{10}$, or -4.3 **15.** 2 **16.** 3

Guided Solution:
13. 8, 8, 8, 7, 2, 11, 8x, 6, 2, 8, 3

Here is a procedure for solving the types of equation discussed in this section.

AN EQUATION-SOLVING PROCEDURE

1. Multiply on both sides to clear the equation of fractions or decimals. (This is optional, but it can ease computations.)
2. If parentheses occur, multiply to remove them using the *distributive laws.*
3. Collect like terms on each side, if necessary.
4. Get all terms with variables on one side and all numbers (constant terms) on the other side, using the *addition principle.*
5. Collect like terms again, if necessary.
6. Multiply or divide to solve for the variable, using the *multiplication principle.*
7. Check all possible solutions in the original equation.

EXAMPLE 11 Solve: $2 - 5(x + 5) = 3(x - 2) - 1$.

$$2 - 5(x + 5) = 3(x - 2) - 1$$

$$2 - 5x - 25 = 3x - 6 - 1 \quad \text{Using the distributive laws to multiply and remove parentheses}$$

$$-5x - 23 = 3x - 7 \quad \text{Collecting like terms}$$

$$-5x - 23 + 5x = 3x - 7 + 5x \quad \text{Adding } 5x$$

$$-23 = 8x - 7 \quad \text{Collecting like terms}$$

$$-23 + 7 = 8x - 7 + 7 \quad \text{Adding 7}$$

$$-16 = 8x \quad \text{Collecting like terms}$$

$$\frac{-16}{8} = \frac{8x}{8} \quad \text{Dividing by 8}$$

$$-2 = x$$

Check:

$2 - 5(x + 5)$	$=$	$3(x - 2) - 1$	
$2 - 5(-2 + 5)$	?	$3(-2 - 2) - 1$	
$2 - 5(3)$		$3(-4) - 1$	
$2 - 15$		$-12 - 1$	
-13		-13	TRUE

The solution is -2.

Do Exercises 17 and 18. ▶

Equations with Infinitely Many Solutions

The types of equations we have considered thus far in Sections 10.1–10.3 have all had exactly one solution. We now look at two other possibilities.

Consider

$$3 + x = x + 3.$$

Let's explore the equation and possible solutions in Margin Exercises 19–22.

Do Exercises 19–22. ▶

Solve.

17. $3(7 + 2x) = 30 + 7(x - 1)$

18. $4(3 + 5x) - 4 = 3 + 2(x - 2)$

Determine whether the given number is a solution of the given equation.

19. 10; $3 + x = x + 3$

20. -7; $3 + x = x + 3$

21. $\frac{1}{2}$; $3 + x = x + 3$

22. 0; $3 + x = x + 3$

Answers

17. -2 **18.** $-\frac{1}{2}$ **19.** Yes **20.** Yes
21. Yes **22.** Yes

We know by the commutative law of addition that the equation $3 + x = x + 3$ holds for any replacement of x with a real number. (See Section 9.7.) We have confirmed some of these solutions in Margin Exercises 19–22. Suppose we try to solve this equation using the addition principle:

$$3 + x = x + 3$$
$$-x + 3 + x = -x + x + 3 \quad \text{Adding } -x$$
$$3 = 3. \quad \text{True}$$

We end with a true equation. The original equation holds for all real-number replacements. Every real number is a solution. Thus the number of solutions is **infinite**.

EXAMPLE 12 Solve: $7x - 17 = 4 + 7(x - 3)$.

$$7x - 17 = 4 + 7(x - 3)$$
$$7x - 17 = 4 + 7x - 21 \quad \text{Using the distributive law to multiply and remove parentheses}$$
$$7x - 17 = 7x - 17 \quad \text{Collecting like terms}$$
$$-7x + 7x - 17 = -7x + 7x - 17 \quad \text{Adding } -7x$$
$$-17 = -17 \quad \text{True for all real numbers}$$

Every real number is a solution. There are infinitely many solutions.

Equations with No Solution

Now consider

$$3 + x = x + 8.$$

Let's explore the equation and possible solutions in Margin Exercises 23–26.

Do Exercises 23–26.

None of the replacements in Margin Exercises 23–26 is a solution of the given equation. In fact, there are no solutions. Let's try to solve this equation using the addition principle:

$$3 + x = x + 8$$
$$-x + 3 + x = -x + x + 8 \quad \text{Adding } -x$$
$$3 = 8. \quad \text{False}$$

We end with a false equation. The original equation is false for all real-number replacements. Thus it has **no** solution.

EXAMPLE 13 Solve: $3x + 4(x + 2) = 11 + 7x$.

$$3x + 4(x + 2) = 11 + 7x$$
$$3x + 4x + 8 = 11 + 7x \quad \text{Using the distributive law to multiply and remove parentheses}$$
$$7x + 8 = 11 + 7x \quad \text{Collecting like terms}$$
$$7x + 8 - 7x = 11 + 7x - 7x \quad \text{Subtracting } 7x$$
$$8 = 11 \quad \text{False}$$

There are no solutions.

Do Exercises 27 and 28.

Determine whether the given number is a solution of the given equation.

23. 10; $3 + x = x + 8$

24. -7; $3 + x = x + 8$

25. $\frac{1}{2}$; $3 + x = x + 8$

26. 0; $3 + x = x + 8$

Solve.

27. $30 + 5(x + 3) = -3 + 5x + 48$

28. $2x + 7(x - 4) = 13 + 9x$

When solving an equation, if the result is:

- an equation of the form $x = a$, where a is a real number, then there is one solution, the number a;
- a true equation like $3 = 3$ or $-1 = -1$, then every real number is a solution;
- a false equation like $3 = 8$ or $-4 = 5$, then there is no solution.

Answers

23. No **24.** No **25.** No **26.** No
27. All real numbers **28.** No solution

10.3 Exercise Set

For Extra Help MyMathLab® MathXL®

Reading Check

Choose from the column on the right the operation that will clear each equation of fractions or decimals.

RC1. $\frac{2}{5}x - 5 + \frac{1}{2}x = \frac{3}{10} + x$

RC2. $0.003y - 0.1 = 0.03 + y$

RC3. $\frac{1}{4} - 8t + \frac{5}{6} = t - \frac{1}{12}$

RC4. $0.5 + 2.15y = 1.5y - 10$

RC5. $\frac{3}{5} - x = \frac{2}{7}x + 4$

a) Multiply by 1000 on both sides.

b) Multiply by 35 on both sides.

c) Multiply by 12 on both sides.

d) Multiply by 10 on both sides.

e) Multiply by 100 on both sides.

a Solve. Don't forget to check!

1. $5x + 6 = 31$

Check: $5x + 6 = 31$?

2. $7x + 6 = 13$

Check: $7x + 6 = 13$?

3. $8x + 4 = 68$

Check: $8x + 4 = 68$?

4. $4y + 10 = 46$

Check: $4y + 10 = 46$?

5. $4x - 6 = 34$

6. $5y - 2 = 53$

7. $3x - 9 = 33$

8. $4x - 19 = 5$

9. $7x + 2 = -54$

10. $5x + 4 = -41$

11. $-45 = 3 + 6y$

12. $-91 = 9t + 8$

13. $-4x + 7 = 35$

14. $-5x - 7 = 108$

15. $\frac{5}{4}x - 18 = -3$

16. $\frac{3}{2}x - 24 = -36$

b Solve.

17. $5x + 7x = 72$

Check: $5x + 7x = 72$?

18. $8x + 3x = 55$

Check: $8x + 3x = 55$?

19. $8x + 7x = 60$

Check: $8x + 7x = 60$?

20. $8x + 5x = 104$

Check: $8x + 5x = 104$?

21. $4x + 3x = 42$

22. $7x + 18x = 125$

23. $-6y - 3y = 27$

24. $-5y - 7y = 144$

25. $-7y - 8y = -15$

26. $-10y - 3y = -39$

27. $x + \frac{1}{3}x = 8$

28. $x + \frac{1}{4}x = 10$

29. $10.2y - 7.3y = -58$

30. $6.8y - 2.4y = -88$

31. $8y - 35 = 3y$

32. $4x - 6 = 6x$

33. $8x - 1 = 23 - 4x$

34. $5y - 2 = 28 - y$

35. $2x - 1 = 4 + x$

36. $4 - 3x = 6 - 7x$

37. $6x + 3 = 2x + 11$

38. $14 - 6a = -2a + 3$

39. $5 - 2x = 3x - 7x + 25$

40. $-7z + 2z - 3z - 7 = 17$

41. $4 + 3x - 6 = 3x + 2 - x$

42. $5 + 4x - 7 = 4x - 2 - x$

43. $4y - 4 + y + 24 = 6y + 20 - 4y$

44. $5y - 7 + y = 7y + 21 - 5y$

Solve. Clear fractions or decimals first.

45. $\frac{7}{2}x + \frac{1}{2}x = 3x + \frac{3}{2} + \frac{5}{2}x$

46. $\frac{7}{8}x - \frac{1}{4} + \frac{3}{4}x = \frac{1}{16} + x$

47. $\frac{2}{3} + \frac{1}{4}t = \frac{1}{3}$

48. $-\frac{3}{2} + x = -\frac{5}{6} - \frac{4}{3}$

49. $\frac{2}{3} + 3y = 5y - \frac{2}{15}$

50. $\frac{1}{2} + 4m = 3m - \frac{5}{2}$

51. $\frac{5}{3} + \frac{2}{3}x = \frac{25}{12} + \frac{5}{4}x + \frac{3}{4}$

52. $1 - \frac{2}{3}y = \frac{9}{5} - \frac{y}{5} + \frac{3}{5}$

53. $2.1x + 45.2 = 3.2 - 8.4x$

54. $0.96y - 0.79 = 0.21y + 0.46$

55. $1.03 - 0.62x = 0.71 - 0.22x$

56. $1.7t + 8 - 1.62t = 0.4t - 0.32 + 8$

57. $\frac{2}{7}x - \frac{1}{2}x = \frac{3}{4}x + 1$

58. $\frac{5}{16}y + \frac{3}{8}y = 2 + \frac{1}{4}y$

C Solve.

59. $3(2y - 3) = 27$

60. $8(3x + 2) = 30$

61. $40 = 5(3x + 2)$

62. $9 = 3(5x - 2)$

63. $-23 + y = y + 25$

64. $17 - t = -t + 68$

65. $-23 + x = x - 23$

66. $y - \frac{2}{3} = -\frac{2}{3} + y$

67. $2(3 + 4m) - 9 = 45$

68. $5x + 5(4x - 1) = 20$

69. $5r - (2r + 8) = 16$

70. $6b - (3b + 8) = 16$

71. $6 - 2(3x - 1) = 2$

72. $10 - 3(2x - 1) = 1$

73. $5(d + 4) = 7(d - 2)$

74. $3(t - 2) = 9(t + 2)$

75. $8(2t + 1) = 4(7t + 7)$

76. $7(5x - 2) = 6(6x - 1)$

77. $5x + 5 - 7x = 15 - 12x + 10x - 10$

78. $3 - 7x + 10x - 14 = 9 - 6x + 9x - 20$

79. $22x - 5 - 15x + 3 = 10x - 4 - 3x + 11$

80. $11x - 6 - 4x + 1 = 9x - 8 - 2x + 12$

81. $3(r - 6) + 2 = 4(r + 2) - 21$

82. $5(t + 3) + 9 = 3(t - 2) + 6$

83. $19 - (2x + 3) = 2(x + 3) + x$

84. $13 - (2c + 2) = 2(c + 2) + 3c$

85. $2[4 - 2(3 - x)] - 1 = 4[2(4x - 3) + 7] - 25$

86. $5[3(7 - t) - 4(8 + 2t)] - 20 = -6[2(6 + 3t) - 4]$

87. $11 - 4(x + 1) - 3 = 11 + 2(4 - 2x) - 16$

88. $6(2x - 1) - 12 = 7 + 12(x - 1)$

89. $22x - 1 - 12x = 5(2x - 1) + 4$

90. $2 + 14x - 9 = 7(2x + 1) - 14$

91. $0.7(3x + 6) = 1.1 - (x + 2)$

92. $0.9(2x + 8) = 20 - (x + 5)$

Skill Maintenance

93. Divide: $-22.1 \div 3.4$. [5.4a], [9.6c]

94. Multiply: $-22.1(3.4)$. [5.3a], [9.5a]

95. Factor: $7x - 21 - 14y$. [9.7d]

96. Factor: $8y - 88x + 8$. [9.7d]

Simplify.

97. $-3 + 2(-5)^2(-3) - 7$ [9.8d]

98. $3x + 2[4 - 5(2x - 1)]$ [9.8c]

99. $23(2x - 4) - 15(10 - 3x)$ [9.8b]

100. $256 \div 64 \div 4^2$ [9.8d]

Synthesis

Solve.

101. $\frac{2}{3}\left(\frac{7}{8} - 4x\right) - \frac{5}{8} = \frac{3}{8}$

102. $\frac{1}{4}(8y + 4) - 17 = -\frac{1}{2}(4y - 8)$

103. $\frac{4 - 3x}{7} = \frac{2 + 5x}{49} - \frac{x}{14}$

104. The width of a rectangle is 5 ft, its length is $(3x + 2)$ ft, and its area is 75 ft^2. Find x.

Formulas

10.4

OBJECTIVES

a Evaluate a formula.

b Solve a formula for a specified letter.

SKILL TO REVIEW

Objective 10.3a: Solve equations using both the addition principle and the multiplication principle.

Solve.

1. $28 = 7 - 3a$

2. $\frac{1}{2}x - 22 = -20$

a EVALUATING FORMULAS

A **formula** is a "recipe" for doing a certain type of calculation. Formulas are often given as equations. When we replace the variables in an equation with numbers and calculate the result, we are **evaluating** the formula. Evaluating was introduced in Section 9.1.

Let's consider a formula that has to do with weather. Suppose you see a flash of lightning during a storm. Then a few seconds later, you hear thunder. Your distance from the place where the lightning struck is given by the formula $M = \frac{1}{5}t$, where t is the number of seconds from the lightning flash to the sound of the thunder and M is in miles.

EXAMPLE 1 *Distance from Lightning.* Consider the formula $M = \frac{1}{5}t$. Suppose it takes 10 sec for the sound of thunder to reach you after you have seen a flash of lightning. How far away did the lightning strike?

We substitute 10 for t and calculate M:

$$M = \frac{1}{5}t = \frac{1}{5}(10) = 2.$$

The lightning struck 2 mi away.

Do Margin Exercise 1.

1. *Storm Distance.* Refer to Example 1. Suppose that it takes the sound of thunder 14 sec to reach you. How far away is the storm?

EXAMPLE 2 *Cost of Operating a Microwave Oven.* The cost C of operating a microwave oven for 1 year is given by the formula

$$C = \frac{W \times h \times 365}{1000} \cdot e,$$

where $W =$ the wattage, $h =$ the number of hours used per day, and $e =$ the energy cost per kilowatt-hour. Find the cost of operating a 1500-W microwave oven for 0.25 hr per day if the energy cost is \$0.13 per kilowatt-hour.

Substituting, we have

$$C = \frac{W \times h \times 365}{1000} \cdot e = \frac{1500 \times 0.25 \times 365}{1000} \cdot \$0.13 \approx \$17.79.$$

The cost for operating a 1500-W microwave oven for 0.25 hr per day for 1 year is about \$17.79.

Do Margin Exercise 2.

2. *Microwave Oven.* Refer to Example 2. Determine the cost of operating an 1100-W microwave oven for 0.5 hr per day for 1 year if the energy cost is \$0.16 per kilowatt-hour.

Answers

Skill to Review:
1. −7 2. 4

Margin Exercises:
1. 2.8 mi 2. \$32.12

3. ***Socks from Cotton.*** Refer to Example 3. Determine the number of socks that can be made from 65 bales of cotton.

EXAMPLE 3 ***Socks from Cotton.*** Consider the formula $S = 4321x$, where S is the number of socks of average size that can be produced from x bales of cotton. You see a shipment of 300 bales of cotton taken off a ship. How many socks can be made from the cotton?

Source: *Country Woman Magazine*

We substitute 300 for x and calculate S:

$$S = 4321x = 4321(300) = 1{,}296{,}300.$$

Thus, 1,296,300 socks can be made from 300 bales of cotton.

◀ **Do Exercise 3.**

b SOLVING FORMULAS

Refer to Example 3. Suppose a clothing company wants to produce S socks and needs to know how many bales of cotton to order. If this calculation is to be repeated many times, it might be helpful to first solve the formula for x:

$$S = 4321x$$

$$\frac{S}{4321} = x. \quad \text{Dividing by 4321}$$

Then we can substitute a number for S and calculate x. For example, if the number of socks S to be produced is 432,100, then

$$x = \frac{S}{4321} = \frac{432{,}100}{4321} = 100.$$

The company would need to order 100 bales of cotton.

EXAMPLE 4 Solve for z: $H = \frac{1}{4}z$.

$$H = \tfrac{1}{4}z \quad \text{We want this letter alone.}$$

$$4 \cdot H = 4 \cdot \tfrac{1}{4}z \quad \text{Multiplying by 4 on both sides}$$

$$4H = z$$

For $H = 2$ in Example 4, $z = 4H = 4(2)$, or 8.

EXAMPLE 5 ***Distance, Rate, and Time.*** Solve for t: $d = rt$.

$$d = rt \quad \text{We want this letter alone.}$$

$$\frac{d}{r} = \frac{rt}{r} \quad \text{Dividing by } r$$

$$\frac{d}{r} = \frac{r}{r} \cdot t$$

$$\frac{d}{r} = t \quad \text{Simplifying}$$

◀ **Do Exercises 4–6.**

4. Solve for q: $B = \dfrac{1}{3}q$.

5. Solve for m: $n = mz$.

6. ***Electricity.*** Solve for I: $V = IR$. (This formula relates voltage V, current I, and resistance R.)

Answers

3. 280,865 socks 4. $q = 3B$
5. $m = \dfrac{n}{z}$ 6. $I = \dfrac{V}{R}$

EXAMPLE 6 Solve for x: $y = x + 3$.

$$y = x + 3 \quad \text{We want this letter alone.}$$
$$y - 3 = x + 3 - 3 \quad \text{Subtracting 3}$$
$$y - 3 = x \quad \text{Simplifying}$$

EXAMPLE 7 Solve for x: $y = x - a$.

$$y = x - a \quad \text{We want this letter alone.}$$
$$y + a = x - a + a \quad \text{Adding } a$$
$$y + a = x \quad \text{Simplifying}$$

Do Exercises 7–9.

Solve for x.

7. $y = x + 5$

8. $y = x - 7$

9. $y = x - b$

EXAMPLE 8 Solve for y: $6y = 3x$.

$$6y = 3x \quad \text{We want this letter alone.}$$
$$\frac{6y}{6} = \frac{3x}{6} \quad \text{Dividing by 6}$$
$$y = \frac{x}{2}, \text{ or } \frac{1}{2}x \quad \text{Simplifying}$$

EXAMPLE 9 Solve for y: $by = ax$.

$$by = ax \quad \text{We want this letter alone.}$$
$$\frac{by}{b} = \frac{ax}{b} \quad \text{Dividing by } b$$
$$y = \frac{ax}{b} \quad \text{Simplifying}$$

Do Exercises 10 and 11.

10. Solve for y: $9y = 5x$.

11. Solve for p: $ap = bt$.

GS **12.** Solve for x: $y = mx + b$.

$$y = mx + b$$
$$y - \_\_\_ = mx + b - b$$
$$y - b = \_\_\_$$
$$\frac{y - b}{m} = \frac{mx}{\_\_\_}$$
$$\frac{y - b}{m} = \_\_\_$$

EXAMPLE 10 Solve for x: $ax + b = c$.

$$ax + b = c \quad \text{We want this letter alone.}$$
$$ax + b - b = c - b \quad \text{Subtracting } b$$
$$ax = c - b \quad \text{Simplifying}$$
$$\frac{ax}{a} = \frac{c - b}{a} \quad \text{Dividing by } a$$
$$x = \frac{c - b}{a} \quad \text{Simplifying}$$

Do Exercises 12 and 13.

13. Solve for Q: $tQ - p = a$.

Answers

7. $x = y - 5$ **8.** $x = y + 7$
9. $x = y + b$ **10.** $y = \frac{5x}{9}$, or $\frac{5}{9}x$
11. $p = \frac{bt}{a}$ **12.** $x = \frac{y - b}{m}$
13. $Q = \frac{a + p}{t}$

Guided Solution:
12. b, mx, m, x

A FORMULA-SOLVING PROCEDURE

To solve a formula for a given letter, identify the letter and:

1. Multiply on both sides to clear fractions or decimals, if that is needed.
2. Collect like terms on each side, if necessary.
3. Get all terms with the letter to be solved for on one side of the equation and all other terms on the other side.
4. Collect like terms again, if necessary.
5. Solve for the letter in question.

EXAMPLE 11 *Circumference.* Solve for r: $C = 2\pi r$. This is a formula for the circumference C of a circle of radius r.

$C = 2\pi r$ We want this letter alone.

$\frac{C}{2\pi} = \frac{2\pi r}{2\pi}$ Dividing by 2π

$\frac{C}{2\pi} = r$

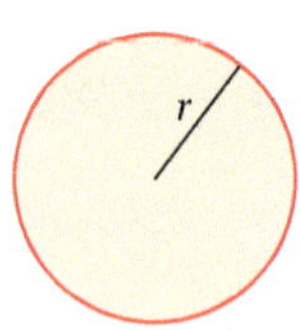

EXAMPLE 12 *Averages.* Solve for a: $A = \frac{a + b + c}{3}$. This is a formula for the average A of three numbers a, b, and c.

$A = \frac{a + b + c}{3}$ We want the letter a alone.

$3 \cdot A = 3 \cdot \frac{a + b + c}{3}$ Multiplying by 3 on both sides

$3A = a + b + c$ Simplifying

$3A - b - c = a$ Subtracting b and c

Do Exercises 14 and 15.

14. *Circumference.* Solve for D:
$$C = \pi D.$$
This is a formula for the circumference C of a circle of diameter D.

15. *Averages.* Solve for c:
$$A = \frac{a + b + c + d}{4}.$$

Answers

14. $D = \frac{C}{\pi}$ 15. $c = 4A - a - b - d$

10.4 Exercise Set

For Extra Help
MyMathLab® MathXL® PRACTICE WATCH READ REVIEW

 Reading Check

Solve each equation for the indicated letter and choose the correct solution from the column on the right.

RC1. Solve $4x = 7y$ for x.

RC2. Solve $y = \frac{1}{4}x - w$ for w.

RC3. Solve $xz + 4y = w$ for x.

RC4. Solve $z = w + 4$ for w.

a) $w = z - 4$

b) $x = \frac{7y}{4}$, or $\frac{7}{4}y$

c) $w = \frac{1}{4}x - y$

d) $x = \frac{w - 4y}{z}$

a Solve.

1. ***Wavelength of a Musical Note.*** The wavelength w, in meters per cycle, of a musical note is given by

$$w = \frac{r}{f},$$

where r is the speed of the sound, in meters per second, and f is the frequency, in cycles per second. The speed of sound in air is 344 m/sec. What is the wavelength of a note whose frequency in air is 24 cycles per second?

2. ***Furnace Output.*** Contractors in the Northeast use the formula $B = 30a$ to determine the minimum furnace output B, in British thermal units (Btu's), for a well-insulated house with a square feet of flooring. Determine the minimum furnace output for an 1800-ft^2 house that is well insulated.

Source: U.S. Department of Energy

3. ***Calorie Density.*** The calorie density D, in calories per ounce, of a food that contains c calories and weighs w ounces is given by

$$D = \frac{c}{w}.$$

Eight ounces of fat-free milk contains 84 calories. Find the calorie density of fat-free milk.

Source: *Nutrition Action Healthletter*, March 2000, p. 9.

4. ***Size of a League Schedule.*** When all n teams in a league play every other team twice, a total of N games are played, where

$$N = n^2 - n.$$

A soccer league has 7 teams and all teams play each other twice. How many games are played?

5. ***Distance, Rate, and Time.*** The distance d that a car will travel at a rate, or speed, r in time t is given by

$$d = rt.$$

a) A car travels at 75 miles per hour (mph) for 4.5 hr. How far will it travel?

b) Solve the formula for t.

6. ***Surface Area of a Cube.*** The surface area A of a cube with side s is given by

$$A = 6s^2.$$

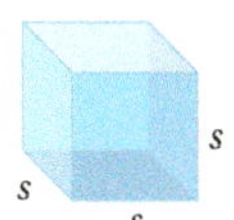

a) Find the surface area of a cube with sides of 3 in.

b) Solve the formula for s^2.

7. *College Enrollment.* At many colleges, the number of "full-time-equivalent" students f is given by

$$f = \frac{n}{15},$$

where n is the total number of credits for which students have enrolled in a given semester.

a) Determine the number of full-time-equivalent students on a campus in which students registered for a total of 21,345 credits.

b) Solve the formula for n.

8. *Electrical Power.* The power rating P, in watts, of an electrical appliance is determined by

$$P = I \cdot V,$$

where I is the current, in amperes, and V is measured in volts.

a) A microwave oven requires 12 amps of current and the voltage in the house is 115 volts. What is the wattage of the microwave?

b) Solve the formula for I; for V.

b Solve for the indicated letter.

9. $y = 5x$, for x

10. $d = 55t$, for t

11. $a = bc$, for c

12. $y = mx$, for x

13. $n = m + 11$, for m

14. $z = t + 21$, for t

15. $y = x - \frac{3}{5}$, for x

16. $y = x - \frac{2}{3}$, for x

17. $y = 13 + x$, for x

18. $t = 6 + s$, for s

19. $y = x + b$, for x

20. $y = x + A$, for x

21. $y = 5 - x$, for x

22. $y = 10 - x$, for x

23. $y = a - x$, for x

24. $y = q - x$, for x

25. $8y = 5x$, for y

26. $10y = -5x$, for y

27. $By = Ax$, for x

28. $By = Ax$, for y

29. $W = mt + b$, for t

30. $W = mt - b$, for t

31. $y = bx + c$, for x

32. $y = bx - c$, for x

33. *Area of a Parallelogram:*

$A = bh$, for h

(Area A, base b, height h)

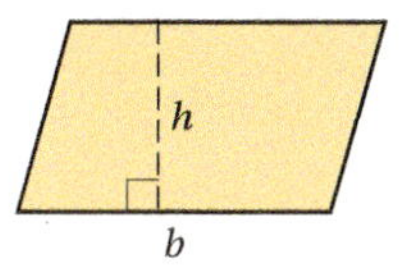

34. *Distance, Rate, Time:*

$d = rt$, for r

(Distance d, speed r, time t)

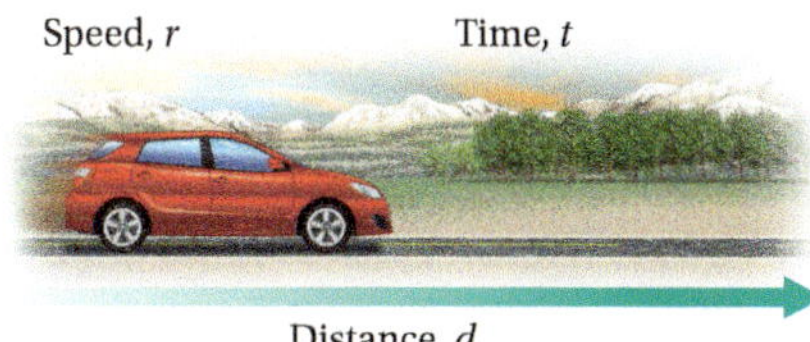

35. *Perimeter of a Rectangle:*

$P = 2l + 2w$, for w

(Perimeter P, length l, width w)

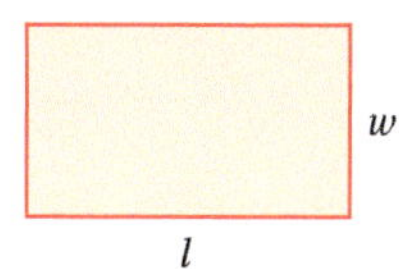

36. *Area of a Circle:*

$A = \pi r^2$, for r^2

(Area A, radius r)

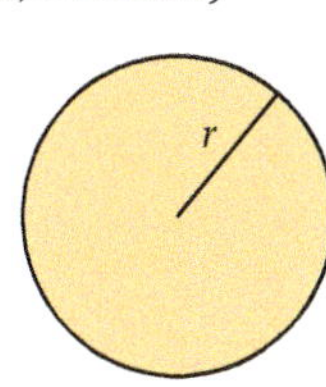

37. *Average of Two Numbers:*

$$A = \frac{a + b}{2}, \text{ for } a$$

a $\quad A = \frac{a+b}{2} \quad$ b

38. *Area of a Triangle:*

$$A = \frac{1}{2}bh, \text{ for } b$$

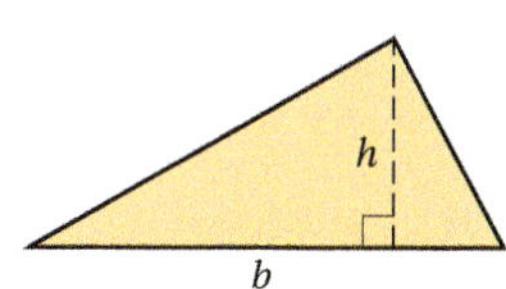

39. $A = \dfrac{a + b + c}{3}$, for b

40. $A = \dfrac{a + b + c}{3}$, for c

41. $A = at + b$, for t

42. $S = rx + s$, for x

43. $Ax + By = c$, for x

44. $Q = \frac{p - q}{2}$, for p

45. *Force:*

$F = ma$, for a

(Force F, mass m, acceleration a)

46. *Simple Interest:*

$I = Prt$, for P

(Interest I, principal P, interest rate r, time t)

47. *Relativity:*

$E = mc^2$, for c^2

(Energy E, mass m, speed of light c)

48. $Ax + By = c$, for y

49. $v = \frac{3k}{t}$, for t

50. $P = \frac{ab}{c}$, for c

Skill Maintenance

51. Evaluate $\frac{3x - 2y}{y}$ when $x = 6$ and $y = 2$. [9.1a]

52. Remove parentheses and simplify:

$4a - 8b - 5(5a - 4b)$. [9.8b]

Subtract. [9.4a]

53. $-45.8 - (-32.6)$

54. $-\frac{2}{3} - \frac{5}{6}$

55. $87\frac{1}{2} - 123$

Add. [9.3a]

56. $-\frac{5}{12} + \frac{1}{4}$

57. $0.082 + (-9.407)$

58. $-2\frac{1}{2} + 6\frac{1}{4}$

Convert to percent notation.

59. 0.034 [6.2b]

60. $\frac{1}{20}$ [6.3a]

61. $\frac{5}{12}$ [6.3a]

62. 0.5 [6.2b]

Synthesis

Solve.

63. $H = \frac{2}{a - b}$, for b; for a

64. $P = 4m + 7mn$, for m

65. In $A = lw$, if l and w both double, what is the effect on A?

66. In $P = 2a + 2b$, if P doubles, do a and b necessarily both double?

67. In $A = \frac{1}{2}bh$, if b increases by 4 units and h does not change, what happens to A?

68. Solve for F: $D = \frac{1}{E + F}$.

Mid-Chapter Review

Concept Reinforcement

Determine whether each statement is true or false.

______ **1.** $3 - x = 4x$ and $5x = -3$ are equivalent equations. [10.1b]

______ **2.** For any real numbers a, b, and c, $a = b$ is equivalent to $a + c = b + c$. [10.1b]

______ **3.** We can use the multiplication principle to divide on both sides of an equation by the same nonzero number. [10.2a]

______ **4.** Every equation has at least one solution. [10.3c]

Guided Solutions

Fill in each blank with the number, variable, or expression that creates a correct statement or solution.

Solve. [10.1b], [10.2a]

5.
$$\begin{aligned} x + 5 &= -3 \\ x + 5 - 5 &= -3 - \square \\ x + \square &= -8 \\ x &= \square \end{aligned}$$

6.
$$\begin{aligned} -6x &= 42 \\ \frac{-6x}{-6} &= \frac{42}{\square} \\ \square \cdot x &= -7 \\ x &= \square \end{aligned}$$

7. Solve for y: $5y + z = t$. [10.4b]
$$\begin{aligned} 5y + z &= t \\ 5y + z - z &= t - \square \\ 5y &= \square \\ \frac{5y}{5} &= \frac{t - z}{\square} \\ y &= \frac{\square}{5} \end{aligned}$$

Mixed Review

Solve. [10.1b], [10.2a], [10.3a, b, c]

8. $x + 5 = 11$

9. $x + 9 = -3$

10. $8 = t + 1$

11. $-7 = y + 3$

12. $x - 6 = 14$

13. $y - 7 = -2$

14. $-\frac{3}{2} + z = -\frac{3}{4}$

15. $-3.3 = -1.9 + t$

16. $7x = 42$

17. $17 = -t$

18. $6x = -54$

19. $-5y = -85$

20. $\frac{x}{7} = 3$

21. $\frac{2}{3}x = 12$

22. $-\frac{t}{5} = 3$

23. $\frac{3}{4}x = -\frac{9}{8}$

24. $3x + 2 = 5$

25. $5x + 4 = -11$

26. $6x - 7 = 2$

27. $-4x - 9 = -5$

28. $6x + 5x = 33$

29. $-3y - 4y = 49$

30. $3x - 4 = 12 - x$

31. $5 - 6x = 9 - 8x$

32. $4y - \frac{3}{2} = \frac{3}{4} + 2y$

33. $\frac{4}{5} + \frac{1}{6}t = \frac{1}{10}$

34. $0.21n - 1.05 = 2.1 - 0.14n$

35. $5(3y - 1) = -35$

36. $7 - 2(5x + 3) = 1$

37. $-8 + t = t - 8$

38. $z + 12 = -12 + z$

39. $4(3x + 2) = 5(2x - 1)$

40. $8x - 6 - 2x = 3(2x - 4) + 6$

Solve for the indicated letter. [10.4b]

41. $A = 4b$, for b

42. $y = x - 1.5$, for x

43. $n = s - m$, for m

44. $4t = 9w$, for t

45. $B = at - c$, for t

46. $M = \frac{x + y + z}{2}$, for y

Understanding Through Discussion and Writing

47. Explain the difference between equivalent expressions and equivalent equations. [9.7a], [10.1b]

48. Are the equations $x = 5$ and $x^2 = 25$ equivalent? Why or why not? [10.1b]

49. When solving an equation using the addition principle, how do you determine which number to add or subtract on both sides of the equation? [10.1b]

50. Explain the following mistake made by a fellow student. [10.1b]

$$x + \frac{1}{3} = -\frac{5}{3}$$
$$x = -\frac{4}{3}$$

51. When solving an equation using the multiplication principle, how do you determine by what number to multiply or divide on both sides of the equation? [10.2a]

52. Devise an application in which it would be useful to solve the equation $d = rt$ for r. [10.4b]

STUDYING FOR SUCCESS *A Valuable Resource—Your Instructor*

- ☐ Don't be afraid to ask questions in class. Other students probably have the same questions you do.
- ☐ Visit your instructor during office hours if you need additional help.
- ☐ Many instructors welcome e-mails with questions from students.

Applications of Percent 10.5

OBJECTIVE

a Solve applied problems involving percent.

SKILL TO REVIEW

Objective 10.2a: Solve equations using the multiplication principle.

Solve.

1. $20 = 0.05a$
2. $0.3z = 327$

a TRANSLATING AND SOLVING

Many applied problems involve percent. Here we begin to see how equation solving can enhance our problem-solving skills.

In solving percent problems, we first *translate* the problem to an equation. Then we *solve* the equation using the techniques discussed in Sections 10.1–10.3. The key words in the translation are as follows.

KEY WORDS IN PERCENT TRANSLATIONS

"**Of**" translates to "·" or "×".
"**Is**" translates to "=".
"**What number**" or "**what percent**" translates to any letter.
"**%**" translates to "$\times \frac{1}{100}$" or "× 0.01".

EXAMPLE 1 Translate:

28% of 5 is what number?
↓ ↓ ↓ ↓ ↓
$28\% \cdot 5 = a$ This is a percent equation.

EXAMPLE 2 Translate:

45% of what number is 28?
↓ ↓ ↓ ↓ ↓
$45\% \times b = 28$

EXAMPLE 3 Translate:

What percent of 90 is 7?
↓ ↓ ↓ ↓ ↓
$n \cdot 90 = 7$

Do Margin Exercises 1–6. ▶

Translate to an equation. Do not solve.

1. 13% of 80 is what number?
2. What number is 60% of 70?
3. 43 is 20% of what number?

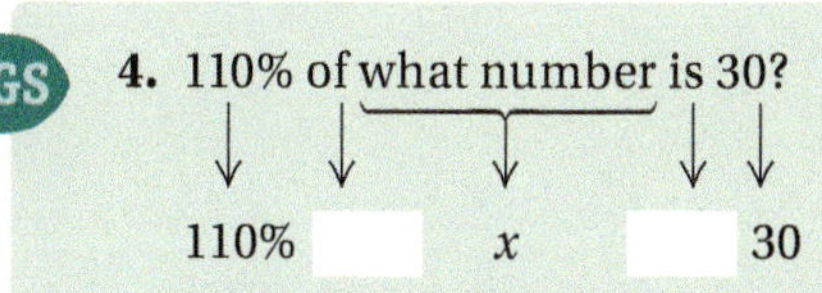

5. 16 is what percent of 80?
6. What percent of 94 is 10.5?

Percent problems are actually of three different types. Although the method we present does *not* require that you be able to identify which type we are studying, it is helpful to know them. Let's begin by using a specific example to find a standard form for a percent problem.

Answers

Skill to Review:
1. 400 2. 1090

Answers to Margin Exercises 1–6 and Guided Solution are on p. 724.

We know that

15 is 25% of 60, or $15 = 25\% \times 60$.

We can think of this as:

Amount = Percent number × Base.

Each of the three types of percent problem depends on which of the three pieces of information is missing in the statement

Amount = Percent number × Base.

1. Finding the ***amount*** (the result of taking the percent)

Example: What number is 25% of 60?

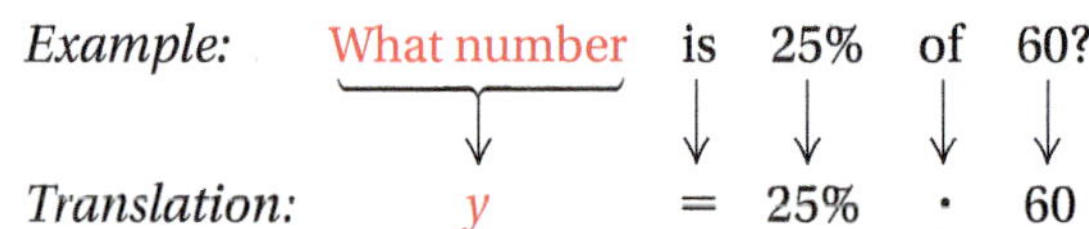

Translation: $y = 25\% \cdot 60$

2. Finding the ***base*** (the number you are taking the percent of)

Example: 15 is 25% of what number?

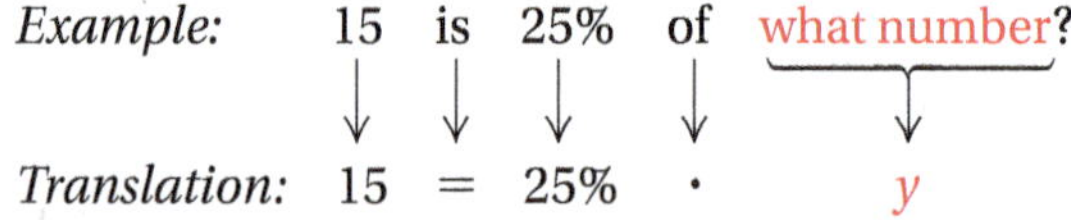

Translation: $15 = 25\% \cdot y$

3. Finding the ***percent number*** (the percent itself)

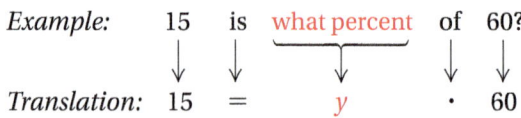

Example: 15 is what percent of 60?

Translation: $15 = y \cdot 60$

Finding the Amount

EXAMPLE 4 What number is 11% of 49?

What number is 11% of 49?

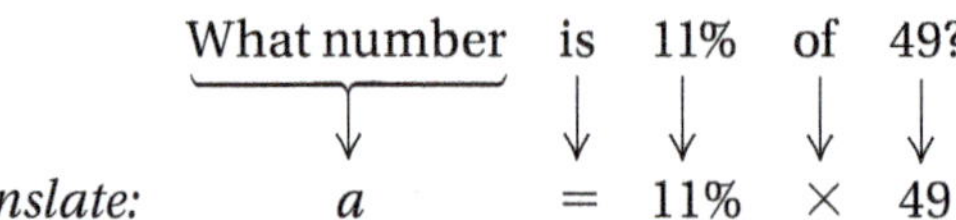

Translate: $a = 11\% \times 49$

Solve: The letter is by itself. To solve the equation, we need only convert 11% to decimal notation and multiply:

$$a = 11\% \times 49 = 0.11 \times 49 = 5.39.$$

Thus, 5.39 is 11% of 49. The answer is 5.39.

7. What number is 2.4% of 80?

◀ Do Exercise 7.

Finding the Base

EXAMPLE 5 3 is 16% of what number?

3 is 16% of what number?

Translate: $3 = 16\% \times b$

$3 = 0.16 \times b$ Converting 16% to decimal notation

Answers

Margin Exercises:
1. $13\% \cdot 80 = a$ 2. $a = 60\% \cdot 70$
3. $43 = 20\% \cdot b$ 4. $110\% \cdot b = 30$
5. $16 = n \cdot 80$ 6. $n \cdot 94 = 10.5$ 7. 1.92

Guided Solution:
4. $\cdot$, $=$

Solve: In this case, the letter is not by itself. To solve the equation, we divide by 0.16 on both sides:

$$3 = 0.16 \times b$$

$$\frac{3}{0.16} = \frac{0.16 \times b}{0.16} \quad \text{Dividing by 0.16}$$

$$18.75 = b. \quad \text{Simplifying}$$

The answer is 18.75.

Do Exercise 8. ▶

Finding the Percent Number

In solving these problems, you *must* remember to convert to percent notation after you have solved the equation.

EXAMPLE 6 $32 is what percent of $50?

$32 is what percent of $50?

Translate: $32 = p \times 50$

Solve: To solve the equation, we divide by 50 on both sides and convert the answer to percent notation:

$$32 = p \times 50$$

$$\frac{32}{50} = \frac{p \times 50}{50} \quad \text{Dividing by 50}$$

$$0.64 = p$$

$$64\% = p. \quad \text{Converting to percent notation}$$

Thus, $32 is 64% of $50. The answer is 64%.

Do Exercise 9. ▶

EXAMPLE 7 ***Donated Girl Scout Cookies.*** Through Operation Cookie Drop, Girl Scout cookies can be donated to all branches of the U.S. military. In 2012, the Girl Scouts of Western Washington sold 3,093,834 boxes of cookies. Of this number, 4.11% were donated to military personnel. How many boxes of cookies did the Girl Scouts of Western Washington donate to Operation Cookie Drop?

Source: Girl Scouts of the USA

To solve this problem, we first reword and then translate. We let $c =$ the number of boxes of cookies donated.

Rewording: What number is 4.11% of 3,093,834?

Translating: $c = 4.11\% \times 3,093,834$

Solve: The letter is by itself. To solve the equation, we need only convert 4.11% to decimal notation and multiply:

$$c = 4.11\% \times 3,093,834 = 0.0411 \times 3,093,834 \approx 127,157.$$

Thus, 127,157 is about 4.11% of 3,093,834, so 127,157 boxes of Girl Scout cookies were donated to Operation Cookie Drop in 2012.

Do Exercise 10. ▶

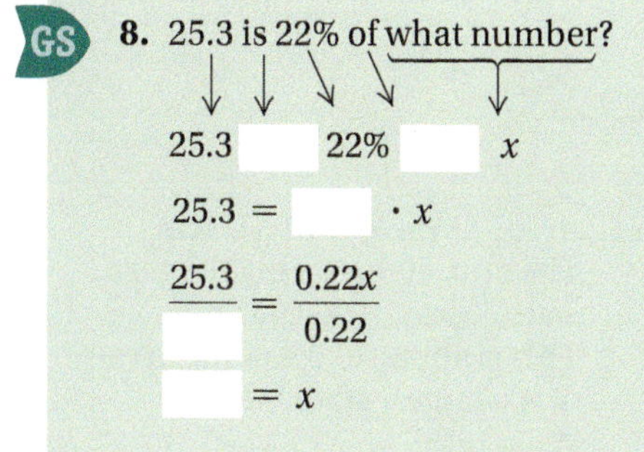

9. What percent of $50 is $18?

10. ***Haitian Population Ages 0–14.*** The population of Haiti is approximately 9,720,000. Of this number, 35.9% are ages 0–14. How many Haitians are ages 14 and younger? Round to the nearest 1000.

Source: Central Intelligence Agency

MCT KidNews 05/05

Answers

8. 115 **9.** 36%
10. About 3,489,000

Guided Solution:
8. =, ·, 0.22, 0.22, 115

EXAMPLE 8 *Motor Vehicle Production.* In 2010, 7.632 million motor vehicles were produced in the United States. This was 10.4% of the world production of motor vehicles. How many motor vehicles were produced worldwide in 2010?

Sources: R. L. Polk, Automotive News Data Center

To solve this problem, we first reword and then translate. We let $P =$ the total worldwide production, in millions, of motor vehicles in 2010.

Rewording: 7.632 is 10.4% of what number?

Translating: $7.632 = 10.4\% \times P$

Solve: To solve the equation, we convert 10.4% to decimal notation and divide by 0.104 on both sides:

$$7.632 = 10.4\% \times P$$
$$7.632 = 0.104 \times P \quad \text{Converting to decimal notation}$$
$$\frac{7.632}{0.104} = \frac{0.104 \times P}{0.104} \quad \text{Dividing by 0.104}$$
$$73.4 \approx P. \quad \text{Simplifying and rounding to the nearest tenth}$$

About 73.4 million motor vehicles were produced worldwide in 2010.

◀ **Do Exercise 11.**

11. *Areas of Texas and Alaska.* The area of the second largest state, Texas, is 268,581 mi^2. This is about 40.5% of the area of the largest state, Alaska. What is the area of Alaska?

EXAMPLE 9 *Employment Outlook.* Jobs at United States auto plants and parts factories (including domestic and foreign-owned) totaled approximately 650,000 in 2012. This number is expected to grow to 756,800 in 2015. What is the percent increase?

Sources: Center for Automotive Research & IHS Global Insight

To solve the problem, we must first determine the amount of the increase:

Jobs in 2015 minus Jobs in 2012 = Increase

$756{,}800 - 650{,}000 = 106{,}800.$

Using the job increase of 106,800, we reword and then translate. We let $p =$ the percent increase. We want to know, "what percent of the number of jobs in 2012 is 106,800?"

Rewording: 106,800 is what percent of 650,000

Translating: $106{,}800 = p \times 650{,}000$

Solve: To solve the equation, we divide by 650,000 on both sides and convert the answer to percent notation:

$$106{,}800 = p \times 650{,}000$$
$$\frac{106{,}800}{650{,}000} = \frac{p \times 650{,}000}{650{,}000} \quad \text{Dividing by 650,000}$$
$$0.164 \approx p \quad \text{Simplifying}$$
$$16.4\% \approx p. \quad \text{Converting to percent notation}$$

The percent increase is about 16.4%.

◀ **Do Exercise 12.**

12. *Median Income.* The U.S. median family income in 2004 was $49,800. This number decreased to $45,800 in 2010. What is the percent decrease?

Source: Federal Reserve's Survey of Consumer Finances, June 2012

Answers

11. About 663,163 mi^2
12. About 8.0% decrease

10.5 Exercise Set

For Extra Help

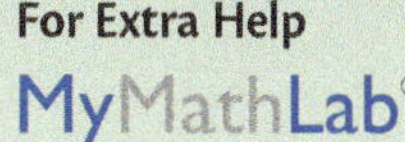

Reading Check

Match each question with the most appropriate translation from the column on the right.

RC1. 13 is 82% of what number?

RC2. What number is 13% of 82?

RC3. 82 is what percent of 13?

RC4. 82 is 13% of what number?

RC5. 13 is what percent of 82?

RC6. What number is 82% of 13?

a) $82 = 13\% \cdot b$

b) $a = 13\% \cdot 82$

c) $a = 82\% \cdot 13$

d) $13 = 82\% \cdot b$

e) $82 = p \cdot 13$

f) $13 = p \cdot 82$

a Solve.

1. What percent of 180 is 36?

2. What percent of 76 is 19?

3. 45 is 30% of what number?

4. 20.4 is 24% of what number?

5. What number is 65% of 840?

6. What number is 50% of 50?

7. 30 is what percent of 125?

8. 57 is what percent of 300?

9. 12% of what number is 0.3?

10. 7 is 175% of what number?

11. 2 is what percent of 40?

12. 16 is what percent of 40?

13. What percent of 68 is 17?

14. What percent of 150 is 39?

15. What number is 35% of 240?

16. What number is 1% of one million?

17. What percent of 575 is 138?

18. What percent of 60 is 75?

19. What percent of 300 is 48?

20. What percent of 70 is 70?

21. 14 is 30% of what number?

22. 54 is 24% of what number?

23. What number is 2% of 40?

24. What number is 40% of 2?

25. 0.8 is 16% of what number?

26. 40 is 2% of what number?

27. 54 is 135% of what number?

28. 8 is 2% of what number?

World Population by Continent. It has been projected that in 2050, the world population will be 8909 million, or 8.909 billion. The following circle graph shows the breakdown of this total population by continent.

World Population by Continent, 2050

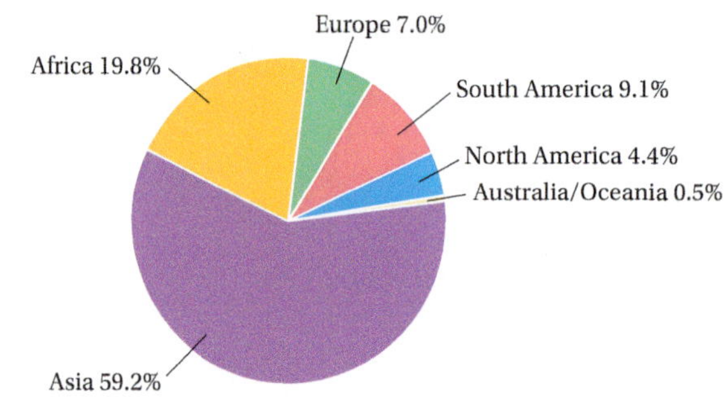

SOURCE: Central Intelligence Agency

Using the data in the figure, complete the following table of projected populations in 2050. Round to the nearest million.

	Continent	Population		Continent	Population
29.	South America		**30.**	Europe	
31.	Asia		**32.**	North America	
33.	Africa		**34.**	Australia/Oceania	

35. ***eBook Revenue.*** Net revenue from 2011 adult book sales totaled approximately $2360 million. eBook revenue accounted for 41% of this amount. What was the net revenue from adult eBook sales in 2011? Round to the nearest million.

Source: Association for American Publishers

36. ***Hardcover Book Revenue.*** Net revenue from 2011 adult book sales totaled approximately $2360 million. Revenue from hardcover books accounted for 54.8% of this amount. What was the net revenue from adult hardcover sales in 2011? Round to the nearest million.

Source: Association for American Publishers

37. ***Student Loans.*** To finance her community college education, Sarah takes out a Stafford loan for $6500. After a year, Sarah decides to pay off the interest, which is 3.4% of $6500. How much will she pay?

38. ***Student Loans.*** Paul takes out a PLUS loan for $5400. After a year, Paul decides to pay off the interest, which is 7.9% of $5000. How much will he pay?

39. *Tattoos.* Of the 237,400,000 adults ages 18 and older in the United States, approximately 49,854,000 have at least one tattoo. What percent of adults ages 18 and older have at least one tattoo?

Sources: U.S. Census Bureau; Harris Poll of 2016 adults; UPI.com

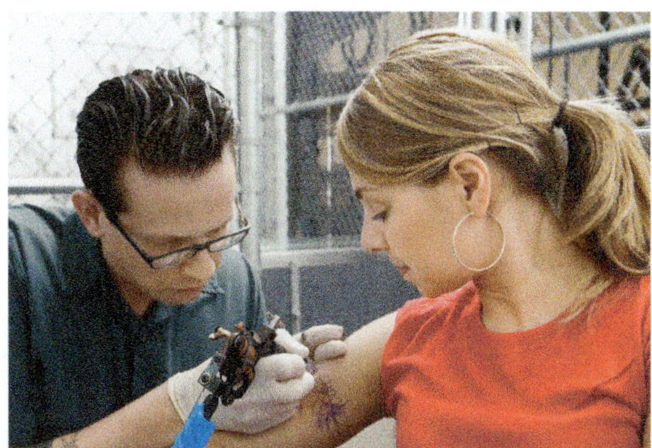

40. *Boston Marathon.* The 2012 Boston Marathon was the 116th running of the race. Since its first race, the United States has won the men's open division 43 times. What percent of the years did the United States win the men's open?

Source: Boston Athletic Association

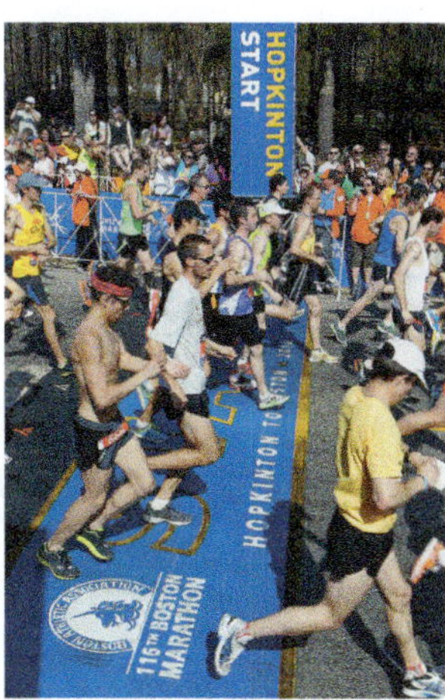

41. *Tipping.* William left a \$1.50 tip for a meal that cost \$12.

a) What percent of the cost of the meal was the tip?
b) What was the total cost of the meal including the tip?

42. *Tipping.* Sam, Selena, Rachel, and Clement left a 20% tip for a meal that cost \$75.

a) How much was the tip?
b) What was the total cost of the meal including the tip?

43. *Tipping.* David left a 15% tip of \$4.65 for a meal.

a) What was the cost of the meal before the tip?
b) What was the total cost of the meal including the tip?

44. *Tipping.* Addison left an 18% tip of \$6.75 for a meal.

a) What was the cost of the meal before the tip?
b) What was the total cost of the meal including the tip?

45. *City Park Space.* Portland, Oregon, has 12,959 acres of park space. This is 15.1% of the acreage of the entire city. What is the total acreage of Portland?

Source: Indy Parks and Recreation Master Plan

46. *Junk Mail.* About 46.2 billion pieces of unopened junk mail end up in landfills each year. This is about 44% of all the junk mail that is sent annually. How many pieces of junk mail are sent annually?

Source: Globaljunkmailcrisis.org

47. *Employment Growth.* In 1980, there were 1.7 million licensed registered nurses in the United States. This number increased to 3.1 million in 2012. What is the percent increase?

Source: Maria Sonnenberg, *Florida Today*, May 14, 2012

48. *Artificial-Tree Sales.* From 2008 to 2012, sales of artificial Christmas trees grew from \$950 million to approximately \$1070 million. What is the percent increase?

Sources: BalsamHill.com and USATradeonline.gov

49. *Newspaper Advertiser Spending.* In 2000, advertisers spent $48.6 billion in newspapers. This number dropped to $22.8 billion in 2010. What is the percent decrease?

Source: Paul Glader, "Black and White, Read No More", *The American Legion Magazine*, August 2012

50. *New Magazines.* In 2010, 301 magazines were launched in North America. In 2011, only 273 were launched. What is the percent decrease?

Source: Media Finder

51. *Dog Bites.* Over one-third of all homeowner insurance liability claims paid in 2011 were for dog bites. The average cost of dog-bite claims in the United States increased from $19,162 in 2004 to $29,396 in 2011. What is the percent increase?

Sources: Adam Belz, Insurance Information Institute, *USA TODAY*

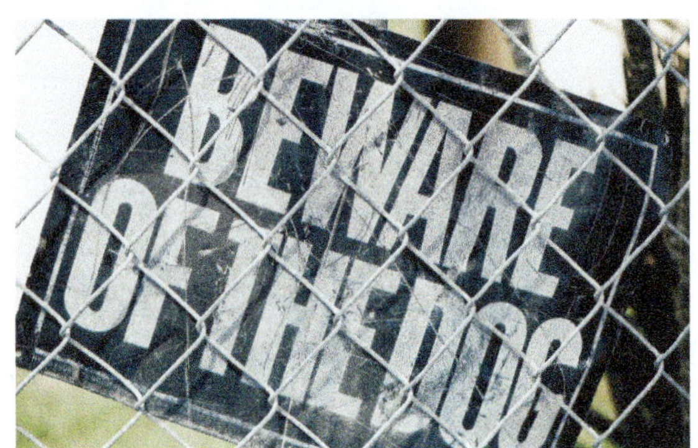

52. *International Students.* In 2001–2002, 582,996 international students were enrolled in U.S. colleges. By 2011–2012, this number had increased to 764,495. What is the percent increase?

Source: Institute of International Education, Open Doors 2012

53. *Little League Baseball.* The number of Little League baseball players worldwide declined from 2.6 million in 1997 to only 2.1 million in 2011. What is the percent decrease?

Source: Little League International

54. *Adoptions from Russia.* The number of Russian child adoptions by Americans has declined from 4381 in 1999 to only 962 in 2011. What is the percent decrease?

Source: U.S. State Department

Skill Maintenance

Multiply. [9.7c]

55. $3(4 + q)$

56. $-\frac{1}{2}(-10x + 42)$

Simplify. [9.8c]

57. $-2[3 - 5(7 - 2)]$

58. $[3(x + 4) - 6] - [8 + 2(x - 5)]$

Simplify. [9.7a]

59. $\frac{75yw}{40y}$

60. $-\frac{18b}{12b}$

Compute.

61. $9.076 \div 0.05$ [5.4a]

62. 9.076×0.05 [5.3a]

Synthesis

63. It has been determined that at the age of 15, a boy has reached 96.1% of his final adult height. Jaraan is 6 ft 4 in. at the age of 15. What will his final adult height be?

64. It has been determined that at the age of 10, a girl has reached 84.4% of her final adult height. Dana is 4 ft 8 in. at the age of 10. What will her final adult height be?

Applications and Problem Solving

10.6

OBJECTIVE

 Solve applied problems by translating to equations.

SKILL TO REVIEW

Objective 9.1b: Translate phrases to algebraic expressions.

Translate each phrase to an algebraic expression.

1. One-third of a number
2. Two more than a number

a FIVE STEPS FOR SOLVING PROBLEMS

We have discussed many new equation-solving tools in this chapter and used them for applications and problem solving. Here we consider a five-step strategy that can be very helpful in solving problems.

FIVE STEPS FOR PROBLEM SOLVING IN ALGEBRA

1. *Familiarize* yourself with the problem situation.
2. *Translate* the problem to an equation.
3. *Solve* the equation.
4. *Check* the answer in the original problem.
5. *State* the answer to the problem clearly.

Of the five steps, the most important is probably the first one: becoming familiar with the problem situation. The box below lists some hints for familiarization.

FAMILIARIZING YOURSELF WITH A PROBLEM

- If a problem is given in words, read it carefully. Reread the problem, perhaps aloud. Try to verbalize the problem as though you were explaining it to someone else.
- Choose a variable (or variables) to represent the unknown and clearly state what the variable represents. Be descriptive! For example, let $L =$ the length, $d =$ the distance, and so on.
- Make a drawing and label it with known information, using specific units if given. Also, indicate unknown information.
- Find further information. Look up formulas or definitions with which you are not familiar. (Geometric formulas appear on the inside back cover of this text.) Consult the Internet or a reference librarian.
- Create a table that lists all the information you have available. Look for patterns that may help in the translation to an equation.
- Think of a possible answer and check the guess. Note the manner in which the guess is checked.

EXAMPLE 1 *Cycling in Vietnam.* National Highway 1, which runs along the coast of Vietnam, is considered one of the top routes for avid bicyclists. While on sabbatical, a history professor spent six weeks biking 1720 km on National Highway 1 from Hanoi through Ha Tinh to Ho Chi Minh City (commonly known as Saigon). At Ha Tinh, he was four times as far from Ho Chi Minh City as he was from Hanoi. How far had he biked and how far did he still need to bike in order to reach the end?

Sources: www.smh.com; *Lonely Planet's Best in 2010*

Answers

Skill to Review:

1. $\frac{1}{3}n$, or $\frac{n}{3}$ 2. $x + 2$, or $2 + x$

1. **Familiarize.** Let's look at a map.

To become familiar with the problem, let's guess a possible distance that the professor is from Hanoi—say, 400 km. Four times 400 km is 1600 km. Since 400 km + 1600 km = 2000 km and 2000 km is greater than 1720 km, we see that our guess is too large. Rather than guess again, let's use the equation-solving skills that we learned in this chapter. We let

$d =$ the distance, in kilometers, to Hanoi, and

$4d =$ the distance, in kilometers, to Ho Chi Minh City.

(We also could let $d =$ the distance to Ho Chi Minh City and $\frac{1}{4}d =$ the distance to Hanoi.)

2. **Translate.** From the map, we see that the lengths of the two parts of the trip must add up to 1720 km. This leads to our translation.

$$\underbrace{\text{Distance to Hanoi}}_{d} \quad \underbrace{\text{plus}}_{+} \quad \underbrace{\text{Distance to Ho Chi Minh}}_{4d} \quad \underbrace{\text{is}}_{=} \quad \underbrace{1720\text{ km.}}_{1720}$$

3. **Solve.** We solve the equation:

$$d + 4d = 1720$$
$$5d = 1720 \quad \text{Collecting like terms}$$
$$\frac{5d}{5} = \frac{1720}{5} \quad \text{Dividing by 5}$$
$$d = 344.$$

4. **Check.** As we expected, d is less than 400 km. If $d = 344$ km, then $4d = 1376$ km. Since 344 km + 1376 km = 1720 km, the answer checks.

5. **State.** At Ha Tinh, the professor had biked 344 km from Hanoi and had 1376 km to go to reach Ho Chi Minh City.

Do Exercise 1.

1. ***Running.*** Yiannis Kouros of Australia holds the record for the greatest distance run in 24 hr by running 188 mi. After 8 hr, he was approximately twice as far from the finish line as he was from the start line. How far had he run?

Source: Australian Ultra Runners Association

Answer

1. $62\frac{2}{3}$ mi

EXAMPLE 2 *Knitted Scarf.* Lily knitted a scarf with shades of orange and red yarn, starting with an orange section, then a medium-red section, and finally a dark-red section. The medium-red section is one-half the length of the orange section. The dark-red section is one-fourth the length of the orange section. The scarf is 7 ft long. Find the length of each section of the scarf.

1. **Familiarize.** Because the lengths of the medium-red section and the dark-red section are expressed in terms of the length of the orange section, we let

$$x = \text{the length of the orange section.}$$

Then $\frac{1}{2}x =$ the length of the medium-red section

and $\frac{1}{4}x =$ the length of the dark-red section.

We make a drawing and label it.

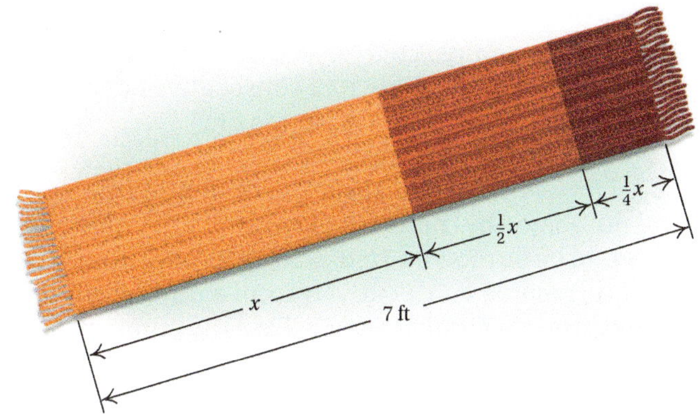

2. **Translate.** From the statement of the problem and the drawing, we know that the lengths add up to 7 ft. This gives us our translation:

Length of orange section	plus	Length of medium-red section	plus	Length of dark-red section	is	Total length
↓	↓	↓	↓	↓	↓	↓
x	$+$	$\frac{1}{2}x$	$+$	$\frac{1}{4}x$	$=$	$7.$

3. **Solve.** First, we clear fractions and then carry out the solution as follows:

$$x + \frac{1}{2}x + \frac{1}{4}x = 7 \qquad \text{The LCM of the denominators is 4.}$$

$$4\left(x + \frac{1}{2}x + \frac{1}{4}x\right) = 4 \cdot 7 \qquad \text{Multiplying by the LCM, 4}$$

$$4 \cdot x + 4 \cdot \frac{1}{2}x + 4 \cdot \frac{1}{4}x = 4 \cdot 7 \qquad \text{Using the distributive law}$$

$$4x + 2x + x = 28 \qquad \text{Simplifying}$$

$$7x = 28 \qquad \text{Collecting like terms}$$

$$\frac{7x}{7} = \frac{28}{7} \qquad \text{Dividing by 7}$$

$$x = 4.$$

2. *Gourmet Sandwiches.* A sandwich shop specializes in sandwiches prepared in buns of length 18 in. Jenny, Emma, and Sarah buy one of these sandwiches and take it back to their apartment. Since they have different appetites, Jenny cuts the sandwich in such a way that Emma gets one-half of what Jenny gets and Sarah gets three-fourths of what Jenny gets. Find the length of each person's sandwich.

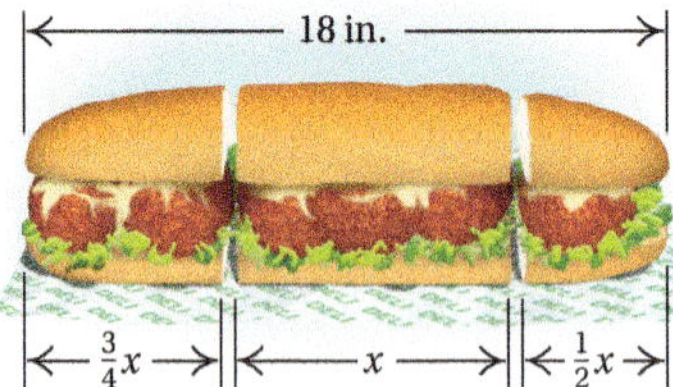

4. **Check.** Do we have an answer to the *original problem*? If the length of the orange section is 4 ft, then the length of the medium-red section is $\frac{1}{2} \cdot 4$ ft, or 2 ft, and the length of the dark-red section is $\frac{1}{4} \cdot 4$ ft, or 1 ft. The sum of these lengths is 7 ft, so the answer checks.

5. **State.** The length of the orange section is 4 ft, the length of the medium-red section is 2 ft, and the length of the dark-red section is 1 ft. (Note that we must include the unit, feet, in the answer.)

Do Exercise 2.

Recall that the set of integers $= \{\ldots, -5, -4, -3, -2, -1, 0, 1, 2, 3, 4, 5, \ldots\}$. Before we solve the next problem, we need to learn some additional terminology regarding integers.

The following are examples of **consecutive integers:** 16, 17, 18, 19, 20; and $-31, -30, -29, -28$. Note that consecutive integers can be represented in the form $x, x + 1, x + 2$, and so on.

The following are examples of **consecutive even integers:** 16, 18, 20, 22, 24; and $-52, -50, -48, -46$. Note that consecutive even integers can be represented in the form $x, x + 2, x + 4$, and so on.

The following are examples of **consecutive odd integers:** 21, 23, 25, 27, 29; and $-71, -69, -67, -65$. Note that consecutive odd integers can be represented in the form $x, x + 2, x + 4$, and so on.

EXAMPLE 3 *Limited-Edition Prints.* A limited-edition print is usually signed and numbered by the artist. For example, a limited edition with only 50 prints would be numbered 1/50, 2/50, 3/50, and so on. An estate donates two prints numbered consecutively from a limited edition with 150 prints. The sum of the two numbers is 263. Find the numbers of the prints.

1. **Familiarize.** The numbers of the prints are consecutive integers. If we let $x =$ the smaller number, then $x + 1 =$ the larger number. Since there are 150 prints in the edition, the first number must be 149 or less. If we guess that $x = 138$, then $x + 1 = 139$. The sum of the numbers is 277. We see that the numbers need to be smaller. We could continue guessing and solve the problem this way, but let's work on developing algebra skills.

Answer

2. Jenny: 8 in.; Emma: 4 in.; Sarah: 6 in.

2. **Translate.** We reword the problem and translate as follows:

Rewording:	First integer	plus	Second integer	is	263
	↓	↓	↓	↓	↓
Translating:	x	$+$	$(x + 1)$	$=$	263.

3. **Solve.** We solve the equation:

$$x + (x + 1) = 263$$
$$2x + 1 = 263 \qquad \text{Collecting like terms}$$
$$2x + 1 - 1 = 263 - 1 \qquad \text{Subtracting 1}$$
$$2x = 262$$
$$\frac{2x}{2} = \frac{262}{2} \qquad \text{Dividing by 2}$$
$$x = 131.$$

If $x = 131$, then $x + 1 = 132$.

4. **Check.** Our possible answers are 131 and 132. These are consecutive positive integers and $131 + 132 = 263$, so the answers check.

5. **State.** The print numbers are 131/150 and 132/150.

Do Exercise 3. ▶

EXAMPLE 4 *Delivery Truck Rental.* An appliance business needs to rent a delivery truck for 6 days while one of its trucks is being repaired. The cost of renting a 16-ft truck is \$29.95 per day plus \$0.29 per mile. If \$550 is budgeted for the rental, how many miles can be driven and stay within budget?

1. **Familiarize.** Suppose the van is driven 1100 mi. The cost is given by the daily charge plus the mileage charge, so we have

6(\$29.95)	+	Cost per mile	times	Number of miles
↓	↓	↓	↓	↓
\$179.70	+	\$0.29	·	1100,

which is \$498.70. We see that the van can be driven more than 1100 mi on the business's budget of \$550. This process familiarizes us with the way in which a calculation is made.

3. ***Interstate Mile Markers.*** The sum of two consecutive mile markers on I-90 in upstate New York is 627. (On I-90 in New York, the marker numbers increase from east to west.) Find the numbers on the markers.

Source: New York State Department of Transportation

Let $x =$ the first marker and $x + 1 =$ the second marker.

Translate and *Solve*:

First marker	+	Second marker	=	627
↓	↓	↓	↓	↓
____	+	(____)	=	627

$$\_\_\_\_ + 1 = 627$$
$$2x + 1 - 1 = 627 - \_\_\_\_$$
$$2x = \_\_\_\_$$
$$\frac{2x}{\_\_\_\_} = \frac{626}{2}$$
$$x = 313.$$

If $x = 313$, then $x + 1 =$ ____. The mile markers are ____ and 314.

Answer

3. 313 and 314

Guided Solution:
3. $x, x + 1, 2x, 1, 626, 2, 314, 313$

We let $m =$ the number of miles that can be driven on the budget of \$550.

2. **Translate.** We reword the problem and translate as follows:

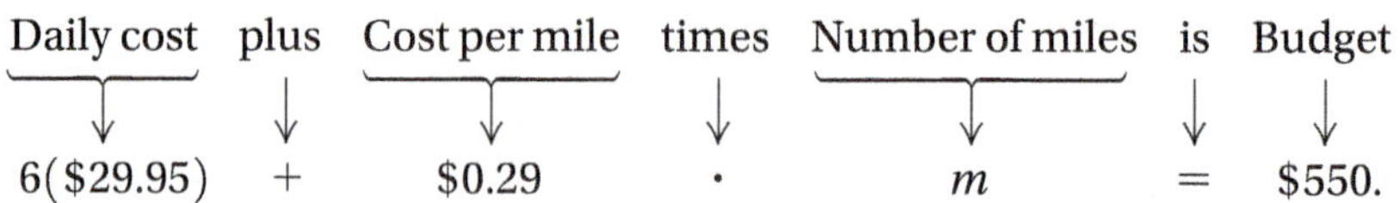

Daily cost plus Cost per mile times Number of miles is Budget
6(\$29.95) + \$0.29 · m = \$550.

3. **Solve.** We solve the equation:

$$6(29.95) + 0.29m = 550$$
$$179.70 + 0.29m = 550$$
$$0.29m = 370.30 \quad \text{Subtracting 179.70}$$
$$\frac{0.29m}{0.29} = \frac{370.30}{0.29} \quad \text{Dividing by 0.29}$$
$$m \approx 1277. \quad \text{Rounding to the nearest one}$$

4. **Check.** We check our answer in the original problem. The cost for driving 1277 mi is $1277(\$0.29) = \$370.33$. The rental for 6 days is $6(\$29.95) = \$179.70$. The total cost is then

$$\$370.33 + \$179.70 \approx \$550,$$

which is the \$550 budget that was allowed.

5. **State.** The truck can be driven 1277 mi on the truck-rental allotment.

4. ***Delivery Truck Rental.*** Refer to Example 4. The business decides to increase its 6-day rental budget to \$625. How many miles can be driven for \$625?

Do Exercise 4.

EXAMPLE 5 ***Perimeter of a Lacrosse Field.*** The perimeter of a lacrosse field is 340 yd. The length is 50 yd longer than the width. Find the dimensions of the field.

Source: www.sportsknowhow.com

1. **Familiarize.** We first make a drawing.

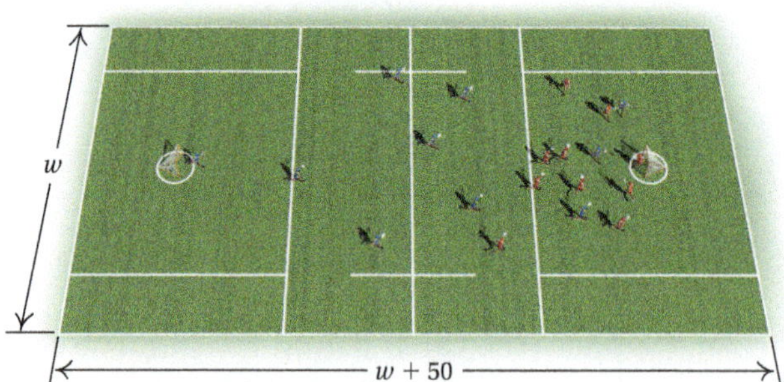

We let $w =$ the width of the rectangle. Then $w + 50 =$ the length. The perimeter P of a rectangle is the distance around the rectangle and is given by the formula $2l + 2w = P$, where

$l =$ the length and $w =$ the width.

Answer

4. 1536 mi

2. **Translate.** To translate the problem, we substitute $w + 50$ for l and 340 for P:

$$2l + 2w = P$$
$$2(w + 50) + 2w = 340.$$

Caution!

Parentheses are necessary here.

3. **Solve.** We solve the equation:

$$2(w + 50) + 2w = 340$$
$$2w + 100 + 2w = 340 \quad \text{Using the distributive law}$$
$$4w + 100 = 340 \quad \text{Collecting like terms}$$
$$4w + 100 - 100 = 340 - 100 \quad \text{Subtracting 100}$$
$$4w = 240$$
$$\frac{4w}{4} = \frac{240}{4} \quad \text{Dividing by 4}$$
$$w = 60.$$

Thus the possible dimensions are

$$w = 60\text{ yd} \quad \text{and} \quad l = w + 50 = 60 + 50, \text{ or } 110\text{ yd}.$$

4. **Check.** If the width is 60 yd and the length is 110 yd, then the perimeter is $2(60\text{ yd}) + 2(110\text{ yd})$, or 340 yd. This checks.
5. **State.** The width is 60 ft and the length is 110 yd.

Do Exercise 5. 

5. *Perimeter of High School Basketball Court.* The perimeter of a standard high school basketball court is 268 ft. The length is 34 ft longer than the width. Find the dimensions of the court.

Source: Indiana High School Athletic Association

Caution!

Always be sure to answer the original problem completely. For instance, in Example 1, we need to find *two* numbers: the distances from *each* city to the biker. Similarly, in Example 3, we need to find two print numbers, and in Example 5, we need to find two dimensions, not just the width.

EXAMPLE 6 *Roof Gable.* In a triangular gable end of a roof, the angle of the peak is twice as large as the angle of the back side of the house. The measure of the angle on the front side is 20° greater than the angle on the back side. How large are the angles?

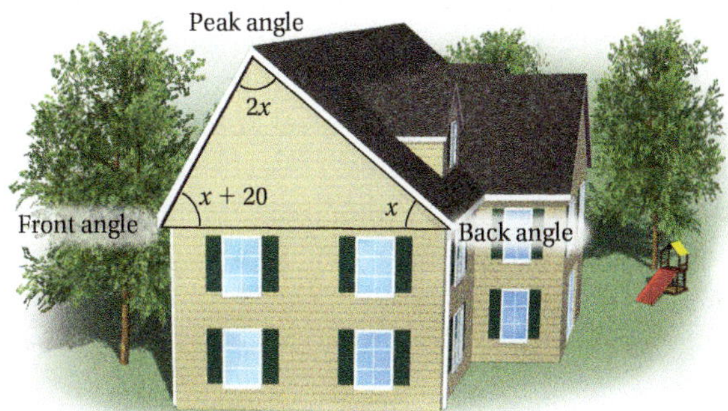

1. **Familiarize.** We first make a drawing as shown above. We let

measure of back angle $= x$.

Then measure of peak angle $= 2x$

and measure of front angle $= x + 20$.

Answer

5. Length: 84 ft; width: 50 ft

2. **Translate.** To translate, we need to know that the sum of the measures of the angles of a triangle is 180°. You might recall this fact from geometry or can look it up in a geometry book or in the list of formulas inside the back cover of this book. We translate as follows:

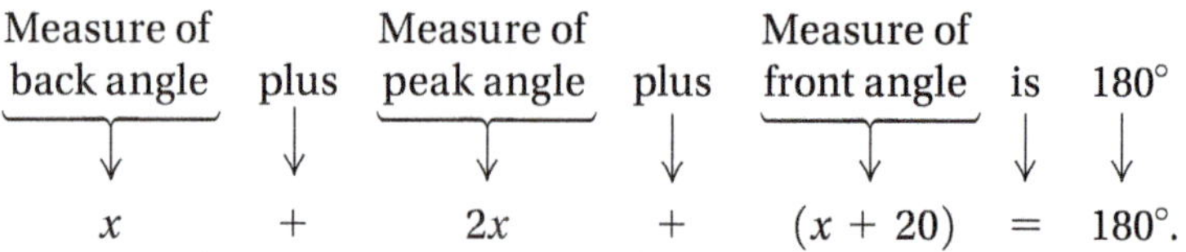

Measure of back angle
plus
Measure of peak angle
plus
Measure of front angle
is
180°

$$x \quad + \quad 2x \quad + \quad (x + 20) \quad = \quad 180°.$$

3. **Solve.** We solve the equation:

$$\begin{aligned} x + 2x + (x + 20) &= 180 \\ 4x + 20 &= 180 \\ 4x + 20 - 20 &= 180 - 20 \\ 4x &= 160 \\ \frac{4x}{4} &= \frac{160}{4} \\ x &= 40. \end{aligned}$$

The possible measures for the angles are as follows:

Back angle: $x = 40°$;
Peak angle: $2x = 2(40) = 80°$;
Front angle: $x + 20 = 40 + 20 = 60°$.

4. **Check.** Consider our answers: 40°, 80°, and 60°. The peak is twice the back, and the front is 20° greater than the back. The sum is 180°. The angles check.

5. **State.** The measures of the angles are 40°, 80°, and 60°.

Caution!

Units are important in answers. Remember to include them, where appropriate.

6. The second angle of a triangle is three times as large as the first. The third angle measures 30° more than the first angle. Find the measures of the angles.

Do Exercise 6.

EXAMPLE 7 ***Fastest Roller Coasters.*** The average top speed of the three fastest steel roller coasters in the United States is 116 mph. The third-fastest roller coaster, Superman: The Escape (located at Six Flags Magic Mountain, Valencia, California), reaches a top speed of 28 mph less than the fastest roller coaster, Kingda Ka (located at Six Flags Great Adventure, Jackson, New Jersey). The second-fastest roller coaster, Top Thrill Dragster (located at Cedar Point, Sandusky, Ohio), has a top speed of 120 mph. What is the top speed of the fastest steel roller coaster?

Source: Coaster Grotto

Answer

6. First: 30°; second: 90°; third: 60°

1. **Familiarize.** The **average** of a set of numbers is the sum of the numbers divided by the number of addends.

We are given that the second-fastest speed is 120 mph. Suppose the three top speeds are 131, 120, and 103. The average is then

$$\frac{131 + 120 + 103}{3} = \frac{354}{3} = 118,$$

which is too high. Instead of continuing to guess, let's use the equation-solving skills we have learned in this chapter. We let $x =$ the top speed of the fastest roller coaster. Then $x - 28 =$ the top speed of the third-fastest roller coaster.

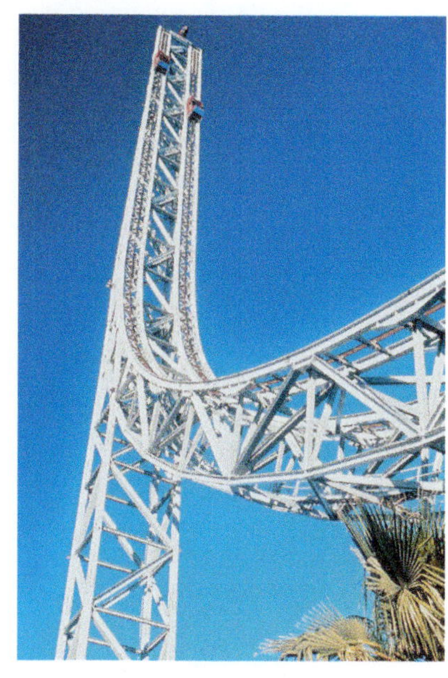

2. **Translate.** We reword the problem and translate as follows:

$$\frac{\text{Speed of fastest coaster} + \text{Speed of second-fastest coaster} + \text{Speed of third-fastest coaster}}{\text{Number of roller coasters}} = \text{Average speed of three fastest roller coasters}$$

$$\frac{x + 120 + (x - 28)}{3} = 116.$$

3. **Solve.** We solve as follows:

$$\frac{x + 120 + (x - 28)}{3} = 116$$

$$3 \cdot \frac{x + 120 + (x - 28)}{3} = 3 \cdot 116 \qquad \text{Multiplying by 3 on both sides to clear the fraction}$$

$$x + 120 + (x - 28) = 348$$

$$2x + 92 = 348 \qquad \text{Collecting like terms}$$

$$2x = 256 \qquad \text{Subtracting 92}$$

$$x = 128. \qquad \text{Dividing by 2}$$

4. **Check.** If the top speed of the fastest roller coaster is 128 mph, then the top speed of the third-fastest is $128 - 28$, or 100 mph. The average of the top speeds of the three fastest is

$$\frac{128 + 120 + 100}{3} = \frac{348}{3} = 116 \text{ mph}.$$

The answer checks.

5. **State.** The top speed of the fastest steel roller coaster in the United States is 128 mph.

Do Exercise 7. ▶

7. ***Average Test Score.*** Sam's average score on his first three math tests is 77. He scored 62 on the first test. On the third test, he scored 9 more than he scored on his second test. What did he score on the second and third tests?

Answer

7. Second: 80; third: 89

EXAMPLE 8 *Simple Interest.* An investment is made at 3% simple interest for 1 year. It grows to \$746.75. How much was originally invested (the principal)?

1. **Familiarize.** Suppose that \$100 was invested. Recalling the formula for simple interest, $I = Prt$, we know that the interest for 1 year on \$100 at 3% simple interest is given by $I = \$100 \cdot 0.03 \cdot 1 = \$3$. Then, at the end of the year, the amount in the account is found by adding the principal and the interest:

$$\begin{array}{ccccc} \text{Principal} & + & \text{Interest} & = & \text{Amount} \\ \downarrow & \downarrow & \downarrow & \downarrow & \downarrow \\ \$100 & + & \$3 & = & \$103. \end{array}$$

In this problem, we are working backward. We are trying to find the principal, which is the original investment. We let $x =$ the principal. Then the interest earned is $3\%x$.

2. **Translate.** We reword the problem and then translate:

$$\begin{array}{ccccc} \text{Principal} & + & \text{Interest} & = & \text{Amount} \\ \downarrow & \downarrow & \downarrow & \downarrow & \downarrow \\ x & + & 3\%x & = & 746.75. \end{array}$$

Interest is 3% of the principal.

3. **Solve.** We solve the equation:

$$\begin{aligned} x + 3\%x &= 746.75 \\ x + 0.03x &= 746.75 && \text{Converting to decimal notation} \\ 1x + 0.03x &= 746.75 && \text{Identity property of 1} \\ (1 + 0.03)x &= 746.75 \\ 1.03x &= 746.75 && \text{Collecting like terms} \\ \frac{1.03x}{1.03} &= \frac{746.75}{1.03} && \text{Dividing by 1.03} \\ x &= 725. \end{aligned}$$

4. **Check.** We check by taking 3% of \$725 and adding it to \$725:

$$3\% \times \$725 = 0.03 \times 725 = \$21.75.$$

Then $\$725 + \$21.75 = \$746.75$, so \$725 checks.

5. **State.** The original investment was \$725.

Do Exercise 8.

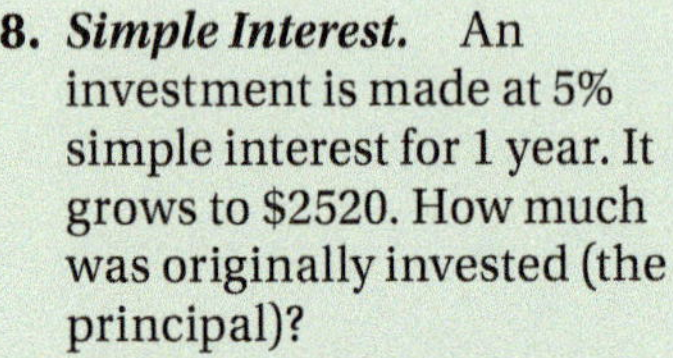

8. ***Simple Interest.*** An investment is made at 5% simple interest for 1 year. It grows to \$2520. How much was originally invested (the principal)?

GS

Let $x =$ the principal. Then the interest earned is $5\%x$.

Translate and *Solve*:

$$\begin{array}{ccccc} \text{Principal} & + & \text{Interest} & = & \text{Amount} \\ \downarrow & \downarrow & \downarrow & \downarrow & \downarrow \\ x & + & \underline{\quad} & = & 2520 \end{array}$$

$$\begin{aligned} x + 0.05x &= 2520 \\ (1 + \underline{\quad})x &= 2520 \\ \underline{\quad}\,x &= 2520 \\ \frac{1.05x}{1.05} &= \frac{2520}{\underline{\quad}} \\ x &= 2400. \end{aligned}$$

EXAMPLE 9 *Selling a House.* The Patels are planning to sell their house. If they want to be left with \$130,200 after paying 7% of the selling price to a realtor as a commission, for how much must they sell the house?

1. **Familiarize.** Suppose the Patels sell the house for \$138,000. A 7% commission can be determined by finding 7% of \$138,000:

$$7\% \text{ of } \$138{,}000 = 0.07(\$138{,}000) = \$9660.$$

Subtracting this commission from \$138,000 would leave the Patels with

$$\$138{,}000 - \$9660 = \$128{,}340.$$

This shows that in order for the Patels to clear \$130,200, the house must sell for more than \$138,000. Our guess shows us how to translate to an equation. We let $x =$ the selling price, in dollars. With a 7% commission, the realtor would receive $0.07x$.

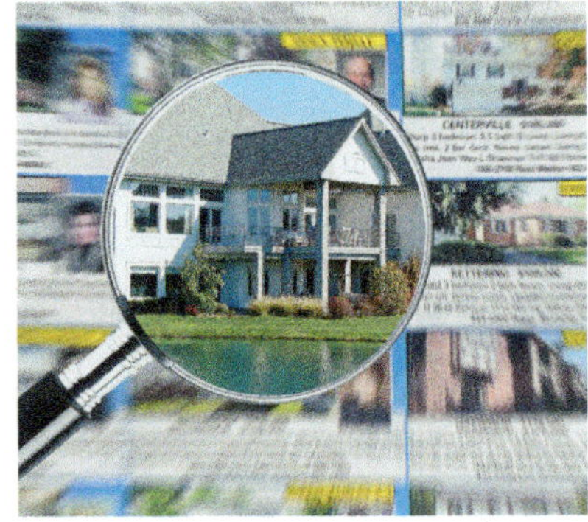

Answer

8. \$2400

Guided Solution:

8. $5\%x$, 0.05, 1.05, 1.05

2. Translate. We reword the problem and translate as follows:

$$\underbrace{\text{Selling price}}_{x} \quad \underbrace{\text{less}}_{-} \quad \underbrace{\text{Commission}}_{0.07x} \quad \underbrace{\text{is}}_{=} \quad \underbrace{\text{Amount remaining}}_{130{,}200.}$$

3. Solve. We solve the equation:

$$x - 0.07x = 130{,}200$$
$$1x - 0.07x = 130{,}200$$
$$(1 - 0.07)x = 130{,}200$$
$$0.93x = 130{,}200$$

Collecting like terms. Had we noted that after the commission has been paid, 93% remains, we could have begun with this equation.

$$\frac{0.93x}{0.93} = \frac{130{,}200}{0.93}$$

Dividing by 0.93

$$x = 140{,}000.$$

4. Check. To check, we first find 7% of \$140,000:

$$7\% \text{ of } \$140{,}000 = 0.07(\$140{,}000) = \$9800.$$

This is the commission.

Next, we subtract the commission to find the remaining amount:

$$\$140{,}000 - \$9800 = \$130{,}200.$$

Since, after the commission, the Patels are left with \$130,200, our answer checks. Note that the \$140,000 selling price is greater than \$138,000, as predicted in the *Familiarize* step.

5. State. To be left with \$130,200, the Patels must sell the house for \$140,000.

Do Exercise 9.

9. ***Selling a Condominium.*** An investor needs to sell a condominium in New York City. If she wants to be left with \$761,400 after paying a 6% commission, for how much must she sell the condominium?

Caution!

The problem in Example 9 is easy to solve with algebra. Without algebra, it is not. A common error in such a problem is to take 7% of the price after commission and then subtract or add. Note that 7% of the selling price ($7\% \cdot \$140{,}000 = \$9800$) is not equal to 7% of the amount that the Patels want to be left with ($7\% \cdot \$130{,}200 = \$9114$).

Answer

8. \$810,000

Translating for Success

1. *Angle Measures.* The measure of the second angle of a triangle is 51° more than that of the first angle. The measure of the third angle is 3° less than twice the first angle. Find the measures of the angles.

2. *Sales Tax.* Tina paid \$3976 for a used car. This amount included 5% for sales tax. How much did the car cost before tax?

3. *Perimeter.* The perimeter of a rectangle is 2347 ft. The length is 28 ft greater than the width. Find the length and the width.

4. *Fraternity or Sorority Membership.* At Arches Tech University, 3976 students belong to a fraternity or a sorority. This is 35% of the total enrollment. What is the total enrollment at Arches Tech?

5. *Fraternity or Sorority Membership.* At Moab Tech University, thirty-five percent of the students belong to a fraternity or a sorority. The total enrollment of the university is 11,360 students. How many students belong to either a fraternity or a sorority?

The goal of these matching questions is to practice step (2), Translate, of the five-step problem-solving process. Translate each word problem to an equation and select a correct translation from equations A–O.

A. $x + (x - 3) + \frac{4}{5}x = 384$

B. $x + (x + 51) + (2x - 3) = 180$

C. $x + (x + 96) = 180$

D. $2 \cdot 96 + 2x = 3976$

E. $x + (x + 1) + (x + 2) = 384$

F. $3976 = x \cdot 11{,}360$

G. $2x + 2(x + 28) = 2347$

H. $3976 = x + 5\%x$

I. $x + (x + 28) = 2347$

J. $x = 35\% \cdot 11{,}360$

K. $x + 96 = 3976$

L. $x + (x + 3) + \frac{4}{5}x = 384$

M. $x + (x + 2) + (x + 4) = 384$

N. $35\% \cdot x = 3976$

O. $2x + (x + 28) = 2347$

Answers on page A-19

6. *Island Population.* There are 180 thousand people living on a small Caribbean island. The women outnumber the men by 96 thousand. How many men live on the island?

7. *Wire Cutting.* A 384-m wire is cut into three pieces. The second piece is 3 m longer than the first. The third is four-fifths as long as the first. How long is each piece?

8. *Locker Numbers.* The numbers on three adjoining lockers are consecutive integers whose sum is 384. Find the integers.

9. *Fraternity or Sorority Membership.* The total enrollment at Canyonlands Tech University is 11,360 students. Of these, 3976 students belong to a fraternity or a sorority. What percent of the students belong to a fraternity or a sorority?

10. *Width of a Rectangle.* The length of a rectangle is 96 ft. The perimeter of the rectangle is 3976 ft. Find the width.

10.6 Exercise Set

For Extra Help MyMathLab® MathXL® PRACTICE WATCH READ REVIEW

Reading Check

Choose from the column on the right the word that completes each step in the five steps for problem solving.

RC1. ______________ yourself with the problem situation.

RC2. ______________ the problem to an equation.

RC3. ______________ the equation.

RC4. ______________ the answer in the original problem.

RC5. ______________ the answer to the problem clearly.

Solve
Familiarize
State
Translate
Check

a Solve. *Although you might find the answer quickly in some other way, practice using the five-step problem-solving strategy.*

1. ***Medals of Honor.*** In 1863, the U.S. Secretary of War presented the first Medals of Honor. The two wars with the most Medals of Honor awarded are the Civil War and World War II. There were 464 recipients of this medal for World War II. This number is 1058 fewer than the number of recipients for the Civil War. How many Medals of Honor were awarded for valor in the Civil War?

Sources: U.S. Army Center of Military History; U.S. Department of Defense

2. ***Milk Alternatives.*** Milk alternatives such as rice, soy, almond, and flax are becoming more available and increasingly popular. A cup of almond milk contains only 60 calories. This number is 89 calories less than the number of calories in a cup of whole milk. How many calories are in a cup of whole milk?

Source: Janet Kinosian, "Nutrition Udder Chaos," *AARP Magazine*, August/September, 2012

3. ***Pipe Cutting.*** A 240-in. pipe is cut into two pieces. One piece is three times the length of the other. Find the lengths of the pieces.

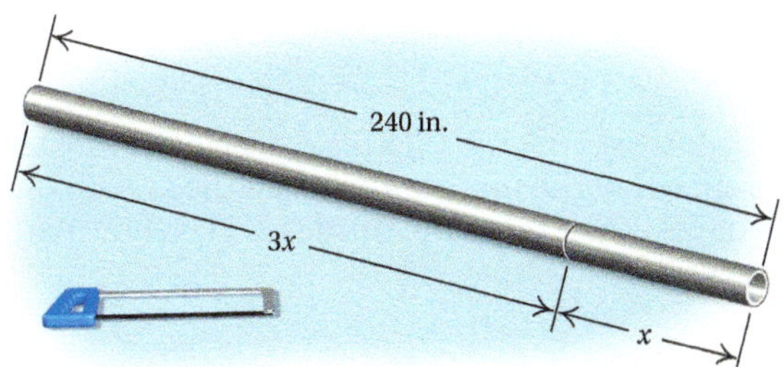

4. ***Board Cutting.*** A 72-in. board is cut into two pieces. One piece is 2 in. longer than the other. Find the lengths of the pieces.

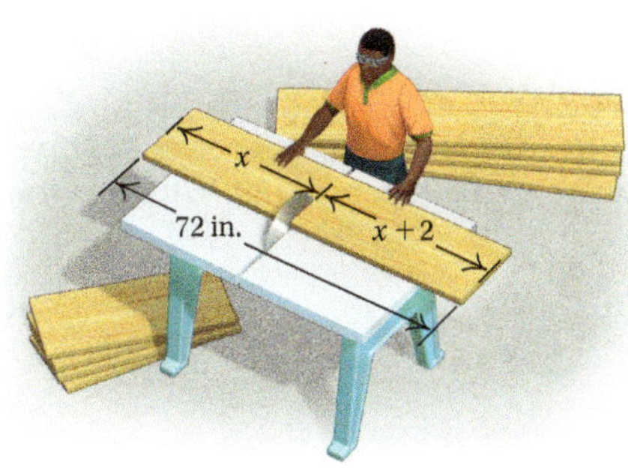

5. ***Public Transit Systems.*** In the first quarter of 2012, the ridership on the public transit system in Boston was 99.2 million. This number is 77.4 million more than the ridership in San Diego over the same period of time. What was the ridership in San Diego during the first quarter of 2012?

Source: American Public Transportation Association

6. ***Home Listing Price.*** In 2011, the average listing price of a home in Hawaii was $72,000 more than three times the average listing price of a home in Arizona. The average listing price of a home in Hawaii was $876,000. What was the average listing price of a home in Arizona?

Source: Trulia

7. ***500 Festival Mini-Marathon.*** On May 4, 2013, 35,000 runners participated in the 13.1-mi One America 500 Festival Mini-Marathon. If a runner stopped at a water station that is twice as far from the start as from the finish, how far is the runner from the finish? Round the answer to the nearest hundredth of a mile.

Source: www.500festival.com

8. ***Airport Control Tower.*** At a height of 385 ft, the FAA airport traffic control tower in Atlanta is the tallest traffic control tower in the United States. Its height is 59 ft greater than the height of the tower at the Memphis airport. How tall is the traffic control tower at the Memphis airport?

Source: Federal Aviation Administration

9. ***Consecutive Apartment Numbers.*** The apartments in Vincent's apartment house are numbered consecutively on each floor. The sum of his number and his next-door neighbor's number is 2409. What are the two numbers?

10. ***Consecutive Post Office Box Numbers.*** The sum of the numbers on two consecutive post office boxes is 547. What are the numbers?

11. ***Consecutive Ticket Numbers.*** The numbers on Sam's three raffle tickets are consecutive integers. The sum of the numbers is 126. What are the numbers?

12. ***Consecutive Ages.*** The ages of Whitney, Wesley, and Wanda are consecutive integers. The sum of their ages is 108. What are their ages?

13. ***Consecutive Odd Integers.*** The sum of three consecutive odd integers is 189. What are the integers?

14. ***Consecutive Integers.*** Three consecutive integers are such that the first plus one-half the second plus seven less than twice the third is 2101. What are the integers?

15. *Photo Size.* A hotel orders a large photo for its newly renovated lobby. The perimeter of the photo is 292 in. The width is 2 in. more than three times the height. Find the dimensions of the photo.

16. *Two-by-Four.* The perimeter of a cross section or end of a "two-by-four" piece of lumber is 10 in. The length is 2 in. more than the width. Find the actual dimensions of the cross section of a two-by-four.

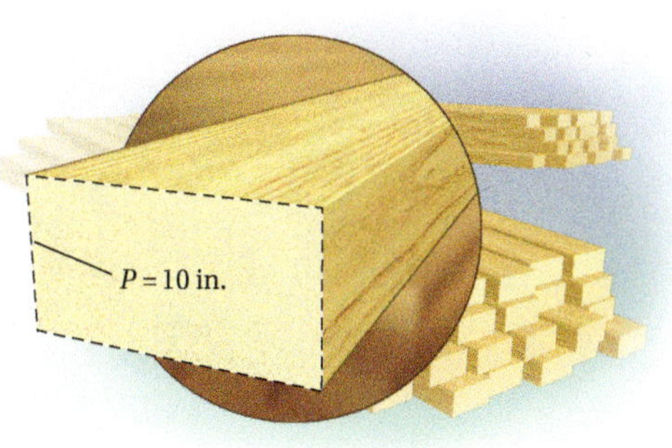

17. *Price of Coffee Beans.* A student-owned and -operated coffee shop near a campus purchases gourmet coffee beans from Costa Rica. During a recent 30%-off sale, a 3-lb bag could be purchased for \$44.10. What is the regular price of a 3-lb bag?

18. *Price of an iPad Case.* Makayla paid \$33.15 for an iPad case during a 15%-off sale. What was the regular price?

19. *Price of a Security Wallet.* Caleb paid \$26.70, including a 7% sales tax, for a security wallet. How much did the wallet itself cost?

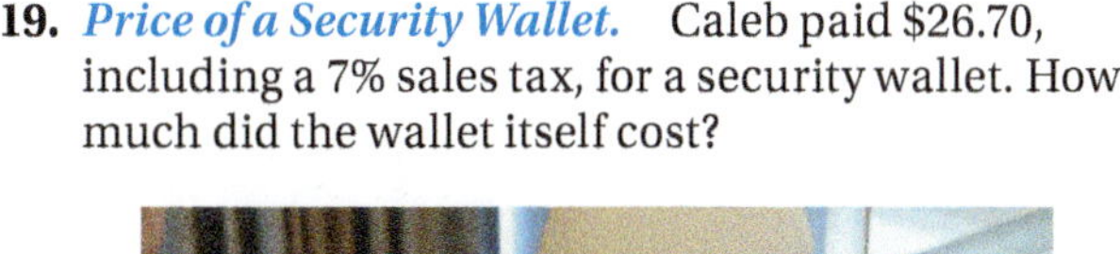

20. *Price of a Car Battery.* Tyler paid \$117.15, including a 6.5% sales tax, for a car battery. How much did the battery itself cost?

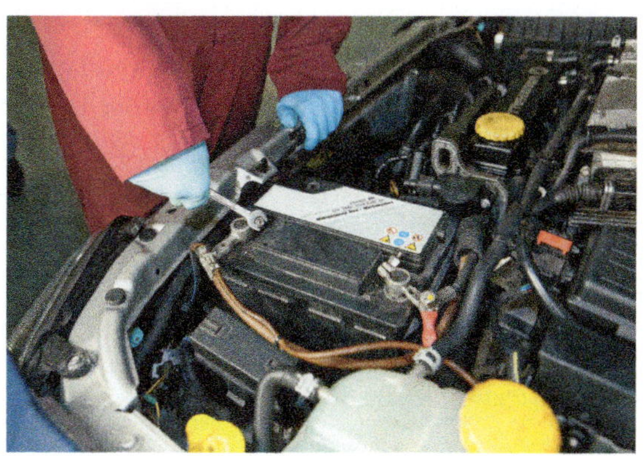

21. *Parking Costs.* A hospital parking lot charges \$1.50 for the first hour or part thereof, and \$1.00 for each additional hour or part thereof. A weekly pass costs \$27.00 and allows unlimited parking for 7 days. Suppose that each visit Hailey makes to the hospital lasts $1\frac{1}{2}$ hr. What is the minimum number of times that Hailey would have to visit per week to make it worthwhile for her to buy the pass?

22. *Van Rental.* Value Rent-A-Car rents vans at a daily rate of \$84.45 plus 55¢ per mile. Molly rents a van to deliver electrical parts to her customers. She is allotted a daily budget of \$250. How many miles can she drive for \$250? (*Hint*: 60¢ = \$0.60.)

23. *Triangular Field.* The second angle of a triangular field is three times as large as the first angle. The third angle is 40° greater than the first angle. How large are the angles?

24. *Triangular Parking Lot.* The second angle of a triangular parking lot is four times as large as the first angle. The third angle is 45° less than the sum of the other two angles. How large are the angles?

25. *Triangular Backyard.* A home has a triangular backyard. The second angle of the triangle is 5° more than the first angle. The third angle is 10° more than three times the first angle. Find the angles of the triangular yard.

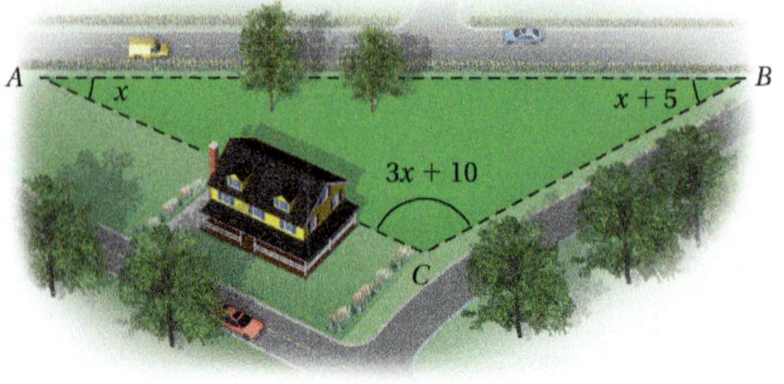

26. *Boarding Stable.* A rancher needs to form a triangular horse pen using ropes next to a stable. The second angle is three times the first angle. The third angle is 15° less than the first angle. Find the angles of the triangular pen.

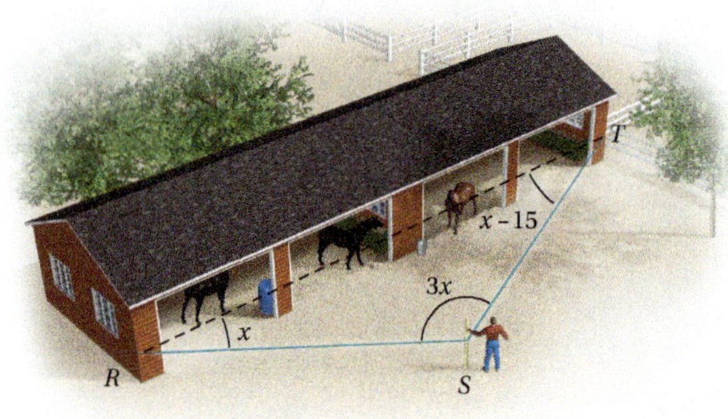

27. *Stock Prices.* Diego's investment in a technology stock grew 28% to \$448. How much did he invest?

28. *Savings Interest.* Ella invested money in a savings account at a rate of 6% simple interest. After 1 year, she has \$6996 in the account. How much did Ella originally invest?

29. *Credit Cards.* The balance on Will's credit card grew 2%, to \$870, in one month. What was his balance at the beginning of the month?

30. *Loan Interest.* Alvin borrowed money from a cousin at a rate of 10% simple interest. After 1 year, \$7194 paid off the loan. How much did Alvin borrow?

31. *Taxi Fares.* In New Orleans, Louisiana, taxis charge an initial charge of \$3.50 plus \$2.00 per mile. How far can one travel for \$39.50?

Source: www.taxifarefinders.com

32. *Taxi Fares.* In Baltimore, Maryland, taxis charge an initial charge of \$1.80 plus \$2.20 per mile. How far can one travel for \$26?

Source: www.taxifarefinders.com

33. *Tipping.* Isabella left a 15% tip for a meal. The total cost of the meal, including the tip, was \$44.39. What was the cost of the meal before the tip was added?

34. *Tipping.* Nicolas left a 20% tip for a meal. The total cost of the meal, including the tip, was \$24.90. What was the cost of the meal before the tip was added?

35. *Average Test Score.* Mariana averaged 84 on her first three history exams. The first score was 67. The second score was 7 less than the third score. What did she score on the second and third exams?

36. *Average Price.* David paid an average of \$34 per shirt for a recent purchase of three shirts. The price of one shirt was twice as much as another, and the remaining shirt cost \$27. What were the prices of the other two shirts?

37. If you double a number and then add 16, you get $\frac{2}{3}$ of the original number. What is the original number?

38. If you double a number and then add 85, you get $\frac{3}{4}$ of the original number. What is the original number?

Skill Maintenance

Calculate.

39. $-\frac{4}{5} - \frac{3}{8}$ [4.3a], [9.4a]

40. $-\frac{4}{5} + \frac{3}{8}$ [4.2b], [9.3a]

41. $-\frac{4}{5} \cdot \frac{3}{8}$ [3.6a], [9.5a]

42. $-\frac{4}{5} \div \frac{3}{8}$ [3.7b], [9.6c]

43. $\frac{1}{10} \div \left(-\frac{1}{100}\right)$ [3.7b], [9.6c]

44. $-25.6 \div (-16)$ [5.4a], [9.6c]

45. $-25.6(-16)$ [5.3a], [9.5a]

46. $-25.6 - (-16)$ [5.2c], [9.4a]

47. $-25.6 + (-16)$ [5.2a], [9.3a]

48. $(-0.02) \div (-0.2)$ [9.6c]

49. Use a commutative law to write an equivalent expression for $12 + yz$. [9.7b]

50. Use an associative law to write an equivalent expression for $(c + 4) + d$. [9.7b]

Synthesis

51. Apples are collected in a basket for six people. One-third, one-fourth, one-eighth, and one-fifth are given to four people, respectively. The fifth person gets ten apples, leaving one apple for the sixth person. Find the original number of apples in the basket.

52. *Test Questions.* A student scored 78 on a test that had 4 seven-point fill-in questions and 24 three-point multiple-choice questions. The student answered one fill-in question incorrectly. How many multiple-choice questions did the student answer correctly?

53. The area of this triangle is 2.9047 in^2. Find x.

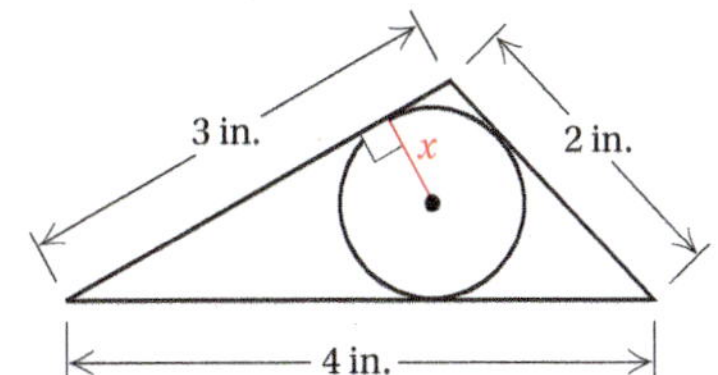

54. Susanne goes to the bank to get \$20 in quarters, dimes, and nickels to use to make change at her yard sale. She gets twice as many quarters as dimes and 10 more nickels than dimes. How many of each type of coin does she get?

10.7 Solving Inequalities

OBJECTIVES

a Determine whether a given number is a solution of an inequality.

b Graph an inequality on the number line.

c Solve inequalities using the addition principle.

d Solve inequalities using the multiplication principle.

e Solve inequalities using the addition principle and the multiplication principle together.

SKILL TO REVIEW

Objective 9.2d: Determine whether an inequality like $-3 \leq 5$ is true or false.

Write true or false.

1. $-6 \leq -8$ **2.** $1 \geq 1$

Determine whether each number is a solution of the inequality.

1. $x > 3$

a) 2 b) 0
c) -5 d) 15.4
e) 3 f) $-\frac{2}{5}$

2. $x \leq 6$

a) 6 b) 0
c) -4.3 d) 25
e) -6 f) $\frac{5}{8}$

Answers

Skill to Review:
1. False **2.** True

Margin Exercises:
1. (a) No; **(b)** no; **(c)** no; **(d)** yes; **(e)** no; **(f)** no
2. (a) Yes; **(b)** yes; **(c)** yes; **(d)** no; **(e)** yes; **(f)** yes

We now extend our equation-solving principles to the solving of inequalities.

a SOLUTIONS OF INEQUALITIES

In Section 9.2, we defined the symbols $>$ (is greater than), $<$ (is less than), $\geq$ (is greater than or equal to), and $\leq$ (is less than or equal to).

An **inequality** is a number sentence with $>$, $<$, $\geq$, or $\leq$ as its verb—for example,

$$-4 > t, \quad x < 3, \quad 2x + 5 \geq 0, \quad \text{and} \quad -3y + 7 \leq -8.$$

Some replacements for a variable in an inequality make it true and some make it false. (There are some exceptions to this statement, but we will not consider them here.)

SOLUTION OF AN INEQUALITY

A replacement that makes an inequality true is called a **solution**. The set of all solutions is called the **solution set**. When we have found the set of all solutions of an inequality, we say that we have **solved** the inequality.

EXAMPLES Determine whether each number is a solution of $x < 2$.

1. -2.7 Since $-2.7 < 2$ is true, -2.7 is a solution.

2. 2 Since $2 < 2$ is false, 2 is not a solution.

EXAMPLES Determine whether each number is a solution of $y \geq 6$.

3. 6 Since $6 \geq 6$ is true, 6 is a solution.

4. $-\frac{4}{3}$ Since $-\frac{4}{3} \geq 6$ is false, $-\frac{4}{3}$ is not a solution.

◄ Do Margin Exercises 1 and 2.

b GRAPHS OF INEQUALITIES

Some solutions of $x < 2$ are $-3, 0, 1, 0.45, -8.9, -\pi, \frac{5}{8}$, and so on. In fact, there are infinitely many real numbers that are solutions. Because we cannot list them all individually, it is helpful to make a drawing that represents all the solutions.

A **graph** of an inequality is a drawing that represents its solutions. An inequality in one variable can be graphed on the number line. An inequality in two variables can be graphed on the coordinate plane. We will study such graphs in Chapter 11.

EXAMPLE 5 Graph: $x < 2$.

The solutions of $x < 2$ are all those numbers less than 2. They are shown on the number line by shading all points to the left of 2. The parenthesis at 2 indicates that 2 *is not* part of the graph.

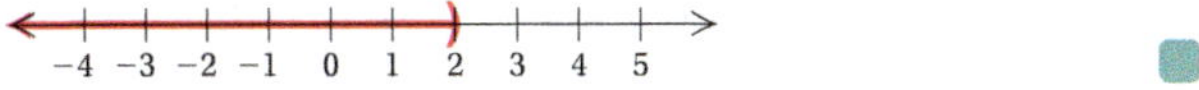

EXAMPLE 6 Graph: $x \geq -3$.

The solutions of $x \geq -3$ are shown on the number line by shading the point for -3 and all points to the right of -3. The bracket at -3 indicates that -3 *is* part of the graph.

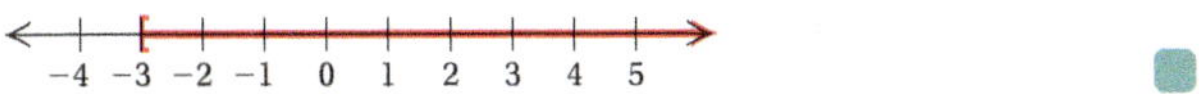

EXAMPLE 7 Graph: $-3 \leq x < 2$.

The inequality $-3 \leq x < 2$ is read "-3 is less than or equal to x *and* x is less than 2," or "x is greater than or equal to -3 *and* x is less than 2." In order to be a solution of this inequality, a number must be a solution of both $-3 \leq x$ and $x < 2$. The number 1 is a solution, as are -1.7, 0, 1.5, and $\frac{3}{8}$. We can see from the following graphs that the solution set consists of the numbers that overlap in the two solution sets in Examples 5 and 6.

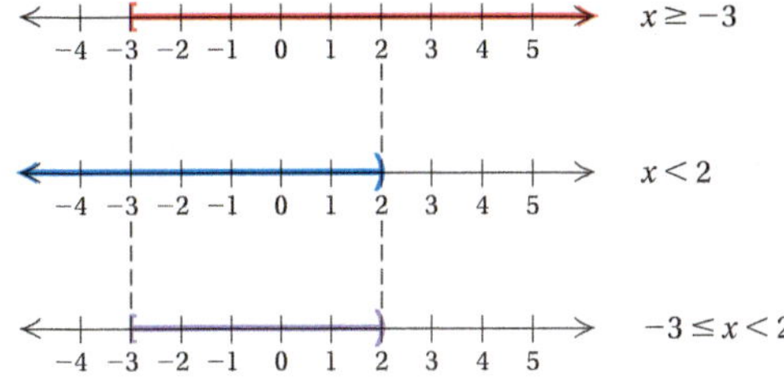

The parenthesis at 2 means that 2 *is not* part of the graph. The bracket at -3 means that -3 *is* part of the graph. The other solutions are shaded.

Do Exercises 3–5. ▶

Graph.

3. $x \leq 4$

−5 −4 −3 −2 −1 0 1 2 3 4 5

4. $x > -2$

−5 −4 −3 −2 −1 0 1 2 3 4 5

5. $-2 < x \leq 4$

−5 −4 −3 −2 −1 0 1 2 3 4 5

C SOLVING INEQUALITIES USING THE ADDITION PRINCIPLE

Consider the true inequality $3 < 7$. If we add 2 on both sides, we get another true inequality:

$$3 + 2 < 7 + 2, \quad \text{or} \quad 5 < 9.$$

Similarly, if we add -4 on both sides of $x + 4 < 10$, we get an *equivalent* inequality:

$$x + 4 + (-4) < 10 + (-4),$$

or $$x < 6.$$

To say that $x + 4 < 10$ and $x < 6$ are **equivalent** is to say that they have the same solution set. For example, the number 3 is a solution of $x + 4 < 10$. It is also a solution of $x < 6$. The number -2 is a solution of $x < 6$. It is also a solution of $x + 4 < 10$. Any solution of one inequality is a solution of the other—they are equivalent.

Answers

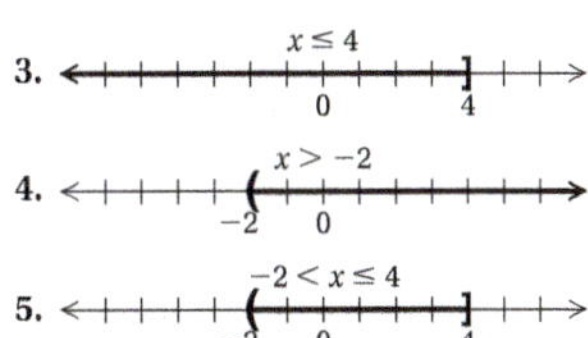

THE ADDITION PRINCIPLE FOR INEQUALITIES

For any real numbers a, b, and c:

$a < b$ is equivalent to $a + c < b + c$;
$a > b$ is equivalent to $a + c > b + c$;
$a \leq b$ is equivalent to $a + c \leq b + c$;
$a \geq b$ is equivalent to $a + c \geq b + c$.

In other words, when we add or subtract the same number on both sides of an inequality, the direction of the inequality symbol is not changed.

As with equation solving, when solving inequalities, our goal is to isolate the variable on one side. Then it is easier to determine the solution set.

EXAMPLE 8 Solve: $x + 2 > 8$. Then graph.

We use the addition principle, subtracting 2 on both sides:

$$x + 2 - 2 > 8 - 2$$
$$x > 6.$$

From the inequality $x > 6$, we can determine the solutions directly. Any number greater than 6 makes the last sentence true and is a solution of that sentence. Any such number is also a solution of the original sentence. Thus the inequality is solved. The graph is as follows:

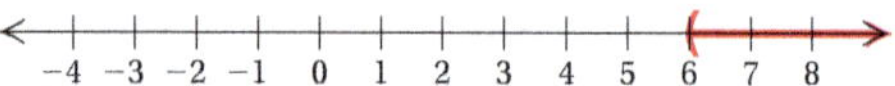

We cannot check all the solutions of an inequality by substitution, as we usually can for an equation, because there are too many of them. A partial check can be done by substituting a number greater than 6—say, 7—into the original inequality:

$x + 2 > 8$	
$7 + 2 \ ?$	8
9	TRUE

Since $9 > 8$ is true, 7 is a solution. This is a partial check that any number greater than 6 is a solution.

EXAMPLE 9 Solve: $3x + 1 \leq 2x - 3$. Then graph.

We have

$$3x + 1 \leq 2x - 3$$
$$3x + 1 - 1 \leq 2x - 3 - 1 \quad \text{Subtracting 1}$$
$$3x \leq 2x - 4 \quad \text{Simplifying}$$
$$3x - 2x \leq 2x - 4 - 2x \quad \text{Subtracting } 2x$$
$$x \leq -4. \quad \text{Simplifying}$$

Any number less than or equal to -4 is a solution. The graph is as follows:

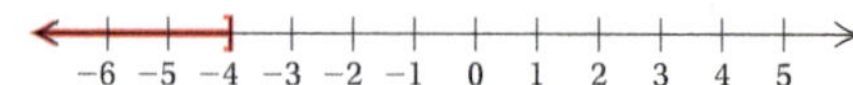

In Example 9, any number less than or equal to -4 is a solution. The following are some solutions:

$$-4, \quad -5, \quad -6, \quad -\frac{13}{3}, \quad -204.5, \quad \text{and} \quad -18\pi.$$

Besides drawing a graph, we can also describe all the solutions of an inequality using **set notation**. We could just begin to list them in a set using roster notation (see p. 615), as follows:

$$\left\{-4, -5, -6, -\frac{13}{3}, -204.5, -18\pi, \ldots\right\}.$$

We can never list them all this way, however. Seeing this set without knowing the inequality makes it difficult for us to know what real numbers we are considering. There is, however, another kind of notation that we can use. It is

$$\{x \mid x \leq -4\},$$

which is read

"The set of all x such that x is less than or equal to -4."

This shorter notation for sets is called **set-builder notation**.

From now on, we will use this notation when solving inequalities.

Do Exercises 6–8. ▶

Solve. Then graph.

6. $x + 3 > 5$

−5 −4 −3 −2 −1 0 1 2 3 4 5

7. $x - 1 \leq 2$

−5 −4 −3 −2 −1 0 1 2 3 4 5

8. $5x + 1 < 4x - 2$

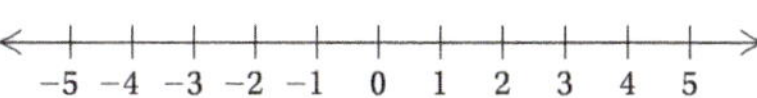

−5 −4 −3 −2 −1 0 1 2 3 4 5

EXAMPLE 10 Solve: $x + \frac{1}{3} > \frac{5}{4}$.

We have

$$x + \tfrac{1}{3} > \tfrac{5}{4}$$

$$x + \tfrac{1}{3} - \tfrac{1}{3} > \tfrac{5}{4} - \tfrac{1}{3} \qquad \text{Subtracting } \tfrac{1}{3}$$

$$x > \tfrac{5}{4} \cdot \tfrac{3}{3} - \tfrac{1}{3} \cdot \tfrac{4}{4} \qquad \text{Multiplying by 1 to obtain a common denominator}$$

$$x > \tfrac{15}{12} - \tfrac{4}{12}$$

$$x > \tfrac{11}{12}.$$

Any number greater than $\frac{11}{12}$ is a solution. The solution set is

$$\left\{x \mid x > \tfrac{11}{12}\right\},$$

which is read

"The set of all x such that x is greater than $\frac{11}{12}$."

When solving inequalities, you may obtain an answer like $\frac{11}{12} < x$. Recall from Chapter 9 that this has the same meaning as $x > \frac{11}{12}$. Thus the solution set in Example 10 can be described as $\left\{x \mid \frac{11}{12} < x\right\}$ or as $\left\{x \mid x > \frac{11}{12}\right\}$. The latter is used most often.

Do Exercises 9 and 10. ▶

Solve.

9. $x + \dfrac{2}{3} \geq \dfrac{4}{5}$

10. $5y + 2 \leq -1 + 4y$

$5y + 2 - \underline{\quad} \leq -1 + 4y - 4y$

$y + 2 \leq -1$

$y + 2 - 2 \leq -1 - \underline{\quad}$

$y \leq \underline{\quad}$

The solution set is $\{y \mid y \underline{\quad} -3\}$.

d SOLVING INEQUALITIES USING THE MULTIPLICATION PRINCIPLE

There is a multiplication principle for inequalities that is similar to that for equations, but it must be modified. When we are multiplying on both sides by a negative number, the direction of the inequality symbol must be changed.

Answers

6. $\{x \mid x > 2\}$;

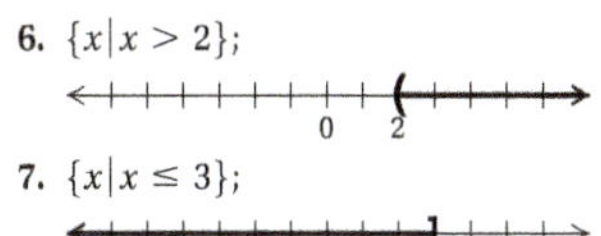

7. $\{x \mid x \leq 3\}$;

8. $\{x \mid x < -3\}$;

9. $\left\{x \mid x \geq \frac{2}{15}\right\}$ 10. $\{y \mid y \leq -3\}$

Guided Solution:

10. $4y, 2, -3, \leq$

Consider the true inequality $3 < 7$. If we multiply on both sides by a *positive* number, like 2, we get another true inequality:

$$3 \cdot 2 < 7 \cdot 2, \quad \text{or} \quad 6 < 14. \qquad \text{True}$$

If we multiply on both sides by a *negative* number, like -2, and we do not change the direction of the inequality symbol, we get a *false* inequality:

$$3 \cdot (-2) < 7 \cdot (-2), \quad \text{or} \quad -6 < -14. \qquad \text{False}$$

The fact that $6 < 14$ is true but $-6 < -14$ is false stems from the fact that the negative numbers, in a sense, mirror the positive numbers. That is, whereas 14 is to the *right* of 6 on the number line, the number -14 is to the *left* of -6. Thus, if we reverse (change the direction of) the inequality symbol, we get a *true* inequality: $-6 > -14$.

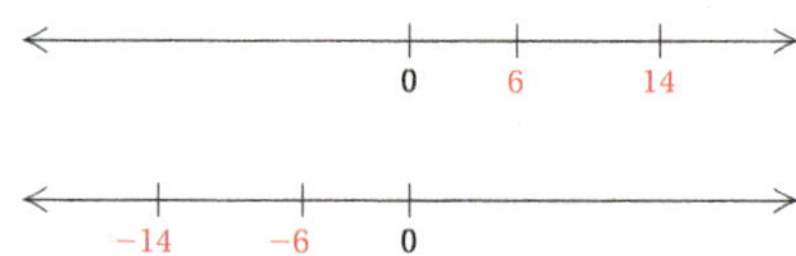

THE MULTIPLICATION PRINCIPLE FOR INEQUALITIES

For any real numbers *a* and *b*, and any *positive* number *c*:

$a < b$ is equivalent to $ac < bc$;

$a > b$ is equivalent to $ac > bc$.

For any real numbers *a* and *b*, and any *negative* number *c*:

$a < b$ is equivalent to $ac > bc$;

$a > b$ is equivalent to $ac < bc$.

Similar statements hold for $\leq$ and $\geq$.

In other words, when we multiply or divide by a positive number on both sides of an inequality, the direction of the inequality symbol stays the same. When we multiply or divide by a negative number on both sides of an inequality, the direction of the inequality symbol is reversed.

EXAMPLE 11 Solve: $4x < 28$. Then graph.

We have

$$4x < 28$$

$$\frac{4x}{4} < \frac{28}{4} \qquad \text{Dividing by 4}$$

The symbol stays the same.

$$x < 7. \qquad \text{Simplifying}$$

The solution set is $\{x \mid x < 7\}$. The graph is as follows:

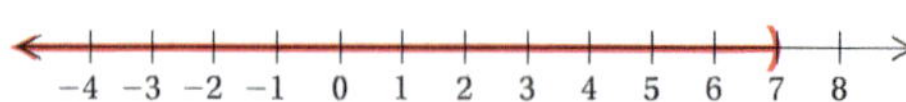

Do Exercises 11 and 12.

Solve. Then graph.

11. $8x < 64$

12. $5y \geq 160$

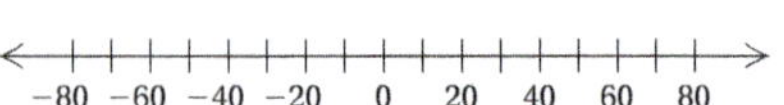

Answers

11. $\{x \mid x < 8\}$;

12. $\{y \mid y \geq 32\}$;

EXAMPLE 12 Solve: $-2y < 18$. Then graph.

$$-2y < 18$$

$$\frac{-2y}{-2} > \frac{18}{-2}$$ Dividing by -2

The symbol must be reversed!

$$y > -9.$$ Simplifying

The solution set is $\{y|y > -9\}$. The graph is as follows:

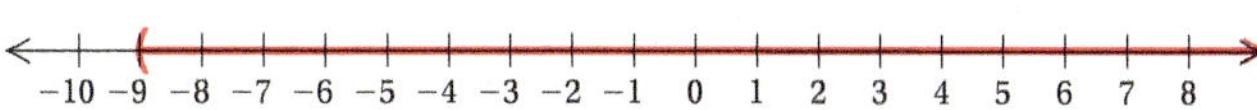

Do Exercises 13 and 14. ▶

Solve.

13. $-4x \leq 24$

14. $-5y > 13$

e USING THE PRINCIPLES TOGETHER

All of the equation-solving techniques used in Sections 10.1–10.3 can be used with inequalities, provided we remember to reverse the inequality symbol when multiplying or dividing on both sides by a negative number.

EXAMPLE 13 Solve: $6 - 5x > 7$.

$$6 - 5x > 7$$

$$-6 + 6 - 5x > -6 + 7$$ Adding -6. The symbol stays the same.

$$-5x > 1$$ Simplifying

$$\frac{-5x}{-5} < \frac{1}{-5}$$ Dividing by -5

The symbol must be reversed because we are dividing by a *negative* number, -5.

$$x < -\frac{1}{5}.$$ Simplifying

The solution set is $\left\{x|x < -\frac{1}{5}\right\}$.

Do Exercise 15. ▶

15. Solve: $7 - 4x < 8$.

EXAMPLE 14 Solve: $17 - 5y > 8y - 9$.

$$-17 + 17 - 5y > -17 + 8y - 9$$ Adding -17. The symbol stays the same.

$$-5y > 8y - 26$$ Simplifying

$$-8y - 5y > -8y + 8y - 26$$ Adding $-8y$

$$-13y > -26$$ Simplifying

$$\frac{-13y}{-13} < \frac{-26}{-13}$$ Dividing by -13

The symbol must be reversed because we are dividing by a *negative* number, -13.

$$y < 2$$

The solution set is $\{y|y < 2\}$.

Do Exercise 16. ▶

16. Solve. Begin by subtracting 24 on both sides.

$$24 - 7y \leq 11y - 14$$

Answers

13. $\{x|x \geq -6\}$ **14.** $\left\{y|y < -\frac{13}{5}\right\}$
15. $\left\{x|x > -\frac{1}{4}\right\}$ **16.** $\left\{y|y \geq \frac{19}{9}\right\}$

Typically, we solve an equation or an inequality by isolating the variable on the left side. When we are solving an inequality, however, there are situations in which isolating the variable on the right side will eliminate the need to reverse the inequality symbol. Let's solve the inequality in Example 14 again, but this time we will isolate the variable on the right side.

EXAMPLE 15 Solve: $17 - 5y > 8y - 9$.

Note that if we add $5y$ on both sides, the coefficient of the y-term will be positive after like terms have been collected.

$$17 - 5y + 5y > 8y - 9 + 5y \quad \text{Adding } 5y$$
$$17 > 13y - 9 \quad \text{Simplifying}$$
$$17 + 9 > 13y - 9 + 9 \quad \text{Adding 9}$$
$$26 > 13y \quad \text{Simplifying}$$
$$\frac{26}{13} > \frac{13y}{13}$$
Dividing by 13. We leave the inequality symbol the same because we are dividing by a positive number.
$$2 > y$$

The solution set is $\{y | 2 > y\}$, or $\{y | y < 2\}$.

17. Solve. Begin by adding $7y$ on both sides.

$24 - 7y \leq 11y - 14$

Do Exercise 17.

EXAMPLE 16 Solve: $3(x - 2) - 1 < 2 - 5(x + 6)$.

First, we use the distributive law to remove parentheses. Next, we collect like terms and then use the addition and multiplication principles for inequalities to get an equivalent inequality with x alone on one side.

$$3(x - 2) - 1 < 2 - 5(x + 6)$$
$$3x - 6 - 1 < 2 - 5x - 30 \quad \text{Using the distributive law to multiply and remove parentheses}$$
$$3x - 7 < -5x - 28 \quad \text{Collecting like terms}$$
$$3x + 5x < -28 + 7 \quad \text{Adding } 5x \text{ and 7 to get all } x\text{-terms on one side and all other terms on the other side}$$
$$8x < -21 \quad \text{Simplifying}$$
$$x < \frac{-21}{8}, \text{ or } -\frac{21}{8}. \quad \text{Dividing by 8}$$

The solution set is $\left\{x | x < -\frac{21}{8}\right\}$.

18. Solve:

GS

$3(7 + 2x) \leq 30 + 7(x - 1)$.

$\square + 6x \leq 30 + 7x - \square$

$21 + 6x \leq \square + 7x$

$21 + 6x - 6x \leq 23 + 7x - \square$

$21 \leq 23 + \square$

$21 - \square \leq 23 + x - 23$

$-2 \leq x$, or

$x \ \square \ -2$

The solution set is $\{x | x \geq \square\}$.

Do Exercise 18.

Answers

17. $\left\{y | y \geq \frac{19}{9}\right\}$ **18.** $\{x | x \geq -2\}$

Guided Solution:

18. 21, 7, 23, $6x$, x, 23, $\geq$, -2

EXAMPLE 17 Solve: $16.3 - 7.2p \leq -8.18$.

The greatest number of decimal places in any one number is *two*. Multiplying by 100, which has two 0's, will clear decimals. Then we proceed as before.

$$16.3 - 7.2p \leq -8.18$$

$$100(16.3 - 7.2p) \leq 100(-8.18)$$ Multiplying by 100

$$100(16.3) - 100(7.2p) \leq 100(-8.18)$$ Using the distributive law

$$1630 - 720p \leq -818$$ Simplifying

$$1630 - 720p - 1630 \leq -818 - 1630$$ Subtracting 1630

$$-720p \leq -2448$$ Simplifying

$$\frac{-720p}{-720} \geq \frac{-2448}{-720}$$ Dividing by –720

The symbol must be reversed.

$$p \geq 3.4$$

The solution set is $\{p | p \geq 3.4\}$.

Do Exercise 19. ▶

19. Solve:

$2.1x + 43.2 \geq 1.2 - 8.4x$.

EXAMPLE 18 Solve: $\frac{2}{3}x - \frac{1}{6} + \frac{1}{2}x > \frac{7}{6} + 2x$.

The number 6 is the least common multiple of all the denominators. Thus we first multiply by 6 on both sides to clear fractions.

$$\frac{2}{3}x - \frac{1}{6} + \frac{1}{2}x > \frac{7}{6} + 2x$$

$$6\left(\frac{2}{3}x - \frac{1}{6} + \frac{1}{2}x\right) > 6\left(\frac{7}{6} + 2x\right)$$ Multiplying by 6 on both sides

$$6 \cdot \frac{2}{3}x - 6 \cdot \frac{1}{6} + 6 \cdot \frac{1}{2}x > 6 \cdot \frac{7}{6} + 6 \cdot 2x$$ Using the distributive law

$$4x - 1 + 3x > 7 + 12x$$ Simplifying

$$7x - 1 > 7 + 12x$$ Collecting like terms

$$7x - 1 - 7x > 7 + 12x - 7x$$ Subtracting $7x$. The coefficient of the x-term will be positive.

$$-1 > 7 + 5x$$ Simplifying

$$-1 - 7 > 7 + 5x - 7$$ Subtracting 7

$$-8 > 5x$$ Simplifying

$$\frac{-8}{5} > \frac{5x}{5}$$ Dividing by 5

$$-\frac{8}{5} > x$$

The solution set is $\left\{x | -\frac{8}{5} > x\right\}$, or $\left\{x | x < -\frac{8}{5}\right\}$.

Do Exercise 20. ▶

20. Solve:

$\frac{3}{4} + x < \frac{7}{8}x - \frac{1}{4} + \frac{1}{2}x$.

Answers

19. $\{x | x \geq -4\}$ **20.** $\left\{x | x > \frac{8}{3}\right\}$

10.7 Exercise Set

For Extra Help
MyMathLab®
MathXL® PRACTICE WATCH READ REVIEW

✓ Reading Check

Classify each pair of inequalities as "equivalent" or "not equivalent."

RC1. $x + 10 \geq 12;\ x \leq 2$

RC2. $3x - 5 \leq -x + 1;\ 2x \leq 6$

RC3. $-\frac{3}{4}y < 6;\ y > -8$

RC4. $2 - t > -3t + 4;\ 2t > 2$

a Determine whether each number is a solution of the given inequality.

1. $x > -4$
a) 4
b) 0
c) -4
d) 6
e) 5.6

2. $x \leq 5$
a) 0
b) 5
c) -1
d) -5
e) $7\frac{1}{4}$

3. $x \geq 6.8$
a) -6
b) 0
c) 6
d) 8
e) $-3\frac{1}{2}$

4. $x < 8$
a) 8
b) -10
c) 0
d) 11
e) -4.7

b Graph on the number line.

5. $x > 4$

−5 −4 −3 −2 −1 0 1 2 3 4 5

6. $x < 0$

−5 −4 −3 −2 −1 0 1 2 3 4 5

7. $t < -3$

−5 −4 −3 −2 −1 0 1 2 3 4 5

8. $y > 5$

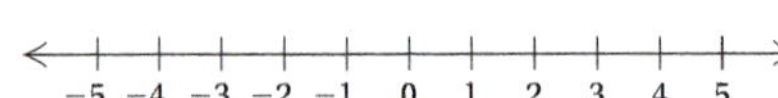

9. $m \geq -1$

−5 −4 −3 −2 −1 0 1 2 3 4 5

10. $x \leq -2$

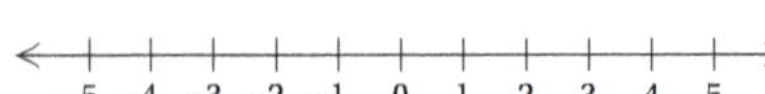

11. $-3 < x \leq 4$

−5 −4 −3 −2 −1 0 1 2 3 4 5

12. $-5 \leq x < 2$

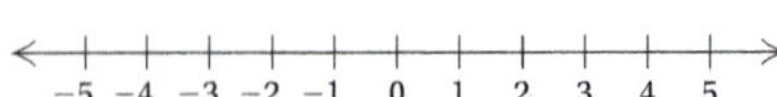

13. $0 < x < 3$

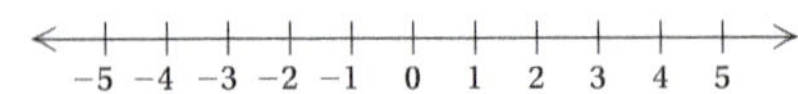

14. $-5 \leq x \leq 0$

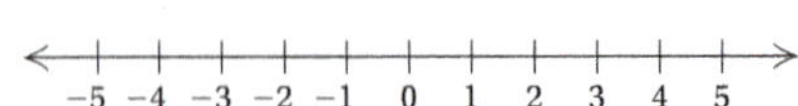

c Solve using the addition principle. Then graph.

15. $x + 7 > 2$

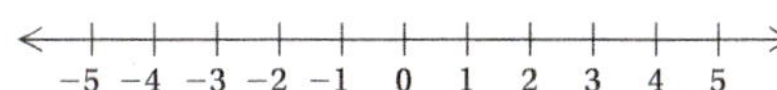

16. $x + 5 > 2$

17. $x + 8 \leq -10$

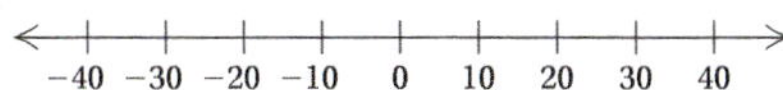

18. $x + 8 \leq -11$

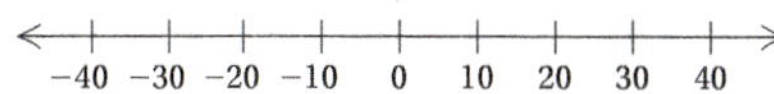

Solve using the addition principle.

19. $y - 7 > -12$

20. $y - 9 > -15$

21. $2x + 3 > x + 5$

22. $2x + 4 > x + 7$

23. $3x + 9 \leq 2x + 6$

24. $3x + 18 \leq 2x + 16$

25. $5x - 6 < 4x - 2$

26. $9x - 8 < 8x - 9$

27. $-9 + t > 5$

28. $-8 + p > 10$

29. $y + \frac{1}{4} \leq \frac{1}{2}$

30. $x - \frac{1}{3} \leq \frac{5}{6}$

31. $x - \frac{1}{3} > \frac{1}{4}$

32. $x + \frac{1}{8} > \frac{1}{2}$

d Solve using the multiplication principle. Then graph.

33. $5x < 35$

34. $8x \geq 32$

35. $-12x > -36$

36. $-16x > -64$

Solve using the multiplication principle.

37. $5y \geq -2$

38. $3x < -4$

39. $-2x \leq 12$

40. $-3x \leq 15$

41. $-4y \geq -16$

42. $-7x < -21$

43. $-3x < -17$

44. $-5y > -23$

45. $-2y > \frac{1}{7}$

46. $-4x \leq \frac{1}{9}$

47. $-\frac{6}{5} \leq -4x$

48. $-\frac{7}{9} > 63x$

e Solve using the addition principle and the multiplication principle.

49. $4 + 3x < 28$

50. $3 + 4y < 35$

51. $3x - 5 \leq 13$

52. $5y - 9 \leq 21$

53. $13x - 7 < -46$

54. $8y - 6 < -54$

55. $30 > 3 - 9x$

56. $48 > 13 - 7y$

57. $4x + 2 - 3x \leq 9$

58. $15x + 5 - 14x \leq 9$

59. $-3 < 8x + 7 - 7x$

60. $-8 < 9x + 8 - 8x - 3$

61. $6 - 4y > 4 - 3y$

62. $9 - 8y > 5 - 7y + 2$

63. $5 - 9y \leq 2 - 8y$

64. $6 - 18x \leq 4 - 12x - 5x$

65. $19 - 7y - 3y < 39$

66. $18 - 6y - 4y < 63 + 5y$

67. $0.9x + 19.3 > 5.3 - 2.6x$

68. $0.96y - 0.79 \leq 0.21y + 0.46$

69. $\frac{x}{3} - 2 \leq 1$

70. $\frac{2}{3} + \frac{x}{5} < \frac{4}{15}$

71. $\frac{y}{5} + 1 \leq \frac{2}{5}$

72. $\frac{3x}{4} - \frac{7}{8} \geq -15$

73. $3(2y - 3) < 27$

74. $4(2y - 3) > 28$

75. $2(3 + 4m) - 9 \geq 45$

76. $3(5 + 3m) - 8 \leq 88$

77. $8(2t + 1) > 4(7t + 7)$

78. $7(5y - 2) > 6(6y - 1)$

79. $3(r - 6) + 2 < 4(r + 2) - 21$

80. $5(x + 3) + 9 \leq 3(x - 2) + 6$

81. $0.8(3x + 6) \geq 1.1 - (x + 2)$

82. $0.4(2x + 8) \geq 20 - (x + 5)$

83. $\frac{5}{3} + \frac{2}{3}x < \frac{25}{12} + \frac{5}{4}x + \frac{3}{4}$

84. $1 - \frac{2}{3}y \geq \frac{9}{5} - \frac{y}{5} + \frac{3}{5}$

Skill Maintenance

Add or subtract.

85. $-\frac{3}{4} + \frac{1}{8}$ [4.2b], [9.3a]

86. $8.12 - 9.23$ [5.2c], [9.4a]

87. $-2.3 - 7.1$ [5.2c], [9.4a]

88. $-\frac{3}{4} - \frac{1}{8}$ [4.3a], [9.3a]

Simplify.

89. $5 - 3^2 + (8 - 2)^2 \cdot 4$ [9.8d]

90. $10 \div 2 \cdot 5 - 3^2 + (-5)^2$ [9.8d]

91. $5(2x - 4) - 3(4x + 1)$ [9.8b]

92. $9(3 + 5x) - 4(7 + 2x)$ [9.8b]

Synthesis

93. Determine whether each number is a solution of the inequality $|x| < 3$.

a) 0
b) -2
c) -3
d) 4
e) 3
f) 1.7
g) -2.8

94. Graph $|x| < 3$ on the number line.

−5 −4 −3 −2 −1 0 1 2 3 4 5

Solve.

95. $x + 3 < 3 + x$

96. $x + 4 > 3 + x$

10.8 Applications and Problem Solving with Inequalities

OBJECTIVES

a Translate number sentences to inequalities.

b Solve applied problems using inequalities.

SKILL TO REVIEW

Objective 10.7d: Solve inequalities using the multiplication principle.

Solve.

1. $-8x \le -512$
2. $-300 > 15y$

Translate.

1. Sara worked no fewer than 15 hr last week.
2. The price of that Volkswagen Beetle convertible is at most \$31,210.
3. The time of the test was between 45 min and 55 min.
4. Camila's weight is less than 110 lb.
5. That number is more than -2.
6. The costs of production of that marketing video cannot exceed \$12,500.
7. At most 1250 people attended the concert.
8. Yesterday, at least 23 people got tickets for speeding.

Answers

Skill to Review:
1. $\{x|x \ge 64\}$ 2. $\{y|y < -20\}$

Margin Exercises:
1. $h \ge 15$ 2. $p \le 31{,}210$ 3. $45 < t < 55$ 4. $w < 110$ 5. $n > -2$ 6. $c \le 12{,}500$ 7. $p \le 1250$ 8. $s \ge 23$

The five steps for problem solving can be used for problems involving inequalities.

a TRANSLATING TO INEQUALITIES

Before solving problems that involve inequalities, we list some important phrases to look for. Sample translations are listed as well.

IMPORTANT WORDS	SAMPLE SENTENCE	TRANSLATION
is at least	Bill is at least 21 years old.	$b \ge 21$
is at most	At most 5 students dropped the course.	$n \le 5$
cannot exceed	To qualify, earnings cannot exceed \$12,000.	$r \le 12{,}000$
must exceed	The speed must exceed 15 mph.	$s > 15$
is less than	Tucker's weight is less than 50 lb.	$w < 50$
is more than	Nashville is more than 200 mi away.	$d > 200$
is between	The film is between 90 min and 100 min long.	$90 < t < 100$
no more than	Cooper weighs no more than 90 lb.	$w \le 90$
no less than	Sofia scored no less than 8.3.	$s \ge 8.3$

The following phrases deserve special attention.

TRANSLATING "AT LEAST" AND "AT MOST"

A quantity x is at least some amount q: $x \ge q$.
(If x is at least q, it cannot be less than q.)

A quantity x is at most some amount q: $x \le q$.
(If x is at most q, it cannot be more than q.)

◀ Do Margin Exercises 1–8.

b SOLVING PROBLEMS

EXAMPLE 1 *Catering Costs.* To cater a company's annual lobster-bake cookout, Jayla's Catering charges a \$325 setup fee plus \$18.50 per person. The cost cannot exceed \$3200. How many people can attend the cookout?

1. **Familiarize.** Suppose that 130 people were to attend the cookout. The cost would then be $\$325 + \$18.50(130)$, or \$2730. This shows that more than 130 people could attend the picnic without exceeding \$3200. Instead of making another guess, we let $n =$ the number of people in attendance.

2. **Translate.** Our guess shows us how to translate. The cost of the cookout will be the \$325 setup fee plus \$18.50 times the number of people attending. We translate to an inequality:

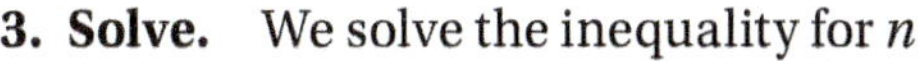

Rewording:	The setup fee	plus	the cost of the meals	cannot exceed	\$3200.
	↓	↓	↓	↓	↓
Translating:	325	+	$18.50n$	$\leq$	3200.

3. **Solve.** We solve the inequality for n:

$$325 + 18.50n \leq 3200$$

$$325 + 18.50n - 325 \leq 3200 - 325 \quad \text{Subtracting 325}$$

$$18.50n \leq 2875 \quad \text{Simplifying}$$

$$\frac{18.50n}{18.50} \leq \frac{2875}{18.50} \quad \text{Dividing by 18.50}$$

$$n \leq 155.4. \quad \text{Rounding to the nearest tenth}$$

4. **Check.** Although the solution set of the inequality is all numbers less than or equal to about 155.4, since $n =$ the number of people in attendance, we round *down* to 155 people. If 155 people attend, the cost will be \$325 + \$18.50(155), or \$3192.50. If 156 attend, the cost will exceed \$3200.

5. **State.** At most, 155 people can attend the lobster-bake cookout.

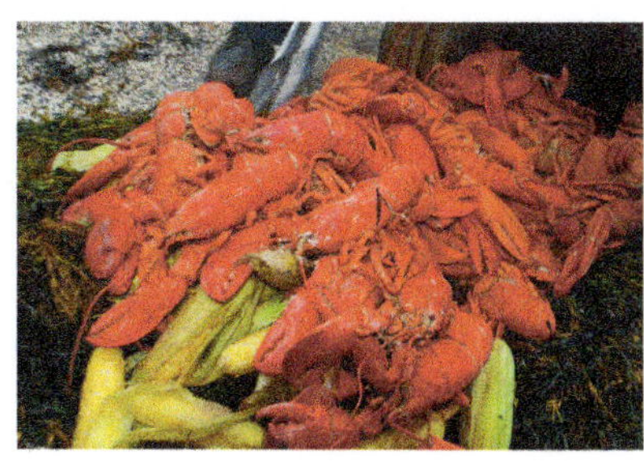

Do Exercise 9. ▶

Caution!

Solutions of problems should always be checked using the original wording of the problem. In some cases, answers might need to be whole numbers or integers or rounded off in a particular direction.

EXAMPLE 2 ***Nutrition.*** The U.S. Department of Agriculture recommends that for a typical 2000-calorie daily diet, no more than 20 g of saturated fat be consumed. In the first three days of a four-day vacation, Ethan consumed 26 g, 17 g, and 22 g of saturated fat. Determine (in terms of an inequality) how many grams of saturated fat Ethan can consume on the fourth day if he is to average no more than 20 g of saturated fat per day.

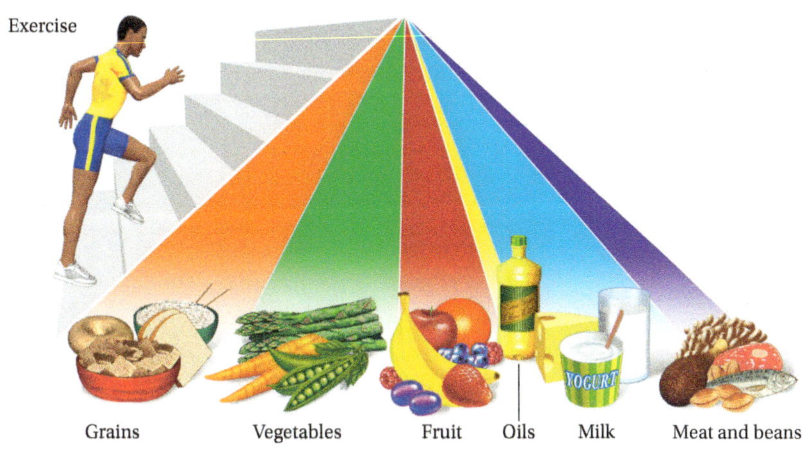

SOURCES: U.S. Department of Health and Human Services; U.S. Department of Agriculture

GS Translate to an inequality and solve.

9. ***Butter Temperatures.*** Butter stays solid at Fahrenheit temperatures below 88°. The formula

$$F = \frac{9}{5}C + 32$$

can be used to convert Celsius temperatures C to Fahrenheit temperatures F. Determine (in terms of an inequality) those Celsius temperatures for which butter stays solid.

Translate and *Solve*:

$$F < 88$$

$$\frac{9}{5}C + 32 < 88$$

$$\frac{9}{5}C + 32 - 32 < 88 - \square$$

$$\frac{9}{5}C < 56$$

$$\square \cdot \frac{9}{5}C < \frac{5}{9} \cdot 56$$

$$C < \frac{\square}{9}$$

$$C < 31\frac{1}{9}.$$

Butter stays solid at Celsius temperatures less than $31\frac{1}{9}°$—that is, $\{C \mid C < 31\frac{1}{9}°\}$.

Answer

9. $\frac{9}{5}C + 32 < 88$; $\{C \mid C < 31\frac{1}{9}°\}$

Guided Solution:

9. $32, \frac{5}{9}, 280$

1. **Familiarize.** Suppose Ethan consumed 19 g of saturated fat on the fourth day. His daily average for the vacation would then be

$$\frac{26\text{ g} + 17\text{ g} + 22\text{ g} + 19\text{ g}}{4} = \frac{84\text{ g}}{4} = 21\text{ g}.$$

This shows that Ethan cannot consume 19 g of saturated fat on the fourth day, if he is to average no more than 20 g of fat per day. We let $x =$ the number of grams of fat that Ethan can consume on the fourth day.

2. **Translate.** We reword the problem and translate to an inequality as follows:

Rewording:	The average consumption of saturated fat	should be no more than	20 g.
Translating:	$\frac{26 + 17 + 22 + x}{4}$	$\leq$	20.

3. **Solve.** Because of the fraction expression, it is convenient to use the multiplication principle first to clear the fraction:

$$\frac{26 + 17 + 22 + x}{4} \leq 20$$

$$4\left(\frac{26 + 17 + 22 + x}{4}\right) \leq 4 \cdot 20 \qquad \text{Multiplying by 4}$$

$$26 + 17 + 22 + x \leq 80$$

$$65 + x \leq 80 \qquad \text{Simplifying}$$

$$x \leq 15. \qquad \text{Subtracting 65}$$

4. **Check.** As a partial check, we show that Ethan can consume 15 g of saturated fat on the fourth day and not exceed a 20-g average for the four days:

$$\frac{26 + 17 + 22 + 15}{4} = \frac{80}{4} = 20.$$

5. **State.** Ethan's average intake of saturated fat for the vacation will not exceed 20 g per day if he consumes no more than 15 g of saturated fat on the fourth day.

Do Exercise 10.

Translate to an inequality and solve.

10. ***Test Scores.*** A pre-med student is taking a chemistry course in which four tests are given. To get an A, she must average at least 90 on the four tests. The student got scores of 91, 86, and 89 on the first three tests. Determine (in terms of an inequality) what scores on the last test will allow her to get an A.

Answer

10. $\frac{91 + 86 + 89 + s}{4} \geq 90$; $\{s \mid s \geq 94\}$

10.8 Exercise Set

For Extra Help

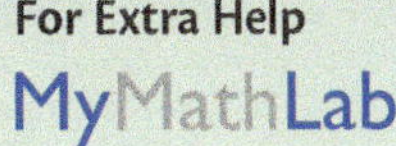

Reading Check

Match each sentence with one of the following.

$q < r \qquad q \leq r \qquad r < q \qquad r \leq q$

RC1. r is at most q.

RC2. q is no more than r.

RC3. r is less than q.

RC4. r is at least q.

RC5. q exceeds r.

RC6. q is no less than r.

a Translate to an inequality.

1. A number is at least 7.

2. A number is greater than or equal to 5.

3. The baby weighs more than 2 kilograms (kg).

4. Between 75 and 100 people attended the concert.

5. The speed of the train was between 90 mph and 110 mph.

6. The attendance was no more than 180.

7. Brianna works no more than 20 hr per week.

8. The amount of acid must exceed 40 liters (L).

9. The cost of gasoline is no less than \$3.20 per gallon.

10. The temperature is at most $-2°$.

11. A number is greater than 8.

12. A number is less than 5.

13. A number is less than or equal to -4.

14. A number is greater than or equal to 18.

15. The number of people is at least 1300.

16. The cost is at most \$4857.95.

17. The amount of water is not to exceed 500 liters.

18. The cost of ground beef is no less than $3.19 per pound.

19. Two more than three times a number is less than 13.

20. Five less than one-half a number is greater than 17.

b Solve.

21. *Test Scores.* Xavier is taking a geology course in which four tests are given. To get a B, he must average at least 80 on the four tests. He got scores of 82, 76, and 78 on the first three tests. Determine (in terms of an inequality) what scores on the last test will allow him to get at least a B.

22. *Test Scores.* Chloe is taking a French class in which five quizzes are given. Her first four quiz grades are 73, 75, 89, and 91. Determine (in terms of an inequality) what scores on the last quiz will allow her to get an average quiz grade of at least 85.

23. *Gold Temperatures.* Gold stays solid at Fahrenheit temperatures below 1945.4°. Determine (in terms of an inequality) those Celsius temperatures for which gold stays solid. Use the formula given in Margin Exercise 9.

24. *Body Temperatures.* The human body is considered to be fevered when its temperature is higher than 98.6°F. Using the formula given in Margin Exercise 9, determine (in terms of an inequality) those Celsius temperatures for which the body is fevered.

25. *World Records in the 1500-m Run.* The formula

$$R = -0.075t + 3.85$$

can be used to predict the world record in the 1500-m run t years after 1930. Determine (in terms of an inequality) those years for which the world record will be less than 3.5 min.

26. *World Records in the 200-m Dash.* The formula

$$R = -0.028t + 20.8$$

can be used to predict the world record in the 200-m dash t years after 1920. Determine (in terms of an inequality) those years for which the world record will be less than 19.0 sec.

27. *Blueprints.* To make copies of blueprints, Vantage Reprographics charges a $5 setup fee plus $4 per copy. Myra can spend no more than $65 for copying her blueprints. What numbers of copies will allow her to stay within budget?

28. *Banquet Costs.* The Shepard College women's volleyball team can spend at most $750 for its awards banquet at a local restaurant. If the restaurant charges an $80 setup fee plus $16 per person, at most how many can attend?

29. *Envelope Size.* For a direct-mail campaign, Hollcraft Advertising determines that any envelope with a fixed width of $3\frac{1}{2}$ in. and an area of at least $17\frac{1}{2}$ in^2 can be used. Determine (in terms of an inequality) those lengths that will satisfy the company constraints.

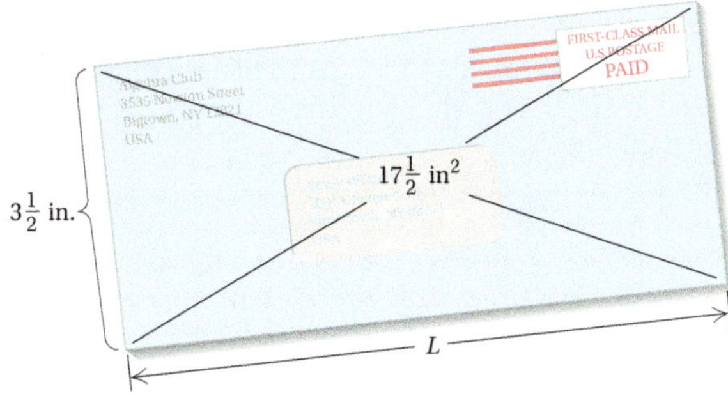

30. *Package Sizes.* Logan Delivery Service accepts packages of up to 165 in. in length and girth combined. (Girth is the distance around the package.) A package has a fixed girth of 53 in. Determine (in terms of an inequality) those lengths for which a package is acceptable.

31. *Phone Costs.* Simon claims that it costs him at least \$3.00 every time he calls an overseas customer. If his typical call costs 75¢ plus 45¢ for each minute, how long do his calls typically last? (*Hint*: 75¢ = \$0.75.)

32. *Parking Costs.* Laura is certain that every time she parks in the municipal garage it costs her at least \$6.75. If the garage charges \$1.50 plus 75¢ for each half hour, for how long is Laura's car generally parked?

33. *College Tuition.* Angelica's financial aid stipulates that her tuition cannot exceed \$1000. If her local community college charges a \$35 registration fee plus \$375 per course, what is the greatest number of courses for which Angelica can register?

34. *Furnace Repairs.* RJ's Plumbing and Heating charges \$45 for a service call plus \$30 per hour for emergency service. Gary remembers being billed over \$150 for an emergency call. How long was RJ's there?

35. *Nutrition.* Following the guidelines of the Food and Drug Administration, Dale tries to eat at least 5 servings of fruits or vegetables each day. For the first six days of one week, he had 4, 6, 7, 4, 6, and 4 servings. How many servings of fruits or vegetables should Dale eat on Saturday in order to average at least 5 servings per day for the week?

36. *College Course Load.* To remain on financial aid, Millie needs to complete an average of at least 7 credits per quarter each year. In the first three quarters of 2013, Millie completed 5, 7, and 8 credits. How many credits of course work must Millie complete in the fourth quarter if she is to remain on financial aid?

37. *Perimeter of a Rectangle.* The width of a rectangle is fixed at 8 ft. What lengths will make the perimeter at least 200 ft? at most 200 ft?

38. *Perimeter of a Triangle.* One side of a triangle is 2 cm shorter than the base. The other side is 3 cm longer than the base. What lengths of the base will allow the perimeter to be greater than 19 cm?

39. *Area of a Rectangle.* The width of a rectangle is fixed at 4 cm. For what lengths will the area be less than 86 cm^2?

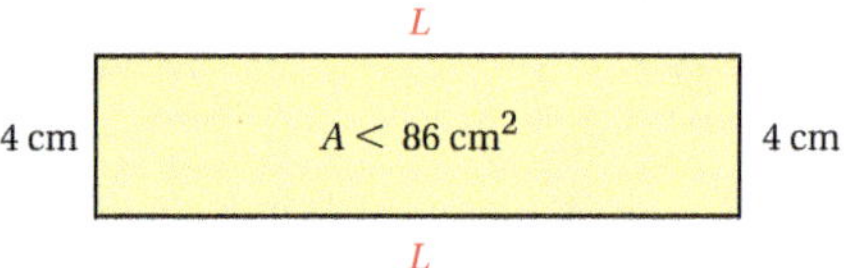

40. *Area of a Rectangle.* The width of a rectangle is fixed at 16 yd. For what lengths will the area be at least 264 yd^2?

41. *Insurance-Covered Repairs.* Most insurance companies will replace a vehicle if an estimated repair exceeds 80% of the "blue-book" value of the vehicle. Rachel's insurance company paid $8500 for repairs to her Toyota after an accident. What can be concluded about the blue-book value of the car?

42. *Insurance-Covered Repairs.* Following an accident, Jeff's Ford pickup was replaced by his insurance company because the damage was so extensive. Before the damage, the blue-book value of the truck was $21,000. How much would it have cost to repair the truck? (See Exercise 41.)

43. *Reduced-Fat Foods.* In order for a food to be labeled "reduced fat," it must have at least 25% less fat than the regular item. One brand of reduced-fat peanut butter contains 12 g of fat per serving. What can you conclude about the fat content in a serving of the brand's regular peanut butter?

44. *Reduced-Fat Foods.* One brand of reduced-fat chocolate chip cookies contains 5 g of fat per serving. What can you conclude about the fat content of the brand's regular chocolate chip cookies? (See Exercise 43.)

45. *Area of a Triangular Flag.* As part of an outdoor education course at Baxter YMCA, Wendy needs to make a bright-colored triangular flag with an area of at least 3 ft^2. What heights can the triangle be if the base is $1\frac{1}{2}$ ft?

46. *Area of a Triangular Sign.* Zoning laws in Harrington prohibit displaying signs with areas exceeding 12 ft^2. If Flo's Marina is ordering a triangular sign with an 8-ft base, how tall can the sign be?

47. *Pond Depth.* On July 1, Garrett's Pond was 25 ft deep. Since that date, the water level has dropped $\frac{2}{3}$ ft per week. For what dates will the water level not exceed 21 ft?

48. *Weight Gain.* A 3-lb puppy is gaining weight at a rate of $\frac{3}{4}$ lb per week. When will the puppy's weight exceed $22\frac{1}{2}$ lb?

49. *Electrician Visits.* Dot's Electric made 17 customer calls last week and 22 calls this week. How many calls must be made next week in order to maintain a weekly average of at least 20 calls for the three-week period?

50. *Volunteer Work.* George and Joan do volunteer work at a hospital. Joan worked 3 more hr than George, and together they worked more than 27 hr. What possible numbers of hours did each work?

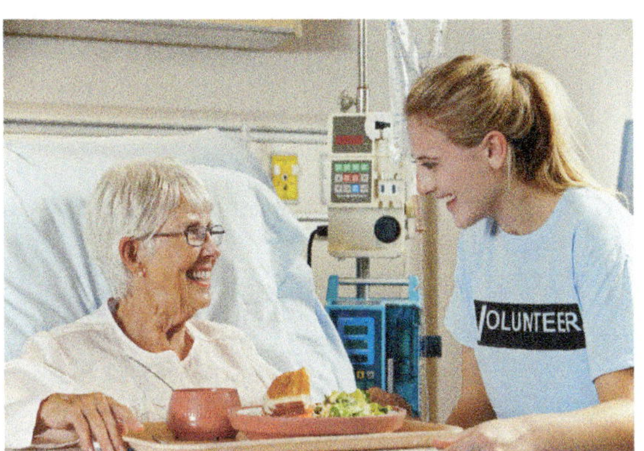

Skill Maintenance

Solve.

51. $-13 + x = 27$ [10.1b]

52. $-6y = 132$ [10.2a]

53. $4a - 3 = 45$ [10.3a]

54. $8x + 3x = 66$ [10.3b]

55. $-\frac{1}{2} + x = x - \frac{1}{2}$ [10.3c]

56. $9x - 1 + 11x - 18 = 3x - 15 + 4 + 17x$ [10.3c]

Solve. [10.5a]

57. What percent of 200 is 15?

58. What is 10% of 310?

59. 25 is 2% of what number?

60. 80 is what percent of 96?

Synthesis

Solve.

61. *Ski Wax.* Green ski wax works best between 5° and 15° Fahrenheit. Determine those Celsius temperatures for which green ski wax works best. Use the formula given in Margin Exercise 9.

62. *Parking Fees.* Mack's Parking Garage charges $4.00 for the first hour and $2.50 for each additional hour. For how long has a car been parked when the charge exceeds $16.50?

63. *Low-Fat Foods.* In order for a food to be labeled "low fat," it must have fewer than 3 g of fat per serving. One brand of reduced-fat tortilla chips contains 60% less fat than regular nacho cheese tortilla chips, but still cannot be labeled low fat. What can you conclude about the fat content of a serving of nacho cheese tortilla chips?

64. *Parking Fees.* When asked how much the parking charge is for a certain car, Mack replies "between 14 and 24 dollars." For how long has the car been parked? (See Exercise 62.)

CHAPTER

10 Summary and Review

Vocabulary Reinforcement

Complete each statement with the correct word or words from the column on the right. Some of the choices may not be used.

1. Any replacement for the variable that makes an equation true is called a(n) ______________ of the equation. [10.1a]
2. The ______________ for equations states that for any real numbers a, b, and c, $a = b$ is equivalent to $a + c = b + c$. [10.1b]
3. The ______________ for equations states that for any real numbers a, b, and c, $a = b$ is equivalent to $a \cdot c = b \cdot c$. [10.2a]
4. An ______________ is a number sentence with $<$, $\leq$, $>$, or $\geq$ as its verb. [10.7a]
5. Equations with the same solutions are called ______________ equations. [10.1b]

addition principle
multiplication principle
solution
equivalent
equation
inequality

Concept Reinforcement

Determine whether each statement is true or false.

______________ 1. Some equations have no solution. [10.3c]

______________ 2. For any number n, $n \geq n$. [10.7a]

______________ 3. $2x - 7 < 11$ and $x < 2$ are equivalent inequalities. [10.7c]

______________ 4. If $x > y$, then $-x < -y$. [10.7d]

Study Guide

Objective 10.3a Solve equations using both the addition principle and the multiplication principle.

Objective 10.3b Solve equations in which like terms may need to be collected.

Objective 10.3c Solve equations by first removing parentheses and collecting like terms.

Example Solve: $6y - 2(2y - 3) = 12$.

$$6y - 2(2y - 3) = 12$$

$$6y - 4y + 6 = 12 \quad \text{Removing parentheses}$$

$$2y + 6 = 12 \quad \text{Collecting like terms}$$

$$2y + 6 - 6 = 12 - 6 \quad \text{Subtracting 6}$$

$$2y = 6$$

$$\frac{2y}{2} = \frac{6}{2} \quad \text{Dividing by 2}$$

$$y = 3$$

The solution is 3.

Practice Exercise

1. Solve: $4(x - 3) = 6(x + 2)$.

Objective 10.3c Solve equations with an infinite number of solutions and equations with no solutions.

Example Solve: $8 + 2x - 4 = 6 + 2(x - 1)$.

$$\begin{aligned} 8 + 2x - 4 &= 6 + 2(x - 1) \\ 8 + 2x - 4 &= 6 + 2x - 2 \\ 2x + 4 &= 2x + 4 \\ 2x + 4 - 2x &= 2x + 4 - 2x \\ 4 &= 4 \end{aligned}$$

Every real number is a solution of the equation $4 = 4$, so all real numbers are solutions of the original equation. The equation has infinitely many solutions.

Example Solve: $2 + 5(x - 1) = -6 + 5x + 7$.

$$\begin{aligned} 2 + 5(x - 1) &= -6 + 5x + 7 \\ 2 + 5x - 5 &= -6 + 5x + 7 \\ 5x - 3 &= 5x + 1 \\ 5x - 3 - 5x &= 5x + 1 - 5x \\ -3 &= 1 \end{aligned}$$

This is a false equation, so the original equation has no solution.

Practice Exercises

2. Solve: $4 + 3y - 7 = 3 + 3(y - 2)$.

3. Solve: $4(x - 3) + 7 = -5 + 4x + 10$.

Objective 10.4b Solve a formula for a specified letter.

Example Solve for n: $M = \dfrac{m + n}{5}$.

$$\begin{aligned} M &= \frac{m + n}{5} \\ 5 \cdot M &= 5\left(\frac{m + n}{5}\right) \\ 5M &= m + n \\ 5M - m &= m + n - m \\ 5M - m &= n \end{aligned}$$

Practice Exercise

4. Solve for b: $A = \dfrac{1}{2}bh$.

Objective 10.7b Graph an inequality on the number line.

Example Graph each inequality: **(a)** $x < 2$; **(b)** $x \geq -3$.

a) The solutions of $x < 2$ are all numbers less than 2. We shade all points to the left of 2, and we use a parenthesis at 2 to indicate that 2 *is not* part of the graph.

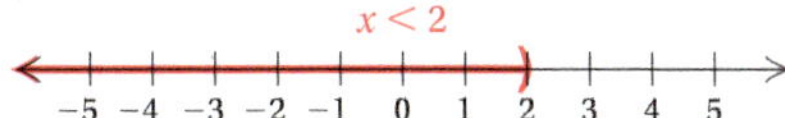

b) The solutions of $x \geq -3$ are all numbers greater than -3 and the number -3 as well. We shade all points to the right of -3, and we use a bracket at -3 to indicate that -3 *is* part of the graph.

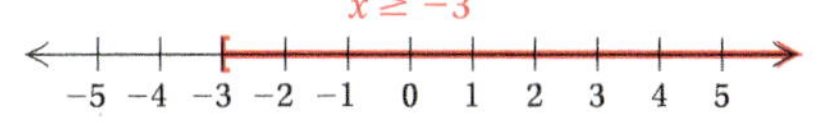

Practice Exercises

5. Graph: $x > 1$.

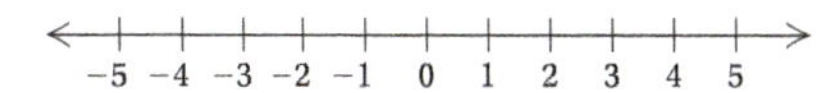

6. Graph: $x \leq -1$.

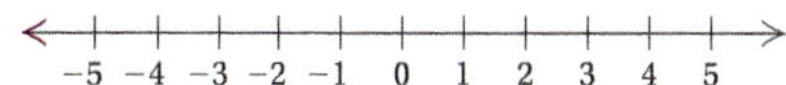

Objective 10.7e Solve inequalities using the addition principle and the multiplication principle together.

Example Solve: $8y - 7 \le 5y + 2$.

$$8y - 7 \le 5y + 2$$
$$8y - 7 - 8y \le 5y + 2 - 8y$$
$$-7 \le -3y + 2$$
$$-7 - 2 \le -3y + 2 - 2$$
$$-9 \le -3y$$
$$\frac{-9}{-3} \ge \frac{-3y}{-3} \quad \text{Reversing the symbol}$$
$$3 \ge y$$

The solution set is $\{y \mid 3 \ge y\}$, or $\{y \mid y \le 3\}$.

Practice Exercise

7. Solve: $6y + 5 > 3y - 7$.

Review Exercises

Solve. [10.1b]

1. $x + 5 = -17$

2. $n - 7 = -6$

3. $x - 11 = 14$

4. $y - 0.9 = 9.09$

Solve. [10.2a]

5. $-\frac{2}{3}x = -\frac{1}{6}$

6. $-8x = -56$

7. $-\frac{x}{4} = 48$

8. $15x = -35$

9. $\frac{4}{5}y = -\frac{3}{16}$

Solve. [10.3a]

10. $5 - x = 13$

11. $\frac{1}{4}x - \frac{5}{8} = \frac{3}{8}$

Solve. [10.3b, c]

12. $5t + 9 = 3t - 1$

13. $7x - 6 = 25x$

14. $14y = 23y - 17 - 10$

15. $0.22y - 0.6 = 0.12y + 3 - 0.8y$

16. $\frac{1}{4}x - \frac{1}{8}x = 3 - \frac{1}{16}x$

17. $14y + 17 + 7y = 9 + 21y + 8$

18. $4(x + 3) = 36$

19. $3(5x - 7) = -66$

20. $8(x - 2) - 5(x + 4) = 20 + x$

21. $-5x + 3(x + 8) = 16$

22. $6(x - 2) - 16 = 3(2x - 5) + 11$

Determine whether the given number is a solution of the inequality $x \le 4$. [10.7a]

23. -3

24. 7

25. 4

Solve. Write set notation for the answers. [10.7c, d, e]

26. $y + \frac{2}{3} \geq \frac{1}{6}$

27. $9x \geq 63$

28. $2 + 6y > 14$

29. $7 - 3y \geq 27 + 2y$

30. $3x + 5 < 2x - 6$

31. $-4y < 28$

32. $4 - 8x < 13 + 3x$

33. $-4x \leq \frac{1}{3}$

Graph on the number line. [10.7b, e]

34. $4x - 6 < x + 3$

−5 −4 −3 −2 −1 0 1 2 3 4 5

35. $-2 < x \leq 5$

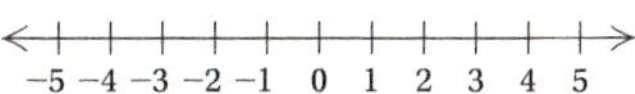

36. $y > 0$

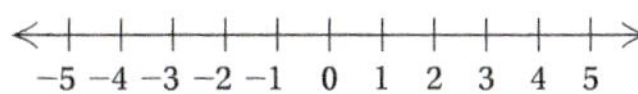

Solve. [10.4b]

37. $C = \pi d$, for d

38. $V = \frac{1}{3}Bh$, for B

39. $A = \frac{a + b}{2}$, for a

40. $y = mx + b$, for x

Solve. [10.6a]

41. ***Dimensions of Wyoming.*** The state of Wyoming is roughly in the shape of a rectangle whose perimeter is 1280 mi. The length is 90 mi more than the width. Find the dimensions.

42. ***Interstate Mile Markers.*** The sum of two consecutive mile markers on I-5 in California is 691. Find the numbers on the markers.

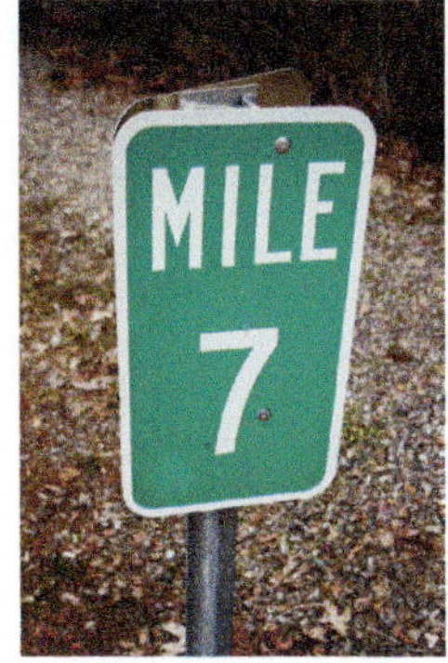

43. An entertainment center sold for \$2449 in June. This was \$332 more than the cost in February. What was the cost in February?

44. Ty is paid a commission of \$4 for each magazine subscription he sells. One week, he received \$108 in commissions. How many subscriptions did he sell?

45. The measure of the second angle of a triangle is 50° more than that of the first angle. The measure of the third angle is 10° less than twice the first angle. Find the measures of the angles.

Solve. [10.5a]

46. What number is 20% of 75?

47. Fifteen is what percent of 80?

48. 18 is 3% of what number?

49. *Black Bears.* The population of black bears in the 26 U.S. states that border on or are east of the Mississippi River increased from 87,872 in 1988 to 164,440 in 2008. What is the percent increase?

Source: Hank Hristienko, Manitoba Conservation Wildlife and Ecosystem Protection Branch

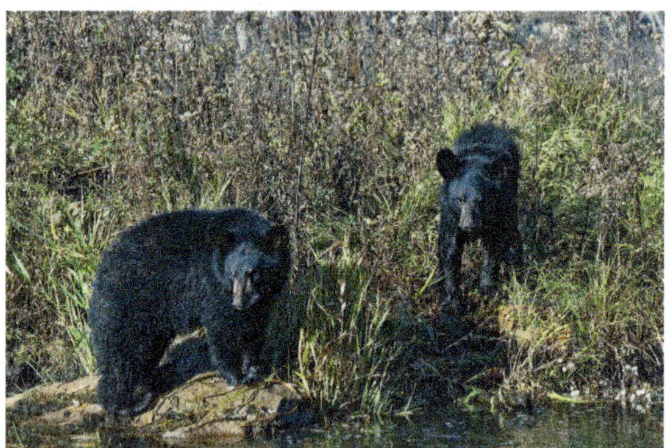

50. *First-Class Mail.* The volume of first-class mail decreased from 102.4 billion pieces in 2002 to only 73.5 billion pieces in 2011. What is the percent decrease?

Source: United States Postal Service

Solve. [10.6a]

51. After a 30% reduction, a bread maker is on sale for \$154. What was the marked price (the price before the reduction)?

52. A restaurant manager's salary is \$78,300, which is an 8% increase over the previous year's salary. What was the previous salary?

53. A tax-exempt organization received a bill of \$145.90 for janitorial supplies. The bill incorrectly included sales tax of 5%. How much does the organization actually owe?

Solve. [10.8b]

54. *Test Scores.* Noah's test grades are 71, 75, 82, and 86. What is the lowest grade that he can get on the next test and still have an average test score of at least 80?

55. The length of a rectangle is 43 cm. What widths will make the perimeter greater than 120 cm?

56. The solution of the equation

$$4(3x - 5) + 6 = 8 + x$$

is which of the following? [10.3c]

A. Less than -1
B. Between -1 and 1
C. Between 1 and 5
D. Greater than 5

57. Solve for y: $3x + 4y = P$. [10.4b]

A. $y = \frac{P - 3x}{4}$
B. $y = \frac{P + 3x}{4}$
C. $y = P - \frac{3x}{4}$
D. $y = \frac{P}{4} - 3x$

Synthesis

Solve.

58. $2|x| + 4 = 50$ [9.2e], [10.3a]

59. $|3x| = 60$ [9.2e], [10.2a]

60. $y = 2a - ab + 3$, for a [10.4b]

Understanding Through Discussion and Writing

1. Would it be better to receive a 5% raise and then an 8% raise or the other way around? Why? [10.5a]

2. Erin returns a tent that she bought during a storewide 25%-off sale that has ended. She is offered store credit for 125% of what she paid (not to be used on sale items). Is this fair to Erin? Why or why not? [10.5a]

3. Are the inequalities $x > -5$ and $-x < 5$ equivalent? Why or why not? [10.7d]

4. Explain in your own words why it is necessary to reverse the inequality symbol when multiplying on both sides of an inequality by a negative number. [10.7d]

5. If f represents Fran's age and t represents Todd's age, write a sentence that would translate to $t + 3 < f$. [10.8a]

6. Explain how the meanings of "Five more than a number" and "Five is more than a number" differ. [10.8a]

CHAPTER 10 Test

For Extra Help For step-by-step test solutions, access the Chapter Test Prep Videos in MyMathLab® or on YouTube (search "BittingerPreIntro" and click on Channels).

Solve.

1. $x + 7 = 15$

2. $t - 9 = 17$

3. $3x = -18$

4. $-\frac{4}{7}x = -28$

5. $3t + 7 = 2t - 5$

6. $\frac{1}{2}x - \frac{3}{5} = \frac{2}{5}$

7. $8 - y = 16$

8. $-\frac{2}{5} + x = -\frac{3}{4}$

9. $3(x + 2) = 27$

10. $-3x - 6(x - 4) = 9$

11. $0.4p + 0.2 = 4.2p - 7.8 - 0.6p$

12. $4(3x - 1) + 11 = 2(6x + 5) - 8$

13. $-2 + 7x + 6 = 5x + 4 + 2x$

Solve. Write set notation for the answers.

14. $x + 6 \le 2$

15. $14x + 9 > 13x - 4$

16. $12x \le 60$

17. $-2y \ge 26$

18. $-4y \le -32$

19. $-5x \ge \frac{1}{4}$

20. $4 - 6x > 40$

21. $5 - 9x \ge 19 + 5x$

Graph on the number line.

22. $y \le 9$

−12 −8 −4 0 4 8 12

23. $6x - 3 < x + 2$

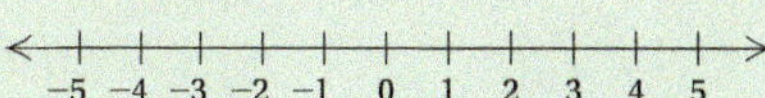

24. $-2 \le x \le 2$

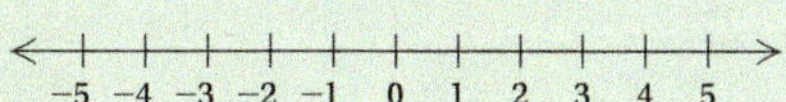

Solve.

25. What number is 24% of 75?

26. 15.84 is what percent of 96?

27. 800 is 2% of what number?

28. *Bottled Water.* Annual bottled water consumption per person in the United States increased from 18.2 gal in 2001 to 29.2 gal in 2011. What is the percent increase?

Source: Beverage Marketing Corporation

29. *Perimeter of a Photograph.* The perimeter of a rectangular photograph is 36 cm. The length is 4 cm greater than the width. Find the width and the length.

30. *Cost of Raising a Child.* It is estimated that $41,500 will be spent for child care and K–12 education for a child to age 17. This number represents approximately 18% of the total cost of raising a child to age 17. What is the total cost of raising a child to age 17?

Sources: U.S. Department of Agriculture

31. *Raffle Tickets.* The numbers on three raffle tickets are consecutive integers whose sum is 7530. Find the integers.

32. *Savings Account.* Money is invested in a savings account at 5% simple interest. After 1 year, there is $924 in the account. How much was originally invested?

33. *Board Cutting.* An 8-m board is cut into two pieces. One piece is 2 m longer than the other. How long are the pieces?

34. *Lengths of a Rectangle.* The width of a rectangle is 96 yd. Find all possible lengths such that the perimeter of the rectangle will be at least 540 yd.

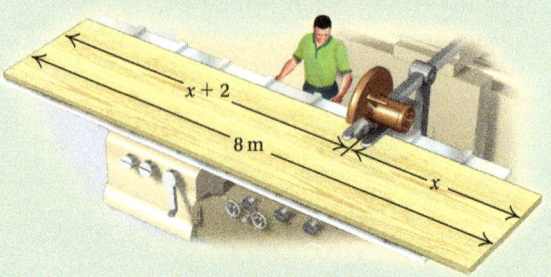

35. *Budgeting.* Jason has budgeted an average of $95 per month for entertainment. For the first five months of the year, he has spent $98, $89, $110, $85, and $83. How much can Jason spend in the sixth month without exceeding his average budget?

36. *Copy Machine Rental.* A catalog publisher needs to lease a copy machine for use during a special project that they anticipate will take 3 months. It costs $225 per month plus 3.2¢ per copy to rent the machine. The company must stay within a budget of $4500 for copies. Determine (in terms of an inequality) the number of copies they can make and still remain within budget.

37. Solve $A = 2\pi rh$ for r.

38. Solve $y = 8x + b$ for x.

39. *Senior Population.* The number of Americans ages 65 and older is projected to grow from 40.4 million to 70.3 million between 2011 and 2030. Find the percent increase.

Source: U.S. Census Bureau

A. 42.5% **B.** 47%
C. 57.5% **D.** 74%

Synthesis

40. Solve $c = \dfrac{1}{a - d}$ for d.

41. Solve: $3|w| - 8 = 37$.

42. A movie theater had a certain number of tickets to give away. Five people got the tickets. The first got one-third of the tickets, the second got one-fourth of the tickets, and the third got one-fifth of the tickets. The fourth person got eight tickets, and there were five tickets left for the fifth person. Find the total number of tickets given away.

CHAPTER

11

Graphs of Linear Equations

STUDYING FOR SUCCESS *Preparing for a Test*

- ☐ Make up your own test questions as you study.
- ☐ Do an overall review of the chapter, focusing on the objectives and the examples.
- ☐ Do the exercises in the mid-chapter review and in the summary and review at the end of the chapter.
- ☐ Take the chapter test at the end of the chapter.

11.1 Graphs and Applications of Linear Equations

OBJECTIVES

a Plot points associated with ordered pairs of numbers; determine the quadrant in which a point lies.

b Find the coordinates of a point on a graph.

c Determine whether an ordered pair is a solution of an equation with two variables.

d Graph linear equations of the type $y = mx + b$ and $Ax + By = C$, identifying the y-intercept.

e Solve applied problems involving graphs of linear equations.

SKILL TO REVIEW

Objective 10.1a: Determine whether a given number is a solution of a given equation.

Determine whether -3 is a solution of each equation.

1. $8(w - 3) = 0$

2. $15 = -2y + 9$

You probably have seen bar graphs like the following in newspapers and magazines. Note that a straight line can be drawn along the tops of the bars. Such a line is a *graph of a linear equation*. In this chapter, we study how to graph linear equations and consider properties such as slope and intercepts. Many applications of these topics will also be considered.

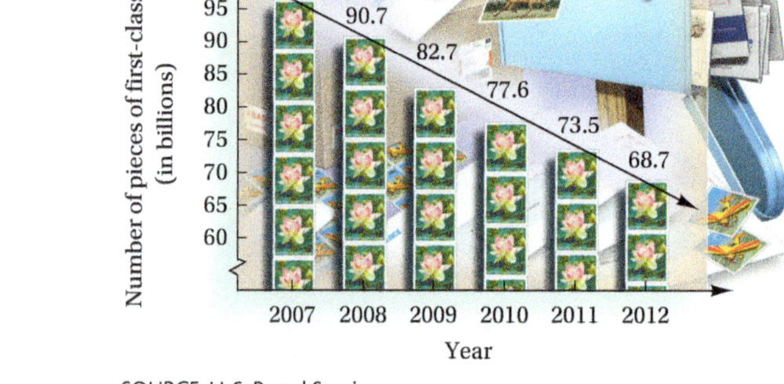

SOURCE: U. S. Postal Service

a PLOTTING ORDERED PAIRS

In Chapter 10, we graphed numbers and inequalities in one variable on a line. To enable us to graph an equation that contains two variables, we now learn to graph number pairs on a plane.

On the number line, each point is the graph of a number. On a plane, each point is the graph of a number pair. To form the plane, we use two perpendicular number lines called **axes**. They cross at a point called the **origin**. The arrows show the positive directions.

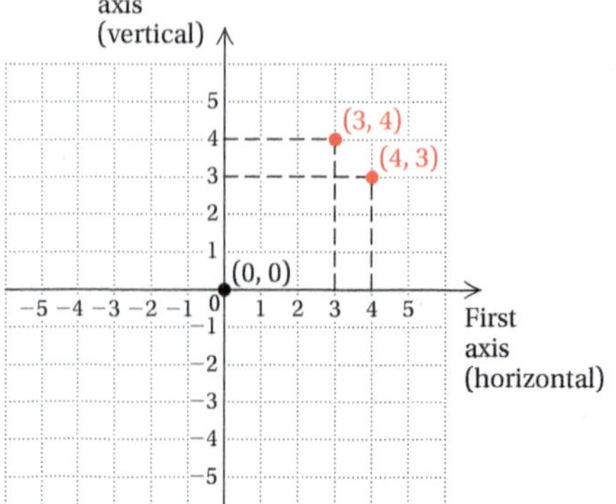

Answers

Skill to Review:
1. -3 is not a solution. **2.** -3 is a solution.

Consider the **ordered pair** $(3, 4)$. The numbers in an ordered pair are called **coordinates**. In $(3, 4)$, the **first coordinate** (the **abscissa**) is 3 and the **second coordinate** (the **ordinate**) is 4. To plot $(3, 4)$, we start at the origin and move *horizontally* to the 3. Then we move up *vertically* 4 units and make a "dot."

The point $(4, 3)$ is also plotted above. Note that $(3, 4)$ and $(4, 3)$ represent different points. The order of the numbers in the pair is important. We use the term *ordered* pairs because it makes a difference which number comes first. The coordinates of the origin are $(0, 0)$.

EXAMPLE 1 Plot the point $(-5, 2)$.

The first number, -5, is negative. Starting at the origin, we move -5 units in the horizontal direction (5 units to the left). The second number, 2, is positive. We move 2 units in the vertical direction (up).

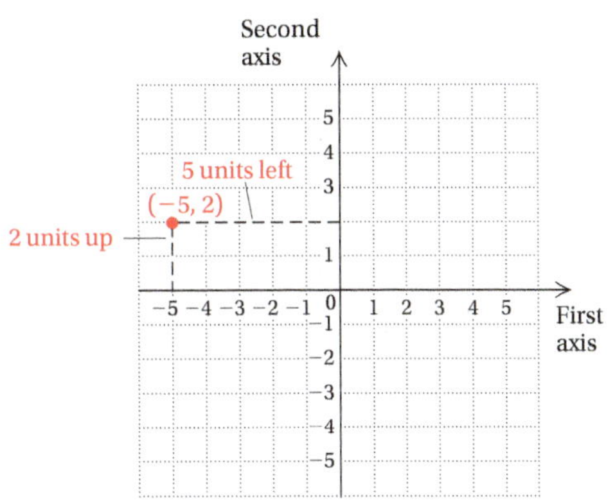

Caution!

The *first* coordinate of an ordered pair is always graphed in a *horizontal* direction, and the *second* coordinate is always graphed in a *vertical* direction.

Do Exercises 1–8. ▶

The following figure shows some points and their coordinates. In region I (the *first quadrant*), both coordinates of any point are positive. In region II (the *second quadrant*), the first coordinate is negative and the second positive. In region III (the *third quadrant*), both coordinates are negative. In region IV (the *fourth quadrant*), the first coordinate is positive and the second is negative.

EXAMPLE 2 In which quadrant, if any, are the points $(-4, 5)$, $(5, -5)$, $(2, 4)$, $(-2, -5)$, and $(-5, 0)$ located?

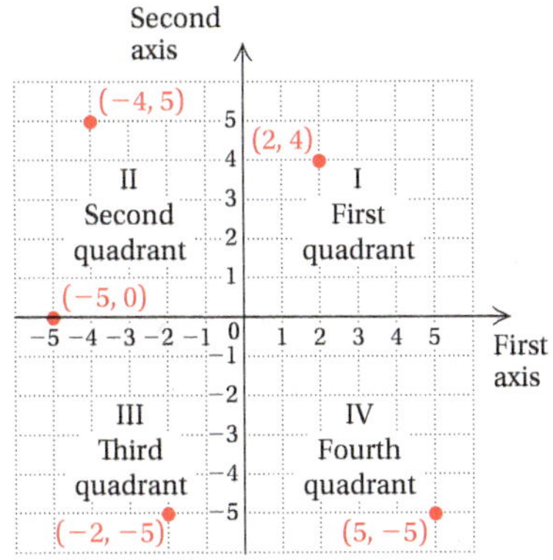

The point $(-4, 5)$ is in the second quadrant. The point $(5, -5)$ is in the fourth quadrant. The point $(2, 4)$ is in the first quadrant. The point $(-2, -5)$ is in the third quadrant. The point $(-5, 0)$ is on an axis and is *not in any quadrant*.

Do Exercises 9–16. ▶

Plot these points on the grid below.

1. $(4, 5)$ **2.** $(5, 4)$

3. $(-2, 5)$ **4.** $(-3, -4)$

5. $(5, -3)$ **6.** $(-2, -1)$

7. $(0, -3)$ **8.** $(2, 0)$

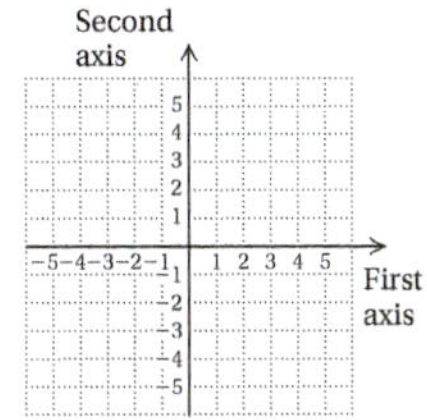

9. What can you say about the coordinates of a point in the third quadrant?

10. What can you say about the coordinates of a point in the fourth quadrant?

In which quadrant, if any, is each point located?

11. $(5, 3)$ **12.** $(-6, -4)$

13. $(10, -14)$ **14.** $(-13, 9)$

15. $(0, -3)$ **16.** $\left(-\frac{1}{2}, \frac{1}{4}\right)$

Answers

1.–8.

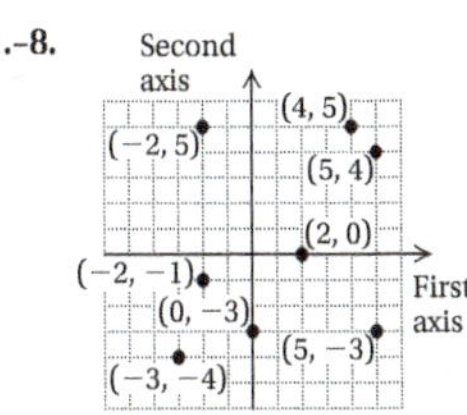

9. First, negative; second, negative **10.** First, positive; second, negative **11.** I **12.** III **13.** IV **14.** II **15.** On an axis, not in any quadrant **16.** II

b FINDING COORDINATES

To find the coordinates of a point, we see how far to the right or to the left of the origin it is located and how far up or down from the origin.

EXAMPLE 3 Find the coordinates of points $A, B, C, D, E, F,$ and G.

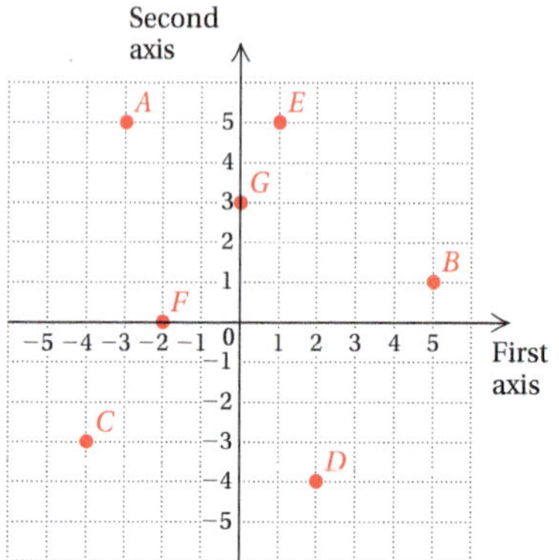

Point A is 3 units to the left (horizontal direction) and 5 units up (vertical direction). Its coordinates are $(-3, 5)$. Point D is 2 units to the right and 4 units down. Its coordinates are $(2, -4)$. The coordinates of the other points are as follows:

B: $(5, 1)$; C: $(-4, -3)$;
E: $(1, 5)$; F: $(-2, 0)$;
G: $(0, 3)$.

◀ Do Exercise 17.

17. Find the coordinates of points $A, B, C, D, E, F,$ and G on the graph below.

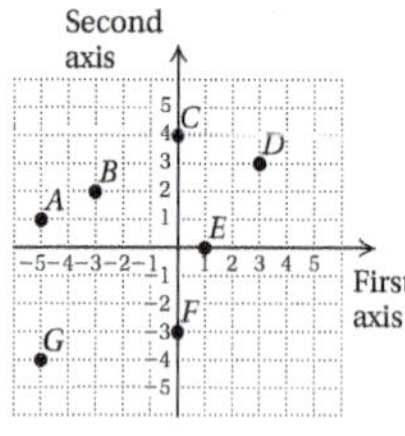

c SOLUTIONS OF EQUATIONS

Now we begin to learn how graphs can be used to represent solutions of equations. When an equation contains two variables, the solutions of the equation are *ordered pairs* in which each number in the pair corresponds to a letter in the equation. Unless stated otherwise, to determine whether a pair is a solution, we use the first number in each pair to replace the variable that occurs first *alphabetically.*

EXAMPLE 4 Determine whether each of the following pairs is a solution of $4q - 3p = 22$: $(2, 7)$ and $(-1, 6)$.

For $(2, 7)$, we substitute 2 for p and 7 for q (using alphabetical order of variables):

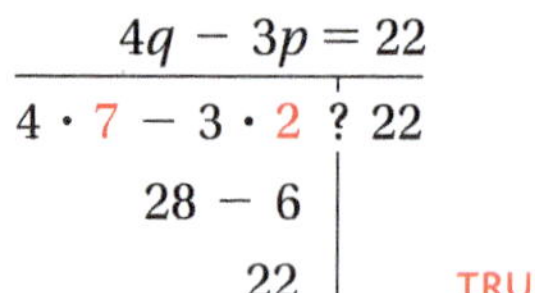

$$\begin{array}{c|l} 4q - 3p & = 22 \\ \hline 4 \cdot 7 - 3 \cdot 2 & ?\ 22 \\ 28 - 6 & \\ 22 & \quad \text{TRUE} \end{array}$$

Thus, $(2, 7)$ *is* a solution of the equation.

For $(-1, 6)$, we substitute -1 for p and 6 for q:

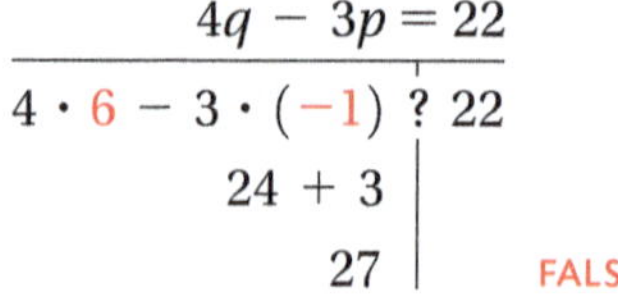

$$\begin{array}{c|l} 4q - 3p & = 22 \\ \hline 4 \cdot 6 - 3 \cdot (-1) & ?\ 22 \\ 24 + 3 & \\ 27 & \quad \text{FALSE} \end{array}$$

Thus, $(-1, 6)$ *is not* a solution of the equation.

◀ Do Exercises 18 and 19.

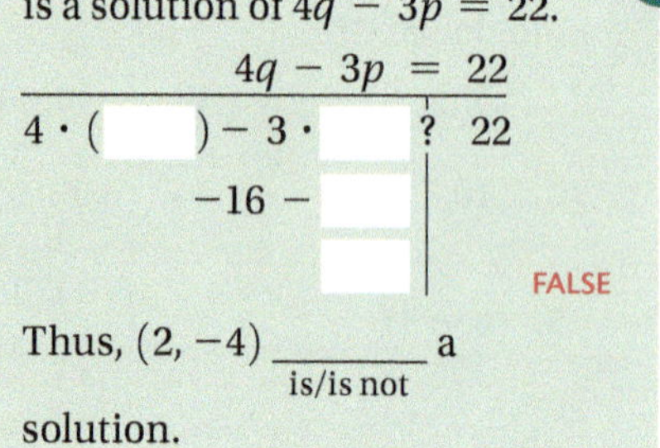

18. Determine whether $(2, -4)$ is a solution of $4q - 3p = 22$. GS

$$\begin{array}{c|l} 4q - 3p & = 22 \\ \hline 4 \cdot (\quad) - 3 \cdot \square & ?\ 22 \\ -16 - \square & \\ \square & \quad \text{FALSE} \end{array}$$

Thus, $(2, -4)$ ______ (is/is not) a solution.

19. Determine whether $(2, -4)$ is a solution of $7a + 5b = -6$.

Answers

17. A: $(-5, 1)$; B: $(-3, 2)$; C: $(0, 4)$; D: $(3, 3)$; E: $(1, 0)$; F: $(0, -3)$; G: $(-5, -4)$ 18. No 19. Yes

Guided Solution:
18. -4, 2, 6, -22, is not

EXAMPLE 5 Show that the pairs $(3, 7)$, $(0, 1)$, and $(-3, -5)$ are solutions of $y = 2x + 1$. Then graph the three points and use the graph to determine another pair that is a solution.

To show that a pair is a solution, we substitute, replacing x with the first coordinate and y with the second coordinate of each pair:

$$\begin{array}{c|l} y & = 2x + 1 \\ \hline 7 & ?\ 2 \cdot 3 + 1 \\ & \ \ 6 + 1 \\ & \ \ 7 \qquad \text{TRUE} \end{array} \qquad \begin{array}{c|l} y & = 2x + 1 \\ \hline 1 & ?\ 2 \cdot 0 + 1 \\ & \ \ 0 + 1 \\ & \ \ 1 \qquad \text{TRUE} \end{array} \qquad \begin{array}{c|l} y & = 2x + 1 \\ \hline -5 & ?\ 2(-3) + 1 \\ & \ \ -6 + 1 \\ & \ \ -5 \qquad \text{TRUE} \end{array}$$

In each of the three cases, the substitution results in a true equation. Thus the pairs are all solutions.

We plot the points as shown below. The order of the points follows the alphabetical order of the variables. That is, x is before y, so x-values are first coordinates and y-values are second coordinates. Similarly, we also label the horizontal axis as the x-axis and the vertical axis as the y-axis.

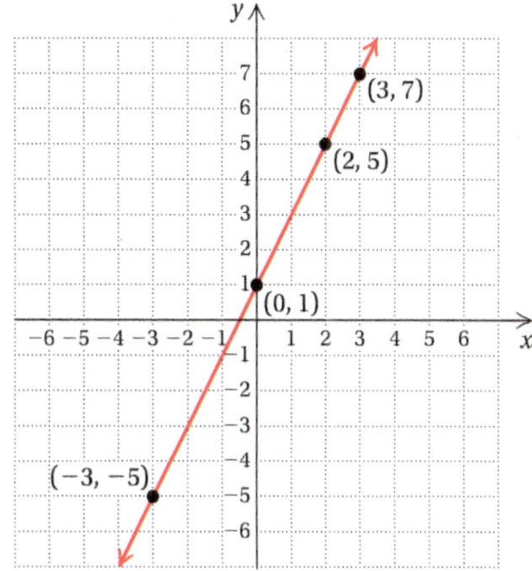

Note that the three points appear to "line up." That is, they appear to be on a straight line. Will other points that line up with these points also represent solutions of $y = 2x + 1$? To find out, we use a straightedge and sketch a line passing through $(3, 7)$, $(0, 1)$, and $(-3, -5)$.

The line appears to pass through $(2, 5)$ as well. Let's see if this pair is a solution of $y = 2x + 1$:

$$\begin{array}{c|l} y & = 2x + 1 \\ \hline 5 & ?\ 2 \cdot 2 + 1 \\ & \ \ 4 + 1 \\ & \ \ 5 \qquad \text{TRUE} \end{array}$$

Thus, $(2, 5)$ is a solution.

Do Exercise 20. ▶

20. Use the graph in Example 5 to find at least two more points that are solutions of $y = 2x + 1$.

Example 5 leads us to suspect that any point on the line that passes through $(3, 7)$, $(0, 1)$, and $(-3, -5)$ represents a solution of $y = 2x + 1$. In fact, every solution of $y = 2x + 1$ is represented by a point on that line and every point on that line represents a solution. The line is the *graph* of the equation.

Answer

20. $(-2, -3)$, $(1, 3)$; answers may vary

GRAPH OF AN EQUATION

The **graph** of an equation is a drawing that represents all of its solutions.

d GRAPHS OF LINEAR EQUATIONS

Equations like $y = 2x + 1$ and $4q - 3p = 22$ are said to be **linear** because the graph of each equation is a straight line. In general, any equation equivalent to one of the form $y = mx + b$ or $Ax + By = C$, where m, b, A, B, and C are constants (not variables) and A and B are not both 0, is linear.

To graph a linear equation:

1. Select a value for one variable and calculate the corresponding value of the other variable. Form an ordered pair using alphabetical order as indicated by the variables.
2. Repeat step (1) to obtain at least two other ordered pairs. Two points are essential to determine a straight line. A third point serves as a check.
3. Plot the ordered pairs and draw a straight line passing through the points.

In general, calculating three (or more) ordered pairs is not difficult for equations of the form $y = mx + b$. We simply substitute values for x and calculate the corresponding values for y.

EXAMPLE 6 Graph: $y = 2x$.

First, we find some ordered pairs that are solutions. We choose *any* number for x and then determine y by substitution. Since $y = 2x$, we find y by doubling x. Suppose that we choose 3 for x. Then

$$y = 2x = 2 \cdot 3 = 6.$$

We get a solution: the ordered pair $(3, 6)$.

Suppose that we choose 0 for x. Then

$$y = 2x = 2 \cdot 0 = 0.$$

We get another solution: the ordered pair $(0, 0)$.

For a third point, we make a negative choice for x. If x is -3, we have

$$y = 2x = 2 \cdot (-3) = -6.$$

This gives us the ordered pair $(-3, -6)$.

We now have enough points to plot the line, but if we wish, we can compute more. If a number takes us off the graph paper, we either do not use it or use larger paper or rescale the axes. Continuing in this manner, we create a table like the one shown on the following page.

CALCULATOR CORNER

Finding Solutions of Equations A table of values representing ordered pairs that are solutions of an equation can be displayed on a graphing calculator. To do this for the equation in Example 6, $y = 2x$, we first access the equation-editor screen. Then we clear any equations that are present. Next, we enter the equation, display the table set-up screen, and set both INDPNT and DEPEND to AUTO.

We will display a table of values that starts with $x = -2$ (TBLSTART) and adds 1 (ΔTBL) to the preceding x-value.

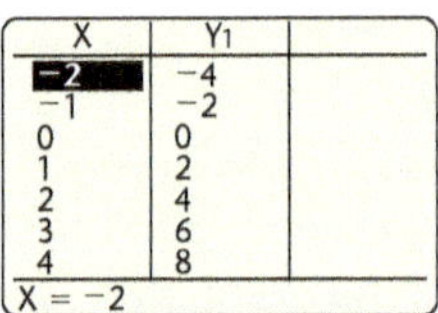

EXERCISE:

1. Create a table of ordered pairs that are solutions of the equations in Examples 7 and 8.

Now we plot these points. Then we draw the line, or graph, with a straightedge and label it $y = 2x$.

x	y $y = 2x$	(x, y)
3	6	(3, 6)
1	2	(1, 2)
0	0	(0, 0)
−2	−4	(−2, −4)
−3	−6	(−3, −6)

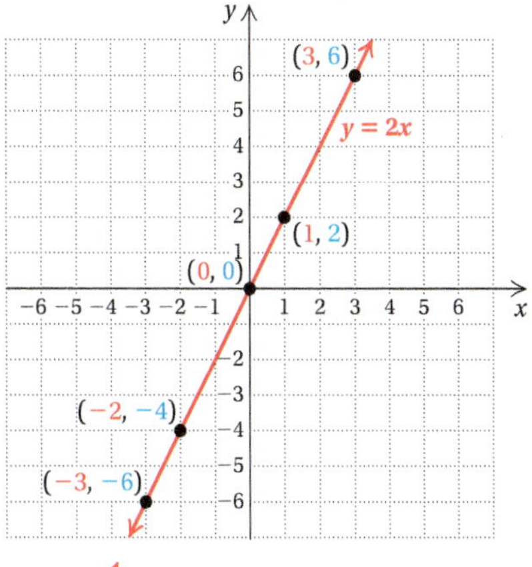

(1) Choose x.
(2) Compute y.
(3) Form the pair (x, y).
(4) Plot the points.

Caution!

Keep in mind that you can choose *any* number for x and then compute y. Our choice of certain numbers in the examples does not dictate those that you must choose.

Do Exercises 21 and 22.

EXAMPLE 7 Graph: $y = -3x + 1$.

We select a value for x, compute y, and form an ordered pair. Then we repeat the process for other choices of x.

If $x = 2$, then $y = -3 \cdot 2 + 1 = -5$, and $(2, -5)$ is a solution.
If $x = 0$, then $y = -3 \cdot 0 + 1 = 1$, and $(0, 1)$ is a solution.
If $x = -1$, then $y = -3 \cdot (-1) + 1 = 4$, and $(-1, 4)$ is a solution.

Results are listed in the following table. The points corresponding to each pair are then plotted.

x	y $y = -3x + 1$	(x, y)
2	−5	(2, −5)
0	1	(0, 1)
−1	4	(−1, 4)

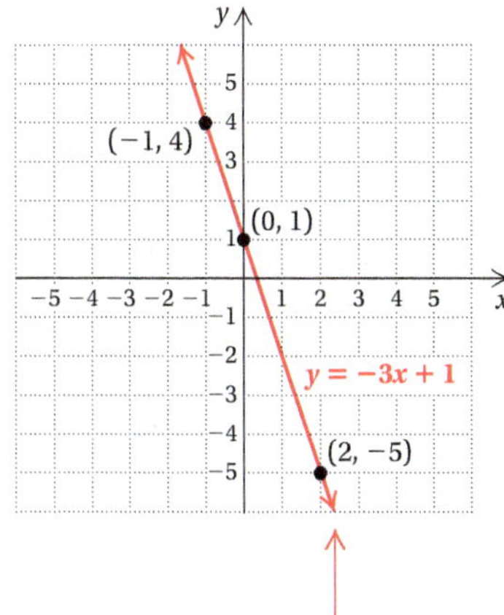

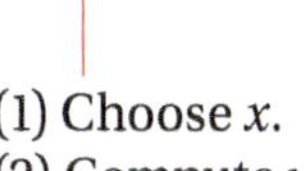

(1) Choose x.
(2) Compute y.
(3) Form the pair (x, y).
(4) Plot the points.

Complete each table and graph.

21. $y = -2x$

x	y	(x, y)
−3	6	(−3, 6)
−1		(−1,)
0	0	(0, 0)
1		(, −2)
3		(3,)

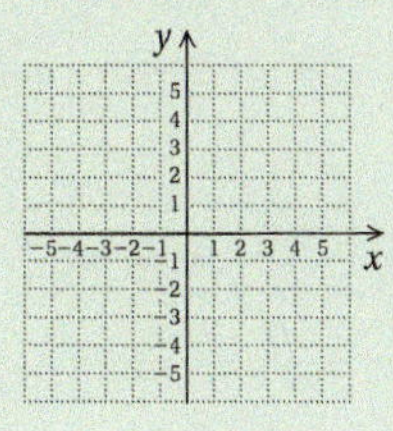

22. $y = \frac{1}{2}x$

x	y	(x, y)
4		
2		
0		
−2		
−4		
−1		

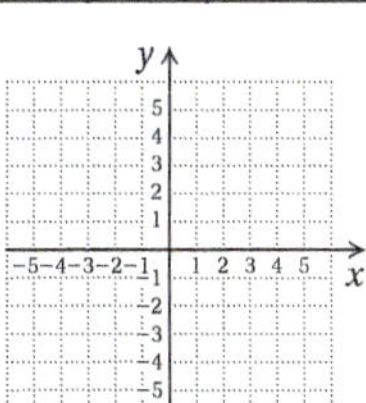

Answers

21.

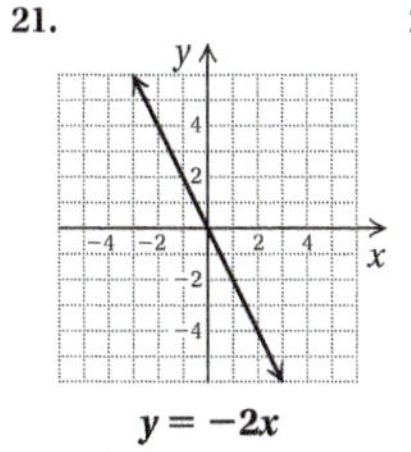

22.

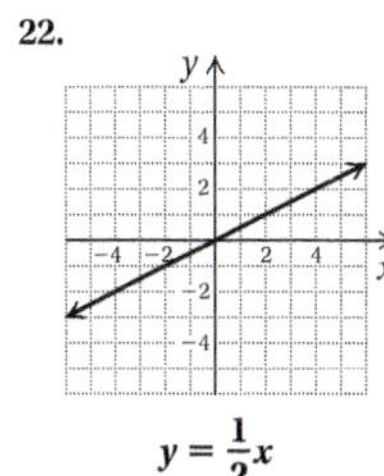

Guided Solution:
21. 2, 2, −2, 1, −6, −6

Complete each table and graph.

23. $y = 2x + 3$

x	y	(x, y)

24. $y = -\frac{1}{2}x - 3$

x	y	(x, y)

Note that all three points line up. If they did not, we would know that we had made a mistake. When only two points are plotted, a mistake is harder to detect. We use a ruler or another straightedge to draw a line through the points. Every point on the line represents a solution of $y = -3x + 1$.

◀ **Do Exercises 23 and 24.**

In Example 6, we saw that $(0, 0)$ is a solution of $y = 2x$. It is also the point at which the graph crosses the y-axis. Similarly, in Example 7, we saw that $(0, 1)$ is a solution of $y = -3x + 1$. It is also the point at which the graph crosses the y-axis. A generalization can be made: If x is replaced with 0 in the equation $y = mx + b$, then the corresponding y-value is $m \cdot 0 + b$, or b. Thus any equation of the form $y = mx + b$ has a graph that passes through the point $(0, b)$. Since $(0, b)$ is the point at which the graph crosses the y-axis, it is called the ***y*-intercept**. Sometimes, for convenience, we simply refer to b as the y-intercept.

y-INTERCEPT

The graph of the equation $y = mx + b$ passes through the ***y*-intercept** $(0, b)$.

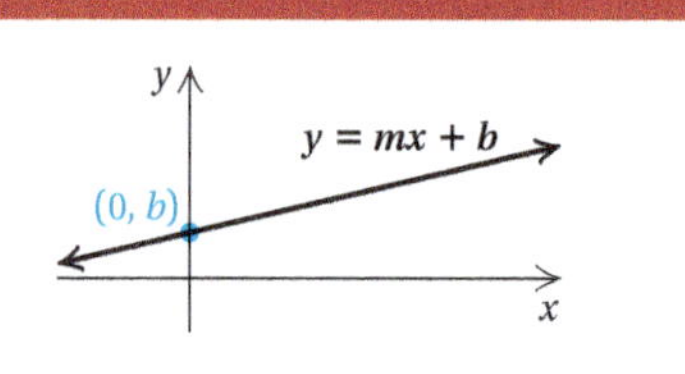

EXAMPLE 8 Graph $y = \frac{2}{5}x + 4$ and identify the y-intercept.

We select a value for x, compute y, and form an ordered pair. Then we repeat the process for other choices of x. In this case, using multiples of 5 avoids fractions. We try to avoid graphing ordered pairs with fractions because they are difficult to graph accurately.

If $x = 0$, then $y = \frac{2}{5} \cdot 0 + 4 = 4$, and $(0, 4)$ is a solution.

If $x = 5$, then $y = \frac{2}{5} \cdot 5 + 4 = 6$, and $(5, 6)$ is a solution.

If $x = -5$, then $y = \frac{2}{5} \cdot (-5) + 4 = 2$, and $(-5, 2)$ is a solution.

The following table lists these solutions. Next, we plot the points and see that they form a line. Finally, we draw and label the line.

x	y $y = \frac{2}{5}x + 4$	(x, y)
0	4	$(0, 4)$
5	6	$(5, 6)$
-5	2	$(-5, 2)$

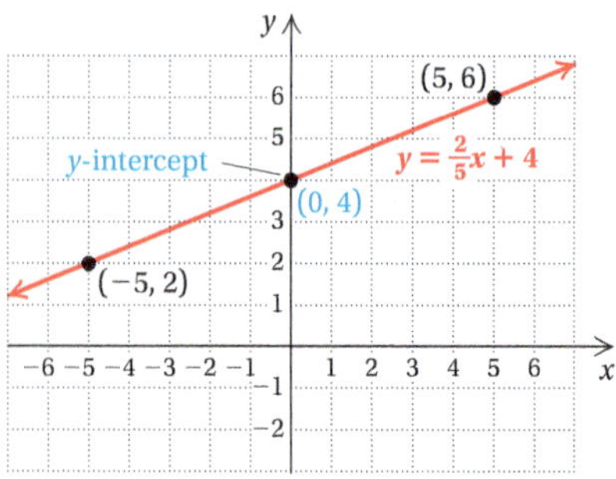

Answers

23.

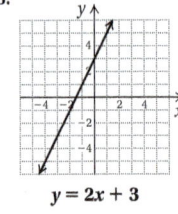

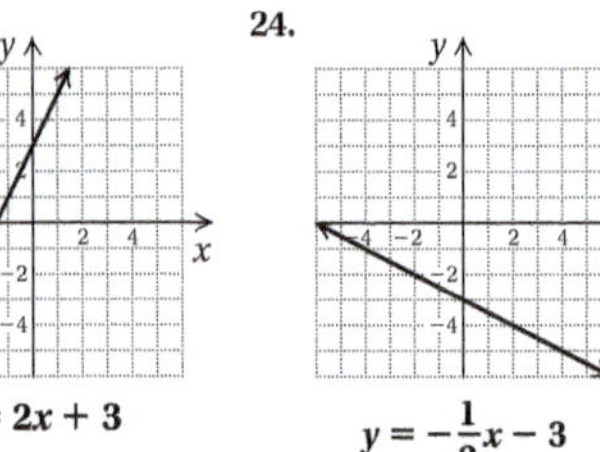

24. $y = -\frac{1}{2}x - 3$

We see that (0, 4) is a solution of $y = \frac{2}{5}x + 4$. It is the y-intercept. Because the equation is in the form $y = mx + b$, we can read the y-intercept directly from the equation as follows:

$$y = \frac{2}{5}x + 4 \qquad (0, 4) \text{ is the } y\text{-intercept.}$$

Do Exercises 25 and 26. ▶

Calculating ordered pairs is generally easiest when y is isolated on one side of the equation, as in $y = mx + b$. To graph an equation in which y is not isolated, we can use the addition and multiplication principles to solve for y. (See Sections 10.3 and 10.4.)

EXAMPLE 9 Graph $3y + 5x = 0$ and identify the y-intercept.

To find an equivalent equation in the form $y = mx + b$, we solve for y:

$$3y + 5x = 0$$
$$3y + 5x - 5x = 0 - 5x \qquad \text{Subtracting } 5x$$
$$3y = -5x \qquad \text{Collecting like terms}$$
$$\frac{3y}{3} = \frac{-5x}{3} \qquad \text{Dividing by 3}$$
$$y = -\frac{5}{3}x.$$

Because all the equations above are equivalent, we can use $y = -\frac{5}{3}x$ to draw the graph of $3y + 5x = 0$. To graph $y = -\frac{5}{3}x$, we select x-values and compute y-values. In this case, if we select multiples of 3, we can avoid fractions.

If $x = 0$, then $y = -\frac{5}{3} \cdot 0 = 0$.

If $x = 3$, then $y = -\frac{5}{3} \cdot 3 = -5$.

If $x = -3$, then $y = -\frac{5}{3} \cdot (-3) = 5$.

We list these solutions in a table. Next, we plot the points and see that they form a line. Finally, we draw and label the line. The y-intercept is (0, 0).

x	y	
0	0	← y-intercept
3	−5	
−3	5	

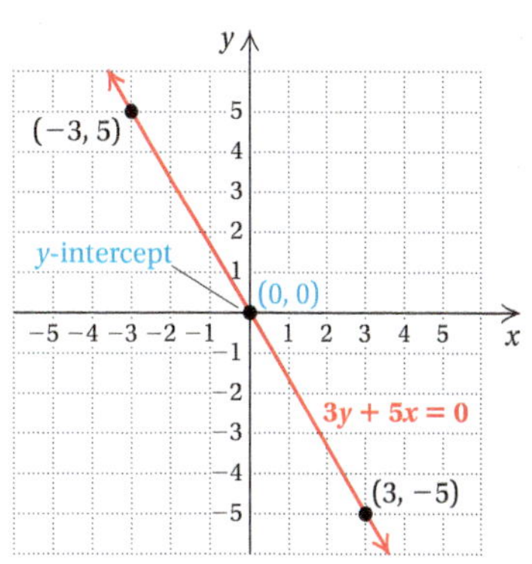

Do Exercises 27 and 28. ▶

Graph each equation and identify the y-intercept.

25. $y = \frac{3}{5}x + 2$

x	y	(x, y)

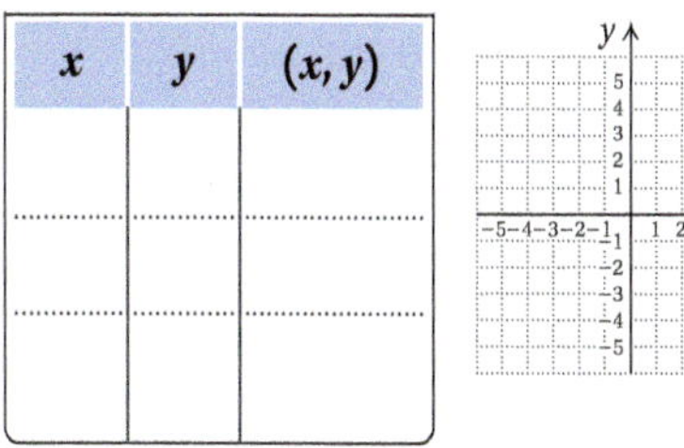

26. $y = -\frac{3}{5}x - 1$

x	y	(x, y)

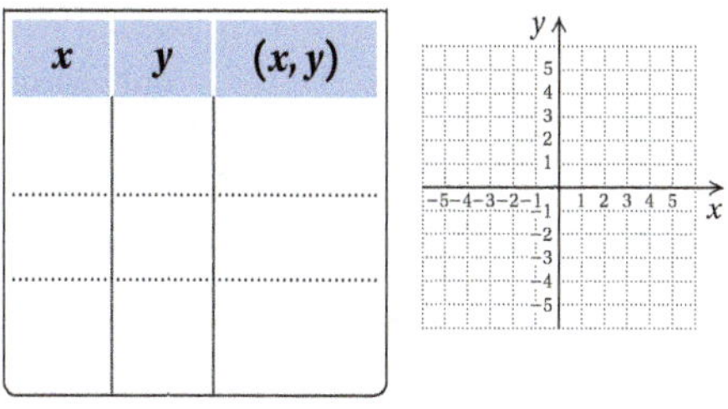

Graph each equation and identify the y-intercept.

27. $5y + 4x = 0$

x	y
0	

28. $4y = 3x$

x	y
0	

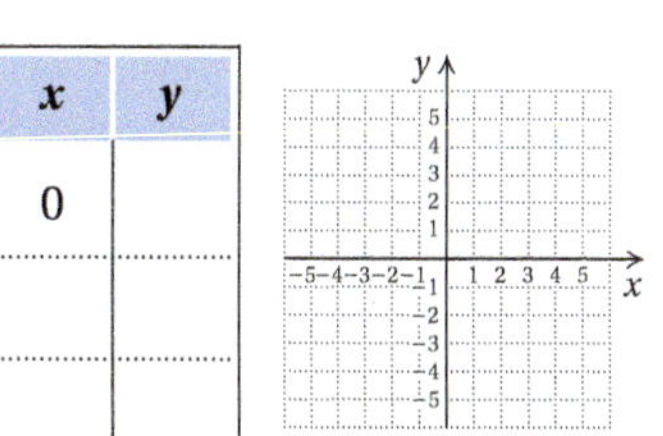

Answers

Answers to Margin Exercises 25–28 are on p. 784.

EXAMPLE 10 Graph $4y + 3x = -8$ and identify the y-intercept.

To find an equivalent equation in the form $y = mx + b$, we solve for y:

$$4y + 3x = -8$$
$$4y + 3x - 3x = -8 - 3x \quad \text{Subtracting } 3x$$
$$4y = -3x - 8 \quad \text{Simplifying}$$
$$\frac{1}{4} \cdot 4y = \frac{1}{4} \cdot (-3x - 8) \quad \text{Multiplying by } \tfrac{1}{4} \text{ or dividing by 4}$$
$$y = \frac{1}{4} \cdot (-3x) - \frac{1}{4} \cdot 8 \quad \text{Using the distributive law}$$
$$y = -\frac{3}{4}x - 2. \quad \text{Simplifying}$$

Thus, $4y + 3x = -8$ is equivalent to $y = -\frac{3}{4}x - 2$. The y-intercept is $(0, -2)$. We find two other pairs using multiples of 4 for x to avoid fractions. We then complete and label the graph as shown.

x	y	
0	-2	← y-intercept
4	-5	
-4	1	

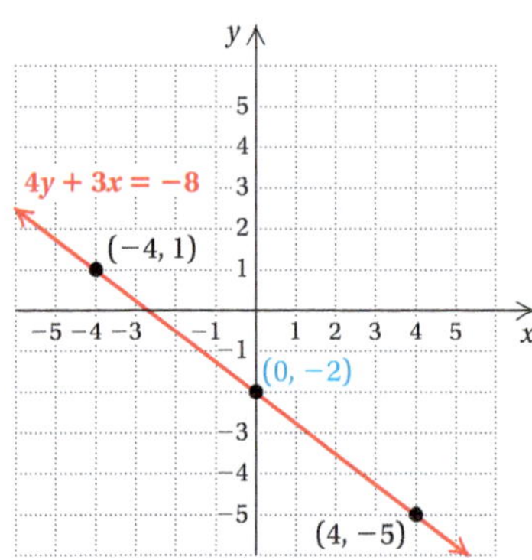

◀ Do Exercises 29 and 30.

Graph each equation and identify the y-intercept.

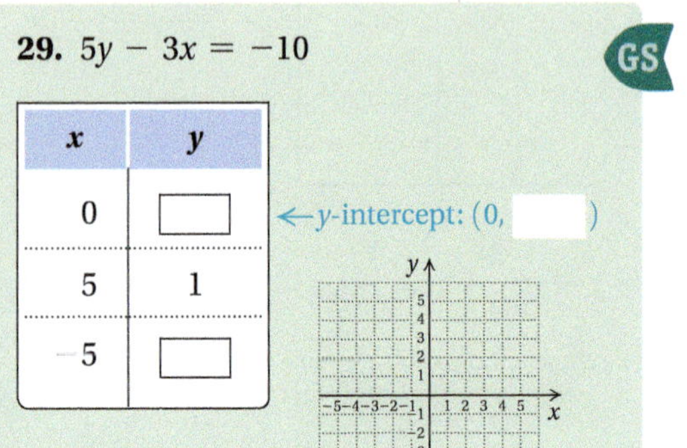

29. $5y - 3x = -10$ GS

x	y	
0	☐	← y-intercept: (0, ☐)
5	1	
-5	☐	

30. $5y + 3x = 20$

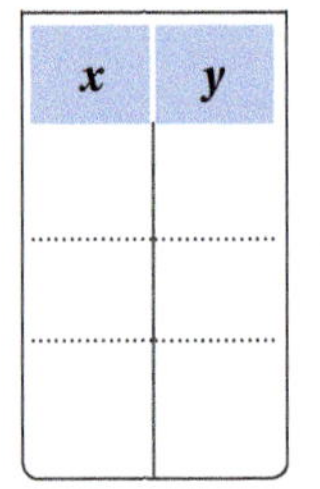

x	y

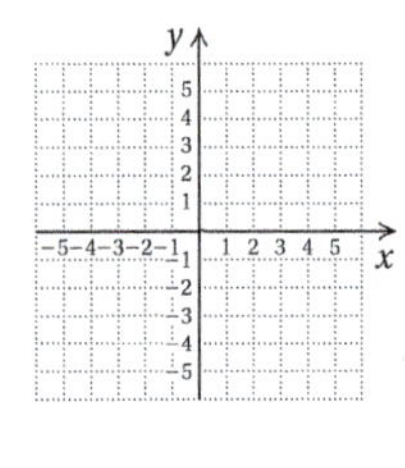

e APPLICATIONS OF LINEAR EQUATIONS

Mathematical concepts become more understandable through visualization. Throughout this text, you will occasionally see the heading Algebraic–Graphical Connection, as in Example 11, which follows. In this feature, the algebraic approach is enhanced and expanded with a graphical connection. Relating a solution of an equation to a graph can often give added meaning to the algebraic solution.

EXAMPLE 11 *World Population.* The world population, in billions, is estimated and projected by the equation

$$y = 0.072x + 4.593,$$

where x is the number of years since 1980. That is, $x = 0$ corresponds to 1980, $x = 12$ corresponds to 1992, and so on.

Source: U.S. Census Bureau

Answers

25.

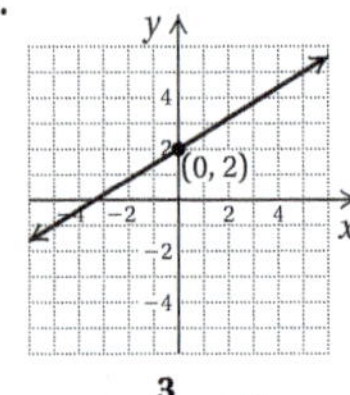

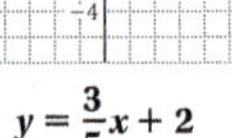

$y = \frac{3}{5}x + 2$

26.

(0, −1)

$y = -\frac{3}{5}x - 1$

27.

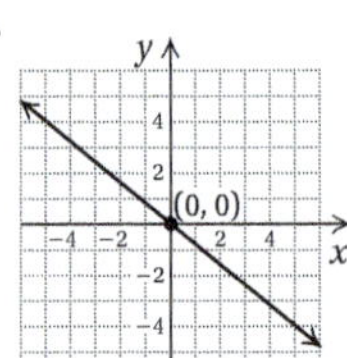

$5y + 4x = 0$

28.

(0, 0)

$4y = 3x$

29.

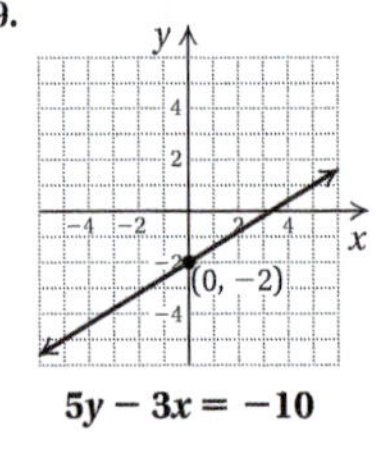

$5y - 3x = -10$

30.

(0, 4)

$5y + 3x = 20$

Guided Solution:
29. $-2, -5, -2$

a) Estimate the world population in 1980 and in 2005. Then project the population in 2030.

b) Graph the equation and then use the graph to estimate the world population in 2015.

c) In what year could we project the world population to be 7.761 billion?

a) The years 1980, 2005, and 2030 correspond to $x = 0$, $x = 25$, and $x = 50$, respectively. We substitute 0, 25, and 50 for x and then calculate y:

$$y = 0.072(0) + 4.593 = 0 + 4.593 = 4.593;$$
$$y = 0.072(25) + 4.593 = 1.8 + 4.593 = 6.393;$$
$$y = 0.072(50) + 4.593 = 3.6 + 4.593 = 8.193.$$

The world population in 1980, in 2005, and in 2030 is estimated and projected to be 4.593 billion, 6.393 billion, and 8.193 billion, respectively.

ALGEBRAIC GRAPHICAL CONNECTION

b) We have three ordered pairs from part (a). We plot these points and see that they line up. Thus our calculations are probably correct. Since we are considering only the year 2015 and the number of years since 1980 ($x \geq 0$) and since the population, in billions, for those years will be positive ($y > 0$), we need only the first quadrant for the graph. We use the three points we have plotted to draw a straight line. (See Figure 1.)

To use the graph to estimate world population in 2015, we first note in Figure 2 that this year corresponds to $x = 35$. We need to determine which y-value is paired with $x = 35$. We locate the point on the graph by moving up vertically from $x = 35$, and then find the value on the y-axis that corresponds to that point. It appears that the world population in 2015 will be about 7.1 billion.

To find a more accurate value, we can simply substitute into the equation:

$$y = 0.072(35) + 4.593 = 7.113.$$

The world population in 2015 is estimated to be 7.113 billion.

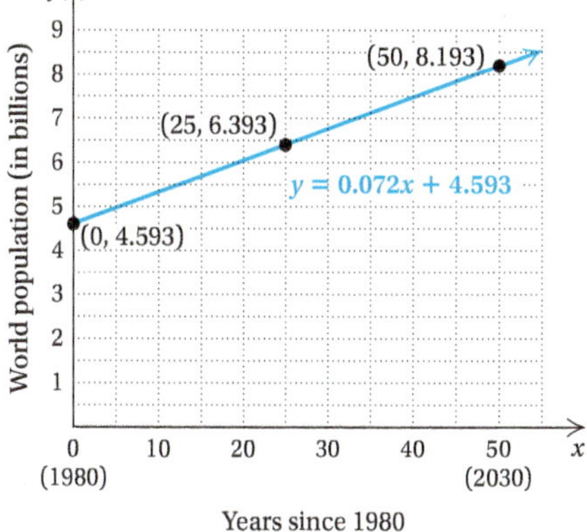

FIGURE 1

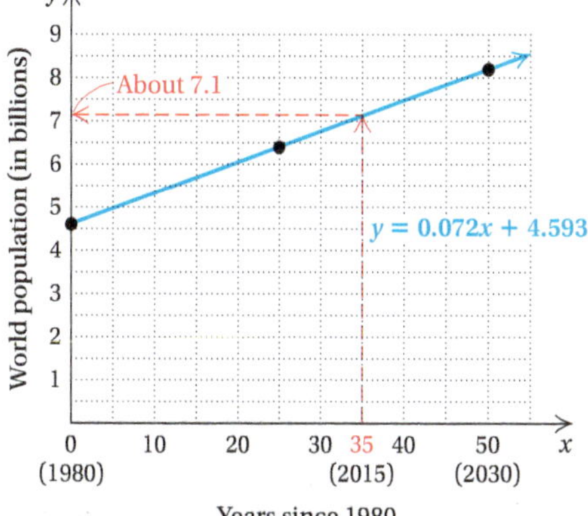

FIGURE 2

c) We substitute 7.761 for y and solve for x:

$$y = 0.072x + 4.593$$
$$7.761 = 0.072x + 4.593$$
$$3.168 = 0.072x$$
$$44 = x.$$

In 44 years after 1980, or in 2024, the world population is projected to be approximately 7.761 billion.

Do Exercise 31 on the following page.

31. ***Milk Consumption.*** Milk consumption per capita (per person) in the United States can be estimated by

$M = -0.183t + 27.776,$

where M is the consumption, in gallons, t years since 1985.

Source: U.S. Department of Agriculture

a) Find the per capita consumption of milk in 1985, in 1995, and in 2015.

b) Graph the equation and use the graph to estimate milk consumption in 2010.

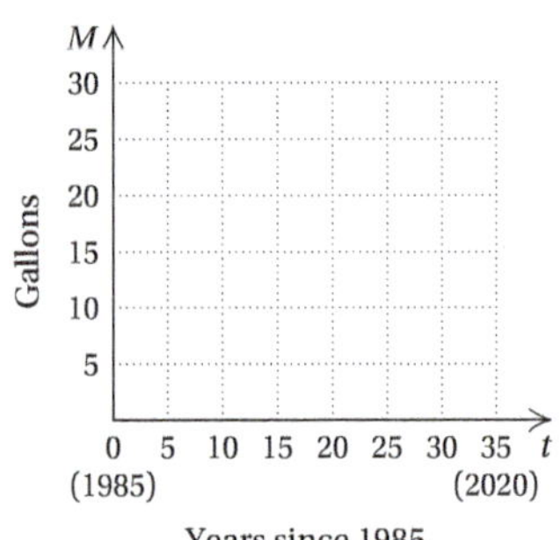

c) In which year would the per capita consumption of milk be 21.737 gal?

Answers

31. (a) 27.776 gal; 25.946 gal; 22.286 gal; **(b)** about 23.2 gal;

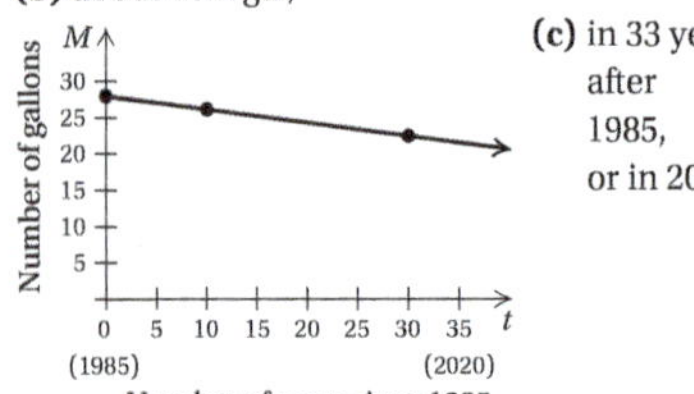

(c) in 33 years after 1985, or in 2018

Many equations in two variables have graphs that are not straight lines. Three such nonlinear graphs are shown below. We will cover some such graphs in the optional Calculator Corners throughout the text and in Chapter 17.

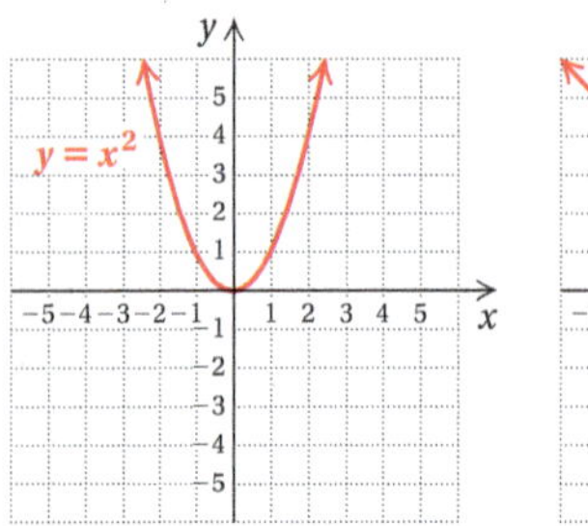

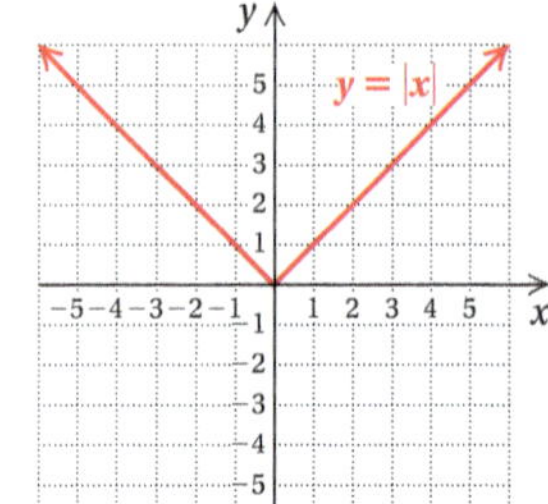

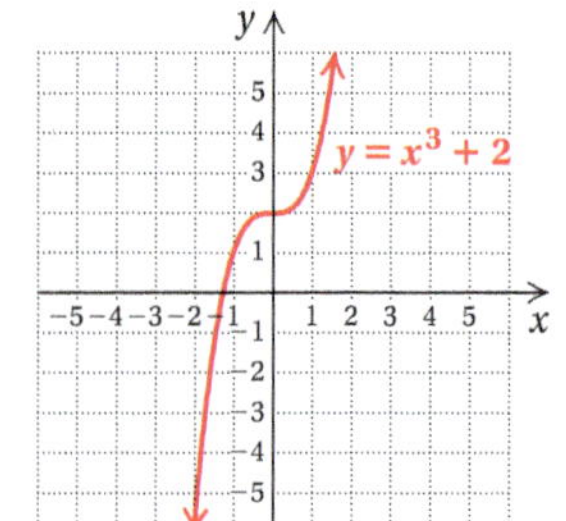

CALCULATOR CORNER

Graphing Equations Equations must be solved for y before they can be graphed on the TI-84 Plus. Consider the equation $3x + 2y = 6$. Solving for y, we get $y = \frac{6 - 3x}{2}$. We enter this equation as $y_1 = (6 - 3x)/2$ on the equation-editor screen. Then we select the standard viewing window and display the graph.

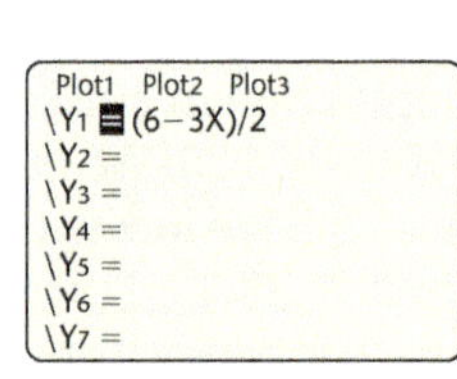

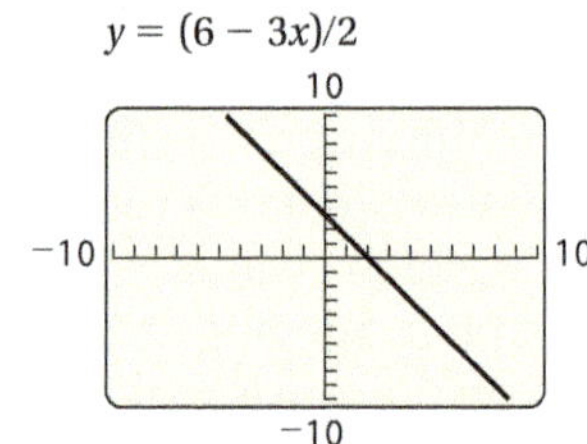

EXERCISES: Graph each equation in the standard viewing window $[-10, 10, -10, 10]$, with Xscl = 1 and Yscl = 1.

1. $y = -5x + 3$

2. $y = 4x - 5$

3. $4x - 5y = -10$

4. $5y + 5 = -3x$

11.1 Exercise Set

For Extra Help

MyMathLab®

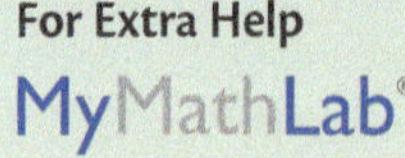

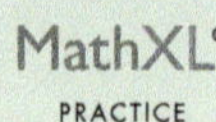

Reading Check

Determine whether each statement is true or false.

RC1. The graph of a linear equation is always a straight line.

RC2. The point $(1, 0)$ is in quadrant I and in quadrant IV.

RC3. The ordered pairs $(4, -7)$ and $(-7, 4)$ name the same point.

RC4. To plot the point $(-3, 5)$, start at the origin and move horizontally to -3. Then move up vertically 5 units and make a "dot."

a

1. Plot these points.

$(2, 5)$ $(-1, 3)$ $(3, -2)$ $(-2, -4)$
$(0, 4)$ $(0, -5)$ $(5, 0)$ $(-5, 0)$

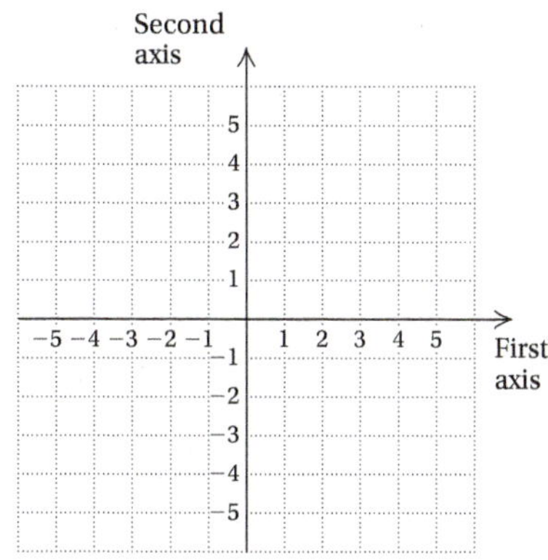

2. Plot these points.

$(4, 4)$ $(-2, 4)$ $(5, -3)$ $(-5, -5)$
$(0, 2)$ $(0, -4)$ $(3, 0)$ $(-4, 0)$

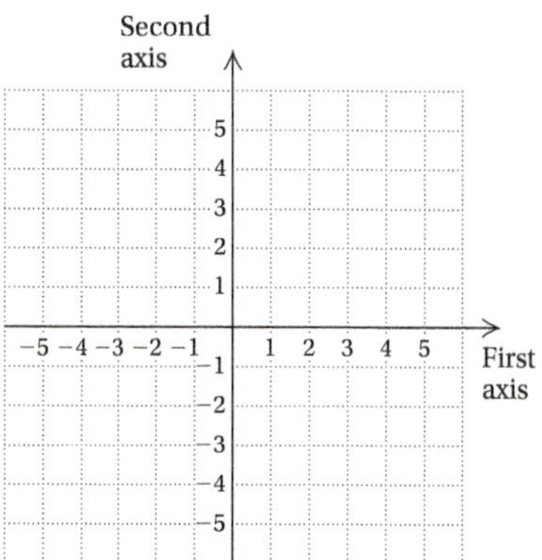

In which quadrant, if any, is each point located?

3. $(-5, 3)$

4. $(1, -12)$

5. $(100, -1)$

6. $(-2.5, 35.6)$

7. $(-6, -29)$

8. $(3.6, 105.9)$

9. $(3.8, 0)$

10. $(0, -492)$

11. $\left(-\frac{1}{3}, \frac{15}{7}\right)$

12. $\left(-\frac{2}{3}, -\frac{9}{8}\right)$

13. $\left(12\frac{7}{8}, -1\frac{1}{2}\right)$

14. $\left(23\frac{5}{8}, 81.74\right)$

In which quadrant(s) can the point described be located?

15. The first coordinate is negative and the second coordinate is positive.

16. The first and second coordinates are positive.

17. The first coordinate is positive.

18. The second coordinate is negative.

19. The first and second coordinates are equal.

20. The first coordinate is the additive inverse of the second coordinate.

b Find the coordinates of points *A*, *B*, *C*, *D*, and *E*.

21.

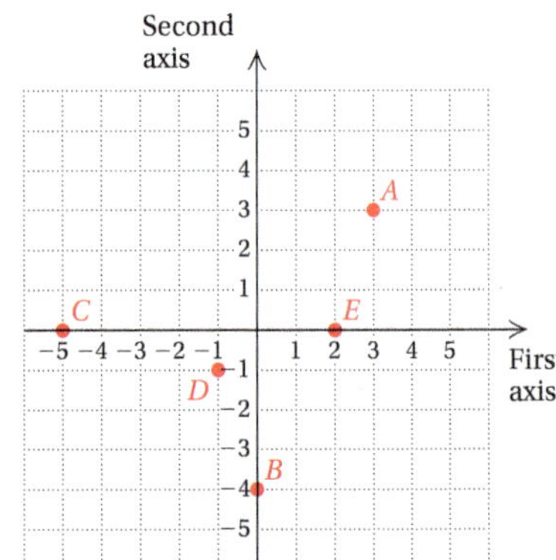

22.

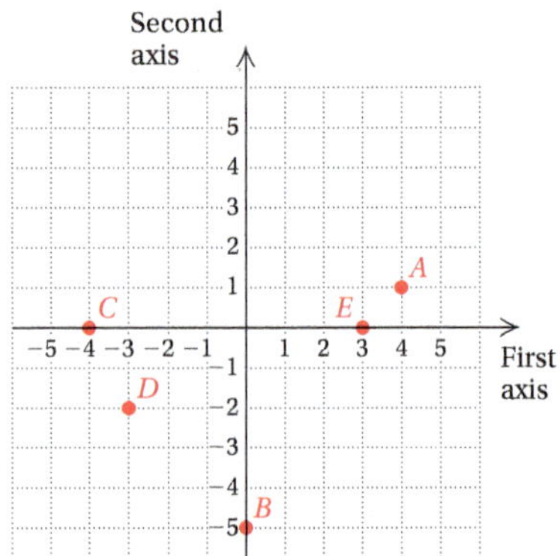

c Determine whether the given ordered pair is a solution of the equation.

23. $(2, 9);\ y = 3x - 1$

24. $(1, 7);\ y = 2x + 5$

25. $(4, 2);\ 2x + 3y = 12$

26. $(0, 5);\ 5x - 3y = 15$

27. $(3, -1);\ 3a - 4b = 13$

28. $(-5, 1);\ 2p - 3q = -13$

In each of Exercises 29–34, an equation and two ordered pairs are given. Show that each pair is a solution of the equation. Then use the graph of the equation to determine another solution. Answers may vary.

29. $y = x - 5;\ (4, -1)$ and $(1, -4)$

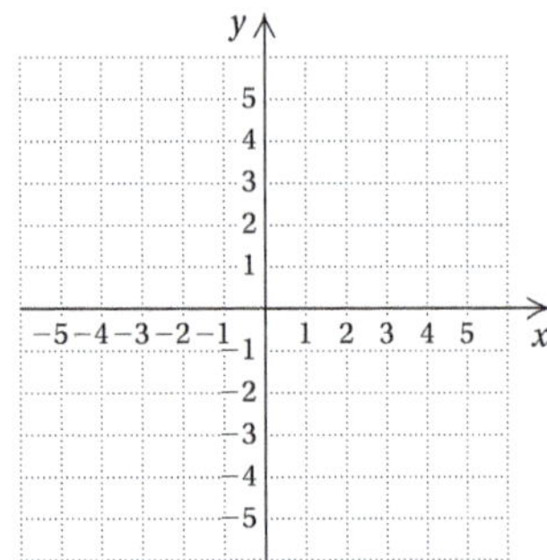

30. $y = x + 3;\ (-1, 2)$ and $(3, 6)$

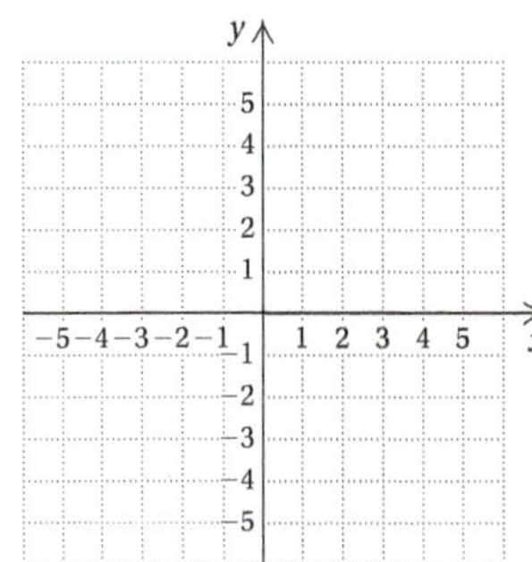

31. $y = \frac{1}{2}x + 3;\ (4, 5)$ and $(-2, 2)$

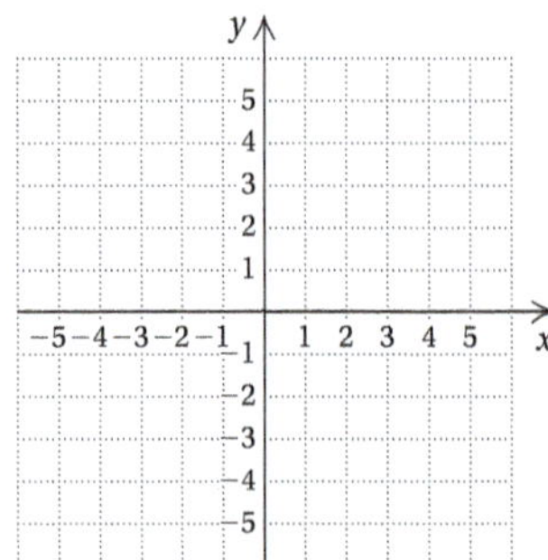

32. $3x + y = 7;\ (2, 1)$ and $(4, -5)$

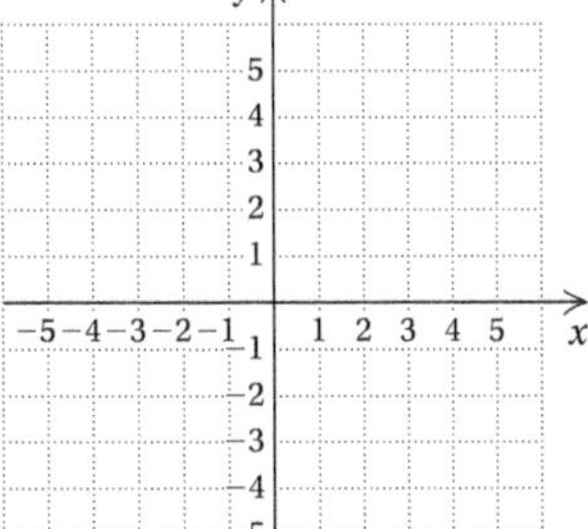

33. $4x - 2y = 10;\ (0, -5)$ and $(4, 3)$

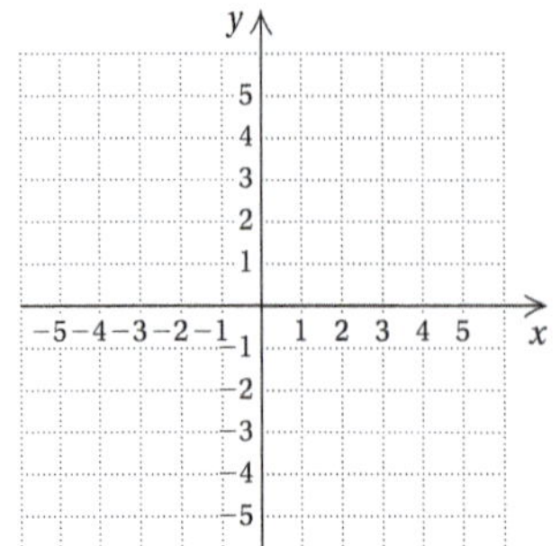

34. $6x - 3y = 3;\ (1, 1)$ and $(-1, -3)$

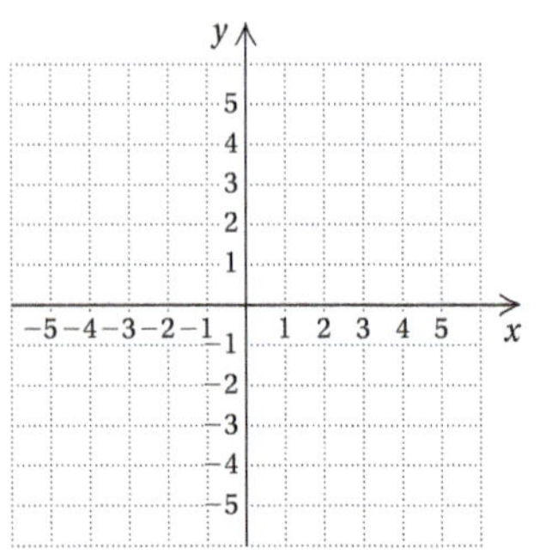

d Graph each equation and identify the y-intercept.

35. $y = x + 1$

x	y
−2	
−1	
0	
1	
2	
3	

36. $y = x - 1$

x	y
−2	
−1	
0	
1	
2	
3	

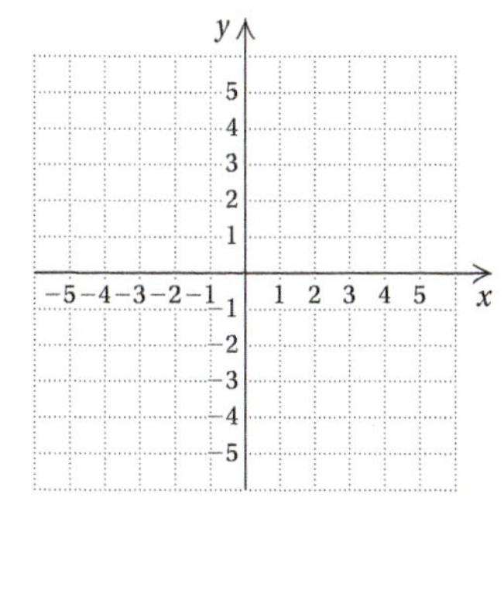

37. $y = x$

x	y
−2	
−1	
0	
1	
2	
3	

38. $y = -x$

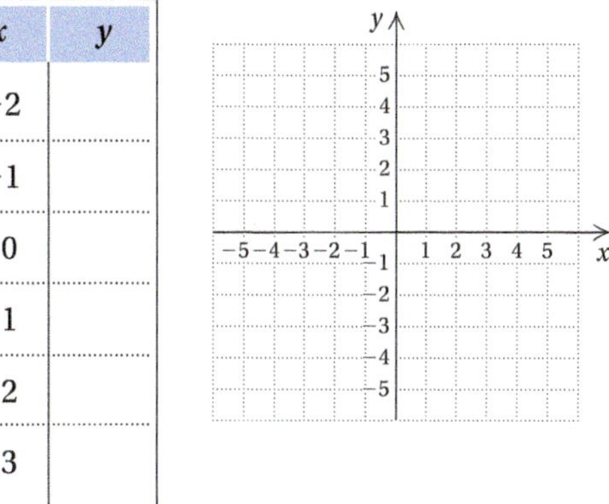

x	y
−2	
−1	
0	
1	
2	
3	

39. $y = \frac{1}{2}x$

x	y
−2	
0	
4	

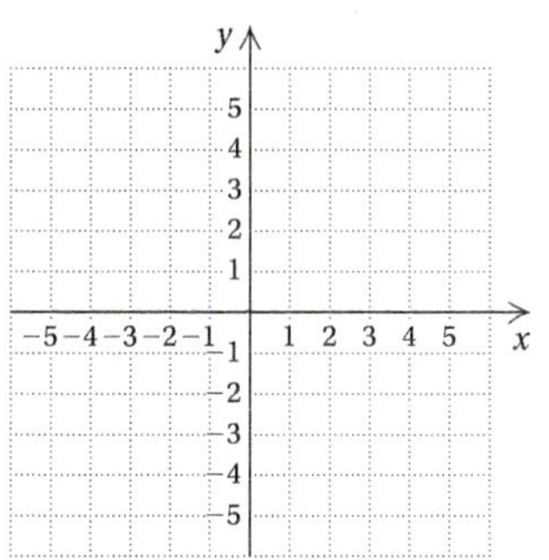

40. $y = \frac{1}{3}x$

x	y
−6	
0	
3	

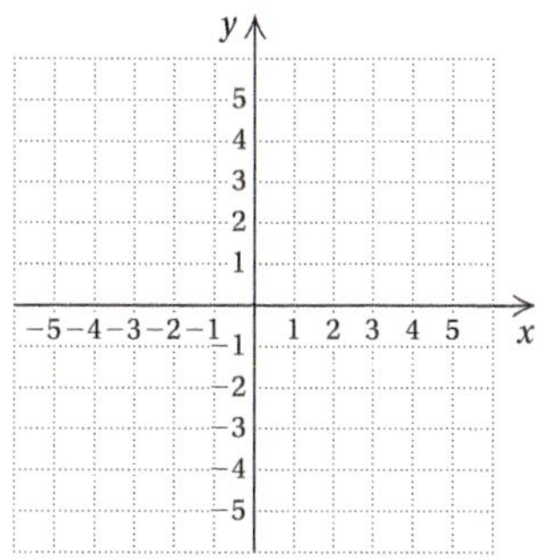

41. $y = x - 3$

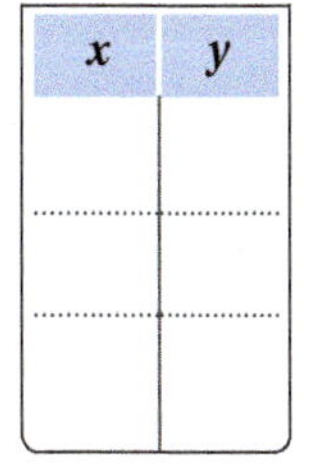

x	y

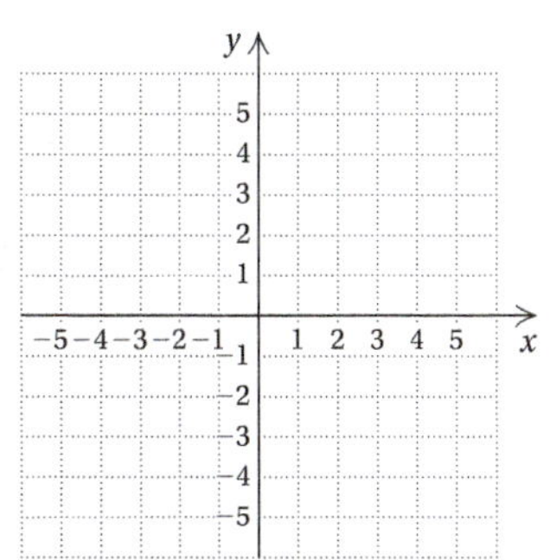

42. $y = x + 3$

x	y

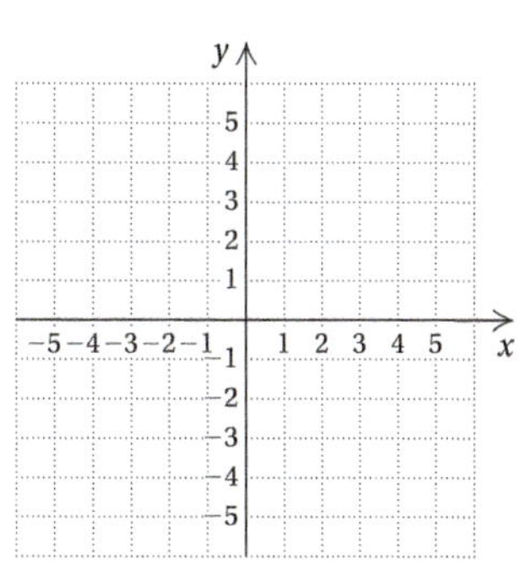

43. $y = 3x - 2$

x	y

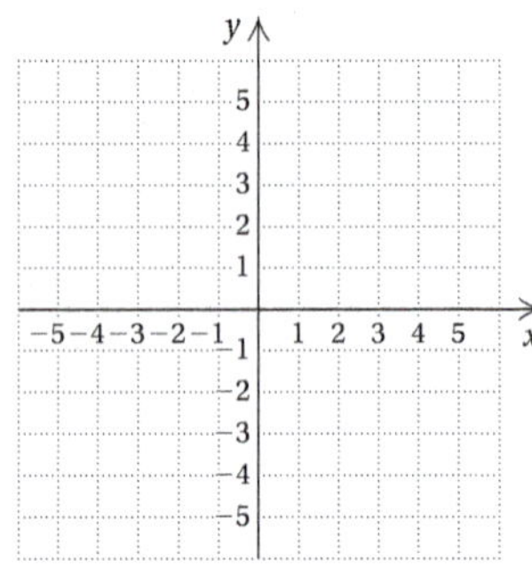

44. $y = 2x + 2$

x	y

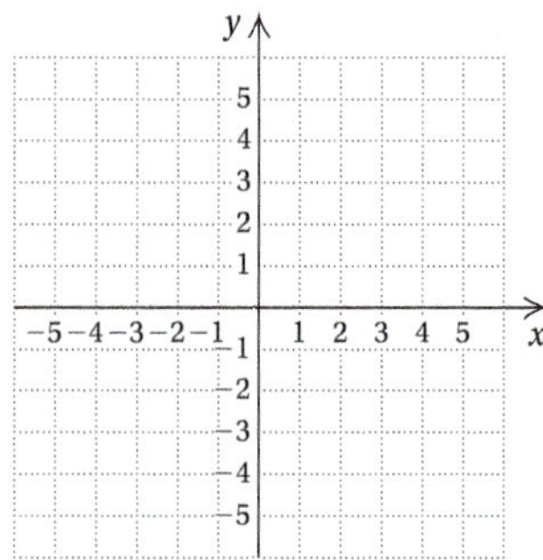

45. $y = \frac{1}{2}x + 1$

x	y

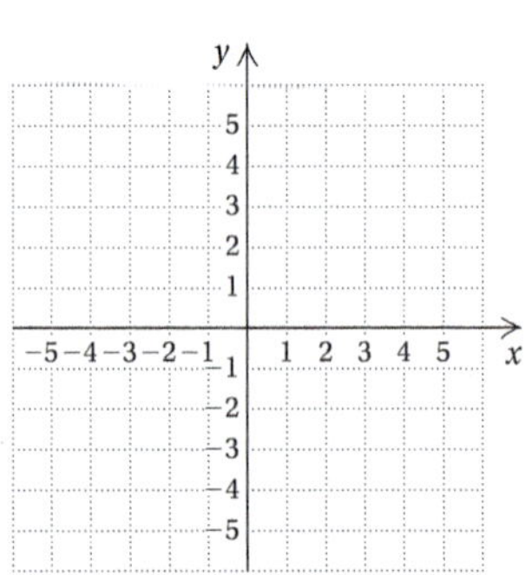

46. $y = \frac{1}{3}x - 4$

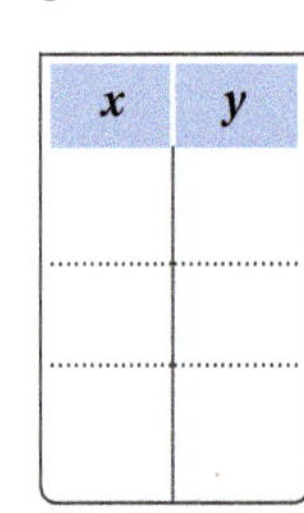

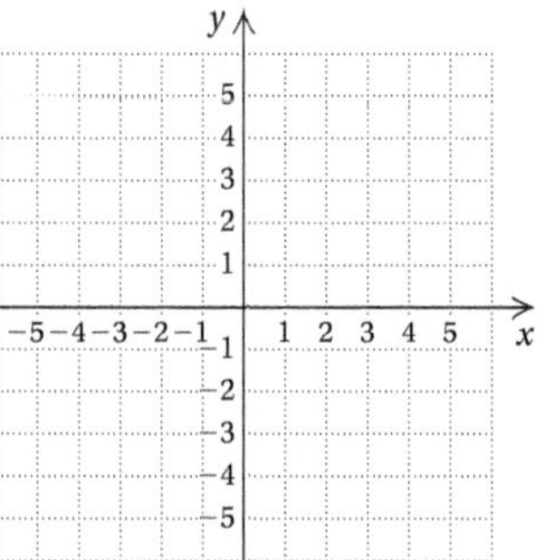

47. $x + y = -5$

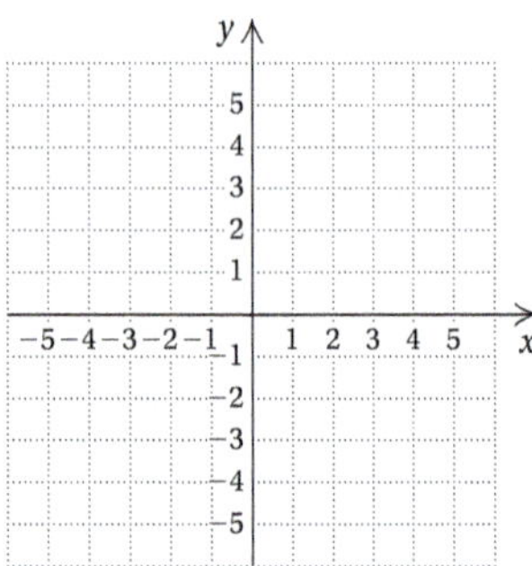

48. $x + y = 4$

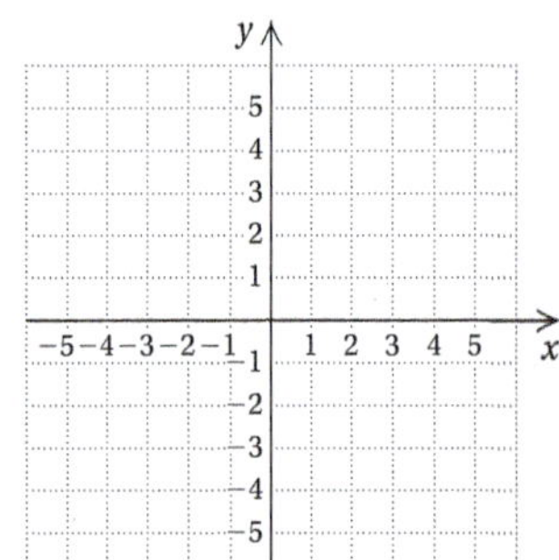

49. $y = \frac{5}{3}x - 2$

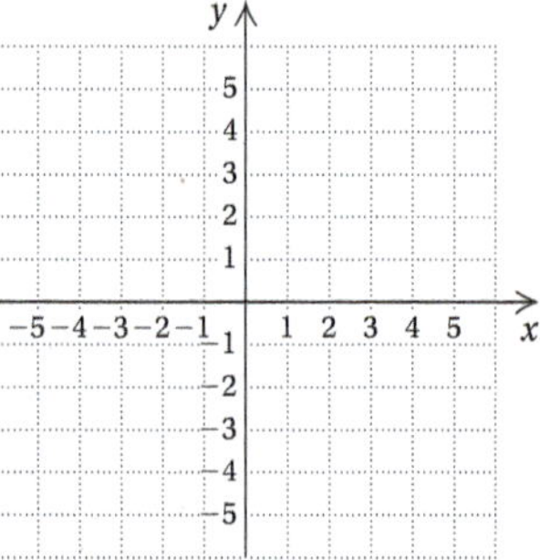

50. $y = \frac{5}{2}x + 3$

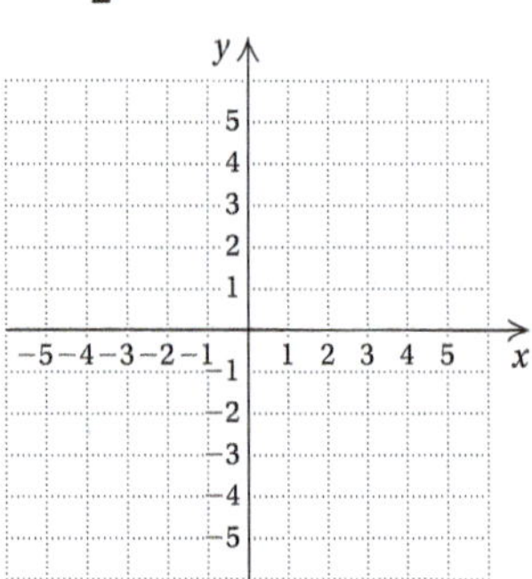

51. $x + 2y = 8$

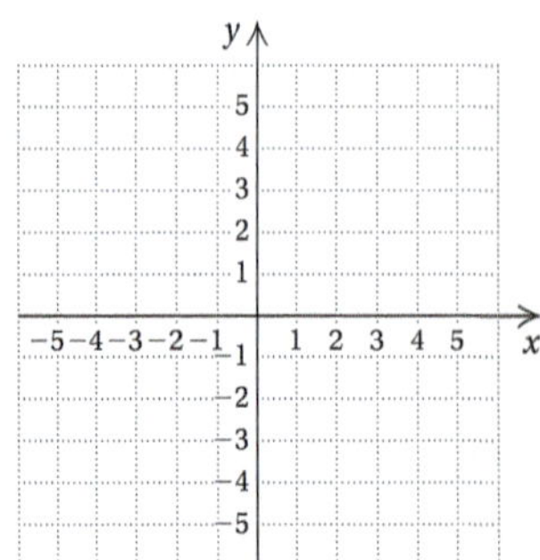

52. $x + 2y = -6$

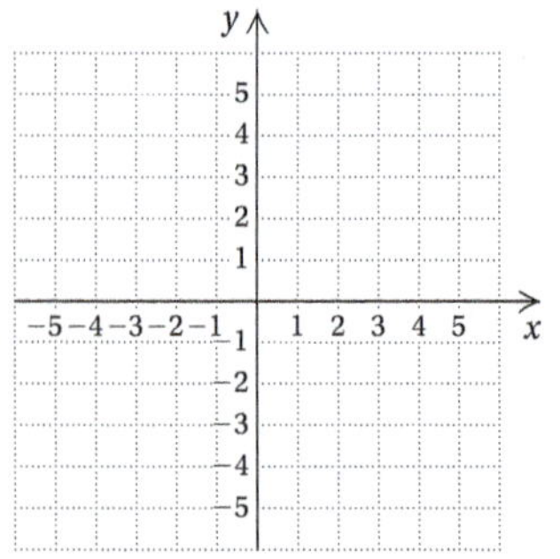

53. $y = \frac{3}{2}x + 1$

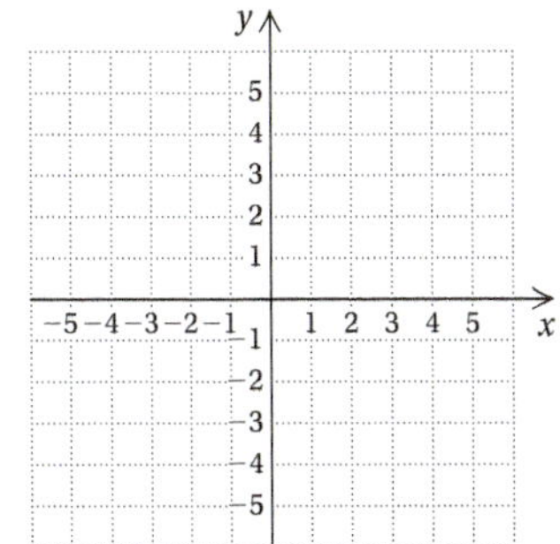

54. $y = -\frac{1}{2}x - 3$

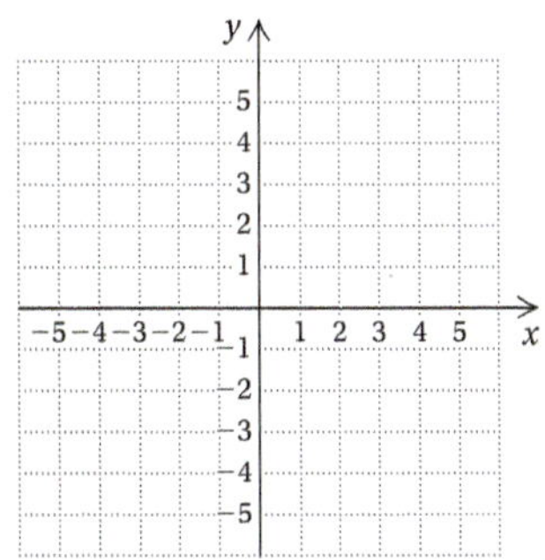

55. $8x - 2y = -10$

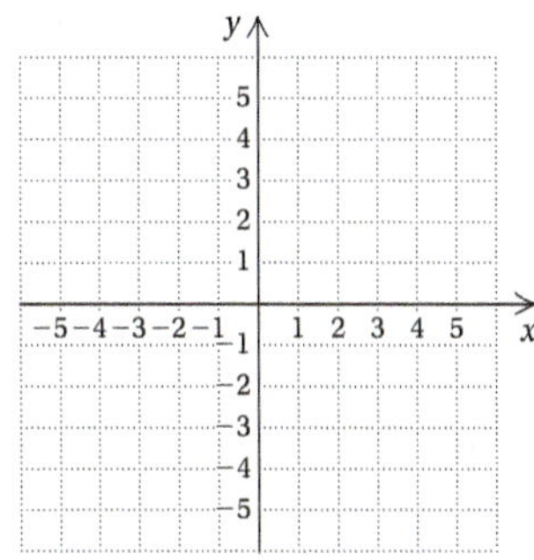

56. $6x - 3y = 9$

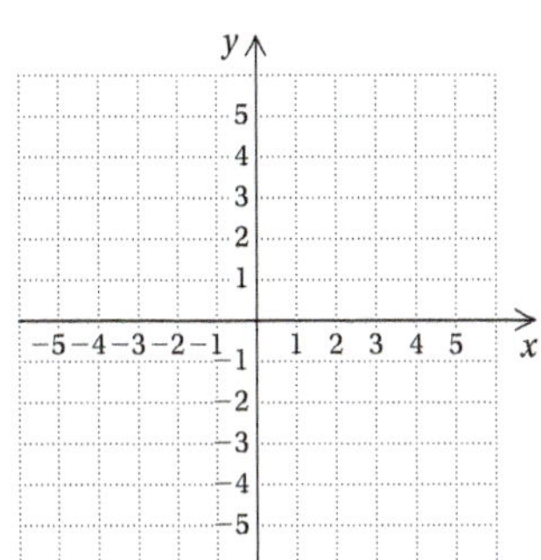

57. $8y + 2x = -4$

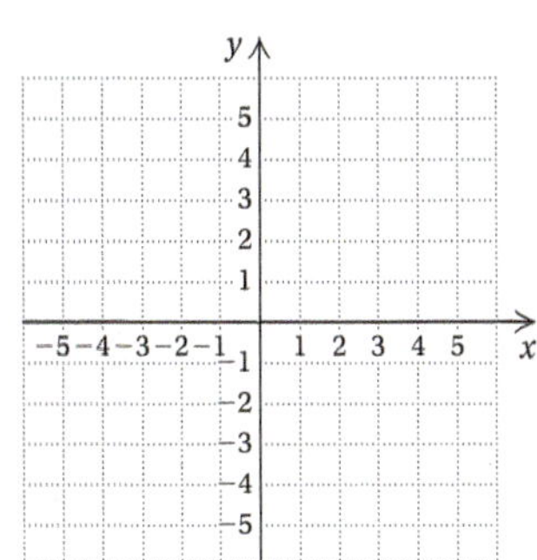

58. $6y + 2x = 8$

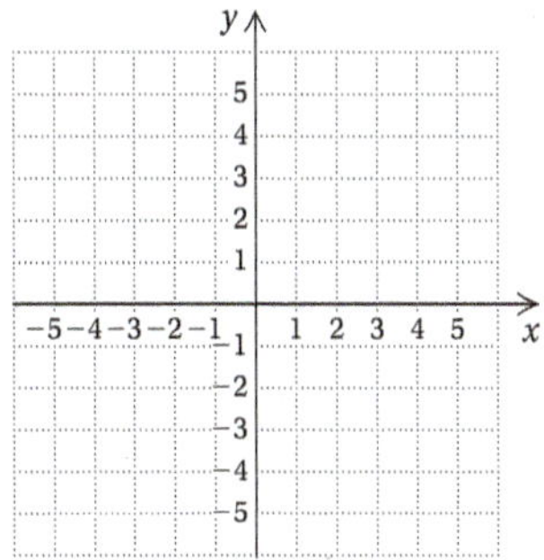

e Solve.

59. *Online Advertising.* Spending for online advertising is increasing. The amount A, in billions of dollars, spent worldwide for online advertising can be approximated and projected by

$$A = 12.83t + 68.38,$$

where t is the number of years since 2010.

a) Find the amount spent for online advertising in 2010, in 2014, and in 2015.

b) Graph the equation and use the graph to estimate the amount spent for online advertising in 2012.

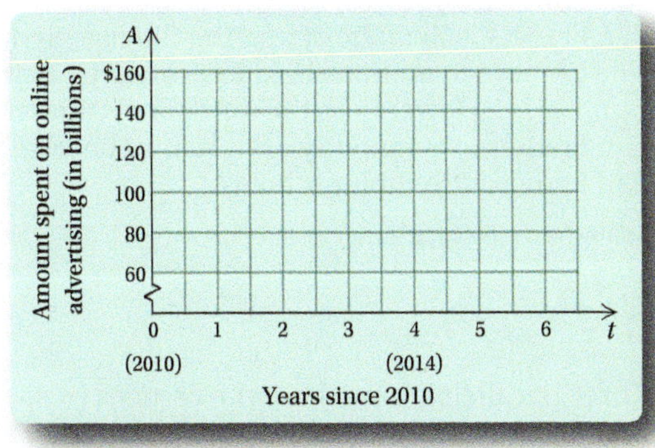

c) In what year will online advertising spending be about \$158.19 billion?

60. *Price of a New Car.* The average price P of a new car, in dollars, can be approximated and projected by

$$P = 426t + 25{,}710,$$

where t is the number of years since 2002.

Source: Edmunds.com

a) Find the average price of a new car in 2002, in 2006, and in 2011.

b) Graph the equation and use the graph to estimate the price of a new car in 2010.

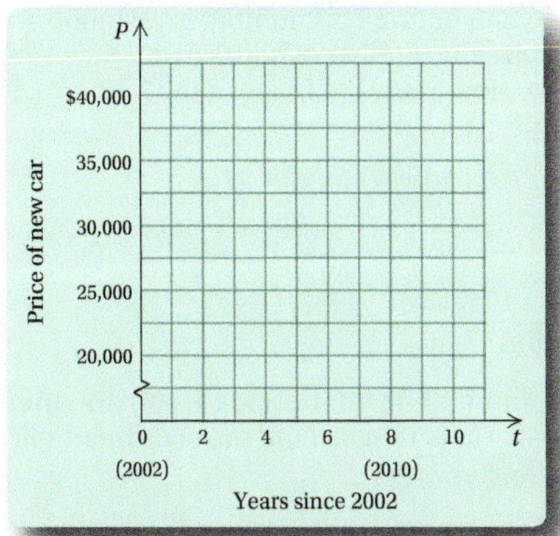

c) In what year will the average price of a new car be approximately \$35,000?

61. *Sheep and Lambs.* The number of sheep and lambs S, in millions, on farms in the United States has declined in recent years and can be approximated and projected by

$$S = -0.125t + 6.898,$$

where t is the number of years since 2000.

Source: U.S. Department of Agriculture

a) Find the number of sheep and lambs in 2000, in 2007, and in 2012.

b) Graph the equation and then use the graph to estimate the number of sheep and lambs in 2010.

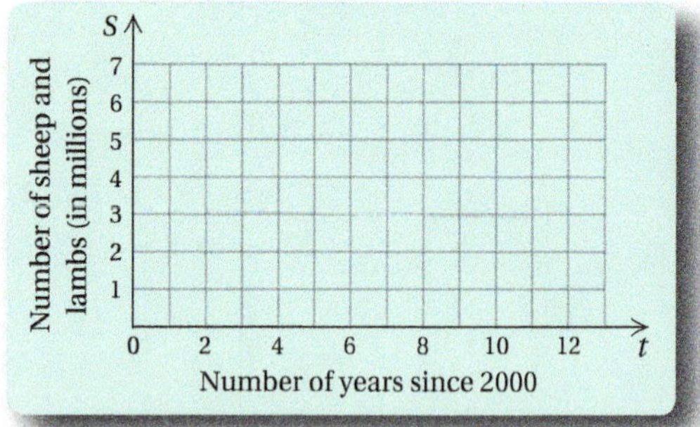

c) At this rate of decline, in what year will the number of sheep and lambs be 4.898 million?

62. *Record Temperature Drop.* On 22 January 1943, the temperature T, in degrees Fahrenheit, in Spearfish, South Dakota, could be approximated by

$$T = -2.15m + 54,$$

where m is the number of minutes since 9:00 that morning.

Source: *Information Please Almanac*

a) Find the temperature at 9:01 A.M., at 9:08 A.M., and at 9:20 A.M.

b) Graph the equation and use the graph to estimate the temperature at 9:15 A.M.

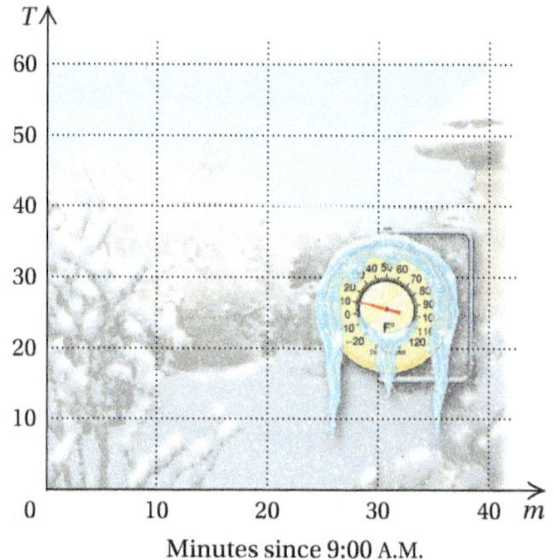

c) The temperature stopped dropping when it reached −4°F. At what time did this occur?

Skill Maintenance

Find the absolute value. [9.2e]

63. $|-12|$

64. $|4.89|$

65. $|0|$

66. $\left|-\frac{4}{5}\right|$

Solve. [10.3a]

67. $2x - 14 = 29$

68. $\frac{1}{3}t + 6 = -12$

69. $-10 = 1.2y + 2$

70. $4 - 5w = -16$

Solve. [10.6a]

71. *Books in Libraries.* The Library of Congress houses 33.5 million books. This number is 0.3 million more than twice the number of books in the New York Public Library. How many books are in the New York Public Library?

Source: American Library Association

72. *Busiest Orchestras.* In 2011, the orchestras who performed the greatest number of concerts were the San Francisco Symphony and the Chicago Symphony. The Chicago Symphony held 133 concerts. This number is 24 fewer than the number of concerts by the San Francisco Symphony. How many concerts did the San Francisco Symphony hold?

Source: www.bachtrack.com

Synthesis

73. The points $(-1, 1)$, $(4, 1)$, and $(4, -5)$ are three vertices of a rectangle. Find the coordinates of the fourth vertex.

74. Three parallelograms share the vertices $(-2, -3)$, $(-1, 2)$, and $(4, -3)$. Find the fourth vertex of each parallelogram.

75. Graph eight points such that the sum of the coordinates in each pair is 6.

76. Graph eight points such that the first coordinate minus the second coordinate is 1.

77. Find the perimeter of a rectangle whose vertices have coordinates $(5, 3)$, $(5, -2)$, $(-3, -2)$, and $(-3, 3)$.

78. Find the area of a triangle whose vertices have coordinates $(0, 9)$, $(0, -4)$, and $(5, -4)$.

More with Graphing and Intercepts

11.2

OBJECTIVES

a Find the intercepts of a linear equation, and graph using intercepts.

b Graph equations equivalent to those of the type $x = a$ and $y = b$.

SKILL TO REVIEW

Objective 10.3a: Solve equations using both the addition principle and the multiplication principle.

Solve.

1. $5x - 7 = -10$

2. $-20 = \frac{7}{4}x + 8$

a GRAPHING USING INTERCEPTS

In Section 11.1, we graphed linear equations of the form $Ax + By = C$ by first solving for y to find an equivalent equation in the form $y = mx + b$. We did so because it is then easier to calculate the y-value that corresponds to a given x-value. Another convenient way to graph $Ax + By = C$ is to use **intercepts**. Look at the graph of $-2x + y = 4$ shown below.

The y-intercept is $(0, 4)$. It occurs where the line crosses the y-axis and thus will always have 0 as the first coordinate. The x-intercept is $(-2, 0)$. It occurs where the line crosses the x-axis and thus will always have 0 as the second coordinate.

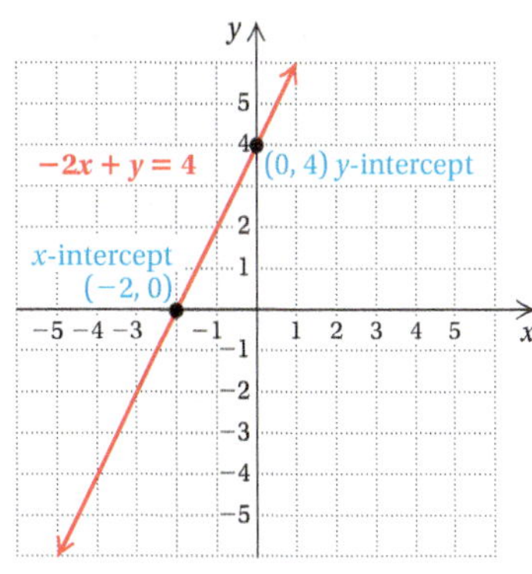

Do Margin Exercise 1. ▶

We find intercepts as follows.

INTERCEPTS

The **y-intercept** is $(0, b)$. To find b, let $x = 0$ and solve the equation for y.

The **x-intercept** is $(a, 0)$. To find a, let $y = 0$ and solve the equation for x.

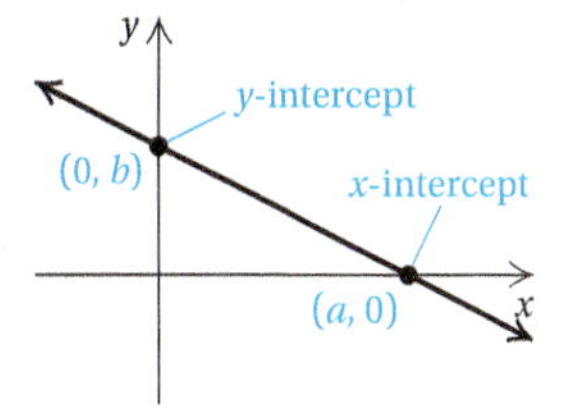

Now let's draw a graph using intercepts.

EXAMPLE 1 Consider $4x + 3y = 12$. Find the intercepts. Then graph the equation using the intercepts.

To find the y-intercept, we let $x = 0$. Then we solve for y:

$$4 \cdot 0 + 3y = 12$$
$$3y = 12$$
$$y = 4.$$

Thus, $(0, 4)$ is the y-intercept. Note that finding this intercept involves covering up the x-term and solving the rest of the equation for y.

To find the x-intercept, we let $y = 0$. Then we solve for x:

$$4x + 3 \cdot 0 = 12$$
$$4x = 12$$
$$x = 3.$$

1. Look at the graph shown below.

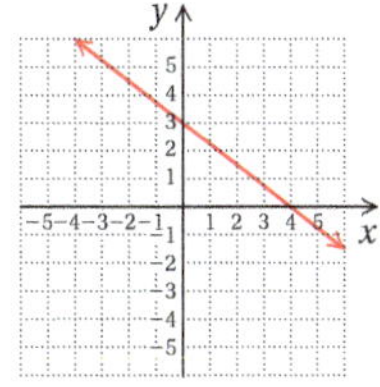

a) Find the coordinates of the y-intercept.

b) Find the coordinates of the x-intercept.

Answers

Skill to Review:

1. $-\frac{3}{5}$ **2.** -16

Margin Exercises:

1. (a) $(0, 3)$; **(b)** $(4, 0)$

Thus, (3, 0) is the x-intercept. Note that finding this intercept involves covering up the y-term and solving the rest of the equation for x.

We plot these points and draw the line, or graph.

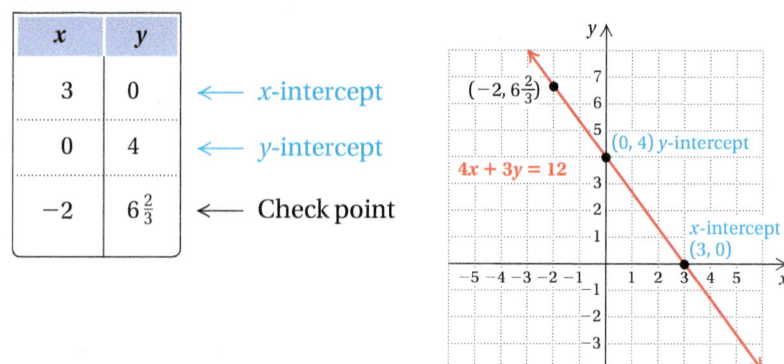

x | y
3 | 0 ← x-intercept
0 | 4 ← y-intercept
−2 | $6\frac{2}{3}$ ← Check point
$(-2, 6\frac{2}{3})$
(0, 4) y-intercept
4x + 3y = 12
x-intercept
(3, 0)

A third point should be used as a check. We substitute any convenient value for x and solve for y. In this case, we choose $x = -2$. Then

$$
\begin{aligned}
4(-2) + 3y &= 12 && \text{Substituting } -2 \text{ for } x\\
-8 + 3y &= 12\\
3y &= 20 && \text{Adding 8 on both sides}\\
y &= \tfrac{20}{3}, \text{ or } 6\tfrac{2}{3}. && \text{Solving for } y
\end{aligned}
$$

It appears that the point $(-2, 6\frac{2}{3})$ is on the graph, though graphing fraction values can be inexact. The graph is probably correct.

Do Exercises 2 and 3.

Graphs of equations of the type $y = mx$ pass through the origin. Thus the x-intercept and the y-intercept are the same, (0, 0). In such cases, we must calculate another point in order to complete the graph. A third point would also need to be calculated if a check is desired.

EXAMPLE 2 Graph: $y = 3x$.

We know that (0, 0) is both the x-intercept and the y-intercept. We calculate values at two other points and complete the graph, knowing that it passes through the origin (0, 0).

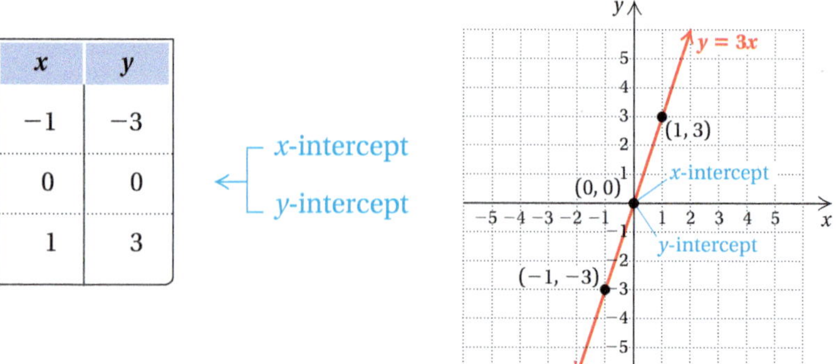

Do Exercises 4 and 5 on the following page.

For each equation, find the intercepts. Then graph the equation using the intercepts.

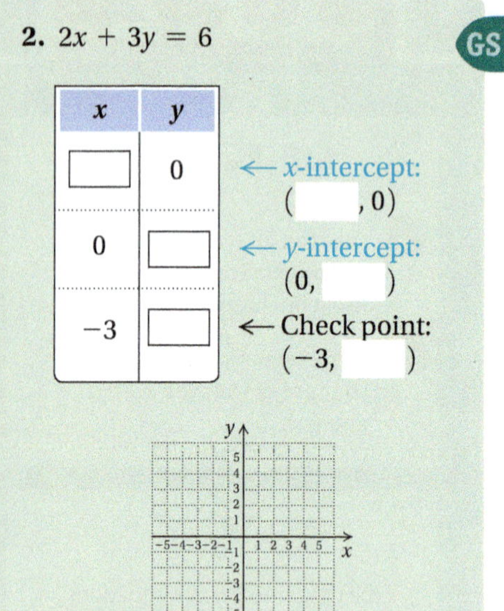

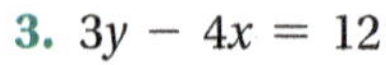

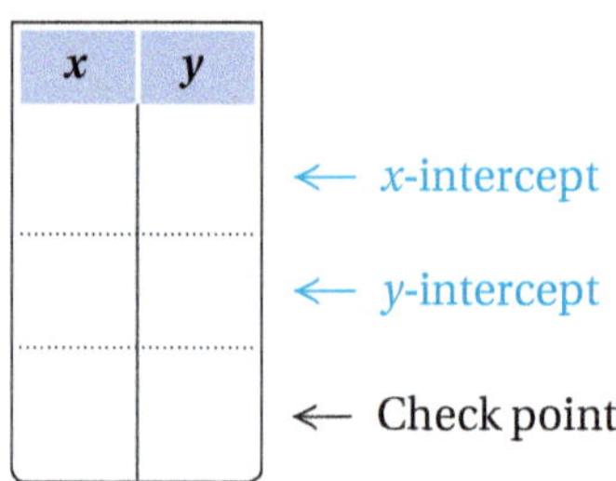

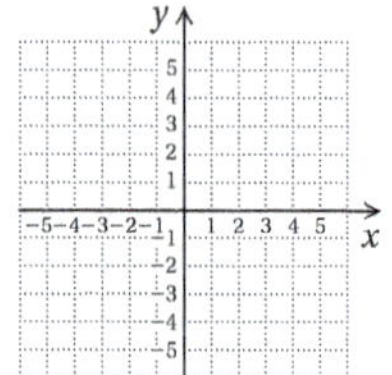

Answers

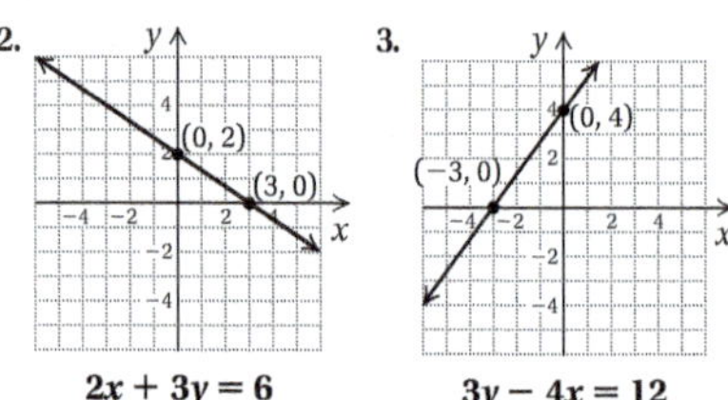

Guided Solution:
2. 3, 2, 4, 3, 2, 4

CALCULATOR CORNER

Viewing the Intercepts Knowing the intercepts of a linear equation helps us to determine a good viewing window for the graph of the equation. For example, when we graph the equation $y = -x + 15$ in the standard window, we see only a small portion of the graph in the upper right-hand corner of the screen, as shown on the left below.

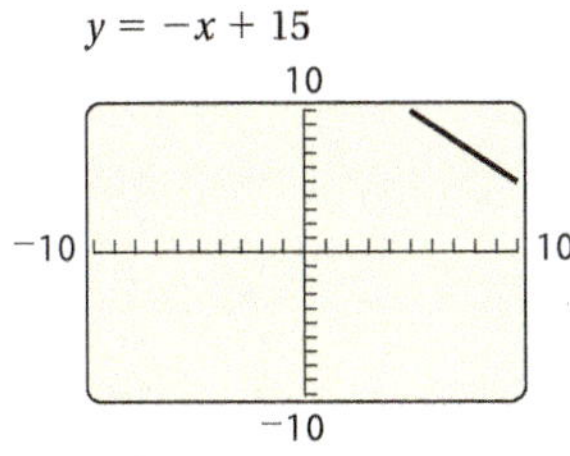

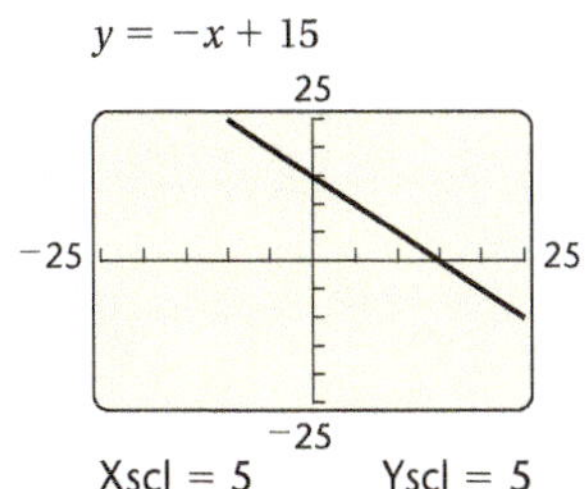

Using algebra, as we did in Example 1, we find that the intercepts of the graph of this equation are $(0, 15)$ and $(15, 0)$. This tells us that, if we are to see more of the graph than is shown on the left above, both Xmax and Ymax should be greater than 15. We can try different window settings until we find one that suits us. One good choice is $[-25, 25, -25, 25]$, with Xscl = 5 and Yscl = 5, shown on the right above.

EXERCISES: Find the intercepts of each equation algebraically. Then graph the equation on a graphing calculator, choosing window settings that allow the intercepts to be seen clearly. (Settings may vary.)

1. $y = -7.5x - 15$

2. $y - 2.15x = 43$

3. $6x - 5y = 150$

4. $y = 0.2x - 4$

5. $y = 1.5x - 15$

6. $5x - 4y = 2$

b EQUATIONS WHOSE GRAPHS ARE HORIZONTAL LINES OR VERTICAL LINES

EXAMPLE 3 Graph: $y = 3$.

The equation $y = 3$ tells us that y must be 3, but it doesn't give us any information about x. We can also think of this equation as $0 \cdot x + y = 3$. No matter what number we choose for x, we find that y is 3. We make up a table with all 3's in the y-column.

x	y
	3
	3
	3

y must be 3.

Choose any number for x. →

x	y
−2	3
0	3
4	3

← y-intercept

Graph.

4. $y = 2x$

x	y
−1	
0	
1	

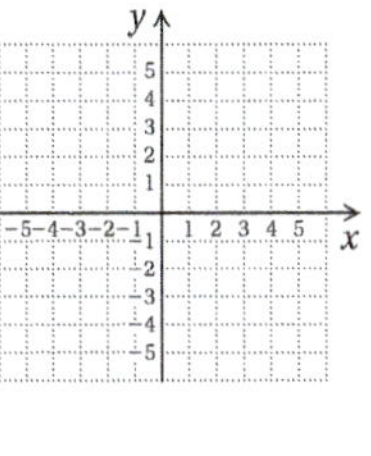

5. $y = -\frac{2}{3}x$

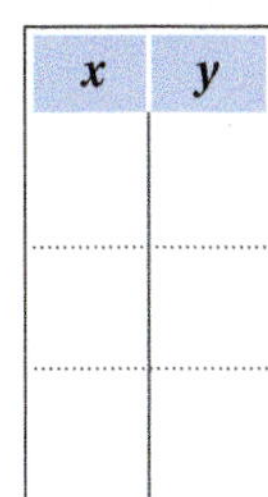

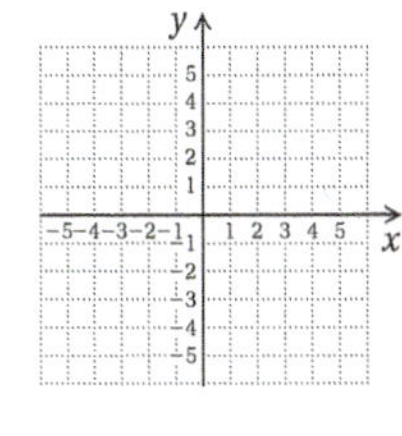

Answers

4.

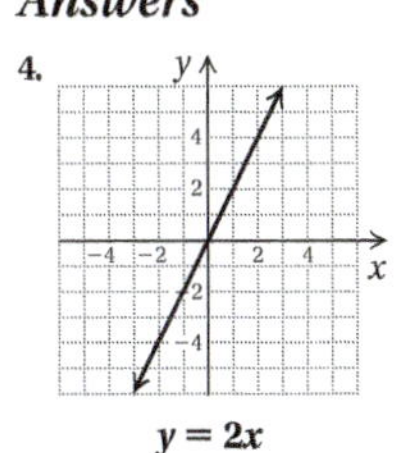

5.

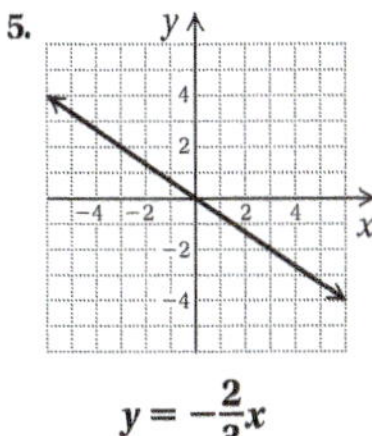

Graph.

6. $x = 5$

GS

x	y
5	-4
☐	0
☐	3

7. $y = -2$

x	y
-1	-2
0	☐
2	☐

8. $x = -3$

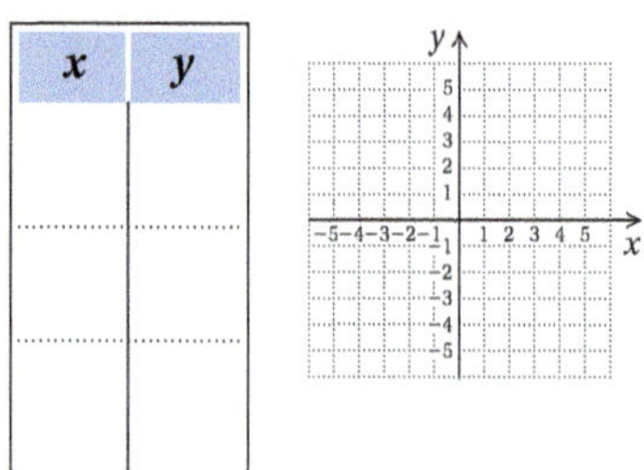

9. $x = 0$

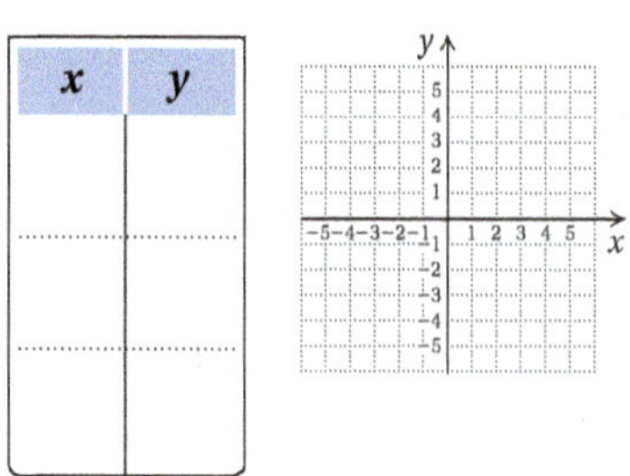

Answers

Answers to Margin Exercises 6–9 and Guided Solutions 6 and 7 are on p. 797.

When we plot the ordered pairs $(-2, 3)$, $(0, 3)$, and $(4, 3)$ and connect the points, we obtain a horizontal line. Any ordered pair $(x, 3)$ is a solution. So the line is parallel to the x-axis with y-intercept $(0, 3)$.

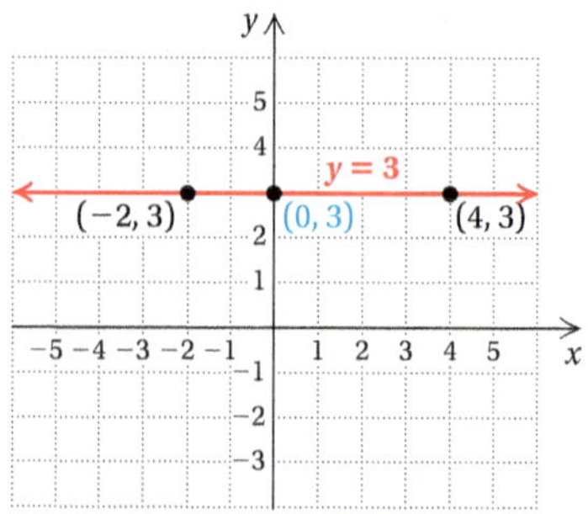

EXAMPLE 4 Graph: $x = -4$.

Consider $x = -4$. We can also think of this equation as $x + 0 \cdot y = -4$. We make up a table with all -4's in the x-column.

x	y
-4	
-4	
-4	
-4	

x must be -4.

x	y
-4	-5
-4	1
-4	3
-4	0

← Choose any number for y.

x-intercept → (row -4, 0)

When we plot the ordered pairs $(-4, -5)$, $(-4, 1)$, $(-4, 3)$, and $(-4, 0)$ and connect the points, we obtain a vertical line. Any ordered pair $(-4, y)$ is a solution. So the line is parallel to the y-axis with x-intercept $(-4, 0)$.

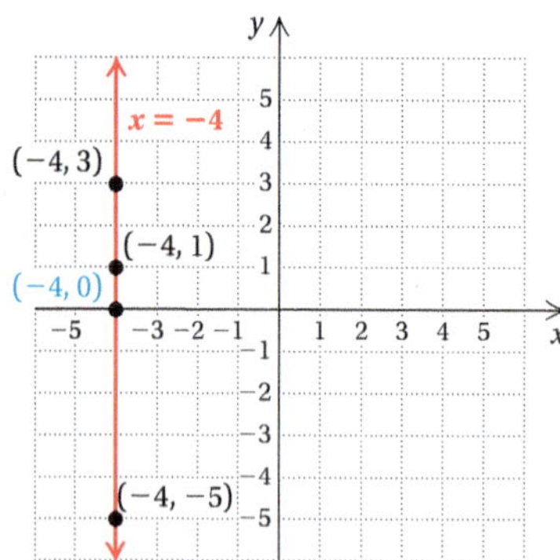

HORIZONTAL LINES AND VERTICAL LINES

The graph of $y = b$ is a **horizontal line**. The y-intercept is $(0, b)$.

The graph of $x = a$ is a **vertical line**. The x-intercept is $(a, 0)$.

Do Exercises 6–9.

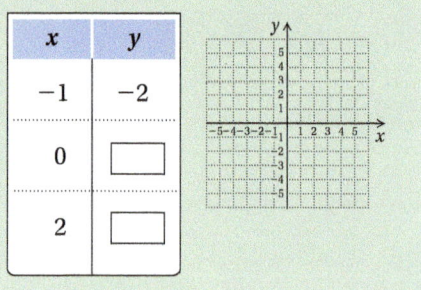

The following is a general procedure for graphing linear equations.

GRAPHING LINEAR EQUATIONS

1. If the equation is of the type $x = a$ or $y = b$, the graph will be a line parallel to an axis; $x = a$ is vertical and $y = b$ is horizontal.

Examples.

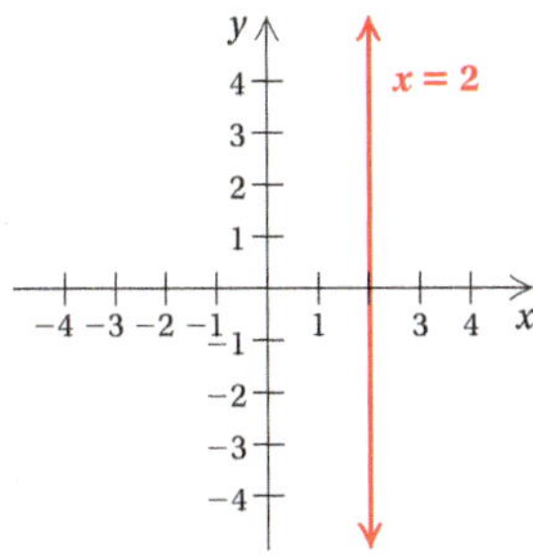

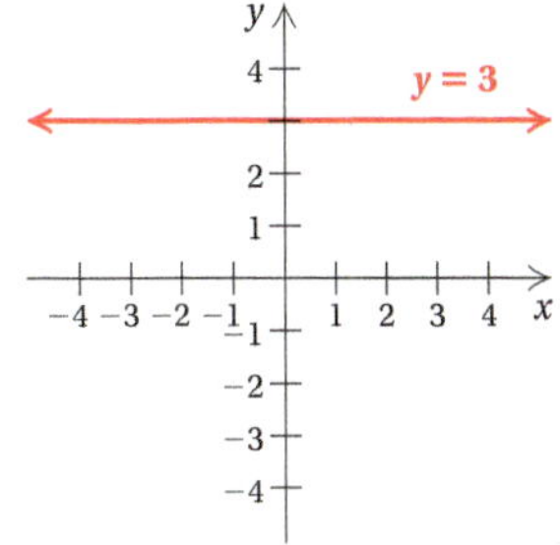

2. If the equation is of the type $y = mx$, both intercepts are the origin, $(0, 0)$. Plot $(0, 0)$ and two other points.

Example.

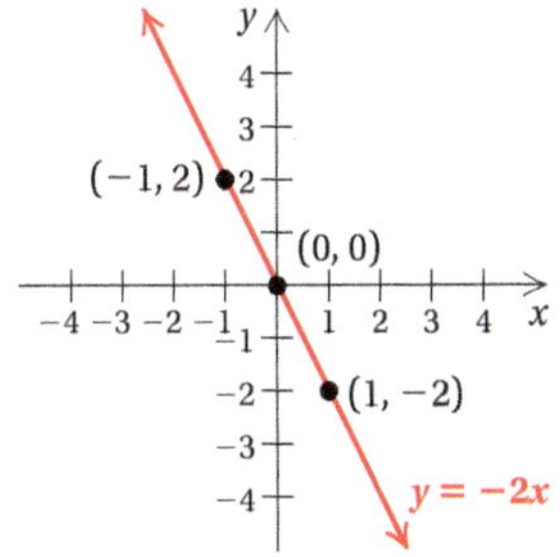

3. If the equation is of the type $y = mx + b$, plot the y-intercept $(0, b)$ and two other points.

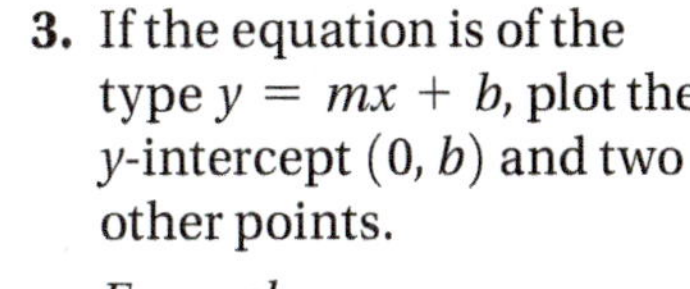

Example.

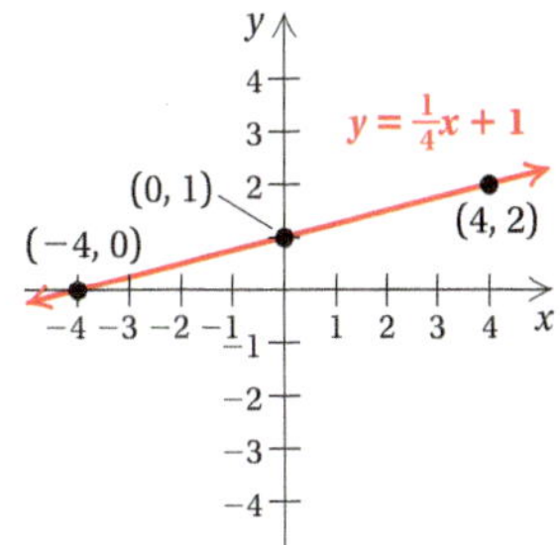

4. If the equation is of the type $Ax + By = C$, but not of the type $x = a$ or $y = b$, then either solve for y and proceed as with the equation $y = mx + b$, or graph using intercepts. If the intercepts are too close together, choose another point or points farther from the origin.

Examples.

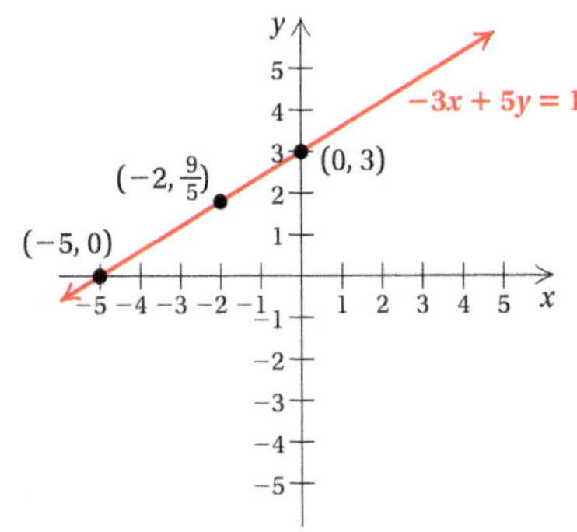

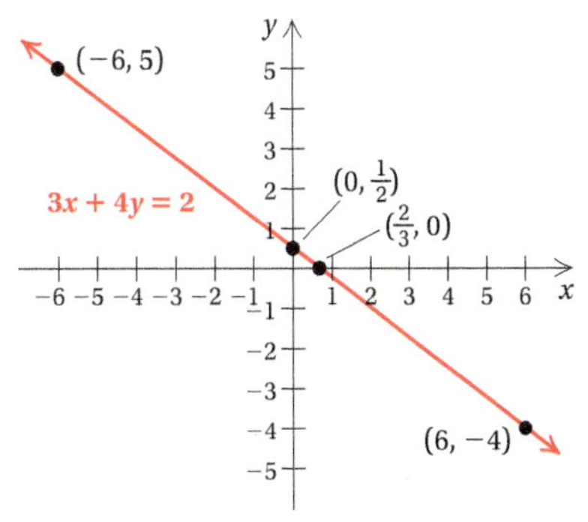

Answers

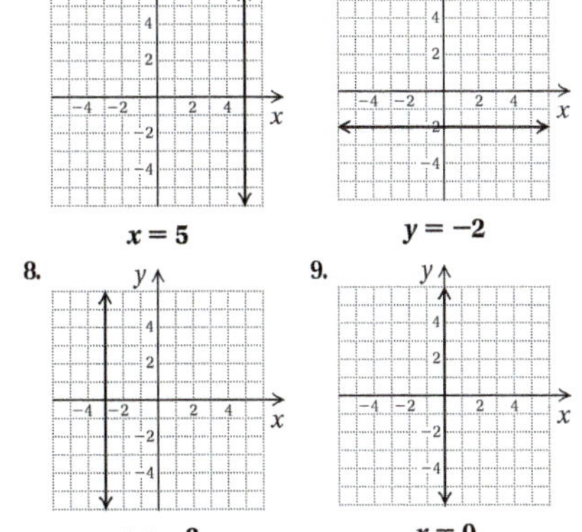

Guided Solutions:
6. 5, 5 **7.** −2, −2

Visualizing for Success

A

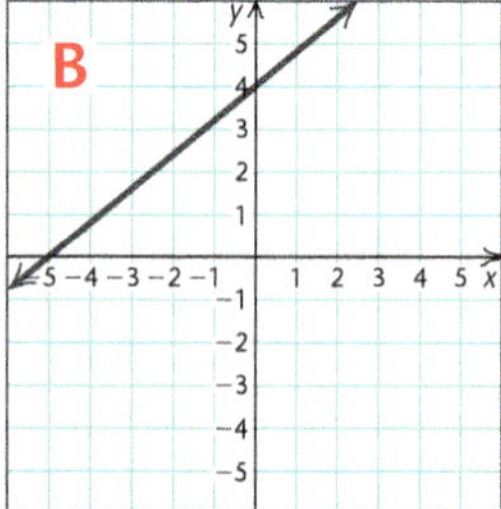

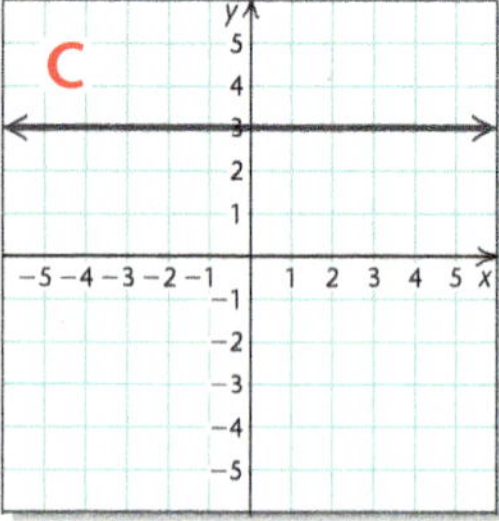

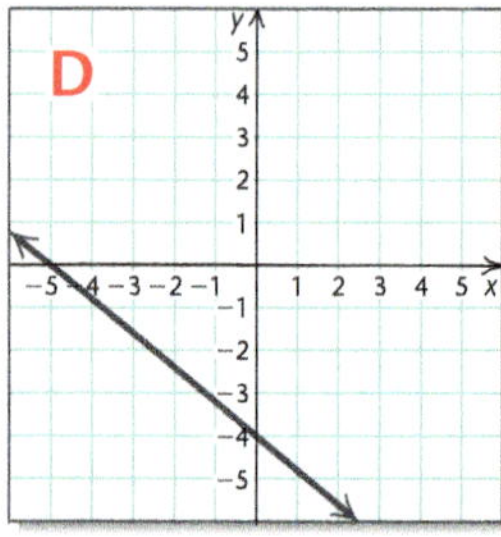

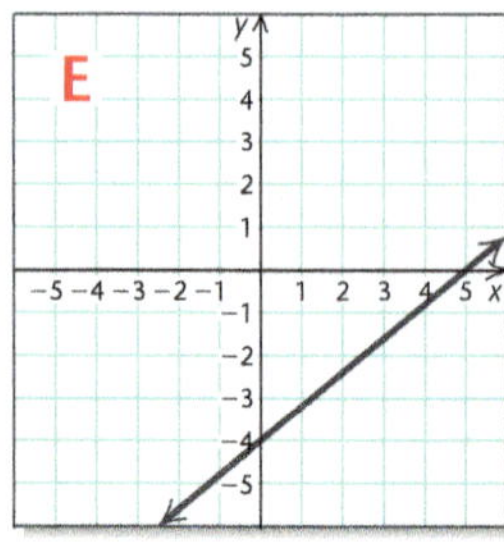

Match each equation with its graph.

1. $5y + 20 = 4x$
2. $y = 3$
3. $3x + 5y = 15$
4. $5y + 4x = 20$
5. $5y = 10 - 2x$
6. $4x + 5y + 20 = 0$
7. $5x - 4y = 20$
8. $4y + 5x + 20 = 0$
9. $5y - 4x = 20$
10. $x = -3$

Answers on page A-22

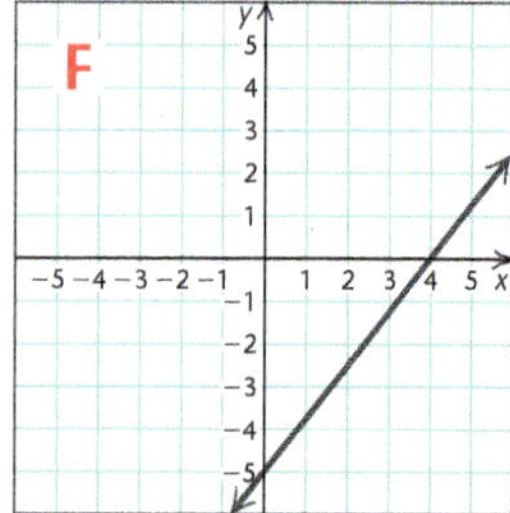

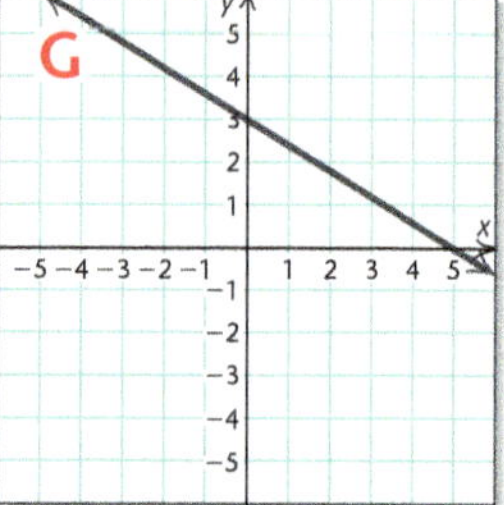

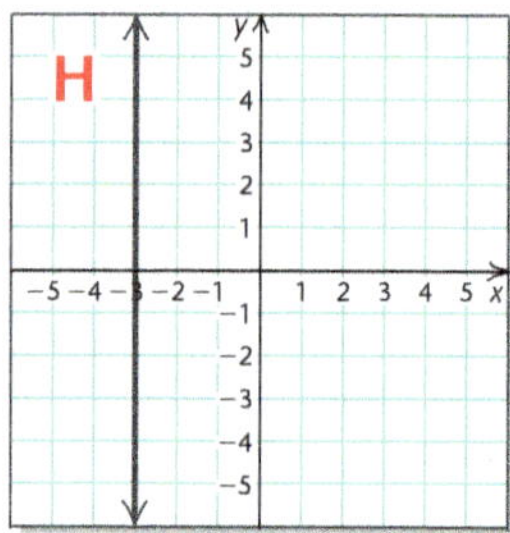

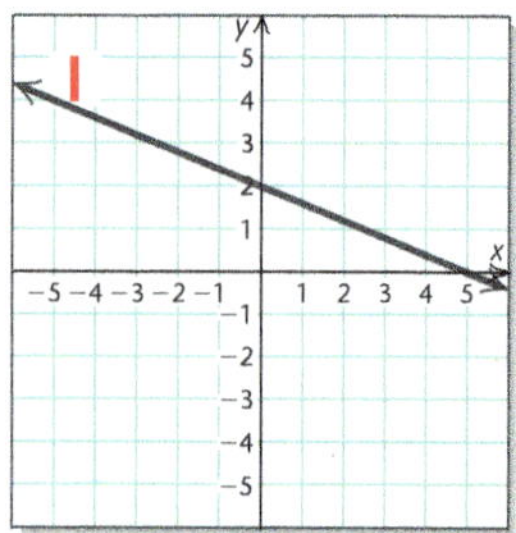

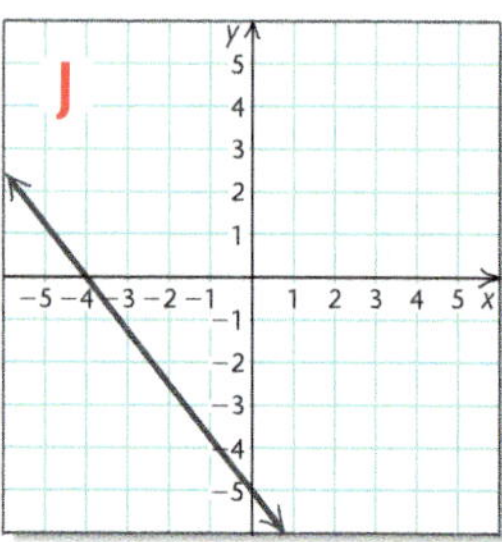

11.2 Exercise Set

For Extra Help
MyMathLab® MathXL® PRACTICE WATCH READ REVIEW

Reading Check

Choose from the column on the right the word or the expression that best completes each statement. Not every choice will be used.

RC1. The graph of $y = -3$ is a(n) ____________ line with $(0, -3)$ as its ____________.

RC2. The x-intercept occurs when a line crosses the ____________.

RC3. To find the x-intercept, let ____________.

RC4. In the graph of $y = 2x$, the point ____________ is both the x-intercept and the y-intercept.

RC5. To find the y-intercept, let ____________.

RC6. The graph of $x = 4$ is a(n) ____________ line with $(4, 0)$ as its ____________.

$(0, 2)$
$(0, 0)$
horizontal
vertical
$x = 0$
$y = 0$
x-intercept
y-intercept
x-axis
y-axis

a For each of Exercises 1–4, find **(a)** the coordinates of the y-intercept and **(b)** the coordinates of the x-intercept.

1.

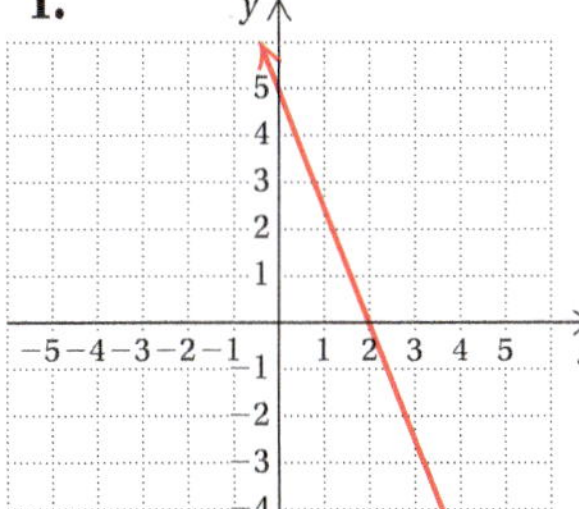

2.

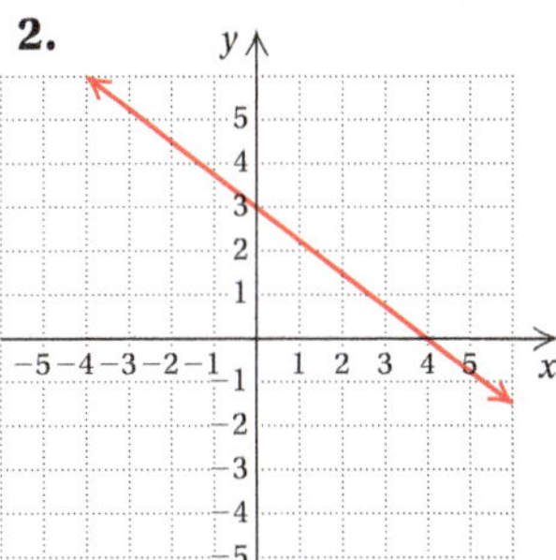

3.

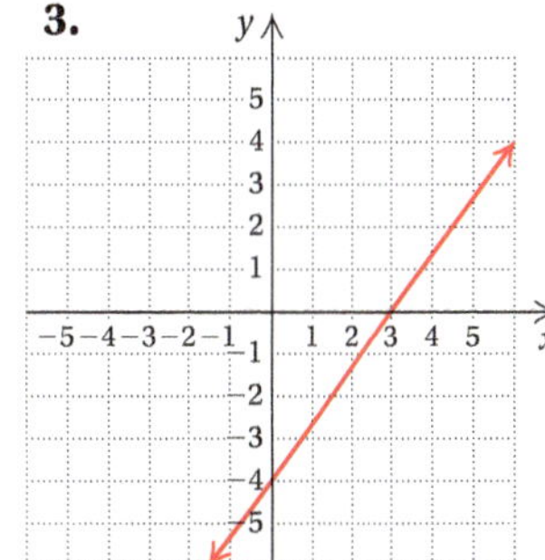

4.

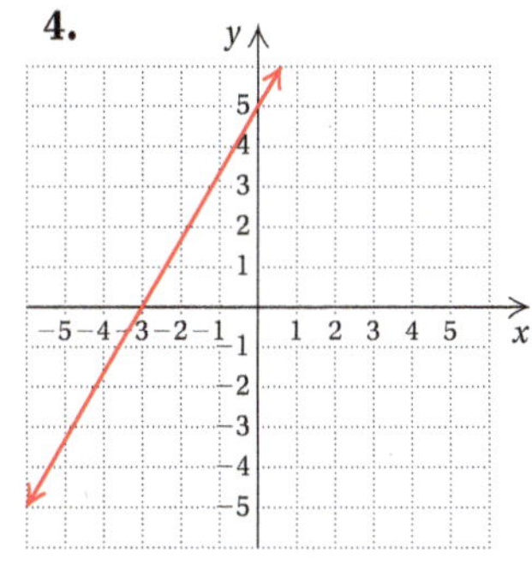

For each of Exercises 5–12, find **(a)** the coordinates of the y-intercept and **(b)** the coordinates of the x-intercept. Do not graph.

5. $3x + 5y = 15$

6. $5x + 2y = 20$

7. $7x - 2y = 28$

8. $3x - 4y = 24$

9. $-4x + 3y = 10$

10. $-2x + 3y = 7$

11. $6x - 3 = 9y$

12. $4y - 2 = 6x$

For each equation, find the intercepts. Then use the intercepts to graph the equation.

13. $x + 3y = 6$

x	y	
0		← y-intercept
	0	← x-intercept

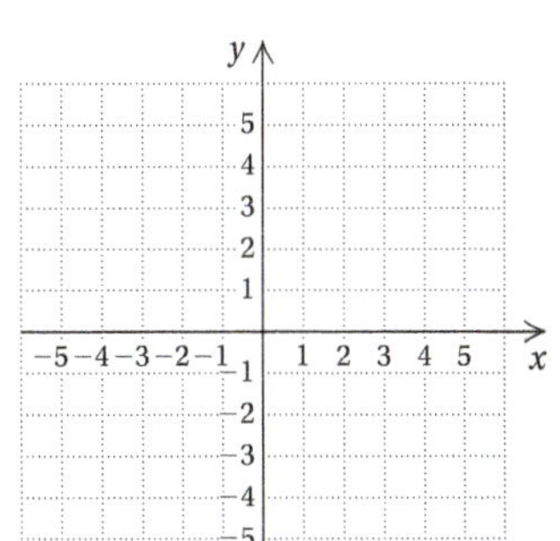

14. $x + 2y = 2$

x	y	
0		← y-intercept
	0	← x-intercept

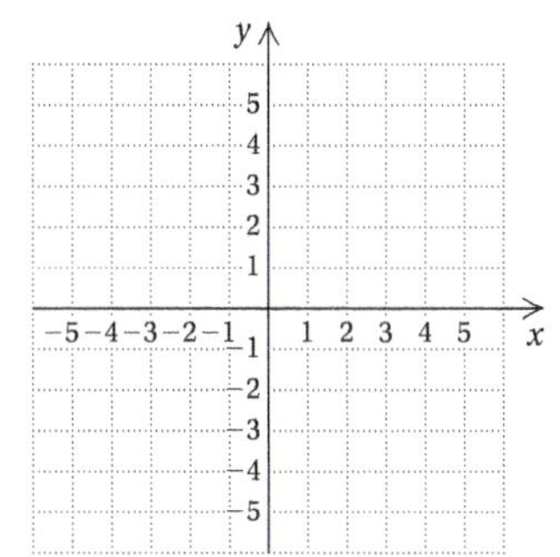

15. $-x + 2y = 4$

x	y	
0		← y-intercept
	0	← x-intercept

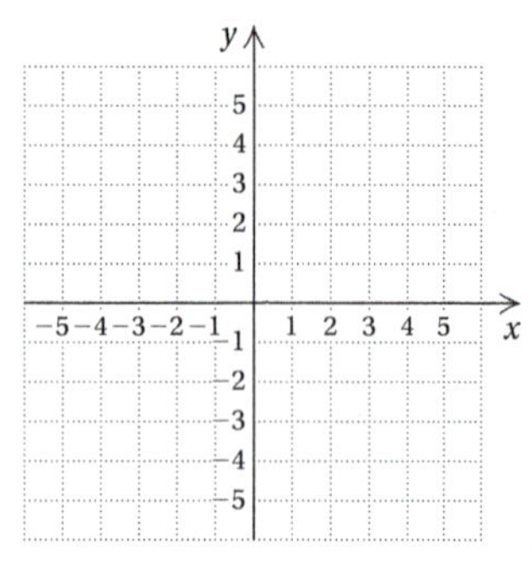

16. $-x + y = 5$

x	y	
0		← y-intercept
	0	← x-intercept

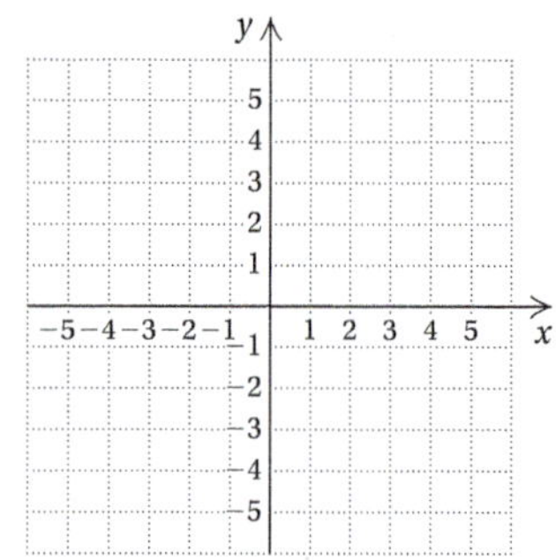

17. $3x + y = 6$

x	y	
0		← y-intercept
	0	← x-intercept

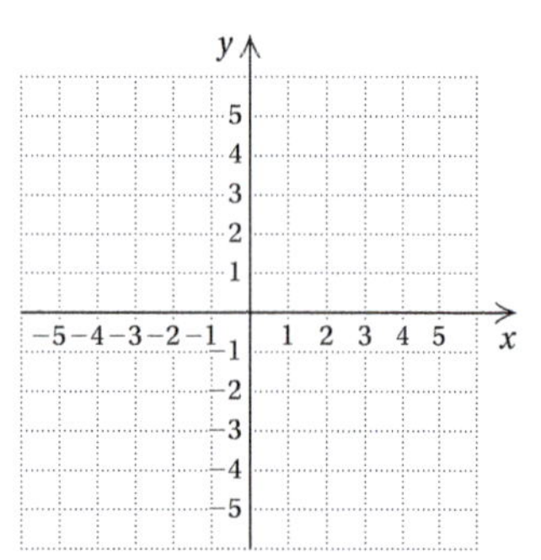

18. $2x + y = 6$

x	y	
0		← y-intercept
	0	← x-intercept

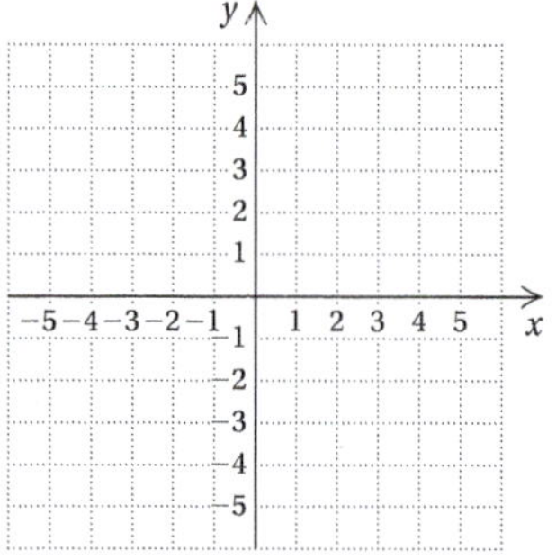

19. $2y - 2 = 6x$

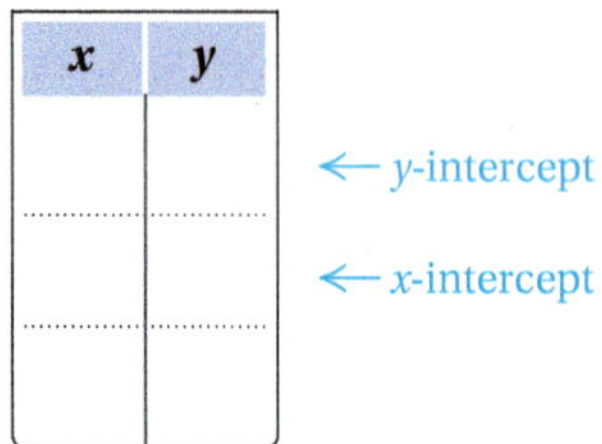

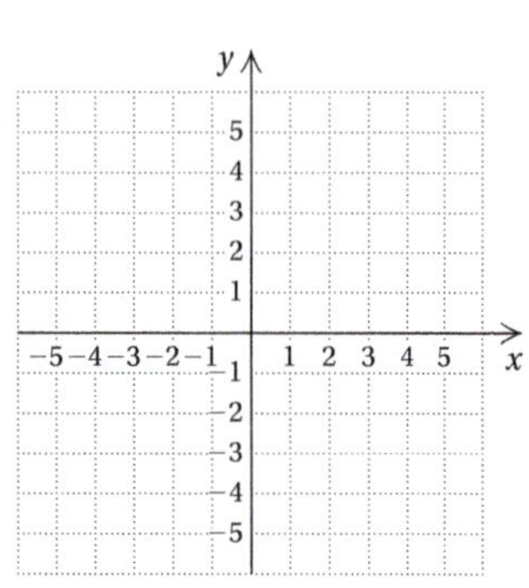

20. $3y - 6 = 9x$

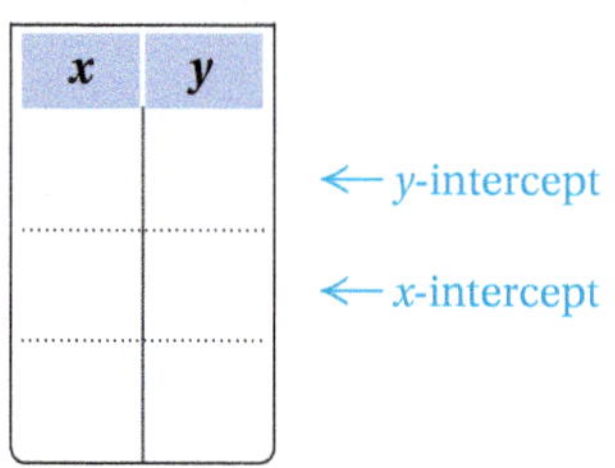

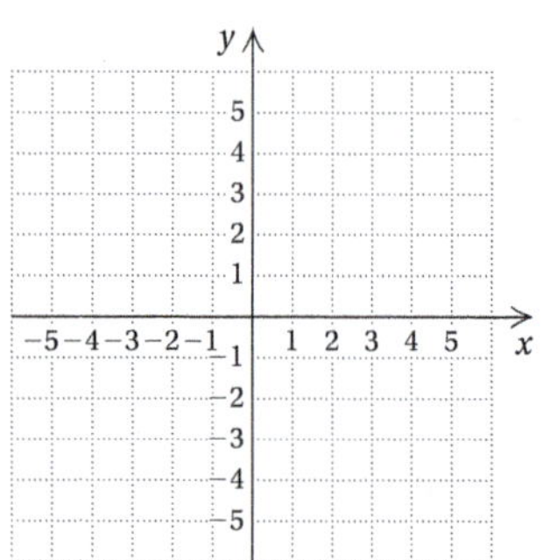

21. $3x - 9 = 3y$

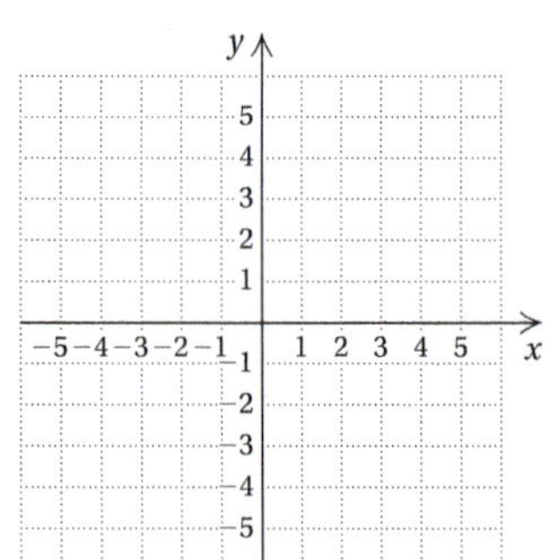

22. $5x - 10 = 5y$

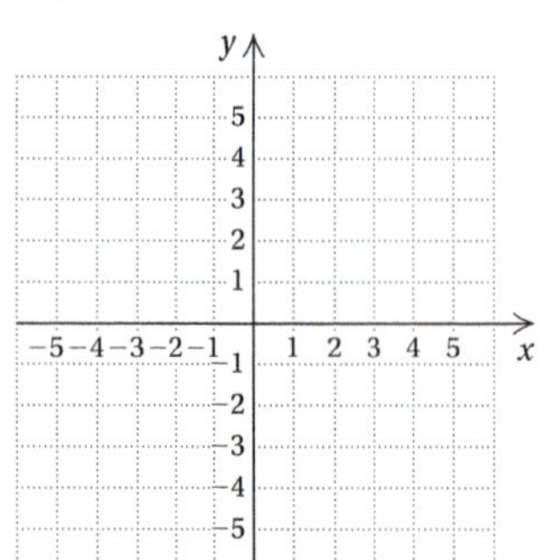

23. $2x - 3y = 6$

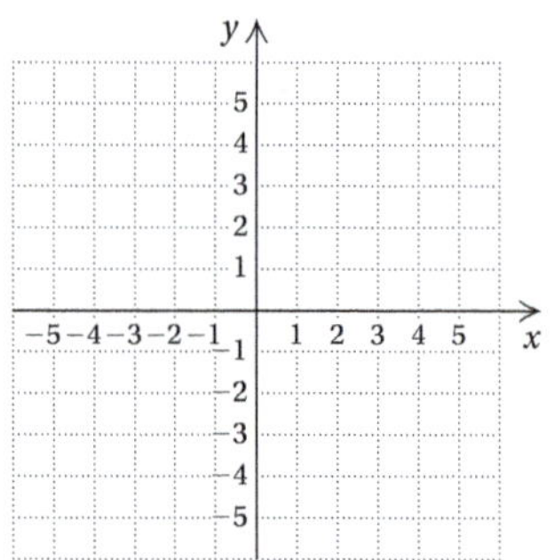

24. $2x - 5y = 10$

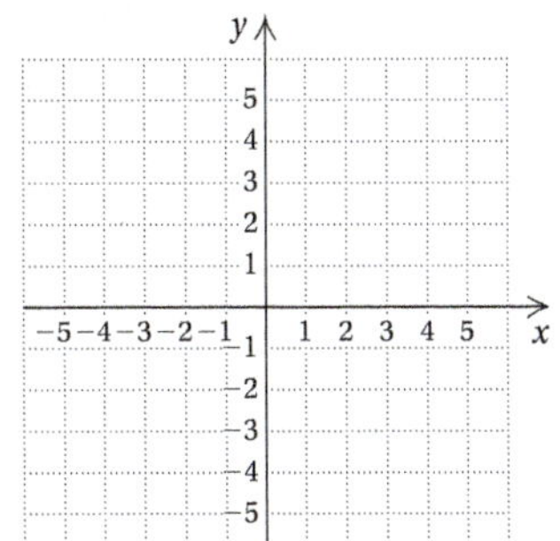

25. $4x + 5y = 20$

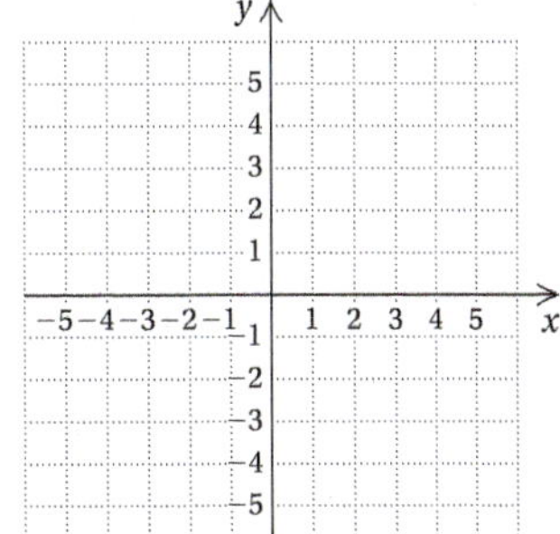

26. $2x + 6y = 12$

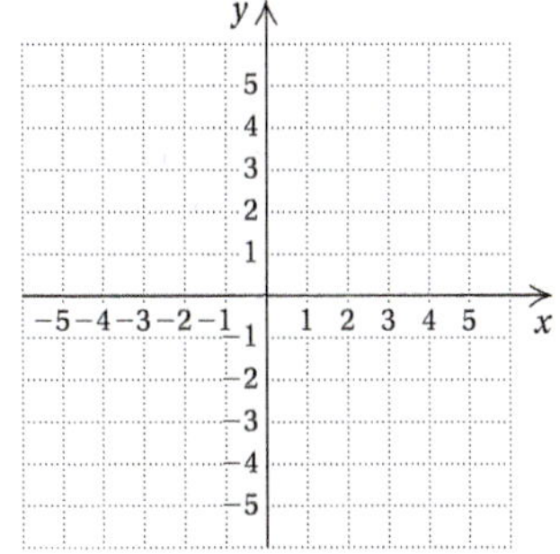

27. $2x + 3y = 8$

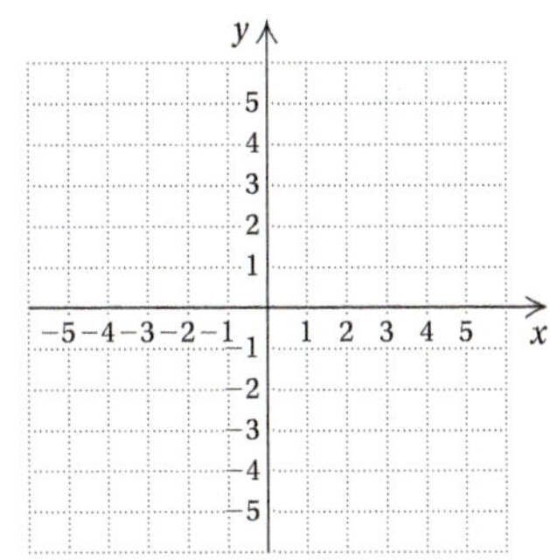

28. $x - 1 = y$

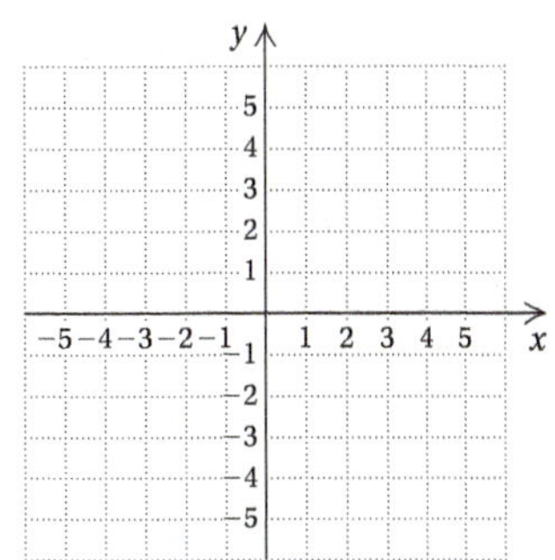

29. $3x + 4y = 5$

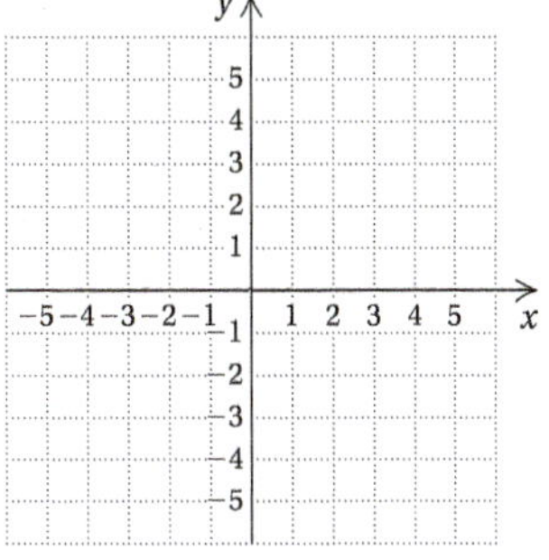

30. $2x - 1 = y$

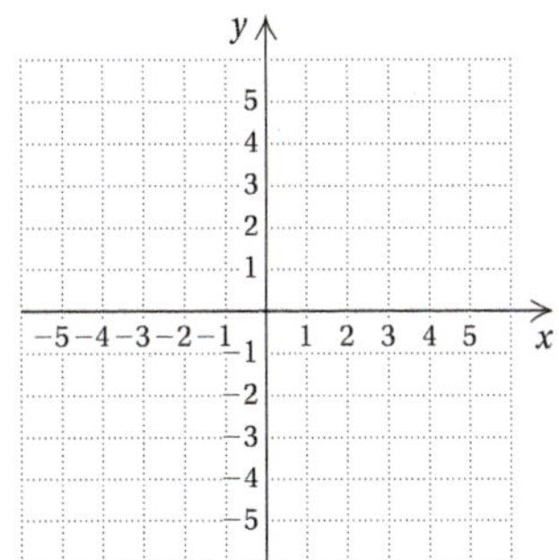

31. $3x - 2 = y$

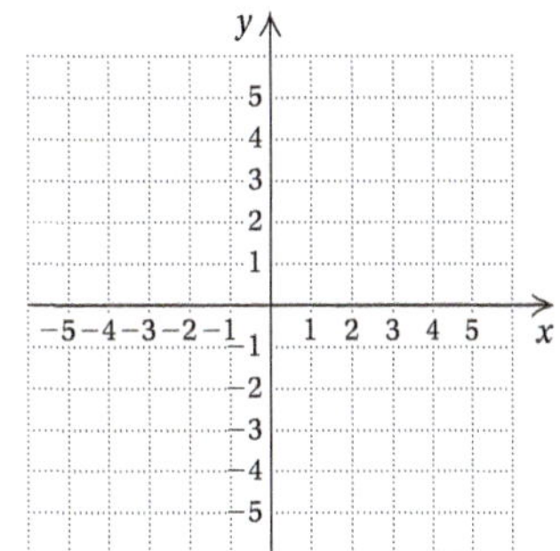

32. $4x - 3y = 12$

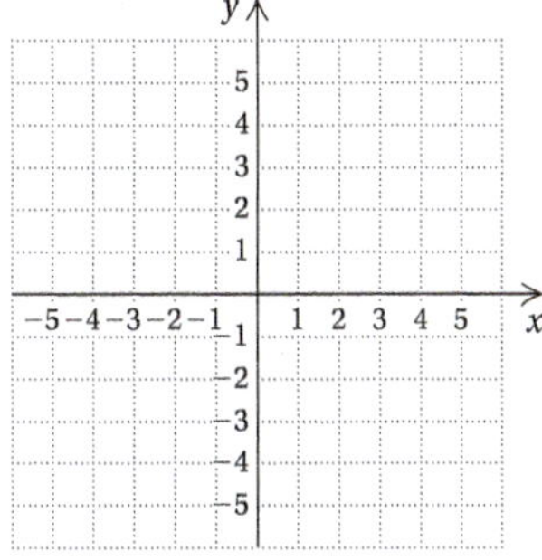

33. $6x - 2y = 12$

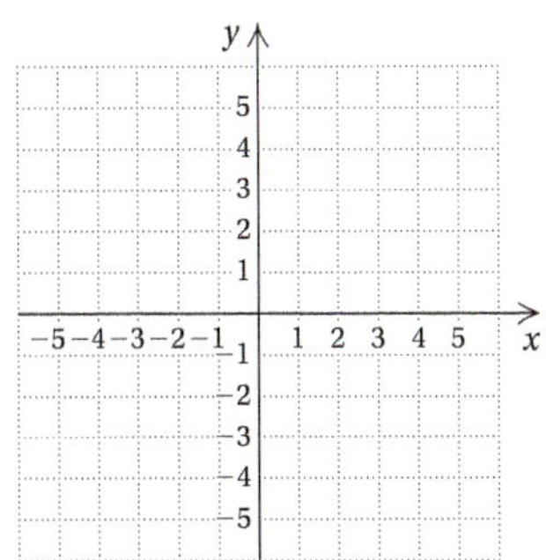

34. $7x + 2y = 6$

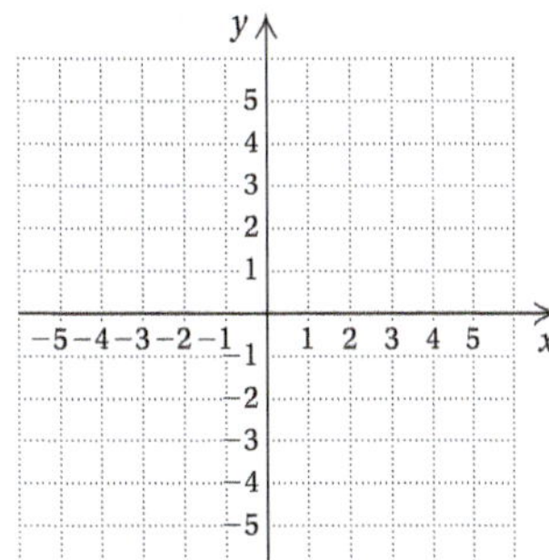

35. $y = -3 - 3x$

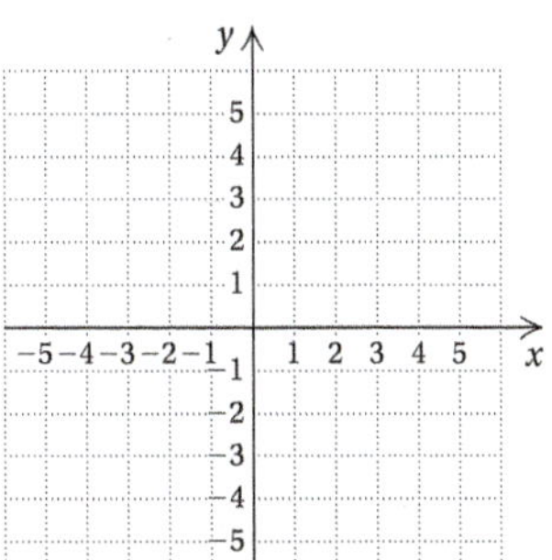

36. $-3x = 6y - 2$

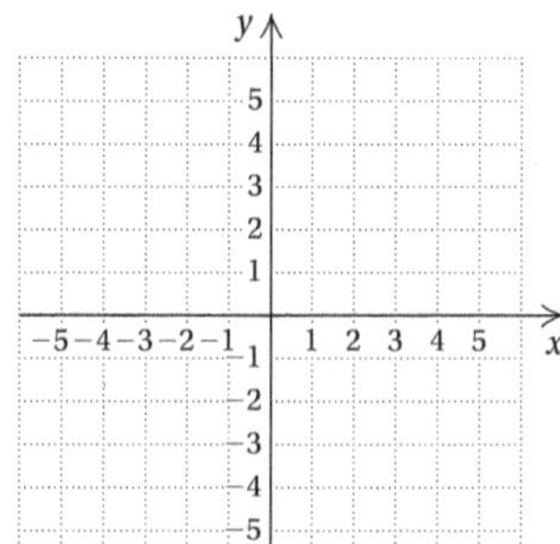

37. $y - 3x = 0$

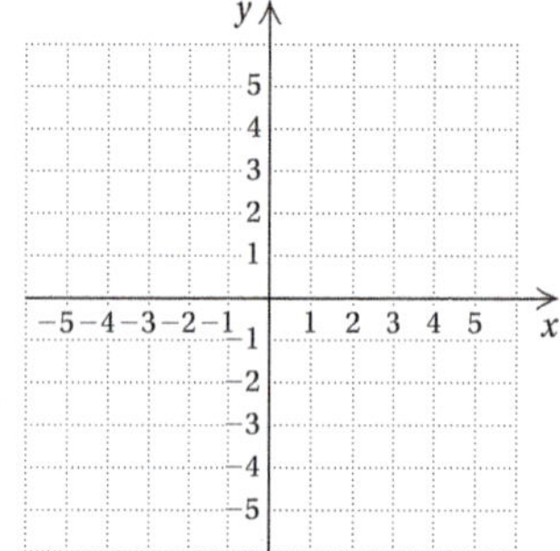

38. $x + 2y = 0$

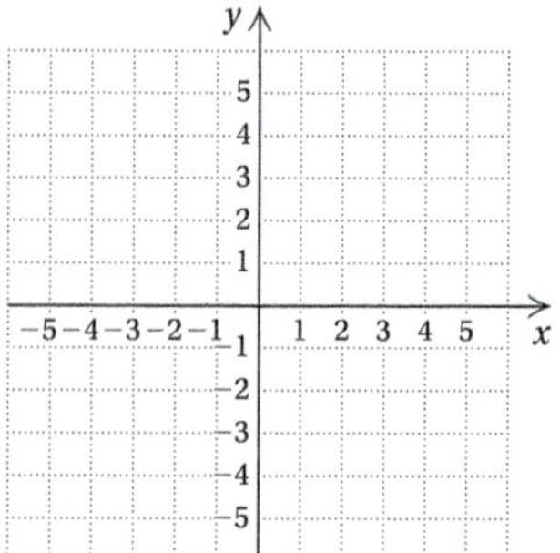

b Graph.

39. $x = -2$

x	y
-2	
-2	
-2	

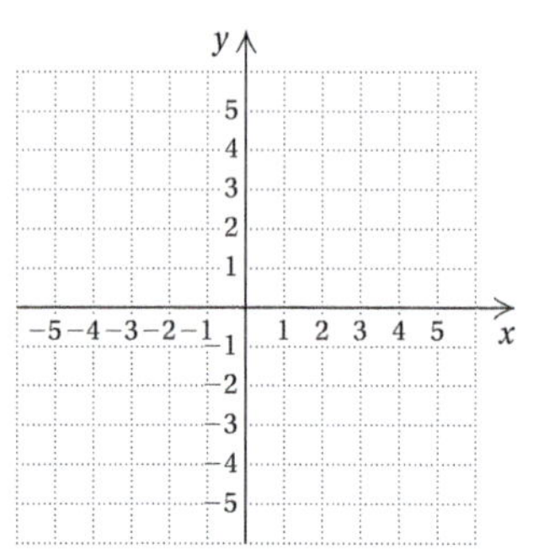

40. $x = 1$

x	y
1	
1	
1	

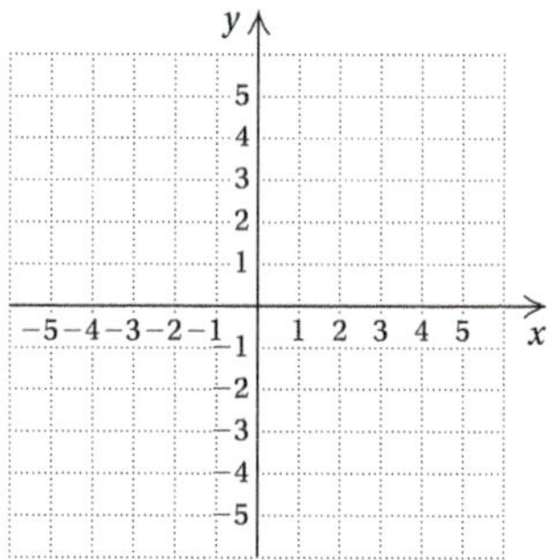

41. $y = 2$

x	y
	2
	2
	2

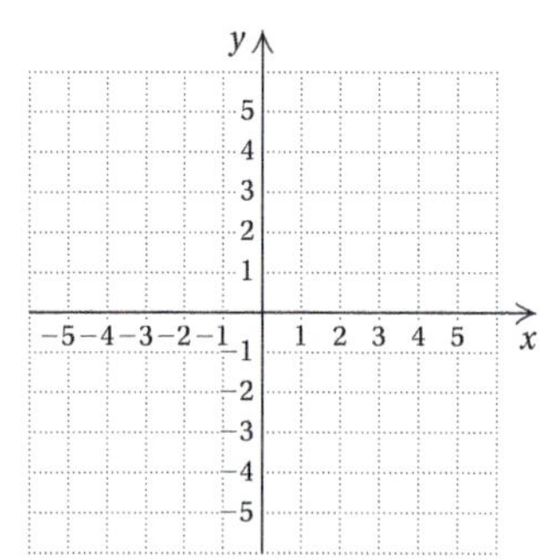

42. $y = -4$

x	y
	-4
	-4
	-4

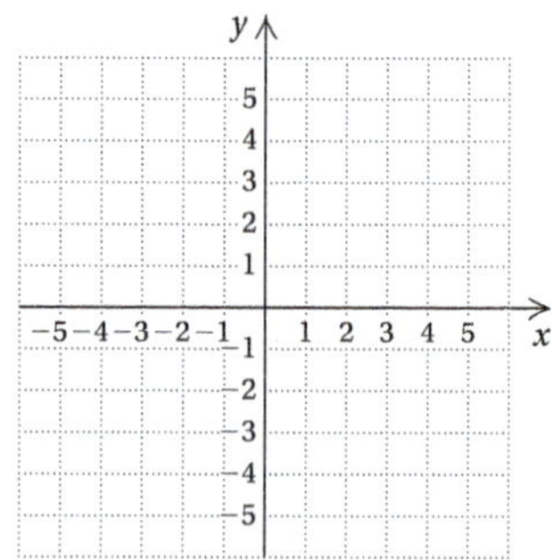

43. $x = 2$

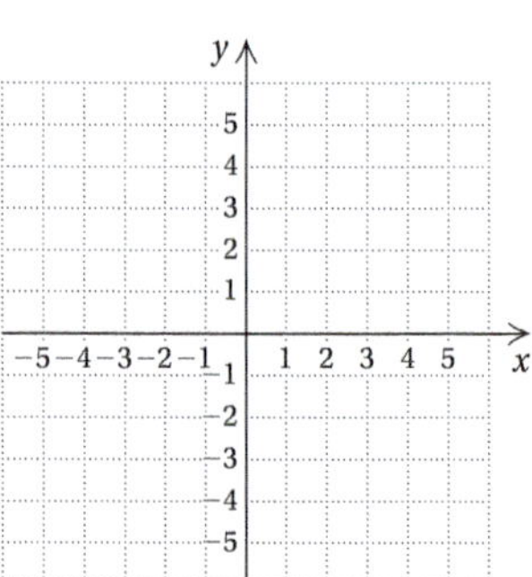

44. $x = 3$

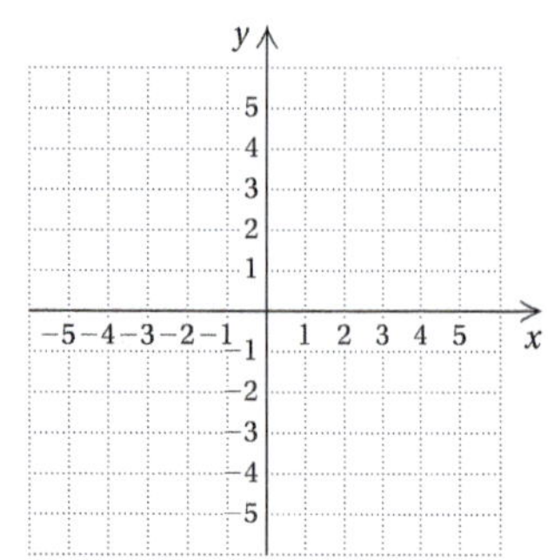

45. $y = 0$

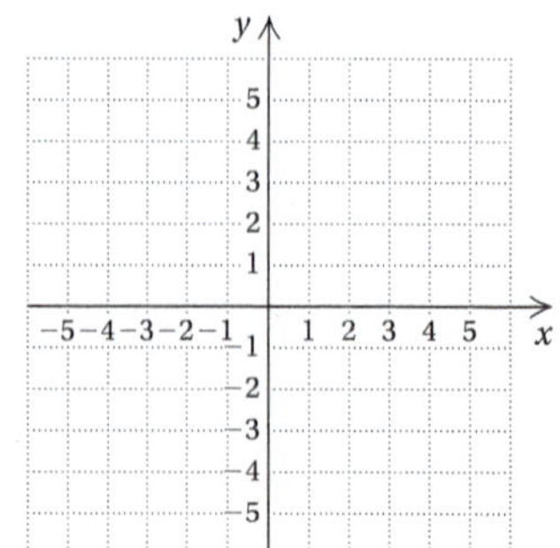

46. $y = -1$

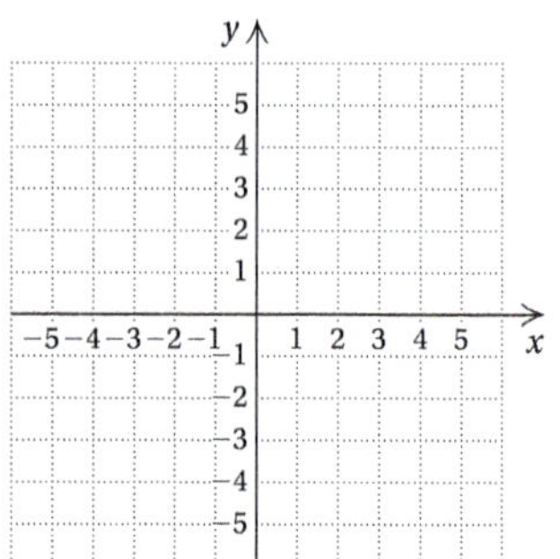

47. $x = \frac{3}{2}$

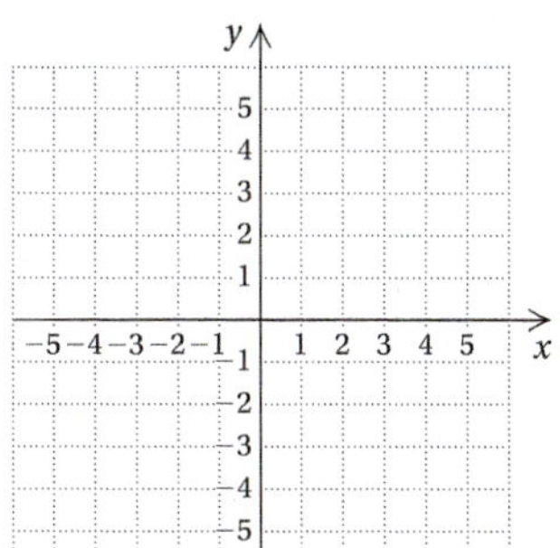

48. $x = -\frac{5}{2}$

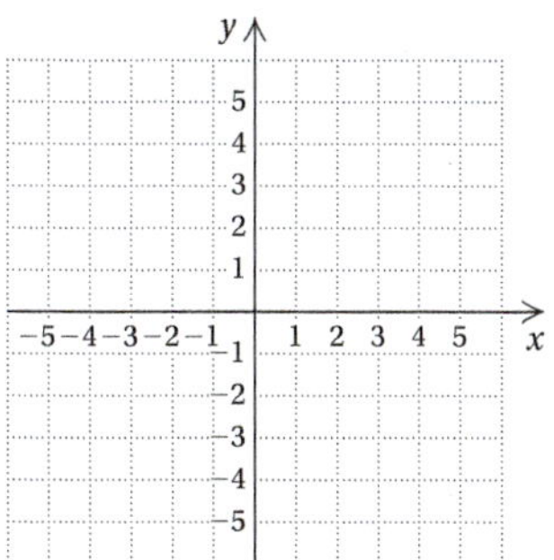

49. $3y = -5$

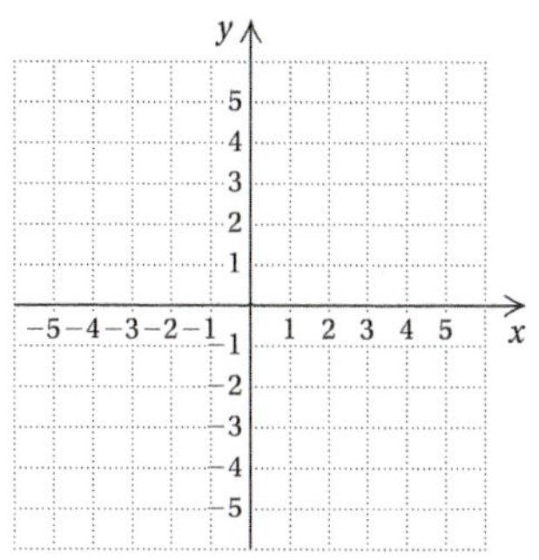

50. $12y = 45$

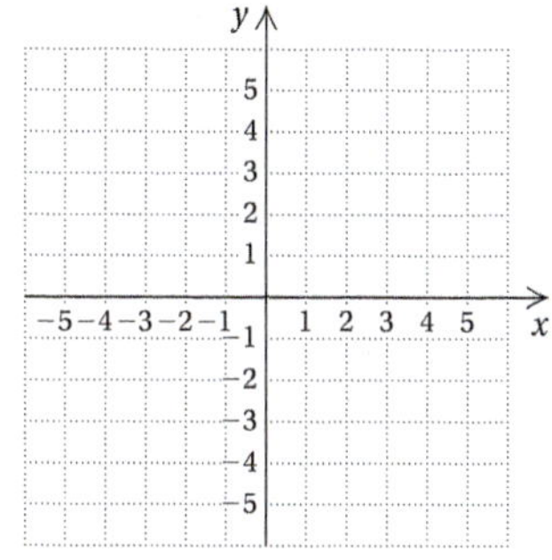

51. $4x + 3 = 0$

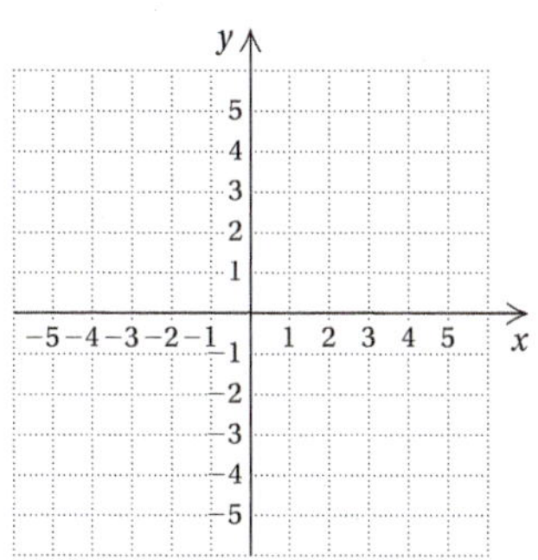

52. $-3x + 12 = 0$

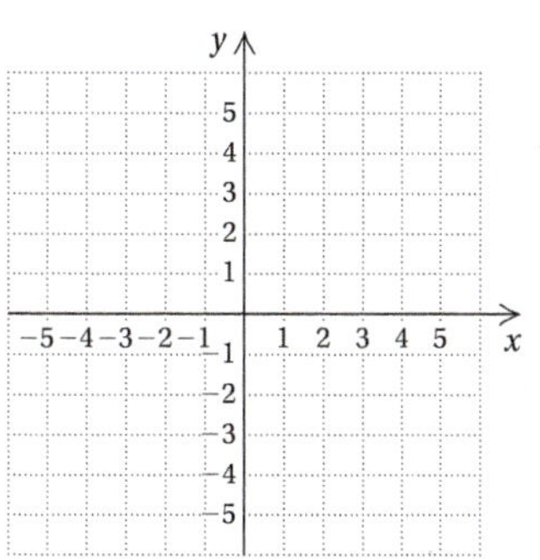

53. $48 - 3y = 0$

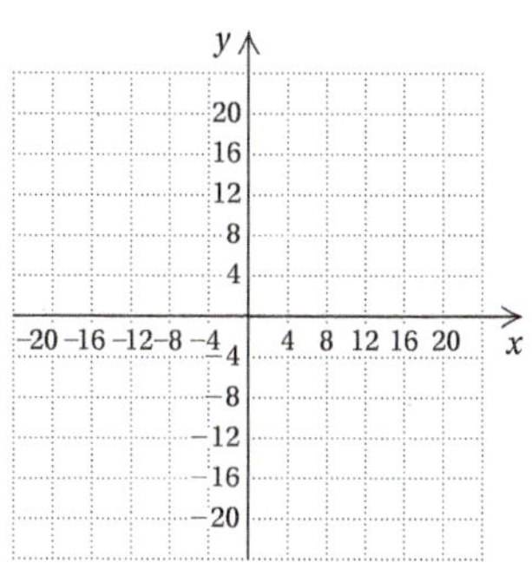

54. $63 + 7y = 0$

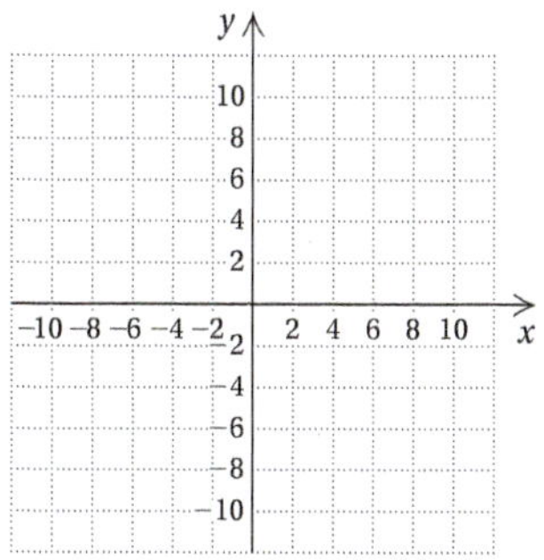

Write an equation for the graph shown.

55.

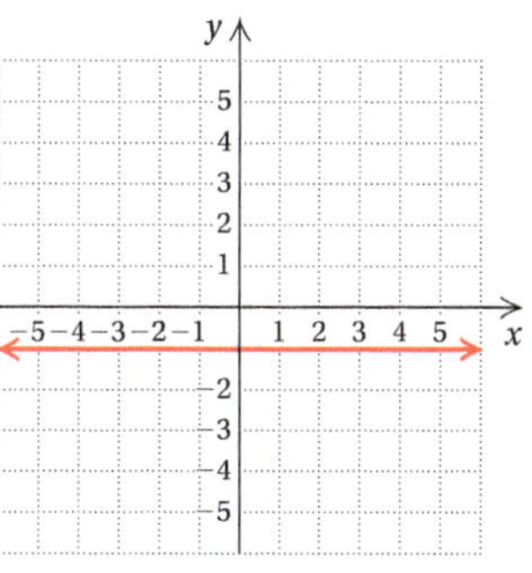

56.

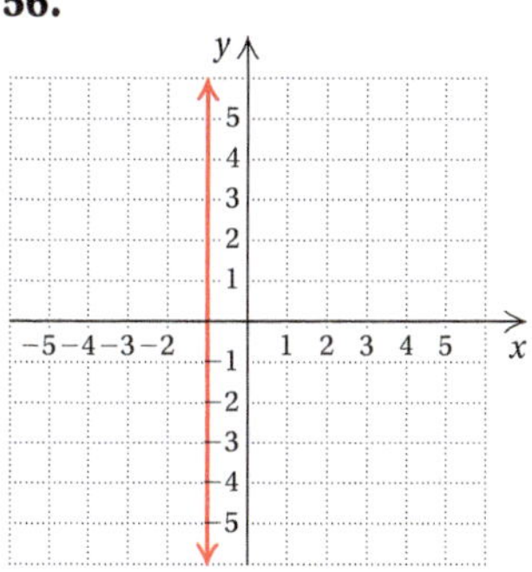

57.

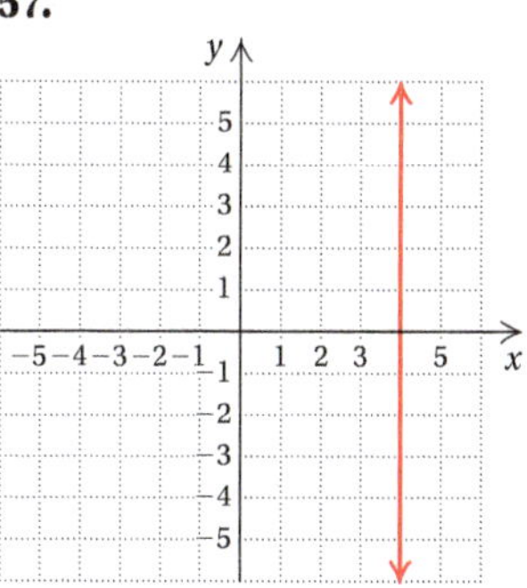

58.

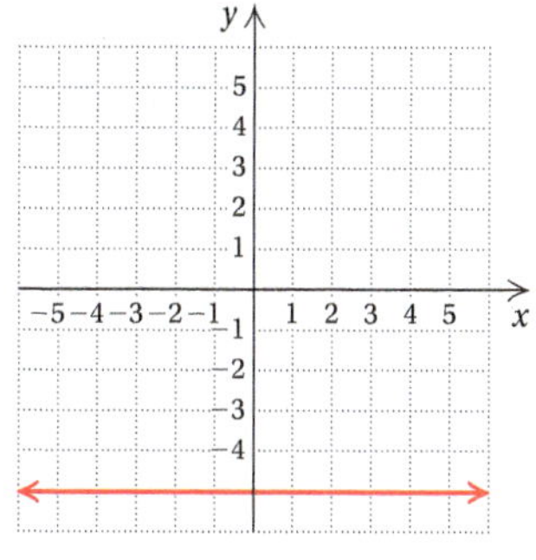

Skill Maintenance

Solve. [10.7e]

59. $x + (x - 1) < (x + 2) - (x + 1)$

60. $6 - 18x \leq 4 - 12x - 5x$

61. $\frac{2x}{7} - 4 \leq -2$

62. $\frac{1}{4} + \frac{x}{3} > \frac{7}{12}$

Synthesis

63. Write an equation of a line parallel to the x-axis and passing through $(-3, -4)$.

64. Find the value of m such that the graph of $y = mx + 6$ has an x-intercept of $(2, 0)$.

65. Find the value of k such that the graph of $3x + k = 5y$ has an x-intercept of $(-4, 0)$.

66. Find the value of k such that the graph of $4x = k - 3y$ has a y-intercept of $(0, -8)$.

11.3 Slope and Applications

OBJECTIVES

a Given the coordinates of two points on a line, find the slope of the line, if it exists.

b Find the slope of a line from an equation.

c Find the slope, or rate of change, in an applied problem involving slope.

SKILL TO REVIEW

Objective 9.4a: Subtract real numbers.

Subtract.

1. $-4 - 20$
2. $-21 - (-5)$

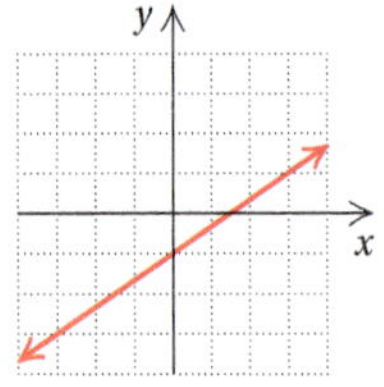

$y = \frac{2}{3}x - 1$

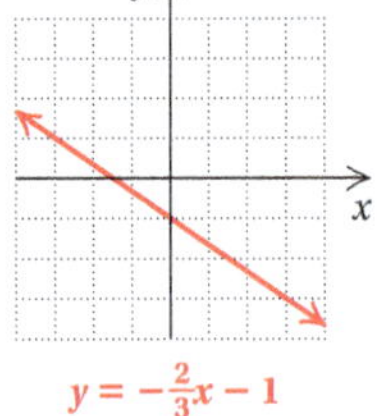

$y = -\frac{2}{3}x - 1$

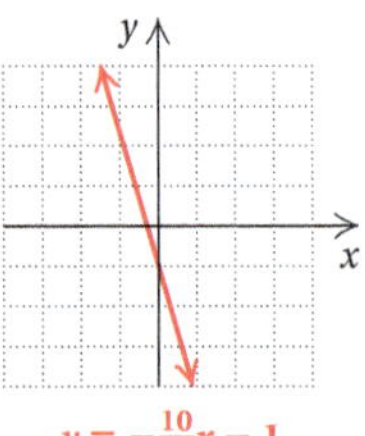

$y = -\frac{10}{3}x - 1$

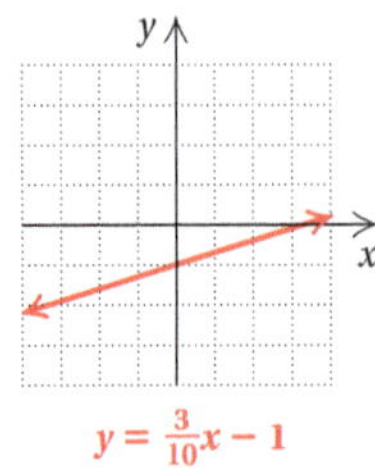

$y = \frac{3}{10}x - 1$

a SLOPE

We have considered two forms of a linear equation, $Ax + By = C$ and $y = mx + b$. We found that from the form of the equation $y = mx + b$, we know that the y-intercept of the line is $(0, b)$.

$$y = mx + b.$$

? (under m) → The y-intercept is $(0, b)$.

What about the constant m? Does it give us information about the line? Look at the graphs in the margin and see if you can make any connection between the constant m and the "slant" of the line.

The graphs of some linear equations slant upward from left to right. Others slant downward. Some are vertical and some are horizontal. Some slant more steeply than others. We now look for a way to describe such possibilities with numbers.

Consider a line with two points marked P and Q. As we move from P to Q, the y-coordinate changes from 1 to 3 and the x-coordinate changes from 2 to 6. The change in y is $3 - 1$, or 2. The change in x is $6 - 2$, or 4.

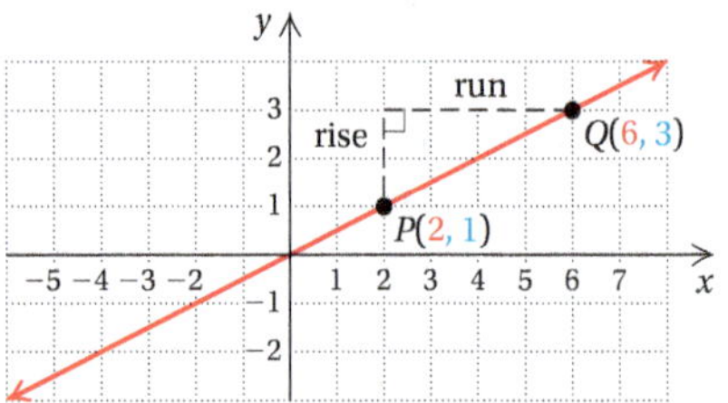

We call the change in y the **rise** and the change in x the **run**. The ratio rise/run is the same for any two points on a line. We call this ratio the **slope** of the line. Slope describes the slant of a line. The slope of the line in the graph above is given by

$$\frac{\text{rise}}{\text{run}} = \frac{\text{the change in } y}{\text{the change in } x}, \text{ or } \frac{2}{4}, \text{ or } \frac{1}{2}.$$

SLOPE

The **slope** of a line containing points (x_1, y_1) and (x_2, y_2) is given by

$$m = \frac{\text{rise}}{\text{run}} = \frac{\text{the change in } y}{\text{the change in } x} = \frac{y_2 - y_1}{x_2 - x_1}.$$

Answers

Skill to Review:
1. -24 2. -16

In the preceding definition, (x_1, y_1) and (x_2, y_2)—read "x sub-one, y sub-one and x sub-two, y sub-two"—represent two different points on a line. It does not matter which point is considered (x_1, y_1) and which is considered (x_2, y_2) so long as coordinates are subtracted in the same order in both the numerator and the denominator:

$$\frac{y_2 - y_1}{x_2 - x_1} = \frac{y_1 - y_2}{x_1 - x_2}.$$

EXAMPLE 1 Graph the line containing the points $(-4, 3)$ and $(2, -6)$ and find the slope.

The graph is shown below. We consider (x_1, y_1) to be $(-4, 3)$ and (x_2, y_2) to be $(2, -6)$. From $(-4, 3)$ and $(2, -6)$, we see that the change in y, or the rise, is $-6 - 3$, or -9. The change in x, or the run, is $2 - (-4)$, or 6.

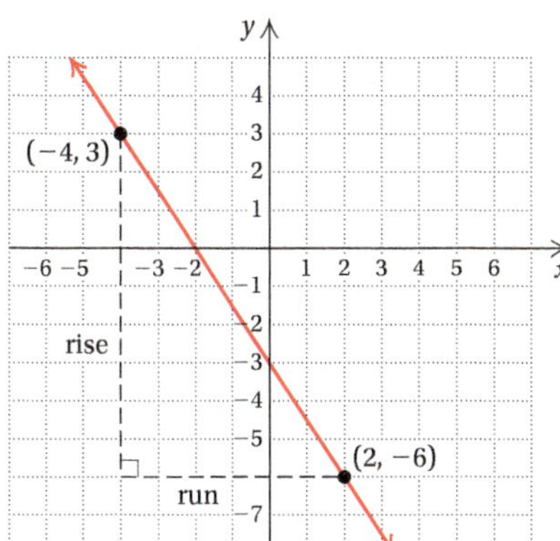

$$\begin{aligned}\text{Slope} = \frac{\text{rise}}{\text{run}} &= \frac{\text{change in } y}{\text{change in } x}\\ &= \frac{y_2 - y_1}{x_2 - x_1}\\ &= \frac{-6 - 3}{2 - (-4)}\\ &= \frac{-9}{6} = -\frac{9}{6}, \text{ or } -\frac{3}{2}\end{aligned}$$

When we use the formula

$$m = \frac{y_2 - y_1}{x_2 - x_1},$$

we must remember to subtract the x-coordinates in the same order that we subtract the y-coordinates. Let's redo Example 1, where we consider (x_1, y_1) to be $(2, -6)$ and (x_2, y_2) to be $(-4, 3)$:

$$\text{Slope} = \frac{\text{change in } y}{\text{change in } x} = \frac{3 - (-6)}{-4 - 2} = \frac{9}{-6} = -\frac{9}{6} = -\frac{3}{2}.$$

Do Exercises 1 and 2.

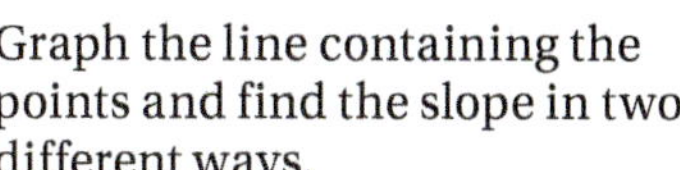
Graph the line containing the points and find the slope in two different ways.

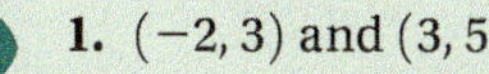
GS **1.** $(-2, 3)$ and $(3, 5)$

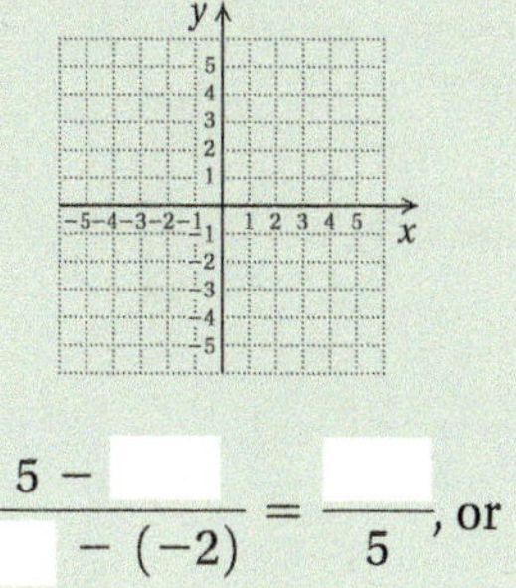

$$\frac{5 - \square}{\square - (-2)} = \frac{\square}{5}, \text{ or}$$

$$\frac{3 - \square}{\square - 3} = \frac{-2}{\square} = \frac{2}{\square}$$

2. $(0, -3)$ and $(-3, 2)$

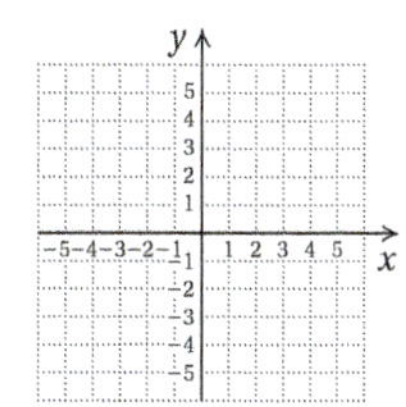

The slope of a line tells how it slants. A line with positive slope slants up from left to right. The larger the slope, the steeper the slant. A line with negative slope slants downward from left to right.

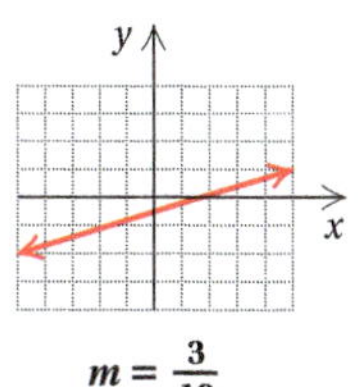
$m = \frac{3}{10}$

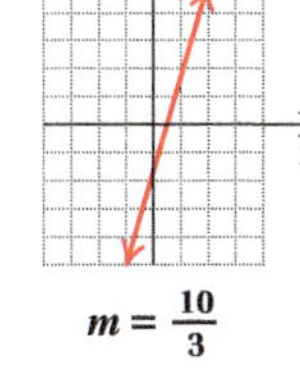
$m = \frac{10}{3}$

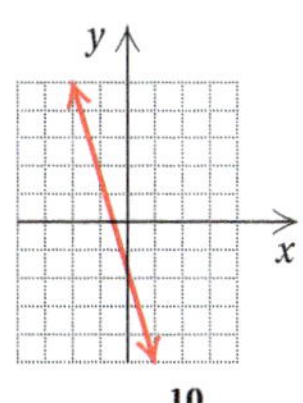
$m = -\frac{10}{3}$

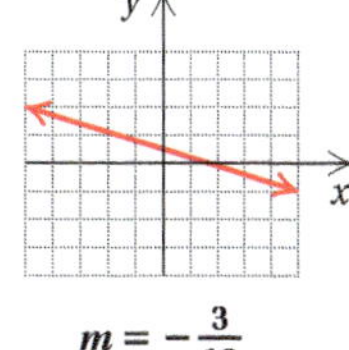
$m = -\frac{3}{10}$

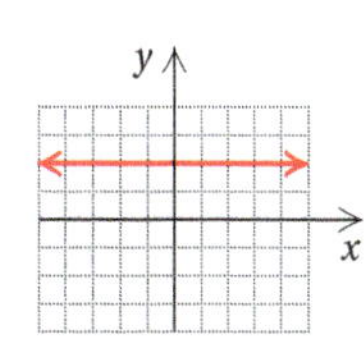
$m = 0$

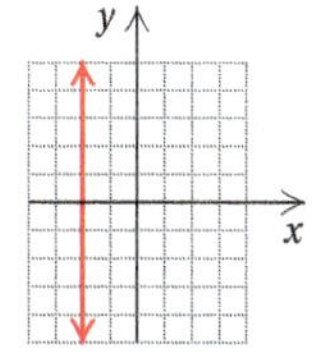
m is not defined.

Later in this section, in Examples 7 and 8, we will discuss the slope of a horizontal line and of a vertical line.

Answers

Answers to Margin Exercises 1 and 2 are on p. 806.

b FINDING THE SLOPE FROM AN EQUATION

It is possible to find the slope of a line from its equation. Let's consider the equation $y = 2x + 3$, which is in the form $y = mx + b$. The graph of this equation is shown at the left. We can find two points by choosing convenient values for x—say, 0 and 1—and substituting to find the corresponding y-values. We find the two points on the line to be $(0, 3)$ and $(1, 5)$. The slope of the line is found using the definition of slope:

$$m = \frac{\text{change in } y}{\text{change in } x} = \frac{5 - 3}{1 - 0} = \frac{2}{1} = 2.$$

The slope is 2. Note that this is also the coefficient of the x-term in the equation $y = 2x + 3$.

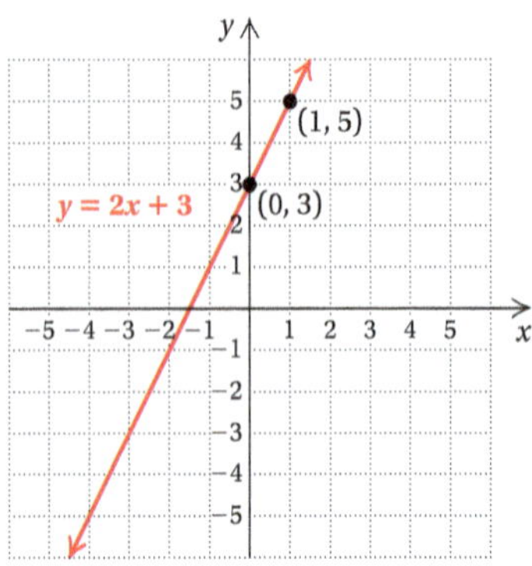

DETERMINING SLOPE FROM THE EQUATION $y = mx + b$

The slope of the line $y = mx + b$ is m. To find the slope of a nonvertical line, solve the linear equation in x and y for y and get the resulting equation in the form $y = mx + b$. The coefficient of the x-term, m, is the slope of the line.

EXAMPLES Find the slope of each line.

2. $y = -3x + \frac{2}{9}$

$\rightarrow m = -3 = \text{Slope}$

3. $y = \frac{4}{5}x$

$\rightarrow m = \frac{4}{5} = \text{Slope}$

4. $y = x + 6$

$\rightarrow m = 1 = \text{Slope}$

5. $y = -0.6x - 3.5$

$\rightarrow m = -0.6 = \text{Slope}$

◀ **Do Exercises 3–6.**

To find slope from an equation, we may need to first find an equivalent form of the equation.

EXAMPLE 6 Find the slope of the line $2x + 3y = 7$.

We solve for y to get the equation in the form $y = mx + b$:

$$2x + 3y = 7$$
$$3y = -2x + 7$$
$$y = \frac{1}{3}(-2x + 7)$$
$$y = -\frac{2}{3}x + \frac{7}{3}. \quad \text{This is } y = mx + b.$$

The slope is $-\frac{2}{3}$.

◀ **Do Exercises 7 and 8.**

Find the slope of each line.

3. $y = 4x + 11$

4. $y = -17x + 8$

5. $y = -x + \frac{1}{2}$

6. $y = \frac{2}{3}x - 1$

Find the slope of each line.

7. $4x + 4y = 7$

8. $5x - 4y = 8$ GS

$5x = \_\_\_ + 8$

$5x - \_\_\_ = 4y$

$\frac{5x - 8}{\_\_\_} = \frac{4y}{4}$

$\_\_\_ \cdot x - 2 = y$, or

$y = \_\_\_ \cdot x - 2$

Slope is $\_\_\_$.

Answers

1. $\frac{2}{5}$ **2.** $-\frac{5}{3}$

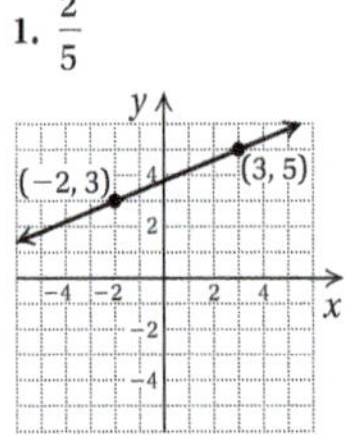

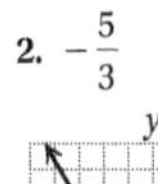

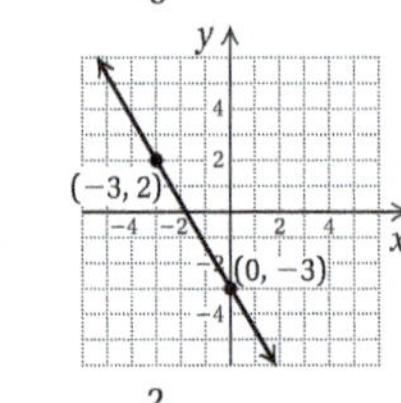

3. 4 **4.** −17 **5.** −1 **6.** $\frac{2}{3}$

7. −1 **8.** $\frac{5}{4}$

Guided Solutions

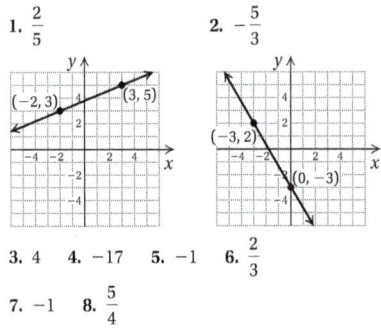

1. 3, 3, 2; 5, −2, −5, 5 **8.** $4y, 8, 4, \frac{5}{4}, \frac{5}{4}, \frac{5}{4}$

What about the slope of a horizontal line or a vertical line?

EXAMPLE 7 Find the slope of the line $y = 5$.

We can think of $y = 5$ as $y = 0x + 5$. Then from this equation, we see that $m = 0$. Consider the points $(-3, 5)$ and $(4, 5)$, which are on the line. The change in $y = 5 - 5$, or 0. The change in $x = -3 - 4$, or -7. We have

$$m = \frac{5 - 5}{-3 - 4} = \frac{0}{-7} = 0.$$

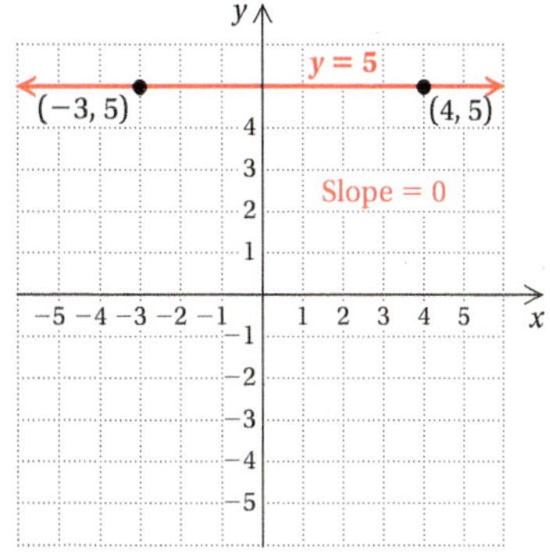

Any two points on a horizontal line have the same y-coordinate. The change in y is 0. Thus the slope of a horizontal line is 0.

EXAMPLE 8 Find the slope of the line $x = -4$.

Consider the points $(-4, 3)$ and $(-4, -2)$, which are on the line. The change in $y = 3 - (-2)$, or 5. The change in $x = -4 - (-4)$, or 0. We have

$$m = \frac{3 - (-2)}{-4 - (-4)} = \frac{5}{0}.$$ Not defined

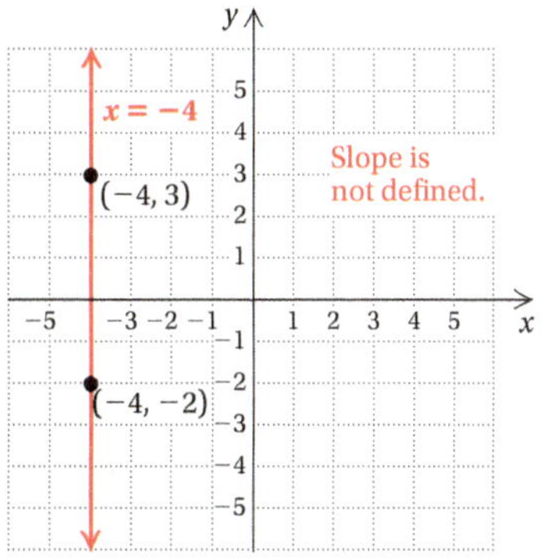

Since division by 0 is not defined, the slope of this line is not defined. The answer in this example is "The slope of this line is not defined."

SLOPE 0; SLOPE NOT DEFINED

The slope of a horizontal line is 0.

The slope of a vertical line is not defined.

Do Exercises 9 and 10.

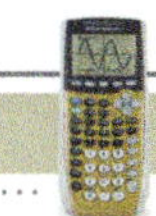

CALCULATOR CORNER

Visualizing Slope

EXERCISES: Graph each of the following sets of equations using the window settings $[-6, 6, -4, 4]$, with Xscl = 1 and Yscl = 1.

1. $y = x$, $y = 2x$, $y = 5x$, $y = 10x$
   What do you think the graph of $y = 123x$ will look like?
2. $y = x$, $y = \frac{3}{4}x$, $y = 0.38x$, $y = \frac{5}{32}x$
   What do you think the graph of $y = 0.000043x$ will look like?
3. $y = -x$, $y = -2x$, $y = -5x$, $y = -10x$
   What do you think the graph of $y = -123x$ will look like?
4. $y = -x$, $y = -\frac{3}{4}x$, $y = -0.38x$, $y = -\frac{5}{32}x$
   What do you think the graph of $y = -0.000043x$ will look like?

Find the slope, if it exists, of each line.

9. $x = 7$

10. $y = -5$

Answers

9. Not defined **10.** 0

C APPLICATIONS OF SLOPE; RATES OF CHANGE

Slope has many real-world applications. For example, numbers like 2%, 3%, and 6% are often used to represent the *grade* of a road, a measure of how steep a road on a hill or mountain is. For example, a 3% grade ($3\% = \frac{3}{100}$) means that for every horizontal distance of 100 ft, the road rises 3 ft, and a -3% grade means that for every horizontal distance of 100 ft, the road drops 3 ft. (Road signs do not include negative signs.)

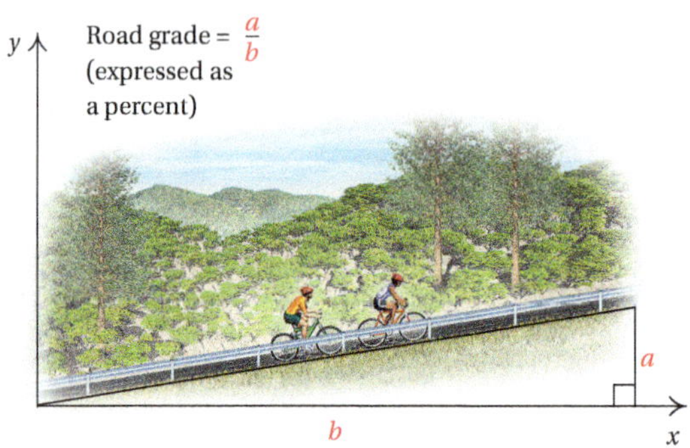

The concept of grade also occurs in skiing or snowboarding, where a 7% grade is considered very tame, but a 70% grade is considered extremely steep.

EXAMPLE 9 *Dubai Ski Run.* Dubai Ski Resort has the fifth longest indoor ski run in the world. It drops 197 ft over a horizontal distance of 1297 ft. Find the grade of the ski run.

197 ft
1297 ft

The grade of the ski run is its slope, expressed as a percent:

$$m = \frac{197}{1297} \begin{array}{l} \longleftarrow \text{Vertical distance} \\ \longleftarrow \text{Horizontal distance} \end{array}$$

$$\approx 0.15$$

$$\approx 15\%$$

◀ **Do Exercise 11.**

11. *Grade of a Treadmill.* During a stress test, a physician may change the grade, or slope, of a treadmill to measure its effect on heart rate (number of beats per minute). Find the grade, or slope, of the treadmill shown below.

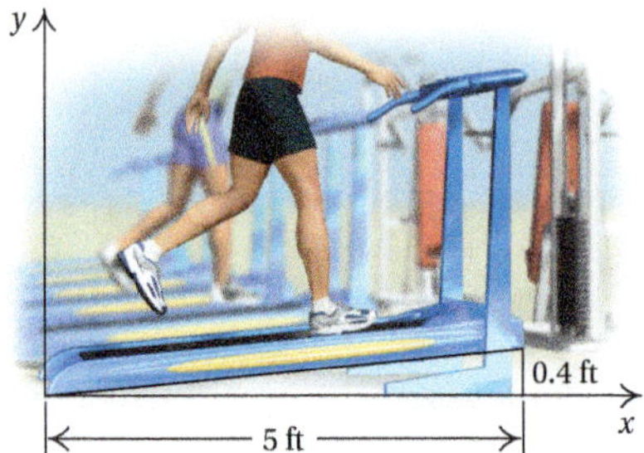

Answer

11. 8%

Slope can also be considered as a **rate of change**.

EXAMPLE 10 *Car Assembly Line.* Cameron, a supervisor in a car assembly plant, prepared the following graph to display data from a recent day's work. Use the graph to determine the slope, or the rate of change of the number of cars that came off an assembly line with respect to time.

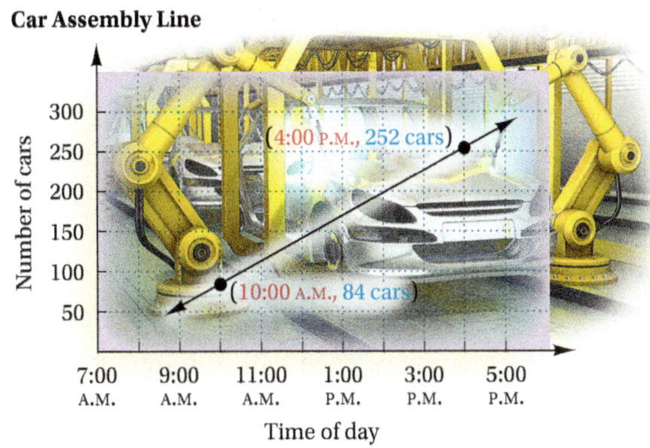

The vertical axis of the graph shows the number of cars, and the horizontal axis shows the time, in units of one hour. We can describe the rate of change of the number of cars with respect to time as

$$\frac{\text{Cars}}{\text{Hours}}, \quad \text{or} \quad \text{number of cars per hour.}$$

This value is the slope of the line. We determine two ordered pairs on the graph—in this case,

(10:00 A.M., 84 cars) and (4:00 P.M., 252 cars).

This tells us that in the 6 hr between 10:00 A.M. and 4:00 P.M., 252 − 84, or 168, cars came off the assembly line. Thus,

$$\begin{aligned}\text{Rate of change} &= \frac{252 \text{ cars} - 84 \text{ cars}}{4{:}00 \text{ P.M.} - 10{:}00 \text{ A.M.}}\\ &= \frac{168 \text{ cars}}{6 \text{ hours}}\\ &= 28 \text{ cars per hour.}\end{aligned}$$

Do Exercise 12. ▶

12. *Masonry.* Daryl, a mason, graphed data from a recent day's work. Use the following graph to determine the slope, or the rate of change of the number of bricks he can lay with respect to time.

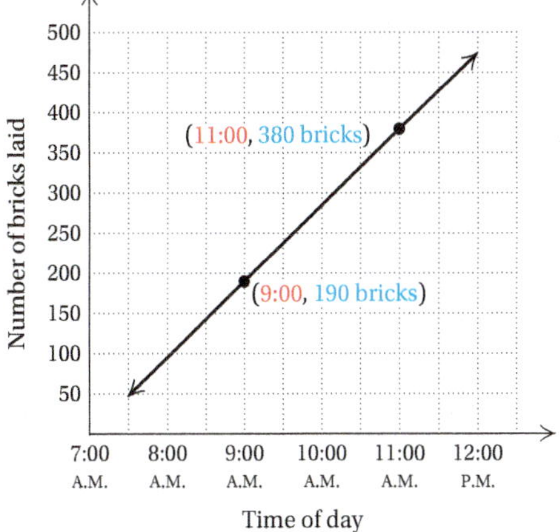

EXAMPLE 11 *Advertising Revenue.* Print-newspaper advertising revenue has been decreasing since 2005. Use the following graph to determine the slope, or rate of change in the advertising revenue with respect to time.

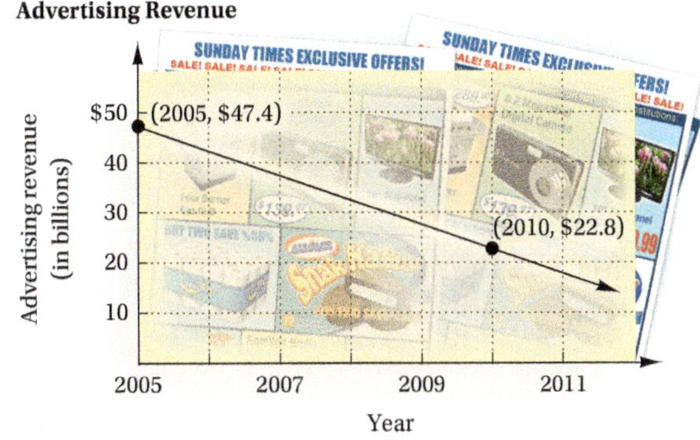

SOURCE: Research Department, Newspaper Association of America

Answer

12. 95 bricks per hour

13. ***Sunday Newspapers.*** Use the following graph to determine the rate of change in the circulation of Sunday newspapers since 2005.

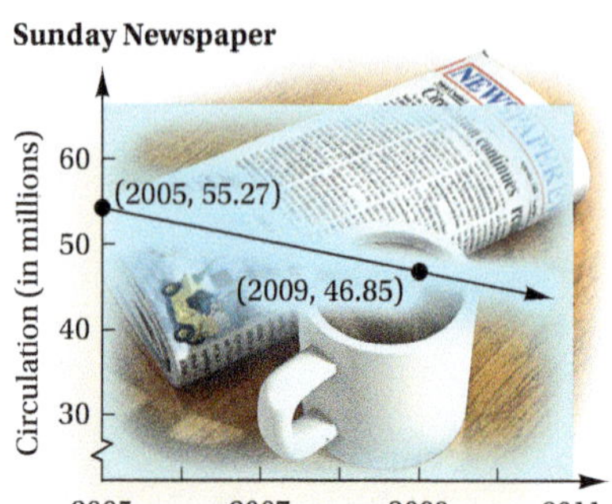

SOURCE: *Editor & Publisher International Yearbook*, 2010

Answer

13. About −2.11 million Sunday newspapers per year

The vertical axis of the graph shows the advertising revenue, in billions of dollars, and the horizontal axis shows the years. We can describe the rate of change in the advertising revenue with respect to time as

$$\frac{\text{Change in advertising revenue}}{\text{Years}}, \quad \text{or} \quad \text{change in advertising revenue per year.}$$

This value is the slope of the line. We determine two ordered pairs on the graph—in this case,

$$(2005, \$47.4) \quad \text{and} \quad (2010, \$22.8).$$

This tells us that in the 5 years from 2005 to 2010, newspaper advertising revenue dropped from \$47.4 billion to \$22.8 billion. Thus,

$$\text{Rate of change} = \frac{\$22.8 - \$47.4}{2010 - 2005}$$
$$= \frac{-\$24.6}{5} \approx -\$4.9 \text{ billion per year.}$$

◀ **Do Exercise 13.**

11.3 Exercise Set

For Extra Help MyMathLab® MathXL® PRACTICE WATCH READ REVIEW

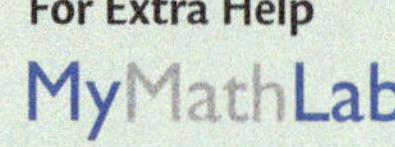

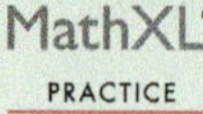

✓ Reading Check

Match each expression with an appropriate description or value from the column on the right.

RC1. Slope of a horizontal line

RC2. y-intercept of $y = mx + b$

RC3. Change in x

RC4. Slope of a vertical line

RC5. Slope

RC6. Change in y

a) Rise
b) Run
c) Rise/run
d) 0
e) Not defined
f) $(0, b)$

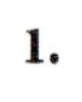

Find the slope, if it exists, of each line.

1.

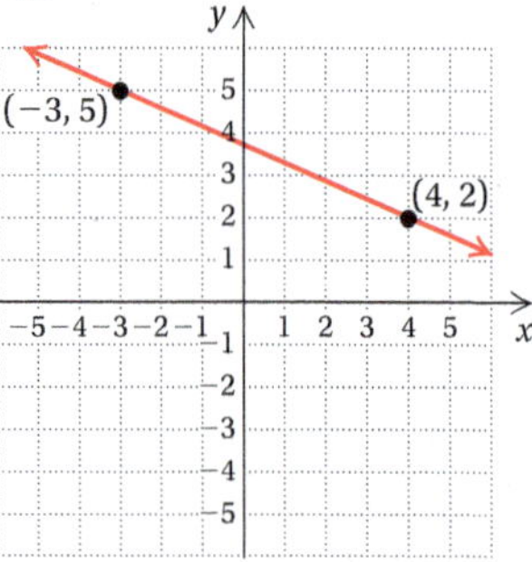

2.

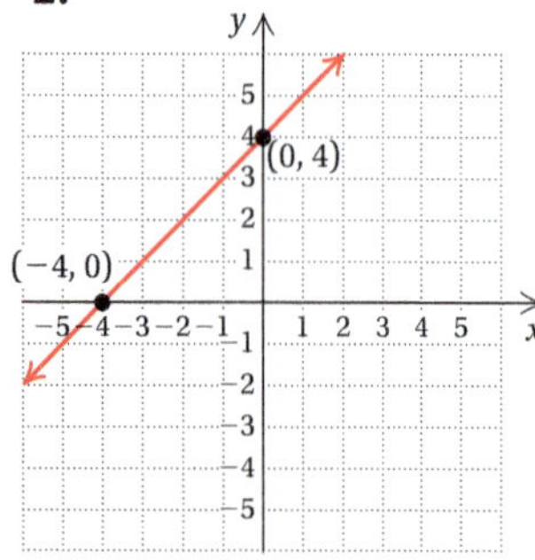

3.

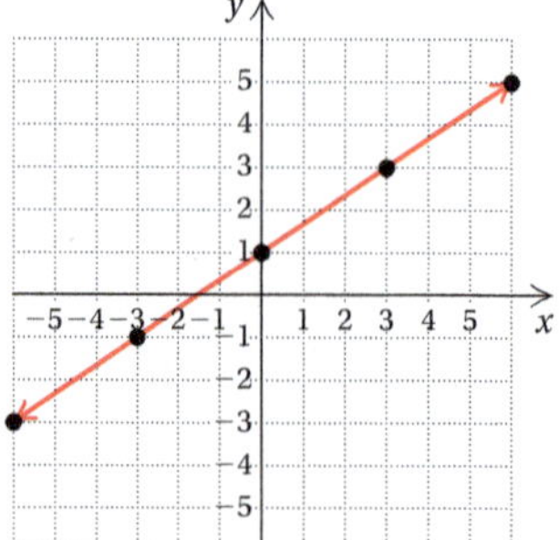

4.

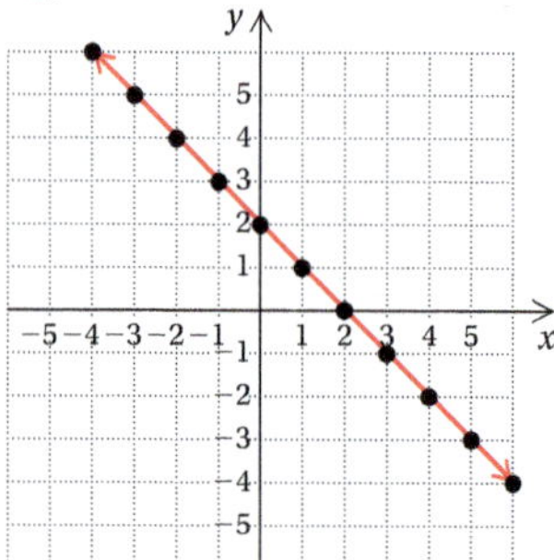

5.

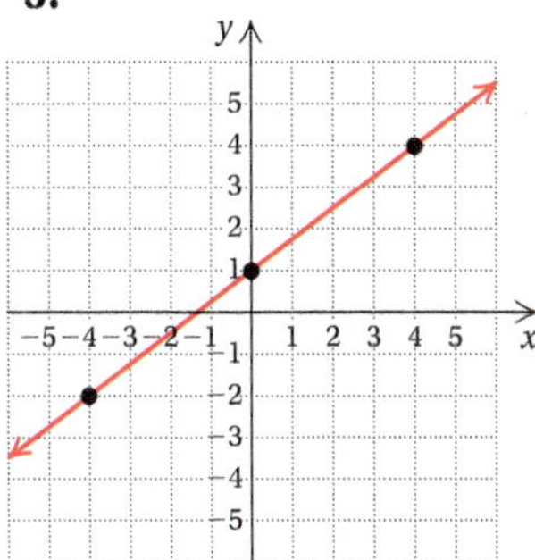

6.

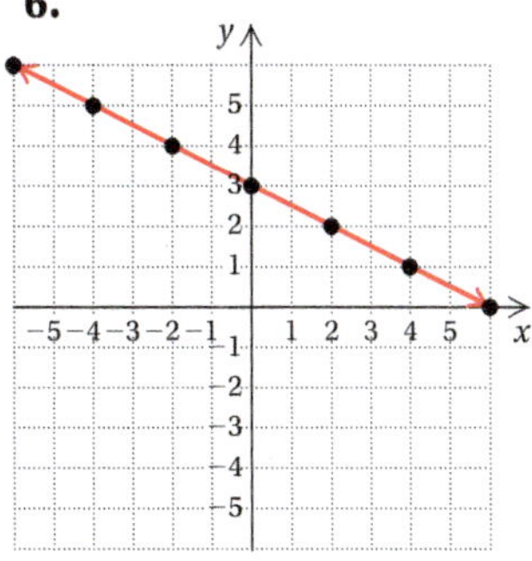

7.

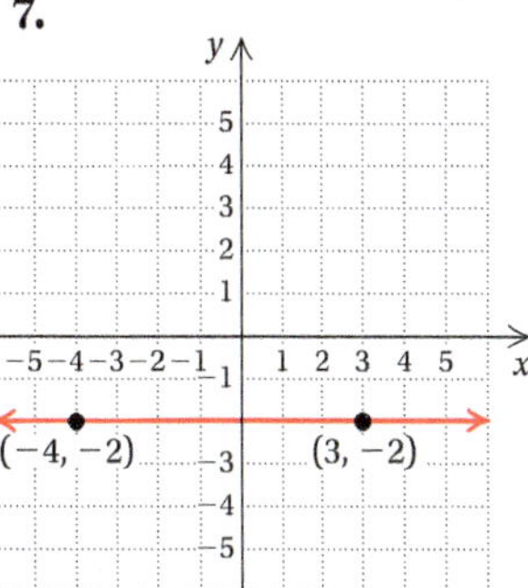

8.

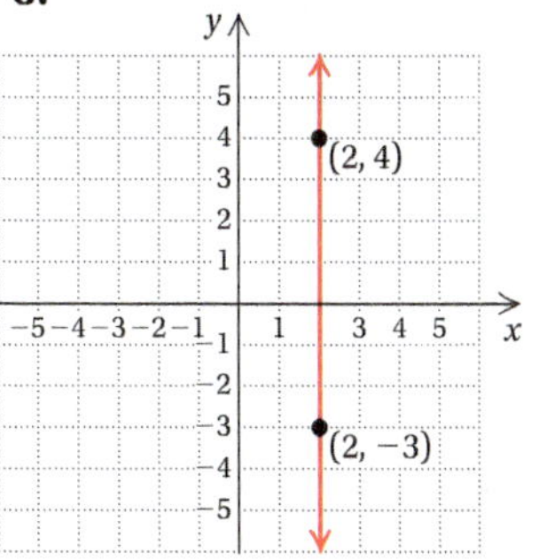

Graph the line containing the given pair of points and find the slope.

9. $(-2, 4), (3, 0)$

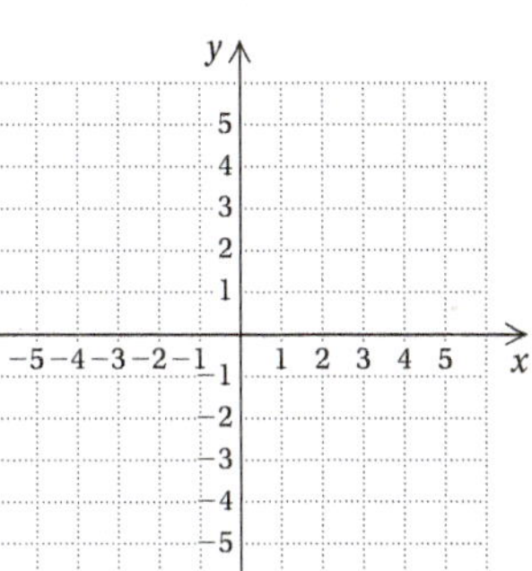

10. $(2, -4), (-3, 2)$

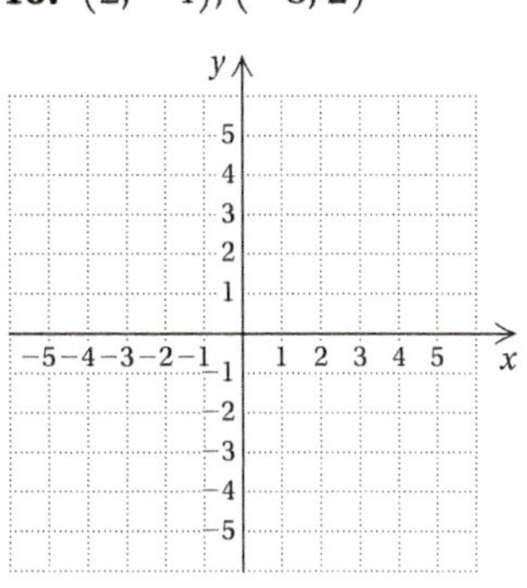

11. $(-4, 0), (-5, -3)$

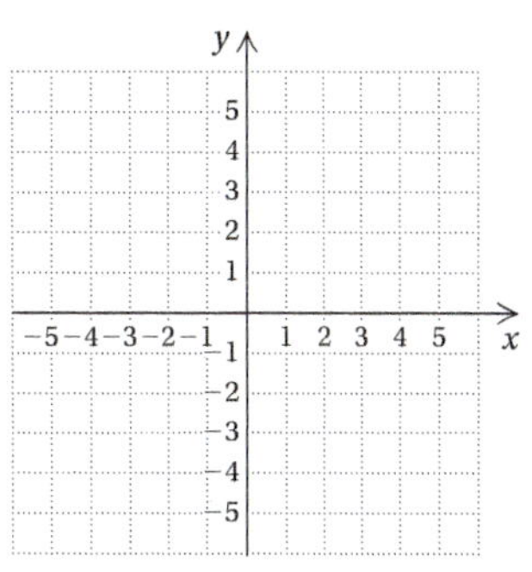

12. $(-3, 0), (-5, -2)$

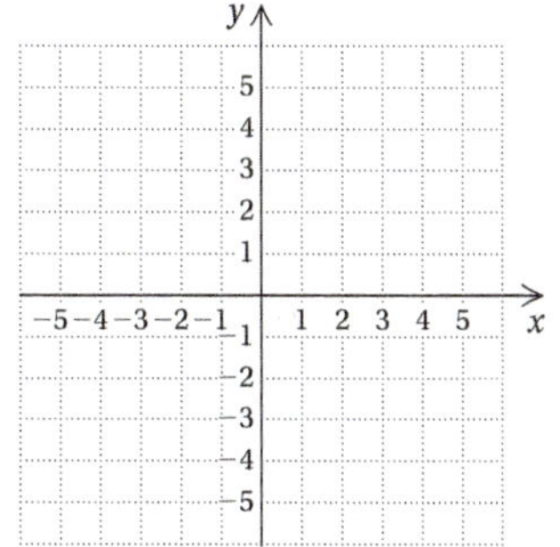

13. $(-4, 1), (2, -3)$

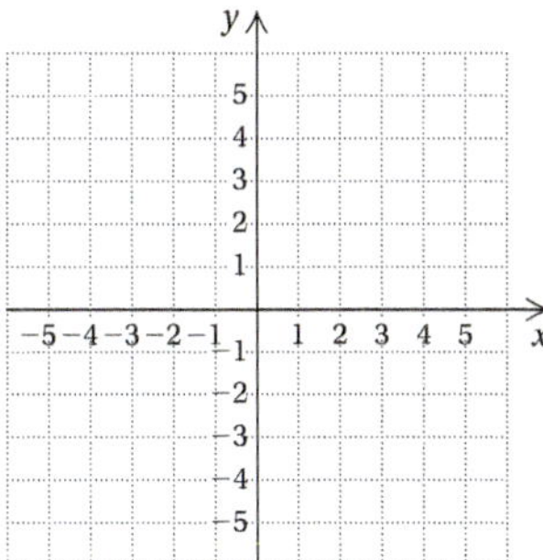

14. $(-3, 5), (4, -3)$

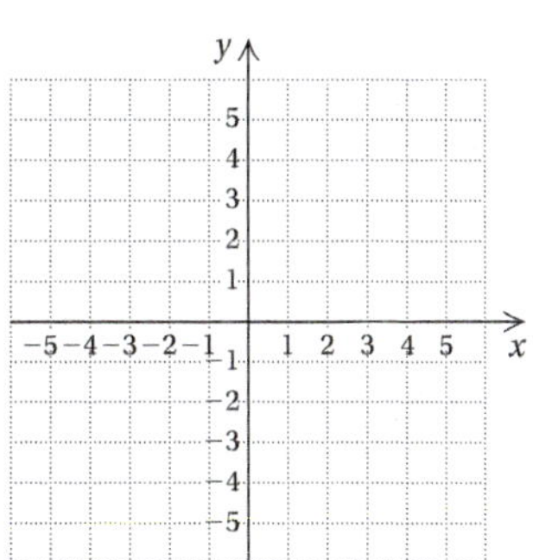

15. $(5, 3), (-3, -4)$

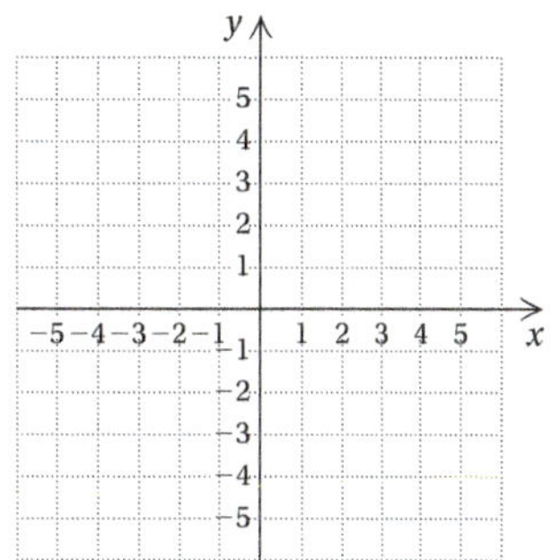

16. $(-4, -3), (2, 5)$

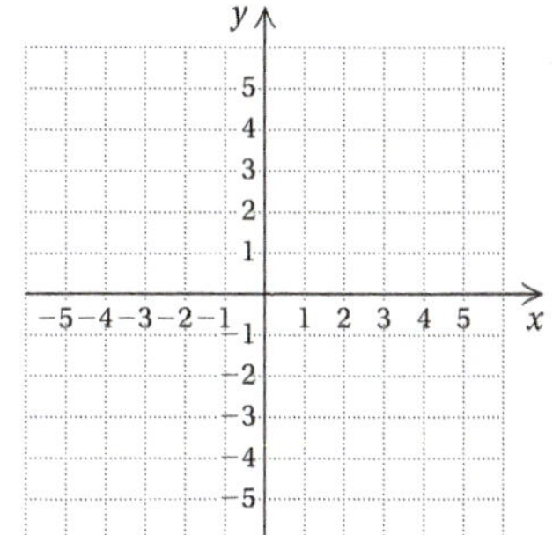

Find the slope, if it exists, of the line containing the given pair of points.

17. $\left(2, -\frac{1}{2}\right), \left(5, \frac{3}{2}\right)$

18. $\left(\frac{2}{3}, -1\right), \left(\frac{5}{3}, 2\right)$

19. $(4, -2), (4, 3)$

20. $(4, -3), (-2, -3)$

21. $(-11, 7), (15, -3)$

22. $(-13, 22), (8, -17)$

23. $\left(-\frac{1}{2}, \frac{3}{11}\right), \left(\frac{5}{4}, \frac{3}{11}\right)$

24. $(0.2, 4), (0.2, -0.04)$

b Find the slope, if it exists, of each line.

25. $y = -10x$

26. $y = \frac{10}{3}x$

27. $y = 3.78x - 4$

28. $y = -\frac{3}{5}x + 28$

29. $3x - y = 4$

30. $-2x + y = 8$

31. $x + 5y = 10$

32. $x - 4y = 8$

33. $3x + 2y = 6$

34. $2x - 4y = 8$

35. $x = \frac{2}{15}$

36. $y = -\frac{1}{3}$

37. $y = 2 - x$

38. $y = \frac{3}{4} + x$

39. $9x = 3y + 5$

40. $4y = 9x - 7$

41. $5x - 4y + 12 = 0$

42. $16 + 2x - 8y = 0$

43. $y = 4$

44. $x = -3$

45. $x = \frac{3}{4}y - 2$

46. $3x - \frac{1}{5}y = -4$

47. $\frac{2}{3}y = -\frac{7}{4}x$

48. $-x = \frac{2}{11}y$

c In each of Exercises 49–52, find the slope (or rate of change).

49. Find the slope (or pitch) of the roof.

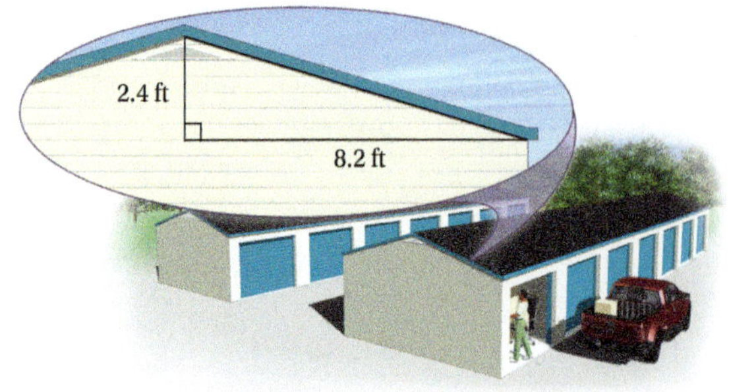

50. Find the slope (or grade) of the road.

51. ***Slope of a River.*** When a river flows, the strength or force of the river depends on how far the river falls vertically compared to how far it flows horizontally. Find the slope of the river shown below.

52. ***Constructing Stairs.*** Carpenters use slope when designing and building stairs. Public buildings normally include steps with 7-in. risers and 11-in. treads. Find the grade of such a stairway.

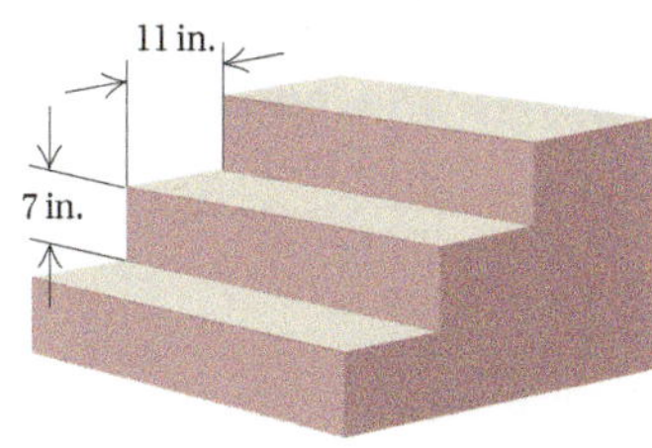

53. ***Grade of a Transit System.*** The maximum grade allowed between two stations in a rapid-transit rail system is 3.5%. Between station A and station B, which are 280 ft apart, the tracks rise $8\frac{1}{2}$ ft. What is the grade of the tracks between these two stations? Round the answer to the nearest tenth of a percent. Does this grade meet the rapid-transit rail standards?

Source: Brian Burell, *Merriam Webster's Guide to Everyday Math*, Merriam-Webster, Inc., Springfield MA

54. ***Slope of Long's Peak.*** From a base elevation of 9600 ft, Long's Peak in Colorado rises to a summit elevation of 14,256 ft over a horizontal distance of 15,840 ft. Find the grade of Long's Peak.

In each of Exercises 55–58, use the graph to calculate a rate of change in which the units of the horizontal axis are used in the denominator.

55. ***Kindergarten in China.*** The number of children enrolled in kindergarten in China is projected to increase from 26.58 million in 2009 to 34 million in 2015. Use the following graph to find the rate of change, rounded to the nearest hundredth of a million, in the number of children enrolled in kindergarten in China with respect to time.

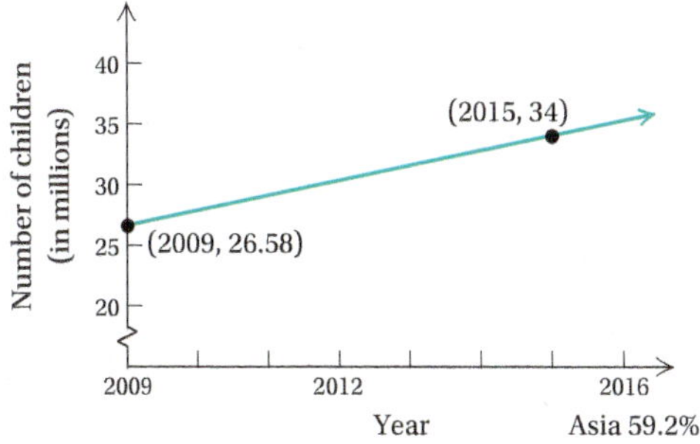

SOURCE: *China Daily*, Lin Qi and Guo Shuhan report, June 23, 2011, p.18

56. ***Injuries on Farms.*** In 1998, there were 37,775 injuries on farms to people under age 20. Since then, this number has steadily decreased. Using the following graph, find the rate of change, rounded to the nearest ten, in the number of farm injuries to people under age 20 with respect to time.

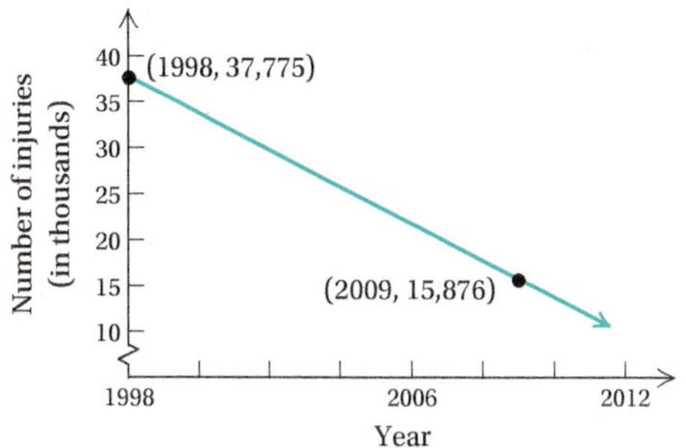

SOURCE: National Institute for Occupational Health and Safety

57. ***Population Decrease of New Orleans.*** The change in the population of New Orleans, Louisiana, is illustrated in the following graph. Find the rate of change, to the nearest hundred, in the population with respect to time.

Population Decrease of New Orleans

Population (in thousands)

New Orleans

Louisiana

(2000, 484,949)

(2010, 343,829)

Year

SOURCE: Decennial Census, U. S. Census Bureau

58. ***Population Increase of Charlotte.*** The change in the population of Charlotte, North Carolina, is illustrated in the following graph. Find the rate of change, to the nearest hundred, in the population with respect to time.

Population Increase of Charlotte

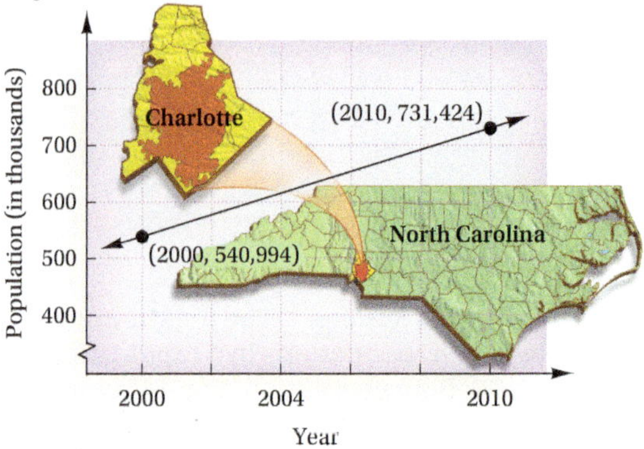

SOURCE: Decennial Census, U. S. Census Bureau

59. ***Production of Blueberries.*** U.S. production of blueberries is continually increasing. In 2006, 358,000,000 lb of blueberries were produced. By 2011, this amount had increased to 511,000,000 lb. Find the rate of change in the production of blueberries with respect to time.

Source: U.S. Department of Agriculture

60. ***Bottled Water.*** Bottled water consumption per person per year in the United States has increased from 16.7 gal in 2000 to 29.2 gal in 2011. Find the rate of change, rounded to the nearest tenth, in the number of gallons of bottled water consumed annually per person per year.

Sources: Beverage Marketing Corporation; International Bottled Water Association

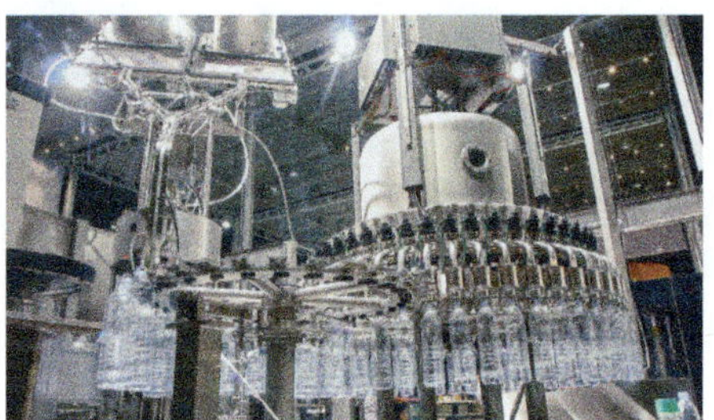

Skill Maintenance

Convert to fraction notation. [6.3b]

61. 16%

62. 37.5%

Collect like terms. [9.7e]

63. $\frac{1}{3}p - p$

64. $t - 6 + 4t + 5$

Synthesis

In each of Exercises 65–68, find an equation for the graph shown.

65.

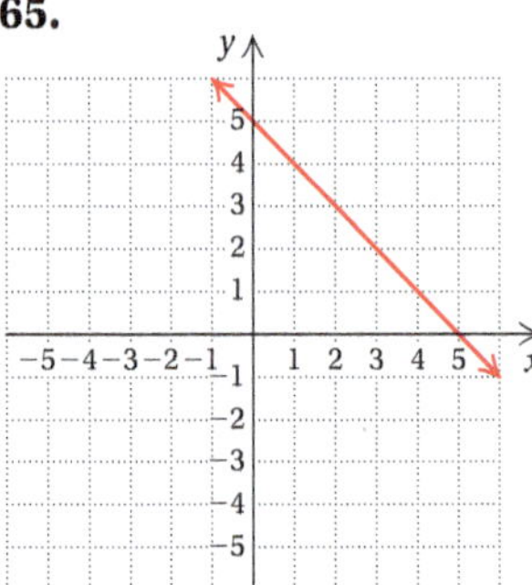

66.

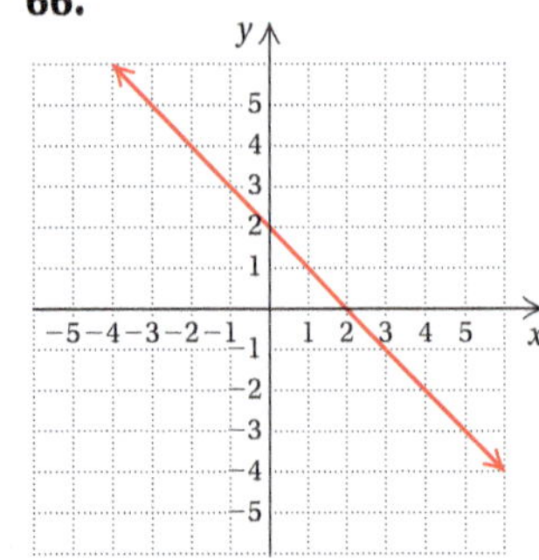

67.

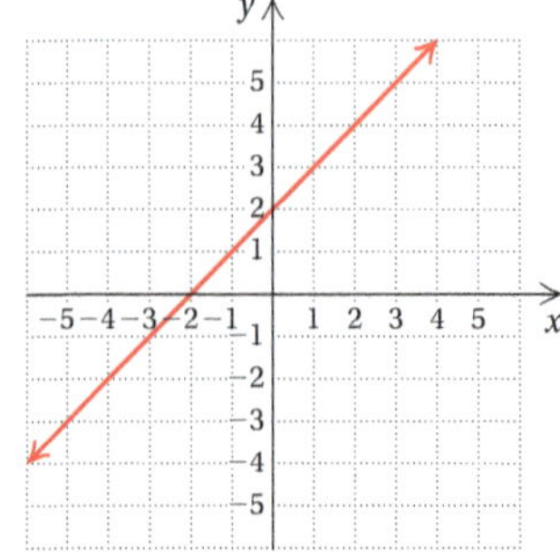

68.

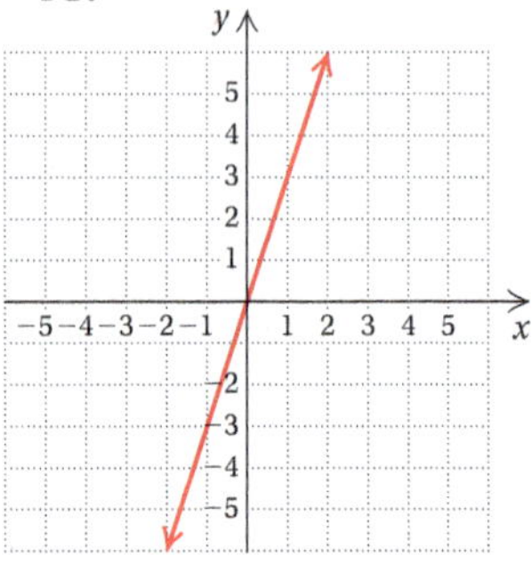

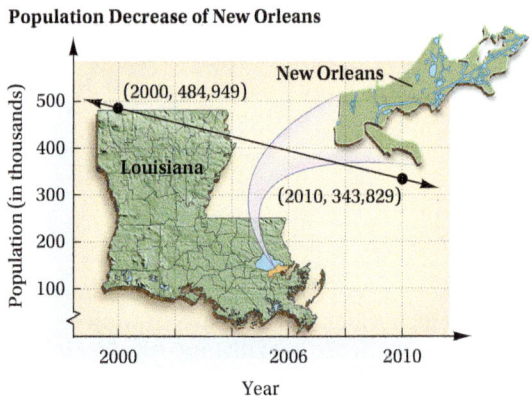

Equations of Lines

11.4

OBJECTIVES

a Given an equation in the form $y = mx + b$, find the slope and the y-intercept; find an equation of a line when the slope and the y-intercept are given.

b Find an equation of a line when the slope and a point on the line are given.

c Find an equation of a line when two points on the line are given.

SKILL TO REVIEW

Objective 11.3a: Given the coordinates of two points on a line, find the slope of the line, if it exists.

Find the slope, if it exists, of the line containing the given pair of points.

1. $(3, 0), (0, 3)$
2. $(-8, 5), (-8, -5)$

We have learned that the slope of a line and the y-intercept of the graph of the line can be read directly from the equation if it is in the form $y = mx + b$. Here we use slope and y-intercept in order to examine linear equations in more detail.

a FINDING AN EQUATION OF A LINE WHEN THE SLOPE AND THE y-INTERCEPT ARE GIVEN

We know from Sections 11.1 and 11.3 that in the equation $y = mx + b$, the **slope** is m and the **y-intercept** is $(0, b)$. Therefore, we call the equation $y = mx + b$ the **slope–intercept equation.***

THE SLOPE–INTERCEPT EQUATION: $y = mx + b$

The equation $y = mx + b$ is called the **slope–intercept equation.** The slope is m and the y-intercept is $(0, b)$.

EXAMPLE 1 Find the slope and the y-intercept of $2x - 3y = 8$.

We first solve for y:

$$2x - 3y = 8$$
$$-3y = -2x + 8 \quad \text{Subtracting } 2x$$
$$\frac{-3y}{-3} = \frac{-2x + 8}{-3} \quad \text{Dividing by } -3$$
$$y = \frac{-2x}{-3} + \frac{8}{-3}$$
$$y = \frac{2}{3}x - \frac{8}{3}.$$

The slope is $\frac{2}{3}$.

The y-intercept is $\left(0, -\frac{8}{3}\right)$.

$2x - 3y = 8$

Do Margin Exercises 1–5.

Find the slope and the y-intercept.

1. $y = 5x$
2. $y = -\frac{3}{2}x - 6$
3. $3x + 4y = 15$
4. $y = 10 + x$
5. $-7x - 5y = 22$

EXAMPLE 2 A line has slope -2.4 and y-intercept $(0, 11)$. Find an equation of the line.

We use the slope–intercept equation and substitute -2.4 for m and 11 for b:

$$y = mx + b$$
$$y = -2.4x + 11. \quad \text{Substituting}$$

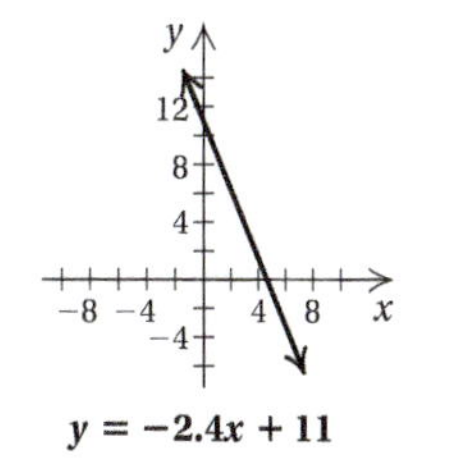

*Equations of lines can also be found using the *point–slope equation*, $y - y_1 = m(x - x_1)$. See Appendix G.

Answers

Skill to Review:
1. -1 2. Not defined

Margin Exercises:
Answers to Margin Exercises 1–5 are on p. 816.

EXAMPLE 3 A line has slope 0 and y-intercept $(0, -6)$. Find an equation of the line.

We use the slope-intercept equation and substitute 0 for m and -6 for b:

$$y = mx + b$$
$$y = 0x + (-6) \quad \text{Substituting}$$
$$y = -6.$$

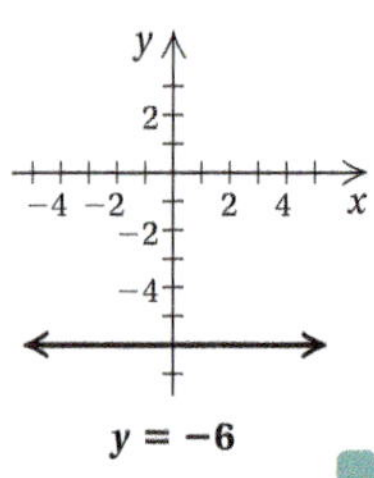

EXAMPLE 4 A line has slope $-\frac{5}{3}$ and y-intercept $(0, 0)$. Find an equation of the line.

We use the slope-intercept equation and substitute $-\frac{5}{3}$ for m and 0 for b:

$$y = mx + b$$
$$y = -\tfrac{5}{3}x + 0 \quad \text{Substituting}$$
$$y = -\tfrac{5}{3}x.$$

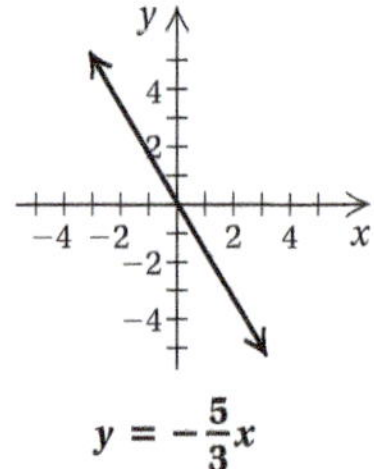

6. A line has slope 3.5 and y-intercept $(0, -23)$. Find an equation of the line.

7. A line has slope 0 and y-intercept $(0, 13)$. Find an equation of the line.

8. A line has slope -7.29 and y-intercept $(0, 0)$. Find an equation of the line.

Do Exercises 6–8.

b FINDING AN EQUATION OF A LINE WHEN THE SLOPE AND A POINT ARE GIVEN

Suppose we know the slope of a line and a certain point on that line. We can use the slope-intercept equation $y = mx + b$ to find an equation of the line. To write an equation in this form, we need to know the slope m and the y-intercept $(0, b)$.

EXAMPLE 5 Find an equation of the line with slope 3 that contains the point $(4, 1)$.

We know that the slope is 3, so the equation is $y = 3x + b$. This equation is true for $(4, 1)$. Using the point $(4, 1)$, we substitute 4 for x and 1 for y in $y = 3x + b$. Then we solve for b:

$$y = 3x + b \quad \text{Substituting 3 for } m \text{ in } y = mx + b$$
$$1 = 3(4) + b \quad \text{Substituting 4 for } x \text{ and 1 for } y$$
$$1 = 12 + b$$
$$-11 = b. \quad \text{Solving for } b \text{, we find that the } y\text{-intercept is } (0, -11).$$

We use the equation $y = mx + b$ and substitute 3 for m and -11 for b:

$$y = 3x - 11.$$

Answers

Margin Exercises:

1. Slope: 5; y-intercept: $(0, 0)$
2. Slope: $-\frac{3}{2}$; y-intercept: $(0, -6)$
3. Slope: $-\frac{3}{4}$; y-intercept: $\left(0, \frac{15}{4}\right)$
4. Slope: 1; y-intercept: $(0, 10)$
5. Slope: $-\frac{7}{5}$; y-intercept: $\left(0, -\frac{22}{5}\right)$
6. $y = 3.5x - 23$ 7. $y = 13$
8. $y = -7.29x$

This is the equation of the line with slope 3 and y-intercept $(0, -11)$. Note that $(4, 1)$ is on the line.

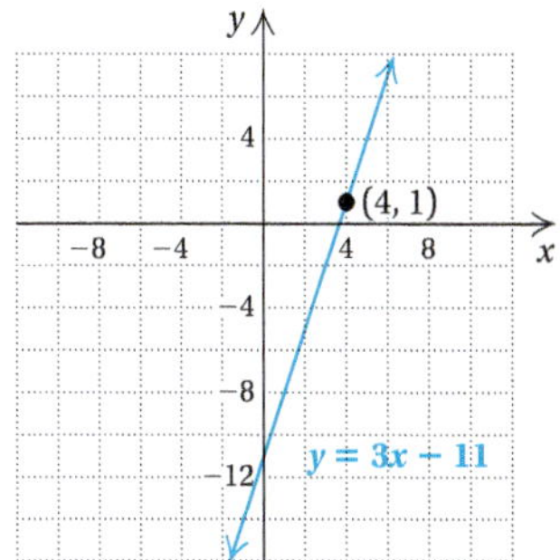

EXAMPLE 6 Find an equation of the line with slope -5 that contains the point $(-2, 3)$.

We know that the slope is -5, so the equation is $y = -5x + b$. This equation is true for all points on the line, including the point $(-2, 3)$. Using the point $(-2, 3)$, we substitute -2 for x and 3 for y in $y = -5x + b$. Then we solve for b:

$$y = -5x + b \qquad \text{Substituting } -5 \text{ for } m \text{ in } y = mx + b$$

$$3 = -5(-2) + b \qquad \text{Substituting } -2 \text{ for } x \text{ and } 3 \text{ for } y$$

$$3 = 10 + b$$

$$-7 = b. \qquad \text{Solving for } b$$

We use the equation $y = mx + b$ and substitute -5 for m and -7 for b:

$$y = -5x - 7.$$

This is the equation of the line with slope -5 and y-intercept $(0, -7)$.

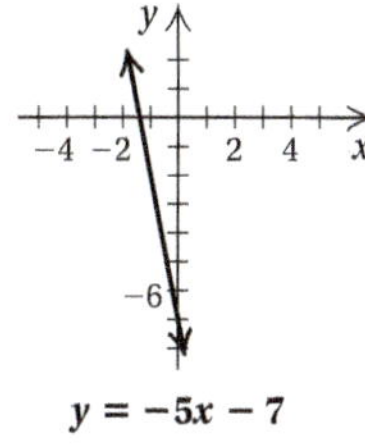

Do Exercises 9–12. ▶

Find an equation of the line that contains the given point and has the given slope.

9. $(4, 2)$, $m = 5$

10. $(-2, 1)$, $m = -3$

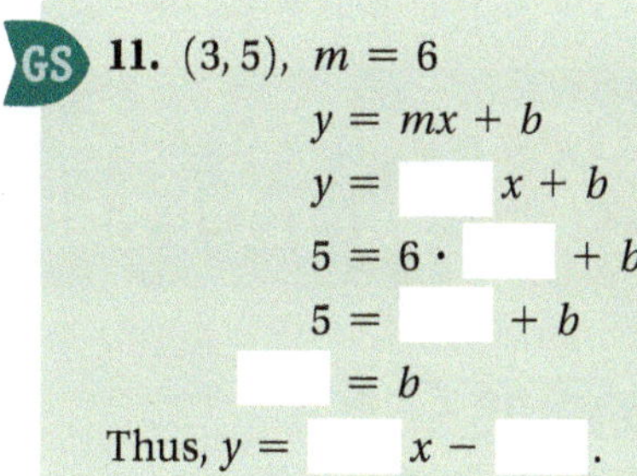

GS **11.** $(3, 5)$, $m = 6$

$y = mx + b$

$y = \square x + b$

$5 = 6 \cdot \square + b$

$5 = \square + b$

$\square = b$

Thus, $y = \square x - \square$.

12. $(1, 4)$, $m = -\frac{2}{3}$

C FINDING AN EQUATION OF A LINE WHEN TWO POINTS ARE GIVEN

We can also use the slope–intercept equation to find an equation of a line when two points are given.

EXAMPLE 7 Find an equation of the line containing the points $(2, 3)$ and $(-2, 2)$.

First, we find the slope:

$$m = \frac{3 - 2}{2 - (-2)} = \frac{1}{4}.$$

Thus, $y = \frac{1}{4}x + b$. We then proceed as we did in Example 6, using either point to find b, since both points are on the line.

Answers

9. $y = 5x - 18$ **10.** $y = -3x - 5$
11. $y = 6x - 13$ **12.** $y = -\frac{2}{3}x + \frac{14}{3}$

Guided Solution:
11. 6, 3, 18, −13, 6, 13

Find an equation of the line containing the given points.

13. $(2, 4)$ and $(3, 5)$

14. $(-1, 2)$ and $(-3, -2)$ GS

First, determine the slope:

$$m = \frac{-2 - 2}{-3 - (\quad)} = \frac{\quad}{-2} = 2;$$

$y = mx + b,$

$y = \quad x + b.$

Use either point to determine b. Let's use $(-3, -2)$:

$\quad = 2 \cdot \quad + b$

$-2 = -6 + b$

$\quad = b.$

Thus, $y = 2x + \quad$.

We choose $(2, 3)$ and substitute 2 for x and 3 for y:

$y = \frac{1}{4}x + b$ Substituting $\frac{1}{4}$ for m in $y = mx + b$

$3 = \frac{1}{4} \cdot 2 + b$ Substituting 2 for x and 3 for y

$3 = \frac{1}{2} + b$

$\frac{5}{2} = b.$ Solving for b

We use the equation $y = mx + b$ and substitute $\frac{1}{4}$ for m and $\frac{5}{2}$ for b:

$$y = \frac{1}{4}x + \frac{5}{2}.$$

This is the equation of the line with slope $\frac{1}{4}$ and y-intercept $\left(0, \frac{5}{2}\right)$. Note that the line contains the points $(2, 3)$ and $(-2, 2)$.

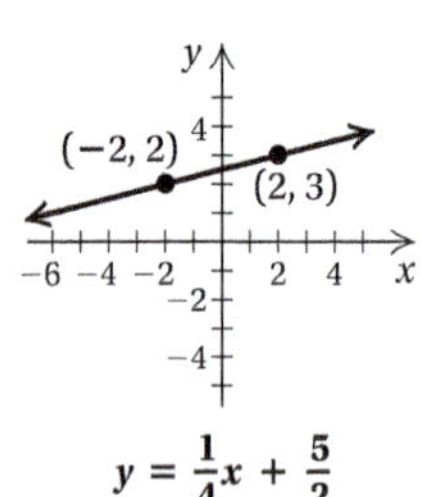

$y = \frac{1}{4}x + \frac{5}{2}$

Answers

13. $y = x + 2$ **14.** $y = 2x + 4$

Guided Solution:

14. $-1, -4, 2, -2, -3, 4, 4$

◀ **Do Exercises 13 and 14.**

11.4 Exercise Set

For Extra Help

MyMathLab®

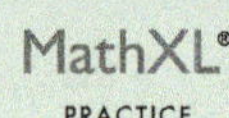

✓ Reading Check

Match each graph with the appropriate equation from the column on the right.

RC1.

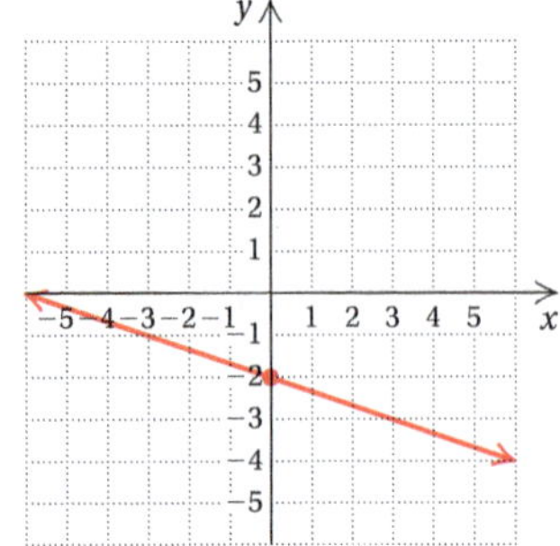

RC2.

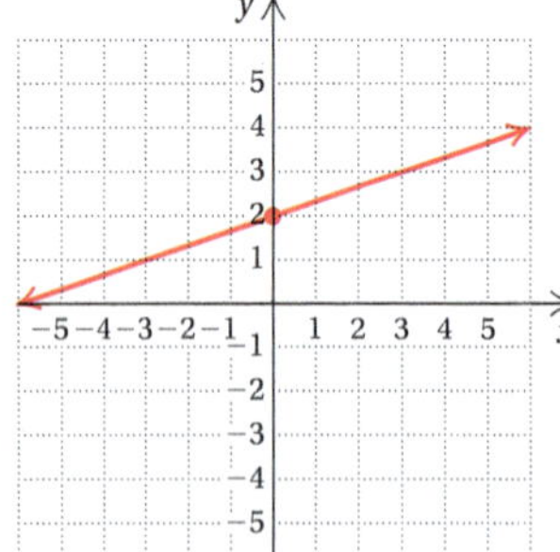

RC3.

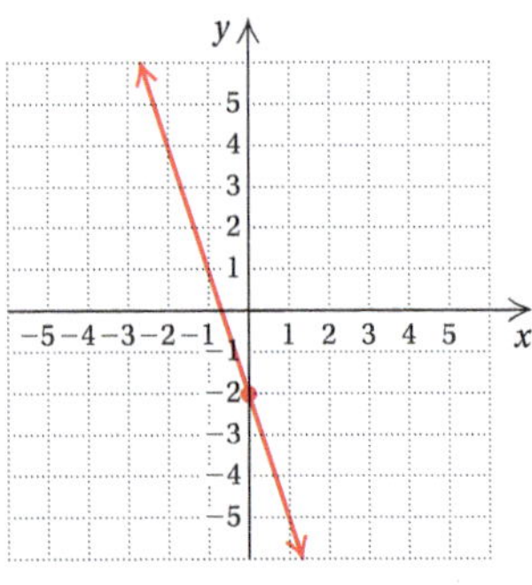

RC4.

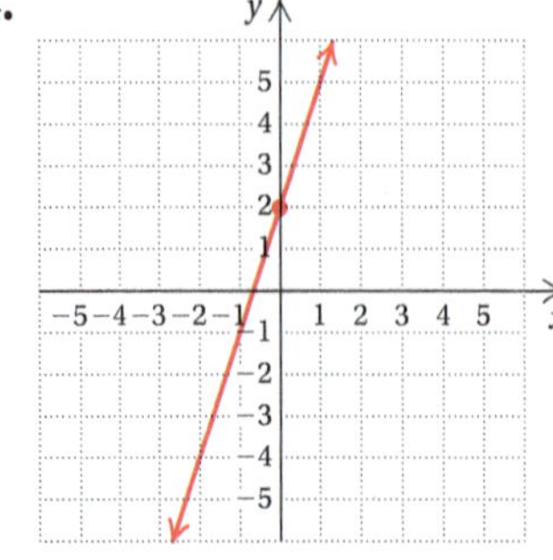

a) $y = -3x - 2$

b) $y = -3x + 2$

c) $y = \frac{1}{3}x - 2$

d) $y = -\frac{1}{3}x + 2$

e) $y = -\frac{1}{3}x - 2$

f) $y = \frac{1}{3}x + 2$

g) $y = 3x + 2$

h) $y = 3x - 2$

a Find the slope and the y-intercept.

1. $y = -4x - 9$

2. $y = 2x + 3$

3. $y = 1.8x$

4. $y = -27.4x$

5. $-8x - 7y = 21$

6. $-2x - 8y = 16$

7. $4x = 9y + 7$

8. $5x + 4y = 12$

9. $-6x = 4y + 2$

10. $4.8x - 1.2y = 36$

11. $y = -17$

12. $y = 28$

Find an equation of the line with the given slope and y-intercept.

13. Slope $= -7$, y-intercept $= (0, -13)$

14. Slope $= 73$, y-intercept $= (0, 54)$

15. Slope $= 1.01$, y-intercept $= (0, -2.6)$

16. Slope $= -\frac{3}{8}$, y-intercept $= \left(0, \frac{7}{11}\right)$

17. Slope $= 0$, y-intercept $= (0, -5)$

18. Slope $= \frac{6}{5}$, y-intercept $= (0, 0)$

b Find an equation of the line containing the given point and having the given slope.

19. $(-3, 0)$, $m = -2$

20. $(2, 5)$, $m = 5$

21. $(2, 4)$, $m = \frac{3}{4}$

22. $\left(\frac{1}{2}, 2\right)$, $m = -1$

23. $(2, -6)$, $m = 1$

24. $(4, -2)$, $m = 6$

25. $(0, 3)$, $m = -3$

26. $(-2, -4)$, $m = 0$

c Find an equation of the line that contains the given pair of points.

27. $(12, 16)$ and $(1, 5)$

28. $(-6, 1)$ and $(2, 3)$

29. $(0, 4)$ and $(4, 2)$

30. $(0, 0)$ and $(4, 2)$

31. $(3, 2)$ and $(1, 5)$

32. $(-4, 1)$ and $(-1, 4)$

33. $\left(4, -\frac{2}{5}\right)$ and $\left(4, \frac{2}{5}\right)$

34. $\left(\frac{3}{4}, -3\right)$ and $\left(\frac{1}{2}, -3\right)$

35. $(-4, 5)$ and $(-2, -3)$

36. $(-2, -4)$ and $(2, -1)$

37. $\left(-2, \frac{1}{4}\right)$ and $\left(3, \frac{1}{4}\right)$

38. $\left(\frac{3}{7}, -6\right)$ and $\left(\frac{3}{7}, -9\right)$

39. *Lottery Sales.* The following line graph shows increasing lottery sales in years t since 2000.

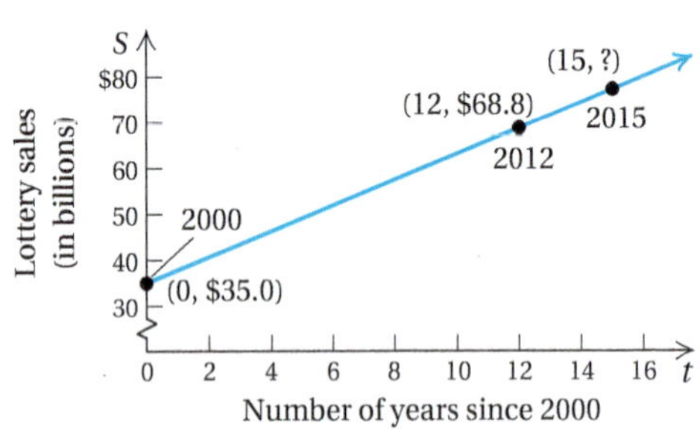

SOURCES: U.S. Census Bureau; National Association of State and Provincial Lotteries

a) Find an equation of the line.
b) What is the rate of change of lottery sales, in billions of dollars, with respect to time?
c) Use the equation to predict lottery sales in 2015.

40. *Aerobic Exercise.* The following line graph describes the *target heart rate T*, in number of beats per minute, of a person of age a, who is exercising. The goal is to get the number of beats per minute to this target level.

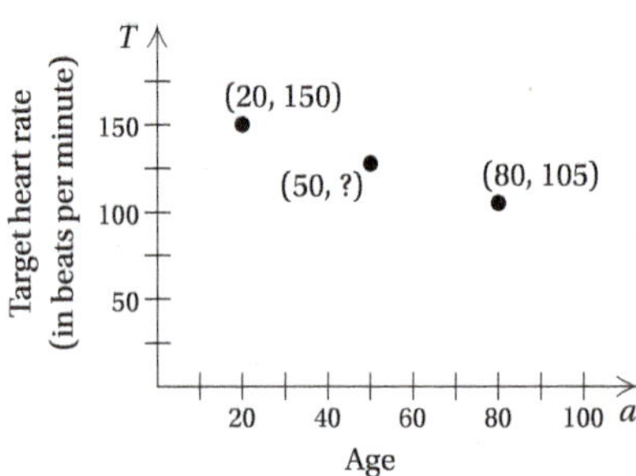

a) Find an equation of the line.
b) What is the rate of change of target heart rate with respect to time?
c) Use the equation to calculate the target heart rate of a person of age 50.

Skill Maintenance

Solve. [10.3c]

41. $3x - 4(9 - x) = 17$

42. $2(5 + 2y) + 4y = 13$

43. $4(a - 3) + 6 = 21 - \frac{1}{2}a$

44. $\frac{2}{3}(t - 3) = 6(9 - t)$

45. $40(2x - 7) = 50(4 - 6x)$

46. $\frac{2}{3}(x - 5) = \frac{3}{8}(x + 5)$

Solve. [10.5a]

47. What is 30% of 12?

48. 15 is 3% of what number?

49. What percent of 50 is 2.5?

50. 4240 is 106% of what number?

Synthesis

51. Find an equation of the line that contains the point $(2, -3)$ and has the same slope as the line $3x - y + 4 = 0$.

52. Find an equation of the line that has the same y-intercept as the line $x - 3y = 6$ and contains the point $(5, -1)$.

53. Find an equation of the line with the same slope as the line $3x - 2y = 8$ and the same y-intercept as the line $2y + 3x = -4$.

Mid-Chapter Review

Concept Reinforcement

Determine whether each statement is true or false.

______________ **1.** A slope of $-\frac{3}{4}$ is steeper than a slope of $-\frac{5}{2}$. [11.3a]

______________ **2.** The slope of the line that passes through (a, b) and (c, d) is $\dfrac{d-b}{c-a}$. [11.3a]

______________ **3.** The y-intercept of $Ax + By = C, B \neq 0$, is $\left(0, \dfrac{C}{B}\right)$. [11.2a]

______________ **4.** Both coordinates of points in quadrant IV are negative. [11.1a]

Guided Solutions

5. Given the graph of the line below, fill in the numbers that create correct statements. [11.2a], [11.3a], [11.4a]

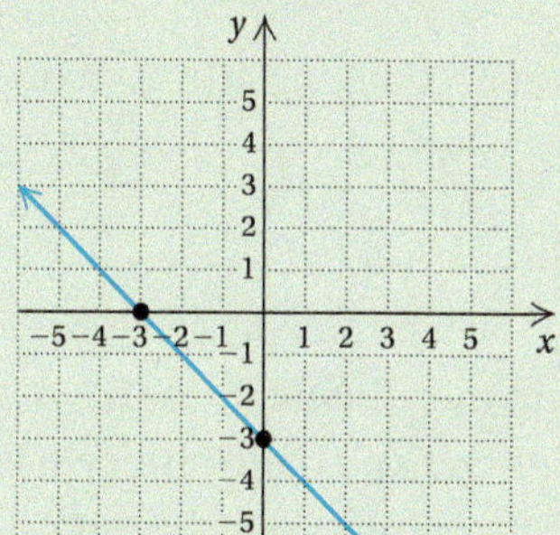

a) The ☐-intercept is (☐, −3).

b) The ☐-intercept is (☐, 0).

c) The slope is $\dfrac{-3 - \square}{\square - \square} = \dfrac{\square}{\square} = \square$.

d) The equation of the line in $y = mx + b$ form is

$y = \square x + \square$, or $y = -x - \square$.

6. Given the graph of the line below, fill in the letters and the number 0 that create correct statements. [11.2a], [11.3a], [11.4a]

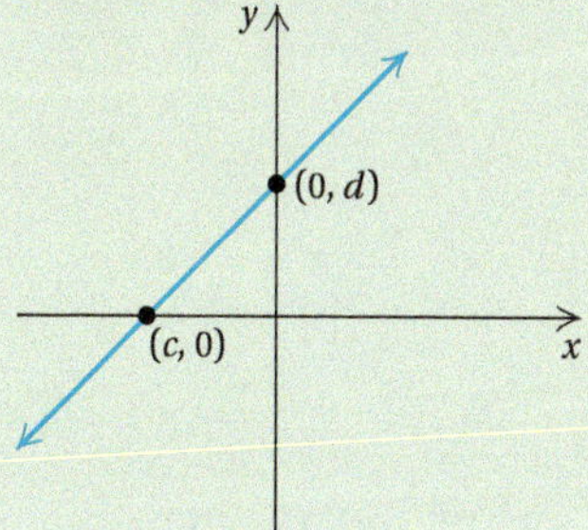

a) The x-intercept is (☐, ☐).

b) The y-intercept is (☐, ☐).

c) The slope is $\dfrac{\square - \square}{\square - c} = \dfrac{\square}{\square} = -\square$.

d) The equation of the line in $y = mx + b$ form is

$y = \square x + \square$.

Mixed Review

Determine whether the given ordered pair is a solution of the equation. [11.1c]

7. $(8, -5)$; $-2q - 7p = 19$

8. $\left(-1, \dfrac{2}{3}\right)$; $6y = -3x + 1$

Find the coordinates of the x-intercept and the y-intercept. [11.2a]

9. $-3x + 2y = 18$

10. $x - \dfrac{1}{2} = 10y$

Graph. [11.1d], [11.2a, b]

11. $-2x + y = -3$

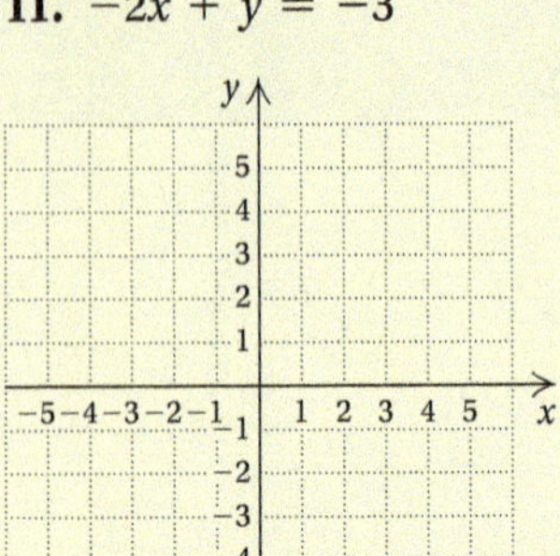

12. $y = -\frac{3}{2}$

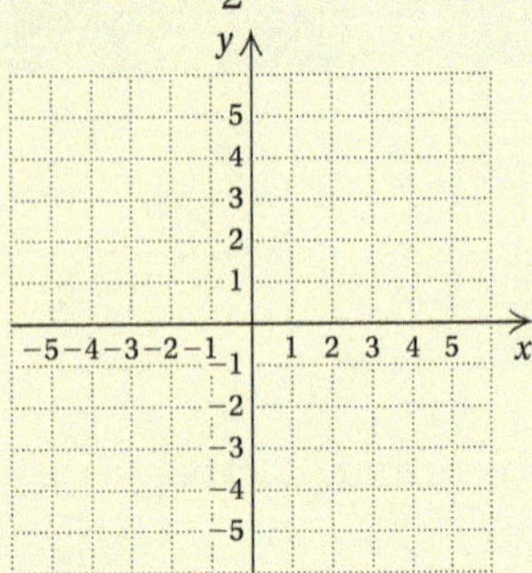

13. $y = -x + 4$

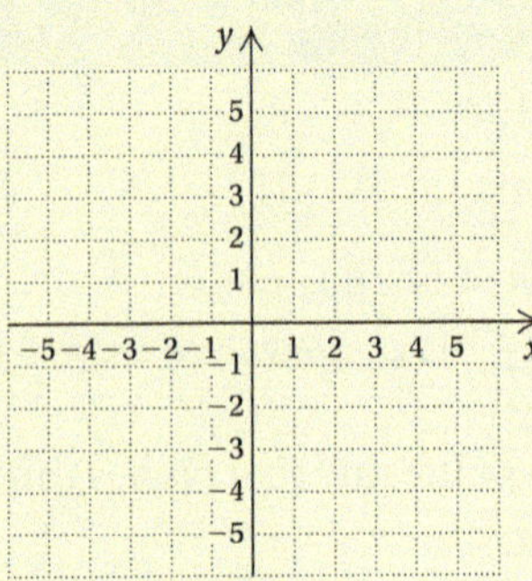

14. $x = 0$

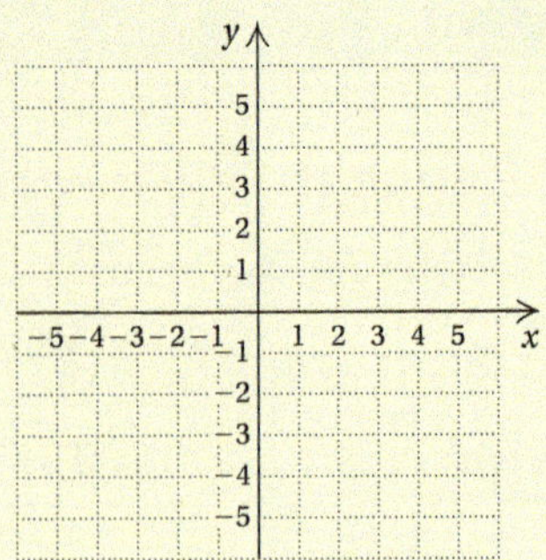

Find the slope, if it exists, of the line containing the given pair of points. [11.3a]

15. $\left(\frac{1}{4}, -6\right), (-2, 4)$

16. $(6, -3), (-6, 3)$

Find the slope, if it exists, of the line. [11.3b]

17. $y = 0.728$

18. $13x - y = -5$

19. $12x + 7 = 0$

20. The population of Texas in 2000 was 20,851,820. The population in 2010 was 25,145,561. Find the rate of change, to the nearest hundred, in the population with respect to time. [11.3c]

Match each equation with the appropriate characteristics from the column on the right. [11.2b], [11.4a, b]

21. $y = -1$

22. $x = 1$

23. $y = -x - 1$

24. $y = x - 1$

25. $y = x + 1$

A. The slope is 1 and the x-intercept is $(-1, 0)$.

B. The slope is -1 and the y-intercept is $(0, -1)$.

C. The slope is not defined and the x-intercept is $(1, 0)$.

D. The slope is 0 and the y-intercept is $(0, -1)$.

E. The slope is 1 and the x-intercept is $(1, 0)$.

26. Find an equation of the line with slope -3 that contains the point $\left(-\frac{1}{3}, 3\right)$. [11.4b]

Find an equation of the line that contains the given pair of points. [11.4c]

27. $\left(\frac{1}{2}, 6\right), \left(\frac{1}{2}, -6\right)$

28. $(3, -4), (-7, -2)$

29. $(3, -4), (2, -4)$

Understanding Through Discussion and Writing

30. Do all graphs of linear equations have y-intercepts? Why or why not? [11.2b]

31. The equations $3x + 4y = 8$ and $y = -\frac{3}{4}x + 2$ are equivalent. Which equation is easier to graph and why? [11.1d]

32. If the graph of the equation $Ax + By = C$ is a horizontal line, what can you conclude about A? Why? [11.2b]

33. Explain in your own words why the graph of $x = 7$ is a vertical line. [11.2b]

STUDYING FOR SUCCESS *Taking a Test*

- ☐ Read each question carefully. Be sure you understand the question before you answer it.
- ☐ Try to answer all the questions the first time through, marking those to recheck if you have time.
- ☐ Write your test in a neat and orderly manner.

Graphing Using the Slope and the y-Intercept

11.5

OBJECTIVE

a Use the slope and the y-intercept to graph a line.

a GRAPHS USING THE SLOPE AND THE y-INTERCEPT

We can graph a line if we know the coordinates of two points on that line. We can also graph a line if we know the slope and the y-intercept.

EXAMPLE 1 Draw a line that has slope $\frac{1}{4}$ and y-intercept $(0, 2)$.

We plot $(0, 2)$ and from there move 1 unit *up* (since the numerator of $\frac{1}{4}$ is *positive* and corresponds to the change in y) and 4 units *to the right* (since the denominator is *positive* and corresponds to the change in x). This locates the point $(4, 3)$. We plot $(4, 3)$ and draw a line passing through $(0, 2)$ and $(4, 3)$. We are actually graphing the equation $y = \frac{1}{4}x + 2$.

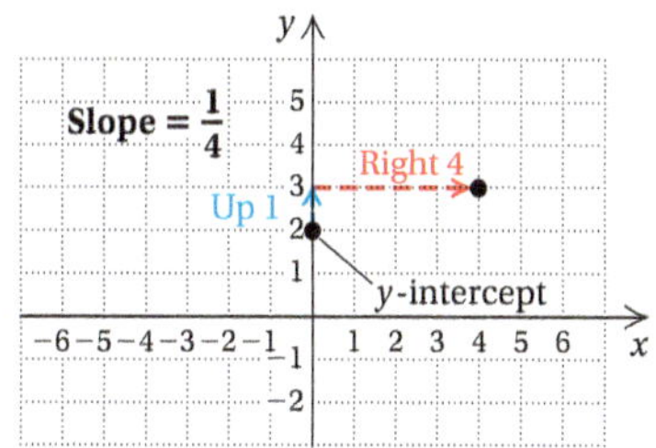

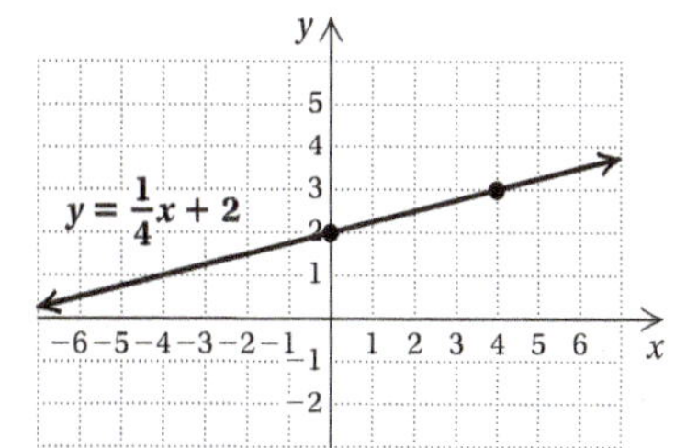

EXAMPLE 2 Draw a line that has slope $-\frac{2}{3}$ and y-intercept $(0, 4)$.

We can think of $-\frac{2}{3}$ as $\frac{-2}{3}$. We plot $(0, 4)$ and from there move 2 units *down* (since the numerator is *negative*) and 3 units *to the right* (since the denominator is *positive*). We plot the point $(3, 2)$ and draw a line passing through $(0, 4)$ and $(3, 2)$. We are actually graphing the equation $y = -\frac{2}{3}x + 4$.

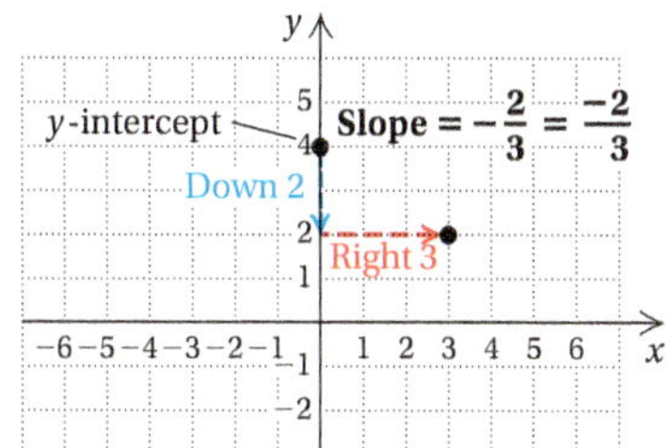

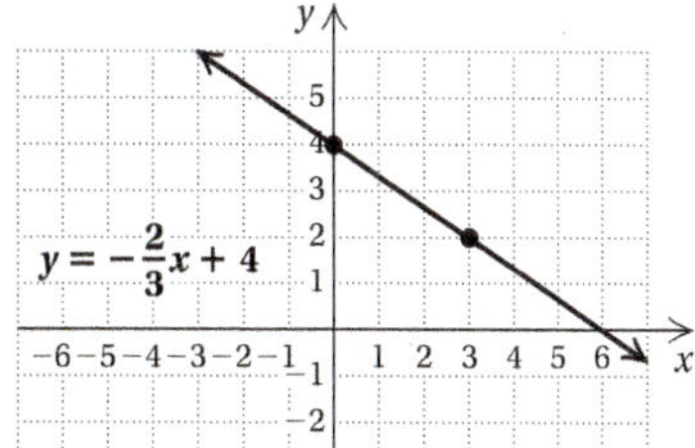

Do Exercises 1–3.

1. Draw a line that has slope $\frac{2}{5}$ and y-intercept $(0, -3)$. What equation is graphed?

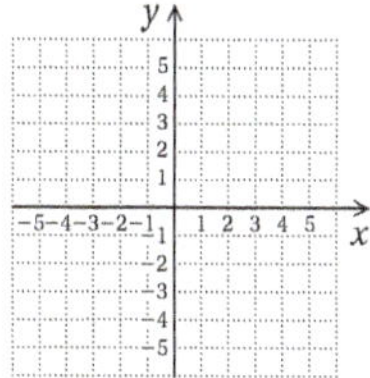

2. Draw a line that has slope $-\frac{2}{5}$ and y-intercept $(0, -3)$. What equation is graphed?

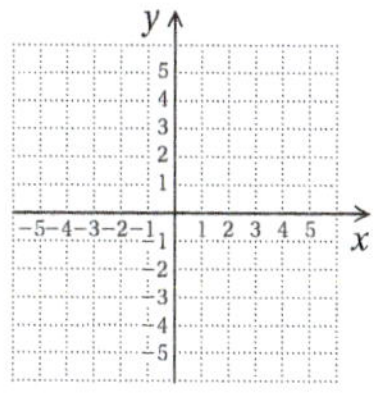

3. Draw a line that has slope 6 and y-intercept $(0, -3)$. Think of 6 as $\frac{6}{1}$. What equation is graphed?

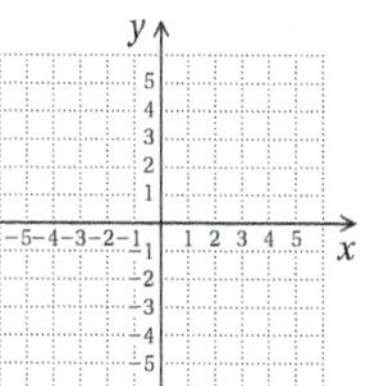

Answers

Answers to Margin Exercises 1–3 are on p. 824.

We now use our knowledge of the slope–intercept equation to graph linear equations.

EXAMPLE 3 Graph $y = \frac{3}{4}x + 5$ using the slope and the y-intercept.

From the equation $y = \frac{3}{4}x + 5$, we see that the slope of the graph is $\frac{3}{4}$ and the y-intercept is $(0, 5)$. We plot $(0, 5)$ and then consider the slope, $\frac{3}{4}$. Starting at $(0, 5)$, we plot a second point by moving 3 units *up* (since the numerator is *positive*) and 4 units *to the right* (since the denominator is *positive*). We reach a new point, $(4, 8)$.

We can also rewrite the slope as $\frac{-3}{-4}$. We again start at the y-intercept, $(0, 5)$, but move 3 units *down* (since the numerator is *negative* and corresponds to the change in y) and 4 units *to the left* (since the denominator is *negative* and corresponds to the change in x). We reach another point, $(-4, 2)$. Once two or three points have been plotted, the line representing all solutions of $y = \frac{3}{4}x + 5$ can be drawn.

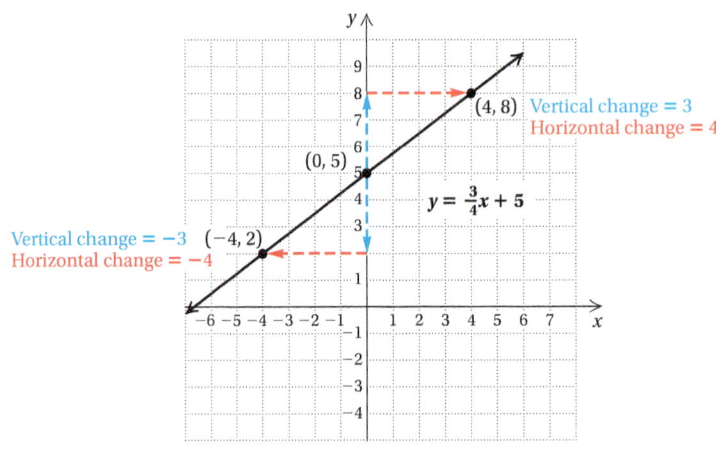

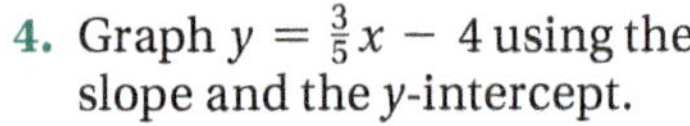

4. Graph $y = \frac{3}{5}x - 4$ using the slope and the y-intercept.

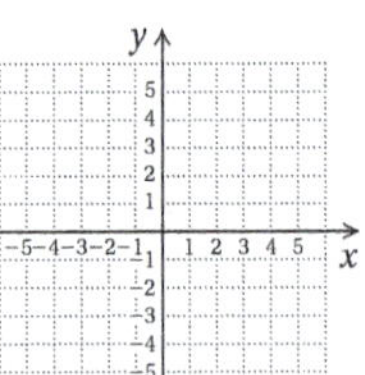

Do Exercise 4.

EXAMPLE 4 Graph $2x + 3y = 3$ using the slope and the y-intercept.

To graph $2x + 3y = 3$, we first rewrite the equation in slope–intercept form:

$$
\begin{aligned}
2x + 3y &= 3 \\
3y &= -2x + 3 && \text{Adding } -2x \\
\tfrac{1}{3} \cdot 3y &= \tfrac{1}{3}(-2x + 3) && \text{Multiplying by } \tfrac{1}{3} \\
y &= -\tfrac{2}{3}x + 1. && \text{Simplifying}
\end{aligned}
$$

To graph $y = -\frac{2}{3}x + 1$, we first plot the y-intercept, $(0, 1)$. We can think of the slope as $\frac{-2}{3}$. Starting at $(0, 1)$ and using the slope, we find a second point by moving 2 units *down* (since the numerator is *negative*) and 3 units *to the right* (since the denominator is *positive*). We plot the new point, $(3, -1)$. In a similar manner, we can move from the point $(3, -1)$ to locate a third point, $(6, -3)$. The line can then be drawn.

Answers

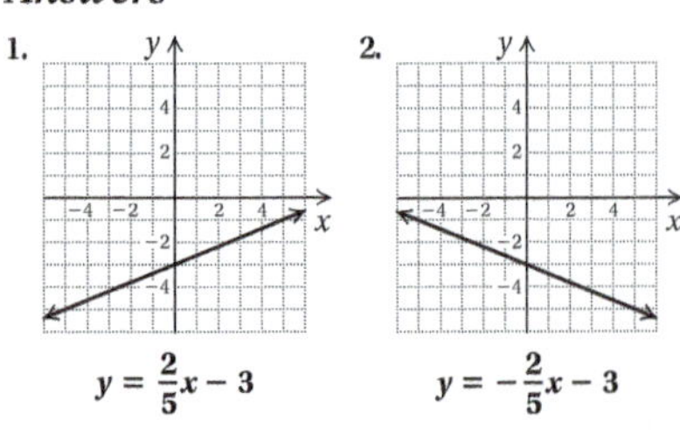

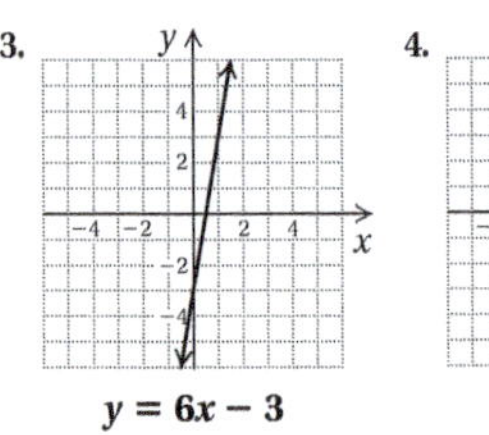

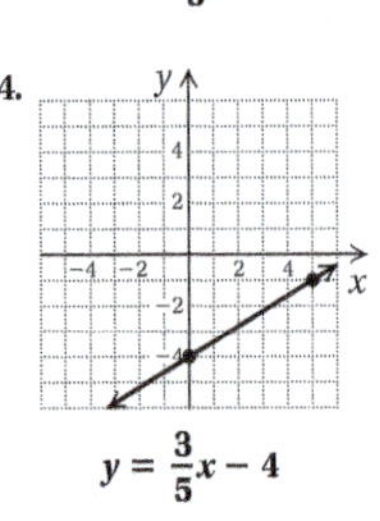

Since $-\frac{2}{3} = \frac{2}{-3}$, an alternative approach is to again plot $(0, 1)$, but this time we move 2 units *up* (since the numerator is *positive*) and 3 units *to the left* (since the denominator is *negative*). This leads to another point on the graph, $(-3, 3)$.

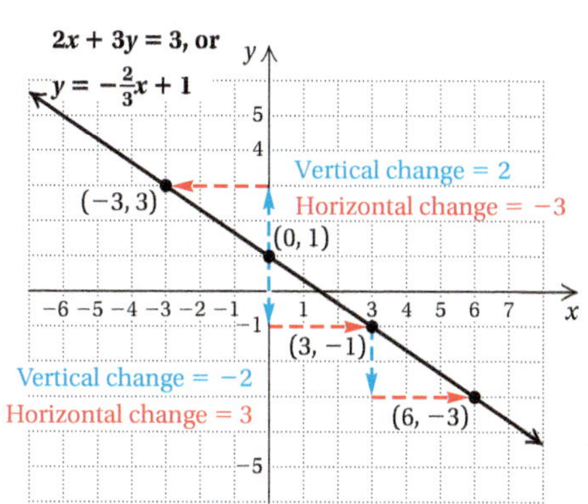

5. Graph: $3x + 4y = 12$.

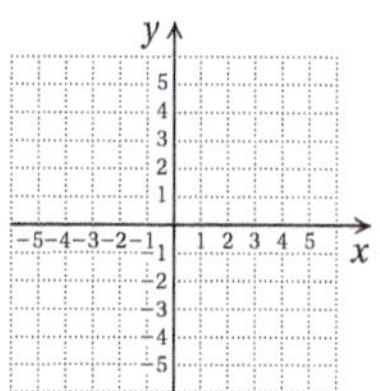

Do Exercise 5. ▶

Answer

5. 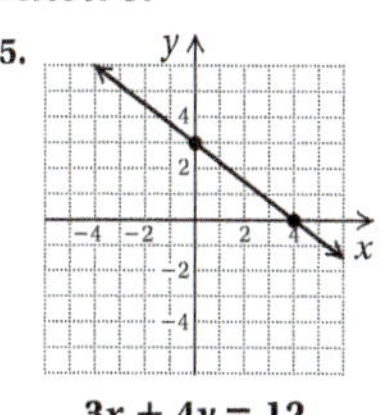

11.5 Exercise Set

For Extra Help

MyMathLab®

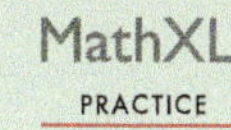

Reading Check

In each exercise, a line has been graphed using its y-intercept and its slope. Choose from the column on the right the appropriate slope of the line.

RC1.

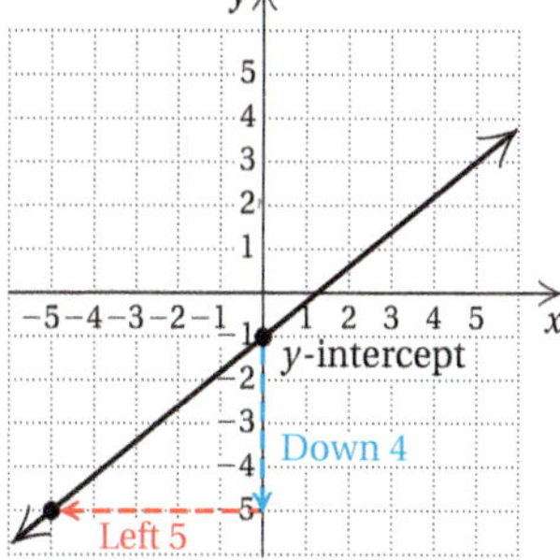

RC2.

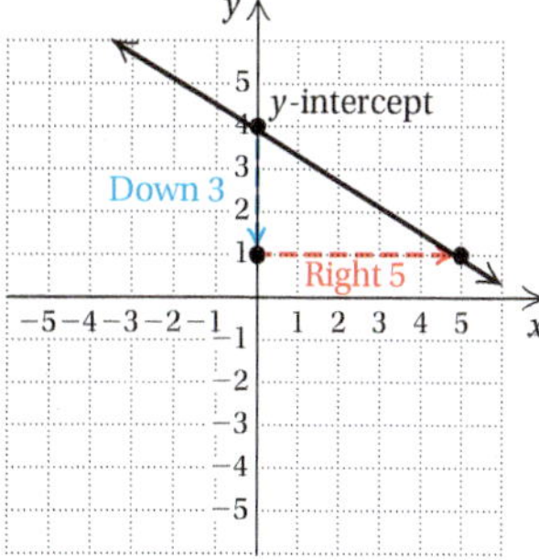

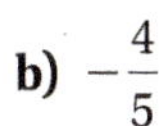

a) $\frac{3}{5}$

b) $-\frac{4}{5}$

c) $\frac{4}{5}$

d) $-\frac{3}{5}$

e) $-\frac{5}{4}$

f) $\frac{5}{3}$

RC3.

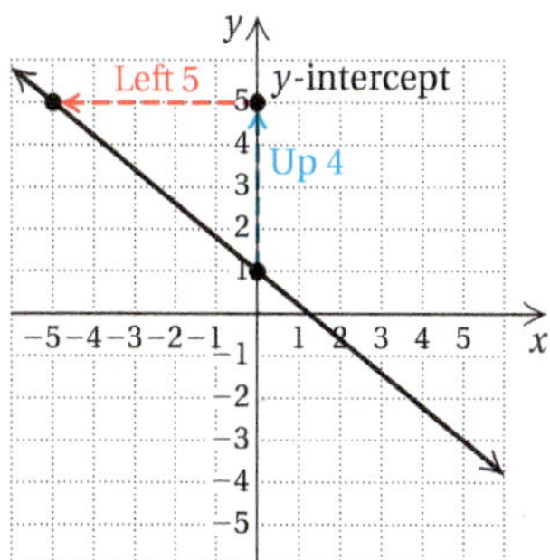

RC4.

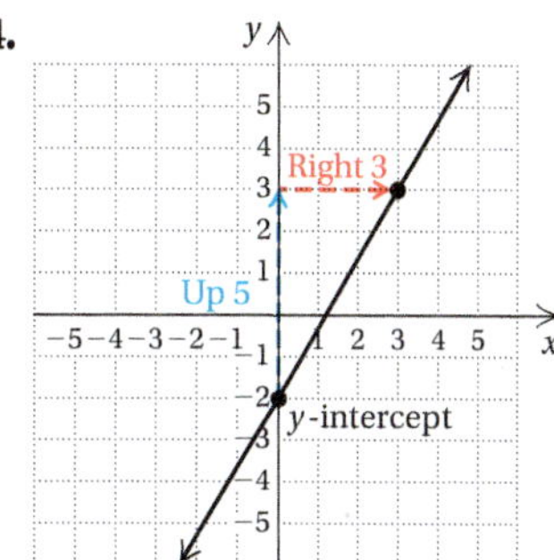

a Draw a line that has the given slope and y-intercept.

1. Slope $\frac{2}{5}$; y-intercept $(0, 1)$

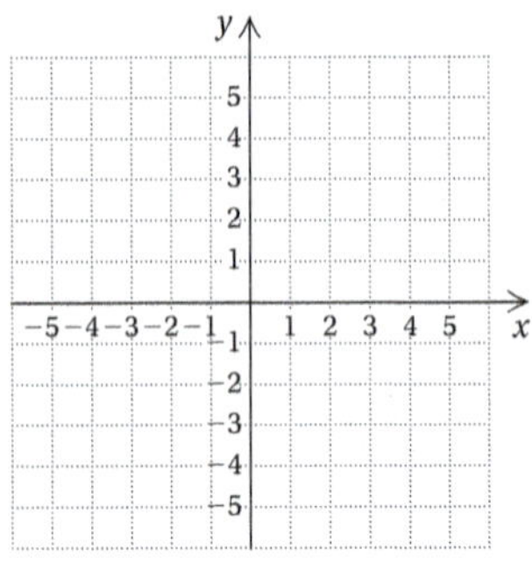

2. Slope $\frac{3}{5}$; y-intercept $(0, -1)$

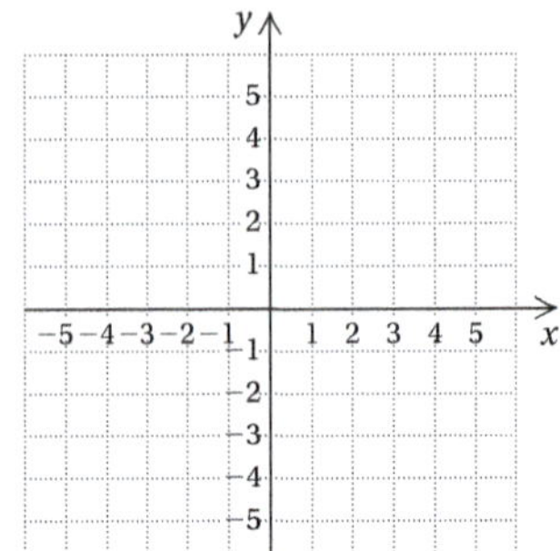

3. Slope $\frac{5}{3}$; y-intercept $(0, -2)$

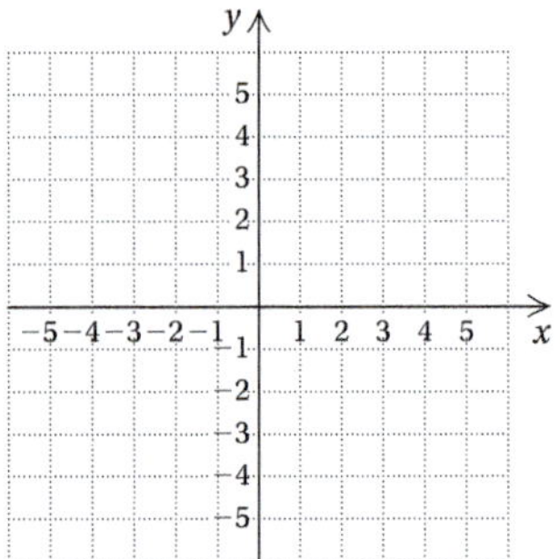

4. Slope $\frac{5}{2}$; y-intercept $(0, 1)$

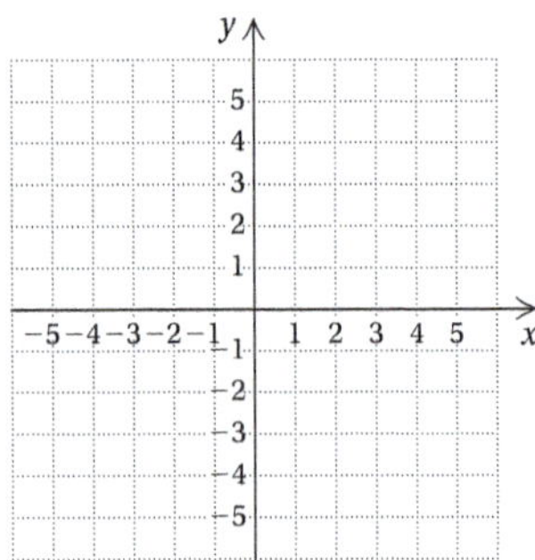

5. Slope $-\frac{3}{4}$; y-intercept $(0, 5)$

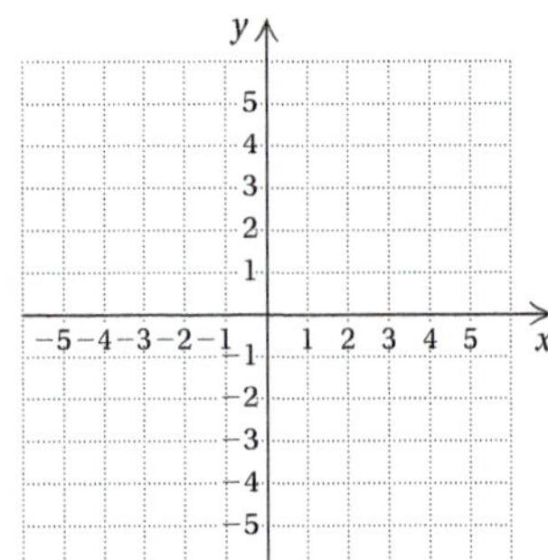

6. Slope $-\frac{4}{5}$; y-intercept $(0, 6)$

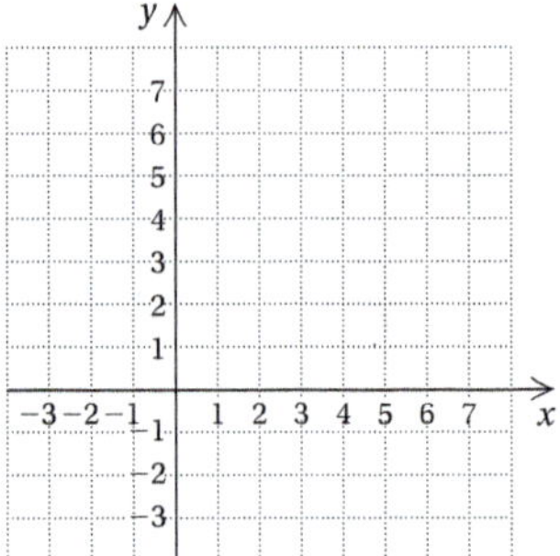

7. Slope $-\frac{1}{2}$; y-intercept $(0, 3)$

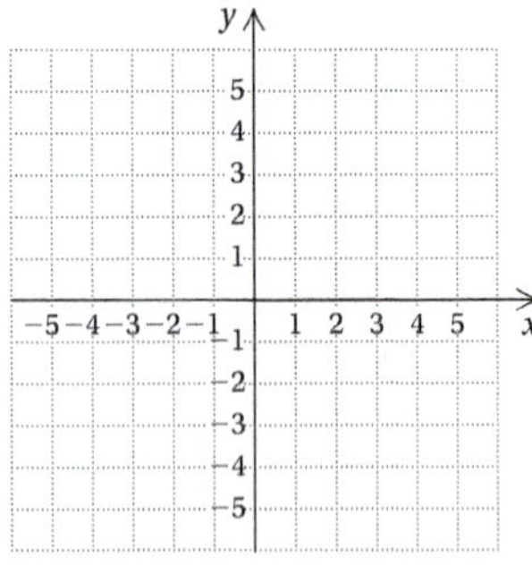

8. Slope $\frac{1}{3}$; y-intercept $(0, -4)$

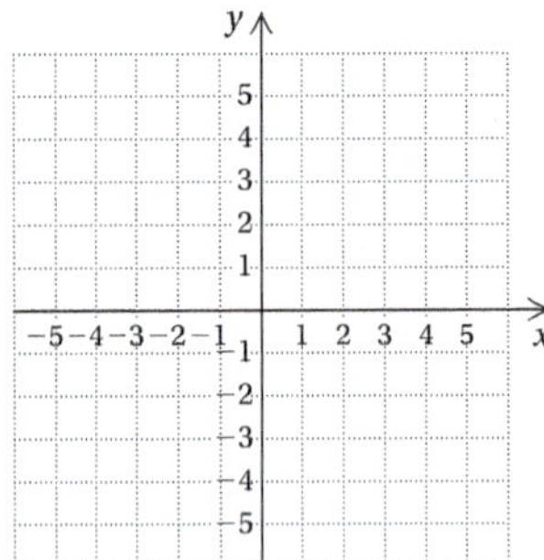

9. Slope 2; y-intercept $(0, -4)$

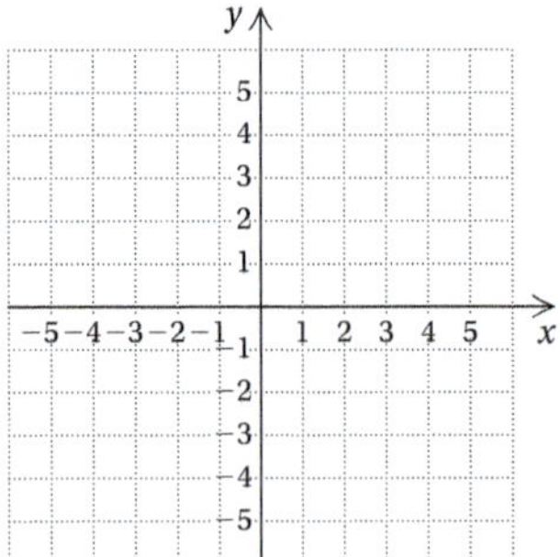

10. Slope -2; y-intercept $(0, -3)$

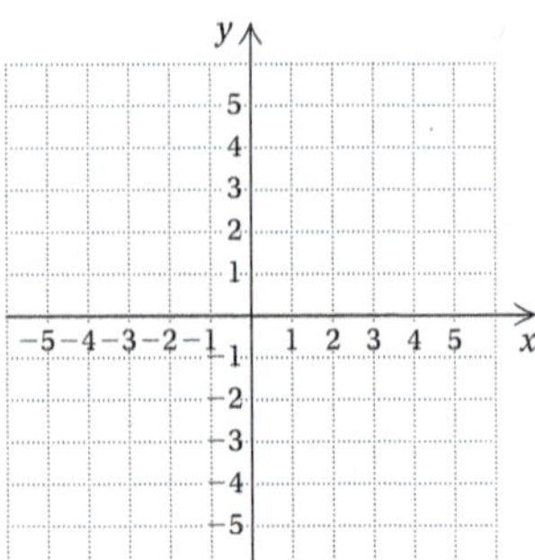

11. Slope -3; y-intercept $(0, 2)$

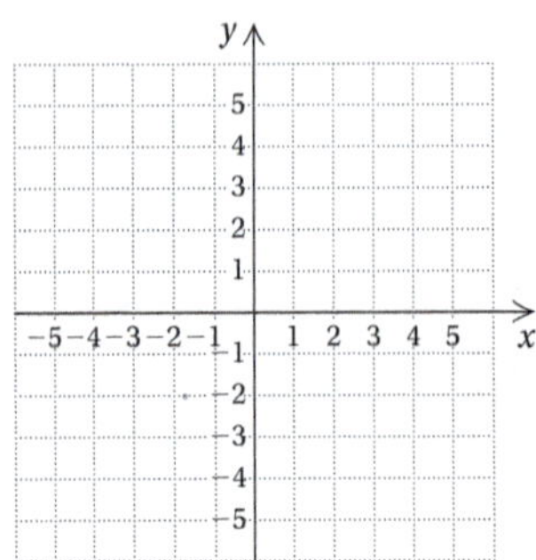

12. Slope 3; y-intercept $(0, 4)$

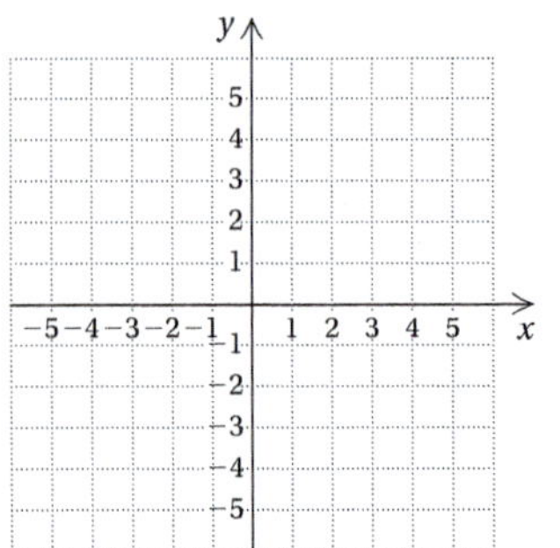

Graph using the slope and the y-intercept.

13. $y = \frac{3}{5}x + 2$

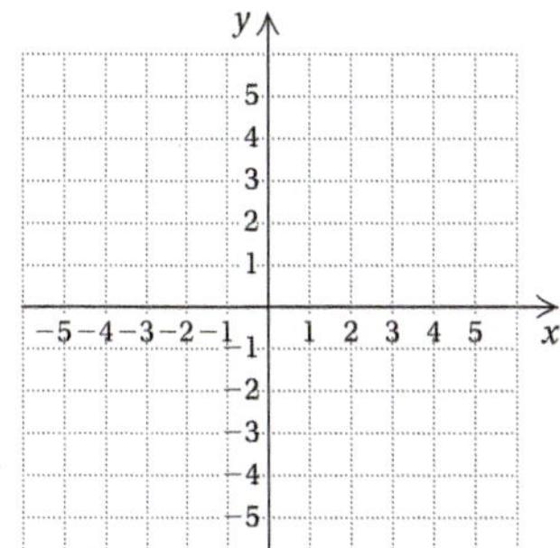

14. $y = -\frac{3}{5}x - 1$

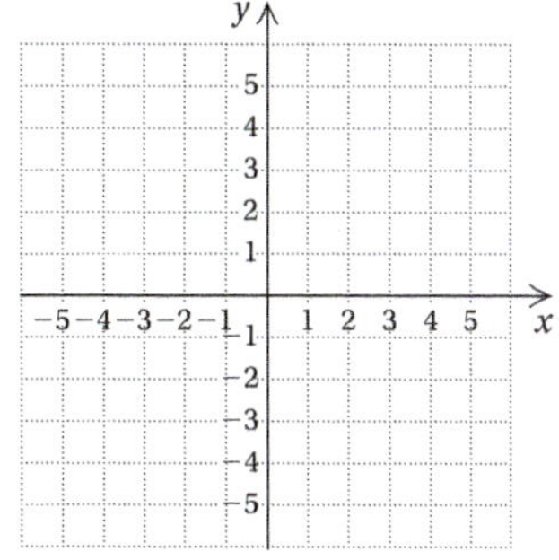

15. $y = -\frac{3}{5}x + 1$

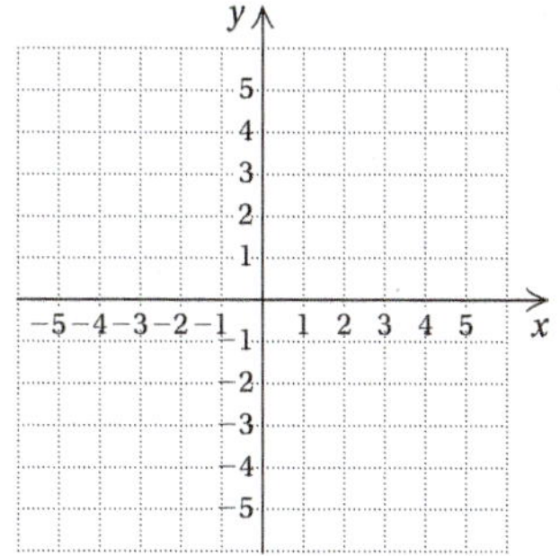

16. $y = \frac{3}{5}x - 2$

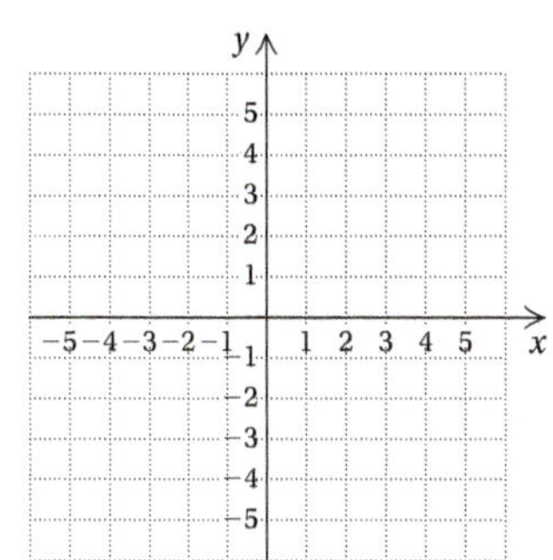

17. $y = \frac{5}{3}x + 3$

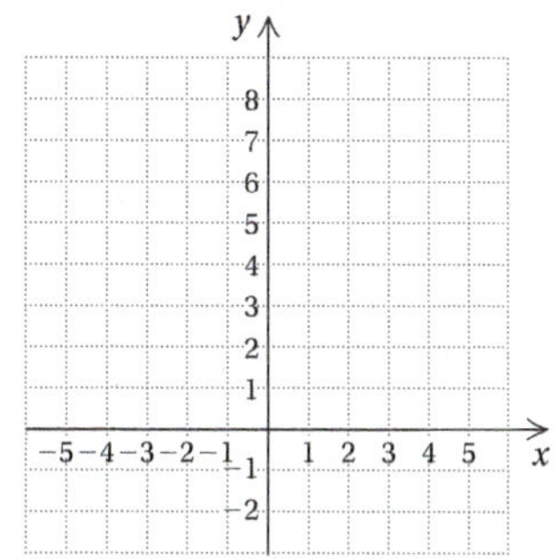

18. $y = \frac{5}{3}x - 2$

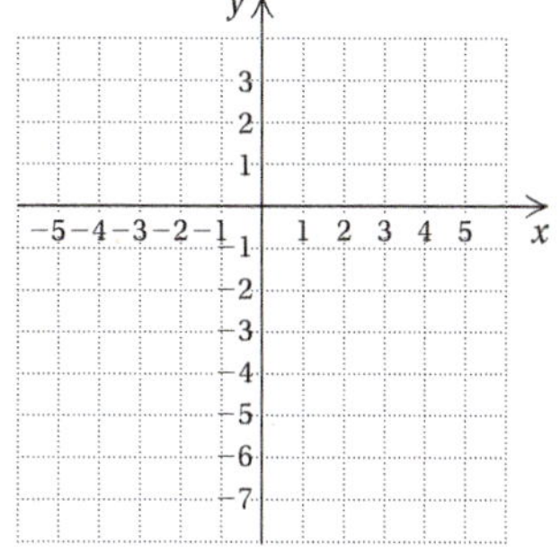

19. $y = -\frac{3}{2}x - 2$

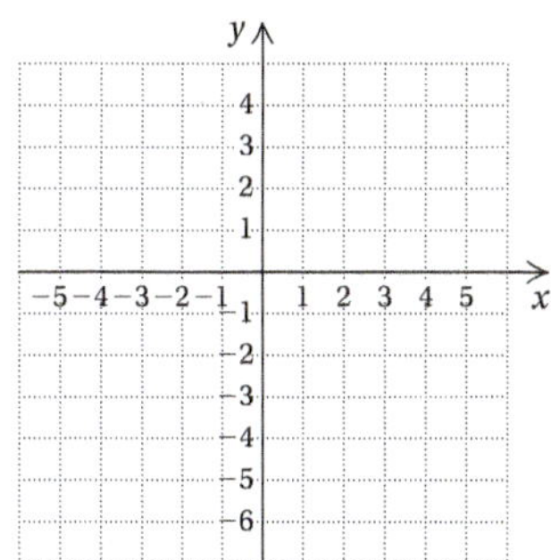

20. $y = -\frac{4}{3}x + 3$

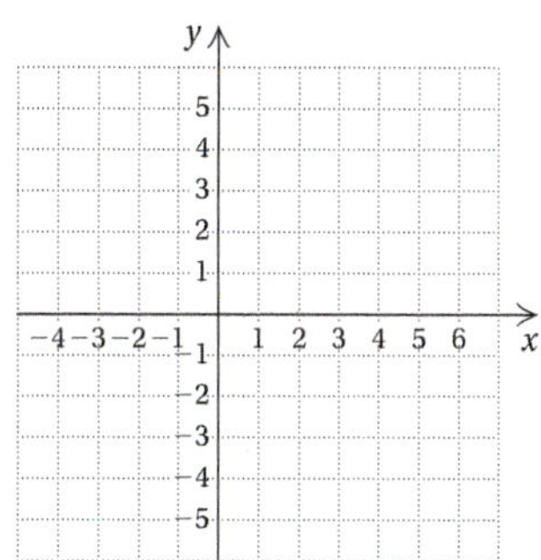

21. $2x + y = 1$

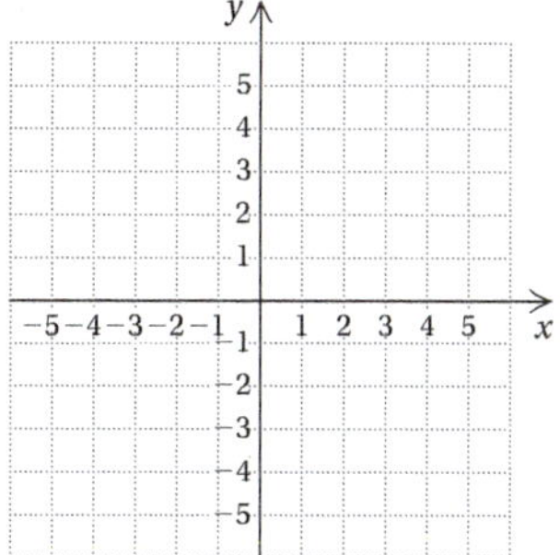

22. $3x + y = 2$

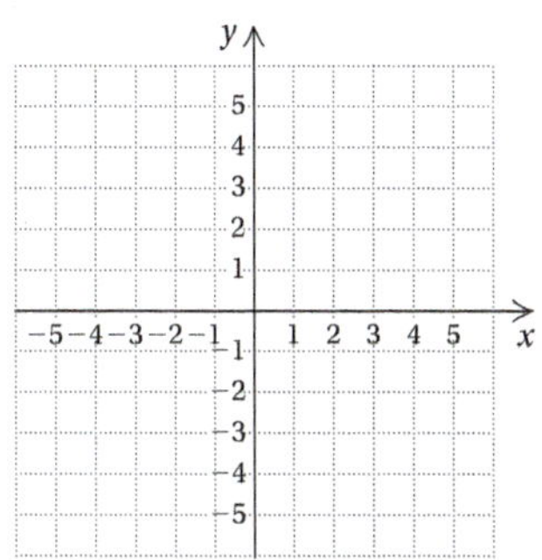

23. $3x - y = 4$

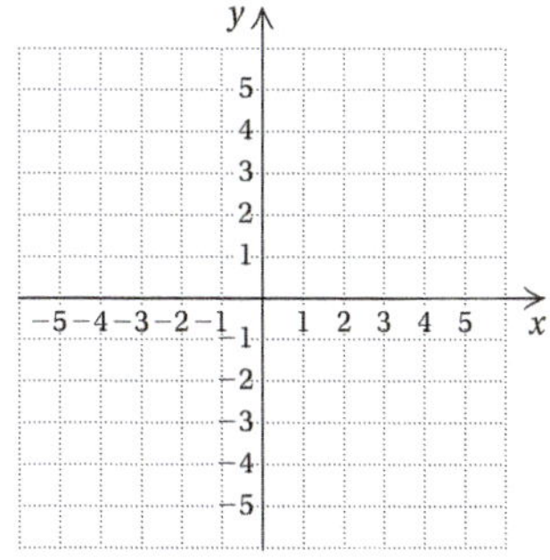

24. $2x - y = 5$

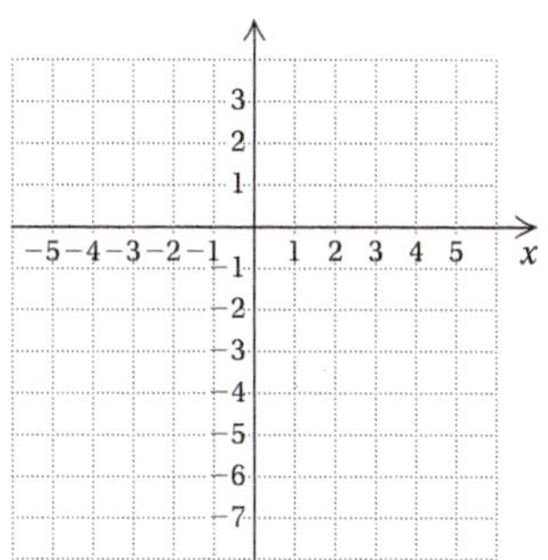

25. $2x + 3y = 9$

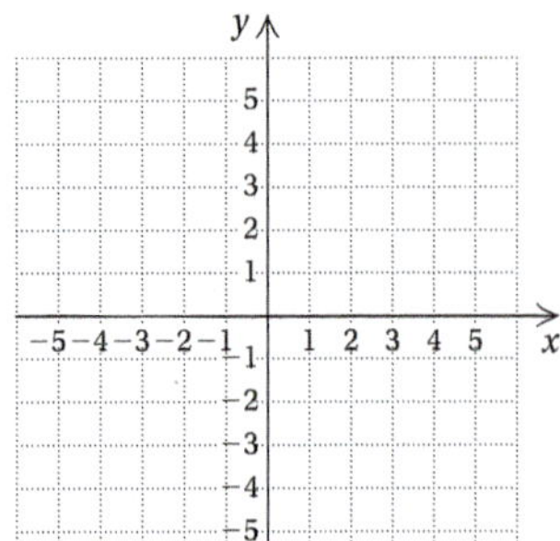

26. $4x + 5y = 15$

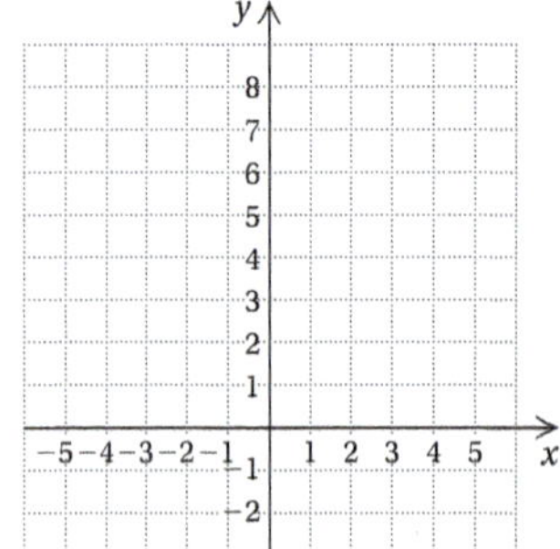

27. $x - 4y = 12$

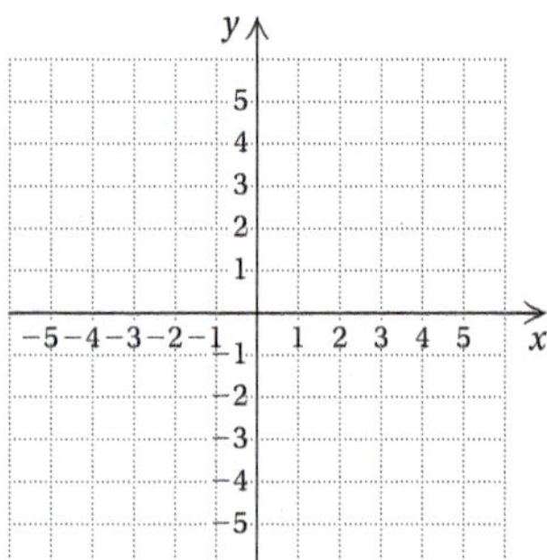

28. $x + 5y = 20$

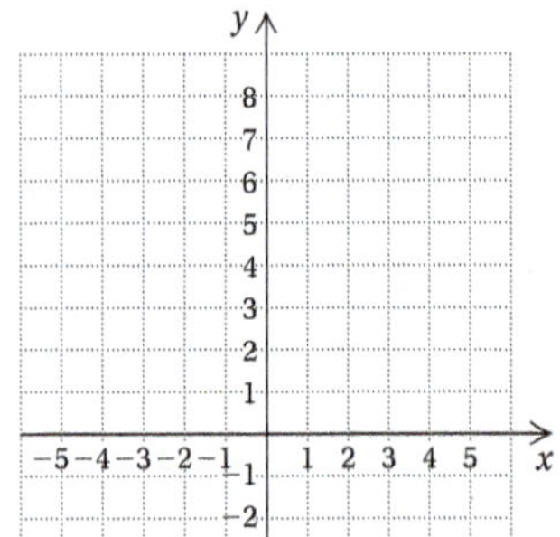

29. $x + 2y = 6$

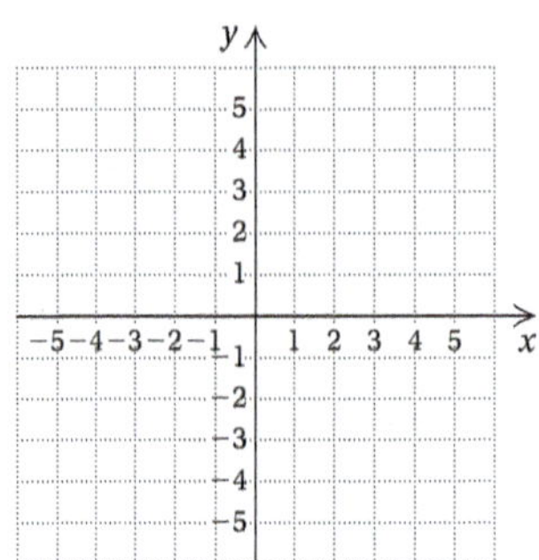

30. $x - 3y = 9$

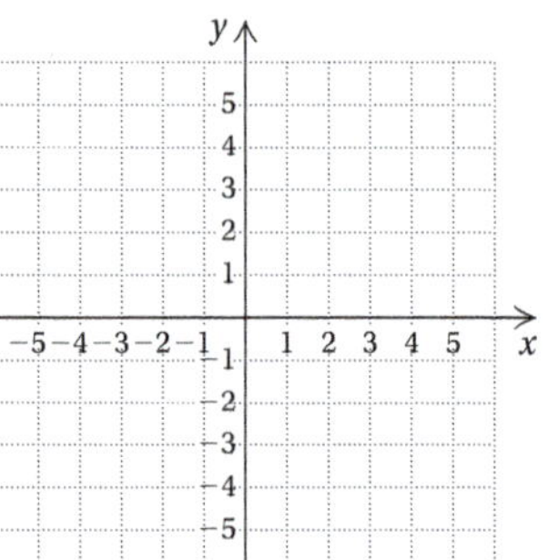

Skill Maintenance

Find the slope, if it exists, of the line containing the given pair of points. [11.3a]

31. $(-2, -6), (8, 7)$

32. $(2, -6), (8, -7)$

33. $(4.5, -2.3), (14.5, 4.6)$

34. $(-0.8, -2.3), (-4.8, 0.1)$

35. $(-2, -6), (8, -6)$

36. $(-2, -6), (-2, 7)$

Simplify. [9.4a]

37. $8 - (-11) + 23$

38. $-200 - 25 + 40$

39. $-10 - (-30) + 5 - 2$

40. $-40 - (-32) + 50 - 1$

Synthesis

41. Graph the line with slope 2 that passes through the point $(-3, 1)$.

42. Graph the line with slope -3 that passes through the point $(3, 0)$.

Parallel Lines and Perpendicular Lines

11.6

OBJECTIVES

a Determine whether the graphs of two linear equations are parallel.

b Determine whether the graphs of two linear equations are perpendicular.

SKILL TO REVIEW

Objective 11.3b: Find the slope of a line from an equation.

Find the slope, if it exists, of each line.

1. $7y - 2x = 10$
2. $x - 13y = -1$

When we graph a pair of linear equations, there are three possibilities:

1. The graphs are the same.
2. The graphs intersect at exactly one point.
3. The graphs are parallel. (They do not intersect.)

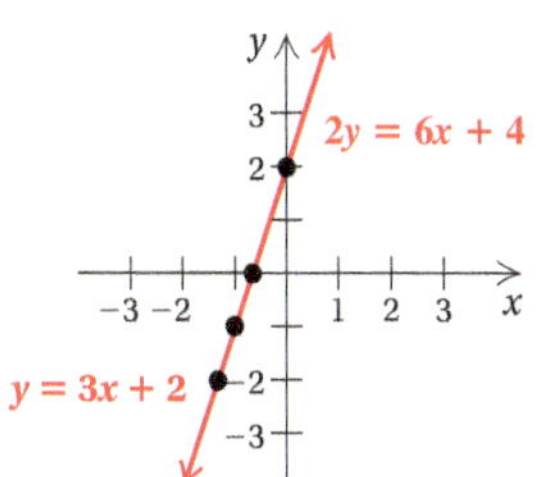

Equations have the same graph.

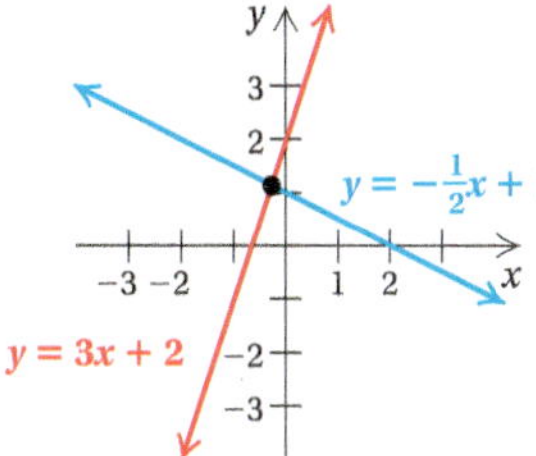

Graphs intersect at exactly one point.

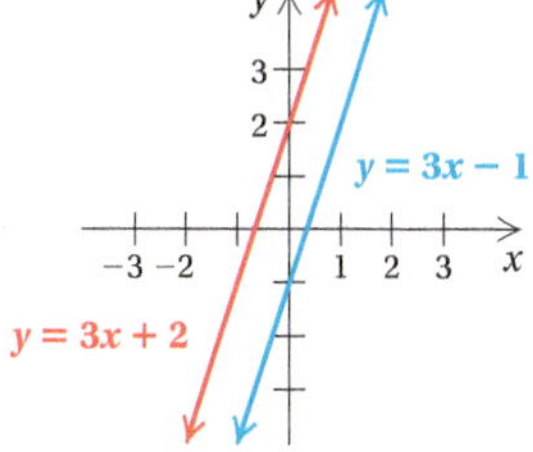

Graphs are parallel.

a PARALLEL LINES

The graphs shown below are of the linear equations

$$y = 2x + 5 \quad \text{and} \quad y = 2x - 3.$$

The slope of each line is 2. The y-intercepts, $(0, 5)$ and $(0, -3)$, are different. The lines do not have the same graph, do not intersect, and are parallel.

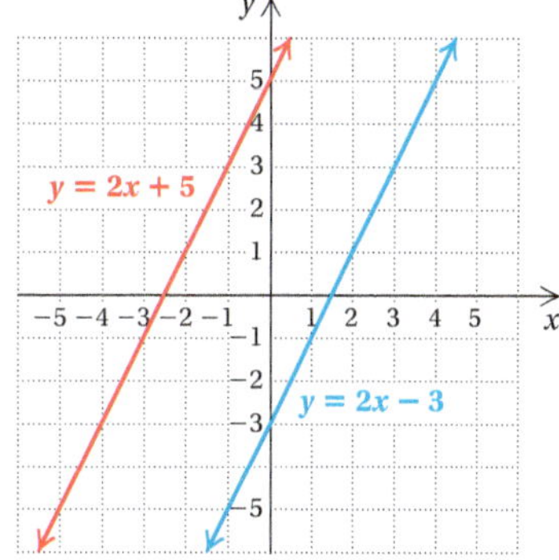

PARALLEL LINES

- Parallel nonvertical lines have the *same* slope, $m_1 = m_2$, and *different* y-intercepts, $b_1 \neq b_2$.
- Parallel horizontal lines have equations $y = p$ and $y = q$, where $p \neq q$.
- Parallel vertical lines have equations $x = p$ and $x = q$, where $p \neq q$.

By simply graphing, we may find it difficult to determine whether lines are parallel. Sometimes they may intersect very far from the origin. We can use the preceding statements about slopes, y-intercepts, and parallel lines to determine for certain whether lines are parallel.

Answers

Skill to Review:

1. $\frac{2}{7}$ 2. $\frac{1}{13}$

Determine whether the graphs of each pair of equations are parallel.

1. $3x - y = -5,$
$y - 3x = -2$

Solve each equation for y and then find the slope.

$3x - y = -5$

$-y = -3x - 5$

$y = \square\, x + 5$

The slope is $\square$.

$y - 3x = -2$

$y = \square\, x - 2$

The slope is 3.

The slope of each line is $\square$.

The y-intercepts, $(0, 5)$ and $(0, \square)$, are different. Thus the lines are parallel.

GS

2. $y - 3x = 1,$
$-2y = 3x + 2$

EXAMPLE 1 Determine whether the graphs of the lines $y = -3x + 4$ and $6x + 2y = -10$ are parallel.

The graphs of these equations are shown below. They appear to be parallel, but it is most accurate to determine this algebraically.

We first solve each equation for y. In this case, the first equation is already solved for y.

a) $y = -3x + 4$

b) $6x + 2y = -10$

$2y = -6x - 10$

$y = \frac{1}{2}(-6x - 10)$

$y = -3x - 5$

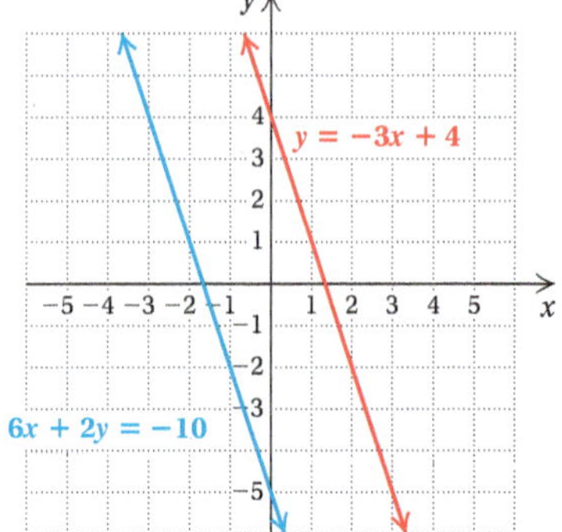

The slope of each line is -3. The y-intercepts are $(0, 4)$ and $(0, -5)$, which are different. The lines are parallel.

◀ Do Exercises 1 and 2.

b PERPENDICULAR LINES

Perpendicular lines in a plane are lines that intersect at a right, or 90°, angle. The lines whose graphs are shown below are perpendicular. You can check this approximately by using a protractor or placing the corner of a rectangular piece of paper at the intersection.

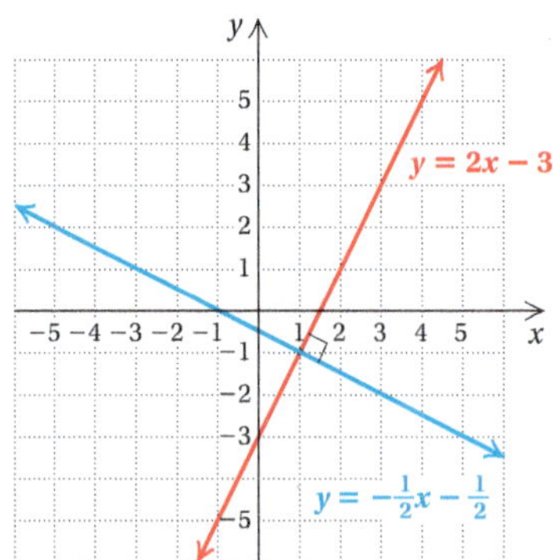

The slopes of the lines are 2 and $-\frac{1}{2}$. Note that $2\left(-\frac{1}{2}\right) = -1$. That is, the product of the slopes is -1.

PERPENDICULAR LINES

- Two nonvertical lines are perpendicular if the product of their slopes is -1, $m_1 \cdot m_2 = -1$. (If one line has slope m, the slope of the line perpendicular to it is $-1/m$.)
- If one equation in a pair of perpendicular lines is vertical, then the other is horizontal. These equations are of the form $x = a$ and $y = b$.

Answers

1. Yes 2. No

Guided Solution:

1. 3, 3, 3, 3, −2

EXAMPLE 2 Determine whether the graphs of the lines $3y = 9x + 3$ and $6y + 2x = 6$ are perpendicular.

The graphs are shown below. They appear to be perpendicular, but it is most accurate to determine this algebraically.

We first solve each equation for y in order to determine the slopes:

a) $3y = 9x + 3$

$y = \frac{1}{3}(9x + 3)$

$y = 3x + 1;$

b) $6y + 2x = 6$

$6y = -2x + 6$

$y = \frac{1}{6}(-2x + 6)$

$y = -\frac{1}{3}x + 1.$

The slopes are 3 and $-\frac{1}{3}$. The product of the slopes is $3\left(-\frac{1}{3}\right) = -1$. The lines are perpendicular.

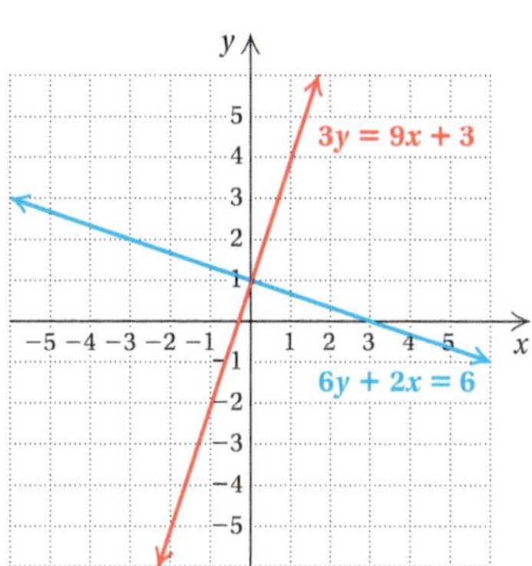

Do Exercises 3 and 4. ▶

CALCULATOR CORNER

Parallel Lines Graph each pair of equations in Margin Exercises 1 and 2 in the standard viewing window, $[-10, 10, -10, 10]$. (Note that each equation must be solved for y so that it can be entered in Y = form on the graphing calculator.) Determine whether the lines appear to be parallel.

Perpendicular Lines Graph each pair of equations in Margin Exercises 3 and 4 in the window $[-9, 9, -6, 6]$. (Note that the equations in Margin Exercise 4 must be solved for y so that they can be entered in Y = form on the graphing calculator.) Determine whether the lines appear to be perpendicular. Note (in the viewing window) that more of the x-axis is shown than the y-axis. The dimensions were chosen to more accurately reflect the slopes of the lines.

Determine whether the graphs of each pair of equations are perpendicular.

3. $y = -\frac{3}{4}x + 7,$

$y = \frac{4}{3}x - 9$

The slopes of the lines are $-\frac{3}{4}$ and ______.

The product of the slopes is $-\frac{3}{4} \cdot \frac{4}{3} =$ ______.

The lines are perpendicular.

4. $4x - 5y = 8,$

$6x + 9y = -12$

Answers

3. Yes **4.** No

Guided Solution:

3. $\frac{4}{3}, -1$

11.6 Exercise Set

For Extra Help

MyMathLab®

MathXL® 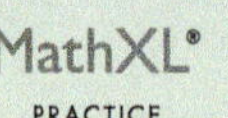   

PRACTICE WATCH READ REVIEW

Reading Check

The graphs of two of the equations listed below are parallel, and the graphs of two of the equations listed below are perpendicular.

RC1. Determine which two lines are parallel.

RC2. Determine which two lines are perpendicular.

a) $y = -4x - 1$

b) $y = -\frac{3}{4}x - 3$

c) $y = -3x + 1$

d) $y = \frac{1}{3}x + 1$

e) $y = -\frac{3}{4}x + 3$

f) $y = -\frac{1}{4}x - 3$

a Determine whether the graphs of the equations are parallel lines.

1. $x + 4 = y,$
$y - x = -3$

2. $3x - 4 = y,$
$y - 3x = 8$

3. $y + 3 = 6x,$
$-6x - y = 2$

4. $y = -4x + 2,$
$-5 = -2y + 8x$

5. $10y + 32x = 16.4,$
$y + 3.5 = 0.3125x$

6. $y = 6.4x + 8.9,$
$5y - 32x = 5$

7. $y = 2x + 7,$
$5y + 10x = 20$

8. $y + 5x = -6,$
$3y + 5x = -15$

9. $3x - y = -9,$
$2y - 6x = -2$

10. $y - 6 = -6x,$
$-2x + y = 5$

11. $x = 3,$
$x = 4$

12. $y = 1,$
$y = -2$

13. $x = -6, y = 6$

14. $x = \frac{1}{3}, y = -\frac{1}{3}$

b Determine whether the graphs of the equations are perpendicular lines.

15. $y = -4x + 3,$
$4y + x = -1$

16. $y = -\frac{2}{3}x + 4,$
$3x + 2y = 1$

17. $x + y = 6,$
$4y - 4x = 12$

18. $2x - 5y = -3,$
$5x + 2y = 6$

19. $y = -0.3125x + 11,$
$y - 3.2x = -14$

20. $y = -6.4x - 7,$
$64y - 5x = 32$

21. $y = -x + 8,$
$x - y = -1$

22. $2x + 6y = -3,$
$12y = 4x + 20$

23. $\frac{3}{8}x - \frac{y}{2} = 1,$
$\frac{4}{3}x - y + 1 = 0$

24. $\frac{1}{2}x + \frac{3}{4}y = 6,$
$-\frac{3}{2}x + y = 4$

25. $x = 0,$
$y = -2$

26. $x = -3,$
$y = 5$

27. $y = 4,$
$y = -\frac{1}{4}$

28. $x = -\frac{7}{8},$
$x = -\frac{8}{7}$

a, **b** Determine whether the graphs of the equations are parallel, perpendicular, or neither.

29. $3y + 21 = 2x,$
$3y = 2x + 24$

30. $3y + 21 = 2x,$
$2y = 16 - 3x$

31. $3y = 2x - 21,$
$2y - 16 = 3x$

32. $3y + 2x + 7 = 0,$
$3y = 2x + 24$

Skill Maintenance

Determine whether the given ordered pair is a solution of the equation. [11.1c]

33. $(1, -1)$; $2x - 15y = -17$

34. $(-14, -6)$; $16 - x = -5y$

Find the intercepts of each equation. [11.2a]

35. $-40x + 5y = 80$

36. $y - 3 = 6x$

Synthesis

37. Find an equation of a line that contains the point $(0, 6)$ and is parallel to $y - 3x = 4$.

38. Find an equation of the line that contains the point $(-2, 4)$ and is parallel to $y = 2x - 3$.

39. Find an equation of the line that contains the point $(0, 2)$ and is perpendicular to $3y - x = 0$.

40. Find an equation of the line that contains the point $(1, 0)$ and is perpendicular to $2x + y = -4$.

41. Find an equation of the line that has x-intercept $(-2, 0)$ and is parallel to $4x - 8y = 12$.

42. Find the value of k such that $4y = kx - 6$ and $5x + 20y = 12$ are parallel.

43. Find the value of k such that $4y = kx - 6$ and $5x + 20y = 12$ are perpendicular.

The lines in the graphs in Exercises 44 and 45 are perpendicular, and the lines in the graph in Exercise 46 are parallel. Find an equation of each line.

44.

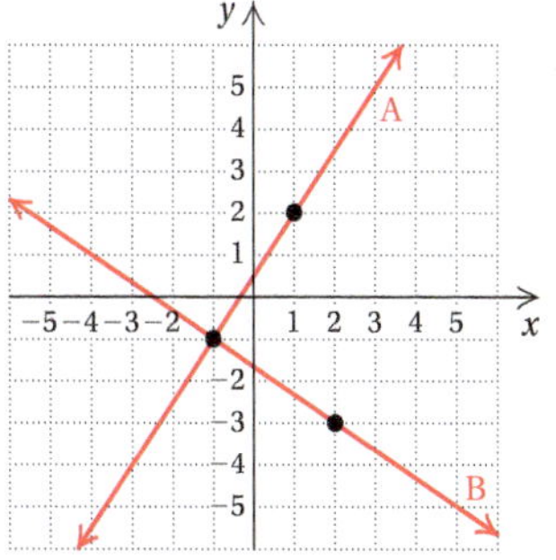

45.

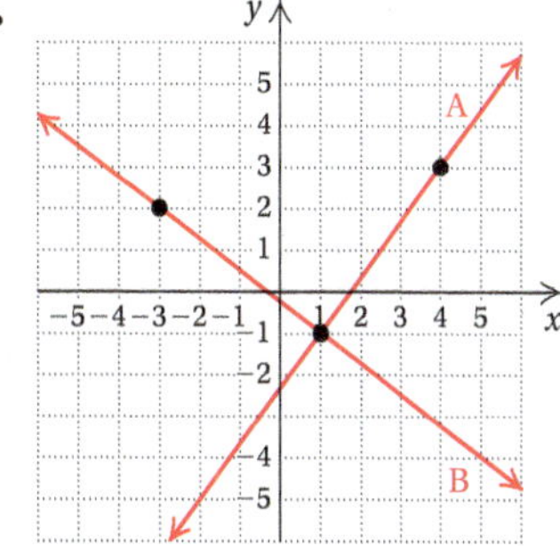

46.

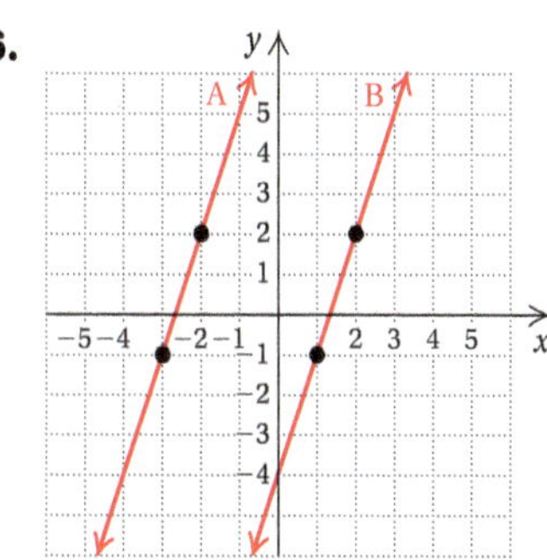

11.7 Graphing Inequalities in Two Variables

OBJECTIVES

a Determine whether an ordered pair of numbers is a solution of an inequality in two variables.

b Graph linear inequalities.

SKILL TO REVIEW

Objective 10.7a: Determine whether a given number is a solution of an inequality.

Determine whether each number is a solution of $x \geq -14$.

1. -26 **2.** -14

A graph of an inequality is a drawing that represents its solutions. An inequality in one variable can be graphed on the number line. An inequality in two variables can be graphed on a coordinate plane.

a SOLUTIONS OF INEQUALITIES IN TWO VARIABLES

The solutions of inequalities in two variables are ordered pairs.

EXAMPLE 1 Determine whether $(-3, 2)$ is a solution of $5x + 4y < 13$.

We use alphabetical order to replace x with -3 and y with 2.

$$\begin{array}{r|l} 5x + 4y & < 13 \\ \hline 5(-3) + 4 \cdot 2 & ?\ 13 \\ -15 + 8 & \\ -7 & \quad \text{TRUE} \end{array}$$

Since $-7 < 13$ is true, $(-3, 2)$ is a solution.

EXAMPLE 2 Determine whether $(6, 8)$ is a solution of $5x + 4y < 13$.

We use alphabetical order to replace x with 6 and y with 8.

$$\begin{array}{r|l} 5x + 4y & < 13 \\ \hline 5(6) + 4(8) & ?\ 13 \\ 30 + 32 & \\ 62 & \quad \text{FALSE} \end{array}$$

Since $62 < 13$ is false, $(6, 8)$ is not a solution.

1. Determine whether $(4, 3)$ is a solution of $3x - 2y < 1$.

2. Determine whether $(2, -5)$ is a solution of $4x + 7y \geq 12$.

Do Margin Exercises 1 and 2.

b GRAPHING INEQUALITIES IN TWO VARIABLES

EXAMPLE 3 Graph: $y > x$.

We first graph the line $y = x$. Every solution of $y = x$ is an ordered pair like $(3, 3)$ in which the first coordinate and the second coordinate are the same. We draw the line $y = x$ dashed (as shown on the left below) because its points are *not* solutions of $y > x$.

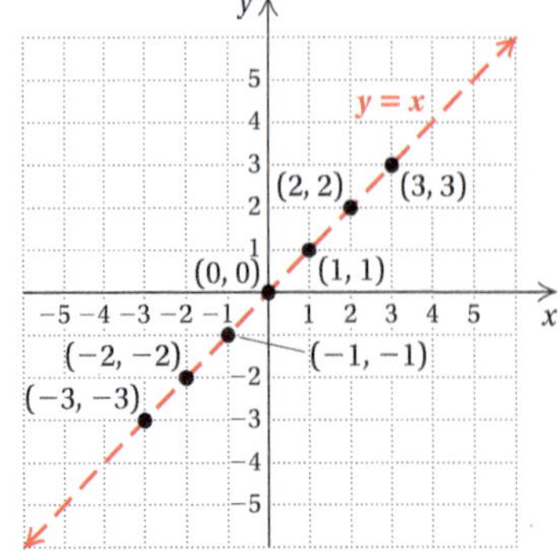

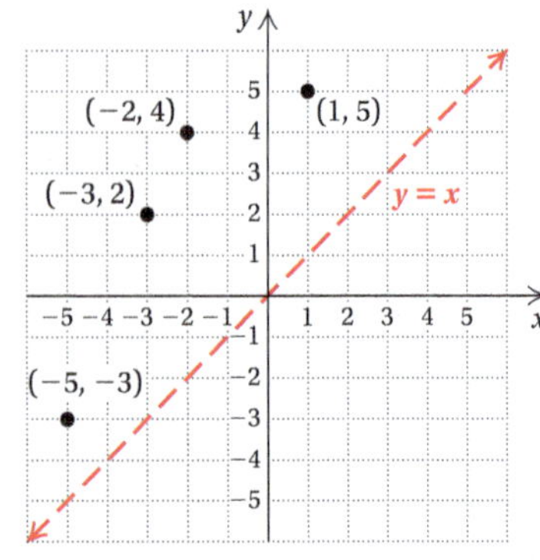

Answers

Skill to Review:
1. No 2. Yes

Margin Exercises:
1. No 2. No

Now look at the graph on the right on the preceding page. Several ordered pairs are plotted in the **half-plane** above the line $y = x$. Each is a solution of $y > x$.

We can check a pair such as $(-2, 4)$ as follows:

$$\begin{array}{c|c} y > x & \\ \hline 4 \; ? \; -2 & \text{TRUE} \end{array}$$

It turns out that any point on the same side of $y = x$ as $(-2, 4)$ is also a solution. *If we know that one point in a half-plane is a solution, then all points in that half-plane are solutions.* We could have chosen other points to check. The graph of $y > x$ is shown below. (Solutions are indicated by color shading throughout.) We shade the half-plane above $y = x$.

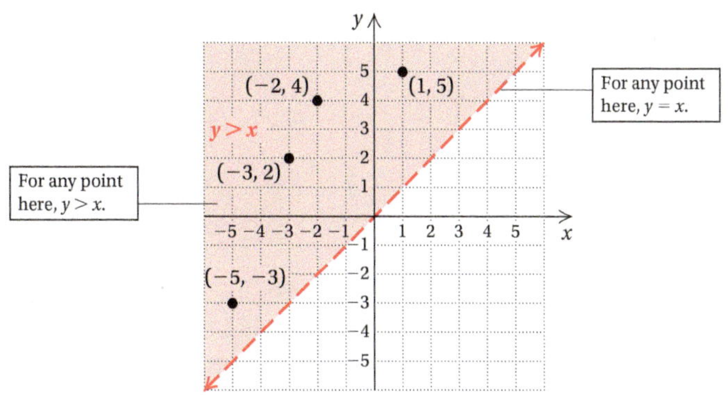

3. Graph: $y < x$.

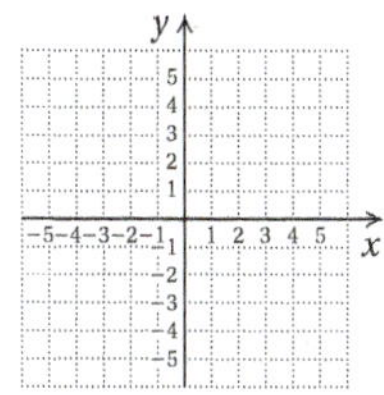

Do Exercise 3. ▶

A **linear inequality** is one that we can get from a linear equation by changing the equals symbol to an inequality symbol. Every linear equation has a graph that is a straight line. The graph of a linear inequality is a half-plane, sometimes including the line along the edge.

To graph an inequality in two variables:

1. Replace the inequality symbol with an equals sign and graph this related linear equation.
2. If the inequality symbol is $<$ or $>$, draw the line dashed. If the inequality symbol is $\leq$ or $\geq$, draw the line solid.
3. The graph consists of a half-plane, either above or below or left or right of the line, and, if the line is solid, the line as well. To determine which half-plane to shade, choose a point *not on the line* as a test point. Substitute to find whether that point is a solution of the *inequality*. If it is, shade the half-plane containing that point. If it is not, shade the half-plane on the opposite side of the line.

Answer

3.

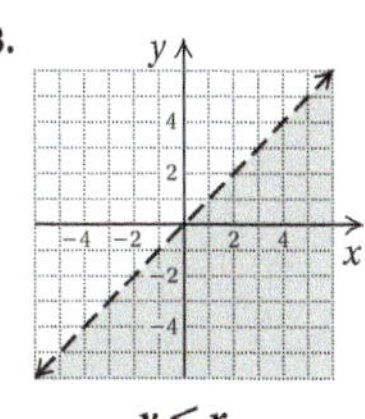

$y < x$

4. Graph: $2x + 4y < 8$.

Related equation:

$2x + 4y$ ______ 8

x-intercept: (______ , 0)

y-intercept: (0, ______)

Draw the line dashed.

Test a point—try $(3, -1)$:

$2 \cdot 3 + 4 \cdot ($ ______ $) < 8$

______ < 8. TRUE

The point $(3, -1)$ is a solution. We shade the half-plane that contains $(3, -1)$.

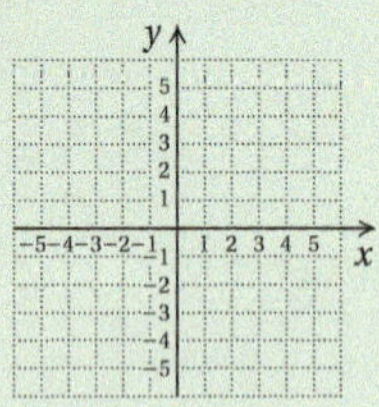

Graph.

5. $3x - 5y < 15$

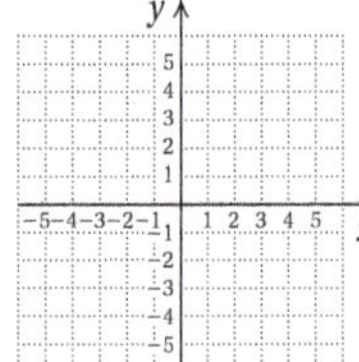

6. $2x + 3y \geq 12$

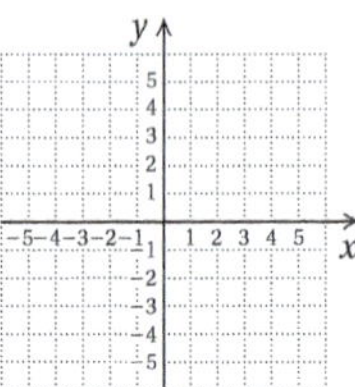

Answers

4.

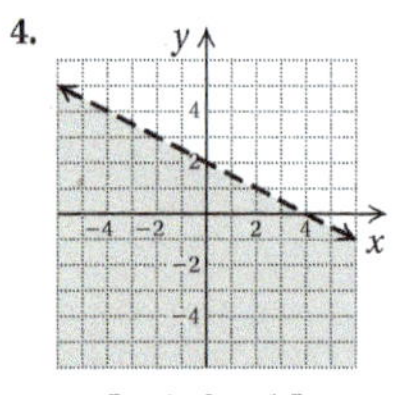

$2x + 4y < 8$

5.

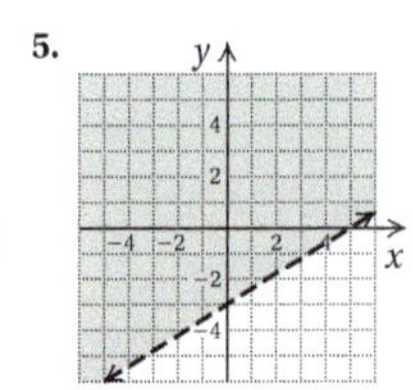

$3x - 5y < 15$

6.

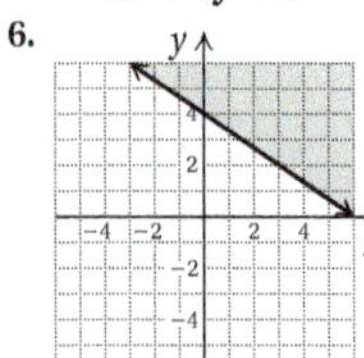

$2x + 3y \geq 12$

Guided Solution:

4. =, 4, 2, −1, 2

EXAMPLE 4 Graph: $5x - 2y < 10$.

1. We first graph the line $5x - 2y = 10$. The intercepts are $(0, -5)$ and $(2, 0)$. This line forms the boundary of the solutions of the inequality.
2. Since the inequality contains the $<$ symbol, points on the line are not solutions of the inequality, so we draw a dashed line.
3. To determine which half-plane to shade, we consider a test point *not* on the line. We try $(3, -2)$ and substitute:

$$\begin{array}{r|l} 5x - 2y & < 10 \\ \hline 5(3) - 2(-2) & ?\ 10 \\ 15 + 4 & \\ 19 & \quad \text{FALSE} \end{array}$$

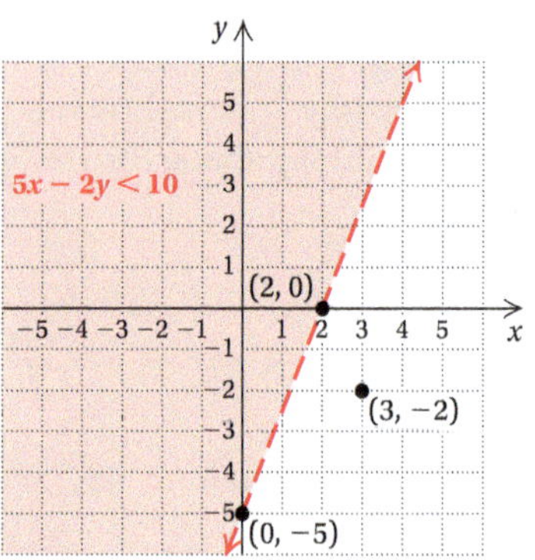

Since this inequality is false, the point $(3, -2)$ is *not* a solution; no point in the half-plane containing $(3, -2)$ is a solution. Thus the points in the opposite half-plane are solutions. The graph is shown above.

◀ **Do Exercise 4.**

EXAMPLE 5 Graph: $2x + 3y \leq 6$.

1. First, we graph the line $2x + 3y = 6$. The intercepts are $(0, 2)$ and $(3, 0)$.
2. Since the inequality contains the $\leq$ symbol, we draw the line solid to indicate that any pair on the line is a solution.
3. Next, we choose a test point that is not on the line. We substitute to determine whether this point is a solution. The origin $(0, 0)$ is generally an easy point to use:

$$\begin{array}{r|l} 2x + 3y & \leq 6 \\ \hline 2 \cdot 0 + 3 \cdot 0 & ?\ 6 \\ 0 & \quad \text{TRUE} \end{array}$$

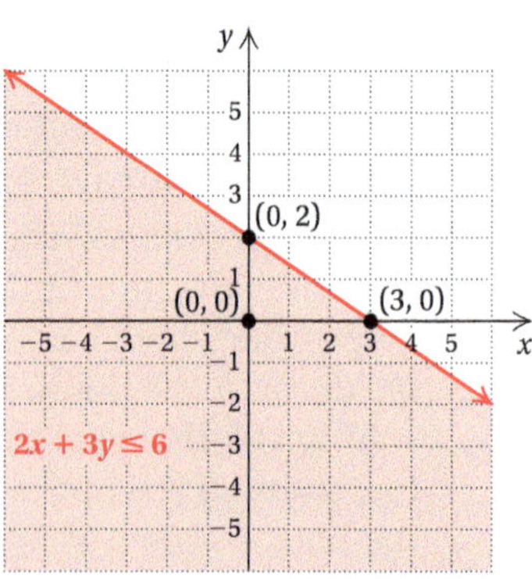

We see that $(0, 0)$ is a solution, so we shade the lower half-plane. Had the substitution given us a false inequality, we would have shaded the other half-plane.

◀ **Do Exercises 5 and 6.**

EXAMPLE 6 Graph $x < 3$ on a plane.

There is no y-term in this inequality, but we can rewrite this inequality as $x + 0y < 3$. We use the same technique that we have used with the other examples.

1. We graph the related equation $x = 3$ on the plane.
2. Since the inequality symbol is $<$, we use a dashed line.
3. The graph is a half-plane either to the left or to the right of the line $x = 3$. To determine which, we consider a test point, $(-4, 5)$:

$$\begin{array}{r|l} x + 0y & < 3 \\ \hline -4 + 0(5) & ?\ 3 \\ -4 & \end{array} \quad \text{TRUE}$$

We see that $(-4, 5)$ is a solution, so all the pairs in the half-plane containing $(-4, 5)$ are solutions. We shade that half-plane.

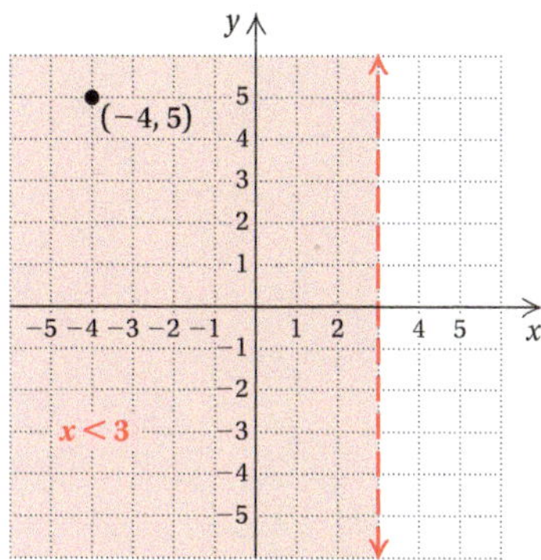

We see from the graph that the solutions of $x < 3$ are all those ordered pairs whose first coordinates are less than 3.

EXAMPLE 7 Graph: $y \geq -4$.

1. We first graph $y = -4$.
2. We use a solid line to indicate that all points on the line are solutions.
3. We then use $(2, 3)$ as a test point and substitute:

$$\begin{array}{r|l} 0x + y & \geq -4 \\ \hline 0(2) + 3 & ?\ -4 \\ 3 & \end{array} \quad \text{TRUE}$$

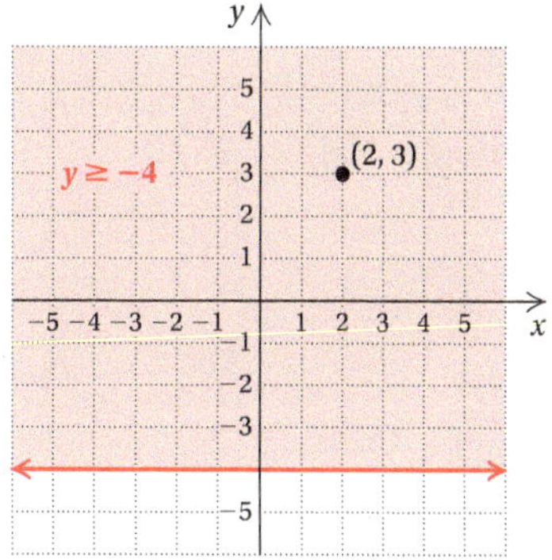

Since $(2, 3)$ is a solution, all points in the half-plane containing $(2, 3)$ are solutions. Note that this half-plane consists of all ordered pairs whose second coordinate is greater than or equal to -4.

Do Exercises 7 and 8. ▶

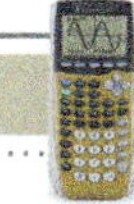

CALCULATOR CORNER

Graphs of Inequalities We can graph inequalities on a graphing calculator, shading the region of the solution set. To graph the inequality in Example 5, $2x + 3y \leq 6$, we first graph the line $2x + 3y = 6$, or $y = (6 - 2x)/3$. After determining algebraically that the solution set consists of all points below the line, we use the graphing calculator's "shade below" graph style to shade this region. On the equation-editor screen, we position the cursor over the graph style icon to the left of the equation and press ENTER repeatedly until the ◣ icon appears. Then we press GRAPH to display the graph of the inequality. Some calculators also have an APP that graphs inequalities.

$y = \dfrac{6-2x}{3}$

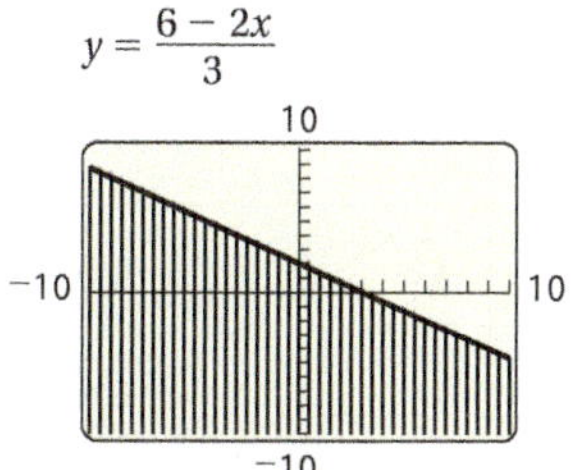

EXERCISE:

1. Use a graphing calculator to graph the inequalities in Margin Exercises 6 and 8 and in Example 7.

Graph.

7. $x > -3$

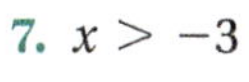

8. $y \leq 4$

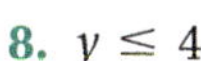

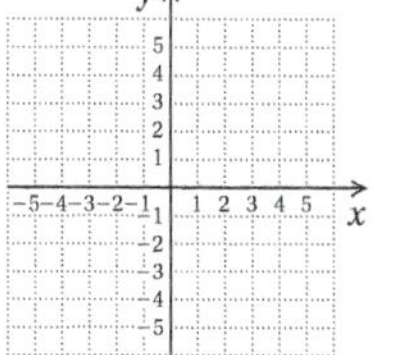

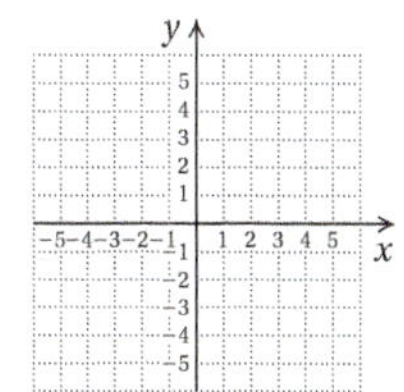

Answers

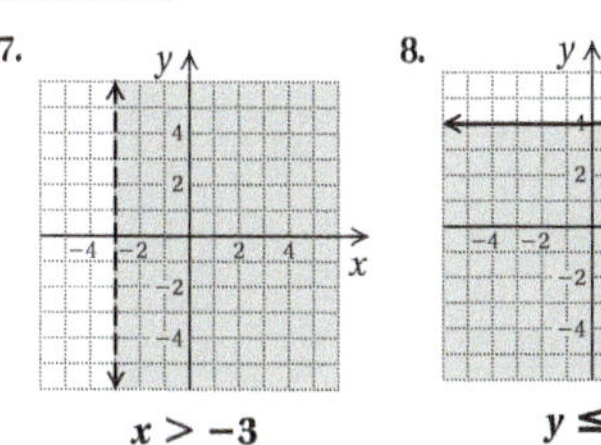

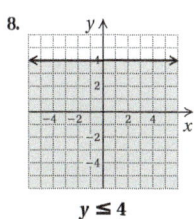

Visualizing for Success

Match each equation or inequality with its graph.

1. $3x - 5y \le 15$

2. $3x + 5y = 15$

3. $3x + 5y \le 15$

4. $3x - 5y \ge 15$

5. $3x - 5y = 15$

6. $3x - 5y < 15$

7. $3x + 5y \ge 15$

8. $3x + 5y > 15$

9. $3x - 5y > 15$

10. $3x + 5y < 15$

Answers on page A-24

A

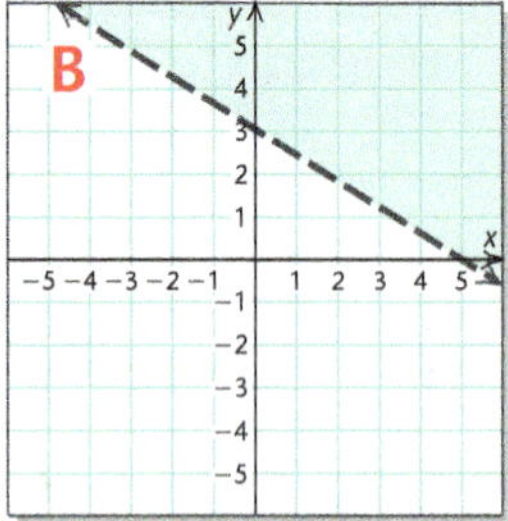

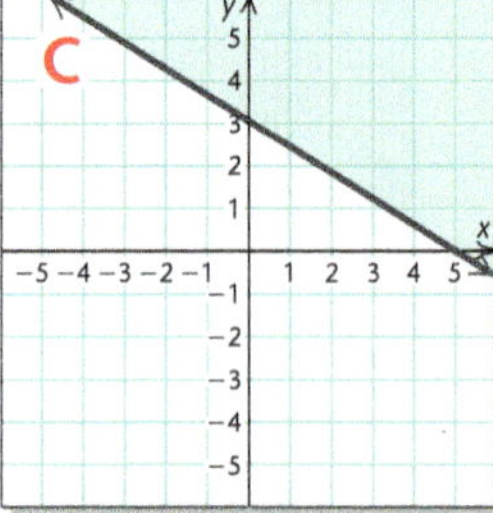

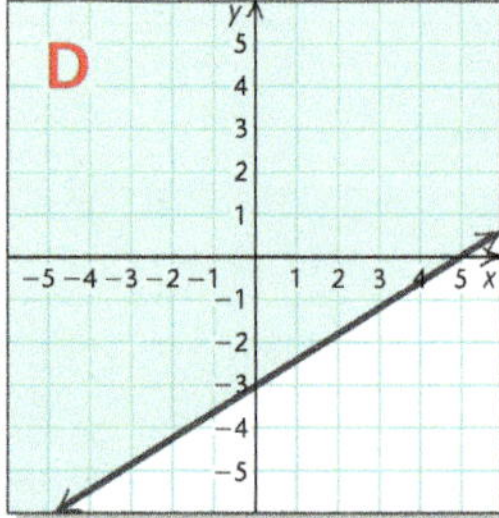

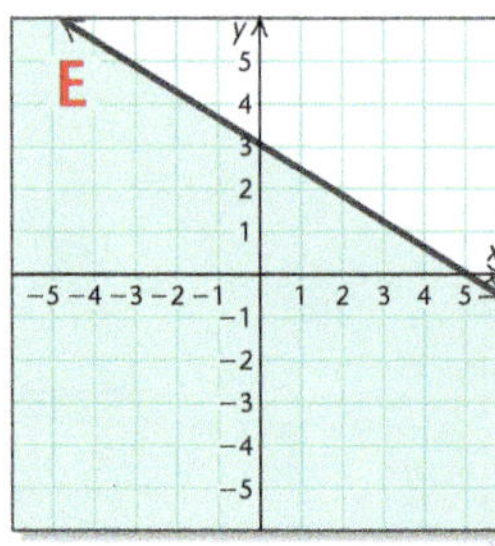

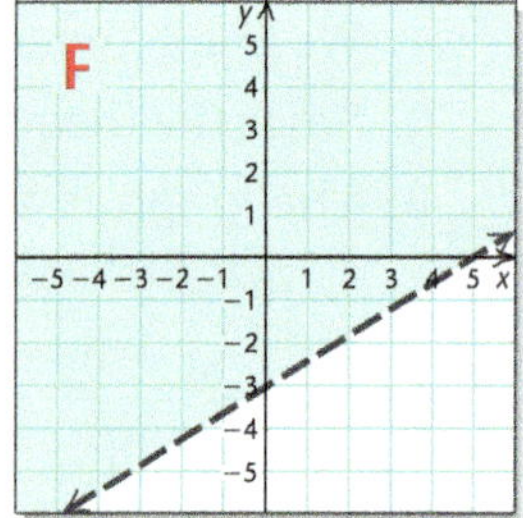

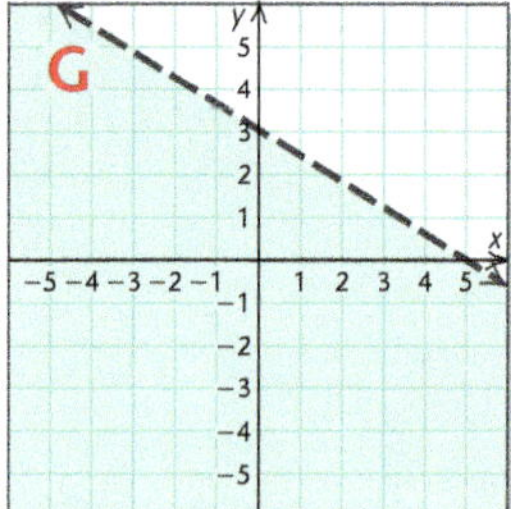

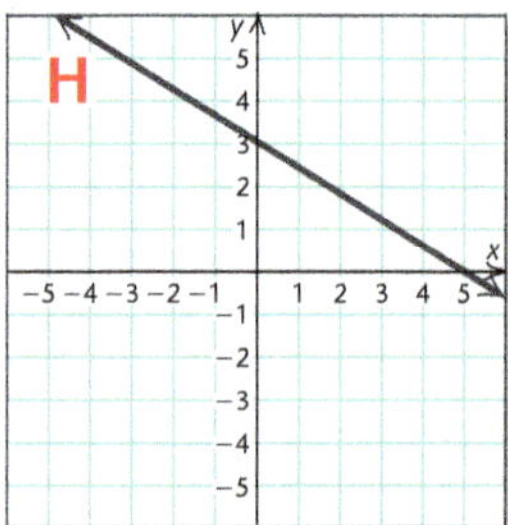

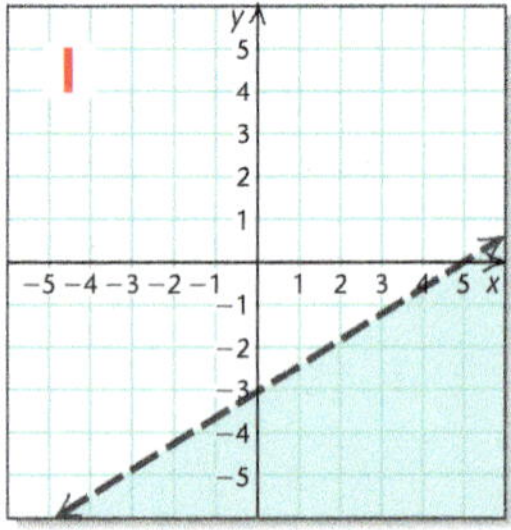

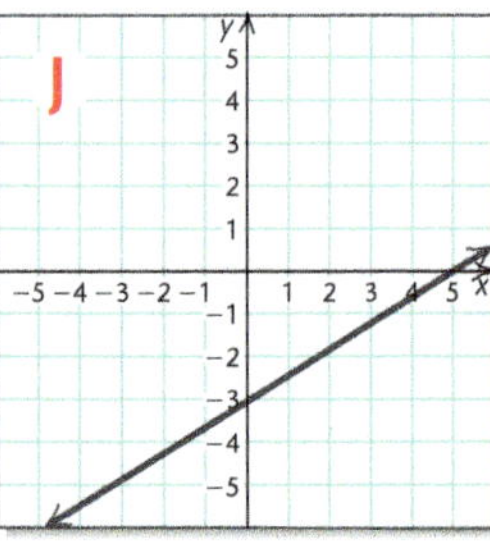

11.7 Exercise Set

For Extra Help MyMathLab® MathXL® PRACTICE WATCH READ REVIEW

Reading Check

The process for graphing the inequality $2x - 4y < -12$ is described in the following paragraph. Choose from the columns on the right the word or the symbol that completes each of Exercises RC1–RC9.

To graph $2x - 4y < -12$, we first replace $<$ with **RC1.** ________ and graph the **RC2.** ________ equation. The **RC3.** ________ is $(-6, 0)$. The **RC4.** ________ is $(0, 3)$. The inequality symbol is **RC5.** ________, so we draw the line **RC6.** ________. The graph consists of a **RC7.** ________, either above or below the line. To determine which half-plane to **RC8.** ________, we test a point not on the line. Let's check $(0, 0)$: $2 \cdot 0 - 4 \cdot 0 < -12$, or $0 < -12$ is **RC9.** ________. We shade the other half-plane.

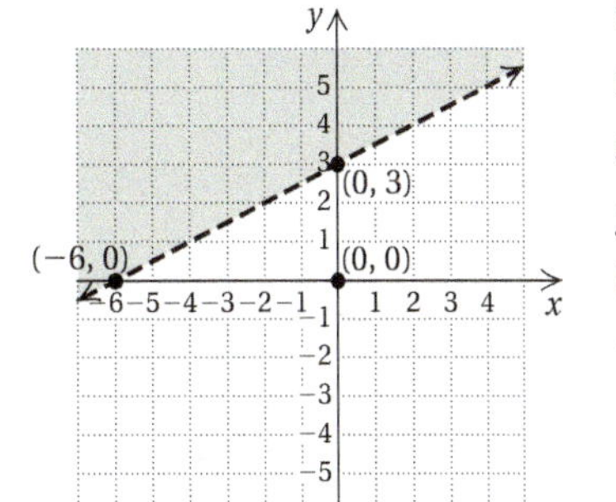

true	$<$
false	$\leq$
x-intercept	solid
y-intercept	dashed
half-plane	related
$=$	shade

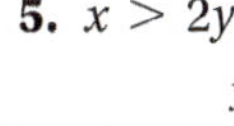

1. Determine whether $(-3, -5)$ is a solution of $-x - 3y < 18$.

2. Determine whether $(2, -3)$ is a solution of $5x - 4y \geq 1$.

3. Determine whether $(1, -10)$ is a solution of $7y - 9x \leq -3$.

4. Determine whether $(-8, 5)$ is a solution of $x + 0 \cdot y > 4$.

b Graph on a plane.

5. $x > 2y$

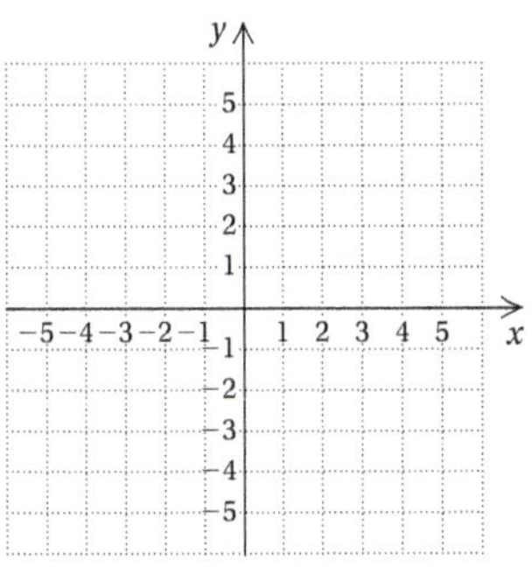

6. $x \geq 3y$

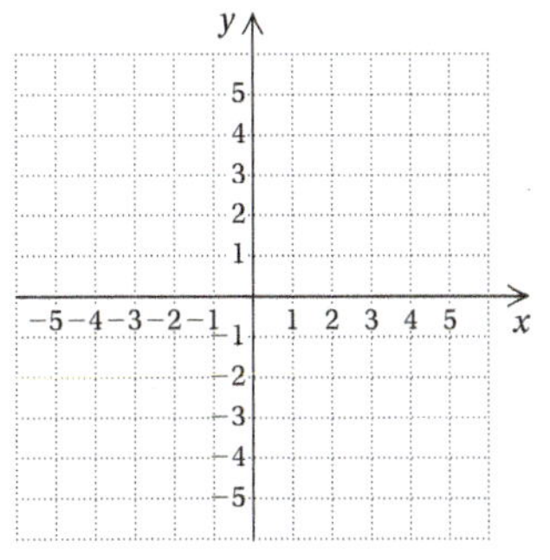

7. $y \leq x - 3$

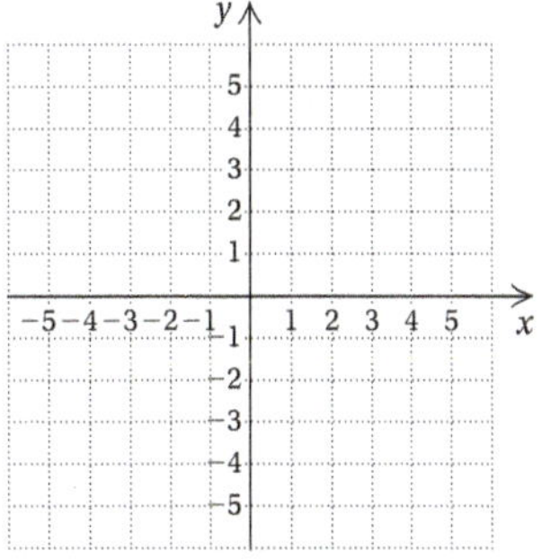

8. $y < x + 4$

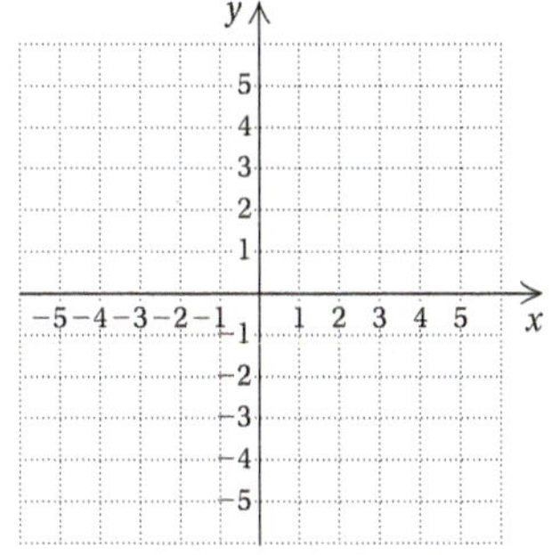

9. $x + y \leq 3$

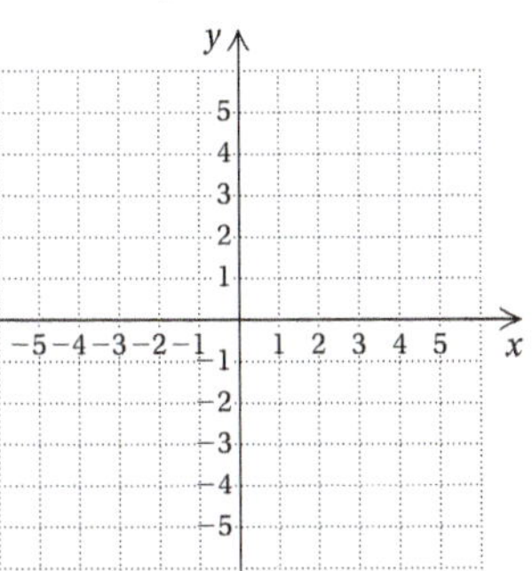

10. $x + y < 4$

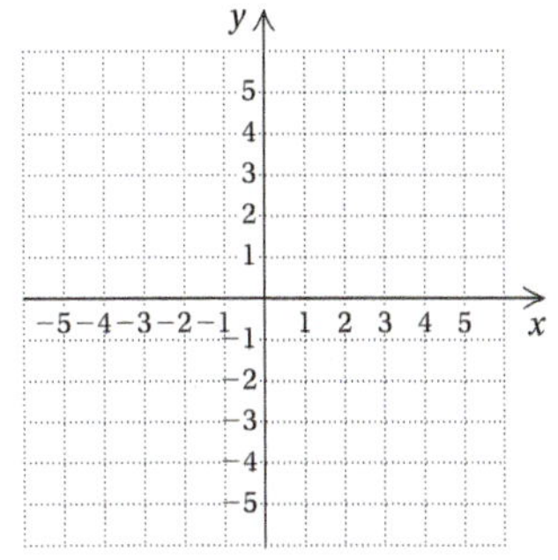

11. $y > x - 2$

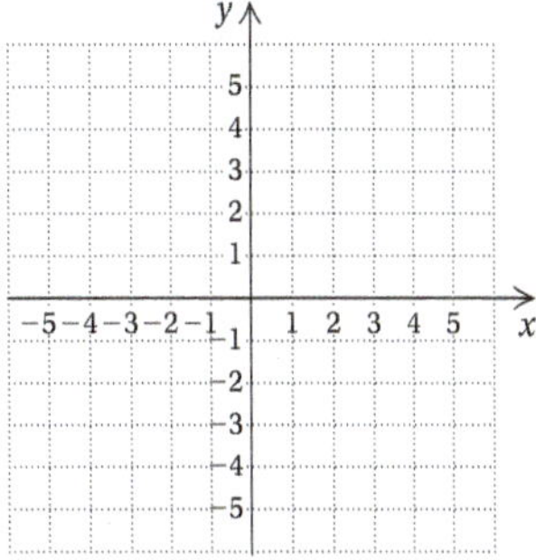

12. $y \geq x - 1$

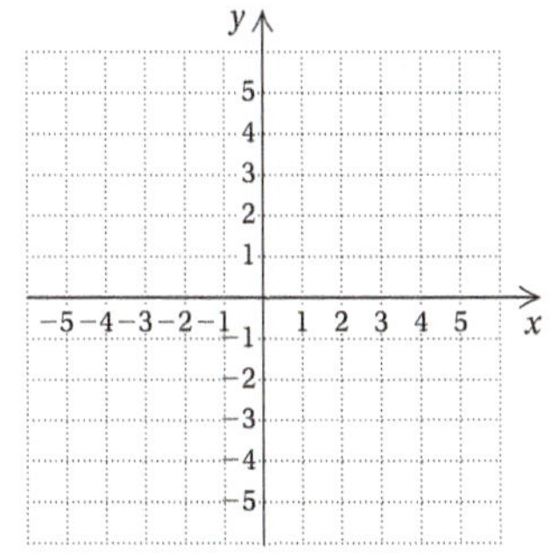

13. $x - y > 7$

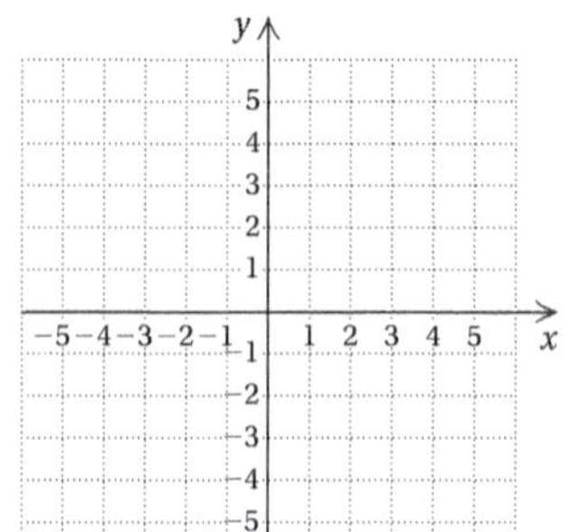

14. $x - y > -2$

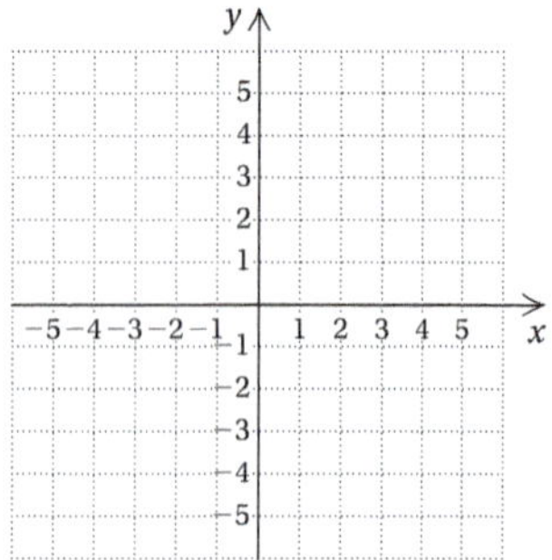

15. $y \geq 4x - 1$

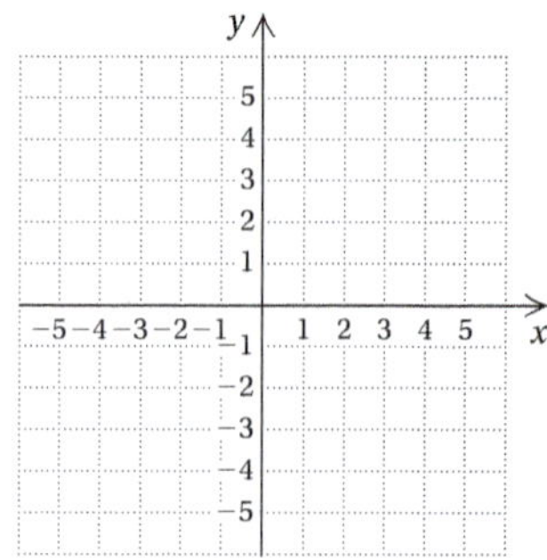

16. $y \geq 3x + 2$

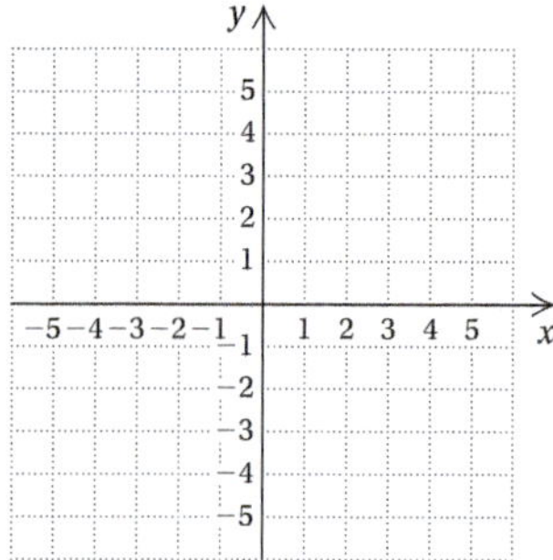

17. $y \geq 1 - 2x$

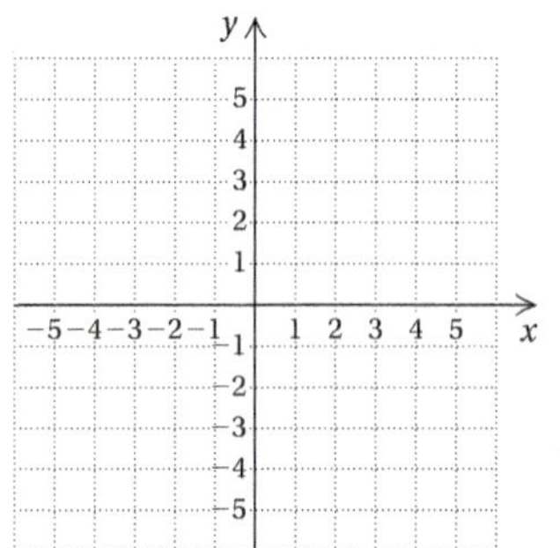

18. $y - 2x \leq -1$

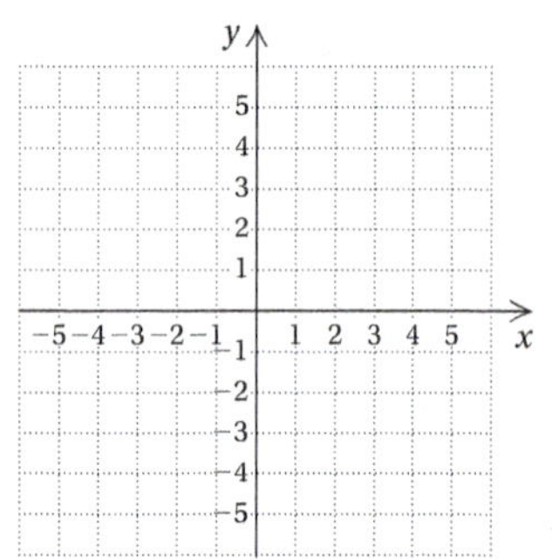

19. $2x + 3y \leq 12$

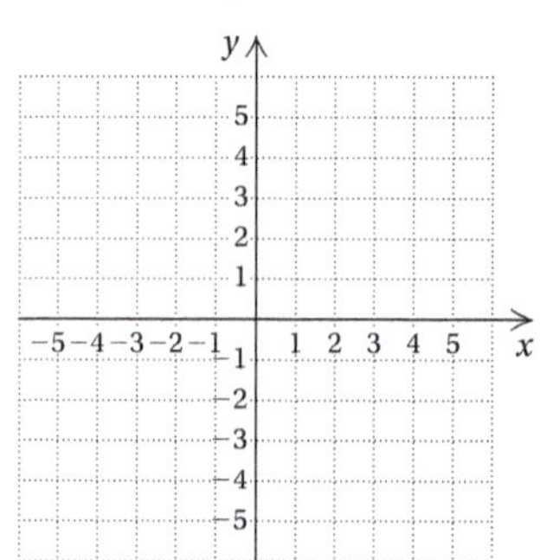

20. $5y - 2x > 10$

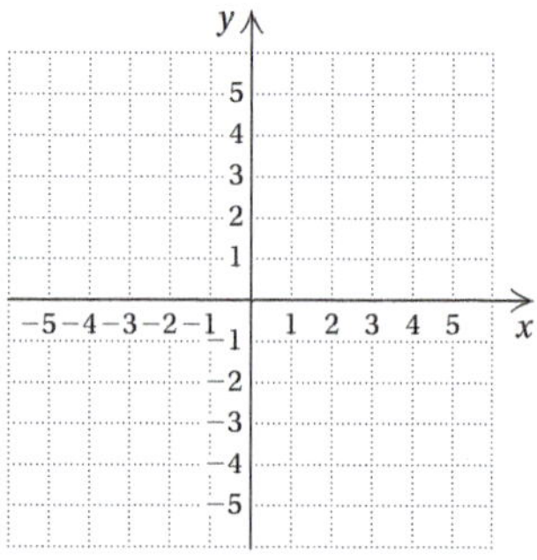

21. $y \leq 3$

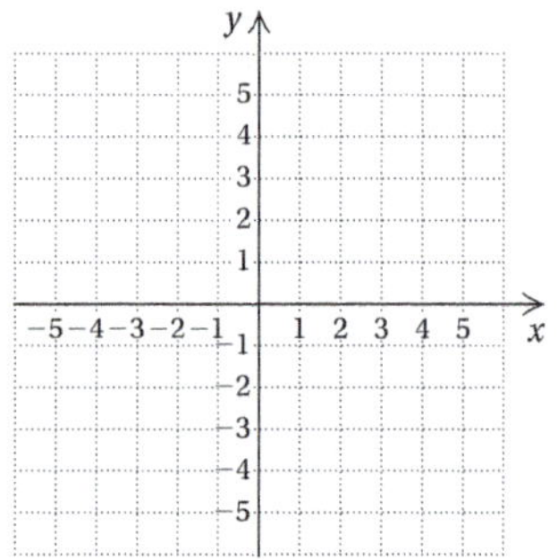

22. $y > -1$

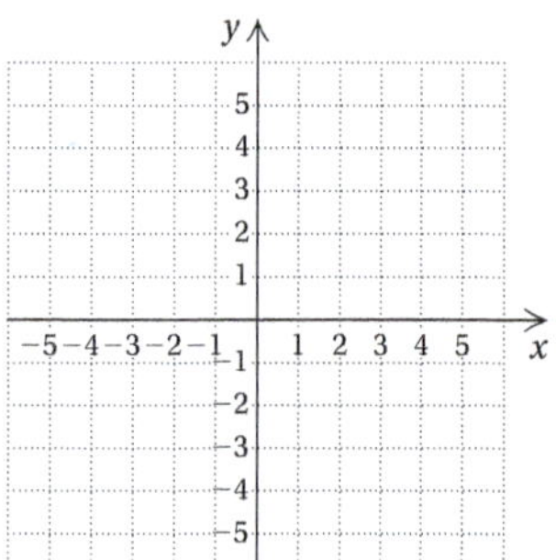

23. $x \geq -1$

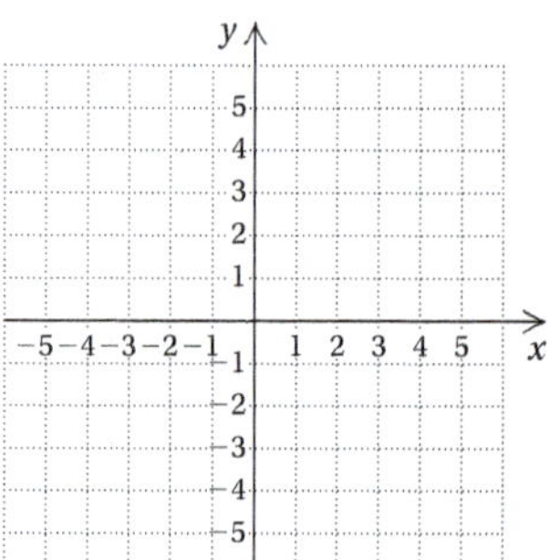

24. $x < 0$

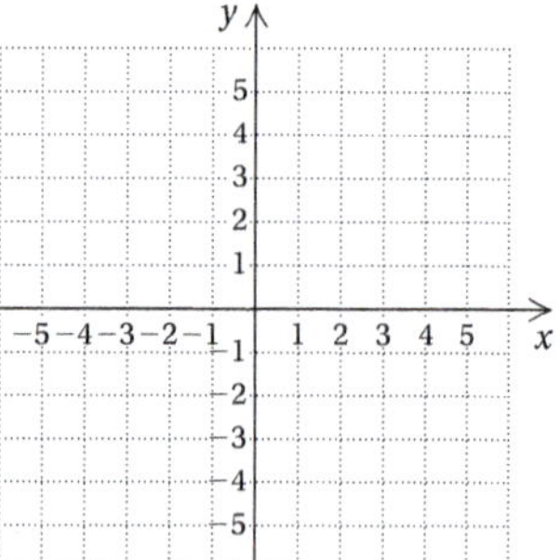

Skill Maintenance

Determine whether the graphs of the equations are parallel, perpendicular, or neither. [11.6a, b]

25. $5y + 50 = 4x,$
$5y = 4x + 15$

26. $5x + 4y = 12,$
$5y + 50 = 4x$

27. $5y + 50 = 4x,$
$4y = 5x + 12$

28. $4x + 5y + 35 = 0,$
$5y = 4x + 40$

Synthesis

29. *Elevators.* Many elevators have a capacity of 1 metric ton (1000 kg). Suppose c children, each weighing 35 kg, and a adults, each weighing 75 kg, are on an elevator. Find and graph an inequality that asserts that the elevator is overloaded.

30. *Hockey Wins and Losses.* A hockey team determines that it needs at least 60 points for the season in order to make the playoffs. A win w is worth 2 points and a tie t is worth 1 point. Find and graph an inequality that describes the situation.

CHAPTER 11

Summary and Review

Vocabulary Reinforcement

Complete each statement with the correct term from the column on the right. Some of the choices may not be used.

1. The equation $y = mx + b$ is called the ______________ equation. [11.4a]
2. ______________ lines are graphs of equations of the type $y = b$. [11.2b]
3. ______________ lines are graphs of equations of the type $x = a$. [11.2b]
4. The ______________ of a line is a number that indicates how the line slants. [11.3a]
5. The ______________ of a line, if it exists, indicates where the line crosses the x-axis, and thus will always have 0 as the ______________ coordinate. [11.2a]
6. The ______________ of a line, if it exists, indicates where the line crosses the y-axis, and thus will always have 0 as the ______________ coordinate. [11.2a]

x-intercept
y-intercept
parallel
perpendicular
vertical
horizontal
slope–intercept
first
second
slope

Concept Reinforcement

Determine whether each statement is true or false.

______________ 1. The x- and y-intercepts of $y = mx$ are both $(0, 0)$. [11.2a]

______________ 2. Parallel lines have the same y-intercept. [11.6a]

______________ 3. The ordered pair $(0, 0)$ is a solution of $y > x$. [11.7a]

______________ 4. The second coordinate of all points in quadrant III is negative. [11.1a]

______________ 5. The x-intercept of $Ax + By = C$, $C \neq 0$, is $\left(\frac{A}{C}, 0\right)$. [11.2a]

Study Guide

Objective 11.1b Find the coordinates of a point on a graph.

Example Find the coordinates of points Q, R, and S.

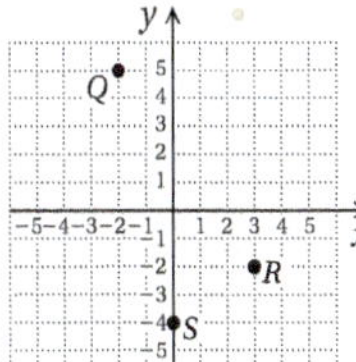

Point Q is 2 units left of the origin and 5 units up. Its coordinates are $(-2, 5)$.

Point R is 3 units right of the origin and 2 units down. Its coordinates are $(3, -2)$.

Point S is 0 units left or right of the origin and 4 units down. Its coordinates are $(0, -4)$.

Practice Exercise

1. Find the coordinates of points F, G, and H.

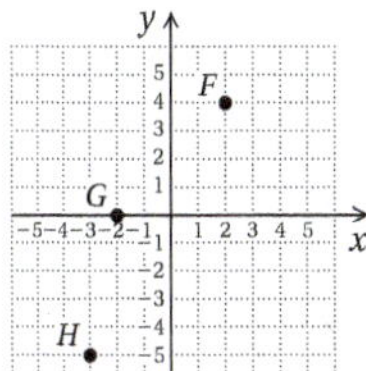

Objective 11.1d Graph linear equations of the type $y = mx + b$ and $Ax + By = C$, identifying the y-intercept.

Example Graph $2y + 2 = -3x$ and identify the y-intercept.

To find an equivalent equation in the form $y = mx + b$, we solve for y: $y = -\frac{3}{2}x - 1$. The y-intercept is $(0, -1)$.

We then find two other points using multiples of 2 for x to avoid fractions.

x	y	
0	−1	← y-intercept
−2	2	
2	−4	

Practice Exercise

2. Graph $x + 2y = 8$ and identify the y-intercept.

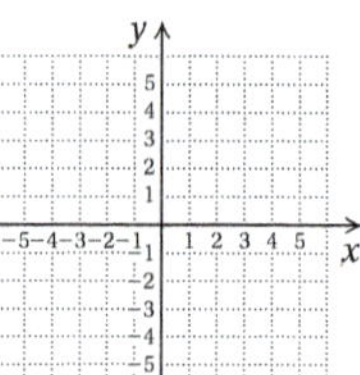

Objective 11.2a Find the intercepts of a linear equation, and graph using intercepts.

Example For $2x - y = -6$, find the intercepts. Then use the intercepts to graph the equation.

To find the y-intercept, we let $x = 0$ and solve for y:

$2 \cdot 0 - y = -6$ and $y = 6$.

The y-intercept is $(0, 6)$.

To find the x-intercept, we let $y = 0$ and solve for x:

$2x - 0 = -6$ and $x = -3$.

The x-intercept is $(-3, 0)$.

We find a third point as a check.

x	y	
0	6	← y-intercept
−3	0	← x-intercept
−1	4	

Practice Exercise

3. For $y - 2x = -4$, find the intercepts. Then use the intercepts to graph the equation.

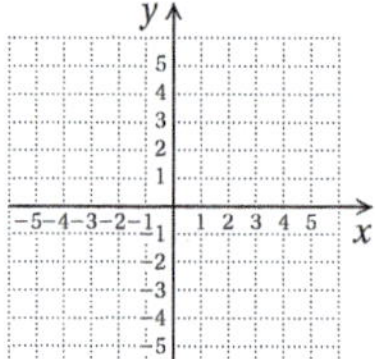

Objective 11.2b Graph equations equivalent to those of the type $x = a$ and $y = b$.

Example Graph: $y = 1$ and $x = -\frac{3}{2}$.

For $y = 1$, no matter what number we choose for x, $y = 1$. The graph is a horizontal line. For $x = -\frac{3}{2}$, no matter what number we choose for y, $x = -\frac{3}{2}$. The graph is a vertical line.

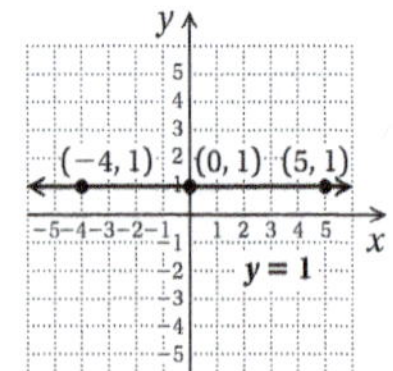

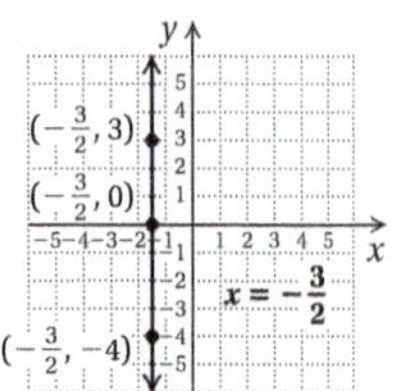

Practice Exercises

Graph.

4. $y = -\frac{5}{2}$

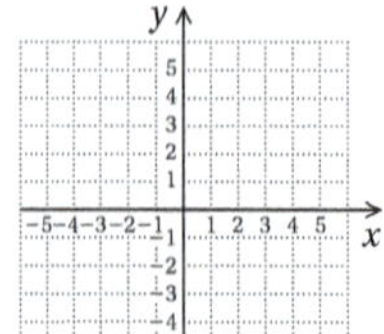

5. $x = 2$

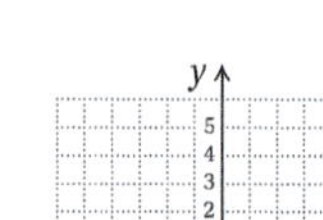

Objective 11.3a Given the coordinates of two points on a line, find the slope of the line, if it exists.

Example Find the slope, if it exists, of the line containing the given points.

$(-9, 3)$ and $(5, -6)$: $m = \frac{-6 - 3}{5 - (-9)} = \frac{-9}{14} = -\frac{9}{14}$;

$\left(7, \frac{1}{2}\right)$ and $\left(-13, \frac{1}{2}\right)$: $m = \frac{\frac{1}{2} - \frac{1}{2}}{-13 - 7} = \frac{0}{-20} = 0$;

$(0.6, 1.5)$ and $(0.6, -1.5)$: $m = \frac{-1.5 - 1.5}{0.6 - 0.6} = \frac{-3}{0}$,

m is not defined.

Practice Exercises

Find the slope, if it exists, of the line containing the given points.

6. $(-8, 20), (-8, 14)$

7. $(2, -1), (16, 20)$

8. $(0.5, 2.8), (1.5, 2.8)$

Objective 11.3b Find the slope of a line from an equation.

Example Find the slope, if it exists, of each line.

a) $5x - 20y = -10$

We first solve for y: $y = \frac{1}{4}x + \frac{1}{2}$. The slope is $\frac{1}{4}$.

b) $y = -\frac{4}{5}$

Think: $y = 0 \cdot x - \frac{4}{5}$. This line is horizontal. The slope is 0.

c) $x = 6$

This line is vertical. The slope is not defined.

Practice Exercises

Find the slope, if it exists, of the line.

9. $x = 0.25$

10. $7y + 14x = -28$

11. $y = -5$

Objective 11.4b Find an equation of a line when the slope and a point on the line are given.

Example Find the equation of the line with slope -2 that contains the point $(3, -1)$.

$y = -2x + b$ Substituting -2 for m in $y = mx + b$

$-1 = -2 \cdot 3 + b$ Substituting 3 for x and -1 for y

$-1 = -6 + b$

$5 = b$ Solving for b

The equation is $y = -2x + 5$.

Practice Exercise

12. Find the equation of the line with slope 6 that contains the point $(-1, 1)$.

Objective 11.4c Find an equation of a line when two points on the line are given.

Example Find an equation of the line that contains $(-10, 5)$ and $(2, -5)$.

Slope $= m = \frac{-5 - 5}{2 - (-10)} = \frac{-10}{12} = -\frac{5}{6}$.

$-5 = -\frac{5}{6}(2) + b$ Substituting $-\frac{5}{6}$ for m, 2 for x, and -5 for y in $y = mx + b$

$-5 = -\frac{5}{3} + b$

$-\frac{10}{3} = b$ Solving for b

The equation is $y = -\frac{5}{6}x - \frac{10}{3}$.

Practice Exercise

13. Find an equation of the line that contains $(7, -3)$ and $(1, -2)$.

Objectives 11.6a, b Determine whether the graphs of two linear equations are parallel, perpendicular, or neither.

Example Determine whether the graphs of the equations are parallel, perpendicular, or neither:

$$2x - y = 8 \quad \text{and} \quad y + \frac{1}{2}x = -2.$$

We solve each equation for y and determine the slope of each:

$$y = 2x - 8 \quad \text{and} \quad y = -\frac{1}{2}x - 2.$$

The slopes are 2 and $-\frac{1}{2}$.

The slopes are not the same. The lines are not parallel. The product of the slopes, $2 \cdot \left(-\frac{1}{2}\right)$, is -1. The lines are perpendicular.

Practice Exercises

Determine whether the graphs of the equations are parallel, perpendicular, or neither.

14. $4y = -x - 12,$
$y - 4x = \frac{1}{2}$

15. $2y - x = -4,$
$x - 2y = -12$

Objective 11.7b Graph linear inequalities.

Example Graph: $3x - y < 3$.

We first graph the line $3x - y = 3$. The intercepts are $(0, -3)$ and $(1, 0)$. Since the inequality contains the $<$ symbol, points on the line are not solutions of the inequality, so we draw a dashed line.

To determine which half-plane to shade, we consider a test point not on the line. We try $(0, 0)$:

$$3 \cdot 0 - 0 < 3$$
$$0 < 3. \quad \text{TRUE}$$

We see that $(0, 0)$ is a solution, so we shade the upper half-plane.

Practice Exercise

16. Graph: $y - 3x \leq -3$.

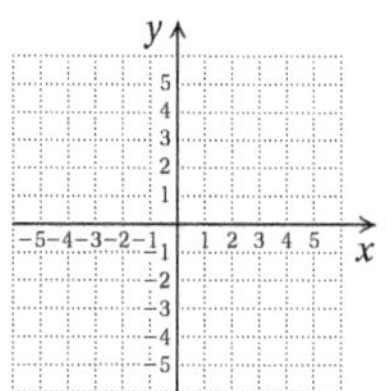

Review Exercises

Plot each point. [11.1a]

1. $(2, 5)$ **2.** $(0, -3)$ **3.** $(-4, -2)$

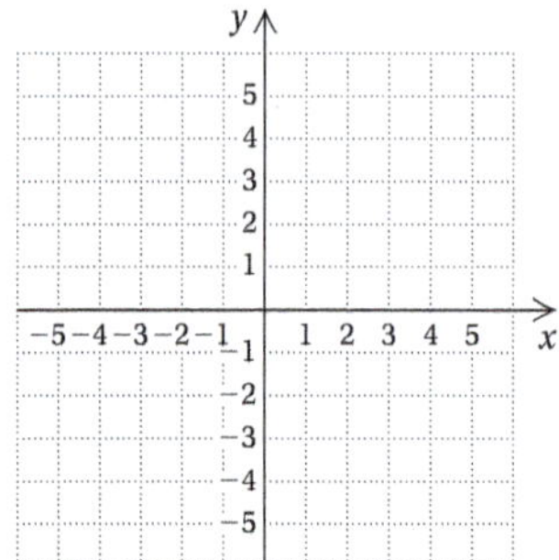

Find the coordinates of each point. [11.1b]

4. A **5.** B **6.** C

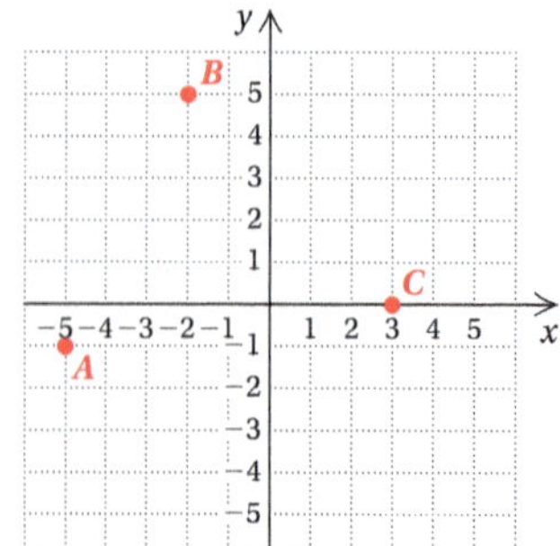

In which quadrant is each point located? [11.1a]

7. $(3, -8)$ **8.** $(-20, -14)$ **9.** $(4.9, 1.3)$

Determine whether each ordered pair is a solution of $2y - x = 10$. [11.1c]

10. $(2, -6)$ **11.** $(0, 5)$

12. Show that the ordered pairs $(0, -3)$ and $(2, 1)$ are solutions of the equation $2x - y = 3$. Then use the graph of the equation to determine another solution. Answers may vary. [11.1c]

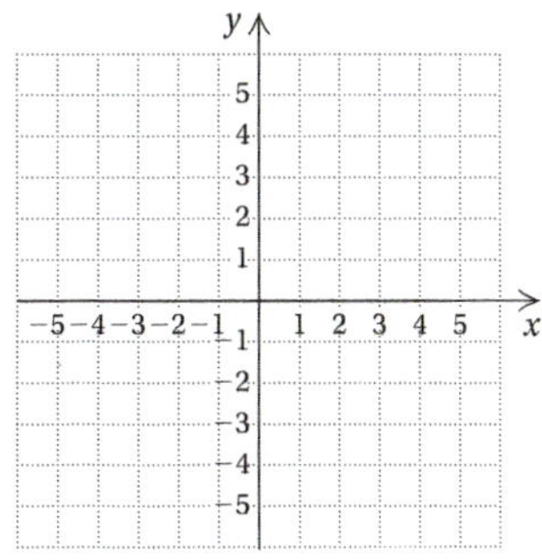

Graph each equation, identifying the y-intercept. [11.1d]

13. $y = 2x - 5$ **14.** $y = -\frac{3}{4}x$

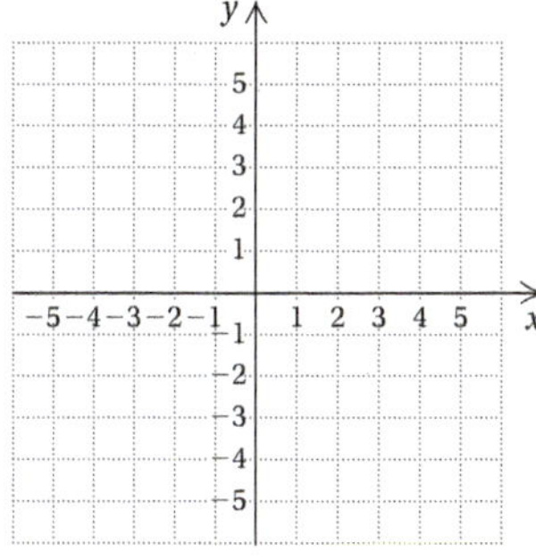

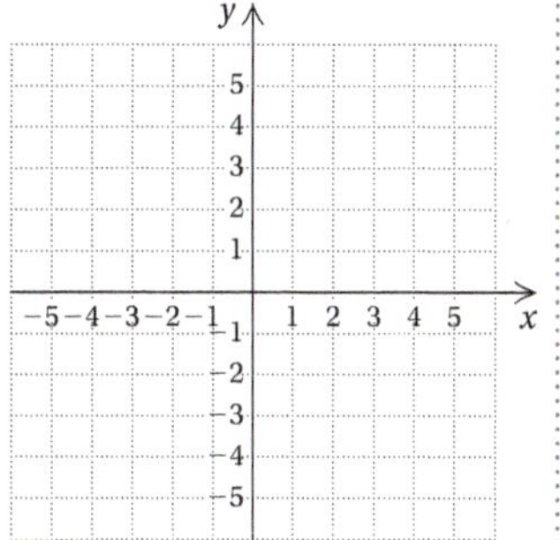

15. $y = -x + 4$ **16.** $y = 3 - 4x$

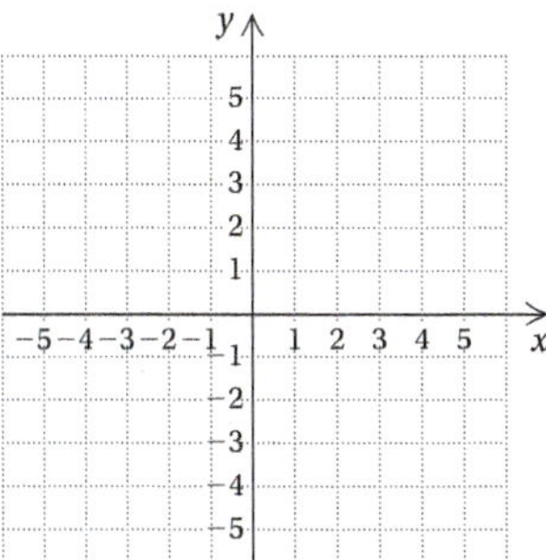

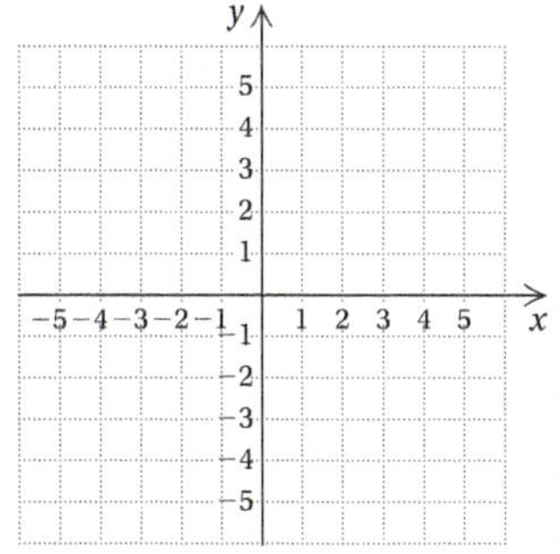

Solve. [11.1e]

17. ***Kitchen Design.*** Kitchen designers recommend that a refrigerator be selected on the basis of the number of people n in the household. The appropriate size S, in cubic feet, is given by

$$S = \frac{3}{2}n + 13.$$

a) Determine the recommended size of a refrigerator if the number of people is 1, 2, 5, and 10.

b) Graph the equation and use the graph to estimate the recommended size of a refrigerator for 4 people sharing an apartment.

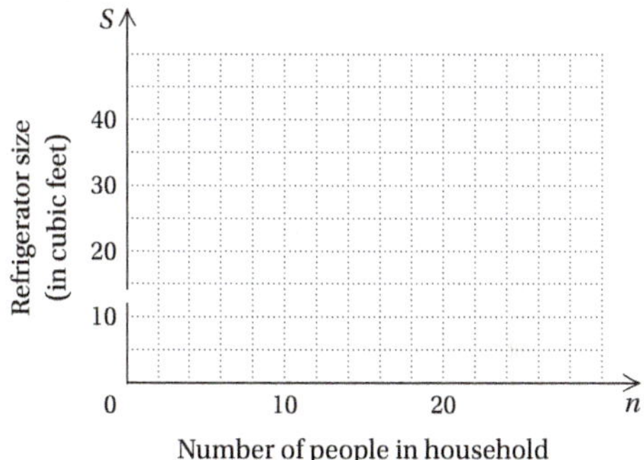

c) A refrigerator is 22 ft^3. For how many residents is it the recommended size?

Find the intercepts of each equation. Then graph the equation. [11.2a]

18. $x - 2y = 6$ **19.** $5x - 2y = 10$

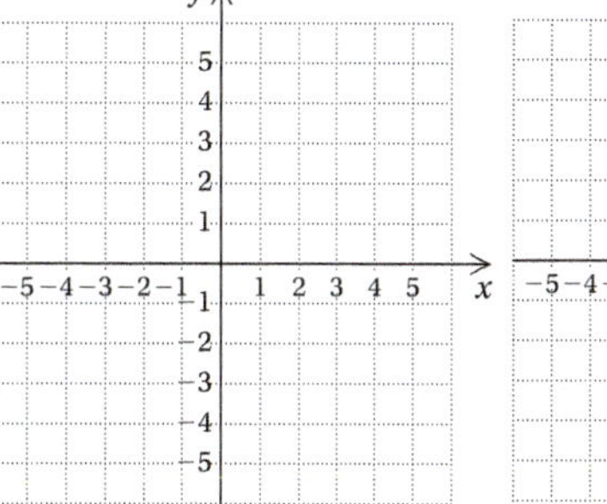

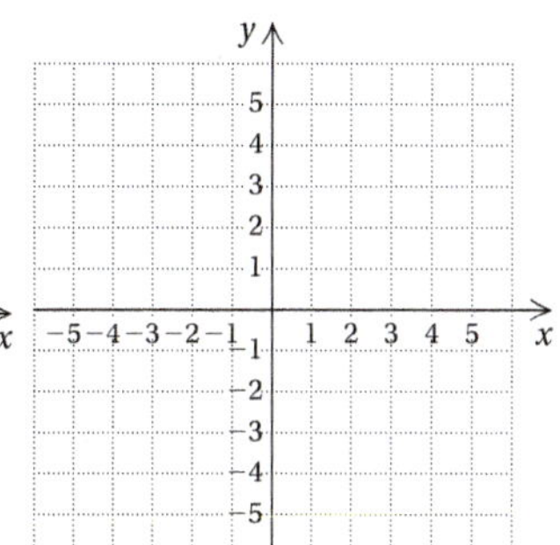

Graph each equation. [11.2b]

20. $y = 3$ **21.** $5x - 4 = 0$

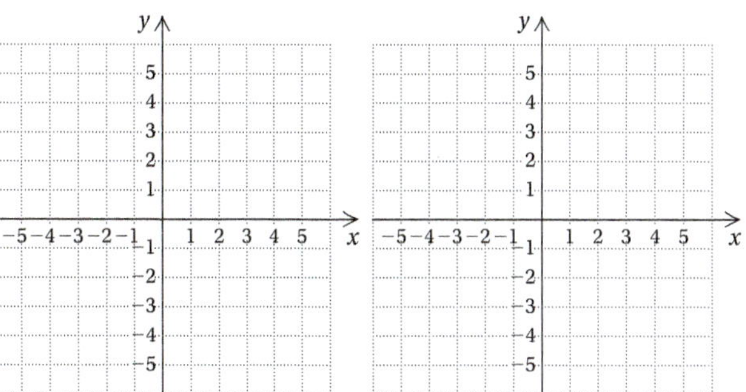

Find the slope. [11.3a]

22. **23.**

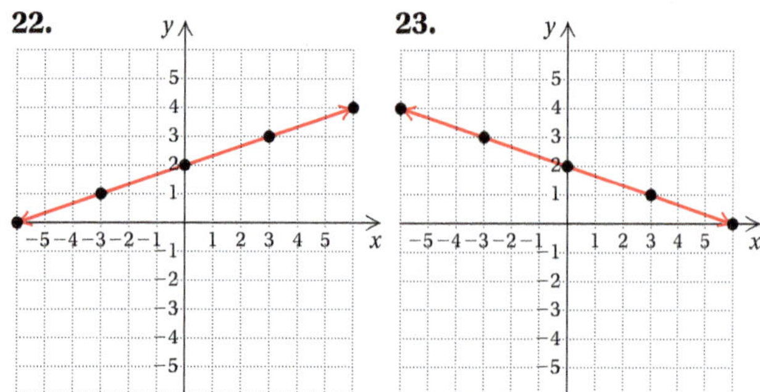

Graph the line containing the given pair of points and find the slope. [11.3a]

24. $(-5, -2), (5, 4)$

25. $(-5, 5), (4, -4)$

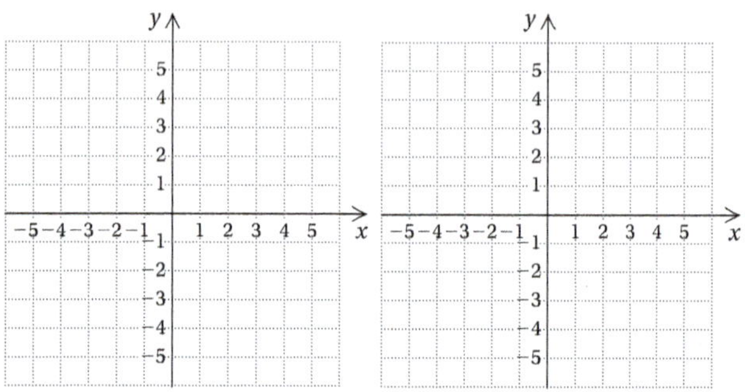

Find the slope, if it exists. [11.3b]

26. $y = -\frac{5}{8}x - 3$

27. $2x - 4y = 8$

28. $x = -2$

29. $y = 9$

30. *Snow Removal.* By 3:00 P.M., Erin had plowed 7 driveways and by 5:30 P.M., she had completed 13. [11.3c]

a) Find Erin's plowing rate, in number of driveways per hour.

b) Find Erin's plowing rate, in number of minutes per driveway.

31. *Road Grade.* At one point, Beartooth Highway in Yellowstone National Park rises 315 ft over a horizontal distance of 4500 ft. Find the slope, or grade, of the road. [11.3c]

32. *Organic Food.* Each year in the United States, the amount of sales in the organic food industry increases. Use the following graph to determine the slope, or rate of change, in the amount of sales, in billions of dollars, of organic food products with respect to time. [11.3a]

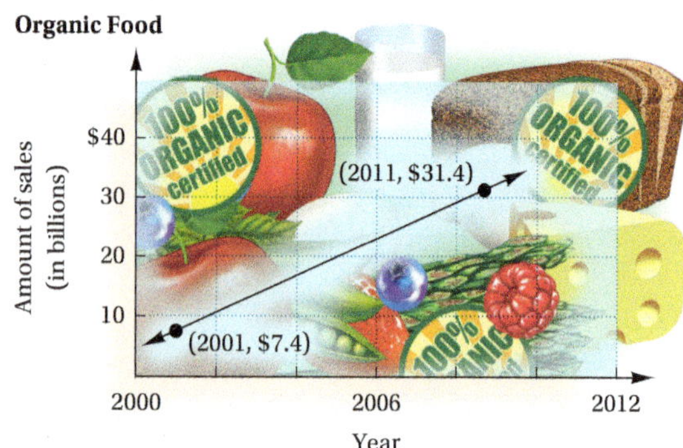

SOURCE: Organic Trade Association

Find the slope and the y-intercept. [11.4a]

33. $y = -9x + 46$

34. $x + y = 9$

35. $3x - 5y = 4$

Find an equation of the line with the given slope and y-intercept. [11.4a]

36. Slope: -2.8; y-intercept: $(0, 19)$

37. Slope: $\frac{5}{8}$; y-intercept: $\left(0, -\frac{7}{8}\right)$

Find an equation of the line containing the given point and with the given slope. [11.4b]

38. $(1, 2),\ m = 3$

39. $(-2, -5),\ m = \frac{2}{3}$

40. $(0, -4),\ m = -2$

Find an equation of the line containing the given pair of points. [11.4c]

41. $(5, 7)$ and $(-1, 1)$

42. $(2, 0)$ and $(-4, -3)$

43. *Taking Cruises.* The following line graph illustrates the number of Americans, in millions, taking cruises for years since 2000. [11.4c]

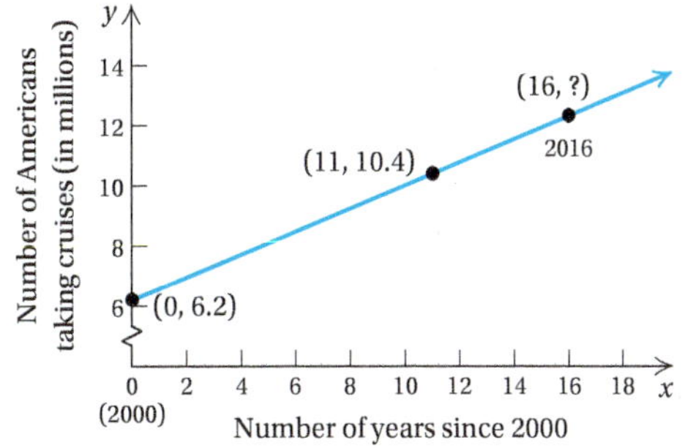

SOURCE: Cruise Lines International Association

a) Find an equation of the line. Let $x =$ the number of years since 2000.
b) What is the rate of change in the number of Americans taking cruises with respect to time?
c) Use the equation to estimate the number of Americans taking cruises in 2016.

44. Draw a line that has slope -1 and y-intercept $(0, 4)$. [11.5a]

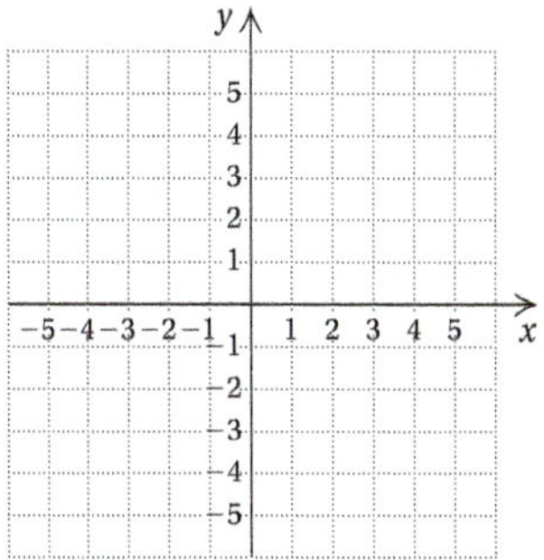

45. Draw a line that has slope $\frac{5}{3}$ and y-intercept $(0, -3)$. [11.5a]

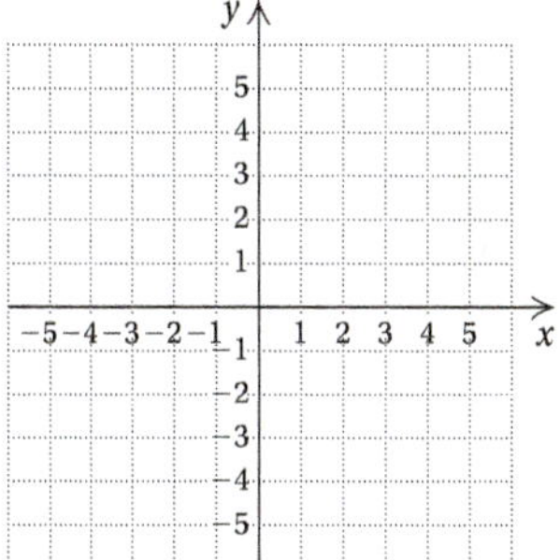

46. Graph $y = -\frac{3}{5}x + 2$ using the slope and the y-intercept. [11.5a]

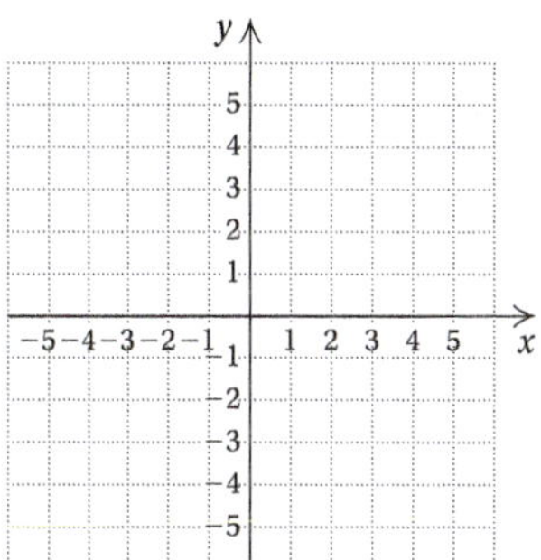

47. Graph $2y - 3x = 6$ using the slope and the y-intercept. [11.5a]

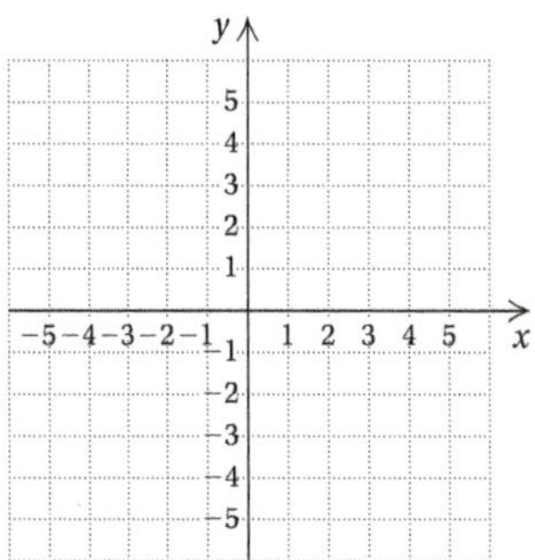

Determine whether the graphs of the equations are parallel, perpendicular, or neither. [11.6a, b]

48. $4x + y = 6,$
$4x + y = 8$

49. $2x + y = 10,$
$y = \frac{1}{2}x - 4$

50. $x + 4y = 8,$
$x = -4y - 10$

51. $3x - y = 6,$
$3x + y = 8$

Determine whether the given point is a solution of the inequality $x - 2y > 1$. [11.7a]

52. $(0, 0)$

53. $(1, 3)$

54. $(4, -1)$

Graph on a plane. [11.7b]

55. $x < y$

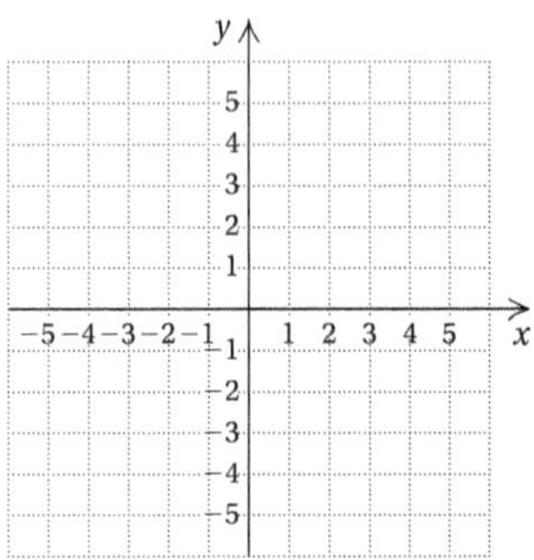

56. $x + 2y \geq 4$

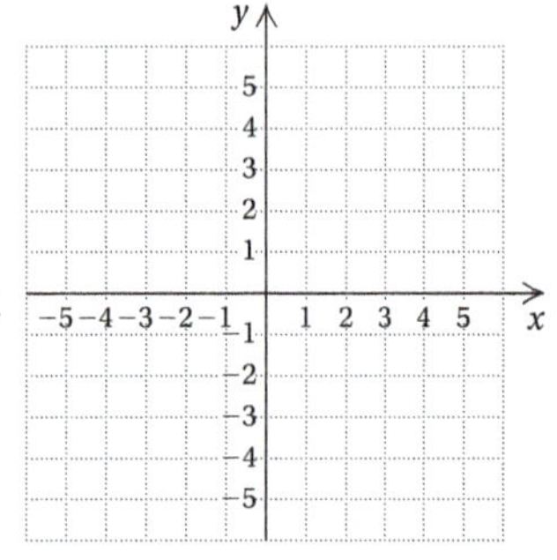

57. $x > -2$

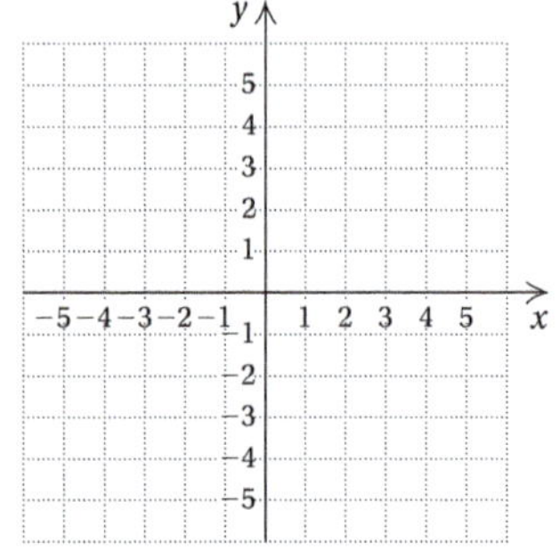

58. Select the statement that describes the graphs of the lines $-x + \frac{1}{2}y = -2$ and $2y + x - 8 = 0$. [11.6a, b]

A. The lines are parallel.
B. The lines are the same.
C. The lines intersect and are not perpendicular.
D. The lines are perpendicular.

59. Find the equation of the line with slope $-\frac{8}{3}$ and containing the point $(-3, 8)$. [11.4b]

A. $y = -\frac{8}{3}x + \frac{55}{3}$

B. $y = -\frac{3}{8}$

C. $y = -\frac{8}{3}x$

D. $8y + 3x = -3$

Synthesis

60. Find the area and the perimeter of a rectangle for which $(-2, 2)$, $(7, 2)$, and $(7, -3)$ are three of the vertices. [11.1a]

61. *Gondola Aerial Lift.* In Telluride, Colorado, there is a free gondola ride that provides a spectacular view of the town and the surrounding mountains. The gondolas that begin in the town at an elevation of 8725 ft travel 5750 ft to Station St. Sophia, whose altitude is 10,550 ft. They then continue 3913 ft to Mountain Village, whose elevation is 9500 ft. [11.3c]

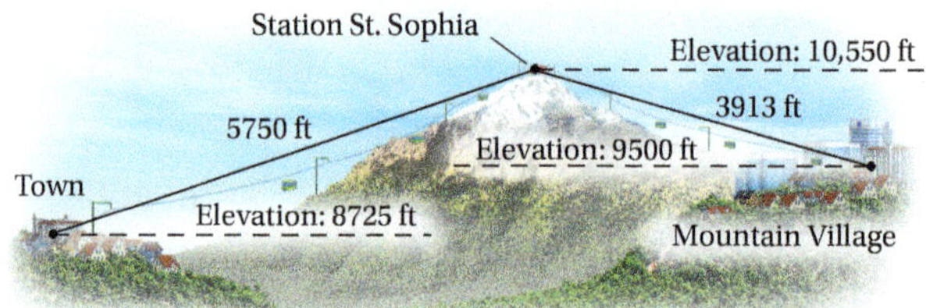

A visitor departs from the town at 11:55 A.M. and with no stop at Station St. Sophia reaches Mountain Village at 12:07 P.M.

a) Find the gondola's average rate of ascent and descent, in number of feet per minute.

b) Find the gondola's average rate of ascent and descent, in number of minutes per foot.

Understanding Through Discussion and Writing

1. Consider two equations of the type $Ax + By = C$. Explain how you would go about showing that their graphs are perpendicular. [11.6b]

2. Is the graph of any inequality in the form $y > mx + b$ shaded *above* the line $y = mx + b$? Why or why not? [11.7b]

3. Explain why the first coordinate of the y-intercept is always 0. [11.1d]

4. Graph $x < 1$ on both the number line and a plane, and explain the difference between the graphs. [11.7b]

5. Describe how you would graph $y = 0.37x + 2458$ using the slope and the y-intercept. You need not actually draw the graph. [11.5a]

6. Consider two equations of the type $Ax + By = C$. Explain how you would go about showing that their graphs are parallel. [11.6a]

CHAPTER 11 Test

For Extra Help For step-by-step test solutions, access the Chapter Test Prep Videos in MyMathLab® or on YouTube (search "BittingerPreIntro" and click on Channels).

In which quadrant is each point located?

1. $\left(-\frac{1}{2}, 7\right)$

2. $(-5, -6)$

Find the coordinates of each point.

3. A

4. B

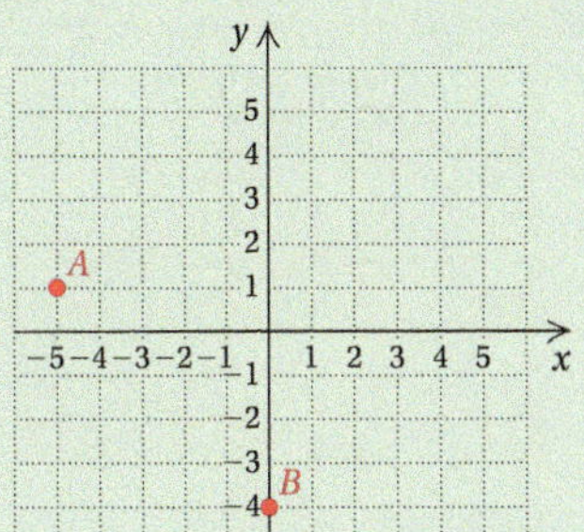

5. Show that the ordered pairs $(-4, -3)$ and $(-1, 3)$ are solutions of the equation $y - 2x = 5$. Then use the graph of the straight line containing the two points to determine another solution. Answers may vary.

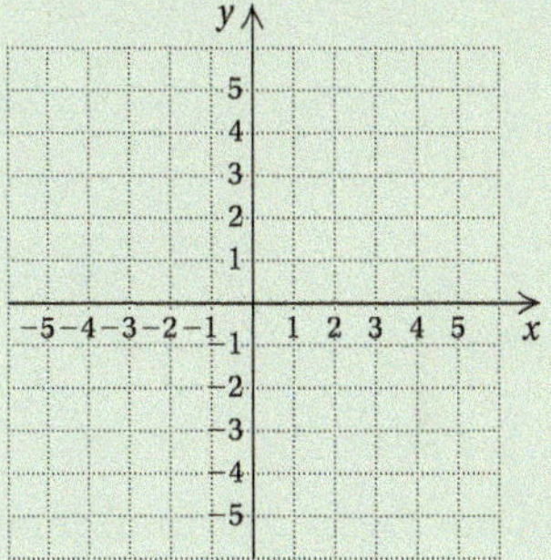

Graph each equation. Identify the y-intercept.

6. $y = 2x - 1$

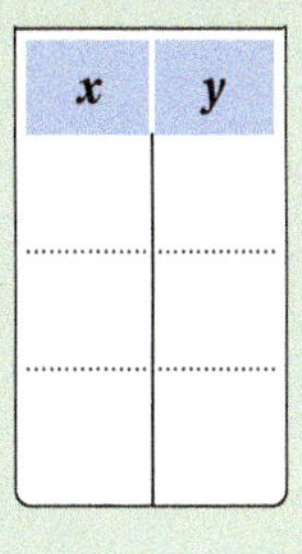

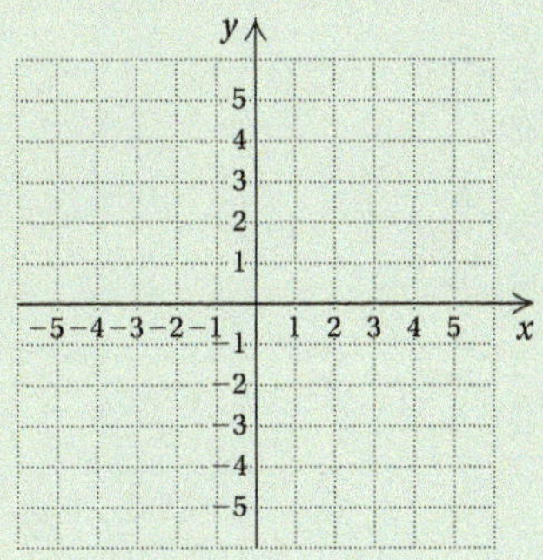

7. $y = -\frac{3}{2}x$

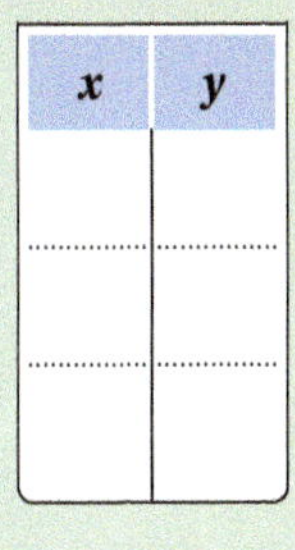

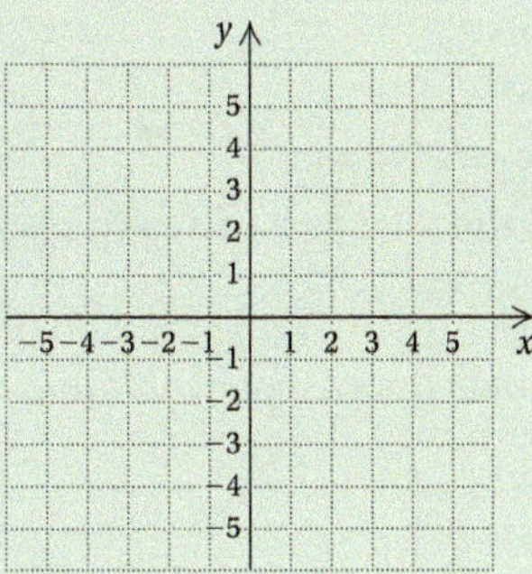

Find the intercepts of each equation. Then graph the equation.

8. $2x - 4y = -8$

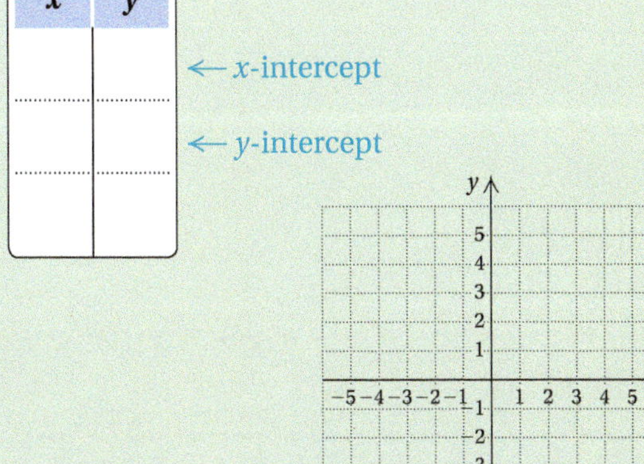

9. $2x - y = 3$

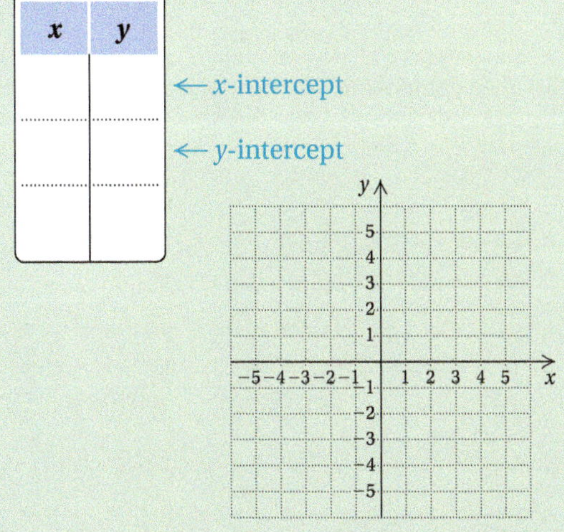

Graph each equation.

10. $2x + 8 = 0$

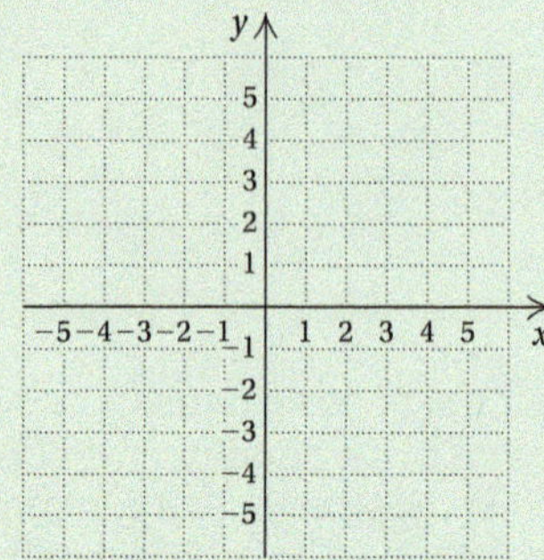

11. $y = 5$

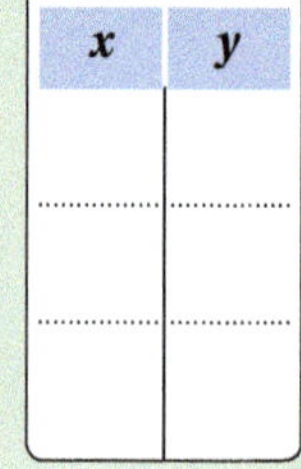

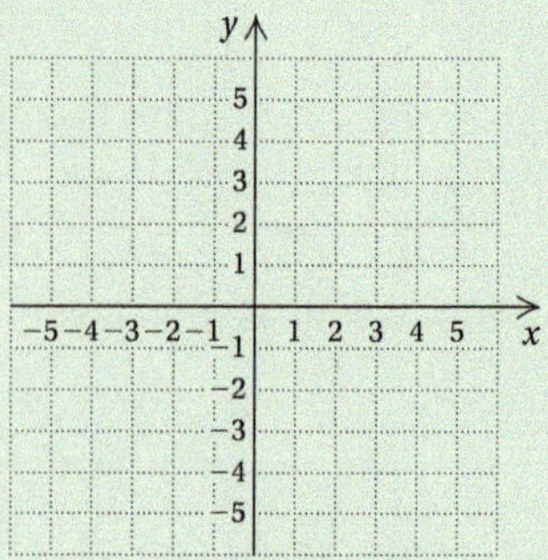

12. *Health Insurance Cost.* The total annual cost, employer plus employee, of health insurance can be approximated by

$$C = 606t + 8593,$$

where t is the number of years since 2007. That is, $t = 0$ corresponds to 2007, $t = 3$ corresponds to 2010, and so on.

Source: Towers Watson

a) Find the total cost of health insurance in 2007, in 2009, and in 2012.

b) Graph the equation and then use the graph to estimate the cost of health insurance in 2016.

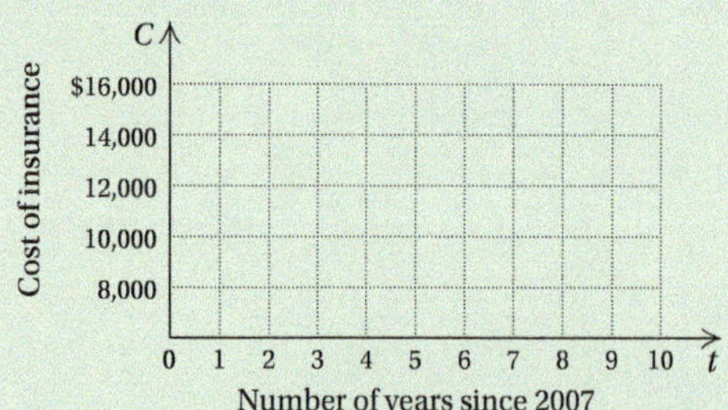

c) Predict the year in which the cost of health insurance will be $15,259.

13. *Train Travel.* The following graph shows data concerning a recent train ride from Denver to Kansas City. At what rate did the train travel?

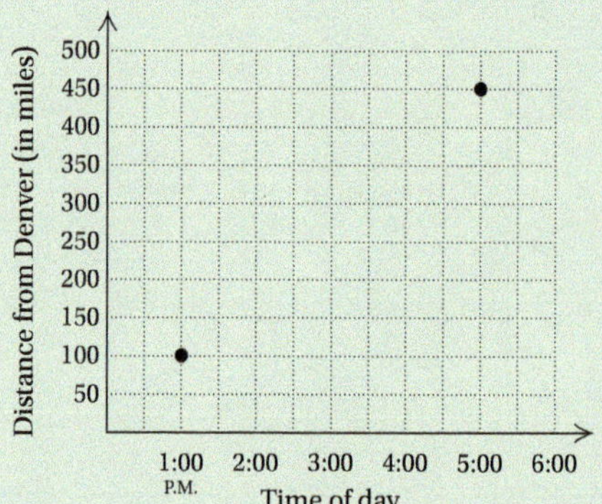

14. Find the slope.

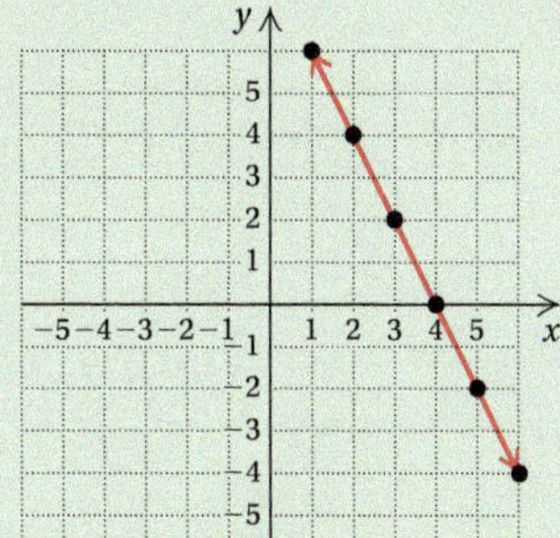

15. Graph the line containing $(-3, 1)$ and $(5, 4)$ and find the slope.

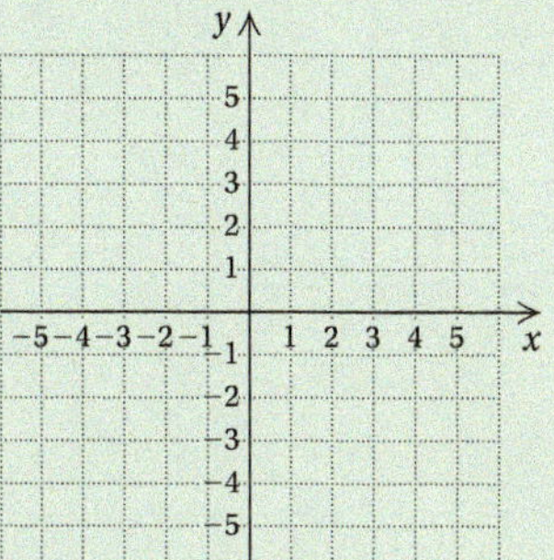

16. Find the slope, if it exists.

a) $2x - 5y = 10$
b) $x = -2$

17. *Navigation.* Capital Rapids drops 54 ft vertically over a horizontal distance of 1080 ft. What is the slope of the rapids?

Find the slope and the y-intercept.

18. $y = 2x - \frac{1}{4}$

19. $-4x + 3y = -6$

Find an equation of the line with the given slope and y-intercept.

20. Slope: 1.8; y-intercept: $(0, -7)$

21. Slope: $-\frac{3}{8}$; y-intercept: $\left(0, -\frac{1}{8}\right)$

Find an equation of the line containing the given point and with the given slope.

22. $(3, 5)$, $m = 1$

23. $(-2, 0)$, $m = -3$

Find an equation of the line containing the given pair of points.

24. $(1, 1)$ and $(2, -2)$

25. $(4, -1)$ and $(-4, -3)$

26. Draw a graph of the line with slope $-\frac{3}{2}$ and y-intercept $(0, 1)$.

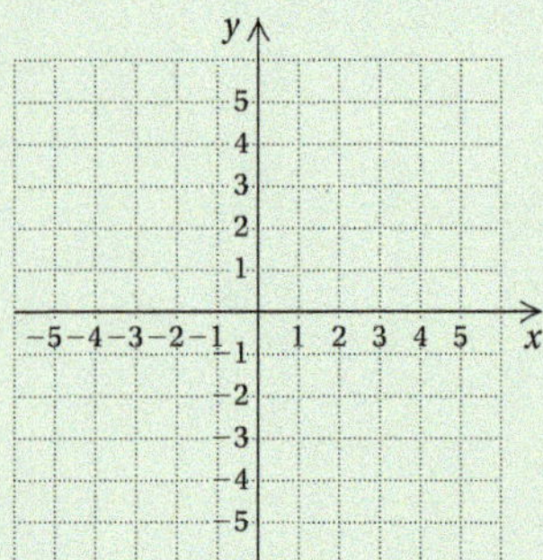

27. Graph $y = 2x - 3$ using the slope and the y-intercept.

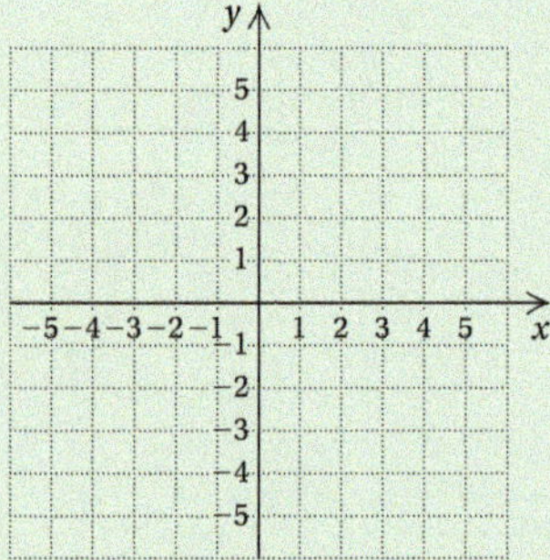

Determine whether the graphs of the equations are parallel, perpendicular, or neither.

28. $2x + y = 8,$
$2x + y = 4$

29. $2x + 5y = 2,$
$y = 2x + 4$

30. $x + 2y = 8,$
$-2x + y = 8$

Determine whether the given point is a solution of the inequality $3y - 2x < -2$.

31. $(0, 0)$

32. $(-4, -10)$

Graph on a plane.

33. $y > x - 1$

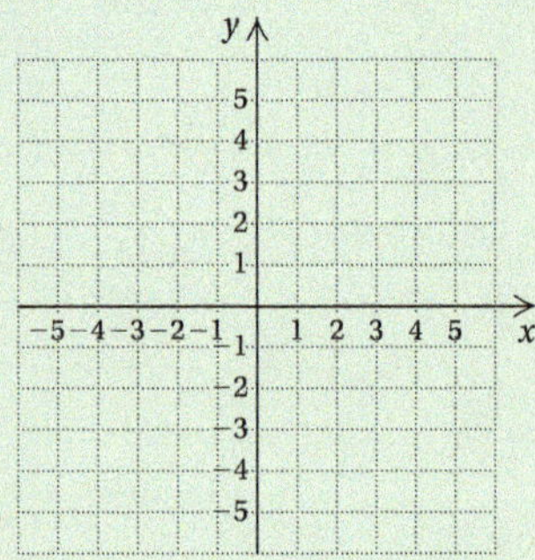

34. $2x - y \geq 4$

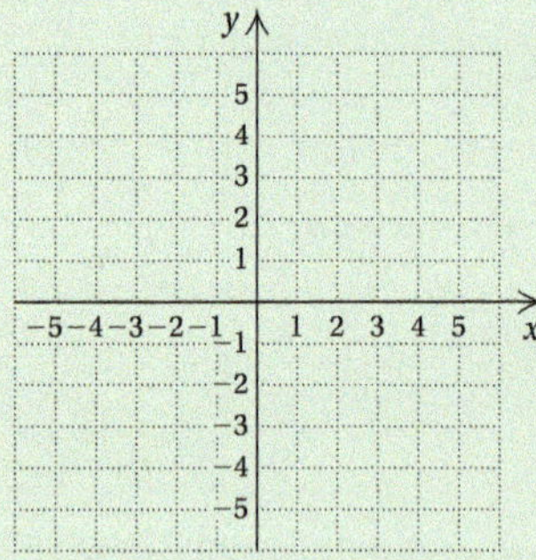

35. Select the statement that best describes the graphs of the lines $15x + 21y = 7$ and $35y + 14 = -25x$.

A. The lines are parallel.
B. The lines are the same.
C. The lines intersect and are not perpendicular.
D. The lines are perpendicular.

Synthesis

36. A diagonal of a square connects the points $(-3, -1)$ and $(2, 4)$. Find the area and the perimeter of the square.

37. Find the value of k such that $3x + 7y = 14$ and $ky - 7x = -3$ are perpendicular.

CHAPTER

12

Polynomials: Operations

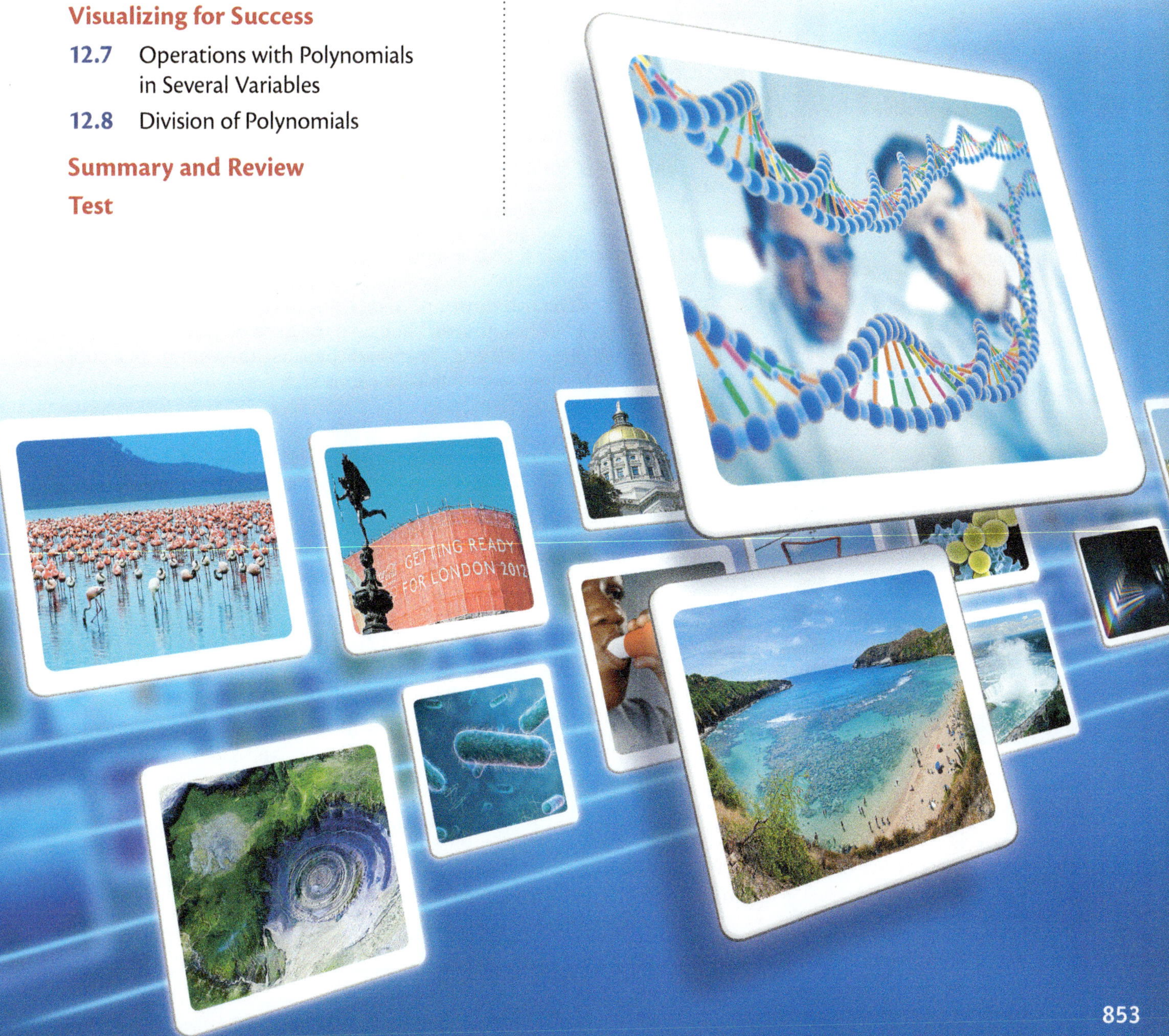

STUDYING FOR SUCCESS *Time Management*

- ☐ As a rule of thumb, budget two to three hours for homework and study for every hour that you spend in class.
- ☐ Make an hour-by-hour schedule of your week, planning time for leisure as well as work and study.
- ☐ Use your syllabus to help you plan your time. Transfer project deadlines and test dates to your calendar.

12.1 Integers as Exponents

OBJECTIVES

- **a** Tell the meaning of exponential notation.
- **b** Evaluate exponential expressions with exponents of 0 and 1.
- **c** Evaluate algebraic expressions containing exponents.
- **d** Use the product rule to multiply exponential expressions with like bases.
- **e** Use the quotient rule to divide exponential expressions with like bases.
- **f** Express an exponential expression involving negative exponents with positive exponents.

SKILL TO REVIEW

Objective 9.1a: Evaluate algebraic expressions by substitution.

1. Evaluate $6y$ when $y = 4$.
2. Evaluate $\frac{m}{n}$ when $m = 48$ and $n = 8$.

a EXPONENTIAL NOTATION

An exponent of 2 or greater tells how many times the base is used as a factor. For example, $a \cdot a \cdot a \cdot a = a^4$. In this case, the **exponent** is 4 and the **base** is a. An expression for a power is called **exponential notation**.

This is the base. $\longrightarrow a^n \longleftarrow$ This is the exponent.

EXAMPLE 1 What is the meaning of 3^5? of n^4? of $(2n)^3$? of $50x^2$? of $(-n)^3$? of $-n^3$?

3^5 means $3 \cdot 3 \cdot 3 \cdot 3 \cdot 3$; n^4 means $n \cdot n \cdot n \cdot n$;
$(2n)^3$ means $2n \cdot 2n \cdot 2n$; $50x^2$ means $50 \cdot x \cdot x$;
$(-n)^3$ means $(-n) \cdot (-n) \cdot (-n)$; $-n^3$ means $-1 \cdot n \cdot n \cdot n$

Do Margin Exercises 1–6 on the following page.

We read a^n as the ***n*th power of *a***, or simply ***a* to the *n*th**, or ***a* to the *n***. We often read x^2 as "***x*-squared**" because the area of a square of side x is $x \cdot x$, or x^2. We often read x^3 as "***x*-cubed**" because the volume of a cube with length, width, and height x is $x \cdot x \cdot x$, or x^3.

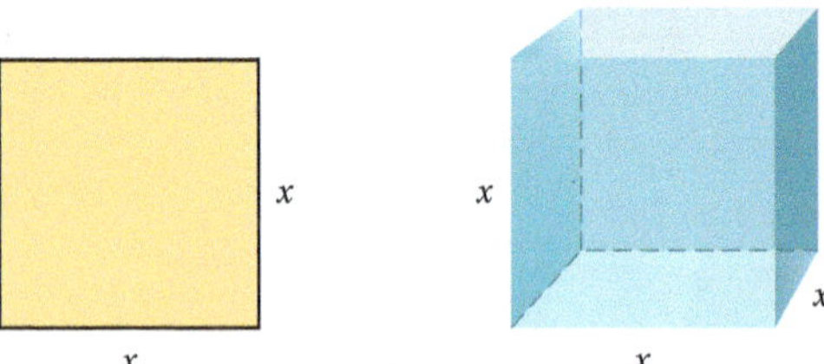

b ONE AND ZERO AS EXPONENTS

Look for a pattern in the following:

On each side, we **divide** by 8 at each step.

$$8 \cdot 8 \cdot 8 \cdot 8 = 8^4$$
$$8 \cdot 8 \cdot 8 = 8^3$$
$$8 \cdot 8 = 8^2$$
$$8 = 8^?$$
$$1 = 8^?.$$

On this side, the exponents **decrease** by 1 at each step.

To continue the pattern, we would say that $8 = 8^1$ and $1 = 8^0$.

Answers

Skill to Review:
1. 24 **2.** 6

We make the following definition.

EXPONENTS OF 0 AND 1

$a^1 = a$, for any number a;
$a^0 = 1$, for any nonzero number a

We consider 0^0 to be not defined. We will explain why later in this section.

EXAMPLE 2 Evaluate 5^1, $(-8)^1$, 3^0, and $(-749.21)^0$.

$5^1 = 5$; $(-8)^1 = -8$;
$3^0 = 1$; $(-749.21)^0 = 1$

Do Exercises 7–12.

What is the meaning of each of the following?

1. 5^4 **2.** x^5

3. $(3t)^2$ **4.** $3t^2$

5. $(-x)^4$ **6.** $-y^3$

Evaluate.

7. 6^1 **8.** 7^0

9. $(8.4)^1$ **10.** 8654^0

11. $(-1.4)^1$ **12.** 0^1

C EVALUATING ALGEBRAIC EXPRESSIONS

Algebraic expressions can involve exponential notation. For example, the following are algebraic expressions:

$$x^4, \quad (3x)^3 - 2, \quad a^2 + 2ab + b^2.$$

We evaluate algebraic expressions by replacing variables with numbers and following the rules for order of operations.

EXAMPLE 3 Evaluate $1000 - x^4$ when $x = 5$.

$1000 - x^4 = 1000 - 5^4$ Substituting
$= 1000 - 625$ Evaluating 5^4
$= 375$

EXAMPLE 4 *Area of a Circular Region.* The Richat Structure is a circular eroded geologic dome with a radius of 20 km. Find the area of the structure.

$A = \pi r^2$ Using the formula for the area of a circle
$= \pi(20\text{ km})^2$ Substituting
$= \pi \cdot 20\text{ km} \cdot 20\text{ km}$
$\approx 3.14 \times 400\text{ km}^2$ Using 3.14 as an approximation for π
$= 1256\text{ km}^2$

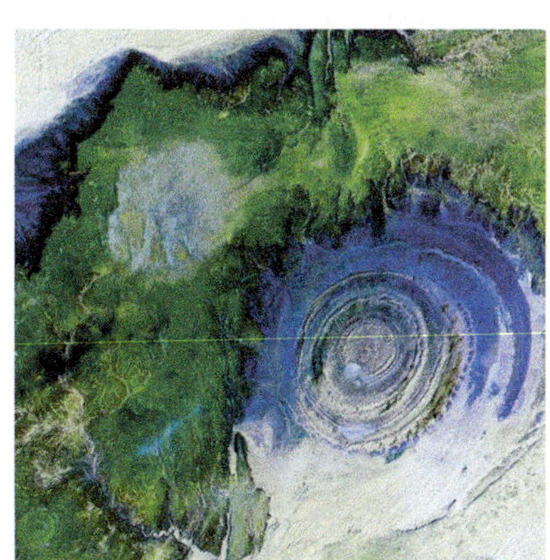

In Example 4, "km²" means "square kilometers" and "≈" means "is approximately equal to."

EXAMPLE 5 Evaluate $(5x)^3$ when $x = -2$.

When we evaluate with a negative number, we often use extra parentheses to show the substitution.

$(5x)^3 = [5 \cdot (-2)]^3$ Substituting
$= [-10]^3$ Multiplying within brackets first
$= [-10] \cdot [-10] \cdot [-10]$
$= -1000$ Evaluating the power

Answers

1. $5 \cdot 5 \cdot 5 \cdot 5$ 2. $x \cdot x \cdot x \cdot x \cdot x$
3. $3t \cdot 3t$ 4. $3 \cdot t \cdot t$
5. $(-x) \cdot (-x) \cdot (-x) \cdot (-x)$ 6. $-1 \cdot y \cdot y \cdot y$
7. 6 8. 1 9. 8.4 10. 1
11. −1.4 12. 0

13. Evaluate t^3 when $t = 5$.

14. Evaluate $-5x^5$ when $x = -2$.

15. Find the area of a circle when $r = 32$ cm. Use 3.14 for π.

16. Evaluate $200 - a^4$ when $a = 3$.

17. Evaluate $t^1 - 4$ and $t^0 - 4$ when $t = 7$.

18. **a)** Evaluate $(4t)^2$ when $t = -3$.
b) Evaluate $4t^2$ when $t = -3$.
c) Determine whether $(4t)^2$ and $4t^2$ are equivalent.

GS

a) $(4t)^2 = [4 \cdot (\quad)]^2$
$= [\quad]^2$
$= \quad$

b) $4t^2 = 4 \cdot (\quad)^2$
$= 4 \cdot (\quad)$
$= \quad$

c) Since $144 \neq 36$, the expressions ________ (are/are not) equivalent.

EXAMPLE 6 Evaluate $5x^3$ when $x = -2$.

$$5x^3 = 5 \cdot (-2)^3 \quad \text{Substituting}$$
$$= 5 \cdot (-2) \cdot (-2) \cdot (-2) \quad \text{Evaluating the power first}$$
$$= 5(-8) \quad (-2)(-2)(-2) = -8$$
$$= -40$$

Recall that two expressions are equivalent if they have the same value for all meaningful replacements. Note that Examples 5 and 6 show that $(5x)^3$ and $5x^3$ are *not* equivalent—that is, $(5x)^3 \neq 5x^3$.

◀ **Do Exercises 13–18.**

d MULTIPLYING POWERS WITH LIKE BASES

We can multiply powers with like bases by adding exponents. For example,

$$a^3 \cdot a^2 = \underbrace{(a \cdot a \cdot a)}_{3\text{ factors}}\underbrace{(a \cdot a)}_{2\text{ factors}} = \underbrace{a \cdot a \cdot a \cdot a \cdot a}_{5\text{ factors}} = a^5.$$

Note that the exponent in a^5 is the sum of those in $a^3 \cdot a^2$. That is, $3 + 2 = 5$. Likewise,

$$b^4 \cdot b^3 = (b \cdot b \cdot b \cdot b)(b \cdot b \cdot b) = b^7, \quad \text{where } 4 + 3 = 7.$$

Adding the exponents gives the correct result.

THE PRODUCT RULE

For any number a and any positive integers m and n,

$$a^m \cdot a^n = a^{m+n}.$$

(When multiplying with exponential notation, if the bases are the same, keep the base and add the exponents.)

EXAMPLES Multiply and simplify.

7. $5^6 \cdot 5^2 = 5^{6+2}$ Adding exponents: $a^m \cdot a^n = a^{m+n}$
$= 5^8$

8. $m^5m^{10}m^3 = m^{5+10+3} = m^{18}$

9. $x \cdot x^8 = x^1 \cdot x^8$ Writing x as x^1
$= x^{1+8}$
$= x^9$

10. $(a^3b^2)(a^3b^5) = (a^3a^3)(b^2b^5)$
$= a^6b^7$

11. $(4y)^6(4y)^3 = (4y)^{6+3} = (4y)^9$

◀ **Do Exercises 19–23.**

Multiply and simplify.

19. $3^5 \cdot 3^5$

20. $x^4 \cdot x^6$

21. $p^4p^{12}p^8$

22. $x \cdot x^4$

23. $(a^2b^3)(a^7b^5)$

Answers

13. 125 14. 160 15. 3215.36 cm^2 16. 119
17. 3; −3 18. **(a)** 144; **(b)** 36; **(c)** no
19. 3^{10} 20. x^{10} 21. p^{24} 22. x^5 23. a^9b^8

Guided Solution:
18. **(a)** −3, −12, 144; **(b)** −3, 9, 36; **(c)** are not

e DIVIDING POWERS WITH LIKE BASES

The following suggests a rule for dividing powers with like bases, such as a^5/a^2:

$$\frac{a^5}{a^2} = \frac{a \cdot a \cdot a \cdot a \cdot a}{a \cdot a} = \frac{a \cdot a \cdot a \cdot a \cdot a}{1 \cdot a \cdot a} = \frac{a \cdot a \cdot a}{1} \cdot \frac{a \cdot a}{a \cdot a}$$
$$= \frac{a \cdot a \cdot a}{1} \cdot 1 = a \cdot a \cdot a = a^3.$$

Note that the exponent in a^3 is the difference of those in $a^5 \div a^2$. That is, $5 - 2 = 3$. In a similar way, we have

$$\frac{t^9}{t^4} = \frac{t \cdot t \cdot t \cdot t \cdot t \cdot t \cdot t \cdot t \cdot t}{t \cdot t \cdot t \cdot t} = t^5, \quad \text{where} \quad 9 - 4 = 5.$$

Subtracting exponents gives the correct answer.

THE QUOTIENT RULE

For any nonzero number a and any positive integers m and n,

$$\frac{a^m}{a^n} = a^{m-n}.$$

(When dividing with exponential notation, if the bases are the same, keep the base and subtract the exponent of the denominator from the exponent of the numerator.)

EXAMPLES Divide and simplify.

12. $\frac{6^5}{6^3} = 6^{5-3}$ Subtracting exponents
$= 6^2$

13. $\frac{x^8}{x} = \frac{x^8}{x^1} = x^{8-1}$
$= x^7$

14. $\frac{(3t)^{12}}{(3t)^2} = (3t)^{12-2}$
$= (3t)^{10}$

15. $\frac{p^5q^7}{p^2q^5} = \frac{p^5}{p^2} \cdot \frac{q^7}{q^5} = p^{5-2}q^{7-5}$
$= p^3q^2$

The quotient rule can also be used to explain the definition of 0 as an exponent. Consider the expression a^4/a^4, where a is nonzero:

$$\frac{a^4}{a^4} = \frac{a \cdot a \cdot a \cdot a}{a \cdot a \cdot a \cdot a} = 1.$$

This is true because the numerator and the denominator are the same. Now suppose we apply the rule for dividing powers with the same base:

$$\frac{a^4}{a^4} = a^{4-4} = a^0.$$

Since $a^4/a^4 = 1$ and $a^4/a^4 = a^0$, it follows that $a^0 = 1$, when $a \neq 0$.

We can explain why we do not define 0^0 using the quotient rule. We know that 0^0 is 0^{1-1}. But 0^{1-1} is also equal to $0^1/0^1$, or $0/0$. We have already seen that division by 0 is not defined, so 0^0 is also not defined.

Do Exercises 24–27.

Divide and simplify.

24. $\frac{4^5}{4^2}$

25. $\frac{y^6}{y^2}$

26. $\frac{p^{10}}{p}$

27. $\frac{a^7b^6}{a^3b^4}$

Answers

24. 4^3 25. y^4 26. p^9 27. a^4b^2

f NEGATIVE INTEGERS AS EXPONENTS

We can use the rule for dividing powers with like bases to lead us to a definition of exponential notation when the exponent is a negative integer. Consider $5^3/5^7$ and first simplify it using procedures we have learned for working with fractions:

$$\frac{5^3}{5^7} = \frac{5 \cdot 5 \cdot 5}{5 \cdot 5 \cdot 5 \cdot 5 \cdot 5 \cdot 5 \cdot 5} = \frac{5 \cdot 5 \cdot 5 \cdot 1}{5 \cdot 5 \cdot 5 \cdot 5 \cdot 5 \cdot 5 \cdot 5}$$

$$= \frac{5 \cdot 5 \cdot 5}{5 \cdot 5 \cdot 5} \cdot \frac{1}{5 \cdot 5 \cdot 5 \cdot 5} = \frac{1}{5^4}.$$

Now we apply the rule for dividing exponential expressions with the same bases. Then

$$\frac{5^3}{5^7} = 5^{3-7} = 5^{-4}.$$

From these two expressions for $5^3/5^7$, it follows that

$$5^{-4} = \frac{1}{5^4}.$$

This leads to our definition of negative exponents.

NEGATIVE EXPONENT

For any real number a that is nonzero and any integer n,

$$a^{-n} = \frac{1}{a^n}.$$

In fact, the numbers a^n and a^{-n} are reciprocals because

$$a^n \cdot a^{-n} = a^n \cdot \frac{1}{a^n} = \frac{a^n}{a^n} = 1.$$

The following is another way to arrive at the definition of negative exponents.

On each side, we **divide** by 5 at each step.

$$5 \cdot 5 \cdot 5 \cdot 5 = 5^4$$
$$5 \cdot 5 \cdot 5 = 5^3$$
$$5 \cdot 5 = 5^2$$
$$5 = 5^1$$
$$1 = 5^0$$
$$\frac{1}{5} = 5^?$$
$$\frac{1}{25} = 5^?$$

On this side, the exponents **decrease** by 1 at each step.

To continue the pattern, it should follow that

$$\frac{1}{5} = \frac{1}{5^1} = 5^{-1} \quad \text{and} \quad \frac{1}{25} = \frac{1}{5^2} = 5^{-2}.$$

EXAMPLES Express using positive exponents. Then simplify.

16. $4^{-2} = \frac{1}{4^2} = \frac{1}{16}$

17. $(-3)^{-2} = \frac{1}{(-3)^2} = \frac{1}{(-3)(-3)} = \frac{1}{9}$

18. $m^{-3} = \frac{1}{m^3}$

19. $ab^{-1} = a\left(\frac{1}{b^1}\right) = a\left(\frac{1}{b}\right) = \frac{a}{b}$

20. $\frac{1}{x^{-3}} = x^{-(-3)} = x^3$

21. $3c^{-5} = 3\left(\frac{1}{c^5}\right) = \frac{3}{c^5}$

Example 20 might also be done as follows:

$$\frac{1}{x^{-3}} = \frac{1}{\frac{1}{x^3}} = 1 \cdot \frac{x^3}{1} = x^3.$$

Caution!

As shown in Examples 16 and 17, a negative exponent does not necessarily mean that an expression is negative.

Do Exercises 28–33. ▶

Express with positive exponents. Then simplify.

28. 4^{-3}

29. 5^{-2}

30. 2^{-4}

31. $(-2)^{-3}$

32. $\frac{1}{x^{-2}}$

GS **33.** $4p^{-3}$

$= 4\left(\frac{1}{\square}\right) = \frac{4}{\square}$

The rules for multiplying and dividing powers with like bases hold when exponents are 0 or negative.

EXAMPLES Simplify. Write the result using positive exponents.

22. $7^{-3} \cdot 7^6 = 7^{-3+6}$ Adding exponents

$= 7^3$

23. $x^4 \cdot x^{-3} = x^{4+(-3)} = x^1 = x$

24. $\frac{5^4}{5^{-2}} = 5^{4-(-2)}$ Subtracting exponents

$= 5^{4+2} = 5^6$

25. $\frac{x}{x^7} = x^{1-7} = x^{-6} = \frac{1}{x^6}$

26. $\frac{b^{-4}}{b^{-5}} = b^{-4-(-5)}$

$= b^{-4+5} = b^1 = b$

27. $y^{-4} \cdot y^{-8} = y^{-4+(-8)}$

$= y^{-12} = \frac{1}{y^{12}}$

Do Exercises 34–38. ▶

Simplify.

34. $5^{-2} \cdot 5^4$

35. $x^{-3} \cdot x^{-4}$

36. $\frac{7^{-2}}{7^3}$

37. $\frac{b^{-2}}{b^{-3}}$

38. $\frac{t}{t^{-5}}$

The following is a summary of the definitions and rules for exponents that we have considered in this section.

DEFINITIONS AND RULES FOR EXPONENTS

1 as an exponent:	$a^1 = a$
0 as an exponent:	$a^0 = 1, a \neq 0$
Negative integers as exponents:	$a^{-n} = \frac{1}{a^n}, \frac{1}{a^{-n}} = a^n; a \neq 0$
Product Rule:	$a^m \cdot a^n = a^{m+n}$
Quotient Rule:	$\frac{a^m}{a^n} = a^{m-n}, a \neq 0$

Answers

28. $\frac{1}{4^3} = \frac{1}{64}$ **29.** $\frac{1}{5^2} = \frac{1}{25}$ **30.** $\frac{1}{2^4} = \frac{1}{16}$ **31.** $\frac{1}{(-2)^3} = -\frac{1}{8}$ **32.** x^2 **33.** $\frac{4}{p^3}$ **34.** 5^2 **35.** $\frac{1}{x^7}$ **36.** $\frac{1}{7^5}$ **37.** b **38.** t^6

Guided Solution:
33. p^3, p^3

12.1 Exercise Set

For Extra Help
MyMathLab®

MathXL® PRACTICE   WATCH  READ  REVIEW

Reading Check

Match each expression with the appropriate value from the column on the right. Choices may be used more than once or not at all.

RC1. _____ y^1

RC2. _____ $y^0, y \neq 0$

RC3. _____ $y^1 \cdot y^1$

RC4. _____ $\dfrac{y^9}{y^8}$

RC5. _____ $\dfrac{y^8}{y^9}$

RC6. _____ $\dfrac{1}{y^{-1}}$

a) 1
b) 0
c) y
d) $\dfrac{1}{y}$
e) y^2

a What is the meaning of each of the following?

1. 3^4 **2.** 4^3 **3.** $(-1.1)^5$ **4.** $(87.2)^6$ **5.** $\left(\frac{2}{3}\right)^4$ **6.** $\left(-\frac{5}{8}\right)^3$

7. $(7p)^2$ **8.** $(11c)^3$ **9.** $8k^3$ **10.** $17x^2$ **11.** $-6y^4$ **12.** $-q^5$

b Evaluate.

13. $a^0, a \neq 0$ **14.** $t^0, t \neq 0$ **15.** b^1 **16.** c^1 **17.** $\left(\frac{2}{3}\right)^0$

18. $\left(-\frac{5}{8}\right)^0$ **19.** $(-7.03)^1$ **20.** $\left(\frac{4}{5}\right)^1$ **21.** 8.38^0 **22.** 8.38^1

23. $(ab)^1$ **24.** $(ab)^0, a \neq 0, b \neq 0$ **25.** $ab^0, b \neq 0$ **26.** ab^1

c Evaluate.

27. m^3, when $m = 3$ **28.** x^6, when $x = 2$ **29.** p^1, when $p = 19$ **30.** x^{19}, when $x = 0$

31. $-x^4$, when $x = -3$ **32.** $-2y^7$, when $y = 2$ **33.** x^4, when $x = 4$ **34.** y^{15}, when $y = 1$

35. $y^2 - 7$, when $y = -10$ **36.** $z^5 + 5$, when $z = -2$ **37.** $161 - b^2$, when $b = 5$ **38.** $325 - v^3$, when $v = -3$

39. $x^1 + 3$ and $x^0 + 3$, when $x = 7$ **40.** $y^0 - 8$ and $y^1 - 8$, when $y = -3$

41. Find the area of a circle when $r = 34$ ft. Use 3.14 for π.

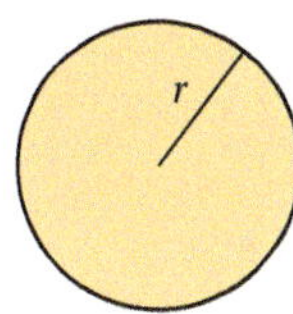

42. The area A of a square with sides of length s is given by $A = s^2$. Find the area of a square with sides of length 24 m.

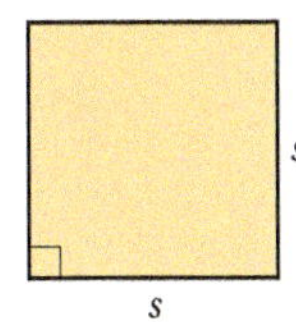

f Express using positive exponents. Then simplify.

43. 3^{-2}

44. 2^{-3}

45. 10^{-3}

46. 5^{-4}

47. a^{-3}

48. x^{-2}

49. $\frac{1}{8^{-2}}$

50. $\frac{1}{2^{-5}}$

51. $\frac{1}{y^{-4}}$

52. $\frac{1}{t^{-7}}$

53. $5z^{-4}$

54. $6n^{-5}$

55. xy^{-2}

56. ab^{-3}

Express using negative exponents.

57. $\frac{1}{4^3}$

58. $\frac{1}{5^2}$

59. $\frac{1}{x^3}$

60. $\frac{1}{y^2}$

61. $\frac{1}{a^5}$

62. $\frac{1}{b^7}$

d, **f** Multiply and simplify.

63. $2^4 \cdot 2^3$

64. $3^5 \cdot 3^2$

65. $9^{17} \cdot 9^{21}$

66. $7^{22} \cdot 7^{15}$

67. $x^4 \cdot x$

68. $y \cdot y^9$

69. $x^{14} \cdot x^3$

70. $x^9 \cdot x^4$

71. $(3y)^4(3y)^8$

72. $(2t)^8(2t)^{17}$

73. $(7y)^1(7y)^{16}$

74. $(8x)^0(8x)^1$

75. $3^{-5} \cdot 3^8$

76. $5^{-8} \cdot 5^9$

77. $x^{-2} \cdot x^2$

78. $x \cdot x^{-1}$

79. $x^{-7} \cdot x^{-6}$

80. $y^{-5} \cdot y^{-8}$

81. $a^{11} \cdot a^{-3} \cdot a^{-18}$

82. $a^{-11} \cdot a^{-3} \cdot a^{-7}$

83. $(x^4y^7)(x^2y^8)$

84. $(a^5c^2)(a^3c^9)$

85. $(s^2t^3)(st^4)$

86. $(m^4n)(m^2n^7)$

e, **f** Divide and simplify.

87. $\frac{7^5}{7^2}$

88. $\frac{5^8}{5^6}$

89. $\frac{y^9}{y}$

90. $\frac{x^{11}}{x}$

91. $\frac{16^2}{16^8}$

92. $\frac{7^2}{7^9}$

93. $\frac{m^6}{m^{12}}$

94. $\frac{a^3}{a^4}$

95. $\frac{(8x)^6}{(8x)^{10}}$

96. $\frac{(8t)^4}{(8t)^{11}}$

97. $\frac{x}{x^{-1}}$

98. $\frac{t^8}{t^{-3}}$

99. $\frac{z^{-6}}{z^{-2}}$

100. $\frac{x^{-9}}{x^{-3}}$

101. $\frac{x^{-5}}{x^{-8}}$

102. $\frac{y^{-2}}{y^{-9}}$

103. $\frac{m^{-9}}{m^{-9}}$

104. $\frac{x^{-7}}{x^{-7}}$

105. $\frac{a^5b^3}{a^2b}$

106. $\frac{s^8t^4}{st^3}$

Matching. In Exercises 107 and 108, match each item in the first column with the appropriate item in the second column by drawing connecting lines. Items in the second column may be used more than once.

107.

5^2	$-\frac{1}{10}$
5^{-2}	$\frac{1}{10}$
$\left(\frac{1}{5}\right)^2$	$-\frac{1}{25}$
$\left(\frac{1}{5}\right)^{-2}$	10
-5^2	25
$(-5)^2$	-25
$-\left(-\frac{1}{5}\right)^2$	$\frac{1}{25}$
$\left(-\frac{1}{5}\right)^{-2}$	-10

108.

$-\left(\frac{1}{8}\right)^2$	16
$\left(\frac{1}{8}\right)^{-2}$	-16
8^{-2}	64
8^2	-64
-8^2	$\frac{1}{64}$
$(-8)^2$	$-\frac{1}{64}$
$\left(-\frac{1}{8}\right)^{-2}$	$-\frac{1}{16}$
$\left(-\frac{1}{8}\right)^2$	$\frac{1}{16}$

Skill Maintenance

Solve.

109. A 12-in. submarine sandwich is cut into two pieces. One piece is twice as long as the other. How long are the pieces? [10.6a]

110. The first angle of a triangle is 24° more than the second. The third angle is twice the first. Find the measures of the angles of the triangle. [10.6a]

111. A warehouse stores 1800 lb of peanuts, 1500 lb of cashews, and 700 lb of almonds. What percent of the total is peanuts? cashews? almonds? [10.5a]

112. The width of a rectangle is fixed at 10 ft. For what lengths will the area be less than 25 ft^2? [10.8b]

Solve.

113. $2x - 4 - 5x + 8 = x - 3$ [10.3b]

114. $8x + 7 - 9x = 12 - 6x + 5$ [10.3b]

115. $-6(2 - x) + 10(5x - 7) = 10$ [10.3c]

116. $-10(x - 4) = 5(2x + 5) - 7$ [10.3c]

Synthesis

Determine whether each of the following equations is true.

117. $(x + 1)^2 = x^2 + 1$

118. $(x - 1)^2 = x^2 - 2x + 1$

119. $(5x)^0 = 5x^0$

120. $\dfrac{x^3}{x^5} = x^2$

Simplify.

121. $(y^{2x})(y^{3x})$

122. $a^{5k} \div a^{3k}$

123. $\dfrac{a^{6t}(a^{7t})}{a^{9t}}$

124. $\dfrac{\left(\frac{1}{2}\right)^4}{\left(\frac{1}{2}\right)^5}$

125. $\dfrac{(0.8)^5}{(0.8)^3(0.8)^2}$

126. $\dfrac{(x - 3)^5}{x - 3}$

Use $>$, $<$, or $=$ for □ to write a true sentence.

127. 3^5 □ 3^4

128. 4^2 □ 4^3

129. 4^3 □ 5^3

130. 4^3 □ 3^4

Evaluate.

131. $\dfrac{1}{-z^4}$, when $z = -10$

132. $\dfrac{1}{-z^5}$, when $z = -0.1$

133. Determine whether $(a + b)^2$ and $a^2 + b^2$ are equivalent. (*Hint*: Choose values for a and b and evaluate.)

12.2 Exponents and Scientific Notation

OBJECTIVES

a Use the power rule to raise powers to powers.

b Raise a product to a power and a quotient to a power.

c Convert between scientific notation and decimal notation.

d Multiply and divide using scientific notation.

e Solve applied problems using scientific notation.

SKILL TO REVIEW

Objective 9.5a: Multiply real numbers.

Multiply.

1. $-5 \cdot 8$ **2.** $(-3)(-5)$

We now consider three rules used to simplify exponential expressions. We then apply our knowledge of exponents to *scientific notation*.

a RAISING POWERS TO POWERS

Consider an expression like $(3^2)^4$. We are raising 3^2 to the fourth power:

$$\begin{aligned}(3^2)^4 &= (3^2)(3^2)(3^2)(3^2)\\ &= (3 \cdot 3)(3 \cdot 3)(3 \cdot 3)(3 \cdot 3)\\ &= 3 \cdot 3 \cdot 3 \cdot 3 \cdot 3 \cdot 3 \cdot 3 \cdot 3\\ &= 3^8.\end{aligned}$$

Note that in this case we could have multiplied the exponents:

$$(3^2)^4 = 3^{2 \cdot 4} = 3^8.$$

THE POWER RULE

For any real number a and any integers m and n,

$$(a^m)^n = a^{mn}.$$

(To raise a power to a power, multiply the exponents.)

EXAMPLES Simplify. Express the answers using positive exponents.

1. $(3^5)^4 = 3^{5 \cdot 4}$ Multiplying exponents
$= 3^{20}$

2. $(2^2)^5 = 2^{2 \cdot 5} = 2^{10}$

3. $(y^{-5})^7 = y^{-5 \cdot 7} = y^{-35} = \dfrac{1}{y^{35}}$

4. $(x^4)^{-2} = x^{4(-2)} = x^{-8} = \dfrac{1}{x^8}$

5. $(a^{-4})^{-6} = a^{(-4)(-6)} = a^{24}$

◀ Do Margin Exercises 1–4.

Simplify. Express the answers using positive exponents.

1. $(3^4)^5$ **2.** $(x^{-3})^4$

3. $(y^{-5})^{-3}$ **4.** $(x^4)^{-8}$

b RAISING A PRODUCT OR A QUOTIENT TO A POWER

When an expression inside parentheses is raised to a power, the inside expression is the base. Let's compare $2a^3$ and $(2a)^3$:

$2a^3 = 2 \cdot a \cdot a \cdot a;$ The base is a.

$$\begin{aligned}(2a)^3 &= (2a)(2a)(2a) && \text{The base is } 2a.\\ &= (2 \cdot 2 \cdot 2)(a \cdot a \cdot a) && \text{Using the associative and commutative laws of multiplication to regroup the factors}\\ &= 2^3a^3\\ &= 8a^3.\end{aligned}$$

We see that $2a^3$ and $(2a)^3$ are *not* equivalent. We also see that we can evaluate the power $(2a)^3$ by raising each factor to the power 3. This leads us to a rule for raising a product to a power.

Answers

Skill to Review:
1. -40 2. 15

Margin Exercises:
1. 3^{20} 2. $\dfrac{1}{x^{12}}$ 3. y^{15} 4. $\dfrac{1}{x^{32}}$

RAISING A PRODUCT TO A POWER

For any real numbers a and b and any integer n,

$$(ab)^n = a^n b^n.$$

(To raise a product to the nth power, raise each factor to the nth power.)

EXAMPLES Simplify.

6. $(4x^2)^3 = (4^1x^2)^3$ — $4 = 4^1$

$= (4^1)^3 \cdot (x^2)^3$ — Raising *each* factor to the third power

$= 4^3 \cdot x^6 = 64x^6$ — Using the power rule and simplifying

7. $(5x^3y^5z^2)^4 = 5^4(x^3)^4(y^5)^4(z^2)^4$ — Raising *each* factor to the fourth power

$= 625x^{12}y^{20}z^8$

8. $(-5x^4y^3)^3 = (-5)^3(x^4)^3(y^3)^3$

$= -125x^{12}y^9$

9. $[(-x)^{25}]^2 = (-x)^{50}$ — Using the power rule

$= (-1 \cdot x)^{50}$ — Using the property of -1: $-x = -1 \cdot x$

$= (-1)^{50}x^{50}$ — Raising each factor to the fiftieth power

$= 1 \cdot x^{50}$ — The product of an even number of negative factors is positive.

$= x^{50}$

10. $(3x^3y^{-5}z^2)^4 = 3^4(x^3)^4(y^{-5})^4(z^2)^4 = 81x^{12}y^{-20}z^8 = \dfrac{81x^{12}z^8}{y^{20}}$

11. $(-x^4)^{-3} = (-1 \cdot x^4)^{-3} = (-1)^{-3} \cdot (x^4)^{-3} = (-1)^{-3} \cdot x^{-12}$

$= \dfrac{1}{(-1)^3} \cdot \dfrac{1}{x^{12}} = \dfrac{1}{-1} \cdot \dfrac{1}{x^{12}} = -\dfrac{1}{x^{12}}$

12. $(-2x^{-5}y^4)^{-4} = (-2)^{-4}(x^{-5})^{-4}(y^4)^{-4} = \dfrac{1}{(-2)^4} \cdot x^{20} \cdot y^{-16}$

$= \dfrac{1}{16} \cdot x^{20} \cdot \dfrac{1}{y^{16}} = \dfrac{x^{20}}{16y^{16}}$

Do Exercises 5–11.

Simplify.

5. $(2x^5y^{-3})^4$

6. $(5x^5y^{-6}z^{-3})^2$

7. $[(-x)^{37}]^2$

8. $(3y^{-2}x^{-5}z^8)^3$

9. $(-y^8)^{-3}$

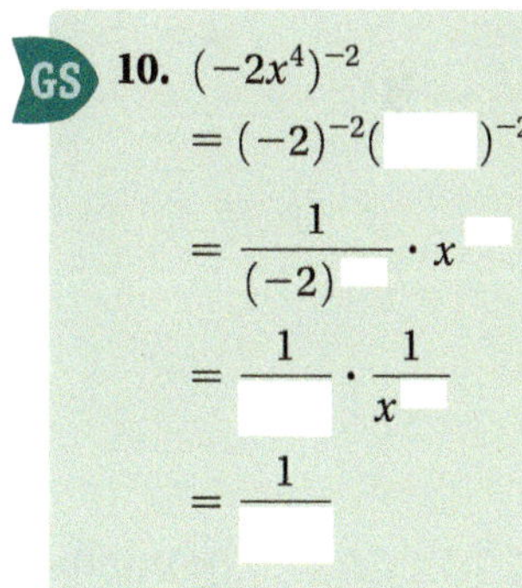

GS **10.** $(-2x^4)^{-2}$

$= (-2)^{-2}(\quad)^{-2}$

$= \dfrac{1}{(-2)^{\square}} \cdot x^{\square}$

$= \dfrac{1}{\square} \cdot \dfrac{1}{x^{\square}}$

$= \dfrac{1}{\square}$

11. $(-3x^2y^{-5})^{-3}$

There is a similar rule for raising a quotient to a power.

RAISING A QUOTIENT TO A POWER

For any real numbers a and b, $b \neq 0$, and any integer n,

$$\left(\frac{a}{b}\right)^n = \frac{a^n}{b^n}.$$

(To raise a quotient to the nth power, raise both the numerator and the denominator to the nth power.)

Answers

5. $\dfrac{16x^{20}}{y^{12}}$ **6.** $\dfrac{25x^{10}}{y^{12}z^6}$ **7.** x^{74} **8.** $\dfrac{27z^{24}}{y^6x^{15}}$ **9.** $-\dfrac{1}{y^{24}}$ **10.** $\dfrac{1}{4x^8}$ **11.** $-\dfrac{y^{15}}{27x^6}$

Guided Solution:
10. x^4, 2, -8, 4, 8, $4x^8$

Simplify.

12. $\left(\frac{x^6}{5}\right)^2$

13. $\left(\frac{2t^5}{w^4}\right)^3$

14. $\left(\frac{a^4}{3b^{-2}}\right)^3$

15. $\left(\frac{x^4}{3}\right)^{-2}$

GS

Do this two ways.

$$\left(\frac{x^4}{3}\right)^{-2} = \frac{(x^4)^{\square}}{3^{-2}} = \frac{x^{\square}}{3^{-2}}$$

$$= \frac{\frac{1}{x^{\square}}}{\frac{1}{3^2}} = \frac{1}{x^8} \div \frac{1}{3^2}$$

$$= \frac{1}{x^8} \cdot \frac{3^2}{\square} = \frac{9}{\square}$$

This can be done a second way.

$$\left(\frac{x^4}{3}\right)^{-2} = \left(\frac{3}{x^4}\right)^{\square}$$

$$= \frac{3^2}{(x^4)^{\square}} = \frac{9}{\square}$$

EXAMPLES Simplify.

13. $\left(\frac{x^2}{4}\right)^3 = \frac{(x^2)^3}{4^3} = \frac{x^6}{64}$ Raising *both* the numerator and the denominator to the third power

14. $\left(\frac{3a^4}{b^3}\right)^2 = \frac{(3a^4)^2}{(b^3)^2} = \frac{3^2(a^4)^2}{b^{3\cdot 2}} = \frac{9a^8}{b^6}$

15. $\left(\frac{y^2}{2z^{-5}}\right)^4 = \frac{(y^2)^4}{(2z^{-5})^4} = \frac{(y^2)^4}{2^4(z^{-5})^4} = \frac{y^8}{16z^{-20}} = \frac{y^8z^{20}}{16}$

16. $\left(\frac{y^3}{5}\right)^{-2} = \frac{(y^3)^{-2}}{5^{-2}} = \frac{y^{-6}}{5^{-2}} = \frac{\frac{1}{y^6}}{\frac{1}{5^2}} = \frac{1}{y^6} \div \frac{1}{5^2} = \frac{1}{y^6} \cdot \frac{5^2}{1} = \frac{25}{y^6}$

The following can often be used to simplify a quotient that is raised to a negative power.

For $a \neq 0$ and $b \neq 0$,

$$\left(\frac{a}{b}\right)^{-n} = \left(\frac{b}{a}\right)^n.$$

Example 16 might also be completed as follows:

$$\left(\frac{y^3}{5}\right)^{-2} = \left(\frac{5}{y^3}\right)^2 = \frac{5^2}{(y^3)^2} = \frac{25}{y^6}.$$

◀ Do Exercises 12–15.

C SCIENTIFIC NOTATION

We can write numbers using different types of notation, such as fraction notation, decimal notation, and percent notation. Another type, **scientific notation**, makes use of exponential notation. Scientific notation is especially useful when calculations involve very large or very small numbers. The following are examples of scientific notation.

The number of flamingos in Africa's Great Rift Valley: $4 \times 10^6 = 4{,}000{,}000$

The length of an *E.coli* bacterium: $2 \times 10^{-6}\,\text{m} = 0.000002\,\text{m}$

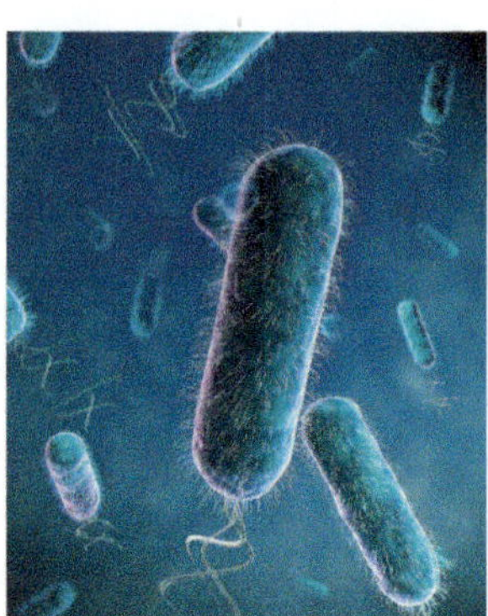

Answers

12. $\frac{x^{12}}{25}$ 13. $\frac{8t^{15}}{w^{12}}$ 14. $\frac{a^{12}b^6}{27}$ 15. $\frac{9}{x^8}$

Guided Solution:

15. $-2, -8, 8, 1, x^8; 2, 2, x^8$

SCIENTIFIC NOTATION

Scientific notation for a number is an expression of the type

$M \times 10^n$,

where n is an integer, M is greater than or equal to 1 and less than 10 ($1 \le M < 10$), and M is expressed in decimal notation. 10^n is also considered to be scientific notation when $M = 1$.

Caution!

Each of the following is *not* scientific notation.

12.46×10^7

This number is greater than 10.

0.347×10^{-5}

This number is less than 1.

You should try to make conversions to scientific notation mentally as often as possible. Here is a handy mental device.

A positive exponent in scientific notation indicates a large number (greater than or equal to 10) and a negative exponent indicates a small number (between 0 and 1).

EXAMPLES Convert to scientific notation.

17. $78{,}000 = 7.8 \times 10^4$

7.8,000. Large number, so the exponent is positive

4 places

18. $0.0000057 = 5.7 \times 10^{-6}$

0.000005.7 Small number, so the exponent is negative

6 places

Do Exercises 16 and 17. ▶

Convert to scientific notation.

16. 0.000517

17. 523,000,000

EXAMPLES Convert mentally to decimal notation.

19. $7.893 \times 10^5 = 789{,}300$

7.89300. Positive exponent, so the answer is a large number

5 places

20. $4.7 \times 10^{-8} = 0.000000047$

.00000004.7 Negative exponent, so the answer is a small number

8 places

Do Exercises 18 and 19. ▶

Convert to decimal notation.

18. 6.893×10^{11}

19. 5.67×10^{-5}

d MULTIPLYING AND DIVIDING USING SCIENTIFIC NOTATION

Multiplying

Consider the product

$400 \cdot 2000 = 800{,}000.$

In scientific notation, this is

$(4 \times 10^2) \cdot (2 \times 10^3) = (4 \cdot 2)(10^2 \cdot 10^3) = 8 \times 10^5.$

Answers

16. 5.17×10^{-4} **17.** 5.23×10^8
18. 689,300,000,000 **19.** 0.0000567

Multiply and write scientific notation for the result.

20. $(1.12 \times 10^{-8})(5 \times 10^{-7})$

21. $(9.1 \times 10^{-17})(8.2 \times 10^{3})$

EXAMPLE 21 Multiply: $(1.8 \times 10^6) \cdot (2.3 \times 10^{-4})$.

We apply the commutative and associative laws to get

$$\begin{aligned}(1.8 \times 10^6) \cdot (2.3 \times 10^{-4}) &= (1.8 \cdot 2.3) \times (10^6 \cdot 10^{-4}) \\ &= 4.14 \times 10^{6+(-4)} \\ &= 4.14 \times 10^2.\end{aligned}$$

We get 4.14 by multiplying 1.8 and 2.3. We get 10^2 by adding the exponents 6 and -4.

EXAMPLE 22 Multiply: $(3.1 \times 10^5) \cdot (4.5 \times 10^{-3})$.

$$\begin{aligned}(3.1 \times 10^5) \cdot (4.5 \times 10^{-3}) &= (3.1 \times 4.5)(10^5 \cdot 10^{-3}) \\ &= 13.95 \times 10^2 && \text{Not scientific notation; 13.95 is greater than 10.} \\ &= (1.395 \times 10^1) \times 10^2 && \text{Substituting } 1.395 \times 10^1 \text{ for } 13.95 \\ &= 1.395 \times (10^1 \times 10^2) && \text{Associative law} \\ &= 1.395 \times 10^3 && \text{The answer is now in scientific notation.}\end{aligned}$$

◀ **Do Exercises 20 and 21.**

CALCULATOR CORNER

To find the product in Example 21 and express the result in scientific notation on a graphing calculator, we first set the calculator in Scientific mode using MODE. Then we go to the home screen and enter the computation by pressing 1 . 8 2ND EE 6 × 2 . 3 2ND EE (−) 4 ENTER. (EE is the second operation associated with the , key.) The decimal portion of a number written in scientific notation appears before a small E and the exponent follows the E.

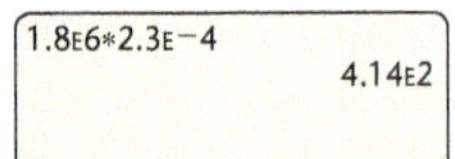

EXERCISES Multiply or divide and express the answer in scientific notation.

1. $(3.15 \times 10^7)(4.3 \times 10^{-12})$
2. $(8 \times 10^9)(4 \times 10^{-5})$
3. $\dfrac{4.5 \times 10^6}{1.5 \times 10^{12}}$
4. $\dfrac{4 \times 10^{-9}}{5 \times 10^{16}}$

Dividing

Consider the quotient $800{,}000 \div 400 = 2000$. In scientific notation, this is

$$(8 \times 10^5) \div (4 \times 10^2) = \frac{8 \times 10^5}{4 \times 10^2} = \frac{8}{4} \times \frac{10^5}{10^2} = 2 \times 10^3.$$

EXAMPLE 23 Divide: $(3.41 \times 10^5) \div (1.1 \times 10^{-3})$.

$$\begin{aligned}(3.41 \times 10^5) \div (1.1 \times 10^{-3}) &= \frac{3.41 \times 10^5}{1.1 \times 10^{-3}} = \frac{3.41}{1.1} \times \frac{10^5}{10^{-3}} \\ &= 3.1 \times 10^{5-(-3)} \\ &= 3.1 \times 10^8\end{aligned}$$

EXAMPLE 24 Divide: $(6.4 \times 10^{-7}) \div (8.0 \times 10^6)$.

$$\begin{aligned}(6.4 \times 10^{-7}) \div (8.0 \times 10^6) &= \frac{6.4 \times 10^{-7}}{8.0 \times 10^6} \\ &= \frac{6.4}{8.0} \times \frac{10^{-7}}{10^6} \\ &= 0.8 \times 10^{-7-6} \\ &= 0.8 \times 10^{-13} && \text{Not scientific notation; 0.8 is less than 1.} \\ &= (8.0 \times 10^{-1}) \times 10^{-13} && \text{Substituting } 8.0 \times 10^{-1} \text{ for } 0.8 \\ &= 8.0 \times (10^{-1} \times 10^{-13}) && \text{Associative law} \\ &= 8.0 \times 10^{-14} && \text{Adding exponents}\end{aligned}$$

◀ **Do Exercises 22 and 23.**

Divide and write scientific notation for the result.

22. $\dfrac{4.2 \times 10^5}{2.1 \times 10^2}$

23. $\dfrac{1.1 \times 10^{-4}}{2.0 \times 10^{-7}}$

Answers

20. 5.6×10^{-15} **21.** 7.462×10^{-13}
22. 2.0×10^3 **23.** 5.5×10^2

e APPLICATIONS WITH SCIENTIFIC NOTATION

EXAMPLE 25 *Distance from the Sun to Earth.* Light from the sun traveling at a rate of 300,000 kilometers per second (km/s) reaches Earth in 499 sec. Find the distance, expressed in scientific notation, from the sun to Earth.

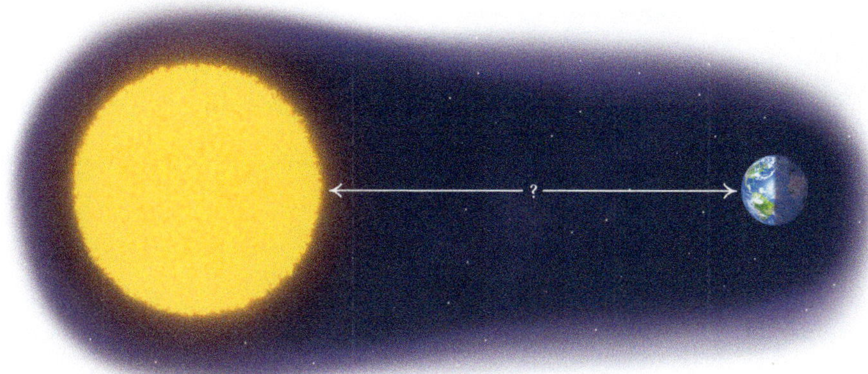

The time t that it takes for light to reach Earth from the sun is 4.99×10^2 sec (s). The speed is 3.0×10^5 km/s. Recall that distance can be expressed in terms of speed and time as

$$\text{Distance} = \text{Speed} \cdot \text{Time}$$
$$d = rt.$$

We substitute 3.0×10^5 for r and 4.99×10^2 for t:

$$\begin{aligned} d &= rt \\ &= (3.0 \times 10^5)(4.99 \times 10^2) && \text{Substituting} \\ &= 14.97 \times 10^7 \\ &= (1.497 \times 10^1) \times 10^7 \\ &= 1.497 \times (10^1 \times 10^7) && \text{Converting to scientific notation} \\ &= 1.497 \times 10^8 \text{ km.} \end{aligned}$$

Thus the distance from the sun to Earth is 1.497×10^8 km.

Do Exercise 24.

24. *Niagara Falls Water Flow.* On the Canadian side, the amount of water that spills over Niagara Falls in 1 min during the summer is about

$$1.3088 \times 10^8 \text{ L.}$$

How much water spills over the falls in one day? Express the answer in scientific notation.

EXAMPLE 26 *Media Usage.* In January 2013, the 800 million active YouTube users viewed 120 billion videos on the site. On average, how many videos did each user view?

Source: YouTube

In order to find the average number of YouTube videos that each user viewed, we divide the total number viewed by the number of users. We first write each number using scientific notation:

$$800 \text{ million} = 800{,}000{,}000 = 8 \times 10^8,$$

$$120 \text{ billion} = 120{,}000{,}000{,}000 = 1.2 \times 10^{11}.$$

Answer

24. 1.884672×10^{11} L

25. *DNA.* The width of a DNA (deoxyribonucleic acid) double helix is about 2×10^{-9} m. If its length, fully stretched, is 5×10^{-2} m, how many times longer is the helix than it is wide?

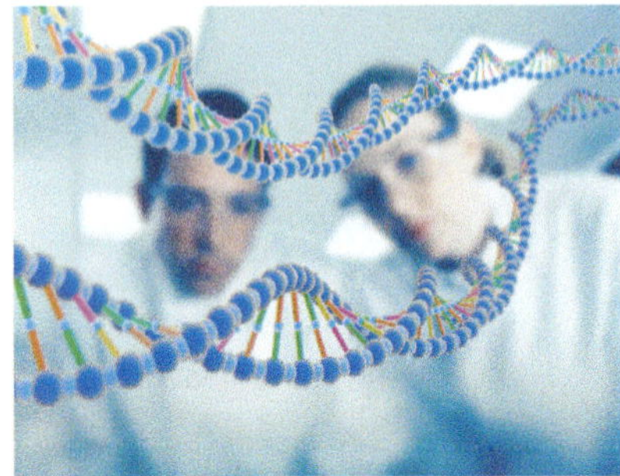

We then divide 1.2×10^{11} by 8×10^{8}:

$$\frac{1.2 \times 10^{11}}{8 \times 10^{8}} = \frac{1.2}{8} \times \frac{10^{11}}{10^{8}}$$
$$= 0.15 \times 10^{3} = (1.5 \times 10^{-1}) \times 10^{3} = 1.5 \times 10^{2}.$$

On average, each user viewed 1.5×10^{2} videos.

Do Exercise 25.

The following is a summary of the definitions and rules for exponents that we have considered in this section and the preceding one.

DEFINITIONS AND RULES FOR EXPONENTS

Exponent of 1:	$a^1 = a$
Exponent of 0:	$a^0 = 1, a \neq 0$
Negative exponents:	$a^{-n} = \frac{1}{a^n}, \frac{1}{a^{-n}} = a^n, a \neq 0$
Product Rule:	$a^m \cdot a^n = a^{m+n}$
Quotient Rule:	$\frac{a^m}{a^n} = a^{m-n}, a \neq 0$
Power Rule:	$(a^m)^n = a^{mn}$
Raising a product to a power:	$(ab)^n = a^n b^n$
Raising a quotient to a power:	$\left(\frac{a}{b}\right)^n = \frac{a^n}{b^n}, b \neq 0;$
	$\left(\frac{a}{b}\right)^{-n} = \left(\frac{b}{a}\right)^n, b \neq 0, a \neq 0$
Scientific notation:	$M \times 10^n$, where $1 \leq M < 10$

Answer

25. The length of the helix is 2.5×10^{7} times its width.

12.2 Exercise Set

For Extra Help
MyMathLab®
MathXL®

PRACTICE

WATCH

READ

REVIEW

Reading Check

Choose from the column on the right the appropriate word to complete each statement. Not every word will be used.

RC1. To raise a power to a power, __________ the exponents.

RC2. To raise a product to the nth power, raise each factor to the _____ power.

RC3. To convert a number less than 1 to scientific notation, move the decimal point to the _________.

RC4. A __________ exponent in scientific notation indicates a number greater than or equal to 10.

add
left
multiply
negative
nth
positive
right

a, b Simplify.

1. $(2^3)^2$

2. $(5^2)^4$

3. $(5^2)^{-3}$

4. $(7^{-3})^5$

5. $(x^{-3})^{-4}$

6. $(a^{-5})^{-6}$

7. $(a^{-2})^9$

8. $(x^{-5})^6$

9. $(t^{-3})^{-6}$

10. $(a^{-4})^{-7}$

11. $(t^4)^{-3}$

12. $(t^5)^{-2}$

13. $(x^{-2})^{-4}$

14. $(t^{-6})^{-5}$

15. $(ab)^3$

16. $(xy)^2$

17. $(ab)^{-3}$

18. $(xy)^{-6}$

19. $(mn^2)^{-3}$

20. $(x^3y)^{-2}$

21. $(4x^3)^2$

22. $4(x^3)^2$

23. $(3x^{-4})^2$

24. $(2a^{-5})^3$

25. $(x^4y^5)^{-3}$

26. $(t^5x^3)^{-4}$

27. $(x^{-6}y^{-2})^{-4}$

28. $(x^{-2}y^{-7})^{-5}$

29. $(a^{-2}b^7)^{-5}$

30. $(q^5r^{-1})^{-3}$

31. $(5r^{-4}t^3)^2$

32. $(4x^5y^{-6})^3$

33. $(a^{-5}b^7c^{-2})^3$

34. $(x^{-4}y^{-2}z^9)^2$

35. $(3x^3y^{-8}z^{-3})^2$

36. $(2a^2y^{-4}z^{-5})^3$

37. $(-4x^3y^{-2})^2$

38. $(-8x^3y^{-2})^3$

39. $(-a^{-3}b^{-2})^{-4}$

40. $(-p^{-4}q^{-3})^{-2}$

41. $\left(\frac{y^3}{2}\right)^2$

42. $\left(\frac{a^5}{3}\right)^3$

43. $\left(\frac{a^2}{b^3}\right)^4$

44. $\left(\frac{x^3}{y^4}\right)^5$

45. $\left(\frac{y^2}{2}\right)^{-3}$

46. $\left(\frac{a^4}{3}\right)^{-2}$

47. $\left(\frac{7}{x^{-3}}\right)^2$

48. $\left(\frac{3}{a^{-2}}\right)^3$

49. $\left(\frac{x^2y}{z}\right)^3$

50. $\left(\frac{m}{n^4p}\right)^3$

51. $\left(\frac{a^2b}{cd^3}\right)^{-2}$

52. $\left(\frac{2a^2}{3b^4}\right)^{-3}$

c Convert to scientific notation.

53. 28,000,000,000

54. 4,900,000,000,000

55. 907,000,000,000,000,000

56. 168,000,000,000,000

57. 0.00000304

58. 0.000000000865

59. 0.000000018

60. 0.00000000002

61. 100,000,000,000

62. 0.0000001

63. *Population of the United States.* It is estimated that the population of the United States will be 419,854,000 in 2050. Convert 419,854,000 to scientific notation.

Source: U.S. Census Bureau

64. *Microprocessors.* The minimum feature size of a microprocessor is the transistor gate length. In 2011, the transistor gate length for a new microprocessor was about 0.000000028 m. Convert 0.000000028 to scientific notation.

65. *Wavelength of Light.* The wavelength of red light is 0.00000068 m. Convert 0.00000068 to scientific notation.

66. *Olympics.* Great Britain spent about $15,000,000,000 to stage the 2012 Summer Olympics. Convert 15,000,000,000 to scientific notation.

Source: CNN.com

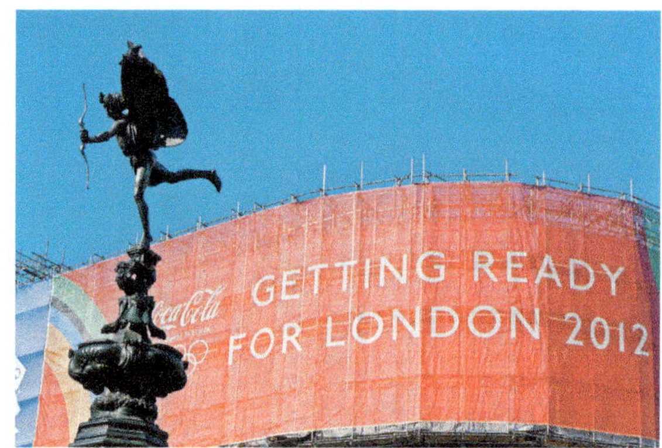

Convert to decimal notation.

67. 8.74×10^{7}

68. 1.85×10^{8}

69. 5.704×10^{-8}

70. 8.043×10^{-4}

71. 10^{7}

72. 10^{6}

73. 10^{-5}

74. 10^{-8}

d Multiply or divide and write scientific notation for the result.

75. $(3 \times 10^{4})(2 \times 10^{5})$

76. $(3.9 \times 10^{8})(8.4 \times 10^{-3})$

77. $(5.2 \times 10^{5})(6.5 \times 10^{-2})$

78. $(7.1 \times 10^{-7})(8.6 \times 10^{-5})$

79. $(9.9 \times 10^{-6})(8.23 \times 10^{-8})$

80. $(1.123 \times 10^{4}) \times 10^{-9}$

81. $\dfrac{8.5 \times 10^8}{3.4 \times 10^{-5}}$

82. $\dfrac{5.6 \times 10^{-2}}{2.5 \times 10^5}$

83. $(3.0 \times 10^6) \div (6.0 \times 10^9)$

84. $(1.5 \times 10^{-3}) \div (1.6 \times 10^{-6})$

85. $\dfrac{7.5 \times 10^{-9}}{2.5 \times 10^{12}}$

86. $\dfrac{4.0 \times 10^{-3}}{8.0 \times 10^{20}}$

Solve.

87. *River Discharge.* The average discharge at the mouths of the Amazon River is 4,200,000 cubic feet per second. How much water is discharged from the Amazon River in 1 year? Express the answer in scientific notation.

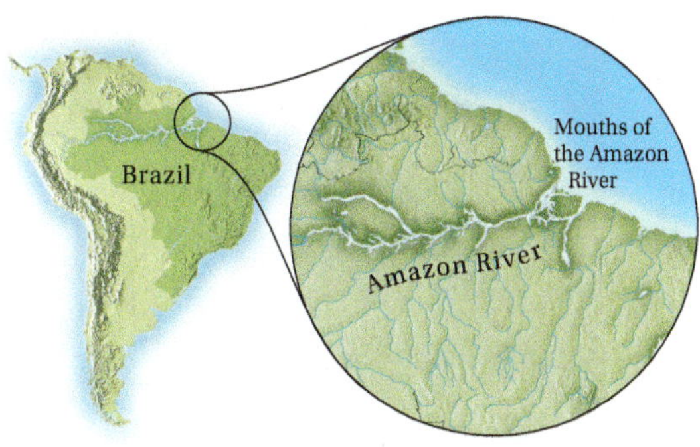

88. *Coral Reefs.* There are 10 million bacteria per square centimeter of coral in a coral reef. The coral reefs near the Hawaiian Islands cover 14,000 km^2. How many bacteria are there in Hawaii's coral reefs?

Sources: livescience.com; U.S. Geological Survey

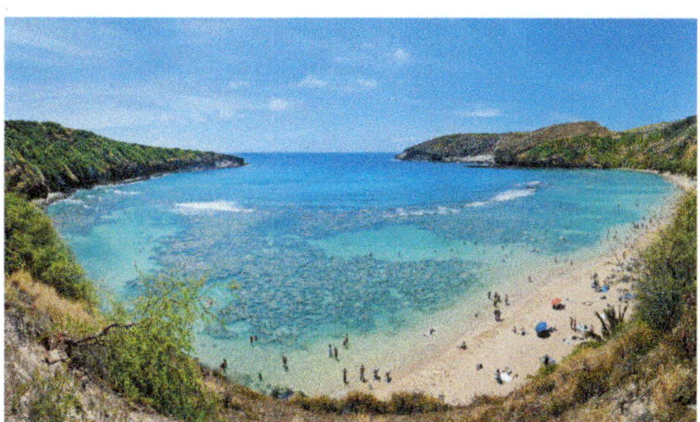

89. *Stars.* It is estimated that there are 10 billion trillion stars in the known universe. Express the number of stars in scientific notation. $(1 \text{ billion} = 10^9; 1 \text{ trillion} = 10^{12})$

90. *Water Contamination.* Americans who change their own motor oil generate about 150 million gallons of used oil annually. If this oil is not disposed of properly, it can contaminate drinking water and soil. One gallon of used oil can contaminate one million gallons of drinking water. How many gallons of drinking water can 150 million gallons of oil contaminate? Express the answer in scientific notation. $(1 \text{ million} = 10^6)$.

Source: *New Car Buying Guide*

91. *Earth vs. Jupiter.* The mass of Earth is about 6×10^{21} metric tons. The mass of Jupiter is about 1.908×10^{24} metric tons. About how many times the mass of Earth is the mass of Jupiter? Express the answer in scientific notation.

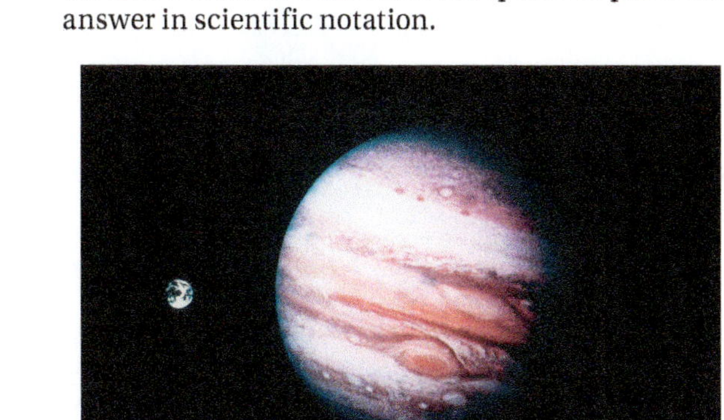

92. *Office Supplies.* A ream of copier paper weighs 2.25 kg. How much does a sheet of copier paper weigh?

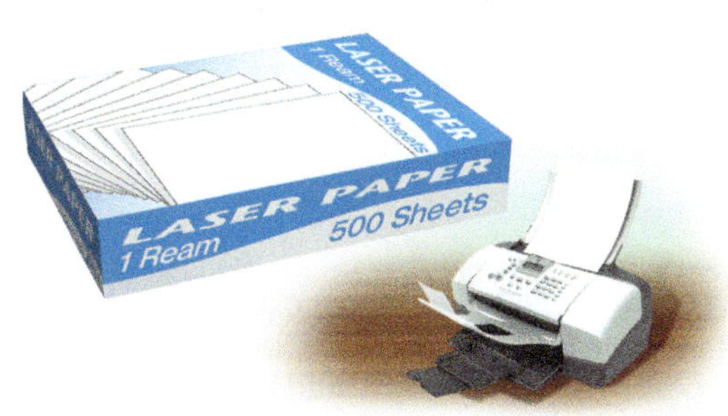

93. *Information Technology.* In 2012, about 2.5×10^{18} bytes of information were generated each day by the worldwide online population of 2×10^9 people. Find the average amount of information generated per person per day in 2012.

Sources: IBM; internetworldstats.com

94. *Computer Technology.* Intel Corporation has developed silicon-based connections that use lasers to move data at a rate of 50 gigabytes per second. The printed collection of the U.S. Library of Congress contains 10 terabytes of information. How long would it take to copy the Library of Congress using these connections? *Note:* 1 gigabyte $= 10^9$ bytes and 1 terabyte $= 10^{12}$ bytes.

Sources: spie.org; newworldencyclopedia.org

95. *Gold Leaf.* Gold can be milled into a very thin film called gold leaf. This film is so thin that it took only 43 oz of gold to cover the dome of Georgia's state capitol building. The gold leaf used was 5×10^{-6} m thick. In contrast, a U.S. penny is 1.55×10^{-3} m thick. How many sheets of gold leaf are in a stack that is the height of a penny?

Source: georgiaencyclopedia.org

96. *Relative Size.* An influenza virus is about 1.2×10^{-7} m in diameter. A staphylococcus bacterium is about 1.5×10^{-6} m in diameter. How many influenza viruses would it take, laid side by side, to equal the diameter of the bacterium?

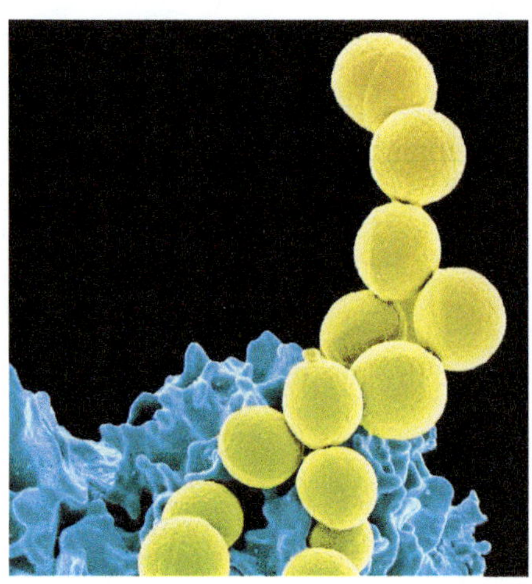

Space Travel. Use the following information for Exercises 97 and 98.

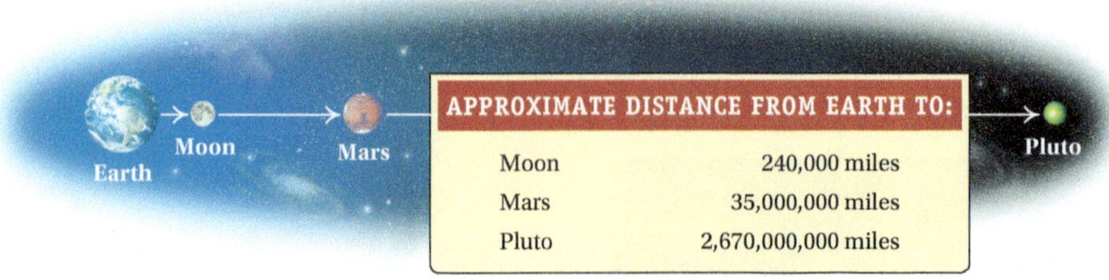

APPROXIMATE DISTANCE FROM EARTH TO:	
Moon	240,000 miles
Mars	35,000,000 miles
Pluto	2,670,000,000 miles

97. *Time to Reach Mars.* Suppose that it takes about 3 days for a space vehicle to travel from Earth to the moon. About how long would it take the same space vehicle traveling at the same speed to reach Mars? Express the answer in scientific notation.

98. *Time to Reach Pluto.* Suppose that it takes about 3 days for a space vehicle to travel from Earth to the moon. About how long would it take the same space vehicle traveling at the same speed to reach the dwarf planet Pluto? Express the answer in scientific notation.

Skill Maintenance

Graph.

99. $y = x - 5$ [11.2a]

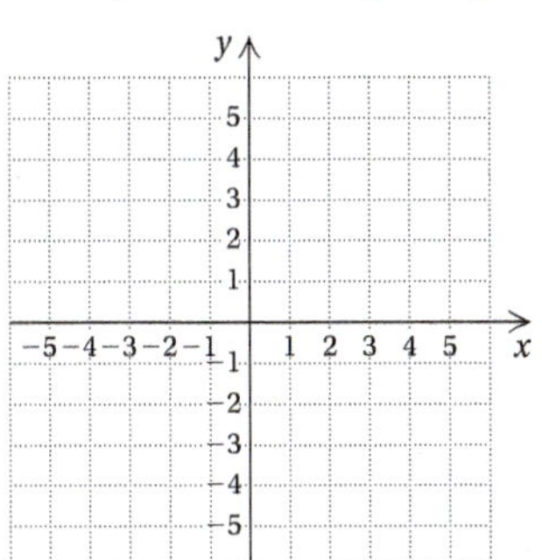

100. $2x + y = 4$ [11.2a]

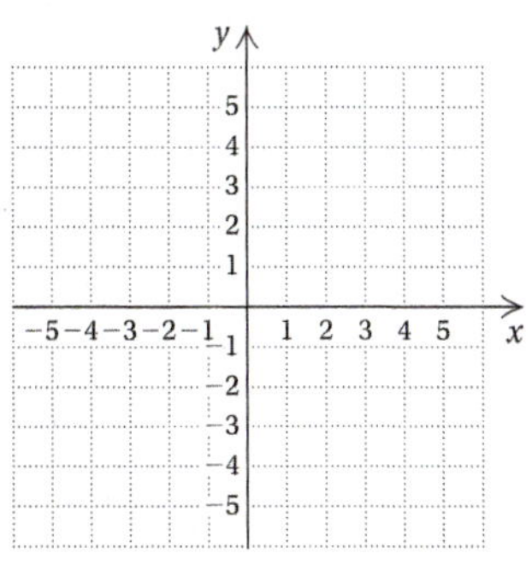

101. $3x - y = 3$ [11.2a]

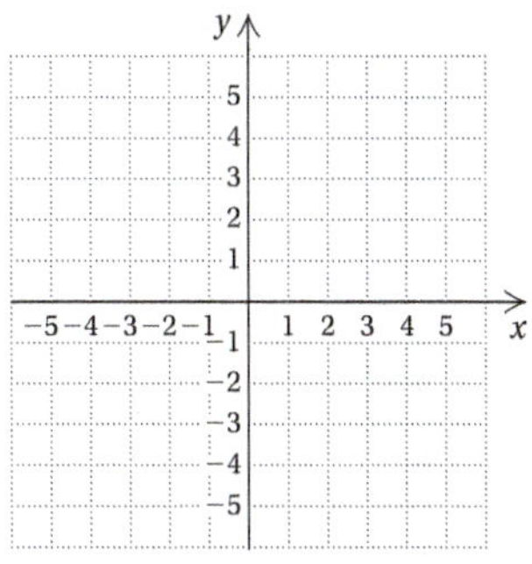

102. $y = -x$ [11.2a]

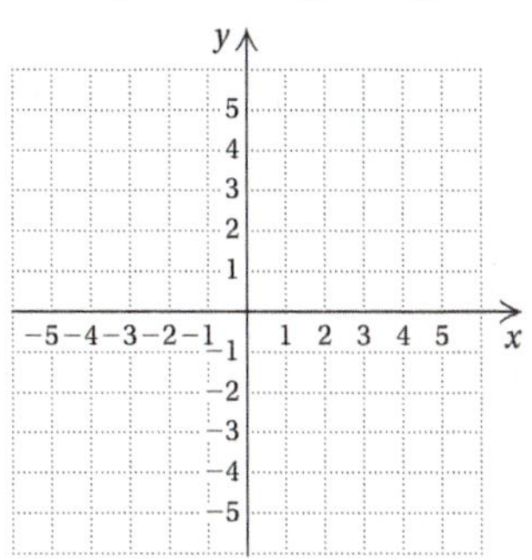

103. $2x = -10$ [11.2b]

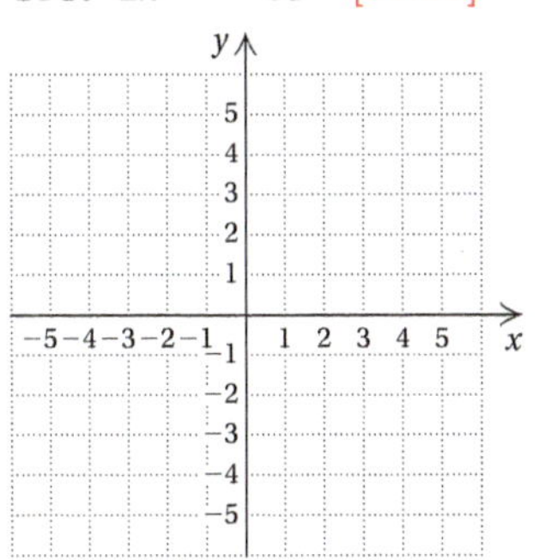

104. $y = -4$ [11.2b]

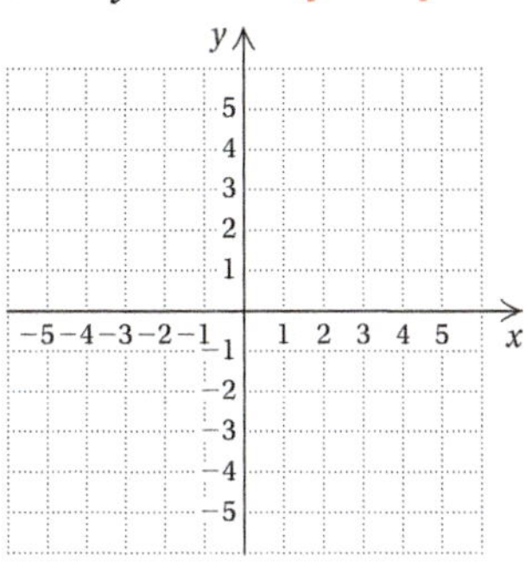

105. $8y - 16 = 0$ [11.2b]

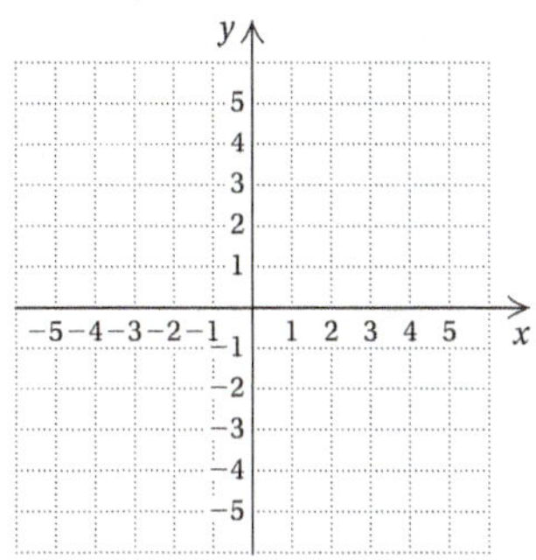

106. $x = 4$ [11.2b]

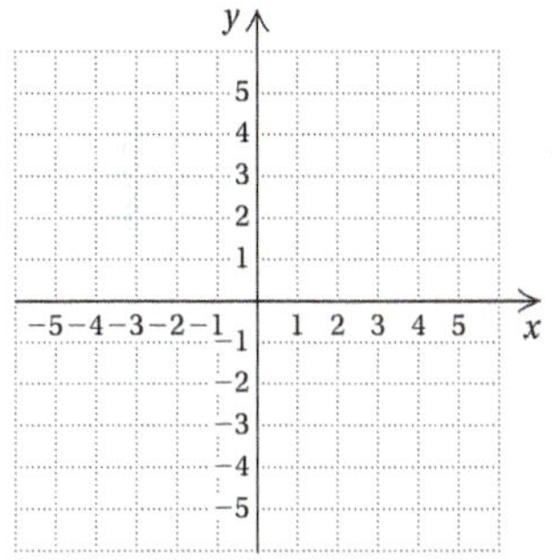

Synthesis

107. Carry out the indicated operations. Express the result in scientific notation.

$$\frac{(5.2 \times 10^{6})(6.1 \times 10^{-11})}{1.28 \times 10^{-3}}$$

108. Find the reciprocal and express it in scientific notation.

$$6.25 \times 10^{-3}$$

Simplify.

109. $\dfrac{(5^{12})^2}{5^{25}}$

110. $\dfrac{a^{22}}{(a^2)^{11}}$

111. $\dfrac{(3^5)^4}{3^5 \cdot 3^4}$

112. $\left(\dfrac{5x^{-2}}{3y^{-2}z}\right)^0$

113. $\dfrac{49^{18}}{7^{35}}$

114. $\left(\dfrac{1}{a}\right)^{-n}$

115. $\dfrac{(0.4)^5}{[(0.4)^3]^2}$

116. $\left(\dfrac{4a^3b^{-2}}{5c^{-3}}\right)^1$

Determine whether each equation is true or false for all pairs of integers m and n and all positive numbers x and y.

117. $x^m \cdot y^n = (xy)^{mn}$

118. $x^m \cdot y^m = (xy)^{2m}$

119. $(x - y)^m = x^m - y^m$

120. $-x^m = (-x)^m$

121. $(-x)^{2m} = x^{2m}$

122. $x^{-m} = \dfrac{-1}{x^m}$

12.3 Introduction to Polynomials

OBJECTIVES

a Evaluate a polynomial for a given value of the variable.

b Identify the terms of a polynomial and classify a polynomial by its number of terms.

c Identify the coefficient and the degree of each term of a polynomial and the degree of the polynomial.

d Collect the like terms of a polynomial.

e Arrange a polynomial in descending order, or collect the like terms and then arrange in descending order.

f Identify the missing terms of a polynomial.

SKILL TO REVIEW

Objective 9.7e: Collect like terms.

Collect like terms.

1. $3x - 4y + 5x + y$
2. $2a - 7b + 6 - 3a + 4b - 1$

We have already learned to evaluate and to manipulate certain kinds of algebraic expressions. We will now consider algebraic expressions called *polynomials.*

The following are examples of *monomials in one variable*:

$$3x^2, \quad 2x, \quad -5, \quad 37p^4, \quad 0.$$

Each expression is a constant or a constant times some variable to a nonnegative integer power.

MONOMIAL

A **monomial** is an expression of the type ax^n, where a is a real-number constant and n is a nonnegative integer.

Algebraic expressions like the following are **polynomials**:

$$\tfrac{3}{4}y^5, \quad -2, \quad 5y + 3, \quad 3x^2 + 2x - 5, \quad -7a^3 + \tfrac{1}{2}a, \quad 6x, \quad 37p^4, \quad x, \quad 0.$$

POLYNOMIAL

A **polynomial** is a monomial or a combination of sums and/or differences of monomials.

The following algebraic expressions are *not* polynomials:

$$\text{(1)}\ \frac{x+3}{x-4}, \qquad \text{(2)}\ 5x^3 - 2x^2 + \frac{1}{x}, \qquad \text{(3)}\ \frac{1}{x^3 - 2}.$$

Expressions (1) and (3) are not polynomials because they represent quotients, not sums or differences. Expression (2) is not a polynomial because

$$\frac{1}{x} = x^{-1},$$

and this is not a monomial because the exponent is negative.

◀ Do Margin Exercise 1.

1. Write three polynomials.

a EVALUATING POLYNOMIALS AND APPLICATIONS

When we replace the variable in a polynomial with a number, the polynomial then represents a number called a **value** of the polynomial. Finding that number, or value, is called **evaluating the polynomial**. We evaluate a polynomial using the rules for order of operations.

EXAMPLE 1 Evaluate the polynomial when $x = 2$.

a) $3x + 5 = 3 \cdot 2 + 5$
$= 6 + 5$
$= 11$

b) $2x^2 - 7x + 3 = 2 \cdot 2^2 - 7 \cdot 2 + 3$
$= 2 \cdot 4 - 7 \cdot 2 + 3$
$= 8 - 14 + 3$
$= -3$

Answers

Skill to Review:
1. $8x - 3y$ 2. $-a - 3b + 5$

Margin Exercise:
1. $4x^2 - 3x + \frac{5}{4}$; $15y^3$; $-7x^3 + 1.1$; answers may vary

EXAMPLE 2 Evaluate the polynomial when $x = -4$.

a) $2 - x^3 = 2 - (-4)^3 = 2 - (-64)$
$= 2 + 64 = 66$

b) $-x^2 - 3x + 1 = -(-4)^2 - 3(-4) + 1$
$= -16 + 12 + 1 = -3$

Do Exercises 2–5.

Evaluate each polynomial when $x = 3$.

2. $-4x - 7$

3. $-5x^3 + 7x + 10$

Evaluate each polynomial when $x = -5$.

4. $5x + 7$

GS 5. $2x^2 + 5x - 4$
$= 2(\quad)^2 + 5(\quad) - 4$
$= 2(\quad) + (\quad) - 4$
$= 50 - \quad - 4$
$= \quad$

ALGEBRAIC ▶◀ GRAPHICAL CONNECTION

An equation like $y = 2x - 2$, which has a polynomial on one side and only y on the other, is called a **polynomial equation**. For such an equation, determining y is the same as evaluating the polynomial. Once the graph of such an equation has been drawn, we can evaluate the polynomial for a given x-value by finding the y-value that is paired with it on the graph.

EXAMPLE 3 Use *only* the given graph of $y = 2x - 2$ to evaluate the polynomial $2x - 2$ when $x = 3$.

First, we locate 3 on the x-axis. From there we move vertically to the graph of the equation and then horizontally to the y-axis. There we locate the y-value that is paired with 3. It appears that the y-value 4 is paired with 3. Thus the value of $2x - 2$ is 4 when $x = 3$. We can check this by evaluating $2x - 2$ when $x = 3$.

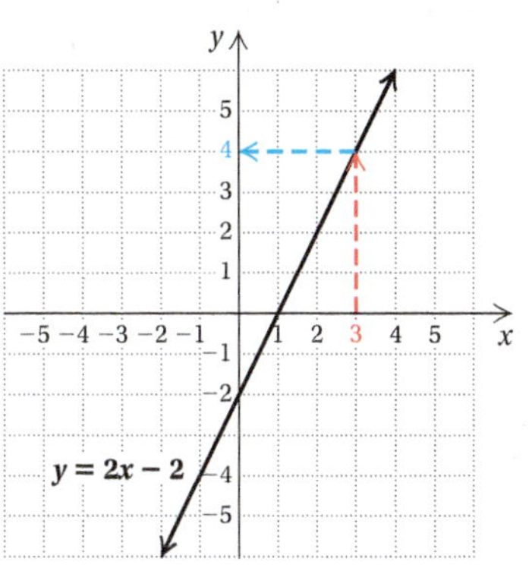

Do Exercise 6.

6. Use *only* the graph shown in Example 3 to evaluate the polynomial $2x - 2$ when $x = 4$ and when $x = -1$.

Polynomial equations can be used to model many real-world situations.

EXAMPLE 4 *Games in a Sports League.* In a sports league of x teams in which each team plays every other team twice, the total number of games N to be played is given by the polynomial equation

$$N = x^2 - x.$$

A women's slow-pitch softball league has 10 teams and each team plays every other team twice. What is the total number of games to be played?

We evaluate the polynomial when $x = 10$:

$$N = x^2 - x = 10^2 - 10 = 100 - 10 = 90.$$

The league plays 90 games.

Do Exercise 7.

7. Refer to Example 4. Determine the total number of games to be played in a league of 12 teams in which each team plays every other team twice.

Answers

2. −19 3. −104 4. −18 5. 21
6. 6; −4 7. 132 games

Guided Solution:
5. −5, −5, 25, −25, 25, 21

EXAMPLE 5 *Medical Dosage.* The concentration C, in parts per million, of a certain antibiotic in the bloodstream after t hours is given by the polynomial equation

$$C = -0.05t^2 + 2t + 2.$$

Find the concentration after 2 hr.

To find the concentration after 2 hr, we evaluate the polynomial when $t = 2$:

$$\begin{aligned} C &= -0.05t^2 + 2t + 2 \\ &= -0.05(2)^2 + 2(2) + 2 && \text{Substituting 2 for } t \\ &= -0.05(4) + 2(2) + 2 && \text{Carrying out the calculation using} \\ &= -0.2 + 4 + 2 && \text{the rules for order of operations} \\ &= 3.8 + 2 \\ &= 5.8. \end{aligned}$$

The concentration after 2 hr is 5.8 parts per million.

ALGEBRAIC GRAPHICAL CONNECTION

The polynomial equation in Example 5 can be graphed if we evaluate the polynomial for several values of t. We list the values in a table and show the graph below. Note that the concentration peaks at the 20-hr mark and after slightly more than 40 hr, the concentration is 0. Since neither time nor concentration can be negative, our graph uses only the first quadrant.

t	C $C = -0.05t^2 + 2t + 2$
0	2
2	5.8 ← Example 5
10	17
20	22
30	17

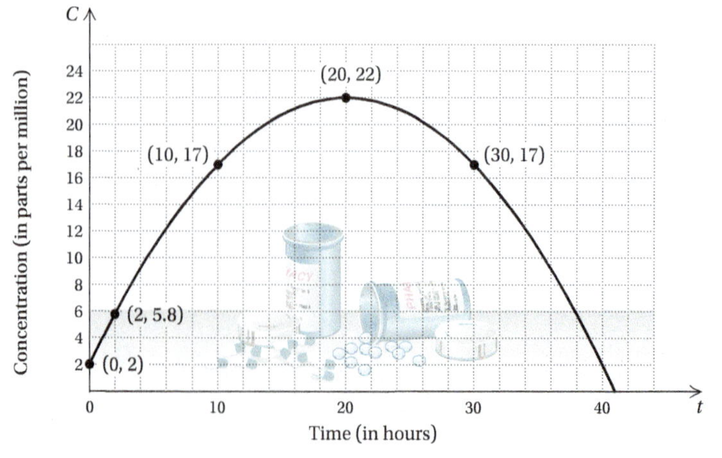

Do Exercises 8 and 9.

CALCULATOR CORNER

To use a graphing calculator to evaluate the polynomial in Example 2(b), $-x^2 - 3x + 1$, when $x = -4$, we can first graph $y_1 = -x^2 - 3x + 1$ in a window that includes the x-value -4. Then we can use the Value feature from the CALC menu, supplying the desired x-value and pressing ENTER to see X = −4, Y = −3 at the bottom of the screen. Thus, when $x = -4$, the value of $-x^2 - 3x + 1$ is -3.

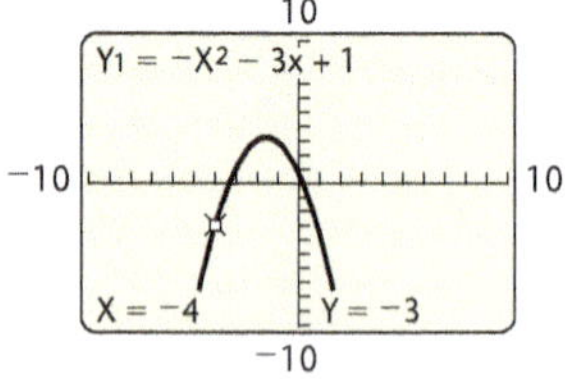

EXERCISES: Use the Value feature to evaluate each polynomial for the given values of x.

1. $-x^2 - 3x + 1$, when $x = -2$, when $x = -0.5$, and when $x = 4$

2. $3x^2 - 5x + 2$, when $x = -3$, when $x = 1$, and when $x = 2.6$

8. *Medical Dosage.* Refer to Example 5.

a) Determine the concentration after 3 hr by evaluating the polynomial when $t = 3$.

b) Use *only* the graph showing medical dosage to check the value found in part (a).

9. *Medical Dosage.* Refer to Example 5. Use *only* the graph showing medical dosage to estimate the value of the polynomial when $t = 26$.

Answers

8. (a) 7.55 parts per million; **(b)** When $t = 3$, $C \approx 7.5$ so the value found in part (a) appears to be correct. **9.** 20 parts per million

b IDENTIFYING TERMS AND CLASSIFYING POLYNOMIALS

For any polynomial that has some subtractions, we can find an equivalent polynomial using only additions.

EXAMPLES Find an equivalent polynomial using only additions.

6. $-5x^2 - x = -5x^2 + (-x)$

7. $4x^5 - 2x^6 + 4x - 7 = 4x^5 + (-2x^6) + 4x + (-7)$

Do Exercises 10 and 11. ▶

When a polynomial is written using only additions, the monomials being added are called **terms**. In Example 6, the terms are $-5x^2$ and $-x$. In Example 7, the terms are $4x^5$, $-2x^6$, $4x$, and -7.

EXAMPLE 8 Identify the terms of the polynomial

$$4x^7 + 3x + 12 + 8x^3 + 5x.$$

Terms: $4x^7$, $3x$, 12, $8x^3$, and $5x$.

If there are subtractions, you can *think* of them as additions without rewriting.

EXAMPLE 9 Identify the terms of the polynomial

$$3t^4 - 5t^6 - 4t + 2.$$

Terms: $3t^4$, $-5t^6$, $-4t$, and 2.

Do Exercises 12 and 13. ▶

Polynomials with just one term are called **monomials**. Polynomials with just two terms are called **binomials**. Those with just three terms are called **trinomials**. Those with more than three terms are generally not specified with a name.

EXAMPLE 10

MONOMIALS	BINOMIALS	TRINOMIALS	NONE OF THESE
$-4x^2$	$2x + 4$	$3x^5 + 4x^4 + 7x$	$4x^3 - 5x^2 + x - 8$
9	$-9x^7 - 6x$	$4x^2 - 6x - \frac{1}{2}$	$z^5 + 2z^4 - z^3 + 7z + 3$

Do Exercises 14–17. ▶

c COEFFICIENTS AND DEGREES

The coefficient of the term $5x^3$ is 5. In the following polynomial, the red numbers are the **coefficients**, 3, -2, 5, and 4:

$$3x^5 - 2x^3 + 5x + 4.$$

Find an equivalent polynomial using only additions.

10. $-9x^3 - 4x^5$

11. $-2y^3 + 3y^7 - 7y - 9$

Identify the terms of each polynomial.

12. $3x^2 + 6x + \frac{1}{2}$

13. $-4y^5 + 7y^2 - 3y - 2$

Classify each polynomial as a monomial, a binomial, a trinomial, or none of these.

14. $3x^2 + x$

15. $5x^4$

16. $4x^3 - 3x^2 + 4x + 2$

17. $3x^2 + 2x - 4$

Answers

10. $-9x^3 + (-4x^5)$
11. $-2y^3 + 3y^7 + (-7y) + (-9)$
12. $3x^2, 6x, \frac{1}{2}$ **13.** $-4y^5, 7y^2, -3y, -2$
14. Binomial **15.** Monomial
16. None of these **17.** Trinomial

EXAMPLE 11 Identify the coefficient of each term in the polynomial

$$3x^4 - 4x^3 + \frac{1}{2}x^2 + x - 8.$$

The coefficient of $3x^4$ is 3.

The coefficient of $-4x^3$ is -4.

The coefficient of $\frac{1}{2}x^2$ is $\frac{1}{2}$.

The coefficient of x (or $1x$) is 1.

The coefficient of -8 is -8.

18. Identify the coefficient of each term in the polynomial $2x^4 - 7x^3 - 8.5x^2 - x - 4$.

◀ **Do Exercise 18.**

The **degree** of a term is the exponent of the variable. The degree of the term $-5x^3$ is 3.

EXAMPLE 12 Identify the degree of each term of $8x^4 - 3x + 7$.

The degree of $8x^4$ is 4.

The degree of $-3x$ (or $-3x^1$) is 1. $x = x^1$

The degree of 7 (or $7x^0$) is 0. $7 = 7 \cdot 1 = 7 \cdot x^0$, since $x^0 = 1$

Because we can write 1 as x^0, the degree of any constant term (except 0) is 0. The term 0 is a special case. We agree that it has *no* degree because we can express 0 as $0 = 0x^5 = 0x^7$, and so on, using any exponent we wish.

The degree of any nonzero constant term is 0.

The **degree of a polynomial** is the largest of the degrees of the terms, unless it is the polynomial 0.

EXAMPLE 13 Identify the degree of the polynomial $5x^3 - 6x^4 + 7$.

$5x^3 - 6x^4 + 7.$ The largest degree is 4.

The degree of the polynomial is 4.

Identify the degree of each term and the degree of the polynomial.

19. $-6x^4 + 8x^2 - 2x + 9$

20. $4 - x^3 + \frac{1}{2}x^6 - x^5$

◀ **Do Exercises 19 and 20.**

Let's summarize the terminology that we have learned, using the polynomial $3x^4 - 8x^3 + x^2 + 7x - 6$.

TERM	COEFFICIENT	DEGREE OF THE TERM	DEGREE OF THE POLYNOMIAL
$3x^4$	3	4	
$-8x^3$	-8	3	
x^2	1	2	4
$7x$	7	1	
-6	-6	0	

Answers

18. 2, −7, −8.5, −1, −4 **19.** 4, 2, 1, 0; 4
20. 0, 3, 6, 5; 6

d COLLECTING LIKE TERMS

When terms have the same variable and the same exponent, we say that they are **like terms**.

EXAMPLES Identify the like terms in each polynomial.

14. $4x^3 + 5x - 4x^2 + 2x^3 + x^2$

Like terms: $4x^3$ and $2x^3$ Same variable and exponent

Like terms: $-4x^2$ and x^2 Same variable and exponent

15. $6 - 3a^2 - 8 - a - 5a$

Like terms: 6 and -8 Constant terms are like terms; note that $6 = 6x^0$ and $-8 = -8x^0$.

Like terms: $-a$ and $-5a$

Do Exercises 21–23. ▶

We can often simplify polynomials by **collecting like terms**, or **combining like terms**. To do this, we use the distributive laws.

EXAMPLES Collect like terms.

16. $2x^3 - 6x^3 = (2 - 6)x^3 = -4x^3$

17. $5x^2 + 7 + 4x^4 + 2x^2 - 11 - 2x^4 = (5 + 2)x^2 + (4 - 2)x^4 + (7 - 11)$
$= 7x^2 + 2x^4 - 4$ ■

Note that using the distributive laws in this manner allows us to collect like terms by adding or subtracting the coefficients. Often the middle step is omitted and we add or subtract mentally, writing just the answer. In collecting like terms, we may get 0.

EXAMPLE 18 Collect like terms: $3x^5 + 2x^2 - 3x^5 + 8$.

$$3x^5 + 2x^2 - 3x^5 + 8 = (3 - 3)x^5 + 2x^2 + 8$$
$$= 0x^5 + 2x^2 + 8$$
$$= 2x^2 + 8$$

Do Exercises 24–29. ▶

Expressing a term like x^2 by showing 1 as a factor, $1 \cdot x^2$, may make it easier to understand how to factor or collect like terms.

EXAMPLES Collect like terms.

19. $5x^8 - 6x^5 - x^8 = 5x^8 - 6x^5 - 1x^8$ Replacing x^8 with $1x^8$

$= (5 - 1)x^8 - 6x^5$ Using a distributive law

$= 4x^8 - 6x^5$

20. $\frac{2}{3}x^4 - x^3 - \frac{1}{6}x^4 + \frac{2}{5}x^3 - \frac{3}{10}x^3$

$= \left(\frac{2}{3} - \frac{1}{6}\right)x^4 + \left(-1 + \frac{2}{5} - \frac{3}{10}\right)x^3$ $-x^3 = -1 \cdot x^3$

$= \left(\frac{4}{6} - \frac{1}{6}\right)x^4 + \left(-\frac{10}{10} + \frac{4}{10} - \frac{3}{10}\right)x^3$

$= \frac{3}{6}x^4 - \frac{9}{10}x^3 = \frac{1}{2}x^4 - \frac{9}{10}x^3$

Do Exercises 30–32. ▶

Identify the like terms in each polynomial.

21. $4x^3 - x^3 + 2$

22. $4t^4 - 9t^3 - 7t^4 + 10t^3$

23. $5x^2 + 3x - 10 + 7x^2 - 8x + 11$

Collect like terms.

24. $3x^2 + 5x^2$

25. $4x^3 - 2x^3 + 2 + 5$

26. $\frac{1}{2}x^5 - \frac{3}{4}x^5 + 4x^2 - 2x^2$

27. $24 - 4x^3 - 24$

28. $5x^3 - 8x^5 + 8x^5$

GS **29.** $-2x^4 + 16 + 2x^4 + 9 - 3x^5$
$= -3x^5 + (-2 + \underline{\quad})x^4 + (16 + \underline{\quad})$
$= -3x^5 + 0x^4 + \underline{\quad}$
$= -3x^5 + 25$

Collect like terms.

30. $5x^3 - x^3 + 4$

31. $\frac{3}{4}x^3 + 4x^2 - x^3 + 7$

32. $\frac{4}{5}x^4 - x^4 + x^5 - \frac{1}{5} - \frac{1}{4}x^4 + 10$

Answers

21. $4x^3$ and $-x^3$ **22.** $4t^4$ and $-7t^4$; $-9t^3$ and $10t^3$
23. $5x^2$ and $7x^2$; $3x$ and $-8x$; -10 and 11
24. $8x^2$ **25.** $2x^3 + 7$ **26.** $-\frac{1}{4}x^5 + 2x^2$
27. $-4x^3$ **28.** $5x^3$ **29.** $-3x^5 + 25$
30. $4x^3 + 4$ **31.** $-\frac{1}{4}x^3 + 4x^2 + 7$
32. $x^5 - \frac{9}{20}x^4 + \frac{49}{5}$

Guided Solution:
29. 2, 9, 25

e DESCENDING ORDER

A polynomial is written in **descending order** when the term with the largest degree is written first, the term with the next largest degree is written next, and so on, in order from left to right.

Arrange each polynomial in descending order.

33. $4x^2 - 3 + 7x^5 + 2x^3 - 5x^4$

34. $-14 + 7t^2 - 10t^5 + 14t^7$

EXAMPLES Arrange each polynomial in descending order.

21. $6x^5 + 4x^7 + x^2 + 2x^3 = 4x^7 + 6x^5 + 2x^3 + x^2$

22. $\frac{2}{3} + 4x^5 - 8x^2 + 5x - 3x^3 = 4x^5 - 3x^3 - 8x^2 + 5x + \frac{2}{3}$

◀ **Do Exercises 33 and 34.**

Collect like terms and then arrange in descending order.

35. $3x^2 - 2x + 3 - 5x^2 - 1 - x$

36. $-x + \dfrac{1}{2} + 14x^4 - 7x - 1 - 4x^4$

EXAMPLE 23 Collect like terms and then arrange in descending order:

$$2x^2 - 4x^3 + 3 - x^2 - 2x^3.$$

$$2x^2 - 4x^3 + 3 - x^2 - 2x^3 = x^2 - 6x^3 + 3 \quad \text{Collecting like terms}$$
$$= -6x^3 + x^2 + 3 \quad \text{Arranging in descending order}$$

◀ **Do Exercises 35 and 36.**

The opposite of descending order is called **ascending order**. Generally, if an exercise is written in a certain order, we give the answer in that same order.

f MISSING TERMS

If a coefficient is 0, we generally do not write the term. We say that we have a **missing term**.

Identify the missing term(s) in each polynomial.

37. $2x^3 + 4x^2 - 2$

38. $-3x^4$

39. $x^3 + 1$

EXAMPLE 24 Identify the missing terms in the polynomial

$$8x^5 - 2x^3 + 5x^2 + 7x + 8.$$

There is no term with x^4. We say that the x^4-term is missing.

◀ **Do Exercises 37–39.**

We can either write missing terms with zero coefficients or leave space.

Write each polynomial in two ways: with its missing term(s) and by leaving space for them.

40. $2x^3 + 4x^2 - 2$

41. $a^4 + 10$

EXAMPLE 25 Write the polynomial $x^4 - 6x^3 + 2x - 1$ in two ways: with its missing term and by leaving space for it.

a) $x^4 - 6x^3 + 2x - 1 = x^4 - 6x^3 + 0x^2 + 2x - 1$ Writing with the missing x^2-term

b) $x^4 - 6x^3 + 2x - 1 = x^4 - 6x^3 \qquad\quad + 2x - 1$ Leaving space for the missing x^2-term

EXAMPLE 26 Write the polynomial $y^5 - 1$ in two ways: with its missing terms and by leaving space for them.

a) $y^5 - 1 = y^5 + 0y^4 + 0y^3 + 0y^2 + 0y - 1$

b) $y^5 - 1 = y^5 \qquad\qquad\qquad\qquad - 1$

◀ **Do Exercises 40 and 41.**

Answers

33. $7x^5 - 5x^4 + 2x^3 + 4x^2 - 3$
34. $14t^7 - 10t^5 + 7t^2 - 14$ **35.** $-2x^2 - 3x + 2$
36. $10x^4 - 8x - \frac{1}{2}$ **37.** x **38.** x^3, x^2, x, x^0
39. x^2, x **40.** $2x^3 + 4x^2 + 0x - 2$; $2x^3 + 4x^2 \quad - 2$
41. $a^4 + 0a^3 + 0a^2 + 0a + 10$; $a^4 \qquad + 10$

12.3 Exercise Set

For Extra Help MyMathLab® MathXL® PRACTICE WATCH READ REVIEW

Reading Check

Choose from the column on the right the expression that best fits each description.

RC1. ____ The value of $x^2 - x$ when $x = -1$

RC2. ____ A polynomial written in ascending order

RC3. ____ A coefficient of $5x^4 - 3x + 7$

RC4. ____ A term of $5x^4 - 3x + 7$

RC5. ____ The degree of one of the terms of $5x^4 - 3x + 7$

RC6. ____ An example of a binomial

a) 0
b) 2
c) 5
d) $-3x$
e) $8x - 9$
f) $y + 6y^2 - 2y^8$

a Evaluate each polynomial when $x = 4$ and when $x = -1$.

1. $-5x + 2$

2. $-8x + 1$

3. $2x^2 - 5x + 7$

4. $3x^2 + x - 7$

5. $x^3 - 5x^2 + x$

6. $7 - x + 3x^2$

Evaluate each polynomial when $x = -2$ and when $x = 0$.

7. $\frac{1}{3}x + 5$

8. $8 - \frac{1}{4}x$

9. $x^2 - 2x + 1$

10. $5x + 6 - x^2$

11. $-3x^3 + 7x^2 - 3x - 2$

12. $-2x^3 + 5x^2 - 4x + 3$

13. *Skydiving.* During the first 13 sec of a jump, the distance S, in feet, that a skydiver falls in t seconds can be approximated by the polynomial equation

$$S = 11.12t^2.$$

In 2009, 108 U.S. skydivers fell headfirst in formation from a height of 18,000 ft. How far had they fallen 10 sec after having jumped from the plane?

Source: www.telegraph.co.uk

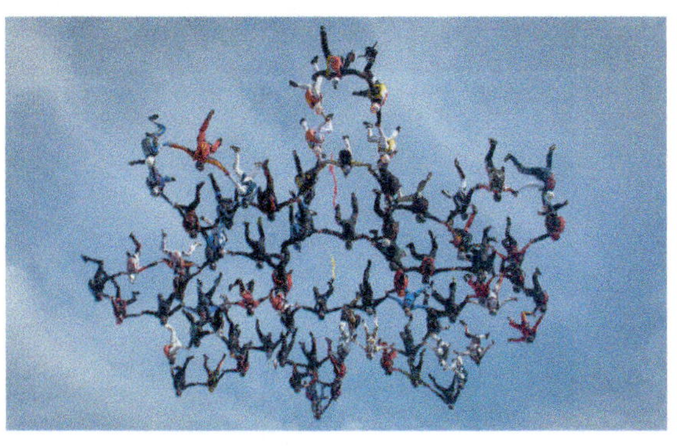

14. *Skydiving.* For jumps that exceed 13 sec, the polynomial equation

$$S = 173t - 369$$

can be used to approximate the distance S, in feet, that a skydiver has fallen in t seconds. Approximately how far has a skydiver fallen 20 sec after having jumped from a plane?

15. ***Stacking Spheres.*** In 2004, the journal *Annals of Mathematics* accepted a proof of the so-called Kepler Conjecture: that the most efficient way to pack spheres is in the shape of a square pyramid. The number N of balls in the stack is given by the polynomial equation

$$N = \frac{1}{3}x^3 + \frac{1}{2}x^2 + \frac{1}{6}x,$$

where x is the number of layers. A square pyramid with 3 layers is illustrated below. Find the number of oranges in a pyramid with 5 layers.

Source: *The New York Times* 4/6/04

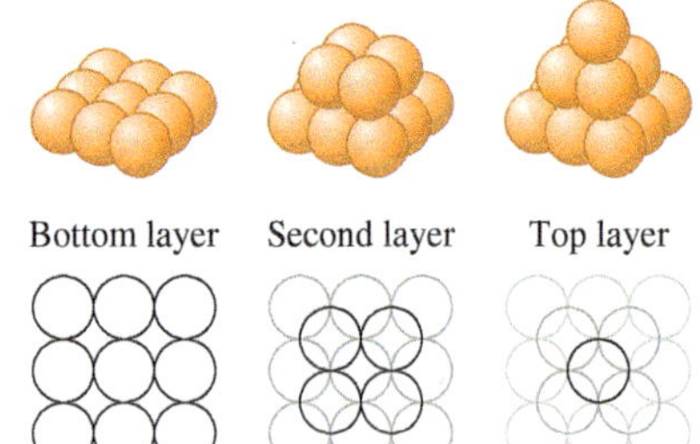

16. ***SCAD Diving.*** The distance s, in feet, traveled by a body falling freely from rest in t seconds is approximated by the polynomial equation

$$s = 16t^2.$$

The SCAD thrill ride is a 2.5-sec free fall into a net. How far does the diver fall?

Source: www.scadfreefall.co.uk

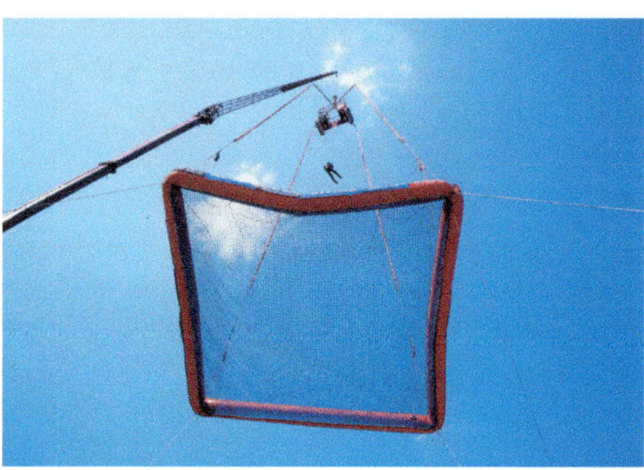

17. The graph of the polynomial equation $y = 5 - x^2$ is shown below. Use *only* the graph to estimate the value of the polynomial when $x = -3$, $x = -1$, $x = 0$, $x = 1.5$, and $x = 2$.

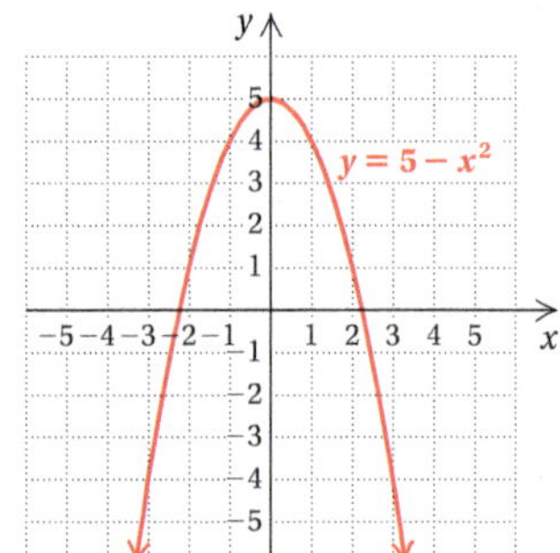

18. The graph of the polynomial equation $y = 6x^3 - 6x$ is shown below. Use *only* the graph to estimate the value of the polynomial when $x = -1$, $x = -0.5$, $x = 0.5$, $x = 1$, and $x = 1.1$.

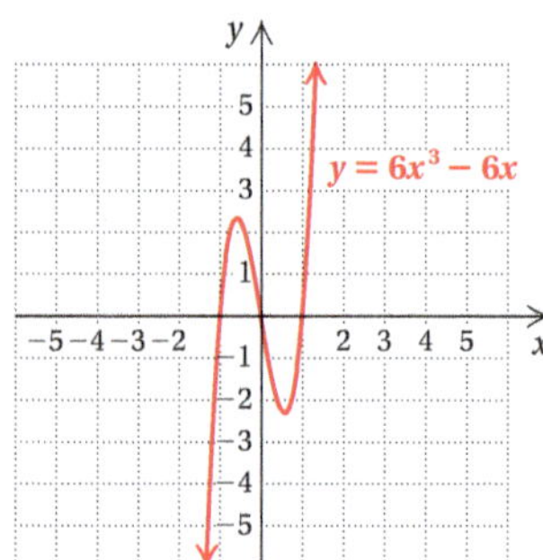

19. ***Solar Capacity.*** The average capacity C, in kilowatts (kW), of U.S. residential installations generating energy from the sun can be estimated by the polynomial equation

$$C = 0.27t + 2.97,$$

where t is the number of years since 2000.

Source: Based on data from IREC 2011 Updates & Trends

a) Use the equation to estimate the average capacity of a solar-energy residential installation in 2010.

b) Check the result of part (a) using the graph below.

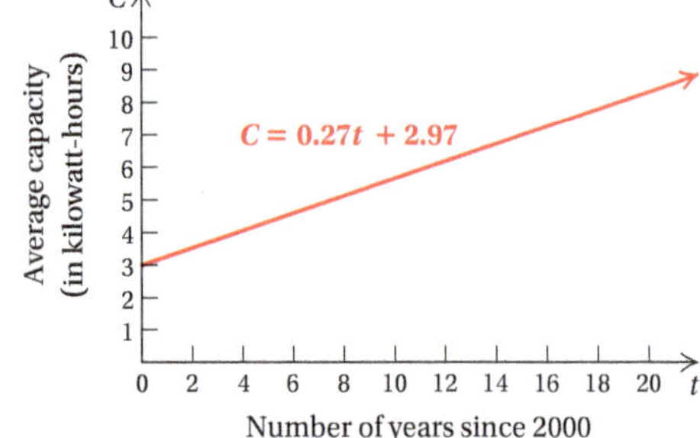

20. ***Senior Population.*** The number N, in millions, of people in the United States ages 65 and older can be estimated by the polynomial equation

$$N = 1.24t + 28.4,$$

where t is the number of years since 2000.

Source: Based on data from U.S. Census Bureau

a) Use the equation to estimate the number of people in the United States ages 65 and older in 2030.

b) Check the result of part (a) using the graph below.

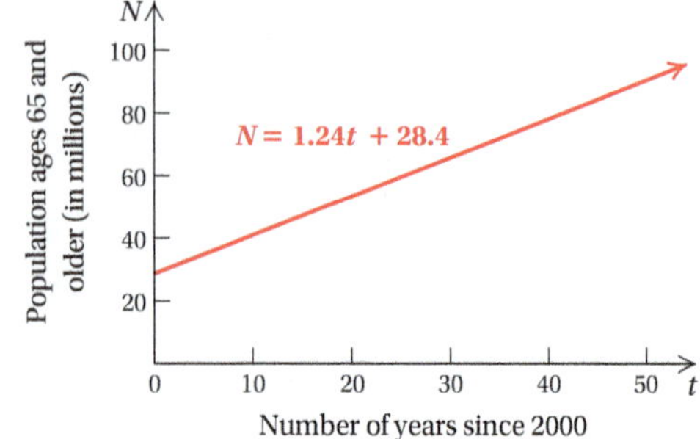

Memorizing Words. Participants in a psychology experiment were able to memorize an average of M words in t minutes, where $M = -0.001t^3 + 0.1t^2$. Use the graph below for Exercises 21–26.

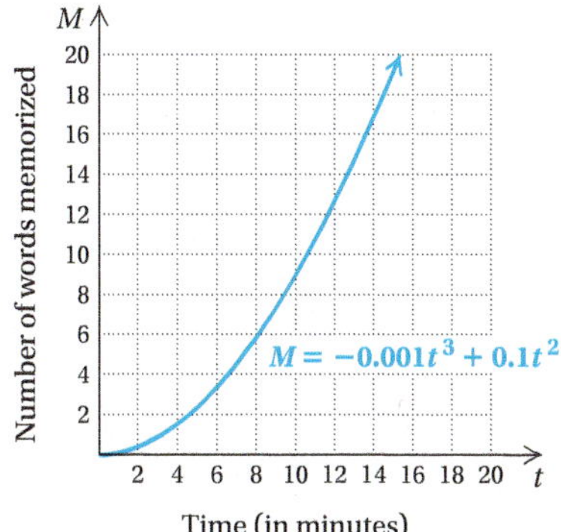

21. Estimate the number of words memorized after 10 min.

22. Estimate the number of words memorized after 14 min.

23. Find the approximate value of M for $t = 8$.

24. Find the approximate value of M for $t = 12$.

25. Estimate the value of M when t is 13.

26. Estimate the value of M when t is 7.

b Identify the terms of each polynomial.

27. $2 - 3x + x^2$

28. $2x^2 + 3x - 4$

29. $-2x^4 + \frac{1}{3}x^3 - x + 3$

30. $-\frac{2}{5}x^5 - x^3 + 6$

Classify each polynomial as a monomial, a binomial, a trinomial, or none of these.

31. $x^2 - 10x + 25$

32. $-6x^4$

33. $x^3 - 7x^2 + 2x - 4$

34. $x^2 - 9$

35. $4x^2 - 25$

36. $2x^4 - 7x^3 + x^2 + x - 6$

37. $40x$

38. $4x^2 + 12x + 9$

c Identify the coefficient of each term of the polynomial.

39. $-3x + 6$

40. $2x - 4$

41. $5x^2 + \frac{3}{4}x + 3$

42. $\frac{2}{3}x^2 - 5x + 2$

43. $-5x^4 + 6x^3 - 2.7x^2 + x - 2$

44. $7x^3 - x^2 - 4.2x + 5$

Identify the degree of each term of the polynomial and the degree of the polynomial.

45. $2x - 4$

46. $6 - 3x$

47. $3x^2 - 5x + 2$

48. $5x^3 - 2x^2 + 3$

49. $-7x^3 + 6x^2 + \frac{3}{5}x + 7$

50. $5x^4 + \frac{1}{4}x^2 - x + 2$

51. $x^2 - 3x + x^6 - 9x^4$

52. $8x - 3x^2 + 9 - 8x^3$

53. Complete the following table for the polynomial $-7x^4 + 6x^3 - x^2 + 8x - 2$.

TERM	COEFFICIENT	DEGREE OF THE TERM	DEGREE OF THE POLYNOMIAL
$-7x^4$			
$6x^3$	6		
		2	
$8x$		1	
	-2		

54. Complete the following table for the polynomial $3x^2 + x^5 - 46x^3 + 6x - 2.4 - \frac{1}{2}x^4$.

TERM	COEFFICIENT	DEGREE OF THE TERM	DEGREE OF THE POLYNOMIAL
		5	
$-\frac{1}{2}x^4$		4	
	-46		
$3x^2$		2	
	6		
-2.4			

d Identify the like terms in each polynomial.

55. $5x^3 + 6x^2 - 3x^2$

56. $3x^2 + 4x^3 - 2x^2$

57. $2x^4 + 5x - 7x - 3x^4$

58. $-3t + t^3 - 2t - 5t^3$

59. $3x^5 - 7x + 8 + 14x^5 - 2x - 9$

60. $8x^3 + 7x^2 - 11 - 4x^3 - 8x^2 - 29$

Collect like terms.

61. $2x - 5x$

62. $2x^2 + 8x^2$

63. $x - 9x$

64. $x - 5x$

65. $5x^3 + 6x^3 + 4$

66. $6x^4 - 2x^4 + 5$

67. $5x^3 + 6x - 4x^3 - 7x$

68. $3a^4 - 2a + 2a + a^4$

69. $6b^5 + 3b^2 - 2b^5 - 3b^2$

70. $2x^2 - 6x + 3x + 4x^2$

71. $\frac{1}{4}x^5 - 5 + \frac{1}{2}x^5 - 2x - 37$

72. $\frac{1}{3}x^3 + 2x - \frac{1}{6}x^3 + 4 - 16$

73. $6x^2 + 2x^4 - 2x^2 - x^4 - 4x^2$

74. $8x^2 + 2x^3 - 3x^3 - 4x^2 - 4x^2$

75. $\frac{1}{4}x^3 - x^2 - \frac{1}{6}x^2 + \frac{3}{8}x^3 + \frac{5}{16}x^3$

76. $\frac{1}{5}x^4 + \frac{1}{5} - 2x^2 + \frac{1}{10} - \frac{3}{15}x^4 + 2x^2 - \frac{3}{10}$

e Arrange each polynomial in descending order.

77. $x^5 + x + 6x^3 + 1 + 2x^2$

78. $3 + 2x^2 - 5x^6 - 2x^3 + 3x$

79. $5y^3 + 15y^9 + y - y^2 + 7y^8$

80. $9p - 5 + 6p^3 - 5p^4 + p^5$

Collect like terms and then arrange in descending order.

81. $3x^4 - 5x^6 - 2x^4 + 6x^6$

82. $-1 + 5x^3 - 3 - 7x^3 + x^4 + 5$

83. $-2x + 4x^3 - 7x + 9x^3 + 8$

84. $-6x^2 + x - 5x + 7x^2 + 1$

85. $3x + 3x + 3x - x^2 - 4x^2$

86. $-2x - 2x - 2x + x^3 - 5x^3$

87. $-x + \frac{3}{4} + 15x^4 - x - \frac{1}{2} - 3x^4$

88. $2x - \frac{5}{6} + 4x^3 + x + \frac{1}{3} - 2x$

f Identify the missing terms in each polynomial.

89. $x^3 - 27$

90. $x^5 + x$

91. $x^4 - x$

92. $5x^4 - 7x + 2$

93. $2x^3 - 5x^2 + x - 3$

94. $-6x^3$

Write each polynomial in two ways: with its missing terms and by leaving space for them.

95. $x^3 - 27$

96. $x^5 + x$

97. $x^4 - x$

98. $5x^4 - 7x + 2$

99. $2x^3 - 5x^2 + x - 3$

100. $-6x^3$

Skill Maintenance

Add. [9.3a]

101. $1 + (-20)$

102. $-\frac{2}{3} + \left(-\frac{1}{3}\right)$

103. $-4.2 + 1.95$

104. $-\frac{5}{8} + \frac{1}{4}$

Subtract. [9.4a]

105. $1 - 20$

106. $\frac{1}{8} - \frac{5}{6}$

107. $\frac{3}{8} - \left(-\frac{1}{4}\right)$

108. $5.6 - 8.2$

Multiply. [9.5a]

109. $(-6)(-3)$

110. $\left(-\frac{1}{2}\right)\left(\frac{2}{3}\right)$

111. $0.5(-1.2)$

112. $(-2)(-3)(-4)$

Divide, if possible. [9.6a, c]

113. $-600 \div (-30)$

114. $\frac{4}{5} \div \left(-\frac{1}{2}\right)$

115. $0 \div (-4)$

116. $\frac{-6.3}{-5 + 5}$

Synthesis

Collect like terms.

117. $6x^3 \cdot 7x^2 - (4x^3)^2 + (-3x^3)^2 - (-4x^2)(5x^3) - 10x^5 + 17x^6$

118. $(3x^2)^3 + 4x^2 \cdot 4x^4 - x^4(2x)^2 + ((2x)^2)^3 - 100x^2(x^2)^2$

119. Construct a polynomial in x (meaning that x is the variable) of degree 5 with four terms and coefficients that are integers.

120. What is the degree of $(5m^5)^2$?

121. A polynomial in x has degree 3. The coefficient of x^2 is 3 less than the coefficient of x^3. The coefficient of x is three times the coefficient of x^2. The remaining coefficient is 2 more than the coefficient of x^3. The sum of the coefficients is -4. Find the polynomial.

Use the CALC feature and choose VALUE on your graphing calculator to find the values in each of the following. (Refer to the Calculator Corner on p. 878.)

122. Exercise 18

123. Exercise 17

124. Exercise 22

125. Exercise 21

Addition and Subtraction of Polynomials

12.4

a ADDITION OF POLYNOMIALS

To add two polynomials, we can write a plus sign between them and then collect like terms. Depending on the situation, you may see polynomials written in descending order, ascending order, or neither. Generally, if an exercise is written in a particular order, we write the answer in that same order.

EXAMPLE 1 Add $(-3x^3 + 2x - 4)$ and $(4x^3 + 3x^2 + 2)$.

$$(-3x^3 + 2x - 4) + (4x^3 + 3x^2 + 2)$$
$$= (-3 + 4)x^3 + 3x^2 + 2x + (-4 + 2) \quad \text{Collecting like terms}$$
$$= x^3 + 3x^2 + 2x - 2$$

EXAMPLE 2 Add:

$$\left(\tfrac{2}{3}x^4 + 3x^2 - 2x + \tfrac{1}{2}\right) + \left(-\tfrac{1}{3}x^4 + 5x^3 - 3x^2 + 3x - \tfrac{1}{2}\right).$$

We have

$$\left(\tfrac{2}{3}x^4 + 3x^2 - 2x + \tfrac{1}{2}\right) + \left(-\tfrac{1}{3}x^4 + 5x^3 - 3x^2 + 3x - \tfrac{1}{2}\right)$$
$$= \left(\tfrac{2}{3} - \tfrac{1}{3}\right)x^4 + 5x^3 + (3 - 3)x^2 + (-2 + 3)x + \left(\tfrac{1}{2} - \tfrac{1}{2}\right) \quad \text{Collecting like terms}$$
$$= \tfrac{1}{3}x^4 + 5x^3 + x.$$

We can add polynomials as we do because they represent numbers. After some practice, you will be able to add mentally.

Do Margin Exercises 1–4. ▶

EXAMPLE 3 Add $(3x^2 - 2x + 2)$ and $(5x^3 - 2x^2 + 3x - 4)$.

$$(3x^2 - 2x + 2) + (5x^3 - 2x^2 + 3x - 4)$$
$$= 5x^3 + (3 - 2)x^2 + (-2 + 3)x + (2 - 4) \quad \text{You might do this step mentally.}$$
$$= 5x^3 + x^2 + x - 2 \quad \text{Then you would write only this.}$$

Do Exercises 5 and 6 on the following page. ▶

We can also add polynomials by writing like terms in columns.

EXAMPLE 4 Add $9x^5 - 2x^3 + 6x^2 + 3$ and $5x^4 - 7x^2 + 6$ and $3x^6 - 5x^5 + x^2 + 5$.

We arrange the polynomials with the like terms in columns.

$$\begin{array}{lllllll}
 & 9x^5 & & -\ 2x^3 & +\ 6x^2 & +\ 3 & \\
 & & 5x^4 & & -\ 7x^2 & +\ 6 & \text{We leave spaces for missing terms.} \\
3x^6 & -\ 5x^5 & & & +\ x^2 & +\ 5 & \\
\hline
3x^6 & +\ 4x^5 & +\ 5x^4 & -\ 2x^3 & & +\ 14 & \text{Adding}
\end{array}$$

We write the answer as $3x^6 + 4x^5 + 5x^4 - 2x^3 + 14$ without the space.

OBJECTIVES

- **a** Add polynomials.
- **b** Simplify the opposite of a polynomial.
- **c** Subtract polynomials.
- **d** Use polynomials to represent perimeter and area.

SKILL TO REVIEW

Objective 9.4a: Subtract real numbers and simplify combinations of additions and subtractions.

Simplify.

1. $-4 - (-8)$

2. $-5 - 6 + 4$

Add.

1. $(3x^2 + 2x - 2) + (-2x^2 + 5x + 5)$

2. $(-4x^5 + x^3 + 4) + (7x^4 + 2x^2)$

3. $(31x^4 + x^2 + 2x - 1) + (-7x^4 + 5x^3 - 2x + 2)$

4. $(17x^3 - x^2 + 3x + 4) + \left(-15x^3 + x^2 - 3x - \dfrac{2}{3}\right)$

Answers

Skill to Review:
1. 4 **2.** −7

Margin Exercises:
1. $x^2 + 7x + 3$
2. $-4x^5 + 7x^4 + x^3 + 2x^2 + 4$
3. $24x^4 + 5x^3 + x^2 + 1$
4. $2x^3 + \dfrac{10}{3}$

Add mentally. Try to write just the answer.

5. $(4x^2 - 5x + 3) + (-2x^2 + 2x - 4)$

6. $(3x^3 - 4x^2 - 5x + 3) + \left(5x^3 + 2x^2 - 3x - \frac{1}{2}\right)$

Add.

7.
$$\begin{array}{r} -2x^3 + 5x^2 - 2x + 4 \\ x^4 \qquad\quad + 6x^2 + 7x - 10 \\ -9x^4 + 6x^3 + x^2 \qquad\quad - 2 \\ \hline \end{array}$$

8. $-3x^3 + 5x + 2$ and $x^3 + x^2 + 5$ and $x^3 - 2x - 4$

Simplify.

9. $-(4x^3 - 6x + 3)$

10. $-(5x^4 + 3x^2 + 7x - 5)$

11. $-(14x^{10} - \frac{1}{2}x^5 + 5x^3 - x^2 + 3x)$

Subtract.

12. $(7x^3 + 2x + 4) - (5x^3 - 4)$

13. $(-3x^2 + 5x - 4) - (-4x^2 + 11x - 2)$

Answers

5. $2x^2 - 3x - 1$ 6. $8x^3 - 2x^2 - 8x + \frac{5}{2}$
7. $-8x^4 + 4x^3 + 12x^2 + 5x - 8$
8. $-x^3 + x^2 + 3x + 3$ 9. $-4x^3 + 6x - 3$
10. $-5x^4 - 3x^2 - 7x + 5$
11. $-14x^{10} + \frac{1}{2}x^5 - 5x^3 + x^2 - 3x$
12. $2x^3 + 2x + 8$ 13. $x^2 - 6x - 2$

◀ Do Exercises 7 and 8.

b OPPOSITES OF POLYNOMIALS

We can use the property of -1 to write an equivalent expression for an opposite. For example, the opposite of $x - 2y + 5$ can be written as

$$-(x - 2y + 5).$$

We find an equivalent expression by changing the sign of every term:

$$-(x - 2y + 5) = -x + 2y - 5.$$

We use this concept when we subtract polynomials.

OPPOSITES OF POLYNOMIALS

To find an equivalent polynomial for the **opposite**, or **additive inverse**, of a polynomial, change the sign of every term. This is the same as multiplying by -1.

EXAMPLE 5 Simplify: $-(x^2 - 3x + 4)$.

$$-(x^2 - 3x + 4) = -x^2 + 3x - 4$$

EXAMPLE 6 Simplify: $-(-t^3 - 6t^2 - t + 4)$.

$$-(-t^3 - 6t^2 - t + 4) = t^3 + 6t^2 + t - 4$$

EXAMPLE 7 Simplify: $-(-7x^4 - \frac{5}{9}x^3 + 8x^2 - x + 67)$.

$$-(-7x^4 - \tfrac{5}{9}x^3 + 8x^2 - x + 67) = 7x^4 + \tfrac{5}{9}x^3 - 8x^2 + x - 67$$

◀ Do Exercises 9–11.

c SUBTRACTION OF POLYNOMIALS

Recall that we can subtract a real number by adding its opposite, or additive inverse: $a - b = a + (-b)$. This allows us to subtract polynomials.

EXAMPLE 8 Subtract:

$$(9x^5 + x^3 - 2x^2 + 4) - (2x^5 + x^4 - 4x^3 - 3x^2).$$

We have

$(9x^5 + x^3 - 2x^2 + 4) - (2x^5 + x^4 - 4x^3 - 3x^2)$

$= 9x^5 + x^3 - 2x^2 + 4 + [-(2x^5 + x^4 - 4x^3 - 3x^2)]$ Adding the opposite

$= 9x^5 + x^3 - 2x^2 + 4 - 2x^5 - x^4 + 4x^3 + 3x^2$ Finding the opposite by changing the sign of *each* term

$= 7x^5 - x^4 + 5x^3 + x^2 + 4.$ Adding (collecting like terms)

◀ Do Exercises 12 and 13.

We combine steps by changing the sign of each term of the polynomial being subtracted and collecting like terms. Try to do this mentally as much as possible.

EXAMPLE 9 Subtract: $(9x^5 + x^3 - 2x) - (-2x^5 + 5x^3 + 6)$.

$(9x^5 + x^3 - 2x) - (-2x^5 + 5x^3 + 6)$

$= 9x^5 + x^3 - 2x + 2x^5 - 5x^3 - 6$ Finding the opposite by changing the sign of each term

$= 11x^5 - 4x^3 - 2x - 6$ Collecting like terms

Do Exercises 14 and 15. ▶

We can use columns to subtract. We replace coefficients with their opposites, as shown in Example 9.

EXAMPLE 10 Write in columns and subtract:

$(5x^2 - 3x + 6) - (9x^2 - 5x - 3)$.

a)
$$\begin{array}{r} 5x^2 - 3x + 6 \\ \underline{-(9x^2 - 5x - 3)} \end{array}$$
Writing like terms in columns

b)
$$\begin{array}{r} 5x^2 - 3x + 6 \\ \underline{-9x^2 + 5x + 3} \end{array}$$
Changing signs

c)
$$\begin{array}{r} 5x^2 - 3x + 6 \\ \underline{-9x^2 + 5x + 3} \\ -4x^2 + 2x + 9 \end{array}$$
Adding ■

If you can do so without error, you can arrange the polynomials in columns and write just the answer, remembering to change the signs and add.

EXAMPLE 11 Write in columns and subtract:

$(x^3 + x^2 + 2x - 12) - (-2x^3 + x^2 - 3x)$.

$$\begin{array}{rrrrl} x^3 & + x^2 & + 2x & - 12 & \\ -(-2x^3 & + x^2 & - 3x & \quad) & \text{Leaving space for the missing term} \\ \hline 3x^3 & & + 5x & - 12 & \text{Changing the signs and adding} \end{array}$$

Do Exercises 16 and 17. ▶

d POLYNOMIALS AND GEOMETRY

EXAMPLE 12 Find a polynomial for the sum of the areas of these four rectangles.

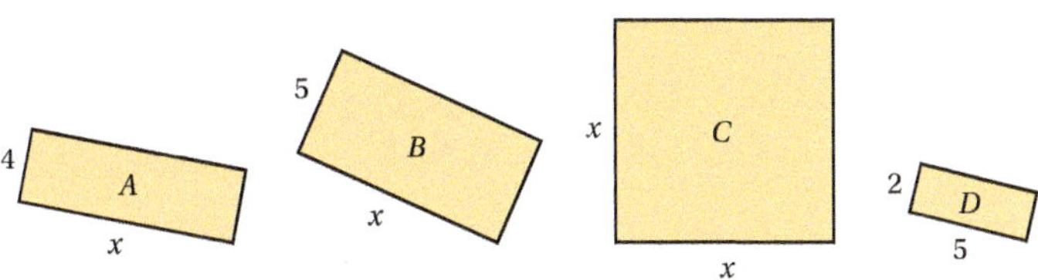

Recall that the area of a rectangle is the product of the length and the width. The sum of the areas is a sum of products. We find these products and then collect like terms.

Subtract.

GS **14.** $(-6x^4 + 3x^2 + 6) - (2x^4 + 5x^3 - 5x^2 + 7)$

$= -6x^4 + 3x^2 + 6 -$

$2x^4$ ___ $5x^3$ ___ $5x^2$ ___ 7

$=$ ___ $x^4 - 5x^3 +$ ___ x^2

$-$ ___

15. $\left(\frac{3}{2}x^3 - \frac{1}{2}x^2 + 0.3\right) - \left(\frac{1}{2}x^3 + \frac{1}{2}x^2 + \frac{4}{3}x + 1.2\right)$

Write in columns and subtract.

16. $(4x^3 + 2x^2 - 2x - 3) - (2x^3 - 3x^2 + 2)$

17. $(2x^3 + x^2 - 6x + 2) - (x^5 + 4x^3 - 2x^2 - 4x)$

Answers

14. $-8x^4 - 5x^3 + 8x^2 - 1$
15. $x^3 - x^2 - \frac{4}{3}x - 0.9$
16. $2x^3 + 5x^2 - 2x - 5$
17. $-x^5 - 2x^3 + 3x^2 - 2x + 2$

Guided Solution:
14. $-, +, -, -8, 8, 1$

18. Find a polynomial for the sums of the perimeters and of the areas of the rectangles.

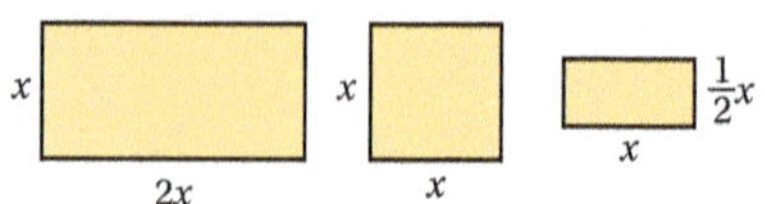

19. *Lawn Area.* An 8-ft by 8-ft shed is placed on a lawn x ft on a side. Find a polynomial for the remaining area.

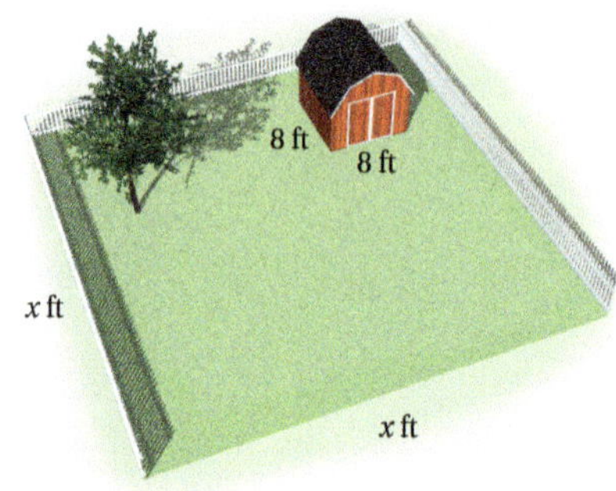

Answers

18. Sum of perimeters: $13x$; sum of areas: $\frac{7}{2}x^2$
19. $(x^2 - 64)$ ft^2

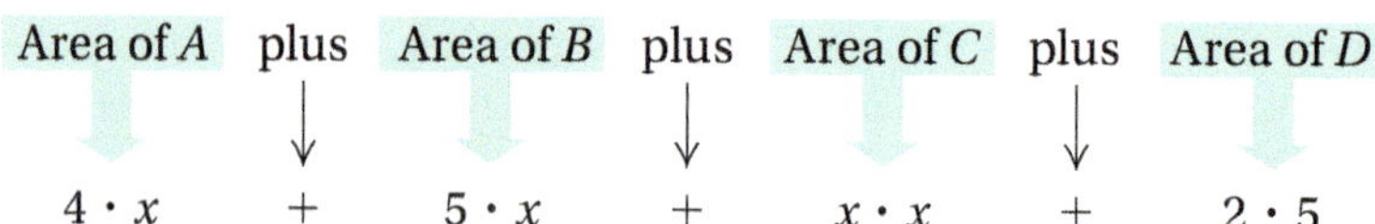

$$4 \cdot x \quad + \quad 5 \cdot x \quad + \quad x \cdot x \quad + \quad 2 \cdot 5$$

We collect like terms:

$$4x + 5x + x^2 + 10 = x^2 + 9x + 10.$$

◀ **Do Exercise 18.**

EXAMPLE 13 *Lawn Area.* A new city park is to contain a square grassy area that is x ft on a side. Within that grassy area will be a circular playground, with a radius of 15 ft, that will be mulched. To determine the amount of sod needed, find a polynomial for the grassy area.

We make a drawing, reword the problem, and write the polynomial.

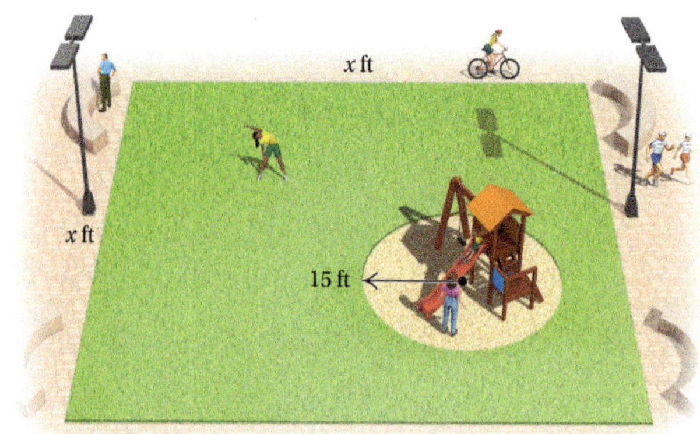

Area of square	−	Area of playground	=	Area of grass
↓	↓	↓	↓	↓
$x \cdot x$ ft^2	−	$\pi \cdot 15^2$ ft^2	=	Area of grass

Then $(x^2 - 225\pi)$ ft^2 = Area of grass.

◀ **Do Exercise 19.**

12.4 Exercise Set

For Extra Help MyMathLab®

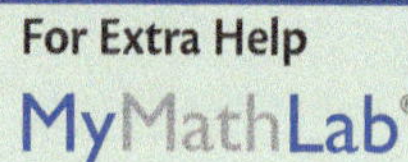

✓ Reading Check

Determine whether each statement is true or false.

RC1. To find the opposite of a polynomial, we need change only the sign of the first term.

RC2. We can subtract a polynomial by adding its opposite.

RC3. The sum of two binomials is always a binomial.

RC4. The area of a rectangle is the sum of its length and its width.

a Add.

1. $(3x + 2) + (-4x + 3)$

2. $(6x + 1) + (-7x + 2)$

3. $(-6x + 2) + \left(x^2 + \frac{1}{2}x - 3\right)$

4. $\left(x^2 - \frac{5}{3}x + 4\right) + (8x - 9)$

5. $(x^2 - 9) + (x^2 + 9)$

6. $(x^3 + x^2) + (2x^3 - 5x^2)$

7. $(3x^2 - 5x + 10) + (2x^2 + 8x - 40)$

8. $(6x^4 + 3x^3 - 1) + (4x^2 - 3x + 3)$

9. $(1.2x^3 + 4.5x^2 - 3.8x) + (-3.4x^3 - 4.7x^2 + 23)$

10. $(0.5x^4 - 0.6x^2 + 0.7) + (2.3x^4 + 1.8x - 3.9)$

11. $(1 + 4x + 6x^2 + 7x^3) + (5 - 4x + 6x^2 - 7x^3)$

12. $(3x^4 - 6x - 5x^2 + 5) + (6x^2 - 4x^3 - 1 + 7x)$

13. $\left(\frac{1}{4}x^4 + \frac{2}{3}x^3 + \frac{5}{8}x^2 + 7\right) + \left(-\frac{3}{4}x^4 + \frac{3}{8}x^2 - 7\right)$

14. $\left(\frac{1}{3}x^9 + \frac{1}{5}x^5 - \frac{1}{2}x^2 + 7\right) +$
$\left(-\frac{1}{5}x^9 + \frac{1}{4}x^4 - \frac{3}{5}x^5 + \frac{3}{4}x^2 + \frac{1}{2}\right)$

15. $(0.02x^5 - 0.2x^3 + x + 0.08) +$
$(-0.01x^5 + x^4 - 0.8x - 0.02)$

16. $(0.03x^6 + 0.05x^3 + 0.22x + 0.05) +$
$\left(\frac{7}{100}x^6 - \frac{3}{100}x^3 + 0.5\right)$

17. $(9x^8 - 7x^4 + 2x^2 + 5) + (8x^7 + 4x^4 - 2x) +$
$(-3x^4 + 6x^2 + 2x - 1)$

18. $(4x^5 - 6x^3 - 9x + 1) + (6x^3 + 9x^2 + 9x) +$
$(-4x^3 + 8x^2 + 3x - 2)$

19.
$$\begin{array}{rrrrr}
0.15x^4 & + 0.10x^3 & - 0.9x^2 & & \\
 & - 0.01x^3 & + 0.01x^2 & + x & \\
1.25x^4 & & + 0.11x^2 & & + 0.01 \\
 & 0.27x^3 & & & + 0.99 \\
\underline{-0.35x^4} & & \underline{+ \quad 15x^2} & & \underline{- 0.03}
\end{array}$$

20.
$$\begin{array}{rrrrr}
0.05x^4 & + 0.12x^3 & - 0.5x^2 & & \\
 & - 0.02x^3 & + 0.02x^2 & + 2x & \\
1.5x^4 & & + 0.01x^2 & & + 0.15 \\
 & 0.25x^3 & & & + 0.85 \\
\underline{-0.25x^4} & & \underline{+ \quad 10x^2} & & \underline{- 0.04}
\end{array}$$

b Simplify.

21. $-(-5x)$

22. $-(x^2 - 3x)$

23. $-\left(-x^2 + \frac{3}{2}x - 2\right)$

24. $-\left(-4x^3 - x^2 - \frac{1}{4}x\right)$

25. $-(12x^4 - 3x^3 + 3)$

26. $-(4x^3 - 6x^2 - 8x + 1)$

27. $-(3x - 7)$

28. $-(-2x + 4)$

29. $-(4x^2 - 3x + 2)$

30. $-(-6a^3 + 2a^2 - 9a + 1)$

31. $-\left(-4x^4 + 6x^2 + \frac{3}{4}x - 8\right)$

32. $-(-5x^4 + 4x^3 - x^2 + 0.9)$

c Subtract.

33. $(3x + 2) - (-4x + 3)$

34. $(6x + 1) - (-7x + 2)$

35. $(-6x + 2) - (x^2 + x - 3)$

36. $(x^2 - 5x + 4) - (8x - 9)$

37. $(x^2 - 9) - (x^2 + 9)$

38. $(x^3 + x^2) - (2x^3 - 5x^2)$

39. $(6x^4 + 3x^3 - 1) - (4x^2 - 3x + 3)$

40. $(-4x^2 + 2x) - (3x^3 - 5x^2 + 3)$

41. $(1.2x^3 + 4.5x^2 - 3.8x) - (-3.4x^3 - 4.7x^2 + 23)$

42. $(0.5x^4 - 0.6x^2 + 0.7) - (2.3x^4 + 1.8x - 3.9)$

43. $\left(\frac{5}{8}x^3 - \frac{1}{4}x - \frac{1}{3}\right) - \left(-\frac{1}{8}x^3 + \frac{1}{4}x - \frac{1}{3}\right)$

44. $\left(\frac{1}{5}x^3 + 2x^2 - 0.1\right) - \left(-\frac{2}{5}x^3 + 2x^2 + 0.01\right)$

45. $(0.08x^3 - 0.02x^2 + 0.01x) - (0.02x^3 + 0.03x^2 - 1)$

46. $(0.8x^4 + 0.2x - 1) - \left(\frac{7}{10}x^4 + \frac{1}{5}x - 0.1\right)$

Subtract.

47. $\begin{array}{r} x^2 + 5x + 6 \\ -(x^2 + 2x) \\ \hline \end{array}$

48. $\begin{array}{r} x^3 \qquad\quad + 1 \\ -(x^3 + x^2 \qquad) \\ \hline \end{array}$

49. $\begin{array}{r} 5x^4 + 6x^3 - 9x^2 \qquad\qquad \\ -(-6x^4 - 6x^3 \qquad\qquad + 8x + 9) \\ \hline \end{array}$

50. $\begin{array}{r} 5x^4 \qquad\quad + 6x^2 - 3x + 6 \\ -(\qquad 6x^3 + 7x^2 - 8x - 9) \\ \hline \end{array}$

51. $\begin{array}{r} x^5 \qquad\qquad\qquad\qquad\quad - 1 \\ -(x^5 - x^4 + x^3 - x^2 + x - 1) \\ \hline \end{array}$

52. $\begin{array}{r} x^5 + x^4 - x^3 + x^2 - x + 2 \\ -(x^5 - x^4 + x^3 - x^2 - x + 2) \\ \hline \end{array}$

Solve.

Find a polynomial for the perimeter of each figure.

53.

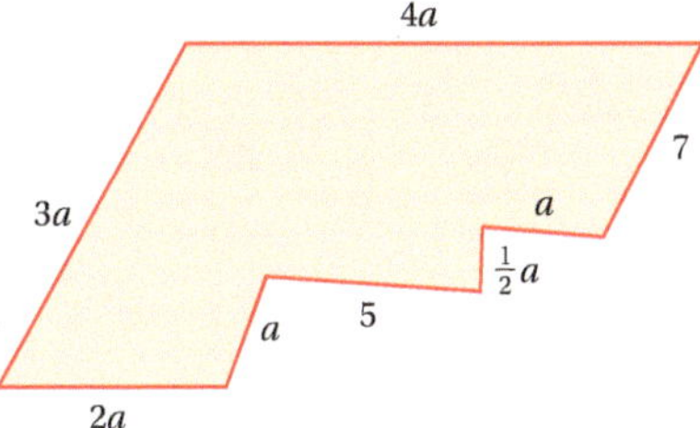

54.

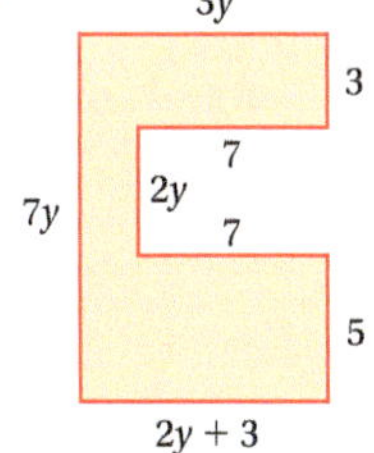

55. Find a polynomial for the sum of the areas of these rectangles.

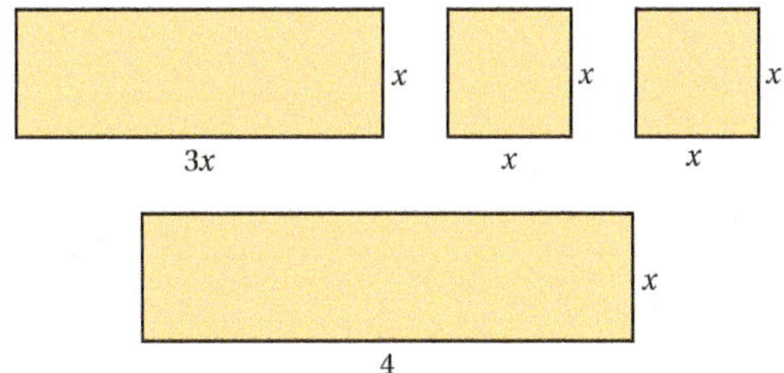

56. Find a polynomial for the sum of the areas of these circles.

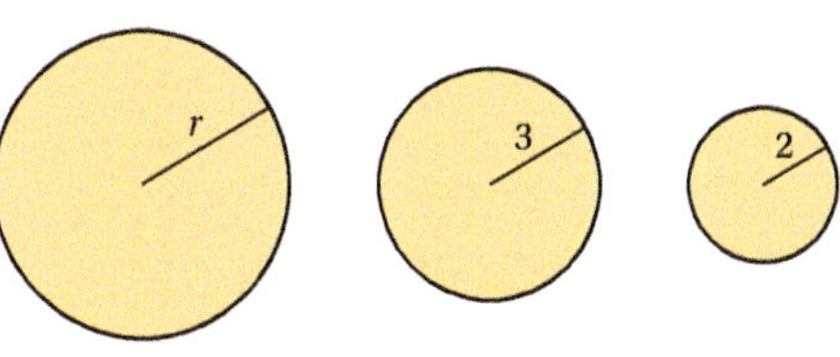

Find two algebraic expressions for the area of each figure. First, regard the figure as one large rectangle, and then regard the figure as a sum of four smaller rectangles.

57.

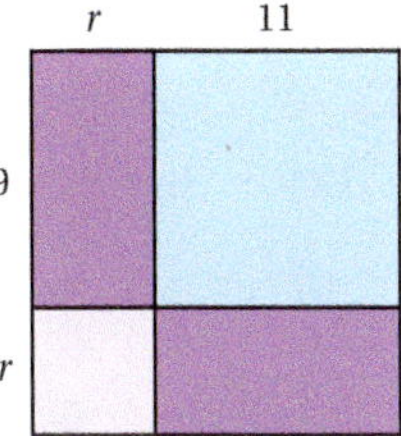

58.

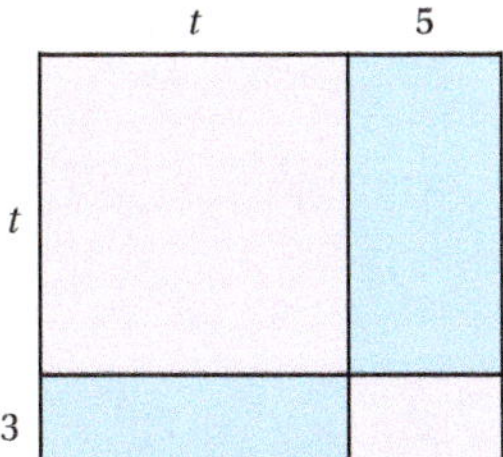

59.

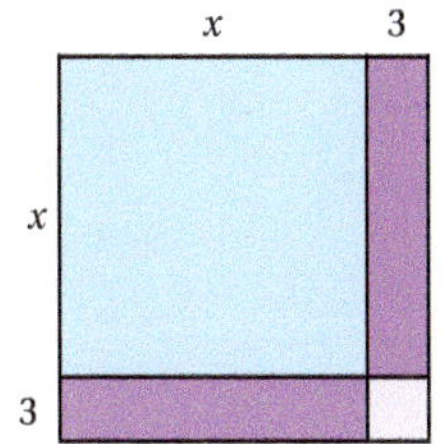

60.

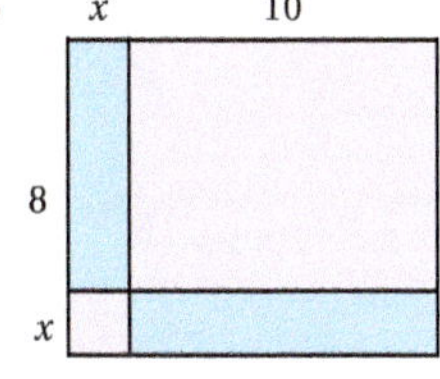

Find a polynomial for the shaded area of each figure.

61.

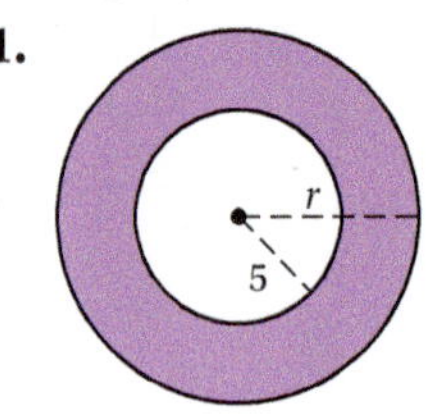

62.

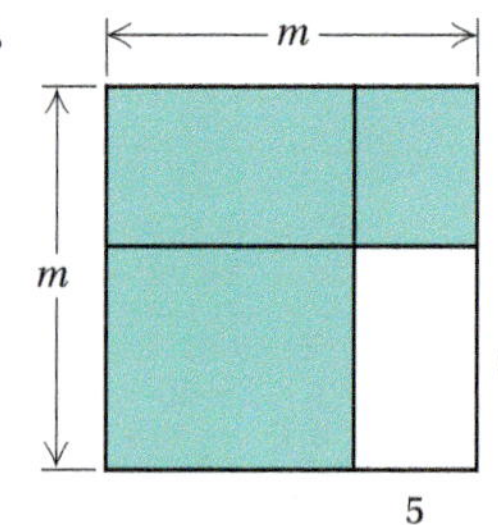

63.

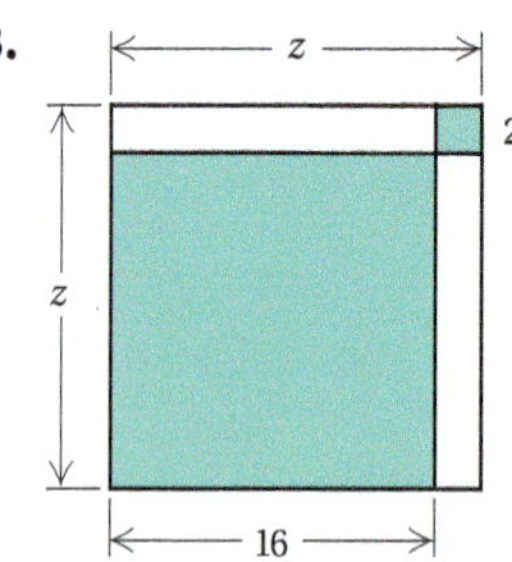

64.

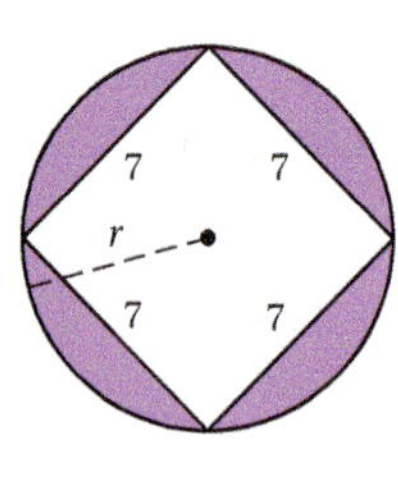

Skill Maintenance

Solve. [10.3b]

65. $8x + 3x = 66$

66. $5x - 7x = 38$

67. $\frac{3}{8}x + \frac{1}{4} - \frac{3}{4}x = \frac{11}{16} + x$

68. $5x - 4 = 26 - x$

69. $1.5x - 2.7x = 22 - 5.6x$

70. $3x - 3 = -4x + 4$

Solve. [10.3c]

71. $6(y - 3) - 8 = 4(y + 2) + 5$

72. $8(5x + 2) = 7(6x - 3)$

Solve. [10.7e]

73. $3x - 7 \leq 5x + 13$

74. $2(x - 4) > 5(x - 3) + 7$

Synthesis

Find a polynomial for the surface area of each right rectangular solid.

75.

76.

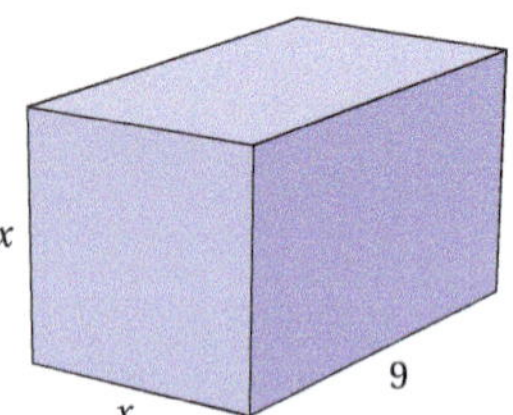

77.

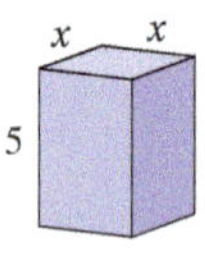

78.

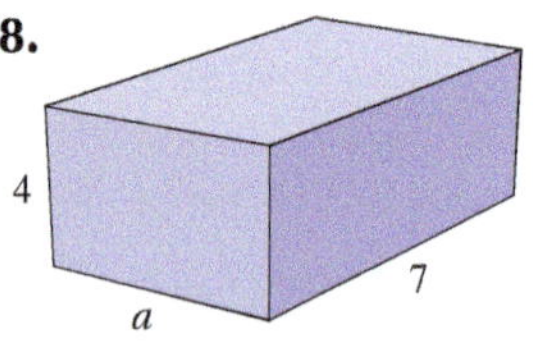

79. Find $(y - 2)^2$ using the four parts of this square.

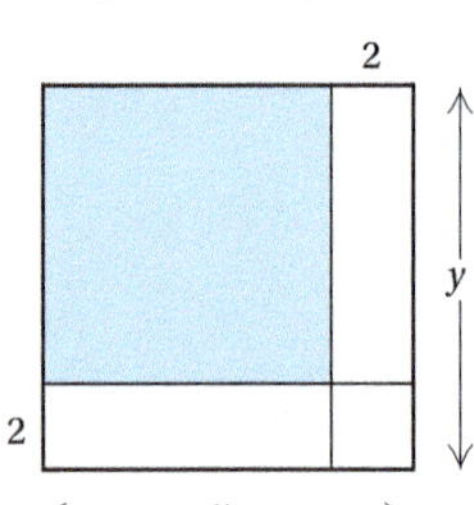

Simplify.

80. $(3x^2 - 4x + 6) - (-2x^2 + 4) + (-5x - 3)$

81. $(7y^2 - 5y + 6) - (3y^2 + 8y - 12) + (8y^2 - 10y + 3)$

82. $(-4 + x^2 + 2x^3) - (-6 - x + 3x^3) - (-x^2 - 5x^3)$

83. $(-y^4 - 7y^3 + y^2) + (-2y^4 + 5y - 2) - (-6y^3 + y^2)$

Mid-Chapter Review

Concept Reinforcement

Determine whether each statement is true or false.

_______________ **1.** a^n and a^{-n} are reciprocals. [12.1f]

_______________ **2.** $x^2 \cdot x^3 = x^6$ [12.1d]

_______________ **3.** $-5y^4$ and $-5y^2$ are like terms. [12.3d]

_______________ **4.** $4920^0 = 1$ [12.1b]

Guided Solutions

 Fill in each blank with the expression or operation sign that creates a correct statement or solution.

5. Collect like terms: $4w^3 + 6w - 8w^3 - 3w$. [12.3d]

$$4w^3 + 6w - 8w^3 - 3w = (4 - 8)\square + (6 - 3)\square$$
$$= \square w^3 + \square w$$

6. Subtract: $(3y^4 - y^2 + 11) - (y^4 - 4y^2 + 5)$. [12.4c]

$$(3y^4 - y^2 + 11) - (y^4 - 4y^2 + 5) = 3y^4 - y^2 + 11 \square y^4 \square 4y^2 \square 5$$
$$= \square y^4 + \square y^2 + \square$$

Mixed Review

Evaluate. [12.1b, c]

7. z^1

8. 4.56^0

9. a^5, when $a = -2$

10. $-x^3$, when $x = -1$

Multiply and simplify. [12.1d, f]

11. $5^3 \cdot 5^4$

12. $(3a)^2(3a)^7$

13. $x^{-8} \cdot x^5$

14. $t^4 \cdot t^{-4}$

Divide and simplify. [12.1e, f]

15. $\frac{7^8}{7^4}$

16. $\frac{x}{x^3}$

17. $\frac{w^5}{w^{-3}}$

18. $\frac{y^{-6}}{y^{-2}}$

Simplify. [12.2a, b]

19. $(3^5)^3$

20. $(x^{-3}y^2)^{-6}$

21. $\left(\frac{a^4}{5}\right)^6$

22. $\left(\frac{2y^3}{xz^2}\right)^{-2}$

Convert to scientific notation. [12.2c]

23. 25,430,000

24. 0.00012

Convert to decimal notation. [12.2c]

25. 3.6×10^{-5}

26. 1.44×10^8

Multiply or divide and write scientific notation for the result. [12.2d]

27. $(3 \times 10^6)(2 \times 10^{-3})$

28. $\dfrac{1.2 \times 10^{-4}}{2.4 \times 10^2}$

Evaluate the polynomial when $x = -3$ and when $x = 2$. [12.3a]

29. $-3x + 7$

30. $x^3 - 2x + 5$

Collect like terms and then arrange in descending order. [12.3e]

31. $3x - 2x^5 + x - 5x^2 + 2$

32. $4x^3 - 9x^2 - 2x^3 + x^2 + 8x^6$

Identify the degree of each term of the polynomial and the degree of the polynomial. [12.3c]

33. $5x^3 - x + 4$

34. $2x - x^4 + 3x^6$

Classify the polynomial as a monomial, a binomial, a trinomial, or none of these. [12.3b]

35. $x - 9$

36. $x^5 - 2x^3 + 6x^2$

Add or subtract. [12.4a, c]

37. $(3x^2 - 1) + (5x^2 + 6)$

38. $(x^3 + 2x - 5) + (4x^3 - 2x^2 - 6)$

39. $(5x - 8) - (9x + 2)$

40. $(0.1x^2 - 2.4x + 3.6) - (0.5x^2 + x - 5.4)$

41. Find a polynomial for the sum of the areas of these rectangles. [12.4d]

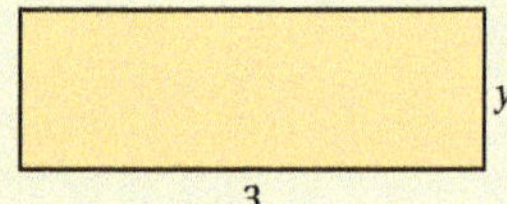

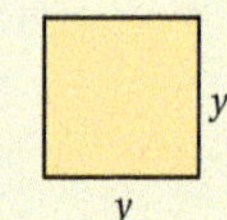

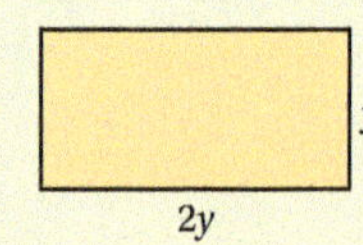

Understanding Through Discussion and Writing

42. Suppose that the length of a side of a square is three times the length of a side of a second square. How do the areas of the squares compare? Why? [12.1d]

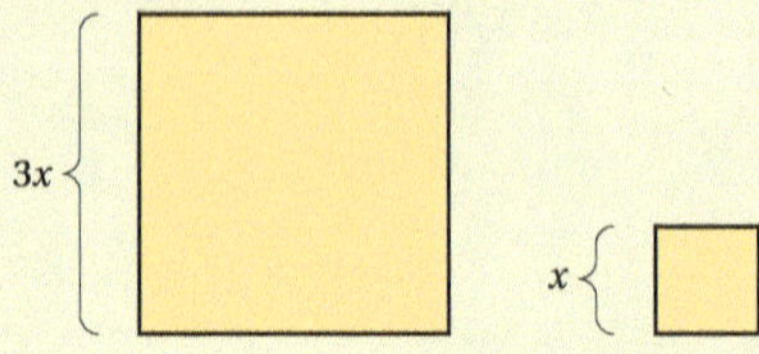

43. Suppose that the length of a side of a cube is twice the length of a side of a second cube. How do the volumes of the cubes compare? Why? [12.1d]

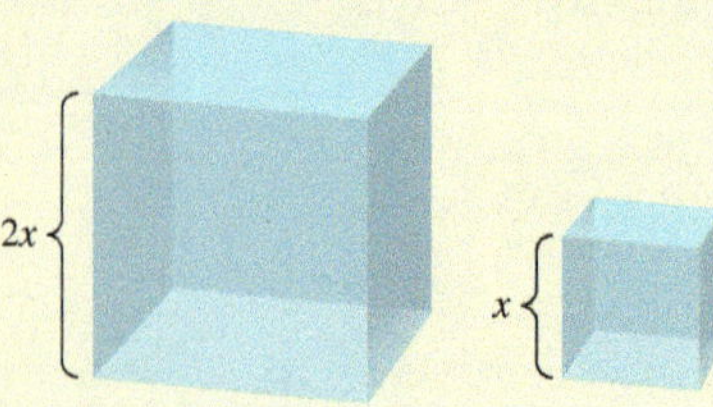

44. Explain in your own words when exponents should be added and when they should be multiplied. [12.1d], [12.2a]

45. Without performing actual computations, explain why 3^{-29} is smaller than 2^{-29}. [12.1f]

46. Is it better to evaluate a polynomial before or after like terms have been collected? Why? [12.3a, d]

47. Is the sum of two binomials ever a trinomial? Why or why not? [12.3b], [12.4a]

STUDYING FOR SUCCESS *As You Study*

- ☐ Find a quiet place to study.
- ☐ Be disciplined in your use of electronics such as video games, the Internet, and television.
- ☐ Pace yourself. It is usually better to study for shorter periods several times a week than to study in one marathon session each week.

Multiplication of Polynomials

12.5

OBJECTIVES

- **a** Multiply monomials.
- **b** Multiply a monomial and any polynomial.
- **c** Multiply two binomials.
- **d** Multiply any two polynomials.

SKILL TO REVIEW

Objective 9.7c: Use the distributive laws to multiply expressions like 8 and $x - y$.

Multiply.

1. $3(x - 5)$
2. $2(3y + 4z - 1)$

We now multiply polynomials using techniques based, for the most part, on the distributive laws, but also on the associative and commutative laws.

a MULTIPLYING MONOMIALS

Consider $(3x)(4x)$. We multiply as follows:

$$\begin{aligned}(3x)(4x) &= 3 \cdot x \cdot 4 \cdot x && \text{By the associative law of multiplication}\\ &= 3 \cdot 4 \cdot x \cdot x && \text{By the commutative law of multiplication}\\ &= (3 \cdot 4)(x \cdot x) && \text{By the associative law}\\ &= 12x^2. && \text{Using the product rule for exponents}\end{aligned}$$

MULTIPLYING MONOMIALS

To find an equivalent expression for the product of two monomials, multiply the coefficients and then multiply the variables using the product rule for exponents.

EXAMPLES Multiply.

1. $5x \cdot 6x = (5 \cdot 6)(x \cdot x)$ By the associative and commutative laws
   $= 30x^2$ Multiplying the coefficients and multiplying the variables
2. $(3x)(-x) = (3x)(-1x) = (3)(-1)(x \cdot x) = -3x^2$
3. $(-7y^5)(4y^3) = (-7 \cdot 4)(y^5 \cdot y^3)$
   $= -28y^{5+3} = -28y^8$ Adding exponents

After some practice, you will be able to multiply mentally.

Do Margin Exercises 1–8. ▶

Multiply.

1. $(3x)(-5)$
2. $(-x) \cdot x$
3. $(-x)(-x)$
4. $(-x^2)(x^3)$
5. $3x^5 \cdot 4x^2$
6. $(4y^5)(-2y^6)$
7. $(-7y^4)(-y)$
8. $7x^5 \cdot 0$

b MULTIPLYING A MONOMIAL AND ANY POLYNOMIAL

To multiply a monomial, such as $2x$, and a binomial, such as $5x + 3$, we use a distributive law and multiply each term of $5x + 3$ by $2x$:

$$\begin{aligned}2x(5x + 3) &= (2x)(5x) + (2x)(3) && \text{Using a distributive law}\\ &= 10x^2 + 6x. && \text{Multiplying the monomials}\end{aligned}$$

Answers

Skill to Review:
1. $3x - 15$ 2. $6y + 8z - 2$

Margin Exercises:
1. $-15x$ 2. $-x^2$ 3. x^2 4. $-x^5$
5. $12x^7$ 6. $-8y^{11}$ 7. $7y^5$ 8. 0

EXAMPLE 4 Multiply: $5x(2x^2 - 3x + 4)$.

$$5x(2x^2 - 3x + 4) = (5x)(2x^2) - (5x)(3x) + (5x)(4)$$
$$= 10x^3 - 15x^2 + 20x$$

MULTIPLYING A MONOMIAL AND A POLYNOMIAL

To multiply a monomial and a polynomial, multiply each term of the polynomial by the monomial.

EXAMPLE 5 Multiply: $-2x^2(x^3 - 7x^2 + 10x - 4)$.

$$-2x^2(x^3 - 7x^2 + 10x - 4)$$
$$= (-2x^2)(x^3) - (-2x^2)(7x^2) + (-2x^2)(10x) - (-2x^2)(4)$$
$$= -2x^5 + 14x^4 - 20x^3 + 8x^2$$

◀ Do Exercises 9–11.

Multiply.

9. $4x(2x + 4)$

10. $3t^2(-5t + 2)$

11. $-5x^3(x^3 + 5x^2 - 6x + 8)$

12. a) Multiply: $(y + 2)(y + 7)$. GS

$(y + 2)(y + 7)$
$= y \cdot (y + 7) + 2 \cdot (\quad)$
$= y \cdot y + y \cdot \quad + 2 \cdot y + 2 \cdot \quad$
$= \quad + 7y + 2y + \quad$
$= y^2 + \quad + 14$

b) Write an algebraic expression that represents the total area of the four smaller rectangles in the figure shown here.

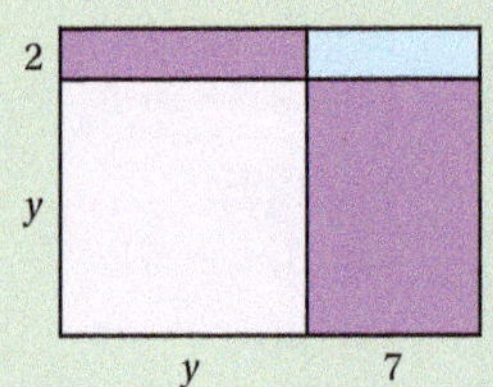

The area is $(y + 7)(y + \quad)$, or, from part (a), $y^2 + \quad + 14$.

C MULTIPLYING TWO BINOMIALS

To find an equivalent expression for the product of two binomials, we use the distributive laws more than once. In Example 6, we use a distributive law three times.

EXAMPLE 6 Multiply: $(x + 5)(x + 4)$.

$(x + 5)(x + 4) = x(x + 4) + 5(x + 4)$ Using a distributive law

$= x \cdot x + x \cdot 4 + 5 \cdot x + 5 \cdot 4$ Using a distributive law twice

$= x^2 + 4x + 5x + 20$ Multiplying the monomials

$= x^2 + 9x + 20$ Collecting like terms

To visualize the product in Example 6, consider a rectangle of length $x + 5$ and width $x + 4$.

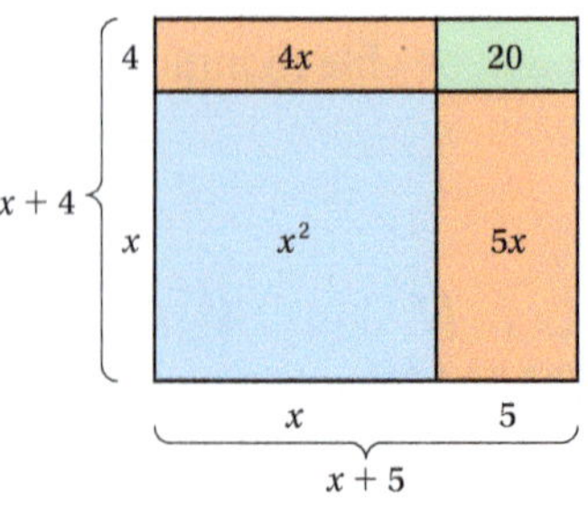

The total area can be expressed as $(x + 5)(x + 4)$ or, by adding the four smaller areas, $x^2 + 4x + 5x + 20$, or $x^2 + 9x + 20$.

◀ Do Exercises 12–14.

Multiply.

13. $(x + 8)(x + 5)$

14. $(x + 5)(x - 4)$

Answers

9. $8x^2 + 16x$ **10.** $-15t^3 + 6t^2$
11. $-5x^6 - 25x^5 + 30x^4 - 40x^3$
12. **(a)** $y^2 + 9y + 14$; **(b)** $(y + 2)(y + 7)$, or $y^2 + 2y + 7y + 14$, or $y^2 + 9y + 14$
13. $x^2 + 13x + 40$ **14.** $x^2 + x - 20$

Guided Solution:
12. **(a)** $y + 7, 7, 7, y^2, 14, 9y$; **(b)** $2, 9y$

EXAMPLE 7 Multiply: $(4x + 3)(x - 2)$.

$$(4x + 3)(x - 2) = 4x(x - 2) + 3(x - 2)$$ Using a distributive law

$$= 4x \cdot x - 4x \cdot 2 + 3 \cdot x - 3 \cdot 2$$ Using a distributive law twice

$$= 4x^2 - 8x + 3x - 6$$ Multiplying the monomials

$$= 4x^2 - 5x - 6$$ Collecting like terms

Do Exercises 15 and 16. ▶

Multiply.

15. $(5x + 3)(x - 4)$

16. $(2x - 3)(3x - 5)$

d MULTIPLYING ANY TWO POLYNOMIALS

Let's consider the product of a binomial and a trinomial. We use a distributive law four times. You may see ways to skip some steps and do the work mentally.

EXAMPLE 8 Multiply: $(x^2 + 2x - 3)(x^2 + 4)$.

$$(x^2 + 2x - 3)(x^2 + 4) = x^2(x^2 + 4) + 2x(x^2 + 4) - 3(x^2 + 4)$$
$$= x^2 \cdot x^2 + x^2 \cdot 4 + 2x \cdot x^2 + 2x \cdot 4 - 3 \cdot x^2 - 3 \cdot 4$$
$$= x^4 + 4x^2 + 2x^3 + 8x - 3x^2 - 12$$
$$= x^4 + 2x^3 + x^2 + 8x - 12$$

Do Exercises 17 and 18. ▶

Multiply.

17. $(x^2 + 3x - 4)(x^2 + 5)$

GS 18. $(3y^2 - 7)(2y^3 - 2y + 5)$
$= 3y^2(2y^3 - 2y + 5) - \_\_\_\_(2y^3 - 2y + 5)$
$= 6y^5 - \_\_\_\_ + 15y^2 - 14y^3 + \_\_\_\_ - 35$
$= 6y^5 - \_\_\_\_ + 15y^2 + 14y - 35$

PRODUCT OF TWO POLYNOMIALS

To multiply two polynomials P and Q, select one of the polynomials—say, P. Then multiply each term of P by every term of Q and collect like terms.

To use columns for long multiplication, multiply each term in the top row by every term in the bottom row. We write like terms in columns, and then add the results. Such multiplication is like multiplying with whole numbers.

3 2 1	300 + 20 + 1	
× 1 2	× 10 + 2	
6 4 2	600 + 40 + 2	Multiplying the top row by 2
3 2 1	3000 + 200 + 10	Multiplying the top row by 10
3 8 5 2	3000 + 800 + 50 + 2	Adding

EXAMPLE 9 Multiply: $(4x^3 - 2x^2 + 3x)(x^2 + 2x)$.

$4x^3 - 2x^2 + 3x$	
$x^2 + 2x$	
$8x^4 - 4x^3 + 6x^2$	Multiplying the top row by $2x$
$4x^5 - 2x^4 + 3x^3$	Multiplying the top row by x^2
$4x^5 + 6x^4 - x^3 + 6x^2$	Collecting like terms
↑ ↑ ↑ ↑	Line up like terms in columns.

Answers

15. $5x^2 - 17x - 12$ 16. $6x^2 - 19x + 15$
17. $x^4 + 3x^3 + x^2 + 15x - 20$
18. $6y^5 - 20y^3 + 15y^2 + 14y - 35$

Guided Solution:
18. $7, 6y^3, 14y, 20y^3$

EXAMPLE 10 Multiply: $(2x^2 + 3x - 4)(2x^2 - x + 3)$.

$$\begin{array}{rrrrrl}
 & & 2x^2 & + 3x & - 4 & \\
 & & 2x^2 & - x & + 3 & \\
\hline
 & & 6x^2 & + 9x & - 12 & \quad \text{Multiplying by 3} \\
 & -2x^3 & - 3x^2 & + 4x & & \quad \text{Multiplying by } -x \\
4x^4 & + 6x^3 & - 8x^2 & & & \quad \text{Multiplying by } 2x^2 \\
\hline
4x^4 & + 4x^3 & - 5x^2 & + 13x & - 12 & \quad \text{Collecting like terms}
\end{array}$$

◀ **Do Exercise 19.**

EXAMPLE 11 Multiply: $(5x^3 - 3x + 4)(-2x^2 - 3)$.

If terms are missing, it helps to leave spaces for them and align like terms in columns as we multiply.

$$\begin{array}{rrrrrl}
 & 5x^3 & & - 3x & + 4 & \\
 & & -2x^2 & & - 3 & \\
\hline
 & -15x^3 & & + 9x & - 12 & \quad \text{Multiplying by } -3 \\
-10x^5 & + 6x^3 & - 8x^2 & & & \quad \text{Multiplying by } -2x^2 \\
\hline
-10x^5 & - 9x^3 & - 8x^2 & + 9x & - 12 & \quad \text{Collecting like terms}
\end{array}$$

◀ **Do Exercises 20 and 21.**

19. Multiply.

$$\begin{array}{r} 3x^2 - 2x - 5 \\ 2x^2 + x - 2 \\ \hline \end{array}$$

Multiply.

20. $$\begin{array}{r} 3x^2 - 2x + 4 \\ x + 5 \\ \hline \end{array}$$

21. $$\begin{array}{r} -5x^2 + 4x + 2 \\ -4x^2 - 8 \\ \hline \end{array}$$

Answers

19. $6x^4 - x^3 - 18x^2 - x + 10$
20. $3x^3 + 13x^2 - 6x + 20$
21. $20x^4 - 16x^3 + 32x^2 - 32x - 16$

CALCULATOR CORNER

Checking Multiplication of Polynomials A partial check of multiplication of polynomials can be performed graphically. Consider the product $(x + 3)(x - 2) = x^2 + x - 6$. We will use two graph styles to determine whether this product is correct. First, we press MODE and select the SEQUENTIAL mode.

Next, on the Y= screen, we enter $y_1 = (x + 3)(x - 2)$ and $y_2 = x^2 + x - 6$. We will select the line-graph style for y_1 and the path style for y_2. To select these graph styles, we use ◀ to position the cursor over the icon to the left of the equation and press ENTER repeatedly until the desired style of icon appears, as shown below. Then we graph the equations.

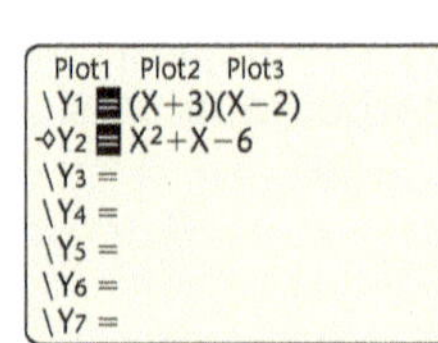

$y_1 = (x + 3)(x - 2)$,
$y_2 = x^2 + x - 6$

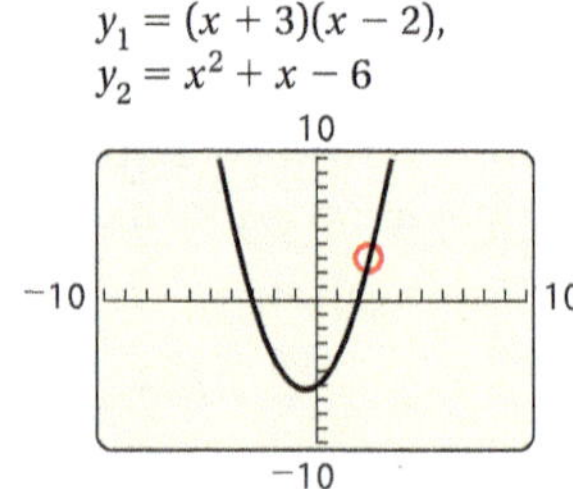

The graphing calculator will graph y_1 first as a solid line. Then it will graph y_2 as the circular cursor traces the leading edge of the graph, allowing us to determine visually whether the graphs coincide. In this case, the graphs appear to coincide, so the factorization is probably correct.

A table of values can also be used as a check.

EXERCISES: Determine graphically whether each product is correct.

1. $(x + 5)(x + 4) = x^2 + 9x + 20$

2. $(4x + 3)(x - 2) = 4x^2 - 5x - 6$

3. $(5x + 3)(x - 4) = 5x^2 + 17x - 12$

4. $(2x - 3)(3x - 5) = 6x^2 - 19x - 15$

12.5 Exercise Set

For Extra Help MyMathLab® MathXL® PRACTICE WATCH READ REVIEW

Reading Check

Match each expression with an equivalent expression from the column on the right. Choices may be used more than once or not at all.

RC1. $8x \cdot 2x$

RC2. $(-16x)(-x)$

RC3. $2x(8x - 1)$

RC4. $(2x - 1)(8x + 1)$

a) $16x^2$
b) $-16x^2$
c) $16x^2 - 1$
d) $16x^2 - 2x$
e) $16x^2 - 6x - 1$

a Multiply.

1. $(8x^2)(5)$ **2.** $(4x^2)(-2)$ **3.** $(-x^2)(-x)$ **4.** $(-x^3)(x^2)$

5. $(8x^5)(4x^3)$ **6.** $(10a^2)(2a^2)$ **7.** $(0.1x^6)(0.3x^5)$ **8.** $(0.3x^4)(-0.8x^6)$

9. $\left(-\frac{1}{5}x^3\right)\left(-\frac{1}{3}x\right)$ **10.** $\left(-\frac{1}{4}x^4\right)\left(\frac{1}{5}x^8\right)$ **11.** $(-4x^2)(0)$ **12.** $(-4m^5)(-1)$

13. $(3x^2)(-4x^3)(2x^6)$ **14.** $(-2y^5)(10y^4)(-3y^3)$

b Multiply.

15. $2x(-x + 5)$ **16.** $3x(4x - 6)$ **17.** $-5x(x - 1)$ **18.** $-3x(-x - 1)$

19. $x^2(x^3 + 1)$ **20.** $-2x^3(x^2 - 1)$ **21.** $3x(2x^2 - 6x + 1)$ **22.** $-4x(2x^3 - 6x^2 - 5x + 1)$

23. $(-6x^2)(x^2 + x)$ **24.** $(-4x^2)(x^2 - x)$ **25.** $(3y^2)(6y^4 + 8y^3)$ **26.** $(4y^4)(y^3 - 6y^2)$

c Multiply.

27. $(x + 6)(x + 3)$ **28.** $(x + 5)(x + 2)$ **29.** $(x + 5)(x - 2)$ **30.** $(x + 6)(x - 2)$

31. $(x - 1)(x + 4)$ **32.** $(x - 8)(x + 7)$ **33.** $(x - 4)(x - 3)$ **34.** $(x - 7)(x - 3)$

35. $(x + 3)(x - 3)$ **36.** $(x + 6)(x - 6)$ **37.** $(x - 4)(x + 4)$ **38.** $(x - 9)(x + 9)$

39. $(3x + 5)(x + 2)$ **40.** $(2x + 6)(x + 3)$ **41.** $(5 - x)(5 - 2x)$ **42.** $(3 - 4x)(2 - x)$

43. $(2x + 5)(2x + 5)$

44. $(3x + 4)(3x + 4)$

45. $(x - 3)(x - 3)$

46. $(x - 6)(x - 6)$

47. $\left(x - \frac{5}{2}\right)\left(x + \frac{2}{5}\right)$

48. $\left(x + \frac{4}{3}\right)\left(x + \frac{3}{2}\right)$

49. $(x - 2.3)(x + 4.7)$

50. $(2x + 0.13)(2x - 0.13)$

Write an algebraic expression that represents the total area of the four smaller rectangles in each figure.

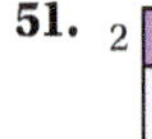

51.
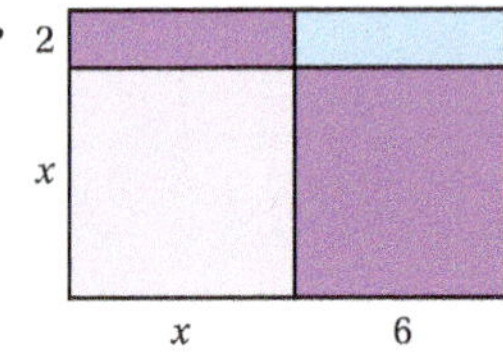

52.
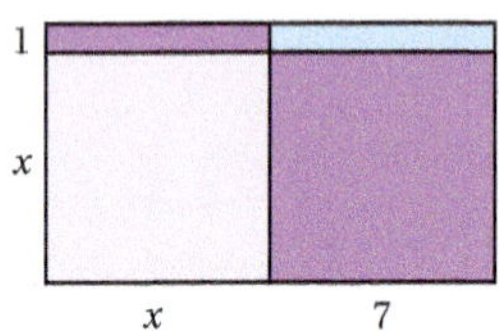

53.
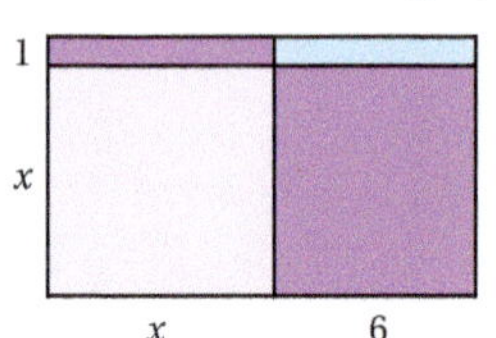

54.
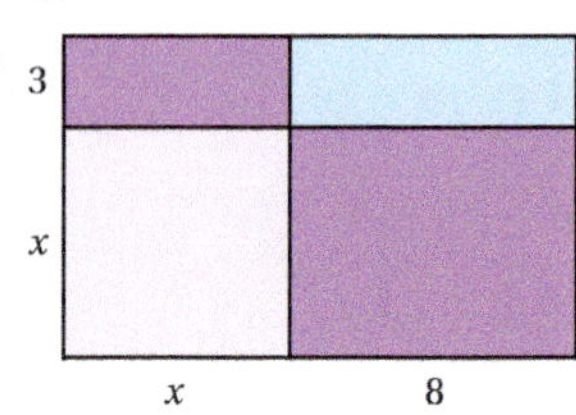

Draw and label rectangles similar to the one following Example 6 to illustrate each product.

55. $x(x + 5)$

56. $x(x + 2)$

57. $(x + 1)(x + 2)$

58. $(x + 3)(x + 1)$

59. $(x + 5)(x + 3)$

60. $(x + 4)(x + 6)$

d Multiply.

61. $(x^2 + x + 1)(x - 1)$

62. $(x^2 + x - 2)(x + 2)$

63. $(2x + 1)(2x^2 + 6x + 1)$

64. $(3x - 1)(4x^2 - 2x - 1)$

65. $(y^2 - 3)(3y^2 - 6y + 2)$

66. $(3y^2 - 3)(y^2 + 6y + 1)$

67. $(x^3 + x^2)(x^3 + x^2 - x)$

68. $(x^3 - x^2)(x^3 - x^2 + x)$

69. $(-5x^3 - 7x^2 + 1)(2x^2 - x)$

70. $(-4x^3 + 5x^2 - 2)(5x^2 + 1)$

71. $(1 + x + x^2)(-1 - x + x^2)$

72. $(1 - x + x^2)(1 - x + x^2)$

73. $(2t^2 - t - 4)(3t^2 + 2t - 1)$

74. $(3a^2 - 5a + 2)(2a^2 - 3a + 4)$

75. $(x - x^3 + x^5)(x^2 - 1 + x^4)$

76. $(x - x^3 + x^5)(3x^2 + 3x^6 + 3x^4)$

77. $(x + 1)(x^3 + 7x^2 + 5x + 4)$

78. $(x + 2)(x^3 + 5x^2 + 9x + 3)$

79. $\left(x - \frac{1}{2}\right)\left(2x^3 - 4x^2 + 3x - \frac{2}{5}\right)$

80. $\left(x + \frac{1}{3}\right)\left(6x^3 - 12x^2 - 5x + \frac{1}{2}\right)$

Skill Maintenance

Simplify.

81. $x - (2x - 3)$ [9.8b]

82. $5 - 2[3 - 4(8 - 2)]$ [9.8c]

83. $(10 - 2)(10 + 2)$ [9.8d]

84. $10 - 2 + (-6)^2 \div 3 \cdot 2$ [9.8d]

Factor. [9.7d]

85. $15x - 18y + 12$

86. $16x - 24y + 36$

87. $-9x - 45y + 15$

88. $100x - 100y + 1000a$

Synthesis

Find a polynomial for the shaded area of each figure.

89.

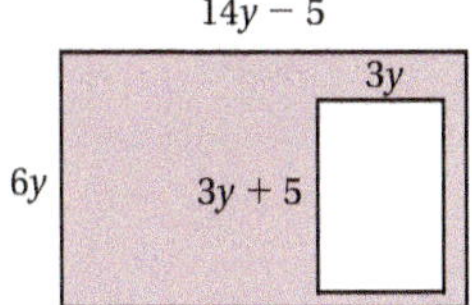

90.

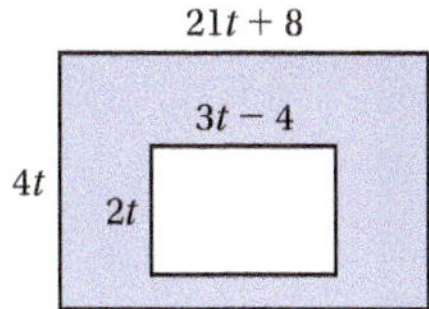

For each figure, determine what the missing number must be in order for the figure to have the given area.

91. Area $= x^2 + 8x + 15$

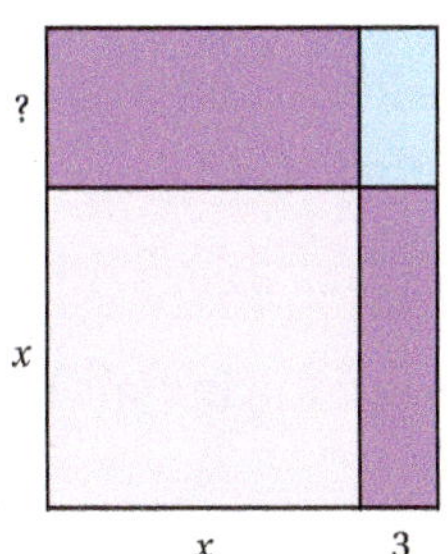

92. Area $= x^2 + 7x + 10$

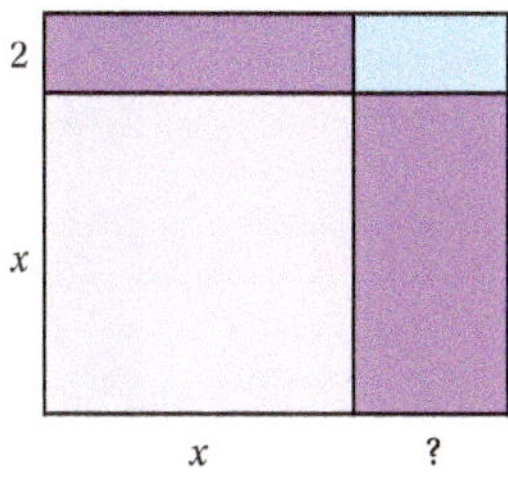

93. Find a polynomial for the volume of the solid shown below.

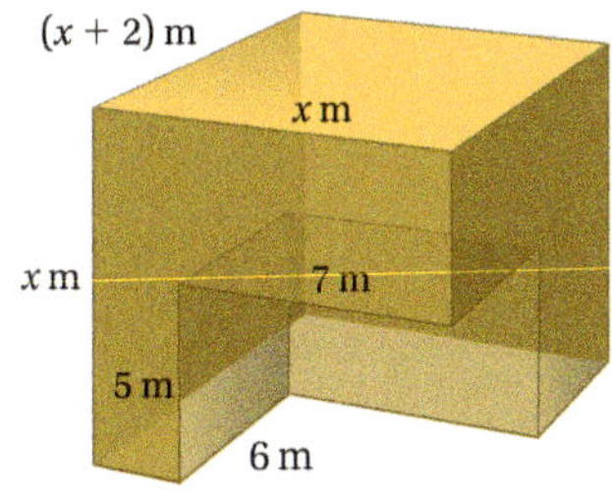

94. An open wooden box is a cube with side x cm. The box, including its bottom, is made of wood that is 1 cm thick. Find a polynomial for the interior volume of the cube.

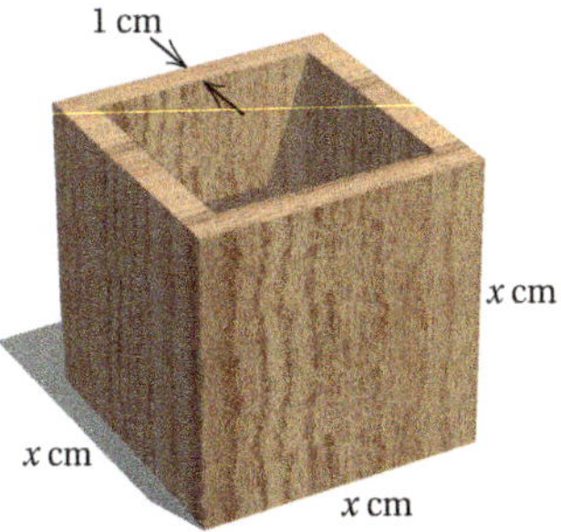

Compute and simplify.

95. $(x - 2)(x - 7) - (x - 7)(x - 2)$

96. $(x + 5)^2 - (x - 3)^2$

97. Extend the pattern and simplify:

$$(x - a)(x - b)(x - c)(x - d) \cdots (x - z).$$

98. Use a graphing calculator to check your answers to Exercises 15, 29, and 61. Use graphs, tables, or both, as directed by your instructor.

12.6 Special Products

OBJECTIVES

a Multiply two binomials mentally using the FOIL method.

b Multiply the sum and the difference of the same two terms mentally.

c Square a binomial mentally.

d Find special products when polynomial products are mixed together.

SKILL TO REVIEW

Objective 9.7e: Collect like terms.

Collect like terms.

1. $\frac{2}{3}x - \frac{2}{3}x$
2. $-12n + 12n$

We encounter certain products so often that it is helpful to have efficient methods of computing them. Such techniques are called *special products.*

a PRODUCTS OF TWO BINOMIALS USING FOIL

To multiply two binomials, we can select one binomial and multiply each term of that binomial by every term of the other. Then we collect like terms. Consider the product $(x + 3)(x + 7)$:

$$\begin{aligned}(x + 3)(x + 7) &= x(x + 7) + 3(x + 7)\\ &= x \cdot x + x \cdot 7 + 3 \cdot x + 3 \cdot 7\\ &= x^2 + 7x + 3x + 21\\ &= x^2 + 10x + 21.\end{aligned}$$

This example illustrates a special technique for finding the product of two binomials:

First terms, Outside terms, Inside terms, Last terms

$$(x + 3)(x + 7) = x \cdot x + 7 \cdot x + 3 \cdot x + 3 \cdot 7.$$

To remember this method of multiplying, we use the initials **FOIL**.

THE FOIL METHOD

To multiply two binomials, $A + B$ and $C + D$, multiply the First terms AC, the Outside terms AD, the Inside terms BC, and then the Last terms BD. Then collect like terms, if possible.

$$(A + B)(C + D) = AC + AD + BC + BD$$

1. Multiply First terms: AC.
2. Multiply Outside terms: AD.
3. Multiply Inside terms: BC.
4. Multiply Last terms: BD.

↓

FOIL

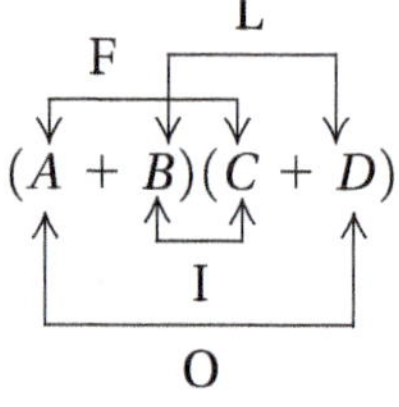

EXAMPLE 1 Multiply: $(x + 8)(x^2 - 5)$.

We have

$$\begin{aligned}(x + 8)(x^2 - 5) &= \overset{F}{x \cdot x^2} + \overset{O}{x \cdot (-5)} + \overset{I}{8 \cdot x^2} + \overset{L}{8(-5)}\\ &= x^3 - 5x + 8x^2 - 40\\ &= x^3 + 8x^2 - 5x - 40.\end{aligned}$$

Since each of the original binomials is in descending order, we write the product in descending order, as is customary, but this is not a "must."

Answers

Skill to Review:
1. 0 2. 0

Often we can collect like terms after we have multiplied.

EXAMPLES Multiply.

2. $(x + 6)(x - 6) = x^2 - 6x + 6x - 36$ Using FOIL
$= x^2 - 36$ Collecting like terms

3. $(x + 7)(x + 4) = x^2 + 4x + 7x + 28$
$= x^2 + 11x + 28$

4. $(y - 3)(y - 2) = y^2 - 2y - 3y + 6$
$= y^2 - 5y + 6$

5. $(x^3 - 1)(x^3 + 5) = x^6 + 5x^3 - x^3 - 5$
$= x^6 + 4x^3 - 5$

6. $(4t^3 + 5)(3t^2 - 2) = 12t^5 - 8t^3 + 15t^2 - 10$

Do Exercises 1–8. ▶

Multiply mentally, if possible. If you need extra steps, be sure to use them.

1. $(x + 3)(x + 4)$
2. $(x + 3)(x - 5)$
3. $(2x - 1)(x - 4)$
4. $(2x^2 - 3)(x^2 - 2)$
5. $(6x^2 + 5)(2x^3 + 1)$
6. $(y^3 + 7)(y^3 - 7)$
7. $(t + 2)(t + 3)$
8. $(2x^4 + x^2)(-x^3 + x)$

EXAMPLES Multiply.

7. $\left(x - \frac{2}{3}\right)\left(x + \frac{2}{3}\right) = x^2 + \frac{2}{3}x - \frac{2}{3}x - \frac{4}{9}$
$= x^2 - \frac{4}{9}$

8. $(x^2 - 0.3)(x^2 - 0.3) = x^4 - 0.3x^2 - 0.3x^2 + 0.09$
$= x^4 - 0.6x^2 + 0.09$

9. $(3 - 4x)(7 - 5x^3) = 21 - 15x^3 - 28x + 20x^4$
$= 21 - 28x - 15x^3 + 20x^4$

(*Note*: If the original polynomials are in ascending order, it is natural to write the product in ascending order, but this is not a "must.")

10. $(5x^4 + 2x^3)(3x^2 - 7x) = 15x^6 - 35x^5 + 6x^5 - 14x^4$
$= 15x^6 - 29x^5 - 14x^4$

Do Exercises 9–12. ▶

Multiply.

9. $\left(x + \frac{4}{5}\right)\left(x - \frac{4}{5}\right)$
10. $(x^3 - 0.5)(x^2 + 0.5)$
11. $(2 + 3x^2)(4 - 5x^2)$
12. $(6x^3 - 3x^2)(5x^2 - 2x)$

We can show the FOIL method geometrically as follows. One way to write the area of the large rectangle below is $(A + B)(C + D)$. To find another expression for the area of the large rectangle, we add the areas of the smaller rectangles.

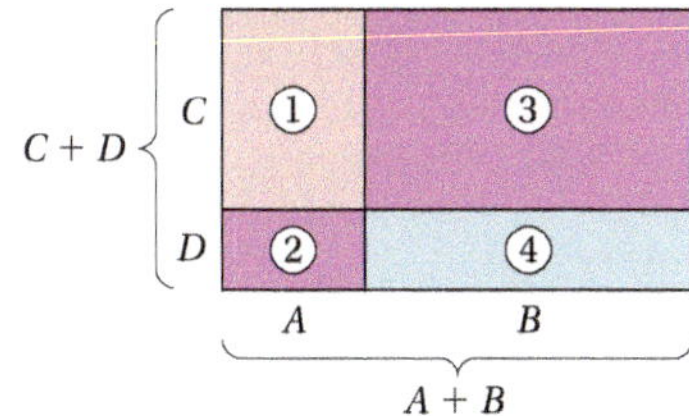

The area of rectangle ① is AC.
The area of rectangle ② is AD.
The area of rectangle ③ is BC.
The area of rectangle ④ is BD.

The area of the large rectangle is the sum of the areas of the smaller rectangles. Thus,

$$(A + B)(C + D) = AC + AD + BC + BD.$$

Answers

1. $x^2 + 7x + 12$ 2. $x^2 - 2x - 15$
3. $2x^2 - 9x + 4$ 4. $2x^4 - 7x^2 + 6$
5. $12x^5 + 10x^3 + 6x^2 + 5$ 6. $y^6 - 49$
7. $t^2 + 5t + 6$ 8. $-2x^7 + x^5 + x^3$
9. $x^2 - \frac{16}{25}$ 10. $x^5 + 0.5x^3 - 0.5x^2 - 0.25$
11. $8 + 2x^2 - 15x^4$ 12. $30x^5 - 27x^4 + 6x^3$

b MULTIPLYING SUMS AND DIFFERENCES OF TWO TERMS

Consider the product of the sum and the difference of the same two terms, such as

$$(x + 2)(x - 2).$$

Since this is the product of two binomials, we can use FOIL. This type of product occurs so often, however, that it would be valuable if we could use an even faster method. To find a faster way to compute such a product, look for a pattern in the following:

a) $(x + 2)(x - 2) = x^2 - 2x + 2x - 4$ Using FOIL
$= x^2 - 4;$

b) $(3x - 5)(3x + 5) = 9x^2 + 15x - 15x - 25$
$= 9x^2 - 25.$

Multiply.

13. $(x + 5)(x - 5)$

14. $(2x - 3)(2x + 3)$

Do Exercises 13 and 14.

Perhaps you discovered in each case that when you multiply the two binomials, two terms are opposites, or additive inverses, which add to 0 and "drop out."

PRODUCT OF THE SUM AND THE DIFFERENCE OF TWO TERMS

The product of the sum and the difference of the same two terms is the square of the first term minus the square of the second term:

$$(A + B)(A - B) = A^2 - B^2.$$

It is helpful to memorize this rule in both words and symbols. (If you do forget it, you can, of course, use FOIL.)

EXAMPLES Multiply. (Carry out the rule and say the words as you go.)

$(A + B)(A - B) = A^2 - B^2$

11. $(x + 4)(x - 4) = x^2 - 4^2$ "The square of the first term, x^2, minus the square of the second, 4^2"
$= x^2 - 16$ Simplifying

12. $(5 + 2w)(5 - 2w) = 5^2 - (2w)^2$
$= 25 - 4w^2$

13. $(3x^2 - 7)(3x^2 + 7) = (3x^2)^2 - 7^2$
$= 9x^4 - 49$

14. $(-4x - 10)(-4x + 10) = (-4x)^2 - 10^2$
$= 16x^2 - 100$

15. $\left(x + \frac{3}{8}\right)\left(x - \frac{3}{8}\right) = x^2 - \left(\frac{3}{8}\right)^2 = x^2 - \frac{9}{64}$

Do Exercises 15–19.

Multiply.

15. $(x + 8)(x - 8)$

16. $(x - 7)(x + 7)$

17. $(6 - 4y)(6 + 4y)$ GS
$(\quad)^2 - (\quad)^2 = 36 - \underline{\quad}$

18. $(2x^3 - 1)(2x^3 + 1)$

19. $\left(x - \frac{2}{5}\right)\left(x + \frac{2}{5}\right)$

Answers

13. $x^2 - 25$ **14.** $4x^2 - 9$ **15.** $x^2 - 64$
16. $x^2 - 49$ **17.** $36 - 16y^2$ **18.** $4x^6 - 1$
19. $x^2 - \frac{4}{25}$

Guided Solution:
17. $6, 4y, 16y^2$

C SQUARING BINOMIALS

Consider the square of a binomial, such as $(x + 3)^2$. This can be expressed as $(x + 3)(x + 3)$. Since this is the product of two binomials, we can use FOIL. But again, this type of product occurs so often that we would like to use an even faster method. Look for a pattern in the following.

a) $(x + 3)^2 = (x + 3)(x + 3)$
$= x^2 + 3x + 3x + 9$
$= x^2 + 6x + 9;$

b) $(x - 3)^2 = (x - 3)(x - 3)$
$= x^2 - 3x - 3x + 9$
$= x^2 - 6x + 9$

Do Exercises 20 and 21.

Multiply.

20. $(x + 8)(x + 8)$

21. $(x - 5)(x - 5)$

When squaring a binomial, we multiply a binomial by itself. Perhaps you noticed that two terms are the same and when added give twice the product of the terms in the binomial. The other two terms are squares.

SQUARE OF A BINOMIAL

The square of a sum or a difference of two terms is the square of the first term, plus twice the product of the two terms, plus the square of the last term:

$$(A + B)^2 = A^2 + 2AB + B^2; \qquad (A - B)^2 = A^2 - 2AB + B^2.$$

It is helpful to memorize this rule in both words and symbols.

EXAMPLES Multiply. (Carry out the rule and say the words as you go.)

$(A + B)^2 = A^2 + 2 \cdot A \cdot B + B^2$

16. $(x + 3)^2 = x^2 + 2 \cdot x \cdot 3 + 3^2$ "x^2 plus 2 times x times 3 plus 3^2"
$= x^2 + 6x + 9$

$(A - B)^2 = A^2 - 2 \cdot A \cdot B + B^2$

17. $(t - 5)^2 = t^2 - 2 \cdot t \cdot 5 + 5^2$
$= t^2 - 10t + 25$

18. $(2x + 7)^2 = (2x)^2 + 2 \cdot 2x \cdot 7 + 7^2 = 4x^2 + 28x + 49$

19. $(5x - 3x^2)^2 = (5x)^2 - 2 \cdot 5x \cdot 3x^2 + (3x^2)^2 = 25x^2 - 30x^3 + 9x^4$

20. $(2.3 - 5.4m)^2 = 2.3^2 - 2(2.3)(5.4m) + (5.4m)^2$
$= 5.29 - 24.84m + 29.16m^2$

Do Exercises 22–27.

Multiply.

22. $(x + 2)^2$

23. $(a - 4)^2$

24. $(2x + 5)^2$

25. $(4x^2 - 3x)^2$

26. $(7.8 + 1.2y)(7.8 + 1.2y)$

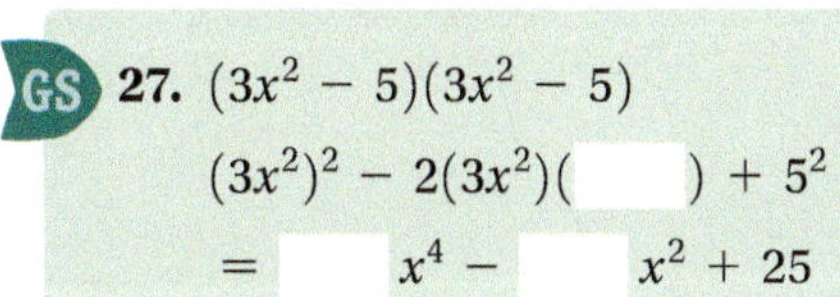

GS 27. $(3x^2 - 5)(3x^2 - 5)$
$(3x^2)^2 - 2(3x^2)(\quad) + 5^2$
$= \quad x^4 - \quad x^2 + 25$

Caution!

Although the square of a product is the product of the squares, the square of a sum is *not* the sum of the squares. That is, $(AB)^2 = A^2B^2$, but

The term $2AB$ is missing.

$$(A + B)^2 \neq A^2 + B^2.$$

To illustrate this inequality, note, using the rules for order of operations, that $(7 + 5)^2 = 12^2 = 144$, whereas $7^2 + 5^2 = 49 + 25 = 74$, and $74 \neq 144$.

Answers

20. $x^2 + 16x + 64$ **21.** $x^2 - 10x + 25$
22. $x^2 + 4x + 4$ **23.** $a^2 - 8a + 16$
24. $4x^2 + 20x + 25$ **25.** $16x^4 - 24x^3 + 9x^2$
26. $60.84 + 18.72y + 1.44y^2$
27. $9x^4 - 30x^2 + 25$

28. In the figure at right, describe in terms of area the sum $A^2 + B^2$. How can the figure be used to verify that $(A + B)^2 \neq A^2 + B^2$?

We can look at the rule for finding $(A + B)^2$ geometrically as follows. The area of the large square is

$$(A + B)(A + B) = (A + B)^2.$$

This is equal to the sum of the areas of the smaller rectangles:

$$A^2 + AB + AB + B^2 = A^2 + 2AB + B^2.$$

Thus, $(A + B)^2 = A^2 + 2AB + B^2$.

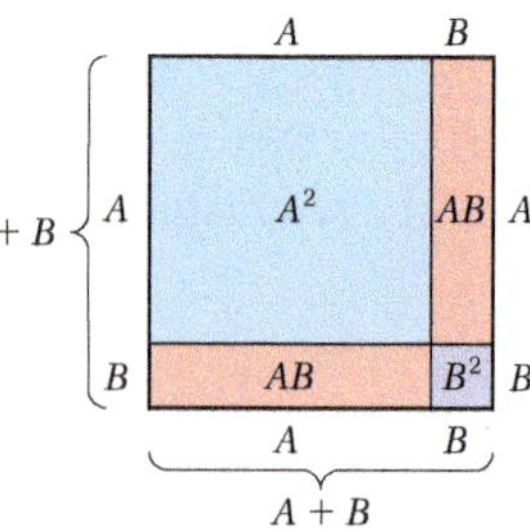

◀ **Do Exercise 28.**

d MULTIPLICATION OF VARIOUS TYPES

Let's now try several types of multiplications mixed together so that we can learn to sort them out. When you multiply, first see what kind of multiplication you have. Then use the best method.

MULTIPLYING TWO POLYNOMIALS

1. Is it the product of a monomial and a polynomial? If so, multiply each term of the polynomial by the monomial.
   Example: $5x(x + 7) = 5x \cdot x + 5x \cdot 7 = 5x^2 + 35x$
2. Is it the product of the sum and the difference of the *same* two terms? If so, use the following:
   $$(A + B)(A - B) = A^2 - B^2.$$
   Example: $(x + 7)(x - 7) = x^2 - 7^2 = x^2 - 49$
3. Is the product the square of a binomial? If so, use the following:
   $$(A + B)(A + B) = (A + B)^2 = A^2 + 2AB + B^2,$$
   or $$(A - B)(A - B) = (A - B)^2 = A^2 - 2AB + B^2.$$
   Example: $$(x + 7)(x + 7) = (x + 7)^2 = x^2 + 2 \cdot x \cdot 7 + 7^2 = x^2 + 14x + 49$$
4. Is it the product of two binomials other than those above? If so, use FOIL.
   Example: $$(x + 7)(x - 4) = x^2 - 4x + 7x - 28 = x^2 + 3x - 28$$
5. Is it the product of two polynomials other than those above? If so, multiply each term of one by every term of the other. Use columns if you wish.
   Example:
   $$\begin{aligned}(x^2 - 3x + 2)(x + 7) &= x^2(x + 7) - 3x(x + 7) + 2(x + 7)\\ &= x^2 \cdot x + x^2 \cdot 7 - 3x \cdot x - 3x \cdot 7 + 2 \cdot x + 2 \cdot 7\\ &= x^3 + 7x^2 - 3x^2 - 21x + 2x + 14\\ &= x^3 + 4x^2 - 19x + 14\end{aligned}$$

Answers

28. $(A + B)^2$ represents the area of the large square. This includes all four sections. $A^2 + B^2$ represents the area of only two of the sections.

Guided Solution:
27. 5, 9, 30

Remember that FOIL will *always* work for two binomials. You can use it instead of either of rules (2) and (3), but those rules will make your work go faster.

EXAMPLE 21 Multiply: $(x + 3)(x - 3)$.

$$(x + 3)(x - 3) = x^2 - 3^2 = x^2 - 9$$

This is the product of the sum and the difference of the same two terms. We use $(A + B)(A - B) = A^2 - B^2$.

EXAMPLE 22 Multiply: $(t + 7)(t - 5)$.

$$(t + 7)(t - 5) = t^2 + 2t - 35$$

This is the product of two binomials, but neither the square of a binomial nor the product of the sum and the difference of two terms. We use FOIL.

EXAMPLE 23 Multiply: $(x + 6)(x + 6)$.

$$(x + 6)(x + 6) = x^2 + 2(6)x + 6^2 = x^2 + 12x + 36$$

This the square of a binomial. We use $(A + B)(A + B) = A^2 + 2AB + B^2$.

EXAMPLE 24 Multiply: $2x^3(9x^2 + x - 7)$.

$$2x^3(9x^2 + x - 7) = 18x^5 + 2x^4 - 14x^3$$

This is the product of a monomial and a trinomial. We multiply each term of the trinomial by the monomial.

EXAMPLE 25 Multiply: $(5x^3 - 7x)^2$.

$$(5x^3 - 7x)^2 = (5x^3)^2 - 2(5x^3)(7x) + (7x)^2 = 25x^6 - 70x^4 + 49x^2$$

$(A - B)^2 = A^2 - 2AB + B^2$

EXAMPLE 26 Multiply: $\left(3x + \frac{1}{4}\right)^2$.

$$\left(3x + \tfrac{1}{4}\right)^2 = (3x)^2 + 2(3x)\left(\tfrac{1}{4}\right) + \left(\tfrac{1}{4}\right)^2 = 9x^2 + \tfrac{3}{2}x + \tfrac{1}{16}$$

$(A + B)^2 = A^2 + 2AB + B^2$

EXAMPLE 27 Multiply: $\left(4x - \frac{3}{4}\right)\left(4x + \frac{3}{4}\right)$.

$$\left(4x - \tfrac{3}{4}\right)\left(4x + \tfrac{3}{4}\right) = (4x)^2 - \left(\tfrac{3}{4}\right)^2 = 16x^2 - \tfrac{9}{16}$$

$(A + B)(A - B) = A^2 - B^2$

EXAMPLE 28 Multiply: $(p + 3)(p^2 + 2p - 1)$.

$$\begin{array}{r} p^2 + 2p - 1 \\ p + 3 \\ \hline 3p^2 + 6p - 3 \\ p^3 + 2p^2 - p \quad \\ \hline p^3 + 5p^2 + 5p - 3 \end{array}$$

Finding the product of two polynomials

Multiplying by 3

Multiplying by p

Do Exercises 29–36.

Multiply.

29. $(x + 5)(x + 6)$

30. $(t - 4)(t + 4)$

31. $4x^2(-2x^3 + 5x^2 + 10)$

32. $(9x^2 + 1)^2$

33. $(2a - 5)(2a + 8)$

34. $\left(5x + \frac{1}{2}\right)^2$

35. $\left(2x - \frac{1}{2}\right)^2$

36. $(x^2 - x + 4)(x - 2)$

Answers

29. $x^2 + 11x + 30$ 30. $t^2 - 16$
31. $-8x^5 + 20x^4 + 40x^2$ 32. $81x^4 + 18x^2 + 1$
33. $4a^2 + 6a - 40$ 34. $25x^2 + 5x + \frac{1}{4}$
35. $4x^2 - 2x + \frac{1}{4}$ 36. $x^3 - 3x^2 + 6x - 8$

Visualizing for Success

In each of Exercises 1–10, find two algebraic expressions for the shaded area of the figure from the list below.

A. $9 - 4x^2$

B. $x^2 - (x - 6)^2$

C. $(x + 3)(x - 3)$

D. $10^2 + 2^2$

E. $x^2 + 8x + 15$

F. $(x + 5)(x + 3)$

G. $x^2 - 6x + 9$

H. $(3 - 2x)^2 + 4x(3 - 2x)$

I. $(x + 3)^2$

J. $(5x + 3)^2$

K. $(5 - 2x)^2 + 4x(5 - 2x)$

L. $x^2 - 9$

M. 104

N. $x^2 - 15$

O. $12x - 36$

P. $25x^2 + 30x + 9$

Q. $(x - 5)(x - 3)$
$+ 3(x - 5) + 5(x - 3)$

R. $(x - 3)^2$

S. $25 - 4x^2$

T. $x^2 + 6x + 9$

Answers on page A-28

1

x 5 x 3

2

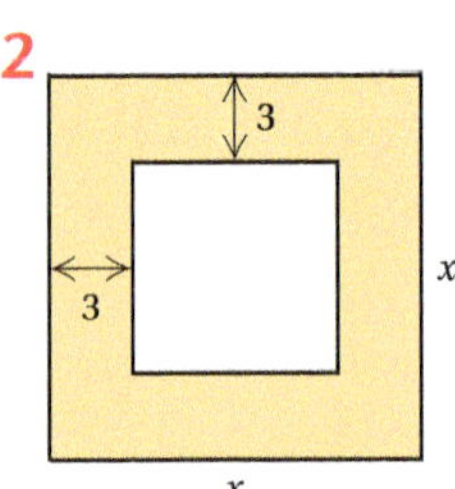

3

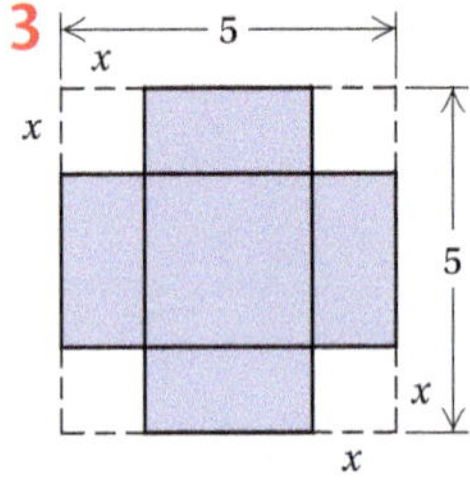

4

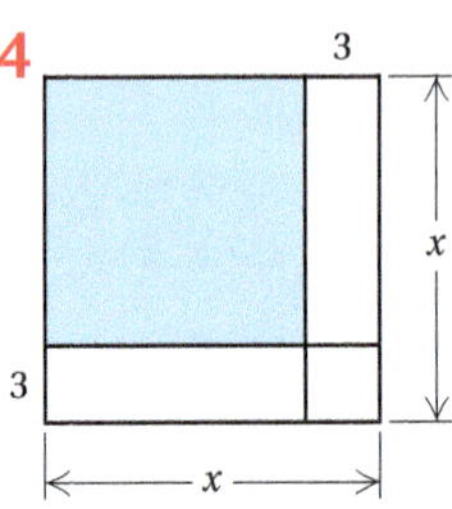

5

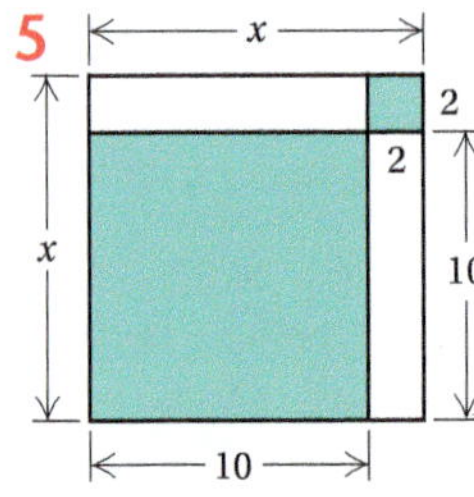

6

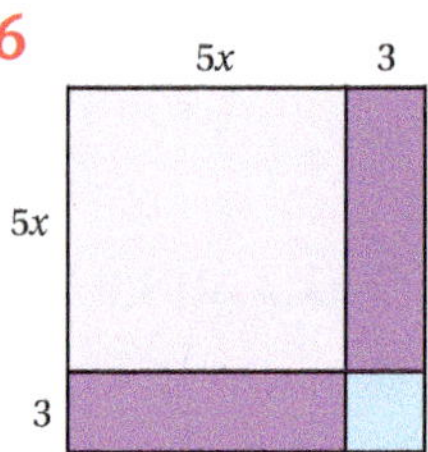

7

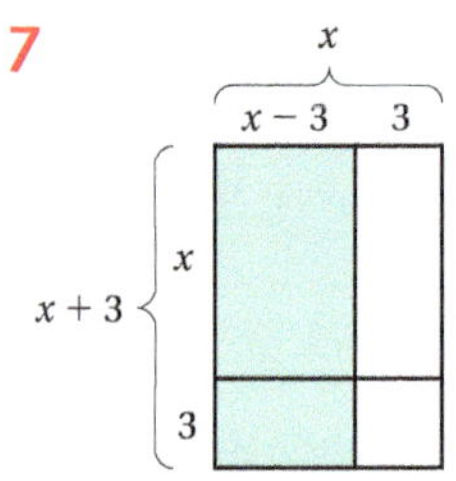

8

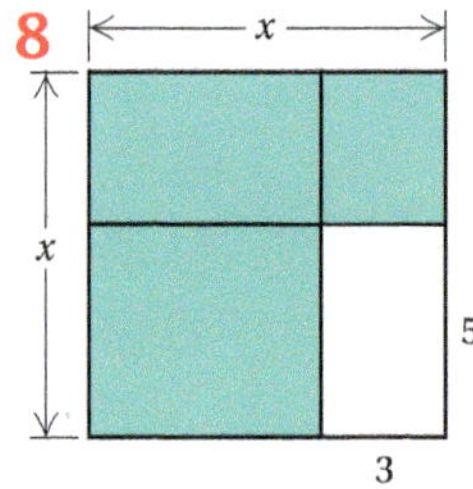

9

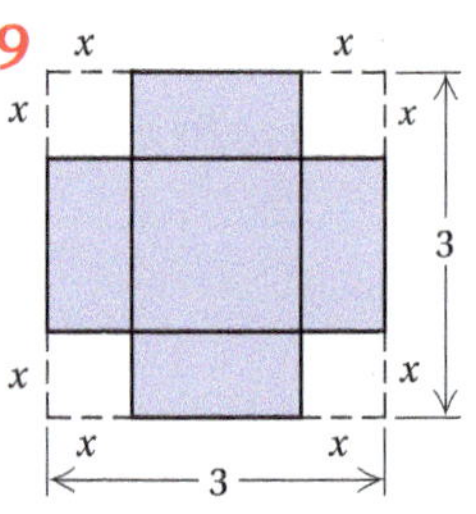

10

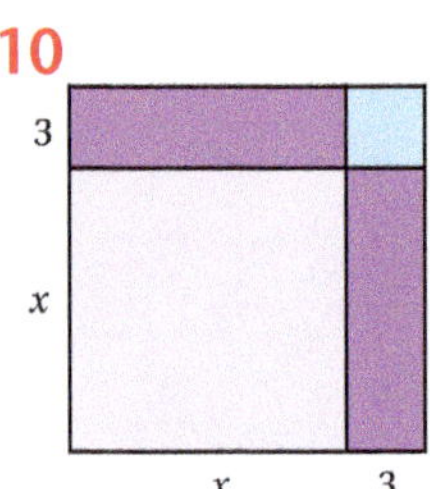

12.6 Exercise Set

For Extra Help MyMathLab® MathXL® PRACTICE WATCH READ REVIEW

Reading Check

Complete each statement with the appropriate word from the column on the right. A word may be used more than once or not at all.

RC1. For the FOIL multiplication method, the initials F O I L represent the words first, ______________, inside, and ______________.

RC2. If the polynomials being multiplied are written in descending order, we generally write the product in ______________ order.

RC3. The expression $(A + B)(A - B)$ is the product of the sum and the ______________ of the same two terms.

RC4. The expression $(A + B)^2$ is the ______________ of a ______________.

RC5. We can find the product of any two ______________ using the FOIL method.

RC6. The product of the sum and the difference of the same two terms is the ______________ of their squares.

ascending
binomial(s)
descending
difference
last
outside
product
square

a Multiply. Try to write only the answer. If you need more steps, be sure to use them.

1. $(x + 1)(x^2 + 3)$

2. $(x^2 - 3)(x - 1)$

3. $(x^3 + 2)(x + 1)$

4. $(x^4 + 2)(x + 10)$

5. $(y + 2)(y - 3)$

6. $(a + 2)(a + 3)$

7. $(3x + 2)(3x + 2)$

8. $(4x + 1)(4x + 1)$

9. $(5x - 6)(x + 2)$

10. $(x - 8)(x + 8)$

11. $(3t - 1)(3t + 1)$

12. $(2m + 3)(2m + 3)$

13. $(4x - 2)(x - 1)$

14. $(2x - 1)(3x + 1)$

15. $\left(p - \frac{1}{4}\right)\left(p + \frac{1}{4}\right)$

16. $\left(q + \frac{3}{4}\right)\left(q + \frac{3}{4}\right)$

17. $(x - 0.1)(x + 0.1)$

18. $(x + 0.3)(x - 0.4)$

19. $(2x^2 + 6)(x + 1)$

20. $(2x^2 + 3)(2x - 1)$

21. $(-2x + 1)(x + 6)$

22. $(3x + 4)(2x - 4)$

23. $(a + 7)(a + 7)$

24. $(2y + 5)(2y + 5)$

25. $(1 + 2x)(1 - 3x)$

26. $(-3x - 2)(x + 1)$

27. $\left(\frac{3}{8}y - \frac{5}{6}\right)\left(\frac{3}{8}y - \frac{5}{6}\right)$

28. $\left(\frac{1}{5}x - \frac{2}{7}\right)\left(\frac{1}{5}x + \frac{2}{7}\right)$

29. $(x^2 + 3)(x^3 - 1)$

30. $(x^4 - 3)(2x + 1)$

31. $(3x^2 - 2)(x^4 - 2)$

32. $(x^{10} + 3)(x^{10} - 3)$

33. $(2.8x - 1.5)(4.7x + 9.3)$

34. $\left(x - \frac{3}{8}\right)\left(x + \frac{4}{7}\right)$

35. $(3x^5 + 2)(2x^2 + 6)$

36. $(1 - 2x)(1 + 3x^2)$

37. $(4x^2 + 3)(x - 3)$

38. $(7x - 2)(2x - 7)$

39. $(4y^4 + y^2)(y^2 + y)$

40. $(5y^6 + 3y^3)(2y^6 + 2y^3)$

b Multiply mentally, if possible. If you need extra steps, be sure to use them.

41. $(x + 4)(x - 4)$

42. $(x + 1)(x - 1)$

43. $(2x + 1)(2x - 1)$

44. $(x^2 + 1)(x^2 - 1)$

45. $(5m - 2)(5m + 2)$

46. $(3x^4 + 2)(3x^4 - 2)$

47. $(2x^2 + 3)(2x^2 - 3)$

48. $(6x^5 - 5)(6x^5 + 5)$

49. $(3x^4 - 4)(3x^4 + 4)$

50. $(t^2 - 0.2)(t^2 + 0.2)$

51. $(x^6 - x^2)(x^6 + x^2)$

52. $(2x^3 - 0.3)(2x^3 + 0.3)$

53. $(x^4 + 3x)(x^4 - 3x)$

54. $\left(\frac{3}{4} + 2x^3\right)\left(\frac{3}{4} - 2x^3\right)$

55. $(x^{12} - 3)(x^{12} + 3)$

56. $(12 - 3x^2)(12 + 3x^2)$

57. $(2y^8 + 3)(2y^8 - 3)$

58. $\left(m - \frac{2}{3}\right)\left(m + \frac{2}{3}\right)$

59. $\left(\frac{5}{8}x - 4.3\right)\left(\frac{5}{8}x + 4.3\right)$

60. $(10.7 - x^3)(10.7 + x^3)$

c Multiply mentally, if possible. If you need extra steps, be sure to use them.

61. $(x + 2)^2$

62. $(2x - 1)^2$

63. $(3x^2 + 1)^2$

64. $\left(3x + \frac{3}{4}\right)^2$

65. $\left(a - \frac{1}{2}\right)^2$

66. $\left(2a - \frac{1}{5}\right)^2$

67. $(3 + x)^2$

68. $(x^3 - 1)^2$

69. $(x^2 + 1)^2$

70. $(8x - x^2)^2$

71. $(2 - 3x^4)^2$

72. $(6x^3 - 2)^2$

73. $(5 + 6t^2)^2$

74. $(3p^2 - p)^2$

75. $\left(x - \frac{5}{8}\right)^2$

76. $(0.3y + 2.4)^2$

d Multiply mentally, if possible.

77. $(3 - 2x^3)^2$

78. $(x - 4x^3)^2$

79. $4x(x^2 + 6x - 3)$

80. $8x(-x^5 + 6x^2 + 9)$

81. $\left(2x^2 - \frac{1}{2}\right)\left(2x^2 - \frac{1}{2}\right)$

82. $(-x^2 + 1)^2$

83. $(-1 + 3p)(1 + 3p)$

84. $(-3q + 2)(3q + 2)$

85. $3t^2(5t^3 - t^2 + t)$

86. $-6x^2(x^3 + 8x - 9)$

87. $(6x^4 + 4)^2$

88. $(8a + 5)^2$

89. $(3x + 2)(4x^2 + 5)$

90. $(2x^2 - 7)(3x^2 + 9)$

91. $(8 - 6x^4)^2$

92. $\left(\frac{1}{5}x^2 + 9\right)\left(\frac{3}{5}x^2 - 7\right)$

93. $(t - 1)(t^2 + t + 1)$

94. $(y + 5)(y^2 - 5y + 25)$

Compute each of the following and compare.

95. $3^2 + 4^2$; $(3 + 4)^2$

96. $6^2 + 7^2$; $(6 + 7)^2$

97. $9^2 - 5^2$; $(9 - 5)^2$

98. $11^2 - 4^2$; $(11 - 4)^2$

Find the total area of all the shaded rectangles.

99.

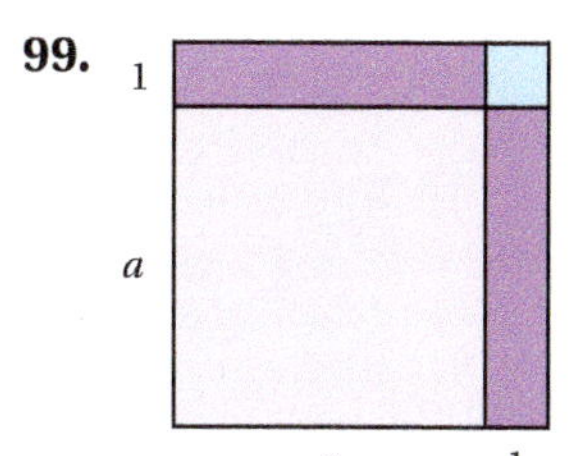

100.

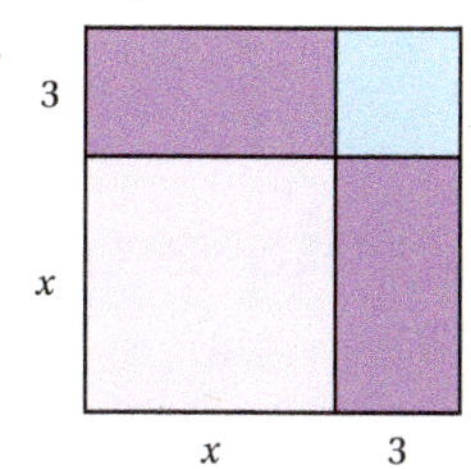

101.

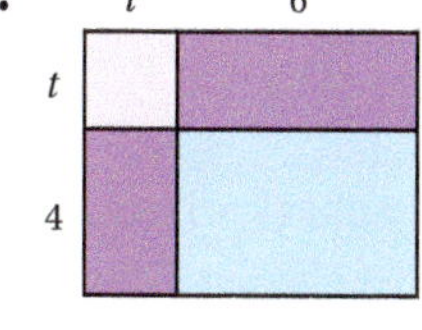

102.

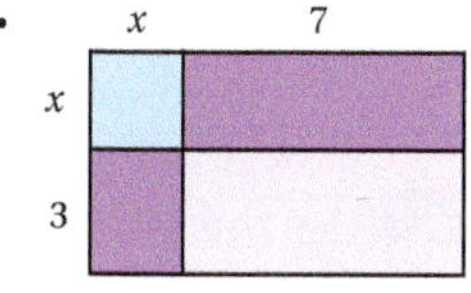

Skill Maintenance

Solve. [10.3c]

103. $3x - 8x = 4(7 - 8x)$

104. $3(x - 2) = 5(2x + 7)$

105. $5(2x - 3) - 2(3x - 4) = 20$

Solve. [10.4b]

106. $3x - 2y = 12$, for y

107. $C = ab - r$, for b

108. $3a - 5d = 4$, for a

Synthesis

Multiply.

109. $5x(3x - 1)(2x + 3)$

110. $[(2x - 3)(2x + 3)](4x^2 + 9)$

111. $[(a - 5)(a + 5)]^2$

112. $(a - 3)^2(a + 3)^2$
(*Hint*: Examine Exercise 111.)

113. $(3t^4 - 2)^2(3t^4 + 2)^2$
(*Hint*: Examine Exercise 111.)

114. $[3a - (2a - 3)][3a + (2a - 3)]$

Solve.

115. $(x + 2)(x - 5) = (x + 1)(x - 3)$

116. $(2x + 5)(x - 4) = (x + 5)(2x - 4)$

117. ***Factors and Sums.*** To *factor* a number is to express it as a product. Since $12 = 4 \cdot 3$, we say that 12 is *factored* and that 4 and 3 are *factors* of 12. In the following table, the top number has been factored in such a way that the sum of the factors is the bottom number. For example, in the first column, 40 has been factored as $5 \cdot 8$, and $5 + 8 = 13$, the bottom number. Such thinking is important in algebra when we factor trinomials of the type $x^2 + bx + c$. Find the missing numbers in the table.

PRODUCT	40	63	36	72	−140	−96	48	168	110			
FACTOR	5									−9	−24	−3
FACTOR	8									−10	18	
SUM	13	16	−20	−38	−4	4	−14	−29	−21			18

118. Consider the rectangle below.

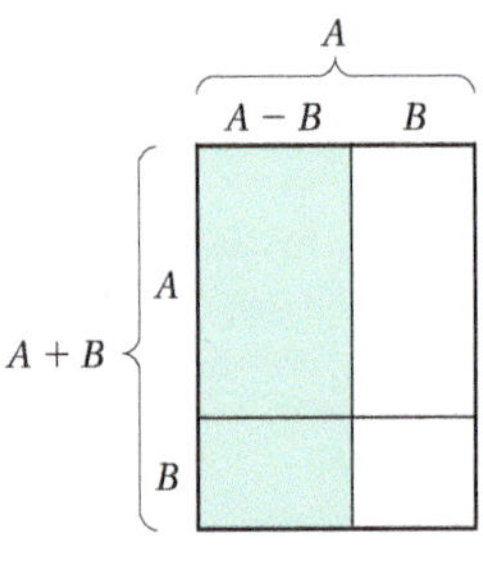

a) Find a polynomial for the area of the entire rectangle.

b) Find a polynomial for the sum of the areas of the two small unshaded rectangles.

c) Find a polynomial for the area in part (a) minus the area in part (b).

d) Find a polynomial for the area of the shaded region and compare this with the polynomial found in part (c).

Use the TABLE or GRAPH feature to check whether each of the following is correct.

119. $(x - 1)^2 = x^2 - 2x + 1$

120. $(x - 2)^2 = x^2 - 4x - 4$

121. $(x - 3)(x + 3) = x^2 - 6$

122. $(x - 3)(x + 2) = x^2 - x - 6$

Operations with Polynomials in Several Variables

12.7

OBJECTIVES

a Evaluate a polynomial in several variables for given values of the variables.

b Identify the coefficients and the degrees of the terms of a polynomial and the degree of a polynomial.

c Collect like terms of a polynomial.

d Add polynomials.

e Subtract polynomials.

f Multiply polynomials.

The polynomials that we have been studying have only one variable. A **polynomial in several variables** is an expression like those you have already seen, but with more than one variable. Here are two examples:

$$3x + xy^2 + 5y + 4, \qquad 8xy^2z - 2x^3z - 13x^4y^2 + 15.$$

a EVALUATING POLYNOMIALS

EXAMPLE 1 Evaluate the polynomial

$$4 + 3x + xy^2 + 8x^3y^3$$

when $x = -2$ and $y = 5$.

We replace x with -2 and y with 5:

$$\begin{aligned} 4 + 3x + xy^2 + 8x^3y^3 &= 4 + 3(-2) + (-2) \cdot 5^2 + 8(-2)^3 \cdot 5^3 \\ &= 4 + 3(-2) + (-2) \cdot 25 + 8(-8)(125) \\ &= 4 - 6 - 50 - 8000 \\ &= -8052. \end{aligned}$$

EXAMPLE 2 *Zoology.* The weight, in kilograms, of an elephant with a girth of g centimeters at the heart, a length of l centimeters, and a footpad circumference of f centimeters can be estimated by the polynomial

$$11.5g + 7.55l + 12.5f - 4016.$$

A field zoologist finds that the girth of a 3-year-old female elephant is 231 cm, the length is 135 cm, and the footpad circumference is 86 cm. Approximately how much does the elephant weigh?

Source: "How Much Does That Elephant Weigh?" by Mark MacAllister on fieldtripearth.org

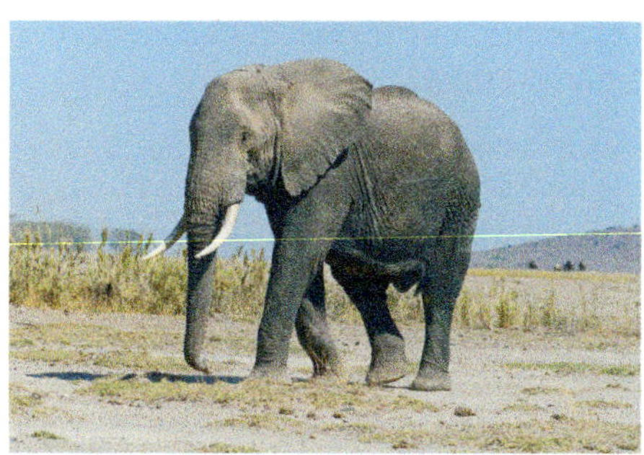

We evaluate the polynomial for $g = 231$, $l = 135$, and $f = 86$:

$$\begin{aligned} 11.5g + 7.55l + 12.5f - 4016 &= 11.5(231) + 7.55(135) + 12.5(86) - 4016 \\ &= 734.75. \end{aligned}$$

The elephant weighs about 735 kg.

Do Exercises 1–3.

1. Evaluate the polynomial $4 + 3x + xy^2 + 8x^3y^3$ when $x = 2$ and $y = -5$.

2. Evaluate the polynomial $8xy^2 - 2x^3z - 13x^4y^2 + 5$ when $x = -1$, $y = 3$, and $z = 4$.

3. *Zoology.* Refer to Example 2. A 25-year-old female elephant has a girth of 366 cm, a length of 226 cm, and a footpad circumference of 117 cm. How much does the elephant weigh?

Answers

1. −7940 **2.** −176 **3.** About 3362 kg

b COEFFICIENTS AND DEGREES

The **degree** of a term is the sum of the exponents of the variables. For example, the degree of $3x^5y^2$ is $5 + 2$, or 7. The **degree of a polynomial** is the degree of the term of highest degree.

EXAMPLE 3 Identify the coefficient and the degree of each term and the degree of the polynomial

$$9x^2y^3 - 14xy^2z^3 + xy + 4y + 5x^2 + 7.$$

TERM	COEFFICIENT	DEGREE	DEGREE OF THE POLYNOMIAL
$9x^2y^3$	9	5	
$-14xy^2z^3$	-14	6	6
xy	1	2	
$4y$	4	1	*Think:* $4y = 4y^1$.
$5x^2$	5	2	
7	7	0	*Think:* $7 = 7x^0$, or $7x^0y^0z^0$.

4. Identify the coefficient of each term:
$-3xy^2 + 3x^2y - 2y^3 + xy + 2.$

5. Identify the degree of each term and the degree of the polynomial
$4xy^2 + 7x^2y^3z^2 - 5x + 2y + 4.$

◀ Do Exercises 4 and 5.

c COLLECTING LIKE TERMS

Like terms have exactly the same variables with exactly the same exponents. For example,

$3x^2y^3$ and $-7x^2y^3$ are like terms;
$9x^4z^7$ and $12x^4z^7$ are like terms.

But

$13xy^5$ and $-2x^2y^5$ are *not* like terms, because the x-factors have different exponents;

and

$3xyz^2$ and $4xy$ are *not* like terms, because there is no factor of z^2 in the second expression.

Collecting like terms is based on the distributive laws.

EXAMPLES Collect like terms.

4. $5x^2y + 3xy^2 - 5x^2y - xy^2 = (5 - 5)x^2y + (3 - 1)xy^2 = 2xy^2$
5. $8a^2 - 2ab + 7b^2 + 4a^2 - 9ab - 17b^2 = 12a^2 - 11ab - 10b^2$
6. $7xy - 5xy^2 + 3xy^2 - 7 + 6x^3 + 9xy - 11x^3 + y - 1$
$= 16xy - 2xy^2 - 5x^3 + y - 8$

Collect like terms.

6. $4x^2y + 3xy - 2x^2y$

7. $-3pt - 5ptr^3 - 12 + 8pt + 5ptr^3 + 4$ GS

The like terms are $-3pt$ and ____, $-5ptr^3$ and ____, and -12 and ____.

Collecting like terms, we have
$(-3 + \_\_\_\_)pt +$
$(-5 + \_\_\_\_)ptr^3 +$
$(-12 + \_\_\_\_)$
$= \_\_\_\_ - 8.$

◀ Do Exercises 6 and 7.

Answers

4. $-3, 3, -2, 1, 2$ 5. $3, 7, 1, 1, 0; 7$
6. $2x^2y + 3xy$ 7. $5pt - 8$

Guided Solution:
7. $8pt, 5ptr^3, 4, 8, 5, 4, 5pt$

d ADDITION

We can find the sum of two polynomials in several variables by writing a plus sign between them and then collecting like terms.

EXAMPLE 7 Add: $(-5x^3 + 3y - 5y^2) + (8x^3 + 4x^2 + 7y^2)$.

$$\begin{aligned}(-5x^3 + 3y - 5y^2) &+ (8x^3 + 4x^2 + 7y^2)\\ &= (-5 + 8)x^3 + 4x^2 + 3y + (-5 + 7)y^2\\ &= 3x^3 + 4x^2 + 3y + 2y^2\end{aligned}$$

EXAMPLE 8 Add:

$$(5xy^2 - 4x^2y + 5x^3 + 2) + (3xy^2 - 2x^2y + 3x^3y - 5).$$

We have

$$\begin{aligned}(5xy^2 - 4x^2y + 5x^3 + 2) &+ (3xy^2 - 2x^2y + 3x^3y - 5)\\ &= (5 + 3)xy^2 + (-4 - 2)x^2y + 5x^3 + 3x^3y + (2 - 5)\\ &= 8xy^2 - 6x^2y + 5x^3 + 3x^3y - 3.\end{aligned}$$

Do Exercises 8–10.

Add.

8. $(4x^3 + 4x^2 - 8y - 3) + (-8x^3 - 2x^2 + 4y + 5)$

9. $(13x^3y + 3x^2y - 5y) + (x^3y + 4x^2y - 3xy + 3y)$

10. $(-5p^2q^4 + 2p^2q^2 + 3q) + (6pq^2 + 3p^2q + 5)$

e SUBTRACTION

We subtract a polynomial by adding its opposite, or additive inverse. The opposite of the polynomial $4x^2y - 6x^3y^2 + x^2y^2 - 5y$ is

$$-(4x^2y - 6x^3y^2 + x^2y^2 - 5y) = -4x^2y + 6x^3y^2 - x^2y^2 + 5y.$$

EXAMPLE 9 Subtract:

$$(4x^2y + x^3y^2 + 3x^2y^3 + 6y + 10) - (4x^2y - 6x^3y^2 + x^2y^2 - 5y - 8).$$

We have

$$(4x^2y + x^3y^2 + 3x^2y^3 + 6y + 10) - (4x^2y - 6x^3y^2 + x^2y^2 - 5y - 8)$$

$$= 4x^2y + x^3y^2 + 3x^2y^3 + 6y + 10 - 4x^2y + 6x^3y^2 - x^2y^2 + 5y + 8$$

Finding the opposite by changing the sign of each term

$$= 7x^3y^2 + 3x^2y^3 - x^2y^2 + 11y + 18.$$

Collecting like terms. (Try to write just the answer!)

Caution!

Do *not* add exponents when collecting like terms—that is,

$7x^3 + 8x^3 \neq 15x^6$; ← Adding exponents is incorrect.

$7x^3 + 8x^3 = 15x^3$. ← Correct

Do Exercises 11 and 12.

Subtract.

11. $(-4s^4t + s^3t^2 + 2s^2t^3) - (4s^4t - 5s^3t^2 + s^2t^2)$

12. $(-5p^4q + 5p^3q^2 - 3p^2q^3 - 7q^4 - 2) - (4p^4q - 4p^3q^2 + p^2q^3 + 2q^4 - 7)$

Answers

8. $-4x^3 + 2x^2 - 4y + 2$
9. $14x^3y + 7x^2y - 3xy - 2y$
10. $-5p^2q^4 + 2p^2q^2 + 3p^2q + 6pq^2 + 3q + 5$
11. $-8s^4t + 6s^3t^2 + 2s^2t^3 - s^2t^2$
12. $-9p^4q + 9p^3q^2 - 4p^2q^3 - 9q^4 + 5$

f MULTIPLICATION

To multiply polynomials in several variables, we can multiply each term of one by every term of the other. We can use columns for long multiplications as with polynomials in one variable. We multiply each term at the top by every term at the bottom. We write like terms in columns, and then we add.

EXAMPLE 10 Multiply: $(3x^2y - 2xy + 3y)(xy + 2y)$.

$$\begin{array}{rrrrl} & 3x^2y & - 2xy & + 3y & \\ & & xy & + 2y & \\ \hline & 6x^2y^2 & - 4xy^2 & + 6y^2 & \text{Multiplying by } 2y \\ 3x^3y^2 & - 2x^2y^2 & + 3xy^2 & & \text{Multiplying by } xy \\ \hline 3x^3y^2 & + 4x^2y^2 & - xy^2 & + 6y^2 & \text{Adding} \end{array}$$

Multiply.

13. $(x^2y^3 + 2x)(x^3y^2 + 3x)$

14. $(p^4q - 2p^3q^2 + 3q^3)(p + 2q)$

Do Exercises 13 and 14.

Where appropriate, we use the special products that we have learned.

EXAMPLES Multiply.

11. $(x^2y + 2x)(xy^2 + y^2) = x^3y^3 + x^2y^3 + 2x^2y^2 + 2xy^2$ Using FOIL

12. $(p + 5q)(2p - 3q) = 2p^2 - 3pq + 10pq - 15q^2$ Using FOIL
$= 2p^2 + 7pq - 15q^2$

$(A + B)^2 = A^2 + 2 \cdot A \cdot B + B^2$

13. $(3x + 2y)^2 = (3x)^2 + 2(3x)(2y) + (2y)^2 = 9x^2 + 12xy + 4y^2$

$(A - B)^2 = A^2 - 2 \cdot A \cdot B + B^2$

14. $(2y^2 - 5x^2y)^2 = (2y^2)^2 - 2(2y^2)(5x^2y) + (5x^2y)^2$
$= 4y^4 - 20x^2y^3 + 25x^4y^2$

$(A + B)(A - B) = A^2 - B^2$

15. $(3x^2y + 2y)(3x^2y - 2y) = (3x^2y)^2 - (2y)^2 = 9x^4y^2 - 4y^2$

16. $(-2x^3y^2 + 5t)(2x^3y^2 + 5t) = (5t - 2x^3y^2)(5t + 2x^3y^2)$ The sum and the difference of the same two terms
$= (5t)^2 - (2x^3y^2)^2 = 25t^2 - 4x^6y^4$

$(A - B)(A + B) = A^2 - B^2$

17. $(2x + 3 - 2y)(2x + 3 + 2y) = (2x + 3)^2 - (2y)^2$
$= 4x^2 + 12x + 9 - 4y^2$

Remember that FOIL will always work when you are multiplying binomials. You can use it instead of the rules for special products, but those rules will make your work go faster.

Do Exercises 15–22.

Multiply.

15. $(3xy + 2x)(x^2 + 2xy^2)$

16. $(x - 3y)(2x - 5y)$

17. $(4x + 5y)^2$

18. $(3x^2 - 2xy^2)^2$

19. $(2xy^2 + 3x)(2xy^2 - 3x)$

20. $(3xy^2 + 4y)(-3xy^2 + 4y)$

21. $(3y + 4 - 3x)(3y + 4 + 3x)$

22. $(2a + 5b + c)(2a - 5b - c)$ GS
$= [2a + (5b + c)][2a - (\quad)]$
$= (2a)^2 - (\quad)^2$
$= \quad - (25b^2 + 10bc + \quad)$
$= 4a^2 - 25b^2 - 10bc - \quad$

Answers

13. $x^5y^5 + 2x^4y^2 + 3x^3y^3 + 6x^2$
14. $p^5q - 4p^3q^3 + 3pq^3 + 6q^4$
15. $3x^3y + 6x^2y^3 + 2x^3 + 4x^2y^2$
16. $2x^2 - 11xy + 15y^2$
17. $16x^2 + 40xy + 25y^2$
18. $9x^4 - 12x^3y^2 + 4x^2y^4$
19. $4x^2y^4 - 9x^2$
20. $16y^2 - 9x^2y^4$
21. $9y^2 + 24y + 16 - 9x^2$
22. $4a^2 - 25b^2 - 10bc - c^2$

Guided Solution:
22. $5b + c, 5b + c, 4a^2, c^2, c^2$

12.7 Exercise Set

For Extra Help MyMathLab® MathXL® PRACTICE WATCH READ REVIEW

Reading Check

Determine whether each sentence is true or false.

RC1. The variables in the polynomial $8x - xy + t^2 - xy^2$ are t, x, and y.

RC2. The degree of the term $4xy$ is 4.

RC3. The terms $3x^2y$ and $3xy^2$ are like terms.

RC4. When we collect like terms, we add the exponents of the variables.

a Evaluate the polynomial when $x = 3$, $y = -2$, and $z = -5$.

1. $x^2 - y^2 + xy$

2. $x^2 + y^2 - xy$

3. $x^2 - 3y^2 + 2xy$

4. $x^2 - 4xy + 5y^2$

5. $8xyz$

6. $-3xyz^2$

7. $xyz^2 - z$

8. $xy - xz + yz$

9. *Lung Capacity.* The polynomial equation

$$C = 0.041h - 0.018A - 2.69$$

can be used to estimate the lung capacity C, in liters, of a person of height h, in centimeters, and age A, in years. Find the lung capacity of a 20-year-old person who is 165 cm tall.

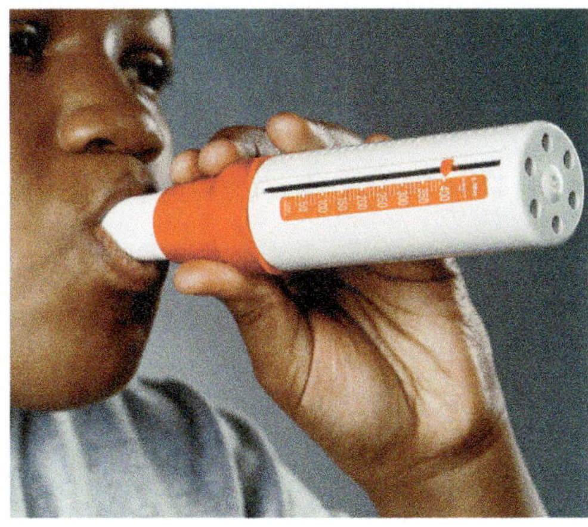

10. *Altitude of a Launched Object.* The altitude h, in meters, of a launched object is given by the polynomial equation

$$h = h_0 + vt - 4.9t^2,$$

where h_0 is the height, in meters, from which the launch occurs, v is the initial upward speed (or velocity), in meters per second (m/s), and t is the number of seconds for which the object is airborne. A rock is thrown upward from the top of the Lands End Arch, near San Lucas, Baja, Mexico, 32 m above the ground. The upward speed is 10 m/s. How high will the rock be 3 sec after it has been thrown?

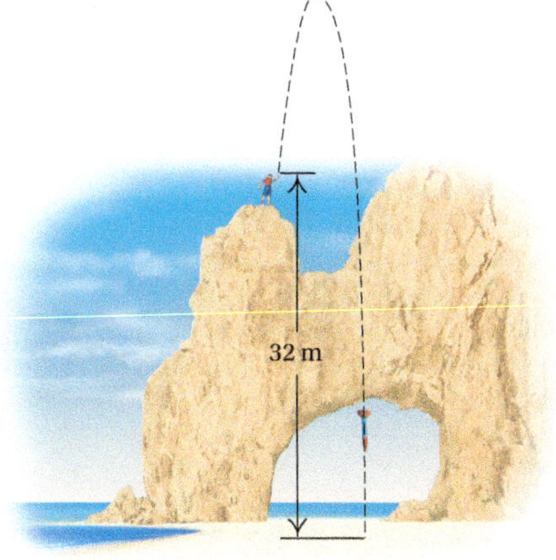

11. *Male Caloric Needs.* The number of calories needed each day by a moderately active man who weighs w kilograms, is h centimeters tall, and is a years old can be estimated by the polynomial

$$19.18w + 7h - 9.52a + 92.4.$$

Steve is moderately active, weighs 82 kg, is 185 cm tall, and is 67 years old. What is his daily caloric need?

Source: Parker, M., *She Does Math.* Mathematical Association of America

12. *Female Caloric Needs.* The number of calories needed each day by a moderately active woman who weighs w pounds, is h inches tall, and is a years old can be estimated by the polynomial

$$917 + 6w + 6h - 6a.$$

Christine is moderately active, weighs 125 lb, is 64 in. tall, and is 27 years old. What is her daily caloric need?

Source: Parker, M., *She Does Math.* Mathematical Association of America

Surface Area of a Right Circular Cylinder. The surface area S of a right circular cylinder is given by the polynomial equation $S = 2\pi rh + 2\pi r^2$, where h is the height and r is the radius of the base. Use this formula for Exercises 13 and 14.

13. A 12-oz beverage can has a height of 4.7 in. and a radius of 1.2 in. Find the surface area of the can. Use 3.14 for π.

14. A 26-oz coffee can has a height of 6.5 in. and a radius of 2.5 in. Find the surface area of the can. Use 3.14 for π.

Surface Area of a Silo. A silo is a structure that is shaped like a right circular cylinder with a half sphere on top. The surface area S of a silo of height h and radius r (including the area of the base) is given by the polynomial equation $S = 2\pi rh + \pi r^2$. Note that h is the height of the entire silo.

15. A coffee grinder is shaped like a silo, with a height of 7 in. and a radius of $1\frac{1}{2}$ in. Find the surface area " of the coffee grinder. Use 3.14 for π.

16. A $1\frac{1}{2}$-oz bottle of roll-on deodorant has a height of 4 in. and a radius of $\frac{3}{4}$ in. Find the surface area of the bottle if the bottle is shaped like a silo. Use 3.14 for π.

b Identify the coefficient and the degree of each term of the polynomial. Then find the degree of the polynomial.

17. $x^3y - 2xy + 3x^2 - 5$

18. $5x^2y^2 - y^2 + 15xy + 1$

19. $17x^2y^3 - 3x^3yz - 7$

20. $6 - xy + 8x^2y^2 - y^5$

c Collect like terms.

21. $a + b - 2a - 3b$

22. $xy^2 - 1 + y - 6 - xy^2$

23. $3x^2y - 2xy^2 + x^2$

24. $m^3 + 2m^2n - 3m^2 + 3mn^2$

25. $6au + 3av + 14au + 7av$

26. $3x^2y - 2z^2y + 3xy^2 + 5z^2y$

27. $2u^2v - 3uv^2 + 6u^2v - 2uv^2$

28. $3x^2 + 6xy + 3y^2 - 5x^2 - 10xy - 5y^2$

d Add.

29. $(2x^2 - xy + y^2) + (-x^2 - 3xy + 2y^2)$

30. $(2zt - z^2 + 5t^2) + (z^2 - 3zt + t^2)$

31. $(r - 2s + 3) + (2r + s) + (s + 4)$

32. $(ab - 2a + 3b) + (5a - 4b) + (3a + 7ab - 8b)$

33. $(b^3a^2 - 2b^2a^3 + 3ba + 4)$
$+ (b^2a^3 - 4b^3a^2 + 2ba - 1)$

34. $(2x^2 - 3xy + y^2) + (-4x^2 - 6xy - y^2)$
$+ (x^2 + xy - y^2)$

e Subtract.

35. $(a^3 + b^3) - (a^2b - ab^2 + b^3 + a^3)$

36. $(x^3 - y^3) - (-2x^3 + x^2y - xy^2 + 2y^3)$

37. $(xy - ab - 8) - (xy - 3ab - 6)$

38. $(3y^4x^2 + 2y^3x - 3y - 7)$
$- (2y^4x^2 + 2y^3x - 4y - 2x + 5)$

39. $(-2a + 7b - c) - (-3b + 4c - 8d)$

40. Subtract $5a + 2b$ from the sum of $2a + b$ and $3a - b$.

f Multiply.

41. $(3z - u)(2z + 3u)$

42. $(a - b)(a^2 + b^2 + 2ab)$

43. $(a^2b - 2)(a^2b - 5)$

44. $(xy + 7)(xy - 4)$

45. $(a^3 + bc)(a^3 - bc)$

46. $(m^2 + n^2 - mn)(m^2 + mn + n^2)$

47. $(y^4x + y^2 + 1)(y^2 + 1)$

48. $(a - b)(a^2 + ab + b^2)$

49. $(3xy - 1)(4xy + 2)$

50. $(m^3n + 8)(m^3n - 6)$

51. $(3 - c^2d^2)(4 + c^2d^2)$

52. $(6x - 2y)(5x - 3y)$

53. $(m^2 - n^2)(m + n)$

54. $(pq + 0.2) \times (0.4pq - 0.1)$

55. $(xy + x^5y^5) \times (x^4y^4 - xy)$

56. $(x - y^3)(2y^3 + x)$

57. $(x + h)^2$

58. $(y - a)^2$

59. $(3a + 2b)^2$

60. $(2ab - cd)^2$

61. $(r^3t^2 - 4)^2$

62. $(3a^2b - b^2)^2$

63. $(p^4 + m^2n^2)^2$

64. $\left(2a^3 - \frac{1}{2}b^3\right)^2$

65. $3a(a - 2b)^2$

66. $-3x(x + 8y)^2$

67. $(m + n - 3)^2$

68. $(a^2 + b + 2)^2$

69. $(a + b)(a - b)$

70. $(x - y)(x + y)$

71. $(2a - b)(2a + b)$

72. $(w + 3z)(w - 3z)$

73. $(c^2 - d)(c^2 + d)$

74. $(p^3 - 5q)(p^3 + 5q)$

75. $(ab + cd^2) \times (ab - cd^2)$

76. $(xy + pq) \times (xy - pq)$

77. $(x + y - 3)(x + y + 3)$

78. $(p + q + 4)(p + q - 4)$

79. $[x + y + z][x - (y + z)]$

80. $[a + b + c][a - (b + c)]$

81. $(a + b + c)(a + b - c)$

82. $(3x + 2 - 5y)(3x + 2 + 5y)$

83. $(x^2 - 4y + 2)(3x^2 + 5y - 3)$

84. $(2x^2 - 7y + 4)(x^2 + y - 3)$

Skill Maintenance

In which quadrant is each point located? [11.1a]

85. $(2, -5)$

86. $(-8, -9)$

87. $(16, 23)$

88. $(-3, 2)$

89. Find the absolute value: $|-39|$. [9.2e]

90. Convert $\frac{9}{8}$ to decimal notation. [9.2c]

91. Use either $<$ or $>$ for $\square$ to write a true sentence: $-17 \square -5$. [9.2d]

92. Evaluate $-(-x)$ when $x = -3$. [9.3b]

Synthesis

Find a polynomial for each shaded area. (Leave results in terms of π where appropriate.)

93.

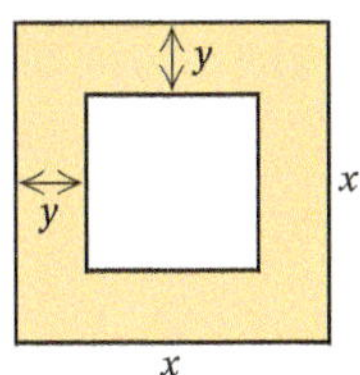

94.

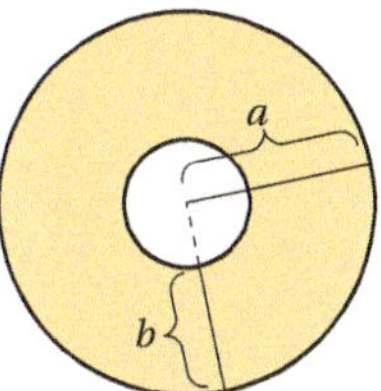

95.

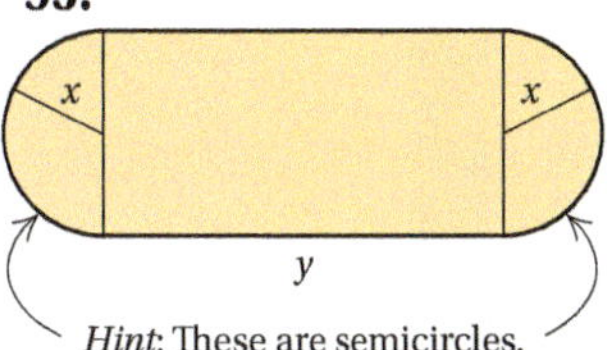

96.

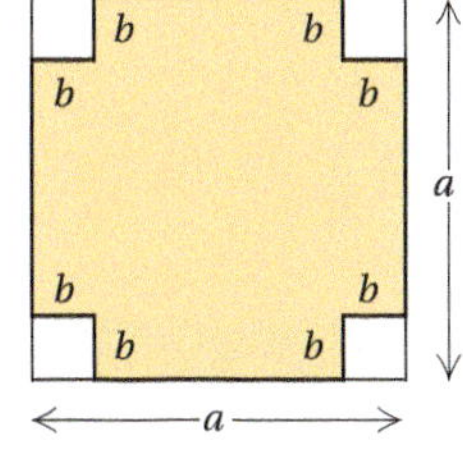

Find a formula for the surface area of each solid object. Leave results in terms of π.

97.

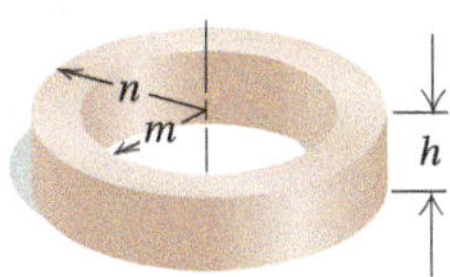

98.

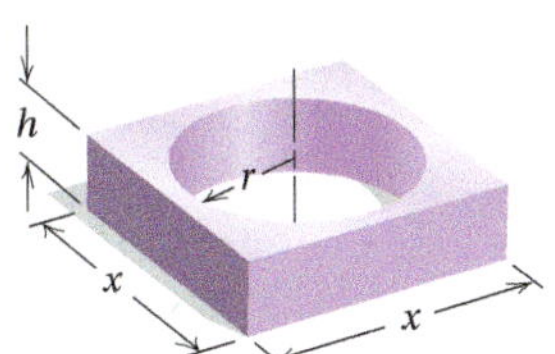

99. *Observatory Paint Costs.* The observatory at Danville University is shaped like a silo that is 40 ft high and 30 ft wide (see Exercise 15). The Heavenly Bodies Astronomy Club is to paint the exterior of the observatory using paint that covers 250 ft^2 per gallon. How many gallons should they purchase?

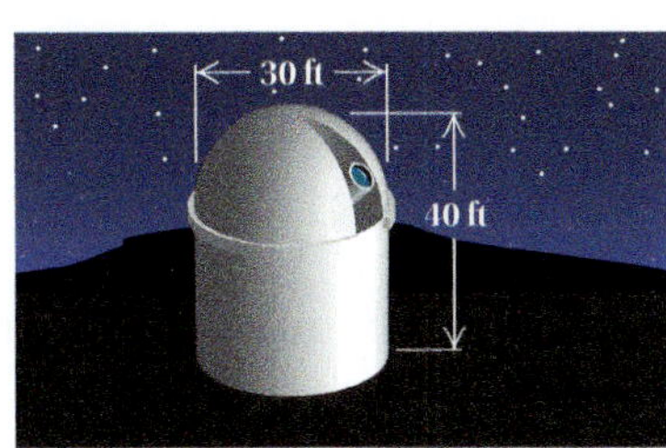

100. *Interest Compounded Annually.* An amount of money P that is invested at the yearly interest rate r grows to the amount

$$P(1 + r)^t$$

after t years. Find a polynomial that can be used to determine the amount to which P will grow after 2 years.

101. Suppose that \$10,400 is invested at 3.5%, compounded annually. How much is in the account at the end of 5 years? (See Exercise 100.)

102. Multiply: $(x + a)(x - b)(x - a)(x + b)$.

12.8 Division of Polynomials

OBJECTIVES

a Divide a polynomial by a monomial.

b Divide a polynomial by a divisor that is a binomial.

SKILL TO REVIEW

Objective 9.6a: Divide integers.

Divide.

1. $\frac{20}{4}$ 2. $\frac{-30}{5}$

In this section, we consider division of polynomials. You will see that such division is similar to what is done in arithmetic.

a DIVIDING BY A MONOMIAL

We first consider division by a monomial. When dividing a monomial by a monomial, we use the quotient rule to subtract exponents when the bases are the same. We also divide the coefficients.

EXAMPLES Divide.

1. $\frac{10x^2}{2x} = \frac{10}{2} \cdot \frac{x^2}{x} = 5x^{2-1} = 5x$

2. $\frac{x^9}{3x^2} = \frac{1x^9}{3x^2} = \frac{1}{3} \cdot \frac{x^9}{x^2} = \frac{1}{3}x^{9-2} = \frac{1}{3}x^7$

3. $\frac{-18x^{10}}{3x^3} = \frac{-18}{3} \cdot \frac{x^{10}}{x^3} = -6x^{10-3} = -6x^7$

4. $\frac{42a^2b^5}{-3ab^2} = \frac{42}{-3} \cdot \frac{a^2}{a} \cdot \frac{b^5}{b^2} = -14a^{2-1}b^{5-2} = -14ab^3$

Caution!

The coefficients are divided, but the exponents are subtracted.

◀ Do Margin Exercises 1–4.

Divide.

1. $\frac{20x^3}{5x}$

2. $\frac{-28x^{14}}{4x^3}$

3. $\frac{-56p^5q^7}{2p^2q^6}$

4. $\frac{x^5}{4x}$

To divide a polynomial by a monomial, we note that since

$$\frac{A}{C} + \frac{B}{C} = \frac{A+B}{C},$$

it follows that

$$\frac{A+B}{C} = \frac{A}{C} + \frac{B}{C}.$$ Switching the left and right sides of the equation

This is actually the procedure we use when performing divisions like $86 \div 2$. Although we might write

$$\frac{86}{2} = 43,$$

we could also calculate as follows:

$$\frac{86}{2} = \frac{80+6}{2} = \frac{80}{2} + \frac{6}{2} = 40 + 3 = 43.$$

Similarly, to divide a polynomial by a monomial, we divide each term by the monomial.

EXAMPLE 5 Divide: $(9x^8 + 12x^6) \div (3x^2)$.

We have

$$(9x^8 + 12x^6) \div (3x^2) = \frac{9x^8 + 12x^6}{3x^2}$$

$$= \frac{9x^8}{3x^2} + \frac{12x^6}{3x^2}.$$ To see this, add and get the original expression.

Answers

Skill to Review:
1. 5 2. −6

Margin Exercises:
1. $4x^2$ 2. $-7x^{11}$ 3. $-28p^3q$ 4. $\frac{1}{4}x^4$

We now perform the separate divisions:

$$\frac{9x^8}{3x^2} + \frac{12x^6}{3x^2} = \frac{9}{3}\cdot\frac{x^8}{x^2} + \frac{12}{3}\cdot\frac{x^6}{x^2}$$
$$= 3x^{8-2} + 4x^{6-2}$$
$$= 3x^6 + 4x^4.$$

Caution!

The coefficients are *divided*, but the exponents are *subtracted*.

To check, we multiply the quotient, $3x^6 + 4x^4$, by the divisor, $3x^2$:

$$3x^2(3x^6 + 4x^4) = (3x^2)(3x^6) + (3x^2)(4x^4) = 9x^8 + 12x^6.$$

This is the polynomial that was being divided, so our answer is $3x^6 + 4x^4$.

Do Exercises 5–7. ▶

Divide. Check the result.

5. $(28x^7 + 32x^5) \div (4x^3)$

$$\frac{28x^7 + 32x^5}{4x^3} = \frac{28x^7}{\square} + \frac{32x^5}{\square}$$
$$= \frac{28}{4}x^{7-\square} + \frac{32}{\square}x^{5-3}$$
$$= 7x^{\square} + \square x^2$$

6. $(2x^3 + 6x^2 + 4x) \div (2x)$

7. $(6x^2 + 3x - 2) \div 3$

EXAMPLE 6 Divide and check: $(10a^5b^4 - 2a^3b^2 + 6a^2b) \div (-2a^2b)$.

$$\frac{10a^5b^4 - 2a^3b^2 + 6a^2b}{-2a^2b} = \frac{10a^5b^4}{-2a^2b} - \frac{2a^3b^2}{-2a^2b} + \frac{6a^2b}{-2a^2b}$$
$$= \frac{10}{-2}\cdot a^{5-2}b^{4-1} - \frac{2}{-2}\cdot a^{3-2}b^{2-1} + \frac{6}{-2}$$
$$= -5a^3b^3 + ab - 3$$

Check: $-2a^2b(-5a^3b^3 + ab - 3) = (-2a^2b)(-5a^3b^3) + (-2a^2b)(ab) - (-2a^2b)(3)$
$$= 10a^5b^4 - 2a^3b^2 + 6a^2b$$

Our answer, $-5a^3b^3 + ab - 3$, checks. ■

To divide a polynomial by a monomial, divide each term of the polynomial by the monomial.

Do Exercises 8 and 9. ▶

Divide and check.

8. $(8x^2 - 3x + 1) \div (-2)$

9. $\dfrac{2x^4y^6 - 3x^3y^4 + 5x^2y^3}{x^2y^2}$

b DIVIDING BY A BINOMIAL

Let's first consider long division as it is performed in arithmetic. When we divide, we repeat the procedure at right.

We review this by considering the division $3711 \div 8$.

```
     4   ←   (1) Divide: 37 ÷ 8 ≈ 4.
 8)3 7 1 1
   3 2   ←   (2) Multiply: 4 × 8 = 32.
     5 1     (3) Subtract: 37 − 32 = 5.
             (4) Bring down the 1.
```

```
       4 6 3
   8 ) 3 7 1 1
       3 2
         5 1
         4 8
           3 1
           2 4
             7
```

To carry out long division:

1. Divide,
2. Multiply,
3. Subtract, and
4. Bring down the next number or term.

Next, we repeat the process two more times. We obtain the complete division as shown on the right above. The quotient is 463. The remainder is 7, expressed as R = 7. We write the answer as

$$463\text{ R }7 \quad\text{or}\quad 463 + \frac{7}{8} = 463\frac{7}{8}.$$

Answers

5. $7x^4 + 8x^2$ **6.** $x^2 + 3x + 2$
7. $2x^2 + x - \frac{2}{3}$ **8.** $-4x^2 + \frac{3}{2}x - \frac{1}{2}$
9. $2x^2y^4 - 3xy^2 + 5y$

Guided Solution:
5. $4x^3, 4x^3, 3, 4, 4, 8$

We check the answer, 463 R 7, by multiplying the quotient, 463, by the divisor, 8, and adding the remainder, 7:

$$8 \cdot 463 + 7 = 3704 + 7 = 3711.$$

Now let's look at long division with polynomials. We use this procedure when the divisor is not a monomial. We write polynomials in descending order and then write in missing terms, if necessary.

EXAMPLE 7 Divide $x^2 + 5x + 6$ by $x + 2$.

$$\begin{array}{r} x \\ x+2\overline{)x^2 + 5x + 6} \\ \underline{x^2 + 2x} \\ 3x \end{array}$$

Divide the first term by the first term: $x^2/x = x$. Ignore the term 2 for this step.

Multiply x above by the divisor, $x + 2$.

Subtract: $(x^2 + 5x) - (x^2 + 2x) = x^2 + 5x - x^2 - 2x = 3x$.

We now "bring down" the next term of the dividend—in this case, 6.

$$\begin{array}{r} x + 3 \\ x+2\overline{)x^2 + 5x + 6} \\ \underline{x^2 + 2x} \\ 3x + 6 \\ \underline{3x + 6} \\ 0 \end{array}$$

Divide the first term of $3x + 6$ by the first term of the divisor: $3x/x = 3$.

The 6 has been "brought down."

Multiply 3 above by the divisor, $x + 2$.

Subtract: $(3x + 6) - (3x + 6) = 3x + 6 - 3x - 6 = 0$.

The quotient is $x + 3$. The remainder is 0. A remainder of 0 is generally not included in an answer.

To check, we multiply the quotient by the divisor and add the remainder, if any, to see if we get the dividend:

$$\underbrace{(x + 2)}_{\text{Divisor}} \cdot \underbrace{(x + 3)}_{\text{Quotient}} + \underbrace{0}_{\text{Remainder}} = \underbrace{x^2 + 5x + 6}_{\text{Dividend}}.$$

The division checks.

◀ **Do Exercise 10.**

10. Divide and check:
$(x^2 + x - 6) \div (x + 3)$.

GS

$$\begin{array}{r} \square - \square \\ x+3\overline{)x^2 + x - 6} \\ \underline{x^2 + \square} \\ \square - 6 \\ \underline{-2x - \square} \\ \square \end{array}$$

EXAMPLE 8 Divide and check: $(x^2 + 2x - 12) \div (x - 3)$.

$$\begin{array}{r} x \\ x-3\overline{)x^2 + 2x - 12} \\ \underline{x^2 - 3x} \\ 5x \end{array}$$

Divide the first term by the first term: $x^2/x = x$.

Multiply x above by the divisor, $x - 3$.

Subtract: $(x^2 + 2x) - (x^2 - 3x) = x^2 + 2x - x^2 + 3x = 5x$.

We now "bring down" the next term of the dividend—in this case, -12.

$$\begin{array}{r} x + 5 \\ x-3\overline{)x^2 + 2x - 12} \\ \underline{x^2 - 3x} \\ 5x - 12 \\ \underline{5x - 15} \\ 3 \end{array}$$

Divide the first term of $5x - 12$ by the first term of the divisor: $5x/x = 5$.

Bring down the -12.

Multiply 5 above by the divisor, $x - 3$.

Subtract: $(5x - 12) - (5x - 15) = 5x - 12 - 5x + 15 = 3$.

Answers

10. $x - 2$

Guided Solution:

10. $$\begin{array}{r} x - 2 \\ x+3\overline{)x^2 + x - 6} \\ \underline{x^2 + 3x} \\ -2x - 6 \\ \underline{-2x - 6} \\ 0 \end{array}$$

The answer is $x + 5$ with $R = 3$, or

$$x + 5 + \frac{3}{x - 3}.$$

Quotient: $x + 5$; Remainder: 3; Divisor: $x - 3$

(This is the way answers will be given at the back of the book.)

Check: We can check by multiplying the divisor by the quotient and adding the remainder, as follows:

$$(x - 3)(x + 5) + 3 = x^2 + 2x - 15 + 3$$
$$= x^2 + 2x - 12.$$

When dividing, an answer may "come out even" (that is, have a remainder of 0, as in Example 7), or it may not (as in Example 8). **If a remainder is not 0, we continue dividing until the degree of the remainder is less than the degree of the divisor.**

Do Exercises 11 and 12.

Divide and check.

11. $x - 2\overline{)x^2 + 2x - 8}$

12. $x + 3\overline{)x^2 + 7x + 10}$

EXAMPLE 9 Divide and check: $(x^3 + 1) \div (x + 1)$.

$$\begin{array}{r} x^2 - x + 1 \\ x + 1\overline{)x^3 + 0x^2 + 0x + 1} \\ \underline{x^3 + x^2} \\ -x^2 + 0x \\ \underline{-x^2 - x} \\ x + 1 \\ \underline{x + 1} \\ 0 \end{array}$$

← Fill in the missing terms. (See Section 12.3.)

Subtract: $x^3 - (x^3 + x^2) = -x^2$.

Subtract: $-x^2 - (-x^2 - x) = x$.

Subtract: $(x + 1) - (x + 1) = 0$.

The answer is $x^2 - x + 1$. The check is left to the student.

EXAMPLE 10 Divide and check: $(9x^4 - 7x^2 - 4x + 13) \div (3x - 1)$.

$$\begin{array}{r} 3x^3 + x^2 - 2x - 2 \\ 3x - 1\overline{)9x^4 + 0x^3 - 7x^2 - 4x + 13} \\ \underline{9x^4 - 3x^3} \\ 3x^3 - 7x^2 \\ \underline{3x^3 - x^2} \\ -6x^2 - 4x \\ \underline{-6x^2 + 2x} \\ -6x + 13 \\ \underline{-6x + 2} \\ 11 \end{array}$$

← Fill in the missing term.

Subtract: $9x^4 - (9x^4 - 3x^3) = 3x^3$.

Subtract: $(3x^3 - 7x^2) - (3x^3 - x^2) = -6x^2$.

Subtract: $(-6x^2 - 4x) - (-6x^2 + 2x) = -6x$.

Subtract: $(-6x + 13) - (-6x + 2) = 11$.

The answer is $3x^3 + x^2 - 2x - 2$ with $R = 11$, or

$$3x^3 + x^2 - 2x - 2 + \frac{11}{3x - 1}.$$

Check: $(3x - 1)(3x^3 + x^2 - 2x - 2) + 11$

$$= 9x^4 + 3x^3 - 6x^2 - 6x - 3x^3 - x^2 + 2x + 2 + 11$$
$$= 9x^4 - 7x^2 - 4x + 13$$

Do Exercises 13 and 14.

Divide and check.

13. $(x^3 - 1) \div (x - 1)$

14. $(8x^4 + 10x^2 + 2x + 9) \div (4x + 2)$

Answers

11. $x + 4$ **12.** $x + 4$ with $R = -2$, or $x + 4 + \frac{-2}{x + 3}$ **13.** $x^2 + x + 1$
14. $2x^3 - x^2 + 3x - 1$ with $R = 11$, or $2x^3 - x^2 + 3x - 1 + \frac{11}{4x + 2}$

For Extra Help
MyMathLab®
MathXL®
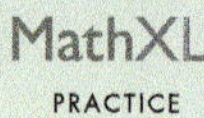
PRACTICE

WATCH

READ

REVIEW

Reading Check

Complete each statement with the appropriate word(s) from the column on the right. A word may be used more than once.

RC1. When dividing a monomial by a monomial, we ________ exponents and ________ coefficients.

RC2. To divide a polynomial by a monomial, we ________ each term by the monomial.

RC3. To carry out long division, we repeat the following process: divide, ________, ________, and bring down the next term.

RC4. To check division, we ________ the divisor and the quotient, and then ________ the remainder.

add
subtract
multiply
divide

a Divide and check.

1. $\dfrac{24x^4}{8}$

2. $\dfrac{-2u^2}{u}$

3. $\dfrac{25x^3}{5x^2}$

4. $\dfrac{16x^7}{-2x^2}$

5. $\dfrac{-54x^{11}}{-3x^8}$

6. $\dfrac{-75a^{10}}{3a^2}$

7. $\dfrac{64a^5b^4}{16a^2b^3}$

8. $\dfrac{-34p^{10}q^{11}}{-17pq^9}$

9. $\dfrac{24x^4 - 4x^3 + x^2 - 16}{8}$

10. $\dfrac{12a^4 - 3a^2 + a - 6}{6}$

11. $\dfrac{u - 2u^2 - u^5}{u}$

12. $\dfrac{50x^5 - 7x^4 + x^2}{x}$

13. $(15t^3 + 24t^2 - 6t) \div (3t)$

14. $(25t^3 + 15t^2 - 30t) \div (5t)$

15. $(20x^6 - 20x^4 - 5x^2) \div (-5x^2)$

16. $(24x^6 + 32x^5 - 8x^2) \div (-8x^2)$

17. $(24x^5 - 40x^4 + 6x^3) \div (4x^3)$

18. $(18x^6 - 27x^5 - 3x^3) \div (9x^3)$

19. $\dfrac{18x^2 - 5x + 2}{2}$

20. $\dfrac{15x^2 - 30x + 6}{3}$

21. $\dfrac{12x^3 + 26x^2 + 8x}{2x}$

22. $\dfrac{2x^4 - 3x^3 + 5x^2}{x^2}$

23. $\dfrac{9r^2s^2 + 3r^2s - 6rs^2}{3rs}$

24. $\dfrac{4x^4y - 8x^6y^2 + 12x^8y^6}{4x^4y}$

b Divide.

25. $(x^2 + 4x + 4) \div (x + 2)$

26. $(x^2 - 6x + 9) \div (x - 3)$

27. $(x^2 - 10x - 25) \div (x - 5)$

28. $(x^2 + 8x - 16) \div (x + 4)$

29. $(x^2 + 4x - 14) \div (x + 6)$

30. $(x^2 + 5x - 9) \div (x - 2)$

31. $\dfrac{x^2 - 9}{x + 3}$

32. $\dfrac{x^2 - 25}{x - 5}$

33. $\dfrac{x^5 + 1}{x + 1}$

34. $\dfrac{x^4 - 81}{x - 3}$

35. $\dfrac{8x^3 - 22x^2 - 5x + 12}{4x + 3}$

36. $\dfrac{2x^3 - 9x^2 + 11x - 3}{2x - 3}$

37. $(x^6 - 13x^3 + 42) \div (x^3 - 7)$

38. $(x^6 + 5x^3 - 24) \div (x^3 - 3)$

39. $(t^3 - t^2 + t - 1) \div (t - 1)$

40. $(y^3 + 3y^2 - 5y - 15) \div (y + 3)$

41. $(y^3 - y^2 - 5y - 3) \div (y + 2)$

42. $(t^3 - t^2 + t - 1) \div (t + 1)$

43. $(15x^3 + 8x^2 + 11x + 12) \div (5x + 1)$

44. $(20x^4 - 2x^3 + 5x + 3) \div (2x - 3)$

45. $(12y^3 + 42y^2 - 10y - 41) \div (2y + 7)$

46. $(15y^3 - 27y^2 - 35y + 60) \div (5y - 9)$

Skill Maintenance

Solve.

47. $-13 = 8d - 5$ [10.3a]

48. $x + \frac{1}{2}x = 5$ [10.3b]

49. $4(x - 3) = 5(2 - 3x) + 1$ [10.3c]

50. $3(r + 1) - 5(r + 2) \geq 15 - (r + 7)$ [10.7e]

51. The number of patients with the flu who were treated at Riverview Clinic increased from 25 one week to 60 the next week. What was the percent increase? [10.5a]

52. Todd's quiz grades are 82, 88, 93, and 92. Determine (in terms of an inequality) what scores on the last quiz will allow him to get an average quiz grade of at least 90. [10.8b]

53. The perimeter of a rectangle is 640 ft. The length is 15 ft more than the width. Find the area of the rectangle. [10.6a]

54. *Book Pages.* The sum of the page numbers on the facing pages of a book is 457. Find the page numbers. [10.6a]

Synthesis

Divide.

55. $(x^4 + 9x^2 + 20) \div (x^2 + 4)$

56. $(y^4 + a^2) \div (y + a)$

57. $(5a^3 + 8a^2 - 23a - 1) \div (5a^2 - 7a - 2)$

58. $(15y^3 - 30y + 7 - 19y^2) \div (3y^2 - 2 - 5y)$

59. $(6x^5 - 13x^3 + 5x + 3 - 4x^2 + 3x^4) \div (3x^3 - 2x - 1)$

60. $(5x^7 - 3x^4 + 2x^2 - 10x + 2) \div (x^2 - x + 1)$

61. $(a^6 - b^6) \div (a - b)$

62. $(x^5 + y^5) \div (x + y)$

If the remainder is 0 when one polynomial is divided by another, the divisor is a *factor* of the dividend. Find the value(s) of c for which $x - 1$ is a factor of the polynomial.

63. $x^2 + 4x + c$

64. $2x^2 + 3cx - 8$

65. $c^2x^2 - 2cx + 1$

CHAPTER

12 Summary and Review

Vocabulary Reinforcement

Complete each statement with the correct word from the column on the right. Some of the choices may not be used.

1. In the expression 7^5, the number 5 is the ______________. [12.1a]
2. The ______________ rule asserts that when multiplying with exponential notation, if the bases are the same, we keep the base and add the exponent. [12.1d]
3. An expression of the type ax^n, where a is a real-number constant and n is a nonnegative integer, is a(n) ______________. [12.3a, b]
4. A(n) ______________ is a polynomial with three terms, such as $5x^4 - 7x^2 + 4$. [12.3b]
5. The ______________ rule asserts that when dividing with exponential notation, if the bases are the same, we keep the base and subtract the exponent of the denominator from the exponent of the numerator. [12.1e]
6. If the exponents in a polynomial decrease from left to right, the polynomial is arranged in ______________ order. [12.3e]
7. The ______________ of a term is the sum of the exponents of the variables. [12.7b]
8. The number 2.3×10^{-5} is written in ______________ notation. [12.2c]

ascending
descending
degree
fraction
scientific
base
exponent
product
quotient
monomial
binomial
trinomial

Concept Reinforcement

Determine whether each statement is true or false.

______________ 1. All trinomials are polynomials. [12.3b]

______________ 2. $(x + y)^2 = x^2 + y^2$ [12.6c]

______________ 3. The square of the difference of two expressions is the difference of the squares of the two expressions. [12.6c]

______________ 4. The product of the sum and the difference of the same two expressions is the difference of the squares of the expressions. [12.6b]

Study Guide

Objective 12.1d Use the product rule to multiply exponential expressions with like bases.

Example Multiply and simplify: $x^3 \cdot x^4$.

$$x^3 \cdot x^4 = x^{3+4} = x^7$$

Practice Exercise

1. Multiply and simplify: $z^5 \cdot z^3$.

Objective 12.1e Use the quotient rule to divide exponential expressions with like bases.

Example Divide and simplify: $\dfrac{x^6y^5}{xy^3}$.

$$\frac{x^6y^5}{xy^3} = \frac{x^6}{x} \cdot \frac{y^5}{y^3}$$
$$= x^{6-1}y^{5-3} = x^5y^2$$

Practice Exercise

2. Divide and simplify: $\dfrac{a^4b^7}{a^2b}$.

Objective 12.1f Express an exponential expression involving negative exponents with positive exponents.

Objective 12.2a Use the power rule to raise powers to powers.

Objective 12.2b Raise a product to a power and a quotient to a power.

Example Simplify: $\left(\dfrac{2a^3b^{-2}}{c^4}\right)^5$.

$$\left(\frac{2a^3b^{-2}}{c^4}\right)^5 = \frac{(2a^3b^{-2})^5}{(c^4)^5}$$
$$= \frac{2^5(a^3)^5(b^{-2})^5}{(c^4)^5} = \frac{32a^{3\cdot5}b^{-2\cdot5}}{c^{4\cdot5}}$$
$$= \frac{32a^{15}b^{-10}}{c^{20}} = \frac{32a^{15}}{b^{10}c^{20}}$$

Practice Exercise

3. Simplify: $\left(\dfrac{x^{-4}y^2}{3z^3}\right)^3$.

Objective 12.2c Convert between scientific notation and decimal notation.

Example Convert 0.00095 to scientific notation.

0.0009.5 (4 places)

The number is small, so the exponent is negative.

$0.00095 = 9.5 \times 10^{-4}$

Example Convert 3.409×10^6 to decimal notation.

3.409000. (6 places)

The exponent is positive, so the number is large.

$3.409 \times 10^6 = 3{,}409{,}000$

Practice Exercises

4. Convert to scientific notation: 763,000.

5. Convert to decimal notation: 3×10^{-4}.

Objective 12.2d Multiply and divide using scientific notation.

Example Multiply and express the result in scientific notation: $(5.3 \times 10^9) \cdot (2.4 \times 10^{-5})$.

$$(5.3 \times 10^9) \cdot (2.4 \times 10^{-5}) = (5.3 \cdot 2.4) \times (10^9 \cdot 10^{-5})$$
$$= 12.72 \times 10^4$$

We convert 12.72 to scientific notation and simplify:

$$12.72 \times 10^4 = (1.272 \times 10) \times 10^4$$
$$= 1.272 \times (10 \times 10^4)$$
$$= 1.272 \times 10^5.$$

Practice Exercise

6. Divide and express the result in scientific notation:

$$\frac{3.6 \times 10^3}{6.0 \times 10^{-2}}.$$

Objective 12.3d Collect the like terms of a polynomial.

Example Collect like terms:

$4x^3 - 2x^2 + 5 + 3x^2 - 12$.

$$\begin{aligned}4x^3 - 2x^2 + 5 + 3x^2 - 12 &= 4x^3 + (-2 + 3)x^2 + (5 - 12)\\ &= 4x^3 + x^2 - 7\end{aligned}$$

Practice Exercise

7. Collect like terms: $5x^4 - 6x^2 - 3x^4 + 2x^2 - 3$.

Objective 12.4a Add polynomials.

Example Add: $(4x^3 + x^2 - 8) + (2x^3 - 5x + 1)$.

$$\begin{aligned}(4x^3 + x^2 - 8) + (2x^3 - 5x + 1) &= (4 + 2)x^3 + x^2 - 5x + (-8 + 1)\\ &= 6x^3 + x^2 - 5x - 7\end{aligned}$$

Practice Exercise

8. Add: $(3x^4 - 5x^2 - 4) + (x^3 + 3x^2 + 6)$.

Objective 12.5d Multiply any two polynomials.

Example Multiply: $(z^2 - 2z + 3)(z - 1)$.

We use columns. First, we multiply the top row by -1 and then by z, placing like terms of the product in the same column. Finally, we collect like terms.

$$\begin{array}{rrrr} & z^2 & -\ 2z & +\ 3\\ & & z & -\ 1\\ \hline & -z^2 & +\ 2z & -\ 3\\ z^3 & -\ 2z^2 & +\ 3z & \\ \hline z^3 & -\ 3z^2 & +\ 5z & -\ 3\end{array}$$

Practice Exercise

9. Multiply: $(x^4 - 3x^2 + 2)(x^2 - 3)$.

Objective 12.6a Multiply two binomials mentally using the FOIL method.

Example Multiply: $(3x + 5)(x - 1)$.

$$\begin{aligned}(3x + 5)(x - 1) &= \overset{F}{3x \cdot x} + \overset{O}{3x \cdot (-1)} + \overset{I}{5 \cdot x} + \overset{L}{5 \cdot (-1)}\\ &= 3x^2 - 3x + 5x - 5\\ &= 3x^2 + 2x - 5\end{aligned}$$

Practice Exercise

10. Multiply: $(y + 4)(2y + 3)$.

Objective 12.6b Multiply the sum and the difference of the same two terms mentally.

Example Multiply: $(3y + 2)(3y - 2)$.

$$\begin{aligned}(3y + 2)(3y - 2) &= (3y)^2 - 2^2\\ &= 9y^2 - 4\end{aligned}$$

Practice Exercise

11. Multiply: $(x + 5)(x - 5)$.

Objective 12.6c Square a binomial mentally.

Example Multiply: $(2x - 3)^2$.

$$\begin{aligned}(2x - 3)^2 &= (2x)^2 - 2 \cdot 2x \cdot 3 + 3^2\\ &= 4x^2 - 12x + 9\end{aligned}$$

Practice Exercise

12. Multiply: $(3w + 4)^2$.

Objective 12.7e Subtract polynomials.

Example Subtract:
$(m^4n + 2m^3n^2 - m^2n^3) - (3m^4n + 2m^3n^2 - 4m^2n^2)$.

$$\begin{aligned}&(m^4n + 2m^3n^2 - m^2n^3) - (3m^4n + 2m^3n^2 - 4m^2n^2)\\&= m^4n + 2m^3n^2 - m^2n^3 - 3m^4n - 2m^3n^2 + 4m^2n^2\\&= -2m^4n - m^2n^3 + 4m^2n^2\end{aligned}$$

Practice Exercise

13. Subtract:
$(a^3b^2 - 5a^2b + 2ab) - (3a^3b^2 - ab^2 + 4ab)$.

Objective 12.8a Divide a polynomial by a monomial.

Example Divide: $(6x^3 - 8x^2 + 15x) \div (3x)$.

$$\begin{aligned}\frac{6x^3 - 8x^2 + 15x}{3x} &= \frac{6x^3}{3x} - \frac{8x^2}{3x} + \frac{15x}{3x}\\&= \frac{6}{3}x^{3-1} - \frac{8}{3}x^{2-1} + \frac{15}{3}x^{1-1}\\&= 2x^2 - \frac{8}{3}x + 5\end{aligned}$$

Practice Exercise

14. Divide: $(5y^2 - 20y + 8) \div 5$.

Objective 12.8b Divide a polynomial by a divisor that is a binomial.

Example Divide $x^2 - 3x + 7$ by $x + 1$.

$$\begin{array}{r} x \;-\; 4 \\ x+1\overline{)\,x^2 - 3x + 7} \\ \underline{x^2 + \;x} \\ -4x + 7 \\ \underline{-4x - 4} \\ 11 \end{array}$$

The answer is $x - 4 + \frac{11}{x + 1}$.

Practice Exercise

15. Divide: $(x^2 - 4x + 3) \div (x + 5)$.

Review Exercises

Multiply and simplify. [12.1d, f]

1. $7^2 \cdot 7^{-4}$

2. $y^7 \cdot y^3 \cdot y$

3. $(3x)^5 \cdot (3x)^9$

4. $t^8 \cdot t^0$

Divide and simplify. [12.1e, f]

5. $\frac{4^5}{4^2}$

6. $\frac{a^5}{a^8}$

7. $\frac{(7x)^4}{(7x)^4}$

Simplify.

8. $(3t^4)^2$ [12.2a, b]

9. $(2x^3)^2(-3x)^2$ [12.1d], [12.2a, b]

10. $\left(\frac{2x}{y}\right)^{-3}$ [12.2b]

11. Express using a negative exponent: $\frac{1}{t^5}$. [12.1f]

12. Express using a positive exponent: y^{-4}. [12.1f]

13. Convert to scientific notation: 0.0000328. [12.2c]

14. Convert to decimal notation: 8.3×10^6. [12.2c]

Multiply or divide and write scientific notation for the result. [12.2d]

15. $(3.8 \times 10^4)(5.5 \times 10^{-1})$

16. $\frac{1.28 \times 10^{-8}}{2.5 \times 10^{-4}}$

17. *Pizza Consumption.* Each man, woman, and child in the United States eats an average of 46 slices of pizza per year. The U.S. population is projected to be about 340 million in 2020. At this rate, how many slices of pizza would be consumed in 2020? Express the answer in scientific notation. [12.2e]

Sources: Packaged Facts; U.S. Census Bureau

18. Evaluate the polynomial $x^2 - 3x + 6$ when $x = -1$. [12.3a]

19. Identify the terms of the polynomial $-4y^5 + 7y^2 - 3y - 2$. [12.3b]

20. Identify the missing terms in $x^3 + x$. [12.3f]

21. Identify the degree of each term and the degree of the polynomial $4x^3 + 6x^2 - 5x + \frac{5}{3}$. [12.3c]

Classify the polynomial as a monomial, a binomial, a trinomial, or none of these. [12.3b]

22. $4x^3 - 1$

23. $4 - 9t^3 - 7t^4 + 10t^2$

24. $7y^2$

Collect like terms and then arrange in descending order. [12.3e]

25. $3x^2 - 2x + 3 - 5x^2 - 1 - x$

26. $-x + \frac{1}{2} + 14x^4 - 7x^2 - 1 - 4x^4$

Add. [12.4a]

27. $(3x^4 - x^3 + x - 4) + (x^5 + 7x^3 - 3x^2 - 5) + (-5x^4 + 6x^2 - x)$

28. $(3x^5 - 4x^4 + x^3 - 3) + (3x^4 - 5x^3 + 3x^2) + (-5x^5 - 5x^2) + (-5x^4 + 2x^3 + 5)$

Subtract. [12.4c]

29. $(5x^2 - 4x + 1) - (3x^2 + 1)$

30. $(3x^5 - 4x^4 + 3x^2 + 3) - (2x^5 - 4x^4 + 3x^3 + 4x^2 - 5)$

31. Find a polynomial for the perimeter and for the area. [12.4d], [12.5b]

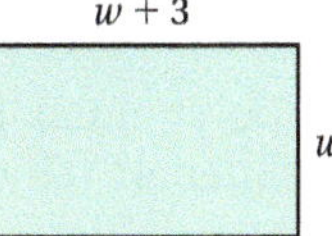

32. Find two algebraic expressions for the area of this figure. First, regard the figure as one large rectangle, and then regard the figure as a sum of four smaller rectangles. [12.4d]

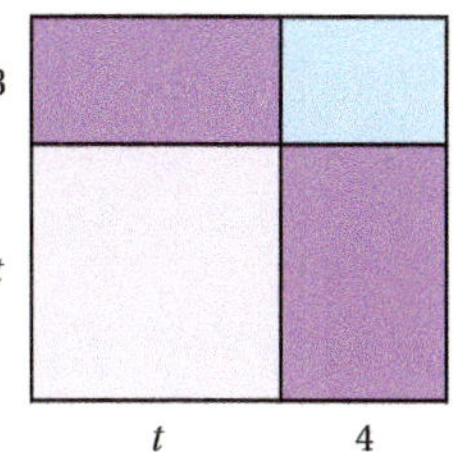

Multiply.

33. $\left(x + \frac{2}{3}\right)\left(x + \frac{1}{2}\right)$ [12.6a]

34. $(7x + 1)^2$ [12.6c]

35. $(4x^2 - 5x + 1)(3x - 2)$ [12.5d]

36. $(3x^2 + 4)(3x^2 - 4)$ [12.6b]

37. $5x^4(3x^3 - 8x^2 + 10x + 2)$ [12.5b]

38. $(x + 4)(x - 7)$ [12.6a]

39. $(3y^2 - 2y)^2$ [12.6c]

40. $(2t^2 + 3)(t^2 - 7)$ [12.6a]

41. Evaluate the polynomial

$$2 - 5xy + y^2 - 4xy^3 + x^6$$

when $x = -1$ and $y = 2$. [12.7a]

42. Identify the coefficient and the degree of each term of the polynomial

$$x^5y - 7xy + 9x^2 - 8.$$

Then find the degree of the polynomial. [12.7b]

Collect like terms. [12.7c]

43. $y + w - 2y + 8w - 5$

44. $m^6 - 2m^2n + m^2n^2 + n^2m - 6m^3 + m^2n^2 + 7n^2m$

45. Add: [12.7d]
$(5x^2 - 7xy + y^2) + (-6x^2 - 3xy - y^2) + (x^2 + xy - 2y^2)$.

46. Subtract: [12.7e]
$(6x^3y^2 - 4x^2y - 6x) - (-5x^3y^2 + 4x^2y + 6x^2 - 6)$.

Multiply. [12.7f]

47. $(p - q)(p^2 + pq + q^2)$

48. $(3a^4 - \frac{1}{3}b^3)^2$

Divide.

49. $(10x^3 - x^2 + 6x) \div (2x)$ [12.8a]

50. $(6x^3 - 5x^2 - 13x + 13) \div (2x + 3)$ [12.8b]

51. The graph of the polynomial equation $y = 10x^3 - 10x$ is shown below. Use *only* the graph to estimate the value of the polynomial when $x = -1$, $x = -0.5$, $x = 0.5$, and $x = 1$. [12.3a]

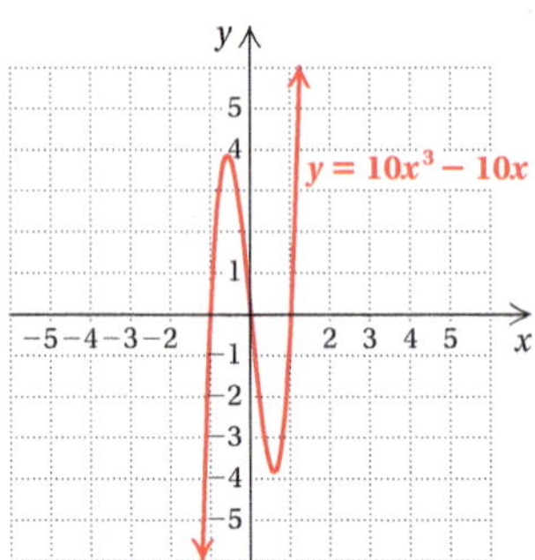

52. Subtract: $(2x^2 - 3x + 4) - (x^2 + 2x)$. [12.4c]

A. $x^2 - 3x - 2$
B. $x^2 - 5x + 4$
C. $x^2 - x + 4$
D. $3x^2 - x + 4$

53. Multiply: $(x - 1)^2$. [12.6c]

A. $x^2 - 1$
B. $x^2 + 1$
C. $x^2 - 2x - 1$
D. $x^2 - 2x + 1$

Synthesis

Find a polynomial for each shaded area. [12.4d], [12.6b]

54.

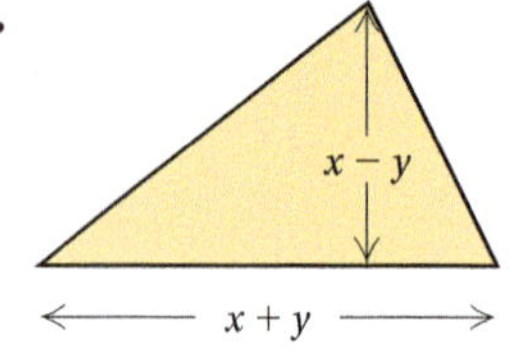

55.

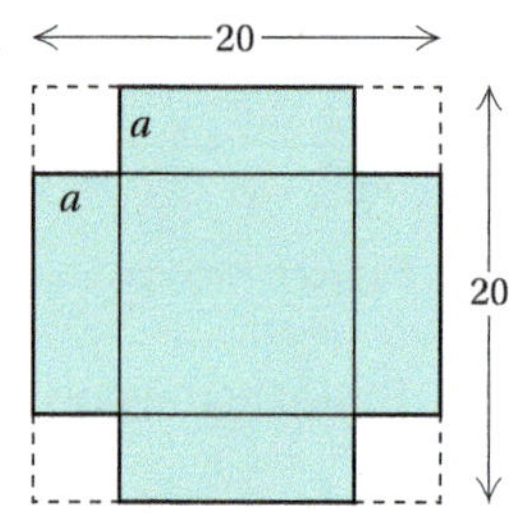

56. Collect like terms: [12.1d], [12.2a], [12.3d]
$-3x^5 \cdot 3x^3 - x^6(2x)^2 + (3x^4)^2 + (2x^2)^4 - 40x^2(x^3)^2.$

57. Solve: [10.3b], [12.6a]
$(x - 7)(x + 10) = (x - 4)(x - 6).$

58. The product of two polynomials is $x^5 - 1$. One of the polynomials is $x - 1$. Find the other. [12.8b]

59. A rectangular garden is twice as long as it is wide and is surrounded by a sidewalk that is 4 ft wide. The area of the sidewalk is 1024 ft^2. Find the dimensions of the garden. [10.3b], [12.4d], [12.5a], [12.6a]

Understanding Through Discussion and Writing

1. Explain why the expression 578.6×10^{-7} is not written in scientific notation. [12.2c]

2. Explain why an understanding of the rules for order of operations is essential when evaluating polynomials. [12.3a]

3. How can the following figure be used to show that $(x + 3)^2 \neq x^2 + 9$? [12.5c]

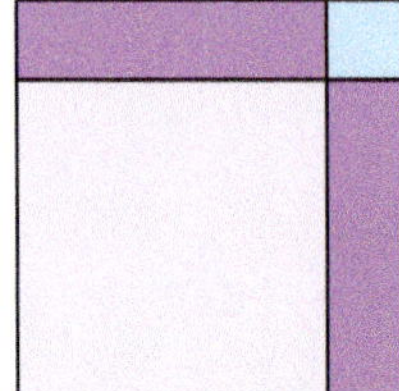

4. On an assignment, Emma *incorrectly* writes

$$\frac{12x^3 - 6x}{3x} = 4x^2 - 6x.$$

What mistake do you think she is making and how might you convince her that a mistake has been made? [12.8a]

5. Can the sum of two trinomials in several variables be a trinomial in one variable? Why or why not? [12.7d]

6. Is it possible for a polynomial in four variables to have a degree less than 4? Why or why not? [12.7b]

CHAPTER 12 Test

For Extra Help For step-by-step test solutions, access the Chapter Test Prep Videos in MyMathLab® or on YouTube (search "BittingerPreIntro" and click on Channels).

Multiply and simplify.

1. $6^{-2} \cdot 6^{-3}$

2. $x^6 \cdot x^2 \cdot x$

3. $(4a)^3 \cdot (4a)^8$

Divide and simplify.

4. $\dfrac{3^5}{3^2}$

5. $\dfrac{x^3}{x^8}$

6. $\dfrac{(2x)^5}{(2x)^5}$

Simplify.

7. $(x^3)^2$

8. $(-3y^2)^3$

9. $(2a^3b)^4$

10. $\left(\dfrac{ab}{c}\right)^3$

11. $(3x^2)^3(-2x^5)^3$

12. $3(x^2)^3(-2x^5)^3$

13. $2x^2(-3x^2)^4$

14. $(2x)^2(-3x^2)^4$

15. Express using a positive exponent: 5^{-3}.

16. Express using a negative exponent: $\dfrac{1}{y^8}$.

17. Convert to scientific notation: 3,900,000,000.

18. Convert to decimal notation: 5×10^{-8}.

Multiply or divide and write scientific notation for the answer.

19. $\dfrac{5.6 \times 10^6}{3.2 \times 10^{-11}}$

20. $(2.4 \times 10^5)(5.4 \times 10^{16})$

21. *Earth vs. Saturn.* The mass of Earth is about 6×10^{21} metric tons. The mass of Saturn is about 5.7×10^{23} metric tons. About how many times the mass of Earth is the mass of Saturn? Express the answer in scientific notation.

22. Evaluate the polynomial $x^5 + 5x - 1$ when $x = -2$.

23. Identify the coefficient of each term of the polynomial $\frac{1}{3}x^5 - x + 7$.

24. Identify the degree of each term and the degree of the polynomial $2x^3 - 4 + 5x + 3x^6$.

25. Classify the polynomial $7 - x$ as a monomial, a binomial, a trinomial, or none of these.

Collect like terms.

26. $4a^2 - 6 + a^2$

27. $y^2 - 3y - y + \dfrac{3}{4}y^2$

28. Collect like terms and then arrange in descending order:

$$3 - x^2 + 2x^3 + 5x^2 - 6x - 2x + x^5.$$

Add.

29. $(3x^5 + 5x^3 - 5x^2 - 3) + (x^5 + x^4 - 3x^3 - 3x^2 + 2x - 4)$

30. $\left(x^4 + \dfrac{2}{3}x + 5\right) + \left(4x^4 + 5x^2 + \dfrac{1}{3}x\right)$

Subtract.

31. $(2x^4 + x^3 - 8x^2 - 6x - 3) - (6x^4 - 8x^2 + 2x)$

32. $(x^3 - 0.4x^2 - 12) - (x^5 + 0.3x^3 + 0.4x^2 + 9)$

Multiply.

33. $-3x^2(4x^2 - 3x - 5)$

34. $\left(x - \frac{1}{3}\right)^2$

35. $(3x + 10)(3x - 10)$

36. $(3b + 5)(b - 3)$

37. $(x^6 - 4)(x^8 + 4)$

38. $(8 - y)(6 + 5y)$

39. $(2x + 1)(3x^2 - 5x - 3)$

40. $(5t + 2)^2$

41. Collect like terms:
$x^3y - y^3 + xy^3 + 8 - 6x^3y - x^2y^2 + 11.$

42. Subtract:
$(8a^2b^2 - ab + b^3) - (-6ab^2 - 7ab - ab^3 + 5b^3).$

43. Multiply: $(3x^5 - 4y^5)(3x^5 + 4y^5).$

Divide.

44. $(12x^4 + 9x^3 - 15x^2) \div (3x^2)$

45. $(6x^3 - 8x^2 - 14x + 13) \div (3x + 2)$

46. The graph of the polynomial equation
$$y = x^3 - 5x - 1$$
is shown at right. Use *only* the graph to estimate the value of the polynomial when $x = -1$, $x = -0.5$, $x = 0.5$, $x = 1$, and $x = 1.1$.

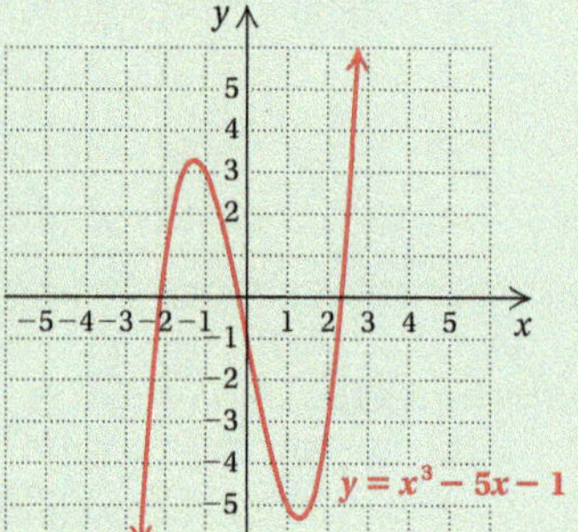

47. Find two algebraic expressions for the area of this figure. First, regard the figure as one large rectangle, and then regard the figure as a sum of four smaller rectangles.

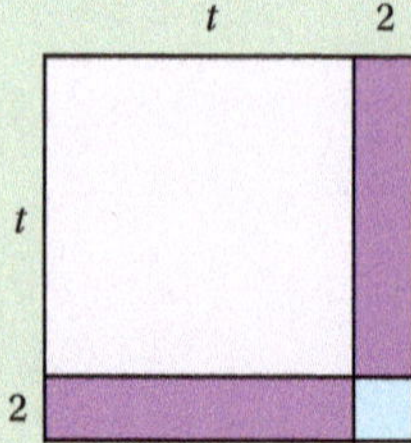

48. Which of the following is a polynomial for the surface area of this right rectangular solid?

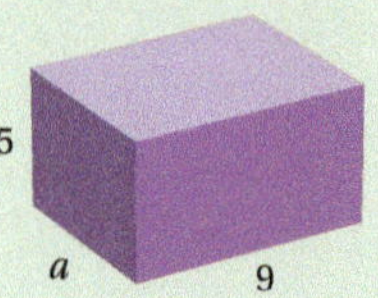

A. $28a$ **B.** $28a + 90$
C. $14a + 45$ **D.** $45a$

Synthesis

49. The height of a box is 1 less than its length, and the length is 2 more than its width. Find the volume in terms of the length.

50. Solve: $(x - 5)(x + 5) = (x + 6)^2.$

CHAPTER

13

Polynomials: Factoring

STUDYING FOR SUCCESS *Working Exercises*

- ☐ Don't begin solving a homework problem by working backward from the answer given at the back of the text. Remember: Quizzes and tests have no answer section!
- ☐ Check answers to odd-numbered exercises at the back of the book.
- ☐ Work some even-numbered exercises as practice doing exercises without answers. Check your answers later with a friend or your instructor.

13.1 Introduction to Factoring

OBJECTIVES

a Find the greatest common factor, the GCF, of monomials.

b Factor polynomials when the terms have a common factor, factoring out the greatest common factor.

c Factor certain expressions with four terms using factoring by grouping.

SKILL TO REVIEW

Objectives 3.2a, 3.2c: Find all the factors of numbers and find prime factorizations of numbers.

Find the prime factorization of each number.

1. 60 **2.** 105

We introduce factoring with a review of factoring natural numbers. Consider the product $3 \cdot 5 = 15$. We say that 3 and 5 are **factors** of 15 and that $3 \cdot 5$ is a **factorization** of 15. Since $15 = 15 \cdot 1$, we also know that 15 and 1 are factors of 15 and that $15 \cdot 1$ is a factorization of 15.

a FINDING THE GREATEST COMMON FACTOR

The numbers 20 and 30 have several factors in common, among them 2 and 5. The greatest of the common factors is called the **greatest common factor, GCF**. We can find the GCF of a set of numbers using prime factorizations.

EXAMPLE 1 Find the GCF of 20 and 30.

We find the prime factorization of each number. Then we draw lines between the common factors.

$$20 = 2 \cdot 2 \cdot 5$$
$$30 = 2 \cdot 3 \cdot 5$$

The GCF $= 2 \cdot 5 = 10$.

EXAMPLE 2 Find the GCF of 180 and 420.

We find the prime factorization of each number. Then we draw lines between the common factors.

$$180 = 2 \cdot 2 \cdot 3 \cdot 3 \cdot 5 = 2^2 \cdot 3^2 \cdot 5^1$$
$$420 = 2 \cdot 2 \cdot 3 \cdot 5 \cdot 7 = 2^2 \cdot 3^1 \cdot 5^1 \cdot 7^1$$

The GCF $= 2 \cdot 2 \cdot 3 \cdot 5 = 2^2 \cdot 3^1 \cdot 5^1 = 60$. Note how we can use the exponents to determine the GCF. There are 2 lines for the 2's, 1 line for the 3, 1 line for the 5, and no line for the 7.

EXAMPLE 3 Find the GCF of 30 and 77.

We find the prime factorization of each number. Then we draw lines between the common factors, if any exist.

$$30 = 2 \cdot 3 \cdot 5 = 2^1 \cdot 3^1 \cdot 5^1$$
$$77 = 7 \cdot 11 = 7^1 \cdot 11^1$$

Since there is no common prime factor, the GCF is 1.

Answers

Skill to Review:
1. $2 \cdot 2 \cdot 3 \cdot 5$ **2.** $3 \cdot 5 \cdot 7$

EXAMPLE 4 Find the GCF of 54, 90, and 252.

We find the prime factorization of each number. Then we draw lines between the common factors.

$$54 = 2 \cdot 3 \cdot 3 \cdot 3 = 2^1 \cdot 3^3,$$

$$90 = 2 \cdot 3 \cdot 3 \cdot 5 = 2^1 \cdot 3^2 \cdot 5^1,$$

$$252 = 2 \cdot 2 \cdot 3 \cdot 3 \cdot 7 = 2^2 \cdot 3^2 \cdot 7^1$$

The GCF $= 2^1 \cdot 3^2 = 18$.

Do Exercises 1–4.

Find the GCF.

1. 40, 100
2. 7, 21
3. 72, 360, 432
4. 3, 5, 22

Consider the product

$$12x^3(x^2 - 6x + 2) = 12x^5 - 72x^4 + 24x^3.$$

To factor the polynomial on the right-hand side, we reverse the process of multiplication:

$$12x^5 - 72x^4 + 24x^3 = 12x^3(x^2 - 6x + 2).$$

This is a *factorization*. The *factors* are $(12x^3)$ and $(x^2 - 6x + 2)$.

FACTOR; FACTORIZATION

To **factor** a polynomial is to express it as a product.

A **factor** of a polynomial P is a polynomial that can be used to express P as a product.

A **factorization** of a polynomial is an expression that names that polynomial as a product.

In the factorization

$$12x^5 - 72x^4 + 24x^3 = 12x^3(x^2 - 6x + 2),$$

the monomial $12x^3$ is called the GCF of the terms, $12x^5$, $-72x^4$, and $24x^3$. The first step in factoring polynomials is to find the GCF of the terms. To do this, we find the greatest positive common factor of the coefficients and the greatest common factors of the powers of any variables.

EXAMPLE 5 Find the GCF of $15x^5$, $-12x^4$, $27x^3$, and $-3x^2$.

First, we find a prime factorization of the coefficients, including a factor of -1 for the negative coefficients.

$$15x^5 = 3 \cdot 5 \cdot x^5,$$

$$-12x^4 = -1 \cdot 2 \cdot 2 \cdot 3 \cdot x^4,$$

$$27x^3 = 3 \cdot 3 \cdot 3 \cdot x^3,$$

$$-3x^2 = -1 \cdot 3 \cdot x^2$$

The greatest *positive* common factor of the coefficients is 3.

Next, we find the GCF of the powers of x. That GCF is x^2, because 2 is the smallest exponent of x. Thus the GCF of the set of monomials is $3x^2$.

Answers

1. 20 2. 7 3. 72 4. 1

EXAMPLE 6 Find the GCF of $14p^2y^3$, $-8py^2$, $2py$, and $4p^3$.

We have

$$14p^2y^3 = 2 \cdot 7 \cdot p^2 \cdot y^3,$$
$$-8py^2 = -1 \cdot 2 \cdot 2 \cdot 2 \cdot p \cdot y^2,$$
$$2py = 2 \cdot p \cdot y,$$
$$4p^3 = 2 \cdot 2 \cdot p^3.$$

The greatest positive common factor of the coefficients is 2, the GCF of the powers of p is p, and the GCF of the powers of y is 1 since there is no y-factor in the last monomial. Thus the GCF is $2p$.

TO FIND THE GCF OF TWO OR MORE MONOMIALS

1. Find the prime factorization of the coefficients, including -1 as a factor if any coefficient is negative.
2. Determine the greatest positive common factor of the coefficients. (If the coefficients have no prime factors in common, the GCF of the coefficients is 1.)
3. Determine the greatest common factor of the powers of any variables. If any variable appears as a factor of all the monomials, include it as a factor, using the smallest exponent of the variable. (If none occurs in all the monomials, the GCF of the variables is 1.)
4. The GCF of the monomials is the product of the results of steps (2) and (3).

Find the GCF.

5. $12x^2$, $-16x^3$

6. $3y^6$, $-5y^3$, $2y^2$

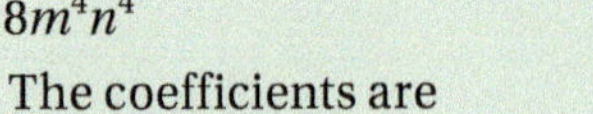

7. $-24m^5n^6$, $12mn^3$, $-16m^2n^2$, $8m^4n^4$ GS

The coefficients are -24, 12, -16, and ____.

The greatest positive common factor of the coefficients is ____.

The smallest exponent of the variable m is ____.

The smallest exponent of the variable n is ____.

The GCF $= 4mn$ ____.

8. $-35x^7$, $-49x^6$, $-14x^5$, $-63x^3$

◀ **Do Exercises 5–8.**

b FACTORING WHEN TERMS HAVE A COMMON FACTOR

To multiply a monomial and a polynomial with more than one term, we multiply each term of the polynomial by the monomial using the distributive laws:

$$a(b + c) = ab + ac \quad \text{and} \quad a(b - c) = ab - ac.$$

To factor, we express a polynomial as a product using the distributive laws in reverse:

$$ab + ac = a(b + c) \quad \text{and} \quad ab - ac = a(b - c).$$

Compare.

Multiply	*Factor*
$3x(x^2 + 2x - 4)$	$3x^3 + 6x^2 - 12x$
$= 3x \cdot x^2 + 3x \cdot 2x - 3x \cdot 4$	$= 3x \cdot x^2 + 3x \cdot 2x - 3x \cdot 4$
$= 3x^3 + 6x^2 - 12x$	$= 3x(x^2 + 2x - 4)$

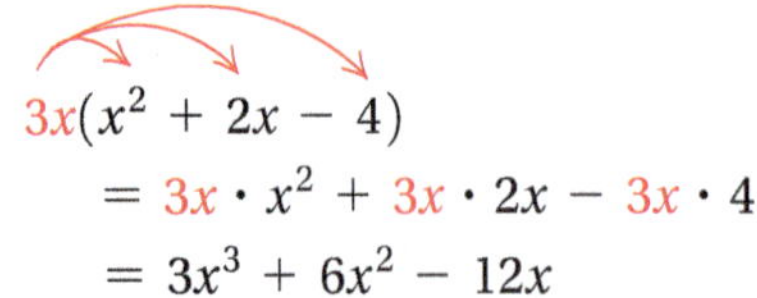

9. a) Multiply: $3(x + 2)$.
   b) Factor: $3x + 6$.

10. a) Multiply: $2x(x^2 + 5x + 4)$.
    b) Factor: $2x^3 + 10x^2 + 8x$.

◀ **Do Exercises 9 and 10.**

Answers

5. $4x^2$ 6. y^2 7. $4mn^2$ 8. $7x^3$
9. (a) $3x + 6$; (b) $3(x + 2)$
10. (a) $2x^3 + 10x^2 + 8x$; (b) $2x(x^2 + 5x + 4)$

Guided Solution:
7. 8, 4, 1, 2, 2

EXAMPLE 7 Factor: $7x^2 + 14$.

We have

$$7x^2 + 14 = 7 \cdot x^2 + 7 \cdot 2 \quad \text{Factoring each term}$$
$$= 7(x^2 + 2). \quad \text{Factoring out the GCF, 7}$$

Check: We multiply to check:

$$7(x^2 + 2) = 7 \cdot x^2 + 7 \cdot 2 = 7x^2 + 14.$$

EXAMPLE 8 Factor: $16x^3 + 20x^2$.

$$16x^3 + 20x^2 = (4x^2)(4x) + (4x^2)(5) \quad \text{Factoring each term}$$
$$= 4x^2(4x + 5) \quad \text{Factoring out the GCF, } 4x^2$$

Although it is always more efficient to begin by finding the GCF, suppose in Example 8 that you had not recognized the GCF and removed only part of it, as follows:

$$16x^3 + 20x^2 = (2x^2)(8x) + (2x^2)(10)$$
$$= 2x^2(8x + 10).$$

Note that $8x + 10$ still has a common factor of 2. You need not begin again. Just continue factoring out common factors, as follows, until finished:

$$= 2x^2(2 \cdot 4x + 2 \cdot 5)$$
$$= 2x^2[2(4x + 5)]$$
$$= (2x^2 \cdot 2)(4x + 5)$$
$$= 4x^2(4x + 5).$$

EXAMPLE 9 Factor: $15x^5 - 12x^4 + 27x^3 - 3x^2$.

$$15x^5 - 12x^4 + 27x^3 - 3x^2 = (3x^2)(5x^3) - (3x^2)(4x^2) + (3x^2)(9x) - (3x^2)(1)$$
$$= 3x^2(5x^3 - 4x^2 + 9x - 1) \quad \text{Factoring out the GCF, } 3x^2$$

Caution!

Don't forget the term -1.

As you become more familiar with factoring, you will be able to spot the GCF without factoring each term. Then you can write just the answer.

EXAMPLES Factor.

10. $24x^2 + 12x - 36 = 12(2x^2 + x - 3)$

11. $8m^3 - 16m = 8m(m^2 - 2)$

12. $14p^2y^3 - 8py^2 + 2py = 2py(7py^2 - 4y + 1)$

13. $\frac{4}{5}x^2 + \frac{1}{5}x + \frac{2}{5} = \frac{1}{5}(4x^2 + x + 2)$

Do Exercises 11–16.

Caution!

Consider the following:

$7x^2 + 14 = 7 \cdot x \cdot x + 7 \cdot 2.$

The terms of the polynomial have been factored, but the polynomial itself has not been factored. This is not what we mean by the factorization of the polynomial. The *factorization* is

$7(x^2 + 2)$. ← A product

The expressions 7 and $x^2 + 2$ are *factors* of $7x^2 + 14$.

Factor. Check by multiplying.

11. $x^2 + 3x$

12. $3y^6 - 5y^3 + 2y^2$

13. $9x^4y^2 - 15x^3y + 3x^2y$

14. $\frac{3}{4}t^3 + \frac{5}{4}t^2 + \frac{7}{4}t + \frac{1}{4}$

15. $35x^7 - 49x^6 + 14x^5 - 63x^3$

16. $84x^2 - 56x + 28$

Answers

11. $x(x + 3)$ **12.** $y^2(3y^4 - 5y + 2)$
13. $3x^2y(3x^2y - 5x + 1)$
14. $\frac{1}{4}(3t^3 + 5t^2 + 7t + 1)$
15. $7x^3(5x^4 - 7x^3 + 2x^2 - 9)$
16. $28(3x^2 - 2x + 1)$

C FACTORING BY GROUPING: FOUR TERMS

Certain polynomials with four terms can be factored using a method called *factoring by grouping.*

EXAMPLE 14 Factor: $x^2(x + 1) + 2(x + 1)$.

The binomial $x + 1$ is a common factor. We factor it out:

$$x^2(x + 1) + 2(x + 1) = (x + 1)(x^2 + 2).$$

The factorization is $(x + 1)(x^2 + 2)$.

Factor.

17. $x^2(x + 7) + 3(x + 7)$

18. $x^3(a + b) - 5(a + b)$

Do Exercises 17 and 18.

Consider the four-term polynomial

$$x^3 + x^2 + 2x + 2.$$

There is no factor other than 1 that is common to all the terms. We can, however, factor $x^3 + x^2$ and $2x + 2$ separately:

$x^3 + x^2 = x^2(x + 1)$; Factoring $x^3 + x^2$

$2x + 2 = 2(x + 1)$. Factoring $2x + 2$

When we group the terms as shown above and factor each polynomial separately, we see that $(x + 1)$ appears in *both* factorizations. Thus we can factor out the common binomial factor as in Example 14:

$$\begin{aligned} x^3 + x^2 + 2x + 2 &= (x^3 + x^2) + (2x + 2) \\ &= x^2(x + 1) + 2(x + 1) \\ &= (x + 1)(x^2 + 2). \end{aligned}$$

This method of factoring is called **factoring by grouping**.

Not all polynomials with four terms can be factored by grouping, but it does give us a method to try.

EXAMPLES Factor by grouping.

15. $6x^3 - 9x^2 + 4x - 6$

$= (6x^3 - 9x^2) + (4x - 6)$ Grouping the terms

$= 3x^2(2x - 3) + 2(2x - 3)$ Factoring each binomial

$= (2x - 3)(3x^2 + 2)$ Factoring out the common factor $2x - 3$

We think through this process as follows:

$$6x^3 - 9x^2 + 4x - 6 = 3x^2(2x - 3) \,\square\, (2x - 3)$$

(1) Factor the first two terms.

(2) The factor $2x - 3$ gives us a hint for factoring of the last two terms.

(3) Now we ask ourselves, "What times $2x - 3$ is $4x - 6$?" The answer is $+ 2$.

Answers

17. $(x + 7)(x^2 + 3)$ **18.** $(a + b)(x^3 - 5)$

Caution!

Don't forget the 1.

16. $x^3 + x^2 + x + 1 = (x^3 + x^2) + (x + 1)$

$= x^2(x + 1) + 1(x + 1)$ — Factoring each binomial

$= (x + 1)(x^2 + 1)$ — Factoring out the common factor $x + 1$

17. $2x^3 - 6x^2 - x + 3$

$= (2x^3 - 6x^2) + (-x + 3)$ — Grouping as two binomials

$= 2x^2(x - 3) - 1(x - 3)$ — *Check:* $-1(x - 3) = -x + 3$.

$= (x - 3)(2x^2 - 1)$ — Factoring out the common factor $x - 3$

We can think through this process as follows.

(1) Factor the first two terms: $2x^3 - 6x^2 = 2x^2(x - 3)$.

(2) The factor $x - 3$ gives us a hint for factoring the last two terms:

$$2x^3 - 6x^2 - x + 3 = 2x^2(x - 3)\square(x - 3).$$

(3) We ask, "What times $x - 3$ is $-x + 3$?" The answer is -1.

18. $12x^5 + 20x^2 - 21x^3 - 35 = 4x^2(3x^3 + 5) - 7(3x^3 + 5)$

$= (3x^3 + 5)(4x^2 - 7)$

19. $x^3 + x^2 + 2x - 2 = x^2(x + 1) + 2(x - 1)$

This polynomial is not factorable using factoring by grouping. It may be factorable, but not by methods that we will consider in this text.

Do Exercises 19–24. ▶

There are two important points to keep in mind when factoring.

> **TIPS FOR FACTORING**
>
> - Before doing any other kind of factoring, first try to factor out the GCF.
> - Always check the result of factoring by multiplying.

Factor by grouping.

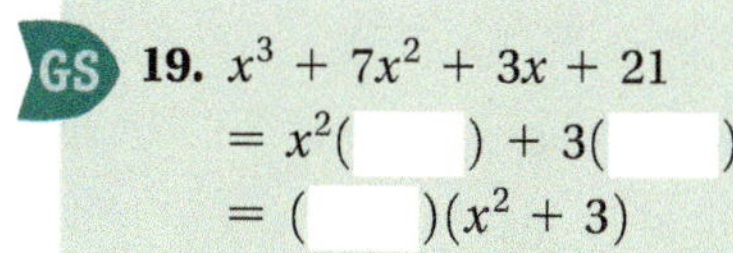

GS **19.** $x^3 + 7x^2 + 3x + 21$

$= x^2(\quad) + 3(\quad)$

$= (\quad)(x^2 + 3)$

20. $8t^3 + 2t^2 + 12t + 3$

21. $3m^5 - 15m^3 + 2m^2 - 10$

22. $3x^3 - 6x^2 - x + 2$

23. $4x^3 - 6x^2 - 6x + 9$

24. $y^4 - 2y^3 - 2y - 10$

Answers

19. $(x + 7)(x^2 + 3)$ **20.** $(4t + 1)(2t^2 + 3)$ **21.** $(m^2 - 5)(3m^3 + 2)$ **22.** $(x - 2)(3x^2 - 1)$ **23.** $(2x - 3)(2x^2 - 3)$ **24.** Not factorable using factoring by grouping

Guided Solution:
19. $x + 7, x + 7, x + 7$

13.1 Exercise Set

For Extra Help MyMathLab® MathXL®

PRACTICE

WATCH

READ

REVIEW

✓ Reading Check

Choose from the column on the right the expression that fits each description.

RC1. ____ A factorization of $36x^2$

RC2. ____ A factorization of $27x$

RC3. ____ The greatest common factor of $36x^2 - 27x$

RC4. ____ A factorization of $36x^2 - 27x$

a) $9x(4x - 3)$
b) $(9x)(4x)$
c) $(9x)(3)$
d) $9x$

a Find the GCF.

1. 36, 42
2. 60, 75
3. 48, 72, 120
4. 90, 135, 225
5. 8, 15, 40
6. 12, 20, 75
7. $x^2, -6x$
8. $x^2, 5x$
9. $3x^4, x^2$
10. $8x^4, -24x^2$
11. $2x^2, 2x, -8$
12. $8x^2, -4x, -20$
13. $-17x^5y^3, 34x^3y^2, 51xy$
14. $16p^6q^4, 32p^3q^3, -48pq^2$
15. $-x^2, -5x, -20x^3$
16. $-x^2, -6x, -24x^5$
17. $x^5y^5, x^4y^3, x^3y^3, -x^2y^2$
18. $-x^9y^6, -x^7y^5, x^4y^4, x^3y^3$

b Factor. Check by multiplying.

19. $x^2 - 6x$
20. $x^2 + 5x$
21. $2x^2 + 6x$
22. $8y^2 - 8y$
23. $x^3 + 6x^2$
24. $3x^4 - x^2$
25. $8x^4 - 24x^2$
26. $5x^5 + 10x^3$
27. $2x^2 + 2x - 8$
28. $8x^2 - 4x - 20$
29. $17x^5y^3 + 34x^3y^2 + 51xy$
30. $16p^6q^4 + 32p^5q^3 - 48pq^2$
31. $6x^4 - 10x^3 + 3x^2$
32. $5x^5 + 10x^2 - 8x$
33. $x^5y^5 + x^4y^3 + x^3y^3 - x^2y^2$
34. $x^9y^6 - x^7y^5 + x^4y^4 + x^3y^3$
35. $2x^7 - 2x^6 - 64x^5 + 4x^3$
36. $8y^3 - 20y^2 + 12y - 16$
37. $1.6x^4 - 2.4x^3 + 3.2x^2 + 6.4x$
38. $2.5x^6 - 0.5x^4 + 5x^3 + 10x^2$
39. $\frac{5}{3}x^6 + \frac{4}{3}x^5 + \frac{1}{3}x^4 + \frac{1}{3}x^3$
40. $\frac{5}{9}x^7 + \frac{2}{9}x^5 - \frac{4}{9}x^3 - \frac{1}{9}x$

c Factor.

41. $x^2(x + 3) + 2(x + 3)$
42. $y^2(y + 4) + 6(y + 4)$
43. $4z^2(3z - 1) + 7(3z - 1)$
44. $2x^2(4x - 3) + 5(4x - 3)$

45. $2x^2(3x + 2) + (3x + 2)$

46. $3z^2(2z + 7) + (2z + 7)$

47. $5a^3(2a - 7) - (2a - 7)$

48. $m^4(8 - 3m) - 3(8 - 3m)$

Factor by grouping.

49. $x^3 + 3x^2 + 2x + 6$

50. $6z^3 + 3z^2 + 2z + 1$

51. $2x^3 + 6x^2 + x + 3$

52. $3x^3 + 2x^2 + 3x + 2$

53. $8x^3 - 12x^2 + 6x - 9$

54. $10x^3 - 25x^2 + 4x - 10$

55. $12p^3 - 16p^2 + 3p - 4$

56. $18x^3 - 21x^2 + 30x - 35$

57. $5x^3 - 5x^2 - x + 1$

58. $7x^3 - 14x^2 - x + 2$

59. $x^3 + 8x^2 - 3x - 24$

60. $2x^3 + 12x^2 - 5x - 30$

61. $2x^3 - 8x^2 - 9x + 36$

62. $20g^3 - 4g^2 - 25g + 5$

Skill Maintenance

Multiply. [12.5b], [12.6d], [12.7f]

63. $(y + 5)(y + 7)$

64. $(y + 7)^2$

65. $(y + 7)(y - 7)$

66. $(y - 7)^2$

67. $8x(2x^2 - 6x + 1)$

68. $(7w + 6)(4w - 11)$

69. $(7w + 6)^2$

70. $(4w - 11)^2$

71. $(4w - 11)(4w + 11)$

72. $-y(-y^2 + 3y - 5)$

73. $(3x - 5y)(2x + 7y)$

74. $(5x - t)^2$

Synthesis

Factor.

75. $4x^5 + 6x^3 + 6x^2 + 9$

76. $x^6 + x^4 + x^2 + 1$

77. $x^{12} + x^7 + x^5 + 1$

78. $x^3 - x^2 - 2x + 5$

79. $p^3 + p^2 - 3p + 10$

80. $4y^6 + 2y^4 - 12y^3 - 6y$

13.2 Factoring Trinomials of the Type $x^2 + bx + c$

OBJECTIVE

a Factor trinomials of the type $x^2 + bx + c$ by examining the constant term c.

SKILL TO REVIEW

Objective 12.6a: Multiply two binomials mentally using the FOIL method.

Multiply.

1. $(x + 3)(x + 4)$

2. $(x - 1)(x + 2)$

a FACTORING $x^2 + bx + c$

We now begin a study of the factoring of trinomials. We first factor trinomials like

$$x^2 + 5x + 6 \quad \text{and} \quad x^2 + 3x - 10$$

by a refined *trial-and-error process*. In this section, we restrict our attention to trinomials of the type $ax^2 + bx + c$, where the **leading coefficient** a is 1.

Compare the following multiplications:

$$\begin{aligned}
& \qquad\qquad\qquad\quad \text{F} \quad \text{O} \quad \text{I} \quad \text{L} \\
(x + 2)(x + 5) &= x^2 + 5x + 2x + 2 \cdot 5 \\
&= x^2 + 7x + 10; \\
(x - 2)(x - 5) &= x^2 - 5x - 2x + (-2)(-5) \\
&= x^2 - 7x + 10; \\
(x + 3)(x - 7) &= x^2 - 7x + 3x + 3(-7) \\
&= x^2 - 4x - 21; \\
(x - 3)(x + 7) &= x^2 + 7x - 3x + (-3)7 \\
&= x^2 + 4x - 21.
\end{aligned}$$

Note that for all four products:

- The product of the two binomials is a trinomial.
- The coefficient of x in the trinomial is the sum of the constant terms in the binomials.
- The constant term in the trinomial is the product of the constant terms in the binomials.

These observations lead to a method for factoring certain trinomials. The first type we consider has a positive constant term, just as in the first two multiplications above.

Constant Term Positive

To factor $x^2 + 7x + 10$, we think of FOIL in reverse. Since $x \cdot x = x^2$, the first term of each binomial is x.

Next, we look for numbers p and q such that

$$x^2 + 7x + 10 = (x + p)(x + q).$$

To get the middle term and the last term of the trinomial, we look for two numbers p and q whose product is 10 and whose sum is 7. Those numbers are 2 and 5. Thus the factorization is

$$(x + 2)(x + 5).$$

Check: $(x + 2)(x + 5) = x^2 + 5x + 2x + 10$
$= x^2 + 7x + 10.$

Answers

Skill to Review:

1. $x^2 + 7x + 12$ **2.** $x^2 + x - 2$

EXAMPLE 1 Factor: $x^2 + 5x + 6$.

Think of FOIL in reverse. The first term of each factor is x:

$$(x + \square)(x + \square).$$

Next, we look for two numbers whose product is 6 and whose sum is 5. All the pairs of factors of 6 are shown in the table on the left below. Since both the product, 6, and the sum, 5, of the pair of numbers must be positive, we need consider only the positive factors, listed in the table on the right.

PAIRS OF FACTORS	SUMS OF FACTORS
1, 6	7
−1, −6	−7
2, 3	5
−2, −3	−5

PAIRS OF FACTORS	SUMS OF FACTORS
1, 6	7
2, 3	**5**

The numbers we need are 2 and 3.

The factorization is $(x + 2)(x + 3)$. We can check by multiplying to see whether we get the original trinomial.

Check: $(x + 2)(x + 3) = x^2 + 3x + 2x + 6 = x^2 + 5x + 6$.

Do Exercises 1 and 2. ▶

Compare these multiplications:

$$(x - 2)(x - 5) = x^2 - 5x - 2x + 10 = x^2 - 7x + 10;$$
$$(x + 2)(x + 5) = x^2 + 5x + 2x + 10 = x^2 + 7x + 10.$$

TO FACTOR $x^2 + bx + c$ WHEN c IS POSITIVE

When the constant term of a trinomial is positive, look for two numbers with the same sign. The sign is that of the middle term:

$$x^2 - 7x + 10 = (x - 2)(x - 5);$$

$$x^2 + 7x + 10 = (x + 2)(x + 5).$$

EXAMPLE 2 Factor: $y^2 - 8y + 12$.

Since the constant term, 12, is positive and the coefficient of the middle term, −8, is negative, we look for a factorization of 12 in which both factors are negative. Their sum must be −8.

PAIRS OF FACTORS	SUMS OF FACTORS
−1, −12	−13
−2, −6	**−8**
−3, −4	−7

The numbers we need are −2 and −6.

The factorization is $(y - 2)(y - 6)$. The student should check by multiplying.

Do Exercises 3–5. ▶

Factor.

1. $x^2 + 7x + 12$

Complete the following table.

PAIRS OF FACTORS	SUMS OF FACTORS
1, 12	13
−1, −12	
2, 6	
−2, −6	
3, 4	
−3, −4	

Because both 7 and 12 are positive, we need consider only the ______ factors in the table above.

$x^2 + 7x + 12$
$= (x + 3)($ ______ $)$

2. $x^2 + 13x + 36$

3. Explain why you would *not* consider the pairs of factors listed below in factoring $y^2 - 8y + 12$.

PAIRS OF FACTORS	SUMS OF FACTORS
1, 12	
2, 6	
3, 4	

Factor.

4. $x^2 - 8x + 15$

5. $t^2 - 9t + 20$

Answers

1. $(x + 3)(x + 4)$ **2.** $(x + 4)(x + 9)$ **3.** The coefficient of the middle term, −8, is negative. **4.** $(x - 5)(x - 3)$ **5.** $(t - 5)(t - 4)$

Guided Solution:
1. −13, 8, −8, 7, −7; positive; $x + 4$

Constant Term Negative

As we saw in two of the multiplications earlier in this section, the product of two binomials can have a negative constant term:

$$(x + 3)(x - 7) = x^2 - 4x - 21$$

and

$$(x - 3)(x + 7) = x^2 + 4x - 21.$$

Note that when the signs of the constants in the binomials are reversed, only the sign of the middle term in the product changes.

EXAMPLE 3 Factor: $x^2 - 8x - 20$.

The constant term, -20, must be expressed as the product of a negative number and a positive number. Since the sum of these two numbers must be negative (specifically, -8), the negative number must have the greater absolute value.

PAIRS OF FACTORS	SUMS OF FACTORS
1, −20	−19
2, −10	**−8**
4, −5	−1
5, −4	1
10, −2	8
20, −1	19

The numbers we need are 2 and −10.

Because these sums are all positive, for this problem all the corresponding pairs can be disregarded. Note that in all three pairs, the positive number has the greater absolute value.

The numbers that we are looking for are 2 and −10. The factorization is $(x + 2)(x - 10)$.

Check: $(x + 2)(x - 10) = x^2 - 10x + 2x - 20$
$= x^2 - 8x - 20.$

TO FACTOR $x^2 + bx + c$ WHEN c IS NEGATIVE

When the constant term of a trinomial is negative, look for two numbers whose product is negative. One must be positive and the other negative:

$$x^2 - 4x - 21 = (x + 3)(x - 7);$$

$$x^2 + 4x - 21 = (x - 3)(x + 7).$$

Consider pairs of numbers for which the number with the larger absolute value has the same sign as b, the coefficient of the middle term.

Do Exercises 6 and 7. (Exercise 7 is on the following page.)

6. Consider $x^2 - 5x - 24$.

a) Explain why you would *not* consider the pairs of factors listed below in factoring $x^2 - 5x - 24$.

PAIRS OF FACTORS	SUMS OF FACTORS
−1, 24	
−2, 12	
−3, 8	
−4, 6	

b) Explain why you *would* consider the pairs of factors listed below in factoring $x^2 - 5x - 24$.

PAIRS OF FACTORS	SUMS OF FACTORS
1, −24	
2, −12	
3, −8	
4, −6	

c) Factor: $x^2 - 5x - 24$.

Answers

6. **(a)** The positive factor has the larger absolute value. **(b)** The negative factor has the larger absolute value. **(c)** $(x + 3)(x - 8)$

EXAMPLE 4 Factor: $t^2 - 24 + 5t$.

We first write the trinomial in descending order: $t^2 + 5t - 24$. Since the constant term, -24, is negative, factorizations of -24 will have one positive factor and one negative factor. The sum of the factors must be 5, so we consider only pairs of factors in which the positive factor has the larger absolute value.

PAIRS OF FACTORS	SUMS OF FACTORS
−1, 24	23
−2, 12	10
−3, 8	**5** ← The numbers we need are −3 and 8.
−4, 6	2

The factorization is $(t - 3)(t + 8)$. The check is left to the student.

Do Exercises 8 and 9. ▶

EXAMPLE 5 Factor: $x^4 - x^2 - 110$.

Consider this trinomial as $(x^2)^2 - x^2 - 110$. We look for numbers p and q such that

$$x^4 - x^2 - 110 = (x^2 + p)(x^2 + q).$$

We look for two numbers whose product is -110 and whose sum is -1. The middle-term coefficient, -1, is small compared to -110. This tells us that the desired factors are close to each other in absolute value. The numbers we want are 10 and -11. The factorization is $(x^2 + 10)(x^2 - 11)$. ■

EXAMPLE 6 Factor: $a^2 + 4ab - 21b^2$.

We consider the trinomial in the equivalent form

$$a^2 + 4ba - 21b^2,$$

and think of $-21b^2$ as the "constant" term and $4b$ as the "coefficient" of the middle term. Then we try to express $-21b^2$ as a product of two factors whose sum is $4b$. Those factors are $-3b$ and $7b$. The factorization is $(a - 3b)(a + 7b)$.

Check: $(a - 3b)(a + 7b) = a^2 + 7ab - 3ba - 21b^2$
$= a^2 + 4ab - 21b^2$. ■

There are polynomials that are not factorable.

EXAMPLE 7 Factor: $x^2 - x + 5$.

Since 5 has very few factors, we can easily check all possibilities.

PAIRS OF FACTORS	SUMS OF FACTORS
5, 1	6
−5, −1	−6

There are no factors of 5 whose sum is -1. Thus the polynomial is *not* factorable into factors that are polynomials with rational-number coefficients. ■

7. Consider $x^2 + 5x - 6$.

a) Explain why you would *not* consider the pairs of factors listed below in factoring $x^2 + 5x - 6$.

PAIRS OF FACTORS	SUMS OF FACTORS
1, −6	
2, −3	

b) Explain why you *would* consider the pairs of factors listed below in factoring $x^2 + 5x - 6$.

PAIRS OF FACTORS	SUMS OF FACTORS
−1, 6	
−2, 3	

c) Factor: $x^2 + 5x - 6$.

Factor.

8. $a^2 - 40 + 3a$

First, rewrite in descending order:
$a^2 + 3a -$ ____.

PAIRS OF FACTORS	SUMS OF FACTORS
−1, 40	☐
−2, 30	☐
−4, 10	☐
−5, 8	☐

The factorization is $(a - 5)($ ____ $)$.

9. $-18 - 3t + t^2$

Answers

7. (a) The negative factor has the larger absolute value. **(b)** The positive factor has the larger absolute value. **(c)** $(x - 1)(x + 6)$
8. $(a - 5)(a + 8)$ **9.** $(t - 6)(t + 3)$

Guided Solution:
8. 40, 39, 28, 6, 3, $a + 8$

Factor.

10. $y^2 - 12 - 4y$

11. $t^4 + 5t^2 - 14$

12. $x^2 + 2x + 7$

In this text, a polynomial like $x^2 - x + 5$ that cannot be factored further is said to be **prime**. In more advanced courses, polynomials like $x^2 - x + 5$ can be factored and are not considered prime.

Do Exercises 10–12.

Often factoring requires two or more steps. In general, when told to factor, we should be sure to *factor completely.* This means that the final factorization should not contain any factors that can be factored further.

EXAMPLE 8 Factor: $2x^3 - 20x^2 + 50x$.

Always look first for a common factor. This time there is one, $2x$, which we factor out first:

$$2x^3 - 20x^2 + 50x = 2x(x^2 - 10x + 25).$$

Now consider $x^2 - 10x + 25$. Since the constant term is positive and the coefficient of the middle term is negative, we look for a factorization of 25 in which both factors are negative. Their sum must be -10.

PAIRS OF FACTORS	SUMS OF FACTORS
$-25, -1$	-26
$-5, -5$	-10

The numbers we need are -5 and -5.

The factorization of $x^2 - 10x + 25$ is $(x - 5)(x - 5)$, or $(x - 5)^2$. The final factorization is $2x(x - 5)^2$. We check by multiplying:

$$\begin{aligned} 2x(x - 5)^2 &= 2x(x^2 - 10x + 25) \\ &= (2x)(x^2) - (2x)(10x) + (2x)(25) \\ &= 2x^3 - 20x^2 + 50x. \end{aligned}$$

Factor.

13. $x^3 + 4x^2 - 12x$

14. $p^2 - pq - 3pq^2$

15. $3x^3 + 24x^2 + 48x$

Do Exercises 13–15.

Once any common factors have been factored out, the following summary can be used to factor $x^2 + bx + c$.

TO FACTOR $x^2 + bx + c$

1. First arrange the polynomial in descending order.
2. Use a trial-and-error process that looks for factors of c whose sum is b.
3. If c is positive, the signs of the factors are the same as the sign of b.
4. If c is negative, one factor is positive and the other is negative. If the sum of two factors is the opposite of b, changing the sign of each factor will give the desired factors whose sum is b.
5. Check by multiplying.

Answers

10. $(y - 6)(y + 2)$ 11. $(t^2 + 7)(t^2 - 2)$
12. Prime 13. $x(x + 6)(x - 2)$
14. $p(p - q - 3q^2)$ 15. $3x(x + 4)^2$

Leading Coefficient −1

EXAMPLE 9 Factor: $10 - 3x - x^2$.

Note that the polynomial is written in ascending order. When we write it in descending order, we get

$$-x^2 - 3x + 10,$$

which has a leading coefficient of -1. Before factoring in such a case, we can factor out a -1, as follows:

$$\begin{aligned} -x^2 - 3x + 10 &= -1 \cdot x^2 + (-1)(3x) + (-1)(-10) \\ &= -1(x^2 + 3x - 10). \end{aligned}$$

Then we proceed to factor $x^2 + 3x - 10$. We get

$$-x^2 - 3x + 10 = -1(x^2 + 3x - 10) = -1(x + 5)(x - 2).$$

We can also express this answer in two other ways by multiplying either binomial by -1. Thus each of the following is a correct answer:

$$\begin{aligned} 10 - 3x - x^2 &= -1(x + 5)(x - 2) \\ &= (-x - 5)(x - 2) && \text{Multiplying } x + 5 \text{ by } -1 \\ &= (x + 5)(-x + 2). && \text{Multiplying } x - 2 \text{ by } -1 \end{aligned}$$

Do Exercises 16 and 17. ▶

Factor.

16. $14 + 5x - x^2$

17. $-x^2 + 3x + 18$

Answers

16. $-1(x + 2)(x - 7)$, or $(-x - 2)(x - 7)$, or $(x + 2)(-x + 7)$
17. $-1(x + 3)(x - 6)$, or $(-x - 3)(x - 6)$, or $(x + 3)(-x + 6)$

13.2 Exercise Set

For Extra Help

MyMathLab®

MathXL®

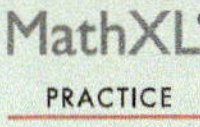

PRACTICE

WATCH

READ

REVIEW

Reading Check

Determine whether each statement is true or false.

RC1. The leading coefficient of $x^2 - 3x - 10$ is 1.

RC2. To factor $x^2 - 3x - 10$, we look for two numbers whose product is -10.

RC3. To factor $x^2 - 3x - 10$, we look for two numbers whose sum is -3.

RC4. The factorization of $x^2 - 3x - 10$ is $(x + 5)(x - 2)$.

a Factor. Remember that you can check by multiplying.

1. $x^2 + 8x + 15$

PAIRS OF FACTORS	SUMS OF FACTORS

2. $x^2 + 5x + 6$

PAIRS OF FACTORS	SUMS OF FACTORS

3. $x^2 + 7x + 12$

PAIRS OF FACTORS	SUMS OF FACTORS

4. $x^2 + 9x + 8$

PAIRS OF FACTORS	SUMS OF FACTORS

5. $x^2 - 6x + 9$

PAIRS OF FACTORS	SUMS OF FACTORS

6. $y^2 - 11y + 28$

PAIRS OF FACTORS	SUMS OF FACTORS

7. $x^2 - 5x - 14$

PAIRS OF FACTORS	SUMS OF FACTORS

8. $a^2 + 7a - 30$

PAIRS OF FACTORS	SUMS OF FACTORS

9. $b^2 + 5b + 4$

PAIRS OF FACTORS	SUMS OF FACTORS

10. $z^2 - 8z + 7$

PAIRS OF FACTORS	SUMS OF FACTORS

11. $t^2 + 3t - 18$

PAIRS OF FACTORS	SUMS OF FACTORS

12. $t^2 + 8t + 16$

PAIRS OF FACTORS	SUMS OF FACTORS

13. $d^2 - 7d + 10$

14. $t^2 - 12t + 35$

15. $y^2 - 11y + 10$

16. $x^2 - 4x - 21$

17. $x^2 + x + 1$

18. $x^2 + 5x + 3$

19. $x^2 - 7x - 18$

20. $y^2 - 3y - 28$

21. $x^3 - 6x^2 - 16x$

22. $x^3 - x^2 - 42x$

23. $y^3 - 4y^2 - 45y$

24. $x^3 - 7x^2 - 60x$

25. $-2x - 99 + x^2$

26. $x^2 - 72 + 6x$

27. $c^4 + c^2 - 56$

28. $b^4 + 5b^2 - 24$

29. $a^4 + 2a^2 - 35$

30. $x^4 - x^2 - 6$

31. $x^2 + x - 42$

32. $x^2 + 2x - 15$

33. $7 - 2p + p^2$

34. $11 - 3w + w^2$

35. $x^2 + 20x + 100$

36. $a^2 + 19a + 88$

37. $2z^3 - 2z^2 - 24z$

38. $5w^4 - 20w^3 - 25w^2$

39. $3t^4 + 3t^3 + 3t^2$

40. $4y^5 - 4y^4 - 4y^3$

41. $x^4 - 21x^3 - 100x^2$

42. $x^4 - 20x^3 + 96x^2$

43. $x^2 - 21x - 72$

44. $4x^2 + 40x + 100$

45. $x^2 - 25x + 144$

46. $y^2 - 21y + 108$

47. $a^2 + a - 132$

48. $a^2 + 9a - 90$

49. $3t^2 + 6t + 3$

50. $2y^2 + 24y + 72$

51. $w^4 - 8w^3 + 16w^2$

52. $z^5 - 6z^4 + 9z^3$

53. $30 + 7x - x^2$

54. $45 + 4x - x^2$

55. $24 - a^2 - 10a$

56. $-z^2 + 36 - 9z$

57. $120 - 23x + x^2$

58. $96 + 22d + d^2$

59. $108 - 3x - x^2$

60. $112 + 9y - y^2$

61. $y^2 - 0.2y - 0.08$

62. $t^2 - 0.3t - 0.10$

63. $p^2 + 3pq - 10q^2$

64. $a^2 + 2ab - 3b^2$

65. $84 - 8t - t^2$

66. $72 - 6m - m^2$

67. $m^2 + 5mn + 4n^2$

68. $x^2 + 11xy + 24y^2$

69. $s^2 - 2st - 15t^2$

70. $p^2 + 5pq - 24q^2$

71. $6a^{10} - 30a^9 - 84a^8$

72. $7x^9 - 28x^8 - 35x^7$

Skill Maintenance

Solve.

73. $2x + 7 = 0$ [10.3a]

74. $-2y + 11y = 108$ [10.3b]

75. $3x - 2 - x = 5x - 9x$ [10.3b]

76. $\frac{1}{2}x - \frac{1}{3}x = \frac{2}{3} + \frac{5}{6}x$ [10.3b]

77. $5(t - 1) - 3 = 4t - (7t - 2)$ [10.3c]

78. $10 - (x - 7) = 4x - (1 + 5x)$ [10.3c]

79. $-2x < 48$ [10.7d]

80. $4x - 8x + 16 \geq 6(x - 2)$ [10.7e]

81. $-6(x - 4) + 8(4 - x) \leq 3(x - 7)$ [10.7e]

Solve. [10.4b]

82. $A = \frac{p + w}{2}$, for p

83. $y = mx + b$, for x

84. $a - c + r = 2$, for c

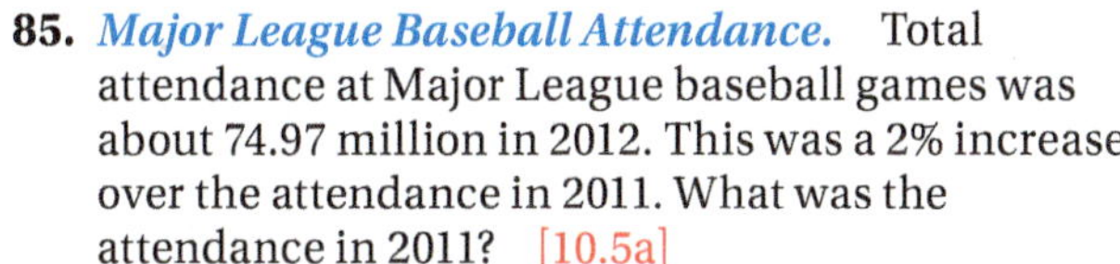

85. *Major League Baseball Attendance.* Total attendance at Major League baseball games was about 74.97 million in 2012. This was a 2% increase over the attendance in 2011. What was the attendance in 2011? [10.5a]

Source: Based on information from baseball-reference.com

86. The first angle of a triangle is four times as large as the second. The measure of the third angle is 30° greater than that of the second. Find the angle measures. [10.6a]

Synthesis

87. Find all integers m for which $y^2 + my + 50$ can be factored.

88. Find all integers b for which $a^2 + ba - 50$ can be factored.

Factor completely.

89. $x^2 - \frac{1}{2}x - \frac{3}{16}$

90. $x^2 - \frac{1}{4}x - \frac{1}{8}$

91. $x^2 + \frac{2}{3}x + \frac{1}{9}$

92. $x^2 - \frac{2}{5}x + \frac{1}{25}$

93. $x^2 + \frac{30}{7}x - \frac{25}{7}$

94. $\frac{1}{3}x^3 + \frac{1}{3}x^2 - 2x$

95. $b^{2n} + 7b^n + 10$

96. $a^{2m} - 11a^m + 28$

Find a polynomial in factored form for the shaded area in each figure. (Leave answers in terms of π.)

97.

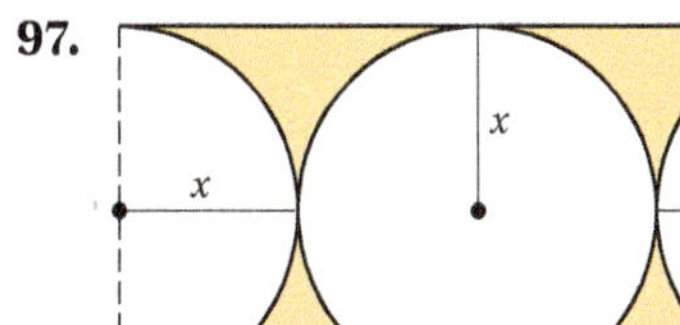

98.

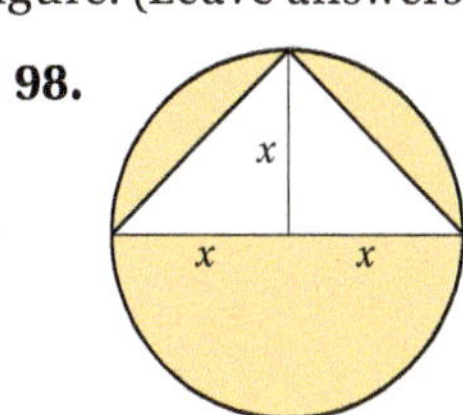

13.3 Factoring $ax^2 + bx + c, a \neq 1$: The FOIL Method

OBJECTIVE

a Factor trinomials of the type $ax^2 + bx + c, a \neq 1$, using the FOIL method.

SKILL TO REVIEW

Objective 12.6a: Multiply two binomials mentally using the FOIL method.

Multiply.

1. $(2x + 3)(x + 1)$
2. $(3x - 4)(2x - 1)$

In this section, we factor trinomials in which the coefficient of the leading term x^2 is not 1. The procedure we use is a refined trial-and-error method.

a THE FOIL METHOD

We want to factor trinomials of the type $ax^2 + bx + c$. Consider the following multiplication:

$$\begin{aligned}(2x + 5)(3x + 4) &= \overset{F}{6x^2} + \overset{O}{8x} + \overset{I}{15x} + \overset{L}{20} \\ &= 6x^2 + 23x + 20\end{aligned}$$

F	O + I	L
$2 \cdot 3$	$2 \cdot 4 \quad 5 \cdot 3$	$5 \cdot 4$

To factor $6x^2 + 23x + 20$, we reverse the above multiplication, using what we might call an "unFOIL" process. We look for two binomials $rx + p$ and $sx + q$ whose product is $(rx + p)(sx + q) = 6x^2 + 23x + 20$. The product of the First terms must be $6x^2$. The product of the Outside terms plus the product of the Inside terms must be $23x$. The product of the Last terms must be 20. We know from the preceding discussion that the answer is $(2x + 5)(3x + 4)$. Generally, however, finding such an answer is a refined trial-and-error process. It turns out that $(-2x - 5)(-3x - 4)$ is also a correct answer, but we generally choose an answer in which the first coefficients are positive.

We will use the following trial-and-error method.

The ac-method in Section 13.4

To the student: In Section 13.4, we will consider an alternative method for the same kind of factoring. It involves factoring by grouping and is called the *ac*-method.

To the instructor: We present two ways to factor general trinomials in Sections 13.3 and 13.4: the FOIL method in Section 13.3 and the *ac*-method in Section 13.4. You can teach both methods and let the student use the one that he or she prefers or you can select just one.

THE FOIL METHOD

To factor $ax^2 + bx + c, a \neq 1$, using the FOIL method:

1. Factor out the largest common factor, if one exists.
2. Find two First terms whose product is ax^2.

   $$(\square x + \quad)(\square x + \quad) = ax^2 + bx + c.$$
   FOIL

3. Find two Last terms whose product is c:

   $$(\quad x + \square)(\quad x + \square) = ax^2 + bx + c.$$
   FOIL

4. Look for Outer and Inner products resulting from steps (2) and (3) for which the sum is bx:

   $$(\square x + \square)(\square x + \square) = ax^2 + bx + c.$$
   I, O — FOIL

5. Always check by multiplying.

Answers

Skill to Review:
1. $2x^2 + 5x + 3$ 2. $6x^2 - 11x + 4$

EXAMPLE 1 Factor: $3x^2 - 10x - 8$.

1) First, we check for a common factor. Here there is none (other than 1 or -1).

2) Find two **F**irst terms whose product is $3x^2$.

   The only possibilities for the **F**irst terms are $3x$ and x, so any factorization must be of the form

   $(3x + \square)(x + \square)$.

3) Find two **L**ast terms whose product is -8.

   Possible factorizations of -8 are

   $(-8) \cdot 1, \quad 8 \cdot (-1), \quad (-2) \cdot 4, \quad \text{and} \quad 2 \cdot (-4).$

   Since the First terms are not identical, we must also consider

   $1 \cdot (-8), \quad (-1) \cdot 8, \quad 4 \cdot (-2), \quad \text{and} \quad (-4) \cdot 2.$

4) Inspect the **O**utside and **I**nside products resulting from steps (2) and (3). Look for a combination in which the sum of the products is the middle term, $-10x$:

Trial	*Product*	
$(3x - 8)(x + 1)$	$3x^2 + 3x - 8x - 8$ $= 3x^2 - 5x - 8$	← Wrong middle term
$(3x + 8)(x - 1)$	$3x^2 - 3x + 8x - 8$ $= 3x^2 + 5x - 8$	← Wrong middle term
$(3x - 2)(x + 4)$	$3x^2 + 12x - 2x - 8$ $= 3x^2 + 10x - 8$	← Wrong middle term
$(3x + 2)(x - 4)$	$3x^2 - 12x + 2x - 8$ $= 3x^2 - 10x - 8$	← **Correct middle term!**
$(3x + 1)(x - 8)$	$3x^2 - 24x + x - 8$ $= 3x^2 - 23x - 8$	← Wrong middle term
$(3x - 1)(x + 8)$	$3x^2 + 24x - x - 8$ $= 3x^2 + 23x - 8$	← Wrong middle term
$(3x + 4)(x - 2)$	$3x^2 - 6x + 4x - 8$ $= 3x^2 - 2x - 8$	← Wrong middle term
$(3x - 4)(x + 2)$	$3x^2 + 6x - 4x - 8$ $= 3x^2 + 2x - 8$	← Wrong middle term

   The correct factorization is $(3x + 2)(x - 4)$.

5) Check: $(3x + 2)(x - 4) = 3x^2 - 10x - 8$.

Two observations can be made from Example 1. First, we listed all possible trials even though we could have stopped after having found the correct factorization. We did this to show that each trial differs only in the middle term of the product. **Second, note that only the sign of the middle term changes when the signs in the binomials are reversed:**

Plus ↓ Minus ↓

$(3x + 4)(x - 2) = 3x^2 - 2x - 8$

Minus ↓ Plus ↓ — Middle term changes sign

$(3x - 4)(x + 2) = 3x^2 + 2x - 8.$

CALCULATOR CORNER

A partial check of a factorization can be performed using a table or a graph. To check the factorization $6x^3 - 9x^2 + 4x - 6 = (2x - 3)(3x^2 + 2)$, for example, we enter $y_1 = 6x^3 - 9x^2 + 4x - 6$ and $y_2 = (2x - 3)(3x^2 + 2)$ on the equation-editor screen. Then we set up a table in **AUTO** mode. If the factorization is correct, the values of y_1 and y_2 will be the same regardless of the table settings used.

X	Y1	Y2
−3	−261	−261
−2	−98	−98
−1	−25	−25
0	−6	−6
1	−5	−5
2	14	14
3	87	87
X = −3		

We can also graph $y_1 = 6x^3 - 9x^2 + 4x - 6$ and $y_2 = (2x - 3)(3x^2 + 2)$. If the graphs appear to coincide, the factorization is probably correct.

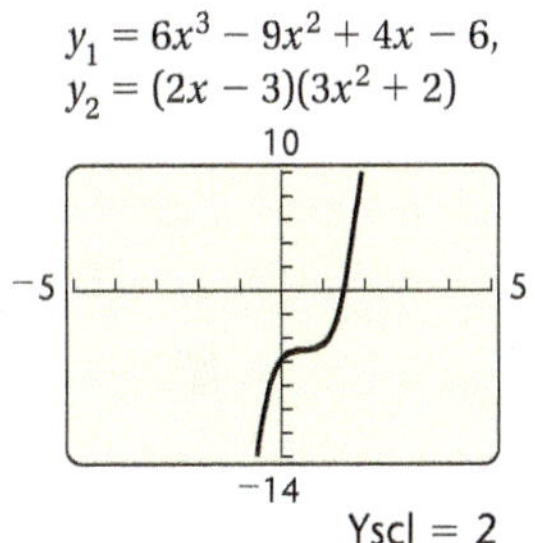

Keep in mind that these procedures provide only a partial check since we cannot view all possible values of x in a table or see the entire graph.

EXERCISES: Use a table or a graph to determine whether each factorization is correct.

1. $24x^2 - 76x + 40 = 4(3x - 2)(2x - 5)$
2. $4x^2 - 5x - 6 = (4x + 3)(x - 2)$
3. $5x^2 + 17x - 12 = (5x + 3)(x - 4)$
4. $10x^2 + 37x + 7 = (5x - 1)(2x + 7)$

Factor.

1. $2x^2 - x - 15$

2. $12x^2 - 17x - 5$

Do Exercises 1 and 2.

EXAMPLE 2 Factor: $24x^2 - 76x + 40$.

1) First, we factor out the largest common factor, 4:

$$4(6x^2 - 19x + 10).$$

Now we factor the trinomial $6x^2 - 19x + 10$.

2) Because $6x^2$ can be factored as $3x \cdot 2x$ or $6x \cdot x$, we have these possibilities for factorizations:

$$(3x + \square)(2x + \square) \quad \text{or} \quad (6x + \square)(x + \square).$$

3) There are four pairs of factors of 10 and each pair can be listed in two ways:

$10, 1 \quad -10, -1 \quad 5, 2 \quad -5, -2$

and

$1, 10 \quad -1, -10 \quad 2, 5 \quad -2, -5.$

4) The two possibilities from step (2) and the eight possibilities from step (3) give $2 \cdot 8$, or 16 possibilities for factorizations. We look for **O**utside and **I**nside products resulting from steps (2) and (3) for which the sum is the middle term, $-19x$. Since the sign of the middle term is negative, but the sign of the last term, 10, is positive, both factors of 10 must be negative. This means only four pairings from step (3) need be considered. We first try these factors with

$$(3x + \square)(2x + \square).$$

If none gives the correct factorization, we will consider

$$(6x + \square)(x + \square).$$

Trial	*Product*	
$(3x - 10)(2x - 1)$	$6x^2 - 3x - 20x + 10$	
	$= 6x^2 - 23x + 10$	← Wrong middle term
$(3x - 1)(2x - 10)$	$6x^2 - 30x - 2x + 10$	
	$= 6x^2 - 32x + 10$	← Wrong middle term
$(3x - 5)(2x - 2)$	$6x^2 - 6x - 10x + 10$	
	$= 6x^2 - 16x + 10$	← Wrong middle term
$(3x - 2)(2x - 5)$	$6x^2 - 15x - 4x + 10$	
	$= 6x^2 - 19x + 10$	← **Correct middle term!**

Since we have a correct factorization, we need not consider

$$(6x + \square)(x + \square).$$

The factorization of $6x^2 - 19x + 10$ is $(3x - 2)(2x - 5)$, but *do not forget the common factor*! We must include it in order to factor the original trinomial:

$$24x^2 - 76x + 40 = 4(6x^2 - 19x + 10)$$
$$= 4(3x - 2)(2x - 5).$$

5) Check: $4(3x - 2)(2x - 5) = 4(6x^2 - 19x + 10) = 24x^2 - 76x + 40$.

Caution!

When factoring any polynomial, always look for a common factor first. Failure to do so is such a common error that this caution bears repeating.

Answers

1. $(2x + 5)(x - 3)$ 2. $(4x + 1)(3x - 5)$

In Example 2, look again at the possibility $(3x - 5)(2x - 2)$. Without multiplying, we can reject such a possibility. To see why, consider the following:

$$(3x - 5)(2x - 2) = (3x - 5)(2)(x - 1) = 2(3x - 5)(x - 1).$$

The expression $2x - 2$ has a common factor, 2. But we removed the *largest* common factor in the first step. If $2x - 2$ were one of the factors, then 2 would have to be a common factor in addition to the original 4. Thus, $(2x - 2)$ cannot be part of the factorization of the original trinomial.

> Given that the largest common factor is factored out at the outset, we need not consider factorizations that have a common factor.

Factor.

3. $3x^2 - 19x + 20$

4. $20x^2 - 46x + 24$

Do Exercises 3 and 4. ▶

EXAMPLE 3 Factor: $10x^2 + 37x + 7$.

1) There is no common factor (other than 1 or -1).

2) Because $10x^2$ factors as $10x \cdot x$ or $5x \cdot 2x$, we have these possibilities for factorizations:

$$(10x + \square)(x + \square) \quad \text{or} \quad (5x + \square)(2x + \square).$$

3) There are two pairs of factors of 7 and each pair can be listed in two ways:

$$1, 7 \quad -1, -7 \qquad \text{and} \qquad 7, 1 \quad -7, -1.$$

4) From steps (2) and (3), we see that there are 8 possibilities for factorizations. Look for **O**uter and **I**nner products for which the sum is the middle term. Because all coefficients in $10x^2 + 37x + 7$ are positive, we need consider only positive factors of 7. The possibilities are

$$(10x + 1)(x + 7) = 10x^2 + 71x + 7,$$
$$(10x + 7)(x + 1) = 10x^2 + 17x + 7,$$
$$(5x + 7)(2x + 1) = 10x^2 + 19x + 7,$$
$$(5x + 1)(2x + 7) = 10x^2 + 37x + 7. \quad \leftarrow \text{Correct middle term}$$

The factorization is $(5x + 1)(2x + 7)$.

5) Check: $(5x + 1)(2x + 7) = 10x^2 + 37x + 7$.

Do Exercise 5. ▶

5. Factor: $6x^2 + 7x + 2$.

TIPS FOR FACTORING $ax^2 + bx + c, a \neq 1$

- Always factor out the largest common factor first, if one exists.
- Once the common factor has been factored out of the original trinomial, no binomial factor can contain a common factor (other than 1 or -1).
- If c is positive, then the signs in both binomial factors must match the sign of b. (This assumes that $a > 0$.)
- Reversing the signs in the binomials reverses the sign of the middle term of their product.
- Organize your work so that you can keep track of which possibilities have or have not been checked.
- Always check by multiplying.

Answers

3. $(3x - 4)(x - 5)$ **4.** $2(5x - 4)(2x - 3)$
5. $(2x + 1)(3x + 2)$

EXAMPLE 4 Factor: $10x + 8 - 3x^2$.

An important problem-solving strategy is to find a way to make new problems look like problems we already know how to solve. The factoring tips on the preceding page apply only to trinomials of the form $ax^2 + bx + c$, with $a > 0$. This leads us to rewrite $10x + 8 - 3x^2$ in descending order:

$$10x + 8 - 3x^2 = -3x^2 + 10x + 8. \quad \text{Writing in descending order}$$

Although $-3x^2 + 10x + 8$ looks similar to the trinomials we have factored, the factoring tips require a positive leading coefficient, so we factor out -1:

$$-3x^2 + 10x + 8 = -1(3x^2 - 10x - 8) \quad \text{Factoring out } -1 \text{ changes the signs of the coefficients.}$$

$$= -1(3x + 2)(x - 4). \quad \text{Using the result from Example 1}$$

The factorization of $10x + 8 - 3x^2$ is $-1(3x + 2)(x - 4)$. Other correct answers are

$$10x + 8 - 3x^2 = (3x + 2)(-x + 4) \quad \text{Multiplying } x - 4 \text{ by } -1$$

$$= (-3x - 2)(x - 4). \quad \text{Multiplying } 3x + 2 \text{ by } -1$$

Factor.

6. $2 - x - 6x^2$

7. $2x + 8 - 6x^2$

Do Exercises 6 and 7.

EXAMPLE 5 Factor: $6p^2 - 13pv - 28v^2$.

1) Factor out a common factor, if any.
   There is none (other than 1 or -1).
2) Factor the first term, $6p^2$.
   Possibilities are $2p, 3p$ and $6p, p$. We have these as possibilities for factorizations:

$$(2p + \square)(3p + \square) \quad \text{or} \quad (6p + \square)(p + \square).$$

3) Factor the last term, $-28v^2$, which has a negative coefficient.
   There are six pairs of factors and each can be listed in two ways:

$-28v, v \quad 28v, -v \quad -14v, 2v \quad 14v, -2v \quad -7v, 4v \quad 7v, -4v$

and

$v, -28v \quad -v, 28v \quad 2v, -14v \quad -2v, 14v \quad 4v, -7v \quad -4v, 7v.$

4) The coefficient of the middle term is negative, so we look for combinations of factors from steps (2) and (3) such that the sum of their products has a negative coefficient. We try some possibilities:

$$(2p + v)(3p - 28v) = 6p^2 - 53pv - 28v^2,$$
$$(2p - 7v)(3p + 4v) = 6p^2 - 13pv - 28v^2. \quad \leftarrow \text{Correct middle term}$$

   The factorization of $6p^2 - 13pv - 28v^2$ is $(2p - 7v)(3p + 4v)$.
5) The check is left to the student.

Factor.

8. $6a^2 - 5ab + b^2$

9. $6x^2 + 15xy + 9y^2$

Do Exercises 8 and 9.

Answers

6. $-1(2x - 1)(3x + 2)$, or $(2x - 1)(-3x - 2)$, or $(-2x + 1)(3x + 2)$ 7. $-2(3x - 4)(x + 1)$, or $2(3x - 4)(-x - 1)$, or $2(-3x + 4)(x + 1)$
8. $(2a - b)(3a - b)$ 9. $3(2x + 3y)(x + y)$

13.3 Exercise Set

For Extra Help
MyMathLab® MathXL® PRACTICE WATCH READ REVIEW

Reading Check

Determine whether each statement is true or false.

RC1. When factoring a polynomial, we always look for a common factor first.

RC2. We can check any factorization by multiplying.

RC3. When we are factoring $10x^2 + 21x + 2$, the only choices for the First terms in the binomial factors are $2x$ and $5x$.

RC4. The factorization of $10x^2 + 21x + 2$ is $(2x + 1)(5x + 2)$.

a Factor.

1. $2x^2 - 7x - 4$

2. $3x^2 - x - 4$

3. $5x^2 - x - 18$

4. $4x^2 - 17x + 15$

5. $6x^2 + 23x + 7$

6. $6x^2 - 23x + 7$

7. $3x^2 + 4x + 1$

8. $7x^2 + 15x + 2$

9. $4x^2 + 4x - 15$

10. $9x^2 + 6x - 8$

11. $2x^2 - x - 1$

12. $15x^2 - 19x - 10$

13. $9x^2 + 18x - 16$

14. $2x^2 + 5x + 2$

15. $3x^2 - 5x - 2$

16. $18x^2 - 3x - 10$

17. $12x^2 + 31x + 20$

18. $15x^2 + 19x - 10$

19. $14x^2 + 19x - 3$

20. $35x^2 + 34x + 8$

21. $9x^2 + 18x + 8$

22. $6 - 13x + 6x^2$

23. $49 - 42x + 9x^2$

24. $16 + 36x^2 + 48x$

25. $24x^2 + 47x - 2$

26. $16p^2 - 78p + 27$

27. $35x^2 - 57x - 44$

28. $9a^2 + 12a - 5$

29. $20 + 6x - 2x^2$

30. $15 + x - 2x^2$

31. $12x^2 + 28x - 24$

32. $6x^2 + 33x + 15$

33. $30x^2 - 24x - 54$

34. $18t^2 - 24t + 6$

35. $4y + 6y^2 - 10$

36. $-9 + 18x^2 - 21x$

37. $3x^2 - 4x + 1$

38. $6t^2 + 13t + 6$

39. $12x^2 - 28x - 24$

40. $6x^2 - 33x + 15$

41. $-1 + 2x^2 - x$

42. $-19x + 15x^2 + 6$

43. $9x^2 - 18x - 16$

44. $14y^2 + 35y + 14$

45. $15x^2 - 25x - 10$

46. $18x^2 + 3x - 10$

47. $12p^3 + 31p^2 + 20p$

48. $15x^3 + 19x^2 - 10x$

49. $16 + 18x - 9x^2$

50. $33t - 15 - 6t^2$

51. $-15x^2 + 19x - 6$

52. $1 + p - 2p^2$

53. $14x^4 + 19x^3 - 3x^2$

54. $70x^4 + 68x^3 + 16x^2$

55. $168x^3 - 45x^2 + 3x$

56. $144x^5 + 168x^4 + 48x^3$

57. $15x^4 - 19x^2 + 6$

58. $9x^4 + 18x^2 + 8$

59. $25t^2 + 80t + 64$

60. $9x^2 - 42x + 49$

61. $6x^3 + 4x^2 - 10x$

62. $18x^3 - 21x^2 - 9x$

63. $25x^2 + 79x + 64$

64. $9y^2 + 42y + 47$

65. $6x^2 - 19x - 5$

66. $2x^2 + 11x - 9$

67. $12m^2 - mn - 20n^2$

68. $12a^2 - 17ab + 6b^2$

69. $6a^2 - ab - 15b^2$

70. $3p^2 - 16pw - 12w^2$

71. $9a^2 + 18ab + 8b^2$

72. $10s^2 + 4st - 6t^2$

73. $35p^2 + 34pt + 8t^2$

74. $30a^2 + 87ab + 30b^2$

75. $18x^2 - 6xy - 24y^2$

76. $15a^2 - 5ab - 20b^2$

Skill Maintenance

Graph.

77. $y = \frac{2}{5}x - 1$ [11.1d]

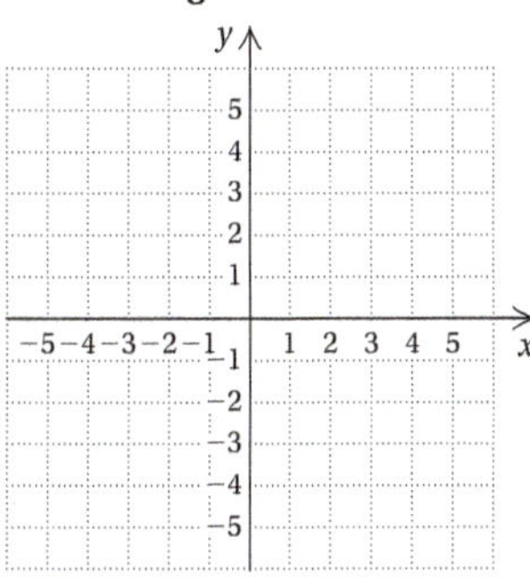

78. $2x = 6$ [11.2b]

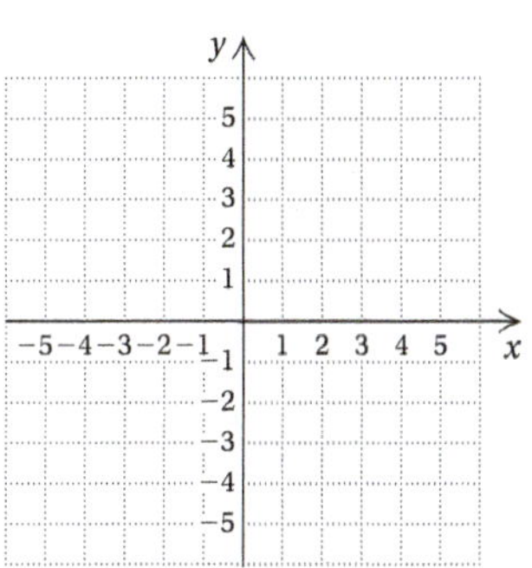

79. $x = 4 - 2y$ [11.1d]

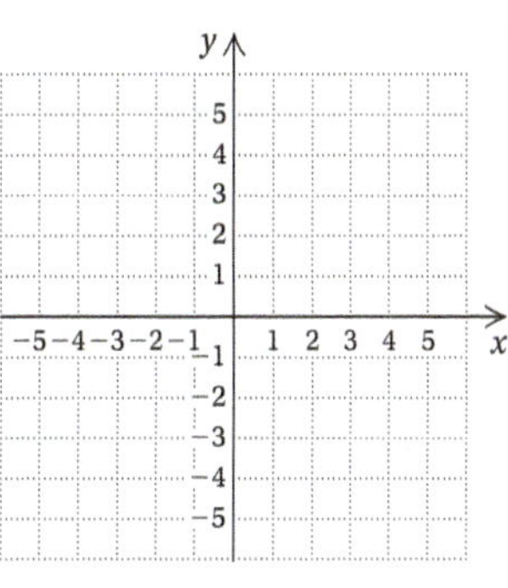

80. $y = -3$ [11.2b]

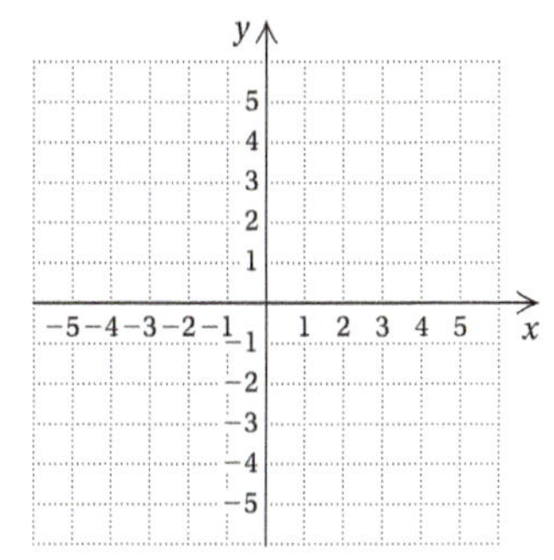

Find the intercepts of each equation. Then graph the equation. [11.2a]

81. $x + y = 4$

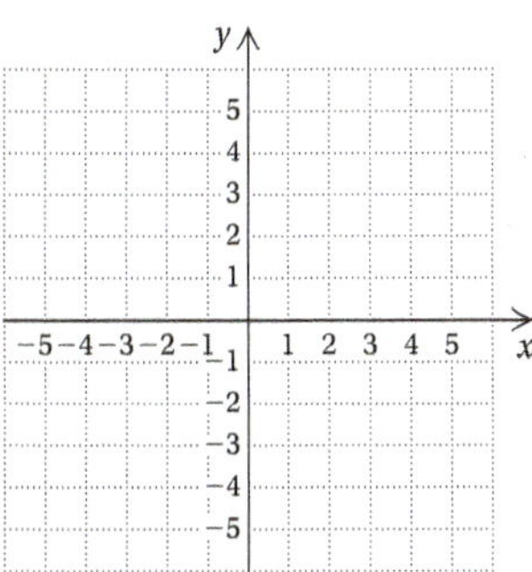

82. $x - y = 3$

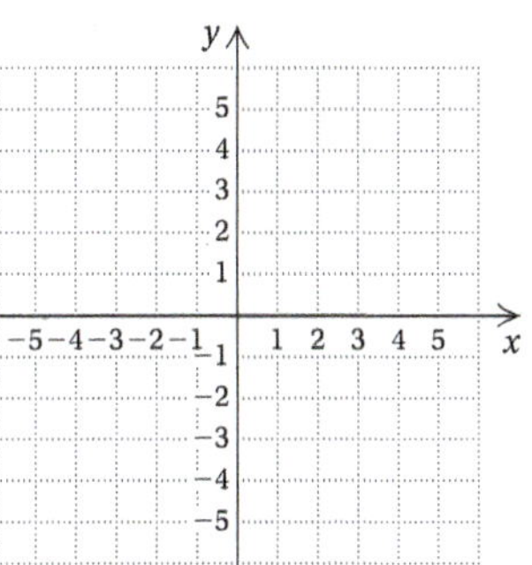

83. $5x - 3y = 15$

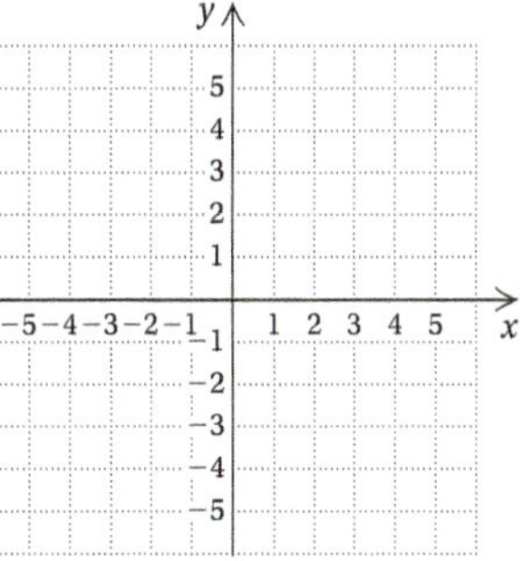

84. $y - 3x = 3$

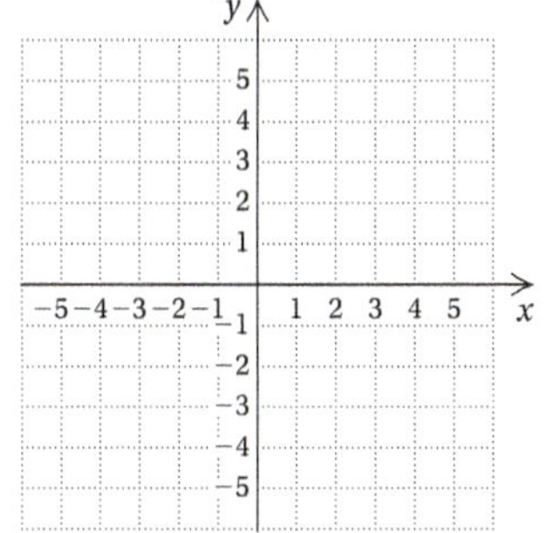

Synthesis

Factor.

85. $20x^{2n} + 16x^n + 3$

86. $-15x^{2m} + 26x^m - 8$

87. $3x^{6a} - 2x^{3a} - 1$

88. $x^{2n+1} - 2x^{n+1} + x$

89.–94. Use the TABLE feature to check the factoring in Exercises 15–20. (See the Calculator Corner on p. 961.)

13.4 Factoring $ax^2 + bx + c, a \neq 1$: The *ac*-Method

OBJECTIVE

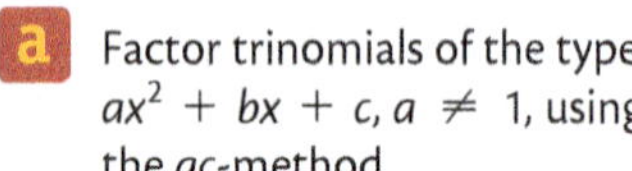

a Factor trinomials of the type $ax^2 + bx + c, a \neq 1$, using the *ac*-method.

SKILL TO REVIEW

Objective 13.1c: Factor certain expressions with four terms using factoring by grouping.

Factor by grouping.

1. $2x^3 - 3x^2 + 10x - 15$

2. $n^3 + 3n^2 + n + 3$

a THE *ac*-METHOD

Another method for factoring trinomials of the type $ax^2 + bx + c, a \neq 1$, involves the product, ac, of the leading coefficient a and the last term c. It is called the ***ac*-method**. Because it uses factoring by grouping, it is also referred to as the **grouping method**.

We know how to factor the trinomial $x^2 + 5x + 6$. We look for factors of the constant term, 6, whose sum is the coefficient of the middle term, 5. What happens when the leading coefficient is not 1? To factor a trinomial like $3x^2 - 10x - 8$, we can use a method similar to the one that we used for $x^2 + 5x + 6$. That method is outlined as follows.

THE *ac*-METHOD

To factor a trinomial using the *ac*-method:

1. Factor out a common factor, if any. We refer to the remaining trinomial as $ax^2 + bx + c$.
2. Multiply the leading coefficient a and the constant c.
3. Try to factor the product ac so that the sum of the factors is b. That is, find integers p and q such that $pq = ac$ and $p + q = b$.
4. Split the middle term, writing it as a sum using the factors found in step (3).
5. Factor by grouping.
6. Check by multiplying.

EXAMPLE 1 Factor: $3x^2 - 10x - 8$.

1) First, we factor out a common factor, if any. There is none (other than 1 or -1).

2) We multiply the leading coefficient, 3, and the constant, -8:

$$3(-8) = -24.$$

3) Then we look for a factorization of -24 in which the sum of the factors is the coefficient of the middle term, -10.

PAIRS OF FACTORS	SUMS OF FACTORS
$-1, \quad 24$	23
$1, \quad -24$	-23
$-2, \quad 12$	10
$2, \quad -12$	-10 ← $2 + (-12) = -10$
$-3, \quad 8$	5
$3, \quad -8$	-5
$-4, \quad 6$	2
$4, \quad -6$	-2

We show these sums for completeness. In practice, we stop when we find the correct sum.

4) Next, we split the middle term as a sum or a difference using the factors found in step (3): $-10x = 2x - 12x$.

Answers

Skill to Review:

1. $(2x - 3)(x^2 + 5)$ **2.** $(n + 3)(n^2 + 1)$

5) Finally, we factor by grouping, as follows:

$$\begin{aligned} 3x^2 - 10x - 8 &= 3x^2 + 2x - 12x - 8 && \text{Substituting } 2x - 12x \text{ for } -10x \\ &= (3x^2 + 2x) + (-12x - 8) \\ &= x(3x + 2) - 4(3x + 2) && \text{Factoring by grouping} \\ &= (3x + 2)(x - 4). \end{aligned}$$

We can also split the middle term as $-12x + 2x$. We still get the same factorization, although the factors may be in a different order. Note the following:

$$\begin{aligned} 3x^2 - 10x - 8 &= 3x^2 - 12x + 2x - 8 && \text{Substituting } -12x + 2x \text{ for } -10x \\ &= (3x^2 - 12x) + (2x - 8) \\ &= 3x(x - 4) + 2(x - 4) && \text{Factoring by grouping} \\ &= (x - 4)(3x + 2). \end{aligned}$$

6) Check: $(3x + 2)(x - 4) = 3x^2 - 10x - 8$.

Do Exercises 1 and 2. ▶

EXAMPLE 2 Factor: $8x^2 + 8x - 6$.

1) First, we factor out a common factor, if any. The number 2 is common to all three terms, so we factor it out: $2(4x^2 + 4x - 3)$.

2) Next, we factor the trinomial $4x^2 + 4x - 3$. We multiply the leading coefficient and the constant, 4 and -3: $4(-3) = -12$.

3) We try to factor -12 so that the sum of the factors is 4.

PAIRS OF FACTORS	SUMS OF FACTORS
$-1, \ 12$	11
$1, -12$	-11
$-2, \ 6$	**4**
$2, -6$	.
$-3, \ 4$	.
$3, -4$	.

$-2 + 6 = 4$

We have found the correct sum, so there is no need to complete the table.

4) Then we split the middle term, $4x$, as follows: $4x = -2x + 6x$.

5) Finally, we factor by grouping:

$$\begin{aligned} 4x^2 + 4x - 3 &= 4x^2 - 2x + 6x - 3 && \text{Substituting } -2x + 6x \text{ for } 4x \\ &= (4x^2 - 2x) + (6x - 3) \\ &= 2x(2x - 1) + 3(2x - 1) && \text{Factoring by grouping} \\ &= (2x - 1)(2x + 3). \end{aligned}$$

The factorization of $4x^2 + 4x - 3$ is $(2x - 1)(2x + 3)$. *But don't forget the common factor!* We must include it to get a factorization of the original trinomial: $8x^2 + 8x - 6 = 2(2x - 1)(2x + 3)$.

6) Check: $2(2x - 1)(2x + 3) = 2(4x^2 + 4x - 3) = 8x^2 + 8x - 6$.

Do Exercises 3 and 4. ▶

Factor.

1. $6x^2 + 7x + 2$

2. $12x^2 - 17x - 5$

1) There is no common factor.

2) Multiplying the leading coefficient and the constant:

$12(-5) = \_\_\_\_$.

3) Look for a pair of factors of -60 whose sum is -17. Those factors are 3 and $\_\_\_\_$.

4) Split the middle term: $-17x = 3x - \_\_\_\_$.

5) Factor by grouping:

$12x^2 + 3x - 20x - 5$

$= 3x(4x + 1) - 5(\_\_\_\_)$

$= (\_\_\_\_)(3x - 5)$.

6) Check:

$(4x + 1)(3x - 5)$

$= 12x^2 - 17x - 5$.

Factor.

3. $6x^2 + 15x + 9$

4. $20x^2 - 46x + 24$

Answers

1. $(2x + 1)(3x + 2)$ **2.** $(4x + 1)(3x - 5)$
3. $3(2x + 3)(x + 1)$ **4.** $2(5x - 4)(2x - 3)$

Guided Solution:

2. $-60, -20, 20x, 4x + 1, 4x + 1$

13.4 Exercise Set

For Extra Help MyMathLab® MathXL® PRACTICE WATCH READ REVIEW

Reading Check

Complete each step in the process to factor $10x^2 + 21x + 2$.

RC1. Using the *ac*-method, multiply the ______________ 10 and the constant 2. The product is 20.

RC2. Find two integers whose ______________ is 20 and whose ______________ is 21. The integers are 20 and 1.

RC3. Split the middle term, $21x$, writing it as the ______________ of $20x$ and x.

RC4. Factor by ______________: $10x^2 + 20x + x + 2 = 10x(x + 2) + 1(x + 2) = (x + 2)(10x + 1)$.

a Factor. Note that the middle term has already been split.

1. $x^2 + 2x + 7x + 14$

2. $x^2 + 3x + x + 3$

3. $x^2 - 4x - x + 4$

4. $a^2 + 5a - 2a - 10$

5. $6x^2 + 4x + 9x + 6$

6. $3x^2 - 2x + 3x - 2$

7. $3x^2 - 4x - 12x + 16$

8. $24 - 18y - 20y + 15y^2$

9. $35x^2 - 40x + 21x - 24$

10. $8x^2 - 6x - 28x + 21$

11. $4x^2 + 6x - 6x - 9$

12. $2x^4 - 6x^2 - 5x^2 + 15$

13. $2x^4 + 6x^2 + 5x^2 + 15$

14. $9x^4 - 6x^2 - 6x^2 + 4$

Factor using the *ac*-method.

15. $2x^2 + 7x + 6$

16. $5x^2 + 17x + 6$

17. $3x^2 - 4x - 15$

18. $3x^2 + x - 4$

19. $5x^2 + 11x + 2$

20. $3x^2 + 16x + 5$

21. $3x^2 - 4x + 1$

22. $7x^2 - 15x + 2$

23. $6x^2 + 23x + 7$

24. $6x^2 + 13x + 6$

25. $4x^2 - 4x - 15$

26. $9x^2 - 6x - 8$

27. $15x^2 + 19x - 10$

28. $6 - 13x + 6x^2$

29. $9x^2 - 18x - 16$

30. $18x^2 + 3x - 10$

31. $3x^2 + 5x - 2$

32. $2x^2 - 5x + 2$

33. $12x^2 - 31x + 20$

34. $35x^2 - 34x + 8$

35. $14x^2 - 19x - 3$

36. $15x^2 - 19x - 10$

37. $49 - 42x + 9x^2$

38. $25x^2 + 40x + 16$

39. $9x^2 + 18x + 8$

40. $24x^2 - 47x - 2$

41. $5 - 9a^2 - 12a$

42. $17x - 4x^2 + 15$

43. $20 + 6x - 2x^2$

44. $15 + x - 2x^2$

45. $12x^2 + 28x - 24$

46. $6x^2 + 33x + 15$

47. $30x^2 - 24x - 54$

48. $18t^2 - 24t + 6$

49. $4y + 6y^2 - 10$

50. $-9 + 18x^2 - 21x$

51. $3x^2 - 4x + 1$

52. $6t^2 + t - 15$

53. $12x^2 - 28x - 24$

54. $6x^2 - 33x + 15$

55. $-1 + 2x^2 - x$

56. $-19x + 15x^2 + 6$

57. $9x^2 + 18x - 16$

58. $14y^2 + 35y + 14$

59. $15x^2 - 25x - 10$

60. $18x^2 + 3x - 10$

61. $12p^3 + 31p^2 + 20p$

62. $15x^3 + 19x^2 - 10x$

63. $4 - x - 5x^2$

64. $1 - p - 2p^2$

65. $33t - 15 - 6t^2$

66. $-15x^2 - 19x - 6$

67. $14x^4 + 19x^3 - 3x^2$

68. $70x^4 + 68x^3 + 16x^2$

69. $168x^3 - 45x^2 + 3x$

70. $144x^5 + 168x^4 + 48x^3$

71. $15x^4 - 19x^2 + 6$

72. $9x^4 + 18x^2 + 8$

73. $25t^2 + 80t + 64$

74. $9x^2 - 42x + 49$

75. $6x^3 + 4x^2 - 10x$

76. $18x^3 - 21x^2 - 9x$

77. $3x^2 + 9x + 5$

78. $4x^2 + 6x + 3$

79. $6x^2 - 19x - 5$

80. $2x^2 + 11x - 9$

81. $12m^2 - mn - 20n^2$

82. $12a^2 - 17ab + 6b^2$

83. $6a^2 - ab - 15b^2$

84. $3p^2 - 16pq - 12q^2$

85. $9a^2 - 18ab + 8b^2$

86. $10s^2 + 4st - 6t^2$

87. $35p^2 + 34pq + 8q^2$

88. $30a^2 + 87ab + 30b^2$

89. $18x^2 - 6xy - 24y^2$

90. $15a^2 - 5ab - 20b^2$

91. $60x + 18x^2 - 6x^3$

92. $60x + 4x^2 - 8x^3$

93. $35x^5 - 57x^4 - 44x^3$

94. $15x^3 + 33x^4 + 6x^5$

Skill Maintenance

Simplify. Express the result using positive exponents.

95. $(3x^4)^3$ [12.2a, b]

96. $5^{-6} \cdot 5^{-8}$ [12.1d, f]

97. $(x^2y)(x^3y^5)$ [12.1d]

98. $\dfrac{a^{-7}}{a^{-8}}$ [12.1e, f]

99. Convert to scientific notation: 30,080,000,000. [12.2c]

100. Convert to decimal notation: 1.5×10^{-5}. [12.2c]

Solve. [10.6a]

101. The earth is a sphere (or ball) that is about 40,000 km in circumference. Find the radius of the earth, in kilometers and in miles. Use 3.14 for π. (*Hint*: 1 km $\approx$ 0.62 mi.)

102. The second angle of a triangle is 10° less than twice the first. The third angle is 15° more than four times the first. Find the measure of the second angle.

Synthesis

Factor.

103. $9x^{10} - 12x^5 + 4$

104. $24x^{2n} + 22x^n + 3$

105. $16x^{10} + 8x^5 + 1$

106. $(a + 4)^2 - 2(a + 4) + 1$

107.–112. Use graphs to check the factoring in Exercises 15–20. (See the Calculator Corner on p. 961.)

Mid-Chapter Review

Concept Reinforcement

Determine whether each statement is true or false.

_______________ **1.** The greatest common factor (GCF) of a set of natural numbers is at least 1 and always less than or equal to the smallest number in the set. [13.1a]

_______________ **2.** To factor $x^2 + bx + c$, we use a trial-and-error process that looks for factors of b whose sum is c. [13.2a]

_______________ **3.** A prime polynomial has no common factor other than 1 and -1. [13.2a]

_______________ **4.** When factoring $x^2 - 14x + 45$, we need consider only positive pairs of factors of 45. [13.2a]

Guided Solutions

Fill in each blank with the number, variable, or expression that creates a correct statement or solution.

5. Factor: $10y^3 - 18y^2 + 12y$. [13.1b]

$$10y^3 - 18y^2 + 12y = \square \cdot 5y^2 - \square \cdot 9y + \square \cdot 6$$
$$= 2y(\square)$$

6. Factor $2x^2 - x - 6$ using the ac-method. [13.4a]

$a \cdot c = \square \cdot \square = -12;$ Multiplying the leading coefficient and the constant

$-x = \square + 3x;$ Splitting the middle term

$$2x^2 - x - 6 = 2x^2 - 4x + \square - 6$$
$$= \square(x - 2) + \square(x - 2)$$
$$= (x - 2)(\square)$$

Mixed Review

Find the GCF. [13.1a]

7. $x^3, 3x$

8. $5x^4, x^2$

9. $6x^5, -12x^3$

10. $-8x, -12, 16x^2$

11. $15x^3y^2, 5x^2y, 40x^4y^3$

12. $x^2y^4, -x^3y^3, x^3y^2, x^5y^4$

Factor completely. [13.1b, c], [13.2a], [13.3a], [13.4a]

13. $x^3 - 8x$

14. $3x^2 + 12x$

15. $2y^2 + 8y - 4$

16. $3t^6 - 5t^4 - 2t^3$

17. $x^2 + 4x + 3$

18. $z^2 - 4z + 4$

19. $x^3 + 4x^2 + 3x + 12$

20. $8y^5 - 48y^3$

21. $6x^3y + 24x^2y^2 - 42xy^3$

22. $6 - 11t + 4t^2$

23. $z^2 + 4z - 5$

24. $2z^3 + 8z^2 + 5z + 20$

25. $3p^3 - 2p^2 - 9p + 6$

26. $10x^8 - 25x^6 - 15x^5 + 35x^3$

27. $2w^3 + 3w^2 - 6w - 9$

28. $4x^4 - 5x^3 + 3x^2$

29. $6y^2 + 7y - 10$

30. $3x^2 - 3x - 18$

31. $6x^3 + 4x^2 + 3x + 2$

32. $15 - 8w + w^2$

33. $8x^3 + 20x^2 + 2x + 5$

34. $10z^2 - 21z - 10$

35. $6x^2 + 7x + 2$

36. $x^2 - 10xy + 24y^2$

37. $6z^3 + 3z^2 + 2z + 1$

38. $a^3b^7 + a^4b^5 - a^2b^3 + a^5b^6$

39. $4y^2 - 7yz - 15z^2$

40. $3x^3 + 21x^2 + 30x$

41. $x^3 - 3x^2 - 2x + 6$

42. $9y^2 + 6y + 1$

43. $y^2 + 6y + 8$

44. $6y^2 + 33y + 45$

45. $x^3 - 7x^2 + 4x - 28$

46. $4 + 3y - y^2$

47. $16x^2 - 16x - 60$

48. $10a^2 - 11ab + 3b^2$

49. $6w^3 - 15w^2 - 10w + 25$

50. $y^3 + 9y^2 + 18y$

51. $4x^2 + 11xy + 6y^2$

52. $6 - 5z - 6z^2$

53. $12t^3 + 8t^2 - 9t - 6$

54. $y^2 + yz - 20z^2$

55. $9x^2 - 6xy - 8y^2$

56. $-3 + 8z + 3z^2$

57. $m^2 - 6mn - 16n^2$

58. $2w^2 - 12w + 18$

59. $18t^3 - 18t^2 + 4t$

60. $5z^3 + 15z^2 + z + 3$

61. $-14 + 5t + t^2$

62. $4t^2 - 20t + 25$

63. $t^2 + 4t - 12$

64. $12 + 5z - 2z^2$

65. $12 + 4y - y^2$

Understanding Through Discussion and Writing

66. Explain how one could construct a polynomial with four terms that can be factored by grouping. [13.1c], [13.4a]

67. When searching for a factorization, why do we list pairs of numbers with the correct *product* instead of pairs of numbers with the correct *sum*? [13.2a]

68. Without multiplying $(x - 17)(x - 18)$, explain why it cannot possibly be a factorization of $x^2 + 35x + 306$. [13.2a]

69. A student presents the following work:

$$4x^2 + 28x + 48 = (2x + 6)(2x + 8)$$
$$= 2(x + 3)(x + 4).$$

Is it correct? Explain. [13.3a], [13.4a]

STUDYING FOR SUCCESS *Applications*

- ☐ If you find applied problems challenging, don't give up! Your skill will improve with each problem that you solve.
- ☐ Make applications real! Look for applications of the math that you are studying in newspapers and magazines.

13.5 Factoring Trinomial Squares and Differences of Squares

OBJECTIVES

- **a** Recognize trinomial squares.
- **b** Factor trinomial squares.
- **c** Recognize differences of squares.
- **d** Factor differences of squares, being careful to factor completely.

In this section, we first learn to factor trinomials that are squares of binomials. Then we factor binomials that are differences of squares.

a RECOGNIZING TRINOMIAL SQUARES

Some trinomials are squares of binomials. For example, the trinomial $x^2 + 10x + 25$ is the square of the binomial $x + 5$. To see this, we can calculate $(x + 5)^2$. It is $x^2 + 2 \cdot x \cdot 5 + 5^2$, or $x^2 + 10x + 25$. A trinomial that is the square of a binomial is called a **trinomial square**, or a **perfect-square trinomial**.

We can use the following special-product rules in reverse to factor trinomial squares:

$$\mathbf{(A + B)^2 = A^2 + 2AB + B^2;}$$
$$\mathbf{(A - B)^2 = A^2 - 2AB + B^2.}$$

How can we recognize when an expression to be factored is a trinomial square? Look at $A^2 + 2AB + B^2$ and $A^2 - 2AB + B^2$. In order for an expression to be a trinomial square:

a) The two expressions A^2 and B^2 must be squares, such as

$$4, \quad x^2, \quad 25x^4, \quad 16t^2.$$

When the coefficient is a perfect square and the power(s) of the variable(s) is (are) even, then the expression is a perfect square.

b) There must be no minus sign before A^2 or B^2.

c) The remaining term is either twice the product of A and B, $2AB$, or its opposite, $-2AB$.

If a number c can be multiplied by itself to get a number n, then c is a **square root** of n. Thus, 3 is a square root of 9 because $3 \cdot 3$, or 3^2, is 9. Similarly, A is a square root of A^2 and B is a square root of B^2.

It would be helpful to memorize this table of perfect squares.

NUMBER, N	PERFECT SQUARE, N^2
1	1
2	4
3	9
4	16
5	25
6	36
7	49
8	64
9	81
10	100
11	121
12	144
13	169
14	196
15	225
16	256
20	400
25	625

EXAMPLE 1 Determine whether $x^2 + 6x + 9$ is a trinomial square.

a) We know that x^2 and 9 are squares.

b) There is no minus sign before x^2 or 9.

c) If we multiply the square roots, x and 3, and double the product, we get the remaining term: $2 \cdot x \cdot 3 = 6x$.

Thus, $x^2 + 6x + 9$ is the square of a binomial. In fact, $x^2 + 6x + 9 = (x + 3)^2$. ■

EXAMPLE 2 Determine whether $x^2 + 6x + 11$ is a trinomial square.

The answer is no, because only one term, x^2, is a square.

EXAMPLE 3 Determine whether $16x^2 + 49 - 56x$ is a trinomial square.

It helps to first write the trinomial in descending order:

$$16x^2 - 56x + 49.$$

a) We know that $16x^2$ and 49 are squares.

b) There is no minus sign before $16x^2$ or 49.

c) We multiply the square roots, $4x$ and 7, and double the product to get $2 \cdot 4x \cdot 7 = 56x$. The remaining term, $-56x$, is the opposite of this product.

Thus, $16x^2 + 49 - 56x$ is a trinomial square.

Do Exercises 1–8. ▶

Determine whether each is a trinomial square. Write "yes" or "no."

1. $x^2 + 8x + 16$
2. $25 - x^2 + 10x$
3. $t^2 - 12t + 4$
4. $25 + 20y + 4y^2$
5. $5x^2 + 16 - 14x$
6. $16x^2 + 40x + 25$
7. $p^2 + 6p - 9$
8. $25a^2 + 9 - 30a$

b FACTORING TRINOMIAL SQUARES

We can use the factoring methods from Sections 5.2–5.4 to factor trinomial squares, but there is a faster method using the following equations.

FACTORING TRINOMIAL SQUARES

$A^2 + 2AB + B^2 = (A + B)^2$;

$A^2 - 2AB + B^2 = (A - B)^2$

We use square roots of the squared terms and the sign of the remaining term to factor a trinomial square.

EXAMPLE 4 Factor: $x^2 + 6x + 9$.

$$x^2 + 6x + 9 = x^2 + 2 \cdot x \cdot 3 + 3^2 = (x + 3)^2$$

$$A^2 + 2\ A\ B + B^2 = (A + B)^2$$

The sign of the middle term is positive.

EXAMPLE 5 Factor: $x^2 + 49 - 14x$.

$x^2 + 49 - 14x = x^2 - 14x + 49$ Changing to descending order

$= x^2 - 2 \cdot x \cdot 7 + 7^2$ The sign of the middle term is negative.

$= (x - 7)^2$

EXAMPLE 6 Factor: $16x^2 - 40x + 25$.

$$16x^2 - 40x + 25 = (4x)^2 - 2 \cdot 4x \cdot 5 + 5^2 = (4x - 5)^2$$

$$A^2 - 2\ A\ B + B^2 = (A - B)^2$$

Do Exercises 9–13. ▶

Factor.

9. $x^2 + 2x + 1$
10. $1 - 2x + x^2$
11. $4 + t^2 + 4t$
12. $25x^2 - 70x + 49$

GS 13. $49 - 56y + 16y^2$

Write in descending order:

$16y^2 - 56y + 49.$

Factor as a trinomial square:

$(4y)^2 - 2 \cdot 4y \cdot \_\_\_ + (\_\_\_)^2$

$= (4y - \_\_\_)^2.$

Answers

1. Yes 2. No 3. No 4. Yes 5. No 6. Yes 7. No 8. Yes 9. $(x + 1)^2$ 10. $(x - 1)^2$, or $(1 - x)^2$ 11. $(t + 2)^2$ 12. $(5x - 7)^2$ 13. $(4y - 7)^2$, or $(7 - 4y)^2$

Guided Solution:
13. 7, 7, 7

EXAMPLE 7 Factor: $t^4 + 20t^2 + 100$.

$$t^4 + 20t^2 + 100 = (t^2)^2 + 2(t^2)(10) + 10^2$$
$$= (t^2 + 10)^2$$

EXAMPLE 8 Factor: $75m^3 + 210m^2 + 147m$.

Always look first for a common factor. This time there is one, $3m$:

$$75m^3 + 210m^2 + 147m = 3m(25m^2 + 70m + 49)$$
$$= 3m[(5m)^2 + 2(5m)(7) + 7^2]$$
$$= 3m(5m + 7)^2.$$

EXAMPLE 9 Factor: $4p^2 - 12pq + 9q^2$.

$$4p^2 - 12pq + 9q^2 = (2p)^2 - 2(2p)(3q) + (3q)^2$$
$$= (2p - 3q)^2$$

Factor.

14. $48m^2 + 75 + 120m$

15. $p^4 + 18p^2 + 81$

16. $4z^5 - 20z^4 + 25z^3$

17. $9a^2 + 30ab + 25b^2$

Do Exercises 14–17.

C RECOGNIZING DIFFERENCES OF SQUARES

The following polynomials are *differences of squares:*

$$x^2 - 9, \quad 4t^2 - 49, \quad a^2 - 25b^2.$$

To factor a difference of squares such as $x^2 - 9$, we will use the following special-product rule in reverse:

$$\mathbf{(A + B)(A - B) = A^2 - B^2.}$$

A **difference of squares** is an expression like the following:

$$A^2 - B^2.$$

How can we recognize such expressions? Look at $A^2 - B^2$. In order for a binomial to be a difference of squares:

a) There must be two expressions, both squares, such as

$$4x^2, \quad 9, \quad 25t^4, \quad 1, \quad x^6, \quad 49y^8.$$

b) The terms must have different signs.

EXAMPLE 10 Is $9x^2 - 64$ a difference of squares?

a) The first expression is a square: $9x^2 = (3x)^2$.
The second expression is a square: $64 = 8^2$.

b) The terms have different signs, $+9x^2$ and -64.

Thus we have a difference of squares, $(3x)^2 - 8^2$.

EXAMPLE 11 Is $25 - t^3$ a difference of squares?

a) The expression t^3 is not a square.

The expression is not a difference of squares.

Answers

14. $3(4m + 5)^2$ 15. $(p^2 + 9)^2$
16. $z^3(2z - 5)^2$ 17. $(3a + 5b)^2$

EXAMPLE 12 Is $-4x^2 + 16$ a difference of squares?

a) The expressions $4x^2$ and 16 are squares: $4x^2 = (2x)^2$ and $16 = 4^2$.

b) The terms have different signs, $-4x^2$ and $+16$.

Thus we have a difference of squares. We can also see this by rewriting in the equivalent form: $16 - 4x^2$.

Do Exercises 18–24.

Determine whether each is a difference of squares. Write "yes" or "no."

18. $x^2 - 25$

19. $t^2 - 24$

20. $y^2 + 36$

21. $4x^2 - 15$

22. $16x^4 - 49$

23. $9w^6 - 1$

24. $-49 + 25t^2$

d FACTORING DIFFERENCES OF SQUARES

To factor a difference of squares, we use the following equation.

FACTORING A DIFFERENCE OF SQUARES

$A^2 - B^2 = (A + B)(A - B)$

To factor a difference of squares $A^2 - B^2$, we find A and B, which are square roots of the expressions A^2 and B^2. We then use A and B to form two factors. One is the sum $A + B$, and the other is the difference $A - B$.

EXAMPLE 13 Factor: $x^2 - 4$.

$$x^2 - 4 = x^2 - 2^2 = (x + 2)(x - 2)$$

$$A^2 - B^2 = (A + B)(A - B)$$

EXAMPLE 14 Factor: $9 - 16t^4$.

$$9 - 16t^4 = 3^2 - (4t^2)^2 = (3 + 4t^2)(3 - 4t^2)$$

$$A^2 - B^2 = (A + B)(A - B)$$

EXAMPLE 15 Factor: $m^2 - 4p^2$.

$$m^2 - 4p^2 = m^2 - (2p)^2 = (m + 2p)(m - 2p)$$

EXAMPLE 16 Factor: $x^2 - \frac{1}{9}$.

$$x^2 - \frac{1}{9} = x^2 - \left(\frac{1}{3}\right)^2 = \left(x + \frac{1}{3}\right)\left(x - \frac{1}{3}\right)$$

EXAMPLE 17 Factor: $18x^2 - 50x^6$.

Always look first for a factor common to all terms. This time there is one, $2x^2$.

$$\begin{aligned} 18x^2 - 50x^6 &= 2x^2(9 - 25x^4) \\ &= 2x^2[3^2 - (5x^2)^2] \\ &= 2x^2(3 + 5x^2)(3 - 5x^2) \end{aligned}$$

Answers

18. Yes **19.** No **20.** No **21.** No **22.** Yes **23.** Yes **24.** Yes

Factor.

25. $x^2 - 9$

26. $4t^2 - 64$

27. $a^2 - 25b^2$
$= a^2 - (\quad)^2$
$= (a + \quad)(a - \quad)$

GS

28. $64x^4 - 25x^6$

29. $5 - 20t^6$
[*Hint*: $t^6 = (t^3)^2$.]

EXAMPLE 18 Factor: $36x^{10} - 4x^2$.

Although this expression is a difference of squares, the terms have a common factor. We always begin factoring by factoring out the greatest common factor.

$$36x^{10} - 4x^2 = 4x^2(9x^8 - 1)$$
$$= 4x^2[(3x^4)^2 - 1^2] \qquad \text{Note that } x^8 = (x^4)^2 \text{ and } 1 = 1^2.$$
$$= 4x^2(3x^4 + 1)(3x^4 - 1)$$

◀ Do Exercises 25–29.

Caution!

Note carefully in these examples that a difference of squares is *not* the square of the difference; that is,

$$A^2 - B^2 \neq (A - B)^2.$$

For example,

$$(45 - 5)^2 = 40^2 = 1600,$$

but

$$45^2 - 5^2 = 2025 - 25 = 2000.$$

Factoring Completely

If a factor with more than one term can still be factored, you should do so. When no factor can be factored further, you have **factored completely**. Always factor completely whenever told to factor.

EXAMPLE 19 Factor: $p^4 - 16$.

$$p^4 - 16 = (p^2)^2 - 4^2$$
$$= (p^2 + 4)(p^2 - 4) \qquad \text{Factoring a difference of squares}$$
$$= (p^2 + 4)(p + 2)(p - 2) \qquad \text{Factoring further; } p^2 - 4 \text{ is a difference of squares.}$$

The polynomial $p^2 + 4$ cannot be factored further into polynomials with real coefficients. ■

Caution!

Apart from possibly removing a common factor, we cannot, in general, factor a sum of squares. In particular,

$$A^2 + B^2 \neq (A + B)^2.$$

Consider $25x^2 + 100$. In this case, a sum of squares has a common factor, 25. Factoring, we get $25(x^2 + 4)$, where $x^2 + 4$ is prime.

Answers

25. $(x + 3)(x - 3)$
26. $4(t + 4)(t - 4)$
27. $(a + 5b)(a - 5b)$
28. $x^4(8 + 5x)(8 - 5x)$
29. $5(1 + 2t^3)(1 - 2t^3)$

Guided Solution:
27. $5b, 5b, 5b$

EXAMPLE 20 Factor: $y^4 - 16x^{12}$.

$$y^4 - 16x^{12} = (y^2 + 4x^6)(y^2 - 4x^6)$$ Factoring a difference of squares

$$= (y^2 + 4x^6)(y + 2x^3)(y - 2x^3)$$ Factoring further. The factor $y^2 - 4x^6$ is a difference of squares.

The polynomial $y^2 + 4x^6$ cannot be factored further into polynomials with real coefficients.

EXAMPLE 21 Factor: $\frac{1}{16}x^8 - 81$.

$$\frac{1}{16}x^8 - 81 = \left(\frac{1}{4}x^4 + 9\right)\left(\frac{1}{4}x^4 - 9\right)$$ Factoring a difference of squares

$$= \left(\frac{1}{4}x^4 + 9\right)\left(\frac{1}{2}x^2 + 3\right)\left(\frac{1}{2}x^2 - 3\right)$$ Factoring further. The factor $\frac{1}{4}x^4 - 9$ is a difference of squares.

Factor completely.

30. $81x^4 - 1$

31. $16 - \frac{1}{81}y^8$

32. $49p^4 - 25q^6$

TIPS FOR FACTORING

- Always look first for a common factor. If there is one, factor it out.
- Be alert for trinomial squares and differences of squares. Once recognized, they can be factored without trial and error.
- Always factor completely.
- Check by multiplying.

Do Exercises 30–32.

Answers

30. $(9x^2 + 1)(3x + 1)(3x - 1)$
31. $\left(4 + \frac{1}{9}y^4\right)\left(2 + \frac{1}{3}y^2\right)\left(2 - \frac{1}{3}y^2\right)$
32. $(7p^2 + 5q^3)(7p^2 - 5q^3)$

13.5 Exercise Set

For Extra Help MyMathLab® MathXL®

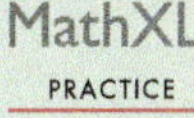

Reading Check

Determine whether each statement is true or false.

RC1. A trinomial can be considered a trinomial square if only one term is a perfect square.

RC2. A trinomial square is the square of a binomial.

RC3. In order for a binomial to be a difference of squares, the terms in the binomial must have the same sign.

RC4. A binomial cannot have a common factor.

a Determine whether each of the following is a trinomial square. Answer "yes" or "no."

1. $x^2 - 14x + 49$

2. $x^2 - 16x + 64$

3. $x^2 + 16x - 64$

4. $x^2 - 14x - 49$

5. $x^2 - 2x + 4$

6. $x^2 + 3x + 9$

7. $9x^2 - 24x + 16$

8. $25x^2 + 30x + 9$

b Factor completely. Remember to look first for a common factor and to check by multiplying.

9. $x^2 - 14x + 49$

10. $x^2 - 20x + 100$

11. $x^2 + 16x + 64$

12. $x^2 + 20x + 100$

13. $x^2 - 2x + 1$

14. $x^2 + 2x + 1$

15. $4 + 4x + x^2$

16. $4 + x^2 - 4x$

17. $y^2 + 12y + 36$

18. $y^2 + 18y + 81$

19. $16 + t^2 - 8t$

20. $9 + t^2 - 6t$

21. $q^4 - 6q^2 + 9$

22. $64 + 16a^2 + a^4$

23. $49 + 56y + 16y^2$

24. $75 + 48a^2 - 120a$

25. $2x^2 - 4x + 2$

26. $2x^2 - 40x + 200$

27. $x^3 - 18x^2 + 81x$

28. $x^3 + 24x^2 + 144x$

29. $12q^2 - 36q + 27$

30. $20p^2 + 100p + 125$

31. $49 - 42x + 9x^2$

32. $64 - 112x + 49x^2$

33. $5y^4 + 10y^2 + 5$

34. $a^4 + 14a^2 + 49$

35. $1 + 4x^4 + 4x^2$

36. $1 - 2a^5 + a^{10}$

37. $4p^2 + 12pt + 9t^2$

38. $25m^2 + 20mn + 4n^2$

39. $a^2 - 6ab + 9b^2$

40. $x^2 - 14xy + 49y^2$

41. $81a^2 - 18ab + b^2$

42. $64p^2 + 16pt + t^2$

43. $36a^2 + 96ab + 64b^2$

44. $16m^2 - 40mn + 25n^2$

c Determine whether each of the following is a difference of squares. Answer "yes" or "no."

45. $x^2 - 4$

46. $x^2 - 36$

47. $x^2 + 25$

48. $x^2 + 9$

49. $x^2 - 45$

50. $x^2 - 80y^2$

51. $-25y^2 + 16x^2$

52. $-1 + 36x^2$

d Factor completely. Remember to look first for a common factor.

53. $y^2 - 4$

54. $q^2 - 1$

55. $p^2 - 1$

56. $x^2 - 36$

57. $-49 + t^2$

58. $-64 + m^2$

59. $a^2 - b^2$

60. $p^2 - v^2$

61. $25t^2 - m^2$

62. $w^2 - 49z^2$

63. $100 - k^2$

64. $81 - w^2$

65. $16a^2 - 9$

66. $25x^2 - 4$

67. $4x^2 - 25y^2$

68. $9a^2 - 16b^2$

69. $8x^2 - 98$

70. $24x^2 - 54$

71. $36x - 49x^3$

72. $16x - 81x^3$

73. $\frac{1}{16} - 49x^8$

74. $\frac{1}{625}x^8 - 49$

75. $0.09y^2 - 0.0004$

76. $0.16p^2 - 0.0025$

77. $49a^4 - 81$

78. $25a^4 - 9$

79. $a^4 - 16$

80. $y^4 - 1$

81. $5x^4 - 405$

82. $4x^4 - 64$

83. $1 - y^8$

84. $x^8 - 1$

85. $x^{12} - 16$

86. $x^8 - 81$

87. $y^2 - \frac{1}{16}$

88. $x^2 - \frac{1}{25}$

89. $25 - \frac{1}{49}x^2$

90. $\frac{1}{4} - 9q^2$

91. $16m^4 - t^4$

92. $p^4t^4 - 1$

Skill Maintenance

Find the intercepts of each equation. [11.2a]

93. $4x + 16y = 64$

94. $x - 1.3y = 6.5$

95. $y = 2x - 5$

Find the intercepts. Then graph each equation. [11.2a]

96. $y - 5x = 5$

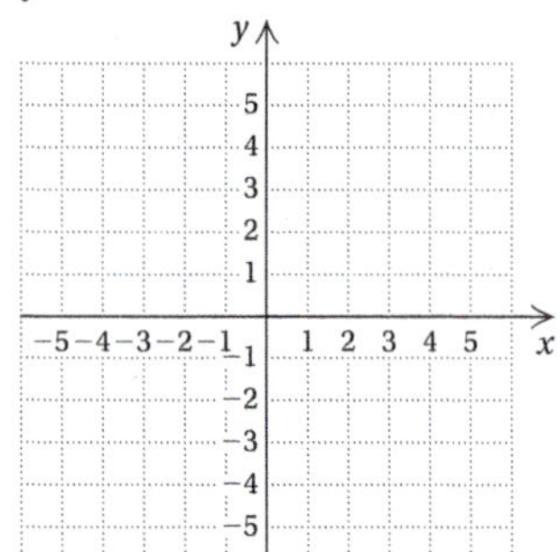

97. $2x + 5y = 10$

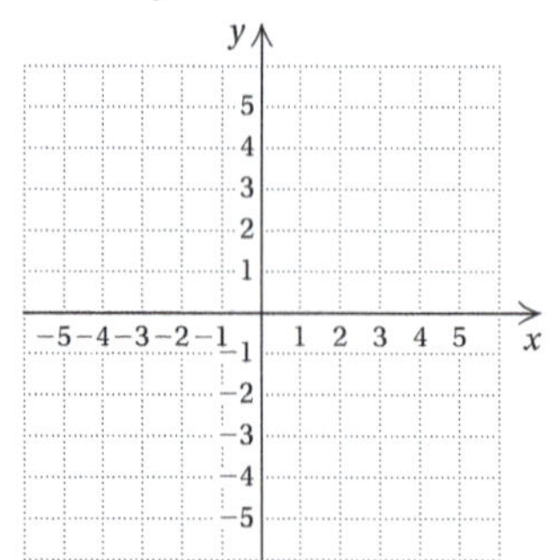

98. $3x - 5y = 15$

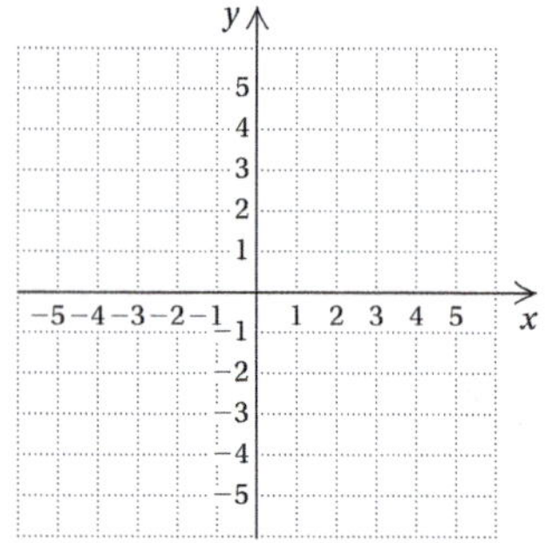

Find a polynomial for the shaded area in each figure. (Leave results in terms of π where appropriate.) [12.4d]

99.

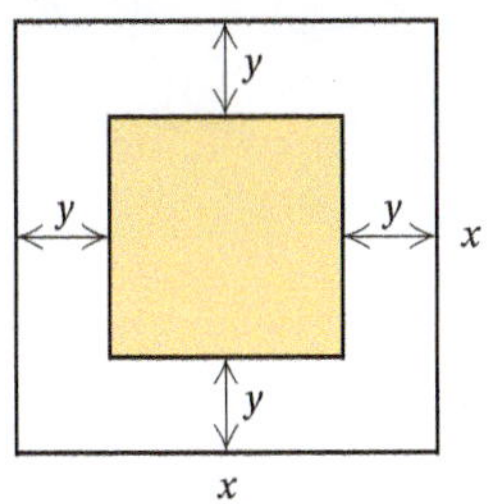

100.

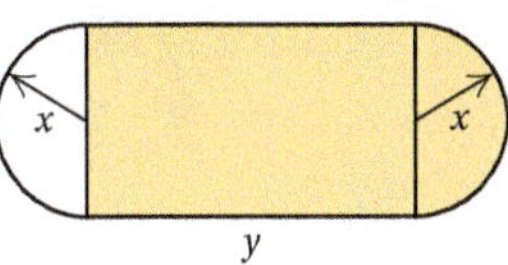

Synthesis

Factor completely, if possible.

101. $49x^2 - 216$

102. $27x^3 - 13x$

103. $x^2 + 22x + 121$

104. $x^2 - 5x + 25$

105. $18x^3 + 12x^2 + 2x$

106. $162x^2 - 82$

107. $x^8 - 2^8$

108. $4x^4 - 4x^2$

109. $3x^5 - 12x^3$

110. $3x^2 - \frac{1}{3}$

111. $18x^3 - \frac{8}{25}x$

112. $x^2 - 2.25$

113. $0.49p - p^3$

114. $3.24x^2 - 0.81$

115. $0.64x^2 - 1.21$

116. $1.28x^2 - 2$

117. $(x + 3)^2 - 9$

118. $(y - 5)^2 - 36q^2$

119. $x^2 - \left(\frac{1}{x}\right)^2$

120. $a^{2n} - 49b^{2n}$

121. $81 - b^{4k}$

122. $9x^{18} + 48x^9 + 64$

123. $9b^{2n} + 12b^n + 4$

124. $(x + 7)^2 - 4x - 24$

125. $(y + 3)^2 + 2(y + 3) + 1$

126. $49(x + 1)^2 - 42(x + 1) + 9$

Find c such that the polynomial is the square of a binomial.

127. $cy^2 + 6y + 1$

128. $cy^2 - 24y + 9$

Use the TABLE feature or graphs to determine whether each factorization is correct. (See the Calculator Corner on p. 961.)

129. $x^2 + 9 = (x + 3)(x + 3)$

130. $x^2 - 49 = (x - 7)(x + 7)$

131. $x^2 + 9 = (x + 3)^2$

132. $x^2 - 49 = (x - 7)^2$

13.6 Factoring: A General Strategy

OBJECTIVE

a Factor polynomials completely using any of the methods considered in this chapter.

SKILL TO REVIEW

Objective 12.7f: Multiply polynomials.

Multiply.

1. $(aw + 5y)(aw - 3y)$
2. $(c + d)(p + t)$

a We now combine all of our factoring techniques and consider a general strategy for factoring polynomials. Here we will encounter polynomials of all the types we have considered, in random order, so you will have the opportunity to determine which method to use.

FACTORING STRATEGY

To factor a polynomial:

a) Always look first for a common factor. If there is one, factor out the largest common factor.

b) Then look at the number of terms.

Two terms: Determine whether you have a difference of squares, $A^2 - B^2$. Do not try to factor a sum of squares: $A^2 + B^2$.

Three terms: Determine whether the trinomial is a square. If it is, you know how to factor. If not, try trial and error, using FOIL or the *ac*-method.

Four terms: Try factoring by grouping.

c) *Always factor completely.* If a factor with more than one term can still be factored, you should factor it. When no factor can be factored further, you have finished.

d) Check by multiplying.

EXAMPLE 1 Factor: $5t^4 - 80$.

a) We look for a common factor. There is one, 5.

$$5t^4 - 80 = 5(t^4 - 16)$$

b) The factor $t^4 - 16$ has only two terms. It is a difference of squares: $(t^2)^2 - 4^2$. We factor $t^4 - 16$ and then include the common factor:

$$5(t^2 + 4)(t^2 - 4).$$

c) We see that one of the factors, $t^2 - 4$, is again a difference of squares. We factor it:

$$5(t^2 + 4)(t + 2)(t - 2).$$

↑ This is a sum of squares. It cannot be factored.

We have factored completely because no factor with more than one term can be factored further.

d) Check:
$$\begin{aligned} 5(t^2 + 4)(t + 2)(t - 2) &= 5(t^2 + 4)(t^2 - 4) \\ &= 5(t^4 - 16) \\ &= 5t^4 - 80. \end{aligned}$$

Answers:

Skill to Review:
1. $a^2w^2 + 2awy - 15y^2$
2. $cp + ct + dp + dt$

EXAMPLE 2 Factor: $2x^3 + 10x^2 + x + 5$.

a) We look for a common factor. There isn't one.

b) There are four terms. We try factoring by grouping:

$$\begin{aligned} &2x^3 + 10x^2 + x + 5 \\ &\quad = (2x^3 + 10x^2) + (x + 5) && \text{Separating into two binomials} \\ &\quad = 2x^2(x + 5) + 1(x + 5) && \text{Factoring each binomial} \\ &\quad = (x + 5)(2x^2 + 1). && \text{Factoring out the common factor } x + 5 \end{aligned}$$

c) None of these factors can be factored further, so we have factored completely.

d) Check: $(x + 5)(2x^2 + 1) = x \cdot 2x^2 + x \cdot 1 + 5 \cdot 2x^2 + 5 \cdot 1$
$= 2x^3 + x + 10x^2 + 5$, or
$2x^3 + 10x^2 + x + 5$. ■

EXAMPLE 3 Factor: $x^5 - 2x^4 - 35x^3$.

a) We look first for a common factor. This time there is one, x^3:

$$x^5 - 2x^4 - 35x^3 = x^3(x^2 - 2x - 35).$$

b) The factor $x^2 - 2x - 35$ has three terms, but it is not a trinomial square. We factor it using trial and error:

$$x^5 - 2x^4 - 35x^3 = x^3(x^2 - 2x - 35) = x^3(x - 7)(x + 5).$$

Don't forget to include the common factor in the final answer!

c) No factor with more than one term can be factored further, so we have factored completely.

d) Check: $x^3(x - 7)(x + 5) = x^3(x^2 - 2x - 35)$
$= x^5 - 2x^4 - 35x^3$. ■

EXAMPLE 4 Factor: $x^4 - 10x^2 + 25$.

a) We look first for a common factor. There isn't one.

b) There are three terms. We see that this polynomial is a trinomial square. We factor it:

$$x^4 - 10x^2 + 25 = (x^2)^2 - 2 \cdot x^2 \cdot 5 + 5^2 = (x^2 - 5)^2.$$

We could use trial and error if we have not recognized that we have a trinomial square.

c) Since $x^2 - 5$ cannot be factored further, we have factored completely.

d) Check: $(x^2 - 5)^2 = (x^2)^2 - 2(x^2)(5) + 5^2 = x^4 - 10x^2 + 25$.

Do Exercises 1–5. ▶

Factor.

1. $3m^4 - 3$

2. $x^6 + 8x^3 + 16$

3. $2x^4 + 8x^3 + 6x^2$

4. $3x^3 + 12x^2 - 2x - 8$

5. $8x^3 - 200x$

a) Factor out the largest common factor:
$8x^3 - 200x$
$= \_\_\_\_(x^2 - 25)$.

b) There are two terms inside the parentheses. Factor the difference of squares:
$8x(x^2 - 25)$
$= 8x(x + \_\_\_\_)(x - \_\_\_\_)$.

c) We have factored completely.

d) Check:
$8x(x + 5)(x - 5)$
$= 8x(x^2 - 25)$
$= 8x^3 - 200x$.

Answers

1. $3(m^2 + 1)(m + 1)(m - 1)$ **2.** $(x^3 + 4)^2$
3. $2x^2(x + 1)(x + 3)$ **4.** $(x + 4)(3x^2 - 2)$
5. $8x(x + 5)(x - 5)$

Guided Solution:
5. $8x$, 5, 5

EXAMPLE 5 Factor: $6x^2y^4 - 21x^3y^5 + 3x^2y^6$.

a) We look first for a common factor:

$$6x^2y^4 - 21x^3y^5 + 3x^2y^6 = 3x^2y^4(2 - 7xy + y^2).$$

b) There are three terms in $2 - 7xy + y^2$. Since only y^2 is a square, we do not have a trinomial square. Can the trinomial be factored by trial and error? A key to the answer is that x is in only the term $-7xy$. The polynomial might be in a form like $(1 - y)(2 + y)$, but there would be no x in the middle term. Thus, $2 - 7xy + y^2$ cannot be factored.

c) Have we factored completely? Yes, because no factor with more than one term can be factored further.

d) The check is left to the student.

EXAMPLE 6 Factor: $(p + q)(x + 2) + (p + q)(x + y)$.

a) We look for a common factor:

$$\begin{aligned}(p + q)(x + 2) + (p + q)(x + y) &= (p + q)[(x + 2) + (x + y)] \\ &= (p + q)(2x + y + 2).\end{aligned}$$

b) The trinomial $2x + y + 2$ cannot be factored further.

c) Neither factor can be factored further, so we have factored completely.

d) The check is left to the student.

EXAMPLE 7 Factor: $px + py + qx + qy$.

a) We look first for a common factor. There isn't one.

b) There are four terms. We try factoring by grouping:

$$\begin{aligned}px + py + qx + qy &= p(x + y) + q(x + y) \\ &= (x + y)(p + q).\end{aligned}$$

c) Since neither factor can be factored further, we have factored completely.

d) Check: $(x + y)(p + q) = px + qx + py + qy$, or

$px + py + qx + qy$.

EXAMPLE 8 Factor: $25x^2 + 20xy + 4y^2$.

a) We look first for a common factor. There isn't one.

b) There are three terms. We determine whether the trinomial is a square. The first term and the last term are squares:

$$25x^2 = (5x)^2 \quad \text{and} \quad 4y^2 = (2y)^2.$$

Since twice the product of $5x$ and $2y$ is the other term,

$$2 \cdot 5x \cdot 2y = 20xy,$$

the trinomial is a perfect square.

We factor by writing the square roots of the square terms and the sign of the middle term:

$$25x^2 + 20xy + 4y^2 = (5x + 2y)^2.$$

c) Since $5x + 2y$ cannot be factored further, we have factored completely.

d) Check: $$\begin{aligned}(5x + 2y)^2 &= (5x)^2 + 2(5x)(2y) + (2y)^2 \\ &= 25x^2 + 20xy + 4y^2.\end{aligned}$$

EXAMPLE 9 Factor: $p^2q^2 + 7pq + 12$.

a) We look first for a common factor. There isn't one.

b) There are three terms. We determine whether the trinomial is a square. The first term is a square, but neither of the other terms is a square, so we do not have a trinomial square. We factor, thinking of the product pq as a single variable. We consider this possibility for factorization:

$$(pq + \square)(pq + \square).$$

We factor the last term, 12. All the signs are positive, so we consider only positive factors. Possibilities are 1, 12 and 2, 6 and 3, 4. The pair 3, 4 gives a sum of 7 for the coefficient of the middle term. Thus,

$$p^2q^2 + 7pq + 12 = (pq + 3)(pq + 4).$$

c) No factor with more than one term can be factored further, so we have factored completely.

d) Check: $(pq + 3)(pq + 4) = (pq)(pq) + 4 \cdot pq + 3 \cdot pq + 3 \cdot 4$
$= p^2q^2 + 7pq + 12.$

EXAMPLE 10 Factor: $8x^4 - 20x^2y - 12y^2$.

a) We look first for a common factor:

$$8x^4 - 20x^2y - 12y^2 = 4(2x^4 - 5x^2y - 3y^2).$$

b) There are three terms in $2x^4 - 5x^2y - 3y^2$. Since none of the terms is a square, we do not have a trinomial square. The x^2 in the middle term, $-5x^2y$, leads us to factor $2x^4$ as $2x^2 \cdot x^2$. We also factor the last term, $-3y^2$. Possibilities are $3y, -y$ and $-3y, y$ and others. We look for factors such that the sum of their products is the middle term. We try some possibilities:

$$(2x^2 - y)(x^2 + 3y) = 2x^4 + 5x^2y - 3y^2,$$
$$(2x^2 + y)(x^2 - 3y) = 2x^4 - 5x^2y - 3y^2.$$

c) No factor with more than one term can be factored further, so we have factored completely. The factorization, including the common factor, is

$$4(2x^2 + y)(x^2 - 3y).$$

d) Check:
$$\begin{aligned}4(2x^2 + y)(x^2 - 3y) &= 4[(2x^2)(x^2) + 2x^2(-3y) + yx^2 + y(-3y)] \\ &= 4[2x^4 - 6x^2y + x^2y - 3y^2] \\ &= 4(2x^4 - 5x^2y - 3y^2) \\ &= 8x^4 - 20x^2y - 12y^2.\end{aligned}$$

EXAMPLE 11 Factor: $a^4 - 16b^4$.

a) We look first for a common factor. There isn't one.

b) There are two terms. Since $a^4 = (a^2)^2$ and $16b^4 = (4b^2)^2$, we see that we have a difference of squares. Thus,

$$a^4 - 16b^4 = (a^2 + 4b^2)(a^2 - 4b^2).$$

c) The last factor can be factored further. It is also a difference of squares.

$$a^4 - 16b^4 = (a^2 + 4b^2)(a + 2b)(a - 2b)$$

d) Check: $(a^2 + 4b^2)(a + 2b)(a - 2b) = (a^2 + 4b^2)(a^2 - 4b^2)$
$= a^4 - 16b^4.$

Do Exercises 6–12. ▶

Factor.

6. $15x^4 + 5x^2y - 10y^2$

7. $10p^6q^2 + 4p^5q^3 + 2p^4q^4$

8. $(a - b)(x + 5) + (a - b)(x + y^2)$

9. $ax^2 + ay + bx^2 + by$

GS **10.** $x^4 + 2x^2y^2 + y^4$

a) There is no common factor.

b) There are three terms. Factor the trinomial square:
$x^4 + 2x^2y^2 + y^4 = (x^2 + \quad)^2$.

c) We have factored completely.

d) Check:
$(x^2 + y^2)^2$
$= (x^2)^2 + 2(x^2)(y^2) + (y^2)^2$
$= x^4 + 2x^2y^2 + y^4$.

Factor.

11. $x^2y^2 + 5xy + 4$

12. $p^4 - 81q^4$

Answers

6. $5(3x^2 - 2y)(x^2 + y)$
7. $2p^4q^2(5p^2 + 2pq + q^2)$
8. $(a - b)(2x + 5 + y^2)$
9. $(x^2 + y)(a + b)$ **10.** $(x^2 + y^2)^2$
11. $(xy + 1)(xy + 4)$
12. $(p^2 + 9q^2)(p + 3q)(p - 3q)$

Guided Solution:
10. y^2

13.6 Exercise Set

For Extra Help
MyMathLab®

MathXL® PRACTICE  WATCH  READ  REVIEW

✓ Reading Check

Complete each step in the following factoring strategy with the appropriate word(s) from the column on the right.

RC1. Always look first for a _______________ factor.

RC2. If there are two terms, determine whether the binomial is a _______________ of squares.

RC3. If there are three terms, determine whether the trinomial is a _______________.

RC4. If there are four terms, try factoring by _______________.

RC5. Always factor _______________.

RC6. Always _______________ by multiplying.

check
completely
grouping
sum
common
product
square
difference

a Factor completely.

1. $3x^2 - 192$

2. $2t^2 - 18$

3. $a^2 + 25 - 10a$

4. $y^2 + 49 + 14y$

5. $2x^2 - 11x + 12$

6. $8y^2 - 18y - 5$

7. $x^3 + 24x^2 + 144x$

8. $x^3 - 18x^2 + 81x$

9. $x^3 + 3x^2 - 4x - 12$

10. $x^3 - 5x^2 - 25x + 125$

11. $48x^2 - 3$

12. $50x^2 - 32$

13. $9x^3 + 12x^2 - 45x$

14. $20x^3 - 4x^2 - 72x$

15. $x^2 + 4$

16. $t^2 + 25$

17. $x^4 + 7x^2 - 3x^3 - 21x$

18. $m^4 + 8m^3 + 8m^2 + 64m$

19. $x^5 - 14x^4 + 49x^3$

20. $2x^6 + 8x^5 + 8x^4$

21. $20 - 6x - 2x^2$

22. $45 - 3x - 6x^2$

23. $x^2 - 6x + 1$

24. $x^2 + 8x + 5$

25. $4x^4 - 64$

26. $5x^5 - 80x$

27. $1 - y^8$

28. $t^8 - 1$

29. $x^5 - 4x^4 + 3x^3$

30. $x^6 - 2x^5 + 7x^4$

31. $\frac{1}{81}x^6 - \frac{8}{27}x^3 + \frac{16}{9}$

32. $36a^2 - 15a + \frac{25}{16}$

33. $mx^2 + my^2$

34. $12p^2 + 24q^3$

35. $9x^2y^2 - 36xy$

36. $x^2y - xy^2$

37. $2\pi rh + 2\pi r^2$

38. $10p^4q^4 + 35p^3q^3 + 10p^2q^2$

39. $(a + b)(x - 3) + (a + b)(x + 4)$

40. $5c(a^3 + b) - (a^3 + b)$

41. $(x - 1)(x + 1) - y(x + 1)$

42. $3(p - q) - q^2(p - q)$

43. $n^2 + 2n + np + 2p$

44. $a^2 - 3a + ay - 3y$

45. $6q^2 - 3q + 2pq - p$

46. $2x^2 - 4x + xy - 2y$

47. $4b^2 + a^2 - 4ab$

48. $x^2 + y^2 - 2xy$

49. $16x^2 + 24xy + 9y^2$

50. $9c^2 + 6cd + d^2$

51. $49m^4 - 112m^2n + 64n^2$

52. $4x^2y^2 + 12xyz + 9z^2$

53. $y^4 + 10y^2z^2 + 25z^4$

54. $0.01x^4 - 0.1x^2y^2 + 0.25y^4$

55. $\frac{1}{4}a^2 + \frac{1}{3}ab + \frac{1}{9}b^2$

56. $4p^2q + pq^2 + 4p^3$

57. $a^2 - ab - 2b^2$

58. $3b^2 - 17ab - 6a^2$

59. $2mn - 360n^2 + m^2$

60. $15 + x^2y^2 + 8xy$

61. $m^2n^2 - 4mn - 32$

62. $p^2q^2 + 7pq + 6$

63. $r^5s^2 - 10r^4s + 16r^3$

64. $p^5q^2 + 3p^4q - 10p^3$

65. $a^5 + 4a^4b - 5a^3b^2$

66. $2s^6t^2 + 10s^3t^3 + 12t^4$

67. $a^2 - \frac{1}{25}b^2$

68. $p^2 - \frac{1}{49}b^2$

69. $x^2 - y^2$

70. $p^2q^2 - r^2$

71. $16 - p^4q^4$

72. $15a^4 - 15b^4$

73. $1 - 16x^{12}y^{12}$

74. $81a^4 - b^4$

75. $q^3 + 8q^2 - q - 8$

76. $m^3 - 7m^2 - 4m + 28$

77. $6a^3b^3 - a^2b^2 - 2ab$

78. $4ab^5 - 32b^4 + a^2b^6$

79. $m^4 - 5m^2 + 4$

80. $8x^3y^3 - 6x^2y^2 - 5xy$

Skill Maintenance

Compute and simplify.

81. $1.2 - 9.87$ [9.4a]

82. $-3 + (-5) + 12 + (-7)$ [9.3a]

83. $\left(-\frac{1}{3}\right)\left(-\frac{3}{5}\right)$ [9.5a]

84. $-3.86 \div 0.5$ [9.6c]

85. $-50 \div (-5)(-2) - 18 \div (-3)^2$ [9.8d]

86. $3(-2) - 2 + |-4 - (-1)|$ [9.8d]

87. Evaluate $-x$ when $x = -7$. [9.3b]

88. Use either $<$ or $>$ for $\square$ to write a true sentence:
$-\frac{1}{3} \square -\frac{1}{2}$. [9.2d]

Synthesis

Factor completely.

89. $t^4 - 2t^2 + 1$

90. $x^4 + 9$

91. $x^3 + 20 - (5x^2 + 4x)$

92. $\frac{1}{5}x^2 - x + \frac{4}{5}$

93. $12.25x^2 - 7x + 1$

94. $x^3 + x^2 - (4x + 4)$

95. $18 + y^3 - 9y - 2y^2$

96. $3x^4 - 15x^2 + 12$

97. $y^2(y - 1) - 2y(y - 1) + (y - 1)$

98. $y^2(y + 1) - 4y(y + 1) - 21(y + 1)$

99. $(y + 4)^2 + 2x(y + 4) + x^2$

13.7 Solving Quadratic Equations by Factoring

OBJECTIVES

a Solve equations (already factored) using the principle of zero products.

b Solve quadratic equations by factoring and then using the principle of zero products.

SKILL TO REVIEW

Objective 10.3a: Solve equations using both the addition principle and the multiplication principle.

Solve.

1. $3x - 7 = 2$

2. $4y + 5 = 1$

Second-degree equations like $x^2 + x - 156 = 0$ and $9 - x^2 = 0$ are examples of *quadratic equations*.

QUADRATIC EQUATION

A **quadratic equation** is an equation equivalent to an equation of the type

$$ax^2 + bx + c = 0, \ a \neq 0.$$

In order to solve quadratic equations, we need a new equation-solving principle.

a THE PRINCIPLE OF ZERO PRODUCTS

The product of two numbers is 0 if one or both of the numbers is 0. Furthermore, *if any product is* 0, *then a factor must be* 0. For example:

If $7x = 0$, then we know that $x = 0$.

If $x(2x - 9) = 0$, then we know that $x = 0$ *or* $2x - 9 = 0$.

If $(x + 3)(x - 2) = 0$, then we know that $x + 3 = 0$ *or* $x - 2 = 0$.

Caution!

In a product such as $ab = 24$, we cannot conclude with certainty that a is 24 or that b is 24, but if $ab = 0$, we can conclude that $a = 0$ or $b = 0$.

EXAMPLE 1 Solve: $(x + 3)(x - 2) = 0$.

We have a product of 0. This equation will be true when either factor is 0. Thus it is true when

$$x + 3 = 0 \quad or \quad x - 2 = 0.$$

Here we have two simple equations that we know how to solve:

$$x = -3 \quad or \quad x = 2.$$

Each of the numbers -3 and 2 is a solution of the original equation, as we can see in the following checks.

Check: For -3:

$$\begin{array}{r|l} (x + 3)(x - 2) & = 0 \\ \hline (-3 + 3)(-3 - 2) & ?\ 0 \\ 0(-5) & \\ 0 & \text{TRUE} \end{array}$$

For 2:

$$\begin{array}{r|l} (x + 3)(x - 2) & = 0 \\ \hline (2 + 3)(2 - 2) & ?\ 0 \\ 5(0) & \\ 0 & \text{TRUE} \end{array}$$

Answers

Skill to Review:

1. 3 **2.** -1

We now have a principle to help in solving quadratic equations.

THE PRINCIPLE OF ZERO PRODUCTS

An equation $ab = 0$ is true if and only if $a = 0$ is true or $b = 0$ is true, or both are true. (A product is 0 if and only if one or both of the factors is 0.)

EXAMPLE 2 Solve: $(5x + 1)(x - 7) = 0$.

We have

$$(5x + 1)(x - 7) = 0$$

$5x + 1 = 0 \quad or \quad x - 7 = 0$ Using the principle of zero products

$5x = -1 \quad or \quad x = 7$ Solving the two equations separately

$x = -\frac{1}{5} \quad or \quad x = 7.$

Check: For $-\frac{1}{5}$:

$$\begin{array}{r|l} (5x + 1)(x - 7) & = 0 \\ \hline (5(-\frac{1}{5}) + 1)(-\frac{1}{5} - 7) & ?\ 0 \\ (-1 + 1)(-7\frac{1}{5}) & \\ 0(-7\frac{1}{5}) & \\ 0 & \text{TRUE} \end{array}$$

For 7:

$$\begin{array}{r|l} (5x + 1)(x - 7) & = 0 \\ \hline (5(7) + 1)(7 - 7) & ?\ 0 \\ (35 + 1) \cdot 0 & \\ 36 \cdot 0 & \\ 0 & \text{TRUE} \end{array}$$

The solutions are $-\frac{1}{5}$ and 7.

When some factors have only one term, you can still use the principle of zero products.

EXAMPLE 3 Solve: $x(2x - 9) = 0$.

We have

$$x(2x - 9) = 0$$

$x = 0 \quad or \quad 2x - 9 = 0$ Using the principle of zero products

$x = 0 \quad or \quad 2x = 9$

$x = 0 \quad or \quad x = \frac{9}{2}.$

Check: For 0:

$$\begin{array}{r|l} x(2x - 9) & = 0 \\ \hline 0 \cdot (2 \cdot 0 - 9) & ?\ 0 \\ 0 \cdot (-9) & \\ 0 & \text{TRUE} \end{array}$$

For $\frac{9}{2}$:

$$\begin{array}{r|l} x(2x - 9) & = 0 \\ \hline \frac{9}{2} \cdot (2 \cdot \frac{9}{2} - 9) & ?\ 0 \\ \frac{9}{2} \cdot (9 - 9) & \\ \frac{9}{2} \cdot 0 & \\ 0 & \text{TRUE} \end{array}$$

When you solve an equation using the principle of zero products, a check by substitution will detect errors in solving.

Do Exercises 1–4.

CALCULATOR CORNER

Solving Quadratic Equations We can solve quadratic equations graphically. Consider the equation $x^2 + 2x = 8$. First, we write the equation with 0 on one side: $x^2 + 2x - 8 = 0$. Next, we graph $y = x^2 + 2x - 8$ in a window that shows the x-intercepts. The standard window works well in this case.

The solutions of the equation are the values of x for which $x^2 + 2x - 8 = 0$. These are also the first coordinates of the x-intercepts of the graph. We use the ZERO feature from the CALC menu to find these numbers. For each x-intercept, we choose an x-value to the left of the intercept as a Left Bound, an x-value to the right of the intercept as a Right Bound, and an x-value near the intercept as a Guess. Beginning with the intercept on the left, we can read its coordinates, $(-4, 0)$, from the resulting screen.

$y = x^2 + 2x - 8$

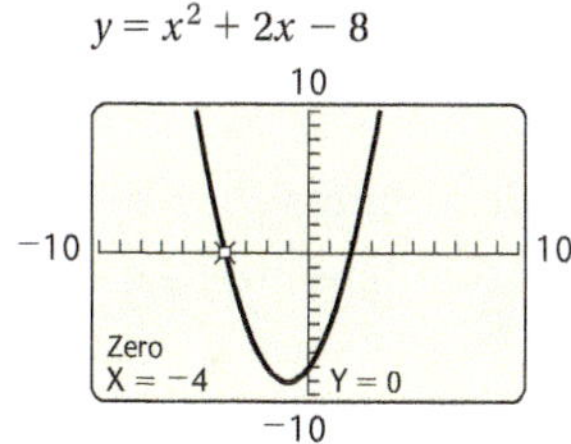

We can repeat this procedure to find the coordinates of the other x-intercept, $(2, 0)$. The solutions of $x^2 + 2x - 8 = 0$ are -4 and 2.

EXERCISE:

1. Solve each of the equations in Examples 4–6 graphically.

Solve using the principle of zero products.

1. $(x - 3)(x + 4) = 0$
2. $(x - 7)(x - 3) = 0$
3. $(4t + 1)(3t - 2) = 0$
4. $y(3y - 17) = 0$

Answers

1. $3, -4$ 2. $7, 3$ 3. $-\frac{1}{4}, \frac{2}{3}$ 4. $0, \frac{17}{3}$

b USING FACTORING TO SOLVE EQUATIONS

Using factoring and the principle of zero products, we can solve some new kinds of equations. Thus we have extended our equation-solving abilities.

EXAMPLE 4 Solve: $x^2 + 5x + 6 = 0$.

There are no like terms to collect, and we have a squared term. We first factor the polynomial. Then we use the principle of zero products.

$$x^2 + 5x + 6 = 0$$

$$(x + 2)(x + 3) = 0 \quad \text{Factoring}$$

$$x + 2 = 0 \quad or \quad x + 3 = 0 \quad \text{Using the principle of zero products}$$

$$x = -2 \quad or \quad x = -3$$

Check: For -2:

$$\begin{array}{c|c} x^2 + 5x + 6 & 0 \\ \hline (-2)^2 + 5(-2) + 6 \ ? & 0 \\ 4 - 10 + 6 & \\ -6 + 6 & \\ 0 & \end{array} \quad \text{TRUE}$$

For -3:

$$\begin{array}{c|c} x^2 + 5x + 6 & 0 \\ \hline (-3)^2 + 5(-3) + 6 \ ? & 0 \\ 9 - 15 + 6 & \\ -6 + 6 & \\ 0 & \end{array} \quad \text{TRUE}$$

The solutions are -2 and -3.

Caution!

Keep in mind that you *must* have 0 on one side of the equation before you can use the principle of zero products. Get all nonzero terms on one side and 0 on the other.

Do Exercise 5.

5. Solve: $x^2 - x - 6 = 0$. GS

$x^2 - x - 6 = 0$

$(x + 2)(\quad) = 0$

$x + 2 = 0 \quad or \quad \quad = 0$

$x = -2 \quad or \quad x = \quad$

Both numbers check.

The solutions are -2 and $\quad$.

EXAMPLE 5 Solve: $x^2 - 8x = -16$.

We first add 16 to get 0 on one side:

$$x^2 - 8x = -16$$

$$x^2 - 8x + 16 = 0 \quad \text{Adding 16}$$

$$(x - 4)(x - 4) = 0 \quad \text{Factoring}$$

$$x - 4 = 0 \quad or \quad x - 4 = 0 \quad \text{Using the principle of zero products}$$

$$x = 4 \quad or \quad x = 4. \quad \text{Solving each equation}$$

There is only one solution, 4. The check is left to the student.

Do Exercises 6 and 7.

Solve.

6. $x^2 - 3x = 28$

7. $x^2 = 6x - 9$

EXAMPLE 6 Solve: $x^2 + 5x = 0$.

$$x^2 + 5x = 0$$

$$x(x + 5) = 0 \quad \text{Factoring out a common factor}$$

$$x = 0 \quad or \quad x + 5 = 0 \quad \text{Using the principle of zero products}$$

$$x = 0 \quad or \quad x = -5$$

The solutions are 0 and -5. The check is left to the student.

Answers

5. $-2, 3$ **6.** $-4, 7$ **7.** 3

Guided Solution:

5. $x - 3, x - 3, 3, 3$

EXAMPLE 7 Solve: $4x^2 = 25$.

$$4x^2 = 25$$

$$4x^2 - 25 = 0 \quad \text{Subtracting 25 on both sides to get 0 on one side}$$

$$(2x - 5)(2x + 5) = 0 \quad \text{Factoring a difference of squares}$$

$$2x - 5 = 0 \quad or \quad 2x + 5 = 0 \quad \text{Using the principle of zero products}$$

$$2x = 5 \quad or \quad 2x = -5 \quad \text{Solving each equation}$$

$$x = \frac{5}{2} \quad or \quad x = -\frac{5}{2}$$

The solutions are $\frac{5}{2}$ and $-\frac{5}{2}$. The check is left to the student.

Do Exercises 8 and 9. ▶

Solve.

GS **8.** $x^2 - 4x = 0$

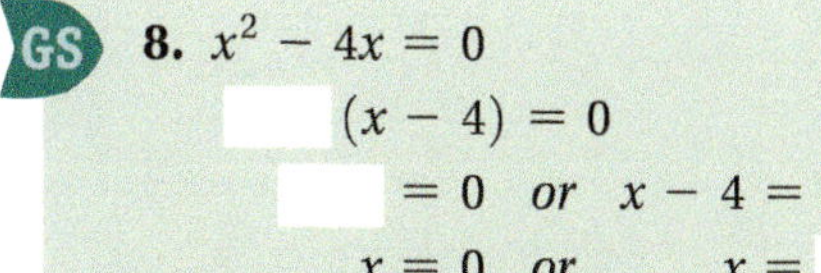

$\underline{\quad}(x - 4) = 0$

$\underline{\quad} = 0 \quad or \quad x - 4 = 0$

$x = 0 \quad or \quad x = \underline{\quad}$

Both numbers check.

The solutions are 0 and $\underline{\quad}$.

9. $9x^2 = 16$

EXAMPLE 8 Solve: $-5x^2 + 2x + 3 = 0$.

In this case, the leading coefficient of the trinomial is negative. Thus we first multiply by -1 and then proceed as we have in Examples 4–7.

$$-5x^2 + 2x + 3 = 0$$

$$-1(-5x^2 + 2x + 3) = -1 \cdot 0 \quad \text{Multiplying by } -1$$

$$5x^2 - 2x - 3 = 0 \quad \text{Simplifying}$$

$$(5x + 3)(x - 1) = 0 \quad \text{Factoring}$$

$$5x + 3 = 0 \quad or \quad x - 1 = 0 \quad \text{Using the principle of zero products}$$

$$5x = -3 \quad or \quad x = 1$$

$$x = -\frac{3}{5} \quad or \quad x = 1$$

The solutions are $-\frac{3}{5}$ and 1. The check is left to the student.

Do Exercises 10 and 11. ▶

Solve.

10. $-2x^2 + 13x - 21 = 0$

11. $10 - 3x - x^2 = 0$

EXAMPLE 9 Solve: $(x + 2)(x - 2) = 5$.

Be careful with an equation like this one! It might be tempting to set each factor equal to 5. **Remember: We must have 0 on one side**. We first carry out the multiplication on the left. Next, we subtract 5 on both sides to get 0 on one side. Then we proceed using the principle of zero products.

$$(x + 2)(x - 2) = 5$$

$$x^2 - 4 = 5 \quad \text{Multiplying on the left}$$

$$x^2 - 4 - 5 = 5 - 5 \quad \text{Subtracting 5}$$

$$x^2 - 9 = 0 \quad \text{Simplifying}$$

$$(x + 3)(x - 3) = 0 \quad \text{Factoring}$$

$$x + 3 = 0 \quad or \quad x - 3 = 0 \quad \text{Using the principle of zero products}$$

$$x = -3 \quad or \quad x = 3$$

The solutions are -3 and 3. The check is left to the student.

Do Exercise 12. ▶

12. Solve: $(x + 1)(x - 1) = 8$.

Answers

8. 0, 4 **9.** $-\frac{4}{3}, \frac{4}{3}$ **10.** $3, \frac{7}{2}$ **11.** $-5, 2$
12. $-3, 3$

Guided Solution:
8. $x, x, 4, 4$

ALGEBRAIC ▶◀ GRAPHICAL CONNECTION

To find the x-intercept of a linear equation, we replace y with 0 and solve for x. This procedure can also be used to find the x-intercepts of a quadratic equation.

The graph of $y = ax^2 + bx + c, a \neq 0$, is shaped like one of the following curves. Note that each x-intercept represents a solution of $ax^2 + bx + c = 0$.

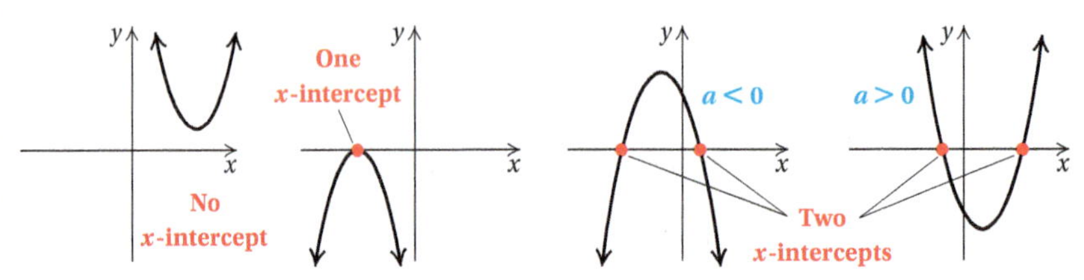

EXAMPLE 10 Find the x-intercepts of the graph of $y = x^2 - 4x - 5$ shown at right. (The grid is intentionally not included.)

To find the x-intercepts, we let $y = 0$ and solve for x:

$$y = x^2 - 4x - 5$$

$$0 = x^2 - 4x - 5 \quad \text{Substituting 0 for } y$$

$$0 = (x - 5)(x + 1) \quad \text{Factoring}$$

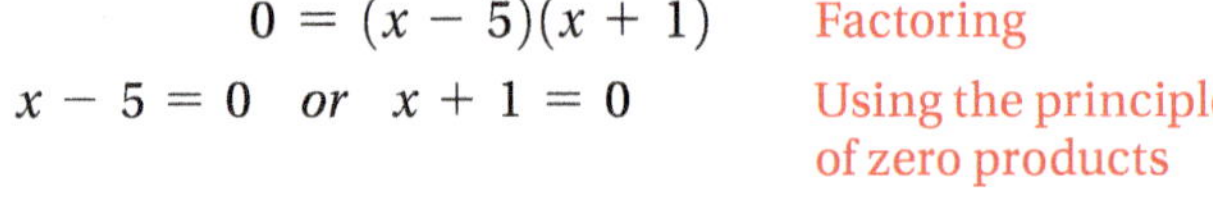

$$x - 5 = 0 \quad or \quad x + 1 = 0 \quad \text{Using the principle of zero products}$$

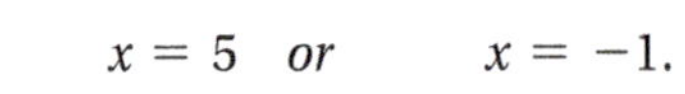

$$x = 5 \quad or \quad x = -1.$$

The solutions of the equation $0 = x^2 - 4x - 5$ are 5 and -1. Thus the x-intercepts of the graph of $y = x^2 - 4x - 5$ are $(5, 0)$ and $(-1, 0)$. We can now label them on the graph.

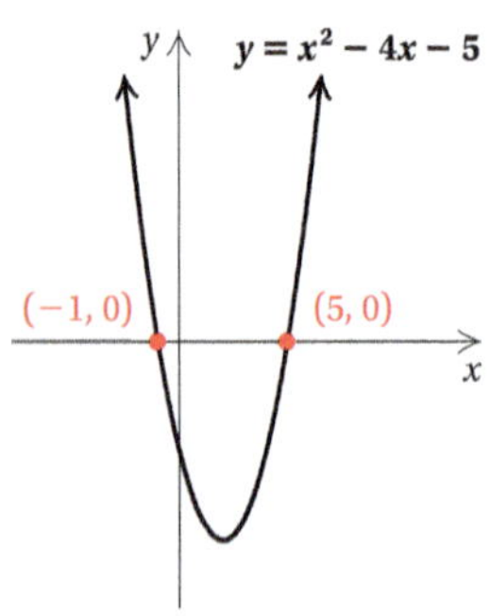

◀ **Do Exercises 13 and 14.**

13. Find the x-intercepts of the graph shown below.

$y = x^2 + 4x - 5$

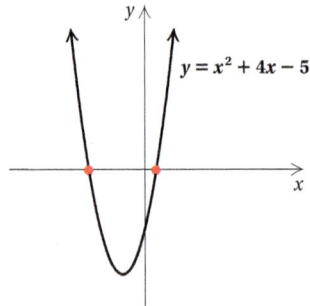

14. Use *only* the graph shown below to solve $3x - x^2 = 0$.

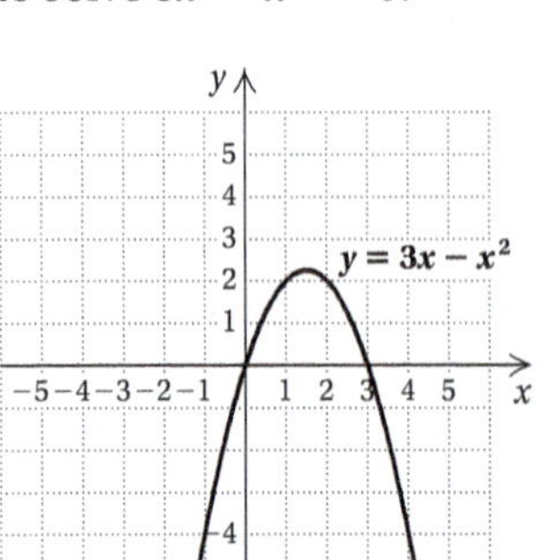

Answers

13. $(-5, 0), (1, 0)$ **14.** 0, 3

13.7 Exercise Set

For Extra Help MyMathLab® MathXL® PRACTICE WATCH READ REVIEW

Reading Check

Determine whether each statement is true or false.

RC1. If $(x + 2)(x + 3) = 10$, then $x + 2 = 10$ *or* $x + 3 = 10$.

RC2. A quadratic equation always has two different solutions.

RC3. The number 0 is never a solution of a quadratic equation.

RC4. If $ax^2 + bx + c = 0$ has no real-number solution, then the graph of $y = ax^2 + bx + c$ has no x-intercept.

a Solve using the principle of zero products.

1. $(x + 4)(x + 9) = 0$

2. $(x + 2)(x - 7) = 0$

3. $(x + 3)(x - 8) = 0$

4. $(x + 6)(x - 8) = 0$

5. $(x + 12)(x - 11) = 0$

6. $(x - 13)(x + 53) = 0$

7. $x(x + 3) = 0$

8. $y(y + 5) = 0$

9. $0 = y(y + 18)$

10. $0 = x(x - 19)$

11. $(2x + 5)(x + 4) = 0$

12. $(2x + 9)(x + 8) = 0$

13. $(5x + 1)(4x - 12) = 0$

14. $(4x + 9)(14x - 7) = 0$

15. $(7x - 28)(28x - 7) = 0$

16. $(13x + 14)(6x - 5) = 0$

17. $2x(3x - 2) = 0$

18. $55x(8x - 9) = 0$

19. $\left(\frac{1}{5} + 2x\right)\left(\frac{1}{9} - 3x\right) = 0$

20. $\left(\frac{7}{4}x - \frac{1}{16}\right)\left(\frac{2}{3}x - \frac{16}{15}\right) = 0$

21. $(0.3x - 0.1)(0.05x + 1) = 0$

22. $(0.1x + 0.3)(0.4x - 20) = 0$

23. $9x(3x - 2)(2x - 1) = 0$

24. $(x + 5)(x - 75)(5x - 1) = 0$

b Solve by factoring and using the principle of zero products. Remember to check.

25. $x^2 + 6x + 5 = 0$

26. $x^2 + 7x + 6 = 0$

27. $x^2 + 7x - 18 = 0$

28. $x^2 + 4x - 21 = 0$

29. $x^2 - 8x + 15 = 0$

30. $x^2 - 9x + 14 = 0$

31. $x^2 - 8x = 0$

32. $x^2 - 3x = 0$

33. $x^2 + 18x = 0$

34. $x^2 + 16x = 0$

35. $x^2 = 16$

36. $100 = x^2$

37. $9x^2 - 4 = 0$

38. $4x^2 - 9 = 0$

39. $0 = 6x + x^2 + 9$

40. $0 = 25 + x^2 + 10x$

41. $x^2 + 16 = 8x$

42. $1 + x^2 = 2x$

43. $5x^2 = 6x$

44. $7x^2 = 8x$

45. $6x^2 - 4x = 10$

46. $3x^2 - 7x = 20$

47. $12y^2 - 5y = 2$

48. $2y^2 + 12y = -10$

49. $t(3t + 1) = 2$

50. $x(x - 5) = 14$

51. $100y^2 = 49$

52. $64a^2 = 81$

53. $x^2 - 5x = 18 + 2x$

54. $3x^2 + 8x = 9 + 2x$

55. $10x^2 - 23x + 12 = 0$

56. $12x^2 + 17x - 5 = 0$

Find the x-intercepts of the graph of each equation. (The grids are intentionally not included.)

57.

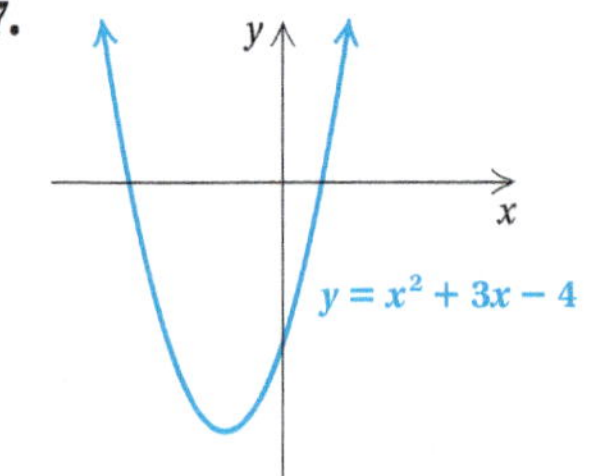

58.

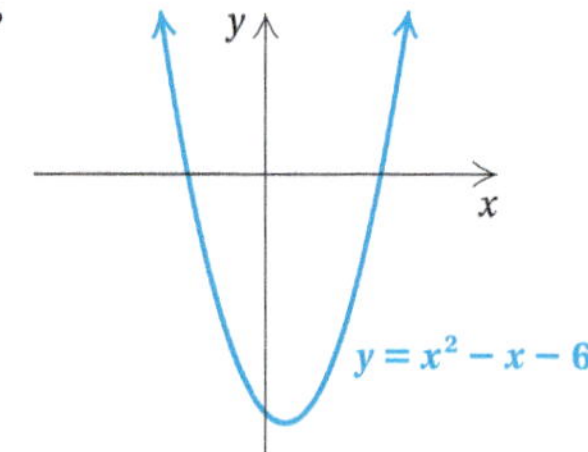

59.

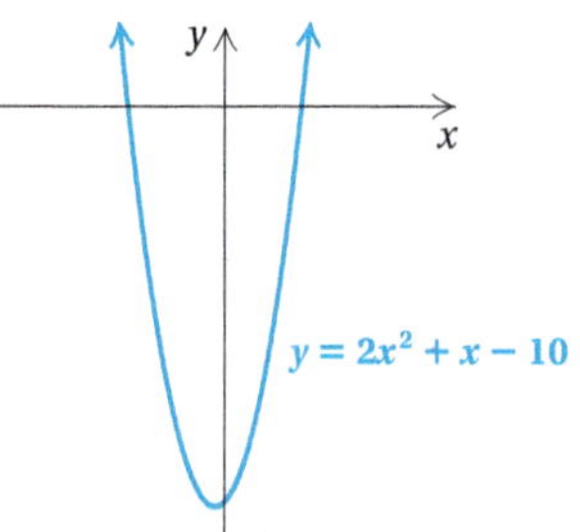

60.

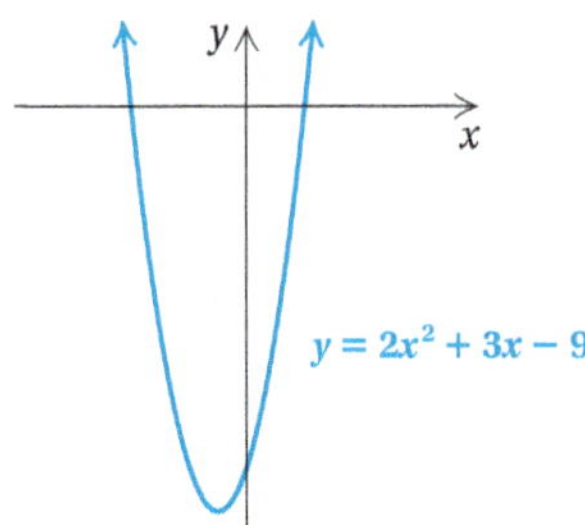

61.

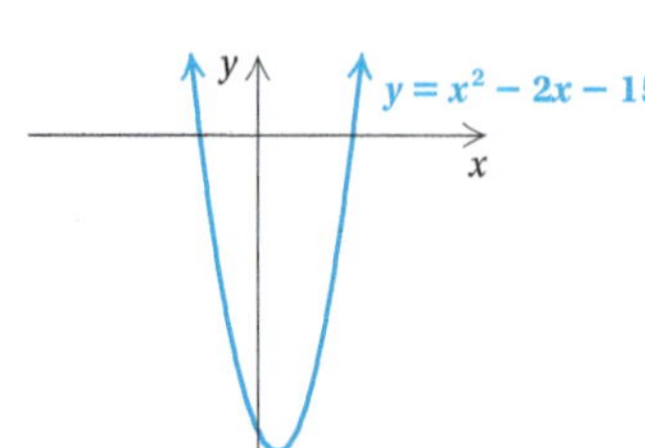

62.

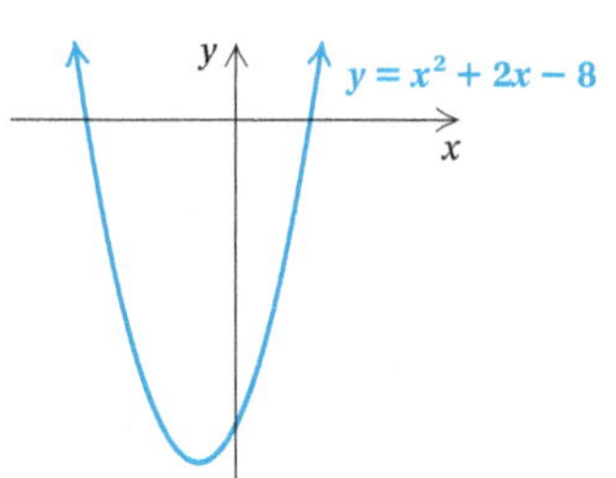

63. Use the following graph to solve $x^2 - 3x - 4 = 0$.

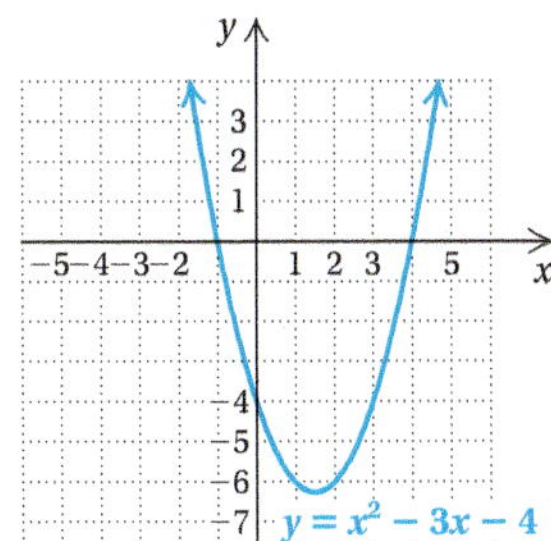

64. Use the following graph to solve $x^2 + x - 6 = 0$.

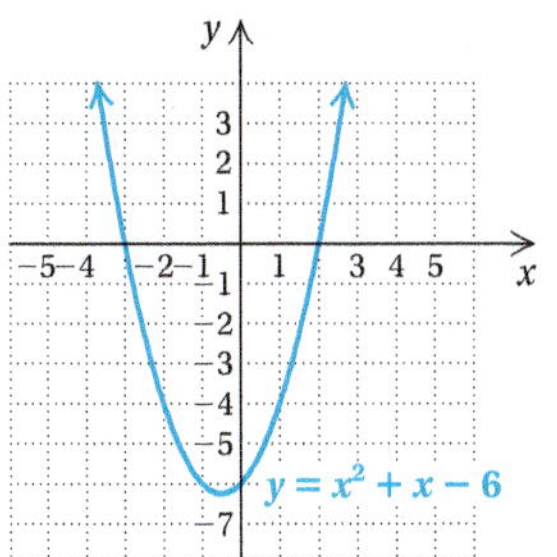

65. Use the following graph to solve $-x^2 + 2x + 3 = 0$.

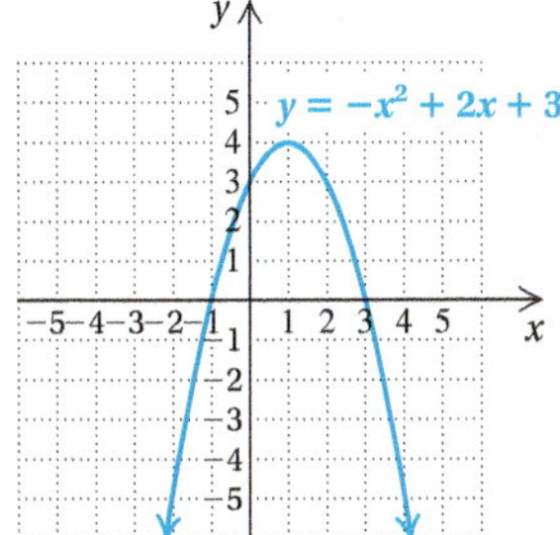

66. Use the following graph to solve $-x^2 - x + 6 = 0$.

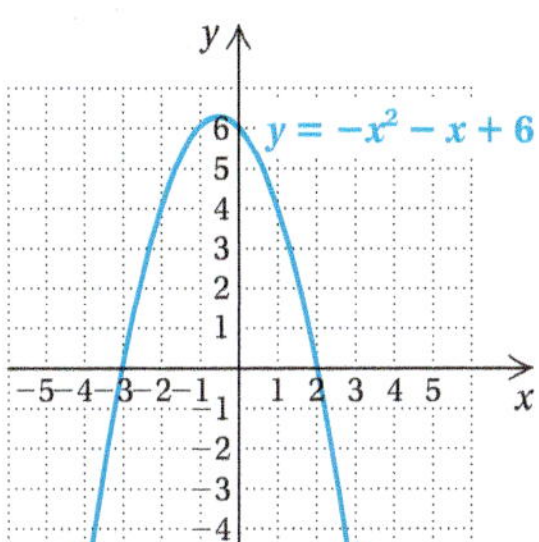

Skill Maintenance

Translate to an algebraic expression. [9.1b]

67. The square of the sum of a and b

68. The sum of the squares of a and b

Solve. [10.7d, e]

69. $-10x > 1000$

70. $6 - 3x \geq -18$

71. $3 - 2x - 4x > -9$

72. $\frac{1}{2}x - 6x + 10 \leq x - 5x$

Synthesis

Solve.

73. $b(b + 9) = 4(5 + 2b)$

74. $y(y + 8) = 16(y - 1)$

75. $(t - 3)^2 = 36$

76. $(t - 5)^2 = 2(5 - t)$

77. $x^2 - \frac{1}{64} = 0$

78. $x^2 - \frac{25}{36} = 0$

79. $\frac{5}{16}x^2 = 5$

80. $\frac{27}{25}x^2 = \frac{1}{3}$

Use a graphing calculator to find the solutions of each equation. Round solutions to the nearest hundredth.

81. $x^2 - 9.10x + 15.77 = 0$

82. $-x^2 + 0.63x + 0.22 = 0$

83. Find an equation that has the given numbers as solutions. For example, 3 and -2 are solutions of $x^2 - x - 6 = 0$.

a) $-3, 4$
b) $-3, -4$
c) $\frac{1}{2}, \frac{1}{2}$
d) $5, -5$
e) $0, 0.1, \frac{1}{4}$

13.8 Applications of Quadratic Equations

OBJECTIVE

a Solve applied problems involving quadratic equations that can be solved by factoring.

SKILL TO REVIEW

Objective 10.6a: Solve applied problems by translating to equations.

Solve.

1. The length of a rectangular garden is twice as long as the width. The perimeter of the garden is 48 ft. Find the length and the width of the garden.
2. The second angle of a triangle is 25° more than the first angle. The third angle is twice as large as the second angle. Find the measures of the angles of the triangle.

a APPLIED PROBLEMS, QUADRATIC EQUATIONS, AND FACTORING

We can solve problems that translate to quadratic equations using the five steps for solving problems.

EXAMPLE 1 *Kitchen Island.* Lisa buys a kitchen island with a butcher-block top as part of a remodeling project. The top of the island is a rectangle that is twice as long as it is wide and that has an area of 800 in^2. What are the dimensions of the top of the island?

1. **Familiarize.** We first make a drawing. Recall that the area of a rectangle is Length · Width. We let x = the width of the top, in inches. The length is then $2x$.

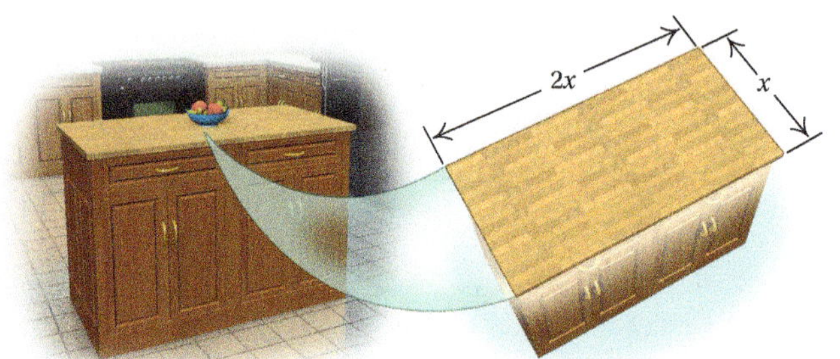

2. **Translate.** We reword and translate as follows:

Rewording:	The area of the rectangle	is	800 in^2
	↓	↓	↓
Translating:	$2x \cdot x$	$=$	800

3. **Solve.** We solve the equation as follows:

$$2x \cdot x = 800$$
$$2x^2 = 800$$
$$2x^2 - 800 = 0 \quad \text{Subtracting 800 to get 0 on one side}$$
$$2(x^2 - 400) = 0 \quad \text{Removing a common factor of 2}$$
$$2(x - 20)(x + 20) = 0 \quad \text{Factoring a difference of squares}$$
$$(x - 20)(x + 20) = 0 \quad \text{Dividing by 2}$$
$$x - 20 = 0 \quad or \quad x + 20 = 0 \quad \text{Using the principle of zero products}$$
$$x = 20 \quad or \quad x = -20. \quad \text{Solving each equation}$$

4. **Check.** The solutions of the equation are 20 and -20. Since the width must be positive, -20 cannot be a solution. To check 20 in., we note that if the width is 20 in., then the length is $2 \cdot 20$ in., or 40 in., and the area is 20 in. · 40 in., or 800 in^2. Thus the solution 20 checks.

5. **State.** The top of the island is 20 in. wide and 40 in. long.

◀ Do Margin Exercise 1.

1. *Dimensions of a Picture.* A rectangular picture is twice as long as it is wide. If the area of the picture is 288 in^2, what are its dimensions?

Answers

Skill to Review:
1. Length: 16 ft; width: 8 ft
2. 26.25°, 51.25°, 102.5°

Margin Exercise:
1. Length: 24 in.; width: 12 in.

EXAMPLE 2 *Butterfly Wings.* The *Graphium sarpedon* butterfly has areas of light blue on each wing. When the wings are joined, the blue areas form a triangle, giving rise to the butterfly's common name, Blue Triangle Butterfly. On one butterfly, the base of the blue triangle is 6 cm longer than the height of the triangle. The area of the triangle is 8 cm^2. Find the base and the height of the triangle.

Source: Based on information from australianmuseum.net.au

1. **Familiarize.** We first make a drawing, letting h = the height of the triangle, in centimeters. Then $h + 6$ = the base. We also recall or look up the formula for the area of a triangle: Area $= \frac{1}{2}$(base)(height).

2. **Translate.** We reword the problem and translate:

Rewording:	$\frac{1}{2}$	times	Base	times	Height	is	8
	↓	↓	↓	↓	↓	↓	↓
Translating:	$\frac{1}{2}$	$\cdot$	$(h+6)$	$\cdot$	h	$=$	8.

3. **Solve.** We solve the quadratic equation using the principle of zero products:

$$\frac{1}{2} \cdot (h + 6) \cdot h = 8$$

$$\frac{1}{2}(h^2 + 6h) = 8 \quad \text{Multiplying } h + 6 \text{ and } h$$

$$2 \cdot \frac{1}{2}(h^2 + 6h) = 2 \cdot 8 \quad \text{Multiplying by 2}$$

$$h^2 + 6h = 16 \quad \text{Simplifying}$$

$$h^2 + 6h - 16 = 16 - 16 \quad \text{Subtracting 16 to get 0 on one side}$$

$$h^2 + 6h - 16 = 0$$

$$(h - 2)(h + 8) = 0 \quad \text{Factoring}$$

$$h - 2 = 0 \quad or \quad h + 8 = 0 \quad \text{Using the principle of zero products}$$

$$h = 2 \quad or \quad h = -8.$$

4. **Check.** The height of a triangle cannot have a negative length, so -8 cannot be a solution. Suppose the height is 2 cm. The base is 6 cm more than the height, so the base is 2 cm + 6 cm, or 8 cm, and the area is $\frac{1}{2}(8)(2)$, or 8 cm^2. The numbers check in the original problem.

5. **State.** The base of the blue triangle is 8 cm and the height is 2 cm.

Do Exercise 2. ▶

2. *Dimensions of a Sail.* The triangular mainsail on Stacey's lightning-styled sailboat has an area of 125 ft^2. The height of the sail is 15 ft more than the base. Find the height and the base of the sail.

Answer

2. Height: 25 ft; base: 10 ft

EXAMPLE 3 *Games in a Sports League.* In a sports league of x teams in which each team plays every other team twice, the total number N of games to be played is given by

$$x^2 - x = N.$$

Maggie's volleyball league plays a total of 240 games. How many teams are in the league?

1., 2. Familiarize and **Translate.** We are given that x is the number of teams in a league and N is the number of games. To find the number of teams x in a league in which 240 games are played, we substitute 240 for N in the equation:

$$x^2 - x = 240. \qquad \text{Substituting 240 for } N$$

3. Solve. We solve the equation as follows:

$$\begin{aligned} x^2 - x &= 240 \\ x^2 - x - 240 &= 240 - 240 && \text{Subtracting 240 to get 0 on one side} \\ x^2 - x - 240 &= 0 \\ (x - 16)(x + 15) &= 0 && \text{Factoring} \\ x - 16 = 0 \quad &or \quad x + 15 = 0 && \text{Using the principle of zero products} \\ x = 16 \quad &or \quad x = -15. \end{aligned}$$

4. Check. The solutions of the equation are 16 and -15. Since the number of teams cannot be negative, -15 cannot be a solution. But 16 checks, since $16^2 - 16 = 256 - 16 = 240$.

5. State. There are 16 teams in the league.

Do Exercise 3.

3. Use $N = x^2 - x$ for each of the following.

   a) *Basketball League.* Amy's basketball league has 19 teams. What is the total number of games to be played if each team plays every other team twice?

   b) *Softball League.* Barry's slow-pitch softball league plays a total of 72 games. How many teams are in the league if each team plays every other team twice?

Answer

3. **(a)** 342 games; **(b)** 9 teams

EXAMPLE 4 *Race Numbers.* Terry and Jody registered their boats in the Lakeport Race at the same time. The racing numbers assigned to their boats were consecutive integers, the product of which was 156. Find the integers.

1. Familiarize. Consecutive integers are one unit apart, like 49 and 50. Let x = the first boat number; then $x + 1$ = the next boat number.

2. Translate. We reword the problem before translating:

Rewording:	First integer	times	Second integer	is	156
	↓	↓	↓	↓	↓
Translating:	x	$\cdot$	$(x + 1)$	$=$	156.

3. **Solve.** We solve the equation as follows:

$$x(x+1) = 156$$
$$x^2 + x = 156 \qquad \text{Multiplying}$$
$$x^2 + x - 156 = 156 - 156 \qquad \text{Subtracting 156 to get 0 on one side}$$
$$x^2 + x - 156 = 0 \qquad \text{Simplifying}$$
$$(x-12)(x+13) = 0 \qquad \text{Factoring}$$
$$x - 12 = 0 \quad or \quad x + 13 = 0 \qquad \text{Using the principle of zero products}$$
$$x = 12 \quad or \quad x = -13.$$

4. **Check.** The solutions of the equation are 12 and -13. Since race numbers are not negative, -13 must be rejected. On the other hand, if x is 12, then $x + 1$ is 13 and $12 \cdot 13 = 156$. Thus the solution 12 checks.
5. **State.** The boat numbers for Terry and Jody were 12 and 13.

Do Exercise 4. ▶

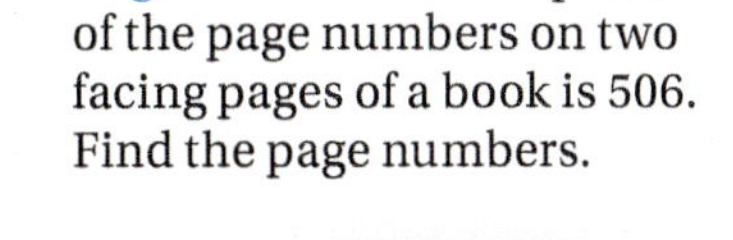
4. ***Page Numbers.*** The product of the page numbers on two facing pages of a book is 506. Find the page numbers.

The Pythagorean Theorem

The Pythagorean theorem states a relationship involving the lengths of the sides of a *right* triangle. A triangle is a **right triangle** if it has a 90°, or *right*, angle. The side opposite the 90° angle is called the **hypotenuse**. The other sides are called **legs**.

THE PYTHAGOREAN THEOREM

In any right triangle, if a and b are the lengths of the legs and c is the length of the hypotenuse, then

$$a^2 + b^2 = c^2.$$

c, a, b

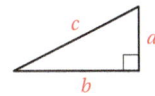

The symbol ⌐ denotes a 90° angle.

EXAMPLE 5 ***Wood Scaffold.*** Jonah is building a wood scaffold to use for a home improvement project. The scaffold has diagonal braces that are 5 ft long and that span a distance of 3 ft. How high does each brace reach vertically?

1. **Familiarize.** We make a drawing as shown above and let h = the height, in feet, to which each brace rises vertically.
2. **Translate.** A right triangle is formed, so we can use the Pythagorean theorem:

$$a^2 + b^2 = c^2$$
$$3^2 + h^2 = 5^2. \qquad \text{Substituting}$$

Answer

4. 22 and 23

3. **Solve.** We solve the equation as follows:

$$
\begin{aligned}
3^2 + h^2 &= 5^2 \\
9 + h^2 &= 25 && \text{Squaring 3 and 5} \\
9 + h^2 - 25 &= 25 - 25 && \text{Subtracting 25 to get 0 on one side} \\
h^2 - 16 &= 0 && \text{Simplifying} \\
(h - 4)(h + 4) &= 0 && \text{Factoring} \\
h - 4 = 0 \quad or \quad h + 4 &= 0 && \text{Using the principle of zero products} \\
h = 4 \quad or \quad h &= -4.
\end{aligned}
$$

4. **Check.** Since height cannot be negative, -4 cannot be a solution. If the height is 4 ft, we have $3^2 + 4^2 = 9 + 16 = 25$, which is 5^2. Thus, 4 checks and is the solution.
5. **State.** Each brace reaches a height of 4 ft.

Do Exercise 5.

5. ***Reach of a Ladder.*** Twila has a 26-ft ladder leaning against her house. If the bottom of the ladder is 10 ft from the base of the house, how high does the ladder reach?

EXAMPLE 6 ***Ladder Settings.*** A ladder of length 13 ft is placed against a building in such a way that the distance from the top of the ladder to the ground is 7 ft more than the distance from the bottom of the ladder to the building. Find both distances.

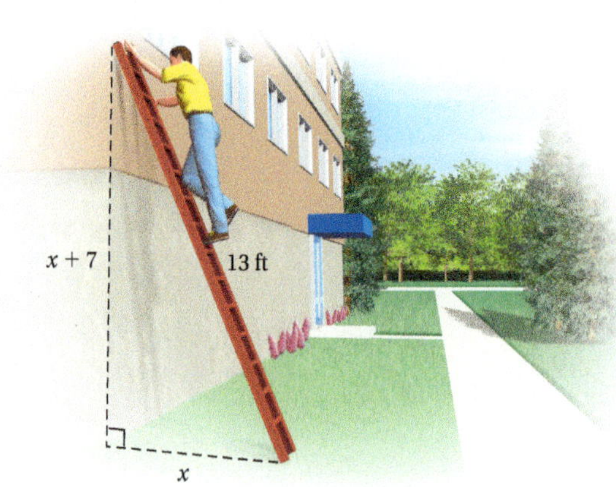

1. **Familiarize.** We first make a drawing. The ladder and the missing distances form the hypotenuse and the legs of a right triangle. We let x = the length of the side (leg) across the bottom, in feet. Then $x + 7$ = the length of the other side (leg). The hypotenuse has length 13 ft.

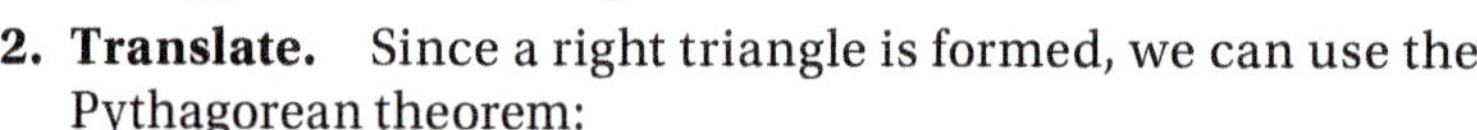

2. **Translate.** Since a right triangle is formed, we can use the Pythagorean theorem:

$$
\begin{aligned}
a^2 + b^2 &= c^2 \\
x^2 + (x + 7)^2 &= 13^2. && \text{Substituting}
\end{aligned}
$$

3. **Solve.** We solve the equation as follows:

$$
\begin{aligned}
x^2 + (x^2 + 14x + 49) &= 169 && \text{Squaring the binomial and 13} \\
2x^2 + 14x + 49 &= 169 && \text{Collecting like terms} \\
2x^2 + 14x + 49 - 169 &= 169 - 169 && \text{Subtracting 169 to get 0 on one side} \\
2x^2 + 14x - 120 &= 0 && \text{Simplifying} \\
2(x^2 + 7x - 60) &= 0 && \text{Factoring out a common factor} \\
x^2 + 7x - 60 &= 0 && \text{Dividing by 2} \\
(x + 12)(x - 5) &= 0 && \text{Factoring} \\
x + 12 = 0 \quad or \quad x - 5 &= 0 && \text{Using the principle of zero products} \\
x = -12 \quad or \quad x &= 5.
\end{aligned}
$$

4. **Check.** The negative integer -12 cannot be the length of a side. When $x = 5, x + 7 = 12$, and $5^2 + 12^2 = 13^2$. Thus, 5 and 12 check.
5. **State.** The distance from the top of the ladder to the ground is 12 ft. The distance from the bottom of the ladder to the building is 5 ft.

Do Exercise 6.

6. ***Right-Triangle Geometry.*** The length of one leg of a right triangle is 1 m longer than the other. The length of the hypotenuse is 5 m. Find the lengths of the legs.

Answers

5. 24 ft **6.** 3 m, 4 m

Translating for Success

1. *Angle Measures.* The measures of the angles of a triangle are three consecutive integers. Find the measures of the angles.

2. *Rectangle Dimensions.* The area of a rectangle is 3599 ft^2. The length is 2 ft longer than the width. Find the dimensions of the rectangle.

3. *Sales Tax.* Claire paid \$40,704 for a new hybrid car. This included 6% for sales tax. How much did the vehicle cost before tax?

4. *Wire Cutting.* A 180-m wire is cut into three pieces. The third piece is 2 m longer than the first. The second is two-thirds as long as the first. How long is each piece?

5. *Perimeter.* The perimeter of a rectangle is 240 ft. The length is 2 ft greater than the width. Find the length and the width.

The goal of these matching questions is to practice step (2), Translate, of the five-step problem-solving process. Translate each word problem to an equation and select a correct translation from equations A–O.

A. $2x \cdot x = 288$

B. $x(x + 60) = 7021$

C. $59 = x \cdot 60$

D. $x^2 + (x + 2)^2 = 3599$

E. $x^2 + (x + 70)^2 = 130^2$

F. $6\% \cdot x = 40{,}704$

G. $2(x + 2) + 2x = 240$

H. $\frac{1}{2}x(x - 1) = 1770$

I. $x + \frac{2}{3}x + (x + 2) = 180$

J. $59\% \cdot x = 60$

K. $x + 6\% \cdot x = 40{,}704$

L. $2x^2 + x = 288$

M. $x(x + 2) = 3599$

N. $x^2 + 60 = 7021$

O. $x + (x + 1) + (x + 2) = 180$

Answers on page A-32

6. *Cell-Phone Tower.* A guy wire on a cell-phone tower is 130 ft long and is attached to the top of the tower. The height of the tower is 70 ft longer than the distance from the point on the ground where the wire is attached to the bottom of the tower. Find the height of the tower.

7. *Sales Meeting Attendance.* PTQ Corporation holds a sales meeting in Tucson. Of the 60 employees, 59 of them attend the meeting. What percent attend the meeting?

8. *Dimensions of a Pool.* A rectangular swimming pool is twice as long as it is wide. The area of the surface is 288 ft^2. Find the dimensions of the pool.

9. *Dimensions of a Triangle.* The height of a triangle is 1 cm less than the length of the base. The area of the triangle is 1770 cm^2. Find the height and the length of the base.

10. *Width of a Rectangle.* The length of a rectangle is 60 ft longer than the width. Find the width if the area of the rectangle is 7021 ft^2.

13.8 Exercise Set

For Extra Help
MyMathLab®
MathXL® PRACTICE WATCH READ REVIEW

Reading Check

Match each statement with an appropriate translation from the column on the right. Choices may be used more than once or not at all.

RC1. The product of two consecutive integers is 20.

RC2. The length of a rectangle is 1 cm longer than the width. The area of the rectangle is 20 cm^2.

RC3. One leg of a right triangle is 1 cm longer than the other leg. The length of the hypotenuse is 20 cm.

RC4. One leg of a right triangle is 1 cm longer than the other leg. The area of the triangle is 20 cm^2.

a) $x + (x + 1) = 20$

b) $x(x + 1) = 20$

c) $\frac{1}{2}x(x + 1) = 20$

d) $x^2 + (x + 1)^2 = 20^2$

a Solve.

1. ***Dimensions of a Painting.*** A rectangular painting is three times as long as it is wide. The area of the picture is 588 in^2. Find the dimensions of the painting.

2. ***Area of a Garden.*** The length of a rectangular garden is 4 m greater than the width. The area of the garden is 96 m^2. Find the length and the width.

3. ***Design.*** The screen of the TI-84 Plus graphing calculator is nearly rectangular. The length of the rectangle is 2 cm more than the width. If the area of the rectangle is 24 cm^2, find the length and the width.

4. ***Construction.*** The front porch on Trent's new home is five times as long as it is wide. If the area of the porch is 320 ft^2, find the dimensions.

5. ***Dimensions of a Triangle.*** A triangle is 10 cm wider than it is tall. The area is 28 cm^2. Find the height and the base.

6. ***Dimensions of a Triangle.*** The height of a triangle is 3 cm less than the length of the base. The area of the triangle is 35 cm^2. Find the height and the length of the base.

7. ***Road Design.*** A triangular traffic island has a base half as long as its height. The island has an area of 64 m^2. Find the base and the height.

8. ***Dimensions of a Sail.*** The height of the jib sail on a Lightning sailboat is 5 ft greater than the length of its "foot." The area of the sail is 42 ft^2. Find the length of the foot and the height of the sail.

Games in a League. Use $x^2 - x = N$ for Exercises 9–12.

9. A Scrabble league has 14 teams. What is the total number of games to be played if each team plays every other team twice?

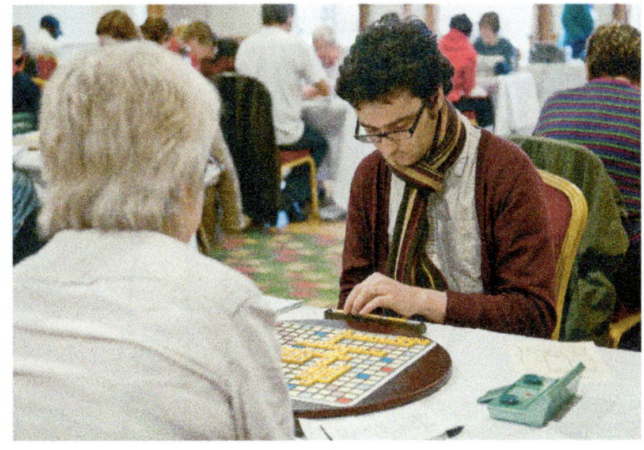

10. A chess league has 23 teams. What is the total number of games to be played if each team plays every other team twice?

11. A slow-pitch softball league plays a total of 132 games. How many teams are in the league if each team plays every other team twice?

12. A basketball league plays a total of 90 games. How many teams are in the league if each team plays every other team twice?

Handshakes. Dr. Benton wants to investigate the potential spread of germs by contact. She knows that the number of possible handshakes within a group of x people, assuming each person shakes every other person's hand only once, is given by

$$N = \tfrac{1}{2}(x^2 - x).$$

Use this formula for Exercises 13–16.

13. There are 100 people at a party. How many handshakes are possible?

14. There are 40 people at a meeting. How many handshakes are possible?

15. Everyone at a meeting shook hands with each other. There were 300 handshakes in all. How many people were at the meeting?

16. Everyone at a party shook hands with each other. There were 153 handshakes in all. How many people were at the party?

17. *Consecutive Page Numbers.* The product of the page numbers on two facing pages of a book is 210. Find the page numbers.

18. *Consecutive Page Numbers.* The product of the page numbers on two facing pages of a book is 420. Find the page numbers.

19. The product of two consecutive even integers is 168. Find the integers. (Consecutive even integers are two units apart.)

20. The product of two consecutive even integers is 224. Find the integers. (Consecutive even integers are two units apart.)

21. The product of two consecutive odd integers is 255. Find the integers. (Consecutive odd integers are two units apart.)

22. The product of two consecutive odd integers is 143. Find the integers. (Consecutive odd integers are two units apart.)

23. *Roadway Design.* Elliott Street is 24 ft wide when it ends at Main Street in Brattleboro, Vermont. A 40-ft long diagonal crosswalk allows pedestrians to cross Main Street to or from either corner of Elliott Street (see the figure). Determine the width of Main Street.

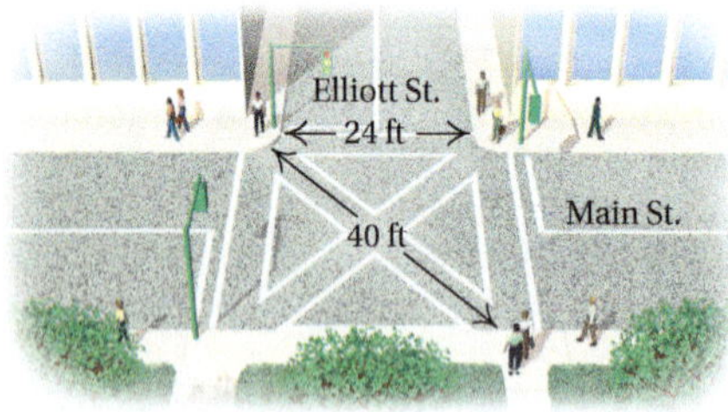

24. *Lookout Tower.* The diagonal braces in a lookout tower are 15 ft long and span a distance of 12 ft. How high does each brace reach vertically?

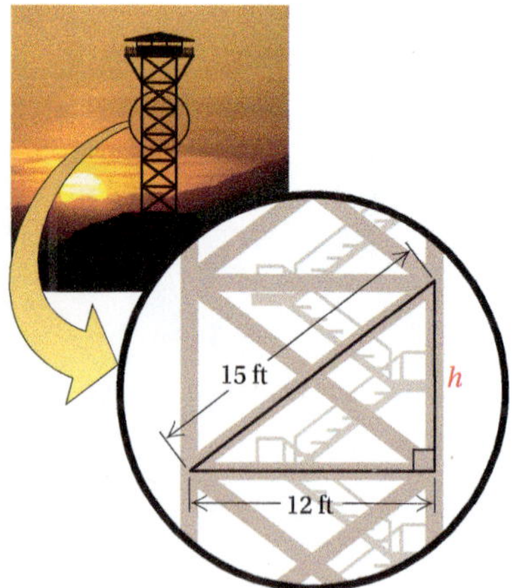

25. *Right-Triangle Geometry.* The length of one leg of a right triangle is 8 ft. The length of the hypotenuse is 2 ft longer than the other leg. Find the lengths of the hypotenuse and the other leg.

26. *Right-Triangle Geometry.* The length of one leg of a right triangle is 24 ft. The length of the other leg is 16 ft shorter than the hypotenuse. Find the lengths of the hypotenuse and the other leg.

27. *Archaeology.* Archaeologists have discovered that the 18th-century garden of the Charles Carroll House in Annapolis, Maryland, was a right triangle. One leg of the triangle was formed by a 400-ft long sea wall. The hypotenuse of the triangle was 200 ft longer than the other leg. What were the dimensions of the garden?

Source: www.bsos.umd.edu

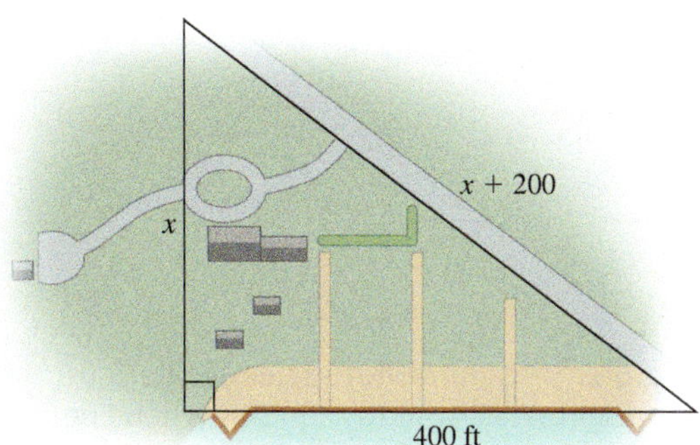

28. *Guy Wire.* The height of a wind power assessment tower is 5 m shorter than the guy wire that supports it. If the guy wire is anchored 15 m from the foot of the antenna, how tall is the antenna?

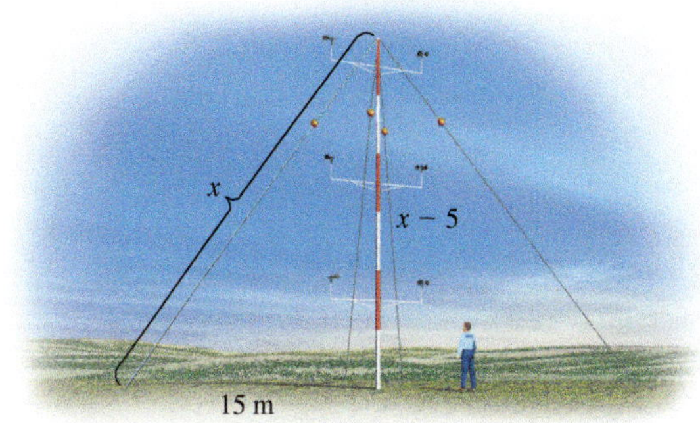

29. *Right Triangle.* The shortest side of a right triangle measures 7 m. The lengths of the other two sides are consecutive integers. Find the lengths of the other two sides.

30. *Right Triangle.* The shortest side of a right triangle measures 8 cm. The lengths of the other two sides are consecutive odd integers. Find the lengths of the other two sides.

31. *Architecture.* An architect has allocated a rectangular space of 264 ft^2 for a square dining room and a 10-ft wide kitchen, as shown in the figure. Find the dimensions of each room.

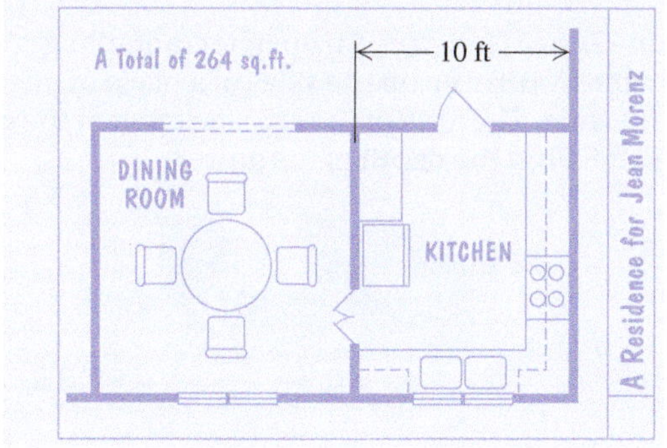

32. *Design.* A window panel for a sun porch consists of a 7-ft tall rectangular window stacked above a square window. The windows have the same width. If the total area of the window panel is 18 ft^2, find the dimensions of each window.

Height of a Rocket. For Exercises 33 and 34, assume that a water rocket is launched upward with an initial velocity of 48 ft/sec. Its height h, in feet, after t seconds, is given by $h = 48t - 16t^2$.

33. When will the rocket be exactly 32 ft above the ground?

34. When will the rocket crash into the ground?

35. The sum of the squares of two consecutive odd positive integers is 74. Find the integers.

36. The sum of the squares of two consecutive odd positive integers is 130. Find the integers.

Skill Maintenance

Compute and simplify.

37. $-3.57 + 8.1$ [9.3a]

38. $-\frac{2}{3} - \frac{1}{6}$ [9.4a]

39. $(-2)(-4)(-5)$ [9.5a]

40. $2 \cdot 6^2 \div (-2) \cdot 3 - 8$ [9.8d]

41. $\frac{2 - |3 - 8|}{(-1 - 4)^2}$ [9.8d]

42. $1.2 + (-2)^3 + 3.4$ [9.8d]

Remove parentheses and simplify.

43. $2(y - 7) - (6y - 1)$ [9.8b]

44. $2\{x - 3[4 - (x - 1)] + x\}$ [9.8c]

Synthesis

45. ***Pool Sidewalk.*** A cement walk of constant width is built around a 20-ft by 40-ft rectangular pool. The total area of the pool and the walk is 1500 ft^2. Find the width of the walk.

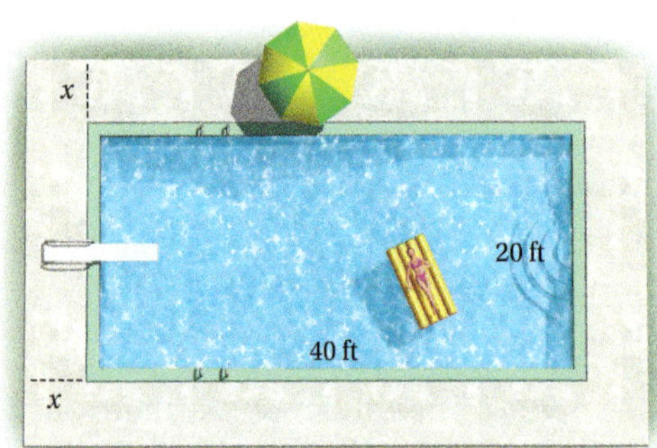

46. ***Roofing.*** A *square* of shingles covers 100 ft^2 of surface area. How many squares will be needed to reshingle the roof of the house shown?

47. ***Dimensions of an Open Box.*** A rectangular piece of cardboard is twice as long as it is wide. A 4-cm square is cut out of each corner, and the sides are turned up to make a box with an open top. The volume of the box is 616 cm^3. Find the original dimensions of the cardboard.

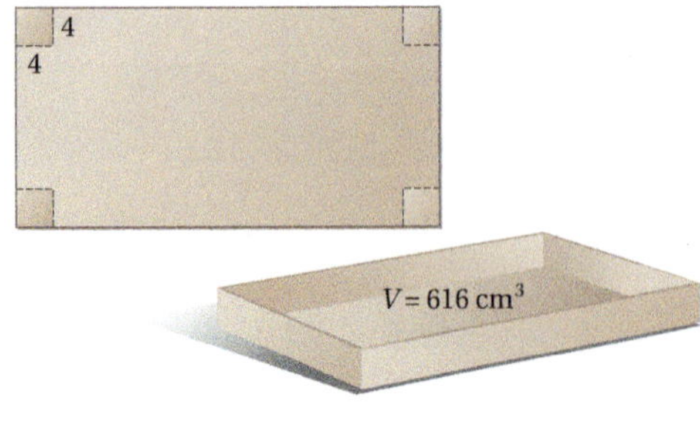

48. ***Rain-Gutter Design.*** An open rectangular gutter is made by turning up the sides of a piece of metal 20 in. wide. The area of the cross-section of the gutter is 50 in^2. Find the depth of the gutter.

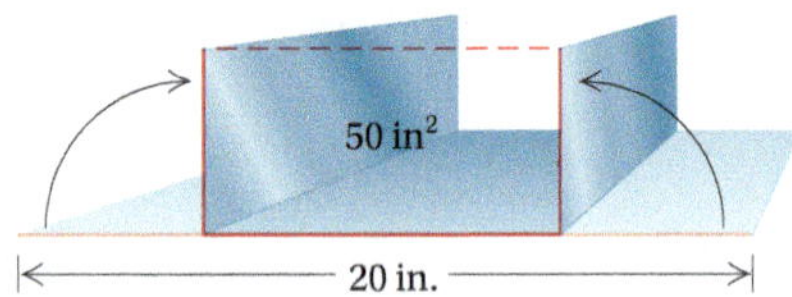

49. ***Right Triangle.*** The longest side of a right triangle is 5 yd shorter than six times the length of the shortest side. The other side of the triangle is 5 yd longer than five times the length of the shortest side. Find the lengths of the sides of the triangle.

50. Solve for x.

60 cm
36 cm
x
63 cm

CHAPTER

13 Summary and Review

Vocabulary Reinforcement

Complete each statement with the correct term from the column on the right. Some of the choices may be used more than once or not at all.

1. To ______________ a polynomial is to express it as a product. [13.1a]
2. A(n) ______________ of a polynomial P is a polynomial that can be used to express P as a product. [13.1a]
3. A(n) ______________ of a polynomial is an expression that names that polynomial as a product. [13.1a]
4. When factoring, always look first for a(n) ______________ factor. [13.1b]
5. When factoring a polynomial with four terms, try factoring by ______________. [13.6a]
6. A trinomial square is the square of a(n) ______________. [13.5a]
7. The principle of ______________ products states that if $ab = 0$, then $a = 0$ or $b = 0$. [13.7a]
8. The factorization of a ______________ of squares is the product of the sum and the difference of two terms. [13.5d]

common
similar
product
difference
factor
factorization
grouping
monomial
binomial
trinomial
zero

Concept Reinforcement

Determine whether each statement is true or false.

______________ 1. Every polynomial with four terms can be factored by grouping. [13.1c]

______________ 2. When factoring $x^2 + 5x + 6$, we need consider only positive pairs of factors of 6. [13.2a]

______________ 3. A product is 0 if and only if all the factors are 0. [13.7a]

______________ 4. If the principle of zero products is to be used, one side of the equation must be 0. [13.7b]

Study Guide

Objective 13.1a Find the greatest common factor, the GCF, of monomials.

Example Find the GCF of $15x^4y^2$, $-18x$, and $12x^3y$.

$$15x^4y^2 = 3 \cdot 5 \cdot x^4 \cdot y^2;$$
$$-18x = -1 \cdot 2 \cdot 3 \cdot 3 \cdot x;$$
$$12x^3y = 2 \cdot 2 \cdot 3 \cdot x^3 \cdot y$$

The GCF of the coefficients is 3. The GCF of the powers of x is x because 1 is the smallest exponent of x. The GCF of the powers of y is 1 because $-18x$ has no y-factor. Thus the GCF is $3 \cdot x \cdot 1$, or $3x$.

Practice Exercise

1. Find the GCF of $8x^3y^2$, $-20xy^3$, and $32x^2y$.

Objective 13.1b Factor polynomials when the terms have a common factor, factoring out the greatest common factor.

Example Factor: $16y^4 + 8y^3 - 24y^2$.

The *largest* common factor is $8y^2$.

$$16y^4 + 8y^3 - 24y^2 = (8y^2)(2y^2) + (8y^2)(y) - (8y^2)(3)$$
$$= 8y^2(2y^2 + y - 3)$$

Practice Exercise

2. Factor $27x^5 - 9x^3 + 18x^2$, factoring out the largest common factor.

Objective 13.1c Factor certain expressions with four terms using factoring by grouping.

Example Factor $6x^3 + 4x^2 - 15x - 10$ by grouping.

$$6x^3 + 4x^2 - 15x - 10 = (6x^3 + 4x^2) + (-15x - 10)$$
$$= 2x^2(3x + 2) - 5(3x + 2)$$
$$= (3x + 2)(2x^2 - 5)$$

Practice Exercise

3. Factor $z^3 - 3z^2 + 4z - 12$ by grouping.

Objective 13.2a Factor trinomials of the type $x^2 + bx + c$ by examining the constant term c.

Example Factor: $x^2 - x - 12$.

Since the constant term, -12, is negative, we look for a factorization of -12 in which one factor is positive and one factor is negative. The sum of the factors must be the coefficient of the middle term, -1, so the negative factor must have the larger absolute value. The possible pairs of factors that meet these criteria are $1, -12$ and $2, -6$ and $3, -4$. The numbers we need are 3 and -4:

$$x^2 - x - 12 = (x + 3)(x - 4).$$

Practice Exercise

4. Factor: $x^2 + 6x + 8$.

Objective 13.3a Factor trinomials of the type $ax^2 + bx + c$, $a \neq 1$, using the FOIL method.

Example Factor $2y^3 + 5y^2 - 3y$.

1) Factor out the largest common factor, y:

$y(2y^2 + 5y - 3)$.

Now we factor $2y^2 + 5y - 3$.

2) Because $2y^2$ factors as $2y \cdot y$, we have this possibility for a factorization:

$(2y +\quad)(y +\quad)$.

3) There are two pairs of factors of -3 and each can be written in two ways:

$3, -1 \qquad -3, 1$

and $\ -1, 3 \qquad 1, -3$.

4) From steps (2) and (3), we see that there are 4 possibilities for factorizations. We look for **O**utside and **I**nside products for which the sum is the middle term, $5y$. We try some possibilities and find that the factorization of $2y^2 + 5y - 3$ is $(2y - 1)(y + 3)$.

We must include the common factor to get a factorization of the original trinomial:

$$2y^3 + 5y^2 - 3y = y(2y - 1)(y + 3).$$

Practice Exercise

5. Factor: $6z^2 - 21z - 12$.

Objective 13.4a Factor trinomials of the type $ax^2 + bx + c, a \neq 1$, using the *ac*-method.

Example Factor $5x^2 + 7x - 6$ using the *ac*-method.

1) There is no common factor (other than 1 or -1).
2) Multiply the leading coefficient 5 and the constant, -6:
   $5(-6) = -30$.
3) Look for a factorization of -30 in which the sum of the factors is the coefficient of the middle term, 7. One number will be positive and the other will be negative. Since their sum, 7, is positive, the positive number will have the larger absolute value. The numbers we need are 10 and -3.
4) Split the middle term, writing it as a sum or a difference using the factors found in step (3):
   $7x = 10x - 3x$.
5) Factor by grouping:
   $$\begin{aligned} 5x^2 + 7x - 6 &= 5x^2 + 10x - 3x - 6 \\ &= 5x(x + 2) - 3(x + 2) \\ &= (x + 2)(5x - 3). \end{aligned}$$
6) Check: $(x + 2)(5x - 3) = 5x^2 + 7x - 6$.

Practice Exercise

6. Factor $6y^2 + 7y - 3$ using the *ac*-method.

Objective 13.5b Factor trinomial squares.

Example Factor: $9x^2 - 12x + 4$.

$9x^2 - 12x + 4 = (3x)^2 - 2 \cdot 3x \cdot 2 + 2^2 = (3x - 2)^2$

Practice Exercise

7. Factor: $4x^2 + 4x + 1$.

Objective 13.5d Factor differences of squares, being careful to factor completely.

Example Factor: $b^6 - b^2$.

$$\begin{aligned} b^6 - b^2 = b^2(b^4 - 1) &= b^2(b^2 + 1)(b^2 - 1) \\ &= b^2(b^2 + 1)(b + 1)(b - 1) \end{aligned}$$

Practice Exercise

8. Factor $18x^2 - 8$ completely.

Objective 13.7b Solve quadratic equations by factoring and then using the principle of zero products.

Example Solve: $x^2 - 3x = 28$.

$$\begin{aligned} x^2 - 3x &= 28 \\ x^2 - 3x - 28 &= 28 - 28 \\ x^2 - 3x - 28 &= 0 \\ (x + 4)(x - 7) &= 0 \\ x + 4 = 0 \quad &or \quad x - 7 = 0 \\ x = -4 \quad &or \quad x = 7 \end{aligned}$$

The solutions are -4 and 7.

Practice Exercise

9. Solve: $x^2 + 4x = 5$.

Review Exercises

Find the GCF. [13.1a]

1. $-15y^2,\ 25y^6$

2. $12x^3,\ -60x^2y,\ 36xy$

Factor completely. [13.6a]

3. $5 - 20x^6$

4. $x^2 - 3x$

5. $9x^2 - 4$

6. $x^2 + 4x - 12$

7. $x^2 + 14x + 49$

8. $6x^3 + 12x^2 + 3x$

9. $x^3 + x^2 + 3x + 3$

10. $6x^2 - 5x + 1$

11. $x^4 - 81$

12. $9x^3 + 12x^2 - 45x$

13. $2x^2 - 50$

14. $x^4 + 4x^3 - 2x - 8$

15. $16x^4 - 1$

16. $8x^6 - 32x^5 + 4x^4$

17. $75 + 12x^2 + 60x$

18. $x^2 + 9$

19. $x^3 - x^2 - 30x$

20. $4x^2 - 25$

21. $9x^2 + 25 - 30x$

22. $6x^2 - 28x - 48$

23. $x^2 - 6x + 9$

24. $2x^2 - 7x - 4$

25. $18x^2 - 12x + 2$

26. $3x^2 - 27$

27. $15 - 8x + x^2$

28. $25x^2 - 20x + 4$

29. $49b^{10} + 4a^8 - 28a^4b^5$

30. $x^2y^2 + xy - 12$

31. $12a^2 + 84ab + 147b^2$

32. $m^2 + 5m + mt + 5t$

33. $32x^4 - 128y^4z^4$

Solve. [13.7a, b]

34. $(x - 1)(x + 3) = 0$

35. $x^2 + 2x - 35 = 0$

36. $x^2 + 4x = 0$

37. $3x^2 + 2 = 5x$

38. $x^2 = 64$

39. $16 = x(x - 6)$

Find the x-intercepts of the graph of each equation. [13.7b]

40.

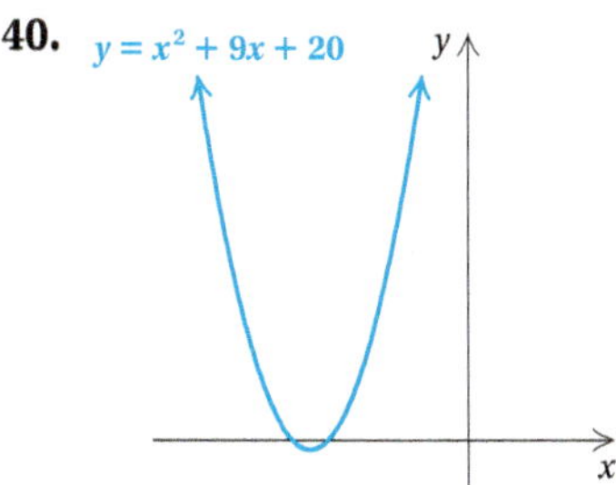

41.

Solve. [13.8a]

42. ***Sharks' Teeth.*** Sharks' teeth are shaped like triangles. The height of a tooth of a great white shark is 1 cm longer than the base. The area is 15 cm^2. Find the height and the base.

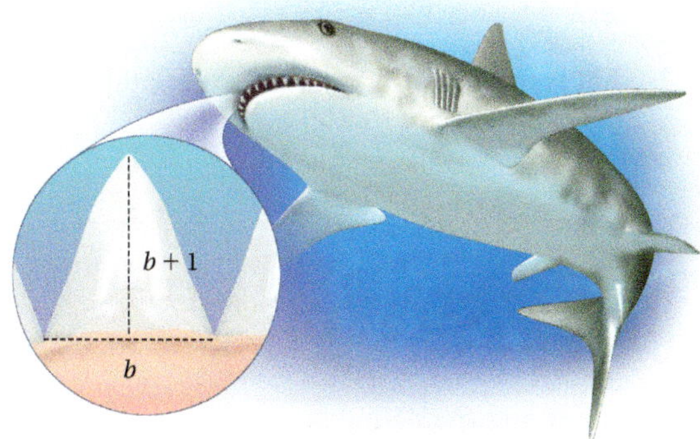

43. The product of two consecutive even integers is 288. Find the integers.

44. ***Zipline.*** On one zipline in a canopy tour in Costa Rica, riders drop 58 ft while covering a distance of 840 ft along the ground. How long is the zipline?

45. ***Tree Supports.*** A duckbill-anchor system is used to support a newly planted Bradford pear tree. Each cable is 5 ft long. The distance from the base of the tree to the point on the ground where each cable is anchored is 1 ft more than the distance from the base of the tree to the point where the cable is attached to the tree. Find both distances.

46. If the sides of a square are lengthened by 3 km, the area increases to 81 km^2. Find the length of a side of the original square.

47. Factor: $x^2 - 9x + 8$. Which of the following is one factor? [13.2a], [13.6a]

A. $(x + 1)$ **B.** $(x - 1)$
C. $(x + 8)$ **D.** $(x - 4)$

48. Factor $15x^2 + 5x - 20$ completely. Which of the following is one factor? [13.3a], [13.4a], [13.6a]

A. $(3x + 4)$ **B.** $(3x - 4)$
C. $(5x - 5)$ **D.** $(15x + 20)$

Synthesis

Solve. [13.8a]

49. The pages of a book measure 15 cm by 20 cm. Margins of equal width surround the printing on each page and constitute one-half of the area of the page. Find the width of the margins.

50. The cube of a number is the same as twice the square of the number. Find all such numbers.

51. The length of a rectangle is two times its width. When the length is increased by 20 in. and the width is decreased by 1 in., the area is 160 in^2. Find the original length and width.

52. Use the information in the figure below to determine the height of the telephone pole.

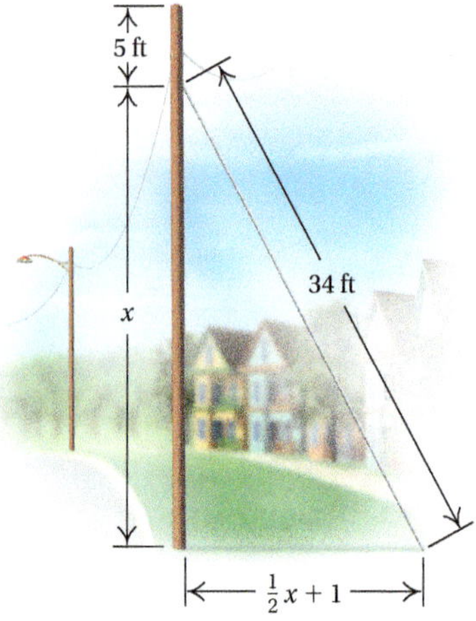

Solve. [13.7b]

53. $x^2 + 25 = 0$

54. $(x - 2)(x + 3)(2x - 5) = 0$

55. $(x - 3)4x^2 + 3x(x - 3) - (x - 3)10 = 0$

Understanding Through Discussion and Writing

1. Gwen factors $x^3 - 8x^2 + 15x$ as $(x^2 - 5x)(x - 3)$. Is she wrong? Why or why not? What advice would you offer? [13.2a]

2. After a test, Josh told a classmate that he was sure he had not written any incorrect factorizations. How could he be certain? [13.6a]

3. Kelly factored $16 - 8x + x^2$ as $(x - 4)^2$, while Tony factored it as $(4 - x)^2$. Evaluate each expression for several values of x. Then explain why both answers are correct. [13.5b]

4. What is wrong with the following? Explain the correct method of solution. [13.7b]

$$(x - 3)(x + 4) = 8$$
$$x - 3 = 8 \quad \textit{or} \quad x + 4 = 8$$
$$x = 11 \quad \textit{or} \quad x = 4$$

5. What is incorrect about solving $x^2 = 3x$ by dividing by x on both sides? [13.7b]

6. An archaeologist has measuring sticks of 3 ft, 4 ft, and 5 ft. Explain how she could draw a 7-ft by 9-ft rectangle on a piece of land being excavated. [13.8a]

Test

For Extra Help For step-by-step test solutions, access the Chapter Test Prep Videos in MyMathLab® or on YouTube (search "BittingerPreIntro" and click on Channels).

1. Find the GCF: $28x^3, 48x^7$.

Factor completely.

2. $x^2 - 7x + 10$

3. $x^2 + 25 - 10x$

4. $6y^2 - 8y^3 + 4y^4$

5. $x^3 + x^2 + 2x + 2$

6. $x^2 - 5x$

7. $x^3 + 2x^2 - 3x$

8. $28x - 48 + 10x^2$

9. $4x^2 - 9$

10. $x^2 - x - 12$

11. $6m^3 + 9m^2 + 3m$

12. $3w^2 - 75$

13. $60x + 45x^2 + 20$

14. $3x^4 - 48$

15. $49x^2 - 84x + 36$

16. $5x^2 - 26x + 5$

17. $x^4 + 2x^3 - 3x - 6$

18. $80 - 5x^4$

19. $6t^3 + 9t^2 - 15t$

Solve.

20. $x^2 - 3x = 0$

21. $2x^2 = 32$

22. $x^2 - x - 20 = 0$

23. $2x^2 + 7x = 15$

24. $x(x - 3) = 28$

Find the x-intercepts of the graph of each equation.

25.

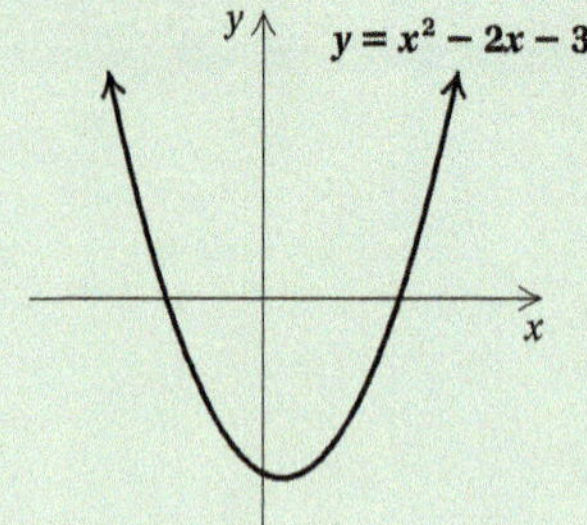

26.

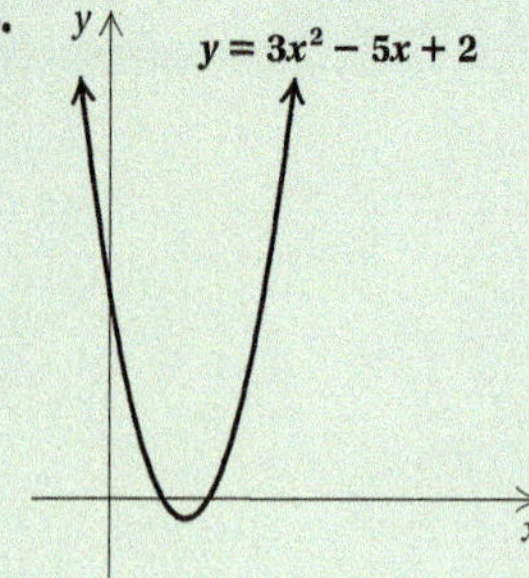

Solve.

27. The length of a rectangle is 2 m more than the width. The area of the rectangle is 48 m^2. Find the length and the width.

28. The base of a triangle is 6 cm greater than twice the height. The area is 28 cm^2. Find the height and the base.

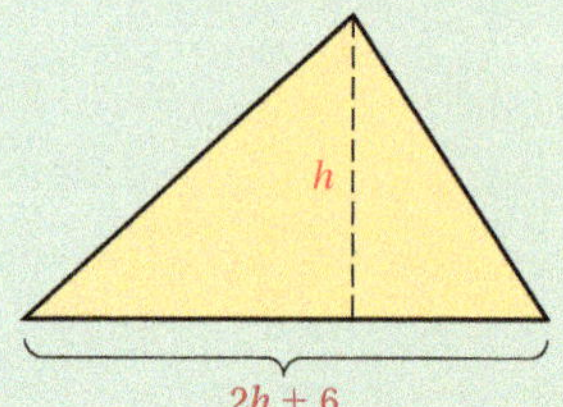

29. *Masonry Corner.* A mason wants to be sure that he has a right-angle corner of a building's foundation. He marks a point 3 ft from the corner along one wall and another point 4 ft from the corner along the other wall. If the corner is a right angle, what should the distance be between the two marked points?

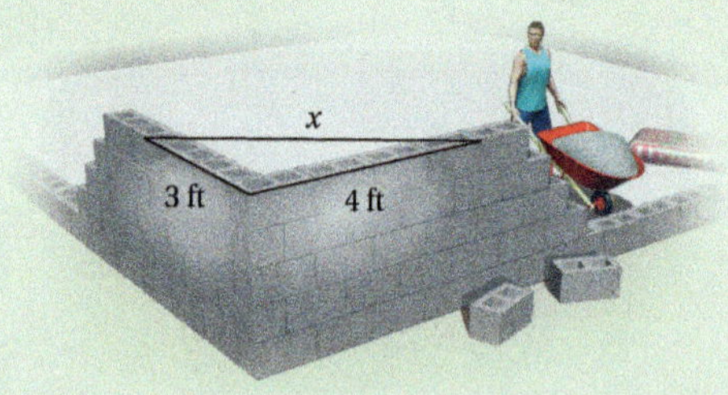

30. Factor $2y^4 - 32$ completely. Which of the following is one factor?

A. $(y + 2)$ **B.** $(y + 4)$
C. $(y^2 - 4)$ **D.** $(2y^2 + 8)$

Synthesis

31. The length of a rectangle is five times its width. When the length is decreased by 3 m and the width is increased by 2 m, the area of the new rectangle is 60 m^2. Find the original length and width.

32. Factor: $(a + 3)^2 - 2(a + 3) - 35$.

33. Solve: $20x(x + 2)(x - 1) = 5x^3 - 24x - 14x^2$.

34. If $x + y = 4$ and $x - y = 6$, then $x^2 - y^2$ equals which of the following?

A. 2 **B.** 10
C. 34 **D.** 24

CHAPTER

14

Rational Expressions and Equations

STUDYING FOR SUCCESS *Quiz and Test Follow-Up*

- ☐ Immediately after completing a chapter quiz or test, write out a step-by-step solution of each question that you missed.
- ☐ Visit your instructor or tutor for help with problems that are still giving you trouble.
- ☐ Keep your tests and quizzes, along with their corrections, to use as a study guide for the final examination.

14.1 Multiplying and Simplifying Rational Expressions

OBJECTIVES

- **a** Find all numbers for which a rational expression is not defined.
- **b** Multiply a rational expression by 1, using an expression such as A/A.
- **c** Simplify rational expressions by factoring the numerator and the denominator and removing factors of 1.
- **d** Multiply rational expressions and simplify.

SKILL TO REVIEW

Objective 3.5b: Simplify fraction notation.

Simplify.

1. $\frac{360}{140}$ **2.** $\frac{189}{252}$

a RATIONAL EXPRESSIONS AND REPLACEMENTS

Rational numbers are quotients of integers. Some examples are

$$\frac{2}{3}, \quad \frac{4}{-5}, \quad \frac{-8}{17}, \quad \frac{563}{1}.$$

The following are called **rational expressions** or **fraction expressions**. They are quotients, or ratios, of polynomials:

$$\frac{3}{4}, \quad \frac{z}{6}, \quad \frac{5}{x+2}, \quad \frac{t^2+3t-10}{7t^2-4}.$$

A rational expression is also a division. For example,

$$\frac{3}{4} \text{ means } 3 \div 4 \quad \text{and} \quad \frac{x-8}{x+2} \text{ means } (x-8) \div (x+2).$$

Because rational expressions indicate division, we must be careful to avoid denominators of zero. When a variable is replaced with a number that produces a denominator equal to zero, the rational expression is not defined. For example, in the expression

$$\frac{x-8}{x+2},$$

when x is replaced with -2, the denominator is 0, and the expression is *not* defined:

$$\frac{x-8}{x+2} = \frac{-2-8}{-2+2} = \frac{-10}{0}. \leftarrow \text{Division by 0 is not defined.}$$

When x is replaced with a number other than -2, such as 3, the expression *is* defined because the denominator is nonzero:

$$\frac{x-8}{x+2} = \frac{3-8}{3+2} = \frac{-5}{5} = -1.$$

EXAMPLE 1 Find all numbers for which the rational expression

$$\frac{x+4}{x^2-3x-10}$$

is not defined.

Answers

Skill to Review:

1. $\frac{18}{7}$ 2. $\frac{3}{4}$

The value of the numerator has no bearing on whether or not a rational expression is defined. To determine which numbers make the rational expression not defined, we set the *denominator* equal to 0 and solve:

$$x^2 - 3x - 10 = 0$$
$$(x - 5)(x + 2) = 0 \quad \text{Factoring}$$
$$x - 5 = 0 \quad or \quad x + 2 = 0 \quad \text{Using the principle of zero products (See Section 13.7.)}$$
$$x = 5 \quad or \quad x = -2.$$

The rational expression is not defined for the replacement numbers 5 and -2.

Do Exercises 1–3. ▶

Find all numbers for which the rational expression is not defined.

1. $\dfrac{16}{x - 3}$

2. $\dfrac{2x - 7}{x^2 + 5x - 24}$

$$x^2 + 5x - 24 = 0$$
$$(x + \square)(x - 3) = 0$$
$$x + 8 = 0 \quad or \quad x - \square = 0$$
$$x = \square \quad or \quad x = 3$$

The rational expression is not defined for replacements -8 and $\square$.

3. $\dfrac{x + 5}{8}$

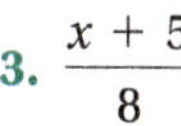

b MULTIPLYING BY 1

We multiply rational expressions in the same way that we multiply fraction notation in arithmetic. For review, we see that

$$\frac{3}{7} \cdot \frac{2}{5} = \frac{3 \cdot 2}{7 \cdot 5} = \frac{6}{35}.$$

MULTIPLYING RATIONAL EXPRESSIONS

To multiply rational expressions, multiply numerators and multiply denominators:

$$\frac{A}{B} \cdot \frac{C}{D} = \frac{AC}{BD}.$$

For example,

$$\frac{x - 2}{3} \cdot \frac{x + 2}{x + 7} = \frac{(x - 2)(x + 2)}{3(x + 7)}. \quad \text{Multiplying the numerators and the denominators}$$

Note that we leave the numerator, $(x - 2)(x + 2)$, and the denominator, $3(x + 7)$, in factored form because it is easier to simplify if we do not multiply. In order to learn to simplify, we first need to consider multiplying the rational expression by 1.

Any rational expression with the same numerator and denominator (except $0/0$) is a symbol for 1:

$$\frac{19}{19} = 1, \quad \frac{x + 8}{x + 8} = 1, \quad \frac{3x^2 - 4}{3x^2 - 4} = 1, \quad \frac{-1}{-1} = 1.$$

EQUIVALENT EXPRESSIONS

Expressions that have the same value for all allowable (or meaningful) replacements are called **equivalent expressions**.

Answers

1. 3 **2.** $-8, 3$ **3.** None

Guided Solution:
2. 8, 3, -8, 3

We can multiply by 1 to obtain an *equivalent expression*. At this point, we select expressions for 1 arbitrarily. Later, we will have a system for our choices when we add and subtract.

EXAMPLES Multiply.

2. $\frac{3x+2}{x+1} \cdot 1 = \frac{3x+2}{x+1} \cdot \frac{2x}{2x}$ Using the identity property of 1. We arbitrarily choose $(2x)/(2x)$ as a symbol for 1.

$= \frac{(3x+2)2x}{(x+1)2x}$

3. $\frac{x+2}{x-7} \cdot \frac{x+3}{x+3} = \frac{(x+2)(x+3)}{(x-7)(x+3)}$ We arbitrarily choose $(x+3)/(x+3)$ as a symbol for 1.

4. $\frac{2+x}{2-x} \cdot \frac{-1}{-1} = \frac{(2+x)(-1)}{(2-x)(-1)}$ Using $(-1)/(-1)$ as a symbol for 1

Multiply.

4. $\frac{2x+1}{3x-2} \cdot \frac{x}{x}$

5. $\frac{x+1}{x-2} \cdot \frac{x+2}{x+2}$

6. $\frac{x-8}{x-y} \cdot \frac{-1}{-1}$

Do Exercises 4–6.

C SIMPLIFYING RATIONAL EXPRESSIONS

Simplifying rational expressions is similar to simplifying fraction expressions in arithmetic. For example, an expression like $\frac{15}{40}$ can be simplified as follows:

$\frac{15}{40} = \frac{3 \cdot 5}{8 \cdot 5}$ Factoring the numerator and the denominator. Note the common factor, 5.

$= \frac{3}{8} \cdot \frac{5}{5}$ Factoring the fraction expression

$= \frac{3}{8} \cdot 1$ $\frac{5}{5} = 1$

$= \frac{3}{8}.$ Using the identity property of 1, or "removing" a factor of 1

Similar steps are followed when simplifying rational expressions: We factor and remove a factor of 1, using the fact that

$$\frac{ab}{cb} = \frac{a}{c} \cdot \frac{b}{b} = \frac{a}{c} \cdot 1 = \frac{a}{c}.$$

In algebra, instead of simplifying

$$\frac{15}{40},$$

we may need to simplify an expression like

$$\frac{x^2 - 16}{x+4}.$$

Just as factoring is important in simplifying in arithmetic, so too is it important in simplifying rational expressions. The factoring we use most is the factoring of polynomials, which we studied in Chapter 13.

Answers

4. $\frac{(2x+1)x}{(3x-2)x}$ 5. $\frac{(x+1)(x+2)}{(x-2)(x+2)}$
6. $\frac{(x-8)(-1)}{(x-y)(-1)}$

To simplify, we can do the reverse of multiplying. We factor the numerator and the denominator and "remove" a factor of 1.

EXAMPLE 5 Simplify: $\frac{8x^2}{24x}$.

$$\frac{8x^2}{24x} = \frac{8 \cdot x \cdot x}{3 \cdot 8 \cdot x}$$ Factoring the numerator and the denominator. Note the common factor, $8x$.

$$= \frac{8x}{8x} \cdot \frac{x}{3}$$ Factoring the rational expression

$$= 1 \cdot \frac{x}{3}$$ $\frac{8x}{8x} = 1$

$$= \frac{x}{3}$$ We removed a factor of 1.

Do Exercises 7 and 8. ▶

Simplify.

7. $\frac{5y}{y}$

8. $\frac{9x^2}{36x}$

EXAMPLES Simplify.

6. $$\frac{5a + 15}{10} = \frac{5(a + 3)}{5 \cdot 2}$$ Factoring the numerator and the denominator

$$= \frac{5}{5} \cdot \frac{a + 3}{2}$$ Factoring the rational expression

$$= 1 \cdot \frac{a + 3}{2}$$ $\frac{5}{5} = 1$

$$= \frac{a + 3}{2}$$ Removing a factor of 1

7. $$\frac{6a + 12}{7a + 14} = \frac{6(a + 2)}{7(a + 2)}$$ Factoring the numerator and the denominator

$$= \frac{6}{7} \cdot \frac{a + 2}{a + 2}$$ Factoring the rational expression

$$= \frac{6}{7} \cdot 1$$ $\frac{a + 2}{a + 2} = 1$

$$= \frac{6}{7}$$ Removing a factor of 1

8. $$\frac{6x^2 + 4x}{2x^2 + 2x} = \frac{2x(3x + 2)}{2x(x + 1)}$$ Factoring the numerator and the denominator

$$= \frac{2x}{2x} \cdot \frac{3x + 2}{x + 1}$$ Factoring the rational expression

$$= 1 \cdot \frac{3x + 2}{x + 1}$$ $\frac{2x}{2x} = 1$

$$= \frac{3x + 2}{x + 1}$$ Removing a factor of 1

Caution!

Note that you *cannot* simplify further by removing the x's because x is not a *factor* of the entire numerator, $3x + 2$, and the entire denominator, $x + 1$.

Answers

7. 5 8. $\frac{x}{4}$

$$9.\quad \frac{x^2+3x+2}{x^2-1} = \frac{(x+2)(x+1)}{(x+1)(x-1)}$$ Factoring the numerator and the denominator

$$= \frac{x+1}{x+1}\cdot\frac{x+2}{x-1}$$ Factoring the rational expression

$$= 1\cdot\frac{x+2}{x-1}$$ $\frac{x+1}{x+1} = 1$

$$= \frac{x+2}{x-1}$$ Removing a factor of 1

Canceling

You may have encountered canceling when working with rational expressions. With great concern, we mention it as a possible way to speed up your work. Our concern is that canceling be done with care and understanding. Example 9 might have been done faster as follows:

$$\frac{x^2+3x+2}{x^2-1} = \frac{(x+2)(x+1)}{(x+1)(x-1)}$$ Factoring the numerator and the denominator

$$= \frac{(x+2)\cancel{(x+1)}}{\cancel{(x+1)}(x-1)}$$ When a factor of 1 is noted, it is canceled, as shown: $\frac{x+1}{x+1} = 1.$

$$= \frac{x+2}{x-1}.$$ Simplifying

Do Exercises 9–12.

Simplify.

9. $\frac{2x^2+x}{3x^2+2x}$

10. $\frac{x^2-1}{2x^2-x-1}$

11. $\frac{7x+14}{7}$

12. $\frac{12y+24}{48}$

Opposites in Rational Expressions

Expressions of the form $a - b$ and $b - a$ are **opposites** of each other. When either of these binomials is multiplied by -1, the result is the other binomial:

$$-1(a-b) = -a+b = b+(-a) = b-a;$$
$$-1(b-a) = -b+a = a+(-b) = a-b.$$
Multiplication by -1 reverses the order in which subtraction occurs.

Consider, for example,

$$\frac{x-4}{4-x}.$$

At first glance, it appears as though the numerator and the denominator do not have any common factors other than 1. But $x - 4$ and $4 - x$ are opposites, or additive inverses, of each other. Thus we can rewrite one as the opposite of the other by factoring out a -1.

EXAMPLE 10 Simplify: $\frac{x-4}{4-x}$.

$$\frac{x-4}{4-x} = \frac{x-4}{-(x-4)} = \frac{1(x-4)}{-1(x-4)}$$ $4 - x = -(x-4)$; $4 - x$ and $x - 4$ are opposites.

$$= \frac{1}{-1}\cdot\frac{x-4}{x-4}$$

$$= -1\cdot 1$$ $1/(-1) = -1$

$$= -1$$

Do Exercises 13–15.

Simplify.

GS

13. $\frac{x-8}{8-x}$

$= \frac{x-8}{-(x-\quad)}$

$= \frac{1(x-8)}{-1(x-8)} = \frac{1}{\quad}\cdot\frac{x-8}{x-8}$

$= -1\cdot\quad = \quad$

14. $\frac{c-d}{d-c}$

15. $\frac{-x-7}{x+7}$

Answers

9. $\frac{2x+1}{3x+2}$ 10. $\frac{x+1}{2x+1}$ 11. $x+2$
12. $\frac{y+2}{4}$ 13. -1 14. -1 15. -1

Guided Solution:
13. 8, -1, 1, -1

d MULTIPLYING AND SIMPLIFYING

We try to simplify after we multiply. That is why we leave the numerator and the denominator in factored form.

EXAMPLE 11 Multiply and simplify: $\frac{5a^3}{4} \cdot \frac{2}{5a}$.

$$\frac{5a^3}{4} \cdot \frac{2}{5a} = \frac{5a^3(2)}{4(5a)}$$ Multiplying the numerators and the denominators

$$= \frac{5 \cdot a \cdot a \cdot a \cdot 2}{2 \cdot 2 \cdot 5 \cdot a}$$ Factoring the numerator and the denominator

$$= \frac{\cancel{5} \cdot \cancel{a} \cdot a \cdot a \cdot \cancel{2}}{\cancel{2} \cdot 2 \cdot \cancel{5} \cdot \cancel{a}}$$ Removing a factor of 1: $\frac{2 \cdot 5 \cdot a}{2 \cdot 5 \cdot a} = 1$

$$= \frac{a^2}{2}$$ Simplifying

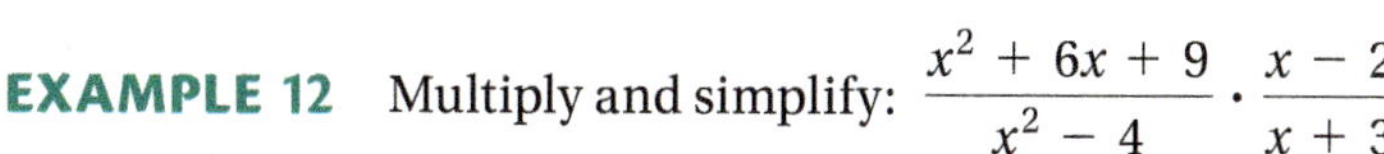

EXAMPLE 12 Multiply and simplify: $\frac{x^2 + 6x + 9}{x^2 - 4} \cdot \frac{x - 2}{x + 3}$.

$$\frac{x^2 + 6x + 9}{x^2 - 4} \cdot \frac{x - 2}{x + 3} = \frac{(x^2 + 6x + 9)(x - 2)}{(x^2 - 4)(x + 3)}$$ Multiplying the numerators and the denominators

$$= \frac{(x + 3)(x + 3)(x - 2)}{(x + 2)(x - 2)(x + 3)}$$ Factoring the numerator and the denominator

$$= \frac{\cancel{(x + 3)}(x + 3)\cancel{(x - 2)}}{(x + 2)\cancel{(x - 2)}\cancel{(x + 3)}}$$ Removing a factor of 1: $\frac{(x + 3)(x - 2)}{(x + 3)(x - 2)} = 1$

$$= \frac{x + 3}{x + 2}$$ Simplifying

Do Exercise 16. ▶

EXAMPLE 13 Multiply and simplify: $\frac{x^2 + x - 2}{15} \cdot \frac{5}{2x^2 - 3x + 1}$.

$$\frac{x^2 + x - 2}{15} \cdot \frac{5}{2x^2 - 3x + 1} = \frac{(x^2 + x - 2)5}{15(2x^2 - 3x + 1)}$$ Multiplying the numerators and the denominators

$$= \frac{(x + 2)(x - 1)5}{5(3)(x - 1)(2x - 1)}$$ Factoring the numerator and the denominator

$$= \frac{(x + 2)\cancel{(x - 1)}\cancel{5}}{\cancel{5}(3)\cancel{(x - 1)}(2x - 1)}$$ Removing a factor of 1: $\frac{(x - 1)5}{(x - 1)5} = 1$

$$= \frac{x + 2}{3(2x - 1)}$$ Simplifying

You need not carry out this multiplication.

Do Exercise 17. ▶

GS **16.** Multiply and simplify:

$$\frac{a^2 - 4a + 4}{a^2 - 9} \cdot \frac{a + 3}{a - 2}.$$

$$\frac{a^2 - 4a + 4}{a^2 - 9} \cdot \frac{a + 3}{a - 2}$$

$$= \frac{(a^2 - 4a + 4)(a + \square)}{(a^2 - \square)(a - 2)}$$

$$= \frac{(a - \square)(a - 2)(a + 3)}{(a + 3)(a - \square)(a - 2)}$$

$$= \frac{\cancel{(a - 2)}(a - 2)\cancel{(a + 3)}}{\cancel{(a + 3)}(a - 3)\cancel{(a - 2)}}$$

$$= \frac{a - \square}{a - \square}$$

17. Multiply and simplify:

$$\frac{x^2 - 25}{6} \cdot \frac{3}{x + 5}.$$

Answers

16. $\frac{a - 2}{a - 3}$ **17.** $\frac{x - 5}{2}$

Guided Solution:
16. 3, 9, 2, 3, 2, 3

14.1 Exercise Set

For Extra Help

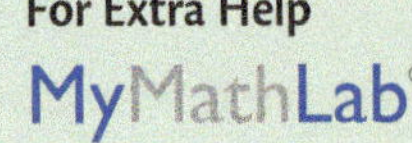
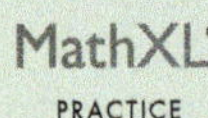

PRACTICE WATCH READ REVIEW

Reading Check

Choose the word below each blank that best completes the statement.

RC1. Expressions that have the same value for all allowable replacements are called ____________ expressions. (rational/equivalent)

RC2. A rational expression is undefined when the ____________ is zero. (denominator/numerator)

RC3. A rational expression can be written as a ____________ of two polynomials. (product/quotient)

RC4. A rational expression is simplified when the numerator and the denominator have no ____________ (other than 1) in common. (factors/terms)

a Find all numbers for which each rational expression is not defined.

1. $\frac{-3}{2x}$

2. $\frac{24}{-8y}$

3. $\frac{5}{x-8}$

4. $\frac{y-4}{y+6}$

5. $\frac{3}{2y+5}$

6. $\frac{x^2-9}{4x-15}$

7. $\frac{x^2+11}{x^2-3x-28}$

8. $\frac{p^2-9}{p^2-7p+10}$

9. $\frac{m^3-2m}{m^2-25}$

10. $\frac{7-3x+x^2}{49-x^2}$

11. $\frac{x-4}{3}$

12. $\frac{x^2-25}{14}$

b Multiply. Do not simplify. Note that in each case you are multiplying by 1.

13. $\frac{4x}{4x} \cdot \frac{3x^2}{5y}$

14. $\frac{5x^2}{5x^2} \cdot \frac{6y^3}{3z^4}$

15. $\frac{2x}{2x} \cdot \frac{x-1}{x+4}$

16. $\frac{2a-3}{5a+2} \cdot \frac{a}{a}$

17. $\frac{3-x}{4-x} \cdot \frac{-1}{-1}$

18. $\frac{x-5}{5-x} \cdot \frac{-1}{-1}$

19. $\frac{y+6}{y+6} \cdot \frac{y-7}{y+2}$

20. $\frac{x^2+1}{x^3-2} \cdot \frac{x-4}{x-4}$

C Simplify.

21. $\frac{8x^3}{32x}$

22. $\frac{4x^2}{20x}$

23. $\frac{48p^7q^5}{18p^5q^4}$

24. $\frac{-76x^8y^3}{-24x^4y^3}$

25. $\frac{4x - 12}{4x}$

26. $\frac{5a - 40}{5}$

27. $\frac{3m^2 + 3m}{6m^2 + 9m}$

28. $\frac{4y^2 - 2y}{5y^2 - 5y}$

29. $\frac{a^2 - 9}{a^2 + 5a + 6}$

30. $\frac{t^2 - 25}{t^2 + t - 20}$

31. $\frac{a^2 - 10a + 21}{a^2 - 11a + 28}$

32. $\frac{x^2 - 2x - 8}{x^2 - x - 6}$

33. $\frac{x^2 - 25}{x^2 - 10x + 25}$

34. $\frac{x^2 + 8x + 16}{x^2 - 16}$

35. $\frac{a^2 - 1}{a - 1}$

36. $\frac{t^2 - 1}{t + 1}$

37. $\frac{x^2 + 1}{x + 1}$

38. $\frac{m^2 + 9}{m + 3}$

39. $\frac{6x^2 - 54}{4x^2 - 36}$

40. $\frac{8x^2 - 32}{4x^2 - 16}$

41. $\frac{6t + 12}{t^2 - t - 6}$

42. $\frac{4x + 32}{x^2 + 9x + 8}$

43. $\frac{2t^2 + 6t + 4}{4t^2 - 12t - 16}$

44. $\frac{3a^2 - 9a - 12}{6a^2 + 30a + 24}$

45. $\dfrac{t^2 - 4}{(t + 2)^2}$

46. $\dfrac{m^2 - 36}{(m - 6)^2}$

47. $\dfrac{6 - x}{x - 6}$

48. $\dfrac{t - 3}{3 - t}$

49. $\dfrac{a - b}{b - a}$

50. $\dfrac{y - x}{-x + y}$

51. $\dfrac{6t - 12}{2 - t}$

52. $\dfrac{5a - 15}{3 - a}$

53. $\dfrac{x^2 - 1}{1 - x}$

54. $\dfrac{a^2 - b^2}{b^2 - a^2}$

55. $\dfrac{6qt - 3t^4}{t^3 - 2q}$

56. $\dfrac{2z - w^5}{5w^{10} - 10zw^5}$

d Multiply and simplify.

57. $\dfrac{4x^3}{3x} \cdot \dfrac{14}{x}$

58. $\dfrac{18}{x^3} \cdot \dfrac{5x^2}{6}$

59. $\dfrac{3c}{d^2} \cdot \dfrac{4d}{6c^3}$

60. $\dfrac{3x^2y}{2} \cdot \dfrac{4}{xy^3}$

61. $\dfrac{x + 4}{x} \cdot \dfrac{x^2 - 3x}{x^2 + x - 12}$

62. $\dfrac{t^2}{t^2 - 4} \cdot \dfrac{t^2 - 5t + 6}{t^2 - 3t}$

63. $\dfrac{a^2 - 9}{a^2} \cdot \dfrac{a^2 - 3a}{a^2 + a - 12}$

64. $\dfrac{x^2 + 10x - 11}{x^2 - 1} \cdot \dfrac{x + 1}{x + 11}$

65. $\dfrac{4a^2}{3a^2 - 12a + 12} \cdot \dfrac{3a - 6}{2a}$

66. $\frac{5v + 5}{v - 2} \cdot \frac{v^2 - 4v + 4}{v^2 - 1}$

67. $\frac{t^4 - 16}{t^4 - 1} \cdot \frac{t^2 + 1}{t^2 + 4}$

68. $\frac{x^4 - 1}{x^4 - 81} \cdot \frac{x^2 + 9}{x^2 + 1}$

69. $\frac{(x + 4)^3}{(x + 2)^3} \cdot \frac{x^2 + 4x + 4}{x^2 + 8x + 16}$

70. $\frac{(t - 2)^3}{(t - 1)^3} \cdot \frac{t^2 - 2t + 1}{t^2 - 4t + 4}$

71. $\frac{5a^2 - 180}{10a^2 - 10} \cdot \frac{20a + 20}{2a - 12}$

72. $\frac{2t^2 - 98}{4t^2 - 4} \cdot \frac{8t + 8}{16t - 112}$

Skill Maintenance

Graph.

73. $x + y = -1$ [11.2a]

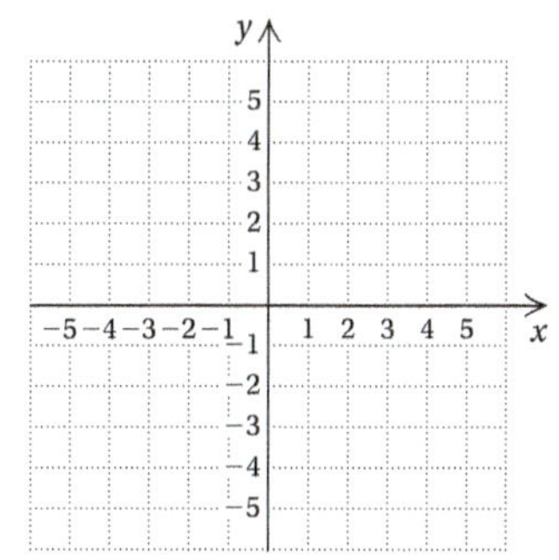

74. $y = -\frac{7}{2}$ [11.2b]

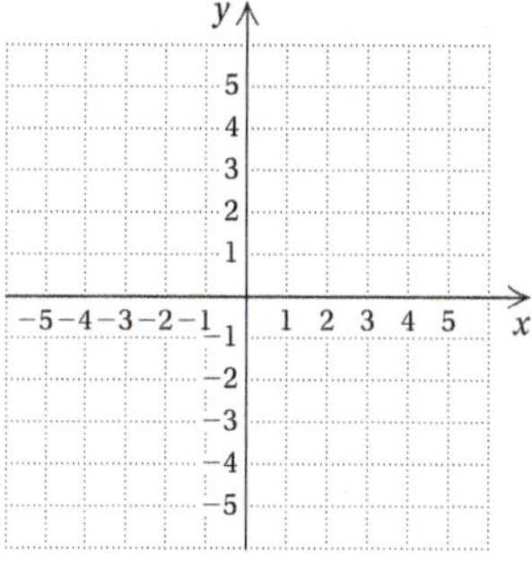

Factor. [13.6a]

75. $x^5 - 2x^4 - 35x^3$

76. $2y^3 - 10y^2 + y - 5$

77. $16 - t^4$

78. $10x^2 + 80x + 70$

Synthesis

Simplify.

79. $\frac{x^2 - y^2}{(x - y)^2} \cdot \frac{x^2 - 2xy + y^2}{x^2 - 4xy - 5y^2}$

80. $\frac{(t + 2)^3}{(t + 1)^3} \cdot \frac{t^2 + 2t + 1}{t^2 + 4t + 4} \cdot \frac{t + 1}{t + 2}$

81. $\frac{x - 1}{x^2 + 1} \cdot \frac{x^4 - 1}{(x - 1)^2} \cdot \frac{x^2 - 1}{x^4 - 2x^2 + 1}$

82. Select any number x, multiply by 2, add 5, multiply by 5, subtract 25, and divide by 10. What do you get? Explain how this procedure can be used for a number trick.

14.2 Division and Reciprocals

OBJECTIVES

a Find the reciprocal of a rational expression.

b Divide rational expressions and simplify.

SKILL TO REVIEW

Objective 13.6a: Factor polynomials.

Factor.

1. $x^2 - 2x$
2. $5y^2 - 11y - 12$

There is a similarity between what we do with rational expressions and what we do with rational numbers. In fact, after variables have been replaced with rational numbers, a rational expression represents a rational number.

a FINDING RECIPROCALS

Two expressions are **reciprocals** of each other if their product is 1. The reciprocal of a rational expression is found by interchanging the numerator and the denominator.

EXAMPLES

1. The reciprocal of $\frac{2}{5}$ is $\frac{5}{2}$. $\left(\text{This is because } \frac{2}{5} \cdot \frac{5}{2} = \frac{10}{10} = 1.\right)$

2. The reciprocal of $\frac{2x^2 - 3}{x + 4}$ is $\frac{x + 4}{2x^2 - 3}$.

3. The reciprocal of $x + 2$ is $\frac{1}{x + 2}$. $\left(\text{Think of } x + 2 \text{ as } \frac{x + 2}{1}.\right)$

Do Margin Exercises 1–4.

Find the reciprocal.

1. $\frac{7}{2}$
2. $\frac{x^2 + 5}{2x^3 - 1}$
3. $x - 5$
4. $\frac{1}{x^2 - 3}$

b DIVISION

We divide rational expressions in the same way that we divide fraction notation in arithmetic.

DIVIDING RATIONAL EXPRESSIONS

To divide by a rational expression, multiply by its reciprocal:

$$\frac{A}{B} \div \frac{C}{D} = \frac{A}{B} \cdot \frac{D}{C} = \frac{AD}{BC}.$$

Then factor and, if possible, simplify.

EXAMPLE 4 Divide and simplify: $\frac{3}{4} \div \frac{9}{5}$.

$$\frac{3}{4} \div \frac{9}{5} = \frac{3}{4} \cdot \frac{5}{9}$$ Multiplying by the reciprocal of the divisor

$$= \frac{3 \cdot 5}{4 \cdot 9} = \frac{3 \cdot 5}{2 \cdot 2 \cdot 3 \cdot 3}$$ Factoring

$$= \frac{\not{3} \cdot 5}{2 \cdot 2 \cdot \not{3} \cdot 3}$$ Removing a factor of 1: $\frac{3}{3} = 1$

$$= \frac{5}{12}$$ Simplifying

Do Margin Exercise 5.

5. Divide and simplify: $\frac{3}{5} \div \frac{7}{10}$.

Answers

Skill to Review:
1. $x(x - 2)$ 2. $(5y + 4)(y - 3)$

Margin Exercises:
1. $\frac{2}{7}$ 2. $\frac{2x^3 - 1}{x^2 + 5}$ 3. $\frac{1}{x - 5}$ 4. $x^2 - 3$ 5. $\frac{6}{7}$

EXAMPLE 5 Divide and simplify: $\frac{2}{x} \div \frac{3}{x}$.

$$\frac{2}{x} \div \frac{3}{x} = \frac{2}{x} \cdot \frac{x}{3}$$ Multiplying by the reciprocal of the divisor

$$= \frac{2 \cdot x}{x \cdot 3} = \frac{2 \cdot \not{x}}{\not{x} \cdot 3}$$ Removing a factor of 1: $\frac{x}{x} = 1$

$$= \frac{2}{3}$$

Do Exercise 6. ▶

6. Divide and simplify: $\frac{x}{8} \div \frac{x}{5}$.

$$\frac{x}{8} \div \frac{x}{5} = \frac{x}{8} \cdot \frac{5}{\square}$$
$$= \frac{x \cdot \square}{8 \cdot x}$$
$$= \frac{\not{x} \cdot 5}{8 \cdot \not{x}}$$
$$= \frac{\square}{8}$$

EXAMPLE 6 Divide and simplify: $\frac{x + 1}{x + 2} \div \frac{x - 1}{x + 3}$.

$$\frac{x + 1}{x + 2} \div \frac{x - 1}{x + 3} = \frac{x + 1}{x + 2} \cdot \frac{x + 3}{x - 1}$$ Multiplying by the reciprocal of the divisor

$$= \frac{(x + 1)(x + 3)}{(x + 2)(x - 1)}$$

We usually do not carry out the multiplication in the numerator or the denominator. It is not wrong to do so, but the factored form is often more useful.

Do Exercise 7. ▶

7. Divide and simplify:

$$\frac{x - 3}{x + 5} \div \frac{x + 5}{x - 2}.$$

EXAMPLE 7 Divide and simplify: $\frac{4}{x^2 - 7x} \div \frac{28x}{x^2 - 49}$.

$$\frac{4}{x^2 - 7x} \div \frac{28x}{x^2 - 49} = \frac{4}{x^2 - 7x} \cdot \frac{x^2 - 49}{28x}$$ Multiplying by the reciprocal

$$= \frac{4(x^2 - 49)}{(x^2 - 7x)(28x)}$$

$$= \frac{2 \cdot 2 \cdot (x - 7)(x + 7)}{x(x - 7) \cdot 2 \cdot 2 \cdot 7 \cdot x}$$ Factoring the numerator and the denominator

$$= \frac{\not{2} \cdot \not{2} \cdot \not{(x - 7)}(x + 7)}{x\not{(x - 7)} \cdot \not{2} \cdot \not{2} \cdot 7 \cdot x}$$ Removing a factor of 1: $\frac{2 \cdot 2 \cdot (x - 7)}{2 \cdot 2 \cdot (x - 7)} = 1$

$$= \frac{x + 7}{7x^2}$$

Do Exercise 8. ▶

8. Divide and simplify:

$$\frac{a^2 + 5a}{6} \div \frac{a^2 - 25}{18a}.$$

EXAMPLE 8 Divide and simplify: $\frac{x + 1}{x^2 - 1} \div \frac{x + 1}{x^2 - 2x + 1}$.

$$\frac{x + 1}{x^2 - 1} \div \frac{x + 1}{x^2 - 2x + 1}$$

$$= \frac{x + 1}{x^2 - 1} \cdot \frac{x^2 - 2x + 1}{x + 1}$$ Multiplying by the reciprocal

Answers

6. $\frac{5}{8}$ **7.** $\frac{(x - 3)(x - 2)}{(x + 5)(x + 5)}$ **8.** $\frac{3a^2}{a - 5}$

Guided Solution:

6. x, 5, 5

Then we multiply numerators and multiply denominators. We have

$$= \frac{(x+1)(x^2-2x+1)}{(x^2-1)(x+1)}$$

$$= \frac{(x+1)(x-1)(x-1)}{(x-1)(x+1)(x+1)}$$ Factoring the numerator and the denominator

$$= \frac{\cancel{(x+1)}\cancel{(x-1)}(x-1)}{\cancel{(x-1)}\cancel{(x+1)}(x+1)}$$ Removing a factor of 1: $\frac{(x+1)(x-1)}{(x+1)(x-1)} = 1$

$$= \frac{x-1}{x+1}.$$

EXAMPLE 9 Divide and simplify: $\frac{x^2-2x-3}{x^2-4} \div \frac{x+1}{x+5}$.

$$\frac{x^2-2x-3}{x^2-4} \div \frac{x+1}{x+5}$$

$$= \frac{x^2-2x-3}{x^2-4} \cdot \frac{x+5}{x+1}$$ Multiplying by the reciprocal

$$= \frac{(x^2-2x-3)(x+5)}{(x^2-4)(x+1)}$$

$$= \frac{(x-3)(x+1)(x+5)}{(x-2)(x+2)(x+1)}$$ Factoring the numerator and the denominator

$$= \frac{(x-3)\cancel{(x+1)}(x+5)}{(x-2)(x+2)\cancel{(x+1)}}$$ Removing a factor of 1: $\frac{x+1}{x+1} = 1$

$$= \frac{(x-3)(x+5)}{(x-2)(x+2)}$$ ← You need not carry out the multiplications in the numerator and the denominator.

◀ Do Exercises 9–11.

Divide and simplify.

9. $\frac{x-3}{x+5} \div \frac{x+2}{x+5}$

10. $\frac{x^2-5x+6}{x+5} \div \frac{x+2}{x+5}$

11. $\frac{y^2-1}{y+1} \div \frac{y^2-2y+1}{y+1}$ GS

$= \frac{y^2-1}{y+1} \cdot \frac{\underline{\quad}+1}{y^2-\underline{\quad}+1}$

$= \frac{(y^2-1)(y+1)}{(y+1)(y^2-2y+1)}$

$= \frac{(y+\underline{\quad})(y-\underline{\quad})(y+1)}{(y+1)(y-\underline{\quad})(y-\underline{\quad})}$

$= \frac{\cancel{(y+1)}\cancel{(y-1)}(y+1)}{\cancel{(y+1)}\cancel{(y-1)}(y-1)}$

$= \frac{y+\underline{\quad}}{\underline{\quad}-1}$

Answers

9. $\frac{x-3}{x+2}$ 10. $\frac{(x-3)(x-2)}{x+2}$ 11. $\frac{y+1}{y-1}$

Guided Solution:

11. $y, 2y, 1, 1, 1, 1, 1, y$

14.2 Exercise Set

For Extra Help

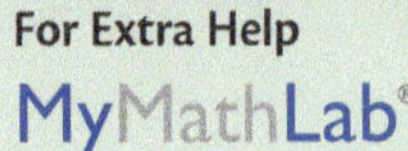

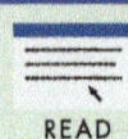

PRACTICE WATCH READ REVIEW

Reading Check

Choose from the choices on the right an equivalent expression.

RC1. The reciprocal of $\frac{2}{x-2}$

RC2. The reciprocal of $x-2$

RC3. $\frac{2}{x} \cdot \frac{x}{2}$

RC4. $\frac{x}{2} \cdot \frac{2}{x+1}$

RC5. $\frac{1}{x} \div \frac{1}{2}$

RC6. $\frac{1}{x} \div \frac{2}{x}$

a) 1

b) $\frac{2}{x}$

c) $\frac{1}{x} \cdot \frac{x}{2}$

d) $\frac{x-2}{2}$

e) $\frac{1}{x-2}$

f) $\frac{x}{2} \div \frac{x+1}{2}$

a Find the reciprocal.

1. $\frac{4}{x}$

2. $\frac{a+3}{a-1}$

3. $x^2 - y^2$

4. $x^2 - 5x + 7$

5. $\frac{1}{a+b}$

6. $\frac{x^2}{x^2-3}$

7. $\frac{x^2+2x-5}{x^2-4x+7}$

8. $\frac{(a-b)(a+b)}{(a+4)(a-5)}$

b Divide and simplify.

9. $\frac{2}{5} \div \frac{4}{3}$

10. $\frac{3}{10} \div \frac{3}{2}$

11. $\frac{2}{x} \div \frac{8}{x}$

12. $\frac{t}{3} \div \frac{t}{15}$

13. $\frac{a}{b^2} \div \frac{a^2}{b^3}$

14. $\frac{x^2}{y} \div \frac{x^3}{y^3}$

15. $\frac{a+2}{a-3} \div \frac{a-1}{a+3}$

16. $\frac{x-8}{x+9} \div \frac{x+2}{x-1}$

17. $\frac{x^2-1}{x} \div \frac{x+1}{x-1}$

18. $\frac{4y-8}{y+2} \div \frac{y-2}{y^2-4}$

19. $\frac{x+1}{6} \div \frac{x+1}{3}$

20. $\frac{a}{a-b} \div \frac{b}{a-b}$

21. $\frac{5x-5}{16} \div \frac{x-1}{6}$

22. $\frac{4y-12}{12} \div \frac{y-3}{3}$

23. $\frac{-6+3x}{5} \div \frac{4x-8}{25}$

24. $\frac{-12+4x}{4} \div \frac{-6+2x}{6}$

25. $\frac{a+2}{a-1} \div \frac{3a+6}{a-5}$

26. $\frac{t-3}{t+2} \div \frac{4t-12}{t+1}$

27. $\frac{x^2-4}{x} \div \frac{x-2}{x+2}$

28. $\frac{x+y}{x-y} \div \frac{x^2+y}{x^2-y^2}$

29. $\frac{x^2-9}{4x+12} \div \frac{x-3}{6}$

30. $\frac{a-b}{2a} \div \frac{a^2-b^2}{8a^3}$

31. $\frac{c^2+3c}{c^2+2c-3} \div \frac{c}{c+1}$

32. $\frac{y+5}{2y} \div \frac{y^2-25}{4y^2}$

33. $\dfrac{2y^2 - 7y + 3}{2y^2 + 3y - 2} \div \dfrac{6y^2 - 5y + 1}{3y^2 + 5y - 2}$

34. $\dfrac{x^2 + x - 20}{x^2 - 7x + 12} \div \dfrac{x^2 + 10x + 25}{x^2 - 6x + 9}$

35. $\dfrac{x^2 - 1}{4x + 4} \div \dfrac{2x^2 - 4x + 2}{8x + 8}$

36. $\dfrac{5t^2 + 5t - 30}{10t + 30} \div \dfrac{2t^2 - 8}{6t^2 + 36t + 54}$

Skill Maintenance

Solve.

37. Camila is taking an astronomy course. In order to receive an A, she must average at least 90 after four exams. Camila scored 96, 98, and 89 on the first three tests. Determine (in terms of an inequality) what scores on the last test will earn her an A. [10.8b]

38. *Triangle Dimensions.* The base of a triangle is 4 in. less than twice the height. The area is 35 in^2. Find the height and the base. [13.8a]

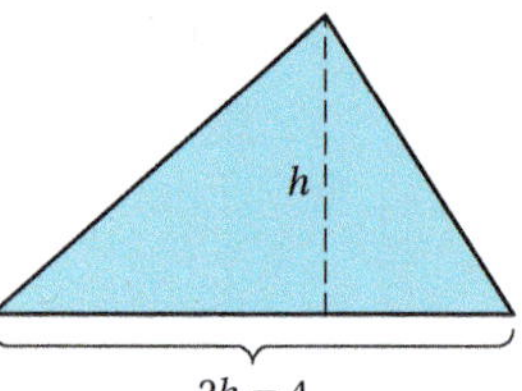

Subtract. [12.4c]

39. $(8x^3 - 3x^2 + 7) - (8x^2 + 3x - 5)$

40. $(3p^2 - 6pq + 7q^2) - (5p^2 - 10pq + 11q^2)$

Simplify. [12.2a, b]

41. $(2x^{-3}y^4)^2$

42. $(5x^6y^{-4})^3$

43. $\left(\dfrac{2x^3}{y^5}\right)^2$

44. $\left(\dfrac{a^{-3}}{b^4}\right)^5$

Synthesis

Simplify.

45. $\dfrac{3a^2 - 5ab - 12b^2}{3ab + 4b^2} \div (3b^2 - ab)$

46. $\dfrac{3x + 3y + 3}{9x} \div \dfrac{x^2 + 2xy + y^2 - 1}{x^4 + x^2}$

47. $\dfrac{a^2b^2 + 3ab^2 + 2b^2}{a^2b^4 + 4b^4} \div (5a^2 + 10a)$

48. The volume of this rectangular solid is $x - 3$. What is its height?

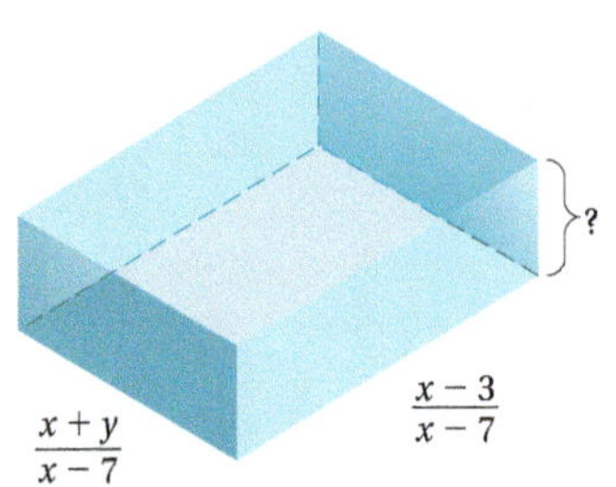

Least Common Multiples and Denominators

14.3

OBJECTIVES

a Find the LCM of several numbers by factoring.

b Add fractions, first finding the LCD.

c Find the LCM of algebraic expressions by factoring.

SKILL TO REVIEW

Objective 3.2a, 3.2c: Find all the factors of numbers and find prime factorizations of numbers.

Find the prime factorization of each number.

1. 750 **2.** 364

a LEAST COMMON MULTIPLES

To add when denominators are different, we first find a common denominator. For example, to add $\frac{5}{12}$ and $\frac{7}{30}$, we first look for the **least common multiple, LCM**, of 12 and 30. That number becomes the **least common denominator, LCD**. To find the LCM of 12 and 30, we factor:

$$12 = 2 \cdot 2 \cdot 3;$$
$$30 = 2 \cdot 3 \cdot 5.$$

The LCM is the number that has 2 as a factor twice, 3 as a factor once, and 5 as a factor once:

$$\text{LCM} = 2 \cdot 2 \cdot 3 \cdot 5 = 60.$$

12 is a factor of the LCM.

30 is a factor of the LCM.

FINDING LCMS

To find the LCM, use each factor the greatest number of times that it appears in any one factorization.

EXAMPLE 1 Find the LCM of 24 and 36.

$$\left.\begin{aligned} 24 &= 2 \cdot 2 \cdot 2 \cdot 3 \\ 36 &= 2 \cdot 2 \cdot 3 \cdot 3 \end{aligned}\right\} \quad \text{LCM} = 2 \cdot 2 \cdot 2 \cdot 3 \cdot 3, \text{ or } 72$$

Do Margin Exercises 1–4.

Find the LCM by factoring.

1. 16, 18

$16 = 2 \cdot 2 \cdot 2 \cdot$ ____

$18 = 2 \cdot$ ____ $\cdot$ ____

$\text{LCM} = 2 \cdot 2 \cdot 2 \cdot$ ____ $\cdot 3 \cdot 3$,

or ____

2. 6, 12

3. 2, 5

4. 24, 30, 20

b ADDING USING THE LCD

Let's finish adding $\frac{5}{12}$ and $\frac{7}{30}$:

$$\frac{5}{12} + \frac{7}{30} = \frac{5}{2 \cdot 2 \cdot 3} + \frac{7}{2 \cdot 3 \cdot 5}.$$

The least common denominator, LCD, is $2 \cdot 2 \cdot 3 \cdot 5$. To get the LCD in the first denominator, we need a 5. To get the LCD in the second denominator, we need another 2. We get these numbers by multiplying by forms of 1:

$$\frac{5}{12} + \frac{7}{30} = \frac{5}{2 \cdot 2 \cdot 3} \cdot \frac{5}{5} + \frac{7}{2 \cdot 3 \cdot 5} \cdot \frac{2}{2}$$ Multiplying by 1

$$= \frac{25}{2 \cdot 2 \cdot 3 \cdot 5} + \frac{14}{2 \cdot 3 \cdot 5 \cdot 2}$$ Each denominator is now the LCD.

$$= \frac{39}{2 \cdot 2 \cdot 3 \cdot 5}$$ Adding the numerators and keeping the LCD

$$= \frac{\not{3} \cdot 13}{2 \cdot 2 \cdot \not{3} \cdot 5}$$ Factoring the numerator and removing a factor of 1: $\frac{3}{3} = 1$

$$= \frac{13}{20}.$$ Simplifying

Answers

Skill to Review:
1. $2 \cdot 3 \cdot 5 \cdot 5 \cdot 5$, or $2 \cdot 3 \cdot 5^3$
2. $2 \cdot 2 \cdot 7 \cdot 13$, or $2^2 \cdot 7 \cdot 13$

Margin Exercises:
1. 144 2. 12 3. 10 4. 120

Guided Solution:
1. 2, 3, 3, 2, 144

Add, first finding the LCD. Simplify if possible.

5. $\frac{3}{16} + \frac{1}{18}$

$= \frac{3}{2 \cdot 2 \cdot 2 \cdot \square} + \frac{1}{2 \cdot \square \cdot 3}$

$= \frac{3}{2 \cdot 2 \cdot 2 \cdot 2} \cdot \frac{3 \cdot 3}{3 \cdot \square}$

$+ \frac{1}{2 \cdot 3 \cdot 3} \cdot \frac{2 \cdot \square \cdot 2}{2 \cdot 2 \cdot 2}$

$= \frac{27 + \square}{2 \cdot 2 \cdot 2 \cdot 2 \cdot 3 \cdot 3}$

$= \frac{35}{\square}$

6. $\frac{1}{6} + \frac{1}{12}$

7. $\frac{1}{2} + \frac{3}{5}$

8. $\frac{1}{24} + \frac{1}{30} + \frac{3}{20}$

Find the LCM.

9. $12xy^2,\ 15x^3y$

10. $y^2 + 5y + 4,\ y^2 + 2y + 1$

11. $t^2 + 16,\ t - 2,\ 7$

12. $x^2 + 2x + 1,\ 3x^2 - 3x,\ x^2 - 1$

Answers

5. $\frac{35}{144}$ 6. $\frac{1}{4}$ 7. $\frac{11}{10}$ 8. $\frac{9}{40}$ 9. $60x^3y^2$
10. $(y + 1)^2(y + 4)$ 11. $7(t^2 + 16)(t - 2)$
12. $3x(x + 1)^2(x - 1)$

Guided Solution:
5. 2, 3, 3, 2, 8, 144

EXAMPLE 2 Add: $\frac{5}{12} + \frac{11}{18}$.

$$\left.\begin{aligned} 12 &= 2 \cdot 2 \cdot 3 \\ 18 &= 2 \cdot 3 \cdot 3 \end{aligned}\right\} \quad \text{LCD} = 2 \cdot 2 \cdot 3 \cdot 3, \text{ or } 36$$

$$\frac{5}{12} + \frac{11}{18} = \frac{5}{2 \cdot 2 \cdot 3} \cdot \frac{3}{3} + \frac{11}{2 \cdot 3 \cdot 3} \cdot \frac{2}{2} = \frac{15 + 22}{2 \cdot 2 \cdot 3 \cdot 3} = \frac{37}{36}$$

Do Exercises 5–8.

C LCMs OF ALGEBRAIC EXPRESSIONS

To find the LCM of two or more algebraic expressions, we factor them. Then we use each factor the greatest number of times that it occurs in any one expression. In Section 14.4, each LCM will become an LCD used to add rational expressions.

EXAMPLE 3 Find the LCM of $12x$, $16y$, and $8xyz$.

$$\left.\begin{aligned} 12x &= 2 \cdot 2 \cdot 3 \cdot x \\ 16y &= 2 \cdot 2 \cdot 2 \cdot 2 \cdot y \\ 8xyz &= 2 \cdot 2 \cdot 2 \cdot x \cdot y \cdot z \end{aligned}\right\} \quad \begin{aligned} \text{LCM} &= 2 \cdot 2 \cdot 2 \cdot 2 \cdot 3 \cdot x \cdot y \cdot z \\ &= 48xyz \end{aligned}$$

EXAMPLE 4 Find the LCM of $x^2 + 5x - 6$ and $x^2 - 1$.

$$\left.\begin{aligned} x^2 + 5x - 6 &= (x + 6)(x - 1) \\ x^2 - 1 &= (x + 1)(x - 1) \end{aligned}\right\} \quad \text{LCM} = (x + 6)(x - 1)(x + 1)$$

EXAMPLE 5 Find the LCM of $x^2 + 4$, $x + 1$, and 5.

These expressions do not share a common factor other than 1, so the LCM is their product:

$$5(x^2 + 4)(x + 1).$$

EXAMPLE 6 Find the LCM of $x^2 - 25$ and $2x - 10$.

$$\left.\begin{aligned} x^2 - 25 &= (x + 5)(x - 5) \\ 2x - 10 &= 2(x - 5) \end{aligned}\right\} \quad \text{LCM} = 2(x + 5)(x - 5)$$

EXAMPLE 7 Find the LCM of $x^2 - 4y^2$, $x^2 - 4xy + 4y^2$, and $x - 2y$.

$$\left.\begin{aligned} x^2 - 4y^2 &= (x - 2y)(x + 2y) \\ x^2 - 4xy + 4y^2 &= (x - 2y)(x - 2y) \\ x - 2y &= x - 2y \end{aligned}\right\} \quad \begin{aligned} \text{LCM} &= (x + 2y)(x - 2y)(x - 2y) \\ &= (x + 2y)(x - 2y)^2 \end{aligned}$$

Do Exercises 9–12.

14.3 Exercise Set

For Extra Help MyMathLab® MathXL® PRACTICE WATCH READ REVIEW

Reading Check

Complete each statement with the best choice from the column on the right. Some choices will not be used.

To add $\frac{5}{16} + \frac{7}{24}$, we begin by finding a **RC1.** ____________ denominator. We first look for the least common **RC2.** ____________ of 16 and 24. That number becomes the least common **RC3.** ____________ of the two fractions. We factor 16 and 24: $16 = 2 \cdot 2 \cdot 2 \cdot 2$ and $24 = 2 \cdot 2 \cdot 2 \cdot 3$. Then to find the LCM of 16 and 24, we use each factor the **RC4.** ____________ number of times that it appears in any one factorization. The LCM is **RC5.** ____________.

$2 \cdot 2 \cdot 2 \cdot 3$
$2 \cdot 2 \cdot 2 \cdot 2 \cdot 3$
common
multiple
numerator
denominator
greatest
least

a Find the LCM.

1. 12, 27
2. 10, 15
3. 8, 9
4. 12, 18
5. 6, 9, 21
6. 8, 36, 40
7. 24, 36, 40
8. 4, 5, 20
9. 10, 100, 500
10. 28, 42, 60

b Add, first finding the LCD. Simplify, if possible.

11. $\frac{7}{24} + \frac{11}{18}$
12. $\frac{7}{60} + \frac{2}{25}$
13. $\frac{1}{6} + \frac{3}{40}$
14. $\frac{5}{24} + \frac{3}{20}$
15. $\frac{1}{20} + \frac{1}{30} + \frac{2}{45}$
16. $\frac{2}{15} + \frac{5}{9} + \frac{3}{20}$

c Find the LCM.

17. $6x^2$, $12x^3$
18. $2a^2b$, $8ab^3$
19. $2x^2$, $6xy$, $18y^2$
20. p^3q, p^2q, pq^2
21. $2(y - 3)$, $6(y - 3)$
22. $5(m + 2)$, $15(m + 2)$
23. t, $t + 2$, $t - 2$
24. y, $y - 5$, $y + 5$
25. $x^2 - 4$, $x^2 + 5x + 6$
26. $x^2 - 4$, $x^2 - x - 2$
27. $t^3 + 4t^2 + 4t$, $t^2 - 4t$
28. $m^4 - m^2$, $m^3 - m^2$

29. $a + 1, (a - 1)^2, a^2 - 1$

30. $a^2 - 2ab + b^2, a^2 - b^2, 3a + 3b$

31. $m^2 - 5m + 6, m^2 - 4m + 4$

32. $2x^2 + 5x + 2, 2x^2 - x - 1$

33. $2 + 3x, 4 - 9x^2, 2 - 3x$

34. $9 - 4x^2, 3 + 2x, 3 - 2x$

35. $10v^2 + 30v, 5v^2 + 35v + 60$

36. $12a^2 + 24a, 4a^2 + 20a + 24$

37. $9x^3 - 9x^2 - 18x, 6x^5 - 24x^4 + 24x^3$

38. $x^5 - 4x^3, x^3 + 4x^2 + 4x$

39. $x^5 + 4x^4 + 4x^3, 3x^2 - 12, 2x + 4$

40. $x^5 + 2x^4 + x^3, 2x^3 - 2x, 5x - 5$

41. $24w^4, w^2, 10w^3, w^6$

42. $t, 6t^4, t^2, 15t^{15}, 2t^3$

Skill Maintenance

Complete the tables below, finding the LCM, the GCF, and the product of each pair of expressions. [12.5a], [13.1a], [14.3a]

	Expressions	LCM	GCF	Product
Example	$12x^3, 8x^2$	$24x^3$	$4x^2$	$96x^5$
43.	$40x^3, 24x^4$			
45.	$16x^5, 48x^6$			
47.	$20x^2, 10x$			

	Expressions	LCM	GCF	Product
44.	$12ab, 16ab^3$			
46.	$10x^2, 24x^3$			
48.	a^5, a^{15}			

Synthesis

49. ***Running.*** Gabriela and Madison leave the starting point of a fitness loop at the same time. Gabriela jogs a lap in 6 min and Madison jogs one in 8 min. Assuming they continue to run at the same pace, when will they next meet at the starting place?

Adding Rational Expressions

14.4

a ADDING RATIONAL EXPRESSIONS

We add rational expressions as we do rational numbers.

ADDING RATIONAL EXPRESSIONS WITH LIKE DENOMINATORS

To add when the denominators are the same, add the numerators and keep the same denominator. Then simplify, if possible.

EXAMPLES Add.

1. $\frac{x}{x+1} + \frac{2}{x+1} = \frac{x+2}{x+1}$

2. $$\frac{2x^2+3x-7}{2x+1} + \frac{x^2+x-8}{2x+1} = \frac{(2x^2+3x-7)+(x^2+x-8)}{2x+1}$$
$$= \frac{3x^2+4x-15}{2x+1}$$
$$= \frac{(x+3)(3x-5)}{2x+1}$$ Factoring the numerator to determine if we can simplify

3. $$\frac{x-5}{x^2-9} + \frac{2}{x^2-9} = \frac{(x-5)+2}{x^2-9} = \frac{x-3}{x^2-9}$$
$$= \frac{x-3}{(x-3)(x+3)}$$ Factoring
$$= \frac{1(x-3)}{(x-3)(x+3)}$$ Removing a factor of 1: $\frac{x-3}{x-3} = 1$
$$= \frac{1}{x+3}$$ Simplifying

Do Margin Exercises 1–3.

When denominators are different, we find the least common denominator, LCD. The procedure we use follows.

ADDING RATIONAL EXPRESSIONS WITH DIFFERENT DENOMINATORS

To add rational expressions with different denominators:

1. Find the LCM of the denominators. This is the least common denominator (LCD).
2. For each rational expression, find an equivalent expression with the LCD. Multiply by 1 using an expression for 1 made up of factors of the LCD that are missing from the original denominator.
3. Add the numerators. Write the sum over the LCD.
4. Simplify, if possible.

OBJECTIVE

a Add rational expressions.

SKILL TO REVIEW

Objective 4.2b: Add using fraction notation when denominators are different.

Add and simplify.

1. $\frac{7}{10} + \frac{11}{15}$

2. $\frac{11}{42} + \frac{5}{14}$

Add.

1. $\frac{5}{9} + \frac{2}{9}$

2. $\frac{3}{x-2} + \frac{x}{x-2}$

3. $\frac{4x+5}{x-1} + \frac{2x-1}{x-1}$

Answers

Skill to Review:
1. $\frac{43}{30}$ 2. $\frac{13}{21}$

Margin Exercises:
1. $\frac{7}{9}$ 2. $\frac{3+x}{x-2}$ 3. $\frac{6x+4}{x-1}$

EXAMPLE 4 Add: $\frac{5x^2}{8} + \frac{7x}{12}$.

First, we find the LCD:

$$\left.\begin{aligned} 8 &= 2 \cdot 2 \cdot 2 \\ 12 &= 2 \cdot 2 \cdot 3 \end{aligned}\right\} \quad \text{LCD} = 2 \cdot 2 \cdot 2 \cdot 3, \text{ or } 24.$$

Compare the factorization $8 = 2 \cdot 2 \cdot 2$ with the factorization of the LCD, $24 = 2 \cdot 2 \cdot 2 \cdot 3$. The factor of 24 that is missing from 8 is 3. Compare $12 = 2 \cdot 2 \cdot 3$ and $24 = 2 \cdot 2 \cdot 2 \cdot 3$. The factor of 24 that is missing from 12 is 2.

We multiply each term by a symbol for 1 to get the LCD in each expression, and then add and, if possible, simplify:

$$\frac{5x^2}{8} + \frac{7x}{12} = \frac{5x^2}{2 \cdot 2 \cdot 2} + \frac{7x}{2 \cdot 2 \cdot 3}$$

$$= \frac{5x^2}{2 \cdot 2 \cdot 2} \cdot \frac{3}{3} + \frac{7x}{2 \cdot 2 \cdot 3} \cdot \frac{2}{2}$$ Multiplying by 1 to get the same denominators

$$= \frac{15x^2}{24} + \frac{14x}{24} = \frac{15x^2 + 14x}{24} = \frac{x(15x + 14)}{24}.$$

Add.

4. $\frac{3x}{16} + \frac{5x^2}{24}$

5. $\frac{3}{16x} + \frac{5}{24x^2}$

GS

$16x = 2 \cdot 2 \cdot 2 \cdot \underline{\quad} \cdot x$

$24x^2 = 2 \cdot 2 \cdot 2 \cdot 3 \cdot \underline{\quad} \cdot x$

$\text{LCD} = 2 \cdot 2 \cdot 2 \cdot \underline{\quad} \cdot 3 \cdot x \cdot x$, or $48x^2$

$\frac{3}{16x} \cdot \frac{3x}{\underline{\quad}} + \frac{5}{24x^2} \cdot \frac{\underline{\quad}}{2}$

$= \frac{\underline{\quad}}{48x^2} + \frac{10}{\underline{\quad}}$

$= \frac{9x + \underline{\quad}}{48x^2}$

EXAMPLE 5 Add: $\frac{3}{8x} + \frac{5}{12x^2}$.

First, we find the LCD:

$$\left.\begin{aligned} 8x &= 2 \cdot 2 \cdot 2 \cdot x \\ 12x^2 &= 2 \cdot 2 \cdot 3 \cdot x \cdot x \end{aligned}\right\} \quad \text{LCD} = 2 \cdot 2 \cdot 2 \cdot 3 \cdot x \cdot x, \text{ or } 24x^2.$$

The factors of the LCD missing from $8x$ are 3 and x. The factor of the LCD missing from $12x^2$ is 2. We multiply each term by 1 to get the LCD in each expression, and then add and, if possible, simplify:

$$\frac{3}{8x} + \frac{5}{12x^2} = \frac{3}{8x} \cdot \frac{3 \cdot x}{3 \cdot x} + \frac{5}{12x^2} \cdot \frac{2}{2}$$

$$= \frac{9x}{24x^2} + \frac{10}{24x^2} = \frac{9x + 10}{24x^2}.$$

◀ **Do Exercises 4 and 5.**

EXAMPLE 6 Add: $\frac{2a}{a^2 - 1} + \frac{1}{a^2 + a}$.

First, we find the LCD:

$$\left.\begin{aligned} a^2 - 1 &= (a - 1)(a + 1) \\ a^2 + a &= a(a + 1) \end{aligned}\right\} \quad \text{LCD} = a(a - 1)(a + 1).$$

We multiply each term by 1 to get the LCD in each expression, and then add and, if possible, simplify:

$$\frac{2a}{(a - 1)(a + 1)} \cdot \frac{a}{a} + \frac{1}{a(a + 1)} \cdot \frac{a - 1}{a - 1}$$

$$= \frac{2a^2}{a(a - 1)(a + 1)} + \frac{a - 1}{a(a - 1)(a + 1)}$$

$$= \frac{2a^2 + a - 1}{a(a - 1)(a + 1)}$$

$$= \frac{(a + 1)(2a - 1)}{a(a - 1)(a + 1)}.$$ Factoring the numerator in order to simplify

Answers

4. $\frac{x(10x + 9)}{48}$ 5. $\frac{9x + 10}{48x^2}$

Guided Solution:

5. $2, x, 2, 3x, 2, 9x, 48x^2, 10$

Then

$$= \frac{\cancel{(a+1)}(2a-1)}{a(a-1)\cancel{(a+1)}} \quad \text{Removing a factor of 1: } \frac{a+1}{a+1} = 1$$

$$= \frac{2a-1}{a(a-1)}.$$

Do Exercise 6. ▶

6. Add:

$$\frac{3}{x^3 - x} + \frac{4}{x^2 + 2x + 1}.$$

EXAMPLE 7 Add: $\dfrac{x+4}{x-2} + \dfrac{x-7}{x+5}$.

First, we find the LCD. It is just the product of the denominators:

LCD $= (x-2)(x+5)$.

We multiply by 1 to get the LCD in each expression, and then add and simplify:

$$\frac{x+4}{x-2} \cdot \frac{x+5}{x+5} + \frac{x-7}{x+5} \cdot \frac{x-2}{x-2}$$

$$= \frac{(x+4)(x+5)}{(x-2)(x+5)} + \frac{(x-7)(x-2)}{(x-2)(x+5)}$$

$$= \frac{x^2 + 9x + 20}{(x-2)(x+5)} + \frac{x^2 - 9x + 14}{(x-2)(x+5)}$$

$$= \frac{x^2 + 9x + 20 + x^2 - 9x + 14}{(x-2)(x+5)}$$

$$= \frac{2x^2 + 34}{(x-2)(x+5)} = \frac{2(x^2 + 17)}{(x-2)(x+5)}.$$

Do Exercise 7. ▶

7. Add:

$$\frac{x-2}{x+3} + \frac{x+7}{x+8}.$$

EXAMPLE 8 Add: $\dfrac{x}{x^2 + 11x + 30} + \dfrac{-5}{x^2 + 9x + 20}$.

$$\frac{x}{x^2 + 11x + 30} + \frac{-5}{x^2 + 9x + 20}$$

$$= \frac{x}{(x+5)(x+6)} + \frac{-5}{(x+5)(x+4)} \quad \text{Factoring the denominators in order to find the LCD. The LCD is } (x+4)(x+5)(x+6).$$

$$= \frac{x}{(x+5)(x+6)} \cdot \frac{x+4}{x+4} + \frac{-5}{(x+5)(x+4)} \cdot \frac{x+6}{x+6} \quad \text{Multiplying by 1}$$

$$= \frac{x(x+4) + (-5)(x+6)}{(x+4)(x+5)(x+6)} = \frac{x^2 + 4x - 5x - 30}{(x+4)(x+5)(x+6)}$$

$$= \frac{x^2 - x - 30}{(x+4)(x+5)(x+6)}$$

$$= \frac{(x-6)\cancel{(x+5)}}{(x+4)\cancel{(x+5)}(x+6)}$$

$$= \frac{x-6}{(x+4)(x+6)}$$

Always simplify at the end if possible: $\dfrac{x+5}{x+5} = 1$.

Do Exercise 8. ▶

8. Add:

$$\frac{5}{x^2 + 17x + 16} + \frac{3}{x^2 + 9x + 8}.$$

Denominators That Are Opposites

When one denominator is the opposite of the other, we can first multiply either expression by 1 using $-1/-1$.

Answers

6. $\dfrac{4x^2 - x + 3}{x(x-1)(x+1)^2}$ 7. $\dfrac{2x^2 + 16x + 5}{(x+3)(x+8)}$

8. $\dfrac{8(x+11)}{(x+16)(x+1)(x+8)}$

Add.

9. $\frac{x}{4} + \frac{5}{-4}$

10. $\frac{2x + 1}{x - 3} + \frac{x + 2}{3 - x}$

$= \frac{2x + 1}{x - 3} + \frac{x + 2}{3 - x} \cdot \frac{-1}{\square}$

$= \frac{2x + 1}{x - 3} + \frac{\square - 2}{x - 3}$

$= \frac{(2x + 1) + (-x - 2)}{x - \square}$

$= \frac{\square - 1}{x - 3}$

GS

11. Add:

$\frac{x + 3}{x^2 - 16} + \frac{5}{12 - 3x}$.

Answers

9. $\frac{x - 5}{4}$ 10. $\frac{x - 1}{x - 3}$ 11. $\frac{-2x - 11}{3(x + 4)(x - 4)}$

Guided Solution:

10. $-1, -x, 3, x$

EXAMPLES

9. $\frac{x}{2} + \frac{3}{-2} = \frac{x}{2} + \frac{3}{-2} \cdot \frac{-1}{-1}$ Multiplying by 1 using $\frac{-1}{-1}$

$= \frac{x}{2} + \frac{-3}{2}$ The denominators are now the same.

$= \frac{x + (-3)}{2} = \frac{x - 3}{2}$

10. $\frac{3x + 4}{x - 2} + \frac{x - 7}{2 - x} = \frac{3x + 4}{x - 2} + \frac{x - 7}{2 - x} \cdot \frac{-1}{-1}$

We could have chosen to multiply this expression by $-1/-1$. We multiply only one expression, *not* both.

$= \frac{3x + 4}{x - 2} + \frac{-x + 7}{x - 2}$ *Note:* $(2 - x)(-1) = -2 + x = x - 2$.

$= \frac{(3x + 4) + (-x + 7)}{x - 2} = \frac{2x + 11}{x - 2}$

◀ Do Exercises 9 and 10.

Factors That Are Opposites

Suppose that when we factor to find the LCD, we find factors that are opposites. The easiest way to handle this is to first go back and multiply by $-1/-1$ appropriately to change factors so that they are not opposites.

EXAMPLE 11 Add: $\frac{x}{x^2 - 25} + \frac{3}{10 - 2x}$.

First, we factor to find the LCD:

$x^2 - 25 = (x - 5)(x + 5);$

$10 - 2x = 2(5 - x).$

We note that $x - 5$ is one factor of $x^2 - 25$ and $5 - x$ is one factor of $10 - 2x$. If the denominator of the second expression were $2x - 10$, then $x - 5$ would be a factor of both denominators. To rewrite the second expression with a denominator of $2x - 10$, we multiply by 1 using $-1/-1$, and then continue as before:

$$\frac{x}{x^2 - 25} + \frac{3}{10 - 2x} = \frac{x}{(x - 5)(x + 5)} + \frac{3}{10 - 2x} \cdot \frac{-1}{-1}$$

$$= \frac{x}{(x - 5)(x + 5)} + \frac{-3}{2x - 10}$$

$$= \frac{x}{(x - 5)(x + 5)} + \frac{-3}{2(x - 5)}$$ LCD $= 2(x - 5)(x + 5)$

$$= \frac{x}{(x - 5)(x + 5)} \cdot \frac{2}{2} + \frac{-3}{2(x - 5)} \cdot \frac{x + 5}{x + 5}$$

$$= \frac{2x}{2(x - 5)(x + 5)} + \frac{-3(x + 5)}{2(x - 5)(x + 5)}$$

$$= \frac{2x - 3(x + 5)}{2(x - 5)(x + 5)} = \frac{2x - 3x - 15}{2(x - 5)(x + 5)}$$

$$= \frac{-x - 15}{2(x - 5)(x + 5)}.$$ Collecting like terms

◀ Do Exercise 11.

14.4 Exercise Set

For Extra Help MyMathLab® MathXL® PRACTICE WATCH READ REVIEW

Reading Check

From the choices on the right, select the names for 1 that result in the same denominator in each addition. Some choices may be used more than once. Some choices may not be used. Do not complete the addition.

RC1. $\frac{2x^2}{15} + \frac{3x}{10} = \frac{2x^2}{3 \cdot 5} \cdot \left(\quad \right) + \frac{3x}{2 \cdot 5} \cdot \left(\quad \right)$

RC2. $\frac{3}{2} + \frac{2x}{x+3} = \frac{3}{2} \cdot \left(\qquad \right) + \frac{2x}{x+3} \cdot \left(\quad \right)$

RC3. $\frac{3x}{x+2} + \frac{2}{x-3} = \frac{3x}{x+2} \cdot \left(\qquad \right) + \frac{2}{x-3} \cdot \left(\qquad \right)$

RC4. $\frac{2}{15x} + \frac{3}{10x^2} = \frac{2}{3 \cdot 5 \cdot x} \cdot \left(\quad \right) + \frac{3}{2 \cdot 5 \cdot x \cdot x} \cdot \left(\quad \right)$

$\frac{x+3}{x+3}$ $\frac{2}{2}$

$\frac{3x}{3x}$ $\frac{x-2}{x-2}$

$\frac{x-3}{x-3}$ $\frac{3}{3}$

$\frac{2x}{2x}$ $\frac{x+2}{x+2}$

a Add. Simplify, if possible.

1. $\frac{5}{8} + \frac{3}{8}$

2. $\frac{3}{16} + \frac{5}{16}$

3. $\frac{1}{3+x} + \frac{5}{3+x}$

4. $\frac{x^2+7x}{x^2-5x} + \frac{x^2-4x}{x^2-5x}$

5. $\frac{4x+6}{2x-1} + \frac{5-8x}{-1+2x}$

6. $\frac{4}{x+y} + \frac{9}{y+x}$

7. $\frac{2}{x} + \frac{5}{x^2}$

8. $\frac{3}{y^2} + \frac{6}{y}$

9. $\frac{5}{6r} + \frac{7}{8r}$

10. $\frac{13}{18x} + \frac{7}{24x}$

11. $\frac{4}{xy^2} + \frac{6}{x^2y}$

12. $\frac{8}{ab^3} + \frac{3}{a^2b}$

13. $\frac{2}{9t^3} + \frac{1}{6t^2}$

14. $\frac{5}{c^2d^3} + \frac{-4}{7cd^2}$

15. $\frac{x+y}{xy^2} + \frac{3x+y}{x^2y}$

16. $\dfrac{2c - d}{c^2d} + \dfrac{c + d}{cd^2}$

17. $\dfrac{3}{x - 2} + \dfrac{3}{x + 2}$

18. $\dfrac{2}{y + 1} + \dfrac{2}{y - 1}$

19. $\dfrac{3}{x + 1} + \dfrac{2}{3x}$

20. $\dfrac{4}{5y} + \dfrac{7}{y - 2}$

21. $\dfrac{2x}{x^2 - 16} + \dfrac{x}{x - 4}$

22. $\dfrac{4x}{x^2 - 25} + \dfrac{x}{x + 5}$

23. $\dfrac{5}{z + 4} + \dfrac{3}{3z + 12}$

24. $\dfrac{t}{t - 3} + \dfrac{5}{4t - 12}$

25. $\dfrac{3}{x - 1} + \dfrac{2}{(x - 1)^2}$

26. $\dfrac{8}{(y + 3)^2} + \dfrac{5}{y + 3}$

27. $\dfrac{4a}{5a - 10} + \dfrac{3a}{10a - 20}$

28. $\dfrac{9x}{6x - 30} + \dfrac{3x}{4x - 20}$

29. $\dfrac{x + 4}{x} + \dfrac{x}{x + 4}$

30. $\dfrac{a}{a - 3} + \dfrac{a - 3}{a}$

31. $\dfrac{4}{a^2 - a - 2} + \dfrac{3}{a^2 + 4a + 3}$

32. $\dfrac{a}{a^2 - 2a + 1} + \dfrac{1}{a^2 - 5a + 4}$

33. $\dfrac{x + 3}{x - 5} + \dfrac{x - 5}{x + 3}$

34. $\dfrac{3x}{2y - 3} + \dfrac{2x}{3y - 2}$

35. $\dfrac{a}{a^2 - 1} + \dfrac{2a}{a^2 - a}$

36. $\dfrac{3x + 2}{3x + 6} + \dfrac{x - 2}{x^2 - 4}$

37. $\frac{7}{8} + \frac{5}{-8}$

38. $\frac{5}{-3} + \frac{11}{3}$

39. $\frac{3}{t} + \frac{4}{-t}$

40. $\frac{5}{-a} + \frac{8}{a}$

41. $\frac{2x + 7}{x - 6} + \frac{3x}{6 - x}$

42. $\frac{2x - 7}{5x - 8} + \frac{6 + 10x}{8 - 5x}$

43. $\frac{y^2}{y - 3} + \frac{9}{3 - y}$

44. $\frac{t^2}{t - 2} + \frac{4}{2 - t}$

45. $\frac{b - 7}{b^2 - 16} + \frac{7 - b}{16 - b^2}$

46. $\frac{a - 3}{a^2 - 25} + \frac{a - 3}{25 - a^2}$

47. $\frac{a^2}{a - b} + \frac{b^2}{b - a}$

48. $\frac{x^2}{x - 7} + \frac{49}{7 - x}$

49. $\frac{x + 3}{x - 5} + \frac{2x - 1}{5 - x} + \frac{2(3x - 1)}{x - 5}$

50. $\frac{3(x - 2)}{2x - 3} + \frac{5(2x + 1)}{2x - 3} + \frac{3(x + 1)}{3 - 2x}$

51. $\frac{2(4x + 1)}{5x - 7} + \frac{3(x - 2)}{7 - 5x} + \frac{-10x - 1}{5x - 7}$

52. $\frac{5(x - 2)}{3x - 4} + \frac{2(x - 3)}{4 - 3x} + \frac{3(5x + 1)}{4 - 3x}$

53. $\frac{x + 1}{(x + 3)(x - 3)} + \frac{4(x - 3)}{(x - 3)(x + 3)} + \frac{(x - 1)(x - 3)}{(3 - x)(x + 3)}$

54. $\frac{2(x + 5)}{(2x - 3)(x - 1)} + \frac{3x + 4}{(2x - 3)(1 - x)} + \frac{x - 5}{(3 - 2x)(x - 1)}$

55. $\dfrac{6}{x - y} + \dfrac{4x}{y^2 - x^2}$

56. $\dfrac{a - 2}{3 - a} + \dfrac{4 - a^2}{a^2 - 9}$

57. $\dfrac{4 - a}{25 - a^2} + \dfrac{a + 1}{a - 5}$

58. $\dfrac{x + 2}{x - 7} + \dfrac{3 - x}{49 - x^2}$

59. $\dfrac{2}{t^2 + t - 6} + \dfrac{3}{t^2 - 9}$

60. $\dfrac{10}{a^2 - a - 6} + \dfrac{3a}{a^2 + 4a + 4}$

Skill Maintenance

Subtract. [12.4c]

61. $(x^2 + x) - (x + 1)$

62. $(4y^3 - 5y^2 + 7y - 24) - (-9y^3 + 9y^2 - 5y + 49)$

Simplify. [12.2a, b]

63. $\left(\dfrac{x^{-4}}{y^7}\right)^3$

64. $(5x^{-2}y^{-3})^2$

Solve.

65. $3x - 7 = 5x + 9$ [10.3b]

66. $x^2 - 7x = 18$ [13.7b]

Graph.

67. $y = \dfrac{1}{2}x - 5$ [11.1d]

68. $2y + x + 10 = 0$ [11.2a]

69. $y = 3$ [11.2b]

70. $x = -5$ [11.2b]

Synthesis

Find the perimeter and the area of each figure.

71.

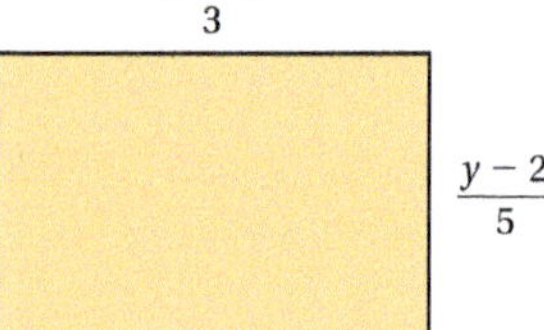

72.

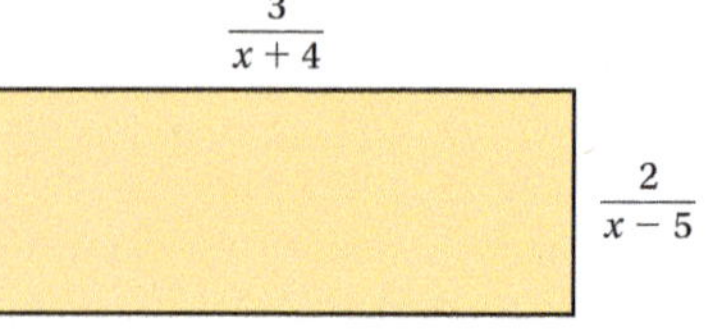

Add. Simplify, if possible.

73. $\dfrac{5}{z + 2} + \dfrac{4z}{z^2 - 4} + 2$

74. $\dfrac{-2}{y^2 - 9} + \dfrac{4y}{(y - 3)^2} + \dfrac{6}{3 - y}$

75. $\dfrac{3z^2}{z^4 - 4} + \dfrac{5z^2 - 3}{2z^4 + z^2 - 6}$

Subtracting Rational Expressions

14.5

a SUBTRACTING RATIONAL EXPRESSIONS

We subtract rational expressions as we do rational numbers.

SUBTRACTING RATIONAL EXPRESSIONS WITH LIKE DENOMINATORS

To subtract when the denominators are the same, subtract the numerators and keep the same denominator. Then simplify, if possible.

EXAMPLE 1 Subtract: $\frac{8}{x} - \frac{3}{x}$.

$$\frac{8}{x} - \frac{3}{x} = \frac{8-3}{x} = \frac{5}{x}$$

EXAMPLE 2 Subtract: $\frac{3x}{x+2} - \frac{x-2}{x+2}$.

$$\frac{3x}{x+2} - \frac{x-2}{x+2} = \frac{3x-(x-2)}{x+2}$$

$$= \frac{3x-x+2}{x+2} \quad \text{Removing parentheses}$$

$$= \frac{2x+2}{x+2} = \frac{2(x+1)}{x+2}$$

Caution!

The parentheses are important to make sure that you subtract the entire numerator.

Do Margin Exercises 1–3.

To subtract rational expressions with different denominators, we use a procedure similar to what we used for addition, except that we subtract numerators and write the difference over the LCD.

SUBTRACTING RATIONAL EXPRESSIONS WITH DIFFERENT DENOMINATORS

To subtract rational expressions with different denominators:

1. Find the LCM of the denominators. This is the least common denominator (LCD).
2. For each rational expression, find an equivalent expression with the LCD. To do so, multiply by 1 using a symbol for 1 made up of factors of the LCD that are missing from the original denominator.
3. Subtract the numerators. Write the difference over the LCD.
4. Simplify, if possible.

OBJECTIVES

a Subtract rational expressions.

b Simplify combined additions and subtractions of rational expressions.

SKILL TO REVIEW

Objective 9.8a: Find an equivalent expression for an opposite without parentheses, where an expression has several terms.

Find an expression without parentheses.

1. $-(3x - 11)$
2. $-(-x + 8)$

Subtract.

1. $\frac{7}{11} - \frac{3}{11}$
2. $\frac{7}{y} - \frac{2}{y}$
3. $\frac{2x^2+3x-7}{2x+1} - \frac{x^2+x-8}{2x+1}$

Answers

Skill to Review:
1. $-3x + 11$ 2. $x - 8$

Margin Exercises:
1. $\frac{4}{11}$ 2. $\frac{5}{y}$ 3. $\frac{(x+1)^2}{2x+1}$

EXAMPLE 3 Subtract: $\frac{x+2}{x-4} - \frac{x+1}{x+4}$.

The LCD $= (x-4)(x+4)$.

$$\frac{x+2}{x-4} \cdot \frac{x+4}{x+4} - \frac{x+1}{x+4} \cdot \frac{x-4}{x-4} \quad \text{Multiplying by 1}$$
$$= \frac{(x+2)(x+4)}{(x-4)(x+4)} - \frac{(x+1)(x-4)}{(x-4)(x+4)}$$
$$= \frac{x^2+6x+8}{(x-4)(x+4)} - \frac{x^2-3x-4}{(x-4)(x+4)}$$

Subtracting this numerator. Don't forget the parentheses.

$$= \frac{x^2+6x+8-(x^2-3x-4)}{(x-4)(x+4)}$$
$$= \frac{x^2+6x+8-x^2+3x+4}{(x-4)(x+4)} \quad \text{Removing parentheses}$$
$$= \frac{9x+12}{(x-4)(x+4)} = \frac{3(3x+4)}{(x-4)(x+4)}$$

◀ **Do Exercise 4.**

4. Subtract:

$$\frac{x-2}{3x} - \frac{2x-1}{5x}.$$

GS

$$\frac{x-2}{3x} - \frac{2x-1}{5x}$$

LCD $= 3 \cdot x \cdot 5 = 15x$

$$= \frac{x-2}{3x} \cdot \frac{5}{\square} - \frac{2x-1}{5x} \cdot \frac{\square}{3}$$
$$= \frac{5x - \square}{15x} - \frac{\square - 3}{15x}$$
$$= \frac{5x - 10 - (6x - \square)}{15x}$$
$$= \frac{5x - 10 - \square + 3}{15x}$$
$$= \frac{\square - 7}{15x}$$

EXAMPLE 4 Subtract: $\frac{x}{x^2+5x+6} - \frac{2}{x^2+3x+2}$.

$$\frac{x}{x^2+5x+6} - \frac{2}{x^2+3x+2}$$
$$= \frac{x}{(x+2)(x+3)} - \frac{2}{(x+2)(x+1)} \quad \text{LCD} = (x+1)(x+2)(x+3)$$
$$= \frac{x}{(x+2)(x+3)} \cdot \frac{x+1}{x+1} - \frac{2}{(x+2)(x+1)} \cdot \frac{x+3}{x+3}$$
$$= \frac{x^2+x}{(x+1)(x+2)(x+3)} - \frac{2x+6}{(x+1)(x+2)(x+3)}$$

Subtracting this numerator. Don't forget the parentheses.

$$= \frac{x^2+x-(2x+6)}{(x+1)(x+2)(x+3)}$$
$$= \frac{x^2+x-2x-6}{(x+1)(x+2)(x+3)} = \frac{x^2-x-6}{(x+1)(x+2)(x+3)}$$
$$= \frac{(x+2)(x-3)}{(x+1)(x+2)(x+3)}$$
$$= \frac{\cancel{(x+2)}(x-3)}{(x+1)\cancel{(x+2)}(x+3)} \quad \text{Simplifying by removing a factor of 1: } \frac{x+2}{x+2} = 1$$
$$= \frac{x-3}{(x+1)(x+3)}$$

◀ **Do Exercise 5.**

5. Subtract:

$$\frac{x}{x^2+15x+56} - \frac{6}{x^2+13x+42}.$$

Denominators That Are Opposites

When one denominator is the opposite of the other, we can first multiply one expression by $-1/-1$ to obtain a common denominator.

Answers

4. $\frac{-x-7}{15x}$ **5.** $\frac{x^2-48}{(x+7)(x+8)(x+6)}$

Guided Solution:

4. 5, 3, 10, 6*x*, 3, 6*x*, $-x$

EXAMPLE 5 Subtract: $\frac{x}{5} - \frac{3x-4}{-5}$.

$$\frac{x}{5} - \frac{3x-4}{-5} = \frac{x}{5} - \frac{3x-4}{-5} \cdot \frac{-1}{-1}$$ Multiplying by 1 using $\frac{-1}{-1}$

This is equal to 1 (not -1).

$$= \frac{x}{5} - \frac{(3x-4)(-1)}{(-5)(-1)}$$

$$= \frac{x}{5} - \frac{4-3x}{5}$$

$$= \frac{x-(4-3x)}{5}$$ Remember the parentheses!

$$= \frac{x-4+3x}{5} = \frac{4x-4}{5} = \frac{4(x-1)}{5}$$

EXAMPLE 6 Subtract: $\frac{5y}{y-5} - \frac{2y-3}{5-y}$.

$$\frac{5y}{y-5} - \frac{2y-3}{5-y} = \frac{5y}{y-5} - \frac{2y-3}{5-y} \cdot \frac{-1}{-1}$$

$$= \frac{5y}{y-5} - \frac{(2y-3)(-1)}{(5-y)(-1)}$$

$$= \frac{5y}{y-5} - \frac{3-2y}{y-5}$$

$$= \frac{5y-(3-2y)}{y-5}$$ Remember the parentheses!

$$= \frac{5y-3+2y}{y-5} = \frac{7y-3}{y-5}$$

Do Exercises 6 and 7. ▶

Subtract.

6. $\frac{x}{3} - \frac{2x-1}{-3}$

7. $\frac{3x}{x-2} - \frac{x-3}{2-x}$

Factors That Are Opposites

Suppose that when we factor to find the LCD, we find factors that are opposites. Then we multiply by $-1/-1$ appropriately to change factors so that they are not opposites.

EXAMPLE 7 Subtract: $\frac{p}{64-p^2} - \frac{5}{p-8}$.

Factoring $64 - p^2$, we get $(8-p)(8+p)$. Note that the factors $8-p$ in the first denominator and $p-8$ in the second denominator are opposites. We multiply the first expression by $-1/-1$ to avoid this situation. Then we proceed as before.

$$\frac{p}{64-p^2} - \frac{5}{p-8} = \frac{p}{64-p^2} \cdot \frac{-1}{-1} - \frac{5}{p-8}$$

$$= \frac{-p}{p^2-64} - \frac{5}{p-8}$$

$$= \frac{-p}{(p-8)(p+8)} - \frac{5}{p-8}$$ LCD $= (p-8)(p+8)$

$$= \frac{-p}{(p-8)(p+8)} - \frac{5}{p-8} \cdot \frac{p+8}{p+8}$$

Answers

6. $\frac{3x-1}{3}$ 7. $\frac{4x-3}{x-2}$

8. Subtract:

$$\frac{y}{16-y^2} - \frac{7}{y-4}.$$

$$\frac{y}{16-y^2} - \frac{7}{y-4}$$
$$= \frac{y}{16-y^2} \cdot \frac{-1}{\square} - \frac{7}{y-4}$$
$$= \frac{-y}{\square - 16} - \frac{7}{y-4}$$
$$= \frac{-y}{(y+4)(y-\square)} - \frac{7}{y-4} \cdot \frac{\square + 4}{y+4}$$
$$= \frac{-y}{(y+4)(y-4)} - \frac{7y+\square}{(y+4)(y-4)}$$
$$= \frac{-y-(7y+28)}{(y+4)(y-4)} = \frac{-y-7y-\square}{(y+4)(y-4)}$$
$$= \frac{-\square - 28}{(y+4)(y-4)} = \frac{\square(2y+7)}{(y+4)(y-4)}$$

Multiplying, we have

$$\frac{-p}{(p-8)(p+8)} - \frac{5p+40}{(p-8)(p+8)}$$

Subtracting this numerator. Don't forget the parentheses.

$$= \frac{-p-(5p+40)}{(p-8)(p+8)}$$
$$= \frac{-p-5p-40}{(p-8)(p+8)} = \frac{-6p-40}{(p-8)(p+8)} = \frac{-2(3p+20)}{(p-8)(p+8)}.$$

◀ Do Exercise 8.

b COMBINED ADDITIONS AND SUBTRACTIONS

Now let's look at some combined additions and subtractions.

EXAMPLE 8 Perform the indicated operations and simplify:

$$\frac{x+9}{x^2-4} + \frac{5-x}{4-x^2} - \frac{2+x}{x^2-4}.$$

$$\frac{x+9}{x^2-4} + \frac{5-x}{4-x^2} - \frac{2+x}{x^2-4}$$
$$= \frac{x+9}{x^2-4} + \frac{5-x}{4-x^2} \cdot \frac{-1}{-1} - \frac{2+x}{x^2-4}$$
$$= \frac{x+9}{x^2-4} + \frac{x-5}{x^2-4} - \frac{2+x}{x^2-4} = \frac{(x+9)+(x-5)-(2+x)}{x^2-4}$$
$$= \frac{x+9+x-5-2-x}{x^2-4} = \frac{x+2}{x^2-4} = \frac{\cancel{(x+2)} \cdot 1}{\cancel{(x+2)}(x-2)} = \frac{1}{x-2}$$

9. Perform the indicated operations and simplify:

$$\frac{x+2}{x^2-9} - \frac{x-7}{9-x^2} + \frac{-8-x}{x^2-9}.$$

◀ Do Exercise 9.

EXAMPLE 9 Perform the indicated operations and simplify:

$$\frac{1}{x} - \frac{1}{x^2} + \frac{2}{x+1}.$$

The LCD $= x \cdot x(x+1)$, or $x^2(x+1)$.

$$\frac{1}{x} \cdot \frac{x(x+1)}{x(x+1)} - \frac{1}{x^2} \cdot \frac{(x+1)}{(x+1)} + \frac{2}{x+1} \cdot \frac{x^2}{x^2}$$
$$= \frac{x(x+1)}{x^2(x+1)} - \frac{x+1}{x^2(x+1)} + \frac{2x^2}{x^2(x+1)}$$

Subtracting this numerator. Don't forget the parentheses.

$$= \frac{x(x+1)-(x+1)+2x^2}{x^2(x+1)}$$
$$= \frac{x^2+x-x-1+2x^2}{x^2(x+1)} \quad \text{Removing parentheses}$$
$$= \frac{3x^2-1}{x^2(x+1)}$$

10. Perform the indicated operations and simplify:

$$\frac{1}{x} - \frac{5}{3x} + \frac{2x}{x+1}.$$

◀ Do Exercise 10.

Answers

8. $\frac{-4(2y+7)}{(y+4)(y-4)}$ **9.** $\frac{x-13}{(x+3)(x-3)}$
10. $\frac{2(3x^2-x-1)}{3x(x+1)}$

Guided Solution:
8. $-1, y^2, 4, y, 28, 28, 8y, -4$

14.5 Exercise Set

For Extra Help
MyMathLab® MathXL® PRACTICE WATCH READ REVIEW

Reading Check

When subtracting rational expressions, parentheses are important to make sure that you subtract the entire numerator. In Exercises RC1–RC3, complete each numerator by **(a)** filling in the parentheses, **(b)** removing the parentheses, and **(c)** collecting like terms.

RC1. $\frac{10x}{x-7} - \frac{3x+5}{x-7} = \frac{10x - (\qquad)}{x-7} = \frac{\qquad}{x-7} = \frac{\qquad}{x-7}$
(a) (b) (c)

RC2. $\frac{7}{4+a} - \frac{4-9a}{4+a} = \frac{7 - (\qquad)}{4+a} = \frac{\qquad}{4+a} = \frac{\qquad}{4+a}$
(a) (b) (c)

RC3. $\frac{9y-2}{y^2-10} - \frac{y+1}{y^2-10} = \frac{9y - 2 - (\qquad)}{y^2-10} = \frac{\qquad}{y^2-10} = \frac{\qquad}{y^2-10}$
(a) (b) (c)

a Subtract. Simplify, if possible.

1. $\frac{7}{x} - \frac{3}{x}$

2. $\frac{5}{a} - \frac{8}{a}$

3. $\frac{y}{y-4} - \frac{4}{y-4}$

4. $\frac{t^2}{t+5} - \frac{25}{t+5}$

5. $\frac{2x-3}{x^2+3x-4} - \frac{x-7}{x^2+3x-4}$

6. $\frac{x+1}{x^2-2x+1} - \frac{5-3x}{x^2-2x+1}$

7. $\frac{a-2}{10} - \frac{a+1}{5}$

8. $\frac{y+3}{2} - \frac{y-4}{4}$

9. $\frac{4z-9}{3z} - \frac{3z-8}{4z}$

10. $\frac{a-1}{4a} - \frac{2a+3}{a}$

11. $\frac{4x+2t}{3xt^2} - \frac{5x-3t}{x^2t}$

12. $\frac{5x+3y}{2x^2y} - \frac{3x+4y}{xy^2}$

13. $\frac{5}{x+5} - \frac{3}{x-5}$

14. $\frac{3t}{t-1} - \frac{8t}{t+1}$

15. $\frac{3}{2t^2-2t} - \frac{5}{2t-2}$

16. $\dfrac{11}{x^2 - 4} - \dfrac{8}{x + 2}$

17. $\dfrac{2s}{t^2 - s^2} - \dfrac{s}{t - s}$

18. $\dfrac{3}{12 + x - x^2} - \dfrac{2}{x^2 - 9}$

19. $\dfrac{y - 5}{y} - \dfrac{3y - 1}{4y}$

20. $\dfrac{3x - 2}{4x} - \dfrac{3x + 1}{6x}$

21. $\dfrac{a}{x + a} - \dfrac{a}{x - a}$

22. $\dfrac{a}{a - b} - \dfrac{a}{a + b}$

23. $\dfrac{11}{6} - \dfrac{5}{-6}$

24. $\dfrac{5}{9} - \dfrac{7}{-9}$

25. $\dfrac{5}{a} - \dfrac{8}{-a}$

26. $\dfrac{8}{x} - \dfrac{3}{-x}$

27. $\dfrac{4}{y - 1} - \dfrac{4}{1 - y}$

28. $\dfrac{5}{a - 2} - \dfrac{3}{2 - a}$

29. $\dfrac{3 - x}{x - 7} - \dfrac{2x - 5}{7 - x}$

30. $\dfrac{t^2}{t - 2} - \dfrac{4}{2 - t}$

31. $\dfrac{a - 2}{a^2 - 25} - \dfrac{6 - a}{25 - a^2}$

32. $\dfrac{x - 8}{x^2 - 16} - \dfrac{x - 8}{16 - x^2}$

33. $\dfrac{4 - x}{x - 9} - \dfrac{3x - 8}{9 - x}$

34. $\dfrac{4x - 6}{x - 5} - \dfrac{7 - 2x}{5 - x}$

35. $\dfrac{5x}{x^2 - 9} - \dfrac{4}{3 - x}$

36. $\dfrac{8x}{16 - x^2} - \dfrac{5}{x - 4}$

37. $\dfrac{t^2}{2t^2 - 2t} - \dfrac{1}{2t - 2}$

38. $\dfrac{4}{5a^2 - 5a} - \dfrac{2}{5a - 5}$

39. $\dfrac{x}{x^2 + 5x + 6} - \dfrac{2}{x^2 + 3x + 2}$

40. $\dfrac{a}{a^2 + 11a + 30} - \dfrac{5}{a^2 + 9a + 20}$

b Perform the indicated operations and simplify.

41. $\dfrac{3(2x + 5)}{x - 1} - \dfrac{3(2x - 3)}{1 - x} + \dfrac{6x - 1}{x - 1}$

42. $\dfrac{a - 2b}{b - a} - \dfrac{3a - 3b}{a - b} + \dfrac{2a - b}{a - b}$

43. $\dfrac{x - y}{x^2 - y^2} + \dfrac{x + y}{x^2 - y^2} - \dfrac{2x}{x^2 - y^2}$

44. $\dfrac{x - 3y}{2(y - x)} + \dfrac{x + y}{2(x - y)} - \dfrac{2x - 2y}{2(x - y)}$

45. $\dfrac{2(x - 1)}{2x - 3} - \dfrac{3(x + 2)}{2x - 3} - \dfrac{x - 1}{3 - 2x}$

46. $\dfrac{5(2y + 1)}{2y - 3} - \dfrac{3(y - 1)}{3 - 2y} - \dfrac{3(y - 2)}{2y - 3}$

47. $\dfrac{10}{2y - 1} - \dfrac{6}{1 - 2y} + \dfrac{y}{2y - 1} + \dfrac{y - 4}{1 - 2y}$

48. $\dfrac{(x + 1)(2x - 1)}{(2x - 3)(x - 3)} - \dfrac{(x - 3)(x + 1)}{(3 - x)(3 - 2x)} + \dfrac{(2x + 1)(x + 3)}{(3 - 2x)(x - 3)}$

49. $\dfrac{a + 6}{4 - a^2} - \dfrac{a + 3}{a + 2} + \dfrac{a - 3}{2 - a}$

50. $\dfrac{4t}{t^2 - 1} - \dfrac{2}{t} - \dfrac{2}{t + 1}$

51. $\frac{2z}{1 - 2z} + \frac{3z}{2z + 1} - \frac{3}{4z^2 - 1}$

52. $\frac{1}{x - y} - \frac{2x}{x^2 - y^2} + \frac{1}{x + y}$

53. $\frac{1}{x + y} - \frac{1}{x - y} + \frac{2x}{x^2 - y^2}$

54. $\frac{2b}{a^2 - b^2} - \frac{1}{a + b} + \frac{1}{a - b}$

Skill Maintenance

Simplify.

55. $(a^2b^{-5})^{-4}$ [12.2a, b]

56. $\frac{54x^{10}}{3x^7}$ [12.1e]

57. $3x^4 \cdot 10x^8$ [12.1d]

Solve. [10.3b]

58. $2.5x + 15.5 = 0.5 + 4x$

59. $\frac{4}{7} + 3x = \frac{1}{2}x - \frac{3}{14}$

60. $6x - 0.5 = 6 - 0.5x$

Find a polynomial for the shaded area of each figure. [12.4d]

61.

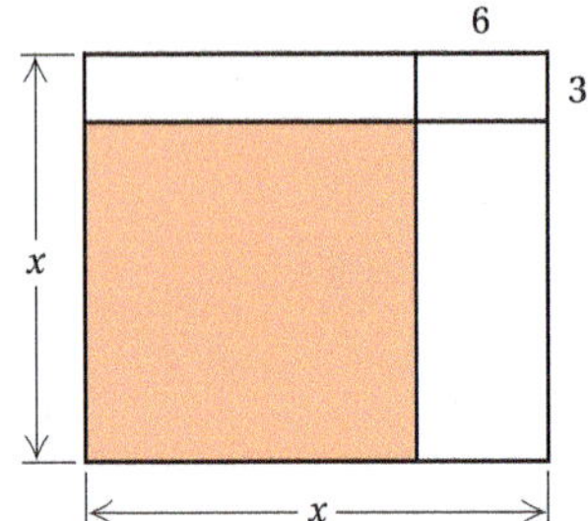

62.

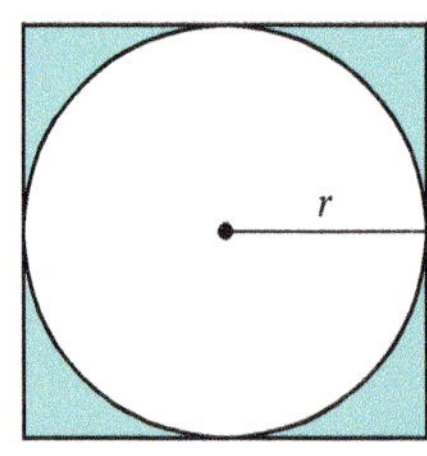

Synthesis

63. The perimeter of the following right triangle is $2a + 5$. Find the missing length of the third side and the area.

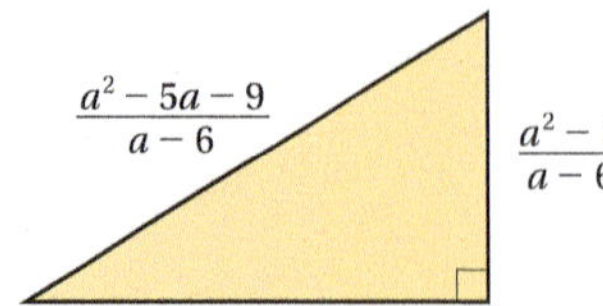

Mid-Chapter Review

Concept Reinforcement

Determine whether each statement is true or false.

__________ **1.** The reciprocal of $\dfrac{3 - w}{w + 2}$ is $\dfrac{w - 3}{w + 2}$. [14.2a]

__________ **2.** The value of the numerator has no bearing on whether or not a rational expression is defined. [14.1a]

__________ **3.** To add or subtract rational expressions when the denominators are the same, add or subtract the numerators and keep the same denominator. [14.4a], [14.5a]

__________ **4.** For the rational expression $\dfrac{x(x - 2)}{x + 3}$, x is a factor of the numerator and a factor of the denominator. [14.1c]

__________ **5.** To find the LCM, use each factor the greatest number of times that it appears in any one factorization. [14.3a, c]

Guided Solutions

Fill in each blank with the number or expression that creates a correct solution.

6. Subtract: $\dfrac{x - 1}{x - 2} - \dfrac{x + 1}{x + 2} - \dfrac{x - 6}{4 - x^2}$. [14.5b]

$$\frac{x-1}{x-2} - \frac{x+1}{x+2} - \frac{x-6}{4-x^2} = \frac{x-1}{x-2} - \frac{x+1}{x+2} - \frac{x-6}{4-x^2} \cdot \frac{\square}{\square} = \frac{x-1}{x-2} - \frac{x+1}{x+2} - \frac{6 - \square}{\square - 4}$$

$$= \frac{x-1}{x-2} - \frac{x+1}{x+2} - \frac{6-x}{(x - \square)(\square + 2)}$$

$$= \frac{x-1}{x-2} \cdot \frac{\square}{\square} - \frac{x+1}{x+2} \cdot \frac{\square}{\square} - \frac{6-x}{(x-2)(x+2)}$$

$$= \frac{x^2 + \square - 2}{(x-2)(x+2)} - \frac{\square - x - 2}{(x-2)(x+2)} - \frac{6-x}{(x-2)(x+2)}$$

$$= \frac{x^2 + x - \square - x^2 + \square + 2 - \square + x}{(x-2)(x+2)}$$

$$= \frac{\square - \square}{(x-2)(x+2)} = \frac{3(\square - \square)}{(x-2)(x+2)} = \frac{\square}{\square} \cdot \frac{\square}{x+2} = \frac{3}{\square}$$

Mixed Review

Find all numbers for which the rational expression is not defined. [14.1a]

7. $\dfrac{t^2 - 16}{3}$

8. $\dfrac{x - 8}{x^2 - 11x + 24}$

9. $\dfrac{7}{2w - 7}$

Simplify. [14.1c]

10. $\dfrac{x^2 + 2x - 3}{x^2 - 9}$

11. $\dfrac{6y^2 + 12y - 48}{3y^2 - 9y + 6}$

12. $\dfrac{r - s}{s - r}$

13. Find the reciprocal of $-x + 3$. [14.2a]

14. Find the LCM of
$x^2 - 100$, $10x^3$, and $x^2 - 20x + 100$. [14.3c]

Add, subtract, multiply, or divide and simplify, if possible.

15. $\dfrac{a^2 - a - 2}{a^2 - a - 6} \div \dfrac{a^2 - 2a}{2a + a^2}$ [14.2b]

16. $\dfrac{3y}{y^2 - 7y + 10} - \dfrac{2y}{y^2 - 8y + 15}$ [14.5a]

17. $\dfrac{x^2}{x - 11} + \dfrac{121}{11 - x}$ [14.4a]

18. $\dfrac{x^2 - y^2}{(x - y)^2} \cdot \dfrac{1}{x + y}$ [14.1d]

19. $\dfrac{3a - b}{a^2 b} + \dfrac{a + 2b}{ab^2}$ [14.4a]

20. $\dfrac{5x}{x^2 - 4} - \dfrac{3}{x} + \dfrac{4}{x + 2}$ [14.5b]

Matching. Perform the indicated operation and simplify. Then select the correct answer from selections A–G listed in the second column. [14.1d], [14.2b], [14.4a], [14.5a]

21. $\dfrac{2}{x - 2} \div \dfrac{1}{x + 3}$

22. $\dfrac{1}{x + 3} - \dfrac{2}{x - 2}$

23. $\dfrac{2}{x - 2} - \dfrac{1}{x + 3}$

24. $\dfrac{1}{x + 3} \div \dfrac{2}{x - 2}$

25. $\dfrac{2}{x - 2} + \dfrac{1}{x + 3}$

26. $\dfrac{2}{x - 2} \cdot \dfrac{1}{x + 3}$

A. $\dfrac{-x - 8}{(x - 2)(x + 3)}$

B. $\dfrac{x - 2}{2(x + 3)}$

C. $\dfrac{2}{(x - 2)(x + 3)}$

D. $\dfrac{x + 8}{(x - 2)(x + 3)}$

E. $\dfrac{2(x + 3)}{x - 2}$

F. $\dfrac{3x + 4}{(x - 2)(x + 3)}$

G. $\dfrac{x + 3}{x - 2}$

Understanding Through Discussion and Writing

27. Explain why the product of two numbers is not always their least common multiple. [14.3a]

28. Is the reciprocal of a product the product of the reciprocals? Why or why not? [14.2a]

29. A student insists on finding a common denominator by always multiplying the denominators of the expressions being added. How could this approach be improved? [14.4a]

30. Explain why the expressions

$$\frac{1}{3 - x} \quad \text{and} \quad \frac{1}{x - 3}$$

are opposites. [14.4a]

31. Explain why 5, −1, and 7 are *not* allowable replacements in the division

$$\frac{x + 3}{x - 5} \div \frac{x - 7}{x + 1}.$$ [14.1a], [14.2a, b]

32. If the LCM of a binomial and a trinomial is the trinomial, what relationship exists between the two expressions? [14.3c]

STUDYING FOR SUCCESS *Make the Most of Your Time in Class*

- ☐ Before class, try to glance at the next section in your text to be discussed, so that you can concentrate on the instruction in class.
- ☐ Get a great seat! Sitting near the front will help you both hear the instruction more clearly and avoid distractions.
- ☐ Let your instructor know in advance if you must miss class, and do your best to keep up with any work that you miss.

Complex Rational Expressions 14.6

a SIMPLIFYING COMPLEX RATIONAL EXPRESSIONS

A **complex rational expression**, or **complex fraction expression**, is a rational expression that has one or more rational expressions within its numerator or denominator. Here are some examples:

$$\frac{1+\frac{2}{x}}{3}, \quad \frac{\frac{x+y}{2}}{\frac{2x}{x+1}}, \quad \frac{\frac{1}{3}+\frac{1}{5}}{\frac{2}{x}-\frac{x}{y}}.$$

These are rational expressions within the complex rational expression.

There are two methods used to simplify complex rational expressions.

METHOD 1: MULTIPLYING BY THE LCM OF ALL THE DENOMINATORS

To simplify a complex rational expression:

1. First, find the LCM of all the denominators of all the rational expressions occurring *within* both the numerator and the denominator of the complex rational expression.
2. Then multiply by 1 using LCM/LCM.
3. If possible, simplify by removing a factor of 1.

EXAMPLE 1 Simplify: $\dfrac{\frac{1}{2}+\frac{3}{4}}{\frac{5}{6}-\frac{3}{8}}$.

$$\frac{\frac{1}{2}+\frac{3}{4}}{\frac{5}{6}-\frac{3}{8}}$$

The denominators *within* the complex rational expression are 2, 4, 6, and 8. The LCM of these denominators is 24. We multiply by 1 using $\frac{24}{24}$. This amounts to multiplying both the numerator *and* the denominator by 24.

$$= \frac{\frac{1}{2}+\frac{3}{4}}{\frac{5}{6}-\frac{3}{8}} \cdot \frac{24}{24}$$ Multiplying by 1

OBJECTIVE

a Simplify complex rational expressions.

SKILL TO REVIEW

Objective 14.3c: Find the LCM of algebraic expressions by factoring.

1. Find the LCM of 2, 4, 6, and 8.
2. Find the LCM of x, x^2, and $5x$.

To the instructor and the student: Students can be instructed either to try both methods and then choose the one that works best for them or to use the method chosen by the instructor.

Answers

Skill to Review:
1. 24 **2.** $5x^2$

Using the distributive laws, we carry out the multiplications:

$$= \frac{\left(\frac{1}{2} + \frac{3}{4}\right)24}{\left(\frac{5}{6} - \frac{3}{8}\right)24} = \frac{\frac{1}{2}(24) + \frac{3}{4}(24)}{\frac{5}{6}(24) - \frac{3}{8}(24)}$$

← Multiplying the numerator by 24

← Multiplying the denominator by 24

$$= \frac{12 + 18}{20 - 9}$$ Simplifying

$$= \frac{30}{11}.$$

Multiplying in this manner has the effect of clearing fractions in both the numerator and the denominator of the complex rational expression.

Do Exercise 1.

1. Simplify: $\dfrac{\frac{1}{3} + \frac{4}{5}}{\frac{7}{8} - \frac{5}{6}}$.

EXAMPLE 2 Simplify: $\dfrac{\frac{3}{x} + \frac{1}{2x}}{\frac{1}{3x} - \frac{3}{4x}}$.

The denominators within the complex expression are x, $2x$, $3x$, and $4x$. The LCM of these denominators is $12x$. We multiply by 1 using $12x/12x$.

$$\frac{\frac{3}{x} + \frac{1}{2x}}{\frac{1}{3x} - \frac{3}{4x}} \cdot \frac{12x}{12x} = \frac{\left(\frac{3}{x} + \frac{1}{2x}\right)12x}{\left(\frac{1}{3x} - \frac{3}{4x}\right)12x} = \frac{\frac{3}{x}(12x) + \frac{1}{2x}(12x)}{\frac{1}{3x}(12x) - \frac{3}{4x}(12x)}$$

$$= \frac{36 + 6}{4 - 9} = \frac{42}{-5} = -\frac{42}{5}$$

Do Exercise 2.

2. Simplify: $\dfrac{\frac{x}{2} + \frac{2x}{3}}{\frac{1}{x} - \frac{x}{2}}$.

GS

$$\frac{\frac{x}{2} + \frac{2x}{3}}{\frac{1}{x} - \frac{x}{2}}$$

LCM of denominators $= 6x$

$$= \frac{\frac{x}{2} + \frac{2x}{3}}{\frac{1}{x} - \frac{x}{2}} \cdot \frac{6x}{\square}$$

$$= \frac{\left(\frac{x}{2} + \frac{2x}{3}\right) \cdot \square}{\left(\frac{1}{x} - \frac{x}{2}\right) \cdot 6x}$$

$$= \frac{3x^2 + \square}{\square - 3x^2} = \frac{7x^2}{3(\square - x^2)}$$

EXAMPLE 3 Simplify: $\dfrac{1 - \frac{1}{x}}{1 - \frac{1}{x^2}}$.

The denominators within the complex expression are x and x^2. The LCM of these denominators is x^2. We multiply by 1 using x^2/x^2. Then, after obtaining a single rational expression, we simplify:

$$\frac{1 - \frac{1}{x}}{1 - \frac{1}{x^2}} \cdot \frac{x^2}{x^2} = \frac{\left(1 - \frac{1}{x}\right)x^2}{\left(1 - \frac{1}{x^2}\right)x^2} = \frac{1(x^2) - \frac{1}{x}(x^2)}{1(x^2) - \frac{1}{x^2}(x^2)} = \frac{x^2 - x}{x^2 - 1}$$

$$= \frac{x(x - 1)}{(x + 1)(x - 1)} = \frac{x}{x + 1}.$$

Do Exercise 3.

3. Simplify: $\dfrac{1 + \frac{1}{x}}{1 - \frac{1}{x^2}}$.

Answers

1. $\frac{136}{5}$ 2. $\frac{7x^2}{3(2 - x^2)}$ 3. $\frac{x}{x - 1}$

Guided Solution:

2. $6x$, $6x$, $4x^2$, 6, 2

METHOD 2: ADDING IN THE NUMERATOR AND THE DENOMINATOR

To simplify a complex rational expression:

1. Add or subtract, as necessary, to get a single rational expression in the numerator.
2. Add or subtract, as necessary, to get a single rational expression in the denominator.
3. Divide the numerator by the denominator.
4. If possible, simplify by removing a factor of 1.

We will redo Examples 1–3 using this method.

EXAMPLE 4 Simplify: $\dfrac{\frac{1}{2}+\frac{3}{4}}{\frac{5}{6}-\frac{3}{8}}$.

The LCM of 2 and 4 in the numerator is 4. The LCM of 6 and 8 in the denominator is 24. We have

$$\frac{\frac{1}{2}+\frac{3}{4}}{\frac{5}{6}-\frac{3}{8}} = \frac{\frac{1}{2}\cdot\frac{2}{2}+\frac{3}{4}}{\frac{5}{6}\cdot\frac{4}{4}-\frac{3}{8}\cdot\frac{3}{3}}$$

Multiplying $\frac{1}{2}$ by 1 to get the common denominator, 4

Multiplying $\frac{5}{6}$ and $\frac{3}{8}$ by 1 to get the common denominator, 24

$$= \frac{\frac{2}{4}+\frac{3}{4}}{\frac{20}{24}-\frac{9}{24}}$$

$$= \frac{\frac{5}{4}}{\frac{11}{24}}$$ Adding in the numerator; subtracting in the denominator

$$= \frac{5}{4} \div \frac{11}{24}$$

$$= \frac{5}{4}\cdot\frac{24}{11}$$ Multiplying by the reciprocal of the divisor

$$= \frac{5\cdot 2\cdot 2\cdot 2\cdot 3}{2\cdot 2\cdot 11}$$ Factoring

$$= \frac{5\cdot 2\cdot \cancel{2}\cdot \cancel{2}\cdot 3}{\cancel{2}\cdot \cancel{2}\cdot 11}$$ Removing a factor of 1: $\frac{2\cdot 2}{2\cdot 2} = 1$

$$= \frac{5\cdot 2\cdot 3}{11}$$

$$= \frac{30}{11}.$$

Do Exercise 4. ▶

4. Simplify. Use method 2.

$$\frac{\frac{1}{3}+\frac{4}{5}}{\frac{7}{8}-\frac{5}{6}}$$

Answer

4. $\frac{136}{5}$

5. Simplify. Use method 2.

$$\frac{\frac{x}{2}+\frac{2x}{3}}{\frac{1}{x}-\frac{x}{2}}$$

$$\frac{\frac{x}{2}+\frac{2x}{3} \leftarrow \text{LCD} = 6}{\frac{1}{x}-\frac{x}{2} \leftarrow \text{LCD} = 2x}$$

$$= \frac{\frac{x}{2}\cdot\frac{3}{\square}+\frac{2x}{3}\cdot\frac{\square}{2}}{\frac{1}{x}\cdot\frac{\square}{2}-\frac{x}{2}\cdot\frac{x}{\square}}$$

$$= \frac{\frac{\square}{6}+\frac{4x}{6}}{\frac{2}{2x}-\frac{\square}{2x}} = \frac{\frac{3x+4x}{\square}}{\frac{2-x^2}{\square}}$$

$$= \frac{7x}{6}\cdot\frac{\square}{2-x^2} = \frac{7\cdot 2\cdot x\cdot x}{2\cdot 3(2-x^2)}$$

$$= \frac{\square}{3(2-x^2)}$$

EXAMPLE 5 Simplify: $\dfrac{\frac{3}{x}+\frac{1}{2x}}{\frac{1}{3x}-\frac{3}{4x}}$.

$$\frac{\frac{3}{x}+\frac{1}{2x}}{\frac{1}{3x}-\frac{3}{4x}} = \frac{\frac{3}{x}\cdot\frac{2}{2}+\frac{1}{2x}}{\frac{1}{3x}\cdot\frac{4}{4}-\frac{3}{4x}\cdot\frac{3}{3}}$$

Finding the LCD, $2x$, and multiplying by 1 in the numerator

Finding the LCD, $12x$, and multiplying by 1 in the denominator

$$= \frac{\frac{6}{2x}+\frac{1}{2x}}{\frac{4}{12x}-\frac{9}{12x}} = \frac{\frac{7}{2x}}{\frac{-5}{12x}}$$

Adding in the numerator and subtracting in the denominator

$$= \frac{7}{2x}\div\frac{-5}{12x}$$

$$= \frac{7}{2x}\cdot\frac{12x}{-5}$$

Multiplying by the reciprocal of the divisor

$$= \frac{7\cdot 6\cdot(2x)}{(2x)(-5)}$$

Multiplying, factoring, and removing a factor of 1: $\frac{2x}{2x} = 1$

$$= \frac{42}{-5} = -\frac{42}{5}$$

◀ **Do Exercise 5.**

EXAMPLE 6 Simplify: $\dfrac{1-\frac{1}{x}}{1-\frac{1}{x^2}}$.

$$\frac{1-\frac{1}{x}}{1-\frac{1}{x^2}} = \frac{1\cdot\frac{x}{x}-\frac{1}{x}}{1\cdot\frac{x^2}{x^2}-\frac{1}{x^2}}$$

Finding the LCD, x, and multiplying by 1 in the numerator

Finding the LCD, x^2, and multiplying by 1 in the denominator

$$= \frac{\frac{x-1}{x}}{\frac{x^2-1}{x^2}}$$

Subtracting in the numerator and subtracting in the denominator

$$= \frac{x-1}{x}\div\frac{x^2-1}{x^2}$$

$$= \frac{x-1}{x}\cdot\frac{x^2}{x^2-1}$$

Multiplying by the reciprocal of the divisor

$$= \frac{(x-1)x\cdot x}{x(x-1)(x+1)}$$

Multiplying, factoring, and removing a factor of 1: $\frac{x(x-1)}{x(x-1)} = 1$

$$= \frac{x}{x+1}$$

◀ **Do Exercise 6.**

6. Simplify. Use method 2.

$$\frac{1+\frac{1}{x}}{1-\frac{1}{x^2}}$$

Answers

5. $\frac{7x^2}{3(2-x^2)}$ **6.** $\frac{x}{x-1}$

Guided Solution:

5. 3, 2, 2, x, $3x$, x^2, 6, $2x$, $2x$, $7x^2$

14.6 Exercise Set

For Extra Help

MyMathLab®

MathXL®

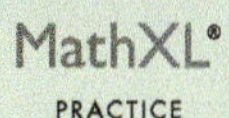
PRACTICE

WATCH

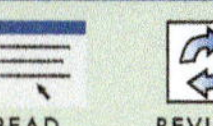
READ

REVIEW

✓ Reading Check

Consider the expression $\dfrac{\frac{8}{x} - \frac{5}{9}}{\frac{2}{x}}$. Complete each statement with the correct word(s) from the column on the right.

RC1. The expression given above is a(n) ______________ rational expression.

RC2. The expression $\frac{8}{x} - \frac{5}{9}$ is the ______________ of the expression.

RC3. The ______________ of the rational expressions $\frac{8}{x}$, $\frac{5}{9}$, and $\frac{2}{x}$ is $9x$.

RC4. To simplify the expression, we can multiply the numerator $\frac{8}{x} - \frac{5}{9}$ by the ______________ of the denominator, $\frac{2}{x}$.

numerator
denominator
opposite
reciprocal
complex
least common denominator

a Simplify.

1. $\dfrac{1 + \frac{9}{16}}{1 - \frac{3}{4}}$

2. $\dfrac{6 - \frac{3}{8}}{4 + \frac{5}{6}}$

3. $\dfrac{1 - \frac{3}{5}}{1 + \frac{1}{5}}$

4. $\dfrac{2 + \frac{2}{3}}{2 - \frac{2}{3}}$

5. $\dfrac{\frac{1}{2} + \frac{3}{4}}{\frac{5}{8} - \frac{5}{6}}$

6. $\dfrac{\frac{3}{4} + \frac{7}{8}}{\frac{2}{3} - \frac{5}{6}}$

7. $\dfrac{\frac{1}{x} + 3}{\frac{1}{x} - 5}$

8. $\dfrac{2 - \frac{1}{a}}{4 + \frac{1}{a}}$

9. $\dfrac{4 - \frac{1}{x^2}}{2 - \frac{1}{x}}$

10. $\dfrac{\frac{2}{y} + \frac{1}{2y}}{y + \frac{y}{2}}$

11. $\dfrac{8 + \frac{8}{d}}{1 + \frac{1}{d}}$

12. $\dfrac{3 + \frac{2}{t}}{3 - \frac{2}{t}}$

13. $\dfrac{\frac{x}{8} - \frac{8}{x}}{\frac{1}{8} + \frac{1}{x}}$

14. $\dfrac{\frac{2}{m} + \frac{m}{2}}{\frac{m}{3} - \frac{3}{m}}$

15. $\dfrac{1 + \frac{1}{y}}{1 - \frac{1}{y^2}}$

16. $\dfrac{\frac{1}{q^2} - 1}{\frac{1}{q} + 1}$

17. $\dfrac{\dfrac{1}{5}-\dfrac{1}{a}}{\dfrac{5-a}{5}}$

18. $\dfrac{\dfrac{4}{t}}{4+\dfrac{1}{t}}$

19. $\dfrac{\dfrac{1}{a}+\dfrac{1}{b}}{\dfrac{1}{a^2}-\dfrac{1}{b^2}}$

20. $\dfrac{\dfrac{1}{x^2}-\dfrac{1}{y^2}}{\dfrac{2}{x}-\dfrac{2}{y}}$

21. $\dfrac{\dfrac{p}{q}+\dfrac{q}{p}}{\dfrac{1}{p}+\dfrac{1}{q}}$

22. $\dfrac{x-3+\dfrac{2}{x}}{x-4+\dfrac{3}{x}}$

23. $\dfrac{\dfrac{2}{a}+\dfrac{4}{a^2}}{\dfrac{5}{a^3}-\dfrac{3}{a}}$

24. $\dfrac{\dfrac{5}{x^3}-\dfrac{1}{x^2}}{\dfrac{2}{x}+\dfrac{3}{x^2}}$

25. $\dfrac{\dfrac{2}{7a^4}-\dfrac{1}{14a}}{\dfrac{3}{5a^2}+\dfrac{2}{15a}}$

26. $\dfrac{\dfrac{5}{4x^3}-\dfrac{3}{8x}}{\dfrac{3}{2x}+\dfrac{3}{4x^3}}$

27. $\dfrac{\dfrac{a}{b}+\dfrac{c}{d}}{\dfrac{b}{a}+\dfrac{d}{c}}$

28. $\dfrac{\dfrac{a}{b}-\dfrac{c}{d}}{\dfrac{b}{a}-\dfrac{d}{c}}$

29. $\dfrac{\dfrac{x}{5y^3}+\dfrac{3}{10y}}{\dfrac{3}{10y}+\dfrac{x}{5y^3}}$

30. $\dfrac{\dfrac{a}{6b^3}+\dfrac{4}{9b^2}}{\dfrac{5}{6b}-\dfrac{1}{9b^3}}$

31. $\dfrac{\dfrac{3}{x+1}+\dfrac{1}{x}}{\dfrac{2}{x+1}+\dfrac{3}{x}}$

32. $\dfrac{x-7+\dfrac{5}{x-1}}{x-3+\dfrac{1}{x-1}}$

Skill Maintenance

Solve. [10.7e]

33. $4-\dfrac{1}{6}x \geq -12$

34. $3(b-8) > -2(3b+1)$

35. $1.5x+19.2 < 4.2-3.5x$

Solve. [13.8a]

36. ***Ladder Distances.*** A ladder of length 13 ft is placed against a building in such a way that the distance from the top of the ladder to the ground is 7 ft more than the distance from the bottom of the ladder to the building. Find these distances.

37. ***Perimeter of a Rectangle.*** The length of a rectangle is 3 yd greater than the width. The area of the rectangle is 10 yd^2. Find the perimeter.

Synthesis

Simplify.

38. $\left[\dfrac{\dfrac{x+1}{x-1}+1}{\dfrac{x+1}{x-1}-1}\right]^5$

39. $1+\dfrac{1}{1+\dfrac{1}{1+\dfrac{1}{1+\dfrac{1}{x}}}}$

40. $\dfrac{\dfrac{z}{1-\dfrac{z}{2+2z}}-2z}{\dfrac{2z}{5z-2}-3}$

Solving Rational Equations

14.7

OBJECTIVE

a Solve rational equations.

SKILL TO REVIEW

Objective 10.3b: Solve equations in which like terms may need to be collected.

Solve. Clear fractions first.

1. $4 - \frac{5}{6}y = y + \frac{7}{12}$

2. $\frac{2}{5}x + \frac{1}{3} = \frac{7}{10}x - 2$

a RATIONAL EQUATIONS

In Sections 14.1–14.6, we studied operations with *rational expressions*. These expressions have no equals signs. We can add, subtract, multiply, or divide and simplify expressions, but we cannot solve if there are no equals signs—as, for example, in

$$\frac{x^2 + 6x + 9}{x^2 - 4} \cdot \frac{x - 2}{x + 3}, \quad \frac{x + y}{x - y} \div \frac{x^2 + y}{x^2 - y^2}, \quad \text{and} \quad \frac{a + 3}{a^2 - 16} + \frac{5}{12 - 3a}.$$

Operation signs occur. There are no equals signs!

Most often, the result of our calculation is another rational expression that has not been cleared of fractions.

Equations *do have* equals signs, and we can clear them of fractions as we did in Section 10.3. A **rational**, or **fraction**, **equation**, is an equation containing one or more rational expressions. Here are some examples:

$$\frac{2}{3} + \frac{5}{6} = \frac{x}{9}, \quad x + \frac{6}{x} = -5, \quad \text{and} \quad \frac{x^2}{x - 1} = \frac{1}{x - 1}.$$

There are equals signs as well as operation signs.

SOLVING RATIONAL EQUATIONS

To solve a rational equation, the first step is to clear the equation of fractions. To do this, multiply all terms on both sides of the equation by the LCM of all the denominators. Then carry out the equation-solving process as we learned it in Chapters 10 and 13.

When clearing an equation of fractions, we use the terminology LCM instead of LCD because we are *not* adding or subtracting rational expressions.

Caution!

We are introducing a new use of the LCM in this section. We previously used the LCM in adding or subtracting rational expressions. *Now* we have equations with equals signs. We clear fractions by multiplying by the LCM on both sides of the equation. This eliminates the denominators. Do *not* make the mistake of trying to clear fractions when you do not have an equation.

EXAMPLE 1 Solve: $\frac{2}{3} + \frac{5}{6} = \frac{x}{9}$.

The LCM of all denominators is $2 \cdot 3 \cdot 3$, or 18. We multiply all terms on both sides by 18:

$18\left(\frac{2}{3} + \frac{5}{6}\right) = 18 \cdot \frac{x}{9}$ Multiplying by the LCM on both sides

$18 \cdot \frac{2}{3} + 18 \cdot \frac{5}{6} = 18 \cdot \frac{x}{9}$ Multiplying each term by the LCM to remove parentheses

$12 + 15 = 2x$ Simplifying. Note that we have now cleared fractions.

$27 = 2x$

$\frac{27}{2} = x.$

The check is left to the student. The solution is $\frac{27}{2}$.

Do Margin Exercise 1.

1. Solve: $\frac{3}{4} + \frac{5}{8} = \frac{x}{12}$.

Answers

Skill to Review:
1. $\frac{41}{22}$ 2. $\frac{70}{9}$

Margin Exercise:
1. $\frac{33}{2}$

ALGEBRAIC ▶◀ GRAPHICAL CONNECTION

We can obtain a visual check of the solutions of a rational equation by graphing. For example, consider the equation

$$\frac{x}{4} + \frac{x}{2} = 6.$$

We can examine the solution by graphing the equations

$$y = \frac{x}{4} + \frac{x}{2} \quad \text{and} \quad y = 6$$

using the same set of axes.

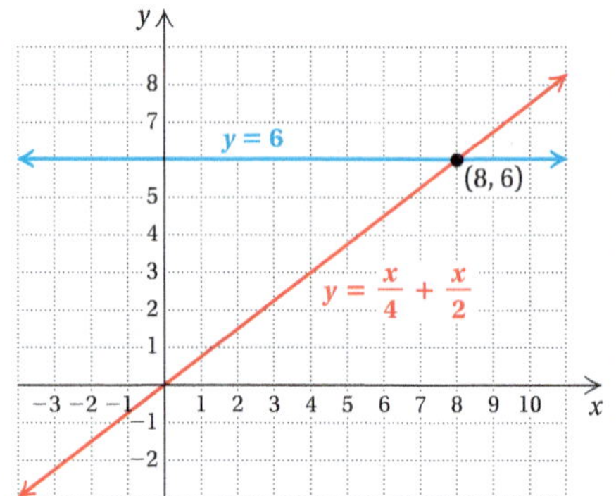

The first coordinate of the point of intersection of the graphs is the value of x for which $\frac{x}{4} + \frac{x}{2} = 6$, so it is the solution of the equation. It appears from the graph that when $x = 8$, the value of $x/4 + x/2$ is 6. We can check by substitution:

$$\frac{x}{4} + \frac{x}{2} = \frac{8}{4} + \frac{8}{2} = 2 + 4 = 6.$$

Thus the solution is 8.

EXAMPLE 2 Solve: $\frac{x}{6} - \frac{x}{8} = \frac{1}{12}$.

The LCM is 24. We multiply all terms on both sides by 24:

$$\frac{x}{6} - \frac{x}{8} = \frac{1}{12}$$

$$24\left(\frac{x}{6} - \frac{x}{8}\right) = 24 \cdot \frac{1}{12} \quad \text{Multiplying by the LCM on both sides}$$

$$24 \cdot \frac{x}{6} - 24 \cdot \frac{x}{8} = 24 \cdot \frac{1}{12} \quad \text{Multiplying to remove parentheses}$$

Be sure to multiply each term by the LCM.

$$4x - 3x = 2 \quad \text{Simplifying}$$

$$x = 2.$$

Check:

$$\frac{x}{6} - \frac{x}{8} = \frac{1}{12}$$

$$\frac{2}{6} - \frac{2}{8} \ ? \ \frac{1}{12}$$

$$\frac{1}{3} - \frac{1}{4}$$

$$\frac{4}{12} - \frac{3}{12}$$

$$\frac{1}{12} \qquad \text{TRUE}$$

This checks, so the solution is 2.

EXAMPLE 3 Solve: $\frac{1}{x} = \frac{1}{4 - x}$.

The LCM is $x(4 - x)$. We multiply all terms on both sides by $x(4 - x)$:

$$\frac{1}{x} = \frac{1}{4 - x}$$

$$x(4 - x) \cdot \frac{1}{x} = x(4 - x) \cdot \frac{1}{4 - x} \quad \text{Multiplying by the LCM on both sides}$$

$$4 - x = x \quad \text{Simplifying}$$

$$4 = 2x$$

$$x = 2.$$

Check:

$$\frac{1}{x} = \frac{1}{4 - x}$$

$$\frac{1}{2} \ ? \ \frac{1}{4 - 2}$$

$$\frac{1}{2} \qquad \text{TRUE}$$

This checks, so the solution is 2.

◀ **Do Exercises 2 and 3.**

Solve.

2. $\frac{x}{4} - \frac{x}{6} = \frac{1}{8}$

3. $\frac{1}{x} = \frac{1}{6 - x}$

Answers

2. $\frac{3}{2}$ 3. 3

EXAMPLE 4 Solve: $\frac{2}{3x} + \frac{1}{x} = 10$.

The LCM is $3x$. We multiply all terms on both sides by $3x$:

$$\frac{2}{3x} + \frac{1}{x} = 10$$

$$3x\left(\frac{2}{3x} + \frac{1}{x}\right) = 3x \cdot 10 \quad \text{Multiplying by the LCM on both sides}$$

$$3x \cdot \frac{2}{3x} + 3x \cdot \frac{1}{x} = 3x \cdot 10 \quad \text{Multiplying to remove parentheses}$$

$$2 + 3 = 30x \quad \text{Simplifying}$$

$$5 = 30x$$

$$\frac{5}{30} = x$$

$$\frac{1}{6} = x.$$

The check is left to the student. The solution is $\frac{1}{6}$.

Do Exercise 4. ▶

GS **4.** Solve: $\frac{1}{2x} + \frac{1}{x} = -12$.

$$\frac{1}{2x} + \frac{1}{x} = -12$$

$$\text{LCM} = 2x$$

$$2x\left(\frac{1}{2x} + \frac{1}{x}\right) = \square(-12)$$

$$2x \cdot \frac{1}{2x} + \square \cdot \frac{1}{x} = 2x(-12)$$

$$1 + \square = \square \cdot x$$

$$\square = -24x$$

$$\frac{3}{\square} = x$$

$$-\frac{\square}{8} = x$$

EXAMPLE 5 Solve: $x + \frac{6}{x} = -5$.

The LCM is x. We multiply all terms on both sides by x:

$$x + \frac{6}{x} = -5$$

$$x\left(x + \frac{6}{x}\right) = x \cdot (-5) \quad \text{Multiplying by } x \text{ on both sides}$$

$$x \cdot x + x \cdot \frac{6}{x} = -5x \quad \text{Note that each rational expression on the left is now multiplied by } x.$$

$$x^2 + 6 = -5x \quad \text{Simplifying}$$

$$x^2 + 5x + 6 = 0 \quad \text{Adding } 5x \text{ to get 0 on one side}$$

$$(x+3)(x+2) = 0 \quad \text{Factoring}$$

$$x + 3 = 0 \quad \text{or} \quad x + 2 = 0 \quad \text{Using the principle of zero products}$$

$$x = -3 \quad \text{or} \quad x = -2.$$

Check: For -3:

$$\begin{array}{c|c} x + \frac{6}{x} = -5 & \\ \hline -3 + \frac{6}{-3} \; ? & -5 \\ -3 - 2 & \\ -5 & \end{array} \quad \text{TRUE}$$

For -2:

$$\begin{array}{c|c} x + \frac{6}{x} = -5 & \\ \hline -2 + \frac{6}{-2} \; ? & -5 \\ -2 - 3 & \\ -5 & \end{array} \quad \text{TRUE}$$

Both of these check, so there are two solutions, -3 and -2.

Do Exercise 5. ▶

CHECKING POSSIBLE SOLUTIONS

When we multiply by the LCM on both sides of an equation, the resulting equation might have solutions that are *not* solutions of the original equation. Thus we must *always* check possible solutions in the original equation.

5. Solve: $x + \frac{1}{x} = 2$.

Answers

4. $-\frac{1}{8}$ **5.** 1

Guided Solution:
4. $2x, 2x, 2, -24, 3, -24, 1$

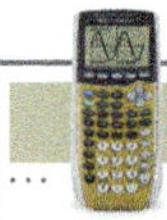

CALCULATOR CORNER

Checking Solutions of Rational Equations Consider the equation in Example 6 and the possible solutions that were found, 1 and −1. To check these solutions, we enter $y_1 = x^2/(x-1)$ and $y_2 = 1/(x-1)$ on the equation-editor screen. Then, with a table set in ASK mode, we enter $x = 1$. The ERROR messages indicate that 1 is not a solution. Next, we enter $x = -1$. Since y_1 and y_2 have the same value, we know that the equation is true, so −1 is a solution.

$y_1 = x^2/(x-1)$,
$y_2 = 1/(x-1)$

X	Y1	Y2
1	ERROR	ERROR
−1	−.5	−.5

EXERCISES: Use a graphing calculator to check the possible solutions.

1. Examples 1, 3, 5, and 7
2. Margin Exercises 1, 3, 6, and 7

Solve.

6. $\dfrac{x^2}{x+2} = \dfrac{4}{x+2}$

7. $\dfrac{4}{x-2} + \dfrac{1}{x+2} = \dfrac{26}{x^2-4}$

LCM $= (x-2)(x+2)$

$(x-2)(x+\underline{\quad})\left(\dfrac{4}{x-2} + \dfrac{1}{x+2}\right)$

$= (x-2)(x+2)\cdot\dfrac{26}{x^2-4}$

$\underline{\quad}(x+2) + \underline{\quad}(x-2) = 26$

$\underline{\quad} + 8 + x - \underline{\quad} = 26$

$\underline{\quad} + 6 = 26$

$5x = \underline{\quad}$

$x = \underline{\quad}$

Answers

6. 2 7. 4

Guided Solution:
7. 2, 4, 1, 4x, 2, 5x, 20, 4

Example 6 illustrates the importance of checking all possible solutions.

EXAMPLE 6 Solve: $\dfrac{x^2}{x-1} = \dfrac{1}{x-1}$.

The LCM is $x - 1$. We multiply all terms on both sides by $x - 1$:

$$\frac{x^2}{x-1} = \frac{1}{x-1}$$

$$(x-1)\cdot\frac{x^2}{x-1} = (x-1)\cdot\frac{1}{x-1} \qquad \text{Multiplying by } x-1 \text{ on both sides}$$

$$x^2 = 1 \qquad \text{Simplifying}$$

$$x^2 - 1 = 0 \qquad \text{Subtracting 1 to get 0 on one side}$$

$$(x-1)(x+1) = 0 \qquad \text{Factoring}$$

$$x - 1 = 0 \quad or \quad x + 1 = 0 \qquad \text{Using the principle of zero products}$$

$$x = 1 \quad or \quad x = -1.$$

The numbers 1 and −1 are possible solutions.

Check: For 1:

$$\frac{x^2}{x-1} = \frac{1}{x-1}$$

$$\frac{1^2}{1-1} \stackrel{?}{=} \frac{1}{1-1}$$

$$\frac{1}{0} \quad \Big| \quad \frac{1}{0} \qquad \text{NOT DEFINED}$$

For −1:

$$\frac{x^2}{x-1} = \frac{1}{x-1}$$

$$\frac{(-1)^2}{(-1)-1} \stackrel{?}{=} \frac{1}{(-1)-1}$$

$$-\frac{1}{2} \quad \Big| \quad -\frac{1}{2} \qquad \text{TRUE}$$

We look at the original equation and see that 1 makes a denominator 0 and is thus not a solution. The number −1 checks and is a solution.

EXAMPLE 7 Solve: $\dfrac{3}{x-5} + \dfrac{1}{x+5} = \dfrac{2}{x^2-25}$.

The LCM is $(x-5)(x+5)$. We multiply all terms on both sides by $(x-5)(x+5)$:

GS

$$(x-5)(x+5)\left(\frac{3}{x-5} + \frac{1}{x+5}\right) = (x-5)(x+5)\left(\frac{2}{x^2-25}\right)$$

Multiplying by the LCM on both sides

$$(x-5)(x+5)\cdot\frac{3}{x-5} + (x-5)(x+5)\cdot\frac{1}{x+5} = (x-5)(x+5)\cdot\frac{2}{x^2-25}$$

$$3(x+5) + (x-5) = 2 \qquad \text{Simplifying}$$

$$3x + 15 + x - 5 = 2 \qquad \text{Removing parentheses}$$

$$4x + 10 = 2$$

$$4x = -8$$

$$x = -2.$$

The check is left to the student. The number −2 checks and is the solution.

Do Exercises 6 and 7.

14.7 Exercise Set

For Extra Help

MyMathLab®

MathXL® PRACTICE WATCH READ REVIEW

Reading Check

One of the common difficulties with this chapter is being sure about the task at hand. Are you combining expressions using operations to get another *rational expression*, or are you solving equations for which the results are numbers that are *solutions* of an equation? To learn to make these decisions, determine for each of the following exercises the type of answer you should get: "Rational expression" or "Solutions." You need not complete the mathematical operations.

RC1. Add: $\frac{5a}{a^2 - 1} + \frac{a}{a^2 - a}$.

RC2. Solve: $\frac{5}{y - 3} - \frac{30}{y^2 - 9} = 1$.

RC3. Subtract: $\frac{4}{x - 2} - \frac{1}{x + 2}$.

RC4. Divide: $\frac{x + 4}{x - 2} \div \frac{6x}{x^2 - 4}$.

RC5. Solve: $\frac{x^2}{x - 1} = \frac{1}{x - 1}$.

RC6. Solve: $\frac{10}{x} + x = -2$.

RC7. Multiply: $\frac{2t^2}{t^2 - 25} \cdot \frac{t^2 + 10t + 25}{t^8}$.

RC8. Solve: $\frac{7}{x - 4} - \frac{2}{x + 4} = \frac{1}{x^2 - 16}$.

a Solve. Don't forget to check!

1. $\frac{4}{5} - \frac{2}{3} = \frac{x}{9}$

2. $\frac{x}{20} = \frac{3}{8} - \frac{4}{5}$

3. $\frac{3}{5} + \frac{1}{8} = \frac{1}{x}$

4. $\frac{2}{3} + \frac{5}{6} = \frac{1}{x}$

5. $\frac{3}{8} + \frac{4}{5} = \frac{x}{20}$

6. $\frac{3}{5} + \frac{2}{3} = \frac{x}{9}$

7. $\frac{1}{x} = \frac{2}{3} - \frac{5}{6}$

8. $\frac{1}{x} = \frac{1}{8} - \frac{3}{5}$

9. $\frac{1}{6} + \frac{1}{8} = \frac{1}{t}$

10. $\frac{1}{8} + \frac{1}{12} = \frac{1}{t}$

11. $x + \frac{4}{x} = -5$

12. $\frac{10}{x} - x = 3$

13. $\frac{x}{4} - \frac{4}{x} = 0$

14. $\frac{x}{5} - \frac{5}{x} = 0$

15. $\frac{5}{x} = \frac{6}{x} - \frac{1}{3}$

16. $\frac{4}{x} = \frac{5}{x} - \frac{1}{2}$

17. $\frac{5}{3x} + \frac{3}{x} = 1$

18. $\frac{5}{2y} + \frac{8}{y} = 1$

19. $\frac{t-2}{t+3} = \frac{3}{8}$

20. $\frac{x-7}{x+2} = \frac{1}{4}$

21. $\frac{2}{x+1} = \frac{1}{x-2}$

22. $\frac{8}{y-3} = \frac{6}{y+4}$

23. $\frac{x}{6} - \frac{x}{10} = \frac{1}{6}$

24. $\frac{x}{8} - \frac{x}{12} = \frac{1}{8}$

25. $\frac{t+2}{5} - \frac{t-2}{4} = 1$

26. $\frac{x+1}{3} - \frac{x-1}{2} = 1$

27. $\frac{5}{x-1} = \frac{3}{x+2}$

28. $\frac{x-7}{x-9}=\frac{2}{x-9}$

29. $\frac{a-3}{3a+2}=\frac{1}{5}$

30. $\frac{x+7}{8x-5}=\frac{2}{3}$

31. $\frac{x-1}{x-5}=\frac{4}{x-5}$

32. $\frac{y+11}{y+8}=\frac{3}{y+8}$

33. $\frac{2}{x+3}=\frac{5}{x}$

34. $\frac{6}{y}=\frac{5}{y-8}$

35. $\frac{x-2}{x-3}=\frac{x-1}{x+1}$

36. $\frac{t+5}{t-2}=\frac{t-2}{t+4}$

37. $\frac{1}{x+3}+\frac{1}{x-3}=\frac{1}{x^2-9}$

38. $\frac{4}{x-3}+\frac{2x}{x^2-9}=\frac{1}{x+3}$

39. $\frac{x}{x+4}-\frac{4}{x-4}=\frac{x^2+16}{x^2-16}$

40. $\frac{5}{y-3}-\frac{30}{y^2-9}=1$

41. $\frac{4 - a}{8 - a} = \frac{4}{a - 8}$

42. $\frac{3}{x - 7} = \frac{x + 10}{x - 7}$

43. $2 - \frac{a - 2}{a + 3} = \frac{a^2 - 4}{a + 3}$

44. $\frac{5}{x - 1} + x + 1 = \frac{5x + 4}{x - 1}$

45. $\frac{x + 1}{x + 2} = \frac{x + 3}{x + 4}$

46. $\frac{x^2}{x^2 - 4} = \frac{x}{x + 2} - \frac{2x}{2 - x}$

47. $4a - 3 = \frac{a + 13}{a + 1}$

48. $\frac{3x - 9}{x - 3} = \frac{5x - 4}{2}$

49. $\frac{4}{y - 2} - \frac{2y - 3}{y^2 - 4} = \frac{5}{y + 2}$

50. $\frac{y^2 - 4}{y + 3} = 2 - \frac{y - 2}{y + 3}$

Skill Maintenance

Add. [12.4a]

51. $(2x^3 - 4x^2 + x - 7) + (4x^4 + x^3 + 4x^2 + x)$

52. $(2x^3 - 4x^2 + x - 7) + (-2x^3 + 4x^2 - x + 7)$

Factor. [13.6a]

53. $50p^2 - 100$

54. $5p^2 - 40p - 100$

Solve.

55. ***Consecutive Even Integers.*** The product of two consecutive even integers is 360. Find the integers. [13.8a]

56. ***Chemistry.*** About 5 L of oxygen can be dissolved in 100 L of water at 0°C. This is 1.6 times the amount that can be dissolved in the same volume of water at 20°C. How much oxygen can be dissolved in 100 L at 20°C? [10.6a]

Synthesis

57. Solve: $\frac{x}{x^2 + 3x - 4} + \frac{x + 1}{x^2 + 6x + 8} = \frac{2x}{x^2 + x - 2}$.

58. Use a graphing calculator to check the solutions to Exercises 13, 15, and 25.

Applications Using Rational Equations and Proportions

14.8

In many areas of study, applications involving rates, proportions, or reciprocals translate to rational equations. By using the five steps for problem solving and the skills of Sections 14.1–14.7, we can now solve such problems.

OBJECTIVES

a Solve applied problems using rational equations.

b Solve proportion problems.

SKILL TO REVIEW

Objective 10.4b: Solve a formula for a specified letter.

Solve for the indicated letter.

1. $x = w \cdot y$, for w
2. $A = c - bt$, for t

a SOLVING APPLIED PROBLEMS

Problems Involving Motion

Problems that deal with distance, speed (or rate), and time are called **motion problems**. Translation of these problems involves the distance formula, $d = r \cdot t$, and/or the equivalent formulas $r = d/t$ and $t = d/r$.

MOTION FORMULAS

The following are the formulas for motion problems:

$d = rt$; Distance = Rate · Time (basic formula)

$r = \frac{d}{t}$; Rate = Distance/Time

$t = \frac{d}{r}$. Time = Distance/Rate

EXAMPLE 1 *Speed of Sea Animals.* The shortfin Mako shark is known to have the fastest speed of all sharks. The sailfish has the fastest speed of all fish. The top speed recorded for a sailfish is approximately 25 mph faster than the fastest speed of a Mako shark. A sailfish can swim 14 mi in the same time that a Mako shark can swim 9 mi. Find the speed of each sea animal.

Sources: International Union for the Conservation of Nature; theshark.dk/en/records.php; thetravelalmanac.com/lists/fish-speed.htm

1. **Familiarize.** We first make a drawing. We let r = the speed of the shark. Then $r + 25$ = the speed of the sailfish.

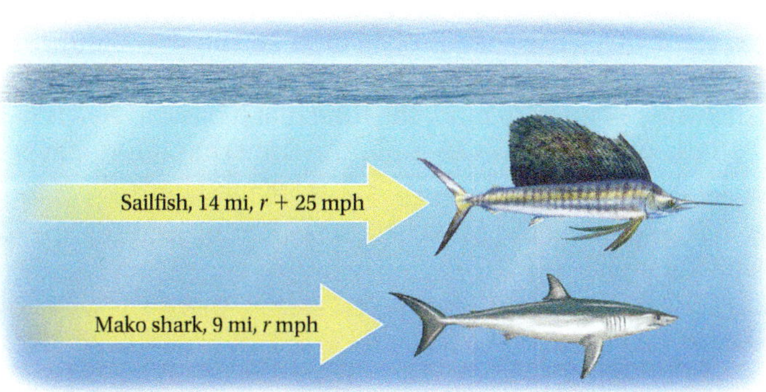

Answers

Skill to Review:

1. $w = \frac{x}{y}$ 2. $t = \frac{c - A}{b}$, or $\frac{A - c}{-b}$

Recall that sometimes we need to use a formula in order to solve an application. As we see above, a formula that relates the notions of distance, speed, and time is $d = rt$, or *Distance* = *Speed* · *Time*.

Since each sea animal travels for the same length of time, we can use just t for time. We organize the information in a chart, as follows.

	DISTANCE (d)	SPEED (r)	TIME (t)	
MAKO SHARK	9	r	t	→ $9 = rt$
SAILFISH	14	$r + 25$	t	→ $14 = (r + 25)t$

2. **Translate.** We can apply the formula $d = rt$ along the rows of the table to obtain two equations:

$$9 = rt \quad \text{and} \quad 14 = (r + 25)t.$$

We know that the sea animals travel for the same length of time. Thus if we solve each equation for t and set the results equal to each other, we get an equation in terms of r.

Solving $9 = rt$ for t: $\quad t = \frac{9}{r}$

Solving $14 = (r + 25)t$ for t: $\quad t = \frac{14}{r + 25}$

Since the times are the same, we have the following equation:

$$\frac{9}{r} = \frac{14}{r + 25}.$$

3. **Solve.** To solve the equation, we first multiply on both sides by the LCM, which is $r(r + 25)$:

$$r(r + 25) \cdot \frac{9}{r} = r(r + 25) \cdot \frac{14}{r + 25}$$ Multiplying on both sides by the LCM, which is $r(r + 25)$

$$9(r + 25) = 14r$$ Simplifying

$$9r + 225 = 14r$$ Removing parentheses

$$225 = 5r$$

$$45 = r.$$

We now have a possible solution. The speed of the shark is 45 mph, and the speed of the sailfish is $r + 25 = 45 + 25$, or 70 mph.

4. **Check.** We check the speeds of 45 mph for the shark and 70 mph for the sailfish. The sailfish does swim 25 mph faster than the shark. If the sailfish swims 14 mi at 70 mph, the time that it has traveled is $\frac{14}{70}$, or $\frac{1}{5}$ hr. If the shark swims 9 mi at 45 mph, the time that it has traveled is $\frac{9}{45}$, or $\frac{1}{5}$ hr. Since the times are the same, the speeds check.

5. **State.** The speed of the Mako shark is 45 mph, and the speed of the sailfish is 70 mph.

Do Exercise 1.

1. ***Driving Speed.*** Catherine drives 20 mph faster than her father, Gary. In the same time that Catherine travels 180 mi, her father travels 120 mi. Find their speeds.

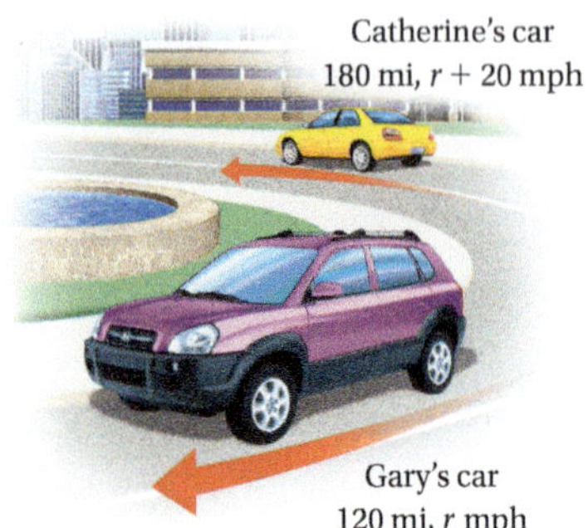

Answer

1. Gary: 40 mph; Catherine: 60 mph

Problems Involving Work

EXAMPLE 2 ***Sodding a Yard.*** Charlie's Lawn Care has two three-person crews who lay sod. Crew A can lay 7 skids of sod in 4 hr, while crew B requires 6 hr to do the same job. How long would it take the two crews working together to lay 7 skids of sod?

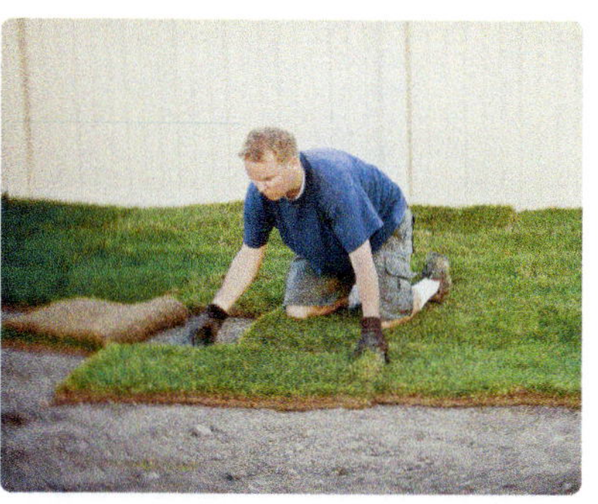

1. **Familiarize.** A common *incorrect* way to translate the problem is to add the two times: $4\text{ hr} + 6\text{ hr} = 10\text{ hr}$. Let's think about this. Crew A can do the job in 4 hr. If crew A and crew B work together, the time that it takes them should be *less* than 4 hr. Thus we reject 10 hr as a solution, but we do have a partial check on any answer we get. The answer should be less than 4 hr.

   We proceed to a translation by considering how much of the job is finished in 1 hr, 2 hr, 3 hr, and so on. It takes crew A 4 hr to do the sodding job alone. Then, in 1 hr, crew A can do $\frac{1}{4}$ of the job. It takes crew B 6 hr to do the job alone. Then, in 1 hr, crew B can do $\frac{1}{6}$ of the job. Working together for 1 hr (see Fig. 1), the crews can do

$$\frac{1}{4} + \frac{1}{6}, \text{ or } \frac{3}{12} + \frac{2}{12}, \text{ or } \frac{5}{12} \text{ of the job in 1 hr.}$$

In one hour:
Crew A Crew B

FIGURE 1

In 2 hr, crew A can do $2\left(\frac{1}{4}\right)$ of the job and crew B can do $2\left(\frac{1}{6}\right)$ of the job. Working together for 2 hr (see Fig. 2), they can do

$$2\left(\frac{1}{4}\right) + 2\left(\frac{1}{6}\right), \text{ or } \frac{6}{12} + \frac{4}{12}, \text{ or } \frac{10}{12}, \text{ or } \frac{5}{6} \text{ of the job in 2 hr.}$$

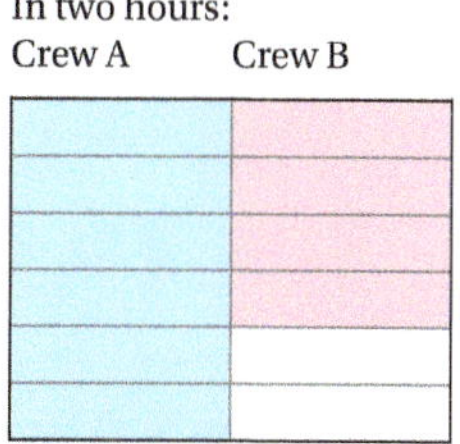

FIGURE 2

TIME	FRACTION OF THE JOB COMPLETED		
	CREW A	CREW B	TOGETHER
1 hr	$\frac{1}{4}$	$\frac{1}{6}$	$\frac{1}{4} + \frac{1}{6}$, or $\frac{5}{12}$
2 hr	$2\left(\frac{1}{4}\right)$	$2\left(\frac{1}{6}\right)$	$2\left(\frac{1}{4}\right) + 2\left(\frac{1}{6}\right)$, or $\frac{5}{6}$
3 hr	$3\left(\frac{1}{4}\right)$	$3\left(\frac{1}{6}\right)$	$3\left(\frac{1}{4}\right) + 3\left(\frac{1}{6}\right)$, or $1\frac{1}{4}$
t hr	$t\left(\frac{1}{4}\right)$	$t\left(\frac{1}{6}\right)$	$t\left(\frac{1}{4}\right) + t\left(\frac{1}{6}\right)$

We see that the answer is somewhere between 2 hr and 3 hr. What we want is a number t such that the fraction of the job that gets completed is 1; that is, the job is just completed.

2. **Translate.** From the table, we see that the time we want is some number t for which

$$t\left(\frac{1}{4}\right) + t\left(\frac{1}{6}\right) = 1, \quad \text{or} \quad \frac{t}{4} + \frac{t}{6} = 1,$$

where 1 represents the idea that the entire job is completed in time t.

3. **Solve.** We solve the equation:

$$12\left(\frac{t}{4}+\frac{t}{6}\right) = 12 \cdot 1 \quad \text{Multiplying by the LCM, which is } 2 \cdot 2 \cdot 3, \text{ or } 12$$

$$12 \cdot \frac{t}{4} + 12 \cdot \frac{t}{6} = 12$$

$$3t + 2t = 12$$

$$5t = 12$$

$$t = \frac{12}{5}, \text{ or } 2\frac{2}{5}\text{ hr.}$$

4. **Check.** In $\frac{12}{5}$ hr, crew A does $\frac{12}{5} \cdot \frac{1}{4}$, or $\frac{3}{5}$, of the job and crew B does $\frac{12}{5} \cdot \frac{1}{6}$, or $\frac{2}{5}$, of the job. Together, they do $\frac{3}{5} + \frac{2}{5}$, or 1 entire job. The answer, $2\frac{2}{5}$ hr, is between 2 hr and 3 hr (see the table), and it is less than 4 hr, the time it takes crew A working alone. The answer checks.
5. **State.** It takes $2\frac{2}{5}$ hr for crew A and crew B working together to lay 7 skids of sod.

2. *Work Recycling.* Emma and Evan work as volunteers at a community recycling center. Emma can sort a morning's accumulation of recyclable objects in 3 hr, while Evan requires 5 hr to do the same job. How long would it take them, working together, to sort the recyclable material?

THE WORK PRINCIPLE

Suppose a = the time that it takes A to do a job, b = the time that it takes B to do the same job, and t = the time that it takes them to do the job working together. Then

$$\frac{t}{a} + \frac{t}{b} = 1.$$

Do Exercise 2.

b APPLICATIONS INVOLVING PROPORTIONS

We now consider applications with proportions. A **proportion** involves ratios. A **ratio** of two quantities is their quotient. For example, 73% is the ratio of 73 to 100, $\frac{73}{100}$. The ratio of two different kinds of measure is called a **rate**. Suppose an animal travels 2720 ft in 2.5 hr. Its **rate**, or **speed**, is then

$$\frac{2720\text{ ft}}{2.5\text{ hr}} = 1088\,\frac{\text{ft}}{\text{hr}}.$$

3. Find the ratio of 145 km to 2.5 liters (L).
4. *Batting Average.* Recently, a baseball player got 7 hits in 25 times at bat. What was the rate, or batting average, in number of hits per times at bat?
5. Impulses in nerve fibers travel 310 km in 2.5 hr. What is the rate, or speed, in kilometers per hour?
6. A lake of area 550 yd^2 contains 1320 fish. What is the population density of the lake, in number of fish per square yard?

Do Exercises 3–6.

PROPORTION

An equality of ratios,

$$\frac{A}{B} = \frac{C}{D},$$

is called a **proportion**. The numbers within a proportion are said to be **proportional** to each other.

Answers

2. $1\frac{7}{8}$ hr 3. 58 km/L
4. 0.28 hit per times at bat 5. 124 km/h
6. 2.4 fish/yd^2

EXAMPLE 3 *Mileage.* A 2013 Fiat 500 Turbo can travel 306 mi in highway driving on 9 gal of gas. Find the amount of gas required for 425 mi of highway driving.

Source: *Car and Driver,* January 2013

1. **Familiarize.** We know that the Fiat can travel 306 mi on 9 gal of gas. Thus we can set up a proportion, letting $x =$ the number of gallons of gas required to drive 425 mi.
2. **Translate.** We assume that the car uses gas at the same rate in all highway driving. Thus the ratios are the same and we can write a proportion. Note that the units of *mileage* are in the numerators and the units of *gasoline* are in the denominators.

$$\begin{array}{r} \text{Miles} \rightarrow \\ \text{Gas} \rightarrow \end{array} \frac{306}{9} = \frac{425}{x} \begin{array}{l} \leftarrow \text{Miles} \\ \leftarrow \text{Gas} \end{array}$$

3. **Solve.** To solve for x, we multiply on both sides by the LCM, which is $9x$:

$$9x \cdot \frac{306}{9} = 9x \cdot \frac{425}{x} \qquad \text{Multiplying by } 9x$$

$$306x = 3825 \qquad \text{Simplifying}$$

$$\frac{306x}{306} = \frac{3825}{306} \qquad \text{Dividing by 306}$$

$$x = 12.5. \qquad \text{Simplifying}$$

We can also use cross products to solve the proportion:

$$\frac{306}{9} = \frac{425}{x} \qquad 306 \cdot x \text{ and } 9 \cdot 425 \text{ are cross products.}$$

$$306 \cdot x = 9 \cdot 425 \qquad \text{Equating cross products}$$

$$\frac{306x}{306} = \frac{3825}{306} \qquad \text{Dividing by 306}$$

$$x = 12.5.$$

4. **Check.** The check is left to the student.
5. **State.** The Fiat will require 12.5 gal of gas for 425 mi of highway driving.

Do Exercise 7. ▶

7. *Mileage.* In city driving, a 2013 Ford Mustang GT can travel 105 mi on 7 gal of gas. How much gas will be required for 243 mi of city driving?

Source: *Motor Trend,* March 2013

EXAMPLE 4 *Fruit Quality.* A company that prepares and sells gift boxes and baskets of fruit must order quantities of fruit larger than what they need to allow for selecting fruit that meets their quality standards. The packing-room supervisor keeps records and notes that approximately 87 pears from a shipment of 1000 do not meet the company standards. Over the holidays, a shipment of 3200 pears is ordered. How many pears can the company expect will not meet the quality required?

Answer

7. 16.2 gal

1. **Familiarize.** The ratio of the number of pears P that do not meet the standards to the total order of 3200 is $P/3200$. The ratio of the average number of pears that do not meet the standard in an order of 1000 pears is $\frac{87}{1000}$.
2. **Translate.** Assuming that the two ratios are the same, we can translate to a proportion:

$$\frac{P}{3200} = \frac{87}{1000}.$$

3. **Solve.** We solve the proportion. We multiply by the LCM, which is 16,000.

$$16{,}000 \cdot \frac{P}{3200} = 16{,}000 \cdot \frac{87}{1000}$$
$$5 \cdot P = 16 \cdot 87$$
$$P = \frac{16 \cdot 87}{5}$$
$$P \approx 278.4$$

4. **Check.** The check is left to the student.
5. **State.** We estimate that in an order of 3200 pears, there are about 278 pears that do not meet the quality standards.

8. *Chlorine for a Pool.* XYZ Pools and Spas, Inc., adds 2 gal of chlorine per 8000 gal of water in a newly constructed pool. How much chlorine is needed for a pool requiring 20,500 gal of water? Round the answer to the nearest tenth of a gallon.

Do Exercise 8.

Similar Triangles

Proportions arise in geometry when we are studying *similar triangles*. If two triangles are **similar**, then their corresponding angles have the same measure and their corresponding sides are proportional. To illustrate, if triangle ABC is similar to triangle RST, then angles A and R have the same measure, angles B and S have the same measure, angles C and T have the same measure, and

$$\frac{a}{r} = \frac{b}{s} = \frac{c}{t}.$$

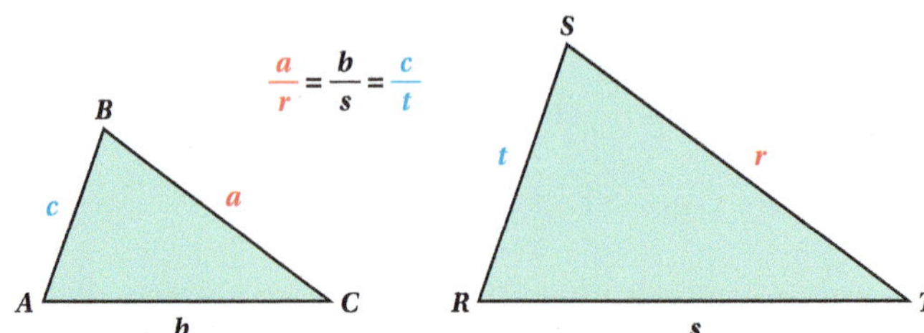

Answer
8. 5.1 gal

SIMILAR TRIANGLES

In **similar triangles**, corresponding angles have the same measure and the lengths of corresponding sides are proportional.

EXAMPLE 5 *Similar Triangles.* Triangles ABC and XYZ below are similar triangles. Solve for z if $a = 8$, $c = 5$, and $x = 10$.

We make a drawing, write a proportion, and then solve. Note that side a is always opposite angle A, side x is always opposite angle X, and so on.

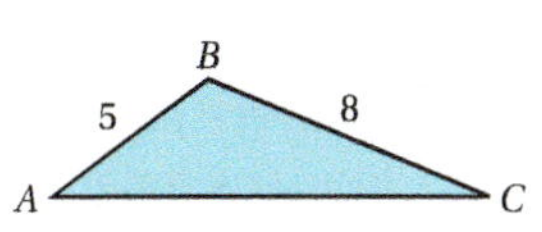

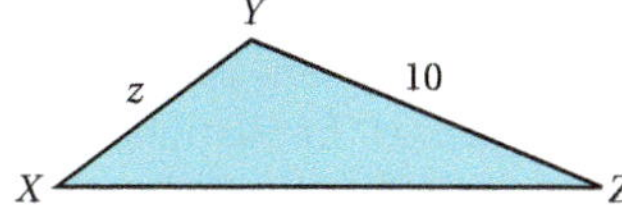

We have

$$\frac{z}{5} = \frac{10}{8} \qquad \text{The proportion } \frac{5}{z} = \frac{8}{10} \text{ could also be used.}$$

$$40 \cdot \frac{z}{5} = 40 \cdot \frac{10}{8} \qquad \text{Multiplying by 40}$$

$$8z = 50$$

$$z = \frac{50}{8} \qquad \text{Dividing by 8}$$

$$z = \frac{25}{4}, \text{ or } 6.25.$$

Do Exercise 9. ▶

9. ***Height of a Flagpole.*** How high is a flagpole that casts a 45-ft shadow at the same time that a 5.5-ft woman casts a 10-ft shadow?

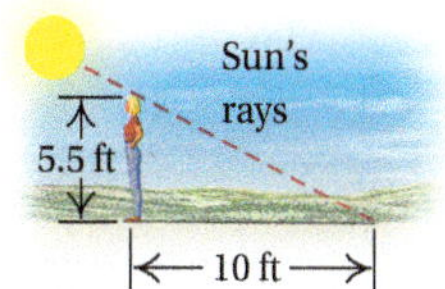

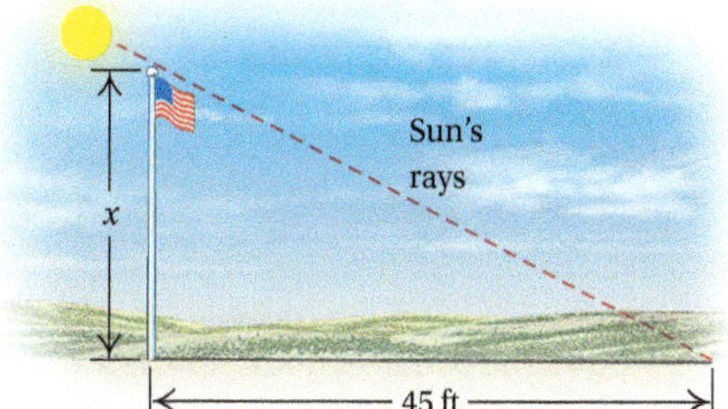

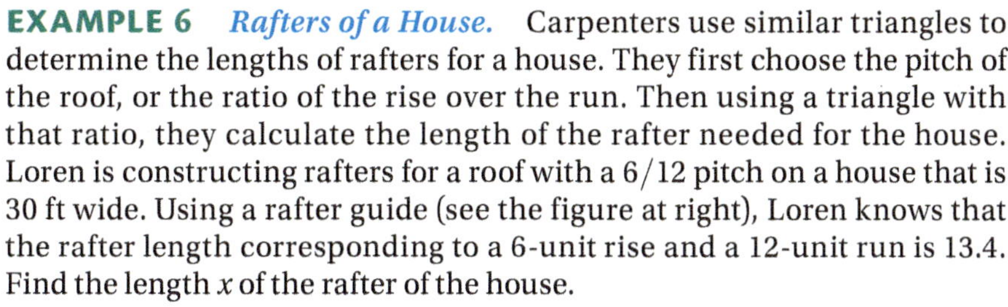
EXAMPLE 6 *Rafters of a House.* Carpenters use similar triangles to determine the lengths of rafters for a house. They first choose the pitch of the roof, or the ratio of the rise over the run. Then using a triangle with that ratio, they calculate the length of the rafter needed for the house. Loren is constructing rafters for a roof with a 6/12 pitch on a house that is 30 ft wide. Using a rafter guide (see the figure at right), Loren knows that the rafter length corresponding to a 6-unit rise and a 12-unit run is 13.4. Find the length x of the rafter of the house.

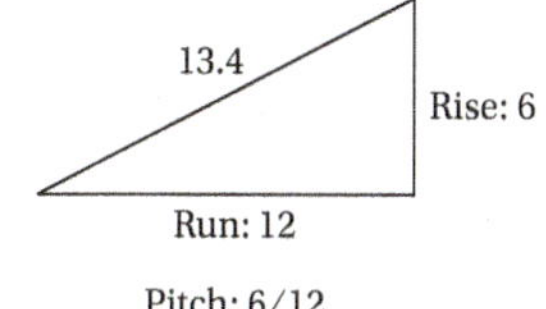

We have the proportion

Length of rafter in 6/12 triangle → 13.4; Length of rafter on the house → x; Run in 6/12 triangle → 12; Run in similar triangle on the house → 15

$$\frac{13.4}{x} = \frac{12}{15}.$$

Solve:

$$13.4 \cdot 15 = x \cdot 12 \qquad \text{Equating cross products}$$

$$\frac{13.4 \cdot 15}{12} = \frac{x \cdot 12}{12} \qquad \text{Dividing by 12 on both sides}$$

$$\frac{13.4 \cdot 15}{12} = x$$

$$16.75 = x$$

The length of the rafter x of the house is about 16.75 ft, or 16 ft 9 in.

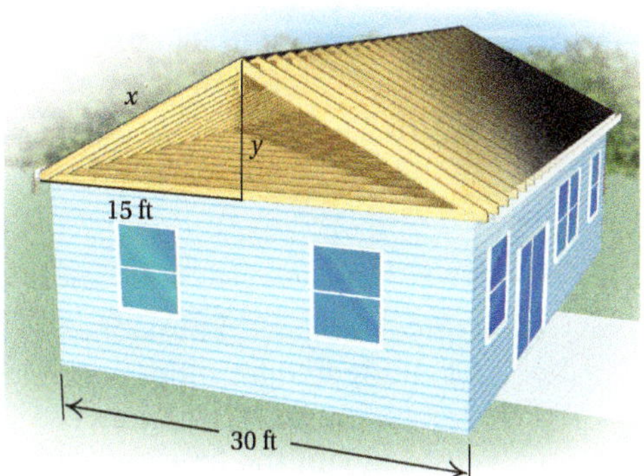

Do Exercise 10. ▶

10. ***Rafters of a House.*** Refer to Example 6. Find the length y in the rafter of the house.

Answers

9. 24.75 ft **10.** 7.5 ft

Translating for Success

1. *Pharmaceutical Research.* In 2013, a pharmaceutical firm spent \$3.6 million on marketing a new drug. This was a 25% increase over the amount spent for marketing in 2012. How much was spent in 2012?

2. *Cycling Distance.* A bicyclist traveled 197 mi in 7 days. At this rate, how many miles could the cyclist travel in 30 days?

3. *Bicycling.* The speed of one bicyclist is 2 km/h faster than the speed of another bicyclist. The first bicyclist travels 60 km in the same amount of time that it takes the second to travel 50 km. Find the speed of each bicyclist.

4. *Filling Time.* A swimming pool can be filled in 5 hr by hose A alone and in 6 hr by hose B alone. How long would it take to fill the tank if both hoses were working?

5. *Office Budget.* Emma has \$36 budgeted for office stationery. Engraved stationery costs \$20 for the first 25 sheets and \$0.08 for each additional sheet. How many engraved sheets of stationery can Emma order and still stay within her budget?

The goal of these matching questions is to practice step (2), Translate, of the five-step problem-solving process. Translate each word problem to an equation and select a correct translation from equations A–O.

A. $2x + 2(x + 1) = 613$

B. $x^2 + (x + 1)^2 = 613$

C. $\dfrac{60}{x + 2} = \dfrac{50}{x}$

D. $20 + 0.08(x - 25) = 36$

E. $\dfrac{197}{7} = \dfrac{x}{30}$

F. $x + (x + 1) = 613$

G. $\dfrac{7}{197} = \dfrac{x}{30}$

H. $x^2 + (x + 2)^2 = 612$

I. $x^2 + (x + 1)^2 = 612$

J. $\dfrac{50}{x + 2} = \dfrac{60}{x}$

K. $x + 25\% \cdot x = 3.6$

L. $t + 5 = 7$

M. $x^2 + (x + 1)^2 = 452$

N. $\dfrac{1}{5} + \dfrac{1}{6} = \dfrac{1}{t}$

O. $x^2 + (x + 2)^2 = 452$

Answers on page A-34

6. *Sides of a Square.* If each side of a square is increased by 2 ft, the area of the original square plus the area of the enlarged square is 452 ft^2. Find the length of a side of the original square.

7. *Consecutive Integers.* The sum of two consecutive integers is 613. Find the integers.

8. *Sums of Squares.* The sum of the squares of two consecutive odd integers is 612. Find the integers.

9. *Sums of Squares.* The sum of the squares of two consecutive integers is 613. Find the integers.

10. *Rectangle Dimensions.* The length of a rectangle is 1 ft longer than its width. Find the dimensions of the rectangle such that the perimeter of the rectangle is 613 ft.

14.8 Exercise Set

For Extra Help MyMathLab® MathXL® PRACTICE WATCH READ REVIEW

Reading Check

Complete each statement with the appropriate word(s) from the column on the right.

same
product
distance
proportion
similar
different
quotient
rate
proportional
corresponding

RC1. If two triangles are similar, then their _______________ angles have the _______________ measures and their corresponding sides are _______________.

RC2. A ratio of two quantities is their _______________.

RC3. An equality of ratios, $\frac{A}{B} = \frac{C}{D}$, is called a(n) _______________.

RC4. Distance equals _______________ times time.

RC5. Rate equals _______________ divided by time.

a Solve.

1. *Car Speed.* Rick drives his four-wheel-drive truck 40 km/h faster than Sarah drives her Kia. While Sarah travels 150 km, Rick travels 350 km. Find their speeds.

Complete this table as part of the *Familiarize* step.

$d = r \cdot t$

	DISTANCE	SPEED	TIME
Car	150	r	
Truck	350		t

2. *Train Speed.* The speed of a CSW freight train is 14 mph slower than the speed of an Amtrak passenger train. The freight train travels 330 mi in the same time that it takes the passenger train to travel 400 mi. Find the speed of each train.

Complete this table as part of the *Familiarize* step.

$d = r \cdot t$

	DISTANCE	SPEED	TIME
CSW	330		t
Amtrak	400	r	

3. *Animal Speeds.* An ostrich can run 8 mph faster than a giraffe. An ostrich can run 5 mi in the same time that a giraffe can run 4 mi. Find the speed of each animal.

Source: www.infoplease.com

4. *Animal Speeds.* A cheetah can run 28 mph faster than a gray fox. A cheetah can run 10 mi in the same time that a gray fox can run 6 mi. Find the speed of each animal.

Source: www.infoplease.com

5. *Bicycle Speed.* Hank bicycles 5 km/h slower than Kelly. In the time that it takes Hank to bicycle 42 km, Kelly can bicycle 57 km. How fast does each bicyclist travel?

6. *Driving Speed.* Kaylee's Lexus travels 30 mph faster than Gavin's Harley. In the same time that Gavin travels 75 mi, Kaylee travels 120 mi. Find their speeds.

7. *Trucking Speed.* A long-distance trucker traveled 120 mi in one direction during a snowstorm. The return trip in rainy weather was accomplished at double the speed and took 3 hr less time. Find the speed going.

8. *Car Speed.* After driving 126 mi, Syd found that the drive would have taken 1 hr less time by increasing the speed by 8 mph. What was the actual speed?

9. *Walking Speed.* Bonnie power walks 3 km/h faster than Ralph. In the time that it takes Ralph to walk 7.5 km, Bonnie walks 12 km. Find their speeds.

10. *Cross-Country Skiing.* Gerard skis cross-country 4 km/h faster than Sally. In the time that it takes Sally to ski 18 km, Gerard skis 24 km. Find their speeds.

11. *Boat Speed.* Tory and Emilio's motorboats travel at the same speed. Tory pilots her boat 40 km before docking. Emilio continues for another 2 hr, traveling a total of 100 km before docking. How long did it take Tory to navigate the 40 km?

12. *Tractor Speed.* Hobart's tractor is just as fast as Evan's. It takes Hobart 1 hr more than it takes Evan to drive to town. If Hobart is 20 mi from town and Evan is 15 mi from town, how long does it take Evan to drive to town?

13. *Gardening.* Nicole can weed her vegetable garden in 50 min. Glen can weed the same garden in 40 min. How long would it take if they worked together?

14. *Harvesting.* Bobbi can pick a quart of raspberries in 20 min. Blanche can pick a quart in 25 min. How long would it take if Bobbi and Blanche worked together?

15. *Shoveling.* Vern can shovel the snow from his driveway in 45 min. Nina can do the same job in 60 min. How long would it take Nina and Vern to shovel the driveway if they worked together?

16. *Raking.* Zoë can rake her yard in 4 hr. Steffi does the same job in 3 hr. How long would it take them, working together, to rake the yard?

17. *Deli Trays.* A grocery needs to prepare a large order of deli trays for Super Bowl weekend. It would take Henry 8.5 hr to prepare the trays. Carly can complete the job in 10.4 hr. How long would it take them, working together, to prepare the trays? Round the time to the nearest tenth of an hour.

18. *School Photos.* Rebecca can take photos for an elementary school with 325 students in 11.5 days. Jack can complete the same job in 9.2 days. How long would it take them working together? Round the time to the nearest tenth of a day.

19. *Wiring.* By checking work records, a contractor finds that Peggyann can wire a home theater in 9 hr. It takes Matthew 7 hr to wire the same room. How long would it take if they worked together?

20. *Plumbing.* By checking work records, a plumber finds that Raul can plumb a house in 48 hr. Mira can do the same job in 36 hr. How long would it take if they worked together?

21. *Office Printers.* The HP Officejet 4215 All-In-One printer, fax, scanner, and copier can print one black-and-white copy of a company's year-end report in 10 min. The HP Officejet 7410 All-In-One can print the same report in 6 min. How long would it take the two printers, working together, to print one copy of the report?

22. *Office Copiers.* The HP Officejet 7410 All-In-One printer, fax, scanner, and copier can make a color copy of a staff training manual in 9 min. The HP Officejet 4215 All-In-One can copy the same report in 15 min. How long would it take the two copiers, working together, to make one copy of the manual?

b Find the ratio of each of the following. Simplify, if possible.

23. 60 students, 18 teachers

24. 800 mi, 50 gal

25. *Speed of Black Racer.* A black racer snake travels 4.6 km in 2 hr. What is the speed, in kilometers per hour?

26. *Speed of Light.* Light travels 558,000 mi in 3 sec. What is the speed, in miles per second?

Solve.

27. *Protein Needs.* A 120-lb person should eat a minimum of 44 g of protein each day. How much protein should a 180-lb person eat each day?

28. *Coffee Beans.* The coffee beans from 14 trees are required to produce 7.7 kg of coffee. (This is the amount that the average person in the United States drinks each year.) How many trees are required to produce 320 kg of coffee?

29. *Hemoglobin.* A normal 10-cc specimen of human blood contains 1.2 g of hemoglobin. How much hemoglobin would 16 cc of the same blood contain?

30. *Walking Speed.* Wanda walked 234 km in 14 days. At this rate, how far would she walk in 42 days?

31. *Mileage.* A 2014 Subaru Forester can travel 242 mi of city driving on 11 gal of gas. Find the amount of gas required for 176 mi of city driving.

Source: *Car and Driver*, January 2013

32. *Mileage.* A 2013 RAM 1500 SLT crew cab can travel 575 mi of highway driving on 23 gal of gas. Find the amount of gas required for 850 mi of highway driving.

Source: *Car and Driver*, February 2013

33. *Estimating Trout Population.* To determine the number of trout in a lake, a conservationist catches 112 trout, tags them, and throws them back into the lake. Later, 82 trout are caught; 32 of them are tagged. Estimate the number of trout in the lake.

34. *Grass Seed.* It takes 60 oz of grass seed to seed 3000 ft^2 of lawn. At this rate, how much would be needed to seed 5000 ft^2 of lawn?

35. *Quality Control.* A sample of 144 firecrackers contained 9 "duds." How many duds would you expect in a sample of 3200 firecrackers?

36. *Frog Population.* To estimate how many frogs there are in a rain forest, a research team tags 600 frogs and then releases them. Later, the team catches 300 frogs and notes that 25 of them have been tagged. Estimate the total frog population in the rain forest.

37. ***Mishandled Baggage.*** The rate of the number of bags mishandled by the airlines fell to 3.39 bags per 1000 passengers in 2011. At this rate, estimate how many bags would be mishandled for 1450 passengers. Round the answer to the nearest hundredth.

Source: U.S. Department of Transportation, Bureau of Transportation Statistics

38. ***Left-Handed People.*** Research shows that 3 in 20 people are left-handed. At a recent NFL game, 63,240 people were in attendance. How many are left-handed?

Source: *Scientific American*

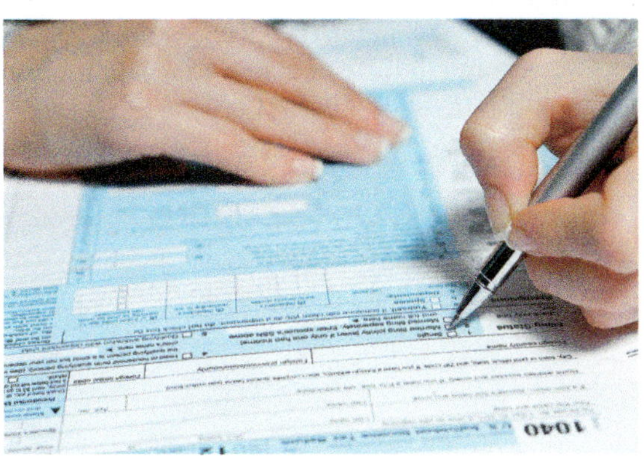

39. ***Honey Bees.*** Making 1 lb of honey requires 20,000 trips by bees to flowers to gather nectar. How many pounds of honey would 35,000 trips produce?

Source: Tom Turpin, Professor of Entomology, Purdue University

40. ***Money.*** The ratio of the weight of copper to the weight of zinc in a U.S. penny is $\frac{1}{39}$. If 50 kg of zinc is being turned into pennies, how much copper is needed?

Geometry. For each pair of similar triangles, find the length of the indicated side.

41. b:

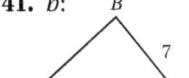

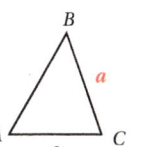

B, 7, A, C, b

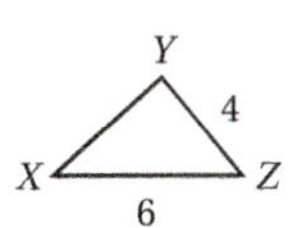

42. a:

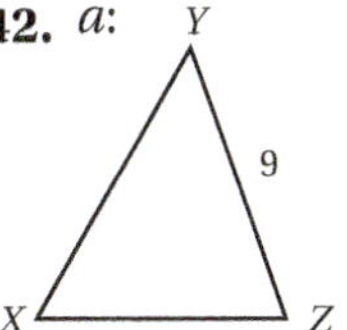

B, a, A, C, 6

43. f:

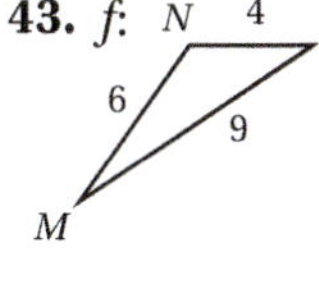

G, f, H, 4, 6, F

44. r:

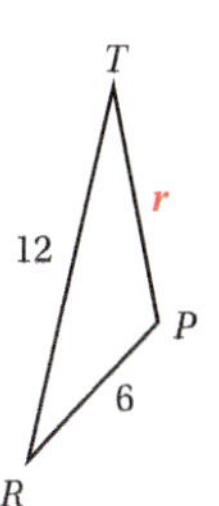

N, 10, 16, O, 8, M

45. h:

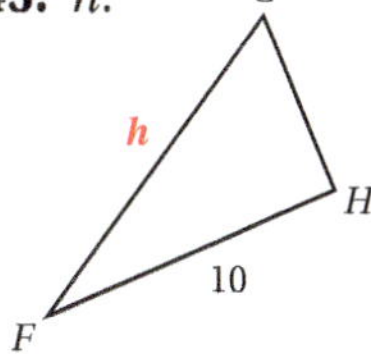

S, 7, T, 6, R

46. n:

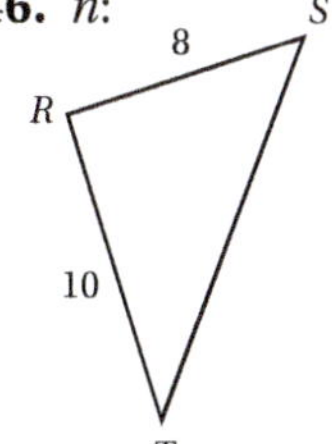

5, N, M, n, P

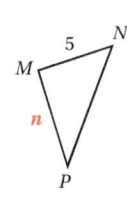

47. ***Environmental Science.*** The Fish and Wildlife Division of the Indiana Department of Natural Resources recently completed a study that determined the number of largemouth bass in Lake Monroe, near Bloomington, Indiana. For this project, anglers caught 300 largemouth bass, tagged them, and threw them back into the lake. Later, they caught 85 largemouth bass and found that 15 of them were tagged. Estimate how many largemouth bass are in the lake.

Source: Department of Natural Resources, Fish and Wildlife Division, Kevin Hoffman

48. ***Environmental Science.*** To determine the number of humpback whales in a pod, a marine biologist, using tail markings, identifies 27 members of the pod. Several weeks later, 40 whales from the pod are randomly sighted. Of the 40 sighted, 12 are from the 27 originally identified. Estimate the number of whales in the pod.

Skill Maintenance

Find the slope, if it exists, of the line containing the given pair of points. [11.3a]

49. $(7, -6), (0, -6)$

50. $(3, -11), (-4, 3)$

Simplify. [12.1d, f]

51. $x^5 \cdot x^6$

52. $x^{-5} \cdot x^6$

53. $x^{-5} \cdot x^{-6}$

54. $x^5 \cdot x^{-6}$

Graph.

55. $y = -\frac{3}{4}x + 2$ [11.1d]

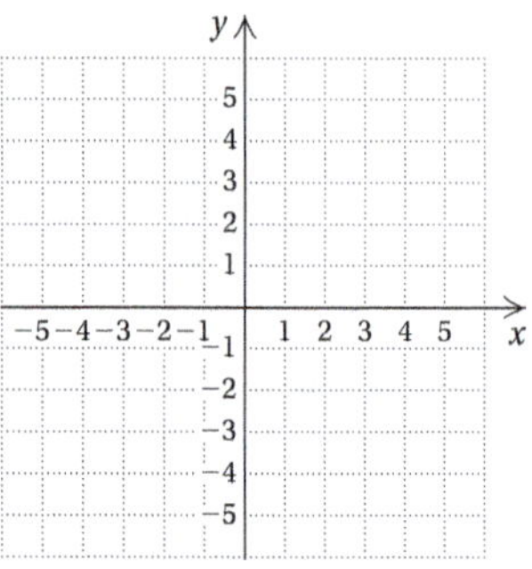

56. $y = \frac{2}{5}x - 4$ [11.1d]

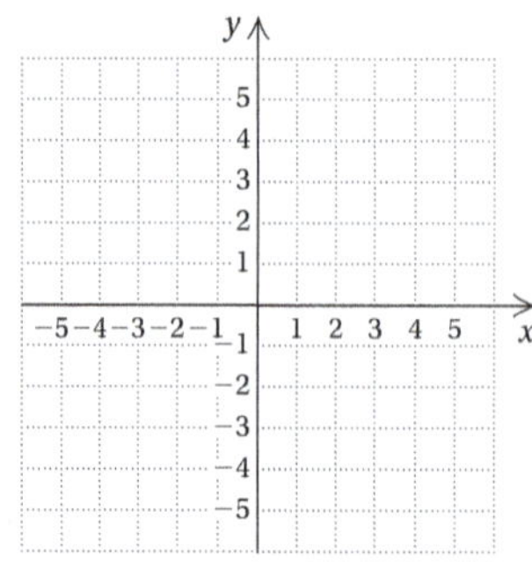

57. $x = -3$ [11.2b]

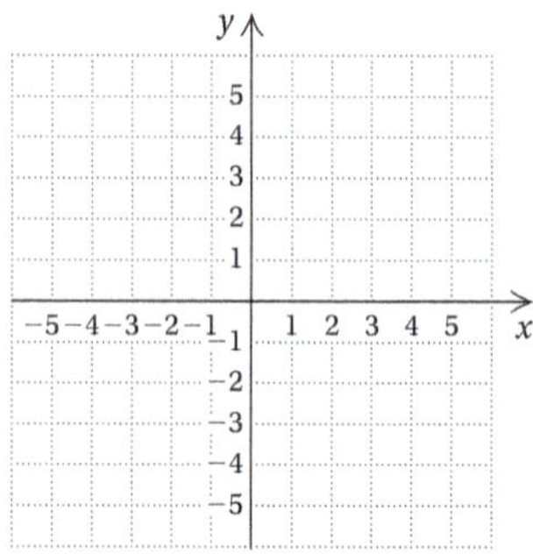

Synthesis

58. Rachel allows herself 1 hr to reach a sales appointment 50 mi away. After she has driven 30 mi, she realizes that she must increase her speed by 15 mph in order to arrive on time. What was her speed for the first 30 mi?

59. How soon, in minutes, after 5 o'clock will the hands on a clock first be together?

Direct Variation and Inverse Variation

14.9

OBJECTIVES

a Find an equation of direct variation given a pair of values of the variables.

b Solve applied problems involving direct variation.

c Find an equation of inverse variation given a pair of values of the variables.

d Solve applied problems involving inverse variation.

a EQUATIONS OF DIRECT VARIATION

A bicycle is traveling at a speed of 15 km/h. In 1 hr, it goes 15 km; in 2 hr, it goes 30 km; in 3 hr, it goes 45 km; and so on. We can form a set of ordered pairs using the number of hours as the first coordinate and the number of kilometers traveled as the second coordinate. These determine a set of ordered pairs:

$$(1, 15), \quad (2, 30), \quad (3, 45), \quad (4, 60), \quad \text{and so on.}$$

Note that the second coordinate is always 15 times the first.

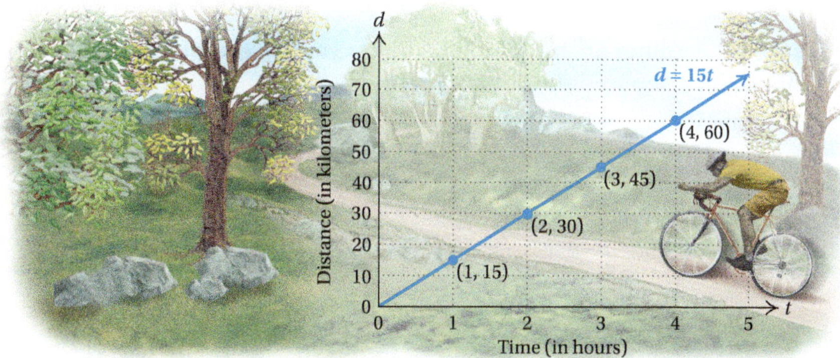

In this example, distance is a constant multiple of time, so we say that there is *direct variation* and that distance *varies directly* as time. The *equation of variation* is $d = 15t$.

DIRECT VARIATION

When a situation translates to an equation described by $y = kx$, with k a positive constant, we say that ***y* varies directly as *x***. The equation $y = kx$ is called an **equation of direct variation**.

In direct variation, as one variable increases, the other variable increases as well. This is shown in the graph above.

The terminologies

"y varies as x,"

"y is directly proportional to x," and

"y is proportional to x"

also imply direct variation and are used in many situations. The constant k is called the **constant of proportionality**, or the **variation constant**. It can be found if one pair of values of x and y is known. Once k is known, other pairs can be determined.

EXAMPLE 1 Find an equation of variation in which y varies directly as x, and $y = 7$ when $x = 25$.

We first substitute to find k:

$$y = kx$$

$$7 = k \cdot 25 \qquad \text{Substituting 25 for } x \text{ and 7 for } y$$

$$\frac{7}{25} = k, \quad \text{or} \quad k = 0.28. \qquad \text{Solving for } k\text{, the variation constant}$$

Then the equation of variation is

$$y = 0.28x.$$

The answer is the equation $y = 0.28x$, *not* simply $k = 0.28$. We can visualize the example by looking at the graph at left.

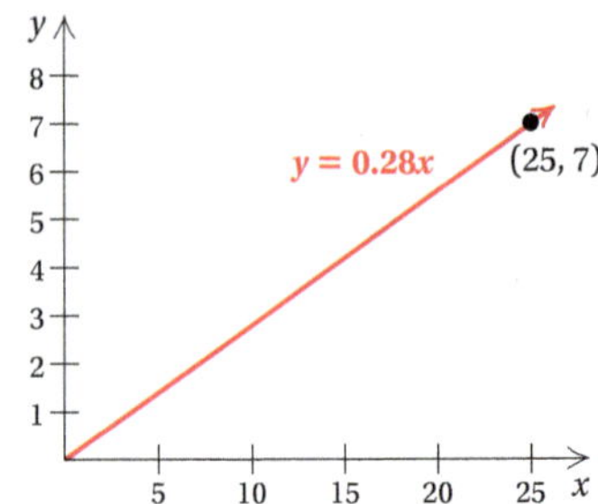

We see that when y varies directly as x, the constant of proportionality is also the slope of the associated graph—the rate at which y changes with respect to x.

1. Find an equation of variation in which y varies directly as x, and $y = 84$ when $x = 12$.

$$y = kx$$
$$84 = k \cdot \square$$
$$k = \frac{\square}{12} = \square$$
$$y = 7 \cdot \square$$
$$y = 7x = 7 \cdot \square$$
$$= \square$$

2. Find an equation of variation in which y varies directly as x, and $y = 50$ when $x = 80$. Then find the value of y when $x = 20$.

EXAMPLE 2 Find an equation in which s varies directly as t, and $s = 10$ when $t = 15$. Then find the value of s when $t = 32$.

We have

$$s = kt \qquad \text{We know that } s \text{ varies directly as } t.$$

$$10 = k \cdot 15 \qquad \text{Substituting 10 for } s \text{ and 15 for } t$$

$$\frac{10}{15} = k, \quad \text{or} \quad k = \frac{2}{3}. \qquad \text{Solving for } k$$

Thus the equation of variation is $s = \frac{2}{3}t$.

$$s = \frac{2}{3}t$$

$$= \frac{2}{3} \cdot 32 \qquad \text{Substituting 32 for } t$$

$$= \frac{64}{3}, \text{ or } 21\frac{1}{3}$$

The value of s is $21\frac{1}{3}$ when $t = 32$.

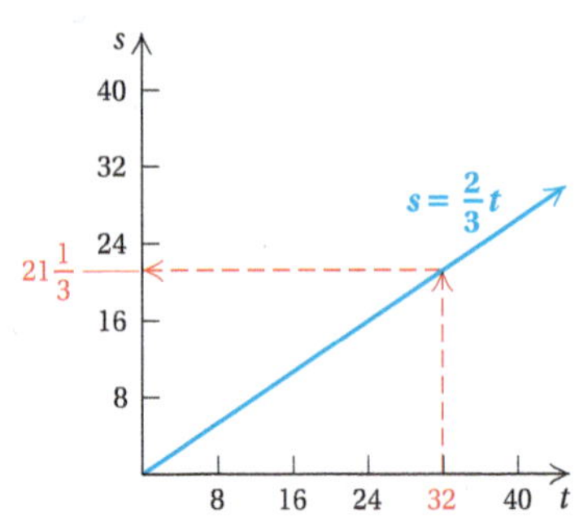

Do Exercises 1 and 2.

b APPLICATIONS OF DIRECT VARIATION

EXAMPLE 3 *Oatmeal Servings.* The number of servings S of oatmeal varies directly as the net weight W of the container purchased. A 42-oz box of oatmeal contains 30 servings. How many servings does a 63-oz box of oatmeal contain?

1., 2. Familiarize and **Translate.** The problem states that we have direct variation between the variables, S and W. Thus an equation $S = kW$, $k > 0$, applies. As the weight of the container increases, the number of servings increases.

Answers

1. $y = 7x$; 287 2. $y = \frac{5}{8}x$; $\frac{25}{2}$

Guided Solution:
1. 12, 84, 7, x; 41, 287

3. **Solve.** The mathematical manipulation has two parts. First, we determine the equation of variation by substituting known values for S and W to find the variation constant k. Second, we compute the number of servings in a 63-oz box of oatmeal.

   a) First, we find an equation of variation:

$$S = kW$$

$$30 = k(42) \qquad \text{Substituting 30 for } S \text{ and 42 for } W$$

$$\frac{30}{42} = k$$

$$\frac{5}{7} = k. \qquad \text{Simplifying}$$

   The equation of variation is $S = \frac{5}{7}W$.

   b) We then use the equation to find the number of servings in a 63-oz box of oatmeal:

$$S = \frac{5}{7}W$$

$$= \frac{5}{7} \cdot 63 \qquad \text{Substituting 63 for } W$$

$$= 45.$$

4. **Check.** The check might be done by repeating the computations. You might also do some reasoning about the answer. The number of servings increased from 30 to 45. Similarly, the weight increased from 42 oz to 63 oz. The answer seems reasonable.

5. **State.** A 63-oz box of oatmeal contains 45 servings.

Do Exercises 3 and 4. ▶

3. ***Gold.*** The karat rating of a gold object varies directly as the percentage of gold in the object. A 14-karat gold chain is 58.25% gold. What is the percentage of gold in a 10-karat gold chain?

4. ***Weight on Venus.*** The weight V of an object on Venus varies directly as its weight E on Earth. A person weighing 165 lb on Earth would weigh 145.2 lb on Venus.

   a) Find an equation of variation.

   b) How much would a person weighing 198 lb on Earth weigh on Venus?

Let's consider direct variation from the standpoint of a graph. The graph of $y = kx$, $k > 0$, always goes through the origin and rises from left to right. Note that as x increases, y increases; and as x decreases, y decreases. This is why the terminology "direct" is used. What one variable does, the other does as well.

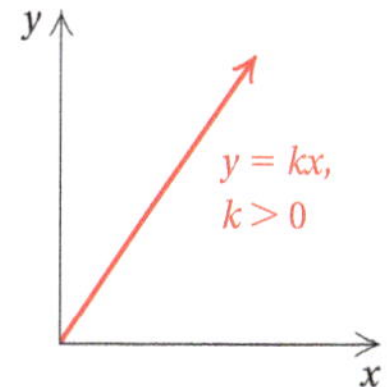

C EQUATIONS OF INVERSE VARIATION

A car is traveling a distance of 20 mi. At a speed of 5 mph, it will take 4 hr; at 20 mph, it will take 1 hr; at 40 mph, it will take $\frac{1}{2}$ hr; and so on. We use the speed as the first coordinate and the time as the second coordinate. These determine a set of ordered pairs:

$$(5, 4), \quad (20, 1), \quad \left(40, \tfrac{1}{2}\right), \quad \left(60, \tfrac{1}{3}\right), \quad \text{and so on.}$$

Note that the product of speed and time for each of these pairs is 20. Note too that as the speed *increases,* the time *decreases.*

In this case, the product of speed and time is constant so we say that there is *inverse variation* and that time *varies inversely* as speed. The equation of variation is

$$rt = 20 \text{ (a constant)}, \quad \text{or} \quad t = \frac{20}{r}.$$

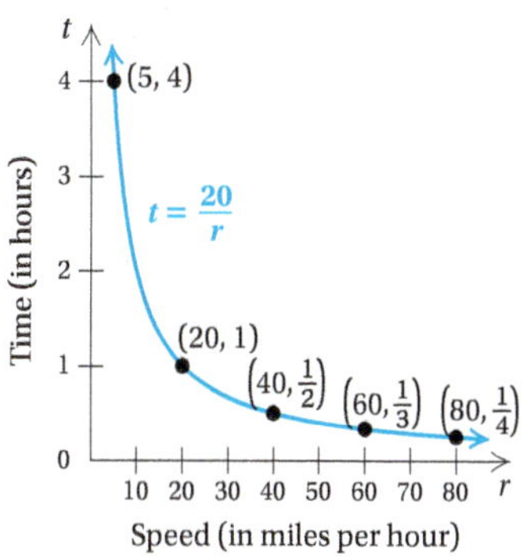

Answers

3. About 41.6%
4. (a) $V = 0.88E$; (b) 174.24 lb

INVERSE VARIATION

When a situation translates to an equation described by $y = k/x$, with k a positive constant, we say that **y varies inversely as x**. The equation $y = k/x$ is called an **equation of inverse variation**.

In inverse variation, as one variable increases, the other variable decreases.

The terminology

"y is inversely proportional to x"

also implies inverse variation and is used in some situations. The constant k is again called the **constant of proportionality**, or the **variation constant**.

EXAMPLE 4 Find an equation of variation in which y varies inversely as x, and $y = 145$ when $x = 0.8$. Then find the value of y when $x = 25$.

We first substitute to find k:

$$y = \frac{k}{x}$$

$$145 = \frac{k}{0.8}$$

$$(0.8)145 = k$$

$$116 = k.$$

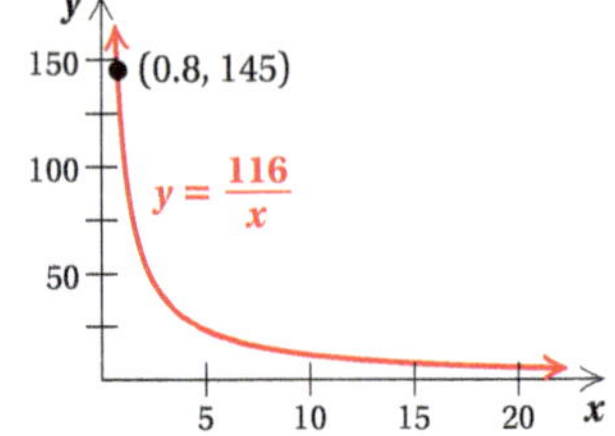

The equation of variation is $y = 116/x$. The answer is the equation $y = 116/x$, *not* simply $k = 116$.

When $x = 25$, we have

$$y = \frac{116}{x}$$

$$= \frac{116}{25} \quad \text{Substituting 25 for } x$$

$$= 4.64.$$

The value of y is 4.64 when $x = 25$.

◀ **Do Exercises 5 and 6.**

The graph of $y = k/x$, $k > 0$, is shaped like the figure at right for positive values of x. (You need not know how to graph such equations at this time.) Note that as x increases, y decreases; and as x decreases, y increases. This is why the terminology "inverse" is used. One variable does the opposite of what the other does.

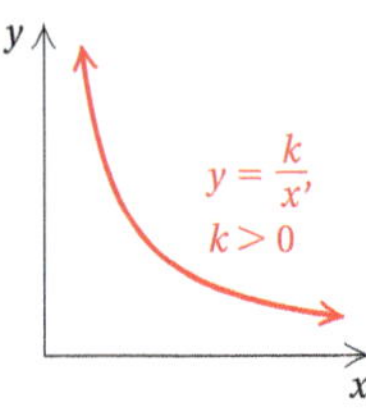

5. Find an equation of variation in which y varies inversely as x, and $y = 105$ when $x = 0.6$. GS

$$y = \frac{k}{x}$$

$$105 = \frac{k}{\square}$$

$$k = \square \cdot 105$$

$$k = \square$$

$$y = \frac{63}{\square}$$

$$y = \frac{63}{x} = \frac{63}{\square}$$

$$= \square$$

6. Find an equation of variation in which y varies inversely as x, and $y = 45$ when $x = 20$. Then find the value of y when $x = 1.6$.

Answers

5. $y = \frac{63}{x}$; 3.15 **6.** $y = \frac{900}{x}$; 562.5

Guided Solution:

5. 0.6, 0.6, 63, x; 20, 3.15

d APPLICATIONS OF INVERSE VARIATION

Often in an applied situation we must decide which kind of variation, if any, might apply to the problem.

EXAMPLE 5 *Trash Removal.* The day after the Indianapolis 500 race, local organizations are assigned the task of cleaning up the grandstands. It takes approximately 8 hr for 30 people to remove the trash from one grandstand. How long would it take 42 people to do the job?

Source: Indianapolis Motor Speedway

1. **Familiarize.** Think about the problem situation. What kind of variation would be used? It seems reasonable that the more people there are working on the job, the less time it will take to finish. Thus inverse variation might apply. We let $T =$ the time to do the job, in hours, and $N =$ the number of people. Assuming inverse variation, we know that an equation $T = k/N$, $k > 0$, applies. As the number of people increases, the time it takes to do the job decreases.

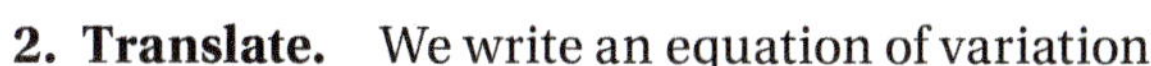

2. **Translate.** We write an equation of variation:

$$T = \frac{k}{N}.$$

Time varies inversely as the number of people involved.

3. **Solve.** The mathematical manipulation has two parts. First, we find the equation of variation by substituting known values for T and N to find k. Second, we compute the amount of time it would take 42 people to do the job.

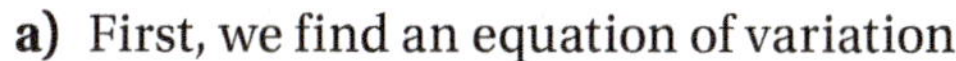

a) First, we find an equation of variation:

$$T = \frac{k}{N}$$

$$8 = \frac{k}{30} \quad \text{Substituting 8 for } T \text{ and 30 for } N$$

$$30 \cdot 8 = k$$

$$240 = k.$$

The equation of variation is $T = \dfrac{240}{N}$.

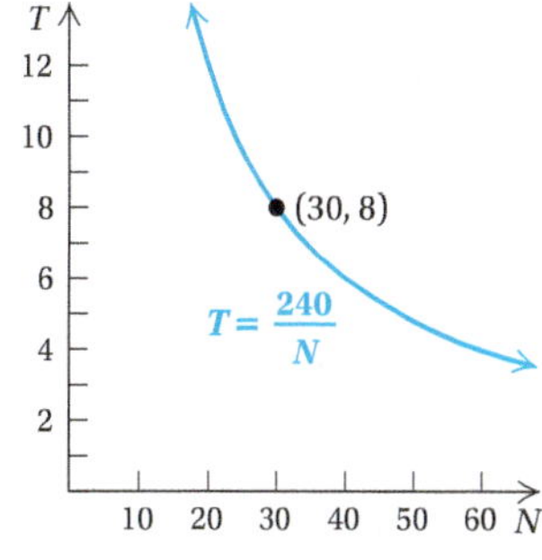

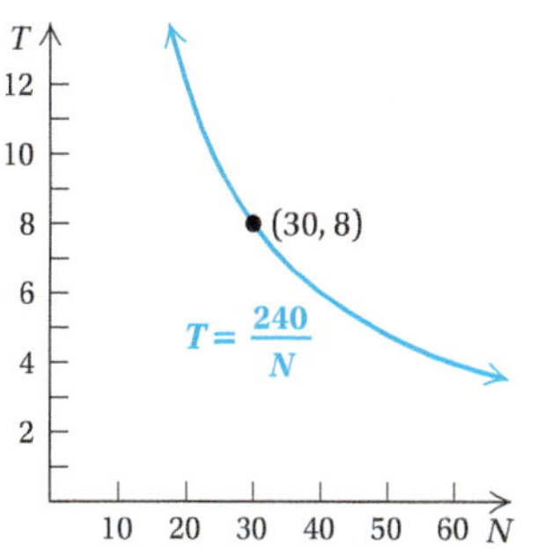

b) We then use the equation to find the amount of time that it takes 42 people to do the job:

$$T = \frac{240}{N}$$

$$= \frac{240}{42} \quad \text{Substituting 42 for } N$$

$$\approx 5.7.$$

4. **Check.** The check might be done by repeating the computations. We might also analyze the results. The number of people increased from 30 to 42. Did the time decrease? It did, and this confirms what we expect with inverse variation.

5. **State.** It should take 5.7 hr for 42 people to complete the job.

Do Exercises 7 and 8. ▶

7. Refer to Example 5. Determine how long it would take 25 people to do the job.

8. *Time of Travel.* The time t required to drive a fixed distance varies inversely as the speed r. It takes 5 hr at 60 km/h to drive a fixed distance.

   a) Find an equation of variation.

   b) How long would it take at 40 km/h?

Answers

7. 9.6 hr

8. (a) $t = \dfrac{300}{r}$; (b) 7.5 hr

14.9 Exercise Set

For Extra Help
MyMathLab® MathXL® PRACTICE WATCH READ REVIEW

✓ Reading Check

Match each description of variation with the appropriate equation of variation from the column on the right.

RC1. r varies directly as s.

RC2. x is inversely proportional to z.

RC3. m varies inversely as n.

RC4. n is directly proportional to m.

RC5. z varies directly as w.

RC6. r varies inversely as s.

a) $r = \frac{k}{s}$

b) $r = ks$

c) $z = kw$

d) $x = \frac{k}{z}$

e) $n = km$

f) $m = \frac{k}{n}$

a Find an equation of variation in which y varies directly as x and the following are true. Then find the value of y when $x = 20$.

1. $y = 36$ when $x = 9$

2. $y = 60$ when $x = 16$

3. $y = 0.8$ when $x = 0.5$

4. $y = 0.7$ when $x = 0.4$

5. $y = 630$ when $x = 175$

6. $y = 400$ when $x = 125$

7. $y = 500$ when $x = 60$

8. $y = 200$ when $x = 300$

b Solve.

9. ***Wages and Work Time.*** A person's paycheck P varies directly as the number H of hours worked. For working 15 hr, the pay is \$180.

a) Find an equation of variation.
b) Find the pay for 35 hr of work.

10. ***Interest and Interest Rate.*** The interest I earned in 1 year on a fixed principal varies directly as the interest rate r. An investment earns \$53.55 at an interest rate of 4.25%.

a) Find an equation of variation.
b) How much will the investment earn at a rate of 5.75%?

11. ***Cost of Sand.*** The cost C, in dollars, to fill a sandbox varies directly as the depth S, in inches, of the sand. The director of Creekside Daycare checks at her local hardware store and finds that it would cost \$67.50 to fill the daycare's box with 6 in. of sand. She decides to fill the sandbox to a depth of 9 in.

a) Find an equation of variation.
b) How much will the sand cost?

12. ***Cost of Cement.*** The cost C, in dollars, of cement needed to pave a driveway varies directly as the depth D, in inches, of the driveway. John checks at his local building materials store and finds that it costs \$1000 to install his driveway with a depth of 8 in. He decides to build a stronger driveway at a depth of 12 in.

a) Find an equation of variation.
b) How much will it cost for the cement?

13. ***Lunar Weight.*** The weight M of an object on the moon varies directly as its weight E on Earth. Jared weighs 192 lb, but would weigh only 32 lb on the moon.

a) Find an equation of variation.
b) Jared's wife, Elizabeth, weighs 110 lb on Earth. How much would she weigh on the moon?
c) Jared's granddaughter, Jasmine, would weigh only 5 lb on the moon. How much does Jasmine weigh on Earth?

14. ***Mars Weight.*** The weight M of an object on Mars varies directly as its weight E on Earth. In 1999, Chen Yanqing, who weighs 128 lb, set a world record for her weight class with a lift (snatch) of 231 lb. On Mars, this lift would be only 88 lb.

Source: *The Guinness Book of Records*, 2001

a) Find an equation of variation.
b) How much would Yanqing weigh on Mars?

15. ***Computer Megahertz.*** The number of instructions N performed per second by a computer varies directly as the speed S of the computer's internal processor. A processor with a speed of 25 megahertz can perform 2,000,000 instructions per second.

a) Find an equation of variation.
b) How many instructions per second will the same processor perform if it is running at a speed of 200 megahertz?

16. ***Water in Human Body.*** The number of kilograms W of water in a human body varies directly as the total body weight B. A person who weighs 75 kg contains 54 kg of water.

a) Find an equation of variation.
b) How many kilograms of water are in a person who weighs 95 kg?

17. ***Steak Servings.*** The number of servings S of meat that can be obtained from round steak varies directly as the weight W. From 9 kg of round steak, one can get 70 servings of meat. How many servings can one get from 12 kg of round steak?

18. ***Turkey Servings.*** A chef is planning meals in a refreshment tent at a golf tournament. The number of servings S of meat that can be obtained from a turkey varies directly as its weight W. From a turkey weighing 30.8 lb, one can get 40 servings of meat. How many servings can be obtained from a 19.8-lb turkey?

C Find an equation of variation in which y varies inversely as x and the following are true. Then find the value of y when $x = 10$.

19. $y = 3$ when $x = 25$

20. $y = 2$ when $x = 45$

21. $y = 10$ when $x = 8$

22. $y = 10$ when $x = 7$

23. $y = 6.25$ when $x = 0.16$

24. $y = 0.125$ when $x = 8$

25. $y = 50$ when $x = 42$

26. $y = 25$ when $x = 42$

27. $y = 0.2$ when $x = 0.3$

28. $y = 0.4$ when $x = 0.6$

b, **d** Solve.

29. ***Production and Time.*** A production line produces 15 CD players every 8 hr. How many players can it produce in 37 hr?

a) What kind of variation might apply to this situation?

b) Solve the problem.

30. ***Wages and Work Time.*** A person works for 15 hr and makes $251.25. How much will the person make by working 35 hr?

a) What kind of variation might apply to this situation?

b) Solve the problem.

31. ***Cooking Time.*** It takes 4 hr for 9 cooks to prepare the food for a wedding rehearsal dinner. How long will it take 8 cooks to prepare the dinner?

a) What kind of variation might apply to this situation?

b) Solve the problem.

32. ***Work Time.*** It takes 16 hr for 2 people to resurface a tennis court. How long will it take 6 people to do the job?

a) What kind of variation might apply to this situation?

b) Solve the problem.

33. ***Miles per Gallon.*** To travel a fixed distance, the number of gallons N of gasoline needed is inversely proportional to the miles-per-gallon rating P of the car. A car that gets 20 miles per gallon (mpg) needs 14 gal to travel the distance.

a) Find an equation of variation.

b) How much gas will be needed for a car that gets 28 mpg?

34. ***Miles per Gallon.*** To travel a fixed distance, the number of gallons N of gasoline needed is inversely proportional to the miles-per-gallon rating P of the car. A car that gets 25 miles per gallon (mpg) needs 12 gal to travel the distance.

a) Find an equation of variation.

b) How much gas will be needed for a car that gets 20 mpg?

35. ***Electrical Current.*** The current I in an electrical conductor varies inversely as the resistance R of the conductor. The current is 96 amperes when the resistance is 20 ohms.

a) Find an equation of variation.

b) What is the current when the resistance is 60 ohms?

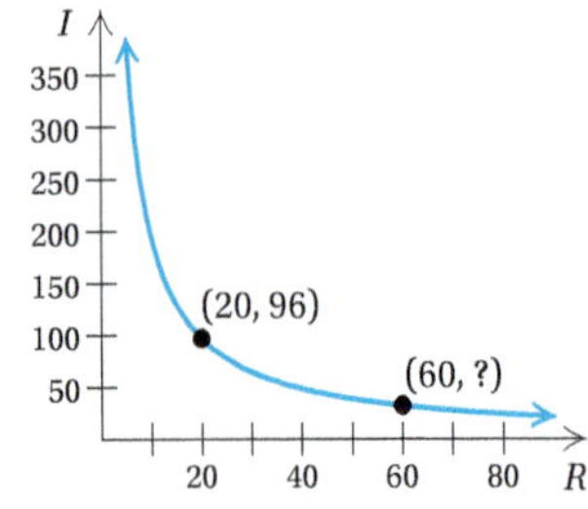

36. ***Gas Volume.*** The volume V of a gas varies inversely as the pressure P on it. The volume of a gas is 200 cm^3 under a pressure of 32 kg/cm^2.

a) Find an equation of variation.

b) What will be its volume under a pressure of 20 kg/cm^2?

37. *Answering Questions.* For a fixed time limit for a quiz, the number of minutes m that a student should allow for each question on a quiz (assuming they are of equal difficulty) is inversely proportional to the number of questions n on the quiz. For a given time limit on a 16-question quiz, students have 2.5 min per question.

a) Find an equation of variation.

b) How many questions would appear on a quiz in which students have the same time limit and have 4 min per question?

38. *Pumping Time.* The time t required to empty a tank varies inversely as the rate r of pumping. A pump can empty a tank in 90 min at a rate of 1200 L/min.

a) Find an equation of variation.

b) How long will it take the pump to empty the tank at a rate of 2000 L/min?

39. *Apparent Size.* The apparent size A of an object varies inversely as the distance d of the object from the eye. A flagpole 30 ft from an observer appears to be 27.5 ft tall. How tall will the same flagpole appear to be if it is 100 ft from the eye?

40. *Driving Time.* The time t required to drive a fixed distance varies inversely as the speed r. It takes 5 hr at 55 mph to drive a fixed distance. How long would it take at 40 mph?

Skill Maintenance

Solve. [14.7a]

41. $\dfrac{x+2}{x+5} = \dfrac{x-4}{x-6}$

42. $\dfrac{x-3}{x-5} = \dfrac{x+5}{x+1}$

Calculate. [9.8d]

43. $3^7 \div 3^4 \div 3^3 \div 3$

44. $\dfrac{37 - 5(4-6)}{2 \cdot 6 + 8}$

45. $-5^2 + 4 \cdot 6$

46. $(-5)^2 + 4 \cdot 6$

Synthesis

Write an equation of variation for each situation.

47. The square of the pitch P of a vibrating string varies directly as the tension t on the string.

48. In a stream, the amount S of salt carried varies directly as the sixth power of the speed V of the stream.

49. The power P in a windmill varies directly as the cube of the wind speed V.

50. The volume V of a sphere varies directly as the cube of the radius r.

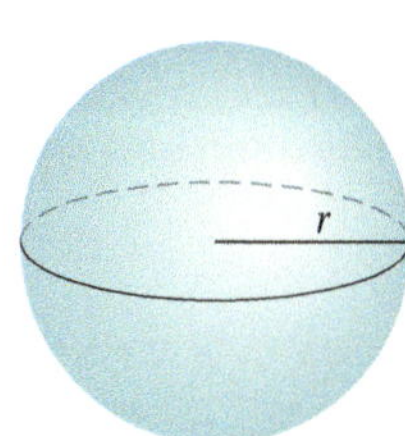

CHAPTER

14 Summary and Review

Vocabulary Reinforcement

Complete each statement with the correct term from the column on the right.

1. A _______________ rational expression is a rational expression that has one or more rational expressions within its numerator or its denominator. [14.6a]
2. An equality of ratios, $\frac{A}{B} = \frac{C}{D}$, is called a(n) _______________. [14.8b]
3. Two expressions are _______________ of each other if their product is 1. [14.2a]
4. Expressions that have the same value for all allowable replacements are called _______________ expressions. [14.1b]
5. Expressions of the form $a - b$ and $b - a$ are _______________ of each other. [14.1c]
6. In _______________ triangles, corresponding angles have the same measure and the lengths of corresponding sides are proportional. [14.8b]
7. When a situation translates to an equation described by $y = \frac{k}{x}$, with k a positive constant, we say that y varies _______________ as x. The equation $y = \frac{k}{x}$ is called an equation of _______________. [14.9c]
8. When a situation translates to an equation described by $y = kx$, with k a positive constant, we say that y varies _______________ as x. The equation $y = kx$ is called an equation of _______________. [14.9a]

reciprocals
proportion
rational
equivalent
directly
complex
direct variation
similar
inversely
opposites
inverse variation

Concept Reinforcement

Determine whether each statement is true or false.

_______________ 1. To determine the numbers for which a rational expression is not defined, we set the denominator equal to 0 and solve. [14.1a]

_______________ 2. The expressions $y + 5$ and $y - 5$ are opposites of each other. [14.1c]

_______________ 3. The opposite of $2 - x$ is $x - 2$. [14.1c]

Study Guide

Objective 14.1a Find all numbers for which a rational expression is not defined.

Example Find all numbers for which the rational expression $\frac{2 - y}{y^2 + 3y - 28}$ is not defined.

$$y^2 + 3y - 28 = 0$$
$$(y + 7)(y - 4) = 0$$
$$y + 7 = 0 \quad or \quad y - 4 = 0$$
$$y = -7 \quad or \quad y = 4$$

The rational expression is not defined for -7 and 4.

Practice Exercise

1. Find all numbers for which the rational expression $\frac{c + 8}{c^2 - 11c + 30}$ is not defined.

Objective 14.1c Simplify rational expressions by factoring the numerator and the denominator and removing factors of 1.

Example Simplify: $\dfrac{6y - 12}{2y^2 + y - 10}$.

$$\frac{6y - 12}{2y^2 + y - 10} = \frac{6(y - 2)}{(2y + 5)(y - 2)}$$
$$= \frac{y - 2}{y - 2} \cdot \frac{6}{2y + 5}$$
$$= 1 \cdot \frac{6}{2y + 5} = \frac{6}{2y + 5}$$

Practice Exercise

2. Simplify:
$$\frac{2x^2 - 2}{4x^2 + 24x + 20}.$$

Objective 14.1d Multiply rational expressions and simplify.

Example Multiply and simplify:
$$\frac{x^2 + 14x + 49}{x^2 - 25} \cdot \frac{x + 5}{x + 7}.$$
$$\frac{x^2 + 14x + 49}{x^2 - 25} \cdot \frac{x + 5}{x + 7} = \frac{(x^2 + 14x + 49)(x + 5)}{(x^2 - 25)(x + 7)}$$
$$= \frac{(x + 7)(x + 7)(x + 5)}{(x + 5)(x - 5)(x + 7)}$$
$$= \frac{x + 7}{x - 5}$$

Practice Exercise

3. Multiply and simplify:
$$\frac{2y^2 + 7y - 15}{5y^2 - 45} \cdot \frac{y - 3}{2y - 3}.$$

Objective 14.2b Divide rational expressions and simplify.

Example Divide and simplify: $\dfrac{a^2 - 9a}{a^2 - a - 6} \div \dfrac{a}{a + 2}$.

$$\frac{a^2 - 9a}{a^2 - a - 6} \div \frac{a}{a + 2} = \frac{a^2 - 9a}{a^2 - a - 6} \cdot \frac{a + 2}{a}$$
$$= \frac{(a^2 - 9a)(a + 2)}{(a^2 - a - 6)a}$$
$$= \frac{a(a - 9)(a + 2)}{(a + 2)(a - 3)a}$$
$$= \frac{a - 9}{a - 3}$$

Practice Exercise

4. Divide and simplify:
$$\frac{b^2 + 3b - 28}{b^2 + 5b - 24} \div \frac{b - 4}{b - 3}.$$

Objective 14.3b Add fractions, first finding the LCD.

Example Add: $\dfrac{13}{30} + \dfrac{11}{24}$.

$$\left.\begin{aligned} 30 &= 2 \cdot 3 \cdot 5 \\ 24 &= 2 \cdot 2 \cdot 2 \cdot 3 \end{aligned}\right\} \text{LCD} = 2 \cdot 2 \cdot 2 \cdot 3 \cdot 5, \text{ or } 120$$

$$\frac{13}{30} + \frac{11}{24} = \frac{13}{2 \cdot 3 \cdot 5} \cdot \frac{2 \cdot 2}{2 \cdot 2} + \frac{11}{2 \cdot 2 \cdot 2 \cdot 3} \cdot \frac{5}{5}$$
$$= \frac{52 + 55}{2 \cdot 2 \cdot 2 \cdot 3 \cdot 5} = \frac{107}{120}$$

Practice Exercise

5. Add: $\dfrac{5}{18} + \dfrac{7}{60}$.

Objective 14.3c Find the LCM of algebraic expressions by factoring.

Example Find the LCM of

$x^2 - 36$ and $x^2 - 5x - 6$.

$$x^2 - 36 = (x + 6)(x - 6)$$
$$x^2 - 5x - 6 = (x - 6)(x + 1)$$
$$\text{LCM} = (x + 6)(x - 6)(x + 1)$$

Practice Exercise

6. Find the LCM of

$x^2 - 7x - 18$ and $x^2 - 81$.

Objective 14.4a Add rational expressions.

Example Add and simplify: $\dfrac{6x - 5}{x - 1} + \dfrac{x}{1 - x}$.

$$\frac{6x - 5}{x - 1} + \frac{x}{1 - x} = \frac{6x - 5}{x - 1} + \frac{x}{1 - x} \cdot \frac{-1}{-1}$$
$$= \frac{6x - 5}{x - 1} + \frac{-x}{x - 1}$$
$$= \frac{6x - 5 - x}{x - 1}$$
$$= \frac{5x - 5}{x - 1}$$
$$= \frac{5(x - 1)}{x - 1} = 5$$

Practice Exercise

7. Add and simplify:

$$\frac{x}{x - 4} + \frac{2x - 4}{4 - x}.$$

Objective 14.5a Subtract rational expressions.

Example Subtract: $\dfrac{3}{x^2 - 1} - \dfrac{2x - 1}{x^2 + x - 2}$.

$$\frac{3}{x^2 - 1} - \frac{2x - 1}{x^2 + x - 2}$$
$$= \frac{3}{(x + 1)(x - 1)} - \frac{2x - 1}{(x + 2)(x - 1)}$$

The LCM is $(x + 1)(x - 1)(x + 2)$.

$$= \frac{3}{(x + 1)(x - 1)} \cdot \frac{x + 2}{x + 2} - \frac{2x - 1}{(x + 2)(x - 1)} \cdot \frac{x + 1}{x + 1}$$
$$= \frac{3(x + 2)}{(x + 1)(x - 1)(x + 2)} - \frac{(2x - 1)(x + 1)}{(x + 2)(x - 1)(x + 1)}$$
$$= \frac{3x + 6 - (2x^2 + x - 1)}{(x + 1)(x - 1)(x + 2)}$$
$$= \frac{3x + 6 - 2x^2 - x + 1}{(x + 1)(x - 1)(x + 2)}$$
$$= \frac{-2x^2 + 2x + 7}{(x + 1)(x - 1)(x + 2)}$$

Practice Exercise

8. Subtract:

$$\frac{x}{x^2 + x - 2} - \frac{5}{x^2 - 1}.$$

Objective 14.6a Simplify complex rational expressions.

Example Simplify $\dfrac{\frac{1}{3}-\frac{1}{x}}{\frac{1}{x}-\frac{1}{2}}$ using method 1.

The LCM of 3, x, and 2 is $6x$.

$$\frac{\frac{1}{3}-\frac{1}{x}}{\frac{1}{x}-\frac{1}{2}} = \frac{\frac{1}{3}-\frac{1}{x}}{\frac{1}{x}-\frac{1}{2}} \cdot \frac{6x}{6x} = \frac{\frac{1}{3}\cdot 6x - \frac{1}{x}\cdot 6x}{\frac{1}{x}\cdot 6x - \frac{1}{2}\cdot 6x}$$

$$= \frac{2x-6}{6-3x} = \frac{2(x-3)}{3(2-x)}$$

Practice Exercise

9. Simplify: $\dfrac{\frac{2}{5}-\frac{1}{y}}{\frac{3}{y}-\frac{1}{3}}$.

Objective 14.7a Solve rational equations.

Example Solve: $12 = \frac{1}{5x} + \frac{4}{x}$.

The LCM of the denominators is $5x$. We multiply by $5x$ on both sides.

$$12 = \frac{1}{5x} + \frac{4}{x}$$

$$5x \cdot 12 = 5x\left(\frac{1}{5x} + \frac{4}{x}\right)$$

$$5x \cdot 12 = 5x \cdot \frac{1}{5x} + 5x \cdot \frac{4}{x}$$

$$60x = 1 + 20$$

$$60x = 21$$

$$x = \frac{21}{60} = \frac{7}{20}$$

This checks, so the solution is $\frac{7}{20}$.

Practice Exercise

10. Solve: $\frac{1}{x} = \frac{2}{3-x}$.

Objective 14.9a Find an equation of direct variation given a pair of values of the variables.

Example Find an equation of variation in which y varies directly as x, and $y = 30$ when $x = 200$. Then find the value of y when $x = \frac{1}{2}$.

$y = kx$ Direct variation

$30 = k \cdot 200$ Substituting 30 for y and 200 for x

$$\frac{30}{200} = k, \text{ or } k = \frac{3}{20}$$

The equation of variation is $y = \frac{3}{20}x$.

Next, we substitute $\frac{1}{2}$ for x in $y = \frac{3}{20}x$ and solve for y:

$$y = \frac{3}{20}x = \frac{3}{20} \cdot \frac{1}{2} = \frac{3}{40}.$$

When $x = \frac{1}{2}$, $y = \frac{3}{40}$.

Practice Exercise

11. Find an equation of variation in which y varies directly as x, and $y = 60$ when $x = 0.4$. Then find the value of y when $x = 2$.

Objective 14.9c Find an equation of inverse variation given a pair of values of the variables.

Example Find an equation of variation in which y varies inversely as x, and $y = 0.5$ when $x = 20$. Then find the value of y when $x = 6$.

$y = \frac{k}{x}$ Inverse variation

$0.5 = \frac{k}{20}$ Substituting 0.5 for y and 20 for x

$10 = k$

The equation of variation is $y = \frac{10}{x}$.

Next, we substitute 6 for x in $y = 10/x$ and solve for y:

$$y = \frac{10}{x} = \frac{10}{6} = \frac{5}{3}.$$

When $x = 6$, $y = \frac{5}{3}$.

Practice Exercise

12. Find an equation of variation in which y varies inversely as x, and $y = 150$ when $x = 1.5$. Then find the value of y when $x = 10$.

Review Exercises

Find all numbers for which the rational expression is not defined. [14.1a]

1. $\frac{3}{x}$

2. $\frac{4}{x - 6}$

3. $\frac{x + 5}{x^2 - 36}$

4. $\frac{x^2 - 3x + 2}{x^2 + x - 30}$

5. $\frac{-4}{(x + 2)^2}$

6. $\frac{x - 5}{5}$

Simplify. [14.1c]

7. $\frac{4x^2 - 8x}{4x^2 + 4x}$

8. $\frac{14x^2 - x - 3}{2x^2 - 7x + 3}$

9. $\frac{(y - 5)^2}{y^2 - 25}$

Multiply and simplify. [14.1d]

10. $\frac{a^2 - 36}{10a} \cdot \frac{2a}{a + 6}$

11. $\frac{6t - 6}{2t^2 + t - 1} \cdot \frac{t^2 - 1}{t^2 - 2t + 1}$

Divide and simplify. [14.2b]

12. $\frac{10 - 5t}{3} \div \frac{t - 2}{12t}$

13. $\frac{4x^4}{x^2 - 1} \div \frac{2x^3}{x^2 - 2x + 1}$

Find the LCM. [14.3c]

14. $3x^2$, $10xy$, $15y^2$

15. $a - 2$, $4a - 8$

16. $y^2 - y - 2$, $y^2 - 4$

Add and simplify. [14.4a]

17. $\frac{x + 8}{x + 7} + \frac{10 - 4x}{x + 7}$

18. $\frac{3}{3x - 9} + \frac{x - 2}{3 - x}$

19. $\frac{2a}{a + 1} + \frac{4a}{a^2 - 1}$

20. $\frac{d^2}{d - c} + \frac{c^2}{c - d}$

Subtract and simplify. [14.5a]

21. $\dfrac{6x-3}{x^2-x-12}-\dfrac{2x-15}{x^2-x-12}$

22. $\dfrac{3x-1}{2x}-\dfrac{x-3}{x}$

23. $\dfrac{x+3}{x-2}-\dfrac{x}{2-x}$

24. $\dfrac{1}{x^2-25}-\dfrac{x-5}{x^2-4x-5}$

25. Perform the indicated operations and simplify: [14.5b]

$$\frac{3x}{x+2}-\frac{x}{x-2}+\frac{8}{x^2-4}.$$

Simplify. [14.6a]

26. $\dfrac{\dfrac{1}{z}+1}{\dfrac{1}{z^2}-1}$

27. $\dfrac{\dfrac{c}{d}-\dfrac{d}{c}}{\dfrac{1}{c}+\dfrac{1}{d}}$

Solve. [14.7a]

28. $\dfrac{3}{y}-\dfrac{1}{4}=\dfrac{1}{y}$

29. $\dfrac{15}{x}-\dfrac{15}{x+2}=2$

Solve. [14.8a]

30. *Highway Work.* In checking records, a contractor finds that crew A can pave a certain length of highway in 9 hr, while crew B can do the same job in 12 hr. How long would it take if they worked together?

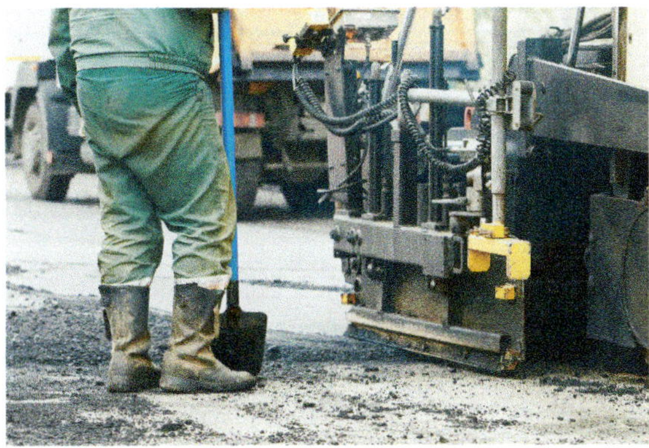

31. *Airplane Speed.* One plane travels 80 mph faster than another. While one travels 1750 mi, the other travels 950 mi. Find the speed of each plane.

32. *Train Speed.* A manufacturer is testing two high-speed trains. One train travels 40 km/h faster than the other. While one train travels 70 km, the other travels 60 km. Find the speed of each train.

Solve. [14.8b]

33. *Quality Control.* A sample of 250 calculators contained 8 defective calculators. How many defective calculators would you expect to find in a sample of 5000?

34. *Pizza Proportions.* At Finnelli's Pizzeria, the following ratios are used: 5 parts sausage to 7 parts cheese, 6 parts onion to 13 parts green pepper, and 9 parts pepperoni to 14 parts cheese.

a) Finnelli's makes several pizzas with green pepper and onion. They use 2 cups of green pepper. How much onion would they use?

b) Finnelli's makes several pizzas with sausage and cheese. They use 3 cups of sausage. How much cheese would they use?

c) Finnelli's makes several pizzas with pepperoni and cheese. They use 6 cups of pepperoni. How much cheese would they use?

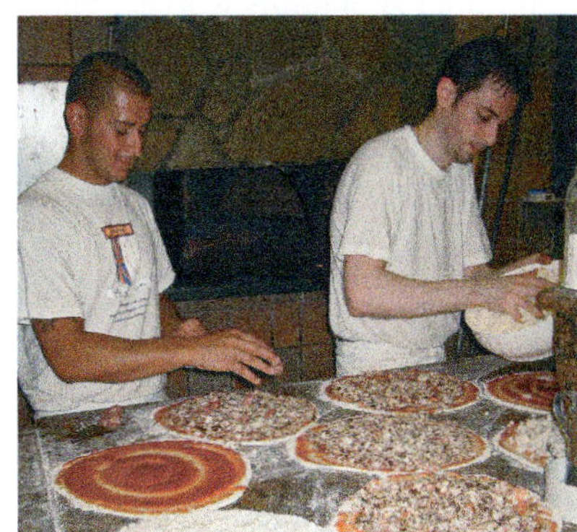

35. *Estimating Whale Population.* To determine the number of blue whales in the world's oceans, marine biologists tag 500 blue whales in various parts of the world. Later, 400 blue whales are checked, and it is found that 20 of them are tagged. Estimate the blue whale population.

36. Triangles ABC and XYZ below are similar. Find the value of x.

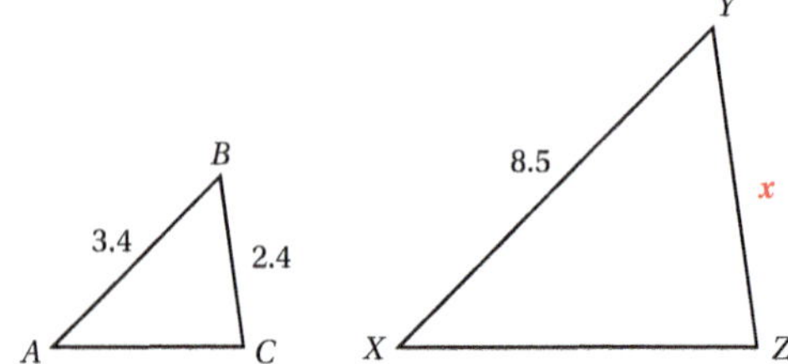

Find an equation of variation in which y varies directly as x and the following are true. Then find the value of y when $x = 20$. [14.9a]

37. $y = 12$ when $x = 4$

38. $y = 0.4$ when $x = 0.5$

Find an equation of variation in which y varies inversely as x and the following are true. Then find the value of y when $x = 5$. [14.9c]

39. $y = 5$ when $x = 6$

40. $y = 0.5$ when $x = 2$

41. $y = 1.3$ when $x = 0.5$

Solve.

42. *Wages.* A person's paycheck P varies directly as the number H of hours worked. The pay is \$165.00 for working 20 hr. Find the pay for 35 hr of work. [14.9b]

43. *WashingTime.* It takes 5 hr for 2 washing machines to wash a fixed amount of laundry. How long would it take 10 washing machines to do the same job? (The number of hours varies inversely as the number of washing machines.) [14.9d]

44. Find all numbers for which

$$\frac{3x^2 - 2x - 1}{3x^2 + x}$$

is not defined. [14.1a]

A. $1, -\frac{1}{3}$ **B.** $-\frac{1}{3}$

C. $0, -\frac{1}{3}$ **D.** $0, \frac{1}{3}$

45. Subtract: $\frac{1}{x - 5} - \frac{1}{x + 5}$. [14.5a]

A. $\frac{10}{(x - 5)(x + 5)}$ **B.** 0

C. $\frac{5}{x - 5}$ **D.** $\frac{10}{x + 5}$

Synthesis

46. Simplify: [14.1d], [14.2b]

$$\frac{2a^2 + 5a - 3}{a^2} \cdot \frac{5a^3 + 30a^2}{2a^2 + 7a - 4} \div \frac{a^2 + 6a}{a^2 + 7a + 12}.$$

47. Compare

$$\frac{A + B}{B} = \frac{C + D}{D}$$

with the proportion

$$\frac{A}{B} = \frac{C}{D}.$$ [14.8b]

Understanding Through Discussion and Writing

1. Are parentheses as important when adding rational expressions as they are when subtracting? Why or why not? [14.4a], [14.5a]

2. How can a graph be used to determine how many solutions an equation has? [14.7a]

3. How is the process of canceling related to the identity property of 1? [14.1c]

4. Determine whether the following situation represents direct variation, inverse variation, or neither. Give a reason for your answer. [14.9a, c]

   The number of plays that it takes to go 80 yd for a touchdown and the average gain per play

5. Explain how a rational expression can be formed for which -3 and 4 are not allowable replacements. [14.1a]

6. Why is it especially important to check the possible solutions to a rational equation? [14.7a]

Test

For Extra Help: For step-by-step test solutions, access the Chapter Test Prep Videos in MyMathLab® or on YouTube (search "BittingerPreIntro" and click on Channels).

Find all numbers for which the rational expression is not defined.

1. $\dfrac{8}{2x}$

2. $\dfrac{5}{x+8}$

3. $\dfrac{x-7}{x^2-49}$

4. $\dfrac{x^2+x-30}{x^2-3x+2}$

5. $\dfrac{11}{(x-1)^2}$

6. $\dfrac{x+2}{2}$

7. Simplify:
$$\frac{6x^2+17x+7}{2x^2+7x+3}.$$

8. Multiply and simplify:
$$\frac{a^2-25}{6a}\cdot\frac{3a}{a-5}.$$

9. Divide and simplify:
$$\frac{25x^2-1}{9x^2-6x}\div\frac{5x^2+9x-2}{3x^2+x-2}.$$

10. Find the LCM:
$$y^2-9,\ y^2+10y+21,\ y^2+4y-21.$$

Add or subtract. Simplify, if possible.

11. $\dfrac{16+x}{x^3}+\dfrac{7-4x}{x^3}$

12. $\dfrac{5-t}{t^2+1}-\dfrac{t-3}{t^2+1}$

13. $\dfrac{x-4}{x-3}+\dfrac{x-1}{3-x}$

14. $\dfrac{x-4}{x-3}-\dfrac{x-1}{3-x}$

15. $\dfrac{5}{t-1}+\dfrac{3}{t}$

16. $\dfrac{1}{x^2-16}-\dfrac{x+4}{x^2-3x-4}$

17. $\dfrac{1}{x-1}+\dfrac{4}{x^2-1}-\dfrac{2}{x^2-2x+1}$

18. Simplify: $\dfrac{9-\dfrac{1}{y^2}}{3-\dfrac{1}{y}}.$

Solve.

19. $\dfrac{7}{y}-\dfrac{1}{3}=\dfrac{1}{4}$

20. $\dfrac{15}{x}-\dfrac{15}{x-2}=-2$

Find an equation of variation in which y varies directly as x and the following are true. Then find the value of y when $x = 25$.

21. $y = 6$ when $x = 3$

22. $y = 1.5$ when $x = 3$

Find an equation of variation in which y varies inversely as x and the following are true. Then find the value of y when $x = 100$.

23. $y = 6$ when $x = 3$

24. $y = 11$ when $x = 2$

Solve.

25. *Train Travel.* The distance d traveled by a train varies directly as the time t that it travels. The train travels 60 km in $\frac{1}{2}$ hr. How far will it travel in 2 hr?

26. *Concrete Work.* It takes 3 hr for 2 concrete mixers to mix a fixed amount of concrete. The number of hours varies inversely as the number of concrete mixers used. How long would it take 5 concrete mixers to do the same job?

27. *Quality Control.* A sample of 125 spark plugs contained 4 defective spark plugs. How many defective spark plugs would you expect to find in a sample of 500?

28. *Zebra Population.* A game warden catches, tags, and then releases 15 zebras. A month later, a sample of 20 zebras is collected and 6 of them have tags. Use this information to estimate the size of the zebra population in that area.

29. *Copying Time.* Kopy Kwik has 2 copiers. One can copy a year-end report in 20 min. The other can copy the same document in 30 min. How long would it take both machines, working together, to copy the report?

30. *Driving Speed.* Craig drives 20 km/h faster than Marilyn. In the same time that Marilyn drives 225 km, Craig drives 325 km. Find the speed of each car.

31. This pair of triangles is similar. Find the missing length x.

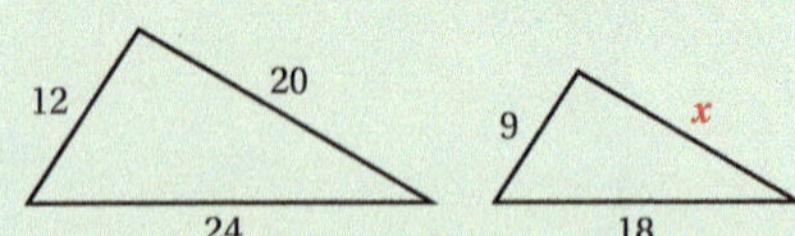

32. Solve: $\dfrac{2}{x-4} + \dfrac{2x}{x^2-16} = \dfrac{1}{x+4}$.

A. -4

B. 4

C. $4, -4$

D. No solution

Synthesis

33. Reggie and Rema work together to mulch the flower beds around an office complex in $2\frac{6}{7}$ hr. Working alone, it would take Reggie 6 hr more than it would take Rema. How long would it take each of them to complete the landscaping working alone?

34. Simplify: $1 + \cfrac{1}{1 + \cfrac{1}{1 + \cfrac{1}{a}}}$.

CHAPTER

15

Systems of Equations

STUDYING FOR SUCCESS *Doing Your Homework*

- ☐ Prepare for your homework by reading explanations of concepts and by following the step-by-step solutions of examples in the text.
- ☐ Include all the steps. This will keep you organized, help you avoid computational errors, and give you a study guide for an exam.
- ☐ Try to do your homework as soon as possible after each class. Avoid waiting until a deadline is near to begin work on an assignment.

15.1 Systems of Equations in Two Variables

OBJECTIVES

a Determine whether an ordered pair is a solution of a system of equations.

b Solve systems of two linear equations in two variables by graphing.

SKILL TO REVIEW

Objective 11.1d: Graph linear equations of the type $y = mx + b$ and $Ax + By = C$.

Graph.

1. $y = 2x + 1$

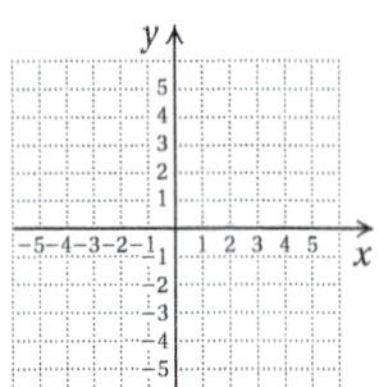

2. $3x - 4y = 12$

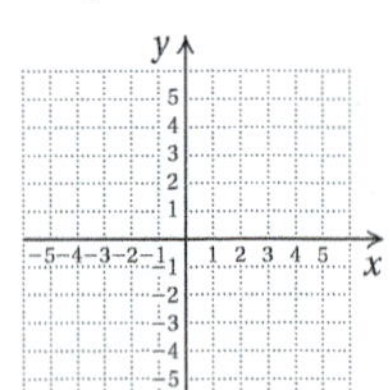

Answers

Answers to Skill to Review Exercises are on p. 1107.

a SYSTEMS OF EQUATIONS AND SOLUTIONS

Many problems can be solved more easily by translating to two equations in two variables. The following is such a **system of equations**:

$$x + y = 8,$$
$$2x - y = 1.$$

SOLUTION OF A SYSTEM OF EQUATIONS

A **solution** of a system of two equations is an ordered pair that makes both equations true.

Look at the graphs shown below. Recall that a graph of an equation is a drawing that represents its solution set. Each point on the graph corresponds to a solution of that equation. Which points (ordered pairs) are solutions of *both* equations?

The graph shows that there is only one. It is the point P where the graphs cross, or intersect. This point looks as if its coordinates are $(3, 5)$. We check to see whether $(3, 5)$ is a solution of *both* equations, substituting 3 for x and 5 for y.

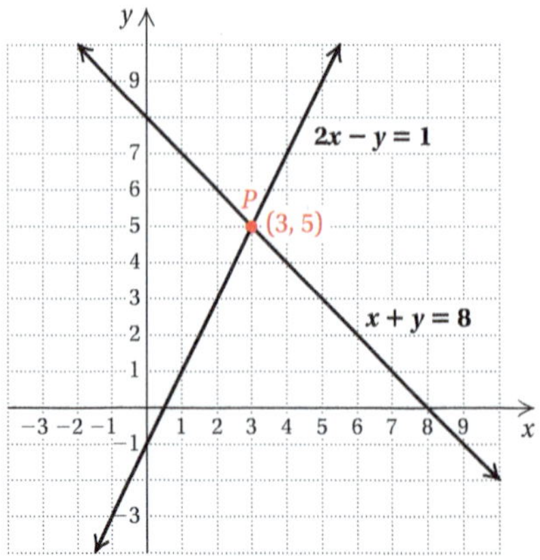

Check:

$$\begin{array}{c|l} x + y = 8 & \\ \hline 3 + 5 \ ? \ 8 & \\ 8 & \text{TRUE} \end{array}$$

$$\begin{array}{c|l} 2x - y = 1 & \\ \hline 2 \cdot 3 - 5 \ ? \ 1 & \\ 6 - 5 & \\ 1 & \text{TRUE} \end{array}$$

There is just one solution of the system of equations. It is $(3, 5)$. In other words, $x = 3$ and $y = 5$.

EXAMPLE 1 Determine whether $(1, 2)$ is a solution of the system

$$y = x + 1,$$
$$2x + y = 4.$$

We check by substituting alphabetically 1 for x and 2 for y.

Check:

$$\begin{array}{c|l} y & = x + 1 \\ \hline 2 & ?\ 1 + 1 \\ & 2 \end{array} \quad \text{TRUE}$$

$$\begin{array}{c|l} 2x + y & = 4 \\ \hline 2 \cdot 1 + 2 & ?\ 4 \\ 2 + 2 & \\ 4 & \end{array} \quad \text{TRUE}$$

This checks, so $(1, 2)$ is a solution of the system of equations.

EXAMPLE 2 Determine whether $(-3, 2)$ is a solution of the system

$$p + q = -1,$$
$$q + 3p = 4.$$

We check by substituting alphabetically -3 for p and 2 for q.

Check:

$$\begin{array}{c|l} p + q & = -1 \\ \hline -3 + 2 & ?\ -1 \\ -1 & \end{array} \quad \text{TRUE}$$

$$\begin{array}{c|l} q + 3p & = 4 \\ \hline 2 + 3(-3) & ?\ 4 \\ 2 - 9 & \\ -7 & \end{array} \quad \text{FALSE}$$

The point $(-3, 2)$ is not a solution of $q + 3p = 4$. Thus it is not a solution of the system of equations.

Do Exercises 1 and 2. ▶

Determine whether the given ordered pair is a solution of the system of equations.

1. $(2, -3)$; $x = 2y + 8$, $2x + y = 1$

Check:

$$\begin{array}{c|l} x & = 2y + 8 \\ \hline & ? \end{array} \qquad \begin{array}{c|l} 2x + y & = 1 \\ \hline & ? \end{array}$$

GS **2.** $(20, 40)$; $a = \frac{1}{2}b$, $b - a = 60$

Check:

$$\begin{array}{c|l} a & = \frac{1}{2}b \\ \hline \_\_\_ & ?\ \frac{1}{2}(\_\_\_) \\ & \_\_\_ \end{array}$$

$$\begin{array}{c|l} b - a & = 60 \\ \hline \_\_\_ - \_\_\_ & ?\ 60 \\ \_\_\_ & \end{array}$$

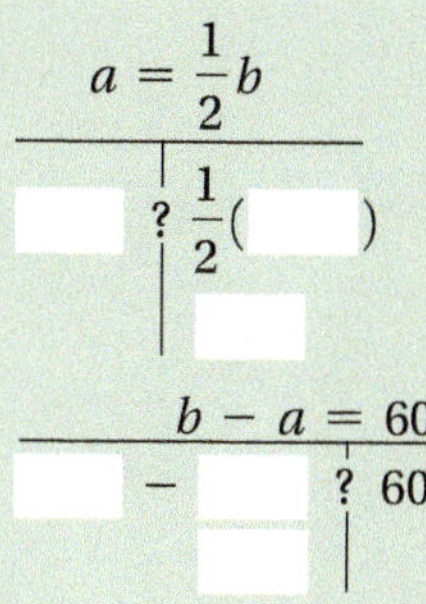

$(20, 40)$ ___ a solution of $a = \frac{1}{2}b$.

$(20, 40)$ ___ a solution of $b - a = 60$.

Therefore, $(20, 40)$ ___ a solution of the system.

b GRAPHING SYSTEMS OF EQUATIONS

When we solve a system of two equations by graphing, we graph both equations and find the coordinates of the points of intersection, if any exist.

EXAMPLE 3 Solve this system of equations by graphing:

$$x + y = 6,$$
$$x = y + 2.$$

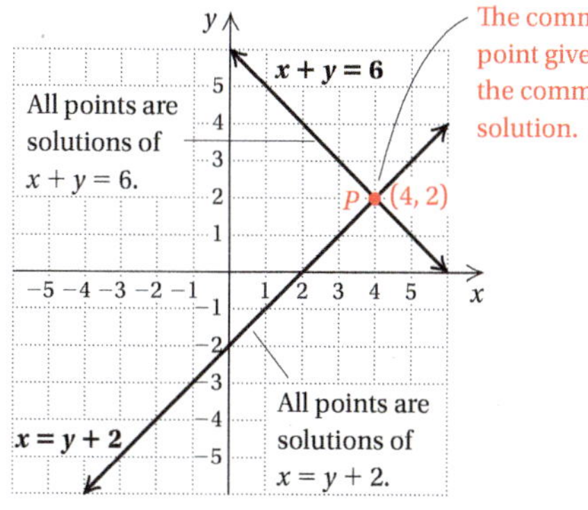

Point P with coordinates $(4, 2)$ looks as if it is the solution. We check the pair as follows.

Check:

$$\begin{array}{c|l} x + y & = 6 \\ \hline 4 + 2 & ?\ 6 \\ 6 & \end{array} \quad \text{TRUE}$$

$$\begin{array}{c|l} x & = y + 2 \\ \hline 4 & ?\ 2 + 2 \\ & 4 \end{array} \quad \text{TRUE}$$

The solution is $(4, 2)$.

Do Exercise 3 on the following page. ▶

Answers

Skill to Review:

1.

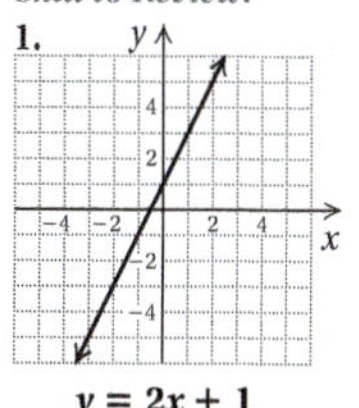

2.

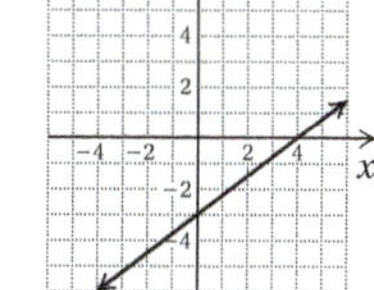

Margin Exercises:

1. Yes 2. No

Guided Solution:

2. $\begin{array}{c|l} a & = \frac{1}{2}b \\ \hline 20 & ?\ \frac{1}{2}(40) \\ & 20 \end{array}$; $\begin{array}{c|l} b - a & = 60 \\ \hline 40 - 20 & ?\ 60 \\ 20 & \end{array}$;

is, is not, is not

3. Solve this system by graphing:

$$2x + y = 1,$$
$$x = 2y + 8.$$

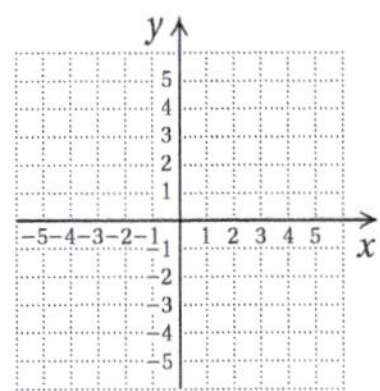

4. Solve this system by graphing:

$$x = -4,$$
$$y = 3.$$

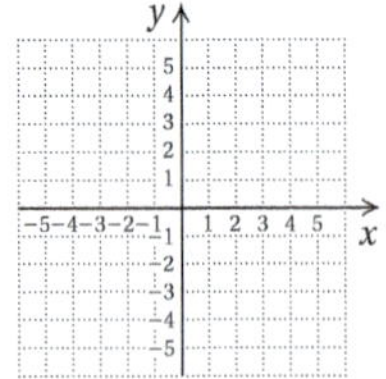

5. Solve this system by graphing:

$$y + 4 = x,$$
$$x - y = -2.$$

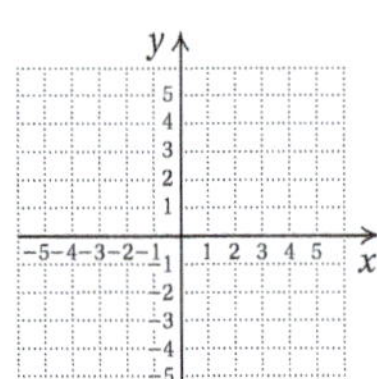

EXAMPLE 4 Solve this system of equations by graphing:

$$x = 2,$$
$$y = -3.$$

The graph of $x = 2$ is a vertical line, and the graph of $y = -3$ is a horizontal line. They intersect at the point $(2, -3)$.

The solution is $(2, -3)$.

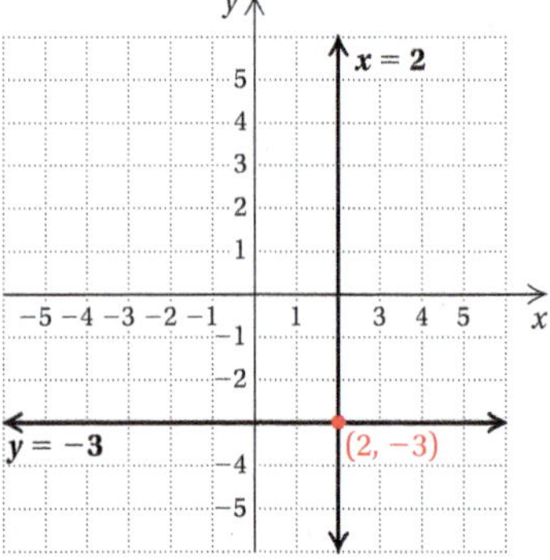

◀ **Do Exercise 4.**

Sometimes the equations in a system have graphs that are parallel lines.

EXAMPLE 5 Solve this system of equations by graphing:

$$y = 3x + 4,$$
$$y = 3x - 3.$$

The lines have the same slope, 3, and different y-intercepts, $(0, 4)$ and $(0, -3)$, so they are parallel.

There is no point at which the lines intersect, so the system has no solution. The solution set is the empty set, denoted $\varnothing$, or $\{\ \}$.

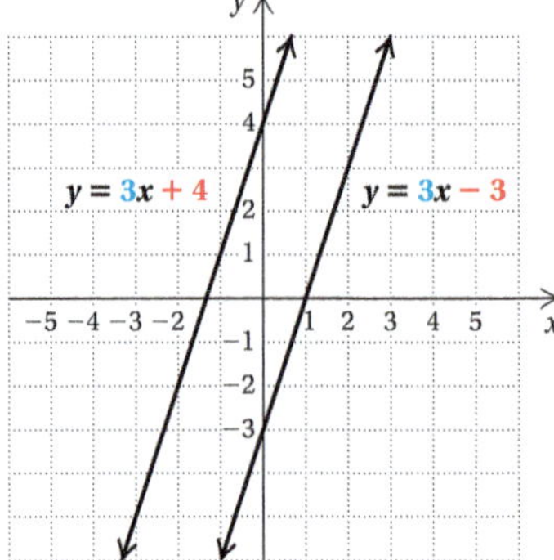

◀ **Do Exercise 5.**

Sometimes the equations in a system have the same graph.

EXAMPLE 6 Solve this system of equations by graphing:

$$2x + 3y = 6,$$
$$-8x - 12y = -24.$$

We graph the equations and see that the graphs are the same. Thus any solution of one of the equations is a solution of the other. Each equation has an infinite number of solutions, some of which are indicated on the graph.

On the following page, we check one such solution, $(0, 2)$, the y-intercept of each equation.

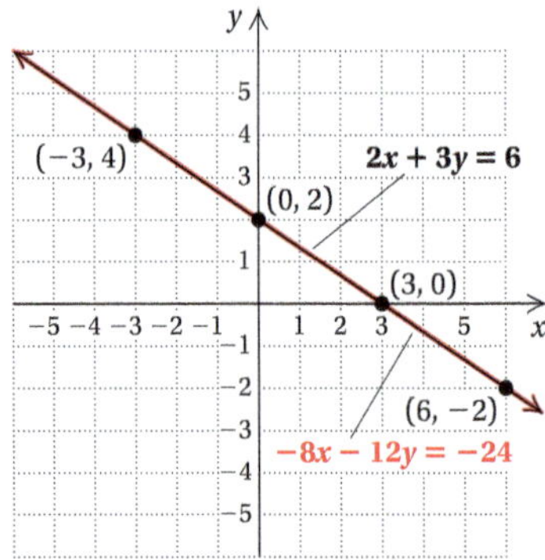

Answers

3. $(2, -3)$ **4.** $(-4, 3)$ **5.** No solution

Check:

$2x + 3y = 6$	
$2(0) + 3(2) \; ? \; 6$	
$0 + 6$	
6	TRUE

$-8x - 12y = -24$	
$-8(0) - 12(2) \; ? \; -24$	
$0 - 24$	
-24	TRUE

We leave it to the student to check that $(-3, 4)$ is also a solution of the system. If $(0, 2)$ and $(-3, 4)$ are solutions, then all points on the line containing them are solutions. The system has an infinite number of solutions.

Do Exercise 6.

6. Solve this system by graphing:

$$2x + y = 4,$$
$$-6x - 3y = -12.$$

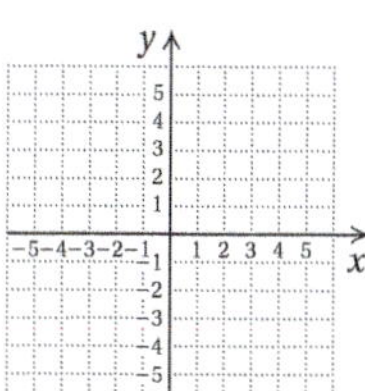

When we graph a system of two equations in two variables, we obtain one of the following three results.

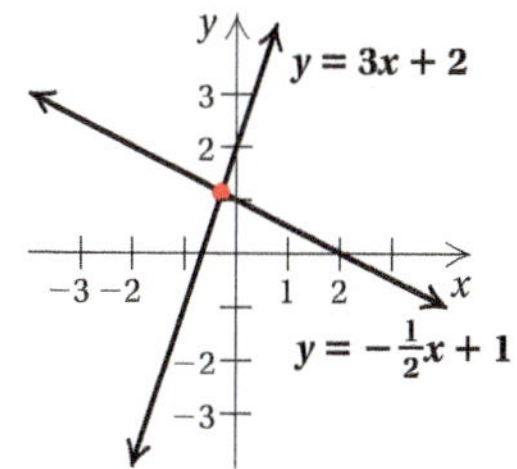

One solution
Graphs intersect.

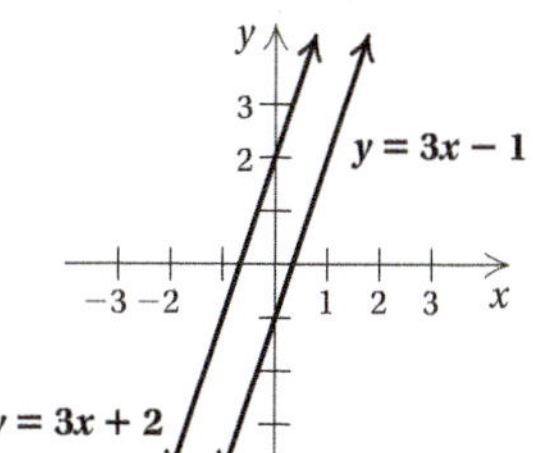

No solution
Graphs are parallel.

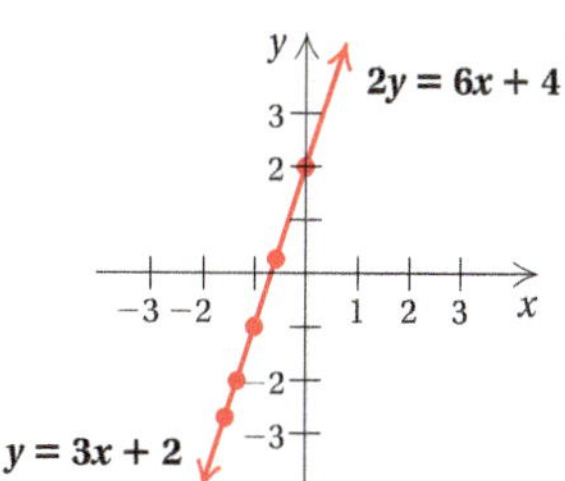

Infinitely many solutions
Equations have the same graph.

ALGEBRAIC ▶◀ GRAPHICAL CONNECTION

Let's take an algebraic–graphical look at equation solving. Such interpretation is useful when using a graphing calculator.

Consider the equation $6 - x = x - 2$. Let's solve it algebraically:

$6 - x = x - 2$	
$6 = 2x - 2$	Adding x
$8 = 2x$	Adding 2
$4 = x.$	Dividing by 2

Can we also solve the equation graphically? We can, as we see in the following two methods.

Method 1. Solve $6 - x = x - 2$ graphically.

We let $y = 6 - x$ and $y = x - 2$. Graphing the system of equations gives us the graph at right. The point of intersection is $(4, 2)$. Note that the x-coordinate of the point of intersection is 4. This value for x is also the *solution* of the equation $6 - x = x - 2$.

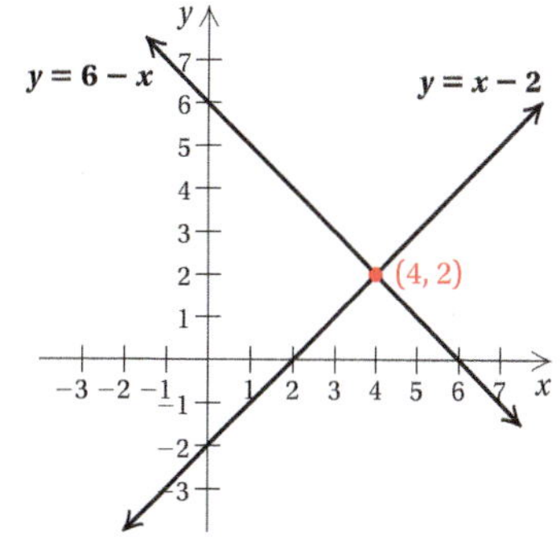

Do Exercise 7.

7. a) Solve $2x - 1 = 8 - x$ algebraically.

b) Solve $2x - 1 = 8 - x$ graphically using method 1.

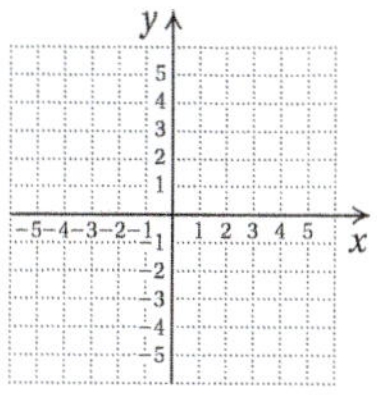

c) Compare your answers to parts (a) and (b).

Answers

6. Infinite number of solutions
7. **(a)** 3; **(b)** 3; **(c)** They are the same.

8. a) Solve $2x - 1 = 8 - x$ graphically using method 2.

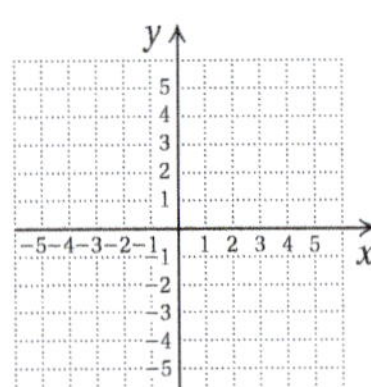

b) Compare your answers to Margin Exercises 7(a), 7(b), and 8(a).

Answers

8. (a) 3; (b) They are the same.

Method 2. Solve $6 - x = x - 2$ graphically.

Adding x and -6 on both sides, we obtain the form $0 = 2x - 8$. In this case, we let $y = 0$ and $y = 2x - 8$. Since $y = 0$ is the x-axis, we need graph only $y = 2x - 8$ and see where it crosses the x-axis. Note that the x-intercept of $y = 2x - 8$ is $(4, 0)$. The x-coordinate of this ordered pair is also the *solution* of the equation $6 - x = x - 2$.

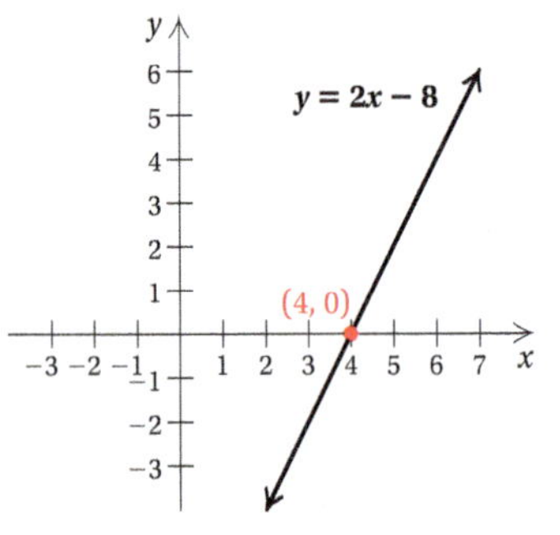

◀ Do Exercise 8.

CALCULATOR CORNER

Solving Systems of Equations We can solve a system of two equations in two variables on a graphing calculator. Consider the system of equations in Example 3,

$$x + y = 6,$$
$$x = y + 2.$$

First, we solve the equations for y, obtaining $y = -x + 6$ and $y = x - 2$. Then we enter $y_1 = -x + 6$ and $y_2 = x - 2$ on the equation-editor screen and graph the equations. We can use the standard viewing window, $[-10, 10, -10, 10]$.

We will use the INTERSECT feature to find the coordinates of the point of intersection of the lines. To use this feature, we select the two graphs, called First curve and Second curve, and then choose a Guess close to the point of intersection. The coordinates of the point of intersection of the graphs, $x = 4, y = 2$, appear at the bottom of the screen. Thus the solution of the system of equations is $(4, 2)$.

$y_1 = -x + 6,\ y_2 = x - 2$

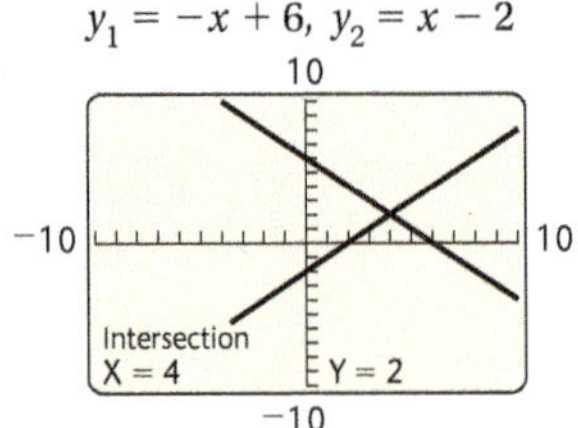

EXERCISES: Use a graphing calculator to solve each system of equations.

1. $x + y = 2,$
   $y = x + 4$

2. $x + 3y = -1,$
   $x - y = -5$

3. $3x + 5y = 19,$
   $4x = 10 + y$

15.1 Exercise Set

For Extra Help
MyMathLab®

MathXL® PRACTICE

WATCH

READ

REVIEW

Reading Check

Determine whether each statement is true or false.

RC1. A solution of a system of two equations in two variables is an ordered pair.

RC2. To check whether $(1, 3)$ is a solution of $y - 3x = 0$, we substitute 1 for x and 3 for y.

RC3. Graphs of two lines may have one point, no points, or an infinite number of points in common.

RC4. Every system of equations has at least one solution.

a Determine whether the given ordered pair is a solution of the system of equations. Use alphabetical order of the variables.

1. $(1, 5)$; $5x - 2y = -5,$ $3x - 7y = -32$

2. $(3, 2)$; $2x + 3y = 12,$ $x - 4y = -5$

3. $(4, 2)$; $3b - 2a = -2,$ $b + 2a = 8$

4. $(6, -6)$; $t + 2s = 6,$ $t - s = -12$

5. $(15, 20)$; $3x - 2y = 5,$ $6x - 5y = -10$

6. $(-1, -5)$; $4r + s = -9,$ $3r = 2 + s$

7. $(-1, 1)$; $x = -1,$ $x - y = -2$

8. $(-3, 4)$; $2x = -y - 2,$ $y = -4$

9. $(18, 3)$; $y = \frac{1}{6}x,$ $2x - y = 33$

10. $(-3, 1)$; $y = -\frac{1}{3}x,$ $3y = -5x - 12$

b Solve each system of equations by graphing.

11. $x - y = 2,$ $x + y = 6$

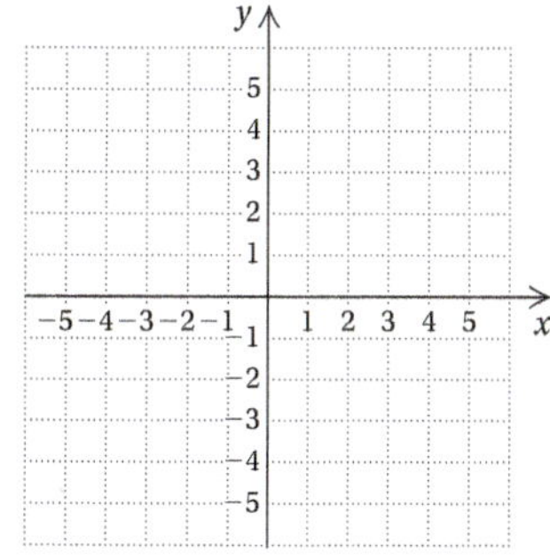

12. 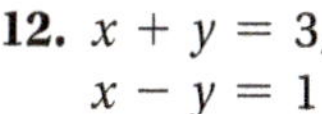
$x + y = 3,$ $x - y = 1$

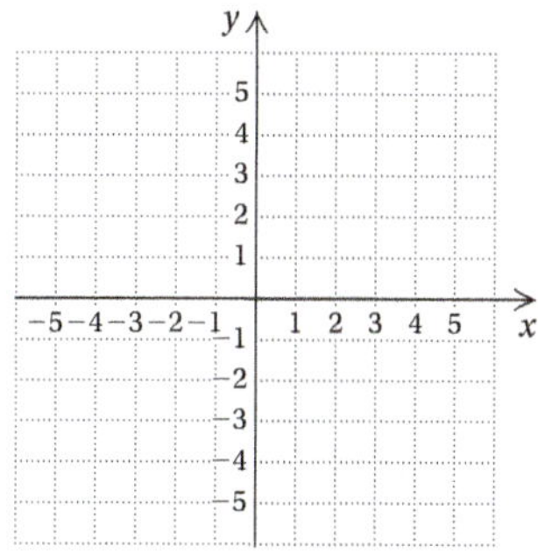

13. 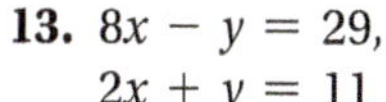
$8x - y = 29,$ $2x + y = 11$

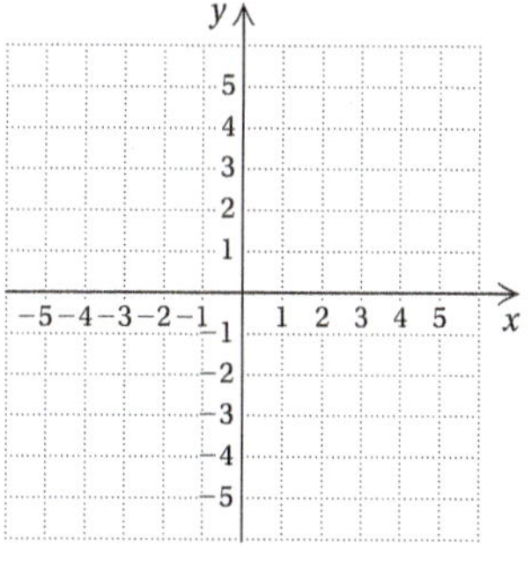

14. 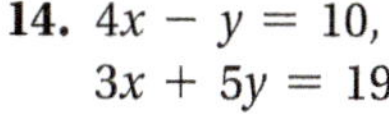
$4x - y = 10,$ $3x + 5y = 19$

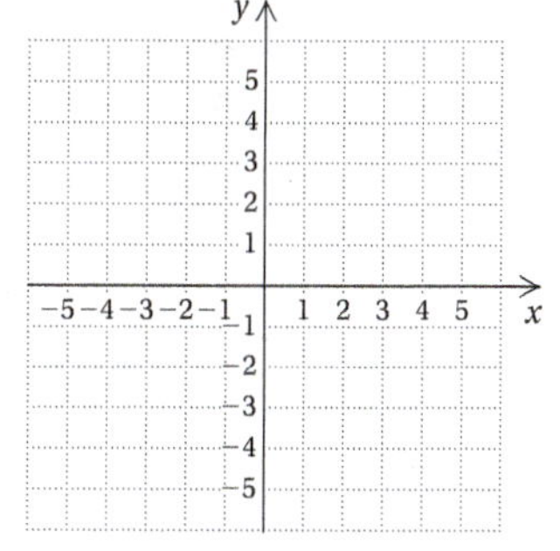

15. $t = v,$ $4t = 2v - 6$

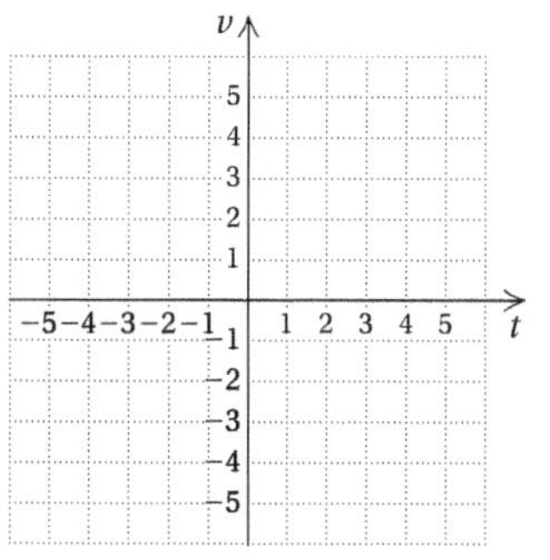

16. 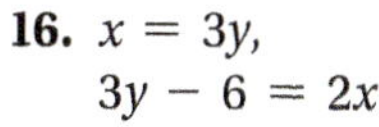
$x = 3y,$ $3y - 6 = 2x$

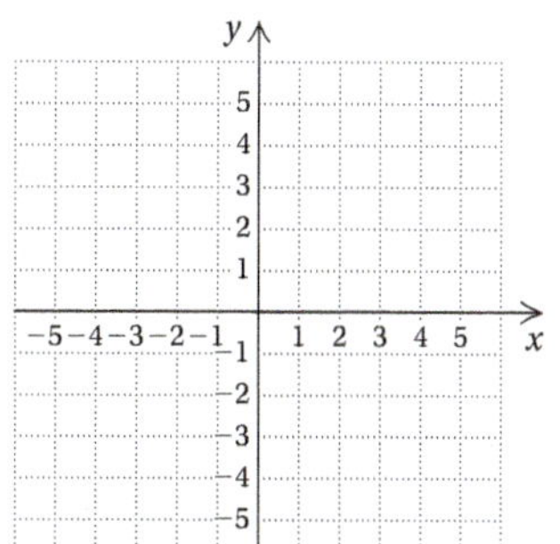

17. 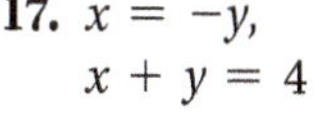
$x = -y,$ $x + y = 4$

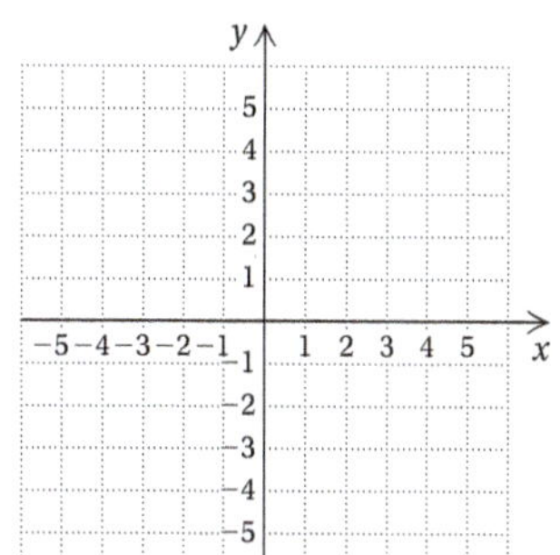

18. 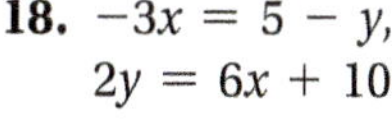
$-3x = 5 - y,$ $2y = 6x + 10$

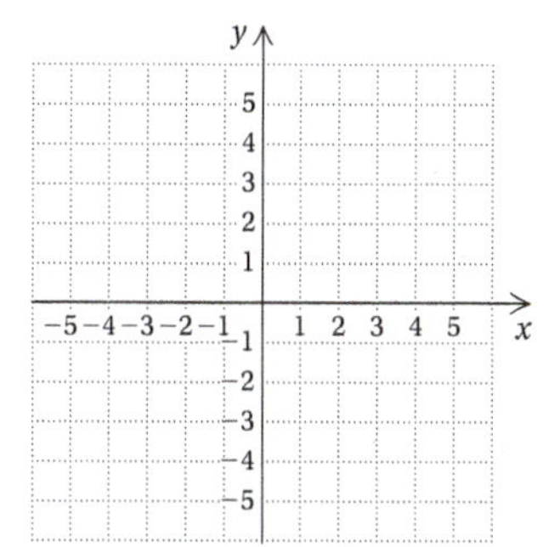

19. $a = \frac{1}{2}b + 1,$
$a - 2b = -2$

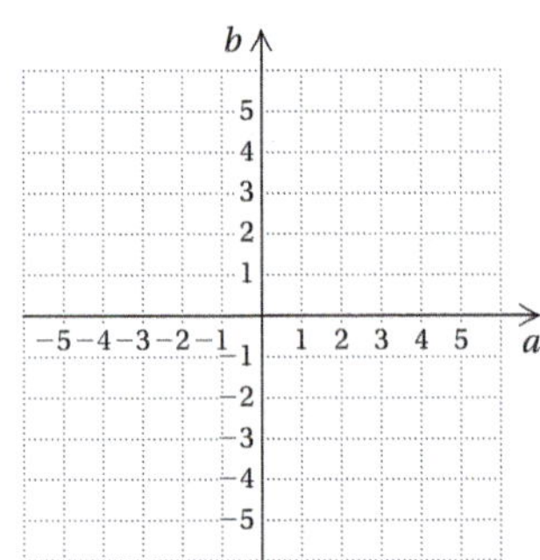

20. $x = \frac{1}{3}y + 2,$
$-2x - y = 1$

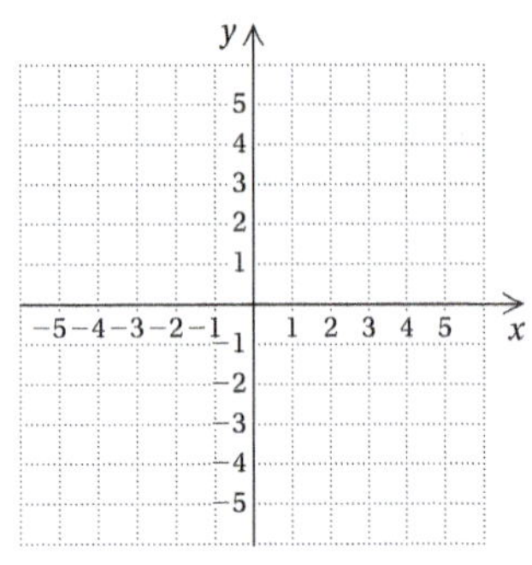

21. $y - 2x = 0,$
$y = 6x - 2$

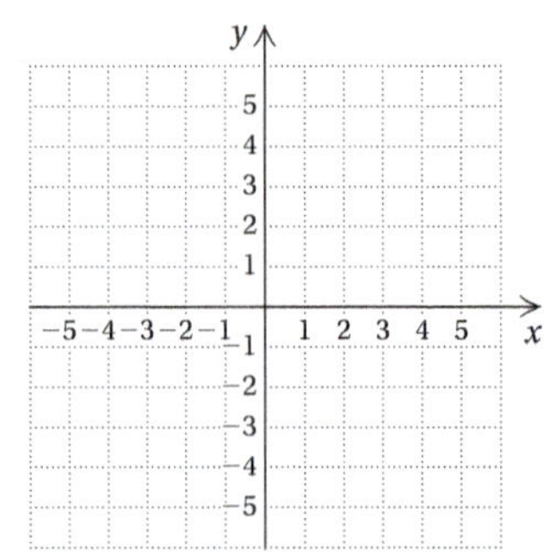

22. $y = 3x,$
$y = -3x + 2$

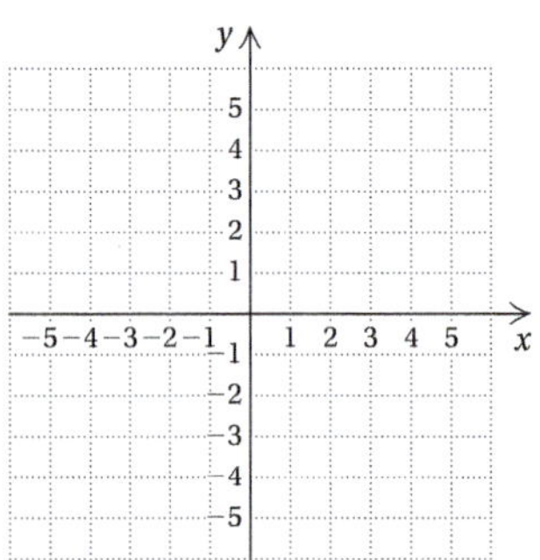

23. $x + y = 9,$
$3x + 3y = 27$

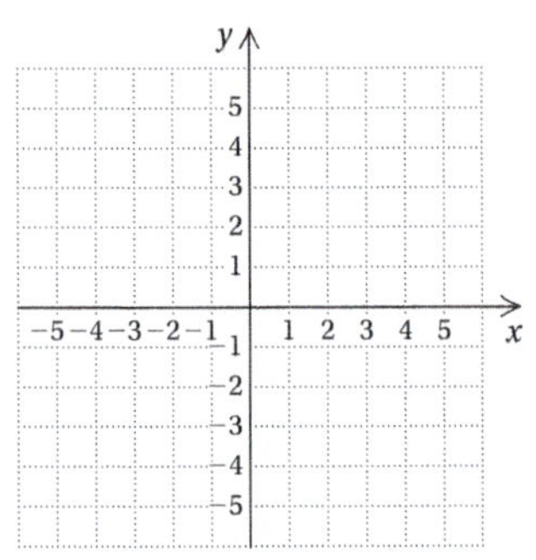

24. $x + y = 4,$
$x + y = -4$

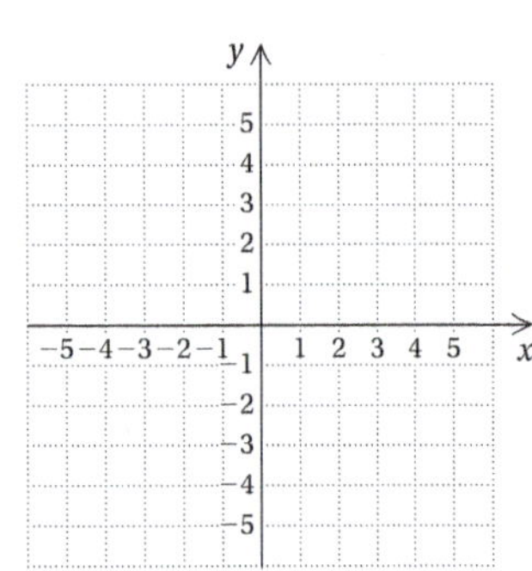

25. $x = 5,$
$y = -3$

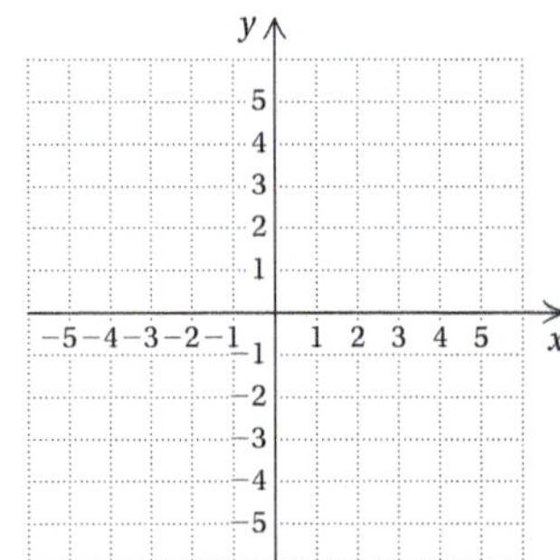

26. $y = 2,$
$y = -4$

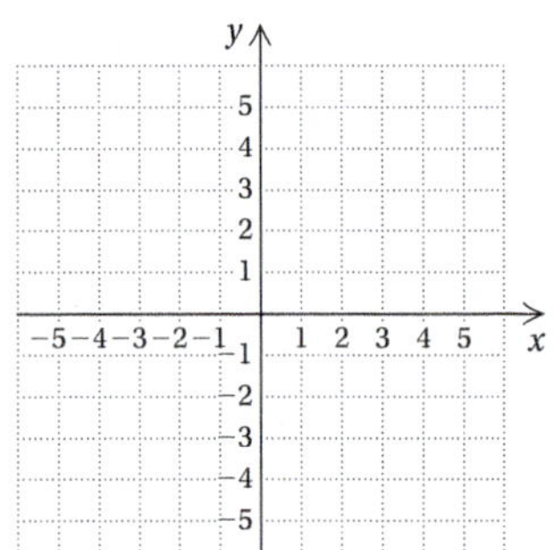

Skill Maintenance

Simplify. [14.1c]

27. $\dfrac{x^2 - 25}{x^2 - 10x + 25}$

28. $\dfrac{8d^2 + 16d}{40d^3 - 8d}$

29. $\dfrac{3y - 12}{4 - y}$

30. $\dfrac{2x^2 - x - 15}{x^2 - 9}$

Classify each polynomial as a monomial, a binomial, a trinomial, or none of these. [12.3b]

31. $5x^2 - 3x + 7$

32. $4x^3 - 2x^2$

33. $1.8x^5$

34. $x^3 + 2x^2 - 3x + 1$

Synthesis

35. The solution of the following system is $(2, -3)$. Find A and B.

$Ax - 3y = 13,$
$x - By = 8$

36. Find an equation to pair with $5x + 2y = 11$ such that the solution of the system is $(3, -2)$. Answers may vary.

37. Find a system of equations with $(6, -2)$ as a solution. Answers may vary.

38.–41. Use a graphing calculator to do Exercises 15–18.

The Substitution Method

15.2

Consider the following system of equations:

$$3x + 7y = 5,$$
$$6x - 7y = 1.$$

Suppose we try to solve this system graphically. We obtain the graph shown at right.

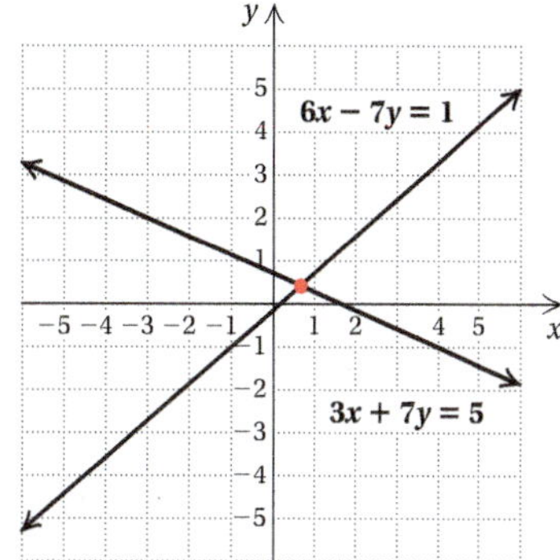

What is the solution? It is rather difficult to tell exactly. It would appear that the coordinates of the point are not integers. It turns out that the solution is $\left(\frac{2}{3}, \frac{3}{7}\right)$. Graphing helps us picture the solution of a system of equations, but solving by graphing, though practical in many applications, is not always fast or accurate. We now learn **algebraic** methods that can be used to determine solutions exactly.

OBJECTIVES

a Solve a system of two equations in two variables by the substitution method when one of the equations has a variable alone on one side.

b Solve a system of two equations in two variables by the substitution method when neither equation has a variable alone on one side.

c Solve applied problems by translating to a system of two equations and then solving using the substitution method.

SKILL TO REVIEW

Objective 10.4b: Solve a formula for a specified letter.

Solve for the indicated letter.

1. $x + 3y = 5$, for x
2. $2x - y = 9$, for y

a SOLVING BY THE SUBSTITUTION METHOD

One nongraphical method for solving systems is known as the **substitution method**.

EXAMPLE 1 Solve the system

$$x + y = 6, \quad (1)$$
$$x = y + 2. \quad (2)$$

Equation (2) says that x and $y + 2$ name the same number. Thus in equation (1), we can substitute $y + 2$ for x:

$$x + y = 6 \qquad \text{Equation (1)}$$
$$(y + 2) + y = 6. \qquad \text{Substituting } y + 2 \text{ for } x$$

This last equation has only one variable. We solve it:

$$y + 2 + y = 6 \qquad \text{Removing parentheses}$$
$$2y + 2 = 6 \qquad \text{Collecting like terms}$$
$$2y + 2 - 2 = 6 - 2 \qquad \text{Subtracting 2 on both sides}$$
$$2y = 4 \qquad \text{Simplifying}$$
$$\frac{2y}{2} = \frac{4}{2} \qquad \text{Dividing by 2}$$
$$y = 2. \qquad \text{Simplifying}$$

We have found the y-value of the solution. To find the x-value, we return to the original pair of equations. Substituting into either equation will give us the x-value.

Answers

Skill to Review:
1. $x = -3y + 5$ **2.** $y = 2x - 9$

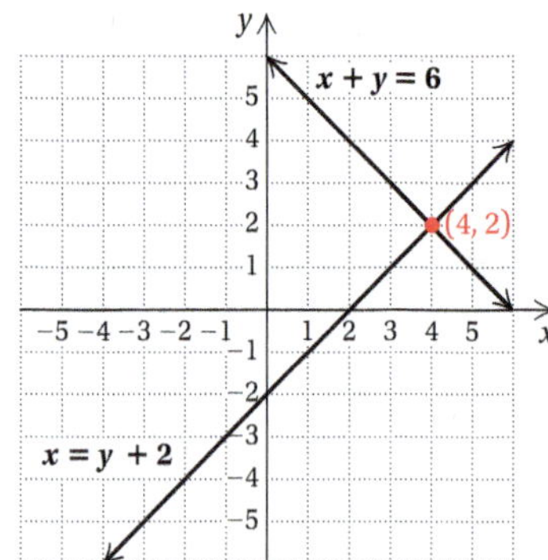

We choose equation (2) because it has x alone on one side:

$$x = y + 2 \quad \text{Equation (2)}$$
$$= 2 + 2 \quad \text{Substituting 2 for } y$$
$$= 4.$$

The ordered pair (4, 2) may be a solution. Note that we are using alphabetical order in listing the coordinates in an ordered pair. That is, since x precedes y alphabetically, we list 4 before 2 in the pair (4, 2).

We check as follows.

Check:

$x + y = 6$			$x = y + 2$		
$4 + 2$	?	6	4	?	$2 + 2$
6		TRUE		4	TRUE

Since (4, 2) checks, we have the solution. The graphical solution shown at left provides another check.

Note in Example 1 that substituting 2 for y in equation (1) will also give us the x-value of the solution:

$$x + y = 6$$
$$x + 2 = 6$$
$$x = 4.$$

Do Exercise 1.

1. Solve by the substitution method. Do not graph. **GS**

$$x + y = 5, \quad \textbf{(1)}$$
$$x = y + 1 \quad \textbf{(2)}$$

Substitute $y + 1$ for x in equation (1) and solve for y.

$$x + y = 5$$
$$(\;\_\_\_\;) + y = 5$$
$$\_\_\_ + 1 = 5$$
$$2y = \_\_\_$$
$$y = \_\_\_$$

Substitute ___ for y in equation (2) and solve for x.

$$x = y + 1$$
$$= \_\_\_ + 1$$
$$= \_\_\_$$

The numbers check. The solution is (___ , ___).

EXAMPLE 2 Solve the system

$$t = 1 - 3s, \quad \textbf{(1)}$$
$$s - t = 11. \quad \textbf{(2)}$$

We substitute $1 - 3s$ for t in equation (2):

$$s - t = 11 \quad \text{Equation (2)}$$
$$s - (1 - 3s) = 11. \quad \text{Substituting } 1 - 3s \text{ for } t$$

Remember to use parentheses when you substitute.

Now we solve for s:

$$s - 1 + 3s = 11 \quad \text{Removing parentheses}$$
$$4s - 1 = 11 \quad \text{Collecting like terms}$$
$$4s = 12 \quad \text{Adding 1}$$
$$s = 3. \quad \text{Dividing by 4}$$

Next, we substitute 3 for s in equation (1) of the original system:

$$t = 1 - 3s \quad \text{Equation (1)}$$
$$= 1 - 3 \cdot 3 \quad \text{Substituting 3 for } s$$
$$= -8.$$

The pair (3, −8) checks and is the solution. Remember: We list the answer in alphabetical order, (s, t). That is, since s comes before t in the alphabet, 3 is listed first and −8 second.

Do Exercise 2.

2. Solve by the substitution method:

$$a - b = 4,$$
$$b = 2 - a.$$

Answers

1. (3, 2) **2.** (3, −1)

Guided Solution:

1. $y + 1$, $2y$, 4, 2, 2, 2, 3, 3, 2

b SOLVING FOR THE VARIABLE FIRST

Sometimes neither equation of a pair has a variable alone on one side. Then we solve one equation for one of the variables and proceed as before, substituting into the *other* equation. If possible, we solve in either equation for a variable that has a coefficient of 1.

EXAMPLE 3 Solve the system

$$x - 2y = 6, \quad (1)$$
$$3x + 2y = 4. \quad (2)$$

We solve one equation for one variable. Since the coefficient of x is 1 in equation (1), it is easier to solve that equation for x:

$x - 2y = 6$ Equation (1)

$x = 6 + 2y.$ Adding $2y$ **(3)**

We substitute $6 + 2y$ for x in equation (2) of the original pair and solve for y:

$3x + 2y = 4$ Equation (2)

$3(6 + 2y) + 2y = 4$ Substituting $6 + 2y$ for x

$18 + 6y + 2y = 4$ Removing parentheses

$18 + 8y = 4$ Collecting like terms

$8y = -14$ Subtracting 18

$y = \frac{-14}{8}$, or $-\frac{7}{4}$. Dividing by 8

To find x, we go back to either of the original equations, (1) or (2), or to equation (3), which is solved for x. It is generally easier to use an equation like equation (3) where we have solved for a specific variable. We substitute $-\frac{7}{4}$ for y in equation (3) and compute x:

$x = 6 + 2y$ Equation (3)

$= 6 + 2\left(-\frac{7}{4}\right)$ Substituting $-\frac{7}{4}$ for y

$= 6 - \frac{7}{2} = \frac{5}{2}.$

We check the ordered pair $\left(\frac{5}{2}, -\frac{7}{4}\right)$.

Check:

$x - 2y = 6$	
$\frac{5}{2} - 2\left(-\frac{7}{4}\right)$	? 6
$\frac{5}{2} + \frac{7}{2}$	
$\frac{12}{2}$	
6	TRUE

$3x + 2y = 4$	
$3 \cdot \frac{5}{2} + 2\left(-\frac{7}{4}\right)$	? 4
$\frac{15}{2} - \frac{7}{2}$	
$\frac{8}{2}$	
4	TRUE

Since $\left(\frac{5}{2}, -\frac{7}{4}\right)$ checks, it is the solution. This solution would have been difficult to find graphically because it involves fractions.

Do Exercise 3. ▶

c SOLVING APPLIED PROBLEMS

Now let's solve an applied problem using systems of equations and the substitution method.

Caution!

A solution of a system of equations in two variables is an ordered *pair* of numbers. Once you have solved for one variable, don't forget the other. A common mistake is to solve for only one variable.

GS **3.** Solve:

$$x - 2y = 8, \quad (1)$$
$$2x + y = 8. \quad (2)$$

Solve for y in equation (2).

$2x + y = 8$

$y = 8 - \_\_\_$ **(3)**

Substitute $8 - 2x$ for y in equation (1) and solve for x.

$x - 2y = 8$

$x - 2(\_\_\_) = 8$

$x - 16 + \_\_\_ = 8$

$\_\_\_ - 16 = 8$

$5x = \_\_\_$

$x = \_\_\_$

Substitute $\_\_\_$ for x in equation (3) and solve for y.

$y = 8 - 2x$

$= 8 - 2(\_\_\_)$

$= \frac{40}{5} - \frac{\_\_\_}{5}$

$= \_\_\_$

The numbers check. The solution is $(\_\_\_, \_\_\_)$.

Answer

3. $\left(\frac{24}{5}, -\frac{8}{5}\right)$

Guided Solution:

3. $2x$, $8 - 2x$, $4x$, $5x$, 24, $\frac{24}{5}$, $\frac{24}{5}$, $\frac{24}{5}$, 48, $-\frac{8}{5}$, $\frac{24}{5}$, $-\frac{8}{5}$

EXAMPLE 4 *Standard Billboard.* A standard rectangular highway billboard has a perimeter of 124 ft. The length is 34 ft more than the width. Find the length and the width.

Source: Eller Sign Company

1. **Familiarize.** We make a drawing and label it. We let l = the length and w = the width.

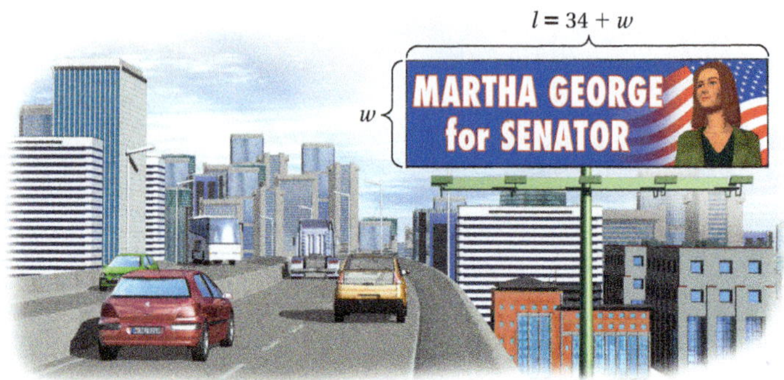

2. **Translate.** The perimeter of a rectangle is given by the formula $2l + 2w$. We translate each statement, as follows.

The perimeter is 124 ft.
$$2l + 2w = 124$$

The length is 34 ft longer than the width.
$$l = 34 + w$$

We now have a system of equations:

$$2l + 2w = 124, \quad (1)$$
$$l = 34 + w. \quad (2)$$

3. **Solve.** We solve the system. To begin, we substitute $34 + w$ for l in the first equation and solve:

$2(34 + w) + 2w = 124$	Substituting $34 + w$ for l in equation (1)
$2 \cdot 34 + 2 \cdot w + 2w = 124$	Removing parentheses
$4w + 68 = 124$	Collecting like terms
$4w = 56$	Subtracting 68
$w = 14.$	Dividing by 4

We go back to one of the original equations and substitute 14 for w:

$l = 34 + w = 34 + 14 = 48.$ Substituting in equation (2)

4. **Check.** If the length is 48 ft and the width is 14 ft, then the length is 34 ft more than the width ($48 - 14 = 34$), and the perimeter is 2(48 ft) + 2(14 ft), or 124 ft. Thus these dimensions check.

5. **State.** The width is 14 ft and the length is 48 ft.

Example 4 illustrates that many problems that can be solved by translating to *one* equation in *one* variable may actually be easier to solve by translating to *two* equations in *two* variables.

Do Exercise 4.

4. *Community Garden.* A rectangular community garden is to be enclosed with 92 m of fencing. In order to allow for compost storage, the garden must be 4 m longer than it is wide. Determine the dimensions of the garden.

Answer

4. Length: 25 m; width: 21 m

15.2 Exercise Set

For Extra Help MyMathLab® MathXL® PRACTICE WATCH READ REVIEW

Reading Check

Determine whether each statement is true or false.

RC1. The substitution method is an algebraic method for solving systems of equations.

RC2. We can find solutions of systems of equations involving fractions using the substitution method.

RC3. When writing the solution of a system, we write the value that we found first as the first number in the ordered pair.

RC4. When solving using substitution, we may have to solve for a variable before substituting.

a Solve using the substitution method.

1. $x = -2y,$ $x + 4y = 2$

2. $r = -3s,$ $r + 4s = 10$

3. $y = x - 6,$ $x + y = -2$

4. $y = x + 1,$ $2x + y = 4$

5. $y = 2x - 5,$ $3y - x = 5$

6. $y = 2x + 1,$ $x + y = -2$

7. $x = y + 5,$ $2x + y = 1$

8. $x = y - 3,$ $x + 2y = 9$

9. $x + y = 10,$ $y = x + 8$

10. $x + y = 4,$ $y = 2x + 1$

11. $2x + y = 5,$ $x = y + 7$

12. $3x + y = -1,$ $x = 2y - 5$

b Solve using the substitution method. First, solve one equation for one variable.

13. $x - y = 6,$ $x + y = -2$

14. $s + t = -4,$ $s - t = 2$

15. $y - 2x = -6,$ $2y - x = 5$

16. $x - y = 5,$ $x + 2y = 7$

17. $r - 2s = 0,$ $4r - 3s = 15$

18. $y - 2x = 0,$ $3x + 7y = 17$

19. $2x + 3y = -2,$ $2x - y = 9$

20. $3x - 6y = 4,$ $5x + y = 3$

21. $x + 3y = 5,$ $3x + 5y = 3$

22. $x + 2y = 10,$ $3x + 4y = 8$

23. $x - y = -3,$ $2x + 3y = -6$

24. $x - 2y = 8,$ $2x + 3y = 2$

C Solve.

25. *Two-by-Four.* The perimeter of a cross section of a "two-by-four" piece of lumber is 10 in. The length is 2 in. more than the width. Find the actual dimensions of a cross section of a two-by-four.

26. *Billboards.* As an advertisement for chocolate candy sales in anticipation of Valentine's Day 2011, the Meiji Seika Kaisha confectionary factory in Takatsuki, Osaka Prefecture, Japan, built a giant billboard in the shape of a chocolate bar. The perimeter of the billboard was 388 m, and the length was 138 m more than the width. Find the length and the width.

Source: www.worldrecordsacademy.org

27. *Dimensions of Wyoming.* The state of Wyoming is roughly a rectangle with a perimeter of 1280 mi. The width is 90 mi less than the length. Find the length and the width.

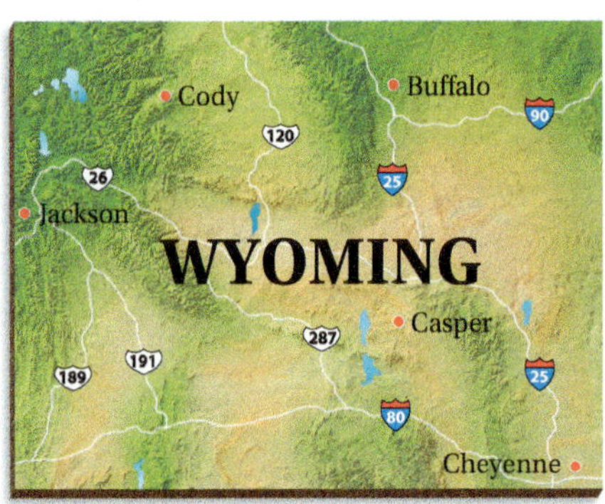

28. *Dimensions of Colorado.* The state of Colorado is roughly a rectangle whose perimeter is 1300 mi. The width is 110 mi less than the length. Find the length and the width.

29. *Lacrosse.* The perimeter of a lacrosse field is 340 yd. The length is 10 yd less than twice the width. Find the length and the width.

30. *Soccer.* The perimeter of a soccer field is 280 yd. The width is 5 more than half the length. Find the length and the width.

31. The sum of two numbers is 37. One number is 5 more than the other. Find the numbers.

32. The sum of two numbers is 26. One number is 12 more than the other. Find the numbers.

33. Find two numbers whose sum is 52 and whose difference is 28.

34. Find two numbers whose sum is 63 and whose difference is 5.

35. The difference of two numbers is 12. Two times the larger is five times the smaller. What are the numbers?

36. The difference of two numbers is 18. Twice the smaller number plus three times the larger is 74. What are the numbers?

37. On average, Americans spend 12.5 hr per month watching videos on both Netflix and Hulu. They spend 7.5 more hours watching videos on Netflix than they do on Hulu. How much time do they spend on each?

Source: *2011 State of Media Report*, Nielsen

38. On average, Americans spend 33.25 hr per week watching both traditional TV and video online. They spend 32.25 more hours watching traditional TV than they do watching video online. How much time do they spend on each?

Source: *2011 State of Media Report*, Nielsen

Skill Maintenance

Graph. [11.1d], [11.2a, b]

39. $2x - 3y = 6$

40. $2x + 3y = 6$

41. $y = 2x - 5$

42. $y = 4$

Factor completely. [13.6a]

43. $6x^2 - 13x + 6$

44. $4p^2 - p - 3$

45. $4x^2 + 3x + 2$

46. $9a^2 - 25$

Simplify. [12.1d, e, f]

47. $\dfrac{x^{-2}}{x^{-5}}$

48. $x^2 \cdot x^5$

49. $x^{-2} \cdot x^{-5}$

50. $\dfrac{a^2b^{-3}}{a^5b^{-6}}$

Synthesis

Solve using the INTERSECT feature on a graphing calculator. Then solve algebraically and decide which method you prefer to use.

51. $x - y = 5,$
$x + 2y = 7$

52. $y - 2x = -6,$
$2y - x = 5$

53. $y - 2.35x = -5.97,$
$2.14y - x = 4.88$

54. $y = 1.2x - 32.7,$
$y = -0.7x + 46.15$

55. *Softball.* The perimeter of a softball diamond is two-thirds of the perimeter of a baseball diamond. Together, the two perimeters measure 200 yd. Find the distance between the bases in each sport.

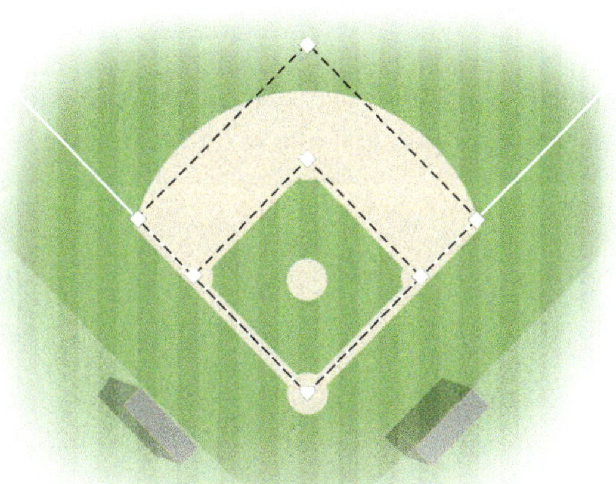

56. Write a system of two linear equations that can be solved more quickly—but still precisely—by a graphing calculator than by substitution. Time yourself using both methods to solve the system.

15.3 The Elimination Method

OBJECTIVES

a Solve a system of two equations in two variables using the elimination method when no multiplication is necessary.

b Solve a system of two equations in two variables using the elimination method when multiplication is necessary.

SKILL TO REVIEW

Objective 10.3c: Solve equations with an infinite number of solutions and equations with no solutions.

Solve.

1. $6(1 - x) = 6 - 6x$
2. $2(x - 1) = 2x + 3$

a SOLVING BY THE ELIMINATION METHOD

The **elimination method** for solving systems of equations makes use of the *addition principle*. Some systems are much easier to solve using this method rather than the substitution method. For example, to solve the system

$$2x + 3y = 13, \quad (1)$$
$$4x - 3y = 17 \quad (2)$$

by substitution, we would need to first solve for a variable in one of the equations. Were we to solve equation (1) for y, we would find (after several steps) that $y = \frac{13}{3} - \frac{2}{3}x$. We could then use the expression $\frac{13}{3} - \frac{2}{3}x$ in equation (2) as a replacement for y:

$$4x - 3\left(\tfrac{13}{3} - \tfrac{2}{3}x\right) = 17.$$

As you can see, although substitution could be used to solve this system, doing so involves working with fractions. Fortunately, another method, elimination, can be used to solve systems and, for problems like this, is simpler to use.

EXAMPLE 1 Solve the system

$$2x + 3y = 13, \quad (1)$$
$$4x - 3y = 17. \quad (2)$$

The key to the advantage of the elimination method for solving this system involves the $3y$ in one equation and the $-3y$ in the other. These terms are opposites. If we add the terms on the left sides of the equations, the y-terms will add to 0, and in effect, the variable y will be eliminated.

We will use the addition principle for equations. According to equation (2), $4x - 3y$ and 17 are the same number. Thus we can use a vertical form and add $4x - 3y$ on the left side of equation (1) and 17 on the right side—in effect, adding the same number on both sides of equation (1):

$$\begin{aligned} 2x + 3y &= 13 && (1) \\ 4x - 3y &= 17 && (2) \\ \hline 6x + 0y &= 30, \text{ or} && \text{Adding} \\ 6x \quad\;\; &= 30. \end{aligned}$$

We have "eliminated" one variable. This is why we call this the **elimination method**. We now have an equation with just one variable that can be solved for x:

$$\begin{aligned} 6x &= 30 \\ x &= 5. \end{aligned}$$

Next, we substitute 5 for x in either of the original equations:

$$\begin{aligned} 2x + 3y &= 13 && \text{Equation (1)} \\ 2(5) + 3y &= 13 && \text{Substituting 5 for } x \\ 10 + 3y &= 13 \\ 3y &= 3 \\ y &= 1. && \text{Solving for } y \end{aligned}$$

Answers

Skill to Review:
1. All real numbers 2. No solution

We check the ordered pair $(5, 1)$.

Check:

$$\begin{array}{c|l} 2x + 3y = 13 & \\ \hline 2(5) + 3(1) \ ? \ 13 & \\ 10 + 3 & \\ 13 & \text{TRUE} \end{array} \qquad \begin{array}{c|l} 4x - 3y = 17 & \\ \hline 4(5) - 3(1) \ ? \ 17 & \\ 20 - 3 & \\ 17 & \text{TRUE} \end{array}$$

Since $(5, 1)$ checks, it is the solution. We can see the solution in the graph shown at right.

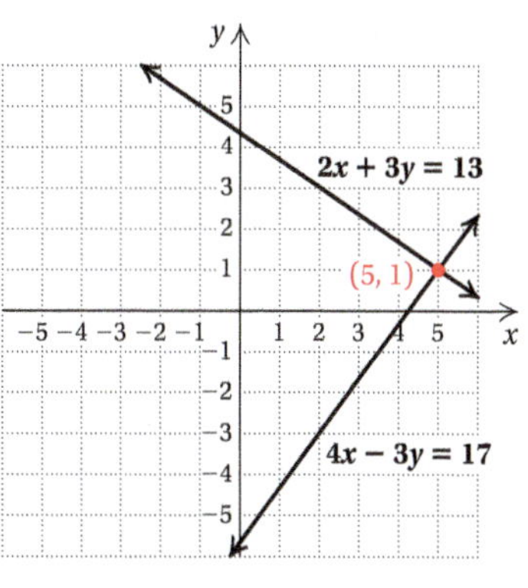

Do Exercises 1 and 2. ▶

Solve using the elimination method.

1. $x + y = 5,$
$2x - y = 4$

2. $-2x + y = -4,$
$2x - 5y = 12$

b USING THE MULTIPLICATION PRINCIPLE FIRST

The elimination method allows us to eliminate a variable. We may need to multiply by certain numbers first, however, so that terms become opposites.

EXAMPLE 2 Solve the system

$$2x + 3y = 8, \quad (1)$$
$$x + 3y = 7. \quad (2)$$

If we add, we will not eliminate a variable. However, if the $3y$ were $-3y$ in one equation, we could eliminate y. Thus we multiply by -1 on both sides of equation (2) and then add, using a vertical form:

$$\begin{array}{rll} 2x + 3y = & 8 & \text{Equation (1)} \\ -x - 3y = & -7 & \text{Multiplying equation (2) by } -1 \\ \hline x \quad\quad = & 1. & \text{Adding} \end{array}$$

Next, we substitute 1 for x in one of the original equations:

$$\begin{array}{rl} x + 3y = 7 & \text{Equation (2)} \\ 1 + 3y = 7 & \text{Substituting 1 for } x \\ 3y = 6 & \\ y = 2. & \text{Solving for } y \end{array}$$

We check the ordered pair $(1, 2)$.

Check:

$$\begin{array}{c|l} 2x + 3y = 8 & \\ \hline 2 \cdot 1 + 3 \cdot 2 \ ? \ 8 & \\ 2 + 6 & \\ 8 & \text{TRUE} \end{array} \qquad \begin{array}{c|l} x + 3y = 7 & \\ \hline 1 + 3 \cdot 2 \ ? \ 7 & \\ 1 + 6 & \\ 7 & \text{TRUE} \end{array}$$

Since $(1, 2)$ checks, it is the solution. We can see the solution in the graph shown at right.

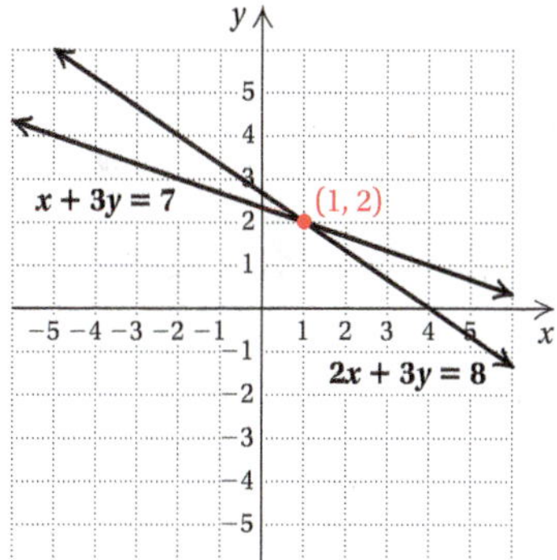

Do Exercises 3 and 4. ▶

3. Solve. Multiply one equation by -1 first.
$5x + 3y = 17,$
$5x - 2y = -3$

4. Solve the system
$3x - 2y = -30,$
$5x - 2y = -46.$

Answers

1. $(3, 2)$ **2.** $(1, -2)$ **3.** $(1, 4)$ **4.** $(-8, 3)$

In Example 2, we used the multiplication principle, multiplying by -1. However, we often need to multiply by something other than -1.

EXAMPLE 3 Solve the system

$$3x + 6y = -6, \quad (1)$$
$$5x - 2y = 14. \quad (2)$$

Looking at the terms with variables, we see that if $-2y$ were $-6y$, we would have terms that are opposites. We can achieve this by multiplying by 3 on both sides of equation (2). Then we add and solve for x:

$$3x + 6y = -6 \qquad \text{Equation (1)}$$
$$15x - 6y = 42 \qquad \text{Multiplying by 3 on both sides of equation (2)}$$
$$18x = 36 \qquad \text{Adding}$$
$$x = 2. \qquad \text{Solving for } x$$

Next, we substitute 2 for x in either of the original equations. We choose the first:

$$3x + 6y = -6 \qquad \text{Equation (1)}$$
$$3 \cdot 2 + 6y = -6 \qquad \text{Substituting 2 for } x$$
$$6 + 6y = -6$$
$$6y = -12$$
$$y = -2. \qquad \text{Solving for } y$$

We check the ordered pair $(2, -2)$.

Check:

$$\begin{array}{c|c} 3x + 6y & -6 \\ \hline 3 \cdot 2 + 6 \cdot (-2) \;?\; & -6 \\ 6 + (-12) & \\ -6 & \end{array} \quad \text{TRUE}$$

$$\begin{array}{c|c} 5x - 2y & 14 \\ \hline 5 \cdot 2 - 2 \cdot (-2) \;?\; & 14 \\ 10 - (-4) & \\ 14 & \end{array} \quad \text{TRUE}$$

Since $(2, -2)$ checks, it is the solution. (See the graph at left.)

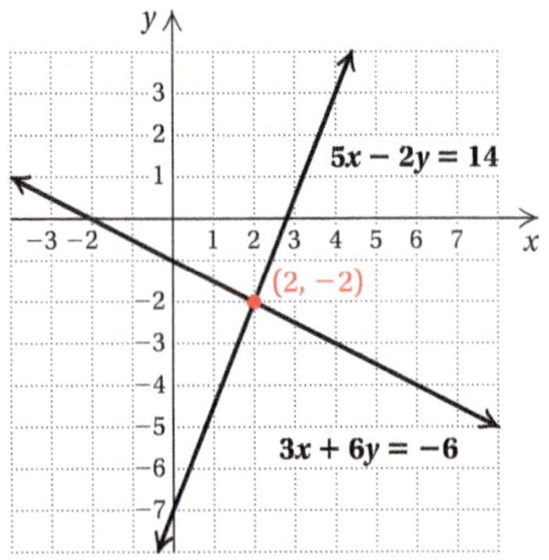

Do Exercises 5 and 6.

Part of the strategy in using the elimination method is making a decision about which variable to eliminate. So long as the algebra has been carried out correctly, the solution can be found by eliminating *either* variable. We multiply so that terms involving the variable to be eliminated are opposites. It is helpful to first get each equation in a form equivalent to $Ax + By = C$.

EXAMPLE 4 Solve the system

$$3y + 1 + 2x = 0, \quad (1)$$
$$5x = 7 - 4y. \quad (2)$$

We first rewrite each equation in a form equivalent to $Ax + By = C$:

$$2x + 3y = -1, \quad (1) \qquad \text{Subtracting 1 on both sides and rearranging terms}$$
$$5x + 4y = 7. \quad (2) \qquad \text{Adding } 4y \text{ on both sides}$$

We decide to eliminate the x-term. We do so by multiplying by 5 on both sides of equation (1) and by -2 on both sides of equation (2). Then we add and solve for y.

Solve each system.

5. $4a + 7b = 11,$
$2a + 3b = 5$

6. $3x - 8y = 2, \quad (1)$
$5x + 2y = -12 \quad (2)$

GS

Multiply equation (2) by 4, add, and solve for x.

$3x - 8y = 2$
$20x +$ ___ $=$ ___
$23x =$ ___
$x =$ ___

Substitute ___ for x in equation (1) and solve for y.

$3x - 8y = 2$
$3($ ___ $) - 8y = 2$
___ $- 8y = 2$
$-8y =$ ___
$y =$ ___

The numbers check. The solution is (___ , ___).

Answers

5. $(1, 1)$ 6. $(-2, -1)$

Guided Solution:
6. $8y, -48, -46, -2, -2, -2, -6, 8, -1, -2, -1$

$$\begin{aligned} 10x + 15y &= -5 && \text{Multiplying by 5 on both sides of equation (1)} \\ -10x - 8y &= -14 && \text{Multiplying by } -2 \text{ on both sides of equation (2)} \\ \hline 7y &= -19 && \text{Adding} \\ y &= \frac{-19}{7}, \text{ or } -\frac{19}{7} && \text{Solving for } y \end{aligned}$$

Next, we substitute $-\frac{19}{7}$ for y in one of the original equations:

$$\begin{aligned} 2x + 3y &= -1 && \text{Equation (1)} \\ 2x + 3\left(-\tfrac{19}{7}\right) &= -1 && \text{Substituting } -\tfrac{19}{7} \text{ for } y \\ 2x - \tfrac{57}{7} &= -1 \\ 2x &= -1 + \tfrac{57}{7} \\ 2x &= -\tfrac{7}{7} + \tfrac{57}{7} \\ 2x &= \tfrac{50}{7} \\ \tfrac{1}{2} \cdot 2x &= \tfrac{1}{2} \cdot \tfrac{50}{7} && \text{Multiplying by } \tfrac{1}{2} \text{ on both sides of the equation} \\ x &= \tfrac{50}{14} \\ x &= \tfrac{25}{7}. && \text{Simplifying} \end{aligned}$$

We check the ordered pair $\left(\frac{25}{7}, -\frac{19}{7}\right)$.

Check:

$3y + 1 + 2x = 0$	
$3\left(-\frac{19}{7}\right) + 1 + 2\left(\frac{25}{7}\right)$	? 0
$-\frac{57}{7} + \frac{7}{7} + \frac{50}{7}$	
0	TRUE

$5x = 7 - 4y$	
$5\left(\frac{25}{7}\right)$	? $7 - 4\left(-\frac{19}{7}\right)$
$\frac{125}{7}$	$\frac{49}{7} + \frac{76}{7}$
	$\frac{125}{7}$ TRUE

The solution is $\left(\frac{25}{7}, -\frac{19}{7}\right)$.

Do Exercise 7. ▶

Let's consider a system with no solution and see what happens when we apply the elimination method.

EXAMPLE 5 Solve the system

$$\begin{aligned} y - 3x &= 2, && (1) \\ y - 3x &= 1. && (2) \end{aligned}$$

We multiply by -1 on both sides of equation (2) and then add:

$$\begin{aligned} y - 3x &= 2 && \text{Equation (1)} \\ -y + 3x &= -1 && \text{Multiplying by } -1 \text{ on both sides of equation (2)} \\ \hline 0 &= 1. && \text{Adding} \end{aligned}$$

We obtain a false equation, $0 = 1$, so there is no solution. The slope–intercept forms of these equations are

$$\begin{aligned} y &= 3x + 2, \\ y &= 3x + 1. \end{aligned}$$

The slopes, 3, are the same and the y-intercepts, $(0, 2)$ and $(0, 1)$, are different. Thus the lines are parallel. They do not intersect. (See the graph at right.)

Do Exercise 8. ▶

Caution!

Solving a *system* of equations in two variables requires finding an ordered *pair* of numbers. Once you have solved for one variable, don't forget the other, and remember to list the ordered-pair solution using alphabetical order.

7. Solve the system

$$\begin{aligned} 3x &= 5 + 2y, \\ 2x + 3y - 1 &= 0. \end{aligned}$$

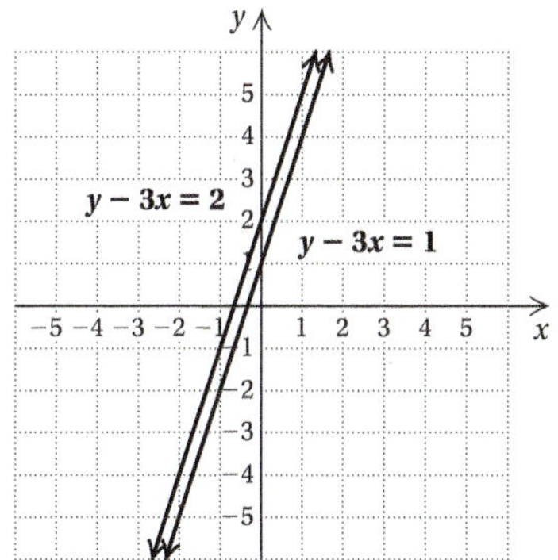

8. Solve the system

$$\begin{aligned} 2x + y &= 15, \\ 4x + 2y &= 23. \end{aligned}$$

Answers

7. $\left(\frac{17}{13}, -\frac{7}{13}\right)$ **8.** No solution

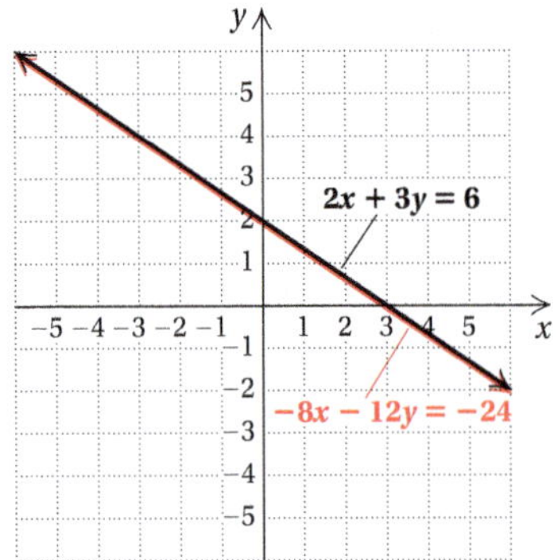

Sometimes there is an infinite number of solutions.

EXAMPLE 6 Solve the system

$$2x + 3y = 6, \quad (1)$$
$$-8x - 12y = -24. \quad (2)$$

We multiply by 4 on both sides of equation (1) and then add the two equations:

$$\begin{aligned} 8x + 12y &= 24 \quad &&\text{Multiplying by 4 on both sides of equation (1)} \\ -8x - 12y &= -24 && \\ \hline 0 &= 0. &&\text{Adding} \end{aligned}$$

We have eliminated both variables, and what remains, $0 = 0$, is an equation easily seen to be true. If this happens when we use the elimination method, we have an infinite number of solutions. The equations in the system have the same graph. (See the graph at left.) Any point on the line gives a solution of the system.

9. Solve the system

$$5x - 2y = 3, \quad (1)$$
$$-15x + 6y = -9. \quad (2)$$

Multiply equation (1) by 3 and add.

$$\begin{aligned} 15x - \_\_\_ &= \_\_\_ \\ -15x + 6y &= -9 \\ \hline 0 &= \_\_\_ \end{aligned}$$

The system has a(n) ___ number of solutions.

Do Exercise 9.

When decimals or fractions appear, we can first multiply to clear them. Then we proceed as before.

EXAMPLE 7 Solve the system

$$\frac{1}{3}x + \frac{1}{2}y = -\frac{1}{6}, \quad (1)$$
$$\frac{1}{2}x + \frac{2}{5}y = \frac{7}{10}. \quad (2)$$

The number 6 is the least common multiple of all the denominators of equation (1). The number 10 is the least common multiple of all the denominators of equation (2). We multiply by 6 on both sides of equation (1) and by 10 on both sides of equation (2):

$$6\left(\frac{1}{3}x + \frac{1}{2}y\right) = 6\left(-\frac{1}{6}\right) \qquad 10\left(\frac{1}{2}x + \frac{2}{5}y\right) = 10\left(\frac{7}{10}\right)$$
$$6 \cdot \frac{1}{3}x + 6 \cdot \frac{1}{2}y = -1 \qquad 10 \cdot \frac{1}{2}x + 10 \cdot \frac{2}{5}y = 7$$
$$2x + 3y = -1; \qquad 5x + 4y = 7.$$

The resulting system is

$$2x + 3y = -1,$$
$$5x + 4y = 7.$$

As we saw in Example 4, the solution of this system is $\left(\frac{25}{7}, -\frac{19}{7}\right)$.

Solve each system.

10. $\frac{1}{2}x + \frac{3}{10}y = \frac{1}{5},$
$\frac{3}{5}x + y = -\frac{2}{5}$

11. $3.3x + 6.6y = -6.6,$
$0.1x - 0.04y = 0.28$

Do Exercises 10 and 11.

Answers

9. Infinite number of solutions
10. $(1, -1)$ **11.** $(2, -2)$

Guided Solution:
9. 6y, 9, 0, infinite

The following is a summary that compares the graphical, substitution, and elimination methods for solving systems of equations.

METHOD	STRENGTHS	WEAKNESSES
Graphical	Can “see” solution.	Inexact when solution involves numbers that are not integers or are very large and off the graph.
Substitution	Works well when solutions are not integers. Easy to use when a variable is alone on one side.	Introduces extensive computations with fractions for more complicated systems where coefficients are not 1 or -1. Cannot “see” solution.
Elimination	Works well when solutions are not integers, when coefficients are not 1 or -1, and when coefficients involve decimals or fractions.	Cannot “see” solution.

15.3 Exercise Set

For Extra Help

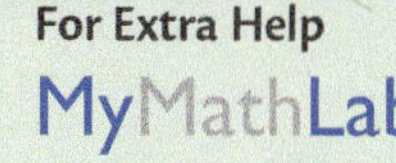

PRACTICE WATCH READ REVIEW

Reading Check

Determine whether each statement is true or false.

RC1. When we are solving a system of equations in x and y using the elimination method, only x can be eliminated.

RC2. Before we add to eliminate a variable, the coefficients of the terms containing that variable should be opposites.

RC3. When solving a system of equations using the elimination method, we may need to multiply before adding in order to eliminate a variable.

RC4. When solving a system of equations using the elimination method, we never need to use substitution.

RC5. Solutions of systems of equations containing fractions cannot be found using the elimination method.

RC6. When we are solving a system of equations algebraically, if we obtain a false equation, the system has infinitely many solutions.

a Solve using the elimination method.

1. $x - y = 7,$
$x + y = 5$

2. $x + y = 11,$
$x - y = 7$

3. $x + y = 8,$
$-x + 2y = 7$

4. $x + y = 6,$
$-x + 3y = -2$

5. $5x - y = 5,$
$3x + y = 11$

6. $2x - y = 8,$
$3x + y = 12$

7. $4a + 3b = 7,$
$-4a + b = 5$

8. $7c + 5d = 18,$
$c - 5d = -2$

9. $8x - 5y = -9,$
$3x + 5y = -2$

10. $3a - 3b = -15,$
$-3a - 3b = -3$

11. $4x - 5y = 7,$
$-4x + 5y = 7$

12. $2x + 3y = 4,$
$-2x - 3y = -4$

b Solve using the multiplication principle first. Then add.

13. $x + y = -7,$
$3x + y = -9$

14. $-x - y = 8,$
$2x - y = -1$

15. $3x - y = 8,$
$x + 2y = 5$

16. $x + 3y = 19,$
$x - y = -1$

17. $x - y = 5,$
$4x - 5y = 17$

18. $x + y = 4,$
$5x - 3y = 12$

19. $2w - 3z = -1,$
$3w + 4z = 24$

20. $7p + 5q = 2,$
$8p - 9q = 17$

21. $2a + 3b = -1,$
$3a + 5b = -2$

22. $3x - 4y = 16,$
$5x + 6y = 14$

23. $x = 3y,$
$5x + 14 = y$

24. $5a = 2b,$
$2a + 11 = 3b$

25. $2x + 5y = 16,$
$3x - 2y = 5$

26. $3p - 2q = 8,$
$5p + 3q = 7$

27. $p = 32 + q,$
$3p = 8q + 6$

28. $3x = 8y + 11,$
$x + 6y - 8 = 0$

29. $3x - 2y = 10,$
$-6x + 4y = -20$

30. $2x + y = 13,$
$4x + 2y = 23$

31. $0.06x + 0.05y = 0.07,$
$0.4x - 0.3y = 1.1$

32. $1.8x - 2y = 0.9,$
$0.04x + 0.18y = 0.15$

33. $\frac{1}{3}x + \frac{3}{2}y = \frac{5}{4},$
$\frac{3}{4}x - \frac{5}{6}y = \frac{3}{8}$

34. $x - \frac{3}{2}y = 13,$
$\frac{3}{2}x - y = 17$

35. $-4.5x + 7.5y = 6,$
$-x + 1.5y = 5$

36. $0.75x + 0.6y = -0.3,$
$3.9x + 5.2y = 96.2$

Skill Maintenance

Solve.

37. $2t - 13 - t = 5t + 7$ [10.3b]

38. $\frac{2}{3}x - \frac{1}{4} = \frac{x}{2}$ [10.3b]

39. $m - (5 - m) = 2(m + 1)$ [10.3c]

40. $-20y \leq 10$ [10.7d]

41. $2x^2 = x$ [13.7b]

42. $x^2 - x - 20 = 0$ [13.7b]

43. $\frac{a - 3}{a - 1} = \frac{2}{5}$ [14.7a]

44. $\frac{1}{x + 1} - \frac{1}{x} = \frac{2}{x - 2}$ [14.7a]

Synthesis

45.–48. Use the INTERSECT feature on a graphing calculator to solve the systems of equations in Exercises 5–8.

49.–54. Use the INTERSECT feature on a graphing calculator to solve the systems of equations in Exercises 25–30.

Solve using the substitution method, the elimination method, or the graphing method.

55. $3(x - y) = 9,$
$x + y = 7$

56. $2(x - y) = 3 + x,$
$x = 3y + 4$

57. $2(5a - 5b) = 10,$
$-5(6a + 2b) = 10$

58. $\frac{x}{3} + \frac{y}{2} = 1\frac{1}{3},$
$x + 0.05y = 4$

59. $y = -\frac{2}{7}x + 3,$
$y = \frac{4}{5}x + 3$

60. $y = \frac{2}{5}x - 7,$
$y = \frac{2}{5}x + 4$

Solve for x and y.

61. $y = ax + b,$
$y = x + c$

62. $ax + by + c = 0,$
$ax + cy + b = 0$

Mid-Chapter Review

Concept Reinforcement

Determine whether each statement is true or false.

__________ **1.** A solution of a system of two equations is an ordered pair that makes at least one equation true. [15.1a]

__________ **2.** Every system of two equations has one and only one ordered pair as a solution. [15.1b]

__________ **3.** The system of equations $y = ax + b$ and $y = ax - b$, $b \neq 0$, has no solution. [15.1b]

__________ **4.** The solution of the system of equations $x = a$ and $y = b$ is (a, b). [15.1b]

Guided Solutions

Fill in each blank with the number or expression that creates a correct solution.

Solve.

5. $x + y = -1,$ **(1)**
$y = x - 3$ **(2)** [15.2a]

$x + \square = -1$ Substituting for y in equation (1)

$\square = -1$ Simplifying

$2x = -1 + \square$

$2x = \square$ Simplifying

$x = \square$

$y = \square - 3$ Substituting for x in equation (2)

$y = \square$ Simplifying

The solution is $(\square, \square)$.

6. $2x - 3y = 7,$ **(1)**
$x + 3y = -10$ **(2)** [15.3a]

$$\begin{aligned} 2x - 3y &= 7 \\ x + 3y &= -10 \\ \hline \square x + \square y &= \square \end{aligned}$$

$\square x + \square y = \square$, or Adding

$\square x = \square$

$x = \square$

$\square + 3y = -10$ Substituting for x in equation (2)

$3y = \square$

$y = \square$

The solution is $(\square, \square)$.

Mixed Review

Determine whether the given ordered pair is a solution of the system of equations. Use alphabetical order of the variables. [15.1a]

7. $(-4, 5)$; $x + y = 1,$
$x = y - 9$

8. $(6, -4)$; $x = y + 10,$
$x - y = 2$

9. $(-1, 1)$; $3x + 5y = 2,$
$2x - y = -1$

10. $(2, -3)$; $2x + y = 1,$
$3x - 2y = 12$

Solve each system of equations by graphing. [15.1b]

11. $x + y = 1,$
$x - y = 5$

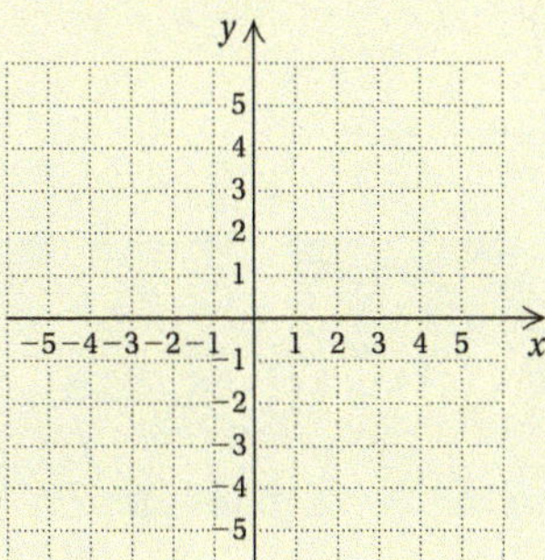

12. $2x + y = -1,$
$x + 2y = 4$

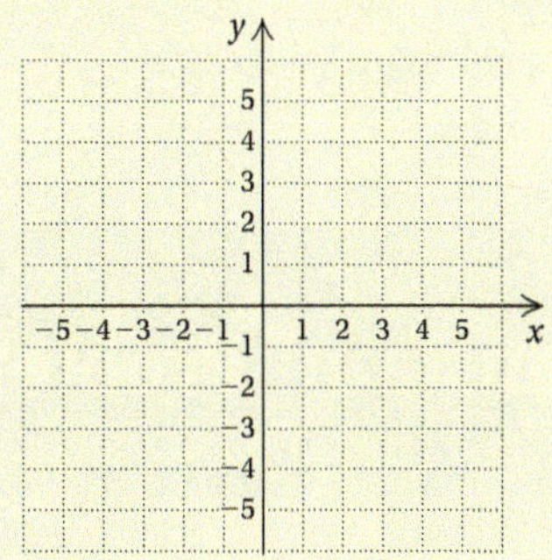

13. $2y = x - 1,$
$3x = 3 + 6y$

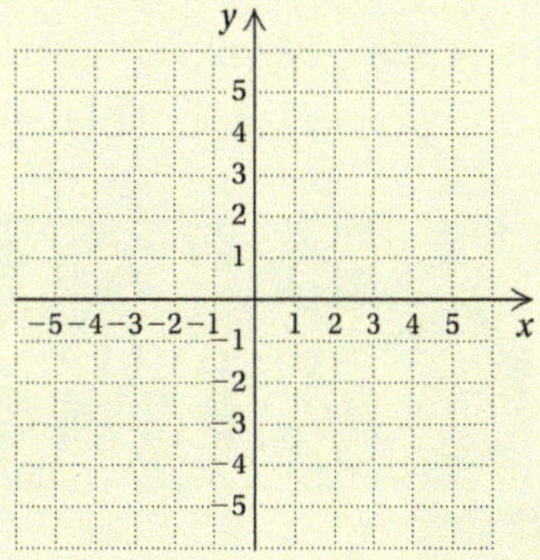

14. $x = y + 3,$
$y = x + 2$

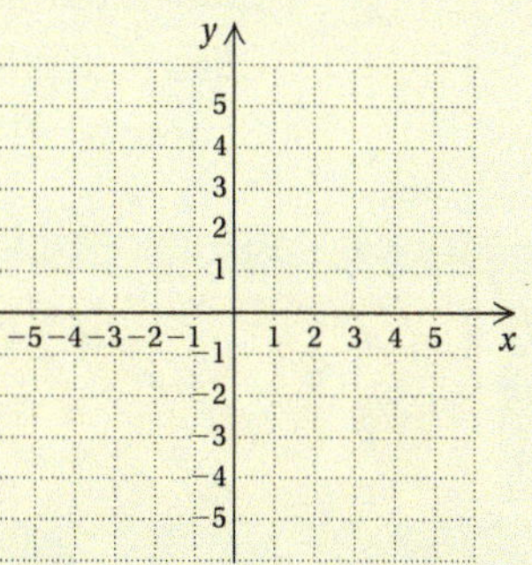

Solve using the substitution method. [15.2a, b]

15. $x + y = 2,$
$y = x - 8$

16. $x = y - 1,$
$2x - 5y = 1$

17. $x + y = 1,$
$3x + 6y = 1$

18. $2x + y = 2,$
$2x - y = -1$

Solve using the elimination method. [15.3a, b]

19. $x + y = 3,$
$-x - y = 5$

20. $3x - 2y = 2,$
$5x + 2y = -2$

21. $2x + 3y = 1,$
$3x + 2y = -6$

22. $2x - 3y = 6,$
$-4x + 6y = -12$

Solve. [15.2c]

23. ***Dimensions of an Area Rug.*** Lily buys an area rug with a perimeter of 18 ft. The width is 1 ft shorter than the length. Find the dimensions of the rug.

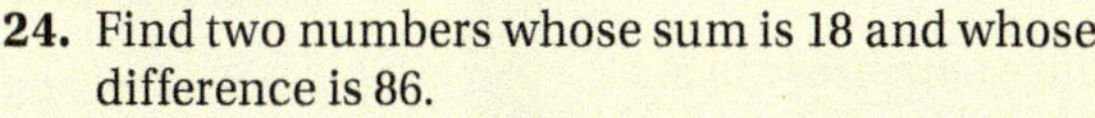

24. Find two numbers whose sum is 18 and whose difference is 86.

25. The difference of two numbers is 4. Two times the larger number is three times the smaller. What are the numbers?

Understanding Through Discussion and Writing

26. Suppose you have shown that the solution of the equation $3x - 1 = 9 - 2x$ is 2. How can this result be used to determine where the graphs of $y = 3x - 1$ and $y = 9 - 2x$ intersect? [15.1b]

27. Graph this system of equations. What happens when you try to determine a solution from the graph? [15.1b]

$$x - 2y = 6,$$
$$3x + 2y = 4$$

28. Janine can tell by inspection that the system

$$x = 2y - 1,$$
$$x = 2y + 3$$

has no solution. How does she know this? [15.1b]

29. Joel solves every system of two equations (in x and y) by first solving for y in the first equation and then substituting into the second equation. Is he using the best approach? Why or why not? [15.2b]

STUDYING FOR SUCCESS *Making Positive Choices*

- ☐ Choose to improve your attitude and raise your goals.
- ☐ Choose to make a strong commitment to learning.
- ☐ Choose to place the primary responsibility for learning on yourself.
- ☐ Choose to allocate the proper amount of time to learn.

15.4 Applications and Problem Solving

OBJECTIVE

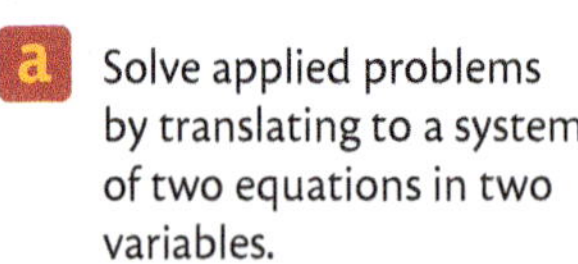

a Solve applied problems by translating to a system of two equations in two variables.

SKILL TO REVIEW

Objective 9.1b: Translate phrases to algebraic expressions.

Translate each phrase to an algebraic expression.

1. 52% of x liters
2. 26 less than a number

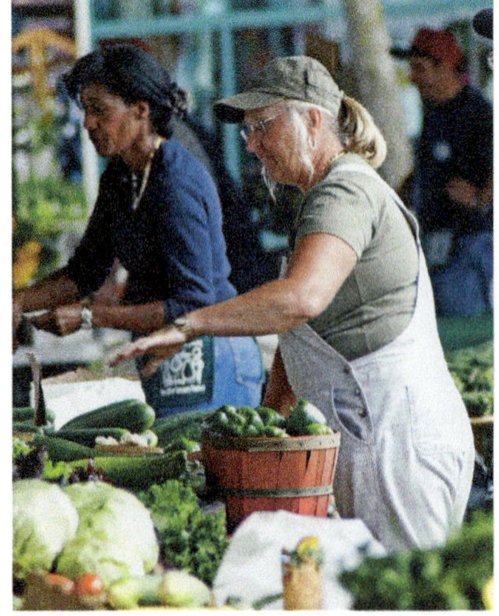

Answers

Skill to Review:
1. 52% x, or $0.52x$
2. $n - 26$

a

We now use systems of equations to solve applied problems that involve two equations in two variables.

EXAMPLE 1 ***Produce Prices.*** Shelby and Donna buy much of their produce at the City Farmer's Market. One Saturday, Shelby bought 3 ears of corn and 2 sweet peppers for \$3.20. Donna bought 8 ears of corn and 1 sweet pepper for \$4.85. Determine the price of one ear of corn and one sweet pepper.

1. **Familiarize.** We let $c =$ the price of one ear of corn and $p =$ the price of one sweet pepper.
2. **Translate.** Each purchase gives us one equation.

Shelby's purchase: 3 ears of corn and 2 sweet peppers cost \$3.20

$$3c + 2p = 3.20$$

Donna's purchase: 8 ears of corn and 1 sweet pepper cost \$4.85

$$8c + p = 4.85$$

3. **Solve.** We solve the system of equations

$$3c + 2p = 3.20 \quad (1)$$
$$8c + p = 4.85. \quad (2)$$

Although we could solve this system using graphing, substitution, or elimination, we decide to use elimination. If we multiply each side of equation (2) by -2 and add, the p-terms can be eliminated and we can solve for c:

$$\begin{aligned} 3c + 2p &= 3.20 && \text{Equation (1)} \\ -16c - 2p &= -9.70 && \text{Multiplying equation (2) by } -2 \\ \hline -13c \quad &= -6.50 && \text{Adding} \\ c &= 0.50. \end{aligned}$$

Next, we substitute 0.50 for c in equation (2) and solve for p:

$$\begin{aligned} 8c + p &= 4.85 \\ 8(0.50) + p &= 4.85 \\ 4 + p &= 4.85 \\ p &= 0.85. \end{aligned}$$

4. **Check.** If one ear of corn cost \$0.50 and one sweet pepper cost \$0.85, then Shelby would have spent 3(\$0.50) + 2(\$0.85) = \$3.20. Donna would have spent 8(\$0.50) + \$0.85 = \$4.85. The prices check.

5. **State.** The price of one ear of corn was \$0.50, and the price of one sweet pepper was \$0.85.

Do Exercise 1. ▶

1. ***Chicken and Hamburger Prices.*** Fast Good Food offers a special two-and-one promotion. The price of one hamburger and two pieces of chicken is \$5.39, and the price of two hamburgers and one piece of chicken is \$5.68. Find the price of one hamburger and the price of one piece of chicken.

EXAMPLE 2 *IMAX Movie Prices.* There were 322 people at a showing of the IMAX 3D movie *To the Arctic.* Admission was \$16.25 each for adults and \$12.00 each for children, and receipts totaled \$4790.50. How many adults and how many children attended?

1. **Familiarize.** To familiarize ourselves with the problem situation, let's make and check a guess.

   The total number of people at the movie was 322, so we choose numbers that total 322. Let's try 242 adults and 80 children. How much money was taken in?

   Money from adults: 242(\$16.25), or \$3932.50
   Money from children: 80(\$12.00), or \$960
   Total receipts: \$3932.50 + \$960, or \$4892.50

   Our guess is not the answer to the problem because the total taken in, according to the problem, was \$4790.50. The steps we have used to see if our guess is correct, however, help us to understand the actual steps involved in solving the problem.

   Let's list the information in a table. We let a = the number of adults and c = the number of children.

	ADULTS	CHILDREN	TOTAL	
ADMISSION	\$16.25	\$12.00		
NUMBER ATTENDING	a	c	322	→ $a + c = 322$
MONEY TAKEN IN	$16.25a$	$12.00c$	\$4790.50	→ $16.25a + 12.00c = 4790.50$

2. **Translate.** The total number of people attending was 322, so

   $$a + c = 322.$$

   The amount taken in from the adults was $16.25a$, and the amount taken in from the children was $12.00c$. These amounts are in dollars. The total was \$4790.50, so we have

   $$16.25a + 12.00c = 4790.50.$$

   We can multiply by 100 on both sides to clear decimals. Thus we have a translation to a system of equations:

   $$a + c = 322, \quad (1)$$
   $$1625a + 1200c = 479{,}050. \quad (2) \quad \text{Multiplying by 100}$$

Answer

1. Hamburger: \$1.99; chicken: \$1.70

2. ***Game Admissions.*** There were 166 paid admissions to a high school basketball game. The price was \$3.10 each for adults and \$1.75 each for children. The amount collected was \$459.25. How many adults and how many children attended?

Complete the following table to aid with the familiarization.

GS

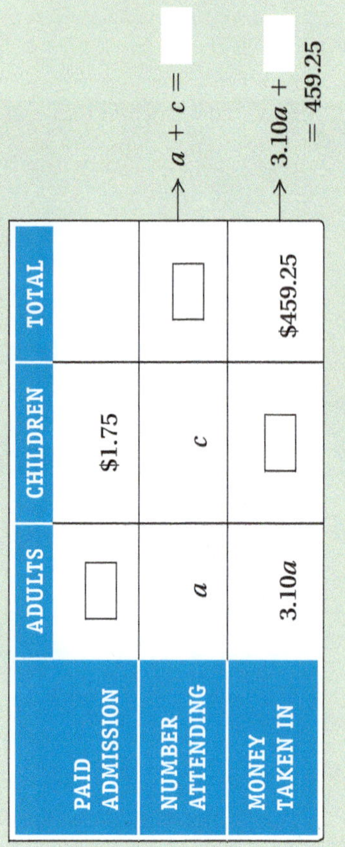

	ADULTS	CHILDREN	TOTAL
PAID ADMISSION	☐	\$1.75	
NUMBER ATTENDING	a	c	☐
MONEY TAKEN IN	$3.10a$	☐	\$459.25

3. **Solve.** We solve the system using the elimination method since the equations are both in the form $Ax + By = C$. (A case can certainly be made for using the substitution method since we can solve for one of the variables quite easily in the first equation. Very often a decision is just a matter of preference.) We multiply by -1200 on both sides of equation (1) and then add and solve for a:

$$\begin{aligned} -1200a - 1200c &= -386{,}400 && \text{Multiplying by } -1200 \\ 1625a + 1200c &= 479{,}050 \\ \hline 425a &= 92{,}650 && \text{Adding} \\ a &= \frac{92{,}650}{425} && \text{Dividing by 425} \\ a &= 218. \end{aligned}$$

Next, we go back to equation (1), substituting 218 for a, and solve for c:

$$\begin{aligned} a + c &= 322 \\ 218 + c &= 322 \\ c &= 104. \end{aligned}$$

4. **Check.** The check is left to the student. It is similar to what we did in the *Familiarize* step.

5. **State.** Attending the showing were 218 adults and 104 children.

◀ Do Exercise 2.

EXAMPLE 3 ***Mixture of Solutions.*** A chemist has one solution that is 80% acid (that is, 8 parts are acid and 2 parts are water) and another solution that is 30% acid. What is needed is 200 L of a solution that is 62% acid. The chemist will prepare it by mixing the two solutions. How much of each should be used?

1. **Familiarize.** We can make a drawing of the situation. The chemist uses x liters of the first solution and y liters of the second solution. We can also arrange the information in a table.

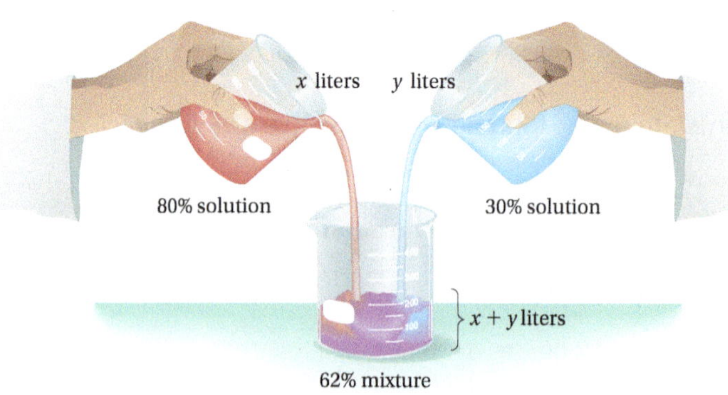

	FIRST SOLUTION	SECOND SOLUTION	MIXTURE	
AMOUNT OF SOLUTION	x	y	200 L	→ $x + y = 200$
PERCENT OF ACID	80%	30%	62%	
AMOUNT OF ACID IN SOLUTION	$80\%x$	$30\%y$	$62\% \times 200$, or 124 L	→ $80\%x + 30\%y = 124$

Answer

2. Adults: 125; children: 41

Guided Solution:

2.

\$3.10	\$1.75	
a	c	166
$3.10a$	$1.75c$	\$459.25

$a + c = 166,$
$3.10a + 1.75c = 459.25$

2. **Translate.** The chemist uses x liters of the first solution and y liters of the second. Since the total is to be 200 L, we have

Total amount of solution: $x + y = 200.$

The amount of acid in the new mixture is to be 62% of 200 L, or 0.62(200 L), or 124 L. The amounts of acid from the two solutions are $80\%x$ and $30\%y$. Thus,

Total amount of acid: $80\%x + 30\%y = 124$

or $0.8x + 0.3y = 124.$

We clear decimals by multiplying by 10 on both sides of the second equation:

$$10(0.8x + 0.3y) = 10 \cdot 124$$
$$8x + 3y = 1240.$$

Thus we have a translation to a system of equations:

$$x + y = 200, \quad (1)$$
$$8x + 3y = 1240. \quad (2)$$

3. **Solve.** We solve the system. We use the elimination method, again because equations are in the form $Ax + By = C$ and a multiplication in one equation will allow us to eliminate a variable, but substitution would also work. We multiply by -3 on both sides of equation (1) and then add and solve for x:

$$-3x - 3y = -600 \qquad \text{Multiplying by } -3$$
$$8x + 3y = 1240$$
$$5x = 640 \qquad \text{Adding}$$
$$x = \frac{640}{5} \qquad \text{Dividing by 5}$$
$$x = 128.$$

Next, we go back to equation (1) and substitute 128 for x:

$$x + y = 200$$
$$128 + y = 200$$
$$y = 72.$$

The solution is $x = 128$ and $y = 72$.

4. **Check.** The sum of 128 and 72 is 200. Also, 80% of 128 is 102.4 and 30% of 72 is 21.6. These add up to 124. The numbers check.

5. **State.** The chemist should use 128 L of the 80%-acid solution and 72 L of the 30%-acid solution.

Do Exercise 3. ▶

3. ***Mixture of Solutions.*** One solution is 50% alcohol and a second is 70% alcohol. How much of each should be mixed in order to make 30 L of a solution that is 55% alcohol?

Complete the following table to aid in the familiarization.

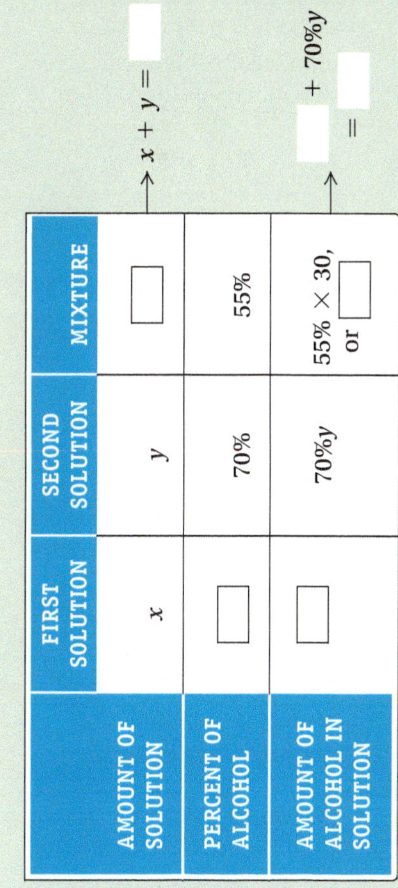

	FIRST SOLUTION	SECOND SOLUTION	MIXTURE
AMOUNT OF SOLUTION	x	y	☐
PERCENT OF ALCOHOL	☐	70%	55%
AMOUNT OF ALCOHOL IN SOLUTION	☐	$70\%y$	$55\% \times 30$, or ☐

EXAMPLE 4 *Candy Mixtures.* A bulk wholesaler wishes to mix some candy worth 45 cents per pound and some worth 80 cents per pound in order to make 350 lb of a mixture worth 65 cents per pound. How much of each type of candy should be used?

1. **Familiarize.** Arranging the information in a table will help. We let x = the amount of 45-cent candy and y = the amount of 80-cent candy.

Answer

3. 50% alcohol: 22.5 L; 70% alcohol: 7.5 L

Guided Solution:

3.

x	y	30
50%	70%	55%
$50\%x$	$70\%y$	$55\% \times 30$, or 16.5

$x + y = 30,$
$50\%x + 70\%y = 16.5$

	INEXPENSIVE CANDY	EXPENSIVE CANDY	MIXTURE	
COST OF CANDY	45 cents	80 cents	65 cents	
AMOUNT (in pounds)	x	y	350	$\rightarrow x + y = 350$
TOTAL COST	$45x$	$80y$	65 cents · (350), or 22,750 cents	$\rightarrow 45x + 80y = 22{,}750$

4. ***Mixture of Grass Seeds.*** Grass seed A is worth $1.40 per pound and seed B is worth $1.75 per pound. How much of each should be mixed in order to make 50 lb of a mixture worth $1.54 per pound?

Complete the following table to aid in the familiarization.

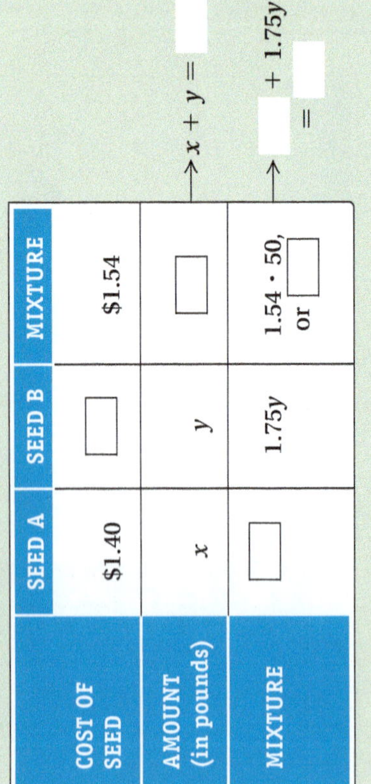

COST OF SEED
AMOUNT (in pounds)
MIXTURE
SEED A
SEED B
MIXTURE
$1.40
$1.54
x
y
1.75y
1.54 · 50, or
x + y =
+ 1.75y
=

Answer

4. Seed A: 30 lb; seed B: 20 lb

Guided Solution:

4.

$1.40	$1.75	$1.54
x	y	50
$1.40x$	$1.75y$	1.54 · 50, or 77

$x + y = 50,$
$1.40x + 1.75y = 77$

Note the similarity of this problem to Example 2. Here we consider types of candy instead of groups of people.

2. Translate. We translate as follows. From the second row of the table, we have

Total amount of candy: $x + y = 350.$

Our second equation will come from the costs. The value of the inexpensive candy, in cents, is $45x$ (x pounds at 45 cents per pound). The value of the expensive candy is $80y$, and the value of the mixture is 65×350, or 22,750 cents. Thus we have

Total cost of mixture: $45x + 80y = 22{,}750.$

Remember the problem-solving tip about dimension symbols. In this last equation, all expressions are given in cents. We could have expressed them all in dollars, but we do not want some in cents and some in dollars. Thus we have a translation to a system of equations:

$$x + y = 350, \quad (1)$$
$$45x + 80y = 22{,}750. \quad (2)$$

3. Solve. We solve the system using the elimination method again. We multiply by -45 on both sides of equation (1) and then add and solve for y:

$$-45x - 45y = -15{,}750 \quad \text{Multiplying by } -45$$
$$45x + 80y = 22{,}750$$
$$35y = 7{,}000 \quad \text{Adding}$$
$$y = \frac{7000}{35}$$
$$y = 200.$$

Next, we go back to equation (1), substituting 200 for y, and solve for x:

$$x + y = 350$$
$$x + 200 = 350$$
$$x = 150.$$

4. Check. We consider $x = 150$ lb and $y = 200$ lb. The sum is 350 lb. The value of the candy is $45(150) + 80(200)$, or 22,750 cents and each pound of the mixture is worth $22{,}750 \div 350$, or 65 cents. These values check.

5. State. The grocer should mix 150 lb of the 45-cent candy with 200 lb of the 80-cent candy.

Do Exercise 4.

EXAMPLE 5 *Coin Value.* A student assistant at the university copy center has some nickels and dimes to use for change when students make copies. The value of the coins is $7.40. There are 26 more dimes than nickels. How many of each kind of coin are there?

1. **Familiarize.** We let d = the number of dimes and n = the number of nickels.
2. **Translate.** There are 26 more dimes than nickels, so we have

$$d = n + 26.$$

The value of the nickels, in cents, is $5n$, since each coin is worth 5 cents. The value of the dimes, in cents, is $10d$, since each coin is worth 10 cents. The total value is given as $7.40. Since we have the values of the nickels and dimes *in cents,* we must use cents for the total value. We express $7.40 as 740 cents. This gives us another equation:

$$10d + 5n = 740.$$

We now have a system of equations:

$$d = n + 26, \quad (1)$$
$$10d + 5n = 740. \quad (2)$$

3. **Solve.** Since we have d alone on one side of one equation, we use the substitution method. We substitute $n + 26$ for d in equation (2):

$10d + 5n = 740$	Equation (2)
$10(n + 26) + 5n = 740$	Substituting $n + 26$ for d
$10n + 260 + 5n = 740$	Removing parentheses
$15n + 260 = 740$	Collecting like terms
$15n = 480$	Subtracting 260
$n = \frac{480}{15}$, or 32.	Dividing by 15

Next, we substitute 32 for n in either of the original equations to find d. We use equation (1):

$$d = n + 26 = 32 + 26 = 58.$$

4. **Check.** We have 58 dimes and 32 nickels. There are 26 more dimes than nickels. The value of the coins is 58($0.10) + 32($0.05), which is $7.40. This checks.
5. **State.** The student assistant has 58 dimes and 32 nickels.

Do Exercise 5. ▶

5. *Coin Value.* On a table are 20 coins, quarters and dimes. Their total value is $3.05. How many of each kind of coin are there?

Look back over Examples 2–5. The problems are quite similar in their structure. Compare them and try to see the similarities. The problems in Examples 2–5 are often called *mixture problems.* These problems provide a pattern, or model, for many related problems.

PROBLEM-SOLVING TIP

When solving a problem, first see if it is patterned or modeled after other problems that you have studied.

Answer

5. Quarters: 7; dimes: 13

15.4 Exercise Set

For Extra Help
MyMathLab®
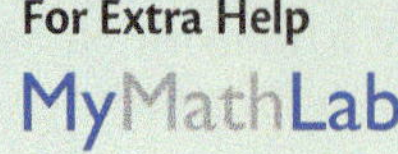

PRACTICE

WATCH

READ

REVIEW

Reading Check

Consider the following mixture problem.

Cherry Breeze is 30% fruit juice and Berry Choice is 15% fruit juice. How much of each should be used in order to make 10 gal of a drink that is 20% fruit juice?

The following table can be used to translate the problem. Choose the expression from the column to the right of the table that best fits each numbered space.

	CHERRY BREEZE	BERRY CHOICE	MIXTURE
NUMBER OF GALLONS	x	y	**RC1.** ___
PERCENT OF FRUIT JUICE	30%	**RC2.** ___%	20%
AMOUNT OF FRUIT JUICE	$0.3x$	**RC3.** ___y	**RC4.** ___

2
10
15
0.15

a Solve.

1. *Assignments.* The professor teaching Introduction to Sociology gives points for each discussion-board post and points for each reply to a post. Ana wrote 3 posts and 10 replies and received 95 points. Jae wrote 8 posts and 1 reply and received 125 points. Determine how many points a discussion post is worth and how many points a reply is worth.

2. *Video Games.* Ethan and Ruth are competing in an online hidden objects game. Ethan found 1 magic ring and 8 four-leaf clovers for a total of 5950 points. Ruth found 4 magic rings and 5 four-leaf clovers for a total of 6250 points. Determine how many points a magic ring is worth and how many points a four-leaf clover is worth.

3. *Butterflies.* Admission prices to the butterflies exhibit at the Brookfield Zoo in Illinois are \$4 each for adults and \$2 each for children. One day, 320 people visited the exhibit, and receipts totaled \$820. How many adults and how many children visited that day?

4. *Zoo Admissions.* The Bronx Zoo charges \$17 admission for each adult and \$13 for each child. One day, a total of \$12,700 was collected from 860 admissions. How many adults and how many children were admitted that day?

5. *Photo Prints.* Lucy paid \$16.50 for 36 prints from Photo World. Some prints were 4×6 and the rest were 5×7, and they were priced as shown in the following table. How many prints of each size did she order?

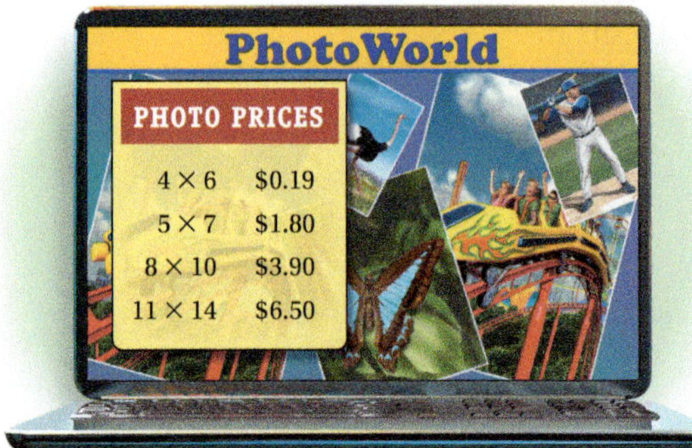

PHOTO PRICES	
4 × 6	\$0.19
5 × 7	\$1.80
8 × 10	\$3.90
11 × 14	\$6.50

6. *Baseball Admissions.* Members of the Benton Youth Club attended a baseball game, buying a total of 29 bleacher and lower reserved seats. Ticket prices are shown in the following table. The total cost of the tickets was \$913. How many of each kind of ticket was bought?

TICKET INFORMATION	
Lower Box	\$36
Upper Box	\$21
Lower Reserved	\$32
Upper Reserved	\$17
Bleacher	\$31

7. ***Basketball Scoring.*** In a game against the Orlando Magic, the Portland Trail Blazers scored 85 of their points on a combination of 40 two- and three-point baskets. How many of each type of shot was made?

Source: National Basketball Association

8. ***Basketball Scoring.*** Tony Parker of the San Antonio Spurs once scored 29 points on 17 shots in an NBA game, shooting only two-point shots and free throws (one point each). How many of each type of shot did he make?

Source: National Basketball Association

9. ***Investments.*** Cassandra has a number of $50 and $100 savings bonds to use for part of her college expenses. The total value of the bonds is $1250. There are 7 more $50 bonds than $100 bonds. How many of each type of bond does she have?

10. ***Commercial Lengths.*** During a football game, a television network aired both 30-sec commercials and 60-sec commercials. The total commercial time was 25 min, and there were 11 more 30-sec commercials shown than 60-sec commercials. How many of each length commercial were shown?

11. ***Mixture of Solutions.*** Solution A is 50% acid and solution B is 80% acid. How many liters of each should be used in order to make 100 L of a solution that is 68% acid? Complete the following table to aid in the familiarization.

	SOLUTION A	SOLUTION B	MIXTURE	
AMOUNT OF SOLUTION	x	y		$\rightarrow x + y = (\quad)$
PERCENT OF ACID	50%		68%	
AMOUNT OF ACID IN SOLUTION		$80\%y$	$68\% \times 100$, or	$\rightarrow 50\%x + (\quad) = (\quad)$

12. ***Production.*** Clear Shine window cleaner is 12% alcohol and Sunstream window cleaner is 30% alcohol. How much of each should be used to make 90 oz of a cleaner that is 20% alcohol?

13. ***Grain Mixtures for Horses.*** Brianna needs to calculate the correct mix of grain and hay to feed her horse. On the basis of her horse's age, weight, and workload, she determines that he needs to eat 15 lb of feed per day, with an average protein content of 8%. Hay contains 6% protein, whereas grain has a 12% protein content. How many pounds of hay and grain should she feed her horse each day?

Source: *Michael Plumb's Horse Journal*, February 1996, pp. 26–29

14. ***Paint Mixtures.*** At a local "paint swap," Kari found large supplies of Skylite Pink (12.5% red pigment) and MacIntosh Red (20% red pigment). How many gallons of each color should Kari pick up in order to mix a gallon of Summer Rose (17% red pigment)?

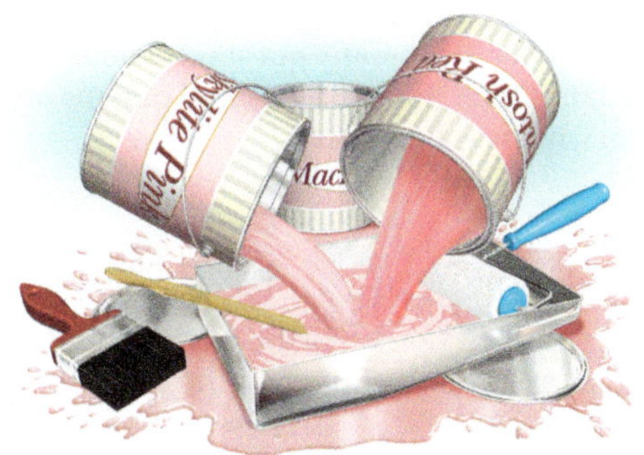

15. *Coin Value.* A parking meter contains dimes and quarters worth \$15.25. There are 103 coins in all. How many of each type of coin are there?

16. *Coin Value.* A vending machine contains nickels and dimes worth \$14.50. There are 95 more nickels than dimes. How many of each type of coin are there?

17. *Coffee Blends.* Carolla's Coffee Shop mixes Brazilian coffee worth \$19 per pound with Turkish coffee worth \$22 per pound. The mixture is to sell for \$20 per pound. How much of each type of coffee should be used in order to make a 300-lb mixture? Complete the following table to aid in the familiarization.

	BRAZILIAN COFFEE	TURKISH COFFEE	MIXTURE	
COST OF COFFEE	\$19		\$20	
AMOUNT (in pounds)	x	y	300	$\rightarrow x + y = (\quad)$
MIXTURE		$22y$	20(300), or \$6000	$\rightarrow 19x + (\quad) = 6000$

18. *Coffee Blends.* The Java Joint wishes to mix organic Kenyan coffee beans that sell for \$7.25 per pound with organic Venezuelan beans that sell for \$8.50 per pound in order to form a 50-lb batch of Morning Blend that sells for \$8.00 per pound. How many pounds of each type of bean should be used to make the blend?

19. *Mixed Nuts.* A customer has asked a caterer to provide 60 lb of nuts, 60% of which are to be cashews. The caterer has available mixtures of 70% cashews and 45% cashews. How many pounds of each mixture should be used?

20. *Mixture of Grass Seeds.* Grass seed A is worth \$2.50 per pound and seed B is worth \$1.75 per pound. How much of each would you use in order to make 75 lb of a mixture worth \$2.14 per pound?

21. *Cough Syrup.* Dr. Zeke's cough syrup is 2% alcohol. Vitabrite cough syrup is 5% alcohol. How much of each type should be used in order to prepare an 80-oz batch of cough syrup that is 3% alcohol?

22. *Mixture of Solutions.* Solution A is 30% alcohol and solution B is 75% alcohol. How much of each should be used in order to make 100 L of a solution that is 50% alcohol?

23. *Test Scores.* Anna is taking a test in which items of type A are worth 10 points and items of type B are worth 15 points. It takes 3 min to complete each item of type A and 6 min to complete each item of type B. The total time allowed is 60 min and Anna answers exactly 16 questions. How many questions of each type did she complete? Assuming that all her answers were correct, what was her score?

24. *Gold Alloys.* A goldsmith has two alloys that are different purities of gold. The first is three-fourths pure gold and the second is five-twelfths pure gold. How many ounces of each should be melted and mixed in order to obtain a 6-oz mixture that is two-thirds pure gold?

25. *Ages.* The Kuyatts' house is twice as old as the Marconis' house. Eight years ago, the Kuyatts' house was three times as old as the Marconis' house. How old is each house?

26. *Ages.* David is twice as old as his daughter. In 4 years, David's age will be three times what his daughter's age was 6 years ago. How old are they now?

27. *Ages.* Randy is four times as old as Marie. In 12 years, Marie's age will be half of Randy's. How old are they now?

28. *Ages.* Jennifer is twice as old as Ramon. The sum of their ages 7 years ago was 13. How old are they now?

29. *Supplementary Angles.* **Supplementary angles** are angles whose sum is 180°. Two supplementary angles are such that one is 30° more than two times the other. Find the angles.

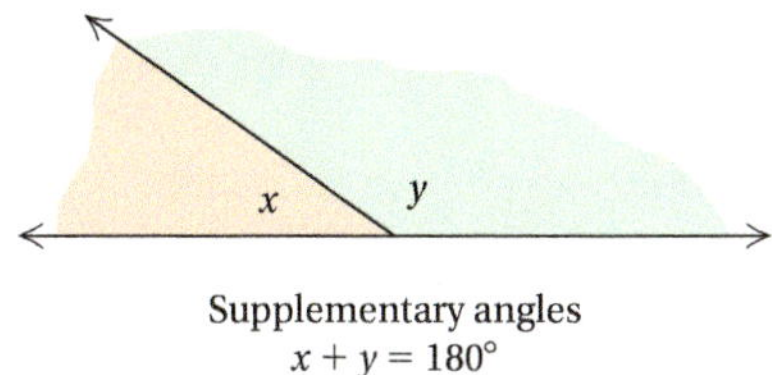

Supplementary angles
$x + y = 180°$

30. *Supplementary Angles.* Two supplementary angles are such that one is 8° less than three times the other. Find the angles.

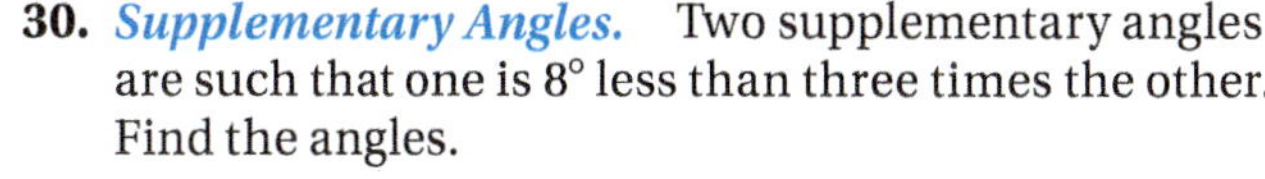

31. *Complementary Angles.* **Complementary angles** are angles whose sum is 90°. Two complementary angles are such that their difference is 34°. Find the angles.

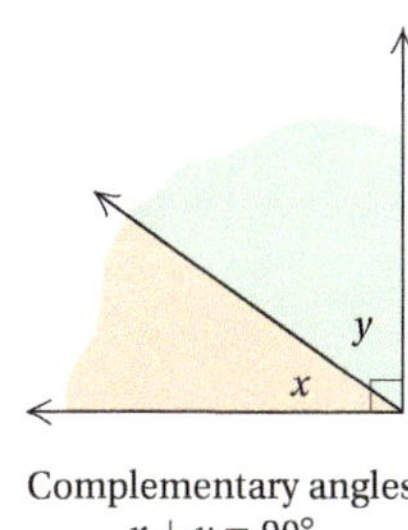

Complementary angles
$x + y = 90°$

32. *Complementary Angles.* Two angles are complementary. One angle is 42° more than one-half the other. Find the angles.

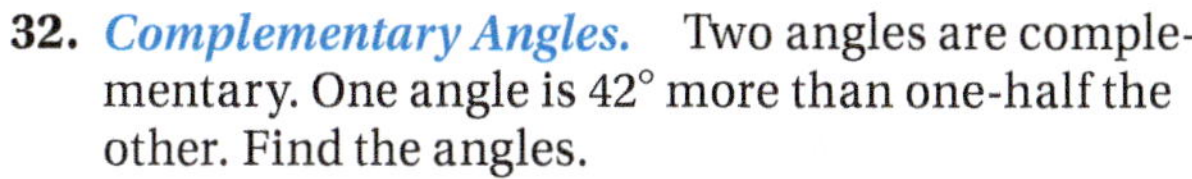

33. *Octane Ratings.* In most areas of the United States, gas stations offer three grades of gasoline, indicated by octane ratings on the pumps, such as 87, 89, and 93. When a tanker delivers gas, it brings only two grades of gasoline, the highest and the lowest, filling two large underground tanks. If you purchase the middle grade, the pump's computer mixes the other two grades appropriately. How much 87-octane gas and 93-octane gas should be blended in order to make 18 gal of 89-octane gas?

Source: Exxon

34. *Octane Ratings.* Refer to Exercise 33. Suppose the pump grades offered are 85, 87, and 91. How much 85-octane gas and 91-octane gas should be blended in order to make 12 gal of 87-octane gas?

Source: Exxon

35. *Printing.* A printer knows that a page of print contains 830 words if large type is used and 1050 words if small type is used. A document containing 11,720 words fills exactly 12 pages. How many pages are in the large type? in the small type?

36. *Paint Mixture.* A merchant has two kinds of paint. If 9 gal of the inexpensive paint is mixed with 7 gal of the expensive paint, the mixture will be worth \$19.70 per gallon. If 3 gal of the inexpensive paint is mixed with 5 gal of the expensive paint, the mixture will be worth \$19.825 per gallon. What is the price per gallon of each type of paint?

Skill Maintenance

Perform the indicated operations and simplify.

37. $(2x^2 - 3) - (x^2 - x - 3)$ [12.4c]

38. $\left(a + \frac{1}{2}\right)\left(a - \frac{1}{2}\right)$ [12.6b]

39. $(t^2 + 1.2)^2$ [12.6c]

40. $(3x + 5)(2x - 7)$ [12.6a]

41. $(x - 1)(x^2 + x + 1)$ [12.5d]

42. $(3mn - m^2n - n^2) + (mn^2 + n^2)$ [12.7d]

43. $\frac{x + 7}{x^2 - 1} - \frac{3}{x + 1}$ [14.5a]

44. $\frac{3 - x}{x - 2} - \frac{x - 7}{2 - x}$ [14.5a]

45. $\frac{1}{x} - \frac{1}{x^2} + \frac{1}{x + 1}$ [14.5b]

46. $\frac{a^2 + a - 20}{a^2 - 4} \cdot \frac{2a^2 + 3a - 2}{a^2 + 10a + 25}$ [14.1d]

47. $\frac{10c^2d}{3x^2} \div \frac{30cd}{x^3}$ [14.2b]

48. $\frac{t^4 - 16}{t^4 - 1} \div \frac{t^2 + 4}{t^2 + 1}$ [14.2b]

Find the intercepts. Then graph the equation. [11.1d], [11.2a]

49. $y = -2x - 3$

50. $y = -0.1x + 0.4$

51. $5x - 2y = -10$

52. $2.5x + 4y = 10$

Synthesis

53. *Milk Mixture.* A farmer has 100 L of milk that is 4.6% butterfat. How much skim milk (no butterfat) should be mixed with it in order to make milk that is 3.2% butterfat?

54. One year, Shannon made \$85 from two investments: \$1100 was invested at one yearly rate and \$1800 at a rate that was 1.5% higher. Find the two rates of interest.

55. *Automobile Maintenance.* An automobile radiator contains 16 L of antifreeze and water. This mixture is 30% antifreeze. How much of this mixture should be drained and replaced with pure antifreeze so that the mixture will be 50% antifreeze?

56. *Employer Payroll.* An employer has a daily payroll of \$1225 when employing some workers at \$80 per day and others at \$85 per day. When the number of \$80 workers is increased by 50% and the number of \$85 workers is decreased by $\frac{1}{5}$, the new daily payroll is \$1540. How many were originally employed at each rate?

57. A two-digit number is six times the sum of its digits. The tens digit is 1 more than the ones digit. Find the number.

Applications with Motion

15.5

OBJECTIVE

a Solve motion problems using the formula $d = rt$.

SKILL TO REVIEW

Objective 10.3c: Solve equations by first removing parentheses and collecting like terms.

Solve.

1. $(x + 3)2 = (x - 1)5$
2. $18(y + 1) = 20y$

a We first studied problems involving motion in Chapter 14. Here we extend our problem-solving skills by solving certain motion problems whose solutions can be found using systems of equations. Recall the motion formula.

THE MOTION FORMULA

Distance = Rate (or speed) · Time

$d = rt$

We use five steps for problem solving. The tips in the margin at right are also helpful when solving motion problems.

As we saw in Chapter 14, there are motion problems that can be solved with just one equation. Let's start with another such problem.

TIPS FOR SOLVING MOTION PROBLEMS

1. Draw a diagram using an arrow or arrows to represent distance and the direction of each object in motion.
2. Organize the information in a chart.
3. Look for as many things as you can that are the same so that you can write equations.

EXAMPLE 1 *Car Travel.* Two cars leave Ashland at the same time traveling in opposite directions. One travels at 60 mph and the other at 30 mph. In how many hours will they be 150 mi apart?

1. **Familiarize.** We first make a drawing.

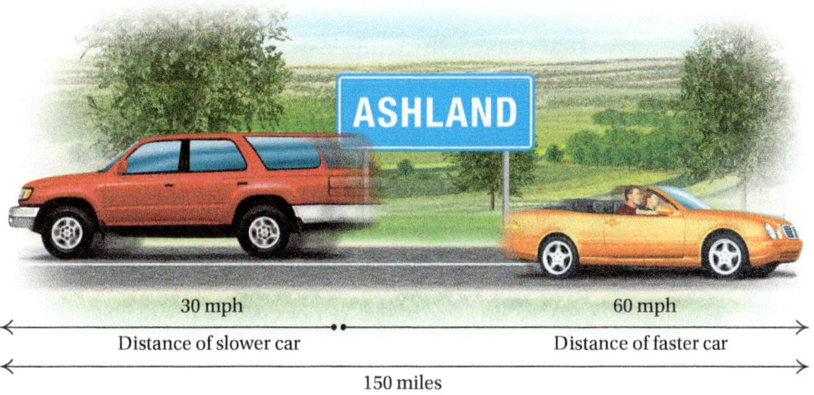

From the wording of the problem and the drawing, we see that the distances may *not* be the same. But the times that the cars travel are the same, so we can use just t for time. We can organize the information in a chart.

$d = r \cdot t$

	DISTANCE	SPEED	TIME
FASTER CAR	Distance of faster car	60	t
SLOWER CAR	Distance of slower car	30	t
TOTAL	150		

Answers

Skill to Review:

1. $\frac{11}{3}$ 2. 9

1. *Car Travel.* Two cars leave town at the same time traveling in opposite directions. One travels at 48 mph and the other at 60 mph. How far apart will they be 3 hr later? (*Hint*: The times are the same. Be *sure* to make a drawing.)

2. *Car Travel.* Two cars leave town at the same time traveling in the same direction. One travels at 35 mph and the other at 40 mph. In how many hours will they be 15 mi apart? (*Hint*: The times are the same. Be *sure* to make a drawing.)

Answers

1. 324 mi 2. 3 hr

2. Translate. From the drawing, we see that

(Distance of faster car) + (Distance of slower car) = 150.

Then using $d = rt$ in each row of the table, we get

$$60t + 30t = 150.$$

3. Solve. We solve the equation:

$$60t + 30t = 150$$

$$90t = 150 \quad \text{Collecting like terms}$$

$$t = \frac{150}{90}, \text{ or } \frac{5}{3}, \text{ or } 1\frac{2}{3}\text{ hr.} \quad \text{Dividing by 90}$$

4. Check. When $t = \frac{5}{3}$ hr,

$$(\text{Distance of faster car}) + (\text{Distance of slower car}) = 60\left(\frac{5}{3}\right) + 30\left(\frac{5}{3}\right)$$

$$= 100 + 50, \text{ or } 150 \text{ mi.}$$

Thus the time of $\frac{5}{3}$ hr, or $1\frac{2}{3}$ hr, checks.

5. State. In $1\frac{2}{3}$ hr, the cars will be 150 mi apart.

Do Exercise 1 and 2.

Now let's solve some motion problems using systems of equations.

EXAMPLE 2 *Train Travel.* A train leaves Stanton traveling east at 35 miles per hour (mph). An hour later, another train leaves Stanton on a parallel track at 40 mph. How far from Stanton will the second (or faster) train catch up with the first (or slower) train?

1. Familiarize. We first make a drawing.

From the drawing, we see that the distances are the same. Let's call the distance d. We don't know the times. We let t = the time for the faster train. Then the time for the slower train $= t + 1$, since it left 1 hr earlier. We can organize the information in a chart.

	d = DISTANCE	r · SPEED	t TIME	
SLOWER TRAIN	d	35	$t + 1$	→ $d = 35(t + 1)$
FASTER TRAIN	d	40	t	→ $d = 40t$

2. **Translate.** In motion problems, we look for quantities that are the same so that we can write equations. From each row of the chart, we get an equation, $d = rt$. Thus we have two equations:

$$d = 35(t + 1), \quad (1)$$
$$d = 40t. \quad (2)$$

3. **Solve.** Since we have a variable alone on one side, we solve the system using the substitution method:

$35(t + 1) = 40t$ Using the substitution method (substituting $35(t + 1)$ for d in equation 2)

$35t + 35 = 40t$ Removing parentheses

$35 = 5t$ Subtracting $35t$

$\frac{35}{5} = t$ Dividing by 5

$7 = t.$

The problem asks us to find how far from Stanton the faster train catches up with the other. Thus we need to find d. We can do this by substituting 7 for t in the equation $d = 40t$:

$$d = 40(7) = 280.$$

4. **Check.** If the time is 7 hr, then the distance that the slower train travels is $35(7 + 1)$, or 280 mi. The faster train travels $40(7)$, or 280 mi. Since the distances are the same, we know how far from Stanton the trains will be when the faster train catches up with the other.

5. **State.** The faster train will catch up with the slower train 280 mi from Stanton.

Do Exercise 3. ▶

3. ***Car Travel.*** A car leaves Spokane traveling north at 56 km/h. Another car leaves Spokane 1 hr later traveling north at 84 km/h. How far from Spokane will the second car catch up with the first? (*Hint*: The cars travel the same distance.)

1. **Familiarize.** Let $t =$ the time for the first car. Then ____ = the time for the second car.
2. **Translate.**
   $d = 56t,$ First car
   $d = 84($ ____ $)$ Second car
3. **Solve.**
   $84(t - 1) = 56t$
   $84t -$ ____ $= 56t$
   $-84 =$ ____
   ____ $= t$
   If $t = 3$, then
   $d = 56($ ____ $) =$ ____.
4. **Check.** The first car travels 168 km in ____ hr, and the second car travels 168 km in ____ hr.
5. **State.** The second car will catch up with the first car in ____ km.

EXAMPLE 3 ***Boat Travel.*** A motorboat took 3 hr to make a downstream trip with a 6-km/h current. The return trip against the same current took 5 hr. Find the speed of the boat in still water.

1. **Familiarize.** We first make a drawing. From the drawing, we see that the distances are the same. Let's call the distance d. We let $r =$ the speed of the boat in still water. Then, when the boat is traveling downstream, its speed is $r + 6$. (The current helps the boat along.) When it is traveling upstream, its speed is $r - 6$. (The current holds the boat back.)

Answer

3. 168 km

Guided Solution:

3. $t - 1, t - 1, 84, -28t, 3, 3, 168, 3, 2, 168$

We can organize the information in a chart. In this case, the distances are the same, so we use the formula $d = rt$.

$$d \quad = \quad r \quad \cdot \quad t$$

	DISTANCE	SPEED	TIME	
DOWNSTREAM	d	$r + 6$	3	→ $d = (r + 6)3$
UPSTREAM	d	$r - 6$	5	→ $d = (r - 6)5$

4. *Air Travel.* An airplane flew for 5 hr with a 25-km/h tail wind. The return flight against the same wind took 6 hr. Find the speed of the airplane in still air. (*Hint*: The distance is the same both ways. The speeds are $r + 25$ and $r - 25$, where r is the speed in still air.)

2. **Translate.** From each row of the chart, we get an equation, $d = rt$:

$$d = (r + 6)3, \qquad \textbf{(1)}$$
$$d = (r - 6)5. \qquad \textbf{(2)}$$

3. **Solve.** Since there is a variable alone on one side of an equation, we solve the system using substitution:

$$(r + 6)3 = (r - 6)5 \qquad \text{Substituting } (r + 6)3 \text{ for } d \text{ in equation (2)}$$
$$3r + 18 = 5r - 30 \qquad \text{Removing parentheses}$$
$$-2r + 18 = -30 \qquad \text{Subtracting } 5r$$
$$-2r = -48 \qquad \text{Subtracting 18}$$
$$r = \frac{-48}{-2}, \text{ or } 24. \qquad \text{Dividing by } -2$$

4. **Check.** When $r = 24$, $r + 6 = 24 + 6$, or 30, and $30 \cdot 3 = 90$, the distance downstream. When $r = 24$, $r - 6 = 24 - 6$, or 18, and $18 \cdot 5 = 90$, the distance upstream. In both cases, we get the same distance so the answer checks.

5. **State.** The speed in still water is 24 km/h.

◀ **Do Exercise 4.**

MORE TIPS FOR SOLVING MOTION PROBLEMS

1. Translating to a system of equations eases the solution of many motion problems.
2. At the end of the problem, always ask yourself, "Have I found what the problem asked for?" You might have solved for a certain variable but still not have answered the question of the original problem. For instance, in Example 2 we solve for t but the question of the original problem asks for d. Thus we need to continue the *Solve* step.

Answer

4. 275 km/h

Translating for Success

The goal of these matching questions is to practice step (2), Translate, of the five-step problem-solving process. Translate each word problem to an equation or a system of equations and select a correct translation from A–O.

1. *Car Travel.* Two cars leave town at the same time traveling in opposite directions. One travels 50 mph and the other travels 55 mph. In how many hours will they be 500 mi apart?

2. *Mixture of Solutions.* Solution A is 20% alcohol and solution B is 60% alcohol. How much of each should be used in order to make 10 L of a solution that is 50% alcohol?

3. *Triangle Dimensions.* The height of a triangle is 3 cm less than the base. The area is 27 cm^2. Find the height and the base.

4. *Fish Population.* To determine the number of fish in a lake, a conservationist catches 85 fish, tags them, and throws them back into the lake. Later, 60 fish are caught, 25 of which are tagged. How many fish are in the lake?

5. *Supplementary Angles.* Two angles are supplementary. One angle measures 36° more than three times the measure of the other. Find the measure of each angle.

A. $20\%x + 60\%y = 50\% \cdot 10,$
$x + y = 10$

B. $18 + 0.35x = 100$

C. $55x + 50x = 500$

D. $11x + 9x = 1$

E. $\frac{85}{x} = \frac{25}{60}$

F. $\frac{x}{11} + \frac{x}{9} = 1$

G. $\frac{1}{2}x(x - 3) = 27$

H. $x^2 + (x + 4)^2 = 8^2$

I. $8^2 + x^2 = (x + 4)^2$

J. $x + (3x + 36) = 180$

K. $20x + 60y = 5,$
$x + y = 10$

L. $x + (3x + 36) + (x - 7)$
$= 180$

M. $18 + 35x = 100$

N. $\frac{x}{85} = \frac{25}{60}$

O. $x + (3x + 36) = 90$

Answers on page A-36

6. *Triangle Dimensions.* The length of one leg of a right triangle is 8 m. The length of the hypotenuse is 4 m longer than the length of the other leg. Find the lengths of the hypotenuse and the other leg.

7. *Costs of Promotional Buttons.* The vice-president of the Spanish club has $100 to spend on promotional buttons for membership week. There is a setup fee of $18 and a cost of 35¢ per button. How many buttons can he purchase?

8. *Triangle Measures.* The second angle of a triangle measures 36° more than three times the measure of the first. The measure of the third angle is 7° less than the measure of the first. Find the measure of each angle of the triangle.

9. *Complementary Angles.* Two angles are complementary. One angle measures 36° more than three times the measure of the other. Find the measure of each angle.

10. *Work Time.* It takes Maggie 11 hr to paint a room. It takes Claire 9 hr to paint the same room. How long would it take to paint the room if they worked together?

15.5 Exercise Set

For Extra Help
MyMathLab®

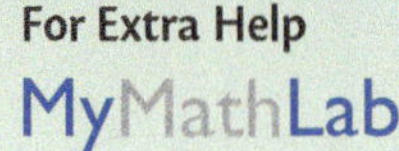

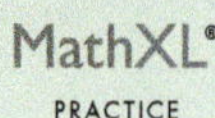

Reading Check

Complete each statement with the appropriate expression from the choices under each blank.

RC1. If Troy drove t hr at 60 mph, he traveled ____________ mi.
$60t, 60/t, t/60$

RC2. Sophia paddles in still water at a rate of r mph. If she is paddling downstream in a river with a current of 2 mph, she is moving at a rate of ____________ mph.
$r + 2, r - 2, 2 - r$

RC3. Rosa's motorboat travels r mph in still water. If she is motoring upstream in a river with a current of 4 mph, she is moving at a rate of ____________ mph.
$r + 4, r - 4, 4 - r$

RC4. Jay's plane travels 125 mph in still air. If he is flying against a head wind of r mph, he is moving at a rate of ____________ mph.
$r + 125, r - 125, 125 - r$

a Solve. In Exercises 1–6, complete the chart to aid the translation.

1. *Car Travel.* Two cars leave town at the same time going in the same direction. One travels at 30 mph and the other travels at 46 mph. In how many hours will they be 72 mi apart?

$d = r \cdot t$

	DISTANCE	SPEED	TIME
SLOWER CAR	Distance of slow car		t
FASTER CAR	Distance of fast car	46	

2. *Car and Truck Travel.* A truck and a car leave a service station at the same time and travel in the same direction. The truck travels at 55 mph and the car at 40 mph. They can maintain CB radio contact within a range of 10 mi. When will they lose contact?

$d = r \cdot t$

	DISTANCE	SPEED	TIME
TRUCK	Distance of truck	55	
CAR	Distance of car		t

3. *Train Travel.* A train leaves a station and travels east at 72 mph. Three hours later, a second train leaves on a parallel track and travels east at 120 mph. When will it overtake the first train?

$d = r \cdot t$

	DISTANCE	SPEED	TIME	
SLOWER TRAIN	d		$t + 3$	→ $d = 72(\quad)$
FASTER TRAIN	d	120		→ $d = (\quad)t$

4. *Airplane Travel.* A private airplane leaves an airport and flies due south at 192 mph. Two hours later, a jet leaves the same airport and flies due south at 960 mph. When will the jet overtake the plane?

$d = r \cdot t$

	DISTANCE	SPEED	TIME	
PRIVATE PLANE	d	192		→ $d = 192(\quad)$
JET	d		t	→ $d = (\quad)t$

5. *Canoeing.* A canoeist paddled for 4 hr with a 6-km/h current to reach a campsite. The return trip against the same current took 10 hr. Find the speed of the canoe in still water.

	d =	r ·	t	
DOWN-STREAM	d	$r + 6$		→ $d = (\quad)4$
UPSTREAM	d		10	→ $\quad = (r - 6)10$

6. *Airplane Travel.* An airplane flew for 4 hr with a 20-km/h tail wind. The return flight against the same wind took 5 hr. Find the speed of the plane in still air.

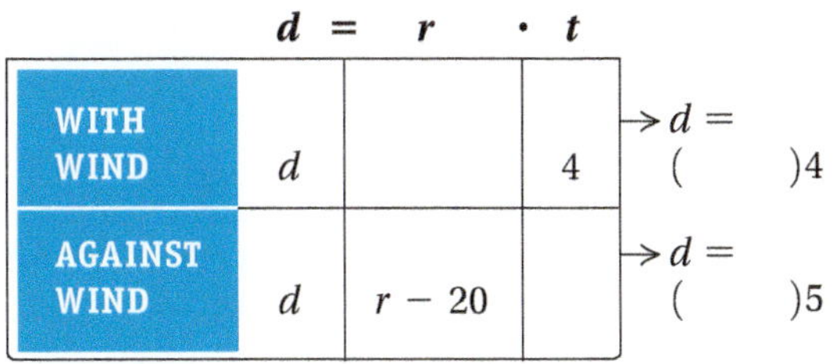

	d =	r ·	t	
WITH WIND	d		4	→ $d = (\quad)4$
AGAINST WIND	d	$r - 20$		→ $d = (\quad)5$

7. *Train Travel.* It takes a passenger train 2 hr less time than it takes a freight train to make the trip from Central City to Clear Creek. The passenger train averages 96 km/h, while the freight train averages 64 km/h. How far is it from Central City to Clear Creek?

8. *Airplane Travel.* It takes a small jet 4 hr less time than it takes a propeller-driven plane to travel from Glen Rock to Oakville. The jet averages 637 km/h, while the propeller plane averages 273 km/h. How far is it from Glen Rock to Oakville?

9. *Motorboat Travel.* On a weekend outing, Antoine rents a motorboat for 8 hr to travel down the river and back. The rental operator tells him to go downstream for 3 hr, leaving him 5 hr to return upstream.

   a) If the river current flows at a speed of 6 mph, how fast must Antoine travel in order to return in 8 hr?

   b) How far downstream did Antoine travel before he turned back?

10. *Airplane Travel.* For spring break, a group of students flew to Cancun. From Mexico City, the airplane took 2 hr to fly 600 mi against a head wind. The return trip with the wind took $1\frac{2}{3}$ hr. Find the speed of the plane in still air.

11. *Running.* A toddler starts running down a sidewalk at 230 ft/min. One minute later, a worried mother runs after the child at 660 ft/min. When will the mother overtake the toddler?

12. *Airplane Travel.* Two airplanes start at the same time and fly toward each other from points 1000 km apart at rates of 420 km/h and 330 km/h. When will they meet?

13. *Motorcycle Travel.* A motorcycle breaks down and the rider must jog the rest of the way to work. The motorcycle was being driven at 45 mph, and the rider jogs at a speed of 6 mph. The distance from home to work is 25 mi, and the total time for the trip was 2 hr. How far did the motorcycle go before it broke down?

14. *Walking and Jogging.* A student walks and jogs to college each day. She averages 5 km/h walking and 9 km/h jogging. The distance from home to college is 8 km, and she makes the trip in 1 hr. How far does the student jog?

Skill Maintenance

Factor completely. [13.6a]

15. $25x^2 - 81$

16. $12a^2 + 16a - 3$

17. $9y^3 - 12y^2 + 4y$

18. $7x^3 + 7x^2 + 14x + 14$

Synthesis

19. *Lindbergh's Flight.* Charles Lindbergh flew the Spirit of St. Louis in 1927 from New York to Paris at an average speed of 107.4 mph. Eleven years later, Howard Hughes flew the same route, averaged 217.1 mph, and took 16 hr and 57 min less time. Find the length of their route.

20. *River Cruising.* An afternoon sightseeing cruise up river and back down river is scheduled to last 1 hr. The speed of the current is 4 mph, and the speed of the riverboat in still water is 12 mph. How far upstream should the pilot travel before turning around?

CHAPTER

15 Summary and Review

Vocabulary Reinforcement

Complete each statement with the correct term from the column on the right. Some of the choices may be used more than once and some may not be used at all.

1. A solution of a system of two equations in two variables is an ordered ______________ that makes both equations true. [15.1a]
2. To solve a system of equations graphically, we graph both equations and find the coordinates of any points of ______________. [15.1b]
3. The substitution method is a(n) ______________ method for solving systems of equations. [15.2a]
4. If, when solving a system algebraically, we obtain a false equation, then the system has ______________. [15.3b]
5. If the graphs of the equations in a system of two equations are parallel, then the system has ______________. [15.1b]
6. If the graphs of the equations in a system of two equations are the same line, then the system has ______________. [15.1b]

algebraic
graphical
intersection
union
infinitely many solutions
no solution
pair
variable

Concept Reinforcement

Determine whether each statement is true or false.

______________ 1. A system of two linear equations can have exactly two solutions. [15.1b]

______________ 2. The solution(s) of a system of two equations can be found by determining where the graphs of the equations intersect. [15.1b]

______________ 3. When we obtain a false equation when solving a system of equations, the system has no solution. [15.3b]

______________ 4. If a system of equations has infinitely many solutions, then *any* ordered pair is a solution. [15.1b]

Study Guide

Objective 15.1a Determine whether an ordered pair is a solution of a system of equations.

Example Determine whether $(2, -3)$ is a solution of the system of equations

$$y = x - 5,$$
$$2x + y = 3.$$

Using alphabetical order of the variables, we substitute 2 for x and -3 for y in both equations.

$$\begin{array}{c|l} y & = x - 5 \\ \hline -3 & ?\ 2 - 5 \\ & -3 \quad \text{TRUE} \end{array} \qquad \begin{array}{r|l} 2x + y & = 3 \\ \hline 2 \cdot 2 + (-3) & ?\ 3 \\ 4 - 3 & \\ 1 & \quad \text{FALSE} \end{array}$$

The pair $(2, -3)$ is not a solution of $2x + y = 3$, so it is not a solution of the system of equations.

Practice Exercise

1. Determine whether $(-2, 1)$ is a solution of the system of equations

$$x + 3y = 1,$$
$$y = x + 3.$$

Objective 15.1b Solve systems of two linear equations in two variables by graphing.

Example Solve this system of equations by graphing:

$$x - y = 1,$$
$$y = 2x - 4.$$

We graph the equations.

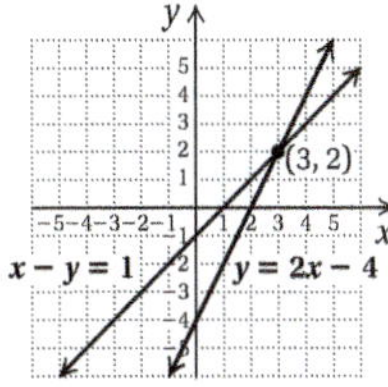

The point of intersection appears to be (3, 2). We check this pair.

$x - y = 1$: $3 - 2 \stackrel{?}{=} 1$; 1 TRUE

$y = 2x - 4$: $2 \stackrel{?}{=} 2 \cdot 3 - 4$; $6 - 4$; 2 TRUE

The pair (3, 2) checks in both equations. It is the solution.

Practice Exercise

2. Solve this system of equations by graphing:

$$2x + 3y = 2,$$
$$x + y = 2.$$

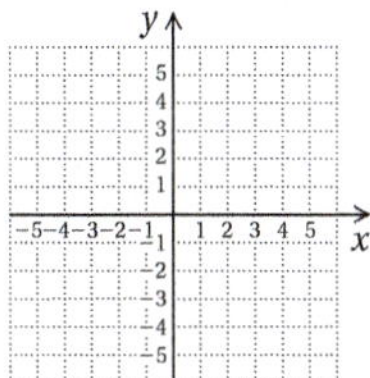

Objective 15.2b Solve a system of two equations in two variables by the substitution method when neither equation has a variable alone on one side.

Example Solve the system

$$x - 2y = 1, \quad \textbf{(1)}$$
$$2x - 3y = 3. \quad \textbf{(2)}$$

We solve equation (1) for x, because the coefficient of x is 1 in that equation:

$$x - 2y = 1$$
$$x = 2y + 1. \quad \textbf{(3)}$$

Now we substitute $2y + 1$ for x in equation (2) and solve for y:

$$2x - 3y = 3$$
$$2(2y + 1) - 3y = 3$$
$$4y + 2 - 3y = 3$$
$$y + 2 = 3$$
$$y = 1.$$

Next, we substitute 1 for y in either equation (1), (2), or (3) and find x. We choose equation (3) since it is already solved for x:

$$x = 2y + 1 = 2 \cdot 1 + 1 = 2 + 1 = 3.$$

We check the ordered pair (3, 1) in both equations.

$x - 2y = 1$: $3 - 2 \cdot 1 \stackrel{?}{=} 1$; $3 - 2$; 1 TRUE

$2x - 3y = 3$: $2 \cdot 3 - 3 \cdot 1 \stackrel{?}{=} 3$; $6 - 3$; 3 TRUE

The pair (3, 1) checks in both equations. It is the solution.

Practice Exercise

3. Solve the system of equations

$$x + y = -1,$$
$$2x + 5y = 1.$$

Objective 15.3b Solve a system of two equations in two variables using the elimination method when multiplication is necessary.

Example Solve the system

$$2a - 3b = 7, \quad (1)$$
$$3a - 2b = 8. \quad (2)$$

We could eliminate either a or b. Here we decide to eliminate the a-terms.

$$6a - 9b = 21 \quad \text{Multiplying equation (1) by 3}$$
$$-6a + 4b = -16 \quad \text{Multiplying equation (2) by } -2$$
$$-5b = 5 \quad \text{Adding}$$
$$b = -1 \quad \text{Solving for } b$$

Next, we substitute -1 for b in either of the original equations:

$$2a - 3b = 7 \quad \text{Equation (1)}$$
$$2a - 3(-1) = 7$$
$$2a + 3 = 7$$
$$2a = 4$$
$$a = 2.$$

The ordered pair $(2, -1)$ checks in both equations, so it is the solution of the system of equations.

Practice Exercise

4. Solve the system of equations

$$3x + 2y = 6,$$
$$x - y = 7.$$

Review Exercises

Determine whether the given ordered pair is a solution of the system of equations. [15.1a]

1. $(6, -1)$; $x - y = 3,$
$2x + 5y = 6$

2. $(2, -3)$; $2x + y = 1,$
$x - y = 5$

3. $(-2, 1)$; $x + 3y = 1,$
$2x - y = -5$

4. $(-4, -1)$; $x - y = 3,$
$x + y = -5$

Solve each system by graphing. [15.1b]

5. $x + y = 3,$
$x - y = 7$

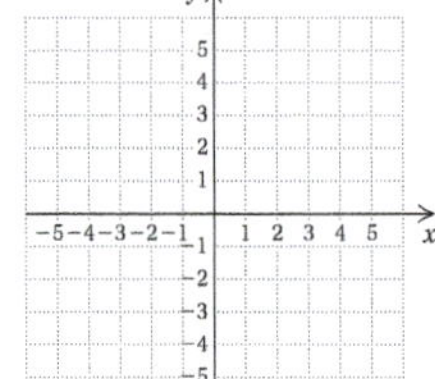

6. $x - 3y = 3,$
$2x - 6y = 6$

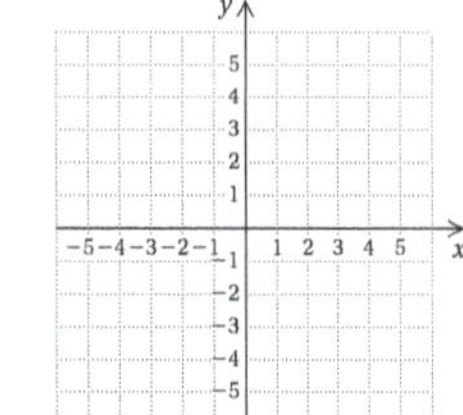

7. $3x - 2y = -4,$
$2y - 3x = -2$

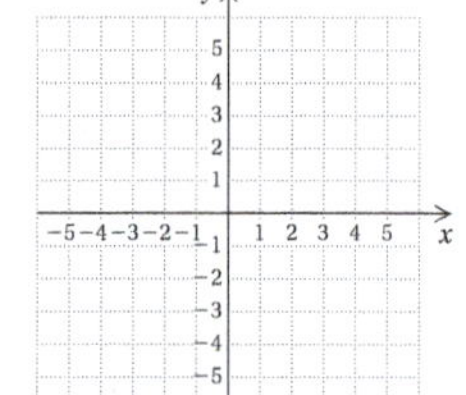

Solve each system using the substitution method. [15.2a, b]

8. $y = 5 - x,$
$3x - 4y = -20$

9. $x + y = 6,$
$y = 3 - 2x$

10. $x - y = 4,$
$y = 2 - x$

11. $s + t = 5,$
$s = 13 - 3t$

12. $x + 2y = 6,$
$2x + 3y = 8$

13. $3x + y = 1,$
$x - 2y = 5$

Solve each system using the elimination method. [15.3a, b]

14. $x + y = 4,$
$2x - y = 5$

15. $x + 2y = 9,$
$3x - 2y = -5$

16. $x - y = 8,$
$2x - 2y = 7$

17. $2x + 3y = 8,$
$5x + 2y = -2$

18. $5x - 2y = 2,$
$3x - 7y = 36$

19. $-x - y = -5,$
$2x - y = 4$

20. $6x + 2y = 4,$
$10x + 7y = -8$

21. $-6x - 2y = 5,$
$12x + 4y = -10$

22. $\frac{2}{3}x + y = -\frac{5}{3},$
$x - \frac{1}{3}y = -\frac{13}{3}$

Solve. [15.2c], [15.4a]

23. *Rectangle Dimensions.* The perimeter of a rectangle is 96 cm. The length is 27 cm more than the width. Find the length and the width.

24. *Paid Admissions.* There were 508 people at a choral concert. Orchestra seats cost $25 each and balcony seats cost $18 each. The total receipts were $11,223. Find the number of orchestra seats and the number of balcony seats sold for the concert.

25. *Window Cleaner.* Clear Shine window cleaner is 30% alcohol, whereas Sunstream window cleaner is 60% alcohol. How much of each is needed to make 80 L of a cleaner that is 45% alcohol?

26. *Weights of Elephants.* A zoo has both an Asian elephant and an African elephant. The African elephant weighs 2400 kg more than the Asian elephant. Together, they weigh 12,000 kg. How much does each elephant weigh?

27. *Mixed Nuts.* Sandy's Catering needs to provide 13 lb of mixed nuts for a wedding reception. The wedding couple has allocated $71 for nuts. Peanuts cost $4.50 per pound and fancy nuts cost $7.00 per pound. How many pounds of each type should be mixed?

28. *Octane Ratings.* The octane rating of a gasoline is a measure of the amount of isooctane in the gas. How much 87-octane gas and 95-octane gas should be blended in order to make a 10-gal batch of 93-octane gas?

Source: Champlain Electric and Petroleum Equipment

29. *Age.* Jeff is three times as old as his son. In 13 years, Jeff will be twice as old as his son. How old is each now?

30. *Complementary Angles.* Two angles are complementary. Their difference is 26°. Find the measure of each angle.

31. *Supplementary Angles.* Two angles are supplementary. Their difference is 26°. Find the measure of each angle.

Solve. [15.5a]

32. *Air Travel.* An airplane flew for 4 hr with a 15-km/h tail wind. The return flight against the wind took 5 hr. Find the speed of the airplane in still air.

	d DISTANCE	$=\ r$ SPEED	$\cdot\ t$ TIME
WITH WIND			
AGAINST WIND			

33. *Car Travel.* One car leaves Phoenix, Arizona, on Interstate highway I-10 traveling at a speed of 55 mph. Two hours later, another car leaves Phoenix traveling in the same direction on I-10 at a speed of 75 mph. How far from Phoenix will the second car catch up to the first?

	d DISTANCE	$=\ r$ SPEED	$\cdot\ t$ TIME
SLOWER CAR			
FASTER CAR			

Solve each system of equations. [15.1b], [15.2a, b], [15.3a, b]

34. $y = x - 2,$
$x - 2y = 6$

A. The y-value is 0.
B. The y-value is -12.
C. The y-value is -2.
D. The y-value is -4.

35. $3x + 2y = 5,$
$x - y = 5$

A. The x-value is 3.
B. The x-value is 2.
C. The x-value is -2.
D. The x-value is -3.

Synthesis

36. The solution of the following system is $(6, 2)$. Find C and D. [15.1a]

$$2x - Dy = 6,$$
$$Cx + 4y = 14$$

37. Solve: [15.2a]

$$3(x - y) = 4 + x,$$
$$x = 5y + 2.$$

38. *Value of a Horse.* Stephanie agreed to work as a stablehand for 1 year. At the end of that time, she was to receive \$2400 and a horse. After 7 months, she quit the job, but still received the horse and \$1000. What was the value of the horse? [15.4a]

Each of the following shows the graph of a system of equations. Find the equations. [11.4c], [15.1b]

39.

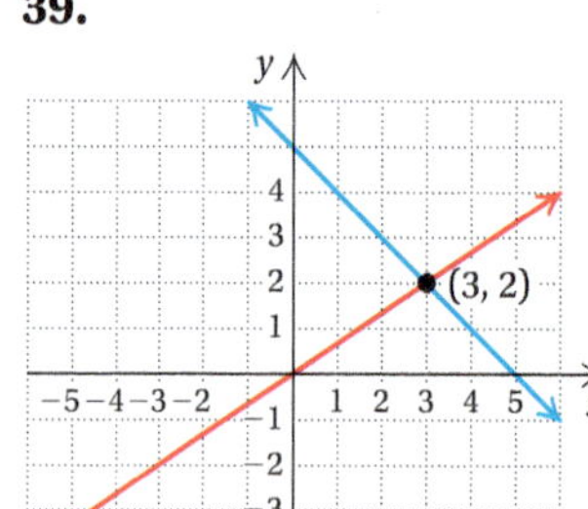

40.

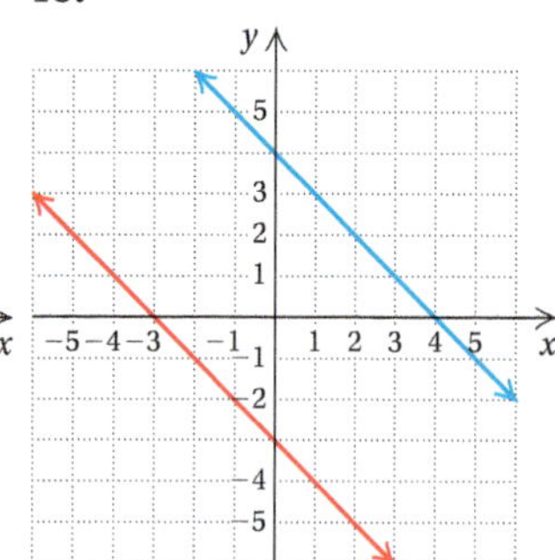

41. *Ancient Chinese Math Problem.* Several ancient Chinese books included problems that can be solved by translating to systems of equations. *Arithmetical Rules in Nine Sections* is a book of 246 problems compiled by a Chinese mathematician, Chang Tsang, who died in 152 B.C. One of the problems is: Suppose there are a number of rabbits and pheasants confined in a cage. In all, there are 35 heads and 94 feet. How many rabbits and how many pheasants are there? Solve the problem. [15.4a]

Understanding Through Discussion and Writing

1. James can tell by inspection that the system

$$2x - y = 3,$$
$$-4x + 2y = -6$$

has an infinite number of solutions. How did he determine this? [15.1b]

2. Explain how the addition and multiplication principles are used to solve systems of equations using the elimination method. [15.3a, b]

3. Which of the five problem-solving steps have you found the most challenging? Why? [15.4a], [15.5a]

4. Discuss the advantages of using a chart to organize information when solving a motion problem. [15.5a]

CHAPTER

15 Test

For Extra Help For step-by-step test solutions, access the Chapter Test Prep Videos in MyMathLab® or on YouTube (search "BittingerPreIntro" and click on Channels).

1. Determine whether the given ordered pair is a solution of the system of equations.

$$(-2, -1);\ 2x - 3y = 4,$$
$$x = 4 + 2y$$

2. Solve this system by graphing. Show your work.

$$x - y = 3,$$
$$x - 2y = 4$$

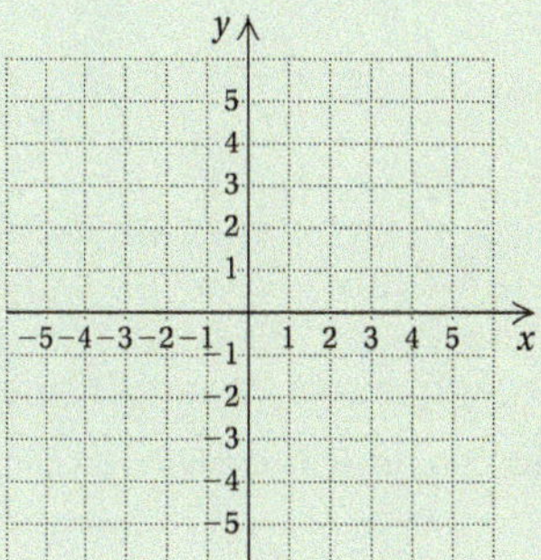

Solve each system using the substitution method.

3. $y = 6 - x,$
$2x - 3y = 22$

4. $x + 2y = 5,$
$x + y = 2$

5. $y = 5x - 2,$
$y - 2 = x$

Solve each system using the elimination method.

6. $x - y = 6,$
$3x + y = -2$

7. $\frac{1}{2}x - \frac{1}{3}y = 8,$
$\frac{1}{3}x - \frac{2}{9}y = 1$

8. $-4x - 9y = 4,$
$6x + 3y = 1$

9. $2x + 3y = 13,$
$3x - 5y = 10$

Solve.

10. *Rectangle Dimensions.* The perimeter of a rectangular field is 8266 yd. The length is 84 yd more than the width. Find the length and the width.

11. *Mixture of Solutions.* Solution A is 25% acid, and solution B is 40% acid. How much of each is needed to make 60 L of a solution that is 30% acid?

12. *Motorboat Travel.* A motorboat traveled for 2 hr with an 8-km/h current. The return trip against the same current took 3 hr. Find the speed of the motorboat in still water.

13. *Carnival Income.* A traveling carnival has receipts of \$4275 one day. Twice as much was made on concessions as on the rides. How much did the concessions bring in? How much did the rides bring in?

14. *Farm Acreage.* The Rolling Velvet Horse Farm allots 650 acres to plant hay and oats. The owners know that their needs are best met if they plant 180 acres more of hay than of oats. How many acres of each should they plant?

15. *Supplementary Angles.* Two angles are supplementary. One angle measures 45° more than twice the measure of the other. Find the measure of each angle.

16. *Octane Ratings.* The octane rating of a gasoline is a measure of the amount of isooctane in the gas. How much 87-octane gas and 93-octane gas should be blended in order to make 12 gal of 91-octane gas?

Source: Champlain Electric and Petroleum Equipment

17. *Phone Rates.* A telephone company offers a domestic calling plan for $2.95 per month plus 10¢ per minute. Another plan charges $1.95 per month plus 15¢ per minute. For what number of minutes will the two plans cost the same?

18. *Ski Trip.* A group of students drove both a car and an SUV on a ski trip. The car left first and traveled at 55 mph. The SUV left 2 hr later and traveled at 65 mph. How long did it take the SUV to catch up to the car?

19. Solve: $x - 2y = 4,$
$2x - 3y = 3.$

A. Both x and y are positive.
B. x is positive; y is negative.
C. x is negative; y is positive.
D. Both x and y are negative.

Synthesis

20. Find the numbers C and D such that $(-2, 3)$ is a solution of the system

$$Cx - 4y = 7,$$
$$3x + Dy = 8.$$

21. *Ticket Line.* Lily is in line at a ticket window. There are two more people ahead of her than there are behind her. In the entire line, there are three times as many people as there are behind her. How many are ahead of Lily in line?

Each of the following shows the graph of a system of equations. Find the equations.

22.

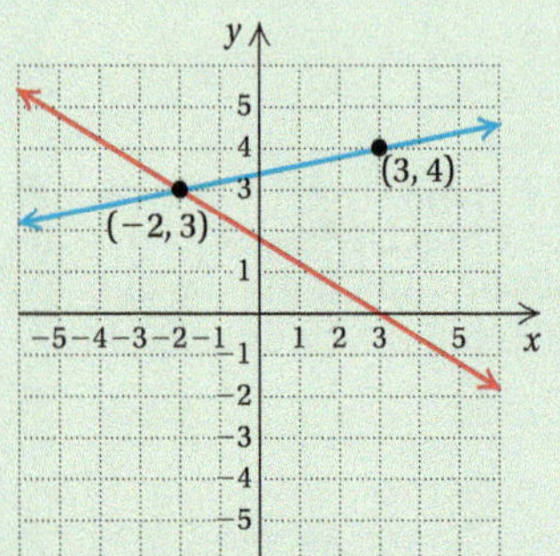

23.

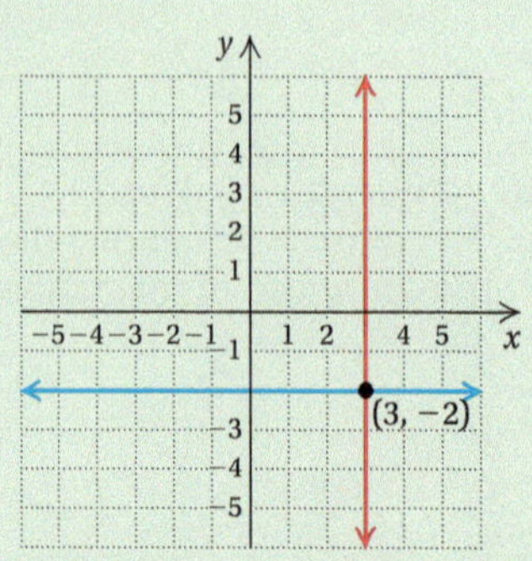

CHAPTER

16

Radical Expressions and Equations

STUDYING FOR SUCCESS *Working with Others*

- ☐ Try being a tutor for a fellow student. You may find that you understand concepts better after you explain them to someone else.
- ☐ Consider forming a study group.
- ☐ Verbalize the math. Often simply talking about a concept with a classmate can help clarify the material.

16.1 Introduction to Radical Expressions

OBJECTIVES

- **a** Find the principal square roots and their opposites of the whole numbers from 0^2 to 25^2.
- **b** Approximate square roots of real numbers using a calculator.
- **c** Solve applied problems involving square roots.
- **d** Identify radicands of radical expressions.
- **e** Determine whether a radical expression represents a real number.
- **f** Simplify a radical expression with a perfect-square radicand.

SKILL TO REVIEW

Objective 1.9b: Evaluate exponential notation.

Evaluate.

1. 7^2 **2.** $\left(\frac{1}{2}\right)^2$

a SQUARE ROOTS

When we raise a number to the second power, we have squared the number. Sometimes we may need to find the number that was squared. We call this process finding a square root of a number.

SQUARE ROOT

The number c is a **square root** of a if $c^2 = a$.

Every positive number has two square roots. For example, the square roots of 25 are 5 and -5 because $5^2 = 25$ and $(-5)^2 = 25$. The positive square root is also called the **principal square root**. The symbol $\sqrt{\ }$ is called a **radical*** (or **square root**) symbol. The radical symbol represents only the principal square root. Thus, $\sqrt{25} = 5$. To name the negative square root of a number, we use $-\sqrt{\ }$. The number 0 has only one square root, 0.

EXAMPLE 1 Find the square roots of 81.

The square roots are 9 and -9. ■

EXAMPLE 2 Find $\sqrt{225}$.

There are two square roots of 225, 15 and -15. We want the principal, or positive, square root since this is what $\sqrt{\ }$ represents. Thus,

$$\sqrt{225} = 15.$$ ■

EXAMPLE 3 Find $-\sqrt{64}$.

The symbol $\sqrt{64}$ represents the positive square root. Then $-\sqrt{64}$ represents the negative square root. That is, $\sqrt{64} = 8$, so

$$-\sqrt{64} = -8.$$

◀ Do Margin Exercises 1–10 on the following page.

*Radicals can be other than square roots, but we will consider only square-root radicals in this chapter. See Appendix J for other types of radicals.

Answers

Skill to Review:

1. 49 **2.** $\frac{1}{4}$

We can think of the processes of "squaring" and "finding square roots" as inverses of each other. We square a number and get one answer. When we find the square roots of the answer, we get the original number *and* its opposite.

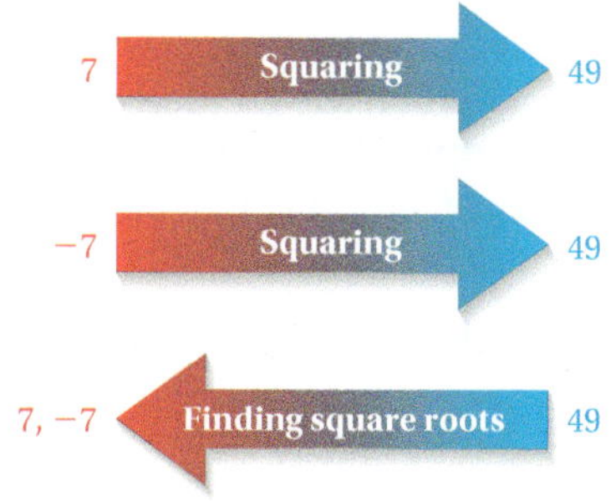

Find the square roots.

1. 36
2. 64
3. 121
4. 144

Find the following.

5. $\sqrt{16}$
6. $\sqrt{49}$
7. $\sqrt{100}$
8. $\sqrt{441}$
9. $-\sqrt{49}$
10. $-\sqrt{169}$

b APPROXIMATING SQUARE ROOTS

We often need to use rational numbers to *approximate* square roots that are irrational. Such approximations can be found using a calculator with a square-root key $\sqrt{\ }$.

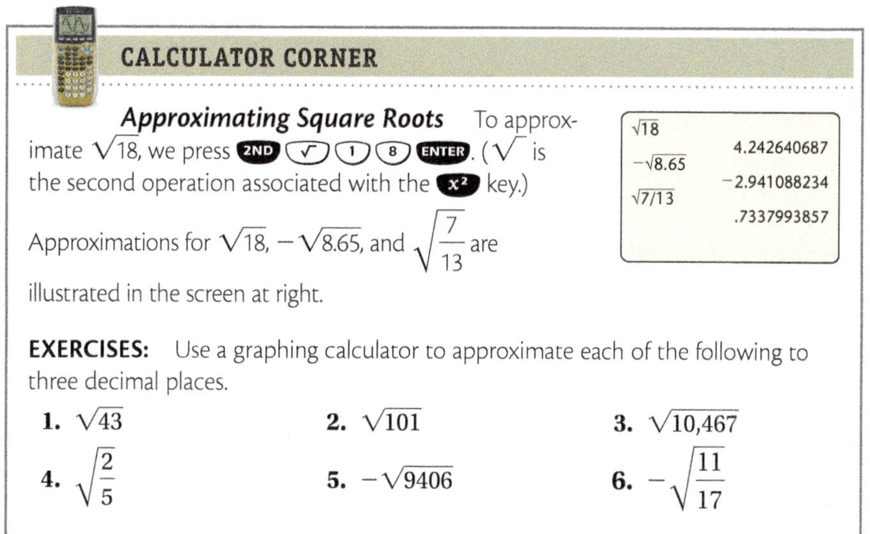

CALCULATOR CORNER

Approximating Square Roots To approximate $\sqrt{18}$, we press 2ND $\sqrt{\ }$ 1 8 ENTER. ($\sqrt{\ }$ is the second operation associated with the x^2 key.)

Approximations for $\sqrt{18}$, $-\sqrt{8.65}$, and $\sqrt{\frac{7}{13}}$ are illustrated in the screen at right.

```
√18
                4.242640687
-√8.65
               -2.941088234
√7/13
                .7337993857
```

EXERCISES: Use a graphing calculator to approximate each of the following to three decimal places.

1. $\sqrt{43}$
2. $\sqrt{101}$
3. $\sqrt{10{,}467}$
4. $\sqrt{\frac{2}{5}}$
5. $-\sqrt{9406}$
6. $-\sqrt{\frac{11}{17}}$

EXAMPLES Use a calculator to approximate each of the following.

	Number	Using a calculator with a 10-digit readout	Rounded to three decimal places
4.	$\sqrt{10}$	3.162277660	3.162
5.	$-\sqrt{583.8}$	−24.16195356	−24.162
6.	$\sqrt{\frac{48}{55}}$	0.934198733	0.934

Do Exercises 11–16. ▶

Use a calculator to approximate each of the following square roots to three decimal places.

11. $\sqrt{15}$
12. $\sqrt{30}$
13. $\sqrt{980}$
14. $-\sqrt{667.8}$
15. $\sqrt{\frac{2}{3}}$
16. $-\sqrt{\frac{203.4}{67.82}}$

Answers

1. 6, −6 **2.** 8, −8 **3.** 11, −11 **4.** 12, −12 **5.** 4 **6.** 7 **7.** 10 **8.** 21 **9.** −7 **10.** −13 **11.** 3.873 **12.** 5.477 **13.** 31.305 **14.** −25.842 **15.** 0.816 **16.** −1.732

c APPLICATIONS OF SQUARE ROOTS

We now consider an application that involves a formula with a radical expression.

17. *Speed of a Skidding Car.* Refer to Example 7. Determine the speed of a car that left skid marks of length **(a)** 40 ft; **(b)** 123 ft.

EXAMPLE 7 *Speed of a Skidding Car.* After an accident, how do police determine the speed at which the car had been traveling? The formula $r = 2\sqrt{5L}$ can be used to approximate the speed r, in miles per hour, of a car that has left a skid mark of length L, in feet. What was the speed of a car that left skid marks of length **(a)** 30 ft? **(b)** 150 ft?

a) We substitute 30 for L and find an approximation:

$$r = 2\sqrt{5L} = 2\sqrt{5 \cdot 30} = 2\sqrt{150} \approx 24.495.$$

The speed of the car was about 24.5 mph.

b) We substitute 150 for L and find an approximation:

$$r = 2\sqrt{5L} = 2\sqrt{5 \cdot 150} \approx 54.772.$$

The speed of the car was about 54.8 mph.

Do Exercise 17.

d RADICANDS AND RADICAL EXPRESSIONS

When an expression is written under a radical, we have a **radical expression**. Here are some examples:

$$\sqrt{14}, \quad \sqrt{x}, \quad 8\sqrt{x^2 + 4}, \quad \sqrt{\frac{x^2 - 5}{2}}.$$

The expression written under the radical is called the **radicand**.

EXAMPLES Identify the radicand in each expression.

8. $-\sqrt{105}$ The radicand is 105.

9. $\sqrt{x} + 2$ The radicand is x.

10. $\sqrt{x + 2}$ The radicand is $x + 2$.

11. $6\sqrt{y^2 - 5}$ The radicand is $y^2 - 5$.

12. $\sqrt{\frac{a - b}{a + b}}$ The radicand is $\frac{a - b}{a + b}$.

Do Exercises 18–21.

Identify the radicand.

18. $\sqrt{227}$

19. $-\sqrt{45 + x}$

20. $\sqrt{\frac{x}{x + 2}}$

21. $8\sqrt{x^2 + 4}$

e EXPRESSIONS THAT ARE MEANINGFUL AS REAL NUMBERS

The square of any nonzero number is always positive. For example, $8^2 = 64$ and $(-11)^2 = 121$. There are no real numbers that when squared yield negative numbers. Thus, $\sqrt{-100}$ does not represent a real number because there is no real number that when squared yields -100. We can try to square 10 and -10, but we know that $10^2 = 100$ and $(-10)^2 = 100$. Neither square is -100. Thus the following expressions do not represent real numbers (they are meaningless as real numbers):

$$\sqrt{-100}, \quad \sqrt{-49}, \quad -\sqrt{-3}.$$

Answers

17. (a) About 28.3 mph; **(b)** about 49.6 mph
18. 227 **19.** $45 + x$ **20.** $\frac{x}{x + 2}$
21. $x^2 + 4$

EXCLUDING NEGATIVE RADICANDS

Radical expressions with negative radicands do not represent real numbers.

Later in your study of mathematics, you may encounter a number system called the **complex numbers** in which negative numbers have defined square roots.

Do Exercises 22–25.

Determine whether each expression represents a real number. Write "yes" or "no."

22. $-\sqrt{25}$ **23.** $\sqrt{-25}$

24. $-\sqrt{-36}$ **25.** $-\sqrt{36}$

f PERFECT-SQUARE RADICANDS

The expression $\sqrt{x^2}$, with a perfect-square radicand, x^2, can be troublesome to simplify. Recall that $\sqrt{\ }$ denotes the principal square root. That is, the answer is nonnegative (either positive or zero). If x represents a nonnegative number, $\sqrt{x^2}$ simplifies to x. If x represents a negative number, $\sqrt{x^2}$ simplifies to $-x$ (the opposite of x), which is positive.

Suppose that $x = 3$. Then

$$\sqrt{x^2} = \sqrt{3^2} = \sqrt{9} = 3.$$

Suppose that $x = -3$. Then

$$\sqrt{x^2} = \sqrt{(-3)^2} = \sqrt{9} = 3, \quad \text{the } \textit{opposite} \text{ of } -3.$$

Note that 3 is the *absolute value* of both 3 and -3. In general, when replacements for x are considered to be *any* real numbers, it follows that

$$\sqrt{x^2} = |x|,$$

and when $x = 3$ or $x = -3$,

$$\sqrt{x^2} = \sqrt{3^2} = |3| = 3 \quad \text{and} \quad \sqrt{x^2} = \sqrt{(-3)^2} = |-3| = 3.$$

PRINCIPAL SQUARE ROOT OF A^2

For any real number A,

$$\sqrt{A^2} = |A|.$$

(That is, for any real number A, the principal square root of A^2 is the absolute value of A.)

EXAMPLES Simplify. Assume that expressions under radicals represent any real number.

13. $\sqrt{10^2} = |10| = 10$

14. $\sqrt{(-7)^2} = |-7| = 7$

15. $\sqrt{(3x)^2} = |3x| = 3|x|$ Absolute-value notation is necessary.

16. $\sqrt{a^2b^2} = \sqrt{(ab)^2} = |ab|$

17. $\sqrt{x^2 + 2x + 1} = \sqrt{(x + 1)^2} = |x + 1|$

Do Exercises 26–31.

Simplify. Assume that expressions under radicals represent any real number.

26. $\sqrt{(-13)^2}$ **27.** $\sqrt{(7w)^2}$

28. $\sqrt{(xy)^2}$ **29.** $\sqrt{x^2y^2}$

30. $\sqrt{(x - 11)^2}$

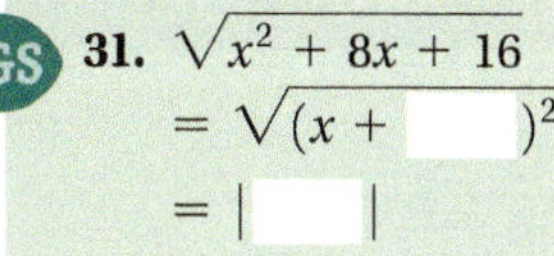

GS **31.** $\sqrt{x^2 + 8x + 16}$

$= \sqrt{(x + \quad)^2}$

$= |\quad|$

Answers

22. Yes **23.** No **24.** No **25.** Yes
26. 13 **27.** $7|w|$ **28.** $|xy|$
29. $|xy|$ **30.** $|x - 11|$ **31.** $|x + 4|$

Guided Solution:
31. $4, x + 4$

Fortunately, in many cases, it can be assumed that radicands that are variable expressions do not represent the square of a negative number. When this assumption is made, the need for absolute-value symbols disappears. Then for $x \geq 0$, $\sqrt{x^2} = x$, since x is nonnegative.

PRINCIPAL SQUARE ROOT OF A^2

For any *nonnegative* real number A,

$$\sqrt{A^2} = A.$$

(That is, for any *nonnegative* real number A, the principal square root of A^2 is A.)

Simplify. Assume that radicands do not represent the square of a negative number.

32. $\sqrt{(xy)^2}$ **33.** $\sqrt{x^2y^2}$

34. $\sqrt{25y^2}$ **35.** $\sqrt{\frac{1}{4}t^2}$

36. $\sqrt{(x-11)^2}$

37. $\sqrt{x^2 + 8x + 16}$
$= \sqrt{(x + \quad)^2}$
$= \quad + 4$

GS

EXAMPLES Simplify. Assume that radicands do not represent the square of a negative number.

18. $\sqrt{(3x)^2} = 3x$ Since $3x$ is assumed to be nonnegative, $|3x| = 3x$.

19. $\sqrt{a^2b^2} = \sqrt{(ab)^2} = ab$ Since ab is assumed to be nonnegative, $|ab| = ab$.

20. $\sqrt{x^2 + 2x + 1} = \sqrt{(x+1)^2} = x + 1$ Since $x + 1$ is assumed to be nonnegative

◀ Do Exercises 32–37.

RADICALS AND ABSOLUTE VALUE

Henceforth, in this text we will assume that no radicands are formed by raising negative quantities to even powers.

We make this assumption in order to eliminate some confusion and because it is valid in many applications. As you study further in mathematics, however, you will frequently need to make a determination about expressions under radicals being nonnegative or positive. This will often be necessary in calculus.

Answers

32. xy **33.** xy **34.** $5y$
35. $\frac{1}{2}t$ **36.** $x - 11$ **37.** $x + 4$

Guided Solution:
37. 4, x

16.1 Exercise Set

For Extra Help
MyMathLab®

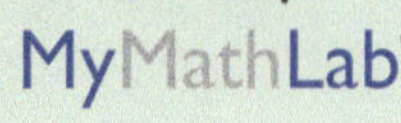

✓ Reading Check

Determine whether each statement is true or false.

RC1. The radical symbol $\sqrt{\ }$ represents only the principal square root.

RC2. For any real number A, $\sqrt{A^2} = A$.

RC3. The radicand in the expression $\sqrt{y} + 3$ is $y + 3$.

RC4. There are no real numbers that when squared yield negative numbers.

RC5. For any nonnegative real number A, $\sqrt{A^2} = A$.

RC6. The number c is a square root of a if $c^2 = \sqrt{a}$.

a Find the square roots.

1. 4 **2.** 1 **3.** 9 **4.** 16 **5.** 100

6. 121 **7.** 169 **8.** 144 **9.** 256 **10.** 625

Simplify.

11. $\sqrt{4}$ **12.** $\sqrt{1}$ **13.** $-\sqrt{9}$ **14.** $-\sqrt{25}$ **15.** $-\sqrt{36}$

16. $-\sqrt{81}$ **17.** $-\sqrt{225}$ **18.** $\sqrt{400}$ **19.** $\sqrt{361}$ **20.** $-\sqrt{441}$

b Use a calculator to approximate each square root. Round to three decimal places.

21. $\sqrt{5}$ **22.** $\sqrt{8}$ **23.** $\sqrt{432}$ **24.** $-\sqrt{8196}$ **25.** $-\sqrt{347.7}$

26. $-\sqrt{204.788}$ **27.** $\sqrt{\frac{278}{36}}$ **28.** $-\sqrt{\frac{567}{788}}$ **29.** $-5\sqrt{189 \cdot 6}$ **30.** $2\sqrt{18 \cdot 3}$

c Solve.

31. *Water Flow of Fire Hose.* The number of gallons per minute discharged from a fire hose depends on the diameter of the hose and the nozzle pressure. For a 2-in. diameter solid bore nozzle, the water flow W, in gallons per minute (GPM), is given by $W = 118.8\sqrt{P}$, where P is the nozzle pressure, in pounds per square inch (psi). Find the water flow, in GPM, when the pressure is **(a)** 650 psi; **(b)** 1500 psi.

Source: www.firetactics.com

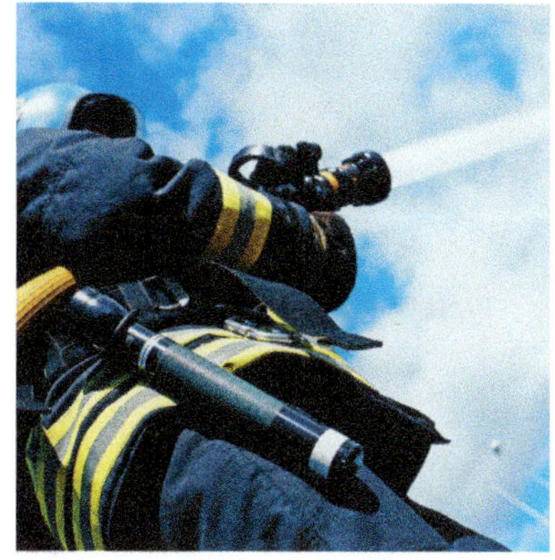

32. *Parking-Lot Arrival Spaces.* The attendants at a parking lot park cars in temporary spaces before the cars are taken to long-term parking spaces. The number N of such spaces needed is approximated by the formula $N = 2.5\sqrt{A}$, where A is the average number of arrivals during peak hours. Find the number of spaces needed when the average number of arrivals is **(a)** 25; **(b)** 62.

Hang Time. An athlete's *hang time* T (the time airborne for a jump), in seconds, is given by $T = 0.144\sqrt{V}$, where V is the athlete's vertical leap, in inches.

Source: Peter Brancazio

33. Vince Carter of the Dallas Mavericks can jump 43 in. vertically. Find his hang time.

34. Jason Richardson of the Philadelphia 76ers can jump 46 in. vertically. Find his hang time.

35. Paul Pierce of the Brooklyn Nets can jump 38 in. vertically. Find his hang time.

36. Shawn Merion of the Dallas Mavericks can jump 42 in. vertically. Find his hang time.

d Identify the radicand.

37. $\sqrt{200}$

38. $\sqrt{16z}$

39. $\sqrt{x} - 4$

40. $\sqrt{3t + 10} + 8$

41. $5\sqrt{t^2 + 1}$

42. $-9\sqrt{x^2 + 16}$

43. $x^2y\sqrt{\frac{3}{x + 2}}$

44. $ab^2\sqrt{\frac{a}{a + b}}$

e Determine whether each expression represents a real number. Write "yes" or "no."

45. $\sqrt{-16}$

46. $\sqrt{-81}$

47. $-\sqrt{81}$

48. $-\sqrt{64}$

49. $-\sqrt{-25}$

50. $\sqrt{-(-49)}$

f Simplify. Remember that we have assumed that radicands do not represent the square of a negative number.

51. $\sqrt{c^2}$

52. $\sqrt{x^2}$

53. $\sqrt{9x^2}$

54. $\sqrt{16y^2}$

55. $\sqrt{(8p)^2}$

56. $\sqrt{(7pq)^2}$

57. $\sqrt{(ab)^2}$

58. $\sqrt{(6y)^2}$

59. $\sqrt{(34d)^2}$

60. $\sqrt{(53b)^2}$

61. $\sqrt{(x+3)^2}$

62. $\sqrt{(d-3)^2}$

63. $\sqrt{a^2-10a+25}$

64. $\sqrt{x^2+2x+1}$

65. $\sqrt{4a^2-20a+25}$

66. $\sqrt{9p^2+12p+4}$

67. $\sqrt{121y^2-198y+81}$

68. $\sqrt{49b^2+140b+100}$

Skill Maintenance

Solve. [15.4a]

69. ***Supplementary Angles.*** Two angles are supplementary. One angle is 3° less than twice the other. Find the measures of the angles.

70. ***Complementary Angles.*** Two angles are complementary. The sum of the measure of the first angle and half the measure of the second is 64°. Find the measures of the angles.

71. ***Food Expenses.*** The amount F that a family spends on food varies directly as its income I. A family making \$39,200 per year will spend \$10,192 on food. At this rate, how much would a family making \$41,000 per year spend on food? [14.9b]

Divide and simplify. [14.2b]

72. $\dfrac{x-3}{x+4} \div \dfrac{x^2-9}{x+4}$

73. $\dfrac{x^2+10x-11}{x^2-1} \div \dfrac{x+11}{x+1}$

74. $\dfrac{x^4-16}{x^4-1} \div \dfrac{x^2+4}{x^2+1}$

Synthesis

75. Use only the graph of $y = \sqrt{x}$, shown below, to approximate $\sqrt{3}$, $\sqrt{5}$, and $\sqrt{7}$. Answers may vary.

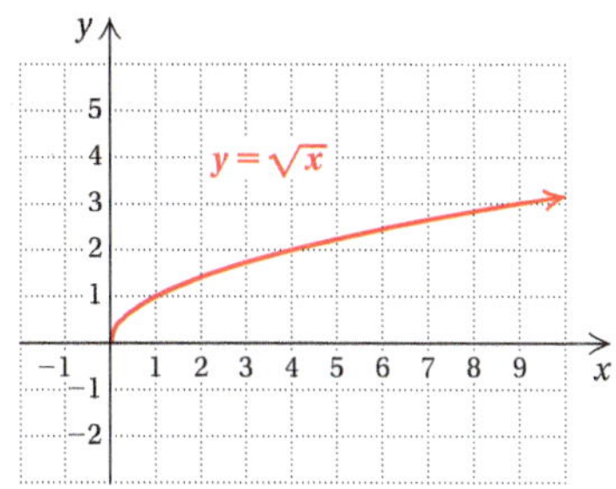

76. ***Wind Chill Temperature.*** When the temperature is T degrees Celsius and the wind speed is V meters per second, the *wind chill temperature,* T_W, is the temperature that it feels like. Here is a formula for finding wind chill temperature:

$$T_W = 13.112 + 0.6215T - 11.37V^{0.16} + 0.3965TV^{0.16}.$$

Estimate the wind chill temperature (to the nearest tenth of a degree) for the given actual temperatures and wind speeds.

a) $T = 7°\text{C},\ V = 8\ \text{m/sec}$
b) $T = -5°\text{C},\ V = 14\ \text{m/sec}$

Solve.

77. $\sqrt{x^2} = 16$

78. $\sqrt{y^2} = -7$

79. $t^2 = 49$

80. Suppose that the area of a square is 3. Find the length of a side.

16.2 Multiplying and Simplifying with Radical Expressions

OBJECTIVES

a Simplify radical expressions.

b Simplify radical expressions where radicands are powers.

c Multiply radical expressions and, if possible, simplify.

SKILL TO REVIEW

Objective 13.5b: Factor trinomial squares.

Factor.

1. $x^2 - 12x + 36$
2. $64x^2 + 48x + 9$

1. Simplify.
   a) $\sqrt{25} \cdot \sqrt{16}$
   b) $\sqrt{25 \cdot 16}$

Multiply.

2. $\sqrt{3}\sqrt{11}$

3. $\sqrt{5}\sqrt{5}$

4. $\sqrt{\frac{5}{11}}\sqrt{\frac{6}{7}}$

5. $\sqrt{x}\sqrt{x+1}$

6. $\sqrt{x+2}\sqrt{x-2}$

a SIMPLIFYING BY FACTORING

To see how to multiply with radical notation, consider the following.

a) $\sqrt{9} \cdot \sqrt{4} = 3 \cdot 2 = 6$ This is a product of square roots.

b) $\sqrt{9 \cdot 4} = \sqrt{36} = 6$ This is the square root of a product.

Note that

$$\sqrt{9} \cdot \sqrt{4} = \sqrt{9 \cdot 4}.$$

Do Margin Exercise 1.

We can multiply radical expressions by multiplying the radicands.

THE PRODUCT RULE FOR RADICALS

For any nonnegative radicands A and B,

$$\sqrt{A} \cdot \sqrt{B} = \sqrt{A \cdot B}.$$

(The product of square roots is the square root of the product of the radicands.)

EXAMPLES Multiply.

1. $\sqrt{5}\sqrt{7} = \sqrt{5 \cdot 7} = \sqrt{35}$
2. $\sqrt{8}\sqrt{8} = \sqrt{8 \cdot 8} = \sqrt{64} = 8$
3. $\sqrt{\frac{2}{3}}\sqrt{\frac{4}{5}} = \sqrt{\frac{2}{3} \cdot \frac{4}{5}} = \sqrt{\frac{8}{15}}$
4. $\sqrt{2x}\sqrt{3x-1} = \sqrt{2x(3x-1)} = \sqrt{6x^2 - 2x}$

Do Margin Exercises 2–6.

To factor radical expressions, we can use the product rule for radicals in reverse.

FACTORING RADICAL EXPRESSIONS

$$\sqrt{AB} = \sqrt{A}\sqrt{B}$$

In some cases, we can simplify after factoring.

A square-root radical expression is simplified when its radicand has no factors that are perfect squares.

Answers

Skill to Review:
1. $(x-6)^2$ 2. $(8x+3)^2$

Margin Exercises:
1. (a) 20; (b) 20 2. $\sqrt{33}$ 3. 5
4. $\sqrt{\frac{30}{77}}$ 5. $\sqrt{x^2 + x}$ 6. $\sqrt{x^2 - 4}$

When simplifying a square-root radical expression, we first determine whether the radicand is a perfect square. Then we determine whether it has perfect-square factors. If so, the radicand is then factored and the radical expression simplified using the preceding rule.

Compare the following:

$$\sqrt{50} = \sqrt{10 \cdot 5} = \sqrt{10}\,\sqrt{5};$$
$$\sqrt{50} = \sqrt{25 \cdot 2} = \sqrt{25}\,\sqrt{2} = 5\sqrt{2}.$$

In the second case, the radicand is written using the perfect-square factor 25. If you do not recognize perfect-square factors, try factoring the radicand into its prime factors. For example,

$$\sqrt{50} = \sqrt{2 \cdot \underbrace{5 \cdot 5}} = 5\sqrt{2}.$$

Perfect square (a pair of the same factors)

Square-root radical expressions in which the radicand has no perfect-square factors, such as $5\sqrt{2}$, are considered to be in simplest form.

EXAMPLES Simplify by factoring.

5. $\sqrt{18} = \sqrt{9 \cdot 2}$ — Identifying a perfect-square factor and factoring the radicand. The factor 9 is a perfect square.

$= \sqrt{9} \cdot \sqrt{2}$ — Factoring into a product of radicals

$= 3\sqrt{2}$ — Simplifying $\sqrt{9}$

The radicand has no factors that are perfect squares.

6. $\sqrt{48t} = \sqrt{16 \cdot 3 \cdot t}$ — Identifying a perfect-square factor and factoring the radicand. The factor 16 is a perfect square.

$= \sqrt{16}\,\sqrt{3t}$ — Factoring into a product of radicals

$= 4\sqrt{3t}$ — Taking a square root

7. $\sqrt{20t^2} = \sqrt{4 \cdot 5 \cdot t^2}$ — Identifying perfect-square factors and factoring the radicand. The factors 4 and t^2 are perfect squares.

$= \sqrt{4}\,\sqrt{t^2}\,\sqrt{5}$ — Factoring into a product of several radicals

$= 2t\sqrt{5}$ — Taking square roots. No absolute-value signs are necessary since we have assumed that expressions under radicals do not represent the square of a negative number.

8. $\sqrt{x^2 - 6x + 9} = \sqrt{(x - 3)^2} = x - 3$ — No absolute-value signs are necessary since we have assumed that expressions under radicals do not represent the square of a negative number.

9. $\sqrt{36x^2} = \sqrt{36}\,\sqrt{x^2} = 6x$, or $\sqrt{36x^2} = \sqrt{(6x)^2} = 6x$

10. $\sqrt{3x^2 + 6x + 3} = \sqrt{3(x^2 + 2x + 1)}$ — Factoring the radicand

$= \sqrt{3(x + 1)^2}$ — Factoring further

$= \sqrt{3}\,\sqrt{(x + 1)^2}$ — Factoring into a product of radicals

$= \sqrt{3}(x + 1)$, or $(x + 1)\sqrt{3}$ — Taking the square root

Do Exercises 7–14. ▶

Simplify by factoring.

7. $\sqrt{32}$ **8.** $\sqrt{92}$

GS **9.** $\sqrt{363q}$

$= \sqrt{\quad \cdot 3 \cdot q}$

$= \sqrt{121}\sqrt{\quad}$

$= \quad \sqrt{3q}$

10. $\sqrt{128t}$ **11.** $\sqrt{63x^2}$

12. $\sqrt{81m^2}$

13. $\sqrt{x^2 + 14x + 49}$

14. $\sqrt{3x^2 - 60x + 300}$

Answers

7. $4\sqrt{2}$ **8.** $2\sqrt{23}$ **9.** $11\sqrt{3q}$ **10.** $8\sqrt{2t}$ **11.** $3x\sqrt{7}$ **12.** $9m$ **13.** $x + 7$ **14.** $\sqrt{3}(x - 10)$, or $(x - 10)\sqrt{3}$

Guided Solution:
9. 121, $3q$, 11

b SIMPLIFYING SQUARE ROOTS OF POWERS

To take the square root of an even power such as x^{10}, we note that $x^{10} = (x^5)^2$. Then

$$\sqrt{x^{10}} = \sqrt{(x^5)^2} = x^5.$$

We can find the answer by taking half the exponent. That is,

$$\sqrt{x^{10}} = x^5. \leftarrow \tfrac{1}{2}(10) = 5$$

EXAMPLES Simplify.

11. $\sqrt{x^6} = \sqrt{(x^3)^2} = x^3 \quad \tfrac{1}{2}(6) = 3$

12. $\sqrt{x^8} = x^4$

13. $\sqrt{t^{22}} = t^{11}$

◀ **Do Exercises 15–18.**

Simplify.

15. $\sqrt{t^4}$ **16.** $\sqrt{t^{20}}$

17. $\sqrt{h^{46}}$ **18.** $\sqrt{x^{100}}$

If an odd power occurs, we express the power in terms of the largest even power. Then we simplify the even power as in Examples 11–13.

EXAMPLE 14 Simplify by factoring: $\sqrt{x^9}$.

$$\begin{aligned}\sqrt{x^9} &= \sqrt{x^8 \cdot x}\\ &= \sqrt{x^8}\sqrt{x}\\ &= x^4\sqrt{x} \longleftarrow\end{aligned}$$

Caution!

Note that $\sqrt{x^9} \neq x^3$.

EXAMPLE 15 Simplify by factoring: $\sqrt{32x^{15}}$.

$\sqrt{32x^{15}} = \sqrt{16 \cdot 2 \cdot x^{14} \cdot x}$ We factor the radicand, looking for perfect-square factors. The largest even power of x is 14.

$= \sqrt{16}\sqrt{x^{14}}\sqrt{2x}$ Factoring into a product of radicals. Perfect-square factors are usually listed first.

$= 4x^7\sqrt{2x}$ Simplifying

◀ **Do Exercises 19 and 20.**

Simplify by factoring.

19. $\sqrt{x^7}$

20. $\sqrt{24x^{11}}$ GS

$= \sqrt{4 \cdot 6 \cdot \underline{\quad} \cdot x}$

$= \sqrt{4}\sqrt{x^{10}}\sqrt{\underline{\quad}}$

$= 2\underline{\quad}\sqrt{6x}$

c MULTIPLYING AND SIMPLIFYING

Sometimes we can simplify after multiplying. We leave the radicand in factored form and factor further to determine perfect-square factors. Then we simplify the perfect-square factors.

EXAMPLE 16 Multiply and then simplify by factoring: $\sqrt{2}\sqrt{14}$.

$\sqrt{2}\sqrt{14} = \sqrt{2 \cdot 14}$ Multiplying

$= \sqrt{2 \cdot 2 \cdot 7}$ Factoring

$= \sqrt{2 \cdot 2}\sqrt{7}$ Looking for perfect-square factors, pairs of factors

$= 2\sqrt{7}$

◀ **Do Exercises 21 and 22.**

Multiply and simplify.

21. $\sqrt{3}\sqrt{6}$ **22.** $\sqrt{2}\sqrt{50}$

Answers

15. t^2 16. t^{10} 17. h^{23} 18. x^{50}
19. $x^3\sqrt{x}$ 20. $2x^5\sqrt{6x}$ 21. $3\sqrt{2}$
22. 10

Guided Solution:
20. x^{10}, $6x$, x^5

EXAMPLE 17 Multiply and then simplify by factoring: $\sqrt{3x^2}\sqrt{9x^3}$.

$$\begin{aligned}\sqrt{3x^2}\sqrt{9x^3} &= \sqrt{3x^2 \cdot 9x^3} && \text{Multiplying}\\ &= \sqrt{3 \cdot x^2 \cdot 9 \cdot x^2 \cdot x} && \text{Looking for perfect-square factors or largest even powers}\\ &= \sqrt{9 \cdot x^2 \cdot x^2 \cdot 3x} && \text{Perfect-square factors are usually listed first.}\\ &= \sqrt{9}\sqrt{x^2}\sqrt{x^2}\sqrt{3x}\\ &= 3 \cdot x \cdot x \cdot \sqrt{3x}\\ &= 3x^2\sqrt{3x}\end{aligned}$$

In doing an example like the preceding one, it might be helpful to do more factoring, as follows:

$$\sqrt{3x^2} \cdot \sqrt{9x^3} = \sqrt{3 \cdot \underline{x \cdot x} \cdot \underline{3 \cdot 3} \cdot \underline{x \cdot x} \cdot x}.$$

Then we look for pairs of factors, as shown, and simplify perfect-square factors:

$$\begin{aligned}&= 3 \cdot x \cdot x\sqrt{3x}\\ &= 3x^2\sqrt{3x}.\end{aligned}$$

EXAMPLE 18 Simplify: $\sqrt{20cd^2}\sqrt{35cd^5}$.

$$\begin{aligned}&\sqrt{20cd^2}\sqrt{35cd^5}\\ &\quad= \sqrt{20cd^2 \cdot 35cd^5} && \text{Multiplying}\\ &\quad= \sqrt{\underline{2 \cdot 2} \cdot 5 \cdot c \cdot \underline{d \cdot d} \cdot 5 \cdot 7 \cdot c \cdot \underline{d \cdot d} \cdot \underline{d \cdot d} \cdot d} && \text{Looking for pairs of factors}\\ &\quad= \sqrt{\underline{2 \cdot 2} \cdot \underline{5 \cdot 5} \cdot \underline{c \cdot c} \cdot \underline{d \cdot d} \cdot \underline{d \cdot d} \cdot \underline{d \cdot d} \cdot 7d}\\ &\quad= 2 \cdot 5 \cdot c \cdot d \cdot d \cdot d\sqrt{7d}\\ &\quad= 10cd^3\sqrt{7d}\end{aligned}$$

Do Exercises 23–25. ▶

Multiply and simplify.

23. $\sqrt{2x^3}\sqrt{8x^3y^4}$

24. $\sqrt{10xy^2}\sqrt{5x^2y^3}$

25. $\sqrt{28q^2r} \cdot \sqrt{21q^3r^7}$

We know that $\sqrt{AB} = \sqrt{A}\sqrt{B}$. That is, the square root of a product is the product of the square roots. What about the square root of a sum? That is, is the square root of a sum equal to the sum of the square roots? To check, consider $\sqrt{A + B}$ and $\sqrt{A} + \sqrt{B}$ when $A = 16$ and $B = 9$:

$$\sqrt{A + B} = \sqrt{16 + 9} = \sqrt{25} = 5;$$

and

$$\sqrt{A} + \sqrt{B} = \sqrt{16} + \sqrt{9} = 4 + 3 = 7.$$

Thus we see the following.

Caution!

The square root of a sum is not the sum of the square roots.

$$\sqrt{A + B} \neq \sqrt{A} + \sqrt{B}$$

CALCULATOR CORNER

Simplifying Radical Expressions

EXERCISES: Use a table or a graph to determine whether each of the following is true.

1. $\sqrt{x + 4} = \sqrt{x} + 2$
2. $\sqrt{3 + x} = \sqrt{3} + x$
3. $\sqrt{x - 2} = \sqrt{x} - \sqrt{2}$
4. $\sqrt{9x} = 3\sqrt{x}$

Answers

23. $4x^3y^2$ **24.** $5xy^2\sqrt{2xy}$ **25.** $14q^2r^4\sqrt{3q}$

16.2 Exercise Set

For Extra Help

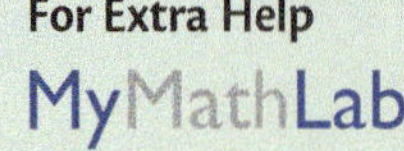
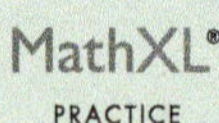

Reading Check

Square-root radical expressions in which the radicand has no perfect-square factors are considered to be in simplest form. Determine whether each radical expression is in simplest form. Answer "yes" or "no."

RC1. $\sqrt{49w^2}$

RC2. $5\sqrt{15}$

RC3. $\sqrt{121q}$

RC4. $4\sqrt{25}$

RC5. $\sqrt{t^2 + 1}$

RC6. $\sqrt{900}$

RC7. $\frac{\sqrt{30}}{4}$

RC8. $\sqrt{221}$

RC9. $\sqrt{x^3 + x^2}$

a Simplify by factoring.

1. $\sqrt{12}$

2. $\sqrt{8}$

3. $\sqrt{75}$

4. $\sqrt{50}$

5. $\sqrt{20}$

6. $\sqrt{45}$

7. $\sqrt{600}$

8. $\sqrt{300}$

9. $\sqrt{486}$

10. $\sqrt{567}$

11. $\sqrt{9x}$

12. $\sqrt{4y}$

13. $\sqrt{48x}$

14. $\sqrt{40m}$

15. $\sqrt{16a}$

16. $\sqrt{49b}$

17. $\sqrt{64y^2}$

18. $\sqrt{9x^2}$

19. $\sqrt{13x^2}$

20. $\sqrt{23s^2}$

21. $\sqrt{8t^2}$

22. $\sqrt{125a^2}$

23. $\sqrt{180}$

24. $\sqrt{320}$

25. $\sqrt{288y}$

26. $\sqrt{363p}$

27. $\sqrt{28x^2}$

28. $\sqrt{20x^2}$

29. $\sqrt{x^2 - 6x + 9}$

30. $\sqrt{t^2 + 22t + 121}$

31. $\sqrt{8x^2 + 8x + 2}$

32. $\sqrt{20x^2 - 20x + 5}$

33. $\sqrt{36y + 12y^2 + y^3}$

34. $\sqrt{x - 2x^2 + x^3}$

b Simplify by factoring.

35. $\sqrt{t^6}$

36. $\sqrt{x^{18}}$

37. $\sqrt{x^{12}}$

38. $\sqrt{x^{16}}$

39. $\sqrt{x^5}$

40. $\sqrt{x^3}$

41. $\sqrt{t^{19}}$

42. $\sqrt{p^{17}}$

43. $\sqrt{(y - 2)^8}$

44. $\sqrt{(x + 3)^6}$

45. $\sqrt{4(x + 5)^{10}}$

46. $\sqrt{16(a - 7)^4}$

47. $\sqrt{36m^3}$

48. $\sqrt{250y^3}$

49. $\sqrt{8a^5}$

50. $\sqrt{12b^7}$

51. $\sqrt{104p^{17}}$

52. $\sqrt{284m^{23}}$

53. $\sqrt{448x^6y^3}$

54. $\sqrt{243x^5y^4}$

C Multiply and then, if possible, simplify by factoring.

55. $\sqrt{3}\sqrt{18}$

56. $\sqrt{5}\sqrt{10}$

57. $\sqrt{15}\sqrt{6}$

58. $\sqrt{3}\sqrt{27}$

59. $\sqrt{18}\sqrt{14x}$

60. $\sqrt{12}\sqrt{18x}$

61. $\sqrt{3x}\sqrt{12y}$

62. $\sqrt{7x}\sqrt{21y}$

63. $\sqrt{13}\sqrt{13}$

64. $\sqrt{11}\sqrt{11x}$

65. $\sqrt{5b}\sqrt{15b}$

66. $\sqrt{6a}\sqrt{18a}$

67. $\sqrt{2t}\sqrt{2t}$

68. $\sqrt{7a}\sqrt{7a}$

69. $\sqrt{ab}\sqrt{ac}$

70. $\sqrt{xy}\sqrt{xz}$

71. $\sqrt{2x^2y}\sqrt{4xy^2}$

72. $\sqrt{15mn^2}\sqrt{5m^2n}$

73. $\sqrt{18}\sqrt{18}$

74. $\sqrt{16}\sqrt{16}$

75. $\sqrt{5}\sqrt{2x-1}$

76. $\sqrt{3}\sqrt{4x+2}$

77. $\sqrt{x+2}\sqrt{x+2}$

78. $\sqrt{x-9}\sqrt{x-9}$

79. $\sqrt{18x^2y^3}\sqrt{6xy^4}$

80. $\sqrt{12x^3y^2}\sqrt{8xy}$

81. $\sqrt{50x^4y^6}\sqrt{10xy}$

82. $\sqrt{10xy^2}\sqrt{5x^2y^3}$

83. $\sqrt{99p^4q^3}\sqrt{22p^5q^2}$

84. $\sqrt{75m^8n^9}\sqrt{50m^5n^7}$

85. $\sqrt{24a^2b^3c^4}\sqrt{32a^5b^4c^7}$

86. $\sqrt{18p^5q^2r^{11}}\sqrt{108p^3q^6r^9}$

Skill Maintenance

Solve. [15.3a, b]

87. $x - y = -6,$
$x + y = 2$

88. $3x + 5y = 6,$
$5x + 3y = 4$

89. $3x - 2y = 4,$
$2x + 5y = 9$

90. $4a - 5b = 25,$
$a - b = 7$

Solve.

91. *Insecticide Mixtures.* A solution containing 30% insecticide is to be mixed with a solution containing 50% insecticide in order to make 200 L of a solution containing 42% insecticide. How much of each solution should be used? [15.4a]

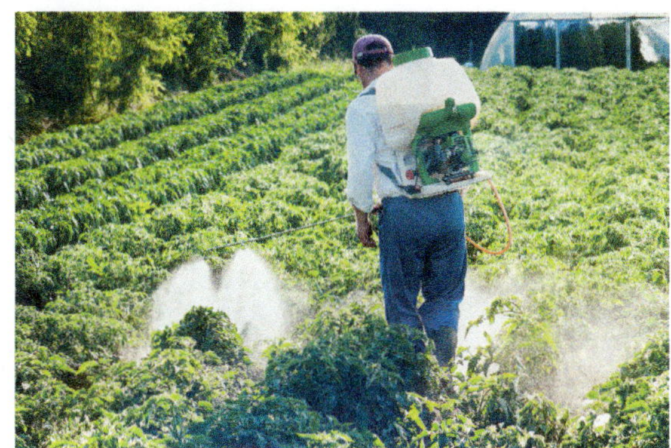

92. *Storage Area Dimensions.* The perimeter of a rectangular storage area is 84 ft. The length is 18 ft greater than the width. Find the area of the rectangle. [15.4a]

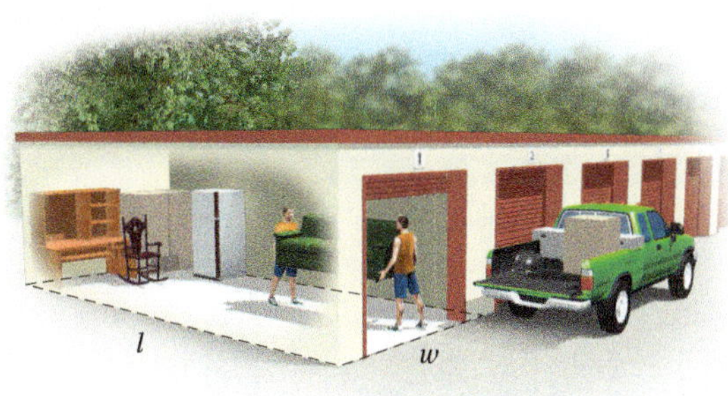

93. *Canoe Travel.* Greg and Beth paddled to a picnic spot downriver in 2 hr. It took them 3 hr to return against the current. If the speed of the current was 2 mph, at what speed were they paddling the canoe? [15.5a]

94. *Fund-Raiser Attendance.* As part of a fund-raiser, 382 people attended a dinner and tour of a space museum. Tickets were \$24 each for adults and \$9 each for children, and receipts totaled \$6603. How many adults and how many children attended? [15.4a]

Synthesis

Factor.

95. $\sqrt{5x - 5}$

96. $\sqrt{x^2 - x - 2}$

97. $\sqrt{x^2 - 36}$

98. $\sqrt{2x^2 - 5x - 12}$

99. $\sqrt{x^3 - 2x^2}$

100. $\sqrt{a^2 - b^2}$

Simplify.

101. $\sqrt{0.25}$

102. $\sqrt{0.01}$

103. $\sqrt{\sqrt{\sqrt{256}}}$

Multiply and then simplify by factoring.

104. $(\sqrt{2y})(\sqrt{3})(\sqrt{8y})$

105. $\sqrt{18(x - 2)}\,\sqrt{20(x - 2)^3}$

106. $\sqrt{27(x + 1)}\,\sqrt{12y(x + 1)^2}$

107. $\sqrt{2^{109}}\,\sqrt{x^{306}}\,\sqrt{x^{11}}$

108. $\sqrt{x}\,\sqrt{2x}\,\sqrt{10x^5}$

109. $\sqrt{a}(\sqrt{a^3} - 5)$

16.3 Quotients Involving Radical Expressions

OBJECTIVES

a Divide radical expressions.

b Simplify square roots of quotients.

c Rationalize the denominator of a radical expression.

SKILL TO REVIEW

Objective 14.1c: Simplify rational expressions by factoring the numerator and the denominator and removing factors of 1.

Simplify.

1. $\dfrac{10x^8}{15x^3}$ **2.** $\dfrac{64a^5b}{24a^2b^6}$

Divide and simplify.

1. $\dfrac{\sqrt{96}}{\sqrt{6}}$ **2.** $\dfrac{\sqrt{75}}{\sqrt{3}}$

3. $\dfrac{\sqrt{x^{14}}}{\sqrt{x^3}}$ **4.** $\dfrac{\sqrt{42x^5}}{\sqrt{7x^2}}$

Answers

Skill to Review

1. $\dfrac{2x^5}{3}$ 2. $\dfrac{8a^3}{3b^5}$

Margin Exercises:

1. 4 2. 5 3. $x^5\sqrt{x}$ 4. $x\sqrt{6x}$

a DIVIDING RADICAL EXPRESSIONS

Consider the expressions

$$\frac{\sqrt{25}}{\sqrt{16}} \quad \text{and} \quad \sqrt{\frac{25}{16}}.$$

Let's evaluate them separately:

a) $\dfrac{\sqrt{25}}{\sqrt{16}} = \dfrac{5}{4}$ because $\sqrt{25} = 5$ and $\sqrt{16} = 4$;

b) $\sqrt{\dfrac{25}{16}} = \dfrac{5}{4}$ because $\dfrac{5}{4} \cdot \dfrac{5}{4} = \dfrac{25}{16}$.

Both expressions represent the same number. This suggests that the quotient of two square roots is the square root of the quotient of the radicands.

THE QUOTIENT RULE FOR RADICALS

For any nonnegative number A and any positive number B,

$$\frac{\sqrt{A}}{\sqrt{B}} = \sqrt{\frac{A}{B}}.$$

(The quotient of two square roots is the square root of the quotient of the radicands.)

EXAMPLES Divide and simplify.

1. $\dfrac{\sqrt{27}}{\sqrt{3}} = \sqrt{\dfrac{27}{3}} = \sqrt{9} = 3$

2. $\dfrac{\sqrt{30a^5}}{\sqrt{6a^2}} = \sqrt{\dfrac{30a^5}{6a^2}} = \sqrt{5a^3} = \sqrt{5 \cdot a^2 \cdot a} = \sqrt{a^2} \cdot \sqrt{5a} = a\sqrt{5a}$

◀ Do Margin Exercises 1–4.

b SQUARE ROOTS OF QUOTIENTS

To find the square root of certain quotients, we can reverse the quotient rule for radicals. We can take the square root of a quotient by taking the square roots of the numerator and the denominator separately.

SQUARE ROOTS OF QUOTIENTS

For any nonnegative number A and any positive number B,

$$\sqrt{\frac{A}{B}} = \frac{\sqrt{A}}{\sqrt{B}}.$$

(We can take the square roots of the numerator and the denominator separately.)

EXAMPLES Simplify by taking the square roots of the numerator and the denominator separately.

3. $\sqrt{\frac{25}{9}} = \frac{\sqrt{25}}{\sqrt{9}} = \frac{5}{3}$ Taking the square root of the numerator and the square root of the denominator

4. $\sqrt{\frac{1}{16}} = \frac{\sqrt{1}}{\sqrt{16}} = \frac{1}{4}$ Taking the square root of the numerator and the square root of the denominator

5. $\sqrt{\frac{49}{t^2}} = \frac{\sqrt{49}}{\sqrt{t^2}} = \frac{7}{t}$

Do Exercises 5–8.

Simplify.

5. $\sqrt{\frac{16}{9}}$

6. $\sqrt{\frac{1}{25}}$

7. $\sqrt{\frac{36}{x^2}}$

8. $\sqrt{\frac{b^2}{121}}$

We are assuming that expressions for numerators are nonnegative and expressions for denominators are positive. Thus we need not be concerned about absolute-value signs or zero denominators.

Sometimes a rational expression can be simplified to one that has a perfect-square numerator and a perfect-square denominator.

EXAMPLES Simplify.

6. $\sqrt{\frac{18}{50}} = \sqrt{\frac{9 \cdot 2}{25 \cdot 2}} = \sqrt{\frac{9}{25} \cdot \frac{2}{2}} = \sqrt{\frac{9}{25} \cdot 1}$

$= \sqrt{\frac{9}{25}} = \frac{\sqrt{9}}{\sqrt{25}} = \frac{3}{5}$

7. $\sqrt{\frac{2560}{2890}} = \sqrt{\frac{256 \cdot 10}{289 \cdot 10}} = \sqrt{\frac{256}{289} \cdot \frac{10}{10}} = \sqrt{\frac{256}{289} \cdot 1}$

$= \sqrt{\frac{256}{289}} = \frac{\sqrt{256}}{\sqrt{289}} = \frac{16}{17}$

8. $\frac{\sqrt{48x^3}}{\sqrt{3x^7}} = \sqrt{\frac{48x^3}{3x^7}} = \sqrt{\frac{16}{x^4}}$ Simplifying the radicand

$= \frac{\sqrt{16}}{\sqrt{x^4}} = \frac{4}{x^2}$

Do Exercises 9–12.

Simplify.

9. $\sqrt{\frac{18}{32}}$

10. $\sqrt{\frac{2250}{2560}}$

GS 11. $\frac{\sqrt{98y}}{\sqrt{2y^{11}}}$

$= \sqrt{\frac{98y}{\square}}$

$= \sqrt{\frac{49}{\square}}$

$= \frac{\sqrt{\square}}{\sqrt{y^{10}}}$

$= \frac{7}{\square}$

12. $\sqrt{\frac{108a^{11}}{3a^{37}}}$

C RATIONALIZING DENOMINATORS

Sometimes in mathematics it is useful to find an equivalent expression without a radical in the denominator. This provides a standard notation for expressing results. The procedure for finding such an expression is called **rationalizing the denominator**. We carry this out by multiplying by 1 in either of two ways.

To rationalize a denominator:

Method 1. Multiply by 1 under the radical to make the denominator of the radicand a perfect square.

Method 2. Multiply by 1 outside the radical to make the radicand in the denominator a perfect square.

Answers

5. $\frac{4}{3}$ 6. $\frac{1}{5}$ 7. $\frac{6}{x}$ 8. $\frac{b}{11}$ 9. $\frac{3}{4}$
10. $\frac{15}{16}$ 11. $\frac{7}{y^5}$ 12. $\frac{6}{a^{13}}$

Guided Solution:
11. $2y^{11}$, y^{10}, 49, y^5

EXAMPLE 9 Rationalize the denominator: $\sqrt{\frac{2}{3}}$.

Method 1: We multiply by 1, choosing $\frac{3}{3}$ for 1. This makes the denominator of the radicand a perfect square:

$$\sqrt{\frac{2}{3}} = \sqrt{\frac{2}{3} \cdot \frac{3}{3}} \quad \text{Multiplying by 1}$$

$$= \sqrt{\frac{6}{9}} = \frac{\sqrt{6}}{\sqrt{9}} \quad \text{The radicand in the denominator, 9, is a perfect square.}$$

$$= \frac{\sqrt{6}}{3}.$$

Method 2: We can also rationalize by first taking the square roots of the numerator and the denominator. Then we multiply by 1, using $\sqrt{3}/\sqrt{3}$:

$$\sqrt{\frac{2}{3}} = \frac{\sqrt{2}}{\sqrt{3}}$$

$$= \frac{\sqrt{2}}{\sqrt{3}} \cdot \frac{\sqrt{3}}{\sqrt{3}} \quad \text{Multiplying by 1}$$

$$= \frac{\sqrt{2} \cdot \sqrt{3}}{\sqrt{3} \cdot \sqrt{3}} = \frac{\sqrt{6}}{\sqrt{9}} \quad \text{The radicand, 9, in the denominator is a perfect square.}$$

$$= \frac{\sqrt{6}}{3}.$$

◀ **Do Exercise 13.**

We can always multiply by 1 to make a denominator a perfect square. Then we can take the square root of the denominator.

EXAMPLE 10 Rationalize the denominator: $\sqrt{\frac{5}{18}}$.

The denominator, 18, is not a perfect square. Factoring, we get $18 = 3 \cdot 3 \cdot 2$. If we had another factor of 2, however, we would have a perfect square, 36. Thus we multiply by 1, choosing $\frac{2}{2}$. This makes the denominator a perfect square.

$$\sqrt{\frac{5}{18}} = \sqrt{\frac{5}{3 \cdot 3 \cdot 2}} = \sqrt{\frac{5}{3 \cdot 3 \cdot 2} \cdot \frac{2}{2}} = \sqrt{\frac{10}{36}} = \frac{\sqrt{10}}{\sqrt{36}} = \frac{\sqrt{10}}{6}$$

EXAMPLE 11 Rationalize the denominator: $\frac{8}{\sqrt{7}}$.

This time we obtain an expression without a radical in the denominator by multiplying by 1, choosing $\sqrt{7}/\sqrt{7}$:

$$\frac{8}{\sqrt{7}} = \frac{8}{\sqrt{7}} \cdot \frac{\sqrt{7}}{\sqrt{7}} = \frac{8\sqrt{7}}{\sqrt{49}} = \frac{8\sqrt{7}}{7}.$$

◀ **Do Exercises 14 and 15.**

13. Rationalize the denominator:

$$\sqrt{\frac{3}{5}}.$$

a) Use method 1.

b) Use method 2.

GS

Rationalize the denominator.

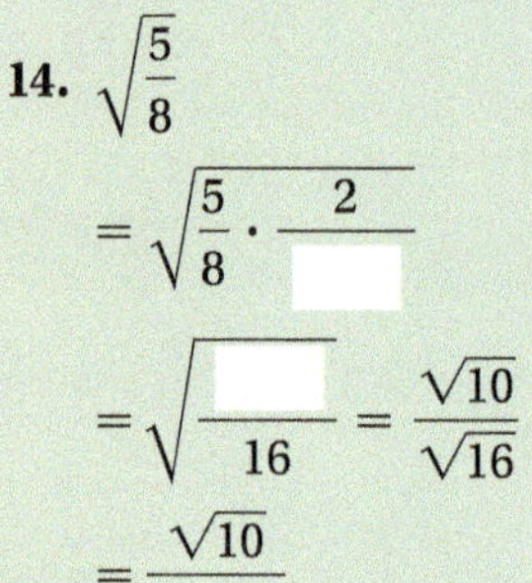

14. $\sqrt{\frac{5}{8}}$

$$= \sqrt{\frac{5}{8} \cdot \frac{2}{\square}}$$

$$= \sqrt{\frac{\square}{16}} = \frac{\sqrt{10}}{\sqrt{16}}$$

$$= \frac{\sqrt{10}}{\square}$$

15. $\frac{10}{\sqrt{3}}$

$$= \frac{10}{\sqrt{3}} \cdot \frac{\square}{\sqrt{3}}$$

$$= \frac{\square\sqrt{3}}{\sqrt{9}}$$

$$= \frac{10\sqrt{3}}{\square}$$

Answers

13. **(a)** $\frac{\sqrt{15}}{5}$; **(b)** $\frac{\sqrt{15}}{5}$ 14. $\frac{\sqrt{10}}{4}$ 15. $\frac{10\sqrt{3}}{3}$

Guided Solutions:
14. 2, 10, 4 **15.** $\sqrt{3}$, 10, 3

EXAMPLE 12 Rationalize the denominator: $\dfrac{\sqrt{3}}{\sqrt{2}}$.

We look at the denominator. It is $\sqrt{2}$. We multiply by 1, choosing $\sqrt{2}/\sqrt{2}$:

$$\frac{\sqrt{3}}{\sqrt{2}} = \frac{\sqrt{3}}{\sqrt{2}} \cdot \frac{\sqrt{2}}{\sqrt{2}} = \frac{\sqrt{3} \cdot \sqrt{2}}{\sqrt{2} \cdot \sqrt{2}} = \frac{\sqrt{6}}{\sqrt{4}} = \frac{\sqrt{6}}{2}, \text{ or } \frac{1}{2}\sqrt{6}.$$

EXAMPLES Rationalize the denominator.

13. $\dfrac{\sqrt{5}}{\sqrt{x}} = \dfrac{\sqrt{5}}{\sqrt{x}} \cdot \dfrac{\sqrt{x}}{\sqrt{x}}$ Multiplying by 1

$= \dfrac{\sqrt{5}\sqrt{x}}{\sqrt{x}\sqrt{x}}$

$= \dfrac{\sqrt{5x}}{x}$ $\sqrt{x} \cdot \sqrt{x} = x$ by the definition of square root

14. $\dfrac{\sqrt{49a^5}}{\sqrt{12}} = \dfrac{\sqrt{49a^5}}{\sqrt{12}} \cdot \dfrac{\sqrt{3}}{\sqrt{3}}$ Factoring 12, we get $2 \cdot 2 \cdot 3$, so we need another factor of 3 in order for the radicand in the denominator to be a perfect square. We multiply by $\sqrt{3}/\sqrt{3}$.

$= \dfrac{\sqrt{49a^5}\sqrt{3}}{\sqrt{12}\sqrt{3}}$

$= \dfrac{\sqrt{49 \cdot a^4 \cdot a \cdot 3}}{\sqrt{36}} = \dfrac{\sqrt{49}\sqrt{a^4}\sqrt{3a}}{\sqrt{36}}$

$= \dfrac{7a^2\sqrt{3a}}{6}$

Do Exercises 16–19.

Rationalize the denominator.

16. $\dfrac{\sqrt{3}}{\sqrt{7}}$ **17.** $\dfrac{\sqrt{5}}{\sqrt{r}}$

18. $\dfrac{\sqrt{64y^2}}{\sqrt{7}}$ **19.** $\dfrac{\sqrt{64x^9}}{\sqrt{15}}$

Answers

16. $\dfrac{\sqrt{21}}{7}$ 17. $\dfrac{\sqrt{5r}}{r}$ 18. $\dfrac{8y\sqrt{7}}{7}$ 19. $\dfrac{8x^4\sqrt{15x}}{15}$

16.3 Exercise Set

For Extra Help

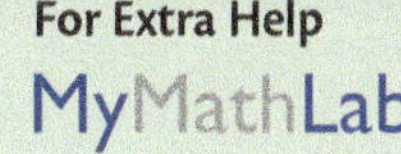

MathXL®

PRACTICE

WATCH

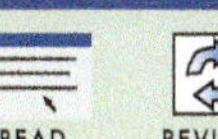

READ REVIEW

Reading Check

Choose from the columns on the right the symbol for 1 that you would use to rationalize the denominator. Some choices may not be used and others may be used more than once.

RC1. $\dfrac{\sqrt{x^3}}{\sqrt{10}}$

RC2. $\dfrac{\sqrt{5}}{\sqrt{3}}$

RC3. $\dfrac{\sqrt{6}}{\sqrt{x^5}}$

RC4. $\dfrac{\sqrt{x}}{\sqrt{5}}$

RC5. $\dfrac{\sqrt{3}}{\sqrt{2}}$

RC6. $\dfrac{\sqrt{3}}{\sqrt{x}}$

a) $\dfrac{\sqrt{2}}{\sqrt{2}}$ b) $\dfrac{\sqrt{x^2}}{\sqrt{x^2}}$

c) $\dfrac{\sqrt{3}}{\sqrt{3}}$ d) $\dfrac{\sqrt{5}}{\sqrt{5}}$

e) $\dfrac{\sqrt{10}}{\sqrt{10}}$ f) $\dfrac{\sqrt{x}}{\sqrt{x}}$

g) $\dfrac{\sqrt{3x}}{\sqrt{3x}}$ h) $\dfrac{\sqrt{6}}{\sqrt{6}}$

a Divide and simplify.

1. $\dfrac{\sqrt{18}}{\sqrt{2}}$

2. $\dfrac{\sqrt{20}}{\sqrt{5}}$

3. $\dfrac{\sqrt{108}}{\sqrt{3}}$

4. $\dfrac{\sqrt{60}}{\sqrt{15}}$

5. $\dfrac{\sqrt{65}}{\sqrt{13}}$

6. $\dfrac{\sqrt{45}}{\sqrt{15}}$

7. $\dfrac{\sqrt{3}}{\sqrt{75}}$

8. $\dfrac{\sqrt{3}}{\sqrt{48}}$

9. $\dfrac{\sqrt{12}}{\sqrt{75}}$

10. $\dfrac{\sqrt{18}}{\sqrt{32}}$

11. $\dfrac{\sqrt{8x}}{\sqrt{2x}}$

12. $\dfrac{\sqrt{18b}}{\sqrt{2b}}$

13. $\dfrac{\sqrt{63y^3}}{\sqrt{7y}}$

14. $\dfrac{\sqrt{48x^3}}{\sqrt{3x}}$

b Simplify.

15. $\sqrt{\dfrac{16}{49}}$

16. $\sqrt{\dfrac{9}{49}}$

17. $\sqrt{\dfrac{1}{36}}$

18. $\sqrt{\dfrac{1}{4}}$

19. $-\sqrt{\dfrac{16}{81}}$

20. $-\sqrt{\dfrac{25}{49}}$

21. $\sqrt{\dfrac{64}{289}}$

22. $\sqrt{\dfrac{81}{361}}$

23. $\sqrt{\dfrac{1690}{1960}}$

24. $\sqrt{\dfrac{1210}{6250}}$

25. $\sqrt{\dfrac{25}{x^2}}$

26. $\sqrt{\dfrac{36}{a^2}}$

27. $\sqrt{\dfrac{9a^2}{625}}$

28. $\sqrt{\dfrac{x^2y^2}{256}}$

29. $\dfrac{\sqrt{50y^{15}}}{\sqrt{2y^{25}}}$

30. $\dfrac{\sqrt{3t^{15}}}{\sqrt{12t}}$

31. $\dfrac{\sqrt{7x^{23}}}{\sqrt{343x^5}}$

32. $\dfrac{\sqrt{125q^3}}{\sqrt{5q^{19}}}$

C Rationalize the denominator.

33. $\sqrt{\frac{2}{5}}$

34. $\sqrt{\frac{2}{7}}$

35. $\sqrt{\frac{7}{8}}$

36. $\sqrt{\frac{3}{8}}$

37. $\sqrt{\frac{1}{12}}$

38. $\sqrt{\frac{7}{12}}$

39. $\sqrt{\frac{5}{18}}$

40. $\sqrt{\frac{1}{18}}$

41. $\frac{3}{\sqrt{5}}$

42. $\frac{4}{\sqrt{3}}$

43. $\sqrt{\frac{8}{3}}$

44. $\sqrt{\frac{12}{5}}$

45. $\sqrt{\frac{3}{x}}$

46. $\sqrt{\frac{2}{x}}$

47. $\sqrt{\frac{x}{y}}$

48. $\sqrt{\frac{a}{b}}$

49. $\sqrt{\frac{x^2}{20}}$

50. $\sqrt{\frac{x^2}{18}}$

51. $\frac{1}{\sqrt{3}}$

52. $\frac{1}{\sqrt{2}}$

53. $\frac{\sqrt{9}}{\sqrt{8}}$

54. $\frac{\sqrt{4}}{\sqrt{27}}$

55. $\frac{\sqrt{11}}{\sqrt{5}}$

56. $\frac{\sqrt{2}}{\sqrt{5}}$

57. $\frac{2}{\sqrt{2}}$

58. $\frac{3}{\sqrt{3}}$

59. $\frac{\sqrt{5}}{\sqrt{11}}$

60. $\frac{\sqrt{7}}{\sqrt{27}}$

61. $\frac{\sqrt{7}}{\sqrt{12}}$

62. $\frac{\sqrt{5}}{\sqrt{18}}$

63. $\frac{\sqrt{48}}{\sqrt{32}}$

64. $\frac{\sqrt{56}}{\sqrt{40}}$

65. $\frac{\sqrt{450}}{\sqrt{18}}$

66. $\frac{\sqrt{224}}{\sqrt{14}}$

67. $\frac{\sqrt{3}}{\sqrt{x}}$

68. $\frac{\sqrt{2}}{\sqrt{y}}$

69. $\frac{4y}{\sqrt{5}}$

70. $\frac{8x}{\sqrt{3}}$

71. $\frac{\sqrt{a^3}}{\sqrt{8}}$

72. $\frac{\sqrt{x^3}}{\sqrt{27}}$

73. $\dfrac{\sqrt{56}}{\sqrt{12x}}$

74. $\dfrac{\sqrt{45}}{\sqrt{8a}}$

75. $\dfrac{\sqrt{27c}}{\sqrt{32c^3}}$

76. $\dfrac{\sqrt{7x^3}}{\sqrt{12x}}$

77. $\dfrac{\sqrt{y^5}}{\sqrt{xy^2}}$

78. $\dfrac{\sqrt{x^3}}{\sqrt{xy}}$

79. $\dfrac{\sqrt{45mn^2}}{\sqrt{32m}}$

80. $\dfrac{\sqrt{16a^4b^6}}{\sqrt{128a^6b^6}}$

Skill Maintenance

Solve. [15.3a, b]

81. $x + y = -7,$
$x - y = 2$

82. $2x - 3y = 7,$
$-4x + 6y = -14$

83. $2x - 3y = 7,$
$2x - 3y = 9$

Divide and simplify. [14.2b]

84. $\dfrac{x - 2}{x + 3} \div \dfrac{x^2 - 4x + 4}{x^2 - 9}$

85. $\dfrac{a^2 - 25}{6} \div \dfrac{a + 5}{3}$

86. $\dfrac{x - 2}{x - 3} \div \dfrac{x - 4}{x - 5}$

Synthesis

Periods of Pendulums. The period T of a pendulum is the time it takes the pendulum to move from one side to the other and back. A formula for the period is

$$T = 2\pi\sqrt{\frac{L}{32}},$$

where T is in seconds and L is the length of the pendulum, in feet. Use 3.14 for π.

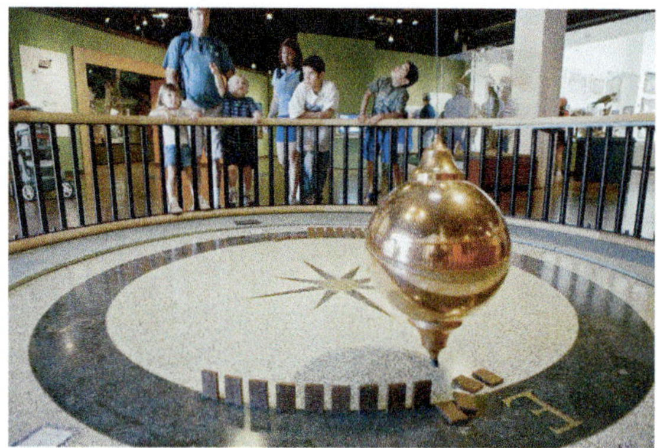

A Foucault pendulum, shown here at the California Academy of Sciences, demonstrates the rotation of the earth. The pendulum knocks down pins at different positions as the earth rotates.

87. Find the periods of pendulums of lengths 2 ft, 8 ft, and 10 in.

88. The pendulum of a grandfather clock is $(32/\pi^2)$ ft long. How long does it take to swing from one side to the other?

89. Rationalize the denominator: $\sqrt{\dfrac{3x^2y}{a^2x^5}}$.

90. Simplify: $\sqrt{\dfrac{1}{x^2} - \dfrac{2}{xy} + \dfrac{1}{y^2}}$.

Mid-Chapter Review

Concept Reinforcement

Determine whether each statement is true or false.

____________ **1.** The radical symbol $\sqrt{\ }$ represents only the principal square root. [16.1a]

____________ **2.** For any nonnegative real number A, the principal square root of A^2 is $-A$. [16.1f]

____________ **3.** Every nonnegative number has two square roots. [16.1a]

____________ **4.** There are no real numbers that when squared yield negative numbers. [16.1e]

Guided Solutions

Fill in each blank with the number or the expression that creates a correct solution.

5. Simplify by factoring:
$\sqrt{3x^2 - 48x + 192}$. [16.2a]

$$\begin{aligned}\sqrt{3x^2 - 48x + 192} &= \sqrt{\square\,(x^2 - 16x + 64)}\\ &= \sqrt{3(\square)^2}\\ &= \sqrt{\square}\,\sqrt{(x-8)^2}\\ &= \sqrt{3}(x - \square)\end{aligned}$$

6. Multiply and simplify by factoring:
$\sqrt{30}\,\sqrt{40y}$. [16.2c]

$$\begin{aligned}\sqrt{30}\,\sqrt{40y} &= \sqrt{30 \cdot \square\, y}\\ &= \sqrt{\square\, y}\\ &= \sqrt{100 \cdot \square \cdot y}\\ &= \sqrt{100 \cdot \square \cdot 3 \cdot y}\\ &= \sqrt{100}\,\sqrt{4}\,\sqrt{\square}\\ &= 10 \cdot \square\,\sqrt{3y}\\ &= \square\,\sqrt{3y}\end{aligned}$$

7. Multiply and simplify by factoring:
$\sqrt{18ab^2}\,\sqrt{14a^2b^4}$. [16.2c]

$$\begin{aligned}\sqrt{18ab^2}\,\sqrt{14a^2b^4} &= \sqrt{18ab^2 \cdot 14\,\square\, b^4}\\ &= \sqrt{2 \cdot 3 \cdot 3 \cdot 2 \cdot 7 \cdot \square \cdot b^6}\\ &= \sqrt{2^2 \cdot 3^2 \cdot 7 \cdot a^2 \cdot \square \cdot b^6}\\ &= \sqrt{2^2}\,\sqrt{3^2}\,\sqrt{a^2}\,\sqrt{b^6}\,\sqrt{\square}\\ &= 2 \cdot 3 \cdot a \cdot \square\,\sqrt{7a}\\ &= 6\,\square\, b^3\sqrt{7a}\end{aligned}$$

8. Rationalize the denominator: $\sqrt{\dfrac{3y^2}{44}}$. [16.3c]

$$\begin{aligned}\sqrt{\frac{3y^2}{44}} &= \sqrt{\frac{3y^2}{2 \cdot \square \cdot 11}}\\ &= \sqrt{\frac{3y^2}{2 \cdot 2 \cdot 11} \cdot \frac{\square}{\square}}\\ &= \sqrt{\frac{33y^2}{\square^2 \cdot 11^2}}\\ &= \frac{\square\,\sqrt{33}}{2 \cdot 11} = \frac{y\sqrt{33}}{\square}\end{aligned}$$

Mixed Review

9. Find the square roots of 121. [16.1a]

10. Identify the radicand: $2x\sqrt{\dfrac{x-3}{7}}$. [16.1d]

11. Determine whether each expression represents a real number. Write "yes" or "no." [16.1e]

a) $\sqrt{-100}$ **b)** $-\sqrt{9}$

Simplify.

12. $\sqrt{128r^7s^6}$ [16.2b]

13. $\sqrt{25(x-3)^2}$ [16.2b]

14. $\sqrt{\frac{1}{100}}$ [16.3b]

15. $-\sqrt{36}$ [16.1a]

16. $-\sqrt{\frac{6250}{490}}$ [16.3b]

17. $\sqrt{225}$ [16.1a]

18. $\sqrt{(10y)^2}$ [16.1f]

19. $\sqrt{4x^2-4x+1}$ [16.2a]

20. $\sqrt{800x}$ [16.2a]

21. $\frac{\sqrt{6}}{\sqrt{96}}$ [16.3a]

22. $\sqrt{32q^{11}}$ [16.2b]

23. $\sqrt{\frac{81}{z^2}}$ [16.3b]

Multiply or divide and, if possible, simplify.

24. $\sqrt{25}\sqrt{25}$ [16.2c]

25. $\frac{\sqrt{18}}{\sqrt{98}}$ [16.3a]

26. $\frac{\sqrt{192x}}{\sqrt{3x}}$ [16.3a]

27. $\sqrt{40c^2d^7}\sqrt{15c^3d^3}$ [16.2c]

28. $\sqrt{24x^5y^8z^2}\sqrt{60xy^3z}$ [16.2c]

29. $\sqrt{2x}\sqrt{30y}$ [16.2c]

30. $\sqrt{21a}\sqrt{35a}$ [16.2c]

31. $\frac{\sqrt{3y^{29}}}{\sqrt{75y^5}}$ [16.3a]

32. Rationalize the denominator and simplify. Match each expression in the first column with an equivalent expression in the second column by drawing connecting lines. [16.3c]

$\frac{x}{\sqrt{3}}$	$\frac{3\sqrt{x}}{x}$
$\sqrt{\frac{3}{x}}$	$\frac{\sqrt{3x}}{3}$
$\frac{3}{\sqrt{x}}$	$\frac{x\sqrt{3}}{3}$
$\frac{3x}{\sqrt{3}}$	$\sqrt{3}$
$\frac{3}{\sqrt{3}}$	$\frac{\sqrt{3x}}{x}$
$\sqrt{\frac{x}{3}}$	$x\sqrt{3}$

Understanding Through Discussion and Writing

33. What is the difference between "**the** square root of 100" and "**a** square root of 100"? [16.1a]

34. Explain why the following is incorrect:

$$\sqrt{\frac{9+100}{25}} = \frac{3+10}{5}.$$ [16.3b]

35. Explain the error(s) in the following:

$$\sqrt{x^2-25} = \sqrt{x^2} - \sqrt{25} = x - 5.$$ [16.2a]

36. Describe a method that could be used to rationalize the *numerator* of a radical expression. [16.3c]

STUDYING FOR SUCCESS *Throughout the Semester*

- ☐ Review regularly. A good way to do this is by doing the Skill Maintenance exercises found in each exercise set.
- ☐ Try creating your own glossary. Understanding terminology is essential for success in any math course.
- ☐ Memorizing is a helpful tool in the study of mathematics. Ask your instructor what you are expected to have memorized for tests.

Addition, Subtraction, and More Multiplication

16.4

OBJECTIVES

a Add or subtract with radical notation, using the distributive laws to simplify.

b Multiply expressions involving radicals, where some of the expressions contain more than one term.

c Rationalize denominators having two terms.

SKILL TO REVIEW

Objective 12.6d: Find special products when polynomial products are mixed together.

Multiply.

1. $(3x - 7)(3x + 7)$
2. $\left(4x - \frac{1}{2}\right)^2$

a ADDITION AND SUBTRACTION

We can add any two real numbers. The sum of 5 and $\sqrt{2}$ can be expressed as $5 + \sqrt{2}$. We cannot simplify this unless we use rational approximations such as $5 + \sqrt{2} \approx 5 + 1.414 = 6.414$. However, when we have *like radicals*, a sum can be simplified using the distributive laws and collecting like terms. **Like radicals** have the same radicands.

EXAMPLE 1 Add: $3\sqrt{5} + 4\sqrt{5}$.

Suppose we were considering $3x + 4x$. Recall that to add, we use a distributive law as follows:

$$3x + 4x = (3 + 4)x = 7x.$$

The situation is similar in this example, but we let $x = \sqrt{5}$:

$$3\sqrt{5} + 4\sqrt{5} = (3 + 4)\sqrt{5} \quad \text{Using a distributive law to factor out } \sqrt{5}$$
$$= 7\sqrt{5}.$$

If we wish to add or subtract as we did in Example 1, the radicands must be the same. Sometimes after simplifying the radical terms, we discover that we have like radicals.

EXAMPLES Add or subtract. Simplify, if possible, by collecting like radical terms.

2. $5\sqrt{2} - \sqrt{18} = 5\sqrt{2} - \sqrt{9 \cdot 2}$ Factoring 18

$= 5\sqrt{2} - \sqrt{9}\sqrt{2}$

$= 5\sqrt{2} - 3\sqrt{2}$

$= (5 - 3)\sqrt{2}$ Using a distributive law to factor out the common factor, $\sqrt{2}$

$= 2\sqrt{2}$

3. $\sqrt{4x^3} + 7\sqrt{x} = \sqrt{4 \cdot x^2 \cdot x} + 7\sqrt{x}$

$= 2x\sqrt{x} + 7\sqrt{x}$

$= (2x + 7)\sqrt{x}$ Using a distributive law to factor out $\sqrt{x}$

Don't forget the parentheses!

Answers

Skill to Review:

1. $9x^2 - 49$ **2.** $16x^2 - 4x + \frac{1}{4}$

Add or subtract and, if possible, simplify by collecting like radical terms.

1. $3\sqrt{2} + 9\sqrt{2}$

2. $8\sqrt{5} - 3\sqrt{5}$

3. $2\sqrt{10} - 7\sqrt{40}$

GS 4. $\sqrt{24} + \sqrt{54}$

$= \sqrt{\quad \cdot 6} + \sqrt{\quad \cdot 6}$

$= \quad \sqrt{6} + 3\sqrt{6}$

$= (2 + \quad)\sqrt{6}$

$= 5\sqrt{\quad}$

5. $\sqrt{9x + 9} - \sqrt{4x + 4}$

Add or subtract.

6. $\sqrt{2} + \sqrt{\frac{1}{2}}$

7. $\sqrt{\frac{5}{3}} + \sqrt{\frac{3}{5}}$

4. $\sqrt{x^3 - x^2} + \sqrt{4x - 4} = \sqrt{x^2(x - 1)} + \sqrt{4(x - 1)}$ Factoring radicands

$= \sqrt{x^2}\sqrt{x - 1} + \sqrt{4}\sqrt{x - 1}$

$= x\sqrt{x - 1} + 2\sqrt{x - 1}$

$= (x + 2)\sqrt{x - 1}$ Using a distributive law to factor out the common factor, $\sqrt{x - 1}$. Don't forget the parentheses!

◀ Do Exercises 1–5.

Sometimes rationalizing denominators enables us to combine like radicals.

EXAMPLE 5 Add: $\sqrt{3} + \sqrt{\frac{1}{3}}$.

$\sqrt{3} + \sqrt{\frac{1}{3}} = \sqrt{3} + \sqrt{\frac{1}{3} \cdot \frac{3}{3}}$ Multiplying by 1 in order to rationalize the denominator

$= \sqrt{3} + \sqrt{\frac{3}{9}} = \sqrt{3} + \frac{\sqrt{3}}{\sqrt{9}}$

$= \sqrt{3} + \frac{\sqrt{3}}{3} = 1 \cdot \sqrt{3} + \frac{1}{3}\sqrt{3}$

$= \left(1 + \frac{1}{3}\right)\sqrt{3}$ Factoring out the common factor, $\sqrt{3}$

$= \frac{4}{3}\sqrt{3}$, or $\frac{4\sqrt{3}}{3}$

◀ Do Exercises 6 and 7.

b MULTIPLICATION

Now let's multiply where some of the expressions may contain more than one term. To do this, we use procedures already studied in this chapter as well as the distributive laws and special products for multiplying with polynomials.

EXAMPLE 6 Multiply: $\sqrt{2}(\sqrt{3} + \sqrt{7})$.

$\sqrt{2}(\sqrt{3} + \sqrt{7}) = \sqrt{2}\sqrt{3} + \sqrt{2}\sqrt{7}$ Multiplying using a distributive law

$= \sqrt{6} + \sqrt{14}$ Using the rule for multiplying with radicals

EXAMPLE 7 Multiply: $(2 + \sqrt{3})(5 - 4\sqrt{3})$.

$(2 + \sqrt{3})(5 - 4\sqrt{3}) = 2 \cdot 5 - 2 \cdot 4\sqrt{3} + \sqrt{3} \cdot 5 - \sqrt{3} \cdot 4\sqrt{3}$ Using FOIL

$= 10 - 8\sqrt{3} + 5\sqrt{3} - 4 \cdot 3$

$= 10 - 8\sqrt{3} + 5\sqrt{3} - 12$

$= -2 - 3\sqrt{3}$

Answers

1. $12\sqrt{2}$ 2. $5\sqrt{5}$ 3. $-12\sqrt{10}$ 4. $5\sqrt{6}$ 5. $\sqrt{x + 1}$ 6. $\frac{3}{2}\sqrt{2}$, or $\frac{3\sqrt{2}}{2}$ 7. $\frac{8}{15}\sqrt{15}$, or $\frac{8\sqrt{15}}{15}$

Guided Solution:
4. 4, 9, 2, 3, 6

EXAMPLE 8 Multiply: $(\sqrt{3} - \sqrt{x})(\sqrt{3} + \sqrt{x})$.

$$(\sqrt{3} - \sqrt{x})(\sqrt{3} + \sqrt{x}) = (\sqrt{3})^2 - (\sqrt{x})^2 \quad \text{Using } (A - B)(A + B) = A^2 - B^2$$
$$= 3 - x$$

EXAMPLE 9 Multiply: $(3 - \sqrt{p})^2$.

$$(3 - \sqrt{p})^2 = 3^2 - 2 \cdot 3 \cdot \sqrt{p} + (\sqrt{p})^2 \quad \text{Using } (A - B)^2 = A^2 - 2AB + B^2$$
$$= 9 - 6\sqrt{p} + p$$

EXAMPLE 10 Multiply: $(2 + \sqrt{5})^2$.

$$(2 + \sqrt{5})^2 = 2^2 + 2 \cdot 2\sqrt{5} + (\sqrt{5})^2 \quad \text{Using } (A + B)^2 = A^2 + 2AB + B^2$$
$$= 4 + 4\sqrt{5} + 5$$
$$= 9 + 4\sqrt{5}$$

Do Exercises 8–12.

Multiply.

8. $\sqrt{3}(\sqrt{5} + \sqrt{2})$
9. $(1 - \sqrt{2})(4 + 3\sqrt{5})$
10. $(\sqrt{2} + \sqrt{a})(\sqrt{2} - \sqrt{a})$
11. $(5 + \sqrt{x})^2$
12. $(3 - \sqrt{7})(3 + \sqrt{7})$

C MORE ON RATIONALIZING DENOMINATORS

Note in Example 8 that the result has no radicals. This will happen whenever we multiply expressions such as $\sqrt{a} - \sqrt{b}$ and $\sqrt{a} + \sqrt{b}$. We see this in the following:

$$(\sqrt{a} + \sqrt{b})(\sqrt{a} - \sqrt{b}) = (\sqrt{a})^2 - (\sqrt{b})^2 = a - b.$$

Expressions such as $\sqrt{3} - \sqrt{x}$ and $\sqrt{3} + \sqrt{x}$ are known as **conjugates**; so too are $2 + \sqrt{5}$ and $2 - \sqrt{5}$. We can use conjugates to rationalize a denominator that involves a sum or a difference of two terms, where one or both are radicals. To do so, we multiply by 1 using the conjugate to form the expression for 1.

Do Exercises 13–15.

Find the conjugate of each expression.

13. $7 + \sqrt{5}$
14. $\sqrt{5} - \sqrt{2}$
15. $1 - \sqrt{x}$

EXAMPLE 11 Rationalize the denominator: $\dfrac{3}{2 + \sqrt{5}}$.

We multiply by 1 using the conjugate of $2 + \sqrt{5}$, which is $2 - \sqrt{5}$, as the numerator and the denominator of the expression for 1:

$$\frac{3}{2 + \sqrt{5}} = \frac{3}{2 + \sqrt{5}} \cdot \frac{2 - \sqrt{5}}{2 - \sqrt{5}} \quad \text{Multiplying by 1}$$
$$= \frac{3(2 - \sqrt{5})}{(2 + \sqrt{5})(2 - \sqrt{5})} \quad \text{Multiplying}$$
$$= \frac{6 - 3\sqrt{5}}{2^2 - (\sqrt{5})^2} \quad \text{Using } (A + B)(A - B) = A^2 - B^2$$
$$= \frac{6 - 3\sqrt{5}}{4 - 5}$$
$$= \frac{6 - 3\sqrt{5}}{-1}$$
$$= -6 + 3\sqrt{5}, \text{ or } 3\sqrt{5} - 6.$$

Answers

8. $\sqrt{15} + \sqrt{6}$ 9. $4 + 3\sqrt{5} - 4\sqrt{2} - 3\sqrt{10}$
10. $2 - a$ 11. $25 + 10\sqrt{x} + x$ 12. 2
13. $7 - \sqrt{5}$ 14. $\sqrt{5} + \sqrt{2}$ 15. $1 + \sqrt{x}$

EXAMPLE 12 Rationalize the denominator: $\dfrac{\sqrt{3}+\sqrt{5}}{\sqrt{3}-\sqrt{5}}$.

We multiply by 1 using the conjugate of $\sqrt{3}-\sqrt{5}$, which is $\sqrt{3}+\sqrt{5}$, as the numerator and the denominator of the expression for 1:

$$\frac{\sqrt{3}+\sqrt{5}}{\sqrt{3}-\sqrt{5}} = \frac{\sqrt{3}+\sqrt{5}}{\sqrt{3}-\sqrt{5}} \cdot \frac{\sqrt{3}+\sqrt{5}}{\sqrt{3}+\sqrt{5}} \quad \text{Multiplying by 1}$$

$$= \frac{(\sqrt{3}+\sqrt{5})^2}{(\sqrt{3}-\sqrt{5})(\sqrt{3}+\sqrt{5})}$$

$$= \frac{(\sqrt{3})^2+2\sqrt{3}\sqrt{5}+(\sqrt{5})^2}{(\sqrt{3})^2-(\sqrt{5})^2} \quad \text{Using } (A+B)^2 = A^2+2AB+B^2 \text{ and } (A+B)(A-B) = A^2-B^2$$

$$= \frac{3+2\sqrt{15}+5}{3-5}$$

$$= \frac{8+2\sqrt{15}}{-2}$$

$$= \frac{2(4+\sqrt{15})}{2(-1)} \quad \text{Factoring in order to simplify}$$

$$= \frac{2}{2} \cdot \frac{4+\sqrt{15}}{-1}$$

$$= \frac{4+\sqrt{15}}{-1}$$

$$= -4-\sqrt{15}.$$

Do Exercises 16 and 17.

Rationalize the denominator.

16. $\dfrac{3}{7+\sqrt{5}}$

$= \dfrac{3}{7+\sqrt{5}} \cdot \dfrac{7-\sqrt{5}}{\square - \sqrt{5}}$

$= \dfrac{\square(7-\sqrt{5})}{7^2-(\square)^2}$

$= \dfrac{\square - 3\sqrt{5}}{49-\square}$

$= \dfrac{21-3\sqrt{5}}{\square}$

GS

17. $\dfrac{\sqrt{5}+\sqrt{7}}{\sqrt{5}-\sqrt{7}}$

EXAMPLE 13 Rationalize the denominator: $\dfrac{5}{2+\sqrt{x}}$.

We multiply by 1 using the conjugate of $2+\sqrt{x}$, which is $2-\sqrt{x}$, as the numerator and the denominator of the expression for 1:

$$\frac{5}{2+\sqrt{x}} = \frac{5}{2+\sqrt{x}} \cdot \frac{2-\sqrt{x}}{2-\sqrt{x}} \quad \text{Multiplying by 1}$$

$$= \frac{5(2-\sqrt{x})}{(2+\sqrt{x})(2-\sqrt{x})}$$

$$= \frac{5 \cdot 2 - 5 \cdot \sqrt{x}}{2^2-(\sqrt{x})^2} \quad \text{Using } (A+B)(A-B) = A^2-B^2$$

$$= \frac{10-5\sqrt{x}}{4-x}.$$

Do Exercise 18.

18. Rationalize the denominator:

$\dfrac{7}{1-\sqrt{x}}$.

Answers

16. $\dfrac{21-3\sqrt{5}}{44}$ 17. $-6-\sqrt{35}$

18. $\dfrac{7+7\sqrt{x}}{1-x}$

Guided Solution:

16. 7, 3, $\sqrt{5}$, 21, 5, 44

16.4 Exercise Set

For Extra Help
MyMathLab® MathXL® PRACTICE WATCH READ REVIEW

Reading Check

Determine whether the two given expressions are conjugates. Answer "yes" or "no."

RC1. $5 - \sqrt{3},\ 5 + \sqrt{3}$

RC2. $\sqrt{3} \cdot \sqrt{5},\ \sqrt{5} \cdot \sqrt{3}$

RC3. $\sqrt{3} - \sqrt{5},\ \sqrt{5} - \sqrt{3}$

RC4. $\sqrt{3} - \sqrt{5},\ \sqrt{3} + \sqrt{5}$

RC5. $-\sqrt{5} + 3,\ \sqrt{5} + 3$

RC6. $\sqrt{5} - 3,\ 3 - \sqrt{5}$

RC7. $\frac{\sqrt{3}}{\sqrt{5}},\ \frac{\sqrt{5}}{\sqrt{3}}$

RC8. $3 - \sqrt{5},\ 5 - \sqrt{3}$

RC9. $5\sqrt{3} + 3\sqrt{5},\ 3\sqrt{5} - 5\sqrt{3}$

a Add or subtract. Simplify by collecting like radical terms, if possible.

1. $7\sqrt{3} + 9\sqrt{3}$

2. $6\sqrt{2} + 8\sqrt{2}$

3. $7\sqrt{5} - 3\sqrt{5}$

4. $8\sqrt{2} - 5\sqrt{2}$

5. $6\sqrt{x} + 7\sqrt{x}$

6. $9\sqrt{y} + 3\sqrt{y}$

7. $4\sqrt{d} - 13\sqrt{d}$

8. $2\sqrt{a} - 17\sqrt{a}$

9. $5\sqrt{8} + 15\sqrt{2}$

10. $3\sqrt{12} + 2\sqrt{3}$

11. $\sqrt{27} - 2\sqrt{3}$

12. $7\sqrt{50} - 3\sqrt{2}$

13. $\sqrt{45} - \sqrt{20}$

14. $\sqrt{27} - \sqrt{12}$

15. $\sqrt{72} + \sqrt{98}$

16. $\sqrt{45} + \sqrt{80}$

17. $2\sqrt{12} + \sqrt{27} - \sqrt{48}$

18. $9\sqrt{8} - \sqrt{72} + \sqrt{98}$

19. $\sqrt{18} - 3\sqrt{8} + \sqrt{50}$

20. $3\sqrt{18} - 2\sqrt{32} - 5\sqrt{50}$

21. $2\sqrt{27} - 3\sqrt{48} + 3\sqrt{12}$

22. $3\sqrt{48} - 2\sqrt{27} - 3\sqrt{12}$

23. $\sqrt{4x} + \sqrt{81x^3}$

24. $\sqrt{12x^2} + \sqrt{27}$

25. $\sqrt{27} - \sqrt{12x^2}$

26. $\sqrt{81x^3} - \sqrt{4x}$

27. $\sqrt{8x + 8} + \sqrt{2x + 2}$

28. $\sqrt{12x + 12} + \sqrt{3x + 3}$

29. $\sqrt{x^5 - x^2} + \sqrt{9x^3 - 9}$

30. $\sqrt{16x - 16} + \sqrt{25x^3 - 25x^2}$

31. $4a\sqrt{a^2b} + a\sqrt{a^2b^3} - 5\sqrt{b^3}$

32. $3x\sqrt{y^3x} - x\sqrt{yx^3} + y\sqrt{y^3x}$

33. $\sqrt{3} - \sqrt{\frac{1}{3}}$

34. $\sqrt{2} - \sqrt{\frac{1}{2}}$

35. $5\sqrt{2} + 3\sqrt{\frac{1}{2}}$

36. $4\sqrt{3} + 2\sqrt{\frac{1}{3}}$

37. $\sqrt{\frac{2}{3}} - \sqrt{\frac{1}{6}}$

38. $\sqrt{\frac{1}{2}} - \sqrt{\frac{1}{8}}$

b Multiply.

39. $\sqrt{3}(\sqrt{5}-1)$

40. $\sqrt{2}(\sqrt{2}+\sqrt{3})$

41. $(2+\sqrt{3})(5-\sqrt{7})$

42. $(\sqrt{5}+\sqrt{7})(2\sqrt{5}-3\sqrt{7})$

43. $(2-\sqrt{5})^2$

44. $(\sqrt{3}+\sqrt{10})^2$

45. $(\sqrt{2}+8)(\sqrt{2}-8)$

46. $(1+\sqrt{7})(1-\sqrt{7})$

47. $(\sqrt{6}-\sqrt{5})(\sqrt{6}+\sqrt{5})$

48. $(\sqrt{3}+\sqrt{10})(\sqrt{3}-\sqrt{10})$

49. $(3\sqrt{5}-2)(\sqrt{5}+1)$

50. $(\sqrt{5}-2\sqrt{2})(\sqrt{10}-1)$

51. $(\sqrt{x}-\sqrt{y})^2$

52. $(\sqrt{w}+11)^2$

c Rationalize the denominator.

53. $\dfrac{2}{\sqrt{3}-\sqrt{5}}$

54. $\dfrac{5}{3+\sqrt{7}}$

55. $\dfrac{\sqrt{3}-\sqrt{2}}{\sqrt{3}+\sqrt{2}}$

56. $\dfrac{2-\sqrt{7}}{\sqrt{3}-\sqrt{2}}$

57. $\dfrac{4}{\sqrt{10}+1}$

58. $\dfrac{6}{\sqrt{11}-3}$

59. $\dfrac{1-\sqrt{7}}{3+\sqrt{7}}$

60. $\dfrac{2+\sqrt{8}}{1-\sqrt{5}}$

61. $\dfrac{3}{4+\sqrt{x}}$

62. $\dfrac{8}{2-\sqrt{x}}$

63. $\dfrac{3+\sqrt{2}}{8-\sqrt{x}}$

64. $\dfrac{4-\sqrt{3}}{6+\sqrt{y}}$

65. $\dfrac{\sqrt{a}-1}{1+\sqrt{a}}$

66. $\dfrac{12+\sqrt{w}}{\sqrt{w}-12}$

67. $\dfrac{4+\sqrt{3}}{\sqrt{a}-\sqrt{t}}$

68. $\dfrac{\sqrt{2}-1}{\sqrt{w}+\sqrt{b}}$

Skill Maintenance

Solve.

69. $3x + 5 + 2(x - 3) = 4 - 6x$ [10.3c]

70. $3(x - 4) - 2 = 8(2x + 3)$ [10.3c]

71. $x^2 - 5x = 6$ [13.7b]

72. $x^2 + 10 = 7x$ [13.7b]

73. Multiply and simplify:

$$\frac{3}{x^2 - 9} \cdot \frac{x^2 - 6x + 9}{12}.$$ [14.1d]

74. The graph of the polynomial equation

$$y = x^3 - 5x^2 + x - 2$$

is shown below. Use either the graph or the equation to estimate or find the value of the polynomial when $x = -1$, $x = 0$, $x = 1$, $x = 3$, and $x = 4.85$. [12.3a]

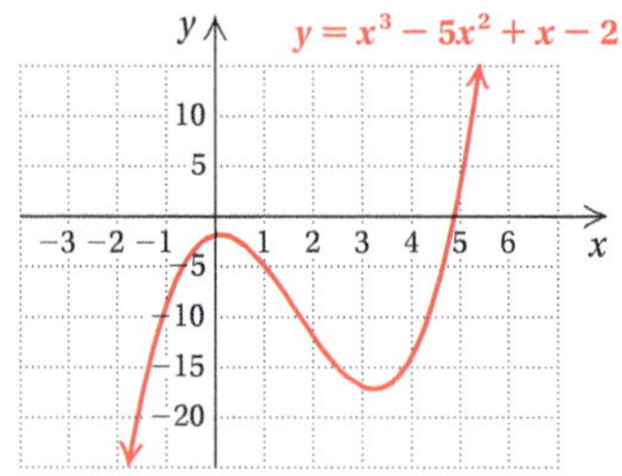

75. ***Continental Divide.*** The Continental Divide in the Americas divides the flow of water between the Pacific Ocean and the Atlantic Ocean. The Continental Divide National Scenic Trail in the United States runs through five states: Montana, Idaho, Wyoming, Colorado, and New Mexico. The Trail's highest altitude is 9990 ft higher than its lowest altitude of 4280 ft. What is the highest altitude of the Trail? [10.6a]

Source: www.continental-divide.net

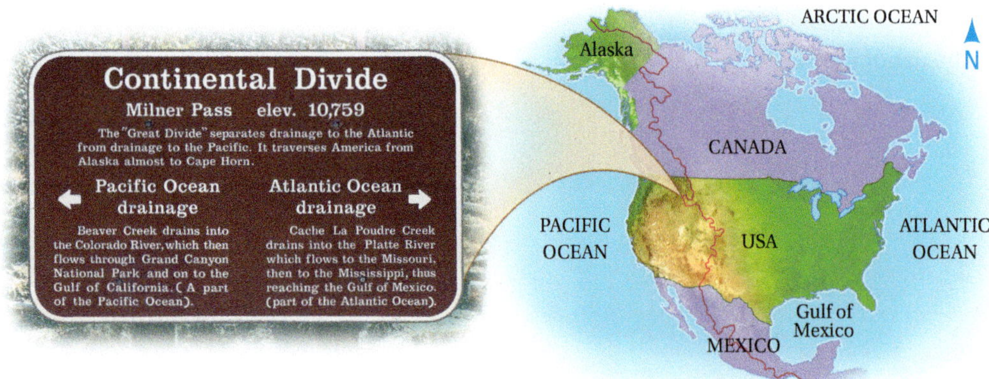

Synthesis

76. Evaluate $\sqrt{a^2 + b^2}$ and $\sqrt{a^2} + \sqrt{b^2}$ when $a = 2$ and $b = 3$. Then determine whether $\sqrt{a^2 + b^2}$ and $\sqrt{a^2} + \sqrt{b^2}$ are equivalent.

Use the TABLE feature to determine whether each of the following is correct.

77. $\sqrt{9x^3} + \sqrt{x} = \sqrt{9x^3 + x}$

78. $\sqrt{x^2 + 4} = x + 2$

Add or subtract as indicated.

79. $\frac{3}{5}\sqrt{24} + \frac{2}{5}\sqrt{150} - \sqrt{96}$

80. $\frac{1}{3}\sqrt{27} + \sqrt{8} + \sqrt{300} - \sqrt{18} - \sqrt{162}$

Radical Equations

16.5

a SOLVING RADICAL EQUATIONS

The following are examples of *radical equations*:

$$\sqrt{2x} - 4 = 7, \qquad \sqrt{x+1} = \sqrt{2x-5}.$$

A **radical equation** has variables in one or more radicands. To solve radical equations, we first convert them to equations without radicals. We do this for square-root radical equations by squaring both sides of the equation, using the following principle.

THE PRINCIPLE OF SQUARING

If an equation $a = b$ is true, then the equation $a^2 = b^2$ is true.

To solve square-root radical equations, we first try to get a radical by itself. That is, we try to isolate the radical. Then we use the principle of squaring. This allows us to eliminate one radical.

EXAMPLE 1 Solve: $\sqrt{2x} - 4 = 7$.

$$\sqrt{2x} - 4 = 7$$

$\sqrt{2x} = 11$ — Adding 4 to isolate the radical

$(\sqrt{2x})^2 = 11^2$ — Squaring both sides

$2x = 121$ — $\sqrt{2x} \cdot \sqrt{2x} = 2x$, by the definition of square root

$x = \frac{121}{2}$ — Dividing by 2

Check:

$\sqrt{2x} - 4 = 7$	
$\sqrt{2 \cdot \frac{121}{2}} - 4$	? 7
$\sqrt{121} - 4$	
$11 - 4$	
7	TRUE

The solution is $\frac{121}{2}$.

Do Margin Exercise 1.

EXAMPLE 2 Solve: $2\sqrt{x+2} = \sqrt{x+10}$.

Each radical is isolated. We proceed with the principle of squaring.

$(2\sqrt{x+2})^2 = (\sqrt{x+10})^2$ — Squaring both sides

$2^2(\sqrt{x+2})^2 = (\sqrt{x+10})^2$ — Raising each factor of the product on the left to the second power

$4(x+2) = x + 10$ — Simplifying

$4x + 8 = x + 10$ — Removing parentheses

$3x = 2$ — Subtracting x and 8

$x = \frac{2}{3}$ — Dividing by 3

OBJECTIVES

a Solve radical equations with one or two radical terms isolated, using the principle of squaring once.

b Solve radical equations with two radical terms, using the principle of squaring twice.

c Solve applied problems using radical equations.

SKILL TO REVIEW

Objective 13.7b: Solve quadratic equations by factoring and then using the principle of zero products.

Solve.

1. $x^2 + 4x - 45 = 0$
2. $1 + x^2 = -2x$

1. Solve: $\sqrt{3x} - 5 = 3$.

Answers

Skill to Review:
1. $-9, 5$ 2. -1

Margin Exercise:
1. $\frac{64}{3}$

Solve.

2. $\sqrt{3x+1} = \sqrt{2x+3}$

$(\sqrt{3x+1})^2 = (\quad)^2$

$\quad + 1 = 2x + \quad$

$\quad = 2$

GS

3. $3\sqrt{x+1} = \sqrt{x+12}$

ALGEBRAIC ▶◀ GRAPHICAL CONNECTION

Consider the equation of Example 3:

$$x - 5 = \sqrt{x+7}.$$

We can visualize the solutions by graphing the equations

$$y = x - 5 \quad \text{and} \quad y = \sqrt{x+7}$$

using the same set of axes.

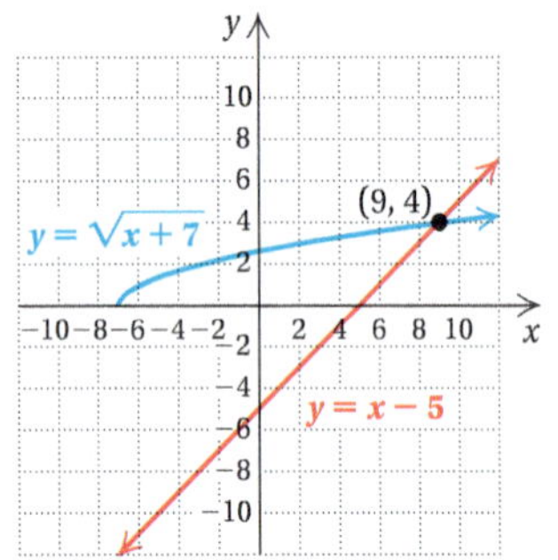

It appears that when $x = 9$, the values of $y = x - 5$ and $y = \sqrt{x+7}$ are the same, 4. We can check this as we did in Example 3. Note also that the graphs *do not* intersect at $x = 2$.

4. Solve: $x - 1 = \sqrt{x+5}$.

Answers

2. 2 3. $\frac{3}{8}$ 4. 4

Guided Solution:

2. $\sqrt{2x+3}$, $3x$, 3, x

Check:

$2\sqrt{x+2}$	$=$	$\sqrt{x+10}$	
$2\sqrt{\frac{2}{3}+2}$	?	$\sqrt{\frac{2}{3}+10}$	
$2\sqrt{\frac{8}{3}}$		$\sqrt{\frac{32}{3}}$	
$2\sqrt{\frac{4\cdot 2}{3}}$		$\sqrt{\frac{16\cdot 2}{3}}$	
$4\sqrt{\frac{2}{3}}$		$4\sqrt{\frac{2}{3}}$	TRUE

The number $\frac{2}{3}$ checks. The solution is $\frac{2}{3}$.

◀ **Do Exercises 2 and 3.**

It is necessary to check when using the principle of squaring. This principle may not produce equivalent equations. When we square both sides of an equation, the new equation may have solutions that the first one does not. For example, the equation

$$x = 1 \qquad (1)$$

has just one solution, the number 1. When we square both sides, we get

$$x^2 = 1, \qquad (2)$$

which has two solutions, 1 and -1. The equations $x = 1$ and $x^2 = 1$ do not have the same solutions and thus are not equivalent. Whereas it is true that any solution of equation (1) is a solution of equation (2), it is *not* true that any solution of equation (2) is a solution of equation (1).

Caution!

When the principle of squaring is used to solve an equation, all possible solutions *must* be checked in the original equation!

Sometimes we may need to apply the principle of zero products after squaring. (See Section 13.7.)

EXAMPLE 3 Solve: $x - 5 = \sqrt{x+7}$.

$$x - 5 = \sqrt{x+7}$$

$$(x-5)^2 = (\sqrt{x+7})^2 \qquad \text{Using the principle of squaring}$$

$$x^2 - 10x + 25 = x + 7$$

$$x^2 - 11x + 18 = 0 \qquad \text{Subtracting } x \text{ and } 7$$

$$(x-9)(x-2) = 0 \qquad \text{Factoring}$$

$$x - 9 = 0 \quad or \quad x - 2 = 0 \qquad \text{Using the principle of zero products}$$

$$x = 9 \quad or \quad x = 2$$

Check: For 9:

$x - 5$	$=$	$\sqrt{x+7}$	
$9 - 5$	?	$\sqrt{9+7}$	
4		4	TRUE

For 2:

$x - 5$	$=$	$\sqrt{x+7}$	
$2 - 5$	?	$\sqrt{2+7}$	
-3		3	FALSE

The number 9 checks, but 2 does not. Thus the solution is 9.

◀ **Do Exercise 4.**

EXAMPLE 4 Solve: $3 + \sqrt{27 - 3x} = x$.

In this case, we must first isolate the radical.

$$3 + \sqrt{27 - 3x} = x$$

$$\sqrt{27 - 3x} = x - 3 \quad \text{Subtracting 3 to isolate the radical}$$

$$(\sqrt{27 - 3x})^2 = (x - 3)^2 \quad \text{Using the principle of squaring}$$

$$27 - 3x = x^2 - 6x + 9 \quad \text{Squaring on each side}$$

$$0 = x^2 - 3x - 18 \quad \text{Adding } 3x \text{ and subtracting 27 to obtain 0 on the left}$$

$$0 = (x - 6)(x + 3) \quad \text{Factoring}$$

$$x - 6 = 0 \quad or \quad x + 3 = 0 \quad \text{Using the principle of zero products}$$

$$x = 6 \quad or \quad x = -3$$

Check: For 6:

$$\begin{array}{c|c} 3 + \sqrt{27 - 3x} & x \\ \hline 3 + \sqrt{27 - 3 \cdot 6} \; ? & 6 \\ 3 + \sqrt{27 - 18} & \\ 3 + \sqrt{9} & \\ 3 + 3 & \\ 6 & \text{TRUE} \end{array}$$

For −3:

$$\begin{array}{c|c} 3 + \sqrt{27 - 3x} & x \\ \hline 3 + \sqrt{27 - 3 \cdot (-3)} \; ? & -3 \\ 3 + \sqrt{27 + 9} & \\ 3 + \sqrt{36} & \\ 3 + 6 & \\ 9 & \text{FALSE} \end{array}$$

The number 6 checks, but −3 does not. The solution is 6.

Do Exercise 5. ▶

5. Solve: $1 + \sqrt{1 - x} = x$.

b USING THE PRINCIPLE OF SQUARING MORE THAN ONCE

Sometimes when we have two radical terms, we may need to apply the principle of squaring a second time.

EXAMPLE 5 Solve: $\sqrt{x} - 1 = \sqrt{x - 5}$.

We have

$$\sqrt{x} - 1 = \sqrt{x - 5}$$

$$(\sqrt{x} - 1)^2 = (\sqrt{x - 5})^2 \quad \text{Using the principle of squaring}$$

$$(\sqrt{x})^2 - 2 \cdot \sqrt{x} \cdot 1 + 1^2 = x - 5 \quad \text{Using } (A - B)^2 = A^2 - 2AB + B^2 \text{ on the left side}$$

$$x - 2\sqrt{x} + 1 = x - 5 \quad \text{Simplifying. Only one radical term remains.}$$

$$-2\sqrt{x} = -6 \quad \text{Isolating the radical by subtracting } x \text{ and 1}$$

$$\sqrt{x} = 3 \quad \text{Dividing by } -2$$

$$(\sqrt{x})^2 = 3^2 \quad \text{Using the principle of squaring}$$

$$x = 9.$$

The check is left to the student. The number 9 checks and is the solution. ■

CALCULATOR CORNER

Solving Radical Equations We can solve radical equations on a graphing calculator. Consider the equation in Example 3: $x - 5 = \sqrt{x + 7}$. We first graph each side of the equation. We enter $y_1 = x - 5$ and $y_2 = \sqrt{x + 7}$ on the equation-editor screen and graph the equations, using the window $[-2, 12, -6, 6]$. Note that there is one point of intersection. Use the INTERSECT feature to find its coordinates. (See the Calculator Corner on p. 1110 for the procedure.)

The first coordinate, 9, is the value of x for which $y_1 = y_2$, or $x - 5 = \sqrt{x + 7}$. It is the solution of the equation. Note that the graph shows a single solution whereas the algebraic solution in Example 3 yields two possible solutions, 9 and 2, that must be checked. The check shows that 9 is the only solution.

$y_1 = x - 5, y_2 = \sqrt{x + 7}$

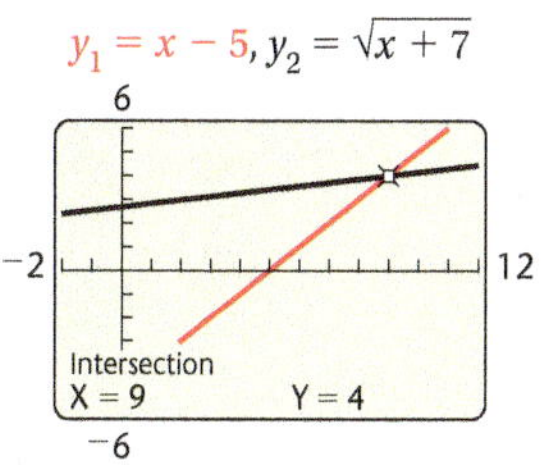

EXERCISES:

1. Solve the equations in Examples 4 and 5 graphically.
2. Solve the equations in Margin Exercises 1–6 graphically.

Answer

5. 1

The following is a procedure for solving square-root radical equations.

SOLVING SQUARE-ROOT RADICAL EQUATIONS

To solve square-root radical equations:

1. Isolate one of the radical terms.
2. Use the principle of squaring.
3. If a radical term remains, perform steps (1) and (2) again.
4. Solve the equation and check possible solutions.

6. Solve: $\sqrt{x} - 1 = \sqrt{x - 3}$.

$$\sqrt{x} - 1 = \sqrt{x - 3}$$
$$(\quad)^2 = (\sqrt{x - 3})^2$$
$$(\quad)^2 - \quad\sqrt{x} + 1 = x - 3$$
$$\quad - 2\sqrt{x} + 1 = x - 3$$
$$-2\sqrt{x} + 1 = -3$$
$$-2\sqrt{x} = \quad$$
$$\sqrt{x} = 2$$
$$(\sqrt{x})^2 = \quad^2$$
$$\quad = 4$$

Do Exercise 6.

C APPLICATIONS

Sighting to the Horizon. How far can you see from a given height? The equation

$$D = \sqrt{2h}$$

can be used to approximate the distance D, in miles, that a person can see to the horizon from a height h, in feet.

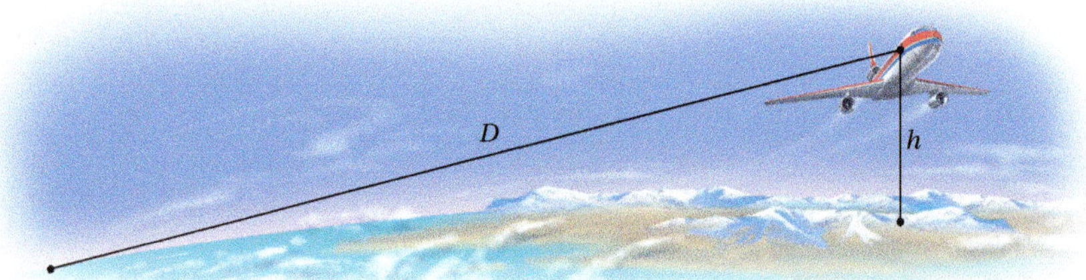

EXAMPLE 6 How far to the horizon can you see through an airplane window at a height, or altitude, of 30,000 ft?

We substitute 30,000 for h in $D = \sqrt{2h}$ and find an approximation using a calculator:

$$D = \sqrt{2 \cdot 30{,}000} \approx 245 \text{ mi}.$$

You can see for about 245 mi to the horizon.

Do Exercises 7 and 8.

7. How far to the horizon can you see through an airplane window at a height, or altitude, of 38,000 ft?

8. A sailor climbs 40 ft up the mast of a ship to the crow's nest. How far can he see to the horizon?

Answers

6. 4 **7.** About 276 mi **8.** About 9 mi

Guided Solution:

6. $\sqrt{x} - 1, \sqrt{x}, 2, x, -4, 2, x$

EXAMPLE 7 *Height of a Ranger Station.* How high is a ranger station if the ranger is able to see out to a fire on the horizon 15.4 mi away?

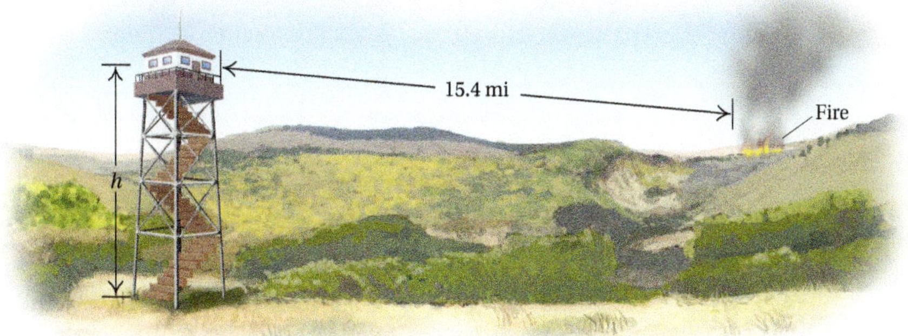

We substitute 15.4 for D in $D = \sqrt{2h}$ and solve:

$$15.4 = \sqrt{2h}$$
$$(15.4)^2 = (\sqrt{2h})^2 \quad \text{Using the principle of squaring}$$
$$237.16 = 2h$$
$$\frac{237.16}{2} = h$$
$$118.58 = h.$$

The height of the ranger tower must be about 119 ft in order for the ranger to see out to a fire 15.4 mi away.

Do Exercise 9. ▶

9. How far above sea level must a sailor climb on the mast of a ship in order to see 10.2 mi out to an iceberg?

Answer

9. About 52 ft

16.5 Exercise Set

For Extra Help

MyMathLab®

MathXL®

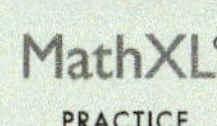
PRACTICE

WATCH

READ

REVIEW

✓ Reading Check

Determine the number of times that the principle of squaring must be used in order to solve the equation. Do not solve the equation.

RC1. $x + 4 = 4\sqrt{x + 1}$

RC2. $\sqrt{x} - 1 = \sqrt{x - 31}$

RC3. $\sqrt{4x - 5} = \sqrt{x + 9}$

RC4. $\sqrt{x + 7} = x - 5$

RC5. $\sqrt{x + 9} = 1 + \sqrt{x}$

RC6. $1 + \sqrt{x + 7} = \sqrt{3x - 2}$

a Solve.

1. $\sqrt{x} = 6$

2. $\sqrt{x} = 1$

3. $\sqrt{x} = 4.3$

4. $\sqrt{x} = 6.2$

5. $\sqrt{y + 4} = 13$

6. $\sqrt{y - 5} = 21$

7. $\sqrt{2x + 4} = 25$

8. $\sqrt{2x + 1} = 13$

9. $3 + \sqrt{x - 1} = 5$

10. $4 + \sqrt{y - 3} = 11$

11. $6 - 2\sqrt{3n} = 0$

12. $8 - 4\sqrt{5n} = 0$

13. $\sqrt{5x - 7} = \sqrt{x + 10}$

14. $\sqrt{4x - 5} = \sqrt{x + 9}$

15. $\sqrt{x} = -7$

16. $\sqrt{x} = -5$

17. $\sqrt{2y + 6} = \sqrt{2y - 5}$

18. $2\sqrt{3x - 2} = \sqrt{2x - 3}$

19. $x - 7 = \sqrt{x - 5}$

20. $\sqrt{x + 7} = x - 5$

21. $x - 9 = \sqrt{x - 3}$

22. $\sqrt{x + 18} = x - 2$

23. $2\sqrt{x - 1} = x - 1$

24. $x + 4 = 4\sqrt{x + 1}$

25. $\sqrt{5x + 21} = x + 3$

26. $\sqrt{27 - 3x} = x - 3$

27. $\sqrt{2x - 1} + 2 = x$

28. $x = 1 + 6\sqrt{x - 9}$

29. $\sqrt{x^2 + 6} - x + 3 = 0$

30. $\sqrt{x^2 + 5} - x + 2 = 0$

31. $\sqrt{x^2 - 4} - x = 6$

32. $\sqrt{x^2 - 5x + 7} = x - 3$

33. $\sqrt{(p + 6)(p + 1)} - 2 = p + 1$

34. $\sqrt{(4x + 5)(x + 4)} = 2x + 5$

35. $\sqrt{4x - 10} = \sqrt{2 - x}$

36. $\sqrt{2 - x} = \sqrt{3x - 7}$

b Solve. Use the principle of squaring twice.

37. $\sqrt{x - 5} = 5 - \sqrt{x}$

38. $\sqrt{x + 9} = 1 + \sqrt{x}$

39. $\sqrt{y + 8} - \sqrt{y} = 2$

40. $\sqrt{3x + 1} = 1 - \sqrt{x + 4}$

41. $\sqrt{x - 4} + \sqrt{x + 1} = 5$

42. $1 + \sqrt{x + 7} = \sqrt{3x - 2}$

43. $\sqrt{x} - 1 = \sqrt{x - 31}$

44. $\sqrt{2x - 5} - 1 = \sqrt{x - 3}$

C Solve.

The speed v, in meters per second, of a wave on the surface of the ocean can be approximated by the formula $v = 3.1\sqrt{d}$, where d is the depth of the water, in meters. Use this formula for Exercises 45–48.

Source: myweb.dal.ca

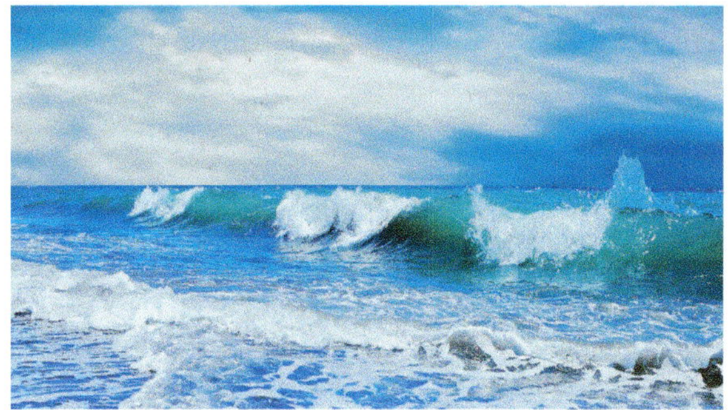

45. What is the speed, in meters per second, of a wave on the surface of the ocean where the depth of the water is 400 m?

46. What is the speed, in meters per second, of a wave on the surface of the ocean where the depth of the water is 225 m?

47. A wave is traveling at a speed of 34.1 m/sec. What is the water depth?

48. A wave is traveling at a speed of 38.75 m/sec. What is the water depth?

Speed of a Skidding Car. How do police determine how fast a car had been traveling after an accident has occurred? The formula

$$r = 2\sqrt{5L}$$

can be used to approximate the speed r, in miles per hour, of a car that has left a skid mark of length L, in feet. (See Example 7 in Section 16.1.) Use this formula for Exercises 49 and 50.

49. How far will a car skid at 65 mph? at 75 mph?

50. How far will a car skid at 55 mph? at 90 mph?

Skill Maintenance

51. Solve $R = \dfrac{s + t}{2}$ for s. [10.4b]

52. Solve: $4 - \frac{1}{4}x < \frac{1}{2}x + 10$. [10.7e]

53. Evaluate $(3x)^3$ when $x = -4$. [12.1c]

54. Divide and simplify: $\dfrac{y^8w^3}{y^3w}$. [12.1e]

Factor.

55. $2x^2 + 11x + 5$ [13.4a]

56. $y^2 - 36$ [13.5d]

57. $9t^2 + 24t + 16$ [13.5b]

58. $1 - x^8$ [13.6a]

Synthesis

Solve.

59. $\sqrt{5x^2 + 5} = 5$

60. $\sqrt{x} = -x$

61. $4 + \sqrt{19 - x} = 6 + \sqrt{4 - x}$

62. $x = (x - 2)\sqrt{x}$

63. $\sqrt{x + 3} = \dfrac{8}{\sqrt{x - 9}}$

64. $\dfrac{12}{\sqrt{5x + 6}} = \sqrt{2x + 5}$

Applications with Right Triangles

16.6

OBJECTIVES

a Given the lengths of any two sides of a right triangle, find the length of the third side.

b Solve applied problems involving right triangles.

SKILL TO REVIEW

Objective 16.1b: Approximate square roots of real numbers using a calculator.

Use a calculator to approximate each square root. Round to three decimal places.

1. $\sqrt{217}$ **2.** $\sqrt{29}$

a RIGHT TRIANGLES

A **right triangle** is a triangle with a $90°$ angle, as shown in the figure below. The small square in the corner indicates the $90°$ angle.

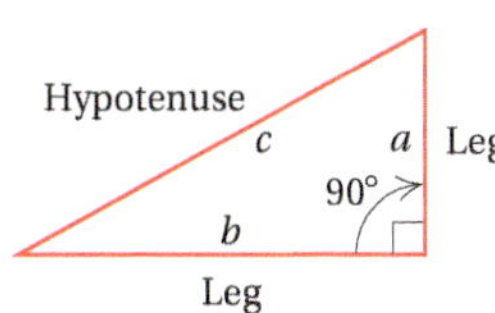

In a right triangle, the longest side is called the **hypotenuse**. It is also the side opposite the right angle. The other two sides are called **legs**. We generally use the letters a and b for the lengths of the legs and c for the length of the hypotenuse. They are related as follows.

THE PYTHAGOREAN THEOREM

In any right triangle, if a and b are the lengths of the legs and c is the length of the hypotenuse, then

$$a^2 + b^2 = c^2.$$

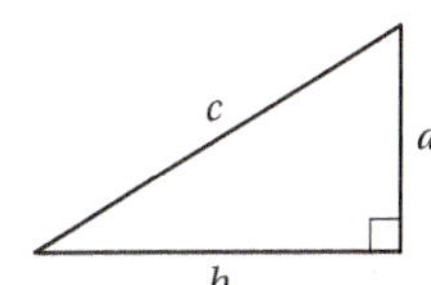

The equation $a^2 + b^2 = c^2$ is called the **Pythagorean equation**.

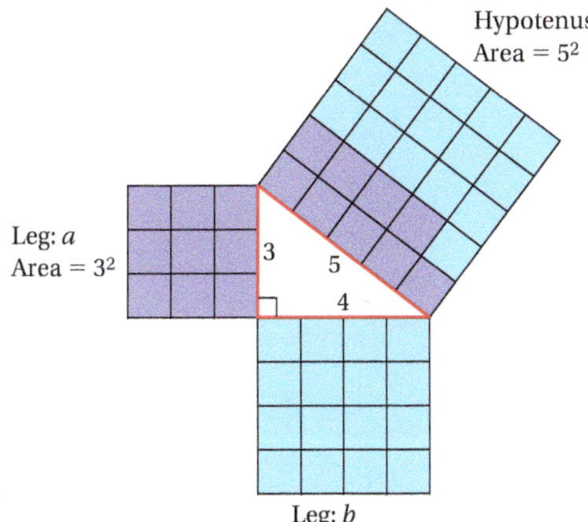

$$a^2 + b^2 = c^2$$
$$3^2 + 4^2 = 5^2$$
$$9 + 16 = 25$$

The Pythagorean theorem is named after the ancient Greek mathematician Pythagoras (569?–500? B.C.). It is uncertain who actually proved this result the first time. A proof can be found in most geometry books.

If we know the lengths of any two sides of a right triangle, we can find the length of the third side.

Answers

Skill to Review:
1. 14.731 **2.** 5.385

1. Find the length of the hypotenuse of this right triangle. Give an exact answer and an approximation to three decimal places.

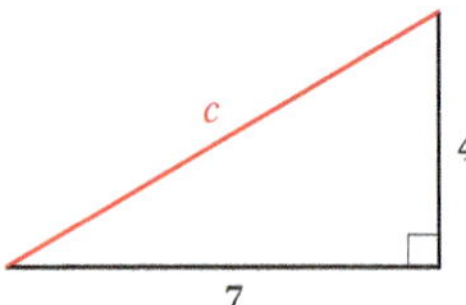

2. Find the length a in this right triangle. Give an exact answer and an approximation to three decimal places.

GS

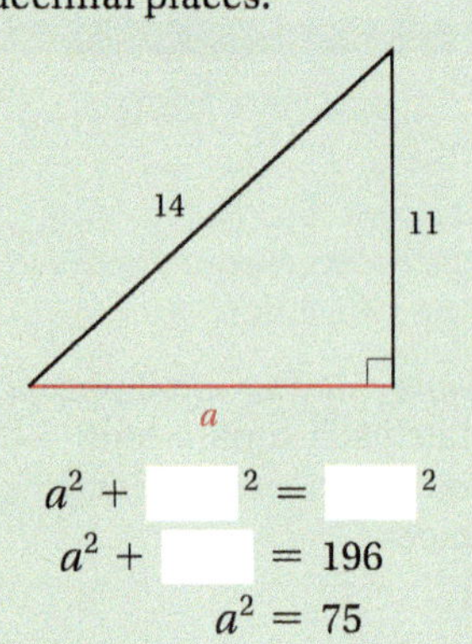

$a^2 + \square^2 = \square^2$

$a^2 + \square = 196$

$a^2 = 75$

$\square = \sqrt{75}$

$a \approx \square$

Find the missing length of a leg in the right triangle. Give an exact answer and an approximation to three decimal places.

3.

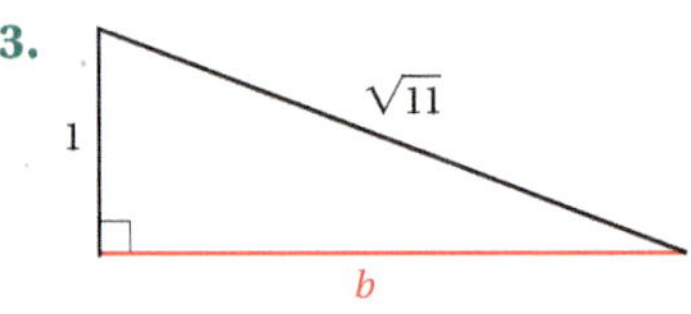

4.

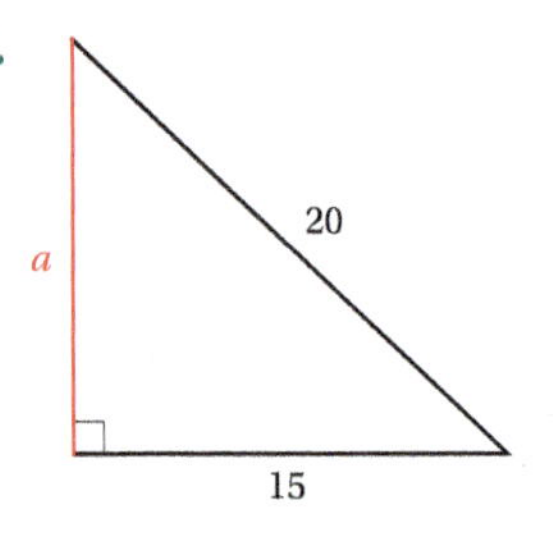

EXAMPLE 1 Find the length of the hypotenuse of this right triangle. Give an exact answer and an approximation to three decimal places.

$4^2 + 5^2 = c^2$ Substituting in the Pythagorean equation

$16 + 25 = c^2$

$41 = c^2$

$c = \sqrt{41}$ Exact answer

$c \approx 6.403$ Using a calculator

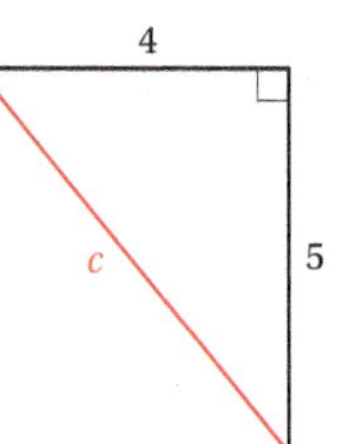

EXAMPLE 2 Find the length b in this right triangle. Give an exact answer and an approximation to three decimal places.

$10^2 + b^2 = 12^2$ Substituting in the Pythagorean equation

$100 + b^2 = 144$

$b^2 = 144 - 100$

$b^2 = 44$

$b = \sqrt{44}$ Exact answer

$b \approx 6.633$ Using a calculator

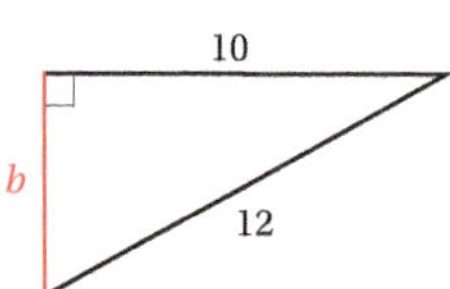

Do Exercises 1 and 2.

EXAMPLE 3 Find the length b in this right triangle. Give an exact answer and an approximation to three decimal places.

$1^2 + b^2 = (\sqrt{7})^2$ Substituting in the Pythagorean equation

$1 + b^2 = 7$

$b^2 = 7 - 1 = 6$

$b = \sqrt{6}$ Exact answer

$b \approx 2.449$ Using a calculator

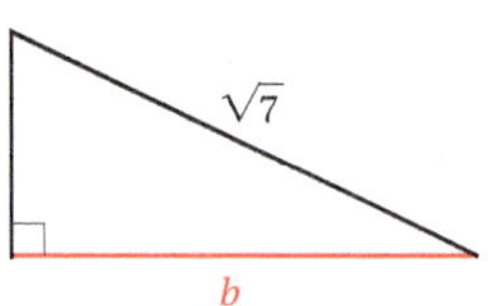

EXAMPLE 4 Find the length a in this right triangle. Give an exact answer and an approximation to three decimal places.

$a^2 + 10^2 = 15^2$

$a^2 + 100 = 225$

$a^2 = 225 - 100$

$a^2 = 125$

$a = \sqrt{125}$ Exact answer

$a \approx 11.180$ Using a calculator

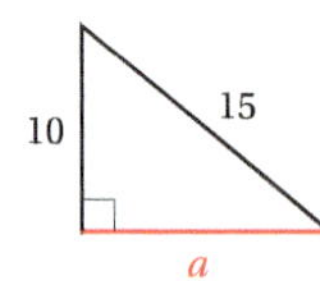

Do Exercises 3 and 4.

Answers

1. $\sqrt{65} \approx 8.062$ 2. $\sqrt{75} \approx 8.660$
3. $\sqrt{10} \approx 3.162$ 4. $\sqrt{175} \approx 13.229$

Guided Solution:
2. 11, 14, 121, a, 8.660

b APPLICATIONS

EXAMPLE 5 *Skateboard Ramp.* Ramps.com of America sells Landwave ramps and decks that can be combined to create a skateboard ramp as high or as wide as one wants. The dimensions of the basic ramp unit are 28 in. wide, 38.5 in. long, and 12 in. high.

Source: www.ramps.com

a) What is the length of the skating surface of one ramp unit?

b) How many ramp units are needed for a 10-ft long skating surface?

a)

1. **Familiarize.** We first make a drawing and label it with the given dimensions. The base and the end of the ramp unit form a right angle. We label the length of the skating surface r.

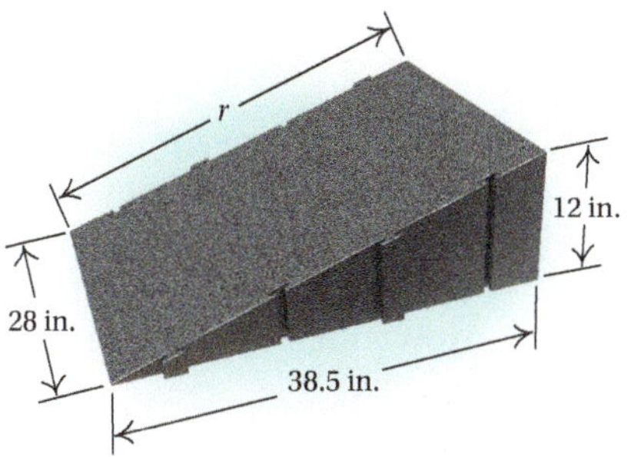

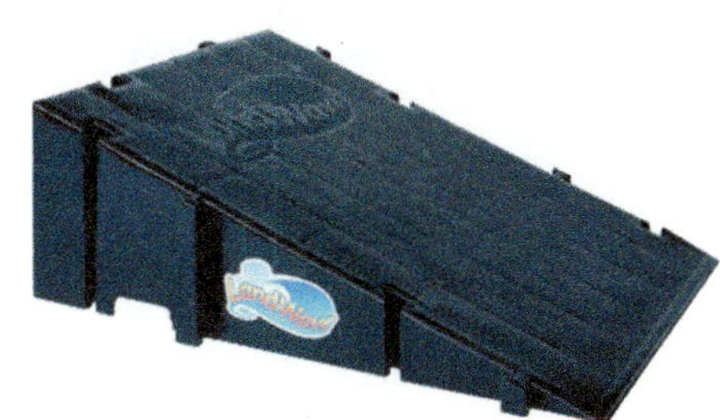

2. **Translate.** We use the Pythagorean equation:

$$a^2 + b^2 = c^2 \quad \text{Pythagorean equation}$$

$$(38.5)^2 + 12^2 = r^2. \quad \text{Substituting 38.5 for } a \text{, 12 for } b \text{, and } c \text{ for } r$$

3. **Solve.** We solve as follows:

$$\begin{aligned}(38.5)^2 + 12^2 &= r^2 \\ 1482.25 + 144 &= r^2 && \text{Squaring} \\ 1626.25 &= r^2 \\ \sqrt{1626.25} &= r && \text{Exact answer} \\ 40.327 &\approx r. && \text{Approximate answer}\end{aligned}$$

4. **Check.** We check the calculations using the Pythagorean equation: $38.5^2 + 12^2 = 1626.25$ and $(40.327)^2 \approx 1626$. The length checks. (Remember that we estimated the value of r.)

5. **State.** The length of the skating surface of a single ramp unit is about 40.327 in.

b) In inches, the length of a 10-ft skating surface is 10×12 in., or 120 in. Each ramp unit is about 40 in. long. Thus it will take $120 \div 40$, or 3, ramp units for a ramp that has a 10-ft long surface.

Do Exercise 5. ▶

5. *Christmas Tree.* Each Christmas season since 1962, the Soldiers' and Sailors' Monument in the center of Indianapolis is decorated as a giant Christmas tree. The tree is composed of 52 strands of lights that are attached at the top of the monument, which is 242 ft tall. How long is each garland of lights if each strand is attached to the ground 39 ft from the center of the monument? Give an exact answer and an approximation to three decimal places.

Source: Indianapolis Downtown Inc.

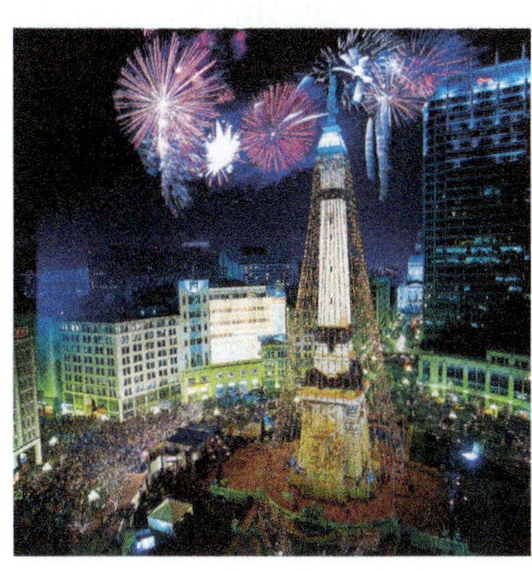

Answer

5. $\sqrt{60{,}085}$ ft $\approx$ 245.122 ft

Translating for Success

The goal of these matching questions is to practice step (2), Translate, of the five-step problem-solving process. Translate each word problem to an equation or a system of equations and select a correct translation from equations A–O.

1. *Coin Mixture.* A collection of nickels and quarters is worth \$9.35. There are 59 coins in all. How many of each coin are there?

2. *Diagonal of a Square.* Find the length of a diagonal of a square whose sides are 8 ft long.

3. *Shoveling Time.* It takes Mark 55 min to shovel 4 in. of snow from his driveway. It takes Eric 75 min to do the same job. How long would it take if they worked together?

4. *Angles of a Triangle.* The second angle of a triangle is three times as large as the first. The third is 17° less than the sum of the other angles. Find the measures of the angles.

5. *Perimeter.* The perimeter of a rectangle is 568 ft. The length is 26 ft greater than the width. Find the length and the width.

6. *Car Travel.* One horse travels 75 km in the same time that a horse traveling 5 km/h faster travels 105 km. Find the speed of each horse.

7. *Money Borrowed.* Emma borrows some money at 5% simple interest. After 1 year, \$8568 pays off her loan. How much did she originally borrow?

8. *TV Time.* The average amount of time per day that TV sets in the United States are turned on is 8 hr. What percent of the time are our TV sets on?

Source: Nielsen Media Research

9. *Ladder Height.* An 8-ft plank is leaning against a shed. The bottom of the plank is 4 ft from the building. How high is the top of the plank?

10. *Lengths of a Rectangle.* The area of a rectangle is 568 ft^2. The length is 26 ft greater than the width. Find the length and the width.

A. $5x + 25y = 9.35,$
$x + y = 59$

B. $4^2 + x^2 = 8^2$

C. $x(x + 26) = 568$

D. $8 = x \cdot 24$

E. $\frac{75}{x} = \frac{105}{x + 5}$

F. $\frac{75}{x} = \frac{55}{x + 5}$

G. $2x + 2(x + 26) = 568$

H. $x + 3x +$
$(x + 3x - 17) = 180$

I. $x + 3x + (3x - 17) = 180$

J. $0.05x + 0.25y = 9.35,$
$x + y = 59$

K. $8^2 + 8^2 = x^2$

L. $x^2 + (x + 26)^2 = 568$

M. $x - 5\% \cdot x = 8568$

N. $\frac{1}{55} + \frac{1}{75} = \frac{1}{x}$

O. $x + 5\% \cdot x = 8568$

Answers on page A-39

16.6 Exercise Set

For Extra Help MyMathLab® MathXL® PRACTICE WATCH READ REVIEW

Reading Check

Choose from the column on the right the equation whose solution is the length of the missing side of the right triangle.

RC1.

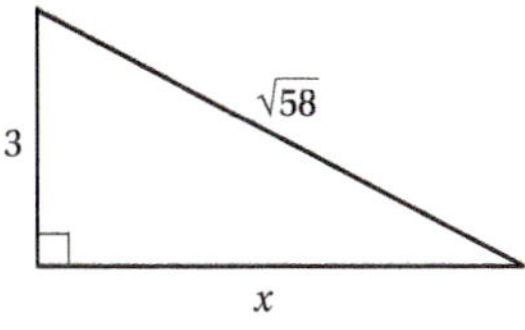

RC2.

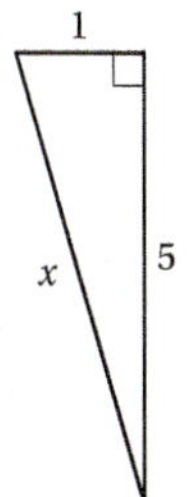

a) $3^2 = x^2 + (\sqrt{58})^2$
b) $1^2 = x^2 + 5^2$
c) $1^2 + 5^2 = x^2$
d) $x^2 + 3^2 = (\sqrt{58})^2$

a Find the length of the third side of each right triangle. Where appropriate, give both an exact answer and an approximation to three decimal places.

1.

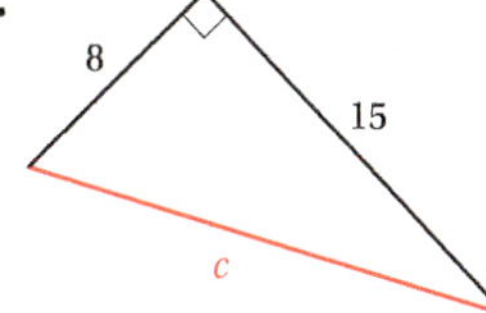

2.

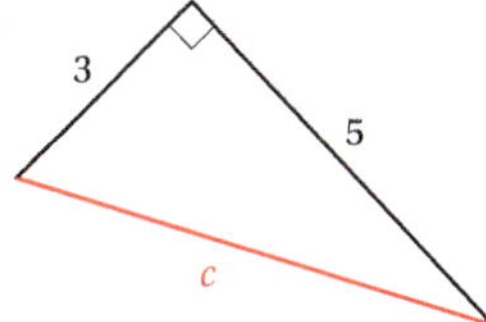

3.

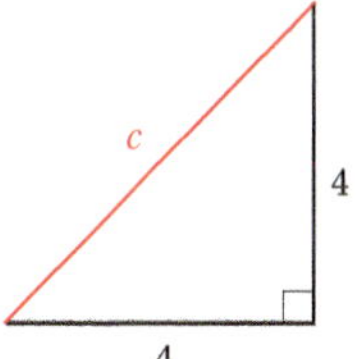

4.

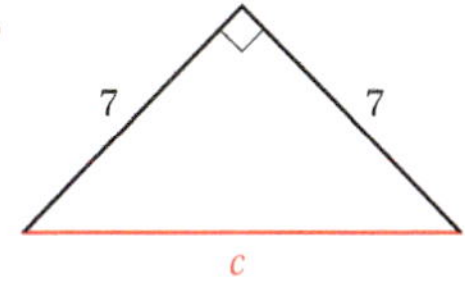

5.

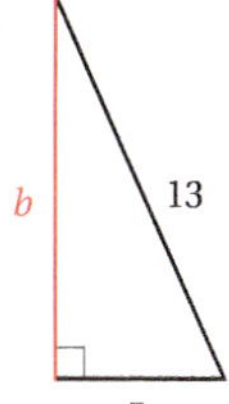

6.

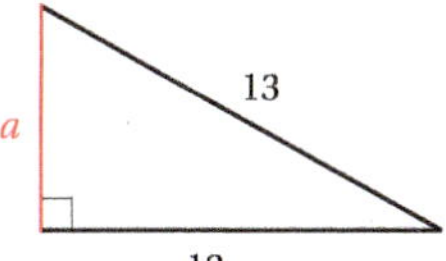

7.

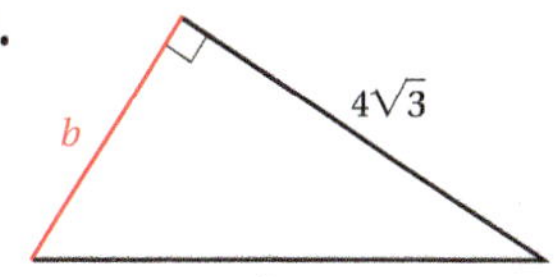

8.

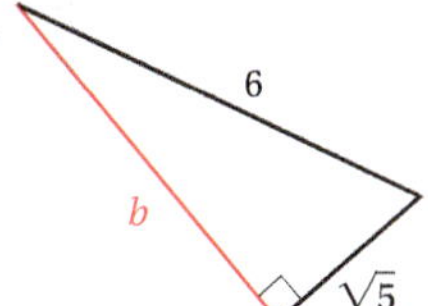

In a right triangle, find the length of the side not given. Where appropriate, give both an exact answer and an approximation to three decimal places. Standard lettering has been used.

9. $a = 10,\ b = 24$

10. $a = 5,\ b = 12$

11. $a = 9,\ c = 15$

12. $a = 18,\ c = 30$

13. $b = 1,\ c = \sqrt{5}$

14. $b = 1,\ c = \sqrt{2}$

15. $a = 1,\ c = \sqrt{3}$

16. $a = \sqrt{3},\ b = \sqrt{5}$

17. $c = 10,\ b = 5\sqrt{3}$

18. $a = 5,\ b = 5$

19. $a = \sqrt{2},\ b = \sqrt{7}$

20. $c = \sqrt{7},\ a = \sqrt{2}$

b Solve. Don't forget to use a drawing. Give an exact answer and an approximation to three decimal places.

21. *Panama Canal.* A photographer is assigned a feature story on a new lock at the Panama Canal and needs to determine the distance between points A and B on opposite sides of the canal. What is the distance between A and B?

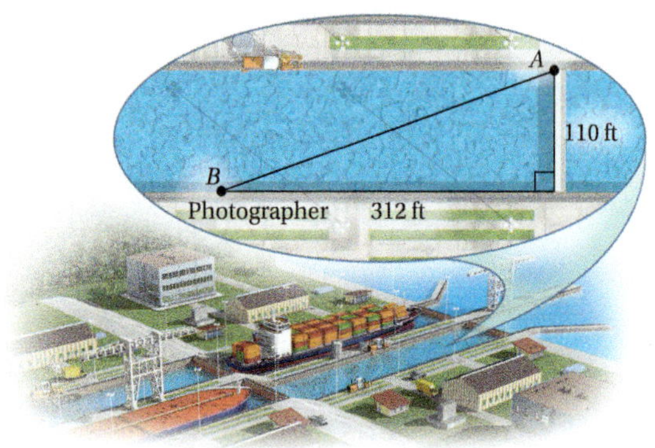

22. *Rope Course.* An outdoor rope course consists of a cable that slopes downward from a height of 37 ft to a resting place 30 ft above the ground. The trees that the cable connects are 24 ft apart. How long is the cable?

23. *Ladder Height.* A 10-m ladder is leaning against a building. The bottom of the ladder is 5 m from the building. How high is the top of the ladder?

24. *Diagonal of a Square.* Find the length of a diagonal of a square whose sides are 3 cm long.

25. *Diagonal of a Soccer Field.* The largest regulation soccer field is 100 yd wide and 130 yd long. Find the length of a diagonal of such a field.

26. *Guy Wire.* How long is a guy wire reaching from the top of a 12-ft pole to a point on the ground 8 ft from the base of the pole?

Skill Maintenance

Solve. [15.3a, b]

27. $5x + 7 = 8y,$
$3x = 8y - 4$

28. $5x + y = 17,$
$-5x + 2y = 10$

29. $3x - 4y = -11,$
$5x + 6y = 12$

30. $x + y = -9,$
$x - y = -11$

31. Find the slope of the line $4 - x = 3y$. [11.4a]

32. Find the slope of the line containing the points $(8, -3)$ and $(0, -8)$. [11.3a]

Synthesis

33. Find x.

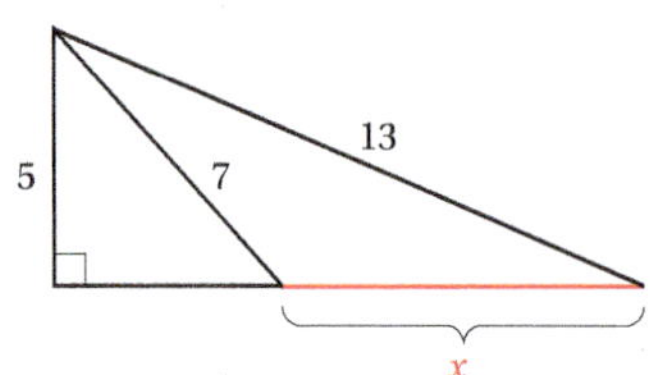

34. *Skateboard Ramp.* Ramps.com of America sells Landwave ramps and decks that can be combined to create a skateboard tower, as shown in Example 5. The dimensions of the Landwave ramp are 28 in. × 38.5 in. × 12 in. The Landwave deck that is a rectangular prism measures 28 in. × 38.5 in. × 12 in. How many ramps and how many decks are needed to build a tower that is 7 ft wide and 7 ft high? (*Hint*: For safety, the tallest column of the tower is built of only decks.)

Source: www.ramps.com

CHAPTER 16

Summary and Review

Vocabulary Reinforcement

Complete each statement with the correct term from the column on the right. Some of the choices may not be used.

principal
Pythagorean
radical
legs
square
radicands
hypotenuse
conjugates
root
rationalizing

1. The positive square root is called the ______________ square root. [16.1a]
2. The symbol $\sqrt{\ }$ is called a(n) ______________ symbol. [16.1a]
3. The number c is a(n) ______________ root of a if $c^2 = a$. [16.1a]
4. The procedure to find an equivalent expression without a radical in the denominator is called ______________ the denominator. [16.3c]
5. Like radicals have the same ______________. [16.4a]
6. Expressions such as $\sqrt{6} - \sqrt{a}$ and $\sqrt{6} + \sqrt{a}$ are known as ______________. [16.4c]
7. In any right triangle, if a and b are the lengths of the legs and c is the length of the ______________, then $a^2 + b^2 = c^2$. The equation $a^2 + b^2 = c^2$ is called the ______________ equation. [16.6a]

Concept Reinforcement

Determine whether each statement is true or false.

______________ 1. When both sides of an equation are squared, the new equation may have solutions that the first equation does not. [16.5a]

______________ 2. The square root of a sum is not the sum of the square roots. [16.2c]

______________ 3. If an equation $a = b$ is true, then the equation $a^2 = b^2$ is true. [16.5a]

______________ 4. If an equation $a^2 = b^2$ is true, then the equation $a = b$ is true. [16.5a]

Study Guide

Objective 16.1d Identify radicands of radical expressions.

Example Identify the radicand in the radical expression.

$$\sqrt{a + 2} + \frac{1}{4}.$$

The radicand is $a + 2$.

Practice Exercise

1. Identify the radicand in the radical expression $10y + \sqrt{y^2 - 3}$.

Objective 16.1e Determine whether a radical expression represents a real number.

Example Determine whether each expression represents a real number.

a) $\sqrt{-11}$ **b)** $-\sqrt{134}$

a) The radicand, -11, is negative; $\sqrt{-11}$ *is not* a real number.

b) The radicand, 134, is positive; $-\sqrt{134}$ *is* a real number.

Practice Exercise

2. Determine whether each expression represents a real number. Write "yes" or "no."

a) $-\sqrt{-(-3)}$ **b)** $\sqrt{-200}$

Objective 16.2a Simplify radical expressions.

Example Simplify by factoring: $\sqrt{162x^2}$.

$$\sqrt{162x^2} = \sqrt{81 \cdot 2 \cdot x^2} = \sqrt{81}\sqrt{x^2}\sqrt{2} = 9x\sqrt{2}$$

Practice Exercise

3. Simplify by factoring: $\sqrt{1200y^2}$.

Objective 16.2b Simplify radical expressions where radicands are powers.

Example Simplify by factoring: $\sqrt{98x^7y^8}$.

$$\sqrt{98x^7y^8} = \sqrt{49 \cdot 2 \cdot x^6 \cdot x \cdot y^8} = \sqrt{49}\sqrt{x^6}\sqrt{y^8}\sqrt{2x} = 7x^3y^4\sqrt{2x}$$

Practice Exercise

4. Simplify by factoring: $\sqrt{175a^{12}b^9}$.

Objective 16.2c Multiply radical expressions and, if possible, simplify.

Example Multiply and then, if possible, simplify: $\sqrt{6cd^3}\sqrt{30c^3d^2}$.

$$\begin{aligned}\sqrt{6cd^3}\sqrt{30c^3d^2} &= \sqrt{6cd^3 \cdot 30c^3d^2}\\ &= \sqrt{2 \cdot 3 \cdot 2 \cdot 3 \cdot 5 \cdot c^4 \cdot d^4 \cdot d}\\ &= \sqrt{4}\sqrt{9}\sqrt{c^4}\sqrt{d^4}\sqrt{5d}\\ &= 2 \cdot 3 \cdot c^2 \cdot d^2 \cdot \sqrt{5d}\\ &= 6c^2d^2\sqrt{5d}\end{aligned}$$

Practice Exercise

5. Multiply and then, if possible, simplify: $\sqrt{8x^3y}\sqrt{12x^4y^3}$.

Objective 16.3a Divide radical expressions.

Example Divide and simplify: $\dfrac{\sqrt{108y^5}}{\sqrt{3y^2}}$.

$$\frac{\sqrt{108y^5}}{\sqrt{3y^2}} = \sqrt{\frac{108y^5}{3y^2}} = \sqrt{36y^3} = \sqrt{36 \cdot y^2 \cdot y} = \sqrt{36}\sqrt{y^2}\sqrt{y} = 6y\sqrt{y}$$

Practice Exercise

6. Divide and simplify: $\dfrac{\sqrt{15b^7}}{\sqrt{5b^4}}$.

Objective 16.3b Simplify square roots of quotients.

Example Simplify: $\sqrt{\dfrac{320}{500}}$.

$$\sqrt{\frac{320}{500}} = \sqrt{\frac{16 \cdot 20}{25 \cdot 20}} = \sqrt{\frac{16}{25} \cdot \frac{20}{20}} = \sqrt{\frac{16}{25} \cdot 1} = \sqrt{\frac{16}{25}} = \frac{\sqrt{16}}{\sqrt{25}} = \frac{4}{5}$$

Practice Exercise

7. Simplify: $\sqrt{\dfrac{50}{162}}$.

Objective 16.3c Rationalize the denominator of a radical expression.

Example Rationalize the denominator: $\dfrac{7x}{\sqrt{18}}$.

$$\begin{aligned}\frac{7x}{\sqrt{18}} &= \frac{7x}{\sqrt{2 \cdot 3 \cdot 3}} \cdot \frac{\sqrt{2}}{\sqrt{2}}\\ &= \frac{7x \cdot \sqrt{2}}{\sqrt{2 \cdot 3 \cdot 3} \cdot \sqrt{2}} = \frac{7x\sqrt{2}}{\sqrt{2 \cdot 3 \cdot 3 \cdot 2}}\\ &= \frac{7x\sqrt{2}}{\sqrt{36}} = \frac{7x\sqrt{2}}{6}\end{aligned}$$

Practice Exercise

8. Rationalize the denominator: $\dfrac{2a}{\sqrt{50}}$.

Objective 16.4a Add or subtract with radical notation, using the distributive laws to simplify.

Example Add and, if possible, simplify.
$\sqrt{9x - 18} + \sqrt{16x^3 - 32x^2}$.

$$\begin{aligned}
&\sqrt{9x - 18} + \sqrt{16x^3 - 32x^2}\\
&\quad = \sqrt{9(x - 2)} + \sqrt{16x^2(x - 2)}\\
&\quad = \sqrt{9}\sqrt{x - 2} + \sqrt{16x^2}\sqrt{x - 2}\\
&\quad = 3\sqrt{x - 2} + 4x\sqrt{x - 2}\\
&\quad = (3 + 4x)\sqrt{x - 2}
\end{aligned}$$

Practice Exercise

9. Add and, if possible, simplify:
$\sqrt{x^3 - x^2} + \sqrt{36x - 36}$.

Objective 16.4b Multiply expressions involving radicals, where some of the expressions contain more than one term.

Example Multiply: $(\sqrt{3} + 4\sqrt{5})(\sqrt{3} - \sqrt{5})$.

$$\begin{aligned}
&(\sqrt{3} + 4\sqrt{5})(\sqrt{3} - \sqrt{5})\\
&\quad = \sqrt{3} \cdot \sqrt{3} - \sqrt{3} \cdot \sqrt{5} + 4\sqrt{5} \cdot \sqrt{3} - 4\sqrt{5} \cdot \sqrt{5}\\
&\quad = 3 - \sqrt{15} + 4\sqrt{15} - 4 \cdot 5\\
&\quad = 3 - \sqrt{15} + 4\sqrt{15} - 20 = 3\sqrt{15} - 17
\end{aligned}$$

Practice Exercise

10. Multiply: $(\sqrt{13} - \sqrt{2})(\sqrt{13} + 2\sqrt{2})$.

Objective 16.4c Rationalize denominators having two terms.

Example Rationalize the denominator: $\dfrac{1 + \sqrt{3}}{5 - \sqrt{3}}$.

$$\begin{aligned}
\frac{1 + \sqrt{3}}{5 - \sqrt{3}} &= \frac{1 + \sqrt{3}}{5 - \sqrt{3}} \cdot \frac{5 + \sqrt{3}}{5 + \sqrt{3}}\\
&= \frac{(1 + \sqrt{3})(5 + \sqrt{3})}{(5 - \sqrt{3})(5 + \sqrt{3})}\\
&= \frac{1 \cdot 5 + 1 \cdot \sqrt{3} + 5 \cdot \sqrt{3} + (\sqrt{3})^2}{5^2 - (\sqrt{3})^2}\\
&= \frac{5 + \sqrt{3} + 5\sqrt{3} + 3}{25 - 3} = \frac{8 + 6\sqrt{3}}{22}\\
&= \frac{2(4 + 3\sqrt{3})}{2 \cdot 11} = \frac{4 + 3\sqrt{3}}{11}
\end{aligned}$$

Practice Exercise

11. Rationalize the denominator: $\dfrac{5 - \sqrt{2}}{9 + \sqrt{2}}$.

Objective 16.5a Solve radical equations with one or two radical terms isolated, using the principle of squaring once.

Example Solve: $x - 3 = \sqrt{x - 1}$.

$$\begin{aligned}
x - 3 &= \sqrt{x - 1}\\
(x - 3)^2 &= (\sqrt{x - 1})^2 \qquad \text{Squaring both sides}\\
x^2 - 6x + 9 &= x - 1\\
x^2 - 7x + 10 &= 0\\
(x - 5)(x - 2) &= 0\\
x - 5 = 0 \quad &or \quad x - 2 = 0\\
x = 5 \quad &or \quad x = 2
\end{aligned}$$

The number 5 checks, but 2 does not. Thus the solution is 5.

Practice Exercise

12. Solve: $x - 4 = \sqrt{x - 2}$.

Objective 16.5b Solve radical equations with two radical terms, using the principle of squaring twice.

Example Solve: $\sqrt{x+20} = 10 - \sqrt{x}$.

$$\sqrt{x+20} = 10 - \sqrt{x}$$
$$(\sqrt{x+20})^2 = (10 - \sqrt{x})^2 \quad \text{Squaring both sides}$$
$$x + 20 = 100 - 20\sqrt{x} + x$$
$$20\sqrt{x} = 80$$
$$\sqrt{x} = 4$$
$$(\sqrt{x})^2 = 4^2 \quad \text{Squaring both sides}$$
$$x = 16$$

The number 16 checks and is the solution.

Practice Exercise

13. Solve: $12 - \sqrt{x} = \sqrt{90 - x}$.

Objective 16.6a Given the lengths of any two sides of a right triangle, find the length of the third side.

Example Find the length of the third side of this right triangle.

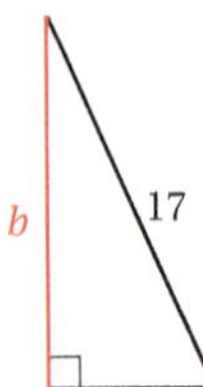

$$a^2 + b^2 = c^2 \quad \text{Pythagorean equation}$$
$$7^2 + b^2 = 17^2 \quad \text{Substituting 7 for } a \text{ and 17 for } c$$
$$49 + b^2 = 289$$
$$b^2 = 240$$
$$b = \sqrt{240} \approx 15.492$$

Practice Exercise

14. Find the length of the third side of this triangle.

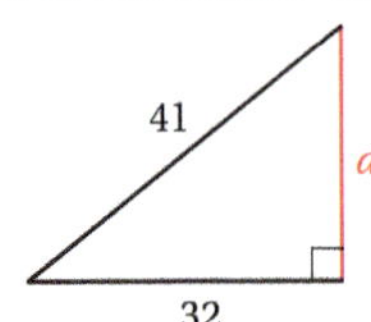

Review Exercises

Find the square roots. [16.1a]

1. 64

2. 400

Simplify. [16.1a]

3. $\sqrt{36}$

4. $-\sqrt{169}$

Use a calculator to approximate each of the following square roots to three decimal places. [16.1b]

5. $\sqrt{3}$

6. $\sqrt{99}$

7. $-\sqrt{320.12}$

8. $\sqrt{\frac{11}{20}}$

9. $-\sqrt{\frac{47.3}{11.2}}$

10. $18\sqrt{11 \cdot 43.7}$

Identify the radicand. [16.1d]

11. $\sqrt{x^2 + 4}$

12. $\sqrt{x} + 2$

13. $3\sqrt{4 - x}$

14. $\sqrt{\frac{2}{y - 7}}$

Determine whether the expression represents a real number. Write " yes" or "no." [16.1e]

15. $-\sqrt{49}$

16. $-\sqrt{-4}$

17. $\sqrt{-36}$

18. $\sqrt{(-3)(-27)}$

Simplify. [16.1f]

19. $\sqrt{m^2}$

20. $\sqrt{(x-4)^2}$

21. $\sqrt{16x^2}$

22. $\sqrt{4p^2 - 12p + 9}$

Simplify by factoring. [16.2a]

23. $\sqrt{48}$

24. $\sqrt{32t^2}$

25. $\sqrt{t^2 - 14t + 49}$

26. $\sqrt{x^2 + 16x + 64}$

Simplify by factoring. [16.2b]

27. $\sqrt{x^8}$

28. $\sqrt{75a^7}$

Multiply. [16.2c]

29. $\sqrt{3}\sqrt{7}$

30. $\sqrt{x-3}\sqrt{x+3}$

Multiply and simplify. [16.2c]

31. $\sqrt{6}\sqrt{10}$

32. $\sqrt{5x}\sqrt{8x}$

33. $\sqrt{5x}\sqrt{10xy^2}$

34. $\sqrt{20a^3b}\sqrt{5a^2b^2}$

Simplify. [16.3a, b]

35. $\sqrt{\frac{25}{64}}$

36. $\sqrt{\frac{49}{t^2}}$

37. $\frac{\sqrt{2c^9}}{\sqrt{32c}}$

Rationalize the denominator. [16.3c]

38. $\sqrt{\frac{1}{2}}$

39. $\frac{\sqrt{x^3}}{\sqrt{15}}$

40. $\sqrt{\frac{5}{y}}$

41. $\frac{\sqrt{b^9}}{\sqrt{ab^2}}$

42. $\frac{\sqrt{27}}{\sqrt{45}}$

43. $\frac{\sqrt{45x^2y}}{\sqrt{54y}}$

Simplify. [16.4a]

44. $10\sqrt{5} + 3\sqrt{5}$

45. $\sqrt{80} - \sqrt{45}$

46. $3\sqrt{2} - 5\sqrt{\frac{1}{2}}$

Simplify. [16.4b]

47. $(2 + \sqrt{3})^2$

48. $(2 + \sqrt{3})(2 - \sqrt{3})$

49. Rationalize the denominator:

$$\frac{4}{2 + \sqrt{3}}. \quad [16.4c]$$

Solve. [16.5a]

50. $\sqrt{x - 3} = 7$

51. $\sqrt{5x + 3} = \sqrt{2x - 1}$

52. $1 + x = \sqrt{1 + 5x}$

53. Solve: $\sqrt{x} = \sqrt{x - 5} + 1$. [16.5b]

Solve. [16.1c], [16.5c]

54. ***Speed of a Skidding Car.*** The formula $r = 2\sqrt{5L}$ can be used to approximate the speed r, in miles per hour, of a car that has left a skid mark of length L, in feet.

a) What was the speed of a car that left skid marks of length 200 ft?

b) How far will a car skid at 90 mph?

In a right triangle, find the length of the side not given. Give an exact answer and an approximation to three decimal places where appropriate. Standard lettering has been used. [16.6a]

55. $a = 15,\ c = 25$

56. $a = 1,\ b = \sqrt{2}$

Solve. [16.6b]

57. ***Airplane Descent.*** A pilot is instructed to descend from 30,000 ft to 20,000 ft over a horizontal distance of 50,000 ft. What distance will the plane travel during this descent?

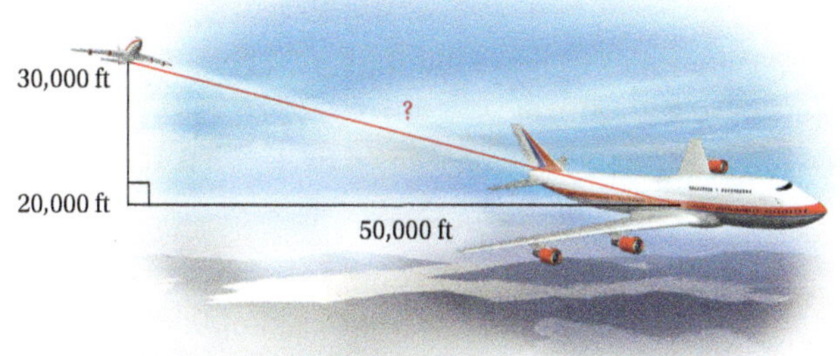

58. ***Lookout Tower.*** The diagonal braces in a lookout tower are 15 ft long and span a distance of 12 ft. How high does each brace reach vertically?

59. Solve: $x - 2 = \sqrt{4 - 9x}$. [16.5a]

A. -5 **B.** No solution
C. 0 **D.** $0, -5$

60. Simplify: $(2\sqrt{7} + \sqrt{2})(\sqrt{7} - \sqrt{2})$. [16.4b]

A. $12 - \sqrt{7}$ **B.** 12
C. $12 - \sqrt{14}$ **D.** $3\sqrt{7} - 2$

Synthesis

61. ***Distance Driven.*** Two cars leave a service station at the same time. One car travels east at a speed of 50 mph, and the other travels south at a speed of 60 mph. After one-half hour, how far apart are they? [16.6b]

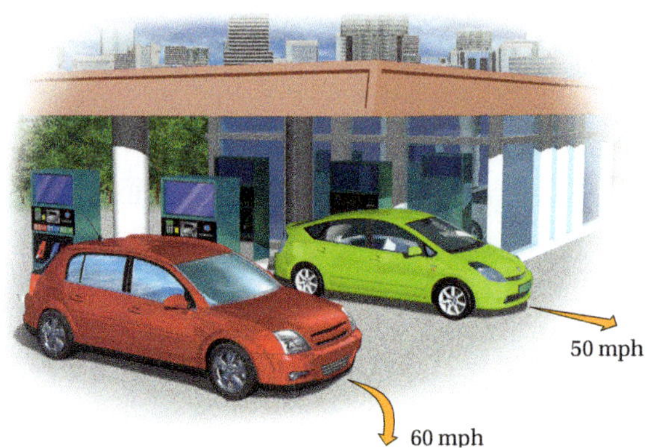

62. Solve $A = \sqrt{a^2 + b^2}$ for b. [16.5a]

63. Find x. [16.6a]

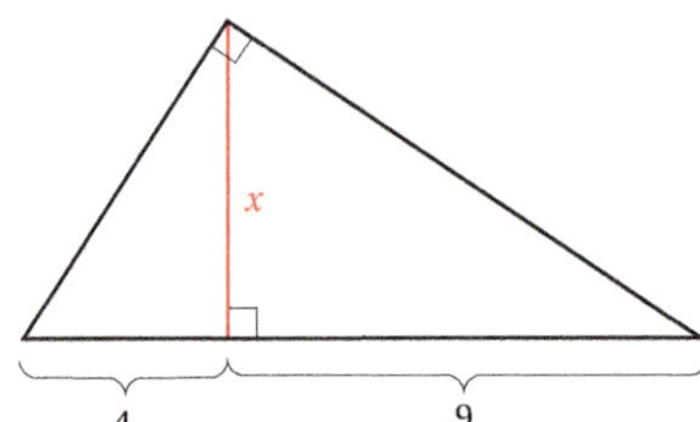

Understanding Through Discussion and Writing

1. Explain why it is necessary for the signs within a pair of conjugates to differ. [16.4c]

2. Determine whether the statement below is true or false and explain your answer. [16.1a], [16.5a]

The solution of $\sqrt{11 - 2x} = -3$ is 1.

3. Why are the rules for manipulating expressions with exponents important when simplifying radical expressions? [16.2b]

4. Can a carpenter use a 28-ft ladder to repair clapboard that is 28 ft above ground level? Why or why not? [16.6b]

5. Explain why possible solutions of radical equations must be checked. [16.5a]

6. Determine whether each of the following is true for all real numbers. Explain why or why not. [16.1f], [16.2a]

a) $\sqrt{5x^2} = |x|\sqrt{5}$
b) $\sqrt{b^2 - 4} = b - 2$
c) $\sqrt{x^2 + 16} = x + 4$

CHAPTER

16 Test

For Extra Help For step-by-step test solutions, access the Chapter Test Prep Videos in MyMathLab® or on YouTube (search "BittingerPreIntro" and click on Channels).

1. Find the square roots of 81.

Simplify.

2. $\sqrt{64}$

3. $-\sqrt{25}$

Approximate the expression involving square roots to three decimal places.

4. $\sqrt{116}$

5. $-\sqrt{87.4}$

6. $4\sqrt{5 \cdot 6}$

7. Identify the radicand in $8\sqrt{4 - y^3}$.

Determine whether each expression represents a real number. Write "yes" or "no."

8. $\sqrt{24}$

9. $\sqrt{-23}$

Simplify.

10. $\sqrt{a^2}$

11. $\sqrt{36y^2}$

Multiply.

12. $\sqrt{5}\sqrt{6}$

13. $\sqrt{x - 8}\sqrt{x + 8}$

Simplify by factoring.

14. $\sqrt{27}$

15. $\sqrt{25x - 25}$

16. $\sqrt{t^5}$

Multiply and simplify.

17. $\sqrt{5}\sqrt{10}$

18. $\sqrt{3ab}\sqrt{6ab^3}$

Simplify.

19. $\sqrt{\dfrac{27}{12}}$

20. $\sqrt{\dfrac{144}{a^2}}$

Rationalize the denominator.

21. $\sqrt{\dfrac{2}{5}}$

22. $\sqrt{\dfrac{2x}{y}}$

Divide and simplify.

23. $\dfrac{\sqrt{27}}{\sqrt{32}}$

24. $\dfrac{\sqrt{35x}}{\sqrt{80xy^2}}$

Add or subtract.

25. $3\sqrt{18} - 5\sqrt{18}$

26. $\sqrt{5} + \sqrt{\frac{1}{5}}$

Simplify.

27. $(4 - \sqrt{5})^2$

28. $(4 - \sqrt{5})(4 + \sqrt{5})$

29. Rationalize the denominator: $\frac{10}{4 - \sqrt{5}}$.

30. In a right triangle, $a = 8$ and $b = 4$. Find c. Give an exact answer and an approximation to three decimal places.

Solve.

31. $\sqrt{3x} + 2 = 14$

32. $\sqrt{6x + 13} = x + 3$

33. $\sqrt{1 - x} + 1 = \sqrt{6 - x}$

34. ***Sighting to the Horizon.*** The equation $D = \sqrt{2h}$ can be used to approximate the distance D, in miles, that a person can see to the horizon from a height h, in feet.

a) How far to the horizon can you see through an airplane window at a height of 28,000 ft?

b) Christina can see about 261 mi to the horizon through an airplane window. How high is the airplane?

35. ***Lacrosse.*** A regulation lacrosse field is 60 yd wide and 110 yd long. Find the length of a diagonal of such a field.

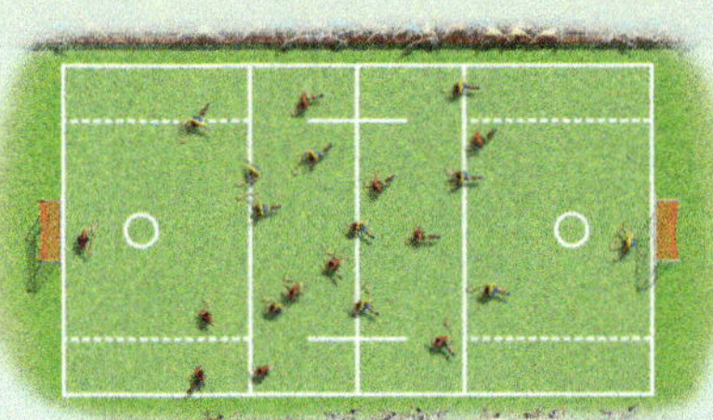

36. Rationalize the denominator: $\sqrt{\frac{2a}{5b}}$.

A. $\frac{\sqrt{10ab}}{5b}$

B. $\frac{a}{b}\sqrt{\frac{2b}{5a}}$

C. $\frac{\sqrt{10}}{5}$

D. $\frac{\sqrt{6a^3b}}{15ab}$

Synthesis

Simplify.

37. $\sqrt{\sqrt{\sqrt{625}}}$

38. $\sqrt{y^{16n}}$

CHAPTER

17

Quadratic Equations

STUDYING FOR SUCCESS *Preparing for the Final Exam*

- ☐ Browse through each chapter, reviewing highlighted or boxed information and noting important formulas.
- ☐ Attend any exam tutoring sessions offered by your college or university.
- ☐ Retake the chapter tests that you took in class, or take the chapter tests in the text.

17.1 Introduction to Quadratic Equations

OBJECTIVES

a Write a quadratic equation in standard form $ax^2 + bx + c = 0, a > 0$, and determine the coefficients a, b, and c.

b Solve quadratic equations of the type $ax^2 + bx = 0$, where $b \neq 0$, by factoring.

c Solve quadratic equations of the type $ax^2 + bx + c = 0$, where $b \neq 0$ and $c \neq 0$, by factoring.

d Solve applied problems involving quadratic equations.

SKILL TO REVIEW

Objective 13.7b: Solve quadratic equations by factoring and then using the principle of zero products.

Solve by factoring and using the principle of zero products.

1. $x^2 + 15x = 0$
2. $x^2 + 8x = 33$

ALGEBRAIC ▶◀ GRAPHICAL CONNECTION

Before we begin this chapter, let's look back at some algebraic–graphical equation-solving connections. In Chapter 11, we considered the graph of a *linear equation* $y = mx + b$. For example, the graph of the equation $y = \frac{5}{2}x - 4$ and its x-intercept are shown below.

If $y = 0$, then $x = \frac{8}{5}$. Thus the x-intercept is $\left(\frac{8}{5}, 0\right)$. This point is also the intersection of the graphs of $y = \frac{5}{2}x - 4$ and $y = 0$.

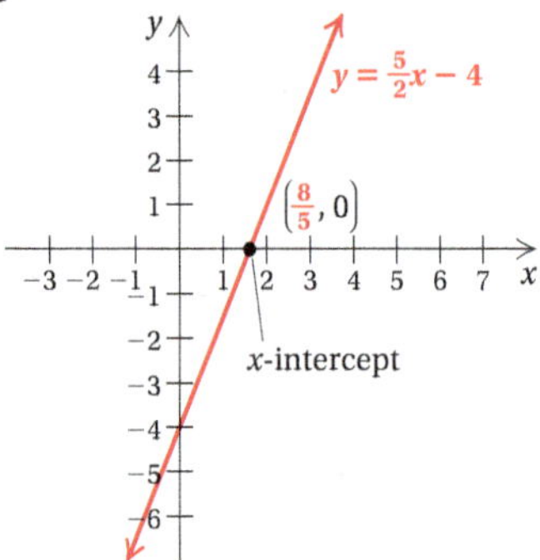

We can solve the linear equation $\frac{5}{2}x - 4 = 0$ algebraically:

$$\frac{5}{2}x - 4 = 0$$
$$\frac{5}{2}x = 4 \quad \text{Adding 4}$$
$$x = \frac{8}{5}. \quad \text{Multiplying by } \frac{2}{5}$$

We see that $\frac{8}{5}$, the solution of $\frac{5}{2}x - 4 = 0$, is the first coordinate of the x-intercept of the graph of $y = \frac{5}{2}x - 4$.

◀ Do Margin Exercise 1 on the following page.

In this chapter, we build on these ideas by applying them to quadratic equations. In Section 13.7, we briefly considered the graph of a *quadratic equation* $y = ax^2 + bx + c, a \neq 0$. For example, the graph of the equation $y = x^2 + 6x + 8$ and its x-intercepts are shown below.

The x-intercepts are $(-4, 0)$ and $(-2, 0)$. We will develop in detail the creation of such graphs in Section 17.6. The points $(-4, 0)$ and $(-2, 0)$ are the intersections of the graphs of $y = x^2 + 6x + 8$ and $y = 0$.

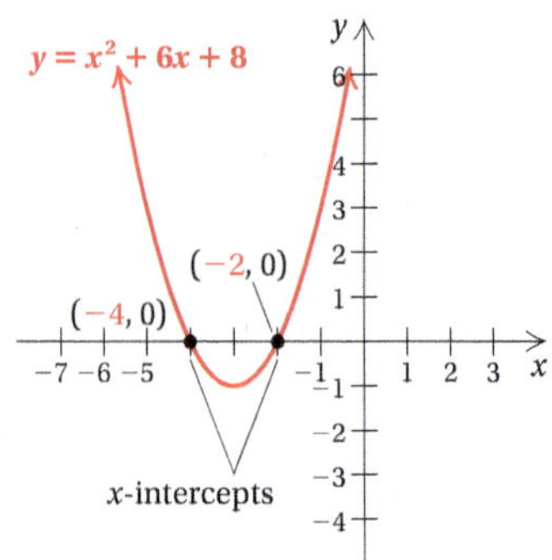

Answers

Skill to Review:
1. $-15, 0$ 2. $-11, 3$

We can solve the quadratic equation $x^2 + 6x + 8 = 0$ by factoring and using the principle of zero products:

$$x^2 + 6x + 8 = 0$$

$$(x + 4)(x + 2) = 0 \qquad \text{Factoring}$$

$$x + 4 = 0 \quad or \quad x + 2 = 0 \qquad \text{Using the principle of zero products}$$

$$x = -4 \quad or \quad x = -2.$$

We see that the solutions of $x^2 + 6x + 8 = 0$, -4 and -2, are the first coordinates of the x-intercepts, $(-4, 0)$ and $(-2, 0)$, of the graph of $y = x^2 + 6x + 8$.

Do Exercise 2. ▶

a STANDARD FORM

The following are **quadratic equations**. They contain polynomials of second degree.

$$4x^2 + 7x - 5 = 0,$$
$$3t^2 - \tfrac{1}{2}t = 9,$$
$$5y^2 = -6y,$$
$$5m^2 = 15$$

The quadratic equation $4x^2 + 7x - 5 = 0$ is said to be in **standard form**. Although the quadratic equation $4x^2 = 5 - 7x$ is equivalent to the preceding equation, it is *not* in standard form.

QUADRATIC EQUATION

A **quadratic equation** is an equation equivalent to an equation of the type

$$ax^2 + bx + c = 0,\ a > 0,$$

where a, b, and c are real-number constants. We say that the preceding is the **standard form of a quadratic equation**.

We define $a > 0$ to ease the proof of the quadratic formula, which we consider later, and to ease solving by factoring. Suppose we are studying an equation like $-3x^2 + 8x - 2 = 0$. It is not in standard form because $-3 < 0$. We can find an equivalent equation that is in standard form by multiplying by -1 on both sides:

$$-1(-3x^2 + 8x - 2) = -1(0)$$

$$3x^2 - 8x + 2 = 0. \qquad \text{Standard form}$$

When a quadratic equation is written in standard form, we can readily determine a, b, and c.

$$ax^2 + bx + c = 0$$
$$3x^2 - 8x + 2 = 0$$

For this equation, $a = 3$, $b = -8$, and $c = 2$.

1. **a)** Consider the linear equation $y = -\frac{2}{3}x - 3$. Find the intercepts and graph the equation.

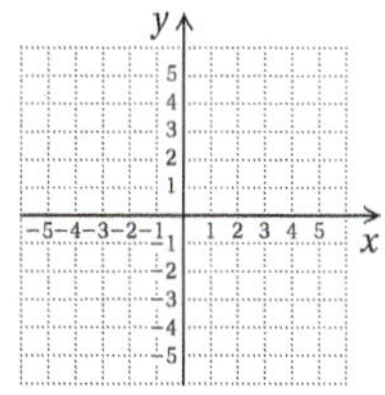

b) Solve the equation

$$0 = -\frac{2}{3}x - 3.$$

c) Complete: The solution of the equation $0 = -\frac{2}{3}x - 3$ is ____. This value is the ____ ____ of the x-intercept, (____, ____), of the graph of $y = -\frac{2}{3}x - 3$.

2. Consider the quadratic equation $y = x^2 - 2x - 3$ and its graph shown below.

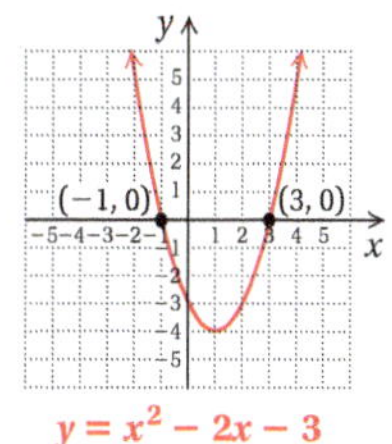

$y = x^2 - 2x - 3$

a) Solve the equation

$$x^2 - 2x - 3 = 0.$$

(*Hint*: Use the principle of zero products.)

b) Complete: The solutions of the equation $x^2 - 2x - 3 = 0$ are ____ and ____. These values are the ____ ____ of the x-intercepts, (____, ____) and (____, ____), of the graph of $y = x^2 - 2x - 3$.

Answers

1. (a) y-intercept: $(0, -3)$; x-intercept: $\left(-\frac{9}{2}, 0\right)$; **(b)** $-\frac{9}{2}$; **(c)** $-\frac{9}{2}$; first coordinate; $\left(-\frac{9}{2}, 0\right)$ **2. (a)** $-1, 3$; **(b)** $-1, 3$; first coordinates; $(-1, 0)$, $(3, 0)$

$y = -\frac{2}{3}x - 3$

EXAMPLES Write in standard form and determine a, b, and c.

1. $4x^2 + 7x - 5 = 0$ The equation is already in standard form.

$a = 4$; $b = 7$; $c = -5$

2. $3x^2 - 0.5x = 9$

$3x^2 - 0.5x - 9 = 0$ Subtracting 9. This is standard form.

$a = 3$; $b = -0.5$; $c = -9$

3. $-4y^2 = 5y$

$-4y^2 - 5y = 0$ Subtracting $5y$
Not positive

$4y^2 + 5y = 0$ Multiplying by -1. This is standard form.

$a = 4$; $b = 5$; $c = 0$

Do Exercises 3–7.

b SOLVING QUADRATIC EQUATIONS OF THE TYPE $ax^2 + bx = 0$

Sometimes we can use factoring and the principle of zero products to solve quadratic equations. In particular, when $c = 0$ and $b \neq 0$, we can always factor and use the principle of zero products.

EXAMPLE 4 Solve: $7x^2 + 2x = 0$.

$$7x^2 + 2x = 0$$
$$x(7x + 2) = 0 \quad \text{Factoring}$$
$$x = 0 \quad or \quad 7x + 2 = 0 \quad \text{Using the principle of zero products}$$
$$x = 0 \quad or \quad 7x = -2$$
$$x = 0 \quad or \quad x = -\tfrac{2}{7}$$

Check: For 0:

$$\frac{7x^2 + 2x = 0}{7 \cdot 0^2 + 2 \cdot 0 \ ?\ 0}$$
$$0 \quad \text{TRUE}$$

For $-\frac{2}{7}$:

$$\frac{7x^2 + 2x = 0}{7\left(-\tfrac{2}{7}\right)^2 + 2\left(-\tfrac{2}{7}\right) \ ?\ 0}$$
$$7\left(\tfrac{4}{49}\right) - \tfrac{4}{7}$$
$$\tfrac{4}{7} - \tfrac{4}{7}$$
$$0 \quad \text{TRUE}$$

The solutions are 0 and $-\frac{2}{7}$.

Caution!

You may be tempted to divide each term in an equation like the one in Example 4 by x. This method would yield the equation

$$7x + 2 = 0,$$

whose only solution is $-\frac{2}{7}$. In effect, since 0 is also a solution of the original equation, we have divided by 0. The error of such division results in the loss of one of the solutions.

Write in standard form and determine a, b, and c.

3. $6x^2 = 3 - 7x$

4. $y^2 = 8y$

5. $3 - x^2 = 9x$

6. $3x + 5x^2 = x^2 - 4 + 2x$

7. $5x^2 = 21$

Solve.

GS

8. $2x^2 + 8x = 0$

$2x(\qquad) = 0$

$2x = 0 \quad or \quad \qquad = 0$

$x = 0 \quad or \quad x = \qquad$

Both numbers check.
The solutions are 0 and ____.

9. $10x^2 - 6x = 0$

Answers

3. $6x^2 + 7x - 3 = 0; a = 6, b = 7, c = -3$
4. $y^2 - 8y = 0; a = 1, b = -8, c = 0$
5. $x^2 + 9x - 3 = 0; a = 1, b = 9, c = -3$
6. $4x^2 + x + 4 = 0; a = 4, b = 1, c = 4$
7. $5x^2 - 21 = 0; a = 5, b = 0, c = -21$
8. $0, -4$ **9.** $0, \frac{3}{5}$

Guided Solution:
8. $x + 4, x + 4, -4, -4$

EXAMPLE 5 Solve: $4x^2 - 8x = 0$.

We have

$$4x^2 - 8x = 0$$
$$4x(x - 2) = 0 \quad \text{Factoring}$$
$$4x = 0 \quad or \quad x - 2 = 0 \quad \text{Using the principle of zero products}$$
$$x = 0 \quad or \quad x = 2.$$

The solutions are 0 and 2.

A quadratic equation of the type $ax^2 + bx = 0$, where $c = 0$ and $b \neq 0$, will always have 0 as one solution and a nonzero number as the other solution.

Do Exercises 8 and 9 on the preceding page.

C SOLVING QUADRATIC EQUATIONS OF THE TYPE $ax^2 + bx + c = 0$

When neither b nor c is 0, we can sometimes solve by factoring.

EXAMPLE 6 Solve: $2x^2 - x - 21 = 0$.

We have

$$2x^2 - x - 21 = 0$$
$$(2x - 7)(x + 3) = 0 \quad \text{Factoring}$$
$$2x - 7 = 0 \quad or \quad x + 3 = 0 \quad \text{Using the principle of zero products}$$
$$2x = 7 \quad or \quad x = -3$$
$$x = \tfrac{7}{2} \quad or \quad x = -3.$$

The solutions are $\frac{7}{2}$ and -3.

EXAMPLE 7 Solve: $(y - 3)(y - 2) = 6(y - 3)$.

We write the equation in standard form and then factor:

$$y^2 - 5y + 6 = 6y - 18 \quad \text{Multiplying}$$
$$y^2 - 11y + 24 = 0 \quad \text{Standard form}$$
$$(y - 8)(y - 3) = 0 \quad \text{Factoring}$$
$$y - 8 = 0 \quad or \quad y - 3 = 0 \quad \text{Using the principle of zero products}$$
$$y = 8 \quad or \quad y = 3.$$

The solutions are 8 and 3.

Do Exercises 10 and 11.

Recall that to solve a rational equation, we multiply both sides by the LCM of all the denominators. We may obtain a quadratic equation after a few steps. When that happens, we solve the quadratic equation, but we must remember to check possible solutions because a replacement may result in division by 0.

ALGEBRAIC ▶◀ GRAPHICAL CONNECTION

Let's visualize the solutions in Example 5.

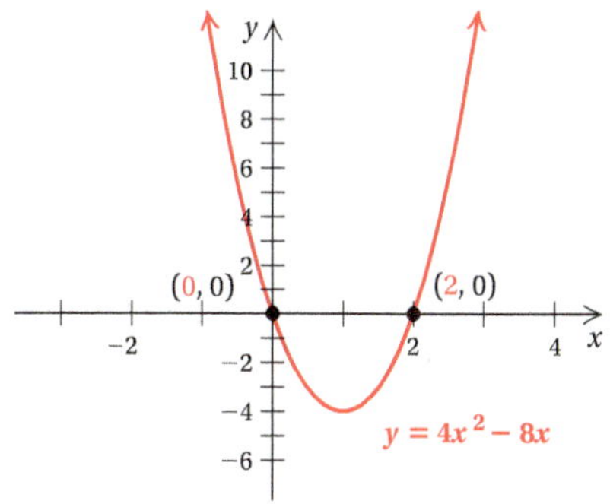

The solutions of $4x^2 - 8x = 0$, 0 and 2, are the first coordinates of the x-intercepts, $(0, 0)$ and $(2, 0)$, of the graph of $y = 4x^2 - 8x$.

ALGEBRAIC ▶◀ GRAPHICAL CONNECTION

Let's visualize the solutions in Example 6.

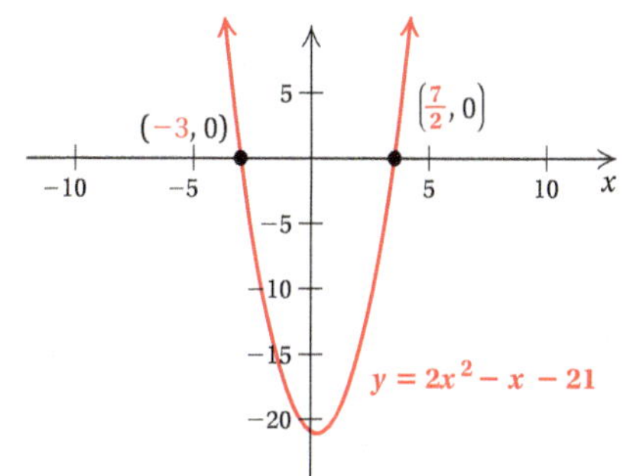

The solutions of $2x^2 - x - 21 = 0$, -3 and $\frac{7}{2}$, are the first coordinates of the x-intercepts, $(-3, 0)$ and $(\frac{7}{2}, 0)$, of the graph of $y = 2x^2 - x - 21$.

Solve.

10. $4x^2 + 5x - 6 = 0$

11. $(x - 1)(x + 1) = 5(x - 1)$

Answers

10. $-2, \frac{3}{4}$ **11.** 1, 4

EXAMPLE 8 Solve: $\dfrac{3}{x-1} + \dfrac{5}{x+1} = 2.$

We multiply by the LCM, which is $(x-1)(x+1)$:

$$(x-1)(x+1) \cdot \left(\frac{3}{x-1} + \frac{5}{x+1}\right) = 2 \cdot (x-1)(x+1).$$

We use the distributive law on the left:

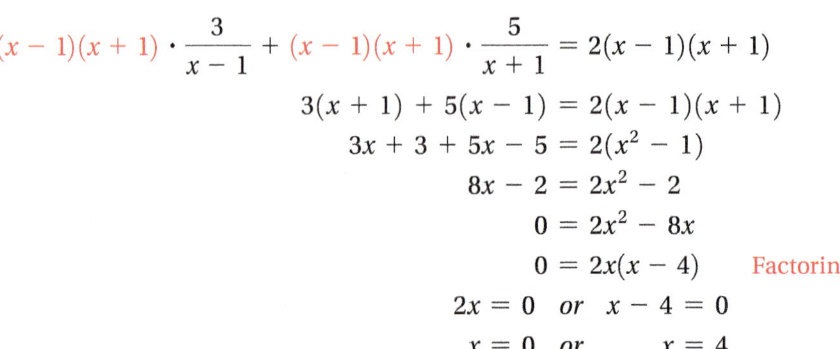

$$(x-1)(x+1) \cdot \frac{3}{x-1} + (x-1)(x+1) \cdot \frac{5}{x+1} = 2(x-1)(x+1)$$

$$3(x+1) + 5(x-1) = 2(x-1)(x+1)$$

$$3x + 3 + 5x - 5 = 2(x^2 - 1)$$

$$8x - 2 = 2x^2 - 2$$

$$0 = 2x^2 - 8x$$

$$0 = 2x(x-4) \quad \text{Factoring}$$

$$2x = 0 \quad \text{or} \quad x - 4 = 0$$

$$x = 0 \quad \text{or} \quad x = 4.$$

Check:

For 0:

$$\frac{3}{x-1} + \frac{5}{x+1} = 2$$

$$\frac{3}{0-1} + \frac{5}{0+1} \; ? \; 2$$

$$\frac{3}{-1} + \frac{5}{1}$$

$$-3 + 5$$

$$2 \qquad \text{TRUE}$$

For 4:

$$\frac{3}{x-1} + \frac{5}{x+1} = 2$$

$$\frac{3}{4-1} + \frac{5}{4+1} \; ? \; 2$$

$$\frac{3}{3} + \frac{5}{5}$$

$$1 + 1$$

$$2 \qquad \text{TRUE}$$

The solutions are 0 and 4.

◀ **Do Exercise 12.**

d SOLVING APPLIED PROBLEMS

EXAMPLE 9 ***Diagonals of a Polygon.*** The number of diagonals d of a polygon of n sides is given by the formula

$$d = \frac{n^2 - 3n}{2}.$$

If a polygon has 27 diagonals, how many sides does it have?

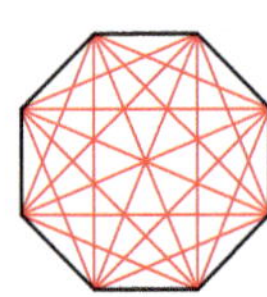

1. **Familiarize.** To familiarize ourselves with the problem, we draw an octagon (8 sides) with its diagonals, as shown at left, and see that there are 20 diagonals. To check this, we evaluate the formula for $n = 8$:

$$d = \frac{8^2 - 3(8)}{2} = \frac{64 - 24}{2} = \frac{40}{2} = 20.$$

2. **Translate.** We substitute 27 for d:

$$27 = \frac{n^2 - 3n}{2}. \qquad \text{The number of diagonals is 27.}$$

---

12. Solve:

GS

$$\frac{20}{x+5} - \frac{1}{x-4} = 1.$$

$$(x+5)(x-4) \cdot \left(\frac{20}{x+5} - \frac{1}{x-4}\right) = (x+5)(x-4) \cdot 1$$

$$(x+5)(x-4) \cdot \frac{20}{x+5} - (x+5)(x-4) \cdot \frac{1}{x-4} = (x+5)(x-4)$$

$$20(x-4) - 1(\quad\quad) = (x+5)(x-4)$$

$$20x - 80 - x - \quad\quad = x^2 + x - \quad\quad$$

$$19x - \quad\quad = x^2 + x - 20$$

$$0 = x^2 - 18x + \quad\quad$$

$$0 = (x-5)(\quad\quad)$$

$$x - 5 = 0 \quad or \quad \quad\quad = 0$$

$$x = 5 \quad or \quad x = \quad\quad$$

Both numbers check in the original equation. The solutions are 5 and ______.

Answer

12. 5, 13

Guided Solution:

12. $x + 5$, 5, 20, 85, 65, $x - 13$, $x - 13$, 13, 13

3. Solve. We solve the equation for n:

$$\frac{n^2 - 3n}{2} = 27$$

$$2 \cdot \frac{n^2 - 3n}{2} = 2 \cdot 27 \quad \text{Multiplying by 2 to clear fractions}$$

$$n^2 - 3n = 54$$

$$n^2 - 3n - 54 = 0 \quad \text{Subtracting 54}$$

$$(n - 9)(n + 6) = 0 \quad \text{Factoring}$$

$$n - 9 = 0 \quad or \quad n + 6 = 0$$

$$n = 9 \quad or \quad n = -6.$$

4. Check. Since the number of sides cannot be negative, -6 cannot be a solution. The answer 9 for the number of sides checks in the formula.

5. State. The polygon has 9 sides. (It is a nonagon.)

Do Exercise 13. ▶

13. A polygon has 44 diagonals. How many sides does it have?

Answer

13. 11 sides

CALCULATOR CORNER

Solving Quadratic Equations We can use the INTERSECT feature to solve a quadratic equation. Consider the equation in Margin Exercise 11, $(x - 1)(x + 1) = 5(x - 1)$. First, we enter $y_1 = (x - 1)(x + 1)$ and $y_2 = 5(x - 1)$ on the equation-editor screen and graph the equations, using the window $[-5, 5, -5, 20]$, Yscl = 2. We see that there are two points of intersection, so the equation has two solutions.

Next, we use the INTERSECT feature to find the coordinates of the left-hand point of intersection. The first coordinate of this point, 1, is one solution of the equation. We use the INTERSECT feature again to find the other solution, 4.

Note that we could use the ZERO feature to solve this equation if we first write it with 0 on one side, that is, $(x - 1)(x + 1) - 5(x - 1) = 0$.

$y_1 = (x - 1)(x + 1)$,
$y_2 = 5(x - 1)$

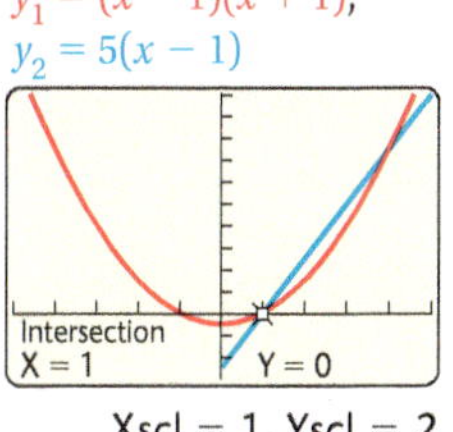

Xscl = 1, Yscl = 2

$y_1 = (x - 1)(x + 1)$,
$y_2 = 5(x - 1)$

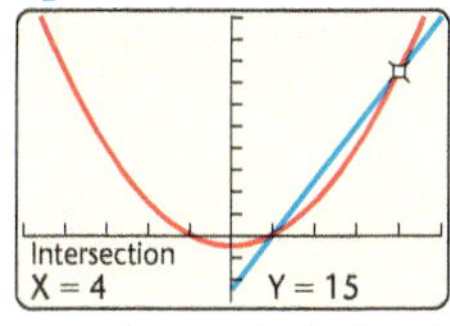

Xscl = 1, Yscl = 2

EXERCISES: Solve.

1. $5x^2 - 8x + 3 = 0$
2. $2x^2 - 7x - 15 = 0$
3. $6(x - 3) = (x - 3)(x - 2)$
4. $(x + 1)(x - 4) = 3(x - 4)$

17.1 Exercise Set

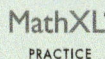

✓ Reading Check

Determine whether each statement is true or false.

RC1. The solutions of $x^2 - x - 12 = 0$ are the second coordinates of the x-intercepts of the graph of $y = x^2 - x - 12$.

RC2. The quadratic equation $14x^2 - 3x - 58 = 0$ is written in standard form.

RC3. One of the solutions of $85x^2 - 96x = 0$ is 0.

RC4. Every quadratic equation can be written in standard form.

a Write in standard form and determine a, b, and c.

1. $x^2 - 3x + 2 = 0$

2. $x^2 - 8x - 5 = 0$

3. $7x^2 = 4x - 3$

4. $9x^2 = x + 5$

5. $5 = -2x^2 + 3x$

6. $3x - 1 = 5x^2 + 9$

b Solve.

7. $x^2 + 5x = 0$

8. $x^2 + 7x = 0$

9. $3x^2 + 6x = 0$

10. $4x^2 + 8x = 0$

11. $5x^2 = 2x$

12. $11x = 3x^2$

13. $4x^2 + 4x = 0$

14. $8x^2 - 8x = 0$

15. $0 = 10x^2 - 30x$

16. $0 = 10x^2 - 50x$

17. $11x = 55x^2$

18. $33x^2 = -11x$

19. $14t^2 = 3t$

20. $6m = 19m^2$

21. $5y^2 - 3y^2 = 72y + 9y$

22. $63p - 16p^2 = 17p + 58p^2$

c Solve.

23. $x^2 + 8x - 48 = 0$

24. $x^2 - 16x + 48 = 0$

25. $5 + 6x + x^2 = 0$

26. $x^2 + 10 + 11x = 0$

27. $18 = 7p + p^2$

28. $t^2 + 14t = -24$

29. $-15 = -8y + y^2$

30. $q^2 + 14 = 9q$

31. $x^2 + 10x + 25 = 0$

32. $x^2 + 6x + 9 = 0$

33. $r^2 = 8r - 16$

34. $x^2 + 1 = 2x$

35. $6x^2 + x - 2 = 0$

36. $2x^2 - 11x + 15 = 0$

37. $3a^2 = 10a + 8$

38. $15b - 9b^2 = 4$

39. $6x^2 - 4x = 10$

40. $3x^2 - 7x = 20$

41. $2t^2 + 12t = -10$

42. $12w^2 - 5w = 2$

43. $t(t - 5) = 14$

44. $6z^2 + z - 1 = 0$

45. $t(9 + t) = 4(2t + 5)$

46. $3y^2 + 8y = 12y + 15$

47. $16(p - 1) = p(p + 8)$

48. $(2x - 3)(x + 1) = 4(2x - 3)$

49. $(t - 1)(t + 3) = t - 1$

50. $(x - 2)(x + 2) = x + 2$

Solve.

51. $\frac{24}{x-2} + \frac{24}{x+2} = 5$

52. $\frac{8}{x+2} + \frac{8}{x-2} = 3$

53. $\frac{1}{x} + \frac{1}{x+6} = \frac{1}{4}$

54. $\frac{1}{x} + \frac{1}{x+9} = \frac{1}{20}$

55. $1 + \frac{12}{x^2-4} = \frac{3}{x-2}$

56. $\frac{5}{t-3} - \frac{30}{t^2-9} = 1$

57. $\frac{r}{r-1} + \frac{2}{r^2-1} = \frac{8}{r+1}$

58. $\frac{x+2}{x^2-2} = \frac{2}{3-x}$

59. $\frac{x-1}{1-x} = -\frac{x+8}{x-8}$

60. $\frac{4-x}{x-4} + \frac{x+3}{x-3} = 0$

61. $\frac{5}{y+4} - \frac{3}{y-2} = 4$

62. $\frac{2z+11}{2z+8} = \frac{3z-1}{z-1}$

d Solve.

63. *Diagonals.* A decagon is a figure with 10 sides. How many diagonals does a decagon have?

64. *Diagonals.* A hexagon is a figure with 6 sides. How many diagonals does a hexagon have?

65. *Diagonals.* A polygon has 14 diagonals. How many sides does it have?

66. *Diagonals.* A polygon has 9 diagonals. How many sides does it have?

Skill Maintenance

Factor completely. [13.6a]

67. $18t - 9p + 9$

68. $2x^3 - 18x^2 + 16x$

69. $x^2 + 14x + 49$

70. $36x^2 - 60xy + 25y^2$

71. $t^4 - 81$

72. $6a^2 - a - 15$

73. $a^2 + ac + a + c$

74. $20a^2b + ab - b$

75. $x^2y^2 - 1$

76. $c^2 - \frac{1}{4}$

77. $3x^4 - 6x^2 + 3$

78. $\frac{1}{4}x^2 - \frac{1}{4}xy + \frac{1}{16}y^2$

Synthesis

Solve.

79. $4m^2 - (m+1)^2 = 0$

80. $x^2 + \sqrt{22}x = 0$

81. $\sqrt{5}x^2 - x = 0$

82. $\sqrt{7}x^2 + \sqrt{3}x = 0$

Use a graphing calculator to solve each equation.

83. $3x^2 - 7x = 20$

84. $x(x-5) = 14$

85. $3x^2 + 8x = 12x + 15$

86. $(x-2)(x+2) = x + 2$

87. $(x-2)^2 + 3(x-2) = 4$

88. $(x+3)^2 = 4$

89. $16(x-1) = x(x+8)$

90. $x^2 + 2.5x + 1.5625 = 9.61$

17.2 Solving Quadratic Equations by Completing the Square

OBJECTIVES

a Solve quadratic equations of the type $ax^2 = p$.

b Solve quadratic equations of the type $(x + c)^2 = d$.

c Solve quadratic equations by completing the square.

d Solve certain applied problems involving quadratic equations of the type $ax^2 = p$.

SKILL TO REVIEW

Objective 13.5b: Factor trinomial squares.

Factor.

1. $x^2 - 18x + 81$
2. $x^2 + 2x + 1$

a SOLVING QUADRATIC EQUATIONS OF THE TYPE $ax^2 = p$

For equations of the type $ax^2 = p$, we first solve for x^2 and then apply the *principle of square roots*, which states that a positive number has two square roots. The number 0 has one square root, 0.

THE PRINCIPLE OF SQUARE ROOTS

- The equation $x^2 = d$ has two real solutions when $d > 0$. The solutions are $\sqrt{d}$ and $-\sqrt{d}$.
- The equation $x^2 = d$ has no real-number solution when $d < 0$.
- The equation $x^2 = 0$ has 0 as its only solution.

EXAMPLE 1 Solve: $x^2 = 3$.

$$x^2 = 3$$

$$x = \sqrt{3} \quad or \quad x = -\sqrt{3} \qquad \text{Using the principle of square roots}$$

Check: For $\sqrt{3}$:

$$\begin{array}{c|c} x^2 & 3 \\ \hline (\sqrt{3})^2 & ?\ 3 \\ 3 & \end{array} \quad \text{TRUE}$$

For $-\sqrt{3}$:

$$\begin{array}{c|c} x^2 & 3 \\ \hline (-\sqrt{3})^2 & ?\ 3 \\ 3 & \end{array} \quad \text{TRUE}$$

The solutions are $\sqrt{3}$ and $-\sqrt{3}$.

1. Solve: $x^2 = 10$.

◀ Do Margin Exercise 1.

EXAMPLE 2 Solve: $\frac{1}{8}x^2 = 0$.

$$\frac{1}{8}x^2 = 0$$

$$x^2 = 0 \qquad \text{Multiplying by 8}$$

$$x = 0 \qquad \text{Using the principle of square roots}$$

The solution is 0.

2. Solve: $6x^2 = 0$.

◀ Do Margin Exercise 2.

EXAMPLE 3 Solve: $-3x^2 + 7 = 0$.

$$-3x^2 + 7 = 0$$

$$-3x^2 = -7 \qquad \text{Subtracting 7}$$

$$x^2 = \frac{-7}{-3} = \frac{7}{3} \qquad \text{Dividing by } -3$$

$$x = \sqrt{\frac{7}{3}} \quad or \quad x = -\sqrt{\frac{7}{3}} \qquad \text{Using the principle of square roots}$$

$$x = \sqrt{\frac{7}{3} \cdot \frac{3}{3}} \quad or \quad x = -\sqrt{\frac{7}{3} \cdot \frac{3}{3}} \qquad \text{Rationalizing the denominators}$$

$$x = \frac{\sqrt{21}}{3} \quad or \quad x = -\frac{\sqrt{21}}{3}$$

Answers

Skill to Review:
1. $(x - 9)^2$ 2. $(x + 1)^2$

Margin Exercises:
1. $\sqrt{10}, -\sqrt{10}$ 2. 0

Check: For $\frac{\sqrt{21}}{3}$:

$$\begin{array}{c|c} -3x^2 + 7 = 0 & \\ \hline -3\left(\frac{\sqrt{21}}{3}\right)^2 + 7 \ ? \ 0 & \\ -3 \cdot \frac{21}{9} + 7 & \\ -7 + 7 & \\ 0 & \text{TRUE} \end{array}$$

For $-\frac{\sqrt{21}}{3}$:

$$\begin{array}{c|c} -3x^2 + 7 = 0 & \\ \hline -3\left(-\frac{\sqrt{21}}{3}\right)^2 + 7 \ ? \ 0 & \\ -3 \cdot \frac{21}{9} + 7 & \\ -7 + 7 & \\ 0 & \text{TRUE} \end{array}$$

The solutions are $\frac{\sqrt{21}}{3}$ and $-\frac{\sqrt{21}}{3}$.

Do Exercise 3. ▶

b SOLVING QUADRATIC EQUATIONS OF THE TYPE $(x + c)^2 = d$

In an equation of the type $(x + c)^2 = d$, we have the square of a binomial equal to a constant. We can use the principle of square roots to solve such an equation.

EXAMPLE 4 Solve: $(x - 5)^2 = 9$.

$$(x - 5)^2 = 9$$

$x - 5 = 3$ *or* $x - 5 = -3$ — Using the principle of square roots

$x = 8$ *or* $x = 2$

The solutions are 8 and 2.

EXAMPLE 5 Solve: $(x + 2)^2 = 7$.

$$(x + 2)^2 = 7$$

$x + 2 = \sqrt{7}$ *or* $x + 2 = -\sqrt{7}$ — Using the principle of square roots

$x = -2 + \sqrt{7}$ *or* $x = -2 - \sqrt{7}$

The solutions are $-2 + \sqrt{7}$ and $-2 - \sqrt{7}$, or simply $-2 \pm \sqrt{7}$ (read "-2 plus or minus $\sqrt{7}$").

Do Exercises 4 and 5. ▶

In Examples 4 and 5, the left sides of the equations are squares of binomials. If we can express an equation in such a form, we can proceed as we did in those examples.

EXAMPLE 6 Solve: $x^2 + 8x + 16 = 49$.

$x^2 + 8x + 16 = 49$ — The left side is the square of a binomial; $A^2 + 2AB + B^2 = (A + B)^2$.

$(x + 4)^2 = 49$

$x + 4 = 7$ *or* $x + 4 = -7$ — Using the principle of square roots

$x = 3$ *or* $x = -11$

The solutions are 3 and -11.

Do Exercises 6 and 7. ▶

GS **3.** Solve: $2x^2 - 3 = 0$.

$2x^2 = \_\_\_$

$x^2 = \frac{3}{2}$

$x = \sqrt{\frac{3}{2}}$ *or* $x = -\sqrt{\_\_\_}$

$x = \sqrt{\frac{3}{2} \cdot \frac{2}{2}}$ *or* $x = -\sqrt{\frac{3}{2} \cdot \_\_\_}$

$x = \frac{\sqrt{6}}{2}$ *or* $x = \_\_\_$

Both numbers check.

The solutions are $\frac{\sqrt{6}}{2}$ and $\_\_\_$.

Solve.

4. $(x - 3)^2 = 16$

5. $(x + 4)^2 = 11$

Solve.

6. $x^2 - 6x + 9 = 64$

7. $x^2 - 2x + 1 = 5$

Answers

3. $\frac{\sqrt{6}}{2}, -\frac{\sqrt{6}}{2}$ **4.** $7, -1$ **5.** $-4 \pm \sqrt{11}$
6. $-5, 11$ **7.** $1 \pm \sqrt{5}$

Guided Solution:
3. $3, \frac{3}{2}, \frac{2}{2}, -\frac{\sqrt{6}}{2}, -\frac{\sqrt{6}}{2}$

C COMPLETING THE SQUARE

We have seen that a quadratic equation like $(x - 5)^2 = 9$ can be solved by using the principle of square roots. We also noted that an equation like $x^2 + 8x + 16 = 49$ can be solved in the same manner because the expression on the left side is the square of a binomial, $(x + 4)^2$. This second procedure is the basis for a method of solving quadratic equations called **completing the square**. *It can be used to solve any quadratic equation.*

Suppose we have the following quadratic equation:

$$x^2 + 10x = 4.$$

If we could add to both sides of the equation a constant that would make the expression on the left the square of a binomial, we could then solve the equation using the principle of square roots.

How can we determine what to add to $x^2 + 10x$ in order to construct the square of a binomial? We want to find a number a such that the following equation is satisfied:

$$x^2 + 10x + a^2 = (x + a)(x + a) = x^2 + 2ax + a^2.$$

Thus, a is such that $2a = 10$. Solving for a, we get $a = 5$; that is, a is half of the coefficient of x in $x^2 + 10x$. Since $a^2 = \left(\frac{10}{2}\right)^2 = 5^2 = 25$, we add 25 to our original expression:

$$x^2 + 10x + 25 \quad \text{is the square of} \quad x + 5;$$

that is,

$$x^2 + 10x + 25 = (x + 5)^2.$$

COMPLETING THE SQUARE

To **complete the square** for an expression like $x^2 + bx$, we take half of the coefficient of x and square it. Then we add that number, which is $(b/2)^2$.

Returning to solve our original equation, we first add 25 on *both* sides to complete the square on the left and find an equation equivalent to our original equation. Then we solve as follows:

$$x^2 + 10x = 4 \qquad \text{Original equation}$$

$$x^2 + 10x + 25 = 4 + 25 \qquad \text{Adding 25: } \left(\tfrac{10}{2}\right)^2 = 5^2 = 25$$

$$(x + 5)^2 = 29$$

$$x + 5 = \sqrt{29} \quad or \quad x + 5 = -\sqrt{29} \qquad \text{Using the principle of square roots}$$

$$x = -5 + \sqrt{29} \quad or \quad x = -5 - \sqrt{29}.$$

The solutions are $-5 \pm \sqrt{29}$.

We have seen that a quadratic equation $(x + c)^2 = d$ can be solved by using the principle of square roots. Any quadratic equation can be put in this form by completing the square. Then we can solve as before.

EXAMPLE 7 Solve: $x^2 + 6x + 8 = 0$.

We have

$$x^2 + 6x + 8 = 0$$
$$x^2 + 6x = -8. \quad \text{Subtracting 8}$$

We take half of 6, $\frac{6}{2} = 3$, and square it, to get 3^2, or 9. Then we add 9 on *both* sides of the equation. This makes the left side the square of a binomial. We have now completed the square.

$$x^2 + 6x + 9 = -8 + 9 \quad \text{Adding 9. The left side is the square of a binomial.}$$
$$(x + 3)^2 = 1$$
$$x + 3 = 1 \quad or \quad x + 3 = -1 \quad \text{Using the principle of square roots}$$
$$x = -2 \quad or \quad x = -4$$

The solutions are -2 and -4.

Do Exercises 8 and 9. ▶

Solve.

8. $x^2 - 6x + 8 = 0$

9. $x^2 + 8x - 20 = 0$

EXAMPLE 8 Solve $x^2 - 4x - 7 = 0$ by completing the square.

We have

$$x^2 - 4x - 7 = 0$$
$$x^2 - 4x = 7 \quad \text{Adding 7}$$
$$x^2 - 4x + 4 = 7 + 4 \quad \text{Adding 4: } \left(\frac{-4}{2}\right)^2 = (-2)^2 = 4$$
$$(x - 2)^2 = 11$$
$$x - 2 = \sqrt{11} \quad or \quad x - 2 = -\sqrt{11} \quad \text{Using the principle of square roots}$$
$$x = 2 + \sqrt{11} \quad or \quad x = 2 - \sqrt{11}.$$

The solutions are $2 \pm \sqrt{11}$.

Do Exercise 10. ▶

GS **10.** Solve: $x^2 - 12x + 23 = 0$.

$$x^2 - 12x = -23$$
$$x^2 - 12x + 36 = -23 + \underline{\quad}$$
$$(\underline{\quad})^2 = 13$$
$$x - 6 = \sqrt{13} \quad or \quad x - 6 = -\underline{\quad}$$
$$x = 6 + \sqrt{13} \quad or \quad x = \underline{\quad}$$

The solutions are $6 \pm \underline{\quad}$.

Example 7, as well as the following example, can be solved more easily by factoring. We solve them by completing the square only to illustrate that completing the square can be used to solve *any* quadratic equation.

EXAMPLE 9 Solve $x^2 + 3x - 10 = 0$ by completing the square.

We have

$$x^2 + 3x - 10 = 0$$
$$x^2 + 3x = 10$$
$$x^2 + 3x + \tfrac{9}{4} = 10 + \tfrac{9}{4} \quad \text{Adding } \tfrac{9}{4}\text{: } \left(\tfrac{3}{2}\right)^2 = \tfrac{9}{4}$$
$$\left(x + \tfrac{3}{2}\right)^2 = \tfrac{40}{4} + \tfrac{9}{4} = \tfrac{49}{4}$$
$$x + \tfrac{3}{2} = \tfrac{7}{2} \quad or \quad x + \tfrac{3}{2} = -\tfrac{7}{2} \quad \text{Using the principle of square roots}$$
$$x = \tfrac{4}{2} \quad or \quad x = -\tfrac{10}{2}$$
$$x = 2 \quad or \quad x = -5.$$

The solutions are 2 and -5.

Do Exercise 11. ▶

11. Solve: $x^2 - 3x - 10 = 0$.

Answers

8. 2, 4 **9.** -10, 2 **10.** $6 \pm \sqrt{13}$
11. -2, 5

Guided Solution:
10. 36, $x - 6$, $\sqrt{13}$, $6 - \sqrt{13}$, $\sqrt{13}$

When the coefficient of x^2 is not 1, we can make it 1, as shown in the following example.

EXAMPLE 10 Solve $2x^2 = 3x + 1$ by completing the square.

We first obtain standard form. Then we multiply by $\frac{1}{2}$ on both sides to make the x^2-coefficient 1.

$$2x^2 = 3x + 1$$

$$2x^2 - 3x - 1 = 0 \qquad \text{Finding standard form}$$

$$\frac{1}{2}(2x^2 - 3x - 1) = \frac{1}{2} \cdot 0 \qquad \text{Multiplying by } \frac{1}{2} \text{ to make the } x^2\text{-coefficient 1}$$

$$x^2 - \frac{3}{2}x - \frac{1}{2} = 0$$

$$x^2 - \frac{3}{2}x = \frac{1}{2} \qquad \text{Adding } \frac{1}{2}$$

$$x^2 - \frac{3}{2}x + \frac{9}{16} = \frac{1}{2} + \frac{9}{16} \qquad \text{Adding } \frac{9}{16}\text{: } \left[\frac{1}{2}\left(-\frac{3}{2}\right)\right]^2 = \left[-\frac{3}{4}\right]^2 = \frac{9}{16}$$

$$\left(x - \frac{3}{4}\right)^2 = \frac{8}{16} + \frac{9}{16}$$

$$\left(x - \frac{3}{4}\right)^2 = \frac{17}{16}$$

$$x - \frac{3}{4} = \frac{\sqrt{17}}{4} \quad \textit{or} \quad x - \frac{3}{4} = -\frac{\sqrt{17}}{4} \qquad \text{Using the principle of square roots}$$

$$x = \frac{3}{4} + \frac{\sqrt{17}}{4} \quad \textit{or} \quad x = \frac{3}{4} - \frac{\sqrt{17}}{4}$$

The solutions are $\frac{3 \pm \sqrt{17}}{4}$.

SOLVING BY COMPLETING THE SQUARE

To solve a quadratic equation $ax^2 + bx + c = 0$ by completing the square:

1. If $a \neq 1$, multiply by $1/a$ so that the x^2-coefficient is 1.
2. If the x^2-coefficient is 1, add so that the equation is in the form

   $x^2 + bx = -c$, or $x^2 + \frac{b}{a}x = -\frac{c}{a}$ if step (1) has been applied.
3. Take half of the x-coefficient and square it. Add the result on both sides of the equation.
4. Express the side with the variables as the square of a binomial.
5. Use the principle of square roots and complete the solution.

Completing the square provides a basis for the quadratic formula, which we will discuss in Section 17.3. It also has other uses in later mathematics courses.

Do Exercise 12.

12. Solve: $2x^2 + 3x - 3 = 0$.

Answer

12. $\frac{-3 \pm \sqrt{33}}{4}$

d APPLICATIONS

EXAMPLE 11 *Falling Object.* The Grand Canyon Skywalk, a horseshoe-shaped glass observation deck, extends 70 ft off the South Rim of the Grand Canyon. This structure, completed in 2007, can support a few hundred people, but the number of visitors allowed on the skywalk at any one time is 120. The Skywalk is approximately 4000 ft above the ground. If a tourist accidentally drops a camera from the observation deck, how long will it take the camera to fall to the ground?

Source: The Grand Canyon Skywalk

1. **Familiarize.** If we did not know anything about this problem, we might consider looking up a formula in a mathematics or physics book. A formula that fits this situation is

   $s = 16t^2,$

   where s is the distance, in feet, traveled by a body falling freely from rest in t seconds. This formula is actually an approximation in that it does not account for air resistance. In this problem, we know the distance s to be 4000 ft. We want to determine the time t that it takes the object to reach the ground.

2. **Translate.** We know that the distance is 4000 ft and that we need to solve for t. We substitute 4000 for s: $4000 = 16t^2$. This gives us a translation.

3. **Solve.** We solve the equation:

$$4000 = 16t^2$$

$$\frac{4000}{16} = t^2 \qquad \text{Solving for } t^2$$

$$\sqrt{\frac{4000}{16}} = t \quad or \quad -\sqrt{\frac{4000}{16}} = t \qquad \text{Using the principle of square roots}$$

$$15.8 \approx t \quad or \quad -15.8 \approx t. \qquad \text{Using a calculator to find the square root and rounding to the nearest tenth}$$

4. **Check.** The number -15.8 cannot be a solution because time cannot be negative in this situation. We substitute 15.8 in the original equation:

   $s = 16(15.8)^2 = 16(249.64) = 3994.24.$

   This answer is close: $3994.24 \approx 4000$. Remember that we rounded to approximate our solution, $t \approx 15.8$. Thus we have a check.

5. **State.** It would take about 15.8 sec for the camera to fall to the ground from the Grand Canyon Skywalk.

Do Exercise 13. ▶

13. *Falling Object.* The CN Tower in Toronto is 1815 ft tall. How long would it take an object to fall to the ground from the top?

Source: *World Almanac*, 2008

Answer

13. About 10.7 sec

17.2 Exercise Set

For Extra Help

MyMathLab®

MathXL® PRACTICE  WATCH  READ  REVIEW 

Reading Check

Determine whether each statement is true or false.

RC1. Every quadratic equation has two different solutions.

RC2. The solutions of $(x + 3)^2 = 5$ are $\sqrt{5}$ and $-\sqrt{5}$.

RC3. To complete the square for $x^2 + 6x$, we need to add 36.

RC4. Every quadratic equation can be solved by completing the square.

a Solve.

1. $x^2 = 121$

2. $x^2 = 100$

3. $5x^2 = 35$

4. $5x^2 = 45$

5. $5x^2 = 3$

6. $2x^2 = 9$

7. $4x^2 - 25 = 0$

8. $9x^2 - 4 = 0$

9. $3x^2 - 49 = 0$

10. $5x^2 - 16 = 0$

11. $4y^2 - 3 = 9$

12. $36y^2 - 25 = 0$

13. $49y^2 - 64 = 0$

14. $8x^2 - 400 = 0$

b Solve.

15. $(x + 3)^2 = 16$

16. $(x - 4)^2 = 25$

17. $(x + 3)^2 = 21$

18. $(x - 3)^2 = 6$

19. $(x + 13)^2 = 8$

20. $(x - 13)^2 = 64$

21. $(x - 7)^2 = 12$

22. $(x + 1)^2 = 14$

23. $(x + 9)^2 = 34$

24. $(t + 5)^2 = 49$

25. $\left(x + \frac{3}{2}\right)^2 = \frac{7}{2}$

26. $\left(y - \frac{3}{4}\right)^2 = \frac{17}{16}$

27. $x^2 - 6x + 9 = 64$

28. $p^2 - 10p + 25 = 100$

29. $x^2 + 14x + 49 = 64$

30. $t^2 + 8t + 16 = 36$

C Solve by completing the square. Show your work.

31. $x^2 - 6x - 16 = 0$

32. $x^2 + 8x + 15 = 0$

33. $x^2 + 22x + 21 = 0$

34. $x^2 + 14x - 15 = 0$

35. $x^2 - 2x - 5 = 0$

36. $x^2 - 4x - 11 = 0$

37. $x^2 - 22x + 102 = 0$

38. $x^2 - 18x + 74 = 0$

39. $x^2 + 10x - 4 = 0$

40. $x^2 - 10x - 4 = 0$

41. $x^2 - 7x - 2 = 0$

42. $x^2 + 7x - 2 = 0$

43. $x^2 + 3x - 28 = 0$

44. $x^2 - 3x - 28 = 0$

45. $x^2 + \frac{3}{2}x - \frac{1}{2} = 0$

46. $x^2 - \frac{3}{2}x - 2 = 0$

47. $2x^2 + 3x - 17 = 0$

48. $2x^2 - 3x - 1 = 0$

49. $3x^2 + 4x - 1 = 0$

50. $3x^2 - 4x - 3 = 0$

51. $2x^2 = 9x + 5$

52. $2x^2 = 5x + 12$

53. $6x^2 + 11x = 10$

54. $4x^2 + 12x = 7$

d Solve.

55. *Burj Khalifa.* The Burj Khalifa in Dubai, The United Arab Emirates, is the tallest structure in the world. It stands at 2723 ft. How long would it take an object to fall from the top?

Source: http://www.skyscrapercenter.com

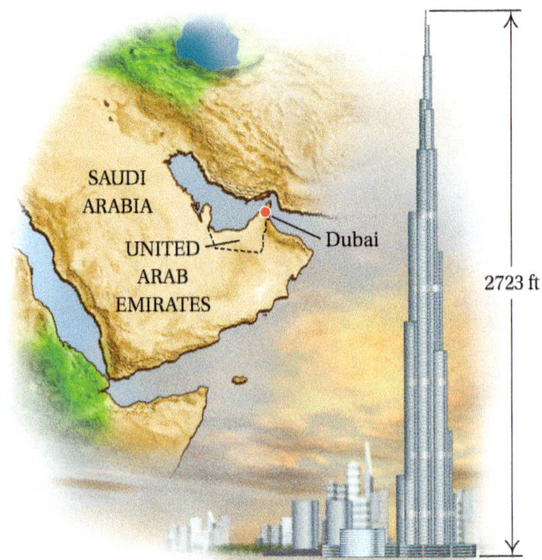

56. *Petronas Towers.* At a height of 1483 ft, the Petronas Towers in Kuala Lumpur, Malaysia, is one of the tallest buildings in the world. How long would it take an object to fall from the top?

Source: *The New York Times Almanac*

57. *Willis Tower.* The Willis Tower in Chicago, formerly called the Sears Tower, is 1451 ft tall. The Willis Tower Skydeck, an observation deck, is 1353 ft above ground. How long would it take an object to fall from the observation deck?

Source: The Willis Tower

58. *Taipei 101.* The Taipei 101 building in Taipei, Taiwan, is 1670 ft tall. How long would it take an object to fall from the top?

Source: *World Almanac*

Skill Maintenance

59. Add: $(3x^4 - x^2 + 1) + (4x^3 - x^2)$. [12.4a]

60. Subtract: $(3x^4 - x^2 + 1) - (4x^3 - x^2)$. [12.4c]

Multiply.

61. $(3x^2 + x)^2$ [12.6c]

62. $(x^2 - x + 1)(2x + 1)$ [12.5d]

Divide.

63. $(12x^4 - 15x^3 - x^2) \div (3x)$ [12.8a]

64. $(5x^2 - x - 3) \div (x - 1)$ [12.8b]

Synthesis

Find b such that the trinomial is a square.

65. $x^2 + bx + 36$

66. $x^2 + bx + 55$

67. $x^2 + bx + 128$

68. $4x^2 + bx + 16$

69. $x^2 + bx + c$

70. $ax^2 + bx + c$

Solve.

71. $4.82x^2 = 12{,}000$

72. $\frac{x}{2} = \frac{32}{x}$

73. $\frac{x}{9} = \frac{36}{4x}$

74. $\frac{4}{m^2 - 7} = 1$

The Quadratic Formula

17.3

We learn to complete the square to prove a general formula that can be used to solve quadratic equations even when they cannot be solved by factoring.

OBJECTIVES

a Solve quadratic equations using the quadratic formula.

b Find approximate solutions of quadratic equations using a calculator.

SKILL TO REVIEW

Objective 9.1a: Evaluate algebraic expressions by substitution.

Evaluate.

1. $-\frac{b}{2a}$, when $a = -2$ and $b = 4$
2. $-x + y$, when $x = -1$ and $y = -10$

a SOLVING USING THE QUADRATIC FORMULA

Each time you solve by completing the square, you perform nearly the same steps. When we repeat the same kind of computation many times, we look for a formula so we can speed up our work. Consider

$$ax^2 + bx + c = 0, \quad a > 0.$$

Let's solve by completing the square. As we carry out the steps, compare them with Example 10 in the preceding section.

$$x^2 + \frac{b}{a}x + \frac{c}{a} = 0 \qquad \text{Multiplying by } \frac{1}{a}$$

$$x^2 + \frac{b}{a}x = -\frac{c}{a} \qquad \text{Adding } -\frac{c}{a}$$

Half of $\frac{b}{a}$ is $\frac{b}{2a}$. The square is $\frac{b^2}{4a^2}$. Thus we add $\frac{b^2}{4a^2}$ on both sides.

$$x^2 + \frac{b}{a}x + \frac{b^2}{4a^2} = -\frac{c}{a} + \frac{b^2}{4a^2} \qquad \text{Adding } \frac{b^2}{4a^2}$$

$$\left(x + \frac{b}{2a}\right)^2 = -\frac{4ac}{4a^2} + \frac{b^2}{4a^2} \qquad \text{Factoring the left side and finding a common denominator on the right}$$

$$\left(x + \frac{b}{2a}\right)^2 = \frac{b^2 - 4ac}{4a^2}$$

$$x + \frac{b}{2a} = \sqrt{\frac{b^2 - 4ac}{4a^2}} \quad or \quad x + \frac{b}{2a} = -\sqrt{\frac{b^2 - 4ac}{4a^2}} \qquad \text{Using the principle of square roots}$$

Since $a > 0$, $\sqrt{4a^2} = 2a$, so we can simplify as follows:

$$x + \frac{b}{2a} = \frac{\sqrt{b^2 - 4ac}}{2a} \quad or \quad x + \frac{b}{2a} = -\frac{\sqrt{b^2 - 4ac}}{2a}.$$

Thus,

$$x = -\frac{b}{2a} + \frac{\sqrt{b^2 - 4ac}}{2a} \quad or \quad x = -\frac{b}{2a} - \frac{\sqrt{b^2 - 4ac}}{2a},$$

so $$x = -\frac{b}{2a} \pm \frac{\sqrt{b^2 - 4ac}}{2a}, \quad \text{or} \quad x = \frac{-b \pm \sqrt{b^2 - 4ac}}{2a}.$$

We now have the following.

THE QUADRATIC FORMULA

The solutions of $ax^2 + bx + c = 0$ are given by

$$x = \frac{-b \pm \sqrt{b^2 - 4ac}}{2a}.$$

Answers

Skill to Review:
1. 1 **2.** −9

The formula also holds when $a < 0$. A similar proof would show this, but we will not consider it here.

EXAMPLE 1 Solve $5x^2 - 8x = -3$ using the quadratic formula.

We first find standard form and determine a, b, and c:

$$5x^2 - 8x + 3 = 0;$$
$$a = 5, \quad b = -8, \quad c = 3.$$

We then use the quadratic formula:

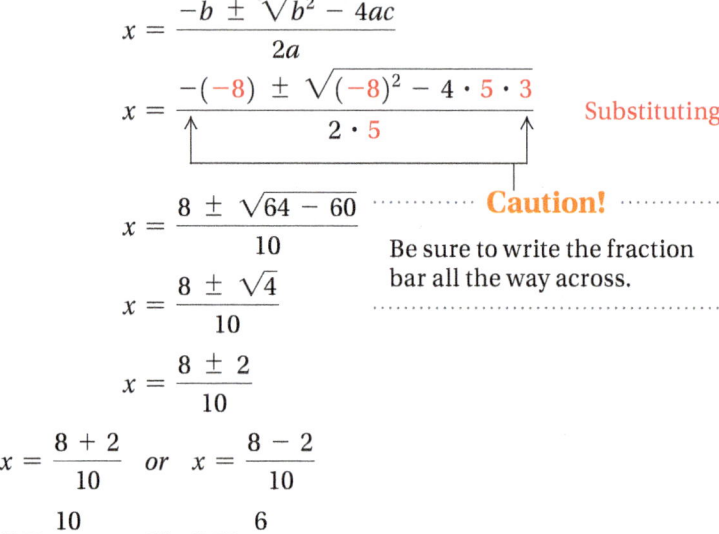

$$x = \frac{-b \pm \sqrt{b^2 - 4ac}}{2a}$$

$$x = \frac{-(-8) \pm \sqrt{(-8)^2 - 4 \cdot 5 \cdot 3}}{2 \cdot 5} \qquad \text{Substituting}$$

$$x = \frac{8 \pm \sqrt{64 - 60}}{10}$$

Caution!

Be sure to write the fraction bar all the way across.

$$x = \frac{8 \pm \sqrt{4}}{10}$$

$$x = \frac{8 \pm 2}{10}$$

$$x = \frac{8 + 2}{10} \quad \text{or} \quad x = \frac{8 - 2}{10}$$

$$x = \frac{10}{10} \quad \text{or} \quad x = \frac{6}{10}$$

$$x = 1 \quad \text{or} \quad x = \frac{3}{5}.$$

The solutions are 1 and $\frac{3}{5}$.

Do Exercise 1.

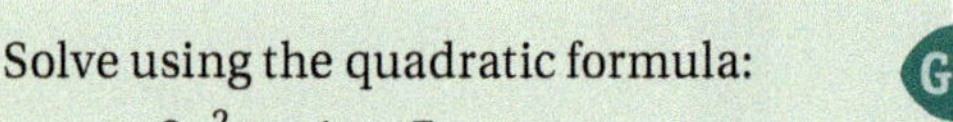

GS

1. Solve using the quadratic formula:

$2x^2 = 4 - 7x.$

Write in standard form:

$2x^2 + 7x - \_\_\_ = 0.$

$a = \_\_\_, b = \_\_\_, c = \_\_\_$

$$x = \frac{-b \pm \sqrt{b^2 - 4ac}}{2a}$$

$$x = \frac{-\_\_\_ \pm \sqrt{\_\_\_^2 - 4 \cdot \_\_\_ \cdot (\_\_\_)}}{2 \cdot \_\_\_}$$

$$x = \frac{-7 \pm \sqrt{49 + \_\_\_}}{4}$$

$$x = \frac{-7 \pm \sqrt{\_\_\_}}{4} = \frac{-7 \pm \_\_\_}{4}$$

$$x = \frac{-7 + 9}{4} \quad \text{or} \quad x = \frac{-7 - 9}{4}$$

$x = \_\_\_$ or $x = \_\_\_$

It would have been easier to solve the equation in Example 1 by factoring. We used the quadratic formula only to illustrate that it can be used to solve any quadratic equation. The following is a general procedure for solving a quadratic equation.

SOLVING QUADRATIC EQUATIONS

To solve a quadratic equation:

1. Check to see if it is in the form $ax^2 = p$ or $(x + c)^2 = d$. If it is, use the principle of square roots.
2. If it is not in the form of (1), write it in standard form, $ax^2 + bx + c = 0$ with a and b nonzero.
3. Then try factoring.
4. If it is not possible to factor or if factoring seems difficult, use the quadratic formula.

The solutions of a quadratic equation can always be found using the quadratic formula. They cannot always be found by factoring. (When the radicand $b^2 - 4ac \geq 0$, the equation has real-number solutions. When $b^2 - 4ac < 0$, the equation has no real-number solutions.)

Answer

1. $-4, \frac{1}{2}$

Guided Solution:

1. $4, 2, 7, -4; \frac{-7 \pm \sqrt{7^2 - 4 \cdot 2 \cdot (-4)}}{2 \cdot 2};$ $32, 81, 9, \frac{1}{2}, -4$

EXAMPLE 2 Solve $x^2 + 3x - 10 = 0$ using the quadratic formula.

The equation is in standard form, so we determine a, b, and c:

$$x^2 + 3x - 10 = 0;$$
$$a = 1, \quad b = 3, \quad c = -10.$$

We then use the quadratic formula:

$$x = \frac{-b \pm \sqrt{b^2 - 4ac}}{2a}$$
$$= \frac{-3 \pm \sqrt{3^2 - 4 \cdot 1 \cdot (-10)}}{2 \cdot 1} \quad \text{Substituting}$$
$$= \frac{-3 \pm \sqrt{9 + 40}}{2}$$
$$= \frac{-3 \pm \sqrt{49}}{2} = \frac{-3 \pm 7}{2}.$$

Thus,

$$x = \frac{-3 + 7}{2} = \frac{4}{2} = 2 \quad or \quad x = \frac{-3 - 7}{2} = \frac{-10}{2} = -5.$$

The solutions are 2 and -5.

Note that when the radicand is a perfect square, as in this example, we could have solved using factoring.

ALGEBRAIC ▶◀ GRAPHICAL CONNECTION

Let's visualize the solutions in Example 2.

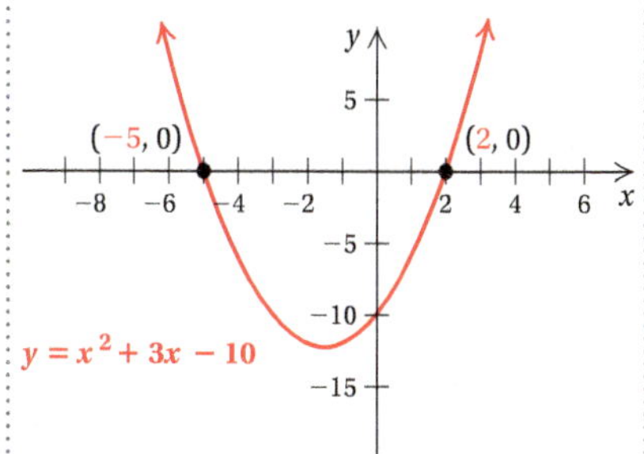

The solutions of $x^2 + 3x - 10 = 0$, -5 and 2, are the first coordinates of the x-intercepts, $(-5, 0)$ and $(2, 0)$, of the graph of $y = x^2 + 3x - 10$.

2. Solve using the quadratic formula:

$x^2 - 3x - 10 = 0.$

Do Exercise 2. ▶

EXAMPLE 3 Solve $x^2 = 4x + 7$ using the quadratic formula. Compare using the quadratic formula here with completing the square as we did in Example 8 of Section 17.2.

We first find standard form and determine a, b, and c:

$$x^2 - 4x - 7 = 0;$$
$$a = 1, \quad b = -4, \quad c = -7.$$

We then use the quadratic formula:

$$x = \frac{-b \pm \sqrt{b^2 - 4ac}}{2a}$$
$$= \frac{-(-4) \pm \sqrt{(-4)^2 - 4 \cdot 1 \cdot (-7)}}{2 \cdot 1} \quad \text{Substituting}$$
$$= \frac{4 \pm \sqrt{16 + 28}}{2} = \frac{4 \pm \sqrt{44}}{2}$$
$$= \frac{4 \pm \sqrt{4 \cdot 11}}{2} = \frac{4 \pm \sqrt{4}\sqrt{11}}{2}$$
$$= \frac{4 \pm 2\sqrt{11}}{2} = \frac{2 \cdot 2 \pm 2\sqrt{11}}{2 \cdot 1}$$
$$= \frac{2(2 \pm \sqrt{11})}{2 \cdot 1} = \frac{2}{2} \cdot \frac{2 \pm \sqrt{11}}{1} \quad \text{Factoring out 2 in the numerator and the denominator}$$
$$= 2 \pm \sqrt{11}.$$

The solutions are $2 + \sqrt{11}$ and $2 - \sqrt{11}$, or $2 \pm \sqrt{11}$.

GS **3.** Solve using the quadratic formula:

$x^2 + 4x = 7.$

Write in standard form:

$x^2 + 4x - \square = 0.$

$a = \square, b = \square, c = \square$

$$x = \frac{-b \pm \sqrt{b^2 - 4ac}}{2a}$$
$$= \frac{-\square \pm \sqrt{\square^2 - 4 \cdot \square \cdot (\square)}}{2 \cdot \square}$$
$$= \frac{-4 \pm \sqrt{16 + \square}}{2} = \frac{-4 \pm \sqrt{\square}}{2}$$
$$= \frac{-4 \pm 2\sqrt{\square}}{2} = \frac{2(-2 \pm \sqrt{11})}{2 \cdot 1}$$
$$= \square \pm \square$$

Do Exercise 3. ▶

Answers

2. $-2, 5$ **3.** $-2 \pm \sqrt{11}$

Guided Solution:

3. 7, 1, 4, -7; $\dfrac{-4 \pm \sqrt{4^2 - 4 \cdot 1 \cdot (-7)}}{2 \cdot 1}$; 28, 44, 11; -2, $\sqrt{11}$

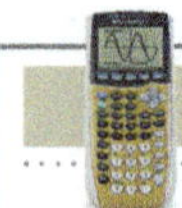

CALCULATOR CORNER

Visualizing Solutions of Quadratic Equations To see that there are no real-number solutions of the equation in Example 4,

$$x^2 + x = -1,$$

we graph $y_1 = x^2 + x$ and $y_2 = -1$.

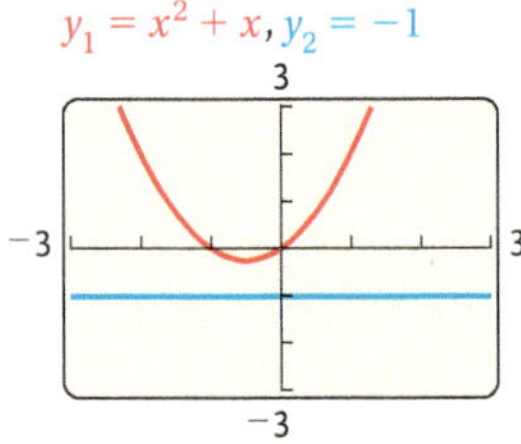

We see that the graphs do not intersect. Thus there is no real number for which $y_1 = y_2$, or $x^2 + x = -1$.

EXERCISE:

1. Explain how the graph of $y = x^2 + x + 1$ shows that the equation in Example 4, $x^2 + x = -1$, has no real-number solutions.

EXAMPLE 4 Solve $x^2 + x = -1$ using the quadratic formula.

We first find standard form and determine a, b, and c:

$$x^2 + x + 1 = 0;$$
$$a = 1, \quad b = 1, \quad c = 1.$$

We then use the quadratic formula:

$$x = \frac{-b \pm \sqrt{b^2 - 4ac}}{2a} = \frac{-1 \pm \sqrt{1^2 - 4 \cdot 1 \cdot 1}}{2 \cdot 1} = \frac{-1 \pm \sqrt{-3}}{2}.$$

Note that the radicand ($b^2 - 4ac = -3$) in the quadratic formula is negative. Thus there are no real-number solutions because square roots of negative numbers do not exist as real numbers.

EXAMPLE 5 Solve $3x^2 = 7 - 2x$ using the quadratic formula.

We first find standard form and determine a, b, and c:

$$3x^2 + 2x - 7 = 0;$$
$$a = 3, \quad b = 2, \quad c = -7.$$

We then use the quadratic formula:

$$x = \frac{-b \pm \sqrt{b^2 - 4ac}}{2a} = \frac{-2 \pm \sqrt{2^2 - 4 \cdot 3 \cdot (-7)}}{2 \cdot 3}$$
$$= \frac{-2 \pm \sqrt{4 + 84}}{2 \cdot 3}$$
$$= \frac{-2 \pm \sqrt{88}}{6} = \frac{-2 \pm \sqrt{4 \cdot 22}}{6}$$
$$= \frac{-2 \pm \sqrt{4}\sqrt{22}}{6} = \frac{-2 \pm 2\sqrt{22}}{6}$$
$$= \frac{2(-1 \pm \sqrt{22})}{2 \cdot 3}$$
$$= \frac{2}{2} \cdot \frac{-1 \pm \sqrt{22}}{3} = \frac{-1 \pm \sqrt{22}}{3}.$$

The solutions are $\frac{-1 + \sqrt{22}}{3}$ and $\frac{-1 - \sqrt{22}}{3}$, or $\frac{-1 \pm \sqrt{22}}{3}$.

Do Exercises 4 and 5.

Solve using the quadratic formula.

4. $x^2 = x - 1$

5. $5x^2 - 8x = 3$

b APPROXIMATE SOLUTIONS

A calculator can be used to approximate solutions of quadratic equations.

EXAMPLE 6 Use a calculator to approximate to the nearest tenth the solutions to the equation in Example 5.

Using a calculator, we have

$$\frac{-1 + \sqrt{22}}{3} \approx 1.230138587 \approx 1.2 \text{ to the nearest tenth, and}$$
$$\frac{-1 - \sqrt{22}}{3} \approx -1.896805253 \approx -1.9 \text{ to the nearest tenth.}$$

The approximate solutions are 1.2 and −1.9.

Do Exercise 6.

6. Approximate the solutions to the equation in Margin Exercise 5. Round to the nearest tenth.

Answers

4. No real-number solutions 5. $\frac{4 \pm \sqrt{31}}{5}$
6. −0.3, 1.9

17.3 Exercise Set

For Extra Help MyMathLab® MathXL® PRACTICE WATCH READ REVIEW

Reading Check

Choose from the column on the right the quadratic equation that has the given values of a, b, and c.

RC1. $a = 1,\ b = -7,\ c = 2$

RC2. $a = 1,\ b = 7,\ c = -2$

RC3. $a = 3,\ b = 7,\ c = -2$

RC4. $a = 3,\ b = -7,\ c = -2$

RC5. $a = 3,\ b = -7,\ c = 0$

RC6. $a = 3,\ b = 0,\ c = -7$

a) $3x^2 + 7x - 2 = 0$

b) $3x^2 - 2 = 7x$

c) $7x - 2 = x^2$

d) $2 - 7x - x^2 = 0$

e) $3x^2 = 7x$

f) $3x^2 = 7$

a Solve. Try factoring first. If factoring is not possible or is difficult, use the quadratic formula.

1. $x^2 - 4x = 21$

2. $x^2 + 8x = 9$

3. $x^2 = 6x - 9$

4. $x^2 = 24x - 144$

5. $3y^2 - 2y - 8 = 0$

6. $3y^2 - 7y + 4 = 0$

7. $4x^2 + 4x = 15$

8. $4x^2 + 12x = 7$

9. $x^2 - 9 = 0$

10. $x^2 - 16 = 0$

11. $x^2 - 2x - 2 = 0$

12. $x^2 - 2x - 11 = 0$

13. $y^2 - 10y + 22 = 0$

14. $y^2 + 6y - 1 = 0$

15. $x^2 + 4x + 4 = 7$

16. $x^2 - 2x + 1 = 5$

17. $3x^2 + 8x + 2 = 0$

18. $3x^2 - 4x - 2 = 0$

19. $2x^2 - 5x = 1$

20. $4x^2 + 4x = 5$

21. $2y^2 - 2y - 1 = 0$

22. $4y^2 + 4y - 1 = 0$

23. $2t^2 + 6t + 5 = 0$

24. $4y^2 + 3y + 2 = 0$

25. $3x^2 = 5x + 4$

26. $2x^2 + 3x = 1$

27. $2y^2 - 6y = 10$

28. $5m^2 = 3 + 11m$

29. $\dfrac{x^2}{x+3} - \dfrac{5}{x+3} = 0$

30. $\dfrac{x^2}{x-4} - \dfrac{7}{x-4} = 0$

31. $x + 2 = \dfrac{3}{x+2}$

32. $x - 3 = \dfrac{5}{x-3}$

33. $\dfrac{1}{x} + \dfrac{1}{x+1} = \dfrac{1}{3}$

34. $\dfrac{1}{x} + \dfrac{1}{x+6} = \dfrac{1}{5}$

b Solve using the quadratic formula. Use a calculator to approximate the solutions to the nearest tenth.

35. $x^2 - 4x - 7 = 0$

36. $x^2 + 2x - 2 = 0$

37. $y^2 - 6y - 1 = 0$

38. $y^2 + 10y + 22 = 0$

39. $4x^2 + 4x = 1$

40. $4x^2 = 4x + 1$

41. $3x^2 - 8x + 2 = 0$

42. $3x^2 + 4x - 2 = 0$

Skill Maintenance

Solve.

43. $2(x + 3) - (x - 5) = 6x$ [10.3c]

44. $-9x \geq \dfrac{1}{2}$ [10.7d]

45. $\dfrac{1}{3}y + \dfrac{1}{2} = \dfrac{2}{3} - \dfrac{5}{6}y$ [10.3b]

46. $t^2 - 7t - 8 = 0$ [13.7b]

47. $3d^2 - 3 = 0$ [13.7b]

48. $\sqrt{3x} = 15$ [16.5a]

49. $\sqrt{2x - 1} = x - 2$ [16.5a]

50. $y^2 = -3y$ [13.7b]

51. $2n^2 - n = 3$ [13.7b]

52. $\dfrac{2}{x} = \dfrac{x}{8}$ [14.7a]

53. $\dfrac{x}{x-1} + \dfrac{1}{x} = 1$ [14.7a]

54. $\dfrac{5x}{x+2} - \dfrac{1}{x-1} = 3$ [14.7a]

Synthesis

Solve.

55. $5x + x(x - 7) = 0$

56. $x(3x + 7) - 3x = 0$

57. $3 - x(x - 3) = 4$

58. $x(5x - 7) = 1$

59. $(y + 4)(y + 3) = 15$

60. $(y + 5)(y - 1) = 27$

61. $x^2 + (x + 2)^2 = 7$

62. $x^2 + (x + 1)^2 = 5$

63. Use a graphing calculator to determine whether the equation $x^2 + x = 1$ has real-number solutions.

64. Use a graphing calculator to determine whether the equation $x^2 = 2x - 3$ has real-number solutions.

65.–72. Use a graphing calculator to approximate the solutions of the equations in Exercises 35–42. Compare your answers with those found using the quadratic formula.

Mid-Chapter Review

Concept Reinforcement

Determine whether each statement is true or false.

__________ **1.** The equation $x^2 = -4$ has no real-number solutions. [17.2a]

__________ **2.** The solutions of $ax^2 + bx + c = 0$ are the first coordinates of the y-intercepts of the graph of $y = ax^2 + bx + c$. [17.1c]

__________ **3.** A quadratic equation of the type $ax^2 + bx = 0$, where $c = 0$ and $b \neq 0$, will always have 0 as one solution and a nonzero number as the other solution. [17.1b]

Guided Solutions

Fill in each blank with the number or the expression that creates a correct solution.

4. Solve $x^2 - 6x - 2 = 0$ by completing the square. [17.2c]

$$x^2 - 6x - 2 = 0$$
$$x^2 - 6x = \square$$
$$x^2 - 6x + \square = 2 + \square$$
$$(x - \square)^2 = \square$$
$$x - \square = \pm\sqrt{\square}$$
$$x = \square \pm \sqrt{11}$$

5. Solve $3x^2 = 8x - 2$ using the quadratic formula. [17.3a]

$$3x^2 = 8x - 2$$
$$3x^2 - \square + \square = 0 \quad \text{Standard form}$$
$$a = \square, \quad b = \square, \quad c = \square$$

We substitute for a, b, and c in the quadratic formula:

$$x = \frac{-b \pm \sqrt{b^2 - 4ac}}{2a} \quad \text{Quadratic formula}$$
$$= \frac{-\square \pm \sqrt{(\square)^2 - 4 \cdot \square \cdot \square}}{2 \cdot \square} \quad \text{Substituting}$$
$$= \frac{\square \pm \sqrt{64 - \square}}{\square} = \frac{8 \pm \sqrt{\square}}{6} = \frac{8 \pm \sqrt{\square \cdot 10}}{6}$$
$$= \frac{8 \pm \square\sqrt{10}}{6} = \frac{2(\square \pm \sqrt{10})}{2 \cdot \square} = \frac{\square \pm \sqrt{10}}{\square}.$$

Mixed Review

Write in standard form and determine a, b, and c. [17.1a]

6. $q^2 - 5q + 10 = 0$

7. $6 - x^2 = 14x + 2$

8. $17z = 3z^2$

Solve by factoring.

9. $16x = 48x^2$ [17.1b]

10. $x(x - 3) = 10$ [17.1c]

11. $20x^2 - 20x = 0$ [17.1b]

12. $x^2 = 14x - 49$ [17.1c]

13. $t^2 + 2t = 0$ [17.1b]

14. $18w^2 + 21w = 4$ [17.1c]

15. $9y^2 - 5y^2 = 82y + 6y$ [17.1b]

16. $2(s - 3) = s(s - 3)$ [17.1c]

17. $8y^2 - 40y = -7y + 35$ [17.1c]

Solve by completing the square. [17.2c]

18. $x^2 + 2x - 3 = 0$

19. $x^2 - 9x + 6 = 0$

20. $2x^2 = 7x + 8$

21. $y^2 + 80 = 18y$

22. $t^2 + \frac{3}{2}t - \frac{3}{2} = 0$

23. $x + 7 = -3x^2$

Solve.

24. $6x^2 = 384$ [17.2a]

25. $5y^2 + 2y + 3 = 0$ [17.3a]

26. $6(x - 3)^2 = 12$ [17.2b]

27. $4x^2 + 4x = 3$ [17.3a]

28. $8y^2 - 5 = 19$ [17.2a]

29. $a^2 = a + 1$ [17.3a]

30. $(w - 2)^2 = 100$ [17.2b]

31. $5m^2 + 2m = -3$ [17.3a]

32. $\left(y - \frac{1}{2}\right)^2 = \frac{5}{4}$ [17.2b]

33. $3x^2 - 75 = 0$ [17.2a]

34. $2x^2 - 2x - 5 = 0$ [17.3a]

35. $(x + 2)^2 = -5$ [17.2b]

Solve and use a calculator to approximate the solutions to the nearest tenth. [17.3b]

36. $y^2 - y - 8 = 0$

37. $2x^2 + 7x + 1 = 0$

For each equation in Exercises 38–42, select from the column on the right the correct description of the solutions of the equation.

38. $x^2 - x - 6 = 0$ [17.3a]

39. $x^2 = -9$ [17.2a]

40. $x^2 = 31$ [17.2a]

41. $x^2 = 0$ [17.2a]

42. $x^2 - x + 6 = 0$ [17.3a]

A. Two real-number solutions
B. No real-number solutions
C. 0 is the only solution.

43. Solve: $(x - 3)^2 = 36$. [17.2b]

A. $-9, 3$
B. $-33, 39$
C. $-3, 9$
D. $\sqrt{6}$

44. Simplify: $\frac{-24 \pm \sqrt{720}}{18}$. [17.3a]

A. $\frac{-8 \pm 4\sqrt{5}}{6}$
B. $\frac{-4 \pm 2\sqrt{5}}{3}$
C. $\frac{-4 \pm \sqrt{20}}{3}$
D. $-2 \pm 2\sqrt{5}$

Understanding Through Discussion and Writing

45. Mark asserts that the solution of a quadratic equation is $3 \pm \sqrt{14}$ and states that there is only one solution. What mistake is being made? [17.2b]

46. Find the errors in the following solution of the equation $x^2 + x = 6$. [17.1c]

$$x^2 + x = 6$$
$$x(x + 1) = 6$$
$$x = 6 \quad or \quad x + 1 = 6$$
$$x = 6 \quad or \quad x = 5$$

47. Explain how the graph of $y = (x - 2)(x + 3)$ is related to the solutions of the equation $(x - 2)(x + 3) = 0$. [17.1a]

48. Under what condition(s) would using the quadratic formula *not* be the easiest way to solve a quadratic equation? [17.3a]

49. Write a quadratic equation in the form $y = ax^2 + bx + c$ that does not cross the x-axis. [17.3a]

50. Explain how you might go about constructing a quadratic equation whose solutions are -5 and 7. [17.1c]

STUDYING FOR SUCCESS *Looking Ahead*

- ☐ As you register for next semester's courses, keep in mind your work and family commitments as well as your course load.
- ☐ If you are registering for another math course, consider keeping your notes, tests, and text from this course as a resource.

Formulas 17.4

OBJECTIVE

a Solve a formula for a specified letter.

SKILL TO REVIEW

Objective 10.4b: Solve a formula for a specified letter.

Solve for the indicated letter.

1. $Q = mx + y$, for x
2. $R = \dfrac{x + y + z}{2}$, for z

a SOLVING FORMULAS

Formulas arise frequently in the natural and social sciences, business, engineering, and health care. The same steps that are used to solve a linear, rational, radical, or quadratic equation can also be used to solve a formula that appears in one of these forms.

EXAMPLE 1 *Intelligence Quotient.* The formula $Q = \dfrac{100m}{c}$ is used to determine the intelligence quotient, Q, of a person of mental age m and chronological age c. Solve for c.

$$Q = \frac{100m}{c}$$

$$c \cdot Q = c \cdot \frac{100m}{c} \qquad \text{Multiplying by } c \text{ on both sides to clear the fraction}$$

$$cQ = 100m \qquad \text{Simplifying}$$

$$c = \frac{100m}{Q}. \qquad \text{Dividing by } Q \text{ on both sides}$$

This formula can be used to determine a person's chronological, or actual, age from his or her mental age and intelligence quotient.

Do Margin Exercise 1.

1. a) Solve for I: $E = \dfrac{9R}{I}$.
   b) Solve for R: $E = \dfrac{9R}{I}$.

EXAMPLE 2 Solve for x: $y = ax + bx - 4$.

$$y = ax + bx - 4 \qquad \text{We want this letter alone on one side.}$$

$$y + 4 = ax + bx \qquad \text{Adding 4. All terms containing } x \text{ are on the right side of the equation.}$$

$$y + 4 = (a + b)x \qquad \text{Factoring out the } x$$

$$\frac{y + 4}{(a + b)} = \frac{(a + b)x}{(a + b)} \qquad \text{Dividing by } a + b \text{ on both sides}$$

$$\frac{y + 4}{a + b} = x \qquad \text{Simplifying. The answer can also be written as } x = \frac{y + 4}{a + b}.$$

Do Margin Exercise 2.

2. Solve for x: $y = ax - bx + 5$.

Answers

Skill to Review:
1. $x = \dfrac{Q - y}{m}$ 2. $z = 2R - x - y$

Margin Exercises:
1. (a) $I = \dfrac{9R}{E}$; (b) $R = \dfrac{EI}{9}$ 2. $x = \dfrac{y - 5}{a - b}$

Caution!

Had we performed the following steps in Example 2, we would *not* have solved for x:

$$y = ax + bx - 4$$

$$y - ax + 4 = bx \quad \text{Subtracting } ax \text{ and adding 4}$$

x occurs on both sides of the equals sign.

$$\frac{y - ax + 4}{b} = x. \quad \text{Dividing by } b$$

The mathematics of each step is correct, but since x occurs on both sides of the formula, *we have not solved the formula for x.* Remember that the letter being solved for should be **alone** on one side of the equation, with no occurrence of that letter on the other side!

EXAMPLE 3 Solve the following work formula for t:

$$\frac{t}{a} + \frac{t}{b} = 1.$$

We clear fractions by multiplying by the LCM, which is ab:

$$ab \cdot \left(\frac{t}{a} + \frac{t}{b}\right) = ab \cdot 1 \quad \text{Multiplying by } ab$$

$$ab \cdot \frac{t}{a} + ab \cdot \frac{t}{b} = ab \quad \text{Using a distributive law to remove parentheses}$$

$$bt + at = ab \quad \text{Simplifying}$$

$$(b + a)t = ab \quad \text{Factoring out } t$$

$$t = \frac{ab}{b + a}. \quad \text{Dividing by } b + a$$

Do Exercise 3.

3. *Optics Formula.* Solve for f:

$$\frac{1}{p} + \frac{1}{q} = \frac{1}{f}.$$

GS

The LCM is ______.

$$pqf\left(\frac{1}{p} + \frac{1}{q}\right) = pqf\left(\frac{1}{f}\right)$$

$$pqf \cdot \frac{1}{p} + pqf\left(\frac{1}{q}\right) = pqf\left(\frac{1}{f}\right)$$

$$qf + \underline{\quad} = pq$$

$$f(\underline{\quad}) = pq$$

$$f = \frac{pq}{\underline{\quad}}$$

EXAMPLE 4 ***Distance to the Horizon.*** Solve $D = \sqrt{2h}$ for h, where D is the approximate distance, in miles, that a person can see to the horizon from a height h, in feet.

This is a radical equation. Recall that we first isolate the radical. Then we use the principle of squaring.

$$D = \sqrt{2h}$$

$$D^2 = (\sqrt{2h})^2 \quad \text{Using the principle of squaring (Section 16.5)}$$

$$D^2 = 2h \quad \text{Simplifying}$$

$$\frac{D^2}{2} = h \quad \text{Dividing by 2}$$

EXAMPLE 5 ***Period of a Pendulum.*** Solve $T = 2\pi\sqrt{\frac{L}{g}}$ for g, where T is the period, in seconds, of a pendulum of length L, in feet, and g is a gravitational constant.

We have

$$\frac{T}{2\pi} = \sqrt{\frac{L}{g}} \quad \text{Dividing by } 2\pi \text{ to isolate the radical}$$

$$\left(\frac{T}{2\pi}\right)^2 = \left(\sqrt{\frac{L}{g}}\right)^2. \quad \text{Using the principle of squaring}$$

Answer

3. $f = \dfrac{pq}{q + p}$

Guided Solution:

3. $pqf, pf, q + p, q + p$

Then

$$\frac{T^2}{4\pi^2} = \frac{L}{g}$$

$$gT^2 = 4\pi^2 L \quad \text{Multiplying by } 4\pi^2 g \text{ to clear fractions}$$

$$g = \frac{4\pi^2 L}{T^2}. \quad \text{Dividing by } T^2 \text{ to get } g \text{ alone}$$

Do Exercises 4–6. ▶

In most formulas, the letters represent nonnegative numbers, so we need not use absolute values when taking square roots.

EXAMPLE 6 *Torricelli's Theorem.* The speed v of a liquid leaving a water cooler from an opening is related to gravity g and the height h of the top of the liquid above the opening by the formula

$$h = \frac{v^2}{2g}.$$

Solve for v.

Since v^2 appears by itself and there is no expression involving v, we first solve for v^2. Then we use the principle of square roots, taking only the nonnegative square root because v is nonnegative.

$$2gh = v^2 \quad \text{Multiplying by } 2g \text{ to clear the fraction}$$

$$\sqrt{2gh} = v \quad \text{Using the principle of square roots. Assume that } v \text{ is nonnegative.}$$

Do Exercise 7. ▶

EXAMPLE 7 Solve $d = \dfrac{n^2 - 3n}{2}$ for n, where d is the number of diagonals of an n-sided polygon.

In this case, there is a term involving n as well as an n^2-term. Thus we must use the quadratic formula.

$$d = \frac{n^2 - 3n}{2}$$

$$n^2 - 3n = 2d \quad \text{Multiplying by 2 to clear fractions}$$

$$n^2 - 3n - 2d = 0 \quad \text{Finding standard form}$$

$$a = 1, \quad b = -3, \quad c = -2d \quad \text{The variable is } n; d \text{ represents a constant.}$$

$$n = \frac{-b \pm \sqrt{b^2 - 4ac}}{2a} \quad \text{Quadratic formula}$$

$$= \frac{-(-3) \pm \sqrt{(-3)^2 - 4 \cdot 1 \cdot (-2d)}}{2 \cdot 1} \quad \text{Substituting into the quadratic formula}$$

$$= \frac{3 + \sqrt{9 + 8d}}{2} \quad \text{Using the positive root}$$

Do Exercise 8. ▶

4. Solve for L: $r = 2\sqrt{5L}$ (the speed of a skidding car).

5. Solve for L: $T = 2\pi\sqrt{\dfrac{L}{g}}$.

6. Solve for m: $c = \sqrt{\dfrac{E}{m}}$.

GS **7.** Solve for r: $A = \pi r^2$ (the area of a circle).

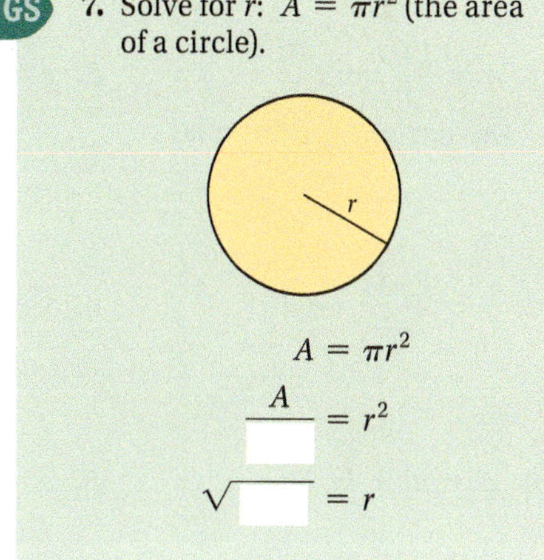

$$A = \pi r^2$$

$$\frac{A}{\square} = r^2$$

$$\sqrt{\square} = r$$

8. Solve for n: $N = n^2 - n$.

Answers

4. $L = \dfrac{r^2}{20}$ **5.** $L = \dfrac{T^2 g}{4\pi^2}$ **6.** $m = \dfrac{E}{c^2}$

7. $r = \sqrt{\dfrac{A}{\pi}}$ **8.** $n = \dfrac{1 + \sqrt{1 + 4N}}{2}$

Guided Solution:

7. $\pi, \dfrac{A}{\pi}$

17.4 Exercise Set

For Extra Help MyMathLab® MathXL® PRACTICE WATCH READ REVIEW

✓ Reading Check

For each formula given, choose from the column on the right the best step to take first when solving the formula for x. Choices may be used more than once or not at all.

RC1. $Q = \dfrac{50x}{n}$

RC2. $t = ax + bx$

RC3. $\dfrac{x}{a} + 2 = \dfrac{x}{b}$

RC4. $N = \sqrt{3x}$

RC5. $x^2 = \dfrac{v}{50n}$

RC6. $p = \dfrac{5x^2 + x}{2}$

a) Factor out the x.
b) Divide both sides by x.
c) Use the principle of squaring.
d) Use the principle of square roots.
e) Clear fractions by multiplying both sides of the equation by the LCM of the denominators.

a Solve for the indicated letter.

1. $q = \dfrac{VQ}{I}$, for I
(An engineering formula)

2. $y = \dfrac{4A}{a}$, for a

3. $S = \dfrac{kmM}{d^2}$, for m

4. $S = \dfrac{kmM}{d^2}$, for M

5. $S = \dfrac{kmM}{d^2}$, for d^2

6. $T = \dfrac{10t}{W^2}$, for W^2

7. $T = \dfrac{10t}{W^2}$, for W

8. $S = \dfrac{kmM}{d^2}$, for d

9. $A = at + bt$, for t

10. $S = rx + sx$, for x

11. $y = ax + bx + c$, for x

12. $y = ax - bx - c$, for x

13. $\dfrac{t}{a} + \dfrac{t}{b} = 1$, for a
(A work formula)

14. $\dfrac{t}{a} + \dfrac{t}{b} = 1$, for b
(A work formula)

15. $\dfrac{1}{p} + \dfrac{1}{q} = \dfrac{1}{f}$, for p
(An optics formula)

16. $\dfrac{1}{p} + \dfrac{1}{q} = \dfrac{1}{f}$, for q
(An optics formula)

17. $A = \dfrac{1}{2}bh$, for b
(The area of a triangle)

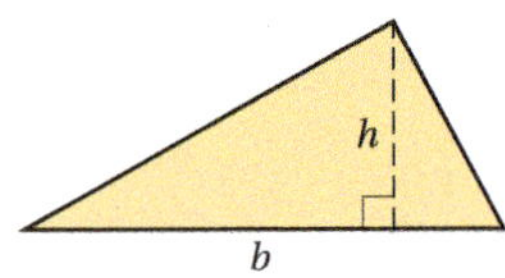

18. $s = \dfrac{1}{2}gt^2$, for g

19. $S = 2\pi r(r + h)$, for h
(The surface area of a right circular cylinder)

20. $S = 2\pi(r + h)$, for r

21. $\frac{1}{R} = \frac{1}{r_1} + \frac{1}{r_2}$, for R
(An electricity formula)

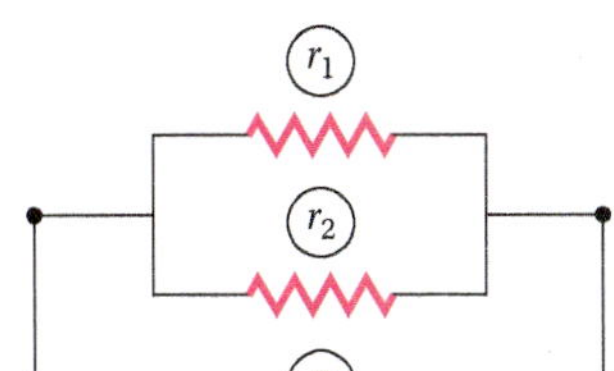

22. $\frac{1}{R} = \frac{1}{r_1} + \frac{1}{r_2}$, for r_1

23. $P = 17\sqrt{Q}$, for Q

24. $A = 1.4\sqrt{t}$, for t

25. $v = \sqrt{\frac{2gE}{m}}$, for E

26. $Q = \sqrt{\frac{aT}{c}}$, for T

27. $S = 4\pi r^2$, for r

28. $E = mc^2$, for c

29. $P = kA^2 + mA$, for A

30. $Q = ad^2 - cd$, for d

31. $c^2 = a^2 + b^2$, for a

32. $c = \sqrt{a^2 + b^2}$, for b

33. $s = 16t^2$, for t

34. $V = \pi r^2 h$, for r

35. $A = \pi r^2 + 2\pi rh$, for r

36. $A = 2\pi r^2 + 2\pi rh$, for r

37. $F = \frac{Av^2}{400}$, for v

38. $A = \frac{\pi r^2 S}{360}$, for r

39. $c = \sqrt{a^2 + b^2}$, for a

40. $c^2 = a^2 + b^2$, for b

41. $h = \frac{a}{2}\sqrt{3}$, for a
(The height of an equilateral triangle with sides of length a)

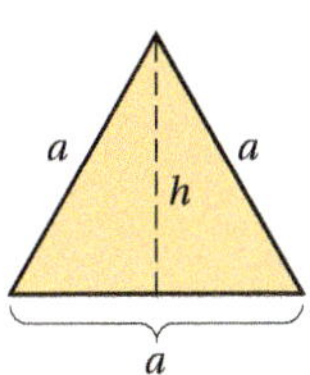

42. $d = s\sqrt{2}$, for s
(The hypotenuse of an isosceles right triangle with s the length of the legs)

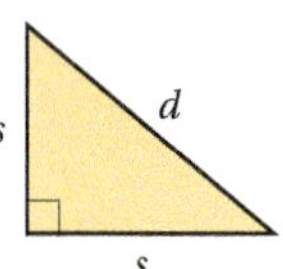

43. $n = aT^2 - 4T + m$, for T

44. $y = ax^2 + bx + c$, for x

45. $v = 2\sqrt{\frac{2kT}{\pi m}}$, for T

46. $E = \frac{1}{2}mv^2 + mgy$, for v

47. $3x^2 = d^2$, for x

48. $c = \sqrt{\frac{E}{m}}$, for E

49. $N = \frac{n^2 - n}{2}$, for n

50. $M = \frac{m}{\sqrt{1 - \left(\frac{v}{c}\right)^2}}$, for c

51. $S = \frac{a + b}{3b}$, for b

52. $Q = \frac{a - b}{2b}$, for b

53. $\frac{A - B}{AB} = Q$, for B

54. $L = \frac{Mt + g}{t}$, for t

55. $S = 180(n - 2)$, for n

56. $S = \frac{n}{2}(a + 1)$, for a

57. $A = P(1 + rt)$, for t (An interest formula)

58. $A = P(1 + rt)$, for r (An interest formula)

59. $\frac{A}{B} = \frac{C}{D}$, for D

60. $\frac{A}{B} = \frac{C}{D}$, for B

Skill Maintenance

In a right triangle, where a and b represent the lengths of the legs and c represents the length of the hypotenuse, find the length of the side not given. Give an exact answer and an approximation to three decimal places. [16.6a]

61. $a = 4, b = 7$

62. $b = 11, c = 14$

63. $a = 4, b = 5$

64. $a = 10, c = 12$

65. $c = 8\sqrt{17}, a = 2$

66. $a = \sqrt{2}, b = \sqrt{3}$

Solve. [16.6b]

67. *Guy Wire.* How long is a guy wire reaching from the top of an 18-ft pole to a point on the ground 10 ft from the pole? Give an exact answer and an approximation to three decimal places.

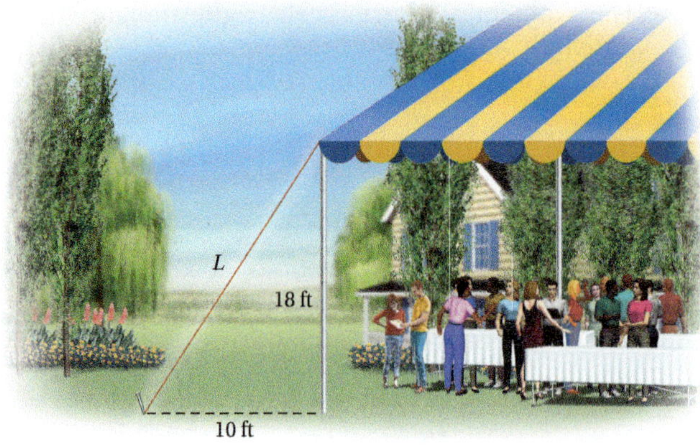

68. *Soccer Fields.* The smallest regulation soccer field is 50 yd wide and 100 yd long. Find the length of a diagonal of such a field.

Simplify.

69. $\sqrt{80}$ [16.2a]

70. $\sqrt{9000x^{10}}$ [16.2b]

Multiply and simplify. [16.2c]

71. $3\sqrt{t} \cdot \sqrt{t}$

72. $\sqrt{8x^2} \cdot \sqrt{24x^3}$

Add or subtract. [16.4a]

73. $\sqrt{40} - 2\sqrt{10} + \sqrt{90}$

74. $\sqrt{18} + \sqrt{50} - 3\sqrt{8}$

Synthesis

75. The circumference C of a circle is given by $C = 2\pi r$.

a) Solve $C = 2\pi r$ for r.

b) The area is given by $A = \pi r^2$. Express the area in terms of the circumference C.

c) Express the circumference C in terms of the area A.

76. Solve $3ax^2 - x - 3ax + 1 = 0$ for x.

Applications and Problem Solving

17.5

OBJECTIVE

a Solve applied problems using quadratic equations.

a USING QUADRATIC EQUATIONS TO SOLVE APPLIED PROBLEMS

EXAMPLE 1 ***Blueberry Farming.*** Kevon is investing in blueberry farming. The area of his rectangular blueberry field is 4800 yd^2. The length is 10 yd longer than five times the width. Find the dimensions of the blueberry field.

1. **Familiarize.** We first make a drawing and label it with both known and unknown information. We let $w =$ the width of the rectangle. The length of the rectangle is 10 yd longer than five times the width. Thus the length is $5w + 10$.

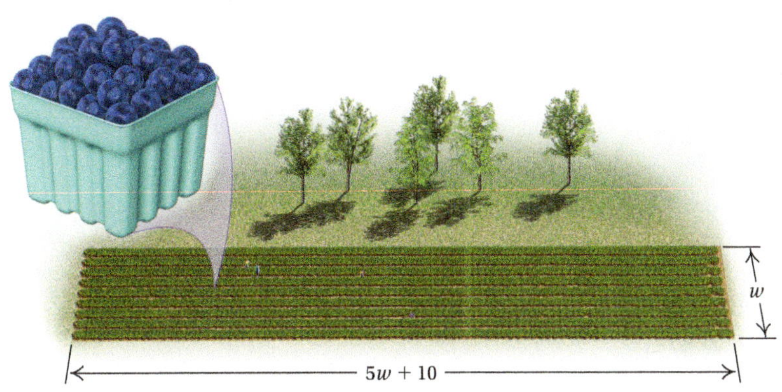

2. **Translate.** Recall that area is length $\times$ width. Thus we have two expressions for the area of the rectangle: $(5w + 10)(w)$ and 4800. This gives us a translation:

$$(5w + 10)(w) = 4800.$$

3. **Solve.** We solve the equation:

$$5w^2 + 10w = 4800$$
$$5w^2 + 10w - 4800 = 0$$
$$w^2 + 2w - 960 = 0 \quad \text{Dividing by 5}$$
$$(w + 32)(w - 30) = 0 \quad \text{Factoring (the quadratic formula could also be used)}$$
$$w + 32 = 0 \quad or \quad w - 30 = 0 \quad \text{Using the principle of zero products}$$
$$w = -32 \quad or \quad w = 30.$$

4. **Check.** We check in the original problem. We know that -32 is not a solution because width cannot be negative. When $w = 30$, $5w + 10 = 160$, and the area is 30×160, or 4800. This checks.

5. **State.** The width of the rectangular blueberry field is 30 yd, and the length is 160 yd.

Do Exercise 1. ▶

GS **1.** ***Mural Dimensions.*** The area of a rectangular mural is 52 ft^2. The length is 5 ft longer than twice the width. Find the dimensions of the mural.

1. **Familiarize.** Let $w =$ the width of the mural. Then the length is $2w +$ ______.
2. **Translate.**
$$(2w + 5)(w) = \_\_\_$$
3. **Solve.**
$$2w^2 + 5w = 52$$
$$2w^2 + 5w - \_\_\_ = 0$$
$$(2w + 13)(\_\_\_) = 0$$
$$2w + 13 = 0 \quad or \quad \_\_\_ = 0$$
$$w = -\tfrac{13}{2} \quad or \quad w = \_\_\_$$
4. **Check.** Only ______ checks. When the width is ______ ft, the length is $2($ ______ $) + 5 =$ ______ ft.
5. **State.** The width is 4 ft, and the length is ______ ft.

Answer

1. Length: 13 ft; width: 4 ft

Guided Solution:

1. 5, 52, 52, $w - 4$, $w - 4$, 4, 4, 4, 4, 13, 13

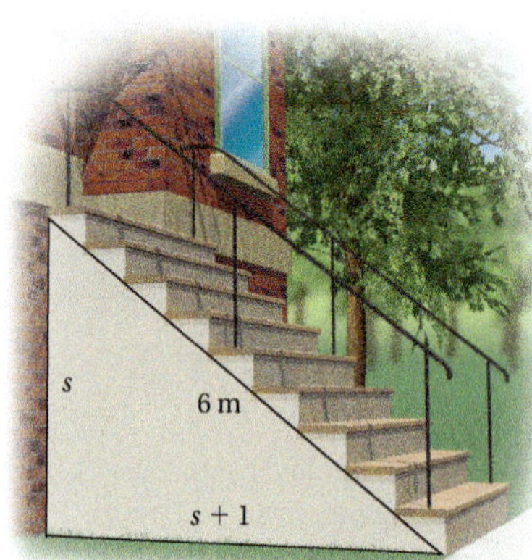

EXAMPLE 2 *Staircase.* A mason builds a staircase in such a way that the portion underneath the stairs forms a right triangle. The hypotenuse is 6 m long. The leg across the ground is 1 m longer than the leg next to the wall at the back. Find the lengths of the legs. Round to the nearest tenth.

1. **Familiarize.** We first make a drawing, letting $s =$ the length of the shorter leg. Then $s + 1 =$ the length of the other leg.
2. **Translate.** To translate, we use the Pythagorean equation:

$$s^2 + (s + 1)^2 = 6^2.$$

3. **Solve.** We solve the equation:

$$s^2 + (s + 1)^2 = 6^2$$
$$s^2 + s^2 + 2s + 1 = 36$$
$$2s^2 + 2s - 35 = 0.$$

Since we cannot factor, we use the quadratic formula:

$$a = 2, \quad b = 2, \quad c = -35;$$
$$s = \frac{-b \pm \sqrt{b^2 - 4ac}}{2a} = \frac{-2 \pm \sqrt{2^2 - 4 \cdot 2(-35)}}{2 \cdot 2}$$
$$= \frac{-2 \pm \sqrt{4 + 280}}{4} = \frac{-2 \pm \sqrt{284}}{4}$$
$$= \frac{-2 \pm \sqrt{4 \cdot 71}}{4} = \frac{-2 \pm 2 \cdot \sqrt{71}}{2 \cdot 2}$$
$$= \frac{2(-1 \pm \sqrt{71})}{2 \cdot 2} = \frac{2}{2} \cdot \frac{-1 \pm \sqrt{71}}{2} = \frac{-1 \pm \sqrt{71}}{2}.$$

Using a calculator, we get approximations:

$$\frac{-1 + \sqrt{71}}{2} \approx 3.7 \quad or \quad \frac{-1 - \sqrt{71}}{2} \approx -4.7.$$

4. **Check.** Since the length of a leg cannot be negative, -4.7 does not check. But 3.7 does check. If the smaller leg s is 3.7, the other leg is $s + 1$, or 4.7. Then

$$(3.7)^2 + (4.7)^2 = 13.69 + 22.09 = 35.78.$$

Using a calculator, we get $\sqrt{35.78} \approx 5.98 \approx 6$, the length of the hypotenuse. Note that our check is not exact because we are using an approximation for $\sqrt{71}$.

5. **State.** One leg is about 3.7 m long, and the other is about 4.7 m long.

Do Exercise 2.

2. *Construction.* Gil and Hal dug a trench across the diagonal of their rectangular backyard in order to install a drainage pipe. If the pipe is 20 m long and the yard is 5 m longer than it is wide, find the dimensions of the yard. Round to the nearest tenth of a meter.

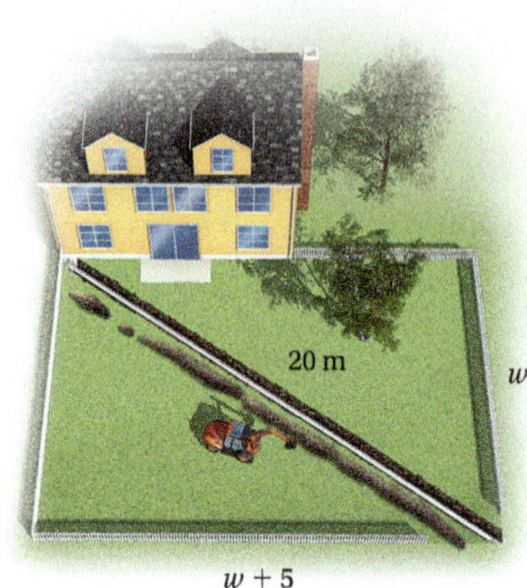

EXAMPLE 3 *Kayak Speed.* The current in a stream moves at a speed of 2 km/h. A kayak travels 24 km upstream and 24 km downstream in a total time of 5 hr. What is the speed of the kayak in still water?

1. **Familiarize.** We first make a drawing. The distances are the same. We let $r =$ the speed of the kayak in still water. Then when the kayak is traveling upstream, its speed is $r - 2$. When it is traveling downstream, its speed is $r + 2$. We let t_1 represent the time it takes the kayak to go upstream and t_2 the time it takes to go downstream. We summarize in a table.

Answer

2. 11.4 m by 16.4 m

	d	r	t	
UPSTREAM	24	$r - 2$	t_1	$\longrightarrow t_1 = \dfrac{24}{r-2}$
DOWNSTREAM	24	$r + 2$	t_2	$\longrightarrow t_2 = \dfrac{24}{r+2}$
TOTAL TIME			5	

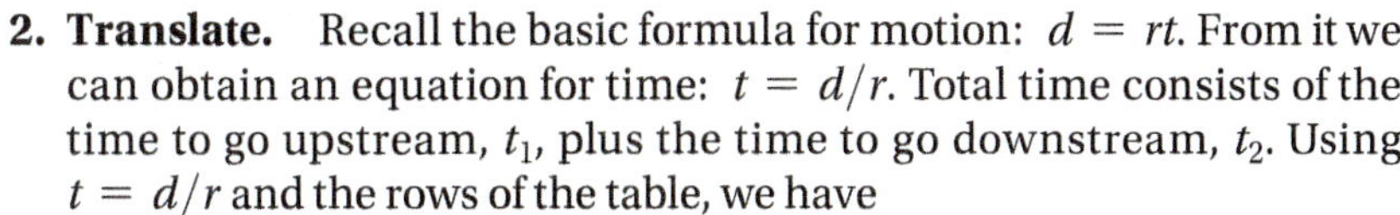

2. **Translate.** Recall the basic formula for motion: $d = rt$. From it we can obtain an equation for time: $t = d/r$. Total time consists of the time to go upstream, t_1, plus the time to go downstream, t_2. Using $t = d/r$ and the rows of the table, we have

$$t_1 = \frac{24}{r-2} \quad \text{and} \quad t_2 = \frac{24}{r+2}.$$

Since the total time is 5 hr, $t_1 + t_2 = 5$, and we have

$$\frac{24}{r-2} + \frac{24}{r+2} = 5. \qquad \text{We have translated to an equation with one variable.}$$

3. **Solve.** We solve the equation. We multiply on both sides by the LCM, which is $(r - 2)(r + 2)$:

$$(r-2)(r+2) \cdot \left(\frac{24}{r-2} + \frac{24}{r+2}\right) = (r-2)(r+2)5$$

$$(r-2)(r+2) \cdot \frac{24}{r-2} + (r-2)(r+2) \cdot \frac{24}{r+2} = (r^2 - 4)5$$

$$24(r+2) + 24(r-2) = 5r^2 - 20$$

$$24r + 48 + 24r - 48 = 5r^2 - 20$$

$$-5r^2 + 48r + 20 = 0$$

$$5r^2 - 48r - 20 = 0 \qquad \text{Multiplying by } -1$$

$$(5r + 2)(r - 10) = 0 \qquad \text{Factoring}$$

$$5r + 2 = 0 \quad \text{or} \quad r - 10 = 0$$

Using the principle of zero products

$$5r = -2 \quad \text{or} \quad r = 10$$

$$r = -\tfrac{2}{5} \quad \text{or} \quad r = 10.$$

4. **Check.** Since speed cannot be negative, $-\frac{2}{5}$ cannot be a solution. But suppose the speed of the kayak in still water is 10 km/h. The speed upstream is then $10 - 2$, or 8 km/h. The speed downstream is $10 + 2$, or 12 km/h. The time upstream, using $t = d/r$, is 24/8, or 3 hr. The time downstream is 24/12, or 2 hr. The total time is 5 hr. This checks.

5. **State.** The speed of the kayak in still water is 10 km/h.

Do Exercise 3. ▶

3. ***Speed of a Stream.*** The speed of a boat in still water is 12 km/h. The boat travels 45 km upstream and 45 km downstream in a total time of 8 hr. What is the speed of the stream? (*Hint*: Let $s =$ the speed of the stream. Then $12 - s$ is the speed upstream and $12 + s$ is the speed downstream.)

1. **Familiarize.**

d	r	t
45	$12 - s$	t_1
$\square$	$12 + s$	t_2
		$\square$

2. **Translate.**

$$t_1 = \frac{45}{12 - s}, \quad t_2 = \frac{45}{\square};$$

$$\frac{45}{12 - s} + \frac{45}{12 + s} = \square$$

3. **Solve.**

$$(12 - s)(12 + s)\left(\frac{45}{12 - s} + \frac{45}{12 + s}\right) = (12 - s)(12 + s)(8)$$

$$45(12 + s) + 45(12 - s) = (144 - s^2)(8)$$

$$1080 = 1152 - \square$$

$$8s^2 - 72 = 0$$

$$s^2 - \square = 0$$

$$s + 3 = 0 \quad \text{or} \quad s - 3 = 0$$

$$s = -3 \quad \text{or} \quad s = \square$$

4. **Check.** The speed of the stream cannot be negative. A speed of 3 km/h checks.

5. **State.** The speed of the stream is $\square$ km/h.

Answer

3. 3 km/h

Guided Solution:

1. 45, 8, $12 + s$, 8, $8s^2$, 9, 3, 3

Translating for Success

1. *Guy Wire.* How long is a guy wire that reaches from the top of a 75-ft cell-phone tower to a point on the ground 21 ft from the pole?

2. *Coin Mixture.* A collection of dimes and quarters is worth \$16.95. There are 90 coins in all. How many of each coin are there?

3. *Wire Cutting.* A 486-in. wire is cut into three pieces. The second piece is 5 in. longer than the first. The third is one-half as long as the first. How long is each piece?

4. *Amount Invested.* Money is invested at 3.2% simple interest. At the end of 1 year, there is \$27,864 in the account. How much was originally invested?

5. *Foreign Languages.* Last year, 3.2% of the 27,864 students at East End Community College took a foreign language course. How many students took a foreign language course?

The goal of these matching questions is to practice step (2), Translate, of the five-step problem-solving process. Translate each word problem to an equation or a system of equations and select a correct translation from A–O.

A. $x^2 + (x - 1)^2 = 7$

B. $\frac{600}{x} = \frac{600}{x + 2} + 10$

C. $13{,}932 = x \cdot 27{,}864$

D. $x = 3.2\% \cdot 27{,}864$

E. $2x + 2(x - 1) = 49$

F. $x + (x + 5) + \frac{1}{2}x = 486$

G. $0.10x + 0.25y = 16.95,$
$x + y = 90$

H. $x + 25y = 16.95,$
$x + y = 90$

I. $3.2x = 27{,}864 - x$

J. $x^2 + (x - 1)^2 = 49$

K. $x^2 + 21^2 = 75^2$

L. $x + 3.2\%x = 27{,}864$

M. $75^2 + 21^2 = x^2$

N. $x + (x + 1) + (x + 2) = 894$

O. $\frac{600}{x} + \frac{600}{x - 2} = 10$

Answers on page A-40

6. *Locker Numbers.* The numbers on three adjoining lockers are consecutive integers whose sum is 894. Find the integers.

7. *Triangle Dimensions.* The hypotenuse of a right triangle is 7 ft. The length of one leg is 1 ft shorter than the other. Find the lengths of the legs.

8. *Rectangle Dimensions.* The perimeter of a rectangle is 49 ft. The length is 1 ft shorter than the width. Find the length and the width.

9. *Car Travel.* Maggie drove her car 600 mi to see her friend. The return trip was 2 hr faster at a speed that was 10 mph greater. Find the time for the return trip.

10. *Literature.* Last year, 13,932 of the 27,864 students at East End Community College took a literature course. What percent of the students took a literature course?

17.5 Exercise Set

For Extra Help
MyMathLab® MathXL®

PRACTICE WATCH READ REVIEW

Reading Check

Determine whether each statement is true or false.

RC1. To find the area of a rectangle, multiply the length of the rectangle by the width.

RC2. The Pythagorean equation is true for all triangles.

RC3. Lengths of sides of rectangles and triangles are positive numbers.

RC4. The speed of a boat moving upstream is the speed of the boat in still water plus the speed of the current in the stream.

a Solve.

1. *Pool Dimensions.* The area of a rectangular swimming pool is 80 yd^2. The length is 1 yd longer than three times the width. Find the dimensions of the swimming pool.

2. The length of a rectangular area rug is 3 ft greater than the width. The area is 70 ft^2. Find the length and the width.

3. *Carpenter's Square.* A *square* is a carpenter's tool in the shape of a right triangle. One side, or leg, of a square is 8 in. longer than the other. The length of the hypotenuse is $8\sqrt{13}$ in. Find the lengths of the legs of the square.

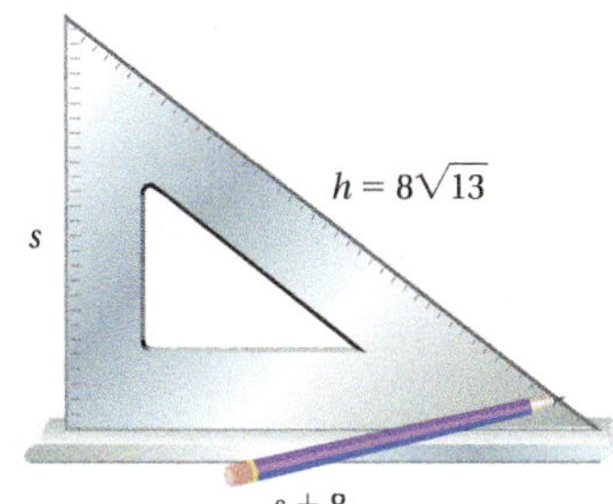

4. *HDTV Dimensions.* When we say that a television is 42 in., we mean that the diagonal is 42 in. For a 42-in. television, the width is 15 in. more than the height. Find the dimensions of a 42-in. high-definition television. Round to the nearest inch.

5. *Rectangle Dimensions.* The length of a rectangular garden is three times the width. The area is 300 ft^2. Find the length and the width of the garden.

6. *Rectangle Dimensions.* The length of a rectangular lobby in a hotel is twice the width. The area is 50 m^2. Find the length and the width of the lobby.

7. *Rectangle Dimensions.* The width of a rectangle is 4 cm less than the length. The area is 320 cm^2. Find the length and the width.

8. *Rectangle Dimensions.* The width of a rectangle is 3 cm less than the length. The area is 340 cm^2. Find the length and the width.

Find the approximate answers for Exercises 9–14. Round to the nearest tenth.

9. *Right-Triangle Dimensions.* The hypotenuse of a right triangle is 8 m long. One leg is 2 m longer than the other. Find the lengths of the legs.

10. *Right-Triangle Dimensions.* The hypotenuse of a right triangle is 5 cm long. One leg is 2 cm longer than the other. Find the lengths of the legs.

11. *Rectangle Dimensions.* The length of a rectangle is 2 in. greater than the width. The area is 20 in^2. Find the length and the width.

12. *Rectangle Dimensions.* The length of a rectangle is 3 ft greater than the width. The area is 15 ft^2. Find the length and the width.

13. *Rectangle Dimensions.* The length of a rectangle is twice the width. The area is 20 cm^2. Find the length and the width.

14. *Rectangle Dimensions.* The length of a rectangle is twice the width. The area is 10 m^2. Find the length and the width.

15. *Picture Frame.* A picture frame measures 25 cm by 20 cm. There is 266 cm^2 of picture showing. The frame is of uniform width. Find the width of the frame.

16. *Tablecloth.* A rectangular tablecloth measures 96 in. by 72 in. It is laid on a tabletop with an area of 5040 in^2, and hangs over the edge by the same amount on all sides. By how many inches does the cloth hang over the edge?

For Exercises 17–24, complete the table to help with the familiarization.

17. *Boat Speed.* The current in a stream moves at a speed of 3 km/h. A boat travels 40 km upstream and 40 km downstream in a total time of 14 hr. What is the speed of the boat in still water?

	d	r	t
UPSTREAM		$r - 3$	t_1
DOWNSTREAM	40		t_2
TOTAL TIME			

18. *Wind Speed.* An airplane flies 1449 mi against the wind and 1539 mi with the wind in a total time of 5 hr. The speed of the airplane in still air is 600 mph. What is the speed of the wind?

	d	r	t
WITH WIND	1539		
AGAINST WIND		$600 - r$	
TOTAL TIME			5

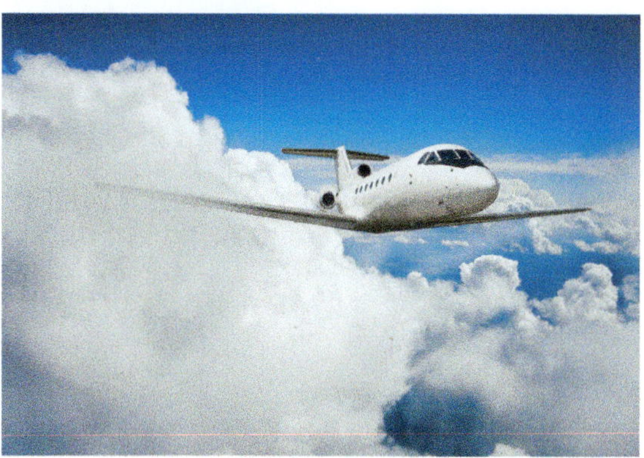

19. *Speed of a Stream.* The speed of a boat in still water is 8 km/h. The boat travels 60 km upstream and 60 km downstream in a total time of 16 hr. What is the speed of the stream?

	d	r	t
UPSTREAM			
DOWNSTREAM			
TOTAL TIME			

20. *Boat Speed.* The current in a stream moves at a speed of 4 mph. A boat travels 5 mi upstream and 13 mi downstream in a total time of 2 hr. What is the speed of the boat in still water?

	d	r	t
UPSTREAM		$r - 4$	t_1
DOWNSTREAM	13		t_2
TOTAL TIME			

21. *Wind Speed.* An airplane flies 520 km against the wind and 680 km with the wind in a total time of 4 hr. The speed of the airplane in still air is 300 km/h. What is the speed of the wind?

	d	r	t
WITH WIND		$300 + r$	
AGAINST WIND	520		
TOTAL TIME			4

22. *Speed of a Stream.* The speed of a boat in still water is 10 km/h. The boat travels 12 km upstream and 28 km downstream in a total time of 4 hr. What is the speed of the stream?

	d	r	t
UPSTREAM			
DOWNSTREAM			
TOTAL TIME			

23. ***Boat Speed.*** The current in a stream moves at a speed of 4 mph. A boat travels 4 mi upstream and 12 mi downstream in a total time of 2 hr. What is the speed of the boat in still water?

	d	r	t
UPSTREAM			
DOWNSTREAM			
TOTAL TIME			

24. ***Boat Speed.*** The current in a stream moves at a speed of 3 mph. A boat travels 45 mi upstream and 45 mi downstream in a total time of 8 hr. What is the speed of the boat in still water?

	d	r	t
UPSTREAM			
DOWNSTREAM			
TOTAL TIME			

25. ***Speed of a Stream.*** The speed of a boat in still water is 9 km/h. The boat travels 80 km upstream and 80 km downstream in a total time of 18 hr. What is the speed of the stream?

26. ***Speed of a Stream.*** The speed of a boat in still water is 10 km/h. The boat travels 48 km upstream and 48 km downstream in a total time of 10 hr. What is the speed of the stream?

Skill Maintenance

Find the coordinates of the y-intercept and of the x-intercept. Do not graph. [11.2a]

27. $8x = 4 - y$

28. $5y - 3x = -45$

Graph.

29. $y = 3x - 5$ [11.5a]

30. $x = -2$ [11.2b]

31. $y = 1$ [11.2b]

32. $2x - y = 2$ [11.2a]

Find an equation of the line with the given slope and y-intercept. [11.4a]

33. Slope: -2; y-intercept: $(0, -5)$

34. Slope: $\frac{1}{2}$; y-intercept: $(0, 1)$

Determine whether each pair of equations represents parallel lines. [11.6a]

35. $y = \frac{3}{4}x - 7$, $3x + 4y = 7$

36. $y = \frac{3}{5}$, $y = -\frac{5}{3}$

Synthesis

37. ***Pizza.*** What should the diameter d of a pizza be so that it has the same area as two 12-in. pizzas? Which provides more servings: a 16-in. pizza or two 12-in. pizzas?

38. ***Golden Rectangle.*** The *golden rectangle* is said to be extremely pleasing visually and was used often by ancient Greek and Roman architects. The length of a golden rectangle is approximately 1.6 times the width. Find the dimensions of a golden rectangle if its area is 9000 m^2.

Graphs of Quadratic Equations

17.6

OBJECTIVES

a Graph quadratic equations.

b Find the x-intercepts of a quadratic equation.

SKILL TO REVIEW

Objective 11.2a: Find the intercepts of a linear equation, and graph using intercepts.

Graph using intercepts.

1. $2x - y = 4$
2. $4y + 20 = -5x$

In this section, we will graph equations of the form

$$y = ax^2 + bx + c, \quad a \neq 0.$$

The polynomial on the right side of the equation is of second degree, or **quadratic**. Examples of the types of equations we are going to graph are

$$y = x^2, \qquad y = x^2 + 2x - 3, \qquad y = -2x^2 + 3.$$

a GRAPHING QUADRATIC EQUATIONS OF THE TYPE $y = ax^2 + bx + c$

Graphs of quadratic equations of the type $y = ax^2 + bx + c$ (where $a \neq 0$) are always cup-shaped. They have a **line of symmetry** like the dashed lines shown in the figures below. If we fold on this line, the two halves will match exactly. The curve goes on forever. The highest or lowest point on the curve is called the **vertex**. The second coordinate of the vertex is either the smallest value of y or the largest value of y. The vertex is also thought of as a turning point. Graphs of quadratic equations are called **parabolas**.

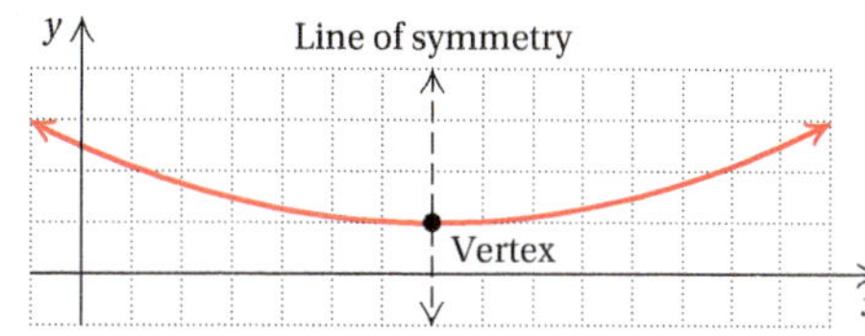

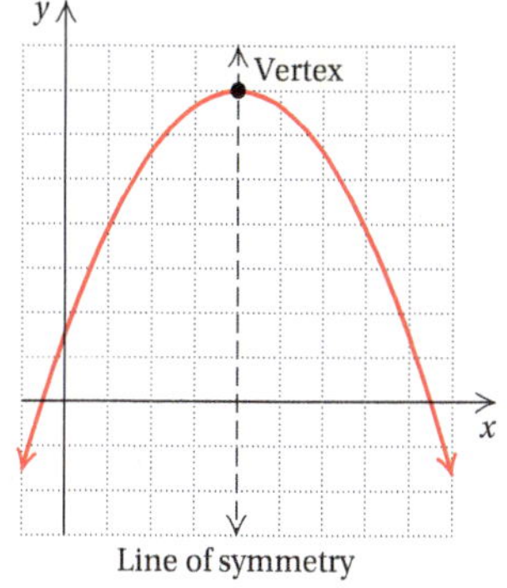

To graph a quadratic equation, we begin by choosing some numbers for x and computing the corresponding values of y.

EXAMPLE 1 Graph: $y = x^2$.

We choose numbers for x and find the corresponding values for y. Then we plot the ordered pairs (x, y) resulting from the computations and connect them with a smooth curve.

For $x = -3$, $y = x^2 = (-3)^2 = 9$.
For $x = -2$, $y = x^2 = (-2)^2 = 4$.
For $x = -1$, $y = x^2 = (-1)^2 = 1$.
For $x = 0$, $y = x^2 = (0)^2 = 0$.
For $x = 1$, $y = x^2 = (1)^2 = 1$.
For $x = 2$, $y = x^2 = (2)^2 = 4$.
For $x = 3$, $y = x^2 = (3)^2 = 9$.

x	y	(x, y)
-3	9	$(-3, 9)$
-2	4	$(-2, 4)$
-1	1	$(-1, 1)$
0	0	$(0, 0)$
1	1	$(1, 1)$
2	4	$(2, 4)$
3	9	$(3, 9)$

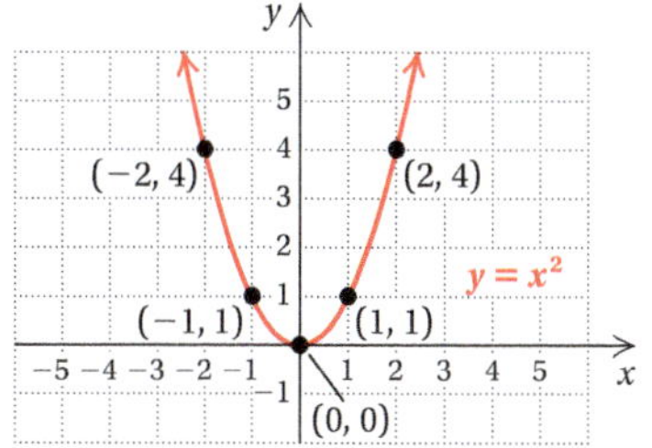

Answers

Skill to Review:

1.

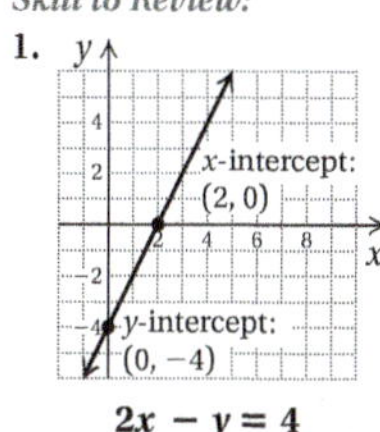

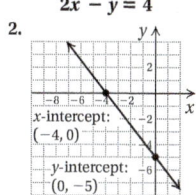

2. x-intercept: $(-4, 0)$; y-intercept: $(0, -5)$; $4y + 20 = -5x$

In Example 1, the vertex is the point $(0, 0)$. The second coordinate of the vertex, 0, is the smallest y-value. The y-axis $(x = 0)$ is the line of symmetry.

Parabolas whose equations are $y = ax^2$ always have the origin $(0, 0)$ as the vertex and the y-axis as the line of symmetry.

A key to graphing a parabola is knowing the vertex. By graphing it and then choosing x-values on both sides of the vertex, we can compute more points and complete the graph.

FINDING THE VERTEX

For a parabola given by the quadratic equation $y = ax^2 + bx + c$:

1. The x-coordinate of the vertex is $-\dfrac{b}{2a}$.

   The line of symmetry is $x = -\dfrac{b}{2a}$.

2. The second coordinate of the vertex is found by substituting the x-coordinate into the equation and computing y.

The proof that the vertex can be found in this way can be shown by completing the square in a manner similar to the proof of the quadratic formula, but it will not be considered here.

EXAMPLE 2 Graph: $y = -2x^2 + 3$.

We first find the vertex. The x-coordinate of the vertex is

$$-\frac{b}{2a} = -\frac{0}{2(-2)} = 0.$$

We next find the second coordinate of the vertex:

$$y = -2x^2 + 3 = -2(0)^2 + 3 = 3. \qquad \text{Substituting 0 for } x$$

The vertex is $(0, 3)$. The line of symmetry is the y-axis $(x = 0)$. We choose some x-values on both sides of the vertex and graph the parabola.

For $x = 1$, $y = -2x^2 + 3 = -2(1)^2 + 3 = -2 + 3 = 1$.
For $x = -1$, $y = -2x^2 + 3 = -2(-1)^2 + 3 = -2 + 3 = 1$.
For $x = 2$, $y = -2x^2 + 3 = -2(2)^2 + 3 = -8 + 3 = -5$.
For $x = -2$, $y = -2x^2 + 3 = -2(-2)^2 + 3 = -8 + 3 = -5$.

x	y
0	3 ← This is the vertex.
1	1
−1	1
2	−5
−2	−5

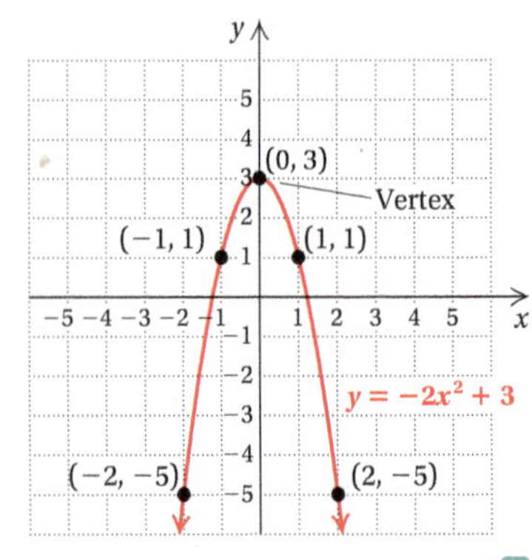

There are two other tips you might use when graphing quadratic equations. The first involves the coefficient of x^2. The a in $y = ax^2 + bx + c$ tells us whether the graph opens up or down. When a is positive, as in Example 1, the graph opens up; when a is negative, as in Example 2, the graph opens down. It is also helpful to plot the y-intercept. It occurs when $x = 0$.

TIPS FOR GRAPHING QUADRATIC EQUATIONS

1. Graphs of quadratic equations $y = ax^2 + bx + c$ are all parabolas. They are *smooth* cup-shaped symmetric curves, with no sharp points or kinks in them.
2. Find the vertex and the line of symmetry.
3. The graph of $y = ax^2 + bx + c$ opens up if $a > 0$. It opens down if $a < 0$.
4. Find the y-intercept. It occurs when $x = 0$, and it is easy to compute.

EXAMPLE 3 Graph: $y = x^2 + 2x - 3$.

We first find the vertex. The x-coordinate of the vertex is

$$-\frac{b}{2a} = -\frac{2}{2(1)} = -1.$$

We substitute -1 for x into the equation to find the second coordinate of the vertex:

$$\begin{aligned} y &= x^2 + 2x - 3 \\ &= (-1)^2 + 2(-1) - 3 \\ &= 1 - 2 - 3 \\ &= -4. \end{aligned}$$

The vertex is $(-1, -4)$. The line of symmetry is $x = -1$.

We choose some x-values on both sides of $x = -1$—say, $-2, -3, -4$ and $0, 1, 2$—and graph the parabola. Since the coefficient of x^2 is 1, which is positive, we know that the graph opens up. Be sure to find y when $x = 0$. This gives the y-intercept.

x	y	
-1	-4	← Vertex
0	-3	← y-intercept
-2	-3	
1	0	
-3	0	
2	5	
-4	5	

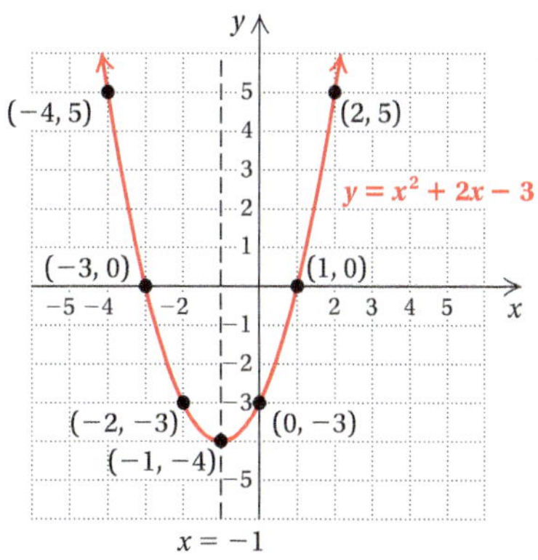

Do Exercises 1–3. ▶

Graph. Label the ordered pairs for the vertex and the y-intercept.

1. $y = x^2 - 3$

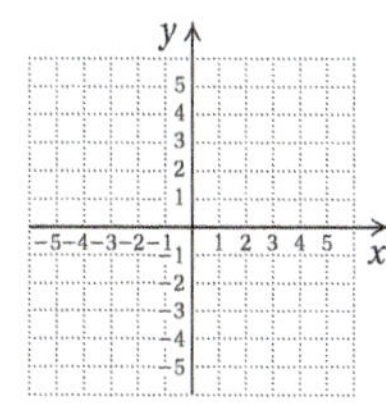

2. $y = -3x^2 + 6x$

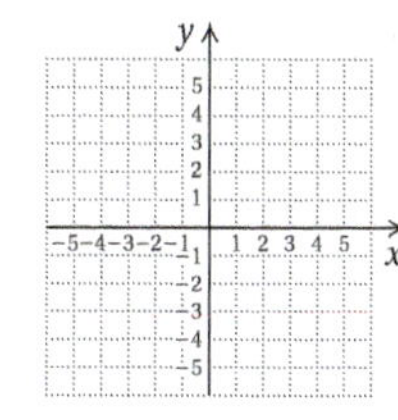

3. $y = x^2 - 4x + 4$

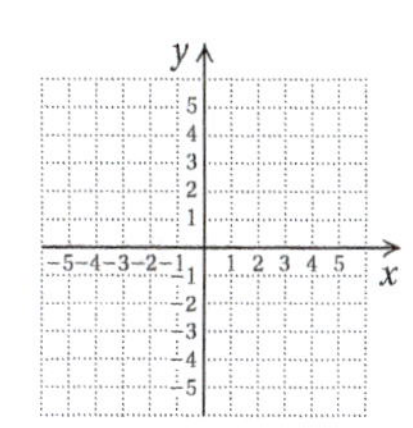

Answers

1.

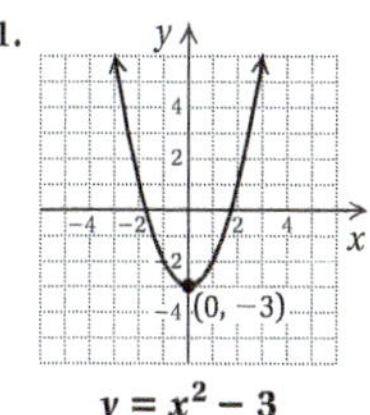

$y = x^2 - 3$

2.

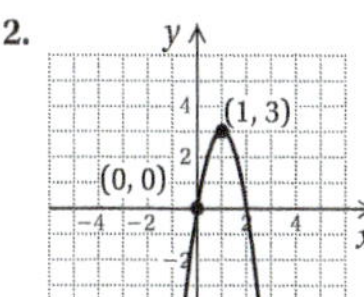

$y = -3x^2 + 6x$

3.

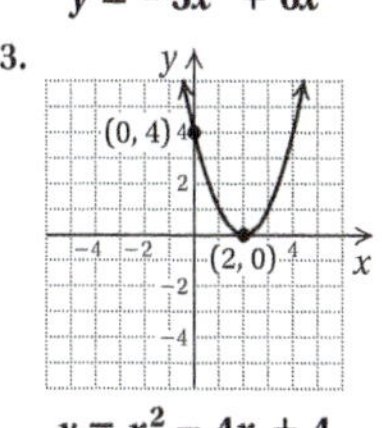

$y = x^2 - 4x + 4$

b FINDING THE x-INTERCEPTS OF A QUADRATIC EQUATION

The x-intercepts of the graph of $y = ax^2 + bx + c$ occur at those values of x for which $y = 0$. Thus the first coordinates of the x-intercepts are solutions of the equation

$$0 = ax^2 + bx + c.$$

We have been studying how to find such numbers in Sections 17.1–17.3.

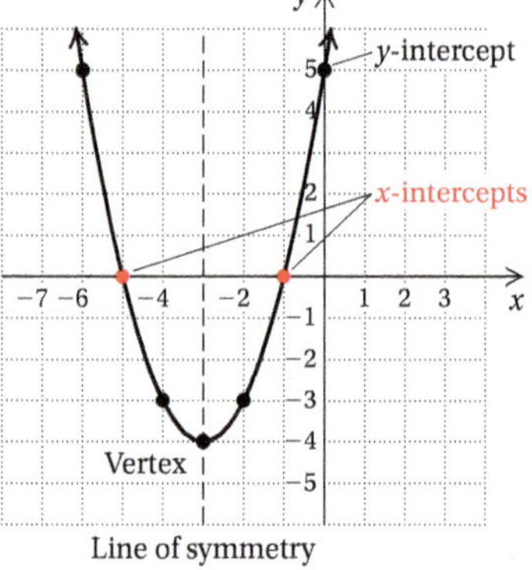

EXAMPLE 4 Find the x-intercepts of the graph of $y = x^2 - 4x + 1$.

We solve the equation $x^2 - 4x + 1 = 0$ using the quadratic formula.

$$a = 1, \quad b = -4, \quad c = 1$$

$$x = \frac{-b \pm \sqrt{b^2 - 4ac}}{2a}$$

$$= \frac{-(-4) \pm \sqrt{(-4)^2 - 4(1)(1)}}{2(1)}$$

$$= \frac{4 \pm \sqrt{16 - 4}}{2}$$

$$= \frac{4 \pm \sqrt{12}}{2} = \frac{4 \pm \sqrt{4 \cdot 3}}{2}$$

$$= \frac{4 \pm 2\sqrt{3}}{2} = \frac{2 \cdot 2 \pm 2\sqrt{3}}{2 \cdot 1}$$

$$= \frac{2}{2} \cdot \frac{2 \pm \sqrt{3}}{1} = 2 \pm \sqrt{3}$$

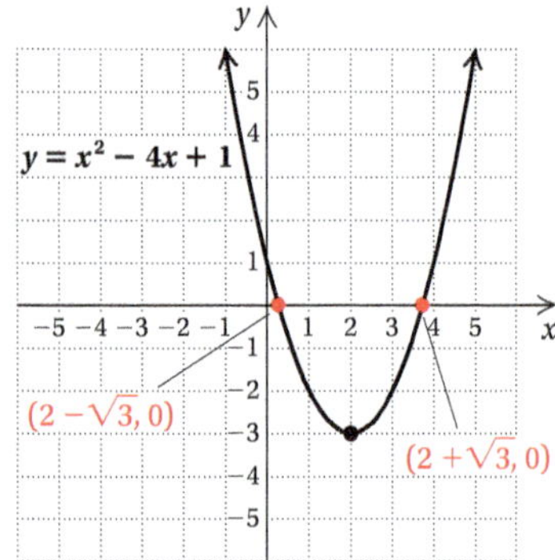

The x-intercepts are $(2 - \sqrt{3}, 0)$ and $(2 + \sqrt{3}, 0)$.

In the quadratic formula $x = \frac{-b \pm \sqrt{b^2 - 4ac}}{2a}$, the radicand $b^2 - 4ac$ is called the **discriminant**. The discriminant tells how many real-number solutions the equation $0 = ax^2 + bx + c$ has, so it also tells how many x-intercepts there are.

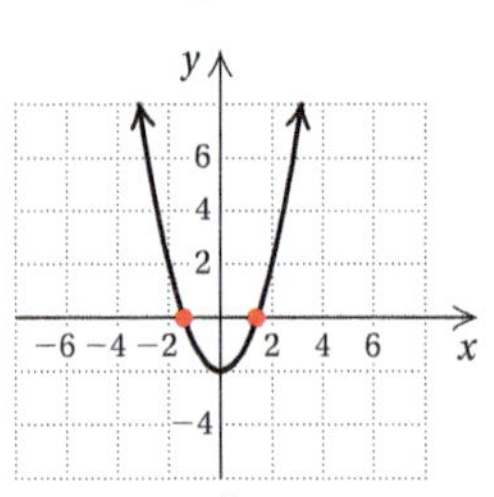

$y = x^2 - 2$
$b^2 - 4ac = 8 > 0$
Two real solutions
Two x-intercepts

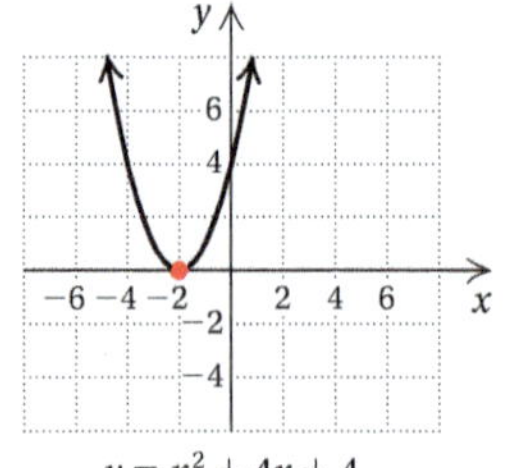

$y = x^2 + 4x + 4$
$b^2 - 4ac = 0$
One real solution
One x-intercept

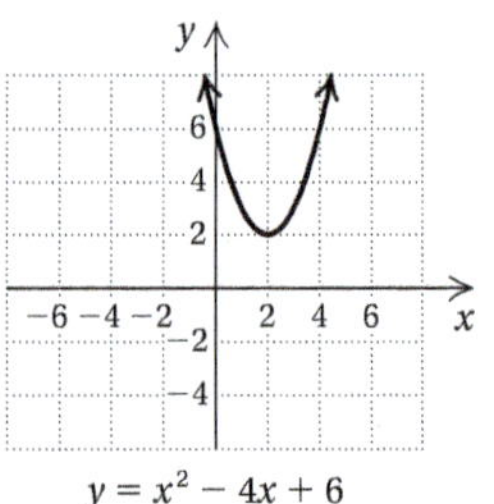

$y = x^2 - 4x + 6$
$b^2 - 4ac = -8 < 0$
No real solutions
No x-intercepts

Do Exercises 4–7.

Find the x-intercepts.

4. $y = x^2 - 3$

5. $y = x^2 + 6x + 8$ GS
$x^2 + 6x + 8 = 0$
$(x + 4)(\quad) = 0$
$x + 4 = 0$ *or* $\quad = 0$
$x = -4$ *or* $x = \quad$
The x-intercepts are $(-4, 0)$ and $(\quad, 0)$.

6. $y = -2x^2 - 4x + 1$

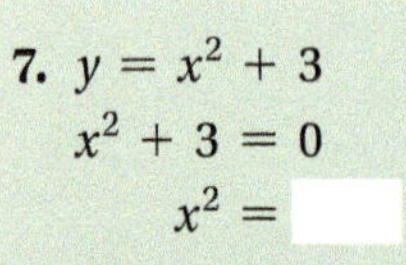

7. $y = x^2 + 3$ GS
$x^2 + 3 = 0$
$x^2 = \quad$
Since -3 is negative, the equation has no real-number solutions. There are no x-intercepts.

Answers

4. $(-\sqrt{3}, 0); (\sqrt{3}, 0)$ 5. $(-4, 0); (-2, 0)$
6. $\left(\frac{-2 - \sqrt{6}}{2}, 0\right); \left(\frac{-2 + \sqrt{6}}{2}, 0\right)$ 7. None

Guided Solutions:
5. $x + 2, x + 2, -2, -2$ 7. -3

Visualizing for Success

Match each equation or inequality with its graph.

1. $y = -4 + 4x - x^2$

2. $y = 5 - x^2$

3. $5x + 2y = -10$

4. $5x + 2y \leq 10$

5. $y < 5x$

6. $y = x^2 - 3x - 2$

7. $2x - 5y = 10$

8. $5x - 2y = 10$

9. $2x + 5y = 10$

10. $y = x^2 + 3x - 2$

Answers on page A-41

A

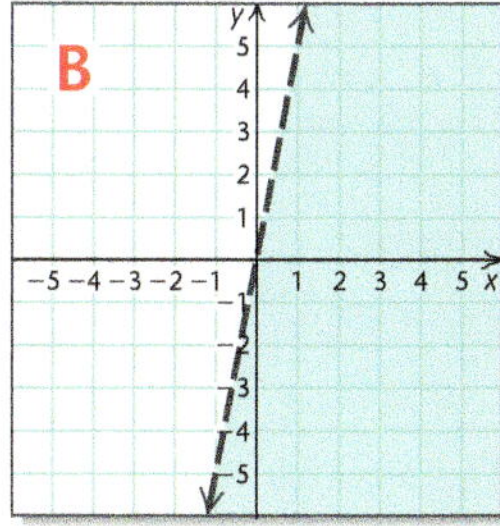

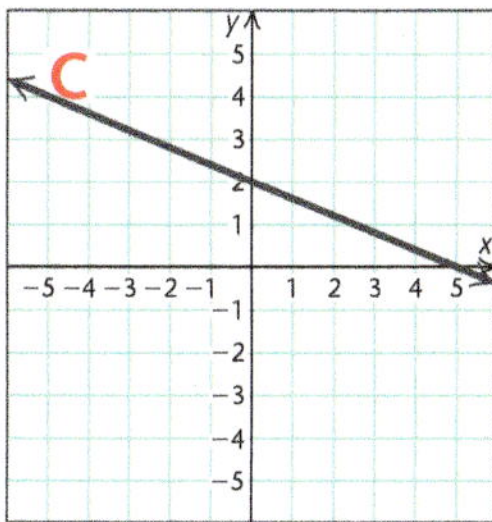

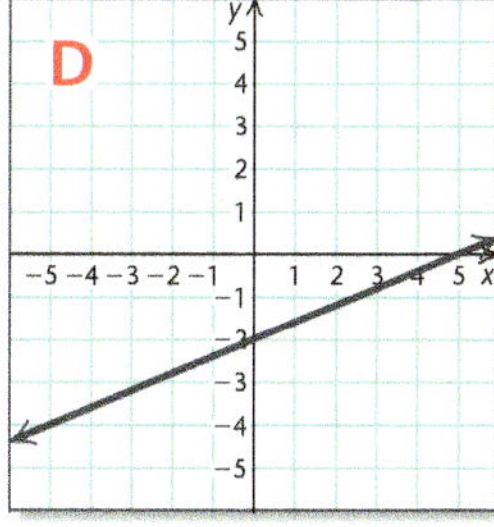

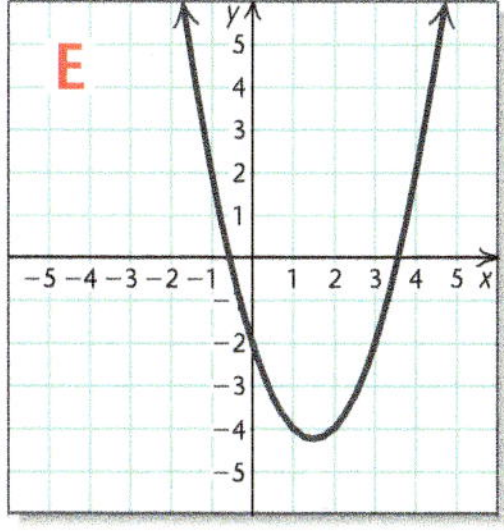

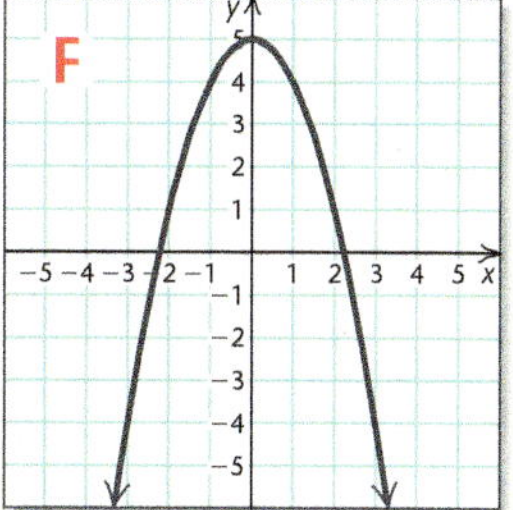

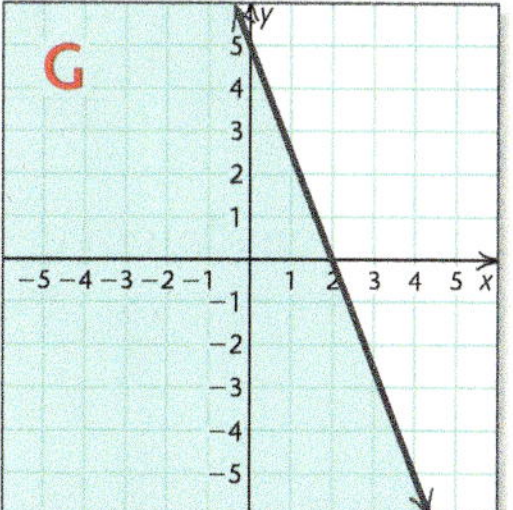

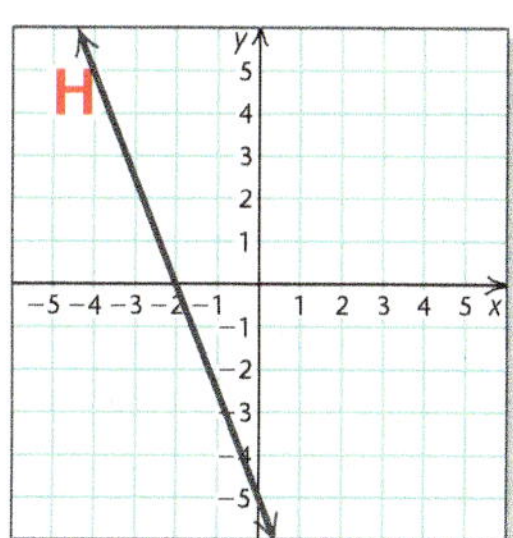

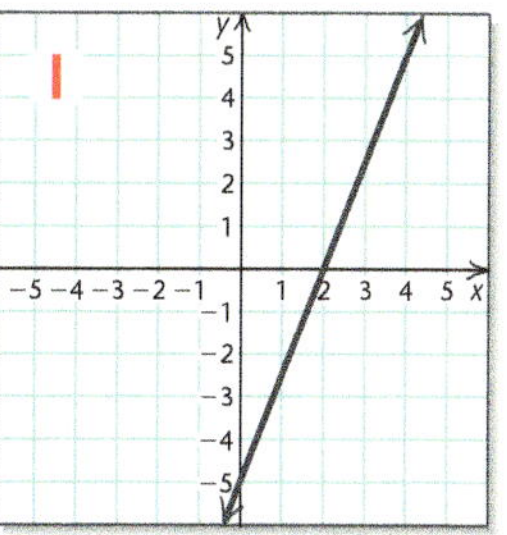

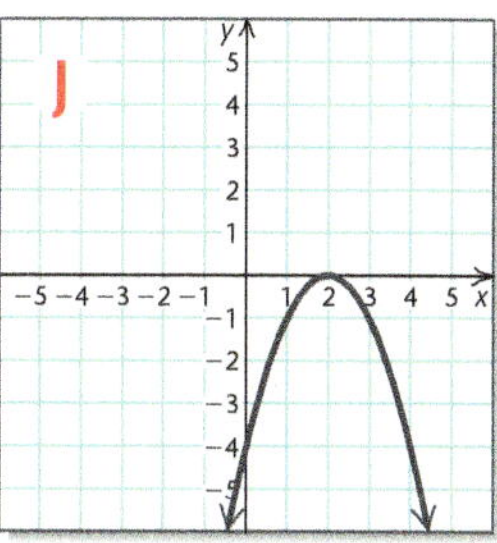

17.6 Exercise Set

For Extra Help

MyMathLab®

MathXL® PRACTICE | WATCH | READ | REVIEW

Reading Check

Choose from the column on the right the word that best completes each statement.

line

parabola

quadratic

vertex

RC1. The equation $x^2 - 9x + 8 = 0$ is an example of a(n) ______________ equation.

RC2. The graph of $y = x^2 - 9x + 8$ is a(n) ______________.

RC3. The turning point of the graph of $y = x^2 - 9x + 8$ is the ______________.

RC4. The graph of $y = x^2 - 9x + 8$ could be folded in half along its ______________ of symmetry.

a Graph the quadratic equation. In Exercises 1–8, label the ordered pairs for the vertex and the y-intercept.

1. $y = x^2 + 1$

x	y
-2	
-1	
0	
1	
2	
3	

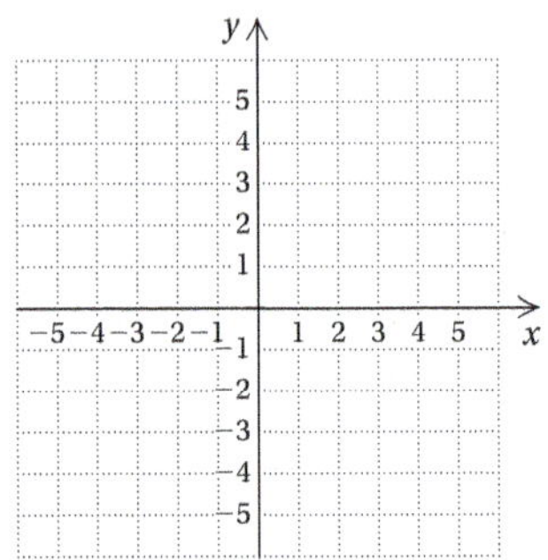

2. $y = 2x^2$

x	y
-2	
-1	
0	
1	
2	
3	

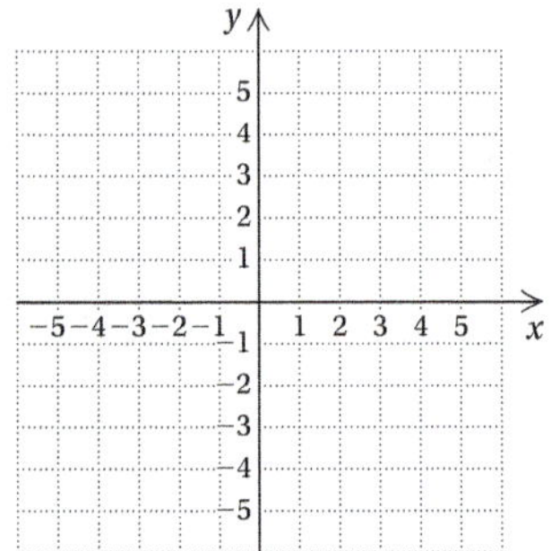

3. $y = -1 \cdot x^2$

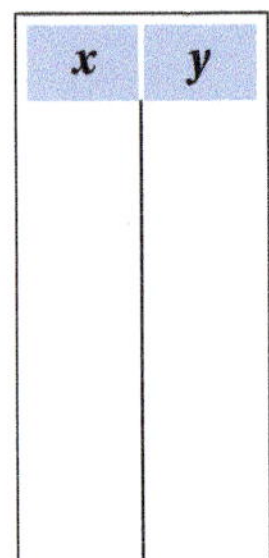

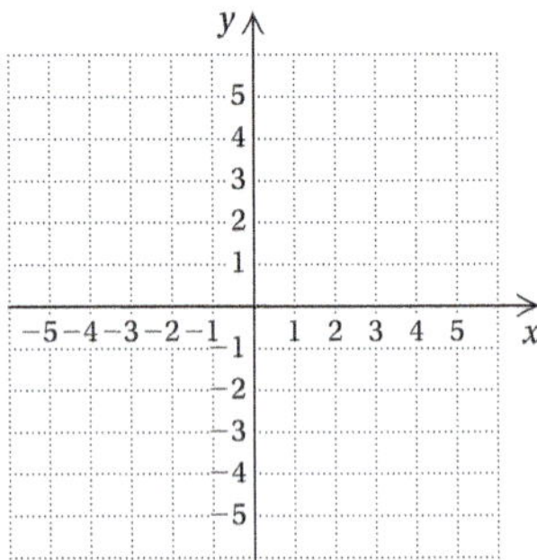

4. $y = x^2 - 1$

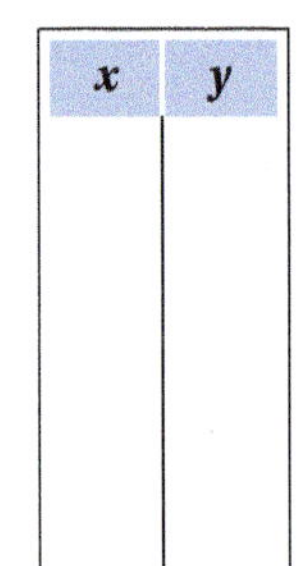

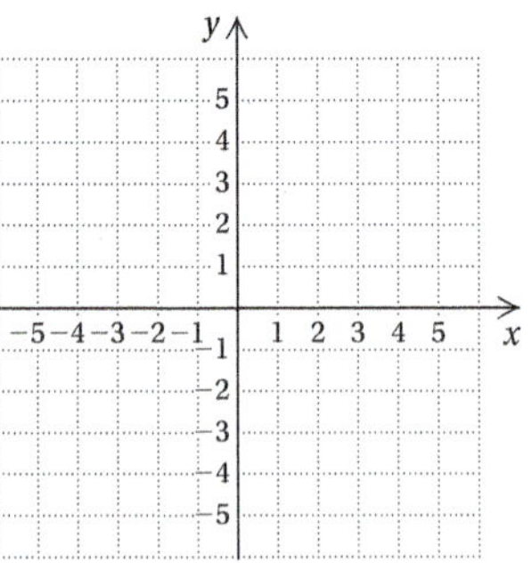

5. $y = -x^2 + 2x$

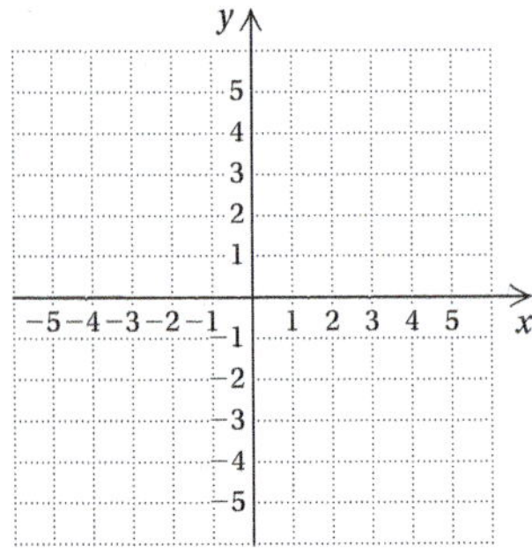

6. $y = x^2 + x - 2$

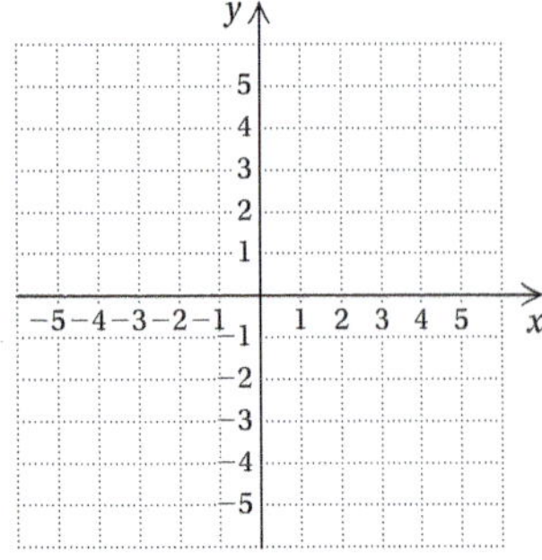

7. $y = 5 - x - x^2$

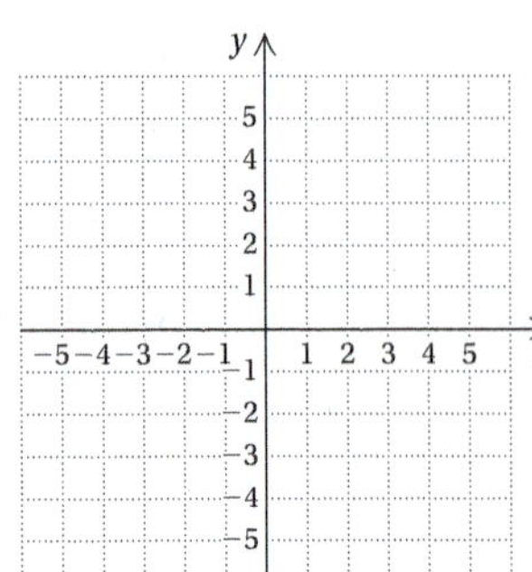

8. $y = x^2 + 2x + 1$

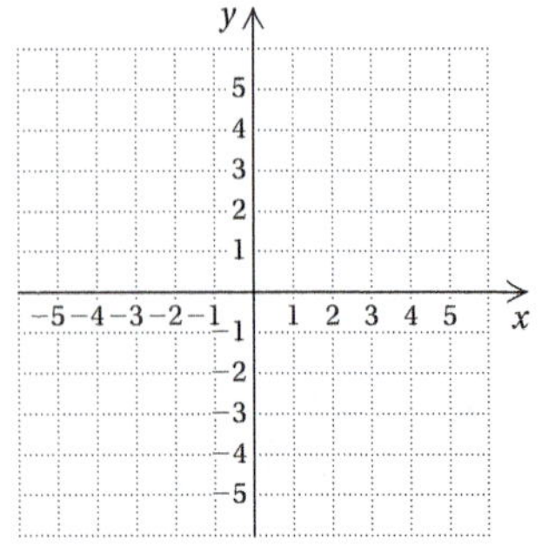

9. $y = x^2 - 2x + 1$

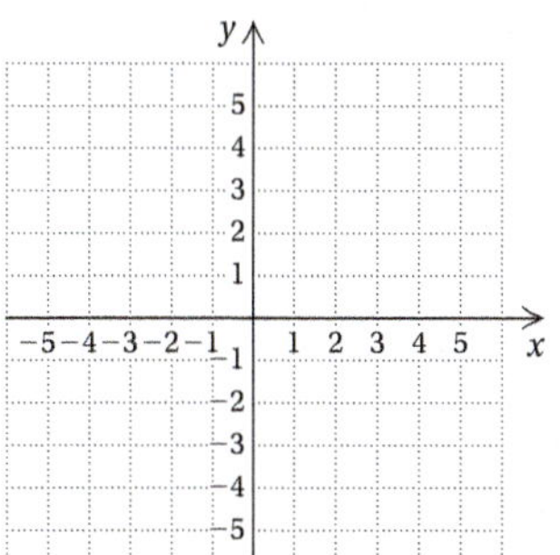

10. $y = -\frac{1}{2}x^2$

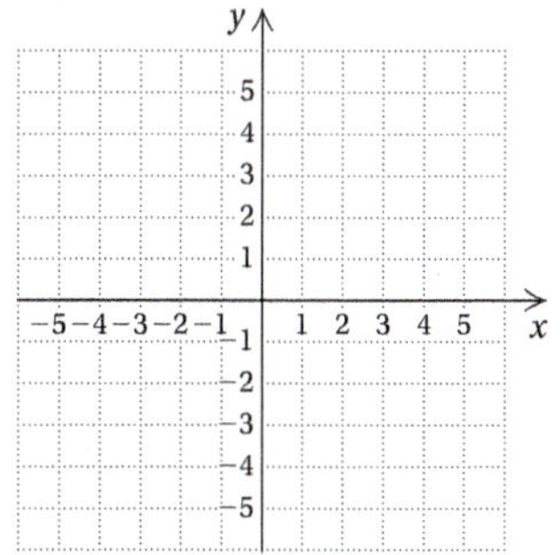

11. $y = -x^2 + 2x + 3$

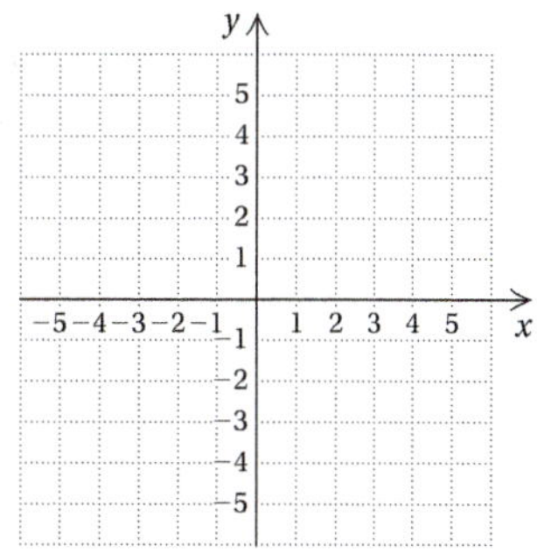

12. $y = -x^2 - 2x + 3$

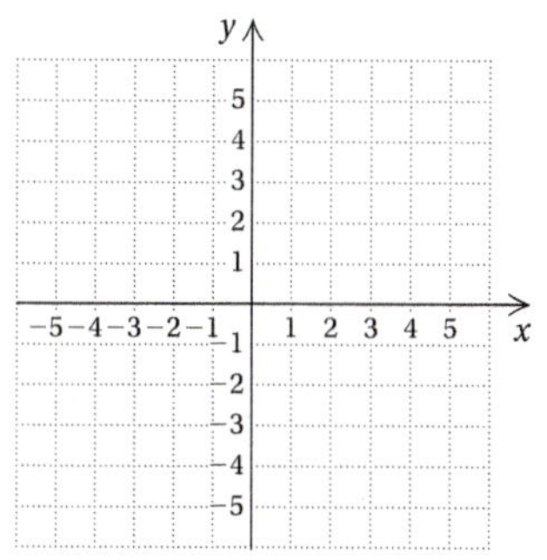

13. $y = -2x^2 - 4x + 1$

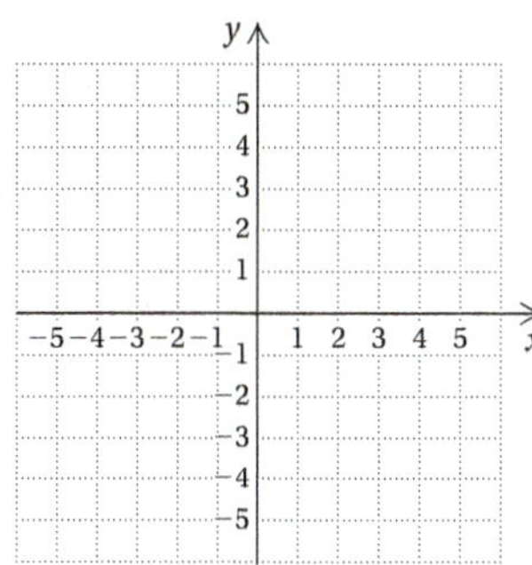

14. $y = 2x^2 + 4x - 1$

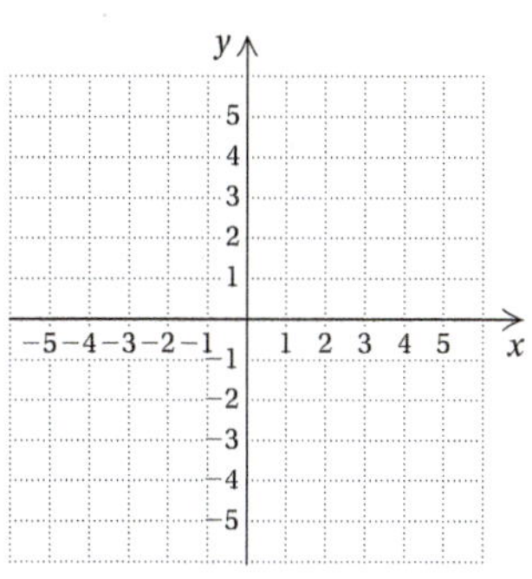

15. $y = 5 - x^2$

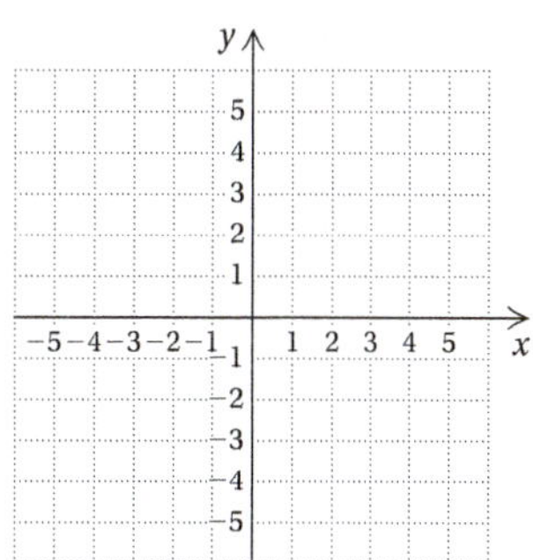

16. $y = 4 - x^2$

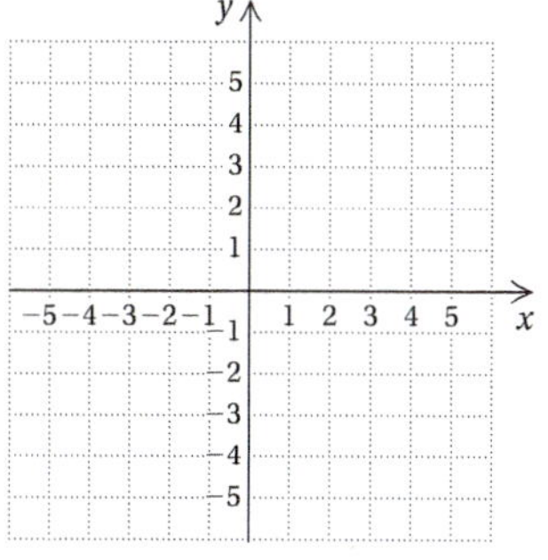

17. $y = \frac{1}{4}x^2$

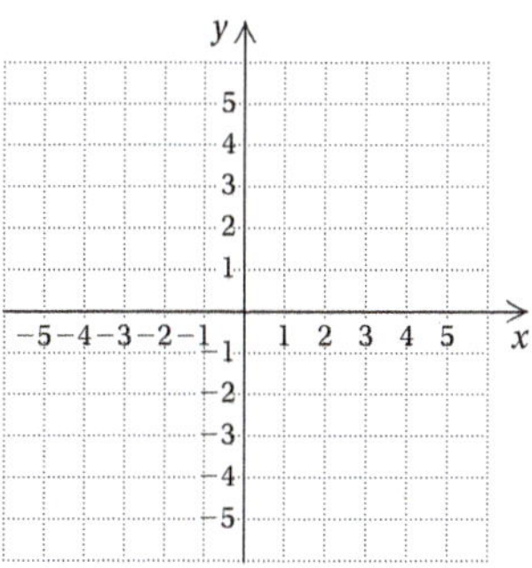

18. $y = -0.1x^2$

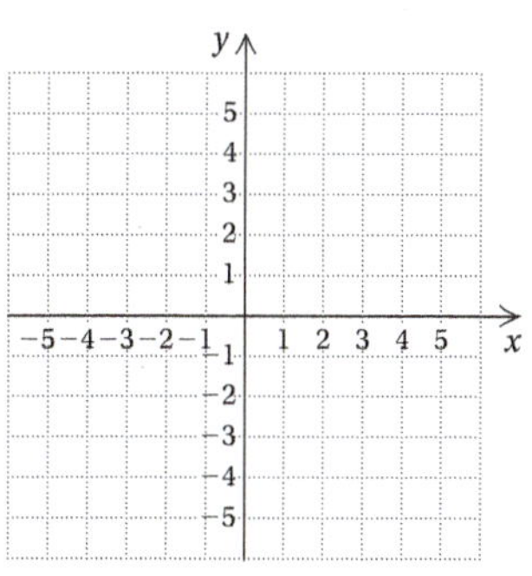

19. $y = -x^2 + x - 1$

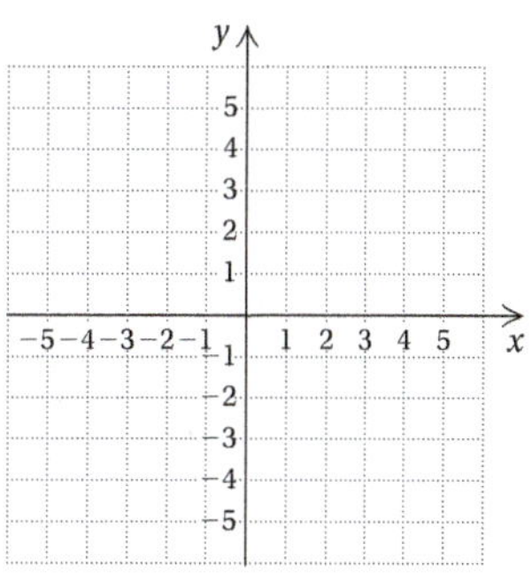

20. $y = x^2 + 2x$

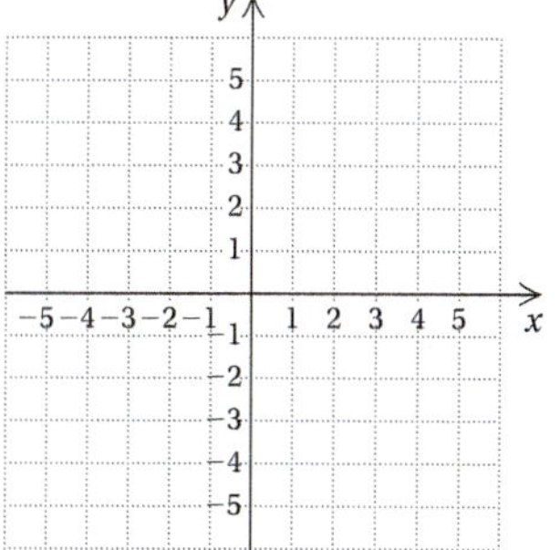

21. $y = -2x^2$

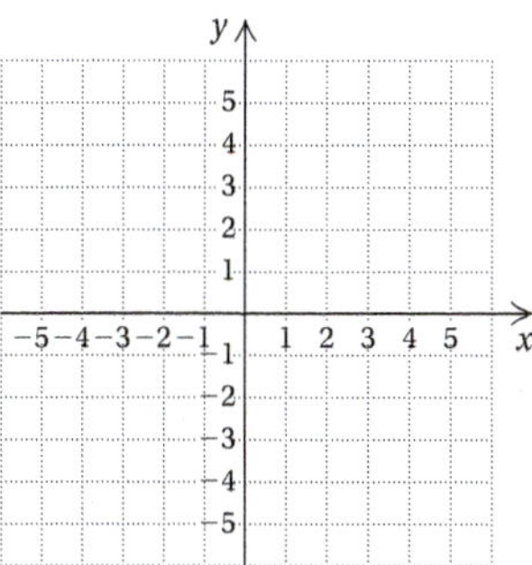

22. $y = -x^2 - 1$

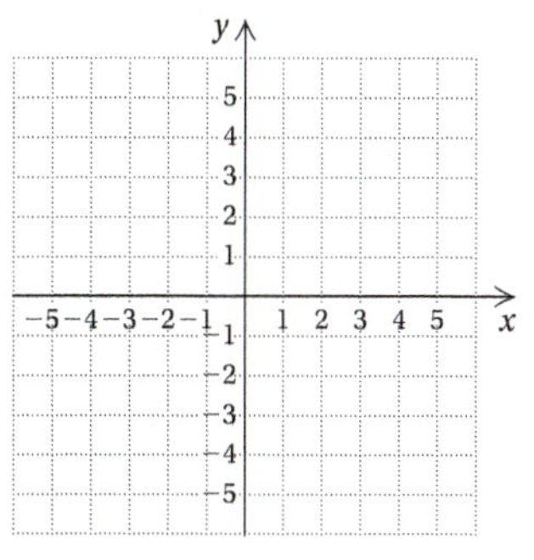

23. $y = x^2 - x - 6$

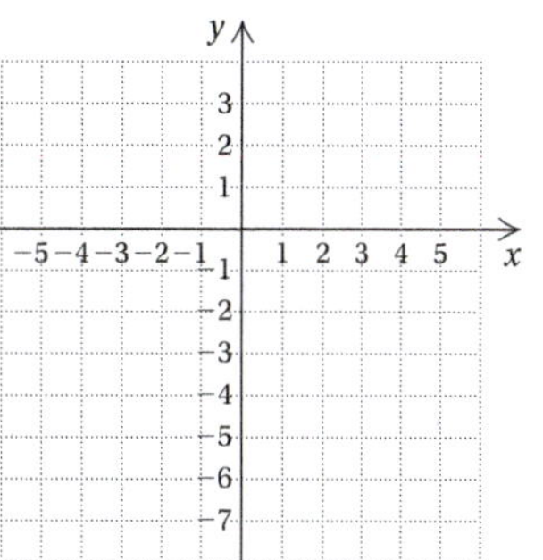

24. $y = 6 + x - x^2$

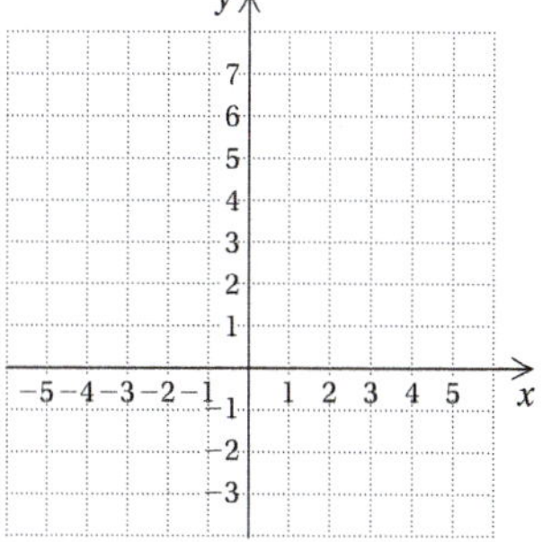

b Find the x-intercepts.

25. $y = x^2 - 2$

26. $y = x^2 - 7$

27. $y = x^2 + 5x$

28. $y = x^2 - 4x$

29. $y = 8 - x - x^2$

30. $y = 8 + x - x^2$

31. $y = x^2 - 6x + 9$

32. $y = x^2 + 10x + 25$

33. $y = -x^2 - 4x + 1$

34. $y = x^2 + 4x - 1$

35. $y = x^2 + 9$

36. $y = x^2 + 1$

Skill Maintenance

37. Add: $\sqrt{8} + \sqrt{50} + \sqrt{98} + \sqrt{128}$. [16.4a]

38. Multiply and simplify: $\sqrt{5y^4}\sqrt{125y}$. [16.2c]

39. Find an equation of variation in which y varies inversely as x, and $y = 12.4$ when $x = 2.4$. [14.9c]

40. Evaluate $3x^4 + 3x - 7$ when $x = -2$. [12.3a]

Simplify. [9.4a]

41. $-\frac{1}{5} + \frac{7}{10} - \left(-\frac{4}{15}\right) + \frac{1}{60}$

42. $-0.63 - 3.4 + 11.08 - (-42.5)$

Synthesis

43. ***Height of a Projectile.*** The height H, in feet, of a projectile with an initial velocity of 96 ft/sec is given by the equation

$$H = -16t^2 + 96t,$$

where t is the time, in seconds. Use the graph of this equation, shown here, or any equation-solving technique to answer the following questions.

a) How many seconds after launch is the projectile 128 ft above ground?

b) When does the projectile reach its maximum height?

c) How many seconds after launch does the projectile return to the ground?

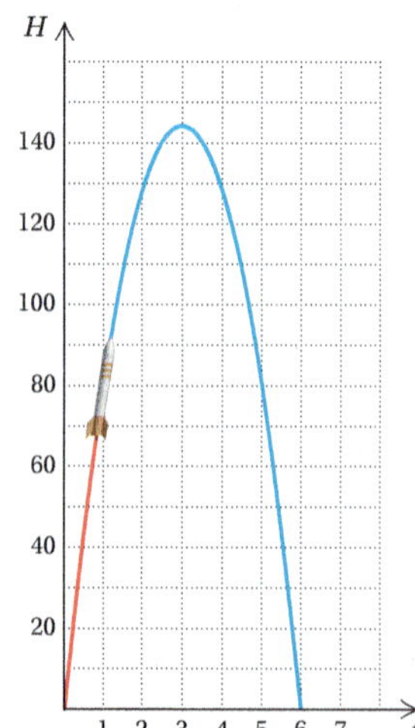

For each equation in Exercises 44–47, evaluate the discriminant $b^2 - 4ac$. Then use the answer to state how many real-number solutions exist for the equation.

44. $y = x^2 + 8x + 16$

45. $y = x^2 + 2x - 3$

46. $y = -2x^2 + 4x - 3$

47. $y = -0.02x^2 + 4.7x - 2300$

Functions

17.7

OBJECTIVES

a Determine whether a correspondence is a function.

b Given a function described by an equation, find function values (outputs) for specified values (inputs).

c Draw a graph of a function.

d Determine whether a graph is that of a function.

e Solve applied problems involving functions and their graphs.

SKILL TO REVIEW

Objective 12.3a: Evaluate a polynomial for a given value of the variable.

Evaluate each polynomial for the indicated value.

1. $10 - \frac{1}{8}x$, when $x = 16$

2. $3x - 5 + x^2$, when $x = -3$

a IDENTIFYING FUNCTIONS

We now develop one of the most important concepts in mathematics: **functions**. We have actually been studying functions all through this text; we just haven't identified them as such. Ordered pairs form a correspondence between first and second coordinates. A function is a special correspondence from one set to another. For example:

To each student in a college, there corresponds his or her student ID number.

To each item in a store, there corresponds its price.

To each real number, there corresponds the cube of that number.

In each case, the first set is called the **domain** and the second set is called the **range**. Given a member of the domain, there is *just one* member of the range to which it corresponds. This kind of correspondence is called a **function**.

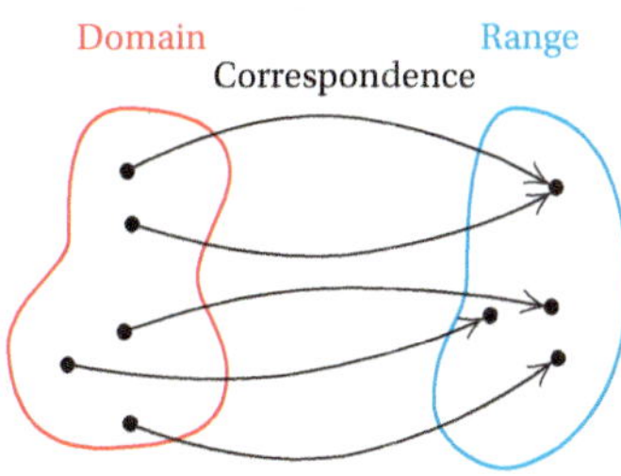

EXAMPLE 1 Determine whether the correspondence is a function.

f:

Domain	Range
1 →	$107.40
2 →	$ 34.10
3 →	$ 29.60
4 →	$ 19.60

g:

Domain	Range
3 →	5
4 →	9
5 →	−7
6 →	−7

h:

Domain	Range
New York →	Mets, Yankees
St. Louis →	Cardinals
San Diego →	Padres

p:

Domain	Range
Mets →	New York
Yankees →	New York
Cardinals →	St. Louis
Padres →	San Diego

The correspondence *f is* a function because each member of the domain is matched to only one member of the range.

The correspondence *g is* also a function because each member of the domain is matched to only one member of the range.

The correspondence *h is not* a function because one member of the domain, New York, is matched to more than one member of the range.

The correspondence *p is* a function because each member of the domain is paired with only one member of the range. Note that a function can pair a number of the range with more than one member of the domain. ■

Answers

Skill to Review:
1. 8 **2.** −5

Determine whether each correspondence is a function.

1. *Domain* — *Range*

Cheetah → 70 mph
Human → 28 mph
Lion → 50 mph
Chicken → 9 mph

2. *Domain* — *Range*

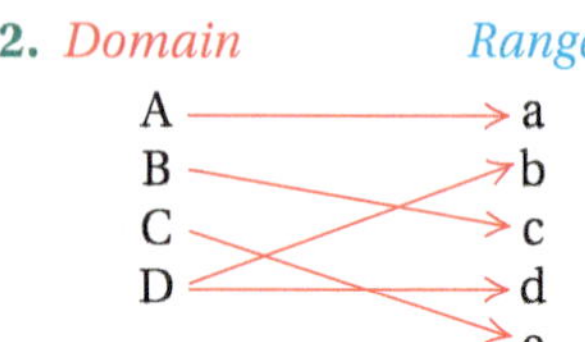

3. *Domain* — *Range*

4. *Domain* — *Range*

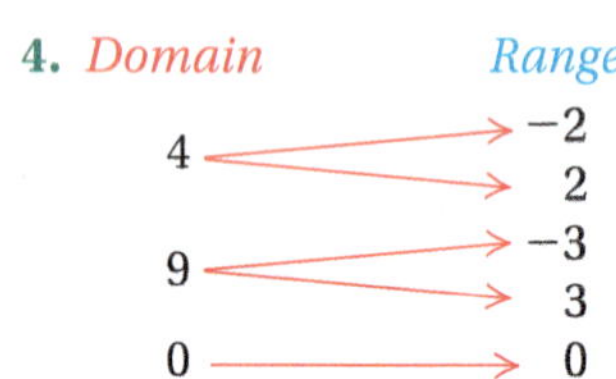

Determine whether each of the following is a function.

5. *Domain*
A set of numbers
Correspondence
10 less than the square of each number
Range
A set of numbers

6. *Domain*
A set of polygons
Correspondence
The perimeter of each polygon
Range
A set of numbers

FUNCTION, DOMAIN, AND RANGE

A **function** is a correspondence between a first set, called the **domain**, and a second set, called the **range**, such that each member of the domain corresponds to *exactly one* member of the range.

◀ **Do Exercises 1–4.**

EXAMPLE 2 Determine whether each correspondence is a function.

	Domain	*Correspondence*	*Range*
a)	A family	Each person's weight	A set of positive numbers
b)	The natural numbers	Each number's square	A set of natural numbers
c)	The set of all states	Each state's members of the U.S. Senate	A set of U.S. Senators

a) The correspondence *is* a function because each person has *only one* weight.

b) The correspondence *is* a function because each natural number has *only one* square.

c) The correspondence *is not* a function because each state has two U.S. Senators.

◀ **Do Exercises 5 and 6.**

When a correspondence between two sets is not a function, it may still be an example of a *relation*.

RELATION

A **relation** is a correspondence between a first set, called the **domain**, and a second set, called the **range**, such that each member of the domain corresponds to *at least one* member of the range.

Thus, although the correspondences of Examples 1 and 2 are not all functions, they *are* all relations. A function is a special type of relation—one in which each member of the domain is paired with *exactly one* member of the range.

b FINDING FUNCTION VALUES

Most functions considered in mathematics are described by equations. A function described by a linear equation like $y = 2x + 3$ is called a **linear function**. A function described by a quadratic equation like $y = 4 - x^2$ is called a **quadratic function**.

Answers

1. Yes **2.** No **3.** Yes **4.** No **5.** Yes **6.** Yes

Recall that when graphing $y = 2x + 3$, we chose x-values and then found corresponding y-values. For example, when $x = 4$,

$$y = 2x + 3 = 2 \cdot 4 + 3 = 11.$$

When thinking of functions, we call the number 4 an **input** and the number 11 an **output**.

The function $y = 2x + 3$ can be named f and described by the equation $f(x) = 2x + 3$. We call the input x and the output $f(x)$. This is read "f of x," or "f at x," or "the value of f at x."

Caution!

The notation $f(x)$ *does not mean* "f times x" and should not be read that way.

It helps to think of a function as a machine; that is, think of putting a member of the domain (an input) into the machine. The machine knows the correspondence and produces a member of the range (the output).

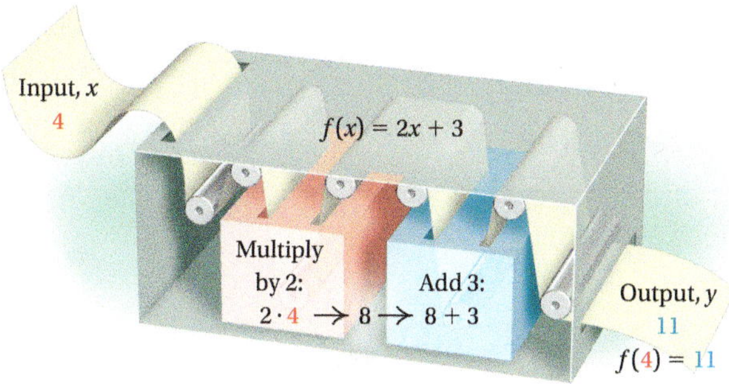

The equation $f(x) = 2x + 3$ describes the function that takes an input x, multiplies it by 2, and then adds 3.

$$f(x) = \underset{\text{Multiply by 2}}{\overset{\text{Input}}{2x}} \quad \underbrace{+\,3}_{\text{Add 3}}$$

To find the output $f(4)$, we take the input 4, double it, and add 3 to get 11. That is, we substitute 4 into the formula for $f(x)$:

$$f(4) = 2 \cdot 4 + 3 = 11.$$

Outputs of functions are also called **function values**. For $f(x) = 2x + 3$, we know that $f(4) = 11$. We can say that "the function value at 4 is 11."

EXAMPLE 3 Find the indicated function value.

a) $f(5)$, for $f(x) = 3x + 2$ **b)** $g(3)$, for $g(z) = 5z^2 - 4$

c) $A(-2)$, for $A(r) = 3r^2 + 2r$ **d)** $f(-5)$, for $f(x) = x^2 + 3x - 4$

a) $f(5) = 3 \cdot 5 + 2 = 15 + 2 = 17$

b) $g(3) = 5(3)^2 - 4 = 5(9) - 4 = 45 - 4 = 41$

c) $A(-2) = 3(-2)^2 + 2(-2) = 3(4) - 4 = 12 - 4 = 8$

d) $f(-5) = (-5)^2 + 3(-5) - 4 = 25 - 15 - 4 = 6$

Do Exercises 7–13. ▶

Find the indicated function value.

7. $f(1)$, for $f(x) = 5x - 3$

8. $g(-4)$, for $g(x) = \frac{1}{2}x + 7$

9. $p(0)$, for $p(x) = x^4 - 5x^2 + 8$

$p(0) = \square^4 - 5 \cdot \square^2 + 8$

$= 0 - 0 + \square$

$= \square$

10. $h\left(\frac{1}{2}\right)$, for $h(x) = 10x$

11. $f(-3)$, for $f(t) = t^2 - t + 1$

12. $g(-94)$, for $g(z) = |z|$

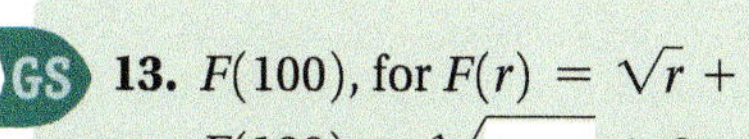

13. $F(100)$, for $F(r) = \sqrt{r} + 9$

$F(100) = \sqrt{\square} + 9$

$= \square + 9$

$= \square$

Answers

7. 2 **8.** 5 **9.** 8 **10.** 5 **11.** 13 **12.** 94 **13.** 19

Guided Solutions:

9. 0, 0, 8, 8 **13.** 100, 10, 19

CALCULATOR CORNER

Finding Function Values We can find function values on a graphing calculator by substituting inputs directly into the formula. After we have entered a function on the equation-editor screen, there are several other methods that we can use to find function values.

Consider the function in Example 3(d), $f(x) = x^2 + 3x - 4$. We enter $y_1 = x^2 + 3x - 4$ and then use a table set in **ASK** mode and enter $x = -5$. We see that the function value, y_1, is 6. We can also use the **VALUE** feature to evaluate the function. To do this, we first graph the function in a window that includes $x = -5$ and then press 2ND CALC to access the **VALUE** feature in order to find the value of y when $x = -5$. A third method uses function notation. We choose Y_1 from the **VARS Y-VARS** function menu, enclose the input in parentheses, and press **ENTER**. All three methods indicate that $f(-5) = 6$.

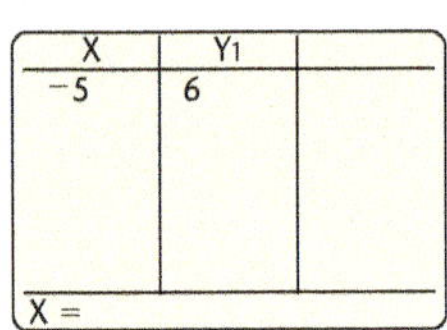

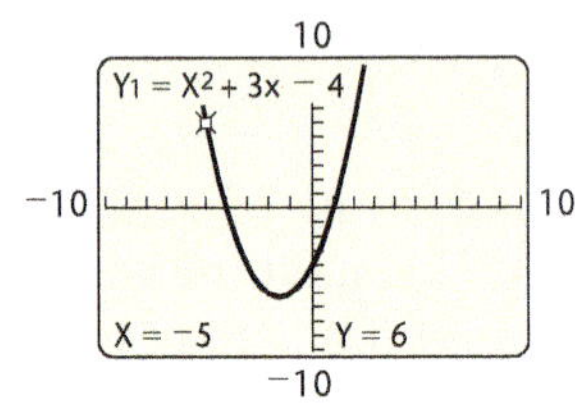

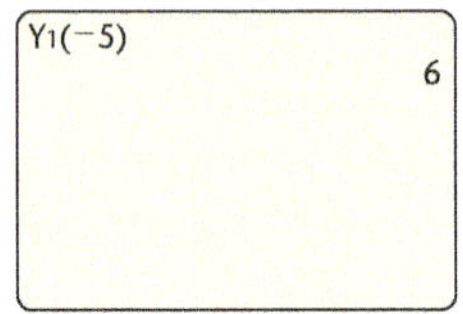

EXERCISES: Find each function value.

1. $f(-3.4)$, for $f(x) = 2x - 6$

2. $f(4)$, for $f(x) = -2.3x$

3. $f(-1)$, for $f(x) = x^2 - 3$

4. $f(3)$, for $f(x) = 2x^2 - x + 5$

C GRAPHS OF FUNCTIONS

To graph a function, we find ordered pairs (x, y) or $(x, f(x))$, plot them, and connect the points. Note that y and $f(x)$ are used interchangeably when we are working with functions and their graphs.

EXAMPLE 4 Graph: $f(x) = x + 2$.

A list of some function values is shown in this table. We plot the points and connect them. The graph is a straight line.

x	$f(x)$
−4	−2
−3	−1
−2	0
−1	1
0	2
1	3
2	4
3	5
4	6

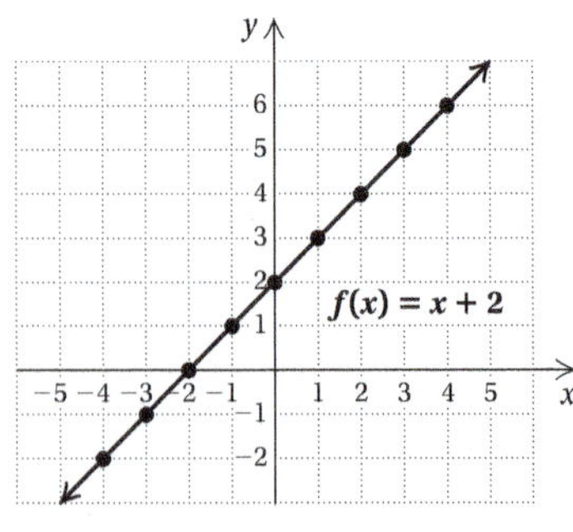

EXAMPLE 5 Graph: $g(x) = 4 - x^2$.

Recall from Section 17.6 that the graph is a parabola. We calculate some function values and draw the curve.

$$g(0) = 4 - 0^2 = 4 - 0 = 4,$$
$$g(-1) = 4 - (-1)^2 = 4 - 1 = 3,$$
$$g(2) = 4 - (2)^2 = 4 - 4 = 0,$$
$$g(-3) = 4 - (-3)^2 = 4 - 9 = -5$$

x	$g(x)$
−3	−5
−2	0
−1	3
0	4
1	3
2	0
3	−5

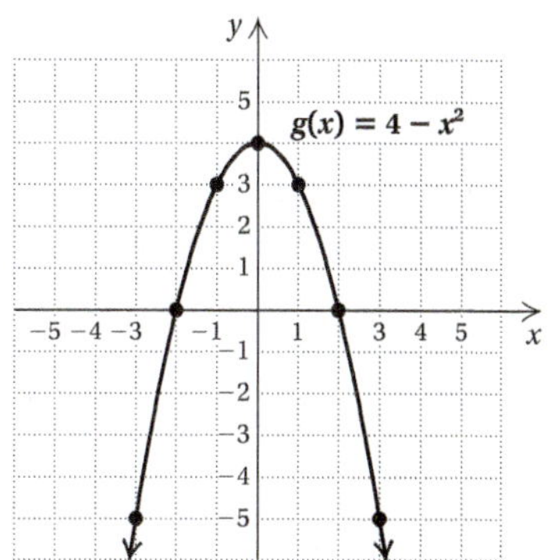

EXAMPLE 6 Graph: $h(x) = |x|$.

A list of some function values is shown in the following table. We plot the points and connect them. The graph is a V-shaped "curve" that rises on either side of the vertical axis.

x	$h(x)$
−3	3
−2	2
−1	1
0	0
1	1
2	2
3	3

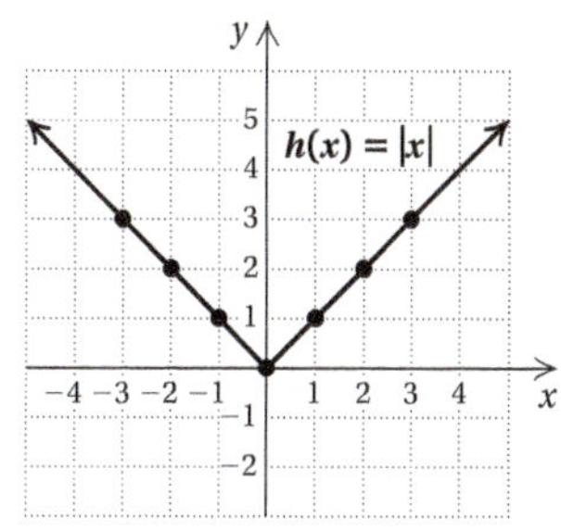

Do Exercises 14–16.

Graph.

14. $f(x) = x - 4$

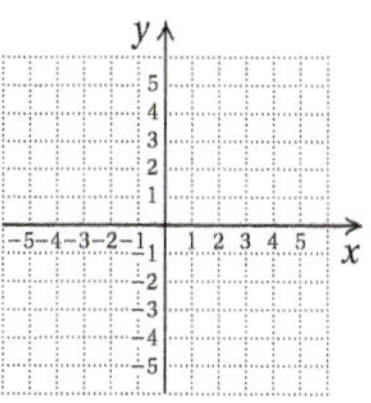

15. $g(x) = 5 - x^2$

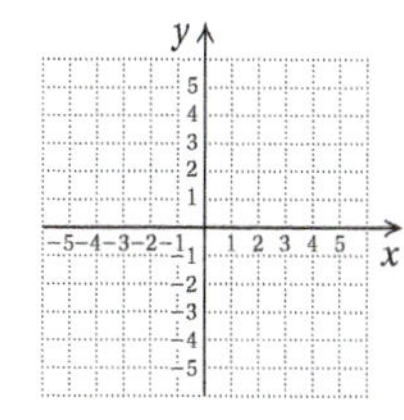

16. $t(x) = 3 - |x|$

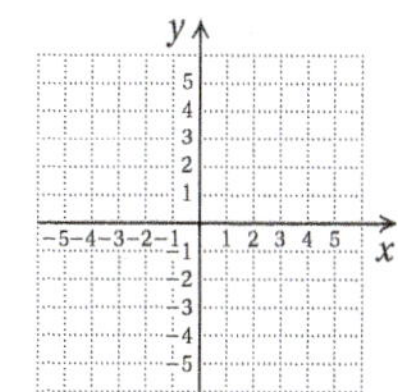

d THE VERTICAL-LINE TEST

Consider the function f described by $f(x) = x^2 - 5$. Its graph is shown at right. It is also the graph of the equation $y = x^2 - 5$.

To find a function value, like $f(3)$, from a graph, we locate the input on the horizontal axis, move vertically to the graph of the function, and then move horizontally to find the output on the vertical axis, where members of the range can be found. As shown, $f(3) = 4$.

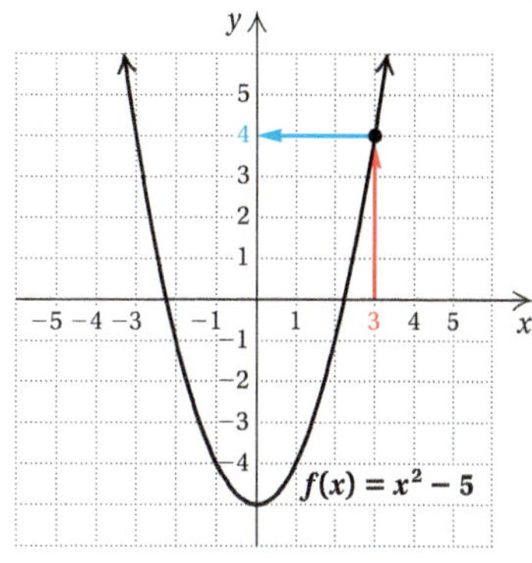

Answers

14.

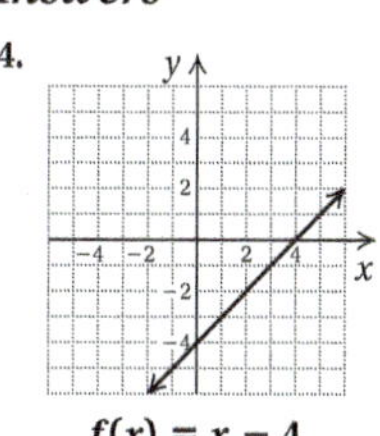

15.

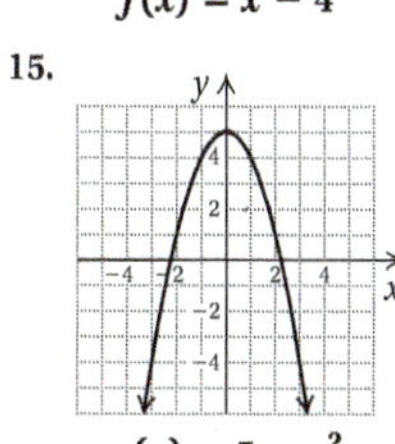

16.

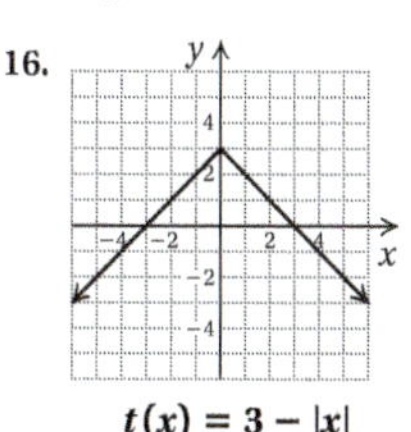

Recall that when one member of the domain is paired with two or more different members of the range, the correspondence is *not* a function. Thus, when a graph contains two or more different points with the same first coordinate, the graph cannot represent a function. Points sharing a common first coordinate are vertically above or below each other. (See the following graph.)

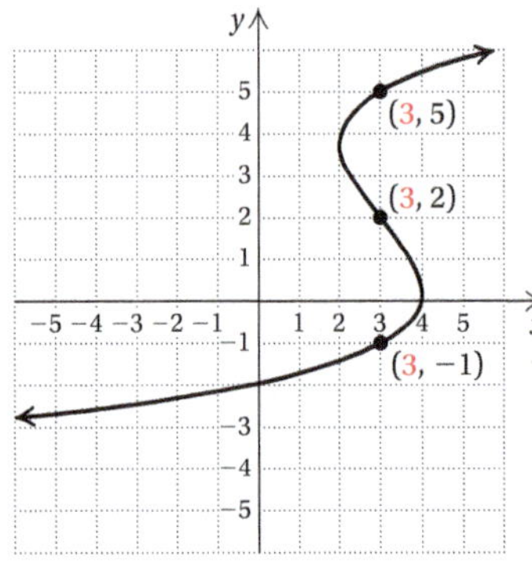

Since 3 is paired with more than one member of the range, the graph does not represent a function.

This observation leads to the *vertical-line test*.

THE VERTICAL-LINE TEST

A graph represents a function if it is impossible to draw a vertical line that intersects the graph more than once.

EXAMPLE 7 Determine whether each of the following is the graph of a function.

a)

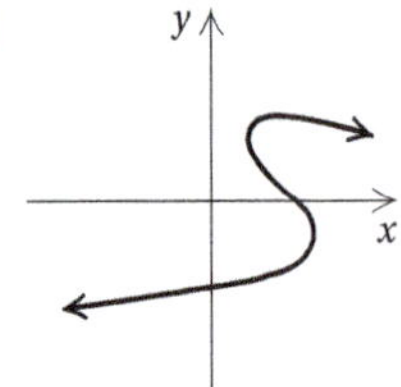

b)

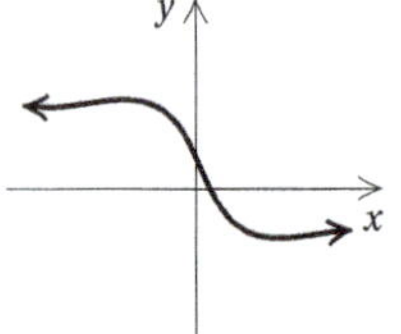

c)

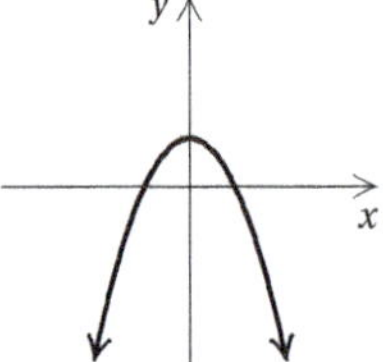

a) The graph *is not* that of a function because a vertical line crosses the graph at more than one point.

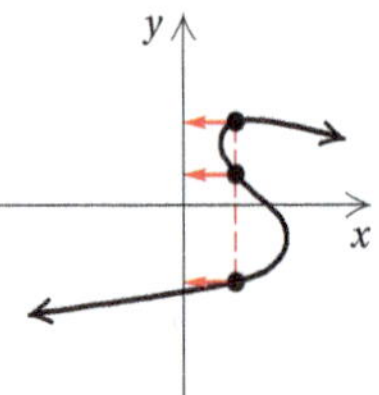

b) The graph *is* that of a function because no vertical line can cross the graph at more than one point. This can be confirmed with a ruler or a straightedge.

c) The graph *is* that of a function.

Do Exercises 17–20.

Determine whether each of the following is the graph of a function.

17.

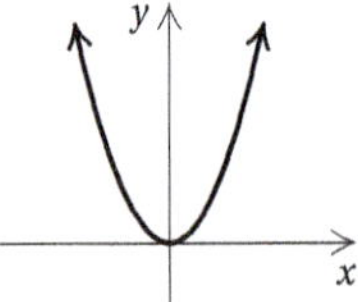

18.

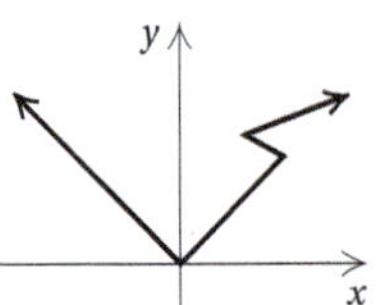

19.

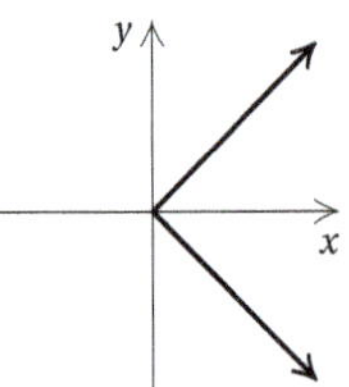

20.

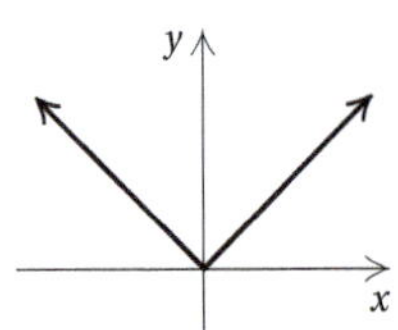

Answers

17. Yes **18.** No **19.** No **20.** Yes

e APPLICATIONS OF FUNCTIONS AND THEIR GRAPHS

Functions are often described by graphs, whether or not an equation is given.

EXAMPLE 8 *Movie Revenue.* The graph shown approximates the weekly revenue, in millions of dollars, from a movie. The revenue is a function of the week, and no equation is given for the function. Use the graph to answer the following.

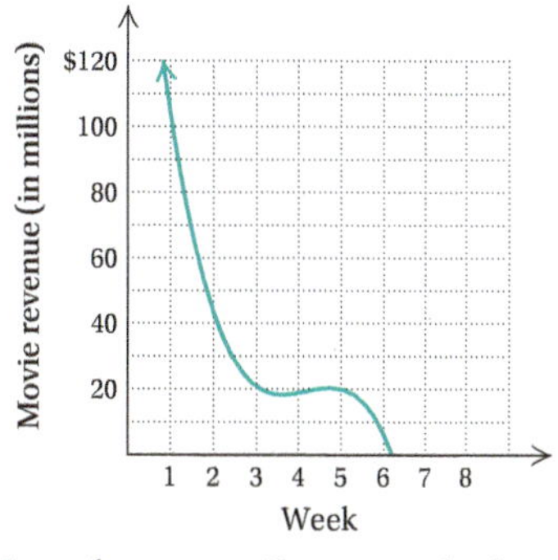

a) What was the movie revenue for week 1?

b) What was the movie revenue for week 5?

a) To estimate the revenue for week 1, we locate 1 on the horizontal axis and move directly up until we reach the graph. Then we move across to the vertical axis. We estimate that value to be about $105 million.

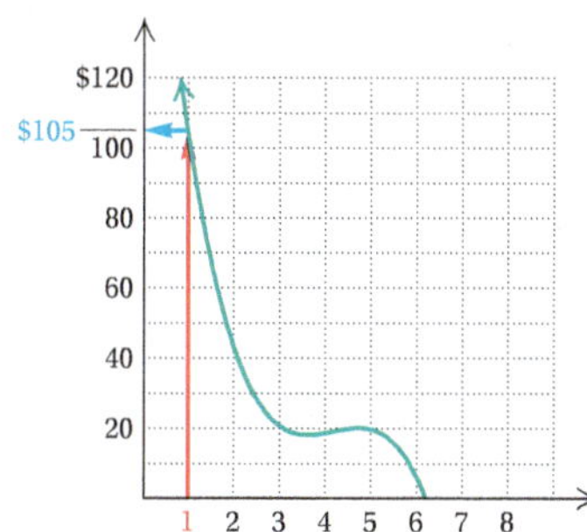

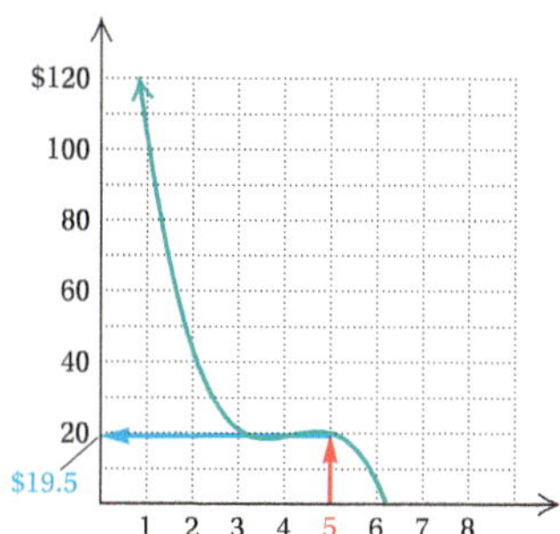

b) To estimate the revenue for week 5, we locate 5 on the horizontal axis and move directly up until we reach the graph. Then we move across to the vertical axis. We estimate that value to be about $19.5 million.

Do Exercises 21 and 22. ▶

Refer to the graph in Example 8.

21. What was the movie revenue for week 2?

22. What was the movie revenue for week 6?

Answers

21. About $43 million **22.** About $6 million

17.7 Exercise Set

For Extra Help MyMathLab® MathXL®

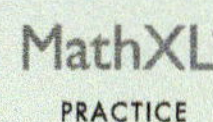
PRACTICE

WATCH

READ

REVIEW

 Reading Check

Choose from the columns on the right the best word or words to complete each sentence. Not all words will be used.

RC1. A function is a special kind of correspondence between a first set, called the ____________, and a second set, called the ____________.

RC2. When we write $f(4) = 9$, we mean that the number 4 is a(n) ____________ of the function.

RC3. We use the ____________ -line test to test whether a graph represents a function.

RC4. The function given by $f(x) = 5x + 7$ is an example of a(n) ____________ function.

domain	linear
range	quadratic
horizontal	input
vertical	output

a Determine whether each correspondence is a function.

1. *Domain* *Range*

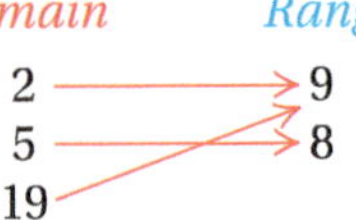

2. *Domain* *Range*

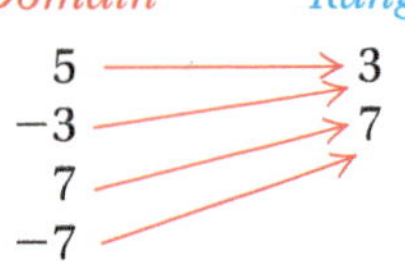

3. *Domain* *Range*

−5 → 1
5
8

4. *Domain* *Range*

5. *Domain* *Range*

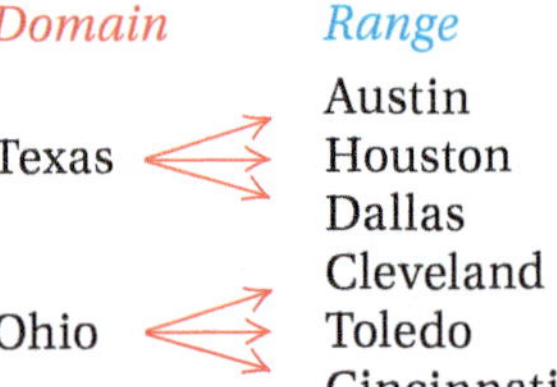

6. *Domain* *Range*

7. *Domain* *Range*

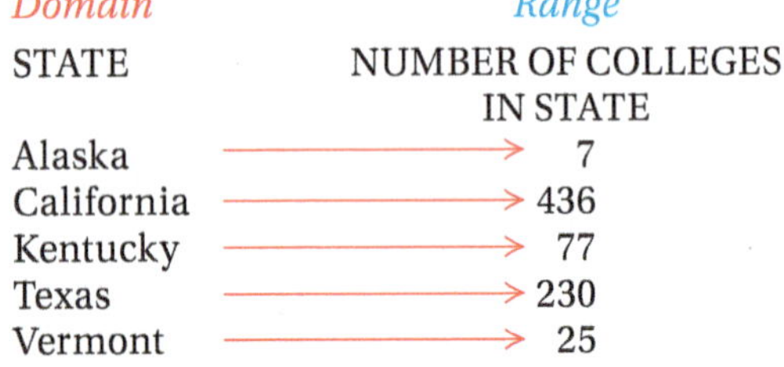

Source: http://www.publicagendaarchives.org

8. *Domain* *Range*

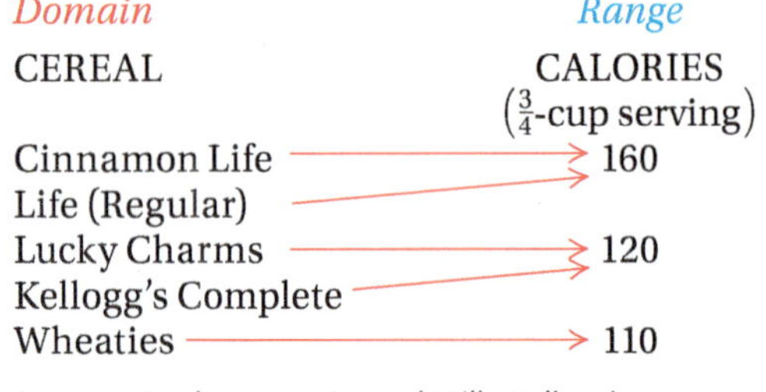

Sources: Quaker Oats; General Mills; Kellogg's

Determine whether each of the following is a function. Identify any relations that are not functions.

	Domain	*Correspondence*	*Range*
9.	A math class	Each person's seat number	A set of numbers
10.	A set of numbers	4 more than the square of each number	A set of numbers
11.	A set of shapes	The area of each shape	A set of numbers
12.	A family	Each person's eye color	A set of colors
13.	The people in a town	Each person's neighbor	A set of people
14.	Students in a college	Each student's classes	A set of classes

d Find the function values.

15. $f(x) = x + 5$

a) $f(4)$ **b)** $f(7)$
c) $f(-3)$ **d)** $f(0)$
e) $f(2.4)$ **f)** $f\left(\frac{2}{3}\right)$

16. $g(t) = t - 6$

a) $g(0)$ **b)** $g(6)$
c) $g(13)$ **d)** $g(-1)$
e) $g(-1.08)$ **f)** $g\left(\frac{7}{8}\right)$

17. $h(p) = 3p$

a) $h(-7)$ **b)** $h(5)$
c) $h(14)$ **d)** $h(0)$
e) $h\left(\frac{2}{3}\right)$ **f)** $h(-54.2)$

18. $f(x) = -4x$

a) $f(6)$ **b)** $f(-\frac{1}{2})$
c) $f(20)$ **d)** $f(11.8)$
e) $f(0)$ **f)** $f(-1)$

19. $g(s) = 3s + 4$

a) $g(1)$ **b)** $g(-7)$
c) $g(6.7)$ **d)** $g(0)$
e) $g(-10)$ **f)** $g(\frac{2}{3})$

20. $h(x) = 19$, a constant function

a) $h(4)$ **b)** $h(-6)$
c) $h(12.5)$ **d)** $h(0)$
e) $h(\frac{2}{3})$ **f)** $h(1234)$

21. $f(x) = 2x^2 - 3x$

a) $f(0)$ **b)** $f(-1)$
c) $f(2)$ **d)** $f(10)$
e) $f(-5)$ **f)** $f(-10)$

22. $f(x) = 3x^2 - 2x + 1$

a) $f(0)$ **b)** $f(1)$
c) $f(-1)$ **d)** $f(10)$
e) $f(2)$ **f)** $f(-3)$

23. $f(x) = |x| + 1$

a) $f(0)$ **b)** $f(-2)$
c) $f(2)$ **d)** $f(-3)$
e) $f(-10)$ **f)** $f(22)$

24. $g(t) = \sqrt{t}$

a) $g(4)$ **b)** $g(25)$
c) $g(16)$ **d)** $g(100)$
e) $g(50)$ **f)** $g(84)$

25. $f(x) = x^3$

a) $f(0)$ **b)** $f(-1)$
c) $f(2)$ **d)** $f(10)$
e) $f(-5)$ **f)** $f(-10)$

26. $f(x) = x^4 - 3$

a) $f(1)$ **b)** $f(-1)$
c) $f(0)$ **d)** $f(2)$
e) $f(-2)$ **f)** $f(10)$

27. *Life Span.* The function given by $l(x) = \dfrac{1700}{x}$ can be used to approximate the life span, in years, of an animal with a pulse rate of x beats per minute.

a) Find the approximate life span of a horse with a pulse rate of 50 beats per minute.
b) Find the approximate life span of a seal with a pulse rate of 85 beats per minute.

28. *Temperature as a Function of Depth.* The function $T(d) = 10d + 20$ gives the temperature, in degrees Celsius, inside the earth as a function of the depth d, in kilometers. Find the temperature at 5 km, 20 km, and 1000 km.

29. *Growth in the Number of Facebook Users.* The function $F(t) = 22t^2 - 41t + 2.7$ can be used to approximate the number of monthly active Facebook users, in millions, t years after 2004. Estimate the number of monthly active Facebook users in 2004 ($t = 0$), in 2010 ($t = 6$), and in 2012 ($t = 8$).

Source: Based on data from Facebook

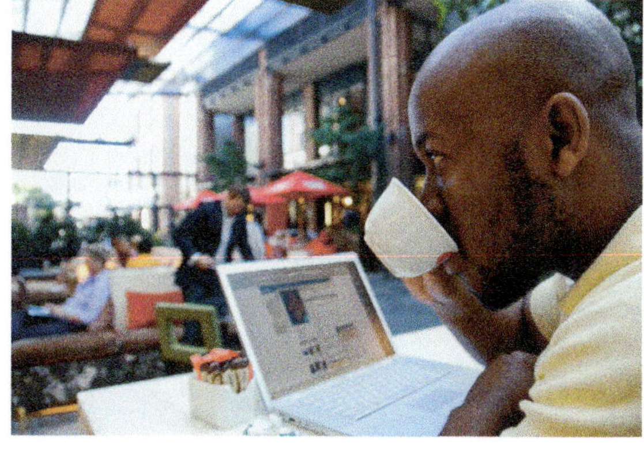

30. *Female Physicians.* The function $p(t) = \frac{1}{8}t^2 + 3t + 35$ can be used to approximate the number of female physicians, in thousands, in the United States t years after 1975. Estimate the number of female physicians in the United States in 1975 ($t = 0$), in 2000 ($t = 25$), and in 2010 ($t = 35$).

Source: Based on data from AMA Physician Characteristics and Distribution in the US, 2012 Edition

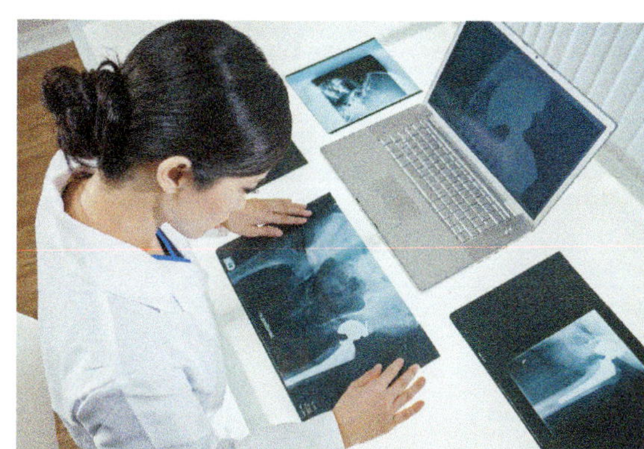

31. ***Pressure at Sea Depth.*** The function $P(d) = 1 + (d/33)$ gives the pressure, in *atmospheres* (atm), at a depth of d feet in the sea. Note that $P(0) = 1$ atm, $P(33) = 2$ atm, and so on. Find the pressure at 20 ft, 30 ft, and 100 ft.

32. ***Temperature Conversions.*** The function $C(F) = \frac{5}{9}(F - 32)$ determines the Celsius temperature that corresponds to F degrees Fahrenheit. Find the Celsius temperature that corresponds to 62°F, 77°F, and 23°F.

Graph each function.

33. $f(x) = 3x - 1$

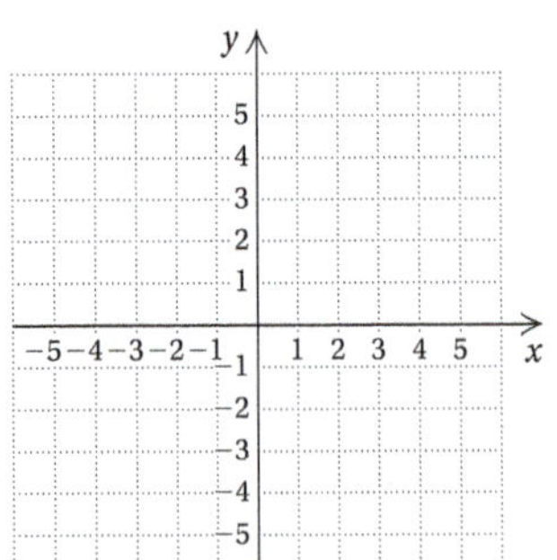

34. $g(x) = 2x + 5$

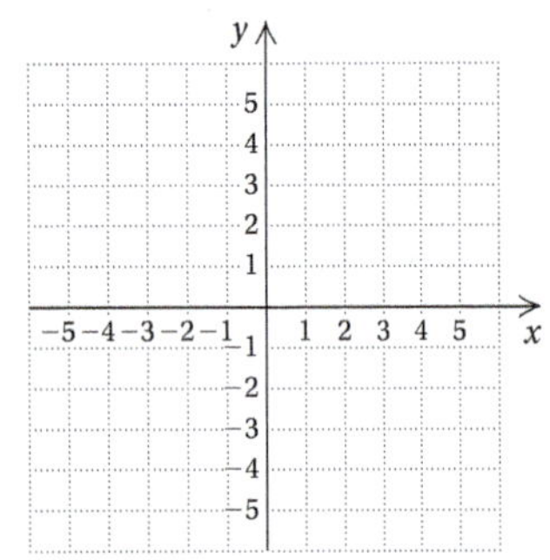

35. $g(x) = -2x + 3$

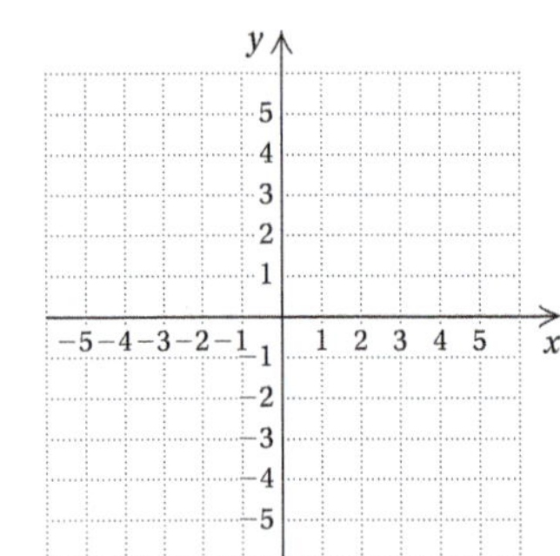

36. $f(x) = -\frac{1}{2}x + 2$

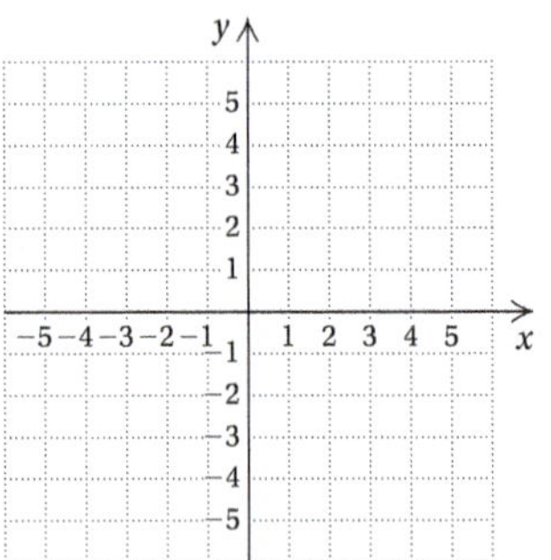

37. $f(x) = \frac{1}{2}x + 1$

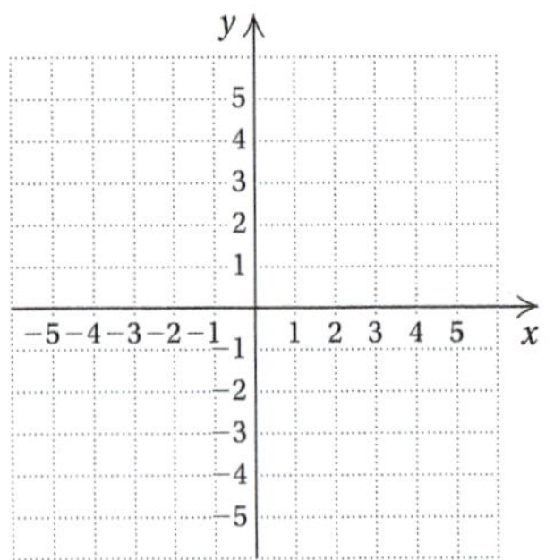

38. $f(x) = -\frac{3}{4}x - 2$

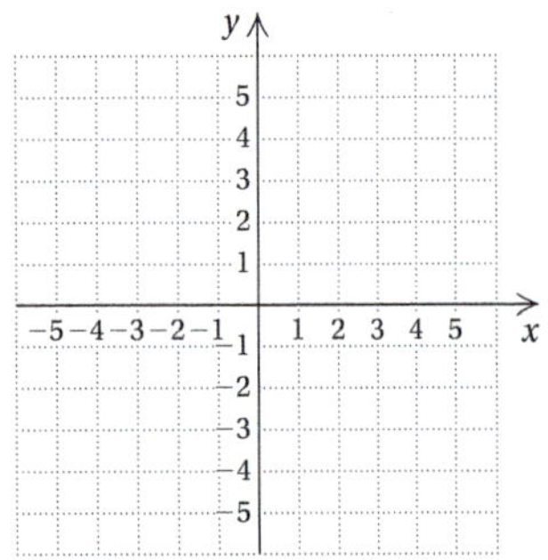

39. $f(x) = 2 - |x|$

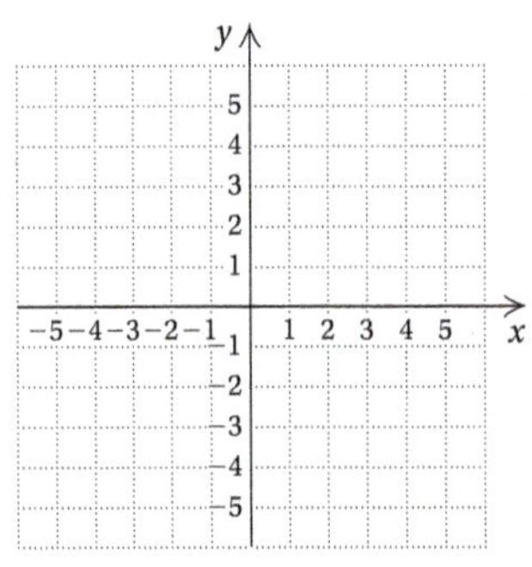

40. $f(x) = |x| - 4$

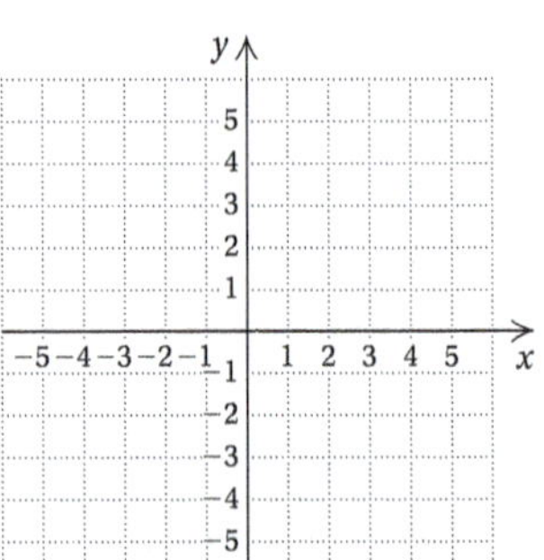

41. $f(x) = x^2$

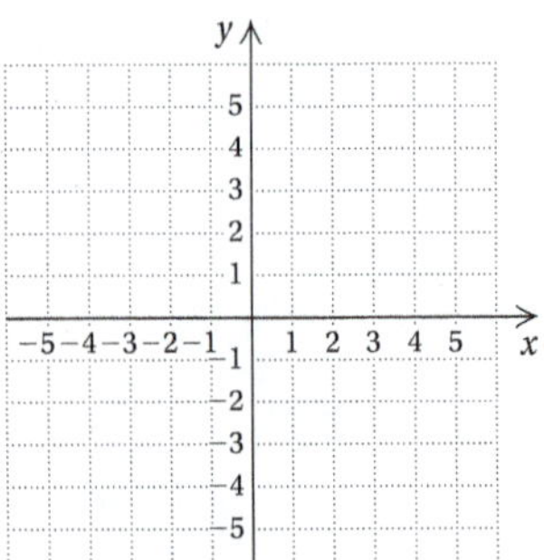

42. $f(x) = x^2 - 1$

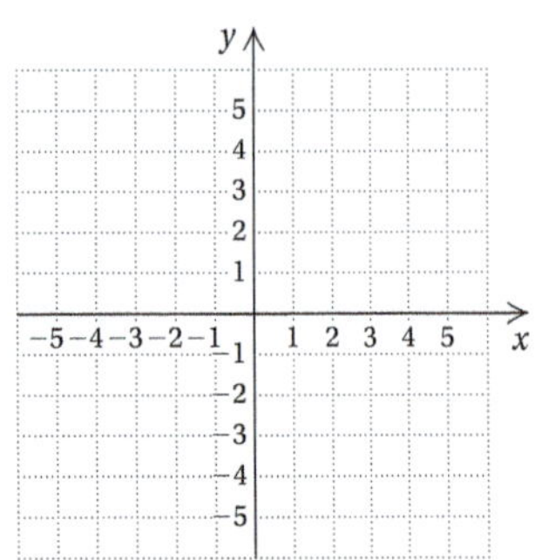

43. $f(x) = x^2 - x - 2$

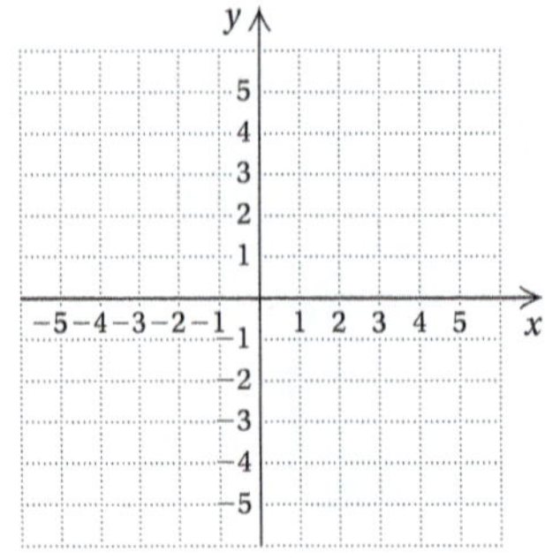

44. $f(x) = x^2 + 6x + 5$

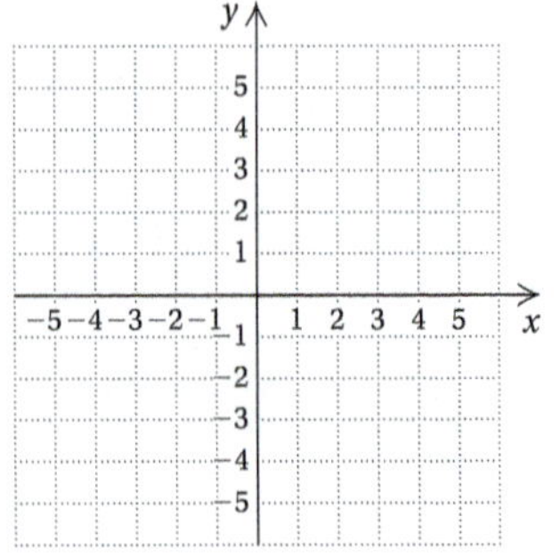

d Determine whether each of the following is the graph of a function.

45.

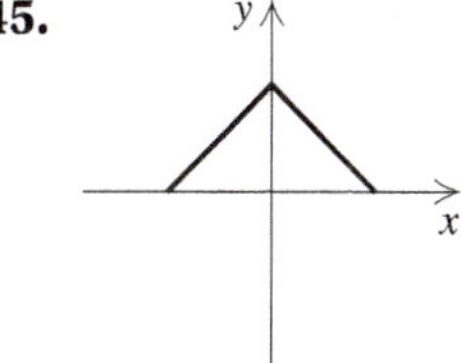

46.

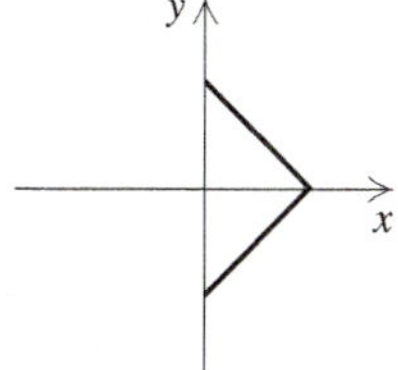

47.

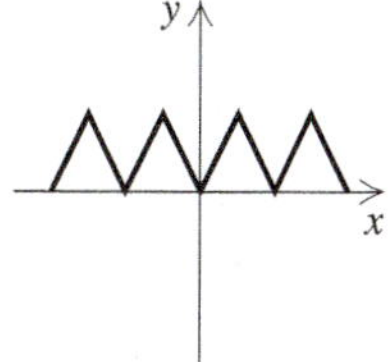

48.

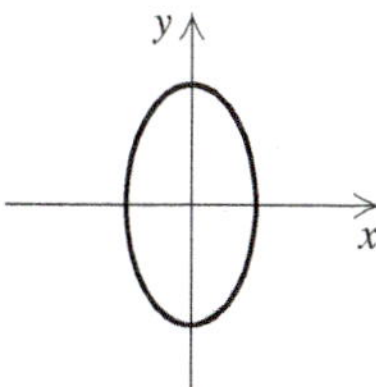

49.

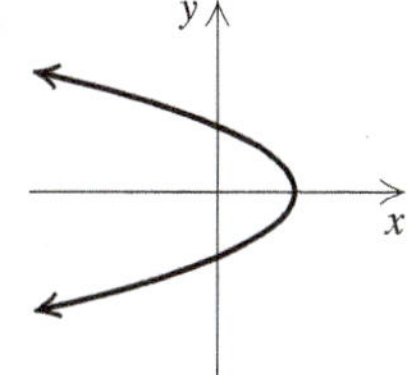

50.

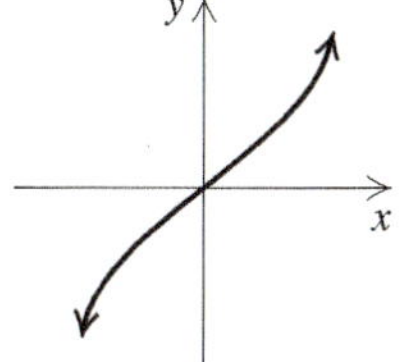

51.

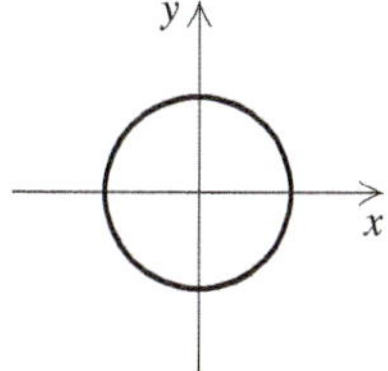

52.

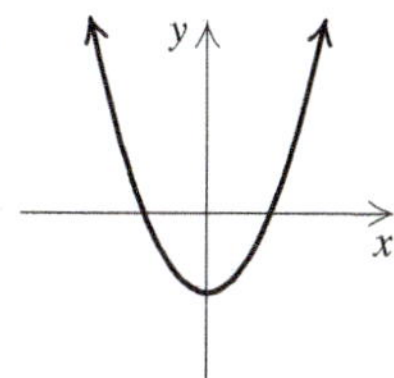

e ***Las Vegas Conventions.*** The number of conventions held in Las Vegas can be modeled as a function of the number of years after 2000. The graph of this function is shown below.

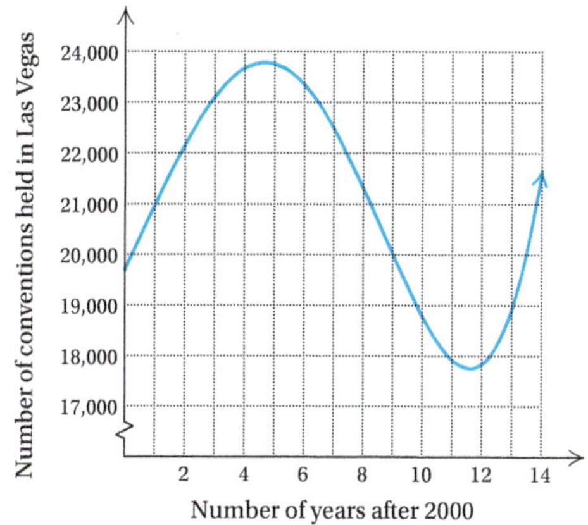

Data from Las Vegas Convention and Visitors Authority

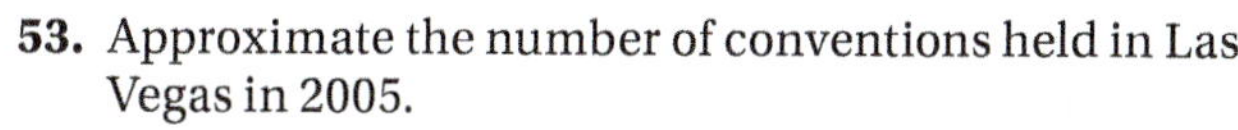

53. Approximate the number of conventions held in Las Vegas in 2005.

54. Approximate the number of conventions held in Las Vegas in 2012.

Skill Maintenance

Solve each system using the substitution method. [15.2b]

55. $x = 2 - y,$
$3x + y = 5$

56. $3x - y = 9,$
$2x = 6 + y$

Solve each system using the elimination method. [15.3a, b]

57. $2x - 5y = 7,$
$x + 5y = 2$

58. $x - 3y = 2,$
$3x - 9y = 6$

Synthesis

Graph.

59. $g(x) = x^3$

60. $f(x) = 2 + \sqrt{x}$

61. $f(x) = |x| + x$

62. $g(x) = |x| - x$

CHAPTER

17 Summary and Review

Key Formulas and Principles

Standard Form of a Quadratic Equation: $ax^2 + bx + c = 0, a > 0$

Principle of Square Roots: The equation $x^2 = d$, where $d > 0$, has two solutions, $\sqrt{d}$ and $-\sqrt{d}$. The solution of $x^2 = 0$ is 0.

The x-coordinate of the vertex of a parabola: $-\frac{b}{2a}$

Quadratic Formula: $x = \frac{-b \pm \sqrt{b^2 - 4ac}}{2a}$

Discriminant: $b^2 - 4ac$

Vocabulary Reinforcement

Complete each statement with the correct term from the column on the right. Some of the choices may be used more than once and some may not be used at all.

1. The equation $ax^2 + bx + c = 0$ is the standard form of a(n) ______________ equation. [17.1a]
2. When we add 25 to $x^2 + 10x$, we are completing the ______________. [17.2c]
3. The expression $b^2 - 4ac$ is called the ______________. [17.6b]
4. The turning point of the graph of a quadratic equation is the ______________. [17.6a]
5. The graph of a quadratic equation is called a(n) ______________. [17.6a]
6. The function given by $f(x) = 7x^2 - 3x$ is an example of a(n) ______________ function. [17.7b]
7. A graph is the graph of a function if it passes the ______________ test. [17.7d]
8. The set of all inputs of a function is called the ______________. [17.7a]

domain
range
function
relation
quadratic
linear
vertical-line
horizontal-line
square
parabola
line of symmetry
vertex
discriminant

Concept Reinforcement

Determine whether each statement is true or false.

______________ 1. A graph represents a function if it is possible to draw a vertical line that intersects the graph more than once. [17.7d]

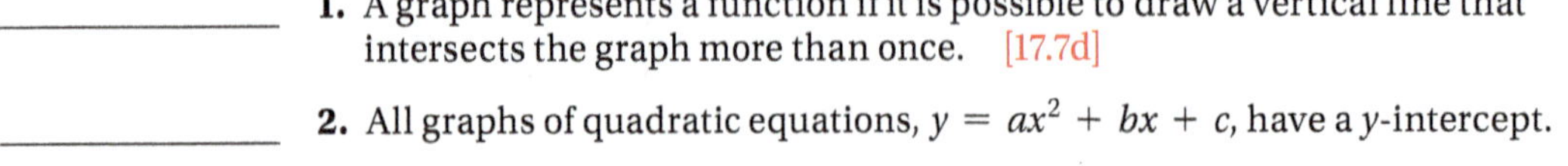

______________ 2. All graphs of quadratic equations, $y = ax^2 + bx + c$, have a y-intercept. [17.6a]

______________ 3. If (p, q) is the vertex of the graph of $y = ax^2 + bx + c, a < 0$, then q is the largest value of y. [17.6a]

______________ 4. If a quadratic equation $ax^2 + bx + c = 0$ has no real-number solutions, then the graph of $y = ax^2 + bx + c$ does not have an x-intercept. [17.6b]

Study Guide

Objective 17.1c Solve quadratic equations of the type $ax^2 + bx + c = 0$, where $b \neq 0$ and $c \neq 0$, by factoring.

Example Solve: $\frac{1}{x} + \frac{2}{x+3} = \frac{3}{2}$.

We multiply by the LCM, which is $2x(x + 3)$.

$$2x(x+3) \cdot \left(\frac{1}{x} + \frac{2}{x+3}\right) = \frac{3}{2} \cdot 2x(x+3)$$

$$2x(x+3) \cdot \frac{1}{x} + 2x(x+3) \cdot \frac{2}{x+3} = 3x(x+3)$$

$$2(x+3) + 2x \cdot 2 = 3x^2 + 9x$$

$$2x + 6 + 4x = 3x^2 + 9x$$

$$0 = 3x^2 + 3x - 6$$

$$0 = 3(x^2 + x - 2)$$

$$0 = 3(x+2)(x-1)$$

$$x + 2 = 0 \quad or \quad x - 1 = 0$$

$$x = -2 \quad or \quad x = 1$$

Both numbers check. The solutions are -2 and 1.

Practice Exercise

1. Solve: $\frac{3}{x+2} + \frac{1}{x} = \frac{5}{4}$.

Objective 17.2a Solve quadratic equations of the type $ax^2 = p$.

Example Solve: $5x^2 - 2 = 12$.

$$5x^2 - 2 = 12$$

$$5x^2 = 14 \qquad \text{Adding 2}$$

$$x^2 = \frac{14}{5} \qquad \text{Dividing by 5}$$

$$x = \sqrt{\frac{14}{5}} \quad or \quad x = -\sqrt{\frac{14}{5}} \qquad \text{Using the principle of square roots}$$

$$x = \sqrt{\frac{14}{5} \cdot \frac{5}{5}} \quad or \quad x = -\sqrt{\frac{14}{5} \cdot \frac{5}{5}} \qquad \text{Rationalizing the denominator}$$

$$x = \frac{\sqrt{70}}{5} \quad or \quad x = -\frac{\sqrt{70}}{5}$$

The solutions are $\frac{\sqrt{70}}{5}$ and $-\frac{\sqrt{70}}{5}$.

Practice Exercise

2. Solve: $7x^2 - 3 = 8$.

Objective 17.2c Solve quadratic equations by completing the square.

Example Solve $x^2 - 10x + 8 = 0$ by completing the square.

$$x^2 - 10x + 8 = 0$$

$$x^2 - 10x = -8 \qquad \text{Subtracting 8}$$

$$x^2 - 10x + 25 = -8 + 25 \qquad \left(\frac{-10}{2}\right)^2 = 25$$

$$(x - 5)^2 = 17$$

$$x - 5 = \sqrt{17} \quad or \quad x - 5 = -\sqrt{17}$$

$$x = 5 + \sqrt{17} \quad or \quad x = 5 - \sqrt{17}$$

The solutions are $5 \pm \sqrt{17}$.

Practice Exercise

3. Solve $x^2 - 4x + 1 = 0$ by completing the square.

Objective 17.3a Solve quadratic equations using the quadratic formula.

Example Solve $6x^2 = 4x + 5$ using the quadratic formula.

$$6x^2 - 4x - 5 = 0 \quad \text{Standard form}$$

$$a = 6, \quad b = -4, \quad c = -5$$

$$x = \frac{-b \pm \sqrt{b^2 - 4ac}}{2a} \quad \text{Quadratic formula}$$

$$= \frac{-(-4) \pm \sqrt{(-4)^2 - 4 \cdot 6 \cdot (-5)}}{2 \cdot 6} \quad \text{Substituting}$$

$$= \frac{4 \pm \sqrt{16 + 120}}{12} = \frac{4 \pm \sqrt{136}}{12}$$

$$= \frac{4 \pm \sqrt{4 \cdot 34}}{12} = \frac{4 \pm 2\sqrt{34}}{12}$$

$$= \frac{2(2 \pm \sqrt{34})}{2 \cdot 6} = \frac{2 \pm \sqrt{34}}{6}$$

The solutions are $\frac{2 \pm \sqrt{34}}{6}$.

Practice Exercise

4. Solve: $4y^2 = 6y + 3$.

Objective 17.6a Graph quadratic equations.

Example Graph: $y = 2x^2 + 4x - 1$.

We first find the vertex. The x-coordinate of the vertex is

$$-\frac{b}{2a} = -\frac{4}{2 \cdot 2} = -1.$$

We substitute -1 for x into the equation to find the second coordinate of the vertex:

$$y = 2(-1)^2 + 4(-1) - 1 = -3.$$

The vertex is $(-1, -3)$. The line of symmetry is $x = -1$. We choose x-values on both sides of $x = -1$ and graph the parabola.

x	y	
-1	-3	← Vertex
-2	-1	
0	-1	
-3	5	
1	5	

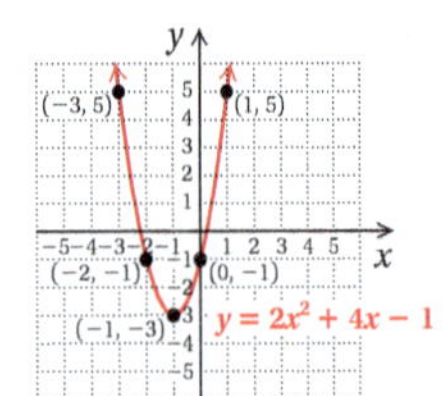

Practice Exercise

5. Graph: $y = x^2 - 4x + 2$.

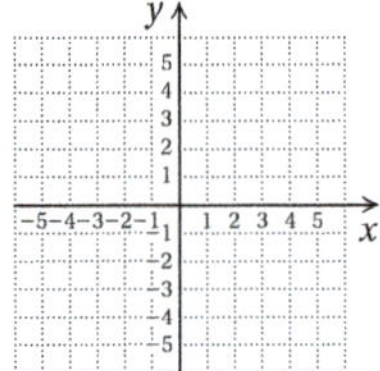

Objective 17.7b Given a function described by an equation, find function values (outputs) for specified values (inputs).

Example Find $f(-2)$, for $f(x) = -\frac{1}{2}x + 7$.

$$f(-2) = -\frac{1}{2}(-2) + 7 = 1 + 7 = 8$$

Practice Exercise

6. Find the indicated function value.

a) $h(5)$, for $h(x) = \frac{1}{5}x^2 + x - 1$

b) $f(0)$, for $f(x) = -3x - 4$

Review Exercises

Solve.

1. $8x^2 = 24$ [17.2a]

2. $40 = 5y^2$ [17.2a]

3. $5x^2 - 8x + 3 = 0$ [17.1c]

4. $3y^2 + 5y = 2$ [17.1c]

5. $(x + 8)^2 = 13$ [17.2b]

6. $9x^2 = 0$ [17.2a]

7. $5t^2 - 7t = 0$ [17.1b]

Solve. [17.3a]

8. $x^2 - 2x - 10 = 0$

9. $9x^2 - 6x - 9 = 0$

10. $x^2 + 6x = 9$

11. $1 + 4x^2 = 8x$

12. $6 + 3y = y^2$

13. $3m = 4 + 5m^2$

14. $3x^2 = 4x$

Solve. [17.1c]

15. $\dfrac{15}{x} - \dfrac{15}{x + 2} = 2$

16. $x + \dfrac{1}{x} = 2$

Solve by completing the square. Show your work. [17.2c]

17. $x^2 - 4x + 2 = 0$

18. $3x^2 - 2x - 5 = 0$

Approximate the solutions to the nearest tenth. [17.3b]

19. $x^2 - 5x + 2 = 0$

20. $4y^2 + 8y + 1 = 0$

21. Solve for T: $V = \dfrac{1}{2}\sqrt{1 + \dfrac{T}{L}}$. [17.4a]

Graph each quadratic equation. Label the ordered pairs for the vertex and the y-intercept. [17.6a]

22. $y = 2 - x^2$

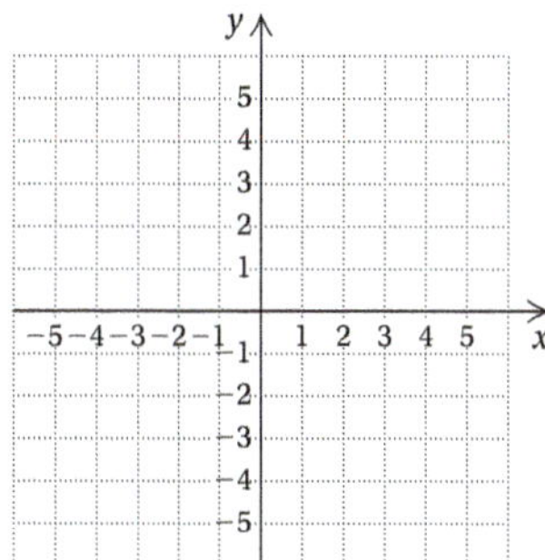

23. $y = x^2 - 4x - 2$

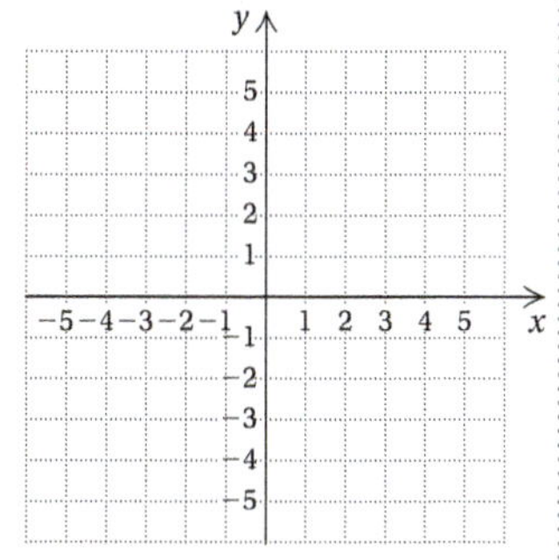

Find the x-intercepts. [17.6b]

24. $y = 2 - x^2$

25. $y = x^2 - 4x - 2$

Solve.

26. *Right-Triangle Dimensions.* The hypotenuse of a right triangle is 5 cm long. One leg is 3 cm longer than the other. Find the lengths of the legs. Round to the nearest tenth. [17.5a]

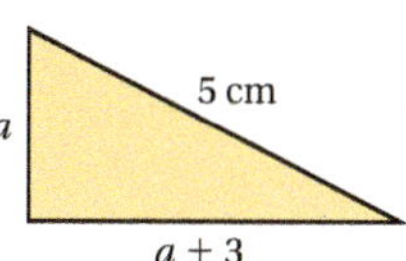

27. *Car-Loading Ramp.* The length of a loading ramp on a car hauler is 25 ft. This ramp and its height form the hypotenuse and one leg of a right triangle. The height of the ramp is 5 ft less than the length of the other leg. Find the height of the loading ramp. [17.5a]

28. *Falling Object.* The Royal Gorge Bridge above Colorado's Arkansas River is the highest suspension bridge in the United States. It hangs 1053 ft above the river. How long would it take an object to fall to the water from the bridge? [17.2d]

Find the function values. [17.7b]

29. If $f(x) = 2x - 5$, find $f(2)$, $f(-1)$, and $f(3.5)$.

30. If $g(x) = |x| - 1$, find $g(1)$, $g(-1)$, and $g(-20)$.

31. *Caloric Needs.* If you are moderately active, you need to consume about 15 calories per pound of body weight each day. The function $C(p) = 15p$ approximates the number of calories C that are needed to maintain body weight p, in pounds. How many calories are needed to maintain a body weight of 180 lb? [17.7e]

Graph each function. [17.7c]

32. $g(x) = 4 - x$

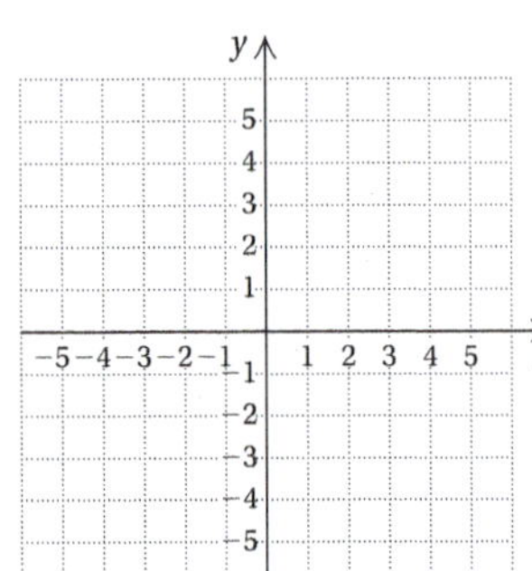

33. $f(x) = x^2 - 3$

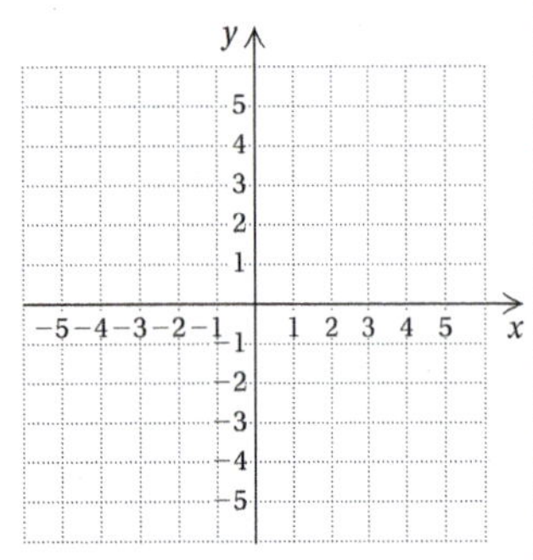

34. $h(x) = |x| - 5$

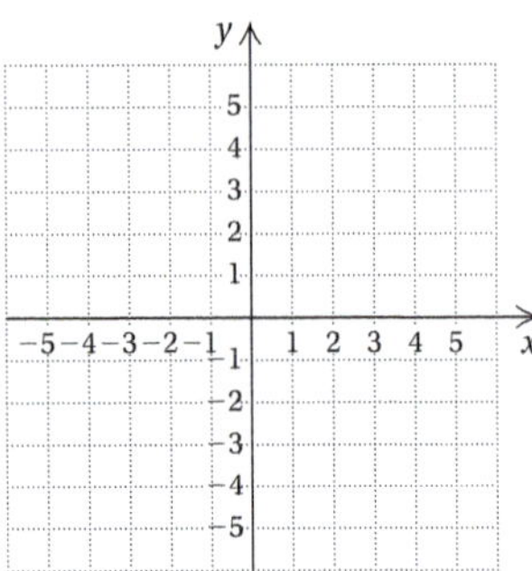

35. $f(x) = x^2 - 2x + 1$

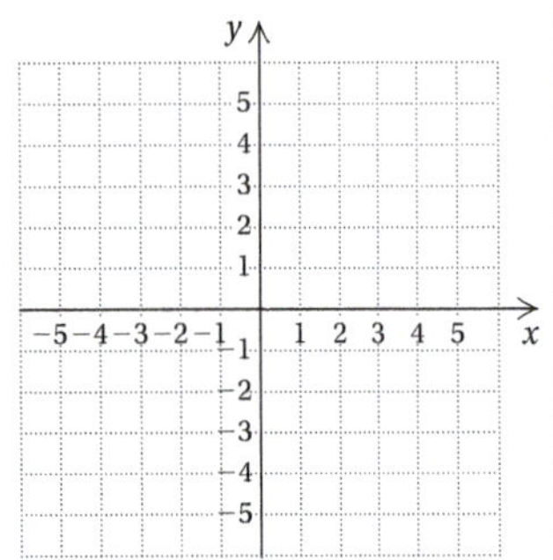

Determine whether each of the following is the graph of a function. [17.7d]

36. 

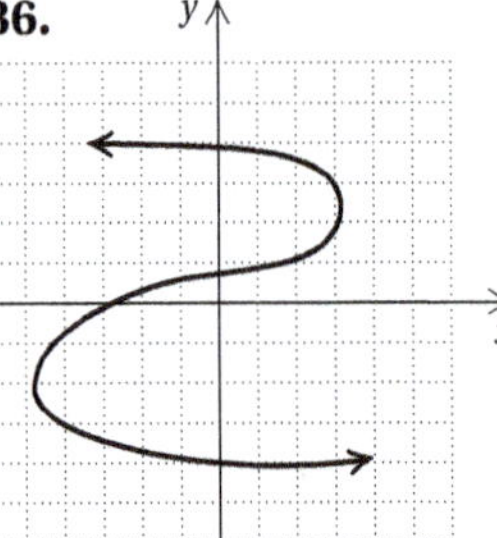

37.

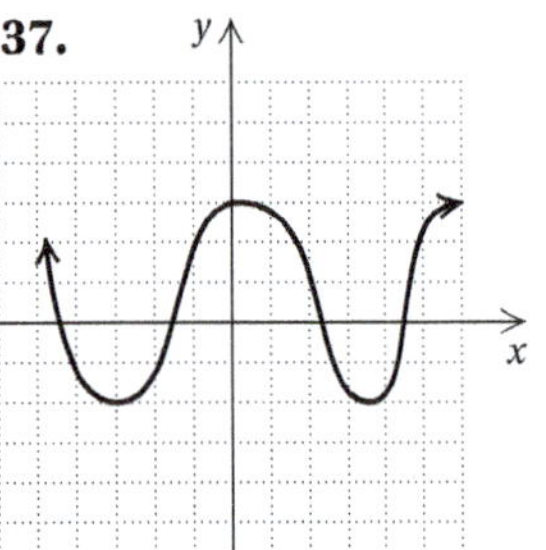

38. Solve: $40x - x^2 = 0$. [17.1b]

A. 40 **B.** $2\sqrt{10}$

C. $-2\sqrt{10}$ **D.** 0, 40

39. Solve: $\frac{1}{2}c^2 + c - \frac{1}{2} = 0$. [17.3a]

A. $-1 \pm \sqrt{2}$ **B.** $-1 \pm \sqrt{5}$

C. $1 \pm \sqrt{2}$ **D.** $-3, 1$

Synthesis

40. Two consecutive integers have squares that differ by 63. Find the integers. [17.5a]

41. A square with sides of length s has the same area as a circle with a radius of 5 in. Find s. [17.5a]

42. Solve: $x - 4\sqrt{x} - 5 = 0$. [17.1c]

Use the graph of

$$y = (x + 3)^2$$

to solve each equation. [17.6b]

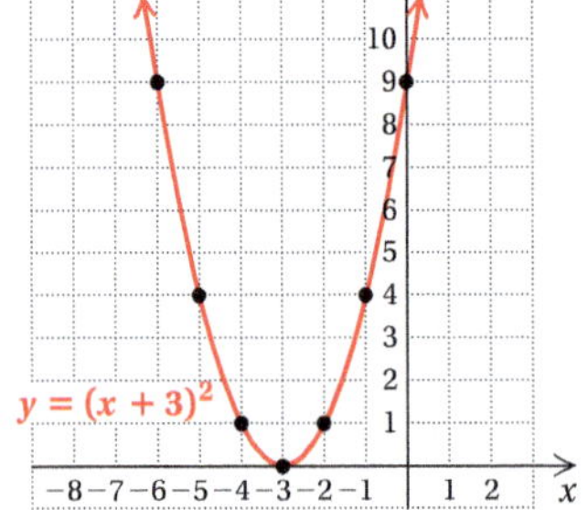

43. $(x + 3)^2 = 1$

44. $(x + 3)^2 = 4$

45. $(x + 3)^2 = 9$

46. $(x + 3)^2 = 0$

Understanding Through Discussion and Writing

1. Find and explain the error(s) in the following solution of a quadratic equation. [17.2b]

$$(x + 6)^2 = 16$$
$$x + 6 = \sqrt{16}$$
$$x + 6 = 4$$
$$x = -2$$

2. Is it possible for a function to have more numbers as outputs than as inputs? Why or why not? [17.7b]

3. Suppose that the x-intercepts of a parabola are $(a_1, 0)$ and $(a_2, 0)$. What is the easiest way to find an equation for the line of symmetry? to find the coordinates of the vertex? [17.6b]

4. Discuss the effect of the sign of a on the graph of $y = ax^2 + bx + c$. [17.6a]

5. If a quadratic equation can be solved by factoring, what type of number(s) will generally be solutions? [17.1c]

CHAPTER 17

Test

For Extra Help: For step-by-step test solutions, access the Chapter Test Prep Videos in MyMathLab® or on YouTube (search "BittingerPreIntro" and click on Channels).

Solve.

1. $7x^2 = 35$

2. $7x^2 + 8x = 0$

3. $48 = t^2 + 2t$

4. $3y^2 - 5y = 2$

5. $(x - 8)^2 = 13$

6. $x^2 = x + 3$

7. $m^2 - 3m = 7$

8. $10 = 4x + x^2$

9. $3x^2 - 7x + 1 = 0$

10. $x - \frac{2}{x} = 1$

11. $\frac{4}{x} - \frac{4}{x + 2} = 1$

12. Solve $x^2 - 4x - 10 = 0$ by completing the square. Show your work.

13. Approximate the solutions to $x^2 - 4x - 10 = 0$ to the nearest tenth.

14. Solve for n: $d = an^2 + bn$.

15. Find the x-intercepts: $y = -x^2 + x + 5$.

Graph. Label the ordered pairs for the vertex and the y-intercept.

16. $y = 4 - x^2$

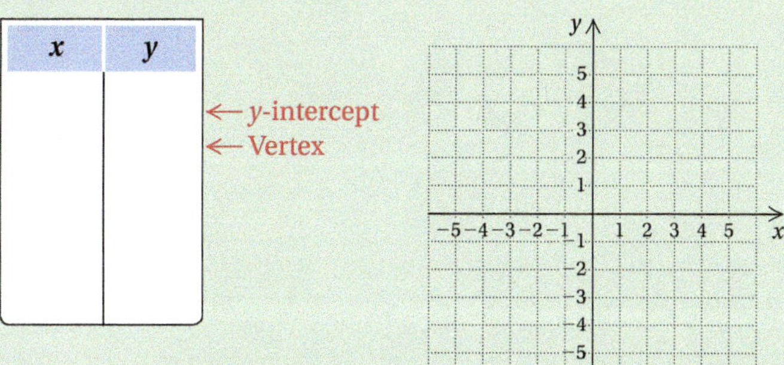

17. $y = -x^2 + x + 5$

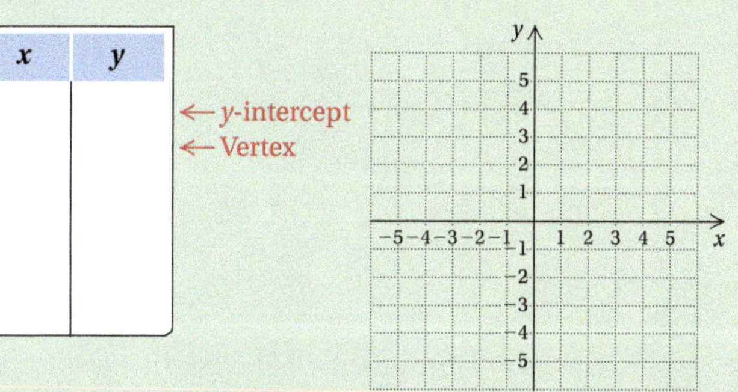

18. If $f(x) = \frac{1}{2}x + 1$, find $f(0)$, $f(1)$, and $f(2)$.

19. If $g(t) = -2|t| + 3$, find $g(-1)$, $g(0)$, and $g(3)$.

Solve.

20. *Rug Dimensions.* The width of a rectangular area rug is 4 m less than the length. The area is 16.25 m². Find the length and the width.

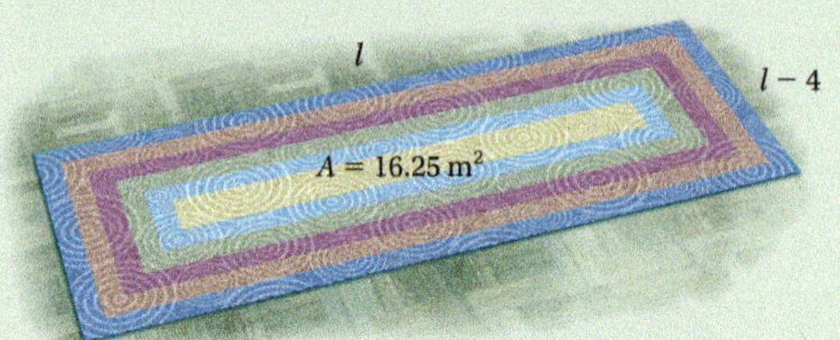

21. *Boat Speed.* The current in a stream moves at a speed of 2 km/h. A boat travels 44 km upstream and 52 km downstream in a total of 4 hr. What is the speed of the boat in still water?

22. *World Record for 10,000-m Run.* The world record for the 10,000-m run has been decreasing steadily since 1940. The function $R(t) = 30.18 - 0.06t$ estimates the record R, in minutes, as a function of t, the time in years since 1940. Estimate the record in 2012.

Graph.

23. $h(x) = x - 4$

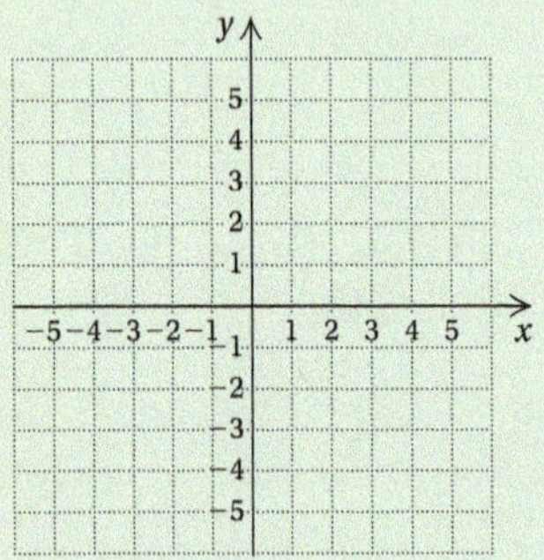

24. $g(x) = x^2 - 4$

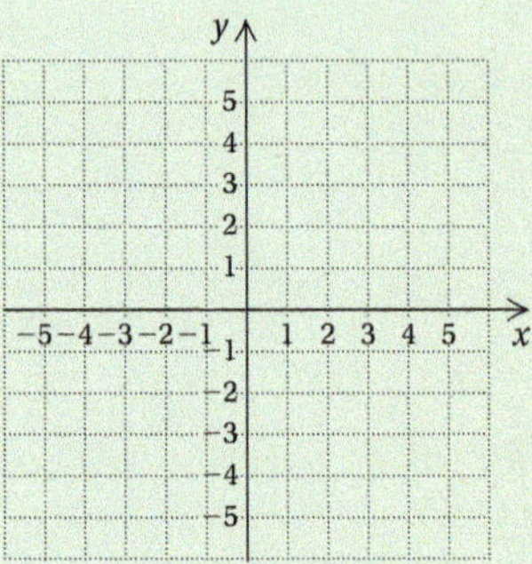

Determine whether each of the following is the graph of a function.

25.

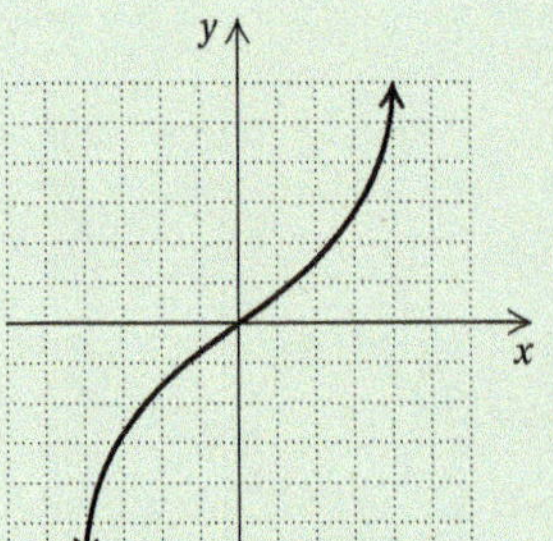

26.

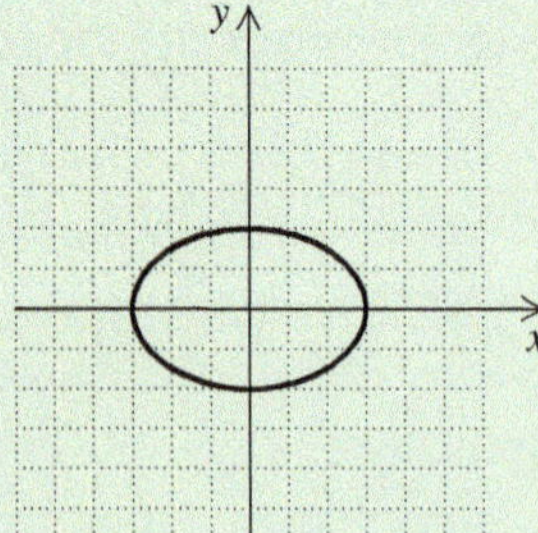

27. Given $g(x) = -2x - x^2$, find $g(-8)$.

A. -80 **B.** 48
C. -66 **D.** -48

Synthesis

28. Find the side of a square whose diagonal is 5 ft longer than a side.

29. Solve:

$$x - y = 2,$$
$$xy = 4.$$

Appendixes

- A LINEAR MEASURES: AMERICAN UNITS AND METRIC UNITS
- B WEIGHT AND MASS; MEDICAL APPLICATIONS
- C CAPACITY; MEDICAL APPLICATIONS
- D TIME AND TEMPERATURE
- E SETS
- F FACTORING SUMS OR DIFFERENCES OF CUBES
- G FINDING EQUATIONS OF LINES: POINT–SLOPE EQUATION
- H EQUATIONS INVOLVING ABSOLUTE VALUE
- I THE DISTANCE FORMULA AND MIDPOINTS
- J HIGHER ROOTS, AND RATIONAL NUMBERS AS EXPONENTS
- K INEQUALITIES AND INTERVAL NOTATION
- L NONLINEAR INEQUALITIES
- M SYSTEMS OF LINEAR INEQUALITIES
- N PROBABILITY
- O THE COMPLEX NUMBERS

A Linear Measures: American Units and Metric Units

OBJECTIVES

a Convert from one American unit of length to another.

b Convert from one metric unit of length to another.

c Convert between American units of length and metric units of length.

Length, or distance, is one kind of measure. To find lengths, we begin with some **unit segment** and assign to it a measure of 1. Suppose $\overline{AB}$ below is a unit segment. Let's measure segment $\overline{CD}$, using $\overline{AB}$ as our unit segment.

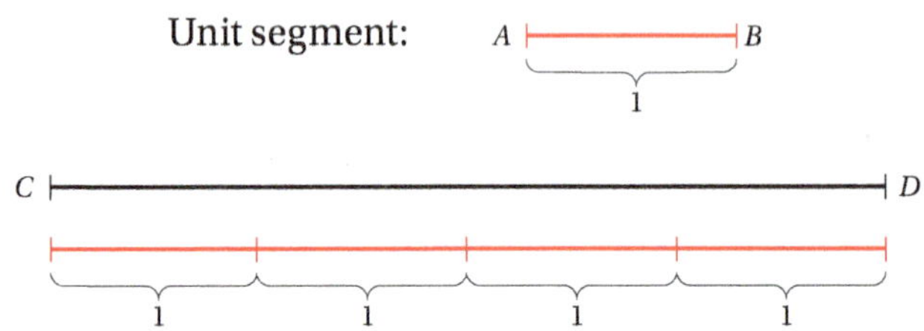

Since we can place 4 unit segments end to end along $\overline{CD}$, the measure of $\overline{CD}$ is 4.

Sometimes we need to use parts of units, called **subunits**. For example, the measure of the segment $\overline{MN}$ below is $1\frac{1}{2}$. We place one unit segment and one half-unit segment end to end.

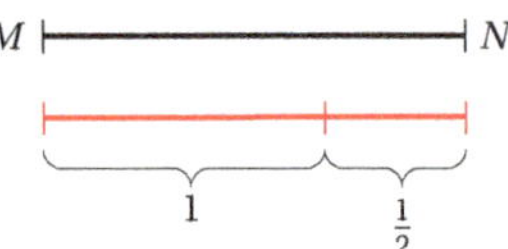

Use the unit below to measure the length of each segment or object.

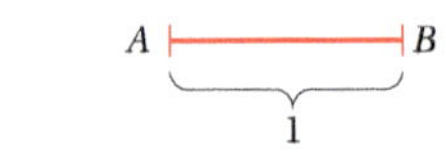

1.

2.

3.

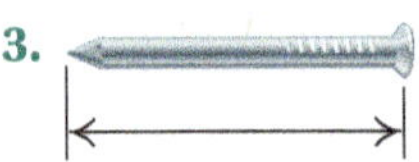

4.

Do Exercises 1–4.

a AMERICAN MEASURES

American units of length are related as follows.

AMERICAN UNITS OF LENGTH	
12 inches (in.) = 1 foot (ft)	3 feet = 1 yard (yd)
36 inches = 1 yard	5280 feet = 1 mile (mi)

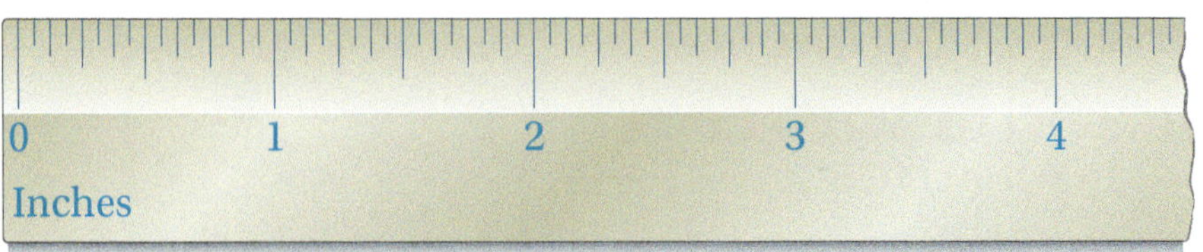

(Actual size, in inches)

Answers

1. 2 2. 3 3. $1\frac{1}{2}$ 4. $2\frac{1}{2}$

We can visualize comparisons of the units as follows:

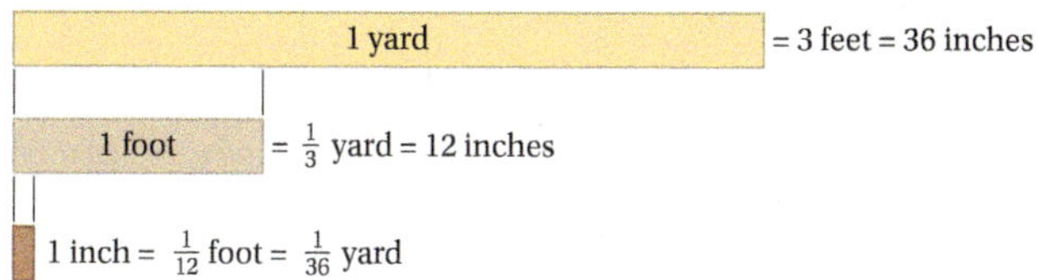

We can also abbreviate the units inches and feet like this: 13 in. = 13″ and 27 ft = 27′. American units have also been called "English," or "British-American," because at one time they were used by both countries. Today, both Canada and England have officially converted to the metric system.

To change from certain American units to others, we make substitutions. Such a substitution is usually helpful when we are converting from a *larger* unit to a *smaller* one.

EXAMPLE 1 Complete: $7\frac{1}{3}$ yd = ________ in.

$$
\begin{aligned}
7\tfrac{1}{3}\text{ yd} &= 7\tfrac{1}{3} \times 1\text{ yd} && \text{We think of } 7\tfrac{1}{3}\text{ yd as } 7\tfrac{1}{3} \times \text{ yd, or } 7\tfrac{1}{3} \times 1\text{ yd.}\\
&= 7\tfrac{1}{3} \times 36\text{ in.} && \text{Substituting 36 in. for 1 yd}\\
&= \frac{22}{3} \times 36\text{ in.}\\
&= 264\text{ in.}
\end{aligned}
$$

Do Exercises 5–7. ▶

Complete.

5. 8 yd = ________ in.

6. $2\frac{5}{6}$ yd = ________ ft

7. 3.8 mi = ________ in.

Sometimes it is helpful to use multiplying by 1 in making conversions. For example, 12 in. = 1 ft, so

$$\frac{12\text{ in.}}{1\text{ ft}} = 1 \quad \text{and} \quad \frac{1\text{ ft}}{12\text{ in.}} = 1.$$

If we divide 12 in. by 1 ft or 1 ft by 12 in., we get 1 because the lengths are the same. Let's first convert from *smaller* units to *larger* units.

EXAMPLE 2 Complete: 48 in. = ________ ft.

We want to convert from "in." to "ft." We multiply by 1 using a symbol for 1 with "in." in the denominator and "ft" in the numerator in order to eliminate inches and to convert to feet:

$$
\begin{aligned}
48\text{ in.} &= \frac{48\text{ in.}}{1} \times \frac{1\text{ ft}}{12\text{ in.}} && \text{Multiplying by 1 using } \frac{1\text{ ft}}{12\text{ in.}} \text{ to eliminate in.}\\
&= \frac{48\text{ in.}}{12\text{ in.}} \times 1\text{ ft}\\
&= \frac{48}{12} \times \frac{\text{in.}}{\text{in.}} \times 1\text{ ft}\\
&= 4 \times 1\text{ ft} && \text{The } \frac{\text{in.}}{\text{in.}} \text{ acts like 1, so we can omit it.}\\
&= 4\text{ ft.}
\end{aligned}
$$

Answers

5. 288 6. $8\frac{1}{2}$ 7. 240,768

We can also look at this conversion as "canceling" units:

$$48 \text{ in.} = \frac{48 \cancel{\text{in.}}}{1} \times \frac{1 \text{ ft}}{12 \cancel{\text{in.}}} = \frac{48}{12} \times 1 \text{ ft} = 4 \text{ ft}.$$

This method is used not only in mathematics, as here, but also in fields such as medicine, chemistry, and physics.

Do Exercises 8 and 9.

Complete.

8. 72 in. = ________ ft

9. 17 in. = ________ ft

EXAMPLE 3 Complete: 25 ft = ________ yd.

Since we are converting from "ft" to "yd," we choose a symbol for 1 with "yd" in the numerator and "ft" in the denominator:

$$25 \text{ ft} = 25 \text{ ft} \times \frac{1 \text{ yd}}{3 \text{ ft}} \qquad 3 \text{ ft} = 1 \text{ yd, so } \frac{3 \text{ ft}}{1 \text{ yd}} = 1 \text{ and } \frac{1 \text{ yd}}{3 \text{ ft}} = 1.$$

We use $\frac{1 \text{ yd}}{3 \text{ ft}}$ to eliminate ft.

$$= \frac{25}{3} \times \frac{\text{ft}}{\text{ft}} \times 1 \text{ yd}$$

$$= 8\frac{1}{3} \times 1 \text{ yd} \qquad \text{The } \frac{\text{ft}}{\text{ft}} \text{ acts like 1, so we can omit it.}$$

$$= 8\frac{1}{3} \text{ yd, or } 8.\overline{3} \text{ yd.}$$

Again, in this example, we can consider conversion from the point of view of canceling:

$$25 \text{ ft} = 25 \cancel{\text{ft}} \times \frac{1 \text{ yd}}{3 \cancel{\text{ft}}} = \frac{25}{3} \times 1 \text{ yd} = 8\frac{1}{3} \text{ yd, or } 8.\overline{3} \text{ yd.}$$

Do Exercises 10 and 11.

Complete.

10. 24 ft = ________ yd

11. 35 ft = ________ yd

EXAMPLE 4 Complete: 23,760 ft = ________ mi.

We choose a symbol for 1 with "mi" in the numerator and "ft" in the denominator:

$$23{,}760 \text{ ft} = 23{,}760 \text{ ft} \times \frac{1 \text{ mi}}{5280 \text{ ft}} \qquad 5280 \text{ ft} = 1 \text{ mi, so } \frac{1 \text{ mi}}{5280 \text{ ft}} = 1.$$

$$= \frac{23{,}760}{5280} \times \frac{\text{ft}}{\text{ft}} \times 1 \text{ mi}$$

$$= 4.5 \times 1 \text{ mi} \qquad \text{Dividing}$$

$$= 4.5 \text{ mi.}$$

Let's also consider this example using canceling:

$$23{,}760 \text{ ft} = 23{,}760 \cancel{\text{ft}} \times \frac{1 \text{ mi}}{5280 \cancel{\text{ft}}}$$

$$= \frac{23{,}760}{5280} \times 1 \text{ mi}$$

$$= 4.5 \times 1 \text{ mi} = 4.5 \text{ mi.}$$

Do Exercises 12 and 13.

Complete.

12. 26,400 ft = ________ mi

13. 2640 ft = ________ mi

Answers

8. 6 **9.** $1\frac{5}{12}$ **10.** 8 **11.** $11\frac{2}{3}$, or $11.\overline{6}$

12. 5 **13.** $\frac{1}{2}$, or 0.5

b METRIC MEASURES

Although the **metric system** is used in most countries of the world, it is used very little in the United States. The metric system does not use inches, feet, pounds, and so on, but its units for time and electricity are the same as those used now in the United States.

An advantage of the metric system is that it is easier to convert from one unit to another within this system than within the American system. That is because the metric system is based on the number 10.

The basic unit of length is the **meter**. It measures just over a yard. In fact, 1 meter $\approx$ 1.1 yd.

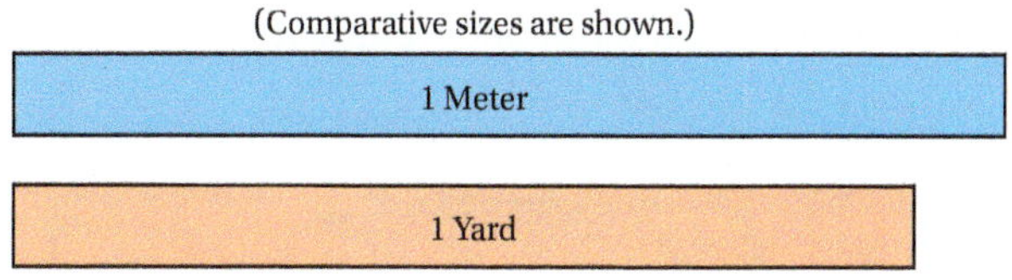

The other units of length are multiples of the length of a meter:

10 times a meter, 100 times a meter, 1000 times a meter, and so on;

or fractions of a meter:

$\frac{1}{10}$ of a meter, $\frac{1}{100}$ of a meter, $\frac{1}{1000}$ of a meter, and so on.

You should memorize the names and abbreviations for metric units of length. Remember *kilo*- for 1000, *hecto*- for 100, *deka*- for 10, *deci*- for $\frac{1}{10}$, *centi*- for $\frac{1}{100}$, and *milli*- for $\frac{1}{1000}$. (The units dekameter and decimeter are not used often.) We will also use these prefixes when considering units of area, capacity, and mass.

METRIC UNITS OF LENGTH

1 *kilo*meter (km) = 1000 meters (m)

1 *hecto*meter (hm) = 100 meters (m)

1 *deka*meter (dam) = 10 meters (m)

1 meter (m)

1 *deci*meter (dm) = $\frac{1}{10}$ meter (m)

1 *centi*meter (cm) = $\frac{1}{100}$ meter (m)

1 *milli*meter (mm) = $\frac{1}{1000}$ meter (m)

Thinking Metric

To familiarize yourself with metric units, consider the following.

1 kilometer (1000 meters)	is slightly more than $\frac{1}{2}$ mile (0.6 mi).
1 meter	is just over a yard (1.1 yd).
1 centimeter (0.01 meter)	is a little more than the width of a paperclip (about 0.3937 inch).

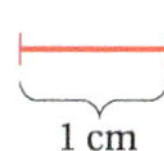

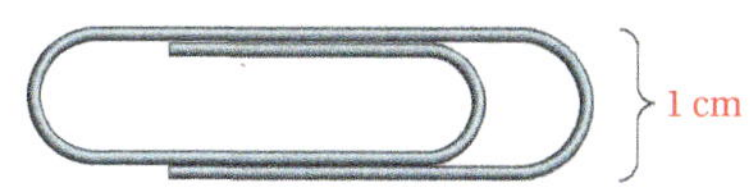

1 inch is 2.54 centimeters.

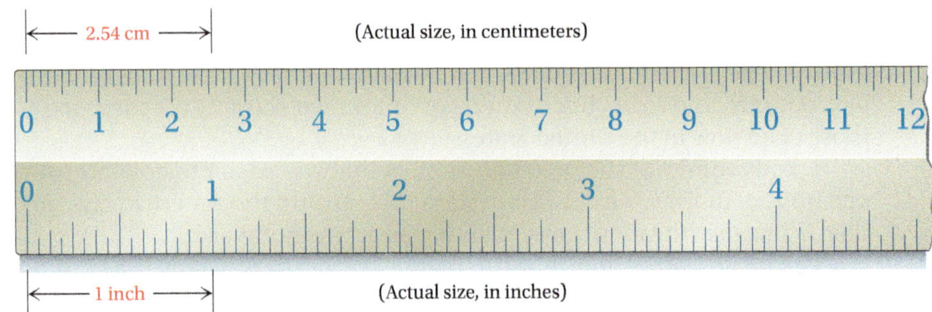

1 millimeter is about the diameter of paperclip wire.

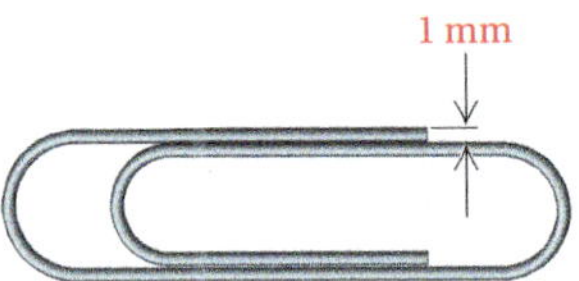

The millimeter (mm) is often used in jewelry making.

In many countries, the centimeter (cm) is used for body dimensions and clothing sizes.

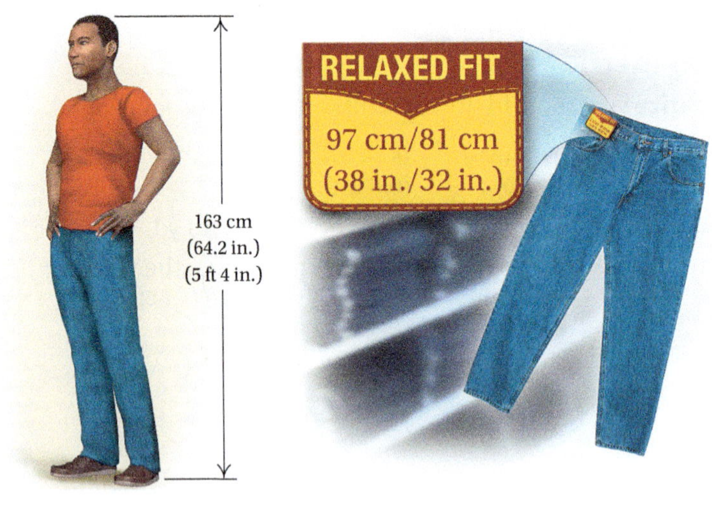

Do Exercises 14–16.

Using a centimeter ruler, measure each object.

14.

15.

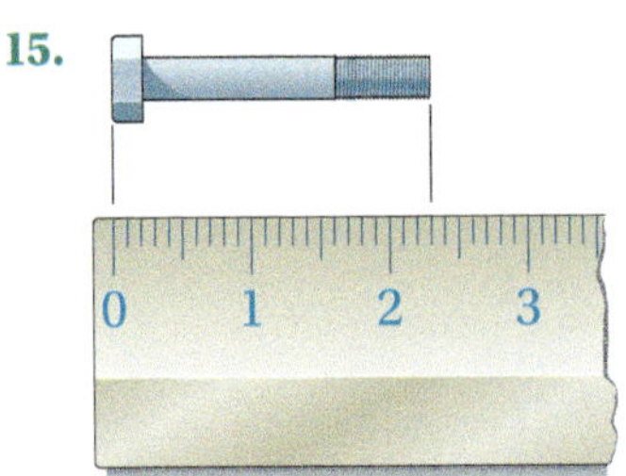

16.

Answers

14. 2 cm, or 20 mm
15. 2.3 cm, or 23 mm
16. 4.4 cm, or 44 mm

The meter (m) is used for expressing dimensions of large objects—say, the height of Hoover dam, 221.4 m, or the distance around a standard athletic track, 400 m—and for expressing somewhat smaller dimensions like the length and the width of an organic vegetable garden.

400 m around

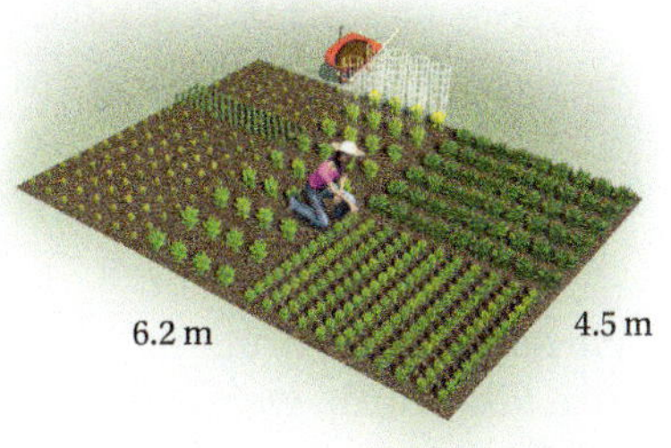

The kilometer (km) is used for longer distances, mostly those that are expressed in miles in American units.

1 mile is about 1.6 km.

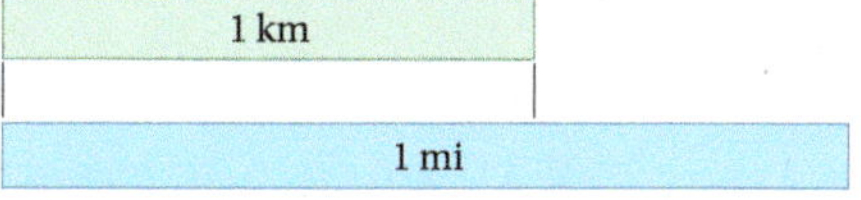

Do Exercises 17–22.

Complete with mm, cm, m, or km.

17. A stick of gum is 7 ______ long.

18. Dallas is 1512 ______ from Minneapolis.

19. A penny is 1 ______ thick.

20. The halfback ran 7 ______.

21. The book is 3 ______ thick.

22. The desk is 2 ______ long.

As with American units, when changing from a *larger* unit to a *smaller* unit, we usually make substitutions.

EXAMPLE 5 Complete: 4 km = ______ m.

Since we are converting from a *larger* unit to a *smaller* unit, we use substitution.

$$\begin{aligned} 4\text{ km} &= 4 \times 1\text{ km} \\ &= 4 \times 1000\text{ m} \qquad \text{Substituting 1000 m for 1 km} \\ &= 4000\text{ m} \end{aligned}$$

Do Exercises 23 and 24.

Complete.

23. 23 km = ______ m

24. 4 hm = ______ m

Since

$$\frac{1}{10}\text{ m} = 1\text{ dm}, \quad \frac{1}{100}\text{ m} = 1\text{ cm}, \quad \text{and} \quad \frac{1}{1000}\text{ m} = 1\text{ mm},$$

it follows that

$$1\text{ m} = 10\text{ dm}, \quad 1\text{ m} = 100\text{ cm}, \quad \text{and} \quad 1\text{ m} = 1000\text{ mm}.$$

EXAMPLE 6 Complete: 93.4 m = ______ cm.

Since we are converting from a *larger* unit to a *smaller* unit, we use substitution. We substitute 100 cm for 1 m:

$$93.4\text{ m} = 93.4 \times 1\text{ m} = 93.4 \times 100\text{ cm} = 9340\text{ cm}.$$

Answers

17. cm **18.** km **19.** mm **20.** m
21. cm **22.** m **23.** 23,000 **24.** 400

EXAMPLE 7 Complete: 0.248 m = ______ mm.

Since we are converting from a *larger* unit to a *smaller* unit, we use substitution.

$$\begin{aligned} 0.248\text{ m} &= 0.248 \times 1\text{ m} \\ &= 0.248 \times 1000\text{ mm} \quad \text{Substituting 1000 mm for 1 m} \\ &= 248\text{ mm} \end{aligned}$$

Do Exercises 25 and 26.

Complete.

25. 1.78 m = ______ cm

26. 9.04 m = ______ mm

We now convert from "m" to "km." Since we are converting from a *smaller* unit to a *larger* unit, we use multiplying by 1. We choose a symbol for 1 with "km" in the numerator and "m" in the denominator.

EXAMPLE 8 Complete: 2347 m = ______ km.

$$\begin{aligned} 2347\text{ m} &= 2347\text{ m} \times \frac{1\text{ km}}{1000\text{ m}} \quad \text{Multiplying by 1 using } \frac{1\text{ km}}{1000\text{ m}} \\ &= \frac{2347}{1000} \times \frac{\text{m}}{\text{m}} \times 1\text{ km} \quad \text{The } \frac{\text{m}}{\text{m}} \text{ acts like 1, so we omit it.} \\ &= 2.347\text{ km} \quad \text{Dividing by 1000 moves the decimal point three places to the left.} \end{aligned}$$

Using canceling, we can work this example as follows:

$$2347\text{ m} = 2347\,\cancel{\text{m}} \times \frac{1\text{ km}}{1000\,\cancel{\text{m}}} = \frac{2347}{1000} \times 1\text{ km} = 2.347\text{ km}.$$

Sometimes we multiply by 1 more than once.

EXAMPLE 9 Complete: 8.42 mm = ______ cm.

$$\begin{aligned} 8.42\text{ mm} &= 8.42\text{ mm} \times \frac{1\text{ m}}{1000\text{ mm}} \times \frac{100\text{ cm}}{1\text{ m}} \quad \text{Multiplying by 1 using } \frac{1\text{ m}}{1000\text{ mm}} \text{ and } \frac{100\text{ cm}}{1\text{ m}} \\ &= \frac{8.42 \times 100}{1000} \times \frac{\text{mm}}{\text{mm}} \times \frac{\text{m}}{\text{m}} \times 1\text{ cm} \\ &= \frac{842}{1000}\text{ cm} = 0.842\text{ cm} \end{aligned}$$

Do Exercises 27–30.

Complete.

27. 7814 m = ______ km

28. 7814 m = ______ dam

29. 9.67 mm = ______ cm

30. 89 km = ______ cm

Mental Conversion

Changing from one unit of length to another in the metric system amounts to the movement of a decimal point. That is because the metric system is based on 10. Let's find a faster way to convert. Look at the following table.

1000 m	100 m	10 m	1 m	0.1 m	0.01 m	0.001 m
1 km	1 hm	1 dam	1 m	1 dm	1 cm	1 mm

Each place in the table has a value $\frac{1}{10}$ that to the left or 10 times that to the right. Thus moving one place in the table corresponds to moving one decimal place.

Answers

25. 178 **26.** 9040 **27.** 7.814 **28.** 781.4
29. 0.967 **30.** 8,900,000

Let's convert mentally.

EXAMPLE 10 Complete: 8.42 mm = _______ cm.

Think: To go from mm to cm in the table is a move of one place to the left. Thus we move the decimal point one place to the left.

1000 m	100 m	10 m	1 m	0.1 m	0.01 m	0.001 m
1 km	1 hm	1 dam	1 m	1 dm	1 cm	1 mm

1 place to the left

8.42 0.8.42 8.42 mm = 0.842 cm

EXAMPLE 11 Complete: 1.886 km = _______ cm.

Think: To go from km to cm in the table is a move of five places to the right. Thus we move the decimal point five places to the right.

1000 m	100 m	10 m	1 m	0.1 m	0.01 m	0.001 m
1 km	1 hm	1 dam	1 m	1 dm	1 cm	1 mm

5 places to the right

1.886 1.88600. 1.886 km = 188,600 cm

EXAMPLE 12 Complete: 3 m = _______ cm.

Think: To go from m to cm in the table is a move of two places to the right. Thus we move the decimal point two places to the right.

1000 m	100 m	10 m	1 m	0.1 m	0.01 m	0.001 m
1 km	1 hm	1 dam	1 m	1 dm	1 cm	1 mm

2 places to the right

3 3.00. 3 m = 300 cm

You should try to make metric conversions mentally as often as possible.

The fact that conversions can be done so easily is an important advantage of the metric system. The most commonly used metric units of length are km, m, cm, and mm. We have purposely used these more often than the others in the exercises.

Do Exercises 31–34. ▶

Complete. Try to do this mentally using the table.

31. 6780 m = _______ km

32. 9.74 cm = _______ mm

33. 1 mm = _______ cm

34. 845.1 mm = _______ dm

Answers

31. 6.78 **32.** 97.4 **33.** 0.1 **34.** 8.451

C CONVERTING UNITS

We can make conversions between American units and metric units by substituting based on the rounded approximations in the following table.

AMERICAN	METRIC
1 in.	2.54 cm
1 ft	0.305 m
1 yd	0.914 m
1 mi	1.609 km
0.621 mi	1 km
1.094 yd	1 m
3.281 ft	1 m
39.370 in.	1 m

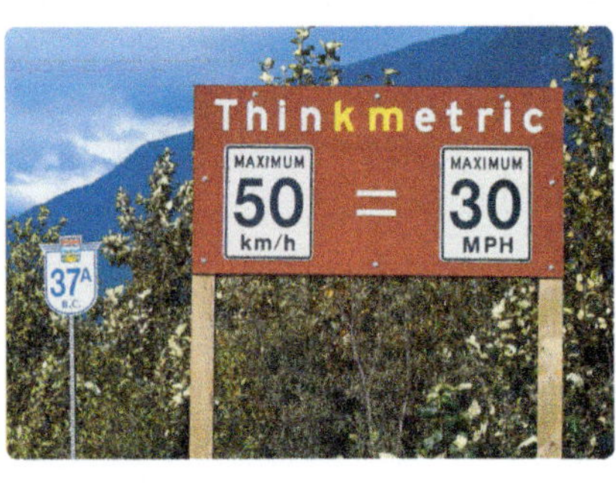

EXAMPLE 13 Complete: 11 in. = ________ cm.

(The wingspan of the world's largest butterfly, the Queen Alexandra)

Source: *Top 10 of Everything 2013*

$$
\begin{aligned}
11 \text{ in.} &= 11 \times 1 \text{ in.} \\
&= 11 \times 2.54 \text{ cm} \qquad \text{Substituting 2.54 cm for 1 in.} \\
&= 27.94 \text{ cm}
\end{aligned}
$$

This answer would probably be rounded to the nearest one: 28 cm.

EXAMPLE 14 Complete: 26.2 mi = ________ km.

(The length of the Olympic marathon)

$$
\begin{aligned}
26.2 \text{ mi} &= 26.2 \times 1 \text{ mi} \\
&\approx 26.2 \times 1.609 \text{ km} \qquad \text{Substituting 1.609 km for 1 mi} \\
&= 42.1558 \text{ km}
\end{aligned}
$$

EXAMPLE 15 Complete: 100 m = ________ ft.

(The length of the 100-m dash)

$$
\begin{aligned}
100 \text{ m} &= 100 \times 1 \text{ m} \\
&\approx 100 \times 3.281 \text{ ft} \qquad \text{Substituting 3.281 ft for 1 m} \\
&= 328.1 \text{ ft}
\end{aligned}
$$

EXAMPLE 16 Complete: 4544 km = ________ mi.

(The distance from New York to Los Angeles)

$$
\begin{aligned}
4544 \text{ km} &= 4544 \times 1 \text{ km} \\
&\approx 4544 \times 0.621 \text{ mi} \qquad \text{Substituting 0.621 mi for 1 km} \\
&= 2821.824 \text{ mi}
\end{aligned}
$$

We would probably round this answer to 2822 mi.

Do Exercises 35–37.

Complete.

35. 100 yd = ________ m
(The length of a football field, excluding the end zones)

36. 2.5 mi = ________ km
(The length of the tri-oval track at Daytona International Speedway)

37. 2383 km = ________ mi
(The distance from St. Louis to Phoenix)

Answers

35. 91.4 **36.** 4.0225 **37.** 1479.843

EXAMPLE 17 *Millau Viaduct.* The Millau viaduct is part of the E11 expressway connecting Paris, France, and Barcelona, Spain. The viaduct has the highest bridge piers ever constructed. The tallest pier is 804 ft high and the overall height including the pylon is 1122 ft, making this the highest bridge in the world. Convert 804 feet and 1122 feet to meters.

Source: www.abelard.org/france/viaduct-de-millau.php

We let P = the height of the pier and H = the overall height of the bridge. To convert feet to meters, we substitute 0.305 m for 1 ft.

$$\begin{aligned} P &= 804 \text{ ft} \\ &= 804 \times 1 \text{ ft} \\ &\approx 804 \times 0.305 \text{ m} \\ &= 245.22 \text{ m} \end{aligned} \qquad \begin{aligned} H &= 1122 \text{ ft} \\ &= 1122 \times 1 \text{ ft} \\ &\approx 1122 \times 0.305 \text{ m} \\ &= 342.21 \text{ m} \end{aligned}$$

Do Exercises 38 and 39. ▶

EXAMPLE 18 Complete: 0.10414 mm = ________ in.
(The thickness of a $1 bill)

In this case, we must make two substitutions or multiply by two forms of 1 since the table on the preceding page does not provide a direct way to convert from millimeters to inches. Here we choose to multiply by forms of 1.

$$\begin{aligned} 0.10414 \text{ mm} &= 0.10414 \times 1 \text{ mm} \times \frac{1 \text{ cm}}{10 \text{ mm}} \\ &= 0.010414 \text{ cm} \\ &= 0.010414 \times 1 \text{ cm} \times \frac{1 \text{ in.}}{2.54 \text{ cm}} \\ &= 0.0041 \text{ in.} \end{aligned}$$

Do Exercise 40. ▶

38. The Pacific Coast Highway, which is part of California State Route 1, is 655.8 mi long. Find this length in kilometers, rounded to the nearest tenth.

39. The height of the Stratosphere Tower in Las Vegas, Nevada, is 1149 ft. Find the height in meters.

40. Complete:
3.175 mm = ________ in.
(The thickness of a quarter)

Answers

38. 1055.2 **39.** 350.445 m **40.** 0.125

A Exercise Set

For Extra Help

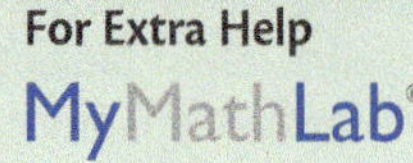

a Complete.

1. 1 ft = ______ in.

2. 1 yd = ______ ft

3. 1 in. = ______ ft

4. 1 mi = ______ yd

5. 1 mi = ______ ft

6. 1 ft = ______ yd

7. 3 yd = ______ in.

8. 10 yd = ______ ft

9. 84 in. = ______ ft

10. 48 ft = ______ yd

11. 18 in. = ______ ft

12. 29 ft = ______ yd

13. 5 mi = ______ ft

14. 5 mi = ______ yd

15. 63 in. = ______ ft

16. 11,616 ft = ______ mi

17. 10 ft = ______ yd

18. 9.6 yd = ______ ft

19. 7.1 mi = ______ ft

20. 31,680 ft = ______ mi

21. $4\frac{1}{2}$ ft = ______ yd

22. 48 in. = ______ ft

23. 45 in. = ______ yd

24. $6\frac{1}{3}$ yd = ______ in.

25. 330 ft = ______ yd

26. 5280 yd = ______ mi

27. 3520 yd = ______ mi

28. 25 mi = ______ ft

29. 100 yd = ______ ft

30. 480 in. = ______ ft

31. 360 in. = ______ ft

32. 720 in. = ______ yd

33. 1 in. = ______ yd

34. 25 in. = ______ ft

35. 2 mi = ______ in.

36. 63,360 in. = ______ mi

b Complete. Do as much as possible mentally.

37. **a)** 1 km = ______ m
b) 1 m = ______ km

38. **a)** 1 hm = ______ m
b) 1 m = ______ hm

39. **a)** 1 dam = ______ m
b) 1 m = ______ dam

40. **a)** 1 dm = ______ m
b) 1 m = ______ dm

41. **a)** 1 cm = ______ m
b) 1 m = ______ cm

42. **a)** 1 mm = ______ m
b) 1 m = ______ mm

43. 6.7 km = ______ m

44. 27 km = ______ m

45. 98 cm = ______ m

46. 0.789 cm = ______ m

47. 8921 m = ______ km

48. 8664 m = ______ km

49. 56.66 m = ______ km

50. 4.733 m = ______ km

51. 5666 m = ______ cm

52. 869 m = ______ cm

53. 477 cm = ______ m

54. 6.27 mm = ______ m

55. 6.88 m = ______ cm

56. 6.88 m = ______ dm

57. 1 mm = ______ cm

58. 1 cm = ______ km

59. 1 km = ______ cm

60. 2 km = ______ cm

61. 14.2 cm = ______ mm

62. 25.3 cm = ______ mm

63. 8.2 mm = ______ cm

64. 9.7 mm = ______ cm

65. 4500 mm = ______ cm

66. 8,000,000 m = ______ km

67. 0.024 mm = ______ m

68. 60,000 mm = ______ dam

69. 6.88 m = ______ dam

70. 7.44 m = ______ hm

71. 2.3 dam = ______ dm

72. 9 km = ______ hm

73. 392 dam = ______ km

74. 0.056 mm = ______ dm

Complete the following table.

	Object	Millimeters (mm)	Centimeters (cm)	Meters (m)
75.	Width of a football field		4844	
76.	Length of a football field			109.09
77.	Length of 4 meter sticks			4
78.	Width of a credit card	56		
79.	Thickness of an index card	0.27		
80.	Thickness of a piece of cardboard		0.23	
81.	Height of One World Trade Center, New York, New York			541.3
82.	Height of The Gateway Arch, St. Louis, Missouri	192,000		

C Complete.

83. 330 ft = ________ m
(The length of most baseball foul lines)

84. 12 in. = ________ cm
(The length of a common ruler)

85. 1171.4 km = ________ mi
(The distance from Cleveland to Atlanta)

86. 2 m = ________ ft
(The length of a desk)

87. 65 mph = ________ km/h
(A common speed limit in the United States)

88. 100 km/h = ________ mph
(A common speed limit in Canada)

89. 180 mi = ________ km
(The distance from Indianapolis to Chicago)

90. 141,600,000 mi = ________ km (The farthest distance of Mars from the sun)

91. 70 mph = ________ km/h
(An interstate speed limit in Arizona)

92. 60 km/h = ________ mph
(A city speed limit in Canada)

93. 10 yd = ________ m
(The length needed for a first down in football)

94. 450 ft = ________ m
(The length of a long home run in baseball)

95. 1.91 m = ________ in.
(The height of Jeremy Lin of the Houston Rockets)

96. 69 in. = ________ m
(The height of Nate Robinson of the Chicago Bulls)

97. 169.41 m = ________ ft
(The height of the Washington Monument)

98. 1671 ft = ________ m
(The height of the Taipei 101 skyscraper)

99. 15.7 cm = ________ in.
(The length of a $1 bill)

100. 7.5 in. = ________ cm
(The length of a pencil)

101. 2216 km = ________ mi
(The distance from Chicago to Miami)

102. 1862 mi = ________ km
(The distance from Seattle to Kansas City)

103. 13 mm = ________ in.
(The thickness of a plastic case for a DVD)

104. 0.25 in. = ________ mm
(The thickness of an eraser on a pencil)

Complete the following table. Answers may vary, depending on the conversion factor used.

	Object	Yards (yd)	Centimeters (cm)	Inches (in.)	Meters (m)	Millimeters (mm)
105.	Width of a piece of typing paper			$8\frac{1}{2}$		
106.	Length of a football field	120				
107.	Width of a football field		4844			
108.	Width of a credit card					56
109.	Length of 4 yardsticks	4				
110.	Length of 3 meter sticks		300			
111.	Thickness of an index card				0.00027	
112.	Thickness of a piece of cardboard		0.23			
113.	Length of the Channel Tunnel connecting France and England				50,500	
114.	Height of Jin Mao Tower, Shanghai	460				

Synthesis

115. Develop a formula to convert from inches to millimeters.

116. Develop a formula to convert from millimeters to inches. How does it relate to the answer for Exercise 115?

B Weight and Mass; Medical Applications

OBJECTIVES

a Convert from one American unit of weight to another.

b Convert from one metric unit of mass to another.

c Make conversions and solve applied problems concerning medical dosages.

There is a difference between **mass** and **weight**, but the terms are often used interchangeably. People sometimes use the word "weight" when, technically, they are referring to "mass." Weight is related to the force of gravity. The farther you are from the center of the earth, the less you weigh. Your mass stays the same no matter where you are.

a WEIGHT: THE AMERICAN SYSTEM

AMERICAN UNITS OF WEIGHT

1 ton (T) = 2000 pounds (lb) 1 lb = 16 ounces (oz)

The term "ounce" used here for weight is different from the "ounce" used for capacity, which we will discuss in Appendix C. We convert units of weight using the same techniques that we use with linear measures.

EXAMPLE 1 A well-known hamburger is called a "quarter-pounder." Find its name in ounces: a "______ ouncer."

Since we are converting from a larger unit to a smaller unit, we use substitution.

$$\frac{1}{4}\text{ lb} = \frac{1}{4}\cdot 1\text{ lb} = \frac{1}{4}\cdot 16\text{ oz} \qquad \text{Substituting 16 oz for 1 lb}$$
$$= 4\text{ oz}$$

A "quarter-pounder" can also be called a "four-ouncer."

EXAMPLE 2 Complete: 15,360 lb = ______ T.

Since we are converting from a smaller unit to a larger unit, we use multiplying by 1.

$$15{,}360\text{ lb} = 15{,}360\text{ lb} \times \frac{1\text{ T}}{2000\text{ lb}} \qquad \text{Multiplying by 1}$$
$$= \frac{15{,}360}{2000}\text{ T} = 7.68\text{ T}$$

Do Exercises 1 and 2.

Complete.

1. 5 lb = ______ oz

2. 8640 lb = ______ T

Answers

1. 80 2. 4.32

b MASS: THE METRIC SYSTEM

The basic unit of mass is the **gram** (g), which is the mass of 1 cubic centimeter ($1\ \text{cm}^3$) of water. Since a cubic centimeter is small, a gram is a small unit of mass.

$1\ \text{g} = 1\ \text{cm}^3$ of water

$$1\ \text{g} = 1\ \text{gram} = \text{the mass of } 1\ \text{cm}^3 \text{ of water}$$

The metric units of mass are listed to the left. The prefixes are the same as those for length.

METRIC UNITS OF MASS

1 metric ton (t) = 1000 kilograms (kg)
1 *kilo*gram (kg) = 1000 grams (g)
1 *hecto*gram (hg) = 100 grams (g)
1 *deka*gram (dag) = 10 grams (g)
1 gram (g)
1 *deci*gram (dg) = $\frac{1}{10}$ gram (g)
1 *centi*gram (cg) = $\frac{1}{100}$ gram (g)
1 *milli*gram (mg) = $\frac{1}{1000}$ gram (g)

Thinking Metric

One gram is about the mass of 1 raisin or 1 package of artificial sweetener. Since 1 kg is about 2.2 lb, 1000 kg is about 2200 lb, or 1 metric ton (t), which is just a little more than 1 American ton (T), which is 2000 lb.

1 gram
1 kilogram of grapes
1 pound of grapes

Small masses, such as dosages of medicine and vitamins, may be measured in milligrams (mg). The gram (g) is used for objects ordinarily measured in ounces, such as the mass of a letter, a piece of candy, or a coin.

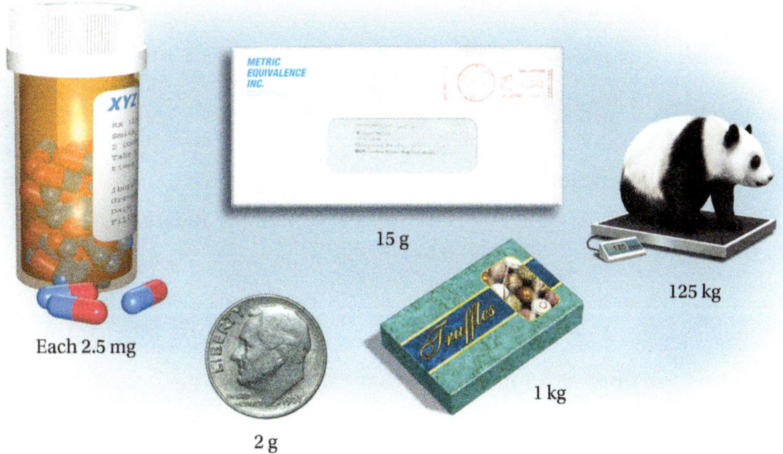

Each 2.5 mg
15 g
125 kg
2 g
1 kg

The kilogram (kg) is used for larger food packages and for body masses. The metric ton (t) is used for very large masses, such as the mass of an automobile, a truckload of gravel, or an airplane.

Do Exercises 3–7. ▶

Complete with mg, g, kg, or t.

3. A laptop computer has a mass of 6 ________.

4. Eric has a body mass of 85.4 ________.

5. This is a 3- ________ vitamin.

6. A pen has a mass of 12 ________.

7. A sport utility vehicle has a mass of 3 ________.

Answers

3. kg **4.** kg **5.** mg **6.** g **7.** t

Changing Units Mentally

As before, changing from one metric unit of mass to another requires only the movement of a decimal point. We use this table.

1000 g	100 g	10 g	1 g	0.1 g	0.01 g	0.001 g
1 kg	1 hg	1 dag	1 g	1 dg	1 cg	1 mg

EXAMPLE 3 Complete: 8 kg = ______ g.

Think: To go from kg to g in the table is a move of three places to the right. Thus we move the decimal point three places to the right.

1000 g	100 g	10 g	1 g	0.1 g	0.01 g	0.001 g
1 kg	1 hg	1 dag	1 g	1 dg	1 cg	1 mg

3 places to the right

8.0 8.000. 8 kg = 8000 g

EXAMPLE 4 Complete: 4235 g = ______ kg.

Think: To go from g to kg in the table is a move of three places to the left. Thus we move the decimal point three places to the left.

1000 g	100 g	10 g	1 g	0.1 g	0.01 g	0.001 g
1 kg	1 hg	1 dag	1 g	1 dg	1 cg	1 mg

3 places to the left

4235.0 4.235.0 4235 g = 4.235 kg

Complete.

8. 6.2 kg = ______ g

9. 304.8 cg = ______ g

Do Exercises 8 and 9.

EXAMPLE 5 Complete: 6.98 cg = ______ mg.

Think: To go from cg to mg is a move of one place to the right. Thus we move the decimal point one place to the right.

1000 g	100 g	10 g	1 g	0.1 g	0.01 g	0.001 g
1 kg	1 hg	1 dag	1 g	1 dg	1 cg	1 mg

1 place to the right

6.98 6.9.8 6.98 cg = 69.8 mg

Answers

8. 6200 **9.** 3.048

The most commonly used metric units of mass are kg, g, cg, and mg. We have purposely used those more than the others in the exercises.

EXAMPLE 6 Complete: 89.21 mg = ______ g.

Think: To go from mg to g is a move of three places to the left. Thus we move the decimal point three places to the left.

1000 g	100 g	10 g	1 g	0.1 g	0.01 g	0.001 g
1 kg	1 hg	1 dag	1 g	1 dg	1 cg	1 mg

3 places to the left

89.21   0.089.21   89.21 mg = 0.08921 g

Do Exercises 10–12.

Complete.

10. 7.7 cg = ______ mg

11. 2344 mg = ______ cg

12. 67 dg = ______ mg

C MEDICAL APPLICATIONS

Another metric unit that is used in medicine is the microgram (mcg). It is defined as follows.

MICROGRAM

$$1 \text{ microgram} = 1 \text{ mcg} = \frac{1}{1{,}000{,}000} \text{ g} = 0.000001 \text{ g}$$

$$1{,}000{,}000 \text{ mcg} = 1 \text{ g}$$

EXAMPLE 7 Complete: 1 mg = ______ mcg.

We convert to grams and then to micrograms:

$$\begin{aligned} 1 \text{ mg} &= 0.001 \text{ g} \\ &= 0.001 \times 1 \text{ g} \\ &= 0.001 \times 1{,}000{,}000 \text{ mcg} && \text{Substituting 1,000,000 mcg for 1 g} \\ &= 1000 \text{ mcg}. \end{aligned}$$

Do Exercise 13.

13. Complete:
1 mcg = ______ mg.

14. ***Medical Dosage.*** A physician prescribes 500 mcg of alprazolam, an antianxiety medication. How many milligrams is this dosage?
Source: Steven R. Smith, M.D.

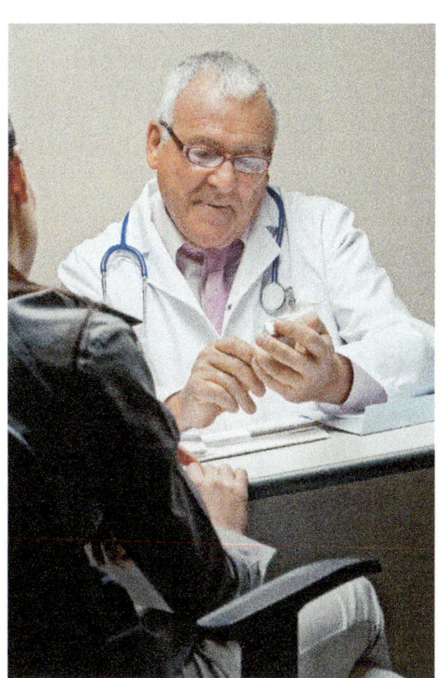

EXAMPLE 8 ***Medical Dosage.*** Nitroglycerin sublingual tablets come in 0.4-mg tablets. How many micrograms are in each tablet?
Source: Steven R. Smith, M.D.

We are to complete: 0.4 mg = ______ mcg. Thus,

$$\begin{aligned} 0.4 \text{ mg} &= 0.4 \times 1 \text{ mg} \\ &= 0.4 \times 1000 \text{ mcg} && \text{From Example 7, substituting 1000 mcg for 1 mg} \\ &= 400 \text{ mcg}. \end{aligned}$$

We can also do this problem in a manner similar to that used in Example 7.

Do Exercise 14.

Answers

10. 77 **11.** 234.4 **12.** 6700 **13.** 0.001
14. 0.5 mg

B Exercise Set

For Extra Help MyMathLab® MathXL® PRACTICE WATCH READ REVIEW

a Complete.

1. 1 T = _______ lb

2. 1 lb = _______ oz

3. 6000 lb = _______ T

4. 8 T = _______ lb

5. 4 lb = _______ oz

6. 10 lb = _______ oz

7. 6.32 T = _______ lb

8. 8.07 T = _______ lb

9. 3200 oz = _______ T

10. 6400 oz = _______ T

11. 80 oz = _______ lb

12. 960 oz = _______ lb

13. *Pecans.* In 2011, U.S. farmers produced 269,700,000 pounds of pecans. How many tons of pecans were produced?

Source: National Agricultural Statistics Service, U.S. Department of Agriculture

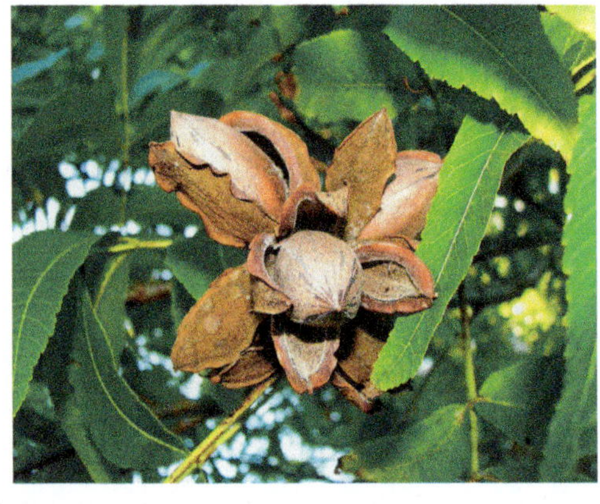

14. *Peaches.* In 2011, U.S. farmers produced 978,260 tons of peaches. How many pounds of peaches were produced?

Source: National Agricultural Statistics Service, U.S. Department of Agriculture

b Complete.

15. 1 kg = _______ g

16. 1 hg = _______ g

17. 1 dag = _______ g

18. 1 dg = _______ g

19. 1 cg = _______ g

20. 1 mg = _______ g

21. 1 g = _______ mg

22. 1 g = _______ cg

23. 1 g = _______ dg

24. 25 kg = _______ g

25. 234 kg = _______ g

26. 9403 g = _______ kg

27. 5200 g = _______ kg

28. 1.506 kg = _______ g

29. 67 hg = _______ kg

30. 45 cg = _______ g

31. 0.502 dg = _______ g

32. 0.0025 cg = _______ mg

33. 8492 g = _______ kg

34. 9466 g = _______ kg

35. 585 mg = _______ cg

36. 96.1 mg = _______ cg

37. 8 kg = _______ cg

38. 0.06 kg = _______ mg

39. 1 t = _______ kg

40. 2 t = _______ kg

41. 3.4 cg = _______ dag

42. 115 mg = _______ g

43. 60.3 kg = _______ t

44. 15.68 kg = _______ t

C Complete.

45. 1 mg = _______ mcg

46. 1 mcg = _______ mg

47. 325 mcg = _______ mg

48. 0.45 mg = _______ mcg

49. 210.6 mg = _______ mcg

50. 8000 mcg = _______ mg

51. 4.9 mcg = _______ mg

52. 0.075 mg = _______ mcg

Medical Dosage. Solve each of the following. (None of these medications should be taken without consulting your own physician.)

Source: Steven R. Smith, M.D.

53. Digoxin is a medication used to treat heart problems. A physician orders 0.125 mg of digoxin to be taken once daily. How many micrograms of digoxin are there in the daily dosage?

54. Digoxin is a medication used to treat heart problems. A physician orders 0.25 mg of digoxin to be taken once a day. How many micrograms of digoxin are there in the daily dosage?

55. Triazolam is a medication used for the short-term treatment of insomnia. A physician advises her patient to take one of the 0.125-mg tablets each night for 7 nights. How many milligrams of triazolam will the patient have ingested over that 7-day period? How many micrograms?

56. Clonidine is a medication used to treat high blood pressure. The usual starting dose of clonidine is one 0.1-mg tablet twice a day. If a patient is started on this dose by his physician, how many total milligrams of clonidine will the patient have taken before he returns to see his physician 14 days later? How many micrograms?

57. Cephalexin is an antibiotic that frequently is prescribed in a 500-mg tablet form. A physician prescribes 2 g of cephalexin per day for a patient with a skin sore. How many 500-mg tablets would have to be taken in order to achieve this daily dosage?

58. Quinidine gluconate is a liquid mixture, part medicine and part water, that is administered intravenously. There are 80 mg of quinidine gluconate in each cubic centimeter (cc) of the liquid mixture. A physician orders 900 mg of quinidine gluconate to be administered daily to a patient with malaria. How much of the solution would have to be administered in order to achieve the recommended daily dosage?

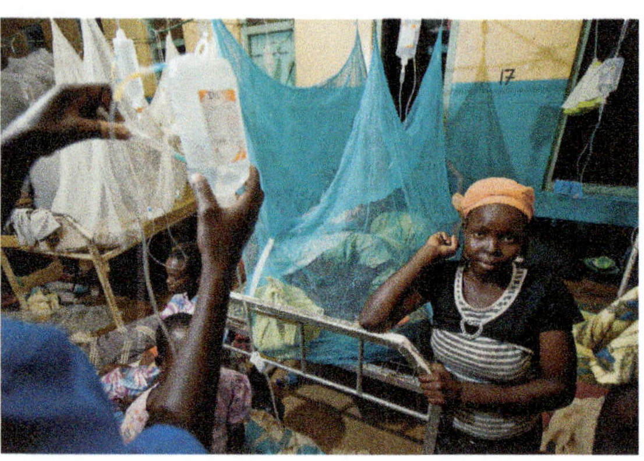

59. Amoxicillin is an antibiotic commonly prescribed for children as a liquid suspension composed of part amoxicillin and part water. In one formulation of amoxicillin suspension, there are 250 mg of amoxicillin in 5 cc of the liquid suspension. A physician prescribes 400 mg per day for a 2-year-old child with an ear infection. How much of the amoxicillin liquid suspension would the child's parent need to administer in order to achieve the recommended daily dosage of amoxicillin?

60. Albuterol is a medication used for the treatment of asthma. It comes in an inhaler that contains 17 mg of albuterol mixed with a liquid. One actuation (inhalation) from the mouthpiece delivers a 90-mcg dose of albuterol.

a) A physician orders 2 inhalations 4 times per day. How many micrograms of albuterol does the patient inhale per day?

b) How many actuations/inhalations are contained in one inhaler?

c) Danielle is leaving for 4 months of college and wants to take enough albuterol to last for that time. Her physician has prescribed 2 inhalations 4 times per day. How many inhalers will Danielle need to take with her for the 4-month period?

Synthesis

61. *Tanzanite.* Tanzanite is a gemstone discovered in 1967 in the East African state of Tanzania, the only place in the world where it has been found. Rarer than diamond, tanzanite ranges in color from ultramarine blue to light violet-blue. The world's biggest piece of tanzanite was unearthed near Tanzania's Mount Kilimanjaro. The gemstone weighed 16,839 carats and was the size of a brick. A **carat** (also spelled **karat**) is a unit of weight for precious stones; 1 carat $=$ 200 mg.

Sources: International Colored Gemstone Association; "World's Biggest Tanzanite Gem Found Near Kilimanjaro," Bloomberg.com, Aug. 3, 2005

a) How many grams is this record tanzanite gemstone?

b) Given that 1 lb $=$ 453.6 g, how many ounces does this gemstone weigh?

Capacity; Medical Applications

C

OBJECTIVES

a Convert from one unit of capacity to another.

b Solve applied problems concerning medical dosages.

a CAPACITY

American Units

To answer a question like "How much soda is in the can?" we need measures of **capacity**. American units of capacity are fluid ounces, cups, pints, quarts, and gallons. These units are related as follows.

AMERICAN UNITS OF CAPACITY

1 gallon (gal) = 4 quarts (qt)	1 pt = 2 cups = 16 fluid ounces (fl oz)
1 qt = 2 pints (pt)	1 cup = 8 fluid oz

Fluid ounces, abbreviated fl oz, are often referred to as ounces, or oz.

EXAMPLE 1 Complete: 9 gal = ______ oz.

Since we are converting from a *larger* unit to a *smaller* unit, we use substitution:

$$\begin{aligned} 9 \text{ gal} &= 9 \cdot 1 \text{ gal} = 9 \cdot 4 \text{ qt} && \text{Substituting 4 qt for 1 gal} \\ &= 9 \cdot 4 \cdot 1 \text{ qt} = 9 \cdot 4 \cdot 2 \text{ pt} && \text{Substituting 2 pt for 1 qt} \\ &= 9 \cdot 4 \cdot 2 \cdot 1 \text{ pt} = 9 \cdot 4 \cdot 2 \cdot 16 \text{ oz} && \text{Substituting 16 oz for 1 pt} \\ &= 1152 \text{ oz.} \end{aligned}$$

EXAMPLE 2 Complete: 24 qt = ______ gal.

Since we are converting from a *smaller* unit to a *larger* unit, we multiply by 1 using 1 gal in the numerator and 4 qt in the denominator:

$$24 \text{ qt} = 24 \text{ qt} \cdot \frac{1 \text{ gal}}{4 \text{ qt}} = \frac{24}{4} \cdot 1 \text{ gal} = 6 \text{ gal.}$$

Do Exercises 1 and 2. ▶

Complete.

1. 5 gal = ______ pt

2. 80 qt = ______ gal

Metric Units

One unit of capacity in the metric system is a **liter**. A liter is just a bit more than a quart. It is defined as follows.

Answers

1. 40 2. 20

METRIC UNITS OF CAPACITY

1 liter (L) = 1000 cubic centimeters (1000 cm^3)

The script letter ℓ is also used for "liter."

The metric prefixes are also used with liters. The most common is **milli-**. The milliliter (mL) is, then, $\frac{1}{1000}$ liter. Thus,

$$1\text{ L} = 1000\text{ mL} = 1000\text{ cm}^3;$$
$$0.001\text{ L} = 1\text{ mL} = 1\text{ cm}^3.$$

Although the other metric prefixes are rarely used for capacity, we display them in the following table as we did for linear measure.

1000 L	100 L	10 L	1 L	0.1 L	0.01 L	0.001 L
1 kL	1 hL	1 daL	1 L	1 dL	1 cL	1 mL (cc)

A preferred unit for drug dosage is the milliliter (mL) or the cubic centimeter (cm^3). The notation "cc" is also used for cubic centimeter, especially in medicine. The milliliter and the cubic centimeter represent the same measure of capacity. A milliliter is about $\frac{1}{5}$ of a teaspoon.

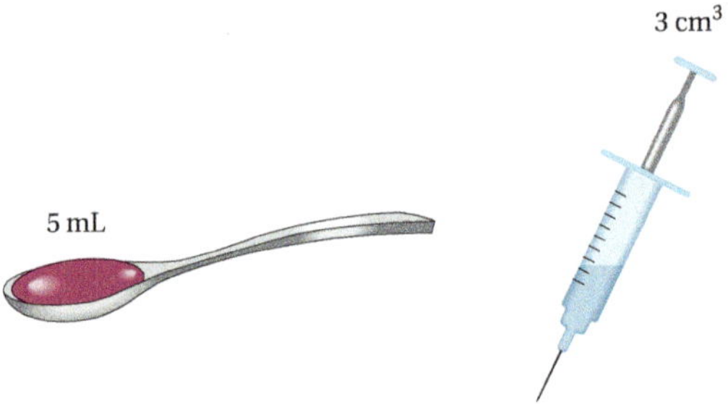

$$1\text{ mL} = 1\text{ cm}^3 = 1\text{ cc}$$

Volumes for which quarts and gallons are used are expressed in liters. Large volumes in business and industry are expressed using measures of cubic meters (m^3).

Do Exercises 3–6.

Complete with mL or L.

3. The patient received an injection of 2 ________ of penicillin.

4. There are 250 ________ in a coffee cup.

5. The gas tank holds 80 ________.

6. Bring home 8 ________ of milk.

EXAMPLE 3 Complete: 4.5 L = ________ mL.

$$4.5\text{ L} = 4.5 \times 1\text{ L} = 4.5 \times 1000\text{ mL} \quad \text{Substituting 1000 mL for 1 L}$$
$$= 4500\text{ mL}$$

1000 L	100 L	10 L	1 L	0.1 L	0.01 L	0.001 L
1 kL	1 hL	1 daL	1 L	1 dL	1 cL	1 mL (cc)

3 places to the right

Answers

3. mL 4. mL 5. L 6. L

EXAMPLE 4 Complete: 280 mL = ______ L.

$$280\text{ mL} = 280 \times 1\text{ mL}$$
$$= 280 \times 0.001\text{ L} \qquad \text{Substituting 0.001 L for 1 mL}$$
$$= 0.28\text{ L}$$

1000 L	100 L	10 L	1 L	0.1 L	0.01 L	0.001 L
1 kL	1 hL	1 daL	1 L	1 dL	1 cL	1 mL (cc)

3 places to the left

We do find metric units of capacity in frequent use in the United States—for example, in sizes of soda bottles and automobile engines.

Do Exercises 7 and 8. ▶

Complete.

7. 0.97 L = ________ mL

8. 8990 mL = ________ L

b MEDICAL APPLICATIONS

The metric system is used extensively in medicine.

EXAMPLE 5 *Medical Dosage.* A physician orders 3.5 L of 5% dextrose in water (abbreviated as D5W) to be administered over a 24-hr period. How many milliliters were ordered?

We convert 3.5 L to milliliters:

$$3.5\text{ L} = 3.5 \times 1\text{ L} = 3.5 \times 1000\text{ mL} = 3500\text{ mL}.$$

The physician ordered 3500 mL of D5W.

Do Exercise 9. ▶

9. *Medical Dosage.* A physician orders 2400 mL of 0.9% saline solution to be administered intravenously over a 24-hr period. How many liters were ordered?

EXAMPLE 6 *Medical Dosage.* Liquids at a pharmacy are often labeled in liters or milliliters. This means that if a physician's prescription is given in ounces, it must be converted. For conversion, a pharmacist knows that 1 fluid oz $\approx$ 29.57 mL.* A prescription calls for 3 fluid oz of theophylline. How many milliliters does the prescription call for?

We convert as follows:

$$3\text{ oz} = 3 \times 1\text{ oz} \approx 3 \times 29.57\text{ mL} = 88.71\text{ mL}.$$

The prescription calls for 88.71 mL of theophylline.

Do Exercise 10. ▶

10. *Medical Dosage.* A prescription calls for 2 oz of theophylline.

a) How many milliliters does the prescription call for?

b) How many liters does the prescription call for?

*In practice, most pharmacists use 30 mL as an approximation to 1 oz.

Answers

7. 970 **8.** 8.99 **9.** 2.4 L
10. (a) About 59.14 mL; **(b)** about 0.059 L

C Exercise Set

For Extra Help MyMathLab® MathXL®  PRACTICE  WATCH  READ  REVIEW

a Complete.

1. 1 L = ________ mL = ________ cm^3

2. ________ L = 1 mL = ________ cm^3

3. 87 L = ________ mL

4. 806 L = ________ mL

5. 49 mL = ________ L

6. 19 mL = ________ L

7. 0.401 mL = ________ L

8. 0.816 mL = ________ L

9. 78.1 L = ________ cm^3

10. 99.6 L = ________ cm^3

11. 10 qt = ________ oz

12. 9.6 oz = ________ pt

13. 20 cups = ________ pt

14. 1 gal = ________ oz

15. 8 gal = ________ qt

16. 1 gal = ________ cups

17. 5 gal = ________ qt

18. 11 gal = ________ qt

19. 56 qt = ________ gal

20. 84 qt = ________ gal

21. 11 gal = ________ pt

22. 5 gal = ________ pt

Complete.

	Object	Gallons (gal)	Quarts (qt)	Pints (pt)	Cups	Ounces (oz)
23.	12-can package of 12-oz sodas					144
24.	Large container of milk			8		
25.	Full tank of gasoline	16				
26.	Dove shampoo					12
27.	Downy fabric softener					51
28.	Williams Lectric Shave					7

Complete.

	Object	Liters (L)	Milliliters (mL)	Cubic Centimeters (cc)	Cubic Centimeters (cm^3)
29.	2-L bottle of soda	2			
30.	Heinz vinegar		3755		
31.	Full tank of gasoline in Europe	64			
32.	Williams Lectric Shave				207
33.	Dove shampoo			355	
34.	Newman's Own salad dressing		473		

b *Medical Dosage.* Solve each of the following.
Source: Steven R. Smith, M.D.

35. An emergency-room physician orders 2.0 L of Ringer's lactate to be administered over 2 hr for a patient in shock. How many milliliters is this?

36. An emergency-room physician orders 2.5 L of 0.9% saline solution over 4 hr for a patient suffering from dehydration. How many milliliters is this?

37. A physician orders 320 mL of 5% dextrose in water (D5W) solution to be administered intravenously over 4 hr. How many liters of D5W is this?

38. A physician orders 40 mL of 5% dextrose in water (D5W) solution to be administered intravenously over 2 hr to an elderly patient. How many liters of D5W is this?

39. A physician orders 0.5 oz of magnesia and alumina oral suspension antacid 4 times per day for a patient with indigestion. How many milliliters of the antacid is the patient to ingest in a day?

40. A physician orders 0.25 oz of magnesia and alumina oral suspension antacid 3 times per day for a child with upper abdominal discomfort. How many milliliters of the antacid is the child to ingest in a day?

41. A physician orders 0.5 L of normal saline solution. How many milliliters are ordered?

42. A physician has ordered that his patient receive 60 mL per hour of normal saline solution intravenously. How many liters of the saline solution is the patient to receive in a 24-hr period?

43. A physician wants her patient to receive 3.0 L of normal saline intravenously over a 24-hr period. How many milliliters per hour must the nurse administer?

44. A physician tells a patient to purchase 0.5 L of hydrogen peroxide. Commercially, hydrogen peroxide is found on the shelf in bottles that hold 4 oz, 8 oz, and 16 oz. Which bottle has a capacity closest to 0.5 L?

Medical Dosage. Because patients do not always have a working knowledge of the metric system, physicians often prescribe dosages in teaspoons (t or tsp) and tablespoons (T or Tbsp). The units are related to the metric system and to each other as follows:

$$5\text{ mL} \approx 1\text{ tsp}, \qquad 3\text{ tsp} = 1\text{ T}.$$

Complete.

45. 45 mL = _____ tsp

46. 3 T = _____ tsp

47. 1 mL = _____ tsp

48. 18.5 mL = _____ tsp

49. 2 T = _____ tsp

50. 8.5 tsp = _____ T

51. 1 T = _____ mL

52. 18.5 mL = _____ T

Synthesis

53. *Wasting Water.* Many people leave the water running while they are brushing their teeth. Suppose that one person wastes 32 oz of water in this way each day. How much water, in gallons, is wasted by one person in a week? in a month (30 days)? in a year (365 days)? Assuming each of the 314 million people in the United States wastes water in this way, estimate how much water is wasted in the United States in a day; in a year.

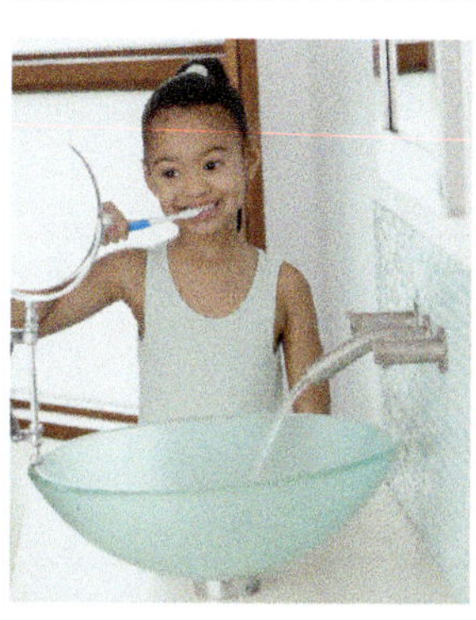

D Time and Temperature

OBJECTIVES

a Convert from one unit of time to another.

b Convert between Celsius and Fahrenheit temperatures using the formulas

$$F = \frac{9}{5} \cdot C + 32$$

and

$$C = \frac{5}{9} \cdot (F - 32).$$

a TIME

A table of units of time is shown below. The metric system uses "h" for hour and "s" for second, but we will use the more familiar "hr" and "sec."

UNITS OF TIME

1 day = 24 hours (hr)	1 year (yr) = $365\frac{1}{4}$ days
1 hr = 60 minutes (min)	1 week (wk) = 7 days
1 min = 60 seconds (sec)	

The earth revolves completely around the sun in $365\frac{1}{4}$ days. Since we cannot have $\frac{1}{4}$ day on the calendar, we give each year 365 days and every fourth year 366 days (a leap year), unless it is a year at the beginning of a century not divisible by 400.

EXAMPLE 1 Complete: 1 hr = _______ sec.

$$\begin{aligned} 1\text{ hr} &= 60\text{ min} \\ &= 60 \cdot 1\text{ min} \\ &= 60 \cdot 60\text{ sec} \qquad \text{Substituting 60 sec for 1 min} \\ &= 3600\text{ sec} \end{aligned}$$

EXAMPLE 2 Complete: 5 years = _______ days.

$$\begin{aligned} 5\text{ years} &= 5 \cdot 1\text{ year} \\ &= 5 \cdot 365\frac{1}{4}\text{ days} \qquad \text{Substituting } 365\frac{1}{4}\text{ days for 1 year} \\ &= 5 \cdot \frac{1461}{4}\text{ days} \\ &= \frac{7305}{4}\text{ days} \\ &= 1826\frac{1}{4}\text{ days} \end{aligned}$$

EXAMPLE 3 Complete: 4320 min = _______ days.

$$4320\text{ min} = 4320\text{ min} \cdot \frac{1\text{ hr}}{60\text{ min}} \cdot \frac{1\text{ day}}{24\text{ hr}} = \frac{4320}{60 \cdot 24}\text{ days} = 3\text{ days}$$

Do Exercises 1–4.

Complete.

1. 2 hr = _______ min
2. 4 years = _______ days
3. 1 day = _______ min
4. 168 hr = _______ wk

Answers

1. 120 2. 1461 3. 1440 4. 1

b TEMPERATURE

Below are two temperature scales: **Fahrenheit** for American measure and **Celsius** for metric measure.

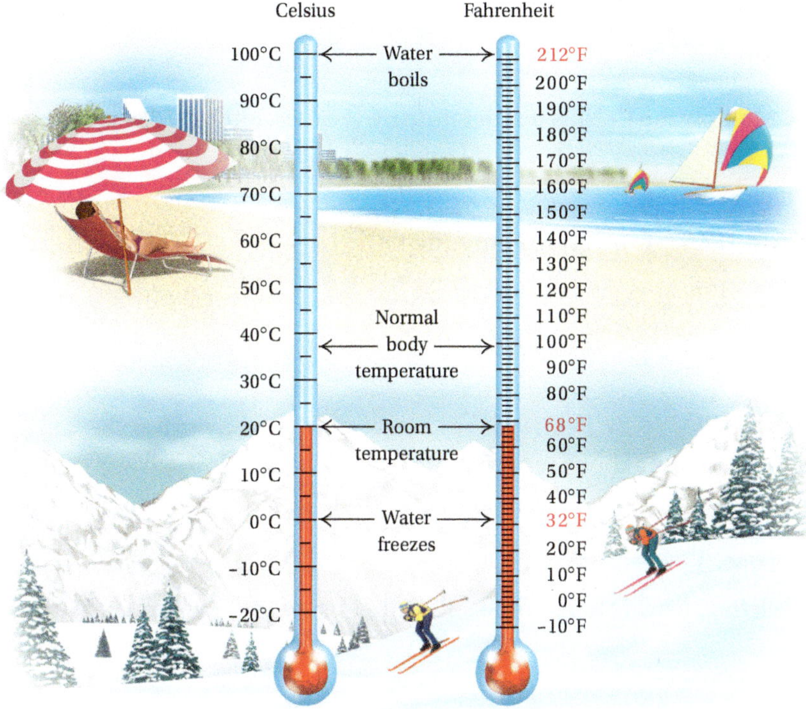

By laying a ruler or other straightedge horizontally between the scales, we can make an approximate conversion from one measure of temperature to the other and get an idea of how the temperature scales compare.

EXAMPLES Convert to Celsius using the scales shown above. Approximate to the nearest ten degrees.

4. 212°F (Boiling point of water) 100°C This is exact.

5. 32°F (Freezing point of water) 0°C This is exact.

6. 105°F 40°C This is approximate.

Do Exercises 5–7.

EXAMPLES Make an approximate conversion to Fahrenheit using the scales shown above.

7. 44°C (Hot bath) 110°F This is approximate.

8. 20°C (Room temperature) 68°F This is exact.

9. 83°C 180°F This is approximate.

Do Exercises 8–10.

Convert to Celsius. Approximate to the nearest ten degrees.

5. 180°F (Brewing coffee)

6. 25°F (Cold day)

7. −10°F (Miserably cold day)

Convert to Fahrenheit. Approximate to the nearest ten degrees.

8. 25°C (Warm day at the beach)

9. 40°C (Temperature of a patient with a high fever)

10. 10°C (Cold bath)

Answers

5. 80°C **6.** 0°C **7.** −20°C **8.** 80°F **9.** 100°F **10.** 50°F

Convert to Fahrenheit.

11. 80°C

12. 35°C

CALCULATOR CORNER

Temperature Conversions Temperature conversions can be done quickly using a calculator. To convert 37°C to Fahrenheit, for example, we press [1] [.] [8] [×] [3] [7] [+] [3] [2] [=]. The calculator displays [98.6], so 37°C = 98.6°F. We can convert 212°F to Celsius by pressing [(] [2] [1] [2] [−] [3] [2] [)] [÷] [1] [.] [8] [=]. The display reads [100], so 212°F = 100°C. Note that we must use parentheses when converting from Fahrenheit to Celsius in order to get the correct result.

EXERCISES Use a calculator to convert each temperature to Fahrenheit.

1. 5°C

2. 50°C

Use a calculator to convert each temperature to Celsius.

3. 68°F

4. 113°F

The following formula allows us to make exact conversions from Celsius to Fahrenheit.

CELSIUS TO FAHRENHEIT

$$F = \frac{9}{5} \cdot C + 32, \text{ or } F = 1.8 \cdot C + 32$$

(Multiply the Celsius temperature by $\frac{9}{5}$, or 1.8, and add 32.)

EXAMPLES Convert to Fahrenheit.

10. 0°C (Freezing point of water)

$$F = \frac{9}{5} \cdot C + 32 = \frac{9}{5} \cdot 0 + 32 = 0 + 32 = 32$$

Thus, 0°C = 32°F.

11. 37°C (Normal body temperature)

$$F = 1.8 \cdot C + 32 = 1.8 \cdot 37 + 32 = 66.6 + 32 = 98.6$$

Thus, 37°C = 98.6°F.

Check the answers to Examples 10 and 11 using the scales on p. 1305.

◀ Do Exercises 11 and 12.

The following formula allows us to make exact conversions from Fahrenheit to Celsius.

FAHRENHEIT TO CELSIUS

$$C = \frac{5}{9} \cdot (F - 32), \text{ or } C = \frac{(F - 32)}{1.8}$$

(Subtract 32 from the Fahrenheit temperature and multiply by $\frac{5}{9}$ or divide by 1.8.)

EXAMPLES Convert to Celsius.

12. 212°F (Boiling point of water)

$$C = \frac{5}{9} \cdot (F - 32)$$
$$= \frac{5}{9} \cdot (212 - 32)$$
$$= \frac{5}{9} \cdot 180 = 100$$

Thus, 212°F = 100°C.

13. 77°F

$$C = \frac{F - 32}{1.8}$$
$$= \frac{77 - 32}{1.8}$$
$$= \frac{45}{1.8} = 25$$

Thus, 77°F = 25°C.

Check the answers to Examples 12 and 13 using the scales on p. 1305.

◀ Do Exercises 13 and 14.

Convert to Celsius.

13. 95°F

14. 113°F

Answers

11. 176°F **12.** 95°F **13.** 35°C **14.** 45°C

D Exercise Set

For Extra Help MyMathLab® MathXL® PRACTICE WATCH READ REVIEW

a Complete.

1. 1 day = _______ hr

2. 1 hr = _______ min

3. 1 min = _______ sec

4. 1 wk = _______ days

5. 1 year = _______ days

6. 2 years = _______ days

7. 180 sec = _______ hr

8. 60 sec = _______ hr

9. 492 sec = _______ min (The amount of time it takes for the rays of the sun to reach the earth)

10. 18,000 sec = _______ hr

11. 156 hr = _______ days

12. 444 hr = _______ days

13. 645 min = _______ hr

14. 375 min = _______ hr

15. 2 wk = _______ hr

16. 4 hr = _______ sec

17. 756 hr = _______ wk

18. 166,320 min = _______ wk

19. 2922 wk = _______ years

20. 623 days = _______ wk

21. ***Actual Time in a Day.*** Although we round it to 24 hr, the actual length of a day is 23 hr, 56 min, and 4.2 sec. How many seconds are there in an actual day?

Source: *The Handy Geography Answer Book*

22. ***Time Length.*** What length of time is 86,400 sec? Is it 1 hr, 1 day, 1 week, or 1 month?

b Convert to Fahrenheit. Use the formula $F = \frac{9}{5} \cdot C + 32$ or $F = 1.8 \cdot C + 32$.

23. 25°C

24. 85°C

25. 40°C

26. 90°C

27. 86°C

28. 93°C

29. 58°C

30. 35°C

31. 2°C

32. 78°C

33. 5°C

34. 15°C

35. 3000°C
(The melting point of iron)

36. 1000°C
(The melting point of gold)

Convert to Celsius. Use the formula $C = \frac{5}{9} \cdot (F - 32)$ or $C = \frac{F - 32}{1.8}$.

37. 86°F

38. 59°F

39. 131°F

40. 140°F

41. 178°F

42. 195°F

43. 140°F

44. 107°F

45. 68°F

46. 50°F

47. 44°F

48. 120°F

49. 98.6°F
(Normal body temperature)

50. 104°F
(High-fever body temperature)

51. ***Highest Temperatures.*** The highest temperature ever recorded in the world is 136°F in the desert of Libya in 1922. The highest temperature ever recorded in the United States is $56\frac{2}{3}$°C in California's Death Valley in 1913.

Source: *The Handy Geography Answer Book*

a) Convert each temperature to the other scale.
b) How much higher in degrees Fahrenheit was the world record than the U.S. record?

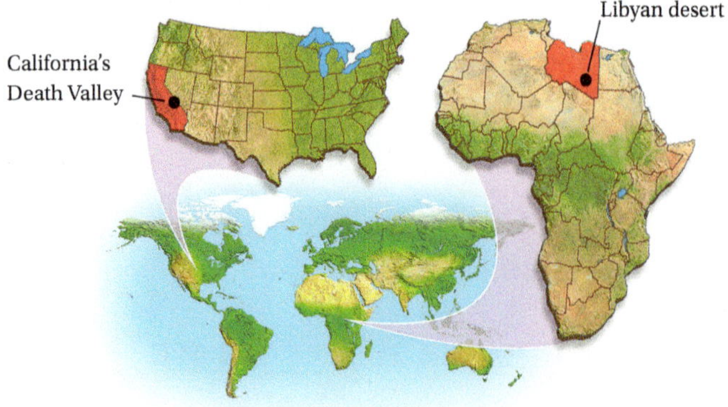

52. ***Boiling Point and Altitude.*** The boiling point of water actually changes with altitude. The boiling point is 212°F at sea level, but lowers about 1°F for every 500 ft that the altitude increases above sea level.

Sources: *The Handy Geography Answer Book; The New York Times Almanac*

a) What is the boiling point at an elevation of 1500 ft above sea level?
b) The elevation of Tucson is 2564 ft above sea level and that of Phoenix is 1117 ft. What is the boiling point in each city?
c) How much lower is the boiling point in Denver, whose elevation is 5280 ft, than in Tucson?
d) What is the boiling point at the top of Mt. McKinley in Alaska, the highest point in the United States, at 20,320 ft?

Synthesis

53. Estimate the number of years in one million seconds.

54. Estimate the number of years in one billion seconds.

Sets

E

OBJECTIVES

a Name sets using the roster method.

b Classify statements regarding set membership and subsets as true or false.

c Find the intersection and the union of sets.

a NAMING SETS

To name the set of whole numbers less than 6, we can use the **roster method**, as follows: $\{0, 1, 2, 3, 4, 5\}$.

The set of real numbers x such that x is less than 6 cannot be named by listing all its members because there are infinitely many. We name such a set using **set-builder notation**, as follows: $\{x | x < 6\}$. This is read "The set of all x such that x is less than 6."

Do Exercises 1 and 2. ▶

Name each set using the roster method.

1. The set of whole numbers 0 through 7

2. $\{x|$ the square of x is $25\}$

b SET MEMBERSHIP AND SUBSETS

The symbol $\in$ means **is a member of** or **belongs to**, or **is an element of**. Thus, $x \in A$ means x is a member of A or x belongs to A or x is an element of A.

EXAMPLE 1 Classify each of the following as true or false.

a) $1 \in \{1, 2, 3\}$
b) $1 \in \{2, 3\}$
c) $4 \in \{x | x \text{ is an even whole number}\}$
d) $5 \in \{x | x \text{ is an even whole number}\}$

a) Since 1 *is* listed as a member of the set, $1 \in \{1, 2, 3\}$ is true.
b) Since 1 *is not* a member of $\{2, 3\}$, the statement $1 \in \{2, 3\}$ is false.
c) Since 4 *is* an even whole number, $4 \in \{x | x \text{ is an even whole number}\}$ is a true statement.
d) Since 5 *is not* even, $5 \in \{x | x \text{ is an even whole number}\}$ is false.

Set membership can be illustrated with a diagram, as shown here.

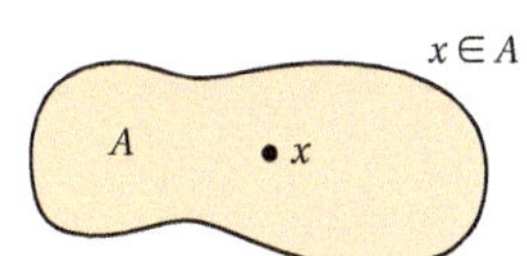

Do Exercises 3 and 4. ▶

Determine whether each of the following is true or false.

3. $8 \in \{x | x \text{ is an even whole number}\}$

4. $2 \in \{x | x \text{ is a prime number}\}$

If every element of A is an element of B, then A is a **subset** of B. This is denoted $A \subseteq B$. The set of whole numbers is a subset of the set of integers. The set of rational numbers is a subset of the set of real numbers.

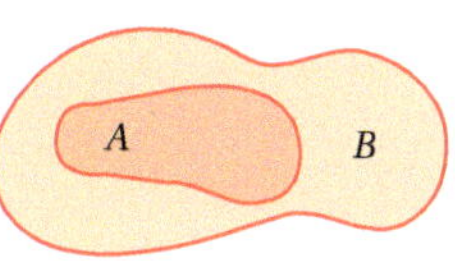

EXAMPLE 2 Classify each of the following as true or false.

a) $\{1, 2\} \subseteq \{1, 2, 3, 4\}$
b) $\{p, q, r, w\} \subseteq \{a, p, r, z\}$
c) $\{x | x < 6\} \subseteq \{x | x \leq 11\}$

a) Since every element of $\{1, 2\}$ is in the set $\{1, 2, 3, 4\}$, the statement $\{1, 2\} \subseteq \{1, 2, 3, 4\}$ is true.

Answers

1. $\{0, 1, 2, 3, 4, 5, 6, 7\}$ **2.** $\{-5, 5\}$
3. True **4.** True

Determine whether each of the following is true or false.

5. $\{-2, -3, 4\} \subseteq \{-5, -4, -2, 7, -3, 5, 4\}$

6. $\{a, e, i, o, u\} \subseteq$ The set of all consonants

7. $\{x \mid x \leq -8\} \subseteq \{x \mid x \leq -7\}$

Find the intersection.

8. $\{-2, -3, 4, -4, 8\} \cap \{-5, -4, -2, 7, -3, 5, 4\}$

9. $\{a, e, i, o, u\} \cap \{m, a, r, v, i, n\}$

10. $\{a, e, i, o, u\} \cap$ The set of all consonants

Find the union.

11. $\{-2, -3, 4, -4, 8\} \cup \{-5, -4, -2, 7, -3, 5, 4\}$

12. $\{a, e, i, o, u\} \cup \{m, a, r, v, i, n\}$

13. $\{a, e, i, o, u\} \cup$ The set of all consonants

Answers

5. True 6. False 7. True
8. $\{-2, -3, 4, -4\}$ 9. $\{a, i\}$ 10. $\{\ \}$, or $\varnothing$
11. $\{-2, -3, 4, -4, 8, -5, 7, 5\}$
12. $\{a, e, i, o, u, m, r, v, n\}$
13. $\{a, b, c, d, e, f, g, h, i, j, k, l, m, n, o, p, q, r, s, t, u, v, w, x, y, z\}$

b) Since $q \in \{p, q, r, w\}$, but $q \notin \{a, p, r, z\}$, the statement $\{p, q, r, w\} \subseteq \{a, p, r, z\}$ is false.

c) Since every number that is less than 6 is also less than or equal to 11, the statement $\{x \mid x < 6\} \subseteq \{x \mid x \leq 11\}$ is true.

Do Exercises 5–7.

C INTERSECTIONS AND UNIONS

The **intersection** of sets A and B, denoted $A \cap B$, is the set of members that are common to both sets.

EXAMPLE 3 Find the intersection.

a) $\{0, 1, 3, 5, 25\} \cap \{2, 3, 4, 5, 6, 7, 9\}$ **b)** $\{a, p, q, w\} \cap \{p, q, t\}$

a) $\{0, 1, 3, 5, 25\} \cap \{2, 3, 4, 5, 6, 7, 9\} = \{3, 5\}$

b) $\{a, p, q, w\} \cap \{p, q, t\} = \{p, q\}$

Set intersection can be illustrated with a diagram, as shown here.

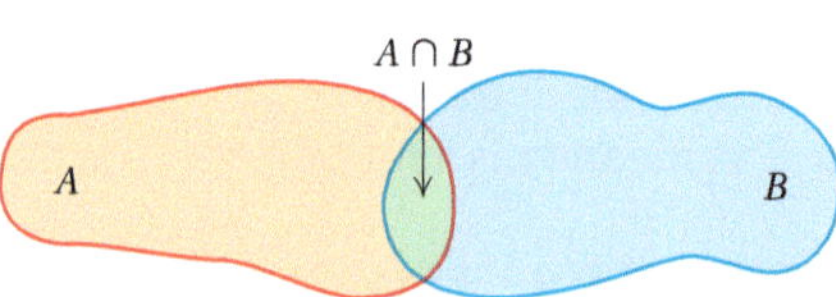

The set without members is known as the **empty set**, and is often named $\varnothing$, and sometimes $\{\ \}$. Each of the following is a description of the empty set:

$\{2, 3\} \cap \{5, 6, 7\}$;

$\{x \mid x \text{ is an even natural number}\} \cap \{x \mid x \text{ is an odd natural number}\}$.

Do Exercises 8–10.

Two sets A and B can be combined to form a set that contains the members of A as well as those of B. The new set is called the **union** of A and B, denoted $A \cup B$.

EXAMPLE 4 Find the union.

a) $\{0, 5, 7, 13, 27\} \cup \{0, 2, 3, 4, 5\}$ **b)** $\{a, c, e, g\} \cup \{b, d, f\}$

a) $\{0, 5, 7, 13, 27\} \cup \{0, 2, 3, 4, 5\} = \{0, 2, 3, 4, 5, 7, 13, 27\}$

Note that the 0 and the 5 are *not* listed twice in the solution.

b) $\{a, c, e, g\} \cup \{b, d, f\} = \{a, b, c, d, e, f, g\}$

Set union can be illustrated with a diagram, as shown here.

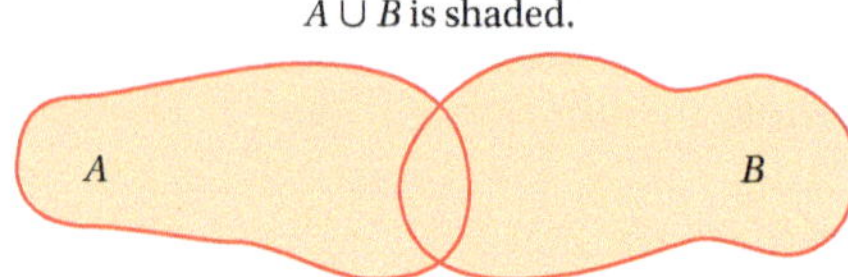

The solution set of the equation $(x - 3)(x + 2) = 0$ is $\{3, -2\}$. This set is the union of the solution sets of $x - 3 = 0$ and $x + 2 = 0$, which are $\{3\}$ and $\{-2\}$.

Do Exercises 11–13.

E Exercise Set

For Extra Help MyMathLab® MathXL® PRACTICE WATCH READ REVIEW

a Name each set using the roster method.

1. The set of whole numbers 3 through 8

2. The set of whole numbers 101 through 107

3. The set of odd numbers between 40 and 50

4. The set of multiples of 5 between 11 and 39

5. $\{x \mid \text{the square of } x \text{ is } 9\}$

6. $\{x \mid x \text{ is the cube of } 0.2\}$

b Classify each statement as true or false.

7. $2 \in \{x \mid x \text{ is an odd number}\}$

8. $7 \in \{x \mid x \text{ is an odd number}\}$

9. Kyle Busch $\in$ The set of all NASCAR drivers

10. Apple $\in$ The set of all fruit

11. $-3 \in \{-4, -3, 0, 1\}$

12. $0 \in \{-4, -3, 0, 1\}$

13. $\frac{2}{3} \in \{x \mid x \text{ is a rational number}\}$

14. Heads $\in$ The set of outcomes of flipping a penny

15. $\{4, 5, 8,\} \subseteq \{1, 3, 4, 5, 6, 7, 8, 9\}$

16. The set of vowels $\subseteq$ The set of consonants

17. $\{-1, -2, -3, -4, -5\} \subseteq \{-1, 2, 3, 4, 5\}$

18. The set of integers $\subseteq$ The set of rational numbers

c Find the intersection.

19. $\{a, b, c, d, e\} \cap \{c, d, e, f, g\}$

20. $\{a, e, i, o, u\} \cap \{q, u, i, c, k\}$

21. $\{1, 2, 5, 10\} \cap \{0, 1, 7, 10\}$

22. $\{0, 1, 7, 10\} \cap \{0, 1, 2, 5\}$

23. $\{1, 2, 5, 10\} \cap \{3, 4, 7, 8\}$

24. $\{a, e, i, o, u\} \cap \{m, n, f, g, h\}$

Find the union.

25. $\{a, e, i, o, u\} \cup \{q, u, i, c, k\}$

26. $\{a, b, c, d, e\} \cup \{c, d, e, f, g\}$

27. $\{0, 1, 7, 10\} \cup \{0, 1, 2, 5\}$

28. $\{1, 2, 5, 10\} \cup \{0, 1, 7, 10\}$

29. $\{a, e, i, o, u\} \cup \{m, n, f, g, h\}$

30. $\{1, 2, 5, 10\} \cup \{a, b\}$

Synthesis

31. Find the union of the set of integers and the set of whole numbers.

32. Find the intersection of the set of odd integers and the set of even integers.

33. Find the union of the set of rational numbers and the set of irrational numbers.

34. Find the intersection of the set of even integers and the set of positive rational numbers.

35. Find the intersection of the set of rational numbers and the set of irrational numbers.

36. Find the union of the set of negative integers, the set of positive integers, and the set containing 0.

37. For a set A, find each of the following.

a) $A \cup \varnothing$ **b)** $A \cup A$
c) $A \cap A$ **d)** $A \cap \varnothing$

38. A set is *closed* under an operation if, when the operation is performed on its members, the result is in the set. For example, the set of real numbers is closed under the operation of addition since the sum of any two real numbers is a real number.

a) Is the set of even numbers closed under addition?
b) Is the set of odd numbers closed under addition?
c) Is the set $\{0, 1\}$ closed under addition?
d) Is the set $\{0, 1\}$ closed under multiplication?
e) Is the set of real numbers closed under multiplication?
f) Is the set of integers closed under division?

39. Experiment with sets of various types and determine whether the following distributive law for sets is true:

$$A \cap (B \cup C) = (A \cap B) \cup (A \cap C).$$

Factoring Sums or Differences of Cubes

F

OBJECTIVE

a Factor sums and differences of two cubes.

a FACTORING SUMS OR DIFFERENCES OF CUBES

We can factor the sum or the difference of two expressions that are cubes.

Consider the following products:

$$\begin{aligned}(A + B)(A^2 - AB + B^2) &= A(A^2 - AB + B^2) + B(A^2 - AB + B^2)\\ &= A^3 - A^2B + AB^2 + A^2B - AB^2 + B^3\\ &= A^3 + B^3\end{aligned}$$

and

$$\begin{aligned}(A - B)(A^2 + AB + B^2) &= A(A^2 + AB + B^2) - B(A^2 + AB + B^2)\\ &= A^3 + A^2B + AB^2 - A^2B - AB^2 - B^3\\ &= A^3 - B^3.\end{aligned}$$

The above equations (reversed) show how we can factor a sum or a difference of two cubes.

FACTORING SUMS OR DIFFERENCES OF CUBES

$A^3 + B^3 = (A + B)(A^2 - AB + B^2)$,
$A^3 - B^3 = (A - B)(A^2 + AB + B^2)$

Note that what we are considering here is a sum or a difference of cubes. We are not cubing a binomial. For example, $(A + B)^3$ is *not* the same as $A^3 + B^3$. The table of cubes in the margin is helpful.

N	N^3
0.1	0.001
0.2	0.008
0	0
1	1
2	8
3	27
4	64
5	125
6	216
7	343
8	512
9	729
10	1000

EXAMPLE 1 Factor: $x^3 - 8$.

We have

$$x^3 - 8 = x^3 - 2^3 = (x - 2)(x^2 + x \cdot 2 + 2^2).$$

$$A^3 - B^3 = (A - B)(A^2 + A\ \ B + B^2)$$

This tells us that $x^3 - 8 = (x - 2)(x^2 + 2x + 4)$. Note that we cannot factor $x^2 + 2x + 4$. (It is not a trinomial square nor can it be factored by trial and error or the *ac*-method.) The check is left to the student.

Do Exercises 1 and 2. ▶

Factor.

1. $x^3 - 27$ **2.** $64 - y^3$

EXAMPLE 2 Factor: $x^3 + 125$.

We have

$$x^3 + 125 = x^3 + 5^3 = (x + 5)(x^2 - x \cdot 5 + 5^2).$$

$$A^3 + B^3 = (A + B)(A^2 - A\ \ B + B^2)$$

Thus, $x^3 + 125 = (x + 5)(x^2 - 5x + 25)$. The check is left to the student.

Do Exercises 3 and 4. ▶

Factor.

3. $y^3 + 8$ **4.** $125 + t^3$

Answers

1. $(x - 3)(x^2 + 3x + 9)$
2. $(4 - y)(16 + 4y + y^2)$
3. $(y + 2)(y^2 - 2y + 4)$
4. $(5 + t)(25 - 5t + t^2)$

EXAMPLE 3 Factor: $x^3 - 27t^3$.

We have

$$x^3 - 27t^3 = x^3 - (3t)^3 = (x - 3t)(x^2 + x \cdot 3t + (3t)^2)$$

$$A^3 - B^3 = (A - B)(A^2 + A\ B + B^2)$$

$$= (x - 3t)(x^2 + 3xt + 9t^2)$$

Factor.

5. $27x^3 - y^3$

6. $8y^3 + z^3$

◀ Do Exercises 5 and 6.

EXAMPLE 4 Factor: $128y^7 - 250x^6y$.

We first look for a common factor:

$$128y^7 - 250x^6y = 2y(64y^6 - 125x^6) = 2y[(4y^2)^3 - (5x^2)^3]$$
$$= 2y(4y^2 - 5x^2)(16y^4 + 20x^2y^2 + 25x^4).$$

EXAMPLE 5 Factor: $a^6 - b^6$.

We can express this polynomial as a difference of squares:

$$(a^3)^2 - (b^3)^2.$$

We factor as follows:

$$a^6 - b^6 = (a^3 + b^3)(a^3 - b^3).$$

One factor is a sum of two cubes, and the other factor is a difference of two cubes. We factor them:

$$(a + b)(a^2 - ab + b^2)(a - b)(a^2 + ab + b^2).$$

We have now factored completely.

In Example 5, had we thought of factoring first as a difference of two cubes, we would have had

$$(a^2)^3 - (b^2)^3 = (a^2 - b^2)(a^4 + a^2b^2 + b^4)$$
$$= (a + b)(a - b)(a^4 + a^2b^2 + b^4).$$

In this case, we might have missed some factors; $a^4 + a^2b^2 + b^4$ can be factored as $(a^2 - ab + b^2)(a^2 + ab + b^2)$, but we probably would not have known to do such factoring.

EXAMPLE 6 Factor: $64a^6 - 729b^6$.

$$64a^6 - 729b^6 = (8a^3 - 27b^3)(8a^3 + 27b^3) \quad \text{Factoring a difference of squares}$$
$$= [(2a)^3 - (3b)^3][(2a)^3 + (3b)^3].$$

Each factor is a sum or a difference of cubes. We factor each:

$$= (2a - 3b)(4a^2 + 6ab + 9b^2)(2a + 3b)(4a^2 - 6ab + 9b^2)$$

Sum of cubes:	$A^3 + B^3 = (A + B)(A^2 - AB + B^2)$;
Difference of cubes:	$A^3 - B^3 = (A - B)(A^2 + AB + B^2)$;
Difference of squares:	$A^2 - B^2 = (A + B)(A - B)$;
Sum of squares:	$A^2 + B^2$ cannot be factored using real numbers if the largest common factor has been removed.

Factor.

7. $m^6 - n^6$

8. $16x^7y + 54xy^7$

9. $729x^6 - 64y^6$

10. $x^3 - 0.027$

◀ Do Exercises 7–10.

Answers

5. $(3x - y)(9x^2 + 3xy + y^2)$
6. $(2y + z)(4y^2 - 2yz + z^2)$
7. $(m + n)(m^2 - mn + n^2) \times (m - n)(m^2 + mn + n^2)$
8. $2xy(2x^2 + 3y^2)(4x^4 - 6x^2y^2 + 9y^4)$
9. $(3x + 2y)(9x^2 - 6xy + 4y^2) \times (3x - 2y)(9x^2 + 6xy + 4y^2)$
10. $(x - 0.3)(x^2 + 0.3x + 0.09)$

F

Exercise Set

For Extra Help
MyMathLab®
MathXL® PRACTICE | WATCH | READ | REVIEW

a Factor.

1. $z^3 + 27$

2. $a^3 + 8$

3. $x^3 - 1$

4. $c^3 - 64$

5. $y^3 + 125$

6. $x^3 + 1$

7. $8a^3 + 1$

8. $27x^3 + 1$

9. $y^3 - 8$

10. $p^3 - 27$

11. $8 - 27b^3$

12. $64 - 125x^3$

13. $64y^3 + 1$

14. $125x^3 + 1$

15. $8x^3 + 27$

16. $27y^3 + 64$

17. $a^3 - b^3$

18. $x^3 - y^3$

19. $a^3 + \frac{1}{8}$

20. $b^3 + \frac{1}{27}$

21. $2y^3 - 128$

22. $3z^3 - 3$

23. $24a^3 + 3$

24. $54x^3 + 2$

25. $rs^3 + 64r$

26. $ab^3 + 125a$

27. $5x^3 - 40z^3$

28. $2y^3 - 54z^3$

29. $x^3 + 0.001$

30. $y^3 + 0.125$

31. $64x^6 - 8t^6$

32. $125c^6 - 8d^6$

33. $2y^4 - 128y$

34. $3z^5 - 3z^2$

35. $z^6 - 1$

36. $t^6 + 1$

37. $t^6 + 64y^6$

38. $p^6 - q^6$

Synthesis

Consider these polynomials:

$$(a + b)^3;\quad a^3 + b^3;\quad (a + b)(a^2 - ab + b^2);$$
$$(a + b)(a^2 + ab + b^2);\quad (a + b)(a + b)(a + b).$$

39. Evaluate each polynomial when $a = -2$ and $b = 3$.

40. Evaluate each polynomial when $a = 4$ and $b = -1$.

Factor. Assume that variables in exponents represent natural numbers.

41. $x^{6a} + y^{3b}$

42. $a^3x^3 - b^3y^3$

43. $3x^{3a} + 24y^{3b}$

44. $\frac{8}{27}x^3 + \frac{1}{64}y^3$

45. $\frac{1}{24}x^3y^3 + \frac{1}{3}z^3$

46. $7x^3 - \frac{7}{8}$

47. $(x + y)^3 - x^3$

48. $(1 - x)^3 + (x - 1)^6$

49. $(a + 2)^3 - (a - 2)^3$

50. $y^4 - 8y^3 - y + 8$

Finding Equations of Lines: Point–Slope Equation

G

OBJECTIVES

a Find an equation of a line when the slope and a point are given.

b Find an equation of a line when two points are given.

We can use the slope–intercept equation, $y = mx + b$, to find an equation of a line. Here we introduce another form, the *point–slope equation*, and find equations of lines using both forms.

a FINDING AN EQUATION OF A LINE WHEN THE SLOPE AND A POINT ARE GIVEN

Suppose we know the slope of a line and the coordinates of one point on the line. We can use the slope–intercept equation to find an equation of the line. Or, we can use what is called a **point–slope equation**. We first develop a formula for such a line.

Suppose that a line of slope m passes through the point (x_1, y_1). For any other point (x, y) to lie on this line, we must have

$$\frac{y - y_1}{x - x_1} = m.$$

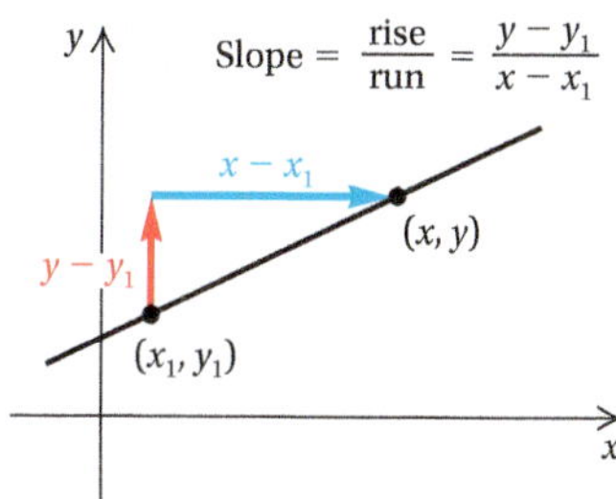

It is tempting to use this last equation as an equation of the line of slope m that passes through (x_1, y_1). The only problem with this form is that when x and y are replaced with x_1 and y_1, we have $\frac{0}{0} = m$, a false equation. To avoid this difficulty, we multiply by $x - x_1$ on both sides and simplify:

$$(x - x_1)\frac{y - y_1}{x - x_1} = m(x - x_1) \qquad \text{Multiplying by } x - x_1 \text{ on both sides}$$

$$y - y_1 = m(x - x_1). \qquad \text{Removing a factor of 1: } \frac{x - x_1}{x - x_1} = 1$$

This is the *point–slope* form of a linear equation.

POINT–SLOPE EQUATION

The **point–slope equation** of a line with slope m, passing through (x_1, y_1), is

$$y - y_1 = m(x - x_1).$$

If we know the slope of a line and a certain point on the line, we can find an equation of the line using either the point-slope equation,

$$y - y_1 = m(x - x_1),$$

or the slope-intercept equation,

$$y = mx + b.$$

EXAMPLE 1 Find an equation of the line with slope -2 and containing the point $(-1, 3)$.

Using the Point-Slope Equation: We consider $(-1, 3)$ to be (x_1, y_1) and -2 to be the slope m, and substitute:

$$\begin{aligned} y - y_1 &= m(x - x_1) \\ y - 3 &= -2[x - (-1)] \qquad \text{Substituting} \\ y - 3 &= -2(x + 1) \\ y - 3 &= -2x - 2 \\ y &= -2x - 2 + 3 \\ y &= -2x + 1. \end{aligned}$$

Using the Slope-Intercept Equation: The point $(-1, 3)$ is on the line, so it is a solution. Thus we can substitute -1 for x and 3 for y in $y = mx + b$. We also substitute -2 for m, the slope. Then we solve for b:

$$\begin{aligned} y &= mx + b \\ 3 &= -2 \cdot (-1) + b \qquad \text{Substituting} \\ 3 &= 2 + b \\ 1 &= b. \qquad \text{Solving for } b \end{aligned}$$

We then use the equation $y = mx + b$ and substitute -2 for m and 1 for b:

$$y = -2x + 1.$$

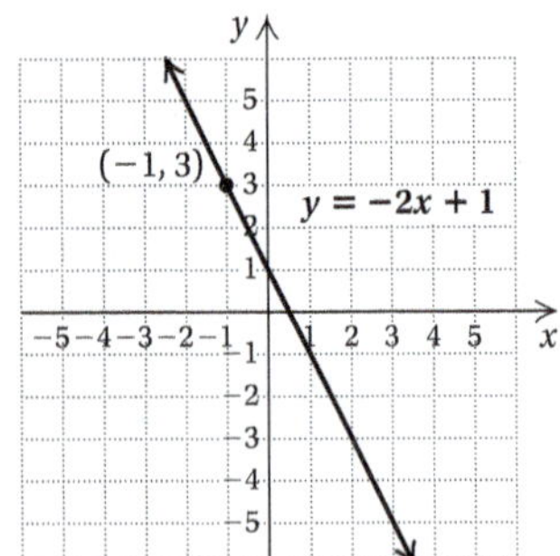

Do Exercises 1–4.

Find an equation of the line with the given slope and containing the given point.

1. $m = -3, \ (-5, 4)$

2. $m = 5, \ (-2, 1)$

3. $m = 6, \ (3, -5)$

4. $m = -\frac{2}{3}, \ (1, 2)$

Answers

1. $y = -3x - 11$ 2. $y = 5x + 11$
3. $y = 6x - 23$ 4. $y = -\frac{2}{3}x + \frac{8}{3}$

b FINDING AN EQUATION OF A LINE WHEN TWO POINTS ARE GIVEN

We can also use the point–slope equation or the slope–intercept equation to find an equation of a line when two points are given.

EXAMPLE 2 Find an equation of the line containing the points $(3, 4)$ and $(-5, 2)$.

First, we find the slope:

$$m = \frac{4 - 2}{3 - (-5)} = \frac{2}{8}, \text{ or } \frac{1}{4}.$$

Now we have the slope and two points. We then proceed as we did in Example 1, using either point, and either the point–slope equation or the slope–intercept equation.

Using the Point–Slope Equation: We choose $(3, 4)$ and substitute 3 for x_1, 4 for y_1, and $\frac{1}{4}$ for m:

$$\begin{aligned} y - y_1 &= m(x - x_1) \\ y - 4 &= \tfrac{1}{4}(x - 3) \qquad \text{Substituting} \\ y - 4 &= \tfrac{1}{4}x - \tfrac{3}{4} \\ y &= \tfrac{1}{4}x - \tfrac{3}{4} + 4 \\ y &= \tfrac{1}{4}x - \tfrac{3}{4} + \tfrac{16}{4} \\ y &= \tfrac{1}{4}x + \tfrac{13}{4}. \end{aligned}$$

Using the Slope–Intercept Equation: We choose $(3, 4)$ and substitute 3 for x, 4 for y, and $\frac{1}{4}$ for m and solve for b:

$$\begin{aligned} y &= mx + b \\ 4 &= \tfrac{1}{4} \cdot 3 + b \qquad \text{Substituting} \\ 4 &= \tfrac{3}{4} + b \\ 4 - \tfrac{3}{4} &= b \\ \tfrac{16}{4} - \tfrac{3}{4} &= b \\ \tfrac{13}{4} &= b. \qquad \text{Solving for } b \end{aligned}$$

Finally, we use the equation $y = mx + b$ and substitute $\frac{1}{4}$ for m and $\frac{13}{4}$ for b:

$$y = \tfrac{1}{4}x + \tfrac{13}{4}.$$

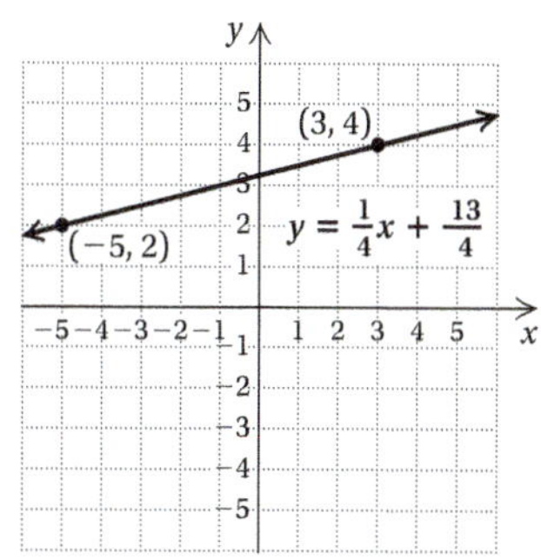

Do Exercises 5 and 6. ▶

5. Find an equation of the line containing the points $(3, -5)$ and $(-1, 4)$.

6. Find an equation of the line containing the points $(-3, 11)$ and $(-4, 20)$.

Answers

5. $y = -\frac{9}{4}x + \frac{7}{4}$ **6.** $y = -9x - 16$

G Exercise Set

For Extra Help
MyMathLab® MathXL® PRACTICE WATCH READ REVIEW

a Find an equation of the line having the given slope and containing the given point.

1. $m = 4,\ (5, 2)$

2. $m = 5,\ (4, 3)$

3. $m = -2,\ (2, 8)$

4. $m = -3,\ (9, 6)$

5. $m = 3,\ (-2, -2)$

6. $m = 1,\ (-1, -7)$

7. $m = -3,\ (-2, 0)$

8. $m = -2,\ (8, 0)$

9. $m = 0,\ (0, 4)$

10. $m = 0,\ (0, -7)$

11. $m = -\frac{4}{5},\ (2, 3)$

12. $m = \frac{2}{3},\ (1, -2)$

b Find an equation of the line containing the given pair of points.

13. $(2, 5)$ and $(4, 7)$

14. $(1, 4)$ and $(5, 6)$

15. $(-1, -1)$ and $(9, 9)$

16. $(-3, -3)$ and $(2, 2)$

17. $(0, -5)$ and $(3, 0)$

18. $(-4, 0)$ and $(0, 7)$

19. $(-4, -7)$ and $(-2, -1)$

20. $(-2, -3)$ and $(-4, -6)$

21. $(0, 0)$ and $(-4, 7)$

22. $(0, 0)$ and $(6, 1)$

23. $\left(\frac{2}{3}, \frac{3}{2}\right)$ and $\left(-3, \frac{5}{6}\right)$

24. $\left(\frac{1}{4}, -\frac{1}{2}\right)$ and $\left(\frac{3}{4}, 6\right)$

Synthesis

25. Find an equation of the line that has the same y-intercept as the line $2x - y = -3$ and contains the point $(-1, -2)$.

26. Find an equation of the line with the same slope as the line $\frac{1}{2}x - \frac{1}{3}y = 10$ and the same y-intercept as the line $\frac{1}{4}x + 3y = -2$.

Equations Involving Absolute Value

H

OBJECTIVE

a Solve equations with absolute-value expressions.

a EQUATIONS WITH ABSOLUTE VALUE

EXAMPLE 1 Solve: $|x| = 4$. Then graph on the number line.

Note that $|x| = |x - 0|$, so that $|x - 0|$ is the distance from x to 0. Thus solutions of the equation $|x| = 4$, or $|x - 0| = 4$ are those numbers x whose distance from 0 is 4. Those numbers are -4 and 4. The solution set is $\{-4, 4\}$. The graph consists of just two points, as shown.

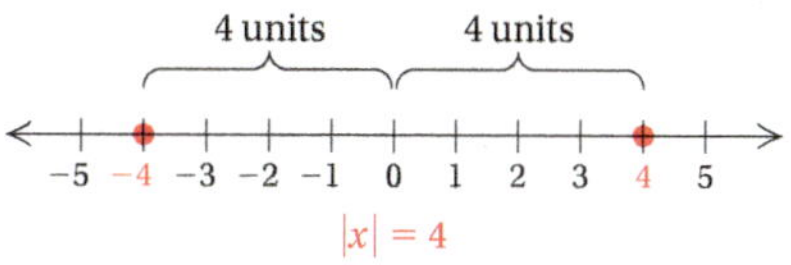

EXAMPLE 2 Solve: $|x| = 0$.

The only number whose absolute value is 0 is 0 itself. Thus the solution is 0. The solution set is $\{0\}$.

EXAMPLE 3 Solve: $|x| = -7$.

The absolute value of a number is always nonnegative. Thus there is no number whose absolute value is -7; consequently, the equation has no solution. The solution set is $\varnothing$.

Examples 1–3 lead us to the following principle for solving linear equations with absolute value.

THE ABSOLUTE VALUE PRINCIPLE

For any positive number p and any algebraic expression X:

a) The solution of $|X| = p$ is those numbers that satisfy $X = -p$ *or* $X = p$.

b) The equation $|X| = 0$ is equivalent to the equation $X = 0$.

c) The equation $|X| = -p$ has no solution.

Do Exercises 1–3. ▶

1. Solve: $|x| = 6$. Then graph on the number line.

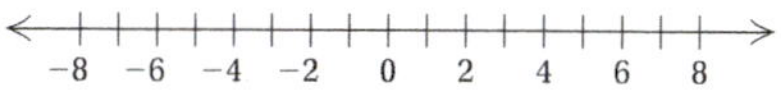

2. Solve: $|x| = -6$.

3. Solve: $|p| = 0$.

We can use the absolute-value principle with the addition and multiplication principles to solve equations with absolute value.

EXAMPLE 4 Solve: $2|x| + 5 = 9$.

We first use the addition and multiplication principles to get $|x|$ by itself. Then we use the absolute-value principle.

Answers

1. $\{6, -6\}$;

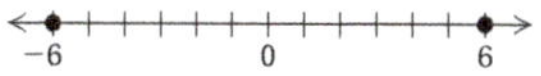

2. $\varnothing$ **3.** $\{0\}$

Solve.

4. $|3x| = 6$

5. $4|x| + 10 = 27$

6. $3|x| - 2 = 10$

We have

$$2|x| + 5 = 9$$
$$2|x| = 4 \quad \text{Subtracting 5}$$
$$|x| = 2 \quad \text{Dividing by 2}$$
$$x = -2 \quad or \quad x = 2 \quad \text{Using the absolute-value principle}$$

The solutions are -2 and 2. The solution set is $\{-2, 2\}$.

◀ **Do Exercises 4–6.**

EXAMPLE 5 Solve: $|x - 2| = 3$.

We can consider solving this equation in two different ways.

Method 1: This allows us to see the meaning of the solutions graphically. The solution set consists of those numbers that are 3 units from 2 on the number line.

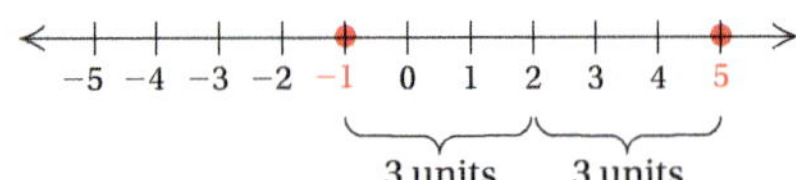

The solutions of $|x - 2| = 3$ are -1 and 5. The solution set is $\{-1, 5\}$.

Method 2: This method is more efficient. We use the absolute-value principle, replacing X with $x - 2$ and p with 3. Then we solve each equation separately.

$$|X| = p$$
$$|x - 2| = 3$$
$$x - 2 = -3 \quad or \quad x - 2 = 3 \quad \text{Absolute-value principle}$$
$$x = -1 \quad or \quad x = 5$$

The solutions are -1 and 5. The solution set is $\{-1, 5\}$.

7. Solve: $|x - 4| = 1$. Use two methods as in Example 5.

◀ **Do Exercise 7.**

EXAMPLE 6 Solve: $|2x + 5| = 13$.

We use the absolute-value principle, replacing X with $2x + 5$ and p with 13:

$$|X| = p$$
$$|2x + 5| = 13$$
$$2x + 5 = -13 \quad or \quad 2x + 5 = 13 \quad \text{Absolute-value principle}$$
$$2x = -18 \quad or \quad 2x = 8$$
$$x = -9 \quad or \quad x = 4.$$

The solutions are -9 and 4. The solution set is $\{-9, 4\}$.

8. Solve: $|3x - 4| = 17$.

◀ **Do Exercise 8.**

EXAMPLE 7 Solve: $|4 - 7x| = -8$.

Since absolute value is always nonnegative, this equation has no solution. The solution set is $\varnothing$.

9. Solve: $|6 + 2x| = -3$.

◀ **Do Exercise 9.**

Answers

4. $\{-2, 2\}$ 5. $\left\{-\frac{17}{4}, \frac{17}{4}\right\}$ 6. $\{-4, 4\}$
7. $\{3, 5\}$ 8. $\left\{-\frac{13}{3}, 7\right\}$ 9. $\varnothing$

H Exercise Set

For Extra Help MyMathLab® MathXL® PRACTICE WATCH READ REVIEW

a Solve.

1. $|x| = 3$

2. $|x| = 5$

3. $|x| = -3$

4. $|x| = -9$

5. $|q| = 0$

6. $|y| = 7.4$

7. $|x - 3| = 12$

8. $|3x - 2| = 6$

9. $|2x - 3| = 4$

10. $|5x + 2| = 3$

11. $|4x - 9| = 14$

12. $|9y - 2| = 17$

13. $|x| + 7 = 18$

14. $|x| - 2 = 6.3$

15. $574 = 283 + |t|$

16. $-562 = -2000 + |x|$

17. $|5x| = 40$

18. $|2y| = 18$

19. $|3x| - 4 = 17$

20. $|6x| + 8 = 32$

21. $7|w| - 3 = 11$

22. $5|x| + 10 = 26$

23. $\left|\frac{2x - 1}{3}\right| = 5$

24. $\left|\frac{4 - 5x}{6}\right| = 7$

25. $|m + 5| + 9 = 16$

26. $|t - 7| - 5 = 4$

27. $10 - |2x - 1| = 4$

28. $2|2x - 7| + 11 = 25$

29. $|3x - 4| = -2$

30. $|x - 6| = -8$

31. $|\frac{5}{9} + 3x| = \frac{1}{6}$

32. $|\frac{2}{3} - 4x| = \frac{4}{5}$

Solve.

33. $|x + 5| > x$

34. $1 - |\frac{1}{4}x + 8| = \frac{3}{4}$

35. $|7x - 2| = x + 4$

36. $|x - 1| = x - 1$

I The Distance Formula and Midpoints

OBJECTIVES

a Use the distance formula to find the distance between two points whose coordinates are known.

b Use the midpoint formula to find the midpoint of a segment when the coordinates of its endpoints are known.

a THE DISTANCE FORMULA

Suppose that two points are on a horizontal line, and thus have the same second coordinate. We can find the distance between them by subtracting their first coordinates. This difference may be negative, depending on the order in which we subtract. So, to make sure that we get a positive number, we take the absolute value of this difference. The distance between two points on a horizontal line (x_1, y_1) and (x_2, y_1) is thus $|x_2 - x_1|$. Similarly, the distance between two points on a vertical line (x_2, y_1) and (x_2, y_2) is $|y_2 - y_1|$.

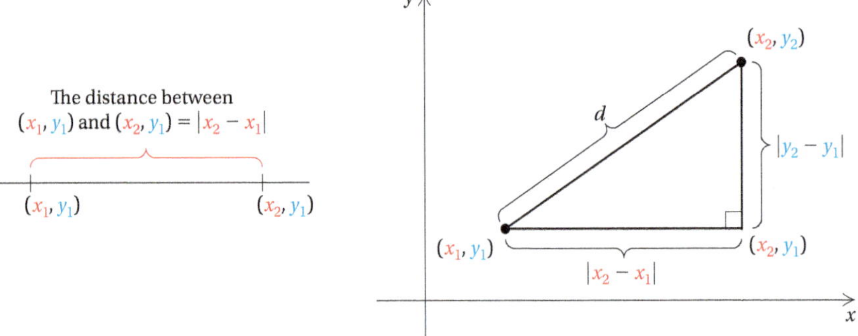

Now consider *any* two points (x_1, y_1) and (x_2, y_2). If $x_1 \neq x_2$ and $y_1 \neq y_2$, these points are vertices of a right triangle, as shown. The other vertex is then (x_2, y_1). The lengths of the legs are $|x_2 - x_1|$ and $|y_2 - y_1|$. We find d, the length of the hypotenuse, by using the Pythagorean equation:

$$d^2 = |x_2 - x_1|^2 + |y_2 - y_1|^2.$$

Since the square of a number is the same as the square of its opposite, we don't need these absolute-value signs. Thus,

$$d^2 = (x_2 - x_1)^2 + (y_2 - y_1)^2.$$

Taking the principal square root, we obtain the formula for the distance between two points.

THE DISTANCE FORMULA

The distance between any two points (x_1, y_1) and (x_2, y_2) is given by

$$d = \sqrt{(x_2 - x_1)^2 + (y_2 - y_1)^2}.$$

This formula holds even when the two points *are* on a vertical line or a horizontal line.

EXAMPLE 1 Find the distance between $(4, -3)$ and $(-5, 4)$. Give an exact answer and an approximation to three decimal places.

We substitute into the distance formula:

$$\begin{aligned} d &= \sqrt{(-5 - 4)^2 + [4 - (-3)]^2} && \text{Substituting} \\ &= \sqrt{(-9)^2 + 7^2} \\ &= \sqrt{130} \approx 11.402. && \text{Using a calculator} \end{aligned}$$

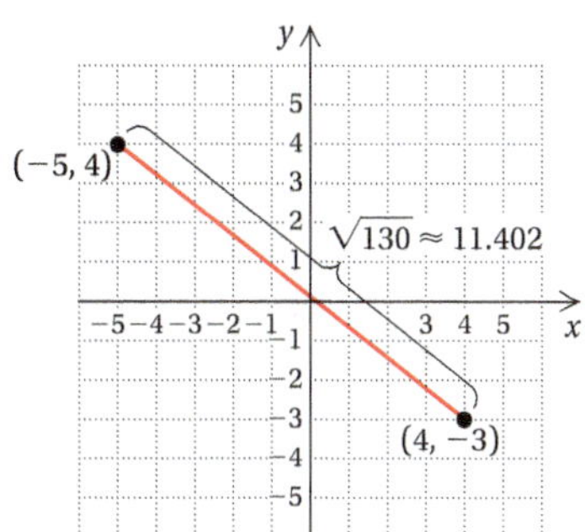

Do Exercises 1 and 2. ▶

Find the distance between each pair of points. Where appropriate, give an approximation to three decimal places.

1. $(2, 6)$ and $(-4, -2)$

2. $(-2, 1)$ and $(4, 2)$

b MIDPOINTS OF SEGMENTS

The distance formula can be used to derive a formula for finding the midpoint of a segment when the coordinates of the endpoints are known.

THE MIDPOINT FORMULA

If the endpoints of a segment are (x_1, y_1) and (x_2, y_2), then the coordinates of the midpoint are

$$\left(\frac{x_1 + x_2}{2}, \frac{y_1 + y_2}{2}\right).$$

(To locate the midpoint, determine the average of the x-coordinates and the average of the y-coordinates.)

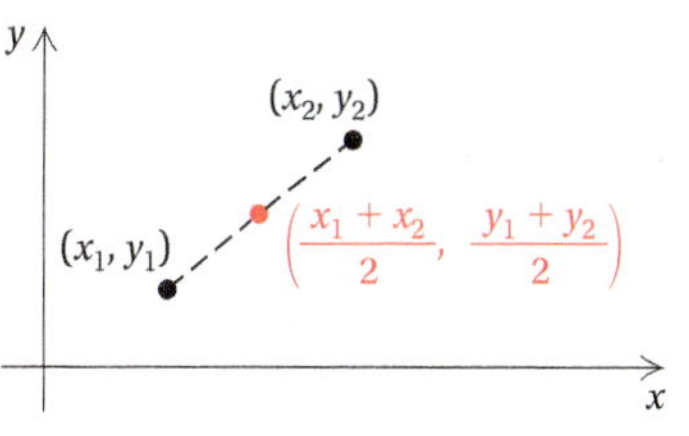

EXAMPLE 2 Find the midpoint of the segment with endpoints $(-2, 3)$ and $(4, -6)$.

Using the midpoint formula, we obtain

$$\left(\frac{-2 + 4}{2}, \frac{3 + (-6)}{2}\right), \text{ or } \left(\frac{2}{2}, \frac{-3}{2}\right), \text{ or } \left(1, -\frac{3}{2}\right).$$

Do Exercises 3 and 4. ▶

Find the midpoint of the segment with the given endpoints.

3. $(-3, 1)$ and $(6, -7)$

4. $(10, -7)$ and $(8, -3)$

Answers

1. 10 2. $\sqrt{37} \approx 6.083$ 3. $\left(\frac{3}{2}, -3\right)$
4. $(9, -5)$

I Exercise Set

For Extra Help MyMathLab® MathXL® PRACTICE WATCH READ REVIEW

a Find the distance between each pair of points. Where appropriate, give an approximation to three decimal places.

1. $(6, -4)$ and $(2, -7)$

2. $(1, 2)$ and $(-4, 14)$

3. $(0, -4)$ and $(5, -6)$

4. $(8, 3)$ and $(8, -3)$

5. $(9, 9)$ and $(-9, -9)$

6. $(2, 22)$ and $(-8, 1)$

7. $(2.8, -3.5)$ and $(-4.3, -3.5)$

8. $(6.1, 2)$ and $(5.6, -4.4)$

9. $\left(\frac{5}{7}, \frac{1}{14}\right)$ and $\left(\frac{1}{7}, \frac{11}{14}\right)$

10. $(0, \sqrt{7})$ and $(\sqrt{6}, 0)$

11. $\left(-5, \frac{3}{4}\right)$ and $\left(-5, -\frac{3}{2}\right)$

12. $\left(-6, \frac{1}{2}\right)$ and $\left(-3, \frac{1}{2}\right)$

13. $(-23, 10)$ and $(56, -17)$

14. $(34, -18)$ and $(-46, -38)$

15. (a, b) and $(0, 0)$

16. $(0, 0)$ and (p, q)

17. $\left(-\frac{2}{5}, 14\right)$ and $(0, 14)$

18. $(0, 7)$ and $(0, -7)$

19. $(\sqrt{2}, -\sqrt{3})$ and $(-\sqrt{7}, \sqrt{5})$

20. $(\sqrt{8}, \sqrt{3})$ and $(-\sqrt{5}, -\sqrt{6})$

21. $(1000, -240)$ and $(-2000, 580)$

22. $(-3000, 560)$ and $(-430, -640)$

 Find the midpoint of the segment with the given endpoints.

23. $(-1, 9)$ and $(4, -2)$

24. $(5, 10)$ and $(2, -4)$

25. $(3, 5)$ and $(-3, 6)$

26. $(7, -3)$ and $(4, 11)$

27. $(-10, -13)$ and $(8, -4)$

28. $(6, -2)$ and $(-5, 12)$

29. $(-3.4, 8.1)$ and $(2.9, -8.7)$

30. $(4.1, 6.9)$ and $(5.2, -6.9)$

31. $\left(\frac{1}{6}, -\frac{3}{4}\right)$ and $\left(-\frac{1}{3}, \frac{5}{6}\right)$

32. $\left(-\frac{4}{5}, -\frac{2}{3}\right)$ and $\left(\frac{1}{8}, \frac{3}{4}\right)$

33. $(\sqrt{2}, -1)$ and $(\sqrt{3}, 4)$

34. $(9, 2\sqrt{3})$ and $(-4, 5\sqrt{3})$

35. $(0, 8)$ and $(-8, -8)$

36. $(-3, 0)$ and $(3, -6)$

Synthesis

Find the distance between the given points.

37. $(-1, 3k)$ and $(6, 2k)$

38. (a, b) and $(-a, -b)$

39. $(6m, -7n)$ and $(-2m, n)$

40. $(-3\sqrt{3}, 1 - \sqrt{6})$ and $(\sqrt{3}, 1 + \sqrt{6})$

If the sides of a triangle have lengths a, b, and c and $a^2 + b^2 = c^2$, then the triangle is a right triangle. Determine whether the given points are vertices of a right triangle.

41. $(-8, -5)$, $(6, 1)$, and $(-4, 5)$

42. $(9, 6)$, $(-1, 2)$, and $(1, -3)$

43. Find the midpoint of the segment with the endpoints $(2 - \sqrt{3}, 5\sqrt{2})$ and $(2 + \sqrt{3}, 3\sqrt{2})$.

44. Find the point on the y-axis that is equidistant from $(2, 10)$ and $(6, 2)$.

J Higher Roots, and Rational Numbers as Exponents

OBJECTIVES

a Find principal square roots and their opposites.

b Simplify radical expressions with perfect-square radicands.

c Find cube roots, simplifying certain expressions.

d Simplify expressions involving odd roots and even roots.

e Write expressions with or without rational exponents, and simplify, if possible.

f Write expressions without negative exponents, and simplify, if possible.

g Use the laws of exponents with rational exponents.

a SQUARE ROOTS AND SQUARE-ROOT FUNCTIONS

When we raise a number to the second power, we say that we have **squared** the number. Sometimes we may need to find the number that was squared. We call this process **finding a square root** of a number.

SQUARE ROOT

The number c is a **square root** of a if $c^2 = a$.

For example:

5 is a *square root* of 25 because $5^2 = 5 \cdot 5 = 25$;

-5 is a *square root* of 25 because $(-5)^2 = (-5)(-5) = 25$.

The number -4 does not have a real-number square root because there is no real number c such that $c^2 = -4$.

PROPERTIES OF SQUARE ROOTS

Every positive real number has two real-number square roots.

The number 0 has just one square root, 0 itself.

Negative numbers do not have real-number square roots.

EXAMPLE 1 Find the two square roots of 64.

The square roots of 64 are 8 and -8 because $8^2 = 64$ and $(-8)^2 = 64$.

Do Exercises 1–3.

Find the square roots.

1. 9 **2.** 36 **3.** 121

PRINCIPAL SQUARE ROOT

The **principal square root** of a nonnegative number is its nonnegative square root. The symbol $\sqrt{a}$ represents the principal square root of a. To name the negative square root of a, we can write $-\sqrt{a}$.

EXAMPLES Simplify.

2. $\sqrt{25} = 5$

Remember: $\sqrt{}$ indicates the principal (nonnegative) square root.

3. $-\sqrt{25} = -5$

4. $\sqrt{\dfrac{81}{64}} = \dfrac{9}{8}$ because $\left(\dfrac{9}{8}\right)^2 = \dfrac{9}{8} \cdot \dfrac{9}{8} = \dfrac{81}{64}$.

Answers

1. $3, -3$ **2.** $6, -6$ **3.** $11, -11$

5. $\sqrt{0.0049} = 0.07$ because $(0.07)^2 = (0.07)(0.07) = 0.0049$.

6. $-\sqrt{0.000001} = -0.001$

7. $\sqrt{0} = 0$

8. $\sqrt{-25}$ Does not exist as a real number. Negative numbers do not have real-number square roots.

Do Exercises 4–13. ▶

Simplify.

4. a) $\sqrt{16}$ b) $-\sqrt{16}$ c) $\sqrt{-16}$

5. a) $\sqrt{49}$ b) $-\sqrt{49}$ c) $\sqrt{-49}$

6. $\sqrt{1}$

7. $-\sqrt{36}$

8. $\sqrt{\frac{81}{100}}$

9. $\sqrt{0.0064}$

10. $-\sqrt{\frac{25}{64}}$

11. $\sqrt{\frac{16}{9}}$

12. $-\sqrt{0.81}$

13. $\sqrt{1.44}$

RADICAL; RADICAL EXPRESSION; RADICAND

The symbol $\sqrt{}$ is called a **radical**.

An expression written with a radical is called a **radical expression**.

The expression written under the radical is called the **radicand**.

These are radical expressions:

$$\sqrt{5}, \quad \sqrt{a}, \quad -\sqrt{5x}, \quad \sqrt{y^2 + 7}.$$

The radicands in these expressions are 5, a, $5x$, and $y^2 + 7$, respectively.

EXAMPLE 9 Identify the radicand in $x\sqrt{x^2 - 9}$.

The radicand is the expression under the radical, $x^2 - 9$.

Do Exercises 14 and 15. ▶

Identify the radicand.

14. $5\sqrt{28 + x}$

15. $\sqrt{\frac{y}{y + 3}}$

b FINDING $\sqrt{a^2}$

In the expression $\sqrt{a^2}$, the radicand is a perfect square. It is tempting to think that $\sqrt{a^2} = a$, but we see below that this is not always the case.

Suppose $a = 5$. Then we have $\sqrt{5^2}$, which is $\sqrt{25}$, or 5.
Suppose $a = -5$. Then we have $\sqrt{(-5)^2}$, which is $\sqrt{25}$, or 5.
Suppose $a = 0$. Then we have $\sqrt{0^2}$, which is $\sqrt{0}$, or 0.

The symbol $\sqrt{a^2}$ never represents a negative number. It represents the principal square root of a^2. Note the following.

SIMPLIFYING $\sqrt{a^2}$

$a \geq 0 \longrightarrow \sqrt{a^2} = a$

If a is positive or 0, the principal square root of a^2 is a.

$a < 0 \longrightarrow \sqrt{a^2} = -a$

If a is negative, the principal square root of a^2 is the opposite of a.

In all cases, the radical expression represents the absolute value of a.

PRINCIPAL SQUARE ROOT OF a^2

For any real number a, $\sqrt{a^2} = |a|$. The principal (nonnegative) square root of a^2 is the absolute value of a.

Answers

4. (a) 4; (b) −4; (c) does not exist as a real number
5. (a) 7; (b) −7; (c) does not exist as a real number
6. 1 7. −6 8. $\frac{9}{10}$ 9. 0.08 10. $-\frac{5}{8}$
11. $\frac{4}{3}$ 12. −0.9 13. 1.2
14. $28 + x$ 15. $\frac{y}{y + 3}$

The absolute value is used to ensure that the principal square root is nonnegative, which is as it is defined.

EXAMPLES Find each of the following. Assume that letters can represent any real number.

10. $\sqrt{(-16)^2} = |-16|$, or 16

11. $\sqrt{(3b)^2} = |3b| = |3| \cdot |b| = 3|b|$

> $|3b|$ can be simplified to $3|b|$ because the absolute value of any product is the product of the absolute values. That is, $|a \cdot b| = |a| \cdot |b|$.

12. $\sqrt{(x-1)^2} = |x-1|$

13. $\sqrt{x^2 + 8x + 16} = \sqrt{(x+4)^2}$
$= |x+4|$

Caution! $|x+4|$ is *not* the same as $|x| + 4$.

Find each of the following. Assume that letters can represent *any* real number.

16. $\sqrt{y^2}$ **17.** $\sqrt{(-24)^2}$

18. $\sqrt{(5y)^2}$ **19.** $\sqrt{16y^2}$

20. $\sqrt{(x+7)^2}$

21. $\sqrt{4(x-2)^2}$

22. $\sqrt{49(y+5)^2}$

23. $\sqrt{x^2 - 6x + 9}$

Do Exercises 16–23.

C CUBE ROOTS

CUBE ROOT

The number c is the **cube root** of a, written $\sqrt[3]{a}$, if the third power of c is a—that is, if $c^3 = a$, then $\sqrt[3]{a} = c$.

For example:

2 is the *cube root* of 8 because $2^3 = 2 \cdot 2 \cdot 2 = 8$;

-4 is the *cube root* of -64 because $(-4)^3 = (-4)(-4)(-4) = -64$.

We talk about *the* cube root of a number rather than *a* cube root because of the following.

> Every real number has exactly one cube root in the system of real numbers. The symbol $\sqrt[3]{a}$ represents *the* cube root of a.

EXAMPLES Find each of the following.

14. $\sqrt[3]{8} = 2$ because $2^3 = 8$.

15. $\sqrt[3]{-27} = -3$

16. $\sqrt[3]{-\frac{216}{125}} = -\frac{6}{5}$

17. $\sqrt[3]{0.001} = 0.1$

18. $\sqrt[3]{x^3} = x$

19. $\sqrt[3]{-8} = -2$

20. $\sqrt[3]{0} = 0$

21. $\sqrt[3]{-8y^3} = \sqrt[3]{(-2y)^3} = -2y$

Answers

16. $|y|$ **17.** 24 **18.** $5|y|$ **19.** $4|y|$
20. $|x+7|$ **21.** $2|x-2|$ **22.** $7|y+5|$
23. $|x-3|$

When we are determining a cube root, no absolute-value signs are needed because a real number has just one cube root. The real-number cube root of a positive number is positive. The real-number cube root of a negative number is negative. The cube root of 0 is 0. That is, $\sqrt[3]{a^3} = a$ whether $a > 0$, $a < 0$, or $a = 0$.

Do Exercises 24–27. ▶

Find each of the following.

24. $\sqrt[3]{-64}$ **25.** $\sqrt[3]{27y^3}$

26. $\sqrt[3]{8(x+2)^3}$ **27.** $\sqrt[3]{-\frac{343}{64}}$

d ODD AND EVEN *k*TH ROOTS

In the expression $\sqrt[k]{a}$, we call k the **index** and assume $k \geq 2$.

Odd Roots

The 5th root of a number a is the number c for which $c^5 = a$. There are also 7th roots, 9th roots, and so on. Whenever the number k in $\sqrt[k]{}$ is an odd number, we say that we are taking an **odd root**.

Every number has just one real-number odd root. If the number is positive, then the root is positive. If the number is negative, then the root is negative. If the number is 0, then the root is 0. For example, $\sqrt[3]{8} = 2$, $\sqrt[3]{-8} = -2$, and $\sqrt[3]{0} = 0$. Absolute-value signs are *not* needed when we are finding odd roots.

> If k is an *odd* natural number, then for any real number a,
>
> $\sqrt[k]{a^k} = a.$

EXAMPLES Find each of the following.

22. $\sqrt[5]{32} = 2$

23. $\sqrt[5]{-32} = -2$

24. $-\sqrt[5]{32} = -2$

25. $-\sqrt[5]{-32} = -(-2) = 2$

26. $\sqrt[7]{x^7} = x$

27. $\sqrt[7]{128} = 2$

28. $\sqrt[7]{-128} = -2$

29. $\sqrt[7]{0} = 0$

30. $\sqrt[5]{a^5} = a$

31. $\sqrt[9]{(x-1)^9} = x - 1$

Do Exercises 28–34. ▶

Find each of the following.

28. $\sqrt[5]{243}$ **29.** $\sqrt[5]{-243}$

30. $\sqrt[5]{x^5}$ **31.** $\sqrt[7]{y^7}$

32. $\sqrt[5]{0}$ **33.** $\sqrt[5]{-32x^5}$

34. $\sqrt[7]{(3x+2)^7}$

Even Roots

When the index k in $\sqrt[k]{}$ is an even number, we say that we are taking an **even root**. When the index is 2, we do not write it. Every positive real number has two real-number kth roots when k is even. One of those roots is positive and one is negative. Negative real numbers do not have real-number kth roots when k is even. When we are finding even kth roots, absolute-value signs are sometimes necessary, as we have seen with square roots. For example,

$$\sqrt{64} = 8, \quad \sqrt[6]{64} = 2, \quad -\sqrt[6]{64} = -2, \quad \sqrt[6]{64x^6} = \sqrt[6]{(2x)^6} = |2x| = 2|x|.$$

Note that in $\sqrt[6]{64x^6}$, we need absolute-value signs because a variable is involved.

Answers

24. -4 **25.** $3y$ **26.** $2(x+2)$ **27.** $-\frac{7}{4}$
28. 3 **29.** -3 **30.** x **31.** y **32.** 0
33. $-2x$ **34.** $3x+2$

EXAMPLES Find each of the following. Assume that variables can represent any real number.

32. $\sqrt[4]{16} = 2$

33. $-\sqrt[4]{16} = -2$

34. $\sqrt[4]{-16}$
Does not exist as a real number.

35. $\sqrt[4]{81x^4} = \sqrt[4]{(3x)^4} = |3x| = 3|x|$

36. $\sqrt[6]{(y+7)^6} = |y+7|$

37. $\sqrt{81y^2} = \sqrt{(9y)^2} = |9y| = 9|y|$

The following is a summary of how absolute value is used when we are taking even roots or odd roots.

SIMPLIFYING

For any real number a:

a) $\sqrt[k]{a^k} = |a|$ when k is an *even* natural number. We use absolute value when k is even unless a is nonnegative.

b) $\sqrt[k]{a^k} = a$ when k is an *odd* natural number greater than 1. We do not use absolute value when k is odd.

Find each of the following. Assume that letters can represent any real number.

35. $\sqrt[4]{81}$

36. $-\sqrt[4]{81}$

37. $\sqrt[4]{-81}$

38. $\sqrt[4]{0}$

39. $\sqrt[4]{16(x-2)^4}$

40. $\sqrt[6]{x^6}$

41. $\sqrt[8]{(x+3)^8}$

42. $\sqrt[7]{(x+3)^7}$

43. $\sqrt[5]{243x^5}$

Do Exercises 35–43.

e RATIONAL EXPONENTS

Expressions like $a^{1/2}$, $5^{-1/4}$, and $(2y)^{4/5}$ have not yet been defined. We will define such expressions so that the general properties of exponents hold.

Consider $a^{1/2} \cdot a^{1/2}$. If we want to multiply by adding exponents, it must follow that $a^{1/2} \cdot a^{1/2} = a^{1/2+1/2}$, or a^1. Thus we should define $a^{1/2}$ to be a square root of a. Similarly, $a^{1/3} \cdot a^{1/3} \cdot a^{1/3} = a^{1/3+1/3+1/3}$, or a^1, so $a^{1/3}$ should be defined to mean $\sqrt[3]{a}$.

$a^{1/n}$

For any *nonnegative* real number a and any natural number index n $(n \neq 1)$,

$$a^{1/n} \quad \text{means} \quad \sqrt[n]{a} \quad \text{(the nonnegative } n\text{th root of } a\text{)}.$$

With rational exponents, we assume that the bases are nonnegative.

EXAMPLES Rewrite without rational exponents, and simplify, if possible.

38. $27^{1/3} = \sqrt[3]{27} = 3$

39. $(abc)^{1/5} = \sqrt[5]{abc}$

40. $x^{1/2} = \sqrt{x}$ An index of 2 is not written.

Rewrite without rational exponents, and simplify, if possible.

44. $y^{1/4}$

45. $(3a)^{1/2}$

46. $16^{1/4}$

47. $(125)^{1/3}$

48. $(a^3b^2c)^{1/5}$

Do Exercises 44–48.

Answers

35. 3 **36.** -3 **37.** Does not exist as a real number **38.** 0 **39.** $2|x-2|$ **40.** $|x|$ **41.** $|x+3|$ **42.** $x+3$ **43.** $3x$ **44.** $\sqrt[4]{y}$ **45.** $\sqrt{3a}$ **46.** 2 **47.** 5 **48.** $\sqrt[5]{a^3b^2c}$

EXAMPLES Rewrite with rational exponents.

41. $\sqrt[5]{7xy} = (7xy)^{1/5}$ We need parentheses around the radicand.

42. $8\sqrt[3]{xy} = 8(xy)^{1/3}$

43. $\sqrt[7]{\frac{x^3y}{9}} = \left(\frac{x^3y}{9}\right)^{1/7}$

Do Exercises 49–52. ▶

Rewrite with rational exponents.

49. $\sqrt[3]{19ab}$

50. $19\sqrt[3]{ab}$

51. $\sqrt[5]{\frac{x^2y}{16}}$

52. $7\sqrt[4]{2ab}$

How should we define $a^{2/3}$? If the general properties of exponents are to hold, we have $a^{2/3} = (a^{1/3})^2$, or $(a^2)^{1/3}$, or $(\sqrt[3]{a})^2$, or $\sqrt[3]{a^2}$. We define this accordingly.

$a^{m/n}$

For any natural numbers m and n ($n \neq 1$) and any nonnegative real number a,

$$a^{m/n} \text{ means } \sqrt[n]{a^m}, \text{ or } (\sqrt[n]{a})^m.$$

EXAMPLES Rewrite without rational exponents, and simplify, if possible.

44. $(27)^{2/3} = \sqrt[3]{27^2} = (\sqrt[3]{27})^2 = 3^2 = 9$

45. $4^{3/2} = \sqrt[2]{4^3} = (\sqrt[2]{4})^3 = 2^3 = 8$

Do Exercises 53–55. ▶

Rewrite without rational exponents, and simplify, if possible.

53. $x^{3/5}$

54. $8^{2/3}$

55. $4^{5/2}$

EXAMPLES Rewrite with rational exponents.

The index becomes the denominator of the rational exponent.

46. $\sqrt[3]{9^4} = 9^{4/3}$

47. $(\sqrt[4]{7xy})^5 = (7xy)^{5/4}$

Do Exercises 56 and 57. ▶

Rewrite with rational exponents.

56. $(\sqrt[3]{7abc})^4$

57. $\sqrt[5]{6^7}$

f NEGATIVE RATIONAL EXPONENTS

Negative rational exponents have a meaning similar to that of negative integer exponents.

$a^{-m/n}$

For any rational number m/n and any positive real number a,

$$a^{-m/n} \text{ means } \frac{1}{a^{m/n}};$$

that is, $a^{m/n}$ and $a^{-m/n}$ are reciprocals.

Answers

49. $(19ab)^{1/3}$ **50.** $19(ab)^{1/3}$ **51.** $\left(\frac{x^2y}{16}\right)^{1/5}$ **52.** $7(2ab)^{1/4}$ **53.** $\sqrt[5]{x^3}$ **54.** 4 **55.** 32 **56.** $(7abc)^{4/3}$ **57.** $6^{7/5}$

Rewrite with positive exponents, and simplify, if possible.

58. $16^{-1/4}$ **59.** $(3xy)^{-7/8}$

60. $81^{-3/4}$ **61.** $7p^{3/4}q^{-6/5}$

62. $\left(\frac{11m}{7n}\right)^{-2/3}$

EXAMPLES Rewrite with positive exponents, and simplify, if possible.

48. $9^{-1/2} = \frac{1}{9^{1/2}} = \frac{1}{\sqrt{9}} = \frac{1}{3}$

49. $(5xy)^{-4/5} = \frac{1}{(5xy)^{4/5}}$

50. $64^{-2/3} = \frac{1}{64^{2/3}} = \frac{1}{(\sqrt[3]{64})^2} = \frac{1}{4^2} = \frac{1}{16}$

51. $4x^{-2/3}y^{1/5} = 4 \cdot \frac{1}{x^{2/3}} \cdot y^{1/5} = \frac{4y^{1/5}}{x^{2/3}}$

52. $\left(\frac{3r}{7s}\right)^{-5/2} = \left(\frac{7s}{3r}\right)^{5/2}$ Since $\left(\frac{a}{b}\right)^{-n} = \left(\frac{b}{a}\right)^{n}$

◀ Do Exercises 58–62.

g LAWS OF EXPONENTS

The same laws hold for rational-number exponents as for integer exponents. We list them for review.

For any real number a and any rational exponents m and n:

1. $a^m \cdot a^n = a^{m+n}$ In multiplying, we add exponents if the bases are the same.
2. $\frac{a^m}{a^n} = a^{m-n}$ In dividing, we subtract exponents if the bases are the same.
3. $(a^m)^n = a^{m \cdot n}$ To raise a power to a power, we multiply the exponents.
4. $(ab)^m = a^m b^m$ To raise a product to a power, we raise each factor to the power.
5. $\left(\frac{a}{b}\right)^n = \frac{a^n}{b^n}$ To raise a quotient to a power, we raise both the numerator and the denominator to the power.

Use the laws of exponents to simplify.

63. $7^{1/3} \cdot 7^{3/5}$

64. $\frac{5^{7/6}}{5^{5/6}}$

65. $(9^{3/5})^{2/3}$

66. $(p^{-2/3}q^{1/4})^{1/2}$

EXAMPLES Use the laws of exponents to simplify.

53. $3^{1/5} \cdot 3^{3/5} = 3^{1/5+3/5} = 3^{4/5}$ Adding exponents

54. $\frac{7^{1/4}}{7^{1/2}} = 7^{1/4-1/2} = 7^{1/4-2/4} = 7^{-1/4} = \frac{1}{7^{1/4}}$ Subtracting exponents

55. $(7.2^{2/3})^{3/4} = 7.2^{2/3 \cdot 3/4} = 7.2^{6/12} = 7.2^{1/2}$ Multiplying exponents

56. $(a^{-1/3}b^{2/5})^{1/2} = a^{-1/3 \cdot 1/2} \cdot b^{2/5 \cdot 1/2}$ Raising a product to a power and multiplying exponents

$= a^{-1/6}b^{1/5} = \frac{b^{1/5}}{a^{1/6}}$

◀ Do Exercises 63–66.

Answers

58. $\frac{1}{2}$ 59. $\frac{1}{(3xy)^{7/8}}$ 60. $\frac{1}{27}$ 61. $\frac{7p^{3/4}}{q^{6/5}}$
62. $\left(\frac{7n}{11m}\right)^{2/3}$ 63. $7^{14/15}$ 64. $5^{1/3}$ 65. $9^{2/5}$
66. $\frac{q^{1/8}}{p^{1/3}}$

Exercise Set

For Extra Help
MyMathLab® MathXL® PRACTICE WATCH READ REVIEW

a Find the square roots.

1. 16 **2.** 225 **3.** 144 **4.** 9 **5.** 400 **6.** 81

Simplify.

7. $-\sqrt{\dfrac{49}{36}}$ **8.** $-\sqrt{\dfrac{361}{9}}$ **9.** $\sqrt{196}$ **10.** $\sqrt{441}$

11. $\sqrt{0.0036}$ **12.** $\sqrt{0.04}$ **13.** $\sqrt{-225}$ **14.** $\sqrt{-64}$

Use a calculator to approximate to three decimal places.

15. $\sqrt{347}$ **16.** $-\sqrt{1839.2}$ **17.** $\sqrt{\dfrac{285}{74}}$ **18.** $\sqrt{\dfrac{839.4}{19.7}}$

Identify the radicand.

19. $9\sqrt{y^2 + 16}$ **20.** $-3\sqrt{p^2 - 10}$ **21.** $x^4y^5\sqrt{\dfrac{x}{y - 1}}$ **22.** $a^2b^2\sqrt{\dfrac{a^2 - b}{b}}$

b Find each of the following. Assume that letters can represent *any* real number.

23. $\sqrt{16x^2}$ **24.** $\sqrt{25t^2}$ **25.** $\sqrt{(-12c)^2}$ **26.** $\sqrt{(-9d)^2}$

27. $\sqrt{(p + 3)^2}$ **28.** $\sqrt{(2 - x)^2}$ **29.** $\sqrt{x^2 - 4x + 4}$ **30.** $\sqrt{9t^2 - 30t + 25}$

c Simplify.

31. $\sqrt[3]{27}$ **32.** $-\sqrt[3]{64}$ **33.** $\sqrt[3]{-64x^3}$ **34.** $\sqrt[3]{-125y^3}$

35. $\sqrt[3]{-216}$ **36.** $-\sqrt[3]{-1000}$ **37.** $\sqrt[3]{0.343(x + 1)^3}$ **38.** $\sqrt[3]{0.000008(y - 2)^3}$

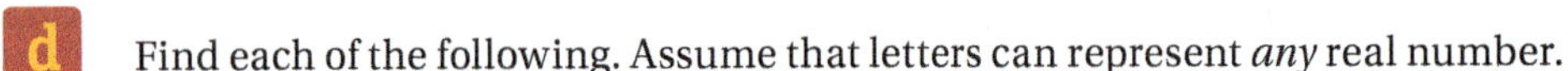

d Find each of the following. Assume that letters can represent *any* real number.

39. $-\sqrt[4]{625}$ **40.** $-\sqrt[4]{256}$ **41.** $\sqrt[5]{-\frac{32}{243}}$ **42.** $\sqrt[5]{-\frac{1}{32}}$

43. $\sqrt[6]{x^6}$ **44.** $\sqrt[8]{y^8}$ **45.** $\sqrt[4]{(5a)^4}$ **46.** $\sqrt[4]{(7b)^4}$

47. $\sqrt[10]{(-6)^{10}}$ **48.** $\sqrt[12]{(-10)^{12}}$ **49.** $\sqrt[414]{(a+b)^{414}}$ **50.** $\sqrt[1999]{(2a+b)^{1999}}$

51. $\sqrt[7]{y^7}$ **52.** $\sqrt[3]{(-6)^3}$ **53.** $\sqrt[5]{(x-2)^5}$ **54.** $\sqrt[9]{(2xy)^9}$

e Rewrite without rational exponents, and simplify, if possible.

55. $y^{1/7}$ **56.** $x^{1/6}$ **57.** $8^{1/3}$ **58.** $16^{1/2}$ **59.** $(a^3b^3)^{1/5}$

60. $(x^2y^2)^{1/3}$ **61.** $16^{3/4}$ **62.** $4^{7/2}$ **63.** $49^{3/2}$ **64.** $27^{4/3}$

Rewrite with rational exponents.

65. $\sqrt{17}$ **66.** $\sqrt{x^3}$ **67.** $\sqrt[3]{18}$ **68.** $\sqrt[3]{23}$ **69.** $\sqrt[5]{xy^2z}$

70. $\sqrt[7]{x^3y^2z^2}$ **71.** $(\sqrt{3mn})^3$ **72.** $(\sqrt[3]{7xy})^4$ **73.** $(\sqrt[7]{8x^2y})^5$ **74.** $(\sqrt[6]{2a^5b})^7$

f Rewrite with positive exponents, and simplify, if possible.

75. $100^{-3/2}$ **76.** $16^{-3/4}$ **77.** $3x^{-1/4}$ **78.** $\frac{1}{a^{-7/8}}$ **79.** $(2rs)^{-3/4}$

80. $5x^{-2/3}y^{4/5}z$ **81.** $\left(\frac{7x}{8yz}\right)^{-3/5}$ **82.** $\left(\frac{2ab}{3c}\right)^{-5/6}$ **83.** $\frac{5a}{3c^{-1/2}}$ **84.** $\frac{2z}{5x^{-1/3}}$

g Use the laws of exponents to simplify. Write the answers with positive exponents.

85. $5^{3/4} \cdot 5^{1/8}$ **86.** $11^{2/3} \cdot 11^{1/2}$ **87.** $\frac{7^{5/8}}{7^{3/8}}$ **88.** $\frac{3^{5/8}}{3^{-1/8}}$ **89.** $\frac{4.9^{-1/6}}{4.9^{-2/3}}$

90. $\frac{2.3^{-3/10}}{2.3^{-1/5}}$ **91.** $(6^{3/8})^{2/7}$ **92.** $(3^{2/9})^{3/5}$ **93.** $a^{2/3} \cdot a^{5/4}$ **94.** $x^{3/4} \cdot x^{2/3}$

Inequalities and Interval Notation

K

OBJECTIVES

a Write interval notation for the solution set or the graph of an inequality.

b Solve an inequality expressing the solution in interval notation and then graph the inequality.

a INEQUALITIES AND INTERVAL NOTATION

The **graph** of an inequality is a drawing that represents its solutions. An inequality in one variable can be graphed on the number line.

EXAMPLE 1 Graph $x < 4$ on the number line.

The solutions are all real numbers less than 4, so we shade all numbers less than 4 on the number line. To indicate that 4 is not a solution, we use a right parenthesis ")" at 4.

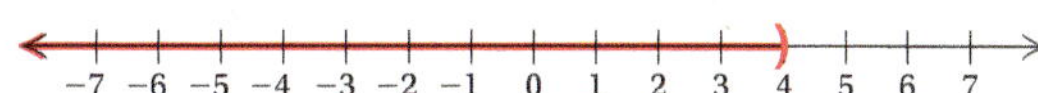

We can write the solution set for $x < 4$ using **set-builder notation**: $\{x \mid x < 4\}$. This is read "The set of all x such that x is less than 4."

Another way to write solutions of an inequality in one variable is to use **interval notation**. Interval notation uses parentheses ( ) and brackets [ ].

If a and b are real numbers such that $a < b$, we define the interval (a, b) as the set of all numbers between but not including a and b—that is, the set of all x for which $a < x < b$. Thus,

$$(a, b) = \{x \mid a < x < b\}.$$

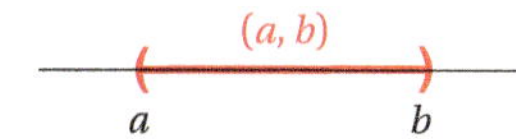

The points a and b are the **endpoints** of the interval. The parentheses indicate that the endpoints are *not* included in the graph.

The interval $[a, b]$ is defined as the set of all numbers x for which $a \le x \le b$. Thus,

$$[a, b] = \{x \mid a \le x \le b\}.$$

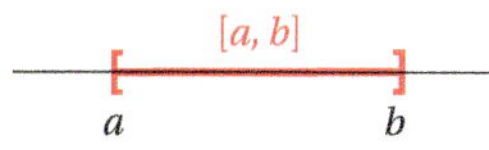

The brackets indicate that the endpoints *are* included in the graph.

The following intervals include one endpoint and exclude the other:

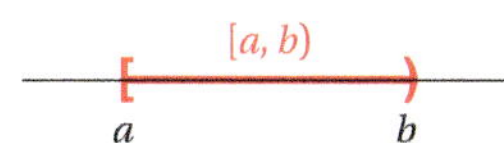

$(a, b] = \{x \mid a < x \le b\}$. The graph excludes a and includes b.

[a, b)

a b

$[a, b) = \{x \mid a \le x < b\}$. The graph includes a and excludes b.

Some intervals extend without bound in one or both directions. We use the symbols ∞, read "infinity," and $-\infty$, read "negative infinity," to name these intervals. The notation (a, ∞) represents the set of all numbers greater than a—that is,

$$(a, \infty) = \{x \mid x > a\}.$$

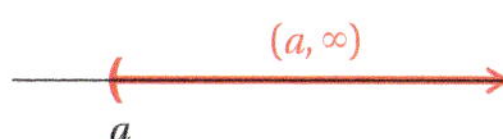

Caution!

Do not confuse the *interval* (a, b) with the *ordered pair* (a, b), which denotes a point in the plane. The context in which the notation appears usually makes the meaning clear.

Similarly, the notation $(-\infty, a)$ represents the set of all numbers less than a—that is,

$$(-\infty, a) = \{x \mid x < a\}.$$

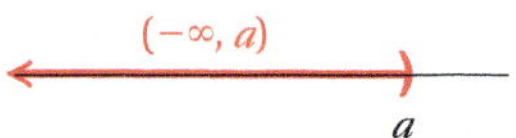

The notations $[a, \infty)$ and $(-\infty, a]$ are used when we want to include the endpoint a. The interval $(-\infty, \infty)$ names the set of all real numbers.

$(-\infty, \infty) = \{x \mid x \text{ is a real number}\}$

Interval notation is summarized in the following table.

Intervals: Notation and Graphs

INTERVAL NOTATION	SET NOTATION	GRAPH
(a, b)	$\{x \mid a < x < b\}$	a b
$[a, b]$	$\{x \mid a \leq x \leq b\}$	a b
$[a, b)$	$\{x \mid a \leq x < b\}$	a b
$(a, b]$	$\{x \mid a < x \leq b\}$	a b
(a, ∞)	$\{x \mid x > a\}$	a
$[a, \infty)$	$\{x \mid x \geq a\}$	a
$(-\infty, b)$	$\{x \mid x < b\}$	b
$(-\infty, b]$	$\{x \mid x \leq b\}$	b
$(-\infty, \infty)$	$\{x \mid x \text{ is a real number}\}$	

Caution!

Whenever the symbol ∞ is included in interval notation, a right parenthesis ")" is used. Similarly, when $-\infty$ is included, a left parenthesis "(" is used.

EXAMPLES Write interval notation for the given set or graph.

2. $\{x \mid -4 < x < 5\} = (-4, 5)$

3. $\{x \mid x \geq -2\} = [-2, \infty)$

4. $\{x \mid 7 > x \geq 1\} = \{x \mid 1 \leq x < 7\} = [1, 7)$

5. −6 −5 −4 −3 −2 −1 0 1 2 3 4 5 6

$(-2, 4]$

6.

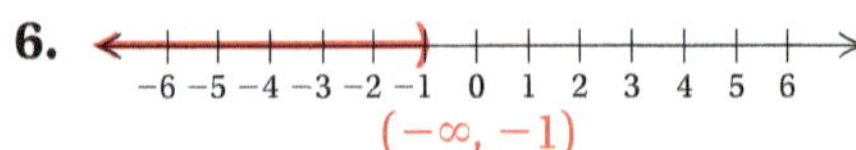

$(-\infty, -1)$

Do Exercises 1–5.

Write interval notation for the given set or graph.

1. $\{x \mid -4 \leq x < 5\}$

2. $\{x \mid x \leq -2\}$

3. $\{x \mid 6 \geq x > 2\}$

4. −40 −30 −20 −10 0 10 20 30 40

5. −40 −30 −20 −10 0 10 20 30 40

Answers

1. $[-4, 5)$ **2.** $(-\infty, -2]$ **3.** $(2, 6]$
4. $[10, \infty)$ **5.** $[-30, 30]$

b SOLVING INEQUALITIES

We now express the solution set in both set-builder notation and interval notation.

EXAMPLE 7 Solve and graph: $x + 5 > 1$.

We have

$$x + 5 > 1$$

$$x + 5 - 5 > 1 - 5 \qquad \text{Using the addition principle: adding } -5 \text{ or subtracting } 5$$

$$x > -4.$$

The solution set is $\{x \mid x > -4\}$, or $(-4, \infty)$. The graph is as follows:

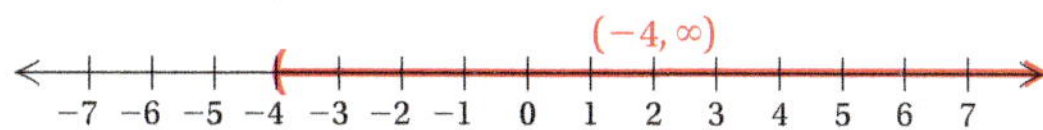

Do Exercises 6 and 7. ▶

Solve and graph.

6. $x + 6 > 9$

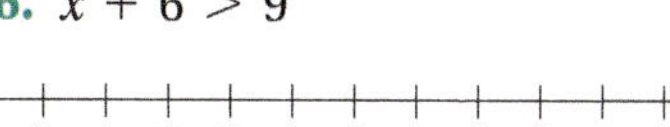

7. $x + 4 \leq 7$

EXAMPLE 8 Solve and graph: $4x - 1 \geq 5x - 2$.

We have

$$4x - 1 \geq 5x - 2$$

$$4x - 1 + 2 \geq 5x - 2 + 2 \qquad \text{Adding 2}$$

$$4x + 1 \geq 5x \qquad \text{Simplifying}$$

$$4x + 1 - 4x \geq 5x - 4x \qquad \text{Subtracting } 4x$$

$$1 \geq x, \text{ or } x \leq 1. \qquad \text{Simplifying}$$

The solution set is $\{x \mid x \geq 1\}$, or $(-\infty, 1]$.

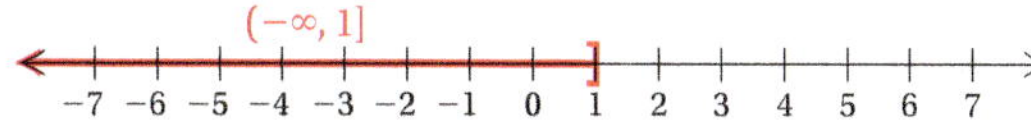

EXAMPLE 9 Solve: $16 - 7y \geq 10y - 4$.

We have

$$16 - 7y \geq 10y - 4$$

$$-16 + 16 - 7y \geq -16 + 10y - 4 \qquad \text{Adding } -16$$

$$-7y \geq 10y - 20 \qquad \text{Collecting like terms}$$

$$-10y + (-7y) \geq -10y + 10y - 20 \qquad \text{Adding } -10y$$

$$-17y \geq -20 \qquad \text{Collecting like terms}$$

$$\frac{-17y}{-17} \leq \frac{-20}{-17} \qquad \text{Dividing by } -17. \text{ The symbol must be reversed.}$$

$$y \leq \frac{20}{17}. \qquad \text{Simplifying}$$

The solution set is $\left\{y \middle| y \leq \frac{20}{17}\right\}$, or $\left(-\infty, \frac{20}{17}\right]$.

Do Exercises 8 and 9. ▶

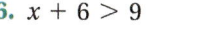

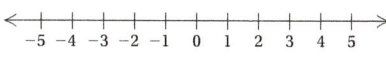

8. Solve and graph:

$2x - 3 \geq 3x - 1$.

9. Solve: $3 - 9x < 2x + 8$.

Answers

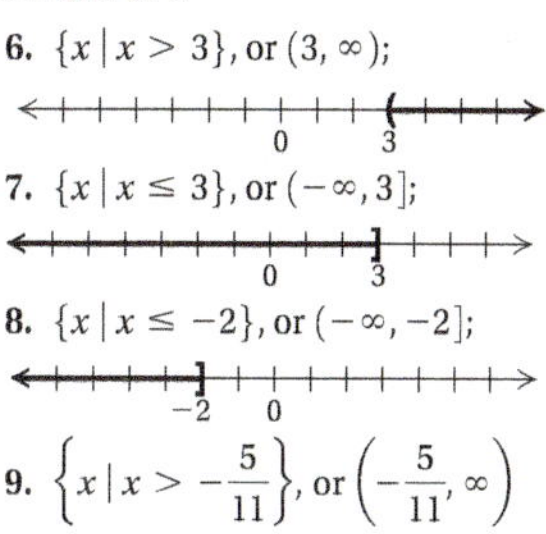

6. $\{x \mid x > 3\}$, or $(3, \infty)$;

7. $\{x \mid x \leq 3\}$, or $(-\infty, 3]$;

8. $\{x \mid x \leq -2\}$, or $(-\infty, -2]$;

9. $\left\{x \middle| x > -\frac{5}{11}\right\}$, or $\left(-\frac{5}{11}, \infty\right)$

K Exercise Set

For Extra Help

MyMathLab®

a Write interval notation for the given set or graph.

1. $\{x \mid x < 5\}$

2. $\{t \mid t \geq -5\}$

3. $\{x \mid -3 \leq x \leq 3\}$

4. $\{t \mid -10 < t \leq 10\}$

5. $\{x \mid -4 > x > -8\}$

6. $\{x \mid 13 > x \geq 5\}$

7. −6 −5 −4 −3 −2 −1 0 1 2 3 4 5 6

8.

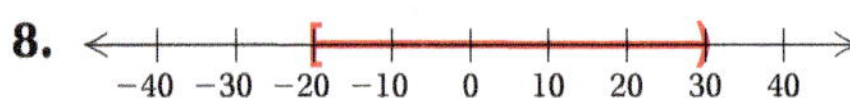

9.

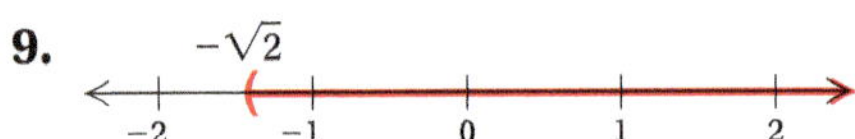

10.

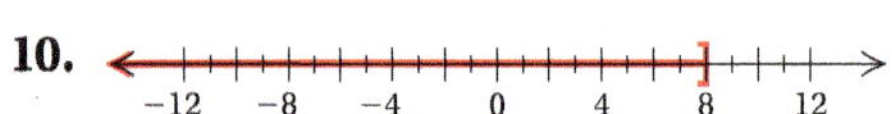

b Solve and graph.

11. $x + 2 > 1$

−6 −5 −4 −3 −2 −1 0 1 2 3 4 5 6

12. $y + 4 < 10$

−6 −5 −4 −3 −2 −1 0 1 2 3 4 5 6

13. $\frac{2}{3}x > 2$

−6 −5 −4 −3 −2 −1 0 1 2 3 4 5 6

14. $0.6x < 30$

−80 −60 −40 −20 0 20 40 60 80

15. $0.3x < -18$

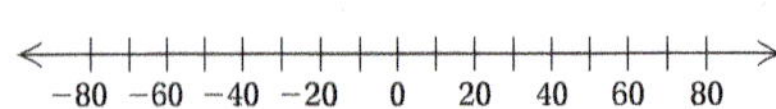

16. $8x \geq 24$

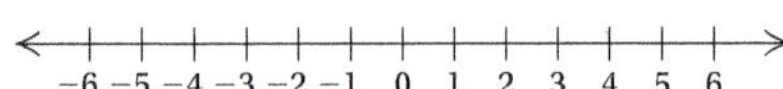

17. $a - 9 \leq -31$

−40 −30 −20 −10 0 10 20 30 40

18. $y - 9 > -18$

−12 −8 −4 0 4 8 12

Solve.

19. $-9x \geq -8.1$

20. $-5y \leq 3.5$

21. $-\frac{3}{4}x \geq -\frac{5}{8}$

22. $-\frac{1}{8}y \leq -\frac{9}{8}$

23. $2x + 7 < 19$

24. $5y + 13 > 28$

25. $5y + 2y \leq -21$

26. $-9x + 3x \geq -24$

27. $2y - 7 < 5y - 9$

28. $8x - 9 < 3x - 11$

Nonlinear Inequalities

L

OBJECTIVES

a Solve quadratic inequalities and other polynomial inequalities.

b Graph quadratic inequalities.

a SOLVING POLYNOMIAL INEQUALITIES

Inequalities like the following are called **quadratic inequalities**:

$$x^2 + 3x - 10 < 0, \qquad 5x^2 - 3x + 2 \geq 0.$$

In each case, we have a polynomial of degree 2 on the left. We can solve a quadratic inequality, such as $ax^2 + bx + c > 0$, by considering the graph of a related equation, $y = ax^2 + bx + c$.

EXAMPLE 1 Solve: $x^2 + 3x - 10 > 0$.

Consider the equation $y = x^2 + 3x - 10$ and its graph. The graph opens up since the leading coefficient $(a = 1)$ is positive. We find the x-intercepts by setting the polynomial equal to 0 and solving:

$$x^2 + 3x - 10 = 0$$
$$(x + 5)(x - 2) = 0$$
$$x + 5 = 0 \quad or \quad x - 2 = 0$$
$$x = -5 \quad or \quad x = 2.$$

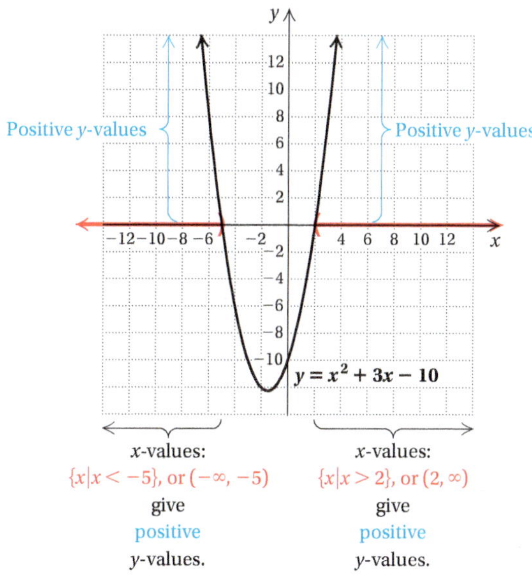

Values of y will be positive to the left and right of the intercepts, as shown. Thus the solution set of the inequality is

$$\{x \mid x < -5 \text{ or } x > 2\}, \quad \text{or} \quad (-\infty, -5) \cup (2, \infty).$$

Do Exercise 1.

We can solve any inequality by considering the graph of a related equation and finding x-intercepts, as in Example 1. In some cases, we may need to use the quadratic formula to find the intercepts.

1. Solve by graphing:
$x^2 + 2x - 3 > 0$.

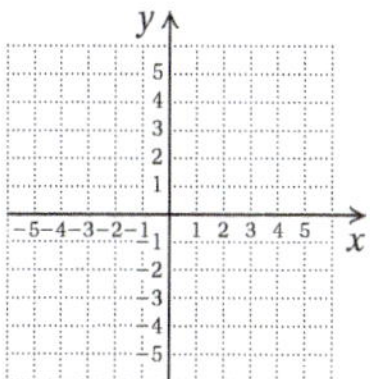

Answer

1. $\{x \mid x < -3 \text{ or } x > 1\}$, or $(-\infty, -3) \cup (1, \infty)$

EXAMPLE 2 Solve: $x^2 + 3x - 10 < 0$.

Looking again at the graph of $y = x^2 + 3x - 10$ or at least visualizing it tells us that y-values are negative for those x-values between -5 and 2.

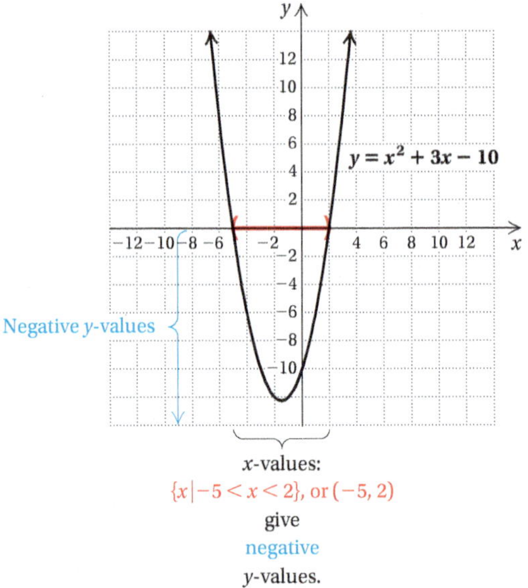

The solution set is $\{x|-5 < x < 2\}$, or $(-5, 2)$.

When an inequality contains $\leq$ or $\geq$, the x-values of the x-intercepts must be included. Thus the solution set of the inequality $x^2 + 3x - 10 \leq 0$ is $\{x|-5 \leq x \leq 2\}$, or $[-5, 2]$.

Do Exercises 2 and 3.

Solve by graphing.

2. $x^2 + 2x - 3 < 0$

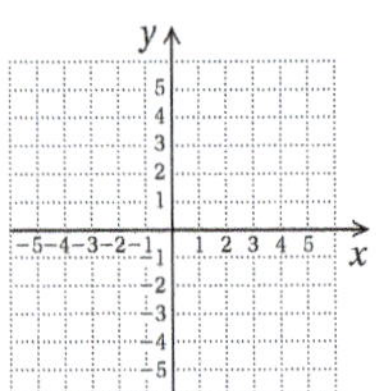

3. $x^2 + 2x - 3 \leq 0$

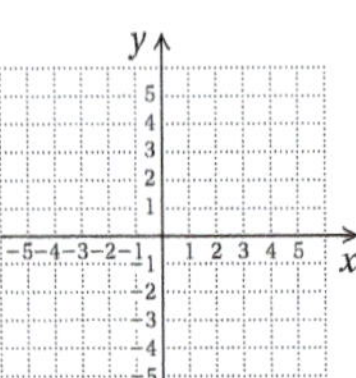

In Examples 1 and 2, we see that the x-intercepts divide the number line into intervals. If a particular equation has a positive output for one number in an interval, it will be positive for all the numbers in the interval. The same is true for negative outputs. Thus we can merely make a test substitution in each interval to solve the inequality. This is very similar to our method of using test points to graph a linear inequality in a plane.

EXAMPLE 3 Solve: $x^2 + 3x - 10 < 0$.

We set the polynomial equal to 0 and solve. The solutions of $x^2 + 3x - 10 = 0$, or $(x + 5)(x - 2) = 0$, are -5 and 2. We locate the solutions on the number line as follows. Note that the numbers divide the number line into three intervals, which we will call A, B, and C.

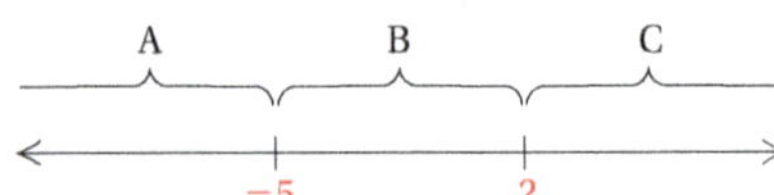

We choose a test number in interval A, say -7, and substitute -7 for x in $y = x^2 + 3x - 10$:

$$y = (-7)^2 + 3(-7) - 10 = 49 - 21 - 10 = 18.$$

Note that since $18 > 0$, the y-values will be positive for any number in interval A.

Next, we try a test number in interval B, say 1, and find the corresponding y-value:

$$y = 1^2 + 3(1) - 10 = 1 + 3 - 10 = -6.$$

Note that since $-6 < 0$, the y-values will be negative for any number in interval B.

Answers

2. $\{x|-3 < x < 1\}$, or $(-3, 1)$
3. $\{x|-3 \leq x \leq 1\}$, or $[-3, 1]$

Next, we try a test number in interval C, say 4, and find the corresponding function value:

$$y = 4^2 + 3(4) - 10 = 16 + 12 - 10 = 18.$$

Note that since $18 > 0$, the y-values will be positive for any number in interval C.

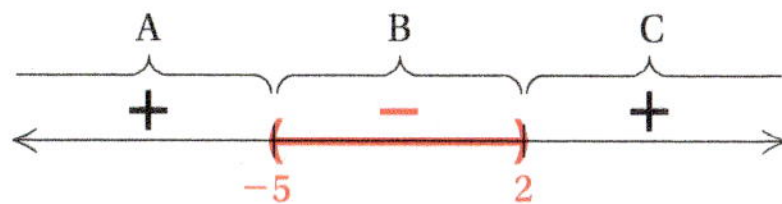

We are looking for numbers x for which $f(x) = x^2 + 3x - 10 < 0$. Thus any number x in interval B is a solution. The solution set is $\{x \mid -5 < x < 2\}$, or the interval $(-5, 2)$. If the inequality had been $\leq$, it would have been necessary to include the endpoints -5 and 2 in the solution set as well.

Do Exercises 4 and 5. ▶

Solve using the method of Example 3.

4. $x^2 + 3x > 4$

5. $x^2 + 3x \leq 4$

EXAMPLE 4 Solve: $5x(x + 3)(x - 2) \geq 0$.

The solutions of $5x(x + 3)(x - 2) = 0$ are 0, -3, and 2. They divide the number line into four intervals, as shown below.

We try test numbers in each interval:

A: Test -5, $\quad y = 5(-5)(-5 + 3)(-5 - 2) = -350 < 0.$

B: Test -2, $\quad y = 5(-2)(-2 + 3)(-2 - 2) = 40 > 0.$

C: Test 1, $\quad y = 5(1)(1 + 3)(1 - 2) = -20 < 0.$

D: Test 3, $\quad y = 5(3)(3 + 3)(3 - 2) = 90 > 0.$

The expression is positive for values of x in intervals B and D. Since the inequality symbol is $\geq$, we need to include the x-intercepts. The solution set of the inequality is

$$\{x \mid -3 \leq x \leq 0 \text{ or } x \geq 2\}, \text{ or } [-3, 0] \cup [2, \infty).$$

Do Exercise 6. ▶

6. Solve: $6x(x + 1)(x - 1) < 0$.

b GRAPHING QUADRATIC INEQUALITIES

Graphing quadratic inequalities involves the same procedure as graphing quadratic equations, but the graph will also include the interior region enclosed by the parabola or the exterior region outside the parabola.

Answers

4. $\{x \mid x < -4 \text{ or } x > 1\}$, or $(-\infty, -4) \cup (1, \infty)$
5. $\{x \mid -4 \leq x \leq 1\}$, or $[-4, 1]$
6. $\{x \mid x < -1 \text{ or } 0 < x < 1\}$, or $(-\infty, -1) \cup (0, 1)$

EXAMPLE 5 Graph: $y \leq x^2 + 2x - 3$.

We first replace the inequality symbol with an equals sign and graph the equation:

$$y = x^2 + 2x - 3.$$

The x-coordinate of the vertex is

$$-\frac{b}{2a} = -\frac{2}{2 \cdot 1} = -1.$$

We substitute -1 for x in the equation to find the second coordinate of the vertex:

$$y = x^2 + 2x - 3 = (-1)^2 + 2(-1) - 3 = 1 - 2 - 3 = -4.$$

The vertex is $(-1, -4)$. The line of symmetry is $x = -1$. We choose some x-values on both sides of the vertex and graph the parabola.

For $x = 0$, $\quad y = x^2 + 2x - 3 = 0^2 + 2 \cdot 0 - 3 = -3$.
For $x = -2$, $\quad y = x^2 + 2x - 3 = (-2)^2 + 2(-2) - 3 = -3$.
For $x = 1$, $\quad y = x^2 + 2x - 3 = 1^2 + 2 \cdot 1 - 3 = 0$.
For $x = -3$, $\quad y = x^2 + 2x - 3 = (-3)^2 + 2(-3) - 3 = 0$.

x	y
-1	-4
0	-3
-2	-3
1	0
-3	0

← This is the vertex.

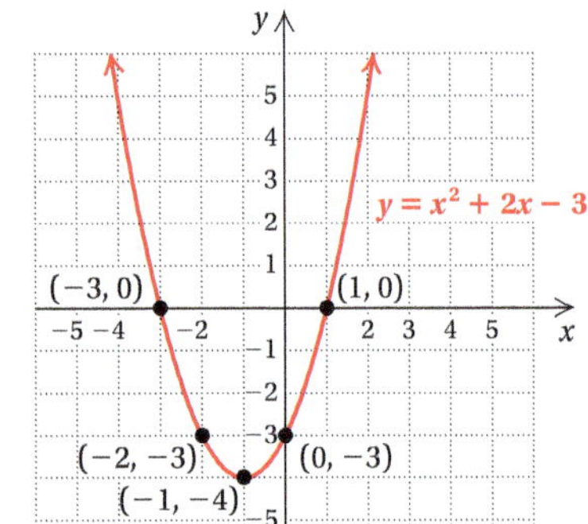

The inequality symbol is $\leq$, so we draw the curve solid.

To determine which region to shade, we choose a point not on the curve as a test point. The origin $(0, 0)$ is usually an easy one to use.

$$\begin{array}{c|l} y & \leq x^2 + 2x - 3 \\ \hline 0 & ?\ 0^2 + 2 \cdot 0 - 3 \\ & \ \ 0 + 0 - 3 \\ & \ \ -3 \qquad \text{FALSE} \end{array}$$

We see that $(0, 0)$ is *not* a solution, so we shade the exterior region. Had the substitution given us a true inequality, we would have shaded the interior.

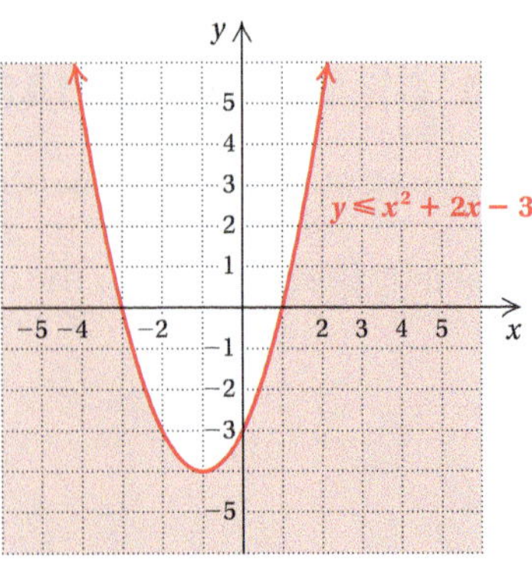

◀ **Do Exercises 7 and 8.**

7. Graph: $y \geq x^2 - 2x - 3$.

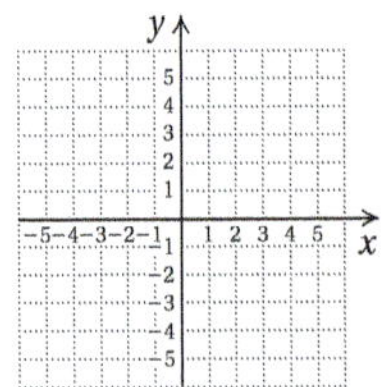

8. Graph: $y > -x^2 + 2x$.

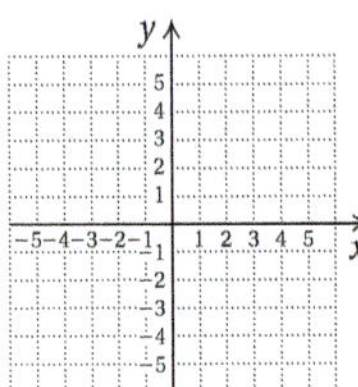

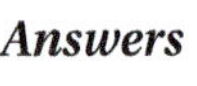

7.
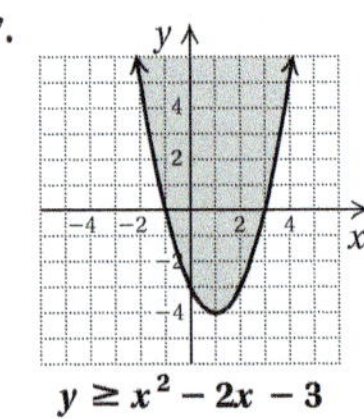

8.
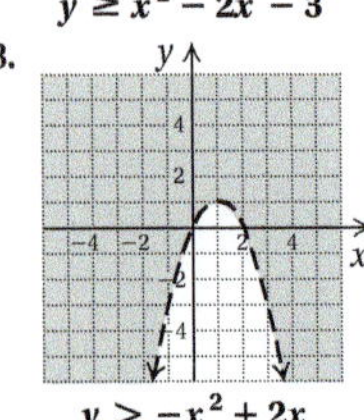

L Exercise Set

For Extra Help MyMathLab® MathXL® PRACTICE WATCH READ REVIEW

a Solve algebraically and verify results from the graph.

1. $(x-6)(x+2)>0$

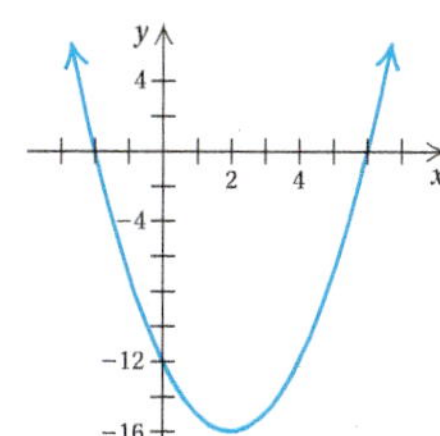

2. $(x-5)(x+1)>0$

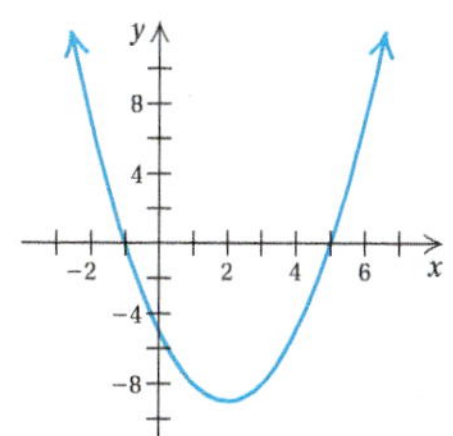

3. $4-x^2\geq 0$

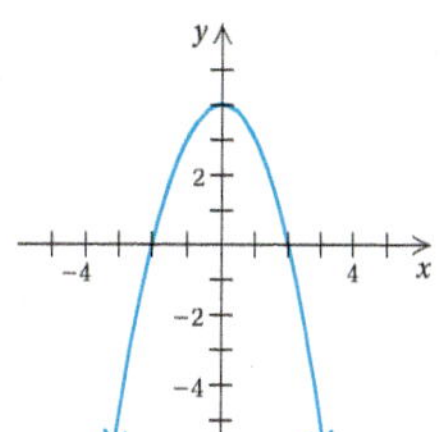

4. $9-x^2\leq 0$

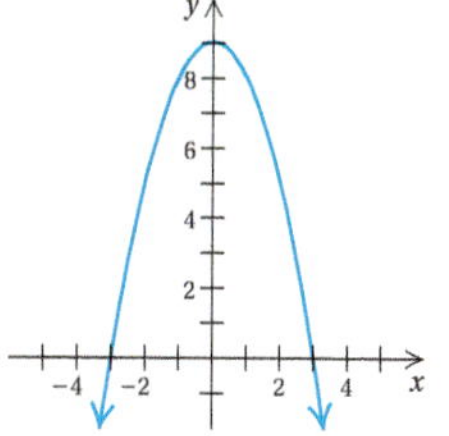

Solve.

5. $3(x+1)(x-4)\leq 0$

6. $(x-7)(x+3)\leq 0$

7. $x^2-x-2<0$

8. $x^2+x-2<0$

9. $x^2-2x+1\geq 0$

10. $x^2+6x+9<0$

11. $x^2+8<6x$

12. $x^2-12>4x$

13. $3x(x+2)(x-2)<0$

14. $5x(x+1)(x-1)>0$

15. $(x+9)(x-4)(x+1)>0$

16. $(x-1)(x+8)(x-2)<0$

17. $(x+3)(x+2)(x-1)<0$

18. $(x-2)(x-3)(x+1)<0$

b Graph.

19. $y \le 3x^2$

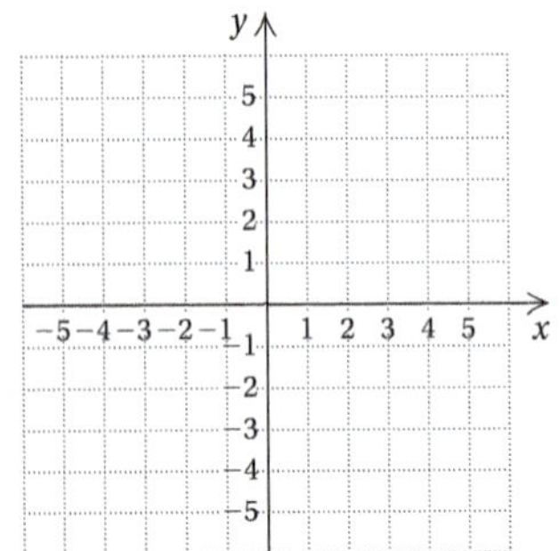

20. $y > -x^2$

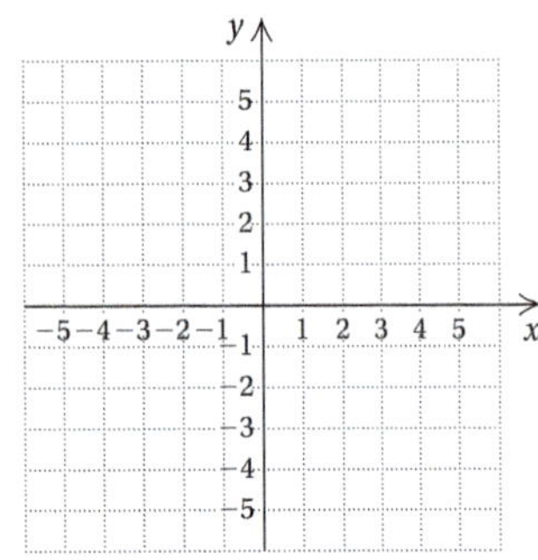

21. $y < 4 - x^2$

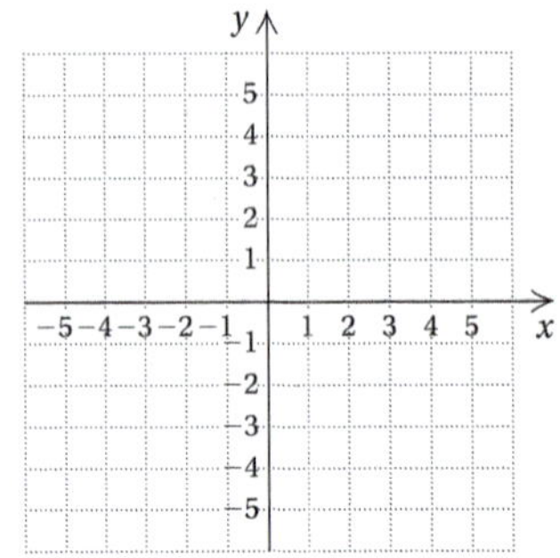

22. $y \le 2 - x^2$

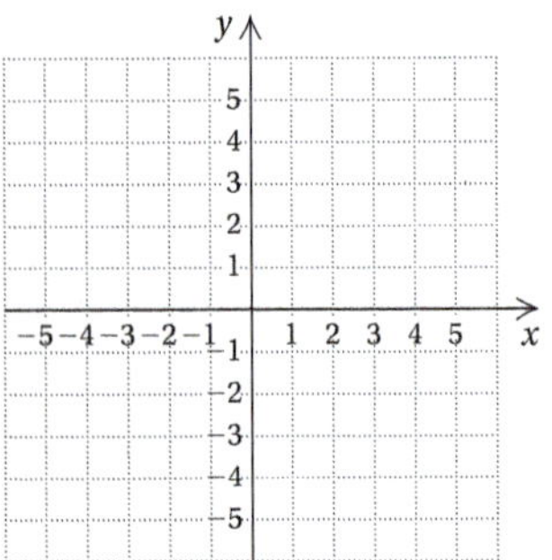

23. $y \le \frac{1}{2}x^2 - 1$

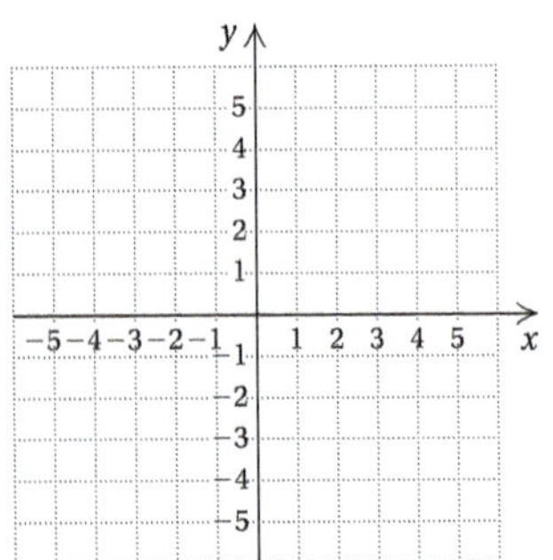

24. $y < -\frac{1}{2}x^2 + 2$

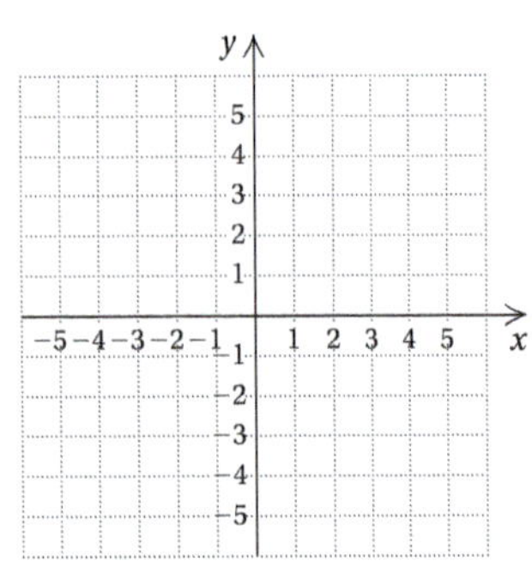

25. $y > x^2 + 4x - 1$

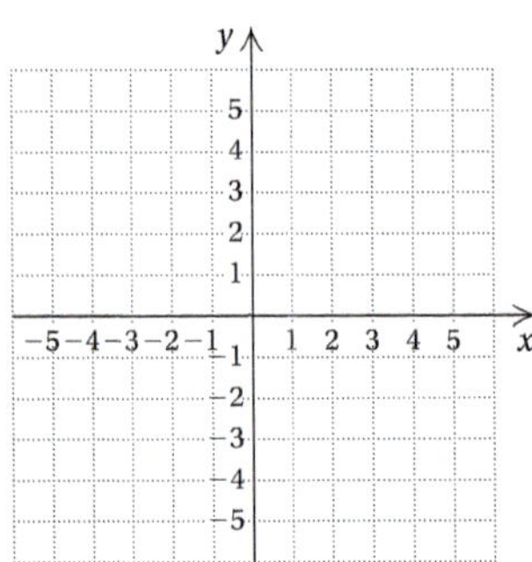

26. $y \le x^2 - 2x - 3$

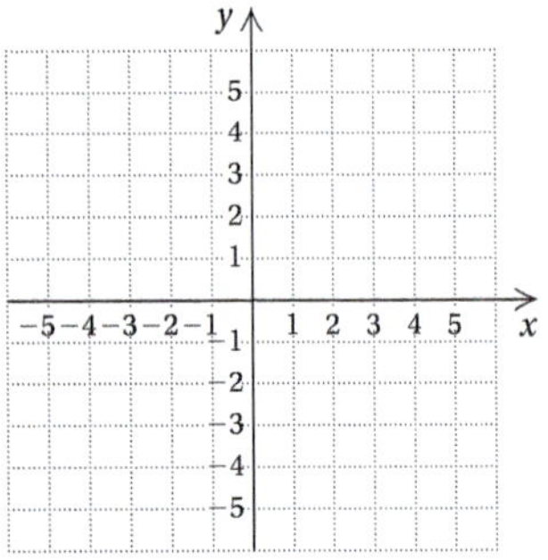

27. $y \ge x^2 + 2x + 1$

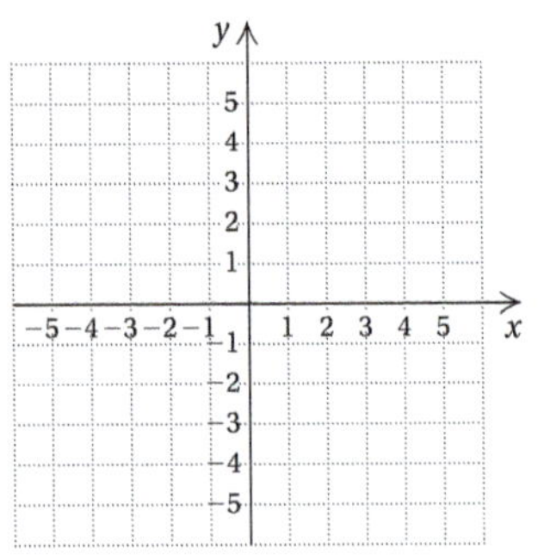

28. $y > x^2 + x - 6$

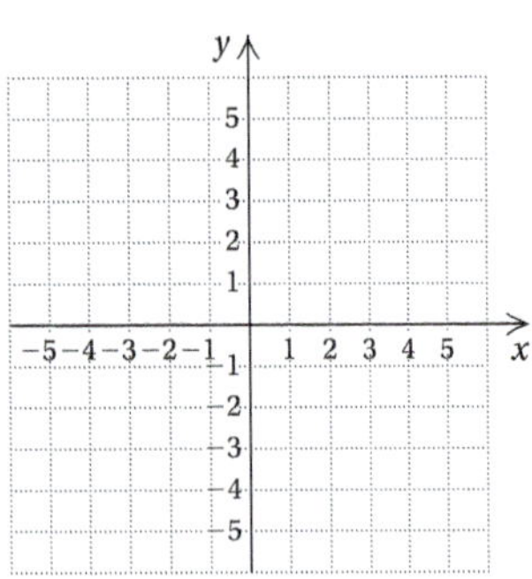

29. $y < 5 - x - x^2$

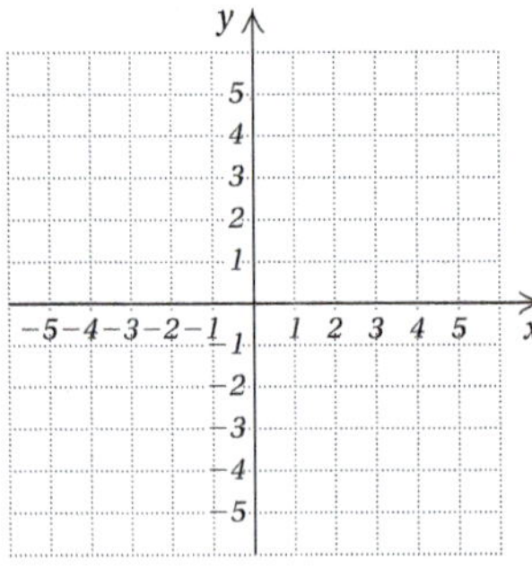

30. $y \ge -x^2 + 2x + 3$

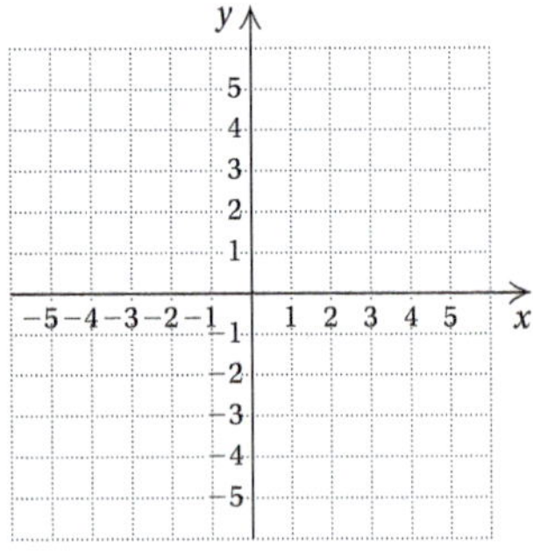

31. $y \le -x^2 - 2x + 3$

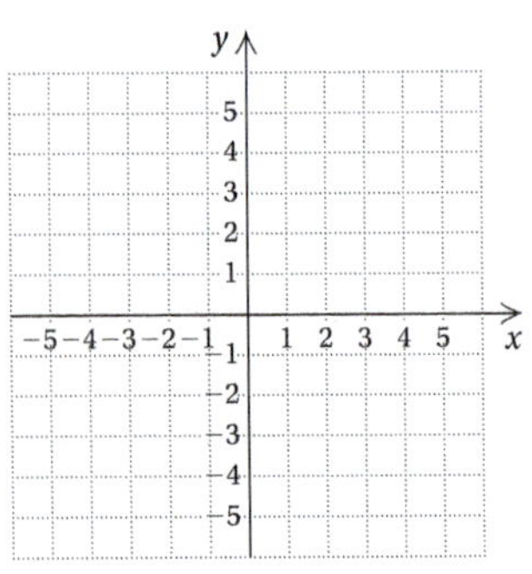

32. $y < -2x^2 - 4x + 1$

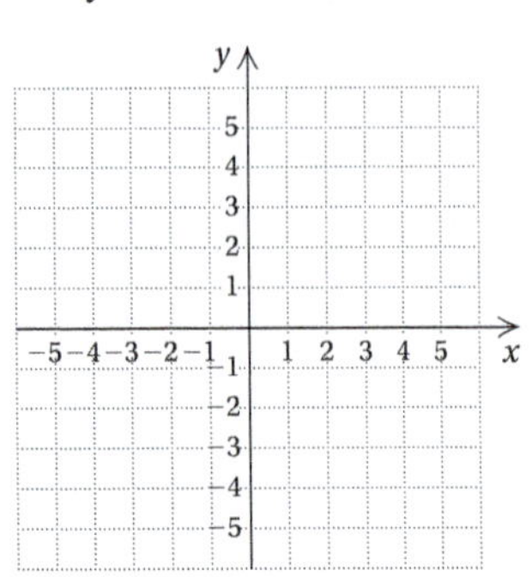

33. $y > 2x^2 + 4x - 1$

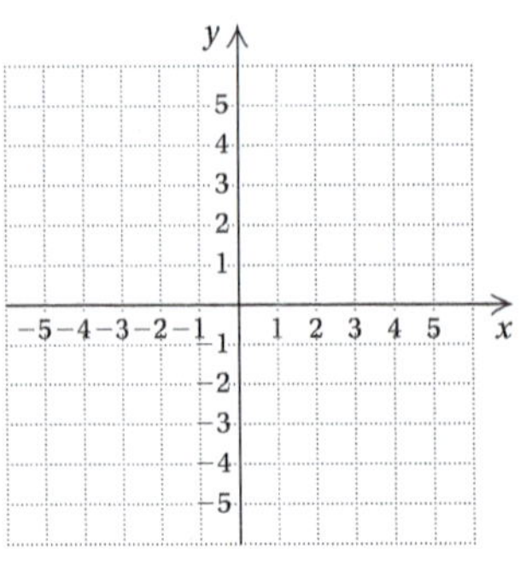

34. $y \le x^2 + 2x - 5$

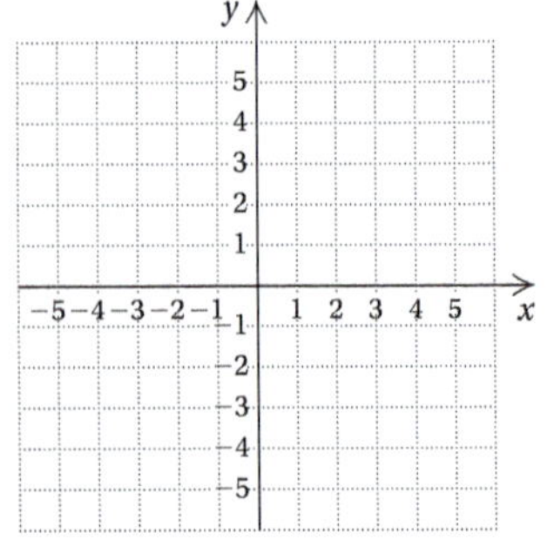

Systems of Linear Inequalities

M

OBJECTIVE

e Graph systems of linear inequalities and find coordinates of any vertices.

In Section 11.7, we studied the graphing of inequalities in two variables. Here we study *systems* of linear inequalities.

a SYSTEMS OF LINEAR INEQUALITIES

The following is an example of a system of two linear inequalities in two variables:

$$x + y \leq 4,$$
$$x - y < 4.$$

A **solution** of a system of linear inequalities is an ordered pair that is a solution of *both* inequalities. To graph solutions of systems of linear inequalities, we graph each inequality and determine where the graphs overlap, or intersect. That will be a region in which the ordered pairs are solutions of both inequalities.

EXAMPLE 1 Graph the solutions of the system

$$x + y \leq 4,$$
$$x - y < 4.$$

We graph $x + y \leq 4$ by first graphing the equation $x + y = 4$ using a solid red line. We consider $(0, 0)$ as a test point and find that it is a solution, so we shade all points on that side of the line using red shading. (See the graph on the left below.) The arrows near the ends of the line also indicate the half-plane, or region, that contains the solutions.

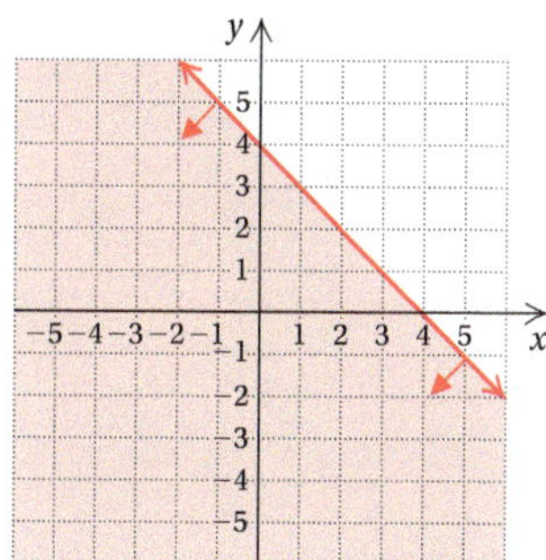

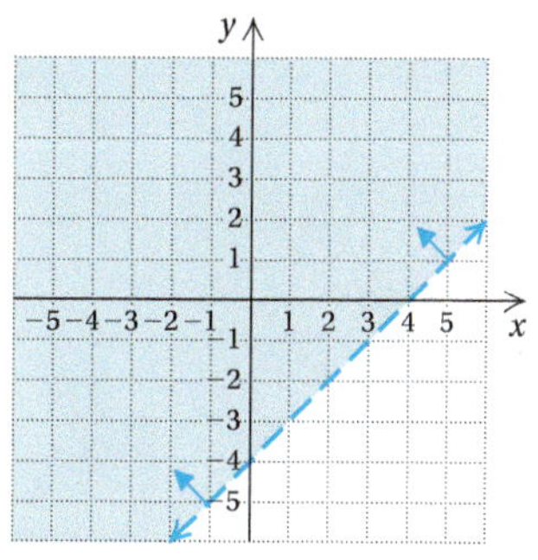

Next, we graph $x - y < 4$. We begin by graphing the equation $x - y = 4$ using a dashed blue line and consider $(0, 0)$ as a test point. Again, $(0, 0)$ is a solution so we shade that side of the line using blue shading. (See the graph on the right at the bottom of the preceding page.) The solution set of the system is the region that is shaded both red and blue and part of the line $x + y = 4$.

1. Graph:

$$x + y \geq 1,$$
$$y - x \geq 2.$$

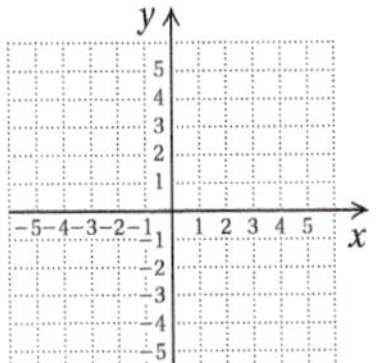

◀ **Do Exercise 1.**

EXAMPLE 2 Graph: $-2 < x \leq 5$.

This is actually a system of inequalities:

$$-2 < x,$$
$$x \leq 5.$$

We graph the equation $-2 = x$ and see that the graph of the first inequality is the half-plane to the right of the line $-2 = x$. (See the graph on the left below.)

Next, we graph the second inequality, starting with the line $x = 5$, and find that its graph is the line as well as the half-plane to the left of it. (See the graph on the right below.) Then we shade the intersection of these graphs.

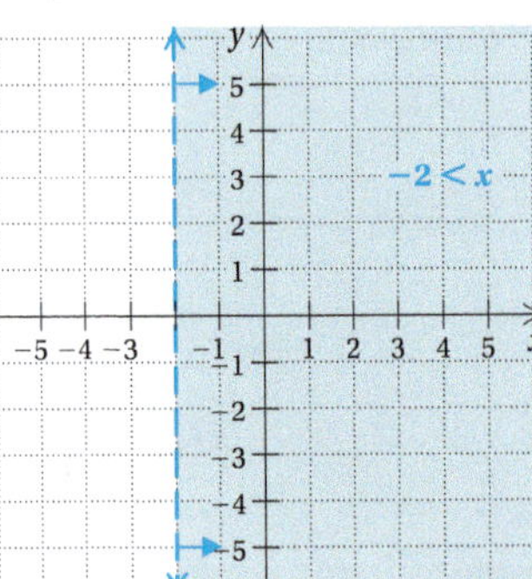

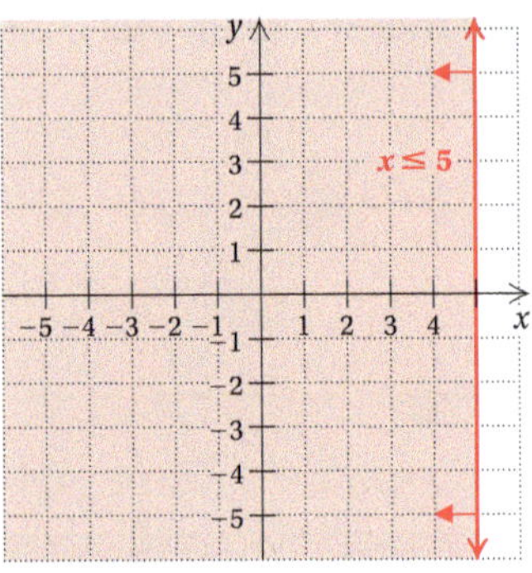

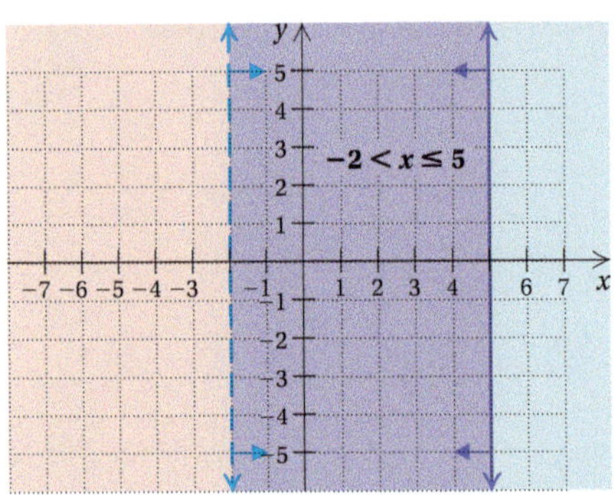

◀ **Do Exercise 2.**

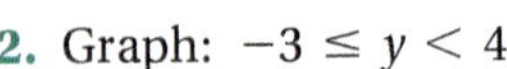

2. Graph: $-3 \leq y < 4$.

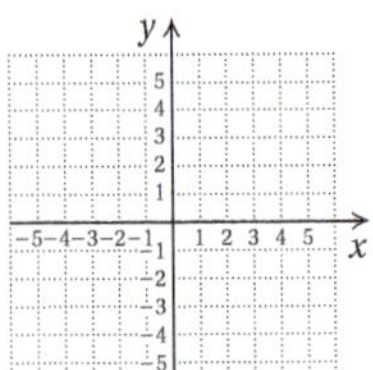

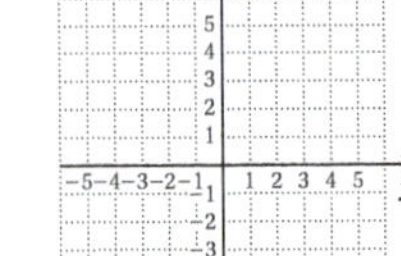

Answers

1.

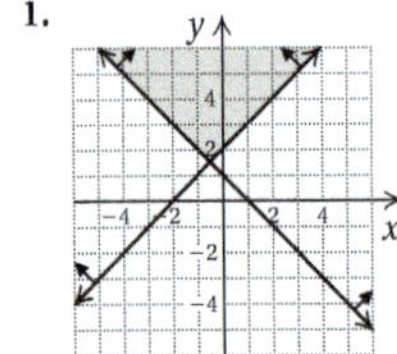

2.

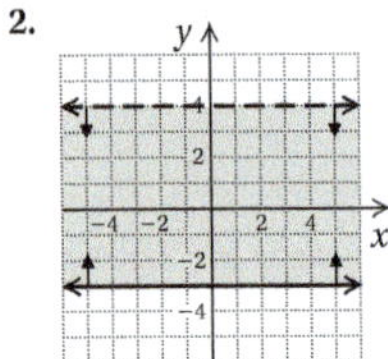

A system of inequalities may have a graph that consists of a polygon and its interior. In *linear programming*, which is a topic rich in application that you may study in a later course, it is important to be able to find the vertices of such a polygon.

EXAMPLE 3 Graph the following system of inequalities. Find the coordinates of any vertices formed.

$$6x - 2y \le 12, \quad (1)$$
$$y - 3 \le 0, \quad (2)$$
$$x + y \ge 0 \quad (3)$$

We graph the lines $6x - 2y = 12$, $y - 3 = 0$, and $x + y = 0$ using solid lines. The regions for each inequality are indicated by the arrows at the ends of the lines. We then note where the regions overlap and shade the region of solutions using one color.

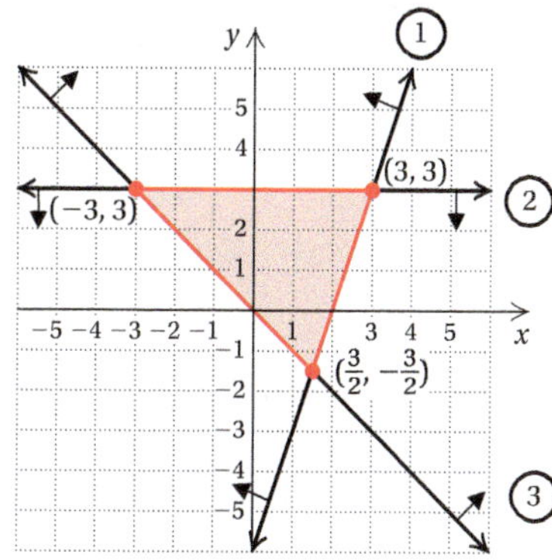

To find the vertices, we solve three different systems of equations. The system of equations from inequalities (1) and (2) is

$$6x - 2y = 12, \quad (1)$$
$$y - 3 = 0. \quad (2)$$

Solving, we obtain the vertex $(3, 3)$.

The system of equations from inequalities (1) and (3) is

$$6x - 2y = 12, \quad (1)$$
$$x + y = 0. \quad (3)$$

Solving, we obtain the vertex $\left(\frac{3}{2}, -\frac{3}{2}\right)$.

The system of equations from inequalities (2) and (3) is

$$y - 3 = 0, \quad (2)$$
$$x + y = 0. \quad (3)$$

Solving, we obtain the vertex $(-3, 3)$.

Do Exercise 3. ▶

3. Graph the system of inequalities. Find the coordinates of any vertices formed.

$$5x + 6y \le 30,$$
$$0 \le y \le 3,$$
$$0 \le x \le 4$$

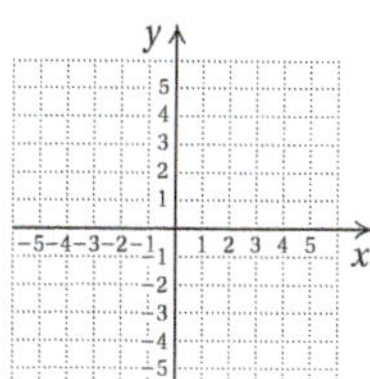

Answer

3. 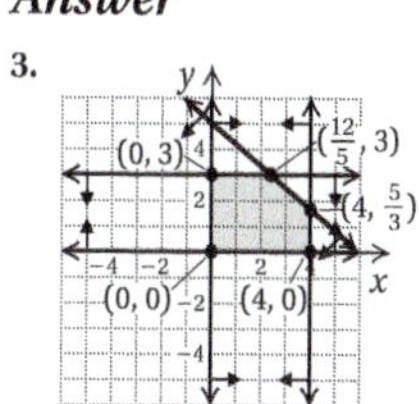

EXAMPLE 4 Graph the following system of inequalities. Find the coordinates of any vertices formed.

$$x + y \le 16, \quad (1)$$
$$3x + 6y \le 60, \quad (2)$$
$$x \ge 0, \quad (3)$$
$$y \ge 0 \quad (4)$$

We graph each inequality using solid lines. The regions for each inequality are indicated by the arrows at the ends of the lines. We then note where the regions overlap and shade the region of solutions using one color.

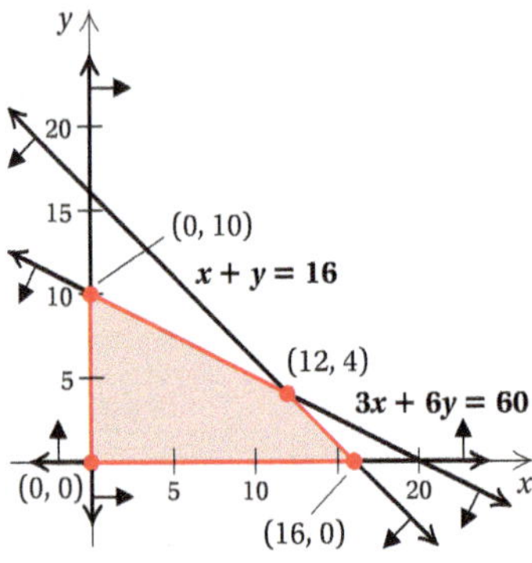

To find the vertices, we solve four different systems of equations. The system of equations from inequalities (1) and (2) is

$$x + y = 16, \quad (1)$$
$$3x + 6y = 60. \quad (2)$$

Solving, we obtain the vertex $(12, 4)$.

The system of equations from inequalities (1) and (4) is

$$x + y = 16, \quad (1)$$
$$y = 0. \quad (4)$$

Solving, we obtain the vertex $(16, 0)$.

The system of equations from inequalities (3) and (4) is

$$x = 0, \quad (3)$$
$$y = 0. \quad (4)$$

The vertex is $(0, 0)$.

The system of equations from inequalities (2) and (3) is

$$3x + 6y = 60, \quad (2)$$
$$x = 0. \quad (3)$$

Solving, we obtain the vertex $(0, 10)$.

◀ **Do Exercise 4.**

4. Graph the system of inequalities. Find the coordinates of any vertices formed.

$$2x + 4y \le 8,$$
$$x + y \le 3,$$
$$x \ge 0,$$
$$y \ge 0$$

Answer

4.

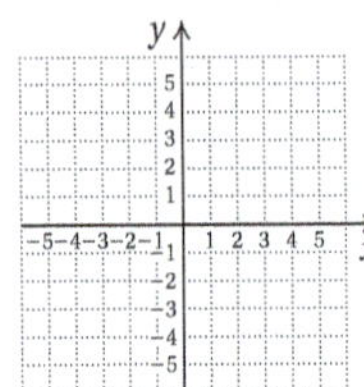

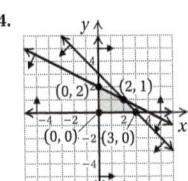

M Exercise Set

For Extra Help
MyMathLab® MathXL® PRACTICE WATCH READ REVIEW

a Graph each system of inequalities. Find the coordinates of any vertices formed.

1. $y \geq x,$
$y \leq -x + 2$

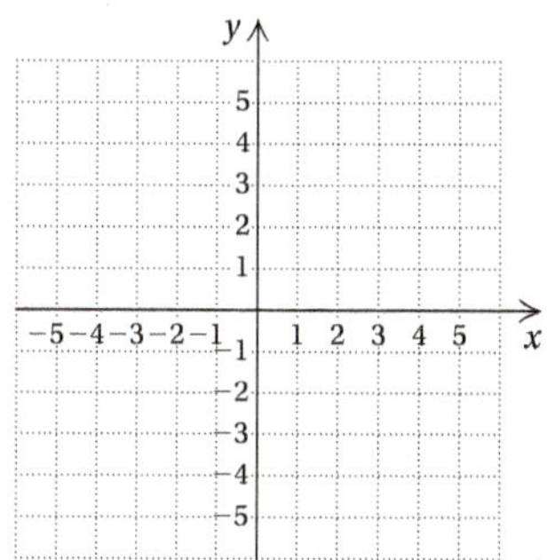

2. $y \geq x,$
$y \leq -x + 4$

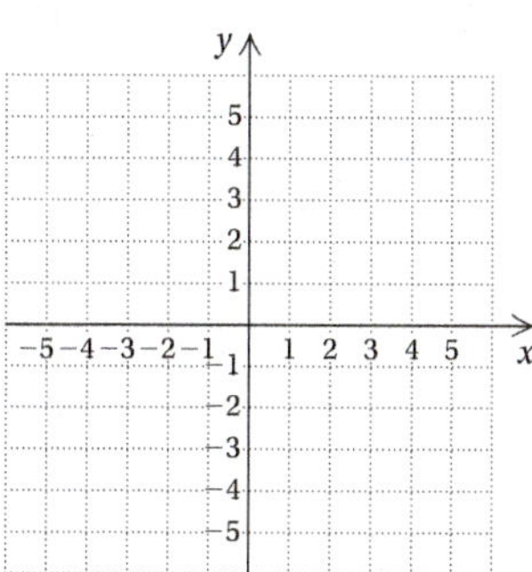

3. $y > x,$
$y < -x + 1$

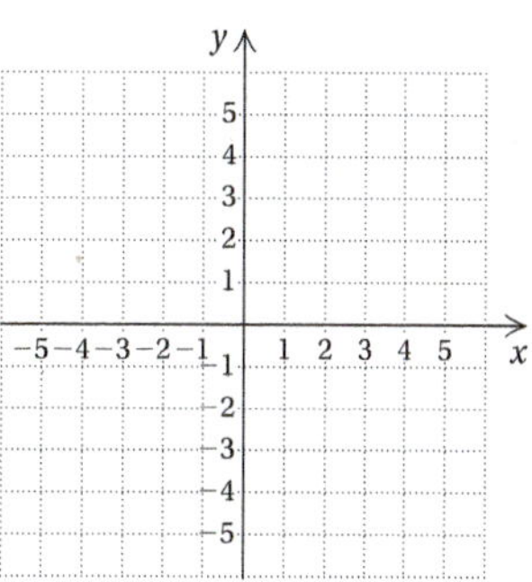

4. $y < x,$
$y > -x + 3$

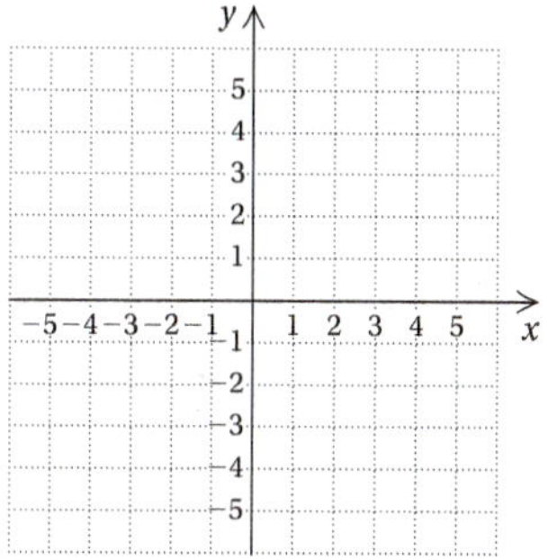

5. $x \leq 3,$
$y \geq -3x + 2$

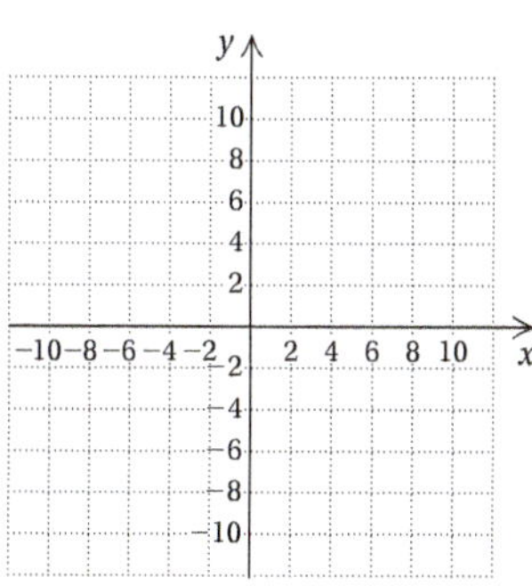

6. $x \geq -2,$
$y \leq -2x + 3$

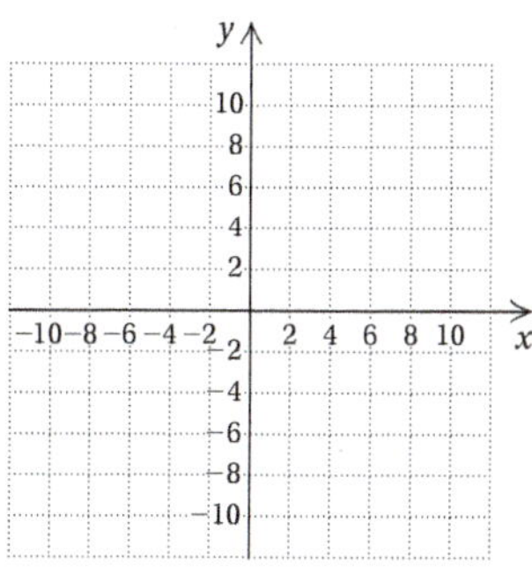

7. $x + y \leq 1,$
$x - y \leq 2$

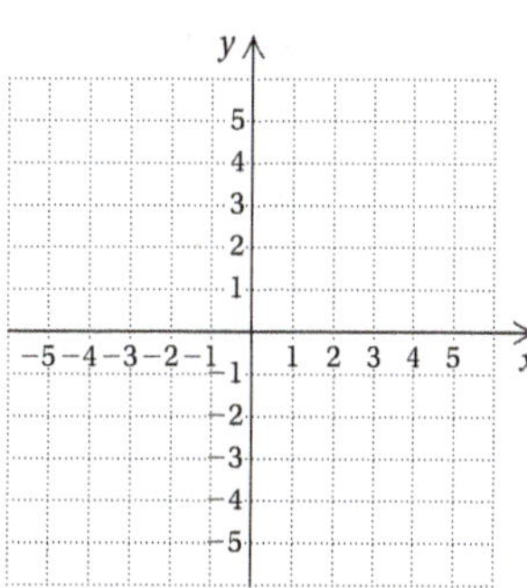

8. $x + y \leq 3,$
$x - y \leq 4$

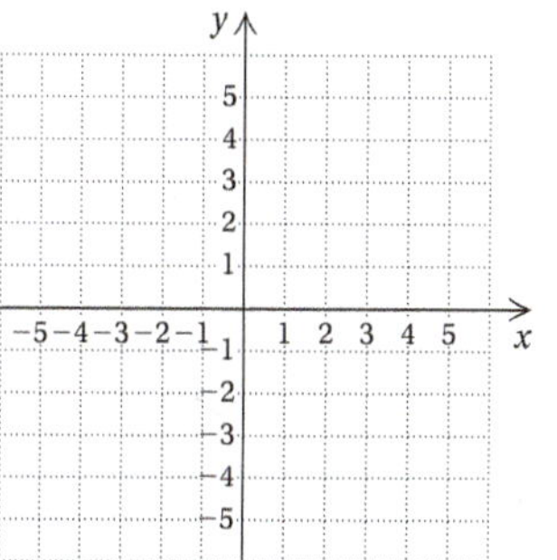

9. $y \leq 2x + 1,$
$y \geq -2x + 1,$
$x \leq 2$

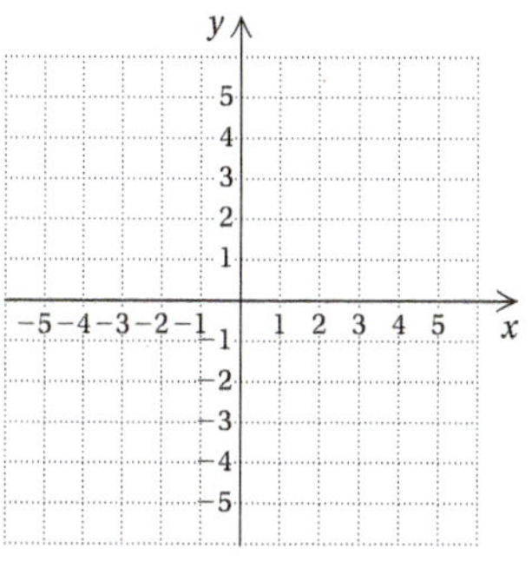

10. $x - y \leq 2,$
$x + 2y \geq 8,$
$y \leq 4$

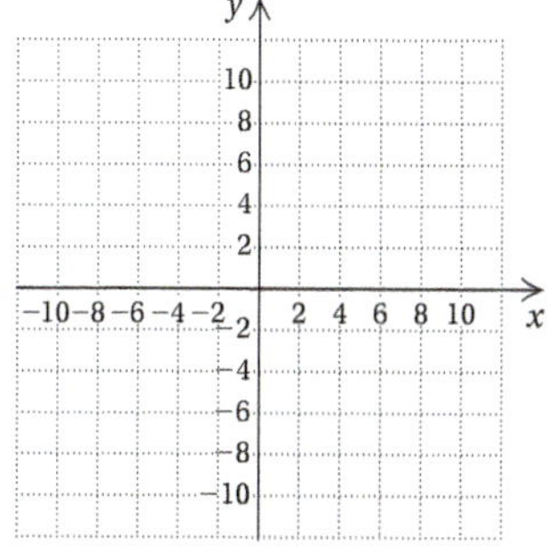

11. $x + 2y \leq 12,$
$2x + y \leq 12,$
$x \geq 0,$
$y \geq 0$

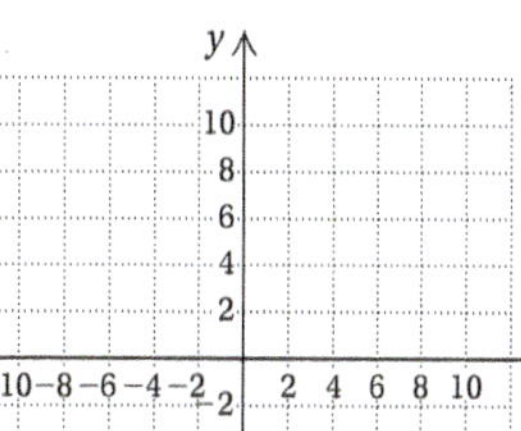

12. $y - x \geq 1,$
$y - x \leq 3,$
$2 \leq x \leq 5$

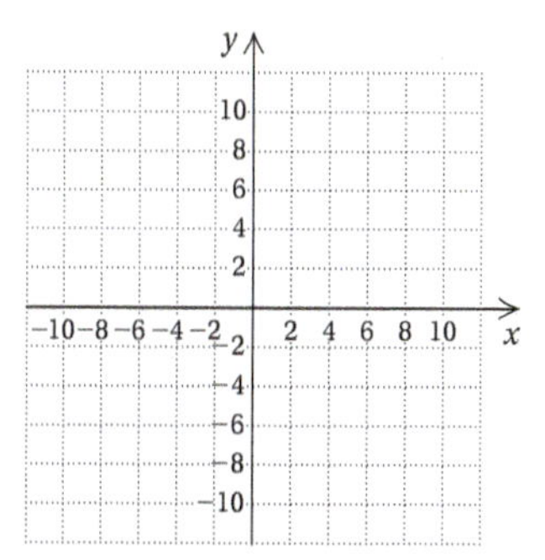

N Probability

OBJECTIVE

a Compute the probability of a simple event.

We say that when a coin is tossed, the chances that it will fall heads are 1 out of 2, or the **probability** that it will fall heads is $\frac{1}{2}$. Of course, this does not mean that if a coin is tossed ten times, it will necessarily fall heads exactly five times. If the coin is tossed a great number of times, however, it will fall heads very nearly half of them.

EXPERIMENTAL AND THEORETICAL PROBABILITY

If we toss a coin a great number of times, say 1000, and count the number of times it falls heads, we can determine the probability of it falling heads. If it falls heads 503 times, we would calculate the probability of the coin falling heads to be

$$\frac{503}{1000}, \quad \text{or} \quad 0.503.$$

This is an **experimental** determination of probability. Such a determination of probability is quite common.

If we consider a coin and reason that it is just as likely to fall heads as tails, we would calculate the probability to be $\frac{1}{2}$. This is a **theoretical** determination of probability. Experimentally, we can determine probabilities within certain limits. These may or may not agree with what we obtain theoretically.

a COMPUTING PROBABILITIES

Experimental Probabilities

We first consider experimental determination of probability. The basic principle we use in computing such probabilities is as follows.

PRINCIPLE *P* (EXPERIMENTAL)

An experiment is performed in which n observations are made. If a situation E, or event, occurs m times out of the n observations, then we say that the **experimental probability** of that event is given by

$$P(E) = \frac{m}{n}.$$

EXAMPLE 1 *Sociological Survey.* An actual survey was conducted to determine the number of people who are left-handed, right-handed, or both. The results are shown in the graph.

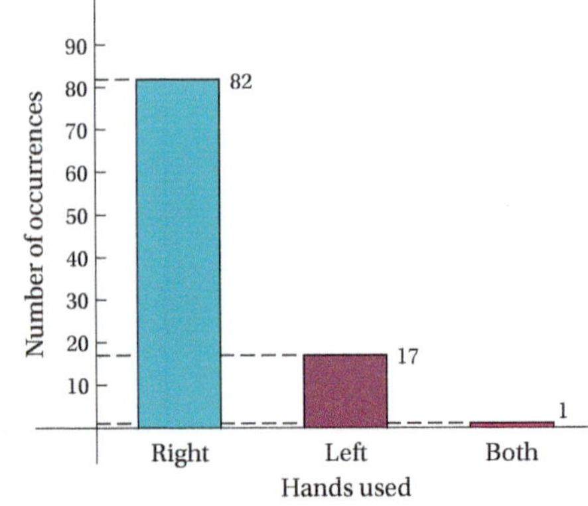

a) Determine the probability that a person is left-handed.

b) Determine the probability that a person is ambidextrous (uses both hands equally well).

a) The number of people who are right-handed was 82, the number who are left-handed was 17, and there was 1 person who is ambidextrous. The total number of observations was $82 + 17 + 1$, or 100. Thus the probability that a person is left-handed is P, where

$$P = \frac{17}{100}.$$

b) The probability that a person is ambidextrous is P, where

$$P = \frac{1}{100}.$$

Do Exercise 1. ▶

1. In reference to Example 1, what is the probability that a person is right-handed?

EXAMPLE 2 *TV Ratings.* The major television networks and others such as cable TV are always concerned about the percentages of homes that have TVs and are watching their programs. It is too costly and unmanageable to contact every home in the country so a sample, or portion, of the homes are contacted. This is done by an electronic device attached to the TVs of about 1400 homes across the country. Viewing information is then fed into a computer. The following are the results of a recent survey.

NETWORK	CBS	ABC	NBC	Other or not watching
NUMBER OF HOMES WATCHING	258	231	206	705

What is the probability that a home was tuned to CBS during the time period considered? to ABC?

The probability that a home was tuned to CBS is P, where

$$P = \frac{258}{1400} \approx 0.184 = 18.4\%.$$

The probability that a home was tuned to ABC is P, where

$$P = \frac{231}{1400} = 0.165 = 16.5\%.$$

Do Exercise 2. ▶

2. In Example 2, what is the probability that a home was tuned to NBC? What is the probability that a home was tuned to a network other than CBS, ABC, or NBC, or was not tuned in at all?

The numbers that we found in Example 2 and in Margin Exercise 2 (18.4 for CBS, 16.5 for ABC, and 14.7 for NBC) are called the *ratings*.

Answers

1. $\frac{82}{100}$ **2.** 14.7%; 50.4%

Theoretical Probabilities

We need some terminology before we can continue. Suppose we perform an experiment such as flipping a coin, throwing a dart, drawing a card from a deck, or checking an item off an assembly line for quality. The results of an experiment are called **outcomes**. The set of all possible outcomes is called the **sample space**. An **event** is a set of outcomes, that is, the subset of the sample space. For example, for the experiment "throwing a dart," suppose the dartboard is as follows.

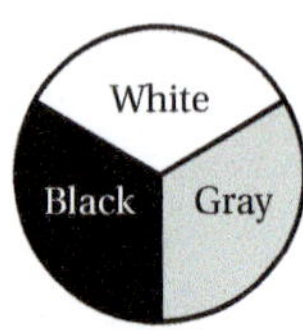

Then one event is

$\{\text{black}\}$, (the outcome is "hitting black")

which is a subset of the sample space

$\{\text{black, white, gray}\}$, (sample space)

assuming that the dart must hit the target somewhere.

We denote the probability that an event E occurs as $P(E)$. For example, "getting a head" may be denoted by H. Then $P(H)$ represents the probability of getting a head. When all the outcomes of an experiment have the same probability of occurring, we say that they are **equally likely**. A sample space that can be expressed as a union of equally likely events can allow us to calculate probabilities of other events.

PRINCIPLE *P* (THEORETICAL)

If an event E can occur m ways out of n possible equally likely outcomes of a sample space S, then the **theoretical probability** of that event is given by

$$P(E) = \frac{m}{n}.$$

A die (pl., dice) is a cube, with six faces, each containing a number of dots from 1 to 6.

EXAMPLE 3 What is the probability of rolling a 3 on a die?

On a fair die, there are 6 equally likely outcomes and there is 1 way to get a 3. By Principle P, $P(3) = \frac{1}{6}$.

3. What is the probability of rolling a prime number on a die?

EXAMPLE 4 What is the probability of rolling an even number on a die?

The event is getting an *even* number. It can occur in 3 ways (getting 2, 4, or 6). The number of equally likely outcomes is 6. By Principle P, $P(\text{even}) = \frac{3}{6}$, or $\frac{1}{2}$.

Do Exercise 3.

Answer

3. $\frac{1}{2}$

We now use a number of examples related to a standard bridge deck of 52 cards. Such a deck is made up as shown in the following figure.

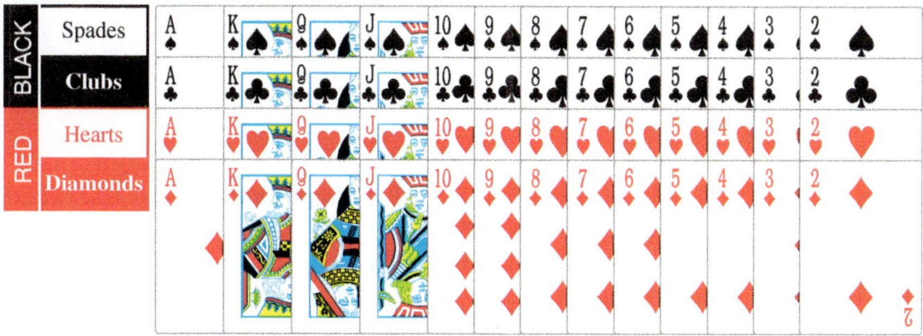

EXAMPLE 5 What is the probability of drawing an ace from a well-shuffled deck of 52 cards?

Since there are 52 outcomes (cards in the deck) and they are equally likely (from a well-shuffled deck) and there are 4 ways to obtain an ace, by Principle *P* we have

$$P(\text{drawing an ace}) = \frac{4}{52}, \quad \text{or} \quad \frac{1}{13}.$$

EXAMPLE 6 Suppose we select, without looking, one marble from a bag containing 3 red marbles and 4 green marbles. What is the probability of selecting a red marble?

There are 7 equally likely ways of selecting any marble, and since the number of ways of getting a red marble is 3,

$$P(\text{selecting a red marble}) = \frac{3}{7}.$$

Do Exercises 4 and 5.

If an event E cannot occur, then $P(E) = 0$. For example, in coin tossing, the event that a coin will land on its edge has probability 0. If an event E is certain to occur (that is, every trial is a success), then $P(E) = 1$. For example, in coin tossing, the event that a coin will fall either heads or tails has probability 1. In general, the probability that an event E will occur is a number from 0 to 1: $0 \leq P(E) \leq 1$.

Do Exercises 6 and 7.

4. Suppose we draw a card from a well-shuffled deck of 52 cards.
   a) What is the probability of drawing a king?
   b) What is the probability of drawing a spade?
   c) What is the probability of drawing a black card?
   d) What is the probability of drawing a jack or a queen?

5. Suppose we select, without looking, one marble from a bag containing 5 red marbles and 6 green marbles. What is the probability of selecting a green marble?

6. On a single roll of a die, what is the probability of getting a 7?

7. On a single roll of a die, what is the probability of getting a 1, 2, 3, 4, 5, or 6?

Answers

4. (a) $\frac{1}{13}$; (b) $\frac{1}{4}$; (c) $\frac{1}{2}$; (d) $\frac{2}{13}$ 5. $\frac{6}{11}$
6. 0 7. 1

Exercise Set

For Extra Help

MyMathLab®

a

1. In an actual survey, 100 people were polled to determine the probability of a person wearing either glasses or contact lenses. Of those polled, 57 wore either glasses or contacts. What is the probability that a person wears either glasses or contacts? What is the probability that a person wears neither?

2. In another survey, 100 people were polled and asked to select a number from 1 to 5. The results are shown in the following table.

NUMBER CHOICES	1	2	3	4	5
NUMBER OF PEOPLE WHO SELECTED THAT NUMBER	18	24	23	23	12

What is the probability that the number selected is 1? 2? 3? 4? 5? What general conclusion might a psychologist make from this experiment?

Linguistics. An experiment was conducted to determine the relative occurrence of various letters of the English alphabet. A paragraph from a newspaper, one from a textbook, and one from a magazine were considered. In all, there was a total of 1044 letters. The number of occurrences of each letter of the alphabet is listed in the following table.

LETTER	A	B	C	D	E	F	G	H	I	J	K	L	M
NUMBER OF OCCURRENCES	78	22	33	33	140	24	22	63	60	2	9	35	30
LETTER	N	O	P	Q	R	S	T	U	V	W	X	Y	Z
NUMBER OF OCCURRENCES	74	74	27	4	67	67	95	31	10	22	8	13	1

Round answers to Exercises 3–6 to three decimal places.

3. What is the probability of the occurrence of the letter A? E? I? O? U?

4. What is the probability of a vowel occurring?

5. What is the probability of a consonant occurring?

6. What letter has the least probability of occurring? What is the probability of this letter not occurring?

Suppose we draw a card from a well-shuffled deck of 52 cards.

7. How many equally likely outcomes are there?

8. What is the probability of drawing a queen?

9. What is the probability of drawing a heart?

10. What is the probability of drawing a 4?

11. What is the probability of drawing a red card?

12. What is the probability of drawing a black card?

13. What is the probability of drawing an ace or a deuce?

14. What is the probability of drawing a 9 or a king?

Suppose we select, without looking, one marble from a bag containing 4 red marbles and 10 green marbles.

15. What is the probability of selecting a red marble?

16. What is the probability of selecting a green marble?

17. What is the probability of selecting a purple marble?

18. What is the probability of selecting a white marble?

Synthesis

19. What is the probability of getting a total of 8 on a roll of a pair of dice? (Assume that the dice are different, say, one red and one black.)

20. What is the probability of getting a total of 7 on a roll of a pair of dice?

21. What is the probability of getting a total of 6 on a roll of a pair of dice?

22. What is the probability of getting a total of 3 on a roll of a pair of dice?

23. What is the probability of getting snake eyes (a total of 2) on a roll of a pair of dice?

24. What is the probability of getting box-cars (a total of 12) on a roll of a pair of dice?

0 The Complex Numbers

OBJECTIVES

- **a** Express imaginary numbers as *bi*, where *b* is a nonzero real number, and complex numbers as $a + bi$, where *a* and *b* are real numbers.
- **b** Add and subtract complex numbers.
- **c** Multiply complex numbers.
- **d** Write expressions involving powers of *i* in the form $a + bi$.
- **e** Find conjugates of complex numbers and divide complex numbers.
- **f** Determine whether a given complex number is a solution of an equation.
- **g** Solve quadratic equations with nonreal solutions.

a IMAGINARY AND COMPLEX NUMBERS

Negative numbers do not have square roots in the real-number system. However, mathematicians have described a larger number system that contains the real-number system, such that negative numbers have square roots. That system is called the **complex-number system**. We begin by defining a number that is a square root of -1. We call this new number *i*.

THE COMPLEX NUMBER *i*

We define the number *i* to be $\sqrt{-1}$. That is,

$$i = \sqrt{-1} \quad \text{and} \quad i^2 = -1.$$

To express roots of negative numbers in terms of *i*, we can use the fact that in the complex-number system, $\sqrt{-p} = \sqrt{-1 \cdot p} = \sqrt{-1}\sqrt{p}$ when *p* is a positive real number.

EXAMPLES Express in terms of *i*.

i is *not* under the radical.

1. $\sqrt{-7} = \sqrt{-1 \cdot 7} = \sqrt{-1} \cdot \sqrt{7} = i\sqrt{7}$, or $\sqrt{7}i$
2. $-\sqrt{-13} = -\sqrt{-1 \cdot 13} = -\sqrt{-1} \cdot \sqrt{13} = -i\sqrt{13}$, or $-\sqrt{13}i$
3. $-\sqrt{-64} = -\sqrt{-1 \cdot 64} = -\sqrt{-1} \cdot \sqrt{64} = -i \cdot 8 = -8i$
4. $\sqrt{-48} = \sqrt{-1 \cdot 48} = \sqrt{-1} \cdot \sqrt{48} = i\sqrt{48}$
   $= i \cdot 4\sqrt{3} = 4i\sqrt{3}$, or $4\sqrt{3}i$

◀ Do Exercises 1–5.

Express in terms of *i*.

1. $\sqrt{-5}$
2. $\sqrt{-25}$
3. $-\sqrt{-11}$
4. $-\sqrt{-36}$
5. $\sqrt{-54}$

IMAGINARY NUMBER

An **imaginary number** is a number that can be named

$$bi,$$

where *b* is some real number and $b \neq 0$.

To form the system of **complex numbers**, we take the imaginary numbers and the real numbers and all possible sums of real and imaginary numbers. These are complex numbers: $7 - 4i$, $-\pi + 19i$, 37, $i\sqrt{6}$.

COMPLEX NUMBER

A **complex number** is any number that can be named

$$a + bi,$$

where *a* and *b* are any real numbers. (Note that either *a* or *b* or both can be 0.)

Answers

1. $i\sqrt{5}$, or $\sqrt{5}i$ 2. $5i$ 3. $-i\sqrt{11}$, or $-\sqrt{11}i$ 4. $-6i$ 5. $3i\sqrt{6}$, or $3\sqrt{6}i$

Since $0 + bi = bi$, every imaginary number is a complex number. Similarly, $a + 0i = a$, so every real number is a complex number. The relationships among various real and complex numbers are shown in the following diagram.

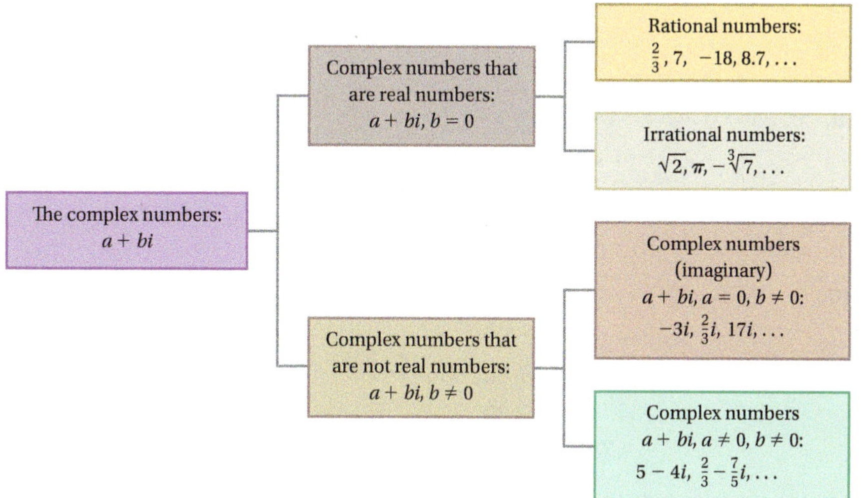

It is important to keep in mind some comparisons between numbers that have real-number roots and those that have complex-number roots that are not real. For example, $\sqrt{-48}$ is a complex number that is not a real number because we are taking the square root of a negative number. *But,* $\sqrt[3]{-125}$ is a real number because we are taking the cube root of a negative number and *any* real number has a cube root that is a real number.

b ADDITION AND SUBTRACTION

The complex numbers follow the commutative and associative laws of addition. Thus we can add and subtract them as we do binomials with real-number coefficients; that is, we collect like terms.

EXAMPLES Add or subtract.

5. $(8 + 6i) + (3 + 2i) = (8 + 3) + (6 + 2)i = 11 + 8i$

6. $(3 + 2i) - (5 - 2i) = (3 - 5) + [2 - (-2)]i = -2 + 4i$

Do Exercises 6–9.

Add or subtract.

6. $(7 + 4i) + (8 - 7i)$

7. $(-5 - 6i) + (-7 + 12i)$

8. $(8 + 3i) - (5 + 8i)$

9. $(5 - 4i) - (-7 + 3i)$

c MULTIPLICATION

The complex numbers obey the commutative, associative, and distributive laws. But although the property $\sqrt{a}\sqrt{b} = \sqrt{ab}$ does *not* hold for complex numbers in general, it does hold when $a = -1$ and b is a positive real number.

To multiply square roots of negative real numbers, we first express them in terms of i. For example,

$$\sqrt{-2} \cdot \sqrt{-5} = \sqrt{-1} \cdot \sqrt{2} \cdot \sqrt{-1} \cdot \sqrt{5} = i\sqrt{2} \cdot i\sqrt{5}$$
$$= i^2\sqrt{10} = -\sqrt{10} \quad \text{is correct!}$$

But $\sqrt{-2} \cdot \sqrt{-5} = \sqrt{(-2)(-5)} = \sqrt{10}$ is wrong!

Caution!

The rule $\sqrt{a}\sqrt{b} = \sqrt{ab}$ holds only for nonnegative real numbers.

Keeping this and the fact that $i^2 = -1$ in mind, we multiply in much the same way that we do with real numbers.

Answers

6. $15 - 3i$ **7.** $-12 + 6i$ **8.** $3 - 5i$
9. $12 - 7i$

EXAMPLES Multiply.

7. $\sqrt{-49} \cdot \sqrt{-16} = \sqrt{-1} \cdot \sqrt{49} \cdot \sqrt{-1} \cdot \sqrt{16}$
$= i \cdot 7 \cdot i \cdot 4$
$= i^2(28)$
$= (-1)(28)$ $i^2 = -1$
$= -28$

8. $\sqrt{-3} \cdot \sqrt{-7} = \sqrt{-1} \cdot \sqrt{3} \cdot \sqrt{-1} \cdot \sqrt{7}$
$= i \cdot \sqrt{3} \cdot i \cdot \sqrt{7} = i^2(\sqrt{21})$
$= (-1)\sqrt{21} = -\sqrt{21}$

9. $-2i \cdot 5i = -10 \cdot i^2 = (-10)(-1) = 10$

10. $(-4i)(3 - 5i) = (-4i) \cdot 3 - (-4i)(5i)$ Using a distributive law
$= -12i + 20i^2$
$= -12i + 20(-1)$ $i^2 = -1$
$= -12i - 20$
$= -20 - 12i$

11. $(1 + 2i)(1 + 3i) = 1 + 3i + 2i + 6i^2$ Using FOIL
$= 1 + 3i + 2i + 6(-1)$ $i^2 = -1$
$= 1 + 3i + 2i - 6$
$= -5 + 5i$ Collecting like terms

Do Exercises 10–17.

Multiply.

10. $\sqrt{-25} \cdot \sqrt{-4}$

11. $\sqrt{-2} \cdot \sqrt{-17}$

12. $-6i \cdot 7i$

13. $-3i(4 - 3i)$

14. $5i(-5 + 7i)$

15. $(1 + 3i)(1 + 5i)$

16. $(3 - 2i)(1 + 4i)$

17. $(3 + 2i)^2$

d POWERS OF *i*

We now want to simplify certain expressions involving powers of i. To do so, we first see how to simplify powers of i. Simplifying powers of i can be done by using the fact that $i^2 = -1$ and expressing the given power of i in terms of even powers, and then in terms of powers of i^2. Consider the following:

$$i,$$
$$i^2 = -1,$$
$$i^3 = i^2 \cdot i = (-1)i = -i,$$
$$i^4 = (i^2)^2 = (-1)^2 = 1,$$
$$i^5 = i^4 \cdot i = (i^2)^2 \cdot i = (-1)^2 \cdot i = i,$$
$$i^6 = (i^2)^3 = (-1)^3 = -1.$$

Note that the powers of i cycle through the values i, -1, $-i$, and 1.

EXAMPLES Simplify. In Examples 16–18, simplify to the form $a + bi$.

12. $i^{37} = i^{36} \cdot i = (i^2)^{18} \cdot i = (-1)^{18} \cdot i = 1 \cdot i = i$

13. $i^{58} = (i^2)^{29} = (-1)^{29} = -1$

14. $i^{75} = i^{74} \cdot i = (i^2)^{37} \cdot i = (-1)^{37} \cdot i = -1 \cdot i = -i$

15. $i^{80} = (i^2)^{40} = (-1)^{40} = 1$

16. $8 - i^2 = 8 - (-1) = 8 + 1 = 9$

17. $17 + 6i^3 = 17 + 6 \cdot i^2 \cdot i = 17 + 6(-1)i = 17 - 6i$

18. $i^{22} - 67i^2 = (i^2)^{11} - 67(-1) = (-1)^{11} + 67 = -1 + 67 = 66$

Do Exercises 18–25.

Simplify.

18. i^{47} **19.** i^{68}

20. i^{85} **21.** i^{90}

22. $8 - i^5$ **23.** $7 + 4i^2$

24. $i^{34} - i^{55}$ **25.** $6i^{11} + 7i^{14}$

Answers

10. -10 **11.** $-\sqrt{34}$ **12.** 42 **13.** $-9 - 12i$ **14.** $-35 - 25i$ **15.** $-14 + 8i$ **16.** $11 + 10i$ **17.** $5 + 12i$ **18.** $-i$ **19.** 1 **20.** i **21.** -1 **22.** $8 - i$ **23.** 3 **24.** $-1 + i$ **25.** $-7 - 6i$

e CONJUGATES AND DIVISION

Conjugates of complex numbers are defined as follows.

CONJUGATE

The **conjugate** of a complex number $a + bi$ is $a - bi$, and the **conjugate** of $a - bi$ is $a + bi$.

EXAMPLES Find the conjugate.

19. $5 + 7i$ The conjugate is $5 - 7i$.

20. $-3 - 9i$ The conjugate is $-3 + 9i$.

21. $4i$ The conjugate is $-4i$.

Do Exercises 26–28. ▶

Find the conjugate.

26. $6 + 3i$

27. $-9 - 5i$

28. $-\frac{1}{4}i$

When we multiply a complex number by its conjugate, we get a real number.

EXAMPLES Multiply.

22. $(5 + 7i)(5 - 7i) = 5^2 - (7i)^2$ Using $(A + B)(A - B) = A^2 - B^2$

$= 25 - 49i^2 = 25 - 49(-1)$

$= 25 + 49 = 74$

23. $(2 - 3i)(2 + 3i) = 2^2 - (3i)^2 = 4 - 9i^2$

$= 4 - 9(-1) = 4 + 9 = 13$

Do Exercises 29 and 30. ▶

Multiply.

29. $(7 - 2i)(7 + 2i)$

30. $(-3 - i)(-3 + i)$

We use conjugates when dividing complex numbers.

EXAMPLE 24 Divide and simplify to the form $a + bi$: $\frac{3 + 5i}{4 + 3i}$.

$$\frac{3 + 5i}{4 + 3i} \cdot \frac{4 - 3i}{4 - 3i} = \frac{(3 + 5i)(4 - 3i)}{(4 + 3i)(4 - 3i)}$$ Multiplying by 1

$$= \frac{12 - 9i + 20i - 15i^2}{4^2 - 9i^2}$$

$$= \frac{12 + 11i - 15(-1)}{16 - 9(-1)}$$ $i^2 = -1$

$$= \frac{27 + 11i}{25} = \frac{27}{25} + \frac{11}{25}i$$

Do Exercises 31 and 32. ▶

Divide and simplify to the form $a + bi$.

31. $\frac{6 + 2i}{1 - 3i}$

32. $\frac{2 + 3i}{-1 + 4i}$

Answers

26. $6 - 3i$ 27. $-9 + 5i$ 28. $\frac{1}{4}i$ 29. 53 30. 10 31. $2i$ 32. $\frac{10}{17} - \frac{11}{17}i$

f SOLUTIONS OF EQUATIONS

The equation $x^2 + 1 = 0$ has no real-number solution, but it has *two* non-real complex solutions.

EXAMPLE 25 Determine whether $1 + i$ is a solution of the equation $x^2 - 2x + 2 = 0$.

We substitute $1 + i$ for x in the equation.

$$\begin{array}{r|l} x^2 - 2x + 2 & 0 \\ \hline (1 + i)^2 - 2(1 + i) + 2 & ?\ 0 \\ 1 + 2i + i^2 - 2 - 2i + 2 & \\ 1 + 2i - 1 - 2 - 2i + 2 & \\ (1 - 1 - 2 + 2) + (2 - 2)i & \\ 0 + 0i & \\ 0 & \quad \text{TRUE} \end{array}$$

The number $1 + i$ is a solution.

Any equation consisting of a polynomial in one variable on one side and 0 on the other has complex-number solutions. (Some may be real.) It is not always easy to find the solutions, but they always exist.

EXAMPLE 26 Determine whether $2i$ is a solution of $x^2 + 3x - 4 = 0$.

$$\begin{array}{r|l} x^2 + 3x - 4 & 0 \\ \hline (2i)^2 + 3(2i) - 4 & ?\ 0 \\ 4i^2 + 6i - 4 & \\ -4 + 6i - 4 & \\ -8 + 6i & \quad \text{FALSE} \end{array}$$

The number $2i$ is not a solution.

Do Exercises 33 and 34.

33. Determine whether $-i$ is a solution of $x^2 + 1 = 0$.

$$\begin{array}{c|c} x^2 + 1 & 0 \\ \hline & ? \end{array}$$

34. Determine whether $1 - i$ is a solution of $x^2 - 2x + 2 = 0$.

$$\begin{array}{c|c} x^2 - 2x + 2 & 0 \\ \hline & ? \end{array}$$

g QUADRATIC EQUATIONS WITH NONREAL SOLUTIONS

In Chapter 17, we solved quadratic equations with real solutions. Some quadratic equations have nonreal solutions, as shown in Examples 27–30.

EXAMPLE 27 Solve: $4x^2 + 9 = 0$.

$$4x^2 + 9 = 0$$

$$x^2 = -\frac{9}{4} \qquad \text{Subtracting 9 and dividing by 4}$$

$$x = \sqrt{-\frac{9}{4}} \quad or \quad x = -\sqrt{-\frac{9}{4}} \qquad \text{Using the principle of square roots}$$

$$x = \frac{3}{2}i \quad or \quad x = -\frac{3}{2}i \qquad \text{Simplifying}$$

We check: $4\left(\pm\frac{3}{2}i\right)^2 + 9 = 4\left(-\frac{9}{4}\right) + 9 = -9 + 9 = 0$. The solutions are $\frac{3}{2}i$ and $-\frac{3}{2}i$, or $\pm\frac{3}{2}i$.

We see that the graph of $f(x) = 4x^2 + 9$ does not cross the x-axis. This is true because the equation $4x^2 + 9 = 0$ has *imaginary* complex-number solutions. Only real-number solutions correspond to x-intercepts.

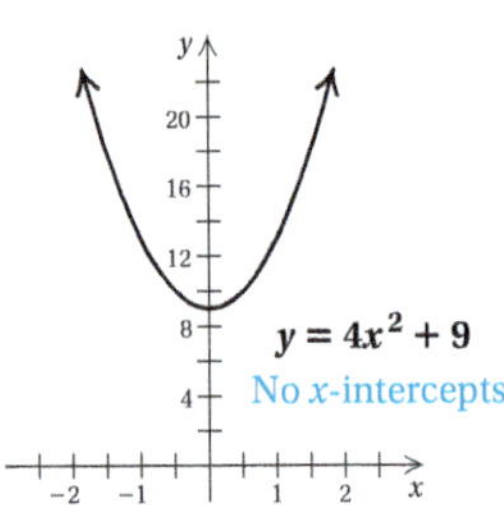

Do Exercise 35.

35. Solve: $2x^2 + 1 = 0$.

Answers

33. Yes **34.** Yes **35.** $\pm\frac{\sqrt{2}}{2}i$

EXAMPLE 28 Solve: $x^2 - 4x + 13 = 0$.

We have $a = 1$, $b = -4$, and $c = 13$. We use the quadratic formula:

$$x = \frac{-(-4) \pm \sqrt{(-4)^2 - 4 \cdot 1 \cdot 13}}{2 \cdot 1} \quad \text{Using the quadratic formula}$$

$$= \frac{4 \pm \sqrt{16 - 52}}{2} = \frac{4 \pm \sqrt{-36}}{2}$$

$$= \frac{4 \pm 6i}{2} = 2 \pm 3i.$$

The solutions are $2 \pm 3i$.

Do Exercise 36. ▶

36. Solve: $x^2 - 8x + 17 = 0$.

EXAMPLE 29 Solve: $2x^2 + 5x = -6$.

We first find standard form and determine a, b, and c:

$$2x^2 + 5x + 6 = 0$$

$$a = 2, \quad b = 5, \quad c = 6.$$

Then we use the quadratic formula:

$$x = \frac{-5 \pm \sqrt{5^2 - 4 \cdot 2 \cdot 6}}{2 \cdot 2}$$

$$= \frac{-5 \pm \sqrt{25 - 48}}{4} = \frac{-5 \pm \sqrt{-23}}{4}$$

$$= \frac{-5 \pm \sqrt{23}i}{4}.$$

The solutions are $\frac{-5 \pm \sqrt{23}i}{4}$, or $-\frac{5}{4} \pm \frac{\sqrt{23}}{4}i$.

Do Exercise 37. ▶

37. Solve: $2x^2 + 7x = -11$.

EXAMPLE 30 Solve: $x^3 + 8 = 0$.

We first factor the difference of cubes and use the principle of zero products:

$$(x + 2)(x^2 - 2x + 4) = 0$$

$$x + 2 = 0 \quad or \quad x^2 - 2x + 4 = 0$$

$$x = -2 \quad or \quad x^2 - 2x + 4 = 0.$$

Then we solve $x^2 - 2x + 4 = 0$ using the quadratic formula with $a = 1$, $b = -2$, and $c = 4$:

$$x = \frac{-(-2) \pm \sqrt{(-2)^2 - 4 \cdot 1 \cdot 4}}{2 \cdot 1} = \frac{2 \pm \sqrt{4 - 16}}{2}$$

$$= \frac{2 \pm \sqrt{-12}}{2} = \frac{2 \pm \sqrt{-1 \cdot 4 \cdot 3}}{2}$$

$$= \frac{2 \pm 2\sqrt{3}i}{2} = 1 \pm \sqrt{3}i.$$

The solutions are -2, $1 + \sqrt{3}i$, and $1 - \sqrt{3}i$, or -2 and $1 \pm \sqrt{3}i$.

Do Exercise 38. ▶

38. Solve: $x^3 - 8 = 0$.

Answers

36. $4 \pm i$ **37.** $\frac{-7 \pm \sqrt{39}i}{4}$, or $-\frac{7}{4} \pm \frac{\sqrt{39}}{4}i$
38. $2, -1 \pm \sqrt{3}i$

0 Exercise Set

For Extra Help MyMathLab® MathXL® PRACTICE  WATCH  READ  REVIEW

a Express in terms of i.

1. $\sqrt{-35}$

2. $\sqrt{-21}$

3. $\sqrt{-16}$

4. $\sqrt{-36}$

5. $-\sqrt{-12}$

6. $-\sqrt{-20}$

7. $\sqrt{-3}$

8. $\sqrt{-4}$

9. $\sqrt{-81}$

10. $\sqrt{-27}$

11. $\sqrt{-98}$

12. $-\sqrt{-18}$

13. $-\sqrt{-49}$

14. $-\sqrt{-125}$

15. $4 - \sqrt{-60}$

16. $6 - \sqrt{-84}$

b Add or subtract and simplify.

17. $(7 + 2i) + (5 - 6i)$

18. $(-4 + 5i) + (7 + 3i)$

19. $(4 - 3i) + (5 - 2i)$

20. $(-2 - 5i) + (1 - 3i)$

21. $(9 - i) + (-2 + 5i)$

22. $(6 + 4i) + (2 - 3i)$

23. $(6 - i) - (10 + 3i)$

24. $(-4 + 3i) - (7 + 4i)$

25. $(4 - 2i) - (5 - 3i)$

26. $(-2 - 3i) - (1 - 5i)$

27. $(9 + 5i) - (-2 - i)$

28. $(6 - 3i) - (2 + 4i)$

c Multiply.

29. $\sqrt{-36} \cdot \sqrt{-9}$

30. $\sqrt{-16} \cdot \sqrt{-64}$

31. $\sqrt{-7} \cdot \sqrt{-2}$

32. $\sqrt{-11} \cdot \sqrt{-3}$

33. $-3i \cdot 7i$

34. $8i \cdot 5i$

35. $-3i(-8 - 2i)$

36. $4i(5 - 7i)$

37. $(3 + 2i)(1 + i)$

38. $(4 + 3i)(2 + 5i)$

39. $(2 + 3i)(6 - 2i)$

40. $(5 + 6i)(2 - i)$

41. $(6 - 5i)(3 + 4i)$

42. $(5 - 6i)(2 + 5i)$

43. $(7 - 2i)(2 - 6i)$

44. $(-4 + 5i)(3 - 4i)$

45. $(3 - 2i)^2$

46. $(5 - 2i)^2$

47. $(1 + 5i)^2$

48. $(6 + 2i)^2$

d Simplify.

49. i^7

50. i^{11}

51. i^{24}

52. i^{35}

53. i^{42}

54. i^{64}

55. i^9

56. $(-i)^{71}$

57. i^6

58. $(-i)^4$

59. $(5i)^3$

60. $(-3i)^5$

Simplify to the form $a + bi$.

61. $7 + i^4$

62. $-18 + i^3$

63. $i^{28} - 23i$

64. $i^{29} + 33i$

65. $i^2 + i^4$

66. $5i^5 + 4i^3$

67. $i^5 + i^7$

68. $i^{84} - i^{100}$

69. $1 + i + i^2 + i^3 + i^4$

70. $i - i^2 + i^3 - i^4 + i^5$

71. $5 - \sqrt{-64}$

72. $\sqrt{-12} + 36i$

73. $\dfrac{8 - \sqrt{-24}}{4}$

74. $\dfrac{9 + \sqrt{-9}}{3}$

e Divide and simplify to the form $a + bi$.

75. $\dfrac{4 + 3i}{3 - i}$

76. $\dfrac{5 + 2i}{2 + i}$

77. $\dfrac{3 - 2i}{2 + 3i}$

78. $\dfrac{6 - 2i}{7 + 3i}$

79. $\dfrac{8 - 3i}{7i}$

80. $\dfrac{3 + 8i}{5i}$

81. $\dfrac{4}{3 + i}$

82. $\dfrac{6}{2 - i}$

83. $\frac{2i}{5 - 4i}$

84. $\frac{8i}{6 + 3i}$

85. $\frac{4}{3i}$

86. $\frac{5}{6i}$

87. $\frac{2 - 4i}{8i}$

88. $\frac{5 + 3i}{i}$

89. $\frac{6 + 3i}{6 - 3i}$

90. $\frac{4 - 5i}{4 + 5i}$

f Determine whether the complex number is a solution of the equation.

91. $1 - 2i$;
$x^2 - 2x + 5 = 0$
?

92. $1 + 2i$;
$x^2 - 2x + 5 = 0$
?

93. $2 + i$;
$x^2 - 4x - 5 = 0$
?

94. $1 - i$;
$x^2 + 2x + 2 = 0$
?

g Solve.

95. $9x^2 + 25 = 0$

96. $36x^2 + 49 = 0$

97. $(x - 7)^2 = -4$

98. $(x + 1)^2 = -9$

99. $2x^2 - 5x + 8 = 0$

100. $2x^2 - 3x + 9 = 0$

101. $x^2 + x + 2 = 0$

102. $x^2 - x + 1 = 0$

103. $x^2 - 4x + 13 = 0$

104. $x^2 - 6x + 13 = 0$

105. $x^2 + 9 = 2x$

106. $x^2 + 7 = 4x$

107. $1 + \frac{2}{x} + \frac{5}{x^2} = 0$

108. $1 + \frac{5}{x^2} = \frac{2}{x}$

109. $x^2 + 5 = 4x$

110. $x^2 + 6 = 4x$

111. $(x - 2)^2 + (x + 1)^2 = 0$

112. $(x + 3)^2 + (x - 2)^2 = 0$

113. $x^3 - 1 = 0$

114. $x^3 + 27 = 0$

Answers

CHAPTER 1

Exercise Set 1.1, p. 6

RC1. digit **RC2.** period **RC3.** expanded **RC4.** standard
1. 5 thousands **3.** 5 hundreds **5.** 1 **7.** 2
9. 2 thousands + 0 hundreds + 5 tens + 8 ones, or 2 thousands + 5 tens + 8 ones **11.** 1 thousand + 5 hundreds + 7 tens + 6 ones **13.** 5 thousands + 7 hundreds + 0 tens + 2 ones, or 5 thousands + 7 hundreds + 2 ones **15.** 9 ten thousands + 3 thousands + 9 hundreds + 8 tens + 6 ones
17. 1 billion + 3 hundred millions + 4 ten millions + 3 millions + 2 hundred thousands + 3 ten thousands + 9 thousands + 9 hundreds + 2 tens + 3 ones **19.** 2 hundred millions + 4 ten millions + 8 millions + 6 hundred thousands + 4 ten thousands + 5 thousands + 0 hundreds + 0 tens + 8 ones, or 2 hundred millions + 4 ten millions + 8 millions + 6 hundred thousands + 4 ten thousands + 5 thousands + 8 ones **21.** Eighty-five **23.** Eighty-eight thousand **25.** One hundred twenty-three thousand, seven hundred sixty-five **27.** Seven billion, seven hundred fifty-four million, two hundred eleven thousand, five hundred seventy-seven
29. Seven hundred one thousand, seven hundred ninety-nine
31. Two million, four hundred seventy-four thousand, two hundred eighty **33.** 632,896 **35.** 50,324 **37.** 2,233,812
39. 8,000,000,000 **41.** 40,000,000 **43.** 30,000,103
45. 64,186,000 **47.** 138

Calculator Corner, p. 10

1. 121 **2.** 1602 **3.** 1932 **4.** 864

Exercise Set 1.2, p. 12

RC1. addends **RC2.** sum **RC3.** 0 **RC4.** perimeter
1. 387 **3.** 164 **5.** 5198 **7.** 100 **9.** 8503 **11.** 5266
13. 4466 **15.** 6608 **17.** 34,432 **19.** 101,310 **21.** 230
23. 18,424 **25.** 31,685 **27.** 132 yd **29.** 1661 ft **31.** 570 ft
33. 8 ten thousands **34.** Nine billion, three hundred forty-six million, three hundred ninety-nine thousand, four hundred sixty-eight **35.** $1 + 99 = 100, 2 + 98 = 100, \ldots, 49 + 51 = 100$. Then 49 100's = 4900 and $4900 + 50 + 100 = 5050$.

Calculator Corner, p. 15

1. 28 **2.** 47 **3.** 67 **4.** 119 **5.** 2128 **6.** 2593

Exercise Set 1.3, p. 17

RC1. minuend **RC2.** subtraction symbol **RC3.** subtrahend
RC4. difference
1. 44 **3.** 533 **5.** 39 **7.** 14 **9.** 369 **11.** 26 **13.** 234
15. 417 **17.** 5382 **19.** 2778 **21.** 3069 **23.** 1089
25. 7748 **27.** 4144 **29.** 56 **31.** 454 **33.** 3749
35. 2191 **37.** 43,028 **39.** 95,974 **41.** 4418 **43.** 1305
45. 9989 **47.** 48,017 **49.** 1345 **50.** 924 **51.** 22,692
52. 10,920 **53.** Six million, three hundred seventy-five thousand, six hundred two **54.** 9 thousands + 1 hundred + 0 tens + 3 ones, or 9 thousands + 1 hundred + 3 ones **55.** 3; 4

Calculator Corner, p. 21

1. 448 **2.** 21,970 **3.** 6380 **4.** 39,564 **5.** 180,480
6. 2,363,754

Exercise Set 1.4, p. 23

RC1. factors **RC2.** product **RC3.** 0 **RC4.** 1
1. 520 **3.** 564 **5.** 1527 **7.** 64,603 **9.** 4770 **11.** 3995
13. 870 **15.** 1920 **17.** 46,296 **19.** 14,652 **21.** 258,312
23. 798,408 **25.** 20,723,872 **27.** 362,128 **29.** 302,220
31. 49,101,136 **33.** 25,236,000 **35.** 20,064,048
37. 529,984 sq mi **39.** 8100 sq ft **41.** 12,685 **42.** 10,834
43. 8889 **44.** 254,119 **45.** 4 hundred thousands **46.** 0
47. 1 ten thousand + 2 thousands + 8 hundreds + 4 tens + 7 ones
48. Seven million, four hundred thirty-two thousand
49. 247,464 sq ft

Calculator Corner, p. 30

1. 28 **2.** 123 **3.** 323 **4.** 36

Exercise Set 1.5, p. 32

RC1. quotient **RC2.** dividend **RC3.** remainder
RC4. divisor
1. 12 **3.** 1 **5.** 22 **7.** 0 **9.** Not defined **11.** 6
13. 55 R 2 **15.** 108 **17.** 307 **19.** 753 R 3 **21.** 74 R 1
23. 92 R 2 **25.** 1703 **27.** 987 R 5 **29.** 12,700 **31.** 127
33. 52 R 52 **35.** 29 R 5 **37.** 40 R 12 **39.** 90 R 22 **41.** 29
43. 105 R 3 **45.** 1609 R 2 **47.** 1007 R 1 **49.** 23
51. 107 R 1 **53.** 370 **55.** 609 R 15 **57.** 304
59. 3508 R 219 **61.** 8070 **63.** 1241 **64.** 66,444
65. 19,800 **66.** 9380 **67.** 40 ft **68.** 99 sq ft
69. 54, 122; 33, 2772; 4, 8 **71.** 30 buses

Mid-Chapter Review: Chapter 1, p. 35

1. False **2.** True **3.** True **4.** False **5.** True
6. False **7.** Ninety-five million, four hundred six thousand, two hundred thirty-seven

8. $\begin{array}{r} \scriptstyle 5\ 9\ 14 \\ \not{6}\,\not{0}\,\not{4} \\ -\ 4\ 9\ 7 \\ \hline 1\ 0\ 7 \end{array}$ **9.** 6 hundreds **10.** 6 ten thousands

11. 6 thousands **12.** 6 ones **13.** 2 **14.** 6 **15.** 5 **16.** 1
17. 5 thousands + 6 hundreds + 0 tens + 2 ones, or 5 thousands + 6 hundreds + 2 ones **18.** 6 ten thousands + 9 thousands + 3 hundreds + 4 tens + 5 ones **19.** One hundred thirty-six
20. Sixty-four thousand, three hundred twenty-five **21.** 308,716
22. 4,567,216 **23.** 798 **24.** 1030 **25.** 7922 **26.** 7534
27. 465 **28.** 339 **29.** 1854 **30.** 4328 **31.** 216
32. 15,876 **33.** 132,275 **34.** 5,679,870 **35.** 253

36. 112 R 5 **37.** 23 R 19 **38.** 144 R 31 **39.** 25 m **40.** 8 sq in. **41.** When numbers are being added, it does not matter how they are grouped. **42.** Subtraction is not commutative. For example, $5 - 2 = 3$, but $2 - 5 \neq 3$. **43.** Answers will vary. Suppose one coat costs \$150. Then the multiplication $4 \cdot \$150$ gives the cost of four coats. Or, suppose one ream of copy paper costs \$4. Then the multiplication $\$4 \cdot 150$ gives the cost of 150 reams. **44.** If we use the definition of division, $0 \div 0 = a$ such that $a \cdot 0 = 0$. We see that a could be *any* number since $a \cdot 0 = 0$ for any number a. Thus we cannot say that $0 \div 0 = 0$. This is why we agree not to allow division by 0.

Exercise Set 1.6, p. 43

RC1. True **RC2.** False **RC3.** False **RC4.** True
1. 50 **3.** 460 **5.** 730 **7.** 900 **9.** 100 **11.** 1000 **13.** 9100 **15.** 32,800 **17.** 6000 **19.** 8000 **21.** 45,000 **23.** 373,000 **25.** $80 + 90 = 170$ **27.** $8070 - 2350 = 5720$ **29.** 220; incorrect **31.** 890; incorrect **33.** $7300 + 9200 = 16{,}500$ **35.** $6900 - 1700 = 5200$ **37.** 1600; correct **39.** 1500; correct **41.** $10{,}000 + 5000 + 9000 + 7000 = 31{,}000$ **43.** $92{,}000 - 23{,}000 = 69{,}000$ **45.** $50 \cdot 70 = 3500$ **47.** $30 \cdot 30 = 900$ **49.** $900 \cdot 300 = 270{,}000$ **51.** $400 \cdot 200 = 80{,}000$ **53.** $350 \div 70 = 5$ **55.** $8450 \div 50 = 169$ **57.** $1200 \div 200 = 6$ **59.** $8400 \div 300 = 28$ **61.** \$800 **63.** \$1200; no **65.** Answers will vary depending on the options chosen. **67. (a)** \$309,600; **(b)** \$360,000 **69.** 90 people **71.** $<$ **73.** $>$ **75.** $<$ **77.** $>$ **79.** $>$ **81.** $>$ **83.** $1{,}335{,}475 < 4{,}134{,}519$, or $4{,}134{,}519 > 1{,}335{,}475$ **85.** $97{,}382 < 98{,}817$, or $98{,}817 > 97{,}382$ **87.** 86,754 **88.** 13,589 **89.** 48,824 **90.** 4415 **91.** 1702 **92.** 17,748 **93.** 54 R 4 **94.** 208 **95.** Left to the student **97.** Left to the student

Exercise Set 1.7, p. 52

RC1. (c) **RC2.** (a) **RC3.** (d) **RC4.** (b)
1. 14 **3.** 0 **5.** 90,900 **7.** 450 **9.** 352 **11.** 25 **13.** 29 **15.** 0 **17.** 79 **19.** 45 **21.** 8 **23.** 14 **25.** 32 **27.** 143 **29.** 17,603 **31.** 37 **33.** 1035 **35.** 66 **37.** 324 **39.** 335 **41.** 18,252 **43.** 104 **45.** 45 **47.** 4056 **49.** 2847 **51.** 15 **53.** 205 **55.** 457 **57.** 142 R 5 **58.** 142 **59.** 334 **60.** 334 R 11 **61.** $<$ **62.** $>$ **63.** $>$ **64.** $<$ **65.** 6,376,000 **66.** 6,375,600 **67.** 347

Translating for Success, p. 62

1. E **2.** M **3.** D **4.** G **5.** A **6.** O **7.** F **8.** K **9.** J **10.** H

Exercise Set 1.8, p. 63

RC1. Familiarize. **RC2.** Translate. **RC3.** Solve. **RC4.** Check.
1. 1962 ft **3.** 1450 ft **5.** 95 milligrams **7.** 18 rows **9.** 43 events **11.** 2054 mi **13.** 2,073,600 pixels **15.** 168 hr **17.** \$273 per month **19.** \$197 **21.** \$7092 **23.** 151,500 **25.** \$78 **27.** \$40 per month **29.** \$24,456 **31.** 35 weeks; 2 episodes **33.** 21 columns **35.** 236 gal **37. (a)** 4200 sq ft; **(b)** 268 ft **39.** 56 cartons **41.** 645 mi; 5 in. **43.** \$247 **45.** 525 min, or 8 hr 45 min **47.** 168,300 jobs **49.** 104 seats **51.** 32 \$10 bills **53.** \$400 **55.** 106 bones **57.** 8273 **58.** 7759 **59.** 806,985 **60.** 147 R 4 **61.** 34 m **62.** 9706 sq ft **63.** $200 \times 600 = 120{,}000$ **64.** 66 **65.** 792,000 mi; 1,386,000 mi

Calculator Corner, p. 71

1. 243 **2.** 15,625 **3.** 20,736 **4.** 2048

Calculator Corner, p. 73

1. 49 **2.** 85 **3.** 36 **4.** 0 **5.** 73 **6.** 49

Exercise Set 1.9, p. 75

RC1. exponent **RC2.** squared **RC3.** multiplication **RC4.** 3
1. 3^4 **3.** 5^2 **5.** 7^5 **7.** 10^3 **9.** 49 **11.** 729 **13.** 20,736 **15.** 243 **17.** 22 **19.** 20 **21.** 100 **23.** 1 **25.** 49 **27.** 5 **29.** 434 **31.** 41 **33.** 88 **35.** 4 **37.** 303 **39.** 20 **41.** 70 **43.** 295 **45.** 32 **47.** 906 **49.** 62 **51.** 102 **53.** 32 **55.** \$94 **57.** 401 **59.** 110 **61.** 7 **63.** 544 **65.** 708 **67.** 27 **69.** 452 **70.** 835 **71.** 13 **72.** 37 **73.** 4898 **74.** 100 **75.** 104,286 sq mi **76.** 98 gal **77.** 24; $1 + 5 \cdot (4 + 3) = 36$ **79.** 7; $12 \div (4 + 2) \cdot 3 - 2 = 4$

Summary and Review: Chapter 1, p. 78

Vocabulary Reinforcement

1. perimeter **2.** minuend **3.** digits; periods **4.** dividend **5.** factors; product **6.** additive **7.** associative **8.** divisor; remainder; dividend

Concept Reinforcement

1. True **2.** True **3.** False **4.** False **5.** True **6.** False

Important Concepts

1. 2 thousands **2.** 65,302 **3.** 3237 **4.** 225,036 **5.** 315 R 14 **6.** 36,000 **7.** $<$ **8.** 36 **9.** 216

Review Exercises

1. 8 thousands **2.** 3 **3.** 2 thousands + 7 hundreds + 9 tens + 3 ones **4.** 5 ten thousands + 6 thousands + 0 hundreds + 7 tens + 8 ones, or 5 ten thousands + 6 thousands + 7 tens + 8 ones **5.** 4 millions + 0 hundred thousands + 0 ten thousands + 7 thousands + 1 hundred + 0 tens + 1 one, or 4 millions + 7 thousands + 1 hundred + 1 one **6.** Sixty-seven thousand, eight hundred nineteen **7.** Two million, seven hundred eighty-one thousand, four hundred twenty-seven **8.** 476,588 **9.** 1,640,000,000 **10.** 14,272 **11.** 66,024 **12.** 21,788 **13.** 98,921 **14.** 5148 **15.** 1689 **16.** 2274 **17.** 17,757 **18.** 5,100,000 **19.** 6,276,800 **20.** 506,748 **21.** 27,589 **22.** 5,331,810 **23.** 12 R 3 **24.** 5 **25.** 913 R 3 **26.** 384 R 1 **27.** 4 R 46 **28.** 54 **29.** 452 **30.** 5008 **31.** 4389 **32.** 345,800 **33.** 345,760 **34.** 346,000 **35.** 300,000 **36.** $>$ **37.** $<$ **38.** $41{,}300 + 19{,}700 = 61{,}000$ **39.** $38{,}700 - 24{,}500 = 14{,}200$ **40.** $400 \cdot 700 = 280{,}000$ **41.** 8 **42.** 45 **43.** 58 **44.** 0 **45.** 4^3 **46.** 10,000 **47.** 36 **48.** 65 **49.** 233 **50.** 260 **51.** 165 **52.** \$502 **53.** \$484 **54.** 1982 **55.** 19 cartons **56.** \$13,585 **57.** 14 beehives **58.** 98 sq ft; 42 ft **59.** 137 beakers filled; 13 mL left over **60.** \$27,598 **61.** B **62.** A **63.** D **64.** 8 **65.** $a = 8, b = 4$ **66.** 6 days

Understanding Through Discussion and Writing

1. No; if subtraction were associative, then $a - (b - c) = (a - b) - c$ for any a, b, and c. But, for example,

$$12 - (8 - 4) = 12 - 4 = 8,$$

whereas

$$(12 - 8) - 4 = 4 - 4 = 0.$$

Since $8 \neq 0$, this example shows that subtraction is not associative. **2.** By rounding prices and estimating their sum, a shopper can estimate the total grocery bill while shopping. This is particularly useful if the shopper wants to spend no more than a certain amount.

3. Answers will vary. Anthony is driving from Kansas City to Minneapolis, a distance of 512 mi. He stops for gas after driving 183 mi. How much farther must he drive? **4.** The parentheses are not necessary in the expression $9 - (4 \cdot 2)$. Using the rules for order of operations, the multiplication would be performed before the subtraction even if the parentheses were not present. The parentheses are necessary in the expression $(3 \cdot 4)^2$; $(3 \cdot 4)^2 = 12^2 = 144$, but $3 \cdot 4^2 = 3 \cdot 16 = 48$.

Test: Chapter 1, p. 83

1. [1.1a] 5 **2.** [1.1b] 8 thousands + 8 hundreds + 4 tens + 3 ones **3.** [1.1c] Thirty-eight million, four hundred three thousand, two hundred seventy-seven **4.** [1.2a] 9989 **5.** [1.2a] 63,791 **6.** [1.2a] 3165 **7.** [1.2a] 10,515 **8.** [1.3a] 3630 **9.** [1.3a] 1039 **10.** [1.3a] 6848 **11.** [1.3a] 5175 **12.** [1.4a] 41,112 **13.** [1.4a] 5,325,600 **14.** [1.4a] 2405 **15.** [1.4a] 534,264 **16.** [1.5a] 3 R 3 **17.** [1.5a] 70 **18.** [1.5a] 97 **19.** [1.5a] 805 R 8 **20.** [1.6a] 35,000 **21.** [1.6a] 34,530 **22.** [1.6a] 34,500 **23.** [1.6b] 23,600 + 54,700 = 78,300 **24.** [1.6b] 54,800 − 23,600 = 31,200 **25.** [1.6b] $800 \cdot 500 = 400,000$ **26.** [1.6c] > **27.** [1.6c] < **28.** [1.7b] 46 **29.** [1.7b] 13 **30.** [1.7b] 14 **31.** [1.7b] 381 **32.** [1.8a] 83 calories **33.** [1.8a] 20 staplers **34.** [1.8a] 1,256,615 sq mi **35.** **(a)** [1.2b], [1.4b] 300 in., 5000 sq in.; 264 in., 3872 sq in.; 228 in., 2888 sq in.; **(b)** [1.8a] 2112 sq in. **36.** [1.8a] 1852 12-packs; 7 cakes left over **37.** [1.8a] $95 **38.** [1.9a] 12^4 **39.** [1.9b] 343 **40.** [1.9b] 100,000 **41.** [1.9c] 31 **42.** [1.9c] 98 **43.** [1.9c] 2 **44.** [1.9c] 18 **45.** [1.9d] 216 **46.** [1.9c] A **47.** [1.4b], [1.8a] 336 sq in. **48.** [1.9c] 9 **49.** [1.8a] 80 payments

CHAPTER 2

Exercise Set 2.1, p. 90

RC1. True **RC2.** True **RC3.** False **RC4.** False **RC5.** True **RC6.** False **RC7.** True **RC8.** True **1.** 820; −541 **3.** 950,000,000; −460 **5.** 1454; −55 **7.** < **9.** > **11.** > **13.** < **15.** < **17.** < **19.** > **21.** 57 **23.** 0 **25.** 24 **27.** 53 **29.** 8 **31.** 7 **33.** −7 **35.** 0 **37.** 21 **39.** −53 **41.** 1 **43.** 7 **45.** −9 **47.** −17 **49.** 23 **51.** −1 **53.** 85 **55.** −345 **57.** 0 **59.** −8 **61.** 825 **62.** 125 **63.** 7106 **64.** 4 **65.** 42 **66.** 69 **67.** > **69.** = **71.** −1, 0, 1 **73.** $-100, -5, 0, |3|, 4, |-6|, 7^2, 10^2, 2^7, 2^{10}$

Calculator Corner, p. 94

1. 13 **2.** −8

Exercise Set 2.2, p. 95

RC1. right **RC2.** left **RC3.** left **RC4.** right **1.** −5 **3.** −4 **5.** 6 **7.** 0 **9.** −4 **11.** −12 **13.** −11 **15.** −15 **17.** 42 **19.** 5 **21.** −4 **23.** 0 **25.** 0 **27.** −9 **29.** 7 **31.** 0 **33.** 45 **35.** −3 **37.** 0 **39.** −10 **41.** −24 **43.** −5 **45.** −21 **47.** −30 **49.** 6 **51.** −21 **53.** 25 **55.** −17 **57.** 6 **59.** −65 **61.** −160 **63.** −62 **65.** −23 **67.** 6681 **68.** 73 **69.** 3 ten thousands + 9 thousands + 4 hundreds + 1 ten + 7 ones **70.** 2352 **71.** 32 **72.** 3500 **73.** −40 **75.** −6483 **77.** All negative **79.** negative **81.** negative

Exercise Set 2.3, p. 100

RC1. (c) **RC2.** (b) **RC3.** (d) **RC4.** (a) **1.** −4 **3.** −7 **5.** −6 **7.** 0 **9.** −4 **11.** −7 **13.** −6 **15.** 0 **17.** 0 **19.** 14 **21.** 11 **23.** −14 **25.** 5 **27.** −7 **29.** −1 **31.** 18 **33.** −10 **35.** −3 **37.** −21 **39.** 5 **41.** −8 **43.** 12 **45.** −19 **47.** −68 **49.** −81 **51.** 116 **53.** 0 **55.** 55 **57.** 19 **59.** −62 **61.** −139 **63.** 6 **65.** 107 **67.** 219 **69.** Profit of $4300 **71.** 3780 m **73.** 17 lb **75.** −3° **77.** 18 points **79.** −10,411 ft **81.** −$83, or $83 in debt **83.** 64 **84.** 4896 **85.** 1 **86.** 4147 **87.** 8 cans **88.** 288 oz **89.** 35 **90.** 3 **91.** 32 **92.** 165 **93.** −309,882 **95.** False; $3 - 0 \neq 0 - 3$ **97.** True **99.** True **101.** 17 **103.** Up 15 points

Calculator Corner, p. 107

1. 148,035,889 **2.** −1,419,857 **3.** −1,124,864 **4.** 1,048,576 **5.** −531,441 **6.** −117,649 **7.** −7776 **8.** −19,683

Exercise Set 2.4, p. 107

RC1. negative **RC2.** positive **RC3.** positive **RC4.** negative **1.** −16 **3.** −60 **5.** −48 **7.** −30 **9.** 15 **11.** 18 **13.** 42 **15.** 20 **17.** −120 **19.** 0 **21.** 72 **23.** −340 **25.** 400 **27.** 0 **29.** 24 **31.** 420 **33.** −70 **35.** 30 **37.** 0 **39.** −294 **41.** 36 **43.** −125 **45.** 10,000 **47.** −16 **49.** −243 **51.** 1 **53.** −121 **55.** −64 **57.** 532,500 **58.** 60,000,000 **59.** 80 **60.** 2550 **61.** 5 **62.** 48 **63.** 40 sq ft **64.** 240 cartons **65.** 5 trips **66.** 4 trips **67.** 243 **69.** 0 **71.** 7 **73.** −2209 **75.** 130,321 **77.** −2197 **79.** 116,875 **81.** −$23 **83.** **(a)** Both m and n must be odd. **(b)** At least one of m and n must be even.

Calculator Corner, p. 112

1. −4 **2.** −2 **3.** 787

Exercise Set 2.5, p. 113

RC1. 0 **RC2.** undefined **RC3.** undefined **RC4.** 0 **1.** −6 **3.** −13 **5.** −2 **7.** 4 **9.** −9 **11.** 2 **13.** −12 **15.** −8 **17.** Undefined **19.** −8 **21.** −23 **23.** 0 **25.** −19 **27.** −41 **29.** −7 **31.** −7 **33.** −334 **35.** 23 **37.** 8 **39.** 12 **41.** −1 **43.** 0 **45.** −10 **47.** −86 **49.** −9 **51.** 18 **53.** −10 **55.** −67 **57.** 10 **59.** −25 **61.** −7988 **63.** −3000 **65.** 60 **67.** 1 **69.** −37 **71.** −22 **73.** 2 **75.** 7 **77.** Undefined **79.** 3 **81.** 2 **83.** 0 **85.** 28 sq in. **86.** 248 rooms **87.** 12 gal **88.** 27 gal **89.** 150 cal **90.** 672 g **91.** 4 pieces; 2 pieces **92.** 4 lozenges; 4 lozenges **93.** 0 **95.** 0 **97.** −2 **99.** 992 **101.** [(] [1] [5] [x^2] [−] [5] [x^y] [3] [)] [÷] [(] [3] [x^2] [+] [4] [x^2] [)] [=] **103.** 5 **105.** Positive **107.** Negative **109.** Positive

Mid-Chapter Review: Chapter 2, p. 116

1. False **2.** False **3.** True **4.** $-x = -(-4) = 4$; $-(-x) = -(-(-4)) = -(4) = -4$ **5.** $5 - 13 = 5 + (-13) = -8$ **6.** $-6 - (-7) = -6 + 7 = 1$ **7.** 450; −79 **8.** −9 **9.** < **10.** < **11.** < **12.** > **13.** 38 **14.** 18 **15.** 0 **16.** 12 **17.** 56 **18.** −3 **19.** 0 **20.** 49 **21.** 19 **22.** 23 **23.** −2 **24.** −16 **25.** 0 **26.** −17 **27.** 1 **28.** 2 **29.** −26 **30.** −4 **31.** −13 **32.** 16 **33.** 6 **34.** −12 **35.** −36 **36.** −54 **37.** 26 **38.** 82 **39.** 81 **40.** −81 **41.** 25 **42.** −5 **43.** 75 **44.** 14 **45.** 42 **46.** −38 **47.** −13 **48.** 2 **49.** 33°C **50.** $44 **51.** Answers may vary. The student may be confusing distance from 0 on the number line with position on the number line. Although −45 is farther from 0 than −21, it is less than −21 because it is to the left of −21 on the number line. **52.** Answers may vary. Subtraction of integers is not associative, as can be illustrated by an example: Compare $3 - (9 - 10) = 4$ with $(3 - 9) - 10 = -16$. **53.** Answers may vary. If we think of the addition on the number line, we start at a negative number and move to the left. This always brings us to a point on the negative portion of the number line. **54.** Yes: consider $m - (-n)$, where both m and n are positive. Then $m - (-n) = m + n$. Now $m + n$, the sum of two positive numbers, is positive.

Calculator Corner, p. 120

1. 243 **2.** 1024 **3.** −32 **4.** −3125

Exercise Set 2.6, p. 122

RC1. Division **RC2.** Multiplication **RC3.** Multiplication **RC4.** Division
1. 20¢ **3.** −2 **5.** 1 **7.** 18 yr **9.** −7 **11.** 14 ft **13.** 14 ft **15.** −21 **17.** −21 **19.** 400 ft **21.** 22 **23.** 0 **25.** 36 **27.** −3 **29.** 11 **31.** −1100 **33.** $\frac{-5}{t}; \frac{5}{-t}$ **35.** $\frac{n}{-b}; -\frac{n}{b}$ **37.** $\frac{-9}{p}; -\frac{9}{p}$ **39.** $\frac{14}{-w}; -\frac{14}{w}$ **41.** −5; −5; −5 **43.** −27; −27; −27 **45.** 36; −12 **47.** 45; 45 **49.** 216; −216 **51.** 1; 1 **53.** 32; −32 **55.** $5a + 5b$ **57.** $4x + 4$ **59.** $2b + 10$ **61.** $7 - 7t$ **63.** $30x - 12$ **65.** $8x + 56 + 48y$ **67.** $-7y + 14$ **69.** $3x + 6$ **71.** $-4x + 12y + 8z$ **73.** $8a - 24b + 8c$ **75.** $4x - 12y - 28z$ **77.** $20a - 25b + 5c - 10d$ **79.** $-3m - 2n$ **81.** $-2a + 3b - 4$ **83.** $-x + y + z$ **85.** Twenty-three million, forty-three thousand, nine hundred twenty-one **86.** 901 **87.** 5280 − 2480 = 2800 **88.** 994 **89.** 5 in. **90.** $63 **91.** 698°F **93.** 4438 **95.** 279 **97.** 2 **99.** 2,560,000 **101.** −32 [×] (88 [−] 29) = −1888 **103.** True **105.** True

Translating for Success, p. 129

1. J **2.** G **3.** A **4.** K **5.** H **6.** N **7.** F **8.** M **9.** D **10.** C

Exercise Set 2.7, p. 130

RC1. closed **RC2.** perimeter **RC3.** perimeter **RC4.** square
1. $2a, 5b, -7c$ **3.** $mn, -6n, 8$ **5.** $3x^2y, -4y^2, -2z^3$ **7.** $14x$ **9.** $-5a$ **11.** $3x + 6y$ **13.** $-13a + 62$ **15.** $-4 + 4t + 6y$ **17.** $6a + 4b - 2$ **19.** $-1 + a - 12b$ **21.** $7x^2 + 3y$ **23.** $7x^4 + y^3$ **25.** $6a^2 - 3a$ **27.** $3x^3 - 8x^2 + 4$ **29.** $9x^3y - xy^3 + 3xy$ **31.** $-4a^6 - 3b^4 + 2a^6b^4$ **33.** 10 ft **35.** 42 km **37.** 8 m **39.** 210 ft **41.** 138 ft **43.** 36 ft **45.** 56 in. **47.** 260 cm **49.** 64 ft **51.** 17 servings **52.** 210 **53.** 16 **54.** 13 **55.** 15 **56.** 25 **57.** 375 **58.** 16 **59.** $7x + 1$ **61.** $-29 - 3a$ **63.** $-10 - x - 27y$ **65.** $45 **67.** 912 mm

Exercise Set 2.8, p. 139

RC1. (c) **RC2.** (a) **RC3.** (d) **RC4.** (b)
1. Equivalent equations **3.** Equivalent expressions **5.** Equivalent expressions **7.** Equivalent equations **9.** Equivalent expressions **11.** Equivalent equations **13.** −3 **15.** −8 **17.** 18 **19.** −15 **21.** 32 **23.** −17 **25.** 17 **27.** 0 **29.** 10 **31.** −14 **33.** 5 **35.** 0 **37.** −11 **39.** −56 **41.** 12 **43.** 390 **45.** 4 **47.** −9 **49.** −15 **51.** −26 **53.** −5 **55.** −4 **57.** −172 **59.** −4 **61.** 7 **63.** 3 **65.** −3 **67.** −7 **69.** −8 **71.** 8 **73.** 24 **75.** 6 **77.** 5 **79.** −8 **81.** 29 **82.** 7 **83.** 8 **84.** 27 **85.** 16 **86.** 26 **87.** 1 **88.** 24 **89.** 8 **91.** 0 **93.** −5 **95.** 29 **97.** −20 **99.** 1027 **101.** −343 **103.** 17

Summary and Review: Chapter 2, p. 142

Vocabulary Reinforcement

1. integers **2.** absolute **3.** opposite **4.** not defined **5.** substituting **6.** constant **7.** distributive law **8.** addition principle **9.** equivalent

Concept Reinforcement

1. True **2.** True **3.** False **4.** False

Study Guide

1. **(a)** 17; **(b)** 300 **2.** 21 **3.** 14 **4.** −90 **5.** −11 **6.** −12 **7.** $30x - 40y - 5z$ **8.** $17a - 7b$ **9.** −6

Review Exercises

1. −45; 72 **2.** > **3.** < **4.** > **5.** 39 **6.** 23 **7.** 0 **8.** 72 **9.** 59 **10.** −9 **11.** −11 **12.** 6 **13.** −24 **14.** −12 **15.** 23 **16.** −1 **17.** 0 **18.** −4 **19.** 12 **20.** 92 **21.** −84 **22.** −40 **23.** −3 **24.** −5 **25.** 0 **26.** −5 **27.** 2 **28.** −180 **29.** −20 **30.** −62 **31.** 7 **32.** −6, −6, −6 **33.** $20x + 36$ **34.** $6a - 12b + 15$ **35.** $-20x - 10y$ **36.** $17a$ **37.** $6x$ **38.** $-3m + 6$ **39.** 36 in. **40.** 100 cm **41.** −8 **42.** −9 **43.** −13 **44.** 11 **45.** 15 **46.** −7 **47.** 8-yd gain **48.** −$130 **49.** A **50.** B **51.** 403 and 397 **52.** **(a)** $-7 + (-6) + (-5) + (-4) + (-3) + (-2) + (-1) + 0 + 1 + 2 + 3 + 4 + 5 + 6 + 7 + 8 = 8$; **(b)** 0 **53.** 662,582 **54.** −88,174 **55.** −240 **56.** $x < -2$ **57.** $x < 0$

Understanding Through Discussion and Writing

1. We know that the product of an even number of negative numbers is positive, and the product of an odd number of negative numbers is negative. Since $(-7)^8$ is equivalent to the product of eight negative numbers, it will be a positive number. Similarly, since $(-7)^{11}$ is equivalent to the product of eleven negative numbers, it will be a negative number. **2.** The expression $-x$ does not always represent a negative number. When $x = 0$, $-x = 0$, and when x is negative, $-x$ is positive. **3.** Jake is expecting the multiplication to be performed before the division. **4.** The expression $-x^2$ represents a negative number, except when $x = 0$. For all other values of x, x^2 is positive, and the opposite of x^2 is negative.

Test: Chapter 2, p. 147

1. [2.1a] −542; 307 **2.** [2.1b] > **3.** [2.1c] 739 **4.** [2.1d] −19 **5.** [2.2a] −11 **6.** [2.2a] −21 **7.** [2.2a] 9 **8.** [2.3a] −12 **9.** [2.3a] −15 **10.** [2.3a] −24 **11.** [2.3a] 19 **12.** [2.3a] 38 **13.** [2.4b] −64 **14.** [2.4a] −270 **15.** [2.4a] 0 **16.** [2.5a] 8 **17.** [2.5a] −8 **18.** [2.5b] −1 **19.** [2.5b] 25 **20.** [2.3b] 14°F higher **21.** [2.6a] −3 **22.** [2.6b] $14x + 21y - 7$ **23.** [2.7a] $4x - 17$ **24.** [2.7b] 20 ft **25.** [2.8b] 5 **26.** [2.8a] −12 **27.** [2.8b] −95 **28.** [2.8d] 4 **29.** [2.6b] C **30.** [2.7b] 66 ft **31.** [2.5b] $35x - 7$ **32.** [2.5b] $-24x - 57$ **33.** [2.5b] 103,097 **34.** [2.5b] 1086

CHAPTER 3

Calculator Corner, p. 151

1. Yes **2.** No **3.** No **4.** Yes

Exercise Set 3.1, p. 154

RC1. (c) **RC2.** (a) **RC3.** (d) **RC4.** (f) **RC5.** (b) **RC6.** (e) **1.** 7, 14, 21, 28, 35, 42, 49, 56, 63, 70 **3.** 20, 40, 60, 80, 100, 120, 140, 160, 180, 200 **5.** 3, 6, 9, 12, 15, 18, 21, 24, 27, 30 **7.** 12, 24, 36, 48, 60, 72, 84, 96, 108, 120 **9.** 10, 20, 30, 40, 50, 60, 70, 80, 90, 100 **11.** 25, 50, 75, 100, 125, 150, 175, 200, 225, 250 **13.** No **15.** Yes **17.** Yes **19.** Yes; the sum of the digits is 12, which is divisible by 3. **21.** No; the ones digit is not 0 or 5. **23.** Yes; the ones digit is 0. **25.** Yes; the sum of the digits is 18, which is divisible by 9. **27.** No; the ones digit is not even. **29.** No; the ones digit is not even. **31.** 6825 is divisible by 3 and 5. **33.** 119,117 is divisible by none of these numbers. **35.** 127,575 is divisible by 3, 5, and 9. **37.** 9360 is divisible by 2, 3, 5, 6, 9, and 10.

39. 555; 300; 36; 45,270; 711; 13,251; 8064 **41.** 300; 45,270 **43.** 300; 36; 45,270; 8064 **45.** 56; 324; 784; 200; 42; 812; 402 **47.** 55,555; 200; 75; 2345; 35; 1005 **49.** 324 **51.** 53 **52.** 5 **53.** −8 **54.** −24 **55.** 45 gal **56.** 4320 min **57.** 125 **58.** 16 **59.** 9^3 **60.** 3^6 **61.** 99,969 **63.** 30 **65.** 210 **67.** 840 **69.** **(a)** $999a + 99b + 9c = 9(111a + 11b + c) = 3(333a + 33b + 3c)$; therefore, $999a + 99b + 9c$ is divisible by both 9 and 3; **(b)** if $a + b + c + d$ is divisible by 9, then $a + b + c + d = 9n$ for some number n. Then $abcd = 999a + 99b + 9c + 9n = 9(111a + 11b + c + n)$, so $abcd$ is divisible by 9. A similar argument holds for 3. **71.** 332,986,412 is divisible by 4 and 11.

Exercise Set 3.2, p. 161

RC1. True **RC2.** False **RC3.** True **RC4.** False **RC5.** False **1.** No **3.** Yes **5.** 1, 2, 3, 6, 9, 18 **7.** 1, 2, 3, 6, 9, 18, 27, 54 **9.** 1, 3, 9 **11.** 1, 13 **13.** 1, 2, 7, 14, 49, 98 **15.** 1, 3, 5, 15, 17, 51, 85, 255 **17.** Prime **19.** Composite **21.** Composite **23.** Prime **25.** Neither **27.** Composite **29.** Prime **31.** Prime **33.** $3 \cdot 3 \cdot 3$ **35.** $2 \cdot 7$ **37.** $2 \cdot 2 \cdot 2 \cdot 2 \cdot 5$ **39.** $5 \cdot 5$ **41.** $2 \cdot 31$ **43.** $2 \cdot 2 \cdot 5 \cdot 5$ **45.** $11 \cdot 13$ **47.** $11 \cdot 11$ **49.** $3 \cdot 7 \cdot 13$ **51.** $5 \cdot 5 \cdot 7$ **53.** $11 \cdot 19$ **55.** $2 \cdot 2 \cdot 2 \cdot 2 \cdot 3 \cdot 5 \cdot 5$ **57.** $3 \cdot 3 \cdot 7 \cdot 11$ **59.** $2 \cdot 2 \cdot 7 \cdot 103$ **61.** $2 \cdot 3 \cdot 11 \cdot 17$ **63.** −26 **64.** 256 **65.** 8 **66.** −23 **67.** 1 **68.** −98 **69.** 0 **70.** 0 **71.** $13 \cdot 19 \cdot 19 \cdot 29$ **73.** $23 \cdot 31 \cdot 61 \cdot 73 \cdot 149$ **75.** Answers may vary. One arrangement is a three-dimensional rectangular array consisting of 2 tiers of 12 objects each, where each tier consists of a rectangular array of 4 rows with 3 objects each.
77.

Product	56	63	36	72	140	96
Factor	7	7	2	2	10	8
Factor	8	9	18	36	14	12
Sum	15	16	20	38	24	20

Product	48	168	110	90	432	63
Factor	6	21	10	9	24	3
Factor	8	8	11	10	18	21
Sum	14	29	21	19	42	24

Exercise Set 3.3, p. 168

RC1. (b) **RC2.** (a) **RC3.** (c) **RC4.** (e) **RC5.** (f) **RC6.** (d) **1.** Numerator: 3; denominator: 4 **3.** Numerator: 7; denominator: −9 **5.** Numerator: $2x$; denominator: $3y$ **7.** $\frac{6}{12}$ **9.** $\frac{1}{8}$ **11.** $\frac{3}{4}$ **13.** $\frac{4}{8}$ **15.** $\frac{12}{12}$ **17.** $\frac{9}{8}$ **19.** $\frac{4}{3}$ **21.** $\frac{11}{9}$ **23.** $\frac{5}{8}$ **25.** $\frac{4}{7}$ **27.** $\frac{12}{16}$ **29.** $\frac{38}{16}$ **31.** **(a)** $\frac{2}{8}$; **(b)** $\frac{6}{8}$ **33.** **(a)** $\frac{3}{8}$; **(b)** $\frac{5}{8}$ **35.** **(a)** $\frac{5}{7}$; **(b)** $\frac{5}{2}$; **(c)** $\frac{2}{7}$; **(d)** $\frac{2}{5}$ **37.** **(a)** $\frac{4}{15}$; **(b)** $\frac{4}{11}$; **(c)** $\frac{11}{15}$ **39.** **(a)** $\frac{1060}{100{,}000}$; **(b)** $\frac{743}{100{,}000}$; **(c)** $\frac{865}{100{,}000}$; **(d)** $\frac{1026}{100{,}000}$; **(e)** $\frac{905}{100{,}000}$; **(f)** $\frac{1728}{100{,}000}$ **41.** $\frac{4}{7}$ **43.** 0 **45.** 7 **47.** 1 **49.** 1 **51.** 0 **53.** $19x$ **55.** 1 **57.** −87 **59.** 0 **61.** Undefined **63.** $7n$ **65.** Undefined **67.** −210 **68.** −322 **69.** 0 **70.** 0 **71.** 451 museums **72.** 71 gal **73.** $\frac{1}{6}$ **75.** $\frac{2}{16}$, or $\frac{1}{8}$

77. 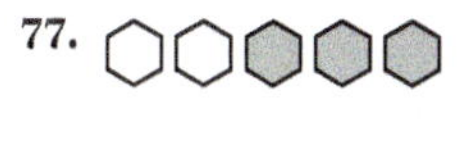 **79.** 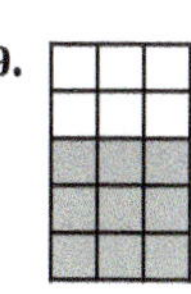 **81.** $\frac{52}{365}$ **83.** $\frac{3}{4}$; $\frac{1}{4}$

Exercise Set 3.4, p. 177

RC1. True **RC2.** True **RC3.** True **RC4.** True **1.** $\frac{3}{8}$ **3.** $\frac{-5}{6}$, or $-\frac{5}{6}$ **5.** $\frac{14}{3}$ **7.** $-\frac{7}{9}$, or $\frac{-7}{9}$ **9.** $\frac{5x}{6}$ **11.** $\frac{-6}{5}$, or $-\frac{6}{5}$ **13.** $\frac{2a}{7}$ **15.** $\frac{-17m}{6}$, or $-\frac{17m}{6}$ **17.** $\frac{6}{5}$ **19.** $\frac{2x}{7}$ **21.** $\frac{4}{15}$ **23.** $\frac{-1}{40}$, or $-\frac{1}{40}$ **25.** $\frac{2}{15}$ **27.** $\frac{2x}{9y}$ **29.** $\frac{9}{16}$ **31.** $\frac{14}{39}$ **33.** $\frac{-3}{50}$, or $-\frac{3}{50}$ **35.** $\frac{7a}{64}$ **37.** $\frac{100}{y}$ **39.** $\frac{-147}{20}$, or $-\frac{147}{20}$ **41.** $\frac{40}{3}$ yd **43.** $\frac{1}{2625}$ **45.** $\frac{1}{24}$ **47.** $\frac{9}{20}$ **49.** 204 **50.** 700 **51.** 13 **52.** −90 **53.** 50 **54.** 6399 **55.** $\frac{71{,}269}{180{,}433}$ **57.** $\frac{-56}{1125}$, or $-\frac{56}{1125}$ **59.** $\frac{1}{80}$ gal

Calculator Corner, p. 182

1. $\frac{14}{15}$ **2.** $\frac{7}{8}$ **3.** $\frac{138}{167}$ **4.** $\frac{7}{25}$

Exercise Set 3.5, p. 184

RC1. Equivalent **RC2.** simplify **RC3.** common **RC4.** cross **1.** $\frac{5}{10}$ **3.** $\frac{-36}{-48}$ **5.** $\frac{35}{50}$ **7.** $\frac{11t}{5t}$ **9.** $\frac{20}{4}$ **11.** $-\frac{51}{54}$ **13.** $\frac{15}{-40}$ **15.** $\frac{-42}{132}$ **17.** $\frac{x}{8x}$ **19.** $\frac{-10a}{7a}$ **21.** $\frac{4ab}{9ab}$ **23.** $\frac{12b}{27b}$ **25.** $\frac{1}{2}$ **27.** $-\frac{2}{3}$ **29.** $\frac{2}{5}$ **31.** 3 **33.** $\frac{3}{4}$ **35.** $-\frac{12}{7}$ **37.** $\frac{3}{4}$ **39.** $\frac{-1}{3}$ **41.** −5 **43.** $\frac{7}{8}$ **45.** $\frac{-2}{3}$ **47.** $\frac{2}{5}$ **49.** $\frac{9}{8}$ **51.** $\frac{12}{13}$ **53.** $\frac{17}{19}$ **55.** $\frac{3}{8}$ **57.** $\frac{3y}{2}$ **59.** $\frac{-9}{10b}$ **61.** $=$ **63.** $\neq$ **65.** $=$ **67.** $\neq$ **69.** $\neq$ **71.** $=$ **73.** $\neq$ **75.** $=$ **77.** $<$ **78.** $<$ **79.** $>$ **80.** $<$ **81.** 3520 **82.** 89 **83.** 6498 **84.** 85 **85.** $\frac{17}{29}$ **87.** $-\frac{29x}{15y}$ **89.** **(a)** $\frac{4}{10} = \frac{2}{5}$; **(b)** $\frac{6}{10} = \frac{3}{5}$ **91.** No; $\frac{162}{489} \neq \frac{193}{555}$ because $162 \cdot 555 \neq 489 \cdot 193$.

Mid-Chapter Review: Chapter 3, p. 186

1. True **2.** False **3.** False **4.** True **5.** $\frac{25}{25} = 1$ **6.** $\frac{0}{9} = 0$ **7.** $\frac{8}{1} = 8$ **8.** $\frac{6}{13} = \frac{18}{39}$ **9.** $\frac{70}{225} = \frac{2 \cdot 5 \cdot 7}{3 \cdot 3 \cdot 5 \cdot 5} = \frac{5}{5} \cdot \frac{2 \cdot 7}{3 \cdot 3 \cdot 5} = 1 \cdot \frac{14}{45} = \frac{14}{45}$ **10.** 84; 17,576; 224; 132; 594; 504; 1632 **11.** 84; 300; 132; 500; 180 **12.** 17,576; 224; 500 **13.** 84; 300; 132; 120; 1632 **14.** 300; 180; 120 **15.** Prime **16.** Prime **17.** Composite **18.** Neither **19.** 1, 2, 4, 5, 8, 10, 16, 20, 32, 40, 80, 160; $2 \cdot 2 \cdot 2 \cdot 2 \cdot 2 \cdot 5$ **20.** 1, 2, 3, 6, 37, 74, 111, 222; $2 \cdot 3 \cdot 37$ **21.** 1, 2, 7, 14, 49, 98; $2 \cdot 7 \cdot 7$ **22.** 1, 3, 5, 7, 9, 15, 21, 35, 45, 63, 105, 315; $3 \cdot 3 \cdot 5 \cdot 7$ **23.** $\frac{8}{24}$, or $\frac{1}{3}$ **24.** $\frac{5}{4}$ **25.** $\frac{7}{9}$ **26.** $\frac{8}{45}$ **27.** $\frac{-40}{11}$, or $-\frac{40}{11}$ **28.** $\frac{2}{5}$ **29.** $\frac{11}{3}$ **30.** 1 **31.** 0 **32.** $\frac{9}{31}$ **33.** $\frac{-9}{5}$ **34.** $\frac{5}{42}$ **35.** $\frac{21}{29}$ **36.** Undefined **37.** $=$ **38.** $\neq$ **39.** $\frac{25}{200}$, or $\frac{1}{8}$ **40.** $\frac{21}{10{,}000}$ mi^2 **41.** Find the product of two prime numbers. **42.** If we use the divisibility tests, it is quickly clear that none of the even-numbered years is a prime number. In addition, the divisibility tests for 5 and 3 show that 2001, 2005, 2007, 2013, 2015, and 2019 are not prime numbers. Then the years 2003, 2009, 2011, and 2017 can be divided by prime numbers to determine whether they are prime. When we do this, we find that 2003, 2011, and 2017 are prime numbers. If the divisibility tests are not used, each of the numbers from 2000 to 2020 can be divided by prime numbers to determine if it is prime. **43.** It is possible to cancel only when identical *factors* appear in the numerator and the denominator of a fraction. Situations in which it is not possible to cancel include the occurrence of identical *addends* or *digits* in the numerator and the denominator.

44. No; since the only factors of a prime number are the number itself and 1, two different prime numbers cannot contain a common factor (other than 1).

Exercise Set 3.6, p. 192

RC1. products **RC2.** Factor **RC3.** 1 **RC4.** Carry out
1. $\frac{1}{3}$ **3.** $-\frac{1}{8}$ **5.** $\frac{4}{7}$ **7.** $\frac{1}{15}$ **9.** $-\frac{27}{10}$ **11.** $\frac{4x}{9}$ **13.** 1
15. 1 **17.** 3 **19.** 7 **21.** 12 **23.** $3a$ **25.** 1 **27.** 1
29. $\frac{1}{5}$ **31.** $\frac{9}{25}$ **33.** $60n$ **35.** $-\frac{10}{3}$ **37.** 1 **39.** 3
41. $\frac{119}{750}$ **43.** $-\frac{20}{187}$ **45.** $-\frac{42}{275}$ **47.** $-\frac{16}{5x}$ **49.** $-\frac{11}{40}$
51. $\frac{5a}{28b}$ **53.** $\frac{5}{8}$ in. **55.** 260 million ounces
57. 480 addresses **59.** $\frac{1}{3}$ cup **61.** \$115,500 **63.** 160 mi
65. Food: \$8400; housing: \$10,500; clothing: \$4200; savings: \$3000; taxes: \$8400; other expenses: \$7500 **67.** 60 in^2
69. $\frac{35}{4}$ mm^2 **71.** $\frac{63}{8}$ m^2 **73.** 92 mi^2 **75.** $\frac{7}{4}$ in^2 **77.** 35
78. 85 **79.** 125 **80.** 120 **81.** 4989 **82.** 8546
83. 8699 **84.** 6407 **85.** $\frac{129}{485}$ **87.** $\frac{1}{12}$ **89.** 13,380 mm^2

Calculator Corner, p. 198

1. $\frac{1}{6}$ **2.** $\frac{20}{9}$ **3.** $-\frac{9}{7}$ **4.** $\frac{3}{2}$

Exercise Set 3.7, p. 200

RC1. True **RC2.** False **RC3.** True **RC4.** True
1. $\frac{3}{7}$ **3.** $\frac{1}{9}$ **5.** 7 **7.** $-\frac{9}{8}$ **9.** $\frac{c}{a}$ **11.** $\frac{m}{-3n}$ **13.** $\frac{-15}{8}$
15. $\frac{1}{7m}$ **17.** $4a$ **19.** $-3z$ **21.** $\frac{4}{7}$ **23.** $\frac{4}{15}$ **25.** 4
27. -2 **29.** $\frac{25}{7}$ **31.** $\frac{1}{8}$ **33.** $\frac{3}{28}$ **35.** -8 **37.** $\frac{x}{2}$
39. $\frac{1}{9x}$ **41.** $35a$ **43.** 1 **45.** $-\frac{2}{3}$ **47.** $\frac{99}{224}$ **49.** $\frac{112a}{3}$
51. $\frac{14}{15}$ **53.** $\frac{7}{32}$ **55.** $-\frac{25}{12}$ **57.** -9 **58.** -81
59. 75 **60.** 81 **61.** 2 **62.** 2 **63.** $6x + 9y - 3$
64. $-4a + 12b - 8c$ **65.** $\frac{100}{9}$ **67.** 36 **69.** $\frac{121}{900}$
71. $\frac{9}{19}$ **73.** $\frac{220}{51}$

Translating for Success, p. 205

1. C **2.** H **3.** A **4.** N **5.** O **6.** F **7.** I
8. L **9.** D **10.** M

Exercise Set 3.8, p. 206

RC1. (b) **RC2.** (c) **RC3.** (d) **RC4.** (a) **1.** 15
3. 9 **5.** -45 **7.** $\frac{2}{17}$ **9.** $\frac{12}{5}$ **11.** $-\frac{16}{21}$ **13.** $-\frac{2}{25}$
15. $\frac{1}{6}$ **17.** -80 **19.** $-\frac{1}{6}$ **21.** $-\frac{7}{13}$ **23.** $\frac{27}{31}$
25. $\frac{6}{7}$ **27.** $\frac{12}{5}$ **29.** $-\frac{7}{15}$ **31.** $\frac{9}{5}$ **33.** 6 **35.** $\frac{10}{7}$
37. 960 extension cords **39.** 32 pairs **41.** 24 bowls
43. 16 L **45.** 20 packages **47.** $\frac{1}{8}$ T **49.** 9 bees
51. 2400 words **53.** 32 laps **55.** 288 km; 108 km
57. $\frac{1}{16}$ in. **59.** 26 **60.** -42 **61.** -67 **62.** -65 **63.** 20
64. 6 **65.** $17x$ **66.** $4a$ **67.** $7a + 3$ **68.** $4x - 7$
69. $\frac{2}{9}$ **71.** $\frac{7}{8}$ lb **73.** 103 slices **75.** 288 bees

Summary and Review: Chapter 3, p. 210

Vocabulary Reinforcement

1. multiplicative **2.** factors **3.** prime **4.** denominator
5. equivalent **6.** reciprocals **7.** factorization **8.** multiple

Concept Reinforcement

1. True **2.** False **3.** True **4.** True

Study Guide

1. 1, 2, 4, 8, 13, 26, 52, 104 **2.** $2 \cdot 2 \cdot 2 \cdot 13$ **3.** 0, 1, 18
4. $\frac{5}{14}$ **5.** $\neq$ **6.** $\frac{70}{9}$ **7.** $\frac{7}{10}$ **8.** $\frac{7}{3}$ cups

Review Exercises

1. 8, 16, 24, 32, 40, 48, 56, 64, 72, 80 **2.** No **3.** No **4.** No
5. Yes **6.** Yes **7.** 1, 2, 3, 4, 5, 6, 10, 12, 15, 20, 30, 60
8. 1, 2, 4, 8, 11, 16, 22, 44, 88, 176 **9.** Prime **10.** Neither
11. Composite **12.** $2 \cdot 5 \cdot 7$ **13.** $2 \cdot 2 \cdot 2 \cdot 3 \cdot 3$
14. $3 \cdot 3 \cdot 5$ **15.** $2 \cdot 3 \cdot 5 \cdot 5$ **16.** $2 \cdot 2 \cdot 2 \cdot 3 \cdot 3 \cdot 3 \cdot 3$
17. $2 \cdot 2 \cdot 2 \cdot 2 \cdot 3 \cdot 5 \cdot 5$ **18.** Numerator: 9; denominator: 7
19. $\frac{3}{5}$ **20.** $\frac{7}{6}$ **21.** (a) $\frac{3}{5}$; (b) $\frac{5}{3}$; (c) $\frac{3}{8}$ **22.** 0 **23.** 1
24. 48 **25.** 1 **26.** $-\frac{2}{3}$ **27.** $\frac{1}{4}$ **28.** -1 **29.** $\frac{3}{4}$
30. $\frac{2}{5}$ **31.** Undefined **32.** $\frac{2}{7}$ **33.** $\frac{32}{225}$ **34.** $\frac{15}{21}$ **35.** $\frac{-30}{55}$
36. $\frac{15}{100} = \frac{3}{20}, \frac{38}{100} = \frac{19}{50}, \frac{23}{100} = \frac{23}{100}, \frac{24}{100} = \frac{6}{25}$ **37.** $\neq$ **38.** $=$
39. $\neq$ **40.** $=$ **41.** $\frac{13}{2}$ **42.** $-\frac{1}{7}$ **43.** 8 **44.** $\frac{5y}{3x}$
45. $\frac{14}{45}$ **46.** $\frac{3y}{7x}$ **47.** $\frac{2}{3}$ **48.** -14 **49.** $\frac{10}{7}$ **50.** 1
51. $24x$ **52.** $\frac{1}{4}$ **53.** $\frac{80}{3}$ **54.** $\frac{1}{15}$ **55.** $6a$ **56.** -1
57. $\frac{3}{4}$ **58.** $\frac{4}{9}$ **59.** 240 **60.** $\frac{-3}{10}$ **61.** 28 **62.** $\frac{2}{3}$
63. 42 m^2 **64.** $\frac{35}{2}$ ft^2 **65.** 9 days **66.** \$32,085
67. 1000 km **68.** $\frac{1}{3}$ cup **69.** $\frac{1}{6}$ mi **70.** 60 bags
71. D **72.** B **73.** $\frac{17}{6}$ **74.** 2, 8
75. $a = 11{,}176$; $b = 9887$ **76.** 13, 11, 101, 37

Understanding Through Discussion and Writing

1. The student is probably multiplying the divisor by the reciprocal of the dividend rather than multiplying the dividend by the reciprocal of the divisor.
2. $9432 = 9 \cdot 1000 + 4 \cdot 100 + 3 \cdot 10 + 2 \cdot 1 = 9(999 + 1) + 4(99 + 1) + 3(9 + 1) + 2 \cdot 1 = 9 \cdot 999 + 9 \cdot 1 + 4 \cdot 99 + 4 \cdot 1 + 3 \cdot 9 + 3 \cdot 1 + 2 \cdot 1$. Since 999, 99, and 9 are each a multiple of 9, $9 \cdot 999$, $4 \cdot 99$, and $3 \cdot 9$ are multiples of 9. This leaves $9 \cdot 1 + 4 \cdot 1 + 3 \cdot 1 + 2 \cdot 1$, or $9 + 4 + 3 + 2$. If the sum of the digits, $9 + 4 + 3 + 2$, is divisible by 9, then 9432 is divisible by 9. **3.** Taking $\frac{1}{2}$ of a number is equivalent to multiplying the number by $\frac{1}{2}$. Dividing by $\frac{1}{2}$ is equivalent to multiplying by the reciprocal of $\frac{1}{2}$, or 2. Thus taking $\frac{1}{2}$ of a number is not the same as dividing by $\frac{1}{2}$.
4. We first consider an object and take $\frac{4}{7}$ of it. We divide the object into 7 parts and take 4 of them, as shown by the shading below in the left figure.

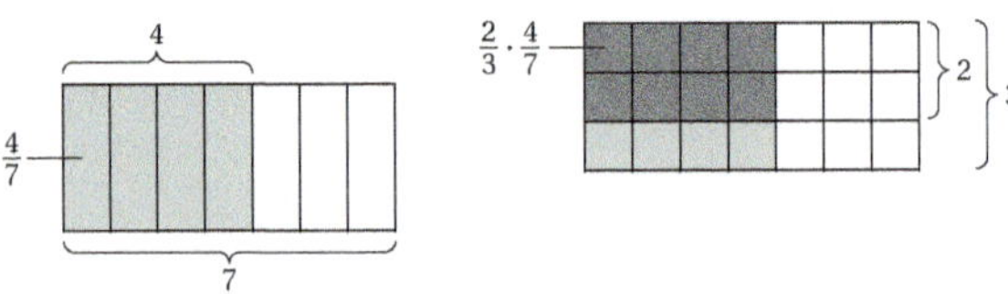

Next, we take $\frac{2}{3}$ of the shaded area in the left figure. We divide it into 3 parts and take two of them, as shown in the right figure. The entire object has been divided into 21 parts, 8 of which have been shaded twice. Thus, $\frac{2}{3} \cdot \frac{4}{7} = \frac{8}{21}$. **5.** Since $\frac{1}{7}$ is a smaller number than $\frac{2}{3}$, there are more $\frac{1}{7}$'s in 5 than $\frac{2}{3}$'s. Thus, $5 \div \frac{1}{7}$ is a greater number than $5 \div \frac{2}{3}$. **6.** No; in order to simplify a fraction, we must be able to remove a factor of the type $\frac{n}{n}$, $n \neq 0$, where n is a factor that the numerator and the denominator have in common.

Test: Chapter 3, p. 215

1. [3.1b] Yes **2.** [3.1b] No **3.** [3.2a] 1, 2, 3, 5, 6, 9, 10, 15, 18, 30, 45, 90 **4.** [3.2b] Composite **5.** [3.2c] $2 \cdot 2 \cdot 3 \cdot 3$
6. [3.2c] $2 \cdot 2 \cdot 3 \cdot 5$ **7.** [3.3a] Numerator: 4; denominator: 9

8. [3.3a] $\frac{3}{4}$ **9.** [3.3a] $\frac{3}{7}$ **10.** [3.3a] **(a)** $\frac{180}{47}$; **(b)** $\frac{47}{93}$
11. [3.3b] 32 **12.** [3.3b] 1 **13.** [3.3b] 0 **14.** [3.5b] $\frac{-1}{3}$
15. [3.5b] 6 **16.** [3.5b] $\frac{1}{5}$ **17.** [3.3b] Undefined **18.** [3.5b] $\frac{2}{3}$
19. [3.5c] $=$ **20.** [3.5c] $\neq$ **21.** [3.5a] $\frac{15}{40}$ **22.** [3.7a] $\dfrac{42}{a}$
23. [3.7a] $-\frac{1}{9}$ **24.** [3.6a] $\frac{5}{2}$ **25.** [3.7b] $\frac{8}{33}$ **26.** [3.4a] $\dfrac{3x}{8}$
27. [3.7b] $-\frac{3}{14}$ **28.** [3.7b] 18 **29.** [3.6a] $\frac{2}{9}$ **30.** [3.8b] $\frac{3}{20}$ lb
31. [3.6b] 125 lb **32.** [3.8a] 64 **33.** [3.8a] $-\frac{7}{4}$
34. [3.6b] $\frac{91}{2}$ m^2 **35.** [3.3a] C **36.** [3.6b] $\frac{7}{48}$ acre
37. [3.6a], [3.7b] $-\frac{7}{960}$

CHAPTER 4

Exercise Set 4.1, p. 223

RC1. True **RC2.** True **RC3.** True **RC4.** False
1. 4 **3.** 50 **5.** 40 **7.** 54 **9.** 150 **11.** 120 **13.** 72
15. 420 **17.** 144 **19.** 180 **21.** 42 **23.** 30 **25.** 72
27. 60 **29.** 36 **31.** 900 **33.** 48 **35.** 210 **37.** 300
39. 60 **41.** abc **43.** $9x^2$ **45.** $4x^3y$ **47.** $24r^3s^2t^4$
49. $a^3b^2c^2$ **51.** Every 60 yr **53.** Every 420 yr **55.** 14 **56.** -27
57. 7935 **58.** $\frac{2}{3}$ **59.** $-\frac{8}{7}$ **60.** -167 **61.** 2592 **63.** 54,033
65. 24 ft **67.** 24 strands **69.** **(a)** Not the LCM because a^2b^5 is not a factor of a^3b^3; **(b)** Not the LCM because a^3b^2 is not a factor of a^2b^5; **(c)** The LCM because both a^3b^2 and a^2b^5 are factors of a^3b^5 and it is the smallest such expression **71.** 2520 **73.** 27 and 2; 27 and 6; 27 and 18

Exercise Set 4.2, p. 231

RC1. True **RC2.** False **RC3.** False **RC4.** True
1. $\frac{5}{9}$ **3.** 1 **5.** $\frac{2}{5}$ **7.** $\dfrac{13}{a}$ **9.** $-\frac{1}{2}$ **11.** $\frac{7}{9}x$ **13.** $\frac{1}{2}t$ **15.** $-\dfrac{9}{x}$
17. $\frac{7}{24}$ **19.** $-\frac{1}{10}$ **21.** $\frac{23}{24}$ **23.** $\frac{83}{20}$ **25.** $\frac{5}{24}$ **27.** $\frac{37}{100}x$ **29.** $\frac{19}{20}$
31. $\frac{7}{8}$ **33.** $-\frac{99}{100}$ **35.** $-\frac{1}{30}x$ **37.** $-\frac{33}{7}t$ **39.** $-\frac{17}{24}$ **41.** $\frac{3}{4}$
43. $\frac{437}{1000}$ **45.** $\frac{5}{4}$ **47.** $\frac{239}{78}$ **49.** $\frac{59}{90}$ **51.** $-\frac{5}{4}$ **53.** $>$ **55.** $<$
57. $>$ **59.** $<$ **61.** $>$ **63.** $>$ **65.** $\frac{4}{15}, \frac{3}{10}, \frac{5}{12}$ **67.** $\frac{37}{12}$ mi
69. $\frac{13}{16}$ lb **71.** $\frac{27}{32}''$ **73.** $\frac{13}{12}$ lb **75.** $\frac{7}{8}$ in. **77.** $\frac{33}{20}$ mi **79.** $\frac{4}{5}$ qt; $\frac{8}{5}$ qt; $\frac{2}{5}$ qt
81. -13 **82.** 4 **83.** -8 **84.** -31 **85.** $\frac{10}{3}$ **86.** 42; 42
87. 35 **88.** 6407 **89.** 11 **90.** 3 **91.** $\frac{9}{7}$ **92.** $-\frac{2}{5}$
93. $\frac{13}{30}t + \frac{31}{35}$ **95.** $7t^2 + \dfrac{9}{a}t$ **97.** $>$ **99.** $\frac{4}{15}$; \$320
101. $a = 2, b = 8$

Translating for Success, p. 239

1. J **2.** E **3.** D **4.** B **5.** I **6.** N **7.** A **8.** C **9.** L
10. F

Exercise Set 4.3, p. 240

RC1. numerators; denominator **RC2.** denominators
RC3. denominators **RC4.** denominators
1. $\frac{2}{3}$ **3.** $-\frac{1}{4}$ **5.** $\dfrac{2}{a}$ **7.** $-\frac{1}{2}$ **9.** $-\dfrac{4}{5a}$ **11.** $\dfrac{2}{t}$ **13.** $\frac{13}{16}$
15. $-\frac{1}{3}$ **17.** $\frac{7}{10}$ **19.** $-\frac{17}{60}$ **21.** $\frac{47}{100}$ **23.** $\frac{26}{75}$ **25.** $-\frac{71}{100}$ **27.** $\frac{13}{24}$
29. $-\frac{29}{50}$ **31.** $-\frac{1}{24}$ **33.** $-\frac{41}{72}$ **35.** $\frac{1}{360}$ **37.** $\frac{2}{9}x$ **39.** $-\frac{7}{20}a$ **41.** $\frac{7}{9}$
43. $\frac{4}{11}$ **45.** $\frac{1}{15}$ **47.** $\frac{9}{8}$ **49.** $\frac{2}{15}$ **51.** $-\frac{7}{24}$ **53.** $\frac{2}{15}$ **55.** $-\frac{5}{4}$
57. $\frac{5}{12}$ hr **59.** $\frac{1}{32}$ in. **61.** $\frac{11}{20}$ lb **63.** $\frac{7}{20}$ hr **65.** $\frac{3}{16}$ in.
67. $\frac{7}{24}$ cup **69.** 1 **70.** Not defined **71.** Not defined **72.** 4
73. $\frac{4}{21}$ **74.** $\frac{3}{2}$ **75.** 21 **76.** $\frac{1}{32}$ **77.** 12 **78.** $\frac{5}{2}$ **79.** $-\frac{3}{10}$
80. $-\frac{9}{28}$ **81.** $\frac{1}{16}$ **83.** $-\frac{11}{10}$ **85.** $-\frac{64}{35}$ **87.** $-\frac{37}{1000}$ **89.** $\frac{1}{4}$ tub
91. $\frac{43}{50}$ **93.** $\frac{1}{8}$ of the dealership **95.** $\frac{14}{3553}$
97. *Day 1*: Cut off $\frac{1}{7}$ of bar and pay the contractor.
Day 2: Cut off $\frac{2}{7}$ of the bar's original length and trade it for the $\frac{1}{7}$.
Day 3: Give the $\frac{1}{7}$ back to the contractor.
Day 4: Trade the $\frac{4}{7}$ remaining for the contractor's $\frac{3}{7}$.
Day 5: Give the contractor the $\frac{1}{7}$ again.
Day 6: Trade the $\frac{2}{7}$ for the $\frac{1}{7}$.
Day 7: Give the contractor the $\frac{1}{7}$ again. This assumes that the contractor does not spend parts of the gold bar immediately.

Exercise Set 4.4, p. 248

RC1. False **RC2.** False **RC3.** False **RC4.** True
1. 3 **3.** $\frac{3}{5}$ **5.** -1 **7.** $\frac{27}{2}$ **9.** $\frac{1}{2}$ **11.** $\frac{3}{2}$ **13.** $-\frac{4}{3}$ **15.** $\frac{1}{2}$
17. $\frac{8}{7}$ **19.** -6 **21.** $\frac{21}{5}$ **23.** 6 **25.** $-\frac{17}{4}$ **27.** $-\frac{3}{4}$ **29.** $\frac{3}{2}$
31. $\frac{9}{2}$ **33.** $-\frac{10}{3}$ **35.** $-\frac{1}{5}$ **37.** $-\frac{1}{6}$ **39.** $\frac{35}{12}$ **41.** -13 **42.** -8
43. 18 **44.** 27 **45.** The balance has decreased \$150.
46. \$1180 profit **47.** $\dfrac{5}{7m}$ **48.** $20n$ **49.** 4 **51.** $-\frac{290}{697}$
53. $\frac{436}{35}$ **55.** 2 cm

Calculator Corner, p. 252

1. $5\frac{4}{7}$ **2.** $8\frac{2}{5}$ **3.** $1476\frac{1}{6}$ **4.** $676\frac{4}{9}$ **5.** $134\frac{1}{15}$ **6.** $12\frac{169}{454}$

Exercise Set 4.5, p. 254

RC1. True **RC2.** True **RC3.** False **RC4.** True
1. $\frac{23}{3}$ **3.** $\frac{25}{4}$ **5.** $-\frac{161}{8}$ **7.** $\frac{51}{10}$ **9.** $\frac{103}{5}$ **11.** $-\frac{100}{3}$ **13.** $\frac{13}{8}$
15. $-\frac{51}{4}$ **17.** $\frac{57}{10}$ **19.** $-\frac{507}{100}$ **21.** $5\frac{1}{3}$ **23.** $7\frac{1}{2}$ **25.** $5\frac{7}{10}$ **27.** $7\frac{2}{9}$
29. $-5\frac{1}{2}$ **31.** $11\frac{1}{2}$ **33.** $-1\frac{1}{2}$ **35.** $61\frac{2}{5}$ **37.** $-8\frac{13}{50}$ **39.** $108\frac{5}{8}$ **41.** $906\frac{3}{7}$
43. $40\frac{4}{7}$ **45.** $-20\frac{2}{15}$ **47.** $-22\frac{3}{7}$ **49.** $275\frac{2}{3}$ backpacks
51. $5504\frac{1}{6}$ hr **53.** $18x + 6y - 24$ **54.** $-3a + 6b - 12c$
55. $237\frac{19}{541}$ **57.** $8\frac{2}{3}$ **59.** $3\frac{2}{3}$ **61.** $52\frac{1}{7}$

Mid-Chapter Review: Chapter 4, p. 256

1. True **2.** True **3.** False **4.** False

5.
$$\begin{aligned}\frac{11}{42} - \frac{3}{35} &= \frac{11}{2 \cdot 3 \cdot 7} - \frac{3}{5 \cdot 7} \\ &= \frac{11}{2 \cdot 3 \cdot 7} \cdot \left(\frac{5}{5}\right) - \frac{3}{5 \cdot 7} \cdot \left(\frac{2 \cdot 3}{2 \cdot 3}\right) \\ &= \frac{11 \cdot 5}{2 \cdot 3 \cdot 7 \cdot 5} - \frac{3 \cdot 2 \cdot 3}{5 \cdot 7 \cdot 2 \cdot 3} \\ &= \frac{55}{2 \cdot 3 \cdot 5 \cdot 7} - \frac{18}{2 \cdot 3 \cdot 5 \cdot 7} \\ &= \frac{55 - 18}{2 \cdot 3 \cdot 5 \cdot 7} = \frac{37}{210}\end{aligned}$$

6.
$$\begin{aligned}x + \tfrac{1}{8} &= \tfrac{2}{3} \\ x + \tfrac{1}{8} - \tfrac{1}{8} &= \tfrac{2}{3} - \tfrac{1}{8} \\ x + 0 &= \tfrac{2}{3} \cdot \tfrac{8}{8} - \tfrac{1}{8} \cdot \tfrac{3}{3} \\ x &= \tfrac{16}{24} - \tfrac{3}{24} \\ x &= \tfrac{13}{24}\end{aligned}$$
The solution is $\frac{13}{24}$.

7.
- 45 and 50 — 450
- 50 and 80 — 400
- 30 and 24 — 120
- 18, 24, and 80 — 720
- 30, 45, and 50 — 450

8. $\frac{16}{45}$ **9.** $\frac{25}{12}$ **10.** $\frac{1}{18}$ **11.** $-\frac{19}{90}$ **12.** $\frac{7}{240}$ **13.** $-\frac{7}{24}x$ **14.** $-\frac{17}{60}$
15. $\frac{6}{91}$ **16.** $\frac{22}{15}$ mi **17.** $\frac{101}{20}$ hr **18.** $\frac{1}{5}, \frac{2}{7}, \frac{3}{10}, \frac{4}{9}$ **19.** $\frac{13}{80}$ **20.** $-\frac{8}{9}$
21. $17\frac{8}{15}$ **22.** C **23.** C **24.** No; if one number is a multiple of the other, for example, the LCM is the larger of the numbers.
25. We multiply by 1, using the notation n/n, to express each fraction in terms of the least common denominator. **26.** Write $\frac{8}{5}$ as $\frac{16}{10}$ and $\frac{8}{2}$ as $\frac{40}{10}$ and since taking 40 tenths away from 16 tenths would give a result less than 0, the answer cannot possibly be $\frac{8}{3}$. You could also find the sum $\frac{8}{3} + \frac{8}{2}$ and show that it is not $\frac{8}{5}$. **27.** No; $2\frac{1}{3} = \frac{7}{3}$ but $2 \cdot \frac{1}{3} = \frac{2}{3}$.

Exercise Set 4.6, p. 264

RC1. (d) **RC2.** (a) **RC3.** (b) **RC4.** (c)
1. $11\frac{2}{5}$ **3.** $9\frac{1}{2}$ **5.** $5\frac{1}{3}$ **7.** $13\frac{7}{12}$ **9.** $12\frac{1}{10}$ **11.** $17\frac{5}{24}$ **13.** $21\frac{1}{2}$ **15.** $27\frac{7}{8}$ **17.** $7\frac{1}{5}$ **19.** $6\frac{1}{10}$ **21.** $1\frac{3}{5}$ **23.** $13\frac{1}{4}$ **25.** $15\frac{3}{8}$ **27.** $7\frac{5}{12}$ **29.** $11\frac{5}{18}$ **31.** $8\frac{3}{4}t$ **33.** $2x$ **35.** $8\frac{3}{10}y$ **37.** $11\frac{8}{9}t$ **39.** $10\frac{5}{24}x$ **41.** $2\frac{1}{3}a$ **43.** $23\frac{3}{20}$ ft **45.** $20\frac{1}{12}$ yd **47.** $\frac{15}{16}$ in. **49.** $134\frac{1}{4}$ in. **51.** $28\frac{3}{4}$ yd **53.** $4\frac{5}{6}$ ft **55.** $7\frac{3}{8}$ ft **57.** $5\frac{7}{8}$ in. **59.** $20\frac{1}{8}$ in. **61.** $3\frac{4}{5}$ hr **63.** $66\frac{5}{6}$ ft **65.** $-\frac{2}{5}$ **67.** $-3\frac{1}{4}$ **69.** $-3\frac{13}{15}$ **71.** $-7\frac{3}{5}$ **73.** $2\frac{7}{12}$ **75.** $-1\frac{8}{9}$ **77.** 3 ten thousands + 8 thousands + 1 hundred + 2 tens + 5 ones **78.** Two million, five thousand, six hundred eighty-nine **79.** 9^4 **80.** 81 **81.** Yes **82.** No **83.** No **84.** Yes **85.** No **86.** Yes **87.** Yes **88.** Yes **89.** $\frac{10}{13}$ **90.** $\frac{1}{10}$ **91.** $8568\frac{786}{1189}$ **93.** $10\frac{7}{12}$ **95.** $-28\frac{3}{8}$ **97.** $55\frac{3}{4}$ in.

Calculator Corner, p. 273

1. $\frac{7}{12}$ **2.** $\frac{11}{10}$ **3.** $\frac{35}{16}$ **4.** $-\frac{3}{10}$ **5.** $10\frac{2}{15}$ **6.** $1\frac{1}{28}$ **7.** $10\frac{11}{15}$ **8.** $2\frac{91}{115}$

Translating for Success, p. 274

1. O **2.** K **3.** F **4.** D **5.** H **6.** G **7.** L **8.** E **9.** M **10.** J

Exercise Set 4.7, p. 275

RC1. True **RC2.** True **RC3.** True **RC4.** True
1. $22\frac{2}{3}$ **3.** $1\frac{2}{3}$ **5.** $-56\frac{2}{3}$ **7.** $16\frac{1}{3}$ **9.** $-10\frac{3}{25}$ **11.** $35\frac{91}{100}$ **13.** $6\frac{1}{4}$ **15.** $1\frac{1}{5}$ **17.** $1\frac{2}{3}$ **19.** $-1\frac{1}{8}$ **21.** $1\frac{8}{43}$ **23.** $\frac{9}{40}$ **25.** $23\frac{2}{5}$ **27.** $15\frac{5}{7}$ **29.** $-27\frac{2}{9}$ **31.** $-1\frac{1}{3}$ **33.** $12\frac{1}{4}$ **35.** $8\frac{3}{20}$ **37.** About \$24 million **39.** About 4,800,000 **41.** About \$42,000,000,000 **43.** 75 mph **45.** $12\frac{4}{5}$ tiles **47.** $5\frac{1}{2}$ cups of flour, $2\frac{2}{3}$ cups of sugar **49.** 15 mpg **51.** $62\frac{1}{2}$ sq ft **53.** 400 cu ft **55.** $16\frac{1}{2}$ servings **57.** $441\frac{1}{4}$ sq ft **59.** $76\frac{1}{4}$ sq ft **61.** $27\frac{5}{16}$ sq cm **63.** 68°F **65.** $13\frac{1}{4}$ in. $\times$ $13\frac{1}{4}$ in.: perimeter = 53 in., area = $175\frac{9}{16}$ sq in.; $13\frac{1}{4}$ in. $\times$ $3\frac{1}{4}$ in.: perimeter = 33 in., area = $43\frac{1}{16}$ sq in. **67.** 22 m **69.** $16\frac{25}{64}$ **71.** $\frac{4}{9}$ **73.** $r = \frac{240}{13}$, or $18\frac{6}{13}$ **75.** 88 gal

Exercise Set 4.8, p. 284

RC1. (c) **RC2.** (d) **RC3.** (a) **RC4.** (b)
1. $\frac{7}{24}$ **3.** $\frac{3}{2}$, or $1\frac{1}{2}$ **5.** $\frac{7}{8}$ **7.** $\frac{59}{30}$, or $1\frac{29}{30}$ **9.** $\frac{7}{16}$ **11.** $\frac{3}{20}$ **13.** -6 **15.** $\frac{1}{36}$ **17.** $-\frac{3}{100}$ **19.** -1 **21.** $\frac{19}{4}$, or $4\frac{3}{4}$ **23.** $\frac{3}{8}$ **25.** $\frac{2}{9}$ **27.** $\frac{3}{11}$ **29.** $-\frac{14}{3}$, or $-4\frac{2}{3}$ **31.** $-\frac{5}{4}$, or $-1\frac{1}{4}$ **33.** $\frac{1}{100}$ **35.** $-\frac{1}{6}$ **37.** $\dfrac{2x}{35}$ **39.** $-\dfrac{3n}{28}$ **41.** $\dfrac{6}{7x}$ **43.** $-\frac{5}{18}$ **45.** $-\frac{7}{12}$ **47.** $\frac{1}{3}$ **49.** $\frac{37}{48}$ **51.** $\frac{25}{72}$ **53.** $\frac{103}{16}$, or $6\frac{7}{16}$ **55.** $16\frac{7}{96}$ mi **57.** $9\frac{19}{40}$ lb **59.** 20 **60.** 2 **61.** 84 **62.** 100 **63.** 1, 2, 3, 6, 7, 14, 21, 42 **64.** No **65.** Prime: 5, 7, 23, 43; composite: 9, 14; neither: 1 **66.** $2 \cdot 3 \cdot 5 \cdot 5$ **67.** $3\frac{11}{14}$ **69.** $\dfrac{8x}{147}$ **71.** 0 **73.** $\frac{1}{2}$ **75.** $\frac{1}{2}$

Summary and Review: Chapter 4, p. 287

Vocabulary Reinforcement

1. least common multiple **2.** mixed numeral **3.** fraction **4.** complex fraction **5.** denominators **6.** least common multiple **7.** greatest **8.** numerators

Concept Reinforcement

1. True **2.** True **3.** True **4.** False

Study Guide

1. 156 **2.** $\frac{28}{45}$ **3.** $<$ **4.** $\frac{4}{35}$ **5.** $-\frac{1}{12}$ **6.** $\frac{26}{3}$ **7.** $7\frac{5}{6}$ **8.** $7\frac{27}{28}$ **9.** $14\frac{14}{25}$ **10.** About 4,500,000 **11.** $\frac{9}{2}$, or $4\frac{1}{2}$

Review Exercises

1. 36 **2.** 90 **3.** 30 **4.** 1404 **5.** $\frac{7}{9}$ **6.** $\dfrac{9}{x}$ **7.** $-\dfrac{7}{15}$ **8.** $\frac{7}{16}$ **9.** $\frac{2}{9}$ **10.** $-\frac{1}{8}$ **11.** $\frac{4}{27}$ **12.** $\frac{1}{18}$ **13.** $>$ **14.** $<$ **15.** $\frac{19}{40}$ **16.** 11 **17.** $-\frac{5}{6}$ **18.** $\frac{12}{25}$ **19.** $\frac{2}{5}$ **20.** $\frac{15}{2}$ **21.** $\frac{67}{8}$ **22.** $\frac{13}{3}$ **23.** $-\frac{12}{7}$ **24.** $2\frac{1}{3}$ **25.** $-6\frac{3}{4}$ **26.** $12\frac{3}{5}$ **27.** $3\frac{1}{2}$ **28.** $-877\frac{1}{3}$ **29.** $82\frac{1}{3}$ **30.** $10\frac{2}{5}$ **31.** $11\frac{11}{15}$ **32.** -9 **33.** $1\frac{3}{4}$ **34.** $7\frac{7}{9}$ **35.** $4\frac{11}{15}$ **36.** $-5\frac{1}{8}$ **37.** $-14\frac{1}{4}$ **38.** $\frac{7}{9}x$ **39.** $3\frac{7}{40}a$ **40.** 16 **41.** $-3\frac{1}{2}$ **42.** $2\frac{21}{50}$ **43.** 6 **44.** -24 **45.** $-1\frac{7}{17}$ **46.** $\frac{1}{8}$ **47.** $\frac{9}{10}$ **48.** $13\frac{5}{7}$ **49.** $2\frac{8}{11}$ **50.** $4\frac{1}{4}$ yd **51.** $3\frac{1}{8}$ pizzas **52.** 24 lb **53.** $63\frac{2}{3}$ pies; $19\frac{1}{3}$ pies **54.** $3\frac{3}{4}$ mi **55.** $177\frac{3}{4}$ sq in. **56.** $50\frac{1}{4}$ sq in. **57.** \$850 **58.** $4\frac{5}{6}$ ft **59.** 1 **60.** $\frac{39}{40}$ **61.** 3 **62.** $\frac{77}{240}$ **63.** $-\frac{8}{9}$ **64.** $-\dfrac{2x}{3}$ **65.** A **66.** D **67.** $\frac{600}{13}$, or $46\frac{2}{13}$ **68.** $\frac{6}{3} + \frac{5}{4} = 3\frac{1}{4}$ **69.** **(a)** 6; **(b)** 10; **(c)** -28; **(d)** -1

Understanding Through Discussion and Writing

1. No; if the sum of the fractional parts of the mixed numerals is n/n, then the sum of the mixed numerals is an integer. For example, $1\frac{1}{5} + 6\frac{4}{5} = 7\frac{5}{5} = 8$. **2.** A wheel makes $33\frac{1}{3}$ revolutions per minute. It rotates for $4\frac{1}{2}$ min. How many revolutions does it make? Answers may vary. **3.** The student is multiplying the whole numbers to get the whole-number portion of the answer and multiplying fractions to get the fraction part of the answer. The student should have converted each mixed numeral to fraction notation, multiplied, simplified, and then converted back to a mixed numeral. The correct answer is $4\frac{6}{7}$. **4.** It might be necessary to find the least common denominator before adding or subtracting. The least common denominator is the least common multiple of the denominators. **5.** Suppose that a room has dimensions $15\frac{3}{4}$ ft by $28\frac{5}{8}$ ft. The equation $2 \cdot 15\frac{3}{4} + 2 \cdot 28\frac{5}{8} = 88\frac{3}{4}$ gives the perimeter of the room, in feet. Answers may vary. **6.** Note that $5 \cdot 3\frac{2}{7} = 5(3 + \frac{2}{7}) = 5 \cdot 3 + 5 \cdot \frac{2}{7}$. The products $5 \cdot 3$ and $5 \cdot \frac{2}{7}$ should be added rather than multiplied together. The student could also have converted $3\frac{2}{7}$ to fraction notation, multiplied, simplified, and converted back to a mixed numeral. The correct answer is $16\frac{3}{7}$.

Test: Chapter 4, p. 293

1. [4.1a] 48 **2.** [4.2a] 3 **3.** [4.2b] $-\frac{5}{24}$ **4.** [4.3a] $\dfrac{2}{t}$ **5.** [4.3a] $\frac{1}{12}$ **6.** [4.3a] $-\frac{1}{12}$ **7.** [4.3b] $\frac{1}{4}$ **8.** [4.4a] $-\frac{12}{5}$ **9.** [4.4a] $\frac{3}{20}$ **10.** [4.2c] $>$ **11.** [4.5a] $\frac{7}{2}$ **12.** [4.5a] $-\frac{75}{8}$ **13.** [4.5a] $-8\frac{2}{9}$ **14.** [4.5b] $162\frac{7}{11}$ **15.** [4.6a] $14\frac{1}{5}$ **16.** [4.6a] $12\frac{5}{12}$ **17.** [4.6b] $4\frac{7}{24}$ **18.** [4.6d] $8\frac{4}{7}$ **19.** [4.6d] $-5\frac{7}{10}$ **20.** [4.3a] $-\frac{1}{8}x$ **21.** [4.6b] $1\frac{54}{55}a$ **22.** [4.7a] 39 **23.** [4.7a] -18 **24.** [4.7b] 6 **25.** [4.7b] 2 **26.** [4.7c] $19\frac{3}{5}$ **27.** [4.7c] $28\frac{1}{20}$ **28.** [4.7d] About 105 kg **29.** [4.7d] 80 books **30.** [4.6c] **(a)** 3 in.; **(b)** $4\frac{1}{2}$ in. **31.** [4.3c] $\frac{1}{16}$ in. **32.** [4.8b] $6\frac{11}{36}$ ft **33.** [4.8a] $3\frac{1}{2}$ **34.** [4.8a] $-\frac{9}{4}$, or $-2\frac{1}{4}$ **35.** [4.8b] $-\frac{2}{3}$ **36.** [4.1a] D **37.** [4.1a] **(a)** 24, 48, 72; **(b)** 24 **38.** [4.3c] Rebecca walks $\frac{17}{56}$ mi farther.

CHAPTER 5

Exercise Set 5.1, p. 303

RC1. 5 **RC2.** 2 **RC3.** 3 **RC4.** 0 **RC5.** 8 **RC6.** 4 **RC7.** 1 **RC8.** 6
1. One hundred nineteen ten-thousandths
3. One hundred thirty-seven and six tenths
5. Five hundred nineteen and twenty-two hundredths
7. Three and seven hundred eighty-five thousandths
9. $\frac{73}{10}$; $7\frac{3}{10}$ **11.** $\frac{2167}{100}$; $21\frac{67}{100}$ **13.** $-\frac{2703}{1000}$; $-2\frac{703}{1000}$ **15.** $\frac{109}{10,000}$ **17.** $-\frac{40,003}{10,000}$; $-4\frac{3}{10,000}$ **19.** $-\frac{207}{10,000}$ **21.** $\frac{7,000,105}{100,000}$; $70\frac{105}{100,000}$ **23.** 0.3 **25.** -0.59 **27.** 3.798 **29.** 0.0078 **31.** -0.00018 **33.** 0.486197 **35.** 7.013 **37.** -8.431 **39.** 2.1739 **41.** 8.953073

43. 0.58 **45.** 0.410 **47.** −5.043 **49.** 235.07 **51.** $\frac{7}{100}$
53. −0.872 **55.** 0.2 **57.** −0.4 **59.** 3.0 **61.** −327.2
63. 0.89 **65.** −0.67 **67.** 1.00 **69.** −0.03 **71.** 0.572
73. 17.002 **75.** −20.202 **77.** 9.985 **79.** 809.5
81. 809.47 **83.** 830 **84.** $\frac{830}{1000}$, or $\frac{83}{100}$ **85.** 182
86. $\frac{182}{100}$, or $\frac{91}{50}$ **87.** $-\frac{12}{55}$ **88.** $-\frac{15}{34}$ **89.** 32,958 **90.** 10,726
91. −1.09, −1.009, −0.989, −0.898, −0.098 **93.** 6.78346
95. 99.99999 **97.** 1998, 2005, 2007, 2010, 2013 **99.** 1976

Calculator Corner, p. 309

1. 317.645 **2.** 49.08 **3.** 33.83 **4.** 0.99 **5.** 242.93
6. −11.692

Exercise Set 5.2, p. 310

RC1. 21.824; 23.7 **RC2.** 146.723; 40.9
1. 464.37 **3.** 1576.015 **5.** 132.56, or 132.560 **7.** 7.823
9. 50.7124 **11.** 10.06 **13.** 771.967 **15.** 20.8649
17. 227.468, or 227.4680 **19.** 41.381 **21.** 49.02 **23.** 3.564
25. 85.921 **27.** 1.6666 **29.** 4.0622 **31.** 29.999 **33.** 3.37
35. 1.045 **37.** 3.703 **39.** 0.9092 **41.** 605.21 **43.** 53.203
45. 161.62 **47.** 44.001 **49.** −3.29 **51.** −2.5 **53.** −7.2
55. 3.379 **57.** −16.6 **59.** 2.5 **61.** −3.519 **63.** 9.601 **65.** 75.5
67. 9.7 **69.** −10.292 **71.** −0.3 **73.** $5.7x$ **75.** $4.86a$
77. $21.1t + 7.9$ **79.** $-2.917x$ **81.** $8.106y - 7.1$ **83.** $-0.9x + 3.1y$
85. $-1.1 - 8.4t$ **87.** $\frac{12}{35}$ **88.** $\frac{14}{45}$ **89.** $\frac{63}{1000}$ **90.** −10
91. −7 **92.** 31 **93.** $-12.001 - 12.2698a + 10.366b$
95. $4.593a - 10.996b - 59.491$ **97.** −138.5 **99.** 2

Calculator Corner, p. 316

1. 142.803 **2.** −0.5076 **3.** 7916.4 **4.** 20.4153

Exercise Set 5.3, p. 320

RC1. (c) **RC2.** (e) **RC3.** (f) **RC4.** (a) **RC5.** (d) **RC6.** (b)
1. 47.6 **3.** 6.72 **5.** 0.252 **7.** 0.522 **9.** 426.3
11. −783,686.852 **13.** −780 **15.** 7.918 **17.** 0.09768
19. −0.287 **21.** 43.68 **23.** 0.030504 **25.** 89.76
27. −322.07 **29.** 55.68 **31.** 3487.5 **33.** 0.1155 **35.** −9420
37. 0.00953 **39.** 5706¢ **41.** 95¢ **43.** 1¢ **45.** \$0.72
47. \$0.02 **49.** \$63.99 **51.** 3,480,000 **53.** 50,960,000,000; 13,410,000,000 **55.** 2,200,000 **57.** 11,000 **59.** 26.025
61. **(a)** 44 ft; **(b)** 118.75 sq ft **63.** **(a)** 37.8 m; **(b)** 88.2 m^2
65. 2.7625 million nurses, or 2,762,500 nurses **67.** −27
68. 36 **69.** 69 **70.** −141 **71.** 1176 R 14 **72.** −27
73. 10^{21} = 1 sextillion **75.** 10^{24} = 1 septillion
77. 6,600,000,000,000 **79.** 366.5488175 **81.** \$61.45

Calculator Corner, p. 324

1. 14.3 **2.** 2.56 **3.** −0.064 **4.** 75.8

Calculator Corner, p. 325

1. 28 R 2 **2.** 116 R 3 **3.** 74 R 10 **4.** 415 R 3

Exercise Set 5.4, p. 329

RC1. Subtract: 2 − 0.04 **RC2.** Divide: 2.06 ÷ 0.01
RC3. Evaluate: 8^3 **RC4.** Add: 4.1 + 6.9 **RC5.** Divide: 9 ÷ 3
RC6. Subtract: 10 − 5
1. 2.99 **3.** 23.78 **5.** 7.08 **7.** 1.2 **9.** −0.9 **11.** −6000 **13.** 140
15. 40 **17.** −0.15 **19.** 3.2 **21.** −3.9 **23.** 0.625 **25.** 0.26
27. 2.34 **29.** −0.3045 **31.** 2.134567 **33.** −2.359 **35.** 1023.7
37. −9236 **39.** 0.08172 **41.** 9.7 **43.** −0.0527
45. −75,300 **47.** −0.0753 **49.** 2107 **51.** −302.997
53. −178.1 **55.** 206.0176 **57.** 0.5 **59.** −400.0108
61. 0.6725 **63.** 5.383 **65.** 10.5 **67.** 13,748.5 ft
69. 6.22 million stays **71.** 31.24 mi **73.** $\frac{19}{73}$ **74.** $\frac{1}{5}$
75. $2 \cdot 2 \cdot 3 \cdot 3 \cdot 19$, or $2^2 \cdot 3^2 \cdot 19$ **76.** $5 \cdot 401$ **77.** $15\frac{1}{8}$
78. $5\frac{7}{8}$ **79.** 343 **80.** 64 **81.** 47 **82.** 41 **83.** −56.6916
85. 6.254194585 **87.** 1000 **89.** 100 **91.** 46.7 points
93. 450 kWh

Mid-Chapter Review: Chapter 5, p. 333

1. False **2.** True **3.** True
4. $P(1 + r) = 5000(1 + 0.045)$
$= 5000(1.045)$
$= 5225$
5. $5.6 + 4.3 \times (6.5 - 0.25)^2 = 5.6 + 4.3 \times (6.25)^2$
$= 5.6 + 4.3 \times 39.0625$
$= 5.6 + 167.96875$
$= 173.56875$
6. Twenty-nine and forty-three hundredths **7.** 9,400,000
8. $\frac{453}{100}$; $4\frac{53}{100}$ **9.** $\frac{287}{1000}$ **10.** 0.13 **11.** −5.09 **12.** 0.7
13. 6.39 **14.** −35.67 **15.** 8.002 **16.** 28.462 **17.** 28.46
18. 28.5 **19.** 28 **20.** 50.095 **21.** 1214.862 **22.** −10.23
23. 18.24 **24.** 272.19 **25.** 5.593 **26.** 15.55 **27.** −19.9
28. 4.14 **29.** 92.871 **30.** 8123.6 **31.** −0.0483 **32.** 5.06
33. 3.2 **34.** 763 **35.** 0.914036 **36.** 2045¢ **37.** \$1.47
38. $-1.22x - 7.1$ **39.** 4.2 **40.** 59.774 **41.** 33.33
42. The student probably rounded over successively from the thousandths place as follows: 236.448 ≈ 236.45 ≈ 236.5 ≈ 237. The student should have considered only the tenths place and rounded down. **43.** The decimal points were not lined up before the subtraction was carried out. **44.** $10 \div 0.2 = \frac{10}{0.2} = \frac{10}{0.2} \cdot \frac{10}{10} = \frac{100}{2} = 100 \div 2.$ **45.** $0.247 \div 0.1 = \frac{247}{1000} \div \frac{1}{10} = \frac{247}{1000} \cdot \frac{10}{1} = \frac{247 \cdot 10}{10 \cdot 100} = \frac{247}{100} = 2.47 \neq 0.0247;$
$0.247 \div 10 = \frac{247}{1000} \div 10 = \frac{247}{1000} \cdot \frac{1}{10} = \frac{247}{10{,}000} = 0.0247 \neq 2.47$

Calculator Corner, p. 337

1. $-0.1\overline{6}$ **2.** $0.\overline{63}$ **3.** $6.\overline{3}$ **4.** $-57.\overline{1}$

Calculator Corner, p. 338

1. 123.150432 **2.** 52.59026102

Exercise Set 5.5, p. 340

RC1. Repeating **RC2.** Terminating **RC3.** Terminating
RC4. Repeating **RC5.** Repeating **RC6.** Terminating
1. 0.375 **3.** −0.5 **5.** 0.12 **7.** 0.225 **9.** 0.52 **11.** −0.85
13. −0.5625 **15.** 1.4 **17.** 1.12 **19.** −1.375 **21.** −0.975
23. 0.605 **25.** $0.5\overline{3}$ **27.** $0.\overline{3}$ **29.** $-1.\overline{3}$ **31.** $1.1\overline{6}$
33. $-1.\overline{27}$ **35.** $-0.41\overline{6}$ **37.** 0.254 **39.** $0.\overline{12}$ **41.** $-0.2\overline{18}$
43. $0.\overline{571428}$ **45.** 0.2; 0.18; 0.182 **47.** 0.3; 0.28; 0.278
49. 0.4; 0.36; 0.364 **51.** −1.7; −1.67; −1.667
53. −0.5; −0.47; −0.471 **55.** 0.6; 0.58; 0.583
57. −0.2; −0.19; −0.193 **59.** −0.8; −0.78; −0.778
61. **(a)** 0.571; **(b)** 1.333; **(c)** 0.429; **(d)** 0.75 **63.** 15.8 mpg
65. 17.8 mpg **67.** 11.06 **69.** 2.736 **71.** $-417.51\overline{6}$
73. 0 **75.** 0.09705 **77.** −1.5275 **79.** 24.375 **81.** 1.08 m^2
83. 5.78 cm^2 **85.** 790.92 in^2 **87.** 21 **88.** $1\frac{1}{2}$ **89.** 10
90. $30\frac{7}{10}$ **91.** $1\frac{1}{24}$ cups **92.** $1\frac{33}{100}$ in. **93.** $0.\overline{142857}$, $0.\overline{285714}$, $0.\overline{428571}$, $0.\overline{571428}$, $0.\overline{714285}$; $0.\overline{857142}$ **95.** 13.86 cm^2
97. 1.76625 ft^2 or 1.767145868 ft^2

Exercise Set 5.6, p. 346

RC1. (b) **RC2.** (d) **RC3.** (c) **RC4.** (e) **RC5.** (a) **RC6.** (f)
1. (d) **3.** (c) **5.** (a) **7.** (c) **9.** 1.6 **11.** 6 **13.** 60 **15.** 2.3 **17.** 180 **19.** (a) **21.** (c) **23.** (b) **25.** (b)
27. $1800 \div 9 = 200$ posts; answers may vary
29. $\$2 \cdot 12 = \$24$; answers may vary **31.** 18,940,000 lb
32. 56 mm^2 **33.** 126 slices **34.** 9 cups
35. $36,500 a year **36.** 16 packages **37.** Yes
39. No **41.** **(a)** $+, \times$; **(b)** $+, \times, -$

Exercise Set 5.7, p. 353

RC1. False **RC2.** True **RC3.** True **RC4.** True
1. 5.4 **3.** 0.2 **5.** −6.24 **7.** −12.6 **9.** −2.55 **11.** 1.97 **13.** 8.38 **15.** −5.9 **17.** 6 **19.** 1.8 **21.** −3.7 **23.** −4.7 **25.** 1.7 **27.** 2.94 **29.** 4.5 **31.** 9 **33.** 3.2 **35.** −1.75 **37.** 30 **39.** 3.2 **41.** 9 **43.** 2.1 **45.** 4.5 **47.** 4.12 **49.** 5.6 **51.** −1.9 **53.** 13 **55.** 0 **57.** −1.5 **59.** 14 m^2
60. 27 cm^2 **61.** $\frac{25}{2}$ in^2 **62.** 24 ft^2 **63.** 5 ft^2
64. 12 m^2 **65.** $-\frac{29}{50}$ **66.** 0 **67.** −2 **68.** 8
69. 3.1 **71.** 36 **73.** 1.1212963

Translating for Success, p. 364

1. I **2.** C **3.** N **4.** A **5.** G **6.** B **7.** D **8.** O **9.** F **10.** M

Exercise Set 5.8, p. 365

RC1. Familiarize **RC2.** Translate **RC3.** Solve **RC4.** Check **RC5.** State
1. Let a = Ron's age: $a + 5$, or $5 + a$ **3.** $b + 6$, or $6 + b$
5. $c - 9$ **7.** Let n = the number; $n - 16$
9. Let s = Nate's speed; $8s$ **11.** $\dfrac{x}{17}$
13. Let x = the number; $\frac{1}{2}x + 20$, or $20 + \frac{1}{2}x$
15. Let x = the number; $4x - 20$
17. Let l = the length and w = the width; $l + w$, or $w + l$
19. Let r = the rate and t = the time; $rt + 10$, or $10 + rt$
21. Let n = the number; $10n + n$, or $n + 10n$
23. Let x and y = the numbers; $5(x - y)$
25. $60.2 billion **27.** $66.28 **29.** $0.51
31. $218,666,666.67 **33.** 9.58 sec **35.** 22,691.5 mi
37. 20.2 mpg **39.** $24.33 **41.** 2.66 cc **43.** 10.8¢
45. $26,900.21 **47.** Area: 268.96 sq ft; perimeter: 65.6 ft
49. 193.04 cm^2 **51.** 2.31 cm **53.** 331.74 ft^2
55. 125 gigabytes **57.** 296,183.55 sq ft **59.** 17 bottles
61. 2.5 million protected acres, 2.9 million unprotected acres
63. $65.30 for food, $195.90 for lodging **65.** $906.50
67. $316,987.20; $196,987.20 **69.** 0 **70.** Undefined
71. 1 **72.** $-\dfrac{1}{10}$ **73.** $-\dfrac{20}{33}$ **74.** $6\dfrac{5}{6}$ **75.** 3999.76 cm^2
77. $0.19 **79.** 25 cm^2. We assume that the figures are nested squares formed by connecting the midpoints of consecutive sides of the next larger square.

Summary and Review: Chapter 5, p. 371

Vocabulary Reinforcement

1. repeating **2.** terminating **3.** billion **4.** million **5.** trillion **6.** rational numbers

Concept Reinforcement

1. True **2.** False **3.** True **4.** False **5.** True

Study Guide

1. $\frac{5093}{100}$ **2.** 81.7 **3.** 42.159 **4.** 153.35 **5.** 38.611 **6.** 207.848 **7.** 19.11 **8.** 0.176 **9.** 60,437 **10.** 7.4 **11.** 0.047 **12.** 15,690

Review Exercises

1. 6,590,000 **2.** 3,100,000,000
3. Three and forty-seven hundredths
4. Thirty-one thousandths
5. Twenty-seven and one ten-thousandth **6.** Nine tenths
7. $\frac{9}{100}$ **8.** $-\frac{4561}{1000}$; $-4\frac{561}{1000}$ **9.** $-\frac{89}{1000}$ **10.** $\frac{30{,}227}{10{,}000}$; $3\frac{227}{10{,}000}$
11. 0.034 **12.** 4.2603 **13.** 27.91 **14.** −867.006 **15.** 0.034 **16.** −0.19 **17.** 0.741 **18.** 1.041 **19.** 17.4 **20.** 17.43 **21.** 17.429 **22.** 17 **23.** 499.829 **24.** 29.148 **25.** 229.1 **26.** 685.0519 **27.** −57.3 **28.** 2.37 **29.** 12.96 **30.** −1.073 **31.** 24,680 **32.** 3.2 **33.** −1.6 **34.** 0.2763 **35.** 0.003056 **36.** −4380 **37.** $2.2x - 9.1y$ **38.** $-2.84a + 12.57$ **39.** 925 **40.** 40.84 **41.** $15.49 **42.** 248.27 **43.** 2.6 **44.** 1.28 **45.** 3.25 **46.** $-1.1\overline{6}$ **47.** 21.08 **48.** −3.2 **49.** −3 **50.** −7.5 **51.** 6.5
52. $2.19 billion **53.** $15.52 **54.** 249.76 ft^2 **55.** $784.47
56. 95 transactions **57.** 195.7 lb of newspaper, 65.7 lb of glass
58. 14.5 mpg **59.** 20.7 million books **60.** $55.50 **61.** $1.33
62. 61.5 ft; 235.625 sq ft **63.** B **64.** A **65.** **(a)** +; **(b)** −
66. $\frac{-13}{15}, \frac{-17}{20}, -\frac{11}{13}, -\frac{15}{19}, \frac{-5}{7}, -\frac{2}{3}$ **67.** 26,260 mi
68. $1 = 3 \cdot \frac{1}{3} = 3(0.33333333\ldots) = 0.99999999\ldots$, or $0.\overline{9}$
69. The rectangular pizza, at $\dfrac{4.4¢}{\text{in}^2}$, is a better buy than the round pizza, which costs $\dfrac{5.5¢}{\text{in}^2}$.

Understanding Through Discussion and Writing

1. Count the number of decimal places. Move the decimal point that many places to the right and write the result over a denominator of 1 followed by that many zeros.
2. $346.708 \times 0.1 = \frac{346{,}708}{1000} \times \frac{1}{10} = \frac{346{,}708}{10{,}000} = 34.6708 \neq 3467.08$
3. When the denominator of a fraction is a multiple of 10, long division is not the fastest way to convert the fraction to decimal notation. Many times this is also the case when the denominator has only 2's or 5's or both as factors. **4.** Multiply by 1 to get a denominator that is a power of 10:

$$\frac{44}{125} = \frac{44}{125} \cdot \frac{8}{8} = \frac{352}{1000} = 0.352.$$

We can also divide to find that $\frac{44}{125} = 0.352$.

Test: Chapter 5, p. 376

1. [5.3b] 2,600,000,000
2. [5.1a] One hundred twenty-three and forty-seven thousandths
3. [5.1b] $-\frac{91}{100}$ **4.** [5.1b] $\frac{2769}{1000}$; $2\frac{769}{1000}$ **5.** [5.1b] 0.074
6. [5.1b] −3.7047 **7.** [5.1b] 756.09 **8.** [5.1b] 91.703
9. [5.1c] 0.162 **10.** [5.1c] 8.049 **11.** [5.1c] −0.09
12. [5.1d] 6 **13.** [5.1d] 5.68 **14.** [5.1d] 5.678 **15.** [5.1d] 5.7
16. [5.2a] 405.219 **17.** [5.3a] 0.03 **18.** [5.3a] 0.21345
19. [5.2b] 44.746 **20.** [5.2a] 356.37 **21.** [5.2c] −2.2
22. [5.2b] 1.9946 **23.** [5.3a] 73,962 **24.** [5.4a] 4.75
25. [5.4a] 30.4 **26.** [5.4a] −0.34682 **27.** [5.4a] 34,682
28. [5.3b] 17,982¢ **29.** [5.2d] $9.8x - 3.9y - 4.6$
30. [5.3c] 11.6 **31.** [5.4b] 7.6 **32.** [5.5c] 1.6 **33.** [5.5c] 5.25
34. [5.5a] −0.4375 **35.** [5.5a] $1.\overline{5}$ **36.** [5.5b] 1.56
37. [5.4b] −22.25 **38.** [5.4b] 1.8045 **39.** [5.5d] 9.72
40. [5.7a] −3.24 **41.** [5.7b] 10 **42.** [5.7b] 1.4
43. [5.8b] $46.69 **44.** [5.8b] 28.3 mpg **45.** [5.8b] $592.45
46. [5.8b] $293.93 **47.** [5.8b] 68.5 yr **48.** [5.6a] C
49. [5.3a] **(a)** Always; **(b)** never; **(c)** sometimes; **(d)** sometimes
50. [5.8b] $35 **51.** [5.8b] **(a)** Fly; **(b)** fly; **(c)** drive

CHAPTER 6

Exercise Set 6.1, p. 391

RC1. ratio **RC2.** proportional **RC3.** proportion **RC4.** cross products

1. $\frac{178}{572}$ **3.** $\dfrac{8\frac{3}{4}}{9\frac{5}{6}}$ **5.** $\frac{29}{75}$ **7.** $\frac{107}{366}$ **9.** $\frac{2}{3}$ **11.** $\frac{7}{9}$ **13.** 40 km/hr **15.** 7.48 mi/sec **17.** 33 mpg **19.** About 393 performances/year **21.** 186,000 mi/sec **23.** 0.623 gal/ft^2 **25.** 25 beats/min **27.** No **29.** Yes **31.** Yes **33.** No **35.** 12 **37.** 20 **39.** 18 **41.** $\frac{28}{3}$, or $9\frac{1}{3}$ **43.** 2.7 **45.** 1.8 **47.** $\frac{3}{8}$ **49.** $\frac{16}{75}$ **51.** 0.7 **53.** $\frac{1}{20}$ **55.** 7680 frames **57.** 60 students **59.** **(a)** 38 euros; **(b)** \$11,368.42 **61.** 90 whales **63.** 232.53 million, or 232,530,000 **65.** 100 oz **67.** 520 calories **69.** 58.1 mi **71.** 64 cans **73.** 162 **74.** 4014 **75.** 12.95 **76.** 0.4 **77.** 14.3 **78.** −0.26 **79.** $-\frac{1}{10}$ **80.** $\frac{5}{4}$ **81.** 1 : 2 : 3

Calculator Corner, p. 398

1. 0.14 **2.** 0.00069 **3.** 0.438 **4.** 1.25

Exercise Set 6.2, p. 399

RC1. 43% **RC2.** 86% **RC3.** 19% **RC4.** 50%

1. $\frac{90}{100}$; $90 \times \frac{1}{100}$; 90×0.01 **3.** $\frac{12.5}{100}$; $12.5 \times \frac{1}{100}$; 12.5×0.01 **5.** 0.67 **7.** 0.456 **9.** 0.5901 **11.** 0.1 **13.** 0.01 **15.** 2 **17.** 0.001 **19.** 0.0009 **21.** 0.0018 **23.** 0.2319 **25.** 0.14875 **27.** 0.565 **29.** 0.13; 0.11 **31.** 0.07; 0.08 **33.** 0.139; 0.4 **35.** 47% **37.** 3% **39.** 870% **41.** 33.4% **43.** 75% **45.** 40% **47.** 0.6% **49.** 1.7% **51.** 27.18% **53.** 2.39% **55.** 65.1% **57.** 34% **59.** 35.9%; 20% **61.** 0.29; 0.3; 0.13; 0.1; 0.18 **63.** 1, 2, 3, 4, 6, 7, 12, 14, 21, 28, 42, 84 **64.** 1, 2, 4, 5, 10, 20, 31, 62, 124, 155, 310, 620 **65.** 180 **66.** $2 \cdot 3 \cdot 3 \cdot 5$ **67.** $\frac{3}{16}$ **68.** 97 **69.** 50% **71.** 70% **73.** 20%

Calculator Corner, p. 403

1. 52% **2.** 38.46% **3.** 110.26% **4.** 171.43% **5.** 59.62% **6.** 28.31%

Exercise Set 6.3, p. 407

RC1. (c) **RC2.** (e) **RC3.** (g) **RC4.** (a) **RC5.** (b) **RC6.** (d) **RC7.** (h) **RC8.** (i)

1. 41% **3.** 5% **5.** 20% **7.** 30% **9.** 50% **11.** 87.5%, or $87\frac{1}{2}\%$ **13.** 80% **15.** $66.\overline{6}\%$, or $66\frac{2}{3}\%$ **17.** $16.\overline{6}\%$, or $16\frac{2}{3}\%$ **19.** 18.75%, or $18\frac{3}{4}\%$ **21.** 81.25%, or $81\frac{1}{4}\%$ **23.** 16% **25.** 5% **27.** 34% **29.** 8%; 59% **31.** 22% **33.** 12% **35.** 15% **37.** $\frac{17}{20}$ **39.** $\frac{5}{8}$ **41.** $\frac{1}{3}$ **43.** $\frac{1}{6}$ **45.** $\frac{29}{400}$ **47.** $\frac{1}{125}$ **49.** $\frac{203}{800}$ **51.** $\frac{176}{225}$ **53.** $\frac{711}{1100}$ **55.** $\frac{3}{2}$ **57.** $\frac{13}{40{,}000}$ **59.** $\frac{1}{3}$ **61.** $\frac{3}{50}$ **63.** $\frac{13}{100}$ **65.** $\frac{19}{25}$ **67.** $\frac{3}{20}$ **69.** $\frac{209}{1000}$

71.

Fraction Notation	Decimal Notation	Percent Notation
$\frac{1}{8}$	0.125	12.5%, or $12\frac{1}{2}\%$
$\frac{1}{6}$	$0.1\overline{6}$	$16.\overline{6}\%$, or $16\frac{2}{3}\%$
$\frac{1}{5}$	0.2	20%
$\frac{1}{4}$	0.25	25%
$\frac{1}{3}$	$0.\overline{3}$	$33.\overline{3}\%$, or $33\frac{1}{3}\%$
$\frac{3}{8}$	0.375	37.5%, or $37\frac{1}{2}\%$
$\frac{2}{5}$	0.4	40%
$\frac{1}{2}$	0.5	50%

73.

Fraction Notation	Decimal Notation	Percent Notation
$\frac{1}{2}$	0.5	50%
$\frac{1}{3}$	$0.\overline{3}$	$33.\overline{3}\%$, or $33\frac{1}{3}\%$
$\frac{1}{4}$	0.25	25%
$\frac{1}{6}$	$0.1\overline{6}$	$16.\overline{6}\%$, or $16\frac{2}{3}\%$
$\frac{1}{8}$	0.125	12.5%, or $12\frac{1}{2}\%$
$\frac{3}{4}$	0.75	75%
$\frac{5}{6}$	$0.8\overline{3}$	$83.\overline{3}\%$, or $83\frac{1}{3}\%$
$\frac{3}{8}$	0.375	37.5%, or $37\frac{1}{2}\%$

75. 70 **76.** 5 **77.** 400 **78.** −18.75 **79.** 23.125 **80.** 25.5 **81.** 4.5 **82.** 8.75 **83.** $-18\frac{3}{4}$ **84.** $7\frac{4}{9}$ **85.** $\frac{203}{2}$ **86.** $-\frac{209}{10}$ **87.** $257.\overline{46317}\%$ **89.** $1.04\overline{142857}$ **91.** $\frac{1}{6}\%$, $\frac{2}{7}\%$, 0.5%, $1\frac{1}{6}\%$, 1.6%, $16\frac{1}{6}\%$, 0.2, $\frac{1}{2}$, $0.\overline{54}$, 1.6

Calculator Corner, p. 414

1. \$5.04 **2.** 0.0112 **3.** 450 **4.** \$1000 **5.** 2.5% **6.** 12%

Exercise Set 6.4, p. 415

RC1. (c) **RC2.** (e) **RC3.** (f) **RC4.** (d) **RC5.** (b) **RC6.** (a)

1. $a = 32\% \times 78$ **3.** $89 = p \times 99$ **5.** $13 = 25\% \times b$ **7.** 234.6 **9.** 45 **11.** \$18 **13.** 1.9 **15.** 78% **17.** 200% **19.** 50% **21.** 125% **23.** 40 **25.** \$40 **27.** 88 **29.** 20 **31.** 6.25 **33.** \$846.60 **35.** 1216 **37.** $\frac{9375}{10{,}000}$, or $\frac{15}{16}$ **38.** $-\frac{125}{1000}$, or $-\frac{1}{8}$ **39.** −0.3 **40.** 0.017 **41.** 7 **42.** 8 **43.** \$10,000 (can vary); \$10,400 **45.** \$1875

Exercise Set 6.5, p. 421

RC1. (b) **RC2.** (e) **RC3.** (c) **RC4.** (f) **RC5.** (d) **RC6.** (a)

1. $\frac{37}{100} = \frac{a}{74}$ **3.** $\frac{N}{100} = \frac{4.3}{5.9}$ **5.** $\frac{25}{100} = \frac{14}{b}$ **7.** 68.4 **9.** 462 **11.** 40 **13.** 2.88 **15.** 25% **17.** 102% **19.** 25% **21.** 93.75%, or $93\frac{3}{4}\%$ **23.** \$72 **25.** 90 **27.** 88 **29.** 20 **31.** 25 **33.** \$780.20 **35.** 200 **37.** 8 **38.** 4000 **39.** 15 **40.** 2074 **41.** $\frac{43}{48}$ qt **42.** $\frac{1}{8}$ T **43.** \$1170 (can vary); \$1118.64

Mid-Chapter Review: Chapter 6, p. 423

1. False **2.** True **3.** False **4.** True **5.** $\frac{1}{2}\% = \frac{1}{2} \cdot \frac{1}{100} = \frac{1}{200}$ **6.** $\frac{80}{1000} = \frac{8}{100} = 8\%$ **7.** $5.5\% = \frac{5.5}{100} = \frac{55}{1000} = \frac{11}{200}$ **8.** $0.375 = \frac{375}{1000} = \frac{37.5}{100} = 37.5\%$

9.
$$\begin{aligned} 15 &= p \times 80 \\ \frac{15}{80} &= \frac{p \times 80}{80} \\ \frac{15}{80} &= p \\ 0.1875 &= p \\ 18.75\% &= p \end{aligned}$$

10.
$$\begin{aligned} \frac{x}{4} &= \frac{3}{6} \\ x \cdot 6 &= 4 \cdot 3 \\ \frac{x \cdot 6}{6} &= \frac{4 \cdot 3}{6} \\ x &= 2 \end{aligned}$$

11. $\frac{1}{3}$ **12.** $\frac{2}{7}$ **13.** 48.67 km/hr **14.** 60.75 mi/hr, or 60.75 mph **15.** 40 **16.** 9 **17.** 4.32 **18.** $\frac{1}{2}$ **19.** 0.28 **20.** 0.0015 **21.** 0.05375 **22.** 2.4 **23.** 71% **24.** 9% **25.** 38.91% **26.** 18.75%, or $18\frac{3}{4}\%$ **27.** 0.5% **28.** 74% **29.** 600% **30.** $83.\overline{3}\%$, or $83\frac{1}{3}\%$ **31.** $\frac{17}{20}$ **32.** $\frac{3}{6250}$ **33.** $\frac{91}{400}$ **34.** $\frac{1}{6}$ **35.** 62.5%, or $62\frac{1}{2}\%$ **36.** 45% **37.** 27 in./day **38.** 58 **39.** $16.\overline{6}\%$, or $16\frac{2}{3}\%$ **40.** 2560 **41.** \$50 **42.** 0.05%, 0.1%, $\frac{1}{2}\%$, 1%, 5%, 10%, $\frac{13}{100}$, 0.275, $\frac{3}{10}$, $\frac{7}{20}$ **43.** B **44.** They all represent the same number. **45.** Since $40\% \div 10 = 4\%$, we can divide 36.8 by 10, obtaining 3.68. Since $400\% = 40\% \times 10$, we can multiply 36.8 by 10, obtaining 368.

Translating for Success, p. 432

1. J **2.** M **3.** N **4.** E **5.** G **6.** H **7.** O **8.** C **9.** D **10.** B

Exercise Set 6.6, p. 433

RC1. \$10, $\frac{\$10}{\$50}$, 20% **RC2.** \$15, $\frac{\$15}{\$60}$, 25% **RC3.** \$120, $\frac{\$120}{\$360}$, $33\frac{1}{3}\%$ **RC4.** \$1600, $\frac{\$1600}{\$4000}$, 40% **1.** United States: about \$104.7 billion; Italy: about \$24.6 billion **3.** \$46,656 **5.** 140 items **7.** About 940,000,000 acres **9.** \$36,400 **11.** 74.4 items correct; 5.6 items incorrect **13.** 261 men **15.** About 27.8% **17.** \$230.10 **19.** About 29.8% **21.** Alcohol: 43.2 mL; water: 496.8 mL **23.** Air Force: 25.2%; Army: 36.5%; Navy: 25.3%; Marines: 13.0% **25.** 5% **27.** 15% **29.** About 34.7% **31.** About 127% **33.** About 49.5% **35.** About 497% **37.** About 9.5% **39.** 34.375%, or $34\frac{3}{8}\%$ **41.** 16,914; 2.8% **43.** 5,130,632; 24.6% **45.** 1,567,582; 21.1% **47.** About 23.8% **49.** $-2.\overline{27}$ **50.** 0.44 **51.** 3.375 **52.** $-4.\overline{7}$ **53.** $\frac{3}{17}$ **54.** $-\frac{9}{34}$ **55.** $\frac{24}{125}$ **56.** $\frac{5}{29}$ **57.** \$42

Exercise Set 6.7, p. 444

RC1. Rate **RC2.** Rate **RC3.** Price **RC4.** Tax **RC5.** Tax **1.** \$9.56 **3.** \$6.66 **5.** \$11.99; \$171.79 **7.** 4% **9.** \$5600 **11.** \$116.72 **13.** \$276.28 **15.** \$194.08 **17.** \$2625 **19.** 12% **21.** \$185,000 **23.** 15% **25.** \$355 **27.** \$30; \$270 **29.** \$2.55; \$14.45 **31.** \$125; \$112.50 **33.** 40%; \$360 **35.** \$300; 40% **37.** 18 **38.** $\frac{22}{7}$ **39.** 4,030,000,000,000 **40.** 5,800,000 **41.** \$17,700

Calculator Corner, p. 450

1. \$16,357.18 **2.** \$12,764.72

Exercise Set 6.8, p. 453

RC1. $\frac{6}{12}$ year **RC2.** $\frac{40}{365}$ year **RC3.** $\frac{285}{365}$ year **RC4.** $\frac{9}{12}$ year **RC5.** $\frac{3}{12}$ year **RC6.** $\frac{4}{12}$ year **1.** \$8 **3.** \$113.52 **5.** \$462.50 **7.** 671.88 **9. (a)** \$147.95; **(b)** \$10,147.95 **11. (a)** \$84.14; **(b)** \$6584.14 **13. (a)** \$46.03; **(b)** \$5646.03 **15.** \$441 **17.** \$2802.50 **19.** \$7853.38 **21.** \$99,427.40 **23.** \$4243.60 **25.** \$28,225.00 **27.** \$9270.87 **29.** \$129,871.09 **31.** \$4101.01 **33.** \$1324.58 **35.** Interest: \$20.88; amount applied to principal: \$4.69; balance after the payment: \$1273.87 **37. (a)** \$98; **(b)** interest: \$86.56; amount applied to principal: \$11.44; **(c)** interest: \$51.20; amount applied to principal: \$46.80; **(d)** At 12.6%, the principal is reduced by \$35.36 more than at the 21.3% rate. The interest at 12.6% is \$35.36 less than at 21.3%. **39.** 800 **40.** $2 \cdot 2 \cdot 3 \cdot 19$ **41.** $-\frac{9}{100}$ **42.** $\frac{32}{75}$ **43.** $\frac{1}{25}$ **44.** $-\frac{7}{300}$ **45.** 55 **46.** $\frac{1}{10}$ **47.** 9.38%

Summary and Review: Chapter 6, p. 456

Vocabulary Reinforcement

1. Percent decrease **2.** Simple **3.** Rate **4.** Proportion **5.** Discount **6.** Sales

Concept Reinforcement

1. True **2.** True **3.** True **4.** False **5.** True

Study Guide

1. $\frac{17}{3}$ **2.** \$7.50/hr **3.** Yes **4.** $\frac{27}{8}$ **5.** 175 mi **6.** 8.2% **7.** 0.62625 **8.** $63.\overline{63}\%$, or $63\frac{7}{11}\%$ **9.** $\frac{17}{250}$ **10.** $4.1\overline{6}\%$, or $4\frac{1}{6}\%$ **11.** 10,000 **12.** About 15.3% **13.** 6% **14.** \$185,000 **15.** Simple interest: \$22.60; total amount due: \$2522.60 **16.** \$6594.26

Review Exercises

1. $\frac{47}{84}$ **2.** $\frac{46}{1.27}$ **3.** $\frac{83}{100}$ **4.** $\frac{0.72}{197}$ **5. (a)** $\frac{12,480}{16,640}$, or $\frac{3}{4}$; **(b)** $\frac{16,640}{29,120}$, or $\frac{4}{7}$ **6.** $\frac{3}{4}$ **7.** $\frac{9}{16}$ **8.** 28 mpg **9.** 6300 revolutions/min **10.** 0.638 gal/ft^2 **11.** Yes **12.** No **13.** 32 **14.** 7 **15.** $\frac{1}{40}$ **16.** 24 **17.** 27 circuits **18. (a)** 247.50 Canadian dollars; **(b)** 50.51 U.S. dollars **19.** 832 mi **20.** 133.5 billion gal, or 133,500,000,000 gal **21.** About 3,774,811 lb **22.** 320 calories **23.** About 12,453 lawyers **24.** 0.04; 0.144 **25.** 0.621; 0.842 **26.** 170% **27.** 6.5% **28.** 37.5%, or $37\frac{1}{2}\%$ **29.** $33.\overline{3}\%$, or $33\frac{1}{3}\%$ **30.** $\frac{6}{25}$ **31.** $\frac{63}{1000}$ **32.** $30.6 = p \times 90$; 34% **33.** $63 = 84\% \times b$; 75 **34.** $a = 38\frac{1}{2}\% \times 168$; 64.68 **35.** $\frac{24}{100} = \frac{16.8}{b}$; 70 **36.** $\frac{42}{30} = \frac{N}{100}$; 140% **37.** $\frac{10.5}{100} = \frac{a}{84}$; 8.82 **38.** 178 students; 84 students **39.** 46% **40.** 2500 mL **41.** 12% **42.** 92 **43.** \$24 **44.** 6% **45.** 11% **46.** About 18.3% **47.** \$42; \$308 **48.** 14% **49.** \$2940 **50.** \$36 **51. (a)** \$394.52; **(b)** \$24,394.52 **52.** \$121 **53.** \$7575.25 **54.** \$9504.80 **55. (a)** \$129; **(b)** interest: \$100.18; amount applied to principal: \$28.82; **(c)** interest: \$70.72; amount applied to principal: \$58.28; **(d)** At 13.2%, the principal is decreased by \$29.46 more than at the 18.7% rate. The interest at 13.2% is \$29.46 less that at 18.7%. **56.** C **57.** C **58.** About 19% **59.** Finishing paint: 11 gal; primer: 16.5 gal

Understanding Through Discussion and Writing

1. A 40% discount is better. When successive discounts are taken, each is based on the previous discounted price rather than on the original price. A 20% discount followed by a 22% discount is the same as a 37.6% discount off the original price. **2.** In terms of cost, a low faculty-to-student ratio is less expensive than a high faculty-to-student ratio. In terms of quality of education and student satisfaction, a high faculty-to-student ratio is more desirable. A college president must balance the cost and quality issues. **3.** No; the 10% discount was based on the original price rather than on the sale price. **4.** Let $S =$ the original salary. After both raises have been given, the two situations yield the same salary: $1.05 \cdot 1.1S = 1.1 \cdot 1.05S$. However, the first situation is better for the wage earner, because $1.1S$ is earned the first year when a 10% raise is given while in the second situation $1.05S$ is earned that year. **5.** For a number n, 40% of 50% of n is $0.4(0.5n)$, or $0.2n$, or 20% of n. Thus, taking 40% of 50% of a number is the same as taking 20% of the number. **6.** The interest due on the 30-day loan will be \$41.10 while that due on the 60-day loan will be \$131.51. This could be an argument in favor of the 30-day loan. On the other hand, the 60-day loan puts twice as much cash at the firm's disposal for twice as long as the 30-day loan does. This could be an argument in favor of the 60-day loan.

Test: Chapter 6, p. 464

1. [6.1a] $\frac{85}{97}$ **2.** [6.1a] $\frac{0.34}{124}$ **3.** [6.1a] $\frac{9}{10}$ **4.** [6.1a] $\frac{25}{32}$ **5.** [6.1b] $1\frac{1}{3}$ servings/lb **6.** [6.1b] 32 mpg **7.** [6.1c] Yes **8.** [6.1c] No **9.** [6.1d] 100 **10.** [6.1d] 360 **11.** [6.1e] 525 mi **12.** [6.1e] **(a)** 13 hats; **(b)** 30 packages **13.** [6.1e] 1512 km **14.** [6.1e] About \$59.17 **15.** [6.2b] 0.613 **16.** [6.2b] 38% **17.** [6.3a] 137.5% **18.** [6.3b] $\frac{13}{20}$ **19.** [6.4a, b] $a = 40\% \cdot 55$; 22 **20.** [6.5a, b] $\frac{N}{100} = \frac{65}{80}$; 81.25% **21.** [6.6a] 16,550 kidney transplants; 5992 liver transplants; 2283 heart transplants **22.** [6.6a] About 611 at-bats **23.** [6.6b] 6.7% **24.** [6.6a] 60.6% **25.** [6.7a] \$25.20; \$585.20 **26.** [6.7b] \$630 **27.** [6.7c] \$40; \$160 **28.** [6.8a] \$8.52 **29.** [6.8a] \$5356 **30.** [6.8b] \$1110.39 **31.** [6.8b] \$11,580.07 **32.** [6.6b] 757,000, 7.4%; 29,000, 40.8%; 21,000, 14.3%; 235,000, 63,000 **33.** [6.7c] \$50; about 14.3% **34.** [6.8c] Interest: \$36.73; amount applied to the principal: \$17.27; balance after payment: \$2687 **35.** [6.4a, b], [6.5a, b] B **36.** [6.1b] C **37.** [6.7b] \$194,600 **38.** [6.7b], [6.8b] \$2546.16

CHAPTER 7

Exercise Set 7.1, p. 473

RC1. Statistic **RC2.** Average **RC3.** Weight **RC4.** Mode **1.** Average: 9.325 million visitors; median: 9.35 million visitors; modes: 9.0 million visitors, 9.5 million visitors **3.** Average: 21; median: 18.5; mode: 29 **5.** Average: 21; median: 20; modes: 5, 20 **7.** Average: 5.38; median: 5.7; no mode exists **9.** Average: 239.5; median: 234; mode: 234 **11.** 36 mpg **13.** 2.7 **15.** Average: \$4.19; median: \$3.99; mode: \$3.99 **17.** 90 **19.** 263 days **21.** **(a)** Jefferson County: \$133,987; Hamilton County: \$146,989; **(b)** Jefferson County **23.** **(a)** 1.455 billion tickets; **(b)** 1.3775 billion tickets; **(c)** 2002 to 2009 **25.** −225.05 **26.** 126.0516 **27.** $\frac{3}{35}$ **28.** $-\frac{14}{15}$ **29.** $a = 30$; $b = 58$ **31.** \$6950

Exercise Set 7.2, p. 482

RC1. False **RC2.** False **RC3.** True **RC4.** False **1.** 92° **3.** 108° **5.** 85°, 60%; 90°, 40%; 100°, 10% **7.** 90° and higher **9.** 30% and higher **11.** 90% − 40% = 50% **13.** 483,612,200 mi **15.** Neptune **17.** 11 Earth diameters **19.** Average: 31,191.75 mi; median: 19,627.5 mi; no mode exists **21.** 300 calories **23.** Yes **25.** 410 mg **27.** White rhino **29.** About 1350 rhinos **31.** About 4100 rhinos **33.** \$1270 per person **35.** 1950 and 1970 **37.** \$1532.70 per person **39.** **(a)** 15% less; **(b)** \$2003.51 more per person **41.** About 24 games **43.** 120–129

Mid-Chapter Review: Chapter 7, p. 486

1. True **2.** True **3.** False **4.** $\frac{60 + 45 + 115 + 15 + 35}{5} = \frac{270}{5} = 54$ **5.** 2.1, 4.8, 6.3, 8.7, 11.3, 14.5; 6.3 and 8.7; $\frac{6.3 + 8.7}{2} = \frac{15}{2} = 7.5$; the median is 7.5. **6.** Average: 83; median: 45; mode: 29 **7.** Average: 18.45; median: 13.895; no mode **8.** Average: $\frac{4}{9}$; median: $\frac{4}{9}$; no mode **9.** Average: 126; median: 116; no mode **10.** Average: \$6.09; median: \$5.24; modes: \$4.96 and \$5.24 **11.** Average: $\frac{27}{32}$; median: $\frac{13}{16}$; no mode **12.** Average: 6; median: 7; modes: 5 and 7 **13.** Average: 38.2; median: 38.2; no mode **14.** 8 oz **15.** 6% **16.** Hershey's Special Dark chocolate bar **17.** 7 oz **18.** Nabisco Chips Ahoy cookies **19.** Tom Brady **20.** About 45 passes **21.** About 22 more touchdown passes **22.** About 39 touchdown passes **23.** Yes. At an average speed of 20 mph, the trip would take $1\frac{1}{2}$ hr (30 mi ÷ 20 mph = $1\frac{1}{2}$ hr). But the driver could have driven at a speed of 75 mph for a brief period during that time and at lower speeds for the remainder of the trip and still have an average speed of 20 mph. **24.** Answers may vary. Some would ask for the average salary since it is a center point that places equal emphasis on all the salaries in the firm. Some would ask for the median salary since it is a center point that deemphasizes the extremely high and extremely low salaries. Some would ask for the mode of the salaries since it might indicate the salary the applicant is most likely to earn.

Exercise Set 7.3, p. 492

RC1. True **RC2.** True **RC3.** False **RC4.** True **1.** Miniature tall bearded **3.** 16 in. to 26 in. **5.** Tall bearded **7.** 25 in. **9.** 190 calories **11.** 1 slice of chocolate cake with fudge frosting **13.** 1 cup of premium chocolate ice cream **15.** About 125 calories **17.** 1950 and 1970 **19.** About 175,000 bachelor's degrees

21.

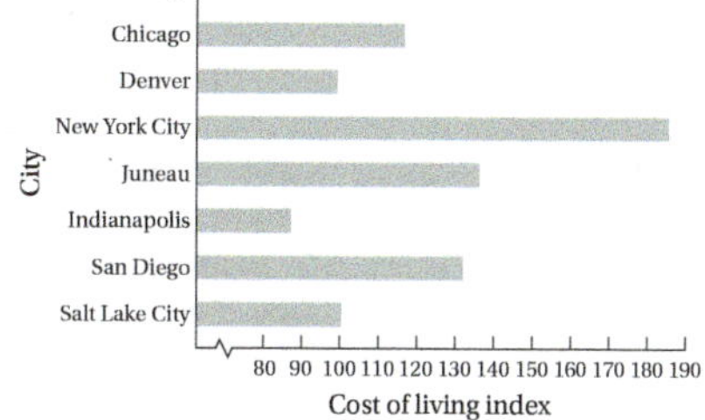

23. Denver and Indianapolis **25.** 68.9 **27.** New York City **29.** 27.4 min **31.** \$42 **33.** June to July, September to October, and November to December **35.** About \$900 **37.** 20 years **39.** About \$440 **41.** About 120(\$1200) = \$144,000

43.

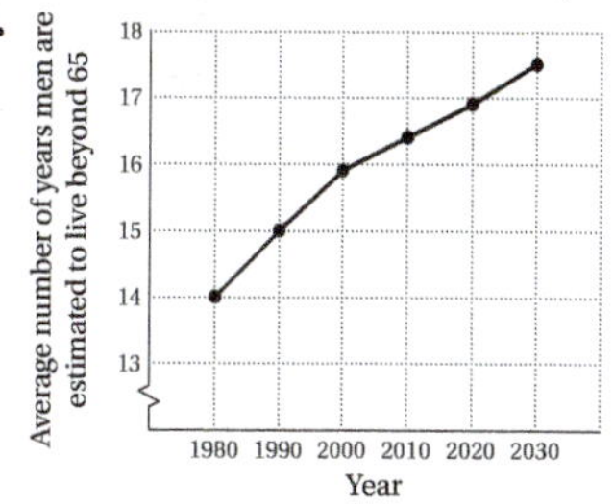

45. 25% **47.** 10.1% **48.** 83 **49.** $-\frac{4}{3}$ **50.** $\frac{7}{24}$ **51.** 2.26 **52.** 6.348 **53.** 7.2 **54.** 80 **55.** 0.9 **56.** 150% **57.** 21 **58.** 17.26 **59.** $\frac{31}{60}$ **60.** 45

Translating for Success, p. 499

1. D **2.** B **3.** J **4.** K **5.** I **6.** F **7.** N **8.** E **9.** L **10.** M

Exercise Set 7.4, p. 500

RC1. True **RC2.** True **RC3.** True **RC4.** False **RC5.** True **RC6.** False **1.** 10% **3.** 98,800 students **5.** Canada **7.**

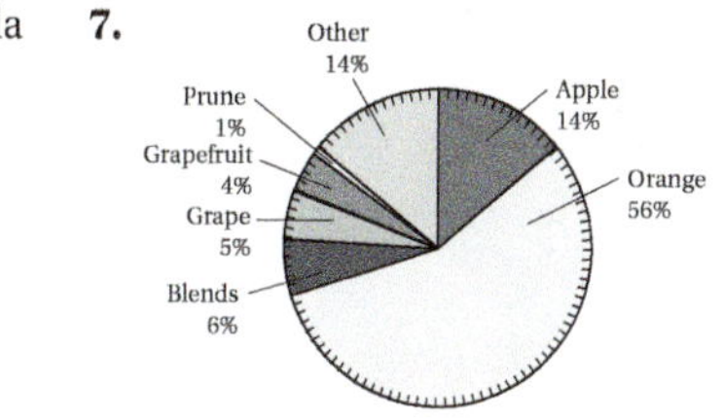

9.

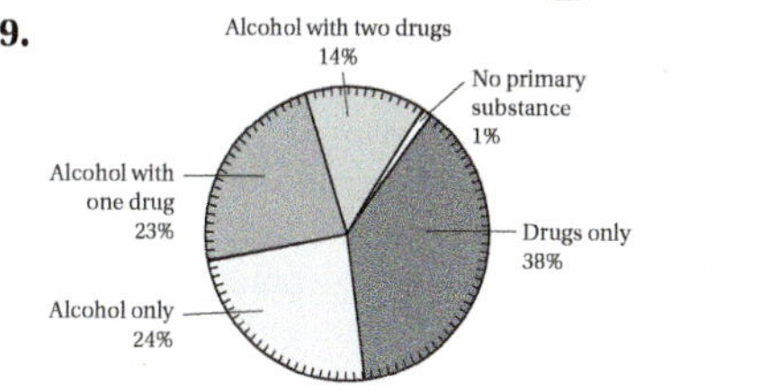

Summary and Review: Chapter 7, p. 502

Vocabulary Reinforcement

1. table **2.** circle graph **3.** pictograph **4.** mode **5.** average **6.** median

Concept Reinforcement

1. False **2.** True **3.** True

Study Guide

1. Average: 8; median: 8; mode: 8 **2.** Quaker Organic Maple & Brown Sugar; \$0.54 per serving **3.** 12 g **4.** Arrowhead Stadium **5.** About \$1350 million more **6.** 80 years and older **7.** 27%

Review Exercises

1. 38.5 **2.** 13.4 **3.** 1.55 **4.** 1840 **5.** $\$16.\overline{6}$ **6.** $321.\overline{6}$ **7.** 96 **8.** 28 mpg **9.** 3.1 **10.** 38.5 **11.** 14 **12.** 1.8 **13.** 1900 **14.** \$17 **15.** 375 **16.** Average: \$260; median: \$228 **17.** 26 **18.** 11 and 17 **19.** 0.2 **20.** 700 and 800 **21.** \$17 **22.** 20 **23.** 52% **24.** Mexico, Germany, and the Philippines **25.** United States **26.** About 21 World Series **27.** 7 games **28.** About 16 more World Series **29.** \$70,000–\$89,000

30. About 2 more governors **31.** About 22 governors **32.** April **33.** About 150 tornadoes **34.** About 150 more tornadoes **35.** In the spring **36.** 1998 **37.** About 10,000 children **38.** 2004 and 2010 **39.** Between 1986 and 1989, between 1995 and 1998, and between 2004 and 2007 **40.** By about 4000 children **41.** 34% **42.** Room and board **43.** 14% **44.** \$4830 **45.** D **46.** A

47.

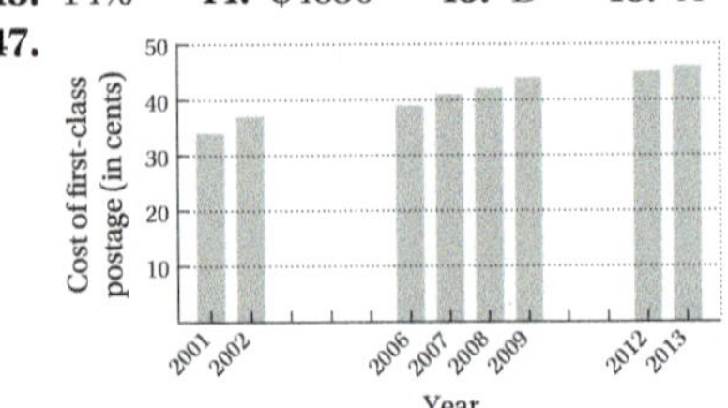

48.

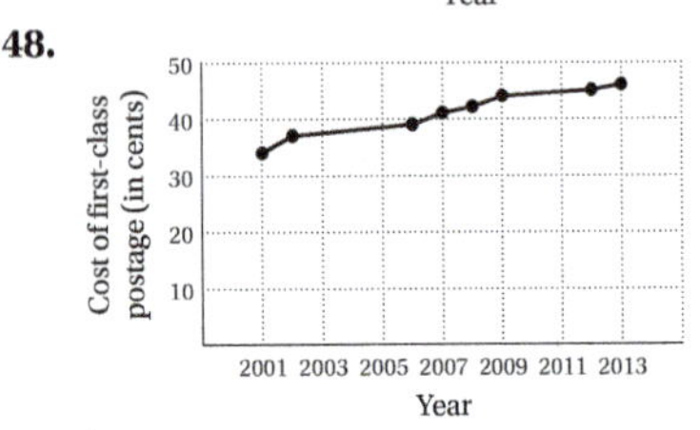

49.

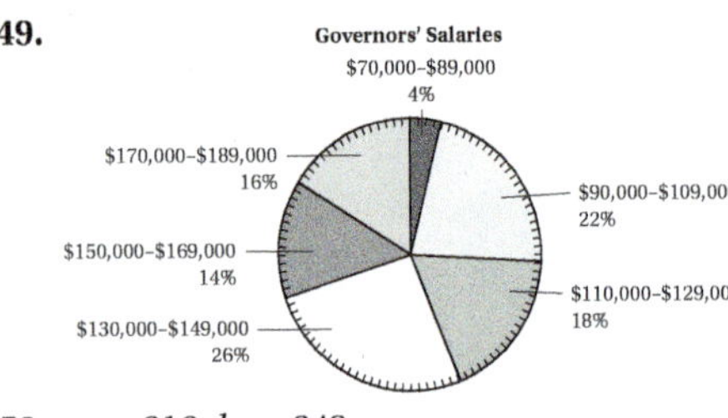

50. $a = 316, b = 349$

Understanding Through Discussion and Writing

1. The equation could represent a person's average income during a 4-year period. Answers may vary. **2.** Bar graphs that show change over time can be converted to line graphs. Other bar graphs cannot be converted to line graphs. **3.** One advantage is that we can use circle graphs to visualize how the numbers of items in various categories compare in size. **4.** A bar graph is convenient for showing comparisons. A line graph is convenient for showing a change over time as well as to indicate patterns or trends. The choice of which to use to graph a particular set of data would probably depend on the type of data analysis desired. **5.** The average, the median, and the mode are "center points" that characterize a set of data. You might use the average to find a center point that is midway between the extreme values of the data. The median is a center point that is in the middle of all the data. That is, there are as many values less than the median as there are values greater than the median. The mode is a center point that represents the value or values that occur most frequently. **6.** Circle graphs are similar to bar graphs in that both allow us to tell at a glance how items in various categories compare in size. They differ in that circle graphs show percents whereas bar graphs show actual numbers of items in a given category.

Test: Chapter 7, p. 508

1. [7.1a] 49.5 **2.** [7.1a] 2.6 **3.** [7.1a] 15.5 **4.** [7.1b, c] 50.5; no mode exists **5.** [7.1b, c] 3; 1 and 3 **6.** [7.1b, c] 17.5; 17 and 18 **5.** [7.1a] 76 **8.** [7.1a] 2.9 **9.** [7.2a] 179 lb **10.** [7.2a] 5 ft 3 in.; medium frame **11.** [7.2a] 9 lb **12.** [7.2a] 32 lb **13.** [7.2b] Spain **14.** [7.2b] Norway and the United States **15.** [7.2b] 900 lb **16.** [7.2b] 1000 lb **15.** [7.3c] 2005 **18.** [7.3c] 2009 **19.** [7.3c] 10 hurricanes **20.** [7.3c] 10 more hurricanes **21.** [7.3c] About 8 hurricanes/year **22.** [7.3c] 2005, 2010, 2012

23. [7.3b]

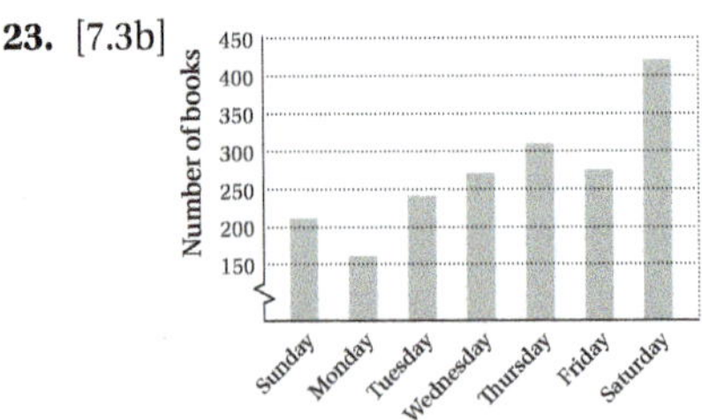

24. [7.3d]

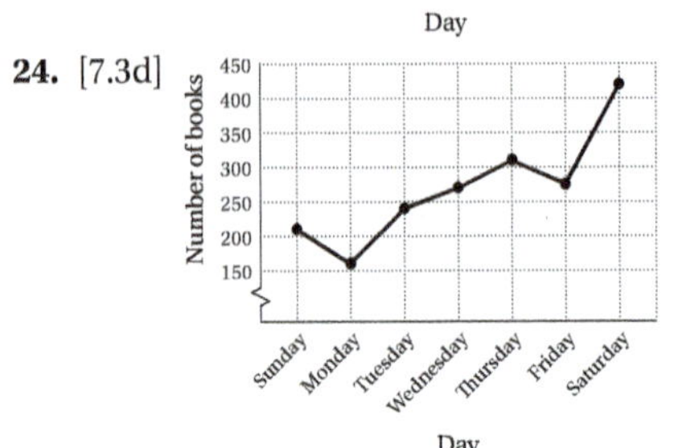

25. [7.4b]

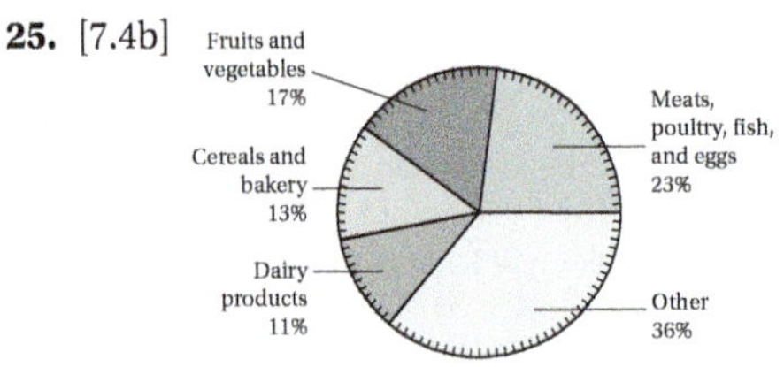

26. [7.4a] C **27.** [7.1a, b] $a = 74, b = 111$

CHAPTER 8

Exercise Set 8.1, p. 519

RC1. (e) **RC2.** (h) **RC3.** (a) **RC4.** (d) **RC5.** (b) **RC6.** (g) **RC7.** (c) **RC8.** (f)

1. G ———— H, $\overline{GH}, \overline{HG}$

3. Q ———— D ——→, $\overrightarrow{QD}$

5. $\overleftrightarrow{DE}, \overleftrightarrow{ED}, \overleftrightarrow{DF}, \overleftrightarrow{FD}, \overleftrightarrow{EF}, \overleftrightarrow{FE}, l$ **7.** Angle GHI, angle IHG, $\angle GHI, \angle IHG$, or $\angle H$ **9.** 10° **11.** 180° **13.** 130° **15.** Obtuse **17.** Acute **19.** Straight **21.** Obtuse **23.** Acute **25.** Obtuse **27.** Not perpendicular **29.** Perpendicular **31.** Scalene; obtuse **33.** Scalene; right **35.** Equilateral; acute **37.** Scalene; obtuse **39.** Quadrilateral **41.** Pentagon **43.** Triangle **45.** Pentagon **47.** Hexagon **49.** 46° **51.** 120° **53.** 43° **55.** 1440° **57.** 900° **59.** 2160° **61.** 3240° **63.** 2.7125 **64.** $8\frac{1}{12}$ **65.** $\frac{7}{60}$ **66.** $\frac{2}{3}$ **67.** −27 **68.** 840 **69.** $\frac{3}{10}$ **70.** $\frac{17}{48}$ **71.** $m\angle ACB = 50°$; $m\angle CAB = 40°$; $m\angle EBC = 50°$; $m\angle EBA = 40°$; $m\angle AEB = 100°$; $m\angle ADB = 50°$

Exercise Set 8.2, p. 526

RC1. closed **RC2.** perimeter **RC3.** perimeter **RC4.** square **1.** 17 mm **3.** 15.25 in. **5.** 18 km **7.** 30 ft **9.** 16 yd **11.** 88 ft **13.** 182 mm **15.** 27 ft **17.** 122 cm **19.** **(a)** 228 ft; **(b)** \$1046.52 **21.** \$19.20 **22.** \$96 **23.** 1000 **24.** 1331 **25.** 225 **26.** 484 **27.** 49 **28.** 64 **29.** 5% **30.** 11% **31.** 64 in.

Exercise Set 8.3, p. 533

RC1. (b) **RC2.** (c) **RC3.** (a) **RC4.** (d) **1.** 15 km^2 **3.** 1.4 in^2 **5.** $6\frac{1}{4}$ yd^2 **7.** 8100 ft^2 **9.** 50 ft^2 **11.** 169.883 cm^2 **13.** $41\frac{2}{9}$ in^2 **15.** 484 ft^2 **17.** 3237.61 km^2 **19.** $28\frac{57}{64}$ yd^2 **21.** 32 cm^2 **23.** 60 in^2 **25.** 104 ft^2 **27.** 45.5 in^2 **29.** 8.05 cm^2 **31.** 297 cm^2 **33.** 7 m^2 **35.** 1197 m^2 **37.** 39,825 ft^2 **39.** **(a)** 630.36 ft^2 **(b)** \$7879.50 **41.** **(a)** 819.75 ft^2; **(b)** 3 gal; **(c)** \$104.85 **43.** 80 cm^2 **45.** 675 cm^2 **47.** 21 cm^2 **49.** 144 ft^2 **51.** $\frac{7}{29}$ **52.** $\frac{1}{1000}$

53. $-\frac{5}{8}$ **54.** $\frac{5}{2}$ **55.** 10% **56.** $16.\overline{6}\%$, or $16\frac{2}{3}\%$ **57.** 91 **58.** 360 **59.** \$30,474.86 **60.** \$413,458.31 **61.** \$641,566.26 **62.** \$429,610.21 **63.** 16,914 in^2

Calculator Corner, p. 541

1. Left to the student **2.** Left to the student

Exercise Set 8.4, p. 543

RC1. radius **RC2.** circumference **RC3.** circumference **RC4.** area **1.** 14 cm; 44 cm; 154 cm^2 **3.** $1\frac{1}{2}$ in.; $4\frac{5}{7}$ in.; $1\frac{43}{56}$ in^2 **5.** 16 ft; 100.48 ft; 803.84 ft^2 **7.** 0.7 cm; 4.396 cm; 1.5386 cm^2 **9.** 94.2 ft **11.** About 55.04 in^2 larger **13.** About 7930.25 mi; about 3965.13 mi **15.** Maximum circumference of barrel: $8\frac{9}{14}$ in.; minimum circumference of handle: $2\frac{86}{133}$ in. **17.** 65.94 yd^2 **19.** 45.68 ft **21.** 26.84 yd **23.** 45.7 yd **25.** 100.48 m^2 **27.** 6.9972 cm^2 **29.** 64.4214 in^2 **31.** 47 mpg **32.** 0.4 oz **33.** 5 lb **34.** \$730 **35.** 43,560 ft^2; 1311.6 ft; \$599.96 **37.** 43,595.47395 ft^2; 739.9724 ft; \$449.97 **39.** 43,560 ft^2; 844 ft; \$449.97

Mid-Chapter Review: Chapter 8, p. 548

1. True **2.** True **3.** False **4.** True

5. $A = \frac{1}{2} \cdot 12\text{ cm} \cdot 8\text{ cm}$
$A = \frac{12 \cdot 8}{2}\text{cm}^2$
$A = \frac{96}{2}\text{cm}^2$, or 48 cm^2

6. $C \approx 3.14 \cdot 10.2\text{ in.}$
$C = 32.028\text{ in.};$
$A \approx 3.14 \cdot 5.1\text{ in.} \cdot 5.1\text{ in.}$
$A = 81.6714\text{ in}^2$

7. 3060° **8.** 15° **9.** Hexagon **10.** Scalene; right **11.** Isosceles; obtuse **12.** Equilateral; acute **13.** 76 mm **14.** $P = 50\frac{2}{3}$ ft; $A = 160\frac{4}{9}$ ft^2 **15.** 800 in^2 **16.** $\frac{9}{16}$ yd^2 **17.** 66 km^2 **18.** $C = 43.96$ in.; $A = 153.86$ in^2 **19.** $C = 27.004$ cm; $A = 58.0586$ cm^2 **20.** Area of a circle with radius 4 ft: $16 \cdot \pi$ ft^2; Area of a square with side 4 ft: 16 ft^2; Circumference of a circle with radius 4 ft: $8 \cdot \pi$ ft; Area of a rectangle with length 8 ft and width 4 ft: 32 ft^2; Area of a triangle with base 4 ft and height 8 ft: 16 ft^2; Perimeter of a square with side 4 ft: 16 ft; Perimeter of a rectangle with length 8 ft and width 4 ft: 24 ft **21.** The area of a 16-in.-diameter pizza is approximately $3.14 \cdot 8\text{ in.} \cdot 8\text{ in.}$, or 200.96 in^2. At \$16.25, its unit price is $\frac{\$16.25}{200.96\text{ in}^2}$, or about \$0.08/in^2. The area of a 10-in.-diameter pizza is approximately $3.14 \cdot 5\text{ in.} \cdot 5\text{ in.}$, or 78.5 in^2. At \$7.85, its unit price is $\frac{\$7.85}{78.5\text{ in}^2}$, or \$0.10/in^2. Since the 16-in.-diameter pizza has the lower unit price, it is a better buy.
22. No; let l and w represent the length and the width of the smaller rectangle. Then $3 \cdot l$ and $3 \cdot w$ represent the length and the width of the larger rectangle. The area of the first rectangle is $l \cdot w$, but the area of the second is $3 \cdot l \cdot 3 \cdot w = 3 \cdot 3 \cdot l \cdot w = 9 \cdot l \cdot w$, or 9 times the area of the smaller rectangle.
23. Yes; let s represent the length of a side of the larger square. Then $\frac{1}{2}s$ represents the length of a side of the smaller square. The perimeter of the larger square is $4 \cdot s$, and the perimeter of the smaller square is $4 \cdot \frac{1}{2}s = 2s$, or $\frac{1}{2}$ the perimeter of the larger square.
24. For a rectangle with length l and width w,

$$\begin{aligned} P &= l + w + l + w \\ &= (l + w) + (l + w) \\ &= 2 \cdot (l + w). \end{aligned}$$

We also have

$$\begin{aligned} P &= l + w + l + w \\ &= (l + l) + (w + w) \\ &= 2 \cdot l + 2 \cdot w. \end{aligned}$$

25. See p. 530 of the text. **26.** No; let $r =$ the radius of the smaller circle. Then its area is $\pi \cdot r \cdot r$, or πr^2. The radius of the larger circle is $2r$, and its area is $\pi \cdot 2r \cdot 2r$, or $4\pi r^2$, or $4 \cdot \pi r^2$. Thus the area of the larger circle is 4 times the area of the smaller circle.

Exercise Set 8.5, p. 556

RC1. (b) **RC2.** (a) **RC3.** (c) **RC4.** (d) **1.** 768 cm^3; 512 cm^2 **3.** 45 in^3; 87 in^2 **5.** 75 m^3; 145 m^2 **7.** $357\frac{1}{2}$ yd^3; $311\frac{1}{2}$ yd^2 **9.** 803.84 in^3 **11.** 353.25 cm^3 **13.** 41,580,000 yd^3 **15.** $4,186,666\frac{2}{3}$ in^3 **17.** 124.72 m^3 **19.** $1950\frac{101}{168}$ ft^3 **21.** 113,982 ft^3 **23.** 24.64 cm^3 **25.** $\frac{33}{40}$ yd^3 **27.** 4747.68 cm^3 **29.** 65,417 m^3 **31.** About 77.7 in^3 **33.** About 8870 m^3 **35.** 152,321 m^3 **37.** 260,579,710,000 mi^3 **39.** 6 cm by 6 cm by 6 cm **41.** **(a)** About 1875.63 mm^3; **(b)** about 11,253.78 mm^3 **43.** 1064 mi^3 **44.** 24,360 mi^2 **45.** 260.4 ft **46.** 1087.8 mi^3 **47.** 3540.68 km^3 **49.** 0.477 m^3

Exercise Set 8.6, p. 568

RC1. True **RC2.** False **RC3.** False **RC4.** True **RC5.** True **RC6.** True **1.** 79° **3.** 23° **5.** 32° **7.** 61° **9.** 177° **11.** 41° **13.** 95° **15.** 78° **17.** Not congruent **19.** Congruent **21.** $m\angle 2 = 67°, m\angle 3 = 33°, m\angle 4 = 80°, m\angle 6 = 33°$ **23.** **(a)** $\angle 1$ and $\angle 3$, $\angle 2$ and $\angle 4$, $\angle 8$ and $\angle 6$, $\angle 7$ and $\angle 5$; **(b)** $\angle 2$, $\angle 3$, $\angle 6$, and $\angle 7$; **(c)** $\angle 2$ and $\angle 6$, $\angle 3$ and $\angle 7$ **25.** $m\angle 6 = m\angle 2 = m\angle 8 = 125°, m\angle 5 = m\angle 3 = m\angle 7 = m\angle 1 = 55°$ **27.** $\angle ABE \cong \angle DCE$, 95°; $\angle BAE \cong \angle CDE$; $\angle AEB \cong \angle DEC$; $\angle BED \cong \angle AEC$ **29.** $\angle AEC \cong \angle DCE$, 50°; $\angle BED \cong \angle EDC$, 41° **31.** $\frac{45}{4}$, or $11\frac{1}{4}$ **32.** $-\frac{129}{8}$, or $-16\frac{1}{8}$ **33.** 118 **34.** $\frac{44}{3}$, or $14\frac{2}{3}$

Exercise Set 8.7, p. 578

RC1. False **RC2.** True **RC3.** True **RC4.** False **1.** $\angle A \cong \angle R, \angle B \cong \angle S, \angle C \cong \angle T; \overline{AB} \cong \overline{RS}, \overline{AC} \cong \overline{RT}, \overline{BC} \cong \overline{ST}$ **3.** $\angle D \cong \angle G, \angle E \cong \angle H, \angle F \cong \angle K; \overline{DE} \cong \overline{GH}, \overline{DF} \cong \overline{GK}, \overline{EF} \cong \overline{HK}$ **5.** $\angle X \cong \angle U, \angle Y \cong \angle V, \angle Z \cong \angle W; \overline{XY} \cong \overline{UV}, \overline{XZ} \cong \overline{UW}, \overline{YZ} \cong \overline{VW}$ **7.** $\angle A \cong \angle F, \angle C \cong \angle D, \angle B \cong \angle E; \overline{AC} \cong \overline{FD}, \overline{AB} \cong \overline{FE}, \overline{CB} \cong \overline{DE}$ **9.** $\angle M \cong \angle Q, \angle N \cong \angle P, \angle O \cong \angle S; \overline{MN} \cong \overline{QP}, \overline{MO} \cong \overline{QS}, \overline{NO} \cong \overline{PS}$ **11.** No **13.** Yes **15.** Yes **17.** No **19.** Yes **21.** Yes **23.** Yes **25.** Yes **27.** Yes **29.** ASA **31.** SAS **33.** SSS or SAS **35.** $\overline{PR} \cong \overline{TR}, \overline{SR} \cong \overline{QR}, \angle PRQ \cong \angle TRS$ (vertical angles); $\triangle PRQ \cong \triangle TRS$ by SAS **37.** $m\angle GLK = m\angle GLM = 90°, \angle GLK \cong \angle GLM, \overline{GL} \cong \overline{GL}, \overline{KL} \cong \overline{ML}; \triangle KLG \cong \triangle MLG$ by SAS **39.** $\overline{AE} \cong \overline{CD}, \overline{AB} \cong \overline{CB}, \overline{EB} \cong \overline{DB}; \triangle AEB \cong \triangle CDB$ by SSS **41.** $\triangle LKH \cong \triangle GKJ$ by SAS; $\angle HLK \cong \angle JGK, \angle LHK \cong \angle GJK, \overline{LH} \cong \overline{GJ}$ **43.** $\triangle PED \cong \triangle PFG$ by ASA. As corresponding parts, $\overline{EP} \cong \overline{FP}$; thus P is the midpoint of $\overline{EF}$. **45.** $m\angle A = 70°, m\angle D = m\angle B = 110°$ **47.** $m\angle M = 71°, m\angle J = m\angle L = 109°$ **49.** $TU = 9, NU = 15$ **51.** $KL = 3\frac{1}{2}, ML = JK = 7\frac{1}{2}$ **53.** $AC = 28, ED = 38$ **55.** 45.2% **56.** $33\frac{1}{3}\%$ **57.** 55% **58.** 88% **59.** $\frac{2.7}{13.1}; \frac{13.1}{2.7}$ **60.** $\frac{1}{4}; \frac{3}{4}$ **61.** 1.75 **62.** 2.34 **63.** −0.234 **64.** 0.0234 **65.** 13.85

Translating for Success, p. 588

1. K **2.** G **3.** B **4.** H **5.** O **6.** M **7.** E **8.** A **9.** D **10.** I

Exercise Set 8.8, p. 589

RC1. 5 **RC2.** 8 **RC3.** a **RC4.** 3 **1.** $\angle R \leftrightarrow \angle A, \angle S \leftrightarrow \angle B, \angle T \leftrightarrow \angle C, \overline{RS} \leftrightarrow \overline{AB}, \overline{RT} \leftrightarrow \overline{AC}, \overline{ST} \leftrightarrow \overline{BC}$ **3.** $\angle C \leftrightarrow \angle W, \angle B \leftrightarrow \angle J, \angle S \leftrightarrow \angle Z, \overline{CB} \leftrightarrow \overline{WJ}, \overline{CS} \leftrightarrow \overline{WZ}, \overline{BS} \leftrightarrow \overline{JZ}$ **5.** $\angle A \cong \angle R, \angle B \cong \angle S, \angle C \cong \angle T; \frac{AB}{RS} = \frac{AC}{RT} = \frac{BC}{ST}$ **7.** $\angle M \cong \angle C, \angle E \cong \angle L, \angle S \cong \angle F; \frac{ME}{CL} = \frac{MS}{CF} = \frac{ES}{LF}$ **9.** $\frac{PS}{ND} = \frac{SQ}{DM} = \frac{PQ}{NM}$ **11.** $QR = 10, PR = 8$ **13.** $EC = 18$ **15.** 36 ft **17.** 100 ft **19.** $\frac{147}{5}$, or $29\frac{2}{5}$ **20.** 0.244 **21.** 78 **22.** −61.1611

Summary and Review: Chapter 8, p. 591

Vocabulary Reinforcement

1. parallel **2.** perimeter **3.** radius **4.** supplementary **5.** scalene **6.** shape

Concept Reinforcement

1. True **2.** False **3.** False **4.** True **5.** True

Study Guide

1. Angle WAQ, angle QAW, $\angle WAQ$, $\angle QAW$, or $\angle A$; 26° **2.** **(a)** Straight; **(b)** acute; **(c)** obtuse; **(d)** right **3.** **(a)** Isosceles, right; **(b)** equilateral, acute; **(c)** scalene, acute; **(d)** scalene, obtuse **4.** 87° **5.** 1260° **6.** 27.8 ft; 46.74 ft^2 **7.** 15.5 m^2 **8.** 8.75 ft^2 **9.** 80 m^2 **10.** 37.68 in. **11.** 616 cm^2 **12.** 1683.3 m^3 **13.** $30\frac{6}{35}$ ft^3 **14.** 1696.537813 cm^3 **15.** 26.49375 ft^3 **16.** Complement: 52°; supplement: 142° **17.** $m\angle 7 = 60°$; $m\angle 9 = 55°$; $m\angle 10 = 60°$; $m\angle 11 = 65°$ **18.** $m\angle 4 = m\angle 1 = m\angle 8 = 105°$; $m\angle 6 = m\angle 3 = m\angle 2 = m\angle 7 = 75°$ **19.** **(a)** SAS; **(b)** ASA **20.** $BC = 45.5$, $AB = DC = 73$; $m\angle A = 30°$, $m\angle B = m\angle D = 150°$ **21.** $ZA = 15$ and $AT = 30$

Review Exercises

1. 54° **2.** 180° **3.** 140° **4.** 90° **5.** Acute **6.** Straight **7.** Obtuse **8.** Right **9.** 60° **10.** Scalene **11.** Right **12.** 720° **13.** 23 ft **14.** 4.4 m **15.** 228 ft; 2808 ft^2 **16.** 36 ft; 81 ft^2 **17.** 17.6 cm; 12.6 cm^2 **18.** 60 cm^2 **19.** 35 mm^2 **20.** 22.5 m^2 **21.** 29.64 yd^2 **22.** 88 m^2 **23.** $145\frac{5}{9}$ in^2 **24.** 840 ft^2 **25.** 8 m **26.** $\frac{14}{11}$ in., or $1\frac{3}{11}$ in. **27.** 14 ft **28.** 20 cm **29.** 50.24 m **30.** 8 in. **31.** 200.96 m^2 **32.** $5\frac{1}{11}$ in^2 **33.** 1038.555 ft^2 **34.** 26.28 ft^2; 20.28 ft **35.** 93.6 yd^3; 150 yd^2 **36.** 193.2 cm^3; 240.4 cm^2 **37.** 31,400 ft^3 **38.** $33.49\overline{3}$ cm^3 **39.** 4.71 in^3 **40.** 942 cm^3 **41.** 49° **42.** 8° **43.** 85° **44.** 147° **45.** 47° **46.** $m\angle 2 = 105°$, $m\angle 3 = 37°$, $m\angle 4 = 38°$, $m\angle 6 = 37°$ **47.** **(a)** $\angle 1$ and $\angle 5$, $\angle 4$ and $\angle 8$, $\angle 3$ and $\angle 7$, $\angle 2$ and $\angle 6$; **(b)** $\angle 4$, $\angle 5$, $\angle 2$, and $\angle 7$; **(c)** $\angle 4$ and $\angle 7$, $\angle 2$ and $\angle 5$ **48.** $m\angle 1 = m\angle 3 = m\angle 7 = m\angle 5 = 45°$, $m\angle 6 = m\angle 2 = m\angle 8 = 135°$ **49.** $\angle D \cong \angle R$, $\angle H \cong \angle Z$, $\angle J \cong \angle K$; $\overline{DH} \cong \overline{RZ}$, $\overline{DJ} \cong \overline{RK}$, $\overline{HJ} \cong \overline{ZK}$ **50.** $\angle A \cong \angle G$, $\angle B \cong \angle D$, $\angle C \cong \angle F$, $\overline{AB} \cong \overline{GD}$, $\overline{AC} \cong \overline{GF}$, $\overline{BC} \cong \overline{DF}$ **51.** ASA **52.** SSS **53.** None **54.** $\overline{IJ} \cong \overline{KJ}$, $\angle HJI \cong \angle LJK$, $\angle HIJ \cong \angle LKJ$; $\triangle JIH \cong \triangle JKL$ by ASA **55.** $m\angle C = 63°$, $m\angle B = m\angle D = 117°$; $BC = 23$, $CD = 13$ **56.** $\angle C \cong \angle F$, $\angle Q \cong \angle A$, $\angle W \cong \angle S$; $\dfrac{CQ}{FA} = \dfrac{CW}{FS} = \dfrac{QW}{AS}$ **57.** $MO = 14$ **58.** B **59.** B **60.** 100 ft^2 **61.** 7.83998704 m^2 **62.** 47.25 cm^2

Understanding Through Discussion and Writing

1. Add 90° to the measure of the angle's complement. **2.** This could be done using the technique in Example 8 of Section 8.5. We could also approximate the volume with the volume of a similarly shaped rectangular solid. Another method is to break the egg and measure the capacity of its contents. **3.** Linear measure is one-dimensional, area is two-dimensional, and volume is three-dimensional. **4.** Divide the figure into 3 triangles.

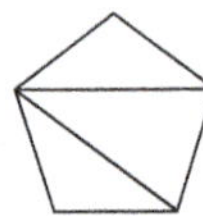

The sum of the measures of the angles of each triangle is 180°, so the sum of the measures of the angles of the figure is $3 \cdot 180°$, or 540°.

5. The volume of the cone is half the volume of the dome. It can be argued that a cone-cap is more energy-efficient since there is less air under it to be heated and cooled. **6.** Volume of two spheres, each with radius r: $2\left(\frac{4}{3}\pi r^3\right) = \frac{8}{3}\pi r^3$; volume of one sphere with radius $2r$: $\frac{4}{3}\pi(2r)^3 = \frac{32}{3}\pi r^3$. The volume of the sphere with radius $2r$ is four times the volume of the two spheres, each with radius r: $\frac{32}{3}\pi r^3 = 4 \cdot \frac{8}{3}\pi r^3$.

Test: Chapter 8, p. 603

1. [8.1b] 90° **2.** [8.1b] 35° **3.** [8.1b] 180° **4.** [8.1b] 113° **5.** [8.1c] Right **6.** [8.1c] Acute **7.** [8.1c] Straight **8.** [8.1c] Obtuse **9.** [8.1f] 35° **10.** [8.1e] Isosceles **11.** [8.1e] Obtuse **12.** [8.1f] 540° **13.** [8.2a], [8.3a] 32.82 cm; 65.894 cm^2 **14.** [8.2a], [8.3a] $19\frac{1}{2}$ in.; $23\frac{49}{64}$ in^2 **15.** [8.3b] 25 cm^2 **16.** [8.3b] 12 m^2 **17.** [8.3b] 18 ft^2 **18.** [8.4a] $\frac{1}{4}$ in. **19.** [8.4a] 9 cm **20.** [8.4b] $\frac{11}{14}$ in. **21.** [8.4c] 254.34 cm^2 **22.** [8.4d] 65.46 km; 103.815 km^2 **23.** [8.5a] 84 cm^3; 142 cm^2 **24.** [8.5e] 420 in^3 **25.** [8.5b] 1177.5 ft^3 **26.** [8.5c] $4186.\overline{6}$ yd^3 **27.** [8.5d] 113.04 cm^3 **28.** [8.6a] Complement: 25°; supplement: 115° **29.** [8.6c] $m\angle 2 = 110°$, $m\angle 3 = 8°$, $m\angle 4 = 62°$, $m\angle 6 = 8°$ **30.** [8.6d] $m\angle 6 = m\angle 2 = m\angle 8 = 120°$, $m\angle 5 = m\angle 3 = m\angle 7 = m\angle 1 = 60°$ **31.** [8.7a] $\angle C \cong \angle A$, $\angle W \cong \angle T$, $\angle S \cong \angle Z$, $\overline{CW} \cong \overline{AT}$, $\overline{WS} \cong \overline{TZ}$, $\overline{SC} \cong \overline{ZA}$ **32.** [8.7a] SAS **33.** [8.7a] None **34.** [8.7a] ASA **35.** [8.7a] None **36.** [8.7b] $m\angle G = 105°$, $m\angle D = m\angle F = 75°$; $EF = 11$, $DE = GF = 20$ **37.** [8.7b] $LJ = 6.4$, $KM = 6$ **38.** [8.8a] $\angle E \cong \angle T$, $\angle R \cong \angle G$, $\angle S \cong \angle F$; $\dfrac{ER}{TG} = \dfrac{RS}{GF} = \dfrac{SE}{FT}$ **39.** [8.8b] $EK = 18$, $ZK = 27$ **40.** [8.5c] D **41.** [8.3a] 2 ft^2 **42.** [8.3b] 1.875 ft^2 **43.** [8.5a] 0.65 ft^3 **44.** [8.5d] 0.033 ft^3 **45.** [8.5b] 0.055 ft^3

CHAPTER 9

Exercise Set 9.1, p. 612

RC1. Division **RC2.** Multiplication **RC3.** Multiplication **RC4.** Division

1. 32 min; 69 min; 81 min **3.** 260 mi **5.** 576 in^2 **7.** 1935 m^2 **9.** 56 **11.** 8 **13.** 1 **15.** 6 **17.** 2 **19.** $b + 7$, or $7 + b$ **21.** $c - 12$ **23.** $a + b$, or $b + a$ **25.** $x \div y$, or $\dfrac{x}{y}$, or x/y **27.** $x + w$, or $w + x$ **29.** $n - m$ **31.** $2z$ **33.** $3m$, or $m \cdot 3$ **35.** $4a + 6$, or $6 + 4a$ **37.** $xy - 8$ **39.** $2t - 5$ **41.** $3n + 11$, or $11 + 3n$ **43.** $4x + 3y$, or $3y + 4x$ **45.** $s + 0.05s$ **47.** $65t$ miles **49.** $\$50 - x$ **51.** $\$12.50n$ **53.** $2 \cdot 2 \cdot 3 \cdot 3 \cdot 3$ **54.** $2 \cdot 2 \cdot 2 \cdot 2 \cdot 2 \cdot 2 \cdot 3$ **55.** $\frac{41}{56}$ **56.** $\frac{31}{54}$ **57.** 0.0515 **58.** 43,500 **59.** 96 **60.** 396 **61.** $\frac{1}{4}$ **63.** 0

Calculator Corner, p. 619

1. 8.717797887 **2.** 17.80449381 **3.** 67.08203932 **4.** 35.4807407 **5.** 3.141592654 **6.** 91.10618695 **7.** 530.9291585 **8.** 138.8663978

Calculator Corner, p. 620

1. −0.75 **2.** −0.45 **3.** −0.125 **4.** −1.8 **5.** −0.675 **6.** −0.6875 **7.** −3.5 **8.** −0.76

Calculator Corner, p. 622

1. 5 **2.** 17 **3.** 0 **4.** 6.48 **5.** 12.7 **6.** 0.9 **7.** $\frac{5}{7}$ **8.** $\frac{4}{3}$

Exercise Set 9.2, p. 623

RC1. H **RC2.** E **RC3.** J **RC4.** D **RC5.** B **RC6.** G **RC7.** True **RC8.** True **RC9.** False **RC10.** False

1. 24; -2 **3.** 7,200,000,000,000; -460 **5.** 1454; -55

7. $\frac{10}{3}$ $-6\ -5\ -4\ -3\ -2\ -1\ 0\ 1\ 2\ 3\ 4\ 5\ 6$ **9.** -5.2 $-6\ -5\ -4\ -3\ -2\ -1\ 0\ 1\ 2\ 3\ 4\ 5\ 6$

11. $-4\frac{2}{5}$ $-6\ -5\ -4\ -3\ -2\ -1\ 0\ 1\ 2\ 3\ 4\ 5\ 6$ **13.** -0.875 **15.** $0.8\overline{3}$

17. $-1.1\overline{6}$ **19.** $0.\overline{6}$ **21.** $0.\overline{1}$ **23.** $-0.\overline{5}$ **25.** $0.\overline{16}$ **27.** $>$ **29.** $<$ **31.** $<$ **33.** $<$ **35.** $>$ **37.** $>$ **39.** $<$ **41.** $>$ **43.** $<$ **45.** $<$ **47.** $x < -6$ **49.** $y \geq -10$ **51.** False **53.** True **55.** True **57.** False **59.** 3 **61.** 11 **63.** $\frac{2}{3}$ **65.** 0 **67.** 2.65 **69.** 1.1 **70.** 0.238 **71.** 52% **72.** 59.375%, or $59\frac{3}{8}\%$ **73.** 81 **74.** 1 **75.** 45 **76.** 0 **77.** $-\frac{2}{3}, -\frac{2}{5}, -\frac{1}{3}, -\frac{2}{7}, -\frac{1}{7}, \frac{1}{3}, \frac{2}{5}, \frac{2}{3}, \frac{9}{8}$ **79.** $\frac{1}{1}$ **81.** $\frac{50}{9}$

Exercise Set 9.3, p. 631

RC1. right; right **RC2.** left; left **RC3.** right; left **RC4.** left; right **1.** -7 **3.** -6 **5.** 0 **7.** -8 **9.** -7 **11.** -27 **13.** 0 **15.** -42 **17.** 0 **19.** 0 **21.** 3 **23.** -9 **25.** 7 **27.** 0 **29.** 35 **31.** -3.8 **33.** -8.1 **35.** $-\frac{1}{5}$ **37.** $-\frac{7}{9}$ **39.** $-\frac{3}{8}$ **41.** $-\frac{19}{24}$ **43.** $\frac{1}{24}$ **45.** $\frac{8}{15}$ **47.** $\frac{16}{45}$ **49.** 37 **51.** 50 **53.** -24 **55.** 26.9 **57.** -8 **59.** $\frac{13}{8}$ **61.** -43 **63.** $\frac{4}{3}$ **65.** 24 **67.** $\frac{3}{8}$ **69.** 13,796 ft **71.** $-3°$F **73.** Profit of \$4300 **75.** He owes \$85. **77.** 0.713 **78.** 0.92875 **79.** 12.5% **80.** 40.625% **81.** $\frac{8}{5}$ **82.** $\frac{1}{4}$ **83.** All positive numbers **85.** B

Exercise Set 9.4, p. 637

RC1. (c) **RC2.** (b) **RC3.** (d) **RC4.** (a) **1.** -7 **3.** -6 **5.** 0 **7.** -4 **9.** -7 **11.** -6 **13.** 0 **15.** 14 **17.** 11 **19.** -14 **21.** 5 **23.** -1 **25.** 18 **27.** -3 **29.** -21 **31.** 5 **33.** -8 **35.** 12 **37.** -23 **39.** -68 **41.** -73 **43.** 116 **45.** 0 **47.** -1 **49.** $\frac{1}{12}$ **51.** $-\frac{17}{12}$ **53.** $\frac{1}{8}$ **55.** 19.9 **57.** -8.6 **59.** -0.01 **61.** -193 **63.** 500 **65.** -2.8 **67.** -3.53 **69.** $-\frac{1}{2}$ **71.** $\frac{6}{7}$ **73.** $-\frac{41}{30}$ **75.** $-\frac{2}{15}$ **77.** $-\frac{1}{48}$ **79.** $-\frac{43}{60}$ **81.** 37 **83.** -62 **85.** -139 **87.** 6 **89.** 108.5 **91.** $\frac{1}{4}$ **93.** 30,383 ft **95.** \$347.94 **97.** 3780 m **99.** 381 ft **101.** 1130°F **103.** $y + 7$, or $7 + y$ **104.** $t - 41$ **105.** $a - h$ **106.** $6c$, or $c \cdot 6$ **107.** $r + s$, or $s + r$ **108.** $y - x$ **109.** False; $3 - 0 \neq 0 - 3$ **111.** True **113.** True

Mid-Chapter Review: Chapter 9, p. 641

1. True **2.** False **3.** True **4.** False **5.** $-x = -(-4) = 4$; $-(-x) = -(-(-4)) = -(4) = -4$ **6.** $5 - 13 = 5 + (-13) = -8$ **7.** $-6 - 7 = -6 + (-7) = -13$ **8.** 4 **9.** 11 **10.** $3y$ **11.** $n - 5$ **12.** 450; -79

13. -3.5 $-6\ -5\ -4\ -3\ -2\ -1\ 0\ 1\ 2\ 3\ 4\ 5\ 6$ **14.** -0.8 **15.** $2.\overline{3}$ **16.** $<$

17. $>$ **18.** False **19.** True **20.** $5 > y$ **21.** $t \leq -3$ **22.** 15.6 **23.** 18 **24.** 0 **25.** $\frac{12}{5}$ **26.** 5.6 **27.** $-\frac{7}{4}$ **28.** 0 **29.** 49 **30.** 19 **31.** 2.3 **32.** -2 **33.** $-\frac{1}{8}$ **34.** 0 **35.** -17 **36.** $-\frac{11}{24}$ **37.** -8.1 **38.** -9 **39.** -2 **40.** -10.4 **41.** 16 **42.** $\frac{7}{20}$ **43.** -12 **44.** -4 **45.** $-\frac{4}{3}$ **46.** -1.8 **47.** 13 **48.** 9 **49.** -23 **50.** 75 **51.** 14 **52.** 33°C **53.** \$54.80 **54.** Answers may vary. Three examples are $\frac{6}{13}$, -23.8, and $\frac{43}{5}$. These are rational numbers because they can be named in the form $\frac{a}{b}$, where a and b are integers and b is not 0. They are not integers, however, because they are neither whole numbers nor the opposites of whole numbers. **55.** Answers may vary. Three examples are π, $-\sqrt{7}$, and 0.31311311131111.... Irrational numbers cannot be written as the quotient of two integers. Real numbers that are not rational are irrational. Decimal notation for rational numbers either terminates or repeats. Decimal notation for irrational numbers neither terminates nor repeats. **56.** Answers may vary. If we think of the addition on the number line, we start at 0, move to the left to a negative number, and then move to the left again. This always brings us to a point on the negative portion of the number line. **57.** Yes; consider $m - (-n)$, where both m and n are positive. Then $m - (-n) = m + n$. Now $m + n$, the sum of two positive numbers, is positive.

Exercise Set 9.5, p. 646

RC1. negative **RC2.** positive **RC3.** positive **RC4.** negative **RC5.** -9 **RC6.** 9 **RC7.** $-\frac{1}{4}$ **RC8.** $-\frac{1}{4}$ **1.** -8 **3.** -24 **5.** -72 **7.** 16 **9.** 42 **11.** -120 **13.** -238 **15.** 1200 **17.** 98 **19.** -72 **21.** -12.4 **23.** 30 **25.** 21.7 **27.** $-\frac{2}{5}$ **29.** $\frac{1}{12}$ **31.** -17.01 **33.** 420 **35.** $\frac{2}{7}$ **37.** -60 **39.** 150 **41.** 50.4 **43.** $\frac{10}{189}$ **45.** -960 **47.** 17.64 **49.** $-\frac{5}{784}$ **51.** 0 **53.** -720 **55.** $-30{,}240$ **57.** 1 **59.** 16, -16; 16, -16 **61.** $\frac{4}{25}, -\frac{4}{25}; \frac{4}{25}, -\frac{4}{25}$ **63.** $-9, -9; -9, -9$ **65.** 441, -147; 441, -147 **67.** 20; 20 **69.** -2; 2 **71.** $-24°$C **73.** -20 lb **75.** \$12.71 **77.** -32 m **79.** 38°F **81.** 180 **82.** $2 \cdot 2 \cdot 2 \cdot 2 \cdot 2 \cdot 2 \cdot 2 \cdot 2 \cdot 2 \cdot 3 \cdot 3$ **83.** $\frac{2}{3}$ **84.** $\frac{8}{9}$ **85.** $\frac{6}{11}$ **86.** $\frac{41}{265}$ **87.** $\frac{11}{32}$ **88.** $\frac{37}{67}$ **89.** $\frac{1}{24}$ **90.** 6 **91.** A **93.** Largest quotient: $10 \div \frac{1}{5} = 50$; smallest quotient: $-5 \div \frac{1}{5} = -25$

Calculator Corner, p. 655

1. -4 **2.** -2 **3.** -32 **4.** 1.4 **5.** 2.7 **6.** -9.5 **7.** -0.8 **8.** 14.44

Exercise Set 9.6, p. 655

RC1. opposites **RC2.** 1 **RC3.** 0 **RC4.** reciprocals **1.** -8 **3.** -14 **5.** -3 **7.** 3 **9.** -8 **11.** 2 **13.** -12 **15.** -8 **17.** Not defined **19.** 0 **21.** $\frac{7}{15}$ **23.** $-\frac{13}{47}$ **25.** $\frac{1}{13}$ **27.** $-\frac{1}{32}$ **29.** -7.1 **31.** 9 **33.** $4y$ **35.** $\frac{3b}{2a}$ **37.** $4 \cdot \left(\frac{1}{17}\right)$ **39.** $8 \cdot \left(-\frac{1}{13}\right)$ **41.** $13.9 \cdot \left(-\frac{1}{1.5}\right)$ **43.** $\frac{2}{3} \cdot \left(-\frac{5}{4}\right)$ **45.** $x \cdot y$ **47.** $(3x + 4)\left(\frac{1}{5}\right)$ **49.** $-\frac{9}{8}$ **51.** $\frac{5}{3}$ **53.** $\frac{9}{14}$ **55.** $\frac{9}{64}$ **57.** $-\frac{5}{4}$ **59.** $-\frac{27}{5}$ **61.** $\frac{11}{13}$ **63.** -2 **65.** -16.2 **67.** -2.5 **69.** -1.25 **71.** Not defined **73.** Percent increase is about 44%. **75.** Percent decrease is about -21%. **77.** 33 **78.** 129 **79.** 1 **80.** 1296 **81.** $\frac{22}{39}$ **82.** 0.477 **83.** 87.5% **84.** $\frac{2}{3}$ **85.** $\frac{9}{8}$ **86.** $\frac{128}{625}$ **87.** $\frac{1}{-10.5}$; -10.5, the reciprocal of the reciprocal is the original number. **89.** Negative **91.** Positive **93.** Negative

Exercise Set 9.7, p. 667

RC1. (g) **RC2.** (h) **RC3.** (f) **RC4.** (e) **RC5.** (d) **RC6.** (a) **RC7.** (b) **1.** $\frac{3y}{5y}$ **3.** $\frac{10x}{15x}$ **5.** $\frac{2x}{x^2}$ **7.** $-\frac{3}{2}$ **9.** $-\frac{7}{6}$ **11.** $\frac{4s}{3}$ **13.** $8 + y$ **15.** nm **17.** $xy + 9$, or $9 + yx$, or $yx + 9$ **19.** $c + ab$, or $ba + c$, or $c + ba$ **21.** $(a + b) + 2$ **23.** $8(xy)$ **25.** $a + (b + 3)$ **27.** $(3a)b$ **29.** $2 + (b + a), (2 + a) + b, (b + 2) + a$; answers may vary **31.** $(5 + w) + v, (v + 5) + w, (w + v) + 5$; answers may vary **33.** $(3x)y, y(x \cdot 3), 3(yx)$; answers may vary **35.** $a(7b), b(7a), (7b)a$; answers may vary **37.** $2b + 10$ **39.** $7 + 7t$ **41.** $30x + 12$ **43.** $7x + 28 + 42y$ **45.** $7x - 21$ **47.** $-3x + 21$ **49.** $\frac{2}{3}b - 4$ **51.** $7.3x - 14.6$ **53.** $-\frac{3}{5}x + \frac{3}{5}y - 6$ **55.** $45x + 54y - 72$ **57.** $-4x + 12y + 8z$ **59.** $-3.72x + 9.92y - 3.41$ **61.** $4x, 3z$ **63.** $7x, 8y, -9z$ **65.** $2(x + 2)$ **67.** $5(6 + y)$ **69.** $7(2x + 3y)$ **71.** $7(2t - 1)$ **73.** $8(x - 3)$ **75.** $6(3a - 4b)$ **77.** $-4(y - 8)$, or $4(-y + 8)$ **79.** $5(x + 2 + 3y)$ **81.** $8(2m - 4n + 1)$ **83.** $4(3a + b - 6)$ **85.** $2(4x + 5y - 11)$ **87.** $a(x - 1)$ **89.** $a(x - y - z)$ **91.** $-6(3x - 2y - 1)$, or $6(-3x + 2y + 1)$ **93.** $\frac{1}{3}(2x - 5y + 1)$ **95.** $6(6x - y + 3z)$ **97.** $19a$ **99.** $9a$ **101.** $8x + 9z$ **103.** $7x + 15y^2$ **105.** $-19a + 88$ **107.** $4t + 6y - 4$ **109.** b **111.** $\frac{13}{4}y$ **113.** $8x$ **115.** $5n$ **117.** $-16y$ **119.** $17a - 12b - 1$ **121.** $4x + 2y$ **123.** $7x + y$ **125.** $0.8x + 0.5y$ **127.** $\frac{35}{6}a + \frac{3}{2}b - 42$

129. 144 **130.** 72 **131.** 144 **132.** 60 **133.** 32 **134.** 72 **135.** 90 **136.** 108 **137.** 180 **138.** $\frac{4}{13}$ **139.** True **140.** False **141.** True **142.** True **143.** Not equivalent; $3 \cdot 2 + 5 \neq 3 \cdot 5 + 2$ **145.** Equivalent; commutative law of addition **147.** $q(1 + r + rs + rst)$

Calculator Corner, p. 676

1. -16 **2.** 9 **3.** 117,649 **4.** $-1{,}419{,}857$ **5.** $-117{,}649$ **6.** $-1{,}419{,}857$ **7.** -4 **8.** -2

Exercise Set 9.8, p. 677

RC1. Multiplication **RC2.** Addition **RC3.** Subtraction **RC4.** Division **RC5.** Division **RC6.** Multiplication
1. $-2x - 7$ **3.** $-8 + x$ **5.** $-4a + 3b - 7c$ **7.** $-6x + 8y - 5$ **9.** $-3x + 5y + 6$ **11.** $8x + 6y + 43$ **13.** $5x - 3$ **15.** $-3a + 9$ **17.** $5x - 6$ **19.** $-19x + 2y$ **21.** $9y - 25z$ **23.** $-7x + 10y$ **25.** $37a - 23b + 35c$ **27.** 7 **29.** -40 **31.** 19 **33.** $12x + 30$ **35.** $3x + 30$ **37.** $9x - 18$ **39.** $-4x - 64$ **41.** -7 **43.** -1 **45.** -16 **47.** -334 **49.** 14 **51.** 1880 **53.** 12 **55.** 8 **57.** -86 **59.** 37 **61.** -1 **63.** -10 **65.** -67 **67.** -7988 **69.** -3000 **71.** 60 **73.** 1 **75.** 10 **77.** $-\frac{13}{45}$ **79.** $-\frac{23}{18}$ **81.** -122 **83.** 18 **84.** 35 **85.** 0.4 **86.** $\frac{15}{2}$ **87.** $-\frac{1}{9}$ **88.** $\frac{3}{7}$ **89.** -25 **90.** -35 **91.** 25 **92.** 35 **93.** $-2x - f$ **95.** **(a)** 52; 52; 28.130169; **(b)** -24; -24; -108.307025 **97.** -6

Summary and Review: Chapter 9, p. 681

Vocabulary Reinforcement

1. integers **2.** additive inverses **3.** commutative law **4.** identity property of 1 **5.** associative law **6.** multiplicative inverses **7.** identity property of 0

Concept Reinforcement

1. True **2.** True **3.** False **4.** False

Study Guide

1. 14 **2.** $<$ **3.** $\frac{5}{4}$ **4.** -8.5 **5.** -2 **6.** 56 **7.** -8 **8.** $\frac{9}{20}$ **9.** $\frac{5}{3}$ **10.** $5x + 15y - 20z$ **11.** $9(3x + y - 4z)$ **12.** $5a - 2b$ **13.** $4a - 4b$ **14.** -2

Review Exercises

1. 4 **2.** 19%x, or $0.19x$ **3.** 620, -125 **4.** 38 **5.** 126
6.
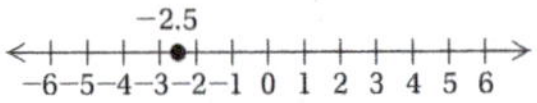

7.
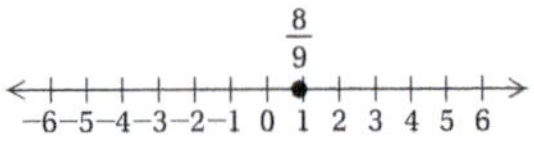

8. $<$ **9.** $>$ **10.** $>$ **11.** $<$ **12.** $x > -3$ **13.** True **14.** False **15.** -3.8 **16.** $\frac{3}{4}$ **17.** $\frac{8}{3}$ **18.** $-\frac{1}{7}$ **19.** 34 **20.** 5 **21.** -3 **22.** -4 **23.** -5 **24.** 1 **25.** $-\frac{7}{5}$ **26.** -7.9 **27.** 54 **28.** -9.18 **29.** $-\frac{2}{7}$ **30.** -210 **31.** -7 **32.** -3 **33.** $\frac{3}{4}$ **34.** 24.8 **35.** -2 **36.** 2 **37.** -2 **38.** 8-yd gain **39.** $-\$360$ **40.** \$4.64 **41.** \$18.95 **42.** $15x - 35$ **43.** $-8x + 10$ **44.** $4x + 15$ **45.** $-24 + 48x$ **46.** $2(x - 7)$ **47.** $-6(x - 1)$, or $6(-x + 1)$ **48.** $5(x + 2)$ **49.** $-3(x - 4y + 4)$, or $3(-x + 4y - 4)$ **50.** $7a - 3b$ **51.** $-2x + 5y$ **52.** $5x - y$ **53.** $-a + 8b$ **54.** $-3a + 9$ **55.** $-2b + 21$ **56.** 6 **57.** $12y - 34$ **58.** $5x + 24$ **59.** $-15x + 25$ **60.** D **61.** B **62.** $-\frac{5}{8}$ **63.** -2.1 **64.** 1000 **65.** $4a + 2b$

Understanding Through Discussion and Writing

1. The sum of each pair of opposites such as -50 and 50, -49 and 49, and so on, is 0. The sum of these sums and the remaining integer, 0, is 0. **2.** The product of an even number of negative numbers is positive, and the product of an odd number of negative numbers is negative. Now $(-7)^8$ is the product of 8 factors of -7 so it is positive, and $(-7)^{11}$ is the product of 11 factors of -7 so it is negative. **3.** Consider $\frac{a}{b} = q$, where a and b are both negative numbers. Then $q \cdot b = a$, so q must be a positive number in order for the product to be negative. **4.** Consider $\frac{a}{b} = q$, where a is a negative number and b is a positive number. Then $q \cdot b = a$, so q must be a negative number in order for the product to be negative. **5.** We use the distributive law when we collect like terms even though we might not always write this step. **6.** Jake expects the calculator to multiply 2 and 3 first and then divide 18 by that product. This procedure does not follow the rules for order of operations.

Test: Chapter 9, p. 687

1. [9.1a] 6 **2.** [9.1b] $x - 9$ **3.** [9.2d] $>$ **4.** [9.2d] $<$ **5.** [9.2d] $>$ **6.** [9.2d] $-2 > x$ **7.** [9.2d] True **8.** [9.2e] 7 **9.** [9.2e] $\frac{9}{4}$ **10.** [9.2e] 2.7 **11.** [9.3b] $-\frac{2}{3}$ **12.** [9.3b] 1.4 **13.** [9.6b] $-\frac{1}{2}$ **14.** [9.6b] $\frac{7}{4}$ **15.** [9.3b] 8 **16.** [9.4a] 7.8 **17.** [9.3a] -8 **18.** [9.3a] $\frac{7}{40}$ **19.** [9.4a] 10 **20.** [9.4a] -2.5 **21.** [9.4a] $\frac{7}{8}$ **22.** [9.5a] -48 **23.** [9.5a] $\frac{3}{16}$ **24.** [9.6a] -9 **25.** [9.6c] $\frac{3}{4}$ **26.** [9.6c] -9.728 **27.** [9.2e], [9.8d] -173 **28.** [9.8d] -5 **29.** [9.3c], [9.4b] Up 15 points **30.** [9.4b] 2244 m **31.** [9.5b] 16,080 **32.** [9.6d] -0.75°C each minute **33.** [9.7c] $18 - 3x$ **34.** [9.7c] $-5y + 5$ **35.** [9.7d] $2(6 - 11x)$ **36.** [9.7d] $7(x + 3 + 2y)$ **37.** [9.4a] 12 **38.** [9.8b] $2x + 7$ **39.** [9.8b] $9a - 12b - 7$ **40.** [9.8c] $68y - 8$ **41.** [9.8d] -4 **42.** [9.8d] 448 **43.** [9.2d] B **44.** [9.2e], [9.8d] 15 **45.** [9.8c] $4a$ **46.** [9.7e] $4x + 4y$

CHAPTER 10

Exercise Set 10.1, p. 694

RC1. (f) **RC2.** (c) **RC3.** (e) **RC4.** (a)
1. Yes **3.** No **5.** No **7.** Yes **9.** Yes **11.** No **13.** 4 **15.** -20 **17.** -14 **19.** -18 **21.** 15 **23.** -14 **25.** 2 **27.** 20 **29.** -6 **31.** $6\frac{1}{2}$ **33.** 19.9 **35.** $\frac{7}{3}$ **37.** $-\frac{7}{4}$ **39.** $\frac{41}{24}$ **41.** $-\frac{1}{20}$ **43.** 5.1 **45.** 12.4 **47.** -5 **49.** $1\frac{5}{6}$ **51.** $-\frac{10}{21}$ **53.** $-\frac{3}{2}$ **54.** -5.2 **55.** $-\frac{1}{24}$ **56.** 172.72 **57.** $\$83 - x$ **58.** $65t$ miles **59.** $-\frac{26}{15}$ **61.** -10 **63.** All real numbers

Exercise Set 10.2, p. 699

RC1. (f) **RC2.** (d) **RC3.** (a) **RC4.** (b)
1. 6 **3.** 9 **5.** 12 **7.** -40 **9.** 1 **11.** -7 **13.** -6 **15.** 6 **17.** -63 **19.** -48 **21.** 36 **23.** -9 **25.** -21 **27.** $-\frac{3}{5}$ **29.** $-\frac{3}{2}$ **31.** $\frac{9}{2}$ **33.** 7 **35.** -7 **37.** 8 **39.** 15.9 **41.** -50 **43.** -14 **45.** $7x$ **46.** $-x + 5$ **47.** $8x + 11$ **48.** $-32y$ **49.** $x - 4$ **50.** $-5x - 23$ **51.** $-10y - 42$ **52.** $-22a + 4$ **53.** $8r$ miles **54.** $\frac{1}{2}b \cdot 10\ \text{m}^2$, or $5b\ \text{m}^2$ **55.** -8655 **57.** No solution **59.** No solution **61.** $\frac{b}{3a}$ **63.** $\frac{4b}{a}$

Calculator Corner, p. 705

1. Left to the student

Exercise Set 10.3, p. 709

RC1. (d) **RC2.** (a) **RC3.** (c) **RC4.** (e) **RC5.** (b)
1. 5 **3.** 8 **5.** 10 **7.** 14 **9.** -8 **11.** -8 **13.** -7 **15.** 12 **17.** 6 **19.** 4 **21.** 6 **23.** -3 **25.** 1 **27.** 6 **29.** -20 **31.** 7 **33.** 2 **35.** 5 **37.** 2 **39.** 10 **41.** 4 **43.** 0 **45.** -1 **47.** $-\frac{4}{3}$ **49.** $\frac{2}{5}$ **51.** -2 **53.** -4 **55.** $\frac{4}{5}$ **57.** $-\frac{28}{27}$ **59.** 6 **61.** 2 **63.** No solution **65.** All real numbers

67. 6 **69.** 8 **71.** 1 **73.** 17 **75.** $-\frac{5}{3}$ **77.** All real numbers **79.** No solution **81.** -3 **83.** 2 **85.** $\frac{4}{7}$ **87.** No solution **89.** All real numbers **91.** $-\frac{51}{31}$ **93.** -6.5 **94.** -75.14 **95.** $7(x - 3 - 2y)$ **96.** $8(y - 11x + 1)$ **97.** -160 **98.** $-17x + 18$ **99.** $91x - 242$ **100.** 0.25 **101.** $-\frac{5}{32}$ **103.** $\frac{52}{45}$

Exercise Set 10.4, p. 717

RC1. (b) **RC2.** (c) **RC3.** (d) **RC4.** (a)

1. $14\frac{1}{3}$ meters per cycle **3.** 10.5 calories per ounce **5.** **(a)** 337.5 mi; **(b)** $t = \frac{d}{r}$ **7.** **(a)** 1423 students; **(b)** $n = 15f$ **9.** $x = \frac{y}{5}$ **11.** $c = \frac{a}{b}$ **13.** $m = n - 11$ **15.** $x = y + \frac{3}{5}$ **17.** $x = y - 13$ **19.** $x = y - b$ **21.** $x = 5 - y$ **23.** $x = a - y$ **25.** $y = \frac{5x}{8}$, or $\frac{5}{8}x$ **27.** $x = \frac{By}{A}$ **29.** $t = \frac{W - b}{m}$ **31.** $x = \frac{y - c}{b}$ **33.** $h = \frac{A}{b}$ **35.** $w = \frac{P - 2l}{2}$, or $\frac{1}{2}P - l$ **37.** $a = 2A - b$ **39.** $b = 3A - a - c$ **41.** $t = \frac{A - b}{a}$ **43.** $x = \frac{c - By}{A}$ **45.** $a = \frac{F}{m}$ **47.** $c^2 = \frac{E}{m}$ **49.** $t = \frac{3k}{v}$ **51.** 7 **52.** $-21a + 12b$ **53.** -13.2 **54.** $-\frac{3}{2}$ **55.** $-35\frac{1}{2}$ **56.** $-\frac{1}{6}$ **57.** -9.325 **58.** $3\frac{3}{4}$ **59.** 3.4% **60.** 5% **61.** $41.\overline{6}\%$, or $41\frac{2}{3}\%$ **62.** 50% **63.** $b = \frac{Ha - 2}{H}$, or $a - \frac{2}{H}$; $a = \frac{2 + Hb}{H}$, or $\frac{2}{H} + b$ **65.** A quadruples. **67.** A increases by $2h$ units.

Mid-Chapter Review: Chapter 10, p. 721

1. False **2.** True **3.** True **4.** False

5.
$$\begin{aligned} x + 5 &= -3 \\ x + 5 - 5 &= -3 - 5 \\ x + 0 &= -8 \\ x &= -8 \end{aligned}$$

6.
$$\begin{aligned} -6x &= 42 \\ \frac{-6x}{-6} &= \frac{42}{-6} \\ 1 \cdot x &= -7 \\ x &= -7 \end{aligned}$$

7.
$$\begin{aligned} 5y + z &= t \\ 5y + z - z &= t - z \\ 5y &= t - z \\ \frac{5y}{5} &= \frac{t - z}{5} \\ y &= \frac{t - z}{5} \end{aligned}$$

8. 6 **9.** -12 **10.** 7 **11.** -10 **12.** 20 **13.** 5 **14.** $\frac{3}{4}$ **15.** -1.4 **16.** 6 **17.** -17 **18.** -9 **19.** 17 **20.** 21 **21.** 18 **22.** -15 **23.** $-\frac{3}{2}$ **24.** 1 **25.** -3 **26.** $\frac{3}{2}$ **27.** -1 **28.** 3 **29.** -7 **30.** 4 **31.** 2 **32.** $\frac{9}{8}$ **33.** $-\frac{21}{5}$ **34.** 9 **35.** -2 **36.** 0 **37.** All real numbers **38.** No solution **39.** $-\frac{13}{2}$ **40.** All real numbers **41.** $b = \frac{A}{4}$ **42.** $x = y + 1.5$ **43.** $m = s - n$ **44.** $t = \frac{9w}{4}$ **45.** $t = \frac{B + c}{a}$ **46.** $y = 2M - x - z$ **47.** Equivalent expressions have the same value for all possible replacements for the variable(s). Equivalent equations have the same solution(s). **48.** The equations are not equivalent because they do not have the same solutions. Although 5 is a solution of both equations, -5 is a solution of $x^2 = 25$ but not of $x = 5$. **49.** For an equation $x + a = b$, add the opposite of a (or subtract a) on both sides of the equation. **50.** The student probably added $\frac{1}{3}$ on both sides of the equation rather than adding $-\frac{1}{3}$ (or subtracting $\frac{1}{3}$) on both sides. The correct solution is -2. **51.** For an equation $ax = b$, multiply by $1/a$ (or divide by a) on both sides of the equation. **52.** Answers may vary. A walker who knows how far and how long she walks each day wants to know her average speed each day.

Exercise Set 10.5, p. 727

RC1. (d) **RC2.** (b) **RC3.** (e) **RC4.** (a) **RC5.** (f) **RC6.** (c)

1. 20% **3.** 150 **5.** 546 **7.** 24% **9.** 2.5 **11.** 5% **13.** 25% **15.** 84 **17.** 24% **19.** 16% **21.** $46\frac{2}{3}$ **23.** 0.8 **25.** 5 **27.** 40 **29.** 811 million **31.** 5274 million **33.** 1764 million **35.** \$968 million **37.** \$221 **39.** 21% **41.** **(a)** 12.5%; **(b)** \$13.50 **43.** **(a)** \$31; **(b)** \$35.65 **45.** About 85,821 acres **47.** About 82.4% **49.** About 53.1% decrease **51.** About 53.4% **53.** About 19.2% decrease **55.** $12 + 3q$ **56.** $5x - 21$ **57.** 44 **58.** $x + 8$ **59.** $\frac{15w}{8}$ **60.** $-\frac{3}{2}$ **61.** 181.52 **62.** 0.4538 **63.** 6 ft 7 in.

Translating for Success, p. 742

1. B **2.** H **3.** G **4.** N **5.** J **6.** C **7.** L **8.** E **9.** F **10.** D

Exercise Set 10.6, p. 743

RC1. Familiarize **RC2.** Translate **RC3.** Solve **RC4.** Check **RC5.** State

1. 1522 Medals of Honor **3.** 180 in.; 60 in. **5.** 21.8 million **7.** 4.37 mi **9.** 1204 and 1205 **11.** 41, 42, 43 **13.** 61, 63, 65 **15.** 36 in. × 110 in. **17.** \$63 **19.** \$24.95 **21.** 11 visits **23.** 28°, 84°, 68° **25.** 33°, 38°, 109° **27.** \$350 **29.** \$852.94 **31.** 18 mi **33.** \$38.60 **35.** 89 and 96 **37.** -12 **39.** $-\frac{47}{40}$ **40.** $-\frac{17}{40}$ **41.** $-\frac{3}{10}$ **42.** $-\frac{32}{15}$ **43.** -10 **44.** 1.6 **45.** 409.6 **46.** -9.6 **47.** -41.6 **48.** 0.1 **49.** $yz + 12$, $zy + 12$, or $12 + zy$ **50.** $c + (4 + d)$ **51.** 120 apples **53.** About 0.65 in.

Exercise Set 10.7, p. 756

RC1. Not equivalent **RC2.** Not equivalent **RC3.** Equivalent **RC4.** Equivalent

1. **(a)** Yes; **(b)** yes; **(c)** no; **(d)** yes; **(e)** yes **3.** **(a)** No; **(b)** no; **(c)** no; **(d)** yes; **(e)** no

5. $x > 4$ 0 4

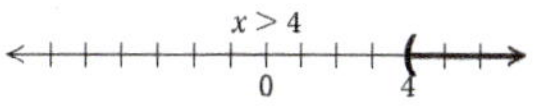

7. $t < -3$ −3 0

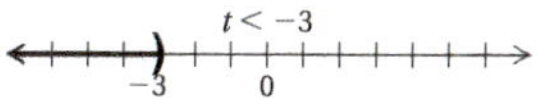

9. $m \geq -1$ −1 0

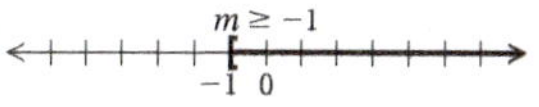

11. $-3 < x \leq 4$ −3 0 4

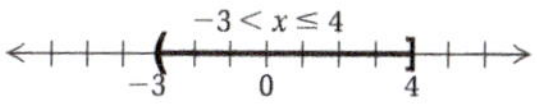

13. $0 < x < 3$ 0 3

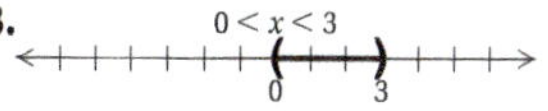

15. $\{x|x > -5\}$ −5 0

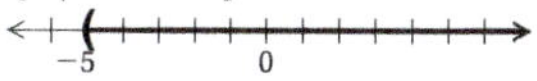

17. $\{x|x \leq -18\}$; −18 −20 0

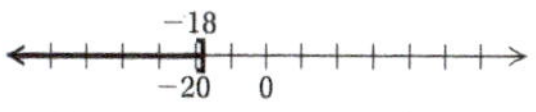

19. $\{y|y > -5\}$

21. $\{x|x > 2\}$ **23.** $\{x|x \leq -3\}$ **25.** $\{x|x < 4\}$ **27.** $\{t|t > 14\}$ **29.** $\{y|y \leq \frac{1}{4}\}$ **31.** $\{x|x > \frac{7}{12}\}$

33. $\{x|x < 7\}$; 0 7

35. $\{x|x < 3\}$; 0 3

37. $\{y|y \geq -\frac{2}{5}\}$ **39.** $\{x|x \geq -6\}$ **41.** $\{y|y \leq 4\}$ **43.** $\{x|x > \frac{17}{3}\}$ **45.** $\{y|y < -\frac{1}{14}\}$ **47.** $\{x|x \leq \frac{3}{10}\}$ **49.** $\{x|x < 8\}$ **51.** $\{x|x \leq 6\}$ **53.** $\{x|x < -3\}$ **55.** $\{x|x > -3\}$ **57.** $\{x|x \leq 7\}$ **59.** $\{x|x > -10\}$ **61.** $\{y|y < 2\}$ **63.** $\{y|y \geq 3\}$ **65.** $\{y|y > -2\}$ **67.** $\{x|x > -4\}$ **69.** $\{x|x \leq 9\}$ **71.** $\{y|y \leq -3\}$

73. $\{y|y < 6\}$ **75.** $\{m|m \geq 6\}$ **77.** $\{t|t < -\frac{5}{3}\}$ **79.** $\{r|r > -3\}$ **81.** $\{x|x \geq -\frac{57}{34}\}$ **83.** $\{x|x > -2\}$ **85.** $-\frac{5}{8}$ **86.** -1.11 **87.** -9.4 **88.** $-\frac{7}{8}$ **89.** 140 **90.** 41 **91.** $-2x - 23$ **92.** $37x - 1$ **93.** **(a)** Yes; **(b)** yes; **(c)** no; **(d)** no; **(e)** no; **(f)** yes; **(g)** yes **95.** No solution

Exercise Set 10.8, p. 763

RC1. $r \leq q$ **RC2.** $q \leq r$ **RC3.** $r < q$ **RC4.** $q \leq r$ **RC5.** $r < q$ **RC6.** $r \leq q$
1. $n \geq 7$ **3.** $w > 2$ kg **5.** 90 mph $< s <$ 110 mph **7.** $w \leq 20$ hr **9.** $c \geq \$3.20$ **11.** $x > 8$ **13.** $y \leq -4$ **15.** $n \geq 1300$ **17.** $W \leq 500$ L **19.** $3x + 2 < 13$ **21.** $\{x|x \geq 84\}$ **23.** $\{C|C < 1063°\}$ **25.** $\{Y|Y \geq 1935\}$ **27.** 15 or fewer copies **29.** $\{L|L \geq 5 \text{ in.}\}$ **31.** 5 min or more **33.** 2 courses **35.** 4 servings or more **37.** Lengths greater than or equal to 92 ft; lengths less than or equal to 92 ft **39.** Lengths less than 21.5 cm **41.** The blue-book value is greater than or equal to \$10,625. **43.** It has at least 16 g of fat. **45.** Heights greater than or equal to 4 ft **47.** Dates at least 6 weeks after July 1 **49.** 21 calls or more **51.** 40 **52.** -22 **53.** 12 **54.** 6 **55.** All real numbers **56.** No solution **57.** 7.5% **58.** 31 **59.** 1250 **60.** $83.\overline{3}\%$, or $83\frac{1}{3}\%$ **61.** Temperatures between $-15°$C and $-9\frac{4}{9}°$C **63.** They contain at least 7.5 g of fat per serving.

Summary and Review: Chapter 10, p. 768

Vocabulary Reinforcement

1. solution **2.** addition principle **3.** multiplication principle **4.** inequality **5.** equivalent

Concept Reinforcement

1. True **2.** True **3.** False **4.** True

Study Guide

1. -12 **2.** All real numbers **3.** No solution **4.** $b = \frac{2A}{h}$

5. 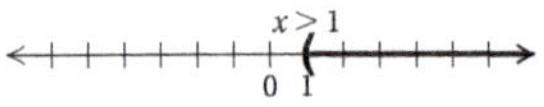

6. 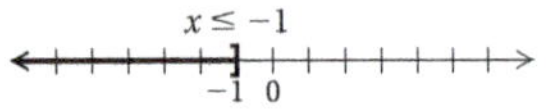

7. $\{y|y > -4\}$

Review Exercises

1. -22 **2.** 1 **3.** 25 **4.** 9.99 **5.** $\frac{1}{4}$ **6.** 7 **7.** -192 **8.** $-\frac{7}{3}$ **9.** $-\frac{15}{64}$ **10.** -8 **11.** 4 **12.** -5 **13.** $-\frac{1}{3}$ **14.** 3 **15.** 4 **16.** 16 **17.** All real numbers **18.** 6 **19.** -3 **20.** 28 **21.** 4 **22.** No solution **23.** Yes **24.** No **25.** Yes **26.** $\{y|y \geq -\frac{1}{2}\}$ **27.** $\{x|x \geq 7\}$ **28.** $\{y|y > 2\}$ **29.** $\{y|y \leq -4\}$ **30.** $\{x|x < -11\}$ **31.** $\{y|y > -7\}$ **32.** $\{x|x > -\frac{9}{11}\}$ **33.** $\{x|x \geq -\frac{1}{12}\}$

34. 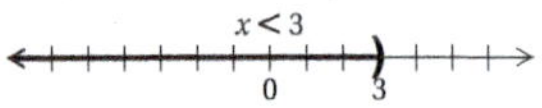

35. 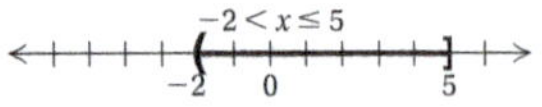

36. 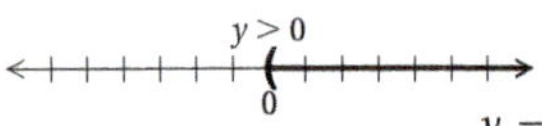

37. $d = \frac{C}{\pi}$ **38.** $B = \frac{3V}{h}$ **39.** $a = 2A - b$ **40.** $x = \frac{y - b}{m}$ **41.** Length: 365 mi; width: 275 mi **42.** 345, 346 **43.** \$2117 **44.** 27 subscriptions **45.** 35°, 85°, 60° **46.** 15 **47.** 18.75% **48.** 600 **49.** About 87.1% **50.** About 28.2% decrease **51.** \$220 **52.** \$53,400 **53.** \$138.95 **54.** 86 **55.** $\{w|w > 17 \text{ cm}\}$ **56.** C **57.** A **58.** 23, -23 **59.** 20, -20 **60.** $a = \frac{y - 3}{2 - b}$

Understanding Through Discussion and Writing

1. The end result is the same either way. If s is the original salary, the new salary after a 5% raise followed by an 8% raise is $1.08(1.05s)$. If the raises occur the other way around, the new salary is $1.05(1.08s)$. By the commutative and associative laws of multiplication, we see that these are equal. However, it would be better to receive the 8% raise first, because this increase yields a higher salary initially than a 5% raise. **2.** No; Erin paid 75% of the original price and was offered credit for 125% of this amount, not to be used on sale items. Now, 125% of 75% is 93.75%, so Erin would have a credit of 93.75% of the original price. Since this credit can be applied only to nonsale items, she has less purchasing power than if the amount she paid were refunded and she could spend it on sale items. **3.** The inequalities are equivalent by the multiplication principle for inequalities. If we multiply on both sides of one inequality by -1, the other inequality results. **4.** For any pair of numbers, their relative position on the number line is reversed when both are multiplied by the same negative number. For example, -3 is to the left of 5 on the number line ($-3 < 5$), but 12 is to the right of -20 ($-3(-4) > 5(-4)$). **5.** Answers may vary. Fran is more than 3 years older than Todd. **6.** Let n represent "a number." Then "five more than a number" translates to the *expression* $n + 5$, or $5 + n$, and "five is more than a number" translates to the *inequality* $5 > n$.

Test: Chapter 10, p. 773

1. [10.1b] 8 **2.** [10.1b] 26 **3.** [10.2a] -6 **4.** [10.2a] 49 **5.** [10.3b] -12 **6.** [10.3a] 2 **7.** [10.3a] -8 **8.** [10.1b] $-\frac{7}{20}$ **9.** [10.3c] 7 **10.** [10.3c] $\frac{5}{3}$ **11.** [10.3b] $\frac{5}{2}$ **12.** [10.3c] No solution **13.** [10.3c] All real numbers **14.** [10.7c] $\{x|x \leq -4\}$ **15.** [10.7c] $\{x|x > -13\}$ **16.** [10.7d] $\{x|x \leq 5\}$ **17.** [10.7d] $\{y|y \leq -13\}$ **18.** [10.7d] $\{y|y \geq 8\}$ **19.** [10.7d] $\{x|x \leq -\frac{1}{20}\}$ **20.** [10.7e] $\{x|x < -6\}$ **21.** [10.7e] $\{x|x \leq -1\}$

22. [10.7b] 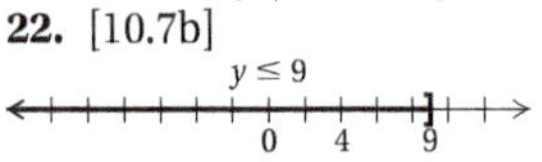

23. [10.7b, e] 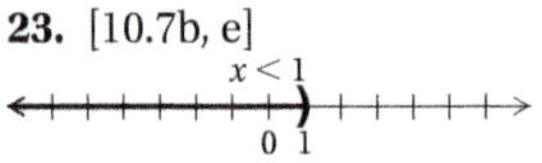

24. [10.7b] 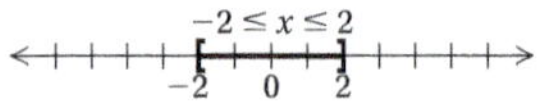

25. [10.5a] 18 **26.** [10.5a] 16.5% **27.** [10.5a] 40,000 **28.** [10.5a] About 60.4% **29.** [10.6a] Width: 7 cm; length: 11 cm **30.** [10.5a] About \$230,556 **31.** [10.6a] 2509, 2510, 2511 **32.** [10.6a] \$880 **33.** [10.6a] 3 m, 5 m **34.** [10.8b] $\{l|l \geq 174 \text{ yd}\}$ **35.** [10.8b] $\{b|b \leq \$105\}$ **36.** [10.8b] $\{c|c \leq 119{,}531\}$ **37.** [10.4b] $r = \frac{A}{2\pi h}$ **38.** [10.4b] $x = \frac{y - b}{8}$ **39.** [10.5a] D **40.** [10.4b] $d = \frac{1 - ca}{-c}$, or $\frac{ca - 1}{c}$ **41.** [9.2e], [10.3a] 15, -15 **42.** [10.6a] 60 tickets

CHAPTER 11

Calculator Corner, p. 780

1. Left to the student

Calculator Corner, p. 786

1. 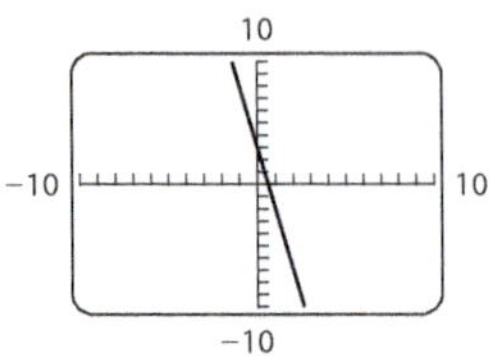

2. 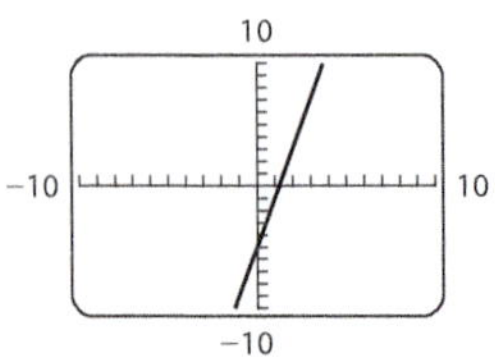

3. $y = \frac{4}{5}x + 2$

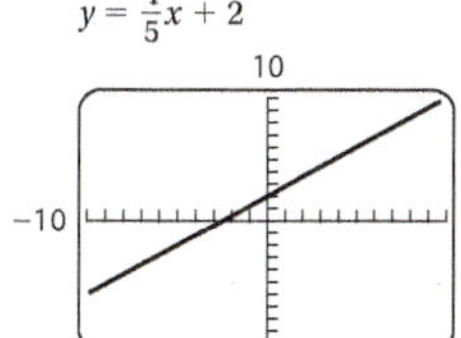

4. $y = -\frac{3}{5}x - 1$

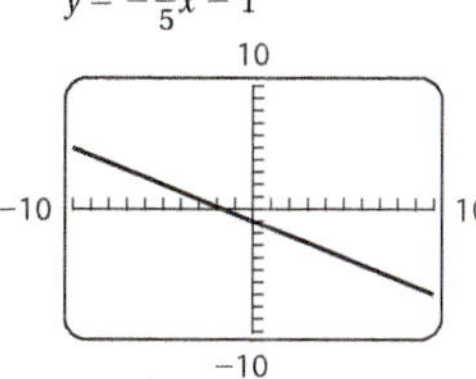

Exercise Set 11.1, p. 786

RC1. True **RC2.** False **RC3.** False **RC4.** True

1.

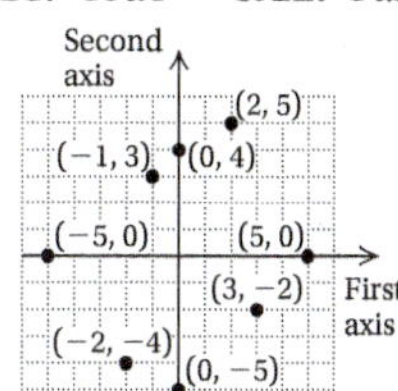

3. II **5.** IV **7.** III
9. On an axis, not in a quadrant
11. II **13.** IV **15.** II
17. I, IV **19.** I, III

21. A: $(3, 3)$; B: $(0, -4)$; C: $(-5, 0)$; D: $(-1, -1)$; E: $(2, 0)$
23. No **25.** No **27.** Yes

29.

$$\begin{array}{c|c} y & x - 5 \\ \hline -1 \; ? & 4 - 5 \\ & -1 \end{array} \quad \text{TRUE}$$

$$\begin{array}{c|c} y & x - 5 \\ \hline -4 \; ? & 1 - 5 \\ & -4 \end{array} \quad \text{TRUE}$$

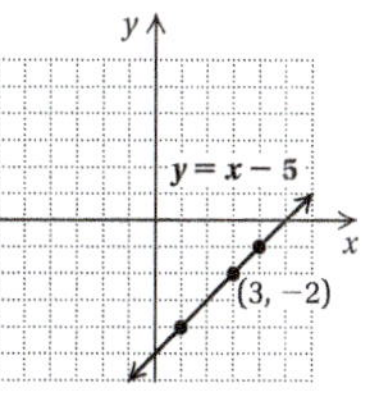

31.

$$\begin{array}{c|c} y & \frac{1}{2}x + 3 \\ \hline 5 \; ? & \frac{1}{2} \cdot 4 + 3 \\ & 2 + 3 \\ & 5 \end{array} \quad \text{TRUE}$$

$$\begin{array}{c|c} y & \frac{1}{2}x + 3 \\ \hline 2 \; ? & \frac{1}{2}(-2) + 3 \\ & -1 + 3 \\ & 2 \end{array} \quad \text{TRUE}$$

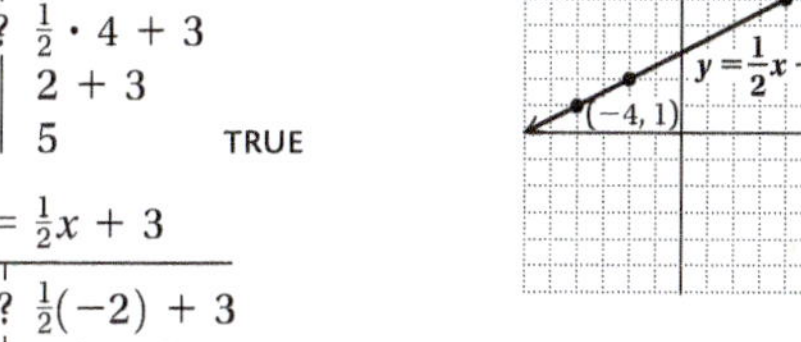

33.

$$\begin{array}{c|c} 4x - 2y & 10 \\ \hline 4 \cdot 0 - 2(-5) \; ? & 10 \\ 0 + 10 & \\ 10 & \end{array} \quad \text{TRUE}$$

$$\begin{array}{c|c} 4x - 2y & 10 \\ \hline 4 \cdot 4 - 2 \cdot 3 \; ? & 10 \\ 16 - 6 & \\ 10 & \end{array} \quad \text{TRUE}$$

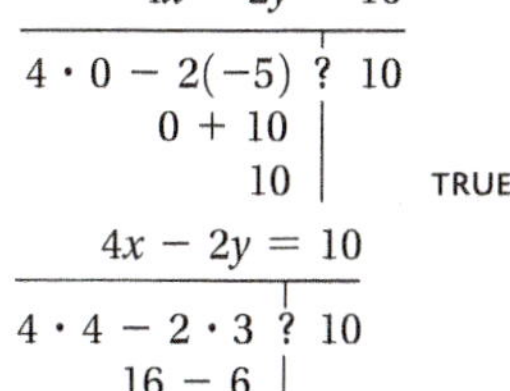

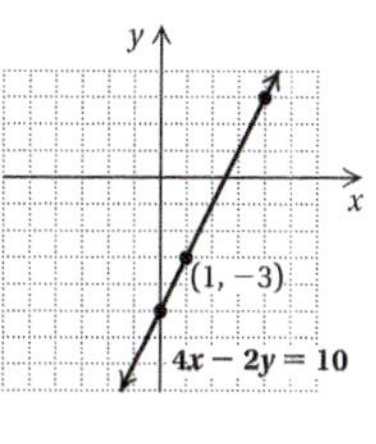

35.

x	y
−2	−1
−1	0
0	1
1	2
2	3
3	4

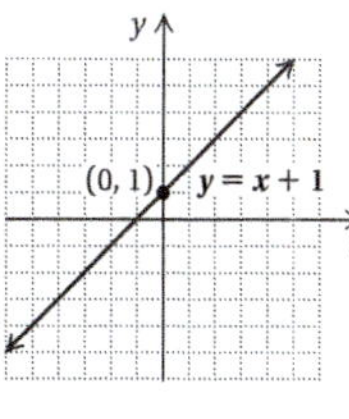

37.

x	y
−2	−2
−1	−1
0	0
1	1
2	2
3	3

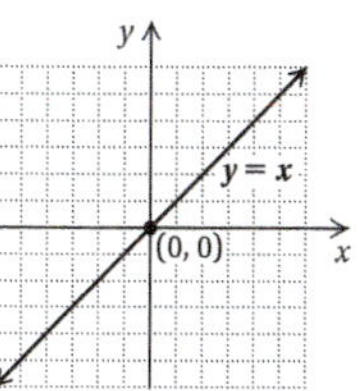

39.

x	y
−2	−1
0	0
4	2

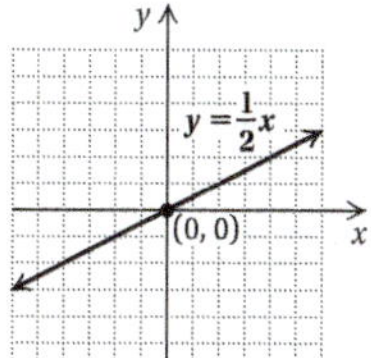

41.

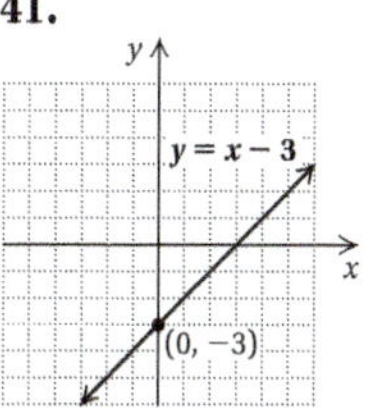

43. **45.**

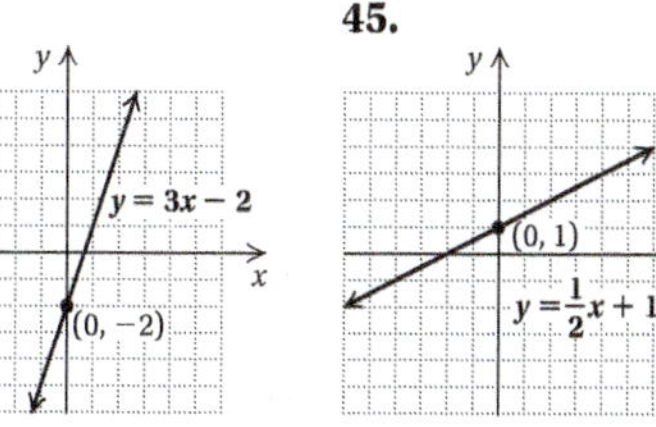

47.

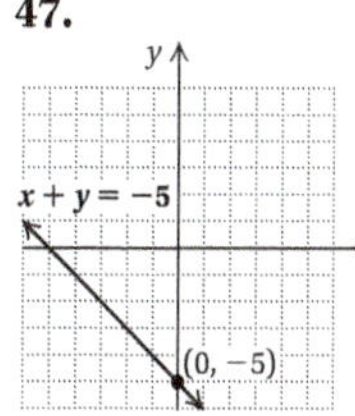

49. **51.**

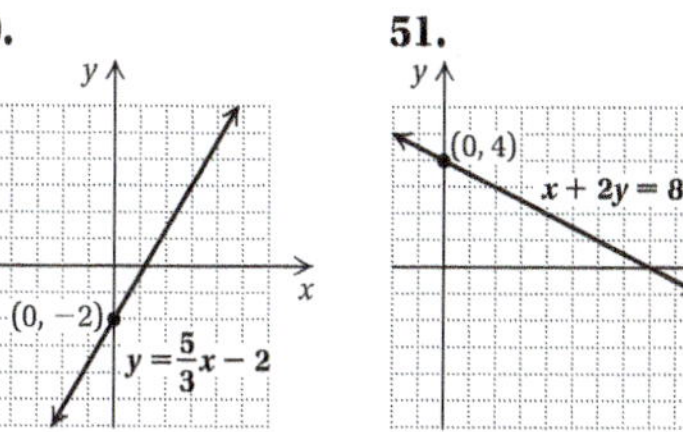

53.

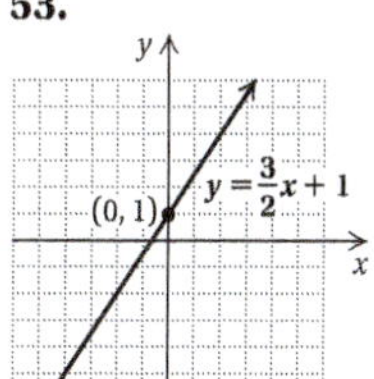

55. **57.**

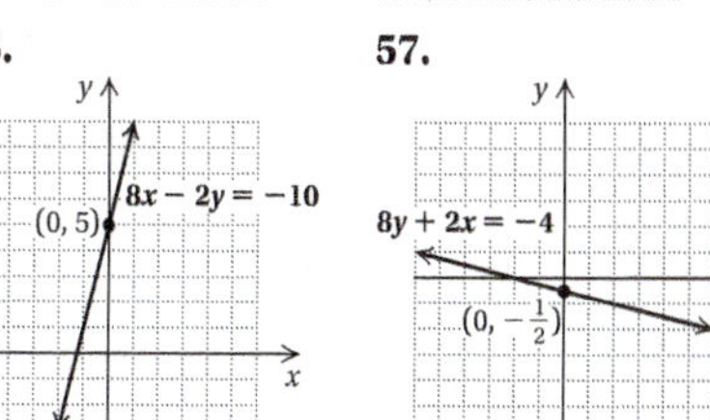

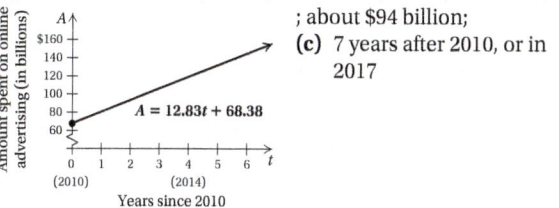

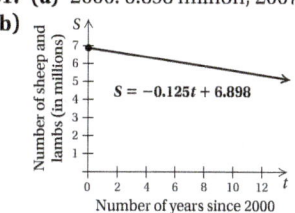

59. **(a)** 2010: \$68.38 billion; 2014: \$119.7 billion; 2015: \$132.53 billion;
(b) [graph]; about \$94 billion;
(c) 7 years after 2010, or in 2017

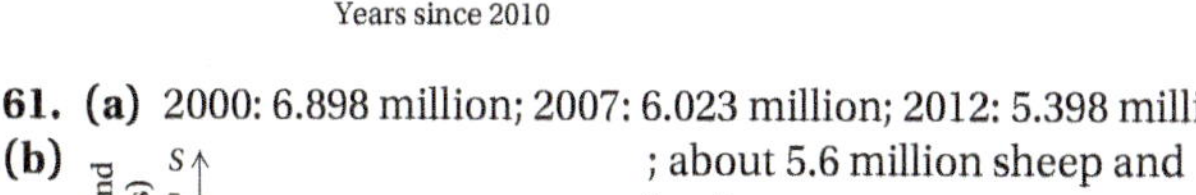

61. **(a)** 2000: 6.898 million; 2007: 6.023 million; 2012: 5.398 million;
(b) [graph]; about 5.6 million sheep and lambs;
(c) 16 years after 2000, or in 2016

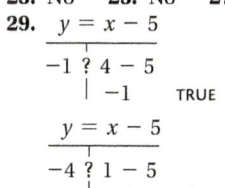

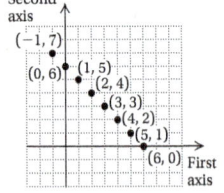

63. 12 **64.** 4.89 **65.** 0 **66.** $\frac{4}{5}$ **67.** $\frac{43}{2}$ **68.** -54 **69.** -10
70. 4 **71.** 16.6 million books **72.** 157 concerts **73.** $(-1, -5)$
75. Second axis, First axis: $(-1, 7)$, $(0, 6)$, $(1, 5)$, $(2, 4)$, $(3, 3)$, $(4, 2)$, $(5, 1)$, $(6, 0)$ **77.** 26 linear units

Calculator Corner, p. 795

1. y-intercept: $(0, -15)$; x-intercept: $(-2, 0)$; $y = -7.5x - 15$

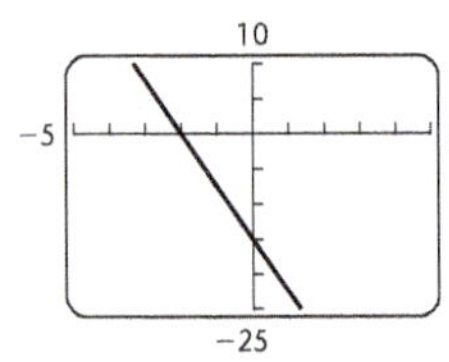

2. y-intercept: $(0, 43)$; x-intercept: $(-20, 0)$; $y = 2.15x + 43$

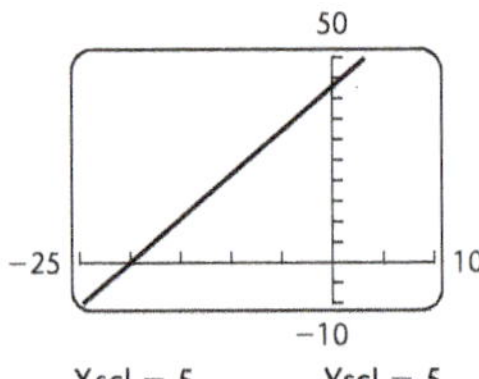

3. y-intercept: $(0, -30)$; x-intercept: $(25, 0)$; $y = (6x - 150)/5$

10
-10 35
-60
Xscl = 5 Yscl = 5

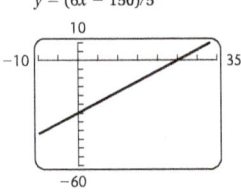

4. y-intercept: $(0, -4)$; x-intercept: $(20, 0)$; $y = 0.2x - 4$

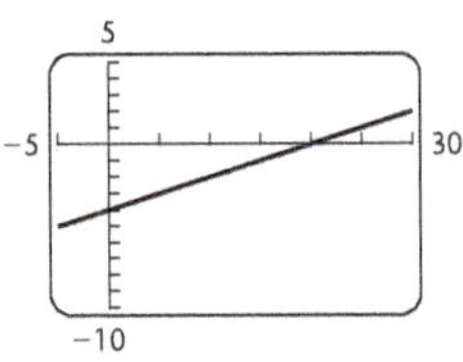

5. y-intercept: $(0, -15)$; x-intercept: $(10, 0)$; $y = 1.5x - 15$

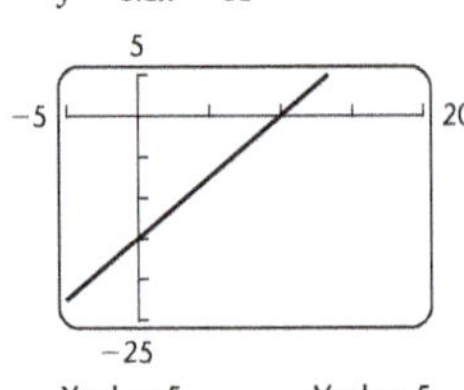

6. y-intercept: $(0, -\frac{1}{2})$; x-intercept: $(\frac{2}{5}, 0)$; $y = (5x - 2)/4$

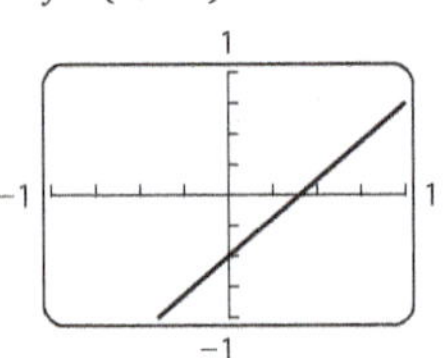

Visualizing for Success, p. 798

1. E **2.** C **3.** G **4.** A **5.** I **6.** D **7.** F **8.** J
9. B **10.** H

Exercise Set 11.2, p. 799

RC1. horizontal; y-intercept **RC2.** x-axis **RC3.** $y = 0$
RC4. $(0, 0)$ **RC5.** $x = 0$ **RC6.** vertical; x-intercept
1. (a) $(0, 5)$; **(b)** $(2, 0)$ **3. (a)** $(0, -4)$; **(b)** $(3, 0)$
5. (a) $(0, 3)$; **(b)** $(5, 0)$ **7. (a)** $(0, -14)$; **(b)** $(4, 0)$
9. (a) $\left(0, \frac{10}{3}\right)$; **(b)** $\left(-\frac{5}{2}, 0\right)$ **11. (a)** $\left(0, -\frac{1}{3}\right)$; **(b)** $\left(\frac{1}{2}, 0\right)$

13.

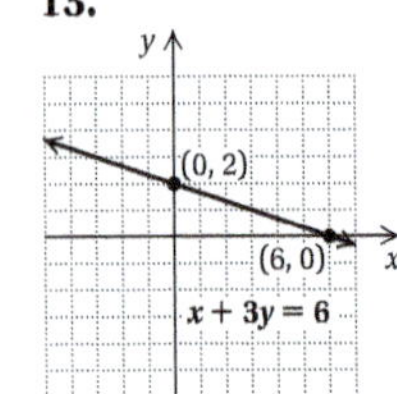

15.

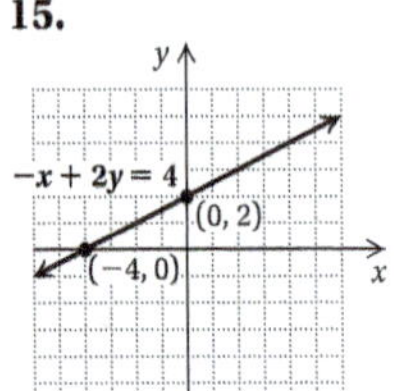

17.

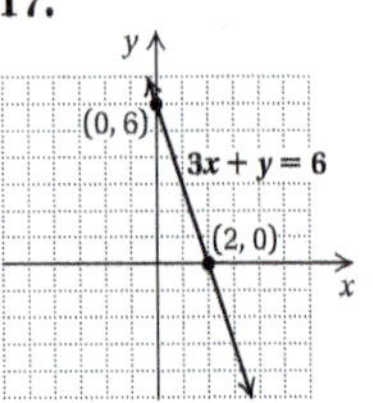

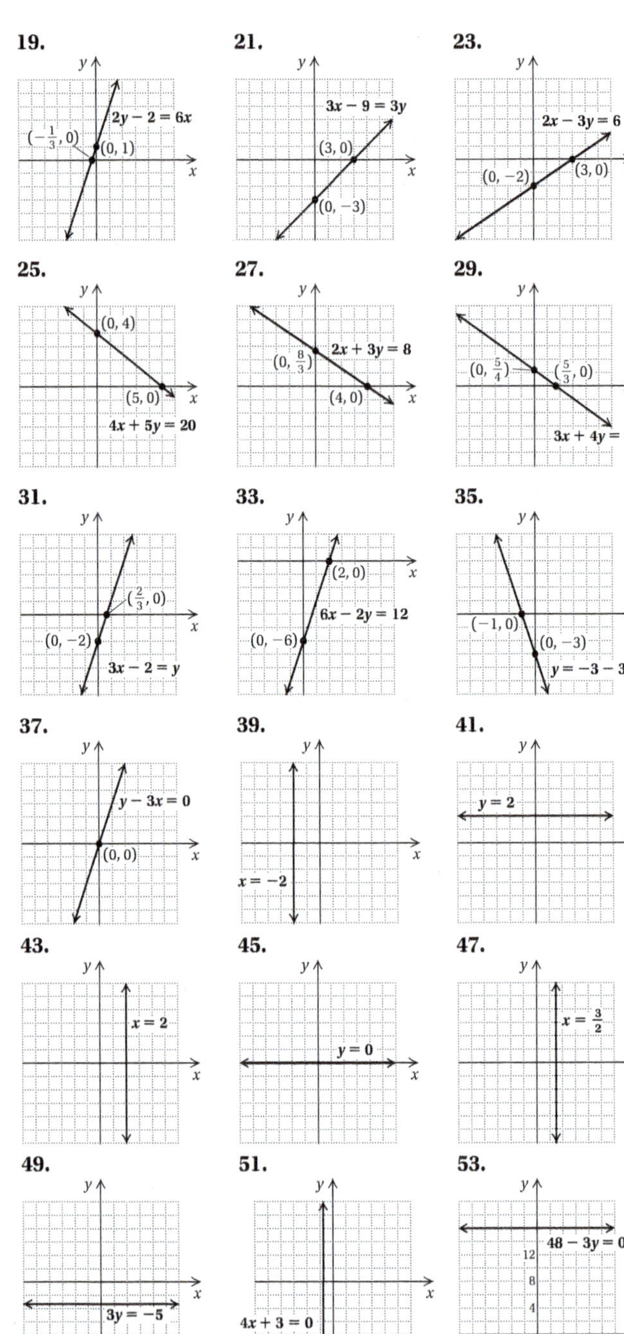

49. **51.** **53.**

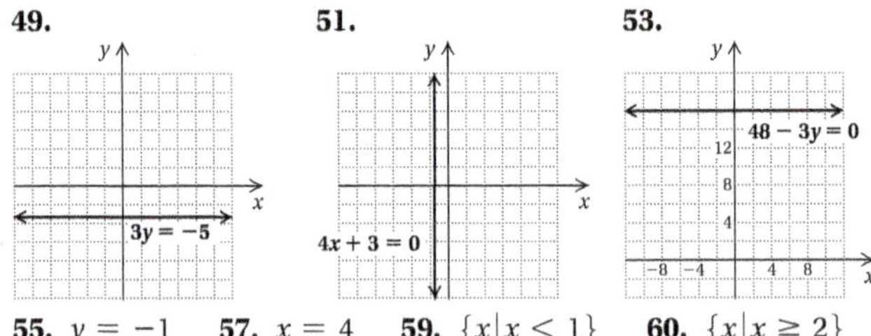

55. $y = -1$ **57.** $x = 4$ **59.** $\{x \mid x < 1\}$ **60.** $\{x \mid x \geq 2\}$
61. $\{x \mid x \leq 7\}$ **62.** $\{x \mid x > 1\}$ **63.** $y = -4$ **65.** $k = 12$

Calculator Corner, p. 807

1. This line will pass through the origin and slant up from left to right. This line will be steeper than $y = 10x$. **2.** This line will pass through the origin and slant up from left to right. This line will be less steep than $y = \frac{5}{32}x$. **3.** This line will pass through the origin and slant down from left to right. This line will be steeper than $y = -10x$. **4.** This line will pass through the origin and slant down from left to right. This line will be less steep than $y = -\frac{5}{32}x$.

Exercise Set 11.3, p. 810

RC1. (d) **RC2.** (f) **RC3.** (b) **RC4.** (e)
RC5. (c) **RC6.** (a)

1. $-\frac{3}{7}$ 3. $\frac{2}{3}$ 5. $\frac{3}{4}$ 7. 0

9. $-\frac{4}{5}$;

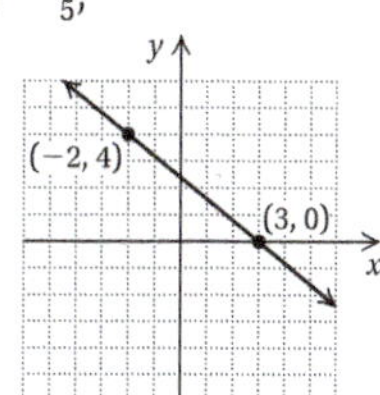

11. 3;

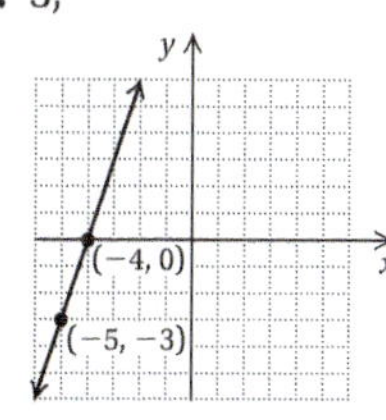

13. $-\frac{2}{3}$;

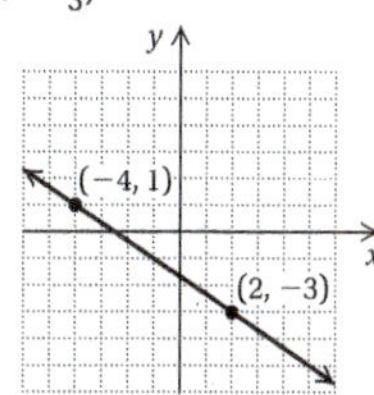

15. $\frac{7}{8}$;

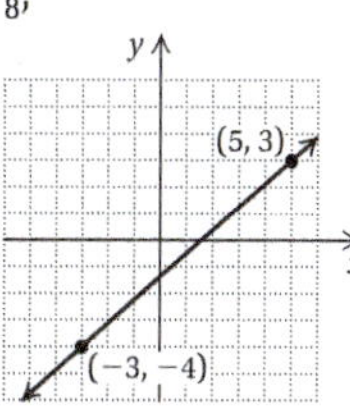

17. $\frac{2}{3}$ 19. Not defined 21. $-\frac{5}{13}$ 23. 0 25. -10 27. 3.78 29. 3 31. $-\frac{1}{5}$ 33. $-\frac{3}{2}$ 35. Not defined 37. -1 39. 3 41. $\frac{5}{4}$ 43. 0 45. $\frac{4}{3}$ 47. $-\frac{21}{8}$ 49. $\frac{12}{41}$ 51. $\frac{28}{129}$ 53. 3.0%; yes 55. About 1.24 million children per year 57. About $-14{,}100$ people per year 59. 30,600,000 lb per year 61. $\frac{4}{25}$ 62. $\frac{3}{8}$ 63. $-\frac{2}{3}p$ 64. $5t - 1$ 65. $y = -x + 5$ 67. $y = x + 2$

Exercise Set 11.4, p. 818

RC1. (e) RC2. (f) RC3. (a) RC4. (g)
1. Slope: -4; y-intercept: $(0, -9)$ 3. Slope: 1.8; y-intercept: $(0, 0)$ 5. Slope: $-\frac{8}{7}$; y-intercept: $(0, -3)$ 7. Slope: $\frac{4}{9}$; y-intercept: $(0, -\frac{7}{9})$ 9. Slope: $-\frac{3}{2}$; y-intercept: $(0, -\frac{1}{2})$ 11. Slope: 0; y-intercept: $(0, -17)$ 13. $y = -7x - 13$ 15. $y = 1.01x - 2.6$ 17. $y = -5$ 19. $y = -2x - 6$ 21. $y = \frac{3}{4}x + \frac{5}{2}$ 23. $y = x - 8$ 25. $y = -3x + 3$ 27. $y = x + 4$ 29. $y = -\frac{1}{2}x + 4$ 31. $y = -\frac{3}{2}x + \frac{13}{2}$ 33. $x = 4$ 35. $y = -4x - 11$ 37. $y = \frac{1}{4}$ 39. (a) $S = 2.82t + 35$; (b) an increase of \$2.82 billion per year; (c) \$77.3 billion 41. $\frac{53}{7}$ 42. $\frac{3}{8}$ 43. 6 44. $\frac{42}{5}$ 45. $\frac{24}{19}$ 46. $\frac{125}{7}$ 47. 3.6 48. 500 49. 5% 50. 4000 51. $y = 3x - 9$ 53. $y = \frac{3}{2}x - 2$

Mid-Chapter Review: Chapter 11, p. 821

1. False 2. True 3. True 4. False
5. (a) The y-intercept is $(0, -3)$. (b) The x-intercept is $(-3, 0)$. (c) The slope is $\frac{-3 - 0}{0 - (-3)} = \frac{-3}{3} = -1$. (d) The equation of the line in $y = mx + b$ form is $y = -1x + -3$, or $y = x - 3$.
6. (a) The x-intercept is $(c, 0)$. (b) The y-intercept is $(0, d)$. (c) The slope is $\frac{d - 0}{0 - c} = \frac{d}{-c} = -\frac{d}{c}$. (d) The equation of the line in $y = mx + b$ form is $y = -\frac{d}{c}x + d$.
7. No 8. Yes 9. x-intercept: $(-6, 0)$; y-intercept: $(0, 9)$
10. x-intercept: $(\frac{1}{2}, 0)$; y-intercept: $(0, -\frac{1}{20})$

11.

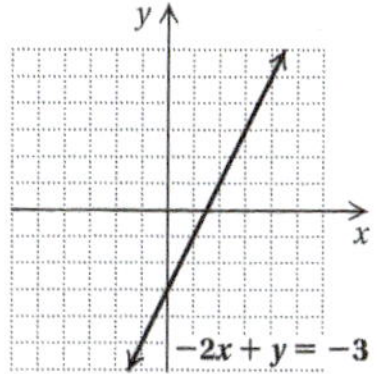

12.

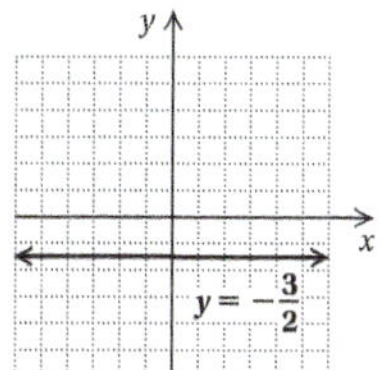

13.

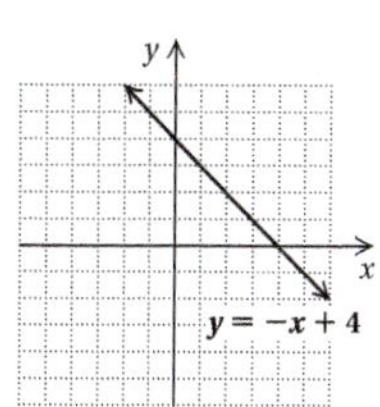

14.

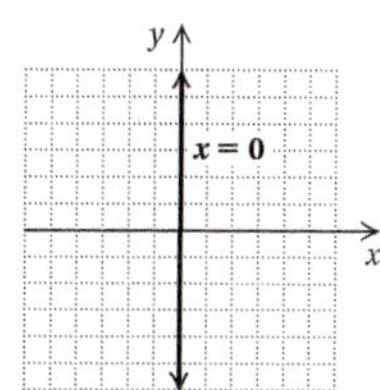

15. $-\frac{40}{9}$ 16. $-\frac{1}{2}$ 17. 0 18. 13 19. Not defined 20. 429,400 people per year 21. D 22. C 23. B 24. E 25. A 26. $y = -3x + 2$ 27. $x = \frac{1}{2}$ 28. $y = -\frac{1}{5}x - \frac{17}{5}$ 29. $y = -4$ 30. No; an equation $x = a$, $a \neq 0$, does not have a y-intercept. 31. Most would probably say that the second equation would be easier to graph because it has been solved for y. This makes it more efficient to find the y-value that corresponds to a given x-value. 32. $A = 0$. If the line is horizontal, then the equation is of the form $y =$ a constant. Thus, Ax must be 0 and, hence, $A = 0$. 33. Any ordered pair $(7, y)$ is a solution of $x = 7$. Thus all points on the graph are 7 units to the right of the y-axis, so they lie on a vertical line.

Exercise Set 11.5, p. 825

RC1. (c) RC2. (d) RC3. (b) RC4. (f)

1.

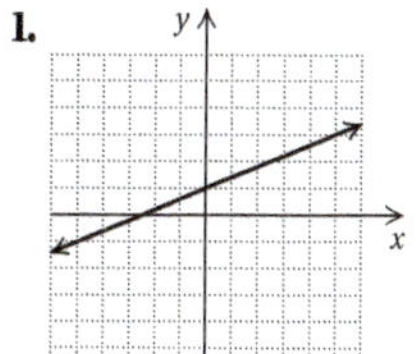

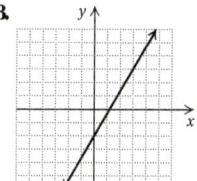

3.

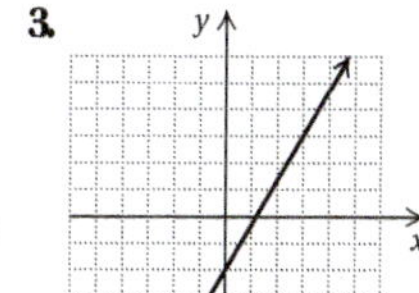

5.

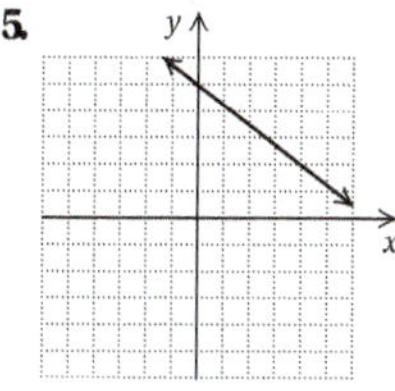

7.

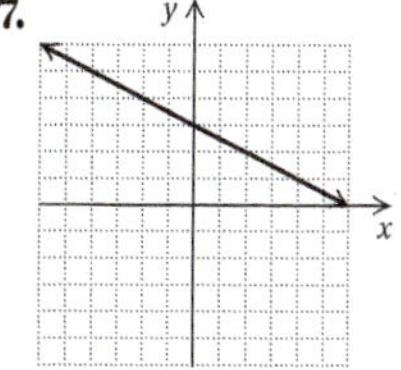

9.

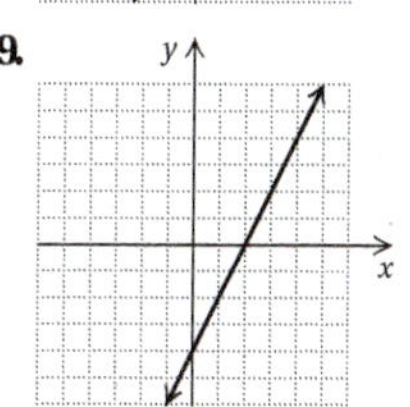

11.

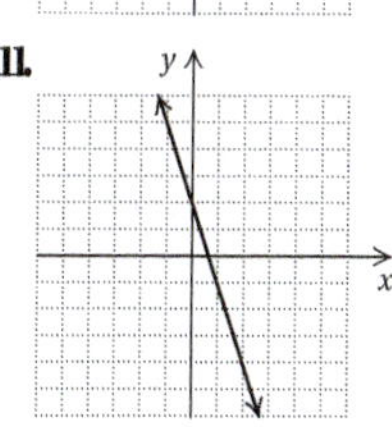

13.

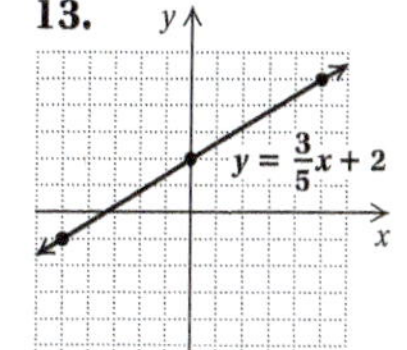

15.

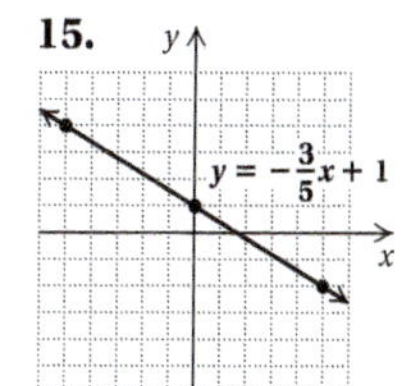

17.

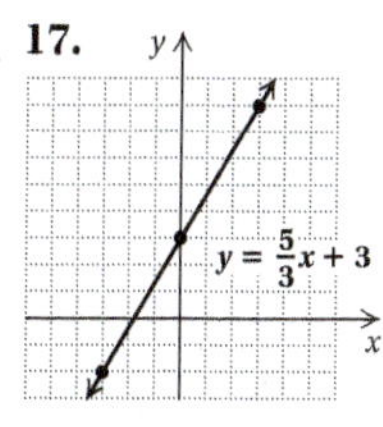

19.

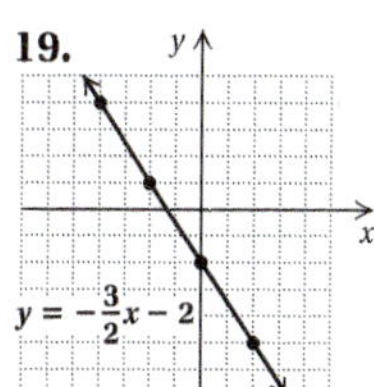

21.

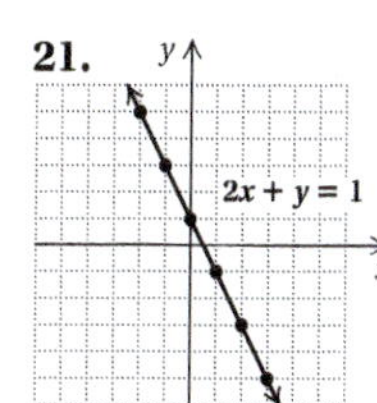

23.

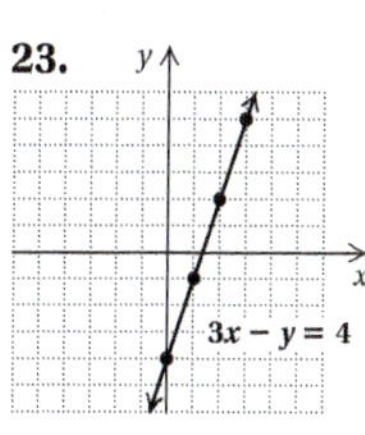

25.

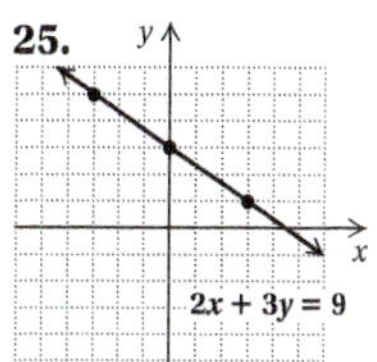

27.

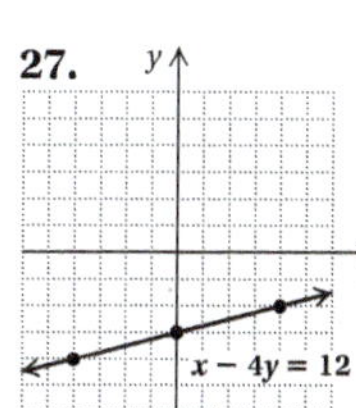

29.

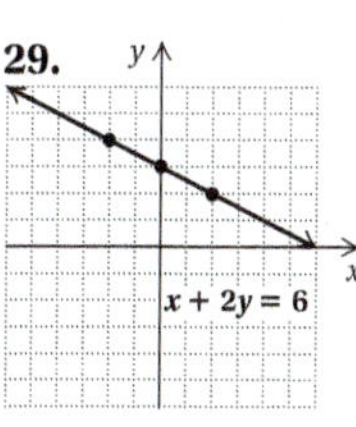

31. $\frac{13}{10}$ 32. $-\frac{1}{6}$ 33. $\frac{69}{100}$, or 0.69 34. $-\frac{3}{5}$, or -0.6 35. 0 36. Not defined 37. 42 38. -185 39. 23 40. 41

41.

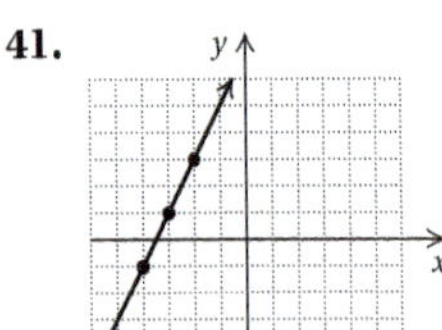

Calculator Corner, p. 831

1. Left to the student

Exercise Set 11.6, p. 832

RC1. (b) and (e) **RC2.** (c) and (d)
1. Yes **3.** No **5.** No **7.** No **9.** Yes **11.** Yes **13.** No **15.** No **17.** Yes **19.** Yes **21.** Yes **23.** No **25.** Yes **27.** No **29.** Parallel **31.** Neither **33.** No **34.** Yes **35.** x-intercept: $(-2, 0)$; y-intercept: $(0, 16)$ **36.** x-intercept: $(-\frac{1}{2}, 0)$; y-intercept: $(0, 3)$ **37.** $y = 3x + 6$ **39.** $y = -3x + 2$ **41.** $y = \frac{1}{2}x + 1$ **43.** 16 **45.** A: $y = \frac{4}{3}x - \frac{7}{3}$; B: $y = -\frac{3}{4}x - \frac{1}{4}$

Calculator Corner, p. 837

1. Left to the student

Visualizing for Success, p. 838

1. D **2.** H **3.** E **4.** A **5.** J **6.** F **7.** C **8.** B **9.** I **10.** G

Exercise Set 11.7, p. 839

RC1. = **RC2.** related **RC3.** x-intercept **RC4.** y-intercept **RC5.** < **RC6.** dashed **RC7.** half-plane **RC8.** shade **RC9.** False
1. No **3.** Yes

5.

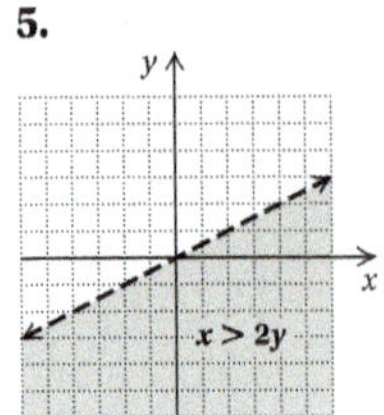

7.

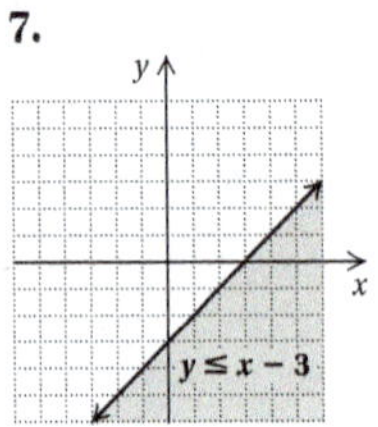

9.

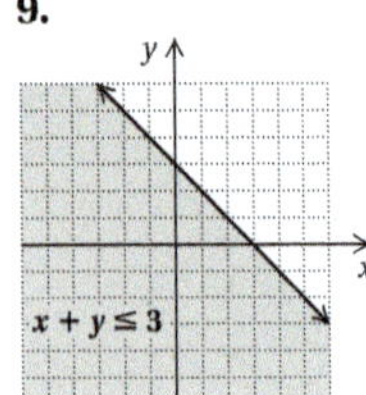

11.

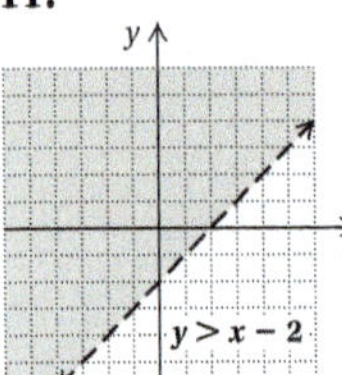

13.

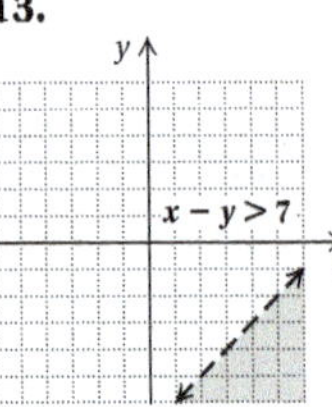

15.

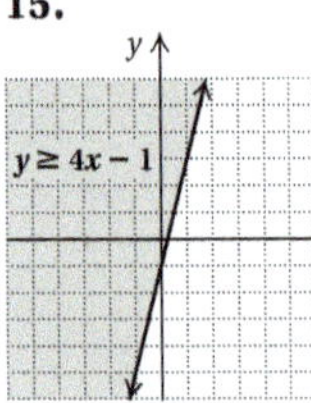

17.

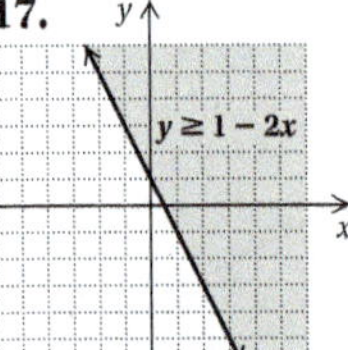

19.

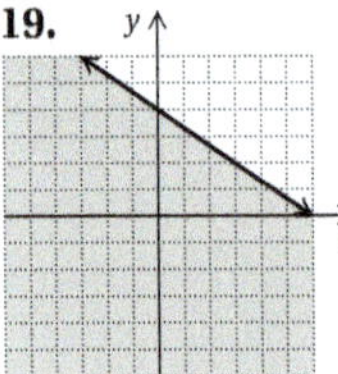

21.

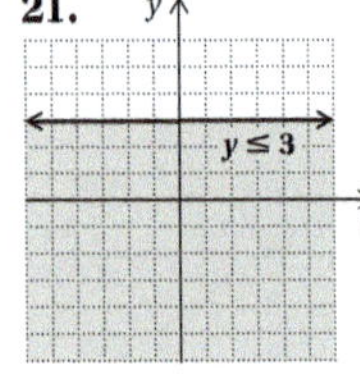

23.

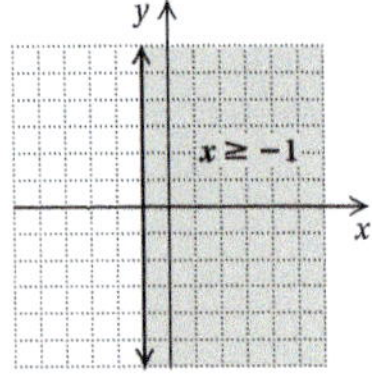

25. Parallel **26.** Perpendicular **27.** Neither **28.** Neither

29. $35c + 75a > 1000$

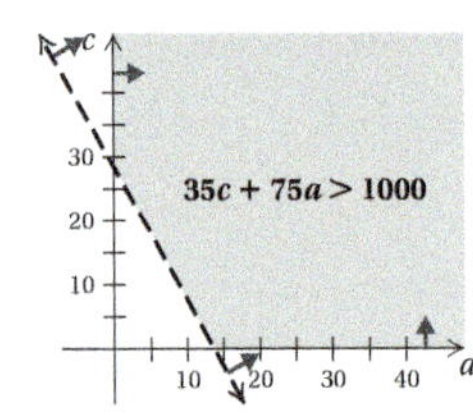

Summary and Review: Chapter 11, p. 841

Vocabulary Reinforcement

1. slope–intercept **2.** Horizontal **3.** vertical **4.** slope **5.** x-intercept; second **6.** y-intercept; first

Concept Reinforcement

1. True **2.** False **3.** False **4.** True **5.** False

Study Guide

1. F: $(2, 4)$; G: $(-2, 0)$; H: $(-3, -5)$

2.

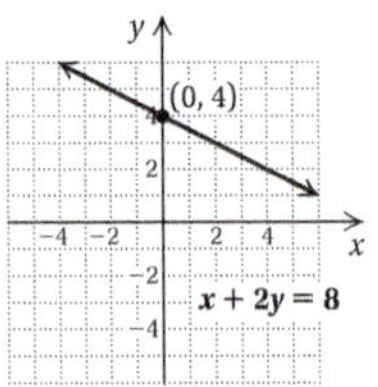

3.

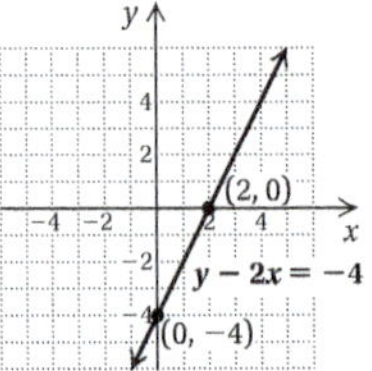

4.

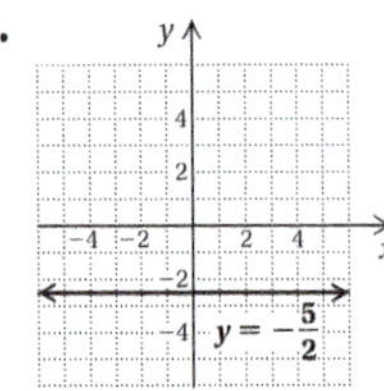

5.

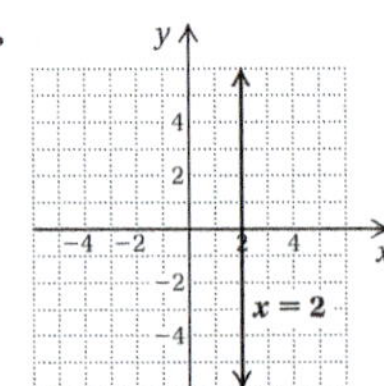

6. m is not defined. **7.** $\frac{3}{2}$ **8.** 0 **9.** m is not defined. **10.** -2 **11.** 0 **12.** $y = 6x + 7$ **13.** $y = -\frac{1}{6}x - \frac{11}{6}$ **14.** Perpendicular **15.** Parallel

16.

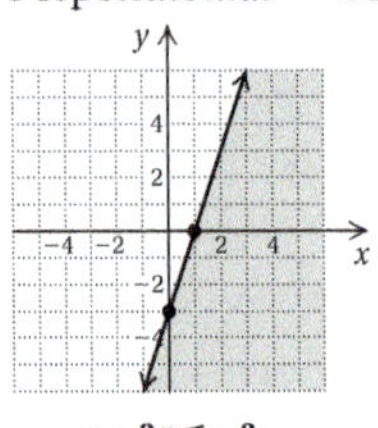

Review Exercises

1.–3.

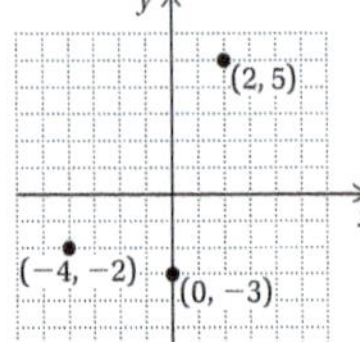

4. $(-5, -1)$ **5.** $(-2, 5)$ **6.** $(3, 0)$ **7.** IV **8.** III **9.** I **10.** No **11.** Yes

12.

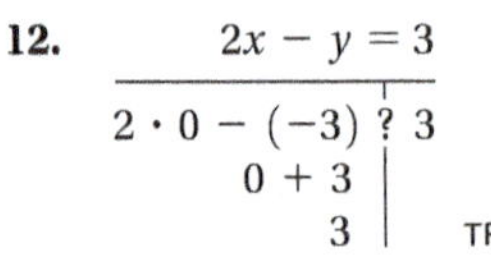

$$\begin{array}{c|c} 2x - y & 3 \\ \hline 2 \cdot 0 - (-3) & 3 \\ 0 + 3 & \\ 3 & \end{array} \quad \text{TRUE}$$

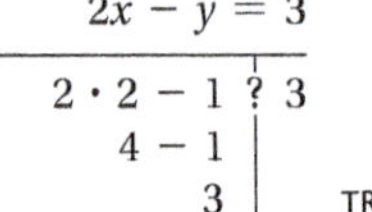

$$\begin{array}{c|c} 2x - y & 3 \\ \hline 2 \cdot 2 - 1 & 3 \\ 4 - 1 & \\ 3 & \end{array} \quad \text{TRUE}$$

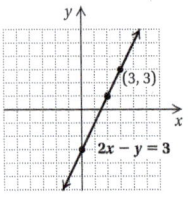

(3, 3)
2x − y = 3

13.

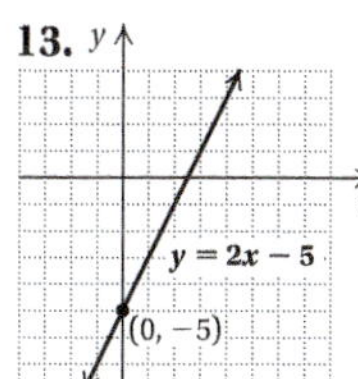

14.

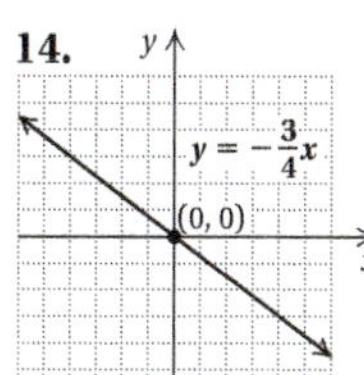

15.

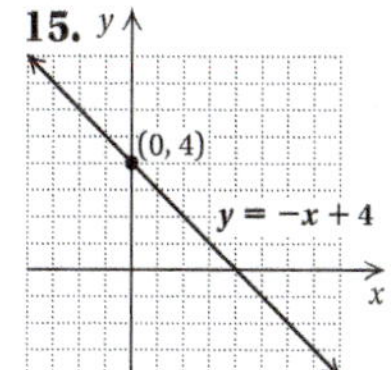

16.

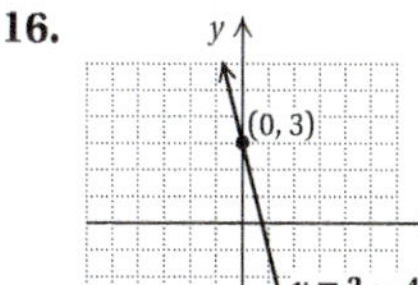
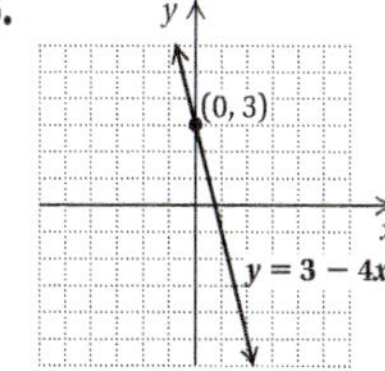

17. **(a)** $14\frac{1}{2}\text{ ft}^3$, 16 ft^3, $20\frac{1}{2}\text{ ft}^3$, 28 ft^3;
(b)

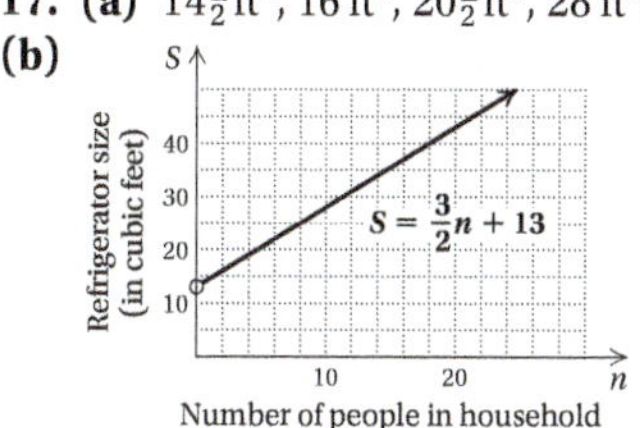
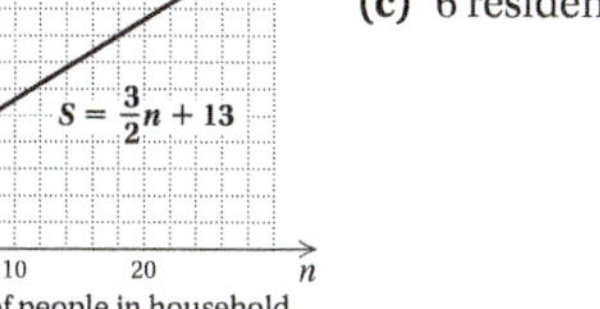

19 ft^3;
(c) 6 residents

18.

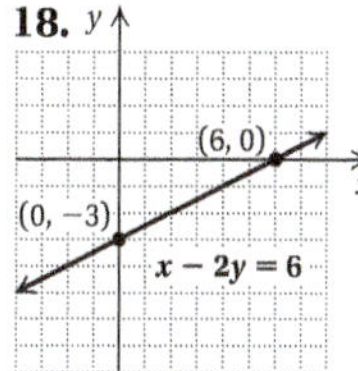

19.

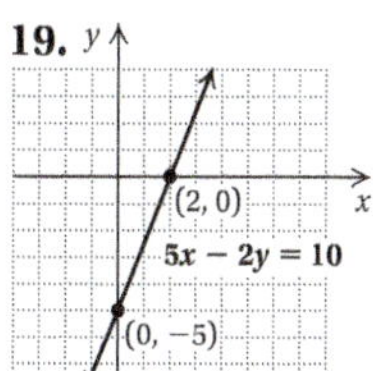

20.

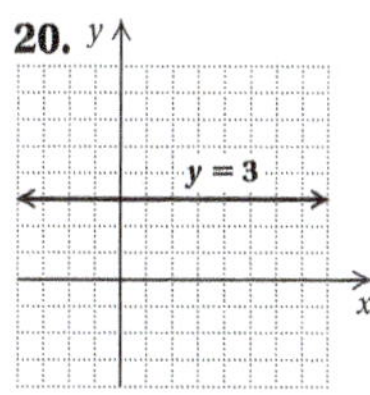

21.

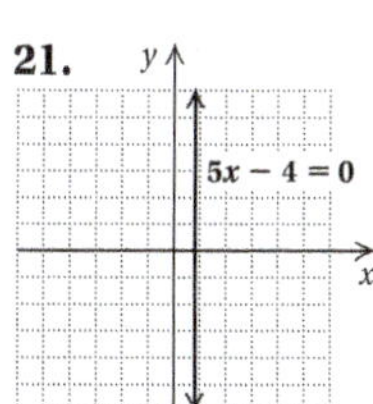

22. $\frac{1}{3}$ **23.** $-\frac{1}{3}$
24. $\frac{3}{5}$;

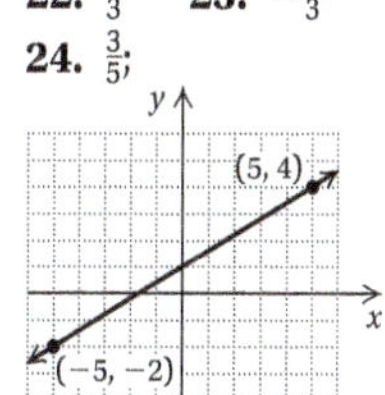

25. -1;

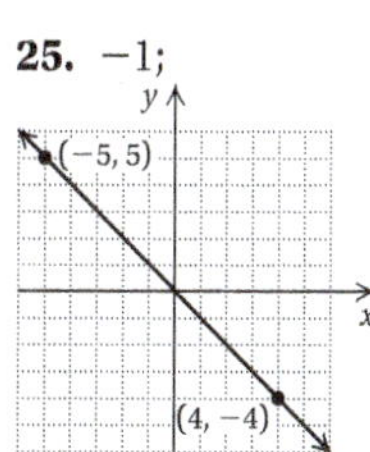

26. $-\frac{5}{8}$ **27.** $\frac{1}{2}$ **28.** Not defined **29.** 0 **30.** **(a)** 2.4 driveways per hour; **(b)** 25 minutes per driveway **31.** 7% **32.** \$2.4 billion per year **33.** Slope: -9; y-intercept: $(0, 46)$ **34.** Slope: -1; y-intercept: $(0, 9)$ **35.** Slope: $\frac{3}{5}$; y-intercept: $\left(0, -\frac{4}{5}\right)$
36. $y = -2.8x + 19$ **37.** $y = \frac{5}{8}x - \frac{7}{8}$ **38.** $y = 3x - 1$
39. $y = \frac{2}{3}x - \frac{11}{3}$ **40.** $y = -2x - 4$ **41.** $y = x + 2$

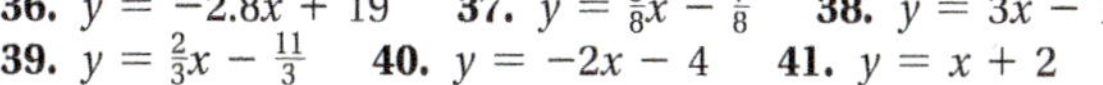

42. $y = \frac{1}{2}x - 1$ **43.** **(a)** $y = 0.38x + 6.2$; **(b)** increase of 0.38 million, or 380,000, people per year; **(c)** 12.28 million, or 12,280,000, people

44.

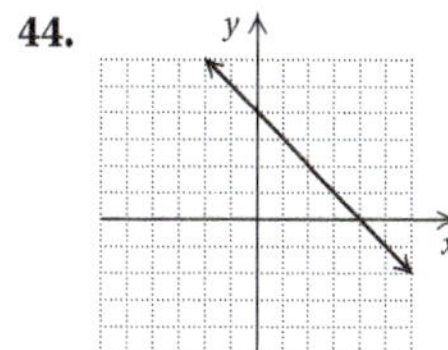

45.

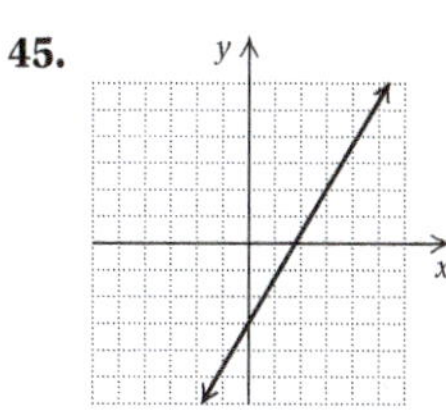

46.

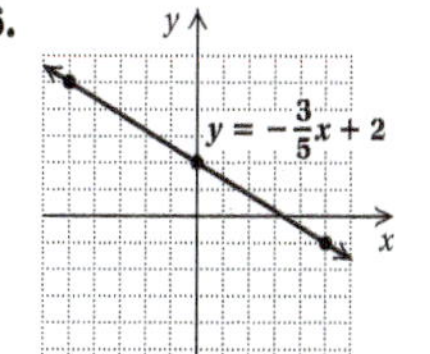

47.

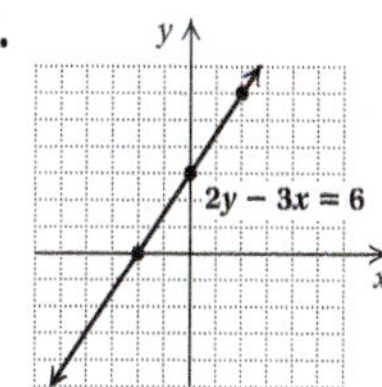

48. Parallel **49.** Perpendicular **50.** Parallel **51.** Neither
52. No **53.** No **54.** Yes
55. **56.** **57.**

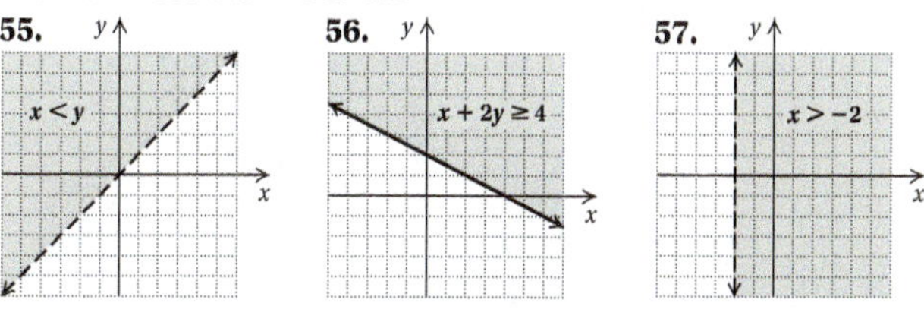

58. D **59.** C **60.** 45 square units; 28 linear units
61. **(a)** $239.58\overline{3}$ ft per minute; **(b)** about 0.004 min per foot

Understanding Through Discussion and Writing

1. If one equation represents a vertical line (that is, it is of the form $x = a$) and the other represents a horizontal line (that is, it is of the form $y = b$), then the graphs are perpendicular. If neither line is of one of the forms above, then solve each equation for y in order to determine the slope of each. Then, if the product of the slopes is -1, the graphs are perpendicular. **2.** If $b > 0$, then the y-intercept of $y = mx + b$ is on the positive y-axis and the graph of $y = mx + b$ lies "above" the origin. Using $(0, 0)$ as a test point, we have the false inequality $0 > b$, so the region above $y = mx + b$ is shaded.

If $b = 0$, the line $y = mx + b$ or $y = mx$ passes through the origin. Testing a point above the line, such as $(1, m + 1)$, we have the true inequality $m + 1 > m$, so the region above the line is shaded.

If $b < 0$, then the y-intercept of $y = mx + b$ is on the negative y-axis and the graph of $y = mx + b$ lies "below" the origin. Using $(0, 0)$ as a test point, we get the true inequality $0 > b$, so the region above $y = mx + b$ is shaded.

Thus we see that in any case the graph of any inequality of the form $y > mx + b$ is always shaded above the line $y = mx + b$.
3. The y-intercept is the point at which the graph crosses the y-axis. Since a point on the y-axis is neither left nor right of the origin, the first or x-coordinate of the point is 0.
4.

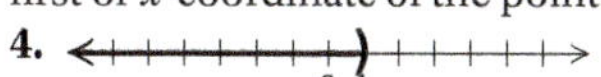

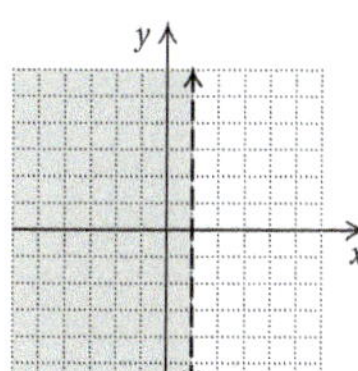

The graph of $x < 1$ on the number line consists of the points in the set $\{x | x < 1\}$. The graph of $x < 1$ on a plane consists of the points, or ordered pairs, in the set $\{(x, y) | x + 0 \cdot y < 1\}$. This is the set of ordered pairs with first coordinate less than 1.

5. First, plot the y-intercept, $(0, 2458)$. Then, thinking of the slope as $\frac{37}{100}$, plot a second point on the line by moving up 37 units and to the right 100 units from the y-intercept. Next, thinking of the slope as $\frac{-37}{-100}$, start at the y-intercept and plot a third point by moving down 37 units and to the left 100 units. Finally, draw a line through the three points. **6.** If the equations are of the form $x = p$ and $x = q$, where $p \neq q$, then the graphs are parallel vertical lines. If neither equation is of the form $x = p$, then solve each for y in order to determine the slope and the y-intercept of each. If the slopes are the same and the y-intercepts are different, the lines are parallel.

Test: Chapter 11, p. 849

1. [11.1a] II **2.** [11.1a] III **3.** [11.1b] $(-5, 1)$
4. [11.1b] $(0, -4)$

5. [11.1c]

$$\begin{array}{c|c} y - 2x & = 5 \\ \hline -3 - 2(-4) & ?\ 5 \\ -3 + 8 & \\ 5 & \end{array} \quad \text{TRUE}$$

$$\begin{array}{c|c} y - 2x & = 5 \\ \hline 3 - 2(-1) & ?\ 5 \\ 3 + 2 & \\ 5 & \end{array} \quad \text{TRUE}$$

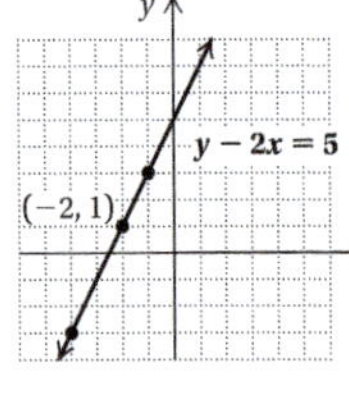

6. [11.1d] **7.** [11.1d] **8.** [11.2a]

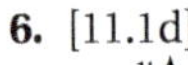

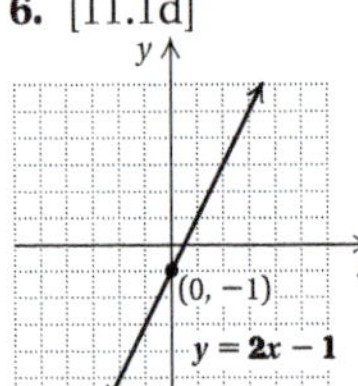

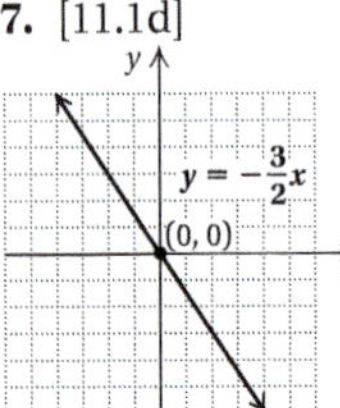

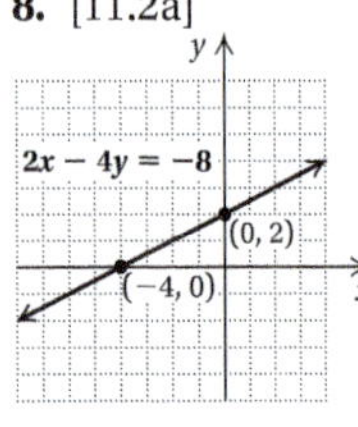

9. [11.2a] **10.** [11.2b] **11.** [11.2b]

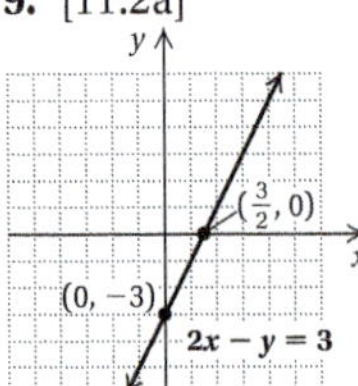

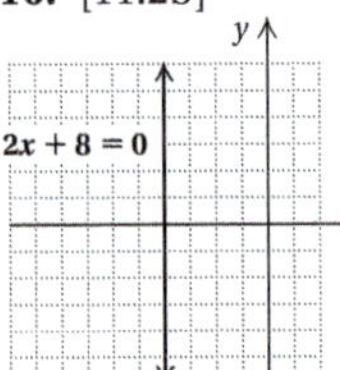

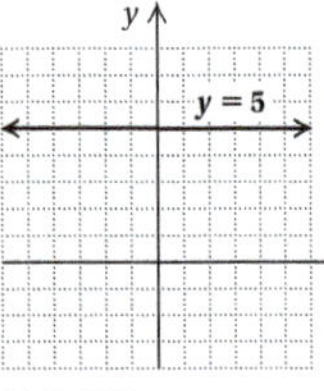

12. [11.1e] **(a)** 2007: \$8593; 2009: \$9805; 2012: \$11,623; **(b)**

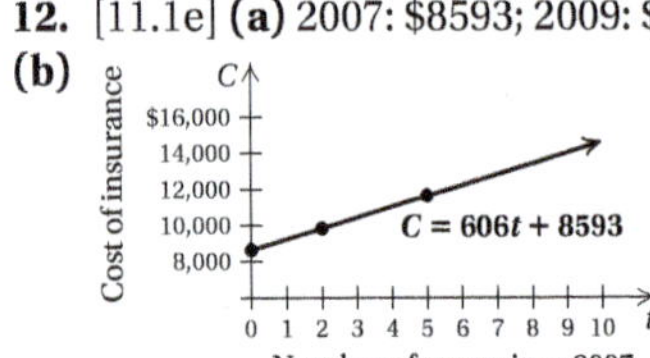

; approximately \$14,000; **(c)** 11 years after 2007, or in 2018

13. [11.3c] 87.5 mph **14.** [11.3a] −2
15. [11.3a] $\frac{3}{8}$;

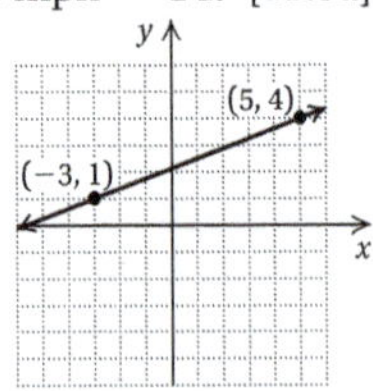

16. [11.3b] **(a)** $\frac{2}{5}$; **(b)** not defined **17.** [11.3c] $-\frac{1}{20}$, or -0.05
18. [11.4a] Slope: 2; y-intercept: $(0, -\frac{1}{4})$ **19.** [11.4a] Slope: $\frac{4}{3}$; y-intercept: $(0, -2)$ **20.** [11.4a] $y = 1.8x - 7$
21. [11.4a] $y = -\frac{3}{8}x - \frac{1}{8}$ **22.** [11.4b] $y = x + 2$
23. [11.4b] $y = -3x - 6$ **24.** [11.4c] $y = -3x + 4$
25. [11.4c] $y = \frac{1}{4}x - 2$
26. [11.5a] **27.** [11.5a]

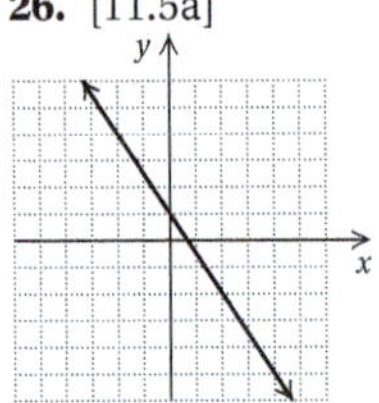

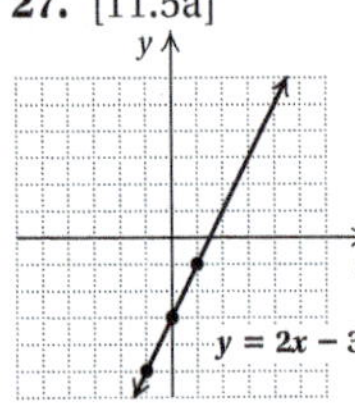

28. [11.6a, b] Parallel **29.** [11.6a, b] Neither
30. [11.6a, b] Perpendicular **31.** [11.7a] No **32.** [11.7a] Yes
33. [11.7b] **34.** [11.7b]

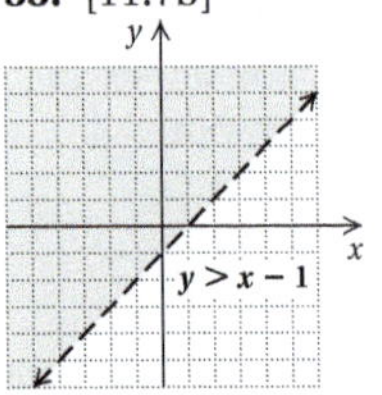

35. [11.6a, b] A **36.** [11.1a] 25 square units; 20 linear units
37. [11.6b] 3

CHAPTER 12

Exercise Set 12.1, p. 860

RC1. (c) **RC2.** (a) **RC3.** (e) **RC4.** (c) **RC5.** (d) **RC6.** (c)
1. $3 \cdot 3 \cdot 3 \cdot 3$ **3.** $(-1.1)(-1.1)(-1.1)(-1.1)(-1.1)$
5. $(\frac{2}{3})(\frac{2}{3})(\frac{2}{3})(\frac{2}{3})$ **7.** $(7p)(7p)$ **9.** $8 \cdot k \cdot k \cdot k$
11. $-6 \cdot y \cdot y \cdot y \cdot y$ **13.** 1 **15.** b **17.** 1 **19.** −7.03
21. 1 **23.** ab **25.** a **27.** 27 **29.** 19 **31.** −81
33. 256 **35.** 93 **37.** 136 **39.** 10; 4 **41.** 3629.84 ft^2
43. $\frac{1}{3^2} = \frac{1}{9}$ **45.** $\frac{1}{10^3} = \frac{1}{1000}$ **47.** $\frac{1}{a^3}$ **49.** $8^2 = 64$
51. y^4 **53.** $\frac{5}{z^4}$ **55.** $\frac{x}{y^2}$ **57.** 4^{-3} **59.** x^{-3} **61.** a^{-5}
63. 2^7 **65.** 9^{38} **67.** x^5 **69.** x^{17} **71.** $(3y)^{12}$
73. $(7y)^{17}$ **75.** 3^3 **77.** 1 **79.** $\frac{1}{x^{13}}$ **81.** $\frac{1}{a^{10}}$ **83.** x^6y^{15}
85. s^3t^7 **87.** 7^3 **89.** y^8 **91.** $\frac{1}{16^6}$ **93.** $\frac{1}{m^6}$ **95.** $\frac{1}{(8x)^4}$
97. x^2 **99.** $\frac{1}{z^4}$ **101.** x^3 **103.** 1 **105.** a^3b^2
107. $5^2 = 25; 5^{-2} = \frac{1}{25}; (\frac{1}{5})^2 = \frac{1}{25}; (\frac{1}{5})^{-2} = 25; -5^2 = -25;$ $(-5)^2 = 25; -(-\frac{1}{5})^2 = -\frac{1}{25}; (-\frac{1}{5})^{-2} = 25$
109. 8 in.; 4 in. **110.** 51°, 27°, 102° **111.** 45%; 37.5%; 17.5%
112. Lengths less than 2.5 ft **113.** $\frac{7}{4}$ **114.** 2 **115.** $\frac{23}{14}$
116. $\frac{11}{10}$ **117.** No **119.** No **121.** y^{5x} **123.** a^{4t} **125.** 1
127. $>$ **129.** $<$ **131.** $-\frac{1}{10{,}000}$ **133.** No; for example, $(3 + 4)^2 = 49$, but $3^2 + 4^2 = 25$.

Calculator Corner, p. 868

1. 1.3545×10^{-4} **2.** 3.2×10^5 **3.** 3×10^{-6} **4.** 8×10^{-26}

Exercise Set 12.2, p. 870

RC1. multiply **RC2.** nth **RC3.** right **RC4.** positive
1. 2^6 **3.** $\frac{1}{5^6}$ **5.** x^{12} **7.** $\frac{1}{a^{18}}$ **9.** t^{18} **11.** $\frac{1}{t^{12}}$ **13.** x^8
15. a^3b^3 **17.** $\frac{1}{a^3b^3}$ **19.** $\frac{1}{m^3n^6}$ **21.** $16x^6$ **23.** $\frac{9}{x^8}$
25. $\frac{1}{x^{12}y^{15}}$ **27.** $x^{24}y^8$ **29.** $\frac{a^{10}}{b^{35}}$ **31.** $\frac{25t^6}{r^8}$ **33.** $\frac{b^{21}}{a^{15}c^6}$
35. $\frac{9x^6}{y^{16}z^6}$ **37.** $\frac{16x^6}{y^4}$ **39.** $a^{12}b^8$ **41.** $\frac{y^6}{4}$ **43.** $\frac{a^8}{b^{12}}$ **45.** $\frac{8}{y^6}$
47. $49x^6$ **49.** $\frac{x^6y^3}{z^3}$ **51.** $\frac{c^2d^6}{a^4b^2}$ **53.** 2.8×10^{10} **55.** 9.07×10^{17}
57. 3.04×10^{-6} **59.** 1.8×10^{-8} **61.** 10^{11} **63.** 4.19854×10^8
65. 6.8×10^{-7} **67.** 87,400,000 **69.** 0.00000005704
71. 10,000,000 **73.** 0.00001 **75.** 6×10^9 **77.** 3.38×10^4
79. 8.1477×10^{-13} **81.** 2.5×10^{13} **83.** 5.0×10^{-4}
85. 3.0×10^{-21} **87.** Approximately 1.325×10^{14} ft^3
89. 1×10^{22} **91.** The mass of Jupiter is 3.18×10^2 times the mass of Earth. **93.** 1.25×10^9 bytes **95.** 3.1×10^2 sheets
97. 4.375×10^2 days
99. **100.**

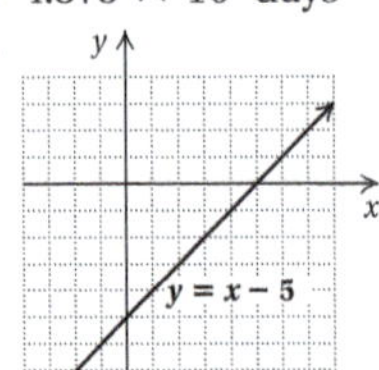

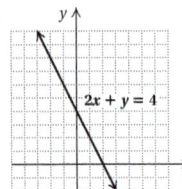

101. 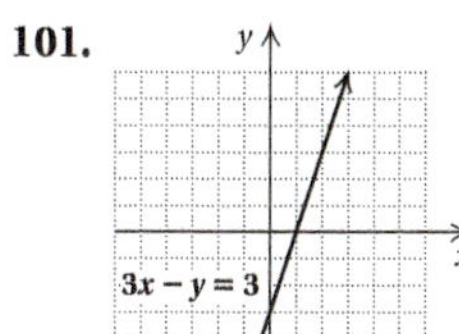

102. 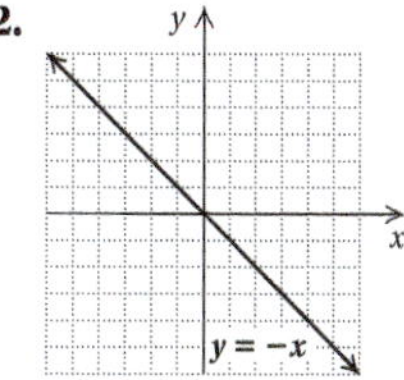

103. 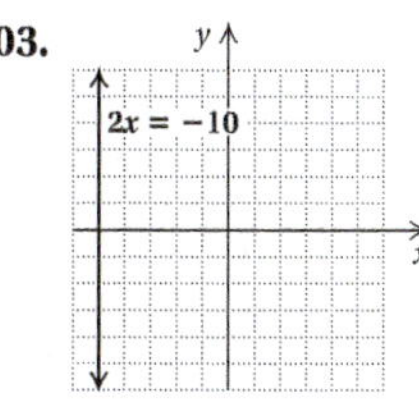

104. 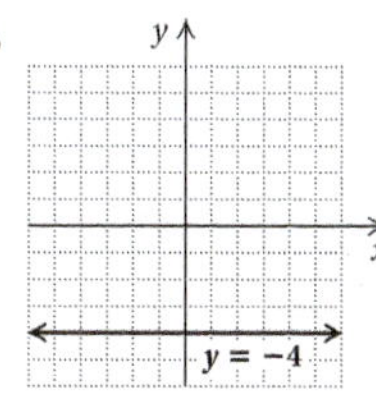

105. 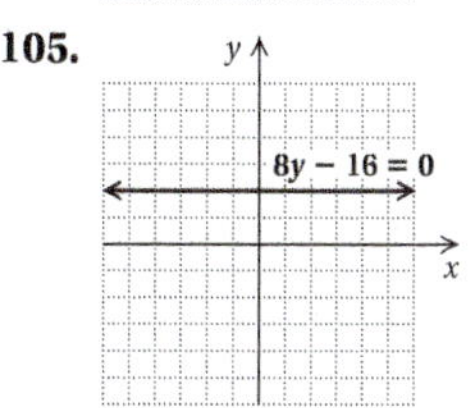

106. 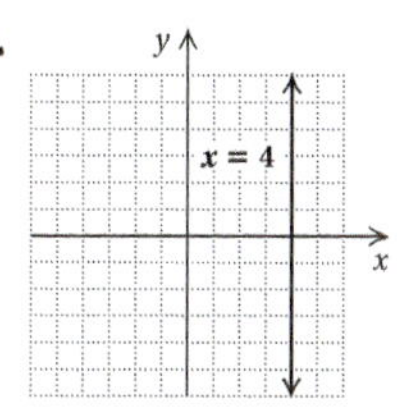

107. 2.478125×10^{-1} **109.** $\frac{1}{5}$ **111.** 3^{11} **113.** 7 **115.** $\frac{1}{0.4}$, or 2.5 **117.** False **119.** False **121.** True

Calculator Corner, p. 878

1. 3; 2.25; −27 **2.** 44; 0; 9.28

Exercise Set 12.3, p. 883

RC1. (b) **RC2.** (f) **RC3.** (c) **RC4.** (d) **RC5.** (a) **RC6.** (e)
1. −18; 7 **3.** 19; 14 **5.** −12; −7 **7.** $\frac{13}{3}$; 5 **9.** 9; 1 **11.** 56; −2 **13.** 1112 ft **15.** 55 oranges **17.** −4, 4, 5, 2.75, 1 **19. (a)** 5.67 kW; **(b)** left to the student **21.** 9 words **23.** 6 **25.** 15 **27.** $2, -3x, x^2$ **29.** $-2x^4, \frac{1}{3}x^3, -x, 3$ **31.** Trinomial **33.** None of these **35.** Binomial **37.** Monomial **39.** −3, 6 **41.** $5, \frac{3}{4}, 3$ **43.** −5, 6, −2.7, 1, −2 **45.** 1, 0; 1 **47.** 2, 1, 0; 2 **49.** 3, 2, 1, 0; 3 **51.** 2, 1, 6, 4; 6
53.

Term	Coefficient	Degree of the Term	Degree of the Polynomial
$-7x^4$	−7	4	
$6x^3$	6	3	
$-x^2$	−1	2	4
$8x$	8	1	
−2	−2	0	

55. $6x^2$ and $-3x^2$ **57.** $2x^4$ and $-3x^4$; $5x$ and $-7x$ **59.** $3x^5$ and $14x^5$; $-7x$ and $-2x$; 8 and −9 **61.** $-3x$ **63.** $-8x$ **65.** $11x^3 + 4$ **67.** $x^3 - x$ **69.** $4b^5$ **71.** $\frac{3}{4}x^5 - 2x - 42$ **73.** x^4 **75.** $\frac{15}{16}x^3 - \frac{7}{6}x^2$ **77.** $x^5 + 6x^3 + 2x^2 + x + 1$ **79.** $15y^9 + 7y^8 + 5y^3 - y^2 + y$ **81.** $x^6 + x^4$ **83.** $13x^3 - 9x + 8$ **85.** $-5x^2 + 9x$ **87.** $12x^4 - 2x + \frac{1}{4}$ **89.** x^2, x **91.** x^3, x^2, x^0 **93.** None missing **95.** $x^3 + 0x^2 + 0x - 27$; $x^3 \qquad - 27$ **97.** $x^4 + 0x^3 + 0x^2 - x + 0x^0$; $x^4 \qquad - x$ **99.** None missing **101.** −19 **102.** −1 **103.** −2.25 **104.** $-\frac{3}{8}$ **105.** −19 **106.** $-\frac{17}{24}$ **107.** $\frac{5}{8}$ **108.** −2.6 **109.** 18 **110.** $-\frac{1}{3}$ **111.** −0.6 **112.** −24 **113.** 20 **114.** $-\frac{8}{5}$ **115.** 0 **116.** Not defined **117.** $10x^6 + 52x^5$ **119.** $4x^5 - 3x^3 + x^2 - 7x$; answers may vary **121.** $x^3 - 2x^2 - 6x + 3$ **123.** −4, 4, 5, 2.75, 1 **125.** 9

Exercise Set 12.4, p. 892

RC1. False **RC2.** True **RC3.** False **RC4.** False
1. $-x + 5$ **3.** $x^2 - \frac{11}{2}x - 1$ **5.** $2x^2$ **7.** $5x^2 + 3x - 30$ **9.** $-2.2x^3 - 0.2x^2 - 3.8x + 23$ **11.** $6 + 12x^2$ **13.** $-\frac{1}{2}x^4 + \frac{2}{3}x^3 + x^2$ **15.** $0.01x^5 + x^4 - 0.2x^3 + 0.2x + 0.06$ **17.** $9x^8 + 8x^7 - 6x^4 + 8x^2 + 4$ **19.** $1.05x^4 + 0.36x^3 + 14.22x^2 + x + 0.97$ **21.** $5x$ **23.** $x^2 - \frac{3}{2}x + 2$ **25.** $-12x^4 + 3x^3 - 3$ **27.** $-3x + 7$ **29.** $-4x^2 + 3x - 2$ **31.** $4x^4 - 6x^2 - \frac{3}{4}x + 8$ **33.** $7x - 1$ **35.** $-x^2 - 7x + 5$ **37.** −18 **39.** $6x^4 + 3x^3 - 4x^2 + 3x - 4$ **41.** $4.6x^3 + 9.2x^2 - 3.8x - 23$ **43.** $\frac{3}{4}x^3 - \frac{1}{2}x$ **45.** $0.06x^3 - 0.05x^2 + 0.01x + 1$ **47.** $3x + 6$ **49.** $11x^4 + 12x^3 - 9x^2 - 8x - 9$ **51.** $x^4 - x^3 + x^2 - x$ **53.** $\frac{23}{2}a + 12$ **55.** $5x^2 + 4x$ **57.** $(r + 11)(r + 9)$; $9r + 99 + 11r + r^2$, or $r^2 + 20r + 99$ **59.** $(x + 3)(x + 3)$, or $(x + 3)^2$; $x^2 + 3x + 9 + 3x$, or $x^2 + 6x + 9$ **61.** $\pi r^2 - 25\pi$ **63.** $18z - 64$ **65.** 6 **66.** −19 **67.** $-\frac{7}{22}$ **68.** 5 **69.** 5 **70.** 1 **71.** $\frac{39}{2}$ **72.** $\frac{37}{2}$ **73.** $\{x|x \geq -10\}$ **74.** $\{x|x < 0\}$ **75.** $20w + 42$ **77.** $2x^2 + 20x$ **79.** $y^2 - 4y + 4$ **81.** $12y^2 - 23y + 21$ **83.** $-3y^4 - y^3 + 5y - 2$

Mid-Chapter Review: Chapter 12, p. 897

1. True **2.** False **3.** False **4.** True **5.** $4w^3 + 6w - 8w^3 - 3w = (4 - 8)w^3 + (6 - 3)w = -4w^3 + 3w$ **6.** $(3y^4 - y^2 + 11) - (y^4 - 4y^2 + 5) = 3y^4 - y^2 + 11 - y^4 + 4y^2 - 5 = 2y^4 + 3y^2 + 6$ **7.** z **8.** 1 **9.** −32 **10.** 1 **11.** 5^7 **12.** $(3a)^9$ **13.** $\frac{1}{x^3}$ **14.** 1 **15.** 7^4 **16.** $\frac{1}{x^2}$ **17.** w^8 **18.** $\frac{1}{y^4}$ **19.** 3^{15} **20.** $\frac{x^{18}}{y^{12}}$ **21.** $\frac{a^{24}}{5^6}$ **22.** $\frac{x^2z^4}{4y^6}$ **23.** 2.543×10^7 **24.** 1.2×10^{-4} **25.** 0.000036 **26.** 144,000,000 **27.** 6×10^3 **28.** 5×10^{-7} **29.** 16; 1 **30.** −16; 9 **31.** $-2x^5 - 5x^2 + 4x + 2$ **32.** $8x^6 + 2x^3 - 8x^2$ **33.** 3, 1, 0; 3 **34.** 1, 4, 6; 6 **35.** Binomial **36.** Trinomial **37.** $8x^2 + 5$ **38.** $5x^3 - 2x^2 + 2x - 11$ **39.** $-4x - 10$ **40.** $-0.4x^2 - 3.4x + 9$ **41.** $3y + 3y^2$ **42.** The area of the smaller square is x^2, and the area of the larger square is $(3x)^2$, or $9x^2$, so the area of the larger square is nine times the area of the smaller square. **43.** The volume of the smaller cube is x^3, and the volume of the larger cube is $(2x)^3$, or $8x^3$, so the volume of the larger cube is eight times the volume of the smaller cube. **44.** Exponents are added when powers with like bases are multiplied. Exponents are multiplied when a power is raised to a power. **45.** $3^{-29} = \frac{1}{3^{29}}$ and $2^{-29} = \frac{1}{2^{29}}$. Since $3^{29} > 2^{29}$, we have $\frac{1}{3^{29}} < \frac{1}{2^{29}}$. **46.** It is better to evaluate a polynomial after like terms have been collected, because there are fewer terms to evaluate. **47.** Yes; consider the following: $(x^2 + 4) + (4x - 7) = x^2 + 4x - 3$.

Calculator Corner, p. 902

1. Correct **2.** Correct **3.** Not correct **4.** Not correct

Exercise Set 12.5, p. 903

RC1. (a) **RC2.** (a) **RC3.** (d) **RC4.** (e)
1. $40x^2$ **3.** x^3 **5.** $32x^8$ **7.** $0.03x^{11}$ **9.** $\frac{1}{15}x^4$ **11.** 0 **13.** $-24x^{11}$ **15.** $-2x^2 + 10x$ **17.** $-5x^2 + 5x$ **19.** $x^5 + x^2$ **21.** $6x^3 - 18x^2 + 3x$ **23.** $-6x^4 - 6x^3$ **25.** $18y^6 + 24y^5$ **27.** $x^2 + 9x + 18$ **29.** $x^2 + 3x - 10$ **31.** $x^2 + 3x - 4$ **33.** $x^2 - 7x + 12$ **35.** $x^2 - 9$ **37.** $x^2 - 16$ **39.** $3x^2 + 11x + 10$ **41.** $25 - 15x + 2x^2$ **43.** $4x^2 + 20x + 25$ **45.** $x^2 - 6x + 9$ **47.** $x^2 - \frac{21}{10}x - 1$ **49.** $x^2 + 2.4x - 10.81$ **51.** $(x + 2)(x + 6)$, or $x^2 + 8x + 12$ **53.** $(x + 1)(x + 6)$, or $x^2 + 7x + 6$

55. 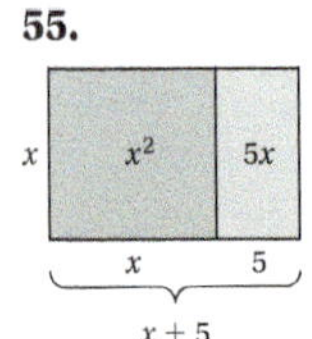

57. 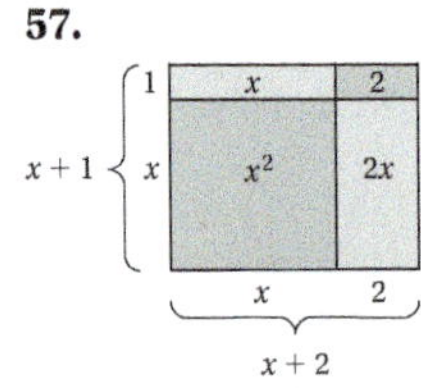

59. 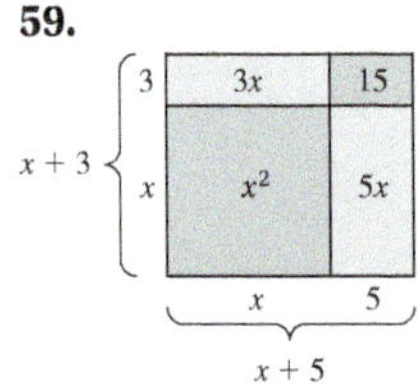

61. $x^3 - 1$ **63.** $4x^3 + 14x^2 + 8x + 1$
65. $3y^4 - 6y^3 - 7y^2 + 18y - 6$ **67.** $x^6 + 2x^5 - x^3$
69. $-10x^5 - 9x^4 + 7x^3 + 2x^2 - x$
71. $-1 - 2x - x^2 + x^4$ **73.** $6t^4 + t^3 - 16t^2 - 7t + 4$
75. $x^9 - x^5 + 2x^3 - x$ **77.** $x^4 + 8x^3 + 12x^2 + 9x + 4$
79. $2x^4 - 5x^3 + 5x^2 - \frac{19}{10}x + \frac{1}{5}$ **81.** $-x + 3$ **82.** 47 **83.** 96
84. 32 **85.** $3(5x - 6y + 4)$ **86.** $4(4x - 6y + 9)$
87. $-3(3x + 15y - 5)$ **88.** $100(x - y + 10a)$ **89.** $75y^2 - 45y$
91. 5 **93.** $(x^3 + 2x^2 - 210)\ \text{m}^3$ **95.** 0 **97.** 0

Visualizing for Success, p. 912

1. E, F **2.** B, O **3.** K, S **4.** G, R **5.** D, M **6.** J, P
7. C, L **8.** N, Q **9.** A, H **10.** I, T

Exercise Set 12.6, p. 913

RC1. outside; last **RC2.** descending **RC3.** difference
RC4. square; binomial **RC5.** binomials **RC6.** difference
1. $x^3 + x^2 + 3x + 3$ **3.** $x^4 + x^3 + 2x + 2$ **5.** $y^2 - y - 6$
7. $9x^2 + 12x + 4$ **9.** $5x^2 + 4x - 12$ **11.** $9t^2 - 1$
13. $4x^2 - 6x + 2$ **15.** $p^2 - \frac{1}{16}$ **17.** $x^2 - 0.01$
19. $2x^3 + 2x^2 + 6x + 6$ **21.** $-2x^2 - 11x + 6$
23. $a^2 + 14a + 49$ **25.** $1 - x - 6x^2$ **27.** $\frac{9}{64}y^2 - \frac{5}{8}y + \frac{25}{36}$
29. $x^5 + 3x^3 - x^2 - 3$ **31.** $3x^6 - 2x^4 - 6x^2 + 4$
33. $13.16x^2 + 18.99x - 13.95$ **35.** $6x^7 + 18x^5 + 4x^2 + 12$
37. $4x^3 - 12x^2 + 3x - 9$ **39.** $4y^6 + 4y^5 + y^4 + y^3$
41. $x^2 - 16$ **43.** $4x^2 - 1$ **45.** $25m^2 - 4$ **47.** $4x^4 - 9$
49. $9x^8 - 16$ **51.** $x^{12} - x^4$ **53.** $x^8 - 9x^2$ **55.** $x^{24} - 9$
57. $4y^{16} - 9$ **59.** $\frac{25}{64}x^2 - 18.49$ **61.** $x^2 + 4x + 4$
63. $9x^4 + 6x^2 + 1$ **65.** $a^2 - a + \frac{1}{4}$ **67.** $9 + 6x + x^2$
69. $x^4 + 2x^2 + 1$ **71.** $4 - 12x^4 + 9x^8$ **73.** $25 + 60t^2 + 36t^4$
75. $x^2 - \frac{5}{4}x + \frac{25}{64}$ **77.** $9 - 12x^3 + 4x^6$ **79.** $4x^3 + 24x^2 - 12x$
81. $4x^4 - 2x^2 + \frac{1}{4}$ **83.** $9p^2 - 1$ **85.** $15t^5 - 3t^4 + 3t^3$
87. $36x^8 + 48x^4 + 16$ **89.** $12x^3 + 8x^2 + 15x + 10$
91. $64 - 96x^4 + 36x^8$ **93.** $t^3 - 1$ **95.** 25; 49 **97.** 56; 16
99. $a^2 + 2a + 1$ **101.** $t^2 + 10t + 24$ **103.** $\frac{28}{27}$ **104.** $-\frac{41}{7}$
105. $\frac{27}{4}$ **106.** $y = \dfrac{3x - 12}{2}$, or $y = \frac{3}{2}x - 6$ **107.** $b = \dfrac{C + r}{a}$
108. $a = \dfrac{5d + 4}{3}$, or $a = \frac{5}{3}d + \frac{4}{3}$ **109.** $30x^3 + 35x^2 - 15x$
111. $a^4 - 50a^2 + 625$ **113.** $81t^{16} - 72t^8 + 16$ **115.** -7
117. First row: 90, −432, −63; second row: 7, −18, −36, −14, 12, −6, −21, −11; third row: 9, −2, −2, 10, −8, −8, −8, −10, 21; fourth row: −19, −6 **119.** Yes **121.** No

Exercise Set 12.7, p. 921

RC1. True **RC2.** False **RC3.** False **RC4.** False
1. −1 **3.** −15 **5.** 240 **7.** −145 **9.** 3.715 L
11. 2322 calories **13.** 44.46 in² **15.** 73.005 in²
17. Coefficients: 1, −2, 3, −5; degrees: 4, 2, 2, 0; 4
19. Coefficients: 17, −3, −7; degrees: 5, 5, 0; 5 **21.** $-a - 2b$
23. $3x^2y - 2xy^2 + x^2$ **25.** $20au + 10av$ **27.** $8u^2v - 5uv^2$
29. $x^2 - 4xy + 3y^2$ **31.** $3r + 7$
33. $-b^2a^3 - 3b^3a^2 + 5ba + 3$ **35.** $ab^2 - a^2b$ **37.** $2ab - 2$
39. $-2a + 10b - 5c + 8d$ **41.** $6z^2 + 7zu - 3u^2$
43. $a^4b^2 - 7a^2b + 10$ **45.** $a^6 - b^2c^2$
47. $y^6x + y^4x + y^4 + 2y^2 + 1$ **49.** $12x^2y^2 + 2xy - 2$
51. $12 - c^2d^2 - c^4d^4$ **53.** $m^3 + m^2n - mn^2 - n^3$
55. $x^9y^9 - x^6y^6 + x^5y^5 - x^2y^2$ **57.** $x^2 + 2xh + h^2$
59. $9a^2 + 12ab + 4b^2$ **61.** $r^6t^4 - 8r^3t^2 + 16$
63. $p^8 + 2m^2n^2p^4 + m^4n^4$ **65.** $3a^3 - 12a^2b + 12ab^2$
67. $m^2 + 2mn + n^2 - 6m - 6n + 9$ **69.** $a^2 - b^2$
71. $4a^2 - b^2$ **73.** $c^4 - d^2$ **75.** $a^2b^2 - c^2d^4$
77. $x^2 + 2xy + y^2 - 9$ **79.** $x^2 - y^2 - 2yz - z^2$
81. $a^2 + 2ab + b^2 - c^2$ **83.** $3x^4 - 7x^2y + 3x^2 - 20y^2 + 22y - 6$
85. IV **86.** III **87.** I **88.** II **89.** 39 **90.** 1.125
91. $<$ **92.** −3 **93.** $4xy - 4y^2$ **95.** $2xy + \pi x^2$
97. $2\pi nh + 2\pi mh + 2\pi n^2 - 2\pi m^2$ **99.** 16 gal
101. $12,351.94

Exercise Set 12.8, p. 930

RC1. subtract; divide **RC2.** divide **RC3.** multiply; subtract
RC4. multiply; add
1. $3x^4$ **3.** $5x$ **5.** $18x^3$ **7.** $4a^3b$ **9.** $3x^4 - \frac{1}{2}x^3 + \frac{1}{8}x^2 - 2$
11. $1 - 2u - u^4$ **13.** $5t^2 + 8t - 2$ **15.** $-4x^4 + 4x^2 + 1$
17. $6x^2 - 10x + \frac{3}{2}$ **19.** $9x^2 - \frac{5}{2}x + 1$ **21.** $6x^2 + 13x + 4$
23. $3rs + r - 2s$ **25.** $x + 2$ **27.** $x - 5 + \dfrac{-50}{x - 5}$
29. $x - 2 + \dfrac{-2}{x + 6}$ **31.** $x - 3$ **33.** $x^4 - x^3 + x^2 - x + 1$
35. $2x^2 - 7x + 4$ **37.** $x^3 - 6$ **39.** $t^2 + 1$
41. $y^2 - 3y + 1 + \dfrac{-5}{y + 2}$ **43.** $3x^2 + x + 2 + \dfrac{10}{5x + 1}$
45. $6y^2 - 5 + \dfrac{-6}{2y + 7}$ **47.** −1 **48.** $\frac{10}{3}$ **49.** $\frac{23}{19}$
50. $\{r \mid r \le -15\}$ **51.** 140% **52.** $\{x \mid x \ge 95\}$
53. 25,543.75 ft² **54.** 228, 229 **55.** $x^2 + 5$
57. $a + 3 + \dfrac{5}{5a^2 - 7a - 2}$ **59.** $2x^2 + x - 3$
61. $a^5 + a^4b + a^3b^2 + a^2b^3 + ab^4 + b^5$ **63.** −5 **65.** 1

Summary and Review: Chapter 12, p. 933

Vocabulary Reinforcement

1. exponent **2.** product **3.** monomial **4.** trinomial
5. quotient **6.** descending **7.** degree **8.** scientific

Concept Reinforcement

1. True **2.** False **3.** False **4.** True

Study Guide

1. z^8 **2.** a^2b^6 **3.** $\dfrac{y^6}{27x^{12}z^9}$ **4.** 7.63×10^5 **5.** 0.0003
6. 6×10^4 **7.** $2x^4 - 4x^2 - 3$ **8.** $3x^4 + x^3 - 2x^2 + 2$
9. $x^6 - 6x^4 + 11x^2 - 6$ **10.** $2y^2 + 11y + 12$ **11.** $x^2 - 25$
12. $9w^2 + 24w + 16$ **13.** $-2a^3b^2 - 5a^2b + ab^2 - 2ab$
14. $y^2 - 4y + \frac{8}{5}$ **15.** $x - 9 + \dfrac{48}{x + 5}$

Review Exercises

1. $\dfrac{1}{7^2}$ **2.** y^{11} **3.** $(3x)^{14}$ **4.** t^8 **5.** 4^3 **6.** $\dfrac{1}{a^3}$ **7.** 1
8. $9t^8$ **9.** $36x^8$ **10.** $\dfrac{y^3}{8x^3}$ **11.** t^{-5} **12.** $\dfrac{1}{y^4}$ **13.** 3.28×10^{-5}
14. 8,300,000 **15.** 2.09×10^4 **16.** 5.12×10^{-5}
17. 1.564×10^{10} slices **18.** 10 **19.** $-4y^5, 7y^2, -3y, -2$
20. x^2, x^0 **21.** 3, 2, 1, 0; 3 **22.** Binomial **23.** None of these
24. Monomial **25.** $-2x^2 - 3x + 2$ **26.** $10x^4 - 7x^2 - x - \frac{1}{2}$
27. $x^5 - 2x^4 + 6x^3 + 3x^2 - 9$ **28.** $-2x^5 - 6x^4 - 2x^3 - 2x^2 + 2$
29. $2x^2 - 4x$ **30.** $x^5 - 3x^3 - x^2 + 8$ **31.** Perimeter: $4w + 6$; area: $w^2 + 3w$ **32.** $(t + 3)(t + 4)$, $t^2 + 7t + 12$
33. $x^2 + \frac{7}{6}x + \frac{1}{3}$ **34.** $49x^2 + 14x + 1$ **35.** $12x^3 - 23x^2 + 13x - 2$

36. $9x^4 - 16$ **37.** $15x^7 - 40x^6 + 50x^5 + 10x^4$ **38.** $x^2 - 3x - 28$ **39.** $9y^4 - 12y^3 + 4y^2$ **40.** $2t^4 - 11t^2 - 21$ **41.** 49 **42.** Coefficients: 1, −7, 9, −8; degrees: 6, 2, 2, 0; 6 **43.** $-y + 9w - 5$ **44.** $m^6 - 2m^2n + 2m^2n^2 + 8n^2m - 6m^3$ **45.** $-9xy - 2y^2$ **46.** $11x^3y^2 - 8x^2y - 6x^2 - 6x + 6$ **47.** $p^3 - q^3$ **48.** $9a^8 - 2a^4b^3 + \frac{1}{9}b^6$ **49.** $5x^2 - \frac{1}{2}x + 3$ **50.** $3x^2 - 7x + 4 + \dfrac{1}{2x + 3}$ **51.** 0, 3.75, −3.75, 0 **52.** B **53.** D **54.** $\frac{1}{2}x^2 - \frac{1}{2}y^2$ **55.** $400 - 4a^2$ **56.** $-28x^8$ **57.** $\frac{94}{13}$ **58.** $x^4 + x^3 + x^2 + x + 1$ **59.** 80 ft by 40 ft

Understanding Through Discussion and Writing

1. 578.6×10^{-7} is not in scientific notation because 578.6 is not a number greater than or equal to 1 and less than 10.
2. When evaluating polynomials, it is essential to know the order in which the operations are to be performed.
3. We label the figure as shown.

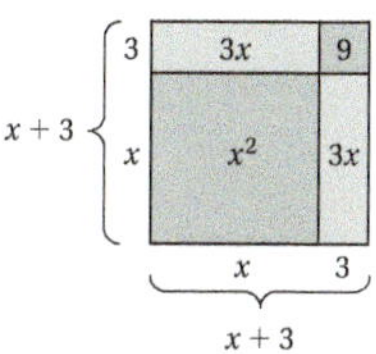

Then we see that the area of the figure is $(x + 3)^2$, or $x^2 + 3x + 3x + 9 \neq x^2 + 9$. **4.** Emma did not divide *each* term of the polynomial by the divisor. The first term was divided by $3x$, but the second was not. Multiplying Emma's "quotient" by the divisor $3x$, we get $12x^3 - 18x^2 \neq 12x^3 - 6x$. This should convince her that a mistake has been made. **5.** Yes; for example, $(x^2 + xy + 1) + (3x - xy + 2) = x^2 + 3x + 3$. **6.** Yes; consider $a + b + c + d$. This is a polynomial in 4 variables but it has degree 1.

Test: Chapter 12, p. 939

1. [12.1d, f] $\dfrac{1}{6^5}$ **2.** [12.1d] x^9 **3.** [12.1d] $(4a)^{11}$ **4.** [12.1e] 3^3 **5.** [12.1e, f] $\dfrac{1}{x^5}$ **6.** [12.1b, e] 1 **7.** [12.2a] x^6 **8.** [12.2a, b] $-27y^6$ **9.** [12.2a, b] $16a^{12}b^4$ **10.** [12.2b] $\dfrac{a^3b^3}{c^3}$ **11.** [12.1d], [12.2a, b] $-216x^{21}$ **12.** [12.1d], [12.2a, b] $-24x^{21}$ **13.** [12.1d], [12.2a, b] $162x^{10}$ **14.** [12.1d], [12.2a, b] $324x^{10}$ **15.** [12.1f] $\dfrac{1}{5^3}$ **16.** [12.1f] y^{-8} **17.** [12.2c] 3.9×10^9 **18.** [12.2c] 0.00000005 **19.** [12.2d] 1.75×10^{17} **20.** [12.2d] 1.296×10^{22} **21.** [12.2e] The mass of Saturn is 9.5×10 times the mass of Earth. **22.** [12.3a] −43 **23.** [12.3c] $\frac{1}{3}$, −1, 7 **24.** [12.3c] 3, 0, 1, 6; 6 **25.** [12.3b] Binomial **26.** [12.3d] $5a^2 - 6$ **27.** [12.3d] $\frac{7}{4}y^2 - 4y$ **28.** [12.3e] $x^5 + 2x^3 + 4x^2 - 8x + 3$ **29.** [12.4a] $4x^5 + x^4 + 2x^3 - 8x^2 + 2x - 7$ **30.** [12.4a] $5x^4 + 5x^2 + x + 5$ **31.** [12.4c] $-4x^4 + x^3 - 8x - 3$ **32.** [12.4c] $-x^5 + 0.7x^3 - 0.8x^2 - 21$ **33.** [12.5b] $-12x^4 + 9x^3 + 15x^2$ **34.** [12.6c] $x^2 - \frac{2}{3}x + \frac{1}{9}$ **35.** [12.6b] $9x^2 - 100$ **36.** [12.6a] $3b^2 - 4b - 15$ **37.** [12.6a] $x^{14} - 4x^8 + 4x^6 - 16$ **38.** [12.6a] $48 + 34y - 5y^2$ **39.** [12.5d] $6x^3 - 7x^2 - 11x - 3$ **40.** [12.6c] $25t^2 + 20t + 4$ **41.** [12.7c] $-5x^3y - y^3 + xy^3 - x^2y^2 + 19$ **42.** [12.7e] $8a^2b^2 + 6ab - 4b^3 + 6ab^2 + ab^3$ **43.** [12.7f] $9x^{10} - 16y^{10}$ **44.** [12.8a] $4x^2 + 3x - 5$ **45.** [12.8b] $2x^2 - 4x - 2 + \dfrac{17}{3x + 2}$ **46.** [12.3a] 3, 1.5, −3.5, −5, −5.25 **47.** [12.4d] $(t + 2)(t + 2)$, $t^2 + 4t + 4$ **48.** [12.4d] B **49.** [12.5b], [12.6a] $V = l^3 - 3l^2 + 2l$ **50.** [12.3b], [12.6b, c] $-\frac{61}{12}$

CHAPTER 13

Exercise Set 13.1, p. 947

RC1. (b) **RC2.** (c) **RC3.** (d) **RC4.** (a) **1.** 6 **3.** 24 **5.** 1 **7.** x **9.** x^2 **11.** 2 **13.** $17xy$ **15.** x **17.** x^2y^2 **19.** $x(x - 6)$ **21.** $2x(x + 3)$ **23.** $x^2(x + 6)$ **25.** $8x^2(x^2 - 3)$ **27.** $2(x^2 + x - 4)$ **29.** $17xy(x^4y^2 + 2x^2y + 3)$ **31.** $x^2(6x^2 - 10x + 3)$ **33.** $x^2y^2(x^3y^3 + x^2y + xy - 1)$ **35.** $2x^3(x^4 - x^3 - 32x^2 + 2)$ **37.** $0.8x(2x^3 - 3x^2 + 4x + 8)$ **39.** $\frac{1}{3}x^3(5x^3 + 4x^2 + x + 1)$ **41.** $(x + 3)(x^2 + 2)$ **43.** $(3z - 1)(4z^2 + 7)$ **45.** $(3x + 2)(2x^2 + 1)$ **47.** $(2a - 7)(5a^3 - 1)$ **49.** $(x + 3)(x^2 + 2)$ **51.** $(x + 3)(2x^2 + 1)$ **53.** $(2x - 3)(4x^2 + 3)$ **55.** $(3p - 4)(4p^2 + 1)$ **57.** $(x - 1)(5x^2 - 1)$ **59.** $(x + 8)(x^2 - 3)$ **61.** $(x - 4)(2x^2 - 9)$ **63.** $y^2 + 12y + 35$ **64.** $y^2 + 14y + 49$ **65.** $y^2 - 49$ **66.** $y^2 - 14y + 49$ **67.** $16x^3 - 48x^2 + 8x$ **68.** $28w^2 - 53w - 66$ **69.** $49w^2 + 84w + 36$ **70.** $16w^2 - 88w + 121$ **71.** $16w^2 - 121$ **72.** $y^3 - 3y^2 + 5y$ **73.** $6x^2 + 11xy - 35y^2$ **74.** $25x^2 - 10xt + t^2$ **75.** $(2x^2 + 3)(2x^3 + 3)$ **77.** $(x^5 + 1)(x^7 + 1)$ **79.** Not factorable by grouping

Exercise Set 13.2, p. 955

RC1. True **RC2.** True **RC3.** True **RC4.** False

1.

Pairs of Factors	Sums of Factors
1, 15	16
−1, −15	−16
3, 5	8
−3, −5	−8

$(x + 3)(x + 5)$

3.

Pairs of Factors	Sums of Factors
1, 12	13
−1, −12	−13
2, 6	8
−2, −6	−8
3, 4	7
−3, −4	−7

$(x + 3)(x + 4)$

5.

Pairs of Factors	Sums of Factors
1, 9	10
−1, −9	−10
3, 3	6
−3, −3	−6

$(x - 3)^2$

7.

Pairs of Factors	Sums of Factors
−1, 14	13
1, −14	−13
−2, 7	5
2, −7	−5

$(x + 2)(x - 7)$

9.

Pairs of Factors	Sums of Factors
1, 4	5
−1, −4	−5
2, 2	4
−2, −2	−4

$(b + 1)(b + 4)$

11.

Pairs of Factors	Sums of Factors
−1, 18	17
1, −18	−17
−2, 9	7
2, −9	−7
−3, 6	3
3, −6	−3

$(t - 3)(t + 6)$
13. $(d - 2)(d - 5)$ **15.** $(y - 1)(y - 10)$ **17.** Prime
19. $(x - 9)(x + 2)$ **21.** $x(x - 8)(x + 2)$
23. $y(y - 9)(y + 5)$ **25.** $(x - 11)(x + 9)$
27. $(c^2 + 8)(c^2 - 7)$ **29.** $(a^2 + 7)(a^2 - 5)$
31. $(x - 6)(x + 7)$ **33.** Prime **35.** $(x + 10)^2$
37. $2z(z - 4)(z + 3)$ **39.** $3t^2(t^2 + t + 1)$
41. $x^2(x - 25)(x + 4)$ **43.** $(x - 24)(x + 3)$
45. $(x - 9)(x - 16)$ **47.** $(a + 12)(a - 11)$ **49.** $3(t + 1)^2$
51. $w^2(w - 4)^2$ **53.** $-1(x - 10)(x + 3)$, or $(-x + 10)(x + 3)$, or $(x - 10)(-x - 3)$
55. $-1(a - 2)(a + 12)$, or $(-a + 2)(a + 12)$, or $(a - 2)(-a - 12)$ **57.** $(x - 15)(x - 8)$
59. $-1(x + 12)(x - 9)$, or $(-x - 12)(x - 9)$, or $(x + 12)(-x + 9)$ **61.** $(y - 0.4)(y + 0.2)$
63. $(p + 5q)(p - 2q)$ **65.** $-1(t + 14)(t - 6)$, or $(-t - 14)(t - 6)$, or $(t + 14)(-t + 6)$ **67.** $(m + 4n)(m + n)$
69. $(s + 3t)(s - 5t)$ **71.** $6a^8(a + 2)(a - 7)$
73. $-\frac{7}{2}$ **74.** 12 **75.** $\frac{1}{3}$ **76.** -1 **77.** $\frac{5}{4}$
78. No solution **79.** $\{x \mid x > -24\}$ **80.** $\{x \mid x \le \frac{14}{5}\}$
81. $\{x \mid x \ge \frac{77}{17}\}$ **82.** $p = 2A - w$ **83.** $x = \dfrac{y - b}{m}$
84. $c = a + r - 2$ **85.** 73.5 million **86.** 100°, 25°, 55°
87. 15, −15, 27, −27, 51, −51 **89.** $\left(x + \frac{1}{4}\right)\left(x - \frac{3}{4}\right)$
91. $\left(x + \frac{1}{3}\right)^2$ **93.** $(x + 5)\left(x - \frac{5}{7}\right)$
95. $(b^n + 5)(b^n + 2)$ **97.** $2x^2(4 - \pi)$

Calculator Corner, p. 961

1. Correct **2.** Correct **3.** Not correct **4.** Not correct

Exercise Set 13.3, p. 965

RC1. True **RC2.** True **RC3.** False **RC4.** False
1. $(2x + 1)(x - 4)$ **3.** $(5x + 9)(x - 2)$ **5.** $(3x + 1)(2x + 7)$
7. $(3x + 1)(x + 1)$ **9.** $(2x - 3)(2x + 5)$ **11.** $(2x + 1)(x - 1)$
13. $(3x - 2)(3x + 8)$ **15.** $(3x + 1)(x - 2)$
17. $(3x + 4)(4x + 5)$ **19.** $(7x - 1)(2x + 3)$
21. $(3x + 2)(3x + 4)$ **23.** $(3x - 7)^2$, or $(7 - 3x)^2$
25. $(24x - 1)(x + 2)$ **27.** $(5x - 11)(7x + 4)$
29. $-2(x - 5)(x + 2)$, or $2(-x + 5)(x + 2)$, or $2(x - 5)(-x - 2)$
31. $4(3x - 2)(x + 3)$ **33.** $6(5x - 9)(x + 1)$
35. $2(3y + 5)(y - 1)$ **37.** $(3x - 1)(x - 1)$
39. $4(3x + 2)(x - 3)$ **41.** $(2x + 1)(x - 1)$
43. $(3x + 2)(3x - 8)$ **45.** $5(3x + 1)(x - 2)$
47. $p(3p + 4)(4p + 5)$ **49.** $-1(3x + 2)(3x - 8)$, or $(-3x - 2)(3x - 8)$, or $(3x + 2)(-3x + 8)$
51. $-1(5x - 3)(3x - 2)$, or $(-5x + 3)(3x - 2)$, or $(5x - 3)(-3x + 2)$ **53.** $x^2(7x - 1)(2x + 3)$
55. $3x(8x - 1)(7x - 1)$ **57.** $(5x^2 - 3)(3x^2 - 2)$ **59.** $(5t + 8)^2$
61. $2x(3x + 5)(x - 1)$ **63.** Prime **65.** Prime
67. $(4m + 5n)(3m - 4n)$ **69.** $(2a + 3b)(3a - 5b)$
71. $(3a + 2b)(3a + 4b)$ **73.** $(5p + 2t)(7p + 4t)$
75. $6(3x - 4y)(x + y)$

77.

$y = \frac{2}{5}x - 1$

78.

$2x = 6$

79.

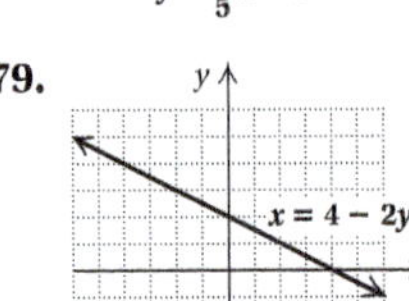

80.

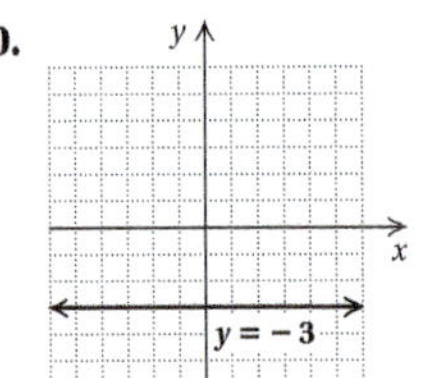

81.

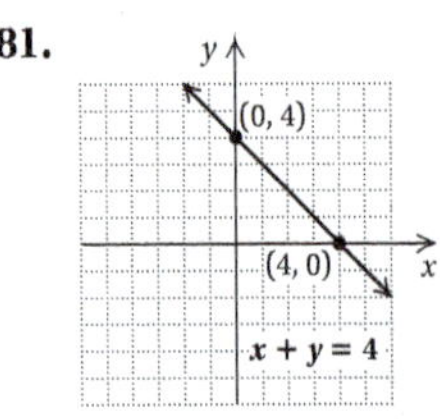

82.

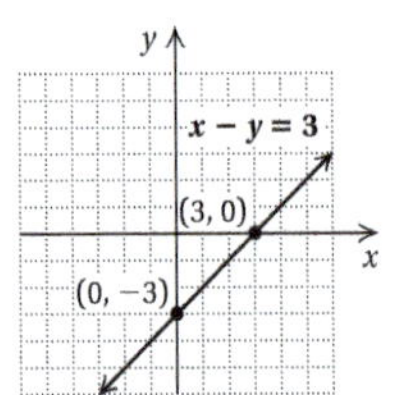

83.

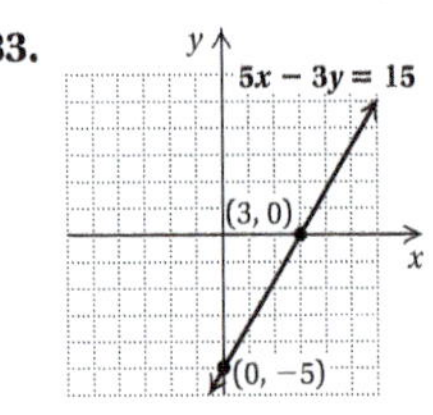

84.

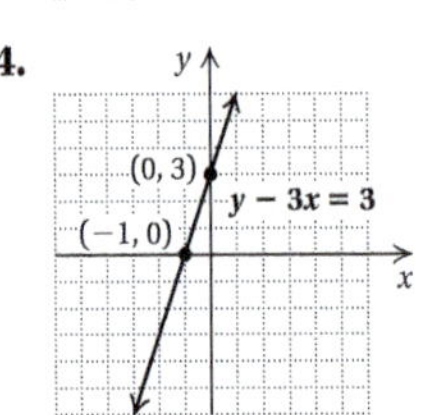

85. $(2x^n + 1)(10x^n + 3)$ **87.** $(x^{3a} - 1)(3x^{3a} + 1)$
89.–93. Left to the student

Exercise Set 13.4, p. 970

RC1. leading coefficient **RC2.** product; sum
RC3. sum **RC4.** grouping **1.** $(x + 2)(x + 7)$
3. $(x - 4)(x - 1)$ **5.** $(3x + 2)(2x + 3)$ **7.** $(3x - 4)(x - 4)$
9. $(7x - 8)(5x + 3)$ **11.** $(2x + 3)(2x - 3)$
13. $(x^2 + 3)(2x^2 + 5)$ **15.** $(2x + 3)(x + 2)$
17. $(3x + 5)(x - 3)$ **19.** $(5x + 1)(x + 2)$ **21.** $(3x - 1)(x - 1)$
23. $(2x + 7)(3x + 1)$ **25.** $(2x + 3)(2x - 5)$
27. $(5x - 2)(3x + 5)$ **29.** $(3x + 2)(3x - 8)$
31. $(3x - 1)(x + 2)$ **33.** $(3x - 4)(4x - 5)$
35. $(7x + 1)(2x - 3)$ **37.** $(3x - 7)^2$, or $(7 - 3x)^2$
39. $(3x + 2)(3x + 4)$ **41.** $-1(3a - 1)(3a + 5)$, or $(-3a + 1)(3a + 5)$, or $(3a - 1)(-3a - 5)$
43. $-2(x - 5)(x + 2)$, or $2(-x + 5)(x + 2)$, or $2(x - 5)(-x - 2)$
45. $4(3x - 2)(x + 3)$ **47.** $6(5x - 9)(x + 1)$
49. $2(3y + 5)(y - 1)$ **51.** $(3x - 1)(x - 1)$
53. $4(3x + 2)(x - 3)$ **55.** $(2x + 1)(x - 1)$
57. $(3x - 2)(3x + 8)$ **59.** $5(3x + 1)(x - 2)$
61. $p(3p + 4)(4p + 5)$ **63.** $-1(5x - 4)(x + 1)$, or $(-5x + 4)(x + 1)$, or $(5x - 4)(-x - 1)$ **65.** $-3(2t - 1)(t - 5)$, or $3(-2t + 1)(t - 5)$, or $3(2t - 1)(-t + 5)$
67. $x^2(7x - 1)(2x + 3)$ **69.** $3x(8x - 1)(7x - 1)$
71. $(5x^2 - 3)(3x^2 - 2)$ **73.** $(5t + 8)^2$ **75.** $2x(3x + 5)(x - 1)$
77. Prime **79.** Prime **81.** $(4m + 5n)(3m - 4n)$
83. $(2a + 3b)(3a - 5b)$ **85.** $(3a - 2b)(3a - 4b)$
87. $(5p + 2q)(7p + 4q)$ **89.** $6(3x - 4y)(x + y)$

91. $-6x(x-5)(x+2)$, or $6x(-x+5)(x+2)$, or $6x(x-5)(-x-2)$ **93.** $x^3(5x-11)(7x+4)$ **95.** $27x^{12}$
96. $\frac{1}{5^{14}}$ **97.** x^5y^6 **98.** a **99.** 3.008×10^{10}
100. 0.000015 **101.** About 6369 km, or 3949 mi **102.** 40°
103. $(3x^5-2)^2$ **105.** $(4x^5+1)^2$ **107.–111.** Left to the student

Mid-Chapter Review: Chapter 13, p. 974

1. True **2.** False **3.** True **4.** False
5. $10y^3 - 18y^2 + 12y = 2y \cdot 5y^2 - 2y \cdot 9y + 2y \cdot 6 = 2y(5y^2 - 9y + 6)$
6. $a \cdot c = 2 \cdot (-6) = -12;$
$-x = -4x + 3x;$
$2x^2 - x - 6 = 2x^2 - 4x + 3x - 6 = 2x(x-2) + 3(x-2) = (x-2)(2x+3)$
7. x **8.** x^2 **9.** $6x^3$ **10.** 4 **11.** $5x^2y$ **12.** x^2y^2
13. $x(x^2-8)$ **14.** $3x(x+4)$ **15.** $2(y^2+4y-2)$
16. $t^3(3t^3-5t-2)$ **17.** $(x+1)(x+3)$ **18.** $(z-2)^2$
19. $(x+4)(x^2+3)$ **20.** $8y^3(y^2-6)$
21. $6xy(x^2+4xy-7y^2)$ **22.** $(4t-3)(t-2)$
23. $(z-1)(z+5)$ **24.** $(z+4)(2z^2+5)$
25. $(3p-2)(p^2-3)$ **26.** $5x^3(2x^5-5x^3-3x^2+7)$
27. $(2w+3)(w^2-3)$ **28.** $x^2(4x^2-5x+3)$
29. $(6y-5)(y+2)$ **30.** $3(x-3)(x+2)$
31. $(3x+2)(2x^2+1)$ **32.** $(w-5)(w-3)$
33. $(2x+5)(4x^2+1)$ **34.** $(5z+2)(2z-5)$
35. $(2x+1)(3x+2)$ **36.** $(x-6y)(x-4y)$
37. $(2z+1)(3z^2+1)$ **38.** $a^2b^3(ab^4+a^2b^2-1+a^3b^3)$
39. $(4y+5z)(y-3z)$ **40.** $3x(x+2)(x+5)$
41. $(x-3)(x^2-2)$ **42.** $(3y+1)^2$ **43.** $(y+2)(y+4)$
44. $3(2y+5)(y+3)$ **45.** $(x-7)(x^2+4)$
46. $-1(y-4)(y+1)$, or $(-y+4)(y+1)$, or $(y-4)(-y-1)$
47. $4(2x+3)(2x-5)$ **48.** $(5a-3b)(2a-b)$
49. $(2w-5)(3w^2-5)$ **50.** $y(y+6)(y+3)$
51. $(4x+3y)(x+2y)$ **52.** $-1(3z-2)(2z+3)$, or $(-3z+2)(2z+3)$, or $(3z-2)(-2z-3)$
53. $(3t+2)(4t^2-3)$ **54.** $(y-4z)(y+5z)$
55. $(3x-4y)(3x+2y)$ **56.** $(3z-1)(z+3)$
57. $(m-8n)(m+2n)$ **58.** $2(w-3)^2$ **59.** $2t(3t-2)(3t-1)$
60. $(z+3)(5z^2+1)$ **61.** $(t-2)(t+7)$ **62.** $(2t-5)^2$
63. $(t-2)(t+6)$ **64.** $-1(2z+3)(z-4)$, or $(-2z-3)(z-4)$, or $(2z+3)(-z+4)$
65. $-1(y-6)(y+2)$, or $(-y+6)(y+2)$, or $(y-6)(-y-2)$
66. Find the product of two binomials. For example, $(2x^2+3)(x-4) = 2x^3 - 8x^2 + 3x - 12$. **67.** There is a finite number of pairs of numbers with the correct product, but there are infinitely many pairs with the correct sum. **68.** Since both constants are negative, the middle term will be negative so $(x-17)(x-18)$ cannot be a factorization of $x^2+35x+306$.
69. No; both $2x+6$ and $2x+8$ contain a factor of 2, so $2 \cdot 2$, or 4, must be factored out to reach the complete factorization. In other words, the largest common factor is 4, not 2.

Exercise Set 13.5, p. 981

RC1. False **RC2.** True **RC3.** False **RC4.** False
1. Yes **3.** No **5.** No **7.** Yes **9.** $(x-7)^2$ **11.** $(x+8)^2$
13. $(x-1)^2$ **15.** $(x+2)^2$ **17.** $(y+6)^2$ **19.** $(t-4)^2$
21. $(q^2-3)^2$ **23.** $(4y+7)^2$ **25.** $2(x-1)^2$
27. $x(x-9)^2$ **29.** $3(2q-3)^2$ **31.** $(7-3x)^2$, or $(3x-7)^2$
33. $5(y^2+1)^2$ **35.** $(1+2x^2)^2$ **37.** $(2p+3t)^2$ **39.** $(a-3b)^2$
41. $(9a-b)^2$ **43.** $4(3a+4b)^2$ **45.** Yes **47.** No **49.** No
51. Yes **53.** $(y+2)(y-2)$ **55.** $(p+1)(p-1)$
57. $(t+7)(t-7)$ **59.** $(a+b)(a-b)$
61. $(5t+m)(5t-m)$ **63.** $(10+k)(10-k)$
65. $(4a+3)(4a-3)$ **67.** $(2x+5y)(2x-5y)$
69. $2(2x+7)(2x-7)$ **71.** $x(6+7x)(6-7x)$
73. $(\frac{1}{4}+7x^4)(\frac{1}{4}-7x^4)$ **75.** $(0.3y+0.02)(0.3y-0.02)$
77. $(7a^2+9)(7a^2-9)$ **79.** $(a^2+4)(a+2)(a-2)$
81. $5(x^2+9)(x+3)(x-3)$ **83.** $(1+y^4)(1+y^2)(1+y)(1-y)$
85. $(x^6+4)(x^3+2)(x^3-2)$ **87.** $(y+\frac{1}{4})(y-\frac{1}{4})$
89. $(5+\frac{1}{7}x)(5-\frac{1}{7}x)$ **91.** $(4m^2+t^2)(2m+t)(2m-t)$
93. y-intercept: $(0, 4)$; x-intercept: $(16, 0)$ **94.** y-intercept: $(0, -5)$; x-intercept: $(6.5, 0)$ **95.** y-intercept: $(0, -5)$; x-intercept: $(\frac{5}{2}, 0)$
96.
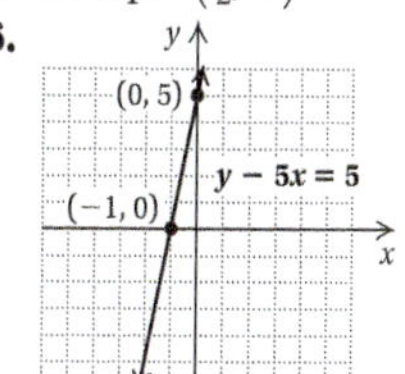

97.
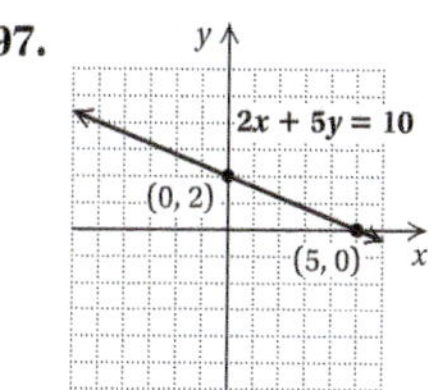

98.
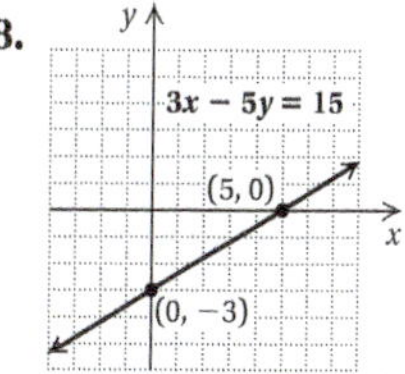

99. $x^2-4xy+4y^2$ **100.** $\frac{1}{2}\pi x^2+2xy$ **101.** Prime
103. $(x+11)^2$ **105.** $2x(3x+1)^2$
107. $(x^4+2^4)(x^2+2^2)(x+2)(x-2)$
109. $3x^3(x+2)(x-2)$ **111.** $2x(3x+\frac{2}{5})(3x-\frac{2}{5})$
113. $p(0.7+p)(0.7-p)$ **115.** $(0.8x+1.1)(0.8x-1.1)$
117. $x(x+6)$ **119.** $\left(x+\frac{1}{x}\right)\left(x-\frac{1}{x}\right)$
121. $(9+b^{2k})(3-b^k)(3+b^k)$ **123.** $(3b^n+2)^2$
125. $(y+4)^2$ **127.** 9 **129.** Not correct **131.** Not correct

Exercise Set 13.6, p. 990

RC1. common **RC2.** difference **RC3.** square
RC4. grouping **RC5.** completely **RC6.** check
1. $3(x+8)(x-8)$ **3.** $(a-5)^2$ **5.** $(2x-3)(x-4)$
7. $x(x+12)^2$ **9.** $(x+3)(x+2)(x-2)$
11. $3(4x+1)(4x-1)$ **13.** $3x(3x-5)(x+3)$
15. Prime **17.** $x(x^2+7)(x-3)$ **19.** $x^3(x-7)^2$
21. $-2(x-2)(x+5)$, or $2(-x+2)(x+5)$, or $2(x-2)(-x-5)$ **23.** Prime **25.** $4(x^2+4)(x+2)(x-2)$
27. $(1+y^4)(1+y^2)(1+y)(1-y)$ **29.** $x^3(x-3)(x-1)$
31. $\frac{1}{9}(\frac{1}{3}x^3-4)^2$ **33.** $m(x^2+y^2)$ **35.** $9xy(xy-4)$
37. $2\pi r(h+r)$ **39.** $(a+b)(2x+1)$ **41.** $(x+1)(x-1-y)$
43. $(n+2)(n+p)$ **45.** $(2q-1)(3q+p)$
47. $(2b-a)^2$, or $(a-2b)^2$ **49.** $(4x+3y)^2$ **51.** $(7m^2-8n)^2$
53. $(y^2+5z^2)^2$ **55.** $(\frac{1}{2}a+\frac{1}{3}b)^2$ **57.** $(a+b)(a-2b)$
59. $(m+20n)(m-18n)$ **61.** $(mn-8)(mn+4)$
63. $r^3(rs-2)(rs-8)$ **65.** $a^3(a-b)(a+5b)$
67. $(a+\frac{1}{5}b)(a-\frac{1}{5}b)$ **69.** $(x+y)(x-y)$
71. $(4+p^2q^2)(2+pq)(2-pq)$
73. $(1+4x^6y^6)(1+2x^3y^3)(1-2x^3y^3)$
75. $(q+8)(q+1)(q-1)$ **77.** $ab(2ab+1)(3ab-2)$
79. $(m+1)(m-1)(m+2)(m-2)$ **81.** -8.67
82. -3 **83.** $\frac{1}{5}$ **84.** -7.72 **85.** -22
86. -5 **87.** 7 **88.** $>$ **89.** $(t+1)^2(t-1)^2$
91. $(x-5)(x+2)(x-2)$ **93.** $(3.5x-1)^2$
95. $(y-2)(y+3)(y-3)$ **97.** $(y-1)^3$ **99.** $(y+4+x)^2$

Calculator Corner, p. 995

1. Left to the student

Exercise Set 13.7, p. 999

RC1. False **RC2.** False **RC3.** False **RC4.** True
1. $-4, -9$ **3.** $-3, 8$ **5.** $-12, 11$ **7.** $0, -3$ **9.** $0, -18$
11. $-\frac{5}{2}, -4$ **13.** $-\frac{1}{5}, 3$ **15.** $4, \frac{1}{4}$ **17.** $0, \frac{2}{3}$ **19.** $-\frac{1}{10}, \frac{1}{27}$
21. $\frac{1}{3}, -20$ **23.** $0, \frac{2}{3}, \frac{1}{2}$ **25.** $-5, -1$
27. $-9, 2$ **29.** $3, 5$ **31.** $0, 8$ **33.** $0, -18$ **35.** $-4, 4$
37. $-\frac{2}{3}, \frac{2}{3}$ **39.** -3 **41.** 4 **43.** $0, \frac{6}{5}$ **45.** $-1, \frac{5}{3}$
47. $-\frac{1}{4}, \frac{2}{3}$ **49.** $-1, \frac{2}{3}$ **51.** $-\frac{7}{10}, \frac{7}{10}$ **53.** $-2, 9$ **55.** $\frac{4}{5}, \frac{3}{2}$
57. $(-4, 0), (1, 0)$ **59.** $(-\frac{5}{2}, 0), (2, 0)$ **61.** $(-3, 0), (5, 0)$
63. $-1, 4$ **65.** $-1, 3$ **67.** $(a + b)^2$ **68.** $a^2 + b^2$
69. $\{x|x < -100\}$ **70.** $\{x|x \leq 8\}$ **71.** $\{x|x < 2\}$
72. $\{x|x \geq \frac{20}{3}\}$ **73.** $-5, 4$ **75.** $-3, 9$ **77.** $-\frac{1}{8}, \frac{1}{8}$
79. $-4, 4$ **81.** 2.33, 6.77 **83.** Answers may vary.
(a) $x^2 - x - 12 = 0$; **(b)** $x^2 + 7x + 12 = 0$; **(c)** $4x^2 - 4x + 1 = 0$;
(d) $x^2 - 25 = 0$; **(e)** $40x^3 - 14x^2 + x = 0$

Translating for Success, p. 1007

1. O **2.** M **3.** K **4.** I **5.** G **6.** E **7.** C **8.** A
9. H **10.** B

Exercise Set 13.8, p. 1008

RC1. (b) **RC2.** (b) **RC3.** (d) **RC4.** (c)
1. Length: 42 in.; width: 14 in. **3.** Length: 6 cm; width: 4 cm
5. Height: 4 cm; base: 14 cm **7.** Base: 8 m; height: 16 m
9. 182 games **11.** 12 teams **13.** 4950 handshakes
15. 25 people **17.** 14 and 15 **19.** 12 and 14; -12 and -14
21. 15 and 17; -15 and -17 **23.** 32 ft **25.** Hypotenuse: 17 ft; leg: 15 ft **27.** 300 ft by 400 ft by 500 ft **29.** 24 m, 25 m
31. Dining room: 12 ft by 12 ft; kitchen: 12 ft by 10 ft **33.** 1 sec, 2 sec **35.** 5 and 7 **37.** 4.53 **38.** $-\frac{5}{6}$ **39.** -40
40. -116 **41.** $-\frac{3}{25}$ **42.** -3.4 **43.** $-4y - 13$ **44.** $10x - 30$
45. 5 ft **47.** 30 cm by 15 cm **49.** 11 yd, 60 yd, 61 yd

Summary and Review: Chapter 13, p. 1013

Vocabulary Reinforcement

1. factor **2.** factor **3.** factorization **4.** common
5. grouping **6.** binomial **7.** zero **8.** difference

Concept Reinforcement

1. False **2.** True **3.** False **4.** True

Study Guide

1. $4xy$ **2.** $9x^2(3x^3 - x + 2)$ **3.** $(z - 3)(z^2 + 4)$
4. $(x + 2)(x + 4)$ **5.** $3(z - 4)(2z + 1)$ **6.** $(3y - 1)(2y + 3)$
7. $(2x + 1)^2$ **8.** $2(3x + 2)(3x - 2)$ **9.** $-5, 1$

Review Exercises

1. $5y^2$ **2.** $12x$ **3.** $5(1 + 2x^3)(1 - 2x^3)$ **4.** $x(x - 3)$
5. $(3x + 2)(3x - 2)$ **6.** $(x + 6)(x - 2)$ **7.** $(x + 7)^2$
8. $3x(2x^2 + 4x + 1)$ **9.** $(x + 1)(x^2 + 3)$ **10.** $(3x - 1)(2x - 1)$
11. $(x^2 + 9)(x + 3)(x - 3)$ **12.** $3x(3x - 5)(x + 3)$
13. $2(x + 5)(x - 5)$ **14.** $(x + 4)(x^3 - 2)$
15. $(4x^2 + 1)(2x + 1)(2x - 1)$ **16.** $4x^4(2x^2 - 8x + 1)$
17. $3(2x + 5)^2$ **18.** Prime **19.** $x(x - 6)(x + 5)$
20. $(2x + 5)(2x - 5)$ **21.** $(3x - 5)^2$
22. $2(3x + 4)(x - 6)$ **23.** $(x - 3)^2$ **24.** $(2x + 1)(x - 4)$
25. $2(3x - 1)^2$ **26.** $3(x + 3)(x - 3)$
27. $(x - 5)(x - 3)$ **28.** $(5x - 2)^2$ **29.** $(7b^5 - 2a^4)^2$
30. $(xy + 4)(xy - 3)$ **31.** $3(2a + 7b)^2$ **32.** $(m + 5)(m + t)$
33. $32(x^2 - 2y^2z^2)(x^2 + 2y^2z^2)$ **34.** $1, -3$ **35.** $-7, 5$
36. $-4, 0$ **37.** $\frac{2}{3}, 1$ **38.** $-8, 8$ **39.** $-2, 8$
40. $(-5, 0), (-4, 0)$ **41.** $(-\frac{3}{2}, 0), (5, 0)$ **42.** Height: 6 cm; base: 5 cm **43.** -18 and -16; 16 and 18 **44.** 842 ft
45. On the ground: 4 ft; on the tree: 3 ft **46.** 6 km **47.** B
48. A **49.** 2.5 cm **50.** 0, 2 **51.** Length: 12 in.; width: 6 in.
52. 35 ft **53.** No solution **54.** $2, -3, \frac{5}{2}$ **55.** $-2, \frac{5}{4}, 3$

Understanding Through Discussion and Writing

1. Although $x^3 - 8x^2 + 15x$ can be factored as $(x^2 - 5x)(x - 3)$, this is not a complete factorization of the polynomial since $x^2 - 5x = x(x - 5)$. Gwen should always look for a common factor first. **2.** Josh is correct, because answers can easily be checked by multiplying.
3. For $x = -3$:

$$(x - 4)^2 = (-3 - 4)^2 = (-7)^2 = 49;$$
$$(4 - x)^2 = [4 - (-3)]^2 = 7^2 = 49.$$

For $x = 1$:

$$(x - 4)^2 = (1 - 4)^2 = (-3)^2 = 9;$$
$$(4 - x)^2 = (4 - 1)^2 = 3^2 = 9.$$

In general, $(x - 4)^2 = [-(-x + 4)]^2 = [-(4 - x)]^2 = (-1)^2(4 - x)^2 = (4 - x)^2$.

4. The equation is not in the form $ab = 0$. The correct procedure is

$$\begin{aligned} (x - 3)(x + 4) &= 8 \\ x^2 + x - 12 &= 8 \\ x^2 + x - 20 &= 0 \\ (x + 5)(x - 4) &= 0 \\ x + 5 = 0 \quad &or \quad x - 4 = 0 \\ x = -5 \quad &or \quad x = 4. \end{aligned}$$

The solutions are -5 and 4.
5. One solution of the equation is 0. Dividing both sides of the equation by x, leaving the solution $x = 3$, is equivalent to dividing by 0.
6. She could use the measuring sticks to draw a right angle as shown below. Then she could use the 3-ft and 4-ft sticks to extend one leg to 7 ft and the 4-ft and 5-ft sticks to extend the other leg to 9 ft.

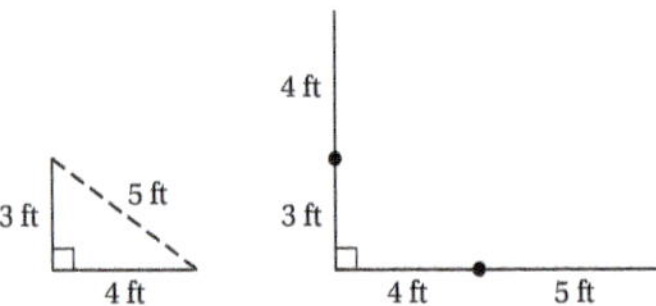

Next, she could draw another right angle with either the 7-ft side or the 9-ft side as a side.

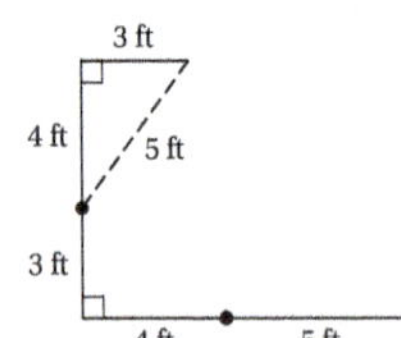

Then she could use the sticks to extend the other side to the appropriate length. Finally, she would draw the remaining side of the rectangle.

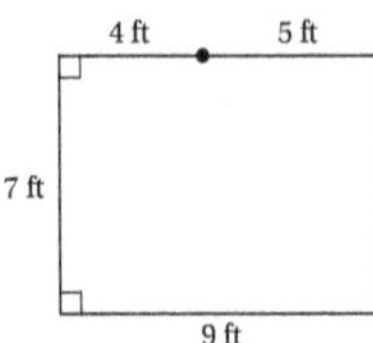

Test: Chapter 13, p. 1019

1. [13.1a] $4x^3$ **2.** [13.2a] $(x - 5)(x - 2)$ **3.** [13.5b] $(x - 5)^2$
4. [13.1b] $2y^2(2y^2 - 4y + 3)$ **5.** [13.1c] $(x + 1)(x^2 + 2)$
6. [13.1b] $x(x - 5)$ **7.** [13.2a] $x(x + 3)(x - 1)$
8. [13.3a], [13.4a] $2(5x - 6)(x + 4)$ **9.** [13.5d] $(2x + 3)(2x - 3)$
10. [13.2a] $(x - 4)(x + 3)$
11. [13.3a], [13.4a] $3m(2m + 1)(m + 1)$
12. [13.5d] $3(w + 5)(w - 5)$

13. [13.5b] $5(3x + 2)^2$ **14.** [13.5d] $3(x^2 + 4)(x + 2)(x - 2)$
15. [13.5b] $(7x - 6)^2$ **16.** [13.3a], [13.4a] $(5x - 1)(x - 5)$
17. [13.1c] $(x + 2)(x^3 - 3)$ **18.** [13.5d] $5(4 + x^2)(2 + x)(2 - x)$
19. [13.3a], [13.4a] $3t(2t + 5)(t - 1)$ **20.** [13.7b] 0, 3
21. [13.7b] $-4, 4$ **22.** [13.7b] $-4, 5$ **23.** [13.7b] $-5, \frac{3}{2}$
24. [13.7b] $-4, 7$ **25.** [13.7b] $(-5, 0), (7, 0)$
26. [13.7b] $(\frac{2}{3}, 0), (1, 0)$ **27.** [13.8a] Length: 8 m; width: 6 m
28. [13.8a] Height: 4 cm; base: 14 cm **29.** [13.8a] 5 ft **30.** [13.5d] A
31. [13.8a] Length: 15 m; width: 3 m **32.** [13.2a] $(a - 4)(a + 8)$
33. [13.7b] $-\frac{8}{3}, 0, \frac{2}{5}$ **34.** [12.6b], [13.5d] D

CHAPTER 14

Exercise Set 14.1, p. 1028

RC1. equivalent **RC2.** denominator **RC3.** quotient
RC4. factors
1. 0 **3.** 8 **5.** $-\frac{5}{2}$ **7.** $-4, 7$ **9.** $-5, 5$ **11.** None
13. $\dfrac{(4x)(3x^2)}{(4x)(5y)}$ **15.** $\dfrac{2x(x - 1)}{2x(x + 4)}$ **17.** $\dfrac{(3 - x)(-1)}{(4 - x)(-1)}$
19. $\dfrac{(y + 6)(y - 7)}{(y + 6)(y + 2)}$ **21.** $\dfrac{x^2}{4}$ **23.** $\dfrac{8p^2q}{3}$ **25.** $\dfrac{x - 3}{x}$
27. $\dfrac{m + 1}{2m + 3}$ **29.** $\dfrac{a - 3}{a + 2}$ **31.** $\dfrac{a - 3}{a - 4}$ **33.** $\dfrac{x + 5}{x - 5}$ **35.** $a + 1$
37. $\dfrac{x^2 + 1}{x + 1}$ **39.** $\frac{3}{2}$ **41.** $\dfrac{6}{t - 3}$ **43.** $\dfrac{t + 2}{2(t - 4)}$ **45.** $\dfrac{t - 2}{t + 2}$
47. -1 **49.** -1 **51.** -6 **53.** $-x - 1$ **55.** $-3t$
57. $\dfrac{56x}{3}$ **59.** $\dfrac{2}{dc^2}$ **61.** 1 **63.** $\dfrac{(a + 3)(a - 3)}{a(a + 4)}$ **65.** $\dfrac{2a}{a - 2}$
67. $\dfrac{(t + 2)(t - 2)}{(t + 1)(t - 1)}$ **69.** $\dfrac{x + 4}{x + 2}$ **71.** $\dfrac{5(a + 6)}{a - 1}$

73.

74.

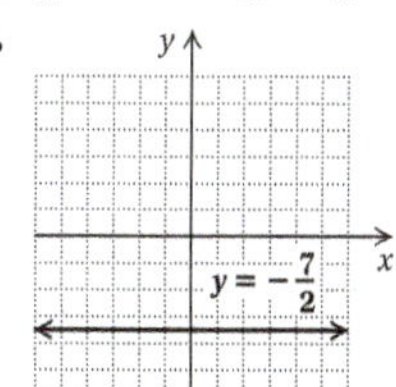

75. $x^3(x - 7)(x + 5)$ **76.** $(2y^2 + 1)(y - 5)$
77. $(2 + t)(2 - t)(4 + t^2)$ **78.** $10(x + 7)(x + 1)$
79. $\dfrac{x - y}{x - 5y}$ **81.** $\dfrac{1}{x - 1}$

Exercise Set 14.2, p. 1034

RC1. (d) **RC2.** (e) **RC3.** (a) **RC4.** (f) **RC5.** (b)
RC6. (c)
1. $\dfrac{x}{4}$ **3.** $\dfrac{1}{x^2 - y^2}$ **5.** $a + b$ **7.** $\dfrac{x^2 - 4x + 7}{x^2 + 2x - 5}$ **9.** $\frac{3}{10}$
11. $\frac{1}{4}$ **13.** $\dfrac{b}{a}$ **15.** $\dfrac{(a + 2)(a + 3)}{(a - 3)(a - 1)}$ **17.** $\dfrac{(x - 1)^2}{x}$ **19.** $\frac{1}{2}$
21. $\frac{15}{8}$ **23.** $\frac{15}{4}$ **25.** $\dfrac{a - 5}{3(a - 1)}$ **27.** $\dfrac{(x + 2)^2}{x}$ **29.** $\frac{3}{2}$
31. $\dfrac{c + 1}{c - 1}$ **33.** $\dfrac{y - 3}{2y - 1}$ **35.** $\dfrac{x + 1}{x - 1}$ **37.** $\{x|x \geq 77\}$
38. Height: 7 in.; base: 10 in. **39.** $8x^3 - 11x^2 - 3x + 12$
40. $-2p^2 + 4pq - 4q^2$ **41.** $\dfrac{4y^8}{x^6}$ **42.** $\dfrac{125x^{18}}{y^{12}}$ **43.** $\dfrac{4x^6}{y^{10}}$
44. $\dfrac{1}{a^{15}b^{20}}$ **45.** $-\dfrac{1}{b^2}$ **47.** $\dfrac{a + 1}{5ab^2(a^2 + 4)}$

Exercise Set 14.3, p. 1039

RC1. Common **RC2.** Multiple **RC3.** Denominator
RC4. Greatest **RC5.** $2 \cdot 2 \cdot 2 \cdot 2 \cdot 3$

1. 108 **3.** 72 **5.** 126 **7.** 360 **9.** 500 **11.** $\frac{65}{72}$ **13.** $\frac{29}{120}$
15. $\frac{23}{180}$ **17.** $12x^3$ **19.** $18x^2y^2$ **21.** $6(y - 3)$
23. $t(t + 2)(t - 2)$ **25.** $(x + 2)(x - 2)(x + 3)$
27. $t(t + 2)^2(t - 4)$ **29.** $(a + 1)(a - 1)^2$ **31.** $(m - 3)(m - 2)^2$
33. $(2 + 3x)(2 - 3x)$ **35.** $10v(v + 4)(v + 3)$
37. $18x^3(x - 2)^2(x + 1)$ **39.** $6x^3(x + 2)^2(x - 2)$ **41.** $120w^6$
43. $120x^4$; $8x^3$; $960x^7$ **44.** $48ab^3$; $4ab$; $192a^2b^4$
45. $48x^6$; $16x^5$; $768x^{11}$ **46.** $120x^3$; $2x^2$; $240x^5$
47. $20x^2$; $10x$; $200x^3$ **48.** a^{15}; a^5; a^{20} **49.** 24 min

Exercise Set 14.4, p. 1045

RC1. $\frac{2}{2}; \frac{3}{3}$ **RC2.** $\dfrac{x + 3}{x + 3}; \dfrac{2}{2}$ **RC3.** $\dfrac{x - 3}{x - 3}; \dfrac{x + 2}{x + 2}$ **RC4.** $\dfrac{2x}{2x}; \dfrac{3}{3}$
1. 1 **3.** $\dfrac{6}{3 + x}$ **5.** $\dfrac{-4x + 11}{2x - 1}$ **7.** $\dfrac{2x + 5}{x^2}$ **9.** $\dfrac{41}{24r}$
11. $\dfrac{2(2x + 3y)}{x^2y^2}$ **13.** $\dfrac{4 + 3t}{18t^3}$ **15.** $\dfrac{x^2 + 4xy + y^2}{x^2y^2}$
17. $\dfrac{6x}{(x - 2)(x + 2)}$ **19.** $\dfrac{11x + 2}{3x(x + 1)}$ **21.** $\dfrac{x(x + 6)}{(x + 4)(x - 4)}$
23. $\dfrac{6}{z + 4}$ **25.** $\dfrac{3x - 1}{(x - 1)^2}$ **27.** $\dfrac{11a}{10(a - 2)}$ **29.** $\dfrac{2(x^2 + 4x + 8)}{x(x + 4)}$
31. $\dfrac{7a + 6}{(a - 2)(a + 1)(a + 3)}$ **33.** $\dfrac{2(x^2 - 2x + 17)}{(x - 5)(x + 3)}$
35. $\dfrac{3a + 2}{(a + 1)(a - 1)}$ **37.** $\frac{1}{4}$ **39.** $-\dfrac{1}{t}$
41. $\dfrac{-x + 7}{x - 6}$, or $\dfrac{7 - x}{x - 6}$, or $\dfrac{x - 7}{6 - x}$ **43.** $y + 3$
45. $\dfrac{2(b - 7)}{(b + 4)(b - 4)}$ **47.** $a + b$ **49.** $\dfrac{5x + 2}{x - 5}$ **51.** -1
53. $\dfrac{-x^2 + 9x - 14}{(x - 3)(x + 3)}$ **55.** $\dfrac{2(x + 3y)}{(x + y)(x - y)}$ **57.** $\dfrac{a^2 + 7a + 1}{(a + 5)(a - 5)}$
59. $\dfrac{5t - 12}{(t + 3)(t - 3)(t - 2)}$ **61.** $x^2 - 1$
62. $13y^3 - 14y^2 + 12y - 73$ **63.** $\dfrac{1}{x^{12}y^{21}}$ **64.** $\dfrac{25}{x^4y^6}$
65. -8 **66.** $-2, 9$

67.

68.

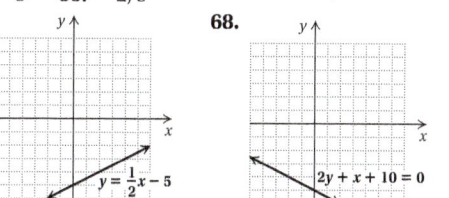

69.

70.

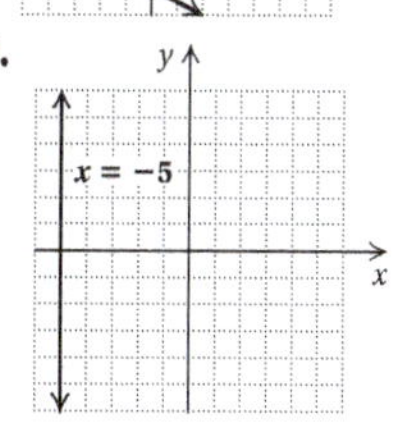

71. Perimeter: $\dfrac{16y + 28}{15}$; area: $\dfrac{y^2 + 2y - 8}{15}$
73. $\dfrac{(z + 6)(2z - 3)}{(z + 2)(z - 2)}$ **75.** $\dfrac{11z^4 - 22z^2 + 6}{(z^2 + 2)(z^2 - 2)(2z^2 - 3)}$

Exercise Set 14.5, p. 1053

RC1. **(a)** $3x + 5$; **(b)** $10x - 3x - 5$; **(c)** $7x - 5$
RC2. **(a)** $4 - 9a$; **(b)** $7 - 4 + 9a$; **(c)** $3 + 9a$
RC3. **(a)** $y + 1$; **(b)** $9y - 2 - y - 1$; **(c)** $8y - 3$

1. $\frac{4}{x}$ 3. 1 5. $\frac{1}{x-1}$ 7. $\frac{-a-4}{10}$ 9. $\frac{7z-12}{12z}$

11. $\frac{4x^2-13xt+9t^2}{3x^2t^2}$ 13. $\frac{2(x-20)}{(x+5)(x-5)}$ 15. $\frac{3-5t}{2t(t-1)}$

17. $\frac{2s-st-s^2}{(t+s)(t-s)}$ 19. $\frac{y-19}{4y}$ 21. $\frac{-2a^2}{(x+a)(x-a)}$

23. $\frac{8}{3}$ 25. $\frac{13}{a}$ 27. $\frac{8}{y-1}$ 29. $\frac{x-2}{x-7}$ 31. $\frac{4}{(a+5)(a-5)}$

33. $\frac{2(x-2)}{x-9}$ 35. $\frac{3(3x+4)}{(x+3)(x-3)}$ 37. $\frac{1}{2}$ 39. $\frac{x-3}{(x+3)(x+1)}$

41. $\frac{18x+5}{x-1}$ 43. 0 45. $\frac{-9}{2x-3}$ 47. $\frac{20}{2y-1}$ 49. $\frac{2a-3}{2-a}$

51. $\frac{z-3}{2z-1}$ 53. $\frac{2}{x+y}$ 55. $\frac{b^{20}}{a^8}$ 56. $18x^3$ 57. $30x^{12}$

58. 10 59. $-\frac{11}{35}$ 60. 1 61. $x^2-9x+18$ 62. $(4-\pi)r^2$

63. Missing length: $\frac{-2a-15}{a-6}$; area: $\frac{-2a^3-15a^2+12a+90}{2(a-6)^2}$

Mid-Chapter Review: Chapter 14, p. 1057

1. False 2. True 3. True 4. False 5. True

6. $$\frac{x-1}{x-2}-\frac{x+1}{x+2}-\frac{x-6}{4-x^2}$$
$$=\frac{x-1}{x-2}-\frac{x+1}{x+2}-\frac{x-6}{4-x^2}\cdot\frac{-1}{-1}$$
$$=\frac{x-1}{x-2}-\frac{x+1}{x+2}-\frac{6-x}{x^2-4}$$
$$=\frac{x-1}{x-2}-\frac{x+1}{x+2}-\frac{6-x}{(x-2)(x+2)}$$
$$=\frac{x-1}{x-2}\cdot\frac{x+2}{x+2}-\frac{x+1}{x+2}\cdot\frac{x-2}{x-2}-\frac{6-x}{(x-2)(x+2)}$$
$$=\frac{x^2+x-2}{(x-2)(x+2)}-\frac{x^2-x-2}{(x-2)(x+2)}-\frac{6-x}{(x-2)(x+2)}$$
$$=\frac{x^2+x-2-x^2+x+2-6+x}{(x-2)(x+2)}$$
$$=\frac{3x-6}{(x-2)(x+2)}$$
$$=\frac{3(x-2)}{(x-2)(x+2)}=\frac{x-2}{x-2}\cdot\frac{3}{x+2}$$
$$=\frac{3}{x+2}$$

7. None 8. 3, 8 9. $\frac{7}{2}$ 10. $\frac{x-1}{x-3}$ 11. $\frac{2(y+4)}{y-1}$ 12. -1

13. $\frac{1}{-x+3}$, or $\frac{1}{3-x}$ 14. $10x^3(x-10)^2(x+10)$ 15. $\frac{a+1}{a-3}$

16. $\frac{y}{(y-2)(y-3)}$ 17. $x+11$ 18. $\frac{1}{x-y}$ 19. $\frac{a^2+5ab-b^2}{a^2b^2}$

20. $\frac{2(3x^2-4x+6)}{x(x+2)(x-2)}$ 21. E 22. A 23. D 24. B

25. F 26. C 27. If the numbers have a common factor, then their product contains that factor more than the greatest number of times it occurs in any one factorization. In this case, their product is not their least common multiple. 28. Yes; consider the product $\frac{a}{b}\cdot\frac{c}{d}=\frac{ac}{bd}$. The reciprocal of the product is $\frac{bd}{ac}$. This is equal to the product of the reciprocals of the two original factors: $\frac{bd}{ac}=\frac{b}{a}\cdot\frac{d}{c}$.

29. Although multiplying the denominators of the expressions being added results in a common denominator, it is often not the *least* common denominator. Using a common denominator other than the LCD makes the expressions more complicated, requires additional simplification after the addition has been performed, and leaves more room for error.

30. Their sum is 0. Another explanation is that $-\left(\frac{1}{3-x}\right)=\frac{1}{-(3-x)}=\frac{1}{x-3}$.

31. $\frac{x+3}{x-5}$ is not defined for $x=5$, $\frac{x-7}{x+1}$ is not defined for $x=-1$, and $\frac{x+1}{x-7}$ $\left(\text{the reciprocal of }\frac{x-7}{x+1}\right)$ is not defined for $x=7$.

32. The binomial is a factor of the trinomial.

Exercise Set 14.6, p. 1063

RC1. complex RC2. numerator RC3. least common denominator RC4. reciprocal

1. $\frac{25}{4}$ 3. $\frac{1}{3}$ 5. -6 7. $\frac{1+3x}{1-5x}$ 9. $\frac{2x+1}{x}$ 11. 8

13. $x-8$ 15. $\frac{y}{y-1}$ 17. $-\frac{1}{a}$ 19. $\frac{ab}{b-a}$ 21. $\frac{p^2+q^2}{q+p}$

23. $\frac{2a(a+2)}{5-3a^2}$ 25. $\frac{15(4-a^3)}{14a^2(9+2a)}$ 27. $\frac{ac}{bd}$ 29. 1

31. $\frac{4x+1}{5x+3}$ 33. $\{x|x\le 96\}$ 34. $\{b|b>\frac{22}{9}\}$

35. $\{x|x<-3\}$ 36. 12 ft, 5 ft 37. 14 yd 39. $\frac{5x+3}{3x+2}$

Calculator Corner, p. 1068

1.–2. Left to the student

Exercise Set 14.7, p. 1069

RC1. Rational expression RC2. Solutions RC3. Rational expression RC4. Rational expression RC5. Solutions RC6. Solutions RC7. Rational expression RC8. Solutions

1. $\frac{6}{5}$ 3. $\frac{40}{29}$ 5. $\frac{47}{2}$ 7. -6 9. $\frac{24}{7}$ 11. $-4, -1$ 13. $-4, 4$
15. 3 17. $\frac{14}{3}$ 19. 5 21. 5 23. $\frac{5}{2}$ 25. -2 27. $-\frac{13}{2}$
29. $\frac{17}{2}$ 31. No solution 33. -5 35. $\frac{5}{3}$ 37. $\frac{1}{2}$
39. No solution 41. No solution 43. 4 45. No solution
47. $-2, 2$ 49. 7 51. $4x^4+3x^3+2x-7$ 52. 0
53. $50(p^2-2)$ 54. $5(p+2)(p-10)$
55. 18 and 20; -20 and -18 56. 3.125 L 57. $-\frac{1}{6}$

Translating for Success, p. 1080

1. K 2. E 3. C 4. N 5. D 6. O 7. F 8. H
9. B 10. A

Exercise Set 14.8, p. 1081

RC1. corresponding; same; proportional RC2. quotient RC3. proportion RC4. rate RC5. distance

1. Sarah: 30 km/h; Rick: 70 km/h
3. Ostrich: 40 mph; giraffe: 32 mph
5. Hank: 14 km/h; Kelly: 19 km/h 7. 20 mph
9. Ralph: 5 km/h; Bonnie: 8 km/h 11. $1\frac{1}{3}$ hr 13. $22\frac{2}{9}$ min
15. $25\frac{5}{7}$ min 17. About 4.7 hr 19. $3\frac{15}{16}$ hr 21. $3\frac{3}{4}$ min
23. $\frac{10}{3}$ students/teacher 25. 2.3 km/h 27. 66 g 29. 1.92 g
31. 8 gal 33. 287 trout 35. 200 duds 37. 4.92 bags
39. 1.75 lb 41. $\frac{21}{2}$ 43. $\frac{8}{3}$ 45. $\frac{35}{3}$
47. About 1700 largemouth bass 49. 0
50. -2 51. x^{11} 52. x 53. $\frac{1}{x^{11}}$ 54. $\frac{1}{x}$

55. 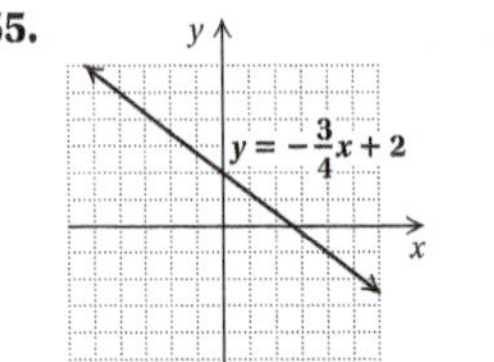

56. 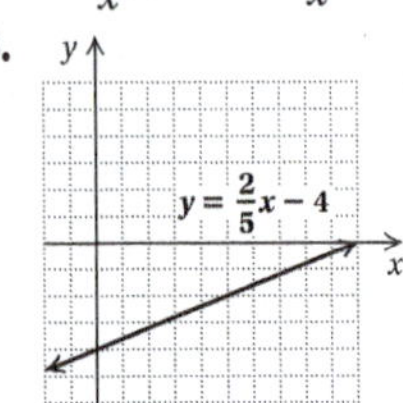

57. 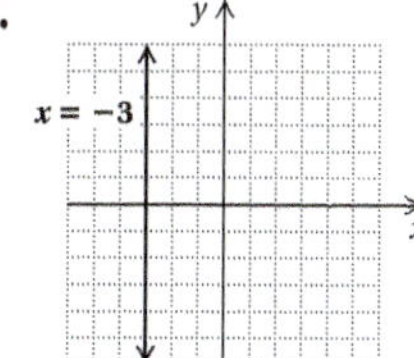

59. $27\frac{3}{11}$ min

Exercise Set 14.9, p. 1092

RC1. (b) **RC2.** (d) **RC3.** (f) **RC4.** (e) **RC5.** (c) **RC6.** (a)
1. $y = 4x; 80$ **3.** $y = 1.6x; 32$ **5.** $y = 3.6x; 72$
7. $y = \frac{25}{3}x; \frac{500}{3}$ **9.** (a) $P = 12H$; (b) \$420
11. (a) $C = 11.25S$; (b) \$101.25 **13.** (a) $M = \frac{1}{6}E$; (b) $18.\overline{3}$ lb; (c) 30 lb **15.** (a) $N = 80{,}000S$; (b) 16,000,000 instructions/sec
17. $93\frac{1}{3}$ servings **19.** $y = \frac{75}{x}; \frac{15}{2}$, or 7.5 **21.** $y = \frac{80}{x}; 8$
23. $y = \frac{1}{x}; \frac{1}{10}$ **25.** $y = \frac{2100}{x}; 210$ **27.** $y = \frac{0.06}{x}; 0.006$
29. (a) Direct; (b) $69\frac{3}{8}$ players **31.** (a) Inverse; (b) $4\frac{1}{2}$ hr
33. (a) $N = \frac{280}{P}$; (b) 10 gal **35.** (a) $I = \frac{1920}{R}$; (b) 32 amperes
37. (a) $m = \frac{40}{n}$; (b) 10 questions **39.** 8.25 ft **41.** $\frac{8}{5}$ **42.** 11
43. $\frac{1}{3}$ **44.** $\frac{47}{20}$ **45.** −1 **46.** 49 **47.** $P^2 = kt$
49. $P = kV^3$

Summary and Review: Chapter 14, p. 1096

Vocabulary Reinforcement

1. complex **2.** proportion **3.** reciprocals **4.** equivalent
5. opposites **6.** similar **7.** inversely; inverse variation
8. directly; direct variation

Concept Reinforcement

1. True **2.** False **3.** True

Study Guide

1. 5, 6 **2.** $\frac{x-1}{2(x+5)}$ **3.** $\frac{y+5}{5(y+3)}$ **4.** $\frac{b+7}{b+8}$ **5.** $\frac{71}{180}$
6. $(x+2)(x-9)(x+9)$ **7.** −1 **8.** $\frac{x^2-4x-10}{(x+2)(x+1)(x-1)}$
9. $\frac{3(2y-5)}{5(9-y)}$ **10.** 1 **11.** $y = 150x; y = 300$
12. $y = \frac{225}{x}; y = 22.5$

Review Exercises

1. 0 **2.** 6 **3.** −6, 6 **4.** −6, 5 **5.** −2 **6.** None
7. $\frac{x-2}{x+1}$ **8.** $\frac{7x+3}{x-3}$ **9.** $\frac{y-5}{y+5}$ **10.** $\frac{a-6}{5}$ **11.** $\frac{6}{2t-1}$
12. $-20t$ **13.** $\frac{2x(x-1)}{x+1}$ **14.** $30x^2y^2$ **15.** $4(a-2)$
16. $(y-2)(y+2)(y+1)$ **17.** $\frac{-3(x-6)}{x+7}$ **18.** −1
19. $\frac{2a}{a-1}$ **20.** $d+c$ **21.** $\frac{4}{x-4}$ **22.** $\frac{x+5}{2x}$ **23.** $\frac{2x+3}{x-2}$
24. $\frac{-x^2+x+26}{(x-5)(x+5)(x+1)}$ **25.** $\frac{2(x-2)}{x+2}$ **26.** $\frac{z}{1-z}$
27. $c-d$ **28.** 8 **29.** −5, 3 **30.** $5\frac{1}{7}$ hr **31.** 95 mph, 175 mph
32. 240 km/h, 280 km/h **33.** 160 defective calculators

34. (a) $\frac{12}{13}$ c; (b) $4\frac{1}{5}$ c; (c) $9\frac{1}{3}$ c **35.** 10,000 blue whales **36.** 6
37. $y = 3x; 60$ **38.** $y = \frac{4}{5}x; 16$ **39.** $y = \frac{30}{x}; 6$ **40.** $y = \frac{1}{x}; \frac{1}{5}$
41. $y = \frac{0.65}{x}; 0.13$ **42.** \$288.75 **43.** 1 hr **44.** C **45.** A
46. $\frac{5(a+3)^2}{a}$ **47.** They are equivalent proportions.

Understanding Through Discussion and Writing

1. No; when we are adding, no sign changes are required, so the result is the same regardless of use of parentheses. When we are subtracting, however, the sign of each term of the expression being subtracted must be changed and parentheses are needed to make sure this is done. **2.** Graph each side of the equation and determine the number of points of intersection of the graphs.
3. Canceling removes a factor of 1, allowing us to rewrite $a \cdot 1$ as a.
4. Inverse variation; the greater the average gain per play, the smaller the number of plays required. **5.** Form a rational expression that has factors of $x+3$ and $x-4$ in the denominator.
6. If we multiply both sides of a rational equation by a variable expression in order to clear fractions, it is possible that the variable expression is equal to 0. Thus an equivalent equation might not be produced.

Test: Chapter 14, p. 1103

1. [14.1a] 0 **2.** [14.1a] −8 **3.** [14.1a] −7, 7 **4.** [14.1a] 1, 2
5. [14.1a] 1 **6.** [14.1a] None **7.** [14.1c] $\frac{3x+7}{x+3}$
8. [14.1d] $\frac{a+5}{2}$ **9.** [14.2b] $\frac{(5x+1)(x+1)}{3x(x+2)}$
10. [14.3a] $(y-3)(y+3)(y+7)$ **11.** [14.4a] $\frac{23-3x}{x^3}$
12. [14.5a] $\frac{2(4-t)}{t^2+1}$ **13.** [14.4a] $\frac{-3}{x-3}$ **14.** [14.5a] $\frac{2x-5}{x-3}$
15. [14.4a] $\frac{8t-3}{t(t-1)}$ **16.** [14.5a] $\frac{-x^2-7x-15}{(x+4)(x-4)(x+1)}$
17. [14.5b] $\frac{x^2+2x-7}{(x-1)^2(x+1)}$ **18.** [14.6a] $\frac{3y+1}{y}$
19. [14.7a] 12 **20.** [14.7a] −3, 5
21. [14.9a] $y = 2x; 50$ **22.** [14.9a] $y = 0.5x; 12.5$
23. [14.9c] $y = \frac{18}{x}; \frac{9}{50}$ **24.** [14.9c] $y = \frac{22}{x}; \frac{11}{50}$
25. [14.9b] 240 km **26.** [14.9d] $1\frac{1}{5}$ hr **27.** [14.8b] 16 defective spark plugs **28.** [14.8b] 50 zebras **29.** [14.7a] 12 min
30. [14.8a] Craig: 65 km/h; Marilyn: 45 km/h
31. [14.8b] 15 **32.** [14.6a] D
33. [14.8a] Rema: 4 hr; Reggie: 10 hr **34.** [14.6a] $\frac{3a+2}{2a+1}$

CHAPTER 15

Calculator Corner, p. 1110

1. (−1, 3) **2.** (−4, 1) **3.** (3, 2)

Exercise Set 15.1, p. 1110

RC1. True **RC2.** True **RC3.** True **RC4.** False
1. Yes **3.** No **5.** Yes **7.** Yes **9.** Yes
11. (4, 2) **13.** (4, 3) **15.** (−3, −3) **17.** No solution
19. (2, 2) **21.** $(\frac{1}{2}, 1)$ **23.** Infinite number of solutions
25. (5, −3) **27.** $\frac{x+5}{x-5}$ **28.** $\frac{d+2}{5d^2-1}$ **29.** −3 **30.** $\frac{2x+5}{x+3}$

31. Trinomial **32.** Binomial **33.** Monomial **34.** None of these **35.** $A = 2, B = 2$ **37.** $x + 2y = 2,$ $x - y = 8$ **39.–41.** Left to the student

Exercise Set 15.2, p. 1117

RC1. True **RC2.** True **RC3.** False **RC4.** True **1.** $(-2, 1)$ **3.** $(2, -4)$ **5.** $(4, 3)$ **7.** $(2, -3)$ **9.** $(1, 9)$ **11.** $(4, -3)$ **13.** $(2, -4)$ **15.** $\left(\frac{17}{3}, \frac{16}{3}\right)$ **17.** $(6, 3)$ **19.** $\left(\frac{25}{8}, -\frac{11}{4}\right)$ **21.** $(-4, 3)$ **23.** $(-3, 0)$ **25.** Length: $3\frac{1}{2}$ in.; width: $1\frac{1}{2}$ in. **27.** Length: 365 mi; width: 275 mi **29.** Length: 110 yd; width: 60 yd **31.** 16 and 21 **33.** 12 and 40 **35.** 20 and 8 **37.** Netfix: 10 hr; Hulu: 2.5 hr

39. $2x - 3y = 6$

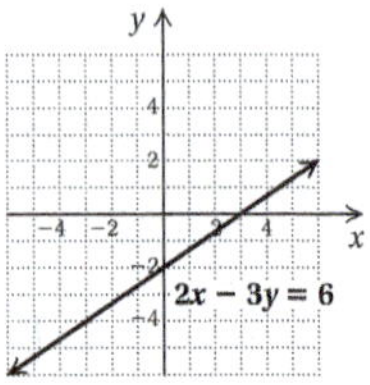

40.

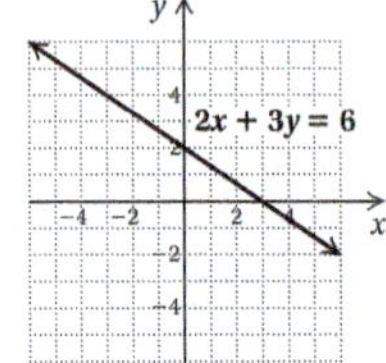

41.

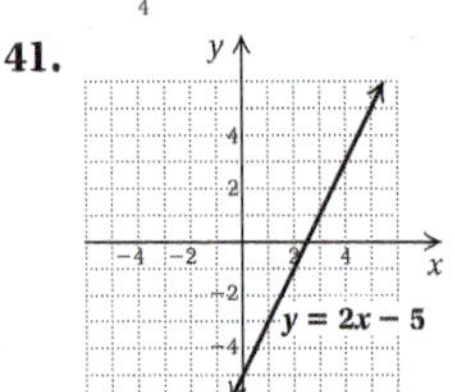

42.

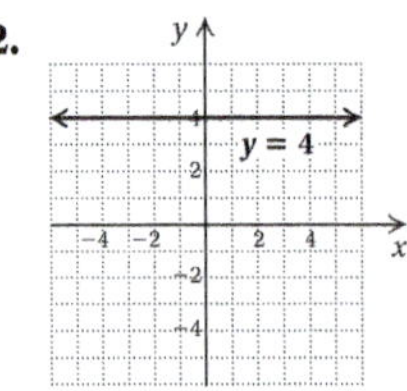

43. $(3x - 2)(2x - 3)$ **44.** $(4p + 3)(p - 1)$ **45.** Not factorable **46.** $(3a - 5)(3a + 5)$ **47.** x^3 **48.** x^7 **49.** $\frac{1}{x^7}$ **50.** $\frac{b^3}{a^3}$ **51.** $(5.\overline{6}, 0.\overline{6})$ **53.** $(4.38, 4.33)$ **55.** Baseball: 30 yd; softball: 20 yd

Exercise Set 15.3, p. 1125

RC1. False **RC2.** True **RC3.** True **RC4.** False **RC5.** False **RC6.** False **1.** $(6, -1)$ **3.** $(3, 5)$ **5.** $(2, 5)$ **7.** $\left(-\frac{1}{2}, 3\right)$ **9.** $\left(-1, \frac{1}{5}\right)$ **11.** No solution **13.** $(-1, -6)$ **15.** $(3, 1)$ **17.** $(8, 3)$ **19.** $(4, 3)$ **21.** $(1, -1)$ **23.** $(-3, -1)$ **25.** $(3, 2)$ **27.** $(50, 18)$ **29.** Infinite number of solutions **31.** $(2, -1)$ **33.** $\left(\frac{231}{202}, \frac{117}{202}\right)$ **35.** $(-38, -22)$ **37.** -5 **38.** $\frac{3}{2}$ **39.** No solution **40.** $\left\{y \mid y \geq -\frac{1}{2}\right\}$ **41.** $0, \frac{1}{2}$ **42.** $-4, 5$ **43.** $\frac{13}{3}$ **44.** $-2, \frac{1}{2}$ **45.–53.** Left to the student **55.** $(5, 2)$ **57.** $(0, -1)$ **59.** $(0, 3)$ **61.** $x = \frac{c - b}{a - 1}, y = \frac{ac - b}{a - 1}$

Mid-Chapter Review: Chapter 15, p. 1128

1. False **2.** False **3.** True **4.** True

5.
$$\begin{aligned} x + x - 3 &= -1 \\ 2x - 3 &= -1 \\ 2x &= -1 + 3 \\ 2x &= 2 \\ x &= 1 \end{aligned}$$

$$\begin{aligned} y &= 1 - 3 \\ y &= -2 \end{aligned}$$

The solution is $(1, -2)$.

6.
$$\begin{aligned} 2x - 3y &= 7 \\ x + 3y &= -10 \\ \hline 3x + 0y &= -3 \\ 3x &= -3 \\ x &= -1 \end{aligned}$$

$$\begin{aligned} -1 + 3y &= -10 \\ 3y &= -9 \\ y &= -3 \end{aligned}$$

The solution is $(-1, -3)$.

7. Yes **8.** No **9.** No **10.** Yes **11.** $(3, -2)$ **12.** $(-2, 3)$ **13.** Infinite number of solutions **14.** No solution **15.** $(5, -3)$ **16.** $(-2, -1)$ **17.** $\left(\frac{5}{3}, -\frac{2}{3}\right)$ **18.** $\left(\frac{1}{4}, \frac{3}{2}\right)$ **19.** No solution **20.** $(0, -1)$ **21.** $(-4, 3)$ **22.** Infinite number of solutions **23.** Length: 5 ft; width: 4 ft **24.** 52 and -34 **25.** 12 and 8 **26.** We know that the first coordinate of the point of intersection is 2. We substitute 2 for x in either $y = 3x - 1$ or $y = 9 - 2x$ and find y, the second coordinate of the point of intersection, 5. Thus the graphs intersect at $(2, 5)$. **27.** The coordinates of the point of intersection of the graphs are not integers, so it is difficult to determine the solution from the graph. **28.** The equations have the same coefficients of x and y but different constant terms. This means that their graphs have the same slope but different y-intercepts. Thus they have no points in common and the system of equations has no solution. **29.** This is not the best approach, in general. If the first equation has x alone on one side, for instance, or if the second equation has a variable alone on one side, solving for y in the first equation is inefficient. This procedure could also introduce fractions in the computations unnecessarily.

Exercise Set 15.4, p. 1136

RC1. 10 **RC2.** 15 **RC3.** $0.15y$ **RC4.** 2 **1.** Discussion post: 15 points; reply: 5 points **3.** Adults: 90; children: 230 **5.** 4×6 prints: 30; 5×7 prints: 6 **7.** Two-pointers: 35; three-pointers: 5 **9.** \$50 bonds: 13; \$100 bonds: 6 **11.** Solution A: 40 L; solution B: 60 L **13.** Hay: 10 lb; grain: 5 lb **15.** Dimes: 70; quarters: 33 **17.** Brazilian: 200 lb; Turkish: 100 lb **19.** 70% cashews: 36 lb; 45% cashews: 24 lb **21.** Dr. Zeke's: $53\frac{1}{3}$ oz; Vitabrite: $26\frac{2}{3}$ oz **23.** Type A: 12 questions; type B: 4 questions; 180 **25.** Kuyatts': 32 years; Marconis': 16 years **27.** Randy: 24; Marie: 6 **29.** 50°, 130° **31.** 28°, 62° **33.** 87-octane: 12 gal; 93-octane: 6 gal **35.** Large type: 4 pages; small type: 8 pages **37.** $x^2 + x$ **38.** $a^2 - \frac{1}{4}$ **39.** $t^4 + 2.4t^2 + 1.44$ **40.** $6x^2 - 11x - 35$ **41.** $x^3 - 1$ **42.** $3mn - m^2n + mn^2$ **43.** $\frac{-2(x - 5)}{(x + 1)(x - 1)}$ **44.** $\frac{-4}{x - 2}$, or $\frac{4}{2 - x}$ **45.** $\frac{2x^2 - 1}{x^2(x + 1)}$ **46.** $\frac{(2a - 1)(a - 4)}{(a - 2)(a - 5)}$ **47.** $\frac{cx}{9}$ **48.** $\frac{(t + 2)(t - 2)}{(t + 1)(t - 1)}$

49.

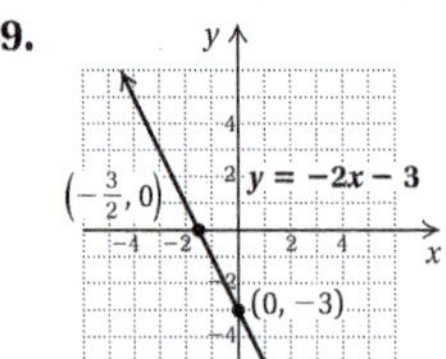

50.

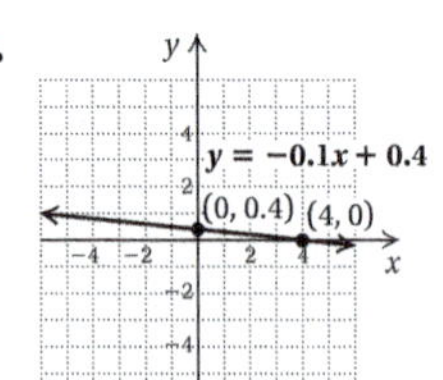

51.

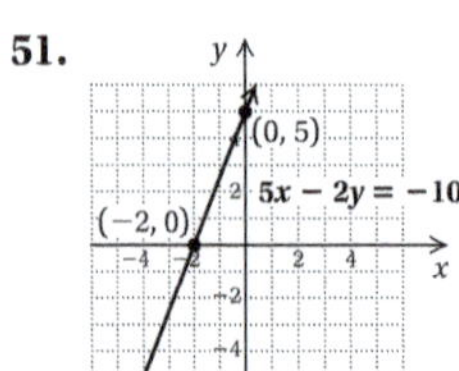

52.

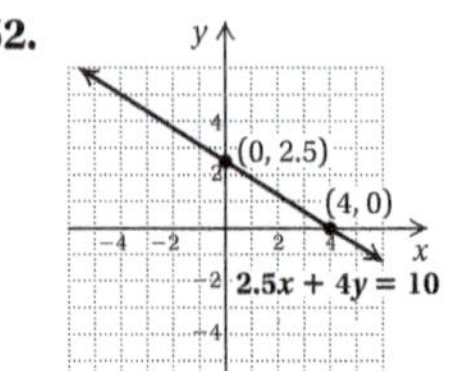

53. 43.75 L **55.** $4\frac{4}{7}$ L **57.** 54

Translating for Success, p. 1145

1. C **2.** A **3.** G **4.** E **5.** J **6.** I **7.** B **8.** L **9.** O **10.** F

Exercise Set 15.5, p. 1146

RC1. $60t$ **RC2.** $r + 2$ **RC3.** $r - 4$ **RC4.** $125 - r$

1.

Speed	Time
30	t
46	t

4.5 hr

3.

Speed	Time	
72	$t + 3$	$\to d = 72(t + 3)$
120	t	$\to d = 120t$

$4\frac{1}{2}$ hr

5.

Speed	Time	
$r + 6$	4	$\to d = (r + 6)4$
$r - 6$	10	$\to d = (r - 6)10$

14 km/h

7. 384 km **9.** **(a)** 24 mph; **(b)** 90 mi **11.** $1\frac{23}{43}$ min after the toddler starts running, or $\frac{23}{43}$ min after the mother starts running **13.** 15 mi **15.** $(5x + 9)(5x - 9)$ **16.** $(6a - 1)(2a + 3)$ **17.** $y(3y - 2)^2$ **18.** $7(x^2 + 2)(x + 1)$ **19.** Approximately 3603 mi

Summary and Review: Chapter 15, p. 1148

Vocabulary Reinforcement

1. pair **2.** intersection **3.** algebraic **4.** no solution **5.** no solution **6.** infinitely many solutions

Concept Reinforcement

1. False **2.** True **3.** True **4.** False

Study Guide

1. Yes **2.** $(4, -2)$ **3.** $(-2, 1)$ **4.** $(4, -3)$

Review Exercises

1. No **2.** Yes **3.** Yes **4.** No **5.** $(5, -2)$ **6.** Infinite number of solutions **7.** No solution **8.** $(0, 5)$ **9.** $(-3, 9)$ **10.** $(3, -1)$ **11.** $(1, 4)$ **12.** $(-2, 4)$ **13.** $(1, -2)$ **14.** $(3, 1)$ **15.** $(1, 4)$ **16.** No solution **17.** $(-2, 4)$ **18.** $(-2, -6)$ **19.** $(3, 2)$ **20.** $(2, -4)$ **21.** Infinite number of solutions **22.** $(-4, 1)$ **23.** Length: 37.5 cm; width: 10.5 cm **24.** Orchestra: 297 seats; balcony: 211 seats **25.** 40 L of each **26.** Asian: 4800 kg; African: 7200 kg **27.** Peanuts: 8 lb; fancy nuts: 5 lb **28.** 87-octane: 2.5 gal; 95-octane: 7.5 gal **29.** Jeff: 39; his son: 13 **30.** 32°, 58° **31.** 77°, 103° **32.** 135 km/h **33.** 412.5 mi **34.** D **35.** A **36.** $C = 1$, $D = 3$ **37.** $(2, 0)$ **38.** $960 **39.** $y = -x + 5, y = \frac{2}{3}x$ **40.** $x + y = 4, x + y = -3$ **41.** Rabbits: 12; pheasants: 23

Understanding Through Discussion and Writing

1. The second equation can be obtained by multiplying both sides of the first equation by -2. Thus the equations have the same graph, so the system of equations has an infinite number of solutions. **2.** The multiplication principle might be used to obtain a pair of terms that are opposites. The addition principle is used to eliminate a variable. Once a variable has been eliminated, the multiplication and addition principles are also used to solve for the remaining variable and, after a substitution, are used again to find the variable that was eliminated. **3.** Answers will vary. **4.** A chart allows us to see the given information and the missing information clearly and to see the relationships that yield equations.

Test: Chapter 15, p. 1153

1. [15.1a] No **2.** [15.1b]

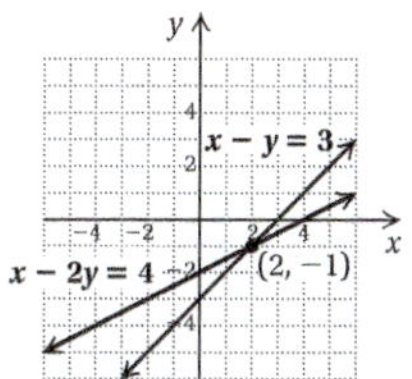

3. [15.2a] $(8, -2)$ **4.** [15.2b] $(-1, 3)$ **5.** [15.2a] $(1, 3)$ **6.** [15.3a] $(1, -5)$ **7.** [15.3b] No solution **8.** [15.3b] $\left(\frac{1}{2}, -\frac{2}{3}\right)$ **9.** [15.3b] $(5, 1)$ **10.** [15.2c] Length: 2108.5 yd; width: 2024.5 yd **11.** [15.4a] Solution A: 40 L; solution B: 20 L **12.** [15.5a] 40 km/h **13.** [15.2c] Concessions: $2850; rides: $1425 **14.** [15.2c] Hay: 415 acres; oats: 235 acres **15.** [15.2c] 45°, 135° **16.** [15.4a] 87-octane: 4 gal; 93-octane: 8 gal **17.** [15.4a] 20 min **18.** [15.5a] 11 hr **19.** [15.1b], [15.2b], [15.3b] D **20.** [15.1a] $C = -\frac{19}{2}$; $D = \frac{14}{3}$ **21.** [15.4a] 5 people **22.** [11.4c], [15.1b] $y = \frac{1}{5}x + \frac{17}{5}$, $y = -\frac{3}{5}x + \frac{9}{5}$ **23.** [11.4c], [15.1b] $x = 3$, $y = -2$

CHAPTER 16

Calculator Corner, p. 1157

1. 6.557 **2.** 10.050 **3.** 102.308 **4.** 0.632 **5.** −96.985 **6.** −0.804

Exercise Set 16.1, p. 1160

RC1. True **RC2.** False **RC3.** False **RC4.** True **RC5.** True **RC6.** False **1.** 2, −2 **3.** 3, −3 **5.** 10, −10 **7.** 13, −13 **9.** 16, −16 **11.** 2 **13.** −3 **15.** −6 **17.** −15 **19.** 19 **21.** 2.236 **23.** 20.785 **25.** −18.647 **27.** 2.779 **29.** −168.375 **31.** **(a)** About 3029 GPM; **(b)** about 4601 GPM **33.** 0.944 sec **35.** 0.888 sec **37.** 200 **39.** x **41.** $t^2 + 1$ **43.** $\dfrac{3}{x + 2}$ **45.** No **47.** Yes **49.** No **51.** c **53.** $3x$ **55.** $8p$ **57.** ab **59.** $34d$ **61.** $x + 3$ **63.** $a - 5$ **65.** $2a - 5$ **67.** $11y - 9$ **69.** 61°, 119° **70.** 38°, 52° **71.** $10,660 **72.** $\dfrac{1}{x + 3}$ **73.** 1 **74.** $\dfrac{(x + 2)(x - 2)}{(x + 1)(x - 1)}$ **75.** 1.7, 2.2, 2.6 **77.** 16, −16 **79.** 7, −7

Calculator Corner, p. 1167

1. False **2.** False **3.** False **4.** True

Exercise Set 16.2, p. 1168

RC1. No **RC2.** Yes **RC3.** No **RC4.** No **RC5.** Yes **RC6.** No **RC7.** Yes **RC8.** Yes **RC9.** No **1.** $2\sqrt{3}$ **3.** $5\sqrt{3}$ **5.** $2\sqrt{5}$ **7.** $10\sqrt{6}$ **9.** $9\sqrt{6}$ **11.** $3\sqrt{x}$ **13.** $4\sqrt{3x}$ **15.** $4\sqrt{a}$ **17.** $8y$ **19.** $x\sqrt{13}$ **21.** $2t\sqrt{2}$ **23.** $6\sqrt{5}$ **25.** $12\sqrt{2y}$ **27.** $2x\sqrt{7}$ **29.** $x - 3$ **31.** $\sqrt{2}(2x + 1)$, or $(2x + 1)\sqrt{2}$ **33.** $\sqrt{y}(6 + y)$, or $(6 + y)\sqrt{y}$ **35.** t^3 **37.** x^6 **39.** $x^2\sqrt{x}$ **41.** $t^9\sqrt{t}$ **43.** $(y - 2)^4$ **45.** $2(x + 5)^5$ **47.** $6m\sqrt{m}$ **49.** $2a^2\sqrt{2a}$ **51.** $2p^8\sqrt{26p}$ **53.** $8x^3y\sqrt{7y}$ **55.** $3\sqrt{6}$ **57.** $3\sqrt{10}$ **59.** $6\sqrt{7x}$ **61.** $6\sqrt{xy}$ **63.** 13 **65.** $5b\sqrt{3}$ **67.** $2t$ **69.** $a\sqrt{bc}$ **71.** $2xy\sqrt{2xy}$ **73.** 18 **75.** $\sqrt{10x - 5}$ **77.** $x + 2$ **79.** $6xy^3\sqrt{3xy}$ **81.** $10x^2y^3\sqrt{5xy}$ **83.** $33p^4q^2\sqrt{2pq}$ **85.** $16a^3b^3c^5\sqrt{3abc}$ **87.** $(-2, 4)$ **88.** $\left(\frac{1}{8}, \frac{9}{8}\right)$ **89.** $(2, 1)$ **90.** $(10, 3)$ **91.** 30% insecticide: 80 L; 50% insecticide: 120 L **92.** 360 ft^2

93. 10 mph **94.** 211 adults and 171 children **95.** $\sqrt{5}\sqrt{x-1}$
97. $\sqrt{x+6}\sqrt{x-6}$ **99.** $x\sqrt{x-2}$ **101.** 0.5 **103.** 2
105. $6(x-2)^2\sqrt{10}$ **107.** $2^{54}x^{158}\sqrt{2x}$ **109.** $a^2 - 5\sqrt{a}$

Exercise Set 16.3, p. 1175

RC1. (e) **RC2.** (c) **RC3.** (f) **RC4.** (d) **RC5.** (a)
RC6. (f) **1.** 3 **3.** 6 **5.** $\sqrt{5}$ **7.** $\frac{1}{5}$ **9.** $\frac{2}{5}$ **11.** 2
13. $3y$ **15.** $\frac{4}{7}$ **17.** $\frac{1}{6}$ **19.** $-\frac{4}{9}$ **21.** $\frac{8}{17}$ **23.** $\frac{13}{14}$ **25.** $\frac{5}{x}$
27. $\frac{3a}{25}$ **29.** $\frac{5}{y^5}$ **31.** $\frac{x^9}{7}$ **33.** $\frac{\sqrt{10}}{5}$ **35.** $\frac{\sqrt{14}}{4}$ **37.** $\frac{\sqrt{3}}{6}$
39. $\frac{\sqrt{10}}{6}$ **41.** $\frac{3\sqrt{5}}{5}$ **43.** $\frac{2\sqrt{6}}{3}$ **45.** $\frac{\sqrt{3x}}{x}$ **47.** $\frac{\sqrt{xy}}{y}$
49. $\frac{x\sqrt{5}}{10}$ **51.** $\frac{\sqrt{3}}{3}$ **53.** $\frac{3\sqrt{2}}{4}$ **55.** $\frac{\sqrt{55}}{5}$ **57.** $\sqrt{2}$ **59.** $\frac{\sqrt{55}}{11}$
61. $\frac{\sqrt{21}}{6}$ **63.** $\frac{\sqrt{6}}{2}$ **65.** 5 **67.** $\frac{\sqrt{3x}}{x}$ **69.** $\frac{4y\sqrt{5}}{5}$ **71.** $\frac{a\sqrt{2a}}{4}$
73. $\frac{\sqrt{42x}}{3x}$ **75.** $\frac{3\sqrt{6}}{8c}$ **77.** $\frac{y\sqrt{xy}}{x}$ **79.** $\frac{3n\sqrt{10}}{8}$ **81.** $\left(-\frac{5}{2}, -\frac{9}{2}\right)$
82. Infinite number of solutions **83.** No solution
84. $\frac{x-3}{x-2}$ **85.** $\frac{a-5}{2}$ **86.** $\frac{(x-2)(x-5)}{(x-3)(x-4)}$
87. 1.57 sec; 3.14 sec; 1.01 sec **89.** $\frac{\sqrt{3xy}}{ax^2}$

Mid-Chapter Review: Chapter 16, p. 1179

1. True **2.** False **3.** False **4.** True

5.
$$\begin{aligned}\sqrt{3x^2 - 48x + 192} &= \sqrt{3(x^2 - 16x + 64)}\\ &= \sqrt{3(x-8)^2}\\ &= \sqrt{3}\sqrt{(x-8)^2}\\ &= \sqrt{3}(x-8)\end{aligned}$$

6.
$$\begin{aligned}\sqrt{30}\sqrt{40y} &= \sqrt{30 \cdot 40y}\\ &= \sqrt{1200y}\\ &= \sqrt{100 \cdot 12 \cdot y}\\ &= \sqrt{100 \cdot 4 \cdot 3 \cdot y}\\ &= \sqrt{100}\sqrt{4}\sqrt{3y}\\ &= 10 \cdot 2\sqrt{3y}\\ &= 20\sqrt{3y}\end{aligned}$$

7.
$$\begin{aligned}\sqrt{18ab^2}\sqrt{14a^2b^4} &= \sqrt{18ab^2 \cdot 14a^2b^4}\\ &= \sqrt{2 \cdot 3 \cdot 3 \cdot 2 \cdot 7 \cdot a^3 \cdot b^6}\\ &= \sqrt{2^2 \cdot 3^2 \cdot 7 \cdot a^2 \cdot a \cdot b^6}\\ &= \sqrt{2^2}\sqrt{3^2}\sqrt{a^2}\sqrt{b^6}\sqrt{7a}\\ &= 2 \cdot 3 \cdot a \cdot b^3\sqrt{7a}\\ &= 6ab^3\sqrt{7a}\end{aligned}$$

8.
$$\begin{aligned}\sqrt{\frac{3y^2}{44}} &= \sqrt{\frac{3y^2}{2 \cdot 2 \cdot 11}} = \sqrt{\frac{3y^2}{2 \cdot 2 \cdot 11} \cdot \frac{11}{11}}\\ &= \sqrt{\frac{33y^2}{2^2 \cdot 11^2}} = \frac{y\sqrt{33}}{2 \cdot 11} = \frac{y\sqrt{33}}{22}\end{aligned}$$

9. $-11, 11$ **10.** $\frac{x-3}{7}$ **11.** **(a)** No; **(b)** yes **12.** $8r^3s^3\sqrt{2r}$
13. $5(x-3)$ **14.** $\frac{1}{10}$ **15.** -6 **16.** $-\frac{25}{7}$ **17.** 15 **18.** $10y$
19. $2x - 1$ **20.** $20\sqrt{2x}$ **21.** $\frac{1}{4}$ **22.** $4q^5\sqrt{2q}$ **23.** $\frac{9}{z}$
24. 25 **25.** $\frac{3}{7}$ **26.** 8 **27.** $10c^2d^5\sqrt{6c}$ **28.** $12x^3y^5z\sqrt{10yz}$
29. $2\sqrt{15xy}$ **30.** $7a\sqrt{15}$ **31.** $\frac{y^{12}}{5}$

32.
$\frac{x}{\sqrt{3}}$ $\qquad$ $\frac{3\sqrt{x}}{x}$
$\sqrt{\frac{3}{x}}$ $\qquad$ $\frac{x}{\sqrt{3x}}$
$\frac{3}{\sqrt{x}}$ $\qquad$ $\frac{x\sqrt{3}}{3}$
$\frac{3x}{\sqrt{3}}$ $\qquad$ $\sqrt{3}$
$\frac{3}{\sqrt{3}}$ $\qquad$ $\frac{\sqrt{3x}}{x}$
$\sqrt{\frac{x}{3}}$ $\qquad$ $x\sqrt{3}$

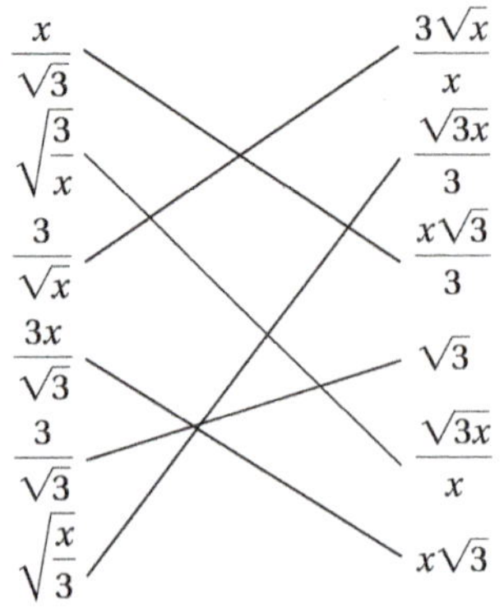

33. The square root of 100 is the principal, or positive, square root, which is 10. **A** square root of 100 could refer to either the positive square root or the negative square root, 10 or -10.
34. It is incorrect to take the square roots of the terms in the numerator individually—that is, $\sqrt{a+b}$ and $\sqrt{a} + \sqrt{b}$ are not equivalent. The following is correct:
$$\sqrt{\frac{9+100}{25}} = \frac{\sqrt{9+100}}{\sqrt{25}} = \frac{\sqrt{109}}{5}.$$
35. In general, $\sqrt{a^2 - b^2} \neq \sqrt{a^2} - \sqrt{b^2}$. In this case, let $x = 13$. Then $\sqrt{x^2 - 25} = \sqrt{13^2 - 25} = \sqrt{169 - 25} = \sqrt{144} = 12$, but $\sqrt{x^2} - \sqrt{25} = \sqrt{13^2} - \sqrt{25} = 13 - 5 = 8$.
36. **(1)** If necessary, rewrite the expression as $\sqrt{a}/\sqrt{b}$. **(2)** Simplify the numerator and the denominator, if possible, by taking the square roots of perfect square factors. **(3)** Multiply by a form of 1 that produces an expression without a radical in the numerator.

Exercise Set 16.4, p. 1185

RC1. Yes **RC2.** No **RC3.** No **RC4.** Yes **RC5.** Yes
RC6. No **RC7.** No **RC8.** No **RC9.** Yes
1. $16\sqrt{3}$ **3.** $4\sqrt{5}$ **5.** $13\sqrt{x}$ **7.** $-9\sqrt{d}$ **9.** $25\sqrt{2}$ **11.** $\sqrt{3}$
13. $\sqrt{5}$ **15.** $13\sqrt{2}$ **17.** $3\sqrt{3}$ **19.** $2\sqrt{2}$ **21.** 0
23. $(2 + 9x)\sqrt{x}$, or $\sqrt{x}(2 + 9x)$ **25.** $(3 - 2x)\sqrt{3}$
27. $3\sqrt{2x+2}$ **29.** $(x+3)\sqrt{x^3 - 1}$ **31.** $(4a^2 + a^2b - 5b)\sqrt{b}$
33. $\frac{2}{3}\sqrt{3}$, or $\frac{2\sqrt{3}}{3}$ **35.** $\frac{13}{2}\sqrt{2}$, or $\frac{13\sqrt{2}}{2}$ **37.** $\frac{1}{6}\sqrt{6}$, or $\frac{\sqrt{6}}{6}$
39. $\sqrt{15} - \sqrt{3}$ **41.** $10 - 2\sqrt{7} + 5\sqrt{3} - \sqrt{21}$ **43.** $9 - 4\sqrt{5}$
45. -62 **47.** 1 **49.** $13 + \sqrt{5}$ **51.** $x - 2\sqrt{xy} + y$
53. $-\sqrt{3} - \sqrt{5}$ **55.** $5 - 2\sqrt{6}$ **57.** $\frac{4\sqrt{10} - 4}{9}$
59. $5 - 2\sqrt{7}$ **61.** $\frac{12 - 3\sqrt{x}}{16 - x}$ **63.** $\frac{24 + 3\sqrt{x} + 8\sqrt{2} + \sqrt{2x}}{64 - x}$
65. $\frac{2\sqrt{a} - a - 1}{1 - a}$ **67.** $\frac{4\sqrt{a} + 4\sqrt{t} + \sqrt{3a} + \sqrt{3t}}{a - t}$ **69.** $\frac{5}{11}$
70. $-\frac{38}{13}$ **71.** $-1, 6$ **72.** 2, 5 **73.** $\frac{x-3}{4(x+3)}$
74. $-9, -2, -5, -17, -0.678375$ **75.** 14,270 ft
77. Not correct **79.** $\frac{-4\sqrt{6}}{5}$

Calculator Corner, p. 1191

1. Left to the student **2.** Left to the student

Exercise Set 16.5, p. 1193

RC1. 1 **RC2.** 2 **RC3.** 1 **RC4.** 1 **RC5.** 2 **RC6.** 2
1. 36 **3.** 18.49 **5.** 165 **7.** $\frac{621}{2}$ **9.** 5 **11.** 3 **13.** $\frac{17}{4}$
15. No solution **17.** No solution **19.** 9 **21.** 12 **23.** 1, 5
25. 3 **27.** 5 **29.** No solution **31.** $-\frac{10}{3}$ **33.** 3
35. No solution **37.** 9 **39.** 1 **41.** 8 **43.** 256
45. 62 m/sec **47.** 121 m **49.** 211.25 ft; 281.25 ft

51. $s = 2R - t$ **52.** $\{x|x > -8\}$ **53.** -1728 **54.** y^5w^2
55. $(2x + 1)(x + 5)$ **56.** $(y - 6)(y + 6)$ **57.** $(3t + 4)^2$
58. $(1 + x^4)(1 + x^2)(1 + x)(1 - x)$ **59.** $-2, 2$ **61.** $-\frac{57}{16}$ **63.** 13

Translating for Success, p. 1200

1. J **2.** K **3.** N **4.** H **5.** G **6.** E **7.** O **8.** D
9. B **10.** C

Exercise Set 16.6, p. 1201

RC1. (d) **RC2.** (c) **1.** 17 **3.** $\sqrt{32} \approx 5.657$ **5.** 12
7. 4 **9.** 26 **11.** 12 **13.** 2 **15.** $\sqrt{2} \approx 1.414$ **17.** 5
19. 3 **21.** $\sqrt{109{,}444}$ ft ≈ 330.823 ft **23.** $\sqrt{75}$ m ≈ 8.660 m
25. $\sqrt{26{,}900}$ yd ≈ 164.012 yd **27.** $\left(-\frac{3}{2}, -\frac{1}{16}\right)$ **28.** $\left(\frac{8}{5}, 9\right)$
29. $\left(-\frac{9}{19}, \frac{91}{38}\right)$ **30.** $(-10, 1)$ **31.** $-\frac{1}{3}$ **32.** $\frac{5}{8}$
33. $12 - 2\sqrt{6} \approx 7.101$

Summary and Review: Chapter 16, p. 1203

Vocabulary Reinforcement

1. principal **2.** radical **3.** square **4.** rationalizing
5. radicands **6.** conjugates **7.** hypotenuse, Pythagorean

Concept Reinforcement

1. True **2.** True **3.** True **4.** False

Study Guide

1. $y^2 - 3$ **2. (a)** Yes; **(b)** no **3.** $20y\sqrt{3}$ **4.** $5a^6b^4\sqrt{7b}$
5. $4x^3y^2\sqrt{6x}$ **6.** $b\sqrt{3b}$ **7.** $\frac{5}{9}$ **8.** $\frac{a\sqrt{2}}{5}$ **9.** $(x + 6)\sqrt{x - 1}$
10. $9 + \sqrt{26}$ **11.** $\frac{47 - 14\sqrt{2}}{79}$ **12.** 6 **13.** 9, 81
14. $a = \sqrt{657} \approx 25.632$

Review Exercises

1. $8, -8$ **2.** $20, -20$ **3.** 6 **4.** -13 **5.** 1.732 **6.** 9.950
7. -17.892 **8.** 0.742 **9.** -2.055 **10.** 394.648 **11.** $x^2 + 4$
12. x **13.** $4 - x$ **14.** $\frac{2}{y - 7}$ **15.** Yes **16.** No **17.** No
18. Yes **19.** m **20.** $x - 4$ **21.** $4x$ **22.** $2p - 3$ **23.** $4\sqrt{3}$
24. $4t\sqrt{2}$ **25.** $t - 7$ **26.** $x + 8$ **27.** x^4 **28.** $5a^3\sqrt{3a}$
29. $\sqrt{21}$ **30.** $\sqrt{x^2 - 9}$ **31.** $2\sqrt{15}$ **32.** $2x\sqrt{10}$ **33.** $5xy\sqrt{2}$
34. $10a^2b\sqrt{ab}$ **35.** $\frac{5}{8}$ **36.** $\frac{7}{t}$ **37.** $\frac{c^4}{4}$ **38.** $\frac{\sqrt{2}}{2}$ **39.** $\frac{x\sqrt{15x}}{15}$
40. $\frac{\sqrt{5y}}{y}$ **41.** $\frac{b^3\sqrt{ab}}{a}$ **42.** $\frac{\sqrt{15}}{5}$ **43.** $\frac{x\sqrt{30}}{6}$ **44.** $13\sqrt{5}$
45. $\sqrt{5}$ **46.** $\frac{1}{2}\sqrt{2}$, or $\frac{\sqrt{2}}{2}$ **47.** $7 + 4\sqrt{3}$ **48.** 1
49. $8 - 4\sqrt{3}$ **50.** 52 **51.** No solution **52.** 0, 3 **53.** 9
54. (a) About 63 mph; **(b)** 405 ft **55.** 20 **56.** $\sqrt{3} \approx 1.732$
57. $\sqrt{2{,}600{,}000{,}000}$ ft $\approx 50{,}990$ ft **58.** 9 ft **59.** B **60.** C
61. $\sqrt{1525}$ mi ≈ 39.051 mi **62.** $b = \pm\sqrt{A^2 - a^2}$ **63.** 6

Understanding Through Discussion and Writing

1. It is necessary for the signs to differ to ensure that the product of the conjugates will be free of radicals. **2.** Since $\sqrt{11 - 2x}$ cannot be negative, the statement $\sqrt{11 - 2x} = -3$ cannot be true for any value of x, including 1. **3.** We often use the rules for manipulating exponents "in reverse" when simplifying radical expressions. For example, we might write x^5 as $x^4 \cdot x$ or y^6 as $(y^3)^2$. **4.** No; consider the clapboard's height above ground level to be one leg of a right triangle. Then the length of the ladder is the hypotenuse of that triangle. Since the length of the hypotenuse must be greater than the length of a leg, a 28-ft ladder cannot be used to repair a clapboard that is 28 ft above ground level. **5.** The square of a number is equal to the square of its opposite. Thus, while squaring both sides of a radical equation allows us to find the solutions of the original equation, this procedure can also introduce numbers that are not solutions of the original equation.
6. (a) $\sqrt{5x^2} = \sqrt{5}\sqrt{x^2} = \sqrt{5} \cdot |x| = |x|\sqrt{5}$. The given statement is correct. **(b)** Let $b = 3$. Then $\sqrt{b^2 - 4} = \sqrt{3^2 - 4} = \sqrt{9 - 4} = \sqrt{5}$, but $b - 2 = 3 - 2 = 1$. The given statement is false. **(c)** Let $x = 3$. Then $\sqrt{x^2 + 16} = \sqrt{3^2 + 16} = \sqrt{9 + 16} = \sqrt{25} = 5$, but $x + 4 = 3 + 4 = 7$. The given statement is false.

Test: Chapter 16, p. 1209

1. [16.1a] $9, -9$ **2.** [16.1a] 8 **3.** [16.1a] -5 **4.** [16.1b] 10.770
5. [16.1b] -9.349 **6.** [16.1b] 21.909 **7.** [16.1d] $4 - y^3$
8. [16.1e] Yes **9.** [16.1e] No **10.** [16.1f] a **11.** [16.1f] $6y$
12. [16.2c] $\sqrt{30}$ **13.** [16.2c] $\sqrt{x^2 - 64}$ **14.** [16.2a] $3\sqrt{3}$
15. [16.2a] $5\sqrt{x - 1}$ **16.** [16.2b] $t^2\sqrt{t}$ **17.** [16.2c] $5\sqrt{2}$
18. [16.2c] $3ab^2\sqrt{2}$ **19.** [16.3b] $\frac{3}{2}$ **20.** [16.3b] $\frac{12}{a}$
21. [16.3c] $\frac{\sqrt{10}}{5}$ **22.** [16.3c] $\frac{\sqrt{2xy}}{y}$ **23.** [16.3a, c] $\frac{3\sqrt{6}}{8}$
24. [16.3a] $\frac{\sqrt{7}}{4y}$ **25.** [16.4a] $-6\sqrt{2}$ **26.** [16.4a] $\frac{6}{5}\sqrt{5}$, or $\frac{6\sqrt{5}}{5}$
27. [16.4b] $21 - 8\sqrt{5}$ **28.** [16.4b] 11 **29.** [16.4c] $\frac{40 + 10\sqrt{5}}{11}$
30. [16.6a] $\sqrt{80} \approx 8.944$ **31.** [16.5a] 48 **32.** [16.5a] $-2, 2$
33. [16.5b] -3 **34.** [16.5c] **(a)** About 237 mi; **(b)** 34,060.5 ft
35. [16.6b] $\sqrt{15{,}700}$ yd ≈ 125.300 yd **36.** [16.3c] A
37. [16.1a] $\sqrt{5}$ **38.** [16.2b] y^{8n}

CHAPTER 17

Calculator Corner, p. 1217

1. 0.6, 1 **2.** $-1.5, 5$ **3.** 3, 8 **4.** 2, 4

Exercise Set 17.1, p. 1217

RC1. False **RC2.** True **RC3.** True **RC4.** True
1. $x^2 - 3x + 2 = 0$; $a = 1, b = -3, c = 2$ **3.** $7x^2 - 4x + 3 = 0$; $a = 7, b = -4, c = 3$ **5.** $2x^2 - 3x + 5 = 0$; $a = 2, b = -3, c = 5$ **7.** $0, -5$ **9.** $0, -2$ **11.** $0, \frac{2}{5}$
13. $0, -1$ **15.** 0, 3 **17.** $0, \frac{1}{5}$ **19.** $0, \frac{3}{14}$ **21.** $0, \frac{81}{2}$
23. $-12, 4$ **25.** $-5, -1$ **27.** $-9, 2$ **29.** 3, 5 **31.** -5
33. 4 **35.** $-\frac{2}{3}, \frac{1}{2}$ **37.** $-\frac{2}{3}, 4$ **39.** $-1, \frac{5}{3}$ **41.** $-5, -1$
43. $-2, 7$ **45.** $-5, 4$ **47.** 4 **49.** $-2, 1$ **51.** $-\frac{2}{5}, 10$
53. $-4, 6$ **55.** 1 **57.** 2, 5 **59.** No solution **61.** $-\frac{5}{2}, 1$
63. 35 diagonals **65.** 7 sides **67.** $9(2t - p + 1)$
68. $2x(x - 1)(x - 8)$ **69.** $(x + 7)^2$ **70.** $(6x - 5y)^2$
71. $(t^2 + 9)(t + 3)(t - 3)$ **72.** $(2a + 3)(3a - 5)$
73. $(a + c)(a + 1)$ **74.** $b(5a - 1)(4a + 1)$
75. $(xy + 1)(xy - 1)$ **76.** $\left(c + \frac{1}{2}\right)\left(c - \frac{1}{2}\right)$
77. $3(x + 1)^2(x - 1)^2$ **78.** $\left(\frac{1}{2}x - \frac{1}{4}y\right)^2$
79. $-\frac{1}{3}, 1$ **81.** $0, \frac{\sqrt{5}}{5}$ **83.** $-1.7, 4$ **85.** $-1.7, 3$
87. $-2, 3$ **89.** 4

Exercise Set 17.2, p. 1226

RC1. False **RC2.** False **RC3.** False **RC4.** True
1. $11, -11$ **3.** $\sqrt{7}, -\sqrt{7}$ **5.** $\frac{\sqrt{15}}{5}, -\frac{\sqrt{15}}{5}$ **7.** $\frac{5}{2}, -\frac{5}{2}$
9. $\frac{7\sqrt{3}}{3}, -\frac{7\sqrt{3}}{3}$ **11.** $\sqrt{3}, -\sqrt{3}$ **13.** $\frac{8}{7}, -\frac{8}{7}$ **15.** $-7, 1$
17. $-3 \pm \sqrt{21}$ **19.** $-13 \pm 2\sqrt{2}$ **21.** $7 \pm 2\sqrt{3}$

23. $-9 \pm \sqrt{34}$ **25.** $\frac{-3 \pm \sqrt{14}}{2}$ **27.** $-5, 11$ **29.** $-15, 1$
31. $-2, 8$ **33.** $-21, -1$ **35.** $1 \pm \sqrt{6}$ **37.** $11 \pm \sqrt{19}$
39. $-5 \pm \sqrt{29}$ **41.** $\frac{7 \pm \sqrt{57}}{2}$ **43.** $-7, 4$ **45.** $\frac{-3 \pm \sqrt{17}}{4}$
47. $\frac{-3 \pm \sqrt{145}}{4}$ **49.** $\frac{-2 \pm \sqrt{7}}{3}$ **51.** $-\frac{1}{2}, 5$ **53.** $-\frac{5}{2}, \frac{2}{3}$
55. About 13.0 sec **57.** About 9.2 sec
59. $3x^4 + 4x^3 - 2x^2 + 1$ **60.** $3x^4 - 4x^3 + 1$
61. $9x^4 + 6x^3 + x^2$ **62.** $2x^3 - x^2 + x + 1$ **63.** $4x^3 - 5x^2 - \frac{x}{3}$
64. $5x + 4 + \frac{1}{x - 1}$ **65.** $-12, 12$ **67.** $-16\sqrt{2}, 16\sqrt{2}$
69. $-2\sqrt{c}, 2\sqrt{c}$ **71.** $49.896, -49.896$ **73.** $-9, 9$

Calculator Corner, p. 1232

1. The equations $x^2 + x = -1$ and $x^2 + x + 1 = 0$ are equivalent. The graph of $y = x^2 + x + 1$ has no x-intercepts, so the equation $x^2 + x = -1$ has no real-number solutions.

Exercise Set 17.3, p. 1233

RC1. (c) **RC2.** (d) **RC3.** (a) **RC4.** (b) **RC5.** (e)
RC6. (f)
1. $-3, 7$ **3.** 3 **5.** $-\frac{4}{3}, 2$ **7.** $-\frac{5}{2}, \frac{3}{2}$ **9.** $-3, 3$ **11.** $1 \pm \sqrt{3}$
13. $5 \pm \sqrt{3}$ **15.** $-2 \pm \sqrt{7}$ **17.** $\frac{-4 \pm \sqrt{10}}{3}$ **19.** $\frac{5 \pm \sqrt{33}}{4}$
21. $\frac{1 \pm \sqrt{3}}{2}$ **23.** No real-number solutions **25.** $\frac{5 \pm \sqrt{73}}{6}$
27. $\frac{3 \pm \sqrt{29}}{2}$ **29.** $-\sqrt{5}, \sqrt{5}$ **31.** $-2 \pm \sqrt{3}$ **33.** $\frac{5 \pm \sqrt{37}}{2}$
35. $-1.3, 5.3$ **37.** $-0.2, 6.2$ **39.** $-1.2, 0.2$ **41.** $0.3, 2.4$
43. $\frac{11}{5}$ **44.** $\{x \mid x \le -\frac{1}{18}\}$ **45.** $\frac{1}{7}$ **46.** $-1, 8$ **47.** $-1, 1$
48. 75 **49.** 5 **50.** $-3, 0$ **51.** $-1, \frac{3}{2}$ **52.** $-4, 4$ **53.** $\frac{1}{2}$
54. $\frac{1}{2}, 4$ **55.** $0, 2$ **57.** $\frac{3 \pm \sqrt{5}}{2}$ **59.** $\frac{-7 \pm \sqrt{61}}{2}$
61. $\frac{-2 \pm \sqrt{10}}{2}$ **63.** Yes **65.–71.** Left to the student

Mid-Chapter Review: Chapter 17, p. 1235

1. True **2.** False **3.** True
4.
$$\begin{aligned} x^2 - 6x - 2 &= 0 \\ x^2 - 6x &= 2 \\ x^2 - 6x + 9 &= 2 + 9 \\ (x - 3)^2 &= 11 \\ x - 3 &= \pm\sqrt{11} \\ x &= 3 \pm \sqrt{11} \end{aligned}$$
5.
$$\begin{aligned} 3x^2 &= 8x - 2 \\ 3x^2 - 8x + 2 &= 0 \quad \text{Standard form} \\ a = 3, \quad b = -8, &\quad c = 2 \end{aligned}$$
We substitute for a, b, and c in the quadratic formula:
$$x = \frac{-b \pm \sqrt{b^2 - 4ac}}{2a} \quad \text{Quadratic formula}$$
$$x = \frac{-(-8) \pm \sqrt{(-8)^2 - 4 \cdot 3 \cdot 2}}{2 \cdot 3} \quad \text{Substituting}$$
$$x = \frac{8 \pm \sqrt{64 - 24}}{6} = \frac{8 \pm \sqrt{40}}{6} = \frac{8 \pm \sqrt{4 \cdot 10}}{6}$$
$$x = \frac{8 \pm 2\sqrt{10}}{6} = \frac{2(4 \pm \sqrt{10})}{2 \cdot 3} = \frac{4 \pm \sqrt{10}}{3}$$
6. $a = 1; b = -5; c = 10$ **7.** $a = 1; b = 14; c = -4$
8. $a = 3; b = -17; c = 0$ **9.** $0, \frac{1}{3}$ **10.** $-2, 5$ **11.** $0, 1$
12. 7 **13.** $-2, 0$ **14.** $-\frac{4}{3}, \frac{1}{6}$ **15.** $0, 22$ **16.** $2, 3$ **17.** $-\frac{7}{8}, 5$
18. $-3, 1$ **19.** $\frac{9 \pm \sqrt{57}}{2}$ **20.** $\frac{7 \pm \sqrt{113}}{4}$ **21.** $8, 10$
22. $\frac{-3 \pm \sqrt{33}}{4}$ **23.** No real-number solutions **24.** $-8, 8$
25. No real-number solutions **26.** $3 \pm \sqrt{2}$ **27.** $-\frac{3}{2}, \frac{1}{2}$
28. $\pm\sqrt{3}$ **29.** $\frac{1 \pm \sqrt{5}}{2}$ **30.** $-8, 12$ **31.** No real-number solutions **32.** $\frac{1 \pm \sqrt{5}}{2}$ **33.** $-5, 5$ **34.** $\frac{1 \pm \sqrt{11}}{2}$
35. No real-number solutions **36.** $-2.4, 3.4$ **37.** $-3.4, -0.1$
38. A **39.** B **40.** A **41.** C **42.** B **43.** C **44.** B
45. Mark does not recognize that the $\pm$ sign yields two solutions, one in which the radical is added to 3 and the other in which the radical is subtracted from 3. **46.** The addition principle should be used at the outset to get 0 on one side of the equation. Since this was not done in the given procedure, the principle of zero products was not applied correctly. **47.** The first coordinates of the x-intercepts of the graph of $y = (x - 2)(x + 3)$ are the solutions of the equation $(x - 2)(x + 3) = 0$. **48.** The quadratic formula would not be the easiest way to solve a quadratic equation when the equation can be solved by factoring or by using the principle of square roots. **49.** Answers will vary. Any equation of the form $ax^2 + bx + c = 0$, where $b^2 - 4ac < 0$, will do. Then the graph of the equation $y = ax^2 + bx + c$ will not cross the x-axis.
50. If $x = -5$ or $x = 7$, then $x + 5 = 0$ or $x - 7 = 0$. Thus the equation $(x + 5)(x - 7) = 0$, or $x^2 - 2x - 35 = 0$, has solutions -5 and 7.

Exercise Set 17.4, p. 1240

RC1. (e) **RC2.** (a) **RC3.** (e) **RC4.** (c) **RC5.** (d)
RC6. (e)
1. $I = \frac{VQ}{q}$ **3.** $m = \frac{Sd^2}{kM}$ **5.** $d^2 = \frac{kmM}{S}$ **7.** $W = \sqrt{\frac{10t}{T}}$
9. $t = \frac{A}{a + b}$ **11.** $x = \frac{y - c}{a + b}$ **13.** $a = \frac{bt}{b - t}$ **15.** $p = \frac{qf}{q - f}$
17. $b = \frac{2A}{h}$ **19.** $h = \frac{S - 2\pi r^2}{2\pi r}$, or $h = \frac{S}{2\pi r} - r$
21. $R = \frac{r_1 r_2}{r_2 + r_1}$ **23.** $Q = \frac{P^2}{289}$ **25.** $E = \frac{mv^2}{2g}$ **27.** $r = \frac{1}{2}\sqrt{\frac{S}{\pi}}$
29. $A = \frac{-m + \sqrt{m^2 + 4kP}}{2k}$ **31.** $a = \sqrt{c^2 - b^2}$ **33.** $t = \frac{\sqrt{s}}{4}$
35. $r = \frac{-\pi h + \sqrt{\pi^2 h^2 + \pi A}}{\pi}$ **37.** $v = 20\sqrt{\frac{F}{A}}$
39. $a = \sqrt{c^2 - b^2}$ **41.** $a = \frac{2h\sqrt{3}}{3}$
43. $T = \frac{2 + \sqrt{4 - a(m - n)}}{a}$ **45.** $T = \frac{v^2 \pi m}{8k}$ **47.** $x = \frac{d\sqrt{3}}{3}$
49. $n = \frac{1 + \sqrt{1 + 8N}}{2}$ **51.** $b = \frac{a}{3S - 1}$ **53.** $B = \frac{A}{QA + 1}$
55. $n = \frac{S + 360}{180}$, or $n = \frac{S}{180} + 2$ **57.** $t = \frac{A - P}{Pr}$
59. $D = \frac{BC}{A}$ **61.** $\sqrt{65} \approx 8.062$ **62.** $\sqrt{75} \approx 8.660$
63. $\sqrt{41} \approx 6.403$ **64.** $\sqrt{44} \approx 6.633$ **65.** $\sqrt{1084} \approx 32.924$
66. $\sqrt{5} \approx 2.236$ **67.** $\sqrt{424}$ ft ≈ 20.591 ft
68. $\sqrt{12{,}500}$ yd ≈ 111.803 yd **69.** $4\sqrt{5}$ **70.** $30x^5\sqrt{10}$
71. $3t$ **72.** $8x^2\sqrt{3x}$ **73.** $3\sqrt{10}$ **74.** $2\sqrt{2}$
75. **(a)** $r = \frac{C}{2\pi}$; **(b)** $A = \frac{C^2}{4\pi}$; **(c)** $C = 2\sqrt{A\pi}$

Translating for Success, p. 1246

1. M **2.** G **3.** F **4.** L **5.** D **6.** N **7.** J **8.** E
9. B **10.** C

Exercise Set 17.5, p. 1247

RC1. True **RC2.** False **RC3.** True **RC4.** False
1. Length: 16 yd; width: 5 yd **3.** 16 in.; 24 in. **5.** Length: 30 ft; width: 10 ft **7.** Length: 20 cm; width: 16 cm **9.** 4.6 m; 6.6 m **11.** Length: 5.6 in.; width: 3.6 in. **13.** Length: 6.4 cm; width: 3.2 cm **15.** 3 cm **17.** 7 km/h **19.** 2 km/h **21.** 0 km/h (no wind) or 40 km/h **23.** 8 mph **25.** 1 km/h **27.** y-intercept: $(0, 4)$; x-intercept: $\left(\frac{1}{2}, 0\right)$ **28.** y-intercept: $(0, -9)$; x-intercept: $(15, 0)$

29.
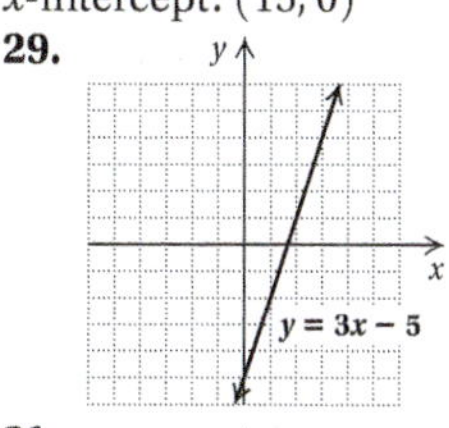

30.
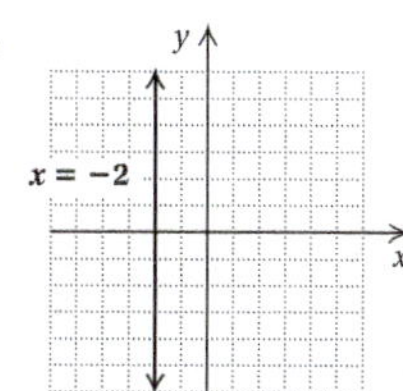

31.
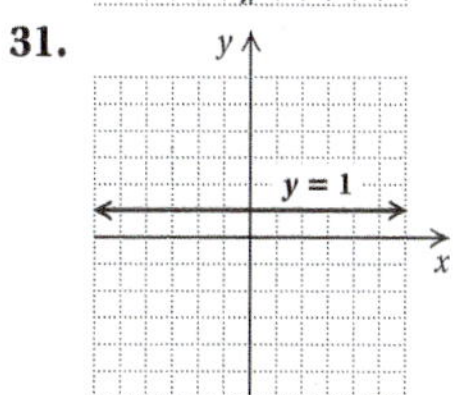

32.
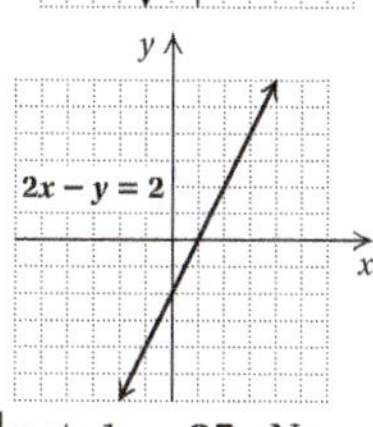

33. $y = -2x - 5$ **34.** $y = \frac{1}{2}x + 1$ **35.** No **36.** Yes
37. $12\sqrt{2}$ in. $\approx$ 16.97 in.; two 12-in. pizzas

Visualizing for Success, p. 1255

1. J **2.** F **3.** H **4.** G **5.** B **6.** E **7.** D **8.** I **9.** C **10.** A

Exercise Set 17.6, p. 1256

RC1. Quadratic **RC2.** Parabola **RC3.** Vertex **RC4.** Line

1.
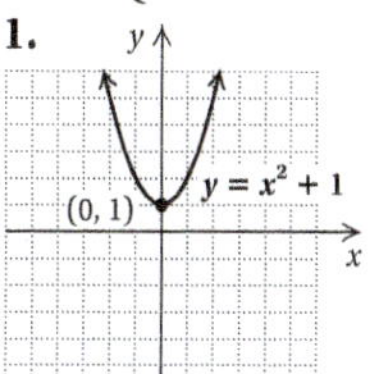

3.
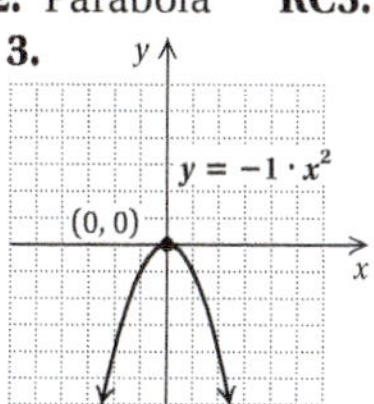

5.
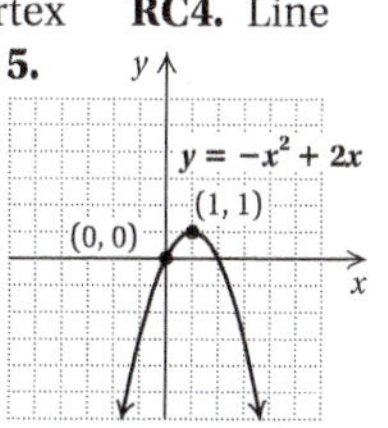

7.
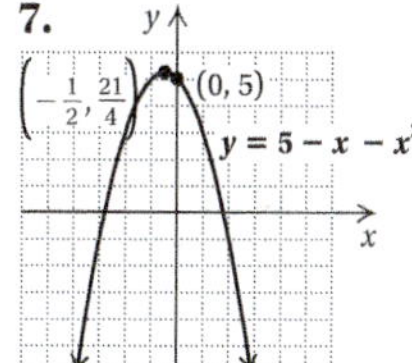

9.
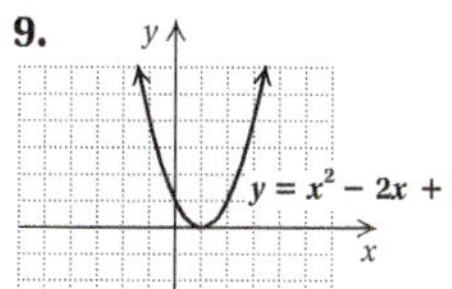

11.
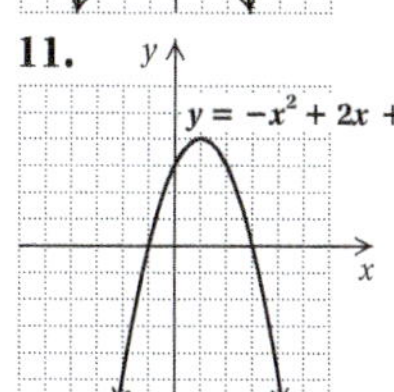

13.
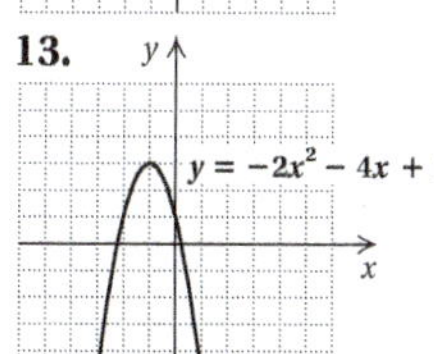

15.
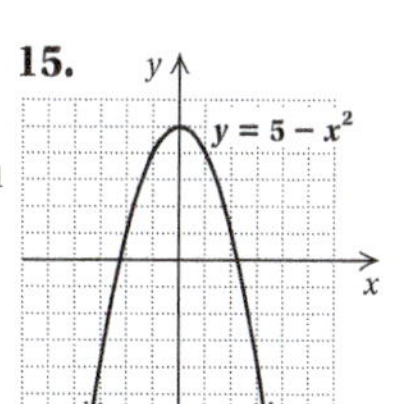

17.
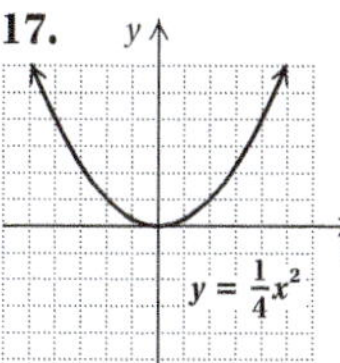

19.
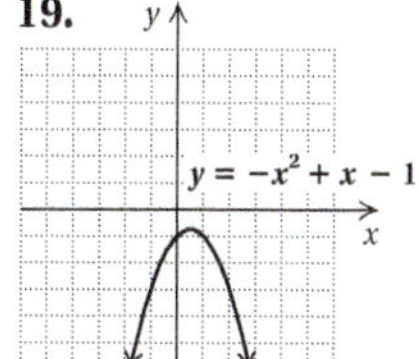

21.
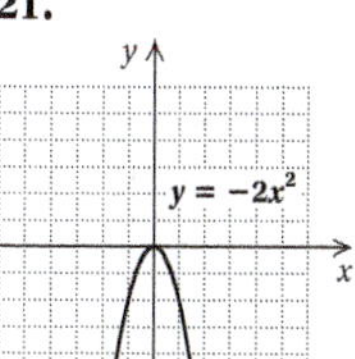

23.
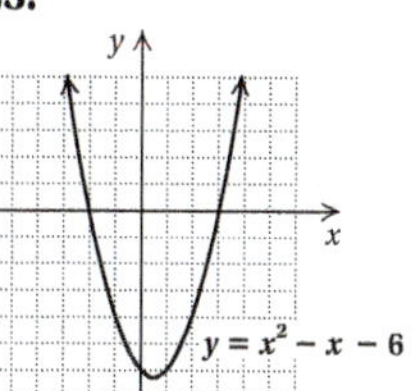

25. $(-\sqrt{2}, 0)$; $(\sqrt{2}, 0)$ **27.** $(-5, 0)$; $(0, 0)$ **29.** $\left(\frac{-1 - \sqrt{33}}{2}, 0\right)$; $\left(\frac{-1 + \sqrt{33}}{2}, 0\right)$ **31.** $(3, 0)$ **33.** $(-2 - \sqrt{5}, 0)$; $(-2 + \sqrt{5}, 0)$ **35.** None **37.** $22\sqrt{2}$ **38.** $25y^2\sqrt{y}$ **39.** $y = \dfrac{29.76}{x}$ **40.** 35 **41.** $\frac{47}{60}$ **42.** 49.55 **43.** **(a)** After 2 sec; after 4 sec; **(b)** after 3 sec; **(c)** after 6 sec **45.** 16; two real solutions **47.** -161.91; no real solutions

Calculator Corner, p. 1262

1. -12.8 **2.** -9.2 **3.** -2 **4.** 20

Exercise Set 17.7, p. 1265

RC1. domain; range **RC2.** input **RC3.** vertical **RC4.** linear
1. Yes **3.** Yes **5.** No **7.** Yes **9.** Yes **11.** Yes **13.** A relation but not a function **15.** **(a)** 9; **(b)** 12; **(c)** 2; **(d)** 5; **(e)** 7.4; **(f)** $5\frac{2}{3}$ **17.** **(a)** -21; **(b)** 15; **(c)** 42; **(d)** 0; **(e)** 2; **(f)** -162.6 **19.** **(a)** 7; **(b)** -17; **(c)** 24.1; **(d)** 4; **(e)** -26; **(f)** 6 **21.** **(a)** 0; **(b)** 5; **(c)** 2; **(d)** 170; **(e)** 65; **(f)** 230 **23.** **(a)** 1; **(b)** 3; **(c)** 3; **(d)** 4; **(e)** 11; **(f)** 23 **25.** **(a)** 0; **(b)** -1; **(c)** 8; **(d)** 1000; **(e)** -125; **(f)** -1000 **27.** **(a)** 34 years; **(b)** 20 years **29.** 2.7 million users; 548.7 million users; 1082.7 million users **31.** $1\frac{20}{33}$ atm; $1\frac{10}{11}$ atm; $4\frac{1}{33}$ atm

33.
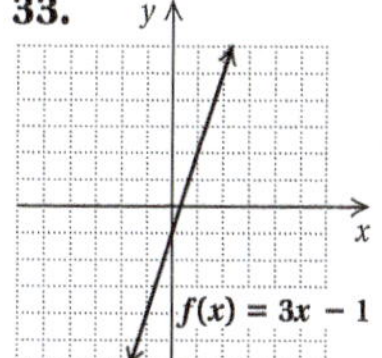

35.
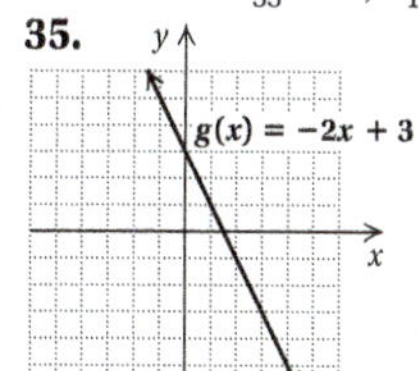

37.
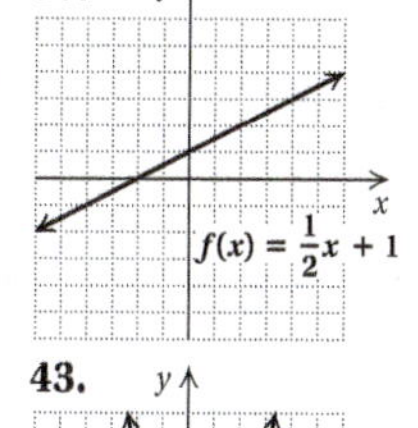

39.
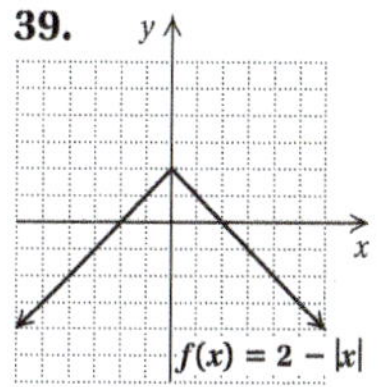

41.
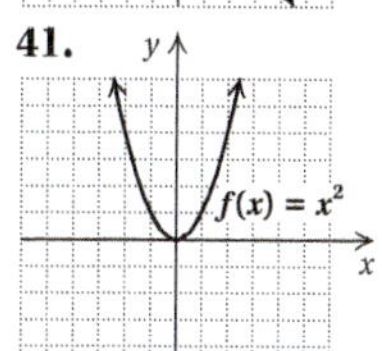

43.
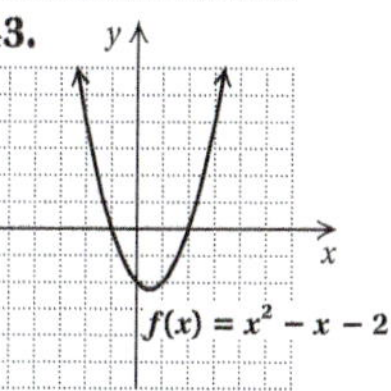

45. Yes **47.** Yes **49.** No **51.** No **53.** About 23,800 conventions **55.** $\left(\frac{3}{2}, \frac{1}{2}\right)$ **56.** $(3, 0)$ **57.** $\left(3, -\frac{1}{5}\right)$ **58.** Infinite number of solutions

59. y or g(x), x, $g(x) = x^3$

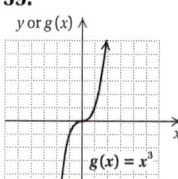

61.
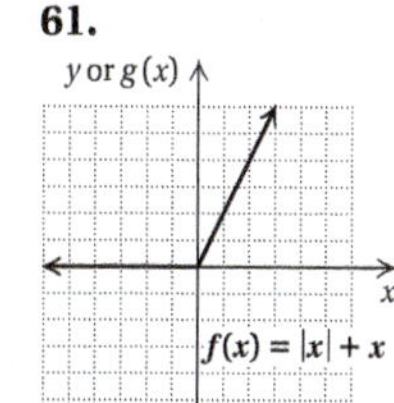

Summary and Review: Chapter 17, p. 1270

Vocabulary Reinforcement

1. quadratic **2.** square **3.** discriminant **4.** vertex **5.** parabola **6.** quadratic **7.** vertical-line **8.** domain

Concept Reinforcement

1. False **2.** True **3.** True **4.** True

Study Guide

1. $-\frac{4}{5}, 2$ **2.** $-\frac{\sqrt{77}}{7}, \frac{\sqrt{77}}{7}$ **3.** $2 \pm \sqrt{3}$ **4.** $\frac{3 \pm \sqrt{21}}{4}$

5.

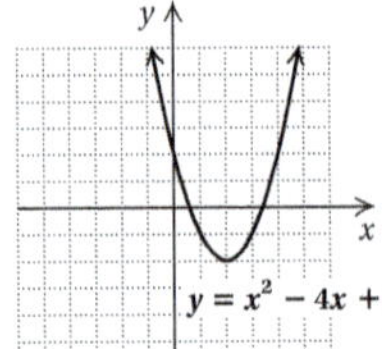

6. **(a)** $h(5) = 9$; **(b)** $f(0) = -4$

Review Exercises

1. $-\sqrt{3}, \sqrt{3}$ **2.** $-2\sqrt{2}, 2\sqrt{2}$ **3.** $\frac{3}{5}, 1$ **4.** $-2, \frac{1}{3}$ **5.** $-8 \pm \sqrt{13}$ **6.** 0 **7.** $0, \frac{7}{5}$ **8.** $1 \pm \sqrt{11}$ **9.** $\frac{1 \pm \sqrt{10}}{3}$ **10.** $-3 \pm 3\sqrt{2}$ **11.** $\frac{2 \pm \sqrt{3}}{2}$ **12.** $\frac{3 \pm \sqrt{33}}{2}$ **13.** No real-number solutions **14.** $0, \frac{4}{3}$ **15.** $-5, 3$ **16.** 1 **17.** $2 \pm \sqrt{2}$ **18.** $-1, \frac{5}{3}$ **19.** 0.4, 4.6 **20.** $-1.9, -0.1$ **21.** $T = L(4V^2 - 1)$

22.

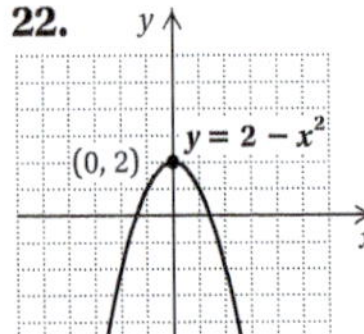

23.

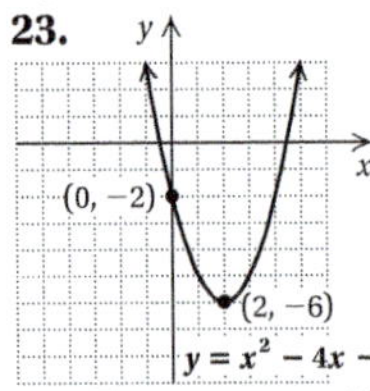

24. $(-\sqrt{2}, 0); (\sqrt{2}, 0)$ **25.** $(2 - \sqrt{6}, 0); (2 + \sqrt{6}, 0)$ **26.** 4.7 cm, 1.7 cm **27.** 15 ft **28.** About 8.1 sec **29.** $-1, -7, 2$ **30.** 0, 0, 19 **31.** 2700 calories

32.

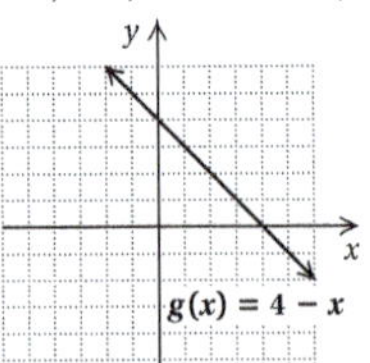

33.

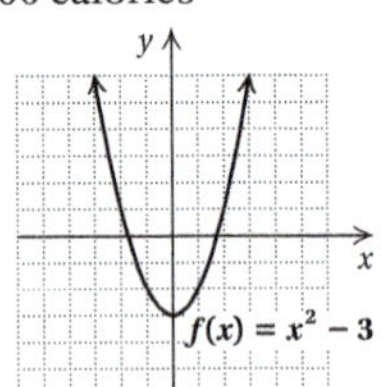

34.

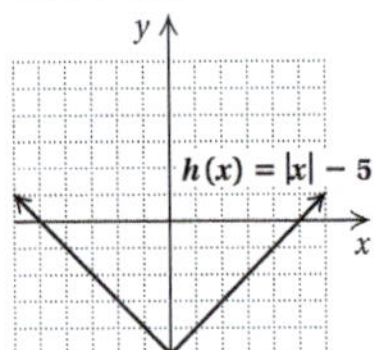

35.

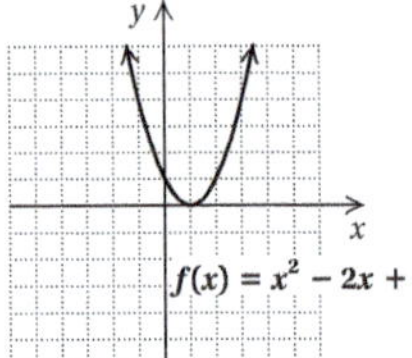

36. No **37.** Yes **38.** D **39.** A **40.** 31 and 32; -32 and -31 **41.** $5\sqrt{\pi}$ in., or about 8.9 in. **42.** 25 **43.** $-4, -2$ **44.** $-5, -1$ **45.** $-6, 0$ **46.** -3

Understanding Through Discussion and Writing

1. The second line should be $x + 6 = \sqrt{16}$ *or* $x + 6 = -\sqrt{16}$. Then we would have

$$x + 6 = 4 \quad \text{or} \quad x + 6 = -4$$
$$x = -2 \quad \text{or} \quad x = -10.$$

Both numbers check so the solutions are -2 and -10. **2.** No; since each input has exactly one output, the number of outputs cannot exceed the number of inputs. **3.** Find the average, v, of the x-coordinates of the x-intercepts, $v = \frac{a_1 + a_2}{2}$. Then the equation of the line of symmetry is $x = v$. The number v is also the first coordinate of the vertex. We substitute this value for x in the equation of the parabola to find the y-coordinate of the vertex. **4.** If $a > 0$, the graph opens up. If $a < 0$, the graph opens down. **5.** The solutions will be rational numbers because each is the solution of a linear equation of the form $mx + b = 0$.

Test: Chapter 17, p. 1275

1. [17.2a] $-\sqrt{5}, \sqrt{5}$ **2.** [17.1b] $-\frac{8}{7}, 0$ **3.** [17.1c] $-8, 6$ **4.** [17.1c] $-\frac{1}{3}, 2$ **5.** [17.2b] $8 \pm \sqrt{13}$ **6.** [17.3a] $\frac{1 \pm \sqrt{13}}{2}$ **7.** [17.3a] $\frac{3 \pm \sqrt{37}}{2}$ **8.** [17.3a] $-2 \pm \sqrt{14}$ **9.** [17.3a] $\frac{7 \pm \sqrt{37}}{6}$ **10.** [17.1c] $-1, 2$ **11.** [17.1c] $-4, 2$ **12.** [17.2c] $2 \pm \sqrt{14}$ **13.** [17.3b] $-1.7, 5.7$ **14.** [17.4a] $n = \frac{-b + \sqrt{b^2 + 4ad}}{2a}$ **15.** [17.6b] $\left(\frac{1 - \sqrt{21}}{2}, 0\right), \left(\frac{1 + \sqrt{21}}{2}, 0\right)$

16. [17.6a]

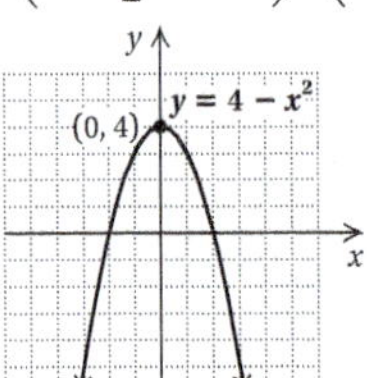

17. [17.6a]

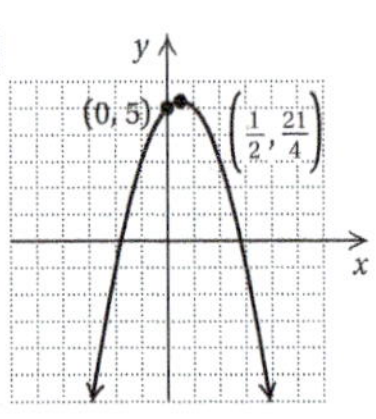

18. [17.7b] 1; $1\frac{1}{2}$; 2 **19.** [17.7b] 1; 3; -3 **20.** [17.5a] Length: 6.5 m; width: 2.5 m **21.** [17.5a] 24 km/h **22.** [17.7e] 25.86 min

23. [17.7c]

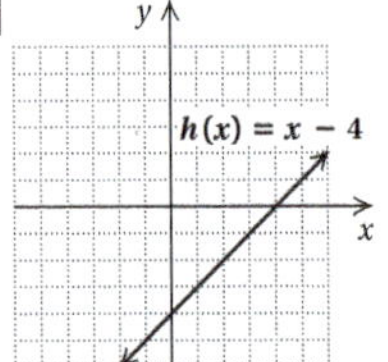

24. [17.7c]

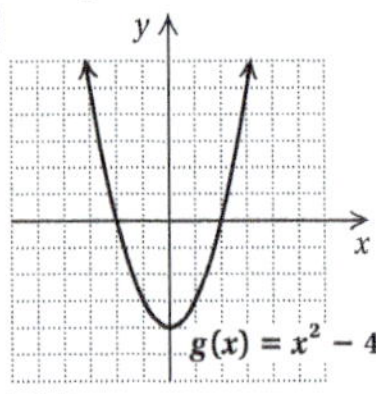

25. [17.7d] Yes **26.** [17.7d] No **27.** [17.7b] D **28.** [17.5a] $(5 + 5\sqrt{2})$ ft **29.** [17.2b], [17.3a] $(1 + \sqrt{5}, -1 + \sqrt{5}), (1 - \sqrt{5}, -1 - \sqrt{5})$

APPENDIXES

Exercise Set A, p. 1288

1. 12 **3.** $\frac{1}{12}$ **5.** 5280 **7.** 108 **9.** 7 **11.** $1\frac{1}{2}$, or 1.5 **13.** 26,400 **15.** $5\frac{1}{4}$, or 5.25 **17.** $3\frac{1}{3}$ **19.** 37,488 **21.** $1\frac{1}{2}$, or 1.5 **23.** $1\frac{1}{4}$, or 1.25 **25.** 110 **27.** 2 **29.** 300 **31.** 30 **33.** $\frac{1}{36}$ **35.** 126,720 **37.** **(a)** 1000; **(b)** 0.001 **39.** **(a)** 10; **(b)** 0.1 **41.** **(a)** 0.01; **(b)** 100 **43.** 6700 **45.** 0.98 **47.** 8.921 **49.** 0.05666 **51.** 566,600 **53.** 4.77 **55.** 688 **57.** 0.1 **59.** 100,000 **61.** 142 **63.** 0.82 **65.** 450 **67.** 0.000024 **69.** 0.688 **71.** 230 **73.** 3.92 **75.** 48,440; 48.44 **77.** 4000; 400 **79.** 0.027; 0.00027 **81.** 541,300; 54,130 **83.** 100.65 **85.** 727.4394 **87.** 104.585 **89.** 289.62 **91.** 112.63 **93.** 9.14 **95.** 75.1967 **97.** 555.83421 **99.** 6.18109 **101.** 1376.136 **103.** 0.51181 **105.–113.** Answers may vary, depending on the conversion factor used.

	yd	cm	in.	m	mm
105.	0.2361	21.59	$8\frac{1}{2}$	0.2159	215.9
107.	52.95934	4844	1907.0828	48.44	48,440
109.	4	365.6	144	3.656	3656
111.	0.000295	0.027	0.0106299	0.00027	0.27
113.	55,247	5,050,000	1,988,892	50,500	50,500,000

115. 1 in. = 25.4 mm

Exercise Set B, p. 1296

1. 2000 **3.** 3 **5.** 64 **7.** 12,640 **9.** 0.1 **11.** 5 **13.** 134,850 tons **15.** 1000 **17.** 10 **19.** $\frac{1}{100}$, or 0.01 **21.** 1000 **23.** 10 **25.** 234,000 **27.** 5.2 **29.** 6.7 **31.** 0.0502 **33.** 8.492 **35.** 58.5 **37.** 800,000 **39.** 1000 **41.** 0.0034 **43.** 0.0603 **45.** 1000 **47.** 0.325 **49.** 210,600 **51.** 0.0049 **53.** 125 mcg **55.** 0.875 mg; 875 mcg **57.** 4 tablets **59.** 8 cc **61.** **(a)** 3367.8 g; **(b)** 118.8 oz

Exercise Set C, p. 1302

1. 1000; 1000 **3.** 87,000 **5.** 0.049 **7.** 0.000401 **9.** 78,100 **11.** 320 **13.** 10 **15.** 32 **17.** 20 **19.** 14 **21.** 88

	gal	qt	pt	cups	oz
23.	1.125	4.5	9	18	144
25.	16	64	128	256	2048
27.	0.3984375	1.59375	3.1875	6.375	51

	L	mL	cc	cm^3
29.	2	2000	2000	2000
31.	64	64,000	64,000	64,000
33.	0.355	355	355	355

35. 2000 mL **37.** 0.32 L **39.** 59.14 mL **41.** 500 mL **43.** 125 mL/hr **45.** 9 **47.** $\frac{1}{5}$ **49.** 6 **51.** 15 **53.** 1.75 gal/week; 7.5 gal/month; 91.25 gal/year; 78.5 million gal/day; 28.6525 billion gal/year

Calculator Corner, p. 1306

1. 41°F **2.** 122°F **3.** 20°C **4.** 45°C

Exercise Set D, p. 1307

1. 24 **3.** 60 **5.** $365\frac{1}{4}$ **7.** 0.05 **9.** 8.2 **11.** 6.5 **13.** 10.75 **15.** 336 **17.** 4.5 **19.** 56 **21.** 86,164.2 sec **23.** 77°F **25.** 104°F **27.** 186.8°F **29.** 136.4°F **31.** 35.6°F **33.** 41°F **35.** 5432°F **37.** 30°C **39.** 55°C **41.** $81.\overline{1}$°C **43.** 60°C **45.** 20°C **47.** $6.\overline{6}$°C **49.** 37°C **51.** **(a)** 136°F = $57.\overline{7}$°C, $56\frac{2}{3}$°C = 134°F; **(b)** 2°F **53.** $53.\overline{3}$°C **55.** About 0.03 year

Exercise Set E, p. 1311

1. $\{3, 4, 5, 6, 7, 8\}$ **3.** $\{41, 43, 45, 47, 49\}$ **5.** $\{-3, 3\}$ **7.** False **9.** True **11.** True **13.** True **15.** True **17.** False **19.** {c, d, e} **21.** $\{1, 10\}$ **23.** { }, or $\varnothing$ **25.** {a, e, i, o, u, q, c, k} **27.** $\{0, 1, 7, 10, 2, 5\}$ **29.** {a, e, i, o, u, m, n, f, g, h} **31.** $\{x \mid x$ is an integer$\}$ **33.** $\{x \mid x$ is a real number$\}$ **35.** { }, or $\varnothing$ **37.** **(a)** A; **(b)** A; **(c)** A; **(d)** { }, or $\varnothing$ **39.** True

Exercise Set F, p. 1315

1. $(z+3)(z^2-3z+9)$ **3.** $(x-1)(x^2+x+1)$ **5.** $(y+5)(y^2-5y+25)$ **7.** $(2a+1)(4a^2-2a+1)$ **9.** $(y-2)(y^2+2y+4)$ **11.** $(2-3b)(4+6b+9b^2)$ **13.** $(4y+1)(16y^2-4y+1)$ **15.** $(2x+3)(4x^2-6x+9)$ **17.** $(a-b)(a^2+ab+b^2)$ **19.** $\left(a+\frac{1}{2}\right)\left(a^2-\frac{1}{2}a+\frac{1}{4}\right)$ **21.** $2(y-4)(y^2+4y+16)$ **23.** $3(2a+1)(4a^2-2a+1)$ **25.** $r(s+4)(s^2-4s+16)$ **27.** $5(x-2z)(x^2+2xz+4z^2)$ **29.** $(x+0.1)(x^2-0.1x+0.01)$ **31.** $8(2x^2-t^2)(4x^4+2x^2t^2+t^4)$ **33.** $2y(y-4)(y^2+4y+16)$ **35.** $(z-1)(z^2+z+1)(z+1)(z^2-z+1)$ **37.** $(t^2+4y^2)(t^4-4t^2y^2+16y^4)$ **39.** 1; 19; 19; 7; 1 **41.** $(x^{2a}+y^b)(x^{4a}-x^{2a}y^b+y^{2b})$ **43.** $3(x^a+2y^b)(x^{2a}-2x^ay^b+4y^{2b})$ **45.** $\frac{1}{3}\left(\frac{1}{2}xy+z\right)\left(\frac{1}{4}x^2y^2-\frac{1}{2}xyz+z^2\right)$ **47.** $y(3x^2+3xy+y^2)$ **49.** $4(3a^2+4)$

Exercise Set G, p. 1320

1. $y = 4x - 18$ **3.** $y = -2x + 12$ **5.** $y = 3x + 4$ **7.** $y = -3x - 6$ **9.** $y = 4$ **11.** $y = -\frac{4}{5}x + \frac{23}{5}$ **13.** $y = x + 3$ **15.** $y = x$ **17.** $y = \frac{5}{3}x - 5$ **19.** $y = 3x + 5$ **21.** $y = -\frac{7}{4}x$ **23.** $y = \frac{2}{11}x + \frac{91}{66}$ **25.** $y = 5x + 3$

Exercise Set H, p. 1323

1. $\{-3, 3\}$ **3.** $\varnothing$ **5.** $\{0\}$ **7.** $\{-9, 15\}$ **9.** $\left\{-\frac{1}{2}, \frac{7}{2}\right\}$ **11.** $\left\{-\frac{5}{4}, \frac{23}{4}\right\}$ **13.** $\{-11, 11\}$ **15.** $\{-291, 291\}$ **17.** $\{-8, 8\}$ **19.** $\{-7, 7\}$ **21.** $\{-2, 2\}$ **23.** $\{-7, 8\}$ **25.** $\{-12, 2\}$ **27.** $\left\{-\frac{5}{2}, \frac{7}{2}\right\}$ **29.** $\varnothing$ **31.** $\left\{-\frac{13}{54}, -\frac{7}{54}\right\}$ **33.** All real numbers **35.** $\left\{1, -\frac{1}{4}\right\}$

Exercise Set I, p. 1326

1. 5 **3.** $\sqrt{29} \approx 5.385$ **5.** $\sqrt{648} \approx 25.456$ **7.** 7.1 **9.** $\frac{\sqrt{41}}{7} \approx 0.915$ **11.** $\frac{9}{4}$ **13.** $\sqrt{6970} \approx 83.487$ **15.** $\sqrt{a^2+b^2}$ **17.** $\frac{2}{5}$ **19.** $\sqrt{17+2\sqrt{14}+2\sqrt{15}} \approx 5.677$ **21.** $\sqrt{9{,}672{,}400} \approx 3110.048$ **23.** $\left(\frac{3}{2}, \frac{7}{2}\right)$ **25.** $\left(0, \frac{11}{2}\right)$ **27.** $\left(-1, -\frac{17}{2}\right)$ **29.** $(-0.25, -0.3)$ **31.** $\left(-\frac{1}{12}, \frac{1}{24}\right)$ **33.** $\left(\frac{\sqrt{2}+\sqrt{3}}{2}, \frac{3}{2}\right)$ **35.** $(-4, 0)$ **37.** $\sqrt{49+k^2}$ **39.** $8\sqrt{m^2+n^2}$ **41.** Yes **43.** $(2, 4\sqrt{2})$

Exercise Set J, p. 1335

1. 4, −4 **3.** 12, −12 **5.** 20, −20 **7.** $-\frac{7}{6}$ **9.** 14 **11.** 0.06 **13.** Does not exist as a real number **15.** 18.628 **17.** 1.962 **19.** $y^2 + 16$ **21.** $\frac{x}{y-1}$ **23.** $4|x|$ **25.** $12|c|$ **27.** $|p+3|$ **29.** $|x-2|$ **31.** 3 **33.** $-4x$ **35.** −6 **37.** $0.7(x+1)$ **39.** −5 **41.** $-\frac{2}{3}$ **43.** $|x|$ **45.** $5|a|$ **47.** 6 **49.** $|a+b|$ **51.** y **53.** $x-2$ **55.** $\sqrt[5]{y}$ **57.** 2 **59.** $\sqrt[5]{a^3b^3}$ **61.** 8 **63.** 343 **65.** $17^{1/2}$ **67.** $18^{1/3}$ **69.** $(xy^2z)^{1/5}$ **71.** $(3mn)^{3/2}$ **73.** $(8x^2y)^{5/7}$ **75.** $\frac{1}{1000}$ **77.** $\frac{3}{x^{1/4}}$ **79.** $\frac{1}{(2rs)^{3/4}}$ **81.** $\left(\frac{8yz}{7x}\right)^{3/5}$ **83.** $\frac{5ac^{1/2}}{3}$ **85.** $5^{7/8}$ **87.** $7^{1/4}$ **89.** $4.9^{1/2}$ **91.** $6^{3/28}$ **93.** $a^{23/12}$

Exercise Set K, p. 1340

1. $(-\infty, 5)$ **3.** $[-3, 3]$ **5.** $(-8, -4)$ **7.** $(-2, 5)$ **9.** $(-\sqrt{2}, \infty)$ **11.** $\{x \mid x > -1\}$, or $(-1, \infty)$ **13.** $\{x \mid x > 3\}$, or $(3, \infty)$ **15.** $\{x \mid x < -60\}$, or $(-\infty, -60)$ **17.** $\{a \mid a \le -22\}$, or $(-\infty, -22]$ **19.** $\{x \mid x \le 0.9\}$, or $(-\infty, 0.9]$ **21.** $\left\{x \mid x \le \frac{5}{6}\right\}$, or $\left(-\infty, \frac{5}{6}\right]$ **23.** $\{x \mid x < 6\}$, or $(-\infty, 6)$ **25.** $\{y \mid y \le -3\}$, or $(-\infty, -3]$ **27.** $\left\{y \mid y > \frac{2}{3}\right\}$, or $\left(\frac{2}{3}, \infty\right)$

Exercise Set L, p. 1345

1. $\{x \mid x < -2 \text{ or } x > 6\}$, or $(-\infty, -2) \cup (6, \infty)$
3. $\{x \mid -2 \le x \le 2\}$, or $[-2, 2]$ **5.** $\{x \mid -1 \le x \le 4\}$, or $[-1, 4]$
7. $\{x \mid -1 < x < 2\}$, or $(-1, 2)$ **9.** All real numbers, or $(-\infty, \infty)$
11. $\{x \mid 2 < x < 4\}$, or $(2, 4)$
13. $\{x \mid x < -2 \text{ or } 0 < x < 2\}$, or $(-\infty, -2) \cup (0, 2)$
15. $\{x \mid -9 < x < -1 \text{ or } x > 4\}$, or $(-9, -1) \cup (4, \infty)$
17. $\{x \mid x < -3 \text{ or } -2 < x < 1\}$, or $(-\infty, -3) \cup (-2, 1)$

19.

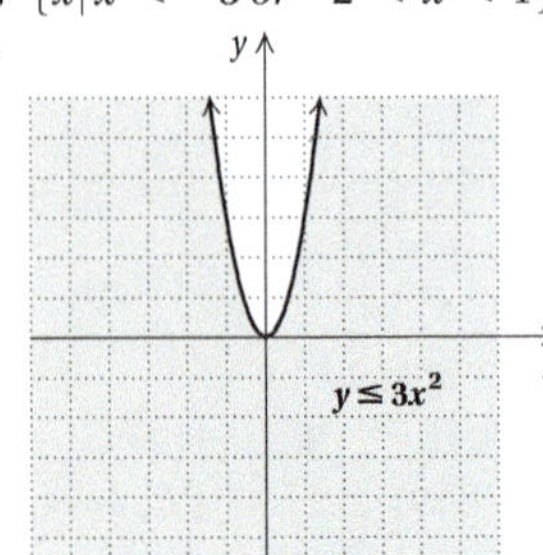

21.

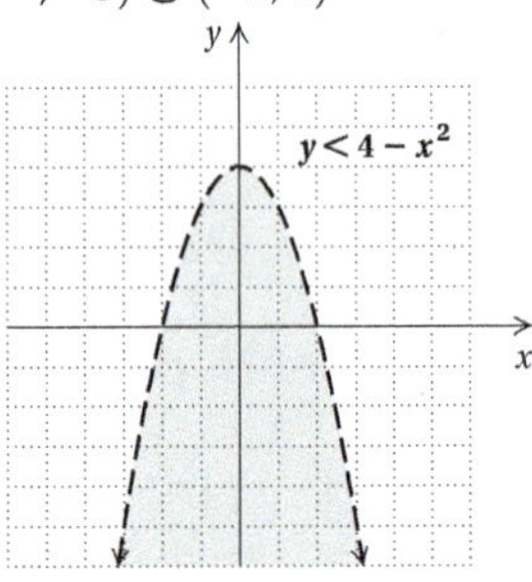

23.

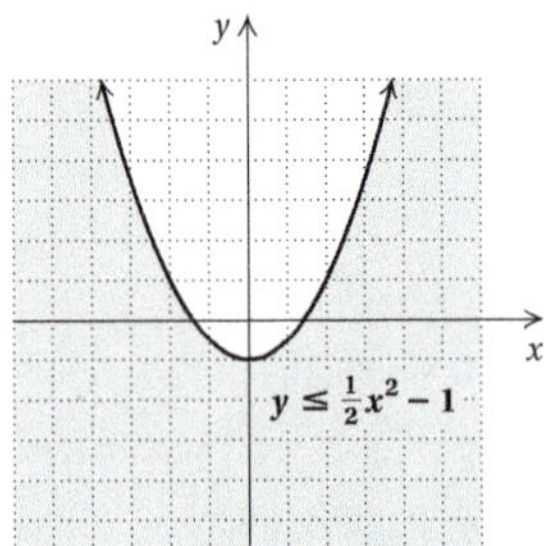

25.

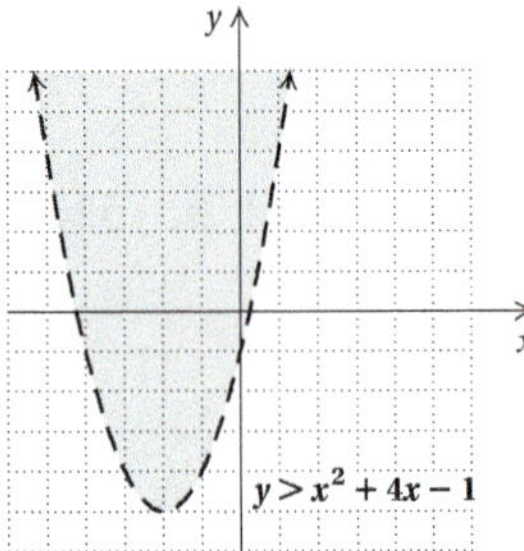

27.

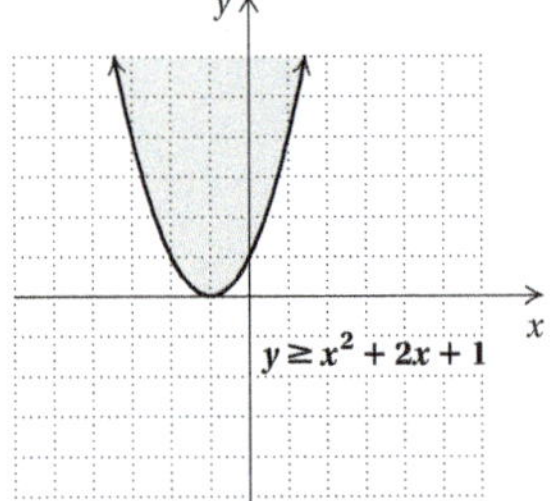

29.

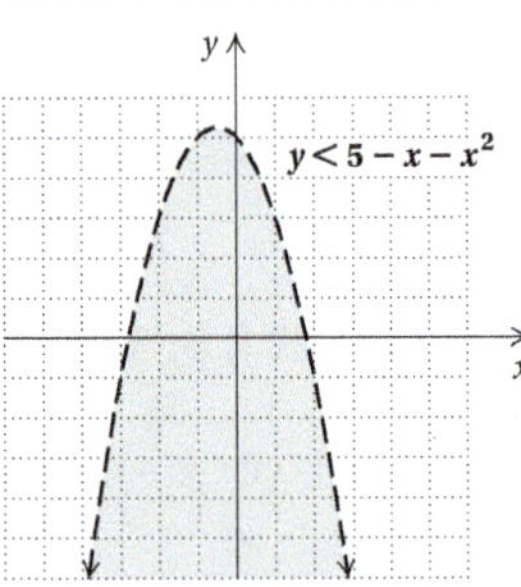

31.

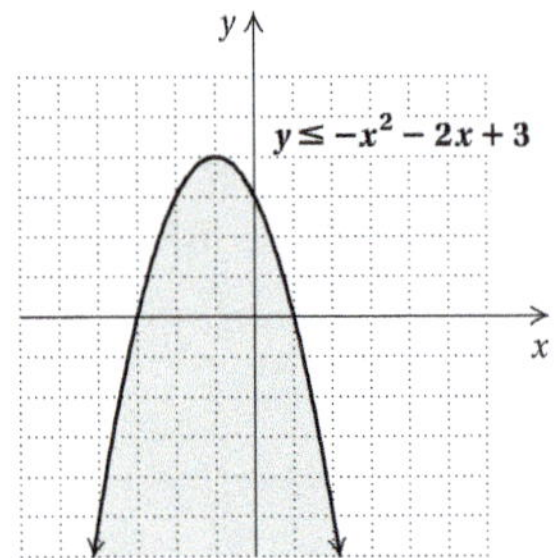

33.

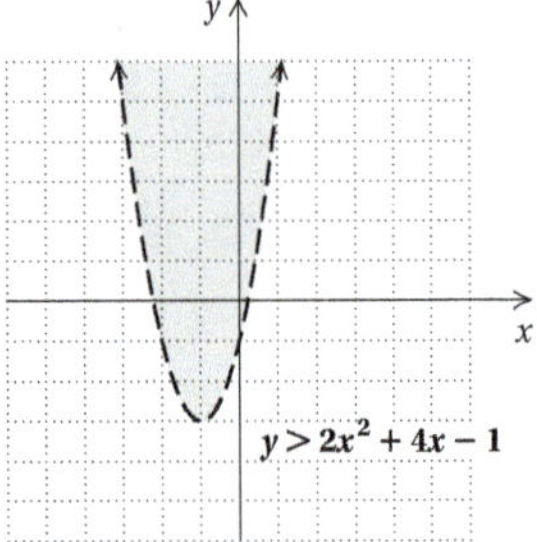

Exercise Set M, p. 1351

1. y, x, (1, 1)

3.

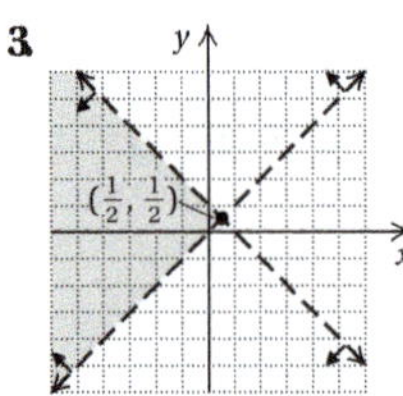

5.

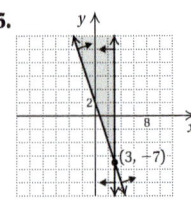

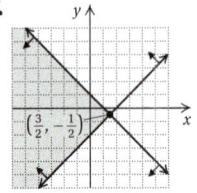

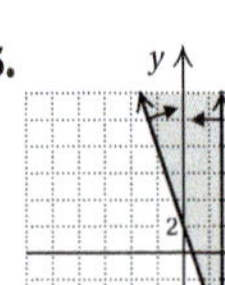

7. y, x, $\left(\frac{3}{2}, -\frac{1}{2}\right)$

9.

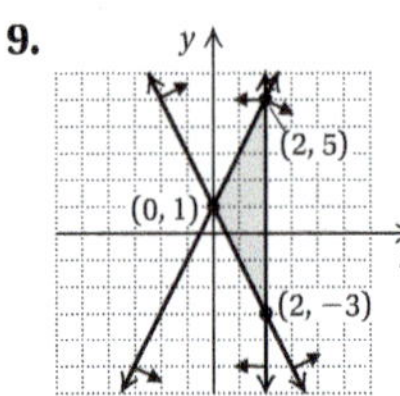

11.

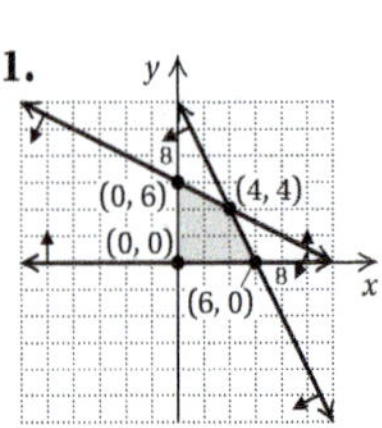

Exercise Set N, p. 1356

1. 0.57, 0.43 **3.** 0.075, 0.134, 0.057, 0.071, 0.030 **5.** 0.633
7. 52 **9.** $\frac{1}{4}$ **11.** $\frac{1}{2}$ **13.** $\frac{2}{13}$ **15.** $\frac{2}{7}$ **17.** 0 **19.** $\frac{5}{36}$
21. $\frac{5}{36}$ **23.** $\frac{1}{36}$

Exercise Set O, p. 1364

1. $i\sqrt{35}$, or $\sqrt{35}i$ **3.** $4i$ **5.** $-2i\sqrt{3}$, or $-2\sqrt{3}i$ **7.** $i\sqrt{3}$, or $\sqrt{3}i$ **9.** $9i$ **11.** $7i\sqrt{2}$, or $7\sqrt{2}i$ **13.** $-7i$ **15.** $4 - 2\sqrt{15}i$, or $4 - 2i\sqrt{15}$ **17.** $12 - 4i$ **19.** $9 - 5i$ **21.** $7 + 4i$
23. $-4 - 4i$ **25.** $-1 + i$ **27.** $11 + 6i$ **29.** -18
31. $-\sqrt{14}$ **33.** 21 **35.** $-6 + 24i$ **37.** $1 + 5i$
39. $18 + 14i$ **41.** $38 + 9i$ **43.** $2 - 46i$ **45.** $5 - 12i$
47. $-24 + 10i$ **49.** $-i$ **51.** 1 **53.** -1 **55.** i **57.** -1
59. $-125i$ **61.** 8 **63.** $1 - 23i$ **65.** 0 **67.** 0 **69.** 1
71. $5 - 8i$ **73.** $2 - \dfrac{\sqrt{6}}{2}i$ **75.** $\frac{9}{10} + \frac{13}{10}i$ **77.** $-i$
79. $-\frac{3}{7} - \frac{8}{7}i$ **81.** $\frac{6}{5} - \frac{2}{5}i$ **83.** $-\frac{8}{41} + \frac{10}{41}i$ **85.** $-\frac{4}{3}i$
87. $-\frac{1}{2} - \frac{1}{4}i$ **89.** $-\frac{3}{5} + \frac{4}{5}i$

91.

$$\begin{array}{r|l} \multicolumn{2}{c}{x^2 - 2x + 5 = 0} \\ \hline (1 - 2i)^2 - 2(1 - 2i) + 5 & ?\ 0 \\ 1 - 4i + 4i^2 - 2 + 4i + 5 & \\ 1 - 4i - 4 - 2 + 4i + 5 & \\ 0 & \quad \text{TRUE} \end{array}$$

Yes

93.

$$\begin{array}{r|l} \multicolumn{2}{c}{x^2 - 4x - 5 = 0} \\ \hline (2 + i)^2 - 4(2 + i) - 5 & ?\ 0 \\ 4 + 4i + i^2 - 8 - 4i - 5 & \\ 4 + 4i - 1 - 8 - 4i - 5 & \\ -10 & \quad \text{FALSE} \end{array}$$

No

95. $\pm\frac{5}{3}i$ **97.** $7 \pm 2i$ **99.** $\dfrac{5}{4} \pm \dfrac{\sqrt{39}}{4}i$ **101.** $-\dfrac{1}{2} \pm \dfrac{\sqrt{7}}{2}i$
103. $2 \pm 3i$ **105.** $1 \pm 2\sqrt{2}i$ **107.** $-1 \pm 2i$ **109.** $2 \pm i$
111. $\frac{1}{2} \pm \frac{3}{2}i$ **113.** $1, -\dfrac{1}{2} \pm \dfrac{\sqrt{3}}{2}i$

Guided Solutions

CHAPTER 1

Section 1.1

8. 2718 = 2 thousands + 7 hundreds + 1 ten + 8 ones

17. One million, eight hundred seventy-nine thousand, two hundred four

Section 1.2

2.
$$\begin{array}{r} {}^{1\ 1\ 1} \\ 7\ 9\ 6\ 8 \\ +\ 5\ 4\ 9\ 7 \\ \hline 1\ 3{,}4\ 6\ 5 \end{array}$$

5. Perimeter = 4 in. + 5 in. + 9 in. + 6 in. + 5 in. = 29 in.

Section 1.3

1.
$$\begin{array}{r} 7\ 8\ 9\ 3 \\ -\ 4\ 0\ 9\ 2 \\ \hline 3\ 8\ 0\ 1 \end{array}$$
Check:
$$\begin{array}{r} 3\ 8\ 0\ 1 \\ +\ 4\ 0\ 9\ 2 \\ \hline 7\ 8\ 9\ 3 \end{array}$$

5.
$$\begin{array}{r} {}^{4\ \ 9\ 13} \\ \not{5}\ \not{0}\ \not{3} \\ -\ 2\ 9\ 8 \\ \hline 2\ 0\ 5 \end{array}$$

Section 1.4

4.
$$\begin{array}{r} {}^{1\ 2\ 4} \\ 1\ 3\ 4\ 8 \\ \times\quad\quad 5 \\ \hline 6\ 7\ 4\ 0 \end{array}$$

20. $A = l \cdot w$
$= 12 \text{ ft} \cdot 8 \text{ ft}$
$= 96 \text{ sq ft}$

Section 1.5

8. $0 \div 2$ means 0 divided by 2.
Since zero divided by any nonzero number is 0, $0 \div 2 = 0$.

9. $7 \div 0$ means 7 divided by 0.
Since division by 0 is not defined, $7 \div 0$ is not defined.

Section 1.6

26. Nearest ten:
$$\begin{array}{r} 8\,4\,0 \\ \times\ 2\,5\,0 \\ \hline 42\ 000 \\ 168\ 000 \\ \hline 210{,}000 \end{array}$$
Nearest hundred:
$$\begin{array}{r} 8\,0\,0 \\ \times\ \ 2\,0\,0 \\ \hline 160{,}000 \end{array}$$

31. Since 8 is to the left of 12 on the number line, $8 < 12$.

Section 1.7

13.
$$\begin{aligned} x + 9 &= 17 \\ x + 9 - 9 &= 17 - 9 \\ x &= 8 \end{aligned}$$
Check:
$$\begin{array}{c|c} x + 9 & 17 \\ \hline 8 + 9 \ ? & 17 \\ 17 & \end{array}$$
Since 17 = 17 is true, the answer checks.
The solution is 8.

19.
$$\begin{aligned} \frac{144}{9} &= \frac{9 \cdot n}{9} \\ 16 &= n \end{aligned}$$
Check:
$$\begin{array}{c|c} 144 & 9 \cdot n \\ \hline 144 \ ? & 9 \cdot 16 \\ & 144 \end{array}$$
Since 144 = 144 is true, the answer checks.
The solution is 16.

Section 1.8

4. **1. Familiarize.** Let p = the number of pages William still has to read.

2. Translate.

Pages already read	plus	Number of pages to read	is	Total number of pages
86	+	p	=	234

3. Solve.
$$\begin{aligned} 86 + p &= 234 \\ 86 + p - 86 &= 234 - 86 \\ p &= 148 \end{aligned}$$

4. Check. If William reads 148 more pages, he will have read a total of 86 + 148 pages, or 234 pages.

5. State. William has 148 more pages to read.

9. **1. Familiarize.** Let x = the number of hundreds in 3500.
Let t = the time it takes to lose one pound.

2. Translate.
$$\begin{aligned} 100 \cdot x &= 3500 \\ x \cdot 2 &= t \end{aligned}$$

3. Solve. From Example 7, we know that $x = 35$.
$$\begin{aligned} x \cdot 2 &= t \\ 35 \cdot 2 &= t \\ 70 &= t \end{aligned}$$

4. Check. Since $70 \div 2 = 35$, there are 35 groups of 2 min in 70 min. Thus you will burn $35 \times 100 = 3500$ calories.

5. State. You must swim for 70 min, or 1 hr 10 min, in order to lose one pound.

Section 1.9

5. $10^4 = 10 \cdot 10 \cdot 10 \cdot 10 = 10{,}000$

15.
$$\begin{aligned} & 9 \times 4 - (20 + 4) \div 8 - (6 - 2) \\ &= 9 \times 4 - 24 \div 8 - 4 \\ &= 36 - 24 \div 8 - 4 \\ &= 36 - 3 - 4 \\ &= 33 - 4 \\ &= 29 \end{aligned}$$

25.
$$\begin{aligned} & [18 - (2 + 7) \div 3] - (31 - 10 \times 2) \\ &= [18 - 9 \div 3] - (31 - 10 \times 2) \\ &= [18 - 3] - (31 - 20) \\ &= 15 - 11 \\ &= 4 \end{aligned}$$

CHAPTER 2

Section 2.1

10. The distance of 18 from 0 is 18, so $|18| = 18$.

22. $-(-x) = -(-(-2))$
$= -(2) = -2$

Section 2.2

19. -12 and 12 have the same absolute value. The answer is 0.

27. Add the positive numbers:

$25 + 10 = 35.$

Add the negative numbers:

$-15 + (-5) + (-9) + (-14) = -43.$

Finally, add the results:

$35 + (-43) = -8.$

Section 2.3

11. $2 - 8 = 2 + (-8) = -6$

17. $-6 - (-2) - (-4) - 12 + 3 = -6 + 2 + 4 + (-12) + 3$
$= -6 + (-12) + 2 + 4 + 3$
$= -18 + 9$
$= -9$

Section 2.4

9. Multiply absolute values:

$3 \cdot 6 = 18.$

The signs are different, so the answer is negative.

$3(-6) = -18$

17. $(-4)(-5)(-2)(-3)(-1) = 20 \cdot 6 \cdot (-1)$
$= 120 \cdot (-1)$
$= -120$

Section 2.5

11. $(-2) \cdot |3 - 2^2| + 5 = (-2) \cdot |3 - 4| + 5$
$= (-2) \cdot |-1| + 5$
$= (-2) \cdot 1 + 5$
$= -2 + 5$
$= 3$

Section 2.6

4. $\frac{-6}{x} = -\frac{6}{x} = \frac{6}{-x}$

17. $6(x + y + z) = 6 \cdot x + 6 \cdot y + 6 \cdot z$
$= 6x + 6y + 6z$

Section 2.7

7. The like terms are

$4m$ and m,

$-2n^2$ and n^2,

5 and -9.

$4m + (-2n^2) + 5 + n^2 + m + (-9)$
$= 4m + m + (-2n^2) + n^2 + 5 + (-9)$
$= 5m + (-n^2) + (-4)$
$= 5m - n^2 - 4$

13. $P = 4s$
$= 4 \cdot 9$ in.
$= 36$ in.

Section 2.8

12. $-t = -3$
$-1 \cdot t = -3$
$\frac{-1 \cdot t}{-1} = \frac{-3}{-1}$
$t = 3$

18. $2x - 9 = 43$
$2x - 9 + 9 = 43 + 9$
$2x = 52$
$\frac{2x}{2} = \frac{52}{2}$
$x = 26$

CHAPTER 3

Section 3.1

4. $2\overline{)16}$ with quotient 8
16
0

Since the remainder is 0, 16 is divisible by 2.

22. Add the digits:
$1 + 7 + 2 + 1 + 6 = 17.$
Since 17 is not divisible by 3, the number 17,216 is not divisible by 3.

Section 3.2

4.

1 is a factor of 45.	$1 \cdot 45$
2 is not a factor of 45.	
3 is a factor of 45.	$3 \cdot 15$
4 is not a factor of 45.	
5 is a factor of 45.	$5 \cdot 9$
6 is not a factor of 45.	
7 is not a factor of 45.	
8 is not a factor of 45.	

Factors of 45: 1, 3, 5, 9, 15, 45.

9. $7\overline{)49}$ with quotient 7
$2\overline{)98}$
$98 = 2 \cdot 7 \cdot 7$

Section 3.3

12. Each gallon is divided into 4 equal parts.

The unit is $\frac{1}{4}$.

There are 7 equal units shaded.

The part that is shaded is $\frac{7}{4}$.

24. $\frac{4 - 4}{567} = \frac{0}{567} = 0$

Section 3.4

9. $\frac{3}{8} \cdot \frac{5}{7} = \frac{3 \cdot 5}{8 \cdot 7}$
$= \frac{15}{56}$

Section 3.5

5. $\frac{4}{3} = \frac{4}{3} \cdot \frac{5}{5}$
$= \frac{4 \cdot 5}{3 \cdot 5}$
$= \frac{20}{15}$

25. $2 \cdot 20 = 40$ $3 \cdot 14 = 42$

$$\frac{2}{3} \square \frac{14}{20}$$

Since $40 \neq 42$, $\frac{2}{3} \neq \frac{14}{20}$.

Section 3.6

1. $\frac{2}{3} \cdot \frac{7}{8} = \frac{2 \cdot 7}{3 \cdot 8}$

$= \frac{2 \cdot 7}{3 \cdot 2 \cdot 2 \cdot 2}$

$= \frac{2}{2} \cdot \frac{7}{3 \cdot 2 \cdot 2}$

$= 1 \cdot \frac{7}{3 \cdot 2 \cdot 2}$

$= \frac{7}{12}$

7. $A = \frac{1}{2} \cdot b \cdot h$

$= \frac{1}{2} \cdot 11 \text{ cm} \cdot \frac{12}{5} \text{ cm}$

$= \frac{1 \cdot 11 \cdot 12}{2 \cdot 5} \text{ cm}^2$

$= \frac{1 \cdot 11 \cdot \not{2} \cdot 2 \cdot 3}{\not{2} \cdot 5} \text{ cm}^2$

$= \frac{66}{5} \text{ cm}^2$

Section 3.7

6. $\frac{6}{7} \div \frac{3}{4} = \frac{6}{7} \cdot \frac{4}{3}$

$= \frac{6 \cdot 4}{7 \cdot 3}$

$= \frac{2 \cdot 3 \cdot 2 \cdot 2}{7 \cdot 3}$

$= \frac{3}{3} \cdot \frac{2 \cdot 2 \cdot 2}{7}$

$= \frac{2 \cdot 2 \cdot 2}{7}$

$= \frac{8}{7}$

Section 3.8

1. $\frac{3}{2} \cdot \frac{2}{3}x = \frac{3}{2} \cdot 8$

$1x = \frac{3 \cdot 8}{2}$

$x = \frac{3 \cdot 2 \cdot 4}{2}$

$x = 12$

4. $-\frac{7}{6} \cdot \left(-\frac{6}{7}a\right) = -\frac{7}{6} \cdot \frac{9}{14}$

$a = -\frac{7 \cdot 3 \cdot 3}{2 \cdot 3 \cdot 2 \cdot 7}$

$a = -\frac{3}{4}$

CHAPTER 4

Section 4.1

12. 1. $18 = 2 \cdot 3 \cdot 3$

$40 = 2 \cdot 2 \cdot 2 \cdot 5$

2. Select the factorization of 40:

$2 \cdot 2 \cdot 2 \cdot 5$.

This is not a multiple of 18. We need two factors of 3.

3. $\text{LCM} = 2 \cdot 2 \cdot 2 \cdot 5 \cdot 3 \cdot 3 = 360$

18. $5a^2 = 5 \cdot a^2$

$a^3b = a^3 \cdot b$

$\text{LCM} = 5 \cdot a^3 \cdot b$, or $5a^3b$

Section 4.2

9. The LCD is 24.

$\frac{3}{8} + \frac{5}{6} = \frac{3}{8} \cdot 1 + \frac{5}{6} \cdot 1$

$= \frac{3}{8} \cdot \frac{3}{3} + \frac{5}{6} \cdot \frac{4}{4}$

$= \frac{9}{24} + \frac{20}{24}$

$= \frac{29}{24}$

20. 1. Familiarize. Let T = the total amount of berries in the salad.

2. Translate. To find the total amount, we add.

$\frac{7}{8} + \frac{3}{4} + \frac{5}{16} = T$

3. Solve. The LCD is 16.

$\frac{7}{8} \cdot \frac{2}{2} + \frac{3}{4} \cdot \frac{4}{4} + \frac{5}{16} = T$

$\frac{14}{16} + \frac{12}{16} + \frac{5}{16} = T$

$\frac{31}{16} = T$

4. Check. The answer is reasonable because it is larger than any of the individual amounts.

5. State. There are $\frac{31}{16}$ qt of berries in the salad.

Section 4.3

6. The LCD is 18.

$\frac{5}{6} - \frac{1}{9} = \frac{5}{6} \cdot \frac{3}{3} - \frac{1}{9} \cdot \frac{2}{2}$

$= \frac{15}{18} - \frac{2}{18}$

$= \frac{13}{18}$

13. $\frac{3}{5} + t = -\frac{7}{8}$

$\frac{3}{5} + t - \frac{3}{5} = -\frac{7}{8} - \frac{3}{5}$

$t + 0 = -\frac{7}{8} \cdot \frac{5}{5} - \frac{3}{5} \cdot \frac{8}{8}$

$t = -\frac{35}{40} - \frac{24}{40}$

$t = \frac{-35 - 24}{40}$

$t = \frac{-35 + (-24)}{40}$

$t = \frac{-59}{40} = -\frac{59}{40}$

Section 4.4

2.
$$\frac{1}{2}x - \frac{1}{5} = \frac{7}{10}$$
$$\frac{1}{2}x - \frac{1}{5} + \frac{1}{5} = \frac{7}{10} + \frac{1}{5}$$
$$\frac{1}{2}x = \frac{7}{10} + \frac{2}{10}$$
$$\frac{1}{2}x = \frac{9}{10}$$
$$2 \cdot \frac{1}{2}x = 2 \cdot \frac{9}{10}$$
$$1x = \frac{2 \cdot 3 \cdot 3}{2 \cdot 5}$$
$$x = \frac{9}{5}$$

6.
$$20 = 6 - \frac{2}{3}x$$
$$3(20) = 3\left(6 - \frac{2}{3}x\right)$$
$$60 = 3 \cdot 6 - \frac{3 \cdot 2}{3}x$$
$$60 = 18 - 2x$$
$$60 - 18 = 18 - 2x - 18$$
$$42 = -2x$$
$$\frac{42}{-2} = \frac{-2x}{-2}$$
$$-21 = x$$

Section 4.5

6.
$$4 \cdot 6 = 24$$
$$24 + 5 = 29$$
$$4\frac{5}{6} = \frac{29}{6}$$

16.
$$5\overline{)17}\ \text{quotient } 3;\quad 15;\quad \text{remainder } 2$$
$$\frac{17}{5} = 3\frac{2}{5}, \text{ so } \frac{-17}{5} = -3\frac{2}{5}$$

Section 4.6

5.
$$8\frac{2}{3} = 8\frac{4}{6}$$
$$-5\frac{1}{2} = -5\frac{3}{6}$$
$$3\frac{1}{6}$$

6.
$$5 = 4\frac{3}{3}$$
$$-1\frac{1}{3} = -1\frac{1}{3}$$
$$3\frac{2}{3}$$

Section 4.7

3.
$$-2 \cdot 6\frac{2}{5} = -\frac{2}{1} \cdot \frac{32}{5}$$
$$= -\frac{64}{5}$$
$$= -12\frac{4}{5}$$

6.
$$2\frac{1}{4} \div 1\frac{1}{5} = \frac{9}{4} \div \frac{6}{5}$$
$$= \frac{9}{4} \cdot \frac{5}{6}$$
$$= \frac{3 \cdot 3 \cdot 5}{2 \cdot 2 \cdot 2 \cdot 3}$$
$$= \frac{3}{3} \cdot \frac{3 \cdot 5}{2 \cdot 2 \cdot 2}$$
$$= \frac{15}{8}$$
$$= 1\frac{7}{8}$$

Section 4.8

2.
$$\frac{1}{3} \cdot \frac{3}{4} \div \frac{5}{8} - \frac{1}{10} = \frac{3}{12} \div \frac{5}{8} - \frac{1}{10}$$
$$= \frac{3}{12} \cdot \frac{8}{5} - \frac{1}{10}$$
$$= \frac{3 \cdot 2 \cdot 2 \cdot 2}{3 \cdot 2 \cdot 2 \cdot 5} - \frac{1}{10}$$
$$= \frac{2}{5} - \frac{1}{10}$$
$$= \frac{4}{10} - \frac{1}{10} = \frac{3}{10}$$

7.
$$\frac{10}{\frac{5}{8}} = 10 \div \frac{5}{8}$$
$$= 10 \cdot \frac{8}{5}$$
$$= \frac{10 \cdot 8}{5}$$
$$= \frac{2 \cdot 5 \cdot 8}{5 \cdot 1}$$
$$= 16$$

CHAPTER 5

Section 5.1

7. 0.896. 3 places

$$0.896 = \frac{896}{1000}$$

10. $\frac{743}{100}$ 7.43.

2 zeros 2 places

$$\frac{743}{100} = 7.43$$

Section 5.2

7.
```
  1 1 1 1
     4 5.7 8 0
   2 4 6 7.0 0 0
+        1.9 9 3
   2 5 1 4.7 7 3
```

8.
```
        13
    6  3 12
   3 7.4 2 8
 − 2 6.6 7 4
   1 0.7 5 4
```

14.
```
   4  9 9  9 10
   5.0 0 0 0
 − 0.0 0 8 9
   4.9 9 1 1
```

Section 5.3

3.
```
       4 2.6 5
   ×   0.8 0 4
     1 7 0 6 0
 3 4 1 2 0 0 0
 3 4.2 9 0 6 0
```

15.
$$\$15.69 = 15.69 \times \$1$$
$$= 15.69 \times 100¢$$
$$= 1569¢$$

Section 5.4

6.
```
         0.0 2 5
 8 6 ) 2.1 5 0
       1 7 2
         4 3 0
         4 3 0
             0
```

7. $\dfrac{0.375}{0.25} = \dfrac{0.375}{0.25} \times \dfrac{100}{100}$

$= \dfrac{37.5}{25}$

$$\begin{array}{r} 1.5 \\ 0.25\overline{)0.375} \\ \underline{25} \\ 125 \\ \underline{125} \\ 0 \end{array}$$

16. $625 \div 62.5 \times 25 \div 6250$

$= 10 \times 25 \div 6250$

$= 250 \div 6250$

$= 0.04$

Section 5.5

3. $\dfrac{1}{6} = 1 \div 6$

$$\begin{array}{r} 0.166 \\ 6\overline{)1.000} \\ \underline{6} \\ 40 \\ \underline{36} \\ 40 \\ \underline{36} \\ 4 \end{array}$$

$\dfrac{1}{6} = 0.1666\ldots = 0.1\overline{6}$

17. Method 1:

$\dfrac{3}{4} \times 0.62 = \dfrac{3}{4} \times \dfrac{0.62}{1}$

$= \dfrac{1.86}{4} = 0.465$

Method 2:

$\dfrac{3}{4} \times 0.62 = 0.75 \times 0.62$

$= 0.465$

Method 3:

$\dfrac{3}{4} \times 0.62 = \dfrac{3}{4} \cdot \dfrac{62}{100}$

$= \dfrac{186}{400}$

$= \dfrac{93}{200} = 0.465$

Section 5.7

6.

$8 + 4x = 9x - 3$

$8 + 4x - 4x = 9x - 3 - 4x$

$8 = 5x - 3$

$8 + 3 = 5x - 3 + 3$

$11 = 5x$

$\dfrac{11}{5} = \dfrac{5x}{5}$

$2.2 = x$

8.

$3(x + 5) = 20 - x$

$3x + 15 = 20 - x$

$3x + 15 + x = 20 - x + x$

$4x + 15 = 20$

$4x + 15 - 15 = 20 - 15$

$4x = 5$

$\dfrac{4x}{4} = \dfrac{5}{4}$

$x = 1.25$

CHAPTER 6

Section 6.1

6. $\dfrac{\text{Length of shortest side}}{\text{Length of longest side}} = \dfrac{38.2}{55.5}$

8. Ratio of 3.6 to 12: $\dfrac{3.6}{12}$

Simplifying:

$\dfrac{3.6}{12} \cdot \dfrac{10}{10} = \dfrac{36}{120} = \dfrac{12 \cdot 3}{12 \cdot 10} = \dfrac{12}{12} \cdot \dfrac{3}{10} = \dfrac{3}{10}$

14. $\dfrac{52\text{ ft}}{13\text{ sec}} = 4\text{ ft/sec}$

18. We compare cross products.

$1 \cdot 39 = 39 \qquad \dfrac{1}{2} \overset{?}{=} \dfrac{20}{39} \qquad 2 \cdot 20 = 40$

Since $39 \neq 40$, the numbers are not proportional.

23. $\dfrac{x}{9} = \dfrac{5}{4}$

$x \cdot 4 = 9 \cdot 5$

$\dfrac{x \cdot 4}{4} = \dfrac{9 \cdot 5}{4}$

$x = \dfrac{45}{4} = 11\dfrac{1}{4}$

28. **1. Familiarize.** Let $p =$ the amount of paint needed, in gallons.

2. Translate. $\dfrac{4}{1600} = \dfrac{p}{6000}$

3. Solve.

$4 \cdot 6000 = 1600 \cdot p$

$15 = p$

4. Check. The cross products are the same.

5. State. For 6000 ft^2, they would need 15 gal of paint.

31. **1. Familiarize.** Let $D =$ the number of deer in the forest.

2. Translate. $\dfrac{153}{D} = \dfrac{18}{62}$

3. Solve.

$153 \cdot 62 = D \cdot 18$

$527 = D$

4. Check. The cross products are the same.

5. State. There are about 527 deer in the forest.

Section 6.3

6. $\dfrac{19}{25} = \dfrac{19}{25} \cdot \dfrac{4}{4}$

$= \dfrac{76}{100} = 76\%$

10. $3.25\% = \dfrac{3.25}{100} = \dfrac{3.25}{100} \times \dfrac{100}{100}$

$= \dfrac{325}{10{,}000} = \dfrac{13 \times 25}{400 \times 25}$

$= \dfrac{13}{400} \times \dfrac{25}{25} = \dfrac{13}{400}$

Section 6.4

9. 20% of what is 45?

$\downarrow \quad \downarrow \quad \downarrow \quad \downarrow \quad \downarrow$

$20\% \cdot b = 45$

$\dfrac{20\% \cdot b}{20\%} = \dfrac{45}{20\%}$

$b = \dfrac{45}{0.2}$

$b = 225$

11. 16 is what percent of 40?

$\downarrow \quad \downarrow \quad \downarrow \quad \downarrow \quad \downarrow$

$16 = p \cdot 40$

$\dfrac{16}{40} = \dfrac{p \cdot 40}{40}$

$\dfrac{16}{40} = p$

$0.4 = p$

$40\% = p$

Section 6.5

8. $\frac{20}{100} = \frac{45}{b}$

$20 \cdot b = 100 \cdot 45$

$\frac{20b}{20} = \frac{100 \cdot 45}{20}$

$b = \frac{4500}{20}$

$b = 225$

9. $\frac{64}{100} = \frac{a}{55}$

$64 \cdot 55 = 100 \cdot a$

$\frac{64 \cdot 55}{100} = \frac{100 \cdot a}{100}$

$\frac{3520}{100} = a$

$35.2 = a$

12. $\frac{12}{40} = \frac{N}{100}$

$12 \cdot 100 = 40 \cdot N$

$\frac{12 \cdot 100}{40} = \frac{40 \cdot N}{40}$

$\frac{1200}{40} = N$

$30 = N$

Thus, \$12 is 30% of \$40.

Section 6.7

2. Sales tax $= 4\% \times 4 \times \$18.95$
$= 0.04 \times \$75.80$
$= \$3.032$
$\approx \$3.03$

Total price $= \$75.80 + \$3.03$
$= \$78.83$

7. $\$2970 = 7.5\% \times S$

$\$2970 = 0.075 \times S$

$\frac{\$2970}{0.075} = \frac{0.075 \times S}{0.075}$

$\$39{,}600 = S$

Section 6.8

1. $I = P \cdot r \cdot t$
$= \$4300 \times 4\% \times 1$
$= \$4300 \times 0.04 \times 1$
$= \$172$

3. a) $I = P \cdot r \cdot t$

$= \$4800 \times 5\frac{1}{2}\% \times \frac{30}{365}$

$= \$4800 \times 0.055 \times \frac{30}{365}$

$\approx \$21.70$

b) Total amount
$= \$4800 + \$21.70$

$= \$4821.70$

CHAPTER 7

Section 7.1

5. Course grade $= \frac{100 \cdot 15 + 92 \cdot 25 + 88 \cdot 40}{15 + 25 + 40}$

$= \frac{7320}{80} = 91.5$

Soha's course grade is 91.5%.

12. Rearrange the numbers in order from smallest to largest:

34, 34, 67, 68, 69, 70.

The middle numbers are 67 and 68.
The average of 67 and 68 is 67.5.
The median is 67.5.

15. Rearrange the numbers in order from smallest to largest.

13, 24, 27, 28, 67, 89.

Each number occurs one time.
There is no mode.

Section 7.2

3. The amount of the decrease in population density is $611 - 603 = 8$.

The percent decrease is $\frac{8}{611} \approx 0.013$, or 1.3%.

8. The graph shows $1\frac{1}{2}$ symbols for South America.

This represents 150 roller coasters.

The graph shows $\frac{1}{2}$ symbol for Africa.

This represents 50 roller coasters.
There are 100 more roller coasters in South America than in Africa.

Section 7.3

7. We look from left to right along a line at \$400 per ounce. The points on the graph that are below this line correspond to the years 1970, 1975, 1985, 1990, 1995, and 2000.

CHAPTER 8

Section 8.1

33. $x + 55° + 61° = 180°$

$x + 116° = 180°$

$x = 180° - 116°$

$x = 64°$

Section 8.2

5. $P = 2 \cdot (l + w)$

$= 2 \cdot (8\frac{1}{4}\text{ in.} + 5\text{ in.})$

$= 2 \cdot (13\frac{1}{4}\text{ in.})$

$= 2 \cdot \frac{53}{4}\text{ in.}$

$= \frac{2 \cdot 53}{2 \cdot 2}\text{ in.}$

$= \frac{53}{2}\text{ in.}$

$= 26\frac{1}{2}\text{ in.}$

8. $P = 4 \cdot s$
$= 4 \cdot 7.8\text{ km}$
$= 31.2\text{ km}$

Section 8.3

6. $A = s \cdot s$

$= 3\frac{1}{2}\text{ yd} \times 3\frac{1}{2}\text{ yd}$

$= \frac{7}{2}\text{ yd} \times \frac{7}{2}\text{ yd}$

$= \frac{49}{4}\text{ yd}^2$

$= 12\frac{1}{4}\text{ yd}^2$

10. $A = \frac{1}{2} \cdot b \cdot h$

$= \frac{1}{2} \times 11\text{ cm} \times 3.4\text{ cm}$

$= 0.5 \times 11 \times 3.4\text{ cm}^2$

$= 18.7\text{ cm}^2$

Section 8.4

3. $C = \pi \cdot d$
$\approx 3.14 \times 18\text{ in.}$
$= 56.52\text{ in.}$

6. $A = \pi \cdot r \cdot r$

$\approx \frac{22}{7} \cdot 5\text{ km} \cdot 5\text{ km}$

$= \frac{22}{7} \cdot 25\text{ km}^2$

$= \frac{550}{7}\text{ km}^2$

$= 78\frac{4}{7}\text{ km}^2$

Section 8.5

6. $V = \pi \cdot r^2 \cdot h$

$\approx 3.14 \times 5\text{ ft} \times 5\text{ ft} \times 10\text{ ft}$

$= 3.14 \times 250\text{ ft}^3$

$= 785\text{ ft}^3$

8. $V = \frac{4}{3} \cdot \pi \cdot r^3$

$\approx \frac{4}{3} \times \frac{22}{7} \times (28\text{ ft})^3$

$= \frac{4}{3} \times \frac{22}{7} \times 21{,}952\text{ ft}^3$

$= \frac{275{,}968}{3}\text{ ft}^3$

$= 91{,}989\frac{1}{3}\text{ ft}^3$

Section 8.6

5. $90° - 67° = 23°$

10. $180° - 71° = 109°$

Section 8.8

5. $\frac{8}{6} = \frac{9}{BT}$

$8(BT) = 6 \cdot 9$

$8(BT) = 54$

$BT = 6\frac{3}{4}$

$\frac{8}{6} = \frac{12}{CT}$

$8(CT) = 6 \cdot 12$

$8(CT) = 72$

$CT = 9$

CHAPTER 9

Section 9.1

5. $A = lw$

$A = (24\text{ ft})(8\text{ ft})$

$= (24)(8)(\text{ft})(\text{ft})$

$= 192\text{ ft}^2$, or 192 square feet

Section 9.3

20. $-\frac{1}{5} + \left(-\frac{3}{4}\right)$

$= -\frac{4}{20} + \left(-\frac{15}{20}\right)$

$= -\frac{19}{20}$

32. $-x = -(-1.6) = 1.6;$

$-(-x) = -(-(-1.6))$

$= -(1.6) = -1.6$

Section 9.4

11. $2 - 8 = 2 + (-8) = -6$

19. $-12 - (-9) = -12 + 9 = -3$

Section 9.6

21. $\frac{4}{7} \div \left(-\frac{3}{5}\right) = \frac{4}{7} \cdot \left(-\frac{5}{3}\right) = -\frac{20}{21}$

25. $\frac{-5}{6} = \frac{5}{-6} = -\frac{5}{6}$

Section 9.7

3. $\frac{3}{4} = \frac{3}{4} \cdot 1 = \frac{3}{4} \cdot \frac{2}{2} = \frac{6}{8}$

4. $\frac{3}{4} = \frac{3}{4} \cdot 1 = \frac{3}{4} \cdot \frac{t}{t} = \frac{3t}{4t}$

8. $\frac{18p}{24pq} = \frac{6p \cdot 3}{6p \cdot 4q}$

$= \frac{6p}{6p} \cdot \frac{3}{4q}$

$= 1 \cdot \frac{3}{4q} = \frac{3}{4q}$

31. $-2(x - 3)$

$= -2 \cdot x - (-2) \cdot 3$

$= -2x - (-6)$

$= -2x + 6$

44. $16a - 36b + 42$

$= 2 \cdot 8a - 2 \cdot 18b + 2 \cdot 21$

$= 2(8a - 18b + 21)$

52. $3x - 7x - 11 + 8y + 4 - 13y$

$= (3 - 7)x + (8 - 13)y + (-11 + 4)$

$= -4x + (-5)y + (-7)$

$= -4x - 5y - 7$

Section 9.8

13. $5a - 3(7a - 6)$

$= 5a - 21a + 18$

$= -16a + 18$

18. $9 - [10 - (13 + 6)]$

$= 9 - [10 - (19)]$

$= 9 - [-9]$

$= 9 + 9$

$= 18$

25. $-4^3 + 52 \cdot 5 + 5^3 - (4^2 - 48 \div 4)$

$= -64 + 52 \cdot 5 + 125 - (16 - 48 \div 4)$

$= -64 + 52 \cdot 5 + 125 - (16 - 12)$

$= -64 + 52 \cdot 5 + 125 - 4$

$= -64 + 260 + 125 - 4$

$= 196 + 125 - 4$

$= 321 - 4$

$= 317$

CHAPTER 10

Section 10.1

8. $x + 2 = 11$

$x + 2 + (-2) = 11 + (-2)$

$x + 0 = 9$

$x = 9$

Section 10.2

1. $6x = 90$

$\frac{1}{6} \cdot 6x = \frac{1}{6} \cdot 90$

$1 \cdot x = 15$

$x = 15$

Check: $6x = 90$

$6x \cdot 15 \ ? \ 90$

90 | TRUE

2. $4x = -7$

$\frac{4x}{4} = \frac{-7}{4}$

$1 \cdot x = -\frac{7}{4}$

$x = -\frac{7}{4}$

6. $\frac{2}{3} = -\frac{5}{6}y$

$-\frac{6}{5} \cdot \frac{2}{3} = -\frac{6}{5} \cdot \left(-\frac{5}{6}y\right)$

$-\frac{12}{15} = 1 \cdot y$

$-\frac{4}{5} = y$

Section 10.3

4. $-18 - m = -57$

$18 - 18 - m = 18 - 57$

$-m = -39$

$-1(-m) = -1(-39)$

$m = 39$

11.
$$\begin{aligned}7x - 17 + 2x &= 2 - 8x + 15\\ 9 \cdot x - 17 &= 17 - 8x\\ 8x + 9x - 17 &= 17 - 8x + 8x\\ 17 \cdot x - 17 &= 17\\ 17x - 17 + 17 &= 17 + 17\\ 17x &= 34\\ \frac{17x}{17} &= \frac{34}{17}\\ x &= 2\end{aligned}$$

13.
$$\begin{aligned}\frac{7}{8}x - \frac{1}{4} + \frac{1}{2}x &= \frac{3}{4} + x\\ 8 \cdot \left(\frac{7}{8}x - \frac{1}{4} + \frac{1}{2}x\right) &= 8 \cdot \left(\frac{3}{4} + x\right)\\ 8 \cdot \frac{7}{8}x - 8 \cdot \frac{1}{4} + 8 \cdot \frac{1}{2}x &= 8 \cdot \frac{3}{4} + 8 \cdot x\\ 7x - 2 + 4x &= 6 + 8x\\ 11x - 2 &= 6 + 8x\\ 11x - 2 - 8x &= 6 + 8x - 8x\\ 3x - 2 &= 6\\ 3x - 2 + 2 &= 6 + 2\\ 3x &= 8\\ \frac{3x}{3} &= \frac{8}{3}\\ x &= \frac{8}{3}\end{aligned}$$

Section 10.4

12.
$$\begin{aligned}y &= mx + b\\ y - b &= mx + b - b\\ y - b &= mx\\ \frac{y - b}{m} &= \frac{mx}{m}\\ \frac{y - b}{m} &= x\end{aligned}$$

Section 10.5

4. 110% of what number is 30?

$110\% \cdot x = 30$

8. 25.3 is 22% of what number?

$$\begin{aligned}25.3 &= 22\% \cdot x\\ 25.3 &= 0.22 \cdot x\\ \frac{25.3}{0.22} &= \frac{0.22x}{0.22}\\ 115 &= x\end{aligned}$$

Section 10.6

3. Let $x =$ the first marker and $x + 1 =$ the second marker.
Translate and *Solve*:

First marker + Second marker = 627

$$\begin{aligned}x + (x + 1) &= 627\\ 2x + 1 &= 627\\ 2x + 1 - 1 &= 627 - 1\\ 2x &= 626\\ \frac{2x}{2} &= \frac{626}{2}\\ x &= 313\end{aligned}$$

If $x = 313$, then $x + 1 = 314$. The mile markers are 313 and 314.

8. Let $x =$ the principal. Then the interest earned is $5\%x$.
Translate and *Solve*:

Principal + Interest = Amount

$$\begin{aligned}x + 5\%x &= 2520\\ x + 0.05x &= 2520\\ (1 + 0.05)x &= 2520\\ 1.05x &= 2520\\ \frac{1.05x}{1.05} &= \frac{2520}{1.05}\\ x &= 2400\end{aligned}$$

Section 10.7

10.
$$\begin{aligned}5y + 2 &\le -1 + 4y\\ 5y + 2 - 4y &\le -1 + 4y - 4y\\ y + 2 &\le -1\\ y + 2 - 2 &\le -1 - 2\\ y &\le -3\end{aligned}$$

The solution set is $\{y \mid y \le -3\}$.

18.
$$\begin{aligned}3(7 + 2x) &\le 30 + 7(x - 1)\\ 21 + 6x &\le 30 + 7x - 7\\ 21 + 6x &\le 23 + 7x\\ 21 + 6x - 6x &\le 23 + 7x - 6x\\ 21 &\le 23 + x\\ 21 - 23 &\le 23 + x - 23\\ -2 &\le x, \text{ or}\\ x &\ge -2\end{aligned}$$

The solution set is $\{x \mid x \ge -2\}$.

Section 10.8

9. *Translate* and *Solve:*
$$\begin{aligned}F &< 88\\ \frac{9}{5}C + 32 &< 88\\ \frac{9}{5}C + 32 - 32 &< 88 - 32\\ \frac{9}{5}C &< 56\\ \frac{5}{9} \cdot \frac{9}{5}C &< \frac{5}{9} \cdot 56\\ C &< \frac{280}{9}\\ C &< 31\frac{1}{9}\end{aligned}$$

Butter stays solid at Celsius temperatures less than $31\frac{1}{9}^\circ$—that is, $\{C \mid C < 31\frac{1}{9}^\circ\}$.

CHAPTER 11

Section 11.1

18. Determine whether $(2, -4)$ is a solution of $4q - 3p = 22$.

$4q - 3p$	$= 22$
$4 \cdot (-4) - 3 \cdot 2$	? 22
$-16 - 6$	
-22	FALSE

Thus, $(2, -4)$ is not a solution.

21. Complete the table and graph $y = -2x$.

x	y	(x, y)
−3	6	(−3, 6)
−1	2	(−1, 2)
0	0	(0, 0)
1	−2	(1, −2)
3	−6	(3, −6)

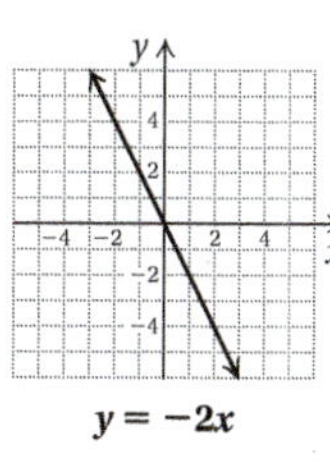

$y = -2x$

29. Graph $5y - 3x = -10$ and identify the y-intercept.

x	y
0	−2
5	1
−5	−5

← y-intercept: (0, −2)

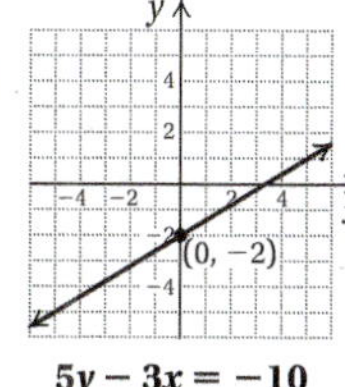

$5y - 3x = -10$

Section 11.2

2. For $2x + 3y = 6$, find the intercepts. Then graph the equation using the intercepts.

x	y
3	0
0	2
−3	4

← x-intercept: (3, 0)
← y-intercept: (0, 2)
← Check point: (−3, 4)

6. Graph: $x = 5$.

x	y
5	−4
5	0
5	3

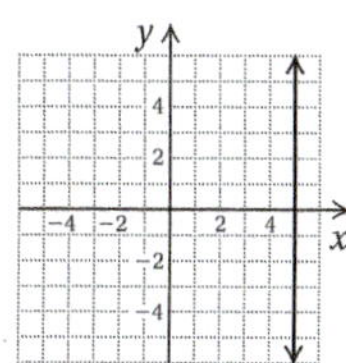

$x = 5$

7. Graph: $y = -2$.

x	y
−1	−2
0	−2
2	−2

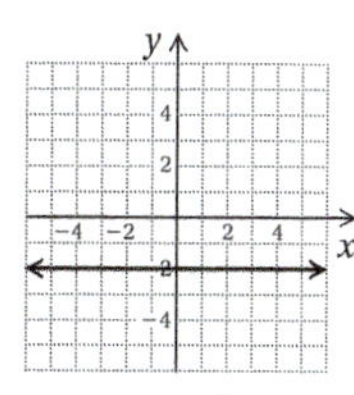

$y = -2$

Section 11.3

1. Graph the line that contains (−2, 3) and (3, 5) and find the slope in two different ways.

$$\frac{5 - 3}{3 - (-2)} = \frac{2}{5}, \text{ or}$$

$$\frac{3 - 5}{-2 - 3} = \frac{-2}{-5} = \frac{2}{5}$$

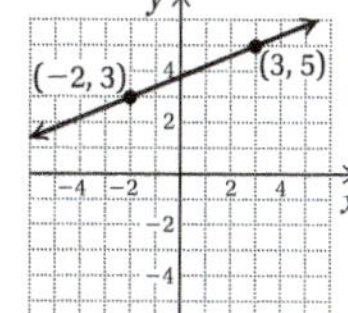

8. Find the slope of the line $5x - 4y = 8$.

$$5x = 4y + 8$$
$$5x - 8 = 4y$$
$$\frac{5x - 8}{4} = \frac{4y}{4}$$
$$\frac{5}{4} \cdot x - 2 = y, \text{ or}$$
$$y = \frac{5}{4} \cdot x - 2$$

Slope is $\frac{5}{4}$.

Section 11.4

11. Find the equation that contains the point (3, 5) and has slope $m = 6$.

$$y = mx + b$$
$$y = 6x + b$$
$$5 = 6 \cdot 3 + b$$
$$5 = 18 + b$$
$$-13 = b$$

Thus, $y = 6x - 13$.

14. Find the equation of the line that contains (−1, 2) and (−3, −2). First, determine the slope:

$$m = \frac{-2 - 2}{-3 - (-1)} = \frac{-4}{-2} = 2;$$
$$y = mx + b,$$
$$y = 2x + b.$$

Use either point to determine b. Let's use (−3, −2):

$$-2 = 2 \cdot -3 + b$$
$$-2 = -6 + b$$
$$4 = b.$$

Thus, $y = 2x + 4$.

Section 11.6

1. Determine whether the graphs of

$$3x - y = -5,$$
$$y - 3x = -2$$

are parallel.

Solve each equation for y and then find the slope.

$$3x - y = -5$$
$$-y = -3x - 5$$
$$y = 3x + 5$$

The slope is 3.

$$y - 3x = -2$$
$$y = 3x - 2$$

The slope is 3.

The slope of each line is 3. The y-intercepts, (0, 5) and (0, −2), are different. Thus the lines are parallel.

3. Determine whether the graphs of

$$y = -\frac{3}{4}x + 7,$$
$$y = \frac{4}{3}x - 9$$

are perpendicular.

The slopes of the lines are $-\frac{3}{4}$ and $\frac{4}{3}$.

The product of the slopes is $-\frac{3}{4} \cdot \frac{4}{3} = -1$.

The lines are perpendicular.

Section 11.7

4. Graph: $2x + 4y < 8$.
Related equation: $2x + 4y = 8$
x-intercept: $(4, 0)$
y-intercept: $(0, 2)$
Draw the line dashed.
Test a point—try $(3, -1)$:
$2 \cdot 3 + 4 \cdot (-1) < 8$
$2 < 8$ TRUE
The point $(3, -1)$ is a solution. We shade the half-plane that contains $(3, -1)$.

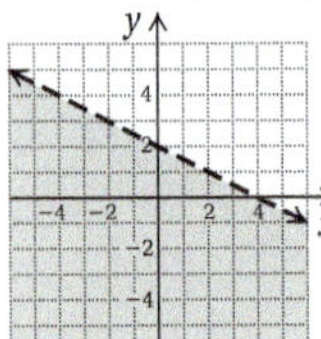

$2x + 4y < 8$

CHAPTER 12

Section 12.1

18. a) $(4t)^2 = [4 \cdot (-3)]^2$
$= [-12]^2$
$= 144$

b) $4t^2 = 4 \cdot (-3)^2$
$= 4 \cdot (9)$
$= 36$

c) Since $144 \neq 36$, the expressions are not equivalent.

33. $4p^{-3} = 4\left(\frac{1}{p^3}\right) = \frac{4}{p^3}$

Section 12.2

10. $(-2x^4)^{-2} = (-2)^{-2}(x^4)^{-2}$
$= \frac{1}{(-2)^2} \cdot x^{-8}$
$= \frac{1}{4} \cdot \frac{1}{x^8}$
$= \frac{1}{4x^8}$

15. $\left(\frac{x^4}{3}\right)^{-2} = \frac{(x^4)^{-2}}{3^{-2}} = \frac{x^{-8}}{3^{-2}}$
$= \frac{\frac{1}{x^8}}{\frac{1}{3^2}} = \frac{1}{x^8} \div \frac{1}{3^2}$
$= \frac{1}{x^8} \cdot \frac{3^2}{1} = \frac{9}{x^8}$

This can be done a second way.
$\left(\frac{x^4}{3}\right)^{-2} = \left(\frac{3}{x^4}\right)^2$
$= \frac{3^2}{(x^4)^2} = \frac{9}{x^8}$

Section 12.3

5. $2x^2 + 5x - 4$
$= 2(-5)^2 + 5(-5) - 4$
$= 2(25) + (-25) - 4$
$= 50 - 25 - 4$
$= 21$

29. $-2x^4 + 16 + 2x^4 + 9 - 3x^5$
$= -3x^5 + (-2 + 2)x^4 + (16 + 9)$
$= -3x^5 + 0x^4 + 25$
$= -3x^5 + 25$

Section 12.4

14. $(-6x^4 + 3x^2 + 6) - (2x^4 + 5x^3 - 5x^2 + 7)$
$= -6x^4 + 3x^2 + 6 - 2x^4 - 5x^3 + 5x^2 - 7$
$= -8x^4 - 5x^3 + 8x^2 - 1$

Section 12.5

12. a) $(y + 2)(y + 7)$
$= y \cdot (y + 7) + 2 \cdot (y + 7)$
$= y \cdot y + y \cdot 7 + 2 \cdot y + 2 \cdot 7$
$= y^2 + 7y + 2y + 14$
$= y^2 + 9y + 14$

b) The area is $(y + 7)(y + 2)$, or, from part (a), $y^2 + 9y + 14$.

18. $(3y^2 - 7)(2y^3 - 2y + 5)$
$= 3y^2(2y^3 - 2y + 5) - 7(2y^3 - 2y + 5)$
$= 6y^5 - 6y^3 + 15y^2 - 14y^3 + 14y - 35$
$= 6y^5 - 20y^3 + 15y^2 + 14y - 35$

Section 12.6

17. $(6 - 4y)(6 + 4y)$
$= (6)^2 - (4y)^2$
$= 36 - 16y^2$

27. $(3x^2 - 5)(3x^2 - 5)$
$= (3x^2)^2 - 2(3x^2)(5) + 5^2$
$= 9x^4 - 30x^2 + 25$

Section 12.7

7. The like terms are $-3pt$ and $8pt$, $-5ptr^3$ and $5ptr^3$, and -12 and 4. Collecting like terms, we have
$(-3 + 8)pt + (-5 + 5)ptr^3 + (-12 + 4)$
$= 5pt - 8$.

22. $(2a + 5b + c)(2a - 5b - c)$
$= [2a + (5b + c)][2a - (5b + c)]$
$= (2a)^2 - (5b + c)^2$
$= 4a^2 - (25b^2 + 10bc + c^2)$
$= 4a^2 - 25b^2 - 10bc - c^2$

Section 12.8

5. $(28x^7 + 32x^5) \div (4x^3)$
$= \frac{28x^7 + 32x^5}{4x^3} = \frac{28x^7}{4x^3} + \frac{32x^5}{4x^3}$
$= \frac{28}{4}x^{7-3} + \frac{32}{4}x^{5-3}$
$= 7x^4 + 8x^2$

10. $(x^2 + x - 6) \div (x + 3)$

$$\begin{array}{r} x - 2 \\ x + 3\overline{)x^2 + x - 6} \\ \underline{x^2 + 3x} \\ -2x - 6 \\ \underline{-2x - 6} \\ 0 \end{array}$$

CHAPTER 13

Section 13.1

7. Find the GCF of $-24m^5n^6$, $12mn^3$, $-16m^2n^2$, and $8m^4n^4$.
The coefficients are -24, 12, -16, and 8.
The greatest positive common factor of the coefficients is 4.
The smallest exponent of the variable m is 1.
The smallest exponent of the variable n is 2.
The GCF $= 4mn^2$.

19. $x^3 + 7x^2 + 3x + 21$
$= x^2(x + 7) + 3(x + 7) = (x + 7)(x^2 + 3)$

Section 13.2

1. Factor: $x^2 + 7x + 12$.
Complete the following table.

Pairs of Factors	Sums of Factors
1, 12	13
−1, −12	−13
2, 6	8
−2, −6	−8
3, 4	7
−3, 4	−7

Because both 7 and 12 are positive, we need consider only the positive factors in the table above.
$x^2 + 7x + 12 = (x + 3)(x + 4)$

8. Factor $a^2 - 40 + 3a$.
First, rewrite in descending order:
$a^2 + 3a - 40$.

Pairs of Factors	Sums of Factors
−1, 40	39
−2, 30	28
−4, 10	6
−5, 8	3

The factorization is $(a - 5)(a + 8)$.

Section 13.4

2. Factor $12x^2 - 17x - 5$.
1) There is no common factor.
2) Multiply the leading coefficient and the constant:
$12(-5) = -60$.
3) Look for a pair of factors of -60 whose sum is -17. Those factors are 3 and -20.
4) Split the middle term: $-17x = 3x - 20x$.
5) Factor by grouping:
$12x^2 + 3x - 20x - 5$
$= 3x(4x + 1) - 5(4x + 1)$
$= (4x + 1)(3x - 5)$.
6) Check: $(4x + 1)(3x - 5) = 12x^2 - 17x - 5$.

Section 13.5

13. Factor $49 - 56y + 16y^2$.
Write in descending order:
$16y^2 - 56y + 49$.
Factor as a trinomial square:
$(4y)^2 - 2 \cdot 4y \cdot 7 + (7)^2$
$= (4y - 7)^2$.

27. $a^2 - 25b^2$
$= a^2 - (5b)^2$
$= (a + 5b)(a - 5b)$.

Section 13.6

5. Factor $8x^3 - 200x$.
a) Factor out the largest common factor:
$8x^3 - 200x = 8x(x^2 - 25)$.
b) There are two terms inside the parentheses. Factor the difference of squares:
$8x(x^2 - 25) = 8x(x + 5)(x - 5)$.
c) We have factored completely.
d) Check: $8x(x + 5)(x - 5) = 8x(x^2 - 25) = 8x^3 - 200x$.

10. Factor $x^4 + 2x^2y^2 + y^4$.
a) There is no common factor.
b) There are three terms. Factor the trinomial square:
$x^4 + 2x^2y^2 + y^4 = (x^2 + y^2)^2$.
c) We have factored completely.
d) Check: $(x^2 + y^2)^2 = (x^2)^2 + 2(x^2)(y^2) + (y^2)^2$
$= x^4 + 2x^2y^2 + y^4$.

Section 13.7

5. $x^2 - x - 6 = 0$
$(x + 2)(x - 3) = 0$
$x + 2 = 0$ *or* $x - 3 = 0$
$x = -2$ *or* $x = 3$
Both numbers check.
The solutions are -2 and 3.

8. $x^2 - 4x = 0$
$x(x - 4) = 0$
$x = 0$ *or* $x - 4 = 0$
$x = 0$ *or* $x = 4$
Both numbers check.
The solutions are 0 and 4.

CHAPTER 14

Section 14.1

2. $\dfrac{2x - 7}{x^2 + 5x - 24}$
$x^2 + 5x - 24 = 0$
$(x + 8)(x - 3) = 0$
$x + 8 = 0$ *or* $x - 3 = 0$
$x = -8$ *or* $x = 3$
The rational expression is not defined for replacements -8 and 3.

13. $\dfrac{x - 8}{8 - x}$
$= \dfrac{x - 8}{-(x - 8)}$
$= \dfrac{1(x - 8)}{-1(x - 8)} = \dfrac{1}{-1} \cdot \dfrac{x - 8}{x - 8}$
$= -1 \cdot 1 = -1$

16. $\dfrac{a^2 - 4a + 4}{a^2 - 9} \cdot \dfrac{a + 3}{a - 2}$
$= \dfrac{(a^2 - 4a + 4)(a + 3)}{(a^2 - 9)(a - 2)}$
$= \dfrac{(a - 2)(a - 2)(a + 3)}{(a + 3)(a - 3)(a - 2)}$
$= \dfrac{\cancel{(a - 2)}(a - 2)\cancel{(a + 3)}}{\cancel{(a + 3)}(a - 3)\cancel{(a - 2)}}$
$= \dfrac{a - 2}{a - 3}$

Section 14.2

6. $\dfrac{x}{8} \div \dfrac{x}{5} = \dfrac{x}{8} \cdot \dfrac{5}{x}$
$= \dfrac{x \cdot 5}{8 \cdot x}$
$= \dfrac{\cancel{x} \cdot 5}{8 \cdot \cancel{x}}$
$= \dfrac{5}{8}$

11. $\dfrac{y^2-1}{y+1} \div \dfrac{y^2-2y+1}{y+1}$

$= \dfrac{y^2-1}{y+1} \cdot \dfrac{y+1}{y^2-2y+1}$

$= \dfrac{(y^2-1)(y+1)}{(y+1)(y^2-2y+1)}$

$= \dfrac{(y+1)(y-1)(y+1)}{(y+1)(y-1)(y-1)}$

$= \dfrac{\cancel{(y+1)}\cancel{(y-1)}(y+1)}{\cancel{(y+1)}\cancel{(y-1)}(y-1)}$

$= \dfrac{y+1}{y-1}$

Section 14.3

1. $16 = 2 \cdot 2 \cdot 2 \cdot 2$
$18 = 2 \cdot 3 \cdot 3$
LCM $= 2 \cdot 2 \cdot 2 \cdot 2 \cdot 3 \cdot 3$,
or 144

5. $\dfrac{3}{16} + \dfrac{1}{18}$

$= \dfrac{3}{2 \cdot 2 \cdot 2 \cdot 2} + \dfrac{1}{2 \cdot 3 \cdot 3}$

$= \dfrac{3}{2 \cdot 2 \cdot 2 \cdot 2} \cdot \dfrac{3 \cdot 3}{3 \cdot 3} + \dfrac{1}{2 \cdot 3 \cdot 3} \cdot \dfrac{2 \cdot 2 \cdot 2}{2 \cdot 2 \cdot 2}$

$= \dfrac{27+8}{2 \cdot 2 \cdot 2 \cdot 2 \cdot 3 \cdot 3}$

$= \dfrac{35}{144}$

Section 14.4

5. $\dfrac{3}{16x} + \dfrac{5}{24x^2}$
$16x = 2 \cdot 2 \cdot 2 \cdot 2 \cdot x$
$24x^2 = 2 \cdot 2 \cdot 2 \cdot 3 \cdot x \cdot x$
LCD $= 2 \cdot 2 \cdot 2 \cdot 2 \cdot 3 \cdot x \cdot x$,
or $48x^2$

$\dfrac{3}{16x} \cdot \dfrac{3x}{3x} + \dfrac{5}{24x^2} \cdot \dfrac{2}{2}$

$= \dfrac{9x}{48x^2} + \dfrac{10}{48x^2}$

$= \dfrac{9x+10}{48x^2}$

10. $\dfrac{2x+1}{x-3} + \dfrac{x+2}{3-x}$

$= \dfrac{2x+1}{x-3} + \dfrac{x+2}{3-x} \cdot \dfrac{-1}{-1}$

$= \dfrac{2x+1}{x-3} + \dfrac{-x-2}{x-3}$

$= \dfrac{(2x+1)+(-x-2)}{x-3}$

$= \dfrac{x-1}{x-3}$

Section 14.5

4. $\dfrac{x-2}{3x} - \dfrac{2x-1}{5x}$
LCD $= 3 \cdot x \cdot 5 = 15x$

$= \dfrac{x-2}{3x} \cdot \dfrac{5}{5} - \dfrac{2x-1}{5x} \cdot \dfrac{3}{3}$

$= \dfrac{5x-10}{15x} - \dfrac{6x-3}{15x}$

$= \dfrac{5x-10-(6x-3)}{15x}$

$= \dfrac{5x-10-6x+3}{15x}$

$= \dfrac{-x-7}{15x}$

8. $\dfrac{y}{16-y^2} - \dfrac{7}{y-4}$

$= \dfrac{y}{16-y^2} \cdot \dfrac{-1}{-1} - \dfrac{7}{y-4}$

$= \dfrac{-y}{y^2-16} - \dfrac{7}{y-4}$

$= \dfrac{-y}{(y+4)(y-4)} - \dfrac{7}{y-4} \cdot \dfrac{y+4}{y+4}$

$= \dfrac{-y}{(y+4)(y-4)} - \dfrac{7y+28}{(y+4)(y-4)}$

$= \dfrac{-y-(7y+28)}{(y+4)(y-4)}$

$= \dfrac{-y-7y-28}{(y+4)(y-4)}$

$= \dfrac{-8y-28}{(y+4)(y-4)} = \dfrac{-4(2y+7)}{(y+4)(y-4)}$

Section 14.6

2. $\dfrac{\dfrac{x}{2}+\dfrac{2x}{3}}{\dfrac{1}{x}-\dfrac{x}{2}}$
LCM of denominators $= 6x$

$= \dfrac{\dfrac{x}{2}+\dfrac{2x}{3}}{\dfrac{1}{x}-\dfrac{x}{2}} \cdot \dfrac{6x}{6x}$

$= \dfrac{\left(\dfrac{x}{2}+\dfrac{2x}{3}\right) \cdot 6x}{\left(\dfrac{1}{x}-\dfrac{x}{2}\right) \cdot 6x}$

$= \dfrac{3x^2+4x^2}{6-3x^2} = \dfrac{7x^2}{3(2-x^2)}$

5. $\dfrac{\dfrac{x}{2}+\dfrac{2x}{3}}{\dfrac{1}{x}-\dfrac{x}{2}}$ ← LCD = 6 (numerator), ← LCD $= 2x$ (denominator)

$= \dfrac{\dfrac{x}{2} \cdot \dfrac{3}{3} + \dfrac{2x}{3} \cdot \dfrac{2}{2}}{\dfrac{1}{x} \cdot \dfrac{2}{2} - \dfrac{x}{2} \cdot \dfrac{x}{x}}$

$= \dfrac{\dfrac{3x}{6}+\dfrac{4x}{6}}{\dfrac{2}{2x}-\dfrac{x^2}{2x}} = \dfrac{\dfrac{3x+4x}{6}}{\dfrac{2-x^2}{2x}}$

$= \dfrac{7x}{6} \cdot \dfrac{2x}{2-x^2} = \dfrac{7 \cdot 2 \cdot x \cdot x}{2 \cdot 3(2-x^2)}$

$= \dfrac{7x^2}{3(2-x^2)}$

Section 14.7

4. $\frac{1}{2x} + \frac{1}{x} = -12$

LCM $= 2x$

$$2x\left(\frac{1}{2x} + \frac{1}{x}\right) = 2x(-12)$$

$$2x \cdot \frac{1}{2x} + 2x \cdot \frac{1}{x} = 2x(-12)$$

$$1 + 2 = -24 \cdot x$$

$$3 = -24x$$

$$\frac{3}{-24} = x$$

$$-\frac{1}{8} = x$$

7. $\frac{4}{x-2} + \frac{1}{x+2} = \frac{26}{x^2-4}$

LCM $= (x-2)(x+2)$

$$(x-2)(x+2)\left(\frac{4}{x-2} + \frac{1}{x+2}\right) = (x-2)(x+2) \cdot \frac{26}{x^2-4}$$

$$4(x+2) + 1(x-2) = 26$$

$$4x + 8 + x - 2 = 26$$

$$5x + 6 = 26$$

$$5x = 20$$

$$x = 4$$

Section 14.9

1. $y = kx$

$$84 = k \cdot 12$$

$$k = \frac{84}{12} = 7$$

$$y = 7 \cdot x$$

Then find y when x is 41.

$$y = 7x = 7 \cdot 41$$

$$= 287$$

5. $y = \frac{k}{x}$

$$105 = \frac{k}{0.6}$$

$$k = 0.6 \cdot 105$$

$$k = 63$$

$$y = \frac{63}{x}$$

Then find y when x is 20.

$$y = \frac{63}{x} = \frac{63}{20}$$

$$= 3.15$$

CHAPTER 15

Section 15.1

2. $(20, 40)$; $a = \frac{1}{2}b$,

$b - a = 60$

Check:

$a = \frac{1}{2}b$	
20 ?	$\frac{1}{2}(40)$
	20

$b - a = 60$	
$40 - 20$	? 60
20	

$(20, 40)$ is a solution of $a = \frac{1}{2}b$.

$(20, 40)$ is not a solution of $b - a = 60$.

Therefore, $(20, 40)$ is not a solution of the system.

Section 15.2

1. $x + y = 5$, **(1)**

$x = y + 1$ **(2)**

Substitute $y + 1$ for x in equation (1) and solve for y.

$$x + y = 5$$

$$(y + 1) + y = 5$$

$$2y + 1 = 5$$

$$2y = 4$$

$$y = 2$$

Substitute 2 for y in equation (2) and solve for x.

$$x = y + 1$$

$$= 2 + 1$$

$$= 3$$

The numbers check. The solution is $(3, 2)$.

3. $x - 2y = 8$, **(1)**

$2x + y = 8$ **(2)**

Solve for y in equation (2).

$$2x + y = 8$$

$$y = 8 - 2x \quad \textbf{(3)}$$

Substitute $8 - 2x$ for y in equation (1) and solve for x.

$$x - 2y = 8$$

$$x - 2(8 - 2x) = 8$$

$$x - 16 + 4x = 8$$

$$5x - 16 = 8$$

$$5x = 24$$

$$x = \frac{24}{5}$$

Substitute $\frac{24}{5}$ for x in equation (3) and solve for y.

$$y = 8 - 2x$$

$$= 8 - 2\left(\frac{24}{5}\right)$$

$$= \frac{40}{5} - \frac{48}{5}$$

$$= -\frac{8}{5}$$

The numbers check. The solution is $\left(\frac{24}{5}, -\frac{8}{5}\right)$.

Section 15.3

6. $3x - 8y = 2$, **(1)**

$5x + 2y = -12$ **(2)**

Multiply equation (2) by 4, add, and solve for x.

$$3x - 8y = 2$$

$$\underline{20x + 8y = -48}$$

$$23x \quad = -46$$

$$x = -2$$

Substitute -2 for x in equation (1) and solve for y.

$$3x - 8y = 2$$

$$3(-2) - 8y = 2$$

$$-6 - 8y = 2$$

$$-8y = 8$$

$$y = -1$$

The numbers check. The solution is $(-2, -1)$.

9. $5x - 2y = 3, \quad (1)$
$-15x + 6y = -9 \quad (2)$

Multiply equation (1) by 3 and add.

$$15x - 6y = 9$$
$$-15x + 6y = -9$$
$$0 = 0$$

The system has an infinite number of solutions.

Section 15.4

2.

\$3.10	\$1.75	
a	c	166
$3.10a$	$1.75c$	\$459.25

$a + c = 166,$
$3.10a + 1.75c = 459.25$

3.

x	y	30
50%	70%	55%
50%x	70%y	55% × 30, or 16.5

$x + y = 30$
$50\%x + 70\%y = 16.5$

4.

\$1.40	\$1.75	\$1.54
x	y	50
$1.40x$	$1.75y$	1.54 · 50, or 77

$x + y = 50,$
$1.40x + 1.75y = 77$

Section 15.5

1. **Familiarize.** Let t = the time for the first car. Then $t - 1$ = the time for the second car.
2. **Translate.**

$$d = 56t,$$
$$d = 84(t - 1)$$

3. **Solve.**

$$84(t - 1) = 56t$$
$$84t - 84 = 56t$$
$$-84 = -28t$$
$$3 = t$$

If $t = 3$ hr, then $d = 56(3) = 168$.

4. **Check.** The first car travels 168 km in 3 hr, and the second car travels 168 km in 2 hr.
5. **State.** The second car will catch up with the first car in 168 km.

CHAPTER 16

Section 16.1

31. $\sqrt{x^2 + 8x + 16}$
$= \sqrt{(x + 4)^2}$
$= |x + 4|$

37. $\sqrt{x^2 + 8x + 16}$
$= \sqrt{(x + 4)^2}$
$= x + 4$

Section 16.2

9. $\sqrt{363q}$
$= \sqrt{121 \cdot 3 \cdot q}$
$= \sqrt{121}\sqrt{3q}$
$= 11\sqrt{3q}$

20. $\sqrt{24x^{11}}$
$= \sqrt{4 \cdot 6 \cdot x^{10} \cdot x}$
$= \sqrt{4}\sqrt{x^{10}}\sqrt{6x}$
$= 2x^5\sqrt{6x}$

Section 16.3

11. $\dfrac{\sqrt{98y}}{\sqrt{2y^{11}}}$
$= \sqrt{\dfrac{98y}{2y^{11}}}$
$= \sqrt{\dfrac{49}{y^{10}}}$
$= \dfrac{\sqrt{49}}{\sqrt{y^{10}}}$
$= \dfrac{7}{y^5}$

15. $\dfrac{10}{\sqrt{3}}$
$= \dfrac{10}{\sqrt{3}} \cdot \dfrac{\sqrt{3}}{\sqrt{3}}$
$= \dfrac{10\sqrt{3}}{\sqrt{9}}$
$= \dfrac{10\sqrt{3}}{3}$

14. $\sqrt{\dfrac{5}{8}}$
$= \sqrt{\dfrac{5}{8} \cdot \dfrac{2}{2}}$
$= \sqrt{\dfrac{10}{16}} = \dfrac{\sqrt{10}}{\sqrt{16}}$
$= \dfrac{\sqrt{10}}{4}$

Section 16.4

4. $\sqrt{24} + \sqrt{54}$
$= \sqrt{4 \cdot 6} + \sqrt{9 \cdot 6}$
$= 2\sqrt{6} + 3\sqrt{6}$
$= (2 + 3)\sqrt{6}$
$= 5\sqrt{6}$

16. $\dfrac{3}{7 + \sqrt{5}}$
$= \dfrac{3}{7 + \sqrt{5}} \cdot \dfrac{7 - \sqrt{5}}{7 - \sqrt{5}}$
$= \dfrac{3(7 - \sqrt{5})}{7^2 - (\sqrt{5})^2}$
$= \dfrac{21 - 3\sqrt{5}}{49 - 5}$
$= \dfrac{21 - 3\sqrt{5}}{44}$

Section 16.5

2. $\sqrt{3x + 1} = \sqrt{2x + 3}$
$(\sqrt{3x + 1})^2 = (\sqrt{2x + 3})^2$
$3x + 1 = 2x + 3$
$x = 2$

6.
$$\sqrt{x} - 1 = \sqrt{x-3}$$
$$(\sqrt{x} - 1)^2 = (\sqrt{x-3})^2$$
$$(\sqrt{x})^2 - 2\sqrt{x} + 1 = x - 3$$
$$x - 2\sqrt{x} + 1 = x - 3$$
$$-2\sqrt{x} + 1 = -3$$
$$-2\sqrt{x} = -4$$
$$\sqrt{x} = 2$$
$$(\sqrt{x})^2 = 2^2$$
$$x = 4$$

Section 16.6

2.
$$a^2 + 11^2 = 14^2$$
$$a^2 + 121 = 196$$
$$a^2 = 75$$
$$a = \sqrt{75}$$
$$a \approx 8.660$$

CHAPTER 17

Section 17.1

8.
$$2x^2 + 8x = 0$$
$$2x(x+4) = 0$$
$$2x = 0 \quad or \quad x + 4 = 0$$
$$x = 0 \quad or \quad x = -4$$

Both numbers check. The solutions are 0 and -4.

12.
$$\frac{20}{x+5} - \frac{1}{x-4} = 1$$
$$(x+5)(x-4) \cdot \left(\frac{20}{x+5} - \frac{1}{x-4}\right) = (x+5)(x-4) \cdot 1$$
$$(x+5)(x-4) \cdot \frac{20}{x+5} - (x+5)(x-4) \cdot \frac{1}{x-4} = (x+5)(x-4)$$
$$20(x-4) - 1(x+5) = (x+5)(x-4)$$
$$20x - 80 - x - 5 = x^2 + x - 20$$
$$19x - 85 = x^2 + x - 20$$
$$0 = x^2 - 18x + 65$$
$$0 = (x-5)(x-13)$$
$$x - 5 = 0 \quad or \quad x - 13 = 0$$
$$x = 5 \quad or \quad x = 13$$

Both numbers check in the original equation. The solutions are 5 and 13.

Section 17.2

3.
$$2x^2 - 3 = 0$$
$$2x^2 = 3$$
$$x^2 = \frac{3}{2}$$
$$x = \sqrt{\frac{3}{2}} \quad or \quad x = -\sqrt{\frac{3}{2}}$$
$$x = \sqrt{\frac{3}{2} \cdot \frac{2}{2}} \quad or \quad x = -\sqrt{\frac{3}{2} \cdot \frac{2}{2}}$$
$$x = \frac{\sqrt{6}}{2} \quad or \quad x = -\frac{\sqrt{6}}{2}$$

Both numbers check. The solutions are $\frac{\sqrt{6}}{2}$ and $-\frac{\sqrt{6}}{2}$.

10.
$$x^2 - 12x + 23 = 0$$
$$x^2 - 12x = -23$$
$$x^2 - 12x + 36 = -23 + 36$$
$$(x-6)^2 = 13$$
$$x - 6 = \sqrt{13} \quad or \quad x - 6 = -\sqrt{13}$$
$$x = 6 + \sqrt{13} \quad or \quad x = 6 - \sqrt{13}$$

The solutions are $6 \pm \sqrt{13}$.

Section 17.3

1. Write in standard form:
$2x^2 + 7x - 4 = 0.$
$a = 2, b = 7, c = -4$
$$x = \frac{-b \pm \sqrt{b^2 - 4ac}}{2a}$$
$$x = \frac{-7 \pm \sqrt{7^2 - 4 \cdot 2 \cdot (-4)}}{2 \cdot 2}$$
$$x = \frac{-7 \pm \sqrt{49 + 32}}{4}$$
$$x = \frac{-7 \pm \sqrt{81}}{4} = \frac{-7 \pm 9}{4}$$
$$x = \frac{-7+9}{4} \quad or \quad x = \frac{-7-9}{4}$$
$$x = \frac{1}{2} \quad or \quad x = -4$$

3. Write in standard form:
$x^2 + 4x - 7 = 0.$
$a = 1, b = 4, c = -7$
$$x = \frac{-b \pm \sqrt{b^2 - 4ac}}{2a}$$
$$= \frac{-4 \pm \sqrt{4^2 - 4 \cdot 1 \cdot (-7)}}{2 \cdot 1}$$
$$= \frac{-4 \pm \sqrt{16 + 28}}{2} = \frac{-4 \pm \sqrt{44}}{2}$$
$$= \frac{-4 \pm 2\sqrt{11}}{2} = \frac{2(-2 \pm \sqrt{11})}{2 \cdot 1}$$
$$= -2 \pm \sqrt{11}$$

Section 17.4

3. Solve for f: $\frac{1}{p} + \frac{1}{q} = \frac{1}{f}$.

The LCM is pqf.
$$pqf\left(\frac{1}{p} + \frac{1}{q}\right) = pqf\left(\frac{1}{f}\right)$$
$$pqf \cdot \frac{1}{p} + pqf\left(\frac{1}{q}\right) = pqf\left(\frac{1}{f}\right)$$
$$qf + pf = pq$$
$$f(q+p) = pq$$
$$f = \frac{pq}{q+p}$$

7. Solve for r: $A = \pi r^2$.
$$\frac{A}{\pi} = r^2$$
$$\sqrt{\frac{A}{\pi}} = r$$

Section 17.5

1. 1. **Familiarize.** Let $w =$ the width of the mural. Then the length is $2w + 5$.
   2. **Translate.**
   $(2w + 5)(w) = 52$
   3. **Solve.**
   $$2w^2 + 5w = 52$$
   $$2w^2 + 5w - 52 = 0$$
   $$(2w + 13)(w - 4) = 0$$
   $$2w + 13 = 0 \quad or \quad w - 4 = 0$$
   $$w = -\tfrac{13}{2} \quad or \quad w = 4$$
   4. **Check.** Only 4 checks. When the width is 4 ft, the length is $2(4) + 5 = 13$ ft.
   5. **State.** The width is 4 ft, and the length is 13 ft.

3. 1. Familiarize.

	d	r	t
Upstream	45	$12 - s$	t_1
Downstream	45	$12 + s$	t_2
Total time			8

2. Translate.

$$t_1 = \frac{45}{12 - s},\ t_2 = \frac{45}{12 + s}$$

$$\frac{45}{12 - s} + \frac{45}{12 + s} = 8$$

3. Solve.

$$(12 - s)(12 + s)\left(\frac{45}{12 - s} + \frac{45}{12 + s}\right) = (12 - s)(12 + s)(8)$$
$$45(12 + s) + 45(12 - s) = (144 - s^2)(8)$$
$$1080 = 1152 - 8s^2$$
$$8s^2 - 72 = 0$$
$$s^2 - 9 = 0$$
$$s + 3 = 0 \quad or \quad s - 3 = 0$$
$$s = -3 \quad or \quad s = 3$$

4. Check. The speed of the stream cannot be negative. A speed of 3 km/h checks.

5. State. The speed of the stream is 3 km/h.

Section 17.6

5.
$$x^2 + 6x + 8 = 0$$
$$(x + 4)(x + 2) = 0$$
$$x + 4 = 0 \quad or \quad x + 2 = 0$$
$$x = -4 \quad or \quad x = -2$$

The x-intercepts are $(-4, 0)$ and $(-2, 0)$.

7.
$$x^2 + 3 = 0$$
$$x^2 = -3$$

Since -3 is negative, the equation has no real-number solutions. There are no x-intercepts.

Section 17.7

9.
$$p(x) = x^4 - 5x^2 + 8$$
$$p(0) = 0^4 - 5 \cdot 0^2 + 8$$
$$= 0 - 0 + 8$$
$$= 8$$

13.
$$F(r) = \sqrt{r} + 9$$
$$F(100) = \sqrt{100} + 9$$
$$= 10 + 9$$
$$= 19$$

CREDITS

p. 3 Tina Manley/Alamy **p. 8** (left) Anton Balazh/Shutterstock, (right) NASA **p. 55** Courtesy of Barbara Johnson **p. 66** (left) Yevgenia Gorbulsky/Fotolia, (right) Courtesy of Geri Davis **p. 82** Sebastian Duda/Fotolia **p. 90** (left) Dave King © Dorling Kindersley, (right) Qingwa/Fotolia **p. 99** Mellowbox/Fotolia **p. 103** Serge Black/Fotolia **p. 146** Melinda Nagy/Fololia **p. 170** Cynoclub/Fotolia **p. 175** Petr84/Shutterstock **p. 176** Larry Roberg/Fotolia **p. 194** (left) Uwimages/Fotolia, (right) Image 100/ Corbis/Glow Images **p. 214** Michaeljung/Fotolia **p. 225** Dr Paulus Gerdes **p. 233** (left) Christophe Fouquin/Fotolia, (right) Jim West/Glow Images **p. 255** Nicholas Piccillo/Fotolia **p. 260** Interfoto/Alamy **p. 261** Ruud Morijn/Fotolia **p. 272** Maridav/ Fotolia **p. 276** Joshua Lott/Reuters **p. 277** epa european pressphoto agency b.v./Alamy **p. 286** Courtesy of Tom Sears **p. 291** Image Source/Glow Images **p. 296** epa european pressphoto agency b.v./Alamy **p. 298** (left) Daily Mail/Rex/Alamy, (right) epa european pressphoto agency b.v./Alamy **p. 303** (left) MSPhotographic/Fotolia, (right) Zai Aragon/Fotolia **p. 317** Estima/ Fotolia **p. 319** Martin Valigursky/Fotolia **p. 321** (left) Michaklootwijk/Fotolia, (right) Santi Visalli/Glow Images **p. 341** Rtimages/Shutterstock **p. 348** Hellen Sergeyeva/Fotolia **p. 349** Ursule/Fotolia **p. 358** TheFinalMiracle/Fotolia **p. 359** Iain Masterton/age fotostock SuperStock **p. 363** Kotangens/Fotolia **p. 366** (left) PCN Photography/Alamy, (right) Christopher Sadowski/Splash News/Newscom **p. 369** (left) Wusuowei/Fotolia, (right) Karrapavan/Fotolia **p. 370** Candan/Fotolia **p. 373** Edward Rozzo/Corbis/Glow Images **p. 383** Carsten Reisinger/Fotolia **p. 387** Ioannis Ioannou/Shutterstock **p. 388** M. Timothy O'Keefe/Alamy **p. 390** Antoni Murcia/Shutterstock **p. 392** (left) Gerard Sioen/Gamma-Rapho/Getty Images, (right) Rudie/Fotolia **p. 393** (left) Pictorial Press, Ltd./Alamy, (right) Fancy Collection/SuperStock **p. 394** (left) Konrad Wothe/Glow Images, (right) ZUMA Press, Inc./Alamy **p. 396** John Dorton/Shutterstock **p. 397** Tina Jeans/Shutterstock **p. 400** (left) Karin Hildebrand Lau/Shutterstock, (right) Susan The/Fotolia **p. 401** (left and right) iStockphoto/Thinkstock **p. 404** Bbbar/Fotolia **p. 412** Nata-Lia/Shutterstock **p. 413** (top) Rehan Qureshi/Shutterstock, (bottom) Studio D/Fotolia **p. 417** Gina Sanders/Fotolia **p. 427** (Barbary macaque) Uabels/Shutterstock, (black rhino) Winfried Wisniewski/AGE Fotostock, (Galapagos seal) William Mullins/Alamy, (Malayan tapir) Roland Seitre/Nature Picture Library, (Cuvier's gazelle) Corbis/ AGE Fotostock, (Darwin's fox) Kevin Schafer/Alamy, (indri) Andy Rouse/Nature Picture Library **p. 429** Visions of America, LLC/Alamy **p. 430** Francis Vachon/Alamy **p. 431** David Taylor/Alamy **p. 433** Daniel Borzynski/Alamy **p. 434** (left) ZUMA Press, Inc./Alamy, (right) Image Source Plus/Alamy **p. 435** iStockphoto/Thinkstock **p. 436** Ashley Cooper Pics/Alamy **p. 437** (left) Rob Wilson/Shutterstock, (right) Wollertz/Shutterstock **p. 442** Flonline Digitale Bildagentur GmbH/Alamy **p. 443** Gmcgill/Fotolia **p. 445** (left) Auremar/Fotolia, (right) Christina Richards/Shutterstock **p. 450** Valua Vitaly/Shutterstock **p. 453** (left) Dmitry Vereshchagin/Fotolia, (right) Andres Rodriguez/Fotolia **p. 460** Imagebroker/Alamy **p. 461** (left) Martin Shields/Alamy, (right) Mark Bonham/Shutterstock **p. 469** YuriArcurs/Shutterstock **p. 471** Mike Wulf/Cal Sport Media/ Newscom **p. 474** (left) Narumol Pug/Fotolia, (right) courtesy of Barbara Johnson **p. 497** All Canada Photos/Alamy **p. 505** Robert Daly/Caia Images/Glow Images **p. 532** Thierry Roge/Reuters **p. 544** (left) Maisna/Fotolia, (right) Wisconsin DNR **p. 558** (left) Luisa Fernanda Gonzalez/Shutterstock, (right) AP Images **p. 559** (top right) Lev1977/Fotolia, (bottom left) Imagebroker/Alamy, (bottom right) Graham Prentice/Shutterstock **p. 610** Veniamin Kraskov/Shutterstock **p. 612** Carlos Santa Maria/Fotolia **p. 616** Ed Metz/Shutterstock **p. 623** Dave King/Dorling Kindersley, Ltd. **p. 630** Comstock/Getty Images **p. 636** Mellowbox/Fotolia **p. 646** Ivan Alvarado/Reuters **p. 649** David Peart/Dorling Kindersley, Ltd. **p. 685** Brian Snyder/ Reuters **p. 713** Leonid Tit/Fotolia **p. 714** London Photos/Alamy **p. 717** SoCalBatGal/Fotolia **p. 718** (left) Sky Bonillo/ PhotoEdit, (right) Anthony Berenyi/Shutterstock **p. 725** Jennifer Pritchard/MCT KRT/Newscom **p. 726** Gjeerawut/Fotolia **p. 729** (left) Iofoto/Shutterstock, (right) Stew Milne/Associated Press **p. 730** (left) Wilson Araujo/Shutterstock, (right) Michael Jung/Shutterstock **p. 739** Magic Mountain/Associated Press **p. 740** Stephen VanHorn/Shutterstock **p. 743** (left) Rodney Todt/Alamy, (right) Corbis/SuperStock **p. 744** (left) Courtesy of Indianapolis Motor Speedway, (right) Lars Lindblad/ Shutterstock **p. 745** (left) Barbara Johnson, (right) Studio 8/Pearson Education Ltd. **p. 746** Elena Yakusheva/Shutterstock **p. 761** (top right) Kike Calvo VWPics/SuperStock, (figure, bottom) U.S. Department of Health and Human Services and U.S. Department of Agriculture **p. 764** Andrey N. Bannov/Shutterstock **p. 767** (left) Reggie Lavoie/Shutterstock, (right) Monkey Business/Fotolia **p. 771** pearlguy/Fotolia **p. 772** Outdoorsman/Fotolia **p. 784** Denis_pc/Fotolia **p. 808** David Pearson/Alamy **p. 809** (figure, bottom) Newspaper Association of America **p. 810** (figure, top) *Editor and Publisher* **p. 813** (middle left) Alex Robinson/Dorling Kindersley, Ltd., (middle right) Evan Meyer/ Shutterstock, (figure, bottom left) *China Daily,* (figure, bottom right) Centers for Disease Control and Prevention **p. 814** (left) Maria Dryfhout/Shutterstock, (right) Onizu3d/Fotolia **p. 846** Arinahabich/Fotolia **p. 855** NASA **p. 866** (left) Anna Omelchenko/Fotolia, (right) Nathan Devery/Science Source/ Photo Researchers, Inc. **p. 869** Lorraine Swanson/Fotolia **p. 870** OJO Images, Ltd./Alamy **p. 872** (left) Terry Leung/Pearson Education Asia Ltd., (right) Lana Rastro/Alamy **p. 873** (upper right) Engine Images/Fotolia, (bottom left) NASA **p. 874** (left) SeanPavonePhoto/Fotolia, (right) Science Source/Photo Researchers, Inc. **p. 883** Brian Buckland **p. 884** Joern Sackermann/ Alamy **p. 917** Linda Whitwam/Dorling Kindersley, Ltd. **p. 921** Coneyl Jay/Science Source/Photo Researchers, Inc. **p. 973** NASA **p. 1004** Hans Neleman/Corbis/Glow Images **p. 1009** Stephen Barnes/Hobbies and Crafts/Alamy **p. 1017** Pattie Steib/ Shutterstock **p. 1040** Igor Normann/Shutterstock **p. 1075** Brocreative/Shutterstock **p. 1078** Ortodox/Shutterstock **p. 1083** (left) Elenathewise/Fotolia, (right) Simon Johnsen/Shutterstock **p. 1084** HandmadePictures/Shutterstock **p. 1085** (left) Lightpoet/Shutterstock, (right) Drazen/Shutterstock **p. 1086** Catmando/Shutterstock **p. 1089** Minadezhda/Fotolia **p. 1091**

Heather Charles/*The Indianapolis Star* **p. 1092** EduardSV/Fotolia **p. 1093** (left and right) NASA **p. 1095** Ibooo7/Shutterstock **p. 1101** (left) Dmitry Kalinovsky/Shutterstock, (right) James McConnachie © Rough Guides/DK Images **p. 1118** Kyod/Landov **p. 1130** Steve Pepple/Shutterstock **p. 1131** Ron Brown/AGE Fotostock **p. 1135** Lightpoet/Shutterstock **p. 1138** Lev Kropotov/Shutterstock **p. 1158** Oskar Eyb/Glow Images **p. 1161** (left) Glow Images, (right) Caro/Alamy **p. 1162** Action Plus Sports Images/Alamy **p. 1171** Fotokostic/Shutterstock **p. 1178** Renaud Visage/AGE Fotostock **p. 1188** Steve Smith/SuperStock **p. 1192** Holger Leue/AGE Fotostock **p. 1196** Andrey Yurlov/Shutterstock **p. 1199** (top) Photo courtesy LandWave Products, Inc., Pioneer, OH, (bottom) Circle of Lights presented by the Contractors of Quality Connection and Electrical Workers of IBEW 481 **p. 1225** National Geographic Image Collection/Alamy **p. 1239** Wavebreakmedia/Shutterstock **p. 1248** EWAStock/Glow Images **p. 1249** Vladimir Sazonov/Fotolia **p. 1265** EditorialByDarrellYoung/Alamy **p. 1267** (left) AfricaMediaOnline/Glow Images, (right) Spotmatik/Shutterstock **p. 1283** Golddc/Shutterstock **p. 1286** (left) The Natural History Museum/Alamy, (right) Klaus Rademaker/Shutterstock **p. 1287** Gaspar Janos/Shutterstock **p. 1295** Tyler Olson/Shutterstock **p. 1296** (left) Viktoriya Field/Shutterstock, (right) Urbanlight/Shutterstock **p. 1298** (top) Jake Lyell/Alamy, (bottom) Greg C. Grace/Alamy **p. 1303** Blend Images/SuperStock

Glossary

A

Abscissa The first coordinate in an ordered pair of numbers

Absolute value The distance that a number is from 0 on the number line

***ac*-method** A method for factoring trinomials of the type $ax^2 + bx + c, a \neq 1$, involving the product, ac, of the leading coefficient a and the last term c; also called the *grouping method*

Acute angle An angle whose measure is greater than 0° and less than 90°

Acute triangle A triangle in which all three angles are acute

Addends In addition, the numbers being added

Additive identity The number 0

Additive inverse A number's opposite; two numbers are additive inverses of each other if their sum is zero.

Additive inverse of a polynomial Two polynomials are additive inverses, or opposites, of each other if their sum is zero.

Algebraic expression A number or variable or a collection of numbers and variables on which operations are performed

Angle A set of points consisting of two rays (half-lines) with a common endpoint (vertex)

Area The number of square units that fill a plane region

Arithmetic mean A center point of a set of numbers found by adding the numbers and dividing by the number of items of data; also called the *average* or the *mean*

Arithmetic numbers The set of whole numbers and positive fractions; also called *nonnegative rational numbers*

Ascending order When a polynomial is written with the exponents of the variable increasing as read from left to right, it is said to be in ascending order.

Associative law of addition The statement that when three numbers are added, regrouping the addends gives the same sum

Associative law of multiplication The statement that when three numbers are multiplied, regrouping the factors gives the same product

Average A center point of a set of numbers found by adding the numbers and dividing by the number of items of data; also called the *mean* or the *arithmetic mean*

Axes Two perpendicular number lines used to identify points in a plane

B

Bar graph A graphic display of data using bars proportional in length to the numbers represented

Base In exponential notation, the number being raised to a power

Binomial A polynomial containing two terms

C

Celsius A temperature scale in which water freezes at 0° and water boils at 100°

Circle The set of all points in a plane that are a given distance (radius) from a given point (center)

Circle graph A graphic display of data using a divided circle to show the percent of a quantity in each of several categories; also called *pie chart*

Circumference The distance around a circle

Coefficient The numeric multiplier of a variable

Commission A percent of total sales paid to a salesperson

Commutative law of addition The statement that when two numbers are added, changing the order in which the numbers are added does not affect the sum

Commutative law of multiplication The statement that when two numbers are multiplied, changing the order in which the numbers are multiplied does not affect the product

Complementary angles Two angles for which the sum of their measures is 90°

Completing the square Adding a particular constant to an expression so that the resulting sum is a perfect square

Complex fraction expression A rational expression that has one or more rational expressions within its numerator and/or denominator

Complex number Any number that can be named $a + bi$, where a and b are any real numbers

Complex-number system A number system that contains the real-number system and is designed so that negative numbers have defined square roots

Complex rational expression A rational expression that has one or more rational expressions within its numerator and/or denominator

Composite number A natural number, other than 1, that is not prime

Compound interest Interest computed on the sum of an original principal and the interest previously accrued by that principal

Congruent angles Two angles that have the same measure

Congruent segments Two line segments that have the same length

Congruent triangles Triangles in which corresponding angles and sides are congruent

Conjugate of a complex number The conjugate of $a + bi$ is $a - bi$, and the conjugate of $a - bi$ is $a + bi$.

Conjugates Pairs of radical terms, like $\sqrt{a} + \sqrt{b}$ and $\sqrt{a} - \sqrt{b}$ or $c + \sqrt{d}$ and $c - \sqrt{d}$, for which the product does not have a radical term

Consecutive even integers Even integers that are two units apart
Consecutive integers Integers that are one unit apart
Consecutive odd integers Odd integers that are two units apart
Constant A number or letter that stands for just one number
Constant of proportionality The constant in an equation of direct or inverse variation
Coordinates The numbers in an ordered pair
Coplanar lines Lines in the same plane
Cross products Given an equation with a single fraction on each side, the products formed by multiplying the left numerator and the right denominator, and the left denominator and the right numerator
Cube root The number c is called a cube root of a, written $\sqrt[3]{a}$, if $c^3 = a$.

D

Decimal notation A representation of a number containing a decimal point
Degree of a polynomial The degree of the term of highest degree in a polynomial
Degree of a term The sum of the exponents of the variables
Denominator The number below the fraction bar in a fraction
Descending order When a polynomial is written with the exponents of the variable decreasing as read from left to right, it is said to be in descending order.
Diagonal of a quadrilateral A line segment that joins two opposite vertices
Diameter A line segment that passes through the center of a circle and has its endpoints on the circle
Difference The result of subtracting one number from another
Difference of cubes An expression that can be written in the form $A^3 - B^3$
Difference of squares An expression that can be written in the form $A^2 - B^2$
Digit A number 0, 1, 2, 3, 4, 5, 6, 7, 8, or 9 that names a place-value location
Direct variation A situation that translates to an equation of the form $y = kx$, with k a positive constant
Discount The amount subtracted from the original price of an item to find the sale price
Discriminant The radicand, $b^2 - 4ac$, from the quadratic formula
Distance formula The equation $d = \sqrt{(x_2 - x_1)^2 + (y_2 - y_1)^2}$ that represents the distance between two points, (x_1, y_1) and (x_2, y_2), on the coordinate plane
Distributive law of multiplication over addition The statement that multiplying a factor by the sum of two numbers gives the same result as multiplying the factor by each of the two numbers and then adding
Distributive law of multiplication over subtraction The statement that multiplying a factor by the difference of two numbers gives the same result as multiplying the factor by each of the two numbers and then subtracting
Dividend In division, the number being divided
Divisible The number b is said to be divisible by another number a if b is a multiple of a.
Divisor In division, the number dividing another number
Domain The set of all first coordinates of the ordered pairs in a function

E

Elimination method An algebraic method that uses the addition principle to solve a system of equations
Empty set The set without members
Equation A number sentence that says that the expressions on either side of the equals sign, =, represent the same number
Equation of direct variation An equation, described by $y = kx$ with k a positive constant, used to represent direct variation
Equation of inverse variation An equation, described by $y = \frac{k}{x}$ with k a positive constant, used to represent inverse variation
Equiangular triangle A triangle in which all angles are congruent
Equilateral triangle A triangle in which all sides are the same length
Equivalent equations Equations with the same solutions
Equivalent expressions Expressions that have the same value for all allowable (or meaningful) replacements
Equivalent fractions Two different fractions that represent the same number
Equivalent inequalities Inequalities that have the same solution set
Evaluate To substitute a value for each occurrence of a variable in an expression
Even root A root with an even index
Event A set of outcomes
Exponent In expressions of the form a^n, the number n is an exponent.
Exponential notation A representation of a number using a base raised to a power

F

Factor *Verb*: to write an equivalent expression that is a product; *noun*: a multiplier
Factoring Writing an expression as a product
Factorization A number expressed as a product of two or more numbers
Factorization of a polynomial An expression that names the polynomial as a product
Fahrenheit A temperature scale in which water freezes at 32° and water boils at 212°
FOIL To multiply two binomials by multiplying the First terms, the Outside terms, the Inside terms, and then the Last terms
Formula An equation that uses numbers or letters to represent a relationship between two or more quantities
Fraction equation An equation containing one or more rational expressions; also called a *rational equation*
Fraction expression A quotient, or ratio, of two polynomials; also called a *rational expression*
Fraction notation A number written using a numerator and a denominator

Function A correspondence between a first set, called the domain, and a second set, called the range, such that each member of the domain corresponds to *exactly one* member of the range

G

Geometric figure A set of points
Grade The measure of a road's steepness
Grade point average (GPA) The average of the grade point values for each credit hour taken
Graph A picture or diagram of the data in a table; a line, curve, or collection of points that represents all the solutions of an equation
Greatest common factor (GCF) The common factor of a polynomial with the largest possible coefficient and the largest possible exponent(s)

H

Hypotenuse In a right triangle, the side opposite the right angle

I

Identity property of 1 The statement that the product of a number and 1 is always the original number
Identity property of 0 The statement that the sum of a number and 0 is always the original number
Imaginary number A number that can be named bi, where b is some real number and $b \neq 0$
Index In the expression $\sqrt[k]{a}$, the number k is called the index.
Inequality A mathematical sentence using $<, >, \leq, \geq$, or $\neq$
Input A member of the domain of a function
Integers The whole numbers and their opposites; $\ldots, -4, -3, -2, -1, 0, 1, 2, 3, 4, \ldots$
Intercept The point at which a graph intersects the x- or y-axis
Interest A percentage of an amount invested or borrowed
Interest rate The percent at which interest is calculated on a principal
Intersecting lines Lines that cross each other at a common point
Intersection of sets *A* and *B* The set of all elements that are common to *both* A and B, denoted $A \cap B$
Inverse variation A situation that translates to an equation of the form $y = \frac{k}{x}$, with k a positive constant
Irrational number A real number that cannot be named as a ratio of two integers
Isosceles triangle A triangle in which two or more sides are the same length

L

Leading coefficient The coefficient of the term of highest degree in a polynomial
Leading term The term of highest degree in a polynomial
Least common denominator (LCD) The least common multiple of the denominators of two or more fractions
Least common multiple (LCM) The smallest number that is a multiple of two or more numbers
Legs In a right triangle, the two sides that form the right angle
Like radicals Radicals that have the same index and the same radicand
Like terms Terms that have exactly the same variable factors
Line Two rays that have common points and continue in opposite directions
Line graph A graph in which quantities are represented as points connected by straight-line segments
Line of symmetry A line that can be drawn through a graph such that the part of the graph on one side of the line is an exact reflection of the part on the opposite side
Linear equation Any equation that can be written in the form $Ax + By = C$, where x and y are variables
Linear function A function that can be described by an equation of the form $y = mx + b$, where x and y are variables
Linear inequality An inequality whose related equation is a linear equation

M

Marked price The original price of an item
Mean A center point of a set of numbers found by dividing the sum of the numbers by the number of items of data; also called the *average* or the *arithmetic mean*
Median In a set of data listed in order from smallest to largest, the middle number if there is an odd number of data items, or the average of the two middle numbers if there is an even number of data items
Metric system A measurement system used in most countries of the world, but very little in the United States
Midpoint formula The formula $\left(\frac{x_1 + x_2}{2}, \frac{y_1 + y_2}{2}\right)$. If the endpoints of a segment are (x_1, y_1) and (x_2, y_2), then this represents the coordinates of the midpoint.
Minuend The number from which another number is being subtracted
Mixed numeral A number represented by a whole number and a fraction less than 1
Mode The number or numbers that occur most often in a set of data
Monomial An expression of the type ax^n, where a is a real-number constant and n is a nonnegative integer
Multiple of a number A product of the number and an integer
Multiplication property of 0 The statement that the product of 0 and any real number is 0
Multiplicative identity The number 1
Multiplicative inverses Reciprocals; two numbers whose product is 1

N

Natural numbers The counting numbers: 1, 2, 3, 4, 5, . . .
Negative integers Integers to the left of zero on the number line
Nonnegative rational numbers The set of whole numbers and positive fractions; also called *arithmetic numbers*
Numerator The number above the fraction bar in a fraction

O

Obtuse angle An angle whose measure is greater than 90° and less than 180°

Obtuse triangle A triangle in which one angle is an obtuse angle
Odd root A root with an odd index
Opposite The opposite, or additive inverse, of a number x is written $-x$. Opposites are the same distance from 0 on the number line but on different sides of 0.
Opposite of a polynomial Two polynomials are opposites, or additive inverses, of each other if their sum is zero.
Ordered pair A pair of numbers of the form (a, b) for which the order in which the numbers are listed is important
Ordinate The second coordinate in an ordered pair of numbers
Origin The point $(0, 0)$ on a graph where the two axes intersect
Original price The price of an item before a discount is deducted
Outcome The result of an experiment
Output A member of the range of a function

P

Palindrome prime A prime number that remains a prime number when its digits are reversed
Parabola The graph of a quadratic equation
Parallel lines Lines in the same plane that never intersect; two lines are parallel if they have the same slope.
Parallelogram A four-sided polygon with two pairs of parallel sides
Percent notation A representation of a number as n parts per 100; $n\%$
Perfect square A rational number p for which there exists a number a for which $a^2 = p$
Perfect-square trinomial A trinomial that is the square of a binomial
Perimeter The distance around an object or the sum of the lengths of its sides
Periods Groups of three digits, separated by commas
Perpendicular lines Two lines that intersect to form a right angle; two lines are perpendicular if the product of their slopes is -1.
Pi (π) The number that results when the circumference of a circle is divided by its diameter; $\pi \approx 3.14$, or $\frac{22}{7}$
Pictograph A graphic means of displaying information using symbols to represent the amounts
Pie chart A graphic display of data using a divided circle to show the percent of a quantity in each of several categories; also called *circle graph*
Point-slope equation The equation $y - y_1 = m(x - x_1)$, where x_1, y_1, and m are real numbers and m is the slope and (x_1, y_1) is a point that lies on the graph of the equation
Polygon A closed geometric figure with three or more line segments as sides
Polynomial A monomial or a combination of sums and/or differences of monomials
Polynomial equation An equation in which two polynomials are set equal to each other
Positive integers Integers to the right of zero on the number line
Prime factorization A factorization of a composite number as a product of prime numbers
Prime number A natural number that has exactly two *different* factors: itself and 1
Prime polynomial A polynomial that cannot be factored using only integer coefficients
Principal An amount of money that is invested or borrowed
Principal square root The nonnegative square root of a number
Principle of zero products The statement that an equation $ab = 0$ is true if and only if $a = 0$ is true or $b = 0$ is true, or both are true
Product The result when one number is multiplied by another
Proportion An equation stating that two ratios are equal
Proportional numbers Two pairs of numbers having the same ratio
Protractor A device used to measure and draw angles
Purchase price The price of an item before sales tax is added
Pythagorean equation The equation $a^2 + b^2 = c^2$, where a and b are lengths of the legs of a right triangle and c is the length of the hypotenuse
Pythagorean theorem In any right triangle, if a and b are the lengths of the legs and c is the length of the hypotenuse, then $a^2 + b^2 = c^2$.

Q

Quadrants The four regions into which the axes divide a plane
Quadratic equation An equation of the form $ax^2 + bx + c = 0$, where $a \neq 0$
Quadratic formula The solutions of $ax^2 + bx + c = 0$, $a \neq 0$, are given by the equation
$$x = \frac{-b \pm \sqrt{b^2 - 4ac}}{2a}.$$
Quadratic function A second-degree polynomial function in one variable
Quadratic inequality An inequality whose related equation is a quadratic equation
Quotient The result when one number is divided by another

R

Radical equation An equation that has variables in one or more radicands
Radical expression An algebraic expression in which a radical symbol appears
Radical symbol The symbol $\sqrt{\ }$
Radicand The expression under the radical
Radius A line segment with one endpoint on the center of a circle and the other endpoint on the circle
Range The set of all second coordinates of the ordered pairs in a function
Rate A ratio used to compare two different kinds of measure
Ratio The quotient of two quantities; the ratio of a to b is $\frac{a}{b}$, also written $a : b$.
Rational equation An equation containing one or more rational expressions; also called a *fraction equation*
Rational expression A quotient, or ratio, of two polynomials; also called a *fraction expression*

Rational numbers Any number that can be written as the ratio of two integers $\frac{a}{b}$, where $b \neq 0$
Rationalizing the denominator A procedure for finding an equivalent expression without a radical in the denominator
Ray A part of a line consisting of one endpoint and all the points on the line on one side of the endpoint
Real numbers All rational and irrational numbers; the set of all numbers corresponding to points on the number line
Reciprocal A multiplicative inverse; two numbers are reciprocals if their product is 1.
Rectangle A four-sided polygon with four 90° angles
Relation A correspondence between a first set, called the domain, and a second set, called the range, such that each member of the domain corresponds to *at least one* member of the range
Repeating decimal A decimal in which a number pattern repeats indefinitely
Right angle An angle whose measure is 90°
Right triangle A triangle in which one angle is a right angle
Rise The change in the second coordinate between two points on a line
Roster notation A way of naming sets by listing all the elements in the set
Rounding Approximating the value of a number; used when estimating
Run The change in the first coordinate between two points on a line

S

Sale price The price of an item after a discount has been deducted
Sales tax A tax added to the purchase price of an item
Sample space The set of all possible outcomes
Scalene triangle A triangle in which all sides are of different lengths
Scientific notation A number written in the form $M \times 10^n$, where n is an integer, $1 \leq M < 10$, and M is expressed in decimal notation
Segment A geometric figure consisting of two points, called *endpoints*, and all points between them
Set A collection of objects
Set-builder notation The naming of a set by describing basic characteristics of the elements in the set
Similar figures Figures with the same shape, but not necessarily the same size
Similar triangles Triangles in which corresponding angles have the same measure and the lengths of corresponding sides are proportional
Simple interest A percentage of an amount P invested or borrowed for t years, computed by calculating principal × interest rate × time in years
Simplify To rewrite an expression in an equivalent, abbreviated form
Slope The ratio of the rise to the run for any two points on a line
Slope–intercept equation An equation of the form $y = mx + b$, where x and y are variables, the slope is m, and the y-intercept is $(0, b)$
Solution of an equation A replacement for the variable that makes the equation true
Solution of a system of equations An ordered pair that makes both equations true
Solution set The set of all solutions of an equation, an inequality, or a system of equations or inequalities
Solve To find all solutions of an equation, an inequality, or a system of equations or inequalities; to find the solution(s) of a problem
Sphere The set of all points in space that are a given distance (radius) from a given point (center)
Square A four-sided polygon with four right angles and all sides of equal length
Square of a number A number multiplied by itself
Square root The number c is a square root of a if $c^2 = a$.
Square-root symbol The symbol $\sqrt{}$
Standard form of a linear equation An equation written in the form $Ax + By = C$
Standard form of a quadratic equation An equation written in the form $ax^2 + bx + c = 0$, $a > 0$, where a, b, and c are real-number constants
Statistic A number that describes a set of data
Straight angle An angle whose measure is 180°
Subsets Sets that are parts of other sets
Substitute To replace a variable with a number
Substitution method An algebraic method for solving systems of equations
Subtrahend In subtraction, the number being subtracted
Sum The result in addition
Sum of cubes An expression that can be written in the form $A^3 + B^3$
Sum of squares An expression that can be written in the form $A^2 + B^2$
Supplementary angles Two angles for which the sum of their measures is 180°
Surface area The sum of the areas of all of the faces of a three-dimensional figure
System of equations A set of two or more equations that are to be solved simultaneously

T

Table A method of presenting data in rows and columns
Term A number, a variable, or a product or a quotient of numbers and/or variables
Terminating decimal A decimal that can be written using a finite number of decimal places
Total price The sum of the purchase price of an item and the sales tax on the item
Transversal A line that intersects two or more coplanar lines in different points
Trapezoid A polygon with four sides, two of which, the bases, are parallel to each other
Triangle A three-sided polygon
Trinomial A polynomial containing three terms
Trinomial square The square of a binomial expressed as three terms

U

Union of sets *A* and *B* The set of all elements belonging to *either* A or B, denoted $A \cup B$

V

Value The numerical result after a number has been substituted into an expression

Variable A letter that represents an unknown number

Variation constant The constant in an equation of direct or inverse variation

Vertex The common endpoint of two rays that form an angle; the point at which the graph of a quadratic equation crosses its axis of symmetry

Vertical angles Two non-straight angles formed by two pairs of opposite rays

Vertical-line test The statement that a graph represents a function if it is impossible to draw a vertical line that intersects the graph more than once

Volume The number of cubic units needed to fill a three-dimensional figure

W

Whole numbers The natural numbers and 0: 0, 1, 2, 3, 4, 5, . . .

X

***x*-intercept** The point at which a graph crosses the x-axis

Y

***y*-intercept** The point at which a graph crosses the y-axis

Index

D

Q

R

S